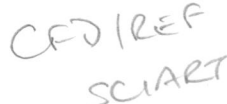

Jane's

RADAR AND ELECTRONIC WARFARE SYSTEMS

Edited by Martin Streetly

Fourteenth Edition

2002-2003

Total number of entries 1,547   New and updated entries 738

Bookmark Jane's homepage on
http://www.janes.com

Jane's award-winning web site provides you with continuously updated news and information.
As well as extracts from our world renowned magazines, you can browse the online catalogue,
visit the Press Centre, discover the origins of Jane's, use the extensive glossary,
download our screen saver and much more.

Jane's now offers powerful electronic solutions to meet the rapid changes in your
information requirements. All our data, analysis and imagery is available on CD-ROM
or via a new secure web service – Jane's Online at http://www.janes.com

Tailored electronic delivery can be provided through Jane's Data Services.
Contact an information consultant at any of our international offices to
find out how Jane's can change the way you work or e-mail us at

info@janes.co.uk *or* **info@janes.com**

ISBN 0 7106 2445 X
"Jane's" is a registered trade mark

Contents

Front cover image: *The 743D Martello air defence radar*
(Alenia Marconi Systems) **2002**/0132385

How to use *Jane's Radar and Electronic Warfare Systems*

Jane's Radar and Electronic Warfare Systems is a one volume, open source reference to the world's military radars and electronic warfare systems. It is divided into the following subject areas: land-based air defence radars; battlefield, missile control and ground surveillance radar systems; naval/coastal surveillance and navigation radars; naval fire-control radars; airborne surveillance, maritime patrol and navigation radars; airborne fire-control radars; Identification Friend-or-Foe (IFF) and Secondary Surveillance Radar (SSR) systems; military Air Traffic Control (ATC), instrumentation and ranging radars; land-based SIGnal INTelligence (SIGINT), electronic support and threat-warning systems; land-based active and passive countermeasures systems and Defensive Aids Suites (DAS); naval SIGINT, electronic support and threat-warning systems; naval active and passive countermeasures systems and DAS; airborne SIGINT, electronic support and threat-warning systems; airborne active and passive countermeasures systems and DAS and radar and electronic warfare simulation and training systems.

Within each subject area, entries are arranged alphabetically by country of manufacture. For each country, entries are attributed (where possible) to a particular contractor. Individual entries are listed alphabetically within each country and are structured as follows:

Title
The service designation and/or commercial name given to the particular piece of equipment.

Type
A brief description of the equipment's role.

Description
A narrative description of the equipment's format and capabilities.

Status
Information on the equipment's production status and/or service use.

Specifications
The equipment's technical specifications.

Contractor
The equipment's manufacturer.

In addition to the main text, *Jane's Radar and Electronic Warfare Systems* includes a glossary, tabulations (detailing in this edition, surface ship and submarine radar and electronic warfare fits), a list of contractors (including names, addresses and (where known) telephone/fax/telex numbers and e-mail and website addresses), a manufacturers index and a general alphabetical index.

To help users of this title evaluate the published data, Jane's Information Group has divided entries into three categories. A full list of all entries indicating their current status is provided in the index.

• **VERIFIED** The editor has made a detailed examination of the entry's content and checked it's relevancy and accuracy for publication in the new edition to the best of his ability.

• **UPDATED** During the verification process, significant changes to content have been made to reflect the latest position known to Jane's at the time of publication.

• **NEW ENTRY** Information on new equipment and/or appearing for the first time in the title.

All new pictures are dated with the year of publication. New pictures this year are dated 2002. Some are followed by a seven digit number for ease of identification by our image library.

A full list of all entries indicating their current status is provided in the index.

Total number of entries ⟨1,547⟩ New and updated entries ⟨738⟩

Copyright enquiries
Contact: Keith Faulkner, Tel/Fax: +44 (0) 1342 305032, e-mail: keith.faulkner@janes.co.uk

British Library Cataloguing-in-Publication Data.
A catalogue record for this book is available from the British Library.

Printed and bound in Great Britain by Biddles Ltd, Guildford and King's Lynn.

Alphabetical list of advertisers

Glossary

The following sets out a selection of the acronyms and abbreviations associated with radar and electronic warfare technology that are used in *Jane's Radar and Electronic Warfare Systems* and within the defence electronics industry and community.

A	Ampère	**DC**	Direct Current	**HI**	HIgh band
AA	Anti-Aircraft	**DECM**	Deceptive Electronic CounterMeasures	**HOJ**	Home On Jam
AAA	Anti-Aircraft Artillery	**DEW**	Distant Early Warning (US)	**HPI**	High-Power Illuminator
AAM	Air-to-Air Missile	**DF**	Direction-Finding	**HPIR**	High Probability of Intercept Receiver
AC	Alternating Current	**DIFM**	Digital Instantaneous Frequency	**HUD**	Head-Up Display
ACET	Automatic Cancellation of Extended		Measurement	**Hz**	Hertz
	Targets	**DINA**	DIrect Noise Amplification		
ADC	Analogue-to-Digital Converter	**DLJ**	DownLink Jamming	**IAGC**	Instantaneous Automatic Gain Control
ADF	Automatic Direction-Finder	**DMA**	Digital Memory Access	**IC**	Integrated Circuit
ADM	Asynchronous Data Modem	**DME**	Distance Measuring Equipment	**ICMS**	Integrated CounterMeasures Systems
AEW	Airborne Early Warning	**DMTI**	Digital Moving Target Indicator	**ICW**	Interrupted Continuous Wave
AEW & C	Airborne Early Warning and Control	**DOA**	Direction Of Arrival	**ID**	IDentification
AF	Audio Frequency	**DPM**	Digital Plotter Map	**IF**	Intermediate Frequency
AFC	Automatic Frequency Control	**DPSK**	Digital Phase Shift Keying	**IFF**	Identification Friend-or-Foe
AFSK	Audio Frequency Shift Keying	**DRFM**	Digital Radio Frequency Memory	**IFM**	Instantaneous Frequency Measurement
AFV	Armoured Fighting Vehicle	**DRTS**	Detection, Ranging and Tracking	**IFRU**	InterFerence Rejection Unit
AGC	Automatic Gain Control		System	**IIR**	Imaging Infra-Red
AGPO	Angle Gate Pull-Off	**DSB**	Double SideBand	**ILS**	Instrument Landing System
Ah	Ampère Hour	**DSP**	Digital Signal Processor	**IMP**	InterModulation Product
AI	(1) Airborne Interception	**DTU**	Data Transmission Unit	**InP**	Indium Phosphide (semiconductor)
	(2) Artificial Intelligence			**INS**	Inertial Navigation System
AJ	Anti-Jamming	**EA**	Electronic Attack (US)	**I/O**	Input/Output
ALARM	Air-Launched Anti-Radiation Missile (UK)	**EAR**	Electronically Agile Radar	**IOC**	Initial Operational Capability (US)
AM	Amplitude Modulation	**EC**	Electronic Combat	**IR**	Infra-Red
ANL	Automatic Noise Levelling	**ECCM**	Electronic Counter-CounterMeasures	**IRCM**	Infra-Red CounterMeasures
AOA	Angle Of Arrival	**ECL**	Emitter Coupled Logic	**IRST**	Infra-Red Search and Track
ARDF	Airborne Radio Direction-Finding	**ECM**	Electronic CounterMeasures	**ISAR**	Inverse Synthetic Aperture Radar
ARGS	Anti-Range Gate Stealing	**EEPROM**	Electronically Erasable Programmable	**ISB**	Independent SideBand
ARM	Anti-Radiation Missile		Read-Only Memory		
ARP	Anti-Rotation Period	**EHF**	Extra High Frequency (30 to 300 GHz)	**JNR**	Jamming-to-Noise Ratio
ARSR	Air Route Surveillance Radar	**EIRP**	Equivalent Isotropically Radiated	**Joint STARS**	Joint Surveillance and Target Attack
ASIC	Application Specific Integrated Circuit		Power		Radar System (US)
ASM	Anti-Shipping Missile	**ELF**	Extremely Low Frequency (0 to 3 kHz)	**JSR**	Jamming-to-Signal Ratio
ASP	Antenna Scan Period	**ELSEC**	ELectronic SECurity		
ASR	(1) Airport Surveillance Radar	**ELINT**	ELectronic INTelligence	**k**	Kilo (one thousand)
	(2) Air Staff Requirement (UK)	**EM**	ElectroMagnetic	**kg**	KiloGram
ASTOR	Airborne STand-Off Radar (UK)	**EMC**	ElectroMagnetic Capability	**kHz**	KiloHertz
ASW	Anti-Submarine Warfare	**EMCON**	EMission CONtrol	**km**	KiloMetre
ASuW	Anti-Surface Warfare	**EMI**	ElectroMagnetic Interference		
ATC	Air Traffic Control	**EMP**	ElectroMagnetic Pulse	**LCD**	Liquid Crystal Display
ATE	Automatic Test Equipment	**EO**	Electro-Optical	**LED**	Light Emitting Diode
ATR	Air Transport Racking	**EOB**	Electronic Order of Battle	**LF**	Low Frequency (30 to 300 kHz)
ATU	Automatic Tuning Unit	**EOCM**	Electro-Optical CounterMeasures	**LIFMOP**	LInear Frequency MOdulated Pulse
AVNL	Automatic Video Noise Limiting	**EP**	Electronic Protection (US)	**LO**	(1) Local Oscillator
AWACS	Airborne Warning And Control	**EPROM**	Erasable Programmable Read-Only		(2) Low band
	Systems (US)		Memory	**LOG**	LOGarithmic
		ERP	Effective Radiated Power	**LORO**	Lobe switching On Receive Only
BD	BauD	**ES**	Electronic Support (US)	**LPA**	Linear Power Amplifier
BFO	Beat Frequency Oscillator	**ESM**	Electronic Support Measures	**LPI**	Low Probability of Intercept
BIT	(1) Binary digIT	**EW**	Electronic Warfare	**LRA**	Line-Replaceable Assembly
	(2) Built-In Test	**EWO**	Electronic Warfare Officer (US)	**LRU**	Line-Replaceable Unit
BITE	Built-In Test Equipment			**LSB**	Lower SideBand
BITS/S	BITs per Second	**FAGC**	Fast Automatic Gain Control	**LSI**	Large Scale Integration
BMEWS	Ballistic Missile Early Warning System	**FDM**	Frequency-Division Multiplex	**LVA**	Log Video Amplifier
	(US)	**FET**	Field Effect Transistor	**LWR**	Laser Warning Receiver
BVR	Beyond Visual Range	**FFT**	Fast Fourier Transform		
B/W	BandWidth	**FLIR**	Forward-Looking Infra-Red	**m**	(1) Milli (one thousand)
BYTE	Combination of 8 BITs	**FM**	Frequency Modulation		(2) Metre
		FML	Frequency Memory Loop	**M**	Mega (one million)
C	Centigrade	**FMOP**	Frequency Modulation On Pulse	**MAW**	Missile Approach Warner
C³	Command, Control and Communications	**FMS**	Foreign Military Sales (US)	**MBAR**	Multiple Beam Acquisition Radar
C³I	Command, Control, Communications	**FOV**	Field Of View	**MCW**	Modulated Continuous Wave
	and Intelligence	**FRESCAN**	FREquency SCANning	**MDS**	Minimum Detectable Signal
CAD	Computer-Aided Design	**FSD**	Full-Scale Development (US)	**MF**	Medium Frequency (300 kHz to 3 MHz)
CCD	Charge-Coupled Device	**FSK**	Frequency Shift Keying	**MHz**	MegaHertz
CDMA	Code Division Multiple Access	**FTC**	Fast Time Constant	**MIPS**	Millions of Instructions Per Second
CdS	Cadmium Sulphide (semiconductor)	**FWE**	Foreign Weapons Evaluation (US)	**mm**	MilliMetre
CFAR	Constant False Alarm Rate			**MMIC**	Monolithic Microwave Integrated
CLOS	Command to Line Of Sight	**g**	Gram		Circuit
cm	CentiMetre	**G**	Giga (one thousand million)	**MMW**	MilliMetre-Wave
CPU	Central Processing Unit	**GaAs**	Gallium arsenide (semiconductor)	**MOP**	Modulation On Pulse
COMINT	COMmunications INTelligence	**GCA**	Ground Controlled Approach	**MPRF**	Modulated Pulse Repetition Frequency
COMSEC	COMmunications SECurity	**GCI**	Ground Controlled Interception	**mrad**	MilliRADian
COSRO	COnical Scan Receive Only	**Ge**	Germanium (semiconductor)	**ms**	MilliSecond
CRT	Cathode Ray Tube	**GHz**	GigaHertz	**MSI**	Medium Scale Integration
CVR	Crystal Video Receiver	**GMR**	Ground Mapping Radar	**MTBF**	Mean Time Between Failure
CW	Continuous Wave	**GPS**	Global Position System (US)	**MTD**	Moving Target Detector
				MTI	Moving Target Indicator
DAS	Defensive Aids Suites			**MTTR**	(1) Mean Time To Repair
dB	DeciBel	**h**	Hour		(2) Multi-Target Tracking Radar
DBF	Digital Beam Forming	**HARM**	High-speed Anti-Radiation Missile (US)	**MUX**	Multiplexer
dBm	Power measurement where the	**HDD**	Head-Down Display	**MW**	MegaWatt
	reference level is 1 milliwatt	**HF**	High Frequency (3 to 30 MHz)		

NADGE	NATO Air Defence Ground Environment
NATO	North Atlantic Treaty Organisation
NBC	Nuclear, Biological and Chemical
Nd:YAG	Neodymium YAG
NEP	Noise Equivalent Power
Ni/Cd	Nickel/Cadmium
NJ	Noise Jamming
ns	NanoSecond
NVG	Night Vision Goggles
O & M	Operations and Maintenance (US)
OTH	Over-The-Horizon
OTHB	Over-The-Horizon Backscatter
PA	Power Amplifier
PCM	Pulse Code Modulation
PLD	Pulse Length Discrimination
PLL	Phase-Lock Loop
PM	Pulse Modulation
PMOP	Phase Modulation On Pulse
POET	Primed Oscillator Expendable Transponder (US)
POI	Probability Of Intercept
PPI	Plan Position Indicator
PPS	Pulses Per Second
PRF	Pulse Repetition Frequency
PRI	Pulse Repetition Interval
PROM	Programmable Read-Only Memory
PSK	Phase Shift Keying
PSP	Programmable Signal Processor
PW	PulseWidth
QPSK	Quadrature Phase Shift Keying
QRC	Quick Reaction Capability (US)
RADAR	Radio Assisted Detection And Ranging
RAM	(1) Random Access Memory (2) Radar Absorbent Material
RATT	RAdioTeleType
RCS	Radar Cross Section
RF	Radio Frequency
RGPI	Range Gate Pull-In
RGPO	Range Gate Pull-Off
RHWR	Radar Homing and Warning Receiver
RINT	Radiation INTelligence
RISC	Reduced Instruction Set Computer
RIU	Radio Interface Unit
RML	Recirculating Memory Loop
RMS	Root Mean Square
ROB	Radar Order of Battle
ROR	Range-Only Radar
RO/RO	Roll-On/Roll-Off
ROTHR	Reloctable Over-The-Horizon Radar
RPD	Random Pulse Discrimination
RPM	Revolutions Per Minute
RSLS	Receiver SideLobe Suppression
RWR	Radar Warning Receiver
RX	Receive
s	Second
SAM	Surface-to-Air Missile
SAR	Synthetic Aperture Radar
SAW	Surface Acoustic Wave
SBB	Single Beam Blanking
SCV	SubClutter Visibility
SDR	Signal-to-Distortion Ratio
Se	Selenium (semiconductor)
SEAD	Suppression of Enemy Air Defences (US)
SHF	Super High Frequency (3 to 30 GHz)
SHORAD	SHOrt Range Air Defence (weapon)
Si	Silicon (semiconductor)
SIGINT	SIGnals INTelligence
SLAR	Side-Looking Airborne Radar
SLB	SideLobe Blanking
SLC	SideLobe Cancellation
SLOR	Swept Local Oscillator Receiver
S/N	Signal-to-Noise ratio
SOJ	Stand-Off Jamming
SORO	Scan On Receive Only
SP	Single Phase

SSA	Solid-State Amplifier
SSB	Single SideBand
SSL	Single Site Location
STANAG	STAndard NATO AGreement
STC	Sensitivity Time Control
STIR	Surveillance and Target Illumination Radar
STRAP	Straight Through Repeater Antenna Performance (US)
SWR	(1) Standing Wave Ratio (2) Surface-Wave Radar
TACDS	Threat Adaptive Countermeasures Dispenser System
TCD	Time Critical Data
TDMA	Time Division Multiple Access
TER	TErrain-following Radar
TIR	Target Illuminating Radar
TIS	Thermal Imaging System
TOA	Time Of Arrival
TOJ	Track On Jam
TSS	Tangential Signal Sensitivity
TTL	Transistor-to-Transitor Logic
TWS	Track-While-Scan
TWT	Travelling Wave Tube
TX	Transmit
UAV	Unmanned Air Vehicle
UHF	Ultra High Frequency (300 MHz-1 GHz)

URE	Unintentional Radiation Exploitation
USB	Upper SideBand
UV	Ultra-Violet
V	Volt
VCO	Voltage Controlled Oscillator
VFO	Variable Frequency Oscillator
VGPO	Velocity Gate Pull-Off
VHF	Very High Frequency (30 to 300 MHz)
VHSIC	Very High-Speed Integrated Circuit
VLF	Very Low Frequency (3 to 30 kHz)
VLSI	Very Large Scale Integration
VOR	VHF Omnidirectional radio Range
VSWR	Voltage Standing Wave Ratio
VVA	Voltage-Variable Attenuator
W	Watt
WLR	Weapon Locating Radar
YIG	Yttrium Indium Garnet

Symbols

°	Degree
>	Greater than
<	Less than
μm	MicroMetre (replaces micron)
μs	MicroSecond
Ω	Ohm
λ	Wavelength

FREE ENTRY IN THIS PUBLICATION

Having your products and services represented in our titles means that they are being seen by the professionals who matter – both by those involved in procurement and those working for the companies that are likely to affect your business. We therefore feel that it is very much in the interests of your organisation, as well as Jane's, to ensure your data is current and accurate.

■ **Don't forget –** You may be missing out on business if your entry in a Jane's book, CD-ROM or Online product is incorrect because you have not supplied the latest information to us.

■ **Ask yourself –** Can you afford not to be represented in Jane's printed and electronic products? And if you are listed, can you afford for your information to be out of date?

■ **And most importantly –** The best part of all is that your entries in Jane's products are TOTALLY FREE OF CHARGE.

Please provide the information on the following categories where appropriate:

1. Organisation name: _____

2. Division name: _____

3. Location address: _____

   _____

4. Mailing address if different: _____

   _____

5. Telephone (please include switchboard and main department contact numbers, e.g. Public Relations, Sales, etc.):

   _____

6. Facsimile: _____

7. E-mail: _____

8. Web sites: _____

9. Contact name and job title: _____

   _____

10. A brief description of your organisation's activities, products and services: _____

    _____

    _____

11. Jane's publications in which you would like to be included: _____

    _____

Please send this information to:
Jacqui Beard, Information Collection, Jane's Information Group,
Sentinel House, 163 Brighton Road, Coulsdon, Surrey, CR5 2YH, UK
Tel: (+44 20) 87 00 38 08
Fax: (+44 20) 87 00 39 59
E-mail: yearbook@janes.co.uk

Copyright enquiries:
Contact: Keith Faulkner
Tel/Fax: (+44 1342) 30 50 32
E-mail: keith.faulkner@janes.co.uk

Please tick this box if you do not wish your organisation's staff to be included in Jane's mailing lists ☐

JREW

EDITORIAL AND ADMINISTRATION

Publishing Director: Alan Condron, e-mail: Alan.Condron@janes.co.uk

Managing Editor: Simon Michell, e-mail: Simon.Michell@janes.co.uk

Global Content Manager: Anita Slade, e-mail: Anita.Slade@janes.co.uk

Content Editing Manager: Jo Fenwick, e-mail: Jo.Fenwick@janes.co.uk

Pre-Press Manager: Christopher Morris, e-mail: Christopher.Morris@janes.co.uk

Team Leaders: Sharon Marshall, e-mail: Sharon.Marshall@janes.co.uk
Neil Grace, e-mail: Neil.Grace@janes.co.uk

Production Editor: Daniel O'Doherty, e-mail: Daniel.O'Doherty@janes.co.uk

Production Controller: Victoria Powell, e-mail: Victoria.Powell@janes.co.uk

Content Update: Jacqui Beard, Information Collection Assistant
Tel: (+44 20) 87 00 38 08 Fax: (+44 20) 87 00 39 59
e-mail: yearbook@janes.co.uk

Jane's Information Group Limited, Sentinel House, 163 Brighton Road,
Coulsdon, Surrey CR5 2YH, UK
Tel: (+44 20) 87 00 37 00 Fax: (+44 20) 87 00 37 88
e-mail: jrew@janes.co.uk

SALES OFFICE

Send EMEA enquiries to: *Group Sales Manager*
Jane's Information Group Limited, Sentinel House, 163 Brighton Road,
Coulsdon, Surrey CR5 2YH, UK
Tel: (+44 20) 87 00 37 00 Fax: (+44 20) 87 63 10 06
e-mail: info@janes.co.uk

Send USA enquiries to: *Robert Loughman – Vice-President Product Sales*
Jane's Information Group Inc, 1340 Braddock Place, Suite 300, Alexandria,
Virginia 22314-1651, USA
Tel: (+1 703) 683 37 00 Fax: (+1 703) 836 02 97 Telex: 6819193
Tel: (+1 800) 824 07 68 Fax: (+1 800) 836 02 971
e-mail: info@janes.com

Send Asia enquiries to: *David Fisher – Group Sales Manager*
Jane's Information Group Asia, 60 Albert Street, #15-01 Albert Complex,
Singapore 189969
Tel: (+65) 331 62 80 Fax: (+65) 336 99 21
e-mail: info@janes.com.sg

Send Australia/New Zealand enquiries to: *David Moden – Business Manager*
Jane's Information Group, PO Box 3502, Rozelle Delivery Centre,
New South Wales 2039, Australia
Tel: (+61 2) 85 87 79 00 Fax: (+61 2) 85 87 79 01
e-mail: info@janes.thomson.com.au

ADVERTISEMENT SALES OFFICES

(Head Office)
Jane's Information Group
Sentinel House, 163 Brighton Road,
Coulsdon, Surrey CR5 2YH, UK
Tel: (+44 20) 87 00 37 00
Fax: (+44 20) 87 00 38 59/37 44
e-mail: defadsales@janes.co.uk

Richard West, Senior Key Accounts Manager
Tel: (+44 1892) 72 55 80 Fax: (+44 1892) 72 55 81
e-mail: richard.west@janes.co.uk

Kate Hamlin, Advertising Sales Manager
Tel: (+44 20) 87 00 38 53 Fax: (+44 20) 87 00 38 59/37 44
e-mail: kate.hamlin@janes.co.uk

Joni Beeden, Advertising Sales Executive
Tel: (+44 20) 87 00 39 63 Fax: (+44 20) 87 00 38 59/37 44
e-mail: joni.beeden@janes.co.uk

Steve Soffe, Advertising Sales Executive
Tel: (+44 20) 87 00 39 43 Fax: (+44 20) 87 00 38 59/37 44
e-mail: steven.soffe@janes.co.uk

(USA/Canada office)
Jane's Information Group
1340 Braddock Place, Suite 300,
Alexandria, Virginia 22314-1651, USA
Tel: (+1 703) 683 37 00
Fax: (+1 703) 836 55 37
e-mail: defadsales@janes.com

USA and Canada
Katie Taplett, US Advertising Sales Director
Tel: (+1 703) 683 37 00 Fax: (+1 703) 836 55 37
e-mail: katie.taplett@janes.com

Northern USA and Eastern Canada
Harry Carter, Northeast Region Advertising Sales Manager
Tel: (+1 703) 683 37 00 Fax: (+1 703) 836 55 37
e-mail: harry.carter@janes.com

South Eastern USA
Kristin D Schulze, Advertising Sales Manager
PO Box 270190, Tampa, Florida 33688-0190
Tel: (+1 813) 961 81 32 Fax: (+1 813) 961 96 42
e-mail: kristin@intnet.net

Western USA and West Canada
Richard L Ayer
127 Avenida Del Mar, Suite 2A, San Clemente, California 92672
Tel: (+1 949) 366 84 55 Fax: (+1 949) 366 92 89
e-mail: ayercomm@earthlink.com

Australia: *Richard West* (see UK Head Office)

Benelux: *Steve Soffe* (see UK Head Office)

Brazil: *Katie Taplett* (see USA address)

Eastern Europe: MCW Media & Consulting Wehrstedt
Dr. Uwe H. Wehrstedt
Hagenbreite 9, D-06463 Ermsleben, Germany
Tel: (+49) 0700/WEHRSTEDT / (+49) 03 47 43/620 90
Fax: (+49) 03 47 43/620 91
e-mail: info@Wehrstedt.org

France: Patrice Février
BP 418, 35 avenue MacMahon,
F-75824 Paris Cedex 17, France
Tel: (+33 1) 45 72 33 11 Fax: (+33 1) 45 72 17 95
e-mail: patrice.fevrier@wandadoo.fr

Germany and Austria: *MCW Media & Consulting Wehrstedt* (see Eastern Europe)

Greece: *Steve Soffe* (see UK Head Office)

Hong Kong: *Joni Beeden* (see UK Head Office)

India: *Joni Beeden* (see UK Head Office)

Israel: Oreet – International Media
15 Kinneret Street, IL-51201 Bene Berak, Israel
Tel: (+972 3) 570 65 27 Fax: (+972 3) 570 65 27
e-mail: admin@oreet-marcom.com
Defence: Liat Shaham
e-mail: liat_s@oreet-marcom.com

Italy and Switzerland: Ediconsult Internazionale Srl
Tel: (+39 010) 58 36 59 Fax: (+39 010) 56 65 78
e-mail: genova@ediconsult.com

Japan: Skynet Media, Inc
748, 1-7 Akasaka 9-chome, Minato-ku, Tokyo 107-0052, Japan
Contact: Mr Osamu Yoneda
Tel: (+81 3) 54 74 78 35
Fax: (+81 3) 54 74 78 37
e-mail: skynetme@wonder.ocn.ne.jp

Middle East: *Steve Soffe* (see UK Head Office)

Pakistan: *Joni Beeden* (see UK Head Office)

Russian Federation: Simon Kay
33 St John's Street, Crowthorne, Berkshire RG45 7NQ, UK
Tel: (+44 1344) 77 71 23 Mobile: (+44 7702) 54 96 84
Fax: (+44 1344) 77 58 85
e-mail: crowkay@msn.com/crowkay@yahoo.com

Scandinavia: The Falsten Partnership
PO Box 21175, London N16 6ZG, UK
Tel: (+44 20) 88 06 23 01 Fax: (+ 44 20) 88 06 81 37
e-mail: sales@falsten.com

Singapore: *Richard West/Joni Beeden* (see UK Head Office)

South Africa: *Richard West* (see UK Head Office)

South Korea: JES Media Inc
2nd Floor, ANA Building, 257-1 Myungil-Dong, Kandong-Gu, Seoul 134-070, Korea
Contact: Mr Young-Seoh Chinn, President
Tel: (+82 2) 481 34 11/34 13
Fax: (+82 2) 481 34 14
e-mail: jesmedia@unitel.co.kr

Spain: Via Exclusivas SL
Contact: Julio de Andres
Viriato 69SC, E-28010 Madrid, Spain
Tel: (+34 91) 448 76 22 Fax: (+34 91) 446 02 14
e-mail: j.a.deandres@viaexclusivas.com

Turkey: *Richard West* (see UK Head Office)

ADVERTISING COPY
Delphine Gandelin (Jane's UK Head Office)
Tel: (+44 20) 87 00 37 42 Fax: (+44 20) 87 00 38 59/37 44
e-mail: delphine.gandelin@janes.co.uk

For North America, South America and Caribbean only:
Shanee Johnson (Jane's USA address)
Alexandria, Virginia 22314-1651, USA
Tel: (+1 703) 683 37 00 Fax: (+1 703) 836 55 37
e-mail: shanee.johnson@janes.com

Information Services & Solutions

Jane's is the leading unclassified information provider for military, government and commercial organisations worldwide, in the fields of defence, geopolitics, transportation and law enforcement.

We are dedicated to providing the information our customers need, in the formats and frequency they require. Read on to find out how Jane's information in electronic format can provide you with the best way to access the information you require.

Jane's Online

Search across the complete portfolio of Jane's Information, via the Internet

Created for the professional seeking specific detailed information, this user-friendly service can be customised to suit your ever-changing information needs. Search across any combination of titles to retrieve the information you need quickly and easily. You set the query — Jane's Online finds the answer!

Key benefits of Jane's Online include:
- the most up to date information available from Jane's
- accessible anytime, anywhere
- saves time — research can be carried out quickly and easily
- archives enable you to compare how specifics have changed over time
- accurate analysis at your fingertips
- site licences available
- user-friendly interface
- high-quality images linked to text

Check out this site today: **http://www.janes.com**

Jane's CD-ROM Libraries

Quickly pinpoint the information you require from Jane's

Choose from nine powerful CD-ROM libraries for quick and easy access to the defence, geopolitical, space, transportation and law enforcement information you need. Take full advantage of the information groupings and purchase the entire library.

Libraries available:
Jane's Air Systems Library
Jane's Defence Equipment Library
Jane's Defence Magazines Library
Jane's Geopolitical Library
Jane's Land and Systems Library
Jane's Market Intelligence Library
Jane's Police and Security Library
Jane's Sea and Systems Library
Jane's Transport Library

Key benefits of Jane's CD-ROM include:
- quick and easy access to Jane's information and graphics
- easy-to-use Windows interface with powerful search capabilities
- online glossary and synonym searching
- search across all the titles on each disc, even if you do not subscribe to them, to determine whether you would like to add them to your library
- export and print out text or graphics
- quarterly updates
- full networking capability
- supported by an experienced technical team

Jane's Data Service

Jane's information on your intranet or controlled military network

Jane's Data Service brings together more than 200 sources of near-realtime and technical reference information serving defence, intelligence, space, transportation and law enforcement professionals. By making Jane's data (HTML) and images (JPEG) available for integration behind Intranet environments or closed networks, this unique service offers you a way to receive information that is updated frequently and works in tune with your organisation. We can also offer a complete management service where Jane's hosts the information and server for you. The most secure way to access Jane's Information.

Jane's Consultancy

A service as individual as your needs
Whether it is research on your competitors' markets, in-depth analysis or customised content that you require, Jane's Consultancy can offer you a tailored, highly confidential personal service to help you achieve your objectives. However large or small your requirement, contact us in confidence for a free proposal and quotation.

Jane's Consultancy will bring you a variety of benefits:
- expert personnel in a wide variety of disciplines
- a global and well-established information network
- total confidentiality
- objective analysis
- Jane's reputation for accuracy, authority and impartiality

The information you require, delivered in a format to suit your needs.

The shield decoy launching system (Alenia Marconi Systems)

2002/0132386

Foreword

Technology Analysis and Forecasts

The 2002-2003 edition of *Jane's Radar and Electronic Warfare Systems* identifies the following trends/developments within its remit as being significant:

- **Active Electronically Scanning Array (AESA) radars**

Jane's Radar and Electronic Warfare Systems believes that AESA radars are poised to become the 'sensor of choice' for future and upgradable fighter aircraft. A late 2001 US Defense Science Board (DSB) report suggests that AESA technology offers 'massive' weight and cost reductions together with 'substantial' increases in capability when compared with mechanically scanning radars. In terms of capability increase, the DSB report claimed that the technology offers a 'factor of 10 to 30 times more net radar capability than competing approaches'. Specific advantages identified include:

- beam agility
- a reduction in radar cross section via the relative simple shape of the AESE radar's transceiver matrix (thereby making it easier to shroud than a conventional mechanical scanner), the ease with which AESA arrays can be canted to direct head-on, main-beam reflections away from the threat emitter's receiver's line-of-sight and the use of frequency selective surface radomes, radar absorbent material shielding and serrated edging patterns
- high reliability levels
- low radio frequency loss levels due to the relationship between the receiver and intermediate frequency amplifier within each AESA transceiver module. Accordingly, the AESA receiver chain minimises spurious noise and delivers a 'very clean' signal for processing
- low probability of intercept via 'smart' power management throughout the engagement sequence (range versus power output), multiple beam search (small sector surveillance with longer dwell times using less power) and pulse-to-pulse parameter variation (including pulsewidth, pulse repetition frequency and beamwidth)
- a 'track-before-detect' capability whereby the radar's detection threshold can be set in such a way as to facilitate scan to scan matching of above threshold returns prior to their confirmation as valid targets in order to provide instantly available track data once they are so confirmed

A key element in the maturation of the technology is the fact that within the US industry at least, the core active transceiver modules needed to power AESA radars are approaching 'commodity status', opening the way for a potential market for some three million such modules for use in the 3,651 AESA radars that are either being planned for, or being integrated aboard, the F-15C (18 radars), F-16 Block 60 (80), F/A-18E/F (258), F-22 (295) and F-35 (approximately 3,000) combat aircraft. The DSB also advocates the rapid integration of AESA technology into the radar sensors carried by the RQ-4A Global Hawk surveillance Unmanned Aerial Vehicle (UAV) and the E-8 Joint Surveillance Target Attack Radar System (Joint STARS) aircraft.

Another major aspect of AESA fighter radar technology *Jane's Radar and Electronic Warfare Systems* believes is worthy of consideration is its potential for retrofit. Of the cited US platforms, the F-15C application is just such a programme. Here, introduction of the AESA (V)2 variant of the aircraft's AN/APG-63 radar is intended to provide the US Air Force with experience of an operational AESA system prior to the widescale introduction of the technology aboard its F-22 and F-35 aircraft. As such, APG-63(V)2 substitutes an AESA array for the sensor's baseline mechanical scanner and integrates it with an 'advanced' identification friend-or-foe capability to provide a fire-control system that is capable of taking 'full advantage' of the latest generation of long-range air-to-air missiles, as exemplified by the AIM-120. Outside the US, *Jane's Radar and Electronic Warfare Systems* notes that the European Airborne Multirole multifunction Solid-state Active array Radar (AMSAR) programme is aimed at producing AESA type equipments for the mid-life updating of combat aircraft, such as the Eurofighter Typhoon, the Rafale C/M and the Mirage 2000-5/-9 series, during the period 2005-2006. Swedish contractor Ericsson Microwave is also involved in the development of an AESA array that can be back fitted to the PS-05/A radar fitted to the Gripen multirole fighter.

Lockheed Martin's F-35 Joint Strike Fighter will be equipped with an AESA radar (Lockheed Martin)

2002/0098624

The Boeing 737/MESA radar AEW solution is on order for the Royal Australian Air Force and has been selected for potential procurement by Turkey (Boeing)

2002/0003970

The US analysis of AEA alternatives has grown out of the need to replace the joint service EA-6B Prowler EA aircraft post 2010 (USN)

2002/0093966

- **Airborne Early Warning**

In the short term, *Jane's Radar and Electronic Warfare Systems* expects Turkey to finalise an Airborne Early Warning (AEW) deal centred around a Boeing 737/Northrop Grumman 1.2 to 1.4 GHz band Multirole Electronically Scanned Array (MESA) radar solution. *Jane's Radar and Electronic Warfare Systems* also expects to see Embraer EMB-145/Ericsson Microwave 2 to 4 GHz Erieye radar solution to build on its success in Brazil, Greece and Mexico. Assuming adequate funding, Mexican acquisition of additional systems seems likely. Elsewhere, both Singapore and Chile are known to be actively considering replacing their existing AEW capabilities with new platforms. Other potential AEW customers include the People's Republic of China (A-50 — see *Jane's Electronic Mission Aircraft* — derivative with Russian radar), India (A-50 variant with either Russian or Israeli radar), Italy, South Korea, Kuwait, Malaysia, Spain, Thailand and the United Arab Emirates. *Jane's Radar and Electronic Warfare Systems* further believes that a UAV-based AEW capability will be demonstrated within the next decade.

- **Future US Airborne Electronic Attack (AEA)**

The US Department of Defense has completed a two year AEA 'analysis of alternatives' in its drive to secure an effective AEA capability in the wake of the withdrawal from service of the EA-6B Prowler (see *Jane's Electronic Mission Aircraft*) circa 2010. Conclusions arrived at by this important study (impacting as it does on both future US and NATO activity) include the following:

- AEA is a 'required capability' for US forces operating in 'defended airspace' against 'current and future threats' and 'enhances/complements' the contributions made by signature reduction ('stealth'), defensive electronic countermeasures and stand-off weapons to platform survival.
- a 'comprehensive' AEA solution requires 'core' and 'stand-in' components with the former providing 'sensing and persistence' and the latter the ability to penetrate heavily defended areas and attack 'advanced frequency agile emitters'. 'Stand-in' components considered included:
  - tactical UAVs equipped with a 'family' of band-specific EA subsystems
  - a family of self-propelled expendables equipped with repeater-based payloads collectively capable of 'full band' coverage
  - a family of 'gravity' expendables equipped with the same payload range as the self-propelled expendables.

A putative 'EA-22' EA version of the USAF's F-22 Raptor air superiority fighter has been proposed as a 'core' component solution within the US analysis of AEA alternatives study (Lockheed Martin)

2002/0048744

Table 1: AEA 'core' option total ownership cost comparison 2010-2040

Option	Cost (US$ billion*)	Remarks
NSHF	20	single land-only platform
BJ	25-26	single land-only platform
HSHF/UAV(CV)	26	land-only/carrier capable platforms
EA-6C with ALQ-99 pods	32	carrier and land-based platform
EA-6C with NT pods	33	carrier and land-based platform
EA-6C/BJ	34	carrier capable/land-only platforms
EA-35(CV)/BJ	35	carrier capable/land-only platforms
EA-35(CV/CTOL)	36-37	carrier and land-based platform
EA-18/BJ	38	carrier capable/land-only platforms
EA-18 with NT pods	39.5	carrier and land-based platform
Boeing 737	41	single land-only platform
EA-6C/Boeing 737	43	carrier capable/land-only platforms
EA-35(CV)/Boeing 737	43	carrier capable/land-only platforms
EA-6C/EA-15	44	carrier capable/land-only platforms
EA-18/EA-16	45	carrier capable/land-only platforms
EA-18/Boeing 737	47	carrier capable/land-only platforms
EA-6C/Boeing 757	47.5	carrier capable/land-only platforms
EA-35(CV)/Boeing 757	47.5	carrier capable/land-only platforms
EA-18/EA-15	47.5	carrier capable/land-only platforms
EA-6C/EB-1	50	carrier capable/land-only platforms
EA-18/Boeing 757	51	carrier capable/land-only platforms
EA-18/EB-1	52	carrier capable/land-only platforms
EA-35(CV)/EB-1	54	carrier capable/land-only platforms
EA-35/EB-52/EA-22	76-77	carrier capable/two land-only platforms
EA-6C/EB-52/EA-22	76-77	carrier capable/two land-only platforms
EA-18/EB-52/EA-22	80-82	carrier capable/two land-only platforms

Key

* Approximate values **BJ** Business Jet **CTOL** Conventional Take-Off and Landing **CV** Aircraft carrier **NSHF** New Start High Flier **NT** New Transmitter **UAV** Unmanned Aerial Vehicle
Source
Airborne Electronic Attack Analysis of Alternatives – Results Briefing (Unclassified) Lieutenant Colonel Steve Sokoly USAF (28 February 2002)

'Stand-in' delivery would be by means of multimission aircraft, dedicated AEA platforms, self-delivery and/or UAVs and Unmanned Combat Air Vehicles (UCAV). The analysis put the total 2010 to 2040 ownership costs of two considered 'stand-in' options (including platform delivery) at between approximately US$7 billion and US$30 billion. 'Core' components considered were as follows:

- *single carrier/land-based capable platforms* = EA-6C, EA-35 (conventional take-off) and EA-18
- *single land-based platforms* = Boeing 737 variant, 'New Start High Flier' (NSHF) and business jet variant
- *carrier-capable and land-based platforms* = UAV/ NSHF, EA-35 (carrier)/business jet variant, EA-6C/ business jet variant, EA-6C/EA-16, EA-6C/EA-15, EA-18/EA-16 and EA-18/EA-15
- *carrier-capable and two land-based platforms* = EA-35 (carrier)/EB-52/EA-22 (two format options), EA-6C/ EB-52/EA-22 (two format options) and EA-18/EB-52/ EA-22 (two format options)
- *carrier-capable and large land-based platforms* = EA-35/Boeing 737 variant, EA-6C/Boeing 737 variant, EA-35 (carrier)/Boeing 757 variant, EA-18/Boeing 737 variant, EA-6C/Boeing 757 variant, EA-6C/EB-1,

EA-35 (carrier)/EB-1, EA-18/Boeing 757 variant and EA-18/EB-1

Total ownership costs for these various 'core' options are given in Table 1.

Provision of a 'complete' AEA capability facilitates:

- denial of situational awareness (via the neutralisation of an enemy's volume search radars and communications links) thereby facilitating the screening of strike forces and the insertion of special operations forces
- the destruction and/or denial of effective fire-control functionality
- the degradation of Surface-to-Air Missile (SAM) systems via neutralisation of their target acquisition sensors and/or their destruction in Destruction of Enemy Air Defences (DEAD) operations

A complete AEA solution requires:

- a 'dedicated' Electronic Support (ES) capability to detect, locate and identify threat emitters
- datalink networking between ES and AEA platforms to increase probability of intercept
- 'human' battle management

- 'full' frequency coverage
- 'preferred' asset placement with regard to the ability to jam (line of sight, range to threat and threat- platform orientation considerations) and AEA platform threat exposure/survivability considerations. Major AEA platform threats were adjudged to be mobile SAMs ('SAM-bush' tactics), anti-radiation missiles, home on jam/track on jam SAM modes and air-to-air missiles.

As of March 2002, a decision as to which AEA approach to follow was scheduled for the Summer of 2002. As of the given date, *Jane's Radar and Electronic Warfare Systems* believed that an EA-18 solution was likely to be implemented.

- **Naval active phased-array radars**

Jane's Radar and Electronic Warfare Systems expects to see the operational deployment of naval active phased-array multimode radars within the short- to medium-term. Identified examples of such technology include:

- the 8 to 20 GHz band MultiFunction Radar (MFR) that is intended for applications such as the US Navy's CVN 77 aircraft carrier and the service's proposed DD(X) Future Surface Combatant (formerly the DD 21 'Zumwalt' class destroyer programme). Here, Raytheon Systems envisages a putative 'SPY-3' radar that would employ between 20,000 and 24,000 active transceiver modules and, complemented by a D-band (1 to 2 GHz) volume search radar, would act as a replacement for the service's Mk 23, Mk 95, AN/ SPQ-9B and AN/SPG-62 fire-control radars. Here, the system would be capable of long-range search, fire-control quality target tracking, surface surveillance, target illumination and missile guidance
- the Alenia Marconi Systems (AMS) SAMPSON multifunction radar that has been selected for installation aboard the UK Royal Navy's next generation Type 45 air defence destroyer. As such, SAMPSON incorporates two back-to-back arrays that are mounted within a spherical radome and are populated by approximately 1,300 four channel transceiver modules. As currently (March 2002) envisaged, AMS will deliver three production standard SAMPSON prototypes by 2004 together with the first of 12 radars for the Type 45 programme itself. *Jane's* sources suggest that a second tranche of 11 radars could be acquired from 2003 onwards. SAMPSON capabilities are understood to include long-range detection of stealthy targets, enhanced fire-control, target classification, multiple target engagement and graceful degradation.
- the multinational APAR system that has been developed by a consortium that includes Thales Nederlands (formerly Signaal – prime contractor), Nortel Networks, European Aeronautic Defence and Space (EADS) Co Deutschland, Euroatlas GmbH, Lockheed Martin Canada, Comdev Ltd, Stork Canada and Thales Canada. Each of APAR's four arrays makes use of 856 four channel transceiver modules that operate in the 8 to 20 GHz frequency band. Each array covers a 120° sector and is capable of ± 60° beam shifting. Delivery of the first seven APAR radars to the German and Dutch navies began during the latter half of 2000.

- **Time critical targeting**

Jane's Radar and Electronic Warfare Systems believes that time critical targeting (that is, the detection, location, tracking, targeting and engagement of targets in real time together with subsequent real-time attack success/failure

The UK's Royal Navy's next generation Type 45 anti-air warfare destroyer will be fitted with the SAMPSON active multifunction radar (BAE Systems)
2002/0084480

The APAR radar has been delivered to the Dutch and German Navies (Thales Nederland)
2002/0080965

assessment) can only grow in importance, particularly in view of America's ongoing 'war on terrorism' and the difficulty inherent in dealing with highly mobile threats, such as the latest generation of surface-to-air missile and theatre ballistic missile systems. To be effective, the described detection-to-engagement sequence against such systems will have to be completed in a time scale of between two and seven minutes. Within the genre, *Jane's Radar and Electronic Warfare Systems* related technologies are likely to include:

- FOliage PENetration (FOPEN) radars. Here, the US Air Force and US Army are known to be working programmes with a Lockheed Martin 30 MHz to 3 GHz dual-band synthetic aperture FOPEN radar demonstrator that are expected to lead to an operational system on platforms such as UAVs.
- the US MultiPlatform Radar Technology Insertion Program (MP-RTIP). MP-RTIP is designed to produce an 8 to 12.5 GHz band scalable synthetic aperture/ Moving Target Indicator (MTI) radar that will be suitable for both manned and UAV applications. Potential platforms include the Block 10 RQ-4A Global Hawk UAV, an E-8 Joint STARS upgrade and/or a new manned host based on aircraft such as the Boeing 737/ 757/767 airliners or one of the latest generation of business jets as exemplified by the Gulfstream V and Global Express.
- the introduction of multiship target ranging in suppression/destruction of enemy air defences (SEAD/ DEAD) operations. Such techniques are known to be under development in the US (the R7 configuration of the AGM-88 High-speed Anti-Radiation Missile (HARM) Targeting System – HTS) and Germany (employing the emitter location system installed aboard Luftwaffe (German Air Force) Electronic Combat and Reconnaissance (ECR) Tornados)
- within the US-UK orbit, the introduction (in the short- to medium-term) of real-time sensor networking using modems and datalinks (particularly the NATO Link 16 system) to fuse surveillance/targeting/damage assessment data from platforms such as the RQ-1/RQ-4

UAVs, the U-2 high altitude reconnaissance aircraft (see *Jane's Electronic Mission Aircraft*), the RC-135, Nimrod R Mk 1, EP-3E and RC-12 'Guardrail' signals intelligence gatherers (see *Jane's Electronic Mission Aircraft*) and ground surveillance platforms such as the E-8, AIP P-3Cs and the UK's forthcoming Airborne STand-Off Radar (ASTOR) system (see *Jane's Electronic Mission Aircraft*). Once acquired and fused, the intention is to hand-off real time, targeting quality information to 'shooters' via datalink

- the introduction by NATO of a multinational airborne Alliance Ground Surveillance (AGS) capability *circa* 2008. Here, the previously cited MP-RTIP is also the basis of the five nation (Belgium, Canada, Denmark, Norway and the USA) NATO Advanced Technology Radar (NATAR) that is being proposed to meet the requirement. As a competitor to this (and as a means of keeping European industry abreast of synthetic aperture/MTI radar technology), contractors in France, Germany, Italy, Netherlands and Spain have formed the SOSTAR (Stand-Off Surveillance and Target Acquisition Radar) joint venture to develop the SOSTAR-X MTI/SAR radar. As currently planned, a SOSTAR-X demonstrator will have been air tested by 2005. With a programme value of then year €85 million (US$78 million), this prototype will be mounted in a Fokker 100 testbed aircraft (see *Jane's Electronic Mission Aircraft*). As outlined at the launch of the joint venture, the SOSTAR technology demonstrator will incorporate active antenna array technology and will feature acquisition, processing and operation control consoles aboard the testbed aircraft, an air-to-ground datalink and associated ground-based data processing, evaluation and dissemination facilities. As a first stage in the trials programme, the SOSTAR testbed will be used to acquire and transmit real-time reconnaissance and targeting data to both stationary and mobile ground stations
- the introduction of the US Navy's Co-operative Engagement Capability (CEC) that will integrate the sensors (including the AN/APS-145 radar carried by

the service's E-2C/Hawkeye 2000 AEW aircraft – see *Jane's Electronic Mission Aircraft*) within a carrier battlegroup to facilitate air defence against targets that are over the radar horizon and/or not 'seen' by particular sensors in particular positions.

Acknowledgements
The production of *Jane's Radar and Electronic Warfare Systems* is a team effort that involves the editor, sources, industry and a dedicated team of production personnel within the *Jane's Information Group*. Accordingly, the editor would like to thank all those who have contributed to the 2002-2003 edition. He also wishes to acknowledge the work of *Jane's Radar and Electronic Warfare Systems'* long-time copy editor Jane Stimson who died suddenly and unexpectedly during December 2001. Mrs Stimson was unflagging in her support of *Jane's Radar and Electronic Warfare Systems* as a project and her attention to detail and skill as a copy editor contributed enormously to the work's success. She is sorely missed as both a valued colleague and as a friend.

Important Notice
While every attempt has been made to keep pace with the changes within the radar and electronic warfare industries during the period March 2001 to March 2002, observant readers will note a number of conflicts over company names within the following text. This is due to the need to adhere to a strict production cycle in order to publish the data in both electronic and hard copy formats on schedule. Accordingly, it should be noted that any such anomalies are addressed as quickly as possible via the medium of *Jane's* online web service.

Martin Streetly
Coulsdon
March 2002

MARTIN STREETLY

Martin Streetly is a full-time defence electronics author and journalist who specialises in the history, technology and application of electronic warfare (EW). He is the editor of both *Jane's Radar and Electronic Warfare Systems* and *Jane's Electronic Mission Aircraft.*

Over the past two decades, Martin Streetly has been a regular contributor to a range of international defense publications including the *Journal of Electronic Defense* (acting as the magazine's European Editor for 12 years up to March 2001), *The Knowles Report, Microwave Journal, Jane's Defence Weekly, Flight International, Naval Forces, International Defence Review, Military Simulation & Training* and the *NAVINT* naval intelligence newsletter. Over the last ten years, he has appeared on the UK's Channel 4 news programme, the BBC and the Discovery Channel. During the 1991 Gulf War, he worked with the UK's Independent Television News Ltd and a range of international newspapers (including the *New York Times* and the *Jerusalem Post*) and has been invited to lecture on EW technology by industry, NATO, the Government of the United Arab Emirates and the *Association of Old Crows*.

Over and above his work on the *Jane's* yearbooks, Martin Streetly has to-date published four books on the history and technology of airborne EW, the details of which are as follows:

Confound & Destroy: 100 Group and The Bomber Support Campaign
Macdonald & Jane's Publishers Ltd, London, 1978 and Jane's Publishing Ltd, London, 1985.
World Electronic Warfare Aircraft
Jane's Publishing Ltd, London, 1983 & 1984.
The Aircraft of 100 Group
Robert Hale Ltd, London, 1984.
Airborne Electronic Warfare: History, Techniques and Tactics
Jane's Publishing Ltd, London, 1988.

Of these, *Confound & Destroy* is considered by many as being the definitive study of the birth of airborne EW within the UK's Royal Air Force while over 5,000 copies of the two editions of *World Electronic Warfare Aircraft* have been sold worldwide.

Users' Charter

This publication is brought to you by Jane's Information Group, a global company with more than 100 years of innovation and an unrivalled reputation for impartiality, accuracy and authority.

Our collection and output of information and images is not dictated by any political or commercial affiliation. Our reportage is undertaken without fear of, or favour from, any government, alliance, state or corporation.

We publish information that is collected overtly from unclassified sources, although much could be regarded as extremely sensitive or not publicly accessible.

Our validation and analysis aims to eradicate misinformation or disinformation as well as factual errors; our objective is always to produce the most accurate and authoritative data.

In the event of any significant inaccuracies, we undertake to draw these to the readers' attention to preserve the highly valued relationship of trust and credibility with our customers worldwide.

If you believe that these policies have been breached by this title, you are invited to contact the editor.

A copy of Jane's Information Group's Code of Conduct for its editorial teams is available from the publisher.

INVESTOR IN PEOPLE

Quality Policy

Jane's Information Group is the world's leading unclassified information integrator for military, government and commercial organisations worldwide. To maintain this position, the Company will strive to meet and exceed customers' expectations in the design, production and fulfilment of goods and services.

Information published by Jane's is renowned for its accuracy, authority and impartiality, and the Company is committed to seeking ongoing improvement in both products and processes, to match the International Standard ISO 9001 1994.

Jane's will at all times endeavour to respond directly to market demands and will also ensure that customer satisfaction is measured and employees are encouraged to question and suggest improvements to working practices.

Jane's will continue to invest in its people through training and development, to meet the Investor in People standards and changing customer requirements.

Jane's

RADAR SYSTEMS

Introduction

Note: *The following overview of radar is intended for the non-specialist reader. It is written in such a way as to be accessible and can in no way be taken as a full description of what is an extremely complex technology.*

Radar (standing for RAdio Direction And Ranging) functions by generating an output of **microwave** (see following) energy which is focused into a beam and illuminates an object in space. A small proportion of this energy is reflected back towards the radar where it is detected by an integral receiver. Microwave energy falls between infra-red radiation and radio waves within the electromagnetic spectrum and has **frequency** and **wavelength** values in the range 0.03 to 100 **Gigahertz** (abbreviated to **GHz**) and 1 m to 3 mm respectively. It should also be noted that the microwave segment of the electromagnetic spectrum is a subset of its **Radio Frequency** (abbreviated to **RF**) segment which has a frequency range of approximately 10^4 to 10^{11} Hertz (see following). Stepping back a stage, frequency may be defined as the number of complete oscillations or cycles of a periodic quantity in unit time and its unit of measurement — the **Hertz** (abbreviated to **Hz**) — as the frequency of a periodic phenomenon that has a period of one second. Within the RF spectrum, frequency values are usually expressed in **Kilohertz** (abbreviated to **kHz**), **Megahertz** (abbreviated to **MHz**) and the already noted GHz. These multiples represent 10^3, 10^6 and 10^9 Hertz respectively. Wavelength is defined as being the distance between two displacements of the same **phase** (that is, the stage or state of development of a regularly recurring quantity such as a radar pulse (see following)) along the direction of propagation. The most widely used military radar frequency/wavelength values are given in Table One.

Because of the low-power level of the reflected or 'echo' signal and its vulnerability to **modulation** (alteration of its characteristics) and interference, the receiver (see following) used in a radar system must be particularly sensitive. The reduction in signal power between transmission and reception is usually defined in terms of a ratio between the two (in the order of 10 to 17:1) and expressed logarithmically in **Decibels** (abbreviated to **dB**). Factors capable of modulating or interfering with the echo signal include:

- **Clutter** obscuration of the main echo signal by additional echo responses generated by surrounding ground features, the sea and rain. Alongside its obscurant effect, clutter can generate **false alarms** in a radar, that is, register what seems to be a valid target but is not. A similar effect can be created by atmospheric effects such as a localised air mass which is a different temperature or pressure from that of the surrounding atmosphere. Such a phenomenon can create phantom targets known as **angels**.
- Electronic **noise** inherent within the radar itself. To be detectable, the echo signal must typically be some 10 dB more powerful than the receiver's own noise level.
- Manmade interference known as **jamming** (see introduction to electronic warfare section).
- Apparent variations in the target's **Radar Cross-Section** (abbreviated to **RCS**) as perceived by the radar due frequency and time modulation effects. Radar cross-section may be defined as a measure of the size of radar response generated by a particular target geometry. This phenomenon is termed **scintillation**.
- Echo signal cancellation caused by reflections produced by flat surface reflectors such as a calm sea. Known as the **multipath** effect.

A list of the major military radar applications is given in Table Two.

Measurements produced by a radar comprise **range**, **bearing** and **elevation**. Target range can be either a **slant** or **ground** value. Slant range is the line of sight distance between the radar and the object illuminated while ground range is the horizontal distance between the emitter and its target and its calculation requires knowledge of the target's elevation. Radar range is established from that time delay experienced between the transmission of the emitter's output and the reception of the echo signal. This is achieved by measuring the particular time value against the known velocity of microwave energy (approximately 300,000,000 m/s). As a rough guide, every microsecond of delay between transmission and reception equals 150 m in range. Target bearing is its direction relative to the radar and is traditionally determined by the mechanical position of the antenna at the moment of reception. Target elevation is identified in a similar manner except for the angular measurement being in the vertical rather than the horizontal plane.

A radar's output can take a number of forms, the most usual of which is termed **pulsed**. In a pulsed radar, the output is transmitted in bursts of energy (individually termed **pulses**) with the echo signal being received during the interval between sequential transmissions. Another form of output is termed **Continuous Wave** (abbreviated to **CW**) which, as its name implies, is continuous. In its basic form, a CW radar is incapable of generating range information and its output must be modulated in some way if such a measurement is to be achieved. A common approach to this problem is **frequency modulation** with **frequency modulated CW** radars being used in applications such as altimeters, tracking radars and instrumentation equipments.

Power sources for modern radar systems most frequently take the form of a **magnetron** or a **Travelling Wave Tube** (abbreviated to **TWT**) or tubes. A basic magnetron consists of a central **cathode** (a negative electrode) surrounded by a cylindrical **anode** (a positive electrode) which is divided into segments or contains a number of **cavity resonators**, that is a space within a closed or substantially closed conductor which will maintain an oscillating electromagnetic field when suitably excited by an external force.

Functionally, a steady electrostatic field is applied between the magnetron's anode and cathode together with a steady magnetic field which is parallel to the device's cylindrical axis and orthogonal to the electrostatic field. Electrons emitted by the cathode are influenced by both fields and their interaction within the gaps or resonating cavities of the device's anode produces microwave frequency

oscillations. Currently, magnetron power sources are available in a number of configurations which include fixed frequency, tunable and **frequency agile**, that is, able to generate an output whose frequency is variable. Within the travelling wave tube, a beam of electrons is generated which continuously interacts with a radio-frequency electromagnetic field to produce amplification or, in some cases, oscillation at microwave frequencies. Like magnetrons, TWTs come in a number of configurations which currently include ring loop, ring bar and coupled cavity types.

A modern, mechanically **scanning** (use of a rotating antenna) pulse radar comprises:

- A **Transmitter** to provide the radar's output energy.
- **Modulation Circuitry** generating the output's operating frequency and **waveform**, that is, the configuration of the pulses used in the output.
- **Timing Circuitry** which manages the **duty cycle** between the radar's transmit and receive functions and **triggers** (activates) the equipment's transmitter and receiver at the appropriate times.
- A **Duplexer** which has the duel function of channelling the transmitter output to the radar's antenna with minimum power loss (see following) and no damage to the receiver and the returning echo signal to the receiver with maximum gain.
- A **Receiver** (usually of superheterodyne type) which is tuned to the frequency of the radar's transmitted signal and which detects the incoming echo response after it has received **Intermediate Frequency** (abbreviated to **IF**) amplification.
- An **Antenna** comprising, in most cases, a back plane reflector (which can be a solid parabolic dish or a shaped, openwork structure) on which a combined transmit and receive element is mounted. The architecture used is optimised to produce a concentrated **beam** of outgoing energy. Antenna movement is controlled by a dedicated drive mechanism and energy is fed to the transmit/receive element by either a coaxial cable or a **waveguide**. A waveguide is a hollow transmission line and can, in some cases, be used as an antenna in its own right (see following). While most modern emitters use dual function transmit/receive elements, the two can be separated with the resultant system being known as a **bistatic** radar.
- A **Signal Processor** which extracts the required data from the received echo signal.
- A **Display Unit** which provides the radar operator with a visual presentation of the data generated by the signal processor.

The concept of the radar beam is something of a misnomer as imperfections in antenna manufacture mean that alongside the desired **main** beam, a number of unwanted **sidelobes** are produced at angles offset from the main illumination. For a reflector antenna (of the type described previously), these effects typically start at about -40 dB near the main beam and fall away as the angular separation increases. Radar designers make strenuous attempts to minimise sidelobes as clutter and jamming returns collected by them can mask the wanted main beam signal. Antenna design is also important in minimising the width of the main beam being generated in order to maximise antenna **gain**. Gain can be defined as a measure of ability to increase the magnitude of a given electrical input parameter against a theoretical perfect system and is measured in dBs. In simpler terms, gain defines an antenna's ability to maximise main beam output in transmit mode and received signal strength in receive mode.

Radar antenna design is a broad church which, alongside the already noted reflector type, includes approaches such as the **Yagi**, **slotted waveguide**, **inverted cassegrain** and **phased-array**. Taking these in order, a Yagi antenna (named after its Japanese inventor, Professor Yagi) comprises (from the rear forwards) a reflector, a **dipole** transmit/receive element and a series of horizontally aligned **directors**. The distance between the dipole and reflector is equal to approximately half the wavelength of the radar's output and, the greater the number of directors used, the better the antenna's directivity. A current application of the Yagi antenna is in the US Navy's AN/APS-130/-139/-145 series of **Airborne Early Warning** (abbreviated to **AEW**) radars. A slotted waveguide antenna takes the form of a folded waveguide into which slots are cut and arranged in such a way as to produce a highly directional main beam output. An example of a slotted waveguide antenna application is the array used with the AN/APY-1/2 radar which is installed in the Boeing E-3 Sentry **Airborne Warning And Control System** (abbreviated to **AWACS**) aircraft.

In the inverted cassegrain design, the radar's output is transmitted from an element mounted in a parabolic reflector into a second reflector mounted in front of it. This bounces the energy back into the main reflector where it is reflected forwards. At this point, the energy's **polarity** (linear or circular) is switched through 90° which makes the forward reflector transparent to it and allows it to pass through it into space. Received energy goes through the same process but in reverse. Although complex, the inverted cassegrain antenna has the advantage of good sidelobe performance (and, therefore, high resistance to jamming and clutter) in a relatively small and lightweight package. An example of the use of such an antenna is in the GEC-Marconi Radar and Defence Systems' Foxhunter radar which is fitted to the Tornado F Mk 3 interceptor.

A phased-array antenna uses a flat plane **planar** arrangement of rows and columns of equally spaced radiating elements. Each radiating element offers a similar output to its neightbours and is designed in such a way as to avoid two elements **coupling** their outputs. The width of the main beam produced by such a system depends on both the spacing of the individual elements within the array (typically, half the wavelength value of the radar's output) and an inverse relationship between beamwidth and the number of elements used (that is, the greater the number of elements, the narrower the beam produced). Beam steering is facilitated through the use of what are termed **phase shifters**. When the phase shifter controlling an individual element or group of elements is set at zero, the output beam is transmitted at right angles to the element or elements. Altering the shifter settings in a particular pattern across the array allows the beam to be 'steered' in the direction required without mechanical movement of the antenna plane. In practice, 'steerability' in phased arrays is restricted to approximately ±60° and requires some mechanical array movement to provide all angle coverage.

One application of this type of technology is used to create what is known as a **passive multifunction radar**. Here, the equipment has a similar function to a conventional surveillance radar but with the mechanical scanned reflector antenna replaced by a planar phased-array. Equally, the equipment can be made to provide coverage in **three dimensions** (abbreviated to **3-D** and representing the ability to deduce target range, bearing and elevation; a **two-dimensional** or **2-D** radar can only provide range and bearing data) by a judicial mix of electronic beam-steering and mechanical movement. Such a capability does away with the need for a supporting height-finding radar but is generally not accurate enough in elevation to act as a fire-control system. Again, the use of electronic beam-steering allows the radar to perform surveillance and multiple target tracking tasks in time sequence within a single unit.

Moving a stage further, an **active array multifunction radar** replaces the common transmitter chain which feeds the radiating elements in the described passive equipment with several thousand individual duplexed transmit/receive modules. Each of these independent units takes the form of a **solid-state Microwave Monolithic Integrated Circuitry** (abbreviated to **MMIC**) which utilises semiconductor material such as **Gallium Arsenide** (chemical symbol **GaAs**) and generates a typical output of around 10 W. Use of such units virtually eliminates loss of microwave energy and further advantage can be gained from applying digital adaptive beam-forming techniques to groups of modules within the array. Using this technique, each sub-array of perhaps 100 transmit/receive units, can be **weighted** in **amplitude** (the peak value of an alternating entity such as a radar pulse in both the positive and negative directions) and phase of the described transmit/receive units, the active array and passive multifunction radars function in a similar manner.

Alongside the traditional form of technology already discussed, the reader will encounter a number of other manifestations which require explanation. Among these, four will probably be encountered most frequently, namely **Synthetic Aperture, Moving Target Indicator, Over-The-Horizon Backscatter** (abbreviated to **SAR, MTI** and **OTH-B** respectively) and **Monopulse** radars. Taking these in order, an SAR radar is an airborne system which utilises the flight path of the aircraft to simulate an extremely large antenna or **aperture**. Over time, individual transmit/receive cycles are completed with the data from each cycle being stored electronically. After a given number of cycles, the stored data is recombined (taking into account the Doppler effects (see following) inherent in the differing transmitter to target geometry experienced in each succeeding cycle) to create a high-**resolution** (that is, a measure of the radar's ability to discern objects of a given size) picture of the terrain being overflown. Using such a technique, radar designers are able to achieve resolutions which would require **real aperture** antennas so large as to be impractical with arrays ranging in size from the 7 m long unit used in the American **Joint Surveillance Target Attack Radar System** (abbreviated to **Joint STARS**) to modified fighter radars. SAR radar is partnered by what is termed **Inverse SAR** (abbreviated to **ISAR**) technology which in the broadest terms, utilises the movement of the target rather than the emitter to create the synthetic aperture. ISAR radars have a significant role aboard maritime patrol aircraft to provide them with radar imagery of a sufficient quality to allow it to be used for target recognition purposes.

Moving target detection is, as its name suggests, the domain of the MTI system which relies for its effect on the **Doppler** shift in frequency which occurs when transmitted radar energy is reflected by a moving target. The Doppler effect is best illustrated by the analogy of the changing pitch of a railway engine (locomotive) whistle as heard by an observer standing by the track. As the engine approaches, the pitch of the whistle appears to be higher than it actually is, correct when it is alongside the observer and lower when it has passed by. Equally, the greater the speed of the engine, the greater is the shift in the pitch of the whistle. In RF terms, the Doppler shift in frequency is directly proportional to the target velocity component towards the radar and the operating frequency of the emitter.

MTI function is essentially that of a filter which removes unwanted low-velocity components of clutter and passes only those returns coming from a moving target. The sensitivity of the technique is a direct function of the radar's **pulse repetition frequency** (abbreviated to **PRF**), that is, the number of pulses occurring in a second. The higher the PRF value, the better is the radar able to reject clutter signals. Within MTI technology, there is a point at which target speed induces a Doppler effect which produces a 360° phase change which prevents the target echo being separated from its clutter background. Known as **blind velocity**, the problem can be overcome by using staggered PRFs where the blind velocity value is different for each transmitted pulse. While some PRF values will still be 'blinded', others will not, allowing the MTI function to continue. Equally, where the target being tracked has a low RCS and generates a minimal Doppler effect, a **range ambiguous** system can be employed which uses a high pulse on target rate but where the time interval between the transmission of each pulse (known as the **pulse repetition interval** and abbreviated to **PRI**) is less than the **radar time delay** (the time for a pulse to reach the target and return to the radar receiver) of the particular target. It is also worth noting here that the length of time the transmitter is switched on to produce a single pulse is known as the **pulse-width** (abbreviated to **PW** and measured in microseconds). The described MTI function can also be taken as being almost synonymous with what are termed **pulse Doppler** radars which use the clutter rejection/target velocity detection capabilities achievable through Doppler filtering to create 'look-down/shoot-down' systems for interceptor aircraft.

OTH-B radars operate in what is known as the 3 to 30 MHz **high frequency** (abbreviated to **HF**) portion of the RF spectrum and make use of the fact that the ionosphere reflects signals transmitted within this part of the spectrum rather than allowing them to pass through and be dissipated in space. Accordingly, such radars are able to detect objects at much longer ranges (perhaps in the order of 2,000 km) than is possible with a microwave radar. To date, OTH-B radars have been deployed or developed by Australia, the USA and Russia and usually take the form of separate transmit and receive arrays located at different sites. While the ability to detect objects at such distances is obviously of considerable value, OTH-B technology is difficult to manage. A key element here is the frequency with which the ionosphere's reflectivity changes and its vulnerability to interference by naturally occurring auroral activity. A second approach to this type of technology utilises the **ground wave** effect whereby an HF band RF frequency signal can attach itself to a conductive surface such as the sea and provide an over-water detection capability at ranges of up to 350 to 400 km. Such technology has been developed in Canada and the UK and can be termed **HF Ground Wave Radar** (abbreviated to **HF-GWR**) or **OTH Surface Wave** (abbreviated to **OTH-SW**) radar. It should also be noted that the described OTH technologies should not be confused with **OTH targeting** where a conventional airborne radar is used to extend the radar horizon of a surface platform beyond that of its onboard sensors. As microwave RF energy (with the exceptions already noted) travels in straight lines, detection range is restricted by the curvature of the earth's surface. Overcoming this requires the elevation of the emitter above the earth's surface, with the increase in detection range being directly proportional to the radar's altitude.

The remaining emitter type cited — monopulse radar — is widely used in precision tracking systems and is currently held to be a key electronic warfare target. Such a system uses a main beam output which is divided into four segments each of which has a different polarity. Target tracking is achieved by comparing the amplitude or phase of the echo signal as perceived by each of the four segments within the transmission beam.

Being a complex technology, radar design requires consideration of a wide range of factors if the end product is to be effective in its role. Considerations (in no particular order of importance) include:

Operating Frequency

Factors here include increased susceptibility to clutter interference, limited detection range and a fall off in transmitter efficiency at the high end of the spectrum as against poor angular resolution and large antenna size at the lower end. As a rough guide, current land-based and shipborne air surveillance radars operate across the B- through G-band section of the spectrum (see Table One for band definitions); ground-based battlefield surveillance and fire-control radars in the B- through M-bands; naval fire-control and navigation radars in the G- through J-bands (with an emphasis on I-band); airborne surveillance radars in the B- through F-band and airborne fire-control radars in the I/J-band. It is also worth noting the increasing use of **frequency-agile radars** which are able to switch frequencies on a pulse-by-pulse or batch-of-pulses by batch-of-pulses basis as a counter jamming measure.

Transmitter Power

Available transmitter power is a major factor in a radar's **aperture product** value (that is, the amount of power radiated multiplied by the effective aperture of the antenna) which is a prime indicator of the emitter's potential target detection performance. As a rough rule of thumb, the smaller the target and the smaller the emitter's antenna, the higher the power level needed to maintain performance. By way of example, a medium-range missile system radar might be able to detect a 1 m² target with a **mean** (average) radiated power of 20 W. When target size is reduced to 0.001 m², this value would have to rise to 20 kW to maintain the same level of detection capability.

Waveform

Selection of an appropriate waveform is crucial in achieving radar performance. By way of example, a pulse Doppler radar, while excellent as a means of extracting a target from a clutter background, cannot resolve range unambiguously because of the high PRF values used (see previously). One way round this is to **code** or **compress** the pulses (by modulating their frequency or phase) so that those pulses forming the echo signal can be readily distinguished from those forming the transmitter output. Equally, **multimode radars** require different waveforms for different tasks and skilful manipulation of the waveform is a major aid to countering jamming. Aside from pulse compression/coding and the already noted CW format, other examples of currently used waveforms include **jitter** (a short duration instability in either the amplitude or phase of the signal), **coherent** (a waveform in which all the pulses have a stable phase relationship to one another) and **non-coherent** (the reverse of coherent).

PRF

PRF is a major factor in determining a radar's detection range and ability to detect targets within clutter. Accordingly, a radar with a low PRF value provides unambiguous range data but is highly susceptible to clutter interference, while an emitter with a high PRF value provides unambiguous velocity information but second order range resolution with inherent inaccuracies. A medium PRF value provides some of the advantages of both the high and low values but requires high-order processing to resolve its inherent ambiguities.

Bandwidth

A radar signal is made up of what is termed a **carrier wave** on to which the desired characteristics are imposed. Accordingly, because the signal is made up of a number of components, it actually contains a range of frequencies grouped around that of the baseline carrier wave. The extent of this range is termed the radar's bandwidth. An example of the impact of bandwidth is the **spread-spectrum radar** which uses an ultra-wide bandwidth (so wide in fact that it frequently appears to be electronic noise rather than a modulated waveform) to minimise detection by a hostile radar warning receiver. Radars which incorporate features such as spread-spectrum, pulse compression or low-output power levels to minimise detectability are termed **Low Probability of Intercept** (abbreviated to **LPI**) emitters.

Scan Pattern

In non-electronically scanning radars, a range of patterns of antenna movement are used to maximise the suitability of the particular radar for its purpose. Commonly used scan patterns include:

- **Circular** in which the antenna rotates through 360° in azimuth continuously.
- **Conical** in which the antenna traces a cone pattern around its central axis. Used in tracking radars with target azimuth and elevation being taken from the mechanical position of the antenna.
- **Conical Scan On Receive Only** in which a conical scan pattern is used while the radar is in receive mode only. Abbreviated to **COSRO**.
- **Frescan** in which the pattern is produced by successive beams (each on a different bearing) which are made to overlap by stepped changes in frequency.
- **Helical** in which the antenna's movement takes the form of a rising and falling helix around the antenna boresight.
- **Lobe On Receive Only** in which the echo signal is sampled at four positions around the antenna boresight while the radar is in receive mode only. Abbreviated to **LORO**.
- **Palmer** in which a circular pattern of movement is imposed on another scan type such as conical or raster.
- **Raster** in which the antenna follows a continuous rectangular pattern of movement which expands and contracts to give area coverage. Used primarily for target acquisition in air-to-air applications.
- **Spiral** in which the antenna follows a spiral pattern in the vertical plane. This approach provides range and relative azimuth/elevation data and is used primarily for target acquisition.
- **Track-While-Scan** in which two unidirectional **sector** (specified area) scan patterns provide simultaneous coverage in the vertical and horizontal planes. Abbreviated to **TWS**.
- **Track-While-Scan On Receive Only** in which TWS is used while the radar is in receive mode only. Abbreviated to **TWSRO**.
- **V-Beam** in which separate radar transmitters feed back-to-back antennas which produce a vertical fan beam and one which has an inclination of approximately 30°. Target elevation is deduced by comparing the differences in echo response experienced by the two antennas. Scan patterns for the individual beams may be circular or over a defined sector.

As a final point in the introductory survey, readers should be aware of **Identification Friend-or-Foe** (abbreviated to **IFF**) or **Secondary Surveillance Radars** (abbreviated to **SSR**). While radars can locate targets in space, they are, for the most part (see note on ISAR systems) unable to distinguish between friendly or hostile contacts. This is achieved by using an IFF/SSR subsystem which transmits an interrogator signal which activates a **transponder** (a combined transmit/receive device which automatically broadcasts a signal when it receives an appropriate trigger signal) aboard the friendly vehicle. The IFF output is coded in some way and shows up as a distinct and recognisable response on either the radar's main display or a co-located dedicated presentation unit. To complete the survey, an explanation of the American 'AN' system of electronic equipment alphanumeric identifiers is given in Table Three.

Table One: Military Radar Frequency/Wavelength Values

Military radar frequencies are described by two concurrent systems of 'bands', the first of which has evolved from the system used during the Second World War. During 1969, NATO (the North Atlantic Treaty Organisation) adopted a second system which originated within the Electronic Warfare (EW) community and is still, somewhat confusingly, referred to as the 'EW' system within some quarters. Both systems are presented here for completeness although the body of *Jane's Radar and Electronic Warfare Systems* uses the NATO/EW values throughout.

Historical Radar Bands			NATO/EW Bands		
Band Designation	Frequency	Wavelength	Band Designation	Frequency	Wavelength
VHF[1]	0.03-0.3 GHz	1,000-100 cm	A	0.03-0.25 GHz	1,000-120 cm
UHF[2]	0.3-1 GHz	100-30 cm	B	0.25-0.5 GHz	120-60 cm
L	1-2 GHz	30-15 cm	C	0.5-1 GHz	60-30 cm
S	2-4 GHz	15-7.5 cm	D	1-2 GHz	30-15 cm
C	4-8 GHz	7.5-3.75 cm	E	2-3 GHz	15-10 cm
X	8-12 GHz	3.75-2.5 cm	F	3-4 GHz	10-7.5 cm
Ku	12-18 GHz	2.5-1.6 cm	G	4-6 GHz	7.5-5 cm
K	18-27 GHz	1.6-1.1 cm	H	6-8 GHz	5-3.75 cm
Ka	27-40 GHz	1.1-0.75 cm	I	8-10 GHz	3.75-3 cm
MM[3]	40-100 GHz	0.75-0.3 cm	J	10-20 GHz	3-1.5 cm
			K	20-40 GHz	1.5-0.75 cm
			L	40-60 GHz	0.75-0.5 cm
			M	60-100 GHz	0.5-0.3 cm

Key
[1] Very High Frequency   [2] Ultra High Frequency   [3] Millimetric

Table Two: Major Military Radar Applications

Air Surveillance	Ground-controlled interception (abbreviated to 'GCI')
	Height-finding
	Target acquisition
	Long-range early warning (including Airborne Early Warning - abbreviated to 'AEW')
	Air traffic management (abbreviated to 'ATM') functions
Tracking and Guidance	Fire control for Anti-Aircraft Artillery (abbreviated to 'AAA'), naval gunfire, land artillery and mortars and missile systems
	Missile guidance
	Precision landing and approach systems
	Range instrumentation
Surface Search and Battlefield Surveillance	Sea search and navigation
	Harbour and waterway management
	Ground mapping
	Mortar and artillery location
	Intruder detection
	Mine detection
	Airfield taxiway management
Space and Missile Surveillance	Ballistic missile warning and acquisition
	Satellite surveillance
Weather Surveillance	Observation/prediction of wind and precipitation conditions
	Weather avoidance (for aircraft)
	Clear-air turbulence detection

Table Three: The US 'AN' Electronic Equipment Identification System

Starting during the Second World War, the US military established the 'Army-Navy' (abbreviated to 'AN') system for the identification of its electronic equipment. Currently, this system is applied to electronic equipment taken into the inventories of the USAF, US Navy, US Army and USMC. 'AN' designations are also increasingly applied to military electronic systems of American origin but which have not been taken up by the country's military. 'AN' identifiers are presented thus: **AN/APG-66(V)3** in which **APG** identifies the equipment's installation type (first letter), the equipment's type (second letter) and its purpose (third letter); **-66** indicates that the particular equipment is the 66th of its type included in the system; **(V)** indicates that the equipment can be configured to suit a number of platforms and/or system applications and **3** indicates that it is the third such variable configuration produced. The initial installation/type/purpose group can be read as follows:

Installation Identifier		Type Identifier		Purpose Identifier	
A	Piloted aircraft	A	Invisible light/heat radiation	B	Bombing
B	Underwater mobile/submarine	C	Carrier	C	Communications
D	Pilotless carrier	D	Radiac	D	Direction-finding/surveillance
F	Fixed ground	G	Telegraph/teletype	E	Release/ejection
G	General ground use	I	Interphone/public address	G	Fire control
K	Amphibious	J	Electromechanical/inertial wire covered	H	Recording/reproduction
M	Ground mobile	K	Telemetry	K	Computing
P	Portable	L	Countermeasures	M	Test/maintenance
S	Water	M	Meteorological	N	Navigation
T	Ground Transportable	N	Sound in air	Q	Special purpose
U	General utility	P	Radar	R	Receiver
V	Ground vehicular	Q	Sonar/underwater sound	S	Search/detection/range bearing
W	Water (surface/subsurface applications combined)	R	Radio	T	Transmitting
Z	Unmanned/piloted air vehicle combination	S	Special/combination of purposes	W	Automatic flight/remote control
		T	Telephone (wire)	X	Identification/recognition
		V	Visible light	Y	Surveillance and control
		W	Armament		
		X	Facsimile/TV		
		Y	Data processing		

Accordingly, **AN/APG-66(V)3** can be identified as the third variable configuration subvariant of the 66th airborne fire-control radar identified within the system. **AN/APY-1** identifies the equipment as being the system's first registered airborne surveillance and control radar while **AN/APX-109(V)** represents its 109th airborne radar identification/recognition system which, like APG-66(V), can be configured for variable applications. The reader should also be aware of the use of a suffix letter to identify succeeding generations of equipment. Thus, **AN/APX-12A** identifies the second-generation of the system's 12th ground mobile radar identification/recognition equipment to be produced.

VERIFIED

LAND-BASED AIR DEFENCE RADARS

AUSTRALIA

Jindalee Operational Radar Network (JORN)

Type
Over-The-Horizon (OTH) backscatter High Frequency (HF-3 to 30 MHz) radar.

Development
Development by the Defence Science and Technology Organisation (DSTO) of the Jindalee system started in the 1970s with the Alice Springs, Northern Territory station. By 1978 the first stage of the project had been completed, using a narrow fixed beam bounced off the ionospheric 'F' layer giving a detection range of 2,000 km. The success of these trials led to Stage B in 1979, to provide a facility for steerable beams over a northwesterly sector. The Stage B radar system used two sites, a transmitting station at Harts Range, near Alice Springs and a receiver at Mount Everard.

During February 1984, Amalgamated Wireless Australia (now British Aerospace Australia) and Computer Sciences of Australia were awarded a contract to examine the feasibility of converting the Jindalee Stage B into an operational system, with the same contractors receiving additional contracts (valued at A$70 million) to undertake the actual modifications between 1986 and 1989. During October 1986, the Australian government announced its intention of establishing an operational OTH radar network which was to be known as the Jindalee Operational Radar Network (JORN). Further testing of the Alice Springs system (involving a variety of airborne target types operating in a variety of environmental and ionospheric conditions) during the period 1987-89 is described as being highly successful. Development and implementation of JORN was put out to tender and in 1991, Telecom Australia (now the Telstra Corporation) was awarded the programme together with a four year maintenance and support contract once the network was up and running. During 1992, the Alice Springs system was commissioned as the Royal Australian Air Force's (RAAF) No 1 Radar Surveillance Unit with the Alice Springs site functioning as both an operational sensor and as a continuation research and development tool.

As prime contractor (see Status), Telstra subcontracted major parts of the necessary work to a range of companies including GEC-Marconi (now Marconi Electronic Systems - digital signal processing software), Radio Frequency Systems (design, manufacture and installation of antennas), DEC (computer hardware) and Telstar Systems (a Telstra/Lockheed Martin joint venture — system software).

Description
Full implementation of JORN involves the construction of two OTH backscatter radars (which are similar in principle to the US ROTHR 2000 system — see separate entry) which will operate in the HF band (3-30 MHz) and will be able to detect airborne and ship targets at ranges of between 1,000 and 3,000 km with a range resolution of 20 to 40 km. Roles envisaged include national defence, the provision of early warning of storm activity, coast watching and the provision of ionospheric data for HF communications. Each JORN radar will incorporate bistatic transmission and reception subsystems geographically separated to prevent mutual interference. Each transmission subsystem will comprise 28 transmitter chains, each of which will incorporate a 20 kW power amplifier. The receiver subsystem near Longreach, Queensland will utilise 480 receiver chains while the one near Laverton, Western Australia will incorporate 960 such chains.

The Longreach site will be equipped with a 0.4 km transmission array while that at Laverton will measure 0.8 km. The Longreach 3 km reception array will be positioned some 100 km from its associated transmitter while the 6 km Laverton receiver will be 85 km distant from its transmission source. The two complexes will offer 90° and 180° coverage respectively. An integral frequency management

system will determine which frequency within the operating band will yield the best signal-to-noise ratio while spectrum and noise monitors will identify clear channels and background noise levels. A backscatter sounder will be used to monitor ionospheric propagation characteristics in the target area and operating frequency selection will be made on the basis of independent data from a low-powered mini radar and a passive channel evaluator. General ionospheric structure characteristics and target ground truths will be obtained via a network of vertical and oblique sounding facilities and transponders located along Australia's northern coastline. A JORN Co-ordination Centre (JCC) will be located at the RAAF base Edinburgh near Adelaide, South Australia. Here, the received data will be processed into usable tracking data. Additionally, the JCC will house a JORN software support and training facility.

Status
As of this issue, all buildings required for the JORN JCC, transmit site and receiver station are reported to have been completed by August 1996. By the same date, the necessary antenna installation and cabling work is noted as having been completed alongside the testing and commissioning of the system's transmitter array at Longreach, Queensland. During 1997, project management of the system was transferred from the Telstra Corporation to RLM Systems (a joint venture between US contractor Lockheed Martin and Australia's Tenix Defence Systems). Sources suggest that RLM received in excess of then year A$600 million as part of the transfer of responsibility. In Febuary 1999, RLM also took over development of the system's software from UK contractor Marconi Electronic Systems, with the latter reported as foregoing then year A$75 million for work already undertaken as part of the deal. As of July 1999, the Longreach array was scheduled to begin air target tracking tests during September 1999. During this test programme, the radar was expected to monitor a region of airspace out to ranges of excess of 3,000 km. As of this edition, final delivery of the JORN system was scheduled for late 2001/ early 2002 (as against an original date of mid 1997).

Contractor
RLM Systems- project manager.

VERIFIED

CHINA, PEOPLE'S REPUBLIC

146-1 target indication radar

Type
UHF band (0.3 to 1 GHz) 3-D mobile Surface-to-Air Missile (SAM) surveillance/ tracking radar.

Description
The 146-1 radar is a mobile target indication radar for use with SAMs. It provides air situation surveillance, multitarget tracking, real-time data processing, optimum target assignment, firing data computation and co-ordinate transformation. The system consists of a trailer for the antenna and transceiver/receiver unit, an electronics van and three vans each with a 75 kW power source. The complete system can be assembled or disassembled in approximately 2 hours. Operation and maintenance personnel strength is given as 12 to 14.

The design features include:
- fully solid-state and electronic phase-scanning system
- array antenna with distributed active solid-state modules
- fully coherent digital moving target indicator with high performance against ground clutter and passive electronic countermeasures
- good electronic counter-countermeasures performance which is achieved by adaptive wideband pulse-to-pulse frequency agility
- low-elevation angle height-finding techniques
- microprocessor-controlled data processing and management
- customer-oriented datalinks and optional interfaces.

Status
As of this edition, 146-1 is thought to be in service with People's Liberation Army (PLA) ground-to-air missile regiments.

Specifications
Frequency: UHF band (0.3-1 GHz)
Antenna: aperture 6.7 × 6.5 m
SCV: >30 dB
Surveillance coverage: 360° (azimuth); 0-65° (elevation); 25 km (altitude)
Range: 15-200 km
Max tracking range: 200 km (2 m² target)
Accuracy: 1° (azimuth); 2° (elevation); 500 m (range)
Range resolution: 120 m (within 90 km)
Data rate: 6 or 12 rpm
Tracked target numbers: 32 batches
ECCM capability: pulse-to-pulse and burst-to-burst frequency agility; DMTI; pulse compression

Prototype JORN array at transmit site, Longreach

The 146-1 surveillance and tracking radar

System reliability: MTBF 120 h (excl computer); MTTR 0.5 h
Antenna operation windspeed: ca 19 m/s

Contractor
Nanjing Research Institute of Electronics Technology, Nanjing.

VERIFIED

HN-401R mobile surveillance radar

Type
UHF band (0.3 to 1 GHz) 2-D mobile air defence radar.

Description
The HN-401R is a mobile long-range air defence 2-D surveillance radar. It consists of a large horizontally polarised dipole planar-array with an aperture of 17 × 8 m, an operations cabin containing all the electronics plus two display positions and a power cabin. The system employs fully coherent pulse compression techniques and uses a highly stable frequency source. The signal processing system is able to adapt to pulse-to-pulse or burst-to-burst frequency agility and moving target indicator. Low-azimuth sidelobes are achieved by feeding unequal power. The HN-401R comprises three units: an operations shelter, an antenna trailer and a prime power shelter and is transportable by road or rail.

Status
As of this edition, HN-401R is no longer in production but is thought to remain in service.

Specifications
Operating frequency: UHF band (0.3-1 GHz)
Antenna
Type: dipole planar-array
Aperture: 17 × 8 m
Gain: 24 dB
Polarisation: horizontal
Transmitter: coherent amplification chain
Output power: 20 kW (peak); 2 kW (average)

The antenna used in the HN-40IR surveillance radar system

Range: 300 km (1.5 m² target)
Resolution: 450 m (range); 9° (azimuth)

Contractor
China National Electronics Import and Export Corporation, Beijing.

VERIFIED

JLG-43 height-finding radar

Type
E-band (2 to 3 GHz) mobile height-finding radar.

Description
The JLG-43 is a nodding height-finder which follows the pattern of the well-known Soviet Cake series of equipments. It is a mobile system in as much as it is readily transportable by military road vehicles and the complete system can probably be carried in a two-truck load.

Operating frequencies are in the E/F-band and a 2 MW transmitter provides for coverage out to a range of 200 km and gives a height coverage of up to 25,000 m.

Status
As of this edition, JLG-43 is no longer in production. Outside China, JLG-43 radars are reported to have been supplied to the Myanmar (Burmese) Air Force.

Specifications
Frequency: E-band (2 GHz)
Pulse-width: 3 μs
Peak power: 2 MW
Nodding frequency: 32.37 cycles/min
Antenna dimensions: 5.5 × 2.13 m
Range: 200 km
Height: 25,000 m
Elevation coverage: 0-30°
Accuracy: 300 m (height); 2° (azimuth)

Contractor
China National Electronics Import and Export Corporation, Beijing.

VERIFIED

The JLG-43 height-finding radar

JLP-40 surveillance radar

Type
E/F- (2 to 4 GHz) and D-band (1 to 2 GHz) tactical air defence radar.

Description
The JLP-40 is a tactical air defence radar designed for use with height-finder radar (s) for Ground Controlled Interception (GCI) or similar applications. It is similar in design to the Russian Federation and Associated States Bar Lock radar from which it has been derived. It features the same arrangement of two large scanners attached to front and rear sides of a rotating cabin that houses the transmitter/receivers.

Operation is in the E/F- and D-bands, each parabolic scanner being illuminated by stacked horn feeds to generate families of multiple beams. There are five E/F-

The JLP-40 surveillance radar

band transmitter/receivers and these are thought to operate with the lower of the two antennas, while the D-band feeds illuminate the upper scanner. An Identification Friend-or-Foe (IFF) interrogator is also incorporated in the system. Three Plan Position Indicator (PPI) displays are provided and one azimuth/range display, plus between two and four additional PPIs. Moving Target Indicator (MTI) facilities are provided in the D-band transmitter/receiver chains.

The whole system is transportable, though hardly mobile despite the use of wheeled trailers for much of the equipment. Clearly, preparing such a large and complex system for operation must occupy several hours from the time of arrival at a surveyed site.

Status
As of this edition, JLP-40 is no longer in production but probably remains in national service. JLP-40 radars are also reported to have been supplied to the Myanmar (Burmese) Air Force.

Specifications
Operating frequencies: E/F- (2-4 GHz) and D- (1-2 GHz) bands
Range: 270 km
Altitude: 20,000 m
Azimuth coverage: 360°
Elevation coverage: 0.5-30°
Accuracy: 500 m (range); 0.5° (bearing)

Contractor
China National Electronics Import and Export Corporation, Beijing.

VERIFIED

JY-8A tactical radar

Type
Short-range 3-D mobile tactical radar.

Description
The JY-8A is a development of the JY-8 system but differs from it in a number of ways. It is designed to provide precise 3-D information on targets and to carry out fire control of missiles and/or anti-aircraft guns. The main differences from the JY-8 system are:
- only one transmitter is employed, the detection coverage is smaller and the height accuracy lower
- frequency diversity techniques are not incorporated
- twelve reception beams are employed, the elevation range being from 0.5 to 30°
- the antenna/transceiver shelter is lighter and more mobile.

The JY-8A 3-D mobile tactical radar

Other than the above differences, the JY-8A operates on the same principles as the JY-8 and is physically identical.

The system is composed of four transportable units: an antenna/transceiver shelter trailer, an operations/reporting shelter truck, a maintenance shelter truck and a tractor truck.

Status
As of this edition, JY-8A is thought to be in service.

Specifications
Antenna beamwidth: 0.55° (horizontal); 0.9° (vertical)
No of beams: 16 (transmit); 12 (receive)
Transmitter output: 800 kW (peak)
Pulse-width: 3-3.3 μs
PRF: 300 Hz
Range: 150 km
Height: 12 km
Elevation: 30°
Detection accuracy: 500 m (range); 0.3° (azimuth); 700 m (height)
Target resolution: 0.6° (azimuth); 1,000 m (range); 0.9° (elevation)
ECCM: frequency agility, CFAR, low sidelobes, blanking of jammed beams
Display: 2 PPIs (31 cm)
MTBF: more than 150 h
MTTR: < 30 min

Contractor
China National Electronics Import and Export Corporation, Beijing.

VERIFIED

JY-9 low-altitude search radar

Type
E-band (2 to 3 GHz) low-altitude search radar.

Description
The JY-9 is a mobile E-band low-altitude search radar intended for use in air defence, gap-filling, airport surveillance and coastal defence. It is designed for the effective detection of targets at low altitude in both electronic countermeasures and natural clutter environments. It consists of an antenna, a radar/operations shelter and a small power trailer and is transportable by air, rail and sea. Setting up or disassembly time with an experienced crew of four men is 20 minutes.

The JY-9 employs a dual beam antenna assembly consisting of a deformed parabolic reflector, two horns and two feed channels. The upper and lower beams cross at 5° elevation. One channel is used for transmit, both are used for receive. The antenna has an aperture of 4.14 × 6.96 m, rotates at 6/12 rpm to provide 360° surveillance in azimuth and can be mounted on the shelter, on the ground or on a steel tower.

The JY-9 has high anti-jamming and anti-clutter capability, including pulse-to-pulse frequency agility, dual channel, pulse compression, a wide operating band, low sidelobes, moving target detection automatic spectrum processing, automatic clutter map and automatic residue map. Mean time between failure is better than 500 hours, a built-in test subsystem is included and automatic fault detection reaches the level of functional modules.

Status
As of this edition, JY-9 is thought to be in service.

Specifications
Frequency: E-band (2-3 GHz)
Antenna: dual beam with cosec^2 pattern
Rotation speed: 6/12 rpm
Beamwidth: 1.5° (horizontal); 40° (vertical)
Polarisation: alternating between linear and circular

The JY-9 low-altitude search radar

Transmitter
Power output: 200 kW (peak); 3.4 kW (average)
Pulse-width: 20 µs
PRF: 850 Hz
Coverage: 360° (azimuth); 0-40° (elevation)
Height: 10 km
Range: 150 km (Swerling I, 2 m²)
Detection accuracy: 80 m (range); 0.4° (azimuth)
Resolution: 200 m; 1.3° (azimuth)
Tracking capacity: ≥ 200 targets or groups of targets

Contractor
ECRIEE (East China Research Institute of Electronic Engineering), Hefei.

VERIFIED

JY-9F surveillance radar

Type
E/F-band (2 to 4 GHz) ground-mobile, low-altitude surveillance radar.

Description
A development of the JY-9 system (see separate entry), JY-9F is a ground-mobile equipment that is designed for national air defence, anti-aircraft artillery and air traffic control applications. The radar is deployed in a two truck, two trailer convoy and features a dual-beam modified CSC² antenna; coherent, four phase coded pulse compression; an electronic counter-countermeasures package that includes low-sidelobe values (less than -34 dB), frequency agility, coded waveform agility, stable frequency synthesis and a coded pulse anti-clutter subsystem and built-in test. Polarisation is selective linear or circular.

Status
As of this edition, JY-9F is understood to be available.

Specifications
Frequency: 2-4 GHz
Range: > 150 km (P_{-d} 80%, P_{-fa} σ X 2)
Altitude coverage: 10,000 m (P_{-d} 80%, P_{-fa} 10^{-6} σ 2 m²)
Accuracy: 0.3° RMS (azimuth); 60 m RMS (range)
Resolution: 1.4° (azimuth at 50%); 120 m (range at 50%)
Antenna rotation speed: 6 rpm
Target processing: > 65 tracks/10 s
MTBF: > 400 h
MTTR: < 0.5 h
Set-up/tear-down time: < 1 h

Contractor
ECRIEE (East China Research Institute of Electrical Engineering), Hefei.

VERIFIED

JY-10F Radar Information Processing Post (RIPP)

Type
RIPP and automatic signal/data processing equipment.

Description
Based on the earlier JY-10 system and customer requirements, the JY-10F RIPP can be accommodated in a command centre or be shelter-mounted and is designed to:
- handle data from up to six radars
- undertake target plot/track processing and data fusion
- function as a radar operations centre for the creation of a local air situation picture

Alongside the baseline configuration, ECRIEE has developed the JY-10(F)B and JY-10F(I) variants. Of these, the single cabinet JY-10F(B) system is designed to provide non-digital radars with an automatic signal/data processing capability and facilitate their integration into an analogue-digital sensor network. The twin cabinet JY-10F(I) configuration is designed to link up to four radar heads that otherwise could not be integrated into a network. The equipment provides automated signal/data processing and is capable of operating in one of four modes. With the addition of an auxiliary height-finding unit (with an integrated height-finding radar), both JY-10F(B) and JY-10F(I) can generate a composite 3-D data output from a 2-D radar network.

Status
As of this edition, JY-10F series equipments are understood to be available.

Specifications
Data fusion capability: 6 radar sources (maximum)
PRF range: 100-1,200 Hz
Pulse-width range: ≥0.5 µs

Antenna rotation rate: 3-24 rpm
Processing capacity: =4,000 plots/10 s
Processing loss: ≤3 dB
False alarm rate: =5 plots/s
MTBF: =5,000 h
MTTR: =30 min

Contractor
ECRIEE (East China Research Institute of Electronic Engineering), Hefei.

VERIFIED

JY-11 surveillance radar

Type
Ground-based, 3-D low-altitude surveillance radar.

Description
JY-11 is a solid-state, ground-mobile surveillance radar that is designed for autonomous function or as a gap filler in a regional air defence network. System features include a low-sidelobe, dual-frequency, frequency scanning antenna; a modular, solid-state transmitter; a large dynamic range receiver; digital waveforming; pulse compression; a constant false alarm rate and automatic set up/tear down.

Status
As of this edition, JY-11 is understood to be available in both ground-mobile and fixed-site configurations.

Specifications
Range: 180 km
Altitude coverage: 12,000 m
Angular coverage: 0-30° (elevation); 360° (azimuth)
Accuracy: 0.5° RMS (azimuth/elevation); 100 m (range)
Antenna rotation speed: 6 rpm
Target processing: 128 tracks/10 s
Anti-clutter performance: ≥ 45 dB (ground clutter); ≥ 30 dB (chaff)
MTBF: ≥ 1,000 h
MTTR: ≥ 0.5 h
Set up/tear down time: ≥ 0.5 h

Contractor
ECRIEE (East China Research Institute of Electrical Engineering), Hefei.

VERIFIED

JY-14 3-D surveillance radar

Type
E/F-band (2 to 4 GHz) long-range 3-D air surveillance radar.

Description
Designed for use in national air defence networks, the JY-14 long-range 3-D radar is described as being suitable for both air surveillance and interceptor control applications. System features include:
- real-time, 3-D target data reporting
- a low side-lobe antenna
- frequency agility and diversity
- a low power 'decoy' mode

The JY-14 long-range radar

- 'advanced' signal processing techniques
- large volume coverage and target processing capabilities.

Status

As of this edition, the JY-14 long-range 3-D air surveillance radar was reported to be in service in 'several' countries.

Specifications

3-D coverage volume (P_4 = 90%, P_{fa} = 10^6, σ = 3 m², Swerling 1)
Range: 450 km (maximum)
Height: 30,000 m (maximum)
Elevation coverage: 20°
Azimuth coverage: 360°
Power output: 10 kW (average); 1 MW (transmit)
ECCM: low side-lobe antenna (-45 dB average); frequency agile (15% of bandwidth); P-to-P, G-to-G or adaptive frequency agility; dual frequency two pulse diversity; digital pulse compression; adaptive moving target indicator
Accuracy: 0.2° (azimuth); 90 m (range); 400 m (height)
Resolution: 0.9° (azimuth); 1° (elevation); 300 m (range)

Contractor

ECRIEE (East China Research Institute of Electrical Engineering), Hefei.

VERIFIED

JY-27 surveillance radar

Type

A-/low B-band (0.2 to 0.3 GHz sub-band) surveillance radar.

Description

The JY-27 surveillance radar is designed for the long-range detection of air targets. System features include a fully coherent, solid-state transmitter; frequency agility; transmission spectrum filtering; digital pulse compression; an unattended operation facility and three channel, airborne moving target indication/constant false alarm rate processing.

Status

As of this edition, JY-27 is understood to be available.

Specifications

Frequency: 0.2-0.3 GHz
Range: ≥330 km (P_{-d} 50%, P_{-fa} 10^{-6} σ 2 m²)
Elevation coverage: 12° (P_{-d} 50%, P_{-fa} 10^{-6} σ 2 m²)
Altitude coverage: 20,000 m (P_{-d} 50%, P_{-fa} 10^{-6} σ 2 m²)
Accuracy: 1° RMS (azimuth); 150 m RMS (range)
Resolution: 5° (azimuth); 300 m (range)
Track processing: 128 tracks/10 s
Antenna rotation speed: 6 rpm

Contractor

ECRIEE (East China Research Institute of Electrical Engineering), Hefei.

VERIFIED

Model 17-C surveillance radar

Type

UHF (0.3 to 1 GHz) band mobile early warning radar.

The Model 17-C surveillance radar

Description

Model 17-C is an intermediate range mobile early warning radar designed for medium- and high-altitude surveillance. The Yagi array-type antenna is mounted on a six-wheel truck which also contains the electronics and displays. Alternatively the antenna can be mounted on the ground. The system is small and lightweight and is normally used in the forward area for aircraft detection. The complete system consists of two vehicles, the second truck being a power vehicle containing two diesel generators.

Status

As of this edition, Model 17-C is thought to be in service with the Chinese defence forces.

Specifications

Antenna: Yagi array with 4 rows, 2 layers and 8 units
Frequency: UHF band
Accuracy: 2 km (range); 2° (bearing)
Resolution: 2 km (range); 15° (bearing)

Contractor

China National Electronics Import and Export Corporation. Beijing.

VERIFIED

Type 408-C long-range air warning radar

Type

VHF band (30 to 300 MHz) air defence radar.

Description

The Type 408-C is a mobile long-range air defence warning radar system operating in the VHF band, at 150 to 180 MHz and 100 to 120 MHz. It employs a dipole planar-array type antenna, mounted on a pedestal, having two back-to-back dipole arrays which can be operated simultaneously at the high and low bands, giving two transmit/receive channels. The equipment has a wide operating frequency range within each band, enabling it to counter active jamming by rapid change of frequency within the bands. By integrating the video signal it is able to suppress asynchronous pulse jamming and by modulating the transmitter pulse repetition frequency, it counters repeater pulse jamming.

The transmitter system employs air-cooled tower triodes and uses Hi-Q coaxial cavity tanks to improve frequency stability. The display system consists of one azimuth/range display and three Plan Position Indicators (PPIs) which can scan in opposite directions and display the signals received at the same time from the two antennas within the two bands. The transmitter and receiver can be controlled remotely from the control cabin. The receiver provides high-sensitivity, high-frequency stability and a wide operating frequency range. It uses both the normal channel and the coherent channel to distinguish moving targets in ground clutter or passive jamming.

The complete system is mobile, with its units installed in eight operation vehicles, forming four transport units; two transmitter/receiver vehicles, one display unit vehicle, one antenna unit vehicle, two transport vehicles and two power supply vehicles (one is a spare).

Status

As of this edition, Type 408-C is thought to be in service with the Chinese armed forces.

The antenna array used in the 408-C radar system

Specifications
Frequency range: 150-180 MHz (high band); 100-120 MHz (low band)
Antenna
Type: 2 back-to-back dipole planar-arrays
Dimensions: 6 × 16 m
Polarisation: horizontal
Horizontal beamwidth: 8° (high band); 12° (low band)
Rotation speed: up to 6 rpm
Transmitter
Peak power: 800 kW
Pulse-length: 10 μs
PRF: 200 Hz

Contractor
China National Electronics Import and Export Corporation, Beijing.

VERIFIED

Type 571 surveillance radar

Type
D-band (1 to 2 GHz) 2-D low-altitude surveillance radar.

Description
The Type 571 is a 2-D D-band radar designed specifically for low-altitude air defence warning. Although apparently operating in a different frequency band, it bears a striking resemblance to the Russian Flat Face target acquisition radar, with its two elliptical paraboloid reflectors mounted one on top of the other and is almost certainly derived from the latter. The Chinese authorities state that it includes both a frequency-hopping and moving target indicator capability.

Two vehicles are required to transport the system, one being the operations cabin and the other the power supply. The antenna system is erected on top of the operations vehicle.

Status
As of this edition, Type 571 is thought to be in service with the Chinese defence forces.

Specifications
Frequency: D-band (1-2 GHz)
Antenna type: two elliptical paraboloid net reflectors and horn-shaped radiators
Antenna size: 5.5 × 2 m (2 reflectors)
Beamwidth: 4.5°
Rotation speed: 3 or 6 rpm
Peak power: 210 kW
Pulse-width: 2 or 4 μs
PRF: 500, 600 or 680 pps

Contractor
China National Electronics Import and Export Corporation, Beijing.

VERIFIED

The Type 571 surveillance radar

Type 581 air warning radar

Type
D-band (1 to 2 GHz) tactical air warning radar.

Description
The Type 581 is a medium- and low-altitude air warning radar for tactical applications. It is a transportable system with the 9.7 m span scanner mounted on turning gear that is carried on a wheeled trailer, which is stabilised for operational use by large folding legs with screw-jacks at their ends. The antenna rotates at up to 6 rpm. Sector scan is 90 or 30° with selectable scan centres.

The operating frequency is in the D-band, giving a detection range of up to 190 km against a fighter aircraft target. There are two units in a Type 581 mobile convoy, one consisting of the scanner and turning gear (and probably the diesel-electric generator unit also) and an electronics vehicle which includes the transmitter/receiver and operators' cabin.

Status
As of this edition, Type 581 is no longer in production but is thought to remain in service.

Specifications
Frequency: D-band (1-2 GHz)
Peak power: 500 kW
Pulse-width: 4 μs; 2 μs
PRF: 300 Hz/600 Hz
Noise figure: 3 dB
Antenna: 9.7 × 3 m
Polarisation: linear
Gain: 34 dB
Range: 190 km (fighter target)
Accuracy: 1 km (range); 1.2° (azimuth)
Resolution: 1 km (range); 4° (azimuth)
MTI: 20 dB (clutter visibility); 34 dB (cancellation ratio)
Displays: PPI (30.5 cm); A/R (18 cm)

Contractor
China National Electronics Import and Export Corporation, Beijing.

VERIFIED

The Type 581 air warning radar

YLC-4 surveillance radar

Type
UHF (0.3 to 1 GHz) band 2-D long-range surveillance radar.

Description
YLC-4 is a UHF 2-D solid-state, fully coherent long-range surveillance radar which features a low-sidelobe antenna; built-in test equipment; pulse-to-pulse, burst-to-burst frequency agility; system redundancy for reliability and airborne moving target indicator. In terms of data manipulation, the radar incorporates semi-automatic or automatic target performance extraction and the ability to synthesise, display and (communicate) data from up to four other radars such as height-finders and gap fillers.

Status
As of this edition, the status of YLC-4 is uncertain.

Specifications
Frequency: UHF band (0.3-1 GHz)
Detection range: 380 km (automatic extraction); 410 km (manual extraction)
Coverage: 360° (azimuth); 0-25° (elevation)
Detection ceiling: 30,000 m
Accuracy: 300 m RMS (range); 0.5° RMS (azimuth)

The antenna array used in the YLC-4 long-range surveillance radar system

Antenna dimensions: 16.5 × 7.12 m
MTI improvement factor: 41 dB
Peak power: 40 kW

Contractor
Nanjing Research Institute of Electronics Technology, Nanjing.

VERIFIED

GERMANY

DR series air defence radars

Type
Family of defence and low-level surveillance radars.

Description
Before its withdrawal from the defence electronics sector, Siemens AG produced a series of air defence and battlefield radars. The range has subsequently been absorbed by DaimlerChrysler Aerospace (now a part of the European Aeronautics, Defence and Space (EADS) company). Details of the air defence models are as follows:

DR 151
The DR 151 is a mobile low-level surveillance D-band (1 to 2 GHz) radar with a range of 45 km. Other improvements have been made in features such as signal

A DR 162 ADV-ER mobile radar deployed in the field

The DR 172S low-altitude surveillance radar

processing, Electronic Counter-CounterMeasures (ECCM) proofing and mechanical design. The system is mounted on one 10-tonne truck and is equipped with an 18 m extensible antenna mast.

DR 162
The DR 162 is basically similar to the DR 151, but with an additional frequency diversity operation mode and with range capability extended to 90 km.

DR 162 ADV-ER
The DR 162 ADV-ER (Air Defence Version-Extended Range) is a mobile D-band (1 to 2 GHz) radar with a detection range of 140 km. The system is based on the DR 172 radar with an added central computer, a large synthetic display and communications facilities. This arrangement permits target tracking and allocation to assigned air defence systems and is highly suitable for both autonomous and integrated air surveillance systems.

DR 172S
DR 172S is the latest member of the DR series radar family and incorporates a 2-D surveillance capacity, solid-state components, frequency diversity and moving target indication filtering. The equipment makes use of a specially shaped antenna reflector and a cosec^2 vertical radiation pattern to enhance subclutter visibility. A secondary surveillance radar capability is integrated into the DR 172S system and employs the same antenna array as the main radar. Other equipment features include:
- two channel frequency diversity
- raster graphic display
- Very/Ultra High Frequency (VHF/UHF) communications subsystem
- two shelter design.

Status
As of this edition, DR series radars are thought to be in operational service.

Specifications
DR 172S
Frequency: D-band (1.2-1.4 GHz sub-band)
Detection range: 140 km
Peak power: 10 kW
Frequency agility: 70 channels

Contractor
European Aeronautic Defense and Space (EADS) Co/SI Sicherungstechnik GmbH & Co KG, Unterschleissheim.

UPDATED

INDIA

GRL 600 air defence radar

Type
D-band (1 to 2 GHz) low-level detection radar.

Description
GRL-600 is a fully coherent low-level radar system catering primarily for the gap filling role in air defence environments. Apart from surveillance, the system can be

The GRL 600 low-level air defence radar

used for local area warning and as a target designation radar. The system is highly mobile and can be deployed in a maximum time of 30 minutes, making it suitable for tactical air defence applications. The complete system is contained in two vehicles, one carrying the antenna and the other the electronics and control system. This radar has been developed by the Electronics Research and Development Establishment, Bangalore in co-operation with Bharat Electronics.

The antenna consists of a shaped parabolic reflector (with a flat-bottom radiation pattern to avoid ground clutter) and turns at 16 rpm. The feed system incorporates a high-performance horn element with integrated Identification Friend-or-Foe (IFF) feed. An auxiliary feed provides improved elevation coverage on reception. The same reflector is used for both primary and IFF beam forming. The antenna is mounted on the vehicle bed and is folded down for transport. Apart from the two vehicles carrying the equipment, there is a standby power supply and a collapsible mast that can raise the antenna to 28 m to provide longer range or look over difficult terrain.

All the electronic subsystems such as the transmitter, receiver, signal processing and operator's station are contained in an air conditioned shelter that can be carried on a standard vehicle. The transmitter operates in D-band with a peak power of 40 kW. Any one of the multiple frequencies can be selected by the operator.

The radar processor is able to track up to 40 targets simultaneously and can work with 12 weapon systems. The display incorporates a multicolour 48 cm plan position indicator catering for both raw and synthetic videos. The operator can interact with the system through an alphanumeric keyboard and other controls. An integral IFF Mk X equipment is included, with the control panel an integral part of the console.

Specifications

Antenna: shaped parabolic (5 × 3.3 m)
Beamwidth: 3° azimuth; 17° elevation
Polarisation: horizontal
Transmitter
Frequency: D-band (1-2 GHz). 6 spot frequencies available
Power output: 40 kW peak
Pulse-width: 3.2 µs
PRF: 3 PRFs staggered scan-to-scan
Range: up to 50 km on a fighter aircraft
Height coverage: 30-1,700 m

Contractor

Bharat Electronics Ltd(an Indian government enterprise), Bangalore.

VERIFIED

GRL 610 air defence radar

Type

D-band (1 to 2 GHz) transportable low-level air defence radar.

Description

GRL 610 is a fully coherent low-level radar system intended for use as a gap filler in the air defence environment. The system is configured as a fully fledged operations

The GRL 610 radar system utilises three vehicles   0006877

centre, together with the radar head. GRL 610 is a transportable and self-contained system with a high mobility and deployment feature. The system is mounted on three vehicles.

The antenna is a shaped parabolic reflector with a flat-bottomed radiation pattern to avoid ground clutter. The system incorporates a high-performance horn element and an integrated Identification Friend-or-Foe (IFF) feed. An auxiliary feed allows improvement in the elevation coverage on reception. The same reflector is used for both primary and IFF beam-forming. The antenna is mounted on the vehicle flat bed and folds flat for transport. For improved coverage the antenna can be mounted on a 28 m high mast.

System electronics are contained in an equipment cabin and an operator's cabin. The equipment cabin contains the transmitter, receiver, signal processing equipment and the IFF interrogator. The operator's cabin contains two displays and their associated controls, together with the data processor and communications facilities.

The transmitter operates at D-band and employs a cross-field amplifier driven by a solid-state chain which delivers an output power of 100 kW. Any one of 12 preset frequencies can be selected by the operator.

The main features of the system are:
- a fully coherent system
- multiple frequencies of operation, selectable by the operator
- full tracking capabilities for manoeuvring targets
- colour plan position indicator display, both raw and synthetic video
- integral IFF
- fast deployment
- advanced signal processing using moving target detection techniques.

Status

As of this edition, GRL 610 is understood to have been procured by the Indian Army.

Specifications

Antenna: shaped parabolic (5 × 3.3 m)
Beamwidth: 3° azimuth; 17° elevation
Polarisation: horizontal
Transmitter
Frequency: D-band (1-2 GHz); 12 spot frequencies available
Power output: 100 kW
Pulse-width: 6.6 µs
PRF: 4 PRFs staggered scan-by-scan
Range: up to 90 km for a fighter aircraft
Height coverage: 30-1,700 m

Contractor

Bharat Electronics Ltd (an Indian government enterprise), Bangalore.

VERIFIED

PSM 33 Mk II mobile radar

Type

E/F-band (2 to 4 GHz) transportable 3-D air defence radar system.

Description

The PSM 33 is a transportable E/F-band 3-D air defence control and reporting radar system, which has been developed in co-operation with Thomson-CSF. It consists of an antenna with a tripod-type mounting plus two cabins, one housing the transmitter and the other containing the receiving and processing units. It can be transported by road, rail or aircraft and can be deployed by eight people in less than one hour.

The antenna consists of a parabolic cylinder reflector fed by a linear array of 40 elements. Each element is preceded by a phase shifter whose phase shifts are varied according to the commands given by a beam steering unit. The secondary

The PSM 33 Mk II air defence radar system

antenna is built in and uses the reflector of the primary antenna. The transmitter shelter houses three E/F-band magnetron transmitters each delivering 1.8 kW average power. The outputs are grouped together in a high-power triplexer. Provision is made for operation, should one or two transmitters fail, by manual or automatic changeover.

Search in azimuth is by mechanical rotation of the antenna and electronic scanning in elevation. A stacked beam pattern in elevation is formed by transmitting three frequencies, using monopulse techniques for accurate height measurement. Height-finding accuracy is claimed as ±3 mrad. Frequency diversity is employed as a protection against electronic countermeasures and a programme to incorporate an area moving target indicator is underway.

The system has several operating modes:

- general surveillance: elevation scanning over 24° with initial elevation adjustable
- single or two sector surveillance: elevation scanning over 6 or 12° as required
- search/designation: accurate height measurement possible from -2 to +29.5° while search continues in the lowest sector
- pure designation mode: accurate height-finding on two closely spaced targets anywhere between -2 and +29.5°.

An integrated interrogator operates in mode 1, 2, 3/A, B, C and D with 3-pulse ISLS 2, 3, or 4 mode interface.

Status
As of this edition, PSM 33 Mk II has been procured by the Indian Air Force.

Specifications
Frequency: E/F-band (2-4 GHz)
Instrumented range: 440 km surveillance mode; 510 km early warning mode
Accuracy (standard deviation up to 200 km): range 50 m, azimuth 0.25°; altitude 460 m
Resolution: azimuth beamwidth 1.5°; elevation beamwidth at low elevations 2°
Antenna: cylindrical paraboloid
Polarisation: circular
Rotation speed: 6 rpm
Elevation coverage: surveillance mode 26°; designation mode 32°
Transmitter
Power output: peak 660 kW; average 10 kW
Pulse duration: 3 × 13 µs
PR period: 930 µs and 3,690 µs according to operating modes and elevation sectors

Contractor
Bharat Electronics Ltd (an Indian government enterprise), Bangalore.

VERIFIED

INTERNATIONAL

Air defence MultiRole Operations Cabin (MROC)

Type
Deployable Command, Control and Communications (C³) facility for air defence networks.

Description
The Alenia Marconi Systems (AMS) MROC is described as being a 'complete' air defence C³ facility that is housed in a deployable, International Standards Organisation (ISO) compliant shelter and is able to integrate primary radar, sensor and tactical data. The use of an open system architecture and modularity facilitate the equipment's use in a variety of applications that include reporting posts, control and reporting posts/centres, sector operations centres and air defence operations centres. MROC utilises 'proven' AMS Systems software products including the company's picture compilation, identification/recognition, threat evaluation,

An interior view of the MROC deployable air defence C³ facility   2000/0080489

resource management, weapons assignment and air mission control packages. The facility's internal layout can be customised to meet specific requirements and makes extensive use of commercial-off-the-shelf platforms. The adaptable human-computer interface used can be implemented in a number of forms, all of which are based on a Windows graphical environment.

Status
As of January 2002, the precise status of MROC was uncertain.

Contractor
Alenia Marconi Systems (an Alenia-BAE Systems joint venture), Chelmsford/Rome.

UPDATED

Argos 45 tactical radar system

Type
G/H-band (4 to 8 GHz) low-altitude air defence radar.

Description
The Argos 45 tactical radar system has been designed to provide co-ordination to forward area low-altitude air defence systems, to defend battalions and selected combat support units which are vulnerable to high-speed, low-multirole aircraft and strike helicopters. The system can take on the role of a low-altitude reporting post, as a gap filler in an air defence network, a tactical command post for the control of air support tactical missions and a designation centre for man-portable missiles and anti-aircraft artillery.

Two configurations are available to meet differing operational requirements. In the compact version, Argos 45C, the system consists of a military shelter housing the primary and secondary radar head, cooling unit, two operator consoles, data handling equipment, communications and a power generator. The operators' positions are in a separate compartment within the shelter. The antenna system consists of a lightweight structure mounted on an erectable tower, stowable in transport configuration on the shelter roof and including an integrated identification friend-or-foe system.

The alternative version, Argos 45T, is suitable for those scenarios where the threat is very high. A new display system and data processing structure (known as MAGICS and MARA respectively) are used in both versions. A hovering helicopter identification capability is incorporated.

Argos 45 is the main sensor for the ARAMIS (Area Multiple Intercept System). This consists of a battery control post carrying an Argos 45 with its antenna mounted on a 10 m high-elevating mast and up to four Aspide fire units.

Status
As of this edition, a number of Argos 45 radars are reported to have been procured by Italy's armed forces.

Specifications
Antenna
Type: planar-array cosec²
Polarisation: horizontal
Scan rate: 30 rpm

Alenia Marconi's Argos 45 tactical radar in operational configuration

Transmitter
Type: coherent chain
Final Stage: TWT
Frequency: G/H-band (4-8 GHz)
Peak power: more than 15 kW
Frequency selection: fixed or random and adaptive frequency selection (option)
IFF: D-band (1-2 GHz), vertically polarised
Range: up to 45 km

Contractor
Alenia Marconi Systems(an Alenia-BAE Systems joint venture), Chelmsford/Rome.

VERIFIED

Argos 73 tactical radar system

Type
E/F-band (2 to 3 GHz sub-band) low/very low altitude mobile air defence radar.

Description
Alenia Marconi's Argos 73 low/very low altitude air defence radar is derived from the American AN/TPS-73 equipment and has been developed in collaboration with the then Unisys (now part of Lockheed Martin). Argos 73 is described as being a 'highly mobile' air defence and coastal surveillance radar that is noted as being able to detect and classify a spread of air target types that range from hovering helicopters to air vehicles flying at speeds of up to Mach 3. The equipment is also claimed to be able to detect targets with small radar cross sections in dense environments polluted by clutter and jamming.

Argos 73 comprises a solid-state primary radar and a monopulse secondary radar with the totality being described as being able to detect fighter sized targets at ranges and altitudes of up to 120 km and 10,000 m respectively. System features are reported to include modular, air-cooled, transistorised, transmitter modules, a high gain cosec^2 antenna and the use of coded waveforms. In addition to its primary roles, the system is noted by its manufacturer as being suitable for use as a gap filler within an air defence network.

Status
As of this edition, Argos 73 radars are understood to have been procured by the German and Pakistani Navies for as coastal surveillance sensors.

Specifications
Frequency: 2-3 GHz
Antenna type: reflector
Polarisation: horizontal/circular (L/R)
Scan rate: 10 (typical), 12 or 15 rpm
Range: up to 120 km
Transmitter: solid state with coded waveforms
Reciever: double frequency conversion; linear and CFAR receivers; I and Q phase detectors; adaptive Doppler filters; matched filters; azimuth correlators
Peak power: 10 kW
PRF: fixed or stagger
Modes: fixed frequency; pulse-to -pulse agility, burst-to-burst agility

Argos 73 coastal radar

Contractor
Alenia Marconi Systems (an Alenia-BAE Systems joint venture), Chelmsford/Rome.

VERIFIED

MARS 402 2-D radar operations centre

Type
2-D radar operations centre.

Description
The MARS (Mobile Automatic Reporting Station) 402 is a mobile radar system providing the detection of low-flying aircraft and surface targets, generation and reporting of the air/surface situation and controlling function for navigational assistance of home-based military aircraft. The system is functionally composed of the following main sections:
- operation section
- Pluto radar section
- power supply.

The Pluto radar section is composed of the radar antenna, transmitter and receiver equipment providing 2-D primary and secondary surveillance radar target detection.

The operation section is composed of the data processing display/communications equipment and provides tracking, reporting and air control functions. The communications subsystem is configured for internal communications and control and is equipped to interface with external systems. The physical components of the MARS 402 in the standard configuration are as follows:
- one antenna module
- one radar shelter
- one operations shelter
- mobiliser(s) (optional)
- tactical tower (optional).

In the compact version of the MARS 402, the radar and operation section equipments are housed in the same shelter to enhance compactness and mobility. Operational roles are:
- 2-D multiple radar input
- early warning surveillance with single or multiple radar input
- navigational assistance
- fixed or transportable applications
- local or remote radar input

Specific functions include:
- automatic target detection and initiation
- automatic tracking
- multiradar processing (with optional multiple radar configuration)
- rate-aided manual tracking
- datalink for exchange of track data, command messages and so on with automatic and manual facilities
- identification by computer-assisted selective identification facility active/passive decoding, automatic datalink, or manual input
- search radar/identification friend-or-foe video display
- synthetic video display.

Status
As of this edition, MARS 402 is thought to be in service with a number of defence forces.

Specifications
Track capacity: 40
Simultaneous intercept vectoring: 2
Datalink: up to 4 sites
Video maps displayed: 2

A view of the MARS-402 radar station showing its Pluto radar sensor and operator shelter

Operator orders: 42
Computer: NDC-160, mobile
Displays: MDU-03

Contractor
Alenia Marconi Systems (an Alenia-BAE Systems joint venture), Chelmsford/
Rome.

VERIFIED

Master series 3-D surveillance radars

Type
Family of E/F-band (2 to 4 GHz) air surveillance radars.

Description
The Thales Raytheon Systems (formerly Thales Air Defence) Master series radars are described as being solid-state, transportable, 2 to 4 GHz band equipments and are designed for air defence applications. As of January 2002, four Master variants had been identified, the known details of which are as follows:

Master A
Master A is a multifunction air defence radar that is capable of surveillance and target tracking and according to Jane's sources, makes use of electronic scanning in both elevation and azimuth. The equipment comprises an antenna assembly (with an Identification Friend-or-Foe (IFF) array mounted above the main antenna face), an operator's shelter and a power supply.

Master M
Master M is an extended range early warning radar and comprises an antenna assembly (with an IFF array mounted above the main antenna face), an operator's shelter and power supply.

Master S
Formerly known as the TRS 2140 Field Low-Altitude Intermediate Range (FLAIR) radar, the Master S equipment is a mobile, E/F-band, solid-state sensor that is designed for the detection of aircraft at medium range/medium, low and very low altitude and which can be used for point and area defence or as a gap filler to complete a coverage volume. It has been developed for easy integration into a sensor network and can be remotely controlled. It is highly mobile, transportable by truck, cargo aircraft or helicopter and can be set up easily in less than 30 minutes by four unskilled personnel. The radar is installed on a self-lifting platform which carries:

- the antenna and identification friend-or-foe array with their rotating gear
- a 2 × 2 m shelter, housing the digital frequency generator, receiver, signal and data processing systems, a technical console, the test and remote-control devices.

Master S has been optimised to detect highly manoeuvrable supersonic aircraft, combat helicopters and missiles, in particular low-flying targets in the most severe environments, such as fixed and moving clutter, and passive and active jamming also in non-standard atmospheric conditions. Nuclear, biological, chemical and electromagnetic pulse protection have been taken into account in the design and can be implemented on request at an optimal cost level.

A solid-state active planar-array antenna, with distributed transmit/receive modules (one per row), has been employed. Simultaneous coherent and non-

A rear view of the Master S antenna showing the system's active elements

coherent processing with map-controlled switching, both in range and azimuth, is particularly effective for clutter suppression. Other clutter rejection features include high-phase stability of the complete coherent chain, very low-sidelobe levels, pencil beam and small radar cell, scanning pattern following the terrain relief, high Pulse Repetition Frequency (PRF), adaptive moving target indicator with Doppler processing and clutter mapping.

Electronic counter-countermeasures performance is provided by frequency agility (burst-to-burst or pulse-to-pulse), very low-sidelobe levels with sidelobe blanking, digital pulse compression and multiple pulse duration, mono pencil beam, PRF staggering, constant false alarm rate, automatic choice of the least jammed frequency and jamming analysis.

Other salient features of the Master S include various operating modes with automatic selection, a low probability of interception which allows the use of very long pulses (high-average power) with low-peak power, unmanned remote command and control operation and high reliability and maintainability. A comprehensive built-in test system for online fault detection is incorporated in the design.

Master T
Master T is a tactical surveillance radar and comprises an antenna assembly (with an IFF array mounted above the main antenna face), an operator's shelter and a power supply.

Status
As of the autumn of 1998, Jane's sources were suggesting that the then Thales Air Defence was working on a Theater Missile Defence (TMD) capability for the Master A radar that would make use of TMD signal processing algorithms developed by Thales Nederland. As of March 2001, Thales is understood to have sold at least one Master A equipment to Singapore. By the following November, Jane's sources were suggesting that Master series radars had been sold to customers in Asia (Master A), Europe (Master A, M and T), the Middle East/Africa (Master T) and the Pacific Rim/Far East (Master A and M).

Specifications
Master A
Frequency: E/F-band (2-4 GHz)
Range: 300 km (surveillance mode - instrumented); 370 km (tracking mode - instrumented)
Master M
Frequency: E/F-band (2-4 GHz)
Range: 460 km (detection - small aircraft target); 470 km (instrumented)
Master S
Frequency: E/F-band (2-4 GHz)
Antenna size: 5 × 2 m
Rotation speed: 12 rpm
Horizontal beamwidth: 1.6°
Vertical beamwidth: 4.2°
Compressed pulse duration: 0.5 µs
Instrumented range: 145 km
Altitude ceiling: 30,000 ft
Max elevation: 45°
Azimuth: 360°
Resolution (after extraction): 210 m (range); 3.5° (azimuth)
Accuracy (after extraction): 30 m (range); 0.2° (azimuth); 2,000 ft at 75 km (altitude)
MTBF: 1,500 h
MTTR: 30 min
Master T
Frequency: E/F-band (2 to 4 GHz)
Range: 370 km (detection - small aircraft target); 440 km (instrumented)

Contractor
Thales Raytheon Systems, Massy, France.

UPDATED

MRCS 403 reporting and control system

Type
Mobile reporting and control system for air defence.

Description
The MRCS (Mobile Reporting and Control System) 403 is a tactical air defence system designed for deployment in the field environment encountered in tactical warfare. It provides all the air defence functions necessary to cope with both friendly and hostile aircraft. It is a modular and flexible mobile system which can operate as an independent air defence operations command and control centre or as an integral part of an extensive air defence network. The functional components of the MRCS 403 consist of a 3-D surveillance, an operations shelter and a prime power source.

The system basic configuration includes the Alenia RAT 31SC radar (see separate entry), which provides the primary function of 3-D surveillance. To increase still further the system flexibility, the radar interface unit has been designed to be adaptable for interfacing requirements with any other 3-D surveillance radars, to provide local integrated networks.

The MRCS-403 reporting and control system with its RAT 31SC radar

The operations shelter includes a dual NDC-160 computer complex, three modular display consoles, a computer peripheral set, a communications control unit, ground-to-air equipment and a datalink interface for the reception of plots.

The MRCS 403 functions include:

- detection, identification, 3-D location and tracking of aircraft
- target designation and control of interceptor aircraft
- target designation to surface-to-air missiles and guns
- reporting of air situation to air defence headquarters
- navigation assistance
- data collection and recording
- simulation.

Status

As of this edition, MRCS 403 is reported to have been procured by Italy (air force) and Austria (two systems).

Specifications

Track capacity: up to 100
Controlled airfields: up to 4
Display synthetic video maps: 2
Simultaneous intercept/recoveries: up to 4
Interceptor aircraft types: 2
Fuel-armament configurations: 8
Operator positions: 3
Display consoles: 3

Contractor

Alenia Marconi Systems (an Alenia-BAE Systems joint venture), Chelmsford/ Rome.

VERIFIED

Pluto low-level coverage surveillance radar

Type

E/F-band (2 to 4 GHz) air surveillance radar.

Description

Pluto is a low-level coverage surveillance radar for air and coastal defence, designed to detect medium, low and very low-flying aircraft and small surface vessels under the most adverse environmental conditions (such as electronic countermeasures pollution, atmospheric and sea clutter). The radar is suitable for fixed and/or mobile installations and can operate either independently or be integrated into a complete defence system. Design features include the following:

Capability to detect and measure the range and azimuth of surface and air targets at medium and low altitudes up to a maximum range of 110 km.

High accuracy and resolution, integral target designation capability and flexibility to interface computer-assisted display and defence systems.

Pulse compression of the radiated waveform with a resultant small-scale radar cell size, resulting in visibility even of very slow moving targets such as small ships. Sophisticated electronic counter-countermeasures processing with visibility maintained even in the presence of heavy active and passive jamming.

Good adaptability and constant false alarm rate characteristics under various environmental conditions, with capability to detect targets and reject interference by automatic evaluation and selection of the best transmission and processing configurations.

Capability of operating in frequency agility from sweep to sweep, with clutter rejection and subclutter visibility of moving targets outgoing or incoming, at speeds up to Mach 3 by means of sophisticated moving target indicator processing techniques.

- Antenna pattern with high-directivity and very low sidelobe levels
- Large transmission bandwidth
- Integrated secondary identification friend-or-foe radar
- High mean time between failure
- High reliability, maintainability and supportability
- Built-in test facilities.

The Pluto radar's antenna mounted on a self-erecting tower

A reduced-range version of Pluto is used to supply surveillance data for the Spada air defence system.

In its basic configuration, Pluto consists of one equipment shelter, one antenna module and a power generator trailer. The transmitter, receiver and processing equipment are all housed in the shelter. In the unmanned configuration, the radar is provided with a datalink interface for connection to an operational shelter or to a control centre.

Status

As of this edition, Pluto radars are understood to have been procured by customers in Italy, the Far East, the Middle East and South America. It should be noted that as of October 1998, Pluto was one of the radars likely to be incorporated into the Alenia Marconi Systems joint venture.

Specifications

Antenna
Type: reflector, feed
Frequency: E/F-band (2-4 GHz)
Polarisation: horizontal or circular, selectable
Beamwidth: 1.5° (horizontal); 4° (vertical)
Beam shape: pencil beam (horizontal); super-cosec^2 matched to a very high-clutter environment (vertical)
Scan rate: 12/15 rpm
Tilt: −2 to +5° (adjustable)
IFF antenna: integrated
Transmitter
Type: coherent chain
Final stage: TWT
Peak power: 135 kW
Bandwidth: 10%
Transmitted waveform: phase-coded (code agility)
Pulse repetition frequency: prefixed or random stagger
Range: up to 110 km on a fighter aircraft
Accuracy: 40 m (range); 0.35° (azimuth)
Operating modes: fixed frequency; random/preprogrammed frequency agility; adaptive frequency selection

Contractor

Alenia Marconi Systems (an Alenia-BAE Systems joint venture), Chelmsford/ Rome.

VERIFIED

RAT 31DL transportable long-range 3-D air defence radar

Type

D-band (1 to 2 GHz) long-range 3-D air defence radar.

Description

RAT 31DL is a D-band, NATO Class 1 transportable air defence radar which is shelterised, adaptable to a range of operating environments and is based on Alenia's experience with the other radars making up the RAT 31 family of equipments. A key element of the RAT 31DL design is the use of a solid-state antenna distributed transmitter which provides for a drastically increased mean time between failure value, a reduction in total weight and volume and the ability to degrade gracefully. Further advantage is gained from the beam-forming technique used which allows the simultaneous independent management of the system's four transmission beams without any break in the frequency band. The approach also allows for reconfiguration of the scanning pattern to take account of any failure in the radar's interchangeable receiver channels. Each transmission beam is configured for monopulse height-finding over the system's entire detection envelope. The specific low-angle monopulse techniques used are claimed to minimise multipath effects and generate height-finding accuracies not available from conventional monopulse systems.

RAT 31DL transmits long pulse bursts which are compressed in the receiver chain to maximise range resolution and resistance to electronic countermeasures (including anti-radiation missiles). Low- and high-pulse repetition frequency waveforms are used to provide long-range and medium-range/Moving Target Indicator (MTI) coverage respectively. Fixed and adaptive notch MTI filtering is used and electronic counter-countermeasures proofing includes sidelobe and peak power control, frequency agility, jam strobe reporting, sidelobe blanking and a dedicated receiver channel for countermeasures environment sampling and automatic, responsive frequency selection. RAT 31DL includes a fully integrated SIR-M secondary monopulse radar.

Status

As of this edition, Jane's sources suggest that three RAT 31DL radars have been supplied to Turkey, with a fourth going to Denmark.

Specifications

Frequency: D-band (1-2 GHz)
Detection envelope: 360° (azimuth); up to 440 km (range); up to 30,480 m (altitude)
System architecture: solid-state antenna distributed transceiver modules
Antenna/transmitter group: planar phased-array with 44 radiating rows made up to 44 transceiver modules per row
Scan rate/mode: 6 rpm/mechanical rotation over 360° in azimuth; squintless electronic phase scanning in elevation
Polarisation: L/H
Number of transmission beams: 4
Transmitted waveform: FM coded pulse
Receiver: 4 identical channels (Σ and Δ); 2 auxiliary channels
Signal processing: fixed and adaptive notch MTI to 3rd order; CFAR and range and azimuth mapping

Contractor

Alenia Marconi Systems (an Alenia-BAE Systems joint venture), Chelmsford/Rome.

VERIFIED

....

RAT 31S and 31SC transportable 3-D air defence radars

Type

E/F-band (2 to 4 GHz) transportable 3-D air defence radar.

Description

The RAT 31S has been designed as a highly mobile 3-D radar, capable of performing a wide range of operational functions associated with air defence, tactical air applications and civil/military air traffic control. The construction method employed makes the system equally suitable for air or land mobile applications. The system comprises an antenna group, a transmitter shelter, a receiver and signal processing shelter and a primary power supply. The antenna group can be moved by a trailer or by a demountable running gear, which is the preferred method when using C-130 aircraft for transportation. The shelters may be moved by any of the methods complying with standard NATO Agreement requirements.

The operating frequency is in the E/F-band and the RAT 31S is intended for aircraft detection and range, elevation and azimuth target measurements at high, medium, and low altitudes. Comprehensive electronic counter-countermeasures facilities include the ability to operate in frequency agility from pulse-to-pulse and/or within pulse-groups, with an automatic adaptive transmission frequency selection. Height measurement of detected aircraft is compatible with frequency

Alenia Marconi's RAT-31SC tactical radar

agility and the subclutter visibility is sustained even when operating in pulse-to-pulse frequency agility.

The RAT 31S uses a novel antenna design that employs a system of three stacked beams with a phased-array control of the elevation of each beam. The configuration is that of a rear-fed planar-array, which avoids aperture blocking, minimises sidelobes and provides good mutual screening characteristics. A coherent chain transmitter with a travelling wave tube final amplification stage is used, which facilitates the frequency-agile features of the radar. A staggered position and phase code is used in the transmitted pulse. Features of the receiver include: monopulse operation with Dicke-fix processing; anti-clutter filtering by means of a double canceller; adaptive clutter attenuator; coherent limiter and moving window azimuth correlator.

The RAT 31S is designed to be an integral element of automated air defence systems. For tactical air defence, Alenia can deploy a mobile command and control centre, the MRCS-403, which is designed to operate with the RAT 31S. This control centre consists of two computers and three control consoles, a computer peripheral set, communications control and ground-to-air equipment. The RAT 31SC variant is a highly mobile tactical 3-D system which is based on RAT 31S technology, with minor electrical changes to comply with battlefield requirements. These include higher rotation rate, higher pulse repetition frequency and circular polarisation.

Status

As of this edition, identified RAT 31S customers are reported to include Austria (six examples), Brazil and Italy (14 examples for the national air defence system). In addition, the radar is noted as having been used in the prototype of the Italian Army's SOATCC low/very low-altitude battlefield air surveillance system as well as being mandated for the service's Forward Area Air Defence System.

Specifications

Antenna
Type: rear-fed planar-array
Polarisation: linear horizontal (RAT-31S); circular (RAT-31S-31S/C)
Beamwidth: 1.5° (azimuth and low-elevation beams)
Azimuth scan mode: mechanical rotation over 360°
Scan rate: 5-10 rpm
Elevation scan mode: electronic by phase control from 0° to 21°
Number of beams: three different groups of beams are radiated on three separate bandwidths
Pattern: multiple pencil beams
Transmitter
Frequency: E/F-band (2-4 GHz)
Type: coherent amplifier chain
Final stage: TWT
Peak power: 135 kW
Max unambiguous range: >300 km
PRF: prefixed or pseudo-random stagger
Transmitted waveform: three separate coded waveforms
MTBF: 450 h
MTTR: 35 min

Contractor

Alenia Marconi Systems (an Alenia-BAE Systems joint venture), Chelmsford/Rome.

VERIFIED

....

RAT 31SL transportable long-range 3-D air defence radar

Type
E/F-band (2 to 4 GHz) long-range 3-D air defence radar.

Description
The RAT 31SL is an E/F-band, NATO Class 1, transportable long-range 3-D air defence radar, providing high performance in heavy clutter and Electronic CounterMeasures (ECM) environments. The radar draws on the design philosophy used in the RAT 31S, taking advantage of such new technologies as the radio frequency networks and phase shifters in the antenna, a new processor where the use of large integration allows a more sophisticated processing and the higher power and radio frequency stability of the transmitter.

The RAT 31SL uses multiple simultaneous independent phase-controlled pencil beams to provide flexible volumetric coverage. Every beam allows monopulse height measurement with excellent accuracy, even in pulse-to-pulse frequency agility. The radar transmits bursts of long pulses, highly compressed in reception, to provide high-range resolution and resistance against both ECM and anti-radiation missiles. Fixed and adaptive moving target indicator filters are enabled by continuously updating clutter maps to enhance target subclutter visibility. High electronic counter-countermeasures performance is provided by very low antenna sidelobes, large operational bandwidth, high-average and peak-limited transmitted power, wideband frequency agility, jam-strobe reporting, sidelobe blanking and by a separate receiver channel for ECM assessment and automatic frequency selection. Detection range of the RAT 31SL is up to 450 km.

The RAT 31SL includes a fully integrated SIR-M secondary monopulse radar. The complete system is ruggedised to withstand adverse weather, solar radiation and various types of electromagnetic interference.

Status
As of this edition, a number of systems are understood to have been procured by the Italian Air Force. Two RAT 31SL systems were ordered by Spain in December 1992, a further two systems in mid-1994. In October 1994, it was announced that the RAT 31SL had been selected for the Norwegian SINDRE II programme (18 examples procured). Most recently, the radar has been successfully bid for a NATO Class 1 system requirement for equipment to be deployed in Turkey.

Specifications
Antenna
Type: corporate rear-feed planar-array
Polarisation: linear/horizontal
Beamwidth: 1.5° azimuth; 1.3° low elevation
Azimuth scan mode: mechanical rotation over 360°
Elevation scan mode: electronic phase control from 0 to 20°
No of beams: 3, independent
Transmitter
Type: TWT amplifier
Frequency: E/F-band (2-4 GHz)
Driver and local oscillators: coherent chain
Transmitted waveform: multiple coded waveform
Receiver
Type: linear and logarithmic
Number of receiving channels: 8

Signal processor
Type: 3 and 4 pulses fixed MTI, adaptive notch 4 pulse MTI, followed by CFAR threshold and post-detection false alarms filtering
Data processor: high-speed modular architecture for real-time applications

Contractor
Alenia Marconi Systems (an Alenia-BAE Systems joint venture), Chelmsford/Rome.

VERIFIED

S711 tactical radar

Type
Mobile tactical low-level and gap filler radar.

Description
The S711 mast-mounted tactical radar has been specifically designed and produced to provide a total capability in low-level and gap filler applications. It is capable of operating in all types of difficult terrain, where the elevated antenna allows unrestricted operation clear of local obstructions. The elevated antenna also allows protective concealment of most of the radar equipment and by operating from unlikely locations, the S711 offers tactical surprise.

The integral display and data handling system and integrated communications facilities can operate in autonomous modes, or can be part of an integrated system interoperating with other control centres and weapon systems.

The double curvature antenna reflector is a lightweight carbon fibre reinforced moulding requiring negligible maintenance and having a highly accurate profile. The antenna horn boom accommodates the vertical to circular polariser and the integral secondary identification friend-or-foe feed elements. The antenna trailer incorporates built-in stabilising legs and a hinged mast with a telescopic upper section. Elevation of the mast is by a built-in, electrically driven, hydraulic system, allowing the use of the antenna at 12 m and 19 m heights.

The electronic units and operator positions are housed in a special purpose container which, together with services modules and a diesel generator, are suitable for use either in the standard configuration mounted on a semi-trailer or for ground deployment. The electronics modules, comprising transmitter/receivers, signal processor, data extractor, and display and data handling system, have been designed to meet the performance and environmental requirements imposed by the operational role of this type of radar.

Status
The S711 is in operational service with a number of armed forces, including NATO. Reliable sources suggest that Jordan has five systems in operational use.

Contractor
Alenia Marconi Systems(an Alenia-BAE Systems joint venture), Chelmsford/Rome.

VERIFIED

Alenia Marconi's RAT 31SL transportable air defence radar

When deployed, the S711 radar's antenna is mast-mounted

S713/723 Martello surveillance radars

Type
D-band (1 to 2 GHz) mobile 3-D air defence surveillance radars.

Description
The Martello radar family consists of a number of mobile 3-D radar systems including the S713, S723, S743-D (see separate entry) equipments. The S713 is the earlier system and differs considerably in both physical appearance and technical characteristics from the later S723. Both systems operate in D-band and are designed to provide long-range complete air defence coverage.

S713
The S713 consists of the antenna, a radar container which houses the high-power transmitter, signal and data processing systems, a cooling/air conditioning unit and a power generation unit. These are all based on the International Standards Organisation freight container concept so that they can be transported by standard prime mover or carried in transport aircraft such as the C-130. The antenna consists of 60 identical horizontal linear array elements, vertically stacked, each with its own receiver. The antenna rotates at 6 rpm and each array receives returns which are assembled into a passive intermediate frequency beam forming network. The transmitter system consists of a 3 MW output coherent transmitter with a frequency synthesiser and transmitter drive.

S723
The later S723 system presents a markedly different appearance, having a wider horizontal but shorter vertical aperture. Technically the S723 planar-array antenna incorporates integral solid-state transmitters giving a fail-soft capability, combined with high-performance receiver modules. Two versions of the S723 are available: S723A and S723C, the latter having six elevation beams as opposed to eight for the S723A.

The S723 is a 3-D stacked beam radar with a parallel receiving system for height-finding. Bearing, range and height are available on every target on each revolution of the antenna. The planar antenna is a vertical stack of array elements each fed from solid-state transmitters housed in the antenna spine. Each element has its own high-performance receiver. The transmitted RF power is fed to each array with appropriate phase and amplitude relationship to obtain a cosec^2 vertical cover with narrow azimuth beamwidth and low sidelobes.

Returns from the targets are received by all arrays. The individual receiver outputs are then combined in a passive beam forming network which synthesises either six or eight elevation beams matched to the required elevation cover. All beams have pulse compression and full adaptive signal processing is provided out to maximum range. Target bearing and range are automatically extracted from a series of individual returns by the plot forming system and monopulse measurement of returns in adjacent beams yields corresponding height data.

The trailer-mounted antenna accommodates the solid-state transmitters, the receivers, the beam forming network, and the Identification Friend-or-Foe/Secondary Surveillance Radar (IFF/SSR) interrogator and its co-mounted antenna. The electronics container houses the six or eight adaptive signal processors, the 3-D plot extraction system, the IFF/SSR data extractor, the plot correlation and digital datalink equipment and a radar management console.

Status
Known S713/S723 Martello customers comprise the UK (five radars for UK Air Defence Ground Environment (UKADGE) including four S723s), Oman (two S713s) and Denmark (two S723s). Of these, the five UKADGE radars are reported to be operational while the Danish equipments are noted as being located on Bornholm and in the Faroes. The Omani S713 sets are quoted as being commissioned during January 1989. Additionally, the Royal Jordanian Air Force is understood to have placed a US$15 million contract for an undisclosed number of an unidentified Martello variant. S713/S723 activity within Marconi Electronic Systems is understood to be centred on support for existing systems rather than new production.

Specifications
S723 System
Frequency: D-band (1-2 GHz)
Bandwidth: 10%
Antenna dimensions: 7.1 m high × 12.2 m wide
Rotation rate: 6 rpm
Azimuth aperture: 1.4°
Elevation aperture: 0-20°
Power output: 132 kW (max); 5 kW (mean)
Instrumented range: 500 km

The S723 Marconi Martello 3-D radar's new planar-array antenna differs noticeably from that of the earlier S713 Martello

Instrumented height: 200,000 ft
Range accuracy: 150 m
Height accuracy: 1,700 ft (S723A); 2,000 ft (S723C)
Azimuth accuracy: 0.13°
Detection range (small aircraft): 500 km (S723A); 425 km (S723C)

Contractor
Alenia Marconi Systems (an Alenia-BAE Systems joint venture), Chelmsford/Rome.

VERIFIED

S743D Martello surveillance radar

Type
D-band (1 to 2 GHz) 3-D long-range transportable or static air defence surveillance radar.

Development
The S743D is a member of the Martello family of D-band (23 cm) transportable radars manufactured by Alenia Marconi Systems for long-range air defence. It is a substantial update of the S723.

Description
The S743D, as with the other members of the Martello family, is a 3-D stacked beam radar with a parallel receiving system for height-finding. Bearing, range and height are available on every target on each revolution of the antenna. The planar-array antenna consists of identical horizontal linear elements vertically stacked, each with its own receiver and transmitter modules. Each array element has the same shaped amplitude distribution, giving narrow azimuth beamwidth and excellent azimuth resolution. Low sidelobes are achieved by precise control of the amplitude and phase, which are fed to each array element. In elevation, the phase of the radio frequency power in the transmitter modules is controlled to give cosec^2 general surveillance cover, or to shape the transit beam in the operational burn through mode.

Returns from the target are received by all arrays. The individual receiver outputs are combined in a passive beam forming network which synthesises eight elevation beams matched to the required elevation cover. All beams have pulse compression and full adaptive processing is provided out to maximum range. Target bearing and range are automatically extracted from a series of individual returns by the plot forming system and monopulse measurement of returns in adjacent beams yields corresponding height data.

The advanced technology planar phased-array antenna and a new fully solid-state transmitter module, combine to give wider bandwidth capability and improved reliability and logistics. The radar antenna and associated electronics antenna are equipped with electromagnetic pulse protection.

A new transputer-based signal processor, using advanced data-farming techniques, provides fail-safe flexibility and has sufficient power to process fully all range cells without reduction of data rate.

The S743D does not rely upon the use of frequency or phase change to achieve height data. Thus the radar can operate on a single spot frequency or in an agile mode to suit prevailing defence requirements. In either case, height data accuracy is preserved. The received signals are processed using pulse compression, Doppler filtering and adaptive thresholding to provide target detection with constant false alarm rate in conditions of heavy clutter, chaff or electronic countermeasures. The signal processor provides a choice of moving target detector or moving target indicator modes, with single or dual clutter notches according to prevailing conditions. The radar cover can be subdivided into a number of range/azimuth cells, and the signal processing automatically or

The S743D Martello transportable radar system

manually optimised to the conditions existing in each cell, thus avoiding compromising the overall output from the radar.

A feature of the radar is the use of Inmos transputers in the signal processor. These devices provide a massive increase in processing power and flexibility. This is important in a system which must react to a rapidly changing radar environment in real time. Each radar uses 4,000 transputers to provide 88 computing modes, giving a massive parallel computing capability. Built-in redundancy enables the system master controller to reconfigure the computer automatically, enabling it to continue to operate if individual units fail.

A secondary surveillance radar/identification friend-or-foe interrogator, responder and plot extractor are incorporated into the S743D, with a large vertical aperture antenna mounted on the top of the planar-array.

The S743D system contains integral data processing, display, simulation and communication facilities and primary and secondary radar data can be correlated to provide radar plot data to a number of operations centres. Remote-control facilities are provided, which enable all important radar parameters to be controlled from an operations centre.

The system provides for transportation and recommissioning on a new site for security purposes. The antenna, with an integral hoist for removal of antenna sections, can be readily broken down for transportation. The radar electronics are housed in a separate container.

Status

As of this edition, known customers for the S743D system comprise Greece (three radars), Malaysia (two), the Oman and Thailand (two). Of these, Jane's sources suggest that Malaysia has established S742D sites at Bukit Puteri, Bukit Ibram, and Labuan, with the latter location understood to be for temporary deployments. The Omani order was first publicised during June 1999 and, as of this edition, was scheduled for completion during 2000/01.

Contractor

Alenia Marconi Systems (an Alenia-BAE Systems joint venture), Chelmsford/
Rome.

VERIFIED

....

S763-LANZA

Type

D-band (1 to 2 GHz) long-range 3-D air surveillance radar.

Description

The S763-LANZA long-range 3-D air surveillance radar is a NATO Class 1 equipment that combines the antenna, transmitter and receivers from the Alenia Marconi Systems' S743SD Martello system (see separate entry) with radar management electronics, displays, an Identification Friend-or-Foe (IFF) subsystem and the 'state-of-the-art' LANZA radar signal processor from the Spanish defence electronics contractor Indra. Looking at some of these elements in more detail, the radar's planar array antenna comprises 40 precision cut horizontal linear elements that are stacked vertically and each of which has its own receiver. The array's sidelobes are controlled via precise adjustment of the phase and amplitude of the signals feed to each of the assembly's elements. A total of 32 distributed, solid-state transmission modules are used to drive the array, with phase control being exercised by a high-speed beam switching system. Here, the number and position of the system's pencil beams are software-controlled to match both the particular threat scenario and to maximise 'beam-on-target' time. Equally, the shape and position of each beam are controlled in range and azimuth (both in transmission and reception modes) to 'step over' and minimise terrain and other clutter. Target height is obtained using monopulse techniques with measurements at particularly low elevation angles being aided by the use of 'special' beam formations.

The S763-LANZA 3-D air surveillance radar

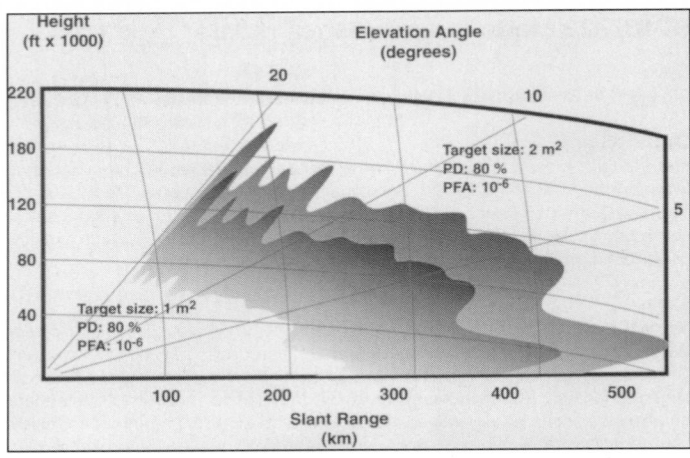

A typical coverage diagram for the S763-LANZA air surveillance radar

2000/0063804

The LANZA signal processor is a fail-safe design that is fully software-controlled and offers adaptive Moving Target Indication/Moving Target Detection (MTI/MTD) modes to overcome all types of clutter out to the radar's maximum instrumented range of 470 km. Digital compression techniques are used to generate 'extremely accurate' target range and height data while the architecture is constantly adapting to the prevailing environmental conditions. LANZA incorporate an 'intelligent' built-in test routine that both recognises failed modules and reconfigures the system to make use of its inherent redundancy.

Other S763-LANZA radar system features include:

- a Mode 4/Mode S integral IFF/monopulse secondary surveillance radar subsystem
- a distributed, solid-state transmitter providing full D-band coverage
- a wide operating bandwidth
- time and frequency processing with adaptive MTI/MTD every three to eight pulses
- 'uncommited' frequency agility
- pulse-to-pulse agility in both elevation and range
- variable Pulse Repetition Frequencies (PRF)
- programmable operating modes including pulse group, pulse coding and power concentration ('burnthrough')
- silent sectors in elevation and azimuth
- automatic radar management with local or remote control and manual overide
- road, rail, sea and air transportable
- an optional radar environment simulator.

Status

As of this edtion, the Spanish Air Force is understood to have procured a total of 10 S763-LANZA radars for use in its Sistema Integrado de Mando y Control Aereo (SIMCA) programme which was launched during 1994.

Specifications

Frequency: 1-2 GHz
Power output: 10-470 kW (instrumented)
Azimuth coverage: 360° (instrumented - 6 rpm scan rate)
Elevation coverage: 0-20° (instrumented)
Altitude: Up to 30,480 m
Accuracy: 0.1° SD (azimuth); 50 m SD (range); 340 m (altitude - at 185 km range)
Resolution: 1° (azimuth - 80% PD); 50 m (range - 80% PD)
MTBF: >1,250 h
MTTR: <30 min
Availability: 99.8%
ECCM features: distributed receivers; fast acting CFAR; jamming strobe processing; moving notch MTI; PRF stagger; pulse compression; sector processing; side-lobe blanking; use of least jammed frequency; very low side-lobe antenna and wide, pulse or burst agile frequency band
Antenna
Azimuth beamwidth: 1.4°
Off-axis side-lobes: -55 dB
Gain: 41 dB (transmit and receive - lower beams)
Dimensions (W x H): 12 x 7 m
Deployment: < 5 h (5 man crew - no crane needed)
Receiver
Number of modules: 40
Number of beams: programmable
Transmitter
Number of modules: 32
Power: 5.35 kW (mean); 53.5 kW (effective peak)
Frquency agility: full NATO D-band spectrum

Contractors

Alenia Marconi Systems (an Alenia-BAE Systems joint venture - antenna, transmitter and receivers), Chelmsford/Rome.
Indra (prime contractor, signal processing, radar management electronics, IFF and displays), Madrid.

VERIFIED

TRM radar series

DaimlerChrysler Aerospace (now a part of the European Aeronautic, Defence and Space (EADS) Co) has developed a family of mobile and shipborne G-band (4 to 6 GHz) medium- and long-range radar systems. Common to all systems are 'excellent' Electronic Counter-CounterMeasures (ECCM) features, high modularity and standardised logistics. Although the land-based versions have been designed primarily for mobile configurations they can also be supplied for operation in fixed stations, as coastal radars for sea and air surveillance and as gap fillers. They are suitable for integration into air defence networks, anti-aircraft weapon systems and for integration with other radar equipments.

TRM-S Area Defence System

TRM-S is a mobile G-band (4 to 6 GHz) 3-D radar that consists of a remote-controlled sensor and an operator shelter (which can be up to 300 m away from the radar), with both system elements being mounted on 10-ton cross-country trucks. On redeployment, setting-up (by six operators) takes less than 30 minutes. The operator shelter is Nuclear/ Biological/Chemical (NBC) warfare-proofed and contains up to three workstations, each of which has Command, Control and Communications (C³) functions.

A selection of operational modes (preprogrammed to user-specific requirements) enables the TRM-S to adapt optimally to all feasible tactical conditions. In all modes, scanning is performed by a high-precision computer-controlled pencil beam. Up to 4,000 targets can be displayed simultaneously and an Identification Friend-or-Foe (IFF) array is integrated with the radar's antenna. The sensor's 'excellent' ECCM performance is 'guaranteed' by the agility of its transmission, pulse repetition and polarisation parameters. Performance monitoring is carried out by an integrated automatic test system and phase-coded pulse-compression 'ensures' long range at low-transmit power together with high resolution. Additional self-protection can be obtained from the TRM-S's ability to remotely control a standoff deception jammer.

The TRM-S mobile area defence radar deployed

The TRM-L point defence system radar deployed

Specifications
Range: 200 km (radar cross-section 2 m², detection probability 60%)
Data renewal time: 3-15 s (depending on mode)
Max antenna height: 12 m (upper edge)
Max wind speed: 100 km/h (operating); 150 km/h (standby)
Power consumption: 45 + 30 kW

TRM-L Point Defence System
The TRM-L is a one-shelter version of a medium-range G-band (4 to 6 GHz) radar that is based on modules from the TRM-S system. The system's NBC proof shelter is mounted on a 15-ton cross-country truck. After relocation the TRM-L can be set up by three operators in less than 15 minutes. TRM-L can be supplied in either 2-D or 3-D versions, with upgrading from 2- to 3-D being possible. System ECCM provision centres on the sensor's frequency agility, pulse repetition frequency, polarisation and phase-coded, pulse compressed sub-pulses. High-probability of detection of anti-radiation missiles and low-level aircraft is achieved via signal processing routines and an additional channel ensures the reliable detection of helicopters. Sidelobe suppression and IFF functions have been integrated into the antenna.

Specifications
Range: 60 km (radar cross-section 1 m, detection probability 80%)
Data renewal time: 2-4 s (depending on mode)
Power consumption: 38 kW
Max antenna height: 12 m (upper edge)
Max wind speed: 100 km/h (operating); 150 km/h (standby)

Status
Over time, the German MoD has (according to Jane's sources) procured seven TRM-S radars for use as the sensor in a German Army mobile anti-aircraft C³ system that integrated the Gepard self-propelled anti-aircraft tank with the Roland Surface-to-Air Missile (SAM). The TRM-L, supplemented by a C³ component, has been used by the German Ministry of Defence as a RAdar Surveillance and Co-Ordination System (RASCOS) centre. Here, the aim was to produce local and wide area situation pictures by means of multiradar tracking. In its Flugabwehr Gefetschtsstand Roland (FGR) format, the TRM-L radar has served as the control sensor for Roland SAM batteries.

Contractor
European Aeronautic, Defence and Space (EADS) Co Systems and Defence Electronics, Ulm, Germany.

UPDATED

....

TRS 2052/TRS 2053/TRS 2056 D-band radars

Type
Family of D-band (1 to 2 GHz) air defence and aircraft recovery radars.

Development
This family of radars stems from models first developed in the mid-1970s and continuously improved since then. New technology versions can now be made available which use an intrinsically modular solid-state transmitter and multiple moving target detection filtering. This gives these later versions (or earlier models refitted with solid-state transmitters) great advantages as regards maintenance, round-the-clock availability, waveform formatting and unattended operation.

A solid-state version of this radar family has been developed and is known as the TRAC 2000/2100. Its main purpose is for use as an approach and terminal area radar and, consequently, the description can be found in the Military ATC, instrumentation and ranging radars section.

Description
The TRS 2052 is a D-band early warning radar which has been developed as a successor to the TRS 2050. Externally and in other general characteristics, the later

The TRS 2052 D-band early warning radar

The TRS 2056 (Centaure) D-band early warning radar

radar is very similar to the TRS 2050. The double curvature antenna is of the same dimensions (9 × 13 m) and independent high- and low-cover beams are radiated. Cosecant beam shaping in elevation extends to over 40° and sharp ground cut-off of the high cover lobe produces a strong contrast between air target echoes and ground returns. High-grade circular polarisation ensures rejection of weather returns.

An Identification Friend-or-Foe (IFF) antenna of the sum/difference type is mounted on the primary radar reflector, as is the sidelobe suppression antenna.

Variations of this successful equipment have led to two more models: the TRS 2053 and TRS 2056. The TRS 2052 air defence long-range surveillance radar and the medium-range and SRE radars TRS 2053 and TRS 2056 use different antennas with the same electronics. A fully equipped cabin (which is transportable by air, rail or road) houses a transceiver unit, which comprises two 2 MW/2 kW class high-stability transceivers (featuring long-life magnetrons, single knob tuning and dual diversity); an IFF interrogator and two reception/processing units (featuring a logarithmic receiver and a digital linear Moving Target Indicator (MTI) filtering assembly). A sidelobe suppression receiver unit can also be included as an option.

In this fashion, a set suitable for both fixed and mobile installations is available to associate with different antennas to constitute the TRS 2052, TRS 2053 and TRS 2056 stations.

With a large antenna (13 × 9 m) featuring high gain and dual coverage, a highly jam-resistant air defence master station, the TRS 2052, is obtained. This has a detection range capability in excess of 350 km for modern fighter aircraft. With a smaller antenna (9 × 5 m), it becomes the TRS 2053, designed for equipping military airfields as recovery stations or as gap filler radar in an air defence network. These radars are widely distributed around the world. The third one of this family, the TRS 2056, developed for the French Air Force as a movable recovery station, uses an 8.5 × 3.5 m antenna which can be dismounted without special tools.

Simultaneous use of operationally variable frequency, frequency diversity, constant false alarm rate receivers, of high performance, compatible MTI and of sidelobe suppression, endows the radar with excellent anti-jamming capabilities.

All three sets are compatible with fully automatic operational use and feature built-in test equipment.

Status
As of March 2001, numerous radars of this family have been supplied to French and overseas customers for air defence, early warning or air traffic control purposes.

Specifications
TRS 2052
Operating frequency range: D-band (100 MHz)
Antenna
Reflector dimensions: 13 × 9 m
Gain (low cover): 36 dB ±0.5 dB
Gain (high cover): 35.5 dB ±0.5 dB
Beamwidth in azimuth (3 dB): 1.2°
Cosecant: >40°
Polarisation: circular, fixed
Rotation speed: 6 rpm
Transmitter
Frequency-tunable magnetron
Peak power: ≥2.2 MW
Pulse duration: 3 μs
PRF: 350 Hz (3-period staggered)
Receiver
Noise figure of RF amplifier: 3 dB
Noise figure at input to reception chain: 4 dB
Intermediate frequency: 30 MHz
Subclutter visibility: ≥25 dB

Digital post-detection integration
Power supply: 3-phase, 220/380 V, 50 Hz
Consumption: 40 kVA (55 kVA for diversity operation)
Performance
Detection range for fluctuating 2 m² target: >333 km
Detection in altitude: >18,288 m

Contractor
Thales Air Defence, Bagneux, France.

VERIFIED

TRS 2054 surveillance radar

Type
D-band (1 to 2 GHz) long-range air surveillance radar.

Description
The TRS 2054 is a coherent, klystron-powered, long-range surveillance radar operating in the D-band. It has been evolved for air traffic surveillance roles and similar applications and is thought to have been based upon experience gained with the TRS 2052 series of radars. It embodies the latest concepts in coherent double-pulse operation and auto-regulating signal processing. This dual-channel 23 cm radar is claimed to be able to detect small aircraft at ranges of up to 370 km and superior clutter see-through capabilities permit detection down to low levels. An integral plot extractor enables the radar to be connected to a data processing system.

The dual-beam antenna uses offset feeds for reduced masking of the energy radiated by the reflector, thereby reducing sidelobes. There is an elongated lower edge to the reflector to intercept radiation from the primary sources so that backlobes are reduced. The feeds are designed to ensure that the reflector edges are only illuminated by low-energy radiation, thus reducing spillover to a level comparable to the background level. A cosec² radiation pattern, boosted at high-elevation angles, is employed, providing good contrast between short-range targets and ground clutter. There is a sharp roll-off at low-elevation angles, which also improves the signal-to-clutter contrast and this is obtained by employing a reflector with a larger than usual vertical dimension. Compact stacking of the main and auxiliary feeds gives good focusing for the high-beam and reduced boresight separation between the two beams, while still maintaining a target-to-clutter enhancement of 16 to 18 dB.

The transmitter uses a TH 2068 klystron which is stated to have a saturation gain of over 50 dB. It is a pulse-driven tube with five cavities giving a peak output of 5 MW, although in the present application it is deliberately under-driven to increase its service life. It is operated at 3.5 MW peak power and 4.7 kW mean power. The transmitter is crystal-controlled. At each pulse repetition interval, the transmitter generates a coherent pulse train consisting of a long duration pulse followed immediately by a short pulse. This provides a means of obtaining simultaneously, in a single radar, the benefits of both long and short pulse operation. The latter gives a precise radar cell and so improves both target resolution and signal-to-clutter contrast, these characteristics being especially valuable at short and medium ranges. At greater ranges, the longer pulse results in an increased useful detection range.

The TRS 2054 (LP23K) long-range surveillance radar

The radar processing system associated with the TRS 2054 radar is called TVD 1000 and is of the Moving Target Detector (MTD) type. It includes a separate processing channel, called weather channel and is designed to improve target detection in various forms of clutter, while providing low-output false alarm rate. The chosen signal processing algorithms are based on the use of a band of eight Doppler filters, on constant false alarm rate thresholding and on an adaptive clutter mapping. The weather channel indicates to the user the areas of dangerous precipitation within the radar coverage. Programmable modular processors are used for the digital signal processing and data processing. Their processing enables complex adaptive algorithms to be carried out. This equipment is incorporated in a radar station which can be unmanned, thanks to the automatic built-in test equipment and the remote maintenance and monitoring system. Results of experiments with the processing system described show clearly that the MTD processing improves the final picture quality. In particular, the detection probability is higher both in clear zones and in clutter, while the false alarm rate remains low.

Other members of the family use the same inherent transmitter/receiver and the antenna of the TRS 2053.

Status
As of March 2001, TRS 2054 is understood to have been supplied to customers in France and a number of other countries.

Contractor
Thales Air Defence, Bagneux, France.

VERIFIED

A TRS Tiger radar equipped with a mast-mounted antenna

TRS 2100 Tiger S radar

Type
E/F-band (2 to 4 GHz) low-altitude surveillance radars.

Description
Under the codename Tiger, Thales Raytheon Systems (formerly Thales Air Defence) has developed a family of lightweight, high-performance radars, the first being the TRS 2100, for the detection of low-flying aircraft. A special version is produced for the French Air Force, designated Aladin. An Improved Tiger version particularly well suited in resistance to high jamming and to non-standard propagation conditions is now being produced. High performance in severe clutter is achieved by the simultaneous use of Moving Target Indicator (MTI) circuits based on Doppler frequency filtering and of pulse compression to limit the volume in which clutter echoes can mix with useful echoes. The radar can be used either as an autonomous detection centre, with local exploitation of radar data, or as a gap filler station linked to an air defence network. Specific operational functions listed by the manufacturer include: control of tactical operations; low-altitude gap filler;

The French Air Force Aladin version of Tiger S - tactical packaging

The Improved Tiger S with electronics housed in the operations cabin

coastal and maritime surveillance; alert and target designation for surface-to-air missile batteries; surveillance radar equipment role and similar mobile or air transportable applications.

The use of E/F-band makes it possible to have an antenna with a high-gain and high-angular resolution but with a relatively small reflector (5 × 2.3 m). The reflector profile is C-shaped with a double curvature and the vertical pattern has a steep slope at low elevations and is super-cosecanted at high elevations. An Identification Friend-or-Foe (IFF) subsystem is incorporated (IFF antenna integral with primary radar antenna) and a built-in test facility monitors the main parameters and locates any faulty functions.

The transmitter uses a coherent amplifier chain. The transmission frequency is obtained from a highly stabilised crystal oscillator and for frequency agility purposes transmission is carried out at different frequencies distributed at random over 250 MHz. The output power is provided by three successive amplifier stages (one transistor stage, one travelling wave tube stage and one cross field amplifier stage). The phase stability of this chain is such as to ensure good clutter rejection performance. Pulse compression is achieved by incorporating crystal dispersive networks, the propagation time of which varies with the frequency in the transmit and receive circuits.

The receiver is preceded by a wideband, low-noise radio frequency amplifier. Rejection of fixed echoes is obtained by a digital linear MTI. A constant false alarm rate chain operates with the MTI processing chain.

The radar head can be linked by either a single coaxial cable or microwave link to a remotely sited operations unit. Utilisation can be manual, semi-automatic or fully automatic.

The radar is transportable by helicopter or by conventional truck or cargo aircraft. The transmitter/receiver can be mounted either on a trailer together with the antenna, or in a technical/operations cabin. The antenna can also be mounted on a mast (14 or 54 m).

Other members of the family are the Tiger G systems (TRS 2105 and TRS 2106) which are described under a separate entry.

Status
As of March 2001, 33 Aladin versions of the Tiger S are known to have been procured by the French Air Force. Elsewhere, more than 10 other countries in the Middle East (including four for Tunisia) and Africa are understood to have procured the Tiger S radar over time. Improved version production commenced in 1986 and deliveries have been made to several customers.

Specifications
Improved Tiger S
Operational performance: on 2 m² (fluctuating target) with P_d 80%
Detection range: 110 km
Altitude detection: >6,000 m
Elevation pattern: cosecanted up to 45°
Subclutter visibility: >40 dB
First blind speed: >1,500 kt
Antenna characteristics
Size: 5 × 2.3 m. IFF antenna integrated with primary radar antenna

Gain: 34 dB
Polarisation: circular, fixed
Azimuth beamwidth: 1.3°
Rotation speed: 12 rpm
Transmitter/receiver characteristics
Mean power: 1 kW
Overall receiver noise figure: <3.5 dB
Frequency agile using synthesiser; dual-pulse PRF operating mode; linear 3 memory DMTI filter; CFAR; post-detection integration circuits.

Contractor
Thales Raytheon Systems, Massy, France.

UPDATED

TRS 2105/TRS 2106 Tiger G low-altitude radars

Type
G-band (4 to 6 GHz) tactical surveillance radars.

Description
The Thales Raytheon Systems (formerly Thales Air Defence) Tiger G radar series (TRS 2105 and 2106) is intended for tactical low-altitude surveillance. These radars are mainly devoted to low-level coverage and gap filling in an air defence network; alert and co-ordination of ground-to-air weapons and control of tactical operations. Both operate in G-band, use coherent transmission, frequency agility, pulse compression, moving target indication, constant false alarm rate and four operating modes. Both are highly resistant to electronic countermeasures and are fitted with a processing unit featuring automated radar adaptivity to the environment and automatic initialisation and tracking of local targets.

TRS 2105
The TRS 2105 is housed in a cabin installed on a two-axle trailer. The cabin supports a foldable 8 m mast bearing the 3 × 1.6 m antenna. The complete system can be carried by road, rail or cargo aircraft (C-130/C-160).

TRS 2106
The TRS 2106, which was a derivative of the TRS 2105 and is particularly suited to battlefield operations, has been replaced by the RAC alert and co-ordination radar (see entry in Battlefield, missile control and ground surveillance radar systems section).

Status
As of November 2001, in excess of 40 Tiger G sensors were reported as having been supplied to customers in Africa and the Middle East. As of the given date, the Tiger G radar series was no longer in production.

Specifications
TRS 2105
Frequency: G-band (550 MHz)
Peak power: 70 kW
Mean power: 1 kW
Detection range: 65/100 km
Rotation rate: 5.5-11 rpm
Automatic tracking capacity: 60 tracks

Contractor
Thales Raytheon Systems, Massy, France.

UPDATED

The Tiger G (TRS 2105) mobile low-altitude surveillance radar

TRS 2201 3-D radar

Type
E/F-band (2 to 4 GHz) 3-D air defence radar.

Description
The Thales Raytheon Systems' (formerly Thales Air Defence) TRS 2201 is a long-range 3-D E/F-band air defence radar that is also known as Palmier. A modern technology version of this radar is used in the NATO Air Defence Ground Environment (NADGE); it ranks among the most powerful radars of its type in the world.
    Major features of this radar are its high-power (20 MW peak) transmitter, its elaborate antenna structure and its range of electronic counter-countermeasures facilities that enable it to function satisfactorily in any environment. The equipment is solid state to the maximum extent possible, is of modular construction and has built-in test facilities.

Status
As of March 2001, TRS 2201 was no longer in production. Over time, a number of countries within NATO/Europe (besides France) and the Middle East/Africa have procured TRS 2201 radars. The equipment was selected by NATO as a primary component of its original NADGE architecture. However, many of these have been replaced by new generation radars such as the TRS 22XX.

Specifications
Frequency: E/F-band (2-4 GHz)
Peak power: 20 MW
Range: >460 km
Accuracy: ±1,500 ft (height)
Data renewal rate: 6/min

Contractor
Thales Raytheon Systems, Massy, France.

UPDATED

The TRS 2201 long-range 3-D air surveillance radar

TRS 2215/2230 3-D radars

Type
Family of E/F-band (2 to 4 GHz) fixed or mobile 3-D air defence radars.

Description
This Thales Raytheon Systems' (formerly Thales Air Defence) family of E/F-band air defence 3-D radars employ electronic phase scanning in elevation and are derived from the French Air Force's height-finding SATRAPE sensor. Within the family, the designation TRS 2230 is given to the fixed-site/relocatable variant, while the equipment's mobile configuration (transportable by road, rail or aircraft (Transall, C-130, or equivalent) and deployable in less than one hour) is termed the TRS 2215. Both radars employ the same space scanning techniques, transmitter, reception and processing circuitry and antenna components married to two sizes of planar phased-array antennas. This commonality (which has 'particular relevance' to maintenance, logistics and training) is claimed to facilitate a 'high degree' of deployment flexibility within air defence systems that make use of both

A TRS 2215 long-range 3-D mobile radar station

A TRS 2230 long-range 3-D radar installed in Brazil

configurations. Functionally, TRS 2215/2230 radars use mechanical rotated antennas to scan in azimuth and the primary aperture used is a direct radiation array of stacked linear sub-arrays fed by two vertical distributors used to form the sum and the elevation difference channels. Each linear sub-array contains elementary sources radiating a circularly polarised wave. The two waveguides allow monopulse operation to perform the height-finding function. Digital electronic phase shift networks inserted between the elementary feeds and the corresponding directive couplers control the pointing of the beams.

The transmitter is an amplifying chain transmitter using a high-efficiency crossed field amplifier as its final tube. It permits the use of the pulse compression technique and that of pulse-to-pulse frequency agility. It is characterised by high efficiency (greater than 70 per cent) and a modular modulator allows repair operations, if required, to be carried out without shutting down the station.

The receiving section groups the receivers, the Moving Target Indicator (MTI) unit, the extractor, the plot processing and altitude computing equipment, the control device for the tilting of the antenna according to the mode of operation imposed and the monitoring desk.

The receiver circuits comprise the anti-jamming chains and three angle error chains. The MTI device processes the signals received from any one of the three lower beams by programming. The extractor performs the extraction of primary and secondary plots and allows them to be associated together. It is followed by a plot processing system which computes altitude and filters plots so as to transmit only useful data. The device generates the data in a digital form permitting the transmission of tracks to a remote centre. The monitoring desk groups the monitor controls of the station's operation.

The high-anti-clutter and anti-jamming capabilities stem from a number of properties, including:

- a high-grade antenna pattern with very low sidelobes
- true random frequency agility in elevation and azimuth, as a result of phase electronic scanning and non-dispersivity respectively
- inhibition of jammed frequencies
- small radar cell (0.2 μs compressed pulse)
- pseudo-random pulse repetition frequency
- intrapulse diversity
- auto-adaptive MTI filtering
- automatic clutter processing
- several operating modes programmable with respect to the operational environment.

Status

As of November 2001, Jane's sources were suggesting that TRS 2215 radars have been sold to customers in Asia, Europe, the Middle East/Africa and the Pacific Rim/Far East, while TRS 2230 sensors have been procured for use in the Middle East/Africa, the Pacific Rim/Far East and South America. In more detail, these procurements are understood to have included the armed forces of Brazil, Cyprus, France, India, Indonesia and Tunisia. Looking at some of these more closely, TRS 2215s are quoted as being supplied into France's *Système de Traitement et de Représentation des Informations de Défense Aérienne (STRIDA)* national air defence network as upgrades, while TRS 2230s are noted as forming part of the CINDACTA I (based on Brasilia), II (Curtiba) and III (Recife) subsets of Brazil's evolving integrated civil and military airspace surveillance and control system. Within the Brazilian system, TRS 2230 radars are thought to be located at (in alphabetical order) Aragarças, Canguçú, Cantanduvas, Chapada Dos Guimares, Couto, Gama, Jaraguari, Morro Da Igreja, Santiago, São Roque, Tanabi, Três Marias and Vitória. Servicing this programme involves Thales working in partnership with a consortium of Brazilian contractors including ESCA SA, ELEBRA Controles SA and Tecnasa Eletronica Profissional SA.

Elsewhere, India is reported to have acquired nine TRS 2215D equipments. Four of these are noted as having been procured directly from Thales with the remaining five being built under licence in India by Bharat Electronics. Indonesia's air defence system is reported to incorporate two TRS 2215D and 12 TRS 2230D radars, while Tunisia is noted as having acquired two TRS 2215 systems. Sources suggest that Pakistan may also have acquired the TRS 2215 radar, a procurement that as of March 2001, had not been confirmed.

Specifications

Detection range: 510 km
Ceiling: >30,500 m
Elevation coverage: −3 to +30°; programmed operating multimode
Data renewal rate: 10 s
TRS 2215 antenna
Size: 5 m span × 5.5 m high
Aperture: 1.5° azimuth; 1.3-3.6° elevation
Gain: 38.5 dB
Beamwidths (3 dB): 1.5° (azimuth); 2-4° (elevation)
TRS 2230 antenna
Gain: 40 dB
Beamwidths (3 dB): 1.5° (azimuth); 1.3-3.6° (elevation)
Polarisation: circular, fixed
Rotation speed: 6 rpm
Peak power: 700 kW
Average power: 10 kW
PRF (average): 380 Hz
Pulse duration: 3 × (13 μs compressed to 0.2 μs)
Reception: CFAR
MTI improvement factor: 40 dB

Contractor

Thales Raytheon Systems, Massy, France.

UPDATED

TRS 22XX long-range 3-D radar

Type

E/F-band (2 to 4 GHz) transportable 3-D air defence radar.

Description

The Thales Raytheon Systems' (formerly Thales Air Defence) TRS 22XX is a mobile 3-D air defence radar complying with the NATO Class 1 radar specifications. It has been developed primarily to replace the long-range 3-D radars operated by the French Air Force. It is designed for efficient operation in conditions of various and simultaneous clutter, heavy jamming, blast and battlefield conditions. It is transportable by road, rail, sea and air.

The TRS 22XX operates in E/F-band with a large frequency bandwidth (15 per cent). Its detection coverage is 470 km range, 100,000 ft in altitude and a 20° elevation pattern. It has a number of advantages including a large operating bandwidth, a high-grade antenna with very low sidelobes and an electronic phase scanning, non-dispersivity single beam operation. Various transmission modes are programmable by narrow azimuthal sectors, including fixed-frequency transmission, pulse-to-pulse or burst-to-burst frequency agility over the complete band and radar silence. Various operating modes programmable by narrow azimuthal sectors are available in order to optimise detection according to the operational environment. Other advantages include auto-adaptive Doppler processing associated with a very short compressed pulse and anti-jamming techniques such as analysis, automatic extraction of jam strobes, selection of least jammed frequency and display of jammed sectors.

The TRS 22XX radar station consists primarily of:

- the aerial assembly; the azimuth coverage results from the mechanical rotation of the antenna while elevation cover stems from beam and phase scanning

A TRS 22XX radar for the Turkish air defence network

techniques. The latter, together with the non-dispersive design of the antenna, allows true frequency agility over the complete bandwidth in all directions
- the high-average power transmitter, located in a cabin, with its ancillaries container
- a reception and signal processing cabin, which houses a local console for the radar station control.

The radar may also be operated from a remote position. It is fitted with built-in test equipment which carries out a permanent check of the operational status and delivers diagnostic information to the local or remote monitoring systems.

Status
Known TRS 22XX contracting activity includes a mid-1990 award from Turkey for 14 such radars for use in its air defence network and a French procurement as part of the upgrading of the country's *Système de Traitement et de Représentation des Informations de Défense Aérienne (STRIDA)* national air defence system. With regard to the Turkish programme, the then Thomson-CSF AIRSYS (subsequently Thales Air Defence) teamed with Turkish contractor Tekfen to produce 10 of the 14 radars in the country. Elsewhere (and as of November 2001), Jane's sources suggest that TRS 22XX has been procured by at least one customer in the Middle East/Africa region.

Specifications
Frequency: E/F-band (2-4 GHz)
Bandwidth: 15%
Antenna
Dimensions: 6 m wide × 4.5 m high
Rotation speed: 6 rpm
Azimuth beamwidth: 1.2°
Elevation beamwidth: 1.6°
Gain: 40 dB
Range: 470 km
Ceiling: 100,000 ft
Elevation: 20°
Accuracy: 50 m (range); 0.25° (azimuth), 2,000 ft (altitude)
Resolution: 120 m (range); 2.5° (azimuth)
MTBF: 500 h

Contractor
Thales Raytheon Systems, Massy, France.

UPDATED

ISRAEL

EL/M-2082 ADAR tactical air defence and air traffic control radar

Type
D-band (390 MHz to 1.55 GHz sub-band) solid-state, three-dimensional (3-D) air defence and air traffic control radar.

Description
Designed to fulfil transportable air defence and fixed-site en route air traffic control roles, Jane's sources suggest that ADAR utilises hybrid electronic/mechanical scanning to provide 3-D coverage. Here, azimuth measurement is provided by mechanically swinging the equipment's active array with elevation data being provided via electronic multibeam (four) steering. A two-dimensional, monopulse, secondary surveillance radar antenna is mounted back-to-back with the primary array to provide interrogation, decoding and tracking in Modes 1, 2, 3/A, C and 4. ADAR's manufacturer claims that the combination of hybrid active scanning, Fast Fourier Transform Doppler processing, variable waveforms and medium- to low-pulse repetition frequencies enables system performance to be maintained in adverse weather conditions and environments containing ground and sea clutter and intentional and unintentional electronic interference. The use of commercial/industrial off-the-shelf components, solid-state technology, multiple transceiver modules and a high redundancy active array are also claimed to ensure high reliability/maintainability and low life-cycle costs.

Status
A transportable ADAR application has been bid to meet a Royal Australian Air Force tactical surveillance radar requirement.

Specifications
Frequency: 390 MHz-1.55 GHz
Azimuth coverage: 360°
Mechanical scan rate: 6 or 12 rpm
Max detection range: 370 km
System weight: 7.5 t (transportable variant)

Contractor
Elta Electronics Industries Ltd (a subsidiary of Israel Aircraft Industries Ltd), Ashdod.

VERIFIED

JAPAN

J/FPS-2 3-D air defence radar

Type
3-D air defence radar.

Description
The J/FPS-2 is a static 3-D air defence radar developed by NEC in Japan. It employs an antenna that embodies a phase/frequency scanning technique of a planar phased-array in elevation, with mechanical rotation in azimuth. A combination of serpentine feed and ferrite phase shifters is used for scanning a pencil beam in elevation.

The radar embodies energy management concepts and signal processing techniques giving high-detection and stable accuracy, with a travelling wave tube and cross field amplifier transmitter chain, chirp pulse compression and amplitude comparison for height measurement. The advanced digital signal processing techniques used for clutter reduction and rejection (and electronic counter-countermeasures) significantly enhance performance for automatic detection and tracking under severe clutter and jamming environments.

The J/FPS-2 is capable of direct interfacing with any automated air defence network and it is linked with air defence ground environment and air defence command post systems. The system has dual channel configuration with high-reliability design features and substantial built-in test equipment to minimise system down time.

The general configuration of the J/FPS-2 can be seen from the photograph. The associated transmitter/receiver, display and other electronics can be housed either in permanent or temporary buildings.

Status
As of this edition, J/FPS-2 is thought to have been procured by the Japan Air Self-Defence Force.

Contractor
NEC Corporation, Tokyo.

VERIFIED

The Planar-array of the J/FPS-2 static 3-D air defence radar

J/FPS-3 air defence radar

Type
Phased-array radar.

Description
The J/FPS-3 air defence radar is a 3-D phased-array radar project, which is the main air defence/interceptor control radar of the Japanese Base Air Defence Ground Environment (BADGE) air defence system. Each J/FPS-3 has a long-range antenna, short-range antenna, false signal emitter, signal processing unit and an underground control system. Flexible detection and tracking is provided through random access beam steering. Resistance to electronic countermeasures is achieved by pulse compression, two wideband frequencies and low-antenna sidelobes.

Status
The Mitsubishi Electric Corporation delivered an advanced phased-array radar to the Japan Defence Agency's research and development institute in December 1986. Communications with control centres was by optical cables.

As of this edition, six J/FPS-3 systems are reported to be operational. While Mitsubishi will neither confirm or deny the reports, Jane's sources suggest that J/FPS-3 operates in the D/E/F-band (1 to 4 GHz), may incorporate an anti-anti-radiation missile decoy subsystem and that approximately 20 such radars are to be acquired.

Contractor
Mitsubishi Electric Corporation, Tokyo.

VERIFIED

J/TPS-102 air defence radar

Type
Mobile air defence radar.

Description
The J/TPS-102 is a short-range, mobile air defence radar which is intended to supplement the fixed-base radars in the Japanese Base Air Defense Ground Environment (BADGE) system. Other than the fact that it employs a cylindrical phased-array antenna for active electronic scanning, details have not been made public.

Status
As of this edition, J/TPS-102 is reported to have been procured by the Japanese Air Self-Defence Force.

Contractor
NEC Corporation, Tokyo.

VERIFIED

NORWAY

Norwegian acquisition radar and control system

Type
Mobile search, acquisition and tracking system.

Development
In order to meet the challenge of low-altitude and electronic warfare threats against ground forces, a joint venture between Hughes Aircraft Company and Kongsberg Gruppen is in operation. The result of this co-operation is a high-performance mobile Acquisition Radar and Control System (ARCS). The system represents further development and adaptation of the Hughes AN/TPQ-36A radar and Kongsberg's multifunction console KMC 9000.

Development of a new medium-range surface-to-air system, known as the Norwegian Advanced Surface-to-Air Missile System (NASAMS) has taken place. It is based on NOAH (NOrwegian Adapted Hawk) and the AN/TPQ-36A radar, with a surface-launched version of the AIM-120 air-to-air missile.

Description
The ARCS system provides: 3-D search, acquisition and tracking with an integrated identification friend-or-foe system; fully automated threat evaluation and weapon assignment; centralised, decentralised and autonomous methods of operation and automatic, semi-automatic and manual modes of operation. Advanced techniques for the rejection of Electronic Counter Measures (ECM) and radar clutter are also included.

The KMC 9000 is designed to meet requirements for military equipment and is based on systems already in use with the Norwegian naval and army forces. The basic concept of the KMC 9000 is modular with an intelligent high-resolution graphic/alphanumeric terminal as the heart of the system.

The KMC 9000 multifunction console

ARCS integrates two additional classes of short-range air defence weapons (anti-aircraft guns and man-portable missiles) to achieve a layered defence system that defies enemy tactics and countermeasures. This highly survivable netted system design means that all elements of the system are utilised most effectively for engagements and that the system remains effective despite loss of elements.

Integrated passive sensor
ARCS utilises an infra-red passive sensor, the Norwegian Tracking Adjunct System (NTAS), which provides for accurate raid-size assessment, hostile-act verification, passive tracking, kill assessment and enhanced system operations in ECM conditions.

Status
The first application of ARCS is in conjunction with the Improved Hawk surface-to-air missile batteries at Norwegian Air Force bases. All 24 ARCS systems have been delivered and the programme has been completed.

Contractor
Kongsberg Gruppen AS, Defence Systems, Kongsberg.

VERIFIED

Radar integration systems

Type
Stationary or mobile systems for integration of primary and secondary radar data.

Description
Radar Integration Systems (RIS) developed by Kongsberg Gruppen are aimed at integration of existing or planned air defence, naval and coastal surveillance radars. They offer advanced functional solutions for radar data processing and displays, providing operators with the facilities required to meet airspace and surface surveillance control.

The REX-2 and REX-300 families of radar extractors provide powerful solutions to automated surveillance radar and secondary surveillance radar plot extraction and tracking. This includes electronic counter measures strobe extraction, clutter-map generation and a serial-link interface. They can be installed as a compact autonomous unit or contained in a KMC-9000 workstation.

The KMC-9000 is a self-contained data processing and radar display system. The family contains ruggedised single or double display workstations with 356 or 508 mm high-resolution raster scan displays. The man/machine interface is based on menu-driven colour display interactions, a programmable entry panel, a tracker-ball, a general keyboard and a selection of dedicated push-buttons. The KMC-9000 can be used independently with an integrated REX-2 or REX-300, or linked together with a suite of other KMC-9000 workstations.

Status
As of this edition, RIS systems are thought to be in service for various naval and air warning applications in Norway and with export customers.

Contractor
Kongsberg Gruppen AS, Defence Systems, Kongsberg.

VERIFIED

The KMC-9000 Workstation

POLAND

N-11 air defence radar

Type
E/F-band (2 to 4 GHz) medium-range 3-D radar system.

Description
The N-11 is a mobile E/F-band air defence surveillance radar giving 3-D information on targets by mechanically rotating the antenna in azimuth and producing a stacked beam pattern in elevation. The system provides the optimum combination of range and height resolution, with excellent performance in electronic countermeasures and clutter. An Identification Friend-or-Foe (IFF) system is built in; alternatively an interface is provided for an autonomous IFF equipment.

The N-11 is self-contained in three operating units, each mounted on a TATRA 815 truck, plus an additional power generator as an option. The operations cabin provides protection for the operators against nuclear/biological/chemical warfare contamination. Cruising speed whilst in transit is given as up to 65 km/h. Typical deployment time when arriving on site is less than 30 minutes and the first radar scan is obtained immediately after deployment. Detected and tracked targets are presented on two raster scan displays. Communication equipment for digital data transmission is also provided. Built-in and self-testing facilities give a fast check of the system after installation and monitor the equipment during operation.

The system can operate in severe environmental conditions, at temperatures ranging from −30 to 50°C, and at humidity of up to 95 per cent.

As of this edition, Jane's sources suggest that the N-11 system has been upgraded under the designation N-11M. N-11M is reported to operate in the E/F-band and as featuring frequency agility; multibeam elevation coverage; digital constant false alarm rate provision and pulse compression. Maximum detection range and altitude coverage values are given as 250 km and 30 km respectively.

Status
As of this edition, N-11 series radars are thought to be available.

Contractor
RADWAR, Warsaw.

VERIFIED

The N-11 radar system antenna

N-21M surveillance radar

Type
E/F-band (2 to 4 GHz) low-altitude surveillance radar.

Description
The N-21M is a 2-D autonomous medium-range battlefield surveillance radar designed primarily for detection of low-level penetration, or as a gap filler in an air defence system. It is particularly suitable for use with tactical operations forces when rapid relocation in difficult terrain and very fast operational readiness are paramount. It is highly mobile and consists of a single armoured, tracked vehicle containing the operations cabin with the radar antenna mounted on top. The vehicle, based on a Type 72 chassis, has two power generators and a land navigation system built in and is able to operate in contaminated terrain. The antenna is mounted on an extendable mast which can be erected to any height up to 8 m and is retracted and stowed for transit. Since all controls are contained

The N-21M low-altitude surveillance radar

within the cabin, the N-21M is capable of single man deployment and is fully operational within five minutes of arriving at site. It is normally operated by a two or three man crew in the cabin, but can also be operated remotely by a single cable link.

The N-21M operates in E/F-band with a fully coherent chirp-pulse transmitter, wideband frequency agility and a low-sidelobe antenna. Frequency diversity is carried out by a number of programmable crystal-controlled frequencies, switching from pulse to pulse within 200 MHz, with automatic selection of the non-jammed carrier frequency. Up to 70 targets can be detected and tracked, with presentation on a raster scan display and automatic remoting by a radio station on each antenna scan. The system also includes microprocessor-aided built-in test equipment and an operator training simulator.

According to Jane's sources, a new 3-D variant of the N-21M radar is under consideration while the basic N-21 design has formed the basis of the N-23 coastal surveillance sensor and the N-25 naval radar. Of these various off-shoots, the 3-D N-21M development is described as featuring a stacked beam phased-array antenna while the N-23 radar is noted as being a fixed site equipment. The surface ship N-25 system is reported to make use of a stabilised antenna platform.

Status
As of this edition, N-21 series radars are thought to have been in service since the mid-1980s.

Specifications
Transmitter: coherent amplifier
Frequency: E/F-band (2-4 GHz)
Staggered pulse repetition frequency: 1,000 Hz average
Pulse duration: 10 µs
Power output: 90 kW peak
Antenna: dual beam, cosec^2
Antenna rotation rate: 12 or 24 rpm
Max range: 100 km
Ceiling: 7,000 m
Elevation coverage: 40°
Accuracy: 100 m range; 1° azimuth
Resolution: 200 m range; 2° azimuth
Target capacity: up to 70 targets can be tracked
IFF: Mark XII compatible
System weight: 35 t (incl vehicle)

Contractor
RADWAR, Warsaw.

VERIFIED

N-22 surveillance radar

Type
E/F-band (2 to 4 GHz) low-altitude surveillance radar.

Description
The N-22 is a 2-D autonomous medium-range battlefield surveillance radar designed primarily for detection of low-level penetration, or as a gap filler in an air defence system. It is particularly suitable for use with tactical operations forces when rapid relocation in difficult terrain and very fast operational readiness are

The N-22 low-altitude surveillance radar

paramount. It is highly mobile and consists of a single armoured vehicle containing the operations cabin with radar antenna mounted on top. The vehicle has two power generators and a land navigation system built in and is able to operate in contaminated terrain. The antenna is mounted on an extendable mast which can be erected to any height up to 8 m and is retracted and stowed for transit. Since all controls are contained within the cabin, the N-22 is capable of single man deployment and is fully operational within five minutes of arriving at site. It is normally operated by a two or three man crew in the cabin, but can also be operated remotely by a single cable link.

The N-22 operates in E/F-band with a fully coherent chirp-pulse transmitter, wideband frequency agility and a low-sidelobe antenna. Frequency diversity is carried out by a number of programmable crystal-controlled frequencies, switching from pulse to pulse within 200 MHz, with automatic selection of the non-jammed carrier frequency. Up to 70 targets can be detected and tracked, with presentation on a raster scan display and automatic remoting by a radio station of each antenna scan. The system also includes microprocessor-aided built-in test equipment and an operator training simulator.

Status
As of this edition, N-22 is thought to be in service.

Specifications
Transmitter: coherent amplifier
Frequency: E/F-band (2-4 GHz)
Staggered pulse repetition frequency: 1,000 Hz average
Pulse duration: 10 µs
Power output: 90 kW peak
Antenna: dual beam, $cosec^2$
Antenna rotation rate: 12 or 24 rpm
Max range: 100 km
Ceiling: 7,000 m
Elevation coverage: 40°
Accuracy: 100 m range; 1° azimuth
Resolution: 200 m range; 2° azimuth
Target capacity: up to 70 targets can be tracked
IFF: Mark XII compatible
System weight: 35 t (incl vehicle)

Contractor
RADWAR, Warsaw.

VERIFIED

N-31M air defence radar system

Type
C/D-band (0.5 to 2 GHz) transportable air defence radar system.

Description
According to Jane's sources, the N-31M system is an air-transportable, ground-mobile (using a Tatra 815 chassis), medium-range, coherent, air defence radar system that was developed during the late 1980s. System features are said to include:
- a low side-lobe antenna
- a coherent, chirp pulse transmitter

- frequency agility
- digital moving target indication filtering
- a 'Dixie Fix' receiver.

A fixed site variant of the system (designed as the N-32M) is noted as utilising the basic N-31M system mated to an RA-83 fixed-site antenna array. So equipped, the N-32M variant is reported to be able to control two E/F-band (2 to 4 GHz) N-41 height-finding radars. So configured, such an architecture is described as offering range accuracy, bearing accuracy, elevation accuracy, range discrimination, azimuth discrimination and elevation discrimination values of 200 m, 0.6 m, 300 m, 200 m, 2.5° and 0.9° respectively. The N-41 height finder itself is reported as having a detection range of up to 360 km at altitudes of up to 30 km.

Status
As of 1998, Jane's sources were suggesting that the N-31M/N-32M radar systems were live programmes. As of this edition, the status of the two equipments is uncertain.

Specifications
(N-31M)
Frequency: 0.5-2 GHz
Detection range: 200 km
Altitude: up to 30 km

Contractor
RADWAR, Warsaw

VERIFIED

ROMANIA

START-1M

Type
E/F-band (3 GHz centre frequency) ground mobile low level detection and gap filler radar.

Description
Ascribed to Romania's Military Equipment and Technology Research Agency, the START-1M radar is described as being designed to provide a 'total capability' in low-level target detection and gap filler air defence applications. The sensor is further noted as being capable of all-terrain deployment and effective operation in countermeasures polluted environments. The architecture is 'two vehicle' portable and can be operated as a stand-alone system or as part of an air defence network.

Status
As of late 2000, the START-1M radar was being actively promoted.

Specifications
Frequency: 3 GHz
Detection range: 150 km
MTI improvement factor: 50 dB
Antenna beamwidth: $cosec^2$ up to 28° (vertical); 1.5° (horizontal)
Gain: 34 dB
Transmitter power: 2 kW (mean); 160 kW (peak)
Pulse-width: 0.313 µs (received - compressed); 12.5 µs (transmitted)
Receiver noise: 3.5 dB
Output: up to 100 tracks and up to 500 plots
Temperature: −20 to +50°C (operating)

Contractor
Romtehnica (export agent), Bucharest.

NEW ENTRY

RUSSIAN FEDERATION AND ASSOCIATED STATES (CIS)

Introductory note: This section deals with land-based air defence radars produced by, exported by or in service with the 21 republics making up the Russian Federation together with the ex-Soviet republics of Armenia, Azerbaijan, Belarus, Georgia, Kazakhstan, Kyrgyzstan, Moldova, Tajikistan, Turkmenistan, Ukraine and Uzbekistan. Collectively, these states are members of the Commonwealth of Independent States (CIS) economic grouping. Despite increasing openness, information on a number of systems remains sketchy. Accordingly, some of the equipments described here are still identified by their NATO reporting names for want of something better.

VERIFIED

1L13-3 surveillance radar

Type
VHF band (30 to 300 MHz) long-range mobile air surveillance radar.

Description
The 1L13-3 radar is a mobile 2-D air surveillance sensor that can be used as part of an air defence system, or can operate autonomously. The complete system appears to be carried on three vehicles and a trailer and can be transported by road, rail or air. Power is supplied by two diesel generators.

The antenna consists of a flat, 18 vertical-dipole array which is mounted on the flat bed of one of the vehicles and folds down for transport. An electric drive provides the power for rotating the antenna at speeds of 10 and 20 rpm. The operations cabin incorporates the transceiver, data processing equipment, the operator position, external telecode communication systems, two radio communication systems and a simulator. Associated with the 1L13-3 radar is the Dog Tail identification friend-or-foe interrogator. Optionally, an additional complete system with a control centre and two operator positions is available.

Status
As of June 2001, the precise status of the1L13-3 air surveillance radar was uncertain.

Specifications
Frequency range: VHF band (30-300 MHz)
Antenna: flat array 16 × 3.24 m
Range: 40 km (altitude); 500 km (azimuth)
Detection range on a fighter: 230/300 km at 10/27 km altitude respectively
Accuracy: 400 m range; 0.7° azimuth
Antenna rotation speed: 10/20 rpm
Switching on time: 3 min
Setting up/closing down time: 45 min max

VERIFIED

The 1L13-3 mobile 2-D surveillance radar

1L117 surveillance radar

Type
E/F-band (3 GHz spot frequency) 3-D early warning, Ground-Controlled Interception (GCI) and Air Traffic Control (ATC) radar.

Description
The ground-mobile 1L117 3-D early warning, GCI and ATC radar is an upgraded version of the P-37 early warning and GCI system (see separate entry) that is 'almost entirely' solid-state (with the exception of its coaxial magnetrons and cathode ray tubes). In standard configuration, the equipment comprises:

- a transceiver cabin (housing the radar and Identification Friend-or-Foe (IFF) subsystem transceivers, the radar and IFF antenna arrays and Moving Target Indicator (MTI) equipment)
- main and standby, diesel-driven electric power generators
- a VIP-117 'remote indicator station' (housing the system's operator workstations).

The operator workstations are understood to incorporate 'two analogue indicators complete with digital displays' and are used to initiate automatic target tracking. Target designation is by means of a joystick-driven cursor and target range, azimuth, altitude, track speed, course and identity are presented to the operators on the noted digital displays. Each workstation is also described as being able to calculate interceptor-target rendezvous points. The 1L117's IFF subsystem comprises an interrogator, primary and secondary radar processors and 'remote' indicators and a combination of computer processing and V-beam determination of the third co-ordinate are claimed to make the radar a 'relatively cheap' 3-D sensor. Each of the 1L117's receiving channels are equipped with their own dedicated digital MTI and signal processors (with the latter being noted as featuring false alarm stabilisation) and the system as a whole is reported to be able to adapt automatically to the particular noise environment of the moment. Here a noise map is employed to establish criteria for switching between the sensor's amplitude and coherent channels. 1L117 is also noted as making use of interscan signal processing.

Status
As of June 2001, the precise status of the 1L117 ground-mobile early warning, GCI and ATC radar was uncertain.

Specifications
Frequency: 3 GHz (spot frequency)
Wavelength: 10 cm
Detection range: 350 km (max)
Vertical scan angle: 28°
Accuracy: 10 min (azimuth); 300 m (range); 400 m (altitude)
Resolution: 1° (azimuth); 500 m (range)
Ground clutter suppression: 25 dB
Information update rate: 10 s
Track capacity: 200
Power requirement: 50 kW
Temperature: −40 to +50°C (operating)
Windspeed: up to 25 m/s
Set up/tear down time: 8 h

Contractor
Lira Design Bureau, Moscow.

VERIFIED

5N69 early warning/Ground Controlled Interception (GCI) radar

Type
D-band (1 to 2 GHz) early warning/GCI radar. NATO reporting name Big Back.

Description
The Big Back early warning/GCI radar made its first appearance during 1981 and is characterised by an antenna array that consists of two large parabolic cylinder antennas, each using an offset linear feed with electronic scanning. The antennas are about 13.5 m high and 16.5 m wide. The linear feeds use electronic scanning to provide 3-D coverage. What appear to be sidelobe suppression antennas are positioned on the lower edges of both reflectors. Big Back is probably a replacement for the Tall King and Back Trap systems and is used at SA-5 'Gammon' surface-to-air missile brigade level to provide early warning. It appears to be primarily a fixed-site system, but with modular construction which would make it transportable. Range is in excess of 500 km.

Status
Over time, the 5N69 early warning/GCI radar is understood to have been procured by the armed forces of the RFAS.

Specifications
Frequency: D-band (1-2 GHz)
Effective range: >500 km

VERIFIED

39N6E Casta-2E2 surveillance radar

Type
UHF band (300 MHz to 1 GHz) surveillance radar.

Description
The 39N6E Casta-2E2 surveillance radar is a multipurpose system that is designed for a range of applications that includes air defence, border surveillance, air traffic control and airport movement control. The equipment comprises hardware, antenna and power supply vehicles, a range that can be augmented by a remote operator post, which can be installed up to 300 m distant from the transmitter site. The radar has been specifically configured to operate in intense clutter environments and system features include:

- a solid-state transmitter
- moving target indicator processing
- 'angel' suppression
- 10 per cent frequency agility (manually selectable from 10 carrier wave frequencies or automatically on a random basis)
- phase modulated transmission pulses (127 or 255 bit M codes with random automatic code change)
- constant false alarm rate
- non-synchronous jamming rejection (via processing and changes in pulse repetition frequency)
- automatic generation of co-ordinated plots
- automatic target lock-on and tracking
- automatic/semi-automatic control of operating modes
- ground/slow moving target tracking and classification
- automatic Identification Friend-or-Foe (IFF) channel and IFF plot/track co-ordination control
- a data dissemination interface and a 50 km range integral voice/data communications subsystem
- built-in test

- a 'flicker' radiation mode for self-defence against anti-radiation missiles
- the ability to accommodate Mk X IFF and/or satellite navigation equipment.

Status

As of June 2001, the precise status of the 39N6E Casta-2E2 surveillance radar was uncertain.

Specifications

Frequency: UHF band (300 MHz-1 GHz)
Range: 5-150 km; 41 km (target at 100 m altitude, antenna phase centre at a height of 14 m); 55 km (target at 100 m altitude, antenna phase centre at a height of 50 m); 95 km (target at 1,000 m altitude)
Azimuth: 360°
Elevation: 25°
Altitude: up to 6 km
Accuracy: 20 m/s (velocity); 40 min (azimuth); 100 m (range); 900 m (target x/y co-ordinates)
Resolution: 5.5° (azimuth); 300 m (range)
MTBF: ≥700 h
MTTR: ≥0.3 h
Uninterrupted operation: ≥20 days
Turn on time: ≤3.3 min
Deployment time: ≤20 min
Power supply: 220 V (3 phase, 400 Hz, internal); 380 V (3 phase, 50 Hz, external)
Power consumption: ≤23 kW
Operator crew: 2
Number of transport vehicles: 3
Temperature: −50 to +50° C

Contractor

All-Russian Radio Engineering Institute, Moscow.

VERIFIED

⋯⋯⋯⋯⋯⋯⋯⋯⋯⋯⋯⋯⋯⋯⋯⋯⋯⋯⋯⋯⋯⋯

51U6 Casta-2E1 surveillance radar

Type

UHF (300 MHz to 1 GHz) surveillance radar.

Description

The 51U6 Casta-2E1 radar is designed for airspace control and is reported to be effective against fixed-wing aircraft, helicopters and small targets such as cruise missiles and unmanned aerial vehicles. The radar is further described as being optimised for the detection of low flying targets in heavy surface and weather clutter environments.

Status

As of June 2001, the precise status of the 51U6 Casta-2E1 surveillance radar was uncertain.

Specifications

Frequency: UHF band (300 MHz-1 GHz)
Range: 5-150 km; 32 km (target at 100 m altitude, antenna phase centre at a height of 7 m); 53 km (target at 100 m altitude, antenna phase centre at a height of 50 m); 95 km (target at 1,000 m altitude, antenna phase centre at a height of 7 m); 105 km (target at 1,000 m altitude, antenna phase centre at a height of 50 m); 32-58 km (target at 100 m altitude); 95-105 km (target at 1,000 m altitude)
Azimuth: 360°
Altitude: up to 6 km
Update rate: 5 and 10/s
Clutter rejection: 53 dB
MTBF: ≥300 h
MTTR: ≤0.5 h
Uninterrupted operation: ≥20 days
Turn on time: ≤3.3 min
Deployment time: ≤20 min
Power consumption: ≤16 kW (max)
Operator crew: 2 (1 shift)
Number of transportation items: 2

Contractors

All-Russian Radio Engineering Institute, Moscow.
Radio Measuring Instrument Plant, Murom.

VERIFIED

⋯⋯⋯⋯⋯⋯⋯⋯⋯⋯⋯⋯⋯⋯⋯⋯⋯⋯⋯⋯⋯⋯

55G6-1 Nebo 3-D surveillance radar

Type

VHF band (30 to 300 MHz) 3-D surveillance radar.

Description

The 55G6-1 Nebo 3-D surveillance radar is a mobile equipment that is designed to provide long-range detection of air targets operating at a wide range of speeds and

The 55G6-1 Nebo 3-D surveillance radar

altitudes. Nebo is further described as being able to perform its mission autonomously or feed its analogue/digital data output to air defence centres. The system's transmission coding and signal processing techniques are optimised for a countermeasures polluted environment and its manufacturer claims that it is effective against low-observability targets. Anti-interference techniques incorporated in the radar include:

- sidelobe blanking control
- optimised sector mode radiation in azimuth
- ground and weather clutter rejection
- moving target indication
- programmable radiation frequencies (adaptive frequency agility over 15 per cent of bandwidth in automatic transmission mode)
- a constant false alarm rate
- an interference direction-finding subsystem.

The 55G6-1 Nebo radar also includes an identification friend-or-foe capability.

Status

As of June 2001, the precise status of the 55G6-1 Nebo surveillance radar was uncertain.

Specifications

Frequency range: VHF band (30-300 MHz)
Detection range: 65 km (fighter-type target at an altitude of 500 m); 300 km (fighter-type target at an altitude of 10,000 m) 400 km (fighter-type target at an altitude of 20,000 m)
Accuracy: 24 min (azimuth); 500 m (range); 850 m (altitude at 200 km range); 2,000 m (altitude at 200-300 km range)
Altitude: 75 km (max)
Interference rejection: 45 dB
Transmitter power: 500 kW (pulse)
Elevation: 16°
Data rate: 10-20/s
Scan pattern: circular
Equipment readiness time: 3.2 min
Uninterrupted operation: ≥300 h
Operation crew: 4
Windspeed: up to 45 m/s
Site altitude: up to 1,000 m above SL
Temperature: −50 to +50° C
MTBF: 150 h

Contractor

NITEL Joint Stock Company, Nizhni Novgorod.

VERIFIED

⋯⋯⋯⋯⋯⋯⋯⋯⋯⋯⋯⋯⋯⋯⋯⋯⋯⋯⋯⋯⋯⋯

55G6-UE Nebo-U 3-D surveillance radar

Type

VHF band (30 to 300 MHz) 3-D surveillance radar.

Description

The 55G6-UE Nebo-U 3-D surveillance radar is a mobile system that is designed for airspace surveillance and air target co-ordinate (range, azimuth, elevation and height) determination. The equipment is further described as featuring built-in test, automatic data readout and the ability to work either autonomously or as part of a co-ordinated air defence system. The radar is also noted as having a detection capability against low observability targets and as being optimised for use in a countermeasures polluted environment.

Status

As of June 2001, the precise status of the 55G6-UE Nebo-U surveillance radar was uncertain.

Specifications

Frequency: VHF band (30-300 MHz)
Detection Range: 70 km (fighter-type target at an altitude of 500 m); 170 km (fighter-type target at an altitude of 3,000 m); 310 km (fighter-type target at an altitude of 10,000 m); 400 km (fighter-type target at an altitude of 20,000 m and above)

Azimuth: 360° (circular scan)
Elevation: 16°
Altitude: 70 km
Accuracy: 15 min (azimuth); 120 m (range); 500 m (altitude at elevation angle of above 1.5°); 800 m (altitude at elevation angle of less than 1.5°)
Interference rejection: 45 dB (localised ground clutter)
Data rate: 10 s
MTBF: 250 h (critical failures)
Power consumption: 100 kW
Deployment time: 22 h
Operator crew: 3
Transport units: 6

Contractor

Nizhegorodsky Radio Engineering Research Institute, Nizhni Novgorod.

VERIFIED

....

55K6-3 air defence radar

Type

VHF band (30 to 300 MHz) mobile 3-D air surveillance radar.

Description

The 55K6-3 radar is a mobile 3-D system that is designed to monitor airspace and to detect and determine range, azimuth, altitude and elevation of a wide range of air targets at long range. It has applications in automatic control systems and will operate efficiently in a severe countermeasures environment. The antenna structure has a very impressive appearance and incorporates an open frame horizontal network, about 15 m wide, surmounted by a vertical open frame which is about 20 m high. The structure is stabilised by a number of guy ropes and takes some 22 hours to bring into operation or break down.

Status

As of June 2001, the precise status of the 55K6-3 air-defence radar was uncertain.

Specifications

Frequency range: VHF band (30-300 MHz)
Operating limits: 1,200 km distance; 75 km altitude, 360° azimuth; 16° elevation
Detection range on a fighter-type aircraft: 65 km at 500 m; 300 km at 10,000 m; 400 km at 27,000 m
Ceiling for fighter detection: 60 km
Accuracy: range 400 m; azimuth 0.4°; altitude 750 m
Data refresh rate: 10 s
Setting up time: 22 h

VERIFIED

....

67N6E Gamma-DE air defence radar

Type

D-band (1 to 2 GHz) phased-array air defence radar.

Description

The 67N6E Gamma-DE phased-array air defence radar is a multipurpose system that is designed for use in automatic/non-automatic control air defence applications and as a tracker in air traffic control systems. The equipment comprises a phased-array antenna (with integral turntable), a processing/control/indicator package, an Identification Friend-or-Foe (IFF) interrogator, a power supply and a spares/calibration package. The antenna used is active in its transmission mode and semi-active when receiving. It comprises:

- transceiver line arrays (each of which contains pattern-forming circuitry)
- a low voltage transmission power amplifier for each emitter element
- phase shift controllers
- low noise reception amplifiers.

Elevation coverage is by means of electronic phase shifting with azimuthal coverage being achieved via electromechanical scanning. Other system features include:

- multichannel Doppler filtering for clutter and 'angel' suppression
- interscan processing
- low radial speed target blanking
- directional power concentration, pulse-to-pulse wideband frequency hopping, signal level limitation (with compression), azimuth sector rejection, multichannel compensation and criteria-based processing to overcome jamming
- built-in test
- six channel (two amplitude, four coherent) space surveillance
- single target classification
- the ability to accommodate Mk X IFF equipment
- the ability to integrate with the Gazetchik electronic warfare system for sensor protection.

Status

As of June 2001, the precise status of the 67N6E Gamma-DE air-defence radar was uncertain.

Specifications

Frequency: D-band (1-2 GHz)
Range: 10-330 km (Iso-Altitude Mode (IAM)); 10-360 km (Iso-Range Mode (IRM)); 220 km (IAM mode max detection range against a 0.13 m² radar cross-section target); 250 km (IRM mode max detection range against a 0.13 m² radar cross-section target); 330 km (IAM mode max detection range against a 1 m² radar cross-section target); 360 km (IRM mode max detection range against a 1 m² radar cross-section target)
Azimuth: 360° (all modes)
Elevation: −2 to +30° (IAM); −2 to +45° (IRM)
Ground clutter suppression: 45 dB
Update rate: 10/s
Target track capacity: 100-200
Accuracy: 10-15 min (azimuth); 15-20 min (elevation); 100 m (range); 600 m (altitude)
Resolution: 1.35° (azimuth); 300 m (range)
Radiated power: 12.5 kW (average)
MTBF: 300 h
MTTR: 0.5 h
Turn on time: 1.5 min
Deployment time: 1.5 h
Crew: 5
Number of transport vehicles: 6
Temperature: −50 to +50° C
Site altitude: up to 2,000 m above SL

Contractor

All-Russian Radio Engineering Institute, Moscow.

VERIFIED

....

96L6E 3-D surveillance radar

Type

Centimetric 3-D air surveillance radar.

Description

The highly automated 96L6E 3-D surveillance radar forms part of the S-300PMU/S-300PMU-1 surface-to-air missile system and is designed to detect, classify and track air targets together with handing-off radar data to users via landline and/or communications radio. Within the S-300 system, the equipment acts as its low-altitude detection and surveillance sensor and as a battery command post. 96L6E is mounted on the MAZ-7930 self-propelled, cross-country chassis and its antenna, if required, can be mounted on the Type 40V6M tower. The system's adaptive radiation and processing techniques are optimised for operation in a countermeasures polluted environment.

Status

As of June 2001, the 96L6E air surveillance radar is understood to have been procured.

Specifications

Frequency: centimetric band
Detection range: 300 km (max)
Azimuth: 360°
Elevation: −3 to 60°
Target velocity: ±30-1,200 m/s (air vehicles at low to high altitude); 50-2,800 m/s (ballistic targets)
Surface clutter suppression: up to 70 dB
Automatic track capacity: up to 100
Start up time: <7 min
MTBF: 300 h
Crew: 2

Contractors

Lira Design Bureau, Moscow.
Lianozovsky Electromechanical Plant, Moscow.

VERIFIED

....

Big Bird early warning radar

Type

F-band (3 to 4 GHz) early warning radar.

Description

Big Bird is an F-band early warning radar which was first deployed with the SA-10 'Grumble' surface-to-air missile in the late 1970s. It is apparently mounted on a tower to assist in long-range detection of very low terrain-following aircraft and cruise missiles. The Big Bird system is also being used to replace older early warning radars.

Status

Over time, Big Bird is thought to have been exported to Syria as part of a Russian commitment to upgrade Syria's air defence network.

Specifications

Frequency: F-band (3.3 GHz)
Scan rate: 5 rpm

VERIFIED

Daryal Anti-Ballistic Missile (ABM) and space vehicle tracking radar

Type

ABM and space vehicle tracking pulsed radar.

Description

The Daryal ABM and space vehicle tracking radar is designed to detect, track and measure the co-ordinates of target vehicles. As such, the equipment makes use of physically separated transmit and receive active phased-array antennas with electronic beam control in both azimuth and elevation. Basic sector coverage is 90° in azimuth by 40° in elevation (expandable to 100° in azimuth by 50° in elevation) and the radar is described as being primarily designed to provide 'barrier beam scanning' of the 'lower portion of the action zone' within an ABM defence system. Other system features include:

- high-output power
- a high level of parameter measurement accuracy
- high-speed target handling
- electronic counter-countermeasures provision
- the ability to handle high-orbit targets
- multichannel signal reception
- high-capacity signal processing and system control computers
- the ability to track at least 20 'composite' targets (with each target having different angular co-ordinates) simultaneously
- ionospheric parameter determination with automatic correction
- a 'round-the-clock' operating life of 10 years.

Status

The Daryal ABM and space vehicle tracking radar is reported to have entered Soviet Russian service during 1983.

Contractor

Mints Radio Engineering Institute, Moscow.

VERIFIED

Delta surveillance radar system

Type

Decimetric surveillance radar system.

Description

The Delta surveillance radar system is designed to detect and identify air targets (including small size vehicles flying at low and medium altitudes) and transmit their co-ordinates to an automatic air defence control system. The equipment is noted as being optimised for use in 'hard-to-reach' areas (desert, polar and forest regions) and comprises two remotely controlled, Unmanned Radar and Communications Stations (URCS) and a Data Processing and Control (DPC) facility. The DPC can be up to 200 km away from its associated URCS sites and an integral tropospheric communications subsystem is used to transmit and receive data within the system. Dependent on terrain features, the RCS antennas can be mast-mounted or installed in radomes, with their phase centres at between 12 and 25 m above ground level.

Status

As of June 2001, the precise status of the Delta surveillance radar was uncertain.

Specifications

Wavelength: decimetric
URCS coverage: 0-12° (elevation); 0-10,000 m (altitude); 360° (azimuth)
Detection range (single URCS, 12 m high phase centre): 40 km (fighter-type target at 100 m); 80 km (fighter-type target at 500 m); 115 km (fighter-type target at 1,000 m); 180 km (fighter-type target at 3,000 to 10,000 m)
Total coverage (2 URCSs, 100 km apart, d × w): 160 × 270 km (target at 500 m); 360 × 460 km (target at 3,000 m)
Accuracy: 60 min (azimuth); 1,000 m (range)
MTI noise-obscured visibility coefficient: 45 dB
Outputs: analogue, co-ordinate points and routes
Data furnished to an automatic air defence control system: 30 (routes); 500 (co-ordinates)
Update rate: 5, 10 and 20 s
MTBF: 800 h (DPC); 1,400 h (URCS)

Power requirement: 16 kW (DPC and URCS)
Crew: 2-3 (1 shift, DPC)

Contractor

Nizhegorodsky Radio Engineering Research Institute, Nizhni Novgorod.

VERIFIED

Desna-M surveillance radar system

Type

Decimetric wavelength 3-D air surveillance radar system.

Description

The Desna-M 3-D air surveillance radar system is designed to automatically detect a 'broad range' of air targets and measure their range, azimuth, altitude and elevation in a heavily countermeasures polluted environment. The system comprises:

- a 6GG rotary antenna assembly
- a 6RR equipment trailer
- a 6GM-M modulator trailer
- two or three power generation trailers
- a spare parts, tools and accessories trailer
- two MAZ-938B 'boxes' and antenna equipment semi-trailers
- an optional Remote Indicator Station (RIS).

The Desna-M antenna assembly is mounted on an artificially created mound that is 6 m high and has a crest angle of no more than 10 minutes. The equipment's RIS can be located at up to 1,000 m from the main site and Desna-M is noted as being transportable by road and rail. In the latter case, the system breaks down into 10 flat car loads. If desired, the radar can take power from a 380 V, 50 Hz industrial supply.

Status

As of June 2001, the precise status of the Desna-M 3-D air surveillance radar was uncertain.

Specifications

Wavelength: decimetric
Range: 300 km (fighter-type target at 10,000 m)
Altitude: 40 km (upper limit against a fighter-type target)
Scan rate: 6 rpm
Accuracy: 15 min (azimuth); 300 m (range); 500 m (altitude)
Clutter suppression factor: 30 dB (min)
Set up/close down time: 10 h (max)
Power requirement: 260 kW
Crew: 16
Temperature: ±50°C (operating)
Windspeed: up to 20 m/s (operating)

Contractor

Pravdinsk Radio Relay Equipment Plant, Nizhni Novgorod.

VERIFIED

Dnepr Anti-Ballistic Missile (ABM) radar

Type

ABM and space vehicle tracking radar system.

Description

A Dnepr ABM and space vehicle tracking radar system comprises two sector radars, each of which is equipped with a fixed antenna array that comprises an elongated double horn with a mouth aperture of 250 by 12 m. Excitation is by means of two rows of slot radiators arranged on two waveguides. A transceiver unit is connected to the waveguides at one end of the array and comprises two transmitters and what is termed a 'receiving-indicating equipment strip'. So configured, the array offers frequency-phase control of the output within a 30 × 30° sector. Overall, Dnepr is quoted as being able to monitor a 120° sector in azimuth with elevation coverage of between 5 and 35°. Functionally, the equipment measures target co-ordinates using bursts of composite coherent signals. It can also make use of detected jamming sources to establish the angular co-ordinates of targets and there is a computer subsystem that is used to automate the equipment's system control, data processing, data reporting and built-in test functions.

Status

As of June 2001, the Dnepr ABM and space vehicle tracking radar system was believed to be the equipment that was given the NATO Reporting Name of Hen House. As such, Jane's sources suggest that, over time, Hen House radars have been set up at Genichesk, Mishelevka, Olenegorsk, Pinsk, Sary Shagan and Skrunda. The Skrunda installation is thought to have been decommissioned during the early to mid-1990s.

Specifications
Wavelength: metric
Polarisation: horizontal
Coverage: 5-35° (elevation); 120° (azimuth)
Pulse power: 1.25 MW (single transmitter)
Main pulse duration: 0.8 µs
Power: 200 kW (mean radiating); 2,100 kW (consumed)
Detection range: 1,900 km (1 m² target radar cross-section)
Accuracy: 1 km (range); 5 m/s (radial speed); 10 min (azimuth); 50 min (elevation)

Contractor
Mints Radio Engineering Institute Joint Stock Company, Moscow.

VERIFIED

Don-2N Anti-Ballistic Missile (ABM) and space vehicle tracking radar

Type
Centimetric band ABM and space vehicle tracking radar.

Description
Designed to detect, track and measure the co-ordinates of ballistic missiles and space vehicles, the Don-2N pulse radar comprises four, fixed, 16 m diameter, phased-array antennas, a digital signal processing subsystem and a 'high-capacity, multiprocessor' system control computer system that is based on the Elbrus-2 equipment. Functionally, the radar automatically detects targets, measures their co-ordinates, determines their trajectories and discriminates re-entry vehicles from decoys in real time. The equipment's system control computer is described as being able to execute up to 1 billion operations per second and countermeasures proofing is provided via frequency agility and adaptive control of the phased-array antennas.

Status
As of this June 2001, the Don-2N ABM and space vehicle tracking radar is understood to form (or to have formed) part of the Moscow ABM defence system.

Specifications
Wavelength: centimetric
Coverage: 1-90° (elevation); 360° (azimuth)
Detection range: 600-1,000 km (space object with a size of 5 cm)
Accuracy: 0.02-0.04° (angular position); 200 m (range)

Contractor
Mints Radio Engineering Institute, Moscow.

VERIFIED

Dunai-3U Anti-Ballistic Missile (ABM) and space vehicle tracking radar

Type
Decimetric wavelength ABM and space vehicle tracking radar.

Description
The Dunai-3U radar is designed to detect, track, measure the co-ordinates of and compute the trajectory of ballistic missiles and satellites. It is described as being an automatic, sectoral, continuous wave equipment that has a scanning sector that measures 51° in azimuth by 48° in elevation. The system utilises physically separated waveguide slot phased transmission and reception arrays and its antenna beam is frequency controlled in azimuth and phase controlled in elevation. Other system features include:
- line-by-line scanning of the sector under surveillance
- a high power output to maximise range
- a low noise target detection capability
- dedicated fine measurement channels
- frequency selective antennas to maximise immunity from interference
- the ability to detect and track fleeting high-speed targets
- a triple redundant, high-capacity, computer control system
- a 90 per cent probability of detecting all components of a re-entering ballistic missile at a range of at least 2,000 km
- the ability to track at least 30 composite ballistic targets simultaneously.

Status
The Dunai-3U ABM and space vehicle tracking radar entered service with the armed forces of the former Soviet Union during 1978. As of June 2001, the radar's status was uncertain.

Specifications
Wavelength: decimetric
Coverage: 48° (elevation); 51° (azimuth)

Range: 4,500 km (designed)
Single target handling capacity: 900 (min)

Contractor
Long Range Radio Communication Research Institute Production Complex JSC, Moscow.

VERIFIED

Furgon early-warning radar

Type
Metric wavelength ground-based early warning radar.

Description
Designed to provide range and azimuth data on air targets at long ranges, the Furgon early-warning radar comprises an antenna array together with one maintenance station, two equipment/operator stations and three power generation units, all of which are housed in semi-trailers. The antenna used is a 32 × 11 m parabolic reflector that is transportable and is mounted on a 'prepared' base. Provision is made for an identification friend-or-foe interrogator subsystem and Furgon can operate in any one of three modes, the known details of which are as follows:
Standard mode
In which, the radar's performance is characterised by target acquisition at maximum range.
High-altitude mode
In which, the radar can detect targets up to its highest elevation angle.
Scanning mode
In which, the radar interleaves its standard and high-altitude modes with a general search mode.
Furgon can be controlled remotely and uses frequency agility to provide a level of active counter-countermeasures protection. Protection against chaff jamming is provided by a 'coherent balancing' subsystem that is based on storage tubes.

Status
As of June 2001, the precise status of the Furgon early-warning radar was uncertain.

Specifications
Wavelength: metric
Coverage: 12° (elevation, standard mode); 17° (elevation, high-altitude mode); 35 km (altitude, standard mode); 360° (azimuth)
Range: 280 km (fighter-type target at 10,000 m, high-altitude mode); 300 km (fighter-type target at 10,000 m, standard mode)
Accuracy: 1.5° (azimuth); 1,500 m (range)
MTI system noise-obscured visibility coefficient: 26 dB
Output: analogue
Update rate: 10 and 20 s
MTBF: 100 h
Power requirement: 74 kW
Crew: 5 (single shift)
Set up time: 62 h
Temperature: ±50° (operating)
Windspeed: up to 3 m/s (operating)

Contractors
Nizhegorodsky Radio Engineering Research Institute, Nizhni Novgorod.
Nitel JSC, Nizhni Novgorod.

VERIFIED

Gamma-S1E

Type
Centimetric band, mobile, 3-D air defence radar.

Description
The Gamma-S1E mobile air defence radar is designed for use in automatic and non-automatic air defence networks or as a sensor in a civilian air traffic control system. The complete equipment comprises:
- an antenna assembly vehicle (designated as 'M1' and equipped with the radar's transceiver equipment and identification friend-or-foe interrogator as well as its antenna)
- a system control, data processing, data presentation and data dissemination vehicle (designated as 'M2')
- a spares, accessories, test equipment and cabling vehicle (designated as 'M3')
- power generation trailers towed by vehicles M1 and M2.

The equipment makes use of a phased-array antenna and can operate as a target acquisition, weapons guidance and/or target designation sensor within an air defence network. The radar is said to feature a high level of automation in its target detection, target parameter measurement and system control/monitoring

A general view of the Gamma-S1E 'M1' antenna vehicle   2000/0063805

A close-up of the phased-array antenna used in the Gamma-S1E air defence radar   2000/0063806

functions and can be operated from a four position Remote Control Facility (RCF). At ranges of up to 1 km, the radar and the RCF are connected by fibre-optic cable with a radio link being used at ranges of between 1 and 15 km.

Status
As of June 2001, the precise status of the Gamma-S1E air defence radar was uncertain.

Specifications
Wavelength: centimetric
Detection range: 10-300 km (10-400 km auxiliary model)
Azimuth coverage: 360°
Elevation coverage: −2 to +30° (−2 to +55° auxiliary model)
Altitude: 30 km
Clutter suppression factor: 45 dB
Scan period: 10s
Tracks per scan: up to 100
Accuracy: 10-15 min (elevation); 15 min (azimuth); 50 m (range); 400 m (altitude)
Resolution: 1.4° (azimuth); 250 m (range)
Power: 10-12 kW (mean generated); 70/90 kW (consumed summer/winter)
MTBF: 500 h
MTTR: 30 min
Start up time: 5 min (3 min in emergency)
Close down time: 40 min
Crew: 3

Contractors
All-Russian Radio Engineering Institute, Moscow.

VERIFIED

Kabina-66 series surveillance radar systems

Type
Family of decimetric wavelength air surveillance radar systems.

Description
Designed for air surveillance applications in the presence of 'heavy' countermeasures pollution, the Kabina-66 and Kabina-66M radars comprise:
- a rotary antenna assembly carried on two trailers designated as 19D-1 and 19D-2 respectively
- a trailer-mounted (designated as 19T) engineering post
- two 'modulation' equipment trailers (designated as MP-1/ML-1 and MP-2/ML-2 respectively)
- an 'indication' equipment trailer (designated as Trailer 1) or multiple VP-87M remote plan position indicator stations
- two to four PRV-13M/PRV-17-2 height-finding radars
- two ground-based 'radio interrogators'
- an RL-30-1M radio broadcast line
- a spare parts and 'measuring equipment' trailer (designated as Trailer Z)
- an 8T210 crane mounted on a URAL-375D truck chassis
- a fully redundant 5E46/5S85 power generation subsystem comprising one 5E97 (two in the Kabina-66M variant), one 5E42 and three ESD-200T/200/4-400 diesel generators; a 5155/TsRP-P main distribution station and one 5P28 and one PPS-P converter units.

Functionally, Kabina-66 series radars scan space using two antennas that rotate at a rate of 6 rpm. Each of these is a parabolic reflector with a horn feed that is capable of forming a beam pattern that covers from 0 to 45° in elevation. Within this range, the radars have a number of operating modes that are designed to optimise their range and altitude coverage. The transmission chain used features a microwave amplifier circuit format that is able to suppress natural clutter and manmade interference while the radar's receiver incorporates travelling wave tube return signal amplifiers to reduce receiver noise and maximise bandwidth. In the Kabina-66M model, the receiver amplifiers used (together with receiver 'protection' devices) are solid-state. Functionally, the Kabina receiver chain processes incoming signals in a multichannel receiver that rejects noise, stores echo returns and 'stabilises' the equipment's false alarm rate. As noted earlier, Kabina systems are completed by separate height-finders that can be located up to 6 km away from the main site.

Status
As of mid 2000, both the Kabina-66 and Kabina-66M air surveillance radar systems were understood to be available.

Specifications
Wavelength: decimetric
Scan rate: 6 rpm
Acquisition range: 330 km (fighter-type target at 10 km altitude); 380 km (fighter-type target at 15 km altitude or higher)
Altitude coverage: 54 km (max, fighter-type target)
Accuracy: 0.8° (azimuth); 300 m (altitude); 1,000 m (range)
Temperature: ±50°C (operating)
Windspeed: up to 25 m/s (operating)
Operating altitude: Up to 1,000 m above SL

Contractor
Pravdinsk Radio Relay Equipment Plant, Nizhni Novgorod.

VERIFIED

Lena early-warning radar

Type
Metric wavelength early-warning radar.

Description
Designed for fixed-site applications, the Lena early-warning radar comprises an antenna array, equipment and power generation buildings and a command post. The antenna used is a 32 × 11 m parabolic reflector and the system as a whole has provision for an identification friend-or-foe interrogator subsystem. Acquired data is displayed at two workstations within the command post (which can be located up to 1,000 m away from the main site) and frequency agility is used to combat active radio frequency jamming. Chaff jamming is handled by a 'coherent balancing' subsystem that is based on storage tubes. The Lena system is understood to be closely related to the mobile Furgon and Oborona-14 radars (see separate entries).

Status
As of June 2001, the precise status of the Lena early-warning radar was uncertain.

Specifications
Wavelength: metric
Coverage: 12° (elevation); 35 km (altitude); 360° (azimuth)
Range: 300 km (fighter-type target at 10,000 m)
Accuracy: 1.5° (azimuth); 1,500 m (range)
MTI noise-obscured visibility coefficient: 26 dB
Output: analogue
Update rate: 10, 15 and 30 s
MTBF: 80 h
Power requirement: 60 kW
Crew: 5 (single shift)

Temperature: ±50°C (operating)
Windspeed: up to 30 m/s

Contractors
Nizhegorodsky Radio Engineering Research Institute, Nizhni Novgorod.
Nitel JSC, Nizhni Novgorod.

VERIFIED

Long Track surveillance radar

Type
E-band (2 to 3 GHz) tactical early warning radar.

Description
Long Track is an E-band radar and was the first highly mobile Russian Federation and Associated States (RFAS) early warning radar to equip the new tactical air defence units. The system has a conventional parabolic mesh reflector with multiple stacked feeds, mounted on a lengthened AT-T heavy tracked transporter. The antenna can be folded, clam-shell style, for transit to protect it from damage. An identification friend-or-foe antenna is fitted on the outer left end of the antenna. The vehicle is fitted with both radio and datalink antennas and can also be connected to control centres by landline. Operation is in E-band with a range in excess of 150 km. It is associated with SA-4, SA-6 and SA-8 surface-to-air missiles. Since it is only a 2-D system it is usually supported by a Thin Skin B nodding height-finder radar.

Status
As of June 2001, the precise status of the Long Track early warning radar was uncertain.

Specifications
Frequency: E-band (2.6 GHz)
Effective range: 150 km
Effective altitude: 30,000 m
Sweep rate: 15 rpm

VERIFIED

Long Track surveillance radar

Oborona-14 early-warning radar

Type
Metric wavelength early-warning radar.

Description
Designed for autonomous or air defence network applications, the Oberona-14 radar is described as being a mobile, 'noise immune' variant of its manufacturer's Lena equipment (see separate entry). The system is mounted on six transportation units (two equipment, two antenna and two power generation semi-trailers) and incorporates a trailer-mounted command post that can be located up to 1 km from the main radar site. The radar has provision for an identification friend-or-foe interrogator subsystem and can be operated in any one of three modes, the known details of which are as follows:
Low-beam mode
In which, the radar's performance is characterised by optimised low- and medium-altitude detection ranges.
High-beam mode
In which, the radar's performance is optimised to provide maximum elevation coverage.
Scanning mode
In which, the radar interleaves its low- and high-beam operating modes.
Frequency agility and automatic, three-channel balancing are used to provide protection against active radio frequency countermeasures while chaff jamming is

handled by what is termed a 'coherent balancing' subsystem based on storage tubes.

Status
As of June 2001, the precise status of the Oborona-14 early-warning radar was uncertain.

Specifications
Wavelength: metric
Coverage: 12° (elevation, low-beam mode); 17° (elevation, high-beam mode); 45 km (altitude, low-beam mode); 360° (azimuth)
Range: 280 km (fighter-type target at 10,000 m, high-beam mode); 300 km (fighter-type target at 10,000 m, low-beam mode)
Accuracy: 1.2° (azimuth); 1,200 m (range)
MTI noise-obscured visibility coefficient: 26 dB
Output: analogue
Update rate: 10 and 20 s
MTBF: 90 h
Power requirement: 100 kW
Crew: 6 (single shift)
Set up time: 24 h
Temperature: ±50°C (operating)
Windspeed: up to 30 m/s

Contractors
Nizhegorodsky Radio Engineering Research Institute, Nizhni Novgorod.
Nitel JSC, Nizhni Novgorod.

VERIFIED

Odd Group height-finding radar

Type
E-band (2 to 3 GHz) height-finder radar system.

Description
The Odd Group radar is a height-finder system, probably operating in E-band (2 to 3 GHz). Virtually no details have been made available. The large elliptical paraboloid antenna appears to be mounted on a pair of lattice trunions cantilevered from the sides of the cabin containing the associated electronics. It is believed to be a successor to the Side Net height-finder. There appears to be a related system called Odd Pair.
Odd Pair is a relatively new height-finder which also operates in E-band. It is very distinctive because of its use of a small parabolic antenna located to the left of the main antenna.

Status
As of June 2001, the status of the Odd Group height-finding radar was uncertain.

VERIFIED

Odd Group height-finder radar

Over-The-Horizon Backscatter (OTH-B) radar

Type
OTH-B radar.

Description
The former Soviet Union had been working on the development and construction of over-the-horizon radar systems since the late 1950s. This was partially a general effort to extend the range of conventional early warning radars and partially

because of the problems with infra-red satellites for missile launch detection. The first OTH radar became functional in the 1970s and concentrated on the backscatter principle to detect missile launches by sensing the disturbances in the ionosphere caused by missile exhaust plumes.

The earlier efforts were beset by problems, mainly because of limitations in computer technology and also by ionospheric instability. These problems were accentuated by the requirement for the system to operate over the north pole region, an area where the ionospheric level and depth is particularly subject to large fluctuations in the earth's magnetic field.

The first OTH-B radar was apparently deployed for trials purposes at Gomel, Belarus facing towards the United States. The second was added a few years later at Komsomolsk-na-Amure in Siberia, also aimed at the United States. A third site was added in the early 1980s at Nikolayev near the Black Sea to monitor Chinese missile tests. In the late 1980s it was reported that a fourth site near Nakhoda, in the coastal region of the Sea of Japan, was under construction to monitor shipping, aircraft and ballistic missile movements in the area between the Chinese coast and the island of Guam. As far as can be ascertained, these radars operate (or operated) in the 4 to 30 MHz frequency band with an output power of 20 to 40 MW.

At one stage, these OTH-B radars were nicknamed 'Russian Woodpeckers' because of the way in which their 10 pps transmissions interfered with short-wave radios. Another nickname given to the system was 'Steel Works' in recognition of their girder-type construction.

It has also been reported that the Russian Federation and Associated States have been working on a forward scatter system but no details are available to confirm or deny this report.

Status

As of June 2001, the precise status of the Russian OTH-B network was uncertain.

VERIFIED

Deployment of a P-12 Spoon Rest A early warning radar

P-10 early warning radar

Type

VHF band (30 to 300 MHz) mobile early warning radars. NATO reporting name Knife Rest B.

Description

The P-10 is a direct derivative of the Knife Rest A, the latter being obsolete. A number of improvements were incorporated, including improved resolution, sensitivity and improved electronic counter-countermeasures provision. The system was made more mobile by fitting the main antenna array on a telescopic mast mounted on the command vehicle. A further improved version, Knife Rest C, was also deployed. Both B and C models are associated with the SA-2 'Guideline' surface-to-air missile. A navalised version was fitted to the 'Sverdlov' class cruisers.

Status

As of June 2001, the P-10 early warning radar must be considered obsolete. This said, as of June 2000, Jane's sources were reporting that the type was still in operational service with the armed forces of Vietnam.

Specifications

Frequency: VHF band (79-93 MHz)
Peak power: 55-100 kW
Effective range: 185-280 km
Range accuracy: 3.7 km
PRF: 50-100 pps
Pulse-width: 4-12 μs
Azimuth accuracy: 36 mils
Range resolution: 2.5 km
Vertical beamwidth: 21°
Horizontal beamwidth: 20-24°

UPDATED

P-12 early warning radar

Type

Very High Frequency (VHF − 30 to 300 MHz) mobile early warning radar. NATO reporting name Spoon Rest.

Description

P-12 (alternative designation 1RL14) is a VHF band early warning radar which has been used in conjunction with the Fan Song fire-control/guidance radar within the SA-2 'Guideline' and SA-3 'Goa' surface-to-air missile systems. The initial version, Spoon Rest A, is mounted on a trailer van, while the improved Spoon Rest B is mounted on a ZiL-157 truck. As in the case of Knife Rest B/C, a full system consists of three elements: the radar/command vehicle, generator truck and NRZ-1 Fish Net identification friend-or-foe antenna. In the 1970s, a modernised version, Spoon Rest C, was developed and is mounted on a van or a Ural 375 truck. A navalised version has been developed and installed on the 'Sverdlov' class cruisers.

A later version, Spoon Rest D is a target acquisition radar derived from Spoon Rest A. The main difference between the A and D models is that the D has eight dipoles instead of six.

Status

As of June 2000, Jane's sources were reporting that the P-12 early warning radar was in operational service with the armed forces of Vietnam (34 examples procured?). Historically, this radar has seen widespread combat service wherever the SA-2 and SA-3 surface-to-air missiles have been used. It is also worth noting that, during late 1998, the Yugoslavian defence exporter Yugoimport - SDPR was offering an upgrade package for the P-12 radar. Package elements included:

- harmonisation of the P-12's capabilities with those of the P-18 equipment
- the introduction of solid-state high frequency and logarithmic amplifiers to improve receiver sensitivity
- an increase in transmitter power
- a reduction in the system's radiation pattern from 12° to 8°
- an increase in antenna tilt (down to 5°)
- installation of a moving target indication capability
- removal of the system's original direction-finder, altitude indicator, Y-transformer, AP-3 automatic device and unit numbers 59 and 75.

Specifications

Frequency: VHF band (147-161 MHz)
Peak power: 180-350 kW
PRF: 310-400 pps
Pulse-width: 4-5 μs
Effective range: 200-275 km
Effective ceiling: 18-20 km
Horizontal beamwidth: 7-9°
Vertical beamwidth: 2.5°

UPDATED

P-14 early warning radar

Type

Very High Frequency (VHF − 30 to 300 MHz) long-range air defence early warning radar. NATO reporting name Tall King.

Description

P-14 is a long-range large early warning radar. The origin of its NATO reporting name stems from the impressive appearance of the structure which incorporates an open frame truncated paraboloid about 30 m wide and 11 m high, mounted on an asymmetric mast. The horizontal wires of the reflector are electrically heated and are connected to transformers for de-icing purposes. The whole reflector assembly is surmounted by a pylon of perhaps another 5 m to which are attached the guy ropes needed to steady this very large array.

The method of illuminating the antenna is not clear but operation is at metric wavelengths. Tall King is reported to have been introduced in the 1950s to provide early warning against high-altitude intruders. The Scoreboard B identification friend-or-foe system is associated with Tall King, as is also the Side Net transportable height-finder radar. Although obviously a fixed-site radar, sectional construction is employed so that, although a lengthy undertaking, Tall King is capable of transportation and re-siting.

Close-up of Tall King early warning radar showing technicians carrying out either final stages of erection or routine maintenance

Status

While the Tall King early warning radar is a dated system (which started to be replaced by more modern systems during the 1970s), as of February 2001, the equipment was still in service as part of Iraq's air defence network. The system is also known to have been deployed in Afghanistan and as of late 1998, the Yugoslavian defence exporter Yugoimport - SDPR was offering an upgrade package for the P-14 radar. Package elements included:

- interfacing the P-14 antenna with the P-12 radar's operator shelter
- installation of the K-14 anti-anti-radiation missile subsystem
- installation of a solid-state high frequency amplifier to enhance receiver sensitivity.

Specifications

Frequency: VHF band (150-180 MHz)
Effective range: 500-600 km
Sweep rate: 2-4 rpm
Altitude range: up to 45,720 m

UPDATED

P-15 early warning/target acquisition radar

Type

UHF band (0.3 to 1 GHz) early warning and target acquisition radar. NATO reporting name Flat Face.

Description

P-15 (alternate designation 1LR13) is a highly mobile 2-D UHF band early warning radar for use with air defence missiles and guns. It is usually deployed co-located

A Flat Face mobile radar unit

with a Side Net height-finder and is most commonly associated with the SA-2 'Guideline' and SA-3 'Goa' Surface-to-Air Missiles (SAM). However, it can also be found in tactical SAM units, including the SA-4 'Ganef', SA-6 'Gainful' and SA-8 'Gecko' in place of the standard Long Track. A later version of Flat Face appears to be the CASTA-2E1 system (see separate entry).

The radar radiates signals from two effective range elliptical paraboloid reflectors, each measuring about 6 × 2.15 m, mounted one on top of the other, with the whole structure mounted on a box-bodied shelter on a ZiL-157 truck. There is a set of mast extensions to elevate the antenna array about 2 m and this is carried on a trailer towed by the truck. The system incorporates basic electronic counter-countermeasures features including frequency selection to counter active jamming and moving target indication to counter chaff.

A Flat Face radar was seized in Chad in 1987 and has been subjected to extensive tests in a French research centre. It has proved to be a 'sturdy system with a high level of resistance to countermeasures and was very good at detecting airborne targets down to low altitudes'.

Status

Operationally, sources suggest that the P-15 radar is generally deployed with 'Guideline' and 'Goa' SAM batteries or as a surveillance sensor with tank and motorised infantry divisions. Outside its service in Eastern Europe, the equipment is also noted as having been associated with export 'Gainful' and 'Gecko' missile systems. As of early 2001, a US Army source has classified the P-15 as one of the 'primary threat radars' faced by NATO during its March-June 1999 'Allied Force' air campaign against Yugoslavia and Serbia. As of February 2001, Jane's sources were reporting that an 'Adapted Radar' variant of the P-15 (designated as the APR-15?) had seen service as part of Yugoslavia's air defence network in the recent past. Here, the radar was reported as incorporating a new antenna array and digital moving target indication and plan position indication. Capabilities are said to have included the ability to track up to 20 targets flying at altitudes and ranges of between 200 and 6,000 m and up to 120 km respectively.

Specifications

Frequency: UHF band (810-850/880-905 MHz)
Effective range: 210-250 km
Peak power: 900 kW
Pulse-width: 2-3 μs
PRF: 200-800; 600-880 pps
Azimuth beamwidth: 10°
Elevation beamwidth: 5°
Range accuracy: 90 m
Angular accuracy: 5°
Antenna dimensions: 6 × 2.15 m

UPDATED

P-15M target acquisition radar

Type

Search and target acquisition radar. NATO reporting name Squat Eye.

Description

P-15M is an air search and target acquisition radar, sometimes used instead of Flat Face and with Low Blow for 'Goa' SA-3 air defence missile batteries. Such applications appear to be those where low-altitude coverage is of particular importance.

The antenna is of relatively large dimensions and of elliptical paraboloid configuration. It is usually mounted on top of a 20 to 30 m slender rectangular lattice tower of sectional construction to increase the effective operating height. The mast is erected alongside a box-bodied ZiL-157 truck which contains the electronics. About 500 kW transmitter power is likely and the radar's role may be essentially that of a gap filler. A maximum range of about 200 km has been estimated. Squat Eye is reported to employ the same electronics as the Flat Face radar.

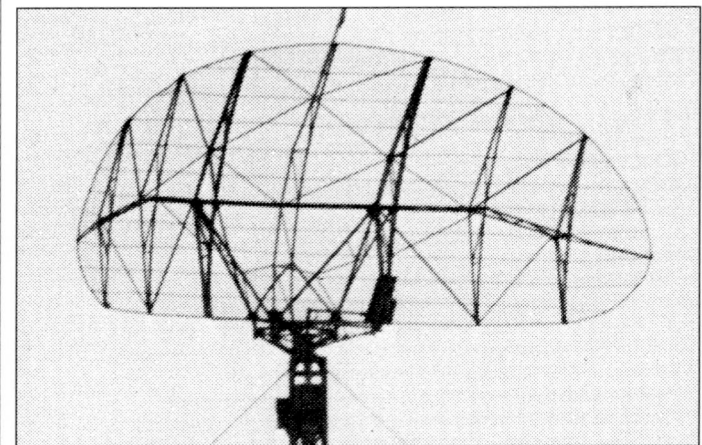

The antenna used in the P-15M target acquisition radar system

Status

As of June 2001, the precise status of the P-15M search and acquisition radar was uncertain.

VERIFIED

P-35/37 series radars

Type

E/F-band (2 to 4 GHz) mobile early warning and Ground Control Interception (GCI) radar. NATO reporting name Bar Lock.

Description

This large 2-D ground-controlled interception and search radar is a trailer-mounted equipment that has two large truncated paraboloid reflectors with clipped corners. A D-band (1 to 2 GHz) identification friend-or-foe antenna is attached to the right corner of the upper array. It is normally associated with the PRV-11 Side Net radar which provides height-finding.

Operating frequencies of the radar are in the E/F-band. Each of the reflectors is fed by three feeds so that six stacked beams result, each beam having a power output in the region of 1 MW. These beams may enable the radar to be used as a height indicator, but it is also likely that they are part of a system for providing surveillance cover over different altitude bands. Both the reflectors can be adjusted in elevation angle by the operator while the radar is operating to permit manual variations of vertical cover. Individual frequency bands for the stacked beams are: 2,695 to 2,715 MHz, 2,715 to 2,750 MHz, 2,815 to 2,835 MHz, 2,900 to 2,990 MHz, 2,990 to 3,025 MHz and 3,080 to 3,125 MHz.

The original Bar Lock has the designation Soviet P-35 or P-37, the latter probably being a slightly improved version. A version known as the P-37M is known to exist but no details are available. In the 1960s an improved version, Bar Lock B, was developed and designated P-50. The differences between the A and B models are not known but almost certainly include improved electronic counter-countermeasures. Bar Lock is widely used as both a GCI and early warning radar. It is closely associated with SA-5 'Gammon' surface-to-air missile units where it is used to provide battalion level surveillance and target acquisition.

Status

P-35/37 series radars are known to have been in operation since the early 1960s and have, over time, been widely deployed in the Russian Federation and Associated States and friendly nations such as North Korea.

Specifications

Frequency bands: see above
Antenna dimensions: 11 × 3.5 m
System rotation: 3 and 6 rpm
Effective range: 390 km
Peak power: 650 kW per beam
Beamwidth: 0.7°
PRF: 375 pps
Pulse-widths: 1.83-3.1 µs
Range accuracy: 900 m
Azimuth accuracy: 0.5°
IFF system: D-band (1-2 GHz), 2.4 µs pulse-width

VERIFIED

The Bar Lock early warning GCI radar

Protivnik-GE surveillance radar

Type

D-band (1 to 2 GHz) 3-D air surveillance radar.

Description

Designed for use as an air traffic control sensor or as part of an automatic ground-controlled interception system, the Protivnik-GE radar is described as being a mobile, high-resolution 3-D sensor that automatically establishes target parameters. Protivnik-GE requires no pre-operation 'adjustment or tuning', features automatic built-in test and can be interfaced with an external, automated control system. The equipment features a secondary surveillance radar subsystem and displays data in digital form on a three-colour, widescope plan position indicator.

Status

As of June 2001, the precise status of the Protivnik-GE 3-D air surveillance radar was uncertain.

Specifications

Frequency: D-band (1-2 GHz)
Range: at least 400 km (circular scan mode)
Altitude coverage: approx 50 m (lower); 120,000 m (upper)
Accuracy: 8 min (elevation); 10 min (azimuth); 50 m (range); 350 m (altitude)
Data rate: 10 s
MTBF: at least 1,000 h
Crew: 2

Contractor

Nizhegorodsky Radio Engineering Research Institute, Nizhni Novgorod.

VERIFIED

PRV-9 height-finding radar

Type

H-band (6 to 8 GHz) nodding height-finder radar. NATO reporting name Thin Skin.

Description

PRV-9 is generally deployed as part of a ground control interception formation. Both trailer- and truck-mounted versions, Thin Skin A and Thin Skin B models, have been observed and the operational role is probably in support of tactical air defence units, as the standard height-finding radar for mobile Surface-to-Air Missiles (SAM) units. In all cases it will be associated with air search and surveillance radars to provide them with complementary height information on targets detected. This radar is reported to be associated with Russian Federation and Associated States (RFAS) 'Ganef', 'Gainful' and 'Gecko' SAM regiments. Range is approximately 240 km.

Status

First reported in 1965, the PRV-9 Thin Skin height-finder is reported as having been supplied to customers in and outside the RFAS.

Specifications

Frequency: H-band (6-8 GHz)
Antenna dimensions: 8.5 × 3.9 m; 6.9 × 3.1 m
Effective range: 240 km
Effective altitude: 30,000 m

VERIFIED

The antenna used in the PRV-9 height-finding radar system

PRV-11 height-finding radar

Type
E-band (2 to 3 GHz) transportable height-finder radar. NATO reporting name Side Net.

Description
PRV-11 is a large transportable nodding height-finder radar, usually employed in conjunction with early warning surveillance radars such as Bar Lock, Back Net and Tall King to provide ground-controlled interception facilities. Operating frequency is in the range 2,560 to 2,710 MHz with an effective range of about 180 km at heights up to 32,000 m. The large narrow elevation elliptical paraboloid antenna is mounted on a pair of lattice trunnions cantilevered from the sides of the cabin containing the associated electronics. The cabin can be rotated in azimuth.

Status
Over time, the PRV-11 height-finder radar has been exported to countries inside and outside the RFAS that have been supplied with 'Goa' and 'Guideline' surface-to-air missile systems.

Specifications
Frequency: E-band (2-3 MHz)
Range: 180 km
Altitude: 30 km
Antenna dimensions: 7.7 × 3 m
Nodding rate: 5-30 cps
Van dimensions: 2.97 × 3.78 × 2.42 m

VERIFIED

The Side Net height-finder radar (Jane's Intelligence Review)

Rezonans surveillance radar

Type
Metric wavelength air surveillance radar.

Description
Designed to detect air targets such as ballistic and cruise missiles, hypersonic vehicles and 'stealth' aircraft, the Rezonans surveillance radar can be operated as an autonomous unit or as part of an automated ground-controlled interception system. The Rezonans radar is housed in five vehicles (three equipment and two power generation semi-trailers) and there is provision for a trailer-mounted, three display console command post that can be located up to 1 km from the main site. The equipment is claimed to feature 'enhanced noise immunity' and 'good' target radial velocity resolution and is able to cover 360° in azimuth. At any time within a single scan, the radar covers a 60° sector, from which, data is received in 16 channels simultaneously. System function is under automatic software control and there is provision for an identification friend-or-foe interrogator subsystem. Frequency agility and an adaptive compensation subsystem (using specifically developed algorithms) are used to provide protection against active radio frequency jamming.

Status
As of June 2001, the precise status of the Rezonans surveillance radar was uncertain.

Specifications
Wavelength: metric
Coverage: 70° (elevation); up to 100 km (altitude); 360° (azimuth)
Range: 350 km (fighter-type target at 10,000 m)

Accuracy: 0.5 m/s (target velocity); 1.5° (azimuth and elevation); 300 m (range)
Noise-obscured visibility factor: 70 dB
Output: numerical data on computer screen
Update rate: 10 s
MTBF: 1,500 h
Power requirement: 50 kW
Crew: 3 (single shift)
Set up time: 24 h
Temperature: –40 to +50°C (operating)
Windspeed: up to 30 m/s

Contractor
Long Range Radio Communication Research Institute Production Complex JSC, Moscow.

VERIFIED

ST-68UM surveillance radar

Type
Centimetric wavelength 3-D air surveillance radar.

Description
Described as being suitable for 'air defence troop, signals unit and air force electronic support' applications, the ST-68UM surveillance radar is noted as being able to detect, identify and track air targets (including long-range, Air Launched Cruise Missiles - ALCM) in an environment polluted by active/passive jamming and ground/weather clutter. The system comprises:
- an equipment and antenna package mounted on the chassis of a MAZ-938B semi-trailer
- a 99X6 power generation system mounted on the chassis of a MAZ-5224V trailer
- an optional 40V6M antenna tower and Type 61103 communications cabin mounted on a KamAZ-4310 truck chassis.

With the 40V6M tower deployed, the ST-68UM's antenna phase centre can be raised to a maximum of 24 m above ground level. The radar's elevation scan makes use of a fractional beam pattern while azimuth coverage is provided by means of mechanical rotation of the array. When incorporated, the system's Type 61103 communications cabin transmits acquired data to external posts up to 35 km away from the main radar site. ST68-UM is also reported to incorporate protection against both passive and active jamming.

Status
As of June 2001, the precise status of the ST-68UM 3-D air surveillance radar was uncertain.

Specifications
Wavelength: centimetric
Range (in the presence of active/passive jamming): 20 km (ALCM at 60 m, without 40V6M antenna tower); 28 km (MiG-21 type target at 50 m, without 40V6M antenna tower); 30 km (ALCM at 100 to 3,000 m, without 40V6M antenna tower); 32 km (ALCM at 60 m, with 40V6M antenna tower); 33 km (MiG-21 type target, with 40V6M antenna tower); 40 km (ALCM at 100 to 3,000 m, with 40V6M antenna tower); 42 km (MiG-21 type target at 100 m, without 40V6M antenna tower); 46 km (MiG-21 type target at 100 m, with 40V6M antenna tower); 80 km (MiG-21 type target at 500 to 6,000 m, with 40V6M antenna tower)
Range (in the presence of ground/weather clutter): 27 km (ALCM at 30 m, with 40V6M antenna tower/ALCM at 60 m, without 40V6M antenna tower); 31 km (MiG-21 type target at 50 m, without 40V6M antenna tower); 33 km (ALCM at 100 m, without 40V6M antenna tower); 40 km (MiG-21 type target at 50 m, with 40V6M antenna tower/ALCM at 60 m, with 40V6M antenna tower); 42 km (MiG-21 type target at 100 m, without 40V6M antenna tower); 48 km (ALCM at 100 m, with 40V6M antenna tower); 51 km (MiG-21 type target at 100 m, with 40V6M antenna tower); 60 km (ALCM at 300 to 3,000 m, with 40V6M antenna tower); 82 km (MiG-21 type target at 500 m, without 40V6M antenna tower); 92 km (MiG-21 type target at 500 m, with 40V6M antenna tower); 147 km (MiG-21 type target at 2,000 to 18,000 m, with 40V6M antenna tower)
Target radial velocity: 60 to 180 km/h (min)
Clutter rejection: 48 dB (min)
MTBF: 100 h
Set up/close down time: 1 h (without 40V6M antenna tower and Type 61103 communications cabin); 2 h (with 40V6M antenna tower and Type 61103 communications cabin)
Crew: 7 (3 shifts)

VERIFIED

Struna-1 surveillance radar

Type
Bistatic surveillance radar system.

Description
Designed to detect 'low signature threats' such as cruise missiles, 'stealth' aircraft, light sporting aircraft and balloons, the bistatic Struna-1 surveillance radar system

is arranged to create a continuous 'barrier' that is several kilometres long by up to 7 km high. Up to 10 individual 'barriers' can be joined together, creating a 'super barrier' with a baseline of up to 400 km. Within the system, acquired data is transmitted via radio relay channels to an operator station that can be up to 400 km away from the particular radar site. Operator station functions include display of acquired data, generation of system control commands, activation of the equipment's built-in test subsystem, data recording and data integration with an external air defence control system. If required, individual Struna-1 radar sites can be unmanned, with control being exercised remotely from the described operator station.

Status

As of June 2001, the precise status of the Struna-1 bistatic surveillance radar system was uncertain.

Specifications

Barrier size (l × d × h): 20-50 km × 1.5-8 km (specific value dependent on target type) × 0-7 km (specific value dependent on target type)
Emitted power: 1-5 W
Data rate: 0.25 s
Power requirement: 750 W (single barrier)
Service life: 40,000 h
Crew: 1 operator
Turn on time: 1 s (max)
Single barrier set up time: 6 h (max)
Single barrier equipment weight: 7 t

Contractor

Nizhegorodsky Radio Engineering Research Institute, Nizhni Novgorod.

VERIFIED

SOUTH AFRICA

ESR 360/360L/380 early warning radar

Type

D-band (1 to 2 GHz) 3-D early warning radar systems.

Description

The ESR 360/360L/380 series radars are high-mobility 3-D early warning systems which are intended for use in the gap filler, airspace control and/or divisional/brigade anti-aircraft asset control roles. ESR 360 series radars are self-contained units which can be transported by fixed-wing aircraft or helicopters and incorporate design features which, according to the manufacturer, maximise reliability and reduce down time. Individual radars can be deployed within 30 minutes and can also be used in fixed site installations.

A typical application consists of an operator cabin which houses the system's controls, full-colour high-resolution raster displays and an integral power supply if no external source is available. The system antenna appears to be integral with the operator shelter and is lowered and stored for transportation. The equipment's design incorporates clutter rejection and an electronic counter-countermeasures suite. The family as a whole can be configured for both single operator usage and

remote-control operation. In this later format, the system design offers master/slave and unmanned operating modes, full radar control and status monitoring via an external datalink. A fully integrated identification friend-or-foe/secondary surveillance radar capability can be incorporated if required.

Specifications

	ESR 360	ESR 360L	ESR 380
Frequency	D-band (1-2 GHz)		
Instrumented range	150 km	300 km	450 km
Detection range (against a 2m² Swerling 1 target)	120 km	200 km	330 km
Azimuth accuracy	0.65°	0.25°	0.25°
Elevation accuracy	±0.6°	±0.4°	±0.2°
Range accuracy		40 m	
Antenna type	low-sidelobe planar-array		
Number of beams	6	6	8
Rotation rate	15 rpm	7.5/15 rpm	6/10 rpm
Transmitter type	high power, wideband, solid state		
Track capability	100 targets/s with track-while-scan		

Contractor

Reunert Defence ESD/Reutech Systems (PTY) Ltd, Stellenbosch.

VERIFIED

SWEDEN

GIRAFFE air defence radar systems

Type

Family of G/H-band (4 to 8 GHz) air defence surveillance radars with Command, Control, Communications and Intelligence (C³I) capabilities.

Description

GIRAFFE is the family designation for a series of combined G/H-band pulse Doppler search radars and combat control centres that are designed for mobile and static, short- and medium-range C³I air defence applications. Within such applications, GIRAFFE's primary purpose is the detection of very low-flying targets in conditions of severe clutter and electronic jamming. Radars within the family employ broadband travelling wave tube final amplifiers and G/H-band operation has been selected as the optimum for an equipment that is designed to provide low-level coverage against small radar cross-section targets operating in a clutter/jamming environment. Countermeasures proofing is provided via frequency agility and the use of low sidelobe antennas, digitial pulse compression, adaptive transmission modes and 'instantaneous' unjammed frequency selection. Digital Doppler processing (with constant false alarm rates) is used to automatically detect, extract and display targets of interest. For communication with associated firing units, the GIRAFFE C³I capability makes use of manpack display units that

The antenna and operator cabin configuration used in the ESR 360 series 3-D early warning radar

The GIRAFFE 75 C³I radar for tactical air defence systems

provide each battery with information such as target bearing and range relative to the firing site, target speed and course, cross-distance, detection time limits and target priorities. As of September 2001, five GIRAFFE variants have been identified, the known details of which are as follows:

GIRAFFE 40

GIRAFFE 40 is a short-range (40 km instrumented range) air defence radar with C³I capability that is designed for use with short-range missile and anti-aircraft gun systems. The equipment employs a folding antenna mast which, when fully extended, gives an array height of 13 m. GIRAFFE 40 can be integrated with an Identification Friend-or-Foe (IFF) subsystem (including Mk XII) and its detection envelope is given as being from ground level to an altitude of 10 km. In Swedish service, the radar has been designated as both the PS-70 and the PS-701 and a 60 kW power output version has been developed under the designation PS-707 (Sweden)/Super GIRAFFE (export). As of September 2001, GIRAFFE 40 was no longer in production.

GIRAFFE 50 AT

Within an instrumented range of 50 km, the GIRAFFE 50 AT has been developed for the Norwegian Army's Low-Level Air Defence System (NALLADS) and is installed on a tracked chassis that is capable of operating over 'severe terrain'. The variant's mast-mounted antenna has an operating height of 7 m and other features include:

- a detection envelope that stretches from ground level to an altitude of 10 km
- fully automatic combat control functions (track initiation, tracking (Kalman filters), target identification, target classification and designation, threat evaluation and 'pop-up' target handling)
- the ability to exercise tactical control over up to 20 firing units
- a radar-to-firing unit target information datalink
- automatic hovering helicopter detection and threat evaluation functions
- the ability to exchange data with GIRAFFE 75 and AMB systems to facilitate radar co-operation and the compilation of a local air picture
- the ability to be integrated with an IFF subsystem (including Mk XII).

GIRAFFE 75

GIRAFFE 75 features a 'complete' C³I capability and is designed for use in medium-range (GIRAFFE 75 has an instrumented range of 75 km) air defence systems or short-range air defence applications where the emphasis is on a high level of electronic counter-countermeasures and C³I performance. Other system features include:

- a detection envelope that stretches from ground level to an altitude of 10 km
- automatic hovering helicopter detection and threat evaluation functions
- a folding antenna mast which, when fully extended, gives an array height of 13 m
- fully automatic combat control functions (track initiation, tracking (Kalman filters), target identification, target classification and designation, threat evaluation and 'pop-up' target handling)
- the ability to exercise tactical control over up to 20 firing units
- a radar-to-firing unit target information datalink
- the ability to exchange data with GIRAFFE 50 AT and AMB systems to facilitate radar co-operation and the compilation of a local air picture
- the ability to be integrated with an IFF subsystem (including Mk XII)
- an optional add-on unit to optimise the radar for coastal surveillance duties.

In Swedish service, GIRAFFE 75 is designated as the PS-90.

GIRAFFE AMB

Developed for use with the Swedish RBS 23 BAMSE and RBS 97 air defence missile systems, GIRAFFE AMB is equipped with an active phased-array antenna and offers a full 3-D capability. An associated, 'next-generation' C³I capability optimised for use with short- to medium-range air defence systems is included. Other system features include:

- multibeam technology
- digital beam shaping
- a detection envelope that stretches from ground level to an altitude of more than 20 km

- automatic hovering helicopter detection and threat evaluation functions
- a folding antenna mast, which, when fully deployed, gives an array height of 13 m
- fully automatic combat control functions (track initiation, tracking (Kalman filters), target identification, target classification and designation, threat evaluation and 'pop-up' target handling)
- the ability to exercise tactical control over up to 20 firing units
- a radar-to-firing unit target information datalink
- the ability to exchange data with GIRAFFE 50 AT, GIRAFFE 75 and other GIRAFFE AMB systems to facilitate radar co-operation and the compilation of a local air picture
- the ability to be integrated with an IFF subsystem (including Mk XII).

In Swedish service, GIRAFFE AMB is designated as UndE 23.

GIRAFFE S

GIRAFFE S is a low-level 'gap filler' and coastal surveillance radar that is designed primarily for remotely controlled applications. System features include:

- a ground level to 6 km altitude detection envelope
- a 180 km instrumented range
- automatic hovering helicopter detection and threat evaluation functions
- fully automatic system control functions (track initiation, tracking (Kalman filters), target identification, target classification and designation, threat evaluation and 'pop-up' target handling)
- simultaneous supervision of air- and sea-space
- fixed-site, transportable and mobile installations options
- a radar-to-firing unit target information datalink.

When used for fixed site operations, GIRAFFE S can be configured for unmanned operation with track and plot data being transmitted to regional or national control centres via narrowband radio link or landline. System control and status monitoring are executed via the control centre and the radar's antenna is mast mounted (up to 30 m high). In mobile applications, GIRAFFE S makes use of a folding 8 m high antenna mast that is integrated with an air conditioned, nuclear/biological/chemical protected, cross-country vehicle or truck-mounted equipment/operator cabin. So configured, GIRAFFE S is fully autonomous, incorporates IFF and communications subsystems and is equipped with two operator workstations. Deployment time is given as beng less than 10 minutes.

Status

As of September 2001, at least 450 GIRAFFE radar/C³I systems are known to have been ordered/procured by customers around the world since 1978. According to Jane's sources (but not confirmed by the equipment's manufacturer), GIRAFFE 40 systems have been supplied to the armed forces of Singapore, Sweden (PS-70 for use with the RBS 70 missile and PS-707 for use with the RBS 90 system) and Thailand. GIRAFFE 50 AT has been supplied to Brazil and Norway (NALLADS programme), while GIRAFFE 75 customers are understood to include Sweden (PS-90) and Venezuela (mounted on MAN LX90 6 × 6 cross-country vehicles). As of the given date, a total of eight GIRAFFE AMB radars had been procured by the Swedish Army (military designated UndE 23, June 2000 contract valued at then year SKr500 million/US$60 million) and the sensor had been ordered by both an unnamed export customer (thought to be Singapore) and the French Air Force (four radars ordered during March 2001). Six 2-D ARTE 740 variants of GIRAFFE AMB (mounted on a MOWAG 10 × 10 Piranha armoured command and control platform, June 2000 contract valued at then year SKr100 million/US$12 million) are also understood to have been ordered during 1999 for use by Sweden's Amphibious Forces (formerly the Coastal Artillery). Jane's sources suggest that ARTE 740 has a range of 30 to 60 km and makes use of the antenna from the GIRAFFE 75 radar mated to the processing technology from the GIRAFFE AMB system. GIRAFFE S is believed to have been acquired by Finland as an air defence 'gap-filler'. Other GIRAFFE (variant unknown) customers are reported to include

The Ericsson GIRAFFE 50 AT search radar and C³I system has been procured for use in the Norwegian Army Low-Level Air Defence System (NALLADS)

The Giraffe AMB radar system is designed for use with medium-range anti-aircraft systems such as the RBS 23 BAMSE air defence missile
0006878

Greece and Yugoslavia and Ericsson notes that 'basic' GIRAFFE systems have been supplied to 'more than' 10 countries worldwide.

Contractor

Ericsson Microwave Systems AB, Surface Surveillance Radar Systems, Mölndal.

UPDATED

HARD/Improved HARD search radar

Type

I/J-band (8 to 12 GHz) 3-D local search and acquisition air defence radar.

Description

The Ericsson Helicopter and Aeroplane Radar Detection (HARD) is a 3-D local search and acquisition radar for ground-based and naval air defence systems. The Swedish Army designation is PS-91. It is a solid-state I/J-band pulse-Doppler radar having a phase-controlled planar-array antenna. It is normally installed in a tracked vehicle for extreme terrain mobility, but can also be used as a free-standing unit. When mounted on a vehicle it automatically compensates for vehicle motion, roll and pitch. Although developed for the Swedish Army's RBS 90 system it can be integrated into any short-range air defence system.

HARD is a 'silent radar', so called because of low-peak power and high-antenna gain, together with low-sidelobes and simultaneous Moving Target Indicator (MTI) and frequency agility. It is stated to be almost immune to electronic jamming because of pulse-to-pulse frequency agility combined with Doppler processing, pulse compression and sidelobes that have been virtually eliminated.

The planar-array, phase-controlled antenna has 16 elements, each fed by a solid-state module consisting of transmitter, receiver and phase shifter. These modules are fed by a fast frequency-shifting radio frequency generator and solid-state driver amplifier.

The system provides target data in elevation by electronic multibeam scanning, azimuth and range with automatic detection, tracking, target evaluation and hovering helicopter detection and analysis. It also has automatic detection and tracking of several targets simultaneously.

The Ericsson HARD 3-D air defence local search radar was developed for use in the Swedish Army's RBS 90 missile system

Improved HARD has been procured for use in Germany's LeFlaSys light air defence system

0022580

Starting in 1992/93, Ericsson began development of a second-generation HARD system known as Improved HARD. The main differences between the two models are the introduction of an alternating pulse repetition frequency mode for the elimination of second-time-around echoes; an improved antenna with lower side-lobe levels; a new adaptive MTI filter which is claimed to improve detection range by 20 per cent and functions automatically in heavy clutter conditions and an increase in power output per module from 6 to 14 W. Improved HARD's detection range against a fighter sized target is quoted as being typically 16 km and the variant replaced the original model from 1995.

Status

As of January 2001, HARD was reported as being in service with the Swedish Army (Swedish military designation PS-91) as part of the service's RBS 90 all-weather air defence system. Here, procurement has included a US$75 million acquisition contract and, within the system, Jane's sources report the radar as being mounted on the roof of a Bv 206 tracked fire-control vehicle. With regard to the Improved HARD configuration, Jane's sources report that 10 such radars have been acquired for use in Germany's LeFlaSys light air defence system were they are installed aboard Mak Wiesel 2 light armoured vehicles. The same sources suggest that Improved HARD entered German service during late 2001.

Specifications
HARD

Frequency: I/J-band (8-12 GHz)
Power output: 65 W (peak); 8 W (average)
Range: 8-10 km against a helicopter; 16-20 km against a fixed-wing aircraft
Weight: 125 kg (antenna/transceiver and processor)

Contractor

Ericsson Microwave Systems AB, Ground Systems Division, Mölndal.

UPDATED

UKRAINE

ST-68UM

Type

E/F-band (2 to 4 GHz) mobile 3-D air surveillance and missile control radar. NATO reporting name Tin Shield.

Description

The ST-68UM (also known as the 36D6 in Russian service) radar is a mobile 3-D system which is designed to detect air targets at low, medium and high altitudes and forms part of the S-300P (NATO reporting name SA-10a 'Grumble') Surface-to-Air Missile (SAM) system. The complete ST-68UM system comprises an antenna array (mounted on an SPP-15 (?) trailer or a 40V6M1 mast), turning gear, an Identification Friend-or-Foe (IFF) subsystem, an electronics/operator cabin and a KP-10 (?) mobile power supply. Of these the antenna trailer is towed by a KrAZ-255V truck, with a KrAZ-255B model being used for the radar's power supply. A MAZ-537 truck is used as to deploy the 40V6M1 antenna mast. System features include:

- multibeam output with independent signal acquisition for each beam
- a high stability transmitter
- a 'unique' automatic Doppler moving target indication facility
- a multiprocessor system for target detection, tracking and trajectory calculation
- built-in test for automatic malfunction detection
- a jamming centre direction-finding capability
- integration of IFF and radar plots

The 40V6M1 mast-mounted antenna configuration used with the ST-68UM radar

The trailer-mounted antenna configuration used with the ST-68UM radar (Martin Mamula/Jane's Intelligence Review)

- 'exceptional' electronic counter-countermeasures provision
- the ability to hand off radar targeting data via 'narrowband communications channels'.

Alongside the described capabilities, Jane's sources suggest that ST-68UM features a remote control operating mode, three operator workstations and can be deployed in one hour (two if the 40V6M1 mast is used).

In the S-300PMU-1 variant of the 'Grumble' SAM system, ST-68UM is replaced by the 3-D phased-array 64N6 (NATO reporting name Tombstone) radar (see separate entry). Developed by the Moscow-based Nauchno Issledovatelskiy Institut Priborostroyeniya (NIIP – Scientific Research Institute for Instrument Engineering), 64N6 features back-to-back antennas and provides a ballistic missile detection capability. Returning to the ST-68UM, the system is also understood to be available for civilian air traffic control applications under the designation Dnepr.

Status

Over time, S-300P SAM systems are understood to have been procured by the armed forces of Bulgaria, Croatia, the Czech Republic, North Korea and the Russian Federation. S-300PMU/PMU-1 systems are quoted as having been supplied to the People's Republic of China (S-300PMU), Cyprus/Greece (S-300PMU-1), Iran (S-300PMU), the Russian Federation and Syria (S-300PMU).

Specifications

Frequency: 2-4 GHz
Instrumented range: 90, 180 or 360 km (dependent on pulse repetition frequency)
Low flying target detection range: 27 km (0.1 m² RCS target flying at an altitude of 50 m); 31 km (1 m² RCS target flying at an altitude of 50 m); 42 km (1 m² RCS target flying at an altitude of 100 m); 110-115 km (1 m² RCS target flying at an altitude of 1,000 m)
Coverage: −0.5 to +30° (elevation); 360° (azimuth)
Antenna scan rate: 6 or 12 rpm
Ground clutter suppression: >48 dB
Accuracy: ≤0.2° (azimuth); ≤50 m (range); 400 m (altitude)
Resolution: ≤3.5° (azimuth); 300, 600 or 1,200 m (range - dependent on PRF)
Track capacity: up to 120 tracks
Environment: −50 to +50°C (temperature); 98% at 35°C (humidity); ≤3,000 m (height)
MTBF: >800 h
MTTR: <30 min
System transportation characteristics:

System element	Body type	Dimensions (l × w × h)	Weight	Vehicle
Radar	SPP-15	13,882 × 2,890 × 3,325 mm	21.54 t	KrAZ-255V
Power supply	KP-10	9,112 × 2,890 × 3,296 mm	14.5 t	KrAZ-255B
Antenna mast		26,110 × 3,200 × 3,950 mm	74.4 t	MAZ-537

Contractor

UKRSPETSEXPORT (export agency), Kiev.

NEW ENTRY

Transportable Over-The-Horizon Surface Wave (OTH-SW) radar

Type

High Frequency (HF - 3 to 30 MHz) OTH-SW radar.

Description

Developed by the Ukrainian Radio Technical Institute, the described 18 to 25 MHz OHT-SW radar is described as being a 'removable' bistatic equipment that is designed to detect aircraft and surface ships and comprises reception and transmission array that are set up at distances of between three and 12 km from each other. As of the second half of 2000, Jane's sources were reporting 200 and 300 km range variants of the architecture. In the first instance, the 200 km range configuration is described as making use of a 330 m long reception array (incorporating 64, 6 m high, 'vibrator' elements) and a transmitter assembly made up of eight, vertical, log periodic antennas. The 300 km range configuration is noted as incorporating a 640 m long, 64 'vibrator' reception aperture.

Status

As of the second half of 2000, UKRSPETSEXPORT was actively promoting the described OTH-SW radar.

Specifications

Frequency: 6-24 MHz (300 km range configuration); 18-25 MHz (200 km range configuration)
Coverage: 60° sector (200 km range configuration)
Range: 60 km (300 km range configuration - 1 dB/m² RCS air target flying at an altitude of 10-100 m); 120 km (300 km range configuration - 1 dB/m² RCS air target flying at an altitude of 100-10,000 m); 180 km (300 km range configuration - surface ship target with an RCS of 20 dB/m²); 200 km (200 km range configuration - max detection range); 300 km (300 km range configuration - 1 dB/m² RCS air target flying at an altitude of above 10,000 m or surface ship target with an RCS of 40 dB/m²)
Detection time: 10-60 s (aircraft); 100-300 s (surface ships)
Track capacity: up to 50 aircraft or 100 surface ships simultaneously

Contractor

UKRSPETSEXPORT (export agency), Kiev.

NEW ENTRY

A general view of part of the transmission array used in the UKRSPETSEXPORT transportable OTH-SW radar (UKRSPETSEXPORT)
2002/0083260

Transportable SkyWave Over-The-Horizon (SkW-OTH) radar

Type

High Frequency (HF - 5 to 28 MHz sub-band) transportable SkW-OTH radar.

Description

Ascribed to the Ukrainian Radio Technical Institute, the cited SkW-OTH radar is reported as being designed to detect and track air, sea surface and ballistic missile targets and as being a bistatic system that incorporates transmission and reception arrays that are set up at distances of between 20 and 200 km from each other. Of the two, the transmission array is noted as incorporating 12 log periodic, vertically polarised antenna elements, each of which is connected to its own 15 kW (mean power) transmitter. The system's reception array is described as being 600 m long. Operated by a 15 man crew (working in shifts), the architecture is further reported as making use of an 'associated' 450 Mflop/s computer system and as having a set up time of 'no more than a month'.

Status

As of the second half of 2000, UKRSPETSEXPORT was actively promoting the described transportable SkW-OTH radar.

Specifications
Frequency: 5-28 MHz
Coverage: 60° sector (to 2,000 km range)
Range: 600 km (min skywave detection range - 15 km surface wave); 2,600 km (max skywave detection range)
Altitude: 10 m-60 km (air targets); 5-100 km (ballistic missiles)
Target velocity: 40-3,600 m/s (ballistic missile targets at an altitude of 5 to 100 km); 18 km/h (threshold for sea surface targets); 100-3,600 km/h (air targets at an altitude of 10 m to 60 km)
Track capacity: more than 50 missiles (within designated 'control areas'); more than 300 sea surface targets (within six 'periodically controlled zones'); up to 1,200 simultaneous air targets

Contractor
UKRSPETSEXPORT (export agent), Kiev.

NEW ENTRY

UNITED KINGDOM

AR-3D air defence radar

Type
E/F-band (2 to 4 GHz) air surveillance radar.

Description
AR-3D is an E/F-band air defence surveillance radar giving 3-D information on targets by mechanically rotating the radar beam in azimuth and electronically scanning in elevation. All elements of the system are trailer mounted for transport by land, sea or air.

The AR-3D antenna, which combines mechanical scanning in azimuth with electronic scanning in elevation, consists of a compact circular polarised linear array positioned at the focus of a simple parabolic cylinder reflector. The reflector has a height of 4.9 m and a width of 7.1 m. The azimuth sidelobes of the antenna are −36 dB in the 5 to 15° sector and −48 dB in the 15 to 90° sector. The transmitter consists of a coherent two stage E/F-band transmitter. Each of the two stages uses a linear beam pulse microwave tube. Both tubes are designed for excellent phase, noise and amplitude performance commensurate with the pulse compression and moving target indicator requirements. The transmitter has a peak pulse of 1.11 MW (mean power 10 kW) and a bandwidth of 200 MHz.

The signals received from the antenna are amplified in a wideband (200 MHz) amplifier. After amplification the signals are separated into channels representing elevation bands of approximately 2° (bandwidth 20 MHz per channel). The signals are then time compressed to 0.1 µs and their frequencies measured to enable the fine elevation within each beam to be obtained.

The function of the plot extractor is to detect targets which meet prescribed criteria. This target information is then fed to tracking and display equipment in the form of digital words containing 3-D positional information as polar co-ordinates. The online processor stores the positional data, carries out azimuth and height calculation, and takes into account various corrections. Plot and track data are then transmitted to the local displays and modems as required.

The varying demands of air defence are met by supplying AR-3D in different operational configurations: as a radar reporting post, control and reporting post, or command and control post. The different requirements are met by variants of the processing and control cabin, with hardware and software organised in modular packages appropriate to the specific function.

As a reporting post the AR-3D provides two operational consoles with automatic plot extraction and reporting to a remote centre of up to 100 combined primary radar and identification friend-or-foe/secondary surveillance radar plots. As a

In some applications, the AR-3D antenna is installed inside a radome as shown here

control and reporting post with five operational consoles, automatic initiation and tracking of up to 100 aircraft is provided with automatic reporting of data to a remote operations centre, while up to 10 simultaneous computer-aided interceptions can be controlled. The command and control post additionally varies in providing for autonomous computer-aided control of own aircraft and missiles.

A number of mid-life upgrade modifications have been developed for the AR-3D which enhance performance and reliability. These include an improved detection performance in clutter, enhanced control of the transmitted beam and an upgrade of the data processing subsystem.

Status
AR-3D has been procured by the UK Ministry of Defence and a number of overseas countries. Two AR-3D systems have been deployed in the Falkland Islands. Although no longer in production, as of mid-2000 BAE Systems was understood to have been providing continuing, long-term support to AR-3D customers. A number of mid-life upgrade modifications have been developed and fitted to the AR-3D which enhance performance, availability and reliability. These include significant improvement in subclutter visibility, improved control of the transmitted beam and upgraded data processing. Readers should note that BAE Systems and Italian contractor Finmeccanica announced on 26 April 2001 that AR-3D manufacturer BAE Systems - Combat and Radar Systems was to be transferred from BAE Systems control to that of their Alenia Marconi Systems joint venture. As of August 2001, said transfer had not been completed.

Specifications
Coverage: adjustable. Typically 333 km on 2 m² target
Radiation: frequency selective within E/F-band. The operating bandwidth is 200 MHz. Independent frequency selective filters for each of 13 elevation channels, each having a bandwidth of approximately 20 MHz
Antenna
Type: linear array and parabolic cylinder
Polarisation: circular
Beamwidth (azimuth): 1°
Beamwidth (elevation): 2°
Peak sidelobe (azimuth): 25 dB
Peak sidelobe (elevation): 23 dB
Scanning angle: 30°
Gain: 41.3 dB
Adjustment: +4, −2°
Rotation rate: 6 rpm
PRF: 250 pps
Frequency: E/F-band (2-4 GHz)
Bandwidth (to 1 dB): 200 MHz
Pulse-length: 36 µs
Power amplifier tube: high-power klystron grid modulated
Transmit modes: continuous or fast on/off
Receiver
Bandwidth: 20 MHz/channel
Compression ratio: 130 :1 (lower channels)
Accuracy
Elevation: 0.15°
Range: 15 m
Range resolution: 40 m

Contractor
BAE Systems - Combat and Radar Systems, Cowes, Isle of Wight.

The antenna used in the AR-3D air surveillance radar system   2000/0079483

UPDATED

AR-320 air defence radar

Type

E/F-band (2 to 4 GHz) 3-D transportable air defence radar (NATO Class 1 long range).

Description

The AR-320 air defence radar is the result of collaboration between the then Plessey Radar Ltd (now BAE Systems - Combat and Radar Systems) and ITT Gilfillan and meets the performance requirements demanded of a NATO standard E/F-band trailer-transportable air defence radar. As such, it combines the ITT Gilfillan Series 320 3-D antenna and transmitter with BAE Systems' AR-3D receiver, signal processor, displays and software. The Series 320 radars represent a family of defence radars based on 3-D radars (such as the AN/SPS-48 and AN/TPS-32) developed for the American Services by Gilfillan and subsequently supplied to a number of offshore customers. The combination of techniques used in the AR-320 provides an inertialess, single pencil beam elevation scan that can be accomplished by either frequency or phase control to steer the beam in space. This technique maintains high resolution throughout the coverage volume, high data rate and accurate target location.

Overall, the AR-320 architecture is self-contained and is housed in a number of vehicles that can be transported by land, sea or air (including as a slung load beneath Chinook-class transport helicopters). It is designed to provide the optimum combination of long range, fine resolution, good performance in electronic countermeasures and clutter, high reliability, ease of maintenance and system survivability. The state-of-the-art digital processing and low-sidelobe antenna provide immunity to active and passive jamming, inter-jammer visibility, detection probability in the presence of clutter and constant low false alarm rate.

The associated radar-mounted Secondary Surveillance Radar (SSR)/ Identification Friend-or-Foe (IFF) interrogator/receiver provides secondary plots which are extracted and correlated with primary plots. The combined plots are then transmitted to an operations centre for use in the tracking process. These capabilities and the common module concept of the radar, ensure that it can be configured to meet a variety of tactical or strategic operational needs and can be used either independently or integrated with a national defence system.

In terms of field deployment, the AR-320's various components are installed in trailer or self-powered vehicle-mounted shelters and (using a trained crew) can be dismantled, set up and brought into service again within six hours, transportation time excluded. Typically, its vehicle group will consist of:
- an antenna assembly which can be transported on a flat bed trailer or rail car. In transit the radar antenna is folded down by means of a hydraulically operated mechanism
- a trailer-installed transmitter cabin containing primary and secondary radar transmitters
- a trailer-mounted processing and display cabin containing signal processing equipment, the radar management suite, operations consoles and the radar environment simulator
- transportable diesel-electric generator(s) able to provide power for the complete system
- a trailer-installed cabin containing workshop, human support and storage facilities.

The entire vehicle group can be deployed on a prepared or unprepared site and can be operated as a field group with radio link communications to a reporting post or sector operations centre.

Where environmental conditions require, the vehicle group can be housed under cover with the radar antenna protected by a rigid radome. In this configuration the antenna is hoisted into the radome and supported on structural cross-members. An alternative arrangement allows for installation of the antenna and associated electronic subsystems within a below-ground shelter or silo. The antenna is permanently mounted on an elevator and, for operational deployment, is raised and unfolded hydraulically. The antenna is capable of rapid retraction into the silo when required. The radar hardware in this configuration is arranged in equipment rooms in conjunction with associated power supplies, heating and ventilation equipment.

In summary, AR-320's main features include:
- long-range, E/F-band (2 to 4 GHz) 'high-cover' performance
- a high-gain, pencil beam output with 'ultra-low' azimuth and elevation sidelobes
- a planar phased-array utilising (within pulse) phase/frequency elevation scan
- wideband frequency agility
- an integral SSR/IFF antenna
- a wideband transmitter with mean power greater than 24 kW
- variable pulse-length and pulse repetition frequency
- multichannel, pulse compressed signal processing
- digital moving target indicator processing with velocity compensation for good anti-clutter and electronic counter-countermeasures performance
- small resolution cell size to facilitate the resolution of multiple targets and the maximisation of signal-to-clutter detectability
- fully automatic, 3-D plot extractor
- 3-D co-ordinates from every plot
- a 'high-performance' Series 9 display system with digital display drive
- bright, synthetic presentation with high-quality characters and excellent registration accuracy
- a radar environment simulation for operator training
- 'high' reliability and availability
- 'wide' environmental tolerance
- a transportable configuration for rapid field deployment.

Status

As of March 2001, six AR-320 radars were understood to be in service as part of the UK's Improved Air Defence Ground Environment. BAE Systems is also reported as having developed a number of AR-320 improvements and upgrades that have enhanced the radar's performance and reliability. Readers should note that BAE Systems and Italian contractor Finmeccanica announced on 26 April 2001 that AR-320 collaborator BAE Systems - Combat and Radar Systems was to be transferred from BAE Systems control to that of their Alenia Marconi Systems joint venture. As of August 2001, said transfer had not been completed.

Contractors

BAE Systems - Combat and Radar Systems, Cowes, Isle of Wight.
ITT Gilfillan Inc, Van Nuys, California, USA.

UPDATED

AR-325 air defence radar

Type

E/F-band (2 to 4 GHz) long-range transportable air surveillance radar.

Description

The AR-325 version of the Commander family is a long-range, 3-D, E/F-band radar intended for air defence purposes. The system is capable of detecting targets at ranges up to 470 km and of operating in the most severe electronic warfare environment. It is frequency-agile and the transmitter operates with a wide range of pulse-lengths and pulse repetition frequencies. This flexibility enables the radar to adapt its transmission pattern to obtain the required coverage. The transmitter provides a high-mean power with a relatively low peak power which contributes to an excellent performance in Electronic CounterMeasures (ECM) conditions. Parallel processing architecture has been used to provide increased operating speeds, using a standard signal processing module which the company has developed for all its large radars. Full moving target detector processing is provided with an automatic switch to moving target indicator.

Elevation coverage is provided by multiple pencil beams for both transmission and reception. As ground clutter is only illuminated in the lowest beam elevations all other elevations are unaffected, compared with stacked beam radars which transmit a fan beam illuminating all elevations simultaneously. Phase monopulse is employed to measure target elevation independently in each of the beam positions.

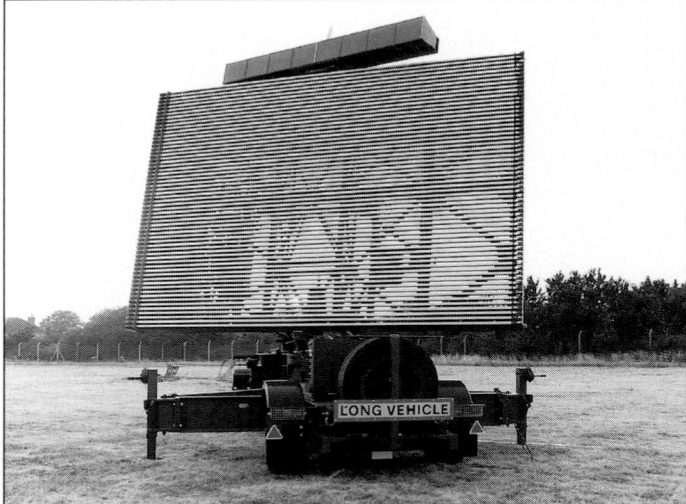

The AR-320 3-D air defence radar
2000/0079482

The antenna array used in the AR-325 Commander transportable air defence radar
2000/0079481

The antenna is an ultra-low sidelobe planar-array unit, giving an azimuth beamwidth of 1.4° and a very narrow vertical beamwidth. These are combined with a compressed pulse-length of 0.4 µs to ensure that the clutter cell is extremely small and the detection probability for wanted target returns is consequently increased.

The transmitter is a wideband two-stage device consisting of a solid-state first stage driving a travelling wave tube output stage. Signal processing, to provide excellent detection characteristics, is achieved by using pulse compression, matched non-linear waveform filter, airborne moving target indication to counter passive ECM and rain and pulse repetition frequency discrimination.

Very low antenna sidelobes, computer controlled energy management of the high-power transmitter operating with full agility over a wide bandwidth and azimuth diversity ensure minimal disruption to detection by ECM. Anti-jamming techniques incorporated within the digital signal processing further enhance the rejection of unwanted interference.

AR-325 operational facilities can include cursive or raster scan display consoles, automatic tracking, multiradar tracking, track repair facilities, voice and data communications, threat assessment, computer aided interception control, fighter recovery data, fighter performance modelling, and resource data catalogue and display. All these facilities can be housed in a BAE Systems' multirole operations cabin. The AR-325 combines the features of the radar sensor and the air defence product. This approach ensures that the size of the configuration can be adapted to meet the exact needs of each customer.

The AR-325 is designed for ease of transport by land, sea or air and is International Standards Organisation compatible.

Status

As of early 2001, AR-325 was reported as having been procured. Readers should note that BAE Systems and Italian contractor Finmeccanica announced on 26 April 2001 that AR-325 manufacturer BAE Systems - Combat and Radar Systems was to be transferred from BAE Systems control to that of their Alenia Marconi Systems joint venture. As of August 2001, said transfer had not been completed.

Specifications

Frequency: 2,700-3,100 MHz (agile band)
PRF: 270 Hz-1.4 kHz
Peak power: 135 kW
Mean power: 7.5 kW
Pulse-length: 200 µs (max, transmitted); 0.4 µs (compressed)
Antenna beamwidth: 1.4° (azimuth); 1.5-3.5° (elevation)
Instrumented range: 470 km
Instrumented height: 30,500 m
Detection range: 430 km against a small target
Data refresh rate: 10 s (6 rpm)

Contractor

BAE Systems - Combat and Radar Systems, Cowes, Isle of Wight.

UPDATED

AR-327 air defence radar

Type

E/F-band (2 to 4 GHz) long-range rapid reaction air surveillance radar.

Description

The AR-327 version of the Commander radar family is a mobile long-range E/F-band system that is designed specifically to meet the growing requirement for air defence radars capable of tactical deployment. By achieving a careful balance between detection performance and rapid mobility, the AR-327 provides full 3-D cover to an instrumented range of 470 km while achieving operational status within 1 hour of deployment. It can be prepared for transportation in the same time scale. The AR-327 radar is installed in standard International Standards Organisation

The antenna used in the AR-327 air surveillance radar

modules for ease of transport by road, rail or air. It is fully compatible with the C-130 aircraft and the CH-47 helicopter for tactical deployment by air.

A modular approach to the design of the AR-327 and the use of subsystems that are common to other radars in the Commander family, ensures that the system is capable of meeting a wide range of operational requirements while achieving high reliability and a low cost of ownership.

Flexible radar energy management allows the radar performance to be tailored to operational requirements. This is achieved by the combination of an electronically steered pencil beam, a highly reliable and efficient driven transmitter and digital pulse compression. Other features include enhanced signal processing, excellent electronic counter-countermeasures including multiple-frequency coded waveforms, auto-tracking with integral operations capability, colour raster displays in the operators' consoles and capability of remote operation. The use of phase monopulse height-finding ensures accurate target height data down to the radar horizon.

In its basic form, the AR-327 operates as a control and reporting post. Its operational capability can be enhanced to the status of a control and reporting centre post by the incorporation of up to four additional control positions in a multirole operations cabin. Digital plot and track exchange with a parent sector operations/command and reporting centre ensures full integration with a national command and control system. All the operational facilities and functions of the AR-325 radar can be incorporated in the AR-327.

Status

As of March 2001, the six AR-327 (service designation - Type 101) radars ordered by the Royal Air Force had all been declared operational. Elsewhere in the world, AR-327 radars are understood to have been procured by a number of offshore customers. Readers should note that BAE Systems and Italian contractor Finmeccanica announced on 26 April 2001 that AR-327 manufacturer BAE Systems - Combat and Radar Systems was to be transferred from BAE Systems control to that of their Alenia Marconi Systems joint venture. As of August 2001, said transfer had not been completed.

Specifications

Frequency: E/F-band (2,700-3,100 MHz - agile band)
Peak power: 150 kW
Pulse-length: digitally compressed to 0.4 µs
Antenna beamwidth: 1.4° azimuth; 2.4° lower elevation
Instrumented range: 470 km
Instrumented height: 30,000 m
Data refresh rate: 10 s (6 rpm)
Range accuracy: 50 m
Azimuth accuracy: 0.16°

Contractor

BAE Systems - Combat and Radar Systems, Cowes, Isle of Wight.

UPDATED

Commander radar family

Type

E/F-band (2 to 4 GHz) 3-D air defence radar family.

Description

Over time, what is now BAE Systems - Combat and Radar Systems has developed a series of 3-D air defence radars that make use of related technology and which are designed to meet 'all' long-range air defence requirements. As such, the Commander family extends from a NATO Class 1 radar to a rapid reaction self-contained system that can function as a basic reporting post. Hardware modularity has facilitated the creation of requirement-specific configurations and flexible energy management is seen as being a key feature of the family. Here, the capability is noted as being achieved via a combination of the use of an electronically scanned pencil beam, what is claimed as being a 'highly reliable and efficient' transmitter design and digital pulse compression. Identified members of the Commander family comprise the following:

- **AR-325** - a long-range E/F-band 3-D air defence radar in transportable and static configurations. See separate entry
- **AR-327** - a mobile rapid reaction E/F-band radar to match the transportation needs of the world's air forces. See separate entry
- **Fixed Air Defence Radar (FADR)** - a long-range, 3-D, E/F-band radar that is designed to meet NATO criteria/standards with enhanced electronic counter-countermeasures and a tactical ballistic missile defence capability

The AR-325 3-D Commander radar   2000/0079480

- **Deployable Air Defence Radar (DADR)** - a long-range mobile rapid reaction 3-D E/F-band radar that is designed to meet NATO criteria/standards with enhanced electronic counter-countermeasures and a tactical ballistic missile defence capability.

Status

Readers should note that BAE Systems and Italian contractor Finmeccanica announced on 26 April 2001 that Commander family manufacturer BAE Systems - Combat and Radar Systems was to be transferred from BAE Systems control to that of their Alenia Marconi Systems joint venture. As of August 2001, said transfer had not been completed.

Specifications

Frequency: 2,700-3,100 MHz
Peak power: up to 200 kW
Transmitted pulse-length: 100 μs
Instrument range: up to 470 km
Instrumented height: 30.5 km
Data refresh rate: 10 s
Beamwidth: 1.4° azimuth; 1.5° at low elevation
Antenna directivity: up to 42 dB
Average sidelobes: −65 dB

Contractor

BAE Systems - Combat and Radar Systems, Cowes, Isle of Wight.

UPDATED

A view of a partly populated array of the type used in BAE Systems' MESAR, SAMPSON and SPECTAR radars   0022582

Multifunction Electronically Scanned Adaptive Radar (MESAR) 2

Type

E/F-band (2 to 4 GHz) electronically scanned adaptive radar technology demonstrator.

Description

The MESAR 2 system is a technology demonstrator that is being used to gather data on the detection of air breathing and ballistic targets in real-time. Launched during 1995, the MESAR 2 programme builds on the earlier MESAR 1 equipment and features a fully populated array that employs 316, 10 W, four channel, line-replaceable transceiver units. Each transceiver unit has an integral dipole and the system as a whole has a 20 per cent bandwidth and is fully agile.

The MESAR 2 system is instrumented as a long-range air defence radar and has an elevation coverage of 90°. In addition to its use of a fully populated array, MESAR 2 differs from its predecessor by way of:
- reduced latency in its high-speed target tracking module
- new range ambiguous and high-resolution waveforms
- pulse-Doppler processing
- new input/outputs for cueing and control
- multiple digital beamforming
- a non co-operative target recognition capability.

The system also features parallel Distributed Array Processing (DAP) signal processing and a real-time control and display subsystem that makes use of commercial-off-the-shelf Intel processors and Ada application Sparc workstations. The radar's digital beamforming capabilities include:
- multiple receive beams
- monopulse angle measurement

- sidelobe blanking to counter pulse jamming
- adaptive suppression of noise jamming in both the main beam and sidelobes.

The technologies demonstrated in the MESAR 1 and 2 systems form the basis for BAE Systems' SAMPSON and SPECTAR naval radars together with proposed land-based sensors for long-range air defence, extended air defence and ballistic missile detection applications.

Status

As of early 2001, the MESAR 2 technology demonstrator is understood to have undergone trials in the UK before its despatch to the White Sands proving ground in the US. The trials conducted with the system are described as having incorporated a 'full spectrum' of target types and as having been undertaken over land and water in conditions of severe clutter and jamming. Readers should note that BAE Systems and Italian contractor Finmeccanica announced on 26 April 2001 that MESAR manufacturer BAE Systems - Combat and Radar Systems was to be transferred from BAE Systems control to that of their Alenia Marconi Systems joint venture. As of August 2001, transfer had not been completed.

Contractor

BAE Systems - Combat and Radar Systems, Cowes, Isle of Wight.

UPDATED

The MESAR 2 technology demonstrator
0022581

Watchman air surveillance radars

Type

E/F band (2 to 4 GHz) family of air surveillance radars.

Description

The Watchman family of E/F-band medium-range radars is designed to provide fighter recovery; helicopter surveillance; drug interdiction and over-land/over-water/coastal/air defence surveillance. The radar can be configured from a number of modules which enables users to match precisely equipments to operational needs. Options include 3-D surveillance, transportable or static aluminium or carbon fibre antennas; adaptive or eight-filter Moving Target Detector (MTD) signal processing and plot or track outputs. The Watchman travelling wave tube-based transmitter is compatible with a range of proven Electronic Counter-CounterMeasures (ECCM) techniques (including frequency agility) and enables a single radar to provide a variety of detection ranges out to a maximum of 222 km. BAE Systems also claims that the use of adaptive MTD and plot processing provides maximum target visibility in severe fixed and moving clutter environments over both land and sea.

The transportable Watchman-T equipment is a military primary surveillance radar which features integral ECCM and offers operational ranges of 111 km, 148 km and 222 km. The system's transmission spectrum and pulse format allows it to be co-located with other E/F-band (2 to 4 GHz) air defence radars, the Mk X/XII Identification Friend-or-Foe systems and monopulse secondary surveillance radars. The equipment can be fitted with Mode 4 encrpytion and its output can be combined into a plot and track message. Watchman-T has an instantaneously available 300 MHz bandwidth and a double conversion receiver which allows for

The carbon fibre antenna array that is used in Watchman series air surveillance radars
2000/0079479

A general view showing the template (left) and arctic tower configurations of the AN/FPS-6 height-finding radar system

frequency agility over the complete bandwidth. Periodic pulse sampling provides for automatic function in the least jammed part of the spectrum while the signal processing format is designed to counter repeater jammers.

Status
Over 90 Watchman radars have been supplied to 19 customers worldwide. Known customers include the UK Royal Air Force (47 radars), NATO (seven), the Finnish Air Force (seven) and Switzerland (three). Watchman-T has been supplied to, among others, the Royal Air Force, NATO and the US Air National Guard. Readers should note that BAE Systems and Italian contractor Finmeccanica announced on 26 April 2001 that Watchman manufacturer BAE Systems - Combat and Radar Systems was to be transferred from BAE Systems control to that of their Alenia Marconi Systems joint venture. As of August 2001, said transfer had not been completed.

Specifications
Watchman-T
Frequency: E/F-band (2,750-3,050 MHz)
Ranges: 111 km, 148 km and 222 km selectable (min 0.9 km)
Coverage: up to 15,240 m at a range of 222 km
Scan rates: 6, 7, 5, 12 or 15 rpm
Accuracy: azimuth: better than 0.15° RMS using plot extracted system; range: better than 30 m RMS
Resolution: azimuth: 90% probability of resolving 2 small targets with 3.7 km separation at a range of 74 km; range: 90% probability of resolving 2 small targets 0.23 m apart at ranges of up to 74 km

Contractor
BAE Systems - Combat and Radar Systems, Cowes, Isle of Wight.

UPDATED

UNITED STATES OF AMERICA

AN/FPS-6 height-finding radar

Type
E-band (2 to 3 GHz) nodding height-finding radar.

Description
For many years the AN/FPS-6 radar has been the principal height-finder used by US armed forces and many allied nations. At least 450 of these systems have been delivered, including the mobile version, the AN/MPS-14.

A high-power E/F-band nodding-beam radar, the AN/FPS-6 is noted for extreme accuracy at long range and three available versions give it wide versatility under a variety of environmental conditions.

The arctic tower installation consists of a 15.24 m radome (either air-supported or rigid) mounted on a 7.62 m two-storey enclosed tower structure. De-icing is provided by a battery of infra-red lights inside the radome. Radome pressurising (when required), Radio Frequency (RF) and other electronic equipment is housed in the tower structure.

The temperate tower installation is designed for moderate or tropical climates. The 7.62 m supporting structure for the antenna includes an enclosure for RF equipment.

The mobile version (AN/MPS-14) is a six-truck, three-trailer system designed for transport at short notice to new strategic or tactical sites.

The reliability and capability of the basic AN/FPS-6 system has benefited from a continuous improvement programme. Sets with various improvements have been designated as AN/FPS-6A, AN/FPS-6B, AN/FPS-6C, AN/FPS-6D, AN/FPS-89 and AN/FPS-90. All these improvements are available in field conversion kit form for the updating of radar sets from earlier production.

Status
As of this edition, FPS-6 radars are reported to be in service in numerous countries around the world.

Specifications
Peak power: 4.5 MW
Average power: 3.6 kW
Frequency: E-band (2,700-2,900 MHz)
PRF: 300-405 pps
Pulse-width: 2 μs
Receiver sensitivity: NF <9.3 dB; MDS −108 dBm
Beam characteristics: pencil beam
Beamwidth: 3.2° (azimuth); 0.9° (elevation)
Antenna azimuth rate: 180° in 4 s
Elevation rate: 20 or 30 nods/min
Antenna gain: 38.5 dB

Contractor
Lockheed Martin Ocean, Radar and Sensor Systems – Syracuse, Syracuse, New York.

UPDATED

AN/FPS-8 search radar

Type
D-band (1 to 2 GHz) early warning and aircraft control radar.

Description
The AN/FPS-8 is a medium power D-band search radar designed for aircraft control and early warning and is installed at commercial airports and military bases both in the US and overseas.

In most installations the antenna is exposed, being mounted on a temperate tower. For severe environmental conditions, the AN/FPS-8 is self-contained in an arctic tower with a protective radome. Over the years improvements have been made to the basic AN/FPS-8, culminating in the present version whose nomenclature is AN/FPS-88 (V). The AN/FPS-8 also has two mobile versions: the AN/MPS-11 and the AN/MPS-11A.

Status
As of this edition, FPS-8 is understood to be in service but no longer in production. Over 200 radars of this type are known to have been produced.

Specifications
Peak power: 1 MW peak
Average power: 1.1 kW
Frequency: D-band (1,280-1,380 MHz)

The AN/FPS-8 search radar

PRF: 360 pps
Pulse-width: 3 μs
Beam characteristics: cosec²
Beamwidth: FPS-88: 1.3° (azimuth), 58° (elevation); FPS-8: 2.5° (azimuth), 30° (elevation)
Antenna azimuth rate: FPS-8: 0-10 rpm (variable); FPS-88: 5, 10 rpm

Contractor
Lockheed Martin Ocean, Radar and Sensor Systems – Syracuse, Syracuse, New York.

UPDATED

AN/FPS-88 surveillance radar

Type
D-band (1 to 2 GHz) surveillance radar.

Description
AN/FPS-88 is an improved version of the FPS-8 radar (see separate entry) which offers enhanced range performance and upgraded signal processing. The radar features a high-gain antenna, circular polarisation, dual channel operation, a parametric amplifier receiver and a radar signal processor with some electronic counter-countermeasures capability. There is also provision for using the main antenna as an identification friend-or-foe radiator.

Status
As of this edition, the FPS-88 surveillance radar is thought to be operational in a number of countries around the world.

Contractor
Lockheed Martin Ocean, Radar and Sensor Systems – Syracuse, Syracuse, New York.

UPDATED

The AN/FPS-88 surveillance radar

AN/FPS-115 Pave Paws radar

Type
UHF band (0.3 to 1 GHz) Submarine Launched Ballistic Missile (SLBM) detection and warning and satellite tracking system.

Description
Pave Paws is the name given to a system of large phased-array AN/FPS-115 UHF solid-state radars to replace an earlier system for ballistic detection and warning. Within the Continental US, FPS-115 radars have been installed at Beale Air Force

The AN/FPS-115 Pave Paws installation at Robins AFB, Georgia

Base (AFB), California; Cape Cod Air Force Station (co-located with Otis AFB), Massachusetts; Eldorado AFB, Texas and Robins AFB, Georgia. Four sites are currently in full operation: at Otis Air Force Base (AFB), Massachusetts; Beale AFB, California; Eldorado AFB, Texas and Robins AFB, Georgia. In addition, versions of the AN/FPS-115 have been used to update the Ballistic Missile Early Warning System (BMEWS) sites at Thule, Greenland and Fylingdales, UK.

The four US-based Pave Paws systems are an important addition to the World Wide Military Command and Control System (WWMCCS) and are operated and maintained by the USAF Space Command. The radar's primary mission of detection and warning of SLBM attack also involves the provision of attack characterisation to the North American Air Defense Command (NORAD) Cheyenne Mountain complex, the Strategic Air Command and the National Command Authorities. The system's secondary role, in support of the USAF Spacetrack system, feeds in positional and velocity data for display of all earth satellites in orbit. Capable of multitarget tracking, it simultaneously detects and discriminates many objects while providing early warning data, launch, impact, position and velocity information as required.

Automated features of the system include detection, tracking initiation and mission decisions. Two standard computers, CYBER 170/865s, which generally serve as the CPU, are programmed for beam steering and the storage and display of data, as well as performing post-mission data reduction and analysis.

Pave Paws consists of a pair of circular planar phased-arrays about 30 m in diameter, each consisting of nearly 2,000 elements. The arrays are inclined from the vertical by 20° and mounted in adjacent sides of a building measuring about 32 m high, forming sloping walls on the seaward side of the structure. Combined coverage of the electronic beams of the two arrays is 85° in elevation and 240° in azimuth. Range is estimated at about 4,800 km. The AN/FPS-115 system being installed at Fylingdales Moor differs from the others in that it has three faces, each with over 2,500 elements, which will provide warning and tracking capabilities over 360° in azimuth.

Status
The Otis AFB Pave Paws was declared operational in 1980 and the Beale AFB system in the following year. The Robins AFB site became operational in 1986 and the Eldorado AFB system in 1987. In March 1988, Raytheon received a US$71 million contract from the USAF which has upgraded the two earlier sites, giving them the same enhanced data processing capabilities as the two later systems.

The BMEWS site at Thule was upgraded with a version of the AN/FPS-115 and became operational in 1987. The Fylingdales Moor site has also been upgraded with a three-face system under a US$167 million contract from the USAF and became operational in late 1992. During November 1998, the FPS-115 radar at Beale AFB was tested to see if it could detect and track a surrogate ballistic missile launched from a potential aggressor state in the Pacific Basin. As of August 1999, the same radar was scheduled to take part in a second round of testing to assess its suitability as a sensor in a potential low-cost, low-risk national missile defence architecture for the US.

Specifications
Frequency: UHF band (420-450 MHz)
Transmitter: solid-state
Module peak power: 322 W
Array type: corporate feed, density tapered
Number of subarrays: 56
Antenna gain: 38.4 (directive gain)
Beamwidth (transmit/receive): 2°/2.2° at boresight
Polarisation (transmit/receive): right hand/left hand circular
Array diameter: 22 m (utilised)
Face tilt: 20°
Azimuth: ±60°; 240° (with 2 faces); 360° (with 3 faces)
Elevation: 3-85°

Contractors
Raytheon Electronic Systems, Marlboro, Massachusetts.
Other contractors include Raytheon Systems Ltd, CDC and United Engineers.

VERIFIED

AN/FPS-117 air defence radar

Type
D-band (1 to 2 GHz) 3-D long-range air defence radar.

Description
The AN/FPS-117 totally solid-state radar system is a D-band, 3-D radar designed to provide long-range accurate aircraft identification and position data for air defence, navigational assistance and tactical control for both close air support and counter air operations. FPS-117 shares a radar architecture with the AN/TPS-59 equipment (see separate entry).

The system provides automatic adaptation to and rejection of, land, sea and weather clutter by using moving target indicator and Doppler processing. Adaptability in siting the radar is achieved by look down beam positions on elevated sites and terrain following. Sidelobe nulling is used to eliminate ground clutter for high-beam positions. Simultaneous monopulse processing is employed for azimuth and elevation position determination. This accounts for the radar's sustained accuracy and target resolution under a variety of conditions.

The planar-array measures 7.32 × 7.32 m and produces a series of pencil beams phased-positioned to scan in elevation up to 20° while the complete antenna rotates in azimuth. The elevation scan consists of 9 to 185 km short-range beams and 185 to 463 km long-range beams. The use of pencil beams is designed to obtain elevation coverage while eliminating clutter problems associated with large transit beamwidths. The transmitter/receiver system consists of 44 row transmitters, each containing up to ten 100 W radio frequency solid-state power modules and 44 receiver modules all mounted on the rear of the array. Identification Friend-or-Foe (IFF) systems, supplied by BAE Systems North America, Raytheon and others are incorporated in the radar.

The data processor controller provides all radar operational management as well as online, automatic performance monitoring to detect possible system failures. The controller provides plot extraction data output and performs sweep-to-sweep and scan-to-scan correlations to reduce false alarms and multiple reports. It also provides automatic radar/IFF correlations as well as monitoring the system for adverse environmental effects. The data processor has recently been updated to utilise commercially available hardware.

In addition to the standard FPS-117, (apparently designed as FPS-117(E)1 in fixed-site applications and FPS-117(E)1T when transportable) there is an RRP 117 model which is being supplied to Germany, a mobile variant (designated as the Type 92) for the UK and a second mobile variant, the TPS-117. TPS-117 utilises the array central enclosure of the FPS-117 combined with the hinged, row feed used in the TPS-59 (see separate entry); an arrangement which allows the radar's full array to be carried on a truck or trailer. Lockheed Martin quotes a set up time of 30 minutes for the equipment and notes its selection by Brazil for its SIVAM (Amazon Surveillance) programme and Australia. Additionally, the company is understood to have developed an FPS-117 upgrade that allows the radar to be used to detect tactical ballistic missiles and cue anti-missile systems. The capability also provides launch and impact point prediction together with improved general air surveillance.

Status
As of May 2001, in excess of 115 AN/TPS-59 - FPS-117 radars are understood to have been procured by customers around the world. The following countries are reported to be operating or to have ordered FPS-117 radars and/or its derivatives:

Australia	Four FPS-117 radars on order
Belgium	FPS-117 is reported to be in service in Belgium as part of NATO's Air Defence Ground Environment (NADGE)
Brazil	Six TPS-117s under contract for use in Brazil's SIVAM programme
Canada	Four FPS-117s for use in the Canadian Coastal Radar programme. Radars installed
Germany	Germany has acquired eight FPS-117 type radars to meet both national and NATO requirements
Iceland	Four relocatable FPS-117s. The Iceland Air Defence system acts as a link between the North American air defence system and NADGE
Italy	Four FPS-117s. Procured under NATO auspices as part of NADGE
Kuwait	One FPS-117
Romania	Five FPS-117(E)1T radars installed
Saudi Arabia	17 FPS-117(V)3s. Form part of the Peace Shield air defence system
South Korea	12 FPS-117s (see Status)
Taiwan	Two Type 592 radars
Turkey	Three FPS-117 radars. Procurement under NATO auspices
UK	Two Type 92 mobile FPS-117 variants. Form part of the Improved UK Air Defence Ground Environment (IUKADGE) and are reported to be located at Buchan and Benbecula in Scotland
US/Canada	16 FPS-117s (11 Canadian and five American) for the North American Northern Warning System (NWS). Thirteen FPS-117s for the US Alaskan Air Defence chain which now forms an adjunct to the NWS. Modernisation has allowed a number of these radars to switch to remote operation under the control of operations centres in Anchorage, Alaska and North Bay, Ontario. During 1995, Lockheed Martin was selected to upgrade all US AF FPS-117 radars to allow full remote operation.

On 22 September 1999, it was announced that the Lockheed Martin Corporation (Syracuse, New York) had been awarded a then year US$6.5 million time and materials contract (F04606-99-D-0106) covering the supply of engineering services in support of North American-based FPS-117 radars. At the time of the announcement, work on the effort was scheduled for completion on 30 September 2002 with the US Air Force's (USAF) Sacramento Air Logistics Center, McClellan Air Force Base, California acting as the programme's contracting activity.

Other 1999/2001 contracting activity includes:
- a then year US$94 million, June 1999 contract covering the supply of an integrated surveillance system (including five FPS 117(E)1T primary and five monopulse secondary surveillance radars) to Croatia
- delivery of the last of five FPS-117 radars to the Romanian Air Force during October 1999. This equipment forms part of an air traffic control and military surveillance system and at the time of delivery was valued at then year US$82 million
- a then year US$2.5 million, May 2000 contract covering the upgrading of the FPS-117/Type 92 radars operated be the Royal Air Force
- a then year US$54.2 million contract covering the supply of four FPS-117 radars to the South Korean Air Force. Scheduled (at the time of the award's announcement) for delivery by 2003, these four radars supplement South Korea's eight existing FPS-117s
- a then year US$30 million (total value with option), March 2001 contract covering the supply of a TPS-117 radar to Estonia for use in the Estonian/Latvian/Lithuanian BALNET air surveillance network. To be operated by the Estonian Air Force in the air surveillance and air traffic control roles, this radar was, at the time of the award's announcement, scheduled for delivery during early 2003. The contract also contained an option covering the supply of a TPS-117 radar to Latvia
- a then year US$47 million, July 2001 contract covering the updating of the 33 FPS-117 radars used in the USAF's Atmospheric Early Warning System (AEWS). Under the award, Lockheed Martin will provide a new Radar Interrogator Set (RIS) subsystem to replace the existing analogue capability. RIS modifications include the introduction of a new large vertical aperture antenna and remote control, monitoring, fault isolation and reliability, maintainability and supportability capabilities that are the equivalent of those of the primary radar. AEWS sites are located in Alaska, Canada, Hawaii, Iceland and Puerto Rico, with Hill Air Force Base, Utah acting as the network's support base. Under the deal, Lockheed Martin will provide system engineering, subsystem integration and software modification services and at the time of the award's announcement, the first radar to be modified (at Fairbanks, Alaska) was scheduled to be finished during the summer of 2002, with the remaining installations being completed by the end of 2006.

Specifications
FPS-117(E)1/FPS-117(E)1T
Antenna: 7.32 × 7.32 m
Range: 5.8 km (min); 288 km (max)
Accuracy: 0.29 km (range); 0.18° (azimuth)
Altitude: 30,480 m
Altitude accuracy: 340 m (at 185 km)
Elevation: 20°
Transmitter power: 24.75 kW (peak)
Frequency: D-band (1,215-1,400 MHz)
Bandwidth: 185 MHz
Agility: >20 frequencies (beam-to-beam selection)
Data rate: 10/12 s
MTBF: >1,000 h

The AN/FPS-117 air defence radar

TPS-117
Antenna: 7.62 × 4.57 m
Transmitter power: 20 kW (peak)
Frequency: D-band (1,215-1,400 MHz)
Beamwidth: 3.4° (azimuth); 2.7° (elevation)
IFF subsystem: modes 1, 2, 3/A and C
MTBF: >1,000 h

Contractor
Lockheed Martin Ocean, Radar and Sensor Systems - Syracuse, Syracuse, New York.

UPDATED

The receiver elements used in one sector of the USA East Coast OTH-B installation

AN/FPS-118 over-the-horizon radar

Type
HF band (3 to 30 MHz) Over-The-Horizon Backscatter (OTH-B) early warning radar.

Development
A contract for a prototype CONUS-OTH-B (CONtinental United States Over-The-Horizon Backscatter) system was awarded to GE Aerospace (now Lockheed Martin Ocean, Radar and Sensor Systems) in 1975 and after the contract was restructured in 1977, an experimental system was developed to demonstrate technical feasibility. This prototype system was situated in Maine, on the eastern seaboard of the US with the transmit site located at Moscow Air Force Station and the receive site at Columbia Air Force Station. The experimental transmissions from the Maine site covered an arc from 016.5 to 076.5° and from 926 to 3,334 km in range. This region was selected to evaluate the radar performance in an area where propagation would be most likely to be affected by the aurora and to provide surveillance of the busy North Atlantic routes where aircraft targets are available at all times. Initial testing took place from June 1980 to June 1981 and, with the success of these trials, GE Aerospace received a contract from the US Air Force in mid-1982 to begin full-scale development of the project.

Description
The Continental US (CONUS)-OTH-B system has been developed to provide electronic surveillance of aircraft at extended ranges of 800 to 2,880 km. Operating in the HF band, where the radar energy is reflected by the ionosphere, this ground-based system is designed for over-the-horizon detection and tracking of aircraft and cruise missiles flying at any altitude. The system operates on frequencies between 5 and 28 MHz and has two separate (bistatic) transmit and receive sites, some 150 to 200 km apart. To adapt continuously in real time to the prevailing ionospheric conditions, both the transmit and receive functions are completely computer controlled. Operation of the transmit and receive sites is synchronised in absolute time to better than 1 µs accuracy by Loran-C.

The complete OTH-B project was intended to consist of a three sector eastern seaboard system, covering 180°, a similar western seaboard system, a two sector system based in Alaska and a four sector central system. The eastern and western systems have been completed but the Alaskan and central systems have been cancelled.

Consideration was also given to the construction of a northwards-looking chain but the wide irregularities in the ionosphere and the effects of the aurora preclude a reliable system, although investigations are still going on. The region is already covered by the new North Warning systems in northern Canada and Alaska.

The transmit antenna array consists of six separate side-by-side 12-element subarrays, each optimised to cover a different portion of the total operating range. Together they provide the capability to operate anywhere between 5 and 28 MHz, and present low-voltage standing wave ratios to the transmitters at all scan angles.

A close-up of the canted dipole elements used in the transmission array of one sector of the USA East Coast OTH-B system

The two highest frequency subarrays (bands E and F) use vertical dipole elements; the other four subarrays use canted dipoles. This arrangement provides elevation patterns which match that required for propagation to the desired ranges via ionospheric reflections. The elements are mounted in front of a common backscreen ranging from 10 to 41 m high and approximately 1,106 m long. A common ground screen extends 230 m in front of the arrays.

Some 12 transmitters operate simultaneously into the 12 elements of a selected subarray. Each transmitter contains six band tuned tank circuits which respectively match the six antenna subarrays. Each transmitter produces up to 100 kW peak power, with very high-spectral purity, at any frequency between 5 and 28 MHz. A 100 kW water-cooled tetrode is used in the final high-power amplifier stage of each transmitter. The 12 transmitters are driven by a beamformer at low power level which causes the transmitter/array combination to collimate the desired 7.5° beam and steer it to the selected positions. In operation the transmit site generates up to 100 MW of effective radiated power.

The receive antenna is a broadside array of 246 monopole elements mounted in front of a backscreen which is 1,517 m long and 20 m high. As in the transmit array a groundscreen extends 230 m in front of the entire receive array. Three bands are used to cover the entire 5 to 28 MHz frequency range. Dividing the array and using variable aperture weights in beam forming results in virtually constant beamwidths of 2.5° over the full 60° azimuth coverage.

The active receive elements are fed via buried coaxial lines to elemental receivers, one for each active element. Each elemental receiver employs 16 RF preselectors to cover the frequency range. In each receiver, the receiver signals are amplified, filtered and digitised before being passed to the beam former/signal processor group. The beam former combines the outputs to form five simultaneous receive beams spatially coincident with the range/azimuth sector. This five beam cluster scans synchronously with the transmit beam. Beamwidth, beam spacing, pointing angle correction for frequency and real-time measured receive subsystem errors are controlled by the receive control computer. The signal processor processes the five receive beams virtually simultaneously. Functions include moving target indicator, interference suppression, range and Doppler resolution processing, non-coherent integration, peak detection and parameter estimation.

At the heart of the system are very high-speed computer, data processing and display systems which are used to control the radar, process the returned signals, monitor the HF propagation environment and display the information in a variety of formats to the operator. These displays are both alphanumeric and graphic types to provide the operator with the maximum amount of information on target speed, track, position, altitude and other special characteristics and also to adjust the transmission frequency to suit the prevailing HF environmental conditions. When the system is fully operational it is designed to give advanced warning of at least one to one and a half hours, even for supersonic aircraft and, since it uses skywave propagation techniques, it can track very low-flying aircraft and cruise missiles at the same ranges as those flying at higher altitudes.

CONUS OTH-B has three modes of operation:

Normal mode
In this mode a surveillance barrier is established up to 60° wide and up to 800 km deep. A step-scan technique is used to illuminate this barrier on a regular, periodic basis by sequential illumination of four contiguous range azimuth sectors. Each sector covers a 7.5° × 800 km area, thus eight sectors form the 60° wide barrier. Five parallel receive beams, with 2.5° centre spacings, are formed to be coincident with each range azimuth sector and collect the signals reflected from the targets in the coverage area. The starting range for each sector in the barrier can be chosen independently.

Interrogate mode
The interrogate mode illuminates a particular range azimuth sector and can be positioned anywhere within the 60° azimuth by 800 to 2,880 km range of the radar. The interrogate range azimuth sector is illuminated by contiguous sector dwells. As with the normal mode, five parallel receive beams, spaced on 2.5° fixed centres, collect the reflected target energy. The interrogate mode is a special mode intended to provide extra detection energy, contiguous illumination, and high-range, velocity and azimuth resolution. It is used to enhance the characterisation of particular targets.

Interleaved mode

This combines the normal and interrogate modes. Interrogate dwell is provided after each full scan of the normal mode barrier. Separate radar operating parameters can be selected independently for the interrogate and normal mode portions of the scan. This mode allows the barrier surveillance to be retained while focusing on special target situations anywhere in the coverage region.

Status

All three eastern sectors achieved limited operational capability at the end of 1988 and were turned over to the USAF in April 1990. An extensive series of tests was carried out during 1988/89 to test the efficiency and detection of the system. These were aimed particularly at the detection of low-flying cruise missile size targets and are understood to have been successful in detecting targets out to ranges of approximately 3,000 km. The results of these tests have also assisted in the development of a number of product improvements which have been incorporated in the western system, as well as being retrofitted into the eastern installation.

For the western seaboard, the OTH radar is located at Mountain Home Air Force Base, Idaho, for the operations centre, Christmas Valley, Oregon, for the transmitter, and Tule Lake, near Alturas, California, for the receiver. GE Aerospace (subsequently absorbed into Lockheed Martin) was awarded a contract valued at US$145 million for the first two sectors of the West Coast system. An additional contract for US$56 million was awarded in November 1987 for the third sector. Total value of the West Coast system contract was expected to be approximately US$313 million. The system was turned over to the USAF at the end of 1990 for operational tests and evaluation.

In March 1991, the US government decided to scrap the complete system, in view of the diminished Russian Federation and Associated States' threat, and dismantle the sites. However, after a few weeks this decision was reversed and the East Coast System has been placed in warm storage and will not be available for daily operational use. Tests and evaluation of the West Coast system have been completed and it has now been closed down, except for maintenance by a skeleton crew. The proposed central and Alaskan systems have been cancelled.

Contractor

Lockheed Martin Ocean, Radar and Sensor Systems – Syracuse, Syracuse, New York (prime contractor).

UPDATED

AN/FPS-124(V) surveillance radar

Type

D-band (1 to 2 GHz) unattended, medium-range air defence surveillance radar.

Description

The AN/FPS-124(V) UnAttended Radar (UAR) is a solid-state, electronic scan, 2-D, D-band system which provides accurate aircraft position and jam strobe data in heavy clutter environments. Coverage is from 3.7 to 130 km, 360° in azimuth, 20° elevation and 4,572 m altitude. No personnel are stationed at the UAR site; the radar automatically isolates faults and reconfigures with redundant elements. Predicted operational mean time between critical failure is in excess of 4,000 hours. Plot data of tracked credible targets (suppression of false reports) and functional status are provided to remotely located regional operational control centres. Data transfer is via a low bit error rate, 2,400 baud communications link.

The AN/FPS-124 unattended surveillance radar

The transmitter/antenna system uses a combination of stripline elements, configured in a fixed cylinder protected by a radome. The stripline elements are dipoles sandwiched between a dielectric and sealed with an aluminium housing. Only the radiating end of the dipole is exposed. The electronics for the transmitter/controller are housed inside the cylinder. Since the dipole elements are scanned electronically there is no mechanical rotation of the antenna. Selective exciting groups of dipoles form and steer the beam around the horizon. This improves reliability and makes adaptive target dwell-and-revisit time possible. A 6 second frame time is employed to provide all surveillance, tracking and self-test functions.

High stability in frequency sources, transmitter, receiver and processing elements provides stationary clutter cancellation in excess of 60 dB. False reports are minimised through data processing algorithms and automatic adaptive constant false alarm rate circuits which include self-generated clutter and bird maps.

The radar command and control functions, system status and target reports are provided by a data processor with operational and maintenance programmes stored in programmable read-only memories. A controller buffers the data for communication purposes, maintains storage of configuration and operational data and provides interactive display and control to the operating and maintenance personnel.

The signal processor incorporates pulse compression, 16 Doppler filters, adaptive residual clutter normalisation, detection threshold control, zero-Doppler detection processing, clutter breakout and bird censor maps, designated range gates for track-while-search and multiple time-around clutter cancellation.

System data processing incorporates:
- adaptive beam scheduling/resource allocation search/verify/track
- control of signal processor detection thresholds, receiver sensitivity time constant/automatic gain control
- burst-to-burst, beam-to-beam, scan-to-scan correlation
- monopulse processing for track
- Kalman filter tracking up to 200 targets
- frequency/pulse repetition frequency waveform selection
- reporting of only tracked credible targets
- remote control of frequencies, credible target velocity reporting thresholds, censor zones and enhanced high data rate search sectors
- prioritisation of detections for allocating verify-and-track resources
- end-to-end performance monitoring, fault isolation and reconfiguration.

Online maintenance consists of replacement of failed line-replaceable units, packaged for periodic transportation to and from unattended sites and replacement by non-technical personnel. Quarterly maintenance visits are anticipated.

Status

The AN/FPS-124(V) is a short- to medium-range surveillance radar which forms part of the upgraded North Warning System. The role of the AN/FPS-124(V) is that of a gap filler in the air defence system, operating in conjunction with the AN/FPS-117 long-range radars.

After extensive testing by the USAF, a US$433 million contract was awarded to Lockheed Martin (formerly Unisys) in 1990 to produce 40 systems, all of which were installed and in operation by the end of 1994.

Specifications

Range: up to 113 km
MTBF: >4,000 h
Coverage: 360° azimuth; 20° elevation; 4.6 km altitude
Antenna dimensions: 3.7 m high × 1.5 m diameter

Contractor

Lockheed Martin Ocean, Radar and Sensor Systems – Syracuse, Syracuse, New York.

UPDATED

AN/TPS-32 surveillance radar

Type

Long-range 3-D tactical surveillance radar.

Description

AN/TPS-32 is a ground-based, long-range, tactical, lightweight radar which automatically provides precise, 3-D position data on multiple targets. The equipment has been designed to operate as the 3-D radar sensor for the marine tactical data system. It also provides two positions for operators to vector interceptor aircraft to assigned targets. It is easily transportable by helicopter, cargo aircraft, or by conventional M35 truck. Automatic target detection, clutter elimination and Identification Friend-or-Foe (IFF) are included.

Air search coverage of the radar extends to 556 km in range to a ceiling of 30,480 m. The search volume is obtained by continuous rotation of the antenna (at 6 rpm) in the azimuth plane and electronic frequency scanning, -1 to +18° in the elevation plane. The elevation scan is accomplished by changing the frequency of the E/F-band RF energy applied to the planar-array, frequency-sensitive antenna. The antenna radiation appears as a series of pencil beams; each beam is 0.84° high and 2.15° wide.

The AN/TPS-32 radar consists of an antenna assembly and three shelters. Two of the shelters contain transmitter stages and the third contains receiver, data

The AN/TPS-32 long-range surveillance radar

processing, including radar and IFF plot extractors; displays and system control equipment. In the transport condition, the radar set consists of two antenna array pallets, one tripod pallet and three shelters. The complete system is helicopter-transportable in six pallets and can be made operational in less than two hours.

To make the most efficient use of transmitter power, the peak power level and interpulse period are changed during the complete elevation scan to fit the coverage volume more precisely. This allows more energy to be spent in the longer range portions of the coverage volume at the lower angles of operation.

Status
Versions of the AN/TPS-32, modified to be NATO Air Defence Ground Environment (NADGE) compatible standard (designated as AN/TPS-64) have been supplied to Kuwait and Turkey.

Specifications
Coverage
Elevation: 19° within interval −1 to +18° (electronically scanned)
Height: 30,480 m
Azimuth: 360° (mechanically scanned)
Displayed range: 4.6-556 km
Data rate: 10 s
Probability of detection/range (1 m² target): 90% to 370 km; 50% to 556 km
False alarm rate: <10 false target messages per azimuth scan
Accuracy (standard deviation)
Range: 229 m
Azimuth: 0.5°
Height: 366 m at 185 km; 914 m at 556 km
Resolution
Range: 457 m (manual); 800 m (automatic)
Azimuth: 2.15°
Elevation: 0.85°
Transmitter
Operating frequency: 2,905-3,080 MHz
Pulse-width: 30 μs
Interpulse period: 1,090-3,772 μs
Peak power output
Final RF amplifier: 2.2 MW (peak) (low elevation, long range)
Radar antenna
Elevation beamwidth: 0.84° (nominal)
Azimuth beamwidth: 2.15° (nominal)
Gain: 41 dB (min)
Sidelobes (azimuth and elevation): −25 dB (min) referenced to peak of main beam
Polarisation: horizontal
Rotational speeds: 6 rpm
Receiver/processor: log FTC, IF double canceller MTI and CPAC receivers with CFAR and fast show AGC; dual 9 channel video processors with 350 kHz 6 W per channel (1.3 MHz independent CPAC); automatic 3-D target detection and clutter censoring; automatic IFF/radar correlation

Contractor
ITT Gilfillan, Van Nuys, California.

VERIFIED

....

AN/TPS-43 tactical 3-D radar

Type
E/F-band (2 to 4 GHz) mobile surveillance radar.

Description
AN/TPS-43 is a lightweight air and ground mobile radar that is designed to control aircraft or surface-to-air missile batteries in a wide variety of tactical environments. It provides complete 3-D coverage out to a range of 447 km on fighter size aircraft and measures heights over the full range by signal amplitude comparisons in six channels. Extensive clutter rejection and electronic counter-countermeasures features are incorporated in the design, including a digital coherent moving target indicator system, pulse-to-pulse frequency agility, Jamming Analysis and Transmission Selection (JATS), coded pulse anti-clutter system and sidelobe blanking. This latest model of the AN/TPS-43 includes extensive equipment refinements and increased operational capability.

For ease of air shipment the equipment divides simply into two pallet loads each of less than 3,400 kg. One load comprises the shelter unit, including processor, radios, transmitter/receiver and displays; the other consists of the antenna assembly and ancillary equipment. The entire equipment can be packed into two M35 trucks for road transport. It can also be airlifted in a single C-130 aircraft.

To minimise weight, light alloys are used wherever possible in the main mechanical structures and micro-miniaturisation techniques are used in the electronics circuits. The feed array features the use of a stripline matrix to form the height-finding beams. The Identification Friend-or-Foe (IFF) antenna is a sum and difference type providing an Interrogator SideLobe Suppression (ISLS) capability. A small printed circuit reference antenna is mounted on the back of the radar feed to act as the radar sidelobe reference antenna. Use of this latter antenna during the dead time between transmitter pulses is available for the JATS function.

Status
The AN/TPS-43 series of radars has been extensively deployed by the armed forces of the USA and more than 20 other countries, with approximately 175 such radars having been delivered in total. Outside the US, countries identified as operating or as having ordered TPS-43 radars are reported to include Argentina (four radars, one of which was lost during the 1982 Falklands (Malvinas) War); Australia (three); Israel (two); Saudi Arabia (up to 28); Spain; Taiwan; Pakistan and Morocco.

The USAF has completed modification of its TPS-43 inventory with the advanced capabilities and features of the AN/TPS-70 system including the Ultra-Low Sidelobe Antenna (ULSA). The upgraded AN/TPS-43 is renamed the AN/TPS-75 (see separate entry).

Specifications
Range: 3-D coverage on fighter aircraft to 447 km
Data rate: 10 s (6 rpm antenna)
Elevation coverage: 0-20°
Accuracy: 107 m (range), 0.35° (azimuth); ±450 m (height)
Electrical characteristics
Power output: 4 MW (peak), 6.7 kW (average)
Frequency: E/F-band (2,900-3,100 MHz in discrete steps (with pulse-to-pulse agility))
Pulse duration: 6.5 μs
Antenna gain: 36 dB (transmit); 40 dB (receive)
Azimuth beamwidth: 1.1°
IFF azimuth beamwidth: 4° (or sum/difference ISLS antenna)
Noise figure: 4.5 dB
Prime power: 400 Hz 3-phase 120/208 V
Mechanical characteristics
Weight: 3,310 kg (shelter module); 2,050 kg (antenna module)
Transport: single C-130 aircraft, 2 M35 trucks, 2 sets of transporters, or 2 helicopter loads

When upgraded with the ULSA array, TPS-43 is redesignated as TPS-75. Note the original TPS-43 antenna in the background of this view

Road speeds: up to 96 km/h
Airlift altitudes: up to 15,000 m
Siting requirements: 6 × 10.5 m clear area on slope of 10% or less
Reaction time: 50 min with a 6-man team
Wind resistance: operate to 96 km/h, survive 170 km/h (tied down)
Operating temperature: −40 to +125°F (−40 to +52°C)
Other
MTBF: 800 h
MTTR: 30 min
Outputs to: 3 × 120 m cables with storage reels
Operations centres plan position indicator: 2 operational positions featuring AN/UYQ-27 (40 cm) displays, digital height readouts and AN/UPA-59A active/passive IFF decoders. Each position also has access to built-in UHF (0.3-1 GHz) ground/air communications and HF (3-30 MHz) point-to-point communications facilities
IFF/selective identification facility equipment: Hazeltine AN/UPX-23 interrogator set and AN/UPA-59A active/passive decoder
Transmitter tube: linear beam klystron

Contractor

Northrop Grumman Corporation, Electronic Sensors and Systems Division, Baltimore, Maryland.

VERIFIED

....

AN/TPS-44 tactical radar (Alert Mk II A/O)

Type

D-band (1 to 2 GHz) transportable air surveillance radar for the US 407L system.

Description

Alert Mk II (AN/TPS-44) is a lightweight, solid-state, transportable D-band air surveillance radar. It was designed for use anywhere in the world as the sensor for the forward air control post of the US 407L Tactical Air Control System.

Alert Mk II is the latest version of the Alert radar series. Information of airborne targets is presented in the form of synchronised radar and Identification Friend-or-Foe (IFF) video on a 0.4 m plan position indicator display console. It also has optional provisions for supplying range and azimuth information to additional tactical operations centres.

The equipment consists of the Alert IIA radar set, a complete IFF/secondary surveillance radar system and the operator display console. All components are housed in two major packages: an equipment shelter houses the electronic components and display console and an antenna pallet carries the folding antenna and feed system, together with the pedestal and antenna drive system. The entire radar can be transported by M35 trucks, C-130 aircraft, helicopters or towed on transporters. It can be converted from transport to operational configuration by a team of four people in less than 40 minutes.

Status

As of this edition, TPS-44 is reported to have been procured by the US Air Force and off-shore customers. Jane's sources suggest that TPS-44 export customers include Argentina, where the radars are said to have been located at naval or air bases in or near Buenos Aires, Port Belgrano, Comodoro Rivadavia, Santa Cruz, Rio Gallegos, Rio Grande and Ushaia.

Specifications

Antenna
Type
Search: modified parabolic
IFF: integral with search radar feed system and reflector
Aperture, search
Horizontal: >45 m
Vertical: >2.7 m
Beamwidth (one-way), search and IFF
Horizontal: 3.8°
Vertical: 8° with cosec^2 7-27°
Polarisation
Search: horizontal
IFF: vertical
Type of feed: search and IFF: horn
Beam pattern search and IFF
Horizontal: conventional fan beam (cosine between 3 dB points), sidelobes 25 dB down, back lobes 30 dB down

The AN/TPS-44 tactical surveillance radar deployed

Vertical: conventional fan beam with cosecant squaring
Scan
Azimuth: 0-15 rpm, clockwise in automatic, or manually searchlighting, clockwise or counter-clockwise. Also available with 6 rpm clockwise fixed-speed option
Elevation: manually adjustable, −3 to +6°
Transmitter
Power
Peak: 1 MW, or greater
Average: 1.12 kW at 800, 0.745 kW at 533, 1.12 kW at 267
Pulse repetition frequency
Manually selectable: 800, 533, 400
Momentary selectable: ±10 µs stagger on 800 PRF. Also available with 4-pulse stagger to eliminate MTI blind speeds
Frequency: 1.25-1.35 GHz, continuously tunable over full range by local or remote tuning control
Tuning rate: 1 MHz/s
Pulse: width automatically selected when PRF is selected:
 1.4 µs for 800 Hz PRF
 1.4 µs for 533 Hz PRF
 4.2 µs for 267 Hz PRF
 2.8 µs for 400 Hz PRF
Rise time: 100 ns (max)
Receiver
Noise figure: typically 3 dB (overall)
Dynamic range: 65 dB
Choice of IF: logarithmic, linear, or wideband limiting
Accuracy
Range: 0.9 km on 400 Hz PRF; 0.19 km on 800 and 533 Hz PRF
Azimuth: ±1°
Type of information displays
- PPI
- A-scope
- output jacks and cabling provided for remote displays.

Contractor

Cardion Inc, Woodbury, New York.

VERIFIED

....

AN/TPS-59(V) tactical radar

Type

D-band (1 to 2 GHz) 3-D long-range air surveillance radar.

Description

The AN/TPS-59(V) is a D-band, long-range 3-D air surveillance phased-array radar. Its transmitters are solid-state and power is generated by direct amplification at the transmission frequency using D-band power transistors. The antenna is a 9.1 m high by 4.9 m rotating planar-array that is located about 2 m above ground level. The three trailer assemblies used as the antenna support structure also carry the array sections during transport. The antenna array itself is made up of 54 identical row feed networks and associated row level transceivers, comprising power supplies, transmitters, preamplifiers, phase shifters, duplexers and logic controls. Each set is housed in a single package directly behind the row feed. The vertical, or row-to-row distribution, is accomplished by three column feeds. The antenna

The AN/TPS-59(V) mobile air defence radar

The AN/TPS-59M/34 3-D air surveillance radar makes use of either a truck or, as shown here, a trailer-mounted antenna

rotates mechanically in azimuth and a pencil beam electronically scans in elevation from 0 to 19° to cover the specified surveillance volume in a raster scan pattern.

Two basic waveforms are used for the surveillance function: one throughout the short-range interval from 5.5 km to 185 km and the other over the long-range interval from 185 km to the limits of coverage. Typically, eight pencil beams are used to scan the long-range interval in elevation and 11 are used to cover the short-range. Provisions are made in the lower long-range and short-range beams to counter the effects of multipath propagation and a special low-angle height-finding technique is employed. There is also a special weather mode of operation which automatically optimises energy management and moving target indicator processing according to prevailing weather conditions. This is performed at computer-controlled intervals to adjust to weather changes.

In addition to its various functions in beam steering, energy management and signal processing, the TPS-59(V) computer stores the position of all detected targets (in three co-ordinates) and correlates the stored positions with fresh data on a scan-to-scan basis. It performs the special range correlation processing associated with second-time-around returns concerned with moving target indicator operation. It also correlates the data from targets in closely spaced beams to avoid the reporting problems associated with multiple detections. All monopulse calculations are performed in the computer.

All controls and indicators for the operation of the TPS-59(V) are located at the display console, of which, there are two operators' positions with two cathode ray tubes. These are plan position indicators with the capability to present an RHI display or for the presentation of alphanumeric data. Other facilities include identification friend-or-foe controls, a performance monitoring status display, a clutter-gating panel and communications controls.

TPS-59(V) is designed primarily to provide long-range surveillance of a tactical airspace to the Tactical Air Operations Centre (TAOC). Its detection capabilities (500 targets per 10 second scan), 3-D positional accuracy and console readouts and controls meet the requirements of an autonomous build-up or back-up ground control intercept role. The system's high-data rate capability supports the additional air traffic control mission. The design specification includes threat criteria such as 1 m² scan-to-scan fluctuating targets flying at speeds up to M4.0. More generally, the modular nature of the sensor's antenna array, coupled with the flexibility inherent in its digital signal processor and its extensive use of software control techniques, have allowed the parallel development of other solid-state radar designs based on its architecture. These include the AN/FPS-117 fixed-site and relocatable radar (see separate entry), the TPS-59M and the Type 92 transportable radar. A reduced size derivative of TPS-59M (designated as the AN/TPS-59M/34) has also been developed, the technical details and specifications of which are given at the end of this entry.

The TPS-59(V)3 configuration represents a significant upgrade of the radar's performance in the areas of reliability, mobility, detection performance against air breathing targets and TBM detection. The (V)3 standard also introduces automatic missile system (such as HAWK) cueing (using several forms of communications links including the Joint Tactical Information Distribution System (JTIDS)), leaves the existing TPS-59(V) antenna array untouched and replaces most of the off-array electronics. All of the new equipment (including display consoles and processing equipment) is housed in a single S-280 shelter rather than in the two shelters currently used. Additionally, the TPS-59(V)'s existing 1970s vintage AN/UYK-7 computer is replaced by a Harris Nighthawk processor.

Status
General Electric (now Lockheed Martin Naval Electronics and Surveillance Systems) was awarded a contract for the development of the AN/TPS-59(V) for the US Marine Corps (USMC) by the US Navy's Naval Electronic Systems Command in 1972. It began USMC acceptance testing in 1976-77. Development of an energy management track-while-scan ground control interface upgrade and an enhanced electronic counter-countermeasures capability began during 1989. An anti-radiation missile defence system is also understood to have been developed. Starting in the early 1990s, TPS-59(V) has been upgraded to enable it to detect and track Tactical Ballistic Missiles (TBM) and has been integrated with the HAWK surface-to-air system, where it provides TBM radar cues. As such, the updated, TBM capable, TPS-59(V) is designated as the TPS-59(V)3. The first preproduction

TPS-59(V)3 upgrade package was completed during December 1995 and in the following August, the TPS-59(V)3 radar/HAWK missile is reported to have completed a live fire test round at the White Sands, New Mexico test range. During these trials, the combination is understood to have successfully engaged and destroyed multiple targets, including simultaneously launched, multiple ballistic and mixed ballistic and cruise missile targets.

At least 16 AN/TPS-59(V) radars have been delivered, with 11 going to the US Marine Corps (USMC) and five to Egypt. The latter are understood to have been linked to 12 operations centres as part of a defence system update. As of late 2000, Jane's sources were indicating that all USMC radars of this type were being upgraded to incorporate the TBM surveillance and tracking capability. Of the various TPS-59 models developed, the TBM (V)3 variant is the latest known to have been in production. Identified TPS-59(V) programme activity during 1999-2000 comprised the following:

29 December 1999
Lockheed Martin Naval Electronics and Surveillance Systems (Syracuse, New York) was awarded a then year US$58.7 million firm, fixed-price contract covering the upgrading of five Egyptian TPS-59(V)2 radars to TBM defence configuration. At the time of its announcement, work on this Foreign Military Sales effort was scheduled for completion by the end of February 2003, with the USMC's Systems Command, Quantico, Virginia acting as the programme's contracting activity.
15 March 2000
Lockheed Martin Naval Electronics and Surveillance Systems (Syracuse, New York) was awarded a then year US$7,498,895 cost-plus, fixed-fee contract covering TPS-59(V)3 radar life cycle system acquisition and support. At the time of its announcement, work on this effort was scheduled for completion by the end of March 2001, with the US Marine Corps' Systems Command, Quantico, Virginia acting as the programme's contracting activity.
8 May 2000
Lockheed Martin Naval Electronics and Surveillance Systems (Syracuse, New York) was awarded a then year US$7.5 million funding increment as part of a continuing programme (valued at then year US$46.8 million) to upgrade 11 TPS-59(V)3 radars operated by the USMC at locations in the continental USA and on Okinawa. Within the context of this specific award, the work undertaken was designed to address cost of ownership, part obsolescence and system performance and was, at the time of the announcement, scheduled for completion by the end of March 2001.

Specifications
AN/TPS-59(V)
Radiated peak power: 46 kW
Average duty cycle: 18%
Transmit gain: 38.9 dB
Effective receive area: 14.5 dBm²
Antenna sidelobe levels: −55 dB
System noise temperature: 540 K
Signal processing losses: 2.4 dB
Frequency: 1,215-1,400 MHz
Antenna beamwidth: 3.4° (azimuth); 1.7° (elevation) (1.4° low angle)
IFF subsystem: modes 1, 2, 3A, C, 4
Surveillance coverage: 360° (azimuth); 5.5-560 km (range); 0-19° (elevation); 30,500 m (altitude)
Frame time: 10 or 5 s
Accuracy: 24 m (range); 3 mrad (azimuth); 300 m (height)
Resolution: 60 m (range); 3.4° (azimuth); 1.7° (elevation)
Reliability: MTBF 1,000 h
Maintainability: MTTR 40 min
Weight: 12,700 kg (system); 2,360 kg (max package)
Prime power: 89 kW (nominal)
Mobility: 1 h (assembly); 30 min (disassembly)

TPS-59M/34 Tactical Air Surveillance Radar
For applications requiring more rapid relocation, a derivation of TPS-59(V), known as TPS-59M/34 has been developed. So configured, TPS-59M/34 provides 3-D air surveillance coverage and a ground controlled intercept capability at ranges of up to 370 km and altitudes of up to 30,500 m. The system can be operational within half an hour of site arrival. The sensor makes use of a 5.8 × 4.6 m, truck or trailer-mounted planar-array antenna that consists of a 34 row feeder network and its associated row transmitters, receivers and power supplies. The structure of this array and its row-to-row signal distribution is similar to that of the baseline TPS-59(V). Again, the TPS-59M/34's system architecture follows that of the TPS-59(V) and its signal/data processing equipment is housed in two S-280 shelters. The architecture's radar control shelter includes two positions for local operation of the system and data acquired by the TPS-59M/34 can be handed-off to remote command and control facilities.

Status
As of July 2001, up to nine TPS-59M/34 radars are understood to have been procured by Egypt, with deliveries starting in late 1991. On 5 April 2001, Lockheed Martin announced to it had been awarded a then year US$16 million Foreign Military Sales contract covering the refurbishment of eight of the Egyptian TPS-59M/34 radars. At the time of the announcement, the refurbishment effort was to be undertaken at a facility in Cairo and the programme's contracting activity was the USMC's Systems Command, Quantico, Virginia.

Specifications
Frequency: D-band (1,215-1,400 MHz)
Peak power: 28 kW

Surveillance coverage: 360° (azimuth); 7-370 km (range); –2 to +20° (elevation); up to 30,500 m (altitude)
Resolution: 43 m (range); 3.5 mrad (azimuth); 460 m (height)
Data update rate: 5 or 10 s
Antenna sidelobes: –55 dB
IFF subsystem modes: modes 1, 2, 3A, C
Mobility: 30 min assembly
Reliability: MTBF >1,000 h
Maintainability: MTTR <40 min
Prime power required: 75 kW

Contractor

Lockheed Martin Naval Electronics and Surveillance Systems - Syracuse, Syracuse, New York.

UPDATED

AN/TPS-63 surveillance radar

Type

D-band (1 to 2 GHz) low-altitude air defence radar.

Development

The AN/TPS-63 was designed and developed under US Navy contracts for use by the US Marine Corps to provide detection, automatic acquisition and target tracking of aircraft targets.

Description

The AN/TPS-63 is a D-band surveillance radar used to provide low-altitude coverage for tactical air defence and air traffic control operations. This radar was built to meet US Marine requirements and features the extensive use of microelectronic and solid-state techniques. An advanced Moving Target Indicator (MTI) subsystem provides more than 60 dB clutter suppression. Extended operator selectable ranges of 222 and 296 km have been added as a USMC processor update programme. The digital MTI uses a four-pulse canceller with variable time intervals between pulses to distinguish moving aircraft from ground returns, while eliminating MTI radar blind speeds. There is also an additional digital three-pulse MTI weather canceller and digital constant false alarm rate processing which is extremely effective against heavy rainfall and chaff. Dual diversity operation is normal and there are built-in test features. Standard features include a solid-state transmitter; a low-sidelobe antenna; an integral automatic primary/secondary surveillance radar tracker; a colour raster scan display; remote control and monitoring key radar functions and digital target reporting (tracks or plots) via standard narrowband communications channels.

Other techniques to detect very small targets in heavy clutter or interference include:

- radio-frequency sensitivity time control which reduces clutter from close-in ground reflections
- processing of both the in-phase and quadrature components of the return signal to improve detection
- transmission of phase-coded pulses and use of coded pulse anti-clutter system signal processing
- digital constant false alarm rate processing to adjust receiver sensitivity
- dual-frequency transmission selectable from 51 frequencies across the frequency band.

Incorporated as part of the AN/TPS-63 is the high-performance AN/UYQ-509 colour raster display which displays simultaneously or individually computer-generated data and radar videos (MTI and normal) mixed with quantised videos such as secondary surveillance radar, gates, symbols and range marks. A state-of-the-art TAC-90 colour raster scan display is being provided in current models or as a direct replacement for the UYQ-27(V) on previous models.

The AN/TPS-63 tactical air surveillance radar

The AN/TPS-63 consists of a single 2.4 m × 2.4 m × 3 m shelter that is suitable for transportable or mobile applications. Fixed site configurations are also available. The 4.9 × 5.5 m antenna is constructed of sections which are stowed within the shelter during transport. The system can be operational within one hour of arrival at a new site. Disassembly of the antenna and preparation for shipment requires only half an hour. A Low Sidelobe Antenna (LSA) for the AN/TPS-63 has been developed and is in use with a number of international customers. The LSA has a stronger mechanical interface and uses dipole radiating elements arranged in 32 interchangeable columns. It is understood to provide greater than a 100 to 1 improvement over the original antenna in electronic counter-countermeasures performance. It also provides an integral sum and difference monopulse or conventional SSR capability. A solid state upgrade of Morocco's TPS-63 radars has been completed.

Status

Over 100 systems have been delivered to the US Marine Corps (22 sets), the US Air Force and nine international customers. These latter comprise Bangladesh, Botswana, Jordan, South Korea, Mexico, Morocco, Taiwan, Venezuela and Yugoslavia. Elsewhere, Egyptian contractor Benha is reported to have co-produced (with Northrop Grumman) 34 of a 42 set Egyptian TPS-63 procurement while (as of 1998) two further examples of the radar were noted as having been ordered by Kuwait. The USAF is using the AN/TPS-63 radar at several of its air-to-air combat ranges for the detection of low-flying fixed-wing aircraft and hovering helicopters in the presence of severe ground and sea clutter.

A version of the AN/TPS-63 has been evolved into the Northrop Grumman E-LASS radar (see separate entry) for use on tethered aerostats which are flown at altitudes up to 4,572 m. Its look down MTI capability provides low-flying small target detection out to ranges of 370 km. It also provides simultaneous maritime and surface detection out to the horizon via a separate independent surface detection channel. As of this edition, the United Arab emirates is understood to have procured at least one E-LASS radar.

Specifications

Frequency: D-band, dual diversity, 80 selectable frequencies between 1,225 and 1,400 MHz
Range: 148, 222, 296 and 370 km (operator selectable)
Data rate: 6, 12, 15 rpm (10, 5, 4 s)
Elevation coverage: 12,000 m shaped coverage to 40°
Power output: 45 kW (peak)
Pulse-width: 64 or 99 μs (dual frequency)
Pulse repetition frequency: 300, 375, 500, 750 average staggered
Antenna gain: 32.5 dB
Angular resolution: 2.9°
Range resolution: 190 m
Accuracy: 150 m (range); 0.35° (azimuth)
ECCM: coded pulse anti-clutter, frequency agility, PRF stagger
MTBF: greater than 2,000 h
MTTR: 30 min
Weight: 3,400 kg
MTI improvement factor: 60 dB

Contractor

Northrop Grumman Corporation, Electronic Sensors and Systems Division, Baltimore, Maryland.

VERIFIED

AN/TPS-70 tactical radar

Type

E/F-band (2 to 4 GHz) mobile/fixed air surveillance radar.

Description

The AN/TPS-70 is a mobile E/F-band radar designed to detect and track hostile aircraft in a variety of environments at ranges out to 450 km. It incorporates clutter rejection and electronic counter-countermeasures features, with a low-sidelobe antenna making it very difficult for enemy countermeasures to detect or jam the system. Also incorporated are advanced signal analysis and processing and a digital coherent moving target indicator system.

The antenna, a flat-slotted array measuring 5.5 m wide by 2.54 m high, is combined with programmable Doppler signal processing and an automated target extraction/tracking computer to provide the total operational scenario capability for automated and manual command centres. A multiple beam architecture with an entire redundant channel is employed, scanning all elevations and ranges simultaneously for targets by means of six beams. This ensures more hits per target for optimum clutter rejection and target extraction.

The AN/TPS-70 provides an automated system interface for digitised target reports in either plot or track format. The Digital Target Extractor (DTE) performs automatic clutter mapping, radar plot extraction, Identification Friend-or-Foe (IFF) decoding/plot extraction, radar/IFF correlation and clutter filtering. A single correlated range/azimuth report is transmitted for each detected target. The forward-tell tracker performs additional scan-to-scan processing on the DTE plot data and automatically initiates and maintains up to 1,000 simultaneous target tracks. The tracker adds target identity, heading, speed and track quality information.

*The AN/TPS-70
mobile tactical radar*

Using a subsystem known as the Missile Launch Warning Subsystem (MLWS), AN/TPS-70 has been given the capacity to detect and track Tactical Ballistic Missiles (TBM) while continuing to detect and track conventional air targets. MLWS automatically alerts the operators that a TBM launch has occurred and forwards the data to appropriate operations centres and/or engagement systems. MLWS provides all relevant TBM data including numbers, types and positions.

All systems controls are entered through a touch-sensitive plasma control terminal. Over 100 separate menus arranged in user-friendly formats allow complete operator control of operating modes, parameters, monitoring of critical functions and automatic fault isolation. Each radar is equipped with two plasma displays. One is located between the operator positions, with the other in the rear of the shelter for maintenance purposes.

The complete system consists of the main antenna (with integral secondary elements) and an operations/electronics cabin standard shelter unit. For ease of transport the AN/TPS-70 is deployed as two modules which can be carried to a forward position by truck, aircraft or helicopter.

The AN/TPS-70 is modular in design, the software being flexible to permit fast reprogramming to meet specific operational requirements. The system can be remotely controlled and monitored and can be operated either autonomously or as part of a larger network. The AN/TPS-70 incorporates extensive built-in redundancy, self-test/diagnosis and solid-state subsystems for virtual 100 per cent system availability. System operation mode and parameters as well as self-test are controlled from a centralised touch-sensitive monitoring and fault isolation panel. Should a system fault occur, the faulty line-replaceable unit is automatically determined and displayed for rapid plug-in replacement. The AN/TPS-70 electronics are very similar to those of the USAF AN/TPS-75 enhancement production programme ensuring the benefits of long-term operational and logistic commonality. In addition to the mobile configuration the system is produced in a fixed-site variant, the FPS-700 and certified for air traffic control as the ARSR-70.

Status

As of this edition, TPS-70 is understood to have been selected by or be in service with the armed forces of at least 15 countries around the world. Specific customers identified by Jane's comprise Abu Dhabi, Bangladesh, Canada, Honduras, Mexico, the Philippines, Saudi Arabia, Switzerland, Thailand, Uganda, the US Air Force, the US National Guard and Yugoslavia. Of these, sources suggest that the Swiss equipments are designated as the Vigilant system and take the form of a TPS-70 variant that has reduced range and 'very high' clutter rejection. The US Air Force radars are used in the Caribbean Basin Radar Network with examples being located in the Cayman Islands, Colombia, Costa Rica, the Dominican Republic, Honduras and Panama. The Saudi systems are designated as AN/TPS-72 equipments and are described as forming the 'backbone' of the Royal Saudi Air Force's Peace Pulse mobile air defence network.

As of this edition, Northrop Grumman is also understood to have developed a new, solid-state TPS-70 variant designated the TPS-70SS. Jane's describes this air-cooled radar as operating in the 2.8 to 3.1 GHz frequency band and as being configurable for both long- (445 km) and short-range (167 km) operations in mobile fixed-site formats. TPS-70SS leverages transmitter technology from Northrop Grumman's ASR-12 air traffic control radar and makes use of solid-state modules mounted in interchangable power panels within its transmission chain. These panels are mounted in the system's operations shelter and the radar is quoted as having a peak power of 32 kW. Here, the use of stacked beams facilitates missile warning and tracking together with fighter control throughout the 360° arc in azimuth. A D-band (1 to 2 GHz) identification friend-or-foe subsystem is incorporated, with the necessary antenna elements being interleaved with those for the primary radar within the main antenna assembly. TPS-70SS weighs less than 6,800 kg in total and can be set up by a four-man crew in 45 minutes in its long-range configuration or in 30 minutes by a two-man crew in its short-range

application. As of this edition, Northrop Grumman was understood to be funding the construction of two prototype TPS-70SS radars for delivery during mid-2001.

Specifications

Frequency: E/F-band (2,900-3,000 MHz)
Antenna
Type: travelling wave planar-array, simultaneous beams, 36 waveguides in array, 98 slots per waveguide
Array size: 5.5 × 2.54 m
Azimuth beamwidth: 1.5°
Elevation beamwidth: 0-20° transmit; 6 simultaneous receive beams (2.3-6°)
Azimuth sidelobes: 50 dB
Rotation rate: 6 rpm
Transmitter
Power output: 3.5 MW (peak); 6.2 kW (average)
Pulse-width: 6.5 μs; 4 state phase coded
Pulse Repetition Frequency: 250/275 pps (average)
Instrumented range: 444 km
Azimuth coverage: 360°
Elevation coverage: 0-20°
Altitude coverage: up to 30,300 m
Accuracy: 107 m (range); 0.22° (bearing); ±457 m (height)
ECCM: discrimination against jamming by low-sidelobe antenna; coded pulse anti-clutter detection; frequency agility (programmed or random); jamming analysis transmission selection; PRF stagger; precision jam strobes for triangulation; cool antenna for reduced IR signature; enhanced resistance to anti-radiation missiles; compatible with decoys; instantaneous radar silence - remote control available
IFF subsystem
Type: interrogation sidelobe suppression
Modes: 1, 2, 3, C
Decode: active/passive; AN/UPA-59

Contractor

Northrop Grumman Corporation, Electronic Sensors and Systems Division, Baltimore, Maryland.

VERIFIED

AN/TPS-71 relocatable Over-The-Horizon Backscatter (OTH-B) radar

Type

HF-band (3 to 30 MHz) tactical OTH-B backscatter radar.

Description

The AN/TPS-71 Relocatable Over-The-Horizon Radar (ROTHR) is an HF frequency, tactical, land-based, bistatic ionospheric backscatter radar designed to provide a wide area over-the-horizon surveillance of both aircraft and ships in support of tactical forces in locations of national interest. At the operator's option the system can be used for overall surveillance and tracking within the coverage area, spotlighting specific regions to handle targets of interest, or the assessment of the number of aircraft of an attacking force. The relocatable aspects of ROTHR apply to the transmitters, receivers and the operational control centre, all of which are in shelters and designed so that they can be moved by land, sea or air to new locations. The transmit and receive antennas are not intended to be relocatable and sites would be preprepared with antennas and cables.

The ROTHR system has to continuously adapt to changing ionospheric conditions. This circumstance is met by the use of a co-located Quasi-Vertical Incidence (QVI) sounder and a backscatter sounder for propagation management assessment. This approach adapts automatically to the ionospheric conditions of the operational environment, whether it be day or night or any season of the year.

The ROTHR transmit system

The system consists of three distinct elements; a transmit site, a receive site and an operations control centre. The transmit and receive sites are separated by about 50 to 100 n miles (92 to 185 km). The transmit site provides radar illumination in accordance with commands from the control centre. It also provides the transmissions for the backscatter and QVI soundings used in the propagation management and assessment function. The solid-state transmitting system power is 200 kW in the frequency band 5 to 28 MHz, over a 64° wide illuminated sector, covering a range from 926 to 2,963 km. In the receive subsystem, the returned beams are formed digitally in the signal processor which also carries out the range and Doppler processing and extracts the target detections. The detections, together with the raw processed data from which they were extracted and signals from the sounder transmissions, are passed to the control centre for further processing.

ROTHR provides surveillance and tracking in a number of sectors called Dwell Illumination Regions (DIRs), which are selected by the operator. The system can scan up to 12 DIRs at the same time and employs beam pointing techniques to concentrate its transmitted energy in a particular DIR for up to 49 seconds. ROTHR can be time shared so that both aircraft and ships can be tracked simultaneously.

The operations control centre is the nerve centre of the system and carries out the processing of the returned target information. Within the centre, the operators interface with the system via displays driven by a data processing subsystem. The system is largely automatic in operation but can be manually overridden if necessary. The operations control centre can be operated either in a co-located or separated (satellite link) configuration. The latter configuration results in significant operation and maintenance cost reduction since it removes operators from remote radar locations.

Status

Raytheon received a contract from the US Navy for full-scale engineering development in 1984. Systems testing began in mid-1987 with operational trials completed at the US Navy radar test facility near Norfolk, Virginia in 1989.

The prototype ROTHR system was relocated from Virginia to Amchitka, Alaska and became operational there in January 1991, providing surveillance along the Aleutian islands chain.

Raytheon, named as the prime contractor for the ROTHR system in 1984, was awarded its first production contract from the US Navy Space and Naval Warfare Systems Command in December that year for three systems. The first of these started operations in Virginia in April 1993, replacing the system that was relocated to Alaska. Another system was scheduled to be operational in April 1995 from a site in Texas. As of the summer of 1999, sources suggest that Raytheon Electronic Systems was awarded a then year US$112.6 million contract by the US Navy covering the operation and maintenance of two ROTHR systems located at Virginia, Texas and in Puerto Rico respectively. At this time, these radars are reported to have been used both to provide early warning for the Service's carrier battle groups operating in the area and as tools in US counter-drug operations. On 28 March 2000, Raytheon was awarded a then year US$10,049,064 cost-plus, fixed-fee, indefinite-quantity/delivery contract covering the supply of support services (including engineering, technical management, training support, maintenance and integrated logistics) for the ROTHR programme. At the time of this contract's announcement, work on the effort was scheduled for completion by the end of March 2001, with the US Navy's Space and Naval Warfare Systems Center, Charleston, South Carolina acting as the programme's contracting activity. The described contract is also known to have included four, one year renewal options which, if taken up, would bring its cumulative value to then year US$45,477,414.

Contractor

Raytheon Electronic Systems, Marlboro, Massachusetts.

UPDATED

AN/TPS-75 tactical 3-D radar

Type

Tactical air defence surveillance radar.

Description

The AN/TPS-75 represents the latest USAF inventory ground-based tactical air defence radar. This system is an upgrade of the AN/TPS-43/70 family of radars with significantly better electronic counter-countermeasures, increased performance and enhanced reliability and maintainability. It employs the Ultra-Low Sidelobe Antenna (ULSA) which decreases sidelobe emission by more than 50 per cent and considerably reduces vulnerability to anti-radiation missiles. The TPS-75 radar is one of three projects in the USAF 'Seek Screen' programme which will provide the USAF with viable surveillance and command, control and communications capability well into the 21st century.

Status

As of May 2001, a total of 57 AN/TPS-75 radars were reported as having been acquired by the USAF. All such systems were reported as operational during 1996, with more than 50 remaining in the inventory as of December 2000. Six (?) radars of this type are understood to have been modified to be able to track tactical ballistic missiles, with the effort centering around the introduction of new AN/UYQ-509 displays and a remote, multisensor input capable correlator. Here, two TPS-75 missile tracking modification kits are known to have been procured by the USAF

under a then year US$12.1 million, fixed-price contract that was announced by the programme contractor (Northrop Grumman) on 11 December 2000. Elsewhere, Jane's sources suggest that the USAF has shown interest in introducing a magnetic modulator to improve TPS-75 reliability and maintainability. An anti-radiation missile decoy for the AN/TPS-75 has also been developed (see AN/TLQ-32 entry).

Contractor

Northrop Grumman Corporation, Electronic Sensors and Systems Sector, Baltimore, Maryland.

UPDATED

ARSR-4 (AN/FPS-130) Air Route Surveillance Radar

Type

D-band (1 to 2 GHz) long-range 3-D air surveillance radar.

Description

The ARSR-4 is a coherent, 3-D radar operating in D-band and is the only radar designed to meet the joint air traffic control and air defence sensor requirements of the Federal Aviation Administration and the US Air Force. The international nomenclature of the ARSR-4 is the AN/FPS-130.

The ARSR-4 generates dual stacks of elevation beams for optimum time energy management. The array-fed aperture provides azimuth sidelobes below -35 dB and circular polarisation to enhance detection of aircraft in poor weather. Eight-pulse-Doppler filters suppress clutter out to 400 km. The modular digital target extractor and tracker is designed to process 800 aircraft plus 200 non-aircraft reports per scan, with a 50 per cent reserve capacity and is expandable to greater capacity. The integrated secondary radar is fully compatible with the US Federal Aviation Administration's Air Traffic Control Radar Beacon System (ATCRBS) and identification friend-or-foe modes S and 4.

The solid-state transmitter is located below the rotary joint, allowing repair to occur while the system is in operation. Built-in automatic reconfiguration, reserve capacity and redundancy contribute to high availability. The system is unmanned with remote control and monitoring and remote fault detection and analysis.

Status

During 1998, ARSR-4 was selected for the joint Federal Aviation Administration (FAA)/USAF radar replacement programme. There are 44 such systems being deployed around the periphery of the United States, providing data for joint use by the FAA for air traffic control and the USAF for air defence. Three AN/FPS-130 radars have been ordered for phase three of the Royal Thai Air Defence System.

Contractor

Northrop Grumman Corporation, Electronic Sensors and Systems Division, Baltimore, Maryland.

VERIFIED

The ARSR-4 air route/air defence radar

Cardion coastal radar system

Type

E/F-band (2 to 4 GHz) automatic coastal surveillance radar.

Description

Cardion has developed and produced a transportable E/F-band coastal defence radar which can rapidly be deployed in remote sites. It includes an E/F-band primary search radar and a D-band (1 to 2 GHz) secondary surveillance radar. A number of these radars are installed around the coasts of Denmark, Norway, Turkey and other unspecified European countries for coastline monitoring and early warning.

*The Cardion coastal
radar system*

The system was designed for use at unattended sites with automatic reporting of data and control signals. Because the equipment is normally unattended, an extensive self-monitoring system has been incorporated into the design. Alerts are sent out immediately upon the first indication of trouble or abnormal situations.

The antenna is a double curvature reflector with integral primary and secondary feed assemblies. An omnidirectional antenna mounted on top of the main reflector is used for sidelobe suppression in the Secondary Surveillance Radar (SSR) system. Since the radar is normally used as a gap filler, the vertical beam shape has been specifically tailored to provide high detection probability of low-level aircraft and surface targets. A tilt mechanism allows adjustment of the beam above and below the horizon.

The transmitter employs a liquid-cooled, coupled cavity travelling wave tube, powered by an all solid-state high-voltage power supply, to produce a highly stable coded waveform. The transmitter is versatile allowing operation in fixed, burst or full frequency agile operation. All operating modes are sector driven, that is a different mode may be selected for any of the four azimuth sectors, each of adjustable size. As the antenna rotates, the preset operating mode will be automatically turned on in each sector. The transmitter may also be silenced in a sector if desired.

Addition of an Automatic Detection and Tracking unit (ADT) provides the system with full automation capability. The (ADT) features coherent and non-coherent signal processing, primary and secondary plot extraction, automatic track initiation and telephone line interface for data remoting. Video processing concepts allow automatic detection of targets in clutter and electronic countermeasures environments, with a well regulated false alarm rate. The data processing portion of (ADT) extracts plot data on both primary and secondary targets, merging the data into a single report when target correlation can be established. An integrated tracking system provides automatic initiation and continuous track update on all plot data.

One of the main functions performed by the (ADT) is the separation and identification of all targets as either surface or air. Either or both types may be selected for transmission. The (ADT) also extracts other non-target information

such as a clutter map and jamming strobes. This data is transmitted, together with target messages, to the operations centre.

The Remote-Control Unit (RCU) is able to control and monitor an entire network of radars using bidirectional communication over telephone lines. Current status information is sent back to the RCU from each of the radar heads where it is displayed. The RCU is equipped with a cathode ray tube overlayed with a touch-sensitive screen which serves as the man/machine interface.

Status
Eight systems have been supplied to Denmark and the system has also been procured by operators in Norway, Turkey and other European countries.

Specifications
Antenna
Type: double curvature, integral with identification friend-or-foe
Polarisation: circular
Rotation rate: 10 rpm
Size: 4.88 × 2.44 m
Transmitter
Frequency: E/F-band (2-4 GHz)
Peak power: >100 kW
Duty cycle: 1%
Waveform: phase-coded pulse for pulse compression
Receiver
Type: superheterodyne, double conversion with image rejection
Moving target indicator: digital

Contractor
Cardion Inc, Woodbury, New York.

VERIFIED

Cobra Dane radar

Type
Ballistic missile tracking radar.

Development
The Cobra Dane programme was authorised in 1971 and requests for proposals were issued in 1972, funding also being approved in that year. A contract for US$39.6 million was awarded to Raytheon in July 1973. System testing was completed in late 1976 and operational capability in 1977.

Under a modernisation programme, obsolete computer hardware and software is being replaced, using the Ada high-level language. The upgraded Cobra Dane will have an expanded catalogue to store data on 12,000 space objects instead of 5,000. It will also provide early warning and space tracking capability simultaneously, a facility not possible in the original configuration. Initial operating capability of the upgraded system was expected to be achieved in early 1994.

Description
The Cobra Dane phased-array radar is designed to detect and track inter-continental and submarine-launched ballistic missiles and satellites. It replaced two radars of the old intelligence variety (AN/FPS-17 and AN/FPS-80), which were themselves pressed into service to undertake Spacetrack duties.

The large fixed phased-array of this radar measures about 30 m in diameter and comprises about 35,000 elements, of which some 15,000 are active elements. The system will survey a 3,220 km corridor to collect data on Russian Federation and Associated States (RFAS) missile development flights, provide early warning of ICBM launches, detect new satellites and update known satellite parameters.

The array provides coverage over a 120° sector of the RFAS missile test range and in its space tracking role the Cobra Dane installation has a range of 46,000 km. In its data collection role on RFAS missile flights, Cobra Dane has the ability to track up to 100 objects simultaneously with precise data on up to 20 targets. The Cobra Dane radar is located at Eareckston Air Force Station (formerly Shemya Air Force Base) in the Aleutian Islands.

Status
Cobra Dane underwent a comprehensive modernisation programme under a contract awarded to Raytheon in 1990.

Contractor
Raytheon Electronic Systems, Marlboro, Massachusetts.

VERIFIED

The Cardion coastal radar system

The USAF's Cobra Dane radar is designed to detect and track ballistic missiles and satellites at very long ranges

FALCON series surveillance radar

Type

G-band (4 to 6 GHz) 2-D air/sea search and acquisition radar.

Description

The FALCON (Frequency Agile Low COverage Netted) radar is a two-dimensional air/sea surveillance radar providing search, acquisition and tracking of low-altitude aircraft and the simultaneous coastal surveillance and tracking of ships, small boats and low-flying helicopters. The automatic tracking of large numbers of aircraft and ship targets can be monitored by operators within the radar control cabin which permits autonomous radar operations. Target data can be transmitted in digital format from a number of FALCON radars, to air and surface control centres at a distant location to provide centralised command and control operations.

The radar equipment can be installed in a protected bunker configuration, in a shelter, or in a mobile configuration and is designed for unattended operation. Integrated Identification Friend-or-Foe (IFF) and radar tracks data can be transmitted by digital datalink to air and surface operations centres.

The antenna is a dual-beam reflector with a cosec^2 coverage to 25° elevation, an integral IFF, switchable horizontal or circular polarisation and low-azimuth sidelobes. The transceiver is crystal controlled; uses a single-stage, high-gain travelling wave tube system; dual channel receiver and quadrature coherent detection.

Frequency agility is employed and the combination of this with the high-gain, low-sidelobe antenna gives considerable protection against active electronic countermeasures. Low-speed surface traffic is detected and tracked using a combination of Moving Target Indicator (MTI) and non-MTI constant false alarm rate. The dual beam antenna, small resolution cell and broadband adaptive MTI filter simultaneously suppress severe ground and sea clutter and moving rain or chaff.

The FALCON operations facility has two consoles with 40 cm circular displays. While both show the same data, the two can be set to display targets moving at different speeds so that one operator can monitor air traffic while the other observes surface traffic.

FALCON-G is a derivative of the standard FALCON system. The main difference between the two is the high-performance antenna of FALCON-G which allows extended height performance while retaining high resolution and very low sidelobes in a configuration that can be rapidly transported and deployed.

Status

Delivery of FALCON radars began during 1986. By 1993 ITT Gilfillan had contracts from a number of countries in Europe, the Middle East, Latin America and Asia for over 50 radars. FALCON coastal defence systems with radar, operations centres and communications have also been delivered.

Specifications

Operating frequency: G-band (5.5 cm wavelength)
Antenna type: reflector, high-gain, dual beam with very low sidelobes
Polarisation: horizontal or circular, operator selectable
Beamwidths: 1.25° (azimuth); cosec^2 to 25° (elevation)
Search scan: 360° at 6 or 12 rpm
Range: 100 km, with a limited capability out of 160 km (FALCON-G 80 km)
Elevation: 25° (FALCON-G 40°)
Accuracy: 20 m (range); 0.2° (azimuth)
Resolution: 85 m (range); 2.4° (azimuth) (FALCON-G 65 m; 1.5°)
Peak power: 55 kW
Mean power: 1,500 kW
Altitude coverage: 0-3,000 m
Simultaneous tracks: up to 100, air-to-surface

Contractor

ITT Gilfillan, Van Nuys, California.

VERIFIED

The G-band dual-feed antenna used in the FALCON radar system

HR-3000 radar (HADR)

Type

E/F-band (2 to 4 GHz) 3-D air defence radar.

Description

HADR is the acronym by which Raytheon Electronic Systems refers to an advanced 3-D multimode radar designed for use in national air defence networks, either to replace ageing equipment, to enhance system performance or to fill gaps in the existing system. The system is also known as the HR-3000. It is understood that many of the advanced features of this radar are a result of experience gained within the AN/TPQ-36 and AN/TPQ-37 mortar and artillery locating radars. Designed to meet new NATO ACE (Allied Command Europe) standard radar specifications, HADR automatically detects, classifies and reports on all targets in its area of coverage. Operation is in the E/F-band of the spectrum and a 4.8 × 6 m planar-array antenna is used, generally protected within a radome.

The HR-3000 primary radar operates over a bandwidth greater than 12 per cent. The radar achieves a detection range of at least 320 km for a 1 m^2 target. The instrumented range coverage is 500 km. Versatile computer control enables radar coverage to 30,000 m altitude and up to 24° elevation. False target reports are controlled to prevent Air Defence Ground Environment (ADGE) system computer overloads.

Target positional data is provided in three dimensions. Target range is measured to the nearest range cell while target azimuth and elevation are obtained by beam splitting. Performance in clutter is achieved by a digital mean velocity clutter tracker in conjunction with a single or double delay Moving Target Indicator (MTI) canceller. A range-gated pulse-Doppler waveform and associated processing implementation is also provided. An antenna rotation period of 10 or 12 seconds is consistent with ADGE system data rate requirements. Primary or secondary radar target report rates can exceed 400 per scan at the radar output. However, transmission rates to the remote command and control site may be limited by datalink capacity.

The HR-3000 antenna subsystem scans a single pencil beam in the elevation plane under computer control by controlling the phase shifters of the antenna. By properly positioning this beam in sequential steps, complete elevation coverage is provided. The elevation scan is completed before the antenna moves one beamwidth in azimuth, providing complete coverage of the surveillance volume with no holes. Antenna beamwidth and the transmitted waveform are controlled at each beam position to optimise management of the radar time and energy resources. Beamwidth is 1.1° in elevation and 1.7° in azimuth.

The operational modes in the HR-3000 system are defined by tables stored in the computer memory. The computer controls the following elements for each beam in the search raster:

- instrumented range
- peak power
- pulse-length
- elevation beamwidth
- waveform
- elevation angle
- frequency
- detection criteria
- plot extraction criteria
- operation changed by software.

The Secondary Surveillance Radar (SSR) subsystem incorporates an Identification Friend-or-Foe (IFF) interrogator and antenna. The HR-3000 antenna makes provisions for mounting the primary IFF antenna as well as the sidelobe blanking IFF antenna and for routing the radio frequency signals through the rotary joint. The beacon video processor which is part of the multifunction processor provides extraction of IFF video and provides active and passive decoding. It also provides SSR target report outputs to the radar controller. The radar controller provides a correlation of these plot reports with primary radar plot reports for a consolidated output to the command control site.

The HR-3000 primary radar utilises several techniques to reduce the effectiveness of Electronic CounterMeasures (ECM). Sophisticated jammers are

The antenna used in the HR-3000 air defence radar system

thwarted by the radar operational characteristics including frequency agility, coded pulses, multiple pulse-widths and multiple pulse repetition frequencies. The primary radar design has low susceptibility to jamming through the use of pencil beams, a low-sidelobe level antenna design, high-receiver dynamic range and sharp channel selectivity. Automatic threshold control, target detection correlation and sidelobe blanking are techniques provided to handle any residual jamming which does get into the system. The burn through mode and automatic frequency selection are special modes available for use in an ECM environment. Single pulse MTI is provided out to 111 km and two-pulse MTI out to 185 km. Seven-pulse range-gated pulse-Doppler is used to suppress rain or chaff clutter.

The system is capable of selecting operating frequency under computer control and each frequency can be selected in about 40 µs and may be changed on pulse-to-pulse or dwell-to-dwell basis. When selected by the operator, the system will automatically measure the interference level in each of the available operational channels and select that which is least affected.

The normal surveillance mode can be interrupted, at the operator's discretion, by a special mode which will intensify the energy on a specific target location. Normal surveillance volume is interrupted and all of the energy available is concentrated in a high-energy waveform with a long dwell time in a sector automatically centred around the jam strobe which has been measured by the system and which indicates jammer direction. At high-elevation angles, full antenna gain will be employed on the target even though the normal surveillance volume might call for a broadened beam.

Status
Four HADR systems have been procured by Germany with two being used in conjunction with the NATO Air Defence Ground Environment (NADGE) and known as GEADGE (German ADGE). So used, these two radars replace AN/FPS-7 equipments within the 412L radar network.

A version of the HR-3000, known as RSRP (Radars for the Southern Region and Portugal), has been selected for the update of NATO's air defence system on its southern flank. Raytheon is understood to have delivered three of these radars to Turkey, two to Greece, two to Italy and one to Portugal. Turkey, Italy and Greece incorporate the equipment into the NADGE system while Portugal is thought to have exercised an option for two more systems. As of November 1994, Raytheon had delivered one RSRP radar to Greece. Nine more were delivered by the end of 1995.

The HR-3000 incorporates a number of modifications to the basic HADR, including increased electronic counter-countermeasures capability, wider receiver bandwidth, improved clutter-rejection waveforms and a faster antenna rotation speed to meet the NATO requirement for a higher data rate. The radars will operate on concrete platforms without supporting towers or radomes.

Raytheon has also supplied three HADR radars for Norwegian air defence. These systems are used in shelters built into the top of mountains and are mounted on elevators so that the antennas can be retracted into environmentally controlled silos for routine maintenance. The silos are designed and manufactured by Kvaerner-Eureka of Norway. The radars have been installed and all are believed to be operational. Norway is now planning to build two new silo-based radars in the northern part of the country, equipped with HADR systems. One HADR is operational in Malaysia as part of the Malaysian Air Defence Ground Environment (MADGE) programme. Raytheon Electronic Systems is prime contractor on this programme.

Contractor
Raytheon Electronic Systems, Fullerton, California.

VERIFIED

....

ITT active aperture radars

Development
ITT Gilfillan has been developing active aperture arrays employing monolithic microwave and very high-speed integrated circuitry and optical and bistatic technologies. When fully developed, the system is expected to provide extended high- and low-altitude coverage, the capability to detect and track very small targets in high-clutter environments, the ability to distinguish and track a variety of targets in a mixed raid and provide multifunction capabilities. One of the keystones in the development of this type of radar is the continuing development and production of gallium arsenide integrated circuit technology.

The development covers all frequency bands from D to J (1 to 20 GHz) to accommodate land-based air defence, tactical, naval and other surveillance requirements, as well as air traffic control, smart skins, hybrid airborne/space sensor systems and for integrated radar/datalink communications requirements.

Status
As of this edition, ITT Gilfillan has a number of solid-state phased-array air traffic control radar systems in production. This range includes the USAF's AN/MPN-25, the PAR-2000 (operational in Brazil and the United Kingdom) and the CGA-2000 equipments. The company is also developing similar technology and applications for airborne and ground-based systems for the US Department of Defense.

Contractor
ITT Gilfillan, Van Nuys, California.

VERIFIED

Low-Altitude Aircraft Detection System (LAADS)

Type
D-band (1 to 2 GHz) transportable low-altitude aircraft detection radar.

Description
LAADS is a transportable air defence radar and command and control system that is designed for the detection and identification of low-flying targets and for the co-ordination of defensive weapons directed against such targets.

The radar employed is a coherent, pulsed Doppler D-band system using a travelling wave tube transmitter. The 12 operating frequencies are generated by highly stable master oscillators. A zero intermediate frequency receiver detects moving targets, rejects fixed clutter and presents analogue data to the digital data processor which drives the local Plan Position Indicator (PPI) and generates digital data messages for broadcast. The radar location is fed into the processor and the target locations are then computed in map co-ordinates. Target data are corrected on each antenna rotation and transmitted to all associated weapons via the datalink. Target track capacity is 64. Instrumented range is 30 or 60 km, with hovering helicopters being detected to 30 km.

An Integrated Weapon Display (IWD - a part of the system's command and control capability) receives digital data messages and displays targets with respect to the weapon location (centre of display). Targets within 16 km from the weapon are displayed in near real time with 1 km display resolution. Targets can be symbolically displayed in either a 'B' scan for forward-looking infra-red or video-equipped weapon systems, or PPI mode for optical systems. Previous positions are scanned to show target direction and to facilitate threat evaluation. The weapons display is mounted directly in front of the gunner, with a cursor showing boresight of the weapon. A hand-held version of the weapons display is available for the unit commander or platoon leader.

The complete radar equipment, with positions for a crew of three, is housed in a standard US S-280 shelter. This is helicopter transportable or land-mobile on an M36 or similar 2½-ton truck. The weight of the complete system is 3,100 kg.

The 188 × 300 cm antenna is mounted on the shelter and is automatically raised and levelled from its folded stowed position. On-mounted antenna facilities are provided for the integrated Identification Friend-or-Foe (IFF) TPX-54.

The general concept of operation is for each radar to provide search, IFF and target assessment for remote weapons positions. Each of the latter is provided with the necessary datalink reception equipment and IWD to permit individual target engagement. The IWDs, displays and the VHF (30 to 300 MHz) datalink provides for an unlimited number of weapon sites within line of sight radio range (normally in excess of 15 km).

LAADS has been designed to include a command and control capability if required. With this option in place, LAADS radars can be deployed as a network of systems which share surveillance responsibilities and exchange radar data. A co-ordinated and automatic 'blinking' capability is incorporated to provide protection from attack by anti-radiation missiles and the command and control display mode includes 1, 2, 4, 8, 15, 30, 60 and 120 km map scaling options, map symbols and lines, weapons location and status, warning and weapon alert messages, engagement and hold fire messages and general text messages.

LAADS is able to serve and provide target acquisition data to a range of SHOrt Range Air Defence (SHORAD) and MAN-Portable Air Defence (MANPAD) equipments including the Chaparral, Vulcan, pedestal-mounted Stinger, Redeye and Blowpipe systems.

Status
As of this edition, approximately 60 LAADS systems have been delivered to Asian (four) and Middle Eastern customers. The latest known LAADS radar delivered was fielded in Thailand during 1998.

The LAADS air defence radar system

Specifications

Frequency: D-band (1-2 GHz)
Power output: 150 W average
PRF: 4.16 kHz, 4.84 kHz (alternating scan-to-scan)
Scan rates: 30; 15; 7.5 rpm
Instrumented range: 30 and 60 km
Detection range: 60 km on 5 m^2 target; 30 km on hovering helicopters
Azimuth accuracy: 1.5° RMS
Range accuracy: ±250 m
IFF system: modified TPX-50

Contractor

Lockheed Martin Ocean, Radar and Sensor Systems, Syracuse, New York.

VERIFIED

Series 320 air defence radars

Type

E/F-band (2 to 4 GHz) 3-D air defence surveillance radar.

Description

The Series 320 family of 3-D air defence radars has been configured to meet various surveillance coverage and siting requirements. The basic system uses a planar-array antenna with frequency steered multiple pencil beam groups to scan in elevation and mechanical rotation to cover 360° of azimuth. Both fixed site and transportable configurations have been produced.

The radar is supplied in either 370 or 556 km range versions. The 370 km version has a peak power output of 150 kW and the 556 km version adds a power amplifier to develop 1.1 MW peak power. Highly stable signal processing techniques maintain target detection and low false alarm rate in high-clutter areas. Operation of the radar is highly automatic with extensive online monitoring of performance and fault isolation. Moving target indication is employed in areas where ground clutter is present. Electronic counter-countermeasures capabilities include jammer frequency analysis, jam-strobe reports and silent sector operation.

Two antenna configurations are available, the 6.4 × 5.18 m antenna pictured provides a nearly circular 1.4° azimuth by 1.6° elevation beam. A 3.65 × 9.75 m antenna, designed for increased elevation accuracy, provides a 2.2° azimuth and a 0.85° elevation beam.

Status

More than 20 systems are understood to have been procured by customers around the world. An agreement with the UK's BAE Systems has resulted in the AR-320 radar which blends ITT 320 Series technology with that from Siemens Plessey's AR-3D radar. AR-320 has been supplied to the UK's Ministry of Defence. Other than this, the Philippines is the only other 320 Series customer which has tentatively been identified.

Specifications

Frequency: E/F-band (2.9-3.1 GHz)
Peak power: 150 kW/1.1 MW
Pulse repetition frequency: variable
Receiver noise figure: 2.5 dB
Antenna: 6.4 × 5.18 m (3.65 × 9.75 m optional)
Gain: 40.6 dB (41.5 dB optional)
Azimuth beamwidth: 1.4° (2.2° optional)
Elevation beamwidth: 1.6° (0.84° optional)
Scan rate: 6 rpm
Polarisation: horizontal
Signal processor: digital
Improvement factor: 40 dB

Detection performance (2 m^2 target): 200/300 n miles (PD ≤ 0.5)
Number of false alarms: 5/scan
Azimuth coverage: 360°
Elevation coverage: 0-20°
Height coverage: 0-30,000 m
Resolution at 185 km: 125 m (range); 2.8° (azimuth); 3.5° (elevation)
Accuracy (1 sigma at 185 km): 18 m (range); 500 m (azimuth); 760 m (height) (400 m with option)

Contractor

ITT Gilfillan, Van Nuys, California.

VERIFIED

Silent Sentry™ detection system

Type

Passive radar-type detection system.

Description

The Lockheed Martin Mission Systems (LMMS) Silent Sentry™ passive detection system utilises a radar-type reception and processing chain to determine air target location from the Doppler shift of echo signals produced when aircraft are illuminated by commercial Frequency Modulated (FM) radio or television transmissions. System architecture comprises Lockheed Martin proprietary, high dynamic range receivers, Silicon Graphic-sourced Challenge L commercial processors (with 12 GP engines), Autometric-sourced software and a phased-array antenna. Functionally, the system is initialised by cataloguing the 50 to 800 MHz band commercial transmitters in its area and establishing their location and parameters. As noted previously, target location is then determined by the Doppler shift in the reflected signal from the object under surveillance.

Status

As of this edition, Silent Sentry™ is understood to have been in development since the early 1980s. By late 1998, the system is reported to have been able to detect 10 m^2 radar cross section targets at ranges of up to 180 km using a single FM illumination source. LMMS also claims that the equipment can track in excess of 200 targets with formation spacings of as little as 15 m. As of this edition, sources suggest that the system has been modified to make use of three FM illuminators and 28 GP engine Challenge XL processors in order to facilitate 3-D target location. An operational Silent Sentry™ architecture is described as offering a 3-D tactical display, as utilising up to six FM illumination sources and as being configurable for fixed-site or deployable applications. The antenna array used in such a system is noted as measuring 2.3 to 2.5 m^2.

Contractor

Lockheed Martin Mission Systems, Gaithersburg, Maryland.

VERIFIED

Tactical ballistic missile radar

Type

Surveillance radar against ballistic missiles.

Description

The tactical ballistic missile radar is another variant of the AN/TPS-70 tactical radar. It is designed to provide high-altitude surveillance against ballistic missiles in

The ITT Gilfillan 320 3-D transportable air defence radar system

Northrop Grumman's tactical ballistic missile radar

intense electronic countermeasures conditions. The flexible antenna design concept allows for virtually instantaneous switch over from this tactical ballistic missile role to the standard air defence mission for which the AN/TPS-70 was originally intended.

The modified processor, including the Missile Launch Warning Subsystem (MLWS), can provide plot data on up to 1,200 targets. Processing capabilities include identification of ballistic missile types, launch points and impact points. For netted operations the processor design allows transmission of digital track data to a command centre or weapons unit. The radar can operate completely unattended. All these capabilities were successfully demonstrated, under US Army contract, in live missile firings and during 1992 under a USAF high-gear programme using AN/TPS-75 radars.

Contractor

Northrop Grumman Corporation, Electronic Sensors and Systems Division, Baltimore, Maryland.

VERIFIED

Theater Missile Defense – Ground Based Radar

Type

I/J-band (8 to 20 GHz) long-range theatre ballistic missile detection.

Development

The Theater Missile Defense – Ground Based Radar (TMD-GBR) will be the surveillance and fire-control radar for the Theater High-Altitude Area Defense (THAAD) system. The US Army Program Executive Office – Missile Defense awarded a contract for the development of the demonstration/validation phase of the TMD-GBR to Raytheon in 1992. This phase consists of the design and fabrication of three radars: one half-aperture system for early testing and concept validation and two full aperture User Operational Evaluation Systems (UOESs) for operational testing with THAAD. As originally planned, the UOES radars were expected to be capable of emergency contingency deployment in early 1997 with the engineering and manufacturing development contract for the objective system scheduled for award in the fourth quarter of FY96. Again as originally planned, this stage of the effort was intended to move the programme towards the fielding of 14 objective system radars supporting two THAAD battalions.

Description

The TMD-GBR is an I/J-band mobile radar system that is designed to detect and track theatre ballistic missiles. The TMD-GBR system includes the following:
- a trailer-mounted, single-faced 9.2 m² wideband phased-array antenna with solid-state transmit/receive modules
- an electronic equipment unit housing radar control and signal/data processing equipment
- a 1.1 MW prime power unit
- a cooling equipment unit which provides cooling for the antenna array
- an operator control unit which contains operator consoles for operations, maintenance and communications monitoring.

The TMD-GBR provides sensor data to the THAAD ballistic missile/command, control, communications and intelligence system via a fibre optic datalink. TMD-GBR provides early warning of theatre ballistic missile launches by detecting and acquiring targets at very long ranges using autonomous search fences and volume search strategies and is capable of cued acquisition modes. The radar performs classification and discrimination to categorise the threat type and identify the target vehicle. It maintains track on the target and provides in-flight updates to the missile, as well as a target object map prior to intercept. The TMD-GBR performs intercept assessment to support the decision to commit additional interceptors or one lower tier system.

Specifications

Aperture: 9.2 m²
Transmitter: solid-state
Detection range: 1,000 km

Contractor

Raytheon Electronic Systems, Waltham, Massachusetts.

VERIFIED

TRACKSTAR low-altitude air defence system

Type

D-band (1 to 2 GHz) low-altitude air surveillance, command and control system.

Description

TRACKSTAR is a self-contained, armoured, fully integrated radar/command and control system. It is designed for low-altitude air defence and provides radar surveillance and real-time automated command datalinks to multiple weapons and gunner displays. The system consists of a radar and a command unit contained in an armoured vehicle, M577, with an antenna system mounted on top of the vehicle.

The D-band radar system employs digital moving target indicator and Doppler processing techniques, with multiple electronic counter-countermeasures features to counter jamming, and has excellent subclutter visibility. It also provides high-order rejection of fixed clutter, ground traffic, inclement weather and chaff. Its digital signal processor employs a unique method of detecting and classifying rotary-wing aircraft even when hovering at ranges up to 30 km. An automated antenna erection, levelling and stowage system is provided and a fully integrated Identification Friend-or-Foe (IFF) interrogator and antenna system is incorporated. The TRACKSTAR system gives 360° surveillance out to a range of 60 km.

Dual command and radar displays are contained in the radar/command unit, which provides fully automated weapon cueing data composed of near real-time target vector and identifying information to ground-based weapons to ensure rapid long-range target acquisition. Each TRACKSTAR has a capacity for 64 internal target tracks and can net tracks received by datalink from up to three other radars. Target track cueing, fire distribution and IFF data are transmitted via VHF (30 to 300 MHz) datalinks to integrated weapon display units which are mounted in or near the supported associated weapons system, such as Blowpipe, Chaparral, Stinger, Vulcan or similar short-range mobile and man-portable air defence weapons.

Status

A number of TRACKSTAR Systems have been sold to Egypt as part of the Chaparral air defence missile system.

Contractor

Lockheed Martin Ocean, Radar and Sensor Systems, Syracuse, New York.

VERIFIED

The TRACKSTAR air defence radar

BATTLEFIELD, MISSILE CONTROL AND GROUND SURVEILLANCE RADAR SYSTEMS

This section includes battlefield surveillance, tracking, fire-control and weapon locating radars whether static, vehicle mounted or man-portable. It includes those systems used for low-level air defence control on the battlefield, weapon guidance for artillery, surface-to-air and surface-to-surface missiles and equipment used for ground surveillance and targeting. Wherever possible full details are given of the associated data processing and displays systems. Comprehensive details of land-based weapon systems, including the missiles, guns, fire-control systems and interface equipments are given in *Jane's Land-Based Air Defence*.

VERIFIED

AUSTRALIA

MVS-470 muzzle velocity system

Type
Muzzle velocity system for artillery and mortars.

Description
The MVS-470 is a small, rugged and lightweight radar system designed to carry out accurate measurements on all known types of artillery and mortars. It has been designed to withstand the rigours of tactical deployment while mounted permanently on the host gun. The system is completely automatic and reduces radar transmission to the absolute minimum while also compensating for errors caused by gun motion. A software-based system provides fundamental flexibility of application for customer supplied algorithms for velocity reduction, as well as enabling the user to add data for new guns and/or munitions. MVS-470 consists of two major assemblies, a radar processor unit which is mounted on the gun and a control display unit. Upon firing the gun, the muzzle velocity of the round and the average round velocity are displayed. An automatic output of average and last round velocities is provided for fire-control systems.

Status
Over time, BAE Systems Australia's muzzle velocity systems have been supplied to the defence forces of Australia, Austria, Brazil, Canada, India, Malaysia, Netherlands, New Zealand, Norway, Singapore, Sweden, Switzerland and Thailand. As of 2001, MVS-470 was a live programme.

Specifications
Types of rounds: calibres down to 20 mm, including conventional, base bleed, flashing propellant and rocket-assisted ammunition
Velocity range: 50-2,400 m/s
Rate of fire: measurement at continuous rates up to 1,000 rds/min
Accuracy: ±0.05% typical

Contractor
BAE Systems Australia, Elizabeth, South Australia.

UPDATED

A close-up of the sensor head used in the MVS-470 system   0006879

BELGIUM

SCB 2130A battlefield radar

Type
I/J-band (8 to 12 GHz) portable battlefield surveillance and target acquisition radar.

Description
The SCB 2130 is a portable, multipurpose, medium range pulse-Doppler radar used for battlefield surveillance and target acquisition. Its main role is to detect,

The SCB 2130A battlefield surveillance radar deployed in the field

locate and classify ground targets and low-flying fixed-wing aircraft and helicopters (both moving and hovering).

The SCB 2130 consists of four units; an operator console unit (or display control console), a transceiver, an antenna unit with its positioning system and a stable quadripod for ground installation. The system can also be deployed on a telescopic mast for vehicle mounting. For remote operation the display control console may be located up to 100 m from the antenna/transceiver. The system can be deployed in 10 minutes by two operators.

The system has four major modes of operation:
- full 360° scan
- limited sector scan between 10 and 360° – user defined sector limits
- zoom mode for focus on a limited area
- artillery spotting.

Different ranges are available, depending on the type of transmitter employed. When a broadband travelling wave tube is used the radar can detect a crawling man up to 10 km, a walking man up to 15 km and armoured vehicles up to 33 km. The solid-state amplifier version gives reduced detection ranges. The SCB 2130 also has a fire deviation measurement function which allows it to observe the point of impact of an artillery shell simultaneously with the target location. Innovative electronic counter-countermeasures features, such as frequency agility and spread spectrum are included.

A 254 mm TV monitor and display processor provide a comprehensive map-like picture of the surveillance area, oriented to the north for easy comprehension. Six types of video display are featured, including high-resolution clutter map, target displays in various colours superimposed with universal transverse Mercator grid and co-ordinates and controllable fading rate for target tracking. Operator controlled, the system's minimal detectable velocity rate can be varied between 1.5 and 48 km/h, reducing false alarms caused by weather conditions and improving target recognition. In addition to the target's audio Doppler signature, recognised by the operator through headphones or loudspeakers, the system can classify between different targets using high-speed pattern recognition algorithms.

A number of options are available for improved target intelligence. These include an optical system for full identification, a plotter for hard copy maps, a modem for communications and a navigation system. All of these can be integrated easily with the radar system.

Status
SCB 2130 has been supplied to the Belgian Army and three export customers.

Specifications
Frequency: I/J-band (8-12 GHz — over 100 frequencies)
Peak power: 70 W (pulse compression) with TWT; 5 W with a solid-state amplifier
Accuracy: ±0.5° azimuth; ±15 m range
Detection ranges (max): crawling man 10 km; walking man 15 km; vehicle 30 km; armoured vehicle 33 km; hovering helicopter 15 km; moving helicopter 20 km
Dynamic range: 45 dB
Antenna
Polarisation: linear or circular
Beamwidth: 1° horizontal; 5° vertical

Contractor
Belgian Advanced Technology Systems SA, Angleur.

VERIFIED

BRAZIL

EDT-FILA fire-control system

Type
I/J/K-band (8 to 40 GHz) anti-aircraft fire-control system.

Description
The EDT-FILA (Fighting Intruders at Low Altitude) has been developed by AVIBRAS to meet the requirements of the Brazilian Army for a mobile fire-control system to be used with 35 or 40 mm anti-aircraft guns and with surface-to-air missile systems.

The basic technology for the FILA was obtained from the Contraves Skyguard anti-aircraft fire-control system, but with a number of improvements such as an additional K-band (20 to 40 GHz) tracking radar, laser range-finder, a new C-2001 computer, a new Search Radar Data Extractor (SRDE) and circular polarisation for the search antenna. The system retains the basic Skyguard I/J-band (8 to 20 GHz) radars for search and tracking, Identification Friend-or-Foe (IFF) and TV tracker.

The pulse-Doppler I/J-band search radar has a high detection capability, high clutter suppression, automatic target alarm and integrated operation with the IFF system.

The K-band and I/J-band monopulse-Doppler tracking radar system has fast pulse repetition frequency and frequency change, automated target exchange, anti-shipping missile detection and alarm, as well as independent (passive) tracking, ensuring a high degree of immunity from noise or disturbances. The high-precision laser range-finder is fully integrated.

The TV tracker allows for automatic/manual high-precision target tracking with automatic zoom. It has an infra-red system to perform accurate parallax measurement of the weapons.

The third-generation digital computer allows real-time data processing and has ample storage capacity for several types of software such as self-test, diagnosis and combat programs.

The control console has a plan position indicator which displays Doppler and raw videos, as well as symbols and markers for presenting the tactical information (including threat evaluation performed by the SRDE). It also provides an electronic counter-countermeasures control panel, a monitor of the TV tracking system, a joystick for manual tracker control and a matrix panel for data input and output.

The complete system is installed in a four-wheel trailer built of fire-resistant glass fibre polyester and its cabin is fully air conditioned.

Status
In operational service with the Brazilian Army since 1985. During 1998, AVIBRAS delivered a number of additional EDT-FILA systems to the Brazilian Army. As of April 2001, total EDT-FILA production for the Brazilian Army was reported as being 13 units.

Contractor
AVIBRAS Industria Aeroespacial SA, Jacareí.

UPDATED

The EDT-FILA fire-control system

BULGARIA

CREDO ground surveillance radar

Type
J-band (10 to 20 GHz) land-based tactical acquisition and surveillance radar.

Description
The CREDO ground surveillance radar has a considerably longer range than the Fara equipment (see next entry). It is designed for search, acquisition, tracking and position finding of moving ground targets, such as men, groups of people and vehicles. It can also be used for surveillance of low-flying aircraft and in a coastal role against surface vessels.

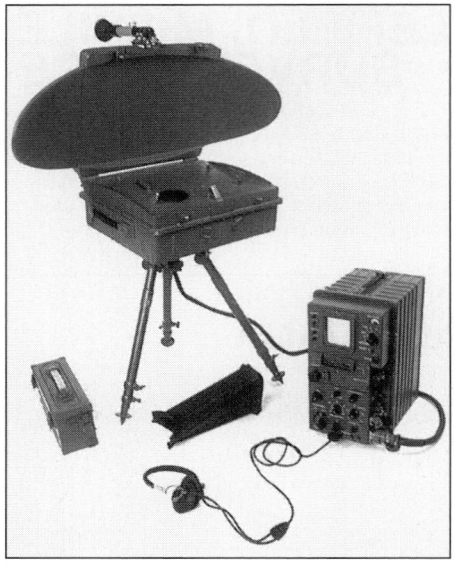

The units making up the CREDO ground surveillance radar

The radar consists of two major units; a receiver/transmitter, an antenna and an indicator unit, plus a tripod, headset, cables and a battery power supply unit. It can be mounted on a variety of fighting vehicles with power provided by the vehicle, or mounted on a tripod with a battery power source. The equipment can be operated by a single person, but two people are required to carry it for portable operation.

The radar is a pulse-Doppler, pseudo-coherent type and features a high degree of natural clutter suppression. It operates in the J-band frequency range and has an option for manual tuning of eight fixed frequencies to ensure better noise suppression. Two modes are available, with or without the moving target selection, for use in varying environmental conditions. Indication of moving targets is effected by visual and audio means. The type of targets can be determined by an experienced operator using headphones.

Status
As of this edition, CREDO is understood to have been procured by the Bulgarian Army.

Specifications
Frequency range: J-band (10-20 GHz)
Range: 4 km (personnel); 10 km (AFV)
Detection probability: 0.8
Min detection range: 200 m
Max moving target velocity: 60 km/h
Max error of moving target positioning
Acoustic indication: 25 m (range); 00-05 mils (azimuth)
Visual indication Type A: 50 m (range); 00-10 mils (azimuth)
Visual indication Type B: 100 m (range); 00-10 mils (azimuth)
Resolution: 100 m (range); 00-50 mils (azimuth)
Sector scan (azimuth): 0-40° (manual control mode); 4-20° (automatic mode)
Battery life: 6 h
Temperature range: −50 to +60°C
Max wind: 15 m/s
Power supply: 24 V
Power consumption: 90 W
MTBF: 250 h
Weight: 50 kg

Contractor
Kintex, Sofia.

VERIFIED

FARA ground surveillance radar

Type
J-band (10 to 20 GHz) short-range, land-based tactical surveillance radar.

Description
This is a portable, short-range battlefield ground surveillance radar designed for acquisition, tracking, determining of co-ordinates and recognition of moving ground targets, such as personnel and vehicles. It is a Doppler type system with continuous emission and phase-code modulation, which ensures qualitative selection of the signals from moving targets and provides a high degree of security and good protection from active countermeasures. Operation of the system is either automatic or manual with indication of target azimuth and range on a digital readout. Recognition of target type depends on operator experience and his use of the headphones for general surveillance and monitoring of return signals.

The complete system consists of an antenna/transceiver, power and control unit, scanning device, battery pack, headphones, cabling, instrumentation set and mounting accessories. The antenna/transceiver can be separated from the power

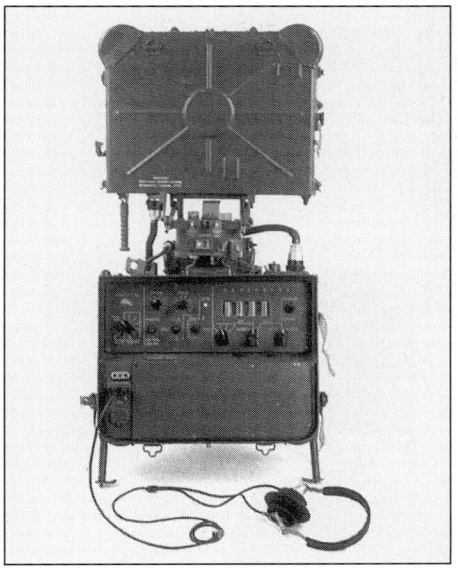

The FARA ground surveillance radar

and control unit to provide remote operation. It is a very compact unit and can be easily carried by one person, contained in a backpack.

Status

As of this edition, FARA is understood to have been procured by Bulgaria's defence forces.

Specifications

Frequency: J-band (10-20 GHz)
Detection range: 0.8-1 km (personnel); 3 km (vehicles)
Resolution: 00-67 mils (azimuth); 100 m (range)
Accuracy: <50 m (range); 00-15 mils (azimuth)
Continuous operating time: ≥ 2 h (at 40°C); 8 h (at 20°C)
Power consumption: 50 W
MTBF: 1,000 h
Weight: 18.3 kg

Contractor

Kintex, Sofia.

VERIFIED

....

NR-100 battlefield radar

Type

Ground-based battlefield surveillance and targeting radar.

Description

NR-100 is a short-range, pulse-Doppler radar that is designed to detect, locate and classify moving targets on the battlefield. The system comprises a transceiver/antenna, a hand-held control unit, a power supply and a set of headphones. The transceiver/antenna can be tripod or vehicle mounted or deployed on a fixed

The units making up the NR-100 battlefield surveillance radar
0006881

object such as a post or tree. NR-100 uses the pulse-Doppler principle to detect personnel and trucks moving with radial speeds of greater than 2 km/h and between 2 and 50 km/h respectively. Bearing and range are displayed on the operator's control unit with target type being identified by audio Doppler signature in their headphones. NR-100 can also be used as an electronic sight for weapon systems such as machine guns and rocket launchers. Power is provided by either a 24 V DC battery or an external 220 V AC supply.

Status

As of this edition, NR-100 is understood to have been procured by the Bulgarian Army.

Specifications

Weight: 9 kg (patrol version); 17 kg
Max detection range: 1,500-1,800 m (erect person moving with a radial speed greater than 2 km/h); 3,500-4,000 m (truck with a radial speed of 2-50 km/h)
Resolution: 6° (angle); 50 m (range)
Scanning sector (azimuth): 30-270°
Max error for target ranging: 50 m
Min spacing for sector presetting: 6°
Range coverage: 6; 50; 200; 1,000; 4,000; 8,000 m
Indicator: Audio, light and digital
Antenna rotation rate: 1 rpm
Consumption: <15 W
Continuous operation: >8 h (1 battery)
Remote control: 25 m
Operating temperature: −20 to +55°C

Contractor

Kintex, Sofia.

VERIFIED

_____

CHINA, PEOPLE'S REPUBLIC

AS901

Type

D-band (1 to 2 GHz) air defence target designation radar.

Description

The China NORth INdustries COrporation (NORINCO) AS901 is described as being a solid-state, pulse-Doppler, D-band 'air defence post target designation' radar that is designed for the acquisition of battlefield attack helicopters and low-flying close support aircraft. As such, the system comprises a pedestal, antenna array, transceiver, identification friend-or-foe interrogator, a terminal unit and associated cabling. Of these, the terminal unit incorporates a light emitting diode colour display and a keyboard. AS901 is normally carried in a light vehicle and can be deployed by its crew in five minutes. NORINCO claims that the radar offers a good electronic counter-countermeasures performance and notes that once AS901 has acquired a target and confirmed it as hostile, it hands off targeting data to an associated weapon system via an automatic datalink.

Status

As of July 2001, Jane's sources were suggesting that the AS901 had been developed under the China People's Liberation Army's 'three new offensives and defensives' battlefield air defence programme.

A general view of the AS901 air defence post target designation radar with its operator's terminal visible to the left of the picture (NORINCO)   2002/0056275

Specifications

Frequency: D-band (1-2 GHz)
Range: 10 km (max helicopter detection with target at an altitude of 100 m); 15 km (max fixed-wing detection with target at 100 m); 25 km (max detection range with target at an altitude of up to 3,500 m)
Accuracy: 2° (azimuth); 400 m (range)
Antenna rotation: 10 rpm
Track-while-scan: up to 10 targets (automatic)
MTBF: 400 h
MTTR: 30 min
Temperature: −40 to +50°C (operating)
Power requirement: 24 V DC (600 W)
Weight: 110 kg

Contractor

China North Industries Corporation, Beijing.

NEW ENTRY

Gin Sling radar

Type

Target acquisition/tracking and missile guidance radar.

Description

Gin Sling is the NATO reporting name for the acquisition and tracking radar employed with the HQ-2 SAM system (Chinese designation HQ-2J). The HQ-2 is a redesign of the HQ-1 — a licence-built version of the RFAS Guideline system. Gin Sling is similar in appearance to the RFAS Fan Song radar and has a matching performance in the ability to track a single target and guide up to three missiles.

The basic elements of Gin Sling are a pair of orthogonal 'trough' antennas, one horizontal and one vertical. The vertical antenna emits a fan-shaped beam which scans from side-to-side, the horizontal antenna scans up and down (for other details see Fan Song entries). Within the HQ-2 system, Gin Sling is understood to be supported by the SJ-202 early warning, acquisition and height-finding radar.

Status

As of this edition, Gin Sling is thought to have been procured by China's armed forces.

Contractor

China Precision Machinery Import and Export Corporation, Beijing.

VERIFIED

The Gin Sling fire-control radar

JY-17 battlefield reconnaissance radar

Type

J-band (10 to 20 GHz) portable battlefield surveillance radar.

Description

The JY-17 is a man-portable, lightweight, solid-state battlefield reconnaissance radar which is designed to detect, recognise, identify and locate moving targets. It can be used to survey the battlefield, detect moving targets, such as armoured vehicles, soft-skinned vehicles and personnel and identify them. Range and azimuth are presented on a control unit and identification is carried out aurally by the operator. Visual and aural alarms are provided. The radar is normally tripod mounted.

The JY-17 is stated to be highly efficient in a number of roles, such as protecting landing fields for helicopters and vertical take-off and landing aircraft, munition storehouses, command posts, oil pipelines, airports, roads and borders. The

The JY-17 battlefield radar

frequency of operation has not been disclosed but it is assumed that it operates in J-band. The main features of the equipment are given as:

- automatic sector scan of antenna with axes and widths selectable through a keyboard
- microstrip antenna
- all solid-state transmitter
- pulse compression techniques employed
- fully coherent transmitter/receiver
- adoption of advanced programmable signal processing techniques
- automatic detection ability
- liquid crystal display or 11.4 cm B-scope (optional)
- aural warning and Doppler monitoring
- remote operation by use of a control unit and keyboard.

Status

As of this edition, JY-17 is thought to be in service.

Contractor

East China Research Institute of Electronic Engineering (ECRIEE), Hefei.

VERIFIED

JY-17A ground surveillance radar

Type

I-band (8 to 10 GHz) ground surveillance radar.

Description

JY-17A is a modular, fully coherent, pulse-Doppler battlefield surveillance radar that is designed to detect, locate and identify moving ground or low-altitude air targets. The equipment can be vehicle mounted or ground deployed and features a solid-state, low probability of intercept transmitter; a high stability frequency synthesiser; selective linear and circular polarisation; digital pulse compression; pulse-Doppler filter bank processing; a raster scan display and automatic target detection and tracking. Display format options comprise terrain profile (ground clutter) and moving target; plot and track; electronic map with moving targets; composite (built-in test, target data and system status); zone enlargement/zone alarm and target data storage/replay.

Status

As of this edition, JY-17A is thought to be available.

Specifications

Frequency: 8 to 10 GHz
Range: (all P_d 80%, P_{fa} 10^{-6}): 10 km (single pedestrian); 15 km (light vehicle); 20 km (helicopter); 25 km (tank/heavy vehicle); 30 km (ship)
Sector axis: 0 to 360° (selectable)
Sector width: 60, 90, 120, 150 or 180° (selectable)
Accuracy: ±10 mils RMS (azimuth); ±10 m RMS (range)
Track processing: 10 tracks (automatic extraction)
Power consumption: < 300 W
MTTR: ≤ 0.5 h

Contractor

ECRIEE (East China Research Institute of Electrical Engineering), Hefei.

VERIFIED

Model 378 battlefield radar

Type
I-band (8 to 12 GHz) portable detection radar.

Description
The Model 378 is a portable battlefield radar operating in I-band which is used for the detection of vehicles and ships. It is a coherent-pulse-Doppler system which consists of an antenna/transceiver mounted on a tripod, a rectifier, a signal processor, a generator and a display. After detecting the target, a visual and audible alarm is activated and the position of the target is displayed on the screen. Although claimed to be a portable system, the total weight of 86 kg would appear to require the use of a vehicle to transport it, particularly with the use of a generator for the power source.

Status
As of this edition, the Model 378 is thought to have been procured by the People's Liberation Army.

Specifications
Frequency: I-band (8-12 GHz)
Max range: 6 km (personnel); 20 km (vehicles)
Accuracy: 30 m (range); 1.5° (azimuth)
Discrimination: 60 m (range); 3° (azimuth)
Power output: 4,800 W peak
Pulse Repetition Frequency: 3,125 ± 5 Hz
Polarisation: horizontal
Beamwidth: 3.2° (horizontal); 5.5° (vertical)
Weights: antenna/transceiver 18 kg; display set 17 kg; rectifier 11 kg; processor 5 kg; generator 30 kg; tripod 5 kg

Contractor
China Fujian Radio Equipments Factory, Taijung.

VERIFIED

The units making up the Model 378 battlefield radar

Model ST-312 ground surveillance radar

Type
Medium-range ground surveillance radar.

Description
ST-312 is a medium-range ground surveillance radar that is designed to detect, identify and locate moving ground targets and low-flying helicopters. System features include a low-noise, wide dynamic range receiver; digital signal processing; coherent pulse-Doppler technology; microprocessor control and a TV scanning, multifunction display. ST-312 can be tripod mounted or deployed from a vehicle using a tower-mounted antenna.

Status
As of this edition, ST-312 is understood to be available.

Specifications
Frequency: 8-12 GHz
Range: 15 km (pedestrian); 25 km (light vehicle); 40 km (tank/truck); 35 km (helicopter)
Accuracy: 0.5° (azimuth); 12 m (range)
Resolution: 2.8° (azimuth); 50 m (range)
Receiver: < −3 dB (noise); 6 MHz (IF bandwidth); 78 dB (dynamic range); < -135 dBW (sensitivity)
Transmitter: 0.33 µs (pulse duration); > 3 kW (peak power); 3.125 kHz (PRF)
Antenna: 2.8° (horizontal beam width); 4° (vertical beam width); > 32 dB (gain)

Antenna polarisation: linear vertical or circular
Environment: −20 to +50°C (operating)
Power consumption: < 200 W
Weight: < 90 kg (tripod mounted)

Contractor
China National Electronics Import and Export Corporation, Beijing.

VERIFIED

MW-7-JB fire-control radar

Type
I/J-band (8 to 20 GHz) mobile anti-aircraft fire-control radar.

Description
The MW-7-JB is a mobile artillery fire-control radar operating in the I/J frequency band. It appears to be a later variant of the MW-5 system. All the main elements are contained in or on a four-wheeled trailer, consisting of a container which also serves as an operations cabin for the crew.

A circular dish antenna for target search and tracking is mounted on the roof of the cabin, using a conical scan technique for target tracking. The antenna can be raised and lowered as required. It is stated to have a diversified countermeasures system, including wide range frequency agility. Other features include moving target indicator processing, automatic change of operation mode and an interface for an identification friend-or-foe system.

Status
As of this edition, MW-7-JB is thought to have been procured by the People's Liberation Army.

Specifications
Frequency: I/J-band (8-20 GHz)
Agility band: 4% of centre frequency
Detection range: 55 km
Tracking range: 35 km
Tracking accuracy: 20 m (range); 1.6 mil (azimuth); 1.8 mil (elevation)
Angular speed of tracking: 600 m (range); 15°/s (elevation); 30°/s (azimuth)
MTBF: 40 h

Contractor
The Huanghe Machine Building Factory.

VERIFIED

The MW-7-JB fire-control radar deployed

SJ-202 target acquisition/tracking radar

Type
Target acquisition/tracking radar.

The SJ-202 target acquisition radar

Description

The SJ-202 is understood to be the acquisition/tracking radar for use with the HQ-2 series of surface-to-air missile systems. The equipment utilises an automated radar van together with a command and control vehicle. Maximum detection range is 115 km with a tracking range of 80 km. The SJ-202 operates in conjunction with the Gin Sling radar.

Status

As of this edition, SJ-202 is thought to have been procured by China's armed forces.

Contractor

China Precision Machinery Import and Export Corporation, Beijing.

VERIFIED

Type 311-A/B/C fire-control radars

Type

I/J-band (8 to 20 GHz) fire-control radar for anti-aircraft guns.

Description

The Type 311 series fire-control radar is for use with anti-aircraft guns and is normally employed with batteries of either 37 mm or 57 mm calibres. It operates on I/J-band frequencies and is capable of both search and target tracking functions and is generally used with a computer and an optical rangefinder.

The Type 311 consists of an operations trailer with the main electronic and mechanical elements of the radar and a towing vehicle in which the operators are carried, together with a power generating set, tools and spare parts. The antenna is mounted on a pedestal on top of the trailer. The complete system can be set up or dismantled to move within about 15 minutes and the weight of the radar trailer is less than four tons. The towing vehicle with equipment weighs under eight tons.

Detection range on a fighter-size aircraft target in the search mode is at least 30 km, with a maximum tracking range of 25 km. Minimum reliable tracking range is about 500 m. The radar can be switched to any one of three preprogrammed operating frequencies by the operator and target position data in the tracking mode can be fed to the computer in either rectangular or spherical co-ordinates. In the search mode, the radar beam is oscillated in the vertical plane at a rate of 4 Hz to broaden the effective beam. A target within this beam that subsequently deviates in either elevation or azimuth by 20 mils or more from the radar boresight axis can then be tracked automatically.

Further development of the Type 311 has led to the 311-B and 311-C. The Type 311-B introduces an integral identification friend-or-foe subsystem, increased frequency coverage and a maximum range of 35 km using a new antenna design. The Type 311-C goes one stage further and has a frequency agile radar with a maximum range of 40 km.

Status

As of this edition, Type 311 series radars are thought to have been procured by the People's Liberation Army and the armed forces of a number of other countries.

Specifications

Operating frequencies: I/J-band (8-20 GHz – 3 switchable sub-bands)
Transmitter
Type: magnetron
Peak power: 200 kW (0.3 μs pulse); 180 kW (0.9 μs pulse)
Pulse-width: 0.3 μs (narrow); 0.9 μs (broad)
PRF: 2,500 Hz (narrow pulse); 833 Hz (broad pulse)
Antenna gain: at least 35 dB

Horizontal beamwidth: 2.6°
Vertical beamwidth: 2.4°
Sidelobe: −18 dB (max)
Receiver sensitivity: −92 dB/mW (CW)
Max range: Type 311-A 30 km; Type 311-B 35 km; Type 311-C 40 km; 25 km (tracking)

Contractor

China National Electronics Import and Export Corporation, Beijing.

VERIFIED

Type 313 fire-control radar

Type

I/J-band (8 to 20 GHz) fire-control radar for anti-aircraft guns.

Description

The Type 313 is a development of the Type 311 described previously. It is intended primarily for use with up to eight 37 mm Type 74 air defence guns, but can also be configured for other applications according to a user's requirements.

The system combines an I/J-band radar with a TV tracker and a laser rangefinder. With a theoretical maximum range of 35 km, a 90 per cent probability of detection is claimed against a target of 2 m² or more at up to 25 km. A version, known as the Type 313A, which has a frequency agile radar has also been developed.

Status

As of this edition, Type 313 is thought to have been procured by the People's Liberation Army.

Contractor

Chuanbei Electronic Industry Co, Beijing.

VERIFIED

Type 702 air defence fire-control system

Type

G/H/I-band (4 to 10 GHz) automatic anti-aircraft fire-control system.

Description

The Type 702 is a mobile automatic all-weather ground-to-air fire-control system consisting of radar, optical unit and computer. Its primary use is against aircraft attacking from low and medium altitude, but it can also be used against ground and sea surface targets. The system will operate with various calibre anti-aircraft guns or surface-to-air missiles. It consists of two wheeled vehicles (a towing van and a trailer) with the antenna mounted on the trailer and folded down for transit. Operating crew is three.

The Type 702 comprises mainly the radar system for both search and tracking roles, TV tracking device, computer and power supply. The fire-control radar is used for fast capture and tracking of the target at low and medium altitude. It determines continuous information on the azimuth angle, elevation angle and slant range of the target and passes this information to the fire-control computer. Automatic target search is in the G/H-band, with tracking in either G/H- or I-bands.

The Type 311-A fire-control radar

The Type 702 air defence fire-control radar

Status

As of this edition, Type 702 is thought to have been procured by the Chinese armed forces, as part of the 37 mm Type 80 weapon system.

Specifications

Frequency: 4-10 GHz
Antenna range: 360° (azimuth); −3 to +87° (elevation)
Detection range: 40 km
Tracking range: 0-40 km
Tracking accuracy: 15 m (range); 1.5 mil (angular)
Tracking speed: 600 m/s (range); 40°/s (elevation); 60°/s (azimuth)
Transmitter: automatic pulse-to-pulse frequency-hopping and self-adaptive frequency agility; G/H-band (4-8 GHz) coaxial magnetron; I-band (8-10 GHz) TWT amplifier chain
Bandwidth: 200 MHz (G/H-band); 1,000 MHz (I-band)
Beamwidth: 1.4° × 1.4° (I-band); 2.2° × 2.2°/2.2° × 22.5° (G/H-band)
MTBF: 100 h
Weight: 5.5 t

Contractor

China North Industries Corporation, Beijing.

VERIFIED

Type 704 artillery locating radar

Type

I/J-band (8 to 20 GHz) weapon location and fire correction radar.

Description

The Type 704 radar is used for the location of hostile artillery (including mortars, guns, howitzers and rocket launchers). It can also be used for tracking friendly artillery fire, calculating the impact error and providing automatic correction parameters. The system features multitarget ability, wide sector scan range, effective electronic counter-countermeasures ability, automatic operation and easy maintenance. The system response time from detection of a projectile to providing location when using automatic altitude correction is given as eight to nine seconds.

The transmitter is a coherent amplifier chain system which uses a high stability frequency synthesiser as its frequency source. A travelling wave tube with an average power of 300 W is used in the end stage to provide range in the I/J-band. The receiver has a field-effect transistor high-frequency amplifier with high sensitivity and low noise characteristics at the front end. Three computer systems are included, one for operations control and data processing; one for antenna beam steering, azimuth elevation and azimuth data extraction and transceiver/antenna vehicle control and the third for failure checking of the operations vehicle.

The Type 704 is located on two vehicles, an antenna/transceiver vehicle and an operations vehicle. The first carries the phase-frequency scanning array, transmitter, receiver and beam steering computer, with the antenna folded flat during transit. The second vehicle consists of two cabins, one for operations control, data processing, displays and so on and the other containing the power supply. Three crew members are needed; a driver, power supply operator and radar operator, although only the latter is required during operation.

Status

As of this edition, Type 704's status is uncertain.

Specifications

Frequency: I/J-band (8-20 GHz)
Antenna
Type: electronic scanning in phase frequency scan mode
Azimuth: ±45° (electronic scan); ±60° (rotation)

The Type 704 artillery locating radar

Elevation: −0.5 to +27° (tilt angle); −0.5 to +90° (rotation); 6° (beamwidth)
Range: 12 km (81 mm mortar); 15 km (120 mm mortar); 16 km (155 mm howitzer)
Location accuracy (CPE): 30 m (mortar); 40 m (flat trajectory gun/registration fire); 60 m (rocket launcher)
Target capacity: 8
Weights: antenna/transceiver vehicle 3,820 kg; operations control vehicle 3,920 kg

Contractor

China North Industries Corporation, Beijing.

VERIFIED

Type 706 (IBIS) ultra-low level search radar

Type

Ultra-low level search radar for air defence.

Description

The Type 706 (IBIS) search radar is a frequency agile air defence system designed for the detection of low-level attack aircraft. It uses fully coherent pulse-Doppler techniques and digital signal processing to enable it to detect and track 10 different types of target which are operating at ultra-low level in difficult topography. It provides multitarget recognition and tracking with a good electronic counter-countermeasures performance and is particularly suitable for use with anti-aircraft weapon systems. It consists of an operations cabin, with the antenna mounted on top, which is transported by a suitable flatbed truck or trailer. The IBIS system can be operated whilst on the vehicle with the legs providing stability, or the vehicle can be removed (as illustrated). The antenna can be raised or lowered to suit the particular terrain.

The transmitter is a coherent type with characteristics of high stability, high purity and low phase noise. It employs a format of main oscillating amplifier chain and a dual-stage grid-control travelling wave tube amplifier. Peak power is approximately 60 kW with a working bandwidth of ±5 per cent. The digital signal processor provides advanced moving target detector and a clutter pattern memory and post-processing system. The data processor has the capability of multitarget adaptive recognition and flight path tracking. It carries out auto-extract and semi-auto-extract of target data recording with a capacity of 200 batches, automatically or semi-automatically as required. Target flight path tracking has a maximum capacity of 20 batches.

The display system consists of two displays, a plan position indicator with a 31 cm cathode ray tube that can display the primary or secondary information of the radar and an alphanumeric display to provide information on target parameters, operation mode, and fault detection and analysis.

Status

As of this edition, Type 706 (IBIS) is thought to have been procured by China's armed forces.

Specifications

Antenna max working height: 12 m
Revolution: 27 rpm
Power output: 60 kW peak
Range: to 40 km
Height coverage: 50-4,000 m

The Type 706 IBIS ultra-low-level search radar

Bearing: 360°
Beamwidth: 3° (horizontal); 45° (vertical)
Resolution: 500 m (range); 3° (azimuth)
Accuracy: 400 m (range); 0.5° (bearing)
Continuous operating time: 12 h
Radar response time: 4-8 s

Contractor
China North Industries Corporation, Beijing.

VERIFIED

GERMANY

BOR-A 550

Type
I-band (9.5 GHz) wide area ground and coastal surveillance radar.

Description
BOR-A 550 is a dual-role solid-state radar that is designed for border/battlefield/coastline surveillance, intruder detection and helicopter navigation aid tasks. It comprises a Radio Frequency (RF) sensor package and an operator unit. The RF sensor package may be broken down into RF and signal processing units while the operator unit comprises a high-resolution, colour, liquid crystal display and a 254 mm fold-down, watertight operator console. Deployment options comprise stand-alone, remote control, vehicle-/shelter-mounted (8 m mast) or trailer-mounted (up to 17 m mast). Technology incorporated in the radar includes pulse compression, spread spectrum, frequency agility, pulse repetition frequency stagger, constant false alarm rate and a neural network for automatic target classification. The system offers automatic tracking and target classification (personnel, wheeled vehicles, tracked vehicles and helicopters on or close to the ground; buoys, inflatables, fishing boats/trawlers, fast patrol boats and large ships at sea) and has four operating modes:
Mode 1: automatic 360° or sector search for moving targets.
Mode 2: antenna controlled in range and bearing by the operator.
Mode 3: small angle search of restricted area to locate, for instance, tracked vehicles which have stopped or disappeared behind an obstacle in the field of view.
Mode 4: automatic tracking in range and bearing of a single moving target.

Alcatel SEL claims that BOR-A 550 has a mean time between failure value of more than 2,000 hours and notes that the equipment is suitable for networked operations within a command, control, communications and intelligence system.

Status
BOR-A 550 was launched during October 1996 and, as of this edition, was understood to have entered series production during the late 1990s.

Specifications
Frequency: I-band (9.5 GHz)
Instrumented range (area): 0-5 km; 0-20 km; 5-25 km; 20-40 km, 0-40 km
Detection range (90% detection probability): 15 km (personnel/small boat); 25 km (light vehicle); 28 km (helicopter); 35 km (fast patrol boat); 38 km (main battle tank); 40 km (vehicle convoy/large ship)
Accuracy: 12 m (range); 0.6° (azimuth)
Resolution: 40 m (range – window); 160 m (range – overview); 2.7° (azimuth)
Deployment time: 5 min
Environment: –33°C to +65°C; up to 4,500 m location altitude; 90 km/h windspeed; waterproof; complies with MIL-STD 810E

Contractor
SEL Defense Systems, Stuttgart.

VERIFIED

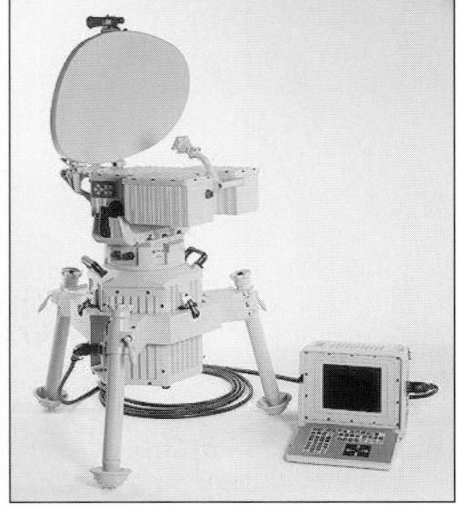

The BOR-A 550 dual-role surveillance radar together with its operator unit

RATAC-S battlefield radar

Type
I-band (8 to 10 GHz) multipurpose battlefield surveillance and target acquisition radar.

Description
RATAC-S is a highly mobile, coherent pulse-Doppler radar with a monopulse antenna configuration used to detect, locate, classify and observe ground targets or low-flying airborne targets. It combines a medium/long-range battlefield surveillance with a high precision artillery fire-control system within one unit. The monopulse principle guarantees the highest angular accuracy, independent of the fluctuation behaviour of the target.

RATAC-S is designed using highly integrated standard components and microprocessors. Considerable progress has been made in a number of fields to give small size, light weight, low power consumption and improved performance. It is a standard equipment for use by army, artillery, border guards, coastguards and paramilitary forces. It combines the functions of surveillance with moving target alarm, target acquisition and classification, automatic target tracking, plotting of target on screen and artillery fire direction into a single portable and simple to operate radar.

The system requires only one operator. Data received by the radar are displayed to the operator in a very comprehensible form, so that an untrained operator can set up and operate RATAC-S after only a short introduction with the simple operating procedures in menu technique, radar map display, automatic observation alarm, target co-ordinate display and built-in test messages in clear words.

RATAC-S consists of only two units, operator unit and Radio Frequency (RF) sensor unit, which are portable and can be deployed quickly for stand-alone operation. Because of its small volume it can also be integrated into the smallest of vehicles. The operator unit contains the console, signal processing, computer and display. The RF sensor unit is subdivided into two units for transport reasons: the antenna compartment containing the antenna, transmitter and receiver and the base compartment with the azimuth drive and control circuitry.

RATAC-S incorporates a compact high-frequency module which consists of a rugged magnetron transmitter with modulator and a high-sensitivity, low-noise monopulse-Doppler receiver with an intermediate frequency of 120 MHz, which avoids interference with the VHF (30 to 300 MHz) communications channels. When jamming or interference is experienced this can normally be overcome by selecting an alternative frequency from the built-in, low-phase-noise coherent 10-channel frequency synthesiser.

The entire system is controlled by a central computer in the operator unit with a co-processor in the antenna compartment. The microprocessor and operational software used enable the elimination of many manual controls, simplify operations and allow a considerable reduction in the number of components. In addition, the system has an extensive growth potential without the need for hardware change.

RATAC-S provides five modes of operation:
Mode 1: automatic scanning with a range of up to 38 km and a sector scan width between 30 and 140°, selectable within 360°. The operator can set direction, range, width, elevation and scan rate of the sector
Mode 2: target acquisition and acoustic target classification up to a range of 38 km, and automatic audio/visual moving target alarm. In this mode the automatic scanning is switched off and the operator can manually position the beam and classify the target by listening to the Doppler tone
Mode 3: precise manual target tracking and deviation measurements for fire control within an area of ±300 m and ±75 mils and automatic audio/visual moving target alarm

The RATAC-S RF sensor unit
0006885

A close-up of the operator unit used in the RATAC-S system 0006886

Mode 4: automatic target tracking and deviation measurement
Mode 5: automatic target tracking and plotting on the display.

The transformation of co-ordinates from polar to UTM, or vice versa, is performed by software in the central computer without the need for additional circuitry. All relevant co-ordinates are automatically presented in allocated text fields on the display.

The use of a graphic display system, consisting of a graphic processor, image storage unit and screen monitor, eliminates the need for a storage cathode ray tube. The graphic processor allows display of various markers and cursors to simplify operation. The digital target image processing enables fixed targets to be stored and redisplayed (radar map), superimposed on the current display at reduced brightness. Target tracking, accurate target positioning and deviation measurements of artillery shell impacts, acoustic classification of targets, continuous and interruptive BITE testing and an alarm facility are provided by the five operating modes.

Status
As of this edition, RATAC-S is understood to have been built under licence in Turkey as the ASKARAD system. In excess of 300 RATAC-S radars have been sold worldwide and the equipment forms part of the German Abra and United Arab Emirates (UAE) Mobile Optronic and Radar surveillance (MORS) systems. Looking at these in more detail, Jane's sources suggest that the Abra system incorporated a total of 38 RATAC-S radars mounted on M113 armoured personnel carriers while the SEL Defense developed MORS is fitted to the UAE AADV armoured artillery observation vehicles.

Specifications
Frequency: I-band (9.5 GHz)
Antenna
Polarisation: vertical or circular
Azimuth scan speed: 7.8°/s (fast); 2°/s (slow)
Elevation angle: ±24°
Range (90% detection probability): up to 8 km (105 mm shell); up to 15 km (155 mm shell); up to 18 km (soldiers); up to 24 km (light vehicle – Jeep); up to 28 km (helicopter); up to 30 km (main battle tank); up to 38 km (vehicle convoy)
Accuracy: ±10 m (range); ±0.6° (azimuth – modes 1/2); ±0.11° (azimuth - mode 3); ±0.17° (azimuth – modes 4/5)
Power output: 7 kW (peak)
System weight: 125 kg
Environment: −32°C to +55°C

Contractor
SEL Defense Systems, Stuttgart.

VERIFIED

INDIA

PIF 518 field artillery radar

Type
I/low J-band (8 to 12 GHz sub-band) mortar location and artillery fire direction radar.

Description
PIF 518 is an integrated radar and computing system which is designed for mortar location and artillery fire direction. Bharat describes the equipment as being similar to the Cymbeline radar it manufactured under licence from UK contractor Thorn EMI (now Racal Thorn Defence). PIF 518 comprises a generator set; power converters; control logic; radar timer; modulator; transceiver; scanner; feed horns; reflector; range and azimuth marker generators; display; resolver transformer; elevation and azimuth resolvers; computer and mortar position indicator. The

The PIF 518 in its trailer-mounted configuration 0006887

equipment is broken down into two major assemblies; a system mounting and a radar head. The mounting assembly (which sits on the ground or is installed on a trailer which is fitted with four levelling/stabilising jacks) incorporates the necessary azimuth drive, ring and computing components. The radar head is made up of transceiver/computer and antenna reflector/scanner/elevation resolution packages. For operator training, PIF 518 includes a target simulation routine (with random delay to overcome operator conditioning) together with built-in test.

Status
As of this edition, PIF 518 is understood to have been procured by the Indian Army.

Specifications
Frequency: 8-12 GHz (adjustable to 1 of 4 discrete sub-bands)
Antenna: Foster scanner with 5 switched elevation beam positions
Peak power: 100 kW
Range: 0-20 km
Azimuth limit: not less than 12,000 mils
Sector scan/sector scan rate: 720 mils/16.6 mils/s
Elevation: −90 to +360 mils
Vertical beam separation: 25, 40, 45, 65 or 90 mils (dependent on mode)
Display: B-type (127 mm^2 CRT display)
Supply voltage: 22-33 V DC (28 V nominal)
Power consumption: 1,200 W (mean); 1,350 W (peak)
Temperature range: −32 to +52°C
Wind velocity: 90 km/h (with tethering)
Altitude: 2,500 m
Dimensions: 1.70 × 1.50 m (unmounted); 1.78 × 2.72 m (trailer mounted)
Weight: 390 kg (unmounted); 980 kg (trailer mounted)

Contractor
Bharat Electronics Ltd, Bangalore.

VERIFIED

PIW 519 weapon control radar

Type
I-band (8 to 10 GHz) and K-band (20 to 40 GHz) anti-aircraft weapon control system.

Description
The PIW 519 is a mobile land-based search and tracking radar which meets the requirement for an anti-aircraft weapon control system. It operates in both I- and K-bands and provides all-weather point and area defence with anti-aircraft guns and surface-to-air missiles against medium- to very low-level air attack. It provides automatic acquisition and tracking while scanning for additional targets. Up to three weapons can be controlled, either three medium-calibre guns, or two guns and one missile. The complete system is mounted on a single trailer, with retractable antennas positioned on top of the operations cabin.

The PIW 519 AA weapon control system

The basic functions of the system are:
- air search and single sweep acquisition in I-band
- moving target indication
- automatic interrogation of the tracked target
- fast and accurate lead angle computation of tracked target
- gun assignment and firing initiation.

The target being tracked is presented to the operator on a TV monitor to assist them in visual target identification, threat assessment and engagement monitoring. PIW 519 is manufactured under licence from Hollandse Signaalapparaten BV.

Specifications

Search antenna
Type: I-band slotted waveguide with IFF dipoles
Horizontal beamwidth: 1.1°
Polarisation: horizontal/circular selectable
Rotating speed: 40 rpm
Tracking antenna
Type: monopulse for I-band; conical scan for K-band
Beamwidth: 2.4° for I-band; 0.6° for K-band
Search coverage: up to 20 km (1 m² target) search-while-track
Search/tracking transmitter
Peak power: 220 kW for I-band
Frequencies: choice of 6 fixed frequencies in I-band
Pulse-length: 0.2 μs
PRF: 4,800-6,000 Hz
K-band transmitter
Peak power: 15 kW
Frequency: 35 GHz approx
Pulse-length: 0.14 μs
PRF: 2,400-3,600 Hz
Weight: 2,750 kg

Contractor
Bharat Electronics Limited, Bangalore.

VERIFIED

INTERNATIONAL

AN/MPQ-64 Sentinel surveillance radar

Type
8 to 12.5 GHz band air defence radar.

Description
The Thales Raytheon Systems' (formerly Raytheon Electronic Systems) AN/MPQ-64 is a new generation, 8 to 12.5 GHz band, 3-D radar for the US Army Forward Area Air Defense System (FAADS). It is used to generate track data to inform FAADS weapons of the location of targets approaching their front-line forces. Based on the TPQ-36A radar, the AN/MPQ-64 is the key to air surveillance and provides target acquisition/tracking information for division and corps weapons.

The system consists of the radar equipment with its prime mover/power source, an integral identification friend-or-foe subsystem and FAAD command, control and information interfaces. The radar uses advanced phased-array technology to detect, track, classify, identify and report targets (fixed- and rotary-wing aircraft, cruise missiles and unmanned aerial vehicles). Targets can be hovering to fast moving, from nap-of-the-earth to the maximum engagements altitude of FAAD weapons.

The AN/MPQ-64 is virtually identical to the TPQ-36A system and has about 90 per cent commonality with the AN/TPQ-36 Firefinder system. The computer control software has been altered from that required to detect weapon-fired projectiles with small radar cross-sections to the requirements of a FAADS system. In addition, non-co-operative target recognition technology has been incorporated. The system can

The AN/MPQ-64 ground-based sensor with its generator mounted on a light truck

operate as a stand-alone radar, or as part of an additional threat location radar giving direct-to-weapons cueing functions. As part of an integrated network the mobility of the AN/MPQ-64 means that it can also be employed as a gap filler.

Clutter rejection is accomplished by combining range-gated pulse-Doppler techniques with moving target indication, so enabling the system to detect and track aircraft travelling at less than 40 kt. In addition, the system is equipped with special Doppler analysis techniques to identify rotor movements and detect hovering or low-speed helicopters. Because stationary clutter is rejected the radar is able to look down from a high vantage point to detect low-flying threats.

The radar computer is able to adjust the waveform of each tracking beam, thereby ensuring that targets can be detected in a clutter free range-gated pulse-Doppler filter. This means that the system can detect low-altitude aircraft or missiles while at the same time producing a very low false plot track reporting rate.

A single AN/MPQ-64 radar is capable of tracking more than 50 targets simultaneously and also of prioritising targets. The operator can override the system manually to examine targets of special interest.

Status
The first production configuration MPQ-64 radar was delivered to the US Army in June 1993 and the first operational 'light' AN/MPQ-64 mounted on a High-Mobility Multipurpose Wheeled Vehicle (HMMWV) was rolled out during July 1994. Over time, the radar has demonstrated the provision of 3-D acquisition, tracking and target engagement data for Bradley Stinger Fighting Vehicles and the Roadrunner (Chaparral), Avenger and Stinger MANPADS weapon systems. Three prototype MPQ-64 systems are also known to have been used to monitor air activity over the venues used to host the 1996 Olympic Games in Atlanta, Georgia.

MPQ-64 has also been integrated with the HAWK and Advanced Medium Range Air-to-Air Missile (in Its ground-launched version) systems as well as air defence guns. The TPQ-36A version of MPQ-64 has been supplied to Norway for use in the Norwegian Adapted HAWK (NOAH) programme. Norway is also understood to have acquired MPQ-64 radars for use in the Norwegian Advanced Surface-to-Air Missile System (NASAMS). NASAMS incorporates a ground-launched AIM-120 missile, the MPQ-64 and a fire direction system developed by Kongsberg Gruppen.

On 26 May 1998, the then Raytheon Electronic Systems was awarded a US Army 10 year US$44 million plus contract covering the supply of 27 MPQ-64 radars and related support activity. Jane's sources suggest that as of 1998, the US Army was procuring at least 100 MPQ-64s against a total regular Army/National Guard requirement of 208 such radars. As of late 2000, MPQ-64 was expected to be able to transmit digitised target data to the US Army's Avenger Block I surface-to-air missile system which was scheduled to enter service during the third quarter of 2001. Returning to 1998, MPQ-64 is also reported to have been under consideration for use in the Dutch Army's air-defence orientated Target Information Command and Control System (TICCS).

Specifications
Frequency band: 8-12.5 GHz
Coverage: 22° (elevation, selectable within −10 to +55°); 360° (azimuth); adjustable horizon mask
Beamwidth: 1.8° (elevation); 2° (azimuth)
Scan rate: 30 rpm (electronic back scan)
Range: 30 km (cruise missile); 40 km (fighter); 75 km (bomber)
Altitude: 15,240 m
Accuracy: 0.2° RMS (azimuth/elevation); 40 m RMS (range)
Mobility: 5 minutes (march order); 15 minutes (emplacement)
Secondary radar: AN/TPX-56
Temperature: − 46°C to + 52°C (operating); − 46°C to + 71°C (non-operating)
Altitude: up to 3,048 m (operating); up to 12,192 m (non-operating)
Wind: 81 km/h (operating - with gusting up to 121 km/h); 105 km/h (non-operating - with gusting up to 161 km/h)
Antenna/transceiver trailer dimensions (h × w × l): 3.4 (antenna erected)/2.4 (antenna stowed) × 2.2 × 3.4 m
Antenna/transceiver trailer weight: 1,742 kg

Contractor
Thales Raytheon Systems, Fullerton, California.

UPDATED

AN/TPQ-36(V) Firefinder weapon locating radar

Type
8 to 12.5 GHz band weapon locating radar.

Description
The Thales Raytheon Systems' (formerly Raytheon Electronic Systems) AN/TPQ-36(V) is an artillery, rocket and mortar locating radar and was originally designed as a replacement for the AN/MPQ-4 and older technology weapon locating radars. Location of artillery at ranges beyond the capability of the AN/TPQ-36(V) is provided by the AN/TPQ-37, the other radar that makes up the Firefinder system. Using only a different computer software program, the same operations shelter can be used for either Firefinder radar.

Location of hostile artillery and mortars by the AN/TPQ-36(V) is completely automatic. The system electronically, scans the horizon over a 90° sector several times a second, intercepting and automatically tracking hostile projectiles, then computing back along the trajectory to the origin. The co-ordinates and altitude of the weapon are then presented to the operator. Automatic location is so rapid that the co-ordinates of the firing weapon are normally with the operator before the enemy round lands.

In addition to its ability to locate artillery, a number of other modifications have been incorporated. The normal 90° sector can be expanded to as much as 360° for use in insurgency operations. The radar can also provide information on the hostile weapon's target by extrapolating the trajectory to the impact point, allowing the information to be used in the priority of return fire.

A very important feature for the modern hostile battlefield is the capability to track and locate weapons firing simultaneously from different locations. Using separate track channels, the radar can track several projectiles at once while continuing to scan for other projectiles in its 90° sector. A further feature allows the radar to ignore subsequent firings from weapon positions already located. As many as 39 different weapon positions simulating an artillery preparation for attack have been used in tests to demonstrate the capability of the system to provide numerous locations in a short period of time.

High mobility allows the system to be used as close as 2 km to the front line. Emplacement takes 15 minutes and displacement 5 minutes or less. The system can be mounted in a single vehicle plus trailer. Since the operations shelter weighs only 2,500 lb (1,136 kg), vehicles as small as 1½ tons can carry the shelter and pull the trailer.

The AN/TPQ-36(V) is a coherent, electronically scanned, range-gated pulse-Doppler radar. Beams are moved electronically in the horizontal dimension by phase shifts and in the vertical dimension by frequency shift. Ground, sky and electronic clutter are filtered by a signal processing software system that discriminates hostile projectiles from aircraft, birds, insects or other interference. The three-dimensional radar then tracks the projectiles, computing the firing point and providing the information to the operator within seconds of the firing. The extent of automation allows a single operator to perform all functions necessary to locate the hostile weapon and inform the counterfire systems.

Fault detection sensors constantly monitor the performance of each subsystem and provide status information. Should a failure occur, built-in test equipment isolates the problem and informs the operator of the location of the problem, frequently down to the specific card that must be replaced. These corrective actions should reveal up to 90 per cent of all problems and can normally be accomplished in 15 minutes.

The AN/TPQ-36(V) has proved to be a reliable system. To date each radar is subjected not only to the normal 'burn-in' at +125°F (+52°C) but is screeened for performance at −25°F (−32°C). The result is a reliability in excess of 125 hours, validating an early field test mean time between failure of 93 hours. In addition to the high and low temperature screening, the performance of each radar is checked using live artillery and mortar firings.

A number of important improvements have been incorporated since the original production models. A more mobile version, mounted in a single vehicle only, is (1995) on field trials. This version is smaller, is more automated with auto-level and other facilities and requires a smaller crew. These improvements enable the system to be deployed much faster.

Status
A full-scale three-year production contract was awarded by the US Army to Hughes (subsequently part of Raytheon Electronic Systems) in August 1978 and since that date over 250 systems have been delivered to the US Army and a number of other defence forces. These have included Australia, Canada, Greece, Jordan, South Korea, Netherlands, Pakistan, Saudi Arabia, Singapore, Spain, Thailand and Turkey. This brings the total number of systems delivered or ordered to over 250 of which in excess of 100 were exported. In August 1990, Hughes received a further US Foreign Military Sale contract to supply four TPS-36(V)3 systems to Saudi Arabia. In December 1992, Hughes received an order for nine more radars from an unnamed customer.

Originally configured using the then standard Gamma Goat articulated vehicle, the TPQ-36(V) subsequently utilised a 2½-ton truck to carry the operations shelter and pull the antenna/transceiver trailer. The system's operations shelter has also been configured for carriage on a militarised 1½-ton pick-up truck and the Mercedes Unimog vehicle (Australian application).

The TPQ-36(V) was deployed in the Gulf during Operation 'Desert Storm'. An air defence variant, the TPQ-36A has been successfully demonstrated to have the capability to track such different targets as pop-up helicopters and high-speed fixed-wing aircraft in high-clutter environments. The first contract for production of this variant was awarded in January 1984. The high percentage of logistic commonality between the two systems should provide benefits in logistic support and life-cycle cost-effectiveness. A further air defence variant of TPQ-36(V) is in

AN/TPQ-36 and MLRS deployed during Operation 'Desert Storm'

production for the US Army's Forward Area Air Defense System (see AN/MPQ-64 entry). As of mid-2000, Raytheon was reported as having delivered more than 340 AN/TPQ-36(V) and TPQ-37(V) Firefinder radars to the US Army, the US Marine Corps and 17 international customers (including the possible sale of two TPQ-36(V) 9 radars to Kuwait during 1998/99). A TPQ-36(V) 8 equipment is understood to have been evaluated in the UK during June 1999 as a possible contender for the British Army's 25 to 30 km range Mamba weapon locating radar programme.

Identified TPQ-36(V) contracting activity during 2000-2001 comprised the following:

30 March 2000
The Northrop Grumman Corporation was awarded a then year US$13,627,629 modification to a firm, fixed-price, time and materials contract (DAAB07-99-C-H004) covering the production of upgrade kits for the TPQ-36(V)8 radar. At the time of the announcement of this contract, work on the effort was to be carried at Northrop Grumman's facilities at Rolling Meadows, Illinois (55% work share) and Fleetville, Pennsylvania (45%) and was scheduled for completion on 1 May 2003. The programme's contracting activity was the US Army's Communications and Electronics Command, Fort Monmouth, New Jersey.

31 March 2000
Raytheon Electronic Systems was awarded a then year US$26,593,476 firm, fixed-price contract covering the supply of six TPQ-36(V)9 radars to Egypt. At the time of the announcement of this contract, work on the effort was to be carried at Raytheon's facilities at Forrest, Mississippi (72% work share) and El Segundo, California (28%) and was scheduled for completion on 30 June 2003. The programme's contracting activity was the US Army's Communications and Electronics Command, Fort Monmouth, New Jersey.

7 July 2000
Raytheon Electronic Systems was awarded a time and materials contract (with an estimated then year potential value of US$10,000,000) covering the provision of depot maintenance support services for a number of radars (including the MPQ-84, TPQ-36(V) and TPQ-37(V) types) that were in service with the US military and the armed forces of allied nations. At the time of the announcement of this contract, work on the effort was to be carried at a number of Raytheon's facilities across the USA, (including Marion, Indiana - 80% work share) and was scheduled for completion on 5 July 2005. The programme's contracting activity was the US Army's Communications and Electronics Command, Fort Monmouth, New Jersey.

9 November 2001
Northrop Grumman Electronic Systems sector (Rolling Meadows, Illinois) was awarded a then year US$20,825,996 contract (as part of a firm, fixed-price contract with an estimated cumulative value of then year US$68,409,697) covering the supply of 11 TPQ-36(V)8 radar kits and associated spares. Three contract options provide for the possible procurement of 28 additional kits and work on the described effort was scheduled for completion by 31 January 2006. The programme's contracting activity was the US Army's Communications and Electronics Command, Fort Monmouth, New Jersey.

29 November 2001
Thales Raytheon Systems was awarded a then year, not-to-exceed US$9,500,000 time and materials contract covering technical, logistical and engineering support services for fielded (Foreign Military Sales and US Army) TPQ-36(V) and AN/TPQ-37(V) radars. At the time of its announcement, work on the effort was scheduled for completion by 29 November 2004 and involved contractor facilities at El Segundo, California (75 per cent workshare); Fullerton, California (15 per cent workshare); Indianapolis, Indiana (five per cent workshare) and Forrest, Mississippi (five per cent workshare). The programme's contracting activity was the US Army's Communications and Electronics Command, Fort Monmouth, New Jersey.

Specifications
Frequency band: 8-12.5 GHz (32 frequencies)
Range: 18 km (effective range against artillery); 24 km (effective range against rockets/maximum detection range)
Azimuth sector: 90°
Prime power: 115/200 V AC (400 Hz, 3-phase, 8 kW)
Transmitted power: 23 kW (minimum peak)

Contractor
Thales Raytheon Systems, Fullerton, California.

UPDATED

AN/TPQ-37(V) Firefinder weapon locating radar

Type
2 to 4 GHz band weapon locating radar.

Description
The Thales Raytheon Systems' (formerly Raytheon Electronic Systems) AN/TPQ-37(V) radar is designed to locate hostile artillery and rocket launchers at their normal firing ranges. It is claimed by its manufacturer to be one of the first tactical radars in the world to make use of phased-array techniques for scanning in both azimuth and elevation. As such, the system utilises miniaturised diode phase shifters distributed throughout the antenna.

The TPQ-37(V) radar programme was initiated in response to the threat of massive concentrations of artillery and rockets in support of both offensive and defensive operations. Before this, there had been no effective means to locate and counter indirect fire, with combat troops having to analyse craters to estimate direction and range as a last resort.

The TPQ-37(V) uses a combination of radar techniques and computer-controlled signal processing for detection, verification, tracking of projectiles and extrapolating the track-data points to the location from which the projectile was fired. Once the origin of a projectile has been identified, that location can be provided by voice or digital datalink to the fire direction centre for initiation of counterfire. These techniques are very similar to those applying to the AN/TPQ-36(V) weapon locating radar described previously.

Like the TPQ-36(V), the TPQ-37(V) is a coherent, electronic-scanned, range-gated pulse-Doppler radar. It incorporates monopulse and its beams are moved in both azimuth and elevation by shifting the inter-element phase using diode phase shifters, the first such application in a tactical system. Significant improvements were incorporated into the TPQ-37 before entering production, ensuring state-of-the-art technology, particularly in the antenna, signal processor and computer. Since then, lightweight Kevlar armour plating has been added to protect the antenna.

Operationally the TPQ-36(V) and TPQ-37(V) will be complementary, the former between 1 and 4 km from the forward line of troops to provide for artillery counterfire against mortars and close-in artillery. Three TPQ-36(V) radars will be deployed in each division sector. The TPQ-37(V) will be sited further behind the front line to locate opposing long-range artillery, with two TPQ-37(V)s to each division sector.

Together, the TPQ-36(V) and TPQ-37(V) are referred to as the Firefinder system. The modification of the programme to incorporate an operations shelter that can be coupled to and operate the antenna of either the TPQ-36(V) or the TPQ-37(V) provides unique operational and maintenance advantages.

The TPQ-37(V) consists of two vehicles and an antenna trailer. A five-ton vehicle carries the 60 kW generator and pulls the antenna trailer. The operations shelter is mounted on a 0.5+ ton vehicle. All components are helicopter transportable. Crew size is between 8 and 12, though once emplaced normal operation can be carried out by a single operator.

Under the TPQ-37(V) Block I Pre-Planned Product Improvement (P3I) programme, the radar's range performance against theatre ballistic missiles has been increased to beyond 100 km. Other upgrades incorporated in the Block I P3I included C-130 transportability, increased mobility in sand and mud with new rubber tracks and reduced false target alarms.

Status
As of mid-2000, Raytheon Electronic Systems was reported as having delivered more than 340 AN/TPQ-36(V) and TPQ-37(V) Firefinder radars to the US Army, the US Marine Corps and 17 international customers. Outside the US, TPQ-37(V) customers are thought to include the People's Republic of China (first two radars delivered during 1989), Egypt, India (12 TPQ-37(V)s for delivery between 1998 and 2002), Israel, Jordan, South Korea, the Netherlands, Saudi Arabia, Singapore (three radars delivered in June 1989), Taiwan and Thailand. The US Army is understood to have procured at least 72 radars of this type.

The AN/TPQ-37 weapon locating radar

Improvements in mobility have received high priority among the evolutionary changes incorporated during production. In addition to placing the operations shelter on a 2½ ton truck in place of a 1.5 ton pick-up, the US Army has replaced the mobilisers initially used for the antenna/transceiver with a tandem-wheel trailer which promises much better stability. Mobility and field tests have shown that the antenna/transceiver can be easily mounted on the FVS carrier, the tracked vehicle chassis used for the Multiple Launch Rocket System in production for the US and several NATO nations. TPQ-37(V) was deployed during Operation 'Desert Storm'. It has also been made available for deployment in Bosnia Herzegovina. Identified TPQ-37(V) contracting activity during 2000-2001 comprised the following:

7 July 2000
Raytheon Electronic Systems was awarded a time and materials contract (with an estimated then year potential value of US$10,000,000) covering the provision of depot maintenance support services for a number of radars (including the AN/MPQ-84, AN/TPQ-36(V) and TPQ-37(V) types) that were in service with the US military and the armed forces of allied nations. At the time of the announcement of this contract, work on the effort was to be carried out at a number of Raytheon's facilities across the USA, (including Marion, Indiana - 80 per cent work share) and was scheduled for completion on 5 July 2005. The programme's contracting activity was the US Army's Communications and Electronics Command, Fort Monmouth, New Jersey.

29 November 2001
Thales Raytheon Systems was awarded a then year, not-to-exceed US$9,500,000 time and materials contract covering technical, logistical and engineering support services for fielded (Foreign Military Sales and US Army) TPQ-36(V) and AN/TPQ-37(V) radars. At the time of its announcement, work on the effort was scheduled for completion by 29 November 2004 and involved contractor facilities at El Segundo, California (75 per cent workshare); Fullerton, California (15 per cent workshare); Indianapolis, Indiana (five per cent workshare) and Forrest, Mississippi (five per cent workshare). The programme's contracting activity was the US Army's Communications and Electronics Command, Fort Monmouth, New Jersey.

Specifications
Frequency band: 2-4 GHz (15 frequencies)
Range: 30 km (effective range against artillery); 50 km (effective range against rockets/maximum detection range)
Azimuth sector: 90°
Prime power: 115/200 V AC (400 Hz, 3-phase, 43 kW)
Transmitted power: 120 kW (minimum peak)

Contractor
Thales Raytheon Systems, Fullerton, California.

UPDATED

AN/TPQ-47 Firefinder weapon locating radar

Type
E/F-band (2 to 4 GHz) weapon locating radar system.

Description
The Thales Raytheon Systems' (formerly Raytheon Electronic Systems) TPQ-47 is an upgrade of the AN/TPQ-37 Firefinder radar that is being undertaken under the auspices of the Firefinder Block II Pre-Planned Product Improvement (P3I) programme. As such, TPQ-47 replaces the existing TPQ-37 Antenna Transceiver Group (ATG) and integrates the US Army's Advanced Field Artillery Tactical Data System (AFATDS) software into the system. Alongside the creation of a more survivable, longer range and less manpower-intensive equipment (when compared with TPQ-37), TPQ-47 also offers the ability to detect theatre tactical ballistic missiles at ranges of up to 300 km and to interface with other US theatre missile defence systems. Physically the system comprises a non-developmental ATG trailer, a 2.5 ton Light/Medium Tactical Vehicle (LMTV) prime mover (equipped with a 60 kW Tactical Quiet Generator) and either a vehicle-mounted Operation Central (OC) or a portable operations suite. Of these, the OC is described as being non-mission essential.

Described as leveraging the electronics upgrade undertaken on the TPQ-37(V)8, other TPQ-47 features include:
- the ability to be driven on/off a C-130 transport aircraft or be delivered by CH-47 or UH-60 helicopter lifts
- first round target location
- the ability to identify boosting projectiles
- antenna auto-calibration (post repair)
- on-line, Level 4 Interactive Electronic Technical Manuals (IETM)
- a search and track facility
- target classification by calibre
- an adaptive threat learning capability
- a high level of system automation.

Status
On 3 June 1998, the then Raytheon Electronic Systems was awarded a then year US$800,000 increment to a then year US$57.3 million competitive Firefinder Block II development contract, under which it was to deliver three engineering development model TPQ-47 radars. This was a precursor to a possible US Fiscal

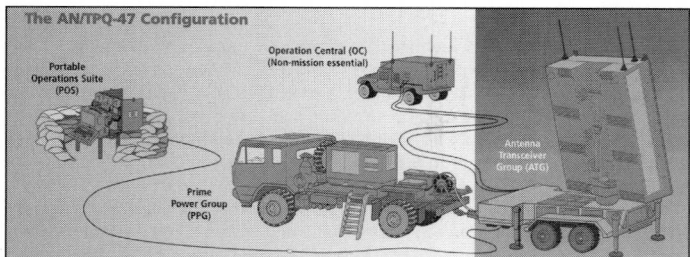

A graphic showing the elements making up the AN/TPQ-47 Firefinder weapon locating radar system   0024877

Year 2002 initial production run of 69 such equipments. TPQ-47 development has been undertaken by a consortium comprising Thales Raytheon Systems (lead contractor), Applied Aerospace Structures Corporation, TRW and the US Army's Tobyhanna Army Depot.

Specifications
Frequency band: 2-4 GHz
Range: 30 km (mortars); 60 km (artillery); 100 km (rockets); 300 km (tactical ballistic missile); 400 km (maximum)
Azimuth sector: 90°
False location rate: one in 12 h
Emplacement time: 15 minutes
Crew: nine
Transport: single C-130 sortie (mission essential equipment) or multiple CH-47 lifts
Prime power: 115/200 V AC (400 Hz, 3-phase, 60 kW)

Contractor
Thales Raytheon Systems, Fullerton, California.

UPDATED

Arabel multifunction radar (land-based application)

Type
I-band (8 to 10 GHz) surveillance, target acquisition, tracking and missile guidance radar.

Description
Arabel is a 3-D, multifunction radar system that is designed for use in either land mobile or naval medium-range, Surface-to-Air Missile (SAM) system applications. The radar performs volumetric surveillance, target tracking, integrated missile datalink guidance (for mid-course correction) and environment mapping. The system's phased-array antenna rotates at 60 rpm to maintain continuous 360° coverage and the data renewal rate necessary for missile guidance. It is also capable of two axis electronic steering to ensure full beam steering flexibility. The combination of frequency agility over a large bandwidth, narrow beamwidth and adaptive steering is claimed to allow Arabel to detect a wide range of threats (including low-flying fixed-wing aircraft, helicopters and small radar cross section missiles) in severe clutter and jamming environments. The system is also noted as having a tactical ballistic missile tracking capability.

Status
As of April 2001, two full-scale engineering development prototypes of the Arabel system had completed acceptance tests in connection with the GIE Eurosam (a consortium of MBDA Missile Systems, Alenia Marconi Systems and Thales Air Defence) medium-range Land Surface-to-Air Anti-Missile (SAAM - also known as

SAMPT) and Naval SAAM systems. The first land-based systems have been ordered by the French and Italian armies where Land SAAM is to replace the HAWK SAM system, while the Naval SAAM system is already in service aboard the French aircraft carrier *Charles de Gaulle*.

Specifications
Frequency: 8-10 GHz
Azimuth coverage: 360°
Elevation coverage: 90°
Detection range: 50 km (missile); 100 km (larger air target)
Track capacity: more than 100
Guidance capacity: up to 16 missiles simultaneously

Contractor
Thales Air Defence, Bagneux, France.

UPDATED

COBRA counterbattery radar

Type
Weapon locating radar.

Description
COBRA (COunterBattery RAdar) is a high-mobility weapon location radar which is being developed for the armies of France, Germany and the UK. It meets all NATO requirements and is designed to detect small cross-section targets across the whole battlefield and to classify ammunition types and firing modes (rockets, swarms, salvos and so on). Integral tracking accuracy combined with a self-positioning system is claimed to enable the system to perform high accuracy location of hostile battery positions and predict impact points.

At the heart of the system is a solid-state, active, modular antenna which is made up of approximately 3,000 gallium arsenide (GaAs) device, low power, transmit/receive modules. Phase to phase steering is employed and sidelobe level gain is controlled by optimising (on command) the system's transmit/receive patterns. Array calibration is automatic.

The COBRA radar is contained in a single cross-country wheeled vehicle which incorporates an operations cabin with the radar antenna mounted on top. The operations cabin houses the system's receiver/processor, command and control subsystem, radar operator's console, command, control and communications console and vehicle intercom. In French service, it is planned to configure COBRA for one-man operation while the German and UK iterations are scheduled to have two operators, enabling the radar to also act as a command post.

System survivability measures include very short transmission times and wideband frequency agility amongst other electronic counter-countermeasures operating modes. The operations cabin is resistant to small arms fire and shell fragments and is provided with nuclear/biological/chemical and electromagnetic pulse protection. Detection range is claimed to be sufficient for the system to function efficiently while being operated outside the reach of enemy batteries.

Status
As of this edition, three COBRA systems have been built and the radar has undergone live fire testing at Meppen in Germany and at Canjuers in France. Prototype equipments have completed British, French and German troop trials and a full-scale production contract for the system was awarded on 6 March 1998. This order is for 29 units, the first of which will be fielded during 2001. Elsewhere, Jane's sources suggest that the system was under test in Switzerland during 1998.

Contractors
Lockheed Martin Naval Electronics and Surveillance Systems - Moorestown, Moorestown, New Jersey, USA.
European Aeronautic, Defence and Space (EADS) Co, Germany.
Thomson-CSF AIRSYS, Bagneux, France.
Thomson Racal Defence, Crawley, UK.

UPDATED

The Arabel radar module antenna   0006989

The COBRA Prototype No 1 in operational mode

Crotale NG surveillance and tracking radars

Type

E/F- (2 to 4 GHz) and J-band (10 to 20 GHz) surveillance, detection and tracking radars for the Crotale New Generation (NG) weapon system.

Description

The Crotale NG radars are an integral part of the Crotale new generation surface-to-air all-weather weapon system.

The surveillance radar is a TRS 2630 Gerfaut system (see separate entry), designed to ensure very short reaction time against low and very low level threats. Its sub-units include:

- a transceiver unit that incorporates a solid-state transmitter
- a digital signal/data processing unit
- a light planar-array antenna that provides enhanced detection range and improved resistance to electronic countermeasures.

The tracking radar is a frequency agile monopulse-Doppler equipment with a high level of anti-clutter and anti-jamming performance. It provides target acquisition and lock on, automatic target range tracking, measurement of the elevation and azimuth deviations between the radar axis and the direction of target and command, line of sight guidance for the system's in-flight VT 1 missiles. Improved electronic counter-countermeasures features include low sidelobes, wideband frequency agility (pulse-to-pulse or burst-to-burst), a constant false alarm rate and jammer tracking.

The monopulse-Doppler radar fitted to the Crotale firing unit   0006991

The latest standard of surveillance unit used in the Crotale system   0006990

Status

As of January 2001, Jane's sources were suggesting that Crotale NG surface-to-air missile systems had been procured by Finland, France, Greece and South Korea (see following). Of these, Finland is reported to have acquired 20 such systems (mounted on Patria XA-180 vehicles) during the early 1990s, while the French procurement is thought to take the form of 12 shelter-mounted systems. The Greek acquisition was announced on 9 October 1998 and covered the purchase of 11 systems at a then year price of US$200 million. During the following year (1999), the South Korean government approved series production of 48 Korean Surface-to-Air Missile (K-SAM) systems, a programme that included the supply and technology transfer of surveillance and fire-control subsystems that were based on those of the Crotale NG. Elsewhere, it has been confirmed that navalised Crotale NG applications have been installed aboard Omani 'Qahir' class corvettes and French 'La Fayette' class frigates.

Specifications

Surveillance radar
Frequency: E/F-band (2-4 GHz)
Antenna: high-gain planar-array
Antenna rotation: 40 rpm
Detection range: >10 km (hovering helicopter); >20 km (combat aircraft)
Automatic tracking: up to 15 tracks with automatic initialisation
Accuracy: 50 m (range, fixed-wing aircraft); 80 m (range, helicopter); 0.4° (azimuth)
Tracking radar
Frequency: J-band (10-20 GHz)
Antenna: high-gain parabolic
Detection range: 20 km
Automatic tracking and missiles guidance (hypervelocity missiles included)

Contractor

Thales Air Defence, Bagneux, France.

VERIFIED

Crotale surveillance and tracking radars

Type

E/F- (2 to 4 GHz) and J-band (10 to 20 GHz) surveillance, detection and tracking radars for the Crotale low-altitude surface-to-air weapon system.

Description

These radars are an integral part of the Crotale low-altitude surface-to-air weapon system that is itself composed of acquisition and firing units. The surveillance radar used is a pulse-Doppler equipment that provides low and very low altitude target acquisition and is mounted on the system's acquisition vehicle. Other system functions include the determination of radar return range and radial speed, target extraction, automatic track-while-scan and the association of identification friend-or-foe responses to interrogated targets. The latest version of the Crotale surveillance radars is based on a large planar antenna and features improved countermeasures resistance and enhanced target detection and designation capabilities.

The Crotale tracking radar is a monopulse-Doppler system that is claimed to have excellent anti-clutter/anti-jamming performance. It is mounted on the firing vehicle and provides target acquisition and lock on and automatic range tracking on a target and up to two in-flight missiles. Here, the system measures the elevation/azimuth deviations between the radar axis, the target and the in-flight missiles.

Status

As of January 2001, Jane's sources were suggesting that over 250 Crotale surface-to-air weapon systems had been procured by a range of countries including Bahrain, Egypt, France, Libya, Pakistan, Saudi Arabia, South Africa and the United Arab Emirates. The same sources were also suggesting that those Crotale systems operated by the French Air Force have been upgraded with a planar antenna, improved radar processing and electronic counter-countermeasures provision and the ability to handle the VT-1 missile. As of the given date, marketing of Crotale was concentrated on logistic support for and upgrades to in-service systems. In this latter context, February 2001 saw Jane's sources reporting that Saudi Arabia had awarded Thales a then year €40 million (US$129 million) contract to upgrade its Crotale systems. Contract elements include system renovation, the supply of spare parts and technical assistance with system upgrading.

Specifications

Surveillance radar
Frequency: E/F-band (2-4 GHz)
Antenna rotation rate: 60 rpm
Target capacity: 30 per antenna revolution
TWS: automatic, 12 targets
Range: 18.5 km on a 1 m² fluctuating target
Altitude: 4,500 m
Tracking radar
Frequency: J-band (10-20 GHz)
Beamwidth: 1.1°

Range: 16 km on a 1 m² fluctuating target
Target speed: 0 to M2.0+

Contractor

Thales Air Defence, Bagneux, France.

....

Cymbeline weapon locating radar

Type

I-/low J-band (8 to 12 GHz sub-band).

Description

Cymbeline is a lightweight, mobile, self-contained radar (including power supply) with a detachable display unit. The radar is mounted on a four-legged structure supported on screw jacks that are fitted with hydraulic absorbers. It is produced in three versions, of which the Mks 1 and 3 are towed on two-wheel trailers and the Mk 2 is mounted on any suitable armoured fighting vehicle or soft-skinned vehicle. The Mk 3 can also be vehicle mounted.

The antenna feed used in the Mks 1 and 2 systems is a mechanical Foster scanner, while that for the Mk 3 is an electronically scanned phased-array. Both illuminate a parabolic cylinder reflector and produce a pencil beam scanning in azimuth. The complete radar head can be rotated to cover any required sector, for example 3,200 mils rotation in 15 seconds. In transit the reflector folds down and the antenna assembly is mounted on top of an equipment box that houses the main electronics unit, the power unit and the display unit during transit. Of these, the main electronics unit contains the transmitter/receiver and the radar timing and computer modules while the radar's display and co-ordinate indicator units can be removed from the equipment box for remote operation for distances of up to 15 m. The display unit consists of a short-persistence 'B' scope and also incorporates all the controls necessary for the operation of the radar. Target location is displayed on the co-ordinate indicator which can be used at distances up to 2 m from the display.

The Cymbeline Mk 3 weapon locating radar

Of the cited variants, the Mk 3 has a number of special features:
- electronically scanned phased antenna system with increased arc of 1,050 mils
- maximum displayed range of 30 km with improved 'B' scope display
- digital computer with liquid crystal display numerical readout
- digital data store as standard fit
- keypad man/machine interface
- greater displacement (30 m) from radar to operator.

The radar enables the operator to plot two points in the trajectory of the projectile and to measure the slant range and bearing to each of these positions. The time taken for the projectile to travel between the two points is also measured and the computer uses this information to extrapolate the firing position. This entire process takes place in about half a minute.

Additional facilities have been provided that are claimed to ensure the maximum accuracy of location and ease of operation over a wide range of operational conditions. For maximum range performance, a switched single beam is used and an additional beam position is available to alert the operator for making the first interception. For short-range work a double-beam mode of operation may be selected to reduce operator reaction time errors. This facility also improves the multiple target capability. Provision has also been made for the internal fitting of an optional digital data storage module. This enables the radar returns to be stored to provide a long-persistence display so that operator concentration can be reduced while improving the marking accuracy. Data storage also improves multiple handling capability.

Status

Cymbeline has been supplied to 18 countries including the UK. As at mid-1994, over 300 Cymbeline systems had been manufactured. Cymbeline Mk 1 is also manufactured under licence by Bharat Electronics, India, under the name MuFAR. As of the mid-1990s, the then Racal (subsequently Racal Defence Electronics and then Thales Sensors) reported continuing export interest in Cymbeline and noted that export sales of the system were then valued at a total of then year £600 million.

A moving target indicator modification has been introduced that is claimed to 'considerably' improve the radar's performance in clutter and electronic countermeasures polluted environments. Over time, Cymbeline systems that have been operated by both the British and New Zealand armies, have been upgraded and UK examples have been deployed to Bosnia to support United Nations operations. In March 1997, the then Racal announced that it had been awarded a then year £4 million contract to support an Egyptian upgrade of 12 of the country's Cymbeline radars. Under the terms of the deal, Egyptian contractor Arab International Optronics provided in-country support and the contract included options to procure additional Cymbelines as well as an integrated logistic support package made up of spares provision and training.

Specifications

Frequency: I-/low J-band (8-12 GHz sub-band)
Display range: 20 km
Measurable range: 500 m (min)
Detection range: 10 km (81 mm finned mortar with better than 40 m CEP location accuracy)
First-time location: 80% probability
Location time: 15 s
Peak transmitter power: 100 kW
Antenna type: triple cone Foster scanner with movable beam switching
Polarisation: circular
Azimuth scan sector: 720 mils
Elevation coverage: –90 to +360 mils
Display: B-scope
Target simulator: incorporated
Power supply: Wankel engine driven 400 Hz 3-phase alternator, full-wave rectified to 28 V DC
Power consumption: 1,200 W (mean); 1,350 W (peak)
Temperature: –32 to +52°C (operating); –36 to +71°C (storage)
Windspeed: 90 km/h (without tethering)
Altitude: 250 m (operating); 1,300 m (storage)
Weight: 390 kg (radar - incl power supply); 980 kg (fully equipped trailer)

Contractor

Thales Sensors, Crawley, UK.

....

DR/MPDR series battlefield radars

Type

Family of battlefield missile control and fire-control radars.

Description

DR 621/622/641/645

These I-band (8 to 10 GHz) pulse-Doppler search radars are multiple application radars for the detection of low- and very low-flying aircraft and helicopters. Because of their construction, the systems can be installed in shelters, vehicles, tanks and ships. The sensors can be used as acquisition and target designation radars (DR 621/622), as part of weapon systems (Wildcat) and as the radars in site defence system battery co-ordination posts (DR 641/645). They can also be expanded to form command and control centres by adding data processing equipment and

The B2 version of 35 mm Flakpanzer Gepard showing Siemens-Albis fire-control radar and MPDR 12 acquisition radar

The Wildcat air defence system with DR 621 radar

display. One version has been fitted to the prototype Daewoo Flying Tiger twin 35 mm anti-aircraft gun in South Korea.

DR 625

The mobile, I-band (8 to 10 GHz) DR 625 radar has been developed to meet modern Very SHOrt- and SHOrt-Range Air Defence (VSHORAD and SHORAD) requirements and is capable of both surveillance and target designation. It is based on the DR621/622/641/645 family. The sensor's range is given as being 20 km, with features such as frequency agility in time on target, pulse compression, very low sidelobe values, a constant false alarm rate and automatic voice advice contributing to its electronic counter-countermeasures capabilities. The compact, low weight DR 625 is mounted on a cross-country vehicle that is air transportable (including as a helicopter slung load).

MPDR 12

The MPDR 12 is the acquisition radar for the Gepard Mks 2 and 3 anti-aircraft tanks. It is an E-band (2 to 3 GHz) pulse-Doppler radar with high subclutter visibility and high data rate (60 rpm). The system is integrated with the MSR 400 Mk XII interrogator (see separate entry) for Identification Friend-or-Foe (IFF). When applied to the Gepard Mk 2 anti-aircraft tank, MPDR 12 is teamed with a J-band (10 to 20 GHz) tracking radar that was developed by Siemens-Albis. Described as being a monopulse pulse-Doppler system, this radar is quoted as having operating range and permanent echo rejection values of 15 km and better than 23 dB respectively.

MPDR 16

This is the search radar used in the Roland mobile surface-to-air missile system and is a pulse-Doppler equipment operating in the D-band (1 to 2 GHz). The minimum and maximum target acquisition ranges are 1.5 and 16.5 km respectively. The 2 × 1 m antenna rotates at a rate of 60 rpm and is of cosec^2 type. An integrated IFF facility is provided, using an MSR 400 Mk XII interrogator and the use of a Doppler processor eliminates ground clutter. The extractor generates radar plots and feeds the computer of the associated tracking radar with azimuth, range and IFF response.

Status

Over time, DR/MPDR series radars have been procured.

Contractor

European Aeronautic, Defence and Space (EADS) Co Systems and Defence Electronics,, Unterschleissheim, Germany.

UPDATED

Flycatcher Mk2 weapon Command and Control (C²) sensor system

Type

Weapon C² sensor system.

Description

The I-band Flycatcher Mk2 weapon C² sensor system is an autonomous, all-weather, hybrid equipment that incorporates an I-band (8 to 10 GHz) radar and a K-band (20 to 40 GHz) radar/Electro-Optical (EO) tracking suite and is applicable to SHOrt Range Air Defence (SHORAD) and Very SHOrt Range Air Defence (VSHORAD) applications. When integrated with VSHORAD missiles, Flycatcher Mk2 provides alerting and targeting cues while for VSHORAD gun systems, it provides three independent weapon channels for 'accurate' all-weather fire-control. The system can be integrated with a variety of SHORAD missile types and supports a number of missile guidance approaches including Command to Line Of Sight (CLOS). Flycatcher Mk 2's I-band radar is a 3-D sensor that covers up to 70° in elevation on every scan. It can be operated in Air Defence (AD) and Command Centre (CC) modes, with the former being the radar's basic operational mode. The CC function sees the Flycatcher Mk2 acting as the co-ordination centre for a group of systems within a SHORAD network. During the deployment phase of an operation, the equipment's I-band radar operates in search mode and detected targets are tracked and interrogated (using an integrated identification friend-or-foe subsystem) automatically. Target identification can be supported by target behavioural analysis (using air space control information) and established tracks can be broadcast to provide VSHORAD missile alerts and cues. Accurate 3-D target tracking during an engagement is performed by either the system's K-band radar or by its EO subsystem. Both approaches are optimised for low-altitude target tracking and the EO subsystem consists of a sensor package that is capable of providing 'round-the-clock' coverage and contains both tracking and observation devices. The K-band radar is used in support of SHORAD missile and VSHORAD gun engagements, generates dedicated target and missile acquisition beams and has a main beam format that can simultaneously track a target and a CLOS missile.

Status

As of August 2001 (and according to Jane's sources), three Flycatcher Mk2 weapon C² sensor systems had been ordered by the Fuerza Aérea Venezolana (Venezuelan Air Force) for integration with a land-based variant of the Israeli Barak-1 vertically launched, point defence, surface-to-air missile. As of the given date, procurement of these system was understood to have been launched during late 1998/early 1999 with delivery scheduled for late 2001/early 2002. More than 150 Flycatcher systems of all types were reported as having been delivered to customers around the world.

Specification

Radar frequency: 8-10 GHz and 20-40 GHz bands
Elevation coverage: 70° (elevation, I-band radar, max)
Range: up to 10 km (EO subsystem); 25 km (I-band radar, AD mode); 50 km (I-band radar, CC mode)

Contractor

Thales Nederland, Hengelo, Netherlands.

UPDATED

Gerfaut ADAS system

Type

E/F-band (2 to 4 GHz) alert and co-ordination radar system.

Description

The Thales Raytheon Systems' (formerly Thales Air Defence) Gerfaut ADAS is an alert and co-ordination system devoted to short-range air defence systems (missiles, anti-aircraft guns and man-portable air defence systems) which are deployed on the battlefield. It is designed to minimise the reaction time and increase weapon efficiency faced with threats such as raids of fixed-wing aircraft and attack helicopters.

Gerfaut ADAS is a modular system which enables the automation of two elements of short-range air defence:

- the radar command post which can be integrated into any type of light battlefield vehicle
- the weapon posts deployed on the field and equipped with terminals enabling automatic datalink with the radar command post. Up to six weapon posts and two types of weapons can be managed.

The system carries out two main functions:

- early warning – fast and forward detection, identification friend-or-foe, track-while-scan and classification of targets flying or hovering at low and very low altitude
- cueing – threat evaluation, selection of the suitable weapons to engage the target(s), computation and transmission of firing data and orders, engagement monitoring.

The system computer carries out system management, data transmission management and alphanumeric display of the tactical situation. The weapon terminal provides display of firing data and orders, transmission of engagement reports, data transmission management and interface to connect the weapon post.

The Gerfaut ADAS system

Data and voice communication is incorporated using fixed and/or hopping frequency VHF (30 to 300 MHz/frequency modulated) radio, and field telephone wires.

Status

As of May 2001, Gerfaut ADAS was understood to be the Gerfaut variant that had been selected for use in the French Army's Martha air defence command and control system. Outside France (and as of November 2001), Jane's sources were suggesting that Gerfaut had been procured by customers in both the Pacific Rim/Far East regions and South America.

Specifications

Radar: Gerfaut E/F-band (2-4 GHz), coherent pulse 2-D Doppler
Instrumented range: 14.1 km
Range detection (in free space): 8.5 km (hovering helicopter); 14.7 km (fixed-wing aircraft)
ECCM: frequency agility; pulse compression; antenna design; Doppler filtering
Track-while-scan: 6 tracks
Data renewal: 1.5 s
Distance between command post and weapon post: 5 km
IFF: Mk X, Mk XII or Secure mode

Contractor

Thales Raytheon Systems, Massy, France.

UPDATED

Integrated radar system for the Royal Netherlands Army's anti-aircraft tank

Type

I- (8 to 10 GHz sub-band) and K-band (27 to 40 GHz sub-band) armoured fighting vehicle-mounted search and tracking radar system.

Description

The Royal Netherlands Army's Pantser Rups Tegen Luchtdoelen (PRTL) 35 mm self-propelled anti-aircraft gun system is equipped with an integrated search and tracking radar system designed and manufactured by Thales. This integrated radar system is mounted on a Leopard tank chassis armed with twin, rapid-firing 35 mm Oerlikon anti-aircraft cannon. The Thales system is based upon the well proven advanced integrated radar technology in which the same I-band transmitter is used for both search, fast acquisition and tracking purposes. A dual-band tracking radar operating in both the 8 to 10 GHz and 27 to 40 GHz sub-bands, with a broad beamwidth compared with that of the search antenna and a very small beamwidth for tracking 'extremely' low-flying targets, is claimed to reduce reaction time to the 'barest minimum' and to facilitate target tracking at 'tree-top level'. Simultaneous search transmissions are continued through the entire tracking phase, permitting the maintenance of complete coverage against further airborne threats. Rapid automatic or manual switching facilities allow the system to cope with a multitarget environment.

The radar system has been designed to detect and identify aircraft at very low to medium altitudes and to track them automatically. It has the following important features:

- integrated all-weather search and tracking systems with a very short reaction time
- search while tracking through 360°
- search on the move, with compensation for vehicle's speed
- good performance in clutter and electronic countermeasures environment
- compact rugged design and module concept to facilitate maintenance
- dual-frequency tracking radar
- integral identification friend-or-foe.

Status

As of January 2001, the updated PRTL search and tracking radar was in service with the Royal Netherlands Army. As of the cited date, the system as a whole was to be updated so that it could be maintained in service until 2015. Here, upgrades are understood to include the introduction of a digital fire-control computer, frangible armour-piercing discarding sabot ammunition, a cooling system, self-test equipment and a datalink that will interface with the Dutch Army's future NEtherlands Short Range Air Defence System (NESRADS). Elsewhere, a 'recent upgrade programme' has replaced the system's existing analogue monochrome display with a full-colour raster scan unit.

Specifications

Antenna system
Search antenna: slotted waveguide type
Length: 1.5 m
Horizontal beamwidth: 1.4°
Vertical beamwidth: 30°
Polarisation: horizontal or circular
Rotational speed: 60 rpm
Range: 15 km
MTI improvement factor: 30 dB
Tracking antenna: parabolic with cassegrain reflector and monopulse feedhorn, plus integrated nozzle motor for the K-band transceiver
Diameter: 0.6 m
Beamwidth: 4.2°
Transmitter/receiver
Transmitter power: 200 mW average
Frequency: I-band (8-10 GHz sub-band)
Search channel: MTI with double canceller (digital)
Tracking channel: pulse-Doppler and built-in anti-image radar in K-band
ECCM: digital video correlator, pulse-length discriminator. Passive tracking in I/low J-band, PRF stagger
Noise figures: search receiver 7 dB, tracking receiver 9 dB
Radar display and control panels
PPI: the PPI provides a tactical airspace picture and incorporates TICCS generated data, alphanumeric data, moving target indication and own radar video among other elements
Power consumption: 200 V 380 Hz approx 2 kVA

Contractor

Thales Nederland, Hengelo, Netherlands.

UPDATED

Man-portable Surveillance and Target Acquisition Radar (MSTAR)

Type

J-band (10 to 20 GHz) man-portable ground and air surveillance radar.

Description

MSTAR is a lightweight coherent J-band pulse-Doppler equipment which is designed for ground surveillance, artillery observation, coast watching and the detection of hovering helicopters. Weighing 55.9 kg, the radar breaks down into four man-portable loads; a 13.8 kg aerial assembly, a 15 kg main electronic assembly, a 16.6 kg control and display assembly and a 10.5 kg mounting tripod. All given weights include ancillaries and carrying cases. Looking at these elements in more detail, the aerial assembly incorporates a Kevlar horn-fed reflector (operator selectable horizontal or circular polarisation) and a solid-state transceiver, while the electronics unit houses the system's signal processor, angulation head, level indicator and an azimuth/elevation local controller. The control and display assembly incorporates an electroluminescent flat panel display, a back illuminated tactile membrane keyboard, a touchbutton cursor control and fault and power low indicators. In addition to being man deployed, MSTAR can be air dropped and vehicle mounted. Power for these various

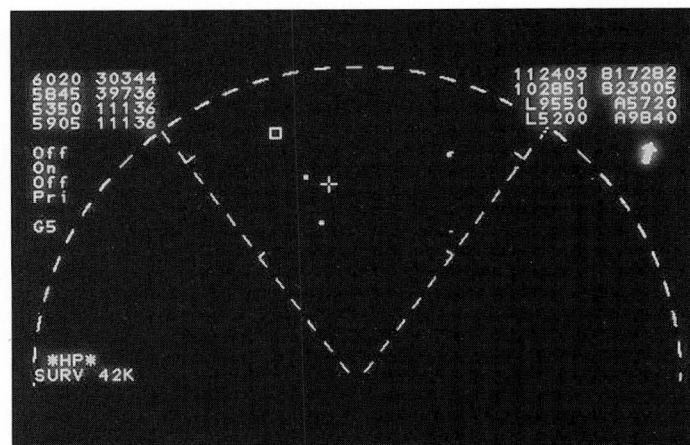

MSTAR's surveillance mode display

The MSTAR battlefield surveillance radar

applications is provided by Ni/Cd or lithium batteries, the host vehicle's supply or an external supply.

System design emphasises ease of use in severe climate and nuclear/biological/chemical warfare environments. Electronic counter-countermeasures features include infra-red reflective paint on the various components, low power output, low sidelobes, narrow beamwidth, pulse compression and operator selectable frequencies, sensitivity and scan arcs. MSTAR is currently the subject of a product enhancement programme which has resulted in the latest variant – the MSTAR 1-10 – featuring selectable power outputs (1 and 10 W) on increase detection range while maintaining low probability of intercept. Other operational enhancements are reported to be in development and Thales is understood to be committed to maintaining MSTAR's viability well into the 21st century.

Status

Over time, more than 400 MSTAR radars are reported as having been sold since 1989. Customers include the British Army (man-portable system and Group B provision on Warrior Mechanised Artillery Observation Vehicles), the Royal Air Force Regiment, Canada, (including installation on Coyote 8 × 8 reconnaissance vehicles), Finland, Netherlands, Spain (licensed technology – see separate ARINE entry) and Saudi Arabia (licence built by the Advanced Electronics Corporation). Canadian use of the system involves licence production of a 4 W variant by System Electronic Incorporated. Elsewhere, eight MSTAR radars were trialled by the US Army (US designation AN/PPS-5C) in Europe during the period April 1994 to April 1995 and a further four have been acquired by an unidentified US agency for border surveillance work. In addition, the USA has combined MSTAR with an electro-optic sensor package and a Global Position System receiver for use in the joint US Army/USMC 'Magic Warrior' proof-of-concept programme which began during 1993. Here, the system was mounted on a high-mobility multipurpose wheeled vehicle. In the UK, MSTAR-type technology has been bid for the sensor suite to be installed in the British Army's next generation Tracer reconnaissance vehicle. Operationally, MSTAR has seen service with the British and Dutch armies in Bosnia and may have been deployed on a limited scale by the US Army in the same theatre during 1996. During mid-1999, it was announced that the then Racal (now Thales Sensors) was under contract to upgrade 139 MSTAR radars operated by the British Army. Enhancements are reported to have included an improved man/machine interface, upgraded signal processing and better target classification. As originally set out, enhanced MSTAR was scheduled to enter service during mid-2000.

The described enhanced MSTAR is also understood to form the basis of an Australian MSTAR CAMSTAR, 61 examples of which were ordered during the second half of 1999 for use in the Australian Army's 'Ninox' night-fighting, surveillance and target acquisition system programme. System elements are said to include a colour man/machine interface and a greater target detection range when compared with baseline MSTAR. In Australian service, a percentage of the AMSTAR procurement will be fitted to ASLAV-5 reconnaissance vehicles and the Australian contractor Vision Abell will undertake system assembly, testing and through life support. As of May 2001, AMSTAR deliveries were scheduled to begin during 2001.

Specifications

MSTAR 1-10
Frequency: J-band (10-20 GHz)
Power output: 1 W and 10 W
Range scales: 50 m-2.5 km; 1.5-6 km; 3-12 km; 6-24 km; 24-42 km
Accuracy: ±6 mils (azimuth) ±10 m (range)
Power consumption: 35 W (surveillance mode); 57 W (fall-of-shot mode)
Power supply: 24 V DC radio battery
Temperature: −32 to +55°C (operating); −33 to +70°C (storage)
Windspeed: up to 22 m/s
Deployment time: < 3 min (man-portable configuration – 2-man crew)
Weight: 46 kg (incl batteries)

Contractor

Thales Sensors, Crawley, UK.

VERIFIED

PAGE low-level air surveillance radar

Type

J-band (10 to 20 GHz) low-level air surveillance radar.

Description

PAGE (Portable Air defence Guard Equipment) is a man-portable/vehicle-mounted J-band low-level air surveillance radar, which is designed for use in very short-range air defence applications involving both anti-aircraft guns and man-portable air defence equipments. The radar makes use of the frequency modulated continuous wave technique which, when combined with its 'very low' power output, is claimed to make the system 'nearly undetectable' by electronic support measures and radar warning receivers. A basic PAGE system is made up of a radar processor, a tripod and an operator unit. Power can be provided by a small motor generator and there is provision for Identification Friend-or-Foe (IFF) integration, automated datalinks and weapon terminals for the supply of real-time pre-warning and cueing data to remote firing units. Primary system functions comprise target detection, target identification (if integrated with IFF), target data display and target tracking. PAGE is a co-development by Thales and Indra of Spain.

Status

As of January 2001, the PAGE low-level air surveillance radar was reported as being in series production for the Spanish Army.

Specifications

Radar front-end
Frequency: J-band (10-20 GHz)
Power output: 20 W
Range: 15-20 km
Resolution: 100 m (range); 2.3° (azimuth)
Antenna beamwidth: 1.7°
Scan rate: 10-20 rpm
Processing: MMI-Doppler Fast Fourier Transform filter, target tracking, strobe on jam and identification by IFF (if integrated)
Operator Unit: ruggedised portable computer
System weight: 85 kg

Contractor

Thales Nederland, Hengelo, Netherlands.

UPDATED

The PAGE air surveillance radar has undergone trials in Spain   0006889

RAC alert and co-ordination 3-D radar

Type

G/H-band (4 to 8 GHz) multipurpose, low-level air defence 3-D radar.

Description

The Thales Raytheon Systems' (formerly Thales Air Defence) RAC alert and co-ordinating radar is designed to provide warning and control for anti-aircraft guns

The RAC radar on a cross-country vehicle

and missiles over very short to medium ranges. The equipment can also act as a gap filler. In more detail, RAC can perform air surveillance tasks such as air situation analysis and target identification together with weapon control functions such as threat evaluation and target assignment.

The concept and design make this radar adaptable to the requirements of anti-aircraft defence systems (stand alone or integrated into a network).

Four main features are provided by the RAC system:
- detection accuracy and short reaction time given by three-dimensional coverage
- the 'see-through' ability in ground and atmospheric clutter
- resistance to jamming
- mobility and serviceability.

The RAC system operates in the G/H-band which offers, within a 100 km range envelope, the best trade-off between atmospheric attenuation, transmitting power resolution and antenna size versus detection of small targets, such as those flying at very low altitude in heavy clutter.

RAC employs a state-of-the-art planar phased-array antenna, featuring a high-grade radiation pattern with ultra-low sidelobe levels. The beam is electronically scanned in elevation to provide three-dimensional target co-ordinates.

The high-stability transmitter is fitted with a single travelling wave tube. A digital pulse compression and moving target detector processes the signal to obtain excellent through-clutter detection, despite ground, rain and chaff echoes, without loss of resolution. Individual echoes of hovering helicopters are catered for by a special processing channel. The high-performance automatic track-while-scan is a result of the use of auto-adaptive Kalman filtering.

A number of basic radar operating modes are available, combining the antenna rotation rate with anti-clutter processing to maintain a high-detection probability in severe environmental conditions. Additional modes allow efficient operation in the face of countermeasures.

Two types of the RAC radar station are available:
- a compact radar station with the radar itself and the combat control system housed on a single vehicle and an antenna which can be elevated up to 13 m by a hydraulic mast
- a two-vehicle radar station in order to separate the radar from the combat control system.

Compact radar station
The compact radar station configuration includes a hydraulic erecting mast (which raises the antenna from 6 m to 13 m above the ground), a hydraulic levelling system and an integral power supply. A cabin shelters the radar and identification friend-or-foe electronics and a two workstation command and control subsystem which is equipped with a computer and communications links. A fully remote command and control format (which features unmanned radar operation) is also available.

Status
Over time, RAC radars are known to have been supplied to the armies of Denmark (as part of the service's Control and Warning System) and Austria together with the Austrian Air Force. Of these, the Austrian Army has procured 16 sets for surface-to-air missile alerting and co-ordination while the country's Air Force has acquired six radars as gap fillers. Elsewhere, RAC is understood to form part of the Spada 2000 surface-to-air missile system procured by Spain and, as of the late 1990s, at least 38 such radars are reported to have been sold to customers in four countries.

Specifications
Transceiver frequency: 4-8 GHz
Transceiver bandwidth: 600 MHz
Compressed pulse: 0.4 μs
Antenna type: 16 row planar-array with mono-pencil beam and electronic tilt
Antenna polarisation: horizontal

Antenna beamwidth: 1.75° (azimuth); 7.9/14° (elevation)
Antenna sidelobe level: −50 dB (average)
Data rate: 2, 3 or 6 s
Instrumented range: 100 km (10 rpm rotation rate, low coverage operating mode)
Min detection range: 1.2 km
Altitude: 9,144 m (normal configuration); 18,288 m (netted configuration)
Elevation coverage: 47°
Accuracy: 0.4/4° (azimuth); 0.5° (elevation); 20/150 m (range)
Signal processing: constant false alarm rate; FIR Doppler filters; high-resolution clutter mapper; moving target detector
Data processing: automatic track initiation, jamming strobes; plot extractor; tracker (more than 100 tracks simultaneously)
IFF provision: co-mounted secondary antenna; fully solid state subsystem; Modes 1, 2, 3/A, C and Secure
ECCM: frequency agility, large operating bandwidth; 'least jammed' frequency selection; low sidelobe antenna
Temperature: −31 to +49°C (operating); −40 to +70°C (storage)
Windspeed: up to 120 km/h (operating); up to 180 km/h (survival)
Prime power: 35 KVA/50 Hz/240 V/400 V, 3-phase + neutral + ground

Contractor
Thales Raytheon Systems, Massy, France.

UPDATED

Rapier Blindfire tracking radar

Type
All-weather tracking and fire-control radar for the Rapier Surface-to-Air Missile (SAM) system.

Description
Historically, the DN 181 Blindfire radar was developed by Marconi Space and Defence Systems, with GEC Avionics as a major subcontractor. Continuing production and enhancement development work is now carried out by Alenia Marconi Systems. This development was carried out on a UK Ministry of Defence (MoD) funded programme with BAE Systems as prime system contractor. The radar tracker, with the Rapier Surface-to-Air Missile (SAM), is designed to meet the threat posed by low-level strike aircraft which can operate at night and in low visibility. The addition of the Blindfire radar gives the system an all-weather/night capability to extend the original 'clear-weather' Rapier devised by BAE Systems with optical tracking to meet this threat. With the radar tracker in the system, the operator has a choice, until the moment of launch, as to which system will be the most efficient for the engagement.

The radar uses differential tracking for simultaneous target and missile tracking. An electronic angle tracking receiver, incorporated within the tracker, detects boresight errors for accurate positioning of its own antenna and the angular difference between the missile and target sightlines for the generation of command guidance signals. The extremely narrow beamwidth, coupled with precisely positioned individual range gates for target and missile, is the key to the accuracy and very low-level capability of the system and also reduces the chances of background clutter, such as ground echoes or rainfall, obscuring the target echo. A number of other electronic measures are also designed into the system to minimise the effect of clutter and other unwanted signals, both spurious and deliberate.

The tracking radar is initially alerted to a hostile aircraft by the system surveillance radar, which indicates the approximate direction of the threat. The tracking radar will immediately swing on to this bearing and rapidly establish the exact bearing, range and height of the target. When the radar is locked on to the target, the missile is launched and this, too, is tracked by the radar. The difference

The Rapier FSC Blindfire tracking radar

between the target and missile angles is instantly derived within the system and commands are automatically transmitted to the missile to guide it on to the target.

The radar tracker is a self-contained, 'plug-in' unit, and requires no extra personnel to operate it.

Status

The initial UK MoD production contract was awarded in early 1972 and the equipment is in widespread operational service with the British Army, the Royal Air Force and seven overseas operators of the Rapier low-level SAM system. The DN181 tracking radar also forms part of the RN Type 911 system ordered by the UK MoD for the lightweight Seawolf programme.

In January 1987, GEC-Marconi Radar and Defence Systems (now Alenia Marconi Systems) was awarded a contract for the full development and production of the new dual tracking radar for the Rapier Field Standard C (FSC) system. This Blindfire FSC radar tracker is a totally new design developed under a £100 million UK MoD-funded contract. The radar tracker is mounted on a common monocoque trailer base fitted with its own power source, and is one of a number of trailers which form a Rapier fire unit.

The target and missile differential angle tracking retains the concepts of the original DN181 radar tracker, resulting in an enhanced low-level capability. A new design angle tracking receiver increases the Blindfire FSC radar tracker's ability to operate in today's complex Electronic CounterMeasures (ECM) environment.

The transmitter/receiver is a completely new design which uses linear frequency-modulated pulse compression techniques to provide increased mean radio frequency power and a high resistance in a hostile ECM environment.

The Blindfire FSC trailer contains a continuous wave Doppler radar gathering unit in place of the optical gathering unit found on the DN181 radar tracker. The radar gathering unit used for initial missile acquisition with a dedicated command transmitter and the onboard computer, allows the Blindfire FSC Rapier system to achieve a unique dual missile firing capability.

The reliability of the electronic circuits is enhanced by the provision of an onboard water glycol cooling system which ensures that the equipment is always operating at optimum temperature and performance.

The Blindfire FSC radar tracker is capable of rapid deployment and uses automatic built-in-test procedures to ensure a minimum into-action time. The design and construction permits ease of maintenance. Built-in-test equipment and continuous monitoring of key parameters results in a high degree of confidence in system availability with rapid fault diagnosis to a first line replacement. The Blindfire radar tracker is self-contained and once deployed operates totally automatically with no increase in the user's workload.

Contractor

Alenia Marconi Systems (an Alenia-BAE Systems joint venture), Chelmsford/Rome.

VERIFIED

Rasit ground surveillance radar

Type

I-band (8 to 10 GHz) multimission ground radar for battlefield surveillance and artillery fire control.

Description

Rasit is a long-range (40 km) ground surveillance radar for the detection, acquisition, localisation and recognition of moving targets, either on or near the ground in all weathers. The Rasit is normally shelter or vehicle mounted, but can be manhandled by breaking it down into four units weighing no more than 30 kg. In French Army service it has been mounted on a shelter-mounted mast.

The equipment is a pulse-Doppler radar. It operates in I-band and uses a coherent receiver and multiple range gates and filters designed for a high probability of detection.

Echoes detected over the surveyed zone are displayed on a daylight display. For each target acquired, the polar and universal transverse Mercator co-ordinates are displayed on the console and the Doppler tone is transmitted to the operator by loudspeaker or by earphone. The operator is thus able to recognise pedestrians, wheeled and tracked vehicles, aircraft and helicopters.

An automatic system can be arranged to trigger an alarm as soon as a target enters into zones defined by the operator within the surveyed area. The operator can select panoramic or sectorial surveillance. The equipment can be operated by a single inexperienced operator in a variety of conditions.

The antenna beam can sweep a sector or can be stopped and then oriented towards a target automatically. The antenna is equipped with a polariser making the radar insensitive to atmospheric perturbations. The radar also has very effective Electronic Counter-CounterMeasures (ECCM) capabilities and performs automatic target tracking.

Rasit is claimed to be the first ground surveillance radar to include surveyed area management. The operator can neutralise certain areas of the radar scanning in order that the alarm should not be triggered by moving targets of no interest in friendly zones. Rasit meets the requirements of border line surveillance as well as battlefield intelligence. Elsewhere, the radar features a fire deviation measurement function that allows it to correct artillery fire by simultaneously observing the target and the point of impact of explosion of the shell. Rasit automatically computes the deviations between these two points. This function is added with three additional plug-in cards into the control console of the radar. The radio frequency unit can be

Rasit can be vehicle mounted, as shown here, where the radar is fitted to a Peugeot P4 light utility truck

mounted on a tripod and remotely controlled (over several hundred metres) from the console. Alternatively, it can be mounted on any of a variety of military vehicles. In addition to built-in tests a tactical test bench can be supplied for third and fourth echelon maintenance. A tactical simulator designed for an efficient and comprehensive training of one or several operators is available.

Status

As of March 2001, in excess of 700 Rasit radars are reported to have been delivered to approximately 20 armies around the world. Customers are understood to include Argentina, Australia, China, Egypt, Estonia (one system for border surveillance), France, Germany, Iraq, South Korea, Libya, Morocco, Nigeria, Norway, Pakistan, Saudi Arabia, Thailand, Tunisia, Venezuela and Zimbabwe. In October 1993, Thomson-CSF (now Thales) was awarded a contract to upgrade those Rasit radars that were in service with the French Army.

Specifications

Frequency: I-band (10 GHz – 10 switchable frequencies)
Polarisation: linear or circular
Peak power: 3 kW
Measurement accuracy: ±0.6° (azimuth); ±10 m (range)
Range (90% probability of detection): 20 km (pedestrian); 40 km (vehicle); 20-30 km (helicopter); 40 km (low-flying aircraft)
Antenna rotation speed: 9°/s (160 mils)
Width of sector: −11-240° (10 widths between 200 and 4,300 mils)
Detected target speed: 150 km/h max
Temperature range: −33 to +70°
Weight: 90 kg (heaviest load 28 kg)
Options: remote slave TV monitor; plotting table; mapping computer
Fire deviation measurement: typical observation range for various shell calibres – 155 mm 25 km; 120 mm 28 km; 105 mm 20 km
Azimuth accuracy: ±0.1° (2 mils)
Range accuracy: ±10 m

Contractor

Thales Air Defence, Bagneux, France.

VERIFIED

RATAC battlefield radar

Type

I-band (8 to 10 GHz) acquisition and tracking radar for ground and low-altitude targets.

Development

Development of RATAC was undertaken by LCT (Laboratoire Central de Télécommunications) under the direction of the French military authority. A subsequent agreement between the French and German governments provided for joint procurement and production for the armies of both countries. Tests were later carried out by the US Army at Fort Sill.

Description

RATAC (RAdar de Tir pour l'Artillerie de Campagne) is a lightweight (eight cabinets of 35 kg, or less, each) battlefield radar providing for detection, acquisition, identification, location and tracking of surface targets such as troops, tanks, vehicles, low-flying helicopters and light aircraft. Other operational functions

The German application of RATAC installed in an M113 armoured vehicle

The French version of the RATAC battlefield and fire director radar (French version)

The units that make up the RB12B surveillance radar

include artillery direction, surveillance of own forces and helicopter control. It comprises six units and is designed for vehicle mounting. Pulse Doppler and monopulse techniques are employed and the operating frequency is in the 3 cm wavelength region. A highly effective fixed target cancellation system is said to make it possible to locate very slow moving targets with a small radar cross-section, such as a single man. Detection ranges are reported to be better than 15 km against vehicles and 8 km against troops, with an accuracy of 10 m in surveillance. Good all-weather performance is claimed.

The system incorporates its own computer and target data can be presented in terms of either polar or grid co-ordinates. An optional plotting board unit permits target positions and movements to be recorded. A loudspeaker is provided to aid target identification by listening to the Doppler characteristics of targets. Automatic target tracking facilities are included and there are provisions for the transmission of data to own artillery. Operation with vehicle installed identification friend-or-foe is a possible option.

Status
As of this edition, RATAC is understood to have been procured by France, Germany, the USA and several other countries. The US nomenclature is AN/TPS-58. The French Army is reported to have carried out a modernisation programme to the RATAC system, which includes a new computer and automation of the manual source message terminal. Within the German Army, RATAC has been replaced by SEL's RATAC-S system (see separate entry).

Contractors
Thomson-CSF AIRSYS, Bagneux, France.
SEL Defense Systems, Stuttgart, Germany.

VERIFIED

....

RB12B short-range ground surveillance radar

Type
J-band (10 to 20 GHz) multimission ground surveillance radar.

Description
The RB12B is a man-portable short-range lightweight coherent pulse-Doppler ground surveillance radar using a solid-state transmitter operating in J-band. It is a ground surveillance system for the detection, acquisition, localisation, tracking and recognition of the nature of moving targets (3 to 100 km/h) on the ground; pedestrians, light vehicles and trucks; as well as low-flying light aircraft, microlight aircraft, unmanned aerial vehicles and helicopters. It can fulfil the requirements of tactical intelligence missions for artillery forward observers, infantry intelligence and paratroopers. The equipment is easily set up and very easy to operate and breaks down into a two-man load for transport.

The RB12B scope displays the azimuth and range co-ordinates for each detected target. The Doppler sound is transmitted to the operator either by loudspeaker or earphones.

The RB12B can also be used for the surveillance of large sensitive areas and borders. To perform this mission the RB12B features surveyed area management, automatic target acquisition and target tracking, surveillance of the perimeter, marker and rallying functions and automatic data transmission.

The operator can select the width of the surveyed zone from 4.5 to 180° (80 to 3,200 mils) and can choose the depth surveyed by operating either in normal mode, overall radar range of 6.4 km, or any zone of 800 m long within the total range in magnified mode.

Display and controls are contained within a unit which can be remote controlled up to a distance of 1,300 m.

Capable of autonomous operation, the RB12B can also be integrated into a surveillance system, using its five switchable frequencies and remote-control facility. It is able to be coupled with other sensors, such as a charge-coupled device and infra-red cameras.

In adapting the RB12B to its Detectit networked static-site defence system, the system's manufacturer (Thales Air Defence) has added an IBM personal computer-based colour display to supplement the standard lightweight monochrome display.

Status
As of May 2001, RB12B was reported as having been procured by the French Air Force. Thales is understood to have developed an extended range version with a 20 km vehicle detection capability.

Specifications
Frequency: J-band (10-20 GHz – 5 switchable frequencies)
Polarisation: vertical or circular
Average power: 25 mW
Measurement accuracy: ±1° (azimuth); ±20 m (range)
Moving target detection speed: 3-100 km/h
Range: 3 km (pedestrian); 6.4 km (vehicle)
Antenna rotation: ≈8°/s (≈143 mils)
Width of sector: preprogrammed 4.5, 90 and 180° (80, 1,600, 3,200 mils); programmable 4.5-180° (≈80 to 3,200 mils)
Control console: liquid crystal display screen 100 × 80 mm; loudspeaker; tactile keyboard; RS-232 plug
Weight: 32 kg incl tripod, battery and 25 m cable
Power supply: 24 V DC, 22 W
Autonomy: 7 h on standard battery
Distance between sensor and operator: 3, 25 and 1,300 m (max)

Contractor
Thales Air Defence, Bagneux, France.

VERIFIED

Reporter tactical early warning system

Type
I-band (8 to 10 GHz) battlefield early warning and designation radar.

Description
Reporter (the latest variant of which is designated as Improved Reporter) is a transportable early warning and cueing radar system designed to search and track air targets at altitudes of 15 to 4,000 m and ranges up to 40 km. It is a compact, I-band radar that has an integrated Identification Friend-or-Foe (IFF) capability for target identification and features a small size, combined radar and IFF antenna. The receiver incorporates digital video processing (moving target indicator) with double canceller techniques ensuring accurate discrimination between fast and slow targets (such as hovering helicopters).

Up to 20 designated targets can be automatically tracked by the built-in video extractor. Target data are transmitted via VHF (30 to 300 MHz) frequency modulated radio, or line communications, to an unlimited number of fire units equipped with a Reporter Target Data Receiver (TDR). At the weapons site, target data from Reporter are processed by the TDR. This is a computer-based, portable, 'highly intelligent' terminal that evaluates target data obtained from the Reporter system. The TDR uses various weapon parameters such as position, priority line and allocated defence sector and with its built-in threat evaluation programme, each TDR displays target data relating to the four highest priority targets with reference to the weapon system position. Various weapon interfaces are available to convert target information into control signals for automatic weapon aiming. These interfaces can be offered separately.

Reporter includes an airspace control function to assist radar operators in discriminating between 'friend', 'unknown' and 'foe' (based on behaviour identification) that is an adjunct to IFF. Airspace information (such as lines, areas and zones) is presented in a synthetic format. The system is further noted as being capable of monitoring an area 40 km in radius. Operated by one man, the system supplies encoded target data to weapon sites, with the operator selecting targets on his colour display for automatic tracking. Selected targets are automatically subjected to IFF interrogation and the radar's computer generates an update on up to 20 targets every antenna revolution. Update messages include target grid co-ordinates, target speed and course and IFF status on each target. Data handover is by means of the standard combat radio net and terminals at each of the weapon sites being served.

Status
As of January 2001, over 40 Reporter systems were reported as having been supplied to customers around the world. To ensure efficient logistic support, radar components are designed for commonality with Flycatcher and the integrated radar for the PRTL anti-aircraft tank. The Reporter system has also been licensed to Indian contractor Bharat Electronics.

Specifications
Frequency: I-band (8-10 GHz)
Power output: 220 kW (peak)
Antenna beamwidth: 6° (horizontal)
Antenna type: parabolic
Scan rate: 40 rpm
Instrumented range: 40 km
PPI: full colour raster scan display

Contractor
Thales Nederland, Hengelo, Netherlands.

UPDATED

Roland tracking radar

Type
J-band (10 to 20 GHz) tracking radar for the Roland weapon system.

Description
This radar system is an integral part of the Roland surface-to-air all-weather weapon system. The tracking radar is used in addition to the periscope sight, their respective axes being maintained in alignment to make it possible for rapid transition from radar to optical sighting, or vice versa.

The radar is a monopulse frequency agile type equipped with a magnetron transmitting valve and a Doppler receiver equipment. The antenna is of the parabolic cassegrain type with circular polarisation.

The radar deals simultaneously with the target and the in-flight missile. Position of the missile relative to the radar beam is established by means of a continuous wave transmission from the missile beacon.

In operation with the Roland 2 system, the target is first detected by the pulse-Doppler European Aeronautic, Defence and Space (EADS) Company Deutschland MPDR 16 D-band (1 to 2 GHz) surveillance radar. The scanner, which can be operated on the move, has an acquisition range of 1.5 to 16.5 km for a 1 m² target operating at speeds from 50 to 450 m/s. The tracking radar, mounted on the front of the turret is a two-channel, monopulse-Doppler microwave Thales Domino 3-D system; one channel tracks the target and the second locks in on the microwave source on the missile.

After launch, an infra-red localiser on the antenna of the tracking radar is used to capture the missile within a distance of 500 to 700 m, at which range the missile has entered the pencil beam of the tracking radar. A second tracking channel follows the missile by means of a transponder carried on it.

Status
As of May 2000, Jane's sources were reporting the delivery of 650 Roland firing units to at least 10 countries worldwide. Among these, the same sources indicate a French procurement of 20 shelterised systems (designated as the Carol variant and mounted on a semi-trailer towed by an ACMAT 6 × 6 vehicle) and a Finnish acquisition of 10 similar equipments mounted on MAN 6 × 6 vehicles. As of the given date, Jane's sources were also reporting France's expectation of maintaining an inventory of 72 upgraded Roland systems (52 AMX-30 and 20 Carol shelter-mounted systems) until 2015 to 2020. During the following November, German contractor LFK GmbH was awarded a then year approximately €20 million contract to develop a Service Life Extension Programme (SLEP) for the German Army's Roland anti-aircraft systems. Here, the intention was to maintain the equipment's operational validity until 2015 via a number of upgrades that included digitisation of the system's fire-control system. At the time of the announcement, LFK was expecting a Roland SLEP production contract during 2003. For its part, the French effort was expected to see delivery of the first upgraded system during mid-2002.

Specifications
Frequency: J-band (10-20 GHz – frequency agile)
Power: 10 kW
Beamwidth: 2° azimuth; 1° elevation
PRF: 5 kHz
Pulse duration: 400 ns
Range: 16 km

Contractor
Thales Air Defence, Bagneux, France.

UPDATED

The operator shelter and radar vehicles used in the Improved Reporter air defence early warning, alerting and weapon cueing system

The Roland surface-to-air weapon system

Shahine surveillance and tracking radars

Type
E/F- (2 to 4 GHz) and J-band (12 to 18 GHz sub-band) surveillance and tracking radars for the Shahine weapon system.

Description
These surveillance and tracking radars are an integral part of the Shahine air defence system (itself composed of acquisition and firing units), which is a more powerful version of Crotale (see separate entry).

The surveillance radar used is a pulse-Doppler equipment that provides target acquisition and is mounted on the Shahine acquisition vehicle. As such, it provides low and very low altitude target detection, a determination of the range and radial speed of radar returns, target extraction, automatic track-while-scan and the association of identification friend-or-foe responses with interrogated targets.

The tracking radar is a monopulse-Doppler system that is claimed to offer 'excellent' anti-clutter and anti-jamming capabilities. It is mounted on the system's firing vehicle and offers target acquisition and lock on, automatic range tracking on the target and up to two in-flight missiles and measurement of the elevation and azimuth deviations between the radar axis, the target and the in-flight missiles.

Status
As of March 2001, Shahine was understood to have been in service with the Royal Saudi Army and Air Force. On 28 March 2001, Thales Air Defence announced that it had been awarded a then year €230 million contract (in co-operation with Sofresa) covering the support and upgrading of Saudi Arabia's Shahine air defence systems. Scheduled to last five years, the effort was described as specifically involving the provision of technical services, 'pyrotechnic elements' and spares together with the renovation of the in-service equipment. Execution of the necessary work was to be split between France and Saudi Arabia.

Specifications
Surveillance radar
Frequency: E/F-band (2-4 GHz)
Antenna rotation rate: 60 rpm
Target capacity: 40 per antenna revolution
TWS: automatic, 12 targets
Range: 19.5 km on a 1 m² fluctuating target
Altitude: 6,000 m
Tracking radar
Frequency: J-band (12-18 GHz)
Beamwidth: 1°
Range: 17 km on a 1 m² fluctuating target

Contractor
Thales Air Defence, Bagneux, France.

VERIFIED

The Shahine surveillance and firing units

SQUIRE ground surveillance radar

Type
J-band (10 to 20 GHz) ground surveillance radar.

Description
SQUIRE (Signaal (now Thales) Quiet Universal Intruder Recognition Equipment) is a man-portable solid-state J-band radar based on frequency modulated continuous wave techniques to ensure low probability of intercept. SQUIRE is used for ground surveillance and target acquisition together with artillery fire adjustment. The system consists of two main elements: a radar and operator units. Total system weight (including tripod, cables, headset, carrying harnesses and battery pack) is less than 45 kg.

SQUIRE's processing is based on Fast Fourier Transform techniques to ensure a high rate of discrimination in both range and speed. Moving targets are presented together with information on speed, location and direction, by means of different colours and sizes. Both audio and video presentation of the target's signature are used for target identification. A walking person is detected at ranges up to 10 km, moving truck or tank-sized targets are detected at ranges up to 24 km. Shell impacts can be detected at ranges of up to 20 km.

The radar unit used in the SQUIRE battlefield surveillance system   0006891

Status
Jane's sources suggest that final development of the SQUIRE radar was completed in June 2000 and that Thales launched an initial series production run of 10 radars during the late summer of that year. As of January 2001, Thales was reporting that SQUIRE radars had been sold to 'several' customers and had been selected by the Netherlands armed forces, who (as of the given date) were expected to place an order for 62 such equipments (55 for the Royal Netherlands Army and seven for the country's Marines) during 2001 for delivery starting in 2002.

Specifications
Frequency: J-band (10-20 GHz)
Power output: 10 mW, 100 mW or 1W
Range: up to 24 km (tanks)
Number of range cells: 512
Antenna beamwidth: 2.8° (horizontal); 8° (vertical)
Scan speed: 0°/s, 7°/s or 14°/s
Display: colour LCD

Contractor
Thales Nederland, Hengelo. Netherlands.

UPDATED

TRS 2620/2630 Gerfaut acquisition radars

Type
Family of E/F-band (2 to 4 GHz) surveillance and acquisition radars.

Description
The TRS 2620 and 2630 Gerfaut radars are designed to operate with all types of short- and very short-range anti-aircraft weapons, for example:
- autonomous weapons – target acquisition and designation for the fire-control unit
- scattered weapons – alert and co-ordination for guns, missiles and man-portable air defence systems.

When engaged in the anti-aircraft defence of battlefield forces or fixed key points, both these weapon systems can be used throughout the theatre of operations, including the front-line contact area, to provide protection for units engaged in combat or on the move. The systems can also be used for point defence of small sensitive areas.

The Gerfaut radars were designed to ensure that these weapon systems have a reaction time that gives them a decisive advantage against the low and very low threats of the 1990s, including anti-tank helicopters. Both the TRS 2620 and TRS 2630 provide target acquisition and classification with the first antenna revolution,

The Clara VSHORAD command and control mobile post with the TRS 2630 Gerfaut radar

A mast-mounted Gerfaut application

localise the target with the accuracy required by the weapons and have a high information renewal rate (40 rpm).

Except for the antenna assembly the TRS 2620 and TRS 2630 include the same sub-units. The overall design includes:

- a light compact transmit/receive antenna unit readily adaptable to a turret or any other support
- a modular processing/operating unit
- a planar-array for the TRS 2630 and a radome enclosed reflector antenna (visual discretion) for the TRS 2620
- all solid-state hardware (except for the display screen)
- a layout designed for maintenance on the battlefield
- built-in-test.

Two salient points of the Gerfaut radars are as follows:

- the TRS 2620 includes an original device to enhance detection performance on a concealed hovering helicopter. This device bears on a twin lobe in azimuth, associated with a specific processing technique
- the TRS 2630 has a larger planar-array antenna providing increased performance in detection range and electronic counter-countermeasures.

Both systems possess a very high resistance to jamming. This is provided by:

- an antenna with very low sidelobes
- a wide transmission band
- true frequency agility associated with a burst mode
- digital pulse compression
- dual frequency change receiver
- velocity filtering
- false alarm regulation.

Status

Over time, Gerfaut radars are understood to have been ordered for a number of French applications and those of other countries. These include the Samantha and Clara Very SHOrt-Range Air Defence (VSHORAD) command and control systems (TRS 2630), the Crotale system, the Swedish CV90 combat vehicle (TRS 2620), and the Blazer air defence weapon system (TRS 2630). In December 1991, a number of Gerfaut radars were ordered by Denmark. Overall, more than 200 Gerfaut radars are reported as having been procured by 10 plus customers around the world.

Specifications

Transceiver frequency: 2-4 GHz
Range: 14-30 km
Altitude: 2.5-5 km
Min detection range: 0.9 km
Accuracy: 7 mrad (azimuth); 25 m range
Data update rate: 1.5/s
Temperature: −40 to +60°C
Power: 28 V DC
IFF: integral IFF antenna, integrated IFF management
Processing: 3 channels (fixed-wing aircraft, helicopters and jammers), 'search on the move'
ECCM: contant false alarm rate; frequency agility; low sidelobes; pulse compression; strobe-on-jam; wideband frequency
MTBF: >1,850 h

Contractor

Thales Air Defence, Bagneux, France.

VERIFIED

ISRAEL

EL/M-2080 anti-tactical ballistic missile early warning and fire-control radar

Type

Search, acquisition and fire-control radar for the Arrow anti-tactical ballistic missile system.

Arrow fire-control radar

Description

The EL/M-2080 is a transportable system made up of a solid-state, phased array radar, a power unit, a cooling system and a control van. Elta describes the radar used as being capable of tracking dozens of surface-to-surface missiles simultaneously over ranges measured in hundreds of kilometres. The system's search, detection, alert, tracking and guidance functions are noted as being performed simultaneously. The EL/M-2080 is quoted as being the 'biggest and most complex radar system' ever developed by the Israeli defence electronics industry.

Status

The EL/M-2080 has been operational with the Israeli Air Force since 1998.

Contractor

Elta Electronics Industries Ltd (a subsidiary of Israel Aircraft Industries Ltd), Ashdod.

VERIFIED

EL/M-2106 and 2106H point defence alert radar

Type

D-band (1 to 2 GHz) portable battlefield radar for air defence early warning, target acquisition and gap filling.

Description

The man-portable EL/M-2106 consists of a transceiver, a 200 × 80 cm antenna assembly, a display unit, a 140 cm quadripod mounting and an optional 2 kg remote-control unit. Deployment time (involving two men) is quoted as being less than 10 minutes and in remote-control mode, the radar and its operator can be up to 100 m apart. Data are displayed in a synthetic sweep memory format using concentric rings of light emitting diodes. The display shows antenna position as well as target data. Up to four display units can be connected to an individual transceiver/antenna assembly so that multiple weapon controllers can have individual access to the incoming radar information.

The EL/M-2106H radar as installed on the M-113 tracked personnel carrier

The EL/M-2106H variant is designed for a variety of fixed-site and mobile applications. For fixed-site use, the radar is configured with a telescopic quadripod mount while soft-skin vehicle applications employ a rear-mounted hydraulic antenna mast. Armoured fighting vehicle installations (such as for the M-113 tracked personnel carrier) make use of a third type of automatically deployed antenna mast. The radiating element of the EL/M-2106H's antenna assembly is slightly smaller than that employed in the M-2106 system and measures 200 × 60 cms. In its vehicular applications, the M-2106H is shock-mounted, draws power from its host's supply and can be operated on the move.

It should also be noted that, according to Jane's sources, Elta licensed the M-2106H radar to US contractor Lear Astronics which produced the system as the AN/UPS-3 air defence surveillance equipment during the 1990s. As far as can be ascertained, UPS-3 was supplied to the US Army (82nd Airborne Division), Navy and Marine Corps (41 examples).

Specifications
(Refers to both M-2106 and M-2106H unless indicated)
Transceiver frequency: D-band (1-2 GHz)
Peak/average power: 150/10 W
Pulse-width: 6.6 or 13.3 µs
PRF: M-2106: 153 µs
M-2106H: 160 or 80 µs
Range gates: 10
MTI improvement factor: M-2106: 50 dB
M-2106H: >50 dB
Range: M-2106: up to 16 km for a 2 m² aerial target
M-2106H: up to 20 km for a 2 m² target (flying aircraft/helicopters); 8-12 km for a 6 m² target (hovering helicopters)
Resolution: M-2106H: 1-2 km (range); 10° (azimuth); 180 km/h (min aircraft target velocity)
Height coverage: M-2106H: up to 2,438 m with an antenna setting of +6°
Weight: M-2106: 26 kg (transceiver); 55 kg (quadripod antenna assembly)
M-2106H: 30 kg (transceiver); 55 kg (quadripod antenna assembly)
Display: synthetic display using concentric circles of red Light Emitting Diodes (LEDs). Azimuth presentation in the form of LEDs at 10° intervals; selectable range presentation in the form of LEDs at 1 km (max 10 km) or 2 km (max 20 km) intervals.

Contractor
Elta Electronics Industries Ltd (a subsidiary of Israel Aircraft Industries Ltd), Ashdod.

VERIFIED

EL/M-2128 miniature intruder detection radar

Type
I/J-band (8 to 20 GHz) miniature intruder detection/point security radar.

Description
The EL/M-2128 is a miniature 350 × 220 × 320 mm radar sensor which is designed for security applications such as perimeter monitoring, gap filling in wider security systems and small point protection of assets such as logistics dumps, government buildings and parked aircraft and trucks. The system provides an all-weather, high-reliability, automatic detection and warning capability and can be installed on watch towers, buildings, fences, the ground and vehicles such as Jeeps or Land Rovers to provide a deployable capability. Clusters of EL/M-2128 sensors can be controlled from a central control station (via an RS-422 port and/or discrete line) and the equipment can be interfaced with an electro-optic surveillance system if required. System control and display are personal computer based.

An EL/M-2128 sensor unit deployed on a perimeter fence

Status
As of this edition, the EL/M-2128 is understood to be available.

Specifications
Frequency: I/J-band (8-12 GHz sub-band)
Transmitter peak power: 1 W
Detection range: 500 m (pedestrians); 1,000 m (vehicles)
Azimuth sector coverage: 120°
Azimuth resolution: 30°
Sensor unit weight: <5 kg

Contractor
Elta Electronics Industries Ltd (a subsidiary of Israel Aircraft Industries Ltd), Ashdod.

VERIFIED

EL/M-2129 Movement Detection and Security Radar (MDSR)

Type
I/J-band (8 to 20 GHz) movement detection and security radar.

Description
The EL/M-2129 MDSR is a ground-based coherent equipment which is designed to detect and monitor movement in restricted border zones in and around sensitive installations such as airfields, industrial areas and power stations. The radar utilises Fast Fourier Transform, Track-While-Scan (TWS) and constant false alarm rate signal processing. It can be installed as either a fixed-site system, a tripod-mounted transportable unit or as a vehicular application. In the latter case, preferred host vehicles are the Jeep and Land Rover, with system function being restricted to when the vehicle is at rest. In TWS mode, the M-2129 is quoted as being able to track automatically up to 60 targets simultaneously. Deployment time is noted as being less than 5 minutes.

Status
As of this edition, Jane's sources suggest that EL/M-2129 is available and that India has procured 56 examples of the equipment as part of a US$14.28 million deal which also includes the acquisition of 200 man-portable radars.

Specifications
Frequency: I/J-band (8-12 GHz sub-band)
Transmitter peak power: 5 W with pulse compression
Detection range: 7 km (pedestrian); 15 km (vehicles and helicopters)
Azimuth sector coverage: up to 359° – selectable by the operator
Resolution: 50 m (range); 2° (azimuth)
Accuracy: 20 m (range); 0.5° (azimuth)
Antenna size: 60 × 40 cm
System weight: approx 30 kg (dependent on PC)

Contractor
Elta Electronic Industries Ltd (a subsidiary of Israel Aircraft Industries Ltd), Ashdod.

VERIFIED

An EL/M-2129 radar deployed to monitor activity around an industrial plant

EL/M-2140 surveillance radar

Type
I/J-band (8 to 20 GHz) battlefield ground surveillance radar.

Description
The EL/M-2140 is a ground surveillance radar system which automatically detects armoured vehicles, light vehicles and personnel. Any sector between 10 and 360° can be covered by this broadband radar at selectable scan rates of up to 4 rpm. Seven area display scales may be explored, including three magnified scales and a

EL/M-2140 surveillance radar

The EL/M-2190 ground penetration radar mounted on a tracked all-terrain chassis   0024844

0 to 30 km full scale. The system's minimal target detectable velocity rate can be varied between 1.5 and 48 km/h to reduce false alarms caused by adverse weather and to improve target recognition.

The EL/M-2140 consists of a transceiver, an antenna assembly and a display and control console. It can be deployed in less than 10 minutes by two personnel. The radar can be operated in the field with the transceiver/antenna assembly mounted on a quadripod, on fixed stations, high towers and similar installations, or installed on armoured vehicles. For remote-control operations, the display and control console may be located up to 50 m away from the antenna and transceiver.

The system operates concurrently on over 100 frequencies in the I/J-band and has a very low probability of intercept radar signature. It is fully coherent and will also detect low-flying aerial targets. Innovative electronic counter-countermeasures facilities radically reduce the probability of intercept. Hostile jamming and interference are defeated by spread spectrum and other techniques.

The 25 cm colour TV monitor and the display processor provide a comprehensive map-like picture of the surveillance area. Six types of video display, including high-resolution clutter map, target displays in colours superimposed with universal transverse Mercator co-ordinates, graphic symbols, a tote menu and controllable fading rate for target tracking are supplied.

Several options are available for improved tactical integration: an optical system for full identification; a plotter for hard-copy maps; a modem for communications and a navigation system. All of these options are easily integrated with the radar system.

Status

As of this edition, Jane's sources report that the EL/M-2140 surveillance radar has been produced under licence in the Czech Republic as the BR2140E. Here, the system forms part of the sensor suite fitted to the BMP-2 based Snezka reconnaissance and observation vehicle. The first Snezka unit is understood to have been delivered to the Czech Army during 1995. Elsewhere in the world, EL/M-2140 is noted as having been evaluated by the Austrian Army.

Specifications

Frequency: I/J-band (8-18 GHz sub-band)
Peak power: 70 W (with pulse compression)
Polarisation: linear vertical or circular
Detection range: 15 km (personnel); 25 km (helicopters); 30 km (vehicle); 33 km (tank)
Azimuth scanning: any sector between 10 and 360°
Scan rate: 5 rates up to 4 rpm
Min detectable velocity: continuous variation between 1-48 km/h
Range gates: 128
Antenna gain: 38 dB
Weight: transceiver 43 kg

Contractor

Elta Electronics Industries Ltd (a subsidiary of Israel Aircraft Industries Ltd), Ashdod.

VERIFIED

EL/M-2190 ground penetration radar

Type

Mobile Ground Penetration Radar (GPR) system.

Description

The EL/M-2190 GPR system is a solid-state digital equipment designed to detect, measure, analyse and classify buried objects such as mines, unexploded ordnance, hazardous waste, cables and pipes together with the mapping of tunnels and bunkers. The radar is mounted on a four-wheeled, low-signature,

A representative EL/M-2190 control station   0024843

remote-controlled vehicle. In the mine clearance role, the system is configured to detect and mark in real time both plastic and metal cased mines and the carrier vehicle's pressure footprint is said to be low enough to allow it to advance through minefields without being destroyed. The radar used is described as having proven capability in a variety of terrains, soils and environments.

Status

As of this edition, an EL/M-2190 robotic mine detection system application has been procured by the Swedish Defence Materiel Administration for use by the Swedish Army in Bosnia.

Specifications

Carrier vehicle's forward velocity: 1 m/s
Radar detection width: 2-3 m
Radar detection depth: 0.3 m
Radar unit weight: 52 kg

Contractor

Elta Electronics Industries Ltd (a subsidiary of Israel Aircraft Industries Ltd), Ashdod.

VERIFIED

ITALY

RASCAL air surveillance radar

Type

Short-range air surveillance radar systems.

Description

RASCAL (*RAdar di Scoperta e Controllo Aereo Locale*) is an autonomous short-range air surveillance radar system for use in the battlefield area. Its main functions are the detection of air targets flying at low and very low level in adverse clutter environments, the automatic recognition on a single scan of hovering and pop-up manoeuvring helicopters, automatic threat evaluation, target designation and assignment to weapons systems.

Rascal air surveillance radar

RASCAL consists of a search radar, a command and control console and a datalink for communications with other air defence systems and data transmission to weapons. The radar is a pulse-Doppler type providing real-time signal and data processing. Detection range is stated to be 20 km on low-flying aircraft and 8 km on hovering helicopters. Other features include automatic 20 target track-while-scan, minimum set up time, adjustable antenna height for optimum operation, command and control of up to 10 weapon systems, search on move and an alphanumeric/graphic colour display.

RASCAL can be integrated into an air defence network and can operate as a stand-alone system. Modular construction is used for ease of vehicle or shelter configuration and the complete system is transportable by cargo aircraft.

Status
As of April 2001, RASCAL is understood to have been procured by the Italian Army for installation aboard M113 armoured personnel carriers.

Contractors
Alenia Difesa Avionic Systems and Equipment Division - Officine Galileo, Florence.

VERIFIED

....

Spada point defence radars

Type
Search/interrogation, tracking and illuminator radars for the Spada air defence system.

Description
Spada is a missile system developed under a contract awarded to Alenia Difesa by the Italian Ministry of Defence (Air Force), for the close-in air defence of relatively small permanent objectives of strategic importance.

The Spada system is organised in two main modules: the Detection Centre (DC) and the Firing Sections (FS).

The DC consists of the Search and Interrogation Radar (SIR), made up of the SIR antenna pedestal and relevant equipment shelter and the Operational Control Centre (OCC) shelter.

The tasks of the DC are target search, detection, identification, evaluation and designation for engagement by the FS. As SIR, the Spada utilises the Alenia Pluto E/F-band (3 to 4 GHz) radar, suitably modified for the specific purpose (use of the available power). The SIR is a low-altitude search radar, particularly suited for operations in dense clutter environments and in the presence of severe electronic countermeasures, which provides data for the OCC. The latter consists of three operational consoles, a data processing system with two NDC-160 interconnected digital computers and a number of communication links (digital data and telephone).

The FS consist of the Fire-Control Centre (FCC), made up of the Tracking and Illumination Radar (TIR) antenna pedestal, inclusive of an on-mounted optronic sensor (TV), the Control Unit (CU) shelter, for TIR and fire-control equipment housing and the Missile Launchers (ML) with six ready-to-fire Aspide missiles each.

The tasks of the FS, which represents the reaction centre of the system, are the acquisition and destruction, through missile intercept, of the targets assigned by the DC. At the FCC the TIR carries out target acquisition, tracking and illumination for missile guidance. As an additional mode of operation the TIR provides for target search (360° or sectorial scan), detection and self-designation. The TV sensor is used both as a back-up to the radar and as an aid for target identification and discrimination as well as for kill assessment. The associated CU supervises these functions which are carried out automatically with the possibility of manual intervention. The ML provide for missile storing, aiming, selection and preparing to fire as well as for automatic launching. As TIR, the Spada uses a pulse-Doppler monopulse radar derived from the Alenia Orion 30X radar. It provides, through the same antenna, for target tracking and illumination for missile semi-active guidance, operating in I- (8 to 10 GHz) and J-bands (10 to 20 GHz) respectively. The TIR

Pluto search and identification radar used with Spada

antenna group is available in two configurations: one mounted on a pallet type dedicated structure and the other, more compact, directly installed on the CU shelter roof.

Status
As of April 2001, the Spada missile system is understood to have been procured by a number of armed services around the world, including the Italian Air Force and the Royal Thai Air Force.

Specifications
Search and Interrogation Radar
Frequency: F-band (3-4 GHz)
Antenna type: supercosec^2
Transmitter: coherent chain, TWT final stage
Waveform: coded
PRF: staggered
Frequency agility: pulse-to-pulse or burst-to-burst
MTI: double canceller, frequency agility compatible
Tracking Radar
Frequency: I-band (8-10 GHz)
Antenna type: monopulse
Transmitter: coherent chain, TWT final stage
PRF: staggered
Frequency agility: pulse-to-pulse or burst-to-burst
Illuminator Radar
Frequency: J-band (10-20 GHz)
Antenna type: monopulse
Transmitter: Klystron oscillator

Contractor
Alenia Difesa Radar Systems Division, Rome.

VERIFIED

....

X-TCP/TAR series short/medium range air defence control systems/radars

Type
Family of air defence control systems that make use of an I-band (8 to 10 GHz) search, acquisition and target designation radar.

Description
The X-TCP/TAR series is a family of air defence control systems that are optimised for very low to medium altitude target detection/designation in support of man-portable air defence systems and small and medium calibre anti-aircraft guns. Identified family members comprise the X-TCP25, X-TCP50, X-TAR25 and X-TAR3D equipments, the known details of which are as follows:
X-TCP25
X-TCP25 is an X-TAR25 radar (see following) equipped tactical control post that is vehicle mounted and optimised for short-range battlefield applications, with an emphasis on the detection and classification of air targets as well as the cueing of dispersed anti-aircraft weapon systems. Here, target data hand-off is by means of direct link or via a target data receiving unit.
X-TCP50
X-TCP50 is described as being a 'more powerful' version of X-TCP25 that has a range of over 40 km and a target capacity of 50.
X-TAR25
X-TAR25 is a fully coherent, I-band pulse-Doppler radar that is optimised for the detection of fixed-wing aircraft and helicopters. System features include a vertical dual-beam configuration, computer control and a 'high' degree of automation (to facilitate ease of operation, multiple target handling and data fidelity), an identification friend-or-foe capability and jammer rejection. Oerlikon Contraves claim that X-TAR25's vertical dual-beam format enhances performance when compared to conventional 2-D radars and that 'several official qualification tests' have shown that the radar is capable of effective operation in heavy clutter and countermeasures polluted environments. Operational applications include acting as a search radar within an Air Defence/Anti-Tank System (ADATS); as a self-propelled acquisition and fire-control sensor when mounted on an armoured personnel carrier such as the M113 or as a high-mobility search and tactical control post (see X-TCP25) for anti-aircraft weapons and man-portable air defence systems.

The X-TCP25 tactical control post

X-TAR3D
X-TAR3D is described as being a 3-D variant of the baseline radar that is equipped with a phased-array antenna.

Status
As of March 2001, X-TCP25 and X-TAR25 were described as being in 'series' production while X-TCP50 and X-TAR3D were noted as being in final development.

Specifications
Frequency: I-band
Scan rate: 30, 40 or 60 rpm
Range: 30-50 km (instrumented)
TWS: 20-50 targets (automatic initialisation)
Total weight: <2 tons (fully equipped shelter with radar)

Contractor
Oerlikon Contraves SpA, Rome.

UPDATED

JAPAN

J/MPQ-N1 (Type 92) radar

Type
Mortar locating radar.

Description
The Type 92 mortar locating radar is apparently a Japanese licence-built version of the American AN/MPQ-4 equipment which was produced by the General Electric Company. The Japanese model, designated J/MPQ-N1, is thought to have essentially the same performance as the US-built radar which is described later in this section.

Japanese Type 92 (J/MPQ-N1) mortar locating radar on the move with its reflector erected (K Ebata)

The Type 92 mortar locating radar's antenna reflector is usually folded when the equipment is being towed (K Ebata)

The purpose of these radars is to detect and locate enemy mortar positions so that counterbattery action can be carried out. The two-beam intercept principle is employed, first-shell acquisition is possible and the system can handle multiple targets.

Status
As of this edition, the status of the J/MPQ-N1 radar was uncertain.

VERIFIED

J/MPQ-P7 radar

Type
I/J-band (8 to 20 GHz) battlefield weapon locating radar.

Description
This artillery locating radar is apparently of Japanese design, but it is not known if there has been any technical collaboration from America or elsewhere. This said, there is an obvious similarity to the Hughes AN/TPQ-36 and TPQ-37 mortar and artillery locating radars. Like the US-designed radars, the J/MPQ-P7 employs electronic scanning of a planar-array antenna to generate the multiple beams necessary for detection and trajectory analysis of incoming projectiles. In this respect the Japanese equipment is so far the only known comparable radar to appear from outside the American defence industry for operational service. The moderately large antenna array is transported on a wheeled trailer towed by a tracked vehicle which houses items of equipment and the radar crew. The antenna is erected when deployed, but folds flat for transit. Electronic beam steering probably obviates any need for the array to rotate in use. It is likely that a sufficiently broad azimuth coverage arc can be achieved by electronic scanning to provide surveillance of the designated threat direction(s).

Few technical details have been obtained, but it is reported that the equipment has a peak output power of 250 kW, operates in the I/J-band and is capable of ranges of more than 30 km.

Status
As of this edition, the status of the J/MPQ-P7 radar was uncertain.

Contractor
Toshiba Electric Company, Tokyo.

VERIFIED

J/MPQ-P7 artillery locating radar equipment in transit configuration (K Ebata)

Tan-SAM engagement radar

Type
Engagement radar for the Tan-SAM missile system.

Description
The engagement radar for the Tan-SAM is a multifunction, multitarget 3-D pulse-Doppler system with three basic modes of operation, using mechanical rotation in azimuth and electronic scanning in the vertical plane. In the omnidirectional search mode, the antenna rotates at 10 rpm and sweeps 360° in azimuth and 15° in elevation during a full rotation. In the sector search/course track mode the antenna

The Tan-SAM fire-control vehicle with the phased-array radar aerial raised ready for use

does not rotate, but automatically sweeps 110° in azimuth and 20° in elevation. Range, azimuth and elevation information on up to six targets can be maintained in this mode and a computer is used to assess and display, on a Cathode Ray Tube (CRT), the degree of threat from each target.

On selection of the most threatening target the radar is switched to the third mode, the fine tracking mode, in which more precise tracking data is obtained and the single shot kill probability is displayed on the CRT and updated second by second. Two targets are selected for engagement and the location data for each is passed automatically to each of two launcher systems.

The antenna of the Tan-SAM system is a phased-array and is mounted on top of the system control cabin. It is 1 m wide and 1.2 m high. The range of the radar is 30 km.

Status
As of this edition, the Tan-SAM engagement radar is thought to have been procured by the Japanese Ground Self-Defence Force (JGSDF) and the Japanese Air Self-Defence Force (JASDF). Japan's total Tan-SAM radar requirement is believed to have been 27 units for the JASDF and 47 for the JGSDF.

Contractor
Toshiba, Chiyoda-ku, Tokyo.

VERIFIED

KOREA, SOUTH

GLAS-830M surveillance radar

Type
I-band (8 to 10 GHz) mobile low-altitude air surveillance radar.

Description
The GLAS-830M is a mobile low-altitude I-band coherent pulse-Doppler air surveillance radar designed to operate with short-range air defence guns and missiles. It is designed to cope with low- and very low-flying threats, such as attacking aircraft and hovering helicopters at tree top level. The GLAS-830M is a versatile sensor system which can be deployed as a gap filler in mountainous terrain, for coastal surveillance and for point defence of a protected fixed asset.

The GLAS-830M employs a travelling wave tube amplifier, pulse compression, a frequency synthesiser and digital raster scan to maximise its effectiveness for low altitude in the forward area. Its mapping function also provides substantial advantages to the operator. For electronic counter-countermeasures purposes, GLAS-830M employs many features, including frequency agility, low peak power and staggered pulse repetition frequency.

Continuous 360° surveillance is provided out to 40 km range. Targets are interrogated by Identification Friend-or-Foe (IFF) and displayed on a 48 cm cathode ray tube. All information with respect to the target is transmitted by frequency modulated radio or cable link to the fire-control data receiver, which is emplaced with the weapon system (anti-aircraft gun or surface-to-air missile).

The system is installed in a shelter mounted on a military vehicle, with a trailer-mounted generator. The antenna, with an IFF dipole, is mounted on top of the shelter and is folded down when in transit. The radar shelter itself consists of a display and control console, IFF interrogator, transceiver console and auxiliary equipment.

Status
As of this edition, GLAS-830M is thought to have been procured by the South Korean defence forces.

Specifications
Transmitter: TWTA
Frequency: I-band (8-10 GHz)

Peak power: 8 kW
Channels: 51
Tracking capacity: 16 targets
Instrumented range: 40 km
Detection range: 36 km
Coverage: 40° (elevation – 0-3 km altitude); 360° (azimuth)
Azimuth resolution: 1.5°
Beamwidth: 1.5° (horizontal); CSC^2 (to 40° vertical)
Antenna rotation: 30 rpm

Contractor
Goldstar Precision, Seoul.

VERIFIED

NORWAY

RD 170BT portable surveillance radar

Type
I-band (8 to 10 GHz) man-portable surveillance radar.

Description
The RD 170 BT is an approximately 60 kg man-portable surveillance radar which breaks down into three approximately 20 kg rucksack loads. Power is supplied from either a battery or a lightweight hand carried generator and the equipment may be operated on-site or by remote control. The radar's magnetron-based transmitter generates short and long pulses to optimise detection performance. The short pulse output is used for three of the set's eight range scales, namely 0.47, 1.39 and 2.8 km. Maximum detection range is quoted as being 89 km. The display used is a 30 cm raster scan daylight viewing unit which shows data in a six colour format. The system was developed specifically for use by the Royal Norwegian Navy who deploy it aboard patrol boats for use as a land-based surveillance sensor when such vessels are using coastal hides and do not wish to reveal their presence through the use of onboard radars. In such circumstances, the RD 170 BT is deployed ashore and is set up at a remote site to provide its host vessel with discrete area surveillance.

Status
Field trials of the RD 170 BT are reported to have taken place during 1986 and an order for 12 such radars for the Royal Norwegian Navy was completed during 1994.

Specifications
Frequency: I-band (9,380-9,440 MHz)
Peak power: 5 kW nominal
Pulse-widths: 0.08 and 0.65 μs
PRF: 1,500 Hz
Scan rate: 25 rpm
Beamwidth: <2.7° at –3 dB
Antenna gain: approx 25 dB
Polarisation: horizontal
Receiver type: linear
Receiver IF bandwidth: 4 MHz (long pulse); 10 MHz (short pulse)
Receiver noise factor: 9.5 dB

Contractor
Racal Norge A/S, Laksevåg.

VERIFIED

POLAND

TRD-1211 surveillance radar

Type
D-band (1 to 2 GHz) air surveillance radar.

Description
The TRD-1211 is a 3-D, beam stacked air surveillance radar which is both transportable (by air or road) and suitable for fixed site applications. The radar can also operate as an unattended equipment. The complete system comprises an antenna unit, a transmitter cabin and an operator cabin. An optional crew rest area/ maintenance shelter can be added if required. The antenna unit incorporates upper transmit and lower receive planar-array elements together with a secondary array mounted above the transmit/receive assembly. The transmission array generates a cosec^2 pattern and eight independent pencil beams are formed for reception. Each beam has its own reception and processing channel. The system's preamplifiers, transmit/receive beam-forming networks and front-end receivers are mounted in a rotary cabin. There is a multichannel rotary joint which carries high and low radio frequency signals.

The antenna assembly used in the TRD-1211 3-D air surveillance radar

The TRD-1211 is noted as incorporating clutter rejection, electronic counter-countermeasures features, built-in test, fault location, an Identification Friend-or-Foe (IFF) subsystem (which meets International Civil Aviation Organisation 10 and Standard NATO Agreement 4193 requirements) and adaptive coherent moving target indicator. The radar automatically initiates and maintains tracks on up to 120 targets and generates target identification, heading and speed data. The system is computer controlled and the operator can call up 100 plus menus (via a keyboard) covering operating modes, system parameters, critical function monitoring and automatic fault isolation. Two colour raster display monitors are provided which can be augmented by a remote operator display which is linked to the system by means of up to 1,000 m of fibre-optic cable. TRD-1211 also has provision for a narrowband datalink to transmit 3-D radar and IFF data to an external command, control communications and intelligence system.

Status
As of the mid-1990s, the TRD-1211 was noted as being in production and in service with the Polish Army.

Specifications
Frequency: D-band (1-2 GHz, 100 MHz bandwidth, frequency diversity)
Detection/instrumented range: 250 km (fighter size targets)/350 km
Altitude coverage: up to 40,000 m
Accuracy (RMS): 0.2°/120 m (bearing); ±600 m (height) at 150 km
Track capacity: 120 targets
ECCM: low-sidelobe antenna; random/programmed frequency agility; jamming analysis transmission selection (dedicated jamming receiver); PRF stagger; cool antenna for reduced IR output; selectable sector operation; remote control; ARM resistance
Signal processor: 8 channels; 4 pulse adaptive MTI (adaptive clutter map control); 38 dB SCV
Antenna
Type: twin planar-array (32 dipole × 16 row transmit; 32 dipole × 40 row receive)
Size: 8.5 × 6.5 m
Azimuth beamwidth/coverage: 2.8/360°
Azimith sidelobes: <35 dB (1st); <45 dB (remaining)
Elevation beamwidth/coverage: 0-30° (1 × cosec² transmit beam; 8 × 3-7.5° simultaneous receive beams)
Directivity: 30 dB (transmit); 35 dB (receive, lower beams)
Rotation rate: 6 rpm
Transmitter
Type: coherent amplifier chain (last stage CFA/solid-state drivers/TWT)
Power output: 0.9 MW (peak); 6.4 kW (average)
Pulse-width: 20 μs LFM (0.4 μs compressed)
PRF: 350 Hz (average)
Receiver
No of receiver channels: 8
Noise figure: 2.5 dB (max)
IFF subsystem
Type: RAWAR SA102M (licensed from Thales)
Modes: 1, 2, 3 & C; 4 (option)
Decode: active/passive; G3b/E3M
Antenna: Σ/Δ patterns: 16 dB gain

Contractor
Przemysłowy Instytut Telekomunikacji, Warsaw.

VERIFIED

RUSSIAN FEDERATION AND ASSOCIATED STATES (CIS)

Introduction

The equipment described in this section is produced by or is in service with the 21 republics making up the Russian Federation together with the independent republics of Armenia, Azerbaijan, Belarus, Georgia, Kazakhstan, Moldova, Tajikistan, Turkmenistan, Ukraine and Uzbekistan. Together, these various states make up the Commonwealth of Independent States (CIS) economic group. Despite increasing openness, information concerning the grouping's defence electronics industry remains less than full. Accordingly, some equipments described here are still referred to by their NATO reporting names.

VERIFIED

1S91 fire-control radar

Type
G/H/I-band (4 to 10 GHz) fire-control radar for the 9M9/9M336 Kub (NATO reporting name SA-6 'Gainful') Surface-to-Air Missile (SAM). NATO reporting name Straight Flush.

Description
The 1S91 is the engagement and command guidance radar used with the 9M9/9M336 SAM. It is believed that the equipment provides the following capabilities:
- limited search
- low-altitude detection/acquisition
- target tracking and illumination
- missile radar command guidance
- secondary radar missile tracking.

Reports of the frequencies used for these functions have not been entirely consistent. It seems, however, that the first three functions are performed in G/H-band, the low-altitude function being performed at about 5 GHz and the high-altitude functions at about 6 GHz. Target tracking (and probably illumination) is an I-band function at around 8 GHz and so probably is the command link, while the secondary radar response for missile tracking is probably at a rather lower frequency. Range is believed to be 60 to 90 km, with a maximum altitude coverage of 10,000 m.

The arrangement of the system can be seen in the accompanying picture. The upper tracking radar antenna assembly is able to rotate independently of the lower antenna. The two antennas and associated apparatus can be rotated relative to the carrying vehicle on a turntable which is presumably located at the top of the circular turret, on which the whole assembly is mounted. Examination of the original photographs suggests that the lower antenna may be pivoted so as to execute a sector scan of some kind and there is some indication that it can be tilted up or down, presumably to compensate for vehicle attitude. These observations are consistent with the functions listed previously. The combined radar superstructure is probably too massive for any kind of continuous circular search process. The arrangement probably provides for a slow circular search into an associated sector scan, the circular motion being halted when a target is located and resumed only if the target starts to move out of the sector.

The feed arrangements of both antennas are interesting. The lower antenna appears to have two feeds, the upper one consisting of a single horn, which presumably produces a low angle pencil beam. The lower feed comprises two or possibly three horns which may well produce a slightly shaped high-cover beam. There could be more sophistication to it than this, but this is not likely. The use of two separate feeds, however, is consistent with the suggestion that the low and high altitude patterns are radiated at different frequencies.

The 1S91 fire-control radar (Tim Ripley)

There is, however, some additional waveguide in this feed which is not explained by what has been described above. It appears to be for a higher radiation frequency (than G/H-band) and could be for one of the other functions (command link or secondary radar – more probably the former). Finally, it looks as though both parts of the main feed have a quarter-wave plate circulariser in front of them. If this is a correct reading of the photograph, it tends to date the technology of the equipment: a radar of this sort built in the West in the late 1960s would have used waveguide circularisers.

The assembly is evidently a conical scanning system using a rotating feed driven by a motor in the housing at the rear of the dish. This housing also almost certainly contains the microwave transmitter/receiver stages to avoid the need for multiple rotating waveguide joints.

Two of the struts supporting the projecting device appear to be simple metal rods or tubes secured to the face of the dish. The third (bottom) strut, however, is different, fairly certainly passing through a hole in the dish and could be a waveguide. The way in which the upper struts meet the device seems a little clumsy if they are no more than supports. While it may be that the device is yet another antenna of some sort, we incline to the view that it is either another quarter-wave plate or a beam-spoiler for some operational purpose and that the bottom strut is merely a kind of push rod (hence the need to go through a hole – a waveguide would not) for moving the device into and out of the way.

Apart from the microwave stages of the tracking radar it seems probable that the bulk of the radar electronics is housed in the bin-shaped structure beneath the tracking radar pedestal. The turret may be assumed to house the displays and control gear. Within SA-6 batteries, the 1S91 is usually teamed with the Scoreboard A, Flat Face, Spoon Rest or Long Track surveillance radar and the Thin Skin B height-finder.

Status

The 9M9/9M336 SAM is understood to have first entered service during 1970 and over time, is understood to have been supplied to Algeria (10 launchers), Angola (six), Azerbaijan, Belarus, Bulgaria (30), Cuba (12), the Czech Republic (180), Egypt (75), Georgia, the former German Democratic Republic (120), Guinea, Guinea-Bissau, Hungary (80), India (30), Iran, Iraq (25), Kazakhstan, Libya, Mozambique, Poland (280), Romania, the Russian Federation, Slovakia, Somalia, Syria, Tanzania, Ukraine, Vietnam, Yemen and the Federal Republic of Yugoslavia. It is perhaps worth noting that a mid-1970s Egyptian SA-6 battery comprised four launch vehicles, one or more reload vehicles, one Straight Flush and one Flat Face radar and four ZU-23 anti-aircraft guns for battery self-defence. More recently, Yugoimport SDPR announced in October 1998 that it was offering an SA-6 upgrade package that included low-noise, solid-state radio amplifiers and a digital moving target indicator capability.

VERIFIED

30N6E/30N6E1 tracking radar

Type

I/J-band (8 to 20 GHz) tracking and missile guidance radar for the S-300PMU and S-300PMU-1 Surface-to-Air Missile (SAM) systems. NATO reporting name Flap Lid.

Description

The 30N6E (30N6E1) equipment is described as being a 'multifunction, four co-ordinating, mono-impulse, impulse Doppler' tracking and missile guidance radar that forms part of the S-300PMU (NATO Reporting Name SA-10a 'Grumble') and S-300PMU-1 (NATO Reporting Name SA-10b 'Grumble' SAM systems. As such, it is reported to be able to track six targets and control up to 12 missiles (two per target) simultaneously. The equipment is further noted as offering 'high' resolution and 'good' countermeasures resistance. A phased-array antenna is used which is either trailer- (designated by NATO as Flap Lid A) or vehicle- (Flap Lid B) mounted and provides 360° coverage in azimuth.

Within the S-300PMU system, 30N6E is used for stand-alone target acquisition/target designation or in conjunction with the 36D6 (NATO Reporting Name Tin Shield) or 76N6 (NATO Reporting Name Shell) radars. In the S-300PMU-1 application, a modified 30N6E radar (the 30N6E1) is used in conjunction with the 83M6E control centre and the ACS types 'Baikal-1Э' and 'Senezh-M1Э'. Here, the 30N6E1 radar provides ballistic missile target acquisition and missile guidance (using a 'missile-as-direction finder' technique).

The vehicle-mounted application of 30N6E is known to NATO as Flap Lid B

Status

As of mid-2000, Jane's sources identify the S-300PMU/PMU-1 SAM system as being in service with the armed forces of the People's Republic of China (S-300PMU), Cyprus/Greece (S-300PMU-1), Iran (S-300PMU), the Russian Federation and Syria (S-300PMU).

Specifications

Missile/target tracking sector limits: –3 to +85° (elevation); 90° (azimuth)
Operating sector: 0-360° (azimuth)
Target tracking range: up to 200 km
Target velocity: up to 2,800 m/s
Number of targets engaged simultaneously: 6
Number of missiles guided simultaneously: 12
Interception altitude: from 25 m upwards

Contractor

Rosoboronexport (export agent), Moscow.

UPDATED

1RL144M fire-control system

Type

E- (2 to 3 GHz) and J-band (10 to 20 GHz) anti-aircraft gun and missile fire-control system. NATO reporting name Hot Shot.

Description

1RL144M is the radar system associated with the Russian Federation and Associated States (RFAS), 2S6 mobile air defence gun missile system. The Hot Shot system consists of two separate radars, one mounted at the front of the turret and used to track the target and direct the gun. This fire-control radar can be independently slewed about 220° for target tracking. The second equipment is positioned at the rear of the turret and is the surveillance and target acquisition radar. This has a full sweep of 360° and can be folded over the rear of the turret during travel to protect the antenna. This technique of using two radars offers a number of advantages, particularly that of allowing the vehicle to continue surveillance during engagements, thus reducing the vulnerability of the vehicle to multiple air attacks.

The complete radar system includes the E-band (2 to 3 GHz) surveillance radar and the J-band (10 to 20 GHz) tracking radar. The system features a special search mode to detect low-flying aircraft (typically 15 m) with the scanning being carried out with an elevation of +1°, together with a strong ground return signal suppression mode. 1RL144M's surveillance radar is credited with being able to detect targets at ranges up to 18 km while its tracking system can handle targets at ranges up to 13 km. The overall system is also known to incorporate a 1RL138 identification friend-or-foe package.

Status

Over time, the 1RL144M fire-control radar has been procured by the armed forces of Russia and a number of East European countries.

VERIFIED

Radars used in the 2S6 air defence system comprise an E-band surveillance set mounted at the rear of the turret and a J-band tracker forward (Christopher F Foss)

9S15MTZ target acquisition radar

Type

3-D, centimetric band target acquisition radar that is designed for use with the S-300V Surface-to-Air Missile (SAM) system's command post or the radar data processing centres of automated air defence command and control system. NATO reporting name Bill Board.

The 9S15MTZ target acquisition radar has the NATO reporting name Bill Board and is shown here deployed for operations (Christopher F Foss)

Description

The 9S15MTZ target acquisition radar is described as being a high-capacity, 3-D, ground-mobile, four-man, coherent pulse equipment that utilises electronic elevation and mechanical azimuth scanning. Target detection zones are noted as being defined by specific pulse-length, pulse repetition frequency, antenna rotation rate and antenna radiation pattern parameter packages. Electronic counter-countermeasures provision comprises:

- low side-lobe levels that are claimed to 'rapidly diminish to the background (noise) level'
- instantaneous 'cycle-to-cycle' frequency hopping based on analysis of the signal strength of detected jamming signals
- 'optimum' filtering and 'limitation' of received echo signals
- three-channel automatic cancellation of detected active jamming signals
- receiver sensitivity time control circuitry
- non-linear moving target indication circuitry with automatic correction for wind velocity
- non-coherent signal accumulation
- channel blanking on those bearings where a specified clutter threshold has been exceeded
- automatic noise jammer bearing calculation and transmission to the S-300V system command post.

The 9S15MTZ radar makes use of two application specific processors that handle system control, target (including jammers) detection, target co-ordinates computation and data transmission to the S-300V missile's command centre. The equipment features a built-in-test routine and is mounted on a tracked chassis that is designated as the Article 832. With an all up weight of 47 tons, 9S15MTZ makes use of a 200 V/400 Hz, 130 kW gas-turbine generator (backed-up by a power unit driven by the vehicle's engine) or can draw power from an external source. A 'coded' datalink is provided for data dissemination, autonomous navigation, topographic survey and orientation and the system's communications suite is completed by radio and voice channels. As yet unconfirmed Jane's sources suggest that the 9S15MTZ radar operates in the F-band (3 to 4 GHz) frequency band.

Status

The S-300V (NATO Reporting Name SA-12a 'Gladiator'/SA-12b 'Giant') SAM system is reported to have entered service during 1986/1987 and as of March 1999, to have been procured by the armed forces of the Russian Federation (100 plus batteries split between the S-300V and S-300V1 systems). During April 1997, India announced its intention of acquiring six to seven S-300V batteries for integration into an air defence/anti-ballistic missile defence system alongside the indigenous Akash low- to medium-altitude SAM.

Specifications

Wavelength: centimetric
Azimuth coverage: 360°
Elevation coverage: 45° (aircraft targets); 55° (ballistic missile targets)
Detection range: 200 km (MiG-21 sized target)

Accuracy: 30 min of arc (elevation); 36 min of arc (azimuth); 250 m (range)
Scan rate: 6-18 s (dependent on selected operating mode)
Number of targets located per scan: 200
Set up/close down time: 5 min

Contractor

NIIP (Measuring Instruments Research Institute), Novosibirsk.

VERIFIED

9S18M1E target acquisition radar

Type

3-D, centimetric band target acquisition radar that is designed for use with the Buk-M1E Surface-to-Air Missile (SAM) system's command post, together with those of associated air defence units. NATO reporting name Snow Drift.

Description

The 9S18M1E target acquisition radar is described as being a 3-D, centimetric band, three-man, coherent pulse equipment that forms part of the Buk-M1E (NATO reporting name SA-11 'Gadfly') SAM system. It should be noted that the designation 'M1E' is thought to delineate the 'Ehksportiynyi' (export) variant of the basic equipment. The radar's antenna takes the form of a flat waveguide array that produces a pencil beam that is scanned electronically in elevation. Azimuthal coverage is provided by mechanical rotation of the antenna. Overall system design incorporates low sidelobes, frequency agility and 'programmed' control of the radar's azimuth and elevation scan patterns to optimise coverage of particular sectors according to their particular threat and jamming environments. System control is by means of an integrated digital computer and pulse-to-pulse cancellation and receiver sensitivity time control circuitry is used to minimise clutter interference. Weighing approximately 35 tons, the 9S18M1E radar is mounted on a GM-567 cross-country tracked chassis. Power is drawn from either a 75 kW gas turbine driven generator, a generator run-off the GM-567's engine or (via converters) a 200 V/50 Hz industrial supply and the equipment features a built-in test routine together with an integral 'telecoded' data communications link with which to disseminate acquired data to the Buk-M1E's command post.

Status

The Buk-M1 SAM system is reported to have entered service during 1982 and, as of March 1999, Jane's sources were suggesting that the missile was in service with the armed forces of Belarus (army), Finland (nine batteries - army), the Russian Federation (250 batteries - army), Syria (air defence command), Ukraine (army) and the Federal Republic of Yugoslavia (army). As of March 2001, Jane's sources were also suggesting that a 9S18M variant formed part of the Russian Ural SAM system (NATO Reporting Name SA-17 'Grizzly') that utilises an upgraded variant of the missile used in the Buk-M1 architecture.

The 9S18M1E target acquisition radar is mounted on a GM-567 tracked cross-country chassis (Christopher F Foss)

Specifications

Wavelength: centimetric
Azimuth coverage: 120° (ballistic missile targets); 360° (aircraft targets)
Elevation coverage: 40° (aircraft targets); 55° (ballistic missile targets)
Detection range: 100 km (MiG-21 sized target)
Accuracy: 30 min of arc (elevation); 38 min of arc (azimuth); 150 m (range)
Scan rate: 6-18 s (deployment on operating mode)
Number of targets located per scan: 100
Set up/close down time: 5 min

Contractor

NIIP (Measuring Instruments Research Institute), Novosibirsk.

UPDATED

9S32 missile guidance station

Type

Missile guidance station for the S-300V Surface-to-Air Missile (SAM) system. NATO reporting name Grill Pan.

Description

Each S-300V (NATO reporting names SA-12a 'Gladiator' (9M83 Type 2 missile) and SA-12b 'Giant' (9M83 Type 1 missile)) battery has a 9S32 missile guidance station attached to it. The station is manned by a crew of three and receives its targeting assignments from the battery's command post. The phased-array radar on top of the station's cabin is carried in the horizontal position while travelling and hydraulically raised for operational use. It is used to provide the final track data of designation targets and to perform horizon searches in its assigned sector from where the low-altitude targets are likely to appear.

When searching for a target with a 2 m² radar cross section the maximum acquisition is 150 km manual mode and 140 km in its automatic mode. Accuracy is understood to be of the order of 7.8 minutes of arc in azimuth, 0.7 to 1.4 m/s in velocity and 10 to 15 m in range. The antenna may be rotated to give 340° in coverage but usually surveys a 42° area in its normal operating mode.

The Grill Pan system is used to remotely control the target illumination radar on the Transport Erector Launchers And Radars (TELAR) by transmitting to them the necessary data for missile launching and guidance. It can track up to 12 targets and control up to six missiles against these targets simultaneously. Within the S-300V system, 9S32 is teamed with the 9S15 (NATO reporting name Bill Board) target acquisition and 9S19 (NATO reporting name High Screen) sector search radars.

Status

According to Jane's sources, the SA-12a SAM system entered service during 1987, with the SA-12b variant becoming operational three years later. Aside from the RFAS, S-300V systems are thought to have been procured by Belarus, Kazakhstan and Ukraine.

VERIFIED

64N6E target acquisition radar

Type

Centimetric target acquisition radar for use with the S-300 PMU-1 Surface-to-Air Missile (SAM) system. NATO reporting name Tombstone.

Description

The centimetric band, ground-mobile, four-man 64N6E target acquisition radar provides 360° coverage and is designed for use with the S-300 PMU-1 (NATO reporting name SA-10d 'Grumble') SAM system's 83M6E Command and Control (C²) facility (see separate entry). As such, the system appears to be divided into two shelterised elements designated as F6E and F8E. Of these, the F6E element houses the radar's phased-array antenna, transmitter and receiver while the F8E section contains its noise protection, data processing and system control equipment. Both elements are mounted on a 60-ton road train that makes use of the Type 74106 tractor and the Type 9988 semi-trailer.

At the heart of the 64N6E radar is a two-faced phased-array antenna that features electronic scanning in azimuth and elevation combined with 'uniform electromechanical' antenna rotation in azimuth. This combination is claimed to facilitate programmed sector coverage, the division of the system's overall coverage into 'regular' scanning and target tracking sub-zones and the optimisation of the radar's operating time via the use of a two-step target detection procedure combined with a Moving Target Indicator (MTI) mode. Within this latter operating mode, MTI limits in elevation and range are set either automatically or by the radar's operator and the mode is used to minimise the effects of chaff jamming. Proofing against spot jamming is by means of carrier wave frequency hopping while false target interference is minimised by the use of time control circuitry (at ranges of up to 65 km) and/or false alarm level stabilisation (at ranges in excess of 65 km). Scan-to-scan blanking is used to eliminate fixed and/or slow-moving clutter detected in a particular sector during successive scans.

System function can be automatic if required and is based on two application specific processors. These, among other tasks, process the received radar data and calculate target co-ordinates. 64N6E also incorporates a built-in test routine

and features a communications suite that includes voice channels and the 5Ya312E wideband communications facility that transmits information to the 83M6E C² centre via radio or landline. Power is provided by a gas turbine powered generator that produces a three phase, 220 V/400 Hz supply. Alternative external power sources include the 63T6A converter-distribution unit and the 5157A diesel-driven generator. As yet unconfirmed Jane's sources suggest that 64N6E operates in the E/F-band (2 to 4 GHz) frequency range.

Status

According to Jane's sources the S-300PMU-1 SAM system entered service during 1992 and as of March 1999 (and outside the Russian Federation) was in service in the People's Republic of China (four to six batteries in total) and Cyprus/Greece (one battery). Jane's sources also suggest that a navalised version of the system (designated as S-300FM Fort-M) is fitted to Russia's Project 1144 'Kirov/Orlan' class cruiser *Pyotr Vealiky*.

Specifications

Wavelength: centimetric
Azimuth coverage: 360°
Elevation coverage: 13.4° (regular scan); 55° (tracking)
Specified sectors: up to 75
Detection range: 260 km (MiG-21 sized target)
Accuracy: 30 min of arc (azimuth); 35 min of arc (elevation); 200 m (range)
Scan rate: 12 s (regular scan); 6 or 12 s (tracking)
Target forwarding rate per scan: 200
Set up/close down time: 5 min

Contractor

NIIP (Measuring Instruments Research Institute), Novosibirsk.

UPDATED

76N6 multifunction radar

Type

Low-altitude search and acquisition radar. NATO reporting name Clam Shell.

Description

The 76N6 frequency modulated continuous wave low-altitude search and acquisition radar is designed for use with the S-300PMU (NATO reporting name SA-10c 'Grumble') and S-300PMU-1 (NATO reporting name SA-10d 'Grumble') Surface-to-Air Missile (SAM) systems. Equipment features include the road-transportable F52MU shelter (that houses the radar's operator console, transceiver, signal processing hardware, track processor, built-in test subsystem and plan position indicator display) and the FA-51MU antenna head assembly. Here, the structure incorporates shielded transmit and receive arrays (each measuring 2.8 m²) and offers full 360° cover in azimuth. The FA-51MU assembly is mast mounted and can be installed on either the 23.8 m high 40V6M or the 38.8 m 40V6M2 hydraulically elevated units. When lowered for travelling, both mast types are carried on a 5T58 trailer that is towed by a MAZ-537 truck. An FA-51MU variant is also available that is directly mounted on a ChMAP trailer (see 76N6S entry). The 76N6 radar's transceiver module is reported to produce a 1.4 kW continuous wave output while its signal processing is claimed to be able to reject terrain, precipitation and chaff clutter together with active jamming. Set up time for the 40V6M mast is given as being 1 hour while that for the 40V6M2 equipment is quoted as being 2 hours.

Status

As of June 2001, Jane's sources were suggesting that outside the Russian Federation, S-300PMU and S-300PMU-1 SAM systems were in service with the armed forces of the People's Republic of China (S-300PMU-1), Cyprus/Greece (S-300PMU1), Iran (S-300PMU) and Syria (S-300PMU).

Specifications

Detection range: 90 km (target at 500 m altitude - up to 0.02 m² RCS, up to 740 m/s velocity); 120 km (target at 1,000 m altitude - up to 0.02 m² RCS, up to 740 m/s velocity)
Coverage: 1° (elevation - single polarisation); 6° (elevation - multiple polarisation); 360° (azimuth)
Chaff rejection: at least 100 dB
False alarm probability: 10^{-8} to 10^{-9}
Data generation rate: 3 s
Track capacity: up to 180
RMS accuracy: 2 km (range); 2.4 m/s (radial speed); 10-20 min (azimuth)
First sidelobe level: <25 dB (radiation pattern); 40-50 dB (lateral spurious)
Power: 1.4 kW (transmitter, average); 55 kW (max consumption)
Turn on time: 3 min
Set up time: 1 h (40V6M mast); 2 h (40V6M2 mast)
MTBF: 100 h
MTTR: <0.5 h

Contractor

Lira Design Bureau, Moscow.

UPDATED

76N6S air defence radar

Type
Self-propelled, ultra-low-altitude air defence radar.

Description
The 76N6S ultra-low-altitude air defence radar is installed on an MAZ-543 eight-wheeled chassis and appears to be a self-propelled 'gap filler' that makes use of the normally mast-mounted F-51MU antenna assembly.

Status
As of June 2001, the precise status of the 76N6S air defence radar was uncertain.

Specifications
Detection range: 90 km (practical radar horizon at 500 m altitude); 120 km (practical radar horizon at 1,000 m)
Target speed: up to 740 m/s
Elevation angle: 1-2°
Data dissemination time: 3 s
Track capacity: up to 180
Accuracy: 2 km (range); 2.4 m/s (radial speed); 10-20 min (azimuth)
Resolution: 1° (azimuth); 4.8 m/s (target velocity)
Power: 1.4 kW (min average transmission power); 55 kW (max power consumption)
Deployment time: 30 min
MTBF: 100 h
MTTR: <30 min
Temperature: –50 to +50°C

Contractors
Lira Design Bureau, Moscow.
Lianozovo Electromechanical Plant, Moscow.

VERIFIED

83M6E Command and Control (C²) system

Type
C² system for use with Surface-to-Air Missile (SAM) systems.

Description
The 83M6E SAM C² system comprises a trailer-mounted command centre and an E/F-band (2 to 4 GHz) 64N6 (NATO reporting name Tombstone) 3-D long range surveillance radar. So configured, the architecture can control up to six S-200VE, (NATO reporting name SA-5 'Gammon') or S-300PMU/PMU-1 (NATO reporting names SA-10c (S-300PMU) and SA-10d (S-300PMU-1) 'Grumble') SAM systems. The centre can be emplaced and readied for action within 5 minutes and aside from the radar, subsystems include a multifunction computer, a battle performance registration capability and communications equipment. System functions (which are described as being automated) comprise control of the radar's scanning mode, tracking of up to 100 targets simultaneously, track identification (using own source data fused with information from neighbouring command centres and high echelons), target identification/selection and target data hand-off. Power requirement is 85 kW and the entire system weighs 39,900 kg of which, 16,000 kg represents the command centre. The radar used is described as offering clutter and electronic countermeasures resistance.

Alongside the 83M6E, the S-300PMU-1 SAM system makes use of the 36N85 (export designation 30N6E1) Illumination and Guidance (IG) radar. Initial target acquisition appears to be by means of a cued 2° elevation × 2° azimuth or 4 × 4° search pattern followed by lock on and automatic tracking within selectable low-altitude (1 × 90°), medium- to high-altitude (5 or 13 × 64°) or ballistic (10 × 32°) sector scans. The radar also generates automatic target priority data and calculates optimum launch windows. Vehicle mounted, the complete IG system is described as weighing 11,500 kg and comprises a transceiver, target identification equipment, radar antenna and an equipment shelter. This latter item houses a multifunction computer system, a signals processor, operator stations and communications equipment. Power consumption is described as being 130 kW and the antenna array used appears to measure 14.5 × 15 × 3.8 m.

Status
Over time, Jane's sources suggest that the S-200 Volga SAM system has seen service with the armed forces of Belarus, Bulgaria, the Czech Republic, Georgia, the former German Democratic Republic, India, Iran, Kazakhstan, North Korea, Libya, Moldova, Poland, the Soviet Union/Russian Federation, Slovakia, Syria and Ukraine. As of early 2001, Bulgaria was reported as intending to keep its S-200 batteries in service until 2008, while Poland was noted as having developed a local upgrade of the system. As of February 2001, Jane's sources were reporting that the Molodechno Guards Air Defence Missile Brigade (based in the Kaliningrad area) was the last Russian air defence unit to be equipped with the S-200 system. As of June 2001, the People's Republic of China, Cyprus/Greece, Iran, the Russian Federation and Syria were noted as having procured S-300PMU/PMU-1 SAM systems. Of these, China, Iran and Syria were understood to be operating the S-300PMU variant, with Cyprus/Greece having acquired the PMU-1 variant.

Contractors
Research and Production Association Almaz (S-300PMU-1).
Research Instrument-Making Institute (83M6E radar).

UPDATED

ABS-1 artillery ballistic station

Type
Muzzle velocity measuring radar.

Description
The ABS-1 serves to determine the muzzle velocity of artillery and mortars within the range 80 to 2,200 m/s under field conditions, with artillery pieces of 100 mm and larger calibres and mortars of 120 mm and larger. The equipment is operated by a single person, weighs approximately 50 kg and can be set up in 5 minutes. Prime source of power is a battery with a rated voltage between 15 and 12.5 V and one battery set can operate the equipment for about 6 hours continuously.

The ABS-1 system provides double measurement of single round muzzle velocities on systems with a firing rate of 5 to 6 rds/min. It is claimed to be very easy to operate. It uses the Doppler effect to measure the velocity of the round according to the time that is required for that round to pass along a measurement base 200 cm long. The mean error in the velocity measured is stated to be less than 0.1 per cent.

Status
As of June 2001, the precise status of the ABS-1 muzzle velocity measuring radar was uncertain.

VERIFIED

Front view of the ABS-1 artillery ballistic station mounted on its tripod. The unit stands about 90 cm high (Jane's Intelligence Review)

ARK-1M Rys artillery reconnaissance and fire-control radar

Type
G/H-band (6 GHz centre frequency) weapon location and fire-control radar. NATO reporting name Small Fred.

Description
The ARK-1M Rys artillery reconnaissance and fire-control radar is a vehicle-mounted 3-D monopulse equipment that has a wavelength of 5 cm and is designed to detect and locate targets for artillery batteries on the battlefield. The radar generates target co-ordinates automatically and, when mounted on an MT-LBu tracked chassis, incorporates a parabolic antenna assembly and a power generation subsystem.

Status
As of May 2001, the status of the ARK-1M Rys artillery reconnaissance and fire-control radar was uncertain.

Specifications
Centre frequency: 6 GHz
Wavelength: 5 cm
Detection range: 7-9 km (howitzers - location mode); 12-13 km (mortars - location mode); 13-15 km (howitzer fire-control); 16-17 km (mortar fire-control); 20-30 km (MRLS fire-control and location); 30 km (battlefield missiles - location); 40 km (battlefield missile fire-control)
Accuracy: 30-90 m (range in fire-control mode)
Track capacity: 30 tracks/min
Sector coverage: 30°
Set up/tear down time: 5 min
Crew: 4

Contractor
Rosoboronexport (export agent), Moscow.

UPDATED

Dog Ear surveillance radar

Type

F/G-band (3 to 6 GHz) surveillance and target acquisition radar.

Description

Dog Ear is the NATO reporting name for an early warning surveillance and target acquisition radar which is used to provide target information for the 9K35 Strela 10 low-altitude Surface-to-Air missile system. The 9K35 makes use of the 9M37 (NATO reporting name SA-13 'Gopher') missile. Dog Ear is normally mounted on an MT-LBu tracked chassis and is allocated on the basis of one system per air defence battery. Frequency is probably in the F/G-band. Acquisition range is 80 km and the tracking range 35 km.

Status

The 9K35 SAM system is understood to have been first deployed during 1978. Jane's sources suggest that over time, the system has been acquired by the armed forces of Afghanistan, Algeria, Angola, Bulgaria, Croatia, Cuba, the Czech Republic, the former German Democratic Republic, Hungary, Iraq, Jordan, Libya, Poland, the Russian Federation, Slovakia and Syria.

VERIFIED

Dog Ear surveillance radar vehicle which is used by SA-13 SAM units to provide target information

Fan Song missile control radars

Type

E/F/G-band (2 to 6 GHz) target detection and missile guidance radars.

Description

Fan Song is the NATO reporting name for a family of target detection and guidance radars that have been developed for use with the S-75 (NATO reporting name SA-2 'Guideline') surface-to-air missile system.

Over time, six of these radars have been identified (by suffix letters A to F) and common to all are the detection and tracking of the target, the command guidance of the missile and tracking of the missile. It is understood that these radars can handle up to six targets at once and guide three missiles at a time. However, the

A Fan Song E target detection and missile guidance radar

missile must pass through the guidance beam within a few seconds of launch if it is to be acquired and steered towards the selected target. There is also a limitation in the amount of steering information that the missile is capable of receiving from the ground station. Nevertheless, the 'Guideline' missile in various versions has remained in service for many years with a number of countries.

Details of the various Fan Song radars appear below. All members of the series contain as a major element a track-while-scan radar which scans a designated sector with two flapping fan beams radiated from two orthogonal antenna systems. The flapping motion has a sawtooth profile and uses the electromechanical system known as the Lewis scanner. The 'Fan' in the codename refers to these beams and the 'Song' to the bird-like sound of the demodulated radiation from these radars.

Fan Song A

As far as is known, Fan Song A was a pre-series type, operating in E/F-band which was intended mainly for initial operating trials. It was quickly superseded by Fan Song B. Fan Song A is reported to have been operated by a four-man crew.

Fan Song B

Fan Song B was an E/F-band member of the family. Radiation from its two Lewis scanners is at 2,965 to 2,990 MHz for one and 3,025 to 3,050 MHz for the other. Peak power is in the region of 600 kW and the equipment has a first-time round range capability of between 60 and 120 km.

The equipment is trailer mounted and the scanned sector can be changed by rotating the trailer and tilting the whole superstructure. The sector scanned is about 10° high and 10° wide, these dimensions being the approximate fan beam width, the beamwidth in the scanning direction being only 2° for each beam.

A small parabolic antenna, mounted on one end of the horizontal Lewis scanner, is used to transmit Ultra High Frequency (UHF – 300 MHz to 1GHz) command guidance signals to the missile. Like Fan Song A, Fan Song B is reported to be operated by a four-man crew.

Fan Song C

Fan Song C is a G-band version of Fan Song B. Very little information is available and it seems that it was very quickly overtaken by another G-band version, Fan Song E. Like Fan Song B, Fan Song C is reported to have been operated by a four-man crew.

Fan Song D

The Fan Song D radar was experimental only and was never deployed operationally. Like its predecessors, Fan Song D is reported to have been operated by a four-man crew.

Fan Song E

Fan Song E is a G-band member of the series and is similar in many respects to Fan Song B. It is normally associated with the SA-2E 'Guideline' Mod 4. The main differences are the frequency of the radiation from the Lewis scanners and the fan beam (and hence the sector dimensions). The beams in Fan Song E are about 7.5° wide in the fan and 1.5° in the scanning direction.

In addition to the antenna for the command guidance signals, which is similar to that of Fan Song B, Fan Song E has two further parabolic dishes mounted on top of the horizontal Lewis scanner (one for each scanner). One of these is vertically and the other horizontally polarised and their purpose is to provide a Lobe-On-Receive-Only (LORO) feature. This is an Electronic Counter-CounterMeasures (ECCM) technique in which the scanner action of the Lewis scanner is restricted to the receive channel by diverting the transmitted signal, which would otherwise be radiated from the scanner, into a dummy load. The signal is then replaced by a signal from one of the parabolic dishes of sufficient power to operate the whole system. Jane's sources suggest that Fan Song E has three operating modes (target acquisition, automatic tracking and low-altitude search and track) and is served by a six-man crew.

Fan Song F

Fan Song F is the latest of the series and reverted to the E/F-band frequency, but with a higher output power. It incorporates scintillation suppression, Moving Target Indicator (MTI) signal processing and manual and mixed mode tracking. The MTI feature allows the radar to pick out aircraft from chaff clouds. Fan Song F is easily identified by a small 'dog-house' structure on top of the horizontal scanner. This houses two crewmen who can guide the missile using optical sights when the radar guidance is suppressed by electronic countermeasures. It replaces the two parabolic dishes on Fan Song E since the LORO feature has been replaced by other ECCM features. Fan Song F is normally associated with the SA-2F 'Guideline' Mod 5 missile. Like Fan Song E, Jane's sources suggest that the six-man Fan Song F radar offers target acquisition, automatic tracking and low-altitude search and track operating modes.

Status

Over time, S-75 missile systems have been acquired by Afghanistan, Albania (estimated at 22 launchers), Angola, (24), Armenia, Belarus, Bosnia-Herzegovina, Bulgaria (132), the People's Republic of China (under the designation HQ-2), Cuba (78), the Czech Republic (282- including reverse-engineered Early Bird missiles), Egypt (282- including reverse-engineered Early Bird missiles), Ethiopia (18), Georgia, Hungary (96), India (96), Iran (60+ – mostly HQ-2 with some missiles adapted for infra-red homing), Iraq, Kazakhstan, Kyrgyzstan, North Korea (270), Libya (108), Mozambique, Pakistan (12+), Peru (18), Poland (240), Romania (120), the Russian Federation, Slovakia, Sudan (18), Syria (138+), Tajikistan, Turkmenistan, Ukraine, Uzbekistan, Vietnam (360), Yemen (20+) and the Federal Republic of Yugoslavia (24). During 1993, the Russian Almaz organisation launched an S-75 modernisation programme while September 2000 saw details of a radar range-finder that had been developed during the early to mid-1970s to support the S-75's primary Fan Song target detection and missile guidance radar emerge. Designated as the RD-75, Jane's sources report that this add-on consisted of a single cabin that incorporated its own antenna and was slaved to the primary radar. Described as a dual-frequency, dual-polarisation, centimetric wavelength equipment, RD-75 was designed to facilitate system function in conditions of 'intense' countermeasures pollution. Here, the primary Fan Song radar would undertake passive tracking of the jamming source and measurement of the target's horizontal and vertical position in space, while the active RD-75 provided target range. As of November 2000, Jane's sources were noting that the RD-75 Fan Song add-on had been procured by the armed forces of both the Russian Federation and Romania.

Specifications

Model	B	C	E	F
Vertical antenna				
Frequency	3,025-3,050 MHz	5,010-5,090 MHz	5,010-5,090 MHz	3,025-3,050 MHz
Beam	10 × 2°	7.5 × 1.5°	7.5 × 1.5°	10 × 2°
Horizontal antenna				
Frequency	2,965 × 2,990 MHz	4,910 × 4,990 MHz	4,910 × 4,990 MHz	2,965 × 2,990 MHz
Beam	2 × 10°	1.5 × 7.5°	1.5 × 7.5°	2 × 10°
Peak power	600 kW	1,500 kW	1,500 kW	600 kW
Search PRF		828-1,440 pps	828-1,440 pps	
Track PRF		1,656-2,880 pps	1,656-2,880 pps	
Scan rate	15.5-17 Hz	15.5-17 Hz	15.5-17 Hz	15.5-17 Hz
Pulse-width		0.4-1.2 ms	0.4-1.2 ms	
Unambiguous range	60-120 km	75-150 km	75-150 km	60-120 km

UPDATED

Flap Wheel fire-control radar

Type
I/J-band (8 to 20 GHz) anti-aircraft gun fire-control radar.

Description
Flap Wheel is the NATO reporting name for a conical scan radar which operates in J-band. The system's antenna is made up of a horizontally polarised parabolic dish and a Yagi array. Flap Wheel is used to provide fire-control data for 57 mm S-60 and 130 mm KS-30 anti-aircraft guns and has been observed in remote positions some 200 m away from the weapon it is supporting. Most recently, the system has been augmented by the installation of a coincident low-light TV tracker.

Status
As of June 2001, the precise status of the Flap Wheel fire-control radar was uncertain.

VERIFIED

A sketch of the antenna array used with the Flap Wheel fire-control radar

GS series radars

Type
I/J-band (8 to 20 GHz) family of battlefield surveillance radars.

Description
The armies of the RFAS employ a number of battlefield surveillance radars, among which are the PSNR-1, GS-11, GS-12 and GS-13.

GS-11 (Alternative designation PSNR-2)
Allocated the NATO reporting name Garpin, GS-11 is a short-range, man-portable battlefield surveillance radar which breaks down into two manpacks and is deployed on a tripod mounting. The equipment weighs approximately 30 kg and is reported to be deployable in approximately 10 minutes. GS-11 operates in the I/J-band (8 to 20 GHz) and has a peak power of 10 kW. Detection ranges against personnel and vehicles are reported to be 1.5 and 4.5 km respectively.

GS-12
GS-12 is understood to be an alternative designation for the PSNR-5M Kredo-M ground surveillance radar (see separate entry).

GS-13
Allocated the NATO reporting name Long Eye, GS-13 is an I/J-band (8 to 20 GHz) trailer-mounted, long-range battlefield surveillance radar which utilises two back-to-back dish antennas. Power output is reported as being 50 kW and it is said to be capable of detecting personnel at ranges of up to 12 km and vehicles at distances of up to 25 km.

Status
Over time, GS series radars are reported as having been procured by the Russian Federation and other customers around the world.

VERIFIED

Low Blow tracking and missile control radar

Type
Family of I-band (8 to 10 GHz) tracking and missile control radars.

Description
Low Blow is the NATO reporting name given to a family of I-band radars used with the land-based S-125 (NATO reporting name SA-3 'Goa') surface-to-air missile. These radars are not the same as those used with shipborne Goa missiles which are known as Peel Group radars. The name Low Blow reflects the ability of the radar to guide the missile towards low-flying targets through heavy clutter.

Like the Fan Song series, the Low Blow radars use pairs of electromechanically scanning trough antennas mounted orthogonally, but to improve low-angle performance the troughs are mounted at 45° from the horizontal. It appears, too, that the troughs are not Lewis scanners but a form of organ-pipe scanner.

The two trough antennas are mounted at a 45° angle in an upside down 'V' shape. This change was introduced to reduce ground clutter. The antennas generate sawtooth fan beams like Fan Song. Inserted into the top of the inverted 'V' is the square parabolic guidance command antenna. Between the two troughs is a Lobe-On-Receive-Only (LORO) mode antenna. When the trough antennas acquire a target the system can be switched to the LORO mode, transmitting from this central parabolic dish antenna. The target tracking antennas scan in 6° swathes, providing a 12° radiated beamwidth.

Carrier frequencies of the Low Blow family lie in the 9,000 to 9,400 MHz band, pulse repetition frequencies between 1,750 and 3,500 pulses/s with unambiguous range correspondingly lying between 40 and 85 km. Radiated beamwidth is in the region of 12° in the fan and 1.5° in the direction of scan. Pulses are between 0.25 and 0.5 μs in duration and peak power output is around 250 kW.

Status
The S-125 system is understood to have entered service during 1961. Over time, Jane's sources suggest that SA-3 batteries have been procured by the armed forces of Afghanistan, Algeria, Angola, Azerbaijan, Belarus, Bulgaria, Cuba, the Czech Republic, Egypt, Ethiopia, Finland, Georgia, the former German Democratic Republic, Hungary, India, Iraq, Kazakhstan, North Korea, Laos, Libya, Mali, Mozambique, Peru, Poland, Slovakia, Somalia, Syria, Tanzania, Ukraine, Vietnam, Yemen, the Federal Republic of Yugoslavia and Zambia.

UPDATED

Monitor point and area guard radar

Type
J-band (10 to 20 GHz) point and area guard radar.

Description
Monitor is a development of the PSNR-5M Kredo-M (NATO Reporting Name Tall Mike - see separate entry) ground surveillance radar that incorporates a tripod-

The elements that make up the baseline Monitor point and area guard radar
0006893

PSNR-5M is understood to form part of the PRP-4M Deytery mobile reconnaissance station that makes use of the BMP tracked chassis

mounted antenna assembly, a power supply, a radar data processing unit, a display/control console (based on an IBM personal computer/AT 386 standard computer) and a wire communications line (see following). It is designed for installation, coastline and border surveillance together with movement control (personnel and vehicles) within designated areas. Functionally, Monitor automatically detects activity within its field of view, generates an alarm signal and displays the activity on a colour map. Data can be transferred to remote recipients via a wire link (up to 4 km long) or radio and the radar is proofed for 24 hour operation in adverse weather conditions. Monitor is further reported as having a service life of 10 years and as being able to be integrated into a surveillance network of up to four radars.

Alongside the baseline Monitor radar described previously, STRELA has subsequently developed the Monitor-M variant which is described as being suitable for area guard and battlefield/border surveillance applications. As such, the variant is noted as featuring:

- an automatic moving target location and tracking capability
- coloured target coding
- a digital map input facility
- restricted zone/control line generation on the radar's display
- an automatic intruder alarm.

Monitor-M is further reported as being suitable for tripod or vehicle mounting or installation as part of an integrated guard system.

Status
Monitor-M was noted as being available during the first half of 2001.

Specifications
Baseline Monitor
Frequency: J-band (10-20 GHz)
Output power: 500 W (pulse)
Coverage: 120° (sector); 48 km² (area)
Resolution: 1.6° (azimuth); 100 m (range)
Instrumented range: 3-15 km
Detection range: 3-4 km (man); 8 km (Jeep-type vehicle); 10 km (helicopter); 12 km (main battle tank); 15 km (vehicles and surface ships)
Target velocity: 2-60 km/h
Max console-to-radar distance: 4,000 m
Power requirement: 220 V AC, 50 Hz
Weight: 70 kg
Monitor-M
Detection range: 8.5 km (man); 20 km ('matériel')
Accuracy: 5 mils (azimuth); 25 m (range)
Coverage: ±18° (elevation); 30-180° (azimuth)
Weight: 50 kg

Contractor
STRELA Scientific Research Institute, Tula.

UPDATED

PSNR-5M Kredo-M ground surveillance radar

Type
H-band (6 to 8 GHz) portable ground surveillance radar. NATO reporting name Tall Mike. Alternative RFAS designation GS-12.

Description
The tripod-mounted PSNR-5M Kredo-M portable ground surveillance radar is described as being a pulse-Doppler, vertically polarised equipment that has a wavelength of 3 cm. Incorporating a magnetron transmitter, a folding parabolic antenna and a display unit with memory, Kredo-M is designed to detect, locate and track moving targets (such as men and vehicles) on the battlefield. A PSNR-5M variant also appears to form part of the PRP-4M Deytery 'mobile reconnaissance station' that combines radar and electro-optic sensors in a package that is installed on a BMP chassis.

Status
As of May 2001, the status of the PSNR-5M Kredo-M portable ground surveillance radar was uncertain.

Specifications
Frequency: H-band (6-8 GHz)
Wavelength: 3 cm
Range: 10,000 m (moving tank - PRP-4M application); 4-5 km (man - tripod mounting); 10-12 km (tank - tripod mounting)
Accuracy: 25 m (range - tripod mounting); 50 m (range - RPR-4M application)
Coverage: 24-120° sector (tripod mounting)
Set up/tear down time: 5 min (tripod mounting)
Crew: 2
Weight: 50 kg (tripod mounting)

Contractor
Rosoboronexport (export agent), Moscow.

UPDATED

Romb tracking radar system

Type
H/I/J-band (6 to 20 GHz) tracking and command guidance radar system for the 9K33 Osa (NATO reporting name SA-8 'Gecko') series of low altitude, Surface-to-Air Missile (SAM) systems. NATO reporting name Land Roll.

Description
Alongside search and tracking arrays, the Romb fire-control radar system features pairs of beacon tracking and transmission antennas for missile guidance. The system's main fire-control array is at the rear of the launch vehicle and folds back through 90° to reduce the overall height for air transport and high-speed road travel. The radar operates in H-band (6 to 8 GHz) with a 360° traverse and a maximum range of about 30 km.

In front of the fire-control radar is the guidance group consisting of a central monopulse target tracking J-band (10 to 20 GHz) radar, with a maximum range of 20 km, using a cassegrain type array and probably employing conical scan and a pulsed transmission waveform and two smaller missile tracking cassegrain antennas which flank the main tracker. They are mounted to provide a limited degree of movement relative to the main array and operate in I-band (8 to 10 GHz). The command link horns for each are mounted beneath the missile tracking antennas. Two rectangular devices, which are believed to assist tracking in an Electronic CounterMeasures (ECM) environment, are mounted to the left and right of the missile guidance radars. Mounted on top of each missile guidance radar is a low-light TV/optical system to assist tracking in low visibility and severe ECM conditions.

Status
Romb is understood to have entered service with the Soviet defence forces during the mid-1970s. Outside Russia (and over time), Jane's sources suggest that the 9K33 system has been exported to Algeria, Angola (15 units), Armenia (20),

The SA-8 SAM system in operating configuration with surveillance radar erected

Azerbaijan, Belarus, Croatia, the Czech Republic, Greece (12), India (48), Iraq, Jordan (50), Libya (60+), Poland (60), Syria (60), Tajikistan, Ukraine and the Federal Republic of Yugoslavia.

Specifications
Frequency: surveillance radar H-band (6-8 GHz); engagement radar J-band (14.5 GHz)
Acquisition range: 30 km
Tracking range: 20 km

VERIFIED

RP-100 Fara-1 ground surveillance radar

Type
J-band (10 to 20 GHz) ground surveillance radar.

Description
The RP-100 Fara-1 ground surveillance radar is designed to provide automatic, all-weather detection of ground-based and seaborne moving targets and can be integrated with night vision devices (such as the NNP-23) and vehicle-mounted weapons (such as the 7.62 mm PKMSN and 12.7 mm NSVS machine guns and the AGS-17 grenade launcher). The equipment comprises an antenna assembly and a display/processor unit. Target identification is thought to be by means of Doppler shift tones (conveyed to the operator via headphones) and Fara-1 is reported to be

The RP-100 Fara-1 surveillance radar in tripod-mounted configuration

0006892

available in man-portable patrol, mobile tripod and vehicle-mounted configurations. When operating in automatic surveillance mode, Fara-1 has provision to transmit threat warnings to off-site recipients via an Aiva radio set. RP-100 is further noted as having a service life of 20 years.

Status
The Fara-1 ground surveillance radar was reported as being available during the first half of 2001.

Specifications
Frequency: J-band (10-20 GHz)
Output power: 0.05 W (continuous wave)
Sector coverage: 24, 45, 90 or 120°
Accuracy: 15 m (range); 15 mrad (bearing)
Detection range: 2 km (man); 5 km ('matériel')
Target velocity: 2-50 km/h
MTBF: 5,000 h
Power consumption: 12 W
Weight: 16 kg

Contractor
STRELA Scientific Research Institute, Tula.

UPDATED

RP-200 Credo-1 ground surveillance radar

Type
J-band (10 to 20 GHz) ground surveillance and intruder detection radar.

Description
As of mid-2001, three Credo-1 ground surveillance and intruder detection radar variants had been identified, the known details of which are as follows:

Credo-1
The RP-200 Credo-1 radar is a multifunction, pulse Doppler sensor that is designed for the automatic, all-weather detection of personnel, vehicles, helicopters and armoured fighting vehicles. It can also be used for artillery fire control, airfield runway control and ship tracking in coastal waters. RP-200 can be tripod-mounted or installed on a wheeled or tracked vehicle, makes use of a klystron power amplifier, can be remotely controlled (at ranges of up to 100 m) and can be integrated with an IBM personal computer if required. Credo-1 is also noted as having a service life of 10 years.

Specifications
Frequency: J-band (10-20 GHz)
Instrumented range: 30 km
Detection range: 7-20 km (artillery fire control); 12-15 km (personnel); 19 km (ship pilotage); 20 km (Jeep-type vehicle); 25 km (main battle tank); 30 km (vehicles and surface ships)

The baseline Credo-1 ground surveillance radar

Target velocity: 3-72 km/h
Accuracy: 0.12° (azimuth); 10 m (range)
Resolution: 1.8° (azimuth); 50 m (range)
Coverage: +18° (elevation); 30-180° (azimuth)
Deployment time: 5 min
Temperature: ± 50°C
Power requirement: 250 W
Weight: 97 kg
Credo-1E
Described as being suitable for tripod, vehicle-mounted, mast-mounted, helicopter and drone applications, the Credo-1E radar is designed for the battlefield surveillance, fire-control and air defence roles. System features include:

• automatic moving target location and tracking
• a target situational display that is overlaid on a topographical terrain background
• target type identification
• alarm signal generation
• a moving target track display
• weather clutter and jamming rejection facilities.

Specifications
Detection range: 15 km (soldier/155 mm shell burst); 30 km (jeep); 40 km (heavy motor vehicle/tank)
Accuracy: 2 mils (target azimuth); 4 mils (shell burst azimuth); 10 m (target range); 20 m (shell burst range)
Coverage: ± 18° (elevation); 30-180 or 360° (azimuth)
Weight: 105 kg

Credo-1S
Described as being a 'reconnaissance complex with [a] mast-lifting facility', Credo-1S is a combined radar and electro-optic (television and thermal imaging) equipment that is vehicle mounted and is designed for the moving target location (on land, on water and in the air) and artillery fire-control roles. System features include:

• the ability to operate in terrain-masked (for example, wooded) locations
• all weather/all light conditions target detection, location and tracking
• automatic moving target co-ordinate determination (with claimed low false alarm rate)
• artillery fire control by means of shell burst co-ordinate determination
• automatic data exchange with fire units and control posts.

Specifications
Detection range: 3 km (TV - soldier identification, thermal imager - soldier detection); 6 km (TV - soldier detection/matériel identification, thermal imager - matériel detection); 10 km (TV - matériel detection); 15 km (radar - soldier/155 mm shell burst); 40 km (radar - motor vehicle/tank); up to 40 km (radar - 'over water target')
Accuracy: 2.5-3 da (azimuth); 10 m (range); 50 m (shell burst)
Mast height: 4, 8 or 12 m

Status
Both the Credo-1E and Credo-1S systems were being reported as being available during the first half of 2001.

Contractor
STRELA Scientific Research Institute, Tula.

UPDATED

....

RPK-2 Tobol fire-control radar

Type
J-band (10 to 20 GHz) Anti-Aircraft (AA) gun and surface-to-air missile fire-control radar. NATO reporting name Gun Dish.

Description
RPK-2 is the fire-control radar that is used in the ZSU-23-4 Shilka quadruple 23 mm cannon-armed, self-propelled AA vehicle. Designed as it is to detect and track low-flying targets, the RPK-2 radar has a short reaction time and is slaved to the ZSU-23-4's armament. To overcome jamming and/or anti-radiation missiles, the radar automatically shuts down when it senses a threat. Fire-control is then undertaken using a co-located optical sighting system.

The RPK-2 operates in the specific 14.6 to 15.6 GHz frequency range. The panoramic search radius extends up to 50 km and the target tracking range up to 18 km. The 1 m diameter antenna can be folded down to reduce overall height of the vehicle for air transport. An identification friend-or-foe system is incorporated. An improved variant of the radar (featuring sidelobe clutter-reduction features) was identified in 1985. Sources suggest that this later version can conduct independent surveillance.

Status
Over time, Jane's sources suggest that the ZSU-23-4 AA systems have been supplied to the armed forces of Afghanistan (20 units), Algeria (210), Angola (20+), Bosnia-Herzegovina, Bulgaria (35), Congo (8), Cuba (36), Egypt (117), Ethiopia (60), Hungary (14), India (100), Iran (100+), Iraq (200+), Israel (60 - captured), Jordan (44), North Korea (100+), Laos (10+), Libya (250), Mongolia, Nigeria (30),

The RPK-2 fire-control radar
(Christopher F Foss)

Peru (35), Poland (87), Russia, Somalia (4), Syria (400), Turkmenistan (28), Ukraine, Vietnam (100+) and Yemen (30+).

During 1996, the Ulyanovsk Mechanical Plant launched a ZSU-23-4 upgrade package. Jane's sources describe this as including:

• radar and fire-control system improvements that increase the system's kill probability factor from 0.3 to 0.6
• improved overall system reliability
• a centralised target designation subsystem
• improved jamming resistance via the introduction of moving target indication
• a built-in test subsystem.

VERIFIED

....

Snap Shot ranging and early warning radar

Type
Ranging and early warning radar for the 9K35 Strela 10 Surface-to-Air Missile (SAM) system.

Description
The 9K35 Strela 10 SAM system (which fires the 9M37 (NATO reporting name SA-13 'Gopher')) is equipped with the Snap Shot range-only radar in order to prevent missile wastage. The complete 9K35 system is mounted on a Transporter-Erector-Launcher And Radar (TELAR) vehicle of which there are three variants: the original TELAR 1, TELAR 2 with the Flat Box electronic support system fitted and the Flat Box-equipped TELAR 3 which appears to have an increased missile storage capacity over its predecessors. Surveillance and acquisition of the target is carried out with the help of a separate Dog Ear battery early warning radar. An

Close-up of one of the 9K35 SAM system's three types of TELAR

identification friend-or-foe system is also fitted which detemines target identification at ranges up to 12 km and altitudes from 100 to 500 m.

Status

As of mid-2000, Jane's sources were suggesting that the 9K35 SAM system was first deployed during 1978 and that by 1987, over 1,000 system TELARs had been produced, 900 of which were in the service of the then Soviet Union. The same sources indicate that apart from the RFAS, the 9K35 system has, over time, been exported to Afghanistan, Algeria, Angola, Bulgaria, Croatia, Cuba, the Czech Republic, the former German Democratic Republic, Hungary, Iraq, Jordan, Libya, Poland, Slovakia and Syria.

VERIFIED

SNAR-10 surveillance radar

Type

K-band (20 to 40 GHz) battlefield surveillance radar. NATO reporting name Big Fred.

Description

SNAR-10 is a widely deployed battlefield surveillance radar system, mounted on a self-propelled, armoured vehicle. It is designed to locate stationary and moving targets at the forward edge of the battle area.

The radar operates at a frequency of about 35 GHz and is capable of scanning the terrain in both a horizontal and vertical mode. This is done by tipping the antenna, thereby overseeing a large area. The dipping antenna is designated as the IRL-127-1.

The radar is a 2-D pulse-Doppler equipment, with the complete system, except the power supply, being installed inside the turret of the vehicle. At the rear of the turret is the rectangular reflector of the antenna system which is almost as wide as the vehicle. On the move the reflector is folded forward on top of the turret and can be covered by a hood. For surveillance work the reflector is erected electromechanically. The narrow radar beam (0.36° wide and 1.3° high) is continuously revolving in a sector of ±13.2°. It can also transmit without scanning a particular sector. In this case the beam sits in the middle of the reflector.

The radar works as either a pulse radar or a Doppler radar, the operator making the selection. In the pulse mode it is impossible to distinguish between stationary and moving targets. To detect moving objects, especially beyond the forward line of troops, the Doppler mode is required. If the entire battlefield is to be searched for targets the sector-search mode is used. The revolving radar beam is activated and this then rotates in the fixed sector of exactly 26.4°. In this mode the operator can make use of all the radar's built-in features, including the elevation search mode.

The radar operator's workstation is equipped with two screens. The main screen, a B-scope, is used for target detection and fire control. The A-scope is for target identification and signal verification. The B-scope provides a display of the distance and azimuth to the target. The A-scope indicates the distance only. In addition, the sound of the radar echo is fed into the earphones of the crew.

Normally, the type of object detected, tank or truck, is not immediately clear. Therefore, the radar echo sound is used for identification. When operating in this mode the antenna has to be traversed and inclined in such a way that the centre of the antenna is pointing at the target. Experienced operators are then able to determine the type of target.

Risk of detection when operating in the search mode can be reduced by selecting the appropriate radar output – power selection is between 14 and 70 kW. Several preprogrammed frequencies can be chosen in the radar working area of 34.55 to 35.25 GHz and can be adjusted manually.

As of mid-2001, *Jane's* sources had identified a second SNAR-10 variant designated as SNAR-10M. Designed for moving target detection and artillery fire-control, SNAR-10M is vehicle mounted and offers:

- automatic over land/over water moving target location and tracking
- automatic co-ordinate 'reading'
- automatic data exchange with fire units and control posts (two independent channels)
- artillery fire-control by means of shell burst tracking
- crew protection from small arms fire and shrapnel
- 'high' mobility.

Status

Over time, the SNAR-10 battlefield surveillance radar is understood to have been procured by Russia and a number of other East European countries. The SNAR-10M variant was reported as being available during the first half of 2001.

Specifications

Baseline SNAR-10

Frequency range: K-band (34.55-35.25 GHz)
Output power: 14-70 kW
PRF: 2,540 Hz (>24 km range); 4,410 Hz (up to 26 km range)
Max range against moving targets
Without MTI: 10 km (100 mm shell impact); 16 km (vehicles); 30 km (ships)
With MTI: vehicles 10 km
Precision locating
Polar co-ordinates: ≤20 m (range); ≤2 mils (traverse)
Grid co-ordinates: radial error ≤30 m
Time for target location: 20 s

The baseline SNAR-10 radar system in operational configuration with its radar reflector raised

Search sector: 26.4°
Radar beam dimensions: 0.36 × 1.3°
Target discrimination: 50 m (range); 6° (traverse)
Time into action: 5 min (prepared position); ≤20 min (unprepared position)
SNAR-10M
Detection range: 15 km (soldier/155 mm shell burst); 40 km (motor vehicle/tank); up to 40 km ('over water target')
Accuracy: 2 mils (target azimuth); 4 mils (shell burst azimuth); 10 m (target range); 20 m (shell burst range)
Coverage: 6° (azimuth - fire-control mode); 30-180 or 360° (azimuth - search mode)

Contractor

STRELA Scientific Research Institute, Tula.

UPDATED

TOR-M1 surface-to-air missile radar system

Type

Surveillance and tracking radar system for the TOR-M1 surface-to-air missile (NATO reporting name SA-15 'Gauntlet').

Description

The TOR-M1 surface-to-air mobile missile system uses vertically launched missiles to engage fixed-wing aircraft, helicopters, unmanned aerial vehicles, smart weapons and guided missiles.

The 3-D pulse-Doppler surveillance radar, mounted on the rear of the system's 9A 331 tracked chassis, provides range, azimuth, elevation and automatic threat information for up to 48 targets to a maximum range of 25 km. It also initiates the tracking of up to 10 targets.

A phased-array pulse-Doppler tracking radar, using pulse compression techniques, is mounted on the front of the turret and is stated to be capable of simultaneous tracking and initiating attacks on two targets travelling at a maximum speed of up to 700 km/h. The antenna assembly can be folded down for transit.

A third dome-type antenna on the vehicle is probably used to gather the missile as it is launched and hand over to the tracking radar.

An autonomous TV system is also fitted for use in battlefield clutter and severe electronic countermeasures environments.

Status

As of January 2001, Jane's sources were suggesting that the SA-15 surface-to-air-missile system had been acquired by the armies of Belarus, the People's Republic of China (15 to 20 systems), Cyprus (six), India, Peru, the Russian Federation (100 plus) and Ukraine together with the Hellenic Air Force (six)

The vehicle-mounted radar system for the SA-15 TOR-M1 surface-to-air missile

Specifications
Surveillance radar
Detection range: 25 km
Elevation coverage: 0-32° (32-64°)
Targets: up to 48 simultaneously
Tracking radar
Tracking range: 25 km
Velocity range: 0-700 m/s
Targets: 2 simultaneously

Contractor
Rosoboronexport (export agent), Moscow.

UPDATED

....

Zoopark-1 reconnaissance and fire-control 'complex'

Type
H-band (6 to 8 GHz) vehicle-mounted artillery location and fire-control radar.

Description
The three-man, MT-LBu-mounted Zoopark-1 radar features electronic scanning and is designed to detect hostile artillery batteries, rocket launchers and tactical missile firing sites and direct counterfire against them. Zoopark-1 can detect and track up to six projectiles simultaneously and calculates projectile trajectories, impact points and the miss-distances of counterfire. Battery/launch site locations are stored in the system's computer memory before being passed to an associated data transmission subsystem or printed out (using a drum type graphic plotter). Alongside its primary role, Zoopark-1 can also be used for air traffic control in an airfield's approach zone (up to a 40 km radius). The complete Zoopark-1 'complex' comprises a 1L259 radar mounted on an MT-LBu tracked chassis, a 1L30 maintenance vehicle (based on the Ural-43203 truck chassis) and a trailer-mounted ED-30-T230P-1RPM generator.

Status
The Zoopark-1 artillery location and fire-control radar was reported as being available during the first half of 2001.

A close up of the Zoopark-1 artillery location radar's antenna
0006894

The Zoopark-1 artillery location radar is mounted on the all-terrain MT-LBu tracked chassis   0006895

Specifications
Frequency: H-band (6-8 GHz)
Detection range: 15 km (field artillery); 25 km (mortars); 40 km (MLRS); 60 km (tactical missiles)
Track capacity: 60-100 tracks/min
Simultaneous scan sector: up to 90°
Simultaneous target tracking limit: up to 12
Crew: 3
Set up time: 5 min
Power consumption: 29 kW (max - all elements of the 'complex')
Guaranteed service life: 8 years

Contractor
STRELA Scientific Research Institute, Tula.

UPDATED

_____

SOUTH AFRICA

EDR110/120/140 target designation radars

Type
Family of air defence surveillance and target designation radars.

Description
EDR110 is the surveillance and target designation radar installed on the South African ZA-35 self-propelled anti-aircraft gun and ZA-HVM surface-to-air missile air defence vehicles. It operates in the D-band (1 to 2 GHz) and is reported to have detection ranges against fixed-winged aircraft and hovering helicopters of 12 and 6 km respectively. An 80 per cent probability of single scan target detection is claimed and a moving ZA-35 is noted as being able to engage a target within 10 seconds of detecting it. EDR110 can track up to 100 targets simultaneously and is credited with range and azimuth accuracies of 40 m and 1° respectively. Target data can be handed-off between vehicles by means of voice radio or datalink and the radar is equipped with a turret-mounted flat screen plasma display. An audio warning of hostile aircraft is provided and the system incorporates electronic counter-countermeasures provision. The set's antenna has two operating positions and is retractable. Maximum elevation is used when the host vehicle is stationary.

The D-band ED120 equipment is derived from EDR110 and incorporates a new signal processor and a larger antenna. Designed to provide target designation for

EDR110 surveillance and target designation radar mounted on a ZA-35 air defence vehicle

The EDR120/140 shelter shown here with its antenna deployed

towed anti-aircraft systems, the radar is quoted as having an instrumented range of 20 km and being able to detect 2 m² fixed-wing aircraft targets at ranges of up to 19 km. Against hovering helicopters, this value reduces to 14 km. Reutech claims that EDR120 offers range and azimuth accuracies of 40 m and 0.6° respectively.

The ED140 is a D-band development of EDR120 which offers longer detection ranges and is, like its predecessor, designed for use with towed anti-aircraft systems. The radar is a solid-state equipment which uses a 23 dB gain, −50 dB sidelobe, planar-array composite antenna. The equipment's signal processor incorporates multiple thresholding, parallel multichannel processing and performance optimisation by sector, range or azimuth gate. The data processor used has a plot extraction/track-while-scan capacity of up to 100 tracks/s. The system's truck- or trailer-mounted shelter (also used in the EDR120 variant) incorporates a full colour display and has a split cover which protects the antenna during transit. Reutech quotes a maximum detection range of 56 km against a 2 m² aircraft target for the system and notes identical range and azimuth accuracies for the radar as those cited for EDR120. EDR140 is also reported to take less than 10 minutes to deploy and can be utilised as an anti-aircraft command centre or as a remotely controlled, unmanned slave system.

Status
As of this edition, EDR110 is reported to have been procured by the South African Army. Development of the EDR120 variant is noted as having been completed and the equipment is described as having been evaluated with the ETS 2400 tracking system and what is termed as an associated anti-aircraft weapon.

Contractor
Reunert Defence ESD/Reutech Systems (PTY) Ltd, Stellenbosch.

VERIFIED

ESR220 local warning radar system

Type
D-band (1 to 2 GHz) local warning system and battery command post.

Description
ESR220 is a 2-D warning radar system and is optimised for the detection of low-flying aircraft. Applications include use as a point defence local warning system, as a gap filler or as a fully integrated divisional/brigade mobile anti-aircraft command post. The radar is self-contained and features a high level of mobility as well as an integral command centre capability. ESR220 can be configured for manual (using a single operator) or remote-control, is transportable by land, sea or air, can be deployed within 10 minutes and is designed for rapid break down. The noted remote-control mode is described as offering 'full' radar control and status monitoring. The ESR220 radar operates in the D-band and utilises a high-power, solid-state, wideband transmitter. The antenna used is a low-sidelobe planar-array and includes an integrated identification friend-or-foe aerial. The system processor unit is described as being fully modular with high pulse compression, adaptable clutter suppression, moving target indication and track-while-scan. A full-colour, high-resolution raster display and a radar control/status display are provided.

Status
As of this edition, ESR220 is understood to have been procured by the South African Army.

Specifications
Frequency: D-band (1-2 GHz)
Instrumented range: 100 km
Detection range: 65 km (fighter-sized target)
Accuracy: 0.65° (azimuth); 40 m (range)

The ESR220 local warning system radar deployed

Scan rate: 15 rpm
Track capacity: 100 simultaneously

Contractor
Reunert Defence ESD/Reutech Systems (PTY) Ltd, Stellenbosch.

VERIFIED

SPAIN

ARINE battlefield radar

Type
J-band (12.5 to 18 GHz sub-band) man-portable battlefield detection radar.

Description
The ARINE battlefield radar has been in development since the early 1990s and uses licensed technology from the then Racal Radar Defence Systems' (now Thales Sensors) MSTAR programme. It is a medium-range man-portable pulse-Doppler battlefield surveillance radar designed to detect, localise and identify vehicles, moving or hovering helicopters and personnel in all weathers. The system is normally tripod mounted but can also be installed on a vehicle. Software changes allow it to be used also for artillery fire correction.

The complete ARINE system is broken down into three modules for transport; the UTRA unit (transmitter, receiver, and parabolic antenna/feed mechanism), the UPS unit (processor and servo-mechanisms) and a display console.

The all solid-state radar elements consist primarily of a transmitter, frequency synthesiser and three very large scale integration application specific integrated circuits which have been specially developed. 'Advanced' signal processing techniques are used to differentiate moving targets (3 to 190 km/h) from static targets, over a maximum 170° arc. For ground use the UPS, surmounted by the UTRA, is mounted on an adjustable tripod with an optical sight provided for alignment. Other add-on options include a light-intensification night sight, a loudspeaker, a battery charger, a track table and an antenna mast.

The system weighs 42 kg (plus 3 kg for batteries) and can be carried by three persons. It can be assembled and brought into operation by one person in less than

The ARINE battlefield radar being deployed

2 minutes. Calibration and adjustment is entirely automatic. The UTRA/UPS assembly is connected to the display console by combined data and power-transmission cable, which is normally 15 to 20 m long but which can be increased to a maximum of 50 m.

The display console incorporates a plan position indicator display, keyboard and an audio alerter. The operator uses menu prompts on the display to inject the radar location (universal transverse Mercator co-ordinates plus north orientation) and the equipment can generate target co-ordinates (including position, range and direction of movement). ARINE's complete range of operating modes are as follows:

- wide area surveillance (plan position indicator display, ranges of up to 24 km divided into three sub-ranges - close surveillance (100 m to 4 km with a resolution of 40 m), medium range (3 to 11 km with a resolution of 80 km) and long range (8 to 24 km with a resolution of 160 m and a zoom facility))
- precision target acquisition (zoom mode, B-scope display, 1 km range)
- clutter map
- auto-alarm (autonomous operation)
- aural identification (using target's Doppler signature)
- digitised map
- joint operation with an electro-optic surveillance system.

Status
As of November 2001, ARINE was understood to be available.

Specifications
Frequency: 12.5-18 GHz
Output: = 7 W
Pulse amplitude: 0.12; 0.33; 10 or 20 µs (according to the operating mode)
Compression ratio: 1:1; 20:1; or 200:1
PRF: 5 or 5.5 kHz
Detection range: 3 km (crawling man); 8 km (hovering helicopter); 10 km (standing man); 20 km (light vehicle/flying helicopter); 24 km (heavy tank)
Ground moving target detection: 3-190 km/h
Accuracy: ± 10 mils (bearing); ± 10 m (range)
Resolution: 45 mils (bearing); 50 m (range)
Antenna type: reflector with horn offset
Polarisation: linear or circular
Beamwidth: 2.1° (azimuth); 5.1° (elevation)
Gain: 34 dB
Azimuth scanning speed: 6.75; 12 or 15°/s
Sector scanned in azimuth: 6-175°
Elevation control: ±20°
Selectable surveillance sector axis: 0-360° (0-6,400 mils)
Operator programmable surveillance sector: up to ±90° (1,600 mils)
Scan rate: 6 and 7.5°/s (zoom); 12 and 15°/s (surveillance)
Tilt adjustment: −300 to +300 mils
Remote control from console: 15-20 m (standard); 50 m (optional)
MTBF: >1,000 h
Power supply: 24 V
Consumption: 85 W
Environmental: −33 to +65°C (operating)
Weight: < 42 kg

Contractor
Indra, Madrid.

UPDATED

SWEDEN

9KA 500 Mk 3 coast artillery fire-control system

Type
Coastal artillery fire-control system for mobile batteries.

Description
The 9KA 500 Mk 3 mobile coastal artillery fire-control system is derived from the 9KA 400 system which has been operational in Sweden and other countries since the mid-1960s.

The 9KA 500 Mk 3 concept additionally emphasises, decreasing the vulnerability by the separation of sensor unit and operations unit by up to 9 km or more.

The 9KA 500 Mk 3 is provided with optronic sensors (laser, TV/(infra-red) and a J-band (10 to 20 GHz), frequency-agile radar giving wideband frequency agility. The set-up of displays and control consoles in the operations unit is similar to that of the earlier 9KA 400. Likewise, the number of consoles and their functions can be easily tailored to suit various operational doctrines and functions. The 9KA 500 Mk 3 is capable of controlling guns or missiles or both.

Status
As of this edition, 9KA 500 Mk 3 is reported to have been procured by the Swedish armed forces.

Contractor
CelsiusTech Systems AB, Järfälla.

VERIFIED

ARTHUR weapon locating radar

Type
G/H-band (4 to 8 GHz) mobile weapon locating and fire-control radar.

Description
ARTHUR (ARTillery HUnting Radar) is a highly mobile, autonomous weapon locating and artillery fire-control radar, capable of operating in a countermeasures polluted environment. It is a fully coherent, broadband system, using a travelling wave tube transmitter and a phased-array, high precision antenna. The antenna is scanned electronically, in both azimuth (over a 90° sector) and elevation. The transmitter is the same as that employed in the Giraffe system, combined with new signal processing. The G/H-band is used to provide an operating range of approximately 15 km against typical targets with cross-sections down to 10 to 5 cm. The latest technologies and special algorithms are employed in the Doppler processor for the suppression of clutter, bird echoes and multiple time around echoes.

ARTHUR will detect automatically, locate and classify artillery, rockets and mortars without any degradation of performance in barrage firing situations and carry out threat assessment based on weapon or impact position. All radar data is transmitted automatically to the combat control centre. The equipment will incorporate its own basic Command, Control and Communications system for direct control of counter-battery fire.

The new system, which is designed for use by both the Swedish and Norwegian armies, are installed in the Bv 208 tracked vehicle, with the radar antenna folded down on the roof for transport. ARTHUR has been developed jointly by Ericsson Microwave Systems and their Norwegian subsidiary Ericsson Radar AS.

Status
As of mid-March 2001, Jane's sources reported sales of the ARTHUR radar at 39 units with 10 (out of 14) delivered to Sweden, 10 (out of 12) to Norway, seven (out of eight) to Denmark and two (out of five) to two unnamed Asian countries. In more detail, the first ARTHUR deliveries to Sweden and Norway were made during February and May 1999 respectively. Within the Norwegian Army, the majority of such radars are understood to be assigned to the service's 6th Division. For its part, Denmark signed a then year US$40 million contract covering the supply of the cited eight ARTHUR sensors during December 1997. Elsewhere in the world, an ARTHUR radar was reported to have been evaluated as a possible contender for the British Army's 25 to 30 km range Mamba weapon locating radar requirement during June 1999 and the equipment is also noted as having been tested by the Swiss during August 2000.

The ARTHUR weapon locating radar mounted on a Bv 208 tracked vehicle

A general view of the radar operator and tactical officer colour-graphic workstations used in the ARTHUR weapon locating radar system   0024878

Specifications
Frequency: G/H-band (5.4-5.9 GHz)
Antenna type: passive phased-array
Antenna size: 1.2 × 2.1 m²
Sector search: 90° (16 subsectors)
Transmitter type: TWT
Receiver type: MTI with adaptive wind compensation
Instrumented range: 20-30 km
Capacity: more than 100 targets/min (sustained rate)

Contractor
Ericsson Microwave Systems AB, Ground Systems Division, Mölndal.

UPDATED

Eagle fire-control radar

Type
K-band (20 to 40 GHz) air defence radar.

Description
The Ericsson Eagle fire-control radar is a 'silent' millimetric system intended for use in mobile ground and naval-based air defence systems. The equipment operates in the K-band (35 GHz), enabling tracking of low-flying targets. The Eagle system has been designed with an extremely low radar signature which has been achieved by pulse compression, high antenna gain and almost no sidelobes, in combination with low output peak power. The radiation pattern and a new transmission technique are claimed to make it impossible for escort or standoff jammers to degrade the radar performance.

The system consists of two units: antenna/transceiver and signal processor, with a total system weight of 200 kg. It is able to track two targets simultaneously, with an angular error of less than 0.2 mrad at 10 km, permitting fast lock-over between targets, air-to-surface missile alert or closed-loop fire control. A special operational mode which allows simultaneous pulse-to-pulse frequency agility and Moving Target Indication (MTI) is provided. Adaptive MTI provides for automatic wind compensation.

Status
As of this edition, Eagle was understood to have been ordered by four customers outside Sweden. An upgraded version of the radar forms part of Sweden's RBS23 BAMSE air defence system, working with the new Giraffe AMB 3-D (see separate entry).

Contractor
Ericsson Microwave Systems AB, Ground Systems Division, Mölndal.

VERIFIED

BAMSE missile control and launch vehicle with the mast-mounted Eagle radar/ TV/IFF and the Giraffe 3-D radar in the background

UAR 1021 search and track radar

Type
I-band (8 to 10 GHz) search and track radar for anti-aircraft systems.

Description
UAR 1021 is a combined, fully coherent, search and track pulse-Doppler radar that employs a single I-band transmitter to feed independent search and track antennas via a power splitter. The tracking subsystem uses a monopulse technique with a cassegrain antenna; the search system uses a cheese antenna or a cosec² antenna with integrated identification friend-or-foe dipoles. Other system features include:

- coherent pulse-Doppler clutter attenuation
- electronic counter-countermeasures provision via bandwidth (900 MHz)
- five selectable pulse-Doppler frequencies (<10 ms switching time)
- random pulse-to-pulse frequency agility across the bandwidth when operating in non-Doppler mode
- monopulse angle tracking
- digital range tracking

Status
As of September 2001, UAR 1021 was reported as having been produced in large numbers for use in the Contraves Skyguard system. The related UAR 1302 equipment (a coherent-on-receive system) is noted as being an integral part of the former Marconi S800 series of anti-aircraft radars and as having entered production during 1975.

Specifications
Frequency: I-band (8.6-9.5 GHz)
MTI improvement factor: 45 dB
Peak power: 25 kW
Noise factor: 9 dB
PRF: 5 different pairs in the 6-10 kHz interval with rapid cyclic switching
Pulse-width: 0.3/1 μs selectable
Tracking antenna gain: 37 dB
Tracking antenna diameter: 1 m
Range coverage: 1-14.7 km (search); 0.3-15 km (track)
Acquisition interval: 1,120 m
Velocity window: 20-450 m/s
Angle tracking accuracy: 1 mrad

Contractor
Ericsson Microwave Systems AB, Ground Systems Division, Mölndal.

VERIFIED

The UAR 1021 combined search and track radar as applied to the Contraves Skyguard system

TAIWAN

ADAR-1 Chang Bei multifunction radar

Type
Target acquisition, tracking and missile guidance radar for the Tien Kung family of Surface-to-Air Missiles (SAM).

Description
The phased-array Chang Bei (Long White) radar is reported to be a joint development between Taiwan's Chung Shan Institute and US contractor Martin Marietta (now Lockheed Martin and providing what is termed as technological

The phased-array ADAR-1 acquisition, tracking and guidance radar used in the Tien Kung SAM system

assistance on the programme). The equipment is paired with the MPG-25 target illumination radar (see separate entry) in the Tien Kung (Sky Bow) SAM system and is reported to have a detection range of up to 500 km. Functionally, each Tien Kung battery is noted as being equipped with one ADAR-1 and two MPG-25 radars to support its three or four four-round firing units.

Status
As of this edition, ADAR-1's status is uncertain.

Contractor
Chung Shan Institute, Taipei.

VERIFIED

MPG-25 target illumination radar

Type
Target illuminator radar for the Tien Kung family of Surface-to-Air Missiles (SAM).

Description
MPG-25 is a continuous wave target illumination radar which is integrated with the ADAR-1 Chang Bei multifunction phased-array radar (see separate entry) and a fire-control processor to create the fire-control suite for Taiwan's Tien Kung (Sky Bow) family of SAMs. Radar integration within this suite is reported to be on a time-share basis and MPG-25 is understood to be a development of the American AN/MPQ-46 HP1 equipment which has been undertaken by Taiwan's Chung Shan Institute. Indigenous features incorporated are said to include a 60 per cent increase in power output together with improvements to the radar's countermeasures resistance and identification friend-or-foe capability. Functionally, two MPG-25 radars are reported to be allocated to each Tien Kung battery which, in turn, is equipped with three or four four-round missile firing units.

Status
As of this edition, MPG-25's status is uncertain.

Contractor
Chung Shan Institute, Taipei.

VERIFIED

The MPG-25 continuous wave illumination radar for the Tien Kung 1 SAM system (DTM)

TURKEY

7941 muzzle velocity radar system

Type
I/low J-band (8 to 12 GHz sub-band) muzzle velocity radar.

Description
ASELSAN's 7941 muzzle velocity radar measures and messages a wide variety of field, anti-aircraft and naval artillery shells and provides accurate muzzle velocity data to enhance first round hit probability. Functionally, the system measures muzzle velocity and calibrates its host weapon on the basis of the measurement of the velocities of multiple rounds. Acquired calibration data can be stored and used in muzzle velocity management. Other system features include:

- estimation of next round velocity
- field reprogrammability to facilitate the system's use with new guns, projectiles and propellants
- built-in test down to printed circuit board or module level
- a printer (for the production of stored data hard-copy)
- the ability to meet specific customer requirements via design modularity.

Status
As of May 2001, ASELSAN were describing the 7941 muzzle velocity radar as being 'in the inventory'.

Specifications
Gun types: all standard weapons
Projectile types: base bleed, extended range, mortar, rocket assisted, standard artillery and sabot discarding
Storage capacity: up to 1,000 gun/projectile/charge/lot combinations (anti-aircraft and naval rounds, non-volatile memory storage)
Trigger mode: acceleration
Signal processing: DSP
Muzzle velocity range: 50-2,000 m/s
Accuracy: better than 0.05%
Supply voltage: 20-30 V DC
Power consumption: 18 W (standby); 27 W (transit)
Frequency: 8-12 GHz
Display: alphanumeric LCD
Temperature: −33 to +55°C (operating); −40 to +70°C (storage)

Contractor
ASELSAN Inc, Microwave and System Technologies Division, Ankara.

NEW ENTRY

ARS-2000 ground surveillance and artillery fire adjustment radar

Type
I/J-band (8 to 20 GHz) ground surveillance and artillery fire adjustment radar.

Description
ARS-2000 is a solid-state, pulse-Doppler radar that makes use of pulse compression, monopulse reception and digital signal processing techniques. The equipment comprises a radio frequency sensor and radar control units, with the

The ARS-2000 ground surveillance and artillery fire adjustment radar
2001/0093595

former breaking down into radar transceiver and radar processing sub-units. Weighing approximately 40 kg each, these latter items are described as being two-man portable. The radar's control unit comes from ASELSAN's graphic data terminal product line and takes the form of a lightweight, personal computer-based, ruggedised terminal that incorporates a 25 cm colour electroluminescent display and a keyboard. ARS-2000's radar processing unit houses the equipment's digital signal processing and azimuth/elevation control circuitry. System applications include:

- battlefield surveillance and intelligence collection
- border surveillance
- coastal surveillance
- precision artillery fire adjustment
- high value facility security.

Display options include:

- B-scope (azimuth on the horizontal axis, range on the vertical axis)
- north-orientated plan position indicator.

Status

As of May 2001, the ARS-2000 ground surveillance and fire adjustment radar was reported as being in service and production. While not confirmed or denied by its manufacturer, Jane's sources suggest that the Turkish Army is operating a mast-mounted ARS-2000 application that is installed on an armoured reconnaissance and surveillance variant of the Otokar Cobra 4 × 4 light armoured personnel carrier.

Specifications

Frequency: I/J-band (8-20 GHz)
Detection range: 40 km (max – ground and water targets)
Accuracy: ± 2 mils (azimuth); ± 10 m (range)
Weights: 5 kg (radar control unit); approx 40 kg (radar transceiver and radar processing units)

Contractor

ASELSAN Inc, Microwave and System Technologies Division, Ankara.

VERIFIED

....

ARS-2001 surveillance and fire adjustment radar

Type

I/J-band (8 to 20 GHz) surveillance and fire adjustment radar.

Description

The ARS-2001 surveillance and fire adjustment radar (a 'complete' ASELSAN design and development programme) is designed to detect, classify and track targets that are moving on or close to the ground and/or sea surface at ranges of up to 40 km together with the provision of all weather, day/night artillery fire adjustment data. System features include:

- a fully solid-state design
- surveillance, detection, tracking and fire adjustment operating modes
- a 'user friendly' man/machine interface
- B-scope or plan position indicator display options
- user defined alarm/friendly zone selection
- a digital map overlay facility
- visual Doppler spectrum target classification
- a remote control facility
- the ability to be integrated into a command and control system
- selective range screens
- adjustable sector widths
- a 25 cm (10 in) electroluminescent colour display
- a hand-held, personal computer-based operator control unit
- a built-in test routine
- survey area map loading
- military standards compliance
- a range of accessories that includes a polariser, a Global Positioning System capability, a TV camera (with integrated display), a printer, an external monitor, fibre-optic remote control cabling, an optical sight, headphones and an external speaker system.

ARS-2001 applications include:

- battlefield, border and coastal surveillance
- artillery and mortar fire adjustment
- 'user defined' roles.

Status

As of May 2001, ARS-2001 was reported as being in production and service.

Specifications

Frequency: I/J-band (8-20 GHz)
Detection range: 40 km (ground and water targets, max)
Accuracy: ± 2 mils (azimuth); ± 10 m (range)
Weight: 5 kg (control unit); approx 40 kg (each, transceiver and processor unit)

Contractor

ASELSAN Inc, Microwave and System Technologies Division, Ankara.

NEW ENTRY

ASKARAD ground surveillance and artillery fire adjustment radar

Type

I/J-band (8 to 20 GHz) battlefield ground surveillance and artillery fire adjustment radar.

Description

The ASKARAD radar is a new-generation equipment that is designed for the surveillance of moving targets and artillery fire adjustment on the battlefield. It uses I/J-band monopulse techniques, has an area observation facility and an audible alarm facility. It combines surveillance, target acquisition and classification, target tracking, and artillery fire adjustment functions within one unit. The configuration of ASKARAD is such that it can be broken down into three man-portable units, each weighing less than 35 kg.

ASKARAD has the following operational roles:

- surveillance
- target acquisition and moving target classification
- precision location of targets
- automatic tracking of targets
- plotting of target trajectories on the display or on a plotter
- adjustment of artillery fire
- guidance of small ground or airborne attack units
- helicopter navigation aid, especially for homing
- border control/intruder detection.

ASKARAD's operation modes are as follows:

- ground surveillance – automatic scanning of ranges up to 38 km with a sector width selectable between 30 and 140°
- acquisition/classification – manual scanning and target classification by means of a Doppler tone up to a range of 38 km
- deviation measurement – precise manual target tracking and deviation measurement within an area of ±300 m and ±75 mils and area observation alarm
- auto tracking – automatic target tracking and deviation measurement
- plotting – automatic target tracking and plotting on the display.

Status

As of May 2001, ASELSAN is understood to have delivered more than 220 ASKARAD radars to customers worldwide.

Specifications

Frequency: 8-12 GHz
Channels: 10
Transmitter power: >5 kW (peak)
Power consumption: <120 W (at 24 V DC, standby); <290 W (at 24 V DC, transmit)
Temperature: −30 to +55° C (operating)
Detection ranges: 8 km (105 mm shell burst); 15 km (155 mm shell burst); 15 km (personnel); 20 km (Jeep-type vehicle); 25 km (helicopter); 30 km (armoured vehicle); 38 km (vehicle convoy)
Target location accuracy: ±10 m (range); ±2 mils (bearing on a moving target)

Contractor

ASELSAN Inc, Microwave and System Technologies Division, Ankara.

VERIFIED

The ASKARAD ground surveillance and artillery fire adjustment radar
2001/0093594

UKRAINE

1L220-U

Type
Battlefield phased-array fire-control radar.

Description
The 1L220-U battlefield phased-array fire-control radar is designed to establish the co-ordinates of hostile artillery, mortars, multiple launch rocket systems and tactical missile batteries, provide targeting data for counter-fire operations and monitor friendly fall of shot. As such, the system's phased-array, electronics, operator workstations (two operators and a crew chief?), navigation and communications subsystems and power supply are installed aboard an armoured tracked chassis, with a second cross-country, wheeled 'spares and accessories' vehicle providing support. Other system functions include threat type identification, impact point forecasting, collection of battlefield reconnaissance data and information hand-off to higher commands and associated weapon systems. The radar's export agency claims that use of the 1L220-U reduces munitions expenditure by a factor of 2.5 to 3 and the time needed to bring targets under accurate counter fire by a factor of 1.5 to 2. The system is also said to increase the area of a specific reconnaissance/destruction zone by a factor of 8 to 10 when compared with 'standard' surveillance techniques.

Status
As of July 2001, the 1L220-U battlefield phased-array fire-control radar was being actively promoted by the Ukraine's UKRSPETSEXPORT export agency.

Specifications
Instrumented range: 48 km ('Scale 1'); 96 km ('Scale 1')
Fire position reconnaissance range: 18-20 km (artillery - not less than 30 km for 115 to 203 mm howitzers); 30 km (mortars); 30-40 km (MLRS); 55 km (tactical missile)
Fire adjustment range: 20-25 km (artillery - ≥30 km demonstrated for 115 to 203 mm howitzers); 40-50 km (MLRS); 80 km (tactical missile)
Throughput: 50 targets/min
Surveillance sector: 60°
Altitude: 3,000 m
Power consumption: 58 kVA
Crew: 3
Deployment/tear down time: 5/3 min

Contractor
UKRSPETSEXPORT (export agent), Kiev.

NEW ENTRY

UNITED KINGDOM

Albacom microwave products

Type
Family of tri-service radar and Electronic CounterMeasures microwave products.

Description
The Albacom family of tri-service radar and ECM microwave products includes the following:

AHPA-12-200
AHPA-12-200 is a medium power I/J-band (7.5 to 18 GHz sub-band) Travelling Wave Tube Amplifier (TWTA) that is designed for ECM applications. The device incorporates the AHTA-12-200 RF solid-state amplifier, a matched power supply, a

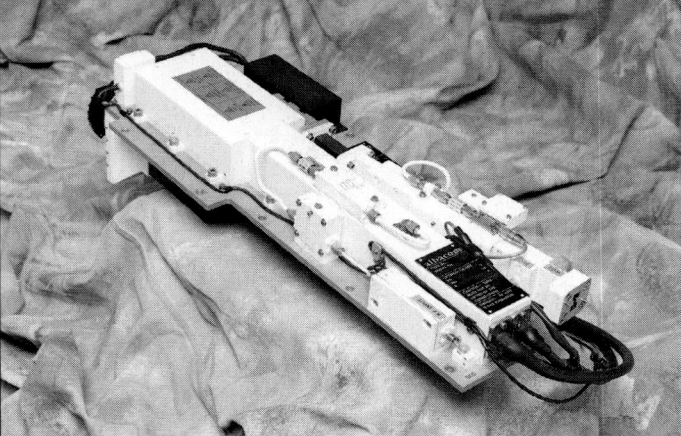

The Albacom AHTA-12-200 naval ECM RF assembly 2000/0085169

broadband helix TWT, a gain equaliser, a harmonic cancellation circuit, remote and local control, output power monitoring, integrated cooling fans, a time elapse indicator and monitoring and protection circuits.

Specifications
Frequency: 7.5-18 GHz
Output power: 200 W
Input power: +1 to +8 dBm
Output power variation: 4 dB
Harmonic power output: -4 dBc (7.5-8.5 GHz); -6 dBc (8.5-18 GHz)
Noise: -14 dBm/MHz (unblanked); -50 dBm/MHz (banked)
Duty ratio: up to 100%
PRF: 20 kHz (maximum)
Warm-up delay: 180 s
Operate delay: 1.5 s
Output power monitor: - 40 dB
Weight: 8.6 kg (RFA assembly); 20.5 kg (PSU)
Dimensions: 427 x 240 x 176 mm (PSU); 570 x 180 x 160 mm (RFA assembly)

AHT-13-70
AHT-13-70 is a medium power I/J-band (8 to 18 GHz sub-band) TWT that is intended for 'multiple' applications and features a beam switching grid, helix circuits, Periodic Permanent Magnet (PPM) focusing, three stage depressed collector operation and conduction cooling.

Specifications
Frequency: 8-18 GHz
Output power: 70 W (minimum peak)
Duty ratio 0 to 100%
PRF: 10 kHz
Gain: 30 dB
Load VSWR: 2:1
Source VSWR: 2:1
Gain flatness: 4 dB (over frequency range)
Harmonic output: -6 dBc
Heater voltage: 6.3 V
Heater power: 6.5 W
Weight: 0.6 kg (approximate)
Dimensions: 200 x 50 x 40 mm (approximate)

AHTA-12-200
AHTA-12-200 is an I/J-band (7.5 to 18 GHz sub-band) Radio Frequency (RF) assembly that is designed for naval ECM applications. The device features the AHT-12-200 TWT, a solid-state amplifier, a gain equaliser, protection circuitry, an integrated cooling fan and harmonic cancellation circuitry.

Specifications
Frequency: 7.5-18 GHz
Output power: 200 W
Input power: +1 to +8 dBm
Output power variation: 4 dB
Harmonic output power: -4 to -6 dBc
Noise: -14 dBm/MHz (unblanked); -50 dBm/MHz (blanked)
Input VSWR: 2:1
Output VSWR: 2:1
Duty ratio: up to 100%
PRF: 20 kHz
Warm-up delay: 180 s
Power monitor: -40 dB
Weight: 8.6 kg (including baseplate)
Dimensions: 570 x 180 x 160 mm (approximate)

Type LY71
Type LY71 is a medium power E/F-band (2.7 to 3.2 GHz sub-band) TWT that is designed for radar applications. The device incorporates a low voltage beam switching grid, ring bar circuits, PPM focusing, depressed collector operation and liquid cooling.

Specifications
Frequency: 2.7-3.2 GHz
Output power: 10 kW (minimum peak)
Duty ratio: 4%
Gain: 45 dB
Load VSWR: 1.5:1
Source VSWR: 1.5:1
Gain flatness: 1 dB (over frequency range)
Harmonic output: -30 dBc (2nd); -40 dBc (3rd and 4th)
Noise figure: 35 dB
Heater voltage: 6.3 V
Heater power: 35 W
Weight: 8.5 kg (approximate)
Dimensions: 630 x 105 x 135 mm (approximate)

Type LY72
Type LY72 is a pulsed E/F-band (2.7 to 3.2 GHz and 3.2 to 3.7 GHz sub-bands) TWTA that is designed for radar applications. The device incorporates built-in test, monitoring, control and protection circuitry, a shadow gridded gun and a single stage depressed collector.

Specifications

Frequency: 2.7-3.2 GHz and 3.2-3.7 GHz (covered by 2 TWTs)
Output power: 10 kW (minimum peak)
Gain: 46 dB
Duty ratio: 2.6% (maximum)
Pulse-length: 20 μs (maximum)
Weight: 46 g (approximate)
Dimensions: 680 x 490 x 180 mm (approximate, including baseplate)

Type LY73

The Type LY73 is a medium power E/F-band (2.7 to 3.2 GHz and 3.2 to 3.7 GHz sub-bands) TWT that is designed for radar applications. The device incorporates a low voltage beam switching grid, ring bar circuits, PPM focusing, depressed collector operation and liquid cooling.

Specifications

Frequency: 2.7-3.2 GHz and 3.2-3.7 GHz
Output power: 10 kW (minimum peak)
Duty ratio: 2.6%
Gain: 46 dB
Load VSWR: 1.5:1
Source VSWR: 1.5:1
Gain flatness: 1 dB (over frequency range)
Harmonic output: -30 dBc (2nd); -40 dBc (3rd and 4th)
Noise figure: 35 dB
Heater voltage: 6.3 V
Heater power: 35 W
Weight: 8.5 kg (approximate)
Dimensions: 630 x 123 x 85 mm (approximate)

Type LY75

Type LY75 is a medium power E/F-band (3.1 to 3.4 GHz sub-band) TWT that is designed for radar applications. The device incorporates a low voltage beam switching grid, ring loop circuits, PPM focusing, depressed collector operation and forced air cooling.

Specifications

Frequency: 3.1-3.4 GHz
Output power: 2.5 kW (minimum peak)
Duty ratio: 10%
Gain: 45 dB
Load VSWR: 1.5:1
Source VSWR: 1.5:1
Gain flatness: 1 dB (over frequency range)
Harmonic output: -25 dBc (2nd); -35 dBc (3rd and 4th)
Noise figure: 35 dB
Heater voltage: 6.3 V
Heater power: 25 W
Weight: 7 kg (approximate)
Dimensions: 560 x 90 x 145 mm (approximate)

Type LY75/1

Type LY75/1 is a medium power E/F-band (3.1 to 3.4 GHz sub-band) TWT that is designed for radar applications. The device incorporates a low voltage beam switching grid, ring loop circuits, PPM focusing, depressed collector operation and conduction cooling.

Specifications

Frequency: 3.1-3.4 GHz
Output power: 2.5 kW (minimum peak)
Duty ratio: 10%
Gain: 45 dB
Load VSWR: 1.5:1
Source VSWR: 1.5:1
Gain flatness: 1 dB (over frequency range)
Harmonic output: -25 dBc (2nd); -35 dBc (3rd and 4th)
Noise figure: 35 dB
Heater voltage: 6.3 V
Heater power: 25 W
Ion pump: 1 mA (s/c, current); 3.5 kW (voltage)
Weight: 6.5 kg (approximate)
Dimensions: 530 x 87 x 115 mm (approximate)

Type LY80

Type LY80 is a medium power D-band (1.2 to 1.8 GHz sub-band) TWT that is designed for radar applications. The device incorporates a low voltage or cathode modulation beam switching grid, ring loop circuits, PPM focusing and conduction cooling.

Specifications

Frequency: 1.2-1.8 GHz (in 2 variants of 300 MHz bandwidth)
Output power: 2 kW (minimum peak)
Duty ratio: 10%
Pulse-length: 15 μs (maximum)
Gain: 40 dB
Load VSWR: 1.5:1 (output); 2:1 (input)
Gain flatness: 1 dB (over frequency range)
Heater voltage: 6.3 V

Heater power: 25 W
Weight: 6 kg (approximate)
Dimensions: 610 x 80 x 100 mm (approximate)

Type LY130

Type LY130 is a medium power I-band (9 to 10 GHz sub-band) TWT that is designed for radar applications. The device incorporates a low voltage beam switching grid, ring loop circuits, PPM focusing, depressed collector operation and conduction cooling.

Specifications

Frequency: 9-10 GHz
Output power: 5 kW (minimum peak)
Duty ratio: 3%
Gain: 60 dB
Load VSWR: 1.5:1
Source VSWR: 1.5:1
Gain flatness: 1 dB (over frequency range)
Harmonic output: -20 dBc
Heater voltage: 6.3 V
Heater power: 20 W
Weight: 2.5 kg (approximate)
Dimensions: 347 x 64 x 83 mm (approximate)

Type LY134

Type LY134 is a medium power I-band (9-10 GHz sub-band) TWT that is designed for radar applications. The device incorporates a low voltage beam switching grid, ring loop circuits, PPM focusing, depressed collector operation and conduction cooling.

Specifications

Frequency: 9-10 GHz
Output power: 8 kW (minimum peak)
Duty ratio: 2%
Gain: 60 dB
Load VSWR: 1.5:1
Source VSWR: 1.5:1
Gain flatness: 1 dB (over frequency range)
Harmonic output: -20 dBc
Heater voltage: 6.3 V
Heater power: 20 W
Weight: 2.5 kg (approximate)
Dimensions: 347 x 64 x 83 mm (approximate)

Status

As of this edition, microwave devices of the types described were reported as being available.

Contractor

Albacom Ltd, Dundee.

VERIFIED

....

Dagger surveillance and target acquisition radar

Type

Air surveillance and acquisition radar for Rapier Field Standard C (FSC) surface-to-air missile and other similar air defence systems.

Description

Dagger provides three-dimensional surveillance, target acquisition and tracking in an autonomous, compact and reliable equipment which, together with the

The Dagger radar as used in the Rapier FSC missile system

launcher, forms the Rapier FSC (export designation Jernas) fire unit. The radar consists of compact sealed modules which are designed for high reliability and ease of maintenance. These fit into the common trailer used in the towed Rapier FSC system.

The radar gives precise range, bearing, elevation and velocity on multiple tracks so that targets can be engaged in order of priority out to the maximum range of the missile system.

Powerful electronic counter-countermeasures capabilities are incorporated and stem from innovative techniques combined with narrow beamwidth, very low sidelobe levels and discrimination against incoherent signals.

Dagger tracks large numbers of targets simultaneously while carrying out automatic identification. It can detect hovering helicopters and very small targets such as unmanned aerial vehicles. By using modern digital filtering techniques it can discriminate small targets in heavy ground clutter from vegetation moving in the wind. Excellent range resolution and the variable pulse repetition frequency help eliminate range ambiguities and mutual interference.

The modular design ensures that the Dagger radar is readily adaptable to other configurations. It can be provided in a self-contained palletised form if a vehicle or trailer-mounted system is required, or the compact modules can be distributed within an armoured vehicle.

Although Dagger performance is optimised for Rapier FSC, other options for range and altitude coverage are available for alternative roles. The accuracy of Dagger makes it an ideal component of a multispectral sensor system and for networking with other radars and weapon systems.

Status
As of March 2001, the Dagger/Rapier FSC combination was reported as having been procured by both the British Army and the Royal Air Force

Contractors
BAE Systems - Combat and Radar Systems, Cowes, Isle of Wight.

UPDATED

....

Rapier search and acquisition radar

Type
F-band (3 to 4 GHz) battlefield air surveillance radar for the Rapier Surface-to-Air Missile (SAM) system.

Description
This system is an integral part of the Rapier SAM system. Little information has been officially released on this part of the system and the following is an assessment based on available data.

This equipment comprises a primary surveillance radar working in conjunction with a computer and a secondary radar (Identification Friend-or-Foe - IFF) interrogator to locate, identify and, where appropriate, initiate the tracking of enemy targets. The whole equipment is located in the fire unit of the Rapier system.

In the search mode the radar (coherent pulsed Doppler equipment) scans continuously through 360° in azimuth. When a target is detected its azimuth range and velocity are measured. On the same scan it is interrogated by the IFF whose antenna is combined with that of the primary radar. If a friendly response is received the information on that target is cancelled from the system and the search continues without interruption.

If the target is adjudged hostile an alarm signal is sent to the remainder of the equipment and the measured co-ordinates are used to direct the tracker (see description of Rapier) and the launcher towards the target.

Racal surveillance radar and missile command transmitter on Rapier fire unit

It is, of course, possible that a friendly target might fail to respond to the IFF interrogation on a single pass of the surveillance system (because of an unfavourable aspect of the airborne antenna, for example) but the system allows for a friendly response to be accepted before the engagement has gone too far.

Status
This equipment forms an integral part of the Rapier system.

Contractors
Thales Sensors, Crawley.
Raytheon Systems Ltd, Harlow (microminiature IFF equipment – including antenna).

UPDATED

UNITED STATES OF AMERICA

Advanced Battlefield Surveillance Radar (ABSR)

Type
I-band (8.75 to 8.95 GHz sub-band) portable ground surveillance radar.

Description
ABSR is described as being a coherent, ground-based, fully encased, all-weather radar that comprises an antenna unit and a remote system control/display unit and is designed for military and paramilitary security applications including border surveillance and the monitoring of activity around sensitive sites. Specifically, the system provides automatic detection and monitoring of moving targets (including people, vehicles and helicopters) at ranges of up to 30 km. Target type identification is by means of an audio signal derived from a Doppler frequency. The system's control/display unit takes the form of a 25 kg ruggedised personal computer-based equipment that can be deployed in less than 5 minutes. A radar levelling subsystem is included and the system's man/machine interface is menu driven and described as being 'intuitive' in order to minimise operator training. ARSS-1 has a range of installation options that include fixed site (buildings or towers), deployable tripod and vehicular (where it is designated as the ARSS-2). In this latter application, the radar functions when its host platform is stationary and is designed to be integrated with an electro-optical viewing system.

Status
As of May 2001, 125 examples (with a further 75 as an option) of the ABSR radar are understood to have been procured by the Israel Defence Forces. ABSR is designed by Israeli contractor Elta Electronics and is manufactured by the Telephonics Corporation.

Specifications
Frequency: 8-12 GHz (50:1, 800 Hz, ±5%)
Transmitter output: 5 kW (peak, pulse compression)
Antenna size: 40 × 60 cm
Antenna gain: 31 dB
Scan rate: 1, 2 and 4 rpm (fixed)
Range: 7-10 km (pedestrian); 10-12 km (pedestrians); 15 km (light vehicle and hovering helicopter); 30 km (heavy vehicle)
Range resolution: 50 m
Accuracy: ±5 mils (azimuth); ±25 m (range)
MTI threshold: 1-3 km/h
Sector coverage: search and detect in any defined sector of up to 359°
Track capacity: automatic tracking and track-while-scan on up to 60 targets
Display options: B-Scope, clutter map, digital map, graphics on colour display and PPI
Outputs: audio and video signals
Power requirement: 24 V source
System weight: 25 kg

Contractor
Telephonics Corporation - Command Systems Division, Farmingdale, New York.

UPDATED

....

AN/MPQ-53 guidance radar

Type
G/H-band (4 to 8 GHz) surveillance and tracking radar for the MIM-104 Patriot surface-to-air missile system.

Description
The AN/MPQ-53 is a frequency-agile multifunction G/H-band radar group which performs all the surveillance, Identification Friend-or-Foe (IFF), tracking and guidance and Electronic CounterMeasures (ECM) functions entailed in the Patriot tactical air defence missile system. The antenna array is a 2.44 m diameter, 5,161-element phased-array planar configuration carried on a semi-trailer chassis. The

Patriot AN/MPQ-53 radar showing main planar-array and supplementary antennas

PSTAR man-portable search and target acquisition radar

antenna unit has separate arrays for target detection and tracking, missile guidance and IFF functions. The last of these tasks is carried out by an AN/TPX-46(V)7 interrogator, using a supplementary array adjacent the main circular search and track array on the antenna unit. Other supplementary arrays are for sidelobe cancellation and missile guidance signal reception.

The Track-Via-Missile (TVM) missile guidance system is unique to Patriot. This technique involves the missile's passive monopulse seeker array being directed by the Engagement Control Station (ECS) to look in the direction of the target which then begins to intercept increasingly precise returns from the reflected electromagnetic energy signals. This in turn triggers the G/H-band onboard downward datalink which is offset in frequency from the target tracks and illumination beam and which transmits target data from the missile guidance package to the ECS computer via the circular 251-element TVM receive-only array at the lower right of the antenna group. The ECS uses this information to calculate guidance instructions which are passed to the missile by the radar's G/H-band command and uplink beam. The phase-coded information is received on the missile by the two sets of guidance antennas, these transmit it to the guidance electronics which in turn use it to move the control surfaces. This procedure is repeated until the point of closest approach when the warhead is detonated.

Radar interrogation of a target is carried out by an AN/TPX-46(V)7 IFF system using a linear antenna array set below the main array position. There are also five diamond-shaped 51-element arrays; two individual ones above the IFF set at the bottom corners of the array and three centred below the level of the TVM receiver array near the lower edge of the front face of the antenna array. These are sidelobe cancellers used to reduce the effect of enemy jamming.

The 3 to 170 km range radar performs its surveillance, tracking, guidance and Electronic Counter-CounterMeasures (ECCM) functions in a time-shared manner by using the system's computer to generate action cycles that last in milliseconds. Up to 32 different radar configurations can be called up with the beams tailored for long range, short range, horizon and clutter, guidance and ECCM functions in terms of their power, waveform and physical dimensions. The data rate for each function can also be selected independently to give 54 different operational modes so that, for example, a long-range search can be conducted over a longer time period than a horizon search for low-altitude pop-up targets. None of the functions requires any given time interval which therefore allows a random sequence of radar actions at any one time, considerably adding to an attacker's ECM problem. The search sector is 90° and the track capability is 120°.

Status

MIM-104 missile systems have or are being supplied to the US Army and a range of export customers including Germany, Israel, Japan, Kuwait, the Netherlands and Saudi Arabia. As of May 1995, a total of 154 Patriot fire units and 7,800 missiles are reported to have been delivered. The system was used operationally during the 1991 Gulf War and as of this edition, has been the subject of a number of upgrade efforts including the introduction of a radar shroud to improve high aircraft density operations, enhanced target detection and discrimination, improved tactical ballistic missile discrimination (PAC-3 upgrade) and an increase in target illumination power.

Contractor

Raytheon Electronic Systems, Bedford, Massachusetts.

VERIFIED

AN/PPQ-2 PSTAR battlefield radar

Type

D-band (1 to 2 GHz) man-portable search and target acquisition air defence radar.

Description

PSTAR is a lightweight, pulse-Doppler search and target acquisition radar and command and control system. It is intended to provide an effective air defence

umbrella for all forward deployed forces and point defence locations. PPQ-2 is noted as being effective in all battle and jamming environments. The radar is billed as offering 'exceptional' subclutter visibility and as being able to detect low-flying aircraft of 'all types'. It can detect and classify helicopters based on the unique signature of their rotor blade returns. PSTAR can classify both friendly and hostile fixed-wing aircraft and helicopters with its Identification Friend-or-Foe (IFF) interrogation capability. The system includes electronic counter-countermeasures features, such as sidelobe cancelling, clear channel search, frequency agility, sector blank and strobe-on-jam. It also has a near-realtime capability to interface directly with a command and control network or weapon systems directly.

PSTAR integrates a mechanically scanned, planar-array antenna; a three-channel superheterodyne receiver; an ultra-low phase noise stable master oscillator, a solid-state power amplifier, advanced digital signal and data processing and a militarised liquid crystal display graphics terminal.

The man-portable unit weighs approximately 135 kg, and breaks down into components of less than 45 kg each. It is designed for deployment from the US Army's new High Mobility Multipurpose Wheeled Vehicle (HMMWV) and can receive power from the vehicle or from a portable generator. Units can be delivered by parachute and can be moved in and out of battle positions quickly (less than 10 minutes). PSTAR is also noted as being able to operate from a moving vehicle in support of manoeuvre operations.

Status

The private venture PSTAR radar has been selected by the US Army as the winner of a competition to build a new battlefield radar for its light infantry forces and is now being fielded as the AN/PPQ-2. Approximately 40 such radars are thought to have been delivered to the US Army. As of May 2001, PSTAR radars are reported to be in service with (or to have been ordered by) six customers on four continents. For the export market, the equipment is fitted with a 20 rpm antenna, an integral IFF interrogator and an enhanced command and control facility. Lockheed Martin also offers an LT PSTAR variant which features reduced range. On 2 February 2000, Lockheed Martin announced that it had signed a Foreign Military Sales (FMS) contract with the US Army to cover the supply of 20 PSTAR radars to the Republic of China (Taiwanese) Army. Potentially valued at then year US$18 million, delivery of the described sensors was scheduled (at the time of the announcement) for completion by mid-2001. Previously, Lockheed Martin has supplied Taiwan with its AN/APS-145 airborne early warning, GD-53 multimode, GE-592 3-D and LAADS radars.

Specifications

Frequency: D-band (1-2 GHz)

Antenna rotation: 10/20 rpm

Radiation pattern: −5 to +30° (nominal); −5 to +5° (adjustable from nominal)

Power output: 50 W (average); 1 kW (peak)

Detection range: out to 20 km on both fixed-wing aircraft and hovering helicopters.

Accuracy: ±200 m (range); ±2° RMS (azimuth).

Contractor

Lockheed Martin Naval Electronics and Surveillance Systems – Syracuse, Syracuse, New York.

UPDATED

AN/PPS-5 combat surveillance radar

Type

I-band (8.8 to 9 GHz sub-band) portable ground surveillance radar.

Description

The AN/PPS-5 is an 8.8 to 9 GHz band portable ground surveillance radar that is designed to detect and locate individuals (at ranges of up to 5,000 m), groups (up to 10,000 m) and small vehicles (up to 10,000 m). An upgrade adds solid-state technology to the baseline model's legacy capabilities. Overall, PPS-5 is claimed to

The AN/PPS-5 ground surveillance radar

*AN/PPS-15A
lightweight ground
surveillance radar*

*AN/PPS-15B base
security radar*

solve the major problems inherent in personnel detection radars by providing automatic sector scanning in azimuth and range coverage that provides a visual display of moving targets within those sectors that are within line of sight. Use of a range-gated filter, moving target indicator processor is deemed to be 'especially suitable' for detecting targets with velocities as low as 1.6 km/h and the radar's operation as a whole is described as being 'silent while scanning'.

A Remote Control Display Unit (RCDU) provides remote control of the radar and takes the form of a ruggedised laptop computer that facilitates operator selection of power and mode. A range mark on the RCDU shows the position of the range gate and a Doppler processor (consisting of a clutter filter and audio amplifier) drives headphones or external speakers. Operator selectable scan speeds are held relatively constant by a feedback motor speed control circuit that supplies only enough power to meet the demand in the presence of varying wind loads; this conserves power. A Windows NT Pentium III operating system allows for 'ease' of operation and 'flexible system integration alternatives'. Telephonics further notes that the system can be 'fully' integrated with other types of surveillance equipment including forward-looking infra-red sensors and day/night cameras.

Status
As of January 2002, the solid-state AN/PPS-5 upgrade was reported as being in development, with first article testing scheduled for Summer 2002.

Specifications
Baseline configuration
Frequency: 8.8-9 GHz
Power output: 1 kW (+30 dBm – average); 1 kW (peak)
PRR: 4,000 pps ±5%
Pulse-width: 0.25 µs
Antenna: parabolic contour with elliptic outline (34 × 107 cm)
Range: 5,000 m (single person); 10,000 m (small vehicle – extended range versions available)
Accuracy: ±20 m (range); ±10 mils (azimuth/elevation)
Weight: 58.9 kg total
Power consumption: 65 W

Contractor
Telephonics Corporation, Command Systems, Farmingdale, New York.

UPDATED

AN/PPS-15A/-15B infantry and base security radar

Type
J-band (10 to 20 GHz) portable ground surveillance radars.

Description
The AN/PPS-15A is a J-band, lightweight ground surveillance radar for the detection and location of moving targets such as men, vehicles, or boats. Range and azimuth of a moving target are read directly from solid-state Light Emitting Diode (LED) displays and the nature of the target is deduced from its aural signature. The radar set consists of the antenna assembly, control indicator, an antenna drive, a headset, a tripod, remote cable and an internal disposable battery which provides power for 12 hours continuous operation. The AN/PPS-15A can be hand held, pintle or tripod mounted and remotely controlled. The control indicator and antenna assembly are integral for most operations but may be separated for remote operation at distances up to 9 m. Automatic sector scan can be centred on a selectable bearing and the scan width also is selectable. Elevation control is manual.

Target returns are fed to an all-range channel and a discrete-range channel. During search, the all-range channel automatically discriminates against clutter and produces both aural and visual alarms when a moving target is detected. During ranging, the discrete-range channel produces a light display and an increased Doppler tone when a moving target enters a range gate positioned by an operator-control. Both channels are active during the search-range mode, enabling an operator to monitor detected targets on the all-range channel as he ranges a selected target by means of the discrete-range gate control. Five frequencies are available for selection by the operator.

AN/PPS-15B is a development of the AN/PPS-15A and provides additional surveillance capability in a 350 m range segment, divided into seven 50 m range gates. The system is designed to detect and locate moving targets, such as vehicles and personnel, to provide perimeter surveillance for virtually any type of installation.

The 350 m segment may be positioned by the operator anywhere within the detection range of the system to provide surveillance without interference from, or alarms caused by, traffic moving inside the base perimeter itself. The seven range gates are individually displayed to provide automatic detection. Sector scan width is variable, by the operator, between limits of 22 and 180°. Five transmitter frequencies are available for selection by the operator. Power sources are supplied with the radar which can be operated from either AC (115 or 230 V) or DC (12 or 24 V).

The AN/PPS-15B can be hand held, pintle or tripod mounted and remotely controlled. Both aural and visual alarms are produced when a moving target is detected and the LED type display is identical to that of the AN/PPS-15A.

Status
Since its introduction in 1975, nearly 2,000 PPS-15 series radars have been sold worldwide. Over 1,000 were delivered to various US military units. Other countries who have taken delivery include Argentina, Canada, Israel, Norway, Spain and Taiwan. PPS-15A is no longer in production. PPS-15B is understood to have been procured by the US Air Force as a security tool for use in its ground-launched cruise missile programme.

Specifications
AN/APS-15A
Frequency: J-band (10-20 GHz)
Peak power: 50 mW
Antenna: 20 × 30 cm
Mode: homodyne
Weight: 10.7 kg (remote operation)
Detection ranges: 1.5 km (personnel); 3 km (vehicles)
Battery: BA4386/PRC-25
AN/APS-15B
Frequency: J-band (10-20 GHz)

Detection range: 3,000 m
Remoting: 35 m
Power source (external): 115/230 V AC, 12/24 V DC
Weight
Radar: 15 kg (incl DC power source and cable)
DC power source: 7 kg
AC power source/battery charger: 9.5 kg
Remote cable and reel: 7 kg

Contractor

Amex Systems Inc, Hawthorne, California.

VERIFIED

....

AN/UPQ-6 (Mark V) velocimeter

Type

J-band (10 to 20 GHz) muzzle velocity radar for tactical weapons.

Description

The AN/UPQ-6 Mk V muzzle velocimeter is a small one piece continuous wave Doppler system that measures the velocity of a projectile at several points along its trajectory. It is intended for use with tactical weapons and can be used with a variety of these, including artillery, small arms and machine guns, air defence weapons, shipborne guns, mortars and fire-control systems. It has been tested with a variety of systems from 5.56 mm to 16 in calibre. The Mk V is also used in experimental testing range applications where its scientific data output can assist in the development of new munitions.

The equipment is mounted either on a tripod or directly on the weapon concerned via a quick disconnect bracket and consists of an antenna which contains an integral transmitter/receiver and data processing circuitry. The unit is fully automatic in all its functions. Built-in test equipment, triggering, false triggering logic, anti-jamming frequency agility and data processing are all conducted without operator input. The system interfaces directly with a fire-control computer, or with an optional display. Computed muzzle velocity is fed directly to the user's fire-control computer, where it can be used as a real-time fire-control input. The system is capable of measuring all types of ammunition, including base bleed, sabot and rocket-assist to an accuracy of ±0.05 per cent. The non-volatile memory will store 1,000 shots. Individual shot data or selected averages can be displayed, together with a variety of scientific data, including waterfall graphs. Velocity output is corrected automatically for non-standard conditions, or is output as an actual velocity.

Status

As of this edition, UPQ-6 (Mark V) is understood to have been procured by US and international armed forces.

Specifications

Frequency: J-band (10.519; 10.522; 10.525; 10.528; 10.531 GHz)
Power output: 250 mW
Velocity range: 20-2,000 m/s
Accuracy: ±0.05% or 0.4 m/s, whichever is greater
Dimensions: 315 × 217 × 100 mm (transceivers);
172 × 79 × 57 mm (display unit)
Weight: 8.2 kg (transceiver); 1 kg (display unit)

Contractor

Lear Astronics Corporation, Santa Monica, California.

VERIFIED

AN/UPQ-6 velocimeter

AN/UPS-3 tactical defence radar

Type

D-band (1 to 2 GHz) short-range air defence tactical radar.

Description

The AN/UPS-3 tactical defence alert radar has been designed to meet the requirement for a short-range air defence command and control sensor for all-weather surveillance and detection of low-flying targets. It operates as a tactical alerting and cueing sensor in the extreme forward area of the battlefield and provides automatic identification of helicopters and fixed-wing aircraft. Target data can be provided to an adjacent fire unit, or transmitted by digital link to a command location for dissemination to a number of units. The system operates in D-band, is all solid state, lightweight, portable and highly reliable. It operates autonomously in the field, breaks down into four transportable kits, has a march order time of less than 10 minutes and has been airdrop qualified. It has also been employed as part of a netted system in the SHORAD C Test Bed by the US Army and can be a low-cost gap filler.

The system consists of a rotating antenna assembly mounted on a pedestal and quadripod, a transceiver/signal processor also on a quadripod and a display which can be remoted up to a kilometre away. Up to four display units can be operated remotely from the equipment, one being the master, by the attachment of the remote-control unit. The radar can be operated with other displays/terminals through an RS-232 output port. The antenna elevation angle can be adjusted −6 to +6° and provides height coverage from ground level to 10,000 ft. The display presents target range and azimuth on a light emitting diode matrix. A low sidelobe antenna and digital signal processor, which give significant improvements in accuracy and Electronic Counter-CounterMeasures (ECCM) capabilities, have been developed, tested and incorporated into the system. This provides a multitarget track-while-scan capability.

ADSR-40/60

Two extended range derivatives of the UPS-3 are being promoted. The ADSR-40 can detect targets with a 2 m² radar cross-section out to 40 km, and has a higher scan rate and improved ECCM features. The ADSR-60 is effective out to 60 km, uses a larger antenna and incorporates identification friend-or-foe, together with a large colour display.

Status

As of this edition, UPS-3 is understood to have been procured by the US Marine Corps (41 radars), US Navy and US Army. The US Army is understood to have operated the system on a high-mobility multipurpose wheeled vehicle to alert and cue Stinger and Chaparral fire units. The US Navy is reported to have used the equipment to provide quick reaction air defence for a classified mission.

Specifications

Frequency: D-band (1,215-1,300 MHz), eight crystal frequencies selectable
Peak power: 210 W
Average power: 18.5 W
MTI improvement factor: 55 dB
Antenna
Beamwidth: 8° (horizontal); 17° (vertical)
Rotation speed: 90°/s
Gain: 21 dB
Track-while-scan: 40 targets simultaneously
Range: 9 km (typical UAV); 14 km (helicopter); 20 km (fighter aircraft)
Accurate velocity data: 18.5 km/h
Accuracy/resolution: 200 m (range); 2° (azimuth)
Weights
Antenna: 55 kg, incl antenna, pedestal and quadripod
Transceiver: 30 kg
Display: 10 kg

Contractor

Lear Astronics Corporation, Santa Monica, California.

VERIFIED

AN/UPS-3 mounted on a High-Mobility Multipurpose Wheeled Vehicle (HMMWV)

Real-Time Velocimeter System (RTVS)

Type
J-band (10 to 20 GHz) mobile artillery tracking radar.

Description
This system is a transportable J-band continuous wave Doppler artillery tracking radar designed to measure and record target radial range, velocity, acceleration and angular position. The equipment is carried in two mobile elements: an antenna trailer which contains the transmitter and receiver plus the steerable antenna system and a control cabin which carries the operator control console and data processing units. The RTVS is intended to allow tracking of targets from 304 to 3,048 m/s and will acquire and track targets down to 0.0032 m². Accuracy at 20 km is stated to be 0.31 m/s for velocity, 0.31 m/s² for acceleration and 6 m in range. Operating frequency of the radar is pretunable between 10 and 10.26 GHz.

Range in the system is determined without modulating the carrier by transmitting three carrier waves simultaneously and measuring the phase shift between transmitted and received beat notes. A high-speed array processor is used to perform digital spectral analysis of the returned signal. The Fast Fourier Transform (FFT) processor extracts target velocity from the raw Doppler signal and spectral details of the transmitted signal resolve the difference.

Unambiguous real-time ranging to 75 km is obtained from the FFTs applied to the Doppler shifted return signal from the three transmitted tones, where the relative shift is proportional to range. The transmitter uses a stable UHF (0.3 to 1 GHz) synthesiser and a low noise master oscillator klystron to drive the final power stage which consists of three klystrons.

Status
As of this edition, RTVS is thought to have been procured for use at the US Army White Sands Missile Range.

Contractor
Datron Systems Inc, Chatsworth, California.

VERIFIED

YUGOSLAVIA, FEDERAL REPUBLIC

IR-3 surveillance radar

Type
J-band (10 to 20 GHz) man-portable battlefield surveillance radar.

Description
The IR-3 is a portable ground surveillance radar that was first shown at the Defendory '90 exhibition in Athens. It is stated to be a fully coherent system with a range of more than 3,000 m on vehicles, and is designed to detect any movement on the battlefield from a crawling man to a heavy vehicle. High sensitivity and low radiated power is claimed, together with portability, high reliability and ease of handling. The antenna/transceiver can be mounted either on a tripod or on the operator's chest. Power is supplied from a battery source.

Antenna/transceiver of the IR-3 surveillance radar

Specifications
Frequency: J-band (10-20 GHz)
Antenna beam accuracy: 5° (azimuth); 7.5° (elevation)
Polarisation: vertical
Detection of mobile targets: 2.5-7.5 km/h (pedestrians); 7.5 km/h (vehicles)
Identification: up to 300 m (crawling man); up to 1,500 m (walking man); up to 2,000 m (light cross-country vehicle); up to 3,000 m (heavy vehicle)
Dimension of antenna/transceiver: 350 × 350 × 125 mm
Weights: 5 kg (transceiver/antenna); 10 kg (system)
MTBF: 5,000 h
Power supply: 24 V Ni/Cd battery; 14 W max consumption
Details have also emerged concerning a new anti-infantry radar designed specifically for a Yugoslav requirement. Designed by the Institute of Microwave Technology and Electronics, the radar appears to be a more powerful system than the IR-3 with a maximum range of over 10 km.

Specific detection ranges are given as 12 km for vehicles and 5 km against personnel. The beamwidth is 3° and the ranging accuracy is claimed to be within 50 m. The peak power output is 3 W. A standard RS-422 computer interface is used for data exchange and control purposes. The complete system weighs about 60 kg. The radar was understood to have been scheduled to enter service with the Federal Yugoslav Army during late 1991.

Contractor
Yugoimport SDPR, Novi Beograd.

VERIFIED

NAVAL/COASTAL SURVEILLANCE AND NAVIGATION RADARS

Introduction

This section contains all shipborne surface search and air surveillance 3-D and 2-D radars, together with height-finders, ballistic tracking radars and their associated data processing and display systems. Since naval navigation radars are frequently used for medium and short search they are also included. Coastal radar systems used for air and surface search, as well as harbour surveillance, also form part of this section.

VERIFIED

AUSTRALIA

CEA-FAR

Type
E/F-band (2 to 4 GHz) multifunction radar.

Description
The CEA-FAR radar is a fixed, active phased-array radar that is available in a number of configurations to meet differing operational, physical and budgetary requirements. Equally, the sensor is claimed to be suitable for use in a range of applications that include:
- self-defence and surveillance for small, medium and large surface ships
- weather monitoring (including wind/precipitation observation and prediction, weather avoidance (for aircraft) and clear-air turbulence detection)
- coastal surveillance (including drug and customs surveillance, border intrusion detection, fisheries/exclusive economic zone protection and search and rescue)
- land-based air surveillance
- surface search and battlefield surveillance
- air traffic control (terminal air surveillance and route monitoring)
- airborne surveillance and command and control

As such, the equipment is modular, programmable and scalable, with the smallest configuration being man-portable and the largest suitable for surface ship applications. The CEA-FAR array is a static, modular, active antenna that utilises discrete antenna tiles as its basic building block. Such tiles are nominally available as an 8 × 8 array that provides a full 3-D capability. Optionally, a 32 × 2 array is available to create a high-resolution, 2-D system for applications such as man-portable ground surveillance and coastal surveillance. Other system features include:
- pulse-to-pulse frequency agility
- electronic scanning and tracking in azimuth and elevation
- high update rate tracking modes that make use of monopulse beam steering in azimuth and elevation
- 'high' levels of system redundancy
- field configurable to meet changing requirements
- integration with a 'full' data fusion track management and display interface
- adaptable frequencies and modes to avoid detected anti-radiation missile threats
- the ability to deploy the sensor's antenna faces remote from the processing system
- an electronic self-leveling capability
- 'low' radar cross section and infra-red signature
- the ability to be configured with non-planar-arrays
- elevation scanning beyond 60°
- multistep 'graceful degradation'
- a remote control facility.

Status
As of late 2000, CEA-FAR was reported as having been one of the four candidate radar systems that were proposed for the now abandoned Australian 'ANZAC' class frigate Warfighting Improvement Programme (WIP). As proposed for ANZAC WIP, a multifacetted main CEA-FAR array would have been installed at the base of the vessel's foremast together with in-duct arrays on either side (?) of its helicopter hangar. During March 2000, CEA Technologies announced that it had been awarded a then year US$9.6 million contract covering the supply of three CEA-FAR radars to the US Government for use as land-based air detection sensors. As of May 2001, CEA Technologies was understood to have been actively promoting the CEA-FAR radar.

Contractor
CEA Technologies Pty Ltd, Canberra, Australian Capital Territory

UPDATED

CEA-MAST

Type
Multisensor track management system.

Description
The CEA-MAST track management system is designed to provide a 'complete' ship-based sensor management system that is optimised for the earliest possible declaration of threat tracks. Centred around sensor interfaces, a multiple hypothesis plot fusion tracker and a track and data fusion processor, the architecture can be integrated with multifunction, long-range surveillance, navigation and search radars, electro-optic sensors (including infra-red search and track systems) and identification friend-or-foe systems. As such, the structure combines radar signal processing, data processing and workstation/display technology to automatically declare surface and air tracks and is further described as featuring 'several' levels of redundancy. Equally, CEA-MAST is noted as offering the option of being 'totally' integrated with a host vessel's combat system (via high-speed interfaces such as Ethernet) or as functioning as a stand-alone system. Here, sensor operation and data display are managed from a dedicated graphical display workstation and CEA Technologies claims that both stand-alone and combat system integrated applications provide 'all' the architecture's facilities.

Status
As of August 2001, the CEA-MAST track management system was noted as being under contract for installation aboard Royal Australian Navy (RAN) 'Adelaide' class frigates as part of their 2002 to 2006 upgrade programme. Here, 'Adelaide' class vessels will receive enhanced AN/SPS-49(V) radars (see separate entry) and Mk 92 weapons control systems, the NTDS combat data system, the Evolved SeaSparrow Missile (ESSM) system, a Link 16 standard datalink capability, improved electronic support and decoy systems and TMS 4350 torpedo defence and TMS 5424 mine avoidance sonars in addition to the CEA-MAST architecture. As of the given date, HMAS *Sydney* was scheduled to be the first of class to receive the described update package.

Specifications
Target capacity: up to 500 tracks/sensor
Target velocity range: programmable up to 1,852 km/h
Range bins: 2.3-58.5 m (programmable)
Surveillance range: 593 km (36.6 m range bin - programmable via range bin selection)
Range accuracy: ±55 m (36.6 m range bin)
Range resolution: 110 m (36.6 m range bin)
False target capacity: up to 1,500 plots (typical value)
MTBF: 10,000 h
MTTR: 15 min

Contractor
CEA Technologies Pty Ltd, Canberra, Australian Capital Territory

NEW ENTRY

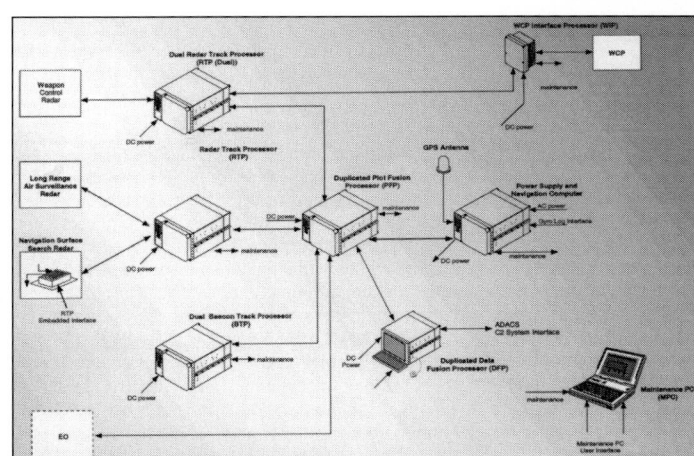

A schematic showing the layout of the CEA-MAST architecture as applied to the RAN's 2002 to 2006 'Adelaide' class upgrade programme (CEA Technologies)
2002/0114374

CEA-Scope

Type
Periscope detection system.

Description
The CEA-Scope periscope detection system is described as being primarily designed to automatically detect and track raised submarine masts. Comprising CEA's Modular Integrated Radar Surveillance System (MIRSS) tracker (see separate entry), a workstation, a monitor and interfaces for a radar and its host vessel's motion sensors, the architecture provides its operator with an auto-alert to any submarine mast within its detection range. This is achieved via real-time pattern recognition of submarine tactical operating modes at periscope depth. Also noted as being effective against ships, small boats and slow moving, low altitude air targets (see following), CEA-Scope can be installed aboard 'any' surface combatant or auxiliary in 'Navy inventory'. Within the system, the MIRSS tracker provides two automatic detection and tracking paths, one of which is for surface and air targets, with the other being dedicated to submarine mast detection. The submarine classification process (performed by a graphic data fusion subsystem) is largely based on the time and motion constraints of submarine operations involving mast exposure and provides a lower false alarm rate and a higher certainty of track declaration than the capability against air and surface targets. Fused sensor data is displayed using CEA's Multisensor Integrated Data fusion Display System (MIDDS), which is described as being a real-time, multisensor, correlation system for MIRSS-based surveillance systems. Here, the correlation capability is provided in a graphical display environment on high resolution computer display terminals. Functionally, all MIRSS derived surveillance data is sent to the MIDDS, where the 'large quantities of incoming sensor data' are preprocessed and displayed using 'easily' recognisable symbology and in 'concise' formats. A communications subsystem reports detections to other interested systems (via high-speed data/communications links, local area networks or RS-232/422 interfaces) and can be modified to provide data on craft other than submarines that operate with definable time, distance and course movements.

Status
CEA-Scope is reported as having 'successfully' completed land-based trials against offshore targets (including surface ships, aircraft and submarine masts) during December 1994 and as having subsequently been installed aboard the Royal Australian Navy frigate *Sydney*. Here, the system was interfaced with the vessel's Mk 92 Mod 2 and CEA Modified Surface Search (MSS - based on a commercial system) radars. Pier side and ocean functional tests of the system's motion compensation capability are described as being successfully completed for both radar types. Range tests against submarine masts were completed on 14 May 1996 and are said to have achieved a 'high success rate' and a 'zero' false alarm rate. During the RIMPAC 98 exercise, a CEA-Scope installation was tested aboard the US Navy's command ship *Coronado*. As of May 2001, CEA Technologies was understood to have been actively promoting the CEA-Scope system.

Contractor
CEA Technologies Pty Ltd, Canberra, Australian Capital Territory

NEW ENTRY

CEA-SHIP

Type
Ship classification radar system.

Description
The CEA-SHIP ship classification radar system is designed to classify targets that are observed during search operations into broad categories of interest. Comprising a compatible surface search/navigation radar (or radars), CEA's Modular Integrated Radar Surveillance System (MIRSS) tracker (see separate entry) and an associated data fusion and display system, CEA-SHIP processes and extracts data from one or more selected radars and utilises it to automatically assess the likely class of ship being observed and present a fused track, radar and assessment picture of the target. Functionally, the system extracts target features and then uses Bayesian and other techniques to classify it on the basis of those features. CEA-SHIP is further noted as employing any available features to derive single look probabilities that a target belongs to a particular category and then as refining the probability set over repeated observations. The utility of the system is claimed to lie in its ability to rapidly focus attention on targets of interest in target rich environments without having to change the observable characteristics of the radar being used.

Status
As of May 2001, a CEA-SHIP ship classification radar system demonstrator was reported as having been 'recently' installed aboard the Royal Australian Navy frigate *ANZAC* as part of a validation programme run by the Australian Department of Defense, CEA Technologies and the Adelaide-based Centre for Sensor Signal and Information Processing (CSSIP).

Contractor
CEA Technologies Pty Ltd, Canberra, Australian Capital Territory

NEW ENTRY

Modular Integrated Radar Surveillance System (MIRSS) tracker

Type
Radar signal processor/tracker.

Description
The CEA Technologies MIRSS is described as being a ruggedised radar signal processor/tracker that provides sensor management, communications and display interface systems. The equipment makes use of 'advanced' signal processing, detection and tracking algorithms and can be integrated with both new and upgraded radar installations. MIRSS is noted as being able to 'seamlessly' acquire, process and display information from multiple sites (each with one or more fixed or mobile sensors) and its radar and video processing subsystems are packaged in a standard 48 cm (19 in) rack-mounted enclosure. Other system features include:

- simultaneous detection and 'high-resolution' tracking of air and surface targets
- automatic detection and track processing to produce tracks, plots and reports
- radar display video extraction (including compression/decompression functions)
- the ability to interface with 'most' radars and a 'wide variety' of other sensors including closed circuit television, direction-finding, Global Positioning System, electronic support and acoustic sensor systems
- automatic detection of and adaptation to a specific type of radar
- a radar control, monitoring and signal digitising interface
- automatic accommodation of a range of communications options including microwave links, fibre-optic cables, data cables and standard telephone lines, at variable data rates down to 9.6 kbps
- the ability to control and monitor co-sited equipments (acting as a communications link multiplexer/demultiplexer)
- built-in local and remote (dial-up) control and diagnostic monitoring facilities
- use of surface mounted component and LSI on multilayered printed circuit board technology
- 'high levels' of module commonality
- remote power up/down.

Status
As of May 2001, the MIRSS radar signal processor/tracker unit was reported as having been procured by the Australian Army, the Sydney Ports Corporation, the US Department of Defense and the US Navy.

Specifications
Target capacity: up to 500 tracks/radar sensor
Target velocity range: programmable up to 1,852 km/h
Range bins: 2.3-58.5 m (programmable)
Surveillance range: 593 km (36.6 m range bin - programmable via range bin selection)
Azimuth accuracy: 0.2 × antenna beamwidth or 0.1°, whichever is the greater (typical value)
Azimuth resolution: 3 × radar beamwidth (typical value)
False target capacity: 1,500 plots (typical value)
Power: 24 V DC/115 V AC
Temperature: 0 to +50°C (operating)
MTBF: 10,000 h
MTTR: <1 h
Inputs: synchronised or azimuth pulses (programmable)
Input video: 4 channels
Dimensions (h × d × w): 222 × 513 × 483 mm
Weight: 22 kg

Contractor
CEA Technologies Pty Ltd, Canberra, Australian Capital Territory

NEW ENTRY

BULGARIA

Automated Radar Data Acquisition, Processing and Display (RDAPD) system

Type
RDAPD system.

Description
The Kintex automated RDAPD system is designed for a range of applications that includes: surveillance and coastguard functions, control of commercial and naval traffic, detection/identification of air targets, surveillance/control of harbour areas, tactical intelligence, weather data processing, forest surveillance and fire-fighting and disaster monitoring/control. RDAPD can be customised to meet customer specific priority levels with 'level one' including coastal/shipborne radars, sonar stations, land-based surveillance stations, meteorological stations and data acquisition sensors. 'Level two' applications comprise regional data processing centres with netted communications for the construction of a general situation picture. RDAPD can interface with a wide range of radar types, older types of which

require dedicated extractor and data transmission units for full integration. RDAPD features multiprocessor computing, can be configured for manual data inputs and can be integrated with digital systems and non-radar sensors such as television trackers.

RDAPD equipped data acquisition and control centres feature multiprocessor operator workstations, high-resolution colour displays, a communications subsystem, a printer, a plotter and a scanner. Functionally, all-source data is synchronised (co-ordinates and time) for distribution and display to multiple or single users. Target trajectories and parameters are defined within the system and the operator is allowed access to any system function by means of text/graphic editing programmes and menus. System software is based on the OS/2 operating system and high order languages are used for programming. Non-authorised system use is prevented by a dedicated guard module. A minimum configuration system using RDAPD comprises a command centre and 15 radars, while a full configuration application incorporates five first level command centres, 75 second level radars (15 per command centre) and three third level command posts.

Status
As of this edition, the Kintex RDAPD system is understood to have been procured by the Bulgarian Army.

Contractor
Kintex, Sofia.

VERIFIED

KALIAKRA coastal surveillance radar

Type
I/J-band (8 to 20 GHz) coastal surveillance and detection radar system.

Description
KALIAKRA is a mobile naval radar designed for detection, identification and tracking of surface vessels and low-flying air targets (fixed-wing aircraft, helicopters, cruise missiles and unmanned air vehicles) in the coastal zone. It is installed in a truck-mounted shelter with the radar antenna mounted on top. An Identification Friend-or-Foe (IFF) antenna is fitted above the main elliptical antenna and the complete antenna structure is lowered during transit. The system electronics, communications equipment, antenna turning gear and the operator's position appear to be installed in the shelter, with a trailer containing power supplies, air conditioning and so on.

The complete system consists of the radar system, IFF equipment, communications components, power supply, test and control instruments, air conditioning, and ancillary equipment.

Status
As of this edition, KALIAKRA is understood to have been procured by the Bulgarian Army.

Specifications
Frequency: 8-20 GHz
Peak pulse power: 25 kW
Detection range: 40 km (surface targets 50-100 m²); 80-90 km (fighter)
Resolution: 70 m (range); 0.8° (azimuth)
Multitarget tracking: 40 pcs
Accuracy: 50-80 m (range); 0.3-0.4° (azimuth)
Direction: <2°
Velocity: <4 km/h

Contractor
Kintex, Sofia.

VERIFIED

A model of the KALIAKRA coastal surveillance radar system

Meduza multipurpose air and sea surface acquisition radar

Type
Mobile multifunctional naval radar for coastal and air surveillance.

Description
Meduza is a multipurpose radar that is designed for sea surface and aerial target detection. System capabilities include:
- detection and positioning of mine splashdown in the course of air minelaying operations
- surface surveillance
- location of small size water targets as part of the coastguard missions
- navigation and ship safety control
- detection and tracking of low-flying air targets
- commercial shipping control.

The main application of the system is mine search and detection. It incorporates a unique anti-mine acquisition facility and features a high-scanning speed enabling detection of transient mine splashdown in the course of several scans. The variable beam pattern of the radar is narrow in the horizontal plane for mine splashdown and wide in the vertical plane for detection of minelaying aircraft.

The radar system is mounted on two ZIL-131 cross-country vehicles, one for the radar equipment and one for the field communications centre. The radar system consists of a rotating antenna assembly, transmitter/receiver, a plan position indicator, a section scan indicator, documentation and recording equipment, automated data acquisition equipment, a simulator, two petrol power supply units mounted on a trailer, a filter/ventilation unit and a conditioner. The system can also be provided in a shipborne configuration, with an interface to the ship's systems and indicators.

Status
As of this edition, Meduza is understood to have been procured by the Bulgarian Army.

Specifications
Weight: 1,300 kg
Detection range: 6-8 km (fishing boat); 8-10 km (floating mine); 12-18 km (submarine periscope); 15-20 km (average sea buoy); 20-25 km (mine splash); 35-45 km (fixed-wing aircraft); 40-50 km (helicopter); 40-59 km (missile boat); 50-55 km (destroyer)
Resolution: 0.7° (bearing); 30 m (range)
Range scales: 0.8, 2, 7, 13, 26, 52 and 103 km
Pulse duration: 0.15, 0.3 and 1.2 µs
Antenna rotation rate: 14.5 rpm
Coverage: 0.7° (horizontal); 15° (vertical)
MTBF: 300 h
Continuous operation: 24 h
Pulse power: 90 kW
Power consumption: 3.5 kW

Contractor
Kintex, Sofia.

VERIFIED

The Meduza multipurpose air and sea surface acquisition radar

Naval traffic control radar system

Type
Sea and riverine traffic control radar system.

Description
Kintex's naval traffic control radar system is software controlled and comprises a control centre which is linked to a number of remote radars. It is designed to monitor and display high density sea and riverine traffic. Acquired data is transmitted to the control centre in a compressed format and overall, the system provides the following:
- a radar picture covering the system's complete surveillance area
- video charts corresponding to coastline features

One of the antenna configurations used in the Kintex naval traffic control radar system   0006896

- automatic tracking of up to 60 targets (extendable to 200 if required)
- definition of the parameters of moving targets
- a radar data archive which can be accessed on request
- picture blanking of operator selected areas
- remote control of system sensors
- operator selectable sector scanning (zoom mode).

Status
As of this issue, Kintex's naval traffic control radar system is understood to have been procured by the Bulgarian Army.

Specifications
Antenna (with reducer)
Antenna length: 2.73 or 5.4 m
Polarisation: linear (5.4 m antenna)
Type: slotted waveguide (5.4 m antenna)
Rotation rate: 24 ± 2 rpm (5.4 m antenna); 32 rpm (2.73 m antenna)
Gain: 32 dB (2.73 m antenna); 36 dB (5.4 m antenna)
Beamwidth: 0.45° (5.4 m antenna); 0.9° (2.73 m antenna)
Transmitter
Carrier frequency: 9,375 ± 30 MHz (5.4 m antenna); 9,410 ± 30 MHz (2.73 and 5.4 m antenna)
Pulse duration: 0.03 µs (≤ 4 km); 0.25 µs (2.73 m antenna); 0.8 µs, 50 ns, 250 ns (5.4 m antenna)
Travel frequency: 1.6, 4 kHz (2.73 m antenna); 0.625, 1.25, 2.5 kHz (5.4 m antenna)
Pulse power: 10 kW ± 50% (2.73 m antenna); 15 kW (5.4 m antenna)
Receiver
Type: linear (2.73 m antenna); LHFE with logarithmic parameters (5.4 m antenna)
Intermediate frequency: 60 MHz (5.4 m antenna)
Bandwidth: 4, 25 MHz (5.4 m antenna)
Data transmission modem
Type: V32 TERBO (compatible with AT)
RS-232C port: 38.4 kbytes/s
Transmission rate: ≥ 19.2 kbytes/s
Computer (min configuration): IBM 486DX2 (40 MHz VESA local databus); 90C31 video controller; 8 Mbytes RAM; 250 Mbytes HDD; 4 pcs RS-232C
Data display: 43 or 51 cm colour; 1,280 × 1,024 pixels
Telephone transmission standard: up to M.1020-CCITT (optional radio communication)
Max cable length: 100 m
Max RS-232C cable length: 2 m

Contractor
Kintex, Sofia.

VERIFIED

CANADA

CMR-90 radar transceiver

Type
I-band (8 to 10 GHz) lightweight surface surveillance system.

Description
The CMR-90 version of the LN66 radar family is designed for use as the primary navigation radar on small craft, or as the secondary radar on larger vessels. It is a modular form-fit-function unit to replace older LN66 10 kW transmitter/receiver units, or for incorporation in new radar systems.

The transceiver is fully solid state, except for the magnetron, providing improved performance, high mean time between failure, reduced power consumption and improved receiver performance. Features include solid-state modulators, low noise receiver and compatibility with existing LN66 10 kW systems.

Status
Over time, CMR-90 radars are understood to have been procured by the US Navy.

Specifications
Frequency: I-band (9,375 ±30 MHz)
Peak power: 12 kW
Pulse-widths: 0.06 and 0.5 µs
PRFs: 2,500 and 1,250 Hz (typical)
Noise figure: 8.5 dB

Contractor
BAE Systems Canada, Kanata, Ontario.

UPDATED

LN66 radar series

Type
I-band (8 to 10 GHz) family of general surveillance and navigation radars.

Description
The LN66 series is a family of modular radars available in a variety of peak power levels from 6 to 90 kW and in a number of system configurations tailored to specific requirements. The system applications include general surface surveillance and navigation, coastal surveillance, gun fire control and anti-submarine warfare roles.

LN66/HP
A rugged five-element radar comprising an antenna, transmitter/receiver, display, north stabilising unit and power supply. The 0.9 m antenna is normally used for airborne applications and is a radome-enclosed end-fed slotted waveguide, horizontally polarised unit. Five and 2.4 m versions are also available for other applications.

Specifications
Frequency: I-band (9,375 ±30 MHz)
Peak power: 75 kW
Rotation speed: 24 ±2.4 rpm
Pulse-width: 0.1 µs or 1 µs
PRF: 500; 2,000 Hz
Weight: antenna 9.7 kg; transmitter/receiver 34 kg; display 22 kg

LN66 ship set
Designed for the short-range navigation role for a large number of surface vessels, the 12 kW ship set is in service with the US Navy in this capacity. The standard configuration consists of a 0.9 m radome, 1.5 or 2.4 m antenna, a solid-state transmitter/receiver, 25 cm display, and a true bearing unit.

Status
Over time, LN66 series radar applications are understood to have included Brazilian 'Ipará' class frigates; Egyptian 'Knox' class frigates; Greek 'Kimon' class destroyers; Iranian 'Babr' class destroyers; Mexican 'Gearing' class destroyers and 'Bronstein' class frigates; Taiwanese 'Knox' class frigates and 'Hai Ou' class guided missile fast patrol craft; Thai 'Knox' class frigates; Turkish 'Tepe' class frigates and US Navy 'Blue Ridge' class amphibious command ships and SH-2 anti-submarine warfare helicopters (75kW variant).

Specifications
Frequency: I-band (9,375 ±30 MHz)
Peak power: 12 kW
Rotation speed: 22 ±2.2 rpm
Pulse-width: 0.5; 0.9 µs
PRF: 800/1,250; 2,500 Hz
Weight: antenna 9.7 kg; transmitter/receiver 13.4 kg; display 22.9 kg

Contractor
BAE Systems Canada, Kanata, Ontario.

UPDATED

An LN66 antenna mounted above the AN/SLQ-32(V)1 antenna on the former US Navy frigate Kirk *(Stefan Terzibaschitsch)*

CHINA, PEOPLE'S REPUBLIC

Chinese naval radar

Information on radars used by the Chinese Navy (other than for five equipments detailed separately) continues to be limited and precludes meaningful entries. Accordingly, the following overviews should be treated with some caution and will be subject to future revision.

Air search and surface search radars

Identified Chinese naval air/surface search radars comprise the I-band (8 to 10 GHz) ESR-1 surface search/fire-control radar; the G-band (4 to 6 GHz) Hai Ying air search radar (see separate entry); the three-dimensional G-band Rice Screen air search radar; the I-band Type 352 surface search/fire-control radar; the G/H-band (4 to 8 GHz) Type 354 (NATO reporting name Eye Shield) air/surface search radar (see separate entry) and the E/F-band (2 to 4 GHz) Type 360 surface search radar. Of these, ESR-1 is reported to be installed aboard Chinese 'Luhu' class destroyers while the three-dimensional Rice Screen radar is noted as being installed aboard Chinese 'Jianghu I/III/IV' class frigates. The Type 352 radar appears to be a Chinese variant of the Russian Square Tie equipment and is described as being fitted to Chinese 'Luda I/II' class destroyers; 'Jianghu I/III/IV' class frigates; 'Huangfen/Hegu/Hoko/Hema' class guided missile fast patrol craft and 'Houijan/Houxin' class guided missile fast attack craft. The remaining cited system – the Type 360 surface search radar – has been noted as being installed aboard Thai 'Naresuan' class frigates. It is also perhaps worth noting that Jane's sources suggest that Chinese 'Jiangwei' class frigates and 'Houijan' class guided missile fast attack craft are fitted with a derivative of the Russian A-band (70 to 73 MHz) P-8 (NATO reporting name Knife Rest) air search radar which may be designated as the Type 765 in Chinese service.

Outside the People's Republic and aside from the already noted Thai use of the Type 360 equipment, Jane's sources suggest that the Type 352 radar is installed aboard the Bangladesh single 'Jianghu I' class frigate and five 'Durdarasha' class guided missile fast patrol craft; the Egyptian pair of Jianghu Is; the North Korean four 'Huangfen' class guided missile fast patrol craft; the Pakistan four 'Huangfens' and four 'Haibat' class guided missile fast patrol craft and the Thai four 'Chao Phraya' class frigates. The Type 765 radar is noted as being fitted to the Egyptian pair of 'Jianghu I' class frigates.

Navigation radars

Sources suggest that China has developed a series of I-band (8 to 10 GHz) navigation/surface search radars (said to include the Types 752, 753, 756 and 757) which are suitable for installation aboard destroyers, escorts, fast patrol craft and submarines. Identified applications include the Chinese 'Houxin' class guided missile fast attack craft (Type 756) and the Bangladesh 'Jianghu I' class frigate (Type 753 – surface search role).

VERIFIED

Hai Ying air surveillance radar

Type
G-band (4 to 6 GHz) air surveillance radar.

Description
Hai Ying (Sea Eagle) is a 3-D air surveillance radar developed by the Chinese Nanjing Marine Institute. The system operates in G-band and utilises a phased scanning, plannar-array antenna. Hai Ying incorporates a stabilised high-power wideband amplifier with a frequency-agile synthesiser and microcomputer. Comprehensive electronic counter-countermeasures facilities are understood to have been included in the design and a maximum target capacity of 25 tracks is quoted.

Status
As of this edition, Jane's sources suggest that the Hai Ying radar has been fitted to a number of the Chinese Navy's 'Luhu' class destroyers.

Specifications
Azimuth coverage: 360°
Altitude coverage: 25,000 m
Range: up to 180 km
Elevation: 0-7.2° or 0-28.8°
Accuracy: ±100 m (range); ±0.8° (azimuth/elevation)

Contractor
Nanjing Marine Institute, Nanjing.

VERIFIED

LR61 coastal defence radar system

Type
I-band (8 to 10 GHz) coastal defence radar.

Description
LR61 is designed automatically to detect and track surface targets and hand-off targeting data to surface-to-surface missile batteries. The system comprises the radar, a data processing station, a communications subsystem and a transportation vehicle. The equipment utilises a 31 cm plan position indicator to display raw video, target counting/plotting and weapon fire-control data together with a 24 × 40 character 23 cm cathode ray tube for tabulated target range, azimuth, velocity and heading information.

Status
As of February 2002, LR61's status was uncertain.

Specifications
Frequency: 8-10 GHz
Range: $\leq$ 5 km (min); $\geq$ 120 km (max), P_{-d} 80%, P_{-f} 10^{-6}, RCS 3,000 m^2, HA $\geq$ 800 m
Azimuth coverage: 360°
Accuracy: $\leq$ 0.2° (azimuth); $\leq$ 6 m (range)
Set up tracking time: $\leq$ 12 s
Target tracking: up to 16 batches (with TWS)
Communication range: > 5 km
Environment: −10 to +40°C (shelter); −20 to +60°C (outside shelter)
Dimensions: 2 × 1.3 × 2.2 m (antenna with pedestal); 5 × 2.4 × 2 m (equipment/operator shelter)
Weight: $\leq$ 1,400 kg

Contractor
China National Electronics Import and Export Corporation, Beijing.

VERIFIED

LR63 splash spotting radar

Type
G/H-band (4 to 8 GHz) splash-spotting radar.

Description
The solid-state, phased-array LR63 is designed to provide a fire correction capability for coastal 130/155 mm gun and surface-to-surface missile batteries and is vehicle mounted for mobility. The radar utilises 36 cm 'B' type tracking, 36 cm 'B' type deviation measurements (with four sub-pictures) and tabular information displays.

Status
As of February 2002, LR63's status was uncertain.

Specifications
Frequency: 4-8 GHz
Range: $\leq$ 1 km (min); $\geq$ 120 km (max), P_{-d} 80%, P_{-f} 10^{-6}, RCS 3,000 m^2, HA 800 m
Azimuth coverage: any 120° sector (automatic tracking); 360° (manual search)
Measuring accuracy: $\leq$ 0.1° RMS (azimuth); $\leq$ 15 m RMS (range)
Environment: −10 to +40°C (shelter); −20 to +60°C (outside shelter)
Dimensions: 2.5 × 1.5 × 0.5 m (antenna); 4 × 2.4 × 2 m (equipment/operator shelter)
Weight: $\leq$ 1,500 kg

Contractor
China National Electronics Import and Export Corporation, Beijing.

VERIFIED

MR33 naval surveillance radar

Type
I-band (8 to 10 GHz) air and surface surveillance radar.

Description
The MR33 surveillance radar is designed for use aboard vessels of more than 200 tonnes and comprises the radar, a data processing station and interfaces to onboard weapon systems and other sensors. MR33 utilises 51 cm colour raster plan position indicator and 640 × 400 pixel 23 cm equal-ion displays.

Status
As of February 2002, MR33's status was uncertain.

Specifications

Frequency: 8-10 GHz
Range: ≤ 500 m (min); ≥ 25 km (aircraft, RCS = 2m²); ≥ 120 km (max) P_{-d} 80%, P_{-f} 10^{-6}, RCS 3,000 m², HA > 10 m
Azimuth coverage: 360°
Tracking accuracy: ≤ 0.3/≤ 05° (azimuth — ship/aircraft); ≤ 50/≤ 150 m (range — ship/aircraft)
Resolution: ≤ 2° (azimuth); ≤ 100 m (range)
Set up tracking time: ≤ 6 s (ship); ≤ 12 s (aircraft)
Tracking capacity: up to 16 batches (with TWS)
Environment: –30 to +60°C (topside equipment)
Dimensions: 1,650 × 674 × 550 mm (equipment cabinet — three installed); 1,800 × 820 mm (antenna with stabilised pedestal)
Weight: 250 kg (antenna); 350 kg (equipment cabinet); 1,200 kg (complete system)
Power consumption: ≤ 10 kW

Contractor

China National Electronics Import and Export Corporation, Beijing.

VERIFIED

....

Type 354 air/surface search radar

Type

G/H-band (4 to 8 GHz) surface and low-altitude search radar.

Description

The Type 354 is a shipborne G/H-band search and target acquisition radar normally employed on missile destroyers. It measures the range and bearing of targets with high accuracy and provides information for the ship's information centre, the fire-control radar and the missile guidance radar. Coherent techniques are employed to combat interference and clutter and the antenna is pitch and roll stabilised. It should also be noted that the Type 354 radar may have the alternative designation MX902.

Status

As of this edition, Type 354 radars are reported to have been installed aboard Chinese 'Luda I/II' class destroyers and 'Jianghu I/III/IV' class frigates together with the single Bangladesh 'Jianghu I' and the four Thai 'Chao Phraya' class frigates.

Specifications

Frequency: G/H-band (4-8 GHz)
Peak power: 500 kW
Pulse-width: 2 μs
PRF: 400 Hz; 800 Hz
Polarisation: linear
Scan rate: 4-10 rpm
Beamwidth: 1.2° (horizontal); 5° (vertical)
Accuracy: 5 m (azimuth); 75 m (range)
Resolution: 800 m (range); 1.3° (azimuth)
Range: more than 93 km on an aircraft with an RCS of 10 m²

Contractor

China National Electronics Import and Export Corporation, Beijing.

VERIFIED

Model 354 shipborne radar

DENMARK

SCANTER radar systems

Type

Family of modular navigation, coastal surveillance and Vessel Traffic Service (VTS) radars.

Description

TERMA describes its SCANTER radar sensor systems as being 'specifically tailored [to meet requirements] where traditional marine radar falls short [of specifications relating to] interference rejection, signal processing, interfacing, data distribution and [environmental toughness]'. As such, the modular equipments are billed as being suitable for ship navigation, short range air and surface search, VTS and coastal surveillance applications. System features include:

- frequency diversity to facilitate the decorrelation of sea clutter, eliminate small target fluctuation, reduce lobing effects and 'improve' long range detection
- auto-adaptive sensitivity control to provide automatic 2-D STC, thereby eliminating the need for operator intervention during 'normal' operations
- pulse-to-pulse integration to enhance signal-to-noise ratio for 'improved' small target detection
- digital FTC and sweep-to-sweep correlation
- horizontal or circular polarisation
- specific transceiver set up (including programmable pulse repetition frequency and pulsewidth and random stagger)
- a remote control and built-in test service tool.

Physically, SCANTER radars comprise an antenna and a transceiver unit. Available antenna options include 2.1 (for navigation applications), 3.7 (harbour/river surveillance), 4, 5.5 (harbour/river surveillance) and 6.4 m ('high performance') SWG assemblies, a cosec² reflector unit (offering 'improved air coverage' for military applications) and a submarine applicable X-band (8 to 12.5 GHz) configuration. TERMA characterises the SCANTER transceiver configurations as incorporating a low noise front-end, 'high' dynamic range receiver and a magnetron transmission chain that makes use of 'special' drive techniques to optimise resolution and range performance and extend magnetron life.

Status

During the period January to September 2001, *Jane's* sources were reporting SCANTER radars as being installed aboard the following warships/submarines and classes of warship/submarine:

Brazil	• the aircraft carrier *Minas Gerais*
	• the corvette *Barroso* (building as of January 2001 - SCANTER or Decca 1226C radar)
	• the anti-submarine warfare capable submarine *Tikuna* (building as of March 2001)
	• 'Niteroi' class frigates (six ships-in-class)
	• 'Tupi' class anti-submarine warfare capable submarines (four boats)
Chile	• the support ship *Merino*
Denmark	• the royal yacht *Dannebrog*
	• the stores ship *Sleipner*
	• 'Niels Juel' (three ships) and 'Thetis' (four ships) class frigates
Lithuania	• the frigates *Aukstaitis* and *Zemaitis*
Sweden	• the coastal patrol craft *Jagaren*
	• the minelayer *Carlskrona*
	• the electronic surveillance ship *Orion*
	• the auxiliary command ship/submarine tender *Visborg*
	• 'Gotland' class anti-submarine warfare capable submarines (three boats)
Venezuela	• 'Sabalo' class anti-submarine warfare capable submarines (two boats)

A TERMA SCANTER X-band antenna

Specifications: Scanter radar systems

Antennas:	2.1 m SWG	3.7 m SWG	4 m SWG	5.5 m SWG **	6.4 m SWG **	2.8 m reflector	1.1 m submarine *
Frequency	8-12.5 GHz	8-12.5 GHz	2-4 GHz	8-12.5 GHz	8-12.5 GHz	8-12.5 GHz	8-12.5 GHz
Vertical beam	fan	fan	Fan	fan	fan	$cosec^2$	$cosec^2$
Polarisation	horiz	horiz	horiz	horiz	circular	horiz	horiz
Gain	= 31 dBi	≥ 33 dBi	≥ 28.5 dBi	≥ 35 dBi	≥ 37 dBi	≥ 32.5 dBi	≥ 27 dBi
Horiz BW at − 3 dB	= 1.1°	≤ 0.6°	≤ 2°	≤ 0.41°	≤ 0.40°	≤ 0.95°	< 2°
Vertical BW at − 3 dB	19°	19°	20°	19°	12°	9°	10°
$Cosec^2$ to						+ 50°	+ 40°
Tilt		optional		optional	− 1.5°		+ 5°
Rotation speed (norm/max)	24/28 rpm	12/24 rpm	12/24 rpm	12/24 rpm	12/24 rpm	12/24 rpm	
Wind speed (op/surv)	50/65 m/s	50/65 m/s	40/50 m/s	45/55 m/s	45/55 m/s	35/50 m/s	
Drive motor	0.4 kW	1.1 kW	1.1 kW	1.1 kW	2.2 kW	2.2 kW	
Weight	= 65 kg	≤ 220 kg	≤ 220 kg	≤ 240 kg	≤ 325 kg	≤ 330 kg	≤ 25 kg

Key

BW Beamwidth **horiz** horizontal **max** maximum **norm** normal **op** operational **surv** survivable

Notes

* The described 1.1 m X–band submarine applicable antenna assembly does not include a drive mechanism

** The 5.5 and 6.4 m SWG antennas are available in reinforced configurations to withstand wind speeds greater than those shown in the table

Transceivers:

	X-band	X-band	S-band
Magnetron	4 kW	25 kW	30 kW
Frequencies	9,375 and 9,410 MHz	9,170, 9,375, 9,410, 9,440 and 9,490 MHz	3,050 MHz
Pulse-widths *	40 ns	40-600 ns	40-600 ns
PRF			
Programming range	800-8,000 Hz	400-4,400 Hz	400-4,400 Hz
Stagger (pseudo random)	0-8%	0-8%	0-8%
Sector transmission	10-360°	10-360°	10-360°

Receiver	
Noise figure (LNFE/system)	2.5 dB (≤4.7 dB - typical)
Dynamic range	=135 dB (total including ≥45 dB STC on RF)
IF bandwidths	3.8, 20 or 40 MHz (optimised for pulse response)
Image rejection on RF mixer	=18 dB for 9,170 MHz (≥20 dB otherwise)
Input/output	
Trigger/video/azimuth	Configured to the individual application
Control	RS-422 serial data lines (3 channels)

Notes

* The pulse-width range cited for the 25 kW X-band and 30 kW S-band configurations are divided into one or two fixed values that are fixed using plug-in modules

Contractor
TERMA Elektronik AS, Lystrup.

UPDATED

GERMANY

Atlas 9500-9800 ARPA series navigation radars

Type
E/F- (2 to 4 GHz) and I/J-band (8 to 20 GHz) family of navigation/surface surveillance radars.

Description
The Atlas 9500-9800 ARPA series is a range of 51 to 74 cm raster scan high-resolution radars characterised by a continuous TV-type display in all ambient conditions. Developed from commercial technology, individual militarised applications within the family comprise compact console-type displays (suitable for stand-alone or combat system integrated operation) linked to either E/F- (10 cm wavelength) or I/J-band (3 cm wavelength) transceiver/antenna assemblies. Systems can be interswitched for cross connection and master-slave function.

New developments within the series include the multifunction 9600M which is a fully integrated system which provides a complete range of radar, ARPA and navigation display functions. Suitable for stand-alone or command and control system applications, 9600M also features route planning, blind pilotage, track control, sector screening and combat system interface capabilities. It can be interfaced with a wide range of target indication or 2-D surveillance radars and has standard outputs for interfacing with combat display systems.

The system's display can be configured in relative and true motion as well as a centred format. Up to 40 targets can be acquired manually or automatically via guard zones together with up to 24 external tracks. Automatic target acquisition is

The display console format used in the Atlas 9500-9800 ARPA series of navigational/surface search radars

by means of limiting lines at up to 200 km/h target speed. With 10 operating ranges in the 0.46 to 178 km band, the equipment also offers an ECDIS option which allows for the real-time superimposition of radar pictures on electronic charts.

Status
Over time, 9600M systems are reported to have been procured for use in the Australasian (Australia and New Zealand) ANZAC and German Navy frigate programmes together with VAMONS patrol and minehunting vessels worldwide.

Specifications

	E/F-band	I/J-band
Transmitter	3,050 MHz ±10 MHz	9,375 MHz ±30 MHz
Peak power	30 kW	25 kW
Antenna	4.3 m slotted array	1.7/2.6 m slotted array
Antenna weight	175 kg	56-57 kg
Horizontal beamwidth	1.7°	1.5/0.9°
Vertical beamwidth	>20°	>20°
Antenna rotation speed	23 rpm	23 rpm (60 rpm option)
Polarisation	horizontal	horizontal

Contractor

STN ATLAS Marine Electronics GmbH, Hamburg.

VERIFIED

INDIA

PFN 513 air and surface surveillance radar

Type

F-band (3 to 4 GHz) medium-range air and surface surveillance radar.

Description

The PFN 513 is an F-band system designed for use on board large- and medium-sized naval vessels or in shore establishments. It is an upgraded version of the Signaal DA05 and provides medium-range air warning, surface warning and target indication to fire-control systems. A high-pulse repetition frequency moving target detection mode facilitates the tracking of sea-skimming targets.

The system uses pulse compression techniques to compress a long transmit pulse, with a relatively low power, to a pulse-length smaller than obtained in a comparable magnetron. An additional short pulse is generated immediately after the main transmitter pulse, with separate receiver and processing circuits to enable short-range coverage used in coastal surveillance of helicopter sorties, and so on. A travelling wave tube provides the high mean output power which determines the required long radar range and high detection probability.

The antenna is a double curvature parabolic reflector, horn fed and radiating a cosec beam. It is stabilised and has provision for mounting a separate identification friend-or-foe array on the primary antenna.

Status

As of this edition, PFN 513 is reported to have been procured by the Indian Navy.

Specifications

Frequency: F-band (3,100-3,500 MHz)
Antenna
Type: double curvature parabolic reflector
Polarisation: vertical or circular
Gain: 31 dB
Beamwidth: 1.55° azimuth; 5.3° elevation
Elevation tilt: 2.5°
Passive high beam elevation: 9°

Rotation speed: 13.5/27 rpm
Transmitter
Peak power: 100 kW
PRF: 500/1,000/4,000 Hz
Pulse-width: 67/35/9 µs
Range: 130 km on 2 m² RCS target

Contractor

Bharat Electronics Limited, Bangalore.

VERIFIED

PIN 523 navigational radar

Type

I-band (8 to 10 GHz) navigation and surveillance radar.

Description

The Bharat navigational radar can also be used for short-range surveillance purposes. It consists of a 1.83 m slotted array, transmitter/receiver, sector control unit and a 51 or 61 cm raster display. Two displays can be operated with one acting as master.

The transmitter is solid state, apart from the thyraton and the magnetron. Sector transmission or blanking is a special feature which enables the operator to shut-off transmission intentionally in an unwanted zone and still have the radar coverage in the zone of interest. Optional features include a radar test set, a remote control for operating the transmitter/receiver and RRA interface.

A choice of two types of Automatic Radar Plotting Aid (ARPA) displays is offered — 51 and 61 cm. Both have identical features and facilities. Other important features include automatic tracking of up to 40 targets, presentation of a combination of navigation data, sea and rain clutter free presentation and interference suppression.

Status

As of this edition, PIN 523 is reported to have been procured.

Specifications

Frequency: I-band (9,375 ± 30 MHz)
Peak output power: 35 kW (short pulse); 45 kW (long pulse)
Pulse-length/PRF: 0.1 µs/4,000 Hz (short pulse); 0.8 µs/1,000 Hz (long pulse)
Antenna
Type: 1.83 m slotted waveguide
Polarisation: horizontal
Beamwidth: 1.4° (horizontal); 22° (vertical)
Gain: 26 dB
Rotation speed: 22 rpm (at 50 Hz) or 27 rpm (at 60 Hz)
Range discrimination: 30 m or better on 1.4 km range
Bearing discrimination: 1.4°
Bearing accuracy: better than 1°

Contractor

Bharat Electronics Limited, Bangalore.

VERIFIED

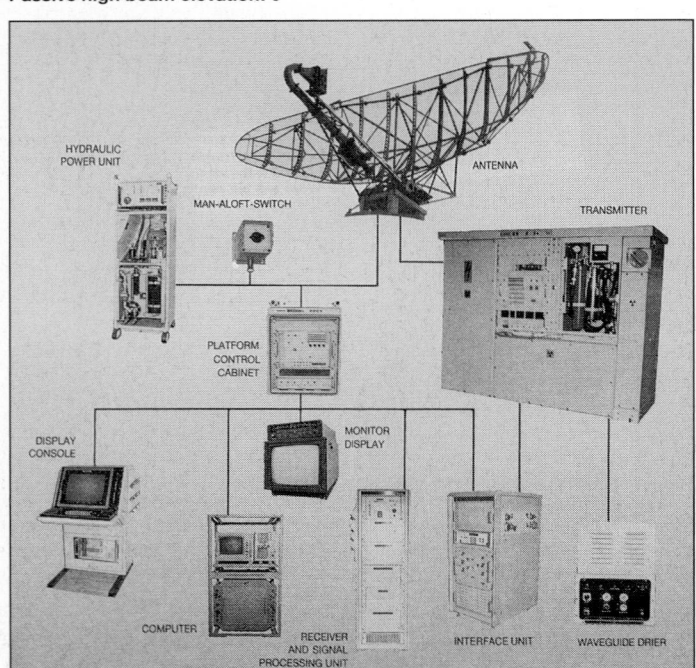

Units of the PFN 513 naval radar

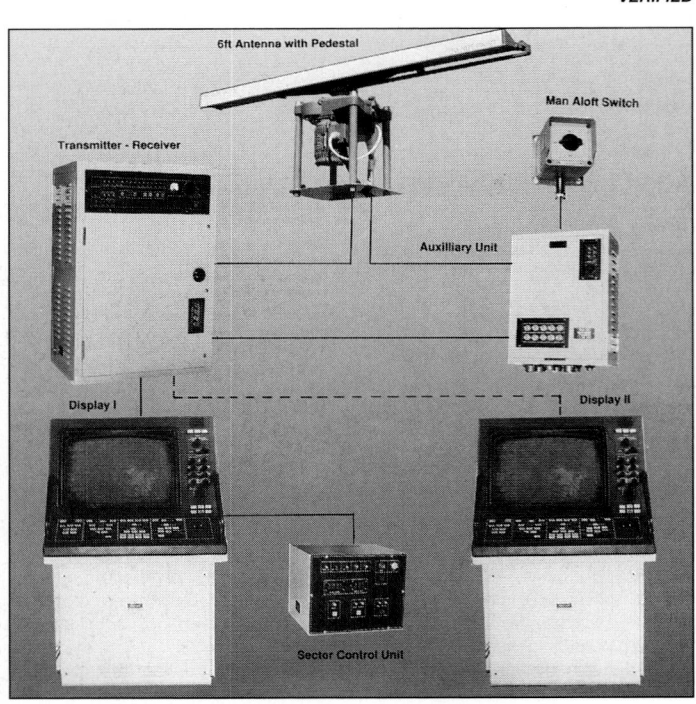

The units that make up the PIN 523 navigation radar   0006899

PIN 524 navigational radar

Type
I-band (8 to 10 GHz) navigation and surveillance radar.

Description
PIN 524 (an upgraded version of the Signaal ZW 06) is a surveillance and navigation radar operating in I-band. It is very similar to the I-band system described in the previous entry, the main difference being in the use of a reflector antenna instead of a slotted waveguide array. This feature gives higher antenna gain and consequently greater range and a narrower beamwidth for better bearing discrimination.

The system consists of a parabolic antenna with pedestal, a transmitter/receiver, a sector control unit, a remote-control unit and a 51 or 61 cm raster scan display. A second display can be incorporated with one being designated as the master.

Sector transmission blanking is a special feature and enables the operator to shut-off transmission intentionally in an unwanted zone while retaining radar coverage in the zone of interest.

The system is suitable for integrating a Rendezvous Radar/Approach (RRA) transponder facility for helicopter control and optional RRA interface components can be supplied if required.

Status
As of this edition PIN 524's status is uncertain.

Specifications
Frequency: I-band (9,410 ± 30 MHz)
Peak output power: 35 kW (short pulse); 45 kW (long pulse)
Pulse-length/PRF: 0.1 μs/4,000 Hz (short pulse); 0.8 μs/1,000 Hz (long pulse)
Antenna
Type: parabolic reflector
Polarisation: horizontal or circular
Beamwidth: 0.9° (horizontal); 19° (vertical)
Gain: 31 dB
Rotation speed: 20 rpm (at 50 Hz); 24 rpm (at 60 Hz)
Range discrimination: 30 m or better on 1.4 km range
Bearing discrimination: 1°
Bearing accuracy: better than 1°

Contractor
Bharat Electronics Limited, Bangalore.

VERIFIED

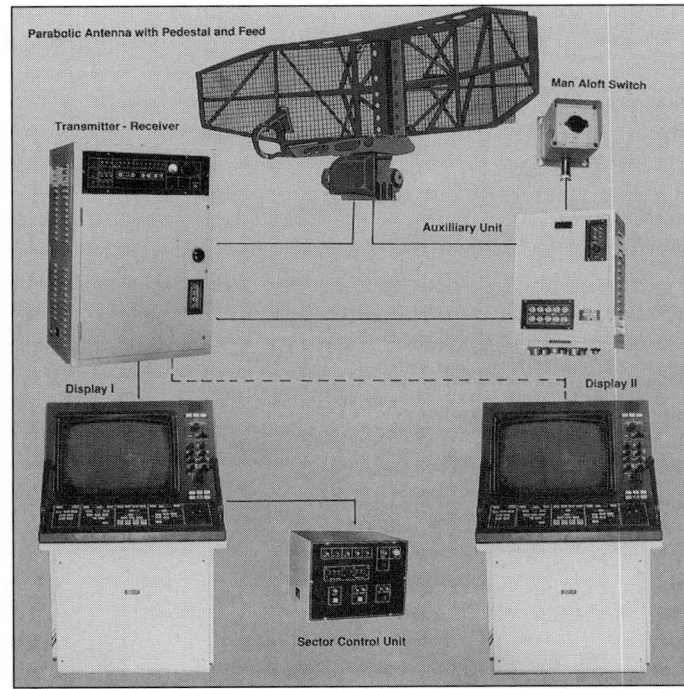

Units of the PIN 524 navigation and surveillance radar

PLN 517 surveillance radar

Type
D-band (1 to 2 GHz) long-range surveillance and air warning radar.

Description
PLN 517 is a long-range D-band radar that is designed for use at shore establishments and aboard medium to large surface ships. It provides air warning and target interception information. PLN 517 is equipped with a high-gain antenna

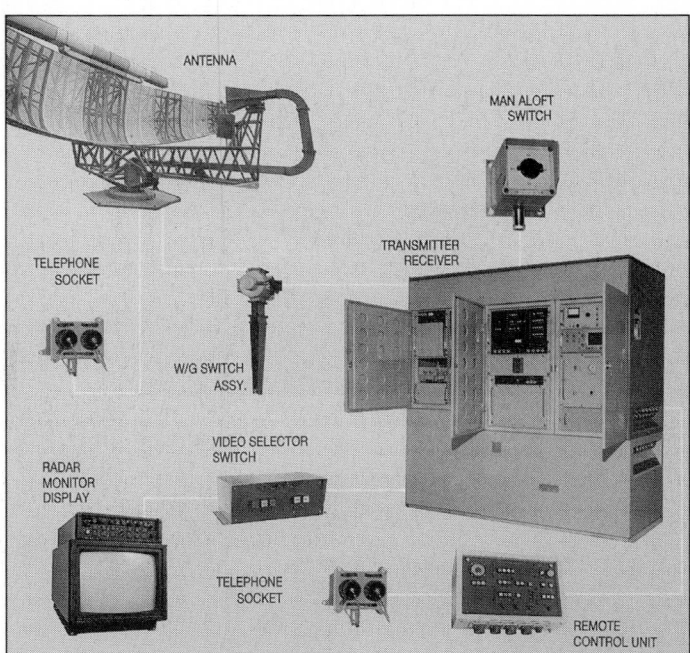

Units of the PLN 517 naval radar

and a high-power transmitter and utilises state-of-the-art signal processing together with solid-state components. The transmitter used is fully solid state except for its magnetron and thyratron. Magnetron drift is corrected by the use of a synthesiser as a stabilised local oscillator. Bharat claims that this technique provides a significantly enhanced moving target indication capability and enables the operator to preselect specific operating frequencies at will. The system incorporates a number of electronic counter-countermeasures, including the ability to vary transmission power in steps and facilities to select a sector for transmission, change the centre of a sector and vary its width.

Status
As of this edition, PLN 517 is reported to have been procured by the Indian Navy.

Specifications
Frequency: 1-2 GHz (tunable to 10 spot frequencies)
Peak pulse power: 1 MW
Peak power options: 500 kW; 750 kW; 1 MW
Pulse-length: 2 ms (short); 4 ms (long)
PRF: 250 Hz (long pulse); 500 Hz (short pulse)
Range coverage: 139-185 km (1 m² RCS target)
Height coverage: 15,240-18,288 m (1 m² RCS target)
Discrimination: 2.2° (azimuth); 300 m (range - short pulse); 600 m (range - long pulse)

Contractor
Bharat Electronics Limited, Bangalore.

VERIFIED

INTERNATIONAL

Active Phased-Array Radar (APAR)

Type
I/J-band (8 to 20 GHz) multifunction naval radar.

Development
Hollandse Signaalapparaten BV (Signaal) and the Royal Netherlands Navy have signed a contract for the product development phase of the APAR multifunction radar. This contract is based on a Memorandum of Understanding (MoU) between Canada, Germany and Netherlands in which the three participants stated their intention to develop, build and test the APAR radar system and proceed with the product development phase.

For project realisation, Signaal is leading a consortium made up of itself; DaimlerChrysler Aerospace AG (now part of the European Aeronautic, Defence and Space (EADS) Co), Germany; Northern Telecom Ltd, Canada and FEL/TNO Physics and Electronics Laboratory, Netherlands. Within this grouping, Signaal is responsible for the overall system design and programme management. FEL/TNO takes part in the overall system design, EADS will design and manufacture the signal generation and processing subsystems, while Northern Telecom will manufacture the transmit/receive modules. All industries will take part in the design of these modules, including monolithic microwave integrated circuits.

A general view of the APAR active phased-array multifunction naval radar
2000/0080965

Description

APAR is a multifunction radar for shipborne applications. The main purpose of the system is to support anti-air warfare and anti-surface warfare tasks of the platform by providing:

- multitarget tracking
- weapon control support
- special search functions.

APAR is the first of a new generation of multifunction radars which have the unique ability to perform multiple capabilities simultaneously; for example multitarget tracking and missile guidance while continuing the search for new targets. This results in many simultaneous engagement channels. Other unique features are the latest adaptive processing and measures against undesired propagation effects.

APAR uses four active phased-array antennas to obtain a 360° coverage in azimuth. Each antenna, with its associated equipment, has the capability to transmit, receive and process radar signals. The four individual channels are linked via the system management part of the system.

The system is compatible with the concept of Signaal's fully distributed architecture. As part of a sensor weapon and command (SEWACO) system, it can fulfil the role of a fire-control support sensor, multitarget tracking radar and dedicated sea-skimmer detection radar.

The antenna of an active phased-array radar such as APAR consists of one or more flat plates, each composed of a large number of transmit/receive elements. Each element is a very small transmitter/receiver. The combination of thousands of these elements in one plate can generate narrow beams that can be pointed in any desired direction within a cone of about ±60°. Pointing from one direction to another can be done rapidly. Each of APAR's four arrays is scheduled to have 3,200 transmit/receive elements grouped in 64 column assemblies.

Status

As of this edition, seven APAR systems are under contract for the German and Royal Netherlands Navies.

Specifications

Frequency: I/J-band (8-20 GHz)
Spatial coverage: 360 × 70°
Instrumented range: 32 km (surface search); 75 km (horizon search); 150 km (air-target/back-up volume search)
Track capacity: more than 150 (surface targets); more than 200 (air targets)
Spacial coverage: 70 × 360° (four-faced system)
Guidance capability for: RIM-7P Sparrow; ESSM (Evolved SeaSparrow Missile); (SM 2) Standard Missile 2

Contractors

Hollandse Signaalapparaten BV, Netherlands.
European Aeronautic, Defence and Space (EADS) Co, Germany.
Northern Telecom Ltd, Canada.
FEL/TNO Physics and Electronics Laboratory, Netherlands.

UPDATED

Arabel multifunction radar (naval application)

Type
I-band (8 to 10 GHz) target acquisition, tracking and missile guidance radar.

Description
Arabel is a 3-D, multifunction radar system that is designed for use in either land-mobile or naval medium-range, Surface-to-Air Missile (SAM) system applications. The radar performs volumetric surveillance, target tracking, integrated missile datalink guidance (for mid-course correction) and environment mapping. The system's phased-array antenna rotates at 60 rpm to maintain continuous 360° coverage and the data renewal rate necessary for missile guidance. It is also capable of two-axis electronic steering to ensure full beam steering flexibility. The combination of frequency agility over a large bandwidth, narrow beamwidth and adaptive steering is claimed to allow Arabel to detect a wide range of threats (including low-flying fixed-wing aircraft, helicopters and small radar cross-section missiles) in severe clutter and jamming environments. As of March 2001, Thales was reporting that the radar had demonstrated 'excellent' performance against low altitude, sea-skimming missiles.

Status
As of March 2001, two full-scale engineering development prototypes of the Arabel system have completed acceptance tests in connection with the GIE Eurosam (a consortium of Aerospatiale Matra Missiles, Alenia Marconi Systems and Thales Air Defence) medium-range Land Surface-to-Air Anti-Missile (SAAM - also known as SAMPT) and Naval SAAM systems. The first land-based systems have been ordered by the French and Italian armies where Land SAAM is to replace the HAWK SAM system, while the Naval SAAM system is already in service aboard the French aircraft carrier *Charles de Gaulle*. This sensor (the first Arabel radar to be produced for the French Navy) was installed aboard the *Charles de Gaulle* during 1997 and forms part of the vessel's SAAM/F system. Elsewhere, three Arabel radars have been ordered for installation aboard Saudi Arabian 'Arriyad' class frigates and according to Jane's sources, the Naval SAAM architecture achieved its sixth successful test firing during early 1999 when the Aster missile/Sylver vertical launcher/Arabel radar package, mounted on the French Navy trials ship *Ile d'Oleron,* intercepted and destroyed a target flying at M1.0 at an altitude of less than 50 m above sea level. As of July 2001, Jane's sources were also suggesting that the five ships making up the French Navy's 'La Fayette' class of frigates would be retrofitted with the Arabel radar as part of a modernisation programme.

Specifications
Frequency: 8-10 GHz
Azimuth coverage: 360°
Elevation coverage: 90°
Detection range: 50 km (missile); 100 km (larger air target)
Track capacity: more than 100
Guidance capacity: up to 16 missiles simultaneously

Contractor
Thales Air Defence, Bagneux, France.

UPDATED

The Arabel radar antenna module as configured for use in the naval SAAM/F SAM system
0007675

Argos 73 coastal radar

Type
E-band (2 to 3 GHz) air and surface surveillance coastal radar.

Description
Reported to have been developed to meet a highly demanding operational requirement, Argos 73 is a fully solid-state coastal radar for combined air and surface surveillance. Its main characteristic is the employment of modular air-cooled active arrays of high-power transistors for the generation of radiated power, so that varying transmitted power can be achieved to cope with any operational requirement. Whatever the number of applied modules the use of active arrays provides an unusually high mean time between failure figure, together with a graceful degradation capability. The result of this application is to make the Argos 73 ideal for applications where high reliability is required, such as in unmanned sites.

In addition to the above feature, the Argos 73 is characterised by the use of a high-gain antenna with cosec^2 pattern, a coded waveform in transmission and a receiver based on the latest electronic counter-countermeasures and anti-clutter facilities.

An identification friend-or-foe antenna with integrated sidelobe suppression can be integrated with the radar.

Status
As of this edition, Argos 73 was reported as being available and as having, over time, been procured by the navies of Germany and Pakistan.

Specifications
Frequency: E-band (2-3 GHz)
Antenna: double curvature parabolic reflector
Rotation rate: typically 10 rpm
Polarisation: linear/circular
Transmitter: frequency agility and fixed frequency, coded pulse waveform
Receiver: double frequency conversion, linear and CFAR receivers, I and Q phase detectors, bank of adaptive Doppler filters, matched filters, azimuth correlators
Range: in excess of 119 km on a fighter aircraft

Contractor
Alenia Marconi Systems (an Alenia-BAE Systems joint venture), Chelmsford/Rome.

VERIFIED

The Argos 73 coastal radar

AWS-6e series naval radars

Type
G-band (4 to 6 GHz) air and surface surveillance radar.

Description
Performance of the well-known series of AWS-6 G-band naval surveillance and target indication radars, now known as AWS-6e, has been considerably enhanced

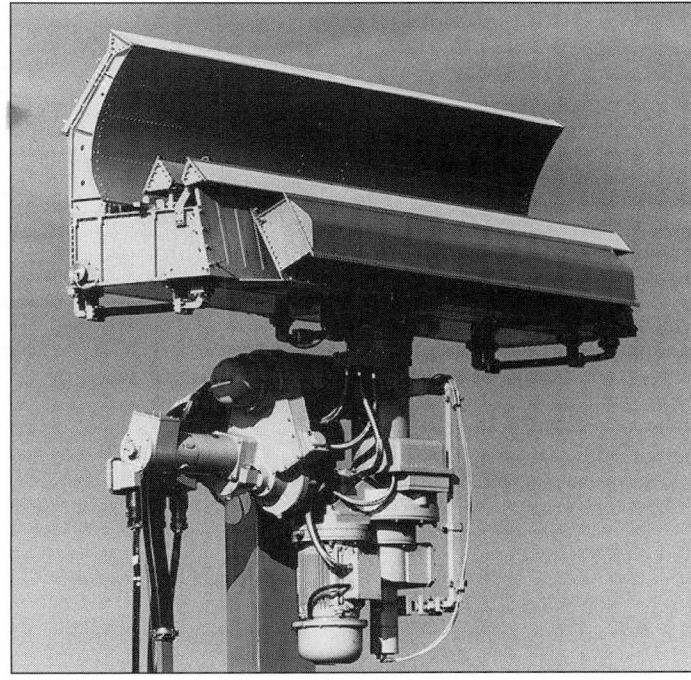

The scanner unit used in the AWS-6e air/surface surveillance radar

by the provision of 3-D data, increased detection and improved Electronic Counter-Countermeasures (ECCM), and offers flexibility to meet customer operational requirements. With low masthead weight and compact below decks equipment, the AWS-6e series can be configured to meet the needs of a wide range of surface ships from 300 tonnes upwards, either as a single multirole radar sensor or integrated as a dedicated radar into a comprehensive multiradar fit.

Variants of the AWS-6 series are:
- a high data rate missile detection radar designed to meet the exacting demands of the surveillance radar module for the Contraves Seaguard Close In Weapon System (CIWS)
- a missile detection and surveillance radar with the antenna rotating at 60 or 30 rpm and designed to meet both the exacting demands of the Seaguard CIWS and to provide general area air and surface surveillance
- a missile detection and surveillance radar with the antenna rotating at 30 or 15 rpm, and designed to meet both the exacting demands of point defence missile systems such as Seawolf and to provide general air and surface surveillance
- a target indication and surveillance radar with the antenna rotating at 30 or 15 rpm and designed as a cost-effective multirole radar for small warships
- a surveillance radar with the antenna rotating at 20 or 10 rpm, designed for maximum surveillance and for use on small warships such as patrol craft and fishery protection vessels.

Dedicated or multirole operation
As a single sensor in offshore patrol vessels, fast strike craft and corvettes the multirole AWS-6e fulfills all the surveillance requirements, providing general area surveillance on air and surface targets together with navigation capability, plus, on demand, the additional target indication performance required for point defence under hostile conditions. High quality plot and track data are available for the weapon systems and command system of the ship. These different operational modes are available within a single radar system and are operator controlled.

As a multisensor in major warships it is practical for more than one sensor to be fitted and for particular radars to be allocated to specified surveillance functions. As the dedicated sensor for point defence, AWS-6e configured as a stabilised, high data rate radar is ideal for missile detection and target indication to a CIWS. The near hemispherical coverage and excellent ECCM performance ensures consistent high quality plot and track data are input to the ship's weapon systems as well as track data to the Command system. The inherent multirole capability of the radar coincidentally provides an additional margin of performance in the event of failure of other shipboard sensors.

The AWS-6e has been designed to operate under severe environmental conditions. Features such as frequency agility, pulse compression and within-pulse coding have minimised the effects of electronic countermeasures. These have been combined with adaptive moving target indicator to give high performance against clutter, chaff and target glint. Improved reaction time and reduction in operator load can be achieved by use of the automatic plot and track extraction facilities.

The equipment is also available in containerised configuration for integration with other weapon systems on ships such as armed merchantmen.

The radars in the AWS-6e series provide a combination of:
- accurate low-flying missile detection out to 35 km
- high diving missile detection (back angle 70°) out to 20 km
- medium-range area air surveillance out to 70 km
- surface surveillance out to the radar horizon
- area surveillance with instrumented range of 200/240 km
- splash spotting out to 20 km
- navigation out to 20 km
- helicopter control.

Status

As of November 2001, Jane's sources were suggesting that AWS-6 radars were in service aboard Danish 'Thetis' class frigates and seven 'Flyvefisken' class multirole combat vessels (*Flyvefisken, Hajen, Havkatten, Laxon, Makrelen, Storen* and *Svaerdfisken*); Omani 'Dhofar' class fast attack craft (three ships?) and Turkish 'Barbaros' and 'Yavuz' class frigates and 'Yildiz' class fast attack craft. As of the given date, the same sources were suggesting that the 'Flyvefisken' class applications are likely to be progressively replaced by the European Aeronautic, Defence and Space (EADS) Co Deutschland's TR-3D radar.

Contractor

Alenia Marconi Systems (an Alenia-BAE Systems joint venture), Chelmsford/ Rome.

UPDATED

AWS-9 series naval surveillance radars

Type

E/F-band (2 to 4 GHz) family of 2/3-D surveillance and target indication radars.

Description

The AWS-9 series of radars has been designed to act as the primary surveillance and target indication sensor aboard major surface warship classes up to and including aircraft carriers. The system's architecture allows the radar's parameters to be adjusted to meet the roles, performance requirements and threat scenarios dictated by the particular platform type to which it is being applied. The radar's three operating modes provide a combination of high veracity target data for point defence missile systems and medium/long-range surveillance for air picture compilation. AWS-9 is supplied with plot and track extraction which allows for automatic and /or manual track initiation of moving air and surface targets as well as generating special alerts on all incoming, fast moving contacts. Operating in the E/F-band, the radar shares pulse transmission, signal processing and electronic counter-countermeasures technology with BAE Systems' 3-D land-based air defence radars. The 2-D member of the family — AWS-9 (2-D) — is intended for applications such as destroyers, frigates, corvettes and offshore patrol vessels and offers two operating modes (Surveillance and Defence); features antenna stabilisation and cosecant beam shaping; incorporates frequency agility, pulse compression and velocity compensated moving target indication to enhance anti-clutter performance and features a frequency coded output and what are termed sophisticated signal processing techniques to overcome hostile jamming.

Status

During the period of March to November 2001, Jane's sources were reporting the Norwegian Navy as operating 2-D AWS-9 radars aboard its 'Oslo' class frigates (installed during the period of 1996 to 1998) while 3-D equipments were identified as being in service with the Royal Navy (as the Type 996 – see separate entry) and the Turkish Navy ('Barbaros' class frigates). Elsewhere, the same sources were suggesting an AWS-9 application for Brunei Darussalam's three 'Brunei' class corvettes, the first two of which was launched during 2001.

Contractor

Alenia Marconi Systems (an Alenia-BAE Systems joint venture), Chelmsford/ Rome.

UPDATED

The AWS-9 radar is the primary air surveillance sensor aboard Turkish 'Barbaros' class frigates

DA05 naval surveillance radar

Type

E/F-band (2 to 4 GHz) air surveillance and target indication radar.

Description

The DA05 is a high-power, E/F-band radar for medium-range surveillance and target indication to weapon control systems. Capable of accurately detecting a jet fighter at approximately 135 km, the DA05 is a stable and reliable system, distinguished by its high-gain antenna, powerful transmitter and sensitive receiver. A narrow beamwidth and short pulse-length result in clear resolution.

A DA05 surveillance radar on a 'Fatahillah' class frigate of the Indonesian Navy

The DA05 has a magnetron-powered transmitter, a receiver with moving target indication, video processing and interference suppression circuits ensuring high reliability in rain, sea and land clutter. The low weight of the antenna system enables installation in smaller ships. A lightweight electro-hydraulic antenna stabiliser is available.

Status

As of the period January to April 2001, Jane's sources were reporting that the DA05 surveillance radar was in service aboard the following warships and classes of warship:

Argentina	• 'Espora' class frigates (five ships in class, one building as of January 2001)
Belgium	• 'Wielingen' class frigates (three ships)
Egypt	• 'Descubierta' class frigates (two ships)
Finland	• the minelayer *Pohjanmaa*
India	• the aircraft carrier *Vikrant*
Indonesia	• 'Amad Yani' (six ships) and 'Fatahillah' (three ships) class frigates
Ireland	• the offshore patrol vessel *Eithne*
South Korea	• 'Ulsan' class frigates (nine ships)
Malaysia	• 'Musytari' class offshore patrol vessels (two ships)
Morocco	• the frigate *Lieutenant Colonel Errhamani*
Netherlands	• 'Jacob Van Heemskerck' class frigates (two ships)
Spain	• 'Descubierta' class corvettes (six ships)
Taiwan	• 'Wu Chin III' class destroyers (seven ships - DA08 radar with DA05 antenna)
Thailand	• the training ship *Makut Rajakumarn*
	• 'Rattanakosin' class corvettes (two ships)

Of these, the DA05 equipments fitted to Belgium's 'Wielingen' class frigates were (as of mid-1998) scheduled for upgrade as part of an effort to keep the ships operationally viable until 2010-12. A version of the radar has also been manufactured under licence by Bharat Electronics (as the PFN 513) in India. As of the mid-1990s, 50 plus radars of this type had been sold worldwide.

Specifications

Antenna: horn fed parabolic reflector
Beamwidth: vertical cosec^2 up to 40°; azimuth 1.7°
Polarisation: linear and circular, selectable
Transmitter: E/F-band (2-4 GHz) tunable coaxial magnetron, with selectable frequencies
Peak power: 1.2 MW
Range: 135 km on a fighter
Pulse-length: 1.3 and 2.6 μs
PRF (staggered): 1,000 and 500 Hz
System Weight: 3,273 kg

Contractor

Thales Nederland, Hengelo, Netherlands.

VERIFIED

DA08 naval surveillance radar

Type

F-band (3 to 4 GHz) surveillance and target indication radar.

Description

The DA08 is a high-power, coherent F-band radar for medium- to long-range surveillance and target indication to weapon control systems. The antenna is mounted on a lightweight hydraulically stabilised platform, providing considerable top weight reduction. The synthesiser-driven travelling wave tube transmitter

The DA08 lightweight antenna

provides for a great flexibility over a wide frequency band, with pulse-to-pulse frequency agility. Together with the low sidelobe level of the antenna, the radar has a high performance in an electronic countermeasures polluted environment. Excellent clutter suppression is the result of the application of stabilisation, a dual beam antenna, dual receiver with a high dynamic range, quadrature digital Moving Target Indicator (MTI) or Fast Fourier Transform (FFT) processing, and circular polarisation.

The DA08 is available in two versions:

- a dual beam MTI version which performs long-range air and surface surveillance as well as target indication. The antenna comprises two feedhorns, one for the active main beam and one for the passive high beam. The high and main beams are processed separately, resulting in a good signal-to-clutter ratio, especially for high-flying, nearby targets. A new lightweight antenna is available
- a single beam medium-range FFT version which has an inherent automatic detection and tracking capability.

By incorporating an optional video extractor the MTI version can perform automatic target tracking of up to 64 air and surface targets. The FFT version has an automatic tracking capability of up to 110 air and 30 surface targets without the need for an additional video extractor.

Status

As of the period January to April 2001, Jane's sources were reporting the DA08 surveillance radar as being in service aboard the following warships and classes of warship:

Argentina	• 'Almirante Brown' class destroyers (four ships in class)
Canada	• 'Iroquois' class destroyers (four ships - DA08 (Canadian designation SPQ-501 - installed as part of the 'TRibal' class Update and Modernisation Programme - TRUMP)
Greece	• 'Hydra' class frigates (four ships)
Malaysia	• 'Leiku' class frigates (two ships)
	• 'Kasturi' class corvettes (two ships)
Netherlands	• the amphibious transport dock/salvage and rescue ship *Rotterdam*
Pakistan	• 'Tariq' class frigates (six ships - DA08 replaces Type 992R radar)
Peru	• the cruiser *Almirante Grau*
Portugal	• 'Vasco da Gama' class frigates (three ships)
Taiwan	• 'Wu Chin III' class destroyers (seven ships - DA08 radar with DA05 antenna)
Turkey	• 'Yavuz' class frigates (four ships)

Specifications

Antenna: horn fed parabolic reflector
Polarisation: linear and circular, selectable
Beamwidth azimuth: 1.55°
Beamwidth elevation: cosec² up to 40°
Rotation speed: 10/20 rpm (version 1); 15 rpm (version 2)
Processing: MTI (version 1); FFT (version 2)
Stabilisation: roll and pitch
Transmitter: F-band coherent (3-4 GHz) TWT system
Power output: 145 kW peak; 5 kW average
PRF: 500/1,000 Hz (version 1); 2,400 Hz (version 2)
Detection range (2 m² RCS target): 125 km (FFT version); 185 km (MTI version)
Resolution: 90 m (range); 1.55° (bearing)
Weights: total top weight 1,100 kg; remainder 3,253 kg

Contractor

Thales Nederland, Hengelo, Netherlands.

VERIFIED

DRBI 10 3-D naval radar

Type

E/F-band (2 to 4 GHz) air surveillance radar.

Description

The DRBI 10 is an E/F-band naval height-finder radar that is designed for interceptor control functions and is related to the land-mobile Picador (TH.D 2200) radar that makes use of electromechanical scanning in elevation and is itself an evolution of the TH.D 1940 equipment.

Status

As of January 2002, Thales Naval France was reporting DRBI 10 as being installed aboard the Brazilian aircraft carrier *Sao Paulo* (formerly the French Navy vessel *Foch* - two radars). As of the given date, the sensor was no longer in production.

Contractor

Thales Naval France, Meudon-la-Forêt, France.

UPDATED

DRBI 23 3-D naval radar

Type

D-band (1 to 2 GHz) 3-D air search and target designation radar.

Description

The DRBI 23 naval radar is a D-band (23 cm) 3-D air search long-range radar providing control of interceptors and target designation to the French Navy Masurca surface-to-air missile system. Transmitter peak power is quoted as several MW and it incorporates a carcinotron oscillator. Six amplification stages are provided, these being capable of virtually identical gain at any frequency within the range of the carcinotron oscillator.

The antenna, which is protected by a large radome, is of the inverse cassegrain type. An array of feed horns directs radiation to a semi-reflective parabolic mirror which in turn returns the energy to a flat plate reflector. This imparts a 45° polarisation change which permits the formed beams to pass through the semi-reflector. Vertical angular information is obtained by the use of stacked beam techniques, these data being processed to provide corrected height information on aircraft targets. This information, together with plan position data, is available for use in the ship's action data system and weapon systems. Stabilisation against ship's motion is incorporated and the antenna assembly includes an IFF antenna unit. The system provides for auto plot extraction and the DRBI 23 is interfaced with the SENIT 1 data handling system. Associated with it are the Masurca missile tracking and guidance radar groups DRBR 51.

Status

As of January 2002, Thales Naval France was reporting DRBI 23 as being installed aboard the French Navy's air defence destroyers *Suffren* and *Duquesne*. As of the given date, the sensor was no longer in production.

Contractor

Thales Naval France, Meudon-la-Forêt, France.

UPDATED

The DRBI 23 radome on the Duquesne

DRBJ 11 naval radar

Type

E/F-band (2 to 4 GHz) 3-D air search and target designation radar.

Description

The DRBJ 11 is a shipboard electronic phase scanned radar developed by Thales Naval France for the French Navy. Few technical details have been released for publication, but some indications can be gathered from the company's previous extensive work in this area of radar technology. The array is composed of numerous separate radiating elements arranged concentrically on a disc surface. The active elements are connected in lines to be fed; while *in-situ* computers determine their phases.

Thales Naval France has revealed a preference for the use of diode phase-shifters for beam steering in electronically scanned arrays. In this technique each source element (or reflector) has an associated module which offers the microwaves several alternative paths corresponding to different propagation times. Diodes controlled by DC voltages perform the switching needed to select required phase shifts. In addition, each source contains its own control logic which receives instructions from the beam steering computer circuitry. The individual elements appear to consist of spirals, supported by ceramic bases, but in the absence of anything from which to obtain an indication of scale it is difficult to deduce the likely frequency, although the probability is E/F-band, with a scan rate of 15 rpm. DRBJ 11 scans electronically in both elevation and bearing, using a single scan to detect and confirm a target and a second scan, repeated after the first, to initiate tracking. Maximum range is understood to be about 366 km.

Data provided by this radar can be processed using Thales Naval France's TAVITAC, TAVITAC 2000 and TAVITAC NT tactical data systems. Of these, TAVITAC 2000 and TAVITAC NT are programmed in Ada and utilise high-resolution, colour TV displays. TAVITAC 2000 systems incorporate an Ethernet Local Area Network (LAN) while TAVITAC NT can be equipped with either an Ethernet LAN or a fibre optic tactical network. Associated sensors and weapons are either connected directly to the network or via an interface unit. TAVITAC systems are suitable for use aboard any size of vessel, make use of ruggedised disk storage and offer database management (map display, ship resources and so on) in addition to a full range of tactical capabilities.

Status

As of January 2002, Thales Naval France was reporting DRBJ 11 as being installed aboard the French Navy's air-defence destroyers *Jean Bart* and *Cassard* and the service's nuclear aircraft carrier *Charles de Gaulle*. According to Jane's sources, *Charles de Gaulle* is fitted with the DRBJ 11B variant of the basic radar that makes use of a circular planar antenna that is inclined at 15° within a protective radome.

Contractor

Thales Naval France, Meudon-la-Forêt, France.

UPDATED

The *Jean Bart* showing the DRBJ 11B radome, with the DRBV 26C forward of the mast

DRBV 22 naval radar

Type

D-band (1 to 2 GHz) long-range air search radar.

Description

The DRBV 22 is a conventional D-band naval search radar. It is widely fitted, both in French Navy vessels and in the ships of other countries. Different versions that have been identified are the DRBV 22 'A', 'C' and 'D' models.

The antenna assembly is very similar in appearance to the US SPS-6 and SPS-12 radars, but the French equipment can be readily distinguished from both US radars by the two supporting stays for the feed horn that are attached to the upper edge of the reflector of the DRBV radar.

Status

As of January 2002, Thales Naval France was reporting the DRBV 22D variant of the baseline DRBV 22 as being installed aboard the French Navy's helicopter carrier *Jeanne d'Arc*. As of the given date, the sensor was no longer in production.

Contractor

Thales Naval France, Meudon-la-Forêt, France.

UPDATED

DRBV 23 naval surveillance radar

Type

D-band (1 to 2 GHz) long-range air surveillance radar.

Description

The DRBV 23 is a D-band (23 cm) long-range naval air search and surveillance radar similar in appearance to the Jupiter naval radar. The most readily discerned external difference is the prominent tubular horizontal supports for the DRBV 23 scanner compared with the lighter structure of the Jupiter.

Status

As of January 2002, Thales Naval France was reporting the DRBV 23B variant of the baseline DRBV 23 as being installed aboard the Brazilian aircraft carrier *Sao Paulo* (formerly the French Navy vessel *Foch*). As of the given date, the sensor was no longer in production.

Contractor

Thales Naval France, Meudon-la-Forêt, France.

UPDATED

DRBV 23 D-band air surveillance radar

DRBV 26 naval air surveillance radar

Type

D-band (1 to 2 GHz) long-range air surveillance radar.

Description

The DRBV 26 is a D-band (23 cm), long-range air surveillance radar with a peak transmitter output power of 2 MW with a 2.5 μs pulse and a pulse repetition frequency of 450 pulses/s. The antenna measures 7.5 × 3 m and has two rotation speeds: 7.5 and 15 rpm. Weight is approximately 1,000 kg. Both primary and secondary (identification friend-or-foe) radar functions are provided by the one antenna system. Beamwidth at half power points of the primary radar pattern is 2.5° or less. The elevation pattern is $cosec^2$ up to 50°.

The antenna array used with the DRBV 26D long-range air surveillance radar installed aboard the French Navy's aircraft carrier Charles de Gaulle   0045036

Solid-state circuitry is used throughout, with the exception of the microwave stages and a fixed frequency, water-cooled magnetron is used in the transmitter. Comprehensive signal processing facilities are incorporated to provide maximum protection from natural interference (such as sea clutter) and electronic countermeasures.

The receiver section provides multiple reception chains, wide dynamic range receiver and anti-clutter circuits. The DRBV 26 is produced in both non-Moving Target Indicator (MTI) and MTI versions. The latter is equipped with a linear digital MTI receiver for operation in clutter environments

Status

As of January 2002, Thales Naval France was reporting DRBV 26 as being installed aboard the following warships and classes of warship:

France	• the aircraft carrier *Charles de Gaulle* (DRBV 26D variant)
	• 'Cassard' (two ships in class - DRBV 26C variant), 'Georges Leygues' (seven ships - DRBV 26A variant) and 'Tourville' (two ships - variant unknown) class destroyers
Saudi Arabia	• 'Arriyad' class frigates (three ships building - DRBV 26C variant)
Taiwan	• 'Kang Ding' class frigates (six ships - DRBV 26D variant)

Of the noted variants, DRBV 26C is understood to differ from DRBV 26A by way of the type of receiver used while DRBV 26D is noted as incorporating a 32 module, solid-state transmitter.

Contractor

Thales Naval France, Meudon-la-Forêt, France.

UPDATED

···

European Multifunction Phased-Array Radar (EMPAR)

Type

G-band (4 to 6 GHz) multimode surveillance and tracking radar.

Description

EMPAR is a G-band phased-array radar being developed by the Alenia Marconi Systems (now an Alenia-BAe Systems joint venture). It produces and provides both surveillance and tracking facilities. The full-phase scan technique employed in the system affords electronic stabilisation of the scanning patterns. A multiplicity of functions (search, correlation, acquisition and tracking), which are usually carried out by individually tasked systems, will now be superseded by the EMPAR system.

The configuration consists of a phased-array antenna with a central high power Travelling Wave Tube (TWT) transmitter, a two-stage superheterodyne receiver, a fully adaptive array signal processor and digital pulse compressor and a comprehensive real-time management computer. The antenna is rotated to achieve full azimuth coverage and at a rate necessary to meet the required update rates. The design incorporates spare capacity providing stretch potential for future

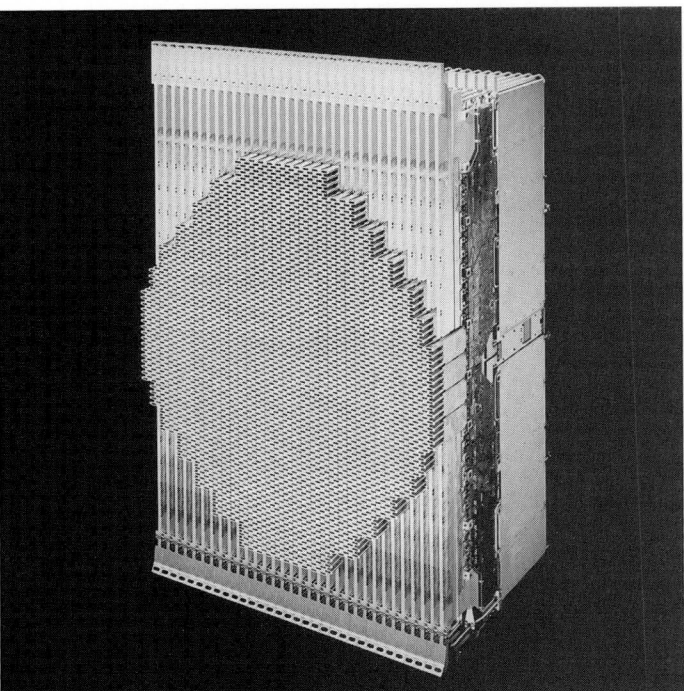

A computer generated image of the antenna configuration used in the EMPAR radar

more stressing requirements. The modular approach allows the system to grow by the use of multiple array faces, thus improving performance still further.

The antenna is a corporate fed, phase-to-phase scanning planar-array. A single pencil beam is formed in space and is electronically scanned by the use of pin diode phase shifters whose phase is controlled from pulse-to-pulse by a beam steering computer. The array is capable of scanning the beam over a wide angle with a low sidelobe level and minimum power loss.

The transmitter is a central TWT providing 120 kW of peak power at a 10 to 12 per cent duty cycle. Operating in a frequency band centred on 5.6 GHz, the radar will provide pulse-to-pulse frequency agility over a range of 10 per cent of the frequency.

The radar management computer is responsible for all real-time task scheduling in EMPAR. The software has been developed jointly by Marconi and Alenia and is hosted on the proven MARA computer developed by Alenia. The radar is one of the building blocks of the FSAF program.

Status

As of this edition, EMPAR field trials are understood to have been completed successfully and extended tests are being carried out on board the Italian Navy test ship *Carabiniere*.

Specifications

Frequency: G-band (5.6 GHz)
Transmitter: driven TWT
Peak power: 120 kW
Antenna: passive phased-array
Radiating aperture: 1.5 × 1.5 m
Scan angles: ±45° azimuth; ±60° elevation
Sidelobes: better than −45 dB
Antenna elements: 2,200
Rotation rate: 60 rpm
Receiver: Superheterodyne; double frequency conversion
Detection of airborne targets: 80 km in normal surveillance mode; 150 km in dedicated surveillance mode
Target tracking accuracy: 3-5 mrad
Total tracks: more than 250
Weights: 2,500 kg above decks; 5,000 kg below decks

Contractors

Alenia Marconi Systems (an Alenia-BAe Systems joint venture), Chelmsford/Rome.

VERIFIED

···

Herakles multifunction 3-D radar

Type

E/F-band (2 to 4 GHz) multifunction 3-D radar.

Description

Thales Naval France's Herakles multifunction 3-D radar is designed for frigate applications where extended self-defence and medium- to long-range surveillance are required. Capable of being integrated with 'all' types of active or semi-active homing missile systems, Herakles is further noted as offering simultaneous 3-D air and surface situation establishment, missile threat detection and weapons deployment. Other system features include:

• confirmation and initiation of tracks during the same antenna scan
• automatic threat and environment adaptation
• a missile command uplink facility
• the use of solid-state technology (including a solid state, coherent transmitter)
• electronic scanning
• a time-on-target/anti-clutter optimised multibeam search mode
• optimisation for littoral warfare
• a target tracking and designation mono-beam mode
• 'very low' sidelobes
• an integrated identification friend-or-foe antenna
• frequency agility
• pulse compression
• electronic stabilisation
• a small topside radar cross section.

Status

As of January 2002, the Herakles multifunction 3-D radar was described as being on order for an export programme. While not confirmed or denied by the manufacturer, as of the given date, Jane's sources were suggesting that the noted 'export programme' involved the supply of equipment for use on six 'La Fayette' class frigates being built for Singapore for delivery during the period 2005-2009. Further, the same sources were reporting that Herakles was being offered to the French Navy for use aboard a new class of multifunction frigates that was being planned for service entry during 2008.

Specifications
Frequency: E/F-band (2-4 GHz)
Angular coverage: up to 70° (elevation); 360° (azimuth)
Range: up to 200 km
Track capacity: >200
Rotation speed: 60 rpm

Contractor
Thales Naval France, Meudon-la-Forêt, France.

UPDATED

LW08 naval surveillance radar

Type
D-band (1 to 2 GHz) surveillance and target indication radar.

Description
The LW08 is a high-power, coherent D-band radar for long-range surveillance and target indication to weapon control systems. The antenna is mounted on a lightweight hydraulically stabilised platform.

The synthesiser-driven travelling wave tube transmitter provides for a great flexibility over a wide frequency band, with pulse-to-pulse frequency agility. Together with the low sidelobe level of the antenna, the radar has a high performance in an electronic countermeasures polluted environment. Excellent clutter suppression is the result of the applications of antenna stabilisation, a receiver with a high dynamic range, quadrature digital moving target indication processing, and circular polarisation.

Status
As of the period March to May 2001, Jane's sources were reporting the LW08 surveillance radar as being in service aboard the following warships and warship classes:

Canada	• 'Iroquois' class destroyers (four ships in class - LW08 (Canadian designation SPQ-502) installed as part of the 'TRibal' class Update and Modernisation Programme - TRUMP)
Germany	• 'Brandenburg' class destroyers (four ships)
Greece	• 'Elli' class frigates (six ships)
India	• 'Delhi' class destroyers (three ships, three proposed as of May 2001 - Bharat 'RAWL' (?) variant of LW08)
	• 'Godavari' (three ships) and 'Nilgiri' (five ships) class frigates
Netherlands	• 'Jacob Van Heemskerck' (two ships), 'Karel Doorman' (eight ships) and 'Kortenaer' (three ships) class frigates
New Zealand	• the frigate *Canterbury*
Peru	• the cruiser *Almirante Grau*
Taiwan	• 'Kang Ding' class frigates (six ships - DRBV-26D Jupiter II radar with LW08 antenna)
Thailand	• 'Naresuan' class frigates (two ships)
United Arab Emirates	• 'Kortenaer' class frigates (two ships)

The LW08 surveillance radar array with SRA-01 IFF antenna mounted on top. In the background is the antenna of a DA08 system with an SRA-02 IFF antenna mounted on top

Specifications
Antenna: horn fed parabolic reflector
Polarisation: linear and circular
Beamwidth azimuth: 2.2°
Beamwidth elevation: cosec² up to 40°
Rotation speed: 7.5/15 rpm
Stabilisation: roll and pitch
Transmitter: D-band (1-2 GHz), TWT or solid state
Power output: 5 kW (average); 150 kW (peak)
PRF: 500/1,000 Hz
Detection range: 260 km on 2 m² RCS target
Resolution: 90 m (range); 2.2° (bearing)
Weights: antenna 3,000 kg; remainder 3,200 kg

Contractor
Thales Nederland, Hengelo, Netherlands.

VERIFIED

LW09 naval surveillance radar

Type
D-band (1 to 2 GHz) surveillance and target indication radar.

Description
The LW09, being a derivative of the LW08 radar, is a high-power coherent D-band radar for long-range surveillance and for providing target indication to weapon control systems. The antenna is mounted on a lightweight hydraulically stabilised platform.

The transmitter is a completely modular solid-state design which ensures a very high system availability and easy maintenance. The transmitter is synthesiser controlled and provides great flexibility over a wide frequency range with pulse-to-pulse frequency agility. Together with a low sidelobe level of the antenna, the radar has a high performance in an electronic countermeasures polluted environment. Excellent performance under various clutter conditions is ensured by wide dynamic range receivers and application of digital video processing supported by circular polarisation.

Status
The solid-state transmitter used in the LW09 radar is understood to have been procured by the Royal Netherlands Navy.

Specifications
Antenna: horn fed reflector
Polarisation: linear and circular
Beamwidth: 2.2° (azimuth); cosec² (up to 40° elevation)
Antenna rotation speed: 7.5/15 rpm
Stabilisation: roll and pitch
Transmitter: D-band (1-2 GHz) solid state
Weight: 3,000 kg above deck; 3,200 kg below deck

Contractor
Thales Nederland, Hengelo, Netherlands.

VERIFIED

Marine ARray EXperimental (MAREX)

Type
G-band (4 to 6 GHz) experimental marine active radar array.

Description
MAREX is an active array demonstrator project that was launched by the then DaimlerChrysler Aerospace (now a part of the European Aeronautic, Defence and Space (EADS) Co). As such, it is a high-duty cycle G-band radar (transmitting over

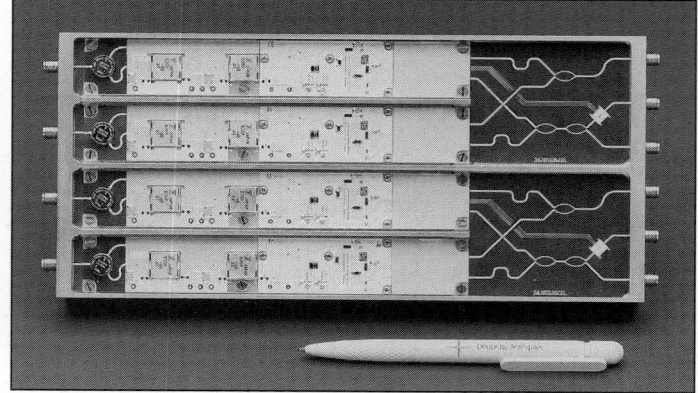

The MAREX transceiver module quadpack

more than 10 per cent of the band) in which the antenna comprises 32 transmit/receive elements that feed a stripline antenna of 32 columns with dipole radiators. A beam steering computer and six-bit phase shifters provide electronic beam steering in one plane (azimuth or elevation). Returns are processed in 16 parallel receivers, analogue/ digital converted and stored for offline analysis.

Status

As of July 2001, a full-scale, functional MAREX test module is reported to have been developed and tested, with the results of a manufacturer's in-house trials programme being incorporated in the development of a new G-band radar technology that features distributed power in the antenna and use of third-generation transceiver modules.

Contractor

European Aeronautic, Defence and Space (EADS) Co Systems and Defence Electronics, Ulm, Germany.

UPDATED

MARS-05 naval surveillance radar

Type

D-band (1 to 2 GHz) medium-range, solid-state surveillance radar.

Description

MARS-05 is a surveillance radar system developed jointly by the Naval Combat Systems Divisions of the then Signaal (now Thales Nederland) and the then Thomson-CSF (now Thales). It is a medium-range air and surface surveillance radar which operates in D-band giving low sensitivity to clutter. MARS-05 is, in fact, a combination of the MARS system developed by the then Thomson-CSF and the MR05 developed by the then Signaal. It is one of the first examples of joint development after the acquisition of the then Signaal by the then Thomson-CSF.

The MARS-05 is provided with a solid-state transmitter and a maintenance-free antenna construction. MARS-05 has several operational features, including very low antenna sidelobes, frequency agility, air detection capability (including moving target indication), surface detection capability, and good clutter performance. For system integration purposes, MARS-05 can be provided with a target extraction and tracking computer. The antenna is equipped with an Identification Friend-or-Foe (IFF) antenna (with an integrated sidelobe suppression capability) inside the antenna radome and in baseline form, MARS-05 provides air and surface surveillance video. Over time, other system configurations have been available.

Status

As of the period February to April 2001, Jane's sources were suggesting that the MARS-05 radar (under its French naval designation of DRBV 21A) was in service aboard the following warships and warship classes:

France	• 'Floreal' class frigates (six ships in class)
	• 'Foudre' landing ship docks (two ships)
Morocco	• 'Floreal' class frigates (two building as of April 2001)

The installation in any or all of the cited vessels may be restricted to below decks equipment.

Specifications

Antenna: stripline array, integrated IFF antenna with ISLS and RSLS
Transmitter type: solid state
Frequency band: D-band (1-2 GHz)
Instrumented range, air: 111 km
Instrumented range, surface: 83 km

Contractor

Thales Nederland, Hengelo, Netherlands.

VERIFIED

MultiRole Radar (MRR)

Type

G-band (4 to 6 GHz) shipboard surveillance and target designation radar.

Description

The Thales Naval France MRR is a combined 3-D air and 2-D surface surveillance and target designation G-band radar that has been developed by Thales Naval France. It is a derivative of the RAC alert and co-ordination radar (see Battlefield, Missile control and ground surveillance radar systems section) and is suitable for installation on fast patrol boats (as the main radar) and on frigates or larger ships (as a complementary radar to the main system). It is optimised to cope with the missile threat, both sea-skimming and diving missiles and to provide 3-D target designation for missile engagement. The coherent, pulse-Doppler, multibeam MRR radar is fitted with a planar antenna that provides high angle coverage of up to 70° and is electronically pitch stabilised and electronically or mechanically stabilised in roll. The equipment is further noted as having detection ranges of up to

180 km against a fighter-sized target and up to 30 km against a sea-skimming missile and as offering jamming analysis, frequency-hopping and digital pulse compression capabilities.

The burst capable MRR transmitter makes use of a frequency-agile Travelling Wave Tube (TWT) and generates long, compressed pulses for the radar's air channel and short pulses for its surface channel. The equipment's receiver has two separate channels, namely a Doppler processing channel with Finite Impulse Response (FIR) for air surveillance and a non-Doppler channel for surface surveillance. MRR is further reported as featuring fully integrated detection and tracking (with initiation of air and surface tracks being automatic and manually selected respectively) and as being fitted with a radar operating mode management computer.

Status

As of January 2002, Thales Naval France was reporting MRR as being in service with the navies of Kuwait (eight equipments for use on 'Um Almaradin' class guided missile fast patrol craft according to Jane's sources) and Qatar (four equipments for use on 'Barzan' class fast patrol craft according to Jane's sources), and as being in production for a third export programme.

Specifications

Antenna: mechanically or electronically roll-stabilised, electronically pitch stabilised
Azimuth beamwidth: 1.75°
Antenna rotation speed: 10 and 30 rpm
Range: 180 km (instrumented)
Polarisation: horizontal
Tracking capacity: 150 (self-defence mode); 300 (standard surveillance mode)

Contractor

Thales Naval France, Meudon-la-Forêt, France.

UPDATED

MW08 surveillance radar

Type

G-band (4 to 6 GHz) air surveillance and target acquisition radar.

Description

The MW08 is an all-weather G-band 3-D short- to medium-range surveillance and target acquisition radar based on the same techniques as SMART-S (see separate entry). It enables very rapid target acquisition and tracking by weapon control systems, by providing target range, elevation, bearing and velocity data for each threat on every radar scan. All system functions, including target detection, air track initiation, target tracking and built-in test equipment, are automatic. Surface track initiation can be performed automatically or manually. Multistripline antennas, with digital Fast Fourier Transform (FFT) beamformer, Doppler FFT processing and tracking, minimise the effects of clutter and jamming.

The MW08 is capable of tracking up to 160 air targets and 40 surface targets simultaneously. Track data is transferred to the command and control system and, if required, to the weapon deployment console for direct surface-to-target engagement via the databus system, or via intercomputer interfaces. With its lightweight roll and pitch hydraulically stabilised platform, the MW08 can be mounted in a high mast position. Complete identification friend-or-foe integration can be provided.

The MW08 radar antenna on the Portuguese Navy frigate Vasco da Gama, *with a Thales STIR fire-control radar below, and the DA08 behind*

Status

As of the period February-March 2001, MW08 surveillance radars were reported as being in service aboard the following warships and warship

Greece	• 'Hydra' class frigates (four ships in class)
South Korea	• 'KDX-2' (three building, three proposed as of February 2001) and 'Okpo' (three ships) class destroyers)
Oman	• 'Qahir' class corvettes (two ships)
Portugal	• 'Vasco da Gama' class frigates (three ships)
Turkey	• 'Kilic' class fast attack craft - missile (three ships, four building as of March 2001)

Specifications

Frequency: G-band (4-6 GHz)
Antenna: stripline array (receive and transmit)
Stabilisation: lightweight; hydraulic in both roll and pitch
Horizontal beamwidth: 2°
Elevation coverage: 70°
Transmitter: coherent amplifier with TWT output stage
Range: 55 km against a fighter aircraft
Weight: 650 kg (above deck); 1,500 kg (below deck)

Contractor

Thales Nederland, Hengelo, Netherlands.

VERIFIED

RAN 20S surveillance radar

Type

E/F-band (2 to 4 GHz) air/surface surveillance radar.

Description

Reported to have been developed for a highly demanding operational application, RAN 20S is a fully solid-state shipborne radar suitable for combined air and surface detection on small and medium tonnage ships. It operates in the E/F-band and is equipped with advanced electronic counter-countermeasures and anti-clutter features in both transmission and reception.

The main characteristic of the RAN 20S is the use of modular air-cooled active arrays of high-power transistors for the generation of radiated power. With this solution it is possible to achieve a modular fully solid-state radar (with power tailored to the operational needs) ensuring an unusually high mean time between failure valve, together with a graceful degradation capability.

The RAN 20S is characterised by the use of different types of antennas, each with very low sidelobes, mounted on a platform fully stabilised in both pitch and roll. This results in the radar being able to supply an accurate and reliable designation of aircraft, missiles and ships to the onboard weapon systems. Examples here include the 5 kW planar - array used in the RAN 20S variant and the 20 kW plus Cosecant squared unit fitted to the RAN 20SEW early warning radar.

An identification friend-or-foe system (including integrated sidelobe suppression) can be incorporated into the system.

Status

Jane's sources suggest that RAN 20S has been selected for the Brazilian 'Niteroi' class frigate update programme.

Specifications

Frequency: E/F-band (2-4 GHz)
Antenna rotation rate: 15 and 30 rpm
Polarisation: linear/circular
Transmitter: frequency agility and fixed frequency, coded pulse waveform
Receiver: double frequency conversion, linear and CFAR receivers, I and Q phase detectors, bank of adaptive Doppler filters, matched filters, azimuth correlators
Range: in excess of 119 km on a fighter aircraft

Contractors

Alenia Marconi Systems (an Alenia-BAe Systems joint venture), Chelmsford/Rome.

VERIFIED

RAN 30X surveillance radar

Type

I/low J-band (8 to 12 GHz sub-band) air and surface surveillance radar.

Description

RAN 30X is a combined surface and air surveillance (low-flying targets) radar operating in I-band and specifically developed for small and medium tonnage ships. The equipment particularly enhances detection of low-altitude threats and steep-diving missiles in clutter and jamming environments and provides high resolution and data rate. RAN 30X utilises the same design philosophy as the RTN 25X and RTN 30X systems with the three radars sharing a number of common components.

The RAN 30X makes extensive use of solid-state techniques to achieve a very compact, reliable and easily maintainable system that incorporates the following features:

• single or multibeam antenna for high coverage
• fully coherent receiver/transmitter architecture with high internal stability to ensure high clutter cancellation and fast frequency jump capability
• pulse compression and adaptive moving target indication
• transmission of high duty ratio coded waveform for improved electronic counter-countermeasures performance and low intercept probability.

An identification friend-or-foe antenna unit including integrated sidelobe suppression can be incorporated into the antenna.

The RAN 30X can be integrated with the RAN 12L radar using a unique antenna. Typical RAN 30X operating roles include air warning against aircraft and missiles; direction-finding against fixed-wing and rotary air targets; surface surveillance; navigation and direction-finding against surface-to-surface missiles. The radar is optimised for gunnery and/or surface-to-air missile target designation.

Status

Jane's sources suggest that a RAN 30X variant has been procured by the Spanish Navy for installation aboard 'Santa Maria' class frigates.

Specifications

Frequency: I/low J-band (8-12 GHz sub-band)
Antenna: cylindrical, parabolic reflector, roll and pitch stabilised
Rotation rate: 15 and 30 rpm
Polarisation: linear (circular optional)
Transmitter: frequency agility and diversity, fixed frequency, coded waveform
Receiver: double frequency conversion, coherent limiting, dual-channel Doppler processing, code matched filtering, azimuth correlation
Range: more than 46 km on a fighter aircraft
Range accuracy: 25 m
Angular accuracy: 0.4°

Contractors

Alenia Marconi Systems(an Alenia-BAe Systems joint venture), Chelmsford/Rome.

VERIFIED

S1810 shipborne radar

Type

I-band (8 to 10 GHz) shipborne radar.

Description

The S1810 is a 3 cm (I-band) frequency-agile radar which has been optimised for shipborne applications. The system gives enhanced detection ranges and precise target indication data to weapon systems and is highly resistant to Electronic CounterMeasures (ECM) polluted environments. It has been designed to detect sea-skimming missiles well beyond the engagement range.

S1810 is a lightweight fully coherent system which provides detection of small aircraft at over 45 km and surface craft out to the radar horizon. The system operates between 8.6 and 9.5 GHz and offers pulse-to-pulse broadband frequency agility. Digital moving target indicator provides improved detection and tracking of low-level air targets in clutter and operates with pulse burst frequency agility or fixed frequency. The S1810 antenna is mounted on a stabilised platform comprising three mutually perpendicular waveguide rotating joints and the azimuth drive. There are three versions:

• the standard version, with 2.44 mm reflector and radome, operating at 24 rpm
• lightweight, with a 2.44 m reflector and radome and operating at 20 or 60 rpm
• lightweight compact, with a 1.2 m reflector and radome, operating at 24 rpm.

All versions of the S1810 provide low-level coverage and angular discrimination. Pulse-to-pulse or short burst frequency agility is achieved using a coherent travelling wave tube transmitter. One of three transmission modes can be selected, including pulse compression to give high effective peak power. A particular transmission mode can be selected within operator defined bearings, with a different mode outside those bearings to achieve, for example, splash spotting within the narrow required sector with full air detection over the rest of the area.

The South Korean 'Po Hang' class frigate Chon An *fitted with an S1810 surface search radar at the top of its foremast*

Status

Over time, S1810 radars are understood to have been installed aboard batch 2 South Korean 'Po Hang' class frigates.

Specifications

Frequency: I-band (8-10 GHz; 3 cm wavelength)
Transmission modes: fixed frequency; pulse burst agility; pulse-to-pulse agility; radar silence; sector transmission
Pulse-length: Mode 1 (normal); 2 µs compressed to 0.1 µs (chirped) low PRF; Mode 2 (air); 1 µs (high PRF); Mode 3 (surface); 0.2 µs (high PRF)
Peak Power: 50 kW
PRF: 2 kHz (low); 4 kHz (high)
Antenna
Type: double curvature GRP metallised coating
Beamshape: $cosec^2$ 5-30°
Polarisation: linear horizontal
Beamwidth: 1.1 ±0.1°
Aperture: 2.44 × 0.84 m or 1.2 × 0.42 m
Rotation rate: 24 or 20/60
Weight: 500 kg

Contractors

Alenia Marconi Systems(an Alenia-BAe Systems joint venture), Chelmsford/Rome.

VERIFIED

S1850 air search radar

Type

D-band (1 to 2 GHz) air surveillance radar.

Description

S1850 is reported to integrate elements of Signaal's SMART-L radar (see separate entry) with transmitter technology from the Alenia Marconi Systems' Martello system (see separate entry). Jane's sources suggest that the equipment utilises a SMART-L antenna modified to accommodate the system's transmission elements.

Status

As of mid-1999, Jane's sources were suggesting that S1850 was likely to form part of the UK variant of the shipboard Principal Anti-Air Missile System (PAAMS) that was being developed jointly by France, Italy and the UK. As of this edition, S1850 is understood to have been selected by the nascent EUROPAAMS consortium for use in its anti-air missile system bid for the three nation 'Horizon' next-generation air defence frigate programme.

Contractors

Alenia Marconi Systems (an Alenia-BAe Sytems joint venture), Chelmsford/Rome. Hollandse Signaalapparaten BV, Hengelo, Netherlands.

VERIFIED

SAMPSON multifunction naval radar

Type

Dual-faced, active array, multifunction naval radar.

Description

The SAMPSON multifunction naval radar is based on the BAE Systems - Combat and Radar Systems/UK Defence Evaluation and Research Agency Multifunction Electronically Scanned Adaptive Radar (MESAR) demonstrator. MESAR incorporates phased-array antenna technology (utilising gallium arsenide transmission modules) and has undergone trials in both the UK and the USA (see Status). The SAMPSON application is designed for point and area defence applications in environments that are heavily polluted with jamming and land and sea clutter. Coverage and radar operation are software controlled with the overall system automatically adapting to the specific operating environment. SAMPSON is compatible with both active and semi-active homing missiles; provides mid-course guidance for such weapons and supports fully automatic operation where rapid reaction is required.

Overall, SAMPSON is a modular system that can be tailored to meet precise operational and weapon system requirements. The radar scans electronically in azimuth and elevation with the system's arrays being rotated mechanically (at a rate of up to 60 rpm) in order to provide continuous 360° cover. Each of SAMPSON's two arrays incorporate between 1,000 and 10,000 transceiver elements, each of which contains a 2 to 20 W power module. According to Jane's sources, when applied to the Type 45 anti-air destroyer (as part of its Principal Anti-Air Missile System (PAAMS) - see following) SAMPSON utilises approximately 2,500 transceivers and has a masthead assembly that weighs 4.6 tonnes and rotates at 30 rpm. Anti-jamming features include adaptive nulling; very low antenna sidelobes; a very wide bandwidth; frequency agility; pulse compression; sidelobe blanking and jammer extraction/tracking. SAMPSON contains no high-voltage, high-power microwave components or associated water-cooling systems (thereby improving

The design of the SAMPSON radar's antenna assembly emphasises low structural weight and a high level of wind tolerance 0024879

reliability and simplifying maintenance and repair); features built-in test routines, solid-state transmission modules and utilises multiple parallel function paths (employing commercial-off-the-shelf technology) to maintain operation in the event of even multiple subsystem failure.

Status

As of March 2001, MESAR technology has been evaluated to provide a multifunction radar for installation aboard future major warships. With regard to the specific SAMPSON application, as of the given date, UK Air Missile System (UKAMS - acting on behalf of PAAMS developer EUROPAAMS) had awarded BAE Systems - Combat and Radar Systems a final development, prototyping and initial production contract covering the supply of three production standard prototypes and a single series production SAMPSON radar for use in the Royal Navy's 'first-of-class' Type 45 anti-air destroyer programme. Valued at 'significantly more' than then year £100 million, the described award is understood to require delivery of all four radars by the end of 2004. A similarly valued Phase 2 award is understood to cover production of a further 11 production standard radars starting in 2003.

Contractor

Alenia Marconi Systems (an Alenia- BAE Systems joint venture), Chelmsford/Rome.

UPDATED

Score coastal radar

Type

I-band (8 to 10 GHz) mobile or fixed coastal and low-altitude surveillance radar.

Description

This improved version of coastal radar uses the transceiver of the Iguane/Agrion/Varan family of airborne radars (see entries in the Airborne surveillance radars section), and the same antenna as the well-proven mobile TRS 3410 radar equipping coastal missile batteries.

For the mobile configuration, the foldable 4.8 m span antenna and the transceiver are installed on a two-wheel trailer and the workstation on a transportable shelter.

The main features of the Score radar are:

- provision of very efficient protection against sea clutter, by means of a very small detection cell (high directivity and pulse-width compressed down to 20 ns)
- high protection against jamming, by means of pulse-to-pulse frequency agility, which also enhances clutter rejection and high power directivity
- very compact equipment complying with military standards
- identification friend-or-foe subsystem compatible
- operational modes best adapted to the prevailing environment enabling the system to benefit from abnormal conditions in particular and to co-operate with other systems fitted with a matched transponder.

Status

As of January 2002, Thales Naval France was reporting Score as having been procured by a European navy.

The Score coastal/low-altitude surveillance radar in mobile configuration
0006897

Specifications
Antenna: 4.8 m parabolic reflector
Gain: 42 dB
Azimuth beamwidth (3 dB): 0.55°
Elevation beamwidth (3 dB): 2.7°
Polarisation: circular, fixed
Rotation speed: 11 rpm
Transmitter
Frequency: I-band
Compressed pulse: 0.02 µs
Range: 93 km (2.5 m² RCS fighter at low altitude); 130 km (fast patrol boat)

Contractor
Thales Naval France, Meudon-la-Forêt, France.

UPDATED

Scout naval radar

Type
I-band (8 to 10 GHz) surveillance/navigation 'quiet' Low Probability of Intercept (LPI) radar.

Description
Scout is an LPI radar designed for surface surveillance tactical navigation and covert operations. It uses Frequency Modulated Continuous Wave (FMCW) techniques and a very low transmitter power which ensures that it is virtually non-detectable by an electronic support measures receiver system. Scout offers the users the possibility of radar-aided pilotage and surface target surveillance in radar silence conditions. It is a complete stand-alone system with a transceiver integrated into the antenna. Scout can also be delivered integrated in pulse radar systems and as of August 2001, a new variant with increased range resolution is under development.

Status
During 1988, the Philips Research Laboratory developed a 'quiet' radar known as Pilot, which was marketed by the then Philips subsidiaries PEAB in Sweden and Signaal in Netherlands. With the sell-off of Philips' defence assets, PEAB was taken over by Bofors (subsequently CelsiusTech and now SaabTech) Electronics and maintained the name of Pilot for this radar. For its part, Signaal (now Thales Nederland) was taken over by the then Thomson-CSF (now Thales) and modified and improved the FMCW Pilot concept and changed the name of the radar to Scout. As of the period January to April 2001, Jane's sources were suggesting that Scout radars were in service aboard the following warships and classes of warship:

Belgium	• 'Wielingen' class frigates (three ships in class)
Finland	• the fast attack craft - missile *Hamina*
Greece	• 'Super Vita' class fast attack craft - missile (three building and four proposed as of February 2001)
Indonesia	• 'Singa' (four ships) and 'Todak' (two ships, two building as of February 2001)

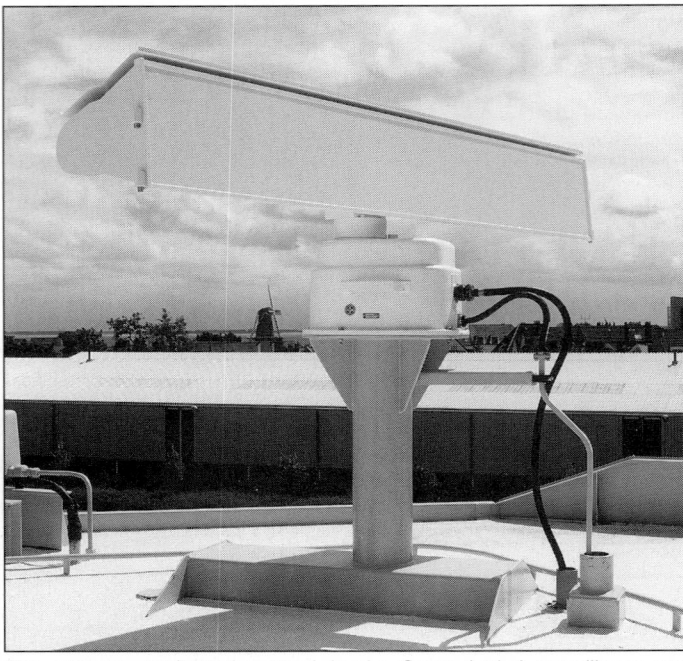

The antenna configuration used in the Scout 'quiet' surveillance and navigation radar
0006903

Netherlands	• the fast combat support ship *Zuiderkruis*
	• 'de Zeven Provinciën' class destroyers (four building as of March 2001)
	• 'Jacob van Heemskerck' (two ships) and 'Karel Doorman' (eight ships) class frigates
	• the coastguard patrol cutters *Jaguar*, *Panter* and *Poema*
United Arab Emirates	• 'Kortenaer' class frigates (two ships)
	• 'Ban Yas' class fast attack craft - missile (six ships)

Elsewhere, the former Magnavox-Signaal Systems Company in the USA is understood to have won a contract from Egypt for 30 mobile Coastal Border Surveillance Systems (CBSS) which use a licence-built Scout radar.

Specifications
Frequency: I-band (8-10 GHz)
Transmitter power: 1 mW, 10 mW, 100 mW or 1 W (operator selectable)
Antenna gain: 30 dB
Antenna rotation speed: 24 rpm
Sweep repetition frequency
Number of range cells: 512
Range scales: 1.4, 2.8, 5.6, 11.1, 22.2 and 44.5 km
Range cell size: 3, 6, 12, 24, 48 and 96 m
Range resolution: 9, 18, 36, 72, 144 or 288 m (range scale dependent)
Power requirements: 115 V (60 Hz, 1-phase, 700 V A approx load)
Dimensions   (w × h × d):   420 × 460 × 300 mm   (processor   unit);
1,933 × 608 × 400 mm (antenna assembly)
Weight: 15 kg (processor unit); 65 kg (antenna assembly)

Contractor
Thales Nederland, Hengelo, Netherlands.

VERIFIED

Sea Tiger surveillance radar

Type
E/F-band (2 to 4 GHz) air and surface surveillance/target designation radar. French Navy designation DRBV 15.

Description
Sea Tiger is a shipborne E/F-band combined surveillance radar that is optimised for the detection and tracking of missile threats, with a particular emphasis on sea-skimming anti-ship weapons. It is designed to perform the following functions in a very severe clutter and jamming environment:
• air surveillance
• surface surveillance
• anti-missile surveillance
• target designation for weapon systems (guns, missiles).
  Visibility in clutter is achieved through the use of complementary techniques such as circular polarisation, pulse compression, Doppler filtering and an anti-clutter reception chain.
  A typical Sea Tiger installation comprises:
• a transmitter with coherent amplifier chain emitting frequency-modulated pulses (pulse compression) over a wide frequency range

The Sea Tiger surveillance radar

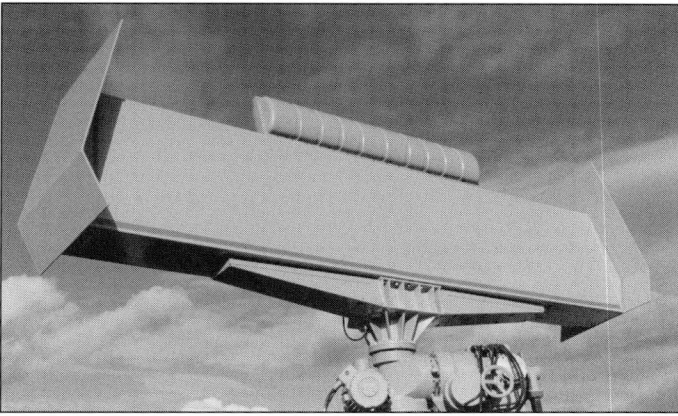

Sea Tiger Mk 2/DRBV 15C antenna

- a receiver with multiple reception chains: Moving Target Indicator (MTI), lin/log, constant false alarm rate. The MTI is a digital linear device whose characteristics, associated with the stability of the amplifier chain, make it possible to detect a missile in the most severe clutter environment
- a roll-stabilised antenna with integrated identification friend-or-foe subsystem.
  Data provided by this radar can be processed using Thales Naval France's TAVITAC, TAVITAC 2000 or TAVITAC NT tactical data systems (see DRBJ 11 entry for more details). Thales has also developed a Mk 2 version of Sea Tiger (French Navy designation DRBV 15C) which is fitted with a low sidelobe planar antenna to improve the equipment's resistance to jamming.

Status
As of January 2002, Thales Naval France was reporting Sea Tiger as being installed aboard the following warships and classes of warship:

China	• 'Luda I/II' (15 ships in class - Sea Tiger as an alternative to Eye Shield) and 'Luhu' (two ships) class destroyers
Columbia	• 'Almirante Padilla' class frigates (four ships)
France	• the aircraft carrier *Charles de Gaulle* (DRBV 15D variant)
	• the research ship *Monge* (DRBV 15C variant)
	• 'Suffren' class destroyers (two ships - DRBV 15A variant)
	• 'La Fayette' class frigates (five ships - DRBV 15C variant)
Saudi Arabia	• 'Madina' class frigates (four ships)
Taiwan	• 'Kang Ding' class frigates (six ships - DRBV 26D variant)

Specifications
Operational performance
Detection range: 111 km on 2 m² (fluctuation) target with Pd ≤50%
Altitude detection: 15,240 m
Elevation pattern: cosec² up to 50°
Subclutter visibility: >40 dB
Antenna
Gain: 30 dB
Polarisation: circular/linear, switchable
Rotation speed: 15 and 30 rpm
Stabilisation: better than ± 25°

Transmitter: frequency agility, pulse compression
Receiver: noise figure better than 5.5 dB. 4 different reception chains: MTI, lin, log, CFAR anti-clutter

Contractor
Thales Naval France, Meudon-la-Forêt, France.

UPDATED

Signaal Coastal Radar System (SCORADS)

Type
G-band (4 to 6 GHz) coastal surveillance radar system.

Description
SCORADS (Signaal COastal RADar System) is a new generation medium-range coastal surveillance radar which has been produced under a turnkey programme for the Greek Navy. The system operates in the G-band for automatic tracking of both surface and low-level air targets up to ranges in excess of 100 km.

SCORADS uses an antenna based on the Signaal (now Thales Nederland) STAR-K air traffic control radar family, with the front end making use of a Signaal's standard travelling wave tube transmitter and 2-D search generator/receiver. The radar processing contains standard Signaal Modular Processing and Common Processing Module units.

Status
The turnkey programme for the Greek Navy noted previously involves the supply of three radar stations and a central control and monitoring centre. In February 1994, the first radar was under test and by August 1996, all three SCORADS stations were fully operational.

Specifications
Frequency: G-band (4-6 GHz)
Peak power: 2,000 W
Antenna gain: 40 dB
Beamwidth: 0.75° (azimuth); 2.75° (elevation)
Scan rate: 10 rpm
Sidelobe levels: −40 dB (peak)
Uncompressed pulse-length: 10 µs (surface mode); 24 µs (air mode)
Compressed pulse-length: 550 ns
PRF: medium (air mode); low (surface mode)
Max range: 105 km (surface); 120 km (air)
Altitude: 3,048 m
Accuracy: 0.25° (azimuth); 50 m (range)
Resolution: 2° (azimuth); 100 m (range)
Capacity: 400 tracks/scan
Dimensions/weight: 4,980 × 4,815 × 3,126 mm/2,950 kg (antenna/transmitter unit); 745 × 1,496 × 445 mm/160 kg (receiver/processor cabinet); 750 × 555 × 650 mm/50 kg (local control workstation)

Contractor
Thales Nederland, Hengelo, Netherlands.

VERIFIED

SMART-L volume search radar

Type
D-band (1 to 2 GHz) 3-D naval volume search radar.

Description
SMART-L is the D-band version of the successful family of SMART multibeam 3-D radars (see separate entries SMART-S and MW08). The radar is a further development of the functional design, incorporating the latest technology and additional advanced radar functions. It is designed according to NATO specifications for a volume search radar. SMART-L is designed to fulfil:
- medium-range detection of the latest generation of small 'stealth' air targets
- long-range detection of conventional aircraft
- high electronic counter-countermeasures performance
- guidance support for patrol aircraft
- surface surveillance up to the radar horizon.

Because of its large power budget, SMART-L is dedicated to the early detection and tracking of very small aircraft and missiles. The accurate 3-D target information (even under multipath conditions), gathered by the SMART-L radar provides an essential contribution to the threat evaluation process, especially in multiple attack scenarios and it allows the weapon control system to perform the fastest lock on.

The antenna consists of a stack of 24 linear arrays. All 24 arrays are used during reception while 16 arrays are used during transmission. By controlling the phase of the radio frequency energy, supplied by the 16 transmission arrays, the shape of the illumination pattern can be controlled and level stabilised. The received radar energy is processed by 24 receiver channels and fed to a beam-forming network in which 14 beams are formed in the elevation coverage of 0 to +70°. The output

A SMART-L radar installed on a Royal Netherlands Navy frigate during the equipment's sea trials   **2000**/0080966

Four SMART-L radars have been ordered for installation aboard the Royal Netherlands Navy's LCF frigates   0024882

signals of the beamformer are further processed by a Doppler processor for the extraction of the target Doppler speed for clutter rejection, and fast but reliable track initiation and maintenance. SMART-L is capable of automatic detection, track initiation and tracking of 1,000 air targets and up to a range of 400 km. Surface surveillance is performed by a D-band surface surveillance channel that is capable of automatic detection and track initiation of 100 surface tracks at ranges of up to 60 km.

Status
As of the period March to April 2001, the SMART-L radar was mandated for Dutch 'de Zeven Provinciën' class destroyers (four building as of March 2001) and German 'Sachsen' class frigates (three building as of April 2001). Production of the seven necessary radars is understood to have began during 1998.

Specifications
Antenna
Type: linear array
Polarisation: vertical
Horizontal beamwidth: 2.2°
Elevation coverage: 0-70°
Illumination pattern: local area, long range, burn-through and customer specific
Vertical beamwidth receiver: 14 beams of 6° approx
Rotation speed: 12 rpm
D-band transmitter
Type: solid state
D-band receiver and processor
Beamformer: FFT
Doppler filtering: FFT
Range: up to 300 km against a fighter aircraft

Contractor
Thales Nederland, Hengelo, Netherlands.

VERIFIED

SMART-S naval surveillance radar

Type
F-band (3 to 4 GHz) 3-D air surveillance radar.

Description
SMART-S (Signaal Multibeam Acquisition Radar for Targeting) is an all-weather 3-D target indication and surveillance radar system intended for all types of naval vessels from fast patrol boats upwards. Its prime application is as the main sensor for data handling and weapon system control and it has a very high performance in the presence of heavy clutter and electronic countermeasures. The equipment has been designed to cope with small high-speed anti-ship missiles with radar cross-sections down to 0.1 m² and approach speeds of M3+, which can be either sea skimmers or arriving from high angles of 70° or more.

The SMART system operates in F-band, where it offers an optimum balance between range, clutter rejection and antenna dimensions. It provides automatic detection, track initiation and track maintenance of both air and surface threats, with gapless coverage over a complete hemisphere from the sea surface upwards. It incorporates anti-clutter and electronic counter-countermeasures features such as multiple reception beams with ultra-low sidelobes in elevation and azimuth, a clutter analysis sensor, broadband transmission, pulse repetition frequency and radio frequency agility per burst and a jamming analysis sensor. SMART is designed to track 160 air targets and 40 surface targets simultaneously.

The system comprises an antenna and three main below-decks units. The hydraulically stabilised antenna consists of a single-element wideband transmitting array, and a multi-element stripline receiving array. The ultra-low sidelobe phased-array allows the formation of multiple receive beams in elevation. To ensure high sensitivity, preprocessing of the received signals takes place in the antenna unit itself. The output of the 16 antennas is fed to a digital beam-forming network in which the 12 independent elevation beams are produced, after which Doppler Fast Fourier Transform processing and automatic tracking is carried out. The transmitter is based on a high-power, pulse-to-pulse coherent travelling wave tube. Integral identification friend-or-foe can be provided.

Status
As of March 2001, Jane's sources were suggesting that SMART-S radars were in service aboard the following warships and warship classes:

Germany   • 'Brandenburg' class frigates (four ships in class)
Netherlands• 'Jacob van Heemskerck' (two ships) and 'Karel Doorman' (eight ships) class frigates

Specifications
Antenna unit
Rotation speed: 27 rpm
Stabilisation: hydraulic in roll and pitch
Transmitting antenna
Type: horn antenna array
Receiving antenna
Type: stripline array
Horizontal beamwidth: 2°
Elevation coverage: 90°
Transmitter
Type: TWT with pulse compression
Power output: 145 kW peak
Frequency: F-band (3-4 GHz)

The SMART-S surveillance radar antenna

Detection range: >100 km against a fighter aircraft
Tracking accuracy: 40 m (range); 0.25° (bearing); 0.6° (elevation)
Weight
Antenna: 1,500 kg
Below-deck units: 2,400 kg

Contractor
Thales Nederland, Hengelo, Netherlands.

VERIFIED

SPECTAR multifunction naval radar

Type
Single-face, active array, multifunction naval radar.

Description
The SPECTAR multifunction, active array radar is designed for installation aboard vessels of corvette size and up and offers weapon system control and long-range air picture compilation capabilities in a system that uses a single masthead antenna. The radar is based on the BAE Systems/UK Defence Evaluation and Research Agency Multifunction Electronically Scanned Adaptive Radar (MESAR) technology demonstrator which has been tested in both the UK and the USA.

Using a combination of electronic scanning in azimuth and elevation and mechanical rotation (at 30 or 60 rpm and providing continuous 360° cover), SPECTAR generates 3-D search and track data for close-in weapons system control and air picture compilation at longer ranges. Radar performance is matched to the detection of current and future low observability targets in adverse environments. Software control of all SPECTAR's major processing and management functions enables it to be readily adapted to future operational needs and evolving threats.

SPECTAR adapts to its operational environment through the automatic selection of appropriate waveforms and processing/management strategies with minimal operation intervention. The system is compatible with active and semi-active missiles and provides mid-course guidance via its primary active antenna. Radar availability is maintained through the use of solid-state components and the incorporation of multiple redundancy in key areas such as the antenna and processing subsystems. SPECTAR incorporates built-in test and utilises commercial-off-the-shelf technology to reduce logistic costs.

Status
As of late 2000, MESAR technology has been evaluated to provide the SPECTAR multifunction radar that is planned for installation on future major warships. Here, SPECTAR applications were a feature of the low radar cross-section Vosper Thornycroft Sea Wraith and BAE Systems Marine Cougar corvette/frigate proposals.

Contractor
Alenia Marconi Systems (an Alenia-BAE Systems joint venture), Chelmsford/Rome.

UPDATED

The antenna configuration proposed for the SPECTAR multifunction naval radar   0006907

Thales Nederland 3-D MultiTarget Tracking Radar (MTTR)

Type
Air search and target tracking radar.

Description
MTTR is an advanced 3-D air search and target tracking radar capable of handling more than 100 aircraft tracks. A very high data rate is achieved by the use of an antenna array consisting of back-to-back pairs of parabolic reflectors and planar electronically scanned antennas, all four of which have common turning gear. Identification friend-or-foe/selective identification facility secondary radars are integrated with the MTTR. The parabolic dishes are used for search and the electronic-scan antennas for target tracking. High and low cover is provided. Beam steering and data extraction are performed by an SMR series digital computer.

Operational functions include: long-range search; search information with high data rate for low-flying aircraft; search information with high resolution of close-in air targets; automatic position and height information; simultaneous tracking of over 100 aircraft targets; target designation facilities for other systems. No performance or other technical details have been obtained. Estimated diameter of the parabolic dishes is in the region of 6 m. Both land and naval uses are likely, and the latter is understood to incorporate full stabilisation.

Status
As of March 2001, Jane's sources were suggesting that the MTTR search and tracking radar was in service aboard the Dutch frigate *de Ruyter*. As of the given date, the frigate *de Ruyter* was scheduled to be decommissioned during January 2003.

Contractor
Thales Nederland, Hengelo, Netherlands.

UPDATED

A view of the Royal Netherlands Navy guided missile frigate Tromp, *showing the large radome housing the Thales 3-D MTTR and the smaller one for the WM25 fire-control system radar*

TRS 3011 Jupiter air surveillance radar

Type
D-band (1 to 2 GHz) long-range air surveillance radar. French Navy designation DRBV 26C (Jupiter) and DRBV 26D (Jupiter-ER).

Description
Jupiter, along with the TRS 3015 Mars surveillance radar (see separate entries), is a member of Thales Naval France's D-band radar family that utilises fully solid-state transmitters. Jupiter itself is a long-range air surveillance sensor for medium and large tonnage vessels. The transmitting power, in D-band and the cosec^2 pattern strengthened at high elevation angles, provide early warning against attacks at all altitudes. An identification friend-or-foe antenna is integrated with the main antenna.

The main feature of Jupiter is its modern, fully solid-state transmitter which, in addition to its advantages related to frequency agility and pulse compression, endows the radar with high maintainability. The transmitter offers an outstanding graceful degradation capability, the possibility of repair without switching off the equipment and drastic reduction of cost.

The reception chain implements the most advanced techniques, using Fast Fourier Transform speed filters, enabling this long-range radar to participate in anti-missile surveillance.

Data provided by this radar can be processed using Thales Naval France's TAVITAC, TAVITAC 2000 or TAVITAC NT tactical data systems (see DRBJ 11 entry for more details). In addition to the basic Jupiter system, Thales has developed the extended range Jupiter-ER variant that employs two solid-state transmitters.

DRBV 26C surveillance radar on the Jean Bart. *The DRBC 33 fire-control radar is on the right of the picture* (Hachiro Nakai)

Status

As of January 2002, Thales Naval France was reporting DRBV 26 as being installed aboard the following warships and classes of warship:

France	• the aircraft carrier *Charles de Gaulle* (DRBV 26D variant)
	• 'Cassard' (two ships in class - DRBV 26C variant), 'Georges Leygues' (seven ships - DRBV 26A variant) and 'Tourville' (two ships - variant unknown) class destroyers
Saudi Arabia	• 'Arriyad' class frigates (three ships building - DRBV 26C variant)
Taiwan	• 'Kang Ding' class frigates (six ships - DRBV 26D variant)

Specifications

Range against incoming fighters: >340 km (Jupiter); >390 km (Jupiter-ER)
Aperture (elevation × azimuth): 10 (cosec 40°) × 2.2° (Jupiter/Jupiter-ER)
Rotation speed: 7.5 and 15 rpm (Jupiter/Jupiter-ER)
Transmitter: solid state
Anti-jamming and anti-clutter reception
Automatic initialisation and automatic tracking of air tracks

Contractor

Thales Naval France, Meudon-la-Forêt, France.

UPDATED

TRS 3015 Mars surveillance radar

Type

D-band (1 to 2 GHz) air and surface surveillance radar. French Navy designation DRBV 21A.

Description

Mars is another member of the D-band TRS 3011 Jupiter/TRS 3015 Mars family of naval surveillance radars (see separate TRS 3011 entry). It is intended for ocean-

The DRBV 21A antenna as installed aboard the French Navy's 'Floréal' class frigates

capable patrol vessels and operates with a simplified transmitter assembled with the receiver in a single electronics cabinet. The antenna weighs approximately 600 kg and rotates at 12 rpm. Instrumented ranges are 110 km for the air channel and 80 km for the optional surface channel. The radar cabinet is fitted for an extraction/tracking unit for automated surveillance. As with Jupiter, the transmitter is solid state with frequency agility, burst-to-burst with Doppler processing, or pulse-to-pulse.

Status

As of January 2002, Thales Naval France was reporting DRBV 21 as being installed aboard the following classes of warship:

France	• 'Floréal' class frigates (six ships in class - DRBV 21A variant)
	• 'Foudre' class landing ships dock (two ships - DRBV 21A variant)

In terms of known contracting activity, the French Navy is known to have ordered three DRBV 21A radars during 1993.

Contractor

Thales Naval France, Meudon-la-Forêt, France.

UPDATED

TRS 3033 Triton S surveillance radar

Type

E/F-band (2 to 4 GHz) air and surface surveillance and target designation radar.

Description

Triton S is an E/F-band air and surface surveillance radar that has been specially developed for the detection of small air targets (aircraft and missiles). It is in a heavy clutter environment, is a fully coherent, digital processing, pulse-Doppler equipment which includes an independent pulse compression and frequency-agile receiver for detection of surface targets.

The antenna is of the non-stabilised type with a broad elevation lobe to counteract the effects of ship motion. The transmitter comprises a travelling wave tube. The radar provides two video signals, an air surveillance video and a surface video, which can be used for display, target designation and track-while-scan processing. Data provided by this radar can be processed using Thales Naval France's TAVITAC, TAVITAC 2000 or TAVITAC NT tactical data systems. (See DRBJ 11 entry for more details).

Status

As of January 2002, Thales Naval France was reporting Triton S as having been procured by the Tunisian Navy for installation (according to Jane's sources) aboard the service's three 'Combattante III M' class fast attack craft.

Specifications

Frequency: E/F-band (2-4 GHz) frequency agile
Two receivers: pulse-Doppler (processed range 1.5-26.2 km); pulse compression
Antenna rotation: 40 rpm
Detection range: air target (fighter), 26.2 km; surface target, radar horizon
SCV: at least 50 dB

Contractor

Thales Naval France, Meudon-la-Forêt, France.

UPDATED

The scanner used in the Triton S surveillance radar system

TRS 3050 Triton G surveillance radar

Type
G-band (4 to 6 GHz) air and search surveillance and target designation radar.

Description
The TRS 3050 is designed for ships of 150 tons upwards to perform air and surface surveillance, low and very low altitude missile detection, target designation to weapon systems and emergency navigation under adverse conditions. It is suitable for either original fitting in new ships, or for use during refits.

It is a G-band radar, fully coherent synthesiser driven, with two receiver channels:
- air: with digital Doppler processing (Fast Fourier Transform), pulse compression, burst-to-burst frequency agility, automatic tracking and initialisation of targets
- surface: with pulse compression, pulse-to-pulse frequency agility and advanced anti-clutter and anti-jamming processing.

As the adjacent photograph shows, Triton G employs the normal radar reflector of the Triton series of radars but can be fitted with other antennas. Most of the operational advantages are the result of advances in transmitter/receiver and processor design. Data provided by this radar can be processed using Thales Naval France's TAVITAC, TAVITAC 2000 or TAVITAC NT tactical data systems. (See DRBJ 11 entry for more details).

Status
As of January 2002, Thales Naval France was reporting Triton G as being installed aboard the following classes of warship:

Germany • 'Tiger' class fast attack craft - missile (five ships in class)
Taiwan   • 'Kang Ding' class frigates (six ships)

Specifications
Frequency: G-band (4-6 GHz)
Max detection range: 19 km (Pd = 0.9, 2 m^2 aircraft target, Doppler operation)
SCV: at least 40 dB
Antenna gain: at least 27 dB
Elevation pattern: 22°

Contractor
Thales Naval France, Meudon-la-Forêt, France.

UPDATED

The scanner used in the Triton G surveillance radar system

TRS 3100 Calypso III submarine radar

Type
I/J-band (8 to 20 GHz) submarine air and surface surveillance radar.

Description
Calypso III is a submarine navigation, surveillance (air and surface) and target designation radar that comprises an antenna (the Calypso II antenna), a non-rotating periscopic mast (hoisting is achieved using the ship's hydraulic system pressure), a transmitter/receiver cabinet (secured at the foot of the mast) and an operator console.

Of these, the radar's transmitter (which is the same as that used in the earlier Calypso II system) is a conventional magnetron unit with a frequency adjustable magnetron. The receiver is fitted with long service life radio frequency elements, a 'modern' mixer and a logarithmic anti-clutter chain. The operator console incorporates a control panel for the radar and its antenna; a 406 mm (16 in) cathode ray tube display (with a digital range-finding facility) and an operator's

panel. Periscope derived bearing electronic support/sonar data can be displayed on the set's plan position indicator and the radar is claimed to be able to determine target range using 'short-time' transmission. Calypso III's detection range against a typical maritime patrol aircraft target flying at an altitude of 2,500 m is given as 33 km, with that for a surface vessel being radar horizon dependent.

Status
As of January 2002, the operational status of the Calypso III radar was uncertain. As of the given date, the sensor was reported as being no longer in production.

Contractor
Thales Naval France, Meudon-la-Forêt, France.

UPDATED

TRS 3110 Calypso IV submarine radar

Type
I/J-band (8 to 20 GHz) submarine air and surface surveillance radar.

Description
The TRS 3110 Calypso IV radar is an I/J-band equipment that is designed to provide surveillance and navigation facilities for submarines. It comprises an antenna assembly, a transmitter/receiver cabinet and an operator's console. Of these, the antenna and operator's console are similar to those used in the Calypso III system (see separate entry). Overall, Calypso IV centres on a navigation radar function that can be role-expanded via the introduction of appropriate circuit cards. Thales describes the radar as using a 1 ms pulse for long range detection (up to at least 19 km in free space) together with a 50 ns configuration for short range (down to approximately 15 m), high resolution target detection. The sensor's receiver is quoted as being optimised for operation in the presence of clutter, while its transmitter is noted as being a 25 kW (peak) klystron-driven device that features continuous, sector, burst and receive only operating modes. The Calypso IV operator's console and associated data processor facilitate the presentation of raw or synthetic radar data (including target echoes, range and bearing labels and transmission sectors).

Status
Over time, a Calypso variant is reported to have been procured for installation aboard three French Navy 'Daphne' class attack and three Argentine 'Santa Cruz' class general purpose submarines (Argentine boats reported as being in service as of January 2001). Calypso technology may also form the basis of the DRUA 33 sensor that has been installed aboard the French Navy's 'Inflexible' class nuclear ballistic missile, 'Rubis' class nuclear attack and 'Agosta' class conventionally powered attack submarines.

Contractor
Thales Naval France, Meudon-la-Forêt, France.

UPDATED

TRS 3405 coastal radar

Type
I-band (8 to 10 GHz) coastal surface surveillance radar.

Description
This equipment is an I-band radar specially intended for the surveillance of maritime traffic in straits, harbour approaches, access channels and harbours.

It is designed to provide high performance in range, resolution and accuracy and protection against the various detection impairments encountered in sea surface surveillance, namely sea and rain clutter and propagation loss in rain and heavy fog. High resolution is achieved by the use of an antenna with a narrow azimuth beamwidth and a very short transmission pulse, which results in a very small radar cell. The use of frequency diversity allows decorrelation of sea clutter peaks, reducing the false alarm rate and improving the detection of small ships or buoys.

The radar operates in the I-band. To achieve a high gain and avoid unwanted rain echoes, which would result from the illumination of clouds at high altitude, the antenna elevation beamwidth is also narrow, and radiation polarisation can be switched from linear to circular. Detection at very close range is ensured by an elevation pattern of the inverted cosec^2 type down to an angle of −25°.

Round-the-clock operating requirements demand two transmitter/receivers for each radar station. In this equipment, they are connected to the antenna through a passive diplexer rather than through a switching device, which enables both units to be operated simultaneously on a frequency diversity basis. As this mode of simultaneous operation is necessary only in heavy weather, the operating cost is not increased significantly. In addition, each reception channel is fitted with an automatic false alarm regulation control slaved to the mean level of sea clutter.

This equipment is fully compatible with extractors and data remoting.

The TRS 3405 maritime surveillance radar installed at Gris-Nez, France

Status
Over time, TRS 3405 is understood to have been procured by the navies of France, Canada, Indonesia and Saudi Arabia.

Specifications
Antenna
Gain: 44 dB
Resolution: 0.24°
Low sidelobes: 25 dB
Inverted cosec: squared pattern in elevation down to −25°
Linear and circular polarisation
Rotation rate: 5 or 10 rpm
Transmitter/receiver
Frequency: I-band (8-10 GHz)
Peak power: 200 kW
High resolution pulse: 0.05 µs
PRF: 1,080-570 Hz
Advanced anti-clutter and anti-jamming processing.

Contractor
Thales Naval France, Meudon-la-Forêt, France.

UPDATED

TRS 3410 mobile coastal radar

Type
I-band (8 to 10 GHz) mobile coastal surveillance radar.

Description
This mobile, cabin-installed station utilises the I-band transmitter/receiver of the TRS 3405 radar operating in frequency diversity for decorrelation of sea clutter peaks, coupled with a 4.8 m antenna located on a trailer. Installed at 40 to 50 m above sea level, this station detects small boats up to 25 km away, whereas for larger ships with superstructures at a height of 10 m or more, the detection range reaches 40 km. Useful range with a coastal antenna at an elevation of 250 m is up to 70 km against patrol boats.

The TRS 3410 radar is fitted with a Track-While-Scan (TWS) device, which refines and maintains the data from a surface track starting from a target designation made by the operator. These data are processed through a computer for the control of various types of anti-ship missiles whose batteries can be located next to the radar station or remotely sited. The compounding of its own accuracy and of the data refinement by the TWS device gives overall target parameters accuracy which is sufficient for the control of coastal artillery fire against surface targets.

Status
Over time, examples of the TRS 3410 radar are understood to have been procured by France, Egypt and Qatar. TRS 3410 is also noted as having been replaced by the Score system within Thales Naval France's product line.

Contractor
Thales Naval France, Meudon-la-Forêt, France.

UPDATED

TRS 3415 SURICATE coastal surveillance radar

Type
I/J-band (8 to 20 GHz) autonomous or remote-controlled coastal surveillance radar.

Description
SURICATE is a coastal surveillance radar which offers the capability of controlling the maritime approaches with an all-weather permanent long-range detection of small units such as dinghies, low-flying aircraft or helicopters.

The antenna is installed on a support or on top of a mast to increase radar horizon range. The transmitter/receiver and processing cabinets are installed inside an electronics room or shelter located at the foot of the antenna mast. The radar can be fitted with three sizes of antenna.

The main features of SURICATE are:

- the latest technology with travelling wave tube, very high pulse compression ratio and very short pulse for clutter rejection
- digital video compression, advanced extractor, automatic initialisation and tracking of surface targets
- unequalled atmospheric clutter resistance because of an ultra-high resolution cell (0.6° × 3 m), pulse-to-pulse frequency agility and circular polarisation.

Status
As of January 2002, Thales Naval France was reporting SURICATE as having been procured by 'several' European and Asian customers.

Specifications
Antenna: 4.40 m
Gain: 40 dB
Azimuth beamwidth: 0.6°
Elevation beamwidth: 4°
Cosec²: −30°
Rotation speed: 11 rpm
Range: 83 km (2.5 m² target); 114 km (10 m² target)
Polarisation: circular or linear (switchable)
Transmitter: I/J-band (8-20 GHz) TWT
Compressed pulse: 20 ns

Contractor
Thales Naval France, Meudon-la-Forêt, France.

UPDATED

TRS series naval radars

Type
Family of G-band (4 to 6 GHz) multifunction naval radars.

Description
The European Aeronautic, Defence and Space (EADS) Co manufactures a range of G-band naval surveillance and multimode radars developed from its land-based systems (see TRM radar series entry in the Land-based air defence radars section for details). Known details of the series are as follows:

TRS-C
TRS-C is a 2-D air and surface surveillance radar for small ship applications. It is the naval version of the TRM-L, available with magnetron or Travelling Wave Tube (TWT) transmitter and moving target indicator or Doppler processing.

TRS-3D
TRS-3D is a G-band, multimode, surveillance and weapons assignment radar that performs Automatic Detection and Tracking (ADT) of surface and air targets in heavy clutter and countermeasures polluted environments. It is derived from the land-based TRM-L and TRM-S radars and because of its modular design, can be readily adapted to meet specific customer range, resolution and accuracy requirements (for example TRS-3D/32 for corvettes and frigates). The TRS-3D transmitter is coherent and incorporates a solid-state driver stage and a travelling wave tube output stage. Electronic counter-countermeasure features inherent in the design include:

- low antenna sidelobes
- high angular resolution in azimuth and elevation against main lobe jamming
- pulse compression for suppression of broadband pulse jammers
- burst-to-burst for frequency, polarisation, pulse repetition frequency, pulse-length and signal coding to reduce deception jamming
- pseudo-random frequency selection over the frequency band to counter spot jammers
- jam detector for continuous monitoring of the RF environment to provide information on jamming activity
- Moving Target Detection (MTD) processing to improve signal to clutter ratio and so reduce the effects of chaff
- automatic jammer avoidance circuits to select the least jammed frequency
- sidelobe blanking and coherent sidelobe cancellation.

The TRS-3D series includes the following variants:

TRS-3D/32 is mandated to replace the DA 08 air/surface search radars aboard German 'Bremen' class frigates such the Emden *shown here* (EADS)
2002/0120324

TRS-3D/16

TRS-3D/16 utilises an antenna that incorporates 16 rows of radiators and a mechanically stabilised platform.

TRS-3D/16ES

TRS-3D/16ES features antenna beam electronic stabilisation and a low mast top antenna weight.

TRS-3D/32 and TRS-3D/32ES

TRS-3D/32 and TRS-3D/32ES feature antennas that utilise 32 rows of radiators and differ from one another in TRS-3D/32's use of a mechanically stabilised antenna platform and TRS-3D/32ES's incorporation of electronic antenna beam stabilisation.

Of the described systems, TRS-3D/16 and /16ES offer the following standard operating modes:

- long range surveillance (90 km instrumented range)
- air and surface surveillance
- target indication (up to 55° elevation in the /16ES model and up to 70° in the /16)
- sea skimmer detection

Additional modes that can be implemented if required comprise:

- extended long range detection (up to 180 km instrumented range)
- gunfire support (for surface fire-control applications).

The antenna group used comprises a primary G-band planar phased-array, an Identification Friend-or-Foe (IFF) antenna and an optional I-band navigation radar antenna. The radar incorporates a full pulse-to-pulse polarisation agility and scanning in elevation is performed using a single, agile pencil beam. Scanning in azimuth is performed mechanically, using a fully stabilised platform that rotates at up to 60 rpm depending on the operating mode selected. MTD Doppler processing facilitates detection of small targets such as sea-skimming and/or diving anti-shipping missiles and there are separate air and surface target channels, each with its own adaptive detection threshold and jamming detector. Helicopter classification is also provided, as is a quick reaction technique to detect pop-up targets.

TRS-CS

A coastal surveillance radar developed from the TRM-L system (see separate entry) with additional equipment for the detection of surface targets.

Status

As of July 2001, Jane's sources report that over time, the TRS-C radar has been deployed aboard ships of the German Navy while TRS-3D/16 has been installed aboard Danish 'Flyvefisken' class multirole warships (14 ships in class). Denmark is also noted as having acquired at least three TRS-3D/16s for retrofit aboard 'Niels Juel' class frigates. Further south, Spanish 'Galicia' class amphibious transport docks (two ships) are reported to be fitted with TRS-3D/16 radars, while the TRS-3D/32 variant has been acquired by the German Navy for retrofit aboard its 'Bremen' class frigates (eight ships). Here, the radar replaces the Thales Nederland (formerly Signaal) DA 08 equipment and is understood to have undergone sea trials (aboard the *Koln*) in the Baltic during February and March of 2000. Elsewhere within the German Navy, an unidentified TRS-3D variant is also understood to have been specified for the service's five 'K130' class corvettes that, as of the given date, were scheduled for delivery during the period 2005 and 2008. During July 2000, Raytheon and EADS announced an agreement whereby the former would produce an unspecified number of TRS-3D/16 radars for use aboard four US built 'Ambassador III' class missile fast attack craft that, as of the given date, were scheduled for delivery to the Egyptian Navy during the period 2004/2005.

Contractor

European Aeronautic, Defence and Space (EADS) Co Systems and Defence Electronics, Ulm, Germany.

UPDATED

Types 967 and 968 radars

Type

D/E-band (1 to 3 GHz) air surveillance and target designation system.

Description

The Royal Navy Type 967 is an air surveillance radar that is integrated with the Type 968 surface surveillance radar to form a very compact medium- to short-range defence radar. They provide high-performance facilities to fulfil the requirements of air and surface warning, search and target designation, from sea level to high elevation angles. These radars, with the Type 910 target tracker and missile guidance radar, form the radar group for the Seawolf surface-to-air missile system, GWS 25. The waveguide antennas for both radars are mounted back-to-back in a common housing carried by a single fully stabilised mount protected by a radome. The Type 967 provides air surveillance with cover up to high elevation angles. Operating frequency is in the D-band (1 to 2 GHz) of the radar spectrum and pulse-Doppler mode is employed. An enhanced version, known as the Type 967M, has been developed. The Type 968 provides low air cover and surface search, and operates in the E-band (2 to 3 GHz). The antenna rotation rate for both is 30 rpm. The Seawolf surveillance radar system is a high-power self-adaptive system, with sophisticated data handling, which resolves both velocity and range ambiguities, initiates track, carries out threat evaluation, takes the engagement decision, and performs attack allocation by assigning a tracker and feeding the track co-ordinates to it. Identification friend-or-foe and electronic counter-countermeasures subsystems are incorporated and the suite is equipped with its own FM 1600 digital processor.

Status

Over time, Type 967/968 radars are understood to have been installed aboard Brazilian 'Broadsword Batch 1' class frigates and Royal Navy Batch 2 and 3 'Broadsword' class frigates. As of January 2002, it was thought likely (but not confirmed) that the Types 967 and 968 radars were being supported by Alenia Marconi Systems.

Contractor

Alenia Marconi Systems (an Alenia-BAE Systems joint venture), Chelmsford/ Rome.

UPDATED

The Type 967/968's antenna assembly as applied to the Brazilian frigate Rademaker *(the former HMS* Battleaxe*). Separate antennas for surface and air search are contained in a common housing* (Stefan Terzibaschitsch)

Type 996 surveillance radar

Type

E/F-band (2 to 4 GHz) 3-D surveillance and target indication radar.

Description

Within the AWS-9 3-D radar series, the Type 996 is the Royal Navy nomenclature for its E/F-band, 3-D surveillance and target indication radar. The equipment utilises a multibeam, phased-array antenna that rotates at 30 rpm and has a co-mounted identification friend-or-foe array. Power is supplied by a travelling wave tube transmitter and the radar incorporates a range of electronic counter-countermeasures features which include low sidelobes, wideband frequency agility, multiple beams and variable transmission patterns. Three operating modes are reported to be available:

The antenna used in the Alenia Marconi Systems Type 996 naval radar

- long-range air target detection
- missile detection
- target indication/long-range surveillance (out to ranges of up to 115 km)

Status

As of March to April 2001, Jane's sources were reporting that the Type 996 surveillance radar was in service aboard the following warships and warship classes:

UK
- the amphibious transport dock *Ocean* (Type 996(2) radar)
- 'Invincible' class aircraft carriers (three ships in class - Type 996 (1) radar)
- 'Type 42 (Batches 1, 2 and 3)' class destroyers (11 (seven Batch 1/2 and four Batch 3) ships - Type 996(1) radar)
- 'Duke' class frigates (15 ships, one building as of April 2001 - Type 996(2) radar)
- 'Albion' class amphibious transport docks (two building as of March 2001 - Type 996(2) radar)
- 'Fort Victoria' class replenishment oilers (two ships - provision for Type 996 radar)

Alenia Marconi Systems is also known to have received a contract to provide a new track extractor (designated as Outfit LFE) for the Type 996 radar together with 'modifications' to the equipment that are designed to improve its maintainability. While neither confirmed nor denied by its manufacturer, Jane's sources suggest that, over time, the RN had procured at least 37 examples of the Type 996 surveillance radar.

Contractor

Alenia Marconi Systems (an Alenia-BAE Systems joint venture), Chelmsford/ Rome.

UPDATED

Type 1022 surveillance radar

Type

D-band (1 to 2 GHz) air and surface surveillance radar.

Description

The Type 1022 is a long-range D-band frequency synthesiser-driven Travelling Wave Tube (TWT) radar with pulse-to-pulse coherence. It was developed as a replacement for the Royal Navy's Type 965 surveillance radar. Long-range performance is aided by the use of a high mean-power TWT in the transmitter. A pulse-compression receiver with Moving Target Indicator (MTI) provides for good performance in rain and in the presence of surface clutter. The radar's antenna uses a squintless feed to provide tapered illumination of the main reflector. This results in good sidelobe performance across the band. Other technical features include: solid-state electronics; frequency agility over a wideband, horizontal and circular polarisation; frequency synthesiser controlled coherent transmitter; MTI with digital quadrature canceller and digital video processing. It is understood that the Type 1022 is a joint Anglo/Dutch project under which Signaal is providing the equipment's transceiver and the then GEC-Marconi (now BAE Systems) the antenna. An automatic radar track extraction unit has been supplied by Racal Defence Electronics.

Status

As of the period April to August 2001, Jane's sources were reporting the Type 1022 surveillance radar as being installed aboard Royal Navy 'Invincible' class aircraft carriers and Batch 1, 2 and 3 'Type 42' class destroyers. As of January 2002, it was thought likely (but not confirmed) that the Type 1022 radar was being supported by Alenia Marconi Systems.

The Type 1022 air surveillance radar (Raymond Cheung)

Contractor

Alenia Marconi Systems (an Alenia-BAE Systems joint venture), Chelmsford/ Rome.

UPDATED

VARIANT naval surveillance radar

Type

G- (4 to 6 GHz) and I-band (8 to 10 GHz) dual-frequency band, short- to medium-range air and surface target surveillance radar.

Description

VARIANT is a compact, lightweight, 2-D surveillance radar specifically intended for small ships, such as fast patrol boats, although it can also be fitted to larger vessels. The general characteristics of VARIANT are particularly suited for use with fast reaction, close-in and point defence weaponry. Excellent all-weather performance and anti-jamming features allow the system to operate in the most hostile detection environment.

The system operates in G and I-bands and will provide automatic air and surface target detection data to a command and control system. Surface target track data can be sent to a weapon system by virtue of VARIANT's track-while-scan channel.

VARIANT comes complete with a fully stabilised platform and integrated identification friend-or-foe antennas. In the low probability of intercept mode, VARIANT performs surface surveillance and gun fire support with a Frequency Modulated Continuous Wave (FMCW) radar component (Scout - see separate entry). The FMCW radar has an excellent surface target detection capability and exhibits high resistance to detection by electronic support measures receiver systems and anti-radiation missile seekers.

The antenna used in the VARIANT naval surveillance radar   0006904

VARIANT is based on a double pill-box antenna, having a horizontal beamwidth of 1.8° and a vertical beamwidth of 14°. The transmitter is a travelling wave tube with an average power of 200 W. Processing is based on digital pulse compression, Fast Fourier Transform processing, plot processing and automatic tracking. Detection range against a typical air target is given as about 30 km in G-band and 28 km in I-band.

Status
As of early 2001, Jane's sources were suggesting three customers for the VARIANT surveillance radar, a total that included Greece (four 'Batch 2 Pirpolitis' class large patrol craft that were building as of March 2001) and Indonesia (two (with two more building as of February 2001) 'Todak' class large patrol craft).

Specifications
Antenna: double pill-box
Beamwidth: elevation 14°; azimuth 1.8° (G-band)
Rotation speed: 14 and 28 rpm
Frequency: G- (4-6 GHz) and I-band (8-10 GHz)
Transmitter: TWT
Average power: 200 W
Operating modes: G-band only; I-band only; G- and I-bands; low probability of intercept
Pulse-lengths: 10 µs (short); 16 and 25 µs (long)
Processing: digital pulse compression; FFT Doppler processing; automatic tracking
Instrumented range: 60 km air; 70 km surface

Contractor
Thales Nederland, Hengelo, Netherlands.

VERIFIED

ZW series naval radars

Type
I-band (8 to 10 GHz) family of surveillance radars.

Description
Thales' ZW series of equipments is a family of general purpose search radars. Representative details of two such sensors - the ZW06 and the ZW07 - are given below.

ZW06
ZW06 is an I-band naval surface search and navigation radar that provides surveillance, navigation, helicopter control and limited air surveillance facilities. Of compact and lightweight construction, it is suitable for installation on very small vessels. The transmitter/receiver is of solid-state design (except for the magnetron) and the antenna is of stainless steel construction to permit siting in adverse environments.

Operational features include surface coverage to radar horizon; air coverage sufficient for helicopter guidance; a high resolution mode for navigation; provision for integration of helicopter transponder systems and a digital video processor. Anti-clutter measures and Electronic Counter-CounterMeasures (ECCM) provisions include circular polarisation, a logarithmic receiver with pulse-length discrimination, sensitivity time control, suppression of non-correlated pulses and a tunable transmitter.

Specifications
Frequency: I-band (8-10 GHz)
Polarisation: linear and circular, selectable
Transmitter: tunable magnetron
Power output: 60 kW
Max range: 22 km against small surface craft

ZW07
ZW07 is an I-band submarine radar that is designed to provide navigation, distance measuring, surface search and limited air warning facilities. A single shot mode has been adapted for ranging of surface targets. Other operational features include

surface coverage up to radar horizon, limited air warning and a high resolution mode for navigation. Anti-clutter measures and ECCM provisions include sector scan facilities, a logarithmic receiver with pulse-length discrimination, suppression of non-correlated pulses and a tunable transmitter.

Specifications
Frequency: I-band (8-10 GHz)
Polarisation: horizontal
Transmitter: tunable magnetron
Power output: 100 W (average); 60 kW (peak)
Max range: 22 km against small surface craft

Status
As of the period January to May 2001, Jane's sources were reporting the ZW06 and ZW07 radars as being in service aboard the following warships and classes of warship:

ZW06
Argentina	• 'Almirante Brown' class destroyers (four ships in class)
Brazil	• 'Niteroi' class frigates (six ships - ZW06 reported as being replaced by Decca TM 1226 as of January 2001)
Egypt	• 'Descubierta' class frigates (two ships)
Greece	• 'Elli' class frigates (six ships)
India	• 'Godavari' (three ships - two ZW06 or two Don Kay radars) and 'Nilgiri' (five ships) class frigates
South Korea	• the 'Ulsan' class frigates *Chung Nam*, *Kyong Buk*, *Masan*, *Seoul* and *Ulsan*
Morocco	• the frigate *Lieutenant Colonel Errhamani*
	• 'Lazaga' class fast attack craft - missile (four ships)
Netherlands	• 'Kortenaer' class frigates (three ships)
Spain	• the 'Descubierta' class corvettes *Cazadora*, *Infanta Cristina*, *Infanta Elena* and *Vencedora*
Taiwan	• 'Hai Lung' class anti-submarine warfare capable submarines (two boats)
Thailand	• the training ship *Makut Rajakumarn*
	• 'Rattanakosin' class corvettes (two ships)
	• 'Chon Buri' class fast attack craft - gun (three ships)

ZW07
Netherlands	• 'Walrus' class anti-submarine warfare capable submarines (four boats)

Contractor
Thales Nederland, Hengelo, Netherlands.

VERIFIED

ISRAEL

EL/M-2226 coastal surveillance radar

Type
I/low J-band (8 to 12 GHz sub-band) coastal surveillance radar.

Description
EL/M-2226 is a digital equipment that is optimised for the detection of small surface targets in adverse sea conditions. The system comprises an antenna assembly (which incorporates a pedestal/turntable), a main electronics package (including the radio frequency front-end, the transmitter, a receiver and a signal processor/system controller) and a remote, personal computer-based colour display/control unit. System features are noted as including optimal small target detection at medium and long ranges; automatic target detection and tracking (without operator intervention); high range resolution and range/azimuth accuracies; 24 hour operation; built-in test and optional interoperability with other sensors.

The antenna used with Thales ZW06 surface search and navigation radar

A general view of the EL/M-2226's tower-mounted radome   0024883

A close-up of the EL/M-2226's antenna assembly   0024884

Status

As of this issue, EL/M-2226 is understood to be available.

Specifications

Frequency: 8-12 GHz (other bands available)
Azimuth beamwidth: 1.5°
Elevation beamwidth: 2.6°
Detection range (Sea State 3): >20 km (rubber boat); >60 km (patrol craft)
Track/target capacity: >100

Contractor

Elta Electronics Industries Ltd (a subsidiary of Israel Aircraft Industries Ltd),
Ashdod.

VERIFIED

EL/M-2228S Automatic Missile Detection Radar (AMDR)

Type

E/F-band (2 to 4 GHz) missile detection and air/surface surveillance radar.

Description

EL/M2228S is a fully coherent E/F-band pulse-Doppler multimode shipboard search radar which was initially developed for the Israeli Navy. The radar is designed to detect sea-skimming missiles, generate automatic threat alerts and provide an automatic, track-while-scan (more than 100 targets simultaneously) air and surface surveillance capability. Elta claims a less than one per day false alarm rate for the system through the use of unique waveforms and signal/data processing techniques. Among these, the use of Doppler filtering with Discrete Fourier Transform and programmable signal/data processing, is noted as allowing the radar to filter out slow moving targets and measure target radial velocity during the course of a single scan, EL/M-2228S can be fitted with an identification friend-or-foe subsystem if required and can be supplied in any one of three system configurations:

The combined reflector/multibeam array antenna assembly used in the 3-D variant of Elta's EL/M-2228S AMDR radar

The antenna used in the 2-D variants of Elta's EL/M-2228S AMDR

- 2-D AMDR: utilises a single equipment rack and a lightweight reflector antenna to provide low/medium elevation coverage
- 2-D high power AMDR: utilises two equipment racks and a lightweight reflector antenna to provide long-range low/medium elevation coverage
- 3-D high power AMDR: provides 70° elevation coverage and height measurement via an integrated antenna system which incorporates the described reflector and a second back-to-back array.

A dual E/F- and I-band (8 to 10 GHz) feed can also be incorporated in the system to create a dual-function AMDR and surveillance/gunnery radar (see EL/M-2228X entry).

Status

Jane's sources suggest that 2-D variants of EL/M-2228S have been procured by the Chilean and Israeli navies and that the 3-D model has been installed on Israeli 'Sa'ar V' class corvettes.

Specifications

Frequency: E/F-band (2-4 GHz)
Transmitter: TWT or TWT + CFA
Range: 20 km (automatic threat alert of incoming missile); 70 km (fighter size air target); 100 km (instrumented); radar horizon (surface targets)
Scan rate: 12 or 24 rpm
Stabilisation: 20° roll and pitch
Antenna type/weight/location: cosec^2 reflector/237 kg/masthead (2-D version); reflector + multibeam array/550 kg/masthead (3-D version)

Contractor

Elta Electronics Industries Ltd (a subsidiary of Israel Aircraft Industries Ltd),
Ashdod.

VERIFIED

EL/M-2228X Surveillance and Gunnery Radar (SGR)

Type

I-band (8 to 10 GHz) air/surface surveillance and gunnery radar.

Description

EL/M-2228X is a lightweight fully coherent I-band pulse-Doppler multimode radar which is designed to perform accurate gunnery spotting while scanning for surface targets. The radar can be applied as a search and fire-control sensor aboard small- and medium-sized vessels while in larger applications it can be used to compliment long-range surveillance and fire-control emitters. EL/M-2228X has a 100 air/surface target track-while-scan capability and can be fitted with an identification friend-or-foe subsystem if required. It can also be configured to incorporate a dual E/F- (2 to 4 GHz) and I-band feed to create a dual-function surveillance/gunnery and AMDR radar (see EL/M-2228S entry).

Status

Jane's sources suggest that a variant designated as EL/M-2228X-1 has been procured for installation aboard Singapore's 'Fearless' class patrol craft.

Specifications

Frequency: I-band (8-10 GHz)
Transmitter: TWT
Range: 50 km (fighter size air target); 100 km (instrumented); radar horizon (surface targets)
Fall of shot measurement accuracy: 10 m (range); 2.5 mrad (bearing)

The antenna used in Elta's EL/M-2228X SGR radar

Scan rate: 12 or 24 rpm
Stabilisation: 20° roll and pitch
Antenna type/weight/location: cosec^2 (up to 30° in elevation) reflector/220 kg/masthead

Contractor
Elta Electronics Industries Ltd (a subsidiary of Israel Aircraft Industries Ltd), Ashdod.

VERIFIED

EL/M-2238 Surveillance and Threat Alert Radar (STAR)

Type
E/F-band (2 to 4 GHz) 3-D volume air surveillance and threat alert radar.

Description
The EL/M-2238 STAR radar is described as being a fully coherent, pulse-Doppler, 3-D multibeam/multimode equipment that is designed to provide 3-D volume air surveillance, surface surveillance and automatic air threat alerts aboard naval vessels ranging in size from corvettes to frigates. Elta claims that the STAR system offers fast detection rates, 'high' spatial resolution and accuracy and what it terms as an 'extremely' low false alarm rate. This latter feature is achieved via the use of 'unique' waveforms and data processing techniques. EL/M-2238 can be configured with medium or large size single-face or dual-face antenna arrays.

One of the antenna arrays used with the EL/M-2238 Star 3-D surveillance and alert radar   0044136

Here, antenna size dictates system range while the dual-face array option features two transmitters, provides twice the update rate of the single-face configuration and offers simultaneous threat alert and long-range detection-while-search functions. Other system features include:

- 3-D elevation coverage
- automatic track-while-scan for both air and surface targets
- identification friend-or-foe correlation
- programmable signal processing
- built-in test
- automatic designation to ship's systems
- multiple interface capabilities.

Status
During the summer of 1999, Jane's sources were reporting that the EL/M-2238 radar had been selected for retrofit aboard Venezuelan 'Lupo' class frigates.

Contractor
Elta Electronics Industries Ltd (a subsidiary of Israel Aircraft Industries Ltd), Ashdod.

VERIFIED

ITALY

ASX-1901 radar antenna

Type
I-/low J-band (8 to 12 GHz sub-band) radar antenna.

Description
The slotted waveguide array ASX-1901 antenna is designed primarily for land-based seaborne traffic and coastal surveillance applications and is available with horizontal or circular polarisation. The array is equipped with a 1,100 W drive motor and can be configured with a variable speed facility if required. GEM Elettronica notes that this last attribute allows the ASX-1901 antenna to be adapted to 'any' I-/low J-band radar transceiver and display system.

Status
As of August 2001, GEM Elettronica was reporting that it had developed the ASX-1901 antenna specifically for use in Lockheed Martin seaborne traffic surveillance systems and that such equipment had been procured by a number of customers around the world.

Specifications
Frequency: 9,260-9,560 MHz
Polarisation: horizontal or circular
Bearing discrimination: 0.4°
Vertical beamwidth: 18°
Gain: ≥35 dB
VSWR: ≤1.20
Beamwidth: ≤0.4° (horizontal); ≤18° (vertical)
Sidelobes: ≤−27 dB (within ±10°); ≤−30 dB (outside ±10°)
Rotation rate: 6, 12 or 22 rpm (0-22 rpm variable option)
Power output: up to 100 kW
Array length: 5.79 m

Contractor
GEM Elettronica Srl, San Benedetto del Tronto.

VERIFIED

CONDO-R naval radar

Type
I-band (8 to 10 GHz) long-range Over-The-Horizon (OTH) air and surface surveillance radar.

Description
The CONDO-R is an OTH surface radar surveillance system designed to exploit electromagnetic anomalous propagation conditions over the sea (ducting).

Specifications: EL/M-2238 Surveillance and Threat Alert Radar (STAR)

	Medium size, single face antenna	Large size, dual face antenna
Frequency	E/F-band (2-4 GHz)	E/F-band (2-4 GHz)
Stabilisation	20° (roll and pitch)	20° (roll and pitch)
Scan rate	12/24 rpm	6/15 rpm
Instrumented range	200 km	350 km
Weight	750 kg (above deck)	2,000 kg (below deck)
	1,300 kg (below deck)	2,400 kg (above deck)
Power consumption	21 kVA	34 kVA

The antenna unit of the CONDO-R OTH radar

It is based on a high-power transmitter and a high-gain antenna in order to ensure strong power density in the layers immediately above the sea surface. CONDO-R provides range performance far exceeding the horizon most of the time, not only against surface ship targets but also low-flying air targets.

The system features modern digital processing and automatic control of beam position inside super-propagation ducts. A CONDO-R variant configured for the coastal surveillance role is designated as the TPS-828 and a retrofit kit is available to upgrade existing I-band radar installation to CONDO-R standard.

Status

Over time, CONDO-R is reported to have been procured by the Italian Navy. A CONDO-R retrofit kit is understood to have been developed for the upgrading of existing I-band radar installations.

Contractor

Alenia Difesa Avionic Systems and Equipment Division - Officine Galileo, Florence.

VERIFIED

Leonardo LD-1500 and LD-1800 series navigation radars

Type

Family of S- (2 to 4 GHz) and X-band (8 to 12.5 GHz) navigation and surface surveillance radars.

Description

The Leonardo LD-1500 and LD-1800 series radars are described as being solid-state, digital equipments that are compliant with International Maritime Organisation (IMO) and ARPA regulations. Of the two, LD-1500 series radars are suitable for small vessel and search and rescue applications while the LD-1800 equipments are configured for use on larger merchant ships, ferries and warships. Both series use colour, flat-screen, liquid crystal display monitors that make use of super thin film technology, offer a greater than 160° field of view in azimuth and elevation and feature pull-down and pop-up menus and a trackball-controlled cursor. LD-1500 series radars incorporate a 38 cm (15 in) monitor (equivalent to a 43 cm (17 in) cathode ray tube display) while the LD-1800 equipments are fitted with a 46 cm (18 in) unit (250 mm PPI display). Both series provide mini or full ARPA facilities and can be fitted with a wide range of antennas. Here, the LD-1500 radars can make use of 1,420 to 2,300 mm arrays while LD-1800 units feature 3,900 mm S-band and 1,980 mm to 3,900 mm X-band antennas.

Status

As of August 2001, LD-1500 and LD-1800 series radars were reported as having been mandated for installation aboard 'several' classes of surface ship operated by Italy's navy, coastguard service and Guardia di Finanza and a number of other customers around the world.

Specifications
LD-1500 series
Antenna
Type: slotted waveguide
Polarisation: horizontal
Beamwidth: ≤1° (horizontal, 2,300 mm open antenna type, at -3 dB), ≤1.2° (horizontal, 1,980 mm open antenna type, at -3 dB); ≤1.9° (horizontal, 1,420 mm open antenna type, at -3 dB); 2.5° (vertical, all antenna types, at -3 dB)
Sidelobes: -25 dB (within ±10°, 2,300 mm open antenna type); -26 dB (within ±10°, 1,420 and 1,980 mm open antenna types); -30 dB (outside ±10°, all antenna types)
Gain: ≥27 dBi (1,420 mm open antenna type); ≥29 dBi (1,980 mm open antenna type); ≥30 dBi (2,300 mm open antenna type)
Rotation speed: 22 rpm (±2 rpm, optional 16, 22 or 40 rpm)
Wind load: 185 km/h (relative)
RF transceiver
Frequency: 9,410 MHz (±30 MHz)
Peak power: 4 or 12 kW (nominal)
Receiver type: solid-state logarithmic
Front end module: microwave integrated circuit
IF: 60 MHz
IF bandwidth: 4 MHz (long/extra long pulses); 8 MHz (medium pulse); 20 MHz (short/medium pulse)
Duplexer: ferrite circular with solid-state limiter diode
Noise figure: ≤4 dB (overall)
Display unit
Type: LCD; super TFT technology; high resolution; colour
Display size: 38 cm (15 in - 43 cm (17 in) CRT equivalent)
Video resolution: 1,024 × 768 pixels (16 levels)
Range discrimination: ±25 m
Range accuracy: ±6 m or less than 0.8% of the scale in use, whichever is the greater
Plot intervals: 15, 30 s, 1, 3 and 6 min
Azimuth discrimination: 0.1°
Bearing accuracy: ±0.5°

Specifications: Leonardo LD-1500 and LD-1800 series navigation radars
Range scale (km):

Pulse-width	0.1	0.2	0.5	0.9	1.4	2.8	5.6	11.1	22.2		44.5	88.9	133.3
80 ns	S	S	S	S	S	S							
300 ns					M	M	M	M	M				
600 ns							L	L	L		L		
1,200 ns											VL	VL	VL	VL
PRF	3,200 Hz				1,600 Hz		800 Hz				500 Hz		

Key
S Short **M** Medium **L** Long **VL** Very Long

LD-1800
Antenna

	X-band (λ = 3 cm)				S-band (λ = 10 cm)
Length	1,980 mm	2,300 mm	3,100 mm	3,900 mm	3,900 mm
Type	slotted waveguide				
Polarisation	H	H	H/C	H/C	V
Horizontal beamwidth*	=1.2°	≤1.1°	≤0.85°	≤0.65°	≤2°
Vertical beamwidth*	~25°	~25°	~25°	~25°	~23°
Sidelobes**	-26 dB	-25 dB	-24 dB	-26 dB	-26 dB
Sidelobes***	-30 dB	-30 dB	-30 dB	-30 dB	-30 dB
Gain	=29 dBi	=30 dBi	=31 dBi	=32 dBi	=28 dBi
Rotation speed	22 rpm (16 or 40 rpm options - selectable via keyboard control)				
Relative wind speed	185 km/h				

Key
* To -3 dB ** Within 10° *** Outside 10° **C** Circular **H** Horizontal **V** Vertical

Specifications: Leonardo LD-1500 and LD-1800 series navigation radars—*continued*

Transmitters

	X-band (λ = 3 cm)		S-band (λ = 10 cm)
Peak power*	10 kW	25 kW	30 kW
Transceiver location	aloft	aloft	below decks
Frequency	9,375 or 9,410 MHz		3,050 MHz
Duplexer	ferrite circulator, solid-state limiter diode		
Dynamic range	= 100 dB		
Min range	better than 15 m on 10 m² target with short pulse		
Discrimination	better than 25 m on 10 m² target with short pulse		

Pulse-length/PRF

	X-band (10 kW)		X-band (25 kW)		S-band (30 kW)	
	Length	*PRF*	*Length*	*PRF*	*Length*	*PRF*
Short pulse	0.08 µs	3,200 Hz	0.08 µs	2,000 Hz	0.08 µs	2,000 Hz
Medium pulse	0.30 µs	1,600 Hz	0.03 µs	1,000 Hz	0.03 µs	1,000 Hz
Long pulse	0.60 µs	800 Hz	0.80 µs	750 Hz	0.80 µs	750 Hz
Extra long pulse	1.20 µs	500 Hz	1.20 µs	500 Hz	1.20 µs	500 Hz

Display

Presentation	colour TV raster
Screen	46 cm (18 in) LCD (51 cm (20 in) standard CRT equivalent)
PPI size	= 270 mm
Screen resolution	1,024 × 1,280 dot/in
Pixel pitch	0.31 mm
Video levels	15
Range scales	0.2, 0.5, 0.9, 1.4, 2.8, 5.6, 11.1, 22.2, 44.5, 88.9 or 133.3 km (177.8 for 25 kW)
Max range PPI	more than 241 km
off centred	
Plot intervals	15, 30 s, 1, 3 and 6 min
Azimuth discrimination	0.1°
Bearing accuracy	±0.5°

Ranges/ring distances	*Scale*	*Ring distance*	*Rings*	*Pulse*
	0.2 km	0.05 km	5	short
	0.5 km	0.09 km	5	short
	0.9 km	0.18 km	5	short
	1.4 km	0.2 km	6	short/medium
	2.8 km	0.5 km	6	short/medium
	5.6 km	0.9 km	6	medium/long
	11.1 km	1.9 km	6	medium/long
	22.2 km	3.7 km	6	long/extra long
	44.5 km	7.4 km	6	long/extra long
	88.9 km	14.8 km	6	extra long
	133.3 km (10 kW)	22.2 km (10 kW)	6	extra long
	177.8 km (25 kW)	29.6 km (25 kW)	6	extra long

General

Temperature: -15 to +55°C (display); -25 to +70°C (scanner)

Power: 18-40 V DC (≤80 W - 4 kW scanner; ≤100 W - 12 kW scanner)

Dimensions (h × w × d): 457 × 400 × 283 mm (display); 450 × 1,420/1,980/2,300 × 390 mm (antenna)

Weight: 17 kg (display); 31 kg (1,420 mm antenna); 33 kg (1,980 mm antenna); 34 kg (2,300 mm antenna)

Presentation mode: course/head/north up (relative); course/north up (true)

Receiver

Type: solid-state logarithmic

IF: 60 MHz

IF bandwidth: 4 MHz (long/extra long pulse); 20 MHz (short/medium pulse)

Noise figure: ≤6 dB

General

Power consumption: <400 W

Weight: 9 kg (X-band 1,980 mm antenna); 10 kg (X-band 2,300 mm antenna); 12 kg (X-band 3,100 mm antenna); 18 kg (interswitch unit); 24 kg (X-band 5 kW and 10 kW scanner unit and display unit); 25 kg (X-band 5, 10 and 25 kW and S-band 10 kW downmast transceiver); 26 kg (X-band 3,900 mm antenna); 32 kg (X-band 25 kW scanner unit); 40 kg (S-band 3,900 mm antenna); 59 kg (X-band 50 kW and S-band 30 kW and 50 kW downmast transceiver); 140 kg (S-band and 50 kW X-band rotation unit)

Contractor

GEM Elettronica Srl, San Benedetto del Tronto.

UPDATED

MM950 navigation radars have been acquired for retrofit aboard 'Alkmaar' class minehunters of the Royal Netherlands Navy   0053499

MM950 series navigation radars

Type

Family of F-band (3,050 MHz centre frequency) and/or I-band (9,375 MHz centre frequency) surface ship navigation radars.

Description

MM950 series navigation radars are automatic target tracking equipments that incorporate rasterscan displays, metal oxide silicon field-effect transistor transceivers and end-fed, slotted waveguide antenna arrays. Operational features include:

- digital manipulation of the display memory to create long duration trails and/or RM true trails
- fixed (10.4 km range) or variable (0.6 to 44.5 km range) automatic acquisition, four sector guard rings
- numerical target track identification

- trackball initiated video mapping (up to 1,800 segments held in non-volatile memories)
- operator adjustable trial manoeuvre display mode
- anchor watch mode
- drift speed calculation (with reference to a selected fixed point)
- on- and off-line diagnostics.

As of this edition, the MM950 family comprises the original MM950 unit and the MM950/A system that features upgraded electronics and software.

Status

MM950 was first introduced during 1994 with the MM950/A variant entering production during 1998. As of this edition, MM950 radars were reported as having been procured by the navies of Finland, India, the Netherlands and Spain together with the Finnish Coast Guard service. Of these various applications, the Finnish systems are described as being dual display I-band equipments that interface with the I-band (8 to 10 GHz), low probability-of-intercept Signaal Scout surveillance/navigation radar. The Indian equipment (installed aboard ten as yet unidentified 'combatants') is described as being a 'modified', dual-display F- and I-band sensor while the Netherlands variant is another dual-display I-band application. Here, 15 radars are understood to have been delivery during 1999 for retrofit aboard that country's 'Alkmaar' class minehunters. The Spanish order involves five single-display I-band radars for installation aboard five 'Angara' class large patrol craft. Delivery of these latter systems was completed during the second half of 1998.

Specifications

Transmit frequency: 3,050 MHz (F-band); 9,375 MHz (I-band)
Peak power: 25 kW (I-band); 30 kW (F-band)
Intermediate frequency: 60 MHz
PRF: 750, 1,500 and 3,000 Hz
IF bandwidth: 1.5, 4.8 and 20 MHz
Pulse-lengths: 60, 250 and 800 ns
Noise figure: W 5 dB

Contractor

Consilium Selesmar Srl, Florence.

VERIFIED

MM/SPN-753/Gemant navigation radar

Type

E/F- (2 to 4 GHz) or I/low J-band (8 to 12 GHz) navigation and surface surveillance radar.

Description

SPN-753 is a ruggedised, military ARPA (Automatic Radar Plotting Aid) radar which can be configured for E/F- or I/low J-band operation and is designed to replace MM/SPN-748 in Italian Navy service. Two versions of the equipment are available, namely SPN-753(V)1 and SPN-753(V)2 with the iterations differing from one another in that SPN-753(V)1 features a data extraction computer for use in target acquisition and tracking. Both variants employ a standard 508 mm (diagonal) high-resolution raster scan display. SPN-753 technology is also used in a second radar designated as Gemant.

Status

As of February 2001, Jane's sources were reporting that SPN-753 and Gemant-type radars were in service aboard the following ships and ship classes:

Greece • 'Etna' class replenishment oiler (one building as of February 2001 - SPN-753 radar)
Italy • the replenishment oiler *Etna* (Italian Navy - SPN-753 radar)
 • the trials ship *Saettia* (Italian Guardia Costiera (Coast Guard) - SPN-753XS(V)2 radar)
 • 'MCC 1101' class water tankers (Italian Navy, four ships in class - SPN-753 radar)
 • 'Antonio Zara' (Italian Guardia di Finanza (Customs Service), three ships - GEM 'ARPA' radar), 'Bigliani (Guardia di Finanza, 12 ships, eight building as of February 2001 - GEM 'ARPA' radar) and 'Corrubia' (Guardia di Finanza, 26 ships - GEM 'ARPA' radar) class patrol ships

Specifications

Frequency: E/F-band (3,050 MHz); I/low J-band (9,410 MHz)
Peak power: 10, 25 or 50 kW
Horizontal beamwidth: 1.1° (2.3 m antenna, I/low J-band operation); 2° (3.7 m antenna, E/F-band operation)
Vertical beamwidth: 22° (E/F-band); 23° (I/low J-band)
Antenna rotation speed: 22 rpm

Contractor

GEM Elettronica Srl, San Benedetto del Tronto.

UPDATED

RAN 3L air search radar

Type

D-band (1 to 2 GHz) long-range air search radar. Italian Navy designation MM/SPS-768.

Description

The RAN 3L is a high-power early warning search radar operating in D-band and suitable for installation on large and medium tonnage vessels. The radar is of advanced design and combines long range with high precision and contemporary electronic counter-countermeasures and anti-clutter facilities.

The system consists of a fully coherent transmitter/receiver chain and a digital processor incorporating a composite coded waveform. It is fully solid-state except for the final radio frequency power amplifier. An identification friend-or-foe antenna with interrogation sidelobe suppression can be mounted on the reflector.

Status

Over time, RAN 3L/SPS-768 radars are understood to have been installed aboard a range of Italian naval vessels including the aircraft carrier *Giuseppe Garibaldi*, the cruiser *Vittorio Veneto* and the four ships making up the 'Audace' class (two) and 'De La Penne' class (two) destroyers. As of this edition, it was thought likely that operational radars of this type were being supported by the Alenia Marconi Systems joint venture.

Specifications

Frequency: D-band (1-2 GHz)
Antenna: double curvature parabolic reflector with hog horn feed
Rotation rate: 6 rpm
Polarisation: linear
Transmitter: frequency agile, frequency diversity and fixed frequency; multipulse coded radiation
Receiver: double frequency conversion, coherent limiting, dual-channel MTI processing, code matched filtering, dual threshold and azimuth correlation
Range: > 204 km on a fighter aircraft
Range accuracy: 70 m
Angular accuracy: 0.4°

VERIFIED

The antenna used in the RAN 3L surveillance radar system

RAN 10S air/surface surveillance radar

Type

F-band (3 to 4 GHz) air and surface surveillance radar. Italian Navy designation MM/SPS-774.

Description

The RAN 10S is a high-power combined air/surface surveillance radar operating in the F-band and suitable for installation on medium tonnage ships, such as destroyers, frigates and corvettes. The design philosophy is the same as that of the RAN 3L and a number of items are common to both systems.

The main features are high elevation coverage, elevated data rate and high precision/resolution. The use of a coded waveform in conjunction with digital processing of the received signal gives excellent capabilities of clutter and jamming rejection, enabling it to operate in full frequency agility while maintaining anti-clutter capabilities.

The RAN 10S antenna on the Italian Navy frigate Fenice *(Stefan Terzibaschitsch)*

The antenna is roll and pitch stabilised and can accommodate an identification friend-or-foe antenna with integrated sidelobe suppression, mounted on top of the reflector.

Typical operating roles include air warning against aircraft and missiles, direction for both fixed- and rotary-wing aircraft, surface surveillance, navigation and direction of surface-to-surface missiles. The radar is optimised for target designation to guns and/or surface-to-air-missile weapon control systems.

Status

Over time, Ran 10S/SPS-774 radars are understood to have been installed aboard a range of Italian naval vessels including the aircraft carrier *Giuseppe Garibaldi*, 'Audace' class (two) and 'De La Penne' class (two) destroyers, 'Maestrale' class (eight) and 'Lupo' class (four) frigates, 'Minerva' class (eight) corvettes and 'Artigliere' class (two) fleet patrol vessels. Outside Italy, Ecuador's six 'Esmeraldas' class corvettes, the Libyan frigate *Dat Assawari*, Peru's 'Meliton Carvajal' class (four) frigates and Venezuela's 'Modified Lupo' class (six) frigates are all thought to have been fitted with this type of radar at some time in their various service lives. As of this edition, it was thought likely that operational RAN 10S equipments were being supported by the Alenia Marconi Systems joint venture.

Specifications

Frequency: F-band (3-4 GHz)
Antenna: roll and pitch stabilised
Antenna rotation: 15 and 30 rpm
Polarisation: linear and circular
Transmitter: frequency agile, fixed frequency, multipulse coded radiation
Receiver: double frequency conversion, hard limiting, dual-channel Doppler processing, code matched filtering, dual threshold azimuth correlation
Range: more than 93 km against a fighter size air target
Range accuracy: 20 m
Angular accuracy: 0.35°

VERIFIED

RAN 11/12 L/X naval radars

Type

D- (1 to 2 GHz) and I-band (8 to 10 GHz) air surveillance and detection radars. Italian Navy designation for RAN 12L/X, MM/SPQ-712.

Description

The RAN 11/12 L/X is a compact integrated radar system which includes an I-band transmitter and receiver for detection of surface and low-flying targets, and a D-band transmitter and receiver for air surveillance. The RAN 12 L/X is the higher power version of the RAN 11 L/X, having a high-power transmitter in the D-band to obtain better air cover. The two individual radars comprising the system can also be supplied as single equipments. (See also RAN 30X entry).

In either of its versions the radar is suitable for installation as the main radar on corvettes, hydrofoils and fast patrol boats, or as a secondary radar in larger vessels.

The D-band transceiver is a modern pulse-Doppler radar executed in solid-state technology which includes final power amplification. This results in compactness combined with no life-limited electronics. The D-band radar incorporates fully coherent processing with high pulse repetition frequency, multipole moving target indicator filtering and pulse compression implemented by digitised hardware. This results in exceptional subclutter visibility together with constant false alarm rate characteristics. An associated identification friend-or-foe system uses the D-band channel.

Status

RAN 11/12 series radars are reported to have entered production during 1995 and are noted as having been installed aboard a range of ships including Italian Navy 'Artigliere' class fleet patrol ships (RAN 12L/X), Iraqi 'Assad' class corvettes (RAN 12L/X), the Libyan frigate *Dat Assawari* (RAN 12), Spanish 'Baleras' class frigates (RAN 12L), Peruvian 'Meliton Carvajal' class frigates (RAN 11L/X) and the Spanish

The RAN 11/12 L/X combined search radar

aircraft carrier *Principe de Asturias* (RAN 12L). Jane's sources also suggest that RAN 12L/X has been acquired by Malaysia (two radars). As of this edition, it was thought likely that operational RAN 11/12 series radars were being supported by the Alenia Marconi Systems joint venture.

Specifications

Frequency: D- (1-2 GHz) and I-band (8-10 GHz)
Antenna: roll and pitch stabilised
Rotation rate: 15 and 30 rpm
Polarisation: linear and circular
I-band transmitter: frequency tunable
I-band receiver: image suppression, mixer, log receiver, Dicke-fix, autogate, automix
D-band transmitter: frequency diversity, waveform coding
D-band receiver: double frequency conversion, coherent limiting, dual-channel Doppler processing, code matched filtering, azimuth correlation.
Range (on a fighter size air target): low altitude (I-band) more than 37 km; medium altitude (D-band RAN 11L) more than 22 km; medium altitude (D-band RAN 12L) more than 26 km
Range accuracy: 45 m (I-band); 160 m (D-band)
Angular accuracy: 0.2° (I-band); 1.5° (D-band)

VERIFIED

RT02 series radar transceivers

Type

Family of E/F- (2 to 4 GHz) and I-/low J-band (8 to 12 GHz sub-band) radar transceivers.

Description

The RT02 series of radar transceivers are designed primarily for land-based seaborne traffic and coastal surveillance applications. System features include:

- internal trigger generation
- PRF random stagger
- blanking on up to 10 angular sectors
- a 'very low' noise figure
- system control via serial link if required
- built-in test (fault detection to 'lowest replaceable item' level)
- optional dual-redundant configuration (one operating, one on 'hot' standby).

Status

As of August 2001, RT02 radar transceivers were reported as having been procured for a number of seaborne traffic and coastal surveillance applications around the world (including customers in Albania, Argentina, China, Georgia and Qatar) and as having been selected for use in a number of forthcoming systems. As of February 2000, GEM Elettronica also noted that it was supplying RT02 transceivers both as prime contractor and as a subcontractor to other manufacturers with, in the latter case, a particular emphasis on collaboration with Lockheed Martin.

Specifications

Frequency: 3,050 or 9,375 MHz
Peak power: 10 kW (E/F- and I-/low J-band); 25 kW (I-/low J-band); 30 kW (E/F-band) or 50 kW (E/F- and I-/low J-band)
Dynamic range: more than 100 dB

Noise figure: ≤3.5 dB
Pulse-width: 4 (operator selectable via serial link)
PRF: 4 (operator selectable via serial link)

Contractor
GEM Elettronica Srl, San Benedetto del Tronto.

VERIFIED

SC-1000 series navigation radars

Type
I/low J-band (8 to 12 GHz) navigation and surface surveillance radar.

Description
The SC-1000 family is a series of digital, solid-state radars which are designed for ship navigation and surface surveillance. The equipments can be fitted with three sizes (254, 305 and 381 mm, measured on the diagonal) of high-resolution, raster scan displays and utilise either a 0.55 m radome enclosed antenna or 1.2 or 1.8 m open air arrays.

Status
SC-1000 series radars are reported as being the primary surface search and navigation radars aboard the Italian Guardia di Finanza's (Customs Service) fleet of coastal patrol vessels and have been supplied to other 'waterborne law enforcement agencies'.

Specifications
Frequency: I/low J-band (9,410 MHz)
Peak power: 5 or 10 kW
Horizontal beamwidth: 1.2° (1.8 m antenna); 1.9° (1.2 m antenna); 4° (radome)
Vertical beamwidth: 23°
Antenna rotation speed: 22 rpm

Contractor
GEM Elettronica Srl, San Benedetto del Tronto.

VERIFIED

An Italian Guardia di Finanza 'Corrubia' class fast patrol boat equipped with GEM Elettronica's 25 kW Gemant (upper antenna) and 10 kW SC-1210 (lower antenna) radars

TPS-755 coastal surveillance radar

Type
Long-range over-the-horizon air and surface surveillance radar.

Description
The TPS-755 is a mobile air and surface surveillance radar which is deployed near shore lines for coastal protection. It is contained in two standard shelters which can be towed easily by trucks and can be airlifted by transport aircraft. Power is provided by a diesel generator, with a battery back-up unit also included. A number of TPS-755 systems can be linked together in a coastal defence network which could include more radar stations and a Command, Control and Communications (C³) system with an operations centre.

The TPS-755 can carry out tracking either automatically or manually, and is capable of local target extraction. Track information can be exchanged with other stations or the C³ centre via a modem system, using standard protocols.

The TPS-755 coastal surveillance radar   0006902

Status
Over time, TPS-755 radars are reported to have been procured by the Italian Ministry of Defence.

Contractor
Alenia Difesa Avionic Systems and Equipment Division - Officine Galileo, Florence.

VERIFIED

JAPAN

OPS series shipborne surveillance and navigation radars

Description
The Japanese Maritime Self-Defence Force (JMSDF) employs a number of air/surface surveillance and navigation radars which have been developed locally from US originals and/or indigenously designed. Examples of such systems are the OPS-9, OPS-11, OPS-12, OPS-14, and OPS-24 air surveillance radars; the OPS-16, OPS-17, OPS-18, OPS-28 and OPS-39 surface search equipments and the OPS-19 and OPS-20 navigation radars. Reported details of these equipments are as follows:

OPS-9
Reported to be an I-band (8 to 10 GHz) surface search radar produced by Fujitsu. Details unconfirmed.
OPS-11
OPS-11 is an air search radar developed by the Mitsubishi Electric Corporation (MELCO). Thought to exist in at least three variants.
OPS-12
Reported to be a D-band 3-D air search radar with a maximum detection range of 119 km. Noted as being developed by the NEC Corporation. Details unconfirmed.
OPS-14
Medium-range air search radar developed by MELCO which is reported to have entered service with the JMSDF during 1971. OPS-14 has been developed in A, B and C models.
OPS-16
Reported to be a D-band surface search radar developed by the Japan Radio Company (JRC). Details unconfirmed.
OPS-17
Reported to be a G/H-band (4 to 8 GHz) surface search radar developed by JRC. Details unconfirmed.

Superstructure of the Japanese destroyer Asagiri showing the OPS-14C air search radar and the OPS-28C surface radar above (W Donko)

Class	Role	Surface Search	Radar Type Air Search	Navigation
'Abukuma'	FFG[1]	OPS-28	OPS-14C	
'Amatsukaza'	DD	OPS-17		
'Asagiri'	DD	OPS-28C	OPS-14C (4 ships)/OPS-24 (4 ships)	
'Chikugo'	FF[1]	OPS-16	OPS-14	
'Haruna'	DD	OPS-28	OPS-11C	
'Hatakaze'	DDG	OPS-28B	OPS-11C	
'Hatsushima/Uwajima'	MHC/MSC[2]	OPS-9 (24 ships)/OPS-39 (7 ships)		
'Hatsuyuki'	DDG	OPS-18	OPS-14B	
'Hayase'	MST/ML	OPS-16	OPS-14	
'Ishikari'	FFG	OPS-28		OPS-19B
'Kongo'	DDG	OPS-28C or D		OPS-20
'Minegumo'	DD	OPS-17	OPS-11B	
'Murasame'	DD	OPS-28D	OPS-24B	
'Shirane'	DD	OPS-28	OPS-12	
'Souya'	MST/ML	OPS-16	OPS-14	
'Tachikaze'	DDG	OPS-16 (2 ships)/OPS-28 (1 ship)	OPS-11C	
'Takatsuki'	DDG	OPS-17	OPS-12	
'Yaeyama'	MSO	OPS-39		
'Yamagumo'	DD	OPS-17	OPS-11	
'Yubari'	FFG	OPS-28C		OPS-19B

Key

[1]The JMSDF is reported to term its FF/FFGs as 'destroyer escorts' [2]Joint class DD Destroyer **DDG** Guided missile destroyer FF Frigate **FFG** Guided missile frigates **MHC** Coastal minehunter **MSC** Coastal minesweeper **MSO** Ocean-going minehunter/sweeper **MST/ML** Mine countermeasure support ship/minelayer.

OPS-18
Reported to be a G/H-band surface search radar which is based on the American AN/SPS-10 equipment and has been developed by JRC. Details unconfirmed.

OPS-19
Reported to be an I-band navigation radar developed by Fujitsu in at least two versions. Details unconfirmed.

OPS-20
Reported to be an I-band navigation radar developed by JRC. Details unconfirmed.

OPS-24
A D-band 3-D air search radar developed by MELCO which incorporates transceiver modules which utilise silicon hybrid integrated circuit and discrete transistor technology. These units are understood to have a peak power of 90 W, 40 and 28 dB transmit and receive gain values; a 3.2 dB noise figure and a 5-bit phase shifter which allows for 11.25° phase shift stages during beam steering. Each 1.23 kg transceiver module is noted as measuring 126 × 253 × 40 mm and the quoted 90 W peak power value points to the radar being equipped with a hybrid active array in which individual transceiver modules feed groups of antenna elements.

OPS-28
Reported to be a G/H-band surface search radar which first entered service during 1980 and which has been developed by JRC. Produced in at least three versions. Details unconfirmed.

OPS-39
Reported to be an I-band surface search radar developed by Fujitsu. Details unconfirmed.

Status
Over time, OPS series radars are reported to have been installed aboard JMSDF vessels including the major surface combatants and mine warfare ships outlined in the accompanying table.

Contractors
Fujitsu Ltd, Tokyo (OPS-9, -19 and -39).
Japan Radio Company, Tokyo (OPS-16, -17, -18, -20 and -28).
Mitsubishi Electric Corporation, Tokyo (OPS-11, -14 and -24).
NEC Corporation, Tokyo (OPS-12).

VERIFIED

KOREA, SOUTH

SPS-95K surface search/navigation radar

Type
G-band (4 to 6 GHz) surface search and navigation radar.

Description
SPS-95K is a lightweight, low-volume surface search and navigation radar suitable for installation aboard destroyers, frigates and corvettes. The equipment comprises an antenna, transceiver and a remote-control unit and is described as making extensive use of solid-state technology. Electronic counter-

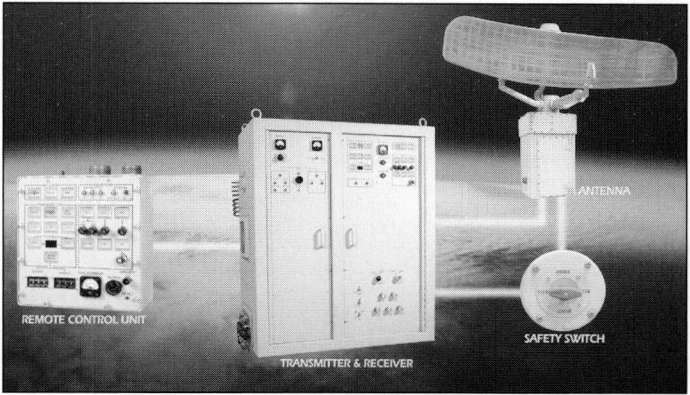

The antenna, transceiver and remote control unit used in the Daewoo SPS-95K surface search and navigation radar

countermeasures provision includes pulse repetition frequency jitter, sector radiation, sensitivity time control, fast time constant, automatic/manual gain and frequency control, pulse interference suppression and constant false alarm rate. Available interfaces comprise video (four outputs), indicator trigger (two), Identification Friend-or-Foe (IFF) trigger (two outputs - SPS-95K incorporates an integral IFF capability), blanking trigger (two), antenna turning signal and gyro.

Status
As of this edition, Jane's sources suggest that SPS-95K may be the navigation radar fitted to South Korean 'Okpo' class frigates.

Specifications
Frequency: G-band (5.45 - 5.825 GHz)
Peak power: 250 kW (nominal)
Pulse-width: 0.12 μs ± 0.01 μs (short); 0.25 μs ± 0.02 μs (medium); 1 μs ± 0.1 μs (long)
PRF: 2,400 Hz ± 1% (short); 1,200 Hz ± 1% (medium); 750 Hz ± 1% (long)
Sector width: 0-360° in 1° steps
Antenna beamwidth: 1.5°
Antenna gain: 30 dB (min)
Sidelobe level: -25 dB (max)
Receiver noise: 5.5 dB (max)
Receiver bandwidth: 11 MHz (short); 2.6 MHz (long)
Power supply: 1.6 kW (transceiver); 1.2 kW (antenna)
MTBF: 600 h (min)
MTTR: 45 min (max)
Instrumented range: 200 km
Free space detection range: 42.6 km (2 m² target)

Contractor
Daewoo Telecom Ltd, Seoul.

VERIFIED

POLAND

CRM-100

Type
I-band (9.3 to 9.5 GHz sub-band) surface surveillance radar.

Description
CRM-100 is described as being a 'quiet', solid-state, frequency modulated continuous wave radar that is designed to detect surface targets, determine their co-ordinates and automatically hand-off tracking data (target number, range (from own position), bearing (from own position), course and speed) to an associated command system. CRM-100's manufacturer claims that it is a low probability of intercept sensor with low power consumption and output values and the ability to 'match' the range coverage provided by 'standard' pulse navigation radars. Additionally, the equipment is noted as being suitable for ground mobile as well as shipboard installations.

Status
As of May 2001, the CRM-100 surface surveillance radar was understood to be available.

Specifications
Carrier frequency: 10 switched frequencies within the 9.3-9.5 GHz range
Frequency deviation: 54 MHz (max)
Output power: 1 mW-1 W (switched according to instrumental range)
Instrumental range: 1.4; 2.8; 11.1; 22.2; 44.5 km
Modulation period: 1 ms
Receiver noise factor: 3 dB
Indicator: 51 cm (20 in) colour
Antenna aperture: 1.4 m
Antenna beamwidth: 1.8° (horizontal, 3 dB); 25° (vertical, 3 dB)
Antenna rotation rate: 30 (±3) rpm
Angular resolution: 0.1°
Bearing accuracy: 1°
Track capacity: up to 40 manually or automatically acquired targets
Tracked target vectors: 1-30 min (adjustable)
Range measurement

Range	1.4 km	3 km	5.6 km	11.1 km	22.2 km	44.5 km
Range cell size	3 m	6 m	12 m	24 m	48 m	96 m
Range resolution	9 m	18 m	36 m	72 m	144 m	288 m

Contractor
Przemysłowy Instytut Telekomunikacji, Warsaw.

NEW ENTRY

RUSSIAN FEDERATION AND ASSOCIATED STATES (CIS)

Introduction

This section deals with those naval surveillance and navigation radars which are produced by, exported by and/or are in service with the navies of the 21 republics which make up the Russian Federation together with the independent states of Armenia, Azerbaijan, Belarus, Georgia, Kazakhstan, Kyrgyzstan, Moldova, Tajikistan, Turkmenistan, Ukraine and Uzbekistan. Taken together, these various republics make up the Commonwealth of Independent States (CIS) economic grouping. Despite increasing openness, many of the entries presented here must be considered incomplete and will be the subject of future revision. Entries primarily reflect equipments which are thought to be currently (2001/2002) operational with the Russian Navy.

VERIFIED

Fregat series surveillance radars

Type
Family of multifunction 3-D naval surveillance radars (NATO reporting name Top Plate).

Description
As of early to mid-2000, a total of five members of the Fregat family of multifunction 3-D naval surveillance radars has been identified, the known details of which are as follows:

Fregat-M2EM
The two channel, E-band (2 to 3 GHz) Fregat-M2EM radar features back-to-back slotted array antennas and adds a higher power transmitter to the Fregat-MAE-3 architecture (see following). Like Fregat-MAE-3, Fregat-M2EM is manufactured in subvariants that have wavelengths of 12 and 15 cm respectively.

Fregat-MAE
The single channel, E-band (2 to 3 GHz) Fregat-MAE radar is described as being the baseline system for the Fregat series and as being manufactured in two subvariants which have wavelengths of 12 and 15 cm respectively.

Fregat-MAE-1
The single channel, E-band (2 to 3 GHz) Fregat-MAE-1 is a variant of Fregat-MAE that weighs less and features electronic beam stabilisation.

Fregat-MAE-3
Fregat-MAE-3 is the baseline two channel, E-band (2 to 3 GHz) Fregat system. As such, it features back-to-back slotted array antennas and is manufactured in subvariants that have wavelengths of 12 and 15 cm respectively.

Fregat-MAE-4K
Fregat-MAE-4K is a lightweight, single channel, H-band (6 to 8 GHz, 4 cm wavelength) variant of the Fregat-MAE-1 system.

'Salyut' notes that the performance of the described Fregat radar variants is optimised by their integration with the company's Poima-E radar data processing system. Configurable for use aboard small, medium and large size ships, Poima-E incorporates initial data input operator and crew commander workstations, each of which is based on an IBM personal computer compatible, open architecture computer equipped with a 53 cm (21 in) high-resolution monitor. System functions include:

- automatic target detection, acquisition, tracking and co-ordinate measurement
- target classification
- primary and secondary data display
- automatic generation of a target allocation plan
- transmission of target tracking data to the host vessel's tactical data and weapon direction systems
- data logging.

The Top Plate radar antenna carried by the 'Udaloy' class destroyer Severomorsk (G Arra)

Specifications: Fregat series surveillance radars

Single-channel variants

Variant	MAE	MAE-1	MAE-4K
Frequency	E-band (2-3 GHz)	E-band (2-3 GHz)	H-band (6-8 GHz)
Wavelength	12 or 15 cm	12 or 15 cm	4 cm
Number of channels	1	1	1
Measured co-ordinates	3	3	3
Coverage	2 km (min range); 30 km (altitude); 45 or 55° (elevation); 150 km (range)	2 km (min range); 30° (elevation); 30 km (altitude); 300 km (range)	1.5 km (min range); 10 km (altitude); 40° (elevation); 150 km (range)
Detection range	27 or 30 km (missile); 125 or 130 (fighter); radar horizon (ship)	27 km (missile); 125 km (fighter); radar horizon (ship)	17 km (missile); 58 km (fighter); radar horizon (ship)
Accuracy	24 min (azimuth); 26 or 40 min (elevation); 120 m (range)	24 min (azimuth); 43 min (elevation); 120 m (range)	14 min (azimuth); 18 min (elevation); 120 m (range)
Track capacity			8
Vision rate	4 s (max)	4 s (max)	2 s (max)
Antenna rotation speed	15 rpm	15 rpm	30 rpm
Number of component units	8	9	9
Area	16 m²	16 m²	20 m²
Power consumption	30 kW	30 kW	30 kW
Weight	2.2 t (antenna); 2.9 t (below decks)	1 t (antenna); 3.1 t (below decks)	0.4 t (antenna); 2.6 t (below decks)

Twin-channel variants

Variant	MAE-3	M2EM
Frequency	E-band (2-3 GHz)	E-band (2-3 GHz)
Wavelength	12 or 15 cm	12 or 15 cm
Number of channels	2	2
Measured co-ordinates	3	3
Coverage	2 km (min range); 30 km (altitude); 55° (elevation); 300 km (range)	2 km (min range); 30 km (altitude); 55° (elevation); 300 km (range)
Detection range	38 km (missile); 180 km (fighter); radar horizon (ship)	50 km (missile); 230 km (fighter); radar horizon (ship)
Accuracy	24 min (azimuth); 30 min (elevation); 120 m (range)	24 min (azimuth); 30 min (elevation); 120 m (range)
Vision rate	2.5 s (max)	2.5 s (max)
Antenna rotation speed	6 or 12 rpm	6 or 12 rpm
Number of component units	15	21
Area	34 m²	48 m²
Power consumption	45 kW	90 kW
Weight	2.5 t (antenna); 6.6 t (below decks)	2.5 t (antenna); 9.6 t (below decks)

Poima-E
Number of channels: 1-3
Track capacity: ≥20
Number of component units: 2-9
Power consumption: 1-4.5 kW
Weight: 300-3,000 kg

Status

As of early to mid-2000, Fregat series radars were reported as being installed aboard Chinese 'Sovremenny' class destroyers; Indian 'Talwar' class frigates; the Russian aircraft carrier *Admiral Kuznetsov*, the destroyer *Admiral Chabanenko*, the frigates *Neustrashimy* and *Novik*, the landing ship *Mitrofan Moskalenko*, the missile range ship *Marshal Krylov* and the same country's Type 956/956A ('Sovremenny/ Sarych' class) destroyers, Type 1135MP ('Krivak III/Nerey' class) frigates (some ships), Type 1144.1/1144.2 ('Kirov/Orlan' class) battle cruisers, Type 1155 ('Udaloy/Fregat' class) destroyers and Type 1164 ('Slava/Atlant' class) cruisers, the Ukrainian frigate *Hetman Sagaidachny* and the proposed Vietnamese Type 2100 (KBO 2000) corvette.

Contractor

State Unitary Enterprise - State Moscow Plant 'Salyut', Moscow.

UPDATED

Gamma-PV

Type

Automatic fixed-site coastguard radar.

Description

The Gamma-PV automatic fixed-site radar is designed for border and coastal surveillance duties together with seaway traffic control 'in the interest of' customs services. System features include:

- automatic target detection in up to Sea States 4
- moving target course and velocity determination
- automatic target tracking in direction and range
- target type identification
- the ability to operate as part of an automated surveillance system.

Status

The Gamma-PV automatic fixed-site radar was reported as being available during the first half of 2001.

Specifications
Detection range: 4.8 km (sea - boat); 11.3 km (sea - cutter, land - person); 22.5 km (sea - medium size ship); 32.2 km (sea - tanker); 40.2 km (land - jeep)
Accuracy: 1° (azimuth); 50 m (range)
Sector coverage: 45, 90, 180 or 360°

Contractor
STRELA Scientific Research Institute, Tula.

UPDATED

MR-212/201 Nyada surface search radar

Type
I/low J-band (8 to 12 GHz sub-band) surface search radar.

Description
MR-212/201 is designed to monitor the close-in surface situation and aid navigation. It can hand-off data to other onboard systems and is described as being applicable to aircraft carriers, missile- and gun-armed surface combatants and anti-submarine warfare ships.

Status
As of June 2001, the precise status of the MR-212/201 search radar was uncertain.

Specifications
Frequency: 8-12 GHz
Bearing/range accuracy: 0.8°/40 m
Bearing/range resolution: 1.2°/25 m
Power consumption: 12 kVA
Weight: 0.8-2.1 t (depending on installation)

Contractor
Rosoboronexport (export agent), Moscow.

UPDATED

MR-300 Angara air and surface surveillance radar

Type
E/F-band (2 to 4 GHz) long-range air and surface surveillance radar. NATO reporting name Head Net A.

Description
The MR-300 Angara air and surface surveillance radar features a large (approximately 6 × 1.5 m) elliptical paraboloid reflector of open lattice construction and a horn feed that is carried on a boom projecting from below the lower edge of the scanner. Angara's principal function is assumed to be air search and surveillance and it can be expected to provide target designation facilities for other fire-control radars for guns and missiles carried by the vessel. It is normally mounted on a mast top location, or other high positions and both single and double installations have been seen on various classes of vessels. Detection range against air targets at medium altitude is given as 111 to 130 km. It is also worth noting that MR-300 is quoted by some sources as operating in the G/H-band (4 to 8 GHz).

The antenna array used in the MR-300 air surveillance radar with a Sun Visor radar director below and to the right

Status
As of early to mid-2000, the MR-300 air surveillance radar was reported as being in service aboard the Russian cruiser *Admiral Golovko.*

UPDATED

MR-310 Angara air and surface search radar

Type
E/F-band (2 to 4 GHz) air and surface search radar. NATO reporting name Head Net B.

Description
MR-310 is an E/F-band air and surface search radar that can generate target co-ordinate data. The equipment forms part of the Angara family of radars, the other members of which are the MR-300 Angara (NATO reporting name Head Net A) and the MR-310U Angara M (NATO reporting name Head Net C).

Status
As of June 2001, the precise status of the MR-310 air and surface search radar was uncertain.

Specifications
Frequency: 2-4 GHz
Range: 180 km
Coverage: 55° (elevation); 360° (azimuth)
Scan rate: 6 or 12 rpm
Co-ordinates measured: 3
Weight: 10 t

VERIFIED

MR-310U Angara M air surveillance radar

Type
E/F-band (2 to 4 GHz) air surveillance radar. NATO reporting name Head Net C.

Description
MR-310U is a back-to-back combination of two MR-300 radars (NATO reporting name 'Head Net A') that are tilted from the horizontal by approximately 30°. This has the effect of displacing the resultant fan-shaped elevation beam by the same amount from the vertical. This, in combination with the vertical beam produced by the companion scanner, provides a height-finding capability using the V-beam technique. In the MR-310U application, the slant and vertical beams are separated in azimuth by 180°. The vertical beam fulfils the search function and the operator selects a target for which height data is required by placing a marker on the target (or similar technique). This places a range gate in the second (slant) beam and excludes other targets. Computation to give a height readout can be performed by analogue or digital methods. Head Net C's maximum detection range is given as being 128 km.

The MR-310U's antenna assembly as installed on the Polish destroyer Warszawa *(Stefan Terzibaschitsch)*

Status

As of early to mid-2000, MR-310U radars were reported as being in service aboard Indian 'Rajput' class destroyers and 'Godavari' and some 'Modified Godavari' class frigates; the Polish destroyer *Warszawa*, the Russian Navy's cruiser *Kerch*, Type 61M ('Modified Kashin' class) destroyers, Type 1135/1135M/1135MP ('Krivak I/II/III' class) frigates and Type 887 ('Smolny' class) training ships and the Ukrainian frigates *Dnipropetrovsk* and *Mikolaiv*.

UPDATED

MR-500 Fut-N search radar

Type

D/E-band (1 to 3 GHz) long-range 3-D air surveillance radar. NATO reporting name Big Net.

Description

MR-500 is a very large, long-range air surveillance radar operating in D/E-band. Behind the reflector are two balance vanes. Range performance is estimated as more than 185 km against an aircraft at medium altitude and up to 370 km at higher altitudes.

Status

As of June 2001, MR-500 radars were reported as having been installed aboard Indian 'Rajput' class destroyers, the Polish destroyer *Warszawa* and Russian Type 1144.1/1144.2 ('Kirov/Orlan' class) battle cruisers, Type 1164 ('Slava/Atlant' class) cruisers and Type 61/61M ('Kashin/Modified Kashin' class) destroyers. When applied to Type 1144 and 1164 vessels, MR-500 is understood to form part of the radar system identified by the NATO reporting name Top Pair.

VERIFIED

Polish destroyer Warszawa *showing its MR-500 radar to the right of its MR-310U equipment* (Stefan Terzibaschitsch)

MR-755 air and surface search radar

Type

Decimetric air and surface search radar. NATO reporting name Half Plate.

Description

MR-755 is described as being able to detect air and surface targets and generate co-ordinate data on them.

Status

As of June 2001, the MR-755 air and surface search radar was reported as having being installed aboard Indian 'Delhi' class destroyers; Russian Type 1124EM ('Grisha V/Albatros' class) frigates (Half Plate B) and the Ukrainian frigate *Lutsk* (Half Plate B).

Specifications

Coverage: 55° (elevation); 360° (azimuth); 150 km (range)
Co-ordinates measured: 3
Scan rate: 15 rpm
Weight: 6.5 t

VERIFIED

Navigation radars

Description

Some 16 types of Russian Federation and Associated States (RFAS) shipboard navigation/short-range surface search radars have been identified to date. These include Eniscy, Mius, Lotsiya and Nayada. The earlier radars such as the Don series, Donets, Stvor and Neptune dating from the 1950s were very similar to the later Okean, Kivach and Lotsiya systems with a 3.2 cm wavelength in the I-band (8 to 10 GHz). It appears that virtually all RFAS navigation radars operate on 9,400 to 9,600 MHz and the Eniscy and Okean on an additional 3,030 to 3,090 MHz band. Maximum range scales vary from 30 km on the Lotsiya to 119 km on Okean and Nayada. The Palm Frond equipment has been replacing the Don series since 1985. Other NATO reporting names given to navigation radars are Ball End, Spin Trough and Sheet Bend. The Don and Nayada radars can have an Alfa equipment connected to them for close in 6 to 22 km navigation assistance. This can project five different type marks on the radar screen to provide the closest point of approach and course to steer. A chart plotter, known as Palma, attaches to the radar. A semi-transparent glass is placed at an angle of 45° between the radar screen and a flat chart; the radar image is projected on to it and superimposed on the chart. The Kirach navigation radar is for small vessels, from 50 to 500 tons and comes in two versions. Kirach 1 is the basic version and becomes Kirach 2 when equipped with a course indicator. The Russian Navy has a complex display of plotting multiple radar inputs called a *Sistema Avtomatichyeskoy Radiolokatsionnoi Prokladki* (SARP), which can be roughly translated as a system of automatic radio location.

Status

As of early to mid-2000, Don-2, Don Kay and Palm Frond navigation radars were reported as being installed aboard the following vessels and ship classes:

Don-2

The Bulgarian survey ship *Admiral Branimir Ormanov* and Russian Type 159A ('Petya II' class) and Type 1135 ('Krivak I' class) frigates.

Don Kay

Indian 'Rajput' class destroyers and 'Godavari' class frigates; the Russian cruiser *Kerch*, the landing ship *Mitrofan Moskalenko*, the intelligence ship *Belomore*, Type 394B ('Primorye' class) intelligence vessels, Type 887 ('Smolny' class) training ships, Type 97P ('Ivan Susanin' class) armed icebreakers/patrol ships, Type 1135/1135M/1135MP ('Krivak I/Krivak II/Krivak III' class) frigates and Type 1595 ('Antonov' class) transports and Ukrainian frigates *Dnipropetrovsk* and *Mikolaiv*.

Palm Frond

Chinese 'Sovremenny' class destroyers; Georgian Type 205P ('Stenka' class) patrol craft; Indian 'Dehli' class destroyers and 'Talwar' class frigates; the Russian aircraft carrier *Admiral Kuznetsov*, the cruiser *Kerch*, the destroyer *Admiral Chabanenko*, the frigate *Neustrashimy*, the landing ship *Mitrofan Moskalenko*, the intelligence ship *Belomore*, the survey and research ships *Leonid Demin* and *Marshal Nedelin*, the submarine rescue ship *Alagez*, Type 05360/1 ('Mikhail Rudnitsky' class) salvage and mooring vessels, Type 97P ('Ivan Susanin' class) armed icebreakers/patrol ships, Type 205P ('Stenka/Tarantul' class) fast attack craft, Type 304/304M ('Amur I/II' class) repair ships, Type 596P ('Vytegrales II' class) supply ships, Type 850 ('Nikolay Zubov' class) and Type 862 ('Yug' class) survey and research ships, Type 956/956A ('Sovremenny' class) destroyers, Type 1041Z ('Svelyak' class) fast patrol craft, Type 1135/1135M/1135MP ('Krivak I/II/III' class) frigates, Type 1144.1/1144.2 ('Kirov' class) battle cruisers, Type 1155 ('Udaloy' class) destroyers, Type 1164 ('Slava' class) cruisers, Type 1453 ('Ingul' class) salvage tugs, Type 1559V ('Boris Chilikin' class) replenishment ships, Type 1595 ('Antonov' class) transports, Type 2020 ('Malina' class) nuclear submarine support ships and Type 12660 'Gorya' class ocean-going minehunters and the Ukrainian auxiliaries *Makivka* and *Slavutich* and Type 205P ('Stenka/Tarantul' class) fast attack craft.

UPDATED

Podberyozovik-E series multifunction radars

Type

Family of 3-D air and surface surveillance and targeting radars (NATO reporting name Flat Screen).

Description

Designed for surface ship air and surveillance and targeting applications, the Podberyozovik-E series of 3-D radars is described as being solid-state equipments that provide a high level of immunity from natural and man-made interference. This is achieved via the use of broadband frequency agility, 'narrow' transmission and reception beams, low sidelobe antennas, non-linear signal processing, 'composite signals' with 'high resolution' and adaptive, digital Moving Target Indication (MTI). Functionally, Podberyozovik-E radars scan the vertical plane with one or two pencil beams with azimuthal coverage being achieved using electromechanical rotation of the antenna. Transmission strategy is governed by matching the radar's Pulse Repetition Frequency (PRF) and pulse-length to the particular target's bearing and elevation. The sensor's field of view is electronically stabilised against the pitching and rolling movements of its host platform while its transmission chain can generate both single and coherent pulse bursts, with the output being

The Podberyozovik-E antenna (the square 'bedstead' array carried amidships) installed on the Russian cruiser Kerch *(Eric Grove)*

characterised by linear frequency intrapulse modulation. Long duration single pulses are used to detect low elevation targets at ranges of up to 500 km with coherent burst of variable repetition frequency pulses being employed to provide 'close-in zone' coverage. The latter approach is said to ensure the reliability of the radar's MTI circuitry.

The antennas used in the Podberyozovik-E series take the form of planar-arrays made up of waveguide strips with variably inclined slots. The arrays are fed by a power divider that takes the form of a serpentine waveguide. In the ET1 model, the antenna used measures 7.16 by 6.26 m while that fitted to the ET2 variant is 7.16 by 2.92 m. These arrays have half-power beamwidths that are 4.2° in the horizontal plane and 5.5° and 8° respectively in elevation. Sidelobe levels do not exceed –25 dB and an identification friend-or-foe interrogator antenna is co-located with the main antenna.

The Podberyozovik-E transmitter takes the form of a three-stage, transistorised amplifier whose input stage incorporates two 'hot' stand-by amplifiers connected in parallel. The second stage contains 16 amplifiers, six of which, form a 'cold' reserve. The final stage employs 128 amplifiers, giving a mean output power of 1.5 kW. A Radio Frequency (RF) signal generator converts Intermediate Frequency (IF) signals to RF using local oscillators arranged in an amplifier-multiplier circuit that incorporates clock crystal oscillators. The signal generator used is made up of microblocks that contain unpackaged micro-assemblies that are mounted on polykor microstrips. Each microblock is sealed and filled with inert gas.

Podberyozovik-E radars are fitted with a superheterodyne receiver that features single-stage frequency conversion and image channel, signal frequency and phase suppression. At its input stage, the receiver is fitted with a low noise, transistorised amplifier (with a semiconductor limiter), a double balanced mixer and an IF amplifier adder. A multichannel digital signal generator/processor creates IF probe bursts, processes interference and switches and multiplies video for external applications. The device employs matched filters, automatic gain control and non-linear signal processing. Protection against chaff and passive clutter is provided by adaptive MTI circuitry, probe burst PRF wobbulation and noise blanking. A beam synchronisation and scanning control unit generates synchronisation signals for both the radar and associated external systems together with appropriate carrier frequencies for particular operating modes. The system's operator console offers on/off control of the radar, high voltage feed/cut-off control of its transmitter and

mode selection facilities. A semi-automatic built-in test and diagnostic subsystem is available and 'Salyut' notes that the performance of the radar can be optimised by integrating it with the company's Poima-E radar data processing system. Configurable for use aboard small, medium and large size ships, Poima-E incorporates initial data input operator and crew commander workstations, each of which is based on an IBM personal computer compatible, open architecture computer equipped with a 53 cm (21 in) high resolution monitor. System functions include:

- automatic target detection, acquisition, tracking and co-ordinate measurement
- target classification
- primary and secondary data display
- automatic generation of a target allocation plan
- transmission of target tracking data to the host vessel's tactical data and weapon direction systems
- data logging.

As a final point, it should be noted that Podberyozovik-E is described by its manufacturer as being a 'C-band' radar. As of this issue, this was thought likely to indicate operation in the 4 to 8 GHz band rather than the 0.5 to 1 GHz frequency range.

Status

As of early to mid-2000, the Russian cruiser *Kerch* was being reported as being equipped with a Flat Screen surveillance radar.

Contractor

State Unitary Enterprise - State Moscow Plant 'Salyut', Moscow.

UPDATED

Pot Drum search radar

Type

H/I-band (6 to 10 GHz) surface search radar.

Description

The H/I-band Pot Drum is a small surface search radar, the scanner for which is mounted in a flat, slightly domed radome from which its NATO reporting name is derived. The diameter of this radome is about 1.5 m and the unit is typically installed at the masthead of small patrol boats. The radar's transceiver is housed below decks while its display unit features an approximately 15 cm diameter cathode ray tube. In addition to surface search and torpedo fire-control, Pot Drum is used for pilotage purposes also and may have a limited air warning capability. The radar's range is given as being approximately 37 km.

'Shershen' class torpedo boat with Pot Drum surface search radar and Drum Tilt fire-control radars for guns

Specifications: Podberyozovik-E series multifunction radars

Variant	ET1	ET2
Number of channels	1	1
Number of measured co-ordinates	3	3
Coverage	5 km (min range);	5 km (min range);
	30° (elevation);	30° (elevation);
	40 km (altitude)	40 km (altitude)
	500 km (range)	500 km (range)
Detection range	55 km (missile);	45 km (missile);
	300 km (fighter);	240 km (fighter);
	radar horizon (ship)	radar horizon (ship)
Accuracy	24 min (azimuth);	24 min (azimuth);
	30 min (elevation)	30 min (elevation)
	150 m (range)	150 m (range)
Track capacity	5	5
Antenna rotation speed	6 or 12 rpm	6 or 12 rpm
Scan rate	5 or 10 s	5 or 10 s
Number of system components	8	8
Power requirement	45 kW	45 kW
Weight	3.2 t (below decks);	2,9 t (antenna);
	4.7 t (antenna)	3.2 t (below decks)

Status

As of June 2001, the Pot Drum surface search radar was reported as having been installed aboard Type 205P ('Stenka/Tarantul' class) fast attack craft operated by the Cuban, Georgian (one example, the *Batumi*), Russian and Ukrainian navies; 'Shershen' class fast attack craft operated by the Egyptian and Vietnamese navies; North Korean 'Najin' class frigates; Romanian 'Naluca' class fast attack craft and Type 206M ('Turya' class) fast attack hydrofoils operated by the Vietnamese navy.

UPDATED

Pot Head search radar

Type

I-band (8 to 10 GHz) surface search radar.

Description

Pot Head utilises an approximately 1.2 m diameter antenna radome and is understood to offer surface target detection, fire control, pilotage and limited air warning capabilities.

Status

As of June 2001, Pot Head search radars (or its Chinese equivalent, the Type 351) were reported as having been installed aboard the Bangladeshi patrol craft *Nirbhoy*, Myanmar 'Hainan' class patrol craft; Chinese 'Haijiu' and 'Haizhui/ Shanghai III' class patrol craft and 'Hainan' and 'Shanghai II' class fast attack craft; Egyptian 'Hainan' class fast attack craft; North Korean 'Najin' class frigates, 'Sariwon' class corvettes, 'Chong-Ju' (Type 351), 'Hainan/SO 1' (Type 351) and 'Taechong I/II' (Type 351) class patrol craft and 'Chaho' (Type 351) and 'Shanghai II' (Type 351) class fast attack craft; the Pakistani patrol craft *Rajshahi*, the Tunisian patrol craft *Utique* and Vietnamese 'SO 1' class patrol craft.

VERIFIED

Pot Head short-range surface radar with identification friend-or-foe antenna

Sky Watch surveillance radar

Type

E/F-band (2 to 4 GHz) air surveillance radar.

Description

Sky Watch is an E/F-band air surveillance radar for the Russian Navy's Project 1143.3 *(Admiral Gorshkov)*/1143.5 *(Admiral Kuznetsov)* aircraft carriers. It utilises four 6 × 6 m planar-array antennas mounted fore and aft and abeam of the superstructure.

Two of the Sky Watch planar-arrays can be seen on the fore part of the Admiral Gorshkov superstructure

Status

As of early to mid-2000, the Sky Watch air surveillance radar was reported as being in service aboard the Russian aircraft carrier *Admiral Kuznetsov*.

UPDATED

Slim Net air and surface search radar

Type

E/F-band (2 to 4 GHz) air and surface search radar.

Description

Slim Net is a high-definition, surface search radar with a secondary air surveillance capability. Its scanner is of open lattice construction and is approximately 5.5 m wide by 1.8 m deep.

Operating status

As of June 2001, Slim Net surface search radars were reported as having been installed aboard the Indian submarine tender *Amba* and the same country's 'Arnala' class frigates; Syrian 'Petya III' class frigates and the Ukrainian auxiliary *Kolomiya*.

VERIFIED

The Slim Net scanner assembly with a Yard Rake rotating Yagi to its right

Snoop class submarine surveillance radars

Type

Submarine surface search radars.

Description

A number of surveillance radars have been fitted to submarines since the Second World War, some of which have been given a NATO reporting name starting with Snoop. The early Soviet submarines were equipped with a small antenna designated Snoop Plate. This is reputed to have a range of 46 km against aircraft and 22 km against surface vessels. Later boats were fitted with a larger and more powerful I-band (8 to 10 GHz) radar known as Snoop Tray that entered service in the early 1960s. Since then, nearly all new Soviet submarines have been fitted with Snoop Tray. Three other Snoop class radars are the Snoop Pair, Head and Half types.

Status

As of early to mid-2000, Snoop series radar installations are thought to have included the following:

Snoop Tray

Algerian Type 877E ('Kilo' class) anti-submarine warfare submarine; the Chinese surface-to-surface missile submarine *351*, Type 033 ('Romeo') general purpose, Type 035 ('Ming' class) general purpose, Type 091 ('Han' class) nuclear attack, Type 092 ('Xia' class) nuclear ballistic missile and Type 877EKM/636 ('Kilo' class) anti-submarine warfare submarines; Indian, Libyan and Polish Type 641 ('Foxtrot' class) general purpose submarines; Indian 'Sindhughosh' class anti-submarine warfare submarines; Iranian Type 877EKM ('Kilo' class) anti-submarine warfare submarines; the Polish anti-submarine warfare submarine *Orzel*; the Romanian anti-submarine warfare submarine *Delfinul 521*; the Russian nuclear-powered attack submarine *K 395*, the nuclear-powered auxiliary submarines *K 403*, *K 411* and *K 433*, Type 641 ('Foxtrot' class) general purpose, Type 641B ('Tango/Som' class) anti-submarine warfare, Type 667B ('Delta I/Murena' class)/667BDR ('Delta III/Kalmar' class)/667BDRM ('Delta IV/Delfin' class) nuclear ballistic missile, Type 671RTM ('Victor III/Schuka' class) nuclear-powered attack and Type 877/ 877K/877M-636 ('Kilo/Kilo 4B/ Vashavyanka' class) anti-submarine warfare submarines and the Ukrainian general purpose submarine *Zaporizya*.

Snoop Plate

The Bulgarian general purpose submarine *Slava 84*; the Chinese ballistic missile submarine *200*, the surface-to-surface missile submarine *351* and Type 033 ('Romeo' class) general purpose submarines and Russian Type 641 ('Foxtrot' class) general purpose submarines - as an alternative to Snoop Plate.

Snoop Pair

The Russian nuclear attack submarine *Tula*, Type 941 ('Typhoon') nuclear ballistic missile, Type 945A ('Sierra II/Kondor' class) nuclear attack, Type /949A ('Oscar II/ Antyey' class) surface-to-surface missile and Type 971/971M ('Akula I/II/Bars' class) nuclear attack submarines. Type 949A and 971/971M boats are also noted as being fitted with Snoop Half radar as an alternative to Snoop Pair.

UPDATED

Square Tie air and surface search radar

Type

I-band (8 to 10 GHz) air and surface search radar.

Description

Square Tie is a small, lightweight, short-range, air and surface search radar that operates in the I-band. The system's scanner is an elliptical paraboloid reflector illuminated by an overhung horn feed. Functions include target detection and tracking for anti-ship missile direction and, possible, target designation for the Drum Tilt gun fire-control radars carried by 'Osa' class craft. Maximum range is believed to be about 130 km.

Status

As of early to mid-2000, Square Tie air and surface search radars (and its Chinese equivalent, the Type 352) were reported as being in service aboard Algerian 'Nanuchka II' class corvettes and 'Osa II' class fast attack craft; the Bangladeshi frigate *Osman* and the same country's 'Durdharsha' and 'Durbar' class patrol craft; Bulgarian 'Osa I/II' class fast attack craft; Chinese 'Luda I/II' class destroyers, 'Jianghu I/II (one ship)/III/IV' class frigates and 'Huangfen', 'Houjian/Huang', 'Houku' and 'Houxim' class fast attack craft and 'Tuzhong' class tugs (one ship); the Croatian fast attack craft *Dubrovnik*; Cuban 'Osa II' class patrol craft; Egyptian 'Jianghu I' class frigates and 'Hegu' class fast attack craft; the Eritrean fast attack craft *FMB 161*; Indian 'Durg' class corvettes and 'Osa II' class fast attack craft; North Korean 'Najin' and 'Soho' class frigates and 'Komar/Sohung', 'Osa I/Huangfen' and 'Soju' class fast attack craft; Latvian 'Osa I' class patrol craft; Libyan 'Nanuchka II' class corvettes and 'Osa II class fast attack craft; Pakistan 'Huangfen' class patrol craft; Polish 'Puck' class patrol craft; Romanian 'Osa I' class patrol craft; Syrian 'Osa II' class patrol craft; Thai 'Chao Phraya' class frigates; Vietnamese 'Osa II' class patrol craft; Yemeni 'Huangfen' class fast attack craft and Yugoslav 'Osa I' class fast attack craft.

UPDATED

Strut Curve/Strut Pair search radars

Type

F-band (3 to 4 GHz) air and surface search radars.

Description

Strut Curve is an F-band medium-range general purpose search radar that provides both air and surface search facilities. An elliptical lattice reflector is used, illuminated by a horn feed carried by a boom projecting from the lower edge of the scanner. Range performance against a 2 m² aircraft target at medium altitude is

approximately 111 km with a likely maximum range of 278 km. A version of this radar, known as Strut Pair, is formed by two Strut Curve antennas mounted back-to-back.

Status

As of early to mid-2000, Strut Curve search radars were reported as being installed aboard Algerian 'Mourad Rais' class frigates; the Bulgarian frigate *Smeli* and the same country's 'Letyashti' class corvettes; Indonesian 'Kapitan Patimura' class corvettes and 'Frosch I/II' class landing ships; Libyan 'Koni' class frigates; Lithuanian 'Grisha III' class frigates; the Polish frigate *Kaszub*; the Romanian destroyer *Marasesti* and the same country's 'Tetal/Improved Tetal' class frigates, 'Cosar' class minelayers and 'Croitor' class logistic support ships; the Russian training ship *Gangut*, the water tanker *Manych*, the submarine depot ship *Volga 879*, the missile support ship *Voronesh 874* and the same country's Type 97P ('Ivan Susanin' class) armed icebreakers/patrol ships, Type 775 ('Ropucha 1' class) landing ships, Type 1124 ('Grisha I' class)/1124K ('Grisha IV' class)/1124M ('Grisha III' class)/1124EM ('Grisha V/Albatros' class) frigates, Type 1559V ('Boris Chilkin' class) replenishment ships and Type 1791 ('Amga' class) missile support ships; the Ukrainian landing ship *Konstantin Olshansky* and Type 1124 ('Grisha I' class)/1124EM ('Grisha V' class)/1124P ('Grisha II class) frigates; Vietnamese Type 159 ('Petya' class) frigates and Yugoslav 'Split/Kotor' class frigates. As of the given date, Strut Pair search radars were noted as being fitted aboard India's Type 1143.4 aircraft carrier (the ex-*Admiral Gorshkov*); Russian aircraft carrier *Admiral Kuznetsov*, the destroyer *Admiral Chabanenko* and the same country's Type 1124EM ('Grisha V' class) frigates (early ships) and Type 1155 ('Udaloy/Fregat' class) destroyers.

UPDATED

Top Pair air surveillance radar

Type

C/D-band (0.5 to 2 GHz) long-range 3-D air surveillance radar.

Description

Top Pair is the name of a back-to-back configuration of two different radar types employed for long-range air surveillance. One application makes use of the MR-600 (NATO reporting name Top Sail) 3-D radar and the MR-500 (NATO reporting name Big Net) long-range search sensor (or a derivative of it). Top Pair is thought to have ranges of about 350 km for larger aircraft and 200 km for fighters.

Status

As of early to mid-2000, Top Pair radars were reported as being installed aboard Russian Type 1144.1/1144.2 ('Kirov/Orlan' class) battle cruisers and Type 1164 ('Slava/Atlant' class) cruisers.

VERIFIED

Strut Pair (upper) and Palm Frond (three lower) antennas in a destroyer application

The cruiser Admiral Ushakov, *showing the Top Pair air surveillance radar group at the very highest position on the ship's superstructure, with another new radar group, Top Steer, to be seen further aft on top of the pyramidal mast*

Top Steer surveillance radar

Type
D/E/F-band (1 to 4 GHz) 3-D air surveillance radar.

Description
Top Steer is a 3-D air search radar that incorporates two back-to-back scanners that are usually mounted high up on the superstructure of the host vessel. The two scanners are not the same, with one being somewhat similar in appearance to that used in the MR-600 (NATO reporting name Top Sail) system. This said, the Top Steer array is noticeably smaller, as can be seen on ships such as the *Kirov* cruiser which has both MR-600 and Top Steer fitted. The other appears to be either a standard MR-300 (NATO reporting name Head Net A) search radar or a derivative thereof. It is not clear whether or not the two scanners serve a single radar system or two separate sensors, the antennas for which being co-located as a matter of structural convenience. Top Steer is likely to operate in the D/E/F-band and it is possible that the larger antenna may incorporate some form of frequency scanning of the beam in elevation to provide a height-finding capability. The probable function of this radar is control of naval helicopters.

Status
As of early to mid-2000, the Top Steer air surveillance radar was reported as being installed aboard Russian Type 1164 ('Slava/Atlant' class) cruisers.

VERIFIED

A SaabTech Systems dual-band antenna

The reflector is of the 'air-porous' type in metal giving low wind forces and high surface accuracy, the latter resulting in ultra-low sidelobes. The antenna is stabilised against ±30° roll and ±10° pitch.

Associated transceiver units
The antenna has been designed to work with a wide variety of Transmit/Receive (T/R) units and associated signal processors. Normally the G-band T/R unit would be a high-power, coherent pulse-Doppler type in order to give maximum performance against small, high-speed air targets. Several such units are available from different manufacturers. The I-band T/R would be normally of the non-coherent type in order to detect primarily slow moving or stationary surface targets. However a navigation type radar T/R unit would be a cost-effective alternative.

Of special interest is the use of the I-band channel of the antenna for the PILOT Mk 2 radar (see separate entry). The higher gain, 37 dB at I-band of the G-I antenna enhances the quietness of PILOT Mk 2 compared to using some 30 dB gain unstabilised conventional type navigation antenna.

Status
Over time, approximately 20 SaabTech Systems dual-band antenna units are understood to have been supplied for a variety of applications including Swedish 'Göteborg' class coastal corvettes; Australian/New Zealand ANZAC class frigates; Bahrain's 'Al Riffa' class fast attack craft; Danish 'Niels Juel' class frigates and 'Willemoes' class fast attack craft; Finnish 'Helsinki' class and 'Rauma' class fast attack craft and Thai 'Handalan' class fast attack craft.

Contractor
SaabTech Systems AB, Järfälla.

UPDATED

Top Steer, Palm Frond navigation radar and Front Door missile guidance radar group on the cruiser Marshal Ustinov *(Steve Zaloga)*

SWEDEN

Dual band (G and I) antenna

Type
Simultaneous G- (4 to 6 GHz) and I-band (8 to 10 GHz) antenna.

Description
SaabTech System's dual band antenna is designed to provide simultaneous, optimised air and surface coverage, and is applicable to small and medium size warships. The capability is achieved by allowing radiation of two wavelengths and waveforms from two separate transmitter/receiver units. Both bands are radiated simultaneously by multiple feed horns illuminating a common reflector. The rotating assembly is stabilised around two axes, roll and pitch, by a conventional gimbal arrangement using well proven hydraulic motors.

The G-band is radiated in three beams in order to obtain long range at low elevation and coverage up to 60° for high dive angle missiles. I-band covers the surface and lowest altitude.

PILOT Mk 2 Low Probability of Intercept (LPI) tactical radar

Type
I-band (8 to 10 GHz) surface surveillance and navigation radar.

Description
PILOT Mk 2 is an LPI radar which can be used as a tactical radar for covert operations. It uses frequency modulated Continuous Wave (CW) transmission for 100 per cent duty cycle and a very low output power which ensures that it is virtually undetectable by tactical electronic support measures systems. Other features of PILOT Mk 2 are high range resolution compared to pulsed radars, high reliability, small lightweight design and ease of installation.

PILOT Mk 2 consists of a transceiver unit, a signal processor unit, a remote-control panel and a waveguide switch unit. The radar has been developed to interface easily to different types of standard I-band antennas and most types of

Specifications: Dual band (G and I) antenna
Antenna aperture: 2.4 × 1 m²

| | G-band | | | I-band |
	Low beam	High beam	Extra high beam	
Frequency range	5.4-5.9 GHz	5.4-5.9 GHz	5.45-5.85 GHz	8.5-9.6 GHz
Gain	34 dB	30.5 dB	26.5 dB	37 dB
Polarisation	Vertical	vertical	vertical	vertical
Horizontal beamwidth	1.9°	1.9°	1.9°	1.1°
Vertical beamwidth	5°	9°	3°	3.7°

displays, including multiple display systems with automatic radar plotting aid trackers, without modifications to the displays.

The main configuration for PILOT Mk 2 is as an add-on to an existing navigation radar, retaining the original antenna, transceiver and display system, where the user can reach tactical advantage by switching between the conventional pulsed radar and PILOT Mk 2. The transceiver design used allows for CW operation using a conventional I-band antenna. Accordingly, visual inspection of the array used will not reveal its connection with PILOT Mk 2.

Status
As of May 2001, PILOT Mk 2 was understood to have been procured by a number of navies around the world.

Specifications
Frequency: I-band (8-10 GHz)
Transmitter: solid state
Output power: 1 mW to 1 W
Instrumented range: up to 44 km
Range cell resolution: down to 2.4 m - equivalent to 16 ns pulse-width

Contractor
SaabTech Systems AB, Järfälla.

UPDATED

Sea Giraffe AMB multimode naval radar

Type
G-band (5.4 to 5.9 GHz sub-band) multimode, phased-array, 3-D surveillance radar.

Description
Sea Giraffe AMB is an air and surface surveillance radar that is applicable to ship classes ranging from patrol vessels to frigates. Above deck equipment comprises a phased-array antenna/receiver assembly and a 'man aloft' switch unit. The below deck suite is made up of a waveguide dryer, a cooling air unit, a transceiver, a signal/data processing cabinet and an operator workstation. Sea Giraffe AMB's transmitter is travelling wave tube based and the system's signal processing chain incorporates Moving Target Indicator (MTI)/non-MTI air, surface and Electronic Counter-CounterMeasures (ECCM) channels together with optional channels for identification friend-or-foe correlation/tracking and helicopter detection/tracking. Within this arrangement, the surface channel has provision for splash spotting. Data processing facilities include automatic air target, surface target and jammer detection and tracking. Other system features include Low Probability of Intercept (LPI), specialised long-range surveillance modes for ducting conditions and littoral operations, together with an automatic detection/tracking override facility that gives priority to high threat, pop-up targets such as sea-skimming anti-shipping missiles. Aside from LPI function and the described dedicated ECCM signal processing channel, Sea Giraffe AMB's ECCM provision includes sector and intermittent transmission, frequency agility and false target cancellation.

Status
As of September 2001, the Sea Giraffe AMB multimode 3-D surveillance radar was reported as being installed aboard Swedish 'Visby' class corvettes and as having been selected for retrofit aboard three 'Orkan' class corvettes of the Polish Navy.

The antenna/receiver assembly used in the Sea Giraffe AMB multimode naval radar   0006905

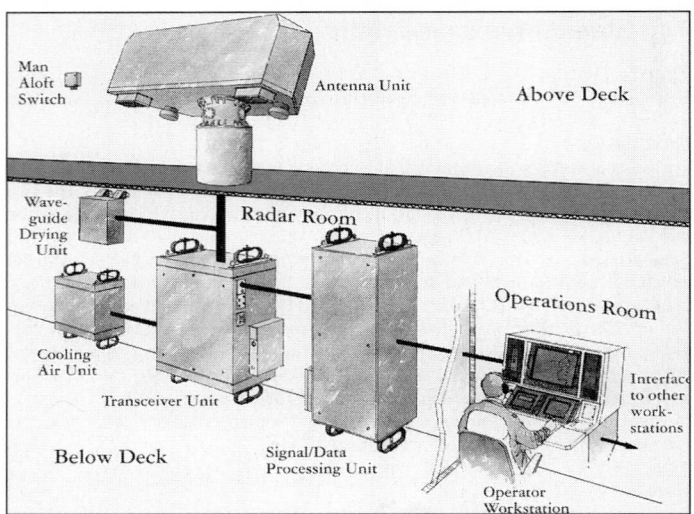

A schematic showing the above/below deck layout of a typical Sea Giraffe AMB application   0006906

Specifications
Frequency: G-band (5.4-5.9 GHz)
Detection range: >60 km (air defence mode against a small air target at an altitude of up to 10 km); 80 km (normal search mode against a small air target at an altitude of up to 10 km)

Contractor
Ericsson Microwave Systems AB, Mölndal.

UPDATED

Sea Giraffe 50/150 naval radars

Type
G-band (4 to 6 GHz) medium-range multipurpose search and target indication radar family.

Description
Sea Giraffe 50/150 radars are described as being suitable for the following roles:

Air/surface search
- long-range detection and tracking of missiles in jammed and highly cluttered environments
- surface target detection and tracking limited only by the radar horizon
- aircraft detection and tracking under severe weather and jamming conditions.

Defence
- accurate air target data to the ship's fire-control systems
- surface target data for gun fire control
- surface target data for surface weapon systems.

Offshore patrol
- long-range navigation
- surface and air target indication and tracking
- guidance for both fixed- and rotary-wing aircraft.

Operational and technical features incorporated in the Sea Giraffe 150 system include:
- long-range detection of anti-ship missiles under both clear and cluttered conditions
- high elevation coverage
- simultaneous suppression of clutter with different velocities
- adapted for ducting environment
- suppression of multiple-time-around clutter
- digital coded pulse compression
- short reaction time
- automatic initiation and tracking of air targets
- automatic tracking of surface targets
- large number of transmit frequencies
- adaptive frequency selection
- prepared for IFF integration.

Status
As of October 2001, Sea Giraffe 50 and 150 were reported as being out of production and sources were suggesting that the Sea Giraffe 150 can accommodate two types of stabilised antenna (an Ericsson-developed three beam elevation array and a SaabTech (formerly CelsiusTech) two beam assembly), both of which can be configured with an optional extreme high beam antenna and an integrated identification friend-or-foe array. During the period February to April 2001, *Jane's* sources were reporting Sea Giraffe 50 and 150 radars as being installed aboard the following warships and classes of warship:

The Sea Giraffe150 high coverage, stabilised antenna used in the Canadian 'Halifax' frigate application

Bahrain	• 'Al Manama' class corvettes (two ships in class - Sea Giraffe 50)
	• 'Ahmad El Fateh' class fast attack craft - missile (four ships - Sea Giraffe 50)
Canada	• 'Halifax' class frigates (12 ships - Sea Giraffe 150)
Kuwait	• the fast attack craft - missile *Al Sanbouk* (Sea Giraffe 50)
Malaysia	• 'Lekiu' class frigates (two ships - Sea Giraffe 150)
Singapore	• 'Victory' class corvettes (six ships - Sea Giraffe 150)
Sweden	• the minelayer *Carlskrona* (Sea Giraffe 50)
	• 'Goteborg' (four ships - Sea Giraffe 150) and 'Stockholm' (two ships - Sea Giraffe 50) class corvettes
	• 'Norrkoping' class fast attack craft - missile (six ships - Sea Giraffe 50)
United Arab Emirates	• 'Muray Jib' class corvettes (two ships - Sea Giraffe 150)
	• 'Ban Yas' (six ships - Sea Giraffe 50) and 'Mubarraz' (two ships - Sea Giraffe 50) class fast attack craft - missile

Specifications

Operating frequency: G-band
Output power: 60 kW; average power 1,200 W
Antenna rotation rate: 60/30 rpm
Instrumented range: 25, 50, 100 and 150 km
Tracking capacity: more than 30 (air targets – auto initiation); more than 50 (surface targets – manual initiation)
Antenna sidelobe levels: –45 dB
Beamwidth, bearing: 1.8°
Elevation coverage: 0-70°
Accuracy: 20 m (range); 0.3° (azimuth)

Contractor

Ericsson Microwave Systems AB, Mölndal.

UPDATED

UKRAINE

Shipboard Over-The-Horizon Surface Wave (OTH-SW) radar

Type:
High Frequency (HF - 15 to 30 MHz sub-band) shipboard OTH-SW radar.

Description

Ascribed to the Ukranian Radio Technical Institute, the cited shipboard OTH-SW radar is described as comprising a 60 m array of vertically polarised reception 'vibrators' mounted along the sides of the host vessel and a two element transmission array located on its mast top. As such, the architecture is designed to detect and track air, anti-shipping missile and surface ship targets.

Status

As of the second half of 2000, UKRSPETSEXPORT was actively promoting the described shipboard OTH-SW radar.

Specifications

Frequency: 15-30 MHz
Coverage: 45° sector (to a range of 170 km)
Range: 50 km (missile at an altitude of 5 m); 80 km (aircraft at an altitude of 10 to 100 m); 130 km (aircraft flying at an altitude in excess of 100 m)

Target velocity: 50 km/h (threshold velocity - air targets)
Track capacity: up to 10 missiles and up to 200 ships simultaneously
Detection-to-tracking time: 10 s (aircraft or missile); 500 s (surface ship)

Contractor

UKRSPETSEXPORT (export agent), Kiev.

NEW ENTRY

UNITED KINGDOM

Introductory note

Alongside the equipments described below, a number of previous generation UK naval radars continue to be used around the world. Examples of such equipment are as follows:

AWS-1
An E/F-band (2 to 4 GHz) search radar developed by the then Plessey company during the late 1950s. AWS-1 is reported to be equipped with a 4.9 × 2 m antenna and to have a peak power of 750 kW. Maximum detection range against a small aircraft is given as 111 km. AWS-1 radars are noted as having been in recent service aboard Iranian 'Alvand' class frigates.

AWS-2
Launched in the early 1970s, AWS-2 is described as being similar to AWS-1 with the addition of a Digital Moving Target Indicator (DMTI) capability. Radars of this type are reported as having been in recent service aboard Portuguese 'Baptista de Andrade' class frigates.

S810/820
With the ability to detect a 5 m^2 RCS air target at ranges of between 25 and 40 km, Marconi Electronic Systems' I-band (8.6 to 9.5 GHz) S810 air/surface surveillance radar is reported to have been in recent service aboard Egyptian 'October' class guided missile fast patrol craft. The company's E/F-band (2.7 to 2.9 or 2.9 to 3.1 GHz) S820 radar is noted as having been installed aboard Egyptian 'Ramadan' class patrol craft.

Type 965
Type 965 is described as being a metric long-range air search radar with a peak power of 450 kW. Two types of antenna array — Outfits AKE(1) and AKE(2) — are noted as being fitted to the radar and it is reported to have seen recent service aboard Argentine 'Hercules' class destroyers (965P) Bangladesh 'Leopard' class frigates and the single frigate *Umar Farooq*; Chilean 'Leander' class frigates and Indonesian 'Khristina Tiyahahu' class frigates.

Type 992
A Marconi product, the Type 992 radar is described as being an E/F-band target indicator radar which, in its earliest version, had a detection range against air targets of 56 km. The later 992Q variant is quoted as being equipped with a fully stabilised, slotted waveguide antenna. Type 992 radars are noted as having seen recent service aboard Chilean 'Prat' class destroyers (992Q or R) and 'Leander' class frigates (992Q) and Pakistan 'Tariq' class frigates (992R). The Pakistan radars of this type are understood to be due to be replaced by Signaal DA08 (see separate entry) units as part of a three ship 'Tariq' class upgrade programme.

Type 993
The E/F-band Type 993 radar is described as being another target indicator which is equipped with a 'Quarter Cheese' antenna and is designed for use aboard frigates. Recent Type 993 installations are thought to include the Bangladesh 'Leopard' class frigates and the single frigate *Umar Farooq*; Indonesian 'Khristina Tiyahahu' class frigates and New Zealand 'Broad Beam Leander' class frigates.

Type 994
The E/F-band Type 994 target indicator radar is described as incorporating the Type 993's antenna array mated to the transceiver from the Plessey (now Siemens Plessey Systems) AWS-4 equipment. The radar is described as having a DMTI facility and is understood to have seen recent service aboard Chilean 'Leander' class frigates; Ecuadorean 'Leander' class frigates and the Royal Navy's aviation training ship *Argus*; 'Castle' class offshore patrol vessels and the assault ships *Fearless* and *Intrepid*.

Type 1006
As of this edition, Jane's sources report that the I-band (8 to 10 GHz) Type 1006 navigation and surface search radar have seen recent service aboard Argentinian 'Hercules' class destroyers; Australian 'Oberon' class attack submarines, 'Bay' class inshore minehunters, 'Freemantle' class patrol craft, the fleet replenishment tanker *Success* and the heavy lift ship *Tobruk*; Bangladeshi 'Island' class training ships; Brazilian 'Broadsword' and 'Humaita' class frigates; Canadian 'Oberon' class attack submarines'; Chilean 'Oberon' class attack submarines and 'Leander' class frigates; Ecuadorian 'Leander' class frigates; the Indian training ship *Krishna* and the aircraft carrier *Viraat*; the Indonesian survey ship *Dewa Kembar* and the replenishment tanker *Arun*; New Zealand 'Broad Beam Leander' class frigates; Pakistan 'Leander' and 'Tariq' class frigates; Philippines 'Jacinto' class corvettes; the Portuguese replenishment tanker *Berrio* and UK 'Swiftsure' class attack submarines, 'Invincible' class aircraft carriers (two ships), 'Hunt' class coastal minesweepers, the assault ships *Fearless* and *Intrepid*, the Antarctic patrol vessel *Endurance*, 'Castle/Island' class offshore patrol vessels, the training ships *Loyal Chancellor* and *Loyal Watcher*, the survey ship *Herald*, the aviation training ship *Argus* and 'Sir Bedivere' class landing ships.

VERIFIED

Naval Chart Display (NCD)

Type
Vector-raster electronic chart display and navigation/tactical workstation for naval applications.

Description
Kelvin Hughes NCD workstation is a commercial-off-the-shelf product that has been ruggedised and electromagnetic compatibility hardened for use in naval applications. As well as being able to display all authorised electronic chart formats (vector ECDIS (IHO S57 Edition 3 - Electronic Navigation Chart), vector DNC (NIMA VPF Digital Navigation Chart data (STANAG 7074) when harmonised with ECDIS), raster ARSC (UK), raster SEAFARER (Australia) and raster BSB (NOAA US national series and Canadian HO NDI product), the unit can be used as a navigation management tool. NCD can output data to most types of ship's command systems, autopilots and DP systems and system options include naval gunfire support and target motion analysis displays. Other unit features include:
- a real-time display of own ship's position
- instant availability of route lines (with alternatives)
- availability of DR/EP positions
- the ability to display navigation construction lines
- the ability to input and store large numbers (hundreds) of preplanned routes
- a continuous display of course/speed over ground data.

Status
As of this edition, NCD is thought to be available.

Contractor
Kelvin Hughes, Hainault.

VERIFIED

The Kelvin Hughes Naval Chart Display workstation   0024880

....

Sperry Marine surveillance and navigation radars

Type
F- (3 to 4 GHz) and I-band (8 to 10 GHz) series of navigation/surveillance radar.

Description
Northrop Grumman Electronic Systems sector - Sperry Marine's UK-developed F/I-band navigation and surveillance radars (formerly the Litton Marine Systems, Decca Division product line) are in service worldwide. Due to their longevity, these equipments have appeared in a number of series, the known details of which are as follows:

20V90 series
A family of modular F- or I-band radars built around the flat screen 20V90 raster scan colour display. An I-band 20V90 variant with a 2.7 m antenna (the 20V90/9) has been given the French Navy designation DRBN 34 while the 20V90TA model adds an air search capability which can detect targets travelling at speeds of up to 1,111 km/h at ranges of up to 74 km. The specific variant is also described as offering track repair, auto-intercept, track labelling, data hand-off to weapon systems, identification friend-or-foe display and helicopter recovery facilities.

Bridgemaster series
Family of F- or I-band radars that employ advanced video processing and are designed to provide high performance at reasonable cost. A Bridgemaster variant has been adopted by the UK Royal Navy as the Type 1008 navigation radar. As of

The Type 2459 F/I radar antenna carried by Netherlands fast combat support ship Zuiderkruis (Stefan Terzibaschitsch)

early 1996, approximately 5,500 Bridgemaster radars had been ordered by customers around the world.

Master series
The Master series includes the Types 1290, 1690 and 2690 equipments. The Types 1290 and 1690 radars are equipped with 30 and 41 cm Plan Position Indicators (PPIs) respectively while the Type 2690 set is fitted with a raster scan display. When fitted with an I-band transceiver, radars of this type employ 1.83 or 2.7 m antennas. F-band applications use a 3.7 m scanner.

Solid State series
The Solid State series includes the Types 1226, 1229 and 1230 radars. Of these, the Types 1226 and 1229 employ 30 cm PPI displays, operate in the I-band, have 20 kW outputs and are equipped with 1.83 and 2.7 m antennas respectively. The F-band Type 1230 utilises a 3.7 m scanner.

Transar series
A parallel development to the Master series, Transar radars are equipped with PPI displays. The series includes equipments such as the Types 916 and 1070.

Type 2459 F/I
A radar which is fitted with a combined F/I-band antenna and which can utilise a variety of transceiver and display units. A variant known as 2459 FID incorporates a D-band (1 to 2 GHz) identification friend-or-foe sub-array.

Status
Sperry Marine's UK-developed navigation and surveillance radar families are very widely used throughout the world. Examples of this usage are reported to include (or, over time, have included) the Australian heavy lift ship *Tobruk* (Type 916) and the survey ships *Flinders* and *Moresby* (both 916); Bahrain's 'Al Riffa', 'Al Manama' and 'Ahmed El Fateh' class Fast Attack Craft (FACs — all fitted with 1226); the Barbadian large patrol craft *Trident* (1226); Belgian 'Aggressive' and 'Flower' class mine warfare vessels (both 1229); Brunei's 'Waspada' class FACs (1229?); the Cameroon's FAC *Bakassi* (1226); the Canadian 'Improved Provider' class fleet replenishment tankers (969); Egyptian 'Osa' class FACs (916); French 'Clemenceau' class carriers, 'Georges Leygues', 'Suffren' and 'Tourville' class destroyers and 'D'Estienne d'Orves', 'Floréal' and 'La Fayette' class frigates (all 1226) together with the helicopter carrier *Jeanne d'Arc* and 'Cassard' class destroyers (both 20V90/9/DRBN 34); Gabon's 'P400' class large patrol craft (1226); Ghanaian 'Lürssen PB57' and FPB 45' class FACs (1226); Greek 'Hydra' class frigates (2690); Indian 'Nilgiri' class frigates (1226); Indonesian 'Fatahillah' and 'Van Speijk' class frigates (both 1229), 'Dagger' class FACs (1226), 'Sibarau' class large patrol craft (916) and 'Singa' class FACs and 'Kakap' class large patrol craft (both 2459); Iranian 'Alvand' class frigates (1226); Kenya's 'Madaraka' class large patrol craft (1226); 'Nyayo' class FACs (1226) Malaysian 'Lekiu' class frigates (20V90TA), 'Kasturi' class corvettes, 'Musytari' class off-shore patrol vessels and 'Mahamiru' class minehunters (all 1226), 'Handalan', 'Kedan', 'Kris', 'Perdana' and 'Sabah' class FACs (616 or 707 — 'Kedan', 'Kris' and 'Sabah') and 'Jerong' class FACs (626); Mexican 'Uribe' class gunships (1226); Netherlands' fast combat support ship *Zuiderkruis* (2459 and 1226) and 'Alkmaar' and 'Dokkum' class mine warfare vessels (1229); New Zealand 'Moa' class inshore patrol craft (916); Nigerian 'Erinomi' class corvettes, 'Ekpe' and 'Siri' class FACs and 'Abeking' and 'Makurdi' class large patrol craft (all 1226); Norwegian 'Oslo' class frigates, 'Storm' and 'Hauk' class FACs and 'Vidar' class coastal minelayers (all 1226); Omani 'Dhofar' class FACs (1226) and 'Al Waafi' class large patrol craft (1226 and 1229); Peruvian 'Velarde' class FACs (1226); Portuguese 'Baptista de Andrade' (316) and 'João Coutinho' (1226) class frigates; Qatar's 'Damsah' class FACs (1226); Saudi Arabian 'Modified La Fayette' and 'Madina' class frigates (1226); Singapore's 'Swift' class coastal patrol craft (1226); Spanish 'Anaga' class large patrol craft (1226); Thai 'Rattansakosin' class corvettes (1226); Turkish 'Barbaros' class (2690)

and 'Yavuz' class (1226) frigates and 'Dogan' and 'Kartal' class FACs (both 1226); United Arab Emirates' 'Muray Jib' class corvettes and 'Ban Yas' and 'Mubarraz' class FACs (all 1226); the UK assault ships *Albion* and *Bulwark* (Type 1008 as an option to the Kelvin Hughes 1007 set) and 'River' class minesweepers/patrol vessels (1226) and Uruguay's 'Commandant Rivière' class frigates (1226).

Contractor

Northrop Grumman Electronic Systems sector - Sperry Marine, New Malden.

UPDATED

....

Type 1007 navigation and search radar

Type

F- (3 to 4 GHz) and I-band (8 to 10 GHz) navigation and surface/air search radar.

Description

The Type 1007 is the standard I-band navigation radar of the Royal Navy. It consists of a range of navigation, surface and air search equipment for naval use. It includes a choice of antennas, I-band and F-band transmitter/receivers and a range of displays. Reduced magnetic signature variants are available for use on mine countermeasures vessels. The Type 1007 radar comprises antenna, transceiver and display subsystems, known details of which are as follows:

Antennas

Three versions of the antenna outfit are available for surface vessels: a 2.4 m single array, a 3.1 m single array, and a 2.4 m dual array for use with helicopter transponders and Outfit RRB. All are horizontally polarised, end-fed slotted line arrays, incorporating vertical polarisation filters to give low sidelobe and back radiation levels. The 3.1 m array has a horizontal beamwidth of 0.75°. Surface ship antenna outfits can operate in winds up to 185 km/h and withstand funnel gas temperatures up to 120°C, as well as gun and shock effects of blast. De-icing systems are available. A 4.1 m antenna is used with the F-band radar and an Identification Friend-or-Foe (IFF) antenna can be surmounted as an option. A fully pressure-tested submarine antenna is also available for fitting to a variety of submarine masts.

Transceivers

The transceiver unit used in the Type 1007 radar is solid state (with the exception of the magnetron) and operates at frequencies of 9,410 and 3,050 MHz with a transmitter power output of 25 kW. A wide dynamic range logarithmic receiver is provided. A built-in monitoring system is included to check that the equipment is operating at peak performance. A low leakage dummy load allows for system testing during periods of radar silence. Centralised emission control circuitry enables command to inhibit transmission immediately. Sector transmission is also incorporated with direct control from the main display (optional for the F-band radar). Blanking pulses are incorporated to safeguard sensitive electronic support measures equipment.

Display

The Colour Tactical Display (CTD) is a highly capable navigation display with a wide selection of operational/tactical facilities. It has a built-in tracking capability of up to 50 Automatically tracked targets and 20 Manually tracked targets. The CTD gives the operator a clear sharp colour tactical picture out to over 300 km with the ability to label tracks with ship's names/numbers. Symbology and a choice of colours are

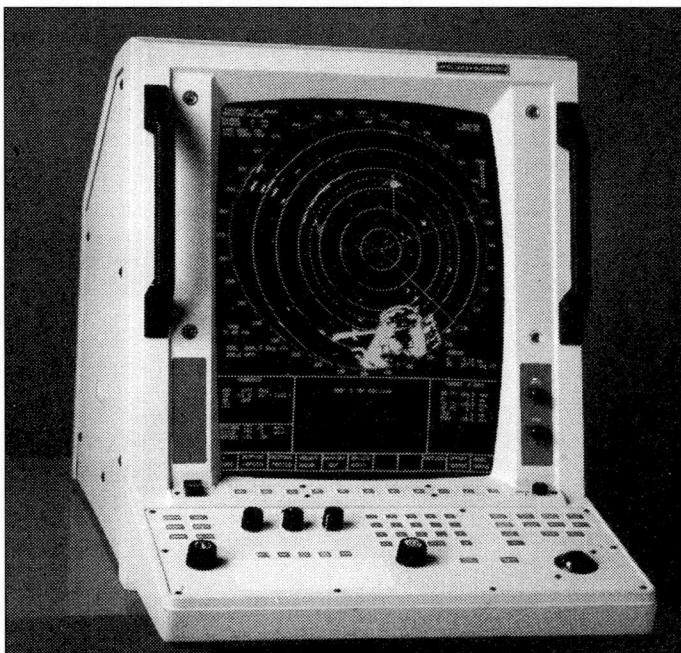

The Type 1007 Colour Tactical Display

used to indicate hostile/friendly/neutral/unknown and whether air/surface or subsurface.

Standard interfacing includes most log, gyro and Global Postioning System equipment. The CTD can receive four radar video inputs and three auxiliary video inputs. Optional facilities include the control and display of frequency modulated continuous wave radars, ability to receive plot extracted data, VESTA, carry out IFF active gate interrogation and interfacing to most combat systems, fire-control systems, infra-red and electronic support measures equipments. A high-speed air tracking facility is available as an option.

The CTD displays can be networked to share track data and form a highly cost-effective command system. A weatherproof auxiliary raster scan display is available as an option. This unit is portable and is designed for use on an open bridge or a submarine fin. A built-in simulator package allows onboard training to be carried out in the minimum of time.

Status

The Kelvin Hughes Type 1007 radar and its associated CTD has, over time, been procured by over 30 navies worldwide. Operational applications are thought to have include Australian 'Collins' class attack submarines; 'Newport' class helicopter support/tank landing ships; 'Huon' class coastal minehunters and the fleet replenishment tanker *Westralia*; the Bangladesh 'Leopard' class frigates; Jordanian 'Al Hussein' class Fast Attack Craft (FAC); Norwegian 'Modernised Kobben' and 'Ula' class attack submarines; Omani 'Qahir' class corvettes; Portuguese 'Albacora' class attack submarines, 'Comandante João Belo' and 'Vasco da Gama' class frigates and 'Cacine' class large patrol craft; Qatari 'Vita' class FACs and UK 'Vanguard' class nuclear-powered ballistic missile submarines. 'Trafalgar' class nuclear-powered attack submarines, 'Duke' class frigates, 'Sandown' class minehunters 'Fort Victoria' class fleet replenishment ships, the proposed assault ships *Albion* and *Bulwark* and the helicopter carrier *Ocean*.

Contractor

Kelvin Hughes, Hainault, UK.

VERIFIED

────────────────────────────

UNITED STATES OF AMERICA

Introductory note

Alongside the equipments described in the following pages, a number of previous generation US surveillance/navigation radars are thought to remain in service around the world. Details of a number of such emitters are as follows:

BPS-15

BPS-15 is described as being a Sperry I-/low J-band submarine navigation and surface search radar with a power output of 35 kW. Variants of BPS-15 are reported to have been installed aboard US Navy 'Ohio' class nuclear-powered ballistic missile submarines (BPS-15A) and 'Los Angeles' class (BPS-15A as an alternative to BPS-16) nuclear-powered attack submarines.

SPS-5

SPS-5 is described as being a Raytheon H-band (6,400 MHz) surface search and navigation radar that was introduced into US Navy service during the early 1950s. SPS-5 was fielded in at least five variants of which, SPS-5A is noted as having introduced a new antenna format and SPS-5C, a peak power of 350 kW. Most recently, SPS-5 radars have been reported as being installed aboard Indonesian 'Samadikun' class frigates; the Myanmar corvettes *Yan Taing Aung* and *Yan Gyi Aung*; Greek 'Patapsco' class support tankers and the dock landing ship *Nafkratoussa*; the Philippines frigate *Rajah Humabon;* Taiwanese 'Crosley' class frigates, 'Auk' class corvettes, 'Cabildo/Ashland' class dock landing ships, the repair ship *Yu Tai* and the frigate *Pin Klao* and Thai 'LSM-1' class landing ships and the frigate *Tachin*.

SPS-10

SPS-10 is described as being a Sylvania G/H-band (4 to 8 GHz) surface search radar that was initially fitted with a 3 m parabolic cylinder antenna. First introduced into US Navy service during late 1953, SPS-10 appeared in at least six variants. Of these, SPS-10B was equipped with a 500 kW transmitter, SPS-10E introduced a new antenna format and SPS-10F had a new pulse repetition frequency of 625 to 660 pps. Over time, the SPS-10 radars have been progressively replaced by the SPS-67 equipment (see entry) but the system has, in recent times, being reported as being installed aboard Brazilian 'Pará' class frigates (SPS-10C); Canadian 'Improved Restigouche' class frigates (possibly now replaced by SPS-67 in at least one case); Egyptian 'Knox' class frigates (as an alternative to SPS-67); German 'Lütjens' class destroyers; Greek 'Kimon' class (SPS-10D/F) and 'Epirus' class frigates (SPS-10F) and 'Terrebonne Parish' class landing ships; Iranian 'Babr' class frigates (SPS-10B); South Korean 'Gearing' class destroyers and 'Ulsan' class frigates (SPS-10C - four ships); Mexican 'Bronstein' class frigates (SPS-10F); Pakistan 'Gearing' class destroyers; Spanish 'Baleares' class frigates and 'Paul Revere' class attack transports; Taiwanese 'Wu Chin I/II' class destroyers, 'Knox' class frigates (alternative to SPS-67) and the amphibious attack flagship *Kao Hsiung*; Thai 'Knox' and 'Tachin' class frigates (one ship) and Turkish 'Gearing' and 'Carpenter' class destroyers and 'Berk' and 'Tepe' class frigates (alternative to SPS-67).

SPS-29

SPS-29 is described as a Westinghouse VHF band (30 to 300 MHz) air search radar with an instrumented range of 500 km. It was produced in six variants. As of this edition, SPS-29 radars were reported as having been in recent service aboard Mexican 'Gearing' class destroyers and Taiwanese 'Wu Chin I/II' class destroyers.

SPS-53

Described as being a Sperry I-/low J-band high resolution search radar which employs a 1.5 m slotted array antenna and has a 40 kW peak power, Most recently, SPS-53 radars have been reported as being installed aboard Greek 'Tolmi' class patrol craft; the Indonesian landing ships *Teluk Bone* and *Teluk Saleh* and the repair ship *Jaya Wijaya*; South Korean 'Diver' and 'Edenton' class salvage ships; the Malaysian landing ships *Raja Jarom* and *Sri Banggi*; the Philippines coastguard tender *Kalinga*; the Taiwanese salvage ship *Ta Hu*; the Turkish patrol craft *Bora*; US Army/Navy 'LCU 1600' class utility landing craft; the Venezuelan patrol craft *Miguel Rodriguez* and Vietnamese 'Admirable' class corvettes.

VERIFIED

....

AN/BPS-16 navigation and search radar

Type

I-band (8 to 10 GHz) submarine navigation and surface search radar.

Description

The AN/BPS-16 (previously known as the AN/BPS-XX) is a submarine radar designed to provide nuclear-powered fast attack and ballistic missile submarines with a navigation and search radar capability when operating on the surface. BPS-16 features a new 50 kW frequency-agile transmitter in I-band and the latest in signal processing techniques to enhance operational performance in heavy weather. Unlike its predecessor, BPS-15, the new system is supplied with a new unique radar mast assembly to raise and retract a modern antenna on a more effective and reliable basis. Range is understood to be approximately 50 km.

Status

As of 1995, the BPS-16 radar was reported as having completed First Article Testing and US Navy (USN) operational evaluation. As of late 2000, the service is understood to have procured at least 10 BPS-16 sensors for use on its 'Seawolf' and 'Los Angeles' class nuclear-powered attack submarines. Recent identified BPS-16 programme activity comprises Litton being awarded a then year US$7,639,942 not-to-exceed letter contract on 11 June 1999 that covered the supply of three (additional ?) BPS-16 radars to the USN. At the time of the contract announcement, work on the effort was scheduled for completion by the end of January 2002 and its contracting activity was the USN's Naval Sea Systems Command, Arlington, Virginia.

Contractor

Northrop Grumman Electronic Systems sector - Sperry Marine, Charlottesville, Virginia.

UPDATED

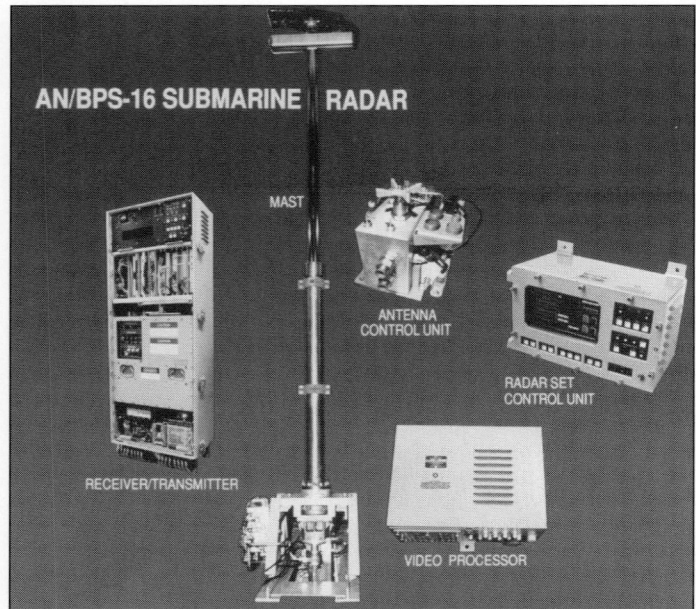

The elements that make up the AN/BPS-16 submarine navigation/surface search radar

....

AN/SPQ-11 Cobra Judy missile tracking radar

Type

E/F- (2 to 4 GHz) and I/J- (8 to 20 GHz) ballistic missile tracking radar.

Description

The Cobra Judy shipborne phased-array radar is a detection and tracking system that is designed to collect data on non-US strategic ballistic missiles tests. The large octagonal array of this radar measures 6.86 m in diameter and is composed of 12,288 antenna elements. The elements are co-ordinated to form transmission patterns by a large scale Control Data CYBER 175-112 computer that allows the radar to detect and track objects at very high rates. After detection and tracking, the information is fed to a computer for storage and display as well as post-mission reduction and analysis.

Cobra Judy is installed on the stern of a former merchant vessel, the USNS *Observation Island*, which has been extensively modified. The system weighs 250 tons (approximately 250,000 kg), stands four storeys high and is integrated into a steel turret that can be rotated mechanically. The octagonal array makes up one wall of the pyramid-like structure. The complete shipborne sensor is operating in the Pacific Ocean, based at Pearl Harbor, Hawaii.

The Cobra Judy system has been upgraded (in a programme managed by the US Air Force Electronic Systems Division) to include an I/J-band radar subsystem (using a parabolic dish antenna) to complement the existing E/F-band phased-array capability. The addition of this new equipment was designed to improve the ability to observe and collect intelligence data on the terminal phase of ballistic missile tests. Since operation in I/J-band offers a better degree of resolution and target separation, the additional system may well improve intelligence considerably by the ability to differentiate between multiple warheads and electronic countermeasures chaff and decoys. The I/J-band subsystem's dish antenna is installed forward of the phased-array, aft of the ship's funnel. The dish and pedestal have been supplied by the USAF and were previously used as a shipborne telemetry radar.

Status

As of this edition, the AN/SPQ-11 missile tracking radar was reported as being in service. The latest identified SPQ-11 contracting activity was a then year US$11,824,227 firm, fixed-price contract that was awarded to the then Raytheon Support Services (Burlington, Massachusetts) on 14 May 1999 and covered the operation and maintenance of the Cobra Judy and Cobra Gemini radars aboard the USNS *Observation Island* during the period 14 May 1999 through 13 May 2000. The US Air Force's 668th Logistics Squadron, Kelly Air Force Base, Texas was the contracting activity for the effort.

Contractor

Raytheon Electronic Systems, Marlboro, Massachusetts.

UPDATED

The Cobra Judy radar installed on the USNS Observation Island, *showing both E/F- and I/J-band radars*

....

AN/SPS-40 surveillance radar

Type

UHF band (300 to 1,000 MHz) air surveillance radar.

Description

The AN/SPS-40 is a two-dimensional naval air search and surveillance radar for detection of targets at long and medium ranges. A nominal range of 320 km has been quoted. SPS-40B baseline (which includes the B, C and D radars) is designed to provide optimum performance capabilities with minimum operator interface. Special features of the SPS-40B include long-range resolution and accuracy, lightweight and flexible packaging for easy shipboard installation, field proven high reliability, maintainability and availability. The UHF(B) band operating frequency provides freedom from weather clutter and low vulnerability to anti-radiation missiles. The system's digital moving target indicator provides excellent subclutter visibility and has solid-state receiver, power supplies and controls. The antenna reflector is of open lattice work construction with an integral identification friend-or-foe antenna.

The antenna used in the AN/SPS-40 surveillance radar (atop the mainmast in this view)

The SPS-40 Solid-State Transmitter (SSTx) replaces system's tube transmitter in the SPS-40E model. To ensure maximum system availability, the SSTx architecture is highly redundant. It is predicted to have a 90 per cent probability of maintenance-free operation for 90 days with no more than 11 per cent projected reduction in radar range performance. Transmitter modules are identical and interchangeable, as also are the power supplies. In the event of component failure, the system undergoes a gradual and graceful degradation in transmitter output. It remains fully operational and capable of detecting targets. The device's solid-state technology offers inherent tactical flexibility. For example, output power is adjustable. As a result, ships can reduce their susceptibility to detection while maintaining substantial air surveillance capability. If the tactical situation requires emission control conditions, the SSTx will respond instantly. Similarly, the transmitter will immediately radiate at full power with just the touch of a push-button. Pulse-to-pulse frequency is also provided. A unique automatic levelling control system greatly reduces the need for maintenance actions. This system automatically senses and compensates for degradations in transmitter module performance.

Status
AN/SPS-40 is reported to have been introduced into US Navy service during the early 1960s and over time, is noted as having been installed aboard a range of the service's ship classes including 'Virginia' class cruisers (SPS-40B); 'Spruance' class destroyers (SPS-40B/C/D in 30 ships); 'Blue Ridge' class amphibious command ships (SPS-40C); 'Tarawa' class amphibious assault ships (SPS-40B/C/D); 'Iwo Jima' class amphibious assault/mine countermeasures support ships; 'Austin' class amphibious transport dock ships (SPS-40B/C) and 'Hamilton/Hero' class US Coast Guard cutters (SPS-40B).

Outside the US, SPS-40 radars are reported to have been fitted to a range of major combatants including Australian 'Perth' class destroyers; the Brazilian aircraft carrier *Minas Gerais* (SPS-40B), the destroyer *Mariz E Barros*, 'Allen M Sumner' class destroyers (three) and 'Para' class frigates (SPS-40B); Egyptian 'Knox' class frigates (SPS-40B); German 'Lütjens' class destroyers; Greek 'Kimon' class (SPS-40B) and 'Gearing' class destroyers and 'Epirus' class frigates (SPS-40B); South Korean 'Gearing' class destroyers; Mexican 'Gearing' class destroyers and 'Bronstein' class frigates (SPS-40D); Pakistan 'Gearing' class destroyers; Saudi Arabian 'Badr' class corvettes (SPS-40B); Taiwanese 'Gearing class (Wu Chin I and II)' and 'Allen M Sumner' class destroyers and 'Knox' class frigates (SPS-40B); and Thai 'Knox' class frigates (SPS-40B); Turkish 'Gearing' and 'Carpenter' class destroyers and 'Berk' and 'Tepe' class (SPS-40B) frigates.

SPS-40 is noted as having been approved for use with the AN/SYS-1 integrated Automatic Detection and Tracking System (ADTS). ADTS correlates inputs from multiple 2/3-D radars and combines them into a single non-duplicated track. Since its introduction, SPS-40 has been upgraded on a number of occasions. One of the system's latest versions — the SPS-40F — includes the solid-state transmitter described above. In May 1992, Northrop Grumman is reported to have been awarded a then year US$15 million US Navy contract for 15 such transmitters, all of which are understood to have been deployed operationally.

Specifications
AN/SPS-40E
Frequency: UHF band (400-450 MHz in 10 channels)
Peak power: 125-225 kW
Average power: 2 kW
Transmitter pulse-width: 60 µs (LRM); 3 µs (SRM)
Scan period: 4/8 s
Track file size: 511
Air surveillance operational capabilities

	Short-range	Long-range
Min range	500 m	3.7 km
Pulse-width	3 µs	60 µs
Pulse compression ratio	-	60:1
Max range	-	370 km

	Short-range	Long-range
Azimuth beamwidth	10°	10°
Vertical beamwidth	19°	19°
Range resolution	457 m	152 m
MTI improvement factor	54 dB	54 dB

Contractors
Northrop Grumman Norden Systems, Melville, New York.
Northrop Grumman Corporation, Electronic Sensors and Systems Division, Baltimore, Maryland.

VERIFIED

AN/SPS-48E air surveillance radar

Type
E/F-band (2 to 3 GHz) long-range 3-D air surveillance radar.

Description
AN/SPS-48E is a 3-D long-range air surveillance radar that can be configured for both ground-based (fixed site and mobile) and shipboard applications. It uses a combination of mechanical scanning in azimuth and electronic beam steering in elevation to provide plan position and height information on air targets. Multiple beams are formed and elevation scanning is accomplished under computer control by multiplexing the E/F-band transmitter radio frequency output to the planar-array frequency-sensitive antenna, thereby simultaneously radiating a series of nine 1.5° pencil beams. These nine beams are overlapped to form a 5.6° elevation group and eight groups typically form a 45° elevation scan. Transmitter power levels and instrumented ranges are adapted to the coverage volume to improve detection probabilities. Stabilisation of the beam position for ship's motion is achieved by electronically changing the transmission frequencies as a function of ship's pitch and roll. Four operational modes are available under computer control and a range of over 400 km at an altitude of 30,500 m for target designation and aircraft direction is reported.

The SPS-48E configuration is understood to have emerged from a 'significant' system redesign that took place during the late 1980s and formed part of the US Navy's (USN) New Threat Upgrade (NTU) programme, with 45 such radars being delivered for use aboard a range of ships including aircraft carriers, amphibious assault ships and amphibious landing docks. The NTU modifications added increased transmitter power, a new low sidelobe antenna and adaptive digital processing to facilitate the detection and tracking of high altitude anti-shipping

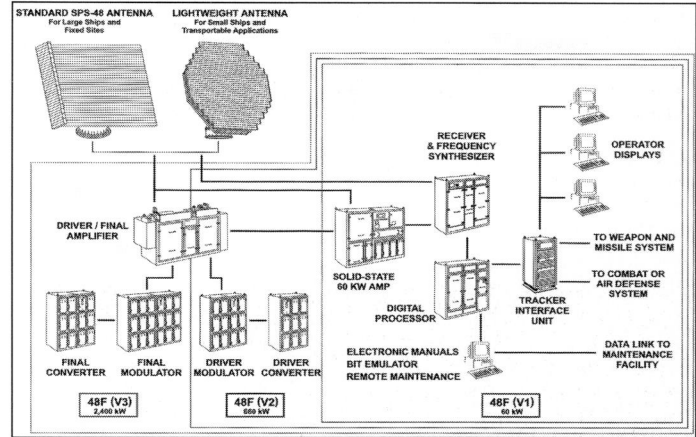

A schematic showing the layout of the three SPS-48F configurations (ITT Gilfillan)
2002/0114010

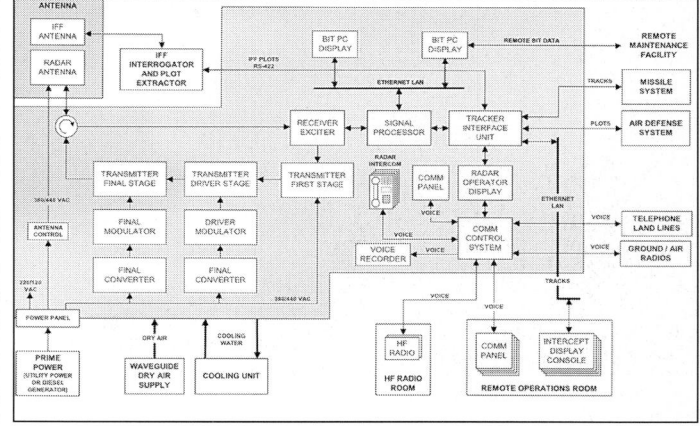

A schematic showing a typical SPS-48 land-based installation (ITT Gilfillan)
2002/0114011

missiles launched in a heavily countermeasures polluted environment from aircraft such as the Tupolev Tu-26. A low elevation field change kit with Digital Moving Target Indicator (DMTI) and Doppler processing was added during 1991 and has been installed in all SPS-48 radars to allow the detection and tracking of low-flying anti-ship cruise missiles. Thus modified, the radar provides positional data for launch on search and mid-course guidance for the SM-2 weapon system together with track data for the Rolling Airframe Missile and Sea Sparrow missiles.

As of 1998, the US Navy was understood to have begun upgrading its SPS-48E radars to include a Co-operative Engagement Capability (CEC) interface that would allow distribution of fire-control quality data to other ships in a battle group. The CEC interface (consisting of VME based commercial-off-the-shelf 68040 processors and interfaces) is also used to interface the radar with the Ship Self Defense System (SSDS) fitted to US Navy 'L' class vessels.

A lightweight antenna variant (less than 1,500 kg) - known as the SPS-48F - has been proposed by ITT Gilfillan for export purposes, subject to US Department of Defense approval. The antenna would be half the weight of the standard SPS-48E unit and provide identical detection range together with a 4 second data rate. SPS-48F is offered in three variant configurations (V1, V2 and V3) depending on the required transmitter power levels. The V1 uses a single stage 60 kW transmitter that changes to a two-stage 660 kW unit in the V2 model. The V3 variant employs a three-stage 2.2 MW transmitter.

Status

Delivery of SPS-48E systems began during mid-1986 and the first such radar was installed aboard the USS *Biddle* during the early part of the following year. The USN is understood to have received a total of 45 SPS-48E radars by July 1994 and as of April 2001, Jane's sources were reporting the sensor as being in service aboard USN 'Nimitz' (eight in service as of the given date) and 'Kitty Hawk/John F Kennedy' (three) class aircraft carriers, 'Tarawa' (five) and 'Wasp' (seven) class amphibious assault ships, 'San Antonio' (four building) class amphibious transport docks and the single unit aircraft carrier USS *Enterprise*. The radars installed aboard the described amphibious warfare vessels are noted as having been recovered from decommissioned cruisers. On 1 March 2000, ITT Gilfillan announced that it had been awarded a then year US$13.7 million contract as part of a continuing SPS-48E refurbishment programme. Here, the specific award covered the supply of three complete upgrade kits (thought to include the described CEC interface, VME-based 68040 processors, Ethernet interfaces and SPARC-10 workstations) the refurbishment of three radars and 'related items' and installation services. Work on the effort was scheduled for completion by the end of June 2002 and at the time of the announcement, the new contract brought the number of USN SPS-48Es to have gone through the programme (which was then scheduled to run until 2025) to 12. As of 2000, a refurbished SPS-48E configuration was also understood to be available for export under the US Foreign Military Sales procedure. On 29 June 2001, the US Marine Corps Systems Command awarded ITT Gilfillan a then year US$143 million contract covering the procurement of six land-based SPS-48E air surveillance radars (plus related logistical services) for Egypt. At the time of the announcement, work on the effort was scheduled for completion by the end of January 2006. In this context, it is also worth noting that the USN is reported as making use of land-based SPS-48E radars as well as those installed aboard the surface ships/classes described previously.

Specifications
SPS-48E/F
Frequency: E/F-band (2.9-3.1 GHz)
Transmitter power: 60 kW (V1); 660 kW (V2); 2.2 MW (V3)

The large ship/fixed land site antenna used in the SPS-48E radar system. Also visible are two AN/SPG-55D fire-control radar antennas immediately below

Range: 407 km
Altitude: up to 30,480 m
Polarisation: horizontal

Contractor
ITT Gilfillan, Van Nuys, California.

UPDATED

AN/SPS-49(V) air surveillance radar

Type
UHF band (300 to 1,000 MHz) long-range air surveillance radar.

Description
AN/SPS-49(V) is a very long-range 2-D air search radar designed for use as the primary detection radar aboard various combatant ships of several countries. In its AN/SPS-49(V)5 iteration, the system features:

- high average power for long-range surveillance and detection of low cross-section threats
- automatic target detection for reliable detection and rapid designation of all target threats
- high electronic counter-countermeasures for assured surveillance in hostile environments
- high bearing accuracy for more accurate target control and designation
- pulse-Doppler processing and clutter maps for weather, land and sea clutter rejection and suppression of chaff
- horizon stabilised antenna for consistent elevation coverage
- digital interface for rapid data transmission
- the ability to track up to 255 targets.

The antenna used is a double-curved reflector which provides 28.5 dB gain with cosec^2 shaping and low sidelobes. Separate auxiliary omni-antennas are installed above and below the main reflector for coherent sidelobe cancellation. Servo-controlled line of sight stabilisation is provided to maintain the antenna beam on the horizon. The SPS-49(V) transmitter consists of a fully coherent solid-state driver and klystron amplifier chain providing 360 kW of peak power. Fully developed and thoroughly tested it provides stable pulse trains for improved Doppler processing with Pulse Repetition Frequency (PRF) stagger and frequency agility. A four loop Coherent SideLobe Canceller (CSLC) is automatically activated to provide significant reduction in the level of electronic countermeasures energy entering the antenna sidelobes. The radar's jamming resistance is further enhanced by sidelobe blanking, frequency agility, single pulse interference reduction, an anti-chaff mode, variance cross field amplification and unspotting of the beam to minimise main lobe jamming.

The AN/SPS-49A (also known as the Medium PRF Upgrade - MPU) variant of the basic design features a new waveform, increased signal processing capability and improved subclutter visibility for increased detection of targets near land. The US Navy is upgrading existing AN/SPS-49 radars with MPU modification kits. SPS-49A radar and MPU modification kit production began during 1994.

Status
By the end of 1994, well over 200 SPS-49 radars are reported to have been sold. During January 1995, Taiwan is noted as having ordered a single SPS-49(V)1 and two SPS-49(V)5 radars together with related medium PRF upgrade kits. Elsewhere (and over time), SPS-49 radars (with sub-types where known) are reported as having been installed aboard or selected for installation aboard a range of ships including Australian 'Adelaide' (V8) and 'ANZAC' (V8) class frigates; Canadian 'Halifax' class frigates (V5); South Korean 'KDX-2' class destroyers (SPS-49(V)5 selected); New Zealand 'ANZAC' class frigates (V8); Spanish 'Santa María' class frigates (V5); Taiwanese 'Cheng Kung' class frigates (V5) and US Navy (USN) 'Kitty Hawk/John F Kennedy' class (V5) and 'Nimitz' class (V5) aircraft carriers (plus the single unit USS *Enterprise* (V5)); 'California' (V5), 'Ticonderoga' and 'Virginia' (V5 as an alternative to SPS-40B) class cruisers; 'Kidd' (V5) and 'Spruance' class (one)

The antenna for the SPS-49 radar installed aboard the 'Spruance' class destroyer USS Hayler (Stefan Terzibaschitsch)

destroyers; 'Oliver Hazard Perry class frigates (V4 and V5) 'Wasp' class multipurpose amphibious assault ships (V9) and 'Whidbey Island/Harpers Ferry' class landing docks/landing docks - aircraft carriers. As of May 2001, identified SPS-49 programme activity during 1999/2001 comprises:

15 December 1999
Raytheon Electronic Systems was awarded a then year US$9,256,165 time and materials award covering the supply of approximately 1,600 parts for the Sidewinder air-to-air missile, the SPS-49 radar, the NATO Sea Sparrow surface-to-air missile and four other weapon systems. At the time of the announcement, work on the effort was to be undertaken at Chula Vista, California (63 per cent workshare), Norfolk, Virginia (23 per cent) and Goleta, California (14 per cent) and was scheduled for completion by the end of December 2000. The contracting activity for this programme was the US Navy's Naval Control Point, Mechanicsburg, Pennsylvania.

30 March 2001
Raytheon Electronic Systems was awarded a then year US$8.2 million firm, fixed-price contract covering the production of one SPS-49A(V)1 Radar FC5 modification kit, four SPS-49(V)5 radar upgrade kits, SPS-49(V)5 INCO spares and SPS-49(V)5 90 days spares for the USN (24.07 per cent of the programme) and the navies of South Korea (58.64 per cent) and Taiwan (17.29 per cent). At the time of the announcement, work on the effort was to have been split between Andover, Massachusetts (84 per cent workshare); Sudbury, Massachusetts (15 per cent) and Waterloo, Ontario, Canada (1 per cent) and was scheduled for completion by the end of April 2004. The programme's contracting activity was the USN's Naval Sea Systems Command, Arlington, Virginia.

Specifications
Frequency band: UHF band (850-942 MHz)
Transmitting power: 360 kW peak; 13 kW average
Frequency selection: fixed or agile
Range: >463 km
Range accuracy: 0.09 km
Azimuth accuracy: 0.5°
Antenna
Dimensions: 7.3 m wide × 4.3 m high
Beamwidth: 3.4° (azimuth); cosec^2 (to 30° elevation)
Gain: 28.5 dB
Scan rate: 6 or 12 rpm
IFF: Type AS-2188 antenna mounted on boom
System weight: 1,425 kg (above decks); 6,325 kg (below decks)

Contractor
Raytheon Electronic Systems, Marlboro, Massachusetts.

UPDATED

....

AN/SPS-52 air surveillance radar

Type
Family of shipboard air surveillance radars.

Description
The SPS-52 series is reported to be a development of the earlier AN/SPS-39A radar which initially featured the SPS-39A architecture combined with a new digital stabilisation computer, planar antenna (see following), parametric amplifier and wide pulse feature to increase range. Over time, the equipment has spawned a succession of variants, the known details of which are as follows:

AN/SPS-52A
Reported to incorporate minor modifications over the original SPS-52 radar. Both SPS-52 and SPS-52A are quoted as being no longer in production.

AN/SPS-52B
Noted as introducing a new solid-state receiver/exciter/processor and a double digital canceller in place of SPS-52A's analogue Moving Target Indicator (MTI), SPS-52B is reported as being no longer in production.

AN/SPS-52C
SPS-52C is described as being an E-band (2 to 3 GHz) 3-D air surveillance radar that utilises mechanical rotation of its antenna to establish target bearing and electronic scanning to determine target elevation. The radar is reported to employ the same AN/SPA-72B antenna as was used in earlier versions of the equipment combined with completely different below-decks electronics. As such, SPS-52C is described as offering significant improvements over its predecessors in terms of detection performance, availability, reliability and maintainability.

The SPA-72B antenna is a backward tilted (25° from the vertical) planar-array made up of rows of slotted waveguide radiators which, in turn, are fed by a sinuous feed system which runs the length of one of the sides of the antenna. Scanning in the vertical plane is achieved by computer controlled variations in transmission frequency. The radar has four operating modes designated as High Angle, Long Range, High Data Rate and MTI. Each mode has a distinct search pattern and is operator selectable according to the particular operational/environmental conditions. Known details of these operating modes are as follows:

The SPS-52C antenna on the destroyer Perth *of the Royal Australian Navy*

High angle
Described as being SPS-52C's primary operating mode, the high-angle mode is noted as offering medium range (up to approximately 280 km instrumented) coverage over a 0 to 45° arc in elevation with a 6 second data refresh rate.

Long range
The long-range mode is described as offering long-range detection out to 450 km over a 0 to 13° arc in elevation with an 8 second data rate.

High data rate
The high data rate mode is noted as providing a detection range of up to approximately 175 km over a 0 to 45° arc in elevation and is described as being particularly effective against close in and pop-up targets such as helicopters.

MTI
The MTI mode is reported to offer coverage out to ranges of 110 km over a 0 to 38° arc in elevation.

Status
Over time, AN/SPS-52 air surveillance radar installations (with sub-variant where known) are reported to have included Australian 'Perth' class destroyers (C); German 'Lütjens' class destroyers; Italian 'Audace' and 'De la Penne' class destroyers together with the cruiser *Vittoria Veneto* and the aircraft carrier *Giuseppe Garibaldi* (all C); Japanese 'Hatakaze' (C) and 'Tachikaze' (B or C (one ship only)) class destroyers; Spanish 'Balearas' class frigates (B) together with the aircraft carrier *Principe de Asturias* (C); the Thai aircraft carrier *Chakri Naruebet* (C) and US Navy 'Wasp' class multipurpose amphibious assault ships (C — one) and 'Tarawa' class general purpose amphibious assault ships (SPS-52 as an alternative to SPS-48E).

Specifications
SPS-52C
Frequency: E-band (2-3 GHz)
Power output: 1 kW
Antenna: tiled (25°) planar-array
Range: 450 km
Altitude coverage: 30,480 m

Contractor
Raytheon Electronic Systems, Fullerton, California.

VERIFIED

....

AN/SPS-55 search and navigation radar

Type
I-band (8 to 10 GHz) surface search and navigation radar.

Description
The AN/SPS-55 is a solid-state I-band surface search and navigation radar developed as a replacement for the AN/SPS-10. It is designed for service on ships of destroyer size or above. Operational uses are: the detection of small surface targets from ranges of less than 50 m to the radar horizon; navigation and pilotage; tracking of low-flying aircraft and helicopters; detection of submarines at snorkel and periscope depth. The system's lightweight antenna (less than 90 kg) has a low profile configuration to minimise installation space requirements and consists of two (selectable) back-to-back, end-fed, slotted arrays, one with circular polarisation

SPS-55 antenna on the US Navy guided missile destroyer Kidd

and the other linear-horizontal polarised. The horizontal beamwidth is 1.5° and beam squint compensation is used to optimise bearing accuracy over the operating frequency range. Vertical beamwidth is 20°. SPS-55's transceiver subsystem is housed below-decks in a single cabinet and is capable of operating at any selected frequency in the band from 9.05 to 10 GHz. Two pulse-widths (1 and 0.12 μs) are provided. The minimum peak transmitter output is 130 kW. Variable sector radiation is also provided. The SPS-55 set does not normally include its own display and a separate control unit is provided to permit remote operation of the transmitter/receiver and scanner subsystems. The same basic modules can be used with a G-band (4 to 6 GHz) antenna to form the AN/SPS-502 radar, which has more than 95 per cent parts commonality with the AN/SPS-55.

Class A antenna kit
Cardion's Class A antenna modification kit provides a lightweight system for I/J-band radar detection for surface search and navigation with Identification Friend-or-Foe (IFF) radar. The D-band (1 to 2 GHz) IFF radar, mounted above the I/J-band radar on a single pedestal, provides low vertical beam tilt and wide vertical beamwidth, allowing both surface search and aircraft detection.

The kit contains the antenna group and receiver/transmitter modification hardware. Mounted below decks and supplied as part of the modification kit to the receiver/transmitter, the bearing circuits perform azimuth squint correction and relative to true bearing correction.

Status
Over time, AN/SPS-55 search and navigation radars are understood to have been installed aboard a range of ship types including Australian 'Adelaide' class frigates; Saudi Arabian 'Badr' class corvettes; the Spanish aircraft carrier *Principe de Asturias* together with the same country's 'Santa Maria' class frigates; Taiwanese 'Cheng Kung' class frigates and the US Navy's 'Ticonderoga' and 'Virginia' class cruisers, 'Kidd' and 'Spruance' class destroyers, 'Oliver Hazard Perry' class frigates and 'Avenger' class mine countermeasures vessels.

Specifications
Antenna
Rotation rate: 16 rpm
Polarisation: circular or linear
Horizontal beamwidth (3 dB): 1.5°
Vertical beamwidth: 20°
Gain: 31 dB
Transmitter
Frequency: I-band (9.05-10 GHz). A G-band (5.45-5.825 GHz) version is also available
Peak power: 130 kW
PRF/pulse-width: 750 pps/1 μs; 2,250 pps/0.12 μs
Receiver
Type: low noise, image-suppression mixer
IF: 60 MHz
Bandwidth: 1.2 MHz (long pulse); 10 MHz (short pulse)
Receiver processors: linear logarithmic, FTC, variable sensitivity time control

Contractor
Cardion Inc, Woodbury, New York.

VERIFIED

AN/SPS-58/65 search and target acquisition radar

Type
D-band (1 to 2 GHz) family of air search and target acquisition radars.

Description
The SPS-58/65 family incorporates at least five variants of the basic SPS-58 D-band air search and target acquisition radar. Known details of these are as follows:

SPS-58
Designed for use with the US Navy's Point Defense Missile System (PDMS), SPS-58 is described as having been optimised for use against low-flying threats by virtue of its low elevation angle performance, data rate and clutter rejection capabilities. In total, the system comprises an antenna, transmitter, receiver/processor, antenna control unit, signal data converter and display unit. The transmitter used is a solid-state, self-contained, air-cooled, coherent amplifier unit with klystron power amplifier stages while the receiver/processor (which is also solid-state) features a low noise superheterodyne receiver and digital Moving Target Indicator (MTI) filtering. Antenna options include a stabilised, horn-fed 4.9 m elliptical unit, a small ship planar-array and a dual feed unit which allows SPS-58 to share the lightweight radar/identification friend-or-foe array used in the G-band (4 to 6 GHz) AN/SPS-10 surface search radar (see following). Alongside the PDMS, SPS-58 type radars are noted as having been successfully integrated with the US Navy's Mk 37 and Mk 68 fire-control systems along with the Mk 5 WDS suite for the Tartar and Standard anti-aircraft missiles. Radars of this type can also interface with the NATO Sea Sparrow missile system.

SPS-58A
SPS-58A comprises a lightweight integral antenna, the SPS-58 transmitter and receiver/processor and a radar control unit. The equipment is designed to feed radar data directly into its host vessel's action/data display system.

SPS-65(V)1
SPS-65(V)1 is an upgraded version of the basic radar which incorporates the SPS-58's transmitter, a high-performance variant of its receiver/processor, a digital interface (to MIL-STD-1397) and a control unit which allows remote control of the system. The interface unit allows the radar to communicate with its host vessel's tactical display system and can be used to facilitate automatic target detection and tracking if its output is fed into a suitable external computer. Antenna options are as described for SPS-58.

SPS-65(V)2
As SPS-65(V)1 but with the addition of what is termed an advanced radar/weapons interface unit.

SPS-65(V)ER
SPS-65(V)ER incorporates a 25 kW, 0.005 duty cycle, solid-state power amplifier and phase-coded pulse compression to increase detection range and target resolution. The equipment also utilises burst-on-burst frequency agility to provide an electronic counter-countermeasures capability. A new three burst transmission strategy is used to enhance further the radar's MTI and range performance as well as provide Doppler coverage. SPS-65(V)ER has a number of antenna options comprising a dedicated planar Doppler array, the ability to function using the existing AN/SPS-6, AN/SPS-12 and SPS-58 arrays and use of a Northrop Grumman developed 1,215-1,400/5,255-5,925 MHz dual feed which allows the SPS-65(V)ER and SPS-10 radars to use the latter's antenna simultaneously, SPS-65(V)ER has an instrumented detection range of 185 km and may also have been known at one time as SPS-58LR. This alternative nomenclature is not confirmed.

Status
As of the period February to April 2001, Jane's were reporting AN/SPS-58 and -65 radars as being installed aboard the following warships and classes of warship:

South Korea	• 'Pae Ku' class fast attack craft - missile (five ships in class - SPS-58)
Taiwan	• the amphibious warfare flagship *Kao Hsiung* (SPS-58)
	• 'Wu Chin III' class destroyers (seven ships - SPS-58)
	• 'Lung Chiang' class fast attack craft - missile (two ships - SPS-58)
USA	• the fast combat support ships *Camden*, *Detroit* and *Sacramento* (SPS-58A)
	• 'Blue Ridge' class amphibious command ships (two ships - SPS-65(V)1)

Contractor
Northrop Grumman Electronic Systems sector, Baltimore, Maryland.

UPDATED

AN/SPS-64 navigation and search radar

Type
F- (3 to 4 GHz) and I-band (8 to 10 GHz) shipborne navigation and surveillance radar.

Description
The AN/SPS-64 Mariners Pathfinder radar is a versatile surface search and navigation radar with raster scan bright display indicators and offers enhanced

Specifications: AN/SPS-64 navigation and search radar

	I-band		F-band			
Frequency range	9,375 ±25 MHz		3,030 ±25 MHz			
Peak power	10, 25 and 50 kW		60 kW			
Wavelength	3 cm		10 cm			
Range	18.3 m-118.5 km		27.4-1,118.5 m			
Pulse-width	0.06, 0.5, 1 µs					
PRF (pps)	3,600, 1,800, 900					
Receiver bandwidth	24, 4, 4 MHz					
Receiver IF	45 MHz		Antenna	1.83 m	2.74 m	3.66 m	3.66 m
Video Amp bandwidth	20 MHz		Horizontal beamwidth	1.25°	0.85°	0.65°	1.85°
Receiver noise	10 dB max		Vertical beamwidth	22°	22°	22°	22°
			Weights	59 kg	61 kg	64 kg	145 kg

Pathfinder/ST displays

operational characteristics. The AN/SPS-64(V) is supplied in I/J-band (9,375-9,420 MHz) or E/F-band (3,030 MHz) versions with a choice of relative, true motion and collision avoidance displays. Plan position indicator displays of 30 and 41 cm are the basic configuration with expansion options and accessories available for the majority of requirements. The RAYCAS V indicator combines the presentation of radar information with collision avoidance, target tracking, navigation and tactical data on a 41 cm display. The RAYPATH indicator combines radar information, collision avoidance and navigation data on either a 30 or 41 cm bright display. These commercial, large ship radar systems have demonstrated high reliability and performance, while offering a most cost-effective radar system for all types of naval vessels. By offering a totally compatible choice of antennas, 25, 50 and 60 kW transmitters and display indicators, over 15 different radar configurations are available. The fully modular design allows duplication of systems when required and intermixing of displays and transceivers in any combination. With an adaptive interface capability, the suite of displays can be used with other radar systems. Interfaces for the AN/SPA-25 and AN/SPA-66 are available, as well as interfaces for fire control and electronic support measures systems.

Status

Over time, AN/SPS-64 navigation and search radars (with sub-variant where known) are reported as having been installed aboard a range of ship types including South Korean 'Dong Hae' class corvettes; Mexican 'Holzinger' class gunships (V6); Thai 'Naresan' class frigates (V5); US Navy 'Kitty Hawk/John F Kennedy' and 'Nimitz' class aircraft carriers (together with the single unit USS *Enterprise* (V9), 'Ticonderoga' and 'Virginia' class cruisers (all V9), 'Arleigh Burke' class (Flights I and II - both V9), 'Kidd' and 'Spruance' class (V9 — to be fitted) destroyers, 'Blue Ridge' class amphibious command ships (V9), 'Wasp' class multipurpose amphibious assault ships (V9), 'Tarawa' class general purpose amphibious assault ships (V9), 'Whidbey Island/Harpers Ferry' class landing docks/landing docks — aircraft carriers (V9) and 'Osprey' class coastal minehunters (V9) and US Coast Guard 'Hamilton/Hero' (V6), 'Famous' (V9) and 'Reliance' class cutters. As of 1995, SPS-64 was reported as being in production. The commercial version of the system is noted as having been fitted to in excess of 5,000 ships worldwide.

Contractor

Raytheon Electronic Systems, Manchester, New Hampshire.

VERIFIED

AN/SPS-67(V) search radar

Type

G-band (5.4 to 5.8 GHz sub-band) surface search radar.

Description

The AN/SPS-67(V) surface search radar is a solid-state G-band equipment that was originally designed to replace the AN/SPS-10 surface search radar, the antenna from which the system was initially used with the new equipment. Below-decks, SPS-67 consists of a transceiver, a video processor, a radar control unit, an antenna controller and an antenna safety switch, all of which are housed in five 'easy-installation' cabinets. System performance is improved via the addition of a very narrow pulse mode (0.1 µs) for better navigation and improved resolution of small targets at short ranges. Long and medium pulse (1 and 0.25 µs) modes are used in open sea for detection of long- and medium-range targets. Performance is further improved by a digital video clutter suppressor and an interference suppressor. Reliability, maintainability and availability are enhanced by the use of standard electronic module technology and an extensive built-in test system.

Alongside the baseline AN/SPS-67(V)1 equipment (described above), the addition of a 'survivable' antenna (in place of the SPS-10 array) creates the AN/SPS-67(V)2 model. Radars fitted with an add-on gunfire support capability (installed aboard US Navy (USN) 'Arleigh Burke' class destroyers) are designated as AN/SPS-67(V)3. APS-67(V)3 provides digital moving target indication, automatic target detection and track-while-scan for surface targets. Data from the (V)3 configuration can be integrated into the AN/SYS-1 Integrated Automatic Detection and Tracking System for automatic correlation with information from several ships' radars to present a combined single tactical track.

Status

During the period February to November 2001, Jane's sources were reporting AN/SPS-67(V) as being installed aboard the following warships and classes of warship:

Chile	• the tank landing ship *Valdivia*
Egypt	• 'Knox' class frigates (two ships in class)
Germany	• 'Lutjens' class destroyers (two ships)
Malaysia	• the tank landing ship *Sri Inderapura*
Mexico	• 'Knox' class frigates (four ships)
Morocco	• the tank landing ship *Sidi Mohammed Ben Abdallah*
Norway	• 'Nordkapp' class offshore patrol vessels (three ships - SPS-67(V)3 or AWS-5)
Spain	• 'Alvaro de Bazan' class frigates (four building as of August 2001)
	• 'Santa Maria' class frigates (six ships - SPS-67(V) or Raytheon 1650/9)
Taiwan	• 'Knox' class frigates (eight ships - SPS-67(V) or SPS-10)
	• 'Newport' class tank landing craft (two ships, one building as of March 2001)
Thailand	• 'Knox' class frigates (two ships - SPS-67(V) or SPS-10)
Turkey	• 'Tepe' class frigates (six ships - SPS-67(V) or SPS-10)
USA	• the aircraft carrier *Enterprise*
	• the tank landing ship *Frederick*
	• the mine countermeasures support ship *Inchon* (SPS-67(V)1)
	• 'Kitty Hawk/John F Kennedy' class aircraft carriers (three ships)
	• 'Nimitz' class aircraft carriers (eight ships with one building and one planned as of April 2001 - SPS-67(V)1)
	• 'Arleigh Burke Flt I/II' (28 ships) and 'Flt IIA' (six ships, nine building or planned as of August 2001) class destroyers (SPS-67(V)3)
	• 'Wasp' class multipurpose amphibious assault ships (seven ships, one building as of August 2001)
	• 'Tarawa' class general purpose amphibious assault ships (five ships - SPS-67(V)3)
	• 'Austin' class amphibious transport docks (11 ships)
	• 'Supply' class fast combat support ships (four ships)
	• 'Mercy' class hospital ships (two ships)

As of February 2002, SPS-67(V) production was being undertaken by DRS Electronic Systems, with the latest identified DRS SPS-67(V) contracting activity being the 1 May 2001 announcement of a then year US$2.1 million award covering the supply of SPS-67(V) radars for installation aboard new build USN 'Arleigh Burke Flt IIA' class destroyers. At the time of this announcement, work on the effort was scheduled for completion by the end of 2004 and DRS was noting that it had al-

The AN/SPS-67(V) antenna on the USN's amphibious assault ship USS Wasp

The AN/SPY-1 phase-scanned array as installed aboard the USS Vincennes (W Donko)

ready supplied the USN with more than 20 SPS-67(V) sets, together with equipments for Spanish 'Alvaro de Bazan' class frigates and Norwegian 'Nordkapp' class offshore patrol vessels. As of the given date, DRS was further reporting that over 100 SPS-67(V)1 radars had been delivered to customers around the world.

Specifications
Frequency: 5.4-5.8 GHz
Azimuth beamwidth: 1.5°
Elevation beamwidth: 12° (-67(V)1); 31° (-67(V)2 & 3)
Pulse-width: 0.1; 0.25; 1.0 µs
Peak power: 280 kW
Scan period: 2/4 s (-67(V)2 & 3); 4 s (-67(V)1)
Instrumented range: 104 km
Track initiation: automatic (SPS-67(V)3)
Track file size: 128 (SPS-67(V)3 - expandable, claimed >24 dB improvement over preceding models, coherent receive auto clutter lock digital moving target indication)
MTBF: >600 h
MTTR: <0.5 h

Contractor
DRS Electronic Systems Inc, Gaithersburg, Maryland.

UPDATED

....

AN/SPY-1 multifunction radar

Type
E/F-band (2 to 4 GHz) search, track and missile direction radar.

Development
The SPY-1 radar was originally developed by RCA (now a part of Lockheed Martin, the prime contractor for the overall Aegis system) under a contract awarded in late 1969. The antenna array began tests in 1972 and by 1973, the transmitter and array were integrated for further tests before embarking on full-scale testing with other elements of the Aegis system. By early 1974, land-based tests of the radar had been completed and the radar moved from RCA's test site to Long Beach Naval Shipyard for installation in the USS *Norton Sound* in preparation for sea trials, including multiple missile firings, which were then carried out successfully.

Description
The AN/SPY-1 multifunction radar is the electronically scanned fixed-array equipment that forms a central part of the US Navy's (USN) Airborne Early warning/ Ground environment Integration Segment (AEGIS) fleet air defence system. It operates in the E/F-band and the output is in several megawatts. The transmitter serves several parallel channels simultaneously. The phase-scanned arrays are mounted in pairs on each AEGIS cruiser, two on the forward deckhouse and two on the aft deckhouse, to provide all-round radar cover. Each array has 4,100 discrete elements and measures 3.65 × 3.65 m. These elements are controlled by AN/ UYK-7 digital computers to produce and steer multiple radar beams for target search, detection and tracking. The SPY-1 also tracks the ship's own missiles fired against hostile targets. It also has the function of providing target tracking data for slaving the target illuminators (up to four on each ship) which support the SM-2 semi-active homing missiles employed in the AEGIS system.

The Aegis system has been fully operational since 1983 and advanced versions of the radar have also been developed. The AN/SPY-1B version is reported as having been fitted to 15 of the 27 'Ticonderoga' class cruisers starting with the USS *Princeton* (CG 59). It uses a new antenna design to give much lower sidelobes, plus an improved signal processor and a new transmitter tube having double the duty cycle with the same peak power. Electronic counter-countermeasures capabilities have also been improved. The AN/SPY-1D version (a virtual twin of the AN/SPY-1B)

has been fitted to the USN's Flights I, II and IIA 'Arleigh Burke' class and Japanese 'Kongou' class destroyers and has been mandated for installation aboard Spanish 'F 100' class frigates. SPY-1D uses a single transmitter with four antennas mounted on a single deckhouse. Considerable space savings have been achieved in the AN/ SPY-1B and AN/SPY-1D by using very large scale integration technology. The AN/ UYK-43B digital computer is used in all the 'Arleigh Burke' class destroyers, as well as the later 'Ticonderoga' class cruisers, starting with USS *Chosin* (CG 65). Lockheed Martin is also understood to have developed an SPY-1D(V) variant (also known as EDM-4B) that addresses both low radar cross section threats (such as sea-skimming anti-shipping missiles) and the needs of littoral warfare. SPY-1D(V) is understood to incorporate a 25 per cent increase in transmitter power and improved subclutter performance. Jane's sources suggest that EDM-4B/SPY-1D (V) was under test during late 1996 with series production originally proposed for US Fiscal Year 1998. Elsewhere, existing SPY-1, SPY-1B and SPY-1D are reported to be being upgraded with improved signal processing, increased average transmission power, improved software and a new track initiation processor. The SPY-1 series is currently completed by the SPY-1F proposal that is intended for the export market.

Status
As of this edition, the AN/SPY-1 multifunction radar was reported as having entered service with the USN during 1983 and was noted as having been installed aboard Japanese 'Kongou' class destroyers (SPY-1D) and USN 'Ticonderoga' class AEGIS cruisers (SPY-1 in 12 ships, SPY-1B in 15 ships — see above) and 'Arleigh Burke' class (Flights I, II and IIA) destroyers (SPY-1D). AN/SPY-1D is further noted as having been selected for use aboard Spanish 'F 100' class frigates. As of the given edition, Jane's sources were also reporting that the US Navy has introduced the Linebacker hardware/software package into SPY-1 in order to enhance its capabilities in the theatre ballistic missile tracking and reporting role. As of October 1998, the Linebacker enhancement is understood to have been installed aboard the USS *Lake Erie* and the USS *Port Royal*. Identified SPY-1 programme activity during the first half of 2000 comprised the following:

25 May 2000
The COLSA Corporation (Huntsville, Alabama) was awarded a then year US$6,667,576 cost-plus, firm, fixed-fee contract covering research and development support for, amongst other things, a real-time digital signal processor (plus associated software) capable of generating simulated Theatre Ballistic Missile radar responses for the SPY-1 radar. At the time of the announcement of this award, its potential cumulative value was estimated at then year US$11,942,913 if all its options were taken up. The effort was scheduled for completion by the end of May 2003 and its contracting activity was the USN's Naval Research Laboratory, Washington DC.

26 May 2000
Raytheon Electronic Systems was awarded a then year US$5,875,424 modification to a previously awarded contract (N00024-98-C-5103) covering the provision of 64,688 man hours of technical production support engineering services for the SPY-1D, SPY-1D(V) and Mk 99 transmitters. At the time of this contract modification's announcement, work on the effort was to be undertaken at Raytheon's facilities at Sudbury, Massachusetts (90% work share) and Andover, Massachusetts (10%) and was scheduled for completion by the end of June 2001. The programme's contracting activity was the USN's Naval Sea Systems Command, Arlington, Virginia.

Contractors
Lockheed Martin Naval Electronics and Surveillance Systems-Moorestown, Moorestown, New Jersey (prime contractor).
Raytheon Electronic Systems, Marlboro, Massachusetts (high-power transmitter).
Computer Sciences Corporation, Moorestown, New Jersey (computer programmes).

UPDATED

Falcon II surveillance radar

Type
Multipurpose air and sea surveillance radar.

Description
Falcon II is a microprocessor-based, multipurpose radar which features bimodal clutter rejection, moving target indication and constant false alarm rate processing. It can be configured for manned or unmanned operation and is designed for single-person maintenance. Electronic counter-countermeasures features include frequency agility and a high-gain, low-sidelobe antenna. An identification friend-or-foe capability can be integrated into the radar that also features built-in test. Falcon II utilises a single-stage, high-gain travelling wave tube transceiver and a dual-beam reflector antenna that offers cosec^2 coverage to 25°. The system is equipped with one or two shelter-mounted operator positions offering graphic, plan position indicator, vector and map display formats. A variant of the basic radar is the Falcon II-G gap filler which is designed to complement long-range surveillance radars. Falcon II-G offers automatic plot extraction and tracking out to ranges and altitudes of 115 km and 12,000 m respectively. Elevation coverage is up to 40° and the radar is noted as being able to provide pop-up target detection.

Status
As of this edition, over 40 Falcon II radars were reported as having been supplied to customers on four continents.

Specifications
Frequency: C-band (0.5-1 GHz)
Power output: 55 kW
Coverage: 100 or 160 km (range); 360° (azimuth); 25° (elevation); 5-10 s (data rate)
Detection range: 100 km (2 m^2 RCS aircraft target); horizon (5 m^2 ship target in Sea State 4)
Accuracy: 20 m (range); 0.2° (azimuth)
Resolution: 85 m (range); 2.4° (azimuth)
MTBF: 600 h (specification); 1,000 h (predicted)
MTTR: 0.5 h

Contractor
Metric Systems Corp, Fort Walton Beach, Florida.

VERIFIED

High-Frequency Surface-Wave Radar (HFSWR)

Type
HF-band (15 to 25 MHz) shipborne surface-wave air and surface surveillance radar.

Description
HFSWR is designed to provide over-the-horizon detection and tracking of low-flying air-to-surface cruise missiles (firm tracks to at least 37 km), aircraft (over 75 km) and surface targets (in excess of 150 km). The approach is also understood to have the potential to detect and track ballistic missile threats at a 500 km plus line of sight range. Jane's sources suggest that the radar has pulse repetition frequency and duty cycle values of 2 kHz and 0.45 respectively and utilises an antenna array which incorporates a vertically stacked, two element transmission antenna together with 24 0.3 × 0.3 × 0.9 m receiver units. These latter antennas are described as being compact, low-profile, meander-line assemblies. The specific frequency range used has been selected because of the good range performance achievable at 15 to 18 GHz and the good ballistic missile detection properties of the 19 to 25 GHz sub-band.

Status
Sanders was awarded a fabrication, integration and test contract for an HFSWR technology demonstration for the US Navy during January 1996. As of this edition, HFSWR trials aboard the service's self-defence test ship (EDDG-31) are

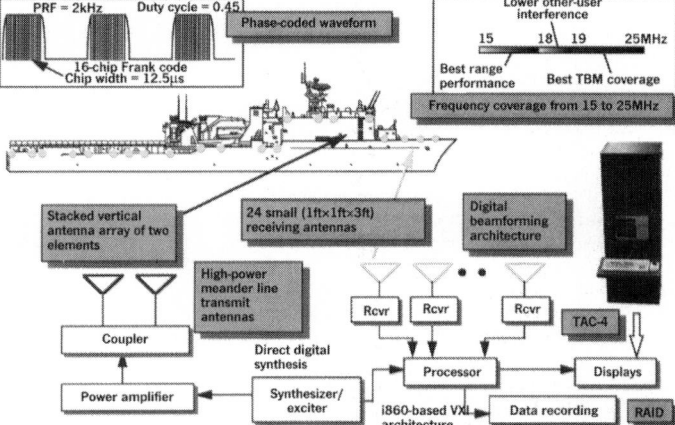

Key elements of the Sanders HFSWR design

understood to have taken place during the period February to April 1998. Thereafter, the system was scheduled to be cross-decked to a 'Whidbey Island' class dock landing ship for a second round of trials starting in mid-1998.

Contractor
Sanders (a Lockheed Martin company), Nashua, New Hampshire.

VERIFIED

MultiMission Surveillance Radar (MMSR)

Type
E-band (2.7 to 2.9 GHz sub-band) coastal surveillance, Air Traffic Control (ATC) and air defence surveillance radar.

Description
Described as being a 'modern, solid-state, commercial-off-the-shelf upgrade' of the AN/TPS-73 ATC radar (see separate entry), the MMSR is noted as being a medium-range, 3-D sensor that is designed for coastal surveillance, ATC and air defence 'gap filler' applications. As such, the radar is designed for 'rapid' tactical deployment, set up and redeployment' and can be 'easily' configured for installation in a High Mobility, Multipurpose Wheeled Vehicle (HMMWV)-mounted shelter, fixed-site location or a transportable format. In this latter context, MMSR is air (C-130 or CH-53), sea (cargo ship) and land (train or truck) transportable and its major system features include:
- 'enhanced' (when compared with the TPS-73) target detection in clutter and countermeasures polluted environments
- a 'soft fail' capability
- hover-to-maximum speed helicopter detection
- selectable polarisations to enhance performance in clutter, rain and/or countermeasures environments
- adaptive moving target detection
- automatic frequency selection
- a tailored STC map to maximise receiver dynamic range
- an integrated radar/beacon antenna array
- Mode S compatibility
- 'comprehensive' built-in test.

Status
As of May 2001, Jane's sources were suggesting that the MMSR was being actively promoted.

Specifications
Frequency: 2.7-2.9 GHz
Peak power: scalable up to 25 kW
Duty cycle: 11%
PRF: fixed or staggered
Pulse-width: 5 and 120 μs
Rotation: 12 or 15 rpm
Signal-to-clutter: ≥60 dB
Detection performance: 110 km (Pd = 0.9, 1 m^2 RCS)
3-D height: 9,144 m-110 km (0°-7°)
Instrumented range: 110 km
Accuracy: 0.18° (azimuth); 30 m (range); <4.572 m-110 km (height)
MTBF: >1,200 h
Availability: >0.99
Deployment: <45 min

Contractor
Lockheed Martin Naval Electronics and Surveillance Systems - Syracuse, Syracuse, New York.

VERIFIED

Orion coastal air defence radar

Type
F-band (3 to 4 GHz) coastal and low-level air defence radar system.

Description
The Orion system is designed to ensure territorial integrity and point defence by providing surveillance coverage over designated land and sea areas. The specific mission applications include coastal surveillance, low-altitude air space surveillance and inshore coastal/harbour traffic monitoring. Coastal surveillance provides early warning and monitoring of naval vessels, commercial vessels and a variety of small craft. The low-altitude air surveillance capability will monitor low-flying aircraft and helicopters.

Orion is a pulse-Doppler system operating in F-band and has been designed for use at unattended sites with automatic remoting of data and control signals. It includes an F-band primary search radar and a D-band (1 to 2 GHz) secondary surveillance radar and can be provided as a fixed-site installation, a transportable or a mobile configuration.

The fixed-site configuration provides permanent surveillance of potential threat approaches. The antenna is tower mounted with or without radome, depending on environmental conditions. The system breaks down for transport by land, sea or air.

In the transportable configuration, the system is designed to operate from semi-permanent sites which may be presurveyed to minimise deployment set up time. A telescopic transportable tower extends the horizon surveillance coverage. The radar system is housed in a military equipment shelter which can be transported by road, C-130 type aircraft, helicopter, ship, rail or towed on mobilisers.

The mobile configuration addresses the requirement for surveillance of sensitive coastlines with a limited number of systems. It is capable of rapid deployment to normally remote, inaccessible locations. The shelterised radar system is installed in a cross-country or tracked vehicle, with the antenna mounted on the vehicle. It is also transportable by C-130 aircraft, ship or rail.

Specifications
Operating frequency: F-band (3.1-3.5 GHz)
IFF frequency: D-band (1,030 and 1,090 MHz)
Coverage: 360° azimuth; 3 km altitude
Detection range: 120 km (1 m^2 target)
Data rate: 6 s
Accuracy: 0.24° azimuth; 15 m range
Resolution: 2.36° azimuth; 90 m range
Antenna: circular/linear polarised direction, shaped for elevation coverage limited to 3,000 m
Transmitter: TWT frequency-agile; 107 kW peak output min
PRF modes: fixed, stagger pulse-to-pulse, stagger burst

Status
Over time, Orion radar systems are reported as having been procured by customers in Denmark, Norway and Turkey.

Contractor
Cardion Inc, Woodbury, New York.

VERIFIED

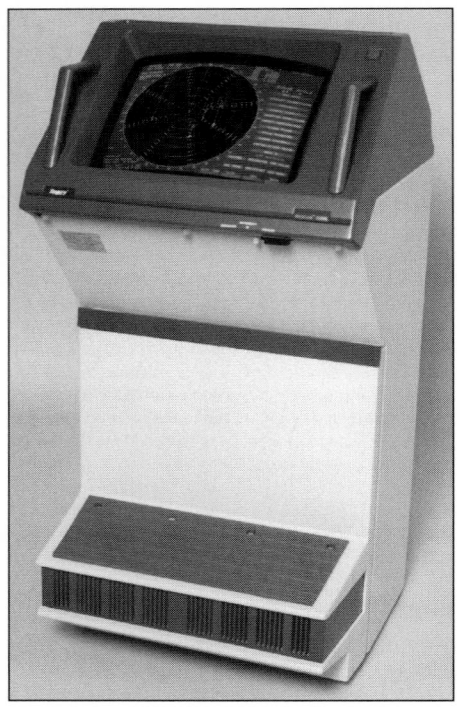

The operator's workstation used with the Rascar navigation and collision avoidance radar

---

Rascar 3400/2500 navigation radar

Type
F- (3 to 4 GHz) or I-band (8 to 10 GHz) shipborne navigation and collision avoidance radar.

Description
The Rascar series equipments are marine radar systems used primarily for navigation and collision avoidance. The series includes three different raster scan displays, a 64 and 48 cm diagonal monochrome, and a 48 cm colour. Each display includes a high-resolution, non-interlaced, high refresh rate presentation. The man/machine interface uses a menu structure and touchscreen. Each display also includes an Automatic Radar Plotting Aid (ARPA) allowing up to 20 targets to be tracked automatically, and their courses, speeds, closest points of approach and times to be monitored continuously against collision threat parameters. The effectiveness and safety of possible 'own ship's' course and speed manoeuvres may be evaluated graphically in a trial manoeuvre display mode. Non-ARPA models are also available.

I-band low-noise transceivers are available in 25 kW and 50 kW peak power, while 30 and 60 kW peak power low-noise transceivers are available in F-band. Slotted array type horizontally polarised I-band antennas in 2.1 and 2.7 m lengths and F-band in a 4.3 m length are available. The antenna range also includes a dual function I-band antenna offering a horizontally polarised array back-to-back with a circularly polarised antenna. The circularly polarised antenna offers improved performance in rain regardless of the density of the storm.

Status
As of January 2002, over 1,000 Rascar series radars are reported to have been delivered to military customers around the world. Within the US military, Rascar type radars are understood to have been installed aboard US Military Sealift Command ships and to have been procured for use aboard US Coast Guard Service ice breakers and buoy tenders (two classes).

Specifications
Frequency: F-band (3,050 MHz); I-band (9,375 MHz)
Peak power: 25 kW (I-band); 30 kW (F-band)
Horizontal beamwidth: 1.1°, 0.9° (I-band); 1.8°(F-band)

Contractor
Northrop Grumman Electronic Systems sector - Sperry Marine, Charlottesville, Virginia.

UPDATED

NAVAL FIRE-CONTROL RADARS

AUSTRALIA

CEA-Mount

Type
I-/low J-band (8 to 12 GHz) illuminator radar.

Description
CEA-Mount is an active phased-array Continuous Wave (CW) radar that is integrated with its host platform's combat system and provides target illumination and guidance uplink commands for surface-to-air missiles such as the NATO SeaSparrow and next-generation Standard Missile systems. Comprising a phased-array transmitter/antenna assembly mounted on an 'agile' director, CEA-Mount is understood to be based on the Solid State Continuous Wave Illuminator (SSCWI) transmitter (see separate entry) that CEA has developed as a form-fit replacement for the Mk 73 Mod 1 system aboard Australian and New Zealand 'ANZAC' class frigates. CEA-Mount's developers (CEA and BAE Systems) also claim that the system features 'novel' phase steering technology that overcomes the effects of phase noise.

Status
Jane's sources suggest that CEA-Mount may have begun a trials programme during February 2000. At this time, CEA-Mount was being offered as part of a surface ship radar package alongside BAE Systems' Sampson system (see separate entry). As of August 2001, the status of the CEA-Mount illuminator was uncertain.

Specifications
Frequency: I-/low J-band (8-12 GHz - including SM-1 and SM-2 bands)
Accuracy: 0.2° (azimuth and elevation)
Elevation coverage: −30 to + 90°
Stabilisation: via electronic beam steering
Mechanical slew rate: 120°/s
Power supply: <10 kVA (Medium Range (MR) model); <30 kVA (Long Range (LR) model)
Peak source level: compatible with full SM-2 coverage requirements
Antenna face size: <1 m aperture (MR model); 1.6 m aperture (LR model)
Antenna RCS: 2 m² RMS at horizon level (≤15 GHz - MR model); <4 m² RMS at horizon level (≤15 GHz - LR model)
Antenna sidelobes: dynamically programmable for sidelobes and auxiliary beam coverage
Uplink: compatible with I-/low J-band (8-12 GHz) uplink modulation requirements
Weight: 1,200 kg (MR model); 2,500 kg (LR model)

Contractors
BAE Systems Australia, Elizabeth, South Australia.
CEA Technologies Pty Ltd, Canberra, Australian Capital Territory

UPDATED

Solid State Continuous Wave Illuminator (SSCWI) transmitter

Type
Surface-to-Air Missile (SAM) target illumination and guidance radar transmitter.

Description
The SSCWI target illumination and guidance radar transmitter is a direct form and functional replacement for the transmitter used in the Mk 73 Mod 1 system that supports the SeaSparrow SAM. CEA Technologies claims that SSCWI provides 'identical' control and functionality with those of the Mk 73's hardware control interface and adds 'performance gains, size reduction and enhanced maintainability and reliability' due to its 'core solid-state technology'. As applied to 'ANZAC' class frigates (see following), SSCWI features include:

- an identical footprint to the existing Mk 73 transmitter
- compatibility with the Evolved SeaSparrow Missile (ESSM) as well as the baseline SeaSparrow weapon
- system specific environmental control unit design
- broadband, solid-state power amplifiers (based on commercial-off-the-shelf discrete devices)
- a remote maintenance access option
- a power supply design based on minimal magnetics
- a programmable master oscillator (with remote control)
- use of a closed circulation, de-ionised water cooling system (with sea water or ship's chilled water (up to an inlet temperature of 42°C) options and excess capability to facilitate higher power output upgrades)
- electrical waveguide dehumidification
- direct power conversion from the host vessel's 440 V AC (50/60 Hz) supply
- built-in test
- touchscreen, 'user friendly' display and control software

- use of data fusion technology
- control message remote frequency switching.

Status
SSCWI has been selected for use aboard 'ANZAC' class frigates of the Royal Australian and New Zealand Navies, with production deliveries starting during 1998.

Contractor
CEA Technologies Pty Ltd, Canberra, Australian Capital Territory

NEW ENTRY

CHINA, PEOPLE'S REPUBLIC

Chinese naval fire-control radars

Identified Chinese naval fire-control radars comprise the I/J-band (8 to 20 GHz) Type 341 (NATO reporting name Rice Lamp), the I/J-band Type 343 (NATO reporting name Sun Visor), the I-band (8 to 10 GHz) Type 347G, the I-band Type 351 and the I-band Type 352 (NATO reporting name Square Tie) equipments. Of these, the Type 351 is reported to be a Chinese version of the Russian Pot Head radar, while the Type 341 equipment may also carry the alternative EFR-1 designation. The Type 343 radar has been linked with the Wok Won designation which, in turn, has also been described as being a G/H-band (6 to 8 GHz) fire-control radar allocated the NATO reporting name Wasp Head. Within the Chinese Navy (and over time), the Type 343 radar is noted as being installed aboard some of the service's 'Luda I/II' class destroyers and the destroyer *Zhuhai*. Outside China, Type 343 radars are also reported as being installed aboard Thai 'Chao Phraya' class frigates. Type 347G is described as having been fitted to Chinese 'Luhu' class destroyers, the destroyer *Zhuhai* and some 'Luda I/II' class vessels. Type 351/Pot Head radars are reported to have been installed on Chinese 'Haizui' coastal patrol craft and on 'Shanghai II' class fast attack craft operated by Bangladesh, China, North Korea, Sierra Leone, Tanzania and Tunisia. Type 352 radars are noted as having been fitted to Chinese 'Jianghu I' class frigates.

VERIFIED

Sun Visor gun fire-control radar

LR62 splash spotting radar

Type
J-band (12 to 18 GHz sub-band) splash spotting radar.

Description
The LR62 electronically scanning splash spotting radar is designed for use with 130 mm and 150 mm surface ship guns and comprises the radar itself and a power generator. LR62 utilises 13 cm 'B' type search, 13 cm 'B' type tracking, 18 cm 'B' type deviation measurement and 5 bit tabular displays.

Status
As of this edition, the status of the LR62 splash spotting radar was uncertain.

Specifications
Frequency: 12-18 GHz
Range: ≤ 0.5 km (min); ≥ 60 km (max) P_d 80%, P_f 10⁻⁶, RCS 1,000 m², HA > 100 m
Azimuth coverage: 360°

Measuring accuracy: ≤ 0.1° (azimuth); ≥ 15 m (range)
Environment: −30 to +60°C (topside equipment)
Dimensions: 2 × 0.66 m (antenna); 4 × 2.4 × 2 m (operator cabin)
Weight: ≤ 3,000 kg (complete system)

Contractor
China National Electronics Import and Export Corporation, Beijing.

UPDATED

MR66 missile defence radar

Type
J-band (10 to 20 GHz) missile defence radar.

Description
MR66 is a shipboard system that is designed to track incoming anti-shipping missiles and hand off targeting data to onboard defensive weapon systems. The equipment is quoted as being fully automatic, having a very short response time and as comprising the radar and fire-control computer and/or gun director interfaces.

Status
As of this edition, the status of the MR66 missile defence radar was uncertain.

Specifications
Frequency: 10-20 GHz
Range: ≤ 600 km (min); ≥ 10 km (max) RCS 0.1 m²
Azimuth coverage: 360°
Elevation coverage: −10 to +85°
Altitude: < 10 m
Environment: −20 to +60°C (topside equipment)
Dimensions: 1,600 × 800 mm (antenna with pedestal); 1,650 × 674 × 550 mm (equipment cabinet – three installed)
Weight: 150 kg (antenna); 300 kg (equipment cabinet); 1,100 kg (complete system)
Power consumption: ≤ 10 kW

Contractor
China National Electronics Import and Export Corporation, Beijing.

UPDATED

Type 88C surveillance and target indication radar

Type
Surveillance and target indication radar.

Description
This is the surveillance, detection and target indication radar for the Type 88C Shipborne Weapon System. It features an elliptical antenna with feed horn, based on a two-axis stabilised mounting and is designed primarily for the detection of low-altitude and surface targets. The system has a track-while-scan capability and is stated to feature frequency agility and moving target indicator modes, with a choice of two rotation rates.

Antenna of the Type 88C radar

The radar acquires the target and designates it for the remotely operated electro-optical sensor. This is a three-axis multisensor tracker with search TV telecamera, tracking TV telecamera, infra-red sensor and a laser range-finder. The sensor complex enables stable and accurate tracking of very low targets in a hostile electronic countermeasures environment.

The command centre of the Type 88C weapon system features the weapon console with three alphanumeric displays showing the ship's location, target data and intercept data. Also in the centre are the radar plan position indicator display and supplementary displays for the other sensors.

Status
As of this edition, Jane's sources suggest that the Type 88C surveillance and target indication radar has been installed aboard Chinese 'Houjian/Huang' class fast attack craft.

Contractor
China North Industries Corporation, Beijing.

VERIFIED

INTERNATIONAL

1802SW fire-control radar

Type
I-band (8 to 10 GHz) naval fire-control radar.

Description
The Alenia Marconi Systems 1802SW is a lightweight fire-control radar that may be supplied as a single- or double-headed configuration for control of both the Seawolf point defence missile and naval gunnery. The system comprises a fully coherent, I-band monopulse frequency-agile target tracking radar, an I-band missile tracking receiver, an electro-optical tracking system and a microwave missile command link. The system can operate with conventionally launched or vertically launched Seawolf missiles and can also be used to control guns from 25 mm to 127 mm calibre used for anti-aircraft, surface and naval gunfire support tasks. The Seawolf system reacts automatically against all allocated closing targets.

Status
As of this edition, two double-headed 1802SW radar systems are understood to have been installed aboard Malaysia's pair of 'Lekiu' class frigates.

Contractor
Alenia Marconi Systems (an Alenia-BAE Systems joint venture), Chelmsford/ Rome.

VERIFIED

The Alenia Marconi Systems 1802SW fire-control radar

Castor 2 series fire-control radars

Type
I/J-band (8 to 20 GHz) family of fire-control radars.

Description
Castor 2 series fire-control radars are described as being 'compact [and] lightweight' equipments that are 'suitable for installation aboard both large and

Castor 2B/2C radar head with TV camera

small vessels'. Details of three of the series (designated as Castor 2B, Castor 2C and Castor 2J/C) are as follows:

Castor 2B

Castor 2B is a broadband, I/J-band, monopulse tracking and moving target indication radar for weapons control applications. As such, it is managed by a microprocessor and is claimed to offer 'high' performance ('even in severe environmental conditions') against low-altitude targets. Castor 2B incorporates anti-clutter filters and an integrated television (TV) tracker and performs the following functions:

* fully automatic target acquisition
* radar, TV and combined radar/TV operating modes
* passive jammer tracking with range data being generated by a dedicated tracking device linked to the surveillance radar
* a splash plotting facility
* autonomous continuous or sector surveillance with set absolute elevation.

Castor 2C

Castor 2C is an I/J-band, Doppler filtered, tracking radar that is designed for weapons control applications where it is required to acquire and track targets in conditions of active/passive jamming and heavy clutter. System performance is achieved via the use of a fully coherent output, frequency agility, analysis of received jamming signals within all the usable bandwidth, auto-adaptive Doppler filtering and central computer management that matches the radar to the environmental conditions at all times. Functionally, Castor 2C provides:

* fully automatic target acquisition
* short-bursts operation (1.5 ms) with simultaneous frequency agility and Doppler processing
* pulse-to-pulse frequency agility

The sensor head assembly used in the Castor 2J/C fire-control system
0045035

* TV tracking or combined TV/radar tracking modes
* passive jammer tracking with range data being generated by a dedicated tracking device linked to the surveillance radar
* shell versus target angular error and jammer tracking angular error displays
* autonomous continuous or sector surveillance.

Castor 2J/C

Castor 2J/C is a multisensor fire-control system that is designed to control guns of all calibres against air and surface (including shore bombardment) targets. As such, the equipment has been optimised to track anti-ship missiles and aircraft flying in a clutter polluted environment and takes the form of a fully automatic, monopulse-Doppler tracking radar operating in the J-band (10 to 20 GHz). The system's overall capabilities can be enhanced by the addition of a TV/infra-red camera package that facilitates passive target tracking. Castor 2J/C is the fire-control sensor for the Thales naval Crotale NG surface-to-air missile system and is designated as the CTM *(Conduite de Tir Modulaire)* sensor when installed aboard 'La Fayette' class frigates.

Status

As of January 2002, Thales Naval France was reporting Castor 2 series radars as being installed aboard the following classes of warship:

China	• 'Luda I/II' (two ships in class - Castor 2J/C) class destroyers
Colombia	• 'Almirante Padilla' class frigates (four ships - Castor 2B)
France	• 'La Fayette' class frigates (five ships - Castor 2J/C)
Germany	• 'Tiger' class fast attack craft - missile (five ships - Castor 2C)
Greece	• 'Laskos' (nine ships - Castor 2 subvariant unknown) and 'Votsis' (six ships - Castor 2 subvariant unknown) fast attack craft - missile
Libya	• 'Combattante IIG' class fast attack craft - missile (seven ships, two building as of February 2001 - Castor 2B)
Nigeria	• 'Combattante IIIB' class fast attack craft - missile (three ships - Castor 2B)
Qatar	• 'Damsah' class fast attack craft - missile (three ships - Castor 2B)
Saudi Arabia	• 'Arriyad' (three ships building as of March 2001 - Castor 2J/C) and 'Madina' (four ships - Castor 2C) class frigates
Taiwan	• 'Kang Ding' class frigates (six ships - Castor 2C)
Tunisia	• 'Combattante IIIM' class fast attack craft - missile (three ships - Castor 2B)

Specifications
Castor 2B and 2C
Frequency: I/J-band (8-20 GHz)
Peak power: more than 30 kW
Range: 0.5-27 km
Acquisition mode: fully automatic
Castor 2J/C
Frequency: J-band (10-20 GHz)
Range: 0.7-30 km
Acquisition mode: fully automatic
Tracking accuracy: 0.4 mrad (5 m)

Contractor
Thales Naval France, Meudon-la-Forêt, France.

UPDATED

DRBC 32 fire-control radar

Type
I/J-band (8 to 20 GHz) fire-control radar.

Description
The DRBC 32 designation embraces a family of I/J-band fire-control radars, of which there are at least five variants (denoted by suffix letters running from A to E) in use in French Navy ships and which are associated with a variety of director units. The latter may or may not include optical direction facilities. Weapons associated with these directors normally are either 100 mm or 57 mm guns. The DRBC family has a peak power of 80 kW and a pulse-width of 4 µs. It can track a 0.1 m² target at up to 15 km range.

Status
As of January 2002, Thales Naval France was reporting DRBC 32 as being installed aboard the following warships and classes of warship:

Argentina	• 'Drummond' class frigates (three ships in class, three proposed as of January 2001 - DRBC 32E)
France	• the helicopter carrier *Jeanne d'Arc* (DRBC 32A)
	• 'Georges Leygues' (four ships ('D 640/641/642/643') - DRBC 32E) and 'Tourville' (two ships - DRBC 32D) class destroyers
	• 'D'Estienne d'Orves' class frigates (nine ships - DRBC 32E)
Uruguay	• 'Commandant Rivière' class frigates (three ships - DRBC 32C)

Contractor
Thales Naval France, Meudon-la-Forêt, France.

UPDATED

DRBC 33 tracking and fire-control radar

Type
I-band (8 to 10 GHz) tracking and fire-control radar.

Description
The DRBC 33 is a fully coherent monopulse tracking and fire-control I-band radar that has been procured by the French Navy. The weapon associated with this equipment is the 100 mm gun, although the system can also be used with both 76 mm and 57 mm weapons. As such, the radar is a version of Thales' Castor 2C equipment and has a peak power of approximately 30 kW.

Status
As of January 2002, Thales Naval France was reporting DRBC 33 as being installed aboard the following classes of warship:

France   • 'Cassard' (two ships in class), 'Georges Leygues' (three ships - 'D 644/645/646') and 'Suffren' (two ships) class destroyers

Contractor
Thales Naval France, Meudon-la-Forêt, France.

UPDATED

The antenna arrangement on the French destroyer Jean Bart *showing the DRBC 33A, DRBV 26C, DIBV 1A and ARBR 17 ESM arrays*

LIROD Mk 2 fire-control system

Type
K-band (20 to 40 GHz) radar/electro-optic fire-control and surveillance system.

Description
LIROD Mk 2 is a sensor subsystem based on a K-band tracking radar and an electro-optic sensor. It is designed primarily to conduct automatic or rate-aided tracking of an air or surface target and provides track data for the control of guns. The system can track targets moving at up to 2,000 m/s and out to a range of 36 km.

LIROD consists of a director which carries a sensor set comprising a K-band monopulse radar antenna and a TV camera. The transmitter is enclosed in the director support. The antenna array is an elliptical paraboloid which offers excellent low-level target performance because of its narrow elevation beam, and at the same time provides easy acquisition with its wider azimuth beam. Radar signal processing for the system is comparable to that applied for STIR and STING. It comprises automatic acquisition and tracking with extensions such as sector search and kill assessment support. Angular optronic tracking is by the optical contrast technique whereby the target is tracked against its background as viewed by a camera; alternatively a correlation tracker, which is capable of automatic target acquisition, can be implemented. Range information is always provided by the radar.

The LIROD sensor subsystem is capable of performing the following functions:
• optical surveillance for air and surface targets
• target acquisition based on 2-D or 3-D handover data
• target tracking using radar information or (for alternative angular tracking) using gated TV-type video either automatic or rate-aided for one air or surface target.
• kill assessment and engagement monitoring
• radar sector surveillance

The LIROD Mk 2 lightweight radar/optronic director   2000/0081437

• system status monitoring
• shell spotting facilities
LIROD's standard interfaces are compatible with Thales' SEWACO FD architecture.

Status
Based on the then Signaal's (now Thales Nederland) LIOD electro-optical fire-control system (but with the emphasis on 'covert' radar), LIROD (LIghtweight Radar/Optronic Director) was developed during the late 1970s, with the first production examples appearing in 1981. LIROD Mk 1 was superseded by LIROD Mk 2 during 1991. As of the period January to March 2001, Jane's sources were reporting LIROD Mk 1 and Mk 2 systems as being in service aboard the following warships and classes of warship:

LIROD variant unknown
Argentina   • 'Almirante Brown' class destroyers (four ships in class - two LIROD per ship)
             • 'Espora' class frigates (five ships, one building as of January 2001 - one LIROD per ship)
Canada   • 'Iroquois' class destroyers (four ships - one LIROD per ship)
Indonesia   • 'Fatahillah' class frigates (three ships - one LIROD per ship)
             • 'Todak' class large patrol craft (two ships, two building as of February 2001 - one LIROD per ship)
South Korea • the 'Ulsan' class frigates *Chung Nam, Kyong Buk, Masan, Seoul* and *Ulsan* (one LIROD per ship)
             • 'Dong Hae' (four ships - one LIROD per ship) and the 'Po Hang' class corvettes *Chung Ju, Jin Ju, Kim Chon, Kun San, Kyong Ju, Mok Po, Po Hang* and *Yo Su* (one LIROD per ship)
Peru   • the cruiser *Almirante Grau* (two LIROD)
Thailand   • 'Rattanakosin' class corvettes (two ships - one LIROD per ship)
             • 'Chon Buri' fast attack craft - gun (three ships - one LIROD per ship)
LIROD Mk 2
Greece   • 'Batch 2 Pirpolitis' class large patrol craft (four building as of March 2001 - one LIROD Mk 2 per ship)
Turkey   • 'Kilic' class fast attack craft - missile (three ships, four building as of March 2001 - one LIROD Mk 2 per ship)

Specifications
Frequency: K-band (35 GHz)
Antenna: elliptical parabolic with monopulse cluster
Polarisation: horizontal
Dimensions: 1 m high × 0.4 m wide
Beamwidth: 15° (horizontal); 55° (vertical)
Transmitter
Power output: 100 W mean
Range on a fighter: 10 km (optronic); >24 km (radar)
Instrumented range: 36 km
Director
Weight: 475 kg
Traverse: 360° unlimited
Elevation: −30° to +85°

Contractor
Thales Nederland, Hengelo, Netherlands.

VERIFIED

Mk91 fire-control system

Type
I-band (8 to 10 GHz) tracking and illumination radar for Sea Sparrow.

Development
Development was initially undertaken in the late 1960s by Raytheon and, subsequently, on a joint basis by companies in six of the seven nations which form the NATO Sea Sparrow group. The six were Canada, Denmark, Italy, Netherlands, Norway and the USA.

Description
This radar set provides search, acquisition, tracking and illumination for the NATO Sea Sparrow point defence missile system Mk57 developed by Belgium, Denmark, Italy, Netherlands, Norway and the USA. As such, the radar forms part of the Mk91 guided missile fire-control system. Mk91 Mod 0 has a single director group, while the Mk91 Mod 1 has dual directors. Each radar has separate antennas for transmission and reception, the dishes of which are protected by radome covers. The radar receiver is carried on the antenna unit. The antennas share a common mounting on the Mk78 Mod 0 missile director. An on-mount TV camera is also provided.

The radar is understood to operate in the I-band and directs the missile to fly a lead angle course to the intercept point. Because of its I-band frequency the Mk91 will detect sea-skimming missiles sooner than lower frequency systems. It can be fully autonomous, conducting search, detection and engagement automatically.

The receiver is the responsibility of the Danish TERMA Elektronik Company and, this unit's functions include: clutter filtering, Electronic CounterMeasures (ECM) channel, signal conversion and amplification. The Mk83 Mod 0 radar set console has search and tracking displays, status and fault indicators, radar operating controls and built-in test equipment. Operating controls provide for such functions as scan pattern selection, ECM facility selection and checkout. The transmitter group Mk73 Mod 0 provides: radio frequency power for search acquisition; tracking in air and surface modes; target illumination and modulation for the Sea Sparrow missile; a reference signal for missile tuning; local oscillator for the receiver; range reference for the radar target data processor which provides target detection, Doppler tracking, range tracking and angle track signals; visual and aural indications to the radar set console; electronic counter-countermeasures and test functions.

Status
Over time, Jane's sources suggest that the Mk91 fire-control system has been installed aboard Norwegian 'Oslo' class frigates and the US Navy aircraft carrier USS *Enterprise* together with the same service's 'Kitty Hawk/John F Kennedy' and 'Nimitz' class aircraft carriers, 'Spruance' class destroyers, 'Wasp' class amphibious assault ships and 'Sacramento' and 'Supply' class fast combat support ships.

Specifications
Frequency: I-band (8-10 GHz)
Range: 25 km

Contractors
Raytheon Electronic Systems, USA (main contractor).
TERMA Elektronik, Denmark (radar receiver).
NEA Lindberg A/S, Denmark (static frequency converter).
Bronswerk-Amersfoort, Netherlands (liquid cooler group).
Kongsberg Gruppen A/S, Norway (radar pedestal and fire-control computer).

VERIFIED

The Mk91 FCS radar installed aboard the 'Spruance' class destroyer USS O'Bannon

NA 25/Orion RTN-25X fire-control radar

Type
I-/low J-band (8 to 12 GHz sub-band) naval fire-control radar.

Description
Before being superseded by the NA 25 system (see Status), the Orion RTN-25X fire-control radar was described as being an I-/low J-band sensor that was to be equipped with a monopulse cassegrain antenna (with polarisation twist) and a high average power, coherent transmission chain. The system was to offer long pulse/coded waveform and single short pulse operating modes while its receiver/data processor was to have incorporated multi-cancellation moving target indication. Tracking accuracy was to have been enhanced through the use of anti-nodding and air defence spotting subsystems. RTN-25X was designed to work with both close-in and longer range weapons in order to provide a single radar, in-depth engagement capability against single threats together with redundancy against multiple threats.

Status
As of this edition, the NA 25 system was reported to be based on the RTN-25X fire-control radar. As of the given edition, the 8 to 12 GHz band NA 25 was noted as being a 'high-accuracy' system that was optimised for use with close-in weapon systems and small and medium calibre surface ship guns.

Specifications
RTN-25X
Frequency: 10-20 GHz
Antenna type: horn cassegrain
Tracking method: monopulse
Transmitter: dual-mode, frequency diverse, pulse compressed coherent chain with TWT
Peak power: 10 kW
Pulse-length: 0.5 μs (long pulse/coded waveform mode); 0.5 μs (single short pulse mode – compressed); 6.5 μs (single short pulse mode – uncompressed)
Receiver: MTI (compatible with frequency diversity and anti-nodding feature)
Range: >45 km (air target)

Contractor
Alenia Marconi Systems (an Alenia-BAE Systems joint venture), Chelmsford/Rome.

VERIFIED

Orion RTN-30X NA 30 fire-control and missile guidance radar

Type
I-band (8 to 10 GHz) fire-control and missile guidance radar. Italian Navy designation SPG-75/76.

Description
The I-band Orion RTN-30X fire-control and missile guidance radar is described as being equipped with a monopulse cassegrain antenna (with polarisation twist) that can also radiate a continuous wave surface-to-air missile guidance signal. The transmission chain is fully coherent and uses a high-average power transmitter. The system features long pulse/coded waveform and short pulse (single pulse) modes and the receiver processor used incorporates multi-cancellation moving target indication and a data processor that enables the radar to perform accurate tracking against 'any' kind of threat ('whatever the flight envelope') in a variety of environments. Accuracy is enhanced by anti-nodding and air defence spotting features. An extended range capability enables RTN-30X to be integrated with a wide range of weapons including close-in weapon systems, guns and missiles.

Status
Over time, RTN-30X is reported to have been installed aboard a range of ships that has included the Italian aircraft carrier *Giuseppe Garibaldi*, 'Audace' and 'De La Penne' class destroyers, 'Maestrale' class frigates and 'Minerva' class corvettes. As of the given edition, it is thought likely that the RTN-30X fire-control and missile guidance radar was being supported by Alenia Marconi Systems (an Alenia-BAE Systems joint venture). It should also be noted that RTN-30X forms the basis of

The antenna assembly used in the Orion RTN-30X fire-control and missile guidance radar

Alenia Marconi's 8 to 12 GHz band NA 30 fire-control radar. NA 30 is noted as being suitable for small and medium calibre ship gun applications together with the guidance of Aspide, ESSM and NSSM missiles.

Specifications
Frequency: I-band (8-10 GHz)
Antenna type: horn cassegrain
Tracking method: monopulse
Transmitter: coherent chain with TWT, dual mode, frequency diversity, pulse compression
Peak power: 12 kW
Pulse-length: long pulse with coded waveform (0.5 µs compressed, 6.5 µs uncompressed); short single pulse (0.5 µs)
Receiver: MTI compatible with frequency diversity, anti-nodding features.
Range: >45 km on an aircraft

Contractor
Alenia Marconi Systems (an Alenia-BAE Systems joint venture), Chelmsford/ Rome.

VERIFIED

STING EO fire-control system

Type
Combined radar (I and K-band - 8 to 10 GHz and 20 to 40 GHz respectively) and Electro-Optical (EO) fire-control system.

Description
STING EO is a multipurpose, lightweight, fire-control radar and electro-optical package which provides air and surface target track data for naval gun and missile systems. Derived from the former Signaal's (now Thales Nederland) STIR family of equipments, STING EO provides automatic target acquisition with subsequent tracking being carried out using both of the equipment's radars. The availability of a K-band capability ensures acquisition and tracking of sea-skimming missile threats. Optimum accuracy is maintained by best data source selection between the two available radar bands supported by parallel TV and Infra-Red (IR) tracking together with laser range-finding. When integrated with a command and control system, STING EO provides capabilities such as sector search (with automatic target detection), missile launch detection, kill assessment and projectile position measurement in addition to its core fire-control function.

Within the equipment's radar subsystem, dual I- and K-band receivers are used together with multichannel Fast Fourier Transforms. Radio frequency counter-countermeasures incorporated include burst-to-burst frequency agility, staggered pulse repetition frequencies and dual-band operation. The availability of the electro-optical subsystem allows the system to function as a passive sensor when required. STING EO can be supplied with or without the IR camera and/or laser range-finder subsystems installed.

Status
As of the period February to March 2001, Jane's sources were reporting that STING fire-control systems were installed aboard the following classes of warship:

Greece • 'Super Vita' class fast attack craft - missile (three building, three proposed as of February 2001)
Oman • 'Qahir' class corvettes (two ships in class)
Qatar • 'Barzan' class fast patrol craft - missile (four ships)
Spain • 'Santa Maria' class frigates (six ships)
Turkey • 'Kilic' fast attack craft - missile (three ships, four building as of March 2001)

The STING EO combined radar and electro-optical naval fire-control system

Specifications
Radar subsystems
Frequency: I- (8-10 GHz) and K-band (20-40 GHz)
Peak power: 2.5 kW (I-band); 30 kW (K-band)
Acquisition performance
Instrumented range: 17 km (K-band); 72 km (I-band)
Director
Elevation: −30 to +120°
Beamwidth: 0.5° (I-band); 2° (K-band)
Dimensions: 2,200 × 2,412 mm
Weight: 750 kg (incl IR camera and laser range-finder)

Contractor
Thales Nederland, Hengelo, Netherlands.

VERIFIED

STIR tracking and illumination radar

Type
I- (8 to 10 GHz) and K-band (20 to 40 GHz) missile and gunfire control radar.

Description
STIR (Signaal Track and Illuminating Radar) performs automatic simultaneous control of both missiles and guns against high-speed sea-skimmers and divers for medium- to very long-range performance. It also includes surface target tracking and full anti-surface weapon control capabilities.

Besides accurate gun control, STIR provides Continuous Wave (CW) illumination guidance for almost all semi-active homing missiles. For enhancement of the low-level performance and elimination of the mirror effect, the basic STIR incorporates integrated dual-frequency (I- and K-band) target tracking, with advanced digital signal processing. As well as target tracking, STIR performs sector search, height measurement, splash spotting and correction for surface target fire and curved path prediction in addition to linear path prediction.

All STIR systems have a number of features, including integrated CW illumination, a variety of moving target indicator and electronic counter-countermeasures facilities, pilot tone injection for online boresight calibration and sector search capability (to search the jammed sector of a surveillance radar). Growth potential and an overcapacity in range are incorporated in the system to cater for future target developments, such as higher speed and lower radar cross-section. Operation can be automatic, semi-automatic or manual, either autonomous or in combination with other sensor and control systems.

STIR is available with an I-band travelling wave tube transmitter (in two power configurations) and with two director types: STIR 1.8 and STIR 2.4. This flexibility matches the various weapons to be controlled and the area and point defence tasks of the ships.

Surface-to-air missiles currently controlled with STIR are: Sea Sparrow (RIM-7H, RIM-7M and RIM-7P); Aspide; Standard missile (SM-1 Block V and VI, SM-2 Block III). The missile launching systems related are trainable (Mk 13, Mk 29) or vertical (Mk 41, Mk 48).

Guns controlled include: FMC 5 in/54-calibre Mk 45 and 3-in/Mk 75; Otobreda 127 mm/54-calibre and 76/62-calibre; Bofors 57 mm Mk 2, 40 mm and 150 mm twin-barrelled; Breda 40 mm twin mount; Oerlikon 25 mm 4-barrelled anti-missile gun and Creusot Loire 100 mm gun mount Mod 68/13.

Status
As of the period January to April 2001, Jane's sources were reporting Thales (formerly Signaal) STIR systems as being installed aboard the following warships and classes of warship:

Argentina • 'Almirante Brown' class destroyers (four ships in class)
Canada • 'Iroquois' class destroyers (four ships - two STIR 1.8 per ship)
 • 'Halifax' class frigates (12 ships - two STIR 1.8 per ship)
Germany • 'Brandenburg' (four ships - two STIR per ship) and 'Bremen' (eight ships) class frigates
Greece • 'Elli' (six ships) and 'Hydra' (four ships - two STIR per ship) class frigates
South Korea • 'KDX-2' (three building, three proposed as of February 2001 - two STIR per ship) and 'Okpo' (three ships - two STIR per ship) class destroyers
Netherlands • 'Jacob van Heemskerck' (two ships - three(?) STIR per ship), 'Karel Doorman' (eight ships - two STIR per ship) and 'Kortenaer' (three ships) class frigates
Nigeria • the frigate *Aradu*
Peru • the cruiser *Almirante Grau*
Portugal • 'Vasco da Gama' class frigates (three ships - two STIR per ship)
Taiwan • 'Wu Chin III' class destroyers (seven ships)
Thailand • the aircraft carrier *Chakri Naruebet*
 • 'Naresuan' class frigates (two ships - two STIR per ship)
Turkey • the 'Barbaros' class frigates *Kemalreis* and *Salihreis* (one or two STIR per ship)
 • 'Yavuz' class frigates (four ships)
United Arab Emirates • 'Kortenaer' class frigates (two ships)

A close-up of the antenna assemblies used in the STIR tracking and illumination system   0006912

Specifications

STIR 1.8 (Basic)
Antenna: single cassegrain 1.8 m
Polarisation: vertical
Beams: I- and K-band pencil beams, widths 1.4 and 0.3° respectively
Peak power: 2.2 kW (I-band); >5 kW (average CW); 30 kW (K-band)
Instrumental range: 17 km (K-band); 72 km (I-band)
Coverage: −30 to +100° (elevation); 360° (azimuth)
Director mass: 1,875 kg
STIR 1.8 (high power)
Antenna: single cassegrain 1.8 m diameter
Polarisation: vertical
Beams: concentric pulse and CW beams, width 1.4°
Average power: 5 kW (I-band); >5 kW (CW)
Instrumented range: 200 km
Director mass: 1,700 kg
STIR 2.4 (high power)
Antenna: single cassegrain 2.4 m diameter
Polarisation: vertical
Beams: concentric pulse and CW beams, width 1°
Average power: 5 kW (I-band); >5 kW (CW)
Instrumented range: 512 km
Director mass: 2,200 kg

Contractor

Thales Nederland, Hengelo, Netherlands.

VERIFIED

TRS 3220 Pollux tracking and fire-control radar

Type

I/J-band (8 to 20 GHz) tracking and fire-control radar.

Description

TRS 3220 Pollux is described as being the shipborne tracker radar that is used in Thales Naval France's Vega series naval fire-control systems. As such, it is usually operated in conjunction with the company's Triton G/H-band (4 to 8 GHz) surveillance and target designation radar. Mainly used in gun control applications, Pollux is further noted as making use of a 'fast' conical scan pattern, as being circularly polarised and as incorporating solid-state circuitry and silicon semi-conductors.

Pollux tracking radar, with opto-electronic sensor

Status

As of January 2002, Thales Naval France was reporting Pollux as being installed aboard the following classes of warship:

Ecuador • 'Manta' (two ships in class) and 'Quito' (three ships) class fast attack craft - missile
Greece • 'Anninos' (four ships) and 'Laskos' (nine ships) class fast attack craft - missile
• 'Jason' class landing ships tank (five ships)
Malaysia • 'Perdana' class fast attack craft - missile (four ships)

Specifications

Frequency: 10 GHz (approx centre frequency, 3 cm wavelength)
Polarisation: circular
Peak power: 200 kW
Range: up to 20 km (target acquisition); up to 30 km (range-finding)
Accuracy: 0.5 mrad (angular); 20 m (range)
Display types: A/R, A/B or A/E
TV tracking: optional
Dimensions: 1,850 × 600 × 600 mm (transmitter/servo and receiver cabinets and fire-control console)
Weight: 450 kg (antenna)

Contractor

Thales Naval France, Meudon-la-Forêt, France.

UPDATED

Type 909 missile guidance radar

Type

G/H-band (4 to 8 GHz) target tracking radar for missiles and guns.

Description

The Royal Navy's Type 909 radar provides target tracking and illuminating facilities for the Sea Dart GWS-30 air defence missile. This weapon has an anti-ship capability and the radar is stated to be suitable for gunlaying also, so that a surface target capability must be assumed. The antenna is of the cassegrain type and has a diameter of 2.44 m. A small dome mounted near the upper edge probably houses an associated identification friend-or-foe antenna. On board ship, the complete antenna assembly will be protected by a cupola radome. The radar head and the office cabin, containing transmitter/receiver unit and associated electronics, are constructed as a single prefabricated assembly to reduce installation and replacement time and to enable functional testing before fitting.

Few technical particulars have been revealed. A G/H-band operating frequency range is likely and a high transmitter power can be expected. It has been stated that elaborate electronic counter-countermeasures is incorporated to counter both active and passive electronic countermeasures. In May 1986 the UK Ministry of Defence awarded a £35 million contract to Marconi (now Alenia Marconi Systems) for update of the Type 909. This update, which is known as the Type 909(I), combines the latest in transmitter and signal processing circuitry to enhance and maintain the system effectiveness in meeting evolving threats.

A Type 909 radar with its protective radome removed, showing its monopulse feed horns in their cassegrain arrangement (Raymond Cheung)

Status

During the period March to August 2001, Jane's sources were reporting the Type 909 fire-control radars as being installed aboard Argentine 'Hercules' class destroyers and UK 'Invincible' class aircraft carriers and 'Type 42' (Batches 1, 2 and 3) class destroyers. The first upgraded Type 909(I) radar is reported to have been delivered to the Royal Navy during December 1988 for installation aboard the Batch 2 Type 42 destroyer HMS Exeter. A further 16 ship sets of Type 909(I) radars (each comprising two equipments mounted fore and aft) are noted as having been procured for installation aboard other Type 42 destroyers and 'Invincible' class aircraft carriers (as of this edition, the 'Invincible' class aircraft carrier HMS Illustrious is reported to have been retrofitted with the Type 909(I) radar). Readers should also note that Type 909(I) radars appear to be fitted with a new dish antenna in place of the existing cassegrain assembly. As of January 2002, it was thought likely, but not confirmed that the Type 909 fire-control radar was being supported by Alenia Marconi Systems.

Contractor

Alenia Marconi Systems (an Alenia-BAE Systems joint venture), Chelmsford/ Rome.

UPDATED

....

Type 911 tracking radar

Type

I-band (8 to 10 GHz) and L/M-band (40 to 100 GHz) tracking radar system for the Seawolf surface-to-air missile.

Description

Type 911 (Alenia Marconi Systems house designation 805SW) is a lightweight tracking radar for use with the Seawolf short-range point defence system. It is designated Type 911 by the Royal Navy and is fitted to its Batch 2 (from HMS Brave onwards) and 3 'Broadsword' and 'Duke' class frigates.

The radar is a dual-frequency (I-band and L/M-band) differential tracking system that includes a command link to control the Seawolf missile in flight. The I-band antenna and transmitter form part of the ST 800 range and the signal processing uses Fast Fourier Transform techniques to ensure effective operation in severe clutter environments. The L/M-band part of the system is a version of the DN 181 Blindfire radar used with the BAE Systems Rapier missile. It provides accurate tracking at low sight angles against targets close to the sea surface, such as sea-skimming missiles and low-flying aircraft. A system of independently illuminating the upper and lower parts of the L/M-band antenna ensures that the missile receives good guidance data while flying at low level.

The radar is fully automatic to provide fast reaction time against small targets and is autonomous, requiring only the allocation of the fire-control channel to the selected target and information on ships' motion. Clutter rejection is provided on both bands for operation in both open and enclosed waters and electronic counter-countermeasures facilities are incorporated to enable use in hostile electronic environments.

Status

Jane's sources note that 'over 50' Type 911 radars have been procured and that, over time, the sensor has been installed aboard Brazilian 'Broadsword' class frigates together with 'Duke' and 'Broadsword' (four Batch 2 and all Batch 3 ships) class frigates of the Royal Navy.

Contractor

Alenia Marconi Systems (an Alenia-BAE Systems joint venture), Chelmsford/ Rome.

VERIFIED

WM20 series radar

Type

I/J-band (8 to 20 GHz) family of fire-control radars.

Description

This radar is the principal subsystem of the Thales WM20 series of naval fire-control systems and in its full form consists of two radars sharing a common stabilised mount and housed in a near-spherical radome. In this installation, the tracker/ illuminator is carried above the search radar, with the gimbal and stabilising assembly in between. The general arrangement can be seen from the accompanying illustration.

Both antennas are fed from a common I/J-band transmitter or from separated transmitters for surveillance and air target tracking.

The search radar carries out the surface and air search functions and automatic tracking of surface targets for weapon control purposes. The tracking radar performs tracking functions for weapon control against air targets.

The typical peak power is 180 kW, but for the WM25 this is increased to 200 kW with the aid of a 1 MW crossed field amplifier. The output power is usually shared by waveguide switching. The performance depends upon the type but generally the search radar has a range of 30 to 32 km against surface targets and the tracker has a range of about 29 km. With the larger search antenna the search radar range is about 46 km. The tracker is capable of monitoring aircraft movement with a 1 m² cross-section at velocities up to 900 m/s, and similar size objects moving at or near the surface at maximum velocities of between 34 and 55 m/s. Most radars are capable of rate-aided manual tracking on four or more targets.

The WM26 variant employs only the search antenna, which is mounted on top of the stabilised platform and enclosed in a hemispherical radome. In all other versions the tracking antenna is fitted in the upper position with the search antenna below.

The WM20 series of weapon control systems carry the following nomenclatures:
WM20 - an integrated gun and torpedo weapon control system for use on small warships against air and surface targets.
WM22 - a simpler version of the WM20 for use on larger vessels, with guns only.
WM24 - a similar system to the WM22 but with increased capability.
WM25 - designed specifically for low-level air defence and uses an integrated system concept with anti-clutter and anti-jamming features.
WM26 - a gun control and navigation system designed to provide continuous air and surface surveillance, radar navigation, combat information, target designation and weapon control.
WM27 - a weapon control system for small vessels using anti-submarine warfare missiles, torpedoes and guns.
WM28 - designed to control one surface-to-surface missile and two light/medium calibre guns.
WM29 - similar to the WM28 but used with a command-guided surface-to-air missile.

Status

As of the period January to April 2001, Jane's sources were reporting WM20 series fire-control radars as being in service aboard the following warships and classes of warship:

WM20
Singapore	• 'Bedok' class coastal minehunters (four ships in class)

WM22
Argentina	• 'Intrepida' class fast attack craft - gun/missile (two ships)
Finland	• the corvette Karjala
Indonesia	• 'Singa' class large patrol craft (four ships)
Malaysia	• 'Kasturi' class corvettes (two ships)

WM24
Nigeria	• 'Vosper Thornycroft Mk 9' class corvettes (two ships)

WM25
Argentina	• 'Almirante Brown' class destroyers (four ships)
Belgium	• 'Wielingen' class frigates (three ships)

The Type 911 tracking radar

WM27 radars on the German Navy fast attack craft Ozelot (Stefan Terzibaschitsch)

Egypt	• 'Descubierta' class frigates (two ships)
Germany	• 'Bremen' class frigates (eight ships)
Greece	• 'Elli' class frigates (six ships)
Japan	• 'Shirane' class destroyers (two ships)
Morocco	• 'Lazaga' class fast attack craft - missile (four ships)
Netherlands	• 'Kortenaer' class frigates (three ships)
Peru	• the cruiser *Almirante Grau*
Spain	• 'Descubierta' class corvettes (six ships)
Thailand	• 'Ratcharit' class fast attack craft - missile (three ships)
Turkey	• 'Yavuz' class frigates (four ships)
United Arab Emirates	• 'Kortenaer' class frigates (two ships)

WM27

Germany	• 'Albatros' (10 ships) and 'Gepard' (10 ships) class fast attack craft - missile

WM28

Argentina	• 'Espora' class frigates (five ships, one building as of January 2001)
Indonesia	• the frigate *Ki Hajar Dewantara*
	• 'Fatahillah' class frigates (three ships)
	• 'Dagger' class fast attack craft - missile (four ships)
Iran	• 'Kaman' class fast attack craft - missile (10 ships)
South Korea	• the patrol ship *Han Kang*
	• the 'Ulsan' class frigates *Chung Nam*, *Kyong Buk*, *Masan*, *Seoul* and *Ulsan*
	• 'Dong Hae' (four ships) and 'Po Hang' (24 ships) class corvettes
Nigeria	• 'Ekpe' class large patrol craft (three ships)
Singapore	• 'Sea Wolf' class fast attack craft - missile (six ships)
Turkey	• 'Doğan' class fast attack craft - missile (eight ships)

Contractor

Thales Nederland, Hengelo, Netherlands.

UPDATED

ISRAEL

EL/M-2221 multifunction radar

Type

I/low J-band (8 to 12 GHz sub-band) or I/low J- and K-band (27 to 40 GHz sub-band) Search, Track and Guidance/gunnery Radar (STGR).

Description

The EL/M-2221 STGR is a single- or dual-band, monopulse, pulse-Doppler radar which offers Surface-to-Air (single round or salvo) and Surface-to-Surface Missile (SA/SSM) guidance; automatic gun fire control against air targets; automatic splash spotting and back up air/surface search together with optional target illumination. System features include multipath elimination; Kalman filtering; electronic counter-countermeasures/clutter elimination; built-in test; programmable signal processing; fast target acquisition; light weight and a composite material antenna designed to resist weathering. As of this edition, EL/M-2221 is available in two configurations, the dual-band EL/M-2221 and the single-band (8 to 12 GHz) EL/M-2221X. Taking these in reverse order, EL/M-2221X is designed to provide automatic gun fire control (against air targets) and splash spotting together with back-up air/surface search. The dual-band EL/M-2221 offers tracking and gunnery control against sea-skimming, anti-shipping missiles

together with guidance for SA/SSMs. However configured, the EL/M-2221 STGR also incorporates a TV or a TV/infra-red passive tracking subsystem.

Status

The EL/M-2221 STGR system was originally developed as part of the BARAK-1 point defence missile system. As of this edition, Jane's sources were suggesting that EL/M-2221 radars had been procured for installation aboard Chilean 'Casma' class patrol craft and 'Prat' class destroyers, Israeli 'Eilat' class corvettes and 'Hetz' class patrol craft and Singaporean 'Victory' class multirole corvettes.

Specifications

Frequency: 8-12 GHz (EL/M-2221X); 8-12 and 27-40 GHz (EL/M-2221)
Transmitter type: TWT
Antenna type: cassegrain
Vertical coverage: −30 to +85°
Scan rate: 12 or 20 rpm
Ranges (typical): up to 4 km (air gunnery − gun dependent); up to 10 km (SAM guidance); 15 km (threat missile acquisition); up to 20 km (surface gunnery − gun dependent); 30 km (fighter aircraft acquisition); up to radar horizon (surface target acquisition)
EL/M-2221 weights: 760 kg (below deck); 765 kg (above deck)
EL/M-2221X weights: 660 kg (above deck); 760 kg (below deck)
Power consumption: 22 kVA (EL/M-2221X); 30 kVA (EL/M-2221)

Contractor

Elta Electronic Industries Ltd (a subsidiary of Israel Aircraft Industries Ltd), Ashdod.

VERIFIED

ITALY

Orion RTN-10X tracking and fire-control radar

Type

I-band (8 to 10 GHz) tracking and fire-control radar. Italian Navy designation SPG-70/73.

Description

RTN-10X is an I-band radar that can be used in conjunction with a number of different fire-control systems to control gun and missile firings. It is based on a conical scan narrowbeam antenna (with a tri-port feed) and uses a magnetron with fast tuning capabilities. The system employs separate signal and data processing techniques to ensure a high-level protection from jamming and interference and also to maintain accurate tracking of both high- and low-flying targets.

Status

Over time, RTN-10X is reported to have been installed aboard a range of ships that has included Brazilian 'Niteroi' class frigates and 'Inhauma' class corvettes; Chilean 'Casma' and 'Iquique' class fast attack craft; Ecuadorean 'Esmeraldas' class corvettes; Israeli 'Aliya' and 'Reshef' class fast attack craft; the Italian cruiser *Vittorio Veneto*, 'Alpino' and 'Lupo' class frigates, 'Artigliere' class fleet patrol craft, 'Cassiopea' class offshore patrol vessels, 'Sparviero' class patrol hydrofoils, 'San Giorgio' class amphibious transport docks and 'Stromboli' class fleet replenishment tankers; Kenyan 'Madaraka' class patrol craft and the single unit *Mamba* and Venezuelan 'Modified Lupo' class frigates, 'Constitucion' class fast attack craft and 'Almirante Clemente' class coastguard cutters. As of this edition, it is thought likely that the Orion RTN-10X tracking and fire-control radar is being supported by Alenia Marconi Systems (an Alenia-BAE Systems joint venture).

The antenna assembly used in Elta's EL/ M-2221 STGR system

The antenna assembly used in the Orion RTN-10X fire-control radar

Specifications
Frequency: I-band (8-10 GHz)
Antenna type: high-gain, tri-port feed
Tracking method: conical scan
Transmitter: frequency changeable, high-power magnetron
Receiver: linear receiver and ECCM receiver capable of withstanding most jammers
Range: >37 km on a fighter aircraft

VERIFIED

Orion RTN-20X fire-control radar

Type
J-band (10 to 20 GHz) fire-control radar for close in weapon systems. Italian Navy designation SPG-74.

Description
RTN-20X is a J-band radar that has been specifically developed for use with close-in weapon systems. It makes use of a monopulse, cassegrain antenna with a polarisation twist and employs a fully coherent transmitter chain, using a travelling wave tube as the final amplifier. Two different pulse-lengths can be selected, both at elevated repetition rates. The radar's transmission frequency can be changed randomly or on a pulse-to-pulse or burst-to-burst basis and the equipment's fully digital receiver uses a multi-cancellation moving target indicator. RTN-20X is also noted as being configured to minimise the effects of multipath when being used against low-flying targets.

Status
Over time, RTN-20X is understood to have been installed aboard a range of ships that has included the Italian aircraft carrier *Giuseppe Garibaldi* and cruiser *Vittorio Veneto*, 'Lupo' and 'Maestrale' class frigates and 'Artigliere' class fleet patrol craft and Venezuelan 'Modified Lupo' class frigates. As of the given edition, it is thought likely that the Orion RTN-20X fire-control radar is being supported by Alenia Marconi Systems (an Alenia-BAE Systems joint venture).

Specifications
Frequency: J-band (10-20 GHz)
Antenna type: cassegrain
Tracking method: monopulse
Transmitter: coherent chain with TWT amplifier, frequency diversity
Receiver: MTI compatible with frequency change, anti-jamming features
Range: >10 km on missiles

VERIFIED

The antenna assembly used in the Orion RTN-20X tracking radar

ST-2 missile seeker

Type
I-band (8 to 10 GHz) active radar homing seeker for ship-to-ship missiles.

Description
The I-band ST-2 seeker is used in the Franco-Italian OTOMAT anti-shipping missile system. ST-2 design features include frequency diversity; a power managed transmitter; a pseudo monopulse receiver; a wide search pattern; automatic target tracking; track on jam and electronic counter-countermeasures provision.

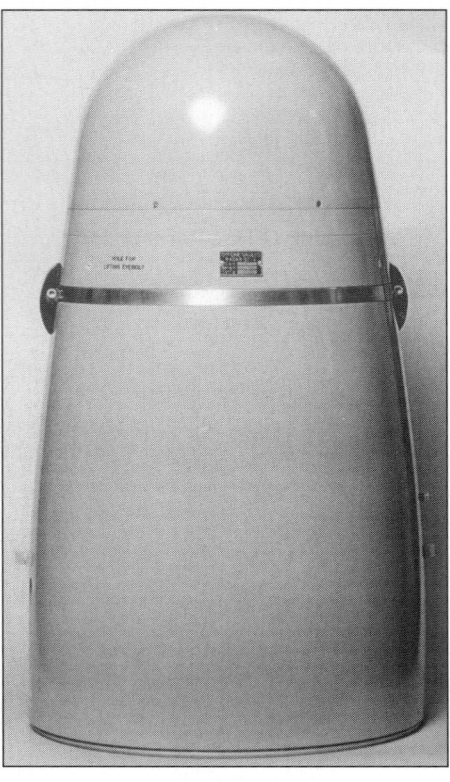

The exterior configuration of the ST-2 active radar missile seeker
0017505

Status
As of April 2001, ST-2 was understood to have been procured by several navies around the world. A millimetric wave variant of the seeker (designated as the DPQ-A07) is noted as having been in development.

Contractor
Alenia Difesa Avionic Systems and Equipment Division - Officine Galileo, Florence.

VERIFIED

RUSSIAN FEDERATION AND ASSOCIATED STATES (CIS)

Introduction
This section details those naval fire-control radars developed by, exported by and/or in service with the navies of the Russian Federation and the independent republics of Armenia, Azerbaijan, Belarus, Georgia, Kazakhstan, Kyrgyzstan, Moldova, Tajikistan, Turkmenistan, Ukraine and Uzbekistan. Together, these 12 entities make up the Commonwealth of Independent States (CIS) economic grouping.

VERIFIED

4R60 Grom missile control radar

Type
G/H/I-band (4 to 10 GHz) family of missile control radars. NATO reporting name Head Light.

Description
A typical Head Light installation consists of four radars, apparently comprising two identical pairs of equipment combined on a common mounting. There is a fifth, smaller dish, possibly fulfilling a command link or identification friend-or-foe function. Three versions of the radar have been identified – Head Light A, B and C. The four main reflectors in the group are open mesh circular dishes (one small and one large to each pair, with the smaller in the upper position) that are disposed symmetrically around a central mounting pillar. Estimated dish sizes are 1.8 and 3.8 m. The electronics for each radar (such as its transceiver) are carried behind the individual reflectors and the whole assembly of four radars is supported from the top of the pedestal turning gear on a yoke of triangular tubular construction. The whole assembly rotates in azimuth and can also move in elevation. The two upper radars also appear to have provision for individual movement in both axes (possibly limited as the dynamic balancing vanes fitted to them suggest). The upper dishes have unusually large feed arrangements (possibly cassegrain) and the larger dishes appear to be front fed. Operating frequencies are reported to be in

The Head Light missile control radar group as fitted to the cruiser Kerch

the G-band for acquisition and H/I-band for tracking. Maximum range of the system is estimated at 60 km with an elevation envelope of between 100 and 25,000 m.

Status

As of early to mid-2000, Head Light B/C missile control radars were reported as being installed aboard the Russian cruiser *Kerch*.

VERIFIED

Cross Swords missile fire-control radar

Type

G- (4 to 6 GHz) and K-band (20 to 40 GHz) surveillance and fire-control radar system for the SA-N-9 missile. Possible Russian designation MR-350.

Description

The Cross Swords system provides surveillance and engagement facilities for the SA-N-9 'Gauntlet' air defence missile. The comparable land-based system (the Tor-M1 - NATO reporting name SA-15 'Gauntlet') has only a single antenna and the multiple array configuration of the naval version is almost certainly designed to provide the high data rate necessary for rapid detection and classification of low-flying aircraft and sea-skimming missiles. Cross Swords operates in G-band and is claimed to be able to provide range, azimuth, elevation and automatic threat evaluation on up to 48 targets at ranges up to 45 km.

The Cross Swords engagement radar employs a single large dish for target tracking (up to a range of 15 km), together with smaller antennas for missile tracking and command uplink. These latter phased-array antennas are mounted on a structure comprising two large electronic boxes disposed about a central pedestal. The large monopulse target tracking dish is angled at 22.5° to the vertical, has a large feed to the front and operates in K-band. The system can track and engage four targets simultaneously and can guide two missiles to each target, provided that all targets are within the same 60 × 60° sector. Of the two enclosed drum-shaped antennas located above the main tracker, the larger presumably tracks the missile while the smaller of the two (which resembles the antenna at the base of the Top Dome fire-control radar) houses a guidance uplink transmitter. Both are mounted at an identical angle to the main tracker. The engagement radar can control both SA-N-9 missiles and gun systems together. The assembly is completed by electro-optical trackers (mounted on either side of the base of the

The Cross Swords engagement radar used with the SA-N-9 SAM system (Steve Zaloga)

main tracking antenna) that can be used independently for missile guidance in a severe electronic countermeasures environment.

Status

As of early to mid-2000, the Cross Swords surveillance and fire-control radar was reported as being installed aboard the Russian destroyer *Admiral Chabanenko*.

UPDATED

Eye Bowl missile fire-control radar

Type

F-band (3 to 4 GHz) fire-control radar for SS-N-14.

Description

The Eye Bowl radar is associated with the SS-N-14 long-range torpedo delivery system. The sensor's main role is the tracking and guidance of the outgoing SS-N-14 missile delivery bus to ensure that it reaches a previously designated target position as determined by the launch vessel's sensors and combat information system. Eye Bowl's reflector dish is comparatively small and there appear to be several auxiliary (possibly electro-optical) sensors mounted immediately behind it. Command signals to the SS-N-14 missile are probably transmitted as coded information within the radar transmissions, as no separate command antennas can be seen. The radar operates in the F-band and is thought to have an operating range of up to 56 km.

Status

As of early to mid-2000, the Eye Bowl fire-control radar was reported as being installed aboard Russian Type 1135 ('Krivak I/Burevestnik' class)/1135M ('Krivak II' class) frigates and Type 1155 ('Udaloy/Fregat' class) destroyers and the Ukrainian frigate *Mikolaiv*.

VERIFIED

A beam view of the two Eye Bowl guidance radars on a 'Krivak' class destroyer, abaft and above the Pop Group radar used with the ship's SA-N-4 anti-aircraft missile system (HMS Londonderry)

Front series missile guidance radars

Type

F- (3 to 4 GHz) and H/I-band (6 to 10 GHz) missile guidance radars.

The Front Door C missile guidance radar on the Russian cruiser Marshal Ustinov *with Palm Frond navigation radar above (Steve Zaloga)*

Description

The Front series are F-band or H/I-band missile guidance radars for shipborne weapons. Front Door provides the mid-course guidance for the SS-N-12 'Sandbox' anti-ship missiles while the H/I-band Front Dome system features a multiple antenna array and provides target illumination for the SA-N-7 and SA-N-17 surface-to-air missiles. It covers from 0 to 70° in elevation and a nominal 360° in azimuth. Each of its designators uses a different frequency and provides one channel of fire-control, although it is believed that in the case of the SA-N-7, three designator channels are used simultaneously against individual high-threat targets. It should also be noted that the Trap Door fire-control radar has been reassigned the NATO reporting name Front Door C. As such, it is associated with the SS-N-12 missile.

Status

As of early to mid-2000, Front series radars were reported as being installed aboard Chinese 'Sovremenny' class destroyers (Front Dome); Indian 'Delhi' class destroyers (Front Dome) and 'Talwar' class frigates (Front Dome) and Russian Type 956/956A ('Sovremenny/Sarych' class) destroyers (Front Dome) and Type 1164 ('Slava' class) cruisers (Front Door).

UPDATED

The radar head used in the Pop Group fire-control system (Steve Zaloga)

Hawk Screech gun fire-control system

Type

Active and passive gun fire-control system.

Description

Hawk Screech is understood to be generally similar to the MR-105 equipment (see separate entry) and is thought to be used to control 45 mm, 57 mm and 76 mm gun mounts.

Status

As of early to mid-2000, Hawk Screech fire-control systems were reported as being installed aboard Algerian 'Mourad Rasi' class frigates; the Bulgarian frigate *Smeli*; Indian 'Arnala' class frigates and the submarine tender *Amba*; Libyan 'Koni' class frigates; the Romanian destroyer *Marasesti* and the same country's 'Tetal' class frigates; Russian Type 97P ('Ivan Susanin' class) armed icebreakers/patrol ships; Syrian 'Petya III' class frigates; the Ukrainian auxiliary *Kolomiya* and Vietnamese 'Petya II/III' class frigates.

UPDATED

A Pop Group fire-control system mounted on a 'Krivak' class destroyer. The Owl Screech gun fire-control radar below the Pop Group is facing the opposite way showing the feed arrangements more clearly

Hot Flash missile control radar

Type

Weapon control system for the CADS-N-1 Kashtan/Kortika (NATO reporting name SA-N-11 'Grissom') air defence system.

Description

The Hot Flash radar is the weapon control system for the CADS-1 (Kashtan/Kortika) close air defence system. It consists of two radar-directed 30 mm guns with a mounting above for four launcher-containers for the SA-N-11 (a navalised SA-19) surface-to-air missile. The Hot Flash radar features two paraboloid antennas with front horn feeds. The central one is probably a search radar to detect and track sea-skimming missiles with the one mounted to starboard probably being used for missile guidance. The system's transceiver electronics are mounted behind the antennas. To the left of the central radar antenna are two electro-optic sensors that are probably a laser range-finder and a remote operated TV camera. The Hot Flash command module can control up to six weapon systems, incorporates a digital frequency control computer and is housed in eight electronics cabinets. One console is likely to be used for sensor control and features a plan position indicator display while the other provides fire control.

Status

As of early to mid-2000, Hot Flash radars were reported as being installed aboard the Russian aircraft carrier *Admiral Kuznetsov* and the same country's Type 1144.1/1144.2 ('Kirov/Orlan' class) battle cruisers.

VERIFIED

MPZ-301 missile fire-control radar

Type

G/J-band (4 to 20 GHz) fire-control radar system for the SA-N-4 Surface-to-Air-Missile (SAM) system. NATO reporting name Pop Group. Alternative Russian designations given (but not confirmed) as Osa-2M or 4P33.

Description

Pop Group is the NATO reporting name for the fire-control radar group associated with the SA-N-4 SAM system and is very similar to the land-based version (NATO

reporting name Land Roll). The Pop Group system incorporates a missile command and two radar antennas and the entire equipment comprises a 2.2 m cube shaped container (likely to house its transceivers, power supplies and turning gear) that is topped off with a trainable radar head. This latter unit is described as featuring a 2 m parabolic reflector that is likely to be used for target search and to rotate independently of the rest of the assembly. On the front face of the radar head are two circular arrays that are assumed to be for target tracking and missile guidance. It appears probable that the Pop Group radar is a monopulse frequency-agile system. Its antenna layout consists of an elliptical, rotating, H-band (6 to 8 GHz) surveillance antenna (30 km acquisition range against 'most' targets), a J-band (10 to 20 GHz) engagement antenna (20 km maximum tracking range), a small J-band missile tracking dish, an I-band (8 to 10 GHz) uplink acquisition antenna and a rectangular command uplink emitter.

Status

As of early to mid-2000, the Pop Group fire-control system is reported as being installed aboard Algerian 'Mourad Rais' class frigates and 'Nanuchka II' class corvettes; the Bulgarian frigate *Smeli*; Indian 'Godavari' class frigates and 'Durg' class corvettes; Libyan 'Koni' class frigates and 'Nanuchka II' class corvettes; Lithuanian 'Grisha III' class frigates; the cruiser *Kerch*, the frigate *Yastreb*, the landing ship *Mitrofan Moskalenko* and the same country's Russian Type 1124 ('Grisha I' class)/1124K ('Grisha IV' class)/1124EM ('Grisha V/Albatros' class)/1124M ('Grisha III' class) frigates, Type 1135MP ('Krivak III/Nerey' class) frigates (one ship?), Type 1144.1/1144.2 ('Kirov/Orlan' class) battle cruisers, Type 1164 ('Slava/Atlant' class) cruisers and Type 1234 ('Nanuchka I/Burya' class)/1234.1 ('Nanuchka III/Veter' class)/1234.2 ('Nanuchka IV/Nakat' class) and Type 1239 ('Dergach/Sivuch' class) corvettes; the Ukrainian frigate *Hetman Sagaidachny* and *Lutsk* the same country's Type 1135 ('Krivak I' class) frigates and Yugoslav 'Kotor/Split' class frigates.

UPDATED

MR-103 Bars gun fire-control radar

Type

H-band (6 to 8 GHz) gun fire-control radar. NATO reporting name Muff Cob.

Description

According to Jane's sources, the MR-103 fire-control radar supports the AK-725 gun system, with each radar controlling one or two mountings. While not

The MR-103 Muff Cob fire-control radar

confirmed, it is thought that MR-103 can be integrated with an electro-optic sighting and tracking system if required.

Status

As of early to mid-2000, MR-103 radars were reported as being installed aboard Algerian 'Nanuchka II' class corvettes; Bulgarian 'Letyashti' class corvettes; Cambodian 'Modified Stenka' class fast attack craft; Cuban 'Stenka/Tarantul' class fast attack craft; Indian 'Godavari' class frigates and 'Durg' class corvettes; Indonesian 'Kapitan Patimura' class corvettes and 'Frosch II' class support ships; Libyan 'Nanuchka II' class corvettes; Romanian 'Cosar' class minelayers and 'Croitor' class logistic support ships; the Russian training ship *Gangut*, the submarine depot ship *Volga 879*, the missile support ship *Voronesh* and the same country's Type 206M ('Turya' class) fast attack hydrofoils, Type 775 ('Ropucha I' class) landing ships tank, Type 1124 ('Grisha I' class) frigates, Type 1234 ('Nanuchka I/Burya' class) corvettes and Type 1559V ('Boris Chilkin' class) replenishment ships (equipped for but not fitted with MR-103); the Ukrainian landing ship *Konstantin Olshansky* and Vietnamese 'Turya' class fast attack craft.

UPDATED

MR-104 Rys gun fire-control radar

Type

I-band (8 to 10 GHz) gun fire-control radar. NATO reporting name Drum Tilt.

Description

MR-104 is an I-band, pedestal-mounted, fire-control radar which provides target acquisition and tracking for the 30 mm AK-230 gun system. Maximum acquisition range against aerial targets is approximately 41 km.

Status

As of early to mid-2000, the MR-104 radar was reported as being installed aboard the Algerian landing ship *471* and the same country's 'Mourad Rais' class frigates and 'Osa II' class fast attack craft; the Bulgarian frigate *Smeli* and the same country's 'Osa I/II' class fast attack craft and 'Polnochny A' class landing ship; Cuban 'Osa II' class fast attack craft; Egyptian 'Osa I' and 'Shershen' class fast attack craft, 'Polnochny A' class landing ships mechanical and 'Yurka' class ocean-going minesweepers; the Eritrean fast attack craft *FMB 161*; the Georgian Type 205P ('Stenka' class) patrol ship *692*; Indian 'Rajput' class destroyers, 'Godavari' class frigates, 'Osa II' class fast attack craft, 'Polnochny C/D' class landing ships mechanical and 'Pondicherry' class ocean-going minesweepers; the Iraqi fast attack craft *Hazirani R15*; North Korean 'Najin' and 'Soho' (one ship) class frigates, 'Huangfen/Osa I' and 'Soju' class fast attack craft and 'Taechong I/II' class patrol craft; Libyan 'Koni' class frigates, 'Osa II' class fast attack craft, 'Natya' class ocean-going minesweepers and 'Polnochny D' class landing ships mechanical; Polish 'Puck' class fast attack craft, 'Modified Obluze' class patrol craft and 'Modified Polnochny C' class landing ships; the Romanian destroyer *Marasesti* and the same country's 'Tetal/Improved Tetal' class frigates, 'Osa I' and 'Naluca' class fast attack craft, 'Cosar' class minelayers, 'Musca' class minesweepers and 'Croitor' class logistic support ships; Russian Type 205P ('Stenka/Tarantul' class) fast attack craft, Type 266M ('Natya I/Akvamaren' class) ocean-going minesweepers (plus one Type 266 ('Yurka/Rubin' class) ocean-going minesweeper), a single Type 771 ('Polnochny B' class) landing ship, Type 887 ('Smolny' class) training ships and Type 1232.1 ('Aist/Dzheyran' class) amphibious warfare hovercraft; the Syrian mine warfare vessel *642* and the same country's 'Osa II' class fast attack craft and 'Polnochny B' class landing ships; the Ukrainian landing ship *Kirovograd* and the same country's 'Stenka/Tarantul' class patrol craft and 'Natya I' class mine warfare vessels; Vietnamese 'Osa II' and 'Shershen' class fast attack craft, 'Yurka/Rubin' class ocean-going minesweepers and 'Polnochny' class landing ships and Yugoslav 'Split/Kotor' class frigates and 'Osa I' class fast attack craft.

UPDATED

MR-105 Turel gun fire-control system

Type

Active and passive gun fire-control system. NATO reporting name Owl Screech.

Description

The MR-105 Turel fire-control system is used with the 76 mm twin AK-726 gun mounting and comprises a target tracking radar (probably H/I-band (6 to 10 GHz), a TV tracker, what is termed 'moving target selection and jamming suppression equipment' and control units. Functionally, Turel provides target co-ordinate and motion parameter measurement, data fusion, gunlaying data, fall of shot measurement and on-system operator training capabilities. The system is classed as a Russian Group 12 fire-control equipment (Class 1230, Complete).

Status

As of early to mid-2000, MR-105 fire-control systems were reported to be installed aboard Indian 'Rajput' class destroyers; the Polish destroyer *Warszawa*; the Russian cruisers *Admiral Golovko* and *Kerch*, the landing ship *Mitrofan Moskalenko*, the missile support ship *Voronesh* and the same country's Type 61/61M ('Kashin/Modified Kashin' class) destroyers and Type 887 ('Smolny' class) training ships; Ukrainian 'Krivak I' class frigates and Yugoslavian 'Split/Kotor' class frigates.

Specifications

Radar subsystem frequency: probably H/I-band (6-10 GHz)
Instrumented range: 55 km
Calibre of gun mounts controlled: 76 mm
Number of gun mounts controlled: up to 2
System weight: 13.5 t

Contractor

Rosoboronexport (export agent), Moscow.

UPDATED

An MR-105 fire-control radar antenna on the Polish Navy destroyer Warszawa. *Above and to the right are the Peel Group antennas (Stefan Terzibaschitsch)*

MR-114 Lev/-145 Drakon/-184 gun fire-control systems

Type

Active and passive series of gun fire-control systems. NATO reporting names Kite Screech (MR-114), Kite Screech A (MR-145) and Kite Screech B (MR-184).

The Kite Screech fire-control radar on the cruiser Marshal Ustinov *(Steve Zaloga)*

The destroyer Otlichnny showing (top to bottom) the antennas used with the vessel's Top Steer surveillance, Kite Screech fire-control and Band Stand radars. Bass Tilt radomes can be seen on either side of the Band Stand installation (Steve Zaloga)

Description

The MR-114/-145/-184 series of active and passive gun fire-control systems is used with 100 mm AK-100 (MR-114 and -145) and AK-130 mm (MR-184) naval gun mounts and comprises a two-band radar (I-band (8 to 10 GHz) for target acquisition and K-band (27 to 40 GHz sub-band) for tracking, a TV tracker, what is termed 'moving target selection and anti-jamming equipment' and control units. System functions comprise measurement of target co-ordinates and motion parameters, data fusion, fall of shot measurement, generation of gunlaying data and on-system operator training. All three versions are classified as Russian Group 12 fire-control equipments (Class 1230, complete).

Status

As of early to mid-2000, MR-114/-145/-184 fire-control systems were reported as being installed aboard Chinese 'Sovremenny' class destroyers; Indian 'Delhi' class destroyers and 'Talwar' class frigates; the Russian destroyer *Admiral Chabanenko*, the frigate *Neustrashimy* and the same country's Type 956/956A ('Sovremenny/Sarych' class) destroyers, Type 1135M/MP ('Krivak II/III-Nerey' class) frigates, Type 1144.1/1144.2 ('Kirov/Orlan' class) battle cruisers, Type 1155 ('Udaloy/Fregat' class) destroyers and Type 1164 ('Slava/Atlant' class) cruisers and the Ukrainian frigate *Hetman Sagaidachny*.

Specifications

Radar subsystem frequency: I- (8-10 GHz) and K- (27-40 GHz sub-band) band
Detection range: 40 km (MR-184, 1 m^2 target)
Instrumented range: 75 km (all models)
Output: 25 kW (K-band, MR-184); 300 kW (I-/low J-band, MR-145)
Calibre of gun mounts controlled: 100 mm (all models); 130 mm
Number of gun mounts controlled: up to 2 (all models)
System weight: 8 t (MR-114/-145/-184)

Contractor

Rosoboronexport (export agent), Moscow.

UPDATED

MR-123 Vympel gun fire-control system

Type

Active and passive gun fire-control system. NATO reporting name Bass Tilt.

Description

MR-123 is a combined active and passive fire-control system designed to control 30 mm AK-630/AK-630M and 76 mm AK-176/AK-176M naval gun mounts. MR-123 is produced in two versions, namely MR-123/76 which can control a single 76 mm weapon and MR-123-02/76 which provides fire-control for one 76 mm or up to two 30 mm weapons. Overall, the MR-123 system is described as incorporating an I-/low J-band (8 to 12 GHz sub-band) target detection and tracking radar, a TV tracker and a control unit. Functionally, the system receives and displays incoming radar and TV information, automatically tracks selected targets and generates firing data for its associated gun mounting/mounts. The antenna for the MR-123 radar subsystem is housed in a drum-shaped radome which is inclined at an angle of 45° and has a diameter of approximately 1.2 m. MR-123 is classified as being a Russian Group 12 fire-control equipment (Class 1230, complete).

Status

As of early to mid-2000, MR-123 gun fire-control systems were reported as being installed aboard the Bulgarian corvette *Mulniya* and the same country's 'Reshitelni' class corvettes; Chinese 'Sovremenny' class destroyers; the Cuban corvette *321*; the Georgian patrol hydrofoil *Tbilisi*; Greek 'Pomornik/Zubr' class amphibious

The radome that houses MR-123 gun fire-control system's target detection and tracking radar scanner (Steve Zaloga)

warfare hovercraft; Indian 'Delhi' and 'Rajput' (two ships) class destroyers, 'Modified Godavari' class frigates and 'Abhay', 'Khukri/Modified Khukri' and 'Veer' class corvettes; Lithuanian 'Grisha III/Albatros' class frigates; the Polish destroyer *Warszawa* and the same country's 'Gornik' and 'Orkan' class corvettes; Romanian 'Zborul' class corvettes; the Russian cruisers *Admiral Golovko* and *Kerch*, the destroyer *Sderzhanny*, the frigates *Novik* and *Yastreb*, the fast attack craft *Vladimirets 060*, the landing ship *Mitrofan Moskalenko*, the research ship *Marshal Krylov* and the same country's Type 61/61M ('Kashin/Modified Kashin' class) destroyers, Type 133 ('Muravey/Antares' class) and Type 1141/1145 ('Babochka/Mukha - Sokol' class) fast attack hydrofoils, Type 206MP ('Matka/Vekhr' class), Type 1041Z ('Svetlyak' class) and Type 1241P ('Pauka I/Molnya' class) fast attack craft, Type 775/775M ('Ropucha I/II' class) landing ships, Type 956/956A ('Sovremenny/Sarych' class) and Type 1155 ('Udaloy/Fregat' class) destroyers, Type 1124M/1124EM ('Grisha III/V - Albatros' class), Type 1135MP ('Krivak III/Nerey' class) and Type 1331 ('Parchim II' class) frigates, Type 1164 ('Slava/Atlant' class) cruisers, Type 1208 ('Yaz/Slepen' class) patrol craft, Type 1232.2 ('Pomornik/Zubr' class) amphibious warfare hovercraft, Type 1234.1 ('Nanuchka III/Veter' class), Type 1239 ('Dergach/Sivuch' class) and Type 1241.1/1241.1M/1241.1MP ('Taruntul I/II/III/Modified III - Molniya' class) corvettes and Type 12660 ('Gorya' class) ocean-going minehunters; the Ukrainian frigates *Hetman Sagaidachny* and *Lutsk*, 'Pauk I' class patrol craft, 'Muravyev/Antares' class patrol hydrofoils, 'Matka/Vekhr' class fast attack hydrofoils and 'Pomornik' class amphibious warfare hovercraft; Vietnamese 'HO-A' and 'Tarantul I' class corvettes and the Yemeni patrol craft *971*.

Specifications

Radar subsystem frequency: I/low J-band (8-12 GHz sub-band)
Instrumented range: 45 km
Calibre of gun mounts controlled: 30 and 76 mm
Number of gun mounts controlled: 1 (76 mm); up to 2 (30 mm)
System weight: 4.5-5 t

Contractor

Rosoboronexport (export agent), Moscow.

UPDATED

MR-352 Positive-E target acquisition and fire-control radar

Type

Target acquisition and fire-control radar. NATO reporting name Cross Dome.

Description

MR-352 is designed for small ship applications and provides air and surface target acquisition, threat assessment and firing data for gun and missile systems. MR-352 is classed as a Russian Group 12 fire-control equipment (Class 1285).

Status

As of early to mid-2000, MR-352 target acquisition and fire-control radars were reported as being installed aboard Greek 'Pomornik/Zubr' class amphibious warfare hovercraft; Indian 'Abhay', 'Khukri' and 'Kora' class corvettes; the Russian frigates *Novik* and *Yastreb* together with the same country's Type 775M ('Ropucha II' class) landing ships, Type 1232.2 ('Pomornik/Zubr' class) amphibious warfare hovercraft, Type 1239 ('Dergach/Sivuch' class) corvettes and Type 1331 ('Parchim II' class) frigates; Ukrainian 'Pomornik/Zubr' class amphibious warfare hovercraft and Vietnamese 'HO-A' class corvettes.

Specifications

Coverage: 128 km (range); 40° (elevation)
System weight: 3.5 t

Contractor

Rosoboronexport (export agent), Moscow.

UPDATED

Top Dome missile guidance radar

Type
J-band (10 to 20 GHz) missile guidance radar.

Description
Top Dome is the NATO reporting name for the J-band radar associated with the vertically launched SA-N-6 surface-to-air missile system fitted aboard Russian Type 1144.1/1144.2 ('Kirov/Orlan' class) battle cruisers and Type 1164 ('Slava/Atlant' class) cruisers. The sensor forms part of the ALTAR Research and Development Corporation's RIF shipborne 'anti-aircraft missile complex for group defence' and utilises a phased-array antenna to provide both long-range target tracking and missile guidance functions for the system. So equipped, Top Dome is reported to be able to track up to six targets and provide guidance for up to 12 missiles simultaneously. Within the RIF complex, initial target detection is via passive sensors or the Top Pair radar. Targets are allocated to either one of two Top Dome director groups within the system according to direction of approach. The sensor is understood to be capable of both tracking the SA-N-6 missile and the designated targets simultaneously and there is also the possibility that it can use a combined mode of guidance that involves using the SA-N-6 seeker for target tracking with its command link relaying data back to the ship. The radar's mounting is fixed in elevation (at a 20° angle), rotates in azimuth and features electronic antenna beam pitch and roll stabilisation. The front face of the mounting carries three semi-cylindrical antenna housings together with a fourth 'thimble' shaped radome. The top antenna tracks both the target and the missile while the 'thimble' shaped antenna is the link for the track-via-missile system. The other antennas probably control two missiles apiece, with tracking of both target and missiles conducted on a time-share basis.

Status
As of early to mid-2000, Top Dome radars were reported to be installed aboard Russian Type 1144.1/1144.2 ('Kirov/Orlan' class) battle cruisers (one aboard *Pyotr Velikiy*, two aboard *Admiral Nakhimov*) and Type 1164 ('Slava/Atlant' class) cruisers.

UPDATED

The Top Dome radar director as installed on the cruiser Marshal Ustinov (Steve Zaloga)

SWEDEN

CEROS 200 tracking radar

Type
12.5 to 18 GHz band fire-control radar.

Description
CEROS 200 is a 12.5 to 18 GHz band, frequency-agile fire-control radar which usually forms part of the 9LV Mk 3 Command and Weapon Control System. It is also available in a stand-alone version. Its primary function is that of target tracking for gun and missile control. The equipment is a monopulse system with a stabilised cassegrain antenna. Use of a multimode feed ensures that this assembly has low-sidelobe values in both sum and difference patterns. The radar uses a helix travelling wave tube transmitter which provides a very wide frequency-agile bandwidth. CEROS 200 can operate in pulse-to-pulse frequency-agile, moving target indicator and pulse-Doppler modes and uses automatic selection of pulse-lengths, pulse repetition frequencies and radar modes to provide a 'high degree' of self-adaption to prevailing target and environment (clutter levels, countermeasures pollution and so on) conditions. The radar also utilises the SaabTech-developed CHASE signal processing technique which allows it to track sea-skimming missiles flying at altitudes of between 3 and 20 m with practically no degradation in performance. The system's pedestal mount can accommodate TV/infra-red cameras and laser range-finders in addition to the radar in order to provide an

The CEROS 200 director unit incorporates tracking radar and 10 GHz illumination capabilities together with TV and infra-red cameras and a laser range-finder

autonomous electro-optic tracking capability. The system makes use of an 'extremely reliable' direct-drive hydraulic motor which has slewing speed and acceleration values of better than 120°/s and 500°/s² respectively. The CEROS 200 director is also available with an integrated 8 to 12.5 GHz band continuous wave antenna for use with the Sea Sparrow missile system. This variant (designated as CEROS CWI) has been procured for installation aboard Australian/New Zealand ANZAC frigate and Danish Standard Flex 300 ('Flyvefisken' class) ship programmes. Alongside the baseline equipment, SaabTech also notes that CEROS 200 is available in an extended range configuration for use with the Extended Sea Sparrow Missile (ESSM) system. A stealth variant (for use on low radar cross-section platforms) forms part of the equipment fit installed aboard the Royal Swedish Navy's 'Visby' class corvettes. Here, use is made of a flat, inclined surface sensor housing, radar absorbent coatings and a frequency selective antenna radome to minimise the overall radar cross-section of the sensor head over a 'multi-octave frequency band'.

Status
SaabTech describes CEROS 200 as being a well proven design which is based on the company's experience in supplying approximately 200 fire-control systems to navies around the world. Such radars are noted as having been operated in a wide variety of environments ranging from arctic to tropical and littoral to open sea. Aside from the noted procurements, 20 December 2001 saw SaabTech Systems being awarded a then year €8.3 million contract covering the supply of CEROS 200 equipments into the Finnish Navy's Squadron 2000 combat vessel programme. At the time of the announcement of this award, deliveries were scheduled to begin during 2004.

Specifications
Baseline version
Type: 12.5-18 GHz frequency-agile MTI monopulse
Frequency range: random pulse-to-pulse in J-band
Transmitter: Helix TWT
Peak power: 1.5 kW
Receiver noise: 5 dB
Antenna diameter: 1 m
Beamwidth (at 3 dB): 1.3°
Gain: 41 dB

Contractor
SaabTech Systems AB, Järfälla.

UPDATED

SWITZERLAND

Seaguard Tracking Module (TMX)

Type
I-band (8 to 10 GHz) tracking and fire-control radar.

Description
Part of the Seaguard range of modular combat systems, the general purpose tracking module (TMX) is used for accurate long-range tracking of air and surface

The sensor head assembly used in the Seaguard Tracking Module (TMX)
0045034

targets for engagement by guns ranging from small (such as Sea Zenith or Millennium) to large calibre, and/or ship-to-air missiles.

The TMX uses a combination of I-band radar, forward-looking infra-red (or optional TV) and laser range-finder sensors. The I-band radar is a fully coherent pulse-Doppler system using pulse compression and a travelling wave tube transmitter. Real-time Fast Fourier Transform Doppler processing is used in both acquisition and tracking modes. The radar incorporates a number of advanced features, including random frequency agility over the entire band, moving target indicator, variable pulse compression ratio, constant false alarm rate, frequency coded transmission, staggered pulse repetition frequency and track-on-jam.

When equipped with electro-optical sensors the combined system processing techniques provide redundancy and make the TMX impervious to electronic countermeasures and multipath effects. Optional second radar receiver and continuous wave injection are available for ship-to-air missile guidance purposes.

Status
As of December 2001, the TMX tracking module was reported as having been procured by a number of customers around the world.

Specifications
Frequency: I-band (8.6-9.5 GHz)
Antenna: cassegrain 1.31 m diameter
Transmitter: TWT
Peak power: 2 or 4 kW
Pulse-length: 0.53-4 µs
Tracking range: 300 m to 70 km
E-O sensors: FLIR spectral bandwidth 8-12 µs, centroid FTT tracking principle; laser spectral wavelength 1.06 µs with a 25 Hz data rate
Tracker: unique 3-axis, high dynamic (3 rad/s; 15 rad/s²); hemispherical tracking (elevation: −35 to +120°)

Contractor
Oerlikon Contraves AG, Zurich.

VERIFIED

UNITED KINGDOM

Naval fire-control radars

As of this edition, a number of earlier generation UK-sourced naval fire-control radars are reported to remain in service around the world. Known details of these are as follows:

ST802
The Marconi/Ericsson ST802 is described as being an I-band (8 to 10 GHz) monopulse tracking radar designed to direct small/medium calibre guns and provide target tracking for missile systems. The equipment is quoted as having a 1 m antenna and generating 'accurate' angular data via a combination of a 2.4° pencil beam and monopulse signal processing. Moving target indication and

constant false alarm rate provision are noted, as is the ability to integrate the radar with a passive electro-optic tracking system. Most recently, Jane's sources suggest that ST802 has been in service aboard Egyptian 'Ramadan' and Kenyan 'Nyayo' class fast attack craft.

ST1802
The Marconi ST1802 is described as being an I-band (specific 8.9 to 9.5 GHz sub-band), travelling wave tube-powered successor to the magnetron-powered ST802 fire-control radar. It is noted as being a monopulse tracker which incorporates switchable moving target indication and an elevation coverage of -30 to +85°. Most recently, Jane's sources suggest that ST1802 has been in service aboard South Korean 'Ulsan' class frigates (four only) and some of the same country's 'Po Hang' class corvettes.

Type 275
Type 275 is described as being a late Second World War vintage fire-control radar which, most recently, is reported to have been in service aboard Bangladeshi 'Leopard' class frigates and the single unit *Umar Farooq*.

Type 903/904
The Plessey (now BAE Systems) Types 903 and 904 radars are described as I/J-band (8 to 20 GHz) fire-control equipments which are used in the MRS 3 gun and GWS 20 series missile control systems respectively. In more detail, MRS 3 is noted as supporting 3 or 4.5 in guns while GWS 20 suites are used with the UK's Seacat surface-to-air missile system. Most recently, Type 903 and 904 radars are reported to have been in service aboard Chilean 'Prat' class destroyers (Type 903 – two only?) and 'Leander' class frigates (Types 903 and 904 – both to be replaced); Ecuadorean 'Leander' class frigates (903 and 904); Indonesian 'Khristina Tiyahahu' class frigates (903) and Pakistani 'Leander' class frigates (904 – likely to be replaced).

VERIFIED

UNITED STATES OF AMERICA

Introductory note

Alongside the equipments described in the following section, readers should be aware of a number of other American naval fire-control radars that are currently in service around the world. Brief details of these are as follows:

AN/SPG-34
AN/SPG-34 is a 50 kW, I/low J-band (8 to 12 GHz) radar that dates from 1953 and was originally designated as the Mk 34. It originally formed part of the Mk 57 and 63 Fire-Control Systems (FCSs). More recently, SPG-34 radars have been reported as having been installed aboard the Brazilian aircraft carrier *Minas Gerais;* the Japanese destroyer *Murakumo;* Portuguese 'Joao Coutinho' class frigates; Spanish 'Paul Revere' class attack transports (as an alternative to AN/SPG-50) and the Turkish fleet support ships *Akar* and *Yarbay Kudret Gungor*.

AN/SPG-50
First delivered to the US Navy during February 1956, the AN/SPG-50 fire-control radar is an updated AN/SPG-34 equipment that features a 1 m antenna dish, a tunable magnetron transmitter, improved signal processing and has a pulse repetition frequency value of 2,000. Range accuracy is given as ± 46 m. More recently, SPG-50 radars have been reported as having been installed aboard Greek 'Tolmi' class patrol ships; South Korean 'Pae Ku' class fast attack craft (as an alternative to W-120); Spanish 'Paul Revere' class attack transports (as an alternative to AN/SPG-34) and the Turkish patrol craft *Bora*.

AN/SPG-52
AN/SPG-52 is a range-only, 50 kW, J-band (12 to 18 GHz) radar which was originally associated with the US Mk 70 GunFire Control System (GFCS). Most recently, the SPG-52 radar was reported as being installed aboard Indonesian 'Samadikum' class frigates.

AN/SPG-53
AN/SPG-53 is a 250 kW, I/low J-band (8 to 12 GHz) radar that was originally associated with the US Mk 68 GFCS. The equipment appeared in a number of variants, known details of which are as follows:
SPG-53A SPG-53 modified to work with missile systems.
SPG-53B Incorporated a Continuous Wave (CW) illuminator for the Tartar Surface-to-Air Missile (SAM). Also noted as being associated with the Mk 68 Mods 9 and 10 GFCSs.
SPG-53D SPG-53A modified to include CW illumination.
SPG-53E SPG-53A modified for monopulse target tracking.
SPG-53F Replacement for SPG-53A which features an embedded electronic warfare simulation subsystem. Most recently, SPG-53 radars are reported as having been installed aboard Egyptian (SPG-53A/D/F), Taiwanese (SPG-53A/D/F) and Thai (SPG-53A/D/F) 'Knox' class frigates; Greek 'Epirus' class frigates; Spanish 'Balearas' class frigates (SPG-53B) and Turkish 'Tepe' class frigates (SPG-53D/F).

AN/SPG-62
AN/SPG-62 is a 10 kW, I/low J-band (8 to 12 GHz) CW illuminator radar that is slaved to the AN/SPY-1 phased-array air search and fire-control radar (see separate entry) in the US Mk 99 FCS. Most recently, SPG-62 radars were reported as being installed aboard Japanese 'Kongou' class destroyers and US Navy 'Ticonderoga' class cruisers and 'Arleigh Burke Flts I/II/IIA' class destroyers.

Separate Target Illumination Radar (STIR)

US STIR was originally developed for use with the US Mk 92 FCS and incorporates continuous wave illumination for the Standard surface-to-air missile system. As of this edition, US STIR was reported as forming part of the FCS fitted to US Navy 'Oliver Hazard Perry' class frigates.

Mk 25

Mk 25 is a 250 kW, I/low J-band (8 to 12 GHz) fire-control radar which forms part of the US Mk 37 GFCS. Most recently, Mk 25 radars were reported as having been installed aboard South Korean (FRAM I/II), Pakistani (FRAM I), Taiwanese (Wu Chin I/II conversions) and Turkish (FRAM I) 'Gearing' class destroyers and the Mexican destroyer *Cuitlahac*. In the Taiwanese application, Mk 25 is noted as being an alternative to RTN-10X.

Mk 35

Mk 35 is a 1950s vintage I/J-band (approximately 10 GHz) anti-aircraft fire-control radar that, most recently, has been reported as being installed aboard Brazilian 'Para' class frigates; the Japanese training ship *Mochizuki* together with the same country's 'Takatsuki' and 'Yamagumo' class destroyers; Mexican 'Bronstein' class frigates and Turkish FRAM I 'Carpenter' class destroyers.

Mk 95

Mk 95 is a 2 kW, I/low J-band (8 to 12 GHz) tracking and illumination radar which features individual continuous wave transmission and reception antenna and forms part of the Mk 91 FCS which is used with the Sea Sparrow SAM. Most recently, Mk 95 radars have been reported as being installed aboard Danish 'Niels Juel' class frigates; Italian 'Lupo' class frigates (Mk 95 Mod 1); Norwegian 'Oslo' class frigates and US Navy 'Kitty Hawk/John F Kennedy' and 'Nimitz' class nuclear-powered aircraft carriers (together with the single unit USS *Enterprise*); 'Spruance' class destroyers; 'Wasp' class multipurpose amphibious assault ships and 'Sacramento' and 'Supply' class fast combat support ships. It should also be noted that Japanese 'Asagiri', 'Haruna', 'Hatsuyuki', 'Murasame' and 'Takatsuki' class destroyers are all armed with the Sea Sparrow system and it may be that the Japanese Type 2 fire-control radar is a licence produced version of Mk 95.

VERIFIED

AN/SPG-51 gun and missile fire-control radar

Type

G- (4 to 6 GHz) and I-band (8 to 10 GHz) missile and gun fire-control radar system.

Description

The AN/SPG-51 pulse-Doppler radar is a major element of the Mk 74 missile/gun fire-control system and provides target tracking and weapon guidance for the Standard surface-to-air missile system. SPG-51's tracking subsystem uses a G-band pulse-Doppler technique to provide automatic target acquisition and track. Peak output power in the G-band is understood to be 30 kW. The I-band continuous wave illuminator provides radar compatibility with the Standard missile.

As part of the US Navy's continuing improvement programme, a number of modifications have been implemented. These include an improved electronic counter-countermeasures capability against jamming, improved surface tracking performance and performance and reliability enhancements to the transmitters. In addition, the Continuous Wave Acquisition and Track (CWAT) has been deployed, adding frequency diversity and high performance against low, small, fast targets in open ocean and land background. This upgrade added the radar target data processor CV-4139/SPG to perform the signal data processing for the CWAT function.

Two SPG-51D antennas aboard the German Navy destroyer Mölders. *Above is the antenna of the AN/ SPS-52 surveillance radar (Stefan Terzibaschitsch)*

The radar line of sight can be rotated continuously in train and can be elevated from −30 to +83° relative to the deck plane. Initial pointing information is supplied to the SPG-51 by a weapon direction function using data from a search radar device. This information is received by the Raytheon CV-3830/SPG radar data processor that generates a search pattern for the SPG-51 until target acquisition is accomplished. Automatic tracking is then initiated and the SPG-51 feeds return signal information to the radar data processor for angle and range tracking. The radar data processor generates the fire-control data for transmission to a weapon direction function, gun mounts and missile launchers. It also provides missile seeker and angle data.

Status

Over time, SPG-51 radar systems are reported as having been installed aboard a range of ship types including Australian 'Perth' class destroyers (SPG-51C); Dutch 'Tromp' class frigates (-51C); French 'Cassard' class destroyers (-51C); German 'Lütjens' class destroyers; Greek 'Kimon' class destroyers (-51D); Italian 'Audace' and 'De La Penne' class destroyers (-51D); Japanese 'Hatakaze' (-51C) and 'Tachikaze' class destroyers; Spanish 'Balearas' class frigates (-51C) and US Navy 'California' and 'Virginia' class cruisers (both -51D) and 'Kidd' class destroyers (-51D).

Contractor

Raytheon Electronic Systems, Marlboro, Massachusetts.

VERIFIED

AN/SPG-60 acquisition and tracking radar

Type

I-band (8 to 10 GHz) fire-control and acquisition radar for the Mk 86 fire-control system.

Description

The AN/SPG-60 is a monopulse, I-band pulse-Doppler tracking radar forming part of the Mk 86 Fire-Control System (FCS). It is used for acquisition and tracking of air targets to ranges of about 111 km, tracking being either automatic or manually aided. The radar's search-to-acquire capability is computer-directed and it accepts two-or three-dimensional target designation co-ordinates in digital or synchro format from either the AN/SPQ-9 radar, which is also part of the Mk 86 FCS, or other shipboard search radars and tactical data systems. The antenna is stabilised. Peak output power is given as 5.5 kW.

Acquisition for SPG-60 is automatic with the radar performing its programmed scan search pattern about the designated point under computer control. The computer resolves the problems of blind range and range rates inherent in Doppler radars, as well as range ambiguity.

The computer also automatically controls and calibrates receiver gain and monopulse channel balance. Other system features include:

- a four-horn monopulse antenna assembly
- continuous azimuth rotation
- a co-located TV camera sighting subsystem
- passive angle tracking
- adaptive scan patterns for automatic target acquisition
- adaptive computer control of pulse repetition frequency and pulse-length.

The Mk 86 FCS uses the SPG-60 to provide tracking data for the SM-1 and SM-2 surface-to-air missile systems and it also provides target illumination by means of continuous wave injection at the SPG-60 antenna.

Status

Over time, SPG-60 radar systems are reported as having been installed aboard a range of ship types including Australian 'Adelaide' class frigates; German 'Lütjens' class destroyers and US Navy 'California' and 'Virginia' class cruisers (SPG-60D),

The antenna for the SPG-60 radar (below the AN/SPQ-9 radome) aboard the German Navy destroyer Lütjens *(Stefan Terzibaschitsch)*

'Spruance' class destroyers, and 'Tarawa' class general purpose amphibious assault ships.

Contractor

Lockheed Martin Naval Electronics and Surveillance Systems-Moorestown, Moorestown, New Jersey.

VERIFIED

AN/SPQ-9B fire-control radar

Type

I/low J-band (8 to 12.5 GHz sub-band) surface search and fire-control radar.

Description

AN/SPQ-9B, which utilises the transmitter from the AN/APG-68 airborne fire-control radar, is designed to function as a stand-alone equipment or as a replacement for SPQ-9A in the Mk 86 gun fire-control system. The system is a pulse-Doppler equipment that detects low flying anti-shipping missiles. SPQ-9B features a low-noise exciter while its receiver and processor are noted as making extensive use of off-the-shelf technology. Other aspects noted include narrow elevation and azimuth beamwidths; variable air channel pulse-widths; a fixed surface channel pulse-width; clutter rejection/improvement (in the region of 90 dB); simultaneous air/surface search and beacon tracking and built-in test.

Status

Northrop Grumman Electronic Systems sector - Norden Systems was awarded a then year US$16 million SPQ-9B development contract in October 1994. The US Navy (USN) began trials with the first preproduction SPQ-9B radar during early 1997. In August 1997, the USN awarded Norden Systems a then year US$9.1 million contract covering first low-rate initial production of two SPQ-9B radar ordnance alteration kits. On 26 August 1999, Norden Systems was awarded a then year US$9,352,040 modification to an existing contract (N00024-94-C-5441) that effected a 'change order' that switched future SPQ-9B antenna production from a 'heavy' to a 'lightweight' configuration and provided engineering change kits to modify existing system antennas to the lightweight standard. At the time of the announcement, work on the effort was scheduled for completion by the end of February 2001 and the programme's contracting activity was the USN's Naval Sea Systems Command, Arlington, Virginia. On 19 January 2001, Norden Systems was awarded a then year US$22,475,000 firm, fixed-price contract covering the fabrication, assembly, inspection, test and delivery of five 'lightweight' configuration, low rate initial production SPQ-9B radars and a single SPQ-9B engineering change kit for converting a radar from 'heavy' to 'lightweight' configuration. At the time of the announcement, work on the effort was split between facilities in Baltimore, Maryland (40 per cent workshare) and Melville, New York (60 per cent workshare), was scheduled for completion by the end of March 2003 and had the USN's Naval Sea Systems Command (Arlington, Virginia) as its contracting activity. When delivered, the five new build SPQ-9Bs procured under this award were noted as being scheduled for installation aboard the aircraft carrier the USS *Eisenhower* and an undisclosed number of USN 'San Antonio' class amphibious transport docks. As of the period April to August 2001, Jane's sources were reporting SPQ-9B as being planned for or installed aboard the following warships and classes of warship:

USA   • the aircraft carrier *Enterprise* (Mk 23 to be replaced by SPQ-9B)
       • the destroyer *Oldendorf*
       • 'Kitty Hawk/John F Kennedy' class aircraft carriers (three ships in class - Mk 23 to be
       replaced by SPQ-9B)
       • 'Nimitz' class aircraft carriers (eight ships, one building and one planned as of April 2001 - Mk 23 or SPQ-9B)
       • 'San Antonio' class amphibious transport docks (four building, eight planned as of April 2001)
       • 'Wasp' class multipurpose amphibious assault ships (seven ships, one building as of August 2001 - SPQ-9B to be retrofitted)

Contractor

Northrop Grumman Electronic Systems sector - Norden Systems, Melville, New York.

UPDATED

Mk 23 target acquisition radar

Type

D-band (1 to 2 GHz) target acquisition radar for the Sea Sparrow missile system.

Description

The Mk 23 Target Acquisition System (TAS) is a two-dimensional D-band rotating fan-beam pulse-Doppler radar intended for use with NATO Sea Sparrow missiles

A Mk 23 TAS radar on the 'Spruance' class destroyer USS Hayler *(Stefan Terzibaschitsch)*

The Mk 23 TAS radar

for protection against sea-skimming, high-diving and pop-up missiles. As a secondary role, the system can also be used for air defence surveillance and aircraft control. The 4.3 m antenna incorporates 26 flared horns which provide 360° azimuth coverage and up to 75° in elevation while rotating at 15 or 30 rpm. It is attached to a 30° stabilised pedestal and is mounted back-to-back with a Mk 12 identification friend-or-foe equipment. The system has a detection range of more than 37 km on a 1 m radar cross-section target and a range of nearly 185 km when used in the secondary role.

The Mk 23 system has four operational roles:
• a normal point defence mode where the equipment is used to detect and track missiles at ranges beyond 32 km and then engage them with anti-missile missiles
• a medium-range mode for surveillance and aircraft control out to nearly 185 km
• a mixed mode combining (a) and (b)
• an emission control mode which allows the operator to scan selected sectors and switch the equipment on and off automatically to avoid detection.

A number of upgrades have been made to the system to improve response times, expand radar coverage and add to the system flexibility. These include a new computer to increase the system speed and expand its memory and a new phased-array antenna using phase shifters instead of horn antennas. According to Raytheon Electronic Systems, this gives the system a four-dimensional capability in bearing, range, elevation angle and range rate to reduce the weapon response time and allow the TAS to be used as a designation radar for 'fire-and-forget' missiles. With changes in the transmitter, the improved system has been designated as the Mk 23M and is reported to have a maximum detection range of over 278 km. Raytheon Electronic Systems is also understood to have undertaken a definition study of an improved Mk 23 TAS (designated as the TAS(I) system). Here, the radar was envisaged as being a rotating phase-to-phase type, capable of steering two 11° wide-focused beams in elevation and azimuth. Such an approach would allow the radar to back-scan and fore-scan while rotating, sending out track confirmation beams and consequently providing a single scan acquisition capability.

Status

Over time, Mk 23 target acquisition radars are reported as having been installed aboard a range of ship types including US Navy 'Kitty Hawk/John F Kennedy' class aircraft carriers (together with the single unit USS *Enterprise*), 'Spruance' class destroyers, 'Blue Ridge' class amphibious command ships, 'Tarawa' class general purpose amphibious assault ships and 'Wasp' class multipurpose amphibious assault ships.

Specifications

Frequency: D-band (1-2 GHz)
Power output: 200 kW (peak)
Scan rate: 2 s
Coverage: 0-75° (elevation); 360° (azimuth)
Antenna
Dimensions: 328 × 815 × 193 cm
Weight: 908 kg

Contractor

Raytheon Electronic Systems, Fullerton, California.

VERIFIED

....

Mk 92 fire-control system

Type

Modular, lightweight missile and gun fire-control system for area and point defence.

Development

The Mk 92 is derived from the WM 25 system (developed by the then Signaal of the Netherlands) and the AN/SPG-60 system (see separate entry). Mk 92 was originally manufactured in 1976 by Sperry Rand (subsequently Lockheed Martin Government Electronic Systems, now Lockheed Martin Naval Electronics and Surveillance Systems) to US Navy requirements. It is currently installed in a variety of ships of the US and other navies. Reliability, maintainability and performance have been progressively upgraded through modifications based on in-service experience. As of 1995, deployed versions of the Mk 92 system are noted as being the Mods 1, 2, 5 and 6.

Description

The fire-control system has a separate search and track Combined Antenna System (CAS), mounted on a stabilised platform and enclosed in a radome. The CAS provides simultaneous volumetric search, monopulse target tracking and illumination functions. A high search data rate and rapid handover to the tracking channels provide fast engagement of pop-up and low elevation surface and shore gunfire engagements. Target designations to an external Separate Tracking and Illumination Radar (STIR) provide a second monopulse and illuminator channel for air engagement with missiles. Operational availability of the Fire-Control System (FCS) is periodically determined by built-in comprehensive testing of overall performance against synthesised target and environmental scenarios. Digital command and control of the FCS and interface with the combat system is through an AN/UYK-7 computer.

Mk 92 Mod 1

The Mk 92 Mod 1 is a lightweight, modular system designed for use in small surface ships and patrol craft. The CAS provides fire-control search radar data at a scan rate of 60 rpm. CAS also provides a radar-tracking channel for control of the Mk 75 gun. Two Track-While-Scan (TWS) channels are available for additional gunfire control. The Mod 1 also provides target data to the Harpoon missile system and can be adapted to control surface-to-air missile control for its host platform with the addition of a Continuous Wave Illumination (CWI) capability.

Mk 92 Mod 2

The Mk 92 Mod 2 is a lightweight system for small surface combatants. It has a combined antenna system with search and tracking radars similar to the Mod 1. Mod 2 CAS also uses CWI for controlling anti-air warfare Standard Missiles (SM-1). A second fire-control channel for SM-1 control is provided by the STIR tracker. The Mod 2 provides Mk 75 gunfire control through the CAS tracker, CAS search TWS and the STIR tracker. Harpoon target information is provided to the SWG-1 launch controller by the Mk 92 Mod 2.

Mk 92 Mod 5

The Mk 92 Mod 5 is a variant of the Mod 1 system with interfaces developed to control the Mk 15 Phalanx close in weapon system gun as well as the Mk 75 gun. It is identical to the Mod 1 in other respects and is deployed aboard Royal Saudi Naval Forces patrol craft.

Mk 92 Mod 6

The Mk 92 Mod 6 is a substantial upgrade of the Mod 2 which incorporates coherent transmitter improvements. It has significantly enhanced the capabilities of the Mod 2 in the areas of inclement weather operations, electronic countermeasures functions, small target detection, reliability and maintainability and improved built-in test.

The Mod 6 has been manufactured as original equipment and as an alteration kit for the Mod 2 system. The Mod 6 has a CAS housed in a radome that includes both a search and tracking antenna that share electronic equipment. A STIR provides a second missile and gunfire channel capability.

At the heart of the Mod 6 upgrade are coherent radar receivers and transmitters using Travelling Wave Tube (TWT) technology. Dual TWTs are provided to ensure high system availability via integral redundancy. Advanced signal processing, increased digital processing capacity and improved software are employed to translate radar data into useful target information. Performance improvements for the Mod 6 relative to the Mod 2 include:

- twice the instrumented radar range (search and track)
- twice the transmitter power
- 100 to 1 improvement in clutter rejection
- coherent cancellation of multiple internal clutter
- order-of-magnitude reduction in radar cross-section detection capability
- improved target velocity handling capability
- improved electronic counter-countermeasures functions.

Two or more simultaneous SAM engagements can be carried out using the CAS and STIR trackers, which are equipped with CWI. Surface targets can be engaged while air engagements are in progress. The Mk 74 Otobreda 76 mm gun engages surface targets and the CAS search radar directs the gun using TWS techniques. The TWS capacity enables the Mk 92 to track two targets for gun engagements and six additional targets for search track management.

Using the Mk 92's modular design, an open architecture version of the Mod 6 STIR tracking and illumination radar has been developed (and is available) for corvette and frigate applications. This stand-alone STIR is described as being capable of directing a variety of short- and long-range semi-active missile types as well as providing gunfire control in air and surface engagements.

A scaled down version of the Mod 6 with CAS search and track is also available. Standard missiles and the Mk 75 gun are controlled by the CAS tracker. Gunfire control is accomplished through two CAS search TWS channels. Six additional targets can be tracked with TWS. Harpoon target information is provided by the SWG-1/1A launch controller. This version is designed for service in small combatants and patrol boats.

Status

Over time, at least 120 Mk 92 fire-control systems are understood to have been manufactured with equipment of the type being reported as having been installed aboard a range of ship types including Australian 'Adelaide' class frigates; Bahrain's 'Oliver Hazard Perry' class frigates; Egyptian 'Oliver Hazard Perry' class frigates; Saudi 'Badr' class corvettes; Spanish 'Santa Maria' class frigates; Taiwanese 'Cheng Kung' class frigates (Mk 92 Mod 6 – then year US$20 million contract placed during January 1994), Turkish 'Gaziantep' class frigates and US Navy 'Oliver Hazard Perry' class frigates and US Coast Guard 'Hamilton/Hero' and 'Reliance' class cutters. On 15 March 2001, Lockheed Martin Naval Electronics and Surveillance Systems - Moorestown was awarded a then year US$45,327,739 cost plus fixed-fee, indefinite delivery/quantity contract covering the provision of materials and engineering/technical services in support of Mk 92 fire-control systems operated by Australia (10 per cent of the programme), Bahrain (1 per cent), Egypt (1 per cent), Saudi Arabia (1 per cent), Spain (4 per cent), Taiwan (82 per cent) and Turkey (1 per cent). At the time of its announcement, work on this Foreign Military Sales programme was to be undertaken at Huntsville, Alabama (52 per cent workshare); Moorestown, New Jersey (44 per cent) and Port Hueneme, California (4 per cent) and was scheduled for completion by the end of March 2006. The contracting activity for the effort was the USN's Naval Surface Warfare Center, Port Hueneme, California.

Contractor

Lockheed Martin Naval Electronics and Surveillance Systems-Moorestown, Moorestown, New Jersey.

UPDATED

The Mk 92 fire-control system aboard the Spanish frigate Victoria
(Stefan Terzibaschitsch)

AIRBORNE SURVEILLANCE, MARITIME PATROL AND NAVIGATION RADARS

CANADA

APS-504(V) series radars

Type
I-band (8 to 10 GHz) family of airborne surveillance radars.

Description
The APS-504(V) series of airborne search radars has been designed for tactical transport and maritime surveillance applications. They can be employed in either a fixed- or rotary-wing type of aircraft. The systems are operationally suitable for use in a search, ground mapping, navigation, station keeping or weather avoidance role. Known details of identified members of the family are as follows:

APS-504(V)3
The APS-504(V)3 is a derivative of the basic APS-504 and has been designed to meet the ever increasing needs of maritime surveillance patrol. Optimised for sea search, it provides digital signal processing and scan conversion allowing the use of standard RS-343A high-resolution 875-line TV displays. This system allows recording and playback of mission radar data and alphanumerics on compatible TV systems.

The APS-504(V)3 system consists of six basic units:
- transceiver
- antenna/pedestal unit
- digital TV displays
- radar control unit with joystick cursor control
- radar processor and converter
- keypad entry unit.

Of these, the antenna is stabilised in pitch and roll and is designed to make use of available radome volume. Features include variable width sector scan, range delay, offset sweep, sweep expansion, true motion display using inputs from the navigation system, and a 1553B interface to the tactical mission computer. Within the transceiver, the transmiter section operates on a frequency of 9,375 MHz with an output of 100 kW peak power. Dual pulse-widths of 0.5 and 2.4 μs are provided. The low noise solid-state receiver provides simultaneous reception of beacon and search signals with the bandwidth automatically matched to the pulse-width.

APS-504(V)5
The APS-504(V)5 is described as 'enhancing' the detection capability of the (V)2 system and adding the digital processing, high definition display of the (V)3. As an anti-submarine warfare class airborne radar it is designed primarily for search and surveillance with a capability for navigation, mapping, weather avoidance and tactical operations use. The (V)5 incorporates wideband frequency agility (500 MHz), pulse compression using advanced surface acoustic wave technology (500:1 and 210:1 pulse compression ratios), scan-to-scan integration and a travelling wave tube amplifier. An SAR derivative of this version is used in the IRIS reconnaissance radar. With frequency agility, APS-504(V)5 has a range of up to 370 km (with selectable range scales) and features pulse compression and scan-to-scan integration. The system's transmit modes include two pulse compression modes and three non-compressed modes for short-range navigation, weather and beacon.

Tactical Data Management System
Although the APS-504(V)5 can be installed as a stand-alone system, its tactical utility is enhanced by the Litton Tactical Data Management System (TDMS). TDMS is a multiprocessor parallel processing system that provides the general purpose processing capability for a maritime patrol system. It contains the tactical application software for control of flight management, provision of tactical aids, geographical/grid co-ordinate conversion, and maintenance of tabular data presentation for the operator. By integrating the radar and navigation outputs through the TDMS, the operator is able to develop tactical and navigation plots with his keyset and joystick. He can also use the TDMS to control some or all of the available sensor systems through his own workstation.

Status
As of this edition, APS-504(V) and APS-504(V)2 were reported as being no longer in production. Jane's sources suggest that a total of 25 APS-504(V)3 and 65 plus APS-504(V)5 radars had been delivered to customers around the world. In US service, APS-504(V)5 is designated as AN/APS-140(V) and is integrated with the AN/APX-76 identification friend-or-foe interrogator. Over time, APS-504(V)5/APS-140(V) radars have been procured by Canada (for use as fishery patrol, ice reconnaissance and search and rescue sensors), Egypt, Indonesia, Ireland (installed aboard two CN.235 patrol aircraft), Taiwan and the US Customs Service (installed aboard six Raytheon Beech 200 surveillance aircraft). Platforms fitted with APS-504(V)/APS-140V range in size from the Raytheon Beech 200 to the Boeing 737. As of October 1998, Jane's sources were also suggesting that Litton was studying the possibility of an upgrade package for the APS-504(V)5 that would add synthetic aperture, moving target indication and range-profiling capabilities to the baseline system.

Specifications
Frequency: I-band (8.9-9.4 GHz sub-band with 2 frequency agility patterns)
Power amplifier: 8 kW peak power pulsed TWT
Antenna: parabolic (various sizes available) mounted on 2/3 axis pedestals
Polarisation: horizontal
Beamwidth: 2.5° (horizontal); 7° (vertical)
Rotation rate: 7-120 rpm in 8 rates automatically selected to match range/mode combinations in turbulence
Scan sector width: 30-120° (operator selectable); continuous
Range: 40 km (periscope); 370 km (mapping/beacon mode/search mode/weather mode)
Display: digital X-Y (in ambient light); multifunction
Interface: MIL-STD-1553B (optional)

Contractor
Litton Systems Canada Ltd, Toronto, Ontario.

VERIFIED

....

Tri-mode Synthetic Aperture Radar (TriSAR) system

Type
Airborne radar image processing system.

Description
TriSAR is a real-time, high-resolution, coherent radar imaging system for military and civilian applications. It uses vector processing to produce detailed, long-range radar cross-section images of both moving and static targets, with the image processing being executed either in the air or post-flight. TriSAR has been designed as a compact, low-weight installation that is applicable to a range of low-cost air platforms. At the heart of its capability is its signal processing subsystem. Here, precision sub-aperture focusing and correlation, target migration tracking and adaptive processing techniques are used. The level of resolution obtained allows the system to perform ship classification with accurate range and bearing.

TriSAR has three imaging modes, the known details of which are as follows:

Spotlight mode
In its spotlight mode, TriSAR directs its associated radar antenna onto a target for a predetermined dwell time. Data is collected and processed to produce a high-

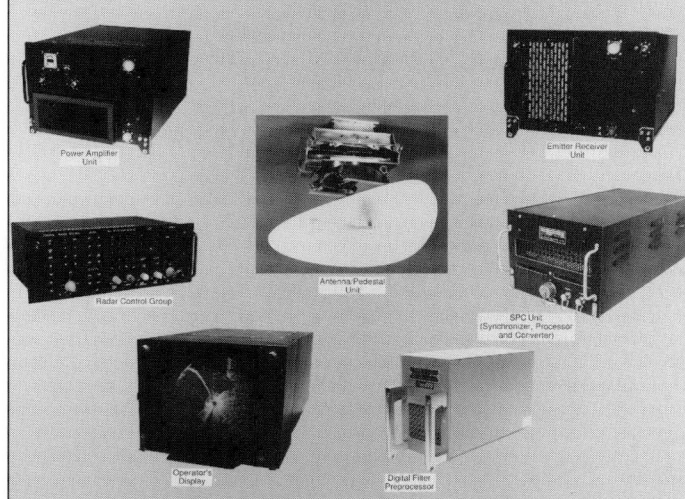

Units of the APS-504(V)5 radar

The type of imagery generated by the TriSAR system

resolution image of the target. Within the mode, TriSAR is able to image both moving and stationary targets. In more detail, moving targets are imaged using an adaptive, sub-aperture focusing algorithm that automatically constructs a finely focused, motion compensated image of the target. Array Systems further notes its belief that the described mode is the 'operator's best choice for imaging potentially threatening maritime targets' and that such a capability is 'unique' to the TriSAR package.

Stripmap mode

The TriSAR stripmap mode provides an 'endless' strip of imagery parallel to the aircraft's line of flight. The TriSAR stripmap mode is further noted as delivering high-resolution imagery in real time and as being effective over both land and sea.

Range Doppler profiling/inverse SAR mode

The TriSAR range Doppler profiling/inverse SAR mode is described as being 'similar' to the 'traditional' inverse SAR function and as producing a continuous series of frames of moving targets such as ships. The target's direction and motion can be determined from the appearance of the image frame and the operator can freeze selected imagery for detailed analysis and classification. During this analysis process, TriSAR continues to produce image frames.

Alongside the previously described modes, the TriSAR system is understood to be able to generate conventional plan position indicator and B-scan displays. An integrated motion compensation subsystem preserves the fidelity of the target data contained in the coherently sampled return signal. The radar also integrates motion data supplied by the host aircraft sensors (such as the Global Positioning System, an Inertial Navigation System and/or a strapdown navigation unit) in order to establish its antenna's phase centre position. This information is used to compensate for high-frequency platform motion which would otherwise smear processed imagery and render high-resolution focusing impossible.

TriSAR produces a 1,280 × 1,024 pixel colour SAR display that consists of a 256 × 1,024 menu area and a 1,024 × 1,024 image area. The command menu used is context sensitive and allows the operator to specify or modify data collection parameters (such as range and cross-range resolution) before and during collection. The system is available as a stand-alone processing suite (complementing an existing conventional coherent onboard radar) or as part of an integrated remote sensing package.

Status

As of early 2001, TriSAR was understood to be a live programme. The radar processor used in the system is reported to make use of algorithms developed for the spot SAR mode in the Canadian Forces' AN/APS-506 maritime patrol aircraft radar. TriSAR technology is also being supplied to the UK for integration into the Searchwater 2000 surveillance radars which are to be fitted to the Royal Air Force's 21 Nimrod MRA Mk 4 patrol aircraft.

Contractor

Array Systems Computing Inc, North York, Ontario.

UPDATED

CHINA, PEOPLE'S REPUBLIC

Type 698 Side-Looking Airborne Radar (SLAR)

Type

I-band (8 to 10 GHz) SLAR system.

Description

The Type 698 is reported to be a side-looking radar designed specifically for the anti-submarine role. It uses coherent moving target detection techniques to detect small moving targets and submarine periscopes in conditions up to Sea State 4. Other features include a parametric amplifier, a high-stability local oscillator, a coherent receiver, IF log amplifier and digital filter. The antenna is a slotted feed double parabolic reflector type.

Status

As of this edition, the Type 698 SLAR is reported as having completed flight testing and as having been procured by one of China's air arms.

Specifications

Frequency: I-band (8-10 GHz)
Detection range: periscope 17 km; ship 60 km
Display ranges: transversal 60 km (normal); longitudinal 30 km (searching)
High resolution: transversal 300 m (searching); longitudinal 50 m
Operational altitude: 50-500 m (searching)
Weight: 230 kg
Volume: 0.8 m³

Contractor

China Leihua Electronic Technology Research Institute, Neijiang, Sichuan.

VERIFIED

DENMARK

TERMA Side-Looking Airborne Radar

Type

I-band (8 to 10 GHz) maritime surveillance SLAR.

Description

Forming part of the TERMA Airborne Surveillance System (TASS), the company's magnetron-powered, I-band SLAR is intended primarily as a pollution/economic zone policing tool and is operated from an associated TASS operator console. Here, the console runs proprietary TERMA Sensor Control System (TSCS) software (implemented on the Windows NT operating system) and features a high resolution, 51 cm (20 in) display monitor. The TSCS software package is described as being able to handle mission administration and all necessary functions for acquisition and data collection/presentation for all onboard TASS sensors (alongside the described SLAR, TASS can incorporate optical and electro-optic cameras, an infra-red/ultra-violet scanner and a microwave radiometer). All acquired data can be post processed at a ground station and the complete airborne TASS package can be configured in a single electronics rack/console unit if required.

Status

As of March 2001, Jane's sources were suggesting that the described TERMA SLAR was in service with a number of customers including Finland's Rajavartiolaitas (Frontier Guard - Dornier Do 228-212 aircraft) and the Royal Danish Air Force (Canadair CL-600 Challenger 604 aircraft).

Specifications

Transceiver
Frequency: 9.375 ± 39 MHz
PRF: 0-2,000 Hz
Pulse-width: 0.25 or 0.60 μs
Magnetron peak power: 25 kW ± 1 dB
IF detection: Lin or Log
Bandwidth: adjustable to pulse-width
Noise figure: < 5.5 dB
Video output: 1.5-5 V (analogue)
Antenna
Configuration: 1 double or 2 single
Beamwidth: < 0.6° (horizontal); 19 ± 3° (vertical)
Gain: > 33 dB
Polarisation: horizontal or vertical
General
Range: 37-46 km (sea surface oil pollution); up to 74 km (ships, ice and land features - both sides of aircraft)
Display modes: port/starboard; expanded; zoom (with imagery north up or track up geocorrected)
Interfaces: ARINC, RS-232, RS-422 or alternatives
Power: 28 V DC (15 A max)
Dimensions (l × w × h): 1,300 × 550 × 1,300 mm (console); 3.66 m (antenna length)
Weight: 130 kg (operator console)

Contractor

TERMA Elektronic AS, Lystrup.

NEW ENTRY

GERMANY

HELLAS Obstacle Warning System (OWS)

Type

Laser radar OWS.

Description

The HELLAS (HELicopter LASer) laser radar OWS is, as its name suggests, designed to detect and warn helicopter pilots of obstacles (such as power and telephone lines) in their line of flight. System componenets include optics, an oscillating mirror, a fibre-optic scanning assembly (made up of two rotating mirrors, optics and two circular/horizontal fibre-optics arrays), an eye-safe laser transmitter, a receiver diode, electronics boxes, a processor and a cockpit display. Functionally, the first of the described fibre-optics is used to transmit laser pulses in a horizontal fan with any reflections being received by the second array. Direction of arrival is determined by the rotation angle of the mirrors within the scanning assembly. Detected obstacles are displayed in a colour-coded, real-time range format in which power lines and the like stand out from their backgrounds.

Status

Work on the laser-based HELLAS OWS began in 1993 under the sponsorship of Germany's Bundesamt für Wehrtechnik und Beschaffung (Federal Office for

Defence Technology and Procurement). Initial flight trials with the system (using Heeresflieger (German Army Aviation) CH-53G and UH-1D helicopters) were carried out during 1994/95. During 1997, HELLAS was used to replicate an accident in which a civilian rescue helicopter had hit a power line and crashed and, as of this edition, the German Federal Border Guard was reported as having procured 25 HELLAS systems for installation aboard its fleet of EC 135 and EC 155 helicopters. Announced in June 1999, this programme was initially reported as being scheduled for completion during 2002. The Heeresflieger was also understood to be intending to flight test an enhanced, all-weather variant of HELLAS during 1999.

Specifications

Wavelength: 1.54 μm (near IR)
Output power: 4 kW
Pulse duration: 5 ns
Detection range: 274 m (10 mm Ø cable, daylight, adverse visibility); 366 m (extended area object, daylight, adverse visibility); 457 m (10 mm Ø cable, daylight, good visibility); 914 m (extended area object, daylight, good visibility). Night-time detection ranges up to 20 % > daylight values due to lack of clutter.
Pixel rate: 100,000/s
Image repetition frequency: 4 Hz (max)
Angular resolution: 0.2° (vertical, 32 × 32° FOV); 0.35° (horizontal , 32 × 32° FOV)

Contractor

Dornier GmbH (a component of the European Aeronautic, Defence and Space (EADS) Co), Friedrichshafen.

UPDATED

INTERNATIONAL

Agrion maritime surveillance radar

Type

I/J-band (8 to 20 GHz) maritime surveillance radar.

Description

Agrion is a member of the Iguane family of maritime radar systems and exists in several versions. It is designed primarily for use aboard helicopters or light aircraft forming a part of task forces, employed for support at sea or for coastal protection. Several types of antenna are available to meet the requirements of various aircraft. The Agrion 15 version allows the guidance of the AS 15TT Aerospatiale air-to-surface missile which has, *inter alia*, demonstrated its efficiency against fast patrol boats. Agrion operates in the I/J-band, using pulse compression and frequency agility to ensure high performance on maritime targets in all combinations of weather, sea state and operating altitude. These same techniques also provide maximum protection against electronic countermeasures. The system provides operational missions such as surface and anti-submarine warfare, over-the-horizon targeting for shipborne surface-to-surface missiles, search and rescue, marine environmental protection, navigation and weather avoidance.

Status

As of April 2001, the Agrion 15 surveillance radar was reported as having been procured for installation aboard 19 Eurocopter AS 565SA Panther anti-submarine/anti-surface ship warfare helicopters of the Royal Saudi Navy.

Contractor

Thales Airborne Systems, Elancourt, France.

VERIFIED

A chin-mounted scanner of an Agrion 15 radar installed on an AS 565SA Panther anti-submarine/anti-ship warfare helicopter

AMASCOS maritime patrol system

Type

Airborne, multisensor, maritime patrol system.

Description

The AMASCOS (Airborne MAritime Situation COntrol System) suite is designed for real-time tactical situation build-up and update, as well as a decision aid to operators. It is a family of maritime patrol systems that use a modular approach to system design to facilitate the integration of AMASCOS architectures aboard 'any' type of fixed- or rotary-wing aircraft. There are three versions of AMASCOS (designated as AMASCOS 100, 200 and 300) that correspond to the three broad categories of mission requirements from simple maritime surveillance to anti-surface and anti-submarine warfare. The typical AMASCOS configuration integrates Thales sourced equipment (a radar (see following), the Chlio Forward-Looking Infra-Red (FLIR) sensor, the DR 3000 Electronic Support (ES) system, the TMS 2000 acoustics chain, the FLASH dipping sonar, a magnetic anomaly detector, a Link Y standard datalink, NATO Link 11, 16 and 22 standard datalinks, a NATO or national standard communications suite, a TOTEM Global Positioning System/inertial navigation system and a TOPDECK cockpit configuration) but its modular architecture makes it possible to tailor each system to specific requirements. The radar used in the described typical configuration is the Ocean Master equipment developed jointly by Thales Airborne Systems and the European Aeronautic, Defence and Space (EADS) Co Deutschland (see separate entry).

AMASCOS 100

The AMASCOS 100 configuration is designed for a range of tasks including Exclusive Economic Zone (EEZ) control, fisheries protection, maritime traffic surveillance, anti-smuggling, anti-piracy and anti-drug patrols, immigration control, pollution monitoring, Search And Rescue (SAR) and sovereignty patrols executed by coastal surveillance services, customs services and/or armed forces. As such, the suite is generally installed aboard lightweight aircraft and is operated by one or two operators working in collaboration with the host aircraft's flight crew. A typical AMASCOS 100 application weighs a maximum of 250 kg and is suitable for installation aboard both fixed-wing aircraft and land or ship-based helicopters.

AMASCOS 200

Designed for Anti-Surface ship Warfare (ASuW), COMmunications and ELectronic INTelligence (COMINT/ELINT), littoral warfare and surveillance and maritime surveillance (including EEZ patrol and SAR) missions, AMASCOS 200 adds an electronic support or radar warning receiver system and an over-the-horizon, anti-shipping missile targeting capability to the AMASCOS 100 architecture. As such, AMASCOS 200 is designed for installation aboard '8-ton class' fixed-wing aircraft and helicopters and requires two to three system operators.

AMASCOS 300

Designed for anti-submarine warfare, ASuW, COMINT/ELINT, littoral warfare and surveillance and maritime surveillance (including EEZ patrol and SAR) missions, AMASCOS 300 adds acoustic equipment and the ability to make use of torpedoes and depth charges to the AMASCOS 200 architecture. As such, AMASCOS 300 is designed for installation aboard '10-ton class' fixed-wing aircraft and helicopters and required three to four system operators (expandable to seven plus operators when configured for long-range maritime patrol aircraft applications).

Status

During 1996, Indonesia awarded the then Thomson-CSF DETEXIS (now Thales Airborne Systems) a contract covering the supply of AMASCOS 100 suites (including Ocean Master radars and Chlio FLIRs) for use aboard six IPTN NC-212 Aviocar maritime patrol aircraft and three NBO 105 helicopters. Elsewhere, an AMASCOS configuration that incorporates Thales' DR 3000 ES system, an Ocean Master radar, Thales' Sadang C1 sonobuoy signal processing system and a Thales Avionics' navigation suite is understood to have been supplied to Pakistan for use in a Dassault Breguet Atlantic 1 maritime patrol aircraft upgrade programme that was completed during 1998. During July 2001, Jane's sources were reporting that Thales had been awarded a then year €50 million contract covering the supply and integration of AMASCOS suites aboard three CN235-220 maritime patrol aircraft being procured by the Indonesian Air Force. Elements making up the specific configuration were noted as including the Ocean Master radar, Chlio FLIR, Gemini navigation computer and an ES system.

Contractor

Thales Airborne Systems, Elancourt, France.

UPDATED

A total of four Pakistani Atlantic 1 maritime patrol aircraft are understood to have been upgraded with an AMASCOS 300 type maritime patrol system during the late 1990s
2001/0102012

Anemone multifunction airborne radar

Type
I/J-band (8 to 20 GHz) search, tracking and ranging radar.

Description
Developed for the French Navy's Super Etendard strike-fighter upgrade programme, the Anemone radar consists of a nose cone (antenna and circuitry), an aircraft/radar interface unit and a control unit and has the following functions:
- air-to-surface (main function)
- air-to-ground (ranging and ground-mapping)
- air-to-air.

The Anemone radar operates in the I/J-band with frequency agility. A wideband monopulse flat slotted array antenna with low-level sidelobes is provided, and reinforced electronic counter-countermeasures are incorporated. In the air-to-surface mode, the system enables a marine target to be detected and tracked. Targets are detected in the search mode with an elevation angle automatically adjusted as a function of the selected range scale. Following target designation and lock on, the change to track-while-scan or continuous tracking is automatic. In the air-to-air mode the radar allows linear scan, search with semi-automatic acquisition, and continuous tracking.

Status
According to Jane's sources, funding for the first ten Super Etendard upgrades was authorised during 1991, with the first Anemone radar for the programme becoming available during mid-1994. As of June 1998, Jane's sources were suggesting that at least 46 Anemone radars had been delivered.

Contractors
Thales Airborne Systems, Elancourt, France.

VERIFIED

Anemone airborne radar

Antilope 5 TC navigation radar

Type
J-band (10 to 20 GHz) terrain-following and navigation radar.

Description
The Antilope 5 TC radar is designed for installation aboard Mirage 2000D and 2000N combat aircraft. Its basic functions are terrain-following, high resolution ground-mapping, interlace (terrain-following and ground-mapping), air-to-air and air-to-surface. Essential characteristics of Antilope 5 TC are:
- J-band transmission providing high-ground reflectivity
- high-speed vertical scanning of the antenna
- asymmetric antenna, providing accurate localisation of obstacles in the path of the aircraft
- an antenna of the flat slotted-array type producing a weighted polar diagram with very low-level diffuse sidelobes
- receiver with a wide dynamic range and image-frequency suppression
- image sharpening of the monopulse type with compression in the elevation plane for highly accurate determination of the height of obstacles
- real-time radar data processing
- continuous and automatic test system
- protection against reception by antenna sidelobes.

Radar information is displayed on a head-up display and on a three-colour multimode cathode ray tube head-down display, as well as being sent to the navigation and weapon system. The system can provide terrain-following commands at 91 m and 1,111 km/h, computing a preset obstacle clearance height, and with a preselected g level.

Status
As of April 2001, more than 180 Antilope 5 TC radars are reported as having been produced for the French Air Force, with the Mirage 2000N application being

Antilope 5 TC terrain-following and navigation radar

reported as having entered service during 1987. First deliveries of radars for use on the Mirage 2000D are thought to have taken place during late 1992. Studies of an Antilope 5 TC synthetic aperture mode are understood to have been launched during 1996.

Contractors
Thales Airborne Systems, Elancourt, France.

VERIFIED

BATTLESCAN battlefield surveillance radar

Type
Airborne standoff battlefield surveillance radar.

Description
The BATTLESCAN battlefield surveillance radar capitalises on Thales Airborne Systems' experience with the Horizon system and takes the form of an architecture that can be mounted on a variety of platforms including light aircraft, helicopters, airships and drones. As such, the radar is designed to detect and locate (in a single scan of a few seconds duration) columns of vehicles, ships and formations of helicopters and is intended to complement and increase the surveillance performance of ground-based radars, with a particular emphasis on the provision of cover in areas that are terrain masked. Functionally, BATTLESCAN provides raw surveillance data that is processed on board the host platform and is then used to amplify and elaborate on alarm and tactical information available to higher commands. BATTLESCAN is interoperable with the US Joint Surveillance Target Attack Radar System (Joint STARS - see separate entries) and incorporates:
- a carrier aircraft motion self-compensating system
- a range of operating modes including moving target indication, sea surveillance and synthetic/inverse synthetic aperture radar
- a flat, very low sidelobe modular antenna
- a wideband travelling wave tube frequency-agile transmitter
- a digital signal processor
- high-performance electronic counter-countermeasures
- operator console with colour display showing target location and speed sorting, on a map with suitable symbols.

The operator console includes automatic operator guiding functions.

Status
As of April 2001, a BATTLESCAN variant was reported as being used in France's Horizon battlefield surveillance system (see separate entry).

Specifications
Radar range: 150 km (in all weathers)
Typical location accuracy: 20 m (range); 5 mrad (azimuth); unambiguous speed sorting for ground targets and helicopters

Contractor
Thales Airborne Systems, Elancourt, France.

VERIFIED

CLARA laser radar

Type
Obstacle avoidance radar.

Description
CLARA is a self-contained, carbon dioxide (CO_2) laser radar that is being developed as a result of an Anglo/French government-to-government initiative. The equipment is housed in an environmentally controlled pod and is applicable to both

The French CLARA demonstrator under test aboard a French Air Force Puma helicopter (CEV)

helicopters and fixed-wing aircraft. CLARA is designed to provide an obstacle avoidance capability against threats such as cables as well as terrain-following, target ranging and designation, short-range true air speed measurement and moving target indication functions.

Status
Under the described government-to-government initiative, two identical demonstrator units have been produced and tested on a fixed-wing aircraft in the UK (under contract to the UK's Defence Evaluation and Research Agency) and on a helicopter in France (under contract to France's *Service des Programmes Aeronautiques*. Flight trials with the device are understood to have begun during 1996.

Contractors
BAE Systems, Milton Keynes, UK.
Thomson-CSF DETEXIS, Elancourt, France.

VERIFIED

DAV warning and surveillance radar

Type
E/F-band (2 to 4 GHz) threat warning and surveillance radar.

Description
The DAV warning and surveillance system, intended for installation in combat helicopters, is an E/F-band pulse-Doppler radar specially designed for detecting air threats that affect flying/hovering helicopters. It enables the crew to detect, identify and designate air threats from a safe distance of up to 9 km. For this purpose the DAV systems antenna assembly is mounted on and above its host's main rotor to facilitate all-round surveillance. This position allows its operational use in tactical flight, especially when the helicopter is partially masked, taking cover behind natural obstacles. DAV consists of three subassemblies as follows:
- a radome enclosed antenna assembly that incorporates the radar's antenna, transceiver, signal processing circuitry (Doppler filtering, signal integration, target extraction, multitarget track-while-scan and target recognition/ identification functions), a low voltage power supply, an antenna angular position encoder and a radar-helicopter rotor interface unit
- a helicopter interface unit that hands off targeting data, generates display imagery and manages control of the radar
- an operator interface that displays target data and allows the operator to control the radar

The system's operational characteristics include:
- all-weather use
- safe distance detection of air threats (radar range suitable for engaging or avoiding threats)
- a 'very short' reaction time
- accurate localisation of threats, allowing the use of weapons (missiles and guns)
- integration into the weapon system
- compatibility with target designation for the missile (launch envelope calculation)
- very low false alarm rate
- low detectability
- little effect on helicopter radar cross-section
- volume and weight compatible with platform helicopter.

DAV can also pick out helicopters in the hover and serve as a mid-air collision warner.

Status
As of April 2001, DAV was reported as having successfully completed tactical and operational tests. Two prototype ('preproduction') examples of the radar are understood to have been delivered to the French Ministry of Defence during November 1995 and April 1996. The DAV programme is understood to have been 50 per cent funded by Thales Airborne Systems (formerly Thomson-CSF DETEXIS) and 50 per cent by the French government.

A computer graphic showing DAV's threat warning envelope

Specifications
Frequency: E/F-band (2-4 GHz)
Transmitter: frequency-agile; transistorised amplifier; solid-state source
Receiver: low noise superheterodyne
Antenna coverage: 24° (elevation - antenna rotating at same speed as main rotor); 360° (azimuth - antenna rotating at same speed as main rotor)
Doppler analysis: digital processing with FFT
Target recognition: helicopter and fixed-wing aircraft
Helicopter recognition: comparison with library signature
Target designation data: azimuth, elevation, range and speed
Power supply: <600 W
Weight: 5 kg (helicopter interface unit); <40 kg (antenna assembly)
Dimensions (Ø × h): 90 × 42 cm (antenna assembly)

Contractor
Thales Airborne Systems, Elancourt, France.

UPDATED

Horizon battlefield surveillance system

Type
I/low J-band (8 to 12.5 GHz sub-band) surveillance and target designation radar-equipped battlefield surveillance system.

Description
Horizon (*Hélicoptère d'Observation Radar et d'Investigation sur Zone*) is a version of the Orchidée battlefield radar surveillance system that was developed in the mid-1980s and consisted of ground-based and airborne elements. Horizon comprises the aircraft itself, a pulse-Doppler 8 to 12.5 GHz band radar, and a datalink for two-way communication with the ground element. Thales Airborne Systems is responsible for the development of the radar installation, the system's datalink and ground station, with Eurocopter handling system integration. Responsibility for the overall programme is the *Délégation Générale pour l'Armement (DGA)*, part of the French Ministry of Defence. The main difference between the Horizon and Orchidée systems is the addition of an onboard operator station and a ground station (with two operator consoles) in the former application.

The radar used in the Horizon system is a very long-range ground surveillance sensor that is mounted aboard a modified Aerospatiale Cougar helicopter and is able to detect and locate (in a single scan of a few seconds duration) vehicles, helicopters and ships, even when they are slow moving. Functionally, the system measures the speed of the detected targets, classifies them as to type (wheeled vehicle, tracked vehicle, helicopter or ship) and processes the raw received data for onboard display and transmission (via datalink) to an associated ground station.

Production Horizon radar systems are installed aboard Eurocopter AS 532UL Cougar helicopters

Range on tanks and helicopters in 90 per cent of European weather conditions is 200 km.

The Horizon radar consists of three modular units (transmitter/receiver, processor and control unit), which are fitted inside the helicopter, and a wide span, mechanically positioned, flat plate antenna that is mounted outside the helicopter. In flight, the complete antenna mechanism is set vertically under the helicopter and the antenna scans over 360°. On landing the antenna is locked crosswise and is raised mechanically under the helicopter's tail boom. Radar data is processed both aboard the aircraft (one workstation) and on the ground (two workstations). The hardware and software used is optimised to operator workload, responsibility level and user organisation. If required, the airborne segment of the system can operate independently of a ground station. Such an operating mode is described as improving reaction time and flexibility.

Status

Orchidée type technology is reported to have been under test since 1986. In June 1990, the French government cancelled the Orchidée programme because of funding difficulties. Despite this, the Orchidée demonstrator was deployed operationally during the 1991 Gulf War where it completed 24 surveillance sorties (under the codename 'Horus') in support of French ground forces and the American XVIII Airborne Corps. In addition to its surveillance work, the Orchidée/Horus aircraft was used to locate hostile radar jammers and provide targeting data for US AH-64 attack helicopters. In October 1992, the Horizon programme was begun and the French army went on to order a total of four operational systems, the first of which was delivered during the summer of 1996. A second system was delivered during July 1997. During 1999, Horizon was used operationally in support of NATO's Operation 'Allied Force' air interdiction campaign against Serbia and by mid-2000, all four Horizon systems had been delivered and declared operational.

Contractor

Thales Airborne Systems, Elancourt, France.

VERIFIED

....

Iguane maritime surveillance radar

Type

I/J-band (8 to 20 GHz) maritime surveillance radar.

Description

Iguane is a maritime surveillance radar that was originally developed as an upgrade for the Breguet Alize carrierborne and anti-submarine warfare aircraft. It has also been fitted to the Dassault-Breguet Atlantique maritime patrol aircraft. Iguane operates in I/J-band and makes use of pulse compression and frequency agility to ensure high performance on maritime targets in all combinations of weather, sea state and operating altitude. These same techniques also provide maximum protection against electronic countermeasures. The system is also capable of providing side-looking radar imagery in real time. Operational missions performed by the Iguane system include surface and anti-submarine warfare, over-the-horizon targeting for shipborne surface-to-surface missiles, search and rescue, marine environmental protection, navigation and weather avoidance.

Status

Over time, the Iguane maritime surveillance radar is reported to have been installed aboard Atlantique Mk 1 maritime patrol aircraft of the Italian Air Force and Atlantique Mk 2 platforms flown by the French Navy. Italian usage of the radar took the form of an upgrade programme that involved the procurement of 18 systems. The Iguane radar is related to Thales' Agrion and Varan sensors.

Contractor

Thales Airborne Systems, Elancourt, France.

VERIFIED

The Iguane airborne radar

Ocean Master radar

Type

I/J-band (8 to 20 GHz) maritime surveillance radar family.

Description

The Ocean Master series is a family of maritime patrol radars that are designed to meet a wide range of mission requirements and which are suitable for installation aboard both fixed- or rotary-wing aircraft. Mission capabilities include anti-surface vessel and anti-submarine warfare, exclusive economic zone surveillance, search and rescue and air-to-air detection. Ocean Master radars provide 'advanced' capabilities for detection, tactical processing, situation display, navigation and weather avoidance and are frequency-agile, compact, lightweight sensors that, in baseline form, make use of either a 100 W (Ocean Master 100) or 400 W (Ocean Master 400) fully coherent travelling wave tube amplifier. Jane's sources suggest that at least three stabilised antenna assemblies are available for use with the sensor and that it is capable of detecting 'all types' of targets including periscopes, small boats and lifeboats. The detection of such small targets is said to rely on the radar's use of pulse compression, programmable radar signal and data processing (with pulse-to-pulse and scan-to-scan integration) and a range of mode selectable antenna rotation rates. Multiple target track-while-scan, radar map memory, target classification and over-the-horizon targeting capabilities are claimed as providing the Ocean Master operator with a 'comprehensive' tactical situation picture and the radar's processing capability is understood to be able to stand in for a dedicated mission computer in onboard mission architectures that lack such provision. Ocean Master interfaces with its host aircraft's navigation and combat system via standard databus types and the radar's target classification capability is noted as being based on both profiling and inverse synthetic aperture radar techniques. In baseline form, Ocean Master comprises an antenna assembly, a transmitter, an exciter/receiver/ processor unit and a Man/Machine Interface (MMI) and the sensor family as a whole forms a central component of Thales' AMASCOS mission system architecture (see separate entry). The system's baseline MMI incorporates a 36 cm (14 in) multifunction colour display together with touch-sensitive, flat panel controls and a trackball. Data displays can be north stabilised or by aircraft heading and incorporate a number of range scales.

Status

The then Thomson-CSF DETEXIS (now Thales Airborne Systems) and the then DaimlerChrysler Aerospace (now a component of the European Aeronautic, Defence and Space (EADS) Co) signed an agreement in January 1992 that covered development, production and marketing of the Ocean Master radar. As of July 2001, Jane's sources were reporting Ocean Master sales as including:

- Ocean Master 100 radars for installation aboard four Dassault Falcon 50 SURMAR aircraft of the Aeronautic Navale (French Naval Air Arm)
- nine Ocean Master 100 radars for installation aboard ShinMaywa US-1A and US-1A Kai seaplanes of the Nihon Kaijyo Jieitai (Japanese Maritime Self-Defence Force) together with an undisclosed number of radars for installation aboard the service's NAMC YS-11T-A navigation trainers.
- unidentified Ocean Master variants for installation aboard six (?) NC-212-200 maritime patrol aircraft and three NBO-105S over-the-horizon targeting helicopters of the Tentara Nasional Indonesia - Angkatan Laut (Indonesian National Defence - Navy)
- unidentified Ocean Master variants for installation aboard four (one subsequently lost) Breguet BR 1150 Atlantic 1 and three Fokker F-27-200 maritime patrol aircraft of the Pakistani Naval Air Arm
- an unidentified Ocean Master variant for installation aboard four CN-235-220 maritime patrol aircraft of the United Arab Emirates Air Force.

The first delivery of a production Ocean Master radar took place during 1994.

Specifications

Frequency: I/J-band (8-20 GHz)
PRF: 300 Hz-125 kHz

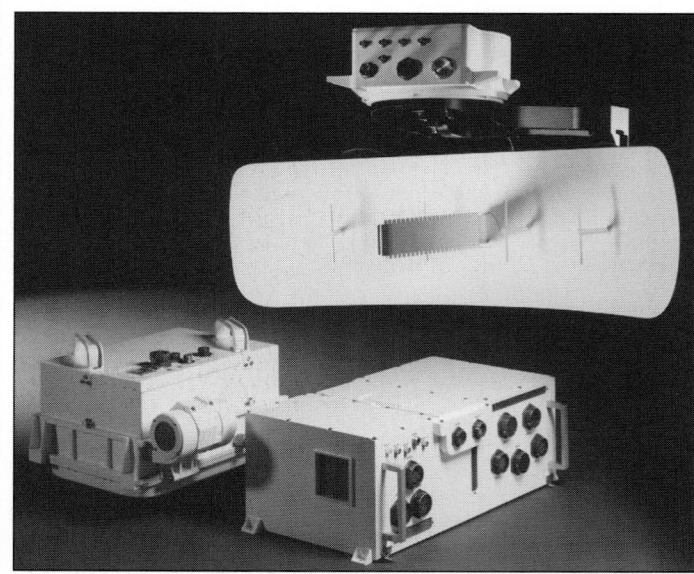

The Ocean Master radar system

Transmitter output: 100 W (Ocean Master 100 - average); 400 W (Ocean Master 400 - average)
Detection range: 200 m (min); 370 km (max)
Range resolution: 3 m (min)
Track-while-scan: up to 32 targets (automatic)
Antenna stabilisation: 2-axis (±20% roll; ±10% pitch)
Antenna tilt: −29 to +4°
Antenna rotation rate: 6-30 rpm (automatic selection by mode)
Antenna gain: 30.5-34 dB
Power requirement: 115 V AC (400 Hz)
Power consumption: 2 kVA (Ocean Master 100); 3.8 kVA (Ocean Master 400)
Interfaces: ARINC 429; MIL-STD-1553B; RS-422; video
Options: 48 cm (19 in) display; IFF compatibility; datalink, ES, IFF, FLIR, sonics and video recorder interfaces; ISAR target classification
Dimensions: 375 × 300 × 415 mm (MMI); 415 × 275 × 320 mm (Ocean Master 100 transmitter); 555 × 200 × 340 mm (exciter/receiver/processor unit); 566 × 208 × 260 mm (Ocean Master 400 transmitter); 660 × 350 mm (antenna option); 940 × 350 mm (antenna option); 1,800 × 300 mm (antenna option)
Weight: 8 kg (MMI control panel); 15 kg (660 × 350 mm antenna option); 16 kg (approx - 940 × 350 mm antenna option); 18 kg (MMI display); 27 kg (exciter/receiver/processor unit and 1,800 × 300 mm antenna option); 29 kg (Ocean Master 100 transmitter); 39 kg (Ocean Master 400 transmitter)

Contractors

Thales Airborne Systems, France.
European Aeronautic, Defence and Space (EADS) Co Systems and Defence Electronics, Germany.

UPDATED

ORB 32 series maritime surveillance/fire-control radars

Type
I-band (8 to 10 GHz) family of maritime surveillance and fire-control radars.

Description
I-band ORB 32 series radars are described as being lightweight, as having low power consumption values and as being compatible with a wide range of fixed- and rotary-wing aircraft. ORB 32 radars are created from modular subassemblies and can be easily upgraded if required. System applications include:

- Exclusive Economic Zone (EEZ) control
- anti-submarine warfare
- anti-surface unit warfare
- active missile fire control
- search and rescue
- radar navigation
- weather avoidance.

ORB 32 series radars are also noted as having a typical peak power of 70 kW and featuring pitch/roll-stabilised antennas; 360° azimuth/30° elevation coverage; 60, 120, 180 or 240° sector scan options and azimuth bearing or true motion stabilisation. Known details of identified family members are as follows:

ORB 3201/ORB 3211
ORB 3201 and 3211 radars are simple, compact and lightweight systems that are suitable for small fixed-wing aircraft or helicopters and have been specifically designed for surface reconnaissance, economic exclusion zone control and search and rescue. They also provide navigation information and weather avoidance.

ORB 3202/ORB 3212
ORB 3202 and 3212 radars are designed for airborne reconnaissance and target designation applications. When integrated into a weapon system, their role is to

The ORB 3203 radar system

detect, designate and accurately track two sea targets. Target co-ordinates may be automatically transmitted to active missiles carried by aircraft, or to a launch vessel.

ORB 3203/ORB 3214
ORB 3203 and 3214 radars are designed for Anti-Submarine Warfare (ASW) applications aboard helicopters or fixed-wing aircraft. They perform helicopter station holding during ASW operations and provide situation information, targeting, navigation, weather avoidance and mapping facilities. ORB 3203/3214 radars offer primary and secondary radar functions and their use of transponders makes it possible to identify helicopters flying at low altitude even if their primary echo is lost in sea clutter.

Status
Over time, ORB 32 series radars are reported as having been supplied to the French Air Force, the French Navy and the air arms of a number of other countries. Platforms fitted with the radar are noted as including Super Frelon, Super Puma and Boeing Vertol 107 helicopters and Nord 262 fixed-wing aircraft.

Contractor
Thales Airborne Systems, Elancourt, France.

UPDATED

ORB 37 multifunction radar

Type
I-band (8 to 10 GHz) surveillance, weather and mapping avoidance radar.

Description
The ORB 37 multifunction radar has been designed to meet French Air Force navigation requirements for the C-160 Transall transport aircraft. It carries out weather avoidance and accurate ground-mapping functions, and is fitted with an interrogation facility for beacon homing. The system consists of a slotted array flat-plate antenna, a transmitter/receiver, a power supply, a Plan Position Indicator (PPI) high definition circular display (for ground-mapping at the navigator's station), a digital PPI display on the flight deck, and flight deck and navigator station control units. For maximum efficiency, the ORB 37 antenna scans at low rate in the weather mode and at high rate when ground-mapping. The corresponding pulse-widths are 2.5 and 0.4 μs.

Status
As of April 2001, the ORB 37 multifunction radar was reported as having been procured by the French Air Force.

Specifications
Frequency: I-band (9,375 MHz)
Power output: 10 kW

Contractor
Thales Airborne Systems, Elancourt, France.

VERIFIED

The ORB 37 multifunction radar

Raphaël-TH/SLAR 2000 surveillance radar

Type
E/F-band (2 to 4 GHz) Side-Looking Airborne Radar (SLAR).

Description
SLAR 2000 (also known as Raphaël-TH) is an all-weather side-looking airborne radar employing synthetic aperture and pulse compression techniques to provide

Thales' Raphaël-TH/SLAR 2000 side-looking, synthetic aperture radar mounted on a French Air Force Mirage F1 reconnaissance aircraft (MS/Thales Airborne Systems)

high-quality mapping. The airborne part of the system is pod mounted for installation on fast jet combat aircraft. It can also be installed in the cargo bay of a commuter or transport aircraft. Radar information is transmitted via datalink to a ground station where it is displayed in real time. Recent upgrades feature moving target detection to be displayed on the ground map. The radar is highly directional and operates in the 3 cm band. It has an effective beamwidth of only a few mrad to provide a sharp and accurate radar map of the ground. Ground echoes are processed in the onboard unit, then transmitted in real time to the ground station. Data are stored both in the pod and in the ground station. The ground station consists of a datalink turret, an operations cabin, and a processing cabin. Imagery is generated of both fixed and moving targets in order to facilitate order of battle updating. The complete ground station is air and land transportable.

Status

As of April 2001, Raphaël-TH/SLAR 2000 was reported as having been procured by the French Air Force and the air arms of a number of other countries. A moving target indicator upgrade is understood to have been launched during 1995 and as of September 1996, Thales announced that it had teamed with the then DaimlerChrysler Aerospace (now part of the European Aeronautic, Defence and Space (EADS) Co Deutschland) on the development of a synthetic aperture radar reconnaissance pod (designated as SARTO) for possible use on German Tornado strike aircraft. The SARTO proposal was understood to have utilised technology from the Raphaël-TH/SLAR 2000 programme.

Contractor

Thales Airborne Systems, Elancourt, France.

VERIFIED

RDN 2000

Type
Doppler velocity sensor for helicopters.

Description

RDN 2000 is designed for installation aboard light, medium and heavy helicopters and provides an autonomous navigation/pilot's aid capability. The equipment utilises a single equipment package that incorporates an antenna, transceiver, power supply and signal processor. When coupled with an inertial navigation or attitude/heading reference system, Thales Airborne Systems claims that RDN 2000 provides a covert navigation capability that is insensitive to active countermeasures. Additionally, the equipment can be operated successfully alongside a co-located self-defence jamming system.

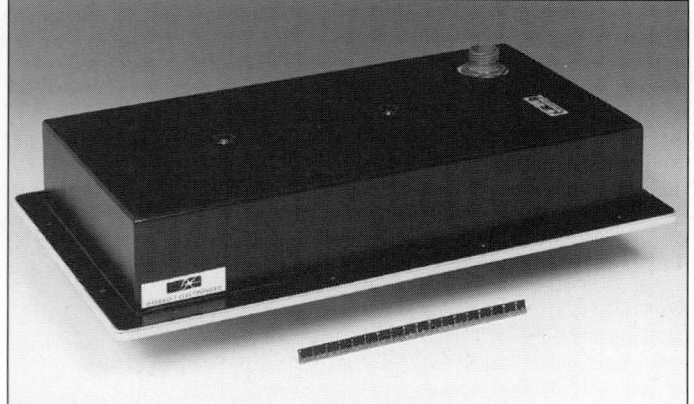

The Thales Airborne Systems RDN 2000 Doppler velocity sensor

Status

Full-scale production of the RDN 2000 radar is reported as having been initiated during 1997 and as of April 2001, the sensor was understood to have been installed aboard French and Saudi Cougar Mk 2 helicopters.

Specifications
Frequency: J-band (10-20 GHz)
MTBF: 8,640 flight h at 60°C ambient
Dimensions: 437 × 240 × 80 mm
Weight: 4.2 kg

Contractor

Thales Airborne Systems, Elancourt, France.

VERIFIED

Searchwater series surveillance radars

Type
Family of I-/low J-band (8 to 12.5 GHz sub-band) maritime surveillance and Airborne Early Warning (AEW) radars.

Description

Searchwater 1 (UK service designation ARI 5980) is a long-range maritime surveillance radar that is capable of target detection, localisation and classification. The equipment's line-replaceable units are described as being functionally self-contained modules which have minimum interconnection with other parts of the system and are designed to make fault location and replacement as easy as possible. Extensive use is made of hybrid and integrated circuitry throughout the system and the radar's transmission chain is noted as including solid-state frequency generators/mixers and cascaded travelling wave tubes. Use of a surface acoustic wave pulse compression technique is noted as is the equipment's use of three axis stabilised, resin-bonded carbon fibre, combined radar and identification friend-or-foe antenna assembly. Other system features include multiple operating modes (including weather avoidance, navigation and beyond-the-surface horizon standoff detection); selectable polarisation; frequency agility; ground-stabilisation of the radar display; integrated scan conversion and plan position indicator A- and B-scope display formats. In a crash programme associated with the 1982 Anglo-Argentine war in the South Atlantic, a number of Searchwater 1 radars were modified to provide the Royal Navy (RN) with airborne AEW and moderate fighter control capabilities. The major developmental items associated with this effort were a deployable antenna and a purpose-built man/machine interface console for the system's chosen host airframe, the Westland Sea King helicopter.

The prospect of replacing the Royal Air Force's (RAF) Nimrod maritime patrol aircraft with a newer type during the early 1990s led the then Thorn EMI (subsequently Thomson Racal Defence and now (May 2001) Thales Sensors) to undertake a largely Private Venture (PV) development effort to produce a modernised radar (designated as Searchwater 2) which could be bid as the new aircraft's primary sensor. This work is understood to have drawn in part on the company's experience with the UK's ASTOR technology demonstration programme (see separate entry) and the 1980s PV Skymaster lightweight AEW/surveillance radar. Cancellation of the Lockheed Martin P-7 (the favoured replacement for the Nimrod) led to the UK rigging the requirement into what became known as the Replacement Maritime Patrol Aircraft (RMPA) programme. After a competition between an upgraded Nimrod (designated as the Nimrod 2000), the new build Lockheed Martin Orion 2000 and a Lockheed Martin UK Government Systems proposal for an upgraded and refurbished P-3A/B Orion (designated as the Valkyrie), the UK Ministry of Defence selected the Nimrod 2000 as its solution to the RMPA requirement during July 1996. Following selection, the upgrade has been officially designated as the Nimrod MRA Mk 4.

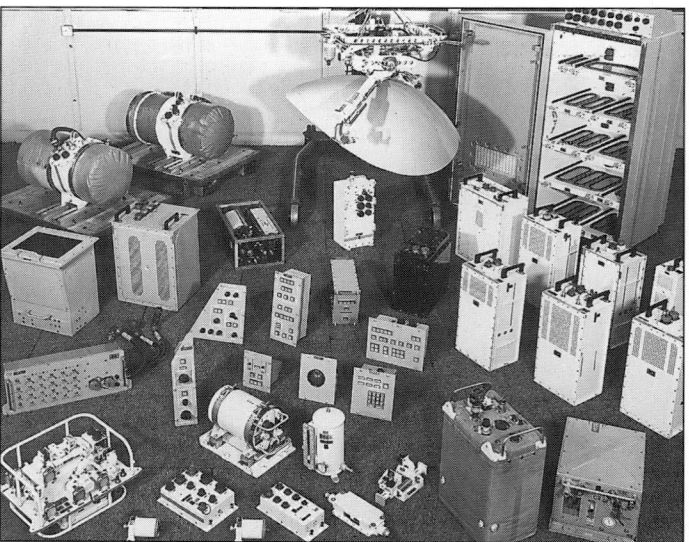

The LRUs making up a Searchwater 1 radar (MS/Thales)

The Searchwater 1 operator's console in a Nimrod MR 2 maritime patrol aircraft (MS/Thales)

Searchwater 2000AEW has been selected for the RN's Sea King AEW Mk 7 programme (MS/Thales)

In addressing the emergence of the UK's RMPA requirement (with its emphasis on value for money rather than absolute compliance with the user's operational/ technical specifications) and the availability of a new generation of lightweight, low-cost maritime surveillance radars, the Searchwater design team radically altered its approach to a follow-on system. Key areas addressed are quoted as including system weight (the original Searchwater 2 proposal had weighed in at more than 400 kg), volume, cost, reliability and maintainability. Criteria set are reported to have included a 60 per cent reduction in weight, a 30 per cent reduction in the required space envelope, a minimum 33 per cent (50 per cent ideal) reduction in cost and three-fold improvement in mean time between failure value. Incorporation of these various elements has resulted in the 165 to 175 kg Searchwater 2000 design melding existing Searchwater/ASTOR/Skymaster-type technology with bought-in components to reduce costs. In this latter area, Thales has an agreement with US contractor Raytheon Systems covering the supply of a scanner for the Searchwater 2000 programme. Thales stresses that the antenna being supplied has been modified to meet the requirements of the application.

As of May 2001, three distinct variants of Searchwater 2000 are being developed or proposed, namely Searchwater 2000MR (for the UK's RMPA programme), Searchwater 2000AEW (for the UK's Sea King AEW Mk 7) and Searchwater 2000MS (for small to medium fixed- and rotary wing maritime patrol applications). All members of the family utilise common generic signal and digital processors. The AEW model utilises the existing AEW Sea King antenna assembly mated to a new transmitter and receiver/exciter. The weight reduction inherent in Searchwater 2000AEW is such as to increase the Sea King AEW Mk 7's endurance by 45 minutes over that of its predecessor. The AEW radar displays via a dedicated man/ machine interface and its operating modes include:

- air-to-air (look up/look down)
- moving target indicator
- surface surveillance (anti-submarine/anti-surface ship warfare)
- littoral
- navigation/ground mapping
- target classification
- weather detection
- beacon.

For its part, Searchwater 2000MR is described as being a multimode, high mean power, travelling wave tube, coherent surveillance radar that incorporates 'all digital' processing and adaptive threshold control. In the Nimrod MRA Mk 4 application, the sensor inputs data directly into its host platform's tactical suite and its system features include:

- selectable frequencies within 8 to 12.5 GHz range
- vertical and horizontal polarisation
- a swath synthetic aperture radar facility
- pulse-Doppler air-to-air, target classification and navigation/weather avoidance operating modes
- use of Array Systems Computing radar imaging software
- full integration with an identification friend-or-foe interrogator
- automatic target tracking (up to 100 plus tracks)
- air cooling
- MIL-STD-1553B databus control and data interfaces
- Two RGB video outputs.

The multimode, Searchwater 2000MS coherent radar is described as combining Thales Sensor's experience with the lightweight 8 to 12 GHz Super Searcher maritime patrol radar (see separate entry) with technology from the Searchwater 2000MR/AEW line. Designed for applications where cost, size and weight are critical, Searchwater 2000MS is a pulse-Doppler radar that features look-up/look-down aircraft detection capabilities; digital processing; adaptive threshold control; synthetic and inverse synthetic aperture functions; fully coherent pulse compression; low probability of intercept and navigation/ground-mapping, weather, beacon and target profile classification modes.

Status

ARI 5980 is reported to have entered service with the RAF in February 1979 aboard British Aerospace Nimrod MR 2 maritime patrol aircraft. More than 40 such radars are understood to have been acquired and have been the subject of a series of updates and reliability improvements. Such work is thought likely to continue until the Nimrod is withdrawn from service circa 2004. AEW configured Searchwater radars are currently (May 2001) installed aboard RN Sea King AEW Mk 2A and three Spanish Navy SH-3 helicopters.

As noted earlier, Searchwater 2000 was ordered for use aboard the UK's Nimrod MRA Mk 4 maritime patrol aircraft during February 1997. In the preceding October, the AEW variant of Searchwater 2000 was selected for use in the RN's Sea King AEW Mk 7 programme. This effort involves the modification of 10 Sea King AEW Mk 2A and three Sea King HAS Mk 5/6 aircraft to the new standard, with a scheduled (as of October 2001) service entry date of March 2002. With regard to the Nimrod MRA Mk 4 application, as of the given date, Jane's sources were suggesting that the Searchwater 2000MR radar completed its company trials programme during May 2001.

Specifications

Searchwater 2000MR
Frequency: 8-12.5 GHz
Elevation coverage: single sweep (cosec^2 beam shape)
Polarisation: vertical and horizontal (selectable)
Surveillance modes: air-to-air (pulse Doppler), ASW, ASuW, littoral waters/open sea and swath SAR
Sub-modes: classification, IFF interrogator and search and rescue
Automatic tracking: >100 tracks
Power consumption: <4 kVA (115 V, 3-phase)
Cooling: air
Interfaces: MIL-STD-1553B (control and data); 2 RGB video outputs
Searchwater 2000MS
Frequency: 8-12.5 GHz
Peak transmitter power: 8 kW (typical)
TWS: >100 tracks
Pulse-length: multiple pulse-widths
PRF: multiple frequencies
Antenna type: planar or reflector
Polarisation: vertical or horizontal
Power consumption: 2.5 kVA (typical)
Cooling: air
Interfaces: ARINC 429, MIL-STD-1553, RS-422 or user defined

Contractor

Thales Sensors, Crawley, UK.

UPDATED

StandOff Surveillance and Target Acquisition Radar (SOSTAR)

Type

I/J-band (8 to 20 GHz) combined Synthetic Aperture Radar (SAR)/Moving Target Indicator (MTI) radar.

Description

The SOSTAR-X combined SAR/MTI radar is a technology demonstrator that is associated with NATO's airborne Alliance Ground Surveillance (AGS) requirement and is being developed by the SOSTAR consortium (made up of European Aeronautic, Defence and Space (EADS) Dornier (Germany - 28 per cent workshare), FIAR (Italy - 28 per cent workshare), Fokker Space (Netherlands - 5 per cent workshare), Indra (Spain - 11 per cent workshare) and Thales Airborne Systems (France - 28 per cent workshare)). As such, the SOSTAR-X radar is reported to be based around an EADS/Thales-developed, scalable (from 1.5 to 6 m plus) active array that incorporates 'existing' Monolithic Microwave Integrated Circuit (MMIC) and gallium-arsenide transceiver module technology. Here, the gallium-arsenide transceiver technology is being derived from work being done within the multinational Airborne Multifunction Solid-state Active array Radar (AMSAR) programme (see separate entry). As initially deployed (aboard a Fokker 100 testbed aircraft), the SOSTAR system will employ a 3 m antenna array and will be equipped with three operator workstations (devoted to target acquisition, processing and operational control respectively) and a datalink system for downlinking data to an associated ground station for information processing, evaluation and distribution. SOSTAR's array scalability is further noted as allowing the system to be tailored to a range of potential platform types including fixed-wing aircraft such as the twin-engined Airbus, high-altitude, long endurance and other types of unmanned aerial vehicle and helicopters such as the NH 90.

Status

On 22 February 2001, EADS Dornier announced the five-nation decision to go ahead with the SOSTAR programme and to form the SOSTAR GmbH joint venture (based at Dornier's Friedrichshafen, Germany facility) to undertake the work At the time of the announcement, the complete programme was valued at then year €85 million, with SOSTAR-X test and demonstration flights scheduled to begin during 2004 and 2005 respectively. Subsequently, EADS entered into an agreement with US contractor Northrop Grumman under which the two companies would promote an AGS radar solution based on the latter's Radar Technology Insertion Programme (RTIP) technology (see separate entry) with 'technology release gaps' (such as a suitable active array antenna) plugged by elements drawn from the SOSTAR effort.

Contractor

SOSTAR GmbH, Friedrichshafen, Germany.

NEW ENTRY

An artist's impression of the Fokker 100 SOSTAR demonstrator aircraft (EADS Dornier)   2002/0083940

Super MAREC maritime surveillance radar

Type

I-band (8 to 10 GHz) airborne maritime surveillance radar.

Description

Super MAREC combines the proven radar sensor components of MAREC II with a scan converted colour TV raster display similar to that developed for the Super Searcher equipment. This saves weight and space, and increases the tactical navigation facilities available to the operator, so reducing his workload on long patrols. Additional facilities include automatic multiple target tracking, and navigation overlay data. Enhancements to Super MAREC include the addition of a full International Standards Organisation contoured weather avoidance display mode, a ground-mapping display and an extra radar mode to give improved close range target detection.

Status

Over time, Super MAREC is reported as having been installed aboard at least 22 Dornier Do 228 patrol aircraft operated by India's Coast Guard service.

Contractor

Thales Sensors, Crawley, UK.

VERIFIED

Super Searcher maritime surveillance radar

Type

I-/low J-band (8 to 12.5 GHz sub-band) maritime surveillance and targeting radar.

Description

Super Searcher is a development of the earlier Sea Searcher radar and is designed for use in multithreat maritime environments. It is a lightweight 8 to 12.5 GHz equipment that, when compared with its predecessor, offers greater detection range, target-tracking capabilities and weapons guidance performance. The system incorporates three selectable pulse-widths, including a short option to give high definition on small targets in bad weather. Contact recognition is also improved by the use of microprocessor techniques and new signal processing algorithms. The visual display unit includes freeze-frame memory storage facilities and graphical overlays of tactical and navigation symbology. The display can either show true or relative motion, with or without offsets. Data are displayed either in latitude and longitude or as a grid reference. The system has a multiple track-while-scan capability and provides a range of navigational features, including ground mapping and waypoint markers. A digital map option is available with extensive graphics overlays. Features such as roads, rivers, railway lines, restricted areas and contours can be individually selected and deselected as required. The map option 'greatly enhances' the capabilities of the radar in the search and rescue and maritime surveillance roles. Additional features can be added (such as sea contours and thermal layers) for anti-submarine warfare operations.

The Super Searcher radar operates in primary and secondary modes, either separately or in combination. A comprehensive symbol library (vectored as required) is claimed to 'ease' the operator's workload, particularly in terms of data storage and extraction. The system is compatible with a range of identification friend-or-foe/secondary surveillance radar interrogators and can be integrated with an electronic support system. A Ground-Mapping Radar (GMR) variant of Super Searcher has been developed to meet a specific Royal Air Force (RAF) requirement for a total avionics update for its fleet of 20 Dominie T1 navigation training aircraft. When compared with the standard Super Searcher, the GMR variant incorporates what have been termed 'significant adaptations' for the ground-mapping role. These include enhanced detection range for both low and medium/high level operations with the 'outstanding display fidelity' required to emulate front-line operational radars in aircraft such as the Tornado and the Jaguar. To meet the radar detection and accuracy requirements inherent in intensive low-level training, the GMR model uses a three-axis platform antenna that is accommodated in a slightly extended airframe nose section. The complete Dominie Avionics Update System (DAUS) also includes a new cabin and flight deck, console layout, integration of navigation sensors, an integrated suite of modern digital radio navigation aids, airframe refurbishment and a contractor logistic support package. Most of the specific-to-role features of the GMR can be incorporated 'as standard' in the maritime version of Super Searcher.

Status

As of May 2001 (and aside from the already noted RAF DAUS programme), Jane's sources were reporting that Super Searcher radars have been installed aboard eight Australian Navy Sikorsky S-70B-2 helicopters, 20 Brazilian Air Force Embraer EMB-III maritime patrol aircraft and 20 Indian Navy Sea King Mk 42B anti-submarine warfare helicopters. In more detail, the RAF DAUS contract was awarded during May 1992 to the then MEL (subsequently Racal Defence Electronics and now Thales Sensors) with Marshall Aerospace as a partner. Super Searcher is also the basis of an upgrade to the ARI 5955 Sea Searcher radars fitted to the RAF's fleet of Sea King HAR Mk 3A search and rescue helicopters. A contract covering this work was awarded in October 1993 and it is understood that a similarly upgraded radar has been offered for use on the service's Sea King HAR Mk 3 helicopters. Elsewhere, Thales has also been contracted to upgrade the ARI 5955 radars fitted to the Royal Australian Navy's Sea King Mk 50 utility helicopters, a programme that appears to form part of a wider upgrade effort (AR 5055) that is intended to bring the helicopters up to Mk 50A standard.

Specifications

Frequency: 8-12.5 GHz
Modes: ground mapping (optional), surveillance and weather
Antenna: planar and reflector, 360° operation with operator selectable sector scanning/blanking
Transmitter: combined transceiver; frequency agile, 4 PRFs; 3 pulse-widths; 8-12.5 GHz band transponder option

The Super Searcher airborne radar

Displays: 525/625 line raster scan; 7.4, 14.8, 29.6, 59.3, 118.5, 237.1 and 474.1 km range scales; 17.8 or 25.4 cm (7 or 10 in) monochrome; 20.3, 25.4, 30.5 or 35.6 cm (8, 10, 12 or 14 in) colour
Signal processing: adaptable menu hierarchy control; automatic track-while-scan (multiple targets); scan-to-scan integration; 'within-beam' integration
Power: 1.4 KVA (typical)
Cooling: ambient air
Missile compatibility: 'all' fire and forget anti-shipping missiles
BIT: yes
Weight: <85 kg (typical)

Contractor
Thales Sensors, Crawley, UK.

VERIFIED

SWIFT Synthetic Aperture Radar (SAR)

Type
G/H-band (4 to 8 GHz) airborne SAR.

Description
Thales Airborne Systems' SWIFT (Standoff all-Weather radar for In-Flight Terrain surveillance) SAR is a miniaturised terrain surveillance radar tailored for integration into unmanned aerial vehicles. SWIFT uses the SAR technique with a fixed side-looking antenna. The image resolution is set on the ground, whatever the flight altitude, thus guaranteeing the same image sharpness throughout all missions. The carrier aircraft's survivability is enhanced since the radar's all-weather capability allows its operation at standoff altitudes. SWIFT is a component of a complete system also including a ground segment consisting of a workstation in the ground control station. Real-time transmission of the processed radar data between the ground and air segments takes place via a datalink.

Status
Developed under a French Ministry of Defence contract, the SWIFT SAR is reported as having been successfully flight-tested during 1994.

Specifications
Frequency: G/H-band (4 to 8 GHz)
Resolution: ground selectable
Swath width: 4 km
Range: 8 km (at 2,438 m altitude)
Volume: 28 litres
Weight: 23 kg
Power requirement: 200 W

Contractor
Thales Airborne Systems, Elancourt, France.

VERIFIED

The SWIFT SAR radar

Varan maritime surveillance radar

Type
I/J-band (8 to 20 GHz) maritime surveillance radar.

Description
Varan is a member of the Iguane family of radars and is essentially an Iguane radar with an antenna type that makes it suitable for virtually all the medium-size maritime patrol aircraft, both fixed- and rotary-wing types. An option with the Varan system is a package giving synthetic aperture capability and/or side-looking radar imagery facilities. The system operates in I/J-band, using pulse compression and frequency agility to ensure high performance on maritime targets in all combinations of weather, sea state and operating altitude. These same techniques also provide maximum protection against electronic countermeasures. The low peak power output level, associated with high receiver sensitivity, increases the difficulty of detection by hostile sensors. Typical detection ranges in Sea State 3/4 are: snorkel

The Varan airborne radar

55 km, fast patrol boat 110 km and freighter 240 km. Overall system weight is 111 kg. Operational missions performed by the Varan radar include surface and anti-submarine warfare, over-the-horizon targeting for shipborne surface-to-surface missiles, search and rescue, marine environmental protection, navigation and weather avoidance. Over time, Thales's DR 2000 electronic support system has been teamed with Varan to create the TRES (Tactical Radar and Electronic support System) wide area maritime surveillance, early warning and weapon control capability.

Status
Over time, Varan radars are reported as having been installed aboard French Navy Falcon 20H Guardian maritime patrol aircraft and as having been procured by Chile (10 systems) for use on both fixed- and rotary-winged platforms.

Contractor
Thales Airborne Systems, Elancourt, France.

VERIFIED

IRAQ

Iraqi Airborne Early Warning (AEW) radar programme

Type
Indigenous AEW radar development programme.

Description
The Iraqi AEW programme centres on two systems: the Adnan 1 and Baghdad 1 derivatives of the Russian Ilyushin Il-76 transport aircraft. Of the two, Baghdad 1 was fitted with a modified Thomson-CSF Tiger ground-based radar while the sensor carried by Adnan 1 remains unknown.
   In more detail, Baghdad 1 had its rear cargo ramp removed and replaced by a radome housing the scanner of an Iraqi built Tiger radar. It is understood that the signal processing of the radar was modified to avoid ground clutter and electronic support measures and modified navigation systems were fitted. Baghdad 1 is reported to have been able to detect, track and identify targets at a maximum range of 350 km. Azimuth coverage is said to have been less than 360°.

Status
As of this edition, one of the three Adnan 1 aircraft is reported to have been destroyed in the 1991 Gulf War with one other being interned in Iran. The remaining Adnan 1 platform may still be in service with the Iraqi Air Force. Nothing is known of the current status of the Baghdad 1 platform.

VERIFIED

ISRAEL

EL/M-2022 series surveillance radars

Type
Family of modular airborne maritime surveillance radars.

Description
The EL/M-2022 equipments are multimode radars which are designed for maritime surveillance, navigation and weather avoidance tasks in both fixed- and rotary-wing applications. The equipment's modularity is claimed to facilitate installation in a wide range of air vehicle types and its capabilities are reported to include: moving

Some of the line-replaceable units which go to make up the EL/M-2022A(V)2 maritime surveillance radar

target indication (land and sea); Doppler beam sharpening; display enlargement and freeze; selectable sector or full scan coverage; Track-While-Scan (TWS) and beacon interrogation. System modules are noted as including low sidelobe antenna assemblies, a front end drive unit, a coherent travelling wave tube-based transmitter and a wide dynamic range, programmable receiver/processor unit. Known details of the various system configurations are as follows:

EL/M-2022A(V)1
A 65 to 75 kg digitally controlled radar which is described as offering medium-range detection capabilities (102 km range against small ships in Sea State 3) for manned and Unmanned Air Vehicle (UAV) applications. TWS and power consumption values are given as up to 50 targets simultaneously and 1 kW respectively.

EL/M-2022A(V)2
An 89 to 98 kg equipment configured for long-range periscope detection. Sea State 3 detection ranges are given as 46 km against a 1 m² target and 148 km against small ships. The radar's power consumption figure is noted as 2 kW.

EL/M-2022A(V)3
A 95 to 103 kg radar which is described as offering long-range periscope detection and target classification capabilities. Sea State 3 high-resolution imaging ranges are given as 56 km against a 1 m² target and 148 km against small ships. TWS and power consumption figures are noted as being up to 100 targets simultaneously and 2.3 kW respectively.

Status
Over time, EL/M-2022A(V)3 is reported to have been selected for installation aboard export Antonov AN-72P/-76 maritime patrol and Australian AP-3C Orion anti-surface warfare aircraft.

Contractor
Elta Electronics Ltd (a subsidiary of Israel Aircraft Industries Ltd), Ashdod.

VERIFIED

EL/M-2055 Synthetic Aperture Radar (SAR)

Type
Modular SAR/Moving Target Indicator (MTI) payload for Unmanned Aerial Vehicle (UAV) applications.

Description
The EL/M-2055 SAR/MTI UAV payload is understood to comprise a SAR/MTI radar, an online signal processor, an Inertial Navigation Unit (INU) and an associated datalink with which to transmit acquired imagery to a Ground Exploitation Segment (GES). Looking at some of these in more detail, the SAR/MTI sensor is equipped with a low sidelobe planar array that features a low-noise receiver front end and is stabilised for drift, roll and vibration. The unit's transmitter is coherent and makes use of miniature travelling wave tube amplification. The radar's processor is noted as incorporating a coherent exciter, a high-speed signal processor and interfaces for the datalink and the host platform's avionics suite. Radar modes comprise strip, spot and MTI, with the latter providing 360° coverage in azimuth. Other system features include squint imagery for flexible area coverage geometry, flexible mission planning and management and real-time re-tasking. For its part, the system's GES makes use of commercial-off-the-shelf equipment and provides computer-aided tools for data exploitation and intelligence reporting. In terms of installation, the GES can be housed in an independent shelter or integrated into the UAV's ground-control station.

Status
As of January/February 2002, EL/M-2055 was understood to be a live programme.

Specifications
Power supply: 28 V DC (MIL-STD-704)
Power consumption: 50 W (INU); 75 W (antenna); 300 W (transmitter); 650 W (processor); 1,075 W (complete system)
Command and control: RS-422A (full duplex)
Downlink output: 2,048 Mbps (10.71 Mbps as option)

Weight: 5 kg (INU); 7.5 kg (transmitter); 16 kg (antenna); 27.5 kg (processor); 56 kg (complete system)

Contractor
Elta Electronics Ltd (a subsidiary of Israel Aircraft Industries Ltd), Ashdod.

UPDATED

EL/M-2060 series Synthetic Aperture Radars (SAR)

Type
Family of airborne surveillance SARs.

Description
As of January/February 2002, two variants of the EL/M-2060 SAR have been identified, the known details of which are as follows:

EL/M-2060P
The EL/M-2060P is described as being a 'field proven' system that is designed for fast jet applications and comprises a detachable pod-mounted SAR, an integral bi-directional datalink and an associated Ground Exploitation Segment (GES). Functionally, EL/M-2060P transmits its imagery (via the built-in datalink) to the GES for exploitation and dissemination. Acquired data is also recorded for non-realtime transmission and/or post-mission analysis. Other system features include strip, spot and ground moving target indication modes (with the latter highlighting moving targets within SAR strip imagery), flexible mission planning and management, real-time re-tasking and 'minimal' pilot workload. The datalink used is drawn from Elta's EL/K-1850 series, with the application being capable of C- (4 to 8 GHz) and X-band (8 to 12.5 GHz) operation, being fully duplex capable and having a range of at least 463 km depending on host aircraft altitude and radio frequency line of sight. The GES is noted as incorporating commercial-off-the-shelf technology, as providing computer-aided tools for data exploitation and intelligence reporting and as being configurable for fixed-site or mobile shelter applications.

EL/M-2060T
EL/M-2060T is described as being a SAR (with GMTI capability) that is suitable for transport aircraft installations and comprises a radar sensor (with an integral online signal processor), airborne and ground exploitation segments and a datalink. Of the cited subsystems, the radar sensor is made up of a low sidelobe planar array antenna (with a stabilised pedestal and a low-noise front-end), a transmitter and a processor that incorporates a coherent exciter, a high-speed signal processor and interfaces to the exploitation elements, the datalink and the host platform's avionics. The transmitter used incorporates coherent travelling wave tube amplification and the system's ground exploitation segment (designated as the GES) makes use of commercial-off-the-shelf technology and is capable of computer-aided mission planning/management, data exploitation and intelligence reporting. Within the host aircraft, EL/M-2060T incorporates up to two exploitation consoles with the GES being configured with up to two (tactical shelter installation), five (mobile S-280 shelter installation) or up to 10 (fixed-site installation) workstations. The datalink used forms part of Elta's EL/K-1850 series (see previously) and radar modes comprise strip, spot and GMTI (with the latter being either a stand-alone mode or one that is integrated with the strip function). There is also a radar squint capability to facilitate flexible area coverage geometry. System options include maritime patrol, inverse SAR and air-to-air modes together with integration with electro-optic and signals intelligence sensors.

Status
As of January/February 2002, both EL/M-2060P and -2060T were understood to be live programmes.

Specifications
Power supply: 4.3 kW (M-2060P - 3.5 kW typical)
Power consumption: 300 W (M-2060T - air data terminal); 750 W (M-2060T - airborne workstation); 3,000 W (M-2050T - radar sensor)
Video output: RS-170 (M-2060P)
C² bus: MIL-STD-1553 (M-2060P)
Weight: 10 kg (M-2060T - air data terminal); 100 kg (M-2060T - airborne workstation); 170 kg (M-2050T - radar sensor); 590 kg (M-2060P - complete system)

Contractor
Elta Electronics Industries Ltd (a subsidiary of Israel Aircraft Industries Ltd), Ashdod.

UPDATED

EL/M-2075 surveillance radar

Type
D-band (1 to 2 GHz) phased-array surveillance radar.

Description
The EL/M-2075 is a D-band phased-array radar which is used for airborne warning and control, tactical surveillance of airborne and surface targets and intelligence gathering.

The first application of Elta's EL/M-2075 AEW radar is as a part of the Boeing 707-based Phalcon surveillance and command and control system

A close-up of the nose and starboard side antenna housings fitted to the Chilean Phalcon Boeing 707 surveillance aircraft (Paul Jackson)

The system, as fitted to the first operational aircraft (a Boeing 707-based system designated as Phalcon), carries three antennas mounted in a 3 m diameter nose radome and two cheek fairings located on either side of the aircraft's forward fuselage. The cheek fairings used measure approximately 12 × 2 m and are mounted on floating beds to prevent air-frame flexing from degrading the radar accuracy. Each side-mounted array has several hundred elements driven by individual liquid-cooled transmit/receive modules, the latter being mounted on racks in spaces behind the cockpit. This three-array configuration provides 280° coverage in azimuth, a figure which can be increased to the full 360° arc via the inclusion of an additional phased array (or arrays) at the rear of the platform.

Each antenna assembly scans a given azimuth sector, scanning being carried out electronically in both azimuth and elevation. Radar modes include high Pulse Repetition Frequency (PRF) search and full track, track-while-scan, a slow-scan detection mode for hovering and low-speed helicopters (using rotor blade returns), and a low PRF ship-detection mode. These modes can be interleaved to provide multimode operation in any scanning sector. It can operate in a passive mode for the detection and tracking of jamming transmitters. Track initiation can be achieved in 2 to 4 seconds and extra long dwell times in selected sectors can extend the radar's detection range.

Effective range of the radar is nominally 375-400 km, although this will vary with the type of target and the mode in which the radar is operating. The system is able to track 100 targets simultaneously. A monopulse IFF system is included, the antennas being incorporated in the primary radar array. Within the specific Phalcon application, the EL/M-2075 radar is teamed with a command and control facility and electronic and communications intelligence subsystems. It should be noted that the designation 'EL/W-2085' has also been associated with the described radar.

Status
As of this edition, the Chilean Air Force had acquired a single Boeing 707-based Phalcon surveillance and command and control aircraft equipped with a three-antenna assembly configured EL/M-2075 radar. Elsewhere, EL/M-2075 variants have been offered to Australia (Airbus A310-300 airframe), the People's Republic of China (Beriev A-50 airframe) and South Korea. In the Chinese context, work on installing an EL/M-2075 type radar in a prototype A-50I airborne early warning aircraft for China's Air Force of the People's Liberation Army was well advanced when, in July 1999, the project was cancelled following intense US pressure on the Israeli government to stop the sale of advanced early warning technology to China. Valued at then year US$250 million, this programme was also noted as including options for three additional aircraft. Elsewhere, Elta and US contractor Raytheon Systems are known to have teamed to offer an Airbus/M-2075 airborne early warning package worldwide.

Contractor
Elta Electronics Ltd (a subsidiary of Israel Aircraft Industries Ltd), Ashdod.

UPDATED

ITALY

APS-705 navigation and search radar

Type
I-band (8 to 10 GHz) airborne navigation and search radar.

Description
The APS-705 family consists of two multipurpose radar systems (APS-705 and APS-705A) that are suitable for installation on both fixed- and rotary-wing aircraft. There are a number of operational functions that include search and rescue, navigation and mapping, surveillance and weapon support. APS-705 operates in the I-band with detection over 360°. The system can be used for target tracking and designation to weapon systems and interfaces with the aircraft's sonar system. The equipment's antenna is tailored to the space and location in the aircraft. For example, on the SH-3D the antenna is placed in the dorsal position on top of the fuselage. Line of sight stabilisation is provided and there are selectable rotation rates. Manual controlled antenna tilt provides ±20° of movement. APS-705 makes use of two transceivers (for frequency diversity operation at 25 kW) and the radar can be integrated with the Officine Galileo's UPX-719 beacon system. The I-band APS-705A is an upgrade of APS-705 that incorporates additional features such as digital processing, a freeze mode, a colour raster scan display, track-while-scan of multiple targets and interfaces for Forward-Looking Infra-Red (FLIR), weapon, electronic support, datalink and sonar systems.

Status
Over time, APS-705A radars are reported to have been installed aboard AB 212 ASW (48 aircraft) and upgraded ASH-3D (19) helicopters of the Italian Navy. APS-705 series radars were also noted as being likely to have been fitted to AB 212 ASW helicopters operated by the Greek, Iranian, Peruvian and Venezuelan navies.

Specifications
Frequency: I-band (8-10 GHz)
Power output: 25 kW
Antenna rotation: 20 or 40 rpm
Pulse-width: 0.05 and 1.5 µs
PRF: 1,600 and 650 Hz
System weight: 80 kg

Contractor
Alenia Difesa Avionic Systems and Equipment Division - Officine Galileo, Florence.

VERIFIED

The units that make up the APS-705A radar

APS-717(V) search and navigation radar

Type
I-band (8 to 10 GHz) airborne search and navigation radar series.

Description
The APS-717 family consists of two search and navigation radar systems (APS-717(V)1 and APS-717(V)2) that are tailored to the needs of individual customers. The family offers many operational features, including search and rescue, surveillance, navigation and target designation. Of the pair, APS-717(V)1 is described as being a lightweight nose-up radar that is suitable for both fixed- and rotary-wing aircraft. It operates in the I-band, providing detection over 180° with automatic stabilisation of line of sight. It can be integrated with the navigation system and a Forward-Looking Infra-Red (FLIR) sensor. Other features include constant false alarm rate, scan-to-scan integrator, pulse-to-pulse integrator, a freeze mode and a colour display with graphics. For its part, the APS-717(V)2 radar is noted as being a high-performance upgrade of the APS-717(V)1 with a number of 'additional' features. These include detection over 360°, integration and automatic initialisation of FLIR/low-light TV, a video recorder output, and an optional track-while-scan facility covering 32 targets.

The units that make up the APS-717(V)2 radar

Status

Over time, APS-717(V)1 is reported as having been installed aboard Italian Air Force AS-61A-4/HH-3F search and rescue helicopters and G222 transport aircraft. The APS-717(V)2 variant is understood to have been procured for installation aboard the AB 412HP utility helicopters operated by the Italian Coastguard service.

Contractor

Alenia Difesa Avionic Systems and Equipment Division - Officine Galileo, Florence.

VERIFIED

APS-784 surveillance radar

Type

I/low J-band (8 to 12 GHz sub-band) airborne maritime surveillance radar.

Description

APS-784 is a maritime surveillance radar suitable for use in both fixed- and rotary-wing aircraft. It can be used in a variety of roles including anti-surface vessel, anti-submarine warfare, weather detection, weapon system support and search and rescue. The radar can be integrated either in the aircraft mission system or as a stand-alone equipment. APS-784 operates in the 8 to 12 GHz frequency band using a coherent travelling wave tube transmitter with pulse-to-pulse frequency agility, and track-while-scan features. Pulse compression techniques are incorporated which enable detection of very small maritime objects. Detection is provided over 360° with sectorial transmission. Other features include integration with an identification friend-or-foe interrogator, constant false alarm rate processing, MIL-STD-1553 interface, and multiple radar video outputs in standard video format. The radar is packaged in four line-replaceable units and has four operational modes featuring track-while-scan with adaptive strategies, anti-surface vessel, anti-submarine warfare, and weather and short range. It has two independent scan converters and scan-to-scan integration.

Status

Development and flight testing of the APS-784 radar is noted as having been completed during 1992. As of October 2001, the APS-784 maritime surveillance radar was reported as being procured for installation aboard eight EH 101 anti-submarine/anti-surface warfare helicopters of the Italian Navy. A modified variant

of this radar (designated as HEW-784) is also known to be under contract for installation aboard four Italian Navy EH 101 airborne early warning helicopters (see separate entry).

Contractor

ELIRADAR consortium (a consortium established by Alenia Difesa Avionic Systems and Equipment Division - Officine Galileo (50 per cent) and FIAR (50 per cent)).

VERIFIED

Creso surveillance radar

Type

I-band (8 to 10 GHz) battlefield surveillance radar.

Description

Creso is one of the sensor systems under development as part of the Italian Army's CATRIN-SORAO command, control, communications and intelligence system and comprises airborne and ground segments. The airborne component is installed aboard an Agusta AB 412 helicopter and incorporates radar, navigation, datalink, defensive countermeasures, Forward-Looking Infra-Red (FLIR) and Electronic Support (ES) subsystems. The radar used is described as being an I-band coherent Doppler moving target indicator unit that is made up of six Line-Replaceable Units (LRUs). Key elements here are a 0.4 × 2 m flat planar slotted-array antenna, a wideband helical travelling wave tube transmitter, a microwave assembly, a double conversion receiver exciter, a signal processor and a MIL-STD-1553B databus compatible data processor. Signal processing is noted as utilising Fast Fourier Transform and constant false alarm rate techniques while the data processor generates display formats, controls antenna function and manages the radar's frequency agility.

As installed in the AB 412, the system utilises two avionics bays built into the rear of the helicopter's main cabin and an operations console mounted centrally behind the flight crew positions. The Creso radar is noted as being optimised for the detection and localisation of moving ground targets at ranges in excess of 100 km. Radar modes are noted as being:

Acquisition
Medium-resolution mode for target detection and localisation.

Count
High-resolution mode for detailed analysis of activity in a specific area.

Memory
Analysis mode which allows the operator to record and recall preceding scans to detect changes over time in a specific geographical area.

Expansion
Digital picture enlargement mode.

The systems's display is described as being divided in two, with the left-hand segment showing graphic information and the right-hand element alphanumeric data. The graphic presentation is noted as showing the Forward Edge of Battle Area (FEBA); platform position, pop-up points (preplanned and usually some 20 km behind the FEBA), speed vector and minimum and maximum detection limits. The alphanumeric display details the system's operating mode, platform position, frequency agility parameters, bandwidth and scan rate. System data is also quoted as being downlinked to a ground-based, shelter-mounted data analysis and distribution centre via a J-band (10 to 20 GHz) 2.5 Mbit/s rate datalink with a range of 70 km.

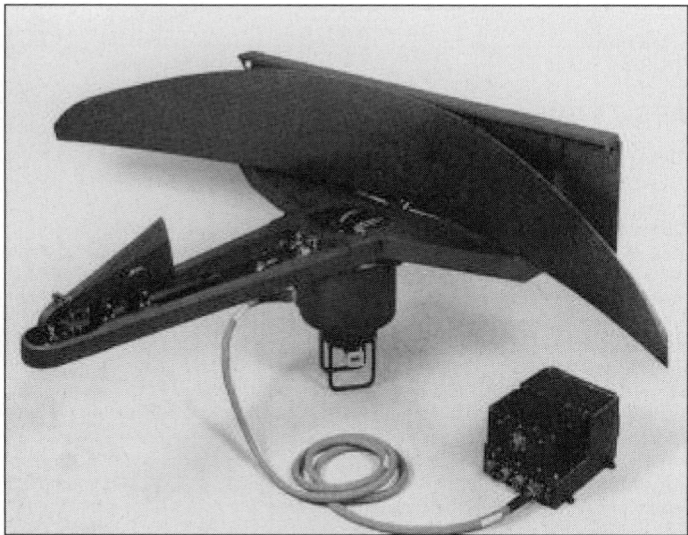

The antenna unit used in the APS-784 radar

A general view of the prototype Creso AB 412 helicopter showing four ports for the ESM system on the helicopter's nose, the Creso antenna housing beneath its nose and the datalink antenna below its tailboom

Status

As of this edition, the Creso radar is reported to have been undergoing environmental and performance trials since 1991. A prototype has been installed aboard Italian Army AB 412 helicopter 'E.I. 453' together with the FLIR, ES and countermeasures subsystems. Additional elements (including the datalink and operator's console) were, as of 1995, scheduled to be installed aboard the testbed during 1996 with follow-on flight tests, inertial navigation system tests and full system test and acceptance flights to be completed by the Spring of 1997. As of this edition, only the prototype Creso system has been procured, with sources suggesting that a full-scale production decision will not now be taken until a decision has been made concerning NATO's multinational Alliance Ground Surveillance (AGS) system programme.

Contractors

Contractors associated with the Creso programme include:
Fabbrica Italiana Apparecchiature Radioelettriche SpA (FIAR - part of Alenia Difesa's Avionic Systems and Equipment Division), Milan (radar).
Elettronica SpA, Rome (ES/defensive countermeasures suite).
Marconi SpA, Genoa (datalink).
Agusta SpA, Rome (airframe/integration).

UPDATED

POLAND

ARS-400

Type

I-band (9.2 to 9.5 GHz sub-band) maritime surveillance radar.

Description

Forming part of the MSC-400 airborne maritime surveillance command and control system, the I-band ARS-400 is described as being a pulse compression radar that features:

- a travelling wave tube transmitter
- multichannel signal processing (sea surface target detection, ground mapping, weather channel and oil pollution channel)
- frequency agility
- scan to scan and digital processing
- electronic counter-countermeasures provision
- a dedicated workstation with colour display.

Status

As of January 2001, Jane's sources were reporting the ARS-400 maritime surveillance radar as being installed aboard the An-28RM Bryza 1RM (*Ratownictwa Morskiego* - Maritime Reconnaissance) aircraft operated by the Polish Navy's 1st and 3rd Naval Air Divisions (based at Gdynia-Babie Doly and Siemirowica respectively).

Specifications

Frequency: 9.2-9.5 GHz
Range: 5-160 km
Transmitter power: 100 W (average)
Antenna beam width: 2° (azimuth); 9° (elevation)
Antenna gain: 32 dB
Pulse-width: 0.2 µs
Noise figure: 2.5 dB
Track capacity: up to 100 targets

Contractor

Przemysłowy Instytut Telekomunikacji, Warsaw.

NEW ENTRY

RUSSIAN FEDERATION AND ASSOCIATED STATES (CIS)

Introduction

The following section details those airborne surveillance, early warning, targeting and navigation radars produced, exported and/or operated by the Russian Federation and the independent republics of Armenia, Azerbaijan, Belarus, Georgia, Kazakhstan, Kyrgyzstan, Moldova, Tajikistan, Turkmenistan, Ukraine and Uzbekistan. Taken together, these 12 entities form the Commonwealth of Independent States (CIS) economic grouping.

VERIFIED

Airborne Early Warning (AEW) radars

The Russian Federation's electronics industry is thought to be the source for a range of AEW radars which are installed aboard the Antonov An-71 (NATO reporting name 'Madcap'), Beriev A-50/A-50U (NATO reporting name 'Mainstay'), Beriev Be-976 and Kamov Ka-31 platforms. Of these, the radar fitted to the A-50U aircraft has been identified as the 3-D, pulse-Doppler Shmel-M equipment. Shmel-M is presumably derived from an earlier model that is fitted to the initial series A-50 aircraft. In turn, this system may be related to the Liana (NATO reporting name 'Flat Jack') radar which was carried by the Soviet Union's first generation AEW platform, the Tupolev Tu-126 (NATO reporting name 'Moss').

As fitted to the A-50U, Shmel-M utilises a 9 m diameter rotodome antenna assembly and is credited with a detection range of up to 230 km against a fighter-sized air target, a figure which rises to 400 km when the radar is observing shipping. The system is manned by 10 operators and is reported to be able to handle 10 simultaneous engagements and up to 50 simultaneous over-land target tracks. The A-50U mission suite is also noted as incorporating VHF band (30 to 300 MHz), UHF band (0.3 to 1 GHz) and satellite communications links, Identification Friend-or-Foe (IFF) and electronic support subsystems and a separate Duran B navigation radar. The type is noted as having provided continuous radar cover over the Black Sea during the 1991 Gulf War and design and production of the A-50U mission suite is credited to the Moscow-based Nauchno Issledovatelskiy Institut Priborostroyeniya (NIIP-Scientific-Research Institute for Instrument Engineering) VEGA organisation. As of May 2001, Jane's sources were suggesting that up to 21 (15 to 16 being a more likely figure) A-50 AEW aircraft were in service with the Russian Air Force and that the type's mission suite was scheduled for upgrading.

The Beriev Be-976 is described as being a 'radar picket' and appears to be generally similar to the A-50 aircraft with the exception that it is fitted with wingtip avionics pods. No details concerning the radar fitted to this aircraft have so far emerged, other than a tenuous report that, despite the use of what appears to be a rotodome, its antenna assembly (shown in the accompanying photograph) is fixed. As of the given date, Jane's sources suggest that at least six Be-976 aircraft have been completed. Of these, four were reported as being non-operational during 1995, with a fifth aircraft being further modified (under the Il-765K designation) to act as an airborne command post for the Russian Burlak space launcher system.

Alongside Shmel-M, NIIP VEGA is also credited with the Kvant radar installed aboard the An-71 AEW aircraft. Sources describe this emitter as being a coherent pulse-Doppler system (featuring pulse compression, variable pulse repetition frequency and digital moving target indication) which, while derived from Shmel-M, utilises a smaller diameter rotodome antenna assembly and operates on different wavelengths. Performance figures suggested for the radar include a maximum detection range of approximately 352 km, a 10 second scan rate, a maximum detection range against a fighter-size air target of 200 km, the ability to detect targets flying at altitudes between sea level and 30,000 m and a simultaneous track capacity of up to 120. Planar co-ordinate accuracy and clutter suppression values of 2.5 km and 50 to 60 dB respectively are also noted. As an option, the An-71's mission suite can be made compatible with Western IFF Mk 12 (modes 1 to 4) and air traffic control (mode A through C and S) transponder systems.

Using a three-man mission crew, the An-71 combined radar, communications and intelligence system was originally developed (in competition with the Yakovlev Yak-44E) to provide an AEW capability for the Russian Navy's aircraft carriers. Despite what has been described as a successful test programme (using two

The Beriev A-50U represents the Russian Air Force's latest generation AEW platform (Paul Jackson)

Flight trials with the Antonov An-71 AEW aircraft are reported to have begun during July 1985

The Ka-31 AEW helicopter is Russia's second attempt to provide its carrier force with an organic early warning capability (Paul Jackson)

A cutaway model of the antenna assembly installed aboard the Beriev Be-976 'radar picket' aircraft (Paul Jackson)

aircraft) during the mid-1980s, the proposal to use the An-71 as a shipboard AEW aircraft foundered on a lack of funds. During the late 1990s, Antonov was promoting the An-71 as a land-based AEW platform and claimed interest in the type from the air forces of Russia and Ukraine together with a number of as yet unidentified potential customers. As of May 2001, the An-71's status was uncertain.

Aside from 10 Tu-126s, which are reported to be still in service with the Russian Air Force, the remaining cited type — the Kamov Ka-31 helicopter — appears to have replaced the An-71 in Russia's search to provide the *Admiral Kuznetsov* with an organic airborne AEW capability. Reported to have been first flown during 1988, the Ka-31 (previously known as the Ka-29RLD) features a 6 m² area antenna which folds through 90° to lie flat against the aircraft's belly for take-off and landing. Designated as the Oko (Eye) equipment, the Ka-31's mission suite is reported to be a product of the Nizhny Novgorod-based Radio Engineering Institute. In terms of performance, Oko is noted as being able to track up to 20 targets simultaneously and detect fighter-sized air targets at ranges of up to 150 km. Surface ships can be detected at ranges of up to 250 km and the system is reported to be operated by a crew of two. Provision for automatic data transfer from the helicopter to its mother ship is also noted. Initial trials with the Ka-31 aboard the *Admiral Kuznetsov* are reported to have begun during 1990 with the type achieving initial operating capability during 1992. As of the given date, a total of three aircraft of this type are thought to have been completed for the Russian Navy with a further four being supplied to the Indian Navy in a then year US$92 million deal that was signed during the second half of 1999. At the time of the deal's announcement (August 1999), the four aircraft involved were scheduled for delivery by the end of 2000 and were intended for initial deployment aboard the Indian Navy's aircraft carrier *Viraat* and the service's three 'Krivak III' class frigates.

VERIFIED

Gukol series radars

Type
I- (8 to 10 GHz) and K-band (20 to 40 GHz) family of weather/navigation radars.

Description
Gukol series radars are modular equipments that are designed to provide a weather/navigation radar capability for lightweight aircraft, helicopters, business aircraft, airliners and military transports and tankers. The equipments' manufacturer claims that Gukol radars provide 'the only proven predictive windshear mode available today and are compatible with a wide range of existing aircraft avionic systems, controls and displays through the use of unique interface modules. As a whole, the Gukol family of weather/navigation radars is quoted as having a mean time between failure value of 5,000 hours and its modularity is said to allow for true two-level maintainability and system growth with limited impact on hardware and software. Gukol series radars also feature built-in test.

As of May 2001, Gukol radars have been identified in four variants, namely the 8 to 10 GHz Gukol-1, -2 and -3 plus the combined 8 to 10 GHz/20 to 40 GHz Gukol-4. Gukol-1 is designed for light and business aircraft applications, while Gukol-2 is optimised for large passenger aircraft use. Gukol-3 is aimed at ultra-light aircraft applications with Gukol-4 being designed for what are termed military transport/tanker requirements.

Gukol radars comprise an antenna and an assembly that incorporates the radar's receiver, transmitter, signal/data processor and power supply/cooling system rack. Operating modes include aircraft, weather and geographical obstacle detection; weather feature direction and velocity determination; real beam and synthetic aperture mapping; map display enlargement and freeze; map display centre stabilisation; aircraft velocity measurement for navigation updating and beacon tracking.

Status
As of June 2001, the precise status of the Gukol radar family was uncertain.

Contractor
Phazotron - NIIR Joint Stock Company, Moscow.

VERIFIED

Kopyo-25 multifunction radar

Type
I/low J-band (8 to 12 GHz) multifunction radar.

Description
Kopyo-25 is a multifunction variant of the Kopyo fire-control radar (see separate entry) which is optimised for use with the Sukhoi Su-25TM/Su-39 close air support aircraft. Pod- or internally mounted, Kopyo-25 comprises a slotted, flat plate antenna, a receiver, an advanced data controller, a signal processor, a synchroniser, an interface unit, a data processor, a power distribution unit, a power supply, a transmitter and an exciter. Kopyo-25 also features built-in test. The equipment is reported to offer air-to-surface, air-to-air and air combat operating modes. Identified capabilities within each of these categories are as follows:

Air-to-surface mode
Real beam ground mapping; Doppler beam sharpening; synthetic aperture with high-resolution mapping for navigation, target location and identification; air-to-ground ranging; display enlargement and freeze; tracking of fixed or moving ground targets and large and small moving target sea search.

Air-to-air mode
Air intercept; range-while-search; track-while-scan and single target tracking.

Air combat mode
Head-up display search; slewable search; boresight and vertical scan.

Status
Kopyo-25 is reported as having begun flight trials during 1998. During June 1999, Jane's sources were noting that the Russian Air Force had initiated a programme to upgrade 80 Su-25S aircraft to Su-25SM standard. Among other elements, this effort involves the introduction of a nose-mounted Kopyo-25 variant.

Specifications: Gukol series radars

	Gukol-1	Gukol-2	Gukol-3	Gukol-4
Frequency	I-band (8-10 GHz)	I-band (8-10 GHz)	I-band (8-10 GHz)	I-band (8-10 GHz) and K-band (20-40 GHz)
Detection range	600 km	400 km	250 km	600 km
Range scale	10/50/100 300/600 km	10/50/100 300/600 km	10/50/100 300/600 km	10/50/100 300/600 km
Antenna size	762 mm	457 mm	380 mm	760 mm
Elevation coverage	±40°	±40°	±15°	±40°
Peak power	0.5 kW	0.5 kW	0.2 kW	1 kW
Weight	28 kg	20 kg	15 kg	60-65 kg
Cooling system		unpressurised air		
System interfaces		MIL-STD-1553B/ARINC-429/RS-170		

On Su-25TM demonstrator aircraft '20', a Kopyo-25 radar pod was carried on the aircraft's centreline (Paul Jackson)

Specifications

Frequency: I/low J-band (8-12 GHz)
Antenna: 500 mm diameter flat slotted array
Angular coverage: ±10 or ±30° (azimuth); 2 or 4 bars (elevation)
Range: 22 km (sea search — small naval vessel); 25 km (air-to-air mode — fighter size target (tail-on aspect) and air-to-surface mode — group of AFVs); 57 km (air-to-air mode — fighter size target (head-on aspect))
Power: 1 kW (average); 5 kW (peak)
MTBF: 120 h
Cooling: Liquid
Weight: 98 kg

Contractor

Phazotron - NIIR Joint Stock Company, Moscow.

VERIFIED

....

NIT Sideways-Looking Airborne Radar (SLAR)

Type

Natural resources managment SLAR.

Description

NIT is a SLAR system for carrying out large scale surveys of the earth's surface regardless of weather conditions. According to the manufacturer, it is intended for use in the survey of natural resources but, as with any SLAR system, use can be made of the information for military purposes.

Investigation of natural resources appears to be the primary aim of the NIT system, covering such tasks as:

• investigation of natural resources (geological mapping, supervision of forestry work and water reservoirs and so on)
• survey of ice situation in the Arctic polar basin
• survey of fires, flooding, and other natural or man-made catastrophes
• supervision of the environment.

The airborne system provides radar information in the coverage zone of up to 80 km, at different polarisations of the transmitted signal. The radar information can be presented on a TV screen in the format required by the operator, and is also transmitted to a ground station via a datalink, for more detailed analysis of the information. Recording of the information can be carried out digitally on a magnetic tape and in analogue form on a photographic film.

Status

Over time, NIT SLARS are reported to have been installed aboard specially modified Antonov An-24, Ilyushin Il-24N fishery observation and Tupolev Tu-134 geophysical survey aircraft. NIT may be related to the putative SLAR that has been associated with the Ilyushin Il-20M signals intelligence platform (NATO reporting name 'Coot-A').

Specifications

Observation band: 40 × 2 km
Resolving power for distance: 15 m
Resolving power along the line of route: 12 min of arc
Weight: 400 kg
Power required: 15 kW

Contractor

RADAR MMS, St Petersburg.

VERIFIED

Sea Dragon maritime patrol radar

Type

I-band (8 to 10 GHz) maritime patrol radar.

Description

This 3 cm wavelength, NIIS Scientific Research Institute-developed radar forms part of the Leninets Holding Company's Sea Dragon maritime patrol aircraft mission suite. System components comprise a planar-array antenna, a power amplifier and an exciter/receiver unit and the radar is described as being both coherent and klystron-based. Radar operating modes include:

• air-to-air detection against a sea clutter background
• synthetic aperture radar
• inverse synthetic aperture radar target classification
• long-range surface search in high sea states and weather/jamming clutter
• beacon tracking
• target acquistion and tracking (range, heading and velocity)
• sonobuoy tracking (linked to the acoustic processing subsystem via a 1553B databus)
• navigation
• weather detection and avoidance.

Within the Sea Dragon suite, the radar displays via one of three multifunction consoles (based on Barco MPRD 134 colour liquid crystal displays) and as of May 2001, has a unit cost of approximately US$2.8 million.

Status

Jane's sources report that trials of the Sea Dragon suite's radar began during late 1998 and that engineering development of the Sea Dragon mission suite was likely to have been completed by the end of 1999. The Sea Dragon mission suite contractor Leninets was also noted as being under contract to the Russian Navy to retrofit a prototype Sea Dragon suite into one of its Ilyushin Il-38 (NATO reporting name 'May') maritime patrol aircraft. As of late 2000, it was also being suggested that the Sea Dragon mission suite had been proposed for the Tupolev Tu-204P maritime patrol concept and was likely to be retrofitted to Russian Tupolev Tu-142M-Z (NATO reporting name 'Bear-F' Mod 4) Anti-Submarine Warfare (ASW) aircraft together with India's fleet of eight Tu-142MK-E (NATO reporting name 'Bear-F' Mod 3), five Il-38 and 14 Ka-28 (NATO reporting name 'Helix-A') ASW platforms.

Specifications

Wavelength: 3 cm
Weight: 260 kg

Contractor

Leninets Holding Company/NIIS Scientific Research Institute, St Petersburg.

VERIFIED

....

Systema millimetric surveillance radar

Type

Millimetric surveillance radar.

Description

The Systema millimetric surveillance radar comprises a combined antenna/transceiver module and processing and system control computer. System applications include helicopter search and rescue, high-resolution terrain imaging, collision avoidance, narrow field of view sensor cueing, automatic targeting of surface targets and as an aircraft landing aid in conditions of low visibility.

Status

Jane's sources report that a prototype of the described Systema millimetric surveillance radar completed its initial trials programme during the period 1995 to 1997.

Specifications

Wavelength: 3 mm
Coverage: −20 to +10° (elevation); ±45° (azimuth)
Range: 3 km (high-voltage cable); 5 km (liferaft/small boat); 6-15 km (automobiles); 15-20 km (small/medium-sized ships and buildings and structures such as bridges)
Range resolution: 8 m (sea targets)
Power requirement: 0.8 kW
Weight: 50 kg

Contractor

Systema (an associate of the Leninets Holding Company), St Petersburg.

VERIFIED

....

Other known RFAS surveillance radars

Alongside the already described systems, the RFAS air forces make use of a number of other surveillance radars about which only limited information is available. Known details of these equipments are as follows:

Kub

Kub (Cube) is described as being the Side-Looking Airborne Radar (SLAR) carried by the Mikoyan MiG-25RBK high-speed reconnaissance bomber (NATO reporting name 'Foxbat-D') which was in production between 1971 and 1980. Alternative Jane's sources suggest the MiG-25RBK was fitted with either the Sabla or Shompol SLAR, both of which are credited to the Moscow-based NIIP VEGA organisation.

Obzor (NATO reporting name Clam Pipe)

Obzor is the missile guidance and navigation radar installed aboard Tupolev Tu-95MS6/MS16 Kh-55 cruise-missile carriers (NATO reporting names 'Bear-H6' and '-H16' respectively) and the same company's Tu-160 strategic bomber ('Blackjack'). Obzor is understood to be a product of the St Petersburg-based Leninets organisation.

PNA-D (NATO reporting name Down Beat)

PNA-D is the I-band (8 to 10 GHz) missile targeting and navigation radar which is fitted to the Tupolev Tu-22M3 strike aircraft (NATO reporting name 'Backfire-C' and may be carried by the earlier Tu-22M2 ('Backfire-B') together with the Tupolev Tu-95K22 ('Bear-G') strike and reconnaissance platform. Possibly also known as Rubin, PNA-D appears to be associated with the Kh-22 air-to-surface missile (AS-4 'Kitchen'), is credited with a maximum range of 300 km and is a product of the Leninets organisation.

Shtyk

Shtyk is the 'multipurpose' SLAR fitted to the Sukhoi Su-24MR reconnaissance aircraft (NATO reporting name 'Fencer-E') and is a product of the NIIP VEGA organisation.

Virazh

Virazh (Turn) is described as being the SLAR installed on the Mikoyan MiG-25RBV high-speed reconnaissance bomber (NATO reporting name 'Foxbat-B') which was produced between 1978 and 1982.

Wet Eye

Wet Eye is generally described as being the I/J-band (8 to 20 GHz) maritime surveillance/Anti-Submarine Warfare (ASW) radar fitted to the Ilyushin Il-38 (NATO reporting name 'May') and Tupolev Tu-142/-142M ('Bear-F') ASW aircraft. An alternative Jane's source suggests that the radars fitted to these platforms are products of the Leninets organisation and are type specific with the Berkut system being fitted to the Il-38 and the Korshum radar to the Tu-142.

VERIFIED

SWEDEN

Coherent All RAdio BAnd Sensing (CARABAS) ultra-wideband Synthetic Aperture Radar (SAR)

Type

20 to 90 MHz ultra-wideband SAR technology demonstrator.

Description

Developed jointly by Sweden's National Defence Research Institute (known by the acronym FOI) and Ericsson Microwave, CARABAS is a low-frequency, ultra-wideband SAR that is designed to validate the technology's ground/foliage-penetration performance in civil and military tactical surveillance and remote sensing applications. As of early 2001, three generations of CARABAS radars have been (or are being) developed, the known details of which are as follows:

CARABAS I

The CARABAS I system made use of Fast Fourier Transform post-collection processing and aft-facing, tubular antennas mounted in 0.3 wide by 5.5 m long inflatable sleeves. Each of these multi-element antennas acted as a transceiver dipole with each antenna receiving and transmitting alternately. Installed aboard a Swedish Air Force TP 86 testbed aircraft, Carabas I is reported to have demonstrated a resolution of 3 by 3m at a range of 10 km during flight trials conducted during the early 1990s.

CARABAS II

The CARABAS II system makes use of post-collection time domain processing and a pair of forward-facing, ultra-light, omnidirectional, 8.2 m long probe antennas.

The aft-facing antenna array used with the CARABAS I ultra-wideband SAR

The forward-facing antenna array used with the CARABAS II ultra-wideband SAR   0006910

Track accuracy is maintained via the use of a ground and air phase-differential application of the Global Positioning System. Again installed aboard a Swedish Air Force TP 86 aircraft, CARABAS II is reported to have achieved a minimum resolution of 2 by 1 m.

CARABAS III

As of early 2001, Jane's sources were reporting that development work on the 20 to 90 MHz CARABAS III radar had been restarted after a programme 'freeze' of several years duration. CARABAS III differs from its predecessors in featuring a two element, phased-array, 5.5 m long antenna assembly (with a receiver for each element) and interferometric processing to resolve the system's returns. Additionally, CARABAS III's design reflects the desire to improve the system's reliability and its power output (to 300 W). One platform option under consideration is a 12 m span medium endurance unmanned aerial vehicle equipped with a pusher engine, wingtip 60 to 90 MHz moving target indication reception antennas, a datalink subsystem and two forward-facing boom antennas for the CARABAS SAR. Potential performance figures for a CARABAS III installation in an air vehicle flying at a speed and altitude of 357 km/h and 9,997 m respectively include the ability to survey a 2 km^2 ground area every second with a resolution of 3 by 3 m.

Status

Flight tests with the CARABAS I system are understood to have begun during the early 1990s with those using the CARABAS II variant beginning in November 1996. As of late 2000, CARABAS trials have been undertaken in Sweden, Pennsylvania and Yuma, Arizona. The Pennsylvania test round was undertaken in conjunction with the US Defense Advanced Research Projects Agency (DARPA) while that at Yuma was intended to demonstrate the system's ability to detect buried objects such as mines. As of early 2001, Jane's sources were suggesting that a 'production ready' CARABAS III configuration would be available by 2004, with roles such as biomass imaging, border surveillance, counter drugs operations and crisis management being proposed for the system.

Contractors

FOI, Linköping.
Ericsson Microwave Systems AB, Mölndal.

UPDATED

....

Erieye Airborne Early Warning (AEW) radar

Type

E/F-band (2 to 4 GHz) AEW radar.

Description

Ericsson has developed a new generation of airborne early warning radar for the Swedish Defence Materiel Administration, with series production for the Swedish Air Force beginning during 1993. The object of the programme was to develop a high-performance, long-range AEW system that would be light enough to be carried by a commercial airframe. This has been achieved primarily through the use of distributed solid-state transmitters, phased-array antennas and extensive employment of composite construction materials.

Erieye is based on an electronic scanning technique that facilitates target tracking in near real-time. The radar's operating frequency is in E/F-band which, together with a large aperture antenna, provides for 'good electronic counter-countermeasures' performance in an 'intense' electronic warfare environment. The dual-sided phased-array antenna is mounted on top of the host aircraft's fuselage in a 9.7 m long fairing. Adaptive waveforms and signal processing facilitate clutter suppression (with 'virtually no degradation' in detection range) and the 'early' detection of air, sea and ground targets. The range performance of the radar against fighter-size targets is approximately 350 km and 150 km against cruise missiles.

The low sidelobe antenna array used contains 200 all solid-state transceiver modules and offers graceful degradation characteristics. E/F-band operation provides narrow beamwidth and yields 'high accuracy with improved target resolution'. Ericsson goes on to characterise Erieye's use of onboard signal processing and waveform generation with adaptive and flexible waveforms, frequency modulation and digital phase coded pulse compression as forming a 'comprehensive' solution to the AEW requirement. The radar also incorporates

In Swedish service, the Erieye AEW radar is installed aboard Saab S100B aircraft
0044137

The Erieye/EMB-145 combination is being produced for the Brazilian SIVAM programme and has been selected for use by the Greek Air Force
2000/0081429

The man/machine interface used in the Greek EMB-145 Erieye-equipped S100B AEW aircraft
0006909

frequency agility and full Doppler processing in low- and medium-pulse repetition frequency modes. When combined with other subsystems such as an onboard Command and Control (C²) capability, identification friend-or-foe/selective identification facility, electronic support measures and a tactical datalink, Erieye-based systems can be configured to provide a variety of AEW capabilities that range from an airborne reporting post to an airborne control and reporting centre.

Status
The Swedish Defence Materiel Administration awarded a SKr1,200 million contract to Ericsson in December 1992 for six Erieye systems. This first series production contract also covered final development work, including the development of a new type of adaptive radar control for increased radar performance. Series production commenced in 1993, with the first delivery of two units being delivered during 1996. As of this edition, final deliveries of the Erieye system to the Swedish armed forces are understood to have taken place during 1999. The platform used for the Swedish application is the Saab 340B (service designation S100B) aircraft. Erieye has also been selected as the airborne surveillance sensor in the SIstema de Vigilância da AMazonia (SIVAM) airspace surveillance and control system for Brazil's Amazonia region. Here, the carrier aircraft will be the Embraer EMB-145 regional jet. Use of the EMB-145 replaces the proposed turboprop EMB-120/Erieye

combination that had generated concerns over take-off weight, speed, range and available space for equipment installation. Jane's sources suggest that Brazil will procure five EMB-145/Erieye combinations for use in the SIVAM programme.

Elsewhere in the world, Ericsson and Thomson-CSF DETEXIS signed a Memorandum of Understanding during March 1998 that covered co-operation on the marketing of a NATO compatible AEW and Control (AEW & C) system that utilises the Erieye radar, DETEXIS's DR 3000 electronic support measures system, Thomson-sourced avionic subsystems (thought to include communications and identification friend-or-foe equipment) and Embraer's EMB-145 airframe. The first success for this package was announced during December 1998, when Greece selected it to meet its AEW&C requirement. Under the terms of the deal, Greece is procuring four EMB-145 based systems that feature five operator stations per aircraft. As of this edition, delivery of the Greek platforms is scheduled to begin three years after contract signature and the programme, as a whole, is understood to involve extensive industrial co-operation and technology transfer between the consortium and the purchaser country.

Contractor
Ericsson Microwave Systems AB, AEW Systems Division, Mölndal.

VERIFIED

UNITED KINGDOM

Airborne STandOff Radar (ASTOR) programme

Description
The UK Ministry of Defence's (MoD) ASTOR requirement (Air Staff Requirement 925) calls for a combined airborne Synthetic Aperture Radar and Moving Target Indicator (SAR/MTI) surveillance system which will be capable of gathering near real-time intelligence in support of air-land and/or tri-service operations. The system is scheduled to have an initial operating capability in 2005. Jane's sources suggest that the system requirement includes a maximum surveillance range of between 250 and 300 km; swath (large area surveillance of fixed targets) and spotlight (small area surveillance of fixed targets with 0.5 m resolution) SAR and MTI (large surveillance of moving targets travelling at less than 10 km/h) operating modes; the ability to detect and track helicopters in flight; the ability to operate off-tether; interoperability with non-national surveillance systems and real-time or near-realtime processing. The ASTOR platform will be required to operate at an altitude of between 12,802 and 15,240 m and the system as a whole is likely to incorporate up to nine ground stations supporting up to six airborne platforms.

Status
On 15 June 1999, the UK MoD announced that its preferred ASTOR bidder was a consortium led by Raytheon Systems (RS) Ltd. At the time of the announcement, the RS-led ASTOR team comprised:
- Bombardier (green airframes)
- Cubic Defense Systems (datalink)
- BAE Systems (SAR antenna technology, signal processing, electronic counter-countermeasures provision and defensive aids suite)
- Marshall Aerospace (airframe logistic support)
- Motorola (image exploitation)
- RS Broughton (system integration on aircraft two to five)
- RS Cossor (logistic support for ground stations)
- RS Greenville Division (airframe modification, design and system integration on first aircraft)
- RS Microelectronics (radar)
- RS Ltd (programme management, system integration, defensive aids suite, logistic support and radar)
- Short Brothers plc (airframe logistic support)
- Thomson-CSF (SAR simulation and radar test facilities)
- DERA (radar processing algorithms)
- Ultra Electronics (datalink).

The consortium's chosen air vehicle is a modified Bombadier BD-700 Global Express long-range corporate transport aircraft with accommodation for a pilot, co-pilot, three-systems operators and a relief operator. The chosen radar is the Raytheon ASTOR Advanced SAR System (ASARS)-2 Plus and the proposal's Tactical Ground Stations are Steyr Pinzgauer 6 × 6 cross-country vehicle-based. At the time of the UK's 'preferred bidder' announcement, the ASTOR Advanced

An artist's impression of Raytheon's Global Express ASTOR air vehicle)

Synthetic Aperture Radar System (ASARS) - 2 Plus dual mode MTI/SAR radar was described as being a 'low risk' development of the US Air Force's (USAF) ASARS-2 equipment that was fully ASTOR specification compliant. As such, the sensor was understood to incorporate the latest USAF standard receiver/exciter (introduced during 1996), upgraded processing and a new BAE Systems developed 4.6 m long antenna. RS went on to note that from an altitude of 15,545 m and a grazing angle of 2°, ASTOR ASARS-2 Plus would offer a detection range in excess of 296 km. As of this edition, an ASTOR-type ground surveillance solution was being promoted worldwide.

UPDATED

Blue Kestrel series maritime surveillance radars

Type
Family of I-band (8 to 10 GHz) long-range surveillance radars.

Description
The Blue Kestrel surveillance radar family consists of three systems, the Blue Kestrel 5000, Blue Kestrel 6000 and the Blue Kestrel 7000. They are allied to the Seaspray radars described elsewhere. In mid-1991, GEC Ferranti (now part of BAE Systems Avionics - Sensor Systems Division) announced a new range of Seaspray and Blue Kestrel systems using commonality of modules throughout the range. This was designed to minimise cost and provides natural growth paths for the customer. Known details of the three cited Blue Kestrel variants are as follows:

Blue Kestrel 5000
Blue Kestrel 5000 is a pulse compression radar sensor developed for the UK Royal Navy's (RN) EH 101 Merlin HM Mk 1 helicopter. It consists of a large flat plate antenna and travelling wave tube transmitter to provide the Merlin with the optimum power aperture product, combined with a high-gain receiver and digital processor. The radar sensor is interfaced with and controlled by the platform's mission avionics management system via the MIL-STD-1553B databus. This highly integrated sensor approach to the radar results in a high-performance pulse compression surveillance radar in a particularly compact and lightweight four line-replacement unit configuration. It is suitable for application in a range of naval helicopters or maritime patrol aircraft.

Status
As of August 2001, delivery of 44 examples of the Blue Kestrel 5000 radar for the Merlin HM Mk 1 naval helicopter was reported as having been completed. As of June 2001, Blue Kestrel was being used to help validate the RN's next-generation anti-surface warfare tactics during NATO exercise 'Blue Game'. Here, two Merlin HM Mk 1 helicopters from 700M Naval Air Squadron were deployed to Aalborg Air Station in northern Denmark and employed their Blue Kestrel radars and Orange Reaper electronic support (see separate entry) as targeting tools during the course of the exercise.

Specifications
Frequency: I-band (8-10 GHz)
Scanner type: planar-array
Transmitter: low peak power, high mean power TWT
Pulse-widths: selectable
PRF: selectable
Coverage: 360°
System weight: 102 kg
Features: pulse compression, CFAR, multiple TWS and operator selectable scan-to-scan integration

Blue Kestrel 6000
Blue Kestrel 6000 is a private venture coherent pulse-Doppler radar sensor. It brings together BAE Systems' experience with the pulse compression Blue Kestrel

A general view of the system operator stations in the Merlin HM Mk 1 anti-submarine warfare helicopter (MS/Westland)

The antenna, transmitter and receiver/processor that make up the Blue Kestrel 7000 radar   0006908

5000, the multimode pulse-Doppler Blue Vixen, and the Inverse Synthetic Aperture Radar work which has been jointly funded by the UK Defence Research Agency (DRA) and the company. Flight trials of a modified Blue Kestrel 5000 in a DRA Sea King testbed have been successfully completed under the latter programme, producing real-time images of co-operative and 'target of opportunity' surface contacts. The resulting radar sensor provides enhanced air-to-surface and air-to-air detection, standoff classification capability and a superior anti-submarine warfare capability with sub-clutter target detection. Additional options include moving target detection and high-resolution synthetic aperture radar ground mapping.

Status
As of August 2001, the status of the Blue Kestrel 6000 radar was uncertain.

Specifications
Frequency: I-band (8-10 GHz)
Type: coherent multimode
Transmitter: TWT fixed frequency/frequency agile
Scanner type: planar-array
Coverage: 360°
LRUs: 4
System weight: 125 kg
Features: CFAR, enhanced TWT, ISAR classification, ASW and air-to-air modes
Options: MTI and SAR ground mapping

Blue Kestrel 7000
Blue Kestrel 7000 is a private venture, lightweight, modular, software driven, multimode maritime surveillance radar that is designed for naval helicopter applications. The equipment features digital pulse compression; constant false alarm rate (for sea clutter rejection); an optimised range of digitally synthesised waveforms and coherent target classification. Blue Kestrel 7000 incorporates a broadband, travelling wave tube-based power amplifier transmitter; switched mode power supplies and a relatively low electrode voltage to enhance reliability; low-noise amplification; spectral purity; commercial-off-the-shelf technology and an integrated data/signal processor and receiver package. The radar's modular software is fully documented and available as source code. Blue Kestrel 7000's operating modes comprise surface surveillance (long and short range and multiple track-while-scan); small target (small target anti-submarine warfare mode); aided target classification (Inverse Synthetic Aperture Radar (ISAR) imaging and covert range profiling; beacon/transponder (detection/interrogation of naval transponders and search-and-rescue beacons); navigation/weather (real beam ground mapping, digital coastline overlay and weather returns); air-to-air and missile (semi-active/fire-and-forget air to surface missile compatible).

Status
Blue Kestrel 7000 development was initiated following BAE Systems' completion of a UK Ministry of Defence sponsored study into coherent helicopter radar in 1991. Related ISAR implementation and motion compensation studies led to ground-based trials during 1992 and 1993/4 respectively. Taken together, the described work enabled the company to develop an interim point design ISAR module with which to upgrade the Blue Kestrel 5000 radar (see previously). During 1996, work was started on the development of a multimode processor line-replaceable unit for Blue Kestrel 7000. Engineering development of the system is understood to have been completed during 1998 and as of August 2001, the status of the Blue Kestrel 7000 radar was uncertain.

Specifications
Frequency: (8-10 GHz)
Pulse-width: selectable (optimised for mode)
PRF: selectable (optimised for mode)
Transmission modes: fixed frequency; frequency agile
Dimensions: 220 × 340 × 150 mm (scanner); 390 × 390 × 195 mm (transmitter); 500 × 390 × 195 (processor/receiver)
Weights: 27.4 kg (scanner/antenna); 28.5 kg (transmitter); 30.1 kg (processor/receiver)

Contractor
BAE Systems Avionics - Sensor Systems Division, Edinburgh.

UPDATED

Seaspray series maritime surveillance radars

Type
I-band (8 to 10 GHz) family of maritime surveillance and targeting radars.

Development
Seaspray is a family of maritime surveillance and targeting radar systems which commenced with the Seaspray Mk 1 in the early 1970s and has continued with Seaspray 2000, Seaspray 3000, Seaspray 4000 and Seaspray 7000. Allied with this family is the range of Blue Kestrel maritime surveillance radars (see separate entry). In mid-1991, GEC Ferranti (now part of BAE Systems Avionics - Sensor Systems Division) announced a new range of these systems using commonality of modules throughout both the Seaspray and Blue Kestrel ranges. This is designed to minimise cost and provides natural growth paths for the customer.

Description
Seaspray Mk 1
Seaspray Mk 1 is an I-band airborne maritime surveillance and targeting radar that was developed in the early 1970s as an integral element of the solution to a Royal Navy Staff Requirement. This requirement was raised as a result of the 1967 sinking of the Israeli destroyer *Eilat* by two Styx missile-armed fast patrol boats. The requirement demanded a light, agile shipborne helicopter that would be able to detect and neutralise long-range surface-to-surface missile-armed fast patrol boats while keeping them outside the engagement range of the helicopter's mother ship. The solution was the Westland Lynx helicopter armed with the British Aerospace Sea Skua missile. To minimise the weight of the missiles and to enable four of them to be carried, semi-active radar homing was selected for their guidance. The link was the helicopter's prime sensor and target illuminator, the Seaspray Mk 1. This lightweight radar provides the high-performance, frequency-agile detection of small targets in adverse conditions and the tenacious monopulse lock-follow target illumination for the Sea Skua. In UK service, Seaspray Mk 1 is designated as ARI 5979.

Seaspray 2000
Seaspray 2000 is an I-band maritime surveillance radar that has been optimised for civil surveillance operations. A range of different sized, full colour combined control and display units provides the man/machine interface. Control of this highly automated radar is via simple on-screen menus. Output ports are provided for hard

A GKN Westland Super Lynx of the South Korean Navy equipped with Seaspray 3000

copy records and video, and one or more displays can be provided as control stations or repeaters. Superior contact detection in adverse conditions is provided by the high peak power, frequency-agile transceiver, constant false alarm rate processing, operator selectable scan-to-scan integration, and a range of optional features. Options include an operator selectable side-looking array radar for ground mapping and pollution control, operator selectable circular polarisation for enhanced performance in precipitation, and multiple target track-while-scan. The display presentation is enhanced by a comprehensive set of synthetic overlay facilities, including digital coastlining, customer defined waypoints and operator defined variable way points.

Seaspray 3000
Seaspray 3000 (previously known as Seaspray Mk 3) introduces significant advances over Seaspray Mk 1, most notably through the provision of a digital signal processor, a revised man/machine interface, and full 360° scan. The system is designed primarily for operation in light, agile, shipborne naval helicopters to provide detection and tracking of small targets in adverse conditions for the Sea Skua missile. For this purpose, it retains the combat-proven monopulse lock-follow target illumination of its predecessor. Consisting of six Line-Replaceable Units (LRUs), Seaspray 3000 is configured to operate on a MIL-STD-1553B databus, with multiple additional standard interfaces being provided. The resultant lightweight, compact and flexible system is applicable to both retrofit and new aircraft. Operating in I-band, Seaspray 3000 uses a high transmitted power and very high speed agility to provide these platforms with high detection performance in sea and weather clutter, and in electronic countermeasures conditions. The digital processor provides advanced features, including constant false alarm rate and track-while-scan facilities to optimise target detection and reduce operator workload. The operator is provided with a comprehensive tactical situation display of scan-converted television format, in monochrome or colour. In addition, the output of other sensors (Forward-Looking Infra-Red (FLIR), electronic support measures, datalink and so on) may also be displayed. The control unit employs a menu structure with soft key operation.

Seaspray 4000
Seaspray 4000 was a private venture proposal for a pulse compression airborne maritime surveillance radar that would be capable of operating as a stand-alone equipment or as the heart of a fully integrated avionics suite. The proposal combined the proven man/machine interface and processing of Seaspray 3000 with a pulse compression front end. Comprising six LRUs, Seaspray 4000 was designed to provide the optimum radar performance for medium-size maritime patrol aircraft and naval helicopters.

The Seaspray Mk1 radar installed in the nose of a Lynx helicopter

Specifications: Seaspray series maritime surveillance radars

	Seaspray Mk 1	Seaspray 2000	Seaspray 3000
Frequency	I-band (8-10 GHz)	I-band (8-10 GHz)	I-band (8-10 GHz)
Peak power	90 kW	90 kW	90 kW
Transmitter	high speed, spin tuned magnetron	frequency agile magnetron	high speed, spin tuned magnetron
Pulse-widths	2 (selectable)	2 (selectable)	2 (selectable)
PRF	3 (selectable)	4 (selectable)	4 (selectable)
Scanner type		front-fed, elliptical section, double curved paraboloid	
Coverage	180°	360° (selectable 60° and 180° sector scans)	360°
LRUs	5	4	6
Weight	75 kg	80 kg	90 kg
Target illumination	monopulse lock-follow		monopulse lock-follow
Features		CFAR and operator selectable scan-to-scan integration	CFAR and operator selectable scan-to-scan integration
Options		operator selectable SLAR; circular polarisation; multiple track-while-scan and coastlining; antenna size (up to 1 m diameter); display size (up to 0.35 m diagonal)	operator selectable circular polarisation; antenna size (up to 1 m diameter); display types

Finnish Frontier Guard Dornier 228 — 212 patrol aircraft are fitted with Seaspray 2000 radars (MS/BAE Systems)

Seaspray 7000

Seaspray 7000 is a lightweight, multimode, surveillance radar that has been specifically designed for installation aboard the GKN Westland Super Lynx 300 naval helicopter. The iteration comprises a two axis stabilised antenna assembly, an I-band (8 to 10 GHz) transmitter, a receiver, a processor and a full colour display. Overall system weight is given as being less than 85 kg and the equipment offers surface surveillance, small target, target classification (inverse synthetic aperture and covert range profiling), beacon/transponder, navigation/weather and missile operating modes. Options include air-to-air detection, synthetic aperture ground mapping and ground moving target indication.

Status

Over time, in excess of 300 Seaspray Mk 1 radars are reported as having been installed aboard British, Brazilian, Danish, Dutch, German and Norwegian GKN Westland Lynx naval helicopters. The Royal Navy is also noted as operating a land-based version of Seaspray Mk 1 as a check on the effectiveness of shipborne electronic warfare equipment. Seaspray Mk 1 is also understood to have been delivered to South Korea (for land-based trials) and to Pakistan in ex-Royal Navy Lynx helicopters. The Lynx/Seaspray Mk 1/Sea Skua combination is noted as having been used operationally in the South Atlantic in 1982 and in the Arabian Gulf during 1991. As of late 2000, Seaspray 2000 was reported as having been procured for installation aboard Dornier 228 patrol aircraft of the Finnish Frontier Guard and a UK Government Agency together with Norwegian Super Puma helicopters. In the Finnish Dornier 228 application, Seaspray 2000 is fully integrated with the BAE Systems' multirole turret system FLIR. For its part, Seaspray 3000 is noted as having been installed aboard (or mandated for installation aboard) GKN Westland Super Lynx helicopters of the German and Republic of Korea navies. Elsewhere, the Brazilian navy is understood to have launched a Seaspray Mk 1 to Seaspray 3000 standard radar upgrade programme and the equipment is also noted as having seen service in a UK government F27 aircraft and as having been proven in Sea Skua coastal battery and fast patrol boat applications (Seaspray 3500 in Kuwait).

Contractor

BAE Systems Avionics - Sensor Systems Division, Edinburgh.

UPDATED

UNITED STATES OF AMERICA

Advanced Synthetic Aperture Radar System 2 (ASARS-2)

Type

High-resolution, ground-mapping, imaging radar.

Description

ASARS-2 is a ground-mapping, imaging radar system that is installed aboard high-altitude U-2S reconnaissance aircraft operated by the US Air Force (USAF). It is described as being able to produce real-time, high-resolution imagery in 'all weathers and at ranges well in excess of those provided by electro-optic systems'. ASARS-2 consists of airborne collection and ground-based processing elements, with the former comprising an antenna array, a liquid cooling system, a heat exchanger, a cockpit 'on-off' system control/status unit, a transmitter, a receiver/exciter, a power control unit and a low-voltage power supply. Aboard the U-2S, this equipment is installed in a dedicated nose configuration (housing the system's antenna, heat exchanger, liquid cooling system, transmitter and receiver/exciter - see accompanying illustration), the aircraft's cockpit (the 'on-off' system control/status unit) and its 'Q' bay (the power control and low-voltage power units). Outside the U-2S application, Raytheon has also developing the ASTOR ASARS-2 Plus

The transmitter unit used in the baseline ASARS-2 radar (MS/Hughes)

The dedicated ASARS-2 nose fairing is 0.76 m longer than the U-2's standard installation. The prominent blister on the front of the structure houses the radar's heat exchanger (Martin Streetly)

variant of ASARS-2 for use in the successful Raytheon Systems Ltd bid for the UK's Airborne STandOff Radar (ASTOR - see separate entry) programme. ASTOR ASARS-2 Plus incorporates the latest USAF standard ASARS-2 receiver/exciter together with upgraded processing and a new, BAE Systems developed, 4.6 m long antenna array. Alongside Raytheon and BAE Systems, the ASTOR ASARS-2 Plus development team is understood to include the UK's Defence Evaluation and Research Agency (contributing processing algorithms). ASARS-2 is understood to be able to download imagery into the USAF's Contingency Airborne Reconnaissance System (CARS) and Tactical RAdar Correlator (TRAC) facilities together with the Anglo-American TR-1 ASARS Data Manipulation System (TADMS).

Status

As of this edition, the ASARS-2 imaging radar system was reported as being in service aboard U-2S aircraft of the USAF's 9th Reconnaissance Wing (Beale Air Force Base, California) and as forming the basis of the radar that is being used in the UK's ASTOR programme. Within the USAF, ASARS-2 is understood to have become operational in 1985 with development of an ASARS-2 Ground Moving Target Indication (GMTI) capability beginning during 1988. In June 1996, the USAF launched an ASARS-2 Improvement Programme (ASARS-2 IP) that Jane's sources suggest will offer real-time processing, enhanced broad area synoptic coverage, an operational GMTI capability (possibly including GMTI search and spot modes) and a measurement intelligence 'complex imaging' facility. The latest standard ASARS-2 receiver/exciter is thought to have been introduced during 1996 and as of the given edition, phase one (validation of legacy modes) of the ASARS-2 IP was reported as having been completed in January 1999. During the following March, Mercury Computer Systems was noted as having received a then year US$11 million contract covering the supply of RACE multi-computer systems for use in the ASARS-2 IP standard equipments. As of September 1999, Jane's sources were suggesting that the first ASARS-2 IP configured radar would be delivered to the USAF during the autumn of 2000.

Contractor

Raytheon Electronic Systems, El Segundo, California.

VERIFIED

AIRborne Surveillance and Target Acquisition Radar (AIRSTAR)

Type
J-band (10 to 20 GHz) tactical surveillance radar for light fixed-wing aircraft and helicopter applications.

Description
AIRSTAR is a lightweight, motion compensated, moving target indicator radar that operates in the 16 to 16.5 GHz sub-band and is intended for battlefield surveillance and target acquisition, mounted primarily aboard UAVs. It uses a pulsed, phase-coherent linear frequency modulated waveform to measure accurately motion in the illuminated region. It automatically separates targets from ground and other clutter and detects them. Sanders states that a UAV equipped with this type of radar is able to report threat targets over a 360° azimuth to an accuracy on the ground within a probable 50 m circle of error. The equipments moving target indicator subsystem has an area coverage of 658×10^5 m^2/s as compared with typical optical or synthetic aperture radar systems for 4×10^5 m^2/s. This means a search pattern of 656 km^2 in 10 seconds (with less than one false alarm) and the ability to record as many as 750 targets per scan whose observed velocity is greater than 1 m/s. Ranges of 6 km against a moving person and 17 km against a large vehicle are claimed.

The radar system is designed for use in light aircraft and helicopters and in addition to battlefield surveillance, can be used for border surveillance, reconnaissance, ground and air interdiction and air search and rescue. As a result of real-time data processing by the radar on board the air vehicle, as many as 100 target reports per second can be downlinked on a narrow data channel of 38.4 kHz.

Status
As of this edition, an AIRSTAR prototype was being reported as having been flight tested aboard a Bell 206C helicopter during July 1992.

Contractor
Lockheed Martin Naval Electronics and Surveillance Systems - Syracuse, Syracuse, New York.

VERIFIED

The AIRSTAR prototype during its initial flight test programme

AN/APN-215(V)/APN-234 multimode radar

Type
I-band (8 to 10 GHz) weather and navigation radar.

Description
AN/APN-215(V) is a lightweight, airborne, digital colour display, multimode radar, which is derived from the commercial RDR-1300B equipment. It is designed to provide weather detection and terrain-mapping for a variety of military aircraft, including rotary- and fixed-wing types from light to heavy twins. When equipped with a sea search function, APN-215(V)'s designation changes to AN/APN-234. In both configurations, the radar consists of three units: receiver/transmitter, combined colour display/control unit and stabilised antenna (antenna drive and antenna array). Weather (levels of precipitation) is shown on the colour display in red, yellow and green. With the use of an auxiliary interface unit, pictorial navigation data from other onboard sensors (such as inertial navigation, Omega or very high frequency omnidirectional radio range systems) can be superimposed on the weather or terrain display. When used with a checklist control unit, the system also provides a pilot-programmable alphanumeric display presentation of operational checklists and other pertinent data.

Status
As of this edition, the AN/APN-215(V) multimode radar was reported as having been procured by the US Army and the US Coast Guard. The AN/APN-234 radar is noted as having been installed aboard US Navy C-2A and EP-3E aircraft.

AN/APN-234 multimode radar

Specifications
APN-215(V)
Frequency: I-band (9,345 ± 40 MHz)
Power output: 10 kW peak (nominal)
Range: 445 km
Blanking pulse output: 5 V into 90 Ω
Antenna: 30.5 cm; 45.7 cm or 30.5 × 45.7 cm (flat plate)
Antenna scan angle: 120° (±60°)
Weight: 16 kg (nominal); 17.8 kg (with auxiliary interface unit)

Contractor
Honeywell, Morristown, New Jersey.

VERIFIED

AN/APN-241 combat aerial delivery radar

Type
I-band (8 to 10 GHz) navigation and weather radar for military tanker/transport.

Description
AN/APN-241 is a lightweight, fully coherent pulse-Doppler radar which is based on the Northrop Grumman AN/APG-66 fire-control radar. It was developed to provide precision airdrop/navigation radar capabilities for military tanker and transport aircraft. The baseline APN-241 configuration has five radar modes: weather/turbulence, predictive windshear, high-resolution ground map, skin paint and beacon. The system also has three display modes: station keeping, flight plan and traffic collision avoidance system. The system can interleave radar modes, allowing the pilot and navigator to view and control separate modes simultaneously. It is capable of accommodating a crew of two or three. The open system architecture of the APN-241 will allow advanced functions such as synthetic aperture, terrain-avoidance and head-up display to be added without hardware modification.

Status
As of February 2001, in excess of 230 AN/APN-241 radars are known to have been procured worldwide. Identified applications include C-130H transport aircraft flown by the air forces of America (entered service during November 1993), Australia (12 aircraft) and Portugal (six) and seven HS.748 navigation trainers operated by the Royal Australian Air Force (APN-241NT variant). Elsewhere, APN-241 is noted as being the baseline radar for the C-130J transport aircraft and as of January 1999, Northrop Grumman was reported as proposing an APN-241B radar (with a 56 cm antenna) for the C-17. On 4 December 2000, Northrop Grumman announced that it had been awarded a then year US$7.7 million contract covering the supply of 24 APN-241 radars for installation aboard USAF C-130H aerial delivery aircraft. At the time of the announcement, the programme was scheduled for completion by 30 June 2002, with the USAF's Warner Robins Air Logistics Center (Robins Air Force Base, Georgia) acting as its contracting activity. As of the given date, Northrop Grumman further noted that it had already supplied the USAF with 120 examples of the APN-241 for use on C-130H aircraft certified for the service's Adverse Weather Aerial Delivery System (AWADS) mission.

Specifications
Frequency: I-band (9.3-9.5 GHz)
Power output: 9.5 W (average); 116 W (peak)
Coverage: −25 to +10° (elevation); ±135° (azimuth in 4 selectable widths)
Weight: 28 kg (electronics unit); 30.8 kg (antenna)
Antenna size: 66 × 80 cm
Range: 515 km
MTBF: >1,000 h

Contractor
Northrop Grumman Electronic Systems sector, Baltimore, Maryland.

UPDATED

AN/APN-242 multimode radar

Type
I-band (8 to 10 GHz) airborne search, weather, mapping and beacon homing radar.

Description
The AN/APN-242 multimode radar is a colour magnetron equipment that is designed as a form, fit and function replacement for the AN/APN-59 sensor aboard (primarily) Lockheed Martin C-130 and Boeing C/KC-135 series aircraft. It comprises a flat-plate antenna assembly, an antenna interface box, a solid-state transceiver, a video processor, a navigator's control panel, a pilot's display and a navigator's azimuth/range indicator. The system's displays are night vision goggles compatible when operating in green mode and its principal operating modes are search, skin painting (aircraft detection), terrain mapping/navigation, weather mapping (at ranges of up to 445 km) and beacon/identification friend-or-foe system interrogation (at ranges of up to 185 km). The operator can choose pencil or fan beams with a variety of pulse-lengths and pulse repetition rates. The system's flat-plate antenna can scan the full 360° arc or a 90° forward facing sector and the equipment as a whole incorporates built-in test.

Status
As of May 2001, the AN/APN-242 multimode radar was reported as being in production for the US Air Force during 2000 and as being applicable to a range of aircraft including the AC-130, C-130E, C-130H, KC-135R, MC-130P and RC-135V/W.

Specifications
Frequency: 9,375 ± 10 MHz (radar); 9,310 MHz (beacon reception)
Transmitted power: 25 kW (nominal peak)
Receiver noise figure: 6.5 dB (nominal)
Ranges: 4.6-37 (in 3.7 km increments), 46, 56, 93, 185 and 445 km
Pulse-length: multiple (0.20, 0.8, 2.35 and 4.5 µm - automatically selected for different ranges/functions)
Scanning features: 90° (forward facing sector); 360° (at 12 rpm for long-range functions and 45 rpm for short-range functions)
Antenna beam selection: instantaneous electronic switching between pencil or fan (both 3° azimuth beamwidth)
Antenna stabilisation: ±15° (pitch); ±30° (roll)
Navigation features: manual electronic cursor (with latitude/longitude or range/bearing readouts); fly-to waypoint (latitude/longitude stabilised)
Interfaces: MIL-STD-1553B; ARINC; synchro (vertical gyro, INS and compass); IFF pulse synchronisation
Power consumption: 800 W
MTBF: > 1,000 h
Dimensions: 152 × 254 × 95 mm (navigator's control unit); 165 × 140 × 272 mm (antenna interface unit); 165 × 165 × 305 mm (pilot's indicator); 197 × 197 × 324 mm (video processor); 216 × 254 × 323 mm (navigator's indicator); 387(Ø) × 384(h) mm (transceiver); 908 × 908 × 864 mm (antenna)
Weight: 2.5 kg (navigator's control unit); 3.2 kg (antenna interface unit); 5.4 kg (pilot's indicator); 9.5 kg (navigator's indicator); 11.3 kg (video processor); 29.5 kg (transceiver); 31.4 kg (antenna)

Contractor
Northrop Grumman Electronic Systems sector - Sperry Marine, Charlottesville, Virginia.

UPDATED

AN/APQ-122(V) multimode radar

Type
I-/low J-band (8 to 12 GHz) and K-band (20 to 40 GHz) navigation and mapping radar.

Description
AN/APQ-122(V) is a dual-requency radar that was originally developed for use in the US Air Force's (USAF) C-130E Adverse Weather Aerial Delivery System (AWADS) programme. It is used for weather avoidance and navigation in supply dropping missions. The equipment provides ground mapping out to more than 370 km, weather information up to 278 km and beacon interrogation up to 445 km when using the I-band frequency radar. K-band frequencies are used when short-range high-resolution performance and target location are required. In the K-band role, the radar provides a high-resolution ground map display to permit target identification and location for position fixing and aerial delivery missions. In this situation, the radar detects and displays a target with a radar cross-section of 50 m² while operating in a rainfall environment of 4 mm/h. In addition to the dual-frequency APQ-122(V)1 AWADS variant, three other APQ-122 configurations have been identified. Of these, APQ-122(V)5 is an I-band (8 to 10 GHz) single frequency system while the APQ-122(V)7 is a navigation training variant of the APQ-122(V)5 that is installed aboard USAF T-43A training aircraft. APQ-122(V)5 was designed as a direct replacement for the AN/APN-59 radar installed aboard C-130E and E-4B aircraft and is described as offering long-range mapping, weather evaluation and avoidance and beacon rendezvous capabilities. The remaining identified family

member - APQ-122(V)8 - is a dual-frequency equipment that incorporates terrain-following and terrain-avoidance capabilities. In addition to these (together with the APQ-122's standard ground-mapping, weather information and beacon interrogation modes), APQ-122(V)8 provides a cross-scan facility. In the terrain-following mode, the radar supplies commands to fly the aircraft at a fixed distance above the terrain, while in the terrain-avoidance mode, the radar displays all terrain at and above the altitude of the aircraft. The cross-scan mode combines the terrain-following and avoidance functions on a time-shared basis. The APQ-122(V)8's K-band subsystem supplies a high-resolution map display for target identification and location. In the terrain-following, terrain-avoidance and cross-scan modes, use is made of a small, circular antenna that is connected to the radar's I-band transceiver. The full set of APQ-122(V)8 operating modes is as follows:

- Terrain-Following (TF)
- Terrain-Avoidance (TA)
- terrain-following/terrain-avoidance (CS)
- long-range ground map
- weather detection
- beacon
- Precision Ground Map (PGM)
- simultaneous TF/PGM, TA/PGM, CS/PGM

Status
Over time, in excess of 300 AN/APQ-122(V) series radars are reported as having been procured by the US Air Force and the air arms of Argentina, Australia, Bolivia, Cameroon, Congo, Denmark, Ecuador, Gabon, Greece, Indonesia, Iran, Israel, Italy, Jordan, Libya, Malaysia, Morocco, Nassau, New Zealand, Niger, Nigeria, Oman, Philippines, Portugal, Saudi Arabia, Singapore, Spain, Sudan, Thailand, Venezuela and Zaïre. Within the USAF, APQ-122(V) radars are noted as having been installed aboard C-130E AWADS, C-130H, T-43A and MC-130E aircraft. With regard to the latter type, Raytheon was awarded a then year US$8.9 million contract during July 1996 that covered the upgrading of the transceivers fitted to 37 MC-130E applicable APQ-122(V)8 radars.

Contractor
Raytheon Electronic Systems, McKinney, Texas.

VERIFIED

AN/APQ-158 terrain-following radar

Type
Terrain-following radar.

Description
The AN/APQ-158 radar is a multimode forward-looking equipment that is used primarily for terrain-following/terrain-avoidance at low altitudes in the MH-53J Pave Low III Night/Adverse Weather Search and Rescue helicopter. The equipment is similar to the AN/APQ-126, but is modified for compatibility with unique helicopter characteristics and Pave Low III mission requirements. The radar contains 15 line-replaceable units that provide the same basic 10 modes of operation as detailed for the AN/APQ-126. System enhancements have provided this radar with the capability to supply updates in all modes except Terrain-Following (TF) and to perform TF missions over very high clutter areas such as cities.

Status
As of this edition, the AN/APQ-158 TF radar was reported as being installed aboard US Air Force MH-53J helicopters.

Contractor
Raytheon Electronic Systems, McKinney, Texas.

VERIFIED

The AN/APQ-158 airborne multimode radar

AN/APQ-164 multimode radar

Type
I-band (8 to 10 GHz) multimode terrain-following, navigation and weapon delivery radar.

Description
The AN/APQ-164 is the radar installed in the US Air Force (USAF) B-1B aircraft. This radar combines technology from the F-16 AN/APG-68 radar and the Electronically Agile Radar (EAR) programme of the US Air Force. As such, the equipment generates data for navigation, penetration, weapon delivery, and for certain other functions such as air refuelling. There are four modes in the AN/APQ-164 system that provide the navigation capability. The primary mode is a high-resolution synthetic aperture radar mapping mode, backed up by a monopulse enhanced real beam ground-mapping mode. The system also detects weather ahead and can display ground beacon returns over a real beam image. The penetration functions of the radar include automatic terrain-following and terrain-avoidance. For weapon delivery, the radar provides four different functions. The first is a velocity update mode, similar to a Doppler navigator, which generates velocity information for the inertial navigation system. Coupled with an accurate Global Positioning System receiver in the avionics system, velocity update produces a dynamic, precision antenna calibration correction. Second, there is a ground moving-target detection and tracking capability for both fast and slow moving vehicles. Third is a high-altitude altimeter function that provides a very accurate measure of local height above the ground. Fourth is a monopulse targeting mode that provides accurate height to the on-scene selected fixed target.

The synthetic aperture mode provides the operator with a high-resolution image of an area of ground that can be chosen by the avionics system or the operator. Long-range maps can be made and five different map scales displayed. The synthetic aperture mapping mode accepts the co-ordinates of a waypoint from the avionics system and makes a map centered on that point. To make an image, the antenna is electronically scanned to the waypoint location. The radar transmits a train of pulses, gathers data for the image, and then switches itself off. At the same time, the image is stored in the radar and presented on the display in a rectangular, ground co-ordinate display.

The radar provides the basic data required for automatic terrain-following. It scans the ground in front of the aircraft and measures the terrain in a range versus height profile out to 19 km and stores that data in the computer. The profile data is sent across the multiplex bus to the terrain-following control unit where the data is used to generate climb/dive commands. This flight profile is then automatically fed into the pilot's flight control system. Since the radar is not continuously scanning in terrain-following, a very low update is used, helping to reduce the risk of detection. This rate is variable and depends on aircraft altitude, manoeuvres, groundspeed and terrain roughness. Under normal conditions, updates are made at 3 to 6 second intervals. However, if the terrain demands it, data can be gathered continuously.

The AN/APQ-164 in the B-1B is a dual-redundant system, with two complete and independent sets of Line-Replaceable Units (LRUs), except for the phased-array antenna. This was the first airborne application of this technology for combat aircraft. Only one set of LRUs is used at a time, the other being maintained on standby.

The phased-array is an outgrowth of the antenna developed on the EAR programme. It contains 1,526 phase control modules and allows virtually instantaneous beam movement to any point in the antenna field of regard. When the radar mission requires a forward, right or left region of regard, the antenna is physically movable to three different positions on a roll detent mount. The radar can, therefore, look off to either side of the aircraft or forward by rolling the antenna about an axis. The normal antenna position is looking forward. However, when the antenna is rolled to one side, the field of view extends from the aircraft nose back to about 115°, permitting a look off to the side of interest without having to change aircraft heading. Once physically moved to one of the three available positions, the antenna is locked into a detent. From the fixed spot, it can be scanned electronically ±60° in azimuth and elevation by means of a unit on the antenna called the beam steering controller, which controls all 1,526 phase control modules.

Status
As of December 2001, the AN/APQ-164 multimode radar was reported as being installed aboard USAF B-1B Lancer strategic bombers.

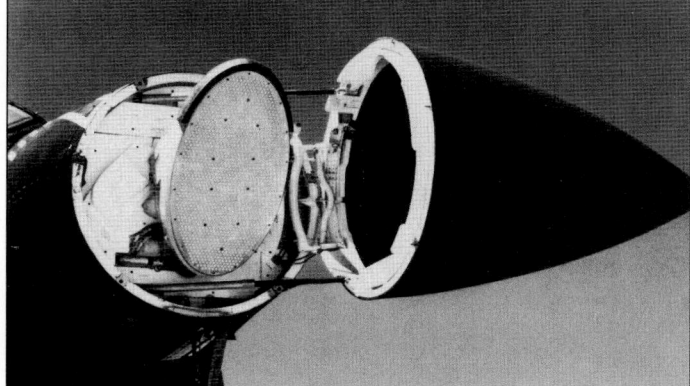

The AN/APQ-164 airborne radar

Specifications
Frequency: I-band (8-10 GHz)
Transmitter: gridded, multiple, peak-power TWT (similar to the AN/APG-68 transmitter)
Antenna: phased-array electronically scanned, 1,118 × 559 mm
Operating modes: (air-to-ground) high-resolution mapping, monopulse enhanced real beam mapping, automatic terrain-following, manual terrain-avoidance, velocity update, ground moving target detection and track, high-altitude calibrate, ground beacon; (air-to-air) weather mapping, air-to-air beacon, rendezvous mode. Growth for full conventional standoff capability and a full air-to-air mode complement is provided.
Weight: 570 kg

Contractor
Northrop Grumman Electronic Systems sector, Baltimore, Maryland.

UPDATED

AN/APQ-170 multimode radar

Type
Multimode terrain-following/avoidance radar.

Description
APQ-170 is the dual-band, forward-looking, terrain-following/avoidance radar that is fitted to the US Air Force's (USAF) MC-130H Combat Talon II special forces support aircraft. The equipment features a high level of redundancy and integrates the core terrain-following/avoidance capability with ground-mapping, weather detection/avoidance and beacon interrogation modes. APQ-170 is claimed to be unique in its ability to provide straight and turning flight cues for climb-limited aircraft and has the capacity to provide radar data for up to four displays in an integrated glass cockpit.

Status
As of October 2001, Jane's sources were reporting APQ-170 as being installed aboard the 24 MC-130H Combat Talon II aircraft assigned to the USAF's Special Operations Command. During December 1997, Systems & Electronics Inc was awarded a US$16,698,561 contract covering interim contractor support of APQ-170 at two US and two offshore bases, system repair services and an upgrade programme to bring APQ-170 up to AN/APQ-425 standard. This latter effort may be aimed at improving the radar's performance in mountainous terrain and enhancing its overall reliability.

Contractor
Systems & Electronics Inc, St Louis, Missouri.

UPDATED

AN/APQ-174/-186 MultiMode Radar (MMR) systems

Type
Terrain-following/avoidance and mapping radar systems.

Description
The AN/APS-174 MMR is a derivative of the radar used in the AN/AAQ-13 Low Altitude Navigation and Targeting Infra-Red for Night (LANTIRN) navigation system and maintains commonality with five of that radar's six Line-Replaceable Units (LRU). It is intended for combat rescue and special operations in aircraft such as the HH/MH-60, CH/MH-47, HH/MH-54 and the V-22. APS-174 provides its host platform with terrain-following, terrain-avoidance, ground mapping, weather direction, beacon interrogation and air-to-ground ranging facilities and its manufacturer describes it as offering increased countermeasures resistance, improved weather penetration, better guidance in turning flight, a power management function for semi-covert operation and low beam reflectivity. MMR also incorporates extensive periodic and initiated built-in test capabilities. The AN/APQ-186 MMR is a derivative of the APQ-174 radar that has been developed for special operations applications on the CV-22 aircraft. As such, APQ-186 is noted as incorporating an enhanced terrain-following capability at low velocity (down to

The pod shell, RIU and pressurisation unit used in the AN/APQ-174 radar

speeds of 9 km/h) together with azimuth monopulse processing to facilitate finer resolution ground mapping. Both APQ-174 and APQ-186 are mounted in a pod that is located on the forward fuselage of the host aircraft and contains a gimbal-mounted antenna, a transmitter, an exciter/receiver and a power supply LRU. The system's radar interface and pressurisation units are separate LRU assemblies.

Status

Raytheon Electronic Systems (formerly Texas Instruments) received funding during 1990 with which to initiate production of the APQ-174B radar for the US Army's Special Operations Aircraft programme. Delivery of 54 systems was completed in August 1994. Flight qualification testing of the equipment began during early 1994 and was completed during 1996. The radar provides the MH-47E and MH-60K helicopters with the ability to operate at low altitudes (down to 30m above ground) by day or night, in adverse weather and high threat environments. In mid-1994, Raytheon Electronic Systems had delivered three APQ-174C radars and two sets of spare LRUs to the US Special Operations Command (USSOCOM) for the US Air Force's CV-22 special operations programme. During 1996, the company was funded for an Engineering and Manufacturing Development (EMD) programme to develop the APQ-186 radar for the CV-22 special operations aircraft.

Specifications

Set clearances: 31, 46, 61, 91 and 152 m
Turning capability: 5.5°/s turn rate
Weather performance: up to 10 mm/h rain
Pod dimensions: 33 × 108 cm
RIU dimensions: 75.4 × 21.5 × 33.7 cm
Weight: 113.4 kg
Power: 115 V AC (400 Hz/2,100 VA); 28 V DC (38 W)
MTBF: 144 h (specified)

Contractor

Raytheon Electronic Systems, McKinney, Texas.

VERIFIED

....

AN/APQ-175 multimode radar

Type
Dual-band (I- (8 to 10 GHz) and K- (20 to 40 GHz) band) multimode radar.

Description
The AN/APQ-175 multifunction radar forms part of the US Air Force (USAF) Adverse Weather Aerial Delivery System (AWADS) that has been installed on a percentage of the service's C-130E transport aircraft fleet. As such, APQ-175 provides a precision navigation capability in conditions of zero visibility and/or darkness and is able to identify terrain features such as rivers, bridges, airfields and mountains. Additionally, the equipment can detect weather cells and provides weather contouring to identify areas of light, moderate or heavy rainfall. Within the AWADS package, APQ-175 operates in conjunction with the AN/APN-169C formation positioning set, the AWADS self-contained navigation system and the PPN-19/SST-181 radar beacon. Basic functional modes are Long-Range Ground Mapping (LRGM), Precision Ground Mapping (PGM), Weather Detection (WD), Beacon Interrogation/Reception (BIR) and Beacon/Precision Ground Map Overlay (BPGMO). Over time, APQ-175 has been enhanced with a covert operating capability and is available in the I-band APQ-175V(X) configuration for applications where there is no requirement for precision ground mapping.

Status
According to Jane's sources, AN/APQ-175 was first fielded during 1988. On 7 November 2001, Engineered Support Systems Inc (of which, Systems & Electronics Inc is a subsidiary) announced that it had been awarded a then year US$18.2 million, five-year duration, depot level repair contract with regard to the USAF's inventory of APQ-175 AWADS radars.

Specifications
Frequency: 8-10 GHz (APQ-175V(X)); 8-10 and 20-40 GHz (APQ-175)
Azimuth coverage: ±145° (APQ-175)
Ranges: up to 56 km (APQ-175 BPGMO and PGM modes); up to 440 km (APQ-175 BIR, LRGM and WD modes)
Field reliability: > 160 h

Contractor
Systems & Electronics Inc, St Louis, Missouri.

UPDATED

....

AN/APQ-181 multimode radar

Type
J-band (10 to 20 GHz) multimode radar for the B-2 strategic bomber.

Development
Hughes (now part of Raytheon Electronic Systems) is reported to have begun work with Northrop (now Northrop Grumman) in the early 1980s to develop radar

concepts for what was then known as the Advanced Technology Bomber. The prototype radar first flew in 1987 and the first AN/APQ-181 production contract was proposed in 1986, authorised in 1987 and fully negotiated in 1988. Ten pre-production radars are understood to have been delivered during the early 1990s and the system is thought to have been in full scale production by 1995.

Description
AN/APQ-181 is a multimode radar system for the B-2 strategic 'stealth' bomber. It operates in J-band (12.5 to 18 GHz) and most of the technical information remains classified. The important and unique feature of the various modes of the AN/APQ-181 is their design for Low Probability-of-Intercept (LPI) operation. The LPI approach used is understood to be a collection of individually effective design and operating techniques that, when integrated, greatly diminish the effectiveness of radio frequency emitter location sensor systems. LPI techniques include unique performance features built-in to the hardware and operated under radar mode control.

Specified radar performance requirements are organised by operating modes. Radar modes enable and support operation of the aircraft in its various missions sequences. For example, the B-2 is capable of autonomous navigation, without the aid of the Global Position System or any other external navigation aid, from base to target and return. The radar has operating modes such as precision position and velocity update measurement to enable high-accuracy inertial navigation. A related mode provides altitude measurement.

AN/APQ-181 also performs vital functions during penetration of defended hostile airspace. The various radar modes available during this phase of a mission provide information about natural (terrain) and manmade hazards to flight. Radar-made images and measurements enable the B-2 to navigate safely over and around hazards while using the same hazards to its advantage to mask defensive systems sensors and to minimise the possible time of observation by these sensors.

During the target search, acquisition, identification and attack phase of a B-2 mission, the radar operates in a variable resolution Synthetic Aperture Radar (SAR) mode, to locate and identify assigned targets precisely. The radar also has other modes and submodes in this mission phase. Typically in the SAR mode, the radar makes a topographic map-like, high-quality image at resolutions that are independent of radar range and provides them to the crew for evaluation. Not only are these radar modes useful for locating targets, they also confirm the accuracy of weapons delivered previously by other vehicles.

APQ-181 has 21 distinct operating modes, including two versions of built-in-test for fault detection and isolation to a Line-Replaceable Unit (LRU), or in the case of the antenna a line-replaceable module. Each mode has its own software programme, which is an assembly of functional code modules that provides detailed radar instructions. In fact each mode has two sets of software; one for the Radar Data Processor (RDP), which is a militarised general purpose computer and one for the Radar Signal Processor (RSP), which is a high-speed special purpose computer. The RSP is programmed in high-efficiency machine language to maximise signal processing throughput.

In addition to the signal and data processors, there are three additional units in a 'radar string': antenna, transmitter and receiver/exciter. There are two of each unit arranged in a string in a B-2 aircraft set. The strings are functionally connected so that any antenna may be connected to either of the four units in a string. This approach to redundancy maximises the radar mission probability of success.

The radar equipment is installed in the aircraft in three zones. Each antenna is mounted in a cavity behind a large radome, approximately 2.4 m outboard of the aircraft centre line and below the leading edge of the wing/body. Each antenna has a large unobstructed field of view forward and to the side of the aircraft fuselage reference line. Six units (two each of the transmitter, receiver and RSP units) are located symmetrically in openings in each side wall of the nose wheel well; the two RDPs are located one above the other in an opening in the aft wall of the nose wheel well.

The antenna is electronically steered in two dimensions and features a monopulse feed design to enable fractional beamwidth angular resolution. It includes a beam steering computer that determines and commands the phase settings of the beam steering phase shifters in response to a pointing direction command from the RDP. The antenna is fitted with a motion sensor subsystem, which is a modified strap-down inertial platform used to measure antenna motion to enable compensation during SAR mode operation. The antenna is equipped with its own power supply and is liquid cooled. It is designed to have carefully controlled and very low scattering performance (low radar cross-section) with respect to both in- and out-of-band Radio Frequency (RF) illumination.

The radar transmitter is a single unit and includes its own high voltage power supplies within its chassis. It employs a gridded travelling-wave tube RF amplifier, similar to other Hughes radars. Because it is such a high-power density package it is liquid cooled.

The receiver/exciter LRU performs several functions usually requiring more than one LRU in contemporary radars. These include generating RF waveforms for amplification by the transmitter (exciter) and amplification, detection and frequency down-conversion to baseband (receiver) of signals received from the antenna. The receiver/exciter also digitises the received signal stream and performs pulse compression to enhance range resolution.

The radar signal processor extracts target images and measurement information from the digitised signal stream and converts this information into a format usable by interfacing avionics or displays. In addition to a digital data output bus, it has also a video output bus that enables direct drive of cockpit displays. The RSP is fully programmable.

The radar signal processor is a dual central processor unit general purpose type computer. It is the command controller for all radar units and serves as the radar terminal on the B-2 avionics databus. It has a throughput of 2.5 Mips and a one million, 16-bit word (16 Mbits) bulk memory.

Specifications
Number of LRUs in one aircraft set: 10
Total weight of an aircraft set: 955 kg
Total volume of an aircraft set: 1.485 m³
Operating prime power demand: 22 kVA AC; 500 W DC

Contractor
Raytheon Electronic Systems, El Segundo, California.

VERIFIED

....

AN/APS-124 search radar

Type
Airborne helicopter search radar.

Description
The AN/APS-124 search radar has been designed and developed specifically for use in the US Navy (USN) SH-60B Seahawk Light Airborne MultiPurpose System (LAMPS) Mk III helicopter. It is specifically designed for helicopter installation featuring a low profile antenna and radome. Optimum detection of surface targets in high sea states is accomplished by several unique features including fast scan antenna and interface with the auxiliary OU/103 digital signal data converter to achieve scan-to-scan integration.

Principal features of the radar include high transmit energy, small surface target detection in high sea states, low profile linear array antenna and a lightweight modular design. These features provide long-range detection in noise, sea clutter decorrelation, installation under the fuselage for full 360° radar coverage and easy installation. APS-124 operates in three modes covering long- and medium-range search and fast scan surveillance. Mode 1 - long-range search - is characterised by long pulse-length, low pulse repetition frequency and slow scan, actual values being 2 μs, 470 pps and 6 rpm. Display ranges are selectable out to 296 km. In the medium-range Mode 2, these values change to 1 μs, 940 pps and 12 rpm. For Mode 3 they become 0.5 μs, 1,880 pps and 120 rpm. The display ranges are selectable up to 74 km and the false alarm rate is adjustable to suit conditions. The radar is integrated with a multipurpose display and with the LAMPS datalink for the transmission of airborne radar video to LAMPS equipped ships.

Status
As of this edition, the AN/APS-124 search radar was reported as being installed aboard USN SH-60B LAMPS Mk III anti-submarine/anti-surface warfare and Spanish Navy S-70B-1 (local designation HS 23) shipboard anti-submarine warfare helicopters. It should be noted that the US Navy plans to remanufacture its SH-60B fleet to -60R standard that will feature the SH-60B's mission suite integrated with the SH-60F's dipping sonar.

Contractor
Raytheon Electronic Systems, McKinney, Texas

VERIFIED

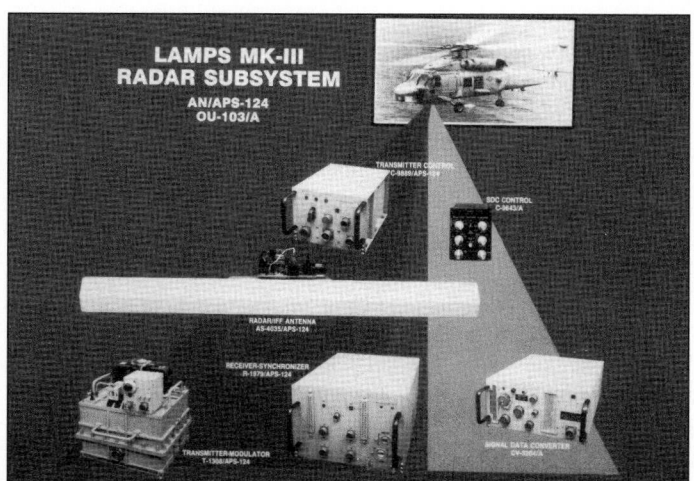

The AN/APS-124 search radar that is installed aboard USN SH-60B LAMPS III helicopter

....

AN/APS-125/-138/-139/-145 Airborne Early Warning (AEW) radars

Type
Family of UHF band (0.3 to 1 GHz) AEW radars.

Description
APS-145 represents the latest development in a line of AEW radars that stretches back to the early 1960s vintage AN/APS-96. Taking the cited radars in

Singapore's E-2C Group O aircraft are fitted with the APS-138 radar (Paul Jackson)

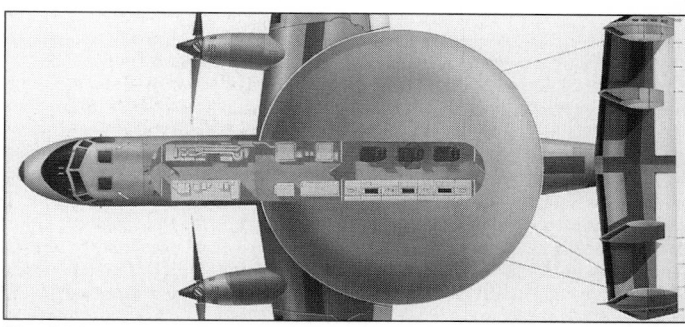

The E-2C Hawkeye AEW aircraft carries a team of three systems operators who are housed in the fuselage beneath the rotodome antenna housing

chronological order, AN/APS-125 emerged as a result of the introduction of what was termed the Advanced Radar Processing Subsystem (ARPS) into the existing APS-120 set (itself a development of AN/APS-111 which, in turn, was evolved from the APS-96) during the 1970s. ARPS is described as incorporating false alarm control and clutter/noise reduction features which taken together, provide the radar with an automatic overland tracking capability. Other enhancements include a switch from analogue to digital moving target indication and improved proofing against hostile jamming. New-build production of APS-125 is noted as having ended during 1982 while the US Navy (USN) is reported to have run an APS-120 to APS-125 upgrade programme during the period 1977 to 1984.

Starting in 1983, APS-125 was replaced in production by the APS-138 model. Alongside what have been termed 'minor improvements' in its circuitry, APS-138 introduced the L-3 Randtron Antenna Systems' (formerly Randtron) AN/APA-171 low-sidelobe 'rotodome' assembly that incorporates the company's Total Radiation Aperture Control Antenna (TRAC-A). So equipped, APS-138 is quoted as offering:

- continuous surface to altitude coverage
- simultaneous automated air and surface detection and tracking
- time difference height-finding
- video type options.

APS-138 was followed by the 1987 vintage APS-139 variant that has been characterised as a 'relatively modest' improvement over its predecessor and appears to have been an attempt to improve the radar's capability against low radar cross section targets such as cruise and anti-shipping missiles.

As of this edition, the latest iteration in the series is the APS-145, which is reported to have entered service with the USN during 1993. APS-145 is described as addressing perceived system shortfalls in the areas of overland clutter rejection, false alarm rate and detection range. Looking at some of these in more detail, the radar's false alarm rate is now managed by an Environmental Management System (EMS) that adjusts its sensitivity according to real-time clutter and target volume conditions. In terms of detection range, the APS-145 value is noted as being 40 per cent greater than that of the APS-138 variant. This increase is reported to have been achieved by means of the introduction of a new Pulse Repetition Frequency (PRF) mode (one of three available) and variable antenna rotation speed (five or six revolutions per minute). PRF variability is also understood to enable the radar to overcome target blind speed problems. To optimise system performance at all times, the sensor's EMS divides its coverage into more than 5,000 range and azimuth related cells per scan, and adapts the equipment's parameters according to a continuously updated assessment of the environmental conditions within each cell.

With regard to its general capabilities, APS-145 is described as offering enhanced performance (when compared with APS-138/-139) over land and sea, when searching in Over-The-Horizon (OTH) mode and at the 'land, sea and OTH interface'. In the first instance, APS-145 is reported to be able to track aircraft overland 'regardless' of the terrain and target density. In terms of ground vehicle detection, the radar is understood to be able to track such targets when densities are 'not too large'. Over sea, APS-145 is noted as being able to track moving and stationary surface and air targets, while at the land, sea and OTH interface, the equipment's EMS automatically modifies its signal processing and tracking algorithms to suit the changing search environments. Other system features include:

- an integral 1 million square mile, fine-grain terrain map that can be customised before or during a mission to include any 'special' environments that may be encountered
- scan-to-scan correlation capable of maintaining in excess of 20,000 tracks
- automatic channel selection and monitoring
- automatic 'clearest' frequency selection

- a continuous surveillance volume of 6 million cubic miles of airspace and more than 388,500 km² of ocean surface when being operated at an altitude of 9,144 m plus.

With regard to platforms, the Northrop Grumman E-2C Hawkeye/Hawkeye 2000 series of carrier-capable AEW aircraft have been the primary recipients of APS-125 through -145 series radars. Other platforms fitted with sensors from the family include the US Coast Guard Service's single EC-130V Hercules surveillance (APS-125 radar and, as of July 1999, serving with the USN's Naval Air Warfare Center's Aircraft Division at Naval Air Station Patuxent River, Maryland) and the US Customs Service's fleet of Lockheed Martin P-3 AEW and Control (AEW & C) aircraft (APS-138/-145). The worldwide E-2C fleet is understood to divide into four categories, namely Group 0 aircraft with APS-125 or APS-138 radars, Group I aircraft with the APS-139 and Group II and Hawkeye 2000 aircraft with the APS-145.

Status

As of January 2002, APS-125/-138/-139/-145 radars are reported to be installed aboard or are scheduled to be installed aboard E-2C Hawkeye aircraft operated by Egypt (five E-2C Group 0s (APS-138 radar to be updated to APS-145 standard) and one Hawkeye 2000E on order); France (two E-2C Group IIs with APS-145 plus one Hawkeye 2000E aircraft on order); Israel (four E-2C Group 0s with APS-125 - withdrawn from service and up for sale); Japan (13 E-2C Group 0s fitted with APS-138 - being progressively upgraded to Hawkeye 2000E standard); Singapore (four E-2C Group 0s with APS-138); Taiwan (four E-2T aircraft (E-2C Group II standard) plus two Hawkeye 2000E aircraft on order) and the USN (72 E-2C Group 0/II aircraft (as of October 2001) plus 21 Hawkeye 2000 platforms on order). Additionally, APS-138/-145 radars are installed aboard the US Customs Service's Lockheed Martin P-3 AEW and C aircraft and APS-145 has been proposed as the primary sensor aboard a putative Lockheed Martin C-130J AEW & C platform.

Contractor

Lockheed Martin Naval Electronics and Surveillance Systems - Syracuse, Syracuse, New York.

UPDATED

AN/APS-128/-128 Model D maritime surveillance radars

Type

I-band (8 to 10 GHz) airborne maritime surveillance radar.

Description

AN/APS-128 Model D is an upgraded version of the AN/APS-128 system. It is an all-digital equipment using a scan converter to present data in TV raster format. It contains target enhancement and clutter reduction circuitry consisting of frequency agility, sensitivity time control, constant false alarm rate and scan-to-scan integration.

The system consists of a rectangular flat-plate antenna and pedestal, a transmitter/receiver, a radar control unit, a digital scan converter, a trackball or joystick cursor and a bright display. Dual display for cockpit weather presentation is available as well as add-on cabin displays. The system uses fully programmable microprocessor operational features with alphanumeric and graphic options to meet varied mission requirements. A number of antenna array sizes are available to suit aircraft and radome configurations. An alternative parabolic antenna features a dual-polarisation capability to provide pencil beam for sea search and shaped beam for mapping.

Target detection of the system is given as:
- 56 km on a snorkel or fishing vessel (assumed 6 m long), 10 m² cross-section, in sea state 3
- 111 km on a trawler (160 m long), 150 m² cross-section, in sea state 5
- 185 km on a freighter (360 m long), 500 m² cross-section, in sea state 5
- 222 km on a tanker (600 m long), 1,000 m² cross-section, in sea state 5

The system also functions as a weather radar with a range of approximately 370 km. With the addition of a keyset control panel and enhanced software, APS-128 Model D becomes the Digital Tactical System (DITACS) surveillance radar. DITACS is reported to offer what is termed 'added tactical sensor capability for enhanced multimission applications'.

Status

As of May 2001, at least 100 APS-128/-128 Model D radars are reported as having been sold worldwide. Over time, users have included the Argentine Coastguard,

APS-128 Model D airborne surveillance radar

Brazilian Air Force, Chilean Navy, Gabon Air Force, Indonesian Air Force, Japanese Maritime Safety Agency, Portuguese Air Force, Royal Malaysian Air Force, the US National Aeronautical and Space Administration (NASA), Singapore Air Force, Spanish Air Force, Spanish Customs, Swedish Coastguard and Venezuelan Navy. APS-128 is quoted as being installed (in both nose-mounted and 360° configurations) aboard a range of aircraft including the C-130, EMB-111, C-212, B-200 T, Short Skyvan, P-3 Orion and Falcon Jet 900. As of the given date, APS-125/-128 Model D radars are no longer in production. However, Telephonics notes that the majority of these radars are still in service and that the company continues to support the APS-128 series with spares.

Specifications

Frequency: I-band (9,375 MHz)
Frequency agility: 85 MHz peak/peak
Power output: 100 kW peak
Pulse-width: 2.4 and 0.5 µs
PRF: 400, 1,200, 1,600 Hz
Noise figure: GAS-FET 3.5 dB low noise RF amplifier
Antenna rotation rate: 15 rpm (sector scan selected from the 60-360° arc); 15-60 rpm (continuous).
Bright display: high ambient light, 875 line high resolution
System weight: 91.8 kg with 20 cm display

Contractor

Telephonics Corporation, Command Systems, Farmingdale, New York.

VERIFIED

AN/APS-130 multimode radar

Type

12.5 to 18 GHz band airborne search and terrain-following radar.

Description

AN/APS-130 is the 12.5 to 18 GHz band multimode radar that is installed aboard the US Navy's (USN) EA-6B Prowler electronic warfare aircraft (see following). Functions performed by the radar include:
- search
- ground mapping
- tracking and ranging of fixed or moving targets
- terrain-avoidance or terrain-following
- beacon detection and tracking

In more detail, APS-130 features a track-while-scan capability that provided simultaneous range, azimuth and elevation data for weapon delivery. Range and azimuth markers are required to be placed on the target response, with elevation data being available on a continuous basis via a separate phase interferometer array that is located below the sensor's main scanner dish. This main dish has a width of about 1 m and is illuminated by a conventional horn-feed to produce a beam that is very narrow in azimuth and has a cosec² profile in elevation. This latter geometry (in combination with the cited interferometric elevation data) eliminates the need for mechanical scanning in the elevation plane. The interferometer array consisted of two adjacent rows of 32 horns and moves with the system's primary scanner dish. Functionally, energy reflected from ground targets arrives at the upper and lower rows with a time difference that is measured using phase comparison techniques and is translated into angular information. APS-130 is further reported to be ± 35° pitch/± 65° roll stabilised.

Status

As of December 2001, the AN/APS-130 multimode radar was reported as being in service aboard USN EA-6B Prowler electronic warfare aircraft. Within the EA-6B programme, Jane's sources suggest that APS-130 was fitted to new build Prowlers from aircraft USN Bureau of Aeronautics number 161120 on and has been retrofitted to those EA-6Bs that have been through the AFC-481 modification programme.

Contractor

Northrop Grumman Electronic Systems sector - Norden Systems, Norwalk, Connecticut.

NEW ENTRY

AN/APS-133(V) multimode radar

Type

I-band (8 to 10 GHz) mapping, weather and beacon homing radar.

Description

The AN/APS-133(V) multimode, digital colour display radar is derived from the commercial RDR-IF equipment and is designed to provide military aircraft with weather avoidance, terrain-mapping and beacon homing facilities. APS-133(V) consists of four units: transceiver, colour display unit, control panel, and a fully stabilised antenna. The colour display presents levels of target signal returns in red,

The AN/APS-133(V) multimode radar

yellow and green (weather mode) and red, yellow and blue (map mode). The display may also be used in conjunction with an auxiliary interface unit to display various types of information from other onboard sensors, such as inertial navigation, Omega, Identification Friend-or-Foe (IFF), station keeping and fuel saving advisory systems. These can be superimposed on the weather or terrain map (except IFF and station keeping), or can be displayed in various combinations.

Status

Over time, the AN/APS-133(V) multimode radar is reported as having been installed aboard a range of aircraft including the C-5, C-141, E-3 and KC-10 types. An improved land mapping version is noted as having been fitted to the USAF's C-25 Air Force One and E-4 National Emergency Airborne Command Post. It is also reported to have been selected for the new US Air Force C-17 airlifter. An improved version is also quoted as being installed aboard KC-130 aircraft.

Specifications

Frequencies: I-band (9,375 ± 5 MHz - transceiver, weather/map modes; 9,310 ± 5 MHz - receiver beacon mode)
Power output: 65 kW peak (nominal)
Range: 560 km
Antenna: 76 cm parabola with 3 pencil beams and CSC cosec^2 fan beam
Antenna scan angle: 180° (±90°)
Weight: 53.6 kg

Contractor

Honeywell, Morristown, New Jersey.

VERIFIED

AN/APS-134(V)/-134 (Plus) maritime surveillance radars

Type

I-band (8 to 10 GHz) maritime surveillance radars.

Description

The AN/APS-134(V) anti-submarine warfare and maritime surveillance radar is the successor to the AN/APS-116 system used in the US Navy (USN) for airborne surveillance. It is designed specifically to detect periscopes under high sea conditions and incorporates all the features of the previous generation with improved performance and a unique maritime surveillance mode. The principal features include a fast scan antenna and associated digital signal processing which form an effective means of eliminating sea clutter. Two other techniques, pulse compression and scan-to-scan processing, are used to combat the inherent problems of detecting small targets in the sea clutter environment.

Three modes of operation are available:

- periscope target detection in sea clutter using fast scan (150 rpm), high Pulse Repetition Frequency (PRF - 2,000 pulses/s) and high resolution with display ranges selectable up to 59 km.
- maritime surveillance out to 278 km with high resolution, low PRF (500 pulses/s) and medium scan speed (40 rpm)
- long-range search and navigation out to 278 km with medium resolution, low PRF and slow scan (6 rpm)

Long-range performance is provided by a high power transmitter (500 kW peak power) coupled with a high antenna gain. A radar control/display gives video presentation on a 25 cm^2 cathode ray tube, with the necessary controls and indicators.

The early- to mid-1990s AN/APS-134 (Plus) concept was designed to add improved periscope detection, digital signal processing, multiple Track-While-Scan (TWS), dual channel digital scan conversion, low probability-of-intercept and built-in diagnostic facilities to the existing APS-134(V) architecture. Long-range,

small target detection was to be achieved through the use of a 500 kW transmitter, a 35 dB high-gain antenna and a preamplifier with a less than 3 dB noise figure. APS-134 (Plus) was also to have incorporated pulse compression and scan-to-scan processing, together with an optional ability to record radar and forward-looking infra-red sensor data. So configured, the equipment was to have had three operating modes, as follows:

- low-altitude periscope detection (fast-scan antenna speed)
- high-altitude maritime surveillance (intermediate scan speed)
- navigation, mapping and weather detection/avoidance (slow scan speed)

A proposed APS-134(LW) variant is reported to have been superseded by the Raytheon Sea Vue family of lightweight, modular maritime surveillance radars (see separate entry).

Status

Over time, AN/APS-134(V) radars are reported as having been installed aboard German Dassault Atlantique Mk 1 (14 examples), South Korean Lockheed Martin P-3C Update III (APS-134(V)6, eight examples), Pakistani P-3C Update 11.75 (two examples) and Singapore's Fokker Maritime Enforcer Mk 2 (APS-134(V)7, five examples) maritime patrol aircraft. As of March 1998, Jane's sources noted that South Korea's APS-134(V)6 radars were to be upgraded to inverse synthetic aperture AN/APS-137(V)6 standard under a then year US$28.7 million US Foreign Military Sales programme. It should also be noted that, as of the given edition, the status of the APS-134 (Plus) concept was uncertain.

Specifications

APS-134(V)
Transmitter frequency: I-band (9.5-10 GHz modes a and b; 9.6-9.9 GHz mode c)
Power output: 500 W (average); 500 kW peak
Antenna beamwidth: 2.4° (azimuth); 4° (elevation)
Polarisation: vertical

Contractor

Raytheon Electronic Systems, McKinney, Texas.

VERIFIED

AN/APS-137(V) maritime surveillance radar

Type

I-band (8 to 10 GHz) multimission maritime surveillance radar.

Description

AN/APS-137(V) is an upgraded version of the AN/APS-116, which forms a part of the Lockheed-Martin S-3A to S-3B Weapon System Improvement Programme (WSIP). The most significant change between the two radars is APS-137's addition of inverse synthetic aperture imaging. This provides it with a long-range ship classification capability. The equipment is also coherent, offering high resolution and featuring multiple track-while-scan, dual channel digital scan conversion, software mode control (for installation versatility), and low probability of intercept capabilities.

As of this edition, the APS-137B(V)5 variant is being procured by the US Navy (USN) for its P-3 Anti-surface warfare Improvement Programme (AIP). APS-137B (V)5 comprises a power supply, a Receiver/Exciter/Synchroniser Processor (RESP), a transmitter, an antenna, a signal data converter, a radar set control-interface unit, a radar set control unit, a radar set computer and a control-indicator. Looking at some of these elements in more detail, the radar's transmission chain incorporates a coherent, air-cooled travelling wave tube amplifier, while its parabolic antenna is stabilised in pitch and roll, incorporates an identification friend-or-foe capability and Synthetic Aperture Radar (SAR) mode motion compensation. Its digital RESP is a fully programmable open architecture (VME and RACEway Interlink), while the radar's operating modes include stripmap and spot SAR, Inverse Synthetic Aperture Radar (ISAR), periscope detection, surface surveillance (land, sea and littoral) and navigation. Display processing options include SAR map, ISAR imagery, plan position indicator, B-scan and range profiling (A-scan). Other system features include:

- precision targeting.
- sector and searchlight scan capabilities.
- on-line/off-line selectable operation.
- multi-resolution SAR function.
- ISAR classification aids.
- video recording.
- a 26 kg auxiliary hardware package made up of a Global Positioning System antenna, fill panel and preamplifier, a video converter and a video tape recorder/control unit

Status

As of this edition, AN/APS-137(V) maritime surveillance radars were reported as being installed aboard 114 USN Lockheed Martin S-3B Viking anti-submarine warfare aircraft. In addition, the service is understood to have procured at least 13 APS-137B(V)5 radars (out of a total potential procurement of 47) for installation on AIP modified Lockheed Martin P-3C Orion maritime patrol aircraft. A further five radars have been procured for US Foreign Military Sales use. During 1993, Jane's sources further suggest that APS-137(V) formed the basis of the USN Naval Research Laboratory's (NRL) Automatic Radar Periscope Detection and Discrimination (ARPDD) technology demonstrator. Working with Raytheon,

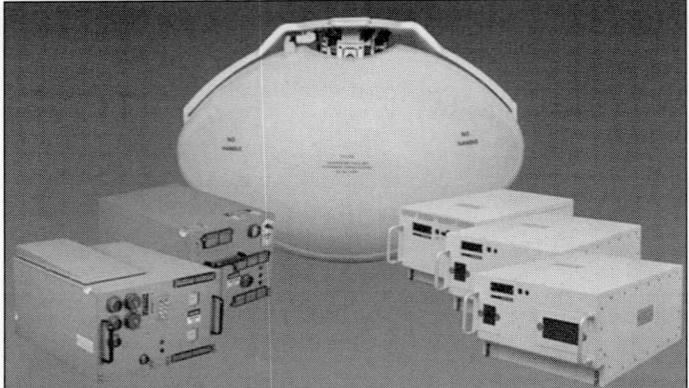

A general view showing the antenna and power supply, receiver/exciter/ synchroniser processor, transmitter, signal data converter and radar set control-interface units used in the APS-137B(V)5 multi-mission surveillance radar

0044140

John Hopkins University and the USN's Naval Air Warfare Centre, the NRL ARPDD teamed ASP-137(V) with new constant false alarm, 'retrospective' (elimination of noise and uncorrelated sea clutter), direct discrimination (automatic target recognition against five second histories of objects with 30 cm range resolution) and indirect discrimination (data fusion for final periscope declarations) processors. Starting in 1998, ARPDD demonstration systems have been tested aboard the destroyer USS *Stump* and an NRL P-3 flying testbed. Identified APS-137(V) contracting activity during the period 1998 to 2000 comprises the following:

29 October 1998
The then Raytheon TI Systems (McKinney, Texas) was awarded a then year US$24,356,155 modification to an existing contract (N00019-95-C-0198) covering the taking up of an option for 11 APS-137B(V)5 radars for installation aboard P-3C maritime patrol aircraft. At the time of the announcement, work on the effort was scheduled for completion by the end of September 2000, with the USN's Naval Air Systems Command, Patuxent River, Maryland acting as the programme's contracting activity.

1 November 1999
The Raytheon Systems Company (McKinney, Texas) was awarded a then year US$5,832,207 modification to an existing contract (N00019-95-C-0198) covering the retrofit of the 'early limited combat identification' capability into 40 APS-137B(V)5 radars, together with the provision of associated technical and administrative data. At the time of the announcement, work on the effort was scheduled for completion by the end of July 2002, with the USN's Naval Air Systems Command, Patuxent River, Maryland acting as the programme's contracting activity.

29 November 1999
The Raytheon Systems Company (McKinney, Texas) was awarded a then year US$30,621,730 modification to an existing contract (N00019-95-C-0198) covering non-recurring efforts associated with the manufacture and support of 12 APS-137(V)6 ISAR radars for the USN, together with a single example for the Royal Norwegian Air Force. The 'efforts' involved were noted as including engineering activities, data prototyping, the conduct of functional/physical audits, generation of technical manuals, source data development, data collection, contractor engineering/ technical support, provision of integrated logistical support, provision of spares, system repair and the provision of support/auxiliary equipment. At the time of the announcement, work on the effort was scheduled for completion by the end of March 2001, with the USN's Naval Air Systems Command, Patuxent River, Maryland acting as the programme's contracting activity.

27 January 2000
Raytheon Electronic Systems (McKinney, Texas) was awarded a then year US$8,112,320 modification to an existing contract (N00019-95-C-0198) covering the taking up of an option to fund the non-recurring support activities associated with the installation of APS-137(V)5 radars aboard five P-3 maritime patrol aircraft. Areas included in the option comprised engineering activities, data prototyping, the conduct of functional/physical configuration audits, generation of technical manuals, source data development, data collection, contractor engineering/technical support, provision of integrated logistical support, provision of spares, system repair and the provision of support/auxiliary equipment. At the time of the announcement, work on the effort was scheduled for completion by the end of April 2001, with the USN's Naval Air Systems Command, Patuxent River, Maryland acting as the programme's contracting activity.

28 January 2000
Raytheon Electronic Systems (McKinney, Texas) was awarded a then year US$16,594,461 firm, fixed-price contract covering the supply of 65 spare parts (including power supplies, receiver processors and antennas) for APS-137B(V)5 radars installed aboard P-3C maritime patrol aircraft. At the time of the announcement, work on the effort was scheduled for completion by the end of December 2001, with the USN's Naval Inventory Control Point, Philadelphia, Pennsylvania acting as the programme's contracting activity.

Specifications
AN/APS-137B(V)5
Frequency: 9.3-10.1 GHz (SAR modes); 9.5-10 GHz (all other modes)
Average power: 200 W (search/navigation modes); 230-500 W (ISAR mode); 350 W (SAR modes); 460 W (periscope detection mode)
Pulse-width: 5 μs (periscope detection mode); 10 μs (all other modes); 12 μs (SAR mode)

Pulse-width compression: 2.5 ns (periscope detection mode); 6 or 13 ns (ISAR mode); 14 ns (search modes); 240 ns (navigation modes); variable (SAR modes)
PRF: 388 Hz (search/navigation modes); 500 or 1,000 Hz (ISAR mode); 1,854 Hz (periscope detection mode); resolution dependent (SAR modes)
Resolution: 0.9 m or 1.8 m (ISAR); R1, R2, 3 m, 9 m or 30 m (SAR)
Scan modes: 6, 60 and 300 rpm
Azimuth coverage: 360°
Antenna stabilisation: ±15° (pitch); ±25° (roll)
Receiver sensitivity: 92 dBm
Receiver noise: 3.5 dB
Power requirement: 28 VDC/115 VAC (400 Hz, 3-phrase)
Video outputs: ASA-70A compatible; E1A-343 RGB raster; E1A-700 monochrome raster; digitised raster

Typical detection performance (Sea State 3):

Target radar cross section	Platform altitude	Detection range
1 m²	305 m	45 km
5 m²	305 m	72 km *
10 m²	305 m	72 km *
200 m²	1,524 m	161 km *
1,000 m²	3,962 m	213 km
10,000 m²	6,401 m	330 km *

* radar horizon limited
MTBF: >500 h (standard)

Dimensions/weights:

Unit	H × W × D (cm)	Weight (kg)
Power supply	22 × 29 × 44	21
Receiver/exciter/synchroniser	33 × 46 × 53	54
Transmitter	28 × 31 × 50	43
Antenna	104 × 92 × 69	35
Signal data converter	22 × 37 × 48	31
Radar set control-interface	22 × 37 × 48	31
Radar set control	28 × 15 × 44	14
Radar set computer	25 × 22 × 48	16
Control indicator	33 × 22 × 11	5

Contractor
Raytheon Electronic Systems, McKinney, Texas.

UPDATED

....

APS-143B(V)3 OceanEye™ sea surveillance radar

Type
I-band (8 to 10 GHz) sea surveillance, imaging and tracking radar.

Description
APS-143B(V)3 is a maritime surveillance and tracking radar that is designed for installation in a variety of fixed-wing aircraft and helicopters. The system uses frequency agility and pulse compression techniques and consists of three units: an antenna, transceiver and signal processor. Control of the radar may be carried out by a dedicated control panel (with on-screen controls), or by a central universal keyset via a MIL-STD-1553B databus. System features include:
- track-while-scan (30, 100 or 200 targets)
- air search with moving target indication (customised for the individual platform)
- integrated electronic support and identification friend-or-foe system interfaces
- electronic counter-countermeasures provision (including sector blanking and staggered pulse repetition frequencies).

The APS-143B(V)3's flat plate antenna can be fitted into any radome and its transmitter is a travelling wave tube type with a peak power output of 8 kW.

Described during June 2001 as being Telephonics' 'flagship' APS-143(V) series product, the I-band APS-143B(V)3 OceanEye™ radar features:
- a Versa Modular Eurocard (VME) - based open architecture (field reprogrammable firmware)
- range profiling
- spotlight and strip map Synthetic Aperture Radar (SAR), Inverse SAR (ISAR) and Side-Looking Airborne Radar (SLAR) functionality
- an air search mode (with moving target indication customised for the specific platform)
- 1:1 to 3,000:1 pulse compression ratios (using surface acoustic wave and digital technology)
- 450 MHz 'ultra wideband' frequency agility
- up to 200 target track-while-scan
- ARINC 407/429/571, MIL-STD-1553B and RS-232/-422 data output and control interfaces
- an air-to-surface missile guidance capability
- optional electronic support system and/or identification friend-or-foe interrogator integration
- a tactical data management system option (controls and manages data presentation for multiple onboard sensor suites)
- built-in test
- field upgradable firmware.

Status
As of January 2002, the APS-143B(V)3 OceanEye™ was the latest known member of the APS-143(V) radar family. Worldwide, APS-143(V) radars are known to be

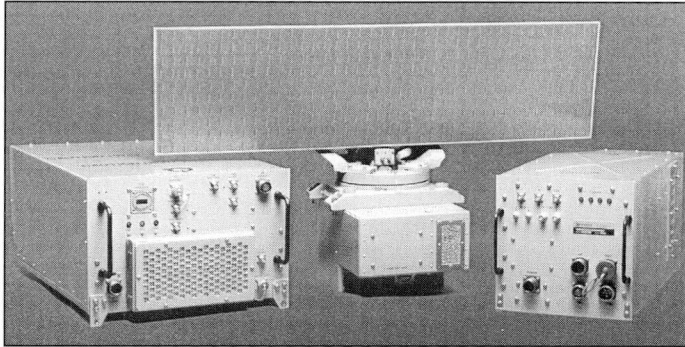

The APS-143B(V)3 OceanEye™ sea surveillance and tracking radar   0017504

installed on or scheduled for installation on a range of platforms, the known details of which are as follows:

- 11 SH-2G(A) helicopters of the Royal Australian Navy (APS-143B(V)3 OceanEye™ variant with imaging capability)
- three CL-604 surveillance aircraft of the Danish Air Force (ISAR capable APS-143B(V)3 OceanEye™ - on order (April 2001))
- 10 S-70B-6 Aegean Hawk helicopters of the Hellenic Navy (potential options for between two and six additional aircraft)
- two Saab 340B Plus SAR-200 search-and-rescue aircraft of the Japanese Maritime Safety Agency
- four Super King Air 200TB maritime patrol aircraft of the Malaysian Air Force (APS-143(V)1)
- five SH-2G(NZ) helicopters of the Royal New Zealand Navy (APS-143(V)3)
- 20 S-70C(M)-1 helicopters of the Taiwanese Navy (APS-143(V)3)
- six S-70B-7 Seahawk helicopters of the Thai Navy
- eight S-70B-28 helicopters of the Turkish Navy (potential options for up to eight additional aircraft)
- two E-9A range surveillance aircraft of the US Air Force (APS-143(V)1)
- 10 US Army TCOM surveillance aerostats (APS-143(V)2 - current (April 2001) status uncertain)
- six HU-25A surveillance aircraft of the US Coast Guard (ISAR capable APS-143B (V)3 OceanEye™ - on order (April 2001))
- two C-26 Metro 'utility' aircraft operated by a US government agency (ISAR capable APS-143B(V)3 OceanEye™ - on order (April 2001))
- three 'fixed-wing, anti-drug surveillance platforms' (APS-143B(V)3 OceanEye™ -on order (June 2001))

Specifications
Frequency: 9.25-9.70 GHz (450 MHz agility)
Transmitter: helix travelling wave tube type
Power output: 260 W (average, nominal); 500 W (average, max); 8 kW (peak, min)
Pulse-width: 0.1, 10, 23.4 or 40 micros
Weighted compressed pulse-width: 100 ns (search modes)
Waveform generation: digital synthesis
PRF: multiple
Receiver noise: 2.5 dB (max)
Receiver bandwidth: matched to pulse-width
Antenna: corporate fed planar-array
Antenna bandwidth: 500 MHz
Antenna gain: 31-34 dB (antenna dependent)
Antenna polarisation: horizontal (vertical option)
Antenna stabilisation: ±30° (pitch and roll)
Sector scan: 45-350° or continuous 360° (operator selectable)
Detection: 1 m² RCS target (typical - range in excess of 43 km, low-altitude operation, Sea State 3)
Range: in excess of 370 km (max)
Display range resolution: 18.5 m (1 m - imaging)
Search range resolution: 15 m
Azimuth accuracy: 0.5° or better
Programmable Video Resolution: CCIR-601, RS-170/-343, VGA and XVGA
Operating modes: ISAR, range profiling, range zoom, return-to-ship, search, SLAR, spotlight SAR, strip map SAR, colour weather and air-to-air
Power requirement: 28 V (12 A); 115 V AC (400 Hz, 3-phase)
MTBF: 650 h (helicopters); 1,164 h (fixed-wing aircraft)
Weight: 82 kg (typical, MIL-STD-1553B based configuration with SAR/ISAR imaging)

Contractor
Telephonics Corporation - Command Systems Division, Farmingdale, New York.

UPDATED

....

AN/APS-147 multimode radar

Type
Maritime Inverse Synthetic Aperture Radar (ISAR).

Description
AN/APS-147 is an ISAR system which uses the latest in high-throughput signal and data processing. Flexibility through programmability provides a product optimised for the maritime surveillance mission. Advanced processing allows the APS-147 to use a collection of waveforms to perform its mission at an output power substantially lower than traditional counterparts in maritime surveillance radars. This results in a radar with an extremely Low Probability of Intercept (LPI) by hostile forces. Using a low peak power waveform with frequency agility, the radar can detect medium- to long-range targets without the threat of electronic support system interception. Radar modes include target imaging, small target (periscope) detection, long-range surveillance, weather detection and avoidance, all-weather navigation, short-range search and rescue, enhanced LPI search and target designation. The AN/APS-147 features:

- flexible modular design which can be tailored to meet specific requirements and can be easily upgraded
- LPI enabling 'see without being seen' performance
- high-resolution images for rapid classification
- light weight by use of composite materials with a unique cooling implementation
- low input power
- simple design for high reliability and maintainability
- fully programmable signal processor with multiple waveform exciter and high-throughput rates
- integrated identification friend-or-foe and SAR option.

Status
During June 1999, Telephonics announced that it had delivered an APS-147 radar for integration and test aboard a US Navy SH-60R helicopter.

Contractor
Telephonics Corporation - Command Systems Division, Farmingdale, New York.

VERIFIED

....

AN/APY-1/2 airborne warning and control radar

Type
E/F-band (2 to 4 GHz) warning and control radar.

Description
APY radars are the core sensors installed aboard the Boeing E-3 Sentry and Boeing 767 (Japanese Air Self-Defence Force designation E-767) Airborne Warning And Control System (AWACS) aircraft. APY-1 and -2 are generally similar with the primary difference being the APY-2's full maritime search capability. Operating within the 10 cm wavelength band, the maritime capable radars are reported to have six operating modes together with a radar technician-controlled test and maintenance format. Known details of these are as follows:

Pulse Doppler Non-Elevation Scan (PDNES)
PDNES is a pulse-Doppler mode that is optimised for target detection out to the radar horizon. By way of example, an APY radar operating in PDNES mode at an altitude of 9,144 m is quoted as being able to detect targets out to a range of approximately 394 km. The PDNES and Maritime (see following) modes can be interleaved.

Pulse Doppler Elevation Scan (PDES)
PDES generates elevation and range data with target elevation being established by electronic scanning of the beam in the vertical plane. The PDES and BTH (see following) modes can be interleaved.

Beyond-The-Horizon (BTH)
BTH utilises a low Pulse Repetition Frequency (PRF), compressed pulse, non Doppler signal for extended-range surveillance in azimuth.

Passive
A receive-only mode designed to localise hostile jamming sources.

Maritime
A low PRF, compressed pulse mode for surface ship detection. The Maritime mode is described as using adaptive digital signal processing to handle real-time variations in sea clutter and to blank land masses.

Standby
In standby mode, the radar is maintained in warmed operational condition ready for immediate use.

While the PDNES/Maritime and PDES/BTH interleaved modes are noted as providing the necessary detection performance for most operational situations, further flexibility is generated by the APY radar being able to divide the surveillance volume of individual azimuth scans into up to 32 subsectors, each of which can be configured with its own operating mode. Selected modes can be commanded on subsequent scans or rearranged to vary the type of coverage in a particular area of interest or to accommodate changes in situation and terrain.

In baseline form, a 3,629 kg maritime capable APY radar divides into 12 component groups mounted in a 9.14 m diameter dorsal rotodome, the aircraft's main cabin and in a bay below the rear of the main cabin floor. Known details of these are as follows:

Radar Control and Maintenance Panel (RCMP)
Located in the main cabin, the RCMP houses the radar's on/off control, a radar status and maintenance data display and the radar technician's system access keyboard. The RCMP controls radar function during maintenance operations.

Radar synchroniser
Mounted in the main cabin, the software controlled synchroniser unit generates all the radar's timing signals and maintains stability. Function is updated once every modulation period.

One of the first-generation SDCs used with the APY-1 radar aboard the Core E-3 Sentry AWACS aircraft (MS/Boeing)

Stable Local Oscillator (STALO)

Located in the main cabin, the STALO assembly generates stable Radio Frequency (RF) signals for the radar transmission and reception (signal conversion) processes as well as acting as the system's central clock. It incorporates a signal generator, phase-locked loops, system clocks and a clutter oscillator.

Transmitter group

Mounted beneath the rear section of the aircraft's main cabin floor, the transmitter group is made up of 21 elements, the known details of which are as follows:

High Voltage Power Supply (HVPS)

The HVPS is made up of five pressurised assemblies (including a 90 kV transformer, filter and regulators) and converts input prime power into a filtered high voltage format.

Transmit electronics

A package made up of two solid-state 2W minimum predrivers providing initial amplification of the STALO's output signal, digitally controlled attenuators to control transmitter power versus elevation angle, intermediate power amplifier travelling wave tubes, two klystron power amplifiers providing pulse modulated RF signals in the form of a high peak power output across the bandwidth and 10 auxiliary units handling power distribution, component and circuit protection against power surges and transmitter group element interfacing.

Rotary Coupler (RC)

The RC provides circuit continuity between the cabin and rotodome electronics. The unit offers one transmission and seven coaxial RF channels and incorporates 105 slip rings. Of these, 16 handle 400 Hz power paths with the remaining 89 being used for radar and non-radar signal activity.

Phase Control Electronics (PCE)

The PCE group is located in the rotodome and comprises two elements, a Phase Shifter Control Unit (PSCU) and a Phase Shift Drive Unit (PSDU). The PSCU houses beam angle control circuits and an antenna tuning data store and accepts commands from the radar's computer to stabilise or scan the beam. The PSDU contains two drive and 28 phase shifter current drive modules and provides current for the system's beam steering and offset phasers.

Antenna Array (AA)

Housed in the rotodome, the 7.32 × 1.5 m AA comprises a stacked array of 28 slotted amplitude weighted waveguides, reflectionless Transmit and Receive Manifolds (TM and RM), 28 ferrite Beam Steering Phase Shifters (BSPS) and 28 low-power non-reciprocal Beam Offset Phase Shifters (BOPS). Of these, the TM accepts the transmitter's RF output and delivers it to the waveguides. The BSPS maintain proper phasing for low-sidelobe transmission, provide phase shifts for vertical beam steering and deliver received energy to the BOPS. The RM accepts signals from the BOPS, combines them in a power divider and delivers the resultant signal to the receiver chain. The remaining element - the BOPS - provide receive/transmit beam offset during elevation scanning and receive/transmit beam space coincidence in non-scanning modes. The radar transmission array is mounted back-to-back with an identification friend-or-foe/secondary surveillance radar aerial with the whole assembly rotating at a speed of 6 rpm.

Microwave Receiver (MR)

Mounted in the rotodome, the MR comprises three channels, each of which is equipped with a five stage receiver protector and a low-noise amplifier.

Analogue Receiver (AR)

Located in the aircraft's main cabin, the AR is made up of a mixer preamplifier, five pulse-Doppler Intermediate Frequency (IF) assemblies, delay line pulse compression circuits and four IF BTH mode units. As a whole, the unit separates and routes the radar's pulse-Doppler, BTH and Maritime receipts into separate receiver channels; coherently detects pulse-Doppler signals in phase and quadrature channels; range gates received pulse-Doppler signals and converts them from analogue to digital format for onward transmission to the digital Doppler processor; compresses, detects and applies Constant False Alarm Rate (CFAR) to BTH pulses and converts BTH data to a digital format for use in the radar data correlator.

Maritime Processor (MP)

Mounted in the aircraft's cabin, the MR is applicable to all APY-2 radars and presumably, those APY-1 sets which have been modified for over-water operations (see following). It is made up of a dedicated receiver incorporating a delay line pulse compressor, sensitivity time control circuitry, an envelope detector, CFAR circuitry, an analogue-to-digital converter and a microprocessor; five IF assemblies and a digital land mass blanking unit which houses a digital map storage memory unit and digital control circuitry. Functionally, the MR compresses, envelope detects and converts analogue Maritime mode signals into a digital format; develops detection criteria based on a CFAR strategy optimised for sea clutter; outputs digitised Maritime mode data to the radar correlator and provides digital land mass blanking.

Digital Doppler Processor (DDP)

Located in the aircraft's main cabin, the DDP notches out mainbeam clutter signals for the radar's pulse-Doppler operating modes; scans matrixed data and compares it with dynamic threshold parameters to provide the CFAR detection facility and outputs digital detection data to the radar data correlator.

Radar Data Correlator (RDC)

Mounted in the aircraft's cabin, the RDC comprises one redundant and two active processors and a core program and semiconductor data memory. The unit accepts commands/input from the system's central computer, the manually operated RCMP, the AR, the DDP and the MP; processes all target and status data; provides target and equipment status data to the system's central computer and RCMP and controls all the radar's internal functions including self-test performance measurement.

As of December 2001, six baseline E-3 Sentry configurations have been identified, the known details of which are as follows:

E-3A Core

Original USAF standard fitted with APY-1 or APY-2 radar. Total procured comprised 24 production aircraft (23 fitted with APY-1 radar and one with APY-2) plus two refurbished EC-137D testbeds. Of these, the two refurbished EC-137Ds and 22 of the APY-1 equipped production aircraft have subsequently been upgraded to E-3B standard (see following). The remaining APY-1 configured Sentry has been retained by Boeing as a trials aircraft and has been brought up to E-3C standard (see following), as has the single APY-2 equipped production aircraft. The Boeing trials aircraft is now known as a JE-3C.

E-3A US/NATO Standard

Combined US and Allied standard which incorporates the APY-2 radar and mission suite enhancement (over the Core aircraft) which include an IBM CC-2 central computer (subsequently upgraded to CC-2E standard on the NATO aircraft - see following); additional HF (3 to 30 MHz) radios; jamming resistant voice communications; a radio teletype capability and provision for defensive countermeasures. Nine such aircraft were delivered to the USAF (all of which have subsequently been upgraded to E-3C standard) together with 18 to NATO and five to Saudi Arabia. The Saudi aircraft are fitted with CFM56 rather than the E-3's standard Pratt & Whitney TF33 engines and have had the computer systems (to CC-2ER standard) and communications upgraded subsequent to delivery. The NATO E-3As have also been the subject of a European Aeronautic, Defence and Space (EADS) Company Deutschland (formerly Daimler-Benz Aerospace) managed Mod Block 1 upgrade programme that adds new colour displays, 'Have Quick' secure radios and Link 16 datalink to the existing mission suite and have been equipped with the AN/AYR-1 (see separate entry) Electronic Support (ES) receiver subsystem.

E-3B/Block 20/30

USAF standard which incorporates APY-1 radar with an austere maritime surveillance capability back-fitted and mission-suite enhancements which include jam resistant voice communications; one additional HF and five additional UHF (0.3 to 1 GHz - with provision for 'Have Quick' anti-jamming circuitry) radios; the IBM CC-2 central computer; five additional Situation Display Consoles (SDC - making a total of 14); the radio teletype capability and provision for defensive countermeasures. The E-3B/Block 20 programme involved the reworking of the two refurbished EC-137Ds and 22 of the APY-1 equipped Core aircraft to the new standard with the first example being redelivered to the USAF on 18 July 1984. An undisclosed number of E-3B aircraft are also reported to have been equipped with the 'Project Snappy' sensor (type unknown) for service during and after the 1991 Gulf War. Alongside this, the E-3B fleet is also understood to be the subject of a Boeing managed upgrade (the ICON programme) which adds the AYR-1 ESM, a Global Positioning System receiver, a Tactical Digital Information Link-J (TADIL-J) capability and the CC-2E computer enhancement to the existing mission suite. So configured, the aircraft are understood to be designated as Block 30s.

E-3C/Block 25/35

USAF standard which incorporates the APY-2 radar and mission suite enhancements, which include five additional SDCs and five more UHF radios with 'Have Quick A-Nets' secure communications provision. The nine USAF US/NATO Standard E-3As plus the single APY-2 equipped Core aircraft have been brought up to this standard. An unknown number of E-3Cs are quoted as having received the 'Project Snappy' sensor system for Desert Storm. In all, a total of seven E-3B/Cs are reported to have been so provisioned during the Gulf War with a further eight being fitted with the device subsequently. In addition, the USAF's E-3C aircraft are included in the ICON upgrade effort, being designated as Block 35 aircraft when so configured.

E-3D Sentry AEW 1

Seven CFM56-powered aircraft equipped with the APY-2 radar and the Loral EW-1017 ESM system which have been procured for the Royal Air Force. The APY-2 radars fitted aboard the RAF's E-3D and Japan's E-767 aircraft feature a unique maritime sub-mode not available in other APY-2 applications.

E-3F

Four CFM56-powered aircraft equipped with APY-2 radar which have been procured for use by the French Air Force.

Following closure of the Boeing 707/E-3/E-6 production line, future airframe activity now centres on an AWACS variant of the Boeing 767 airliner. As currently (2001) described, this platform is equipped with the APY-2 radar with an enhanced maritime surveillance capability and a mission suite that is generally similar to that used in the E-3. A full crew complement is described as being two pilots, a mission director, tactical director, a fighter allocator, two weapon controllers, a link manager, seven surveillance operators, a communications operator, a radar technician, a communications technician and computer display technician.

Status

As of December 2001, APY radar equipped E-3 AWACS aircraft are operated by, or are on order for, the air forces of France (four E-3F aircraft operated by the 36e Escadre de Détection Aéroportée's (EDA) Escadrons 1/36 'Berry' and 2/36 'Nivernais'. The 36e EDA was formed on 1 March 1990), Japan (four Boeing E-767 aircraft delivered during 1998/1999 and operated by the Japanese Air Self-Defence Force's Misawa-based 601 Hikotai), Luxembourg (17 US/NATO Standard E-3A aircraft with Luxembourg acting as the required registration agency for NATO's multinational Airborne Early Warning Force (AEWF). The AEWF's aircraft were delivered between 1982 and 1985), Saudi Arabia (five E-3A aircraft to an equivalent US/NATO Standard E-3A configuration that were delivered during 1986-1987 and are operated by the Royal Saudi Air Force's Riyadh-based No 18 Squadron), the UK (seven E-3D/Sentry AEW 1 aircraft operated by Nos 8 and 23 Squadrons, RAF Waddington. Of these, No 8 Squadron was formed at Waddington on 1 July 1991 and was declared to NATO as the AEWF's E-3D Component on 1 July 1992) and the USAF (33 E-3B/C aircraft operated by the 552nd Air Control Wing, Tinker Air Force Base (AFB), Oklahoma, the 961st Airborne Air Control Squadron (AACS), Kadena Air Base, Japan and the 962nd AACS based at Elmendorf AFB, Alaska).

In more detail, the NATO AEWF is reported to have its central operating base at Geilenkirchen, Germany with forward operating locations (housing up to six deployed E-3s at a time) at Konya, Turkey; Previza, Greece; Trapani/Birgi, Italy and Orland, Norway. NATO AEWF personnel are drawn from Belgium, Canada, Denmark, Greece, Italy, Luxembourg (non-flying), Netherlands, Norway, Portugal, Turkey and the US. Within the USAF, the E-3 entered service during March 1977 with the last of the 34 aircraft ordered being delivered in June 1984. As of 1996, at least 71 APY radars had been produced.

Looking at the APY radar specifically, Northrop Grumman received a US$223 million contract in mid-1989 covering the full-scale development phase of an E-3 Radar System Improvement Programme (RSIP). RSIP is designed to improve the radar's effectiveness against manned aircraft and cruise missiles that have low radar cross sections. RSIP modifications include introduction of a new pulse compressed waveform type which will increase sensitivity; improvements to the system's man/machine interfaces including the introduction of a Fast Fourier Transform signals analyser into the RCMP; replacement of the RDC and DDP by new general purpose and adaptable signal processors respectively with the two forming a surveillance radar group using Ada software and improvements to the system's reliability and maintainability. As of May 2001, Boeing (in collaboration with Northrop Grumman) is reported as having delivered 18 RSIP kits to the European Aeronautic, Defence and Space (EADS) Company Deutschland (formerly DaimlerChrysler Aerospace) for installation aboard NATO E-3s. RSIP installation across the NATO E-3A and RAF E-3D fleets was completed during 2000, with that for the USAF continuing into 2001. As of December 2001, Northrop Grumman was reporting that RSIP had been implemented on 17 NATO E-3A aircraft, seven RAF E-3D aircraft and nine USAF E-3B/C aircraft and that is implementation aboard France's four E-3F aircraft was under contract. The company further noted that as of the given date, total production of RSIP kits stood at 46, with all 46 upgrades scheduled to be in service by 2005. Within the US, RSIP installation is undertaken at Tinker AFB, with BAE Systems having undertaking the work for the RAF. Specifically identified APY-1/2 contracting activity during the period 1998-2001 comprised the following:

8 October 1998

The Boeing Defense and Space Group was awarded a then year US$44,066,955 face value increase to an existing firm, fixed-price contract covering the provision of RSIP kits, related sub-kits, material and installation support for five E-3 aircraft. At the time of the increase's announcement, work on the effort was scheduled for completion on 31 May 2001 and the programme's contracting activity was the USAF's Electronic Systems Center, Hanscom AFB, Massachusetts.

10 November 1999

Litton (San Carlos, California) was awarded a then year US$20,952,239 requirements contract (F04606-00-D-0008) covering the repair of approximately 240 klystrons used to control operating frequency within the APY-1/2 radar. At the time of the award's announcement, work on the effort was scheduled for completion by 9 November 2004 and the programme's contracting activity was the USAF's Sacramento Air Logistics Center, McClellan AFB, California.

19 May 2000

Raytheon Electronic Systems (McKinney, Texas) was awarded a then year US$6,575,500 firm, fixed-price contract covering the supply of 435 Common Large Area Display Sets (CLADS) with which to replace the displays used in the existing SDC units aboard E-3 aircraft. At the time of the noted announcement, this contract was divided into 45 set tranches (with each tranche being delivered 120 days after the exercising of the particular tranche option and the order being completed by a final tranche of 30 equipments) and its contracting activity was the USAF's Air Force Research Laboratory, Kirtland AFB, New Mexico.

A CFM56-powered E-3D Sentry AEW 1 operated by the Royal Air Force's No 8 Squadron ()

9 June 2000

Boeing Defense and Space Group (Seattle, Washington State) was awarded a then year US$45 million firm, fixed-price, not-to-exceed contract covering support for the APY-2 RSIP programme; provision of a partial APY-2 Group B kit and 'special' test equipment for the USAF's Avionics Integration Support Facility (Tinker Air Force Base, Oklahoma) and the modification of two APY-2 'transmit angel controls' and ten thermal assemblies to RSIP configuration. At the time of the award's announcement, work on the effort was scheduled for completion by the end of March 2003 and the programme's contracting activity was the USAF's Electronic Systems Center, Hanscom AFB, Massachusetts.

19 October 2001

Boeing Space and Communications (Seattle, Washington State) was awarded a then year US$27,686,851 option to provide four Group A and B radar kits and four Group A and B in-flight maintenance spares packages in support of the USAF's APY-2 RSIP programme. At the time of the announcement, work on the effort was scheduled for completion by the end of December 2005, with the USAF's Electronic Systems Center, Hanscom AFB, Massachusetts, acting as the programme's contracting activity.

Contractor

Northrop Grumman Electronic Systems sector, Baltimore, Maryland.

UPDATED

....

AN/APY-3 multimode Side-Looking Airborne Radar (SLAR)

Type

I-band (8 to 10 GHz) multimode phased-array SLAR.

Description

APY-3 is the radar sensor used in the USAF/US Army Joint Surveillance Target Attack Radar System (Joint STARS). Functionally, Joint STARS is a long-range, long-endurance, air-to-ground surveillance and battle management system which is capable of looking deep behind hostile borders to detect and track movement in forward and rear echelon areas. The system is further intended to provide air and ground commanders with intelligence and targeting data that is consistent with optimal use of the forces under their control. Joint STARS comprises an airborne segment together with an unlimited number of Ground Station Modules (GSM) which receive real-time processed radar data from the airborne platform. The airborne element of the system is designated as the E-8C and comprises the host airframe, the APY-3 radar and Operations and Control (O & C) and communications subsystem.

The airframe used is an extensively modified and remanufactured Boeing 707-300 series airliner. The system's prime contractor, Northrop Grumman notes that the 707's ' robust design' offers 'many advantages that are not available in more recent designs which are optimised for greater efficiency within very narrow parameters'. In E-8C configuration, the aircraft is equipped with 18 operator consoles (17 of which are devoted to O & C with the remaining station being manned by a navigation/self-defence operative) which, according to Northrop Grumman, is sufficient to give the platform a 'battle management capability, especially (relating to) air operations'. The E-8C is noted as having an endurance of up to 20 hours with air-to-air refuelling and there is a crew rest area within the aircraft to maintain crew performance on long missions.

An E-8A Joint STARS aircraft

A central feature of the Joint STARS system is its 7.32 m long radar antenna which is carried beneath the forward fuselage of the E-8 aircraft

The APY-3 radar subsystem utilises a side-looking phased-array antenna which is housed in a 7.32 long 'canoe' radome mounted beneath the E-8's forward fuselage and is electronically steered and mechanically elevated. APY-3 is noted as being extremely agile electronically and able to provide interleaved battlefield management data to system operators located in the aircraft and GSMs. While not confirmed, Jane's sources suggest that an APY-3 equipped E-8 flying at an altitude of between 9,144 and 12,192 m can survey a 1 million km² area during the course of an 8 hour sortie and that the radar has a maximum detection range which is in excess of 250 km. APY-3 has a number of operating modes, the known details of which are as follows:

Moving Target Indicator/Wide Area Surveillance (MTI/WAS)
MTI/WAS is described as APY-3's basic operating mode and is designed to detect, locate and classify slow-moving vehicles throughout the radar's Field Of View (FOV). The MTI technique used is noted as allowing the system to differentiate between wheeled and tracked vehicles.

Moving Target Indicator/Sector Search (MTI/SS)
As its name suggests, The MTI/SS mode focuses on selected areas of the radar FOV and is optimised for enhanced resolution and attack guidance. MTISS's rapid revisit time also provides an automatic tracking facility against operator selected targets.

Synthetic Aperture Radar/Fixed Target Indication (SAR/FTI)
SAR/FTI provides a black and white photographic-like ground imaging capability against selected geographic areas in the radar FOV. This is enhanced by the FTI submode that highlights the largest fixed targets visible in the area under surveillance. SAR/FTI can be interleaved with both the MTI/WAS and MTI/SS modes and as such, is noted as being particularly useful in the post-attack assessment role.

The E-8's O & C subsystem controls the radar and is seen as being the key to the successful exploitation of the acquired sensor data. As a whole, the subsystem is built around a real-time, VAX-based, distributed processing architecture that includes individual DEC ALPHA-based digital processors at each of the aircraft's 18 Raytheon AXP-3000/500 workstations. The subsystem architecture is so configured as to allow all operators simultaneous access to all available data. The individual consoles manage the exchange and display of data and offer a range of display formats. In this area, as well as being integrated into digitally stored map data bases, moving target data can be superimposed on SAR imagery or enhanced with information provided by offboard sensors. All display formats can be recorded and replayed at the optimum speed at which it is easiest to detect patterns of movement and/or changes in such patterns. Other elements of the O & C subsystem are thought to include five Raytheon 920/866 'supermini' computers, three programmable signal processors, Interstate Electronics high-resolution workstation colour displays, Orbit International operator keyboards, Telephonics intercom control units and Miltope message printers.

As of the mid-1990s, the E-8's communications subsystem was noted as incorporating two datalinks and three types of voice radio. An encrypted, highly jam resistant Cubic Surveillance and Control DataLink (SCDL) allows the E-8 to communicate with the system's GSMs, provide them with the same still or moving imagery as is being seen aboard the aircraft and provide a radar service request uplink for a network of up to 15 GSMs which gives the ground operators the same level control over the radar system as their airborne counterparts. For communications with air command elements, the aircraft has two Joint Tactical Information Distribution System (JTIDS) terminals while voice communication needs are catered for by a suite of 12 encrypted UHF band (0.3 to 1 GHz), three encrypted VHF band (30 to 300 MHz) and two encrypted HF band (3 to 20 MHz) radios. The E-8 may also be equipped with Single Channel Ground and Airborne Radio System (SINCGARS) provision in its VHF radio net but this is not confirmed.

Status
The current operational Joint STARS capability is a combined USAF and US Army development which fuses the former's mid-1970s 'Pave Mover' programme with the latter's 1979 vintage 'Assault Breaker' effort. Work on the Joint STARS programme began as a joint service programme during 1983. Aimed at providing an airborne surveillance and anti-tank radar system, three industrial teams bid for

the programme with that led by the then Grumman (now Northrop Grumman) being awarded a US$657 million contract for its airborne element during September 1985. To handle the work, the company created a new Joint STARS division - the Melborne Systems Division (MSD) - to design, develop and integrate the system. This included responsibility for all the radar signal processing, post processing and battle management functions inherent in the concept. Development of the APY-3 radar subsystem was in the hands of the then Norden Systems (now Northrop Grumman Norden Systems) while Motorola began work on the Joint STARS GSM under a separate US Army contract.

Two Boeing 707-328C airliners were acquired from American Airlines and the Australian flag carrier Quantas to act as what were then known as E-18C Joint STARS testbeds. After modification by Boeing, the two airframes were delivered to MSD for fitting out during 1986-1987. The first fully configured prototype (now known as an E-8A) made its maiden flight on 22 December 1988. The MSD Joint STARS development programme was interrupted by the 1991 Gulf War which saw both development aircraft brought up to an interim operational capability and deployed to Saudi Arabia under the control of the USAF's 4411th Joint STARS Squadron (Provisional). The first operational E-8A sortie was flown on 14 January 1991 and by the end of the conflict, the two aircraft are reported to have flown a total of 534 hours in the course of 49 surveillance missions.

In April 1992, Grumman was awarded a US$125 million low rate E-8 initial production advanced procurement contract. Initially, the production standard aircraft was to have been the new-build E-8B model but this was abandoned after the delivery of the first airframe in favour of reworking existing Boeing 707 airframes. Such aircraft are designated as E-8Cs, the first example of which was completed in March 1994. A favourable programme review in May 1993 led to the authorised construction of six E-8Cs at the rate of two per year. As of this edition, a total of 16 E-8C aircraft are expected to be procured by the USAF. The service activated the 93rd Air Control Wing at Robins Air Force Base, Georgia during January 1996 to operate the type. As of late 1996, the 93rd had received at least one production E-8C. Prior to the establishment of this new unit, the USAF formed the 4500th Joint STARS Squadron (Provisional) to support the NATO 'Joint Endeavour' operation in Bosnia. Equipped with one E-8A and an E-8C, the 4500th deployed to Rhein-Main Air Base, Germany on 14 December 1995 and remained on station until the end of the following March. During this deployment, the Squadron's aircraft are reported to have completed 97 missions. On 7 November 1996, the 93rd Air Control Wing detached to Rhein-Main for a 30 to 60 day serial of operations in support of 'Joint Endeavour'. This was the unit's first operational deployment since its formation.

Outside the US, the Joint STARS system has been proposed to meet both the UK's Airborne STand Off Radar (ASTOR - see separate entry) and NATO's Alliance Ground Surveillance (AGS) system requirements. In both instances, the offers were not taken up. Elsewhere, Jane's sources have, over time, also suggested interest in the system from Japan, South Korea and Saudi Arabia.

Alongside the continuing production effort, the E-8 is also to be the subject of a MultiStage Improvement Programme (MSIP), the first phase of which (introduction of Tactical Digital Intelligence Link - Joint (TADIL-J) datalink and infra-red countermeasures) is noted as having been funded. Phase 2 of the MSIP is quoted as involving the integration of satellite communications, an improved data modem and an automatic target recognition facility and as of 1995, was expected to be launched during 1996. Trials of the countermeasures provision took place at Eglin Air Force Base, Florida during late 1995 using an E-8A fitted with five AN/ALE-47 infra-red decoy flare launchers (two on each side of the rear fuselage plus a fifth pointing aft) and an AN/AAR-47 missile approach warning (360° sensor units mounted above and below the aircraft's fuselage) system.

As of this edition, APY-3 may be the subject of the Joint STARS Radar Technology Insertion Programme (RTIP) which looks to significantly upgrade the radar's capability. According to Jane's sources, the Joint STARS RTIP is effectively a new radar that features an electronically steered antenna array that is approximately half the size of the APY-3's 7.62 x 0.6 m unit and incorporates 15 phase centres as against the APY-3's three; simultaneous SAR and MTI functions; improved MTI and SAR resolution and an MTI revisit rate of between six and ten times per second. As of the given edition, the USAF had ordered its 15th E-8

The phased-array antenna used in the APY-3 radar

The production standard E-8C features multiple operator stations of the type shown here

aircraft. Elsewhere, it was announced on 15 June 1999 that NATO's National Armament Directors Committee had endorsed an RTIP based solution to meet the Alliance's Ground Surveillance (AGS) requirement. Here, the radar would be installed aboard an Airbus A321 airframe. At the time of the endorsement, the AGS requirement was for six aircraft with an in-service date of 2006. At the same time, a two year window was allocated for consideration of the four nation (France, Germany, Italy and the Netherlands) Stand-Off Surveillance Target Acquisition Radar (SOSTAR) technology demonstrator as a sensor for AGS. Subsequent to its endorsement, Northrop Grumman has gone on to establish the NATO Transatlantic Advanced Radar (NATAR) consortium to develop the RTIP- Airbus A321 AGS concept. As of this edition, NATAR consortium members included Belgium, Canada, Denmark, Luxembourg, Norway and the USA.

Contractors

Rockwell Collins (flight management system).
Cubic Defense Systems (SCDL datalink).
Intererstate Electronics (graphic displays).
Litton (inertial navigation system).
Magnavox (UHF radios).
Miltope Corporation (message printers).
Motorola (ground stations modules).
Northrop Grumman, Melbourne Systems Division (E-8 prime).
Norden Systems (a unit of Northrop Grumman's Electronic Sensors and Systems Division _ APY-3 radar).
Orbit International (workstation keyboards).
Raytheon Electronic Systems (processors and AXP-3000/500 workstations).
RF Products (VHF co-location filter).
Telephonics Corporation (intercom control sets).

UPDATED

AN/APY-6 surface surveillance radar

Type

8 to 12.5 GHz band Synthetic Aperture Radar (SAR)/Ground Moving Target Indicator (GMTI) surface surveillance radar.

Description

The Northrop Grumman AN/APY-6 SAR/GMTI surface surveillance radar is described as being able to image surface-ship targets and provide coverage of seaborne and land-based moving and fixed targets in littoral warfare scenarios. Jane's sources suggest that the system makes use of a travelling wave tube amplified transmission chain and incorporates a 'flexible' planar-array antenna that can be configured to provide forward hemisphere, side-looking or 360° coverage from a nose- or belly-mounted location. The same sources report the use of a four channel receiver (one GMTI and three SAR) and a system architecture that is based on an Ethernet with fibre channel interconnectivity and industry-standard programming in 'C'. Identified APY-6 operating modes include strip SAR, spotlight SAR, inverse SAR (ISAR), GMTI and rotating antenna detection. Of these, the GMTI mode provides precision geolocation of moving targets (such as theatre ballistic missile Transporter/Erector/Launcher (TEL) vehicles) which can be overlayed on SAR imagery.

Status

Understood to have been developed at the behest of the US Navy's (USN) Office of Naval Research (ONR), the AN/APY-6 SAR/MTI radar is reported as having begun

flight trials aboard the USN's Naval Air Warfare Center Aircraft Division's Naval Force Aircraft Test Squadron's 'Hairy Buffalo' NP-3C testbed aircraft during 1999. Jane's sources suggest that the 'Hairy Buffalo' testbed is primarily intended to investigate airborne sensors and communications networks that could enhance the USN's ability to engage time critical targets using organic assets. As of December 2001, APY-6 was noted as having 'successfully' participated in the Affordable Moving Surface Target Engagement (AMSTE) programme, Fleet Battle Experiments-Golf and -Hotel, and a 'major' NATO exercise during US Fiscal Year 2000. As of May 2001, Jane's sources were noting that the ONR had awarded Northrop Grumman a then year US$390,000 contract covering the integration of an Active Electronically Scanned Array (AESA) antenna into the APY-6 system.

Specifications

Frequency: 8-12.5 GHz
Coverage: 200° (initial 'Hairy Buffalo' installation with mechanical scanner)
Swath width: 6-48 km (overland strip SAR mode)
Resolution: 0.3-1 m (surface ship ISAR imaging and overland spot SAR mode); 1-8 m (overland strip SAR mode)
Range: up to 200 km (surface ship ISAR imaging and overland strip SAR mode)
Weight: 186 kg

Contractor

Northrop Grumman Electronic Systems sector - Norden Systems, Norwalk, Connecticut

NEW ENTRY

AN/ZPQ-1 Tactical Endurance Synthetic Aperture Radar (TESAR)

Type

J-band (12.5 to 18 GHz sub-band) SAR for Unmanned Aerial Vehicle (UAV) applications.

Description

The AN/ZPQ-1 TESAR is a high-resolution, J-band (12.5 to 18 GHz sub-band) radar that is designed for use aboard UAVs and manned aircraft. The system weighs 74.9 kg, occupies a volume of 0.12 m³ and draws 1,050 W of power. Within the UAV domain, the radar is installed on the General Atomics Predator UAV which is capable of operating at altitudes of up to 7,620 m, a range of up to 741 km and of undertaking missions of up to 20 hours duration when equipped with an electro-optical/radar/datalink payload. As applied to the Predator, ZPQ-1 comprises an antenna assembly, an exciter/transceiver and a commercial-off-the-shelf processor. Of these, the antenna assembly features single axis electronic scanning, a two axis mechanical gimbal and has a maximum ±135° field of regard (selectable ±22.5°, 45°, 60°, 90° or 135° sectors at ranges of between 5 and 25 km and elevations of 1, 2 or 3 bars). The exciter/transceiver is described as being a single channel equipment that incorporates digital waveform generation, microwave power module transmitter technology and can be configured with a second channel for moving target indication if required. The system processor is noted as being capable of 540 Mops and as featuring a MILitary STandard (MIL-STD)-1553 interface. As installed on the Predator, ZPQ-1 is understood to be able to process imagery onboard the UAV and datalink it to a ground station to give the operator a real-time strip map. The ground control station used also incorporates Northrop Grumman sourced high-resolution image equipment that is reported as being derived from the US Army's Enhanced Radar Correlator equipment.

Status

As of July 2001, the AN/ZPQ-1 radar was reported as having completed its US Army Communications and Electronics Command Advanced Concept Technology Demonstration (ACTD) programme (during which, it was operationally deployed for mission and user evaluation) and as having transitioned to operational system status. Initial field use of ZPQ-1 was in support of NATO's 'Joint Endeavour' operation in Bosnia and was deemed as being 'highly successful'. The equipment has continued in use in the peacekeeping role since March 1996. During 1999, a then year US$38.2 million follow-on award covering the supply of 34 radars brought total ZPQ-1 procurement to 60 and extended production of the system until November 2001.

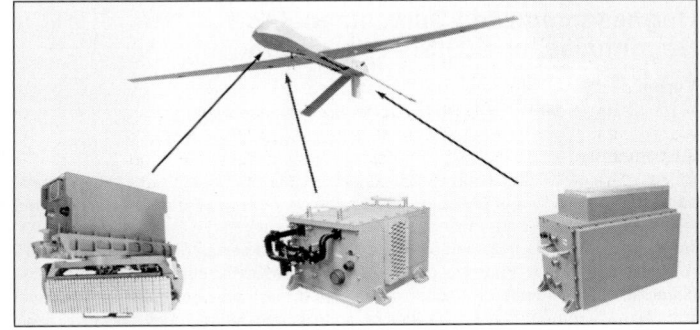

The components making up the ZPQ-1 radar as applied to the Predator UAV (Northrop Grumman)   2002/0114012

Specifications

Frequency: 12.5-18 GHz
MTI mode min detectable velocity: 1.5 m/s (forward hemisphere); 3.5 m/s (±135° azimuth)
MTI mode FAR: <2/min
MTI mode frame time: <1 min
SAR spot mode size/resolution: 800 × 800 m/0.3 m, 1.6 × 1.6 km/0.6 m or 2.6 × 2.6 km/1 m
SAR spot mode CEP: 25 m
SAR spot mode range: 18 km (0.3 m resolution), 25 km (0.6 m resolution) or 28 km (1 m resolution)
SAR strip mode swath width/resolution: 800 m/0.3 m, 1.6 km/0.6 m or 2.6 km/1 m
SAR strip mode range: 10 km (0.3 m resolution), 18 km (0.6 m resolution) or 25 km (1 m resolution)
SAR strip mode rate: 100.8 km^2/h (0.3 m resolution), 201.6 km^2/h (0.6 m resolution) or 336 km^2/h (1 m resolution)
Groundspeed: 93-167 km/h
Altitude: 7,620 m
Power: 1,050 W
Cooling: ambient air
MTBF: >700 h
Volume: 0.12 m^3
Dimensions (l × w × h): 453 mm (antenna assembly total height); 457 × 188 × 190 mm (antenna assembly box); 470 mm (antenna Ø); 482 × 326 × 261 mm (exciter/transceiver); 528 × 233 × 378 mm (processor)
Weight: 74.9 kg

Contractor

Northrop Grumman Electronic Systems sector, Baltimore, Maryland.

UPDATED

B-52 terrain avoidance and ground mapping radar

Type

8 to 12.5 GHz band terrain avoidance and ground mapping radar.

Description

This 8 to 12.5 GHz band equipment is the terrain avoidance and ground mapping radar fitted to the US Air Force's (USAF) B-52H heavy bombers. The system has been upgraded to enhance its reliability, maintainability and operational capabilities. As a part of this process, analogue components have, in many cases, been replaced with digital devices. The current configuration incorporates 16 distinct units, eight of which replace 50 elements in the original radar. Functionally, the system provides a +45° azimuth elevation contour for terrain avoidance and features a 360° scan mode for ground mapping. Boeing (the prime contractor) has developed the equipment's antennas, interface electronics and displays, while Northrop Grumman's Norden Systems facility is responsible for its transceiver/modulator, radar processor and display generator.

Status

As of December 2001, this Boeing/Northrop Grumman Norden radar was in service aboard USAF B-52H heavy bomber aircraft and was expected to remain operational until at least 2010.

Contractors

Boeing Aerospace Support Maintenance and Modification Center, Wichita, Kansas.
Northrop Grumman Electronic Systems sector - Norden Systems, Norwalk, Connecticut.

UPDATED

Hughes Integrated Surveillance And Reconnaissance (HISAR™) system

Type

Modular, multisensor surveillance and reconnaissance system.

Description

HISAR™ is a surveillance and reconnaissance suite that is built around a synthetic aperture/Moving Target Indicator (MTI) radar that derives from the AN/APQ-181 (see separate entry) and Advanced Synthetic Aperture Radar System 2 (ASARS-2 - see separate entry) equipments that Raytheon Systems has developed for the US Air Force's (USAF) B-2 bomber and U-2S reconnaissance aircraft respectively. The baseline HISAR™ suite is a radar only system that can be customised to meet specific requirements via the addition of signals intelligence, forward-looking infra-red and long-range optical sensors together with communications, datalink and ground station packages. The HISAR™ radar is described as being an all-weather, day or night equipment that offers wide area moving target, combined synthetic aperture/MTI strip, synthetic aperture spot, sea surveillance and air-to-air operating modes. It is noted as being suitable for installation aboard a wide range of airframe types which includes turboprop/jet business aircraft through to larger platforms such as the Lockheed Martin P-3 Orion.

Status

As of this edition, HISAR™ radar variants are in service aboard the US Army's RC-7B Airborne Reconnaissance Low - Multisensor (ARL-M) reconnaissance aircraft and have been test flown in a company Raytheon Beech King Air 2000 demonstrator in connection with the USAF's Global Hawk unmanned aerial vehicle programme. The US State Department is reported to have granted export approval for HISAR™ systems to more than 60 countries and in October 1997, Raytheon Systems announced its first international HISAR™ sale (thought to be a German customer). It should be noted that despite Hughes' absorption by Raytheon, this radar appears to retain its original HISAR designation and that synthetic aperture radar sensor installed on the Global Hawk long endurance reconnaissance unmanned aerial vehicle is reported to make use of HISAR radar hardware.

Specifications

Area coverage: 10 km^2 (SAR spot mode); 10,000 km^2 in 30-120 km sectors (wide area MTI mode)
Swath: 37 km (combined SAR/MTI strip mode)
Resolution: 1.8 m (SAR spot mode); 6 m (combined SAR/MTI strip mode)
Detection range: 20-110 km (combined SAR/MTI strip mode)

Contractor

Raytheon Electronic Systems, El Segundo, California.

VERIFIED

Lockheed Martin Advanced Imaging Radar System (LAIRS)

Type

Synthetic Aperture Radar (SAR) system.

Description

LAIRS is a development of the Advanced SAR System 1 (ASARS-1) which Lockheed Martin produced for use in the USAF's Mach 3 SR-71A reconnaissance aircraft. It uses the latest SAR techniques and is composed of both airborne and ground-based subsystems. The ground element supports mission planning, radar data collection, airborne system maintenance, image processing and image exploitation.

The design of LAIRS is such that the specific system configuration can be tailored to the user's requirement, or individual mission requirements. The airborne radar equipment can include an onboard high-density magnetic tape recorder, a real time datalink and an in-flight processor and display. The datalink can include an optional narrowband up-link to allow ground control of the collection of data by the radar system. The ground processing equipment can be configured to include electronic image outputs for a soft copy exploitation system, image hard copy generation and generation of computer compatible image products. The recommended image exploitation equipment consists of soft copy image exploitation stations that are designed to allow interpreters to make the maximum use of the high-quality imagery generated by LAIRS. The number of workstations, the management of imagery data and hard copy image exploitation equipment are a selection the user can specify to tailor the exploitation equipment to his needs.

LAIRS provides three imaging modes which produce imagery for exploitation purposes and one mode for navigation or target detection purposes. The Swath mode is a wide swath, high-resolution imaging mode intended for area surveillance and is similar to the imaging modes provided by other synthetic aperture radar systems. This mode generates a 19 km swath image with an along track extent that is determined by the mission plan. The Fixed/Moving Target Indicator (F/MTI) mode is similar to the Swath mode except that the range extent of the swath is set at 9 km and provides simultaneous detection of moving targets and the generation of high-quality images. The Navigation mode is similar to the Swath mode, except the image data is generally not used for exploitation purposes. Instead, the radar information is processed into a moderate resolution image and displayed to the flight crew when the radar system includes the optional in-flight processor. When the Navigation mode is used to image a target of known location, the flight crew can interact with the radar equipment to determine the current navigation error. The measured aircraft position location error can then be used to correct the aircraft navigation system (provided that the navigation system has the update capability), or the radar system can account for the error to ensure that subsequent collection will cover the desired targets. The Navigation mode can be used to search large areas and allow the flight crew to designate specific targets for imaging using the high-resolution mode, as is usually required in a sea surveillance mission.

Status

As of this edition, a LAIRS variant was reported as having been installed aboard a US Customs Service P-3 surveillance aircraft and the sensor may have been selected by South Korea for use on four Hawker 800XP radar surveillance aircraft.

Specifications

	Swath mode	F/MTI mode	Navigation mode
Min slant range	<19 km	<37 km	<19 km
Max slant range	185 km	185 km	185 km
Angle swath width	19 km	9 km	19 km
Azimuth swath width	continuous	continuous	continuous
3 dB impulse response	3 m	3 m	3 m

Contractor

Lockheed Martin Tactical Defense Systems, Litchfield Park, Arizona.

VERIFIED

Lockheed Martin aerostat-borne radar systems

Type

Family of D-band (1 to 2 GHz) aerostat-borne surveillance radar.

Description

Lockheed Martin Ocean, Radar and Sensor Systems supports the L-88 radars which have been retrofitted to the USAF's 'Seek Skyhook' surveillance aerostats and installed aboard a further balloon located at Lajas in Puerto Rico. Available in a number of variants, L-88 is a solid-state, dual-channel, fully coherent D-band radar which is described as being able to simultaneously track air and surface targets out to ranges of 370 km. As of this edition, identified family members comprise the L-88(V)2 and L-88A. Both based on the AN/DPS-5 (see following), AN/FPS-117 and AN/TPS-59 radars, the cited equipments differ from one another in the L-88A incorporating eight fewer row antennas than the L-88(V2) and weighing 317.5 kg less than the latter. General L-88 series system features include:

- automatic clutter locking to enhance performance consistency and clutter cancellation
- continuous, automatic monitoring and adjustment of system performance to facilitate real-time adjustment to changes in target dynamics, environment and weather conditions
- use of the radar's antenna pattern and sensitivity time control to maintain uniform coverage over the sensor's 'entire' range
- dual Moving Target Indicator (MTI) channels
-]an integral Litton Data Systems digital target extractor
- MTI and non-MTI video options to eliminate blind spots and facilitate manual maximum range adjustments and automatic compensation for changes in altitude, weather and terrain
- automatic target report processing
- automatic compensation for aerostat blow-down
- continuous performance assessment
- integration with the Apple Macintosh OS-based Lockheed Martin Radar Display Unit.

Status

The L-88 aerostat radar traces its history back to experimental work undertaken by the then General Electric company (subsequently Martin Marietta and Lockheed Martin Government Electronic Systems) during the late 1960s. This resulted in the 1974, four aerostat 'Seek Skyhook' programme. As initially delivered, the 'Seek Skyhook' vehicles were fitted with the AN/DPS-5 radar that is described as being a single channel E/F-band (2 to 4 GHz) surveillance sensor. Equipped with a 6.7 × 3.8 m parabolic antenna (rotating at a rate of 5 rpm), DPS-5 is quoted as being able to detect targets out to a range of 275 km when operated from an altitude of 3,658 m. Three of the four 'Seek Skyhook' systems were located at Cudjoe Key and Cape Canaveral (both in Florida) during the early stages of their service life and as noted above, have been upgraded with the L-88 radar. Alongside 'Seek Skyhook', Lockheed Martin also produced a shipborne surveillance aerostat equipped with I-/low J-band (8 to 12 GHz) AN/APS-128J radars which were used by the US Army to monitor sea and air traffic at five selected 'choke' points in the Caribbean during 1985-86. Readers should note that as of this edition, all US aerostat radar systems appear to have been gathered into the USAF's Tethered Aerostat Radar System programme which has operating sites (from west to east) at Yuma, Arizona; Fort Huachuca, Arizona; Deming, New Mexico; Marfa, Texas; Eagle Pass, Texas; Rio Grande, Texas; Matogorda, Texas; Morgan City, Louisiana; Horseshoe Beach, Florida; Cudjoe Key, Florida and Lajas, Puerto Rico.

Specifications

Coverage
Range: 19.5-278 km or 19-370 km (L-88(V2) and L-88A).
Azimuth: 360° (with four operator-defined inhibit sectors - L-88(V2) and L-88A).
Altitude: ground level to at least 610 m above aerostat flight altitude (L-88(V2) and L-88A)
Antenna
Type: shaped paraboloid (L-88(V2) and L-88A)
Rows: 8 (L-88A); 16 (L-88(V2))
Dimensions: 5.2 x 8.8 m (L-88(V2) and L-88A)
Polarisation: vertical (L-88(V2) and L-88A)
Stabilisation: gravity dampened (L-88(V2) and L-88A)
Azimuth/elevation beamwidth*: 1.9°/3.5° (L-88(V2) and L-88A)

US Army 'choke' point monitoring aerostat

Tilt: -1.5° (adjustable - L-88(V2) and L-88A)
Gain*: 35.5 dB (L-88(V2) and L-88A)
Azimuth sidelobes*: -22 dB or greater (L-88(V2) and L-88A)
Rotation rate: 5 rpm (L-88(V2) and L-88A)
Transmitter
Type: solid-state (L-88(V2) and L-88A)
Frequency: 1,215-1,400 MHz (L-88(V2) and L-88A)
Frequency diversity/agility: 15 MHz (L-88(V2) and L-88A)/pulse-to-pulse (L-88A - two pairs, L-88(V2) - five pairs)
Pulse-width: two sequential pulses (each 130 μs or 160 μs - L-88(V2) and L-88A)
PRF: 369, 304 Hz (average - L-88(V2) and L-88A)
Peak/average power*: 9 kW (measured at output of circulator)/801, 875 W (L-88A); 16.6 kW (measured at output of circulator)/1,478, 1,614 W (L-88(V2))
Modulation: FM (L-88(V2) and L-88A)
BT product: 162,200 (L-88(V2) and L-88A)
Receiver
STC: digital, eight functions of range, eight operator-definable sectors, totalling 31 dB (L-88(V2) and L-88A)
Dynamic range: 90 dB (including STC - L-88(V2) and L-88A)
IF frequencies: 75 and 390 MHz (L-88(V2) and L-88A)
Channels: two (L-88(V2) and L-88A)
System noise temperature*: 418°K (includes antenna and line loss - L-88(V2) and L-88A)
Sensitivity: -111 dBm, S=N (uncompressed - L-88(V2) and L-88A)
Signal processing
Type: digital, two independent channels, each with MTI and normal video (L-88(V2) and L-88A)
Modes: 278 km, 370 km (L-88(V2) and L-88A)
Sample rate: 1.67 MHz (L-88(V2) and L-88A)
Pulse compression: 162:1 and 200:1 (L-88(V2) and L-88A)
Weighting: linear FM with hamming weighting or non-linear FM operator selection (L-88(V2) and L-88A)
MTI filters: 8 (L-88(V2) and L-88A)
MTI filter notch width: 9, 14, 19, 23, 28, 37, 46 or 56 km/h, depending on radar mode and operator selection (L-88(V2) and L-88A)
Clutter lock: approximately 3,000 independent areas (L-88(V2) and L-88A)
Video integrator: sliding window, 2, 8, 16 and 32 IPPs (L-88(V2) and L-88A)
A/D converters: 12-bit (including sign - L-88(V2) and L-88A)
DC offsets: automatic correction (L-88(V2) and L-88A)
CFAR: sliding window, 60 cells (L-88(V2) and L-88A)
Log/FTC: Lin/Log function plus operator-selectable FTC (L-88(V2) and L-88A)
Non-MTI/MTI selection: automatic or manually selected range (L-88(V2) and L-88A)
MTI improvement factor: greater than 65 dB mountain clutter (L-88(V2) and L-88A)

* *system parameters that vary over the transmission band are specified at band centre of 1,300 MHz*

Contractor

Lockheed Martin Naval Electronics and Surveillance Systems - Syracuse, Syracuse, New York.

VERIFIED

Lockheed Martin Small Tactical Synthetic Aperture Radar (STacSAR™)

Type

J-band (10 to 20 GHz) Synthetic Aperture Radar (SAR) system.

Description

The Small Tactical Synthetic Aperture Radar (STacSAR™) is a lightweight (less than 29.5 kg) SAR system produced for use on unmanned aerial vehicles or light

The radar electronics package used in STacSAR™

The antenna/gimbal assembly used in STacSAR™

The STacSAR™ radome configuration as used in a light aircraft application

manned aircraft. It uses the latest radar technology and processing techniques. STacSAR™ provides fixed target imagery in both spotlight and search modes, as well as moving target indications. Onboard processing is included, allowing real-time exploitation. STacSAR™ is a modular design with inherent flexibility. System parameters are programmable under software control. SAR modes include both a strip map search mode and a higher resolution spotlight mode along with a moving target indicator mode. The user can specify mode parameters in selecting range and azimuth resolutions and swath widths. Transmitter pulse-width and repetition frequency can also be varied to achieve maximum average power for selected modes. The system is designed for datalink operation where imagery can be downlinked in real time to user ground stations. STacSAR™ consists of two subsystems: an antenna/gimbal assembly and an electronics package.

Status
As of this edition, the STacSAR™ radar was reported as having begun a trials programme during 1996.

Specifications
Wavelength: J-band (17 GHz)
Resolution: 0.5-50 m programmable

Swath: 500-5,000 m programmable
Slant range: 10 km (with 12 W transmitter)
Weight: < 29.5 kg

Contractor
Lockheed Martin Tactical Defense Systems, Litchfield Park, Arizona.

VERIFIED

....

Lynx Synthetic Aperture Radar (SAR)

Type
J-band (15.2 to 18.2 GHz sub-band) SAR.

Description
The Lynx SAR is a lightweight, high-resolution equipment that is designed for Unmanned Aerial Vehicle (UAV) applications and comprises Radar Electronics Assembly (REA) and Sensor Front-End/Gimbal Assembly (SFE/GA) packages. Of these, the REA houses the system's radar control, waveform generation, up-conversion, receiver, video, Analogue-to-Digital Conversion (ADC) and signal processing elements, all of which are housed in a custom VME chassis. In more detail, the SAR's Radio Frequency (RF)/microwave functions are contained in a set of five (STABLO, up-converter, frequency, receiver and RF interconnect) VME modules. Digital waveform synthesis is accomplished on a custom VME board that generates a chirp with 42-bit precision at 1 GHz. Another custom board (operating at 125 MHz and providing 8-bit data) facilitates ADC, while the equipment's signal processor comprises 16 nodes of Mercury Computer Systems RACEway connected, 200 MHz, Power personal computers. These implement a scalable architecture for image formation. Four additional nodes are used for other radar functions including motion measurement, radar control and optional data recording. The Lynx SFE/GA contains the system's antenna, motion measurement hardware and its front-end components (including transmitter travelling wave tube and receiver low-noise amplifiers). The gimbal itself is a three axis, custom designed assembly while the antenna is a vertically polarised, horn-fed unit. Motion measurement is undertaken by a carrier-phase, Global Positioning System (GPS) - assisted, inertial navigation system that centres on a Litton LN-200 fibre-optic inertial measurement unit (augmented by an Interstate Electronics Corporation GPS receiver).

Image formation within the Lynx system is by means of stretch processing and the radar has four primary operating modes as follows:

Geo-Referenced Stripmap (GRS) mode
In GRS mode, the radar's operator specifies a precise strip on the ground that is to be imaged. The sensor then patches together a continuous and seamless strip of images to yield the required coverage. Image strips can be formed on either side of the host air vehicle, which is not constrained to fly parallel to the strip area during image collection.

SAR Transit Stripmap (STS) mode
In STS mode, the radar's operator specifies a range from the air vehicle to a target line and the sensor forms a stripmap parallel to the line-of-flight. The radar patches together a continuous and seamless strip of images until commanded to do otherwise or until the air vehicle deviates too far from the required flight track. In the event of such an occurrence, the radar immediately starts building a new stripmap.

SAR Spotlight (SS) mode
In SS mode, the radar's operator specifies the co-ordinates of a target and the sensor dwells on that location until commanded to do otherwise or until its imaging geometry limits are exceeded. Lynx can produce spot images on both sides of its host vehicle's line-of-flight and incorporates an automatic zoom feature that can generate sequential images at ever-finer resolutions until the system's limits are reached.

Ground Moving Target Indication (GMTI) mode
In GMTI mode, the Lynx SAR is reported as being able to scan over a 270° sector. Alongside the foregoing, the Lynx SAR also incorporates coherent change

The Lynx SAR as installed aboard the I-GNAT UAV
(Sandia National Laboratories)   2001/0093004

Imagery generated by the Lynx SAR (Sandia National Laboratories)

2001/0093005

detection in order to allow the detection of 'subtle' scene changes between one image and the next.

Status

As of this edition, the Lynx SAR was reported as being available. Lynx is a collaborative venture between UAV builder General Atomics (GA) and the Sandia National Laboratories, with work starting on the project during 1996. Flight testing of the radar (with the sensor mounted in a Twin-Otter manned testbed aircraft) is reported to have taken place during the period July 1998 to February 1999. The sensor's first flight aboard a UAV (a GA I-GNAT vehicle) occurred during March 1999. As of January 2000, the Sandia National Laboratories had produced two Lynx radars with GA building a third. As of the given date, series production of the system was to be undertaken by GA. Despite its use of custom boards, Lynx is noted as being essential a commercial-off-the-shelf equipment. During May 2000, Jane's sources were reporting that the Lynx SAR was one of two radar options being studied by Italy for use on the six Predator surveillance UAVs that it was intending to procure. The radar is also known to have undergone operational testing aboard a US Army C-12 aircraft in the Balkans.

Specifications

Frequency: 15.2-18.2 GHz
Power output: 320 W (35% duty cycle)
Antenna beamwidth: 3.2° (azimuth); 7° (elevation)
System noise figure: 4.5 dB (approximately)
System weight: 52 kg
Stripmap SAR mode
Resolution: 0.3-3 m (azimuth and slant range)
Range: 7-30 km (slant range, 3-60 km with reduced performance)
Ground swath: 2,600 pixels (16 node system, 3,500 pixels at coarser resolutions)
View size: 934 m (0.3 m resolution, 45° depression)
Squint angle: ± 45°-135° (scene centreline versus aircraft velocity vector)
Spotlight SAR mode
Resolution: 0.1-3 m (five selectable scales)
Range: 4-25 km (slant range, 3-60 km with reduced performance)
Patch size: 2 x (640 x 480) pixels
View size: 640 x 480 pixels (over NTSC video link)
Squint angle: ± 45°-135° (0.15 m resolution or coarser, scene centreline versus aircraft velocity vector); ± 50°-135° (0.1-0.15 m resolution, scene centreline versus aircraft velocity vector)
GMTI mode
Minimum detectable velocity: 11 km/h
Range: 4-25 km (slant range)
Angular coverage: ± 135° (total possible swept angle)
Ground swath: 10 km
Minimum detectable target: +10 dBsm
Maximum clutter: -10 dBsm/m² (average distribution)

Contractors

Sandia National Laboratories, Albuquerque, New Mexico.
General Atomics, San Diego, California.

UPDATED

Multirole Electronically Scanned Array (MESA) radar

Type

D-band (1.2 to 1.4 GHz sub-band) Airborne Early Warning (AEW) radar.

Description

As applied to the Boeing 737-700 AEW and Control (AEW & C) platform, the MESA radar makes use of a static, electronically scanned antenna array that is mounted above the aircraft's rear fuselage. Within this assembly's radome, the array is divided into lateral (two) and 'top hat' sub-arrays that are made-up of 'hundreds' of silicon carbide, solid-state, transceiver modules. The sensor's operating modes are noted as being spot, high Pulse Repetition Frequency (PRF) and low PRF. Of these, the spot mode is used for tracking high priority targets and 'enhanced' long-range performance, while the high PRF option being aimed at air-to-air targets. MESA is noted as being able to interleave modes during a single scan period, the duration of which can be varied between three and 40 seconds. Sector width is constrained by the sensor's bandwidth and the spot mode incorporates an increase in dwell period when compared with the radar's other operating options. Each of the sensor's sub-array is reported as incorporating an Identification Friend-or-Foe (IFF)

capability that is used when the radar function is at 'off'. Accordingly, the 'top hat' array can operate as an IFF sensor when the lateral arrays are transmitting radar energy and vice versa.

Status

As of this edition, the MESA AEW radar was reported as having been bid for both Australia's Project 'Wedgetail' and an outstanding Turkish airborne early warning and control requirement.

Specifications

Frequency: 1.2-1.4 GHz
Antenna coverage: 60° (azimuth, fore and aft, 'top hat' array); 120° (azimuth, lateral arrays)
Detection range: 350 km (max)
Scan duration: 3-40 s (operator variable)
Beamwidth: 2-8° (operator variable)

Contractor

Northrop Grumman Electronic Systems sector, Baltimore, Maryland.

UPDATED

RDR-1400C search-and-rescue and weather avoidance radar

Type

I/low J-band (8 to 12.5 GHz sub-band) search-and-rescue and weather radar.

Description

The RDR-1400C search-and-rescue and weather radar is designed for both rotary- and fixed-wing applications and aside from its primary modes, can be used for oil slick detection. As such, the system incorporates three specialised search modes, designated as 'Search 1' (incorporating sea clutter rejection to facilitate small boat and buoy detection down to a minimum range of 274 m), 'Search 2' (high resolution, precision ground mapping) and 'Search 3' (ground mapping and the detection and tracking of sea surface phenomena such as oil slicks). The sensor also offers a beacon mode that when operating with a graphics interface unit, automatically displays target longitude and latitude with navigation sensor data on a moving map background. A white course line (rotatable through 360° around the centre of a target beacon) offers a digital flightpath to the beacon from any direction. Left and right deviation cues are provided by a deviation control bar that is located on the radar's indicator. In 'OBS trac' mode, RDR-1400C provides an additional course selection option. Here, with the radar operating in weather or search mode, a course bearing cursor is generated from the host aircraft's position and controlled by its horizontal situation indicator course selector.

Status

As of May 2001, Telephonics was reporting sales of more than 8,000 RDR series radars worldwide. As of the given date, RDR-1400C radars were noted as being installed aboard C-130, C-141, CASA C-212, C-295 and CN-235, Dornier Do 228, F-27, L-410 and Raytheon Beech King Air fixed-wing aircraft together with A-109, AB-212, AB-412, AS 332 Super Puma, Bell 214, BK 117, Ecureuil/Fennec, EH 101 Cormorant, Eurocopter 145, 155 and 365, HH-60, MBB 105, S-76, Sea King and Super Lynx helicopters.

Specifications

Frequency: 8-12.5 GHz
Power output: 10 kW
Antenna: 254, 305, 457 or 305 × 457 mm
Scan angle: 60 or 120°
Scan rate: 28°/s
Display range/marks: 0.9/0.2 km, 1.8/0.5 km, 4/0.9 km, 9/2 km, 19/5 km, 37/9 km, 74/19 km, 148/37 km, 296/74 km or 444/111 km
Min tracking range: 274 m
Beacon range: LOS to 296 km
Power requirement: 3 A at 115 V AC (400 Hz); 4.2 A at 28 V DC
Temperature: –40 to +55°C (indicator); –50 to +55°C (transceiver)
Dimensions (w × h × d): 127 × 159 × 352 mm (transceiver); 159 × 159 × 276 mm (indicator)
Weight: 5.22 kg (indicator); 6.58 kg (transceiver)

Contractor

Telephonics Corporation - Command Systems Division, Farmingdale, New York.

UPDATED

RDR-1500B multimode surveillance radar

Type

I-band (8 to 10 GHz) airborne search and surveillance radar.

Description

RDR-1500B is a development of the RDR-1300C radar fitted to US Coast Guard HH-65A helicopters. It is designed to provide multimode radar capability for

The RDR-1500B search and surveillance radar

helicopters and fixed-wing aircraft engaged in low- and medium-altitude maritime missions. The primary mission of the system is for airborne search and surveillance on sea search operations. Secondary missions include terrain-mapping, weather avoidance, beacon navigation and the display navigation information from the aircraft navigation system.

The RDR-1500B is a lightweight, I-band (8 to 10 GHz), 360° digital colour radar system consisting of six line-replaceable units: the RT-1501A transceiver, the IN-1502A or -1502B multifunction colour indicator, a CN-1506A control panel, an AA-1504A antenna, a DA-1203A or DA-1503A antenna drive and an IU-1507A interface unit. Of these, the RT-1501A transceiver functions as a short-range pulse radar for high-resolution sea search and terrain-mapping and as a long-range pulse radar for long-range sea search, terrain-mapping and conventional weather avoidance. The 216 × 241 mm IN-1502A indicator provides a continuous, three-colour display of radar targets (including ground mapping and beacon modes) and VHF omnidirectional radio ranging and navigation data. The 127 × 152 mm IN-1502B indicator provides similar capabilities and is designed for limited space applications. Standard display modes include aircraft heading reference, north oriented and ground reference displays. The system also can offset the sweep centre to any location on the display. Additionally, the radar has target marker capability, allowing the operator to determine range and bearing (or latitude/longitude) of a target from the aircraft, and also relative range and bearing between targets.

The 991 × 227 mm AA-1504A antenna is a flat-plate, phased-array equipment that is the radar's baseline. Other antenna formats are available for installations if the AA-1504A assembly is not suitable. Of the antenna drive options, the pitch, roll and tilt line of sight stabilised DA-1503A assembly covers 360° in azimuth and ±25° of true vertical in elevation. The DA-1203A assembly is designed for applications requiring 120° sector scan functionality. System options include track-while-scan, a forward-looking infra-red sensor pointing output, a target datalink and a video link interface. Overall, RDR-1500B provides surface search, terrain mapping, beacon navigation and weather avoidance operating modes. System options include sector scan, navigation interfaces (area and inertial navigation, Decca hyperbolic, Doppler, GPS, LORAN C, VLF/Omega and VOR/DME), forward-looking infra-red sensor (with pointing capability) and low-light television/video interfaces, multifunction colour indicators (127 × 152 mm or 216 × 241 mm), ANVIS compatibility and track-while-scan (up to 20 targets).

Status

Over time, RDR-1500B radars are reported as having been installed aboard a range of aircraft and helicopters that includes the AgustaWestland 109, Lynx and Sea King, Antonov An-32, Bell 230 and 412, British Aerospace Jetstream 31, CASA 212, Convair T-29, deHavilland Canada DHC-7, Dornier Do 228, Eurocopter BO-105, Dauphin, Fennec and Super Puma, Fokker F27, Heli-Dyne Sentinel, Piaggio P166, Raytheon Beech King Air C90 and 200, Reims Caravan 2 and Shorts Sherpa. The RDR-1500B variant is noted as having been co-developed with FIAR of Italy.

Specifications

Frequency: 9,375 MHz (transmit/receive, surface search/weather mode); 9,310 MHz (receive, beacon mode)
Power output: 10 kW peak (nominal)
Ranges: 1.16 km; 2.32 km; 4.63 km; 9.26 km; 18.52 km; 37.04 km; 74.08 km; 148.16 km; 296.32 km
PRF: 200, 800 or 1,600 Hz
Pulse-width: 0.1, 0.5 or 2.35 µs
Antenna gain: 31 dBi (AA-1504A)
Antenna beamwidth: 2.6° (azimuth); 10.5° (elevation)
Scan angle: 120° (sector scan); 360°
Scan rate: 28°/s (120° sector scan); 45-90°/s
Interfaces: ARINC 407, 419 and 429; SIN COS; XYZ
Weight: 34.4 kg (complete system)

Contractor

Telephonics Corporation - Command Systems Division, Farmingdale, New York.

UPDATED

RDR-1600 search-and-rescue and weather avoidance radar

Type

I/low J-band (8 to 12.5 GHz sub-band) search-and-rescue and weather avoidance radar.

Description

RDR-1600 addresses the needs of modern 'glass cockpit' fixed-wing aircraft and helicopters and is described by its manufacturer as being a 'state-of-the-art' digital equipment that is '25 per cent lighter in weight and 25 per cent lower in power consumption than the industry standard radar'. As such, the radar has five primary operational modes and its system features include:

- suitability for weather detection/target alerting, search and rescue, surveillance and oil slick detection and mapping tasks
- built-in test circuitry
- normal and precision ground mapping
- a beacon tracking mode
- a narrow pulse precision approach mode (137 m minimum range) to meet offshore oil platform and other precision landing requirements
- ARINC 429 and 453 interfaces
- enhanced clutter detection.

Status

As of June 2001, Telephonics was actively promoting the RDR-1600 search-and-rescue and weather avoidance radar.

Specifications

Frequency: 8-12.5 GHz
RF power output: 10 kW
Antenna size: 254, 305, 457 or 305 × 457 mm
Scan angle: 60 or 120°
Scan rate: 28°/s
Display range/marks: 0.9/0.2, 1.9/0.5, 3.7/0.9, 9.3/2.3, 37/9.3, 74/19, 148/37, 296/74 or 445/111 km
Min detection range: 457 m (weather mode); 137 m (search mode)
Beacon range: LOS or up to 296 km
Power requirement: 3 A at 28 V DC; 3 V A at 115 V AC (400 Hz); + 5 V at 4 A for panel light (when used)
Temperature: −50 to +55°C (transceiver and control panel)
Dimensions (w × h × d): 127 × 159 × 352 mm (transceiver); 159 × 159 × 276 mm (control panel); 168(Ø) × 195(D) mm (254 mm antenna); 194(Ø) × 195(D) mm (305 mm antenna); 270(Ø) × 195(D) mm (457 mm antenna)
Weight: 0.8 kg (control panel); 3.4 kg (254 mm antenna); 3.5 kg (305 mm antenna); 4 kg (457 mm antenna); 7.2 kg (transceiver)

Contractor

Telephonics Corporation - Command Systems Division, Farmingdale, New York.

UPDATED

RDR-1700

Type

I-band (8 to 10 GHz) search, surveillance and weather avoidance radar.

Description

Intended to be a successor to Telephonics' RDR-1500B radar (see separate entry), the I-band RDR-1700 search, surveillance and weather avoidance radar is specifically designed for integration with 'glass cockpit' configured aircraft. Comprising an antenna/pedestal unit, a transceiver and an interface unit, RDR-1700 makes use of an ARINC-429 interface to facilitate system control via 'glass cockpit' multifunction displays and has, as its primary role, search and surveillance. Secondary roles include terrain mapping, weather avoidance, beacon navigation and oil slick detection and the radar can be configured to scan over the full 360° arc or a 120° sector. Antenna options comprise 737 × 229 mm, 838 × 229 mm and 991 × 229 mm 360° scan units together with 254 mm Ø, 305 mm Ø, 457 mm Ø and 457 × 305 mm 120° scan assemblies. System options include multiple interfaces for navigation (area and inertial navigation, Decca hyperbolic, Doppler, GPS, LORAN C, VLF/Omega and VOR/DME), forward-looking infra-red (including pointing) sensors and low-light television/video cameras; a 20 target track-while-scan mode and a joystick controller.

Status

On 11 June 2001, Telephonics announced that AgustaWestland was procuring a maximum of 23 RDR 1700 surveillance and weather avoidance radar for installation aboard naval helicopters.

Specifications

Frequency: 9,310 MHz (beacon mode receive); 9,375 MHz (transmit and search/weather mode receive)
Transmitter power output: 10 kW (nominal)
PRF: 200, 800 or 1,600 Hz
Pulse-width: 0.1, 0.5 or 2.35 ms
Antenna gain: 26-31 dBi (antenna dependent)

The line-replaceable units that make up the RDR-1700 surveillance and weather avoidance radar (Telephonics)   2002/0114013

Scan angle: 120° sector or full 360°
Scan rate: 28°/s (120°), 45°/s (360°) or 90°/s (360°)
Ranges: 1.16, 2.32, 4.63, 9.26, 18.52, 37.04, 74.08, 148.16 and 296.32 km
Operating modes: beacon navigation, surface search (primary), terrain mapping, and weather avoidance
Situation display: aircraft heading, beacon location/identification, fixed ground reference (true motion), navigation waypoint overlay, north reference, search pattern, target marker (range/bearing or latitude/longitude) and variable display offset
Integral interface: ARINC 429
Weight: 23 kg (120° scan configuration); 27 kg (360° scan configuration)

Contractor
Telephonics Corporation - Command Systems Division, Farmingdale, New York.

UPDATED

Sea Vue surveillance radar

Type
Family of lightweight, modular I-band (8 to 10 GHz) surveillance radars.

Description
The Sea Vue (SV) family is made up of a series of lightweight, modular surveillance radars that combine technology from Raytheon Systems' AN/APS-134(V) and -137(V) radars with an open-architecture Receiver/Exciter/Synchroniser Processor (RESP), an air-cooled transmitter and a range of fixed-tilt or tilt-stabilised antennas. Such radars are suitable for fixed-wing and helicopter applications. Sea Vue radars have three major operating modes, namely:
Search 1
Optimised for the detection of small targets in high sea clutter using pulse compression and fast scan-to-scan processing.
Search 2
Optimised for long-range detection in clutter with coastline/landmass mapping facilities.
Weather
Optimised for the detection and avoidance of adverse weather.

Looking at the various elements making up a Sea Vue radar in more detail, the equipment's transmitter chain incorporates a coherent, solid-state, grid modulator travelling wave tube amplifier and generates linear frequency modulated, fixed, frequency agile or biphase coded waveforms. Sea Vue antennas offer sector and spotlight search options and have an identification friend-or-foe capability. The RESP makes use of linear frequency modulated pulse compression, digital pulse compression, sensitivity time control and automatic gain control. System signal processing features include a constant false alarm rate, pulse-to-pulse integration, high resolution scan-to-scan processing, multitarget track-while-scan, Kalman filter tracking, adaptive aperture Inverse Synthetic Aperture Radar (ISAR) processing and Synthetic Aperture Radar (SAR) processing. The processing chain also includes Raytheon Systems' proprietary SeaVision™ technology that utilises a mix of system parameters and signal processing techniques to enhance target detection in all sea states. Display processing options include plan position indicator and B-scan, multiple high-resolution video formats, ISAR imagery, range profiling (A-scan as an addition to ISAR imagery) and SAR map. Display options include colour, monochrome or flat panel. System growth options include moving target discrimination, an ISAR classification aid, digital map coastline overlays, Doppler beam sharpening, coherent air target look-down, customer specific antenna systems and controls and displays and video recording. Sea Vue radars also include built-in-test.

Status
As of this edition, known Sea Vue radar applications were reported as including Australian Coastwatch Dash 8 and Reims Caravan II patrol aircraft (SV-1022 variant) and Japanese Air Self-Defence Force U-125A search and rescue aircraft (SV-1011 variant). The SV-1021-IS variant (featuring both synthetic and inverse synthetic aperture functions) is also understood to be the preferred maritime/land surveillance radar sensor for the Pilatus PCXII Eagle multimission surveillance aircraft.

Specifications
Frequency: 9.4-9.8 GHz
Power output: 8, 15 or 50 kW (transmitter model dependent)
Detection range: 50 km (15 kW transmitter, 1 m² RCS target, Sea State 3, aircraft at an altitude of 183 m); 352 km (15 kW transmitter, 10,000 m² RCS target, aircraft at an altitude of 7,955 m)
Azimuth coverage: 360°
Scan rate: 6, 60 or 120 rpm
Power requirement: 28 VDC/115 VAC (400 Hz, 3-phase)
MTBF: >500 h
Interfaces: ARINC 429; MIL-STD-1553; RS-170 Video; RS-232; RS-343 Video; RS-422; STANAG 3350B
Nominal Dimensions/Weights:

Unit	W × H × D (cm)	Weight (kg)
Antenna	124* × 65 × -	23
15 kW transmitter	33 × 28.9 × 49.8	30
RESP	39.1 × 25.7 × 49.8	37

* swept diameter with smaller sizes available

Contractor
Raytheon Electronic Systems, McKinney, Texas.

VERIFIED

Side-Looking Airborne Modular Multimission Radar (SLAMMR)

Type
Side-looking long-range surveillance radar.

Description
SLAMMR is a long-range surveillance radar intended for use in maritime patrol, border surveillance and mapping. It can also be used for ice patrol missions, oil pollution detection, search and rescue and fishing area patrol. It is a member of the same side-looking airborne radar family as the AN/APS-131 and the AN/APS-135 (see separate entry) and is identical in many respects.

The aircraft installation consists of seven main subassemblies: antennas, antenna switching unit, receiver/transmitter, display processor, display, control unit and signal processor. The antenna consists of two yaw stabilised, horizontally or vertically polarised, slotted waveguide arrays. These can be mounted back-to-back within a single pod, or individually mounted on either side of the aircraft to provide an unobstructed view.

The remaining units are either pallet- or rack-mounted inside the aircraft for easy removal. The display processor provides radar timing and control functions and, based on inputs from the aircraft inertial navigation system, creates latitude and longitude references for display with the radar imagery. The receiver/transmitter contains the magnetron transmitter and the low noise receiver. The antenna switching unit directs the radiated power and received signal to either the left or right antenna.

The data can also be recorded for future use as well as transmitted to ground stations, in real time, via a radio link. The signal processor is a moving target indicator option that is available for the detection of moving targets at long range for border surveillance. By the use of this processor, surveillance can be carried out at ranges up to 148 km. A high-quality fixed target map is also available for radar mapping and geological exploration.

Status
As of this edition, SLAMMR radars were reported as having been installed aboard Lockheed Martin C-130 and Raytheon Beechcraft 1900 aircraft together with the Indonesian Air Force's three Boeing 737-2X9 Surveiller combined maritime patrol/transports. During the early 1990s, Boeing is reported to have been awarded a four year, US$117 million contract to upgrade the mission suites installed aboard the three Indonesian Surveiller aircraft. Work undertaken is noted as including the introduction of a Boeing developed digital processing and display system, a real-time display for the SLAMMR radar, a nose-mounted search radar and an identification friend-or-foe system. Work on the first aircraft is reported to have been completed by Boeing during the Summer of 1993, with the remaining two machines being modified at the Bandung facility of Indonesian aerospace contractor BPPT/ITPN.

Contractor
Motorola, Scottsdale, Arizona.

VERIFIED

Tactical Unmanned Aerial Vehicle Radar (TUAVR)

Type
Synthetic Aperture Radar (SAR)/Moving Target Indicator (MTI) radar for tactical Unmanned Aerial Vehicle (UAV) applications.

Description
The TUAVR sensor is a SAR/MTI radar that is designed for use aboard the US Army's RQ-7A Shadow 200 tactical surveillance and target acquisition UAV. As

such, the radar offers high resolution SAR strip and spot map modes together with MTI. In its Advanced Technology Demonstration Programme (ATDP - see Status) configuration, TUAVR is described as being mounted in a single chassis and as having a weight and volume of 29 kg and 0.04 m³ respectively. In its production form, the system's overall weight and volume are set to drop to 26 kg and 0.03 m³ respectively. As applied to the RQ-7A, TUAVR is intended to be an 'all-weather, day or night source of battlefield information [that is] organic to the brigade level' and will provide a 'line of sight all-weather precision targeting' capability.

Status

Northrop Grumman was awarded a TUAVR development contract during April 1998, with a prototype system making its first flight (aboard a company Islander testbed aircraft) during November 2000. In March 2001, the sensor was flight tested aboard a US Army RQ-5A Hunter at Fort Huachuca, Arizona. Here, the flight trials amounted to 2.7 hours of air time spread over several days and the radar was integrated with L-3 Communications' Tactical Common DataLink (TCDL) to facilitate the transmission of imagery to a ground station. On 17 July 2001, Northrop Grumman announced that it had delivered two TUAVR radars and spares to the US Army's Communications and Electronics Command's Intelligence and Information Warfare Directorate following the completion of formal US government acceptance testing. In late July 2001, a formal demonstration of TUAVR's capabilities was accomplished at Fort Huachuca, Arizona. Again installed aboard an RQ-5A UAV, the test highlighted sensor performance and resulted in the development of suitable concepts of operation.

Contractor

Northrop Grumman Electronic Systems sector, Baltimore, Maryland.

UPDATED

....

TCOM aerostat-borne surveillance systems

Type

Family of tethered aerostat vehicles designed to support surveillance radars and other electronic equipment at high altitude.

Description

TCOM LP (formerly a subsidiary of Westinghouse and a privately owned partnership since 1989) has constructed and operated a number of aerostats ranging in size from the model 15M (15 m long) to the large model 71M® with a volume of 16,700 m³. Older model aerostats (now out of production) were known as the Systems 250 and 365. As of this edition, several new designs in the 80 to 110 m long class are under consideration in order to carry heavier payloads to higher altitudes. When used in Low Altitude Surveillance Systems (LASS®), model 71M® and System 365 aerostats are used to support large surveillance radars which are able to detect and track small, low-flying aircraft and moving surface targets. Current LASS® applications make use of two radars - variants of the Northrop Grumman AN/TPS-63 (see E-LASS entry) and the Lockheed Martin L-88 (see Lockheed Martin balloon-borne Radar Systems entry) - which both have similar detection ranges and are both dual channel equipments in which one channel is used against air targets and the other for sea or land search.

Status

As of this edition, the following TCOM aerostat surveillance systems have been identified:

Cariball I

TCOM was awarded a contract for the CARIbbean BALLoon (CARIBALL) programme in 1984 and installed a first LASS® system on Grand Bahama Island during the following year. Equipped (according to Jane's sources) with a TPS-63 variant radar, this LASS® application utilised a System 365 aerostat to elevate the sensor to an altitude of 3,050 m from where, it had a 90 per cent single scan probability of detection against a 5 m² air target flying at an altitude of 150 m and at ranges of up to 260 km. CARIBALL I was the first of a chain of aerostat surveillance systems set up around the Caribbean rim to detect illicit smuggling activities.

The E-LASS equipped TCOM Model 71M® surveillance aerostat deployed at Matagorda, Texas as part of the USA's anti-narcotics LASS® system

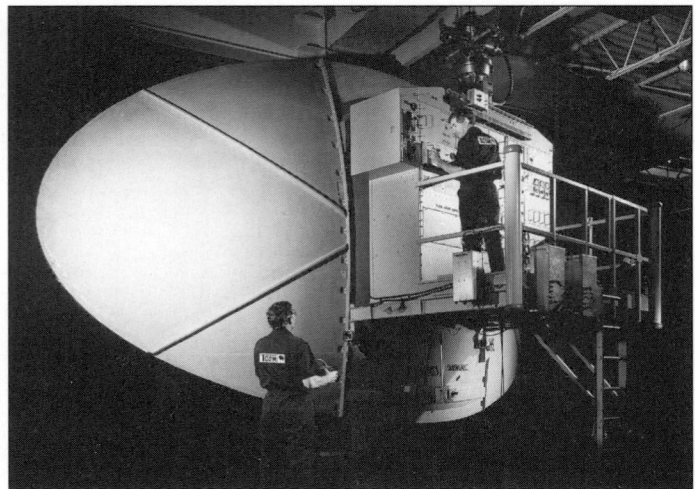

A TPS-63 radar variant is used in both TCOM's CARIBALL and SOWRBALL surveillance aerostats

Israeli Systems

Israel has been operating a mix of System 365 and model 71M® aerostats in southern Israel for what TCOM terms 'several years'. Payloads vary and it is understood that a 365 LASS® system made up of a System 365 aerostat and a dual channel air and ground surveillance radar payload is included in the inventory.

Kuwait LASS®

A TCOM LASS® (using a model 71M® aerostat) was installed and operating in Kuwait at the time of the 1990 Iraqi invasion and was instrumental in the escape of the Emir and his family. This original system was destroyed by the Iraqis and has subsequently been replaced by a new aerostat/radar combination. According to TCOM, this new system has 'recently' made an uninterrupted flight of more than 30 days.

Small Aerostat Surveillance System (SASS)

TCOM reports that it has developed a family of aerostats (ranging in size from the model 25M to the model 32M®) for use with specialised radar and electronic warfare payloads. Using such vehicles, the SASS programme was sponsored by the US Army and involved one land- (located in South Korea) and three sea-based systems (located, according to Jane's sources, in the Caribbean), all of which were equipped with a variant of the AN/APG-66 fire-control radar (see separate entry). So applied, the modified APG-66 is described as having a 3 × 1 m scanner which is capable of 360° rotation and the ability to detect a 2 m² target out to the line-of-sight horizon when operated from an altitude of 915 m. As originally deployed, the SASS systems utilised a 31M model aerostat which was then replaced by the larger 32M® model to provide a greater safety factor for sea-based operations in tropical climates.

SOWRBALL

In 1986, the US Customs Service expanded its aerostat drug barrier to include four new LASS® systems in an effort designated as the SOuthWest Region BALLoon (SOWRBALL) programme. Three systems were based within the continental US (located at Fort Huachuca, Arizona (TPS-63 radar according to Jane's sources); Yuma, Arizona (TPS-63) and Deming, New Mexico (TPS-63)) with the fourth being set up on Great Exuma Island in the Bahamas. This Bahamian system was later redesignated as CARIBALL II. All SOWRBALL/CARIBALL II systems make use of TCOM's model 71M® aerostat which is designed to lift its radar payload to an altitude of 4,575 m.

Tethered Aerostat Antenna Programme (TAAP)

During 1984, TCOM developed a very low frequency communications aerial system which used the tether for a model 25M aerostat as a high-efficiency, long-wire antenna. The TAAP system was successfully tested by both the USAF and the US Navy and has gone on to form the basis of export systems which employ model 32M® to model 50M aerostats to support their antennas.

A sea-based TCOM SASS application

Tethered Aerostat Radar System (TARS)

The TARS programme installed four LASS® systems (using 71M® model aerostats) at Matagorda, Texas (E-LASS radar according to Jane's sources); Morgan City, Louisiana (E-LASS); Horseshoe Beach, Florida (E-LASS) and on Great Inagua in the Bahamas (E-LASS). Each TARS aerostat is designed to carry its radar payload to an altitude of 4,575 m where line of sight range to the horizon is 277 km and air targets can be detected out to 368 km (the maximum instrumented range of the radar). Readers should note that as of this edition, the term TARS appears to have widened and, according to Jane's sources, now describes a USAF run system which utilises both TCOM and Lockheed Martin (see separate entry) aerostat systems located (from west to east) at Yuma, Arizona; Fort Huachuca, Arizona; Deming, New Mexico; Marfa, Texas; Eagle Pass, Texas; Rio Grande, Texas; Matagorda, Texas; Morgan City, Louisiana; Horseshoe Beach, Florida; Cudjoe Key, Florida and Lajas, Puerto Rico.

United Arab Emirates LASS®

A TCOM LASS® (using a model 71M® aerostat) has been installed in the United Arab Emirates. The use of a large volume aerostat is necessary to maintain operational altitude in the hot and humid climate of the Persian Gulf. According to Jane's sources, the radar used in this system is an E-LASS equipment.

Contractor

TCOM,LP, Columbia, Maryland.

VERIFIED

AIRBORNE FIRE-CONTROL RADARS

Introduction

Airborne weapon aiming and fire-control radars are covered in this section, including details of the associated processing and display equipment, together with any interfaces with head-up and head-down display systems and other electro-optical systems. In some instances, particularly in the case of multimode radars, a certain amount of search and surveillance capabilities has been included.

VERIFIED

FRANCE

RBE2 multifunction radar

Type

D- (1 to 2 GHz) through M- (60 to 100 GHz) band multifunction airborne radar.

Description

The RBE2 multifunction radar has been designed for installation aboard the Rafale multirole combat aircraft and comprises an antenna, a travelling wave tube transmitter, a receiver and a processor unit and makes use of electronic scanning via what its manufacturer describes as a 'unique' beam steering lens arrangement. In more detail, a broadband illuminator (consisting of a 'snakeline' waveguide flat plate array) produces and defines the focusing of the radar beam. Two deflector units (consisting of a honeycomb array of diodes that can delay propagation in a selected line of channels) are positioned ahead of the illuminator array and a phase controller induces a progressive delay in each consecutive line of channels, causing the beam to swing through the full scanning angle in a single plane. The approach is analogous to an optical prism, but with a variable angle.

In the air defence role, RBE2 offers four operating modes, as follows:

- search in look-up/look-down mode (whatever the target aspect or altitude)
- identification by an integrated IFF
- automatic multitarget tracking
- dogfight

Target detection is optimised via the use of automatic waveform selection and high (for targets at low altitude and in ground clutter), medium (relatively low speed targets) and low (medium to high altitude targets) pulse repetition frequency modes. In the tracking role, RBE2 offers Track-While-Scan (TWS) and Single Target Tracking (STT) modes. Of these, TWS takes the form of an automatic, multitarget tracking mode that assesses the tactical environment, identifies targets and establishes target priorities. The TWS function is automatic and establishes target numbers even if the volume is outside the radar display's resolution. At any given moment, the eight most dangerous threats are logged and displayed in terms of target, range and closing speed symbology. The pilot can change priorities and the system updates its display as the tactical situation develops. RBE2's TWS mode is designed for use with active radar homing missiles while its STT function covers those weapons that are fitted with semi-active seekers.

In the air-to-ground role, the RBE2 supports the Rafale's all-weather deep strike, close support, battlefield interdiction and maritime strike capabilities. System air-to-ground modes comprise:

- automatic Terrain-Following/Terrain-Avoidance/Threat-Avoidance (TF/TA^2)
- high-resolution mapping
- fixed target search and tracking
- ground moving target search and tracking
- air-to-ground ranging

In the maritime strike role, RBE2 provides long-range target detection, multitarget discrimination and target recognition and assessment facilities. The system is also noted as having 'low observability', as being fully integrated with the overall Rafale weapon system and as interfacing in real time with the platform's electronic warfare suite.

The RBE2 multifunction radar

Status

RBE2 has been developed by a *Groupement d'Intérêt Economique* (Economic Interest Group) made up from elements of the then Thomson-CSF (now part of Thales - two thirds share) and the former Dassault Electronique (subsequently part of Thomson-CSF DETEXIS and now Thales Airborne Systems - one third share). Designed for installation aboard France's next generation Rafale multirole combat aircraft, the prototype RBE2 radar is reported to have been delivered at the end of 1991, with flight tests (aboard a Mystère trials aircraft) starting during July 1992. Thereafter, the RBE2 test programme is noted as involving five Mystère 20s, four Mirage 2000s and three of the Rafale prototypes (starting with aircraft B01). According to Jane's sources, the first production standard RBE2 radar was delivered during 1997, with operational qualification (to F-1 standard) following in 1998.

Contractor

GIE radar ACT. ACM Rafale, Elancourt.

UPDATED

INTERNATIONAL

4A series active radar missile seekers

Type

Family of active radar missile seekers.

Description

4A series active radar seekers are designed for use in air-to-air, surface-to-air and ground-to-air missile systems and feature integral growth potential to match future platform, weapon system and threat environment developments. 4A seekers consist of 10 interchangeable subassemblies that include an agile scanning antenna, multiple receivers, a high-power coherent transmitter, a high spectral purity frequency synthesiser and multiple processor boards. All told, the architecture makes use of 77 application specific integrated circuits of 37 different types. Total system weight (including radome) is given as being below 12 kg.

Status

As of March 2001, 4A missile seeker variants were reported as having been selected by the French Ministry of Defence for use in the MICA air-to-air and EUROSAM ASTER surface-to-air missile systems. A further variant has been proposed for the MBDA Missile Systems (formerly Matra BAe Dynamics) Meteor weapon.

Contractor

Thales Airborne Systems, Elancourt, France.

UPDATED

ADAC Mk 2 active missile seeker

Type

Active anti-shipping missile seeker.

Description

ADAC Mk 2 is the latest version of the active seeker that is used in the MBDA Missile Systems Exocet anti-shipping missile. The new model is electronically and mechanically interchangeable with its predecessor (ADAC Mk 1) and incorporates enhanced electronic counter-countermeasures provision. Functionally, ADAC Mk 2 becomes live after a passive flight phase, carries out a target scan and initiates automatic tracking if an assigned or preprogrammed criteria matching target is detected. Having locked on to a target, ADAC Mk 2 then generates steering commands for its host missile. The final part of the attack profile is conducted at surface skimming altitudes.

Status

As of March 2001, the ADAC Mk 2 anti-shipping missile seeker was reported as having been procured.

Contractor

Thales Airborne Systems, Elancourt, France.

UPDATED

Airborne Multirole multifunction Solid-state Active-array Radar (AMSAR)

Type
Advanced active-array radar.

Description
GTDAR (originally standing for GEC/Thomson/DASA Airborne Radar) is a joint venture between BAE Systems Avionics - Sensor Systems Division in the UK, Thales Airborne Systems in France and the European Aeronautic, Defence and Space (EADS) Company's Systems and Defence Electronics business unit in Germany. As such, its purpose is to develop technology for a future Airborne Multirole multifunction Solid-state Active-array Radar (AMSAR) and it is under contract to the British, French and German Ministries of Defence. All three partners will participate in supplying complete production equipment to meet domestic requirements through GTDAR. The UK's Defence Procurement Executive handles British government involvement, the *Délégation Générale pour l'Armement* looks after the French interests and the *Bundesant fur Wehrtechnik und Beschaffung* acts for Germany. Funding is divided between all three nations and has initially focused on the development of gallium arsenide (GaAs) Monolithic Microwave Integrated Circuits (MMIC), active-array technologies.

Status
Work on the AMSAR programme began during 1993 and is aimed at producing operational radars for use on Eurofighter 2000/Typhoon, the UK's Future Offensive and Future Carrier-Based Aircraft (FOA/FCBA), Mirage 2000-5/-9, Rafale C/M and Tornado F Mk 3 post 2000. As of August 2001, Phase 1 of the AMSAR programme (feasibility/definition studies and the construction of the 200 transceiver module Partial Antenna Demonstrator) was reported as having been completed, with Phase 2 (construction and testing of the 60 cm diameter, 1,000 transceiver module Core Antenna Radar (CAR) demonstration) launched. As of May 2001, flight trials of the CAR demonstrator were scheduled to begin during 2002-2003, with production AMSAR-type equipments coming on stream during 2005-2006. The potential market for AMSAR-type radars has been put at up to 1,000.

Contractors
GTDAR, Elancourt, France (prime).
BAE Systems Avionics - Sensor Systems Division, Edinburgh, UK.
Thales Airborne Systems, Elancourt, France.
EADS - Systems and Defence Electronics, Ulm, Germany.

UPDATED

An impression of the AMSAR radar   0006992

....

Captor multimode radar

Type
I/J-band (8 to 20 GHz) multimode radar.

Description
The Captor sensor (formerly the European Collaborative Radar (ECR) - 90) is a third-generation, coherent I/J-band, pulse-Doppler multimode radar that has been designed specifically for installation aboard the Eurofighter/Typhoon combat aircraft and is the product of the four-nation Euroradar consortium (see following). As such, the equipment is optimised to detect, identify, prioritise and engage targets at ranges greater than the effective range of enemy weapon systems and in conditions of severe electronic jamming. Using BAE Systems' Blue Vixen radar as a starting point, Captor is described as incorporating 'significant' increases (when compared with Blue Vixen) in processing power to 'fully exploit' the high data content within its wideband, spread spectrum transmitter waveforms and to rapidly switch between modes and functions. The equipment also incorporates built-in test routines (to maximise availability) together with 'intelligent automation' in order to optimise it for single seat operation. It is further noted as being compatible with the latest generation of aircraft weaponry (particularly the MBDA Missile Systems

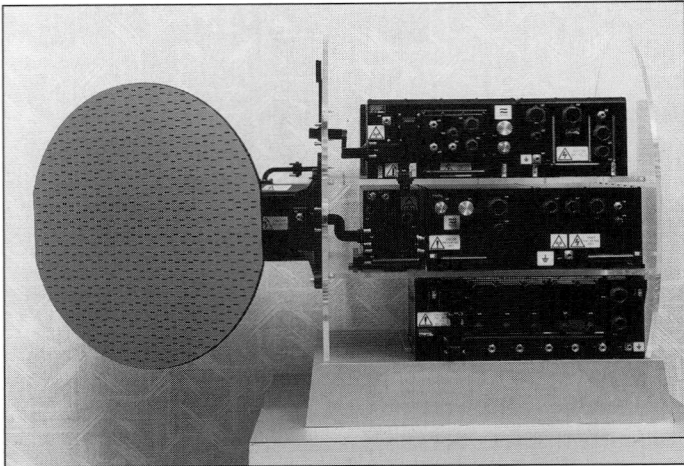

The Euroradar Captor multimode radar that has been designed specifically for use aboard the Eurofighter/Typhoon multirole combat aircraft

Meteor Beyond Visual Range (BVR) air-to-air missile). Other system features include:

- a transducer-computer architecture (using software to achieve complexity)
- all aspect air target detection (look-up/look-down in all conditions
- use of multiple waveforms (high, medium and low pulse repetition frequencies) that are optimally selected for specific roles
- a very high-power aperture with minimal loss
- 'powerful and flexible' electronic-countermeasures provision with inherent growth
- 'life cycle support' design
- automated tracking (multiple target Track-While-Scan (TWS), automatic track initiation and transition, TWS to combat and a data adaptive scan pattern)
- tactical functions including RAID assessment and non-co-operative target recognition
- four selectable combat scan functions
- Optimised for use with the Eurofighter/Typhoon weapons load (including the AIM-120 Advanced Medium Range Air-to-Air Missile (AMRAAM), Advanced Short Range Air-to-Air Missile (ASRAAM), AIM-9 Sidewinder and Meteor BVR missiles)

Alongside those operating modes noted previously, Captor offers

- multimode air search and track
- single air target tracking
- sea surface search
- ground moving target indication
- air-to-surface tracking
- real beam mapping
- high-resolution mapping (scanning and spotlight digital beam sharpening)
- terrain avoidance (contour mapping)
- air-to-surface ranging
- precision velocity update

A Captor installation is described as weighing 193 kg and as incorporating 61 shop-replaceable items packaged into six line-replaceable units. System cooling is by means of air and liquid and the radar is compatible with STANAG 3910C and MIL-STD-1553B databusses for integration into an overall avionics system. Captor is also noted as being able to support the 'latest' airborne interrogation equipment. Of the various elements making up the radar, the system's processor is described as being an 'evolutionary extension' of the architecture used in the Blue Vixen equipment and as making use of surface-mounted device packaging, Very Large Scale Integration (VLSI) and Application Specific Integrated Circuit (ASIC) technologies. The radar's receiver features I-band (8 to 10 GHz) to serial digital output; multiple radio frequency channels; wideband frequency agility; waveform diversity; surface acoustic wave dispersive and non-dispersive filters and processor controlled calibration and built-in test.

Captor's waveguide and power monitoring components are noted as being packaged in such a way as to enhance maintainability and 'cost of ownership' while the equipment's transmitter comprises a power amplifier and an auxiliary unit. Features here include an embedded computer; electronically commutated direct current motor drives; hybridised switch mode supplies and power amplifier and a double-depressed, collector-coupled, cavity travelling wave tube. The system's scanner takes the form of a slotted waveguide antenna with an integrated identification friend-or-foe array and is mechanically driven in azimuth and elevation.

Within the Euroradar consortium, the UK's BAE Systems Avionics - Sensor Systems Division is the programme's prime contractor and has part responsibility for the equipment's processor and receiver. The consortium's German partner - the European Aeronautic, Defence and Space (EADS) Company's Systems and Defence Electronics business unit - has part responsibility for the system's scanner assembly, processor and receiver, together with full responsibility for its waveguide unit. Italian partner FIAR handles the radar's transmitter auxiliary and power units while Spanish partner ENOSA has part responsibility for its scanner assembly.

Status
According to Jane's sources, the A model Captor multimode radar was first flown aboard a BAC One-Eleven testbed aircraft during January 1993 and had, by late 1998, made at least 700 test and data collection flights. The C Model Captor was

the first to be configured for installation aboard the Eurofighter/Typhoon combat aircraft and is reported to have begun flight trials aboard the BAC One-Eleven testbed during July 1996. The first C Model radar for installation on a Eurofighter/Typhoon was delivered to DaimlerChrysler Aerospace (now part of EADS) during June 1996 and began flight trials aboard development aircraft DA 5 on 24 February 1997. Subsequently, two-seat development aircraft DA 4 (first flight 14 March 1997) and DA 6 (first flight 31 August 1996) are reported as having been fitted with the system and as of May 2001, Euroradar was understood to have received an order for 147 production examples of the sensor for installation aboard the Tranche 1 production batch of Eurofighter/Typhoons. During February 2001, BAE Systems delivered the first two production-standard Captor radars to the Eurofighter/Typhoon production facility at Warton, Lancashire in the UK and during the period 15 to 29 March 2001, development aircraft DA 5 undertook a seven sortie evaluation of Captor's look-up/look-down and track-while-scan capabilities in head-on and tail-chase engagements in a 'clutter and interference rich' environment. During this effort, DA 5 was pitted against 16 F-4F and four MiG-29 'target' aircraft from the Luftwaffe's (German Air Force) Jagdgeschwader 73 'Steinhoff'. By the following May, Jane's sources noted that a total of 16 preproduction C model Captor radars had been delivered to the Eurofighter construction consortium and that the sensor had amassed 400 hours of 'radar on' trials aboard the BAC One-Eleven testbed. Elsewhere, the radar was further described as having built up a total of 2,000 'reliability growth' test hours at the EADS - Systems and Defence Electronics facility at Ulm, Germany. On 19 June 2001, Euroradar announced that it had formerly delivered the software package required to facilitate Captor's full range of air-to-air and air-to-ground operating modes. As of the given date, the then current Captor software (delivered during 1999) is reported as facilitating the radar's initial operating capability, that is, a 'comprehensive' air-to-air capability and an 'initial' air-to-surface one.

Contractors

BAE Systems Avionics - Sensor Systems Division, Edinburgh, UK.
EADS - Systems and Defence Electronics, Ulm, Germany.
ENOSA, Madrid, Spain.
FIAR, Milan, Italy.

UPDATED

Cyrano IV/IVM/IVMR/IVM3 multimode radars

Type

I/J-band (8 to 20 GHz) family of airborne multimode radars.

Description

The Cyrano IV is a multimode airborne radar that in its baseline form offers:

- air-to-air search
- automatic tracking
- interception and fire domain computations
- dogfight engagements
- home-on-jam
- ground-mapping

Options include contour mapping and terrain-avoidance, blind let-down and air-to-ground ranging modes. In its Cyrano IVM variant, the radar incorporates track-while-scan and is described as adding air-to-sea search and tracking modes to the baseline equipment's capabilities. Cyrano IV series radars present data to the pilot via a Type 196 gunsight or a head-up display and can input data directly to onboard weapons if required. Other aircraft systems that are interfaced with the sensor include an inertial or gyro platform and the host platform's air data computer. The Cyrano IVMR subvariant of the IVM sensor is reported to have been developed for use on the Mirage F.1CR tactical reconnaissance aircraft and is understood to offer ground mapping, contour mapping, air-to-surface ranging and blind let-down modes. A further IVM subvariant - Cyrano IVM3 - is noted as having been developed for use on the Mirage 50 and as a retrofit item for the Mirage III and V. Incorporating technology from the RDM and RDI programmes, Cyrano IVM3 is understood to provide air-to-air, air-to-surface and maritime modes. The subvariant is also noted as having a built-in test facility and maintainability features.

Status

Over time, Cyrano IV series radars are reported as having been procured for installation aboard Mirage F.1 aircraft operated by the air forces of Ecuador, France, Greece, Iraq, Jordan, Kuwait, Libya, Morocco, South Africa and Spain.

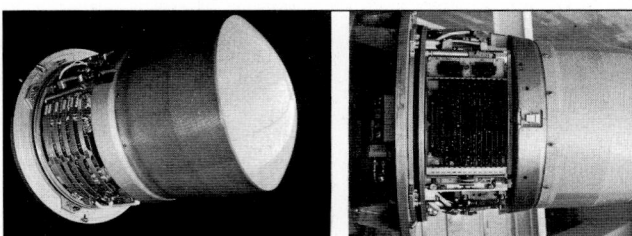

Cyrano IV and Cyrano IVM attack multifunction radars pictured side-by-side for comparison. The later Cyrano IVM (right) features a track-while scan mode and improved reliability and maintainability when compared with its predecessor

The Cyrano IVM3 subvariant is noted as having been used in Venezuela's eight aircraft Mirage 50EV/DV upgrade programme, while a range-finding mode is understood to have been implemented on those Cyrano radars fitted to Spanish F.1C/CE aircraft.

Contractor

Thales Airborne Systems, Elancourt, France.

VERIFIED

PARAD anti-radar seekers

Type

Family of passive anti-radar seekers.

Description

PARAD passive anti-radar seekers are designed for use in air-to-surface anti-radiation missiles and anti-radar drones and smart munitions. PARAD technology is based on a wideband, four-arm spiral antenna design that is described as offering a wide instantaneous field of view, phase and amplitude measurement and 'excellent' signal discrimination and identification.

Status

As of March 2001, prototype and form-factored preproduction PARAD seekers were reported as having been produced for test purposes.

Contractor

Thales Airborne Systems, Elancourt, France.

VERIFIED

RC series multifunction radars

Type

Family of multifunction fire-control radars.

Description

Derived from experience gained on the RDY radar programme (see separate entry), the RC (Radar Compact) product line is a family of lightweight, cost-effective, high-performance, multimission radars that are intended for combat aircraft and advanced trainer applications. Their baseline architecture is made up of four autonomous line-replaceable units (an antenna unit; exciter/receiver; transmitter and processor) and the system's modularity allows it to be repackaged to fit a wide range of space volumes. RC radars feature simple cooling (non-filter air), have a low power requirement and incorporate a number of antenna configurations. Here, the use of low-inertia, flat slotted plate technology is claimed to offer high levels of agility and low sidelobes. The transmitter employed is noted as being able to optimise power output over a broad spectrum of frequencies and waveforms, irrespective of radar operating mode. System radio frequency design is optimised for high sensitivity, low noise and loss levels, high spectral purity and a wide dynamic range. The signal processor used is a programmable, high throughput, greater than one gigaflop unit.

The RC radar's track-while-scan function features smart, automatic scan management with target prioritisation. Such facilities provide what is termed a 'reliable and accurate, fire-and-forget, multifiring capability' that meets the datalink requirements of active missile types and provides electronic counter-countermeasures proofing. Multiple Pulse Repetition Frequencies (PRF - low, medium and high) are used to maximise performance in both air-to-air and air-to-

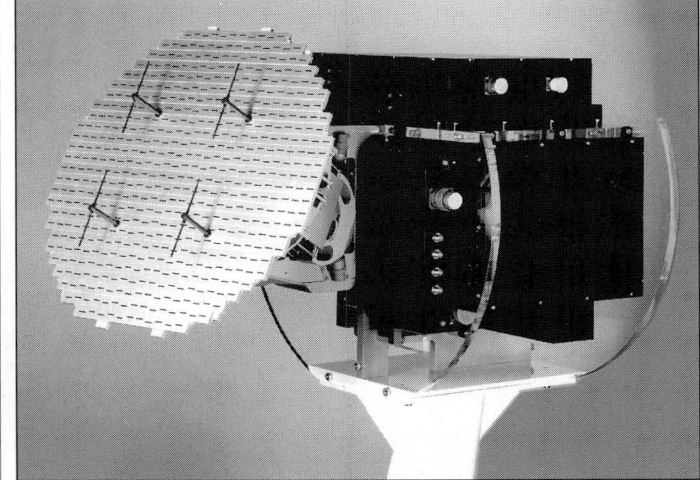

A general view of an RC 400 radar showing its antenna unit (front), transmitter (top rear) and exciter/receiver (lower rear)
0007318

surface modes. Within a given mode, the PRF is automatically managed from bar to bar in the scan pattern (with respect to attack geometry) to reduce pilot workload.

Status

As of April 2001, RC series radars were being promoted for multirole fighter upgrade, new build lightweight fighter and role flexible advanced trainer applications. As of the given date, the RC 400 radar is known to have been proposed for installation aboard the Mikoyan MiG-29 SMT fighter (potential retrofit programme/s), the Mirage F.1CH/EH fighter (potential retrofit programme for Morocco) and the European Aeronautic, Defence and Space (EADS) Co MAKO advanced trainer.

Specifications

RC 400

Frequency: I/low J-band (8-12 GHz sub-band)
Detection range: 102 km class on 5 m² target (look-up/look-down)
Transmitter: air-cooled spectrally pure chain
Antenna: elliptical or circular, monopulse, flat slotted array with IFF dipoles
Receiver/exciter: wide dynamic range, 2 channel receiver
Air-to-air modes: all aspect look-up/look-down; multipriority automatic lock on; simultaneous multitarget firing; combat; IFF interrogation
Air-to-ground modes: azimuth beam compressed ground-mapping; Doppler beam sharpening; range-finding; moving target indication with track-while-scan; high resolution mapping (option)
Air-to-sea modes: search and twin target track-while-scan; target calibration

Contractor

Thales Airborne Systems, Elancourt, France.

VERIFIED

RDI and RDM multimode radars

Type

Multimode radar for the Mirage 2000.

Description

The high Pulse Repetition Frequency (PRF) pulse-Doppler RDI and the low PRF pulse-Doppler RDM radars are fitted aboard Mirage 2000 combat aircraft. Known details of the two equipments are as follows:

RDI

RDI is optimised for the interception role and incorporates an 8 to 12 GHz coherent travelling wave tube transmitter together with a flat slotted plate antenna. The antenna has an integrated Identification Friend-or-Foe (IFF) antenna and the overall system incorporates an IFF transmitter/receiver. The equipment is optimised for air superiority in all weather conditions at any height, and range is stated as being in the region of 111 km against a head-on fighter target. The RDI is a combat-proven radar that features high operational performance and availability in the most severe conditions.

The 100 kHz + PRF guarantees accurate target speed assessment, with a Thales patented process to give range data at maximum range in search mode (as well as tracking). Track-While-Scan (TWS) facilities are available and the RDI radar incorporates an internal digital databus to facilitate sophisticated radar signal processing. Other facilities include digital Doppler filtering circuits and higher resistance to electronic countermeasures. Total weight of the 11 line-replaceable units is 255 kg. The available air-to-air modes will allow the following functions to be performed:
- air-to-air search at all altitudes
- long-range TWS or continuous tracking and missile guidance
- automatic short-range tracking for missiles or guns

Air-to-ground modes are for:
- ground mapping

RDM Doppler radar for the multirole version of the Mirage 2000

RDI pulse-Doppler radar for the air superiority versions of the Mirage 2000

- contour mapping
- air-to-ground ranging

Status

As of April 2001, the RDI radar was reported as having become fully operational with the French Air Force during 1986. According to Janes's sources, RDI radars are (or have been) installed on Mirage 2000B aircraft numbers 515 and 518 to 530 and Mirage 2000C aircraft numbers 38 to 124.

RDM

RDM is a low PRF Doppler radar that employs a travelling wave tube coherent transmitter and an inverse cassegrain antenna. Features emphasised in the design are operational versatility, extensive options and growth potential. The radar is capable of accommodating a continuous wave illuminator for the guidance of missiles that use a Doppler homing head, an IFF interrogator, Doppler beam-sharpening circuits for high-resolution mapping and ground target identification, as well as new high-reliability subsystems for low-level penetration and sea surface search and tracking. Upgraded versions of RDM offer automatic modes, new look-down processing, advanced man/machine interface and enhanced resistance to jamming. The basic operational functions provided are:
- air-to-air search (all altitude, all sectors) and interception
- low-level strike (mapping, terrain-avoidance, blind let-down)
- air-to-ground attack
- sea surface search and attack

For air-to-air combat, the RDM offers a 120° cone of coverage, the antenna scanning at either 50°/s or 100°/s, with ±60, ±30 or ±15° of scan. A 5 m² target can be detected at up to 111 km range. For air-to-air gun attacks, the 3.5° beamwidth radar can be locked on to the target at 19 km range, with automatic tracking within the head-up display's field of view or in a 'super-search' area, or in vertical search. In a look-down air-to-air situation, a 5 m² target can be detected at 46 km range. TWS facilities are available and the design incorporates an internal digital databus and built-in test equipment. Both the RDI and RDM radars display data on a head-up display and on a trichromatic, multimode, cathode ray tube head-down display that is linked to an interception and firing computer.

Status

As of April 2001, more than 250 RDM radars are reported as having been delivered. According to Jane's sources, radars of this type have been fitted to Mirage 2000B aircraft numbers 501 to 514, 516 and 517; Mirage 2000C aircraft numbers 1 to 37 together with most (if not all) Mirage 2000E single-seat and Mirage 2000ED two seat export aircraft. As of the given date, Jane's sources were reporting E model Mirage 2000 aircraft (with subvariant and numbers flying where known) as having been procured by Abu Dhabi (20 EAD and 5 EAD), Egypt (14 EM), Greece (32 EG and 4 BG) and India.

Contractor

Thales Airborne Systems, Elancourt, France.

VERIFIED

RDY multifunction radar

Type

I-/low J-band (8 to 12 GHz sub-band) multifunction fire-control radar.

Description

RDY is a multifunction radar that was originally designed for the Mirage 2000-5 fighter. In the air-to-air mode, the baseline system offers multiple range-gate high, medium and low pulse repetition frequency operating modes and is described as being capable of the long-range detection of low- and high-altitude targets, irrespective of their angle of approach. RDY's track-while-scan function facilitates multitarget tracking and engagement, presents the pilot with tactical situation analysis data and identifies targets through an 'interactive dialogue' with an

The RDY radar is a standard fit on Mirage 2000 series combat aircraft

onboard identification friend-or-foe interrogator. In the air-to-ground role, RDY provides mapping functions (with Doppler beam-sharpening for navigation update), a 'blind' penetration capability, terrain-avoidance and air-to-ground ranging. In the maritime environment, RDY is noted as being capable of detecting targets in high seas at ranges of up to 296 km, performing multitarget tracking and target dissemination and providing anti-shipping missile (such as the AM39 or Kormoran II weapons) fire-control.

RDY is equipped with a flat plate, low-inertia aperture antenna that is claimed as offering 'excellent agility and very low levels of secondary lobes' and executes an all aspect 60° conical scan pattern. For future upgrading, RDY is understood to be potentially compatible with an electronically scanning array. The radar is equipped with a compact, dual peak power I/J-band transmitter (that is capable of generating an output across a broad spectrum of frequencies, irrespective of mode) and a one gigaflop programmable signal processor. Thales also notes that it is able to offer an RDY synthetic aperture reconnaissance capability if required.

Status

As of April 2001, the RDY multifunction radar was reported as being the standard fit on the Mirage 2000-5, -5 Mk 2 ('enhanced' RDY) and -9 ('enhanced' RDY) aircraft and as being retrofitted aboard 37 French Air Force Mirage 2000Cs (aircraft to Mirage 2000-5F standard; 11 aircraft redelivered during 1998, 22 during 1999), 25 Greek Mirage 2000EG s ('enhanced' RDY radar, aircraft to Mirage 2000-5 Mk 2 standard) and 62 United Arab Emirates' Mirage 2000EAD/DADs ('enhanced' RDY, aircraft to Mirage 2000-9 standard). As of the given date, other identified offshore customers for the Mirage 2000-5 comprised Qatar (9 Mirage 2000-5EDA and 3 -5DDA aircraft) and Taiwan (48 Mirage 2000-5Ei and 12 -5Di aircraft).

Contractor

Thales Airborne Systems, Elancourt, France.

VERIFIED

Scipio radar family

Type

Multimode attack and targeting radar.

The SCP-01 multimode radar

Description

The Scipio family is a series of coherent, lightweight and compact frequency-agile radars. The radar is specially designed to suit one-man operation on fighter aircraft, with multirole capabilites following Hands On Throttle And Stick (HOTAS) philosophy. It has a total weight of less than 75 kg and is characterised by a high mean time between failure and a low mean time to repair value. An extensive built-in-test system provides fault detection and location.

The Scipio family offers a range of operational modes including air-to-air (with look-down); automatic detection, designation and tracking for air combat; maritime target detection and tracking and ground ranging and mapping.

The SCP-01 Scipio variant is an I-band (8 to 10 GHz) equipment which is described as featuring pulse compression/pulse-Doppler techniques; a frequency agile travelling wave tube transmitter; a software reconfigurable signal processor; monopulse tracking; track-while-scan and a colour video output with graphics.

Status

As of this edition, the SCP-01 multimode radar was reported as having been installed aboard Brazilian AMX strike aircraft.

Contractors

Alenia Difesa Avionic Systems and Equipment Division - Officine Galileo, Florence, Italy.
TECTELCOM, Sao Jose Dos Campos, Brazil.

VERIFIED

ISRAEL

EL/M-2001B radar

Type

I/J-band (8 to 20 GHz) pulse-Doppler radar for combat aircraft.

Description

The EL/M-2001B dual mode I/J-band pulse-Doppler radar for combat aircraft is a range-only equipment for air-to-air and air-to-ground operations. Of compact dimensions and weighing less than 50 kg, the radar is described as being suitable for installation in the nose of small new combat aircraft or for retrofitting to earlier types. M-2001B is constructed of six line-replaceable units (transmitter, receiver, power supply, Radio Frequency (RF) exciter and servo unit), all of which (together with the equipment's beam directing unit, gear box, RF head and travelling wave tube) are attached to the radar's chassis/mounting assembly. The sensor's travelling wave tube is its only non solid-state component. Modes of operation are air-to-air (all altitudes and all aspects) and air-to-ground (manual or computer controlled). Target detection is by visual means and the radar will automatically acquire and track any target viewed in the Head-Up Display (HUD) by the pilot. It operates in heavy ground clutter and remains clutter-free even (according to Elta) in 'very low level' air combat. Target acquisition is also automatic in the air-to-ground mode, although in both modes there is provision for pilot intervention for selection of targets of interest. Information provided by the EL/M-2001B can either be displayed on the aircraft HUD or fed into the weapons' control computer where it may be used for continuously computed impact point and continuously computed release point weapon delivery.

Status

As of this edition, Jane's sources were suggesting that over time, EL/M-2001B has been fitted to upgraded Chilean Mirage 50C/CN aircraft and the 75 close air support and 10 two-seat training MiG-21s included in Romania's Lancer upgrade programme. EL/M-2001B is also understood to be the basis for the FIAR manufactured Pointer ranging radar fitted to Italian Air Force AMX strike aircraft.

Specifications

Frequency: I/J-band (8-20 GHz)
Form factor: frustum of a cone
Length: 490 mm (incl connectors)
Base diameter: 450 mm

EL/M-2001B is the radar installed aboard the close support and training MiG-21s included in Romania's Lancer upgrade programme
(Michael J Gething/Jane's)

Antenna aperture: 195 mm
Weight: <50 kg
Power requirements: 115 V, 400 Hz, 3-phase, 1 kVA, DC 30 W

Contractor
Elta Electronics Industries Ltd, (a subsidiary of Israel Aircraft Industries Ltd), Ashdod.

VERIFIED

EL/M-2032 fire-control radar

Type
Multimode fire-control radar.

Description
EL/M-2032 is a software-controlled, pulse-Doppler radar designed for multimission fighter aircraft applications and is reported to feature all aspect look-up/look-down detection capabilities; a coherent travelling wave tube transmitter; a low-sidelobe planar-array antenna; a two axes monopulse guard channel and a programmable signal processor. System weight is given as being between 75 and 100 kg dependent on the size of antenna used and the radar is reported to have 14 operating modes divided between air-to-air and air-to-surface functions as follows:
Air-to-Air Modes: Range-while-search; single target tracking; track-while-scan; air combat vertical scan; air combat slewable; air combat head-up display search and boresight.
Air-to-Surface Modes: High-resolution mapping (synthetic aperture); real beam mapping; Doppler beam sharpening; air-to-ground ranging; sea search; beacon and terrain avoidance.

Status
As of this edition, Jane's sources were suggesting that EL/M-2032 formed part of Israel Aircraft Industries' F-5 upgrade programme for the Chilean Air Force and that it was the radar being installed in the 25 air defence MiG-21s that were included in Romania's Lancer upgrade programme. A variant of the radar was also understood to have been selected by India for use on Jaguar M maritime strike aircraft. This latter programme is thought to have involved the procurement of ten operational radars plus an additional set for integration trials.

Contractor
Elta Electronics Industries Ltd (a subsidiary of Israel Aircraft Industries Ltd), Ashdod.

VERIFIED

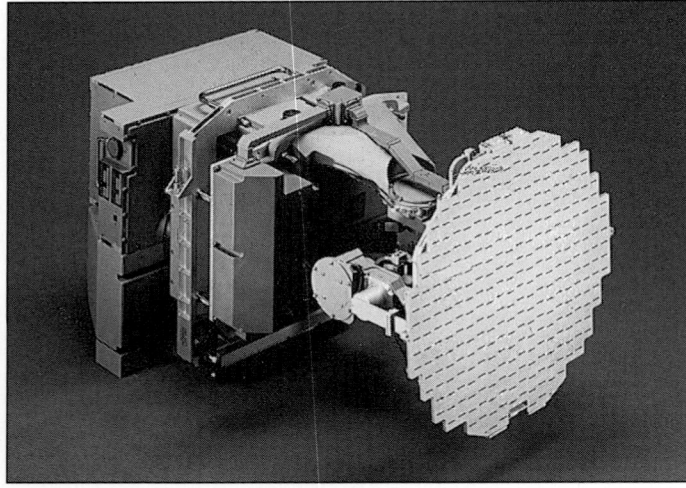

The EL/M-2032 multimode fire-control radar

ITALY

Grifo series fire-control radars

Type
Family of I/low J-band (8 to 12.5 GHz sub-band) airborne pulse-Doppler fire-control radars.

Description
Grifo series fire-control radars are lightweight, multimode pulse-Doppler equipments that are designed for new build and retrofit programmes. They are 8 to 12.5 GHz band systems that offer a range of air-to-air, air combat and air-to-surface modes. They employ travelling wave tube, pulse compression/frequency agile transmitters, a programmable Fast Fourier Transform processor and a monopulse flat plate array antenna. Aside from the Grifo 7 (which is described separately), as of

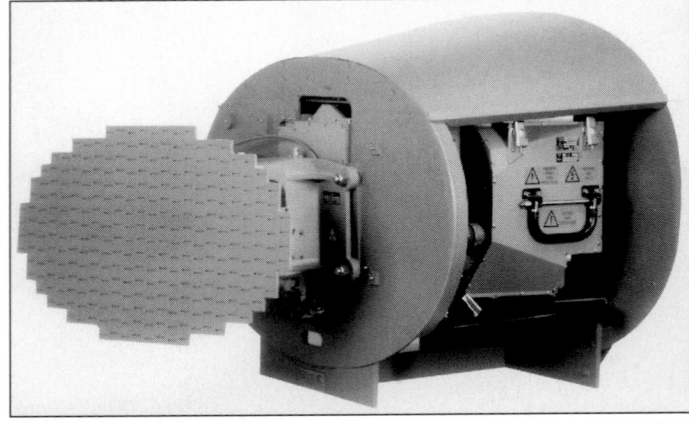

The Grifo F multimode radar   2001/0077469

October 2001, the Grifo family comprises the Grifo F, Grifo L and Grifo M equipments, the known details of which are as follows:

Grifo F
Grifo F is an 8 to 12.5 GHz sub-band multimode radar that is designed for installation aboard F-5E and F-5F aircraft. System features include:
- low, medium and high pulse repetition frequencies
- digital pulse compression
- use of a dual-channel receiver
- a 'fully processed' guard channel
- 'extensive' built-in test
- 'full' electronic counter-countermeasures provision
- compatibility with infra-red, semi-active and active missile types

Grifo M
Grifo M is an 8 to 12.5 GHz sub-band multimode radar that is designed for installation aboard Mirage III aircraft and incorporates the system features noted previously. In terms of functionality, Grifo F and Grifo M share a common set of operating modes, the known details of which are as follows:

Air-to-air
- boresight acquisition (air combat)
- dual target track
- head-up display acquisition (air combat)
- range-while-search (adaptive mode)
- range-while-search (normal mode)
- single target track
- situational awareness
- slew (air combat)
- spot
- track-while-scan
- velocity search
- vertical acquisition (air combat)

Air-to-surface
- air-to-ground ranging
- Doppler beam-sharpening 1 (8:1)
- Doppler beam-sharpening 2 (64:1)
- ground moving target indicator
- ground/sea moving target track
- real beam map
- sea (low)
- sea (high)
- sea (single target track)

Navigation
- beacon
- terrain avoidance
- weather

The elements making up the Grifo M multimode radar   2001/0077468

Status

As of October 2001, FIAR is reported as having received an upgrade contract to equip 40 of Singapore's fleet of Northrop F-5E/F and RF-5E aircraft with the Grifo-F radar variant during February 1992. So configured, such aircraft have subsequently been redesignated as F-5S/T and RF-5F platforms respectively. The same basic radar (but with a different flat plate antenna and designated as the Grifo-M) is further understood to have been used to upgrade 35 Pakistani Mirage IIIs in a programme that was launched during 1995. During September 1997, FIAR announced that it was to supply The Boeing Company with 70 Grifo-L radars and install them aboard Czech L-159 light combat aircraft in a deal worth then year US$56 million. By the end of 1997, in excess of 150 Grifo fire-control radars were reported as having been procured.

Specifications

	Grifo F	Grifo M
Frequency	8-12.5 GHz	8-12.5 GHz
Dissipation	<1.5 kW	<1.5 kW
Power	2 kVA	2 kVA
Cooling	forced air	forced air
MTBF	>200 h	>200 h
Weight	85 kg	87 kg

Contractor

Fabbrica Italiana Apparecchiature Radioelettriche SpA (FIAR - part of Alenia Difesa's Avionic Systems and Equipment Division), Milan.

UPDATED

··

Grifo 7 fire-control radar

Type

I/low J-band (8 to 12.5 GHz sub-band) airborne fire-control radar.

Description

The Grifo 7 is an 8 to 12.5 GHz sub-band, pulse compressed airborne fire-control radar that features Hands On Throttle And Stick (HOTAS) control and 'full' electronic counter-countermeasures provision. Other system features include a dual-channel receiver, a 'fully processed' guard channel and 'extensive' built-in test. Grifo 7 offers 'super search' and boresight air-to-air combat modes (both of which feature automatic transition to single target tracking) together with air-to-ground ranging.

Status

As of June 2001, Jane's sources were reporting that the Pakistan Avionics and Radar Factory (PA & RF - Kamra, Pakistan) had delivered its first licence-built, production standard Grifo 7 radar for use in Pakistan's F-7PG and ex-Royal Australian Air Force Mirage IIIO upgrade programmes. Here, the intention was to equip 46 F-7PG aircraft with the radar together with approximately 45 Mirage IIIOs. PA & RF is understood to have delivered a prototype preproduction Grifo 7 during October 2000 and as of the given date, to have been offering the radar for Mirage III upgrade programmes 'worldwide'.

Specifications

Frequency: 8-12.5 GHz
Dissipation: 850 W
Power: 500 V A/450 W
Cooling: fan
MTBF: >200 h (flight guaranteed)

Contractor

Fabbrica Italiana Apparecchiature Radioelettriche SpA (FIAR - part of Alenia Difesa's Avionic Systems and Equipment Division), Milan.

UPDATED

The elements making up the Grifo 7 multimode radar   2001/0077470

SM-1S active missile seeker

Type

I-band (8 to 10 GHz) active homing radar seeker for air-launched anti-shipping missiles.

Description

SM-1S is a version of the SM-1 active missile seeker that is designed for use on the Marte Mk 2S anti-shipping missile. Improvements inherent in the new model include enhanced target detection range; enhanced target selection; a littoral attack capability; improved Electronic Counter-CounterMeasures (ECCM) provision; power management and a low probability of intercept operating mode. Other system features include frequency diversity; pseudo monopulse function; a low-power transmitter; a programmable search pattern; a target selection capability; automatic home on target/home on jam tracking; digital processing and embedded ECCM.

Status

As of April 2001, the SM-1S active missile seeker was understood to have completed its development cycle. The earlier SM-1 seeker was noted as having been procured by the Italian Navy.

Contractor

Alenia Difesa Avionic Systems and Equipment Division - Officine Galileo, Florence.

VERIFIED

The SM-1S active missile seeker's external configuration
0017506

JAPAN

Mitsubishi airborne radar

Type

Airborne fire-control radar.

Description

The Mitsubishi Electric Corporation (under the management of the Japan Defence Agency's Technical Research and Development Institute) has developed an active, phased-array radar for use on the Mitsubishi F-2 support fighter. The sensor is reported as featuring a 61 × 72 cm, 808 transceiver module phased-array antenna, with extensive use being made of gallium arsenide and microwave integrated circuit technology in the production of the antenna modules. According to sources, the new radar offers air-to-air, dog fight, air-to-surface and navigation modes. In more detail, air-to-air sub-modes are said to include medium and short range missile engagement, gun engagement and visual identification. In dog fight mode, the sensor is noted as providing automatic target tracking and engagement, while its air-to-surface functions are reported as including continuously computed release point (for 'dumb' munitions), dive-toss release, continuously computed impact point, manual aiming, pre-planned anti-shipping missile launch and visual anti-shipping missile launch. Navigation modes comprise ground mapping, terrain-avoidance and position update. Other noted system features include track-while-scan, multiple target track and engage and look-down/shoot-down.

Status

As of this edition, at least five test models of the described radar are reported as having been delivered to the Japanese Defence Agency. Installed in XF-2 aircraft, these equipments are noted as having been undergoing flight trials since 1996.

As of the given edition, Jane's sources were reporting the inclusion of nine production F-2 support fighters in Japan's Fiscal Year 2000 defence budget.

Contractor
Mitsubishi Electric Corporation, Tokyo.

VERIFIED

RUSSIAN FEDERATION AND ASSOCIATED STATES (CIS)

Russian Federation

The radars described here are produced by, exported by and/or are in service with the armed forces of the Russian Federation and the independent republics of Armenia, Azerbaijan, Belarus, Georgia, Kazakhstan, Kyrgyzstan, Moldova, Tajikistan, Turkmenistan, Ukraine and Uzbekistan. Taken together, these 12 entities make up the Commonwealth of Independent States economic grouping.

VERIFIED

9B-1348E active radar seeker

Type
Multifunction monopulse-Doppler active radar seeker for air-to-air missile applications.

Description
9B-1348E is a monopulse-Doppler active radar seeker that is used in air-to-air missiles such as the RVV-AE. This equipment incorporates what are termed active and correction datalink channel units and provides target acquisition search, lock-on and tracking (cued by the launch aircraft's fire-control radar); missile-to-target angle and closing speed measurement; reception and decoding of control update signals and generation of missile guidance commands. The seeker is fully autonomous, provides a fire-and-forget capability, updates the missile's inertial guidance system during flight and can be customised to meet specific customer requirements. Other active missile seekers identified as being AGAT products comprise the 14.5 kg, 200 mm diameter 9B-1103M, the 10 kg, 200 mm diameter updated 9B-1103M, the 40 kg, 380 mm diameter 9B-1388 and the 28.5 kg, 310 mm diameter 9E50M1E. Of these, the 9B-1103M is reported to be able to detect a 5 m² target at a maximum range of 20 km while the equivalent values for the updated 9B-1103M are 5 m² and 25 km respectively. The 9B-1388 device is quoted as being able to 'defeat long- and super long-range air-to-air missiles at the far boundary of the engagement envelope' while the 9E50M1E seeker is noted as being able to detect a 1 m² target at a maximum range of 40 km. Readiness times for the 9B-1103M, updated 9B-1103M, 9B-1388 and 9E50M1E units are given as 1.5 s (after preliminary warm-up), 1 s (after preliminary warm-up) or 7 to 8 s (from cold start), 5 s (after preliminary warm-up) and 14 s (after application of power) respectively. Recent seeker technology developed by the contractor includes:
- miniature 'super high-frequency' transmitter and receiver units that feature accelerated klystron and travelling wave tube heating
- an antenna unit that incorporates a pair of fibre-optic gyros
- a high-speed, digital signal processor and computer that is capable of multichannel signal processing, features a 'large' memory volume and executes more than 50 operations per second.

Status
As of this edition, all the noted seeker units were understood to be available. The 9B-1348E unit is reported to have undergone live firing trials.

Specifications
Range: 16 km (locked-on to a 5 m² RCS target); up to 50 km (datalinked as part of the MIG-29's fire-control system); up to 70 km (installed on an RVV-AE missile fired at a 5 m² RCS target)
Time-to-ready: <5 s (following 2 min warm-up)
Dimensions (Ø × L): 200 × 604 mm
Weight: <16 kg (excl radome)

Contractor
AGAT Research Institute, Moscow.

VERIFIED

ARGS-35

Type
Active radar seeker for anti-shipping missile applications.

Description
ARGS-35 is the active, coherent, multichannel, multifunctional active radar seeker which is fitted to the Kh-35 Uran anti-shipping missile. The seeker is described as

The ARGS-35 seeker forms part of the Kh-35 Uran anti-shipping missile system

being able to discriminate specific targets and using detected target azimuth, elevation and range (plus closing speed), generate guidance commands for its host missile. The seeker is also noted as being able to function within a bracket of missiles. According to Jane's sources, ARGS-35 operates in the I-/low J-band (8 to 12 GHz sub-band) and is equipped with a 420 mm diameter, slotted, flat-plate antenna.

Status
As of this edition, Jane's sources were suggesting that the Kh-35 Uran anti-shipping missile entered full-scale production during 1996.

Specifications
Frequency range: 8-12 GHz
Angular coverage: -20° to +10° (elevation); ± 45° (azimuth)
Range: 20 km (max)
Environmental envelope: ± 50°C (temperature); 4 mm/s (precipitation); Sea States 5-6 (day and night)
Dimensions (Ø x L): 420 × 700 mm
Weight: 40 kg (excluding radome)

Contractor
RADAR MMS, Moscow.

VERIFIED

Komar/Super-Komar multimode radars

Type
I/low J-band (8 to 12 GHz) multirole airborne radar.

Description
The Komar (Gnat) fire-control radar appears to have been designed primarily for the upgrade market with proposals for installation aboard Chinese F-7 II and A5 and Indian MiG-27 combat aircraft being touted during the mid-1990s. In more detail, the radar has a built-in test system and is described as offering 15 air-to-air, air-to-ground and strike mission operating modes, as follows:
Air-to-air modes Target detection and ranging; single target tracking; multiple-target (up to eight) tracking; vertical air combat manoeuvre; head-up display search; boresight and slewable.
Air-to-ground modes Ground mapping; real beam; Doppler beam sharpening; synthetic aperture and air-to-ground ranging.
Strike mission modes Target designation for radar-guided missiles; gun ranging and bomb aiming.
    A variant of the basic design designated as Super-Komar features digital signal and data processing and was originally aimed the Chinese FC-1 fighter. Super-Komar offers 14 operating modes, the known details of which are as follows:
Air-to-air modes Look-up/look-down range-while-search; track-while-scan (up to eight targets with simultaneous engagement of two); vertical air combat manoeuvre; head-up display search; wide angle search and boresight.

Air-to-ground modes Ground mapping; real beam; Doppler beam sharpening; synthetic aperture; enlargement/freeze; track-while-scan (up to four targets); moving target indication and air-to-ground ranging.

Status

As of this edition, the status of the Komar and Super-Komar multimode radars was uncertain.

Specifications

Super-Komar
Frequency: I/low J-band (8-12 GHz)
Detection range: 45 km (rear hemisphere); 75 km (forward hemisphere)
Angular coverage: ±10 to ±30° in azimuth
Peak power: 5 kW
Cooling: air and liquid
Weight: 90 kg

Contractor

Phazotron-NIIR Joint Stock Company, Moscow.

VERIFIED

....

Kopyo series multimode radars

Type

I/low J-band (8 to 12 GHz) family of multimode search and track radars.

Description

As of this edition, four radars have been identified as belonging to this series, namely Kopyo (spear), Super Kopyo, Super Kopyo-PH and Kopyo-25. Kopyo-25 is a pod-mounted, multipurpose radar used on the Sukhoi Su-25SM, -25TM and -39 ground-attack aircraft and is described in a separate entry. Known details of the remaining variants are as follows:

Kopyo

Aimed specifically at the Mikoyan MiG-21 upgrade market, the 8 to 12 GHz Kopyo radar is a coherent pulse-Doppler multimode equipment which comprises a flat plate antenna assembly, a receiver, an advanced data controller, signal and data processing units, a transmitter, a synchroniser unit, an exciter and a power supply. Kopyo is described as employing digital processing and being compatible with a range of missiles and smart weapons that includes the R-27R1, R-27T, R-60MK and Kh-31A types. It is noted as having 13 operating modes which are divided between air-to-air and air-to-surface functions as follows:
Air-to-Air Modes: Look-up/look-down range-while-search; Track-While-Scan (TWS) of up to eight targets with simultaneous engagement of two; air combat vertical search; air combat head-up display search; air combat wide angle search and air combat boresight.
Air-to-Surface Modes: Real beam ground-mapping; Doppler beam sharpening; synthetic aperture; display enlargement/freeze; TWS of up to four targets; ground moving target indicator and air-to-surface ranging.

Super Kopyo

Super Kopyo is described as being a modernised Kopyo radar in which new high throughput signal and data processors are used. Aside from this, trade show display material suggests that the only major differences between the two variants is a reduction in overall weight to 90 kg and increase in forward and rear hemisphere maximum detection ranges. In all other respects, the two equipments appear identical.

Super Kopyo-PH

Super Kopyo-PH is an ultra lightweight phased-array radar that is intended to provide a first look, first shot, first kill capability. Component units of the radar appear to include a phased-array antenna assembly, digital signal and data processors and a 12 bit analogue-digital converter unit. Super Kopyo-PH is also

billed as incorporating a built-in test facility (down to printed circuit board level) and features air-to-air and air-to-surface operating modes as follows:
Air-to-Air Modes: Look-up/look-down range-while-search; track-while-scan; single target tracking and air combat.
Air-to-Surface Modes: Real beam ground-mapping; Doppler beam sharpening and synthetic aperture target identification. Aside from those described above, Phazotron notes the availability of other operating modes for this radar.

Status

As of this edition, the baseline Kopyo multimode radar was reported as forming part of Mikoyan's MiG-21-93 upgrade package for up to 125 Indian MiG-21bis aircraft. The prototype Indian MiG-21-93 made its maiden flight on 6 October 1998. As of the given edition, the status of the Super Kopyo and Super Kopyo-PH radars was uncertain.

Specifications

Kopyo
Frequency: I/low J-band (8-12 GHz)
Transmitter type: liquid-cooled TWT
Detection range: 45 km (rear hemisphere); 75 km (forward hemisphere)
Angular coverage: ±10/±30° (azimuth); 2/4 bars (elevation)
Power output: 1 kW (average); 5 kW (peak)
Cooling: air and liquid
MTBF: 120 h
Antenna: mechanically scanned planar-array
Weight: 105 kg
Volume: 185 dm^3

Contractor

Phazotron-NIIR Joint Stock Company, Moscow.

VERIFIED

....

MOSKIT multimode radar

Type

I/low J-band (8 to 12 GHz) multimode radar.

Description

MOSKIT is a modular, coherent, multimode radar designed for combat and latest generation training aircraft applications. The equipment comprises four line-replaceable units, namely a flat, slotted-array antenna; an air-cooled, travelling wave tube transmitter; a combined receiver/exciter and a signal/data processor. Available operating modes are as follows:
Air-To-Air: Range-while-search in look-up/look-down and eight target Track-While-Scan (TWS) with two target simultaneous engagement.
Close Air Combat Modes: Head-up-display search; slewable scan; boresight and vertical scan.
Air-To-Ground: Real beam ground mapping; Doppler beam sharpening; synthetic aperture; enlargement; freeze; beacon; two target TWS; air-to-ground ranging and ground moving target tracking.
　MOSKIT is described as being compatible with the KAB-500KR, Kh-31A, R-27R, R-27T, R-73E and RVV-AE munitions.

Status

As originally marketed, the MOSKIT multimode radar was aimed at the MiG-21M/MF retrofit (MiG-21-98) and MiG-AT trainer markets. As of this edition, the status of the radar was uncertain.

Specifications

Frequency: I/low J-band (8-12 GHz)
Power output: 0.15 kW (average); 2.5 kW (peak)
Detection range: 30 km (rear hemisphere); 45 km (forward hemisphere)
Angular coverage: ±10/±30° (azimuth); 2/4 bars (elevation)
Cooling: air
Weight: 60 kg
Volume: 90 dm^3

Contractor

Phazotron -NIIR Joint Stock Company, Moscow.

VERIFIED

....

MOSQUITO multimode maritime attack radar

Type

I/low J-band (8 to 12 GHz) multimode, maritime attack radar.

Description

MOSQUITO is a multimode, maritime attack radar that is designed to act as what is termed a radar sensor when used with Indian and Western munitions and a fire-control system when used with Russian weapons. The radar comprises four line-

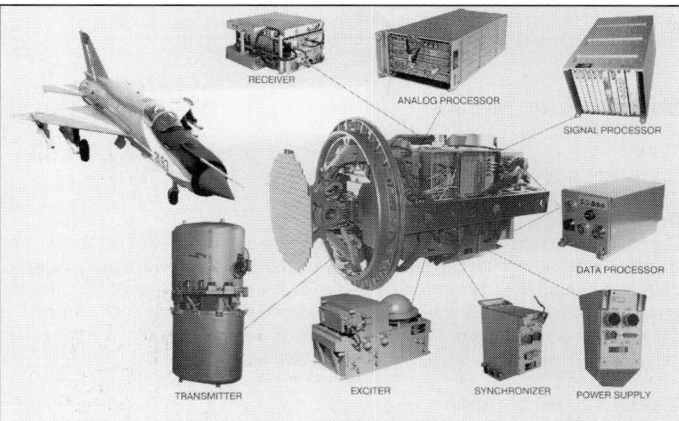

The component units that make up the Kopyo multimode pulse-Doppler radar

replaceable units, namely a flat, slotted array antenna; an air-cooled, travelling wave tube transmitter; a combined receiver/exciter and a signal/data processor. A range of operating modes are offered, as follows:

Air Combat: Detection, lock-on and tracking of air targets (using both Russian and Western air-to-air missiles).

Close Air Combat Modes: Head-up display search; slewable scan; boresight and vertical scan.

Air-To-Surface: Detection and co-ordinate measurement of port and fleet anchorage targets for attack with unguided munitions.

Air-To-Sea: Detection and co-ordinate measurement of surface ship targets using the Sea Eagle anti-shipping missile.

Status

MOSQUITO appears to be a variant of Phazotron's MOSKIT radar (see separate entry). As of this edition, the radar's status was uncertain.

Specifications

Frequency: I/low J-band (8-12 GHz)
Power output: 0.15 kW (average); 2.5 kW (peak)
Detection range: 30 km (rear hemisphere); 45 km (forward hemisphere); 100 km (300 m² RCS ship target in Sea State 4-6)
Angular coverage: ±10/±30° (azimuth); 2/4 bars (elevation)
Cooling: air
Weight: 60 kg
Volume: 90 dm³

Contractor

Phazotron-NIIR Joint Stock Company, Moscow.

VERIFIED

Nauchno-Issledovatelskiy Institut Priborostroyeniya (NIIP) fire-control radars

Alongside the Osa and Zaslon systems described separately, the NIIP organisation is credited as being the source for the N001, N007, N011, N011M and N014 fire-control radars. Known details of these various equipments are as follows:

N001
The NIIP N001 fire-control radar is understood to be a mechanically scanned, air-to-air system that is capable of engaging one target at a time. Jane's sources suggest that the radar has been installed aboard (or proposed for installation aboard) Sukhoi Su-27, Su-30 and Su-33 aircraft. As of this edition, NIIP was understood to have proposed adding a ground mapping function to the N001 radars installed on Chinese Su-27s.

N007
N007 is an alternative designation for the Zaslon electronically-steered radar fitted to the Mikoyan MiG-31 interceptor (see separate entry).

N011
Apparently a development of the N001 equipment, NIIP's N011 radar is reported as featuring a slotted, flat-plate antenna and as offering both air-to-air and air-to-surface (ground mapping and terrain-following/terrain-avoidance) operating modes. Jane's sources suggest that N011 is applicable to the Sukhoi Su-27M aircraft and has maximum head-on and rear-on detection ranges against 3 m² targets of 80 to 100 km and 30 to 40 km respectively. The equipment is quoted as being able to track up to 15 targets and engage up to six of them simultaneously.

N011M
NIIP's N011M radar is a multimode, multi-frequency (D- (1 to 2 GHz) and I-/low J- (8 to 12 GHz sub-band) band), phased-array equipment that has forward and aft maximum detection ranges against a 2 m² target of 80 to 100 km and 30 to 40 km respectively. Its maximum search range against 'large' air targets (such as airborne early warning and control aircraft) is noted as being 400 km with surface targets being detectable at ranges of up to 200 km. The radar's air-to-surface operating modes are reported to include ground mapping and terrain-following/terrain-avoidance. N011M is known to have been flight-tested on Sukhoi Su-27M/-35 aircraft and is noted as being applicable to Su-30MK/MKI, -35 and -37 platforms.

N014
As of this edition, the N014 radar was reported as being an electronically-scanned radar, the development of which had been abandoned by the late summer of 1998. N014 may have been originally intended for installation on Mikoyan's MiG 1-42 fifth generation combat aircraft.

VERIFIED

Osa multimode radar

Type

I-/low J-band (8 to 12 GHz) multimode radar

Description

The 8 to 12 GHz Osa (Wasp) radar is a lightweight, phased array equipment that is designed for new build and retrofit applications. System features include:

- track-while-scan on up to 16 air targets
- simultaneous engagement of up to four air targets

- simultaneous tracking of up to two ground targets
- a surface mapping capability that includes image freeze and zoom options

Status

As of this edition, the Osa radar was reported as having been test flown aboard a Mikoyan MiG-29UBT two-seat combat trainer/ground-attack aircraft and as having been proposed for MiG-21M/MF series upgrades and use aboard combat capable variants of the MiG-AT and Yak-130 trainers.

Specifications

Frequency: I-/low J-band (8-12 GHz)
Range: 40 km (5 m² RCS target, tail-on position); 85 km (5 m² RCS target, head-on position)
Coverage: 360° (azimuth)
Weight: 22 kg (460 mm diameter antenna assembly); 120 kg (total system)

Contractor

Nauchno-Issledovatelskiy Institut Priborostroyeniya (NIIP), Zhukovsky.

VERIFIED

RP-35 multimode radar

Type

I/low J-band (8 to 12 GHz) multimode radar.

Description

RP-35 is a coherent, multimode, digital fire-control sensor that comprises a high gain, low sidelobe, electronically scanning phased-array antenna; a liquid-cooled travelling wave tube transmitter; an exciter; a three channel microwave receiver and programmable signal and data processors. The radar is designed for one-man operation with all combat critical radar controls integrated into the host aircraft's throttle grip and stick controller. Data is displayed to the pilot via head-up and head-down displays. The radar has a built-in test system. Identified operating modes are as follows:

Air-To-Air: Range-while-search; velocity search; 24 target Track-While-Scan (TWS) with simultaneous engagement of four targets; single target track; raid cluster resolution; automatic terrain avoidance; head-up display search; vertical scan; boresight and wide angle.

Air-To-Surface: Real beam ground mapping; Doppler beam sharpening; synthetic aperture; enlargement; freeze; four target TWS; air-to-ground ranging and ground moving target indication and track.

RP-35 is described as being compatible with a wide range of weapons including KAB-500KR, Kh-29T, Kh-31A, Kh-35U, Kh-38, R-27ER1, R-27ET1, R-27R1, R-27T, R-73E and RVV-AE munitions.

Status

As of this edition, the RP-35 multimode radar was reported as having been designed to 'meet the operational requirements of new-generation single-seat, multirole combat aircraft'.

Specifications

Frequency: I/low J-band (8-12 GHz)
Power output: 2 kW (average); 8 kW (max)
Detection range: 65 km (rear hemisphere); 140 km (forward hemisphere)
Angular coverage: ±20/±60° (azimuth); 2/4 bars (elevation)
MTBF: >150 h
Cooling: air and liquid
Weight: 220 kg
Volume: 350 dm³

Contractor

Phazotron-NIIR Joint Stock Company, Moscow.

VERIFIED

Sapfir series radars

Type

Series of airborne fire-control radars.

Description

The codename Sapfir (Sapphire) has been applied to a series of Soviet and Russian Federation airborne radars that stretch back to the 1960s. Radars within the series all appear to have been developed by the entity now known as Phazotron and have additional alphanumeric N, R, RP, S and TsD service designations and/or design bureau codenames. The known details of Sapfir radars that are thought to be currently in service or under development are as follows:

TsD-30/R-2L/RP-21 Sapfir
RP-21 (known as TsD-30 in its preproduction series) is a 1960s vintage fire-control radar that can detect a 16 m² target at a range of 20 km and begin tracking it at

10 km. The set uses a single dish antenna and covers a 60° arc in azimuth. In terms of applications, the TsD-30 radar is reported to have been first fitted to the 1960 vintage MiG-21P-13. The series production RP-21 radar made its début aboard the MiG-21MF. The designation R-2L (NATO reporting name Spin Scan?) was applied to an RP-21 variant which was installed on the MiG-21FL export aircraft which, alongside being built in the Soviet Union between 1965 and 1968, was produced under licence by India's Hindustan Aeronautics Ltd (HAL) starting in 1966.

A modified version of RP-21, known as RP-21M was fitted to export MiG-21PFMs built between 1966 and 1968, while the TsD-30 variant reappeared on the 1965 vintage MiG-21R tactical reconnaissance aircraft. A modified RP-21M radar (the RP-21MA) was installed on the MiG-21M that was both exported by the Soviet Union between 1968 and 1971 and built in India by HAL between 1973 and 1981. As March 2000, MiG-21FL, -21M, -21PF, -21PFM and -21Rs were reported as being in service with a number of air forces including those of Algeria (35 x MiG-21PFM); Bulgaria (15 x -21R); Cuba (35 x -PFM); Egypt (? x -21PF/PFM and 10 x -21R); India (20 x -21FL and ? x -21M - see Kopyo entry); Iraq (? x -21PF); North Korea (130 x -21PF/PFM); Laos (? x -21PFM), Poland (14 x -21M, 20 x -21PFM and 12 x -21R), Romania (? x -21M - see EL/M-2001B and EL/M-2032 entries), Syria (? x -21PF) and the Federal Republic of Yugoslavia (6 x -21R).

RP-22 Sapfir-21

RP-22 (NATO reporting name Jay Bird) is reported to be capable of detecting a 16 m² target at a range of 30 km and begin tracking it at 15 km. The set utilises a two-reflector Cassegrain antenna and has angular coverages of ±30° in azimuth and ±20° in elevation. It is also noted as offering a fire sequencing display. The equipment operates in the 12.88 to 13.2 GHz frequency band and has three pulse repetition frequency ranges; 1,592 to 1,792 Hz, 2,042 to 2,048 Hz and 2,716 to 2,724 Hz. In terms of usage, the radar appears to have been first fitted to the 1965 vintage MiG-21S under the designation RP-22S. As the RP-22 Saphir-21, it was applied to the MiG-21SM (produced between 1968 and 1974), the MiG-21MF (1970 to 1975), the MiG-21SMT (1971 to 72) and the MiG-21bis (mass produced in the Soviet Union from the early 1970s and built under licence by HAL in India starting 1974). As of March 2000, MiG-21MF and -21bis aircraft were reported as serving with a number of air forces including those of Algeria (23 x MiG-21MF/bis); Angola (19 x -21bis); Croatia (28 x -21bis); Cuba (80 x -21bis and 20 x -21MF); the Czech Republic (36 x -21MF - some in storage); Egypt (? x -21MF); Hungary (36 x -21bis - 22 in service at any one time); India (230 -21bis (125 to be upgraded to MiG-21-93 standard - see Kopyo entry) and ? x -21MF); Libya (42 x -21bis); Nigeria (12 x -21MF); Poland (55 (25 naval) x -21bis and 60 x -21MF); Romania (? -21MF - see EL/M-2001B and EL/M-2032 entries); the Slovak Republic (c.12 (active) x -21MF); Syria (? x -21MF/bis), Vietnam (125 x -21MF/bis); the Yemen (58 x -21MF/bis) and Zambia (13 x -21MF). It is also worth noting that Sapfir-21 was also fitted to the 50 MiG-23S aircraft between mid-1969 and the end of 1970.

S-23 Sapfir-23

Sapfir-23 is reported to be a fire-control radar featuring Doppler detection, a fire sequencing display and a two-reflector Cassegrain antenna. The radar is noted as being able to detect a 16 m² target at a range of 70 km and begin tracking it at 55 km. Angular coverage is given as ±45° in azimuth and ±30° in elevation. The radar also features free space look-up modes capable of functioning in clutter backgrounds. In production form, the radar was first applied (under the designation Sapfir-23-SL) to the MiG-23M that was first flown during June 1972. The MiG-23M formed the basis of the MF and MS export models which featured less advanced systems than those carried by the M. In the 1976 vintage MiG-23ML, the Sapfir-23-SL radar was replaced by the upgraded Sapfir-23ML variant. As of March 2000, MiG-23MF, -23MS and -23ML aircraft were reported as being in service with a number of air forces including those of Cuba (20 x MiG-23MF/MS); Iraq (60 x -23ML); Poland (30 x -23MF); Romania (24 x -23MF) and Syria (80 x -23MF/MS/ML). It is also worth noting that at least one of the Sapfir-23 variants may have the alternative designation N003E. Sapfir-23 (or a variant thereof) is also likely to be the J-band (10 to 20 GHz) radar which has been given the NATO reporting name High Lark.

RP-25 Sapfir-25

Sapfir-25 is described as featuring a two-reflector Cassegrain antenna and as providing a true look-down/shoot-down capability for the 1978 vintage MiG-25PD Mach 2 interceptor. As such, it replaced the Smerch-A radar installed in the preceding MiG-25P model. Smerch-A is described as being able to detect a 16 m² target at a range of 100 km and begin tracking it at 50 km. Offering a degree of low-altitude capability, Smerch-A is noted as having angular coverages of 120° in azimuth and ±30° in elevation. The Sapfir-25 equipment is quoted as having the same maximum detection range against a 16 m² target as Smerch-A together with the ability to track targets at a range of 75 km. The set's angular coverages are given as ±60° in azimuth and ±30° in elevation. During 1979, all operational MiG-25P aircraft were upgraded to -25PD standard and given the designation MiG-25PDS. Outside the Russian Federation (which, as of March 2000, was reported to be operating approximately 80 MiG-25PDS aircraft), MiG-25 interceptors have been supplied to the air forces of Algeria (as of March 2000, 13 x -25PD aircraft according to Jane's sources), Iraq (12 x -25PD), Kazakhstan (10 x -25PD - some in storage), Libya (40 x -25PD) and Syria (30 x -25PD).

N019 Sapfir-29

Sapfir-29 (which is also known in some instances as the RP-29 or Topaz and has been given the NATO reporting name Slot Back) is a look-up/look-down pulse-Doppler fire-control radar which operates in the I/low J-band (8 to 12 GHz) frequency range and can track up to 10 targets simultaneously. The radar's angular coverage values are ±70° in azimuth and -40° to +60° in elevation. Sapfir-29 is understood to be installed on the initial production model of the Mikoyan MiG-29. Subsequent to this, Jane's sources suggest that the MiG-29S variant has been fitted with either the N019M or N019ME variants of Sapfir-29; the MiG-29SE with the N019ME and the MiG-29N with the N019ME.

Phazotron describes the basic Topaz variant as comprising an antenna assembly, receiver, transmitter, antenna control unit/synchroniser, exciter and its associated power supply, conversion unit, digital computer, input/output unit, test signal conversion unit and built-in test unit. It is quoted as providing detection and covert automatic tracking of targets in free space and/or ground clutter; track-while-scan against 10 targets with simultaneous engagement of two; a study/ training mode; a high-speed vertical scan/lock on mode for visual close combat and compatibility with a range of weapons including the R-27R1, R-29R1E, R-27T1, R-27T1E, R-60MK, R-73E and RVV-AE missiles. The radar's mean time between failure figure is given as 100 hours while forward hemisphere detection ranges are reported as 70 km in air-to-surface mode and 80 km in free space against small fighter-type targets. The equivalent rear hemisphere figures are 30 and 40 km respectively. Jane's sources suggest that the N019M variant of the basic radar has a maximum detection range of 100 km and features revised hard- and software, a built-in test facility and a 15 m resolution synthetic aperture ground mapping capability. The N019ME model is understood to be the export version of the N019M.

As of March 2000, MiG-29 aircraft were reported as being in service with a number of air forces including those of Belarus (58 x MiG-29S and 8 x -29UB according to Jane's sources); Bulgaria (17 x -29 and 4 x -29UB); Cuba (14 x -29 and 2 x -29UB); Germany (19 x -29 and 4 x -29UB); Hungary (20 x -29 and 6 x -29UB); India (70 x -29 and 8 -29UB); Iran (14 x -29/UB); Iraq (? x -29/UB - placed in storage during 1995); Kazakhstan (33 (estimate) x -29 and 9 (estimate) x -29UB); North Korea (25 x -29 and 5 x -29UB); Malaysia (14 x -29N and 2 x -29NUB); Peru (10 (estimate) x -29 and 2 x -29UB); Poland (19 x -29 and 4 x -29UB); Romania (15 x -29 and 3 x -29UB); Russia (460 x -29 (up to 180 being upgraded to -29SMT standard) and 150 x -29UB (approximately 30 being converted to -29UBT standard); the Slovak Republic (20 x -29 and 3 x -29UB), Syria (42 x -29 and 6 x -29UB); the Ukraine (215 x -29/-29UB); Uzbekistan (33 x -29); the Yemen (8 x -29 and 2 x -29UB) and the Federal Republic of Yugoslavia (1 to 2 x -29 (local designation L-18) and 2 x -29UB (NL-18)).

Contractor

Phazotron-NIIR Joint Stock Company, Moscow.

VERIFIED

....

SOKOL multimode fire-control radar

Type

I/low J-band (8 to 12 GHz) multimode fire-control radar.

Description

SOKOL is a multiple pulse repetition frequency (low, high and medium), multimode fire-control radar that comprises a low sidelobe, 1 m diameter, phased-array antenna; a liquid-cooled, travelling wave tube transmitter; a software-controlled, high stability exciter; a multichannel receiver; a programmable signal processor and a high-order language data processor. The radar features a built-in test system and offers the following operating modes:

Air-To-Air: Look-up/look-down range-while-search; velocity search; 24 target Track-While-Scan (TWS) with simultaneous engagement of up to six targets; single target track; raid cluster resolution and automatic terrain avoidance.

Close Air Combat Modes: Vertical scan; head-up display search; boresight and wide angle.

Air-To-Ground: Enhanced real beam mapping; Doppler beam sharpening; synthetic aperture; enlargement; freeze; four target TWS; precision velocity update; air-to-ground and air-to-sea ranging and ground moving target indication and tracking.

SOKOL is also described as being compatible with the Kh-31A, R-27R1, R-27T1, R-27ER1, R-27ET1, R-73E and RW-AE munitions while the system's antenna makes use of non-equidistant field distribution rather than the traditional linear approach. Alongside the baseline system, Phazotron is reported as having developed 75 kg and 45 kg variants of the SOKOL radar designated as Pharaon and Pharaon-M respectively. The baseline Pharaon system is noted as having a 75 km detection range and the two variants are noted as being designed for use on aircraft such as the Sukhoi Su-27IB and Su-33KUB.

Status

As of this edition, the status of the SOKOL/Pharaon radar family was uncertain.

Specifications

SOKOL

Frequency: I/low J-band (8-12 GHz)
Power output: 2 kW (average); 8 kW (peak)
Detection range: 80 km (rear hemisphere); 180 km (forward hemisphere, 3 m² target)
Angular coverage: ±70° (azimuth)
MTBF: >150 h
Cooling: air and liquid
Weight: 250-270 kg class
Volume: 600 dm³

Contractor

Phazotron-NIIR Joint Stock Company, Moscow.

VERIFIED

Zaslon fire-control radar

Type

I-band (8 to 10 GHz) airborne fire-control radar. NATO reporting name Flash Dance.

Description

Also referred to as the N007 and/or S-800 (Shield), Zaslon is reported to be an electronically steered, look-down/shoot-down passive phased-array radar that operates in the 9 to 9.5 GHz frequency band. The baseline equipment is noted as offering a 200 km maximum detection range against a 16 m² target with tracking beginning at 120 km. Zaslon's angular coverages are given as 140° in azimuth and -60/+70° in elevation. Western sources suggest that the radar can track up to 10 targets and engage four simultaneously and has rear hemisphere look-up/look-down ranges of 89 and 69 km respectively.

Status

As of March 2000, the Zaslon fire-control radars are reported as having been installed aboard approximately 320 MiG-31 interceptors of the Russian Air Force. As of the given date, Jane's sources suggest that this number was divided between MiG-31B aircraft fitted with the updated Zaslon-A weapon system and radar and MiG-31BS platforms (existing MiG-31s that have been bought up to MiG-31B standard). A further radar upgrade (designated as Zaslon-M) is understood to have been proposed for a definitive second-generation MiG-31BM aircraft.

Contractor

Nauchno-Issledovatelskiy Institut Priborostroyeniya (NIIP), Zhukovsky.

VERIFIED

The Zaslon phased-array radar carried on the MiG-31 was the world's first use of an electronically steered array on a fighter aircraft

....

Zhuk fire-control radar

Type

I/low J-band (8 to 12 GHz) airborne multimode radar family.

Description

The look-up/look-down Zhuk (Beetle) radar has appeared (or has been proposed) in a number of variants including the N010 Zhuk-29, N010MP Zhuk-29, Zhuk-27, Zhuk-811 and Zhuk-PH. The baseline radar comprises an antenna assembly, receiver, advanced data controller, data and signal processors, synchroniser, power supply, exciter, transmitter and what is termed a TV-former unit. The radar features a built in test capability and is credited with 15 operating modes divided between air-to-air and air-to-surface functions as follows:

Air-to-Air Modes: Look-up/look-down range-while-search and Track-While-Scan (TWS) of 10 targets with simultaneous engagement of up to four.

Air Combat Modes: Vertical search; head-up display search; wide-angle search; boresight and automatic terrain avoidance for low-altitude combat operations.

Air-to-Surface Modes: Real beam ground-mapping; Doppler beam sharpening; synthetic aperture; display enlargement/freeze; TWS on four targets; ground target Moving Target Indicator (MTI)/tracker; air-to-surface ranging and navigation update. Weapons compatibility includes the Kh-31A, R-27R1, R-27T1, R-37E and RW-AE munitions.

The baseline Zhuk-29 radar is fitted with a 680 mm antenna and offers uniform scale ground-mapping; navigation upgrade, air-to-surface target designation; look-up/look-down air-to-air modes (including a dedicated visual close combat mode) and automatic terrain avoidance. Detection range (against up to 10 targets) is noted as being greater than 100 km with TWS on up to four targets. Zhuk-27 is generally similar to the baseline system with the exception that forward and rear hemisphere detection ranges are given as 140 and 55 km respectively while angular coverage in azimuth is noted as being 40, 120 or 170° selectable sectors.

Zhuk-PH adds a phased-array antenna and optionally, a new high-performance signal processor to the basic radar and offers 17 operating modes divided between air-to-air and air-to-surface operating modes as follows:

Air-to-Air Modes: Target detection with velocity measurement; look-up/look-down range-while-search and TWS on eight targets (selected from 24) with simultaneous engagement of up to eight.

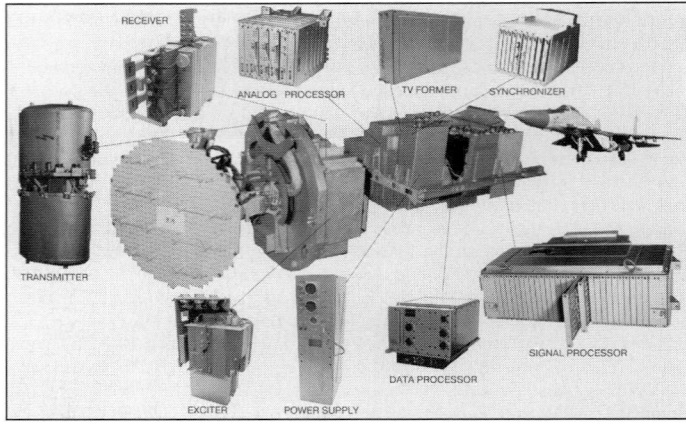

The component units which go to make up the Zhuk fire-control radar

Air Combat Modes: Vertical search; head-up display search; wide-angle search; boresight and individual target localisation within a formation.

Air-to-Surface Modes: Real beam ground mapping; Doppler beam sharpening; synthetic aperture; display enlargement/freeze; TWS on up to eight targets; ground target MTI/tracking; air-to-surface ranging; terrain avoidance and navigation update.

System weight is given as 275 kg with peak and average power values being noted as 10 and 2 kW respectively. Against a 500 m² surface ship target, Zhuk-27 is quoted as having a maximum detection range of 250 km. With the improved signal processor installed, the radar is described as being able to detect air targets and measure their velocities at a maximum range of 245 km while standard forward and rear hemisphere detection ranges are in the order of 183 and 60 km respectively. Angular coverages are given as 120° in azimuth and ±60° in elevation. Weapons compatibility is said to include the Kh-31A, Kh-35A, R-27R1 and R-27T1 munitions.

Status

As of this edition, Jane's sources were suggesting that the N010 Zhuk-29 radar was installed aboard the MiG-29M, the N010MP Zhuk-29 variant (N010 with a synthetic aperture radar mode and 70° (azimuth) by −40° to +50° (elevation) angular coverage) aboard the MiG-29MST and that Zhuk PH was a radar option for a putative MiG-29SMT-II upgrade. Elsewhere, the same sources were suggesting that the Chinese signed a procurement contract covering the supply of up to 200 Zhuk-811 radars for use on J-8IIM air superiority fighter during 1996. As yet unconfirmed sources suggested that Zhuk-811 offered a six fold increase in data handling and signal processing power over the baseline Zhuk radar, offered an 80 km detection range and a 24 target track-while-scan capability and the ability to engage up to four targets simultaneously.

Specifications

Zhuk-29

Frequency: I/low J-band (8-12 GHz)
Power output: 1 kW (average); 5 kW (peak)
Detection range: 40 km (rear hemisphere); 80 km (forward hemisphere)
Angular coverage: ±20/±60/±90° (azimuth); 2/4 bars (elevation)
MTBF: 150 h
Cooling: air and liquid
Weight: 250 kg
Volume: 500 dm³

Contractor

Phazotron-NIIR Joint Stock Company, Moscow.

VERIFIED

SWEDEN

Active Electronically Scanned Array (AESA)

Type

AESA radar system technology demonstrator.

Description

Sweden's AESA programme is aimed at developing an active, electronically scanned array that can be back fitted into existing radars (such as the PS-05/A - see separate entry) as well as forming the basis for the next generation of such sensors. The AESA technology demonstration features:

- an array of transceiver modules mounted on a moving platform to maximise the equipment's azimuthal search limits
- instantaneous beam shifting
- improved detection range and jamming resistance (when compared with existing radars)
- a multibeam capability in which beams can be controlled individually and simultaneously

A model of the Ericsson AESA array showing both the layout of its transceiver modules and the modules themselves   0044143

- simultaneous fire-control and obstacle warning functions
- the ability to facilitate concurrent radar, datalink, radar warning and jamming functions within the overall avionics system
- graceful system degradation (via the use of 1,000 plus transceiver modules)

Status

During 1994, the Swedish Defence Materiel Administration (FMV) awarded Ericsson an AESA study contract. By 1997, the company had begun AESA laboratory tests using a breadboard array fitted with approximately 100 transceiver modules. 'Successful' trials with this array were completed during 1998 and were followed by a contract covering full-scale development of an AESA radar system demonstrator. Here, the aim was to demonstrate both the technology and the benefits of an AESA radar system to the Swedish Air Force and to provide Ericsson with the necessary experience to develop a next-generation AESA radar for the JAS 39 Gripen multirole combat aircraft's post 2010 mid-life upgrade. As of September 2001, the described AESA radar system demonstrator was scheduled to be installed in a Viggen testbed aircraft, with flight trials beginning during 2004.

Contractor

Ericsson Microwave Systems AB, Airborne Radar Division, Mölndal.

UPDATED

PS-05/A multimode radar

Type

I/J-band (8 to 20 GHz) multimode search and fire-control radar.

Description

PS-05/A is the multimode search and fire-control radar installed in the Swedish JAS-39 Gripen combat aircraft and comprises a 25 kg antenna/platform assembly, a 73 kg power liquid-cooled, travelling wave tube power amplifier/transmitter unit, a 32 kg software-controlled exciter/receiver unit and a 23 kg signal/data processor. Within this latter unit, the signal processing function is programmable while the D80 data processor uses software written in Pascal and incorporates a built-in test routine. To provide the required operational flexibility, low, medium and high Pulse Repetition Frequency (PRF) functions are incorporated together with a range of waveform modes which include the following:

HPD: A high PRF mode that utilises Doppler processing for clutter rejection and which is designed primarily for use against nose aspect air targets.

MPD: A medium PRF mode that utilises Doppler processing and is designed for use against nose and tail aspect air targets. A high-resolution submode is incorporated to facilitate target tracking.

LPD: A Doppler processing mode that is used against moving surface targets.

LPRF: A low PRF mode with pulse-to-pulse frequency agility that is used for real-beam mapping and surface target detection.

AGR: A ground target ranging mode.

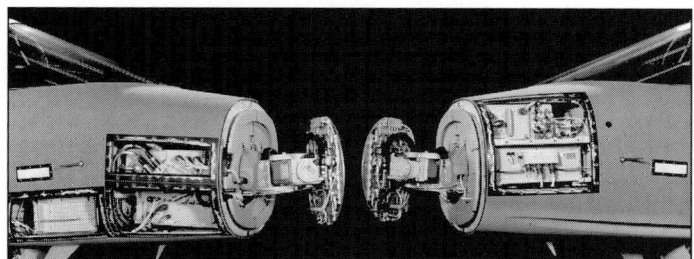

A composite view showing the location of PS-05/A components in the JAS-39 combat aircraft   0006994

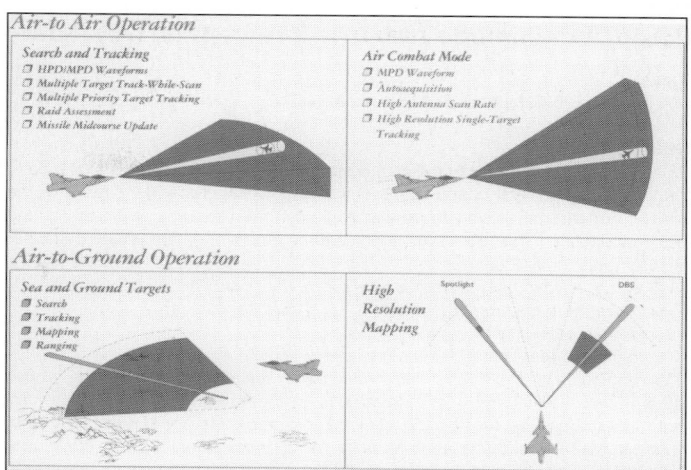

A graphic showing PS-05/A's air-to-air and air-to-ground operating modes

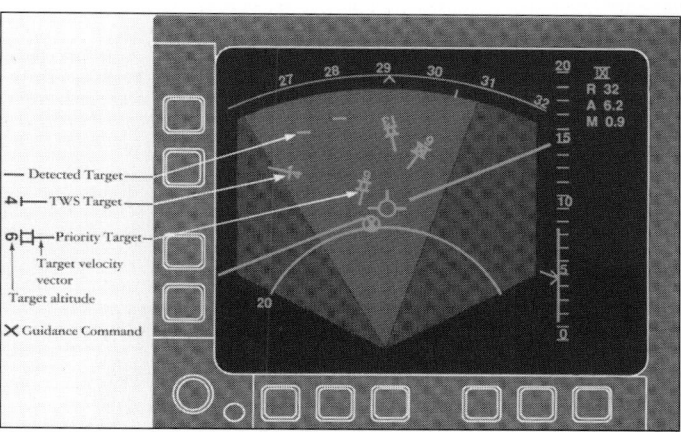

A representation of a typical display format generated by the PS-05/A radar

DBS: A synthetic aperture mode that utilises Doppler processing and is used for high-resolution ground mapping with high angular coverage obtained via continuous antenna scanning.

SLM: A synthetic aperture spotlight mode that utilises Doppler processing for very high-resolution mapping.

Use of these various waveform/PRF combinations allows the radar to offer the following operational modes:

Air-To-Air: Long-range search; multiple target track-while-scan; multiple priority target tracking; short-range, wide angle search and track (air combat mode); single target tracking and raid assessment.

Air-To-Surface: Long-range search; ground and sea priority target tracking; mapping and air-to-surface ranging.

Clutter rejection features include transmitter spectral purity; wide receiver dynamic range and low sidelobe levels while electronic counter-countermeasures features include use of monopulse; high output power; low sidelobes; a wide radio frequency bandwidth and frequency agility. As a whole, the radar makes use of a range of modern component technologies including gallium arsenide field-effect chip transistors; thin film on aluminium oxide substrate technology; high density analogue/digital packaging; multilayer thick-film technology; microwave integrated circuitry and application specific integrated circuitry. Acquired data can be displayed to the pilot via head-up and head-down displays and the system incorporates built-in test.

Status

As of September 2001, over 100 PS-05/A multimode radars were reported as having been delivered for installation aboard single- and two-seat JAS-39 Gripen combat aircraft. During 1997, Ericsson began work on a PS-05/A upgrade that incorporates new processors and components. A key area within this effort was the replacement of the D80 systems computer with a new data processor (designated as the Modular Airborne Computer System-MACS) in Batch 3 Gripens. The upgrade also included an enhanced signal processing capability that was noted as 'dramatically' improving the type's air-to-air and air-to-ground mission modes.

Specifications

Frequency: I/J-band (8-20 GHz)
Power output: 1 kW (average)
MTBF: 250 h (air operations - predicted)
Antenna type/size: Planar-array/600 mm (diameter)
Total system weight: 156 kg

Contractor

Ericsson Microwave Systems AB, Airborne Radar Division, Mölndal.

UPDATED

PS-46/A multimode Airborne Interception (AI) radar

Type
I/J-band (8 to 20 GHz) AI and fire-control radar.

Description
PS-46/A is the multimode AI radar fitted to the JA 37 fighter version of the Saab Viggen. To the general operational requirements of an all-weather capability, the ability to operate effectively in an electronic countermeasures environment and high availability, the PS-46/A specification adds the following particular requirements for air-to-air roles:

- capability against high-performance aircraft, transports, and helicopters
- all-hemisphere coverage
- look-down capability
- air-to-ground ranging.

In considering the transmitter waveform options to meet the above requirements, a number of techniques are available. The air-to-ground requirements can be met by the conventional non-coherent pulsed waveform, whereas the air-to-air requirement calls for a more sophisticated waveform. Waveforms that provide look-down capability are high Pulse Repetition Frequency (PRF) pulse-Doppler (HPD), medium PRF pulse-Doppler (MPD) and low PRF pulse-Doppler (LPD). These waveforms are mainly characterised by their pulse repetition frequencies since this parameter is the determining factor on overall performance. HPD and MPD require an internal radar frequency reference and subsequently coherent transmission, whereas LPD in addition can use a pseudo-coherent transmission (coherent-on-receiver) or a non-coherent transmission, resulting in auto-coherent or non-coherent airborne moving target indicator. A careful review of the pros and cons of these waveforms led to the selection of the two waveforms:

- low and medium PRF waveforms for look-up modes
- medium PRF waveform with Doppler filters and signal processing to resolve range ambiguities for look-down modes.

The radar is designed as a 'cartridge' fixed to the forward aircraft bulkhead with four bolts. The aircraft systems supply the necessary electric power, cooling air, and hydraulic power for the hydraulic vane motors that drive the two-axis antenna. The radar consists of 10 Line-Replaceable Units (LRUs) and these are supported or housed in a lightweight frame that is in itself an LRU. The construction provides for simple removal and replacement of units.

The main design characteristics of the PS-46/A radar are summarised as:

- I/J-band
- medium PRF pulse-Doppler operation
- coherent travelling wave tube transmitter
- large antenna with low sidelobes
- digital signal and data processing
- integrated continuous wave target illumination
- test.

Target data extraction and smoothing are carried out by Kalman filtering in stabilised Cartesian co-ordinates.
The PS-46/A radar can be configured to provide the following modes of operation:

- target search
- sea surface search
- passive emitter search
- passive emitter track (track-while-scan or continuous track)
- target acquisition (automatic using Head-Up Display (HUD) or semi-automatic using HUD)
- target tracking (track-while-scan or continuous track mode)
- target illumination
- air-to-ground ranging
- Advanced Medium Range Air-to-Air Missile (AMRAAM) datalink.

Single and multibar, wide or narrow scan patterns are provided together with raster scan for HUD lock on in dogfights.

Status
As of September 2001, the PS-46/A radar was reported as being no longer in production, but still in service aboard the Swedish Air Force's fleet of JA 37 Viggen interceptors.

Specifications
Frequency: I/J-band (8-20 GHz)
Antenna: 700 mm diameter
Power: 200 W (CW illuminator); 500 W (average)
Detection range: >50 km (look-down mode)
Processor: digital signal processor (32-bit floating point, 500 kword programme memory)
Weight: 300 kg
MTBF: 180 h

Contractor
Ericsson Microwave Systems AB, Airborne Radar Division, Mölndal.

UPDATED

UAP 13 series fire-control radars

Type
I/J-band (8 to 20 GHz) fire-control radars for interceptor aircraft.

Description
The UAP 13 equipments are a series of nose-mounted, forward-looking radars for search and fire-control applications in interceptor aircraft. The radars in this series are pulse radars operating in the I/J-band. The basic model is the UAP 13102, which is also designated PS-01/A. It is fitted to the Saab Draken J 35F and, possibly, J. The UAP 13103 (PS-011/A) is similar to the UAP 13102 but is designed to be used with the S71N infra-red search and track set. Both these models are intended for use in aircraft having a fire-control computer.

The following operating modes are provided: search, acquisition, lock on and tracking. An additional feature is the provision for slaving of missile radar and infra-red guidance systems to the aircraft radar. A two-bar scanned search pattern is used in the elevation plane for the UAP 13102, and a choice of two- or four-band in the UAP 13103.

Status
Over time, UAP 13102 and 13103 (integrated with the S71N infra-red search and track system) radars are reported to have been fitted to Swedish Air Force Drakens. As of October 2001, Jane's sources were reporting that a UAP 13 series radar was in service aboard J 35Ö aircraft of the Austrian Air Force.

Contractor
Ericsson Microwave Systems AB, Airborne Radar Division, Mölndal.

UPDATED

UNITED KINGDOM

Blue Fox radar

Type
I-band (8 to 10 GHz) airborne interception and fire-control radar.

Description
Blue Fox is a lightweight radar designed to fulfil the dual role of airborne interception and air-to-surface search and strike. It was developed to form part of a fully integrated weapon system. It operates in the I-band and uses frequency agility to enhance the radar's immunity to electronic countermeasures and improve its ability to detect small targets in bad weather and rough sea states. For air-to-air

The Ericsson PS-46/A multimode AI radar installed in a Viggen

The flat aperture radar used in the Blue Fox radar system

interception, Blue Fox can be used for lead-pursuit or chase attacks and incorporates a transponder mode for identifying friendly aircraft or ships. The equipment is built on the line-replaceable unit principle, each component part of the radar (transmitter, receiver, processor, amplifier) can be easily checked or removed independently for servicing. The antenna is a flat aperture slotted array, stabilised in pitch and roll. The display used provides a bright, digital scan-converted picture. Superimposed on the display are the flight symbols showing aircraft altitude, speed, heading and so on, so that the pilot can monitor and control the aircraft's manoeuvres while using the radar.

Status

As of April 2001, the Blue Fox radar was understood to be in service aboard the 16 Sea Harrier FRS Mk 51 aircraft assigned to the Indian Navy's Air Squadron 300 (shore based at Dabolim, India).

Contractor

BAE Systems Avionics - Sensor Systems Division, Edinburgh.

UPDATED

Blue Vixen radar

Type

I-band (8 to 10 GHz) airborne interception multimode radar.

Description

Blue Vixen is a lightweight, coherent, pulse-Doppler radar that operates in the I-band. It is a true multimode equipment that maintains full power in all modes; has all-weather look-up/look-down capabilities; can detect targets over land and sea and offers high-resolution air-to-air ranging.

Blue Vixen operates in low, medium and high Pulse Repetition Frequency (PRF) modes. Selection of the appropriate PRF for optimum detection is automatic and depends on background clutter and target density. Low PRF is in look-down mode to provide accurate range and velocity with all aspect detection. High PRF provides for the look-down detection of targets approaching at high speed in a high clutter environment.

Automatic track-while-scan and single target track are available in air-to-air modes. It is claimed to be the first AI radar in the world to be designed from the outset with full Advanced Medium Range Air-to-Air Missile (AMRAAM) compatibility. It is also compatible with other medium range air-to-air missiles, the AIM-9 Sidewinder and the Sea Eagle anti-shipping missile.

Blue Vixen has been designed with a flexible line-replaceable unit configuration. The equipment's antenna, receiver and transmitter are co-located in the aircraft's nose bay with the radar power supply and processor being housed elsewhere in the airframe. System weight is given as 145 kg.

Status

As of April 2001, Jane's sources were reporting that the Blue Vixen radar was installed aboard the 49 Sea Harrier F/A Mk 2 air defence/strike aircraft assigned to the Royal Navy's Nos 800 and 801 Squadrons (both shore-based at Yeovilton, Somerset, UK) and the Sea Harrier Operational Evaluation Unit (based at Boscombe Down, Wiltshire, UK). A successful series of Sea Harrier AMRAAM firing trials using the radar are noted as having taken place during 1993-94. With Blue Vixen fitted, the Sea Harrier is reported to be able to ripple fire up to four AMRAAMs (AIM-120s) while maintaining guidance datalink and target tracking functions.

Contractor

BAE Systems Avionics - Sensor Systems Division, Edinburgh.

UPDATED

Foxhunter Airborne Interception (AI) radar

Type

I-band (8 to 10 GHz) AI radar.

Description

The design of Foxhunter (RAF designation AI-24) provides a multimode system compatible with the size and weight limits of the air defence variant of the Tornado and the operational requirements of the next two decades. A substantial part of the signal processing is performed digitally in addition to digital radar data handling. The equipment anticipates trends in offensive tactics, (such as low-level penetration and the use of electronic countermeasures) and the latest improvements give it the potential to support active missiles. Additionally, Foxhunter has the flexibility to operate as part of ground or Airborne Early Warning (AEW)-based control environments while retaining the ability to operate autonomously. The radar is designed to detect and track subsonic and supersonic targets at ranges in excess of 185 km at both low and high level. Although it is intended to form part of an overall air defence system, the integration of the radar into the Tornado Air Defence Variant's avionics system permits effective independent operation.

Foxhunter operates in I-band (3 cm) and in its primary mode, uses a pulse-Doppler technique known as Frequency Modulated Interrupted Continuous Wave (FMICW). At the heart of the radar is a master timing and synchronising unit (employing phase-locked loop techniques) which generates the transmission waveforms for amplification by the transmitter and the reference signals and timing pulses employed within the receiver and signal processing circuits. Compact and lightweight surface acoustic wave devices provide signal waveforms for the pulse compression modes of the radar. The antenna uses the 'Elliott' twist-reflecting cassegrain principle which combines rigidity and light weight with low-level spurious radiation lobes; a key feature in the rejection of ground clutter and jamming signals. This type of antenna has been developed to an advanced level of performance. The scanner employs a hydraulic drive mechanism and servos to achieve the speed and precision of beam pointing and stabilisation demanded while the aircraft is manoeuvring.

The radar's travelling wave tube transmitter has been designed with the inherent flexibilty to handle the differing waveforms used in the various modes of operation. At the heart of the signal processing system is a digital processor that uses Fast Fourier Transform frequency analysis techniques to filter signal returns into narrow frequency channels. Target echoes are segregated from clutter returns and a uniform detection threshold is achieved against all target velocities. Foxhunter also provides target illumination for the Sky Flash medium-range semi-active air-to-air missiles and incorporates extensive anti-jamming provision. Since heavy jamming is virtually certain, the system incorporates strong electronic counter-countermeasures facilities.

Foxhunter Product Improvement Programme (PIP)

Since Foxhunter was introduced into service, a two-stage PIP has been implemented. Stage 1 introduces unspecified system modifications, radar functions related to the aircraft's Hands On Throttle And Stick (HOTAS) modes and refined Track-While-Scan (TWS) and anti-jamming algorithms. Stage 2 updates include a redesigned data processor, improved data processing software and modified transmitter and receiver paths. Taken together, these modifications improve TWS performance in high radio-frequency noise environments. The new processor also allows automatic and manual scan management techniques to be employed in TWS. Further, the capability can handle a significantly greater number of tracks than its predecessor and provides the basis for further growth potential (including the ability to support active radar missiles).

Status

As of the period October 2000 to April 2001, Jane's sources were reporting the Foxhunter AI radar as being installed aboard approximately 164 Tornado F Mk 3/Air Defence Variant (ADV) aircraft that were in service with the Italian (24 Tornado F Mk 3 aircraft (October 2000 figure) assigned to the 12 and 21° *Gruppo Caccia*

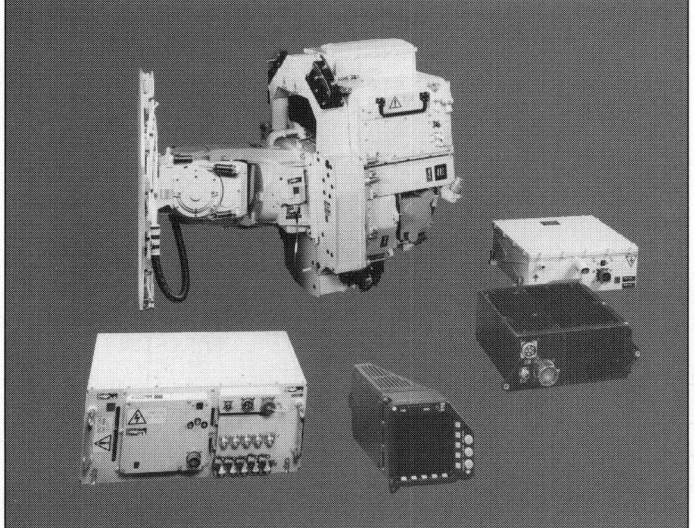

Units of the Blue Vixen radar

A Foxhunter AI radar mounted in the nose of a UK Tornado ADV variant (courtesy of RAF Coningsby)

Intercettori Ognitempo (All-weather Interceptor Squadron) based at Gioia del Colle, Italy), Saudi (22 Tornado ADV aircraft (October 2000 figure) assigned to No 29 Squadron based at Dharan, Saudi Arabia) and UK (118 Tornado F Mk 3 aircraft (April 2001 figure) assigned to the Air Defence Operational Evaluation Unit (based at Coningsby, Lincolnshire, UK), Nos 5 (Coningsby), 25 (Leeming, Yorkshire, UK), 43 (Leuchars, Fife, UK), 56(R) (Coningsby) and 111 (Leuchars) Squadrons and the Tornado F Mk 3 detachment at Al Kharj, Saudi Arabia) air forces. The Foxhunter Stage 1 PIP is noted as having been introduced into service during 1989, with Stage 2 being implemented during 1996.

Contractor
BAE Systems Avionics - Sensor Systems Division, Edinburgh.

UPDATED

Skyranger airborne radar

Type
I-band (8 to 10 GHz) airborne weapon control radar.

Description
Skyranger is a lightweight airborne weapon control radar developed for retrofit to light fighter and attack aircraft. It consists of three main units: antenna, transmitter/receiver and signal processor/power supply and, since the amount of space available for retrofit programmes can often be limited and irregular in shape, the modularity of Skyranger has been established at printed circuit card level. The individual cards can, therefore, be packaged into housings designed for the space available. The equipment has been designed as part of an integrated avionics suite, the other systems being an air data computer, radar altimeter, Head-Up Display (HUD), weapon aiming computer and secure communications.

Skyranger accepts discrete digital commands from a cockpit-mounted control panel and provides output data in the form of a digital serial link (ARINC 429) to an HUD and other weapon aiming systems. It has two main modes; guns and missiles, the former having a shorter range, wide-angle beam. In the missile mode the radar energy is fed from the feed horn and reflected back from the parabolic antenna in a 6° beam with a maximum range of 15 km. For gun attacks, the radar energy is fed directly out from the antenna through a polarised window. This results in an 18° beamwidth and a range of 5 km. Minimum range is 300 m for guns and 150 m for missiles, with ranging accuracy in the order of ±15 m below 3 km and ±30 m above. Target relative velocities of from −500 to +1,000 m/s may be handled.

The equipment operates in I-band and has 5 per cent pulse-to-pulse agility. Mean time between failure value is given as 200 hours and the equipment contains built-in test systems. The current version has a fixed antenna.

Status
Over time, more than 300 Skyranger radars are reported to have been produced. The sensor is known to be fitted to the Chinese F-7M/P series Airguard fighter aircraft. As of April 2001, Jane's sources were reporting that somewhere in the region of 78 F-7M/P series aircraft equipped with Skyranger radars had been exported to the air forces of Bangladesh (16 (original figure, 13 as of October 2000) F-7MB aircraft assigned to No 5 Squadron based at Dhaka-Tejgaon/Chittagong, Bangladesh), Iran (at least 18 (original figure, 12 as of April 2001) F-7M aircraft), Myanmar (24 (original figure, 20 as of April 2000) F-7M aircraft assigned to three squadrons based at Hmawbi (two units) and Moulmein, Myanmar) and Pakistan (20 (original figure) F-7P aircraft). As of the given date, Jane's sources were suggesting the Skyranger radars aboard Pakistan's F-7Ps could be replaced with a variant of the FIAR Grifo radar (see separate entry) to bring them closer to the sensor fit of the F-7MP aircraft, of which, Pakistan has procured 100 examples. Readers are cautioned that the foregoing F-7 quantity/assignment data should be regarded as provisional and treated accordingly.

The Skyranger radar configuration used on the Chinese F-7P/M series fighter aircraft

Specifications
Frequency: I-band (8-10 GHz)
Range: 5 km (gun operation); 15 km (missiles)
Range resolution: 150 m
Pulse-to-pulse agility: 5%
Power requirements: 27 V DC, <50 W; 115 V 400 Hz single phase, <400 VA
Weight
Antenna: 4 kg
Transmitter/receiver: 25 kg
Signal processor/power supply: 8 kg
Total installed weight: 40 kg

Contractor
BAE Systems Avionics - Sensor Systems Division, Edinburgh.

UPDATED

UNITED STATES OF AMERICA

AN/APG-63(V) multimode airborne radar

Type
I/J-band (8 to 20 GHz) multimode airborne radar.

Description
The AN/APG-63(V) multimode radar is a pulse-Doppler system that operates over a number of selectable frequencies in the I/J-band. A gridded travelling wave tube transmitter is employed, with digital Doppler signal processing and digital mode/data management. These features permit operation over a wide range of pulse repetition frequencies, pulse-widths and processing modes. The I-band antenna is a planar-array type, carried on a three axis gimbal system.

Radar information is digitally processed and there are two types of display employed: one is a small cathode ray tube located at the upper left-hand corner of the instrument panel and called the Vertical Situation Display (VSD); the other is the Head-Up Display (HUD). In general, the VSD is employed for the longer range, initial stages of an interception, while the HUD is for use during actual engagements or close-in encounters. The VSD presents the pilot with a 'cleaned' synthetic display of computer-processed radar video data, together with alphanumerics and symbols.

Controls for the APG-63(V) are located at three positions in the cockpit. The main control panel is on the console on the left side of the pilot. This console also carries the two throttles and key radar operating controls are located on them. The third location for radar controls is the aircraft control stick. Typical alphanumeric information that can be presented on the VSD are target altitude, groundspeed, heading, range, aspect angle, closure rate, and g-force. Groundspeed and g-force data give valuable indications of the kind of target being tracked. A satisfactory response to an identification friend-or-foe interrogation is indicated by a symbol displayed on the VSD.

An automatic acquisition switch on the aircraft control stick enables the pilot to lock the radar on to targets within close-in ranges. There are three modes: in the boresight mode, the radar locks on to the first target that enters the aircraft boresight, as designated by the gun reticule on the HUD; the second, called supersearch, locks the radar on to the first target that comes within the HUD field of view; the third, vertical scan, locks the radar on to the first target that enters an elevation scan pattern normal to the aircraft's lateral axis.

Alongside the baseline radar, Raytheon has developed two upgraded variants of the basic design under the designations APG-63(V)1 and APG-63(V)2. Of these, the APG-63(V)1 has been designed for field retrofit to the F-15 and focuses on enhancing system performance, reliability (120 hours mean time between failure), maintainability and supportability. System features include improved built-in test (frequently to module level); compatibility with both mechanically and

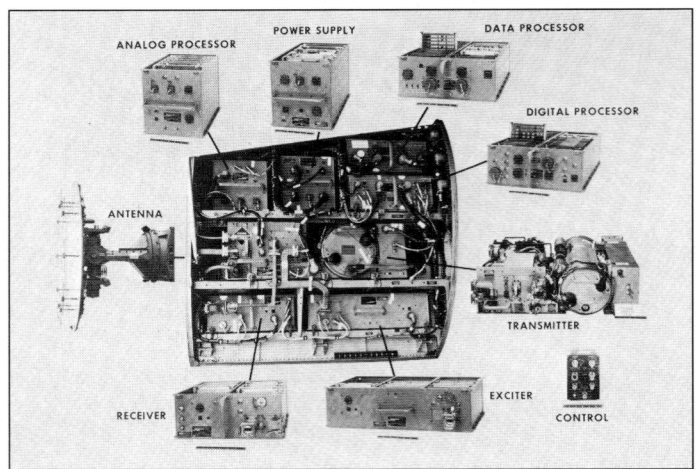

A view of the baseline AN/APG-63(V) radar that shows its subsystems and their location in the complete equipment

A general view of the AESA antenna that is used in the APG-63(V)2 radar (Boeing)

2002/0103897

electronically scanned antennas; increased receiver/exciter signal-to-noise ratio; increased transmitter bandwidth; increased processing throughput and increased processor memory. System operating modes include boresight; supersearch; vertical scan; auto gun; track-while-scan; range-while-scan; precision velocity update (navigation); ground moving target detection and high resolution ground mapping. The APG-63(V)2 features an Active Electronically Scanned Array (AESA) antenna, an improved Honeywell environmental control system and integration with an 'advanced' BAE Systems North America identification friend-or-foe subsystem. APG-63(V)2 is also noted as taking 'full advantage' of the AIM-120 air-to-air missile's capabilities.

Status

By September 1986, Raytheon Electronic Systems is reported as having delivered approximately 1,000 APG-63(V) radars, including those produced under licence in Japan. As such, the sensor has been installed aboard F-15A, B, C, D and J fighter/combat training aircraft operated by the US (102 F-15As, 20 F-15Bs, 345 F-15Cs and 53 F-15Ds as of October 2000) Israeli (38 F-15As, six F-15Bs, 16 F-15Cs and 11 F-15Ds as of April 2001), Japanese (158 F-15J as of April 2001)and Saudi Arabian (87 F-15C/Ds as of October 2000) air forces and four P-3 'interceptor' aircraft of the US Customs Service. Of these, the first US Customs Service Orion 'interceptor' was introduced into service during 1984, with the remaining three having been delivered by the end of 1986. Production of the APG-63(V)1 retrofit package for USAF F-15C/D aircraft was sanctioned in July 1999, when Raytheon was awarded a then year US$93,496,863 firm, fixed-price contract covering the supply of 22 equipments. At the time of this award, the USAF was understood to have a total APG-63(V)1 requirement of 162 units, with production being planned for US Fiscal Years 1999 through 2003. According to Jane's sources, the AESA antenna-equipped APG-63(V)2 has been fitted to 18 F-15C interceptors based at Elmendorf Air Base, Alaska with installation being originally scheduled for completion by the end of 2000. Raytheon is also noted as having introduced a new programmable signal processor into APG-63 during the mid-1980s with the resultant radar being able to undertake Doppler beam sharpened ground mapping. As of July 2001, identified APG-63(V) programme activity during 2000/2001 was as follows:

14 January 2000
The former McDonnell Douglas Corporation (St Louis, Missouri) was awarded a then year US$11,000,416 option to an existing firm, fixed-incentive contract (F33657-97-C-2044-P00010) that covered contractor support for the APG-63(V)1 radars installed on F-15 aircraft. At the time of the award, work on the effort was scheduled for completion on 14 January 2001 and the programme's contracting activity was the USAF's Aeronautical Systems Center, Wright-Patterson Air Force Base (AFB), Ohio.

31 May 2000
The former McDonnell Douglas Corporation (St Louis, Missouri) was awarded a then year US$98,758,455 option to an existing firm, fixed-price contract covering the provision of 25 APG-63(V)1 radar upgrade kits for installation aboard F-15 aircraft. At the time of the award, work on the effort was scheduled for completion on 31 December 2002 and the programme's contracting activity was the USAF's Aeronautical Systems Center, Wright-Patterson AFB, Ohio.

20 June 2000
The former McDonnell Douglas Corporation (St Louis, Missouri) was awarded a then year US$21,090,309 option to an existing firm, fixed-price contract covering the supply of eight APG-63(V)1 radars and associated spares packages for use on F-15 aircraft. At the time of the award, work on the effort was scheduled for completion on 31 December 2002 and the programme's contracting activity was the USAF's Aeronautical Systems Center, Wright-Patterson AFB, Ohio.

28 June 2001
McDonnell Douglas Training Systems (St Louis, Missouri) was awarded a then year US$146,433,030 firm, fixed-price/cost plus contract covering the upgrading of 36 APG-63(V) radars applicable to F-15s and the provision of associated contractor support. At the time of the announcement, work on the effort was to be carried out by the Raytheon Systems Company at El Segundo, California (55 per cent workshare); Forest, Mississippi (29 per cent) and 'other locations' (16 per cent) and

was scheduled for completion by the end of December 2003. The programme's contracting activity was the USAF's Aeronautical Systems Center, Wright-Patterson AFB, Ohio.

Specifications
Frequency: selectable in the I/J-band (8-20 GHz)
Range: 161 km
Antenna: 3 gimbal axes, mechanical scan
No of LRUs: 9
Weight: 221 kg
Volume: 0.25 m³
MTBF: 60 h

Contractor
Raytheon Electronic Systems, El Segundo, California.

UPDATED

AN/APG-65 multimode airborne radar

Type
I/low J-band (8 to 12.5 GHz sub-band) multimode fire-control radar.

Description
The AN/APG-65 radar was developed by Hughes (now part of Raytheon Electronic Systems) for use on the US Navy's (USN) F/A-18 Hornet strike fighter aircraft. It is an I/low J-band equipment designed as an all-digital, multimode system suitable for both air-to-air combat and air-to-ground weapon delivery missions. It provides radar information for the control of the F/A-18's 20 mm gun, Sparrow, Sidewinder and AIM-120 missiles in aerial combat and a full range of conventional and precision guided weapons in the ground attack role. During Operation Desert Storm two USN F/A-18s, en route to attack ground targets, found and destroyed two Iraqi MiG-21s in air-to-air combat. They then switched back to the air-to-ground mode and completed their primary mission by delivering their bombs accurately. This dramatically demonstrated the versatility and multirole capability of the APG-65 radar. In its air-to-air role, the APG-65 presents a 'clean-scope', synthetic scan-converted display against airborne targets in all aspects, all altitudes and through all target manoeuvres. It incorporates complete search, track and air combat mode variations.

The radar includes: a velocity search mode to provide maximum detection range capability against nose aspect targets; a range-while-search mode to detect all-aspect targets; a track-while-scan mode which, combined with an autonomous missile such as AIM-120, gives the aircraft a launch-and-leave capability; a single target track mode; a gun director mode and a rapid assessment mode which enables the operator to expand the region centred on a single tracked target, permitting radar separation of closely spaced targets.

Three air combat manoeuvring modes provide automatic target acquisition in various search volumes: the gun acquisition mode to scan the entire head-up display volume and to lock onto the first target located within a specified range; the vertical acquisition mode in which the radar scans vertically in a narrow width volume and automatically acquires the first target found within a specific range and a boresight acquisition mode to allow the pilot to point the aircraft at the desired target and acquire it automatically. The pilot can step through successive targets until he acquires the one he wants.

Surface attack modes include long-range, high-resolution surface mapping which, combined with other modes, gives the pilot the ability to detect and track fixed or moving targets on land or sea. The APG-65 radar includes: a precision velocity update feature to improve navigational accuracy; a terrain-avoidance mode for low-level penetration missions in limited visibility conditions; ground moving target indication/track or fixed target modes which the pilot may select depending on the target tactical situation; air-to-surface ranging and a sea surface mode which enables the radar to detect ship targets regardless of sea condition.

An upgrade designated the APG-73 (see separate entry) is currently underway which will change the bandwidth, increase the internal operating rates of the receiver/exciter and increase the processing speed of the radar signal processor. New radar data processing hardware will increase throughput speed and expanded memory.

Status
As of November 2001, the APG-65 multimode radar was reported as being installed (or mandated for installation) aboard F/A-18A/B/C (early production)/D (early production) aircraft operated by the USN and US Marine Corps and the air forces of Australia (55 × F/A-18A and 16 × F/A-18B), Canada (83 × CF-188A and 39 × CF-188B), Kuwait (32 × F/A-18C and 8 × F/A-18D) and Spain (91 × EF-18A Plus/C.15 and 12 × EF-18B Plus/CE.15); AV-8B Plus aircraft of the US Marine Corps and the Italian (16 × AV-8B Plus) and Spanish (8 × EAV-8B Plus/VA.2) navies; 39 F-4E Avionics Upgrade Programme (AUP) aircraft of the Greek Air Force (see following) and 110 F-4F Improved Combat Efficiency (ICE) aircraft (also known as the F-4F KWS) of the German Air Force (see following). With regard to the F-4 AUP and ICE efforts, selection of APG-65 for use in the latter programme was announced by the German government in early 1985 and initial deliveries were made in early 1988. The radars used in the ICE upgrade have been manufactured under licence by DaimlerChrysler Aerospace (now part of the European Aeronautic, Defence and Space (EADS) Co) and during July 1997, the same company announced that it had been selected (in collaboration with Israeli

Italian Navy Harrier II Plus aircraft fitted with the APG-65 fire-control radar

contractor Elbit) to undertake the Greek F-4E AUP programme. The prototype F-4E AUP made its maiden flight on 28 April 1999 and Jane's sources were suggesting that the in-country programme contractor (Hellenic Aerospace Industries) would have delivered four AUP aircraft to the Greek Air Force by the end of 2000.

Other programme highlights include the Spring 1990 signature of a US-Canada agreement to jointly develop an APG-65 upgrade designated as the AN/APG-73. At the time of signature, this effort had an estimated cost of then year US$260 million and in April and November of 1990, Boeing is understood to have been awarded F/A-18 APG-73 integration contracts valued at then year US$221 million. During January 1991, the same contractor (together with BAE Systems) was awarded a then year US$181 million contract to integrate and flight test the APG-65 radar in the AV-8B Harrier II Plus aircraft. The prototype AV-8B Plus made its maiden flight in September 1992 and in the following November, the Italian government authorised the purchase of 16 such aircraft. The Italian aircraft are noted as incorporating a 'slightly modified' APG-65 variant. In March 1993, the Spanish government authorised the purchase of eight EAV-8B Plus aircraft and on 29 January 2001, Boeing (in the form of its McDonnell Douglas subsidiary at St Louis, Missouri) was awarded a then year US$35,304,098 funding modification that turned an existing advanced acquisition contract (N00019-00-C-0363) into a firm, fixed-price contract covering the re-manufacture of two Spanish EAV-8B aircraft to EAV-8B Plus standard. At the time of the announcement, work on the effort was to be divided between St Louis (63 per cent work share) and Brough in the UK (37 per cent work share) and was scheduled for completion by the end of July 2003. The programme's contracting activity was the USN's Naval Air Systems Command, Patuxent River, Maryland. As of November 2001, the US Marine Corps' AV-8B to AV-8B Plus conversion programme was understood to be continuing.

Specifications
Frequency: I/low J-band (8-12.5 GHz)
PSP: 7.2 million complex operations/s
No of LRUs: 5
Weight: 154 kg
Volume: 0.126 m³ (excl antenna)
Transmitter: liquid-cooled, software controlled TWT
Antenna type/size: planar-array/71 cm (approx)
Processors: 250 k memory (radar); 7.2 Mops speed (signal)
MTBF: 120 h

Contractor
Raytheon Electronic Systems, El Segundo, California.

UPDATED

AN/APG-66 multimode airborne radar

Type
H- through low J-band (6 to 10 GHz sub-band) multimode weapon control radar.

Description
The AN/APG-66 multimode airborne radar is a 6 to 10 GHz band, coherent, pulse-Doppler sensor that was originally designed for use on the F-16 aircraft and is capable of supporting AIM-7F/M Sparrow, AIM-9L/M Sidewinder, AIM-120A, Penguin, MICA and Skyflash E missiles. The baseline radar is of modular design and comprises six functional Line-Replaceable Units (LRUs), each of which has its own power supply. As such, the baseline configuration is noted as containing 9,500 component parts and in its production form, has no associated hydraulics or rate/roll gyros. So configured, the radar offers 10 operating modes, some of which are associated with frequency agility-based countermeasures resistance. All system functions (including self-test) are computer controlled via a serial digital databus. LRU production has involved a European consortium that has included companies in Belgium, Denmark, Netherlands and Norway. Over time, the following APG-66 variants have been identified:

APG-66(T47)
APG-66(T47) is the APG-66 radar variant installed in the Cessna OT-47B surveillance aircraft. In this installation, the radar is integrated with the Northrop Grumman WF-360 Forward-Looking Infra-Red (FLIR) sensor. As originally scheduled, delivery of the OT-47B aircraft was scheduled for completion by March 1997.

APG-66(V)2
APG-66(V)2 is an upgrade of the basic radar that is matched to the US/NATO F-16A/B Mid-Life Update (MLU) programme and is applicable to Block 10/15 and

ADF configuration aircraft. According to Jane's sources, the enhanced radar incorporates a new signal processor, a higher power transmitter and faster antenna phase shifting and offers greater detection range (increased from 65 to 83 km in the presence of heavy clutter and jamming), improved reliability (an MTBF figure of 210 hours), improved protection against electromagnetic interference, enhanced operational performance (including an improved mapping capability and a reduced false alarm rate) and the ability to support a colour display. Northrop Grumman further notes that the APG-66(V)2 system contains 'significant' growth capabilities and offers combined search and track, track-while-scan, air combat search and automatic tracking air-to-air modes together with real beam mapping, 64:1 Doppler beam sharpening, scan freeze, ranging, beacon homing, combined ground mapping/air target tracking and sea search air-to-surface options. As of June 2001, Northrop Grumman was claiming that APG-66(V)2 offered 'significant operational and sustainment advantages at a third of the cost of a new radar' and noted that it had teamed with the US Air Force's Ogden Logistics Center to offer F-16A/B users a general APG-66(V)2 upgrade package. Here, options comprised a full F-16 MLU avionics upgrade/sensor combination or a radar kit with software only changes to the host platform's avionics system.

APG-66(V)3
Closely related to APG-66(V)2, the APG-66(V)3 is configured for use aboard Block 20 F-16A/B aircraft. While neither confirmed nor denied by its manufacturer, Jane's sources suggest that APG-66(V)3 incorporates a continuous wave illumination capability and is the radar selected by Taiwan for use on its 150 new-buy, AIM-7 air-to-air missile capable F-16A/Bs. As of June 2001, in excess of 550 APG-66(V)2 and (V)3 radars were reported as being operational with more than 150 being new build equipments.

APG-66(V)X
As of June 2001, the APG-66(V)X configuration was understood to be an upgrade option for F-16A/B (ADF, Block 15/20 and MLU standard) aircraft that makes use of the antenna, modular receiver/exciter and common radar processor from the AN/APG-68(V)9 radar (see separate entry) mated with a 'smart interface' for the processor and a modified APG-68(V)9 medium duty transmitter. The 'smart interface' allows the radar to automatically sense and adapt to its host avionics configuration and the sensor incorporates all the capabilities of the APG-66(V)2/(V)3 variants together with an increase in detection range and tracking accuracy and a fourfold improvement in reliability. If its host platform has not received the F-16 MLU avionics upgrade, the radar is 'improved Synthetic Aperture Radar (SAR) ready' rather than incorporating the high resolution SAR capability inherent in the combined MLU/APG-66(V)X package.

APG-66H
APG-66H is an APG-66 variant that is optimised for use on the BAE Systems Hawk 200 single-seat multirole combat aircraft. To fit the space constraints of the Hawk installation, APG-66H features a slightly smaller antenna than the standard equipment and a new signal data processor. This latter unit is reported to offer most of the modes available in the APG-68 radar (see separate entry) including TWS.

APG-66J
APG-66J is an APG-66 variant that has been configured for use in the Japanese Air Self-Defence Force's F-4EJ Kai upgrade programme. As of January 1995, 83 out of a planned total of 86 F-4EJ aircraft are reported to have been brought up to Kai standard.

APG-66NT
APG-66NT is the APG-66 variant installed in the 17 civilian-operated T-39N aircraft that are used in the US Navy's Undergraduate Naval Flight Officer training programme.

APG-66NZ
APG-66NZ is an APG-66 variant that is reported to have incorporated a maritime tracking mode and to have been installed aboard New Zealand's fleet of 14 A-4K and five TA-4K Kahu Skyhawks.

APG-66SR
APG-66SR is an extended range variant of the basic radar that is used in Northrop Grumman's MultiSensor Surveillance Aircraft (MSSA) system. Here, the radar is teamed with a WF-360 FLIR, a navigation subsystem which includes a Litton LTN-92 ring laser gyro inertial navigation unit and a Global Positioning System receiver and a communications suite comprising dual VHF/UHF band (30 MHz to 1 GHz) radios. APG-66SR is fitted with a larger aperture antenna and, as installed in the first MSSA platform (the Pilatus Britten-Norman BM2T-4R Defender), offers 360° coverage in azimuth. As of May 2001, two MSSA Defender aircraft had been delivered to Northrop Grumman by January 1995.

APG-66 Small Aerostat Surveillance System (SASS) variant
APG-66 SASS is a minimally modified baseline APG-66 fire-control radar that is rack-mounted on the underside of the aerostat's hull to provide a range of airborne early warning and surveillance capabilities. So used, the radar installation includes two gyro systems to maintain accurate elevation pointing and north referenced azimuth-pointing angles. Data accquired by the radar is passed to an associated ground station via a fibre-optic link contained within the aerostat's tether. In the SASS application, APG-66 has seven operating modes, as follows:
- medium Pulse Repetition Frequency (PRF) air surveillance with 19, 37, 74 and 148 km range scales
- non-coherent, frequency agile, low PRF sea surveillance

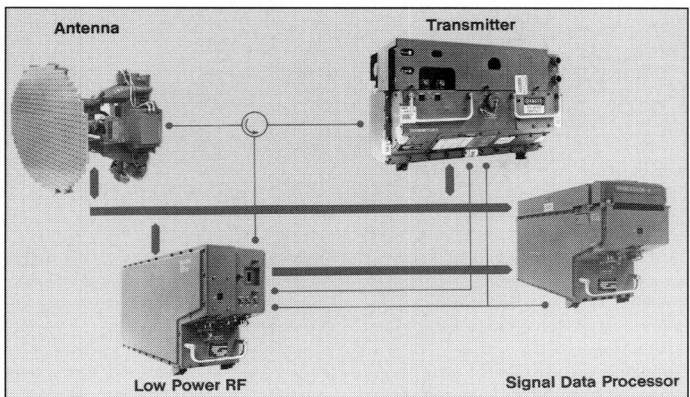

A schematic showing the elements that make up the APG-66(V)2 multimode radar   0044144

- coherent Doppler sea surveillance with clutter cancellation
- ground moving target indicator
- 8 to 12 GHz band beacon interrogation
- weather surveillance
- real beam mapping.

APG-66T
APG-66T is an APG-66 variant that is described as offering multitarget track-while-scan, four dogfight operating modes and situational awareness mode. APG-66T is noted as having completed flight tests aboard a US Navy testbed aircraft during October 1991.

ARG-1
ARG-1 is an APG-66(V)2 variant that has been developed for installation aboard Argentina's 32 A-4AR Fighting Hawk multirole combat aircraft.

Status
As of May 2001, the multimode APG-66 radar was reported as having been procured (or mandated for procurement) by the air forces of Argentina, Belgium, Denmark, Egypt, Indonesia, Israel, Malaysia, Mexico, Netherlands, Norway, Oman, Pakistan, Portugal, Singapore, Taiwan, Thailand, USA and Venezuela together with the US Coast Guard, US Customs Service and the US Navy (USN). Within the F-16 MLU context, Jane's sources suggest that Northrop Grumman received a then year US$106 million contract for 301 APG-66(V)2 update kits (for use on Belgian, Danish, Dutch and Norwegian sensors) during September 1994, with a further 49 firm orders having been placed by December 1995. Identified programme activity during 1999 comprised the following:

27 May 1999
The Northrop Grumman Corporation (Baltimore, Maryland) was awarded a then year US$16,919,556 modification to an existing contract (F33657-98-C-2015) that covered the provision of 49 APG-66(V)2 radar kits and associated spares for use on Belgian (24 radar kits), Dutch (radar spares), Norwegian (radar spares) and Portuguese (25 radar kits) F-16 aircraft. At the time of the modification's announcement, work on the effort was scheduled for completion on 29 March 2002, with the USAF's Aeronautical Systems Division, Wright-Patterson Air Force Base, Ohio acting as the programme's contracting activity.

1 June 1999
Northrop Grumman's Electronic Sensors and Systems Sector (Baltimore, Maryland) was awarded an estimated then year US$15,942,335 requirements contract (with firm fixed prices) covering the repair of APG-66 radars fitted aboard USN P-3 Orions and 'various' aircraft that were in service with the US Coast Guard (USCG) and the US Customs Service, together with WF-360 FLIRs fitted to aircraft operated by the USCG and the USN's Navy Test Pilot School. At the time of the contract's announcement, work on the effort was scheduled for completion by the end of October 2002, with the USN's Naval Inventory Control Point, Philadelphia, Pennsylvania acting as the programme's contracting activity.
As of June 2001, Northrop Grumman noted that it had produced in excess of 1,500 APG-66 radars.

Specifications
APG-66 general
Frequency: H- through low J-band (6.2-10.9 GHz sub-band (baseline APG-66); 8-10 GHz sub-band (APG-66(V)2)
Transmitter: TWT
Search angle: 120° (azimuth and elevation)
Angular coverage: ±10/±30/±60° (azimuth); 1/2/4 bars (elevation)
Range scales: 19, 37, 74 and 148 km
Electronic protection: EMI pulse editor; fast phase shifting; frequency agility
MTTR: 5 min
MTBF: >210 h (over 300 h demonstrated)
Weight: 98.4 kg (APG-66T); 107.7 kg (APG-66H); 115.9 kg (APG-66(V)2); 134.3 kg (APG-66(V)1)
Volume: 0.08 m³ (APG-66T); 0.082 m³ (APG-66H); 0.097 m³ (APG-66(V)2); 0.102 m³ (APG-66(V)1)
APG- 66 SASS
Frequency: 9.7-9.9 GHz

Power: 200 W (average); 17.5 kW (peak)
Beamwidth: 0.75° (azimuth); 2.25° (elevation)
PRF: 500 Hz-15 kHz
Pulse-width: 0.284-4 µs
Emission type: uncoded waveform
Subclutter visibility: 60 dB
Receiver noise figure: 4.3 dB
Range gates: 128
Range: 59 km (GMTI, 10 m² target, 9 km/h min velocity); 74 km (sea surveillance, 10 m² target, 0 km/h min velocity, Sea State 2); 74 km (sea surveillance, 50 m² target, 19 km/h min velocity, Sea State 2); 83 km (air surveillance, 2 m² target, 102 km/h min velocity); 148 km (instrumented and air surveillance, 10 m² target, 102 km/h min velocity)
Angular coverage: ±20° (elevation); continuous 360° or sector scan (azimuth)
Antenna size: 1 × 3 m
Antenna gain: 42 dB
Antenna polarisation: vertical
Antenna pointing accuracy: 5 mrad
Antenna rotation rate: 0.5-3 rpm
Antenna sidelobes: >–40 dB (azimuth)

Contractor
Northrop Grumman Electronic Systems sector, Baltimore, Maryland.

UPDATED

AN/APG-67 multimode radar

Type
I-/low J-band (8 to 12 GHz sub-band) multimode airborne radar.

Description
The AN/APG-67 multimode radar is described as being a modern, digital, computer-controlled system that provides a full range of air-to-air capabilities at ranges of up to 148 km. Air-to-air modes include look-up and look-down range while search, situation awareness, velocity search, track-while-scan, manual and automatic target designation and single target track. Air-to-surface modes include ground map with expand, Doppler beam sharpening, freeze over land and sea and surface moving target indication and track capabilities. Overall, APG-67 is designed to meet the sensor requirements of fighter aircraft and is billed as having a small volume, 'high' reliability and 'low' initial and life cycle costs. An optional Synthetic Aperture Radar (SAR) capability (see following) is noted as being an 'easy' addition to new-build or follow-on radars of this type. APG-67 also features programmable signal and data processors, a software-controlled antenna servo system, full software implementation of the radar's detection, tracking and display functions and MIL-STD-1553 databus compatibility. In terms of the cited SAR capability, Jane's sources suggest that modifications to the baseline architecture include upgraded software and a real-time, commercial-off-the-shelf signal processor and motion measurement subsystem. SAR parameters are reported to include all-weather search out to ranges of 150 km, strip map swath (up to 40 km wide and with 1 m resolution), spotlight mode (with 1 m resolution) and slow target indication.

Status
As of January 2002, a version of APG-67 (designated as the Golden Dragon 53 - GD-53) was reported as being in service aboard Taiwan's F-CK-1A Ching-Kuo interceptor (also known as the Indigenous Defense Fighter) and the type further described as having been selected for use on South Korea's KTX-2 advanced training aircraft. A low-cost commercial variant of the system has been developed under the designation GSAR. As of late 2000, GSAR (which may offer a 3-D interferometric SAR capability) was noted as being used 'worldwide' in the geological survey role.

Contractor
Lockheed Martin Naval Electronics and Surveillance Systems - Syracuse, Syracuse, New York.

VERIFIED

AN/APG-68(V) airborne fire-control radar

Type
I/low J-band (8 to 12.5 GHz sub-band) fire-control radar.

Description
AN/APG-68(V) is the fire-control radar fitted to F-16C/D multirole combat aircraft and is claimed to be the most reliable radar within the US Air Force's (USAF) inventory. Since 1988, the model installed in Block 30/40 aircraft (APG-68(V)1 to (V)4 on Block 30/40s, (V)6 on the Block 40s) has demonstrated a mean time between failure figure of 160 hours. This is reported to have increased to 300 hours in the Block 50 aircraft installation (APG-68(V)5 for USAF F-16s, (V)7 and (V)8 on export aircraft). The system was developed for use in the F-16 Multinational Staged Improvement Programme (known as MSIP). As such, APG-68(V) offers an AIM-120

based, multitarget, beyond visual range capability together with enhanced situational awareness, increased detection range and improved map resolution. Within the architecture, all programme instructions are stored in a 1,024 kword, non-volatile, electronically erasable, programmable, read-only memory and the sensor's computer is programmed in Jovial J73, the USAF approved high order language. Initial operating capability for the F-16C/D aircraft (the first Fighting Falcon models to incorporate APG-68(V)) was in December 1984.

APG-68(V)'s capabilities are primarily facilitated by the introduction of a Programmable Signal Processor (PSP) and a dual-mode transmitter. Of these, the PSP makes use of an 'advanced' modular processing architecture and a high density, solid-state memory, while the transmitter selects the best waveform for each mode of operation. Here, the range includes a low Pulse Repetition Frequency (PRF) option for air-to-surface engagements and medium and high PRF modes for long-range air interception. Looking at these in more detail, a high PRF velocity search mode increases target detection range where the target is travelling at a relatively high velocity. Once a target is detected, the radar switches to a medium PRF range-while-search mode that can be employed against targets (with any aspect) to gain additional range and angle information. In the track-while-scan mode, the sensor can track up to 10 targets simultaneously, assess the degree of threat from each and launch missiles as appropriate. By using high-resolution Doppler techniques, closely spaced targets can be distinguished. In air combat mode, APG-68(V) scans selected airspace and automatically acquires the nearest target. In look-down situations, land-based targets such as moving vehicles or vessels are ignored because the radar rejects returns with less than a specified threshold. This ground-mapping terrain radar technique ensures that only airborne targets are portrayed on the radarscope.

Starting with the APG-68(V)5 configuration (designed for use on the F-16C/D Block 50/52 aircraft), the radar features the insertion of Very High Speed Integrated Circuit (VHSIC) technology into the equipment's PSP, a move that is claimed to improve system mean time between failure to greater than 300 hours. So configured, the unit is termed the Advanced PSP (APSP). Northrop Grumman is also reported as having developed an upgraded radar software suite for the Block 50 F-16 that includes dual target situation awareness, enhanced monopulse ground-map, electronic boresight, enhanced long-range target tracking and enhanced air combat manoeuvring acquisition modes.

Elsewhere in the programme, May 2000 saw the launch of the APG-68(V)9 (formerly APG-68(V)XM) configuration for new build F-16 Block 50 aircraft. Designed to provide a range of capability/system improvements (including a high resolution Synthetic Aperture Radar (SAR) mode, an increase in air-to-air range, a twofold improvement in reliability and improved cost of ownership to combat capability ratio), APG-68(V)9 features:

- a modified antenna assembly that incorporates a strapdown inertial measurement unit to facilitate the introduction of the previously noted SAR mode
- a new, modular, receiver/exciter that is a form-fit replacement for the radar's existing low-power radio frequency unit. The new line-replaceable unit incorporates open architecture technology, wideband WFG for SAR and 'high' bandwidth A/Ds
- a modified transmitter that converts the existing dual-mode architecture to a medium duty one that provides improved average and peak power values
- a programmable, commercial-off-the-shelf, open architecture, common radar processor that replaces the existing PSP/APSP. Here, the insertion is described as offering 10 times the throughput of the PSP/APSP and as incorporating specific APG-68(V)9 radar mode software
- the introduction of 'some' new rack cables and waveguide microwave components.

The APG-68(V)9 configuration is further described as:

- eliminating the need for intermediate level maintenance
- offering enhanced fault isolation and a new extended range search mode (±60° scan) that replaces the radar's heritage air-to-air modes and can perform multi-target scans that facilitate the acquisition of up to four targets simultaneously
- being compatible with the AIM-120 and AIM-9X air-to-air missiles, the Joint Helmet Mounted Cueing System, 'J'-series 'smart' munitions, the Lightening II and 'other' electro-optical targeting pods and the ALQ-131, ALQ-165 and 'other' electronic warfare systems.

Alongside APG-68(V)9, Northrop Grumman is also offering the APG-68(V)X configuration for upgrade applications. As such, APG-68(V)X makes use of the (V)9's antenna, receiver/exciter and common radar processor (with 'smart interface' - see following) and a modified (V)9 transmitter when applied to F-16A/B models. When applied to F-16C/D airframes, (V)X makes use of an unmodified (V)9 transmitter. The 'smart interface' allows the (V)X architecture to automatically sense and adapt to its host's avionic environment. Here, the configuration is compatible with both existing F-16 and Mid-Life Update (MLU) avionic configurations. When applied to F-16A/B aircraft, (V)X provides a baseline capability that is the equivalent of the AN/APG-66(V)2/(V)3 (see separate entry) together with increases in detection range/tracking accuracy and a fourfold improvement in reliability. When applied to F-16C/D airframes, the (V)X capability closely matches that of the previously described (V)9 configuration. In the air-to-surface domain, (V)X offers a high resolution SAR capability if its host platform has received the F-16 MLU avionics update or 'improved SAR readiness' if it has not. The extended range search mode is also included in the V(X) release and Northrop Grumman further notes that APG-68(V)9 and (V)X configurations are available for aircraft other than the F-16.

Status

As of May 2001, in excess of 2,900 APG-68(V) radars were reported as having been procured by the US Air Force (USAF) and seven US Foreign Military Sales (FMS)

customers. As of late 2000/early 2001, identified APG-68(V) FMS clients were as follows:

- Bahrain (18 × F-16C and 4 × F-16D - October 2000 data)
- Egypt (116 (out of an on-order total of 136) × F-16C and 28 (42) × F-16D - April 2001 data)
- Greece (62 × F-16CG and 14 × F-16DG - 34 × F-16C and 16 × F-16D on order - October 2001 data)
- Israel (71 × F-16C and 48 × F-16D - 102 F-16I on order - June 2001 data)
- South Korea (107 (125) × F-16C and 49 (55) × F-16D - April 2001 data)
- Turkey (189 × F-16C and 35 × F-16D - October 2001 data)

In more detail, some of the Greek, South Korean and Turkish radars formed part of a June 1994 then year US$115 million contract that covered the supply of 93 USAF and FMS APG-68(V) systems. Identified APG-68(V) contracting activity during the period September 1999 to September 2001 comprised the following:

3 September 1999

The Northrop Grumman Corporation (Baltimore, Maryland) was awarded a then year US$40,833,152 firm, fixed-price contract covering the supply of 24 APG-68(V)8 radars for installation aboard F-16 aircraft of the Egyptian Air Force. At the time of the announcement, work on this FMS effort was scheduled for completion by 31 December 2001, with the programme's contracting activity being the USAF's Aeronautical Systems Center, Wright-Patterson Air Force Base (AFB), Ohio.

16 March 2000

CPU Technology (Pleasanton, California) was awarded a then year US$6 million 'other transaction' agreement covering the development of a prototype modernisation kit for the array processor used in the APG-68(V) fire-control radar. At the time of the announcement, work on the effort was scheduled for completion by 23 February 2000, with the programme's contracting activity being the USAF's Ogden Air Logistics Center, Hill AFB, Utah. An 'other transaction' agreement is a 'unique' contracting authority that allows for greater flexibility to negotiate 'special circumstance' contract terms than the US Department of Defense's normal way of doing business.

10 April 2000

The Northrop Grumman Corporation (Baltimore, Maryland) was awarded a then year US$52,179,753 modification to an existing firm, fixed-price contract that covered the supply of 24 APG-68(V)8 radars and 11 associated spares packages for use on Egyptian F-16 aircraft. At the time of the announcement, work on this FMS effort was scheduled for completion by 31 December 2001, with the programme's contracting activity being the USAF's Aeronautical Systems Center, Wright-Patterson AFB, Ohio.

15 May 2000

Honeywell International (Teterboro, New Jersey) was awarded a then year US$21,822,000 firm, fixed-price contract covering the provision of test and repair support for APG-68(V) radars installed aboard Turkish F-16 fighters. At the time of the announcement, work on this FMS effort was scheduled for completion by the end of May 2002, with the USAF's Ogden Air Logistics Center, Hill AFB, Utah acting as the programme's contracting activity.

29 May 2001

Litton Systems Inc (a Northrop Grumman subsidiary - San Carlos, California) was awarded an estimated then year US$8,939,423 firm, fixed requirement contract covering the provision of repair services for travelling wave tubes used in the APG-68(V) radars installed aboard F-16C/D aircraft. At the time of the announcement, work on the effort was scheduled for completion by the end of May 2004, with the programme's contracting activity being the USAF's Ogden Air Logistics Center, Hill AFB, Utah.

27 September 2001

The Northrop Grumman Corporation (Baltimore, Maryland) was awarded a then year US$5,680,000 firm, fixed-price contract covering the supply of APG-68(V) radar sets and 'main attack radar units' for installation aboard Block 50 F-16 combat aircraft. At the time of the announcement, the effort was scheduled for completion by the end of May 2003 and the programme's contracting activity was the USAF's Ogden Air Logistics Center, Hill AFB, Utah.

As of July 2001, Northrop Grumman was reporting orders for 125 APG-68(V)9 radars with Jane's sources suggesting Greece and Israel as being among the variant's customers.

Specifications

Frequency: 8-12.5 GHz
Transmitter: gridded, dual peak power TWT (APG-68)
Antenna: planar-array 480 × 720 mm
Search range: 296 km (APG-68(V))
Operating modes: (APG-68(V) air-to-air) range-while-search, track-while-scan (10 targets), velocity search, uplook search, raid cluster resolution, situation awareness mode, auto-acquisition air combat; (APG-68(V) air-to-ground) real-beam mapping with Doppler beam sharpening and scan freeze, sea surface search, beacon homing, ground moving target indication and tracking, fixed-target tracking and air-to-ground ranging.
Cooling: 9.9 kg/min at 27°C (APG-68(V)9)
Power: 5,606 V A (APG-68(V)9)
Volume: 0.13 m³ (APG-68(V)9)
Weight: 172 kg (APG-68(V)); 164 kg (APG-68(V)9)

Contractor

Northrop Grumman Electronic Systems sector, Baltimore, Maryland.

UPDATED

AN/APG-70 airborne fire-control radar

Type
I/J-band (8 to 20 GHz) airborne fire-control radar.

Description
The AN/APG-70 fire-control radar is an upgrade of the AN/APG-63 sensor that makes use of four new 'advanced design, higher performance' Line-Replaceable Units (LRUs - replacing five units in its predecessor) and two existing LRUs that have been modified. The radar's transmitter features a high average power gridded travelling wave tube, multiple pulse repetition frequencies and increased stability (giving increased dynamic range). Its newly designed receiver/exciter has increased bandwidth, improved tracking in jamming conditions, greater sensitivity and longer range detection. Other new assemblies include the equipment's data processor, programmable signal processor and the analogue signal converter. The antenna and power supply are the same as those used in the AN/APG-63. The APG-70 signal processor employs modular parallel processing (controlled by a MIL-STD-1750A central processor unit) and operates at speeds in excess of 30 million complex operations/s (potentially upgradable to 40 Mops). The sensor's radar data processor performs general purpose computations and has been upgraded to 1,024 k of memory. This is over 10 times greater than that available in the APG-63 and the unit operates at between four and five times faster. Approximately 220 k is devoted to air-to-air modes, 110 k to air-to-ground, 200 k to the built-in test feature and 64 k to scratchpad memory. The remainder is spare memory that is available for future enhancements.

Status
As of May 2001, AN/APG-70 variants were reported as being installed aboard Israeli F-15I (25 examples ordered), Saudi F-15S (72 ordered) and US Air Force (USAF) F-15C (36 examples), F-15D (six) and F-15E (205) strike aircraft. As of the given date, Jane's sources were suggesting that the APG-70 radars fitted to Saudi F-15S aircraft feature an application-specific processor, antenna drive rate and processor memory. Another APG-70 variant is understood to be the AN/APQ-180 radar that is installed aboard USAF AC-130U gunships.

Specifications
Frequency: selectable in the I/J-band (8-20 GHz)
Antenna: 3 gimbal axes, mechanical scan
Range: 15 m-18.5 km (automatic acquisition); +92 km (ground mapping); 185 km (air-to-air)
Resolution: 2.6 m at 37 km (ground mapping)
Weight: 251 kg
Volume: 0.25 m³
MTBF: 80 h

Contractor
Raytheon Electronic Systems, El Segundo, California.

VERIFIED

The AN/APG-70 radar

AN/APG-71 airborne fire-control radar

Type
I/low J-band (8 to 12 GHz) fire-control radar.

Description
The AN/APG-71 is an upgraded version of the radar portion of the AN/AWG-9 airborne weapon control system (see separate entry) that is installed aboard US Navy F-14D Tomcat aircraft. It is essentially a digital version of the AN/AWG-9, the key elements in the upgrade being a fully programmable processor and a companion radar data processor that replace the AN/AWG-9's four analogue

processors. Digital radar displays are incorporated and the system includes 'sophisticated' electronic counter-countermeasures provisions. Detection and tracking modes have also been improved over those of the AN/AWG-9.

Status
Testing of APG-71 engineering development models began in late 1986, with flight trials beginning during March 1988. Operational evaluation of the F-14D/APG-71 combination was completed during December 1990 and APG-71 production (55 examples) continued through until 1993. As of this edition, Jane's sources reported 46 F-14D aircraft as being in service. Identified recent APG-71 contracting activity comprises a then year US$8,062,657 not-to-exceed ceiling price contract that was awarded to the then Raytheon Training and Services Co (Indianapolis, Indiana) in respect of the supply of four equipment applicable antenna assemblies and two equipment applicable microwave processors. At the time of the contract announcement, work on the effort was scheduled for completion by the end of October 2001, with the USN's Naval Inventory Control Point, Philadelphia, Pennsylvania acting as the programme's contracting activity.

Specifications
Frequency: I/low J-band (8-12 GHz)
Power: 500 W (average - pulse mode); 7 kW (average - pulse-Doppler mode); 10 kW (peak)
Range: 213 km across a 213 km wide front
Scan rate: 80°/s (horizontal); 2 scans/s (vertical)
Tracking capacity: up to 24 targets simultaneously
Antenna: slotted planar-array
Weight: 590 kg
Volume: 0.78 m³

Contractor
Raytheon Electronic Systems, El Segundo, California.

UPDATED

AN/APG-73 multimode airborne radar

Type
I/low J-band (8 to 12 GHz) multimode fire-control airborne radar.

Description
The AN/APG-73 multimode radar is an upgraded, all digital sensor that is based on the earlier AN/APG-65 equipment. It has a new multifunction data/signal processor, power supply and receiver/exciter. The upgrade gives increased memory, bandwidth, frequency agility and higher analogue/digital sampling rates. 'Advanced' technology is used to enhance the radar's electronic counter-countermeasures capability, with a strong emphasis on flexible software that allows the sensor to adapt quickly to differing threats. The programmable data/signal processor takes the form of a general purpose dual 1750A computer that provides mode and antenna control, target tracking and display processing and allows the system to be adapted for new weapons or tactics via software rather than hardware changes. Faster analogue-to-digital air-to-air conversion improves the radar resolution cell and the new data/signal processor improves Doppler resolution. This enhances the radar's ability to discriminate between closely spaced targets. As such, it operates at more than 2 Mips and is equipped with a 2 million word firm memory and a 256,000 16-bit working memory. The radar's signal processing element has 4 Mbytes of bulk memory and its throughput has been increased to 60 million complex operations/s through the use of multichip gate arrays. The travelling wave tube transmitter and antenna from the APG-65 are retained in the APG-73, while the sensor's receiver/exciter features the circuitry and input/output interfaces required for integration of an active electronically scanned antenna array, if required. When fitted with a motion-sensing subsystem and stretch waveform generator and special test equipment, instrumentation and Reconnaissance (SIR) modules, APG-73 can generate high resolution ground maps and make use of 'advanced' image correlation algorithms to enhance weapon designation accuracy. The full capability APG-73's operating modes are as follows:

Air-to-air modes
- high Pulse Repetition Frequency (PRF)
- velocity search (for maximum detection range against head-on aspect targets)
- high/medium PRF range-while-search
- four short-range, automatic acquisition modes
- track-while-scan (providing the AIM-120 missile with a fire-and-forget capability when integrated with the F/A-18)
- gun director
- raid assessment/situation awareness
- single target track

Air-to-surface modes
- Doppler beam-sharpened sector and patch mapping
- medium resolution synthetic aperture radar imaging
- radar navigation ground mapping
- real beam ground mapping
- fixed and moving ground target indication and tracking
- air-to-surface ranging
- terrain avoidance

The AN/APG-73 fire-control radar as installed aboard a US Navy F/A-18 aircraft

- precision velocity update
- inverse range angle
- sea surface search (with clutter suppression)

Reconnaissance modes
- strip map
- spotlight map.

Status

As of May 2001, the AN/APG-73 multimode radar was reported as being installed aboard F/A-18C/D/E/F aircraft operated by Finland (57 × F-18C and 7 × F-18D), Malaysia (8 × F-18D), Switzerland (26 × F-18C and 7 × F-18D) and the US Navy/US Marine Corps (412 × F/A-18C, 57 × F/A-18D (procurement continuing), 8 × F/A-18E (initial procurement of 35 aircraft) and 2 × F/A-18F (initial procurement of 34 aircraft)). During April 1999, the US Marine Corps announced its wish to upgrade its F/A-18A aircraft to F/A-18C avionic standard (Engineering Change Programme -583, including the introduction of the APG-73 radar) and made a US Fiscal Year 2000 request for authorisation to modify a first tranche of 24 airframes (see following). In September 1999, Australia formerly requested a total of 71 APG-73 radars with which to upgrade its F/A-18A and B platforms (Australia's Project Air 5376 - see following). Canada also wishes to introduce the APG-73 on its CF-188s as part of its CF-188 Hornet UpGrade (HUG) programme. Identified APG-73 contracting activity during 1999-2001 comprised the following:

22 April 1999
The then Raytheon Sensors and Electronic Systems (El Segundo, California) was awarded a then year US$9,905,550 firm, fixed-price contract covering the supply of five retrofit kits (comprising a power supply, a receiver, a data processor and an electrical rack) in support of APG-73 radars installed aboard F/A-18 aircraft. At the time of the announcement, work on the effort was scheduled for completion by the end of November 2000 and the programme's contracting authority was the US Navy's (USN) Inventory Control Point, Philadelphia, Pennsylvania.

27 May 1999
The then Raytheon Sensors and Electronic Systems (El Segundo, California) was awarded a then year US$6,600,000 firm, fixed-price contract covering the supply of five retrofit kits (comprising a power supply, a receiver, a data processor and an electrical rack) in support of APG-73 radars installed aboard US Marine Corps F/A-18 aircraft. At the time of the announcement, work on the effort was scheduled for completion by the end of December 2000 and the programme's contracting authority was the USN's Inventory Control Point, Philadelphia, Pennsylvania.

16 December 1999
Raytheon Electronic Systems (El Segundo, California) was awarded a then year US$10,545,000 firm, fixed-price contract covering the supply of 21 transmitter units in support of APG-73 radars slated for use aboard F/A-18 aircraft. At the time of the announcement, work on the effort was scheduled for completion by the end of May 2001 and the programme's contracting authority was the USN's Inventory Control Point, Philadelphia, Pennsylvania.

12 January 2000
Raytheon Electronic Systems (El Segundo, California) was awarded a then year US$200,251,988 firm, fixed price order (against a previously awarded basic ordering agreement) covering the supply of 111 APG-73 radar kits for installation aboard F/A-18 aircraft operated by the Royal Australian Air Force (RAAF - 71 radars), the US Marine Corps (21) and the US Navy (19). At the time of the announcement, work on the effort was to be undertaken at Forest, Mississippi (45 per cent workshare), El Segundo, California (35 per cent), Andover, Massachusetts (11 per cent), Dallas, Texas (8 per cent) and Newport Beach, California (1 per cent) and was scheduled for completion by the end of October 2002. The programme's contracting activity was the USN's Naval Air Systems Command, Patuxent River, Maryland.

30 March 2000
Raytheon Electronic Systems (El Segundo, California) was awarded a then year US$10,643,252, firm, fixed-price order covering the supply of seven receivers and eight radar data processors for use in APG-73 radars fitted to F/A-18 aircraft. At the time of the announcement, work on the effort was to be undertaken at El Segundo, California and was expected to be completed by the end of March 2003. The programme's contracting activity was the USN's Inventory Control Point, Philadelphia, Pennsylvania.

2 May 2000
Raytheon Electronic Systems (El Segundo, California) was awarded a then year US$8,851,788 firm, fixed-price Foreign Military Sales order for nine APG-73 radars for installation aboard F/A-18 aircraft of the RAAF. At the time of the announcement, work on the effort was to be undertaken at El Segundo, California and was scheduled for completion by the end of December 2002. The programme's contracting activity was the USN's Inventory Control Point, Philadelphia, Pennsylvania.

15 March 2001
Raytheon Electronic Systems (El Segundo, California) was awarded a then year US$8,383,905 firm, fixed-price Foreign Military Sales order for 54 APG-73 components (including antenna assemblies and transmitters) for use on F/A-18 aircraft of the RAAF. At the time of the announcement, work on the effort was to be undertaken at El Segundo, California and was scheduled for completion by the end of April 2003. The programme's contracting activity was the USN's Naval Inventory Control Point, Philadelphia, Pennsylvania.

29 June 2001
Raytheon Electronic Systems (El Segundo, California) was awarded a then year US$39,250,000 firm, fixed-price delivery order (against the previously agreed basic ordering agreement N00383-01-G-100A) covering the supplies and services required to manufacture 26 APG-73 Phase I radar upgrade kits for use on F/A-18 aircraft. Work on the effort was to be undertaken at El Segundo, California (47 per cent workshare); Forest, Mississippi (30 per cent); Andover, Massachusetts (15 per cent) and Dallas, Texas (8 per cent). At the time of the announcement, the programme was scheduled for completion by the end of November 2003 and its contracting activity was the USN's Naval Air Systems Command, Patuxent River, Maryland.

21 September 2001
Raytheon Electronic Systems (El Segundo, California) was awarded a then year US$6,083,724 firm, fixed-price contract covering the procurement of an APG-73 radar test generator/simulator by the government of Finland. At the time of the announcement of this US Foreign Military Sales programme, work on the effort was scheduled for completion by the end of October 2003, with the USN's Naval Air Systems Command, Aircraft Division, Lakehurst, New Jersey acting as its contracting activity.

25 October 2001
Raytheon Electronic Systems (El Segundo, California) was awarded a then year US$27,090,128 firm, fixed-price contract covering the procurement of 107 line items (consisting of 15 various components including transmitters, receivers, antennas and power supplies) for use in APG-73 radars installed aboard F/A-18 aircraft. At the time of the announcement, work on the effort was scheduled for completion by the end of October 2005, with the programme's contracting activity being the USN's Naval Inventory Control Point, Philadelphia, Pennsylvania.

Specifications
Frequency: 8-12 GHz
Number of LRUs: 5 plus antenna
Volume: 0.126 m³ (excl antenna)
Weight: 154 kg

Contractor
Raytheon Electronic Systems, El Segundo, California.

UPDATED

AN/APG-76 MultiMode Radar System (MMRS)

Type
J-band (10 to 20 GHz) multimode radar.

Description
The AN/APG-76 MMRS is designed to provide the F-4 Phantom with enhanced air-to-air and air-to-ground capabilities. Air-to-air capabilities include look-up, look-down and beacon mode, as well as air track and air combat modes. Air-to-ground performance is enhanced through the inclusion of real-beam ground map, high-

The AN/APG-76 multimode radar

resolution Synthetic Aperture Radar (SAR) and Doppler beam sharpening, ground mapping with simultaneous ground moving target indication and beacon modes. Software enables the multimode radar system to provide guidance for future standoff air-to-ground weapons. A powerful clutter suppression interferometer provides a clutter-free resolution SAR map, as well as multitarget tracking capability. The radar's air-to-ground capabilities have also been used in the Gray Wolf technology demonstration. Here, the radar was mounted in a pod and was described as offering a number of operating modes including real beam, Doppler sharpening and spotlight. In real beam, the Gray Wolf application was noted as having a maximum detection range in excess of 185 km while the Doppler beam sharpening capability allowed 26 × 26 km sectors to be scanned with high-resolution values. The equipment's spotlight mode was thought to incorporate three submodes, the most sensitive of which provides 0.3 m resolution on targets at ranges of up to 130 km. The Gray Wolf application is also noted as having been being able to track up to 75 moving surface targets simultaneously.

Status

As of December 2001, Northrop Grumman's Norden Systems was reported as having supplied Israel with 50 APG-76 radars for use in Israel Aircraft Industries' F-4-2000 upgrade programme.

Contractor

Northrop Grumman Electronic Systems sector - Norden Systems, Norwalk, Connecticut.

UPDATED

AN/APG-77 multimode airborne radar

Type

Multimode radar for the F-22 Advanced Tactical Fighter.

Description

The AN/APG-77 multimode radar incorporates a 'low-observability', active aperture, electronically scanned array (incorporating approximately 2,000 transceiver modules) and is described as offering long-range, multitarget, all-weather, stealth vehicle detection, electronic intelligence gathering and multiple missile engagement capabilities. The sensor is noted as making use of Monolithic Microwave Integrated Circuit (MMIC) and Very High-Speed Integrated Circuit (VHSIC) technologies in order to enhance reliability and achieve a mean time between failure value of more than 400 hours. Alongside its air-to-air capabilities, the radar is understood to incorporate 'advanced' dogfight and air-to-surface operating modes that are all aspect and effective in a heavy clutter environment. As yet unconfirmed sources suggest that APG-77 has a 'typical' operating range of 193 km and is specified to achieve an 86 per cent probability of intercept against a 1 m^2 target at its maximum detection range using a single radar paint. In surveillance mode, the same sources describe the radar as using an air-to-air moving target indicator mode with target recognition being handled by spot and/or Ultra High-Resolution (UHR) modes. This latter mode is described as offering 31 cm resolution at ranges in excess of 161 km and it has been suggested that the returns generated by the UHR mode are matched to an integral signature library to facilitate non co-operative target recognition. In terms of intelligence gathering, it is suggested that APG-77 has an approximately 2 GHz bandwidth when it is functioning in a forward-looking, high-gain, passive listening mode.

Status

The AN/APG-77 radar is being developed by a Northrop Grumman/Raytheon Systems Joint Venture and an early version was demonstrated in a flying testbed as part of a demonstration/validation phase. As of January 2001, 11 Engineering and Manufacturing Development (EMD) APG-77 radars are reported to have been delivered and as having been used for systems integration and checkout as part of the sensor's EMD process. On 15 November 2000, F-22 aircraft 4004 made its maiden flight with an operational, APG-77 type, active, electronically scanning array antenna installed. So equipped, the aircraft's radar is noted as having successfully tracked multiple targets 'almost immediately' after the platform left the runway. As of January 2001, Northrop Grumman was expecting APG-77 low-rate initial production to begin during the first quarter of 2001.

Contractors

Northrop Grumman Corporation, Electronic Sensors and Systems Sector, Baltimore, Maryland.
Raytheon Electronic Systems, McKinney, Texas.

UPDATED

AN/APG-78 Longbow fire-control radar

Type

Millimetric fire-control radar.

Description

The AN/APG-78 radar forms part of the Longbow fire-and-forget anti-armour system that is fitted to AH-64D Apache battlefield attack helicopters. The radar subsystem comprises a low probability of intercept millimetric (35 GHz frequency) radar mounted on top of the helicopter's main rotor mast and a millimetric radar for

AH-64D Longbow Apache helicopter equipped with a fire-control radar

the Radio Frequency (RF) Hellfire missile. This latter item uses the weapon's standard bus. The Longbow radar is designed to interface with the AH-64's fire-control system and provides rapid target area search; automatic target detection, classification and prioritisation; fixed and moving target detection (at maximum standoff range) and missile fire-and-forget capabilities within a 55 km^2 area. As such, the radar is reported to have shown a 28-fold improvement in battlefield effectiveness over the non-radar AH-64A during US government operational tests. An integrated passive radar band interferometer subsystem (designated as the AN/APR-48A) is used to locate and identify radiating targets while the Longbow system as a whole identifies and ranks targets in priority order for attack by RF Hellfire missiles. Here, up to 16 of the highest priority targets can be displayed simultaneously, at ranges of up to 8 km for moving targets.

Status

The Longbow system is produced by the Longbow Limited Liability Company (LLLC - a joint venture between Northrop Grumman's Electronic Systems sector and Lockheed Martin Missiles and Fire-Control). Initial Operational Test and Evaluation of the equipment was completed in March 1995 and the system completed full-scale development (on schedule and to cost) during the fourth quarter of 1995. In April 1996, Longbow was selected for application to the British Army Air Corps' 67 WAH-64 battlefield attack helicopter, and the first US Army attack helicopter unit to be equipped with the AH-64D Apache Longbow system was fielded during 1998. As of December 2001, the US Army was reported as planning to convert at least 501 of its AH-64A helicopters to AH-64D and as having selected an AN/APG-78 variant for installation aboard its RAH-66 Comanche scout helicopter. As of the given date, other Longbow radar orders include units for Israel (nine AH-64D ordered during the spring of 2001), Japan (the AH-64D being selected to meet Japan's AH-X attack helicopter requirement on 27 August 2001, with a 'percentage' of up to 60 aircraft to be acquired being fitted with the Longbow radar) and Singapore (20 AH-64D helicopters). Identified Longbow radar contracting activity is as follows:

March 1996
US Lot 1 production contract let.
April 1996
Longbow International (LLLC's international business unit) was awarded a contract covering the supply of Longbow radars into the UK's WAH-64D programme. The first UK sensor was delivered during June 1998.
February 1997
US Lot 2 production contract let. According to Jane's sources, US Lot 1 and 2 Longbow radar production totaled 20 equipments.
November 1997
LLLC was awarded a multi-year contract (valued at then year US$565 million) covering the production and supply of 207 Longbow radar systems. As part of this, the joint venture was contracted to supply 44 radars and procure long lead items for 57 more during February 2001, an award that brought the value of the multi year work to then year US$442.9 million.

Contractors

Lockheed Martin Missiles and Fire-Control, Orlando, Florida.
Northrop Grumman Electronic Systems sector, Baltimore, Maryland.

UPDATED

AN/APG-80 Active Electronically Scanned Array (AESA) radar

Type
Multimode AESA airborne radar.

Description
Developed originally for installation aboard the United Arab Emirates' F-16 Block 60 multirole fighter, the AN/APG-80 AESA is described by Jane's sources as offering 'advanced' mode interleaving that enable the host platform's pilot to maintain situational awareness and weapons quality air-to-air target tracking while prosecuting an air-to-ground attack. APG-80 is further noted as offering:
- an expanded (when compared with the APG-68 equipment - see separate entry) bandwidth
- a greater (when compared with APG-68) detection range
- a 140° track volume
- a 20 target multitrack capability (with growth potential to up to 50 to enhance situational awareness)
- up to six target simultaneous target tracking capability with single target tracking accuracy
- a low radar cross-section
- an automatic terrain following capability
- an 'ultra' high resolution synthetic aperture radar mode
- 'extensive' electronic counter-countermeasures provision
- a planned mean time between failure value of 500 hours

System options are reported as including:
- an enhanced SAR/automatic target cueing mode
- a ground moving target indicator mode

Status
As of January 2002, AN/APG-80 was understood to be in development.

Contractor
Northrop Grumman Electronic Systems sector, Baltimore, Maryland.

NEW ENTRY

AN/APQ-126(V) airborne fire-control radar

Type
J-band (10 to 20 GHz) fire-control and navigation radar.

Description
AN/APQ-126(V) is a forward-looking, variable configuration, airborne navigation and attack radar for A-7D/E series strike aircraft. Operating in the J-band, the radar's primary functions are ground mapping, air-to-ground ranging and safety of flight. The APQ-126(V)'s operating modes are as follows:
- ground mapping - pencil-shaped beam
- air-to-ground ranging
- air-to-air boresight ranging
- terrain-avoidance
- terrain-following
- cross-scan terrain-following/terrain-avoidance
- cross-scan terrain-following/ground-map pencil
- beacon
- television
- radar homing and warning.

AN/APQ-126(V) also features adverse weather 'look-through' using selectable circular polarisation, slaved antenna pointing in air-to-air ground ranging and variable antenna tilt control which allows the system operator to optimise ground-map displays and highlight targets of interest.

Status
As of this edition, APQ-126(V) radars were reported as being in service aboard Greek A-7E/H and TA-7C/H aircraft. An APQ-126(V) variant (designated as the

AN/APQ-126 airborne radar

AN/APQ-158) is noted as having been developed for use on the USAF's MH-53J special operations helicopter. As of the given edition, total APQ-126(V) production was given as being in excess of 1,000 units.

Contractor
Raytheon Electronic Systems, McKinney, Texas.

VERIFIED

AN/APQ-153/-157 fire-control radars

Type
I-band (8 to 10 GHz) fire-control and tracking radars.

Description
The APQ-153 and APQ-157 Airborne Search Target Attack Radars (ASTAR) are lightweight search and range tracking equipments that are installed on F-5E fighter and F-5F combat training aircraft respectively. The equipments provide stabilised search, automatic target acquisition/illumination, automatic ranging/boresight missile steering and gunnery facilities In APQ-153 form the equipment comprises: an antenna assembly, a transceiver, a radar processor, an indicator unit and a control unit. APQ-157 differs in being configured for use by the F-5F's two-man crew and adds rear cockpit indicator and control units and a coupler power supply to the basic APQ-153 architecture. The antenna assembly used incorporates a 30 × 40 cm parabolic dish (with horizontal polarisation) and the indicator unit used features a 127 mm direct view tube and provides 'B' type search and missile lock-on displays. In terms of operating modes, APQ-153/-157 radars offer search, boresight missile lock-on and air-to-air gunnery options. Of these, search utilises a 7° elevation beamwidth that is stepped up 3° and down 3° at its respective starboard and port azimuth limits to create two-bar coverage at 10° in elevation. The missile lock-on mode is designed to work with the AIM-9 Sidewinder weapon and provides aircraft steering information to align the missile's acquisition envelope with a target. The gunnery options are dogfight and AA1/AA2.

Status
Over time, more than 1,400 APQ-153/-157/-159(V) (see separate entry) radars are reported to have been produced for installation aboard F-5 series aircraft around the world.

Specifications
Frequency: 8-10 GHz
Range: 20 km (missile lock on mode); 37 km (search mode)
Coverage: ±45° (search mode)
Weight: 50 kg (APQ-153); 60 kg (APQ-157)
MTBF: 62 h

Contractor
Systems & Electronics Inc, St Louis, Missouri.

UPDATED

AN/APQ-159(V) fire-control radar

Type
I/J-band (8 to 20 GHz) multimode fire-control radar.

Description
APQ-159(V) is designed to provide a 'state-of-the-art' radar upgrade for F-5E fighter aircraft currently equipped with the AN/APQ-153 equipment. Installation of the system is noted as requiring minimum aircraft modification and as not impacting on platform weight and centre of gravity. When compared with APQ-153, APQ-159(V) incorporates frequency agility, improved reliability components, a new processor power supply transformer, a redesigned servo amplifier, environmental stress screening for the receiver and modulator, the introduction of a gallium arsenide, field-effect transistor, low noise receiver and a new high gain, low sidelobe, planar-array antenna. Using this basic architecture, a number of APQ-159(V) variants have been developed, the known details of which are as follows:

APQ-159(V)1 and 2
APQ-159(V)1 and 2 incorporate a scan converter and television display that are compatible with the AGM-65 Maverick air-to-surface missile.

APQ-159(V)5
APQ-159(V)5 was installed aboard US Air Force F-5 dissimilar combat training aircraft.

Specifications
Frequency: 8-20 GHz
Range: 22 km (APQ-159(V)5 - lock on against an F-16 sized target); 34 km (APQ-159(V)5 - max detection range against an F-16 sized target)
MTBF: 127 h (APQ-159(V)5)

Contractor
Systems & Electronics Inc, St Louis, Missouri.

VERIFIED

AN/AWG-9 Airborne Weapon Control System (AWCS)

Type
Airborne fire-control system.

Description
The AN/AWG-9 AWCS has been developed to support the AIM-54 Phoenix air-to-air missile aboard US Navy (USN) F-14 interceptors. As such, it comprises a fire-control and target illuminating radar, digital computer and displays. Provision is also included for the automatic exchange of datalink information between the AWG-9 and the Naval Tactical Data System (NTDS) and Airborne Tactical Data System (ATDS) for target designation and other functions.

For onboard target acquisition, the AWCS includes a long-range, high-power pulse-Doppler radar. This system has a look-down capability that enables it to pick moving targets out of the ground clutter that normally obscures targets in a conventional radar. In addition to its long range, the AWCS provided the US Navy's first operational airborne multiple target tracking and prioritisation capability.

AWG-9 is designed to consecutively launch and simultaneously guide up to six AIM-54 missiles that have been launched against individual targets and makes use of a planar, slotted-plate radar antenna array. Alongside AIM-54, the system is able to handle fire control of the AIM-7 Sparrow and AIM-9 Sidewinder missiles together with the 20 mm M-61 Vulcan rotary cannon. In terms of system architecture, AWG-9 incorporates a pulse-Doppler radar as its primary target sensor, a multipurpose digital computer and associated control and display subsystems. The radar is optimised for long-range target acquisition, a function that is assisted by the presence of a low noise parametric amplifier in its receiver section. The use of a pulse-Doppler modes facilitates look-down target acquisition and the radar can also function in conventional pulse modes. A separate travelling wave tube transmitter provides Continuous Wave (CW) illuminating energy for use with the semi-active AIM-7F/M Sparrow missiles and time-sharing techniques allow for up to six AIM-54 missiles to be given pulse-Doppler mid-course guidance simultaneously.

Data processing in the AWG-9 AWCS is performed by a general purpose digital computer that features high-speed operation and a large memory capacity in an extremely compact package. The central computer keeps track of targets detected by the radar while the radar continues to search. Based on preprogrammed logic, the computer evaluates threats, generates steering information for the pilot and paints a complete tactical situation for the system operator in standard Naval Tactical Data System symbology, all based on data generated either internally or obtained through external datalinks.

For control of the AIM-54 weapon system, the aircraft's system operator is provided with two cathode ray tube display units. One, a 12.7 cm diameter unit, is used as a multimode display for the presentation of raw radar derived target information and identification friend-or-foe returns, while a larger (25.4 cm diameter) unit is used for the display of processed data. The latter includes target track information, alphanumeric and symbolic data obtained via datalink from other units of a naval force. The display is also used as a computer readout device for the presentation of computer-generated missile/target assignments.

Status
Hughes (now part of Raytheon Electronic Systems) delivered the last AN/AWG-9 AWCS during 1988, although production of spares continued into the following year. As of this edition, AWG-9 was reported as being installed aboard 111 USN F-14A interceptors. Over time, the USN has completed a retrofit programme to double the core memory of the AWG-9's central processor and an upgraded version of the AWG-9's radar subsystem has been developed under the designation AN/APG-71 (see separate entry).

Contractor
Raytheon Electronic Systems, El Segundo, California.

VERIFIED

The AN/AWG-9 weapon control system, with an AIM-54 Phoenix missile in foreground

F-35 Active Electronically Scanned Array (AESA) radar

Type
Electronically scanned array multipurpose airborne radar.

Description
Building on its experience with the APG-68 and APG-77 radars, Northrop Grumman's Electronic Systems sector is developing a multipurpose (air-to-air, air-to-ground and electronic warfare capable) AESA radar for installation aboard the Lockheed Martin F-35 Joint Strike Fighter (JSF). The contractor further notes that it is a partner (with Lockheed Martin Missiles and Fire-Control) in the development of the F-35's Electro-Optical (EO) distributed aperture and EO targeting systems and is involved in the design, development, test and integration of a section of the aircraft's passive electronic warfare/countermeasures system. Here, equipment/software being developed includes wide band acquisition receivers and air-to-air/air-to-ground emitter location algorithms.

Status
As of December 2001, the F-35 AESA radar was in its Systems Development and Demonstration (SDD) phase and had been test flown aboard Northrop Grumman's BAC 1-11 JSF avionics testbed aircraft.

Contractor
Northrop Grumman Electronic Systems sector, Baltimore, Maryland.

NEW ENTRY

IDENTIFICATION FRIEND-OR-FOE (IFF) AND SECONDARY SURVEILLANCE RADAR (SSR) SYSTEMS

Introduction

The IFF and SSR section covers ground-based, shipborne, airborne and weapon-mounted interrogators and transponders, and their associated equipments. Since *Jane's Radar and Electronic Warfare Systems* is devoted to military equipment,

secondary radar for civil air traffic control purposes is not included, except where necessary because of system or operational integration reasons. Details of civil secondary radar systems can be found in the *Jane's Air Traffic Control* and *Jane's Avionics* yearbooks.

VERIFIED

INDIA

Bharat Electronics Identification Friend-or-Foe (IFF) systems

Type
D-band (1 to 2 GHz) IFF/Secondary Surveillance Radar (SSR) system.

Description
Ground IFF/SSR Mk X systems manufactured by Bharat Electronics are designed to operate in association with various types of primary radars, both military and civilian. The primary radars include different types of Russian origin (ex-Soviet Union), shipborne radars, and also the Flycatcher radar. These systems are designed to meet International Civil Aviation Organisation specifications and are available in a wide range of configurations that can be tailored to meet user specifications. The configurations are:
- on-mounted where the IFF/SSR antenna is physically mounted on top of the primary antenna
- off-mounted where the IFF/SSR antenna is on a separate mast. Its rotation is synchronised with the primary radar antenna using a servo-control system
- integrated type where the IFF/SSR function is realised by using the primary antenna reflector itself for beam forming. The IFF feed is specially designed for optimum performance with a primary reflector, and is mounted on the antenna structure.

The electronics cabinet associated with the Bharat Mk X IFF/SSR system
0006997

The on/off-mounted IFF antenna is a linear array type using horn radiators. The antenna generates integrated interrogation and control patterns with three pulse sidelobe suppression. Three types of antenna are available, these being, respectively, 1.7, 3.5 and 9 m long. New version lightweight printed dipole array antennas are also available.

Status
As of this edition, Bharat IFF/SSR systems were reported as having been procured.

Specifications
Transmitter frequency: 1,030 ± 0.1 MHz
Receiver frequency: 1,090 MHz nominal
Power output: 200 W-4 kW peak
Pulse-width: 0.8 ± 0.1 µs
Modes: 1, 2, 3/A, B, C, D

Contractor
Bharat Electronics Limited (a Government of India Enterprise), Bangalore.

VERIFIED

Type 405A Identification Friend-or-Foe (IFF) transponder

Type
D-band (1 to 2 GHz) airborne IFF/Secondary Surveillance Radar (SSR) transponder.

Description
The Type 405A is an airborne solid-state IFF transponder which provides automatic replies to appropriate ground or airborne interrogators operating on the IFF Mk X system. It includes Mode C altitude coding and an interrogator sidelobe suppression function that inhibits replies to sidelobe signals from the interrogator. The basic equipment consists of a transmitter/receiver unit, a panel-mounted control unit, and a switching unit.

Status
Over time, more than 1,800 Type 405A transponders are reported as having been installed aboard fighter aircraft and helicopters of the Indian Air Force, Navy, Army and Coastguard.

Specifications
Transmitter frequency: 1,090 MHz nominal
Stability: ± 3 MHz
Power output: not less than 350 W peak at 1% duty cycle
Receiver frequency: 1,030 MHz
Interrogation modes: 1, 2, 3A and C
Number of codes: 4,096
Dimensions: 122 × 400 × 201 mm (transponder); 146 × 81.5 × 70 mm (control unit)
Weight: 0.6 kg (control unit); 9 kg (transponder)

Contractor
Hindustan Electronics Limited, Hyderabad.

VERIFIED

The antenna array used with the Bharat Mk X IFF/SSR system is shown here mounted above the scanner of an air traffic control radar
0006998

The Type 405A IFF transponder

INTERNATIONAL

MSR 200/400/2000 Identification Friend-or-Foe (IFF) interrogators

Type
D-band (1 to 2 GHz) ground-based IFF interrogators.

Description
Known details of the MRS 200/400/2000 equipments are as follows:

MSR 200/MSR 2000
MSR 200 and MSR 2000 form a family of lightweight IFF interrogators that make use of fully solid-state technology. Of the two, MSR 200 is described as being a compact, mobile IFF interrogator that is designed for use in short- and medium-range radar systems. As such, it employs passive decoding (including civil and military emergency decoding as options) in compliance with STAndard NATO AGreement (STANAG) 4193. For example it is compatible with Mk X-A and, if applicable, with Mk XII techniques. Applications of the MSR 200 so far are in the DR 600 Siemens radar and short/medium-range radars of different radar manufacturers. The MSR 2000 IFF/secondary surveillance radar interrogator reported as being suitable for 'all' civil and military air traffic control applications and is described (within the family context) as being 'a more powerful' interrogator for use ground-based and shipborne longer-range radars and weapon systems.

MSR 400
MSR 400 is an IFF interrogator that is designed for use in mobile, short- and medium-range radar systems. The equipment employs passive decoding and is compatible with Mk X selective identification feature and Mk XII techniques.

Status
As of July 2001, MSR 200/400/2000 series equipments were reported as having been procured with MSR 400 applications being understood to include the Gepard Mk 2 anti-aircraft tank's MPDR-12 radar (see separate entry), the Roland surface-to-air missile system and a number of as yet unidentified short- to medium-range radars.

Specifications
Transmitter frequency: 1,030 ± 0.2 MHz
Receiver centre frequency: 1,090 MHz
Peak power: 200-600 W (MSR 200); 400 W (MSR 400); > 1,500 W (MSR 2000)
Modes: 1, 2, 3/A, 4 (MSR 200/400); 1, 2, 3/A, B, C, D, 4 (MSR 2000)
Duty cycle: 1%

Contractor
European Aeronautic, Defence and Space (EADS) Co Systems and Defence Electronics, Unterschleissheim, Germany.

UPDATED

New-Generation Identification Friend-or-Foe (NGIFF) system

Type
Multiplatform question and answer identification system.

Description
NGIFF is a multinational attempt to create a common IFF system that builds on the mid-1980s STAndard NATO AGreement (STANAG) 4162 requirement for a common equipment that was to be known as the NATO Identification System or NIS. While NIS foundered, the work done on the system has allowed the 'definition and development' of equipments which are compliant with the STANAG 4193 IFF Mk XII requirement, the International Civil Aviation Organisation's (ICAO) Annex 10 on Mode S (a mandatory capability on military aircraft from January 1999) and can be upgraded to a NGIFF-type standard in the future.

Status
France, Italy and Germany are understood to have agreed to proceed with NGIFF, with the work to be done in two phases. Phase 1 involves the development and procurement of Mk X, XII, Mode S and NGIFF upgrade compatible transponders, the definition of Mode S NGIFF upgrade compatible interrogators and studies into encryption, STANAG and frequency supportability. Phase 2 is described as involving the development of NGIFF modules for use in transponders, the development/procurement of suitable interrogators (including Mode S) and the development/ procurement of NGIFF encryption modules. NGIFF work is to be synchronised in the three participating countries, with one country eventually being designated as NGIFF leader. A combined development company will then be created by the three national contractors, with the designated lead country acting as the programme's procurement agency.

Contractors
European Aeronautic, Defence and Space (EADS) Co Systems and Defence, Germany.
Thales Communications, France.
Marconi Italia Defence, Italy.

UPDATED

NRAI-7(.)/SC10(.) Identification Friend-or-Foe (IFF) transponder

Type
D-band (1 to 2 GHz) airborne IFF transponder.

Description
The NRAI-7 is a solid-state Mk XII diversity transponder which inhibits replies to interrogator sidelobe transmissions, and automatically codes special replies to provide assistance in the position identification of particular aircraft and in emergencies. The diversity function is provided by a dual receiver with inputs connected to upper and lower antennas, a system for comparison of received signals and an antenna switch that directs the response to the antenna that has received the strongest interrogation signal. This allows more accurate identification, particularly during aircraft manoeuvres which can blanket or interrupt signals. The pilot may also insert codes such as radio failure alert and warning of hijackers aboard. It is available in one- or two-box housing (see NRAI-9A). The single-box version is claimed to be one of the smallest transponders in the world. A naval version is also available.

Status
Over time, NRAI-7 transponders are reported as having been installed on Mirage 2000, Mirage F1, Mirage III, Mirage IV, AS 332/335, C-130, C-160 Transall and other aircraft types together with ships of the French and other world navies. The unit has also been integrated into a number of Polish platforms including MiG fighters. Overall, between 2,000 and 3,000 NRAI-7 systems have been delivered to the French defence forces and those of other countries.

Specifications
Peak power: 500 W
Sensitivity: −77 dBm
Frequency: 1,030 MHz (receive); 1,090 MHz (transmit)
Modes available: 1, 2, 3A/C and Mode 4 capability
No of codes: 32 (Mode 1); 4,096 (Modes 2 and 3/A); 2,048 (Mode C)
Dimensions: 130 × 127 × 145 mm
Weight: 3 kg

Contractor
Thales Communications, Colombes Cedex, France.

VERIFIED

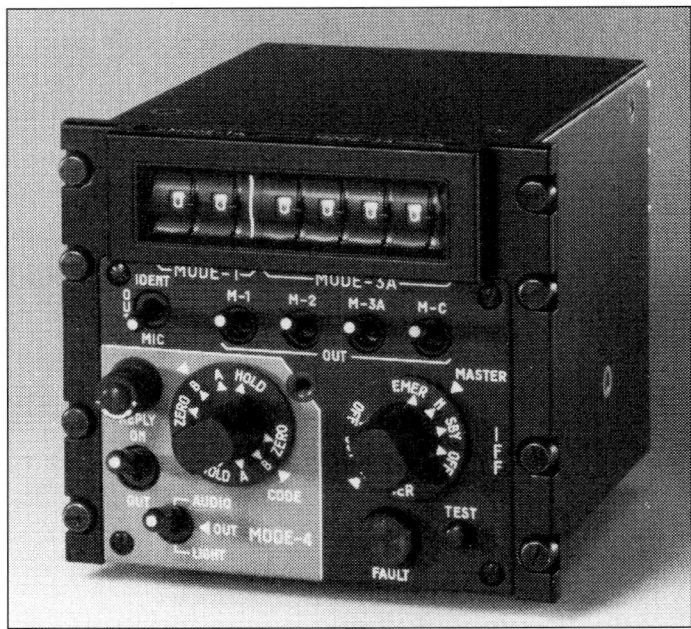

The NRAI-7(.)/SC10(.) airborne IFF transponder

NRAI-9(.)/SC15(.) Identification Friend-or-Foe (IFF) transponder

Type
D-band (1 to 2 GHz) airborne IFF transponder.

Description
The NRAI-9A is essentially a two-box version of the NRAI-7 which incorporates a number of improvements. These include the elimination of sidelobe response, and automatic special-code referral, with positive identification permitting a ground operator to locate a particular aircraft. Special emergency codes may also be employed, chosen by the pilot, such as radio failure. Dual receiver channels connected to upper and lower antenna and comparison circuits provide a diversity function.

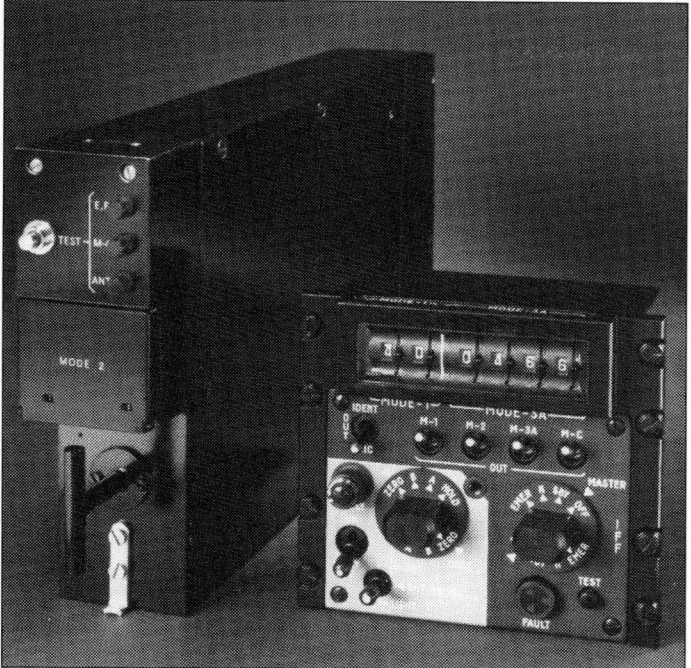

The NRAI-9(.)/SC15(.) IFF transponder

Status

Over time, the NRAI-9 transponder is reported as having been installed aboard ATL2 maritime patrol aircraft.

Specifications

Frequency: 1,030 MHz (receive); 1,090 MHz (transmit)
Power output: 500 W peak
Modes available: 1, 2, 3A/C and Mode 4 capability
Codes: 32 (Mode 1); 4,096 (Modes 2 and 3/A); 2,048 (Mode C)
Dimensions: 58 × 193 × 361 mm (transmitter/receiver); 127 × 130 × 80 mm (control unit)
Weight: 2.5 kg (transmitter/receiver); 1.4 kg (control unit)

Contractor

Thales Communications, Colombes Cedex, France.

VERIFIED

NRAI-11(.)/SB13(.) Identification Friend-or-Foe (IFF) interrogator-decoder

Type

D-band (1 to 2 GHz) airborne IFF interrogator.

Description

The NRAI-11A is an interrogator-decoder for identifying and determining the range and bearing of any friendly mobile unit within the surveillance area of the IFF system with which it is operating. The equipment consists of an encoder, a transmitter, an antenna switch, two receivers, an analogue processing unit, a defruiter, a passive decoder, an evaluator, an extractor, a wobbulator and an automatic self-test. Peak output power of the system is 1 kW with a duty cycle of 1 per cent.

The NRAI-11(.)/SB13(.) IFF interrogator-decoder

The dual receiver system determines bearing and elevation of the target by monopulse techniques and an antenna switch is incorporated for suppression of sidelobes during interrogation sidelobe suppression. Interrogation mode is 1, 2, 3 without interlacing, with passive decoding on Mode 1, 2 or 3 and 4,096 combinations. The extractor has a memory capacity of 255 plots.

Volume and power consumption has been reduced substantially by the use of a solid-state transmitter, low-power integrated and monolithic high-density circuits, hybrid and custom-built large-scale integration circuits, and switching power supply.

Status

Over time, the NRAI-11 interrogator-decoder is reported as having been installed aboard Mirage 2000 variants.

Specifications

Peak power: 1 kW
Modes: 1, 2, 3/A and 4
Frequency: 1,030 MHz (transmit); 1,090 MHz (receive)
Receiver sensitivity: −79 dBm (1,087-1,093 MHz)
Dimensions: 124 × 80 × 194 mm
Weight: 12 kg

Contractor

Thales Communications, Colombes Cedex, France.

VERIFIED

SB 14 Identification Friend-or-Foe (IFF) interrogator

Type

Ground-based IFF interrogator.

Description

The SB 14 IFF is a small, lightweight, battery-operated ground-to-air interrogator for very short-range weapon systems. A Mode 4 internal memory and a reply evaluator model eliminate the need for a cryptographic computer during operation. A programmer loads the Modes 1, 2, 3/A, 4 memory with the appropriate code catalogue. The programmer connector and test connector are located on the interrogator. A single cable interfaces the interrogator to the weapon system.

The interrogator includes both interrogation and receiver sidelobe suppression circuits. It can be programmed to operate in each mode sequentially, according to the mission requirements. The modes and codes are controlled by the contents of the memory unit, the interrogation procedure being predetermined at the loading of the memory unit by the programmer. Codes are loaded for a multiday mission and in the secure mode the interrogation challenges are randomly chosen to avoid compromise.

The IFF challenge is initiated by action of the operator on the weapon trigger before firing and is discrete since a challenge sequence consists of a short burst of interrogations. A positive outcome from a self-test sequence allows the interrogation process to start.

Status

Over time, the SB 14 interrogator is understood to have been procured by a number of armed forces around the world (including the French Army) for use with missile systems such as IGLA, Mistral and Stinger.

Specifications

Antenna: 3-element dual Yagi
Modes: 1, 2, 3/A and 4
Dimensions: 182 × 256 × 124 mm
Weight: 5.6 kg, incl antenna
Range: 10 km

Contractor

Thales Communications, Colombes Cedex, France.

VERIFIED

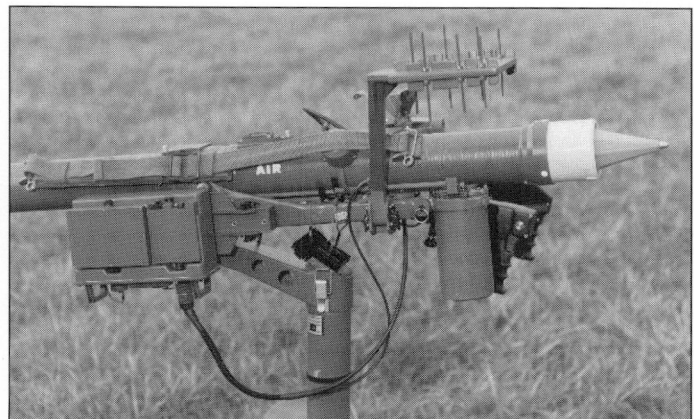

SB 14 IFF interrogator for short-range weapon systems

SB 16 Identification Friend-or-Foe (IFF) interrogator

Type
Ground-based IFF interrogator.

Description
The SB 16 is a modular ground-to-air IFF interrogator that is designed for use with short-range (up to 100 km) radar-controlled weapon systems and features considerable commonality with the SB 14 Very SHOrt-Range Air Defence System (VSHORADS) unit (see separate entry). Capable of interfacing and synchronising with 'various' kinds of radars, SB 16 comprises interrogator and processor units. Of these, the interrogator is described as being small enough and light enough to be mounted on a radar scanner (thereby saving up and down link losses) while the processor is noted as being mounted 'near to' the host radar's display. In order to eliminate the need for a permanent link to a communications security computer, SB 16 stores Modes 1, 2, 3A and S interrogations in a removable memory module that is mounted on the front panel of its processor. The available memory capacity provides for up to four days worth of Selective Identification Feature (SIF) and crypto-secure operations and is also noted as taking into account code tolerance during transition periods.

When fitted with an optional control box, SB-16 is provided with local control and display facilities. Functionally, the equipment transmits interrogations in the appropriate azimuth sector before weapon system initiation and according to programmed IFF sequences. Replies are processed within the host radar's range gate and the interrogator unit's spatial coverage is controlled and matched to the particular weapon system using ISLS, RSLS and reply gating techniques together with appropriate antennas. Where two targets are close in range within the antenna beam, SB 16 utilises two different interrogation patterns to optimise discrimination. An integral built-in test facility performs a complete system test during each interrogation.

Status
As of May 2001, the status of the SB 16 IFF interrogator was uncertain.

Contractor
Thales Communications, Colombes Cedex, France.

VERIFIED

STR 700 Identification Friend-or-Foe (IFF) transponder

Type
D-band (1 to 2 GHz) airborne IFF transponder.

Description
Over time, the STR 700 IFF transponder has been installed aboard a large number of German military aircraft. It features high reliability, small dimensions, low weight and diversity of operations. The equipment meets all aircraft identification monitoring system specifications and consists of two basic units: the receiver transmitter and the control unit with logic section. The control unit has interfaces for the Mode 4 decoder/coder facility and for an altitude encoder. The STR 700 has been specifically designed for applications where diversity of operation is required and is provided with two receiving channels. However, it can be equipped with a single receiving channel only. The receiver/transmitter is modular in design, accommodated in a single ½ Air Transport Racking short case for both models. Normally the control unit consists of the actual control section, with switches and lamps, plus the logic section that decodes the interrogation signals. If the available space in the cockpit is too narrow, the logic section can be accommodated separately from the control section. Reliable identification of the target aircraft depends largely on the correct action of the transponder. Consequently the STR 700 transponder is provided with many test circuits, to reveal the functional condition at any time automatically by internal interrogations. For system checking on the ground the receiver/transmitter of the two-channel version is provided with 12 light-emitting diodes to indicate proper performance of the individual subassemblies.

Status
As of early 2001, Jane's sources were reporting that STR 700 was in production and service aboard aircraft of the German armed forces.

Specifications
Frequencies: 1,030 ± 1.5 MHz (receive); 1,090 ± 3 MHz (transmit)
Power supply: 16-32 V DC (up to 70 W)
Dimensions: 124 × 193 × 382 mm (transceiver); 146 × 134 × 155 mm (control unit/logic section)
Weight: 3.2 kg (control unit/logic section); 9.8 kg (transceiver - single channel); 10.8 kg (transceiver - two channel)

Contractor
European Aeronautic, Defence and Space (EADS) Co Systems and Defence Electronics, Unterschleissheim, Germany.

UPDATED

STR/TSC 2000

Type
Airborne Identification Friend-or-Foe (IFF) transponder.

Description
Developed by European Identification Systems (EIS - a 50-50 joint venture between the European Aeronautic, Defence and Space (EADS) Co Systems and Defence Electronics business unit in Germany and Thales Communications in France), the STR 2000 (German designation)/TSC 2000 (French designation) airborne IFF transponder is described by Jane's sources as incorporating a full Mk XII capability (as defined in STAndard NATO AGreement (STANAG) 4193) and as offering Mode S level 3 (as defined in International Civil Aviation Organisation (ICAO) Annex 10. As such, the equipment comprises a transponder and a dedicated Control and Display Unit (CDU). To facilitate Mode S levels 2 and 3 functionality (with all available communications protocols), STR/TSC 2000 makes use of its host platform Air Datalink Processor (ADP), with the unit being connected to the transponder's MILitary STanDard (MIL-STD) -1553B serial databus interface. So configured, STR/TSC 2000 provides remote terminal databus functions to facilitate bidirectional Mode S data exchange with the ADP. When not so organised, STR/TSC 2000 (together with its CDU) can provide an 'elementary' Mode S level 2 capability. The transponder also provides all the interfaces necessary for either an external or an appliqué Mode 4 cryptographic computer as standard. Upgrades of the system facilitates traffic collision avoidance system (as defined in Air Radio INCorporated (ARINC) 429) and Mode 5 (New Generation IFF (NGIFF) - see separate entry) functionality.

Status
According to Jane's sources, Germany's BWB procurement agency awarded EIS a then year €160 million STR/TSC 2000 contract during late 2000/early 2001, with the EIS partners working in the proportion of 2:1 as a reflection of the numbers of transponders required by Germany and France respectively. Initial STR/TSC 2000 applications are believed to include 16 German Navy Breguet Atlantic I maritime patrol/signals intelligence aircraft, French and German Transall C.160 transport aircraft, French and German Eurocopter AS 532 and Tiger helicopters and French Air Force C-130H transport and C-135FR/KC-135R tanker aircraft. STR/TSC 2000 is also understood to be mandated for the multinational NH-90 helicopter.

Specifications
Dimensions: 146 × 133 × 81 mm (CDU); 318 × 124 × 194 mm (transponder)
Weight: 1.5 kg (CDU); 6.8 kg (transponder without crypto appliqué)

Contractor
European Identification Systems, France/Germany.

NEW ENTRY

TSA 1010 Identification Friend-or-Foe (IFF) interrogator

Type
D-band (1 to 2 GHz) ground-based or shipboard IFF interrogator.

Description
Derived from Thales' SA 10 system, the TSA 1010 IFF interrogator is designed for LOng Range Air Defence System (LORADS) applications and can be installed in fixed stations, mobile land-based radars and ships. The equipment is packaged in a 48 cm × 6U rack that houses interrogator, extractor (target data processor), cooling fan and cryptocomputer (KIR format or miniaturised) units. If required, the cryptocomputer can be external to the main rack. TSA 1010 is compliant with STAndard NATO AGreement (STANAG) 4193 and International Civil Aviation Organisation (ICAO) Annex 10. It is capable of IFF/secondary surveillance radar Modes 1, 2, 3/A, C and, optionally, Mode 4/Secure. As a further option, the unit can

The TSA 1010 IFF interrogator with KIR, ACC memory module and SR19 control box
0006995

be configured to incorporate Automatic Code Changing (ACC) in Modes 1 and 3/A and there is provision for upgrading to Mode S and New Generation IFF (NGIFF) if required.

TSA 1010 is an easy to reconfigure equipment that increases baseline performance and functionality via the addition of modules. The main system configurations are as follows:

- Interrogator only (baseline configuration)
- Interrogator with video and reply preprocessing. Thales notes that this configuration is a cost-effective means of improving IFF system performance without changing the existing processing and exploitation subsystems
- Interrogator-extractor
- Interrogator-extractor with preprocessed video output.

Status
Over time, TSA 1010 is reported to have been integrated with a 'broad spread' of radar systems and has been selected for global retrofit on a range of Romanian sensors.

Specifications
Frequency: 1,030 ±0.2 MHz; 1,090 ±3 MHz (receive)
Modes: 1, 2, 3/A, 4/Secure, ACC (option for Modes 1 and 3/A), NGIFF (upgrade) and S (upgrade)
Peak power: 33 dBW (typical)
Range: up to 482 km
Target capacity: up to 65/beam; >1,500/scan
Dimensions (w × h × d): 483 × 266 × 594 mm
Weight: 39 kg

Contractor
Thales Communications, Colombes Cedex, France.

VERIFIED

....

TSB 2500 Identification Friend-or-Foe (IFF) interrogator-transponder

Type
D-band (1 to 2 GHz) combined airborne IFF interrogator-transponder.

Description
TSB 2500 (which exists in both interrogator-transponder and interrogator only configurations) comprises two Line-Replaceable Units (LRU), a Combined Interrogator-Transponder (CIT) and an Antenna Control or Antenna Adaptor Unit (ACU or AAU). The ACU is used with electronically scanning antennas while the AAU is designed to interface with mechanically scanning arrays. The equipment is modular in design and functions as a Mk XII and Mode S level 2/3 transponder and a Mk XII interrogator. It also incorporates provision for the integration of Mode 5 New-Generation IFF (NGIFF) transponder and interrogator functions. Looking at the equipment's LRU in more detail, the CIT handles all the necessary interrogator-transponder transmission, reception and signal/data processing functions and interfaces with its host platform via a US MILitary STanDard (MIL-STD)-1553B databus. As noted previously, the ACU controls electronically scanning antennas while the AAU acts primarily as a booster and a Radio Frequency (RF) front end for mechanically scanned arrays. The packaging of the ACU and AAU in separate LRU allows them to be installe d close to the antenna systems they are serving in order to minimise RF losses. In terms of encryption, TSB 2500 is STAndard NATO AGreement (STANAG) 4193 compliant and can interface with any US National Security Agency (NSA) Mode 4 or Secure (for non-NATO applications) cryptocomputer. In the same context, it can also accommodate a front panel, dual KIT/KIR appliqué cryptocomputer.

Status
Over time, the TSB 2500 combined IFF interrogator-transponder is understood to have been selected for installation on France's next-generation Rafale combat aircraft, the Swedish S 100B Argus airborne early warning system and the multinational NH90 helicopter.

Specifications
	Interrogator	Transponder
Frequency	1,030 ±0.2 MHz	1,090 ±0.5 MHz
Power	>32 dBW	500 W (±2 dB)
Operating modes	1, 2, 3/A, C and 4 (Mode 5 upgradable)	1, 2, 3/A, 4 and S (Mode 5 upgradable)
Dimensions	32 × 193 × 290 mm (AAU); 230 × 115 × 105 mm (ACU); 228.6 × 157.2 × 193.5 mm (CIT)	
Weight	<4 kg (AAU); 5.5 kg (ACU); <10 kg (CIT)	

Contractor
Thales Communications, Colombes Cedex, France.

VERIFIED

TSC 2050 Identification Friend-or-Foe (IFF) transponder

Type
D-band (1 to 2 GHz) airborne IFF transponder.

Description
TSC 2050 is a true diversity Mk XII and Mode S level 3, Traffic alert and Collision Avoidance System (TCAS) compatible IFF transponder that fully meets the requirements of the International Civil Aviation Organisation (ICAO) Annex 10, STAndard NATO AGreement (STANAG) 4193 and US Department of Defense Aircraft Identification Monitoring System (AIMS) 65-100B specifications. Additionally, the equipment incorporates provision for the future integration of the New-Generation IFF (NGIFF) Mode 5 function.

TSC 2050 is designed for shelf mounting in a non-pressurised section of its host airframe and is reported to be suitable for both latest generation combat aircraft and retrofit applications. System control can be by means of a US MILitary STanDard (MIL-STD)-1553B interface or via a dedicated control box. In terms of encryption, TSC 2050 can interface with any US National Security Agency (NSA) Mode 4 or Secure (for non-NATO applications) cryptocomputer. A further option is the use of a mechanical adaptor to install the cryptocomputer as an appliqué.

Status
Over time, the TSC 2050 IFF transponder is understood to have been selected for use on the UK's Nimrod MRA Mk 4 maritime patrol aircraft and as a global retrofit on a range of as yet unidentified Romanian platforms.

Specifications
Dimensions: 136 × 124 × 212 mm
Weight: <5.2 kg
Altitude: 21,336 m

	Receiver	Transmitter
Type	dual channel superheterodyne	all solid-state
Frequency	1,030 ±0.5 MHz	1,090 ±0.5 MHz
Peak power		500 W (typical)
Bandwidth	6-9 MHz (6 dB bandwidth)	
Dynamic range	55 dB	
Sensitivity	−77 dBm	
Reply rate		1,000 Hz (Mode 4); 1,200 Hz (SIF)
VSWR		output protected against mismatch
Droop		1 dB
Output channels		2 with built-in PIN switches
Modes		1, 2, 3/A, C, 4 and S level 3 (B, D and test when operated through a -1553B databus)

Contractor
Thales Communications, Colombes Cedex, France.

VERIFIED

ISRAEL

EL/M-2099 Identification Friend-or-Foe (IFF) transponder interrogator

Type
Airborne IFF transponder-interrogator.

Description
EL/M-2099 is a solid-state, airborne IFF transponder-interrogator that is Mk 10 to 12 compatible and is designed for installation in a wide range of combat aircraft types. System features include compact packaging; spatial coverage commensurate with

The EL/M-2099 IFF transponder-interrogator (1998)
0009907

onboard fire-control radars: a 32 to 64-bar target capacity; integral electronic counter-countermeasures provision; software controlled target extraction and processing and programmable interrogation and reply codes.

Status
As of this edition, the EL/M-2099 IFF transponder-interrogator was understood to be available.

Specifications
Transmitter
Type: solid-state
Frequency: 1,030 (±0.2) MHz; 1,090 (±0.2) MHz
Peak power: 59 ± 1 dBm
Receiver
Type: dual logarithmic
Frequency: 1,030 MHz (diversity); 1,090 MHz (RSLS incl GTC)
90% decoding sensitivity: −77 dBm (1,030 MHz); −80 dBm (1,090 MHz)
Coder: M1, M2 and M3/A (1,030 MHz); emergency, M1, M2, M3/A and MC (1,090 MHz); user-specified pulse pattern (option)
Decoder/pulse processing: bracket decoding, CFAR circuits and pulse-width discrimination (interrogator function); AOC, diversity processor, echo rejection, ISLS, Modes 1/2/3/C decoding and pulse-width discrimination (transponder functions); user-specified pulse pattern (option)
Extractor function: software controlled (by dedicated DSP)
System management: microprocessor controlled
Altitude: up to 18,288 m
Temperature: −40 to +71°C (operating)
Power requirement: 115 V AC; 400 Hz
Dimensions (H × W × D): 200 × 147 × 350 mm
Weight: 12.2 kg
Volume: 10,290 cm³

Contractor
Elta Electronics Industries Ltd (a subsidiary of Israel Aircraft Industries Ltd), Ashdod.

VERIFIED

ITALY

AN/UPA-59A(V) decoder

Type
Identification friend-or-foe decoder group for shipboard use.

Description
The basic decoder selects modes for interrogation and accepts reply video inputs for processing. The video inputs may be mode-separated video, composite video (all modes of reply video and mode tags with reset on a single input), or tagged video (all modes of reply video on one input and mode tags on another input). Passive decoding of all codes is provided in Modes 1, 2 and 3/A for any condition of interface. A Selected Altitude Layer (SAL) allows the display of Mode C altitude replies for air traffic control purposes. Four channel active decoding and readout is provided for Modes 1, 2, 3/A and C. Reply codes are read out directly in Modes 1, 2 and 3/A and are translated in Mode C to display the corresponding altitude in 30 m increments.

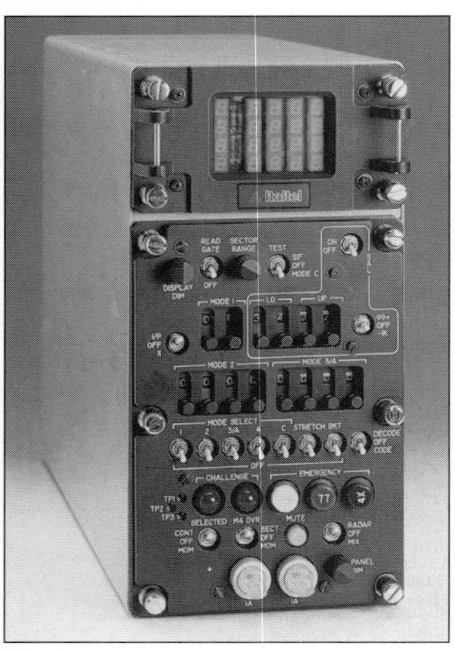

The AN/UPA-59A(V) decoder group

Active readout is supplied for targets within a range/azimuth gate, initiated by a gate signal from the indicator. A plan position indicator modification kit is available to generate the active gating signal if one is not already generated. Light gun activation can also be employed as an option.

The AN/UPA-59A(V) provides a number of features, including:
- solid-state decoding and degarbling delay lines
- improved degarbler action
- active readout using a light-emitting diode display.

Contractor
ITALTEL, Defence Telecommunications Division, Milan.

VERIFIED

SIT 421T Identification Friend-or-Foe (IFF) transponder

Type
D-band (1 to 2 GHz) airborne IFF transponder.

Description
The SIT 421T is a one-box airborne IFF transponder, suitable for fitting in fixed-wing aircraft, helicopters and naval vessels. It is a licence-built version of a BAE Systems North America transponder. It operates in Modes 1, 2, 3/A, 4 and C. The receiver/transmitter includes a 500 W solid-state transmitter, dual-channel receiver and RF interface module. The first of these comprises a delay line oscillator, modulator, driver and power amplifier.

The dual-channel receiver handles space diversity operation to ensure the maximum and most reliable transponder response. The receiver may also be set for operation with one antenna only (single channel operation).

The controls for operation of the transponder, code and mode selection and so on are mounted on the front of the equipment, which is designed for cockpit mounting.

Two remote-mounted versions are in production:
- SIT 421T controlled via multiwire cable from the SIT 901 control unit
- SIT 421T/1553 controlled via databus MIL-STD-1553B.

Status
Over time, the SIT 421T IFF transponder is reported to have been installed aboard Italian Mangusta anti-tank/escort helicopters (Mode 4 encryption unit added on Lot 2 aircraft), frigates and hydrofoils.

Specifications
Frequency: 1,030 ± 0.3 MHz (receiver); 1,090 MHz (transmitter)
Sensitivity: −77 dBm (adjustable 69-77)
Dynamic range: 55 dB
Output power: 500 W
Operation modes: 1, 2, 3/A, 4 and C
Dimensions: 136.5 × 136.5 × 213 mm
Weight: SIT 421T 4.1 kg (transponder) 0.8 kg (controller)

Contractor
ITALTEL, Defence Telecommunications Division, Milan.

VERIFIED

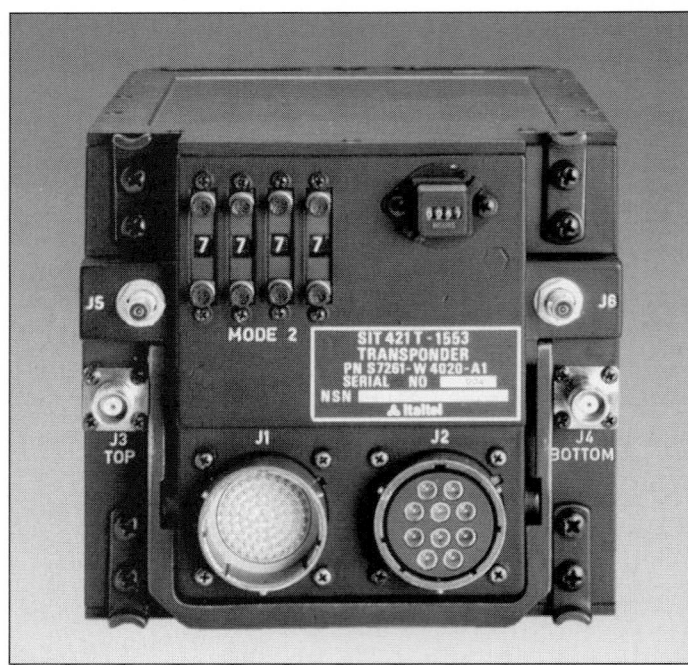

The SIT 421T IFF transponder

SIT 422 Identification Friend-or-Foe (IFF) interrogator

Type
D-band (1 to 2 GHz) weapon systems IFF interrogator.

Description
The ITALTEL short-range air defence SIT 422 interrogator is a Mark XII equipment designed to operate and survive in tactical battlefield environments.

It provides weapon systems with an all-weather fire capability enabling rapid engagement of hostile aircraft at maximum weapon system range. It is particularly suitable for installation in land mobile systems, small ships, and for all those applications where a need for short-range identification exists, and where small size, lightweight and low power consumption are especially important. The SIT 422 interrogator can be controlled directly from an associated fire-control system computer, or by means of a control unit. Associated equipment is a SIT 904 or SIT 906 control unit (where applicable), and a Model AS-131 75 cm antenna.

Interrogator codes are stored in a removable memory module that has an internal clock that extracts codes for the interrogator. The SIT 422 is also suitable for direct interfacing to a KIR/1A-TSEC crypto-computer for Mode 4 operation.

The interrogator consists of a receiver-transmitter, video processor, reply processor/sequential observer, timing and control, self-test, memory module or KIR interface and power supply. Each function is modular in design for ease of maintenance.

Status
SIT 422 was developed in conjunction with BAE Systems North America. Over time, the device is reported as having been used by the Italian army with the Skyguard/Aspide missile system, the LPD-20 radar and the SIDAM air defence gun. It is also noted as having been specified for the Otobreda OTOMATIC 76 mm SP anti-aircraft gun.

Specifications
Frequency: 1,090 ± 0.2 MHz (receiver); 1,030 ± 0.2 MHz (transmitter)
Sensitivity: −76 dBm
Output power: 200 W
Duty cycle: 1% max
Weight: 12.71 kg

Contractor
ITALTEL, Defence Telecommunications Division, Milan.

VERIFIED

....

SIT 434 Identification Friend-or-Foe (IFF) interrogator

Type
D-band (1 to 2 GHz) airborne IFF interrogator.

Description
The SIT 434 is a modular architecture IFF interrogator, intended for both fixed- and rotary-wing aircraft and has considerable growth potential. In its basic configuration, the equipment operates under control of the SIT 905 control box.

It is capable of operating in each of the Selective Identification Facility (SIF) Modes 1, 2, 3/A and 4, used either separately or interlaced (1 SIF mode + Mode 4). For Mode 4 operation an associated crypto unit, with its mounting and key-loading devices, must be used.

Challenge control is possible by means of an enabling signal coming from the control box or from the radar system. Video output signals are generated as the result of the processing relevant to the last interrogation cycle. Different symbols are provided for target, SIF friend or Mode 4 friend.

SIT 434 interrogator and SIT 905 controller

Optional features include:
SIT 434B
- pulse-to-pulse 2/2 defruiter, replacing the internal decoding function
- interface for an external active/passive decoder, replacing the digital symbol generation

SIT 434C
- control of interrogator and target information interface by means of an embedded dual redundant RTU per MIL-STD-1553B
- challenge management, including interrogation sector(s) capability (based on target expected position), interface pattern and interrogation pulse repetition frequency optimisation, and power management
- antenna synchro interface to associate azimuth data to SIF/Mode 4 plot; both rotating and sector scanning antenna can be accommodated.

Specifications
Transmitter power (peak): 1,200 W/300 W selectable
Transmit frequency: 1,030 ± 0.2 MHz
Receiver frequency: 1,090 ± 0.2 MHz
Duty cycle: 1% max
Decoding: passive decoding (4,096 codes for SIF modes; dual code capacity for M3/A overlap period)
Size: 1 ATR short
Weight: 15 kg including controller and mounting

Contractor
ITALTEL, Defence Telecommunications Division, Milan.

VERIFIED

RUSSIAN FEDERATION AND ASSOCIATED STATES (CIS)

RFAS Identification Friend-or-Foe (IFF) systems

Description
Known RFAS IFF systems comprise the following:
Ground-based
Score Board
Shipborne
Dead Duck
Dog Tail
End Curve
Gin Pole
High Pole A
High Pole B
Salt Pot A
Salt Pot B
Ski Pole
Square Head
Yard Rake
Airborne
Beacon
Cross Up
Slap Shot
SOD series (ATC/SIF)
SRO-2/2M (NATO Odd Rods)
SRZ-2
Odd Rods
This is the NATO reporting name for the SRO-2/2M (and sometimes for the SRZ-2 and SRZO systems). It appears to be the latest transponder for airborne applications and is the standard fit on most RFAS military aircraft. The antennas visible on RFAS fighter aircraft indicate that this system is a conventional type of diversity IFF.

The End Curve IFF system (Christopher F Foss)

The Dog Tail IFF system

Score Board

This is the NATO reporting name for a transportable ground interrogator used by RFAS forces. It is very early equipment since it has an antenna array consisting of two horizontal rows of two radiators per element. The two double rows are mounted one above the other to give a four-array of what appears to be a total of 32 dipole radiators. The whole system is mounted on a rectangular rotatable framework.

Square Head Naval System

The antenna of this shipborne system consists of a broadside array of dipoles with a rectangular reflector/support frame measuring about 1.4 m high by 2 m wide. It was originally thought to be a long wavelength search radar of modest power, but later evidence confirms that its function is that of an IFF interrogator antenna, or directional array for transmission of guidance signals to surface-to-surface missiles. It is most frequently seen on the older types of RFAS missile-carrying boats, and these carry two or four such arrays on sponsons fore and aft of the mainmast. Recent information strongly suggests that the system operates in G-band (4 to 6 GHz).

VERIFIED

SPAIN

IRS-10M Secondary Surveillance Radar (SSR) system

Type

Ground-based SSR system.

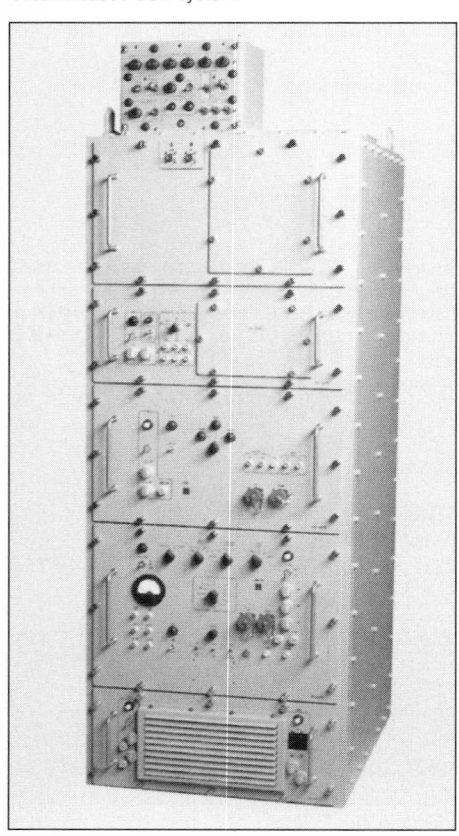

The electronics cabinet used in the IRS-10M secondary surveillance radar

Description

The IRS-10M is a Mark X secondary surveillance radar with Mark XII operating capability. The standard configuration of the IRS-10M is dual channel, including supervision units providing automatic changeover of the operational channel in the event of failure. The military version has the capability of generating Mark X selective identification feature, Modes 1, 2, 3/A and C interrogations and Mark XII Mode 4.

The system consists of an antenna, two receiver/transmitters, two coder-extractors, a digital switching unit, a radio frequency switching unit and an antenna control unit. All units other than the antenna are designed for mounting in a standard 48 cm rack. In order to operate the system from a site remotely located from the radar site, a remote-control unit is available, and communicates via a telephone line. The CSL-10 antenna is an open array developed according to Federal Aviation Authority specifications, and mounted on a pedestal. Three channels provide capabilities for interrogation and receiver sidelobe suppression functions and make it suitable for monopulse processors.

Status

As of this edition, IRS-10M is thought to be in both military and civil service.

Specifications

Frequency: 1,030 ± 3.5 MHz (transmit); 1,090 ± 5 MHz (receive)
Polarisation: vertical
Transmitter peak power: 33 dBW min
Weights: 350 kg (antenna); 260 kg (IRS-10)

Contractor

INDRA DTD, Madrid.

VERIFIED

IRS-20MP monopulse Secondary Surveillance Radar (SSR)

Type

Ground-based SSR system.

Description

The IRS-20MP monopulse SSR has been developed for air defence and air traffic control applications. The system is an improved SSR that can be used in a stand-alone surveillance role or in association with a primary/search surveillance radar. It provides accurate plot data on each pulse of an airborne transponder reply. It also provides smoother tracking and better plot data from garbled and defective replies. Because of its inherent lower pulse repetition frequency, less interference (fruit) is generated, and substantially less load to airborne transponders, thereby reducing lock ups. Its planar open array antenna utilises sum and difference radiation patterns developed to improve azimuthal accuracy. The vertical radiation pattern is designed with a bottom edge that reduces the amount of ground incident energy. No changes to transponders are necessary with the monopulse SSR.

Features of the IRS-20MP include improved accuracy with minimal interrogations, high efficiency with the large vertical aperture antenna, protection against reflections, an all solid-state modular interrogator, a high-speed databus and multiple microprocessors, improved sidelobe suppression, improved degarbling and phantom reply rejection, and programmable output formats.

Specifications

Sum pattern beamwidth: 2.35°
Antenna gain: 28 dB
Sidelobes: −27 dB
Interrogator peak power: 2 kW
GTC: sector and range programmable
OBA accuracy: < 6 min of arc
Range: 463 km
Target capacity: 400 at 7.5 rpm

Contractor

INDRA DTD, Madrid.

VERIFIED

The IRS-20MP monopulse SSR system

IRS-M/IFF-25FR identification system

Type
Military Identification Friend-or-Foe (IFF) package.

Description
The IRS-M/IFF-25FR IFF package is able to generate raw and synthetic video simultaneously and is designed to be operated locally or by remote control. The equipment is solid state and modular in design; utilises 68000 (IRS-M subsystem) and 68020 (IFF-25SR subsystem) processors and is STAndard NATO AGreement (STANAG) 4193/5017 and International Civil Aviation Organisation (ICAO) compliant. System features include: built-in test (fault isolation to line-replaceable unit level); programmable test targets and power outputs; degarbling, dephantoming and defruiting capabilities; active and passive decoding and interrogation and rear sidelobe suppression facilities.

Status
As of this edition, the IRS-M/IFF-25FR package is understood to be available.

Specifications
Instrumented range: 474 km
Detection probability: ≥98.7%
Code validation probability: ≥99.7%
Azimuth accuracy: 0.022°
MTTR: <30 mins
MTBF: >1,000 h
Operating temperature: −28 to +65°C
Transmitter
Frequency: 1,030 ± 1 MHz
Peak Power: ≥ 2,300 W
Duty cycle: ≥ 1%
Interrogation modes: SIF and Mode 4
Receiver
Frequency: 1,090 MHz
Sensitivity: −87 dBm
Dynamic range: ≥70 dB
Noise figure: ≤7 dB
Bandwidth: 8 MHz
Extractor
Fruit rejection: 99.8%
Interrogation format: SIF and Mode 4 (as per ICAO and STANAG)
Decoding: Modes 1, 2, 3/A and emergencies

Contractor
INDRA DTD, Madrid.

VERIFIED

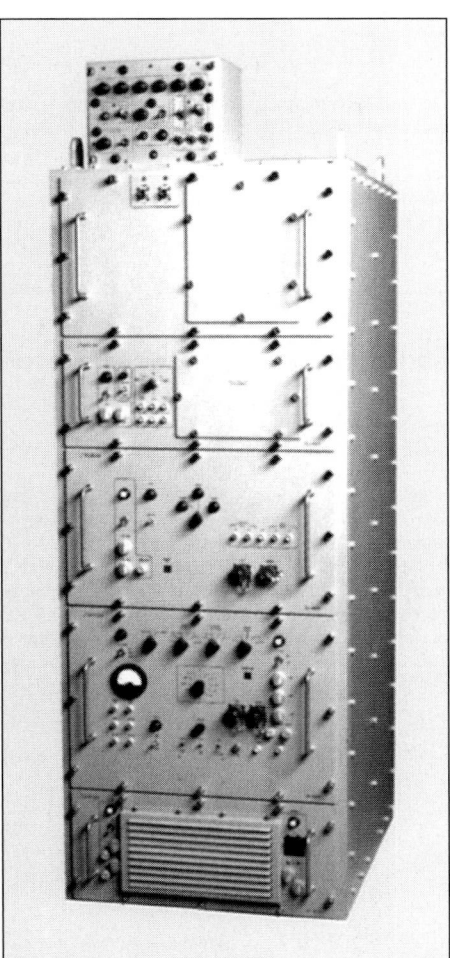

The electronics cabinet used in the IRS-M/IFF-25FR identification package
0006999

UNITED KINGDOM

ARI 5983 Identification Friend-or-Foe (IFF) transponder system

Type
I-band (8 to 10 GHz) airborne IFF transponder.

Description
ARI 5983 is designed to provide a means of locating, identifying and providing navigational assistance for aircraft operating outside the limits of normal radar coverage. The system is interrogated by a primary radar and gives an edge of band response. Response codes are selected via a control unit that can also be used to initiate a built-in test routine and switch between the system's two antennas for optimum coverage.

ARI 5983 receives interrogator signals from pulse radars in two bands, each of which has a bandwidth of 100 MHz wide. When interrogated, the transponder responds with either a single radio frequency pulse or a coded group of up to six pulses. Sixteen different reply code formats are available. Transponder function is automatically suppressed during the operation of other onboard I-band equipment. An output pulse is also generated to trigger I-band blanking when the transponder is operating. Correct transponder function is indicated by a light emitting diode display on the equipment's control unit. System options include double or multiple pulse interrogation (to minimise false triggering) and ARI 5983 is fully NATO codified.

Status
Over time, ARI 5983 is reported as having been installed aboard Royal Navy Sea Harrier F/A Mk 2 aircraft and Lynx, Merlin HM Mk 1 and Sea King helicopters. The system is also noted as having been fitted to off-shore Do 228, Lynx and Dauphin aircraft. As of October 2000, it was understood that ARI 5983 was to be replaced aboard the Sea Harrier F/A Mk 2 by a BAE Systems combined interrogator/transponder as part of the UK's Successor IFF programme (see separate entry).

Specifications
Receiver frequencies: 9,190-9,290 and 9,360-9,460 MHz
Sensitivity: −93 dBw
Transmitter frequency: 9,310 ± 7 MHz
Output power: 135-300 W max peak
Pulse duration: 0.45 ± 0.1 µs
Reply code: single-pulse or 6-pulse code, 16 settings
Pulse spacing: 2.9 µs nominal
Duty cycle: 0.005 max
Power supply: 28 V DC, 40 W max
Dimensions: 160 × 217 × 87 mm (transponder); 147 × 117 × 48 mm (control unit)

Contractor
M/A-COM Ltd, Dunstable.

UPDATED

Cossor™ Condor 2 Monopulse Secondary Surveillance Radar (MSSR) antennas

Type
Family of MSSR specific antennas.

Description
As of January 2001, Raytheon Systems Ltd was offering two types of MSSR specific antenna, the known details of which are as follows:

Flat Plate Antenna (FPA)
Raytheon's MSSR FPA is described as being a lightweight structure that is suitable for use in ground (stand-alone or on-mounted), ground mobile (wheeled vehicle) and shipboard applications. As such, it incorporates two identical (mirror image) outer radiating sections and a central element that incorporates front and rear radiators. The described sections can be attached directly to a primary radar array or supported by a 'simple' spine in stand-alone applications.

Large Vertical Aperture (LVA) antenna
Raytheon's LVA MSSR antenna is designed for installation above primary radar arrays or as part of an autonomous MSSR system. As such, it comprises 35 vertical radiating columns that are mounted on a central aluminium spine. Each radiating element incorporates an array of 12 dipole aerials and bonded glass/PTFE stripline circuitry, all of which are housed in an aluminium jacket that incorporates a polycarbonate/ glass radome. The individual radiating element dipole arrays are phase and amplitude controlled to minimise ground illumination and maximise air vehicle coverage. The LVA's support spine provides the rigidity needed to survive wind and icing loads and provides environmental protection for system cabling and distribution networks. Overall, the Raytheon LVA is noted as producing a ground roll off for the vertical path of at least 2 dB/° at the −6 dB point.

Status
As of January 2001, Raytheon Condor 2 FPA arrays are reported as having been supplied for transportable applications in Australia and Brazil. In the LVA context

(and as of the given date), more than 250 LVA units were noted as being operational worldwide.

Specifications

	FPA	LVA
Antenna type	linear patch array	open dipole planar-array
Operating frequencies	SSR/IFF bands	SSR/IFF bands
Azimuth patterns	sum; difference; SLS	sum; difference; SLS
Elevation pattern	fan beam	shaped beam
Gain	19.3 dBi	27 dBi
Azimuth beamwidth	5°	2.4°
Azimuth sidelobe level	−24 dB	−26 dB
Operating temperature	−50 to +70°C	−50 to +70°C
Dimensions (l × h × d)	4.3 × 0.5 × 0.35 m	8 × 1.85 × 0.93 m
Weight	50 kg	480 kg

Contractor

Raytheon Systems Ltd, Harlow.

NEW ENTRY

....

Cossor™ Condor 2 Monopulse Secondary Surveillance Radar (MSSR) systems

Type
Family of ground-based MSSR systems.

Description
The International Civil Aviation Organisation (ICAO) Annex 10 compliant Raytheon Systems' Condor 2 series MSSRs centre on the company's solid-state Condor 2 interrogator and an associated antenna (see separate entry). As such, the Condor 2 interrogator is described as incorporating 'full' transmission, reception, processing and control capabilities within a single cabinet and as making use of very large scale integrated chips and microprocessors. In the type of MSSR application being described, the Condor 2 interrogator is co-located with a power supply, with the whole being regarded as a single interrogator channel. Systems are typically supplied in dual channel configurations to ensure system availability, with cross coupling facilitating the transfer of processed targets (before track processing) to the standby channel in the event of a primary channel failure. Use of this procedure ensures that both channels have the current track history if a changeover is required. Condor 2 series MSSRs incorporate 'comprehensive' monitoring systems to provide 'complete and accurate' assessment of their operational performance. Any failures detected by the equipment's self-test routine is reported locally to a Local Maintenance Terminal and remotely to a Remote System Control Terminal within 'seconds' of an event. Automatic channel reconfiguration maintains system function in order to minimise manual intervention. As of January 2001, three Condor 2 configurations had been identified, the known details of which are as follows:

Condor 2
The baseline MSSR with provision for a 'simple' module upgrade to Mode S.

Condor 2 Mode S Level 2
Condor 2 Mode S Level 2 offers 'enhanced' surveillance to ICAO Level 2 standard together with a 'comprehensive' air traffic management capability. Designed for Mode S multisite operations, Condor 2 Mode S Level 2 can be supplied with the Mode S capability incorporated or as a field upgrade.

Condor 2 Mode S Level 5
The Condor 2 Mode S Level 5 ground network system is a 'full' Mode S performance variant that has been developed within the PreOperational European Mode S (POEMS) programme.

In addition, all Condor 2 configurations can include a Mode 4 military cryptographic identification friend-or-foe capability and can be supplied with interlaced or sector Mode S and Mode 4 provision if required.

Status
As of January 2001, in excess of 220 Condor 2 systems were reported as having been commissioned worldwide. As of the given date, current Condor 2 programmes comprised:

- 213 Condor 2 MSSR systems ordered for incorporation in the US department of Defense/Federal Aviation Administration's (FAA) DASR11 programme
- an unknown number of Condor 2 MSSR systems with the Mode 4 capability ordered by the Royal Australian Air Force
- 127 Condor 2 Mode S Level 2 systems ordered for use in the US FAA's ATCBI-6 programme
- selection of Eurocontrol POEMS Mode S Level 5 standard systems by the UK's NATS and Germany's DFS
- 21 Condor 2 Mode S Level 5 systems ordered under the Common Mode S procurement programme for Germany, Swisscontrol and LVNL.

Jane's sources suggest that the first Condor 2 systems were delivered during 1993 and that a Mode S configuration was supplied to India during 1995.

While Condor 2 represents the Raytheon Systems' early 2001 standard MSSR, readers should note that as of May 2001, Jane's sources were suggesting that examples of the company's earlier generation MSSR systems remained in service.

Here (and over time), the UK's Ministry of Defence is known to have installed more than 180 Raytheon MSSRs at its military airfields in the British Isles and Germany, while the country's Civil Aviation Authority is noted as having acquired examples of the Cossor™ SSR 950 system. Elsewhere in the world, the Canadian Department of Transport is understood to have procured 41 dual Cossor™ SSR 955 systems (from which the Condor 9600 system evolved) for deployment throughout Canada. Earlier generation Cossor™ MSSR systems are also noted as having been procured by customers in Greece, Trinidad and Tobago and several Middle East countries including Saudi Arabia, the United Arab Emirates and Bahrain. For its part, the already noted Condor 9600 system is known to have been acquired by the Swedish Civil Aviation Administration for the country's national air traffic control radar replacement programme, with the first such equipment becoming operational at Romele near Malmo, in August 1989.

Specifications
Interrogator
Transmitter type: solid-state plug-in module
Frequency: 1,030 (± 0.01) MHz
Output power: 33 dBW
Duty cycle: 4.2% (average); up to 6% (peak); 63.7% (high duty cycle transmitter, peak)
Operating modes: 1, 2, 3/A, 3/B, 3/C, 4 and S
Log receivers
Frequency: 1,090 (± 0.2) MHz
Sensitivity: −90 dBm (tangential)
Dynamic range: −16 to −86 dBm
MTBF: 25,000 h/channel (field proven data)

Contractor
Raytheon Systems Ltd, Harlow.

UPDATED

....

Cossor™ Interrogation and Reply Cryptographic Equipment (CIRCE)

Type
Identification Friend-or-Foe (IFF) interrogation and reply cryptographic equipment.

Description
CIRCE is a modular encryption system for military IFF applications where NATO KIT/KIR is unavailable. It can be integrated into existing IFF units and incorporates a customer specific cryptographic algorithm to provide a specific, independent and secure national IFF capability. CIRCE is also noted as featuring a STAndard NATO AGreement (STANAG) 4193 interface.

Status
Over time, CIRCE is reported as having been procured by the defence forces of at least three Gulf and Asian countries.

Contractor
Raytheon Systems Ltd, Harlow.

VERIFIED

....

Outfit RRB Identification Friend-or-Foe (IFF) receiver

Type
C-band (0.5 to 1 GHz) shipboard IFF receiver.

Description
Designed to work with the ARI 5893 IFF transponder (see separate entry), the Outfit RRB receiver features a swept gain facility (that can be controlled by an external extractor/processor); an interface for a Video Code Suppression Unit (VCSU); selectable coded or uncoded outputs; compatibility with a wide range of I-band (8 to 10 GHz) radars; built-in test and natural cooling. Functionally, transponder replies are filtered from the received radar echo and are channelled to the RRB receiver. As noted previously, the equipment's output can be either coded or quantised uncoded video pulses. These may be used for presentation on a display or be utilised for further signal processing and plot correction.

Status
As of this issue, Outfit RRB is understood to have been procured by the Royal Navy.

Specifications
Frequency: 9,310 ± 7 MHz
Sensitivity: 65 dBm (pulse)
Noise figure: 10 dB
Swept gain variation: 42 dB
Output threshold: + 0.5 V
Output noise level: 0.0725 V (into matched load)
Power supply: 115 VAC; 115 VA

Dimensions (w × d × h): 455 × 200 × 405 mm
Weight: 15.5 kg
Temperature: 0°C to +45°C

Contractor
M/A-COM Ltd, Dunstable.

VERIFIED

PA6150 airborne Identification Friend-or-Foe (IFF)/ Secondary Surveillance Radar (SSR) transponder

Type
D-band (1 to 2 GHz) IFF/SSR transponder for military and civil aircraft.

Description
The PA6150 is an airborne transponder which has been designed for both military and civil use. The equipment has a 500 W solid-state transmitter and a logarithmic receiver that is based on a BAE Systems' semi-conductors monolithic integrated log amplifier. The remaining circuitry uses application-specific integrated circuit gate technology to allow all the high-speed signal processing to be achieved digitally. The transponder is fully microprocessor-controlled and system control is achieved by the use of either a microprocessor-based cockpit-mounted control unit via an RS-422 link or via a MIL-STD-1553B databus. An interface to a MIL-STD-1553B or ARINC 429 databus is required for operation at Mode S Level 2 and 3. IFF/SSR modes and codes, including Mode S selection, are entered and selected using key switches with light-emitting diode status indicators. Provision is made for automatic code changing. For military identification a secure cryptographically encoded transponder reply option is available. A small appliqué cryptographic computer module is provided which is accessible from the front of the unit. The cryptographic key may be electronically loaded using a preloaded fill gun. PA6150 is packaged in an ARINC 600 4MU case. The minimum standard transponder satisfies the basic IFF/SSR requirements for Modes 1, 2, 3/A and Mode S Level 1 operation. Mode S Level 2 or 3 operation is available when either a MIL-STD-1553B or ARINC 429 databus is connected to the transponder to allow data exchange with the aircraft Mode S Air DataLink Processor.

Status
As of this edition, the status of PA6150 is uncertain.

Specifications
Transmitter frequency: 1,090 ± 0.5 MHz
Power output: 500 W (nominally 27 dBW)
Diversity isolation: 20 dB min
Receiver frequency: 1,030 MHz
Dimensions: 95 × 146 × 100 mm (control unit); 194 × 124 × 318 mm (transponder)
Weight: 2.2 kg (control unit); 9.5 kg (transponder)

Contractor
BAE Systems, Portsmouth.

VERIFIED

PA6817 Identification Friend-or-Foe (IFF)/Secondary Surveillance Radar (SSR) transponder

Type
Shipboard IFF/SSR transponder and control unit.

Description
PA6817 is a modular IFF/SSR transponder system which is compatible with the standard NATO agreement 4193 IFF Mk XII requirement and operates in conjunction with the PA6807 control unit. The two equipments are mounted in what is termed a versatile console and there is a built-in test facility for the transponder which includes self mode interrogation and fault finding to module level. The PA6807 unit houses the system's Mode 1 and 3/A code and KIT-1C Mode 4 cryptographic computer controls. Mode 2 code, power and transmit/standby/remote safe-to-transmit controls are mounted on the transponder assembly. The PA6817 unit uses a solid-state power amplifier with all necessary IFF signal processing being handled by a 113,000 gate digital application specific integrated circuit. General transponder operation is controlled by microprocessor running software resident in a programmable read-only memory.

Status
As of this edition, PA6817 is understood to have been procured by the Royal Navy.

Specifications
Power output: 27 dBW (nominal); 25 dBW (min)
Reply rate: 1,200 replies/s (with up to 15 pulses per reply)
Triggering sensitivity: −72 to −80 dB (min)
Transponder dimensions/weight: 324 × 272 × 487 mm/14 kg

Control unit dimensions/weight: 152 × 152 × 200 mm/1.8 kg
Operational temperature/relative humidity: 15-45°/30-85% rh (transponder); 15-40°C/30-70% rh (control unit)

Contractor
BAE Systems, Portsmouth.

VERIFIED

PTR 283 Mk 1/PVS 1280 Mk II Identification Friend-or-Foe (IFF) interrogators

Type
D-band (1 to 2 GHz) airborne IFF interrogators.

Description
These interrogator systems have been designed to meet the requirements of in-flight secondary radar interrogations. The transmitter/receiver uses pulsed oscillator techniques employing automatic frequency control, and a logarithmic receiver using silicon integrated circuits. The equipment is designed to interrogate on Modes 1, 2 and 3/A, the pulses driving the modulator being generated by the encoder/decoder.

The PTR 283 Mk 1 equipment consists of a lightweight D-band transmitter/receiver unit, together with an associated encoder/decoder unit and a control unit. Other units associated with the system are an antenna switch, and dual antenna system together with an L-trace radar display.

The PVS 1280 Mk II system consists of a D-band lightweight transmitter/receiver, together with an associated encoder/decoder that offers the facility of active decoding and defruiting. This equipment is designed to integrate into an airborne primary radar system and the control of the system is performed by the primary radar controller. The system offers Interrogation SideLobe Suppression (ISLS) operation and an ISLS switch is available which enables the transmitted power to be distributed equally to two antennas, and also enables the antennas to be fed alternatively in phase and anti-phase for ISLS operation.

Status
Over time, equipments of the types described are reported as having been procured.

Specifications
Transmitter
Frequency: 1,030 ± 0.5 MHz
Power output: 5 kW (peak nominal); 5.5 kW (PVS 1280 peak nominal)
Duty cycle: 0.11%
Pulse length: 0.8 ± 0.2 μs
Receiver
Frequency: 1,090 ± 0.2 MHz
Decoder: 496 codes

Contractor
BAE Systems, Portsmouth.

VERIFIED

PTR 446A Identification Friend-or-Foe (IFF)/ Secondary Surveillance Radar (SSR) transponder

Type
Lightweight IFF/SSR transponder.

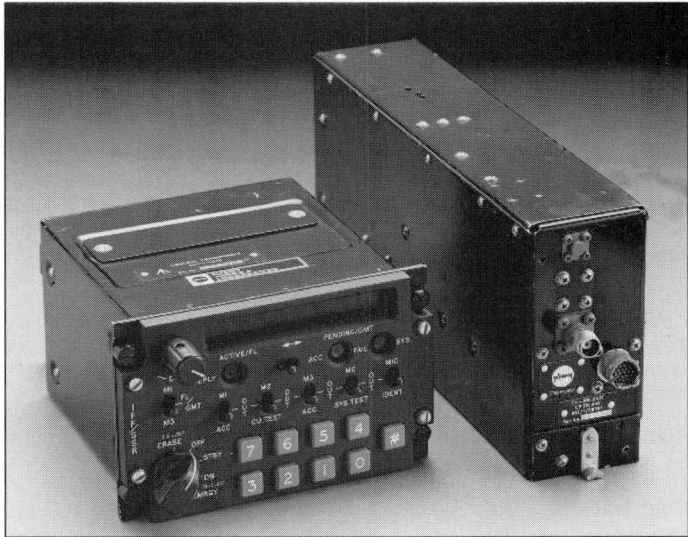

PTR 446A transponder and control unit

Description

The PTR 446A is a lightweight, micro-miniaturised transponder that provides a means of aircraft identification in response to ground radar interrogation. Both military and civil modes are covered by the equipment. To achieve the reduction in size and weight, extensive use has been made of specially designed microcircuits. A digital shift register replaces the conventional delay lines in the encoder and decoder circuits thus providing time delays independent of temperature. Decoder, encoder and associated switches are located in the control unit. In operation the reply codes are either set up on the control unit on Modes 1 and 3/A and B, or external to the control unit for Mode C or Mode 2. The operational capability of the equipment is determined by the control unit in use, ranging from a three mode facility given by the control unit PV 447 to a full mode capability given by the control unit PV 1447. Three-pulse sidelobe suppression is used. The PV 1447 control unit has been developed to provide full standard NATO agreement 5017 facilities for the PTR 446. Capable of operating simultaneously on Modes 1, 2, 3/A and C, the unit incorporates special features for reply code selection.

Status

Over time, PTR 446A is reported to have been installed on UK military helicopters and BAE Systems Hawk fixed-wing aircraft.

Contractor

BAE Systems, Portsmouth.

UPDATED

....

PTR 461 Identification Friend-or-Foe (IFF)/Secondary Surveillance Radar (SSR) transponder

Type

Shipborne and mobile IFF/SSR transponder.

Description

The PTR 461 operates on Modes 1, 2 and 3/A and provides a means of ship identification in response to interrogation by ground, shipborne or airborne secondary radars. It is designed in accordance with the military Mk 10A requirements defined in standard NATO agreement 5017. The transponder uses a solid-state transmitter and operates in conjunction with the control unit PV 462 which may be installed remotely from the transponder. The code controls for Modes 1 and 3/A are housed on the control unit and the code control for Mode 2 is situated on the transponder.

The PTR 461 is derived from the lightweight transponder PTR 446, and uses common modules which give a logistic compatibility. It includes its own monitoring facilities to allow fault finding down to module level.

Status

Over time, PTR 461 is reported to have been procured by the Royal Navy.

Contractor

BAE Systems, Portsmouth.

VERIFIED

....

PV 846 Identification Friend-or-Foe (IFF) decoder

Type

Shipborne IFF automatic decoder.

Description

The PV 846 is a shipborne IFF automatic decoder designed to meet the requirements for a decoding interface between an IFF interrogator/responser and a computer data handling system. The PV 846 accepts instructions from the computer which specify a particular radar target for interrogation in terms of its range and bearing. On receipt of this information, the decoder instructs the interrogator to transmit on a particular mode over the given bearing, the mode being selected from Modes 1, 2 or 3/A. Reply signals received by the interrogator are gated in range and azimuth by the decoder and passed to the computer. Facilities are provided for an occasional distress search over the full interrogation range. At present rotation intervals (between 8 and 12 antenna rotations), the output from the interrogator is accepted over a full 360° at maximum range. The decoder also accepts bearing information from digitisers on the antenna. Each time an octant boundary is crossed by the antenna, information is passed to the computer, enabling it to prepare data defining the next targets. PV 846 is fully solid state and possesses a built-in test system either operated automatically by the computer, or manually by the use of a self-test button on the front of the equipment. Weight of the decoder is 28.1 kg.

Status

As of this edition, PV 846 is noted as having been procured.

Contractor

BAE Systems, Portsmouth.

VERIFIED

Raytheon IFF 860/890 series Identification Friend-or-Foe (IFF) interrogators

Type

Family of D-band (1 to 2 GHz) ground-based and naval IFF interrogators.

Description

Raytheon Systems' IFF 860 series IFF interrogators are designed to operate in conjunction with an integral or external radio frequency switch to provide an Interrogator SideLobe Suppression (ISLS) function. Received transponder replies are processed by the interrogator and are delivered in the form of video signal or processed (various serial and parallel data formats) outputs. Raytheon Systems claims that IFF 860's performance matches that of civil Air Traffic Control (ATC) systems and notes that the system incorporates a built-in test routine. Options include an azimuth data interface and conventional or monopulse operation. IFF 890 series equipments are solid-state systems which utilise a series of common modules and include ISLS, receiver sidelobe suppression and receiver gain time control to minimise the number of unwanted replies passed to the decoder processor. The decoder modules used provide full decoding and IFF reply evaluation facilities (for any selected mode) together with defruiting and degarbling. Current key variants of the combined family are as follows:

Raytheon IFF 863

The Raytheon IFF 863 interrogator is a long-range (up to 500 km depending on antenna type) equipment with a modular architecture which allows it to be configured for a range of applications including military and civil ATC and ground and naval air defence.

Raytheon IFF 891

The Raytheon IFF 891 is a single box interrogator which provides short-range surveillance (typically 12 km) using a miniature flat plate antenna. It is battery powered, operates on Modes 1, 2, 3/A and 4 and features automatic code change and stored code crypto as standard. IFF 891 is described as being suitable for a range of man-portable and short-range air defence applications.

Raytheon IFF 892

The Raytheon IFF 892 is a two-box interrogator which has modular commonality with IFF 891. Depending on antenna gain, it offers surveillance out to 20 km and is suitable for short-range air defence missile and gun fire-control applications.

Raytheon IFF 896

The Raytheon IFF 896 is a medium-range (up to 45 km depending on antenna type) interrogator which is suitable for forward alerting and other radar applications. The equipment's mode and code capabilities are similar to those of IFF 891.

Status

As of March 2001, Raytheon IFF 860 variants are reported to have been first delivered (for air defence system applications) during 1992, with the IFF 863 model being selected for use in the Australia-New Zealand ANZAC frigate programme (first deliveries during 1993). Raytheon notes that 860 series interrogators are suitable for land-based air defence radars such as the AN/FPS-117, Commander and Martello equipments. IFF 890 series interrogators are noted as being in service on the British Aerospace Rapier, Contraves Skyguard and Shorts Starburst air defence systems and as having been ordered for use on Ericsson Giraffe air defence radars. Raytheon Systems further notes that the IFF 891 is suitable for systems such as Starburst and RBS 70 while the IFF 892 is applicable to equipments such as Rapier and Skyguard. The IFF 896 equipment is the likely variant for use with Giraffe.

A Raytheon IFF 891 interrogator mounted on a Starburst air defence system

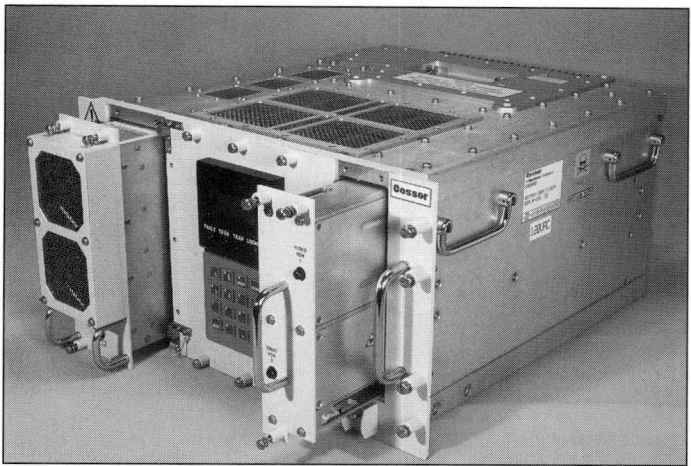

The Raytheon IFF 863 interrogator unit

Specifications

	Raytheon IFF 860	Raytheon IFF 891	Raytheon IFF 892	Raytheon IFF 896
Transmitter frequency	1,030 (± 0.2) MHz	1,030 (± 0.2) MHz	1,030 (± 0.2) MHz	1,030 (± 0.2) MHz
Receiver f requency	1,090 (± 0.2) MHz	1,090 (± 0.2) MHz	1,090 (± 0.2) MHz	1,090 (± 0.2) MHz
Interrogator modes	1, 2, 3/A, 4	1, 2, 3/A, 4	1, 2, 3/A, 4	1, 2, 3/A, 4
Peak power (nominal)	33 dBW	20 dBW	20 dBW	27 dBW
Duty cycle	up to 1%			
Sensitivity (each channel)	−85 dBm	−65 dBm	−65 dBm	−65 dBm
Receiver bandwidth (nominal)	10 MHz (to −3 dB)			
Dynamic range	70 dB			
Input voltage (nominal)	120/240 V	12-18 V DC	24 V DC	12-18 V DC/115 V/400 Hz
Operating temperature	−40 to +71°C	−40 to +75°C	−40 to +75°C	−20 to +50°C
Dimensions	123 ×50 × 129 mm	425 × 180 × 125 mm[1]	355 × 180 × 100 mm[2] 308 × 180 × 90 mm[3]	530 × 280 ×260 mm

Key

[1] incl batteries   [2] transceiver unit   [3] decoder unit

Contractor

Raytheon Systems Ltd, Harlow.

UPDATED

Raytheon IFF 4500 Identification Friend-or-Foe (IFF) interrogator

Type

Monopulse IFF system interrogator.

Description

Raytheon's IFF 4500 series monopulse interrogators provides a comprehensive IFF Mk XII identification facility when used to identify co-operative target platforms suitably equipped with IFF Mk XII transponders and Mode 4 cryptographic units (NATO Applications) or CIRCE or similar cryptographic units (non-NATO Applications). However, a Mode 4 cryptographic unit is not necessary if a transponder is required only to respond with IFF Mk XA replies. Therefore, civilian aircraft and those military aircraft not yet equipped to the IFF Mk XII, Mode 4 standard can still be identified using IFF Mk XA interrogation protocols. IFF 4500 series interrogators are a one Air Transport Racking (ATR) short replacement for existing Raytheon IFF 3500 series equipments, providing IFF Mk XII operation on Modes 1, 2, 3/A, C and 4, with selective and all-code decoding using either manual or Automatic Code Change (ACC). The optional ACC facility is provided within the interrogator unit. This utilises a highly modular construction within the 1 ATR short form factor. Alternatively, a ½ ATR medium form factor is available.

IFF 4500 interrogator units provide all necessary IFF processing including azimuth degarble/defruiting processing. IFF target reports are available via a MIL-STD-1553B databus interface for display on the host radar display. Control of the units can be provided by the host radar system over the MIL-STD-1553B databus interface or suitable discrete alternative. Additional discrete lines provide interrogator on/off control, antenna azimuth information, sector blanking and KV/ACC Erase. Standard discrete interfaces are provided for D-band (1 to 2 GHz)

suppression in/out and the Mode 4 weight on wheels line (if used). Sum and difference radio frequency outputs interface with the IFF antenna system via the existing feeders. Extensive Built-In-Test (BIT) facilities are also included. These facilities include power-up BIT, continuous BIT and manually initiated BIT. The BIT identifies failure to module level. The highly adaptable modular architecture of the baseline IFF 4500 interrogator can be configured to meet with a wide variety of airborne weapon system and surveillance applications with the following key features:

- full IFF Mk XII capability in accordance with standard NATO agreement 4193
- enhanced monopulse azimuth measurement capability, yielding high accuracy in high target density and long-range applications
- high reliability, all solid-state technology
- monopulse reply processor capable of handling over 500 target reports within each 360° antenna scan
- highly effective degarble processing (4 overlapped replies).

Status

As of May 2001, the Raytheon IFF 4500 interrogator was understood to have been selected for retrofit to the Royal Air Force's Tornado F Mk 3 interceptors as part of the UK's Successor IFF (SIFF) programme.

Contractor

Raytheon Systems Ltd, Harlow.

UPDATED

Raytheon IFF 4700 series Identification Friend-or-Foe (IFF) transponders

Type

Family of D-band (1 to 2 GHz) IFF transponders.

Description

Raytheon IFF 4700 series equipments are built from a series of six basic modules (three printed circuit boards, a power supply unit, a chassis assembly and a radio frequency unit) which are mixed and matched to provide the required form factors. The various models provide a full diversity IFF facility and can be interrogated by Mk XA or Mk XII interrogators conforming to STAndard NATO AGreements (STANAG) 4193 and 5017 together with civil secondary surveillance radars built to International Civil Aviation Organisation Annex 10 standard. Responses are generated for Mode 1, 2, 3A, C and 4 interrogations with Mode 4 function being available only on equipments fitted with a cryptographic unit conforming to STANAG 4193. IFF 4700 equipments can also be fitted with Raytheon Systems' CIRCE encryption module (see separate entry). Identified members of the IFF 4700 family include the following:

Raytheon IFF 4720

The 4.7 kg, 90 × 194 × 321 mm Raytheon IFF 4720 is an equipment bay mounted unit that is packaged in an Air Transport Radio (ATR) compliant housing and is described as being suitable for a range of naval and airborne applications. The equipment can interface with a MIL-STD-1553B databus if required.

Raytheon IFF 4740

The 5 kg, 145 × 132 × 172 mm Raytheon IFF 4740 model is an aircraft cockpit mounted variant that can be configured to meet specific lighting and/or Night Vision Goggle (NVG) compatibility requirements. The unit provides full IFF and control facilities in a single package.

Raytheon IFF 4760

The 4.5 kg, 136 × 136 × 212 mm Raytheon IFF 4760 is another equipment bay mounted model which can operate in ambient temperatures of up to 95°C and interfaces with a MIL-STD-1553B databus and the Raytheon IFF 4770 control unit.

Cossor™ IFF 4760 remote-mounted transponder

Cossor™ IFF 4720 for air and naval use

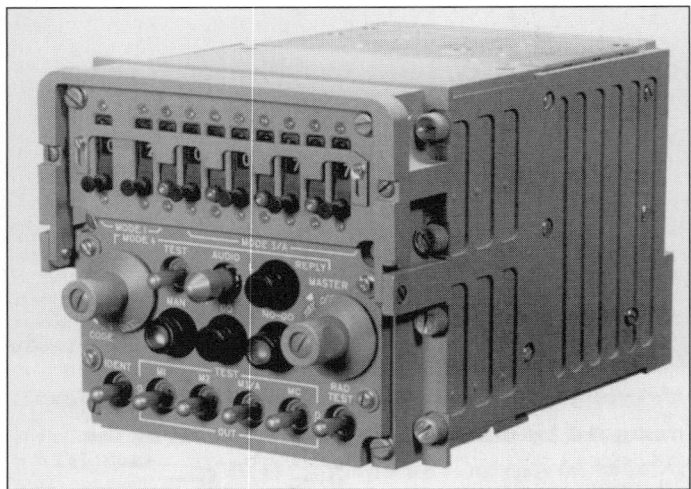

Cossor™ IFF 4740 for high-performance aircraft and helicopters

Raytheon IFF 4770

The 1.2 kg, 146 × 95 × 76 mm Raytheon IFF 4770 is the operator control unit which supports the Raytheon IFF 4700 series transponders. It provides full Mk XII controllability and is available in a range of lighting colours and voltages with full NVG compatibility if required.

Status

As of May 2001, at least 260 Raytheon IFF 4760 transponders were reported as having been procured by the Royal Air Force for use aboard Harrier GR. Mk 5/7 close support and, possibly, other aircraft types. Additionally, the Raytheon IFF 4720 transponder is understood to have been supplied for use in a MiG-29 application and Raytheon Systems describe the Raytheon IFF 4740 unit as being suitable for installation in the Tornado strike aircraft. According to Jane's sources, other IFF 4700 series applications include the CN-235 transport and Hawk 100 training aircraft.

Specifications

Transmit frequency: 1,090 ± 0.5 MHz
Receiver frequency: 1,030 ± 0.5 MHz
Input voltage: +22 to +29 V (full performance)
Input power (normal operation): 35 W (nominal); 70 W (max)
Peak power: 500 W (nominal)
Transmit duty cycle: 1% (max)
Receiver bandwidth: 7.5 MHz (3 dB down); 24 MHz (30 dB down); 50 MHz (60 dB down)
Sensitivity: −75 ± 3 dBm
Dynamic range: 55 dB (min)
BIT: receiver frequency/sensitivity; transmitter frequency/power; mode decoding; diversity decision logic; antenna voltage standing-wave ratio; Mode 4 interface; altitude digitiser interface
MTBF: 1,500-2,000 h (MIL-STD-781)
Temperature range: −40 to +70°C (normal operation); +95°C (short term)

Contractor

Raytheon Systems Ltd, Harlow.

UPDATED

Raytheon IFF 4800 series Identification Friend-or-Foe (IFF) transponders

Type

Family of D-band (1 to 2 GHz) IFF transponders.

Description

Raytheon IFF 4800 series equipments are of modular construction and are housed in a single line-replaceable unit. The transponders within the family provide complete dual-antenna diversity IFF facilities in fixed-wing, helicopter and shipboard applications and can be interrogated by Mk XA or Mk XII interrogators (conforming to STAndard NATO AGreements (STANAG) 4193 and 5017) together with civil secondary surveillance radars built to International Civil Aviation Organisation Annex 10 (including Mode S) standard. They respond to Mode 1, 2, 3/A, C, 4 and S challenges with automatic code change being provided for Modes 1 and 3/A. An appliqué Mode 4 cryptographic unit is incorporated (which can be removed where Mode 4 responses are not required) and there are interfaces for a Mode S air datalink processor, an airborne collision avoidance system and Global Positioning System reporting. The equipments feature built-in test routines and have provision for growth to Next-Generation IFF standard (STANAG 4193 Part V). System control can be by means of a discrete control or MIL-STD-1553B databus. IFF 4800 series equipments (including the IFF 4810 and 4830 transponders) integrate with the Raytheon IFF 4870 control unit.

Status

As of January 2002, the status of the Raytheon IFF 4800 series transponder family was uncertain.

Specifications

Frequency: 1,030 (±0.5) MHz (receive); 1,090 (±0.5) MHz (transmit)
Input voltage: 28 V (nominal)
Input power: 90 W (nominal); 120 W (max)
Peak power: 500 W ± 2 dB (all conditions)
Reply rate: 50 Hz (Mode S - 16 replies averaged over 1 s); 1,200 Hz (selective identification facility/Mode 4 - averaged over 1 s)
Receiver bandwidth: 8 MHz (3 dB down); 50 MHz (60 dB down)
Sensitivity (MTL): −75 ± 3 dBm
Dynamic range: 55 dB (min)
MTBF: 1,500-2,000 h
Temperature range: −40 to +70°C (normal operation); +90°C (short term)
Altitude: to 21,336 m
Weight: 8.5 kg
Cooling: convection

Contractor

Raytheon Systems Ltd, Harlow.

UPDATED

UNITED STATES OF AMERICA

Air traffic control radar beacon system/ Identification friend-or-foe Mk XII System (AIMS)

Type

Electronically steered antenna.

Description

The AIMS OE-120/UPX group, has been developed and improved by Sanders to support Identification Friend-or-Foe (IFF) and Air Traffic Control Radar Beacon Systems (ATCRBS). The system is also suitable for land-based operations. It consists of an electronically scanned circular antenna that offers multifunction performance and provides military IFF in all interrogation modes, as well as operation with military and commercial air traffic control equipment. It is particularly suited for operation in systems where rapid multitarget identification is necessary, such as track-while-scan, phased-array or multiple search radars.

The antenna group consists of an antenna, antenna position programmer and a control unit. The antenna itself measures 3.75 m diameter × 40.6 cm high and has an array of 64 Radio Frequency (RF) radiators on its perimeter. For shipborne applications, it is mounted concentrically around the mast to avoid any blockage inherent with mechanical rotating systems. Interrogation SideLobe Suppression (ISLS) is provided by selectively energising radiating elements to produce the required omnidirectional and directional beam patterns. The system can be operated in a continuous, sector or jump-scan mode. In the latter mode, switching from one beam to another can be accomplished within 50 μs.

The antenna position programmer is located within the perimeter of the antenna. On command from the control unit it distributes RF power to the radiators which, in turn, form and steer the beam. The remotely positioned control unit receives either digital or synchro beam position data and translates these into digital antenna beam-steering commands. In the shipborne application, the control unit also provides compensation for roll, pitch and ship's heading. In addition, internally generated beam-steering commands may be selected by means of front panel controls.

AIMS shipborne IFF antenna system

Sum and difference mode execution can be employed simultaneously when required for special operational conditions such as Mode S and receiver sidelobe suppression. ISLS queuing is electronically commanded for sum/omni or sum/difference pulse transmissions. Boresight peak of sum beam to null of the difference ratios typically exceed 35 dB, providing high monopulse azimuth angle resolution and received pulse amplitude sorting potentials to meet the needs of most IFF and ATC.

System maintenance is enhanced by a built-in diagnostic test system that automatically monitors more than 200 operational test points. A front panel digital display provides fast indication of the position of any malfunction within the system.

Status

As of this edition, at least 60 antennas of this type have been delivered to the navies of Japan and the US. Six antennas configured for use on land have been delivered to the US Air Force. Outstanding US Navy orders during 1997 totalled eight units. Four antennas plus a spare set have been ordered for the Spanish F-100 frigate programme with delivery scheduled for 1998/99. On 9 June 2000, Sanders was awarded a then year US$5,702,254 firm, fixed-price contract covering the supply of five OE-120/UPX antenna assemblies, associated 'installation and checkout spare' kits and technical data. At the time of the award's announcement, work on the effort was scheduled for completion by the end of September 2002 and its contracting activity was the US Navy's Naval Air Systems Command, Patuxent River, Maryland.

Contractor

Sanders (a Lockheed Martin company), Nashua, New Hampshire.

UPDATED

....

AN/APX-76 Identification Friend-or-Foe (IFF) interrogator

Type

D-band (1 to 2 GHz) airborne IFF interrogator.

Description

The AN/APX-76 airborne interrogator exists in three main versions: the electron-tube APX-76A, the solid-state transmitter APX-76B and the Technically Improved Product (TIP) APX-76C. Known details of these variants are as follows:

AN/APX-76A

APX-76A is an air-to-air interrogator for all-weather interceptor and other tactical aircraft, and has full Aircraft Identification Monitoring System (Air Traffic Control Radar Beacon System/IFF Mk XII system) capability in Modes 1, 2, 3/A and 4. It achieves a narrow antenna beamwidth, and a reduction in clutter and 'fruit' through interrogation and receiver sidelobe suppression circuits, in conjunction with special antennas having sum (main lobe) and difference (sidelobe suppression) patterns. Bracket-decoded video and discrete coded-decoded video are displayed on the radar screen to provide unambiguous correlation between IFF and primary radar targets. The equipment is fully IFF Mk XII compatible.

AN/APX-76B

GEC-Marconi Hazeltine produces an all solid-state transmitter and power supply unit for the APX-76, which provides both operational advantages and improved reliability and maintainability. Mean time between failure increases to 400 hours, compared with 225 hours for the APX-76A. APX-76B is directly interchangeable with its predecessor, and no realignment is necessary; the complete operation requires only 15 minutes.

AN/APX-76C model

The nomenclature APX-76C designates a system with the TIP package incorporated. TIP takes the form of modification kits for the receiver/transmitter and synchroniser portions of the system. TIP eliminates dated circuitry and offers a dramatic reduction in power supply current which enables operation at lower

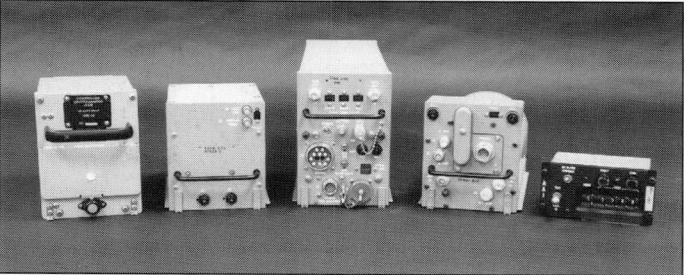

The AN/APX-76 IFF interrogator system

temperatures and improves reliability. Other technical features include an improved triple pulse decoder for Mode 4 use, an additional defruiter function, a multipath ditch circuit, and adaptive threshold anti-jam circuitry.

Status

APX-76 entered production during 1967 and some 7,000 such systems are reported to have entered service to date. Within the US military, the interrogator is noted as having been fitted (over time) to E-1B, E-2B/C, F-4B/C/D/E/G/J, F-14A, F-15A/B/C/D, F-111D, KC-130F, P-3, S-3A/B and SH-60B aircraft. Outside the US, APX-76 systems have been supplied to air arms of Australia, Canada, Greece, Israel, Japan, South Korea, Portugal, Saudi Arabia and the UK. As of this edition, GEC-Marconi Hazeltine notes that it is producing APX-76C. The US Navy is known to have upgraded its APX-76As with solid-state transmitters/power supplies while large numbers of the TPI (receiver/transmitter/synchroniser) package have been sold to both the US Navy and the US Air Force.

Specifications

Frequency: 1,030 MHz (transmit); 1,090 MHz (receive)
Duty cycle: 1% max
Power output: 2 kW
Sensitivity: –83 dBm
MTBF: 225 h (APX-76A); 400 h (APX-76B)
Weight: 16.8 kg

Contractor

BAE Systems North America, Greenlawn, New York.

VERIFIED

....

AN/APX-100(V) Identification Friend-or-Foe (IFF) transponder

Type

Airborne IFF transponder.

Description

The 'next-generation' AN/APX-100 (V) transponder can be either panel mounted or installed in an equipment bay and uses microminiature technology in both its digital and radio frequency circuitry. The equipment includes integrated crypto, full level 3 Mode S and Global Positioning System position reporting in the same space envelope as the original APX-100 set. The system is currently in production for use on a range of US, NATO and international military aircraft applications. The equipment's manufacturer also notes the availability of upgrade kits to bring older systems up to the latest standard.

The system is a completely solid-state, modular equipment with a complete dual channel diversity system, comprehensive built-in test, digital coding and encoding and high anti-jamming capability. Two antennas form part of the equipment and the diversity system receives signals from each and switches the transmitter output to the antenna that received the strongest interrogation signal. This system is designed to cure the problems of poor coverage with a single antenna that suffers on occasions from aircraft manoeuvre and antenna 'shadowing'.

The transmitter is all solid-state with its 500 W peak power output obtained from four parallel microwave transistors. Two additional transistors complete the transmitter oscillator and driver stages. The diversity system provides improved antenna coverage and allows improvement in performance in overloaded, jamming and multipath environments. Automatic overload control and anti-jamming features are also incorporated.

For aircraft that have no cockpit space available, an equipment bay-mounted configuration, RT-1157/APX-100(V), is available. In addition, the RT-1471/APX-100(V), a databus version, operating in accordance with MIL-STD-1553B, is available. There are also Night Vision Goggle (NVG) compatible units available which operate at the low-light levels needed for NVG operations. These latter models, together with top- and bottom-mounted antennas form a complete diversity transponder system.

Status

Over time, AN/APX-100(V) transponders are reported as having been installed aboard a range of US military aircraft including the AH-1S/T, AH-64, AV-8B, C-5, C-12, C-17, C-20, C-21, C-23A, C-130, CH/MH-47, E-2C, EC-130, F-14D, F-22, F/A-18, HH-60, HH-65, OH-58, OV-10, RAH-66, SH-60, T-45A, UH-60, UH/MH-60, V-22 and VC-6. In December 1993, AlliedSignal (now Honeywell) is understood to

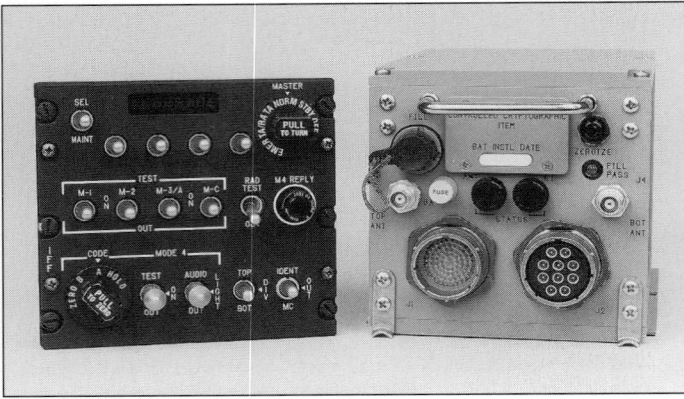

The AN/APX-100(V) IFF transponder

have delivered 113 APX-100 Mk XII transponders to Hungary for use on MiG-21 and -23 aircraft. This was followed in October 1994 by the Czech Republic's selection of the equipment for installation aboard its Mil-8 and -24 helicopters and L-159 fixed-wing aircraft. The equipment is also understood to have been produced under licence in Japan by Toyocom. On 9 August 2001, the Raytheon Company was awarded a then year US$9.5 million (maximum) modification to an existing firm, fixed-price contract (DAAB07-00-D-N406) in respect of a three year indefinite delivery/indefinite quantity type effort that covered the production of two types of A1 radio frequency module for the APX-100(V) transponder. At the time of the announcement, it was expected that a minimum quantity of 30 modules would be required in each of the three years, with overall production being capped at a maximum of 300. Work on the effort was to be carried out in Baltimore, Maryland and was scheduled for completion by 29 February 2004. The programme's contracting activity was the US Army's Communications and Electronics Command, Fort Monmouth, New Jersey.

Specifications
Frequency: 1,030 ±0.5 MHz (receiver); 1,090 ±0.5 MHz (transmitter)
Peak power output : 500 W ±3 dB
Duty cycle: 1% max
Diversity operation: full diversity capability
Dimensions: 136.5 × 136.5 × 212.7 mm
Weight: 4.53 kg

Contractor
Honeywell, Morristown, New Jersey.

UPDATED

AN/APX-101(V) Identification Friend-or-Foe (IFF) transponder

Type
D-band (1 to 2 GHz) airborne Mk XII diversity transponder for military use.

Description
The AN/APX-101(V) is a completely solid-state Mark XII IFF diversity transponder, including the 500 W transmitter and dual-tuned radio frequency receiver and

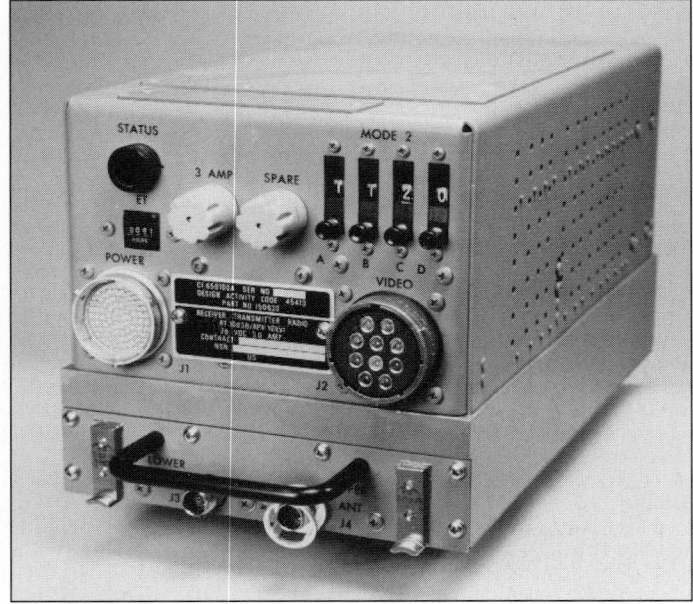

The AN/APX-101 airborne transponder

consists of replaceable modules. The transponder is housed in a single line-replaceable unit of standard dimensions and is mounted in the airframe without the use of shockmounts. Extensive use of stripline techniques and micro-electronic configurations in the transmitter and receiver has eliminated the need for space-consuming Radio Frequency (RF) coaxial cavities. Crystal-controlled pulse-width discrimination, decoding and encoding ensure positive and accurate responses to interrogations. Unique channel selection circuitry ensures positive antenna channel control for proper diversity operation.

The AN/APX-101(V) operates in Modes 1, 2, 3/A, 4 (Secure) and C. It receives RF interrogations from two antenna systems, decodes the interrogations into the proper mode, encodes the selected reply and transmits the coded RF reply through the correct antenna. The transponder operates with the aid of an air data computer for Mode C, with a transponder computer for Mode 4 and with the Type C-6280(P)/APX Control, Transponder Set.

Status
As of this edition, in excess of 11,000 AN/APX-101(V) transponders were reported as having been delivered to customers in 30 countries (including Belgium, Brazil, Israel, Taiwan, Tunisia and Turkey) around the world. Within the US military, the APX-100(V) is noted as having been installed aboard A-10, E-3A, F-5E/F, F-16, F-15 and KC-10A aircraft.

Specifications
MTBF: 1,200 h
Volume: 0.006 m³
Weight: 6.53 kg
Dimensions (w × h × d): 15.24 × 14.73 × 27.80 cm
Transmitter
Type: solid-state
Output power: 500 W (min at 1% duty cycle)
Frequency: 1,090 MHz (± 1.5 MHz stability)
Reply rate: 1,200/s (for 15 pulse-coded reply)
VSWR: no effect (antenna may be shorted or open)
Altitude: up to 30,480 m
Temperature: −54 to +70°C (operating)
Input power: 28 V DC (65 W)
Receiver
Type: 2-channel TRF
Frequency: 1,030 MHz
Bandwidth: −6 dB points (>7 MHz); −60 dB points (> ± 25 MHz)
Sensitivity: −77 dBm
Dynamic range: >50 dB

Contractor
Litton Guidance and Control Systems Division, Woodland Hills, California.

VERIFIED

AN/APX-103 Identification Friend-or-Foe (IFF) interrogator

Type
Airborne IFF interrogator.

Description
AN/APX-103 is the standard Airborne Warning And Control System (AWACS) Mark XII capable IFF interrogator and is used to selectively identify and locate civil and military aircraft when they are equipped with suitable transponders. In the military environment, the system can identify friendly aircraft while, simultaneously, providing instantaneous range, azimuth, altitude and identification data from high density target environments that are within its surveillance volume. The acquired information is used in conjunction with the AWACS AN/APY-1/2 radar to perform air traffic control and airborne early warning and control missions. The standard system configuration incorporates two redundant transceivers and a signal processor. APX-103 is configured as a digital beam split system, while the later AN/APX-103B and -103C models make use of monopulse processing and the 'advanced' target detection and code processing algorithms.

Status
As of November 2001, the APX-103 IFF interrogator family was reported as being in service aboard American E-3B/C, French E-3F, Japanese E-767, NATO E-3A, Saudi E-3A and UK E-3D AWACS aircraft. As of the given date, Telephonics noted that two APX-103 upgrades were to be deployed. In the first instance, a NATO Mid-term Upgrade programme has introduced a Mode S interrogation capability into those equipments installed aboard the Alliance's E-3As. As of January 2002, this upgrade is reported as having completed qualification and flight testing. The second effort is described as being a reliability upgrade of the APX-103's transceiver that utilises a new, high-power, solid-state transmitter, new low-noise receivers and a new low-voltage power supply. This second programme is noted as having been implemented across 'most' of the combined French, Japanese, Saudi, UK and US E-3/E-767 fleet. In terms of specifically identified contracting activity, Telephonics is known to have been awarded a then year US$9,499,117 modification to an existing APX-103 time and materials contract on 31 October 2001. Here, the baseline effort involved the provision of 123 APX-103 receiver assemblies and 123 transmitters for use aboard E-3 aircraft. At the time of the announcement, work on

the programme was scheduled for completion by the end of April 2003 and its contracting activity was the US Air Force's Warner Robins Air Logistics Center, Warner Robins Air Force Base, Georgia.

Contractor
Telephonics Corporation, Command Systems Division, Farmingdale, New York.

UPDATED

AN/APX-109(V) combined Identification Friend-or-Foe (IFF) interrogator/transponder system

Type
Combined airborne interrogator/transponder system.

Description
AN/APX-109(V) is a solid-state design that uses advanced microwave packaging techniques, surface mount devices, and complementary metal-oxide semiconductor large scale integration gate array circuits to reduce significantly the component count, weight and volume of the system.

Using digital signal processing the APX-109(V) offers: a prioritised four channel assumption of control message, a first in IFF design; adaptive thresholding of the received video to improve performance in high noise and jamming environments; and monopulse processing of received target video for enhanced azimuth accuracy.

The MIL-STD-1750A processor supports the built-in defruiter and statistical reply evaluator. It provides operational flexibility for the system and allows optimisation of system parameters based upon mission requirements. At the organisational level of maintenance, the 1750A processor supports the built-in test functions of the equipment, and allows it to report up to 97 per cent of its critical internal failures over the 1553B MUX bus. The APX-109(V) can be interfaced to either a mechanically or electronically scanned antenna.

Status
As of May 2001, the APX-109(V) combined interrogator/transponder system was reported as having been procured for installation aboard South Korean and Turkish F-16s. As of the given date, the latest identified APX-109(V) related activity was Litton's (a Northrop Grumman subsidiary) receipt of a US$19 million modification to a firm, fixed-price contract covering the supply of 20 APX-109(V)3 systems for installation aboard South Korean F-16s. At the time of the announcement, work on this Foreign Military Sales effort was scheduled for completion by the end of March 2004, with the programme's contracting activity being the US Air Force's Aeronautical Systems Center, Wright-Patterson Air Force Base, Ohio.

Specifications
AN/APX-109(V)3
Altitude: 21,336 m
Temperature: −40 to +71°C (operating)
Force air cooling: −40 to +49°C
Input power: +28 V DC (150 W long term, 220 W peak)
MTBF: 2,500 h
Weight: 15.5 kg (incl COMSEC applique)
Dimensions (w × h × d): 15.24 × 21.28 × 36.83 cm
Interrogator subsystem
Detection range: > 185 km
Coverage: ± 60° (azimuth and elevation)
Azimuth accuracy: 2°
Range resolution: 152 m
In-beam targets: 32
Number of antenna elements: 4
Bottom antenna: required for platforms with space-based co-ordinate displays

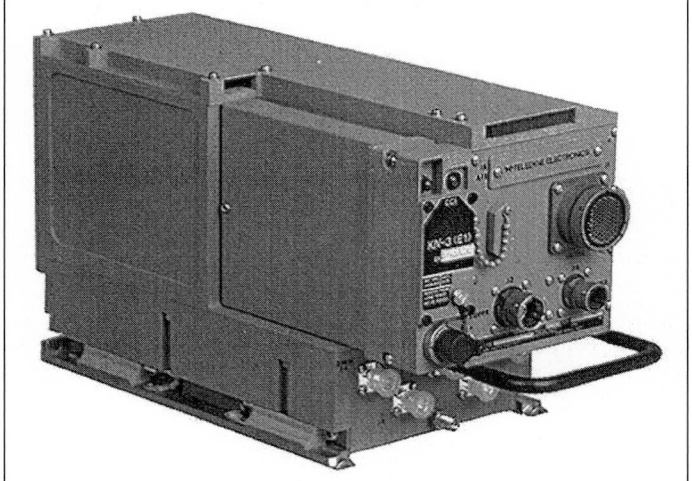

The AN/APX-109(V)3 interrogator/transponder

Receiver

	Interrogator	Transponder
Type	superheterodyne	superheterodyne
Frequency	1,090 MHz (±0.2 MHz)	1,030 MHz (±0.5 MHz)
Bandwidth	>±4 MHz (3 dB);	>±4 MHz (3 dB);
	<±25 MHz (60 dB)	<±25 MHz (60 dB)
Dynamic range	>55 dB	>55 dB
Channels	4	2

Transmitter

	Interrogator	Transponder
Type	Solid-state	Solid-state
Frequency	1,030 MHz (±0.2 MHz)	1,090 MHz (±0.5 MHz)
Output power	2,000 W (nominal)	500 W (±2 dB) at 1% DC
VSWR	1.5:1 (max)	1.5:1 (max)
Droop	<1 dB	<1 dB
Modes of operation	1, 2, 3/A, C and 4	1, 2, 3/A, C and 4
Interrogation rate	450 Hz	
Reply rate limit		1,500/s

Contractor
Litton Guidance and Control Systems Division, Woodland Hills, California.

UPDATED

AN/APX-111 combined Identification Friend-or-Foe (IFF) interrogator/transponder system

Type
Airborne IFF interrogator/transponder system.

Description
The AN/APX-111 interrogator/transponder system is one of the most recent in the family of airborne IFF products developed and produced by Marconi Aerospace Systems. It consists of the RT-1679(V) Combined Interrogator/Transponder (CIT), the C-12222(V) beam forming network, the AS-4267(V) antenna array and, possibly, two associated cryptographic computers. The packaging approach used provides space-constrained platforms with a full IFF equipment suite in a volume of 9,700 cm³. For retrofit applications, APX-III can replace as many as four to six boxes making up the existing interrogator and transponder.

APX-III's architecture enables it to support Mk X (selective identification facility), Mk XII (Mode 4) and 'custom-crypto' systems. The crypto module is an integral unit that is removable at the front panel. Mode S transponder capability and growth to next-generation IFF are also part of the system architecture.

APX-III meets international standards for IFF and air traffic control including US Department of Defense Aircraft Identification Monitoring Systems 65-1000B and STAndard NATO AGreement (STANAG) 4193. AN/APX-111 utilises miniature low-profile fuselage-mounted electronically scanned antenna arrays. It provides 185 km full interrogator range capability without the need for external amplifiers.

The design is all solid-state with a unique approach to thermally efficient cooling; a MIL-STD-1750 processor and 1553 databus are included. The latest IFF techniques are provided, including digital target reports for a clear operator display, monopulse processing for accurate target azimuth, a statistical reply evaluator for high confidence identification, a defruiter for dense environments and cryptographic coding for security. In addition, both continuous and operator-initiated built-in test is provided with reporting via the 1553 databus.

Status
As of this edition, APX-111 is understood to have been selected for retrofit to US Navy and Foreign Military Sales F/A-18 aircraft and for installation aboard new build F/A-18E/Fs.

Specifications
Dimensions/weights

Unit	h × w × d (mm)	weight (kg)
RT-1679(V)	208 × 146 × 320	15
C-12222(V)	178 × 249 × 76	5.4
AS-4267(V)*	29 × 55 × 237	0.2

* Individual elements

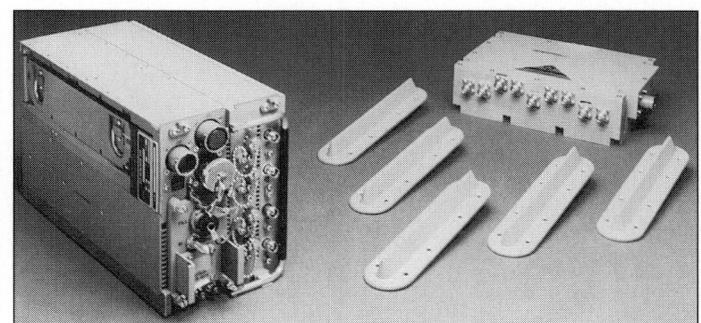

The AN/APX-111 IFF interrogator/transponder

Interrogator subsystem
Detection Range: 185 km
Sector coverage: ±60° (elevation); ±70° (azimuth)
Azimuth accuracy: ±2°
Range accuracy: 152 m
In-beam targets: 32
Modes: 1, 2, 3/A, C, 4
Power output: 1.4 kW
Transponder subsystem
Power output: 400-460 W
Receive sensitivity: −77dBm
Modes: 1, 2, 3/A, C, 4
MTBF: 1,600 h (system); 1,800 h (RT-1679(V) CIT)
MTTR: 0.25 h
Prime power: 28 V DC (115 V AC option); 180 W

Contractor
BAE Systems North America, Greenlawn, New York.

VERIFIED

AN/APX-113(V) combined Identification Friend-or-Foe (IFF) transponder-interrogator system

Type
Combined IFF transponder-interrogator system.

Description
The solid-state AN/APX-113(V) system comprises an interrogator, transponder and two cryptographic computers packaged in a single, low-volume, lightweight housing. The equipment provides Mk X (selective identification facility) and Mk XII (Mode 4) interrogator/transponder functions together with a level 3 Mode S transponder facility. The system's cryptographic module is a removable front panel appliqué unit and as a whole, APX-113(V) incorporates a 'custom-crypto' capability. The equipment utilises miniature, low-profile, fuselage-mounted, electronically scanning antennas which can be replaced, if required, by dipoles mounted on the host platform's radar antenna. Other features include: digital target reports for clear operator display; a MIL-STD-1553B databus interface; Ada software; digital video processing (as an anti-jamming measure); monopulse azimuth processing; a defruiter and statistical reply evaluation for high confidence identification. Marconi Aerospace Systems also notes that APX-113(V) is designed to be upgraded to Next-Generation IFF standard and meets STAndard NATO AGreement (STANAG) 4193 and US Department of Defense Aircraft Identification Monitoring System 65-1000B specifications.

Status
As of this edition, APX-113(V) transponder-interrogator systems have been delivered for use in the Belgian, Danish, Netherlands and Norwegian F-16 Mid-Life Upgrade (MLU) programme; Greek F-4 and F-16 Block 50 aircraft; Japanese F-2 aircraft; Taiwanese F-16s (plus other aircraft types); and UK Sea Harrier FA Mk 2, Sea King AEW Mk 7 helicopters and Nimrod MRA Mk 4 maritime patrol aircraft. In addition, as of the given edition, APX-113(V) was also mandated for installation aboard US Air Force F-16 Block 50/52 aircraft.

Specifications
Weight: 13.6 kg
Volume: 9,785 cm³
Interrogator subsystem
Detection range: 185 km
Sector coverage: ±60° (electronic scan - azimuth/elevation)
Accuracy: 152 m (range); ±2° (azimuth)
In-beam targets: 32
Modes: 1, 2, 3/A, C, 4
Power output: 2,400 W
Transponder subsystem

Power output: 400-600W
Modes: 1, 2, 3/A, 4, S
MTBF: 2,000 h

Contractor
BAE Systems North America, Greenlawn, New York.

VERIFIED

AN/GYQ-51 Advanced Tracking System (ATS)

Type
Ground-based, surveillance radar extractor/tracker.

Development
Developed by Litton Data Systems as an enhancement of previous generation extractor/tracker equipment, namely the Remote Radar Tracking System (RRTS) '77 and the Radar Beacon Digitizer (RBD) CV-3682/UPX '80. In 1982 Litton developed an advanced parallel processor and integrated it into the RBD as a prototype of the ATS. In 1984 the USAF selected the system for use with the AN/TPS-43E and AN/TPS-75 radars as part of the Module Control Equipment (MCE) AN/TYQ-23(V)2 programme. In 1987 the USAF procured the equipment for use in the Seek Skyhook aerostat radar programme at which time current nomenclature was assigned.

Description
The ATS is a stand-alone, unattended extractor/tracker. It performs automatic search and beacon target extraction, jamming detection and high-capacity adaptive tracking for virtually all pulse search radars and co-located beacons including Mode 4. The system is capable of continuous, unattended extraction and tracking in the presence of ground, sea, weather, and electronic countermeasures clutter. Sensor inputs are processed in the form of receive video or digital plot report data. Noisy area tracking is employed to provide automatic track initiation in clutter regions with low false track rates. Output data consists of beacon target data, filtered search target data, track data, strobes and status information in a variety of formats and protocols. Operator adaptation data and continuous data readout is provided by an accompanying PC-based Remote Controller. The system is noted as being able to handle up to 1,000 beacon and 5,000 search targets and up to 10,000 trial tracks per 10 second scan at ranges of up to 463 km.

Status
AN/GYQ-51 is noted as having been fielded in a variety of military and civil air surveillance applications. Over time, the USAF and US Air National Guard are thought to have procured 65 units (designated as the MCE Interface Group) to

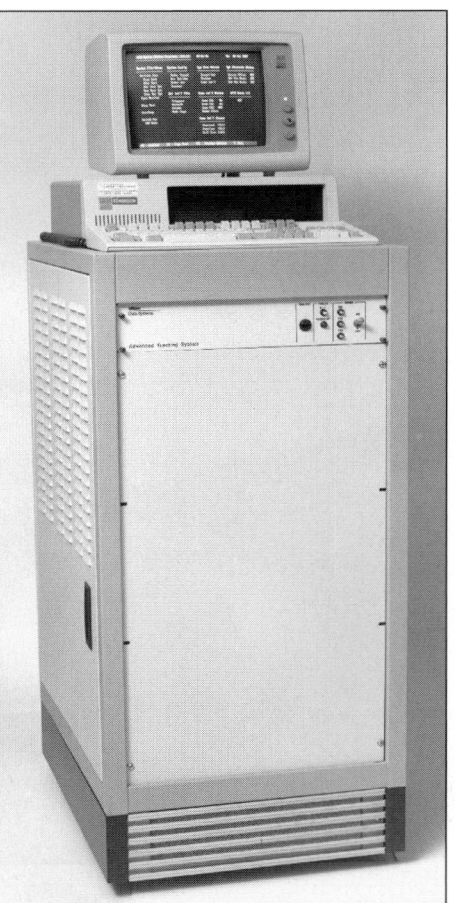

The various subsystems making up the APX-113(V) combined IFF transponder-interrogator system

The AN/GYQ-51 ATS

provide the interface between the AN/TPS-75 radar and the AN/TYQ-23(V)2 equipment. Such units are noted as having been modified to incorporate a two-way fibre optic communications link for radar/IFF control. ATS systems are also noted as having been applied to the Seek Skyhook aerostat, feeding filtered plots and track data to the North American Air Defense command and the US Customs Service command, control, communications and intelligence centre in Florida. Other ATS systems have been fitted to US Customs Service aerostats (in the southern United States and the Caribbean) and the Pave Paws radar.

Contractor

Litton Data Systems, Agoura Hills, California.

VERIFIED

AN/MPX-12A Identification Friend-or-Foe (IFF) interrogator

Type

D-band (1 to 2 GHz) weapon systems short-range IFF interrogator.

Description

AN/MPX-12A is a compact, solid-state IFF interrogator which is designed to operate and survive in the battlefield. The unit provides weapon systems with an all-weather fire capability permitting rapid engagement of enemy aircraft at maximum system range. It is particularly suitable for land-mobile systems, helicopters and small ships, where the need is for short-range identification, small size, light weight and low power consumption. The equipment includes both interrogator sidelobe suppression and receiver sidelobe suppression capabilities for good azimuth accuracy while using small antennas. The interrogator can be controlled directly from an associated fire control system computer or an associated control box.

MPX-12A consists of a receiver/transmitter, video processor, reply processor (sequential observer), timing and control, self-test, one of three modular options (Mk XII memory, coder interface or Mk X) and power supply. It has operator controls that allow mode selection, initiation of challenges and target identification.

It should be noted that when fitted with either the coder interface or Mk X modules, this equipment is known as the Model 2697 Short-Range Interrogator with the MPX-12A designation only applying to those units equipped with the Mk XII memory module. In the MPX-12A configuration, an AN/GSX-3 programmer is used to load Mk XII codes into the memory module.

Status

As of this edition, MPX-12A is reported to have been selected for use in the US Army's Forward Area Air Defense System (FAADS) programme.

Specifications

Frequency: 1,030 ± 0.2 MHz (transmitter); 1,090 ± 0.2 MHz (receiver)
Peak power output: 200 W
Duty cycle: up to 1%

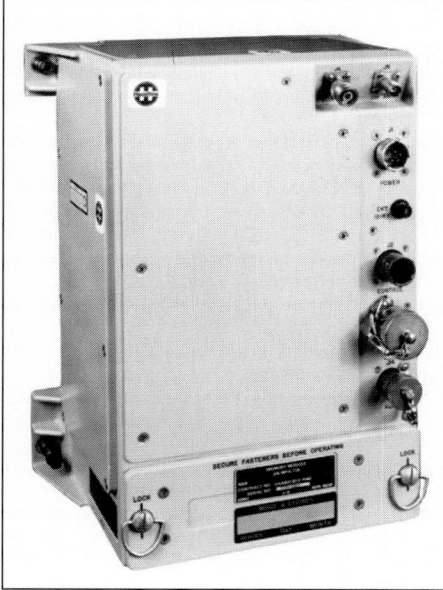

The MPX-12A short-range IFF interrogator
0007313

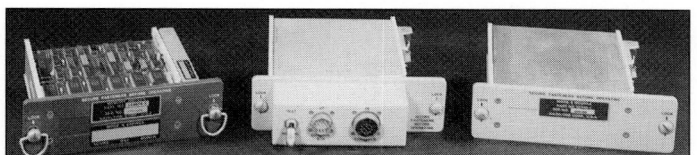

AN/MPX-12A modules showing (left to right) secure mode code storage module, coder interface module, Mk X module

---

Sensitivity: −76 dBm
Weight: 12.7 kg
Dimensions (h × w × d): 350 × 273 × 178 mm
Operating temperature: −46 to +71°C
MTBF: 6,000 h (predicted in ground mobile application at 25°C)
Power: 115 V AC (400 Hz, single phase)

Contractor

BAE Systems North America, Greenlawn, New York.

VERIFIED

AN/TPX-42(V) Secondary Surveillance Radar (SSR) system

Type

Ground-based SSR system.

Description

The AN/TPX-42(V) variants of the basic TPX-42 design now incorporate either the militarised -42(V)13 configuration (which is suitable for shipboard installations) or ruggedised upgrade packages for existing TPX-42 equipments. The upgrade kit consists of replacement displays and enhancements to the SSR system. The display upgrade kits centre on the ruggedised ADP-2000 class display. The ADP-2000 class unit combines the radar data processing and display processor functions in a single chassis and features a built-in network capabilty. The equipment provides full air traffic control facilities and can be configured in a standard embedded computer processor format or in an A/T configuration that consists of a UNIX workstation and a graphical user interface. The SSR upgrade kit incorporates the Telephonics BTE-2000 beacon target extractor as a replacement for the existing video signal processor. The TPX-42(V)13 makes use of a militarised display indicator and Telephonics beacon and search radar extractors.

Status

As of March 2001, Telephonics was reported as having produced at least 322 baseline TPX-42s for deployment at sites in the US and overseas bases. Over time, special versions of the system have been installed on US Navy aircraft carriers and basic and programmable versions of the system are noted as having been supplied to the civil/military authorities of Brazil, Bulgaria, Canada, Czech Republic and Slovakia, Iran, Netherlands, Norway, Philippines, Poland, Spain and Taiwan.

Contractor

Telephonics Corporation, Command Systems Division, Farmingdale, New York.

VERIFIED

AN/TPX-46(V) Identification Friend-or-Foe (IFF) interrogator

Type

D-band (1 to 2 GHz) weapon system IFF interrogator.

Description

The Marconi Aerospace Systems AN/TPX-46(V)1 to (V)6 interrogators are compatible with a wide range of missile systems including the HAWK, Improved HAWK and Nike-Hercules. Interface capabilities for other radar systems are incorporated. The AN/TPX-46(V)7 equipment (which is used in the Patriot missile system) is different from other versions in that the interrogator is computer-directed for single target evaluation and utilises an IFF antenna that is integrated into the main radar antenna. This IFF antenna is not supplied by Hazeltine. Components and modules from the AN/TPX-46(V)1 to (V)6 receiver/transmitter and processor are used extensively in the AN/TPX-46(V)7.

Functionally, TPX-46(V) interrogates radar targets and, when the aircraft replies correctly, provides indications of aircraft identity for display on the associated radar Plan Position Indicator (PPI). The interrogator also has the ability, when its associated radar has been disabled - except for the PPI and its power supplies - to generate its own main trigger, to rotate its own antenna producing antenna synchronising signals, and to display on the radar indicator IFF replies from suitably equipped aircraft. Even when subjected to environmental extremes, the interrogator antenna can synchronise with the radar antenna at speeds up to 25 rpm. In addition, the interrogator transmitter power can be reduced, making it suitable for use with ground control approach radar systems.

The interrogator consists of four basic operating units: the receiver/transmitter, the coder-decoder group, the control box and the antenna group. The antenna consists of an array capable of a beamwidth of 5.5° (nominal). This is achieved by means of sum and difference antenna pattern techniques providing interrogator sidelobe suppression and receiver sidelobe suppression. Until now, production of so narrow a beamwidth has required the use of arrays approximately twice as long.

AN/TPX-46(V) provides a clean display by means of the defruiting action of its processor unit. Other circuits make possible the automatic countdown of the repetition rate of the radar trigger, when necessary, to a frequency suitable for use by the interrogator set. Also provided is a gain time control circuit.

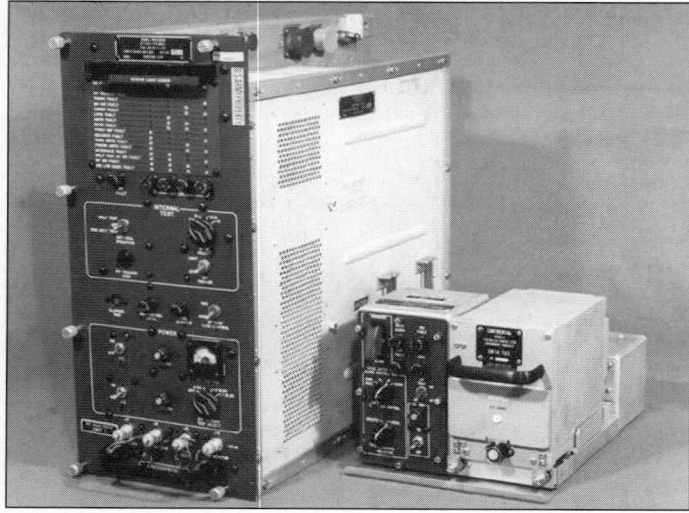

The TPX-46(V)7 IFF interrogator as used in the Patriot missile system

A solid-state transmitter module is now available as a direct replacement for the original electron tube version, as well as technology upgrade printed circuit cards, resulting in reduced maintenance costs. The AN/TPX-46(V) operates in Modes 1, 2, 3/A and C or in the Mk XII configuration on Mode 4 in conjunction with a KIR-1A/TSEC computer.

Status

Over time, TPX-46(V) is reported to have been procured for use with US Army Improved HAWK missile systems, AN/TSQ-38 radars, AN/TSQ-51 AADCPs and US Marine Corps Improved HAWK missile batteries. It is also in service with a number of foreign states. The AN/TPX-46(V)7 configuration is understood to be in service with the US Army Patriot air defence system, with at least 130 plus examples being known to have been produced.

Specifications

Frequency: 1,030 MHz (transmit); 1,090 MHz (receive)
Sensitivity: −80 dBm
Peak power: 1,000 or 2,000 W, selectable; lower power model available
Antenna: 2.23 m (long); 22.7 kg (weight - TPX-46(V)7)

Contractor

BAE Systems North America, Greenlawn, New York.

VERIFIED

AN/TPX-54(V) Identification Friend-or-Foe (IFF)/Secondary Surveillance Radar (SSR) interrogator

Type

D-band (1 to 2 GHz) ground-based and shipborne IFF/SSR interrogator.

Description

Using a modular building block approach, the all solid-state AN/TPX-54(V) provides a Mk XII IFF/SSR radar target data acquisition system which meets military and civil needs of secondary surveillance radar users, and positive friendly identification, in ground and shipborne applications. AN/TPX-54(V) is designed to military specifications and operates in up to four selective identification feature modes from Modes 1 or B, 2 or D, 3/A and C and also in Mode 4. The unit is capable of operating with up to a five mode interface when used with a KIR-1A/TSEC computer in Mode 4. The mode selection may be made remotely at an external decoder or at a remote-control box.

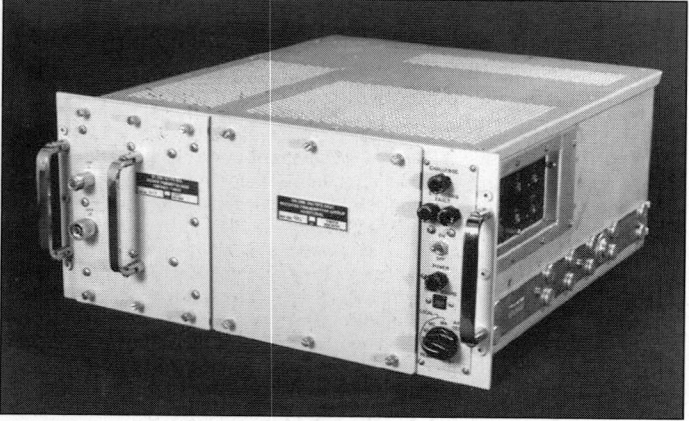

The AN/TPX-54(V) IFF/SSR interrogator

The basic system comprises five plug-in printed circuit boards which provide transmit/receive, synchronisation, coding video processing and built-in test functions. Eleven additional boards are available as options covering defruiting, computer interface, national secure operation, automatic radar trigger countdown and target data extraction (with reply correlation) capabilities. This latter feature provides digital data for the host radar's computer and allows for radar computer control of the system.

Status

Over time, TPX-54(V) has been procured by customers in the US and in excess of 21 other countries around the world.

Specifications

Frequency: 1,030 ± 0.2 MHz (transmit); 1,090 ± 0.5 MHz (receive)
Peak power: 30.5 dBW (min at antenna port)
Duty cycle: 1% max
Receiver bandwidth: −3 dB, 8 MHz (nominal)
Sensitivity: −82 dBm (min 90% decode, at antenna port)
Extractor range: 878 km (instrumented)
Altitude: up to 3,048 m (operating)
Temperature: −40°C to +71° (operating)
Capacity: 60 in-process targets (min, simultaneously)
Range accuracy: ±0.1 km
Range resolution: 0.1 km
Azimuth resolution: effective beamwidth plus 7 pulse recurrence periods, all modes responding
MTBF: 10,200 h
MTTR: 15 min
Dimensions (w × h × d): 48.3 × 22.9 × 53.3 cm
Weight: 27 kg
Power input: 115 or 230 V AC (47 - 440 Hz)

Contractor

BAE Systems North America, Greenlawn, New York.

VERIFIED

AN/UPX-24(V) Identification Friend-or-Foe (IFF) processor

Type

Shipborne IFF processor.

Description

The AN/UPX-24(V) is described as being the 'core' IFF processor within the AN/UPX-29(V) IFF interrogator system that is installed aboard US Navy (USN) 'Arleigh Burke', 'Ticonderoga' and 'Wasp' class surface combatants. As such, the equipment's main function is the interrogation and identification of surface platforms and aircraft that are equipped with Selective Identification Feature (SIF) Modes 1, 2, 3A/C and 4 (specific US release only) transponders. UPX-24(V) is compatible with radar rotation rates of up to 60 rpm; is made up of a central processor controller, a remote control monitor and up to 22 remote control indicators; operates with the OE-120/UPX electronically steered antenna array (which can be re-directed to interrogate a target at any azimuth within less than 50 microseconds) and can be configured to operate with a mechanically rotating antenna if required.

Functionally, the UPX-24(V)'s processor controller receives mode selection and interrogation commands from both operators (using its remote control indicators) and via a Navy Tactical Data System (NTDS) MILitary STanDard (MIL-STD) - 1397 combat system interface. Working from these inputs, the processor controller generates steering commands for the OE-120/UPX antenna and interrogation commands to the UPX-29(V) interrogator. In turn, the unit receives IFF video and mode tags from the UPX-29(V) and performs target detection, decoding, code validation, defruiting and degarbling, using established criteria for target start, continuation and end, verification and code validation. For its part, the UPX-24(V) control monitor allows the user to define up to three manually entered azimuth sectors with individual control of interrogation frequencies, pulse repetition frequencies and radio frequency power. It also provides remote alarms, processor controller master reset, NTDS maximum/minimum range control and Mode 4 control and encryption zeroise.

The UPX-24(V) remote control indicators are co-located with the host vessel's plan position indicators or NTDS consoles. Functionally, they enable the console operators to interrogate selected modes and request active and 'pop-up' (see following) evaluation of both SIF and Mode 4 targets. They also allow passive decode of operator-selected targets, thereby providing unique target of operator interest symbology for display. For its part, the shipboard weapon system can request modes of interrogation and all target information in specified sectors (or full scans), test targets at specified ranges and azimuths and display 'pop-up' target information in a range-azimuth window that is centered at a specified point. UPX-24(V) provides target reports/replies for the host combat data system and has three operating modes, as follows:

Pop-up

The UPX-24(V)'s pop-up mode (which requires the availability of the OE-120/UPX antenna) responds to a identification request by immediately re-directing the antenna to interrogate an area of interest, thereby permitting immediate acquisition of IFF data relative to a specified target or area.

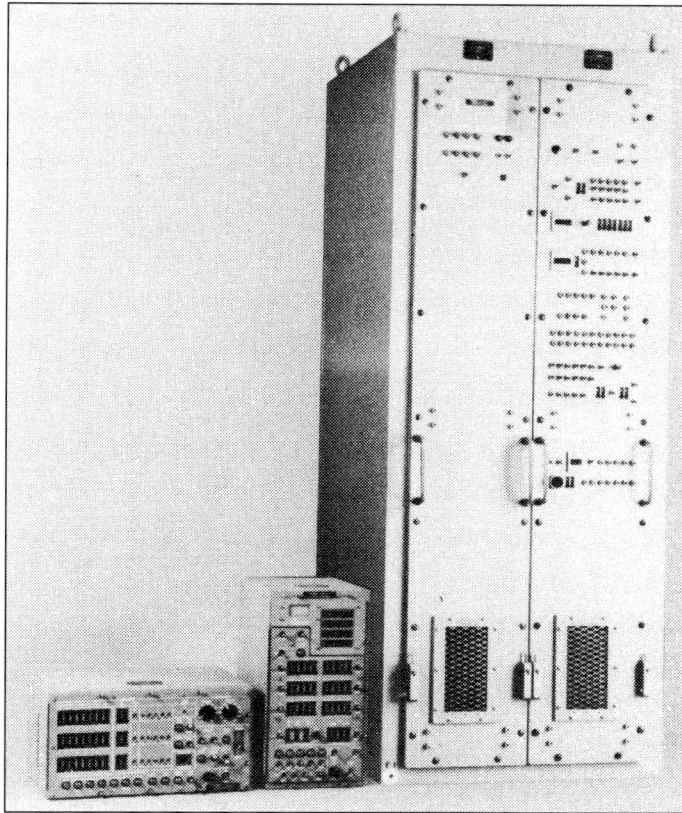

The AN/UPX-24(V) shipborne IFF system

Active
In active mode, UPX-24(V) responds to an interrogation request by gathering data when its associated antenna (electronically or mechanically scanned) is pointing in the direction of interest.

Passive
In passive mode, UPX-24(V) draws 360° scan data from its memory.
Other system features include:

- 40 k steady-state and 100 k peak defruiters
- the ability to process up to 100 targets simultaneously during a single sweep
- a jittered pulse repetition time capability
- separate processing files for active and 'pop-up' processing
- full sweep and window-gated Mode 4 evaluation in override and interlace modes
- detection and recombination of target splits in range and/or azimuth
- effective separation of targets that have the same range and overlap in azimuth
- indication-of-position and military/civilian emergency detection
- a distributed processing architecture
- compatibility with MIL-STD-167, -461, -740, -901 and -16400.

Specifications
Primary power: 115 V AC (10%, 60-40 Hz, 1,500 W max)
Weight: 544 kg (max)
Dimensions (H × W × D): 1.8 × 0.9 × 0.7 m

Status
As of November 2001, the AN/UPX-24(V) IFF system was reported as having been installed aboard a range of American ('Ticonderoga' class cruisers, 'Arleigh Burke' class destroyers and 'Wasp' class amphibious assault ships) and Japanese warships. Engineering change proposals are also noted as having been funded in order to improve the system's display processor, operator interface, data throughput and radar type compatibility. Identified UPX-24(V) contracting activity during 2000/2001 is as follows:

9 June 2000
Litton Systems was awarded a then year US$17,099,103 firm, fixed-price contract covering the procurement of 10 UPX-24(V) IFF systems, four system retrofit kits, seven 'installation and checkout spares' and associated technical data. At the time of the award's announcement, the effort was scheduled for completion by the end of May 2002 and its contracting activity was the US Navy's (USN) Naval Air Systems Command, Patuxent River, Maryland.

15 November 2001
Litton Integrated Systems (Northridge, California) was awarded a then year US$6,230,000 firm, fixed-price contract covering the US Fiscal Year 2001 procurement of four UPX-24(V) IFF interrogators. At the time of the announcement, work on the effort was to be split between the company's facilities at San Diego, California (85 per cent work share) and Northridge (15 per cent work share) and was scheduled for completion by the end of June 2003. The programme's contracting activity was the USN's Naval Air Systems Command, Patuxent River, Maryland.

Contractor
Litton Integrated Systems (a part of Northrop Grumman Electronic Systems' Navigation Systems Division), Northridge, California.

UPDATED

AN/UPX-30(V) Identification Friend-or-Foe (IFF) system

Type
Shipborne IFF system.

Description
The AN/UPX-30(V) is a central IFF equipment processor with remote-control displays for sensor information. The design is based on a standard VME modular data fusion system that is capable of interfacing the C and D computer with three MIL-STD-1397 data control interfaces, two Mk XII IFF interrogators, six shipboard primary radars, 29 command and control displays, two SARTIS UPX-34, one SLQ-20 upgrade, and an optional Doctrine Processor.

The CIFF/AutoID system associates real-time targets' IDentification (ID) information and provides an automated quick reaction, high-confidence ID assessment of radar and IFF tracks through the use of multisensor integration and optional ID Doctrine processing. The system also operates with the ACDS Block 0 or Block 1 combat systems. The CIFF/AutoID system has comprehensive built-in-test for fault detection. The system is intended to replace the AN/UPX-24 on AEGIS class ships.

Status
During 1992, the US Navy was reported as having awarded the then AlliedSignal (now Honeywell) a design and production contract that covered the supply of three engineering development models of the UPX-30(V) system. Intended for use aboard the US Navy's 'Arleigh Burke' class destroyers, UPX-30(V) is further noted as having undergone a critical design hardware review during November 1994.

Contractor
Honeywell, Morristown, New Jersey.

VERIFIED

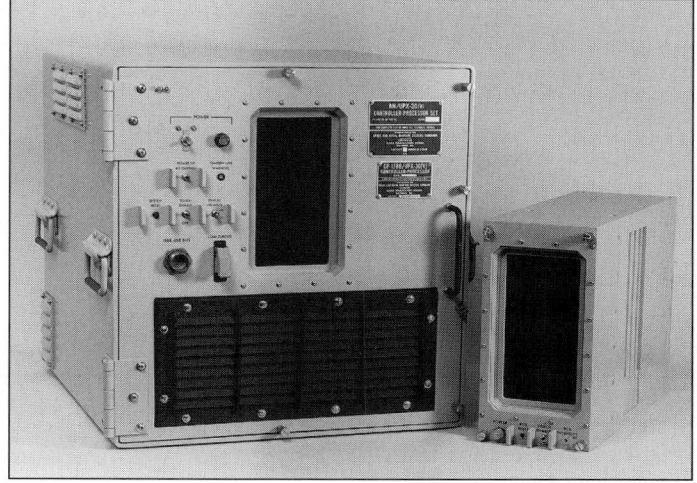

The AN/UPX-30(V) IFF system

AN/UPX-37 Identification Friend-or-Foe (IFF) interrogator

Type
D-band (1 to 2 GHz) digital IFF interrogator.

Description
The AN/UPX-37 digital IFF interrogator is an open architecture, VME-based modular equipment that is designed for naval, land-based air defence, airborne surveillance and air traffic control applications. It is used for Mk XII and Next Generation IFF processing (including Modes S and 5) and operates either autonomously or in conjunction with a host radar. The system conforms to US Department of Defense, NATO, International Civil Aviation Organisation and US Federal Aviation Administration requirements. UPX-37's modular/digital architecture allows for customised configurations and performance optimisation for 'most' applications. Digital target reports can be provided in addition to wideband video for subsequent passive/active decoding. The equipment also provides amplitude monopulse functionality for 'significant' azimuth accuracy improvement over conventional interrogators.

Status
Over time, the UPX-37 IFF interrogator has been procured by the US Navy as a replacement for the AN/UPX-27 system.

Specifications
Frequency: 1,030 ± 0.01 MHz (transmit); 1,090 ± 0.5 MHz (receive)
Peak power: 33 dBW (1 transmitter module, output at antenna ports, adjustable by −9 dB in 3 dB steps); 36 dBW (2 transmitter modules, output at antenna ports, adjustable by −9 dB in 3 dB steps)

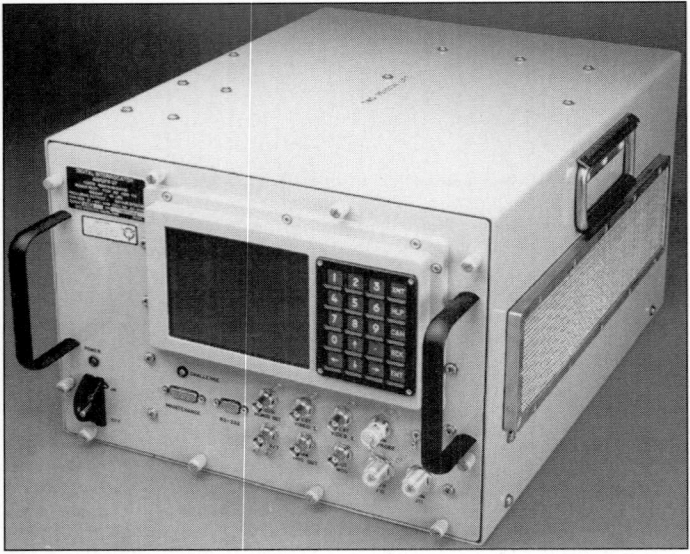

The AN/UPX-37 digital IFF interrogator **2000**/0081430

Duty cycle: 2% (max)
Receiver bandwidth: −3 dB, 8 MHz (nominal)
Sensitivity: −84 dBm (min, 90% decode, measured at antenna port)
Extractor instrumented range: >556 km
Altitude: up to 3,048 m
Temperature: −28°C to +65°C (operating)
Capacity: 60 in-beam targets, 1,000 targets per scan
Range accuracy: 1,152 m
Resolution: effective beamwidth plus 7 RPRs (azimuth, all modes responding); 1,152 m (range)
MTBF: >5,000 h
MTTR: 19 min
Power input: 115/230 V AC (<100 V A, 47 to 440 Hz)
Weight: 36.3 kg
Dimensions (w × h × d): 37.5 × 26.8 × 45.7 cm

Contractor
BAE Systems North America, Greenlawn, New York.

VERIFIED

AT100 radar transponder

Type
G-band (5.4 to 5.9 GHz sub-band) radar transponder.

Description
AT100 is a miniature precision radar augmentation device that operates within the 5.4 to 5.9 GHz frequency band and is designed to enhance the tracking capabilities of G/H-band radars. Utilised primarily as a range safety tool, AT100 is described as being suitable for installation aboard manned and unmanned vehicles, a range that includes aircraft, missiles and air- and seaborne target drones. The device is noted as being all solid-state with the exception of its triode oscillator. Other system features include:
- a 'long life' triode cavity transmitter
- a single antenna connector
- open and short circuit protection
- reverse polarity protection
- adjustable code and delay selection.

Status
As of this edition, the AT100 radar transponder is understood to be available.

Specifications
Transmitter
Frequency range: 5.4-5.9 GHz (factory set, field adjustable)
Frequency stability/selection: ± 3 MHz (± 50 kHz/°C)/continuous manual screw
Modulation type: AM pulse
Output device/power: triode cavity oscillator/20-100 W (factory set)
Pulse-width/pulse-width jitter: 0.5 ± 0.1 μs/0.01 μs
Pulse rise/fall time: 0.1 μs/0.2 μs
Power spectrum: < 3 MHz/0.5 μs at ¼dB points
PRF: up to 2,600 pps or pgps
Receiver
Frequency range/selection: 5.4-5.9 GHz (factory set, field adjustable)/screw adjustable poles
Frequency stability: ± 5 MHz

Sensitivity: −65 dBm (min)
Dynamic range: −65 dBm to 20 dBm at 99% reply
Bandwidth: 11 ± 3 MHz (3 dB level)
Pulse decoder: single or double
Pulse-width: 0.25-5 μs (single and double, factory set)
Double/second pulse spacing: 3-12 μs (factory set, field adjustable)/± 0.15 μs (accepts), ± 0.3 μs (rejects)
Random triggering: 10 pps (max)
Mechanical/electrical/environmental
Temperature: −40°C to +80°C (operating); −65°C to +95°C (non-operating)
Altitude: sea level to 70,104 m
Input voltage: 24-32 V DC
Power consumption: 0.25 A at 28 V DC, 1,000 pps
Volume: 0.000174 m³
Weight: 0.34 kg
Dimensions: 9.6 × 5.7 × 3.2 cm

Contractor
L-3 Telemetry-East, Newtown, Pennsylvania.

VERIFIED

AT1000 radar transponder

Type
G-band (5.4 to 5.9 GHz sub-band) radar transponder.

Description
AT1000 is a miniature, precision radar augmentation device that operates within the 5.4 to 5.9 GHz frequency range and is designed to enhance the tracking capabilities of G/H-band radars. Utilised primarily as a range safety tool, AT1000 is described as being suitable for installation aboard manned and unmanned vehicles, a range that includes aircraft, missiles and air- and seaborne target drones. The device is noted as being all solid-state with the exception of its triode oscillator. Other system features include:
- a 'long-life' triode cavity transmitter
- single or double pulse interrogations
- a single antenna connector
- open and short circuit protection
- reverse polarity protection
- adjustable code and delay selection.

Status
As of this edition, the AT1000 radar transponder is understood to be available and as having been procured.

Specifications
Transmitter
Frequency range: 5.4-5.9 GHz (factory set, field adjustable)
Frequency stability/selection: ± 3 MHz (± 50 kHz/°C)/continuous manual screw
Modulation type: AM pulse
Output device/power: triode cavity oscillator/100-500 W (factory set)
Pulse-width/pulse-width jitter: 0.5 ± 0.1 μs/0.01 μs
Pulse rise/fall time: 0.1 μs/0.2 μs
Power spectrum: < 3 MHz/0.5 μs at ¼dB points
PRF: up to 2,600 pps or pgps
Delay variation/jitter: ± 0.1 μs (max from −65 dBm to +20 dBm)/± 0.03 μs (max from −65 dBm to +20 dBm)
Receiver
Type: TRF
Frequency range/selection: 5.4-5.9 GHz (factory set, field adjustable)/screw adjustable poles
Frequency stability: ± 5 MHz
Sensitivity: −65 dBm (min)
Dynamic range: −65 dBm to 20 dBm at 99% reply
Bandwidth: 11 ± 3 MHz (3 dB level)
Pulse decoder: single or double
Pulse-width: 0.25-5 μs (single and double, factory set)
Double/second pulse spacing: 3-12 μs (factory set, field adjustable)/± 0.15 μs (accepts), ± 0.3 μs (rejects)
Random triggering: 10 pps (max)
Mechanical/electrical/environmental
Temperature: −40°C to +80°C (operating); −65°C to +95°C (non-operating)
Altitude: sea level to 70,104 m
Input voltage: 24-32 V DC
Power consumption: 0.25 A at 28 V DC, 1,000 pps
Volume: 0.000208 m³
Weight: 0.3969 kg
Dimensions: 9.6 × 5.7 × 3.8 cm

Contractor
L-3 Telemetry-East, Newtown, Pennsylvania.

VERIFIED

AV12X radar transponder

Type
I-band (9 to 9.5 GHz sub-band) radar transponder.

Description
AV12X is a miniature, precision radar augmentation device that is designed to augment the tracking capabilities of radars operating in the 8 to 12 GHz frequency band. Utilised primarily for range safety functions, AV12X is described as being suitable for use aboard a range of manned and unmanned vehicles that includes aircraft, missiles and air- and seaborne target drones. Other system features include:

- a 'long-life' Gunn cavity transmitter
- single and double pulse interrogations
- a single antenna connection
- open and short circuit protection
- reverse polarity protection
- adjustable code and delay selection.

Status
As of this edition, the AV12X radar transponder is understood to be available.

Specifications
Transmitter
Frequency range/selection: 9-9.5 GHz/continuous mechanical by single screw
Frequency stability: ± 5 MHz (all conditions)
Modulation type: AM pulse
Output device/power: Gunn cavity oscillator
Pulse-width/jitter: 0.4 ± 0.15 μs /0.01 μs (max)
Pulse rise/fall time: 0.1 μs (max)/0.2 μs (max)
PRF: up to 2,500 pps
Reply delay/variation: 1-6 μs/0.05 μs (max for signal levels between 0 and 45 dBm)
Delay jitter: 0.02 μs (max 0 to −46 dBm)
Fixed delay: 3-12 μs (internally adjustable)
Receiver
Type: direct (TRF)
Frequency range/selection: 9-9.5 GHz/continuous mechanical, 2 cavity preselector
Sensitivity: -46 dBm (min)
Dynamic range: 99% (min for +20 dBm to −46 dBm input levels)
Stability: ± 3 MHz
Bandwidth: 26 ± 7 MHz
Pulse decoder: single or double (internally selectable)
Pulse width: 0.25-1 μs (double pulse); 0.25-5 μs (single pulse)
Random triggering: 10 pps (max)
Double pulse decoder: 3-12 μs (adjustable)
Decoder accept/reject limits: ± 0.5 μs (accept); ± 0.8 μs (reject)
Mechanical/electrical/environmental
Temperature: −35°C to +60°C (operating); −40°C to +75°C (non operating)
Altitude: sea level to 45,720 m
Input voltage: 22-32 V DC (return connected to ground)
Power consumption: 0.2 A nominal at 28 V DC, 1,000 pps
Weight: 0.51 kg (nominal)
Dimensions: 8.7 × 9 × 3.5 cm

Contractor
L-3 Telemetry-East, Newtown, Pennsylvania.

VERIFIED

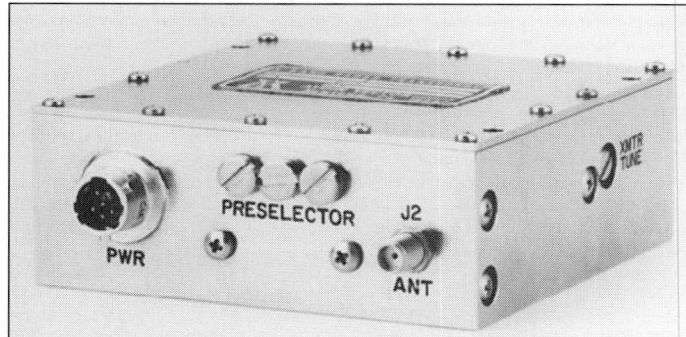

The 9 to 9.5 GHz frequency band AV12X radar transponder

Battlefield Combat Identification Systems (BCIS)

Type
All-weather, question and answer battlefield identification system.

Description
BCIS is an all-weather battlefield identification system that operates in the 38 GHz frequency band and comprises a universal transponder unit and an interrogator

The BCIS interrogator antenna (rectangular unit, lower right) as fitted to a Bradley Fighting Vehicle

that is fitted to 'shooter' systems such as tanks. Functionally, the operator initiates the identification process by activating the BCIS interrogator that, in turn, transmits a query (made up of five individual pulses) via a narrowbeam, directional antenna. The transponder mounted in the target vehicle receives the interrogation signal (via an omnidirectional antenna), validates it and if valid, responds with its own identification signal together with that of its interrogator (again, transmitted in the form of a string of five pulses). At the same time, the unit informs its host vehicle's crew that it has been interrogated. The particular BCIS interrogator receives the incoming response, validates it, displays the results (as a 'friend' or 'foe' icon) in its host vehicle's sighting subsystem and generates an appropriate voice cue in its crew's headphones. Overall system response time is approximately a half to one second and the response signal is time synchronised using Global Positioning System (GPS) clock data. A BCIS digital datalink add-on is also understood to have been developed. Here, a software enhancement allows the system to use its existing GPS input to create geographic co-ordinates that can be added to a platform-unique identification code for onward transmission via datalink. Such an approach allows the BCIS system to be expanded into a situational awareness tool. Trials of this add-on have been undertaken (using a High-Mobility Multipurpose Wheeled Vehicle (HMMWV) and an Abrams battle tank analogue) and the package

The BCIS symbology (circle, lower right) indicating a 'friend' in the weapon sight of a Bradley Fighting Vehicle

has been found to be able to hand-off data at ranges of up to 1.2 km, with the transmitted information being displayed on laptop computers in both vehicles. Trials on airborne applications (see Status) are reported to have demonstrated ranges in excess of 8 km.

Status

Jane's sources suggest that TRW's Space and Electronic Group was awarded a US$16 million contract covering the design and manufacture of 46 BCIS systems in August 1993. In the following November, the effort was reported to have been put on hold for budgetary reasons. Following programme restructuring (which may have included TRW teaming with Raytheon and Electromagnetic Sciences Inc rather than its original BCIS partner Magnavox), TRW is understood to have delivered prototype BCIS ship sets to the US Army during October 1994. As of this edition, BCIS is understood to have been installed on a range of platforms (including the M1A1 and M1A2 Abrams main battle tanks, the M2A1 and M2A2 Bradley infantry fighting vehicles, an M7 Bradley Fire Support Team (BFIST) vehicle, an M9 ACE armoured earthmover, an M93 Fox nuclear, biological and chemical reconnaissance vehicle, an M113 armoured personnel carrier, an M981 Fire Integrated Support Team Vehicle (FISTV), an HMMWV, an AV-8B close-support aircraft and an AH-64 battlefield attack helicopter) for trials. BCIS was the US entrant in the four nation (France, Germany, the UK and the US) NATO battlefield identification demonstration that took place in Germany during May 1997 and has demonstrated waveform compatibility with the French Battlefield Identification Friend-or-Foe (BIFF) system (itself based on Thomson-CSF's (now Thales) *Dispositif d'Identification au Combat* demonstrator). As of August 2000, Jane's sources are suggesting that TRW had delivered 115 BCIS systems to the US Army, 111 of which were in engineering and manufacturing development configuration, with the remaining four having been built to Initial Operational Test and Evaluation (IOT&E) standard. Here, the IOT&E sets are described as having featured 'improved' waveforms and 'enhanced' frequency and digital electronics. The same sources go on to report that BCIS low-rate initial production was sanctioned during the third quarter of 2000, with the first of up to 952 production units (646 units over three years, with a further 306 being on a fourth year option) being scheduled for delivery to the US Army on 1 May 2001. As of the given date, the first unit to receive BCIS was mandated as the US Army's 4th Infantry Division (Mechanised).

Specifications

Frequency: 38 GHz
Range: 7 km (dust, demonstrated); 7.3 km (4 mm of rain/h, demonstrated); >8 km (air-to-ground, demonstrated); 15 km (clear sky, demonstrated)
Probability of correct ID: >99% (demonstrated)

Contractor

TRW Space and Electronics Group (prime), Redondo Beach, California.

UPDATED

Northrop Grumman Monopulse Secondary Surveillance Radar (MSSR) system

Type

Monopulse interrogator/receiver subsystem.

Description

The Northrop Grumman MSSR is available as a monopulse interrogator/receiver subsystem of the Federal Aviation Agency's Mode S system. It is the only SSR designed from the beginning for risk-free upgrading to full Mode S capabilities, including selective identification and datalink. Monopulse processing provides aircraft detection and tracking with up to 10 times the accuracy of current Air Traffic Control Radar Beacon System (ATCRBS)/SSR systems. Because the MSSR system can vary its transmit power and receiver sensitivity, it performs equally well

in terminal and en route applications. It can be co-located with any present or future primary radar and is fully compatible with current ATCRBS, Mk XII selective identification feature/IFF and Mode 4 systems. MSSR is billed as being a 'highly reliable', with a single channel system mean time between failure value of 2,000 hours. It is dual channel and operates in a simplex mode. If failure occurs in the operating channel, switch over to the back-up channel is automatic, ensuring safe, non-interrupted operations at all times.

Mode S Beacon System Sensor

The Mode Select system is replacing the ATCRBS currently in use throughout the United States. With a range coverage of over 450 km, the Mode S transmitter/modulator interrogates each aircraft individually, eliminating transponder interference from closely spaced aircraft and enabling higher air traffic densities. A multichannel monopulse receiver provides extremely accurate aircraft position data, even when used with existing aircraft transponders. The monopulse subsystem described above is available separately, permitting phased acquisition of the Mode S system. For aircraft equipped with Mode S transponders, an integral two-way datalink enables pilots to request and receive weather and other data without using ATC voice channels.

Status

As of this edition, the Mode S Beacon System Sensor is reported to have been selected for use by Aruba, India and the US. Northrop Grumman MSSRs have been procured by Belgium, Bosnia-Herzegovina, the People's Republic of China, Egypt, El Salvador Georgia, Mexico, Morocco, Panama, Peru, Poland, Saudi Arabia, Taiwan and Tunisia.

Contractors

Northrop Grumman/Lockheed Martin Joint Venture, Baltimore, Maryland (Mode S).
Northrop Grumman Corporation, Electronic Sensors and Systems Sector, Baltimore, Maryland (MSSR).

VERIFIED

OX-60/FPS-117 Identification Friend-or-Foe (IFF) interrogator

Type

D-band (1 to 2 GHz) IFF interrogator for the AN/FPS-117 radar.

Description

The Marconi Aerospace Systems IFF interrogator OX-60/FPS-117 is the beacon interrogator group for the AN/FPS-117 minimally attended radar. The OX-60 is a 370 km system with receiver sidelobe suppression and interrogator sidelobe suppression. It is capable of interlacing Modes 2, 3/A, C, and 4 and parallel processing the replies. The system is all solid state and to meet the requirements of a minimally attended site, it is designed in a redundant, automatic switch over configuration. The equipment consists of a control processor, two processors, signal data and two receiver/transmitters which provide a mean time between failure value of in excess of 118,000 hours. The automatic fault test and isolation is extensive and under the control of an 8086 microprocessor. Self-test of the beacon system is carried out on a continuous basis and consists of both a local circuits self-check as well as a full test accomplished by the generation of simulated targets. This function occurs in parallel to normal operation and is completely transparent to it.

Once a fault has been detected the microprocessor goes into a fault-isolation function to isolate the fault automatically to a single or group line-replaceable unit. A status message is then passed to the computer indicating which module is at fault and needs to be replaced. This automatic fault isolation reduces mean time to repair and simplifies maintenance procedures.

Status

As of this edition, OX-60 is understood to have been procured for use with the FPS-117 radars deployed as part of the US/Canadian North American air defence system, as well as those operated by a number of international customers.

A Mode S beacon sensor array mounted on an ASR-9 airport surveillance radar

OX-60/FPS-117 is the beacon interrogator subsystem for the AN/FPS-117 surveillance radar

Specifications

Frequency: 1,030 ± 0.2 MHz (transmitter); 1,090 ± 3 MHz (receiver)
Sensitivity: −83 dBm decoding
Peak transmitter power: 2 kW
Duty cycle: 1%
Range: 370 km
Effective beamwidth: 2.75-0.5°
Power input: 115/200 V AC 50-400 Hz, 3-phase. Single channel 350 W, dual channel 700 W
Specification: designed to MIL-E-4158

MTBF: 118,765 h (redundant); 2,969 h (single channel)
MTTR: 20 min
Dimensions: 482 × 266.5 × 520.5 mm (signal processor); 482 × 222 × 609.4 mm (receiver/transmitter); 482 × 133.3 × 381 mm (control unit)
Weights: 29.7 kg (signal processor); 16 kg (receiver/transmitter); 12.3 kg (control unit)

Contractor

BAE Systems North America, Greenlawn, New York.

VERIFIED

MILITARY AIR TRAFFIC CONTROL (ATC), INSTRUMENTATION AND RANGING RADARS

This section includes ATC surveillance, approach and landing radars where used primarily for military purposes. It does not cover those radars that are used mainly in the civil ATC networks. However, it should be remembered that many countries use their ATC radars for military surveillance as well and in some cases have integrated these systems with the national air defence network to provide an overall air defence surveillance capability. This section also covers instrumentation and test range radars, both land-based and shipborne, and meteorological radars used in military applications.

VERIFIED

BRAZIL

RMT 0100A meteorological radar

Type
E-band (2 to 3 GHz) meteorological/multipurpose radar.

Description
The RMT 0100A operates in E-band to detect weather/precipitation over 400 km, and pluviometric capacity over 150 km. Applications of this radar include meteorological research, short-term weather forecasting, river basin control, agricultural planning, civil defence and air traffic protection.

Structured to be used in a meteorological information network, the system has a master operations deck, remote or local, through which the radar is operated. It can be connected to a meteorological data processing system with larger capacity, or through a network to a remote viewing station. Information can be presented in a number of formats with the option being menu selected.

The RMT 0100A includes a powerful digital signal processor and uses commercial microprocessors in both the master operations deck and the remote viewing station. Given intrinsic flexibility in hardware and software, the RMT 0100A can be adapted to a wide variety of systems to match the user's operational requirements.

Status
As of this edition, RMT 0100A is understood to have been procured.

Specifications
Frequency: E-band (2.7-2.9 GHz)
Antenna diameter: 3.7 m
Beamwidth: 2°
Output power: 600 kW
Pulsewidth: 2 µs
PRF: 250 Hz
Interface: RS-232C, 9,600 b/s

Contractor
TECTELCOM, São José dos Campos.

VERIFIED

Antenna assembly of the RMT 0100A radar (Ronaldo S Olive)

BULGARIA

ROZA meteorological system

Type
Mobile radar-based meteorological system.

Description
ROZA is a self-contained, computer-based, mobile meteorological system and is designed to measure actual/relative wind speed/direction in the near-ground atmosphere. It employs the technique of automated radar tracking of a balloon carrying a corner reflector. The information obtained is available for use by missile, artillery and aviation units. The complete system is mounted on a ZIL-131 wheeled vehicle, and consists of a radar assembly, a special computer unit with a digital printing device, a portable optical unit, a hydrogen generator, communications equipment, power supply and non-designated special equipment. The instrument cabin contains the receiver/transmitter unit and accessory equipment.

Status
As of this edition, ROZA is understood to have been procured by the Bulgarian Army.

Specifications
Sounding height: 5 km (without extrapolation); 12 km (with extrapolation)
Number of layers: 14
Type of meteobulletin: accurate (with round off; up to 0.1 m/s (speed); up to 00-10 mils (direction)); coarse (with round off; up to 0.1 m/s (speed); 0.1 to 10 mils (direction)
Max slope position: 6°
Set up time: not more than 3 min (3-man crew)
Power consumption: 2 kW
Operating temperature range: −40 to +50°C
Power supply: 220 V (self-contained)
Max tracking distance: 20 km
Average error: 0.7 m/s (velocity); 00-50 mils (direction)
Sounding cycle: 20 min
MTBF: 600 h
Weight: 10,500 kg

Contractor
Kintex, Sofia.

VERIFIED

The ROZA meteorological system   0007315

CHINA, PEOPLE'S REPUBLIC

HN-C03-M precision instrumentation radar

Type
G-band (4 to 6 GHz) mobile precision instrumentation radar.

Description
A vehicle-mounted precision instrumentation radar, HN-C03-M is a G-band long-range, high-accuracy measuring equipment. The system consists of an antenna vehicle, an electronics shelter vehicle, a power supply vehicle and a calibration vehicle. It uses microcomputer circuits to process radar data in real time, and the multifunction console can be operated by one person. Short-range and low-elevation targets are also tracked by TV so that low-elevation tracking performance is improved.

Status
As of this edition, the status of the HN-CO3-M instrumentation radar was uncertain.

Specifications
Frequency: G-band (5.5-5.7 GHz)
Range: 300 km for reflecting target
Transmitter
Power: 1 MW (peak)
Pulsewidth: 0.8 or 1.7 µs
PRF: 585.5 Hz
Tracking accuracy: 0.2 mil (azimuth and elevation); 5 m (range)

Contractor
Jiangsu Huaning Electronics Corporation, Nanjing.

VERIFIED

HN-C03-M mobile instrumentation radar

....

Model 793 Precision Approach Radar (PAR) system

Type
E/F- (2 to 4 GHz) and I-band (8 to 10 GHz) PAR system for military and civil applications.

Description
The Model 793 radar is an air surveillance and ground-controlled PAR system that can be supplied in both static and mobile versions. It includes an E/F-band surveillance radar and an I-band PAR to provide an independent airport terminal area surveillance and approach system. The system is equipped with VHF (30 to 300 MHz) and UHF (0.3 to 1 GHz) direction-finding equipment.
   Features of the precision approach radar include:
• rejection of weather clutter by using variable circular polarisation techniques and log IF amplifier
• Digital Moving Target Indicator (DMTI) techniques to eliminate ground clutter and asynchronous interference
• with the advanced digital techniques in deviation measurement of range and bearing, the radar can automatically track a single aircraft, give a dynamic display of range, altitude and altitude deviation of the aircraft

Antenna of the Model 793 PAR system

   Features of the surveillance radar include:
• with its hyperboloidal reflector, the antenna dual-beam arrangement and variable circular polarisation the radar is able to reject close ground clutter and weather interference
• DMTI technology eliminates ground clutter and asynchronous interference
• a low-noise parametric amplifier in the receiver
• a digital dual-pole feedback video accumulative technique to get the echoes with full content and a clear display picture.

Status
Over time, the Model 793 PAR is understood to have been procured for both civil and military applications within the Chinese mainland.

Specifications
PAR
Coverage: 35 km for fighter-type aircraft in clear weather, 15 km in moderate rain
Azimuth accuracy: not more than ±0.5° but not less than 12 m
Elevation accuracy: not more than ±0.35° but not less than 7 m
Range accuracy: not more than 2.5% of range but not less than 60 m
Surveillance
Range: 100 km for small aircraft in normal weather
Azimuth coverage: 360°
Vertical coverage: 40° cosec^2
Max altitude: 10,000 m
Azimuth accuracy: ±1.5°
Range accuracy: 2.5% of range

Contractor
China National Electronics Import and Export Corporation, Beijing.

VERIFIED

....

REL-1 airport surveillance radar

Type
D-band (1 to 2 GHz) guidance and surveillance radar for military and civil use.

Description
REL-1 is a transportable or static D-band radar for both guidance and airport surveillance. It employs a four-channel frequency diversity system to provide a high

The antenna used with the REL-1 airport surveillance radar

probability of detection and high reliability. It can detect targets in different space area and distance by changing the radiative angle of the antenna. Digital moving target indicator technology provides good clutter rejection. The transmitter employs a tunable magnetron oscillator so that the operating frequency can be changed rapidly. A radio link is incorporated in the system for long-distance transmission of the video images. REL-1 has been designed for easy erection and dismantling in the transportable version. The complete system is conveyed by five trucks and five trailers.

Status
As of this edition, the status of the REL-1 guidance and surveillance radar was uncertain.

Specifications
Operating frequency: D-band (1,220-1,350 MHz)
Frequency diversity operation: 4 channels
Antenna
Dimensions: 15.5 m span; 8 m high
Polarisation: linear
Gain: 34 dB; 38 dB
Azimuth dB beamwidth: 1.1°
Sidelobe level: <−20 dB
Rotation speed: 3 or 6 rpm

Contractor
China National Electronics Import and Export Corporation, Beijing.

VERIFIED

Type 791-A Precision Approach Radar (PAR)

Type
I/J-band (8 to 20 GHz) military PAR.

Description
The Type 791-A PAR is a typical military mobile radar equipment. It is mounted on a six-wheeled truck, although the manufacturers state that both mobile and fixed installations are available.

This Chinese-manufactured system closely resembles the former Soviet RSP-7 PAR (NATO reporting name Two Spot) and it could well be a licence-made version although there are slight differences. For example, in the former Soviet version the PAR antenna heads are located between the driver's cab and the radar cab, whereas in the Chinese model the antennas are at the rear end of the whole assembly. However, this could be nothing more significant than the fact that the equipment was designed as a cabin-housed equipment meant to be carried on a flatbed truck, and capable of being loaded either way round. There are similar slight differences in the scanner outlines.

Operation is in the I/J-band and dual transmitter/receivers are provided to ensure continuity of service; for the same reason the display indicators are also duplicated. The two scanners, for search and azimuth guidance and for elevation guidance, are co-located on a mounting at the rear of the vehicle. There is an operator's cab in front of this which also contains communications facilities. Circular polarisation is provided to combat weather clutter and the receiver includes a logarithmic intermediate frequency amplifier. Double B scope displays

Type 791-A PAR

are used to improve the accuracy at short range. When operating the antenna is able to change direction rapidly to cope with aircraft landing on different runways. Dual transceivers, indicators and generators are supplied.

Status
As of this edition, the status of the Type 791-A PAR was uncertain.

Specifications
Frequency: I/J-band (8-20 GHz)
Range: 35 km (15 km in rain)
Coverage: 1-8° (elevation); 20° (azimuth)
Accuracy: 0.35° (elevation); 0.5° (azimuth)
Range accuracy: approx 60 m
Resolution: 200 m (range); 1.2° (angular)

Contractor
China National Electronics Import and Export Corporation, Beijing.

VERIFIED

CZECH REPUBLIC

OPRM-71 Air Traffic Control (ATC) radar

Type
D- (1 to 2 GHz) and I-band (8 to 10 GHz) ATC radar system.

Description
The mobile, three-man OPRM-71 ATC system is optimised for military use and is designed for terminal zone air traffic navigation and landing control applications on temporary or permanent airfields. As such, the equipment incorporates Airport Surveillance Radar (ASR), Secondary Surveillance Radar (SSR) and Precision Approach Radar (PAR) functions in a single unit. Operating modes comprise ASR/SSR, ASR/SSR/PAR (with ASR as primary function) and PAR/ASR/SSR (with PAR as primary function). Systems components include a coherent transmitter, a direction-finder, a transceiver subsystem and a line communications interface. Other features include:
- built-in test
- digital signal processing
- clutter attenuation
- duplex data transmission between the radar and a central operations post via fibre optic cable (up to 1 km) or microwave transceiver (up to 10 km)
- digital recording of the air situation and voice communications.

Status
As of this edition, the status of the OPRM-71 ATC system was uncertain.

Specifications
Pulse power: 0.5 kW (SSR function); 50 kW (ASR/PAR functions)
Range: 25 km (PAR function); 60 km (ASR function); 120 km (SSR function)
Altitude (max): 2,100 m (PAR function); 4,000 m (ASR function); 10,000 m (SSR function)
AMTI coefficient: 40 dB (ASR/PAR functions)

Contractor
Omnipol Ltd, Prague.

VERIFIED

DENMARK

TERMA ballistic instrumentation products

Type
Family of ballistic instrumentation products.

Description
Danish contractor TERMA produces a range of ballistic instrumentation products that are designed for projectile velocity determination, in-bore ballistic measurement and terminal ballistic measurement applications. As of this edition, the product family is known to include:

Antennas
TERMA produces a range of E/F-band (2 to 4 GHz), I/low J-band (8 to 12 GHz) and high-frequency Doppler radar antennas that are designed for short-, medium- and long-range projectile and missile velocity determination applications.

Data units
TERMA produces the DR 5000 general signal analyser that offers a digital ballistic instrumentation data acquisition and processing capabilities and can be integrated with any of the previously described antenna units. The software used in the DR 5000 analyser is described as being optimised for ease of data acquisition and presentation as well as for comprehensive data reduction and modelling tasks.

Systems
TERMA's DR 6700 automatic Doppler tracking radar is designed for simultaneous measurement of a projectile's velocity and position in space. DR 6700 is further described as integrating digital signal processing, microwave technology and an 'advanced' user interface into a signal unit.

TERMA also indicates that its range of ballistic measurement products can be mixed and matched into customer specific systems that are complete with all the necessary mechanical structures, electronics and software.

Status
As of this edition, the described TERMA ballistic instrumentation products were thought to be available.

Contractor
TERMA Elektronik AS, Lystrup.

VERIFIED

Danish contractor TERMA produces a range of ballistic instrumentation products   0044146

Antenna of the PCS 514 tracking radar

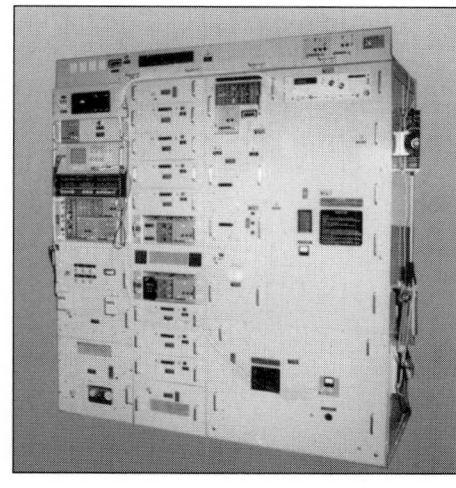

The transmitter cabinet used in the PCS514 tracking radar
0007316

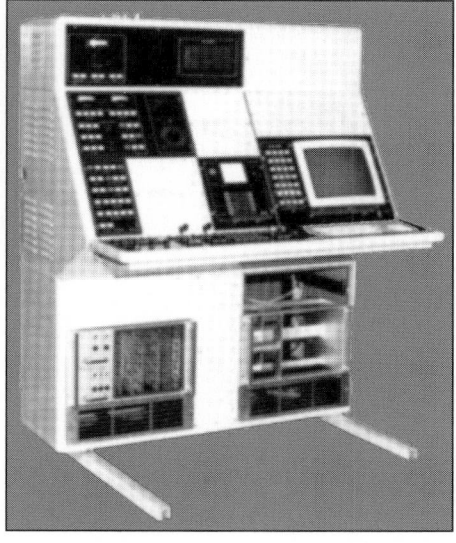

The display/control console used with the PCS 514 tracking radar
0007317

INDIA

PCS 514 tracking radar

Type
G-band (4 to 6 GHz) instrumentation and tracking radar.

Description
PCS 514 is a G-band precision coherent monopulse radar that is designed for range safety and rocket/space vehicle performance analysis applications. It can acquire and track rockets and launch vehicles up to a range of 3,200 km, providing the range, bearing, height and velocity of detected targets. PCS 514 comprises a transmitter, an antenna (with associated drives), a multipolarisation radio frequency feed, a range tracking subsystem, a multiCoho subsystem, a Doppler tracking subsystem, an angle tracking subsystem, a display/control console and data processors.

Status
As of this edition, the status of the PCS 514 instrumentation and tracking radar was uncertain.

Specifications
Frequency: G-band (5.45-5.85 GHz)
Tracking range: 200 km (1 m² skin echo); 3,200 km (transponder)
Peak power: 1 MW
Pulse-width: 0.25, 0.5 and 1 μs (selectable)
PRF: 585.5, 292.75 and 146.375 pps
Angle coverage: −5 to +185° (elevation); continuous (azimuth)

Contractor
Bharat Electronics Ltd, Bangalore.

VERIFIED

INTERNATIONAL

Adour II instrumentation/tracking radar

Type
G-band (4 to 6 GHz) instrumentation and tracking radar.

Description
Adour II instrumentation radar is an improved version of the previous Adour and is a G-band tracking system using the conical scanning principle. It is intended for acquisition and automatic tracking and is specifically designed for making measurements in:
- flight test centres, for the calibration and evaluation of prototype aircraft, of airborne equipment such as altimeters and autopilots, and of ground equipment such as surveillance radars
- rocket and missile test centres and launching centres, for measurements relating to missile, sounding rockets, and satellite launchers.

  In addition, the radar is used at such centres for wind-finding and similar measurements.

  Adour II is available in fixed or mobile form. It has been modified recently to improve its class of accuracy by using more extensive digital techniques and improving the operating mode. Adour II can be provided with a TV and/or infra-red camera and with an optical designation station to facilitate target acquisition and identification.

Status
Over time, Adour radars are reported as having been acquired for use on test ranges in Australia, Brazil, France, Switzerland and other countries.

Specifications
Antenna
Type: scanning, cassegrain feed system, 3 m diameter
Polarisation: vertical (circular optional)
Beamwidth: 1.3° (3 dB)
Gain: 41 dB
Transmitter
Type: magnetron
Frequency: G-band (5,450-5,825 MHz)
Peak power: 250 kW
Pulse-length: 0.5-1.7 μs (double pulse optional)
Tracking range: 170 km (1 m² skin echo); 2,000 km (50 W transponder)
Tracking accuracy: 3 m (peak range); 0.2 mrad (angular)
Turret max speed and acceleration: 0.6 rad/s and 3 rad/s² (elevation); 1 rad/s and 4 rad/s² (azimuth)

Contractor
Thales Air Defence, Bagneux, France.

VERIFIED

Adour II instrumentation and tracking radar

Armor instrumentation/tracking radar

Type
G-band (4 to 6 GHz) instrumentation and tracking radar.

Description
Armor is a high-precision G-band instrumentation and tracking radar designed for adjustment and performance checking of ballistic missiles. It is intended to be installed on an instrumentation ship. Facilities offered by the radar include:
- electromagnetic analysis of target cross-section (by transmitting various waveforms), and offline analysis

Armor instrumentation and tracking radar

- slipstream analysis (plasma)
- automatic acquisition of a target designation by the rendezvous method
- automatic tracking of radar echoes or transponder signals up to 4,000 km.

  Armor is able to track simultaneously three targets in the antenna beam, thus allowing the study of the ballistic swarm at the approach of the impact point.

Status
Over time, two Armor radars are reported as having been installed aboard the French range ship *Monge*. For its own part, *Monge* is noted as having been commissioned during 1992.

Specifications
Antenna: cassegrain feed system, diameter 10 m
Reception channels: 3 real time, 2 analysis
Beamwidth (3 dB): 0.37°
Transmitter: solid state + TOP + klystron
Frequency: G-band (5,400-5,900 MHz)
Peak power: 2 MW
Pulse duration: 0.5-100 μs
Tracking range: 6,000 km (1 m² skin echo); 20,000 km (transponder)
Tracking accuracy: 1 m (peak range); 0.1 mrad (angular RMS)

Contractor
Thales Air Defence, Bagneux, France.

VERIFIED

Artois instrumentation/tracking radar

Type
Instrumentation and tracking radar.

Description
Developed in 1973, Artois is a 3-D electronic scanning radar used to track several targets simultaneously during complex experiments with weapon systems, and perform high-accuracy differential trajectory measurements. The principal features of this radar are electronic scanning in a cone having a vertex angle of 10° minimum, instantaneous deflection from pulse-to-pulse and multitarget tracking with controlled variable interlace.

Artois instrumentation radar

Status

Over time, an Artois radar is noted as having been installed at France's *Centre d'Essais des Landes* on the country's Atlantic coast for ballistic and tactical missile trials and training.

Contractor

Thales Air Defence, Bagneux, France.

VERIFIED

....

Atlas instrumentation/tracking radar

Type

G-band (4 to 6 GHz) satellite tracking and instrumentation system.

Description

Atlas is a very high-precision tracking radar designed for trajectography and satellite tracking. A monopulse G-band radar, Atlas automatically tracks radar echoes or transponder signals at distances up to about 4,000 km. The equipment comprises an antenna turret (housing the equipment's radio and intermediate frequency receiver circuitry), a 1 MW peak power transmitter (able to be coded and tuned across the 5,450 to 5,825 MHz frequency band), an operating console, an interconnection cabinet and a mains supply cabinet. A console-mounted range-finder is fitted with a synchronising device that allows the radar to be used in series with other tracking radars of the same or different types. Other features include 'very accurate' optical angular encoding and a built-in processor for data preprocessing, display and output.

The Atlas radar is available in fixed or transportable form. Recent improvements include more extensive use of digital techniques and operating mode. Operational optional facilities include TV monitoring, TV tracking and infra-red tracking.

Status

As of May 2001, production of the Atlas instrumentation and tracking radar was understood to have been completed.

Specifications

Antenna
Type: monopulse cassegrain feed system
Diameter: 4 m
Polarisation: vertical/circular
Gain: 44 dB
Beamwidth (3 dB): 0.9°
Transmitter
Type: magnetron, codable and tunable
Frequency: G-band (5,450-5,825 MHz)
Peak power: 1 MW
Pulse-length: 0.25-1.7 µs
Tracking range: 400 km (1 m² skin echo); 5,000 km (50 W transponder)
Tracking accuracy: 3 m (peak range); 0.075 mrad (angular RMS)
Transmitter frequency controlled by a high-precision standard cavity

Contractor

Thales Air Defence, Bagneux, France.

VERIFIED

The antenna array used with the Atlas tracking radar

Battlefield METeorological System (BMETS)

Type

Passive, independent, mobile meteorological system.

Description

BMETS is intended to replace the UK's existing in-service radar-based Artillery METeorological System (AMETS). As such, BMETS is claimed to offer 'significant advantages' over AMETS, notably:
- reduced manpower – BMETS needs only two operators to conduct atmospheric soundings
- no radiated energy – BMETS will use radio direction-finding theodolites to track the radiosonde

A complete BMETS section comprises a Meteorological Command Post (MCP) and a stores vehicle, each of which tows a trailer. The MCP houses ground meteorological instrumentation and data processing and communications equipment. The MCP trailer is used to carrying the system's radio theodolite and electrical generator. The stores vehicle houses the necessary upper-air consumables (radiosondes and balloons) with its trailer being used to carry the hydrogen cylinders used for balloon inflation. When the BMETS is deployed, a hydrogen filled balloon carries a radiosonde to a height of over 20 km. The radiosonde (which can be of any commercially available type) transmits data on ambient pressure, temperature and humidity during its ascent and is tracked by the radio theodolite in bearing and elevation. These data are received by the MCP's data processing equipment and converted to meteorological messages, which are then transmitted to the various users of such information.

Status

As of this edition, BMETS was noted as being fully developed.

Contractor

Alenia Marconi Systems (an Alenia-BAE Systems joint venture), Chelmsford/Rome.

VERIFIED

The MCP vehicle used in the BMETS system

....

Béarn instrumentation/tracking radar

Type

G-band (4 to 6 GHz) instrumentation and tracking radar.

Description

Béarn is a high-precision G-band automatic tracking radar, using the conical scanning principle, designed for measuring the trajectory of high-speed missiles at long range. Facilities offered by the radar include:
- manual or automatic acquisition of a target dynamically or statically designated by the rendezvous method
- automatic tracking of radar echoes or transponder signals up to 4,000 km
- elevation, azimuth, and range co-ordinate readout as numerical data in real time
- automatic changeover to memory tracking in case of signal loss
- polarisation switching without interruption of tracking
- synchronisation of a chain of radars interrogating the same transponder
- autonomous and automatic stabilisation of the pointing axis (for shipboard installations).

Status

Over time, the Béarn instrumentation and tracking radar has been procured by a number of customers in France and around the world. As of May 2001, Béarn was noted as being no longer in production.

Specifications
Antenna
Type: cassegrain feed
Diameter: 4 m
Gain: 44 dB
Beamwidth: 0.9° at 3 dB
Polarisation: 3 (at operator choice)
Transmitter
Type: magnetron
Frequency: G-band (5,450-5,825 MHz tunable)
Peak power: 1 MW
Pulse-length: 1.7 µs
PRF: 585.5 Hz

Contractor
Thales Air Defence, Bagneux, France.

VERIFIED

ESTEREL instrumentation/tracking radar

Type
E-band (2 to 3 GHz) instrumentation and tracking radar.

Description
ESTEREL is a mobile instrumentation radar used to monitor air-to-air, surface-to-air or sea-to-sea missiles launched from moving carriers against moving targets. It automatically performs the acquisition and tracking of the carrier, acquisition and tracking of the launched missile and miss-distance measurement with respect to the target. Tracking can be carried out against co-operative and non-co-operative targets (including low altitude) using the system's moving target indicator capability. ESTEREL's daylight display includes:
- a scope (double tracks, markers and magnifier) for raw and processed video data monitoring
- an interactive alphanumeric display for radar parameters, status, input and output data.

Other system features include automatic radar mode selection and sequencing, TV angular tracking and X and Y input/output interfaces.

Status
Over time, the French Ministry of Defence is understood to have procured the ESTEREL instrumentation and tracking radar for a number of test range applications.

Specifications
Antenna: cassegrain systems, 2.1 m diameter, horizontal polarisation
Beamwidth (3 dB): 3°
Transmitter: tunable, coaxial magnetron
Frequency: E-band (2.7-2.9 GHz)
Peak power: 800 kW
Pulse-length: 0.7 µs
Tracking range: ±10 mrad (field angle); ±11 km (field range); 130 km (1 m² skin echo)
Tracking accuracy: 5 m (peak range); 0.7 mrad (angular RMS)

Contractor
Thales Air Defence, Bagneux, France.

VERIFIED

The ESTEREL tracking radar

MDR 2700 miss-distance indicator

Type
Airborne miss-distance indicator.

Description
The MDR 2700 miss-distance indicator provides accurate measurement of the true distance by which missiles or other projectiles miss scaled targets during live air defence training and can be configured with instrumentation features for weapon system evaluation applications. Thales Sensors also notes that its scoring systems are suitable for installation in a range of remotely piloted and towed targets produced by 'leading manufacturers'. The specific MDR 2700 system comprises a Doppler radar mounted in the target and a receiving ground station that contains a processor that is used to analyse the Doppler signal transmitted from the target in order to obtain a real-time miss-distance value. Time of projectile flight and closest approach to the target are also computed and a post-trial analysis system is available to provide additional data such as warhead/fuze performance. MDR 2700 can also be rigged for vector scoring.

Status
As of late 2000, the MDR 2700 miss-distance indicator was reported as being in service.

Specifications
Airborne unit
RF power: 2 W max
RF Frequency: D-band (1,382.5 ±1 MHz – other frequencies available)
Telemetry: UHF band (0.3-1 GHz – other options available)
Bandwidth: 12 MHz
Dimensions: 63 × 115 × 215 mm (standard); 50 × 50 × 200 mm (miniature)
Weight: 1.7 kg (excl antenna)
Ground processor
Display: 7 segment LCD, printer
Performance: 0-25 m ±10% (typical)
Angular performance: ±10° RMS
Dimensions: 560 × 300 × 520 mm
Weight: 25 kg nominal

Contractor
Thales Sensors, Crawley, UK.

VERIFIED

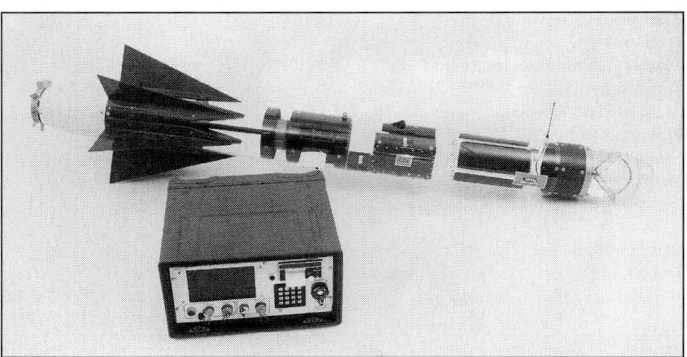

MDR 2700 miss-distance indicators fitted to a Hayes towed target, with the ground processor unit in the foreground

---

S500 series primary airfield surveillance radars

Type
E-band (2 to 3 GHz) airfield surveillance radars.

Description
The S500 series of radars has been designed for airfield surveillance at military and civil airports. Good subclutter visibility in fixed and moving clutter combined with an 'excellent' electronic counter-countermeasures capability are claimed to make S500 series equipments suitable for low cover and gap filler applications alongside their primary air traffic control role. The series consists of three systems, (S511 Surveyor, S511C and S512), all of which utilise the same antenna design, signal processing techniques and displays. Model differences are to be found in the type of transceiver used.

The S500 series antenna array consists of a 5 m aperture, double cosec², carbon fibre reflector and a double horn system for dual-beam operation. Variable polarisation between linear (vertical) and circular is provided on the lower beam and fixed circular polarisation on the upper beam. The antenna is turned at 15 rpm by a turning gear that is designed for a 100,000 hours working life without major component replacement. A secondary surveillance radar antenna can be mounted on top of the reflector and series radars come complete with a modular mounting tower that incorporates an enclosed scanner turning gear housing. The series has provision for remote control and monitoring and two display systems (the S361 multiprocessor, colour or monochrome raster unit and the ASTRID cursive system)

*The S511 airfield
surveillance radar*

are applicable to such radars. Of the three cited models, S511 makes use of the S2022 transceiver that incorporates a tunable, high-stability magnetron (with a solid-state modulator) and a 'high-stability' Coho-Stalo receiver section.

The long-range S511C model is fitted with a high-power coaxial magnetron and larger modulator while the S512 variant employs the S2062 wideband, cathode modulated travelling wave tube transceiver. This coherent device makes use of pulse compression and provides full frequency flexibility (block-by-block and pulse-by-pulse frequency agility).

Frequency diversity using two transceivers is noted as being available with all the cited systems and a three-channel signal processor is used to eliminate moving and fixed clutter and reduces tangential fading. Three temporal threshold subsystems (clutter maps) are fitted to ensure a low constant false alarm rate. Optional equipment includes a separate weather processor for controllable display weather contouring and the S7204 plot extractor that provides the necessary interface between the radar's signal processing and a data handling system.

Status
Over time, at least 30 S500 series equipments are understood to have been procured by customers around the world. Of these, six examples are known to have been acquired by Canada, where they were designated as the AN/FPS-509.

Specifications
Frequency: E-band (2.7-2.9 GHz)
Power output: 550 W (mean); 650 kW (peak)
Pulse-width: 0.85 µs
PRF: 1,000 pps (max)
PRF stagger: 6 period up to ±14%
Antenna gain: 31.5 dB (auxiliary beam); 34 dB (main beam)
MTBF: 1,714 h (single channel); 6,000 h (dual diversity, 1 channel failed)

Contractor
Alenia Marconi Systems (an Alenia-BAE Systems joint venture), Chelmsford/Rome.

VERIFIED

....

Savoie instrumentation radar

Type
Ground or shipborne long-range instrumentation radar.

Description
Savoie is a multitracking radar, intended for automatic tracking at long ranges. The equipment is designed to be shipborne or ground-based.

Status
Over time, the Savoie instrumentation radar is reported as having been installed aboard the French Navy's range ship *Monge* and as having been procured for use on a number of French land ranges. The data processing capability of the Savoie radar is understood to have been updated at least once.

The French Navy range ship Monge *is equipped with Savoie, Gascoigne and Armor missile tracking radars*

Specifications
Type: monopulse tracking radar
Antenna
Diameter: 8 m
Gain: 28 dB
Beamwidth (3 dB): 6°
Polarisation: linear
Turret
Weight: 12 t
Azimuth rotation: +110°
Elevation rotation: 0-90°
Speed: 0.25 rad/s
Acceleration: 0.5 rad/s^2
Accuracy of analysis axis: 5×10^{-4} rad
Transmitter
Peak power: 150 kW
Mean power: 22 kW
Pulse duration: pulse compression with several modes
Receiver
Pulse compression: receiver controlled by computer
Noise figure: 2.5 dB

Contractor
Thales Air Defence, Bagneux, France.

VERIFIED

....

Sirocco meteorological radar station

Type
I-band (8 to 10 GHz) tracking radar system for meteorological purposes.

Description
Sirocco is an I-band tracking radar for use with meteorological sounding balloons, associated telemetry systems, and processing equipment, with provision for the transmission of meteorological information for use by artillery units.

The complete system is built as a self-contained land-mobile unit, the tracking radar being mounted on a two-wheeled trailer, and an air conditioned SH 17 shelter

*Sirocco
meteorological radar
system*

is carried on the towing vehicle. The latter houses the operating crew, the radar and telemetry operating console, a radio-telegraphy system with built-in modem, and stowage for the sondes and radar reflectors. In addition to the radar, the carrier holds the telemetry receiver and two electric generators.

The Sirocco system produces standard meteorological messages, online, for transmission by teleprinter or a radio-telegraphy link for use by artillery units. The messages provide windspeed, wind direction, and air temperatures at a series of standard altitudes. Maximum range is in excess of 130 km and angular accuracy is better than 0.05°. Range can be measured to within 5 m and windspeed can be determined with an accuracy of better than 1 kt (0.5 m/s). In its trajectography version, Sirocco establishes the trajectories of a wide range of moving objects, projectiles (radar returns), drones and missiles (transponder).

Status

Over time, Sirocco is noted as having been procured by a number of countries including France, Jordan, Morocco, Netherlands, Saudi Arabia and Venezuela.

Specifications

Frequency: I-band (8-10 GHz)
Peak power: 80 kW
Receiver: 2-axis monopulse. Optional transmitted level control
Ranging: digital in 5 m increments
Angular tracking limits: −8 to +160° (elevation); unlimited (azimuth)
Max range: 130 km
Angular accuracy: better than 1 mil
Range accuracy: better than 5 m
Windspeed accuracy: better than 1 kt (0.5 m/s)

Contractor

Thales Airborne Systems, Elancourt, France.

VERIFIED

Sirocco Mk 2 meteorological station

Type

Dual-mode tracking system for meteorological purposes.

Description

Sirocco Mk 2 is an upgrade of the Sirocco system (see separate entry) which is designed to supply accurate meteorological data for use by artillery units. This upgrade is based on a dual-mode system (discrete radar mode and passive mode) by addition of a Marwin System link to the Omega network, a PC ATD 486 computer and an automatic ground parameter system.

The complete system is built as a self-contained land-mobile unit, with the tracking unit being mounted on a two-wheeled trailer and an air conditioned SH 17 shelter carried on the towing vehicle. The latter houses the operating crew, the radar and telemetry operating console, the PC ATD 486 computer, the Marwin system to receive the Omega network, a radio-telegraphy system (with built-in modem) and stowage for the sondes and radar reflectors. One power generator (10

Sirocco Mk 2 dual-mode meteorological station

kVA) mounted on the truck is possible. In addition to the tracking unit, the trailer carries the telemetry receiver.

The Sirocco Mk 2 system produces standard meteorological messages, online, for transmission by teleprinter or a radio-telegraphy link for use by artillery units. The messages provide windspeed, wind direction and air temperatures at a series of standard altitudes.

The sounding can be initiated with the radar (in low power) and carried on with the Omega system. Sounding can also be carried out while the station is moving. In radar mode, the maximum range is in excess of 150 km and angular accuracy is better than 0.05°. Range can be measured to within 5 m and windspeed can be determined with an accuracy of better than 0.5 m/s.

In the trajectography version, Sirocco establishes the trajectories of a wide range of moving objects such as projectiles, drones and missiles.

Specifications

Active mode - radar
Frequency: I-band (8-10 GHz)
Peak power: 80 kW
Low power (undetectable): −52 dB of max power
Receiver: 2-axis monopulse. Optional transmitted level control
Ranging: digital in 5 m increments
Angular tracking limits: −8 to +160° (elevation); unlimited (azimuth)
Max range: 130 km
Accuracy: better than 5 m (range); better than 1 mil (angular); better than 0.5 m/s (windspeed)
Passive mode
Receiver: omega
Frequencies: 11.1-13.8 kHz
Max range: 100 km
Resolution: speed 0.1 m/s; direction 1°
Windspeed accuracy: 0.5 m/s

Contractor

Thales Airborne Systems, Elancourt, France.

VERIFIED

STAR-2000 surveillance/terminal approach radar

Type

E-band (2 to 3 GHz) area surveillance and terminal approach control radar.

Description

STAR-2000 is a dedicated, solid-state, modular terminal approach radar that is suitable for both civilian and military air traffic control applications. The equipment incorporates a dedicated weather channel and its overall range capability can be extended from 111 to 167 km through the use of incremental power increases. STAR-2000 configurations exist for stand-alone, Monopulse Secondary Surveillance Radar (MSSR)/Identification Friend-or-Foe (IFF) associated or Mode S operation with the radar's data output format being configurable to match all transmission formats. In summary, the radar's manufacturer cites its main features as being the following:

- fixed, shelter-mounted and transportable configurations
- full coherence and clutter-driven adaptive processing for improved target detection in severe clutter conditions
- independent dual-polarisation weather channel
- modular, fail-safe, online maintainable, solid-state, frequency diverse/agile transmitter
- digital frequency synthesiser and pulse compression with low sidelobes
- auto-adaptive moving target detection with clutter rejection techniques
- false alarm free plot extraction and tracking of up to 1,000 targets
- MSSR/IFF beacon and Mode S reinforcement
- programmable output data formatting
- full built-in test and remote monitoring
- automatic reconfiguration.

Status

Over time, STAR-2000 is reported as having been procured by a number of customers around the world.

Specifications

Frequency: E-band (2,700-2,900 MHz)
Peak power: 9/18/34 kW
Noise figure: 1.6 dB
Signal processor: adaptive FIR filters
Target tracking capacity: to 1,000
Azimuth accuracy/resolution: 0.15/2.8° RMS
Range accuracy/resolution: 60/230 m
MTBF: 24,000 h
MTTR: 30 min
Availability: 99.998%

Contractor

Thales Air Defence, Bagneux, France.

VERIFIED

Thales Defence instrumented radar systems

Type
Family of instrumented radar systems for radar signature research applications.

Description
As part of Thales Defence's radar signature management business, the company's Radar and Systems Research Group (RSRG) has developed three high-resolution instrumented radar systems, the known details of which are as follows:

Airborne Data Acquisition System (ADAS)
The RSRG developed and operated ADAS package is an airborne version of Thales' MIDAS multimode, multiband, polarimetric instrumented radar system (see following) that is mounted in an Ecureuil (Squirrel) helicopter. ADAS can be fitted with a range of application specific antenna arrays and is understood to be both demountable and operated by a single individual.

Specifications
Frequencies: F- (3.15 GHz), I- (9-11.25 and 9.75 GHz), J- (15.75 GHz), K- (35 GHz) and M-band (94 GHz)
Polarisation: linear (vertical and horizontal)
Agility: band and polarisation agile, pulse-by-pulse
Receiver: simultaneous 2-channel, co and cross-polar
PRF: 1-40 kHz (variable)
Bandwidth: 100 or 500 MHz; 2.25 GHz (I-band only)
Data gathering modes: high-range resolution (2.1 m using linear chirp modulation over 100 MHz bandwidth, 256 or 1,024 contiguous 1.5 m range cells); super high resolution (0.36 m using DPDPS) linear chirp modulation over 500 MHz bandwidth, 256 to 4,096 contiguous 0.3 m range cells); ultra-high resolution (0.1 m using DPDPS linear chirp modulation over 2.25 GHz bandwidth (8 × 500 MHz, 50% overlap), 256 to 4,096 contiguous range cells).

Maritime Clifftop Radar (MCR)
The UK former Defence Evaluation and Research Agency owned, RSRG developed MCR system is a deployable multimode, multiband, polarimetric instrumented radar system that is designed for open air radar measurement of ship targets. The equipment is installed in two trucks (one carrying an operator's shelter and the other the system's antenna assembly) and can be configured with a variety of application specific antenna arrays. These include a high gain dish assembly and a 9.75 GHz high gain, scanning array that has a scan rate of up to 300°/s.

Specifications
Frequencies: F- (3.15 GHz), I- (9-11.25 and 9.75 GHz), J- (15.75 GHz), K- (35 GHz) and M-band (94 GHz)
Polarisation: circular (left and right hand) and linear (vertical and horizontal)
Agility: band and polarisation agile, pulse-by-pulse
Receiver: simultaneous 2-channel, co and cross-polar
PRF: 1-40 kHz (variable)
Bandwidth: 100 or 500 MHz; 2.25 GHz (I-band only)
Data gathering modes: high-range resolution (2.1 m using linear chirp modulation over 100 MHz bandwidth, 256 or 1,024 contiguous 1.5 m range cells); super high resolution (0.36 m using DPDPS linear chirp modulation over 500 MHz bandwidth, 256 to 4,096 contiguous 0.3 m range cells); ultra-high resolution (0.1 m using DPDPS linear chirp modulation over 2.25 GHz bandwidth (8 × 500 MHz, 50% overlap), 256 to 4,096 contiguous range cells).

Mobile Instrumented Data Acquisition System (MIDAS)
MIDAS is a van-mounted, multimode, multiband, polarimetric instrumented radar system that can be operated with its host vehicle either static or moving. MIDAS is owned and operated by Thales' RSRG and as with the group's other instrumented radar systems, MIDAS can be fitted with a range of application specific antenna arrays.

Specifications
Frequency: A/B/C- (100-600 MHz, short range only), F- (3.15 GHz), I- (9-11.25 and 9.75 GHz), J- (15.75 GHz), K- (35 GHz) and M-band (94 GHz)
Polarisation: circular (left and right hand) and linear (vertical and horizontal)
Agility: band and polarisation agile, pulse-by-pulse
Receiver: simultaneous 2-channel, co and cross-polar
Bandwidth: 100 or 500 MHz; 2.25 GHz (I-band only)
Data gathering modes: high-range resolution (2.1 m using linear chirp modulation over 100 MHz bandwidth, 256 or 1,024 contiguous 1.5 m range cells); super high resolution (0.36 m using DPDPS linear chirp modulation over 500 MHz bandwidth, 256 to 4,096 contiguous 0.3 m range cells); ultra-high resolution (0.1 m using DPDPS linear chirp modulation over 2.25 GHz bandwidth (8 × 500 MHz, 50% overlap), 256 to 4,096 contiguous range cells)

Operational status
As of May 2001, the ADAS, MCR and MIDAS systems were understood to be in service.

Contractor
Thales Defence, Wells, UK.

VERIFIED

TRAC 2000 series aircraft recovery radars

Type
Family of aircraft recovery radars.

Description
The Thales Raytheon Systems' (formerly Thales Air Defence) TRAC 2000 is a family of radar systems which has both military and civil applications. When used in the military application, protection from electronic countermeasures is provided.

The basis of the radar is a solid-state D-band (1 to 2 GHz) transmitter, having a redundant configuration consisting of 20 individual modules, each of which can be replaced in the event of failure without interrupting the radar service. Each module has a peak power output of 600 W, the assembly of 20 modules producing 10 kW of peak power. In the event of a single module failure only 3 per cent of range is lost; transmitter operation is ensured for up to three module failures. The transmitter can be pulsed by long or short pulses in frequency diversity transmission, and sectorised transmission. The transmitter is structured to be fully redundant, not only in the transmission and reamplification stages, but with regard to the power supplies, cooling and system protection devices. The transmitter, when operating in frequency diversity, is modulated by two 1 µs pulses, then by two 60, 100 or 200 µs pulses, depending on the application.

Two receivers are supplied to provide the diversity function, and the necessary redundancy in terms of system reliability. The receivers feature:
- a low-noise Radio Frequency (RF) amplifier preceded by a limiter and incorporating a sensitivity time constant function
- a frequency changer using the dual-frequency changer technique to enable rapid transmitter frequency switching without time consuming receiver realignment
- a short/long pulse adaptive filter. Thales claims that TRAC 2000's fully adaptive Doppler filtering and thresholding provides the equipment with a significantly improved false alarm rate. The equipment's signal and data processing is undertaken in a single unit.

A common RF amplifier is connected to the high beam used for reception only and which feeds the high beam input to the HI/LO beam switch in each receiver.

The radar types covered in this family are as follows:

Approach/Terminal Radar TRAC 2000/2100
This uses one or two transmitter units together with an antenna THD 286. This antenna has been designed for the terminal area surveillance role and is able to rotate at 10, 12 or 15 rpm, with an adapted beamwidth of 1.7°. The antenna dimensions are 9 m span by 5 m high. The TRAC 2000 with one transmitter gives a range of 185 km, the TRAC 2100 has two transmitters and a range of 250 km. The radar can be supplied with a choice of signal processing units:
- the TVD 900, four-filter Moving Target Indication (MTI) using conventional three and four canceller filter techniques. These filters provide a zero velocity channel, a moving target channel, and two meteorological effects suppression filters. Each filter is followed by a super clutter visibility mass memory
- the TVD 1000 eight-filter MTI using Fast Fourier filters coupled with high-performance processors for calculating the position, speed, range and barycentre of detected targets.

Area Surveillance Radar TRAC 2300/2400
This may accommodate two or four transmitter units, and for long-range detection uses the antenna AT432. This antenna has been specifically designed for the long-range role, having reflector dimensions of 13 m span by 7 m high. The beamwidth is 1.3° for a turning speed of 5 or 6 rpm.

The TRAC 2300, with two transmitters provides a range of 296 km. The TRAC 2400, which has four transmitters, has a range exceeding 370 km.

Modular solid-state transmitter of the TRAC series radar

As with the approach/terminal radars, the TRAC 2300/2400 are available with a choice of the TVD 900 or TVD 1000 signal processing equipments.

Status
As of May 2001, TRAC 2000 series aircraft recovery radars were reported as having been 'field proven' around the world.

Contractor
Thales Raytheon Systems, Massy, France.

UPDATED

TRS 2310 precision approach radar

Type
I-band (8 to 10 GHz) mobile approach radar for military use.

Description
The TRS 2310 is a mobile approach radar facility designed for military aircraft guidance on a chosen glide path. The radar is part of a ground-controlled approach system comprising the surveillance radar, the associated identification friend-or-foe and the control cabin.

The TRS 2310 features:

- excellent performance in the presence of ground clutter and weather returns through the use of a coherent transmitter with travelling wave tube amplifier, a Doppler moving target indicator with four to eight velocity filters, and transmission at two or three frequencies in bursts of six to 10 pulses at variable pulse repetition frequency
- a choice between two operating modes (bad weather or fine weather) to adapt processing to the environmental conditions. Optimised Doppler filtering makes recovery possible under the most severe precipitation conditions
- duplicated signal transmission, reception and processing chains
- the elevation scanning antenna is cosec² mainly on one side to ensure that the target once captured is not lost near touchdown
- wide azimuth coverage is compatible with use of the system for parallel runway operation
- with a wide beam antenna assembly several landings can be controlled simultaneously
- the system is capable of four QFU (direction/designation of runway to be used) positions with very fast change of QFU (less than 30 seconds).

Status
As of May 2001, TRS 2310 was no longer in production. Over time, it has been supplied to the French armed forces and to various overseas customers.

Specifications
Frequency: I-band (9-9.2 GHz)
Azimuth antenna
Reflector size: 2.6 × 0.9 m
Polarisation: circular
Sector scan: 20° (−13 to +7° or −7 to +13°)
Elevation antenna
Reflector size: 4.4 × 0.7 m
Sector scan: 10° (−1 to +9°)
Polarisation: circular

The TRS 2310 PAR system

Max displayed range: 17.5 km (bad weather); 36 km (fine weather)
Peak power: 30 kW
Average power (tube output): 50 W (fine weather); 100 W (bad weather)
Pulse duration: 0.5 µs

Contractor
Thales Air Defence, Bagneux, France.

VERIFIED

TRS 2505 Picardie instrumentation radar

Type
I/J-band (8 to 20 GHz) acquisition and tracking radar for test centres.

Description
TRS 2505 Picardie radar is an acquisition and tracking radar that is designed for use at artillery, missile and flight-test centres. The equipment has been designed primarily for use in the development of modern weapon systems and operational doctrines, as well as in training procedures. The capabilities of the system include all essential phases of tactical missile experimentation, whether surface-to-air, air-to-air, air-to-surface or surface-to-surface. Picardie is available in both mobile and fixed configurations and consists of a cassegrain antenna (with a monopulse feed), a fully coherent travelling wave tube transmitter (facilitating frequency agility and burst operation), a Doppler signal processor (with Fast Fourier Transform filtering) and a mission management processor. A range-finder unit allows simultaneous tracking on skin echoes from two targets in the radar beam. Depending on the particular mission, a number of automatic acquisition and tracking modes can be used to match its characteristics.

Status
Over time, the Picardie instrumentation radar is reported as having been procured for range use in France.

Specifications
Antenna
Type: cassegrain, monopulse feed
Diameter: 1.5 m
Transmitter
Type: TWT, frequency and PRF agile, burst operation
Frequency: I/J-band (8-20 GHz)
Signal and data processing: fully digital
Tracking range: 60 km (1 m² skin echo)
Tracking accuracy: 4 m (peak range); 0.4 mrad (angular RMS)

Contractor
Thales Air Defence, Bagneux, France.

VERIFIED

The antenna assembly used with the Picardie instrumentation radar

ISRAEL

Elta Super Instrumentation System (SIS)

Type
G-band (5.4 to 5.9 GHz sub-band).

Description
The Elta Super Instrumentation System (SIS) is an instrumentation and tracking system designed for a variety of applications, and serves as an integrated

Elta Super Instrumentation System

evaluation centre which can be installed in various test ranges. These applications include general test range work, weapons development, ammunition evaluation, aircraft training test ranges, and space activities.

The system consists of: a coherent, precise, high-performance instrumentation radar; an integrated electro-optical system for acquisition, angle tracking and star calibration; and a powerful computer system for online and offline signal and data processing. These capabilities are fully integrated to provide precise measurement and real-time evaluation of various types of airborne targets, simultaneous measurement of all the relevant trajectory parameters, real-time data analysis, recording of the obtained data, and computer processing and display of the results.

The pedestal-mounted sensor system is an accurate elevation-over-azimuth type designed to withstand large loads without affecting accuracy. The receiving system is a low-noise superheterodyne, two axes monopulse system with high dynamic range. The computer is a general purpose unit with a very high processing rate. It shares memory with an array processor, and has powerful input-output controllers. A command and control subsystem includes a computer terminal; an assemblies control panel; target/trajectory status and tracking (range, Doppler and signal) displays and range and angle stick controllers.

Main features of the SIS include:
- operation in all adverse environmental conditions and clutter
- direct measurement of azimuth, elevation, range, radial velocity and absolute power
- pulse-to-pulse amplitude and phase data recording
- ballistic, aerodynamic, tracking analysis algorithms
- graphical and numerical display and printing means
- user-friendly software packages
- online built-in test of system and assemblies
- computer-aided operation and calibration.

Specifications
Frequency: G-band (5.4-5.9 GHz)
Transmitter peak output: 60 kW (low power); 0.4 MW (high power)
Instrumented range (for 1 m² skin tracking): 200 km (full power); 80 km (low power)
Nominal tracking precision: (RMS for S/N = 20 dB) angles 100 μrad; range 1.5 m; radial velocity 5 cm/s
Dynamic properties: 42 m/s (max angular velocity); 300 m/s (max radial acceleration); 10,000 m/s (max radial velocity); 20 m/s² (max angular acceleration)
Pedestal performance: 50 μrad (precision); −5 to +185° (elevation); 360° (azimuth - continuous)

Contractor
Elta Electronics Industries Ltd (a subsidiary of Israel Aircraft Industries Ltd), Ashdod.

VERIFIED

NETHERLANDS

Vesta/Vesta-VC transponder system

Type
VHF band (30 to 300 MHz) helicopter transponder and datalink system.

Description
Vesta is a landing, safety, identification and positioning aid primarily intended for ship-based helicopters. Helicopters are difficult to locate because of their slow and low-level operations and small radar cross-section. Vesta has been designed to solve this problem.

Vesta consists of a helicopter-mounted transponder which transmits a VHF signal when the aircraft is illuminated by the ship's search radar. This reply is picked up by the Vesta shipborne component to produce a clear echo on the radar screen.

The helicopter is equipped with two spherical antennas for reception of the ship's search radar signals. The transponder generates a VHF reply containing a number of code pulses for identification purposes. The airborne component can also be used on light aircraft.

The ship component consists of two simple rod antennas to receive the transponder signals, which are then routed via the receiver (and extractor unit) to the ship's radar display. The Vesta system is controlled by either a control unit and a radar indicator, or a data handling system.

The Vesta system can indicate unambiguously any number of Vesta-equipped helicopters on the ship's radar display. A limited number of helicopters can be identified by means of preselected codes. As an option the Vesta system can be extended with an extractor. This provides the control unit or data handling system with a maximum of five tracks. The extractor is incorporated in the receiver unit.

Vesta-VC
Vesta-VC is designed to extend the ship's radar horizon by using the helicopter transponder system described above as a datalink. The data part makes use of existing voice channels. Information is transmitted in short bursts, initiated by the pilot. Both the transponder and datalink signals are processed by the shipborne Vesta component.

Specifications
Transmission frequency: VHF (30 to 300 MHz)
Radar reception frequency: D- through I-band (1-10 GHz)
Transmission power: 10 W
Pulse-length: 2 μs
PRF (helicopter part): 20 kHz (without code)
Sensitivity (ship part): −78 dBm
Range: 0-230 km
Number of codes: 5 (optionally 64)
Number of co-operating radar systems: 4
Weight of helicopter part: 2.3 kg (2.7 kg with datalink)
Weight of shipborne part: 58 kg

Contractor
Hollandse Signaalapparaten, Hengelo.

VERIFIED

RUSSIAN FEDERATION AND ASSOCIATED STATES (CIS)

This section deals with equipment produced by, exported by and/or in service in the Russian Federation and the independent states of Armenia, Azerbaijan, Belarus, Georgia, Kazakhstan, Kyrgyzstan, Moldova, Tajikistan, Turkmenistan, Ukraine and Uzbekistan. Taken together, these 12 entities form the Commonwealth of Independent States economic grouping.

VERIFIED

Kama-N trajectory-measuring radar

Type
Mobile trajectory-measuring radar.

Description
The Kama-N trajectory-measuring radar is designed to establish the co-ordinates of air and space vehicles such as satellites, missiles, projectiles and 'deep penetration' balloons and comprises:
- a KN-1 antenna station
- a KN-2 equipment cabin
- a K22M primary voltage converter
- an ED-60 diesel generator set.
Of these, the antenna station houses the system's directional antenna and drives, while the equipment cabin accommodates radar control consoles, the radar's receiver, a synchronisation subsystem, range and altitude measuring equipment, an antenna control unit, a data transceiver and data recording unit, the radar's transmitter, a built-in test subsystem and a spares holding. The architecture's antenna and equipment cabin are connected by cables and a waveguide line. The integral generator set can operate for up to 6 hours continuously and produces a three-phase, 380 V, 50 Hz output. This is converted to a three-phase, 220 V, 400 Hz format by the K22M unit. Functionally, Kama-N can operate in response to both transponder and echo signals.

Status
As of this edition, the status of the Kama-N trajectory-measuring radar was uncertain.

Specifications

Range: 0-2,880 km (in response to a transponder signal); up to 50 km (skin tracking on a 1 m² RCS target)
Radial speed: up to 11,000 m/s
Radial acceleration: up to 300 m/c²
Antenna rate: up to 9°/s (elevation); up to 18°/s (azimuth)
Acceleration: up to 3°/s (azimuth and elevation)
Accuracy: up to 5 min (azimuth and elevation); up to 8 m (range)
Set-up/close-down time: up to 12 h (close-down); up to 24 h (set-up)
Crew: 4
Power requirement: 56 kW
Temperature: −40 to +50°C (operating)
Weight: 3,500 kg (K22M); 6,000 kg (ED-60); 8,000 kg (KN-1); 16,000 kg (KN-2)
Dimensions (l × w × h): 5.7 × 2.3 × 2.93 m (K22M); 6.24 × 2.35 × 2.72 m (ED-60); 6.86 × 2.53 × 4.36 m (KN-1); 11 × 2.57 × 3.64 m (KN-2)

Contractors

Kuntsevo Design Bureau, Moscow.
Moscow Radio Engineering Plant, Moscow.

UPDATED

Vehicle-mounted version of Two Spot PAR

Long Talk airfield search/surveillance radar

Type

Mobile airfield search and surveillance radar.

Description

Long Talk is the air search and surveillance radar element of the RFAS standard ground-controlled approach and aircraft recovery system used at RFAS air bases. The other elements of the system are the precision azimuth and elevation tracking radars (NATO reporting name Two Spot), that form the precision approach radar portion. In some installations all three radars are mounted on one long four-wheeled trailer vehicle, which may also house the two- or three-person operating crew, but the Two Spot installation has been observed mounted on a platform above the driving cab of a standard military truck.

The Long Talk antenna has a moderately large elliptical paraboloid reflector, illuminated by radiating elements carried on a pyramid-shaped strut built up of four tubes. At the horizontal extremities of the reflector, one or two vertically positioned supplementary antennas (which could be identification friend-or-foe arrays) are located. Some installations have one such item on each end of the primary radar reflector; others have only one altogether. A radar similar in general appearance to Long Talk and with one vertical supplementary antenna attached to it is noted under the reporting name One Eye.

Status

As of this edition, the status of the Long Talk airfield search and surveillance radar was uncertain.

VERIFIED

The Long Talk airfield surveillance radar

Two Spot Precision Approach Radar (PAR)

Type

I/J-band (8 to 20 GHz) precision approach/landing guidance radar.

Description

Two Spot is the radar group employed by RFAS air forces for use in PAR aircraft recovery and landing guidance. It is usually associated with the Long Talk air search radar, which provides aircraft with air traffic control and homing directions to the airfield and positioning for the PAR 'talk-down' landing itself. In some installations, the Two Spot and Long Talk systems are both mounted on the same trailer vehicle, but Two Spot has also been observed mounted on a platform above the driving cab of a truck.

There are two antennas, both of parabolic section: one for elevation tracking and the other for azimuth. They are carried on a combined mounting, which permits limited rotation of the whole assembly enabling the system to be aligned to the bearing of the runway in use. The antennas and other items can be folded for transit. The operating frequency is in the I/J-band. Two prominent discone Very High Frequency (VHF - 30 to 300 MHz) or Ultra High Frequency (UHF - 0.3 to 1 GHz) communications antennas for the ground/air link used for talk-down are carried on a tubular mast and these may be the origin of the NATO reporting name.

Status

As of this edition, the status of the Two Spot precision approach and landing radar was uncertain.

VERIFIED

UNITED KINGDOM

Litton Marine Systems surveillance radars

Type

Range, coastal surveillance and harbour radars.

Description

Litton Marine Systems' Decca Division maintains a comprehensive capability in the design, manufacture and installation and support of surveillance radar for a wide range of applications. These include firing range safety, coastal surveillance for intruder detection and transportable applications. Such systems are modular and can be configured to meet a particular requirement. They range in size from a single radar and display to a multiradar installation covering a large area. The data in this case would be relayed from remote sites to one or more operations centres. The sensor data can be relayed by microwave links, fibre optic cable, or narrowband links. If required, equipments at various sites can be supplied in purpose-built containers fully fitted out for use. Standard displays are available that use colour digital scan conversion with a wide range of operational and processing facilities. These include the capability to output track table data. The radar may be interfaced to specialised display systems.

A typical Litton Marine Systems firing range surveillance system

Status

As of this edition, Litton Marine Systems sourced surveillance radars were thought to be available.

Contractor

Litton Marine Systems BV, Decca Division, New Malden.

VERIFIED

SIGMA Radar Cross Section (RCS) measurement system

Type

RCS measuring and analysis system.

Description

In conjunction with the UK's former Defence Evaluation and Research Agency, BAE Systems Combat and Radar Systems and partner company Roke Manor Research have developed a range of RCS measuring systems, the latest of which (designated as SIGMA) has been developed for the export market. Overall, the SIGMA system incorporates a tracking radar, a measuring radar and an electro-optic system to provide real-time RCS measurement of ships, aircraft, land vehicles and chaff decoys. The architecture incorporates comprehensive facilities to allow post trials analysis for further characterisation of targets. Different configurations are available dependent upon required frequency coverage and the SIGMA system is packaged on an International Standards Organisation (ISO) compliant pallet to allow easy transportation between measurement sites. Being fully self-contained, it can be used at unprepared locations. Associated with SIGMA is an RCS prediction programme called Epsilon that predicts target RCS via analysis of its physical shape and surface properties.

Status

Over time, the SIGMA RCS measuring and analysis system is understood to have been procured by at least two customers, one of whom was located in a NATO country.

Contractor

BAE Systems - Combat and Radar Systems, Cowes, Isle of Wight.

UPDATED

Type 282 ranging radar

Type

K-band (20 to 40 GHz) tracking and ranging radar for test ranges.

Description

The Type 282 ranging radar has been developed to meet the requirements for a highly accurate sensor to form part of a new generation of tracking equipment at UK Defence Evaluation and Research Agency ranges. It is used in conjunction with a television tracker to provide high accuracy range co-ordinates and replaces the kine-theodolites originally used. It can measure the range of a target (single artillery shells, missiles, aircraft, and so on) to a very high accuracy. Within the system, the ranging radar used is a K-band, short pulse coherent equipment that employs digital signal processing for clutter rejection and filtering for range accuracy. It has achieved static ranging accuracies of better than 0.25 m and dynamic accuracies

of 0.4 m and has exceeded the design accuracy requirement of 0.5 m at 10 km (0.005 per cent of range out to 35 km). As a result, trajectories can be measured in real time from a single station, without the requirement for three or more triangulation stations. The system consists of two components: a radar head comprising the antenna, microwave injection test equipment, transmitter and receiver and a signal processing cabinet containing the processing circuits and the electrical interface with the electro-optical equipment. Type 282 can be integrated with four types of electro-optical theodolite (Contraves Goertz, SFIM Mount Stem 600, MBB 2000 series and Photosonics compact tracking mount), giving each instrument 3-D, real-time, target trajectory tracking capabilities.

Status

Twelve systems of this type are noted as having been supplied to the UK's Ministry of Defence, with the first example being handed over in August 1988. Additional Type 282 radars are reported as having been supplied to companies in France, Germany, Spain and the US. As of this edition, it was thought likely that operational Type 282 radars were being supported by Alenia Marconi Systems (an Alenia-BAE Systems joint venture). The Type 282 ranging radar was originally developed by GEC-Marconi Radar and Defence Systems Ltd.

Specifications

Range accuracy: <10 km is 0.5 m (1 sigma); >10 km is 0.005% (1 sigma)
Tracking performance: 18 km (0.1 m^2 RCS target); 35 km (1.7 m^2 RCS target)
Transmitter: short pulse K-band magnetron
Frequency: 34-34.3 GHz with 4 frequency settings
Beamwidth: typically 0.65° (1 m diameter antenna)
Receiver: coherent with digital MTI

VERIFIED

UNITED STATES OF AMERICA

AN/FPN-62 Precision Approach Radar (PAR)

Type

Airfield PAR system.

Description

The I-band (8 to 10 GHz) AN/FPN-62 is an updated version of earlier equipment in which new components have been substituted for older, more difficult to maintain hardware. The original, electronically scanned azimuth and elevation antennas are retained and guidance signals are normally passed back to the aircraft voice radio link.

Status

As of this edition, the precise status of the FPN-62 PAR was uncertain. Historically, the equipment was developed by Raytheon in a competitive USAF programme aimed at updating of fixed PAR capabilities of the AN/CPN-4, AN/MPN-13 and AN/FPN-16 equipments. This resulted in a contract for the upgrading and supply of 40 systems for the USAF and the prototype was produced in 1976. In 1979, Raytheon's UK subsidiary Cossor Electronics (now Raytheon Systems Ltd), won a similar contract for the upgrading of RAF SLA-3 PARs and produced an Anglicised version of the FPN-62 designated as 'Falconer'. The UK concern has supplied 43 static and two mobile systems for use at home and overseas RAF bases and has also sold six systems to Norway. A further equipment was supplied to Zimbabwe.

Contractor

Raytheon Electronic Systems, Marlboro, Massachusetts.

VERIFIED

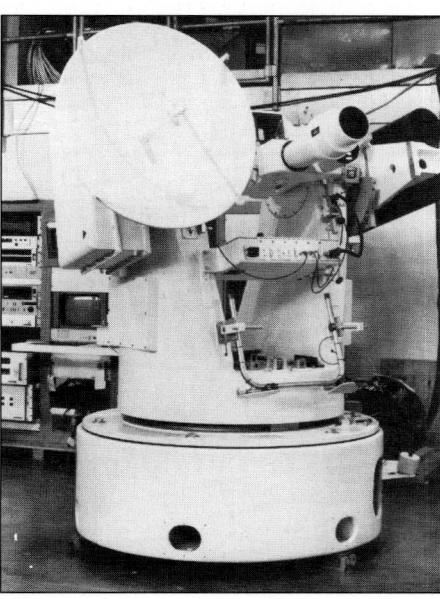

A Type 282 radar mounted on a Contraves Goertz electro-optical tracking system mount

The AN/FPN-62 PAR

AN/FPN-66 terminal approach radar station

Type
Terminal approach radar station.

Description
The AN/FPN-66 terminal approach radar station consists of the Thomson-CSF/Wilcox Electric Inc TA 10 MTD primary radar and the RSM 970 secondary surveillance radar. Of these, the TA 10 MTD operates in the E-band (2,700 to 2,900 MHz) and has been designed for the surveillance of the close terminal areas and airport approaches, as well as range control. The RSM 970 is a monopulse secondary surveillance radar that is entirely solid-state and fully compatible for Mode S.

Status
As of this edition, at least 80 AN/FPN-66 terminal approach radar stations were reported as having been sold worldwide. A major customer has been the US Army, who, over time, is known to have installed the system at Forts Bragg, Drum, Huachuca, Lewis, Polk, Riley and Stewart in the continental US; at Grafenwöhr and Heidelberg in Germany and at Kwajalein Atoll in the South Pacific.

Contractors
Wilcox Electric Inc, Kansas City, Missouri.
Thomson-CSF AIRSYS, Bagneux, France.

VERIFIED

The AN/FPN-66 approach radar station

AN/FPS-16 instrumentation radar

Type
G/H-band (4 to 8 GHz) range instrumentation radar.

Description
The AN/FPS-16 tracking radar is designed specifically for range instrumentation purposes and is claimed to be the first to be so designed. Its transportable version is the AN/MPS-25. Both systems are G/H-band units. The antenna pedestal can be rotated continuously in azimuth and from −10 to +190° in elevation. First introduced in the mid-1950s, these units have been extensively modified in the intervening years; almost all now use integrated circuit electronics. Most of the 60 FPS-16 and seven AN/MPS-25 radars have additional modifications including larger diameter antennas, three microwave transmitters, low-noise receivers and pulse-Dopplers. An upgrade has been carried out by BAE Systems North America (see RIR 716 entry).

Status
Over time, the AN/FPS-16 instrumentation radar is known to have been procured for use in both the UK and the US.

Specifications
Angle tracking precision (20 dB S/N): 0.1 mils RMS
Range tracking precision (20 dB S/N): 4.7 m RMS

The FPS-16 instrumentation radar

Azimuth tracking rate: 750 mils/s
Azimuth tracking acceleration: 1,020 mils/s^2
Elevation tracking rate: 400 mils/s
Elevation tracking acceleration: 1,020 mils/s^2
Ro, 1 m^2, 0 dB S/N: 235 km

Contractors
BAE Systems North America, Fort Walton Beach, Florida (upgrade).

VERIFIED

AN/GPN-22 Precision Approach Radar (PAR)

Type
I/J-band (8 to 20 GHz) PAR.

Description
The AN/GPN-22 PAR is intended specifically for unattended use in fixed base high-density air traffic control system operations under all weather conditions. It operates in I/J-band and is a development of the high-performance AN/TPN-25 PAR used in the US Air Force (USAF) AN/TPN-19 landing control system. Both these radars employ phased-array antennas to perform simultaneous scanning and tracking modes over the full radar coverage volume without any mechanical motion of the antenna. In the case of the TPN-25, up to six aircraft targets can be monopulse tracked simultaneously; in the case of GPN-22, the same number of targets can be handled but the coverage pattern measures 20 × 8° compared with the 20 × 15° of the earlier equipment. The range is about 35 km in both cases. Provisions are made for the display of landing aircraft position data in the operations centre up to 3,500 m from the radar head. The only significant difference between the GPN-22 and the TPN-25 is in the physical arrangements of the antenna. The GPN-22 feed assembly is supported by a boom structure projecting over the upper edge of the main reflector, whereas in the TPN-25 it is mounted beside the reflector.

The AN/GPN-22 precision approach radar

Status

As of this edition, 39 examples of the GPN-22 PAR radar are known to have been delivered to the USAF, with a further two going to Austria, six to South Korea and 11 to Netherlands. Under a then year US$50 million USAF contract, Raytheon has upgraded the AN/GPN-22 with the same dual-channel receivers used to upgrade the AN/TPN-25. Deliveries of the upgrade kit began in 1994 and continued into 1995. Austria is also known to have upgraded its two equipments with these receivers.

Contractor

Raytheon Electronic Systems, Marlboro, Massachusetts.

VERIFIED

AN/MPN-14K mobile Air Traffic Control (ATC) radar

Type

Mobile military ATC radar.

Description

The AN/MPN-14K is a mobile tactical RAPCON (Radar APproach CONtrol) designed for operations anywhere in the world. It is a virtually solid-state equipment that provides for control of aircraft within a 110 km terminal manoeuvring area, guidance during precision approach and control during departure through the terminal area. The Precision Approach Radar (PAR) radiates in I-band (9 to 9.6 GHz sub-band) with dual 80 kW transmitters, using solid-state modulators and coaxial magnetrons, with a gallium arsenide field-effect transistor, low-noise amplifier and a solid-state moving target indicator processor. The Approach Surveillance Radar (ASR) is an E-band (2.7 to 2.9 GHz sub-band) equipment with dual 1 MW transmitters. This radar provides enhanced target detection capability and significantly improved range performance of the PAR to nearly 40 km for a 1 m² target. The dual redundancy technology provides over 2,000 hours Mean Time Between Failure (MTBF) for the PAR function and more than 3,500 hours for the ASR function. The AN/MPN-14K is packaged in a two-trailer configuration that provides a complete tactical radar and communications facility that is air deliverable in two C-130 transport aircraft loads. As such, the US Air Force's (USAF) MPN-14Ks are claimed to offer a latest generation, fixed site RAPCON capability in a mobile package.

MPN-14K system configurations range from two four-wheel trailers with integrated turntables, to fixed installation ASRs with turntable-mounted PARs, to a single trailer ASR. Upgrades for the MPN-14K ASR and PAR radars to improve the subclutter visibility to >51 dB have been developed together with new solid-state transmitters and state-of-the-art signal and data processors (including all digital, high-performance, universal displays). These upgrades are currently available and are intended significantly to improve the equipment's MTBF and MTTR values together with a reduction in its overall cost of ownership.

Status

As of this edition, the AN/MPN-14K RAPCON radar is reported as having been supplied to the US Air Force, US Air National Guard and several Pacific Rim customers and as having been marketed in Asia and Europe. Here, the system has been offered with enhanced computer controlled data processing equipment and large digital ASR and ITT proprietary digital PAR displays for flight plan data presentation. In more detail, known MPN-14K contracting activity includes 1995 awards covering the supply of MPN-14K systems to a number of international customers including Turkey (two contracts in July and October 1995 with a total value of then year US$40 million) and countries in the Asia-Pacific region. Turkey placed a third MPN-14K contract (valued at then year US$30 million) during the summer of 1996. During the summer of 1997, South Korea placed two contracts for MPN-14K systems with a total value of then year US$40 million, with the new radars being used to augment the country's existing inventory of such equipments.

Specifications
ASR
Frequency: E-band (2.7-2.9 GHz)
Range: 111 km

MPN-14K mobile RAPCON

Azimuth: 360°
Elevation: 7,620 m or 30°
Accuracy: azimuth ±1°; range 4% or 152 m
PAR
Frequency: I-band (9-9.6 GHz sub-band)
Range: 37 km
Azimuth: 20°
Elevation: 8°
Accuracy: 7 m or 0.2° (whichever greater - elevation) 9 m or 0.2° (whichever greater - azimuth)

Contractor

ITT Gilfillan, Van Nuys, California.

VERIFIED

AN/MPN-25 tactical area surveillance and precision approach landing system

Type

All weather Airport Surveillance Radar (ASR)/Precision Approach Radar (PAR)/ Secondary Surveillance Radar (SSR) system.

Description

The AN/MPN-25 tactical area surveillance and precision approach landing system is a high mobility version of ITT Gilfillan's GCA/PAR-2000 equipment (see separate entry). As such, MPN-25 adds a secondary surveillance radar capability to the baseline GCA/PAR-2000 configuration. MPN-25 is an active aperture radar that features gallium arsenide transceiver modules integrated into both its azimuth and elevation antennas. ITT Gilfillan claims that the use of air as the cooling medium for these modules enhances their maintainability and reliability. MPN-25 also includes multiple displays and a 'full' communications subsystem housed in an S-280 type Operations Shelter (OS). The system's OS can be located more than a kilometre away from the radar and system set up and tear down times are given as both being less than 2 hours. Each element within MPN-25 has its own power and environmental control subsystems and ITT Gilfillan further suggests that the equipment is the only one of its type to be able to change runways (to any one of the six preset touchdown points) in less than a minute. The MPN-25 system is transportable as a single C-130 load.

Status

As of July 2001, the AN/MPN-25 tactical area surveillance and precision approach landing system is understood to have been procured by the US Air Force.

Specifications

Target and detection definition: 1-1,000 m² (RCS); ±74 to ±463 km/h (velocity - PAR); ±74 to ±741 km/h (velocity - ASR); Swerling 1 (fluctuation model)
Instrumented ASR coverage: 0-20° (elevation); 0-2,438 m (altitude); every 5 s (update rate - 60 rpm antenna rotation); 35 km (range - rain mode); 56 km (range - clear mode); 360° (azimuth)
Instrumented PAR coverage: −1 to +7° (elevation); 1/s (update rate); 28 km (range - rain mode); 30° (azimuth); 31 m (min altitude above ground intercept point); 37 km (range - clear mode)
Instrumented SSR coverage: every 4.8 s (update rate - 12.5 rpm antenna rotation); 111-463 km (range - dependent on interrogator selected); 360° (azimuth)
Aircraft target processing: 22 plots/scan (PAR targets - elevation); 50 plots/scan (PAR targets - azimuth); 250 plots/scan (ASR and SSR targets)
Weather processing: entire radar coverage area - 3 levels
Reliability: 2,212 h (MTBCF)
Maintainability: 0.25 h (MTTR)

Contractor

ITT Gilfillan, Van Nuys, California

UPDATED

An artist's impression of the MPN-25 tactical area surveillance and precision approach landing system
0044147

AN/MPS-39 multiple object tracking radar

Type
G-band (5.4 to 5.9 GHz sub-band) general purpose, multiple object tracking instrumentation radar.

Description
The AN/MPS-39 Multiple Object Tracking Radar (MOTR) uses a phased-array with a ±30° electronic beam steering, mounted on a high accuracy pedestal that may be moved while electronic beam steering is in effect. The instrument operates in the 5.4 to 5.9 GHz band and is transportable. It was designed by Lockheed Martin to US Department of Defense specifications under a contract with the US Army White Sands Missile Range, New Mexico. MOTR provides digital data output of position versus time on up to 40 simultaneously tracked objects. Tracking accuracies on all tracked objects are equal to or better than that provided by the predecessor class of instrumentation radars represented by the AN/FPS-16. The radar incorporates an n-pulse code that facilitates transponder tracking. Power output is 1 MW peak, 5 kW average.

Status
The first AN/MPS-39 MOTR was accepted at the White Sands Missile Range in December 1988 with a second example going to the US Air Force's Eastern Space and Missile Center (Patrick Air Force Base, Florida) during June 1990. A third radar of this type was deployed at the White Sands Missile Range towards the end of 1990 with a fourth example going to the Western US test range during 1992. The UK's Ministry of Defence acquired the fifth MPS-39 MOTR to be produced during 1994.

Specifications
Antenna
Type: space fed lens array, linear vertical polarisation
Directive gain on broadside: 45.9 dB
Scan angles: 60° cone plus grating lobe cusps
Aperture: 3.66 m
Beamwidth: 1° broadside
Transmitter
Type: coherent, TWT driven CFA
Frequency: G-band (5.4-5.9 GHz tunable)
Power: 5 kW (average); 1 MW (peak)
Tracking accuracies
Angles: 0.15 mils RMS (relative); 0.2 mils RMS (absolute)
Range: 0.3 m RMS (relative); 0.73 m RMS (absolute)

Contractor
Lockheed Martin Naval Electronics and Surveillance Systems - Surface Systems, Moorestown, New Jersey.

UPDATED

The MPS-39 multiple object tracking radar

AN/SPN-35 and AN/SPN-43 Air Traffic Control (ATC) radars

Type
F-band (3 to 4 GHz) ATC radar systems for surface ship applications.

Description
AN/SPN-35 is an approach radar that provides precise range, altitude and heading data to assist in the guidance of fixed-wing aircraft and helicopters during their final approach to landing aboard aircraft carriers and other types of platform. SPN-35 provides landing control up to the final transition point from which recovery is made

The SPN-43 radar system

and is mounted on a gyrostabilised platform that keeps it horizon aligned at all times. For its part, the 3.5 to 3.7 GHz band AN/SPN-43 ATC radar is described as forming a 'rugged part' of the US Navy (USN) carrier ATC system that is designed to marshal in-bound aircraft before their negotiating final landings. Mounted on a pitch and roll stabilised platform, SPN-43 has a maximum detection range of 93 km and a peak power output value of 850 kW.

Status
Over time, the AN/SPN-35 and AN/SPN-43 ATC radars are reported as having been installed aboard the USN's aircraft carrier USS *Enterprise* (SPN-43A) and the same service's 'Tarawa' (SPN-35A and SPN-43B) and 'Wasp' (SPN-35A and SPN-43B) class amphibious assault ships and 'Kitty Hawk/John F Kennedy' (SPN-43A) and 'Nimitz' (SPN-43B) class aircraft carriers. ITT Gilfillan is also noted as having developed upgrades for both systems.

Contractor
ITT Gilfillan, Van Nuys, California.

VERIFIED

AN/SPN-46(V) approach radar

Type
I- (8 to 10 GHz) and K- (20 to 40 GHz) band precision approach and landing system.

Description
Developed during the early 1980s, AN/SPN-46(V) provides simultaneous and automatic control for up to two aircraft during final approach and landing aboard aircraft carriers and other landing platforms. The system is a precision dual-band automatic acquisition/tracking radar and features cross-band beacon and aircraft skin tracking (with Mode I, II and III capabilities).

The equipment consists of a radar/ship motion sensor subsystem, a central computer, a display unit and ancillary equipment. Designed primarily as a navigation aid for use during CASE III recovery periods, SPN-46(V) offers three modes of operation as follows:

Mode I
A closed-loop, hands-off automatic landing mode.
Mode II
An instrument landing system approach mode, in which, data is displayed on board the in-bound aircraft.
Mode III
A carrier-controlled, talk-down approach mode.
Other system features include:
• pilot selectable mode control
• high-accuracy aircraft track information
• ship motion compensation synchronised to aircraft motion and computation of optimum flight path
• an 18 km operating range
• a 7 km approach control zone under all weather and deck conditions.

Status
Over time, the AN/SPN-46(V) precision approach and landing system is reported as having been installed aboard the US Navy's (USN) aircraft carrier USS *Enterprise* together with the same service's 'Kitty Hawk/John F Kennedy' and 'Nimitz' class aircraft carriers. The system was also noted as being a possible retrofit item for the USN's 'Wasp' class amphibious warfare ships.

Specifications
Frequencies: I- (9,310 ±35 MHz) and K- (33.2 ±0.2 GHz) band
Peak power: 50 kW

Pulse-width: 0.2 µs
PRF: 2,000 pps
Automatic search: 366 m (range); 1° (elevation); 25° (azimuth)
Scan rate: 12 scans/min
Radar coverage: −15 to +30° (elevation); ±150° (azimuth)
Antenna: 1.22 m dual-band parabolic; K-band conical scan; I-band monopulse cassegrain feed; circular and linear polarisation

Contractor
Textron Defense Systems, Wilmington, Massachusetts.

VERIFIED

AN/TPN-18A Ground-Controlled Approach (GCA) radar

Type
I-band (8 to 10 GHz) transportable GCA system.

Description
Originally developed for the US Army, the AN/TPN-18A GCA radar system is a lightweight, tactical ground-based radar that provides precise three-function (terminal area surveillance, precision approach for control of landings, and height-finding for aircraft monitoring) information to airport/heliport controllers. TPN-18A can be transported by helicopter, cargo aircraft, or truck and is suitable for truck/trailer mounting for full mobility in combat-related operations. The earlier AN/TPN-8 version, designed for the US Marine Corps' military all-weather tactical operations, has since been adapted by other military services. TPN-18A is a component of the AN/TSQ-71A and AN/TSQ-72 landing control central systems.

Status
As of this edition, the status of the AN/TPN-18A GCA radar was uncertain.

Specifications
Frequency: I-band (9-9.6 GHz)
Peak power: 200 kW nominal
PRF: 1,200 pps
Pulse-width: 0.2 or 0.8 µs
Azimuth beam: 1.3° (horizontal); 3.5° (vertical)
Elevation beam: 1.1° (vertical); 3.4° (horizontal)

Contractor
ITT Gilfillan, Van Nuys, California.

VERIFIED

The TPN-18A GCA radar system

AN/TPN-19 landing control central

Type
Transportable surveillance radar, Precision Approach Radar (PAR) and operations centre.

Description
AN/TPN-19 is a terminal air traffic control system that comprises an Airport Surveillance Radar (ASR - AN/TPN-24), a PAR, (AN/TPN-25) and an Operations Centre (OC). The radar systems are linked to the OC by microwave links (with a maximum range of 19 km) and the system as a whole is designed for both rapid worldwide deployment and fixed base installation. Packaging techniques make the system completely transportable by aircraft, helicopter or road vehicle. The TPN-24

ASR is a dual-channel, 2 × 500 kW, E-band (2 to 3 GHz), Moving Target Indicator (MTI) radar. System features include:
- a multiple (12) horn antenna feed that produces lobe-free coverage up to 12,000 m and 111 km on a 1 m² target
- a low angle pattern which is electrically variable and programmable in range and azimuth to reduce clutter returns
- a digital coherent and non-coherent MTI system to eliminate weather returns
- a staggered pulse repetition frequency to eliminate blind speeds below 2,037 km/h.

The TPN-25 PAR operates in I-band (8 to 10 GHz) and has a range of 37 km, even with precipitation at 5 cm/h. A monopulse system, it can track up to six targets simultaneously. For multiple runway coverage, the antenna assembly can be slewed through 270°. In the ground-controlled approach configuration, TPN-19 features three dual-mode displays, a figure that increases to seven when a full Radar APproach CONtrol (RAPCON) capability is required.

Status
Over time, the complete TPN-19 system and the separate ASR and PAR sets incorporated in it, have been supplied to US and off-shore customers. Here, Australia is understood to have procured one TPN-19 and six TPN-25s; Austria, two GPN-22s (an improved, fixed-site version of TPN-25); South Korea, six GPN-22s; Netherlands, a fixed-site TPN-25 and 11 GPN-22s and the USAF, 11 TPN-19s and 39 GPN-22s.

Contractor
Raytheon Electronic Systems, Marlboro, Massachusetts.

VERIFIED

AN/TPN-22 precision approach radar

Type
Precision approach radar.

Description
AN/TPN-22 is a precision track-while-search radar that uses phase/frequency scanning to provide high data rates for automatic detection and tracking while simultaneously searching a 46 × 8° sector. TPN-22 can execute Mode 1 remote-controlled, hands-off landings, Mode II cross-hair landings and Mode III conventional ground control approach landings. The radar consists of antenna, transmitter, receiver, digital processor and subsystems. The equipment's planar-array antenna operates at I-band and radiates a computer-controlled pencil beam using phase/frequency scanning. The antenna consists of 94 electronic digitally controlled diode phase shifters, feeding 94 serpentine arrays and incorporates circular polarisation for improved performance in bad weather. TPN-22's transmitter final stage consists of an air-cooled, high-power Travelling Wave Tube (TWT) using a solid-state regulated high-voltage power supply and line-type modulator. The transmitter first stage is also a TWT amplifier.

Status
Over time, the AN/TPN-22 precision approach radar has been procured by the US Marine Corps.

Contractor
ITT Gilfillan, Van Nuys, California.

VERIFIED

AN/TPN-24 airfield surveillance radar

Type
E-band (2 to 3 GHz) mobile airfield surveillance radar.

Description
The E-band, dual-diversity, multiple beam AN/TPN-24 is the airfield surveillance radar employed in the AN/TPN-19 landing control central system (see separate entry) and was originally designed for that system. It has subsequently been widely delivered separately for use on its own or with other equipment. The original version was designed and packaged for use as a highly mobile, tactical radar equipment, but other versions have been configured for different applications, including a fixed site variant, a tower-mounted model (designated as the ASR-910) and a semi-static version (ASR-909). The antenna and transceiver units are the same in each version as those of the basic TPN-24. Other system features include:
- two receiver/transmitter units, diplexed in two frequency diversity operations
- a 12 horn feed antenna
- vertical coverage from 0.5 to 30° above the horizon
- remotely selectable circular or linear polarisation
- staggered pulse repetition frequency (four different intervals)
- selectable coherent moving target indicator signal processing with double cancellation.

Status
In addition to those supplied for use in TPN-19 systems (a total of 12), some 58 separate TPN-24s are understood to have been procured over time. Customers are

*The TPN-24 airfield
surveillance radar*

noted as including Australia, Austria, Germany (44 fixed-site installations), South Korea and Netherlands.

Specifications

Radar: dual-channel, dual-frequency, diversity. Redundant or single channel and spectrum filter options
Frequency: E-band (2,700-2,900 MHz)
Transmitter type: magnetron
Peak power: 500 kW (per channel)
Pulse-length: 1 μs
Pulse rate: 1,050 pps, average
PRF: 12 staggers
Antenna size: 4.26 × 2.43 m
Azimuth beamwidth: 1.6°
Elevation beamwidth: 4°
Rotation speed: 13 rpm
Beam elevation tilt: +5 to −2°
Detection range: 110 km (clear weather, 1 m² RCS target)

Contractor

Raytheon Electronic Systems, Marlboro, Massachusetts.

VERIFIED

AN/TPN-25 Precision Approach Radar (PAR)

Type
I-band (8 to 10 GHz) mobile PAR system.

Description
AN/TPN-25 is the PAR used in the AN/TPN-19 landing control central tactical airfield control system. Designed to fulfil a requirement for a highly mobile system capable of assisting the landing of fixed-wing aircraft and helicopters in adverse weather conditions, TPN-25 can also be deployed at fixed bases. It employs a unique phased-array antenna that constantly scans the entire service volume of 20° azimuth × 15° elevation and 35 km in range, as well as tracking up to six approaching aircraft in elevation and azimuth. No antenna movement is required except when it is necessary to reorientate it to allow a different runway direction to be used. This can be performed under remote control to permit any one of four runways on an airfield to be served and TPN-25 is capable of automatic remote feeding of guidance information to the landing aircraft via a datalink. Operation is in the I-band and no mechanical scanning is necesary. A four-lobe monopulse feed horn is directed at the main antenna reflector, radiation having to pass through a digitally controlled microwave 'lens' consisting of an array of 824 phase-shifting elements. The latter are controlled to produce sharp, digitally steered, step-scanned pencil beams.

Status
Over time, 11 AN/TPN-25 radars are noted as having been supplied to the USAF for use in the AN/TPN-19 system (see separate entry), together with a further eight going to Australia for use in both fixed-site and mobile applications. With enhanced performance, TPN-25 is designated as the AN/GPN-22; a variant that is reported to have been supplied in quantity to the USAF and off-shore customers. Under a US$50 million contract awarded by the USAF, Raytheon has upgraded the AN/TPN-25 with redundant transmitters and dual-channel receivers capable of automatic alignment, fault detection, isolation and switch over. Deliveries of the upgraded equipment began in 1994.

Specifications
Radar: dual channel, dual function (scan/track)
Frequency: I-band (9,000-9,200 MHz)
Transmitter type: 2 solid-state oscillators with TWT
Peak power: 12.5 kW min each transmitter

Pulse-length: 1 μs
PRF: 3,500 pps (average)
Stagger ratio: 3:2
Antenna beamwidth: 1.4° (azimuth); 0.75° (elevation)
Detection range: 37 km (clear weather, trainer aircraft target)

Contractor
Raytheon Electronic Systems, Marlboro, Massachusetts.

VERIFIED

AN/TPS-73 Air Traffic Control (ATC) radar

Type
E-band (2 to 3 GHz) tactical ATC and surveillance radar.

Description
AN/TPS-73 is a highly mobile, rapidly deployable, 2-D, solid-state, E-band, primary and monopulse secondary radar system that is designed for detection and identification of airborne targets. A highly stable solid-state transmitter is used to provide a signal-to-clutter improvement ratio greater than 56 dB for stationary clutter. A 'negligible' false alarm rate is achieved via the use of:
- adaptive thresholds tailored to the changing clutter
- burst-to-burst, beam-to-beam and scan-to-scan correlation
- tracking parameters
- asynchronous interference blanking.

Although designed as an ATC radar, TPS-73's performance in clutter and countermeasures environments makes it suitable for other applications including gap filling and coastal surveillance. The radar is configured in a single 2.44 × 2.44 × 3.05 m International Standards Organisation (ISO) compliant shelter that houses the transmitter, auto-tracker, radar electronics, beacon equipment, a power distribution unit and a display console. It also provides storage during transport for the antenna group components (reflector, polariser, support collar and arm), the beacon omni-antenna, a beacon test set, a remote-control panel, air conditioning ducts, levelling jacks and tie-down cables/anchors. It is transportable by land, sea or air (C-130, CH-53). Other system features include:
- a solid-state transmitter with full band frequency agility, giving enhanced target detection in clutter/ECM environments, and a fail-soft capability
- dual E-band antenna beams for improved performance in clutter
- linear polarisation with selectable circular polarisations for improved target detection in clutter, rain and countermeasures environments
- adaptive moving target detection
- independent range/azimuth adaptive thresholding of Doppler filters for maximum target visibility
- high-resolution clutter maps for each range/azimuth/Doppler bin for improved detection
- a variable sensitivity time constant map to maintain receiver linearity at short ranges
- automatic, rapid selection of 'least interfered with' transmission frequencies
- an integrated radar/beacon antenna
- a monopulse beacon system for improved accuracy/resolution and Mode S compatibility
- high reliability/maintainability/availability provided by standby redundant processing channels and fail-soft features
- pulse compression to maintain resolution and use of low peak power.

The TPS-73 radar

Status

Over time, 18 TPS-73 radars are understood to have been delivered to the US military as part of the US Marine Corps' Marine ATC And Landing System (MATCALS) programme where they are teamed with the AN/TPN-22 and AN/TSQ-131 equipments.

Specifications

Frequency: E-band (2,700-2,900 MHz)
Power output: 1.1 kW (average at 12 rpm); 10 kW (peak)
PRF: burst-to-burst stagger
Modes: fixed frequency, frequency agility (random or automatic selection of least interfered with frequencies)
Antenna: reflector fed, horizontal or circular (L/R) polarisation
Beamwidth: 1.5° (azimuth); 5° (elevation)
Scan rate: 12 or 15 rpm
Coverage: 0.9-111 km on 1 m² RCS target; 40° elevation
Accuracy: 60 m (range); 0.18° (azimuth)
MTBF: 5,000 h system

Contractor

Lockheed Martin Naval Electronics and Surveillance Systems - Syracuse, Syracuse, New York.

VERIFIED

··

ASR-9 airport surveillance radar

Type

Airfield surveillance radar with military and civil applications.

Description

Northrop Grumman claims that ASR-9 is the first airport surveillance radar to display weather and aircraft simultaneously. ASR-9 combines circular polarisation with moving target detection for vastly improved aircraft detection in weather. A separate weather channel generates six weather levels, any two of which may be selected at one time by the controller. For detection of small targets in severe clutter, ASR-9 employs a dual-beam antenna, advanced digital processing, sophisticated constant false alarm rate circuitry, and a scan-to-scan tracker. A super clutter processor with a fine grain clutter map enhances returns from tangentially flying aircraft.

ASR-9 provides coverage out to 110 km, with availability of greater than 99.9 per cent. The system is completely unattended, incorporating a remote maintenance and monitoring system and has a dual-channel mean time between failure of over 3,500 hours. Should a fault occur, a built-in test detects and isolates the problem, a capability that can be controlled from a central facility. Redundancy has also been incorporated in the antenna subsystem by the use of dual-drive motors and dual-azimuth pulse generators.

Status

Over time, the ASR-9 airfield surveillance radar has been part of a joint Federal Aviation Administration (FAA)/US Air Force/US Army national airspace systems programme that encompassed the procurement of 134 examples of the radar. Three transportable systems have been produced for the US Army while Aruba, Belgium, the People's Republic of China, Georgia, India, Morocco, Panama, Philippines, Poland, Taiwan and Tunisia are all known to have acquired or ordered the ASR-9 radar.

Contractor

Northrop Grumman Corporation, Electronic Sensors and Systems Division, Baltimore, Maryland.

VERIFIED

The ASR-9 airport surveillance radar

ASR-12 airport surveillance radar

Type

Solid-state terminal airspace surveillance radar.

Description

ASR-12 is a modular, fully solid-state radar that features moving target detection; a separate weather channel; a soft failure transmitter and dual-beam reception in elevation. The equipment is available in fixed site or transportable configurations and is described as meeting all relevant International Civil Aviation Organisation requirements. ASR-12 is reported to be able to provide accurate aircraft position data in poor weather, ground clutter, interference (natural and man-made) and vehicular traffic conditions. The system's signal and data processor design is based on commercial-off-the-shelf hardware with a software implementation of the moving target detection capability. Particular attention is noted as having been paid to maximising azimuth resolution.

Status

Over time, the ASR-12 terminal airspace surveillance radar is reported as having been selected for air traffic control applications in El Salvador, Egypt, Mexico, Peru and Saudi Arabia. The first such system was delivered to Peru during January 1998.

Contractor

Northrop Grumman Corporation, Electronic Sensors and Systems Division, Baltimore, Maryland.

VERIFIED

··

Compact Tracking Radar (CTR)

Type

I-band (8 to 10 GHz) tracking radar for electro-optics tracking platforms.

Description

The CTR is a full angle and range-tracking sensor in a configuration that is easily adaptable to a variety of electro-optical tracking platforms. This I-band radar adds the flexibility of a full angle and range-tracking sensor, significantly improving the capabilities of both new and existing optical tracking systems. The CTR provides long-range, all-weather target acquisition; centroid tracking of clustered targets and a target Doppler data output. The equipment uses a digital moving target indicator receiver to provide coherent signal processing. Recorded data may be used for post-mission pulse-Doppler analysis, providing an additional dimension to the angle and range data available from typical radar and optical tracking systems. The CTR transmitter (with a magnetron and solid-state modulator) provides 85 kW minimum peak output power. Standard metric pulse repetition frequencies to 1,144 pps are provided. The 1 m antenna has a gain of 36 dB, low sidelobes, and a −3 dB beamwidth of 2.5°. The CTR can skin track aircraft beyond 65 km and munitions to 20 km. Beacon tracking capability is included.

Status

As of this edition, an example of the CTR system was reported as having been delivered to an 'international test range' customer during 1996.

Specifications

Frequency: I-band (8-10 GHz)
Transmitter output: 85 kW
Antenna size: 1 m
Antenna gain: 36 dB
PRF: 286, 572, 1,144
Angle accuracy: 0.25 mil RMS
Range accuracy: 3 m RMS

Contractor

BAE Systems North America, Fort Walton Beach, Florida.

VERIFIED

··

Digital Airport Surveillance Radar (DASR)

Type

Digital air traffic control radar and beacon system.

Description

DASR is a combined solid-state primary and monopulse secondary radar system, which is designed to replace existing civil and military AN/ASR-78 equipments. Within its primary radar subsystem, DASR incorporates a modular solid-state transmitter, a dual-beam antenna, digital pulse compression and programmable digital signal processors, while the monopulse secondary radar incorporates the potential for a Mode S field upgrade. The system as a whole features fully redundant electronic and data communication systems; automatic primary/secondary data combination; intelligent workstation-based remote control and remote maintenance monitoring facilities; automatic fault isolation (90 per cent for one replaceable circuit card/module or 95 per cent for three); a performance analysis and diagnostic routine; graceful degradation of transmitter function and a mean time between failure value of better than 1,300 hours.

Status
DASR is a joint US Department of Defense/Federal Aviation Administration procurement that was awarded to Raytheon in August 1996. As then structured, the procurement was in two phases; a preproduction phase running from August 1996 to July 1998 and a production period running from October 1997 to September 2006. In the preproduction phase, Raytheon was to deliver three test radars, develop suitable interfaces for existing automation systems, conduct design qualification tests and support a six month government test and evaluation programme. The follow-on production phase was to cover the manufacture and delivery of 213 DASRs, site survey and preparation work, system installation and test, the provision of site spares and test equipment, operator training and interim contractor/organic depot support for the programme.

Specifications
Range: 111 km (primary surveillance); 222 km (secondary surveillance)
MTBF: >1,300 h

Contractor
Raytheon Electronic Systems, Marlboro, Massachusetts.

VERIFIED

....

GCA/PAR-2000/PAR-2000 Airport Surveillance Radar (ASR)/Precision Approach Radar (PAR) system

Type
All-weather surveillance and landing radar.

Description
GCA/PAR-2000 is an International Civil Aviation Organisation (ICAO) compliant, all-weather surveillance and landing radar that is suitable for both civil and military applications and provides remotely controlled multirunway coverage and a height-finding capability. Based on a proven active array architecture that utilises gallium arsenide transceiver modules, GCA/PAR-2000 is designed to provide simplicity of operation in a high-reliability/availability equipment. A dedicated PAR variant (designated as PAR-2000) makes use of the same subsystems that are used in the GCA/PAR-2000 configuration.

Remotely monitored, each GCA/PAR-2000 site is unmanned. Built-in test routines isolate faults down to replaceable unit level and the system's modularity allows for graceful degradation with continued operation before and during maintenance. The equipment's mean time to repair value is given as 20 minutes. ITT notes that the elimination of high-voltage circuitry, oil reservoirs and system/subsystem alignments offers dramatic improvements in equipment reliability together with reduced training and manning requirements. GCA/PAR-2000 offers four modes of operation (Air Surveillance Radar (ASR) - PAR, ASR, PAR and PAR height-finder) and uses fibre optic links for real-time system control. PAR-2000 offers PAR and PAR height-finding operating modes. Both equipments incorporate high-resolution digital colour workstations with raster-scan colour displays. Data rate for both equipments in height-finder mode is 1 second.

Status
As of July 2001, GCA/PAR-2000 is reported as having completed its formal test programme, with the PAR-2000 configuration being noted as being in service with the air forces of Brazil and the UK. The US Air Force has selected a mobile version of GCA/PAR-2000 (designated AN/MPN-25 - see separate entry) for use in its Tactical Landing System programme. The first MPN-25 delivery is understood to have taken place during 1998.

Specifications
Target and detection definition: 1-1,000 m² (RCS); ±74 to ±463 km/h (velocity - PAR); ±74 to ±741 km/h (velocity - ASR); Swerling 1 (fluctuation model)

An unmanned GCA/PAR-2000 site   0007605

Instrumented ASR coverage: 0-20° (elevation); 0-2,438 m (altitude); every 5 s (update rate - 60 rpm antenna rotation); 28 km (range - rain mode); 56 km (range - clear mode); 360° (azimuth)
Instrumented PAR coverage: −1 to +7° (elevation); 30° (azimuth); 28 km (range - rain mode); 31 m (min altitude above ground intercept point); 37 km (range - clear mode); 1/s (update rate)
Aircraft target processing: 22 plots/scan (PAR targets - elevation); 50 plots/scan (PAR targets - azimuth); 250 plots/scan (ASR targets)
Weather processing: entire radar coverage area - 3 levels
Reliability: 2,212 h (MTBCF)
Maintainability: 2 h (routine maintenance - once/quarter)

Contractor
ITT Gilfillan, Van Nuys, California.

UPDATED

....

Laser instrumentation radar

Type
Laser-based instrumentation radar.

Description
The BAE Systems series of laser-based instrumentation radars has been designed for use in very high-accuracy tracking applications. The initial system produced is understood to have been used by the Federal Aviation Administration to validate Global Positioning System-based navigation and landing systems. The company's laser radar systems are capable of tracking both augmented and unaugmented targets. Highest accuracy is achieved when tracking a target augmented with a small array of passive retro-reflectors. Equipment is available in high-power 1.06 μm or eye-safe 1.5 μm versions and the systems use a pulsed, high repetition rate laser and an intelligent software-based receiver as their primary tracking sensor. Target angle and range information is derived from a quadrant detection technique similar to a conventional monopulse tracking radar system. Automatic calibration techniques are used to measure anomalies in the atmosphere, tracking mount, and optics. Real-time data correction is used during target tracking to compensate for errors that would otherwise degrade accuracy.

Status
As of this edition, the first BAE Systems laser instrumentation radar was reported as having been delivered during 1998.

Specifications
Frequency range: 1.06 μm or 1.5 μm (eye-safe)
Tracking accuracy: 0.3 m RMS (range); 5 arc/s RMS (angular)
Transmitter output: 10-100 mJ at 1.06 μm or 2-10 mJ at 1.5 μm
Receiver: monopulse, quadrant detector with intelligent digital receiver

Contractor
BAE Systems North America, Fort Walton Beach, Florida.

VERIFIED

....

MR-710 modified Nike tracking radar system

Type
I-band (8 to 10 GHz) tracking radar.

Description
The modified Nike-Hercules radar system retains the best features of the original pedestal, while upgrading the electronics to modern technologies. The complete upgrade provides a cost-effective precision tracking radar for the 1990s and beyond. New equipment installed on the completely refurbished BAE Systems- or customer-supplied pedestal includes a three-channel monopulse receiver, transmitter with pulse-width improvements, a totally new console, solid-state high-power servo amplifiers and a level sensor. A television tracker with optics is optional. The radar is designed for installation in a building, or as a mobile or transportable system.

Status
Over time, the MR-710 tracking radar system is reported as having been procured for use on 'several' US test ranges.

Specifications
Frequency range: 8.5-9.6 GHz
Peak power: 250 kW
RF source: Varian VMX1077A coaxial magnetron
Pulse-widths: 0.25, 0.5 and 1 μs, selectable from the console
Antenna polarisation: vertical
Antenna beamwidth: 1°

Contractor
BAE Systems North America, Fort Walton Beach, Florida.

VERIFIED

PAR-80 Precision Approach Radar (PAR)

Type
I-band (8 to 10 GHz) PAR system.

Description
PAR-80 is a fixed site, turntable-mounted PAR that features a pencil beam planar-array antenna employing ferrite phase shifters for azimuth beam displacement and mechanical actuators for vertical beam positioning. Radar scan angles of 30° in azimuth and 7° in elevation when combined with the 19 to 37 km range range provide large service volume coverage for all aircraft types and traffic pattern variations. PAR-80 utilises circular polarisation for operation in adverse weather, with optional constant false alarm rate circuitry. The radar's components (antenna, scan programmer and its dual transceivers) are pallet mounted on a turntable that is remotely controlled for coverage of multiple runway approaches. Control/indicators (with an accompanying video processor) may be remotely controlled from operations centres or control towers up to 3,048 m distant. Other system features include:

- solid-state design
- digital clutter-reduction subsystems
- dual-channel radar transceivers
- a high-gain low-sidelobe antenna
- remote control of PAR landing direction with four sets of preset course and glideslope cursors.

Status
Over time, the PAR-80 PAR system is reported as having been procured by the German Air Force and Navy and the Belgian Air Force.

Specifications
Frequency: I-band (9,000-9,160 MHz)
Pulse-width: 0.2 µs
PRF: 3,450 Hz
Peak power: 150 kW
Beamwidth: 1.1° (horizontal) × 0.6° (vertical)
Antenna gain: 45 dB
Sidelobes: −22 dB max
Max range: 37 km
Resolution: 70 m (range); 0.6° (elevation); 1.1° (azimuth)
Receiver noise: 11 dB max
Signal processing: lin/log FTC and linear; log CFAR (optional)

Contractor
ITT Gilfillan, Van Nuys, California.

VERIFIED

The ITT Gilfillan PAR-80 precision approach radar

RIR 716 instrumentation radar

Type
G-band (4 to 6 GHz) range instrumentation radar.

Description
The RIR 716 range instrumentation radar is an upgraded AN/FPS-16 system that can be configured for mobile, transportable or static installations. It is designed for

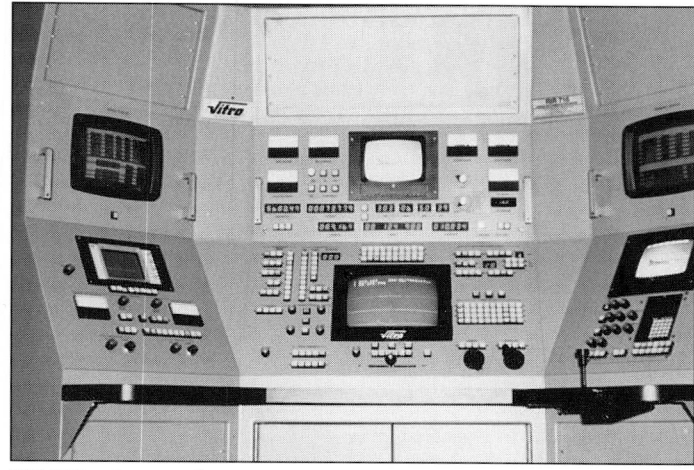

RIR 716 radar console

a variety of test range applications. The equipment retains the better features of the AN/FPS-16 antenna/pedestal assembly integrated with the latest electronics technology. An electro-optical system provides angle tracking, acquisition aid and star calibration capabilities. Highly accurate time/space/position information is provided on various types of airborne targets. The interactive computer and system software enables precise calibration, data correction, data analysis, vector track and timely reconfiguration to meet changing test requirements. RIR 716 is a three-channel monopulse system that includes an interactive computer and modular software and hardware that enables extensive flexibility in configuration to meet unique requirements. Star calibration techniques are used to identify mechanical anomalies such as mislevel, radio frequency skew, encoder bias, droop and non-orthogonality. The output data is corrected in real time for anomalies. Extensive software such as trajectory acquisition to aid in acquiring track on various types of projectiles and satellite acquisition has been developed for the system. Software and hardware is periodically updated to meet new requirements and these updates are made available to all users. A three-channel digital moving target indicator has been added to recent systems.

Status
Over time, at least eight RIR 716 range instrumentation radars are reported to have been procured.

Specifications
Frequency: G-band (5.4-5.9 GHz sub-band)
Tracking accuracy: 2.74 m (range); 0.1 mil RMS (angular)
Azimuth coverage: 360° continuous
Elevation coverage: −10 to +180°
Transmitter output: 1 MW
Antenna size: varies from 4-5 m diameter

Contractor
BAE Systems North America, Fort Walton Beach, Florida.

VERIFIED

RIR 772 range instrumentation radar

Type
G- (4 to 6 GHz) or I-band (8 to 10 GHz) range tracking radar.

Description
The RIR 772 range instrumentation radar is an active radio frequency tracking system designed for precision tracking of rockets, missiles, aircraft and other types of airborne targets. It is available in a mobile, transportable or fixed configuration. It provides continuous and highly accurate time/space/position information to support test range programmes. Standard configurations are G-band (5.4 to 5.9 GHz sub-band) or I-band (8.5 to 9.6 GHz sub-band) frequencies with 2.44 m or 3.05 m parabolic reflectors (RIR 772A or RIR 772B). The transmitter is a magnetron type with hard tube modulator, giving a peak output power of 250 kW. As an option, a 1 MW output G-band transmitter is available. In its standard configuration, the RIR 772 consists of an antenna/pedestal assembly with all additional equipment housed in the electronics shelter. Included in the shelter are the operator's console, transmitter, receiver, digital range tracker and other integral and ancillary equipment.

Status
As of this edition, the status of the RIR 772 instrumentation radar was uncertain.

Specifications
Transmitter: magnetron with hard tube modulator
Power output: 250 kW
Pulse-width: 0.25, 0.5 and 1 µs
Duty cycle: 0.001 max

Pulse coding: 2 0.25 or 0.5 μs pulses spaced 3-12 μs
Range: RIR 772A 115 km on I-band, 102 km on G-band; RIR 772B 93 km on I-band, 78 km on G-band (these figures are based on the range that a 1 m² RCS target can be skin tracked with a 12 dB above noise return signal)

Contractor
BAE Systems North America, Fort Walton Beach, Florida.

VERIFIED

RIR 778 and RIR 779 instrumentation radars

Type
G- (4 to 6 GHz) and I-band (8 to 10 GHz) range instrumentation radars.

Description
The RIR 778 and RIR 779 range instrumentation radars were designed to meet the constantly changing requirements of test range work. These three-channel monopulse systems include an interactive computer and modular software and hardware that enables extensive flexibility in configuration to meet unique requirements. Star calibration techniques are used to identify mechanical anomalies such as mislevel, Radio Frequency (RF) skew, encoder bias, droop and non-orthogonality. The output data is corrected in real time for all anomalies. Extensive software such as trajectory acquisition to aid in acquiring track on various types of projectiles and satellite acquisition has been developed. Software and hardware are periodically updated to meet new requirements and these updates are made available to all users. A three-channel digital moving target indicator has been added to recent systems. In addition to automatic Radio Frequency (RF) track, a TV tracker enables optical tracking in angles. Vector track can be accomplished in either RF or optical track modes of operation. Systems have been manufactured for mobile, transportable and permanent installations and G- (5.4 to 5.9 GHz sub-band) and I-band (8.5 to 9.6 GHz sub-band) variants.

Status
Over time, BAE Systems is reported as having delivered at least 42 RIR 778/779 radars to US and international customers.

Specifications
Frequency range: G-band (5.4-5.9 GHz sub-band); I-band (8.5-9.6 GHz sub-band)
Tracking accuracy: 2.75 m RMS (range); 0.1 mil RMS (angular)
Transmitter output: up to 1 MW
Antenna size: antenna sizes vary from 2.44-5 m diameter

Contractor
BAE Systems North America, Fort Walton Beach, Florida.

VERIFIED

The antenna/pedestal assembly used in the BAE Systems RIR 779 instrumentation radar

---

ROR 721 range-only radar

Type
I-band (8 to 10 GHz) range-only radar.

Description
The ROR 721 is designed for integration with electro-optical tracking mounts to provide real-time range data. The system uses an I/low J-band transceiver unit with a 0.4 to 1.0 m cassegrain antenna. Range tracking and system control are accomplished via a ruggedised PC-AT bus computer with data storage and RS-232/422 interface capability. The system requires minimal operator interface and is capable of tracking a 0.1 m² radar cross-section target to at least 12 km. Tracking range can be increased by incorporation of a larger antenna and a higher power transmitter. Addition of the ROR 721 enables a single-station solution for electro-optical tracking mounts. The radio frequency transceiver operates at approximately 9.4 GHz with an intermediate frequency of 60 MHz. Transmitter power is 10 or 25 kW with pulse-widths of 0.25 and 0.5 μs, and a pulse repetition frequency of 1,280 pps. The noise figure of the system is 6 dB nominal.

Status
Over time, at least two ROR 721 range-only radars are reported as having been procured by customers in the USA.

Specifications
Frequency: 9.4 GHz
Transmitter output: 10 or 25 kW
Antenna size: 0.4-1.0 m
Pulse-width: 0.25, 0.50 μs
PRF: 1,280 pps

Contractor
BAE Systems North America, Fort Walton Beach, Florida.

VERIFIED

Series 52 Precision Approach Radar (PAR) system

Type
I-band (8 to 10 GHz) PAR system.

Description
The ITT Gilfillan Series 52 is a turntable-mounted PAR. The radar data is remotely fed to control centres or control towers. Air traffic controllers provide precise approach guidance to aircraft down to the runway threshold through voice communications with the pilot. The system's turntable is remotely controlled for aligning to cover multiple runway approaches. Up to six radar displays may be remotely controlled a maximum 3,636 m from the equipment shelter. Radar ranges of 19, 28, and 37 km provide coverage for all types of aircraft and all traffic pattern variations. For transportation, the Series 52 is packaged into two major subassemblies: the shelter and turntable pallet. Radar scan angles of 20° in azimuth and 8° in elevation are developed by the delta-A scanning antennas. These antennas feature minimum sidelobe performance and circular polarisation, coupled with a digital moving target indicator and integrated video. Dual-channel radar transmitters and receivers and high-gain, low-sidelobe antennas, remote control of PAR landing direction and four sets of preset course and glideslope cursors are operational features of the Series 52 PAR. Over time, military versions of the Series 52 included the US Navy's AN/FPN-63 and Royal Navy's MPN-23. Another derivative of the Series 52 has been sold to the Canadian Forces as the FPN-503.

Status
As of this edition, the status of the Series 52 PAR system was uncertain.

A Series 52 PAR system

Specifications
Frequency: I-band (9,000-9,160 MHz)
Peak power: 80 kW
Azimuth antenna gain: 39.7 dB
Elevation antenna gain: 40.3 dB
Max range: 37 km
Resolution: 48 m (range); 0.55° (elevation); 0.85° (azimuth)

Contractor
ITT Gilfillan, Van Nuys, California.

VERIFIED

SPC instrumentation radar systems

Type
Range of instrumentation radar systems.

Description
System Planning Corporation (SPC) has been a pioneer in the design and manufacture of instrumentation radar systems and ranges for the measurement of very low Radar Cross-Sections (RCS). In addition to designing, developing and fabricating the hardware, SPC designs radar measurement ranges, develops advanced measurement and processing techniques and software and provides 24-hour support to field operations. A wide range of standard and customised ultra-wideband imaging radars is available, together with data processing, workstation and component products, measurement and range system design service, and turnkey systems. SPC instrumentation radars have been used effectively in quality control tests of low-observable systems such as the SR-71 reconnaissance aircraft, the F-117 stealth fighter, the B-2 bomber and the Advanced Cruise Missile. The company's mobile Mk IV range radar is easily adapted for employment on truck and airborne platforms and is a pulsed, coherent, frequency-agile measurement system that collects data for processing and display in a variety of formats, including RCS versus azimuth/frequency/downrange, phase versus azimuth, and two-way inverse synthetic aperture radar images.

Status
Over time, at least 30 SPC instrumentation radars are reported as having been procured for use on US government and aerospace industry test ranges.

Specifications
Mark IV system
Frequency coverage: UHF through J-band (0.35-18 GHz - other bands available)
Output power: 2 W-1 kW (peak)
Max PRF: 1 MHz
Min pulse-width: 5 μs
Frequency: agile pulse-to-pulse, enabling multiple frequencies to be collected during 1 target rotation
Antenna: 0.45 m offset-fed dual-polarised 8-18 GHz; 1.22 m offset-fed polarised 2-18 GHz; 1.83 m centre-fed single polarisation I-band (system can be adapted to customer provided antennas)

Contractor
System Planning Corporation, Arlington, Virginia.

VERIFIED

Terminal Airport Surveillance Radar (TASR)

Type
E-band (2 to 3 GHz) airport surveillance radar.

Description
TASR is an all-weather air traffic control airport surveillance radar. It is a modular system that can be configured to meet any operational requirement. Configurations include a trailer-mounted mobile system for tactical operations or a fixed-based system. The system has a dual redundant transceiver and is also adaptable to preferred secondary surveillance radars. TASR radiates at E-band (2.7 to 2.9 GHz sub-band) with a digital solid-state transmitter at the heart of the system. The all solid-state radar also includes a dual-beam antenna and high-resolution digital colour workstation displays. The TASR is suitable for both commercial or military air traffic applications. Based on a proven modular design, the TASR provides simplified operations with improved reliability and availability. The solid-state system is claimed to provide significant benefits over existing tube-type radar technology.

Status
As of this edition, the civilian TASR variant was reported as having completed its integration and testing programme while at least one military configured system is understood to have been supplied to a 'classified' customer during the late 1990s.

Specifications
Air surveillance coverage: 111 km (range); 7,620 m (altitude); 360° (azimuth)
Air surveillance data rate: 4.8-5 s

Contractor
ITT Gilfillan, Van Nuys, California.

VERIFIED

The antenna array used in the TASR system

ELECTRONIC WARFARE SYSTEMS

ELECTRONIC WARFARE SYSTEMS

Introduction

Electronic Warfare (EW) is one of the key elements of the modern battle scenario, protecting one's own forces from attack, denying information to the enemy and intercepting and disrupting an enemy's voice communication and datalinks. In effect, EW is a continuing war between active systems that 'attack' and defensive systems which protect. A widespread network of electronic intelligence stations is operated by many countries by land, sea and air, not only to monitor the electromagnetic spectrum, but also to disrupt hostile transmissions by jamming in a number of ways.

In the land-based role, much of EW equipment is dedicated to the passive electronic intelligence role and to disrupting the communications links and surveillance systems of the other side. The naval scenario is somewhat different and, although considerable use is made of intelligence gathering, the prime role of naval EW is the protection of the unit or fleet from aircraft or missile attack. In the airborne applications, EW is employed both for intelligence gathering and for protection of the aircraft from surface and air-launched missiles. In space the major powers have developed a number of ELINT satellites for overall surveillance of the radar and radio frequencies. Although much of EW is in the electromagnetic spectrum, the increasing use of Infra-Red (IR) guidance systems for missiles and lasers for target marking and range-finding, has meant that these wavelengths are now the subject of both jamming and decoy equipments.

Description

There are three basic types of electronic warfare:

- passive-Electronic Support Measures (ESM)
- active-Electronic CounterMeasures (ECM)
- anti-ECM Electronic Counter-CounterMeasures (ECCM).

One of the problems with the last two types is the continuous development of one equipment to counter the other. As the ECM specialist produces systems to provide jamming and decoy methods so the ECCM engineer develops equipment to overcome these methods. Frequently the same manufacturers are doing both!

Electronic Support Measures (ESM)

There are two basic types of ESM: ELectronic INTelligence (ELINT) and COMmunications INTelligence (COMINT). The latter, as its name implies, is intended for the interception of communications, whether by voice or datalink. The former is primarily dedicated to the interception and analysis of radar emissions from surveillance, fire-control or missile guidance radars and is often allied to an ECM system to provide protection from these. The combination of ELINT and COMINT work is known as SIGnals INTelligence (SIGINT).

COMINT provides both interception, direction-finding and analysis of hostile transmissions, primarily to assess the movements and intentions of the opposing forces. Analysis of the signals provides much valuable information of the intentions for command and control purposes. The most recent systems provide the operator with the ability to detect and analyse unusual and complex signals, as well as the normal interception and Direction-Finding (DF) facilities. The receiving equipment is frequently allied to a computer-based processing and display system so that automatic position fixing in the land-based role can be carried out by the use of remote-controlled DF stations. Spectra and/or time waveforms are normally provided, together with alphanumeric readouts which include type of transmission, frequency, modulation and other signal parameters. These parameters are used to determine the types of communication and radar systems in use, whether they are mobile or static, the direction of any movement and so on. Multisignal detection and analysis is provided in nearly all equipment. A map display, overprinted with the intercepted information can be incorporated to give the battlefield commander an overall picture of both the tactical and the electromagnetic situation.

In the airborne surveillance role, information is often passed to the ground by a datalink and is also recorded for subsequent detailed analysis. Increasing use is being made of remotely piloted vehicles for this purpose.

For platform protection ELINT is vital, particularly in the airborne and shipborne roles, in that it provides not only direction-finding but also analysis of the incoming signals to provide immediate warning of threat radars, including surveillance, fire control, targeting and missile guidance systems. Signals from radar systems are intercepted by a warning receiver and are analysed by an associated processor to give a wide range of parameters, including direction, type of radar, frequency, frequency agility, Pulse Repetition Frequency (PRF) and PRF type. These parameters are usually sufficient to characterise the type of emitter and complete identification is then carried out by comparing the analysed signal with parameters of hostile and friendly emitter characteristics stored in a library within the computer memory. Analysis of the signals and warning of a threat is virtually instantaneous and enables countermeasures of jamming and/or decoys to be initiated. These can be carried out either automatically or manually.

For aircraft, ships and armoured fighting vehicles effective warning systems are essential for survival in the electromagnetic threat environment of the modern battlefield. The warning receivers mentioned in the previous paragraph are being continuously updated to cope with the latest threats. These receivers are normally either crystal video or superheterodyne-based equipments, both of which have their own advantages. Crystal video receivers, either narrowband or wideband, can operate over a frequency range from 0.5 to 40 GHz, covering all radar transmissions except those in the 94/95 GHz millimetric waveband. They are effective against pulsed, frequency-agile, PRI-agile, spread spectrum and continuous-wave transmitters. Superheterodyne receivers are more expensive but provide a coverage from 0.01 to 40 GHz with a high level of sensitivity, plus long pick-up ranges and sidelobe penetration. More recently, the use of Very High-Speed Integrated Circuit (VHSIC) technology has been introduced to provide much smaller equipments with very high scan speeds. Another interesting system is that of acousto-optic receivers performing signal processing using Bragg cells. In these cells, RF signals are converted into acoustic waves which are then sampled with light beams. This is a type of technology which provides the capability of handling large instantaneous bandwidths, fast response times, 100 per cent probability of interception and the ability to detect low-level signals in heavy clutter. A number of companies in the USA and Western Europe have been working on this technology for several years.

Both ELINT and COMINT systems are heavily dependent on the digital computer to provide all the analysis functions. The software program upon which these functions are based is keyed in before any operational mission and will carry out the necessary analysis on multiple signals. The library of the average processor will contain the parameters of 2,000 or more radar systems and, in many cases, can be reprogrammed by the operator to store unidentified signals for later processing and analysis. The processing involved consists of three stages in series: sorting of the radar pulses as they come in, segregation of the pulse trains and identification of the emitters. The system must be able to operate in a dense electromagnetic environment, discriminating between actual signals and clutter. Advanced processing techniques to de-interleave complex radar signals and minimise false alarms are essential. A key feature of all these processors is speed and accuracy, so that defensive measures can be taken before it is too late.

The operator's display varies from a simple radar warning receiver in the aircraft cockpit, giving the direction of an incoming radar-guided missile, to a graphic and/or alphanumeric readout on board a ship which provides complete information, together with a threat priority identification, when multiple signals have been analysed.

Of major importance also is the antenna system and these vary widely depending on the type of ESM equipment and the platform on which they are mounted. For obvious reasons aircraft antennas have to be small and lightweight and consequently provide less facilities than their ground-based or shipborne contemporaries. For the latter applications a wide variety of passive and active dipoles and DF antennas, both directional and omnidirectional, are employed. Submarines have a particular need for antennas that give the least possible radar cross-section to avoid detection and are sufficiently compact to be mounted in the restricted space on the fin. More recently, the use of millimetric wave radars has presented a major threat, although countermeasures systems are now coming into service.

One of the most important developments in recent years has been the use of infra-red homing systems as the 'seekers' in both surface-to-air and air-to-air missiles. They are already in widespread use and this has meant that warning systems have had to be developed to provide indication of missile lock on, with the indicator normally combined with that of the radar warning receiver. One of the more recent systems to become operational is the Missile Approach Warning (MAW) equipment, using pulse-Doppler radar techniques, to alert aircrews of infra-red guided missiles homing in from the rear of the aircraft and to provide range and closing speed of the missile. However, radar-based MAW systems are active and many defence forces prefer passive systems which give no emissions. Laser-based systems have also come to the fore in recent years and are used for target designation and marking, as well as illuminators for 'smart' weapons and as the guidance and homing for beam-riding missiles. Several companies have developed systems to warn of laser 'painting'. Millimetric wave radar and laser threats are particularly applicable to battlefield helicopters, ground attack aircraft and armoured fighting vehicles. As far as the latter is concerned, there has been considerable development in the warning systems for Armoured Fighting Vehicles (AFVs) but very little effort to provide effective countermeasures.

Electronic CounterMeasures (ECM)

ECM is the active part of EW and is intended to disrupt the surveillance systems of the enemy, whether by radar or radio communications and also to counter any of his weapons which use electromagnetic, infra-red or laser systems for guidance or aiming. There are two main methods of achieving this: by jamming, or by the use of decoys, both of which are effective when used properly. Many modern ECM equipments, particularly in the naval scenario, employ both methods in an integrated system.

Noise jamming is the use of transmissions to disrupt the enemy's communications channels or to saturate his radar to obscure its target. Although this denies the enemy his information channels, it also means that the jamming source cannot read the signals for intelligence purposes. Apart from this, modern frequency-agile communication systems are no longer easy to jam effectively. In addition, what if the military communications network or at least some of it, is 'hidden' in the civil network? This makes it very difficult to detect and even more difficult to jam without disrupting the civil communications network. As far as radar jamming is concerned, whilst noise jamming may be effective, it does have the drawback of bringing your presence to the attention of the enemy, particularly in the case of naval forces. It can also act as a homing aid to missiles, most of which have a home-on-jam capability. For this reason the latest techniques of anti-missile jamming employ noise and deception on a complementary basis, using noise initially and then providing electronically generated false targets, such as chaff clouds infra-red flares and offboard decoys (see later). Simple noise jamming is still in widespread use in the land warfare scenario, one important application being in remotely operated expendable jammers. These can be hand-emplaced, artillery-delivered, dropped from aircraft or used in unmanned aerial vehicles and serve as short term jammers for a particular operation.

To combat infra-red guided missiles a jammer, normally employing a pulsed radiation transmitter, is used and transmits modulated infra-red energy. This is an omnidirectional device which has the effect of degrading the ability of the guidance head and rendering it ineffective. These devices are largely employed on aircraft and are normally externally mounted, either in pods or in the engine nacelles.

The second method of ECM is the use of decoys, either chaff in the case of electromagnetic threats or flares to combat infra-red devices. The use of chaff goes back over 50 years to the Second World War and the material itself has changed very little. What has changed has been the method of dispersal and this varies according to the type of platform. For aircraft it takes the form of an internally or externally mounted dispenser which ejects chaff-filled cartridges, either manually or automatically, in combination with a radar warning receiver. In some cases a chaff cutter dispenser is employed so that the operator can select various tuned dipole lengths to meet changing threats in frequency bands. These dispensers are used both for aircraft self-protection and for 'seeding' selected flight paths for large scale operations. For infra-red countermeasures, flare cartridges are ejected from the dispensers and most dispensers have a dual role of carrying both chaff and flares. It should be noted, however, that several techniques that enable infra-red missile seekers to detect and reject the IR emissions from flares have now been developed. Most flare rejection techniques use two phases - 'switch', which basically measures a sharp rise in energy level as being caused by a flare and 'response', which rejects the flare emission and reacquires the original target.

During the past few years a few manufacturers have been working on active off-board airborne decoys. These are normally towed decoys, although free-flight decoys are also now available, either 'glider' types or powered decoys which emulate an aircraft's flight path. However, they are much more expensive than chaff or flares!

For naval applications, both chaff and flares are deployed by means of rockets to present an alternative target to a missile radar during its search phase. The decoy must have similar characteristics to that of the intended target to be of any use, although this usually means that the decoy only needs a radar cross-section larger than the intended target. To be most effective it should also be deployed in time and position, so that the missile radar locks on to it before it can see its intended target. This is known as the 'distraction' mode and provides a number of separate decoys deployed in a pattern around the ship to present several alternative targets to hostile missiles. An alternative mode, known as 'seduction', is used when the missile has already locked on to its target. In this mode, the ship deploys a decoy above or by the side of itself and then moves away while the missile remains locked on to the bigger echo. In both the distraction and seduction modes considerable use is made of the wind to allow the decoy to drift downwind with the missile locked on to it. Yet another alternative is the use of 'offboard' decoys, either floating or parachute-suspended, which emit the same signal as the ship's radar, but more strongly.

Electronic Counter-CounterMeasures (ECCM)

ECCM is the method by which you endeavour to combat the ECM systems of the enemy by either making your equipment ECM-resistant or by using techniques to nullify his jamming and/or decoy systems. It is an extremely sensitive area in that any disclosure of ECCM measures designed into a system are likely to inform the enemy of its vulnerability to ECM.

Against jamming systems, the most commonly used method is frequency agility, whereby the transmissions are made to 'hop' over a large frequency band in a random fashion. This means that either the jammer has to spread its power over the entire band with the inevitable loss of strength on any particular frequency, or it must attempt to follow the signal as it hops randomly. The latter is a difficult and costly method, often employing the Bragg Cell technique.

In the past decade, a number of anti-radiation missiles have been developed and these pose a considerable threat to both land-based and shipborne surveillance radars. During the 1990-91 Gulf War, nearly 1,000 anti-radiation missiles were launched at Iraqi air defence and tactical radar sites with a very high success rate. The missile is passive in operation so that it cannot be picked up by ESM systems and normally locks on to the sidelobes of the radar transmission. The main countermeasures against this type of missile are low sidelobes, frequency agility and the use of decoy transmitters, which must be positioned close enough to the surveillance radar to 'seduce' the missile, but not so close as to endanger the main system.

A great deal of development effort has recently gone into the use of passive surveillance systems for detection. These are either optical or infra-red in design and, because they do not transmit, are not vulnerable to the anti-radiation missile. They are, however, very limited in range and are still vulnerable to infra-red or optically guided missiles. A number of the latter type have been in operational use for some years, using a television camera in the nose. The defence against these is normally by the use of a smoke screen and several systems which combine the deployment of smoke, chaff and infra-red flares are in current use.

Finally, the latest advance is the use of 'stealth' techniques to combat the radar system. This is beginning to be employed in both ships and aircraft and consists of a number of methods to reduce the radar echoing area of the vehicle. The main techniques employed are: firstly, to design the airframe itself to avoid sharp corners and flat surfaces which act as radar reflectors and secondly, the use of radar absorbent material which minimises the amount of energy reflected back to the radar. In the case of aircraft (and two of the latest American aircraft, the F-117 and the B-2, are examples of this) the most important parts of the fuselage can be covered in radar absorbent material to make it extremely difficult to detect. Ships, with their very large radar reflecting area, are a different proposition with the main efforts being centred on the reduction of radar cross-section and the optimisation of offboard decoy performance. During the past few years a great deal of effort has been aimed at the theoretical analysis of vehicle radar signatures, particularly those of warships, aircraft and armoured fighting vehicles, to provide mathematical modelling for signature reduction purposes. These efforts, of course, are of no consequence against infra-red guided missiles and the only recourse there is to reduce the vehicle's 'hot spots' to present a more difficult target for homing.

EW in the 1990-91 Gulf War

In some respects, the Gulf War was a bonus for the various electronic warfare manufacturers in that many of the equipments supplied for aircraft, ships and land forces underwent their first tests in actual warfare. On the whole the ESM surveillance systems, the jammers and the self-protection devices operated as planned. The initial stages of the battle were fought entirely in the air with the allied aircraft using their electronic suppression equipment to jam the Iraqi communications networks and air defence radar systems. The F-4G Wild Weasels, EA-6B Prowlers, EF-111As, Compass Call EC-130s, Tornados and other dedicated electronic warfare aircraft flew thousands of sorties to deny use of the Iraqi command, control and communication networks. In addition, the use of the HARM and ALARM anti-radiation missiles against radar sites virtually destroyed the Iraqi systems. ESM systems were used extensively to locate Iraqi air defence sites and armour and, in particular, the mobile Scud missile sites. Communications and electronic intelligence aircraft included RC-135s, TR-1s, EP3s and EA-3Bs.

Comments from allied commanders stated that the electronic combat force closed down the Iraqi air defence network completely during the first 24 hours of the air war. In addition to the radar and command and control systems, Iraq was reported to have several hundred Soviet SAMs and over 100 Roland SAMs. However, the command and control system was highly centralised and the communications jamming by the EC-130 aircraft disrupted the communication links from the central command to the point where there was no control of the radars, missiles and artillery batteries.

One factor that did occur during the Gulf War was the very high signal density in the electromagnetic spectrum. It had always been assumed that the 'Cold War' scenario presented the greatest threat in terms of signal density and 'local' wars would present little or no problem. Much to everyone's surprise this was not the case and although the extreme density across the spectrum by the use of radars and EW systems by both sides did not jeopardise the battle, it did cause a certain amount of rapid rethinking, particularly in the ESM field.

Although the net result of the electronic warfare side of the battle was to enable the allied aircraft to operate with almost complete safety, it should be remembered that the allied force was overwhelming. This, together with the virtual lack of opposition by the Iraqi Air Force, makes it rather difficult to assess the value of EW against a highly sophisticated air defence system and a strong opposition air force.

The Future

What of the future in this era of glasnost, with relations between East and West improving? Is electronic warfare all that necessary, is the development of better and better systems a real requirement? Only time can tell. The era of glasnost is only a few years old and the break-up of the Soviet Union into its constituent republics poses a difficult scenario for the West. Although the Russian Federation and Associated States has an overall Commander-in-Chief for the armed forces, there is already a heated argument as to the disposition of some of these forces. In addition to any problems of the former Soviet Union, there are many flashpoints in the world which could escalate into more serious conflicts.

Despite wars and rumours of wars, there has been a move towards cutting back on defence procurement and electronic warfare is likely to be just as hard hit as any other sector. An example of this is the AN/ALQ-165 ASPJ programme which has had considerable problems before being officially cancelled, although it has now made a comeback in the export market. The INtegrated Electronic Warfare System (INEWS) is another area which is being closely looked at by the US Government. If the world situation really quietens down, then there is bound to be a major cutback in arms procurement and the development of new techniques, new equipment and advanced technology in electronic warfare will slow down. In this case, the probability is for more and more updates of current equipment by way of improved software and hardware, with less money being spent on the forward edge of technology projects. There will also be a greater demand to update existing platforms by retrofitting currently available EW equipment, rather than developing new platforms.

If one looks back to the mid-1980s when the 'Cold War' was still in existence, the attitude towards a new EW requirement was to design and develop a totally new system. Technology was improving rapidly, new threats were appearing and there seemed little problem in obtaining Research and Development (R&D) funding. Times have changed and the latest technique is known as P^3I (PrePlanned Product Improvement). This can be defined in two ways: long-range plans for the upgrading of systems already in operational service and building into new equipments the capacity and capability for future improvement packages without large scale and expensive modification programmes. An example of this is the US Navy's next airborne EW procurement, the Integrated Defensive Electronic CounterMeasures (IDECM) system, an upgrade to the F/A-18 E/F ECM capability. It is intended that it will build largely on existing EW equipment, including the AN/ALR-67(V)2/(V)3, a towed, active radar decoy, AN/ALE-47 and a missile warning system.

Electronic warfare is probably fortunate in that it represents the one area of defence which is so important that even defence ministries are reluctant to cut it back too far. If one considers the value of capital equipment, such as a ship, an aircraft or an AFV, to say nothing of the personnel operating the platform, the provision of efficient warning and countermeasures systems is really only an insurance policy.

New technologies such as data fusion are emerging to provide an overall picture of information from a variety of active and passive sensors. Combat identification is an area which is attracting much attention. There is an increasing use of lasers for both protection and offense, although their usage as 'blinding weapons' against troops raises justifiable humanitarian considerations. Directed energy weapons, as a means of jamming, are also being investigated.

However, the key to many EW systems, or as one manufacturer put it, 'what separates the leading EW suppliers from the second string' is the database and its ability to be large and fast enough to carry all the parameters that are likely to be encountered and provide rapid analysis. It must be capable of being reprogrammed quickly, preferably in the air in the case of airborne equipment, with its library tailored to the current operational scenario and constructed to the needs of the user.

Conclusions

The electronic warfare scene is a continuously evolving battle between the various aspects of ESM, ECM and ECCM. With the complexity of modern weapons and the speed of reaction necessary to combat them, the weak link in the chain would appear to be the human being who has to make the decision! This is not necessarily the case, because in some instances an operator is far better than an automatic processor. He can interpret situations based on previous experience more readily and can alter his thresholds easily to perform basic functions such as detecting a signal in heavy background clutter, whereas a machine can only operate at the threshold for which it has been programmed. Nevertheless, great strides have been made in artificial intelligence systems and, although there is still a long way to go, the era of complete automation will eventually arrive. The amount of raw information from modern sensor systems is so vast that better and better processors, employing highly complex software programs, are vital to analyse the inputs. This speed and complexity is such that the operators must be highly efficient and an extensive business in providing EW training and simulation systems has grown up over the past years. Even so, this does not help, say the pilot of a single-seat fighter who is being presented with a vast amount of electronic information and, perhaps, only a second or so to react against a missile attack. All that this means, in that type of situation, is that the warning system must be fully automatic in its countermeasures role, with an overriding manual facility as a safety measure.

Another vital component is the provision of programmable software so that the system program can be changed easily. It is interesting to note that EW systems used in the Gulf War were designed to cope with Soviet missiles and radars but, in some cases, found themselves faced with Western systems. Fortunately the most up-to-date radar warning receivers and jammers are software controlled and were able to be reprogrammed to meet the threats.

VERIFIED

LAND-BASED SIGNALS INTELLIGENCE (SIGINT), ELECTRONIC SUPPORT AND THREAT WARNING SYSTEMS

AUSTRALIA

CELTIC High Frequency Direction-Finding (HF/DF) systems

Type
Communications band DF system.

Description
CELTIC is an integrated rack-mounted, modular HF/DF system which may be operated independently or as part of a SIGnals INTelligent (SIGINT) system. It is capable of being configured to suit many diverse roles in strategic and transportable environments. Established methods and techniques have been applied to the communications-surveillance field, where 'processing gain' at the receiver is used to detect, acquire, DF and analyse communication signals.

Current requirements to extend the ground and sky wave detection range against weak emitters (−10 dB signal to noise ratio in a 3 kHz bandwidth) have set the normal threshold of processing gain at about 20 dB. The technology employed in CELTIC permits this to be increased to at least 40 dB, if the situation demands, allowing acquisition of signals well into the noise levels.

Superior signal selection is achieved in CELTIC by a combination of spectral filtering in the frequency domain and windowing in the spatial domain. Proprietary algorithms also allow the accurate labelling of each frame captured data with a figure-of-merit quality indicator. These algorithms permit reliable bearing estimates on ground waves out to 80 to 100 km, on surface waves over the sea out to 250 to 300 km and on sky waves out to beyond 3,000 km. CELTIC also employs algorithms to deduce the position of an emitter in the HF sky wave mode from a Single Site Location (SSL) to an accuracy normally better than 10 per cent of range.

The CELTIC system has been designed to operate efficiently in areas about the geomagnetic equator where ionospheric conditions normally affect the accuracy of azimuth and elevation measurements. CELTIC operates at 40 frames/s and measures both elevation and azimuth angles of arrival on ground wave and sky wave signals. Overall, the system's key features include interferometry; SSL Function; speed and accuracy; commercial-off-the-shelf hardware; configurability; effectiveness against short signals; interference rejection; the ability to be networked; SIGINT support and PUSHER upgradeability.

In terms of SIGINT system applications, CELTIC is noted as being 'easily' incorporated into architectures such as BAE Systems ASIS. ASIS has been designed around a modular architecture which allows a variety of system configurations to be purpose built, using common modules, to meet a range of different operational requirements for defence and non-defence applications in war and peace. Major functions undertaken by the system include: search; monitor; analysis; location (Celtic HF/DF); report generation and system supervision.

To provide the required intelligence, ASIS enables the following tasks to be undertaken:
- receive tasks from a higher authority
- detect and intercept target transmissions
- identify the location of the transmitter
- demodulate the transmission
- extract intelligence from the message being intercepted
- record content as appropriate for message type (text, graphics, voice and so on)
- log intercepts on reporting database
- prepare and issue intelligence reports.

ASIS normally comprises a Master (or Control) Station and a number of Slave (or Remote) Stations linked by a communications system. The open system architecture design of the system enables easy configuration by adding or removing modules to meet the specific operational requirements.

Status
Over time, the CELTIC HF/DF system is reported as having been procured by customers in Pakistan and Papua New Guinea.

Specifications
Frequency: 2-30 MHz - extension to 500 kHz
No channels: 3-7
Frame rate: 5/s (4 channel); 40/s (7 channel)
No antennas: 3-7 dipole or loop elements
DF method: cross spectral interferometer
Effective bandwidth: 24.4 Hz
Frequency resolution: 10 Hz
Min sig duration: >5 m/s
S/N threshold: −10 dB in 3 kHz
System sensitivity: better than −120 dBm
Instrumental DF accuracy: 0.2° at 30 MHz
Typical Sky wave accuracy: <2°
Co-channel rejection: PSD windowing
Multimode rejection: elevation windowing
Co-channel interference: azimuth windowing
SSL: yes (5 options)
Wavefront testing: yes (standard)
Super resolving: yes (optional)
Remote operation: yes (via modem)

Contractor
BAE Systems Australia, Elizabeth, South Australia.

UPDATED

PRISM AD Electronic Support (ES) receiver

Type
Land-based microwave ES surveillance receiver.

Description
PRISM AD (standing for Air Defence) is a land-based version of the naval PRISM ES receiver (see separate entry) and is for air defence and early warning applications. As such, the system can be deployed independently as a stand-alone sensor or as part of a network complementing the information supplied by surveillance radars.

PRISM AD uses the amplitude comparison technique coupled with a distributed processor architecture to provide passive detection, direction-finding and classification of pulsed radar systems in a multi-emitter environment. It has been

The operator's console used in the CELTIC HF/DF system

The PRISM AD ES system mounted in a Land Rover

designed specifically as a transportable system to allow quick and simple installation in a variety of vehicles. PRISM AD may be deployed in a number of ways, including:

- a single PRISM AD can be co-located with a ground-based air search radar
- the system can be used as a remote sensor to provide effective long-range early warning and identification
- it can be mounted in a light vehicle for rapid deployment.

Status
As of August 2001, the PRISM AD system is understood to have been procured by the Royal Australian Air Force.

Contractor
BAE Systems Australia, Elizabeth, South Australia.

UPDATED

BULGARIA

Kintex COMmunications INTelligence (COMINT) systems

Type
Range of COMINT equipments.

Description
Kintex of Bulgaria provides a range of communications equipment that can be used for COMINT and monitoring purposes. Two examples of such units are:

R-55P Receiver
This is a programmable, digitally tuned receiver covering the frequency range from 100 kHz to 30 MHz. It is of modular construction, weighing 25 kg and is designed for continuous operation in both stationary and mobile applications. The fully remote control of the receiver allows it to be built into a complex, computerised system. The receiver functions include a digitally tuned frequency synthesiser with 10 Hz spacing, presetting of 90 channels, automatic scanning in preset frequency range, automatic channel scanning and a non-volatile channel memory.

Jasmine P HF Receiver
Jasmine P is a programmable receiver operating within the 100 kHz to 30 MHz frequency range. It is microprocessor-controlled and designed for use in stationary and mobile services. The receiver functions include a digitally tuned frequency synthesiser of 50 Hz spacing, 99 operating channels memory, scanning capability for a preset frequency range, frequency and channel editing during normal operation and tuning time of 50 ms.

Status
As of this edition, the status of the Jasmine P and R-55P receivers was uncertain.

Specifications
Frequency range: 0.1-30 MHz
Emission types: continuous wave (AAA, A2A); amplitude modulation (A3E, H2A, H2B, HBE); single sideband (R2A, R3E, J2A, J3E, B8E); frequency shift keying (F1A, F1B 125/170/250/500 Hz, F7B 250 Hz)
Sensitivity: (at 12 dB SINAD): ≤0.5 μV (AAA, F1A); =1 μV (J3E, B8E)
Selectivity: 0.3-3.1 kHz (for 3 dB); 1.2-5.6 kHz (for 80 dB)
Automatic gain: >90 dB
AGC delay time: 0.4 (2) s (switchable)
Remote control: RS-232 (1,200 bit/s)
Temperature range: −10 to +50°C (operating)
Humidity at 40°C: 95%
Power supply: 20-27 V (DC); 220 V (AC)
Dimensions: 470 × 177 × 480 mm (19 in model); 480 × 197 × 480 mm (stationary model)

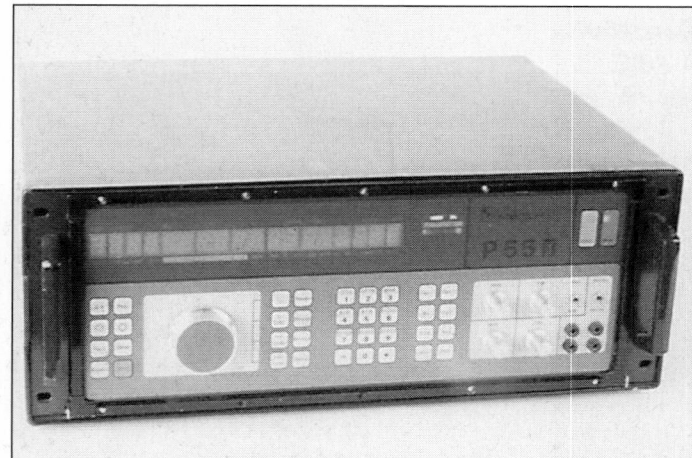

The Kintex R-55P high-frequency receiver

Contractor
Kintex, Sofia.

VERIFIED

Radon SIGnals INTelligence (SIGINT) receiver

Type
C- through J-band (0.65 to 18 GHz sub-band) SIGINT receiver.

Description
The Radon receiver is designed to intercept and convert signals in the 0.65 to 18 GHz frequency band and is intended for use in mobile of fixed site SIGINT systems. In its fixed site application, Radon can be remotely controlled and the complete equipment comprises a base frequency module, six frequency expansion modules, converters, an antenna illumination unit and a packing case.

Status
As of this edition, the Radon receiver was reported as having been procured by the Bulgarian Army.

Specifications
Overall frequency range: 0.65-18 GHz
Base module frequency range: 1.4-3.4 GHz
Individual frequency expansion module range: >2 GHz
Sensitivity: <1E-13 W (at 3 MHz bandwidth)
Noise: 2 dB (up to 6 GHz); 5 dB (over 6 GHz)
Frequency fluctuation: 1E-7
Dynamic range: >50 dB
Video amplifier bandwidth: 0,0003-18
Power supply: 12 V; 220 V/50 Hz
Power consumption: 80 W

Contractor
Kintex, Sofia.

VERIFIED

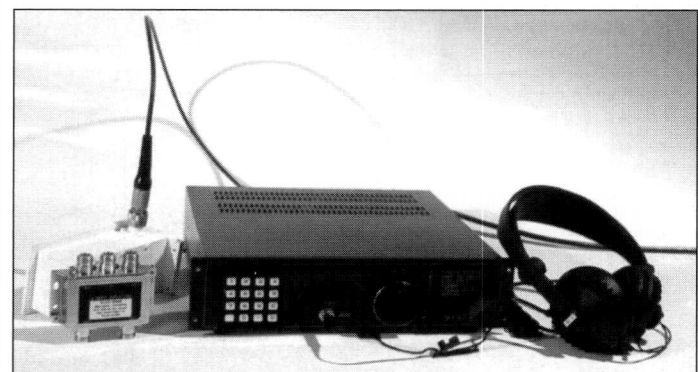

The 0.65-18 GHz Radon SIGINT receiver   0007606

CHINA, PEOPLE'S REPUBLIC

DZ9001 ELectronic INTelligence (ELINT) system

Type
D- through J-band (1 to 18 GHz) ground-mobile ELINT system.

Description
The DZ9001 ground-mobile ELINT system comprises a three-truck convoy in which two vehicles carry deployable (scissors lift) antenna radomes with the third being configured as a control centre. The system is completed by what appears to be a generator or logistics trailer.

Status
As of this edition, the status of the DZ9001 ELINT system was uncertain.

Specifications
Frequency range: 1-18 GHz
Dynamic range: >50 dB
Sensitivity: =70 dBm
Direction-finding accuracy: >3° RMS
Frequency accuracy: 2 MHz

Contractor
China National Electronics Import and Export Corporation, Beijing.

VERIFIED

ZJ9301-1 radar Electronic Support (ES) system

Type
D- through H-band (1 to 8 GHz) or I- through J-band (8 to 18 GHz sub-band) man-portable radar ES system.

Description
The ZJ9301 man-portable, battlefield ES system is available in 1 to 8 GHz and 8 to 18 GHz configurations and includes an operator display/control/processing unit and a tripod-mounted antenna. The equipment is described as being able to handle between three and five detected radars simultaneously.

Status
As of this edition, the status of the ZJ9301-1 ES system was uncertain.

Specifications
Frequency range: 1-8 GHz or 8-18 GHz
Receiver sensitivity: −60 dBm
Direction-finding accuracy: >4° RMS
Frequency accuracy: 15 MHz RMS

Contractor
China National Electronics Import and Export Corporation, Beijing.

VERIFIED

CZECH REPUBLIC

Battlefield radar Electronic Support (ES) station

Type
C- through E-band (0.8 to 2.8 GHz sub-band) battlefield radar monitoring system.

Description
Mounted on a Tatra 815 × 6 × 6 truck, this battlefield radar ES system is designed to detect ground surveillance and artillery radars operating in the 0.8 to 2.8 GHz band. Frequency coverage is divided into three system specific sub-bands with the necessary reception arrays being mounted on a trailer. Direction-finding and parametric workstations are incorporated and system function is described as being fully automatic. The equipment utilises tropospheric reflection to maximise detection range and utilises an integral threat library for emitter identification. Two systems working together are claimed to be able to generate threat location data that is accurate enough for weapons targeting.

Status
As of this edition, equipment of the type described is reported as having been evaluated by the Czech Army with production examples being delivered to the service during 1998.

Specifications
Frequency range: 0.8-2.8 GHz
Detection range: up to 500 km

Contractors
RAMET, Slovak Republic.
Vojensky Technicky Ustav Ochrany (VTUO), Czech Republic.

VERIFIED

BORAP 'radio-technical' reconnaissance system

Type
SIGnals INTelligence (SIGINT) system.

Description
The BORAP 'radio-technical' SIGINT system is designed to detect, identify, locate and track tri-service emitters and comprises two receiver assemblies (mounted on 7 m deployable masts), two processing/control units and a command, control and data display station. In its mobile configuration, BORAP is housed in three 'mid-size' trucks and two trailers and the system as a whole is based on wideband interferometric direction-finding and a 'high-performance' electronic intelligence analysis subsystem. Alongside radars (including continuous wave and pulse-Doppler types), the architecture can handle TACtical Air Navigation (TACAN)/ Distance Measuring Equipment (DME), Identification Friend-or-Foe (IFF)/Selective Identification Feature (SIF) and other types of 'communication' emitters. It is also noted as being able to provide 'full analysis' of recorded signals, a data dissemination capability and simultaneous functionality.

Status
As of February 2002, BORAP was understood to be a live programme.

Specifications
Frequency coverage: 1,025-1,150 MHz (TACAN/DME interrogators); 1,090 MHz (IFF/SIF transponders); 0.1-1 GHz (option); 1-18 GHz (standard coverage); 18-40 GHz (option)
Range: up to 400 km
Coverage: 120° (minimum azimuth)
Track capacity: up to 200 targets simultaneously
Output data: IFF/SIF modes (1, 2, 3/A and C), IFF Mode 4 flag, Mode C barometric altitude, Mode S address/altitude, radar parameters, radar type/mode, TACAN/DME (channel, frequency and mode), target identification and X–Y co-ordinates,
Crew: one operator (single station)
Environmental: −30 to +55°C (operating)

Contractor
Omnipol a.s. (export agent), Prague.

UPDATED

MCS-90 series monitoring and reconnaissance systems

Type
C- through J-band (0.82 to 18 GHz sub-band) family of radar monitoring and reconnaissance systems.

Description
The MCS-90 series of monitoring and reconnaissance systems are designed to provide information on all types of tri-service pulse emitters (including radars, Selective Identification Facility (SIF) transponders, TACtical Air Navigation (TACAN) interrogators and noise/pulse jammers) operating in the 0.82 to 18 GHz frequency band. Within the family, the MCS-90/M is a mobile equipment while the MCS-90/S is configured for static site applications. MCS-90/B is designed for integration into a customer specific facility, while the MCS-90/C designation is used to describe a combined architecture made up of two or three MCS-90/M, /S or /B equipments. The baseline MCS-90 architecture consists of three receiving stations and a processing and control centre that is normally co-located with one of the receiving stations. Distance between the two outer receiving stations and the centre is from 10 to 35 km. Signals from targets are received and processed by the receiving assemblies and are sent, together with the measured parameters, via microwave link to the centre processing and control station. Further parameters are measured at the centre and the main signal processing is carried out to identify the target type by comparison with information in the computer library. The target type, location, frequency, operation mode, track and other parameters are displayed to the operator in the control cabin. Up to 24 radar and 48 SIF air targets can be processed simultaneously on selected portions of the received frequency band. Full analysis of the signal is carried out automatically, except for intrapulse modulation that is performed manually. Detection range is up to 450 km (for targets above the radar horizon for all three receiving stations). The manufacturer states that the system could control a jammer but this has not yet been developed. Of the cited variants, the fixed-site MCS-90/S is considered to be the basic configuration.

The MCS-90/M mobile variant of the MCS-90 radar monitoring and reconnaissance system

It comprises a central shelter-mounted processing and control station and the three receiving stations, each consisting of an antenna assembly contained in a radome and mounted on a mast, and a shelter-mounted receiving control station. In the MCS-90/M mobile model, the equipment's reception antenna assemblies are mounted in special radomes on top of truck-mounted telescopic antenna carriers. Shelters containing the electronics are mounted on TATRA 6 × 6 trucks. The telescopic mast is 8 m high and folds flat when the vehicle is moving. When erect it can be extended to any height from 12.5 to 25 m. Total weight of the mobile version, including vehicle, is 28 tonnes.

Status
As of this edition, equipment of the types described were reported as having been procured.

Contractor
Tesla Pardubice, U Zámecku 26, Pardubice.

VERIFIED

SDD 'radioelectronic' signals monitoring station

Type
SIGnals INTelligence (SIGINT) system.

Description
The SDD 'radioelectronic' SIGINT system is designed to automatically detect, monitor, identify, locate and analyse signals from, predominantly, ground-based radars. System range is achieved via the use of the troposcater effect and the architecture as a whole comprises two monitoring and direction-finding stations, each of which incorporates a multi-antenna/receiver assembly and a processing/control unit. SDD is further noted as being a modular equipment that is 'easily' upgraded and as being available in static and mobile configurations. Here, the mobile version is transportable by truck, helicopter or fixed-wing cargo aircraft. Other system features include automatic azimuth and frequency scanning, automatic emitter identification (including a finger printing capability), built-in test and nuclear, biological and chemical protection.

Status
As of February 2002, SDD was understood to be a live programme.

Specifications
Frequency coverage: 0.5-18 GHz
Azimuth coverage: 360°
DF accuracy: better than 0.5° (automatic co-ordinate location)
Bandwidth: 4,000 MHz (instantaneous frequency measurement)
Range: up to 700 km
Crew: one operator (single station)

Contractor
Omnipol a.s. (export agent), Prague.

UPDATED

TAMARA MCS-93 ELectronic INTelligence (ELINT) system

Type
C- through J-band (0.82 to 18 GHz sub-band) ELINT system.

Description
According to Jane's sources, the TAMARA ELINT system is designed to detect, identify, locate and track tri-service emitters, Selective Identification Feature (SIF) transponders, identification friend-or-foe interrogators, TACtical Air Navigation (TACAN) systems, distance measuring units, jammers and data transmission networks operating in the 0.82 to 18 GHz frequency range. The equipment is described as being available in mobile, fixed-site and platform integrated configurations with the mobile application being considered as the baseline for the system as a whole.

A mobile TAMARA system is described as comprising three RS-AJ/M receiver subsystems, an RS-KB receiver control station, an RS-KM processing station and, if required, a local commander's ZZP-5 monitoring and plotting station. Taking these in the order given, each RS-AJ/M subsystem takes the form of a mast-mounted antenna/receiver assembly installed on a Tatra 815 8 × 8 truck chassis. To fit it for the role, the vehicle has been modified to incorporate four hydraulic jacks for levelling purposes and a front-mounted bulldozer blade for limited excavation and terrain adjustment work. The rear-mounted, hydraulically driven, self-support antenna/receiver assembly mast can be set at a height of 8.5 m or between 12.5 and 25 m depending on the terrain. The antenna/receiver assembly itself takes the form of a cylindrical radome that houses the necessary antennas and receivers, a microwave link for communication between individual antenna/receiver assemblies and built-in diagnostics. Operationally, three RS-AJ/M units are networked at distances of between 10 and 35 km from one another.

The RS-KB receiver control station is shelter mounted and houses an operator workstation from which RS-AJ/M function can be monitored. The RS-KM processing station is described as exercising overall system control, undertaking data processing and providing the interface to the user's communications or control network. RS-KM subsystems are reported to include an R15 main control computer, an MAK high-speed measurement unit, a signals recorder and an MCA communications adaptor. The ZZP-5 monitoring and plotting station is noted as featuring two 50 cm colour monitoring displays, a plotter, a printer, a link modem, radio communications and an RS-232C interface. Data generated by the system is said to include target identity, time of transmission for data signals, X and Y target co-ordinates, SIF response modes, TACAN channel identification together with ground-based target monitoring and frequency surveillance reports. Parametric information produced includes frequency, pulse repetition interval, pulse-width and antenna scan period values for individual emitters.

Status
During the late 1990s, Omnipol was reported to be seeking export customers for the TAMARA MCS-93 ELINT system and as of this edition, the VERA-E passive surveillance system had been promoted as the MCS-93's 'successor'.

Specifications
Frequency: 0.82-18 GHz
Detection range: 450 km
Azimuth coverage: 100° (per RS-AJ/M unit)
Set up time: 1 h (RS-AJ/M unit)
Track capacity: more than 72 (automatic, simultaneous function)

Contractor
HTT-Tesla Pardubice, Pardubice (equipment manufacturer).
Omnipol Ltd, Prague (export agent).

VERIFIED

VERA-E passive surveillance system

Type
SIGnals INTelligence (SIGINT) system.

Description
The VERA-E SIGINT system is described as being a long-range equipment that is designed to detect, identify, locate and track tri-service emitters. It is billed as being suitable for early warning, field intelligence and air defence applications and can be configured in 2-D and 3-D configurations. Here, the 'standard' 2-D arrangement comprises three receiver assemblies (designated as left, right and central), a central processing station and microwave links for intercommunication between the various system elements. Of these, the receiver assemblies are mounted on 17 m masts and the 'right' and 'left' units can be located at up to 50 km from the central station. The central processing station is housed in an International Standards Organisation compliant transport container and the complete architecture is deployed in a three 'mid-size' truck/single trailer convoy. In its 3-D configuration, the architecture incorporates three receiver assemblies together with a fourth unit that is co-located with the central processing station. As such, 3-D VERA-E provides 360° coverage in azimuth and is deployed aboard a four-truck/single-trailer convoy. In all cases, VERA-E functionality is based on the time difference of arrival technique. The system is further noted as providing a frequency activity survey and technical analysis (including inter- and intra-pulse analysis and emitter finger printing) of a range of emitter types including radars, jamming transmitters, Identification Friend-or-Foe (IFF)/Selective Identification Feature (SIF) transponders, TACtical Air Navigation (TACAN)/Distance Measuring Equipment (DME) interrogators, datalinks and other types of pulsed equipments.

Status
As of February 2002, VERA-E was understood to be a live programme.

Specifications
Frequency range: 1,025-1,150 MHz (TACAN/DME interrogators); 1,090 MHz (IFF/SIF transponders); 0.1-1 GHz (option); 1-18 GHz (standard coverage); 18-40 GHz (option)
Range: up to 450 km
Coverage: 140° (2-D configuration); 360° (3-D configuration)
Tracking capacity: up to 200 targets simultaneously
Location accuracy: 20 m RMS (2-D*, side error - radar and IFF/SIF); 80 m RMS (2-D*, side error - TACAN/DME); 150 m (3-D**, range error (D = 15 km) - radar and IFF/SIF); 200 m (3-D**, altitude error - radar and IFF/SIF); 200 m RMS (2-D*, range error - radar and IFF/SIF); 400 m (3-D**, range (D = 15 km) and altitude errors - radar and IFF/SIF); 800 m RMS (2-D*, range error - TACAN/DME); 1,500 m (3-D**, range error (D = 30 km) - TACAN/ DME); 4,000 m (3-D**, range error (D = 15 km) - TACAN/DME)
Crew: three drivers, three technicians and a system operator (2-D configuration)
* 2 × 25 km geometry, target range of 150 km and 25 pulses being processed
** 3 × D km geometry, target range of 150 km, target height of 5,000 m and 25 pulses being processed

Contractor
Omnipol s.a. (export agent), Prague.

NEW ENTRY

GERMANY

AMPLUS 14 Direction-Finding (DF) system

Type
Communications band radio DF receiver.

Description
AMPLUS 14 is a high-performance communications band DF receiver that is suitable for mobile, semi-mobile and stationary applications, over the frequency range 20 to 500 MHz. It is designed for optimum linking into systems andconsists of a BAP387 operation and display unit and a DDF7106 direction-finding receiver unit. The two units can be operated using an RS-422 interface when separated by up to 1,200 m.

The BAP387 contains a powerful microprocessor which controls the display, the DDF7106 and the input and output interfaces. Some 64 memories are available for frequencies and associated operating parameters. This enables preset device settings to be called up if required.

The Watson-Watt three-channel DF technique is used in the receiver section. This has the advantages of extensive immunity to common-channel interference and high-speed direction-finding of short-term signals.

Status
As of this edition, the AMPLUS 14 DF system is reported to be in service.

Specifications
Frequency range: 20-500 MHz
Frequency scanning: up to 64 channels
Bearing accuracy: ±1°
Weights: 2.5 kg (control/display unit); 16 kg (receiver)

Contractor
C Plath GmbH, Hamburg.

VERIFIED

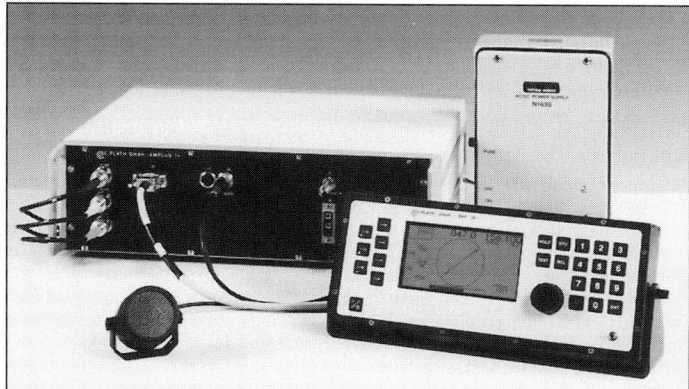

The AMPLUS 14 DF receiver

Antenna matrix 16/256 and 32/256

Type
1 to 30 MHz band antenna matrix for monitoring station applications.

The C Plath 1 to 30 MHz band antenna matrix   2000/0068852

Description
C Plath's 1 to 30 MHz antenna matrix is designed for use in monitoring station applications and can be customised to accommodate customer specific combinations of antenna inputs and connectable receivers. An example of such a configuration might be 16 antenna inputs and 256 outputs (designated as 16/256) or 32 antenna inputs with 256 outputs (32/256). Within the matrix, individual antennas can be switched between individual monitoring receivers independently without antenna/receiver distortion. De-coupling between input/input, output/output and input/output over the matrix's total frequency range is noted as being 'easily' accomplished with a 'high degree' of efficiency. Typical IP2 and IP3 intermodulation points are in the order of 76 dBm and 40 dBm respectively. An automatic programme switching to stand-by mode is reported to lower overall power consumption and there is a built-in test facility to optimise mean time between failures and mean time to repair values.

Status
As of this edition, the described antenna matrix is reported as having been in service since 1998.

Contractor
C Plath GmbH, Hamburg.

UPDATED

APF 1050 Direction-Finding (DF) system

Type
Automatic radio DF system.

Description
The APF 1050 system covers the 10 kHz to 30 MHz frequency range. Each DF-Station consists of eight DF receivers (Type SFP 5200) which can be operated at the same time and independently of each other. The system can be used in both semi-mobile and in stationary applications. Either independent DF networks with a number of APFs can be set up or existing DF networks can be extended. Since the APF 1050 is easy to relocate, DF networks can be set up to suit requirements and matched to the relevant situation. The remote control feature of the equipment makes unattended operation a possibility.

Status
As of this edition, the APF 1050 DF system was reported as having been procured.

Contractor
C Plath GmbH, Hamburg.

VERIFIED

The APF 1050 automatic DF system

DDF0xM series direction-finders

Type
Family of digital communications band direction-finders.

Description
DDF0xM series equipments cover the 0.3 to 3,000 MHz frequency range and can be operated in Watson-Watt or correlative interferometer modes depending on the type of direction-finding antenna used. A wide range (including units from manufacturers other than Rohde and Schwarz) of such antennas is available for

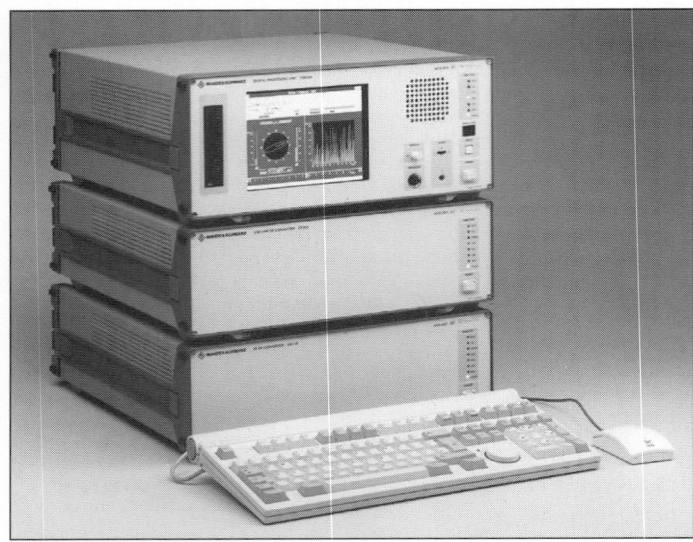

The 0.3 to 1,300 MHz DDF06M digital direction-finder

both mobile and stationary applications. The equipments can intercept signals of very short duration (>10 µs in Watson-Watt mode, >500 µs for correlative interferometer) and are capable of operating in scan mode where they make use of the Fast Fourier Transform technique (25 kHz real-time bandwidth for the HF band and 200 kHz for the VHF/UHF bands). DDF0xM direction-finders are computer controlled (external or internal as an available option) and can use AC or DC power supplies. The equipments can also be remotely controlled when networked and the family includes the following variants:

DDF01M

DDF01M covers the 0.3 to 30 MHz band and when operated in the correlative interferometer mode (using a nine crossed-loop or monopole antenna array with a diameter of 50 m), can deliver elevation and azimuth data and thereby provide a single station location capability. For mobile applications, a 1 to 30 MHz, 1.1 m diameter antenna can be used.

DDF05M

DDF05M covers the 20 to 1,300 MHz or 20 to 3,000 MHz bands and utilises a range of Watson-Watt or correlative interferometer antenna arrays. It can be used in fixed-site and mobile applications and has a typical scanning speed in interferometer mode of 45 MHz/s with 25 kHz resolution. Provision is made for the interception of GSM signals.

DDF06M

DDF06M covers the 0.3 to 1,300 MHz band (extendable to 3,000 MHz) in a single equipment.

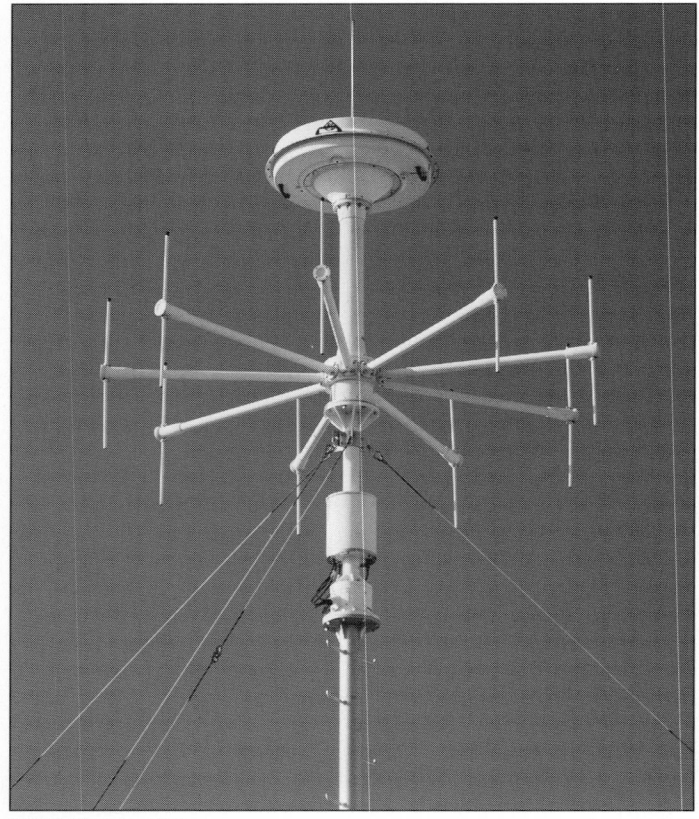

DDF0xM series D/F   2000/0056270

A close-up of the ADD 150 direction-finding antenna

Status
As of this edition, DDF0xM series digital communications band direction-finders are reported as having been procured.

Contractor
Rohde & Schwarz, Munich.

VERIFIED

DDF0xS scanning direction-finder

Type
Intercept and Direction-Finding (DF) system for frequency-hopping and burst signals.

Description
To cope with rapidly changing situations such as frequency-hopping transmitters or burst transmissions, the DDF0xS has been developed to satisfy the requirements for new concepts of electronic reconnaissance systems. In the past, if separate units were used for detection and direction/position-finding, it often happened that a newly discovered frequency activity could not be passed on to the DF system because of the short signal duration and thus the signal could not be processed by the DF system. Advanced systems must be capable of performing simultaneous measurements of frequency, level and bearing of such signals and DDF0xS incorporates a fast-scanning DF receiver to realise this concept and examine very wide frequency bands within a short time.

The DDF0xS makes extensive use of digital signal processing, especially of Fast Fourier Transform which allows simultaneous analysis of a wide frequency band with selectable channel resolutions within the analyser bandwidth.

Applications of the DDF0xS include:
- automatic position finding systems with high intercept probability
- interception and direction-finding of frequency-hopping and burst signals
- optimised use in automatic interception systems through data reduction so that results are limited to sources of interest
- SSL (Single Station Location) systems by additional determination of elevation angle using interferometric evaluation in the HF band (0.5 to 30 MHz)
- highly flexible in stationary and mobile applications (land-based vehicles, ships and aircraft) by selectable use of Watson-Watt or interferometric evaluation algorithms and thus various antenna configurations, especially with wide aperture characteristics.

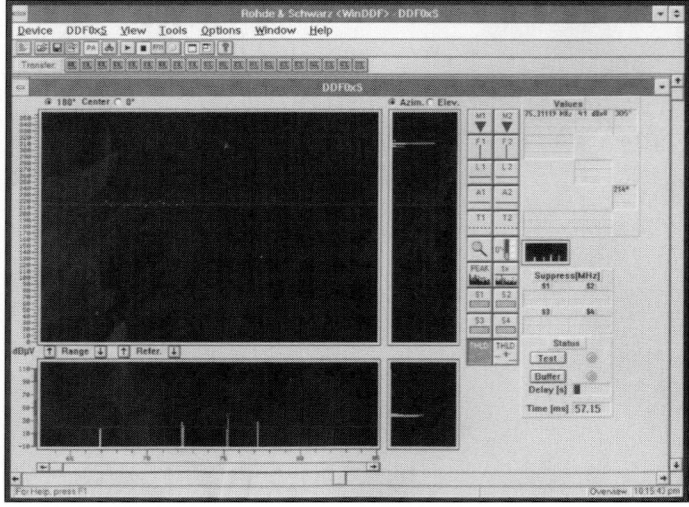

The DDF0xS man/machine interface in scan mode

Status

As of this edition, the DDF0xS scanning direction-finder was reported as having been procured.

Specifications

Frequency range: 0.5-1,300 MHz
Analysis window: 200 kHz

Contractor

Rohde & Schwarz, Munich.

VERIFIED

DDF190 direction-finder

Type

Communications band digital direction-finder.

Description

DDF190 covers the 20 to 3,000 MHz frequency range, makes use of digital processing techniques and operates in the correlative interferometer mode. The equipment consists of a Direction-Finding (DF) processor (connected to the intermediate frequency output (10.7 to 21.4 MHz) of a suitable receiver) and a DF antenna. Mobile and fixed-site application DF antennas covering the 20 to 1,300 MHz and 1,300 to 3,000 MHz bands are available for use with the equipment. DDF190 offers a range of DF modes and bearing results can be displayed in histogram form if required. Minimum signal duration is less than 50 ms and the system can be remotely controlled and make use of either an AC or DC power supply.

The 1.3 to 3 GHz ADD 071 (bottom) and 20 to 1,300 MHz ADD 190 (top) antennas used with the DDF190 direction-finder

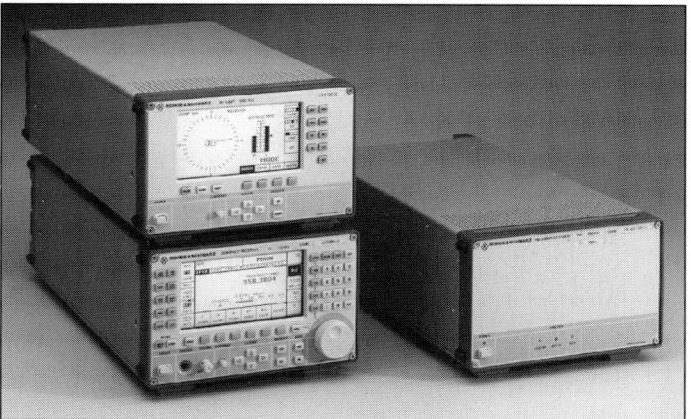

The DDF190 direction-finder (top left) shown here with Rohde & Schwarz's VHF/UHF ESMC receiver (bottom left) and ESMC-FE frequency extension unit (right)

Status

As of this edition, the DDF190 digital direction-finder was reported as having been procured.

Contractor

Rohde & Schwarz, Munich.

VERIFIED

DFP 5300 radio Direction-Finder (DF)/analyser

Type

Broadband LF/HF band (0.3 to 30 MHz) radio DF and analysis equipment.

Description

As a broadband direction-finding equipment, the DFP 5300 offers real-time processing, fast signal acquisition, 100,000 bearings/s, data reduction by sector mode, replay capability and the detection and bearing of all transmitters over the frequency range 0.3 to 30 MHz. The DFP 5300 has an independent monitoring receiver which can be controlled within a surveillance window of up to 1.6 MHz by means of a cursor. It provides reliable detection and bearing of all transmitters whether conventional, frequency-hopping, spread spectrum, chirpsounder or similar. Display modes include frequency over amplitude, frequency over time, bearing over frequency, bearing over time and histogram indication. A new post-processor has been integrated to reduce the data rates. Commanding a DF-net equipped with DFP 5300 and the retransmission of results can now be done via conventional telephone links.

Status

DFP 5300 is reported to have been in service since 1994.

Specifications

Frequency range: 0.3-30 MHz
Synthesiser resolution: 10 kHz
Frequency span: 10 kHz-1.6 MHz
Resolution width: 800 channels
Demodulators: A1A, A3E, F3E, J3E

Contractor

C Plath GmbH, Hamburg.

VERIFIED

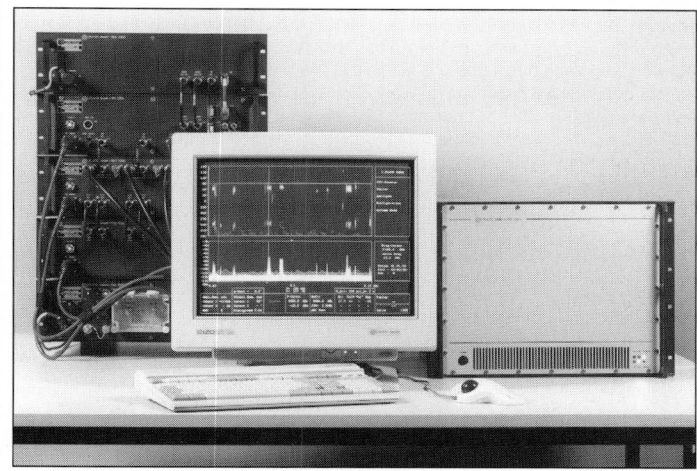

The DFP 5300 broadband direction-finder

DFP 7107 direction-finder

Type

Communications band direction-finder.

Description

DFP 7107 is a compact, broadband direction-finder which features multichannel, time-frequency transformation and direction-finding evaluation according to Watson-Watt or interferometer principles. System features include large band coverage (where there is high transmitter congestion) in fast time; broadband detection of transmitters (including frequency hoppers and spread spectrum types); anticipated threat bearing recognition; transmission activity warning; fast signal acquisition, suitability for mobile and semi-mobile applications and low power consumption.

Status

As of this edition, DFP 7107 is reported to be in production.

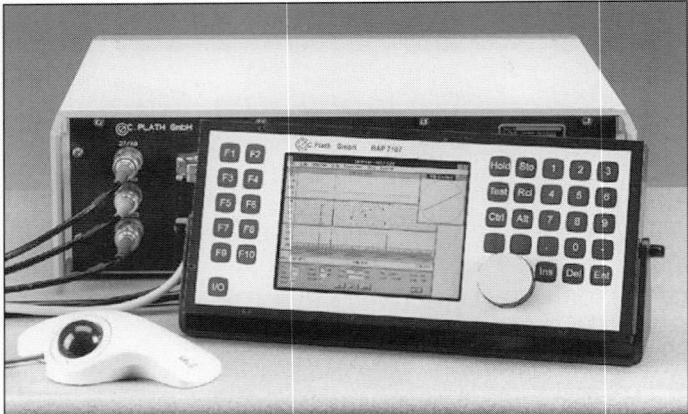

The DFP 7107 communications band direction-finder

Specifications
Frequency: 20-1,350 MHz
Resolution: 500/250/40 lines
Scanning speed: =100 MHz/s (at 250 line resolution)
Acquisition time: =10 ms
Accuracy: ±2° (up to 500 MHz); ±3° (above 500 MHz)

Contractor
C Plath GmbH, Hamburg.

VERIFIED

EB 200 miniport receiver

Type
10 kHz to 3 GHz portable Electronic Support (ES) receiver.

Description
EB 200 is a low battery power, miniport receiver that intercepts signals, locates transmitters and generates graphic overviews of frequency bands. The equipment's design is such as to allow one hand on body operation while its full graphic display provides an optimised format for each task undertaken. For close range direction-finding, EB 200 is teamed with the hand-held, active HE 200 antenna. HE 200 covers the 10 kHz to 3,000 MHz frequency range. As a full feature High/Very High/Ultra High Frequency (HF/VHF/UHF) equipment, Rohde & Schwarz suggests that EB 200 is equally at home in fixed-site applications. Here, remote control via a local area network allows the receiver to generate fast spectra and parametric readouts; undertake master-slave functions and be subjected to built in test. The equipment's intermediate frequency panoramic liquid crystal display is noted as facilitating rapid detection, monitoring and assessment of received signals.

Status
As of this edition, the EB 200 ES receiver was reported as having been procured.

Specifications
Frequency range: 10 kHz-3 GHz
Dynamic range: −10 to 110 dBµV
Scanning modes: frequency scanning, frequency spectrum and memory scanning
Demodulation: AM; CW; FM; LSB; none; USB
Bandwidths: 15 (150 Hz-1 MHz)

The 10 KHz to 3 GHz EB 200 miniport receiver shown here with an associated active hand-held directional antenna   0044245

Display: full graphics (240 × 64 pixels)
Operator controls: numerical keypad; softkeys; controls; rollkey
RF connectors: for antenna; IF (10.7 ± 1 MHz); int/ext reference
Audio connectors: balanced; unbalaned; 600 Ω mono output; stereo output; ext loudspeakers; headphones
Digital connectors: IF as I/Q signal; AF
Remote control: via RS-232C or LAN interface; SCPI conform
Case dimensions: 88 × 210 × 270 mm
Weight: 4 kg; 5.5 kg (with battery pack)
Power consumption: <22 W

Contractor
Rohde & Schwarz, Munich.

VERIFIED

ESMA search receiver

Type
Radio frequency scanning receiver for COMmunications INTelligence (COMINT) purposes.

Description
The ESMA search receiver intelligently scans the frequency spectrum at a rate of 3,000 channels/s over the frequency range 20 to 1,300 MHz. The receiver is designed for COMINT and radio monitoring and, in conjunction with other equipment (such as direction-finders, monitoring receivers and printers), general purpose radio monitoring installations can be established which provide high detection probability with a low false alarm rate. The receiver is controlled from a standard PC-AT which also allows control of the other equipment.

Status
As of this edition, the ESMA search receiver was reported as having been procured.

Specifications
Frequency range: 20-1,300 MHz
Tuning time: 0.15 ms

Contractor
Rohde & Schwarz, Munich.

VERIFIED

The ESMA search receiver

ESMC compact receiver

Type
Communications band communications intelligence receiver.

Description
ESMC is only half the size of a customary 48 cm multipurpose receiver and has been optimised for radio monitoring applications over the frequency range 20 to 650 MHz (extendable to 500 kHz to 3,000 MHz) with the following features:
* fast frequency scan for hopper detection
* compact design and low weight
* simple operation via a liquid crystal display
* suitable for receiving any signal in the designated range
* wide dynamic range and high overload capacity
* 1 Hz frequency resolution
* low phase noise
* accurate measurement of signal level
* offset display for channel frequency
* AC/DC supply without exchanging the power supply unit
* fully remote controllable via IEEE-488 or RS-232C interface using SCPI protocol.

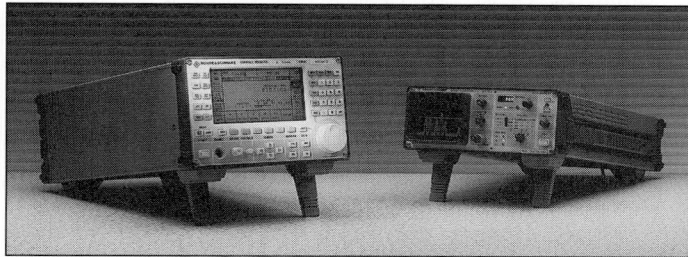

The ESMC receiver

The operating concept of ESMC meets the requirements of modern radio monitoring receivers with all main functions (type of demodulation, bandwidth and so on) set directly via labelled keys. A hot key allows returning to the main menu from any sub-menu. Menu control is organised in priority levels so that signal processing is not interrupted by menu changes and the user never loses sight of what is going on.

Status
As of this edition, the ESMC communications band communications intelligence receiver was reported as having been procured.

Contractor
Rohde & Schwarz, Munich.

VERIFIED

GIGATUNE microwave receiver system

Type
Suite of microwave receiver tuning systems.

Description
The microwave receiver system GIGATUNE is a suite of approximately 20 different devices allowing anyone to set up the correct microwave receiver system for any radio monitoring application in the frequency range from 1 to 18 GHz.

A range of microwave antennas with dish diameters between 0.9 and 3 m is available for the GIGATUNE system. They can be fitted with log-periodic broadband feeds. These active feeds already contain low-noise preamplifiers and selectors.

The microwave band from 1 to 18 GHz is divided into octave bands on five tuners which are tuned with crystal accuracy in 50 MHz steps by the associated synthesisers. Subsequent selection takes place at the first intermediate frequency in the GHz band, the bandwidth being 100 MHz; this considerably reduces intermodulation. Both frequency outputs from the first and second oscillator are derived from a highly accurate master oscillator. The first oscillator is Yttrium indium garnet tuned, the second contains a dielectric resonator.

Status
As of this edition, GIGATUNE type receivers were reported as having been procured.

Contractor
Rohde & Schwarz, Munich.

VERIFIED

Microwave antenna system for GIGATUNE

GX 200 signal analyser

Type
Frequency/phase shift keying data signal analyser.

Description
The GX200 signal analyser allows automatic assignment of detected emissions to their sources by means of parameters such as type of modulation, data rate, frequency shift, centre frequency and content related parameters, bit pattern, and coding. The analyser can be used:
• for data analysis in any HF (3 to 30 MHz) radio monitoring test system
• for interception and identification of short time emissions (bursts)
• for triggering an alarm when specific targets are detected
• for determining source-specific parameters (with built-in Fast Fourier Transform optional).

Status
As of this edition, the GX200 frequency/phase shift keying data signal analyser was noted as having been procured.

Specifications
Baud rate: 10-1,200 Baud
Shift: 50-2,000 Hz
Min measurement duration: 30 symbol changes

Contractor
Rohde & Schwarz, Munich.

VERIFIED

The GX200 signal analyser   0044246

GA950/GA960 satellite monitoring systems

Type
Family of satellite monitoring systems.

Description
The GA950 and GA960 equipments are designed to monitor satellite communications traffic carried by the INMARSAT M, mini M and B communications systems. Key GA 9XX series features include:
• compact system (mobile 48 cm rack) and antenna design
• optimisation for tactical applications
• a scan mode to detect all communications links
• a monitoring mode to obtain link details and content
• a search mode for the monitoring of dedicated targets
• a find mode to detect active mobiles in the area of coverage
• voice monitoring (via speaker and/or headphones) and recording (on hard disc)
• facsimile decoding and storage on hard disc
• data/AST monitoring and storage on hard disc.

Status
As of October 1999, the GA950 system was reported to be available as a commercial-off-the-shelf item, while the GA960 application is understood to have become available during the first half of 2000.

Contractor
Rohde & Schwarz, Munich.

VERIFIED

HF 2000 computer supported intercept position

Type
HF band (3 to 30 MHz) general purpose radio reconnaissance system.

Description
HF 2000 is a general purpose HF band radio reconnaissance system which is based on digital signal processing receiving techniques and processor modules which are application configured (software downloads) for functions such as signal classification, automatic acquisition of data transmissions, parameter measurement and the decoding of amplitude, Frequency and Phase Shift Keying (FSK and PSK) signals. A digital demodulator is used to demodulate amplitude and frequency modulated, continuous wave, signal sideband, FSK, PSK and quadrature amplitude modulated signals while all system control functions (including that of the built in DAT-recorder and audio matrix) are executed by a Windows-based graphic user interface. Technical specifications relating to this system may be found in the entry for the E 2000 LH receiver.

Status
As of this edition, the HF 2000 general purpose radio reconnaissance system was reported as having been procured. As of December 2000, it was thought likely that the described equipment was being supported by European Aeronautic, Defence and Space (EADS) EWATION (Ulm, Germany).

UPDATED

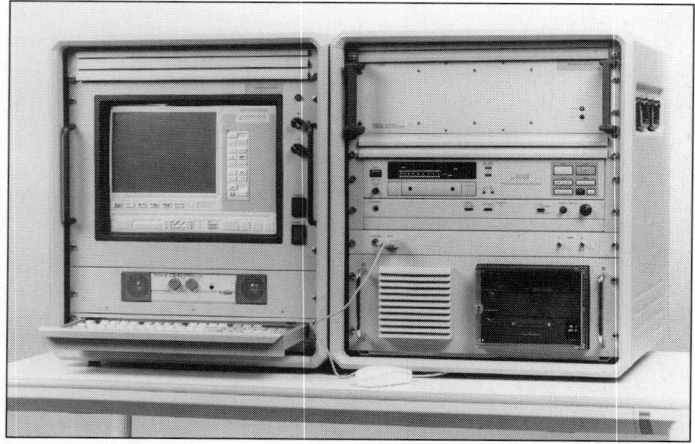

The HF 2000 computer supported intercept position

HPS 8000 interception post controller

Type
Interception post control.

Description
The HPS 8000 is a special development for the control of equipment in an interception station using a commercially available PC. The HPS 8000 software enables the configuration of the workstations as interception or supervisory posts. Data communication takes place over a local area network. Information obtained in the interception station can be saved and managed. Hard disks or other storage media can be used for storage depending on the PC employed. An interception station normally consists of a supervisory post and a number of interception posts. The task of the interception is the acquisition, analysis and documentation of radio signals and, where applicable, the initation of a position-fixing process.

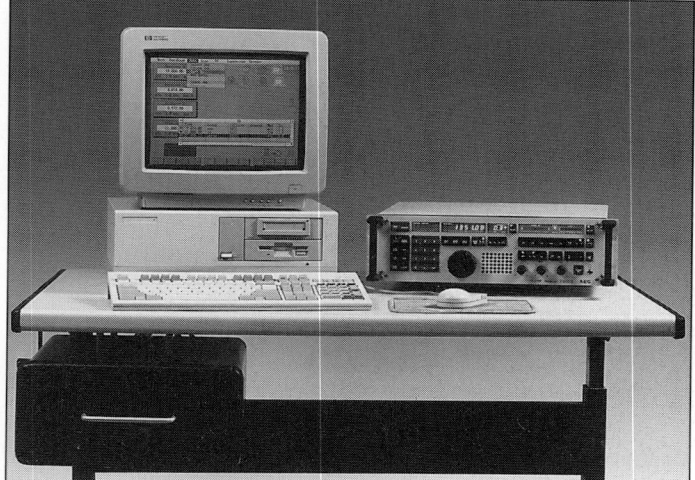

The HPS 8000 interception post controller

Status
As of this edition, the HPS 8000 interception post control is reported to have first appeared during 1994 and as having been procured.

Contractor
C Plath GmbH, Hamburg.

VERIFIED

MCP 8000

Type
Monitoring control position for Radio Direction-Finding (RDF) applications.

Description
The MCP 8000 monitoring control position is designed to control independently all the operating functions (including antenna selection) of up to four monitoring receivers and to direct the operations of an RDF net. Here, location results are displayed on a map background on the equipment's display in such a way as to allow the operator to recognise immediately if a detected transmission is in an area of interest. Results obtained can be printed out on customised reports. Each MCP 8000 workstation includes its own database with all frequencies and location results (including all necessary receiver settings and 'additional' data) being stored for subsequent use. Audio information can be recorded and linked to the database if required. Additional frequency lists (detailing such things as a particular frequency range, type of modulation or target area) can be generated from the database. All lists can be sorted by entry and there is a list entry search facility. MCP 8000 also manages scanning within a given frequency range, using programmed dwell times and frequency steps. The system can also execute frequency list scanning. Such lists are generated from the database, are storable and can be activated as required. MCP 8000 also incorporates a 'job list' facility that allows preprogrammed frequency lists to be called up automatically in particular time slots. RDF net control is by means of conventional telephone line or via a datalink and C Plath notes that 'powerful' monitoring stations can be constructed via the combination of a number of MCP 8000 systems.

Status
As of this edition, MCP 8000 is noted as having been in service since December 1998.

Contractor
C Plath GmbH, Hamburg.

UPDATED

NETTRAP mobile location system

Type
Mobile communications band Direction-Finding (DF)/emitter location system.

Description
NETTRAP is a mobile DF/emitter location system that covers the 20 to 1,000 MHz frequency band. Its compact design enables the equipment to be installed in what is termed a small car or carried as a manpack. Operationally, NETTRAP can be configured to contain one central DF post and up to three remote stations. The system includes full-duplex radio links for data/voice communications between the central station and each of the remotes. At the heart of the equipment is the PA 1555 DF unit (see separate entry) which is able to detect short time emissions, as well as handling intermittent and duplex signals. In the central NETTRAP station, the DF unit is connected to the communications subsystem and a ruggedised PC that displays received data within a geographical map format. The remote DF stations are connected to the communications subsystem only. The DF antenna used can be folded for transport and can be mounted on a plug-in mast if required.

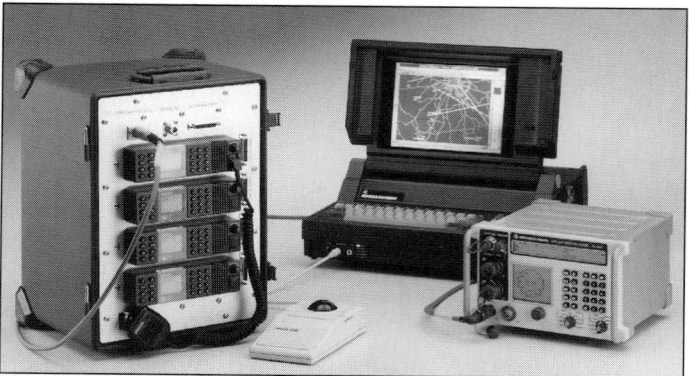

The NETTRAP mobile location system

Status
As of this edition, the NETTRAP DF/emitter location system was reported as having been procured.

Specifications
Frequency: 20-1,000 MHz
DF accuracy: 2° RMS
Min signal duration: 50 ms
Power supply: 12 V DC
Weight: 7.5 kg (PA 1555 DF unit); 10 kg (remote communications unit); 11 kg (antenna); 16.7 kg (central communications unit)
Antenna dimensions: 1 × 2 m

Contractor
Rohde & Schwarz, Munich.

VERIFIED

PA1555 direction-finder

Type
Portable Direction-Finding (DF) system for tactical applications.

Description
Because of its compact design and foldable DF antenna system, the PA1555 Direction-Finder is easily transportable and highly suitable for special missions, as well as for sites with difficult access. This is helped by a DC power supply of low power consumption. For use in vehicles, a DF antenna is available which covers the entire frequency range and which will be housed in a round plastic radome of 1.1 m diameter and approximately 25 cm height. Since the PA1555 permits the selection of several DF modes, it is possible to match to different types of transmission and thus achieve optimum DF results. Among these, for example, is the presentation of the bearings in the form of a histogram that allows the identification of different stations involved in a radio communication network. The PA1555 also offers the possibility to search within specified frequency limits for activities by using the frequency scan mode. In a similar way up to 100 preprogrammed frequency channels can be scanned. A serial data interface (RS-232C) allows the remote control of the PA1555 and thus its integration into radiolocation stations.

Status
As of this edition, the PA1555 DF system was reported as having been procured.

Specifications
Frequency range: 20-200/1,000 MHz
Bearing accuracy: 2° RMS
Weight: 7.5 kg

Contractor
Rohde & Schwarz, Munich.

VERIFIED

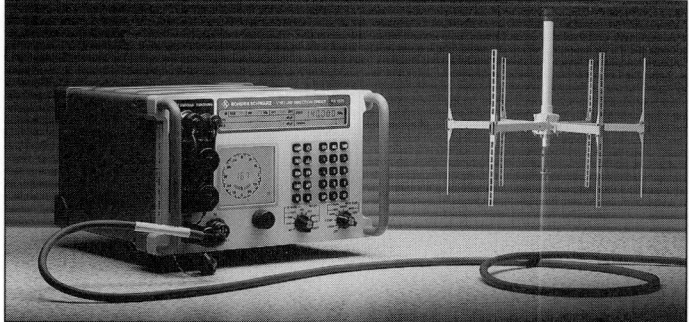

The PA1555 VHF/UHF direction-finder

RAMON COMmunications INTelligence (COMINT) system

Type
Modular radio communications intercept and Direction-Finding (DF) system.

Description
The modular RAMON COMINT system intercepts and locates emissions in the 10 kHz to 40 GHz frequency range and can be configured to meet customer requirements. The system is suitable for land-based, naval and airborne applications. Configurations available range from a single-operator 'compact' system through to one that incorporates dedicated supervisor, search, monitoring, DF and analysis positions. In multiposition iterations, all positions are connected via a local or wide area network to facilitate the exchange of orders and reports.

A general view of the RAMON COMINT system in shelter-mounted configuration
0007618

Generated reports are analysed and condensed to supply information about the overall situation and to form part of an electronic order of battle. Due to its modular design, RAMON can be upgraded in what Rohde & Schwarz terms 'easy steps'. Individual positions within the system are fitted with the 'necessary' equipment, a range that includes the ESMA fast scan receiver, the High Frequency (HF - 3 to 30 MHz) band EK890 receiver, the Very/Ultra High Frequency (V/UHF - 30 MHz to 1 GHz) band ESMC receiver or the 1 to 18 GHz band GIGATUNE receiver; signal analysers; classifiers and HF through UHF band digital direction-finders.

Status
As of this edition, the RAMON COMINT system is reported as having been procured in airborne, shipborne and fixed station configurations.

Contractor
Rohde & Schwarz, Munich.

VERIFIED

Rohde & Schwarz specialist antennas

Type
Range of specialised antennas for transmitter and receiver applications.

Description
Over time, Rohde & Schwarz has produced a range of specialist antennas, the known details of which are as follows:

AC008 microwave directional antenna
AC008 is a manually adjustable, directional dish (Ø 0.9 m) and feed antenna that is both collapsible and designed for mobile radio monitoring and field strength

The AC008 microwave directional antenna

urement in the 1 to 18 GHz frequency band and up to 26.5 GHz respectively. he equipment can be directed towards geostationary satellites and utilises three types of feed (designated as HL024A1, HL024S2 and HL025) to enable it to handle any type of signal polaristion.

AC090/120/180/300 microwave directional antennas

The AC090/120/180/300 series of microwave antennas is used in outdoor, multi-element assemblies and is designed for the detection and monitoring of terrestrial and satellite signals in the 1 to 18 GHz frequency band (expandable to 1 to 40 GHz if required). Azimuth coverage is ±180° with that in elevation being −5 to +95°. Antenna positioning is controlled from a detached control unit that also incorporates a computer interface to allow a particular assembly to be operated as part of a remotely controlled monitoring network. Antenna positioning in azimuth and elevation is simultaneous. When manually controlled, assemblies of the type can be rotated clockwise and counter-clockwise and tilted up and down at a constant speed. In automatic mode, auto-tracking (to preset azimuth and elevation limits) and optional, controller directed sector scanning are available.

Functionally, an active linear or dual-linear polarised log periodic antenna with an LNA follower is located near the focus of the reflector in order to minimise antenna to receiver loss. Preamplifier gain is such as to optimise the relationship between system noise and dynamic range despite cable losses. If the 1 to 40 GHz band is being covered, polarisation and antenna type are selected via an RS-485 control board connection, with the necessary commands being listed in an accompanying manual. The man/machine interface used is based on Windows 95 with Windows NT 4.0 as an available option.

HE016 active antenna system

HE016 is a combination of the active, high frequency HE010 rod antenna and two crossed horizontal, high frequency dipole antennas combined via a 90° coupler.

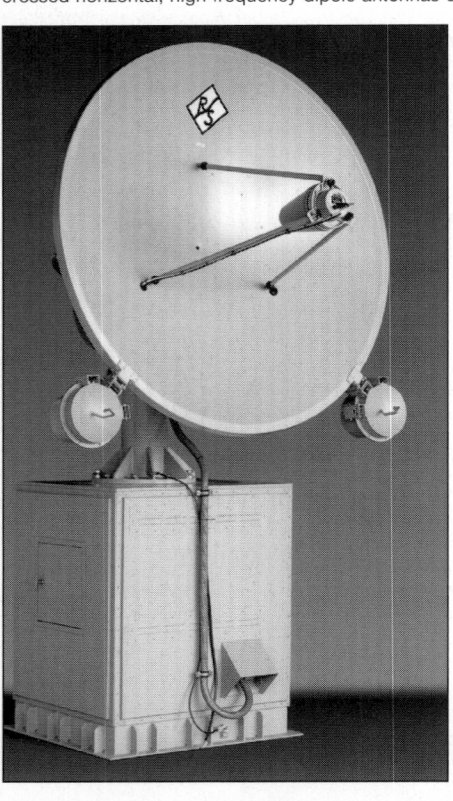

A multiple AC090/ 120/180/300 microwave antenna assembly
0007619

The HE 016 active antenna   0044247

HE016 provides omnidirectional reception of horizontally and vertically polarised waves in the ranges of 600 kHz to 30 MHz (horizontal) and 10 kHz to 80 MHz (vertical).

HE202 and 302 receiving dipoles

The HE202 and 302 active receiving dipoles cover the 200 MHz to 1 GHz and 20 to 500 MHz frequency bands respectively . They are designed to replace antenna systems that feature multiple, passive, switch selected elements for the reception of horizontally or vertically polarised signals. The two active dipole types are also noted as offering comparable electrical characteristics to the systems they are intended to replace.

According to Rohde & Schwarz, the non-linear distortion and noise values exhibited by the HE 202 and 302 dipoles provide a dynamic range that is 'seldom attained' by passive antennas equipped with quality preamplifiers. The units are also noted as providing near frequency independent radiation patterns over their respective frequency ranges. Further, the input and output of the single stage, symmetrical push-pull amplifiers incorporated in these units are protected against the exclusive voltage produced by atmospheric discharge or proximity to transmitting antennas.

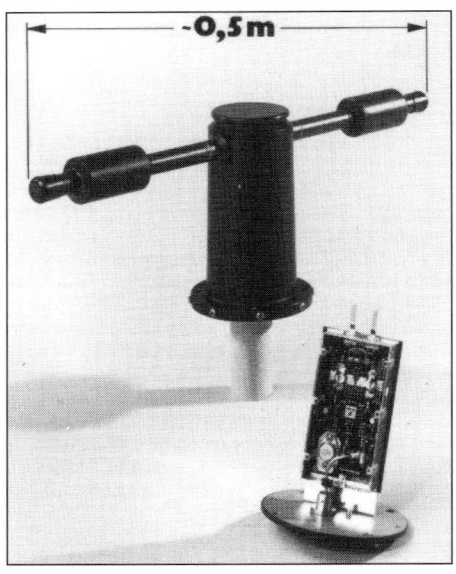

The HE 202 active receiving dipole

The HE 309 active vertical dipole

The HE 302 active receiving dipole

The HK 001 UHF coaxial dipole

The HK 014 coaxial dipole

HE309 active vertical dipole

The 20 to 1,300 MHz HE309 antenna is a small size (1 m radiator), high signal-to-noise ratio, active dipole that requires minimum distribution and switching support and is claimed to offer a very large bandwidth, a wide dynamic range and high sensitivity. Designed for radio communications reception, detection and monitoring applications, a single HE309 antenna is noted as being able to replace multiple installations of passive antennas while as a whole, the unit is described as offering high immunity to close proximity lightning strikes and non-linear distortion (with the latter being of a similar level to that offered by a passive antenna with a high grade preamplifier) together with a unit weight of approximately 3 kg.

HK001 225 to 400 MHz coaxial dipole

HK001 is a vertically polarised, omnidirectional antenna for use in fixed and mobile applications with a particular emphasis on shipboard installations. The dipole has been designed to be lightweight and to have a low wind load and as such, is noted as being suitable for installation at a considerable height above the ground. Such an installation regime is noted as offering increased field strength in transmission, higher sensitivity in reception and reduced environmental influences (particularly those caused by neighbouring parasitic elements).

HK014 coaxial dipole

HK014 is a vertically polarised, broadband, omnidirectional antenna that covers the 100 MHz to 1.3 GHz frequency range and is suitable for both transmission and reception applications. As a transmission antenna, HK014 is suitable for continuous wave operation at 400 W maximum over the 100 to 400 MHz band and

at 200 W maximum over the 400 MHz to 1.3 GHz spread. The VSWR of this coaxial is described as being better than 2 dB over its complete frequency range with a typical maximum value being 1.8 dB. HK014 operates as an electrically shortened, unbalanced fed dipole in its lower frequency range and as a double cone antenna in its upper range. The radiator construction used is noted as ensuring continuous transition between modes of radiation, resulting in vertical diagrams with relatively low frequency dependence. Rohde & Schwarz characterise HK014 as particularly suitable for shipboard and monitoring applications.

Status

As of this edition, antennas of the types described were thought to be available.

Contractor

Rohde & Schwarz, Munich.

VERIFIED

SCANLOC Direction-Finding (DF) and location system

Type

Scanning radio DF and emitter location system.

Description

The SCANLOC DF and emitter location system operates across the 0.5 MHz to 1,300 MHz frequency band and is designed for use against short time, frequency-agile (hopping) and intermittent radio emissions using multiple direction-finders in synchronised scanning mode. At the heart of the capability is the fast-scanning (up to 200 MHz/s) DDF0xS digitally scanning DF equipment (see separate entry) with up to four such units being integrated into an individual SCANLOC system. SCANLOC features include instantaneous signal interception, direction-finding and emitter location; automatic recognition of emission type; mobile emitter tracking and the storage of location results in a database.

Status

As of this edition, the SCANLOC DF and emitter location system was described as being a fully developed, 'off-the-shelf' system.

Contractor

Rohde & Schwarz, Munich.

VERIFIED

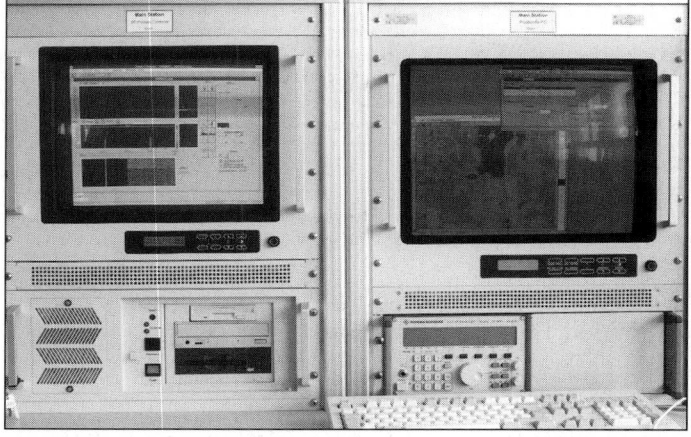

The SCANLOC operator's console   0007620

SFP 5200 Direction-Finding (DF) system

Type

Communications band DF system.

Description

The SFP 5200 is a visual direction-finder, operating over the frequency range 0.01 to 30 MHz, which is particularly characterised by its high acquisition rate. A maximum of 8 ms is required from the start of the DF command to the presentation of a digitally evaluated bearing (with quality criteria indication). This time also includes the setting time for the synthesiser at 2 ms and channel tuning with a new system at 1 ms. With a suitable drive this DF receiver is able to operate against frequency-hopping radios and to determine their location with DF networks. Different bandwidths can be selected for direction-finding and audio information. The complete contents of a message can be received with the narrow bandwidths most favourable for direction-finding. The digital filters that are used also ensure absolute synchronisation of the receiver during the transient period of tuning or tuning to the filter edge.

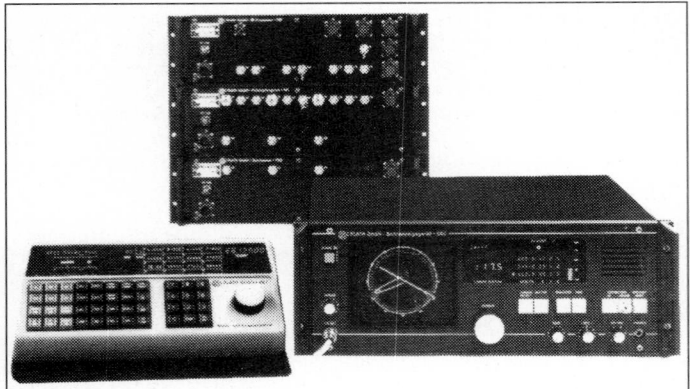

The SFP 5200 visual direction-finder

Status

As of this edition, the SFP 5200 DF system was reported as having been procured.

Contractor

C Plath GmbH, Hamburg.

VERIFIED

SIGMA communications band Electronic Support (ES) system

Type

Automatic communications ES system for interception, goniometry, monitoring and analysis.

Description

SIGMA (System for Interception, Goniometry, Monitoring and Analysis) is an automatic communications intelligence ES system designed for long-term and tactical reconnaissance activities. It copes with both conventional fixed frequency emitters and frequency-hoppers and is designed for stationary and mobile/semi-mobile deployment on various ground-based and airborne platforms. SIGMA may be operated as either a stand-alone system or with other stations in a netted surveillance system. The mechanical configuration is adaptable to the platform requirements. As a standard, SIGMA is installed in 48 cm (19 in) racks with an integrated operating position. The whole system is controlled by the operator via the window-oriented man/machine interface of the position computer. If required the operating position can be remoted. For unmanned operation, full remote control is possible via a 64 kbits/s link (including satellite communications). SIGMA, extended by measuring receivers, can carry out tasks of spectrum monitoring according to *Comité Consultatif International des Radiocommunications* recommendations.

Status

As of early 2001, Jane's sources were reporting the SIGMA ES system as having been procured by national and international customers.

Specifications

Frequency range: 1-1,000 MHz or sub-bands (extendable to 3 GHz)
Search speed: 125 channels/s (HF band max); 40,000 channels/s (V/UHF band max)
Probability of intercept: 99%
False alarm rate: 10^{-6}

The SIGMA ES system's operator position showing activity on a polar cartesian display   0044248

DF processing: automatic
Computer workstation: PC, AT compatible, Pentium

Contractor

European Aeronautic, Defence and Space (EADS) Co Systems and Defence Electronics, Ulm.

UPDATED

TACINT radio monitoring and Direction-Finding (DF) system

Type

Mobile radio communications intercept and DF system.

Description

The 20 MHz to 1.3/3 GHz TACINT mobile radio communications intercept and DF system is described as making use of a transportation box for mobility and as requiring either a 12 V DC or 110/220 V AC power supply. System features include:
- the ability to link individual TACINT stations into a DF/location network using an optional communications box
- real-time spectrum scanning with a scanning speed of up to 13 GHz/s
- target-orientated tasking, recording and evaluation of electromagnetic scenarios
- bearing line and emitter displays on an electronic map for processing situation data
- pre-evaluation of intelligence and the generation of contributions to situation reports.

Status

As of the late 1990s, the TACINT mobile radio communications intercept and DF system was described as being an available, commercial-off-the-shelf item.

Contractor

Rohde & Schwarz, Munich.

VERIFIED

WinLoc radiolocation software

Type

Software for direction-finding and radiolocation.

Description

The WinLoc radiolocation software offers:
- a graphic user interface under Windows 3.1
- display of direction-finding and radiolocation results on a digitised map
- support of well proven Rohde & Schwarz direction-finders
- support of different direction-finders in one system
- use as a single operating position
- use as a networked operating position in a master monitoring system
- frequency scanning for automatic monitoring of frequency bands and frequency lists.

Digitised maps can be generated with the MapEdit Windows software. With MapEdit, bit and vector maps can be imported or generated and edited with the aid of scanner or digitiser.

Status

As of the late 1990s, the WinLoc software package was reported as being available.

Contractor

Rohde & Schwarz, Munich.

VERIFIED

INTERNATIONAL

ARAMIS radio acquisition and monitoring system

Type

0.3 kHz to 3,000 MHz automated radio acquisition and monitoring system.

Description

The modular ARAMIS acquisition and monitoring system is designed to detect, gather and analyse radio communications traffic within the 0.3 kHz to 3,000 MHz frequency band and is made up of a number of mobile, semi-mobile and fixed intelligence gathering stations. The individual equipments and workstations making up a particular ARAMIS system are matched to the type and quantity of data to be gathered from a designated geographical area. Depending on customer requirements, system function can be manual, automatic or a mixture of the two, while the architecture's upper frequency limit can be extended if required. ARAMIS systems are further configured to facilitate data interchange between the national control centres, regional centres, equipments and databases within an overall communications intelligence network. ARAMIS's operational functions comprise listening-in, decoding and/or recording and emitter location, with mode selection being made on the basis of frequency range to be covered, the number of transmitters to be monitored, the size of the geographical area in which the target net is operating and the technical nature of the targeted traffic. Database updating within the architecture is automatic and ARAMIS's modularity allows it to be grown to accommodate evolving requirements.

Status

Over time, the ARAMIS acquisition and monitoring system is reported as having been procured.

Contractor

Thales Communications, RadioSurveillance and COMINT Systems Unit, Gennevilliers, France.

VERIFIED

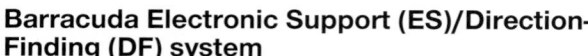

Barracuda Electronic Support (ES)/Direction-Finding (DF) system

Type

Communications band ES/DF system.

Description

Barracuda is a mobile ES/DF system which covers the 20 to 1,000 MHz frequency range in 1 Hz steps and makes use of advanced radio frequency, digital signal processing and microprocessor technologies. The system's operator PC control/display unit utilises Windows-based software while its receivers are based on commercial-off-the-shelf hardware (Thales' RDF 3725 equipment). Barracuda's Very/Ultra High Frequency (V/UHF) subsystem is integral with its host vehicle (wheeled platforms such as the Land Rover through to tracked armoured personnel carriers) while the necessary High Frequency (HF) antenna array is deployed on site.

Status

Thales Defence is reported as having received a then year £3 million UK Ministry of Defence contract covering the supply of eight Barracuda systems for use by the British Army during the mid-1990s. As of September 1996, four of the eight systems had been delivered (mounted in converted 436 Series armoured personnel carriers).

The British Army's Barracuda ESM/DF systems are mounted in modified 436 Series armoured personnel carriers

The interior of the British Army's Barracuda ESM/DF system vehicle

Specifications

Frequency: 20-1,000 MHz (baseline system)
Channel memory: up to 500
Demodulation modes: AM, CW, FM, FSK, ISB, LSB, USB (up to 30 kHz bandwidth)
Operating temperature: –10 to +55°C
Dimensions: 133 × 483 × 520 mm

Contractor

Thales Defence Information Systems, Wells, UK.

UPDATED

Beady Eye radar intercept and analysis system

Type

Radar intercept and analysis system.

Description

Beady Eye is a 1 to 18 GHz band, 360° cover, single-operator, tactical, automated radar intercept, bearing, classification and analysis system that is designed to detect battlefield threat radar emitters. A typical operational deployment would consist of three or more Electronic Support (ES) stations, linked by voice and digital datalinks to a Ferranti Computer Systems-developed control station. Other system elements include:

- a high-gain Direction-Finding (DF) antenna
- an automatic signal logger
- a digital instantaneous frequency measuring receiver
- a digital random access display (with multiple display options)
- a visual display message formatting; compilation and presentation capability
- a hard-copy printer
- microprocessor controlled system switching
- video and audio outputs for operator signal analysis
- a digital datalink.

Of these various elements, the Beady Eye control station provides system control and co-ordination, communications, the location and classification of emitters and a data management facility. Overall, the equipment is configured for forward-edge-of-battle deployments and offers 'high' sensitivity and probability of intercept values, accurate DF, single pulse frequency measurement and automatic parameter analysis.

Status

Over time, the British Army is reported to have procured at least four Beady Eye sensor and two Beady Eye control stations. As of September 2001, a variant of Thales Sensors' Meerkat-S ES system (designated as INCE - see separate entry) was being procured as a replacement for Beady Eye within the service.

Contractors

Thales Sensors, Crawley, UK.

UPDATED

Corvus III ELectronic INTelligence (ELINT) system

Type

Mobile or fixed-site 0.4 to 18 GHz ELINT system.

Description

The Corvus III ELINT system is described as providing an integrated approach to the gathering, analysis and management of data pertaining to intercepted radar signals. The architecture consists of a network of sensor stations and a Command

A shelter-mounted Corvus III sensor station

and Control (C²) facility (designated as the Electronic Warfare Operations Centre - EWOC) that tasks the sensor stations and analyses their reports in real time in order to generate and monitor a dynamic electronic order of battle. A typical Corvus III sensor station comprises:

- an ELINT sensor that incorporates a channelised receiver and 'comprehensive' pulse analysis and data recording subsystems
- a commercially based workstation that features a colour operator interface, 'extensive' database functions and data recording/playback facilities
- an air-conditioned operator shelter
- communications and power generation subsystems
- an optional 0.6 to 18 GHz wideband Electronic Support (ES) subsystem to facilitate 100 per cent probability of intercept.

The associated EWOC provides facilities for:

- generating an electronic order of battle
- generating higher control level reports
- correlating, managing and analysing data from its associated sensor stations and other network assets
- communications frequency management
- sensor station management and tasking.

Looking at some of the architecture's technology in more detail, the channelised receiver used has a 1 GHz instantaneous bandwidth while the de-interleaver used is derived from Thales' work on naval ES systems. A relational emitter identification database format is used and the colour, 'point-and-click' display options available to the operator include tactical (including maps with plotted emitter overlays) and tabular formats. The architecture has the ability to designate a system element as its master station and analysis tools are available to provide pulse-by-pulse analysis of a previously unseen emitter mode. The Corvus III system also incorporates Thales' Electronic Warfare Support System (EWSS) emitter and platform database.

Status

Over time, Corvus-type systems are reported as having been procured by customers in the UK and the Middle East.

Specifications

Frequency range: 0.4-18 GHz (optional extension to 40 GHz)
Mean sensitivity: −76 dBmi (0.4-2 GHz sub-band); −90 dBmi (2-18 GHz sub-band)
Dynamic range: 60 dB
DF accuracy: 2° RMS
Receiver type: scanning dish with 1 GHz IF, 29 channel channelised receiver
Bearing accuracy: 1.5° RMS (2-18 GHz sub-band)
Azimuth coverage: 360° (scanning)

Contractor

Thales Sensors, Crawley, UK.

VERIFIED

Corvus Electronic Warfare (EW) products (EP4220 and EP4410 modules)

Type

Range of equipments for EW systems.

Description

Over time, Thales Defence Information Systems' EP EW product range has included the EP4410 pulse analyser, EP4220 channelised receiver and the EP4250 tuner. Of these, EP4220 is a wideband channelised receiver that has been specifically designed for electronic intelligence, electronic support and signal monitoring applications that require high sensitivity and wide instantaneous

bandwidth. Operating in the 0.4 to 18 GHz frequency, it is described as being capable of detecting complex signals in a dense electromagnetic environment. It has a wide bandwidth (1 GHz compared with the much narrower bandwidths of swept superheterodyne receivers) which is claimed to enable it to 'fully examine' frequency-agile transmitters. Further, the detection bandwidth within each of the receiver's channels is reported to be sufficiently narrow as to give measurement values that are comparable with narrowband electronic intelligence receivers.

EP4410 is a pulse analyser that is designed to work with the EP4220 or conventional superheterodyne receivers. For each incoming pulse, it produces a set of pulse descriptor parameters. Qualification based upon these parameters allows pulses of the required signal of interest to be isolated and their pulse descriptors recorded together with correlated intrapulse data. EP4410 can analyse up to 64 k pulse descriptors together with 2 Mbytes of amplitude intrapulse data and 2 Mbytes of frequency intrapulse data. The equipment also includes options for precision modulation-on-pulse demodulation and growth potential for multichannel input operation. System control can be via the equipment's colour graphics front panel display (using a point and click device), Ethernet or an IEEE-488 interface.

Status

Over time, equipments of the types described are reported as having been procured. For its part, the EP4410 pulse analyser is known to have been acquired by the UK's Ministry of Defence.

Specifications

EP 4410
Input range: −1 to +1 V (nominal input range)
Frequency measurement:

Encoding range	Scale factor	Resolution
±3 MHz	333 mV/MHz	6 kHz
±10 MHz	100 mV/MHz	20 kHz
±20 MHz	50 mV/MHz	40 kHz
±40 MHz	25 mV/MHz	80 kHz
±250 MHz	4 mV/MHz	500 kHz

Amplitude measurement: 0-1.7 V nominal input range; 0.33 dB resolution; 20 mV/dB scale factor
Pulse-width measurement: 35 ns-10 ms
TOA measurement: 24 h clock rollover period; 5 ns resolution; 1, 2, 5 or 10 MHz optional external time reference
Threshold detection: 0-85 dB range, 1 dB step size (programmable threshold); 0-30 dB threshold offset, selectable time constant (noise riding threshold)
Data acquisition modes: 8 independent qualification windows (each defined in frequency, amplitude and pulse-width)
Intrapulse sampling: high-speed sampling channels and buffer stores for the measurement of amplitude and frequency intrapulse characteristics
Amplitude: 0.85 dB encoding range; 0.33 dB resolution
Frequency:

Encoding range	Resolution
±3 MHz	24 kHz
±10 MHz	80 kHz
±20 MHz	160 kHz
±40 MHz	320 kHz
±250 MHz	2 MHz

Sampling rate: 1.56-200 MHz (programmable)
Storage capacity: 2 Mbytes (amplitude and frequency)
Control interface: Ethernet local area network
Power: 115/240 V (40-60 Hz, 300 W)
Weight: 20 kg
Dimensions (w × h × d): 483 × 178 × 575 mm

Contractor

Thales Defence Information Systems, Wells, UK.

UPDATED

DFS2000

Type

Man/Machine Interface (MMI) for Direction-Finding (DF) systems.

Description

The MRCM (a joint venture between European Aeronautic, Defence and Space (EADS) EWATION and South Africa's Grintek EWATION) DFS2000 MMI provides an 'integrated and modular environment' for DF applications (including MRCM's MRD30W3, MRD1920 and MRD2000 equipments) and running Windows NT, allows the user 'freedom' to construct his desired working environment. Each element of the MMI (map, histogram, tune DF and such like) is displayed in separate windows, most of which can be resized and/or moved at will. DFS2000 requires (as a baseline) a 1,280 × 1,024 display screen and can use the maximum display area allowed by the particular graphics card. No specific graphic driver is required, with all graphic information going through the standard Windows NT graphic API. If desired, the architecture can make use of multiple monitors and can be operated as a stand-alone workstation with a single DF system or as part of a network, with or without direct coupling to a particular DF equipment. Configuration options include master-slave, multimaster and no predefined

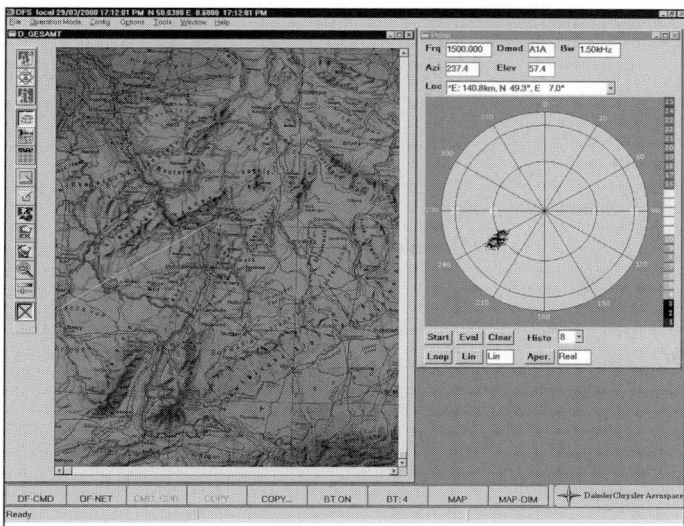

The interferometric (with single station location) version of DFS2000 (MRCM)
2002/0098312

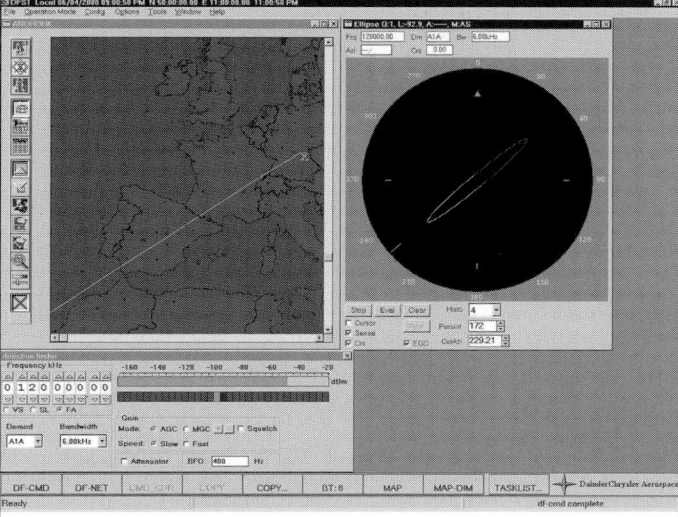

The Watson Watt version of DFS2000 (MRCM)   2002/0098311

master. In this latter case, individual workstations exchange result lists and can work in pairs or larger subnet groups. Overall, DFS2000's MMI functions comprise:

- histogram
- memory and frequency scan
- zoom (operator is able to zoom in and out of a particular frequency range)
- map
- record/replay (all scenario (audio and DF) data)
- result list management (save, recall, exchange and such like)
- mobile emitter tracking
- reception of external commands from non DFS systems (here, current activity is stopped for the duration of the higher priority command and is then resumed)
- automatic redirection of DF results to an external output
- Global Positioning System input (for the establishment of individual station location and network synchronisation)
- course input (for shipboard applications)
- optional connection to monitoring or search receivers.

Status
As of late 2000, the DFS2000 MMI was reported as being in service.

Contractor
MRCM (an EADS/Grintek EWATION joint venture), Ulm, Germany/Pretoria, South Africa.

NEW ENTRY

....

E 2000 series digital receivers

Type
Family of communications band digital receivers.

Description
The E 2000 receiver family is described as being a 'state-of-the-art suite of digital receiver channels and additional attachments' that may be configured with up to

five separate receiving channels in a single three unit high 48 cm (19 in) rack mounting unit. Within the family, there are three receiver variants, the known details of which are as follows:

E 2000 LH
Communications band receiver covering the frequency range from 0.3 kHz to 30 MHz. In addition to the common features noted in the following, E 2000 high frequency receivers set up priorities in universal demodulator functions. Capabilities include amplitude, frequency and phase shift keying; amplitude and frequency modulation; frequency/pulse shift keying decode (including method and codes); Morse decode and auxiliary functions.

E 2000 LU
Communications band receiver covering the frequency range from 10 kHz to 3,000 MHz.

E 2000 VU
Communications band receiver covering the frequency range from 20 to 3,000 MHz. The system exploits digital signal processing technology to provide the same technical performance as the E 2000 LH receiver.
E 2000 system features include:

- digital signal processing and filtering for 'optimum' noise suppression
- 'numerous' intermediate frequency filters
- notch filter and passband-tuning for the elimination of interference
- ancillary slot card processors with software modules for phase/frequency shift keying demodulation, signal classification or recognition of procedures in data transmissions, measuring units for spectrum monitoring systems (following International Telecommunications Union regulations), all of them optionally
- a computer controlled operator interface or operational BF 2000 control panel for manual operation.

Known details of the system's optional ancillaries are as follows:

BF 2000 control panel
BF 2000 is a manual tuning and function control module that is available in either full rack (48 cm (19 in) rack mounting and including power supply) or front panel configuration. Frequency tuning is by means of a control knob and all receiver functions and displays are shown on an integral light emitting diode display.

PI 2000 panoramic display
PI 2000 is a plug in module that displays received signal spectra or modulation parameters. Narrow (PS 2000 — 10 kHz bandwidth at high frequency and 100 kHz at very/ultra high frequency) and wide (PB 2000 — 5 MHz bandwidth, 42 MHz intermediate frequency) band display funtions are available and if the display is

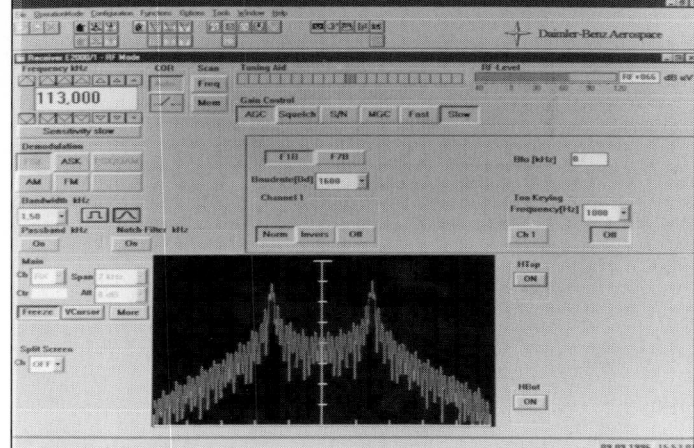

The E 2000's user interface showing a narrowband signal display   0007611

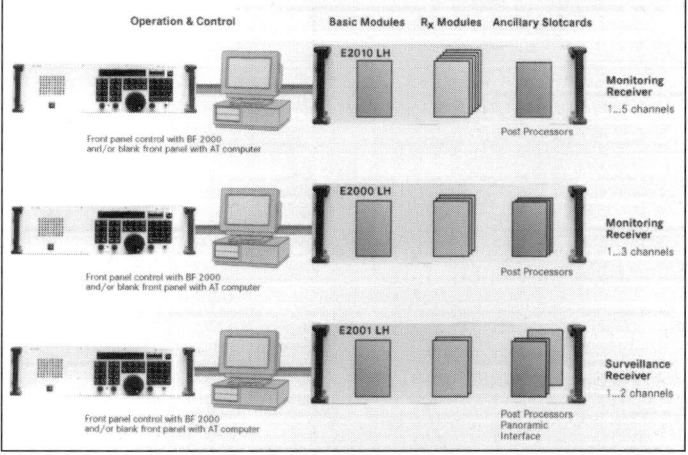

A graphic of the E 2000 receiver family showing the types of system combinations that are available   0007612

The BF front control panel with light emitting diode display   0007613

being controlled from a personal computer, it is integrated into the user's graphic interface. Dual channel display formats are used offering a range of modes that include detail, maximum, average or frozen panorama; simultaneous display of two spectra and phase shift keying vector diagram, constellation, eye pattern or spectrum.

Alongside the described ancillaries and variants, the former DaimlerChrysler Aerospace (now part of the European Aeronautic, Defence and Space (EADS) Co) developed the ADAS 2000 data acquisition system as an adjunct to the E 2000 series. Designed to automatically acquire and analyse radio data signals, ADAS 2000 is operator programmed and consists of additional hardware and DSP software for the E 2000 receiver together with a number of personal computer modules. ADAS 2000 can accommodate new functions and/or operations via the addition/exchange of modules and is divided into three subsystems as follows:

Acquisition
Tasked with the interception and decoding of known radio data signals (bit streams) and the interception of unknown signals for subsequent offline fine analysis.

Bit stream analysis
This function supports the operator in offline analysis of stored unknown signals. Here, the goal is to determine the used transmission protocol and coding procedure. Identified structures and bit patterns are described by the operator in decoding instructions that include the applicable code table.

Programming
Facilitates the user's conversion of analysis results into programmes that describe signals for the E 2000's modules.

Status
As of early 2001, Jane's sources were reporting the E 2000 receiver family as having been procured 'worldwide'.

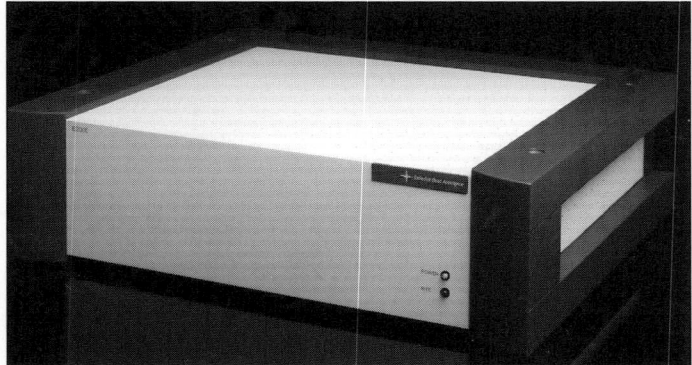

The E 2000 receiver mounted in a desktop cabinet   0007614

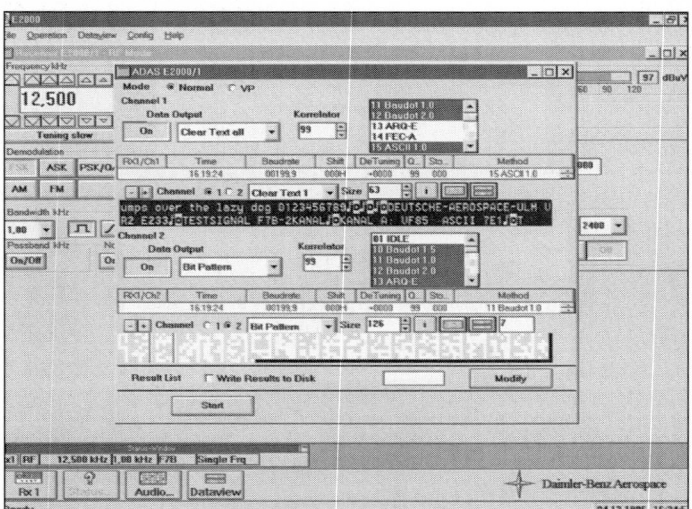

The ADAS 2000 acquisition module showing an F7B signal and a display of clear text and bit patterns on an E 2000 graphic user interface   0044244

Specifications

	E 2000 LH	E 2000 VU	E 2000 LU*
Frequency range	0.3 kHz-30 MHz	20 MHz-3 GHz	10 kHz-3 GHz
Frequency increments	1 Hz (min)		
No of receiving channels (per unit)	1-5	1-4	up to 2
Typical setting time	20 ms (1 ms option)	8 ms (100 μs option)	
Notch filters	100, 200 and 400 Hz (10 Hz resolution)	100, 200, 400, 800, 1,600 and 3,200 Hz	
Passband tuning	5 kHz (asymmetrical setting)	±1/2 selected bandwidth (B≤15 kHz)	
Demodulation types	all types used in HF range		
Modes		A1A, A1B, A2A, J2A, J2B, A3E, R3E, H3E, J3E, J7B, F3E and G3E (standard); 2/4/8/16 PSK, 2/4/8/16 DPSK, F1A, F1C and F7B (option with UD 2000)	
Interfaces	RS-232C, RS-485 and Ethernet	RS-232, RS-485 and Ethernet	

Key
* Except where noted, technical data for E 2000 LU as E 2000 LH/VU

ADAS 2000
Radio frequency: as E 2000 (see previously)
Demodulation: FSK, PSK and QAM
Modulation speed measurement range: 1-600 Bd (F7B, automatic, separately for 2 channels); 1-4,800 Bd (PSK/QAM, automatic within window around setting rate)
Line spacing: 20-8,000 Hz (F1B)
Measurement range: 40-2,400 Hz (F7B)
Library: unlimited scope (max of 100 code/bit structures can be loaded simultaneously, approximately 24 standard code/bit structures are supplied with the equipment)
Standard codes and (code tables): ARQ-242/-342/-E/-E3/-M2/-M4 (CCITT 3); ASCII 1.0/2.0 (CCITT 5); AUTOSPEC (Bauer); AX 25 (CCITT 5); Baudot 1.5/2.0 (CCITT 2); CIS-11/-14/-27 (M2); FEC-A (CCITT 2); HNG-FEC (CCITT 2); IDLE 2/7/8/28/56; POL-ARQ (Sitor 7); RUM-FEC (RUM 16); SI-ARQ (CCITT 3); SI-FEC (CCITT 3); SITOR-A/B (Sitor 7); SPREAD 11/21/51 (Bauer)

Contractor
European Aeronautic, Defence and Space (EADS) Co Systems and Defence Electronics, Ulm, Germany.

UPDATED

···

European Aeronautic, Defence and Space (EADS) Electronic Warfare (EW) antennas

Type
Family of reception and direction-finding antennas.

The A1228VU2 interferometer antenna

Description

Tactical electronic support and countermeasures systems developed by the former DaimlerChrysler Aerospace (now part of EADS) are fitted with antennas designed and manufactured by the company. These sensor elements consist of omnidirectional and logarithmic periodic receiving antennas and various Adcock Direction-Finding (DF) antennas. The variety of types covers the complete frequency range required for communications and EW systems. Many types are designed for both tactical semi-mobile and fixed operation. Some examples of such systems are as follows:

A1294/2 VLF (3 to 30 kHz) receiving antenna
A1146 HF (3 to 30 MHz) log-periodic directional receiving antenna
AK 1200 HF Adcock antenna
A1228VU2 VHF/UHF (30 to 1,000 MHz) interferometer antenna combination
A1284/1 HF/VHF/UHF (3 to 1,000 MHz) DF Adcock antenna combination
AK 1296 HF interferometer antenna combination.

Status

As of late 2000, EW antennas of the types described were reported as having been procured.

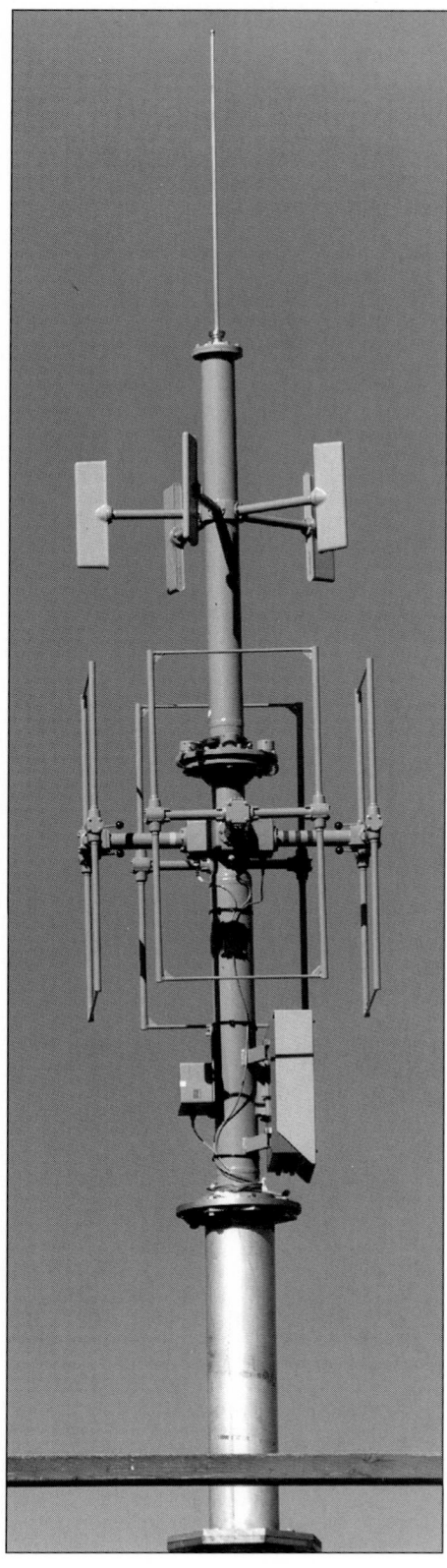

The A1228 DF multimode antenna (interferometer and/or Watson Watt)
0007610

The A1284/1 DF antenna combination

Contractor

European Aeronautic, Defence and Space (EADS) Co Systems and Defence Electronics, Ulm, Germany.

UPDATED

LWD 2 laser warning system

Type

Laser warning system for Armoured Fighting Vehicles (AFV).

Description

LWD 2 is an AFV laser warning system that comprises a control unit, a junction box and between one and 12 detector arrays. In the single detector head configuration, the junction box is replaced by a mechanical/electrical interface unit. In more detail, the system's control unit houses the equipment's processing electronics, the operator controls and a display panel. Both visual and audible threat warnings are provided together with threat identification (single-pulse range-finders or multiple-pulse designators). Threat bearing is indicated on a 24 light emitting diode ring display and is additionally shown as a four-digit bearing readout calibrated in degrees or mils as preferred. An overhead threat display is incorporated when overhead detectors are in use and there are two control switches – power and audio alarm select. An RS-422 interface is provided for external system control or communication with an onboard countermeasures suite. The LWD 2 junction box interconnects the detector arrays with the system control unit and collates received data before transmitting it for processing. The box can be configured to recognise any arrangement of detectors, can be reset for alternative detector layouts and can be mounted internally or externally. The detector arrays used have an azimuth field of view of more than 180° with a resolution of ±7.5°. As noted earlier, the LWD 2 system can accommodate up to 12 detector arrays. When a single-pair array is used, it is configured as a low-profile head with a 360° field of view. Available system options include incorporation of overhead detectors into the standard detector arrays; configuration of arrays to be corner mounted at 45° to the host vehicle's axis; an eight array plus junction box; gunner or driver remote displays; software to allow standard and non-standard detector mountings to be mixed and a programme for calculating directional data on the turret/hull relationship.

Status

As of October 2001, at least 500 Thales Optronics (formerly Avimo) - sourced laser warning systems were reported as having been procured by customers around the world.

Specifications

FOV: −12 to +47° (standard vertical FOV); −12 to +90° (vertical FOV with overhead option); 195° (azimuth FOV)

The elements that make up the LWD 2 AFV laser warning system

Azimuth resolution: ±7.5°
Spectral response: 400-1,600 nm
Response time: <0.1-0.2 s
Supply voltage: 18-32 V DC
Operating temperature: −40 to +55°C

Contractor

Thales Optronics, Taunton, UK.

UPDATED

Meerkat-S

Type

0.4 to 18 GHz band ground mobile Electronic Support (ES) and ELectronic INTelligence (ELINT) system.

Description

Meerkat-S manufacturer Thales describes it as being a 'high-performance, very high-reliability ES-ELINT system' that is designed to detect, identify and locate hostile and 'other' types of radar emission within battlefield and worldwide out-of-area tactical applications. As such, it consists of a number of mobile sensor vehicles that feed a similarly mobile Non-communications Control Centre (NCC). Here, an Electronic Order of Battle (EOB) picture is created and maintained. Equally, the architecture is scalable, with each NCC being capable of operating with two, three or four sensor stations. As an alternative, the Meerkat sensor station can function as a stand-alone equipment, with communication between the various system elements being by means of voice and data combat net radio.

Looking at the Meerkat-S sensor station in more detail, the system incorporates a mast-mounted Antenna-Receiver Assembly (ARA), a Receiver Digitiser Unit (RDU), a Pulse Train Analyser Unit (PTAU), main and auxiliary mode libraries and a single person operator's console. Of these, the ARA is noted as containing a rotating dish antenna/log video receiver subsystem for intercept and direction-finding and an omnidirectional antenna/receiver package for sidelobe suppression. The RDU performs frequency measurement, establishes relevant signal parameters and transmits a digitised pulse descriptor to the PTAU. Here, the received pulse stream is de-interleaved and accurate and stable individual emitter data sets are generated. Emitter identification is by means of library comparison. In this context, the station's main mode library has a capacity of 'at least' 10,000 modes, while its auxiliary library holds up to 1,000 emitter modes that have been operator loaded. Main library loading is achieved via an integral disk unit. The operator's console features a high-resolution, colour, flat panel display that offers tabular and graphic display formats. Processed data is transmitted to the NCC or is recorded to disk for further analysis or archiving. Other system features include:

- fully automatic ES functionality
- automatic and manual ELINT analysis as standard
- shelter mounting aboard a wheeled Pinzgauer all-terrain chassis
- air transportable by C-130 aircraft
- claimed 100 per cent probability of intercept
- an integral Global Positioning System sensor location capability
- optional precision measurement and intrapulse analysis ELINT receivers.

For its part, the Pinzgauer-mounted NCC receives parametric data from each of its associated sensor stations via the already noted combat net radio provision and correlates it to produce position fixes. The various individual fixes are overlaid on a

A general view of the sensor station configuration used in the INCE Meerkat-S ES/ELINT system application (Thales Sensors) 2002/0116876

digital mapping system to create a real-time EOB. A rearward reporting capability is available as an option.

Status

On 11 September 2001, Thales Sensors announced that it had been awarded a then year £6 million contract to supply the British Army with a three sensor station/NCC application of the Meerkat-S architecture to meet its Interim Non-Communications Electronic support (INCE) requirement. At the time of supply, INCE was intended to bridge the gap between the existing Beady Eye and Pinemartin ES systems and the 'Soothsayer' electronic warfare architecture (then scheduled to come on stream during 2006) within the service's 14 Signals Regiment. At the time of the announcement, INCE was described as being effective against airborne, anti-aircraft, artillery location, battlefield surveillance and meteorological radars together with 'civil' emitters such as air traffic control radars, identification friend-or-foe transponders, television transmitters and cellphone base stations. In addition to system supply, the INCE award was also understood to include operator training and a five-year integrated logistics support package. As of October 2001, the baseline Meerkat-S architecture was further reported as being available.

Specifications

Meerkat-S

Frequency range: 0.4-18 GHz (40 GHz option)
Bearing accuracy: 1.5° RMS (>2 GHz, average)
Sensitivity: 76 dBW/m² (−85 dBmi at 9 GHz)
Dynamic range: 60 dB
Coverage: cosec² (elevation); 360° (azimuth)
Mode tracks: up to 500 emitters
Pulse density: up to 1 Mpps
Library: up to 1,000 modes (auxiliary); up to 10,000 modes (main)

Contractor

Thales Sensors, Crawley, UK.
Thales Airborne Systems, Elancourt, France.

NEW ENTRY

Meerkat-SA

Type

2 to 40 GHz band fixed-site Electronic Support (ES) and ELectronic INTelligence (ELINT) system.

Description

Meerkat-SA manufacturer Thales describes it as being a 'high-performance, very high-reliability ES-ELINT system' that is designed to detect, identify and locate hostile and 'other' types of radar emission and can function as an unattended sensor architecture within passive air defence and strategic intercept applications. As such, it consists of a number of remote, fixed-site, sensor stations that feed into a central Control Site (CS). Here, multiple operators control the architecturer's remote sensor stations and control processors that identify intercepts, fuse data and maintain a 'dynamic' Electronic Order of Battle (EOB). Equally, the system is scalable, with each CS being capable of operating with up to 12 sensor stations. Communication between the architecture's various elements is by means of trunk radio or cable connection into a wide area network.

Looking at the Meerkat-SA sensor station in more detail, the system incorporates a mast-mounted Antenna-Receiver Assembly (ARA), a Receiver Digitiser Unit (RDU) and a Pulse Train Analyser Unit (PTAU). Of these, the ARA is noted as containing a rotating dish antenna/receiver subsystem for 'high-sensitivity' intercept and Direction-Finding (DF), an omnidirectional antenna/receiver package for sidelobe suppression and a multiport instantaneous DF receiver for all-round surveillance. A dual redundant RDU performs frequency measurement, establishes relevant signal parameters and transmits digitised pulse descriptors to the MINERVA processor-based PTAU. Although the Meerkat-SA sensor station's primary mode of operation is unattended, an operator workstation can be connected for use at any accessible site.

For its part, the CS receives parametric data from each of its associated sensor stations and identifies and correlates it to produce emitter position fixes. Each established fix is then overlaid on an electronic mapping system to form a real-time EOB. Emitter/platform identification is by means of library comparison, using both main and auxiliary libraries. Of these, the main library features an integral disk unit for data loading and has a capacity of up to 10,000 emitter modes and up to 1,000 platform types. The architecture's operator loaded auxiliary library has a capacity of up to 1,000 emitter modes. Workstations are provided for sensor station control, emitter/platform identification and database management, data fusion and EOB maintenance and detailed technical analysis of detected emitters. Each workstation is equipped with a high-resolution, colour, flat screen monitor that can handle tabular and graphical display formats. Data may also be recorded to disk for subsequent analysis. Other system features include:

- claimed 100 per cent probability of intercept
- precision pulse repetition interval measurement for specific emitter identification
- a Global Positioning System facility for time referencing
- remote sensor station equipment failure tolerance.

Status
As of October 2001, the Meerkat-SA architecture was reported as being available.

Specifications
Frequency range: 2-40 GHz (0.4 GHz option)
Bearing accuracy: 1.5° RMS (>2 GHz, average)
Sensitivity: 76 dBW/m² (−85 dBmi at 9 GHz)
Dynamic range: 60 dB
Coverage: cosec² (elevation); 360° (azimuth)
Mode tracks: up to 500 emitters
Pulse density: up to 1 Mpps
Library: up to 1,000 emitter modes (auxiliary); up to 1,000 platform types/up to 10,000 emitter modes (main)

Contractor
Thales Sensors, Crawley, UK.
Thales Airborne Systems, Elancourt, France.

NEW ENTRY

....

MRD2000 LH

Type
1.6 (0.5) to 30 MHz band Direction-Finding (DF) system.

Description
The MRCM (a joint venture between European Aeronautic, Defence and Space (EADS) EWATION and South Africa's Grintek EWATION) MRD2000 LH is an interferometric, 'high precision' DF system that covers the 1.6 (0.5) to 30 MHz frequency range and comprises a multichannel DF receiver, an antenna assembly and a standard personal computer running the DFS2000 Man/Machine Interface

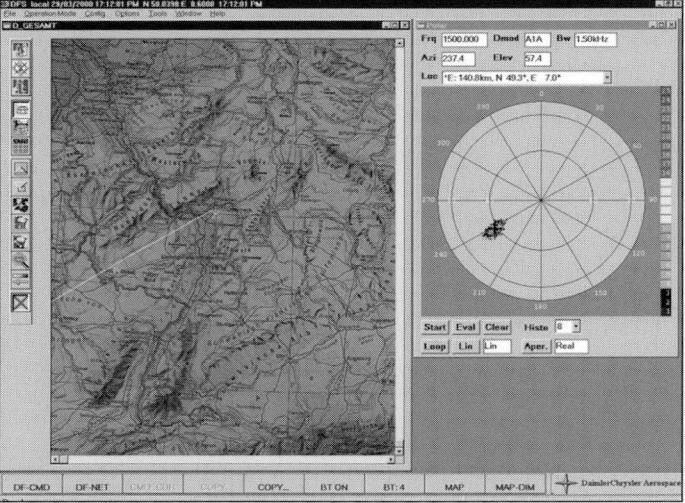

The MRD2000 LH DF system MMI (MRCM)   2002/0098313

(MMI - see separate entry) under Windows NT. The architecture is packaged in a three unit, 48 cm (19 in) rack configuration and system features include:

- parallel signal processing (with the five channel P2000 LH DF receiver)
- 'reliable' interception and bearing measurement of burst and frequency-hopping signals
- 'absolute' phase and amplitude synchronisation of the DF channels
- use of intermediate frequency/variable notch filters and passband-tuning to 'eliminate' interference
- a wide aperture antenna system
- a Single Station Location (SSL) capability
- stand-alone (with SSL capability) or networked functionality
- suitability for fixed-site or semi-mobile applications
- local or remote control (fully automatic or operator supported functionality)

Status
As of late 2000, the MRD2000 LH DF system was reported as being in production and service.

Specifications
Frequency range: 1.6 (0.5)-30 MHz
Number of DF channels: 5
Demodulation: AM, CW, FM, FSK and SSB
Intermodulation: +40 dBm (IPIP 3); +70 dBm (IPIP 2)
Min signal duration/DF time: 2 ms
DF sensitivity: −15 dBµV/m type
DF accuracy: 1°/cos e (skywave); <1° RMS (groundwave - 0.5° typical)
SSL accuracy: 15% RMS (typical - distances from >100 km); 15 km RMS (distances of up to 100 km)

Contractor
MRCM (an EADS/Grintek EWATION joint venture), Ulm, Germany/Pretoria, South Africa

NEW ENTRY

....

MRD4008

Type
1.6 (0.5) to 30 MHz Direction-Finding (DF) system.

Description
The MRCM (a joint venture between European Aeronautic, Defence and Space (EADS) EWATION and South Africa's Grintek EWATION) MRD4008 system is described as being a wideband, multichannel, DF equipment that provides continuous monitoring and surveillance of 'all' radio activity within the 1.6 (0.5 - see Specifications) to 30 MHz frequency band. It is optimised for the interception of low probability of intercept signals (burst, frequency hopping and chirp) and can be used to detect and analyse unknown signals as well as searching for and retrieving transmissions from known communications nets. Within its 800 kHz instantaneous bandwidth, MRD4008 is claimed to be able to detect all active transmitters and execute a DF algorithm on each of them. The equipment's Man/Machine Interface (MMI) presents frequency band occupation (as a time versus frequency waterfall display) and shows either the azimuth or amplitude of the signals observed. System options include:

- a single station location capability (interferometric mode)
- a 700 s to 30 minutes raw data buffer

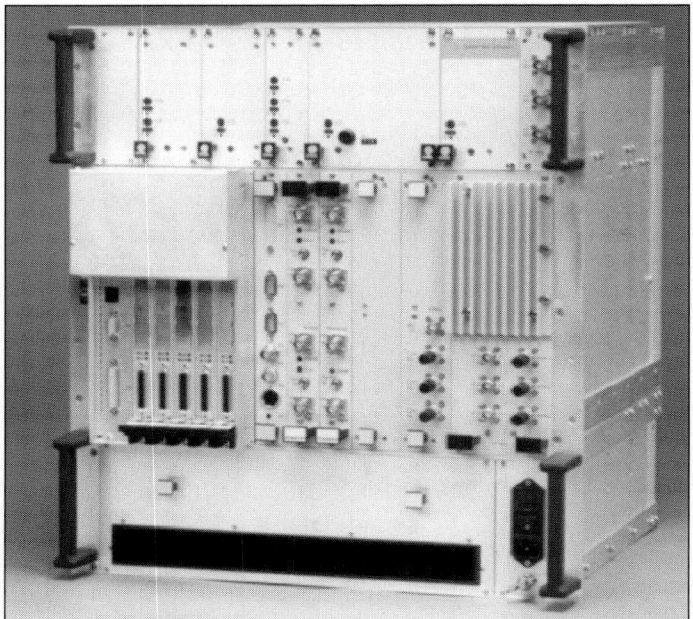

The MRD4008 DF system hardware (MRCM)   2002/0098329

The MRD4008 DF system MMI (MRCM) 2002/0098327

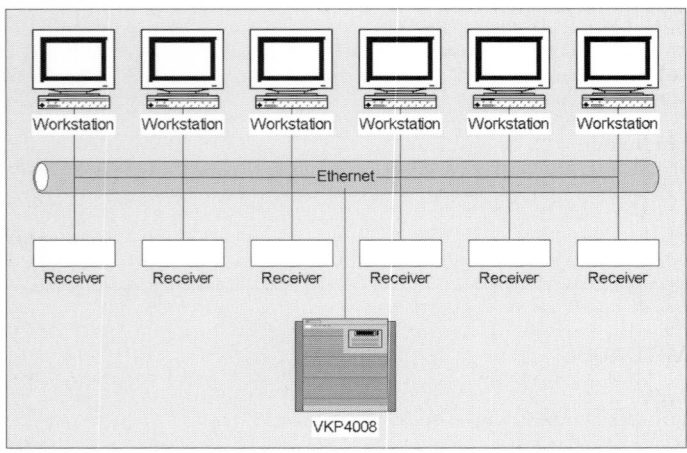

A schematic showing the MRD4008 DF system in a multi-user configuration (MRCM) 2002/0098330

- connection to the MCL4000W classification system (segmentation, parameter measurement, mode of operation classification, data protocol classification, demodulation and decoding)
- connection to the MRR4000 wideband receiver system (simultaneous interception of signals, digital signal conversion for signal extraction, scaleable (up to 30 MHz) bandwidth and a signal buffer for predetection recording)
- multistation networking for triangulation

Status
As of late 2000, the MRD4008 DF system was reported as being in production and service.

Specifications
Frequency range: 0.5-30 MHz (Adcock antenna); 1.6-30 MHz (interferometric antenna)
DF mode: interferometric (small and wide aperture mode) or Watson Watt
Noise figure: <13 dB
Sensitivity: >−115 dBm (15 dB S/N, 1 kHz bandwidth)
Dynamic range: >85 dB (typical - 800 kHz band); >110 dB (125 Hz sub-channel)
Real-time bandwidth: 800 kHz with 6,400 FFT channels
Scanning speed: up to 110 MHz/s
DF rate: up to 100,000 DFs/s
Bearing accuracy: <1° RMS (with DF antenna)
Min signal duration: 1 ms (down to 125 µs)

Contractor
MRCM (an EADS/Grintek EWATION joint venture), Ulm, Germany/Pretoria, South Africa.

NEW ENTRY

MSD 2000 signals display unit

Type
Multifunction signals display unit.

Description
The MSD 2000 is a dual-channel panoramic display unit that is microprocessor controlled and, in combination with former DaimlerChrysler Aerospace (now part of

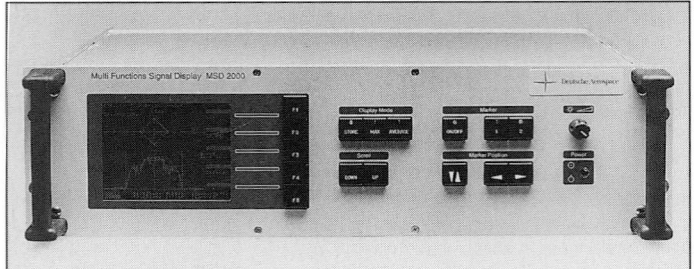

The multifunction MSD 2000 signal display unit

the European Aeronautic, Defence and Space (EADS) Co) developed receivers, is suitable for panorama monitoring and narrowband analysis. Up to 1 MHz in the HF range or 5 MHz in the VHF/UHF range can be displayed and real-time panorama (as well as maximum, mean value and frozen panorama) displays are shown on a high-resolution, controllable brightness, electroluminescent screen. MSD 2000 is claimed to be 'highly suited' for presentation of demodulated phase shift keying-specific signals such as phase vectors or eye patterns. Two serial interfaces RS-232C or IEC/IEEE-488 for command and control facilitate system integration.

Status
As of mid 2000, the MSD 2000 multifunction signals display unit was reported as having been procured.

Specifications
IF inputs: 2 inputs, optionally 10.7 MHz, 21.4 MHz or 42.2 MHz; 2 inputs, optionally 200 kHz or 10 kHz
Frequency resolution: 1 Hz (at 250 Hz display width)
Display ranges: 250 Hz-10 kHz (IF 10 kHz, 200 kHz); 1 kHz-5 MHz (IF 21.4 MHz); 250 Hz-1 MHz (IF 10.7 MHz, 42.2 MHz)

Contractor
European Aeronautic, Defence and Space (EADS) Co Systems and Defence Electronics, Ulm, Germany.

UPDATED

Odette communications band intercept and Direction-Finding (DF) system

Type
High/Very High/Ultra High Frequency (HF/VHF/UHF) signals intercept and DF system.

Description
The Odette HF/VHF/UHF signals intercept and DF system incorporates commercial-off-the-shelf software and receivers and is designed for installation in a range of platform types. These include the wheeled Land Rover utility vehicle, armoured fighting vehicles and the tracked Hägglund BV206 all-terrain vehicle. The system's manufacturer stresses its ability to locate frequency-hopping emitters and notes that it is equipped with a communications subsystem built around Thales Defence's Panther enhanced digital tactical radio.

Status
During November 1998, the then Racal Radio (now Thales Defence Information Systems) announced that it had been awarded a then year £26 million plus contract to deliver an undisclosed number (thought to be in excess of 50) of Odette signals intercept and DF systems to the British Army and the Royal Marines. Supply of the Odette system was competitively tendered in the UK, Europe and the USA and it is understood that initial deliveries of the equipment were made during the first quarter of 1999. As originally scheduled, programme completion was set for the end of 1999 and the cited contract included a full in-service support and training package.

Contractor
Thales Defence Information Systems, Wells, UK.

UPDATED

Polygon MRD30w3/n

Type
1.5 to 30 MHz band Direction-Finding (DF) system.

Description
The MRCM (a joint venture between European Aeronautic, Defence and Space (EADS) EWATION and South Africa's Grintek EWATION) Polygon MRD30w3/n is a wideband, fast Fourier transform DF equipment that is designed for wideband spectrum surveillance of and DF against fixed frequency and low probability of

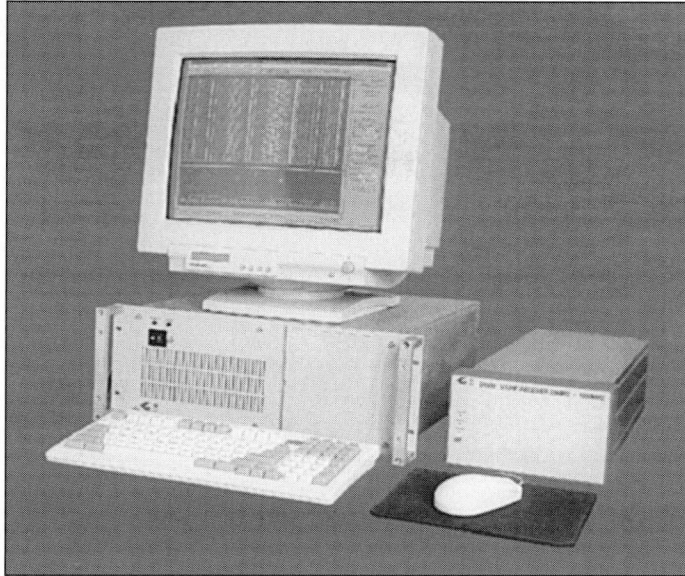

The Polygon MRD30w3/n DF system hardware (MRCM)   2002/0098332

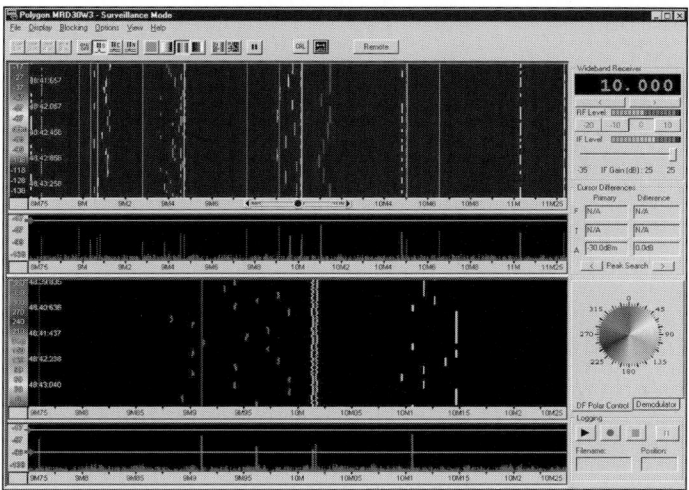

The Polygon MRD30w3/n man/machine interface (MRCM)   2002/0098331

intercept (including frequency hoppers, burst, chirp and direct sequence spread spectrum) signals within the 1.5 to 30 MHz frequency range. As such, the system is claimed to offer a 'high' probability of intercept and 'accurate' DF results, capabilities that are noted as being vested in the use of sensitive, large aperture antenna systems and parallel signal processing. The baseline MRD30w architecture can be configured with three (MRD30w3/n) or six (MRD30w6/n) channels.

Status
As of late 2000, the Polygon MRD30w3/n DF system was reported as being in production and service.

Specifications
Frequency range: 1.5-30 MHz (interferometric antenna)
Instrumented accuracy: <1° RMS
Sensitivity: −110 dBm signals are detected (3 kHz, 10 dB (S+N)/N)
IF rejection: 80 dB
Instantaneous bandwidth: 500 kHz
Frequency resolution: 625 Hz
Noise figure: <13 dB
DF algorithm: interferometric and Watson Watt
Search scanning rate: 150 MHz/s (without DF)
DF rate: 8,000 DFs/s (interferometric, scanning, 500 kHz bandwidth, 25% occupancy)
Min signal duration: 3.5 ms (interferometric)
Polyphase filters: filter factors 1, 2, 4 and 8
Zoom: zoom factors 1, 2, 4 and 8
Display windows: scan and zoom windows with time versus frequency waterfall, angle of arrival versus frequency waterfall and time DF waterfall
User interface: Windows NT (XGA colour, 1,024 × 768)
Dimensions: 48 cm × 3U (MRR 2010 LH1); 48 cm × 4U (digital signal processor); 48 cm × 5U (radio frequency unit - incl calibration unit)

Contractor
MRCM (an EADS/Grintek EWATION joint venture), Ulm, Germany/Pretoria, South Africa.

NEW ENTRY

Polygon MRD3000w5

Type
20 to 3,000 MHz band Direction-Finding (DF) system.

Description
The MRCM (a joint venture between European Aeronautic, Defence and Space (EADS) EWATION and South Africa's Grintek EWATION) Polygon MRD3000w5 is a wideband, fast Fourier transform DF system that is designed for wide spectrum surveillance of and DF against fixed and frequency-agile signals within the 20 to 3,000 MHz frequency band. It is claimed to combine the 'utmost' in probability of intercept with high accuracy DF, capabilities that are noted as being achieved via the use of sensitive, large aperture, interferometric antenna systems and parallel signal processing in five DF channels. MRD3000w5 can be configured for fixed-site, semi-mobile and mobile operations in land-based, shipboard and airborne applications.

Status
As of late 2000, the Polygon MRD3000w5 DF system was reported as being in production and service.

Specifications
Frequency range: 20-3,000 MHz
Number of DF channels: 5
Instantaneous bandwidth: 1.25 or 10 MHz (selectable)
Frequency resolution: 1.531, 3.063, 6.125 or 12.5 kHz
Noise figure: 10 dB
Spurious free dynamic range: 70 dB
Scanning speed: up to 10 GHz/s (up to 25 GHz/s as an option)
Scanning acquisition time: 80 µs
Bearing accuracy: <1° RMS (typical fixed-site application, 20-1,000 MHz frequency range, 0.5° typical); <1.5° RMS (typical fixed-site application, 1,000-3,000 MHz frequency range, 1° typical)
Bearing sensitivity: 1,0 µV/m (typical fixed-site application, 20-1,000 MHz frequency range); 1,5 µV/m (typical fixed-site application, 1,000-3,000 MHz frequency range)
MMI operating system: Windows NT

Contractor
MRCM (an EADS/Grintek EWATION joint venture), Ulm, Germany/Pretoria, South Africa.

NEW ENTRY

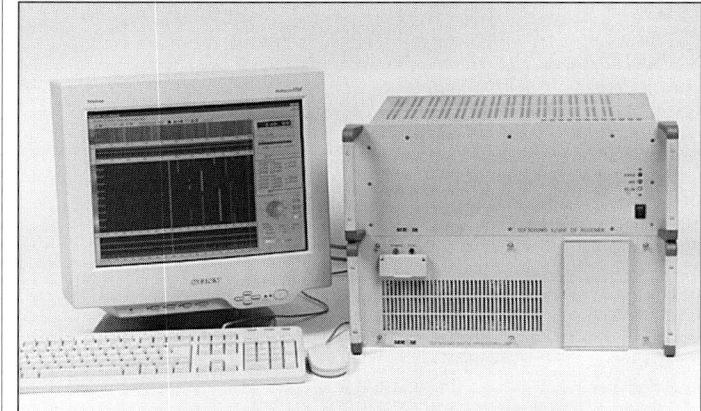

The Polygon MRD3000w5 DF system hardware (MRCM)   2002/0098324

PSI 2000 Direction-Finding (DF) system

Type
Communications band interferometer DF system.

Description
The high-frequency Interferometer PSI 2000 is a large aperture, digital DF system that is designed to provide high accuracy bearings by means of parallel digital signal processing in five DF channels. In addition to use of the interferometric DF technique, PSI 2000 employs digital signal processing and filtering to provide 'absolute' phase and amplitude synchronisation within the DF channels so as to ensure 'reliable' interception of and bearings on short duration signals such as burst transmissions. The equipment's control and bearing display unit takes the form of a personal computer compatible workstation with a high-resolution colour monitor. Bearing results are displayed in a polar co-ordinate system as an azimuth/elevation histogram. Map display as well as a single station location capability are available options.

Status
As of late 2000, Jane's sources were reporting the PSI 2000 DF system as having been procured.

The AK 2000 interferometer antenna system is made up of seven combined loop/linear units of the type shown here

The operator's console (including monitoring position) used in the shelterised version of the PSI 2000 DF system   0007617

Specifications
Frequency range: 1.6-30 MHz
DF accuracy: 1° RMS typical
Min signal duration: 2 ms
Power supply: 115/230 V AC
Dimensions: 48 cm unit 3 U height

Contractor
European Aeronautic, Defence and Space (EADS) Co Systems and Defence Electronics, Ulm, Germany.

UPDATED

RA3720 series surveillance receivers

Type
Family of communications band surveillance receivers.

Description
The RA3720 series of high-performance receivers covers the 15 kHz to 1 GHz band and are suitable for use in many types of surveillance and monitoring systems. They include built-in features that will enhance the performance of such systems. 'Advanced' radio frequency, digital signal processing and microprocessor technologies are combined to produce receivers with high performance, thereby allowing the system designer to produce a surveillance or monitoring system which is effective even when operated in a crowded electromagnetic environment. The flexible design and extensive user facilities are claimed to make the RA3720 series ideal for use in modern systems. The receivers are fitted with an operator's control panel providing a wide range of facilities. Operation is simplified by the use of single function push-buttons for the basic receiver functions, while a menu system is used to control the many special facilities.

Comprehensive channel and frequency scanning facilities are built into the basic receiver. These include the ability to scan multiple frequency ranges and to set up mixed frequency/channel scan routines. Specific channels or frequency ranges with a scan range may be omitted from the scan. During scan operation, automatic

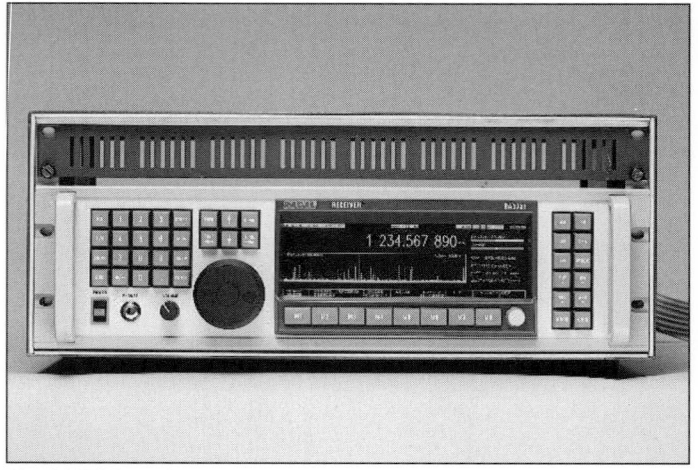

The RA3721 receiver

'stop-no-signal' is possible using the built-in signal activity detector which is designed to provide a high probability of interception combined with a low false alarm rate. Manual control of the scan is also possible either from the front panel or by means of external inputs. When the receiver detects signal activity, the receiver frequency may be sent automatically to another equipment via the remote-control interface. The receiver may be set to resume the scan automatically after a preset time, or on receipt of an external command.

Status
Over time, RA3720 series receivers were reported as having been procured.

Contractor
Thales Defence Information Systems, Wells, UK.

UPDATED

RA3790 series surveillance receivers

Type
Family of digital communications band surveillance receivers.

Description
RA3790 series surveillance receivers are modular in construction, cover the 10 kHz to 30 MHz frequency range and have been designed with particular attention being given to sensitivity, intermodulation, reciprocal mixing and spurious responses. Equal thought has been given to the ergonomics of the front panel used. Here, a menu system is provided to handle system special features and single function push-buttons are incorporated to activate the most commonly used modes. Elsewhere, key system features include 'high-performance' radio frequency circuitry; digital signal processing; digital Intermediate Frequency (IF) filtering (providing up to 100 bandwidths); tunable IF notch (plus passband tuning); automatic channel and frequency scanning; remote controllability; provision of a controller (the MA3790 unit) for slave equipments and built-in test. All RA3790 series receivers include a serial ASCII interface with a built-in multi-address capability of up to 100 units and remotely controlled and operator manned equipments can be utilised in the same network. Identified family members are as follows:
RA3791: Single receiver with operator's front panel.
RA3792: Dual receiver with operator's front panel.
RA3793: Single receiver configured for remote control.
RA3794: Dual receiver configured for remote control.

Status
Over time, RA3790 series receivers are reported as having been procured.

Specifications
Frequency: 10 kHz-30MHz (in 1 Hz steps)
Modes: A1A, A1B, (CW); A2A, A2B (MCW); A3E (AM); F3E (FM); F1B (FSK); H2A, H2B, H3E, J2A, J2B, J3E, R2A, R2B, R3E (USB/LSB); B7B, B8E, B9W (ISB option)
Channel store: 100 frequencies in EEPROM memory
Scan modes: Channel (0.1-9.99 s dwell time per channel); frequency (100 Hz-999.9 kHz step size, 10 Hz/s-999.9 kHz/s sweep rate)
Sensitivity: 0.5-30 MHz (for full frequency range)
Power supply: 90-132 V and 175-264 V AC (with automatic range selection 47-63 Hz)
Power consumption: 50 W (RA3791)
Temperature: −10 to +55°C
Dimensions: 133 × 483 × 450 mm

Contractor
Thales Defence Information Systems, Wells, UK.

UPDATED

RDF3210 Direction-Finding (DF) system

Type
HF band (3 to 30 MHz) interferometer DF system.

Description
The fixed-site or transportable RDF3210 DF system is noted as being able to handle high-angle skywave and low-angle/ground wave signals and as featuring 'easy' antenna deployment and a 'comprehensive range' of DF displays. The system is capable of position fixing by single site location techniques when suitable ionospheric data are available. RDF3210 measures both the azimuth and elevation of received signals and is particularly effective against high angle skywave signals. In fixed-site applications, remote control of the equipment and displays permits unattended operation over suitable communications links. Remote operation includes diagnosis of faults down to unit level. To assist in optimum fixed-site selection the antenna array can be situated over 500 m from the main equipment. When configured for transportable deployment, RDF3210 is mounted in light-alloy, shockproof cases with deployment being minimised through the use of composite cables between the antenna system and main equipment. The antenna system used is a 'unique' five element, 46 m diameter circular array with an identical central reference element. The antenna elements comprise active crossed loops, the outputs of which are combined through a 90° hybrid network to provide omnidirectional coverage at all elevations. A Fast Fourier Transform is used to provide a high-resolution signal spectrum display. The operator is able to discriminate large interfering signals (including within the specified signal bandwidth) and to exclude them from bearing calculations. The equipment's menu-driven display provides the operator with a choice of presentations, including:

- readouts of averaged bearings and quality control factors
- the elevation angle of the target signal. This may be used in conjunction with the known height of the ionosphere to determine the range of the target transmitter
- a polar display of instantaneous azimuth bearings. This option is automatically chosen when the unit is switched on
- histogram display of azimuth and elevation
- a record of target activity within the previous 30 minutes
- position fixing results when the station is used in the single location role.

Status
Over time, the RDF3210 DF system is reported as having been procured.

Contractor
Thales Defence Information Systems, Wells, UK.

UPDATED

The RDF3210 antenna system during deployment

SEEKER series Electronic Support (ES) systems

Type
Family of communications band ES systems.

Description
As of May 2001, two members of the SEEKER communications band ES system family have been identified, the known details of which are as follows:
SEEKER-4
The 20 to 1,000 MHz SEEKER-4 ES system is described as being suitable for transportable (wheeled vehicles and shelter) or fixed-site applications. It incorporates a Racal RA3726 dual receiver (48 cm (19 in) racking compatible), a Personal Computer (PC) and AE3007 High Frequency (HF - 3 to 30 MHz) or AE3020 Very/Ultra High Frequency (VHF/UHF - 30 MHz to 1 GHz) Adcock Direction-Finding (DF) antennas. SEEKER-4 makes use of digital signal processing techniques and provides a number of operator control functions in order to facilitate its use against a 'wide variety' of signal types. The architecture's PC can be deployed remotely from its receiver and antennas, while the system as a whole is noted as being suitable for use in an automated position fixing system that shows target locations on a map display. Functionally, the system's receiver is tuned to the target frequency and the four elements of its antenna array are switched sequentially to produce four beam patterns. This modulates the target emitter's incident waveform with bearing information and the resultant output is fed to the receiver where it is amplified, demodulated and digitised. The resultant digital data stream is applied to a microprocessor unit that calculates the signal's bearing at a rate of approximately 50 bearings/s. This information is integrated to provide average bearings over periods of time.
SEEKER-5
Derived from equipment supplied to the British Army and Royal Marines, the 20 to 1,000 MHz (upgradable) SEEKER-5 ES system is described as combining a 'number of proprietary equipments' to provide a search, DF, intercept and analysis capability against fixed- and agile-frequency communications radios. SEEKER-5 elements are mounted in shockproof cabinets that can be installed in a variety of vehicle types including armoured fighting vehicles, Land Rovers and the Hägglunds Bv206 all-terrain vehicle. The mast that carries the equipment's DF and monitoring antennas is co-located on the host vehicle and, as an alternative, the SEEKER-5 equipment can be demounted and operated from a building. System control is by means of a ruggedised computer (using the Windows NT operating system) and an Ethernet connection is used to share data between operator positions and vehicles. Remote data communication between sensors and sensors and a headquarters is by means of a VHF radio subsystem (based on Thales' Panther enhanced digital VHF radio). In terms of functionality, SEEKER-5 makes use of Butler Matrix DF processing to automatically produce Lines Of Bearing (LOB) on all detected transmissions within its operating bandwidth. LOBs from multiple DF/intercept units can be fed back to a DF tasking position where all transmissions are automatically geolocated. SEEKER-5 can be deployed in a number of configurations with a typical deployment consisting of a DF baseline (made up of four sensors) operating under the overall control of a command vehicle. An individual SEEKER-5 station provides one or more of the following operator facilities:
Search/DF
- automatic scanning for and detection of radio frequency signals within the target band
- automatic DF against fixed- and agile-frequency emitters
- automatic position fixing using an integral VHF link to communicate with other SEEKER sensors.
Intercept
- interception of fixed-frequency emitters
- generation of a local target activity database
- demodulation and recording of intercepted signal audio contents.
Analysis
- reception of data from search/DF and intercept positions
- sorting and correlation of target information within an intelligence database
- display of data in text and digital colour map formats
- generation of intelligence reports for tasking headquarters.

Status
As of May 2001, SEEKER series communications band ES systems were reported as being available.

Specifications
SEEKER-4
Frequency: 20-1,000 MHz in 1 Hz steps (HF option down to 10 kHz, DF option down to 1.6 MHz with appropriate antennas)
Channel memory: up to 500 (non-volatile memory)
Demodulation modes: AM, CW, FM, FSK, LSB and USB (bandwidth of up to 30 kHz)
Bandwidth: selectable from the range 75 Hz-30 kHz (12 kHz max HF bandwidth option, nominal 280 kHz wideband FM bandwidth mode option)
Intermodulation: +10 dBm (20-1,000 MHz, 3rd order intercept point); +22 dBm (1.6-30 MHz option, 3rd order intercept point, RF amp on); +30 dBm (1.6-30 MHz option, 3rd order intercept point, RF amp off); +36 dBm (1.6-30 MHz option, 2nd order intercept point, RF amp on); +46 dBm (1.6-30 MHz option, 2nd order intercept point, RF amp off); +50 dBm (20-1,000 MHz, 2nd order intercept point)
Synthesiser phase noise: −115 dBc/Hz (20 kHz offset, 20-1,000 MHz); −131 dBc/Hz (20 kHz offset, 1.6-30 MHz option)
AF outputs: 600 Ω balanced line output, external loudspeaker, internal loudspeaker and headphones output
Remote control: serial ASCII interface (receiver, RS-232C compatible); IEEE802.3 compliant Ethernet high-speed serial link (option)
DF displays: bearing log, histogram, map, polar and waterfall (displayed on an associated Pentium PC)
Power supply: 20-30 V DC (negative earth); 100-240 V AC (47-440 Hz)
Temperature: −10 to +55°C (operating); −40 to +70°C (storage)
Weight: 25 kg (RA3726 receiver)
Dimensions (h × w × d): 133 × 483 × 520 mm

Contractor
Thales Defence Information Systems, Wells, UK.

UPDATED

SIGMA II

Type
20 to 3,000 MHz band signals intercept and analysis system.

Description
The MRCM (a joint venture between European Aeronautic, Defence and Space (EADS) EWATION and South Africa's Grintek EWATION) SIGMA (System for Interception, Goniometry, Monitoring and Analysis) II signals intercept and analysis system is designed for 'long-term and tactical reconnaissance activities' centered around the interception and analysis of signals within the 20 to 3,000 MHz frequency band. As such, the system can monitor and Direction-Find (DF) against communications nets using a fast scanning search receiver/direction-finder with a 'short bearing time'. Claimed to be effective against frequency hoppers, SIGMA II features a processor that automatically controls system function (using menu driven application software) and processes acquired data. The system can be operated as a stand-alone unit or with other stations within a netted configuration when position fixing is required. It is designed for stationary and mobile deployment in a range of ground-based and airborne applications.

Status
As of late 2000, the SIGMA II signals intercept and analysis system was reported as being in production and service.

Specifications
Frequency range: 20-3,000 MHz (search, DF and monitoring functions - 10 kHz-30 MHz as an option for monitoring purposes)
DF accuracy: <1° RMS (1,000-3,000 MHz sub-band); <1.5° RMS (20-1,000 MHz sub-band); <2.5° RMS (20-100 MHz sub-band)
Instantaneous bandwidth: 1.25-10 MHz
Effective scan rate: up to 1 GHz/s

Contractor
MRCM (an EADS/Grintek EWATION joint venture), Ulm, Germany/Pretoria, South Africa.

NEW ENTRY

The strategic ES system vehicle
2000/0085240

A general view of a ground mobile application of the SIGMA II signals intercept and analysis system (MRCM)
2002/0098333

STRATEGIE electronic surveillance system

Type
Vehicle-mounted tactical Electronic Support (ES)/ELectronic INTelligence (ELINT) system.

Description
STRATEGIE is a vehicle-mounted tactical ES/ELINT system that consists of a network of highly mobile stations that can be operational within 15 minutes of deployment. It is designed for deployment in the battlefield area where it is used to intercept, identify and localise ground-based radar emitters. Each STRATEGIE station consists of the following:
- a fixed interferometer antenna array with an instantaneous wide field of view (no spin), mounted at the top of an erectable mast
- a high-sensitivity/high-scan speed, broadband superheterodyne receiver
- a high-speed processor for signal sorting and accurate measurement of the direction of arrival of signals
- a parameter analysis device
- a computer for the classification and identification of targets, threat library management and triangulation
- an operator's console with a high-resolution screen, for tactical situation monitoring, mapping and radar interception analysis (histograms and so on)
- a system for the inter-station transmission of data by radio

- a printer
- Global Positioning System to establish precise threat location.
   Parameters measured and processed in the baseline version include bearing, frequency, signal level, pulse repetition frequency, pulse-width and antenna scan period. An ELINT function is available to allow the detailed analysis of pulse data, signal modulation and recording of these parameters. Automatic identification of the target radar is accomplished by comparison with a threat library, which can be updated by the user. STRATEGIE may be associated with an electronically scanned, multithreat jammer. The system may be operated in a number of modes. Normally it is used for tactical ELINT, generating an ESM situation picture. This information can be used to command adjoining electronic countermeasures stations. Basically, the system is able to operate in three modes, tactical ELINT (automatic surveillance) with two submodes of stand-alone mode and remote mode, technical ELINT mode and training mode. In the tactical ELINT mode the system runs automatically under computer control. It collects information about adverse deployment of threat emitters in the interception area. Tactical information is presented on a high-resolution colour graphic monitor. The system is also equipped with sophisticated equipment for technical ELINT operation. With this equipment, the operator is able to analyse and store the parameters of highly complex signals under manual control, even when intercepted in a dense signal environment. The equipment's training mode allows operators to exercise missions without being deployed to a specific operation area. For this purpose the signal environment is fed into the system by a simulator.

Status
The first example of the STRATEGIE ES/ELINT system was delivered to the French Army in June 1995 and the architecture is reported as having been deployed to Bosnia in support of NATO operations. In French service, the equipment is designated as STAIR.

Specifications
Frequency band: 0.8-18 GHz (possible extension to 0.5-40 GHz)
Field of view: 180° (360° extension in 90° sectors available)

Contractor
Thales Airborne Systems, Elancourt, France.

VERIFIED

SURICATE passive radar detection system

Type
C- to J-band (0.5 to 20 GHz) passive radar detection system.

Description
SURICATE is a passive detection system that is designed to intercept, identify, localise and track airborne radars in ground air defence applications. The equipment takes the form of a mobile station and is deployed in groups of two or three. In multiple applications, slave equipments are remotely controlled. The system uses fixed antennas to provide 360° coverage, fast search and update rate capabilities and uses interferometry to maximise direction-finding accuracy and the equipment's instantaneous field of view. Compressive receiver technology is employed to provide long range detection of difficult emitters such as airborne terrain-following radars and the active sensors fitted aboard battlefield helicopters. SURICATE may be teamed with the land-based variant of Thales' SALAMANDRE

The SURICATE passive radar detection system
2000/0085241

naval radar jammer (see separate entry) to provide electronic cover for high value ground sites.

Status

Over time, the SURICATE passive radar detection system is reported as having been procured by the French Army.

Specifications

Frequency: C- to J-band (0.5-20 GHz); C- to K-band (0.5-40 GHz) optional
Field of view: 360°
Accuracy: 2 MHz class (frequency); 0.5° RMS (bearing)
Data processing: Real time; high order software; self-adapting in search and track to site environment

Contractors

Thales Airborne Systems, Elancourt, France.

VERIFIED

Tac-Weasel Electronic Support (ES)/ELectronic INTelligence (ELINT) system

Type

Ground mobile ES/ELINT collection and analysis system.

Description

Described by its manufacturer as a 'lightweight, high-performance' ground mobile ES/ELINT collection and analysis system, Tac-Weasel covers the 0.7 to 18 GHz frequency range (extendable to up to 40 GHz if required). It can be installed in a range of military vehicles (including the Land Rover and High-Mobility Multipurpose Wheeled Vehicle (HMMWV) types) and is air portable (including

A Land Rover mounted Tac-Weasel application with its antenna array deployed

slung-loaded beneath a helicopter). Functionally, Tac-Weasel stations are operated in groups of two or more (to facilitate emitter triangulation), with one station being designated as 'master'. 'Slave' stations (which are identical to the 'master' configuration) in the net are remotely controlled and do not need to be manned. The single operator in the 'master' station is able to view all data generated by his 'slave' stations and is provided with facilities to compute emitter location on selected targets. The results are displayed in digital map format and are used to create a local Electronic Order of Battle (EOB). If required, the architecture can accommodate additional equipment to allow for deeper ELINT analysis of received signals, a package that can include a dedicated, fine grain ELINT receiver. Tac-Weasel features a rotating antenna array, which outputs (in a common, 2 to 6 GHz band intermediate frequency format) to a five channel, multiplexed receiver. The frequency, pulse-width and other characteristics (see Specifications) of each detected emitter are established and digitised by a Parameter Measurement Unit (PMU). The PMU time-stamps the data before passing to a data processing unit that incorporates Thales' proprietary SADIE de-interleaving architecture. So equipped, Tac-Weasel can derive the pulse train characteristics of up to 250 emitters simultaneously. Emitter identification is undertaken at the 'master' operating workstation, which also maintains intercepted radar and emitter library databases and incorporates the necessary software for emitter location and EOB compilation purposes.

Status

As of March 2001, the Tac-Weasel ES/ELINT collection and analysis system was reported as being available. Jane's sources suggest that the British Army's Pinemartin equipment was a variant of the Tac-Weasel architecture. As of September 2001, Pinemartin was being replaced by a variant of Thales Sensors' Meerkat-S ES system (designated as INCE - see separate entry) within the British military.

Specifications

Frequency range: 0.7-18 GHz (extendable up to 40 GHz if required)
Sensitivity: –80 dBW/m²
Dynamic range: 60 dB
Accuracy: 1° RMS (2-18 GHz sub-band); 3° RMS (0.7-2 GHz sub-band)
Measured parameters: amplitude, bearing, frequency, pulse repetition interval, pulse-width and time of arrival
Active intercepts: 250 (expandable)
Main emitter library: 10,000 modes
Auxiliary library: 100 modes (operator entered)

Contractor

Thales Sensors Information Systems, Crawley, UK.

UPDATED

Telegon 10 Direction-Finding (DF) system

Type

Radio communications band DF system.

Description

The PGS 1720 Telegon 10 DF system is a communications direction-finder which covers the frequency range 10 kHz to 1,000 MHz. Configurations are also available with smaller frequency ranges. The equipment employs the Watson-Watt method and is able to obtain a bearing on signals with a duration as low as 1 ms. The Telegon 10 consists of the DF receiver P 1720, control unit BP 1620 and the cathode ray tube display unit SG 1620. Other system features include:

- three-channel DF with non-ambiguous sense indication
- microprocessor-controlled DF procedures
- effective against burst signals and frequency hopping emitters when appropriately configured
- an electronic bearing cursor with digital bearing readout
- resolution of non-coherent co-channel interference via real-time DF displays
- an integrated automatic bearing processor
- 100 memory channels for direction-finder settings

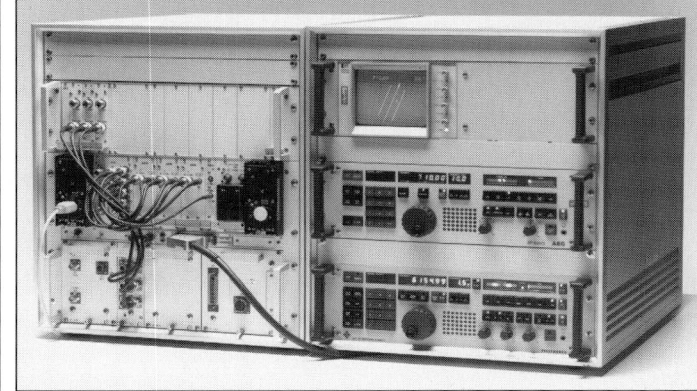

The Telegon 10 DF equipment with search receiver

- a remote-control facility
- built-in test.

Telegon 10 can be employed in stationary and mobile systems and, because of its integral 'intelligence', is suitable for integration into electronic warfare systems. An example of a Telegon 10-based mobile application is the armoured personnel carrier-mounted PGS 3601 DF system that entered service with the German Army and another 'NATO user' during the mid-1990s. Here, three independent Telegon 10 subsystems are used and the architecture is equipped with a folding antenna array that is raised to a height of 10 m on a hydraulic mast. Bearings are displayed on an eight digit cathode ray tube display and the system as a whole consumes approximately 150 W of power.

Status

Aside from the previously noted PGS 3601 application (and as of February/March 2001), Jane's sources were reporting that Telegon 10 DF systems were installed aboard Australian and New Zealand 'ANZAC' class frigates, Greek 'Hydra' class frigates and Saudi Arabian 'Arriyad' class frigates.

Specifications

Frequency range: 10 kHz-1,000 MHz
Frequency resolution: 10 Hz
Equipment error: <1°
Operational modes: A1A, A2A, A3E, F2A, F3E, J3E

Contractor

European Aeronautic, Defence and Space (EADS) Co Systems and Defence Electronics, Ulm, Germany.

UPDATED

Telegon 12 Direction-Finding (DF) system

Type
Digital multimode DF system.

Description

The digital Telegon 12 DF system incorporates the P 1920 DF receiver/processor and a computer workstation (Personal Computer (PC) - compatible) as the control/display element. Depending on the variant adopted, the equipment covers the frequency range from 10 kHz to 1,000 MHz. Other system features include:

- multiple DF control and processing modes including Watson-Watt and interferometer. The three-channel DF receiver with digital DF processing enables both conventional and advanced signals, including burst signals, to be intercepted and measured for direction of arrival
- identical digital filters in the DF channels ensure maximum accuracy and synchronism; they are free from tolerance, need no maintenance and do not age
- system control via an external PC-based computer with graphic user interface for menu-driven control of the DF functions, and for display of various graphic data which simplifies the interpretation of reconnaissance information.

Status

As of July 2001, the Telegon 12 DF system was reported as having been procured and as forming part of the MAIGRET communications band electronic support system (see separate entry).

Specifications

Frequency range: 10 kHz-1,000 MHz (optionally 3,000 MHz)
Direction-finder: 3-channel DF equipment with digital IF processing
DF methods: Watson-Watt or interferometer, or both in quasi-simultaneous operation

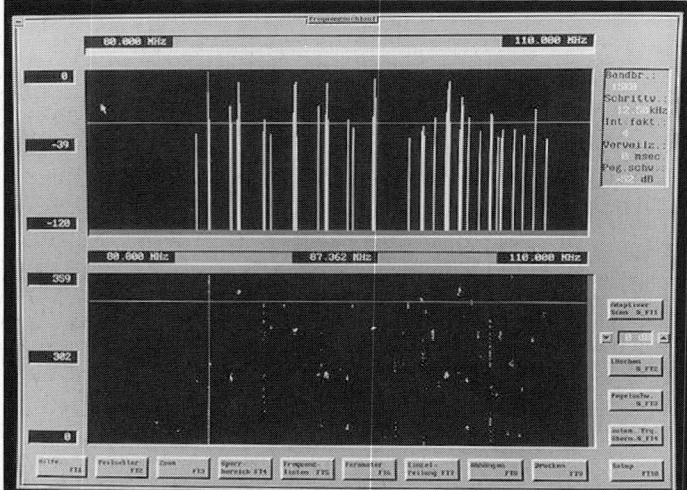

The Telegon 12 display

Contractor

European Aeronautic, Defence and Space (EADS) Co Systems and Defence Electronics, Ulm, Germany.

UPDATED

Telegon 111 Direction-Finding (DF) system

Type
Communications band DF system.

Description

The Telegon 111 communications band DF system consists of a DF processor and integrated receiver and is controlled by a personal computer (lap- or desktop) - compatible computer. Because of its small dimensions and weight, the Telegon 111 is suited for mobile, portable or stationary applications. A serial data interface facilitates the integration into netted DF systems for spectrum monitoring and security tasks.

Status

As of 2000, Jane's sources were reporting the Telegon 111 DF system as being in production and in service with a number of customers around the world.

Specifications

Frequency range: 25-1,000 (1,999) MHz
Frequency resolution: 100 Hz
DF method: ADF with electronic rotating goniometer
Demodulation modes: AM, FM, SSB
Min signal duration: 50 ms
Power supply: 12-24 V DC (mains power as an option)
Dimensions: 276 × 136 × 327 mm (Telegon 111); 280 × 40 × 170 mm (control unit)
Weights: 8 kg (Telegon 111); 1 kg (control unit)

Contractor

European Aeronautic, Defence and Space (EADS) Co Systems and Defence Electronics
Ulm, Germany.

UPDATED

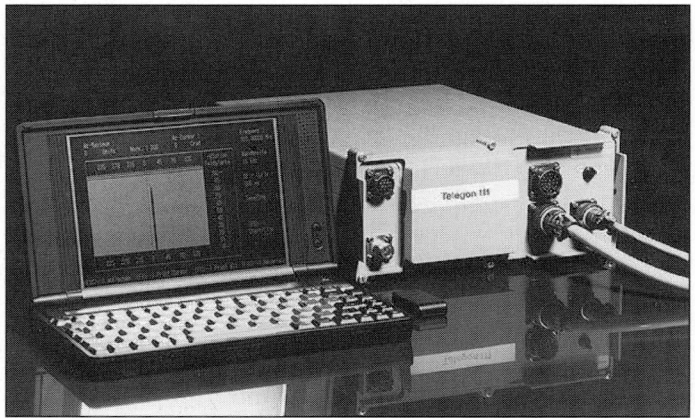

The Telegon 111 direction-finder

Telegon 112 Direction-Finding (DF) system

Type
Communications band DF system.

Description

The Telegon 112 communications band DF system is described as being a compact DF receiving module that covers the 20 to 3,000 MHz frequency band and incorporates a standard personal computer with which to control its bearing functions and display data. The system's DF module is based on the digital E 2000 VU receiver and the equipment as a whole is suitable for both fixed-site and mobile applications.

Status

As of 2000, the Telegon 112 DF system was reported as having been procured by national and international customers.

Specifications

Frequency: 20-3,000 MHz
DF method: ADF with electronic rotating goniometer
Frequency resolution: 10 Hz

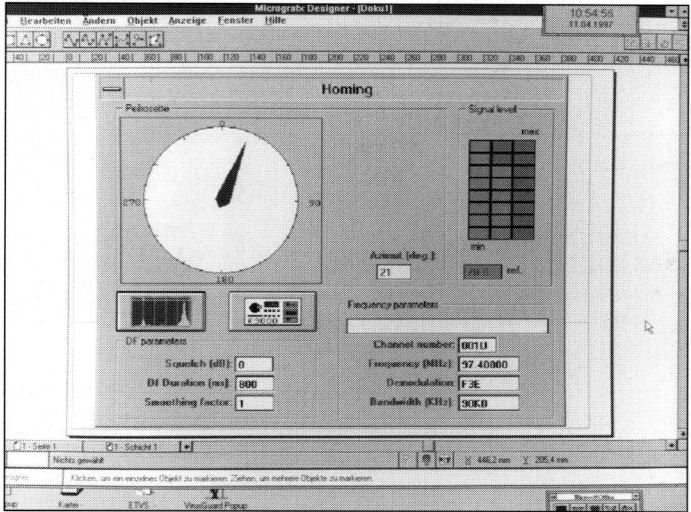

The Telegon 112 DF display   0009342

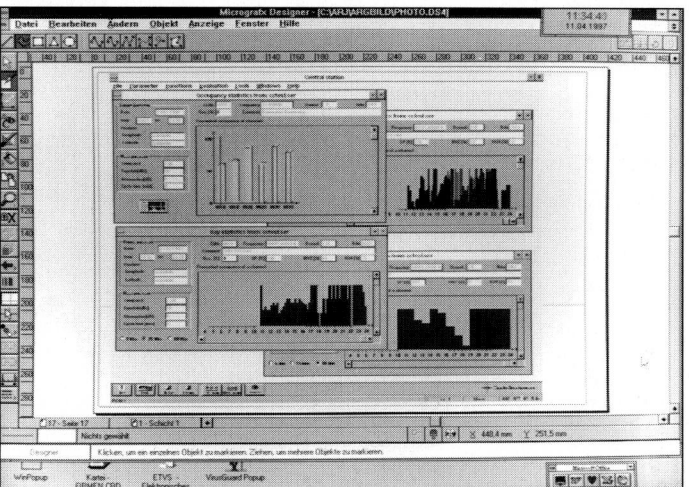

The static evaluation (frequency occupancy, density) display generated by Telegon 112   0009343

Min signal duration: 50 ms
Power supply: 115/230 V AC (optional 12 V or 24 V DC battery supply)

Contractor

European Aeronautic, Defence and Space (EADS) Co Systems and Defence Electronics, Ulm, Germany.

UPDATED

Telegon MRD 1920

Type
Multimode Direction-Finder.

Description
The MRCM (a joint venture between European Aeronautic, Defence and Space (EADS) EWATION and South Africa's Grintek EWATION) Telegon MRD 1920 multimode DF system is quoted as being able to handle 'several' DF techniques (including interferometric and Watson Watt) and as operating within the 10 kHz to 1 GHz frequency range. To achieve this coverage, the equipment comes in three variants designated as MRD 1920 LH (10 kHz to 30 MHz band), MRD 1920 LU (10 kHz to 1 GHz band) and MRD 1920 VU (20 MHz to 1 GHz band). MRD 1920

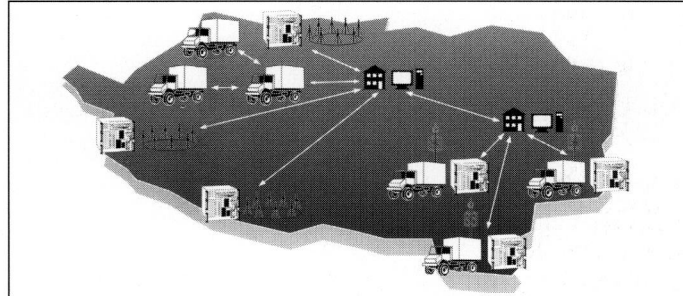

A schematic showing MRD 1920 functionality within a network of fixed and mobile stations (MRCM)   2002/0098326

series equipment feature 'state-of-the-art' analogue/digital receivers that are coupled to a digital signal processing unit that runs 'robust' DF algorithms. System performance is claimed to include a 'very high' dynamic range (up to 140 dB with automatic gain control), 'high' selectivity, 'very high' sensitivity and 'very good' DF accuracy. The equipment's dynamic range, selectivity and sensitivity are further claimed to facilitate the detection and bearing measurement of weak signals that are 'almost buried' in ambient noise or intermodulation from a nearby, more powerful transmitter. 'Special' DF algorithms are quoted as producing a bearing error of 'less than a degree' against continuous wave and frequency-hopping signals. In terms of 'complex' signals, MRD 1920 also offers 'special' DF modes for use against pulse or burst transmissions together with pulse analysis of such signals. The equipment can be remotely controlled by means of a modem, radio link and/or a local area network.

Status
As of late 2000, MRD 1920 series equipments were reported as being available and in service in mobile, fixed-site and shipboard applications.

Specifications
Frequency range: 10 kHz-30 MHz (MRD 1920 LH); 10 kHz-1 GHz (MRD 1920 LU); 20 MHz-1 GHz (MRD 1920 VU)
Number of channels: 3
Bearing accuracy: 1°
DF algorithms: interferometric or Watson Watt (frequency and antenna dependent)
Frequency scan: 10 MHz/s (linear - 25 kHz channel spacing); 1.5 GHz/s (adaptive)
Memory scan: 400 channels/s (30 kHz bandwidth)
Min signal duration: 1 ms
Sensitivity: 1 µV/m (with MRA 1228, 1° bearing fluctuation, T=200 ms integration time and 15 kHz bandwidth); − 115 dB (3 kHz, 10 dB S/N ratio)
Dynamic range: >92 dB (without gain control, 0.6 kHz bandwidth); >140 dB (with gain control, 0.6 kHz bandwidth)
Noise figure: 10-13 dB (frequency dependent)
Temperature: −25 to +55°C (operating); −40 to +70°C (storage)
User interface: DFS2000 (Windows NT); 1,280 × 1,024

Contractor
MRCM (an EADS/Grintek EWATION joint venture), Ulm, Germany/Pretoria, South Africa.

NEW ENTRY

Thales Communications high-performance intercept/direction-finding products

Type
Family of High Frequency (HF) and Very/Ultra High Frequency (V/UHF) high-performance combined intercept and direction-finding equipments.

Description
Thales Communications has developed a range of high-performance HF and V/UHF combined intercept and direction-finding equipments that are designed to meet what the company terms the 'most severe' requirements within electronic support and communications intelligence systems. Such equipments feature fast scanning and a processed synthetic data format that is optimised for ease of operator situational assessment and any additional processing that may be required. System functions comprise:
- search and detection
- direction-finding on all detected emitters (azimuth and elevation angle for the HF band)
- emitter location (position fixing with one single site location station in HF or triangulation using several sensors)
- technical characterisation of emitters.

Other system features include:
- real-time processing
- automatic detection with noise estimate (low false alarm rate)
- automatic extraction processing
- automatic recognition of signal type (including frequency hopping, fixed frequency and burst)
- automatic tracking (periodic update of synthetic processed data)
- specific frequency-hopping emitter position fixing modes
- real-time display of raw and synthetic data (various graph formats)
- optional storage of raw and synthetic data.

Status
As of May 2001, Thales Communications HF and V/UHF combined intercept and direction-finding equipments were understood to be available.

Specifications
Instantaneous bandwidth: several MHz
Scanning speed: several GHz/s (V/UHF)

Contractor
Thales Communications RadioSurveillance and COMINT Systems Unit, Gennevilliers, France.

VERIFIED

TRC 197 direction-finding system

Type
HF band (3 to 30 MHz) interceptor/direction-finder.

Description
The TRC 197 is a very compact interception/monitoring and direction-finding workstation able to process very short signals. Based on an interferometry principle it consists of a simple, modular broad base antenna array. This enables the effects of multiple paths, a source of high bearing error, to be minimised and the azimuth and elevation of the received signal to be determined with very high accuracy.

The crossed loop antenna ensures precise sensitivity, even to nearby transmissions received from a high elevation angle and regardless of polarisation. Operation by means of a computer, including operation of the receiver, makes the equipment user-friendly and open-ended. Various optional software enables the radio direction-finder to access new functions (location, remote control, offline sorting), or to be tailored to any operational mission.

The TRC 197 direction-finding function can work either autonomously if the Single Station Location (SSL) option is implemented, or in a system involving several radio direction-finders if the triangulation localisation option is implemented.

The TRC 197 consists of:
- an antenna array including a reference antenna and five to seven antennas located on the two sides of an equilateral triangle. Each antenna is connected to the switch through a single coaxial cable carrying both the HF signal and the power supply
- a TRC 1974 antenna switch enabling one antenna out of five or seven to be selected
- a TRC 1973 receiver, including two HF receiving channels, one of which is used for permanent monitoring
- a PC-AT type computer, including a phase and amplitude measuring printed circuit board to ensure the acquisition of two intermediate frequency signals and the direction-finding computation.

Status
Over time, the TRC 197 direction-finding system is noted as having been produced.

Contractor
Thales Communications, RadioSurveillance and COMINT Systems Unit, Gennevilliers, France.

VERIFIED

The TRC 197 HF interception/direction-finder

TRC 297 direction-finding system

Type
VHF/UHF band (30 MHz to 1 GHz) intercept/direction-finding system.

Description
The TRC 297 is a very compact interception/monitoring and direction-finding workstation able to process very short signals. Based on an interferometry principle, it consists of broad-based antennas. This enables the effects of multipaths, a source of high bearing errors, to be minimised and the azimuth of the received signals to be determined with very high accuracy.

Operation by means of a PC, including operation of the receiver, makes the equipment user-friendly and open-ended. Various options of software enable the

The TRC 297 V/UHF interceptor/direction-finder

radio direction-finder to access new functions (location, remote control, offline sorting and so on) or to be tailored to any operational mission.

The TRC 297 can work in a system involving several direction-finders if the triangulation localisation option is implemented.

Status
Over time, the TRC 297 direction-finding system is understood to have been procured.

Contractor
Thales Communications, RadioSurveillance and COMINT Systems Unit, Gennevilliers, France.

VERIFIED

TRC 297D direction-finding system

Type
VHF/UHF band (30 MHz to 1 GHz) mobile radio direction-finder.

Description
The TRC 297D is a mobile version of the TRC 297 equipped with a lightweight antenna. It uses the same interferometry principle to provide a high degree of accuracy. Designed for spectrum control mission and radio monitoring, it can be fitted on board any land-based vehicle, ship or aircraft. It may also be used in a fixed station when a non-compromising antenna network is required. The TRC 297D can also be integrated in a system involving different direction-finders if the triangulation localisation option is implemented.

Status
Over time, TRC 297D is understood to have been procured.

Contractor
Thales Communications, RadioSurveillance and COMINT Systems Unit, Gennevilliers, France.

VERIFIED

The TRC 297D mobile radio direction-finder

TRC 610 series direction-finding equipment

Type
Family of communications band radio interceptor direction-finders for fixed station, semi-mobile or mobile applications.

Description
The TRC 610 family includes high-performance radio interceptor direction-finders covering all or part of the 0.3 to 2,700 MHz frequency range. The three main equipments in this family are:

- TRC 611 covering the 0.3 to 30 MHz range
- TRC 612 covering the 20 to 2,700 MHz range
- TRC 613 covering the 0.3 to 2,700 MHz range.

These products are basically designed and developed for fast interception and direction-finding, accuracy and ease of integration.

The measuring principle is interferometry on a broad base antenna. The angle of arrival is determined from the measurement of instantaneous phase differences induced by the incoming signal. This principle provides a high accuracy and sensitivity, even in severe operational conditions. Phase measurements are performed simultaneously on the antenna networks and provide the equipments with fast interception and direction-finding capability, allowing direction-finding of frequency-hopping transmissions.

The TRC 610 family of equipments is designed for integration into radio surveillance and electronic warfare systems. In the basic version they include a high-speed remote control. The system modular design permits selection of the best solution for adaptation to the various operational configurations (fixed, semi-mobile or mobile) and to the various frequency ranges (10 ranges from 0.3 to 2,700 MHz). Because of their automation the direction-finders are easy to operate and with their powerful processor, they include many DF data processing possibilities.

In local operation, the terminal with its keyboard and colour display provides a complete service for easy operator direction-finding dialogue. For each operating phase of mode, the data are displayed on the screen. In remote-controlled operation, the direction-finder receives remote-control messages from an external computer and retransmits bearing results.

Three operating modes are available:

- preset scanning where the direction-finder is tuned to a preselected fixed frequency and performs successive bearings. In this mode the operator can use two types of display — polar histogram display and, in HF, azimuth/elevation display
- memory scanning where the TRC 610 is tuned successively to selected frequencies and performs bearings when activity is detected. In this mode, the TRC 610's 'extraction' capability is used to perform automatic threat classification (fixed frequency, burst, frequency hopping and so on)
- frequency scanning on frequency sub-band where the interceptor direction-finder performs automatic bearings on active frequencies.

This scanning mode is particularly applicable to electronic counter-countermeasures radiocommunication processing. A specific display enables the operator to discriminate between frequency-hopping and fixed frequency transmissions.

Status
Over time, TRC 610 series direction-finders are reported as having been procured by the armed forces of France and those of a number of other countries around the world.

Specifications
Scanning rate: 2 GHz/s
DF bearing duration: 500 µs (1 µs option)
Accuracy: = 0.5° RMS (azimuth); 2° RMS (elevation in the 2-30 MHz band)
Sensitivity: 0.7-15 µV/m

Contractor
Thales Communications, RadioSurveillance and COMINT Systems Unit, Gennevilliers, France.

VERIFIED

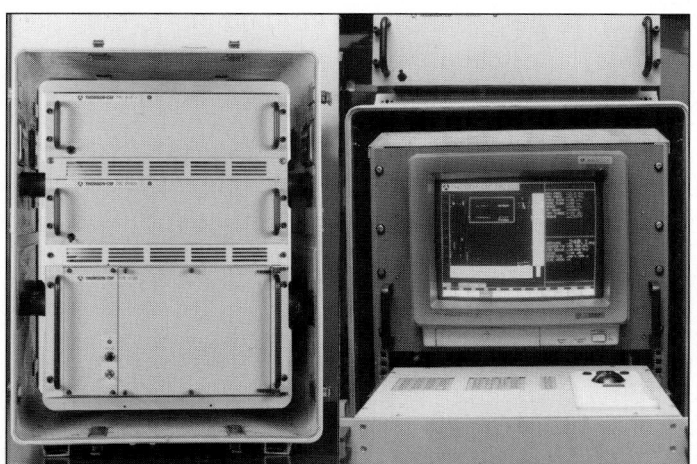

The TRC 612 radio direction-finder

TRC 641 technical analysis equipment

Type
Family of High/Very High/Ultra High Frequency (HF/VHF/UHF – 3 MHz to 1 GHz) technical analysis equipments.

Description
The TRC 641 series of technical analysers is designed to handle digitised voice and data transmissions and, according to their manufacturer, utilise 'new' processing techniques. As such, they make use of Fast Fourier Transform (FFT), digital filtering and demodulation processing algorithms and high speed, specialised processors to provide modulation recognition, demodulation, code recognition, decoding and signal characteristic measurement capabilities. TRC 641 analysers are further quoted as being able to operate with a wide range of receiver types, as being able to be integrated with 'all' types of electronic warfare and electronic support measures systems and as being suitable for use in both fixed-site and mobile stations. Thales also states that the TRC 641 series is suitable for civilian and paramilitary applications as well as strategic and tactical military applications.

Status
Over time, TRC 641 series technical analysers are reported as having been procured.

Contractor
Thales Communications, RadioSurveillance and COMINT Systems Unit, Gennevilliers, France.

VERIFIED

The operator's keyboard and display that is used with TRC 641 series technical analysers   0044241

TRC 2000 digital receiver

Type
Very Low to Ultra High Frequency (VLF to UHF band – 300 Hz to 1 GHz) digital receiver.

Description
The TRC 2000 is a VLF to High Frequency (HF), Very High Frequency (VHF) to UHF or VLF to UHF receiver that is described by its manufacturer as having 'outstanding' technical performance and capabilities (such as the easy addition of channel, notch and passband filters) that are not available in analogue equipments. The receiver is designed for general search, surveillance and monitoring applications and can be integrated into radio surveillance and intelligence gathering systems. An individual TRC 2000 receiver can be controlled from a single front panel and/or by an external Personal Computer (PC) running Windows NT, while control of a number of such receivers is by means of a single front panel, a PC or system software. Up to four receivers (mounted in a 3U19 rack) can be controlled by the same computer. TRC 2000 operates in listening, technical analysis and scanning modes and is configured to allow its operator to monitor activity on single or multiple frequencies (up to 100 table of frequencies) or sub-bands (up to 20). Frequency masking for up to 800 frequencies and sub-bands is available, thereby allowing the equipment to monitor only those ranges deemed to be operationally relevant.

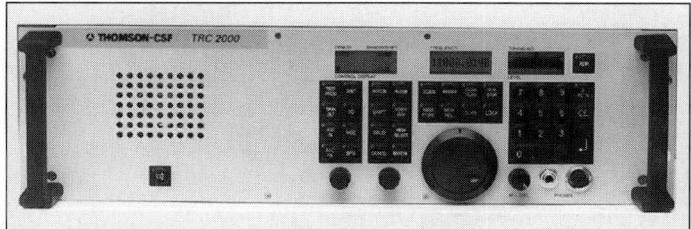

The TRC 2000 front panel

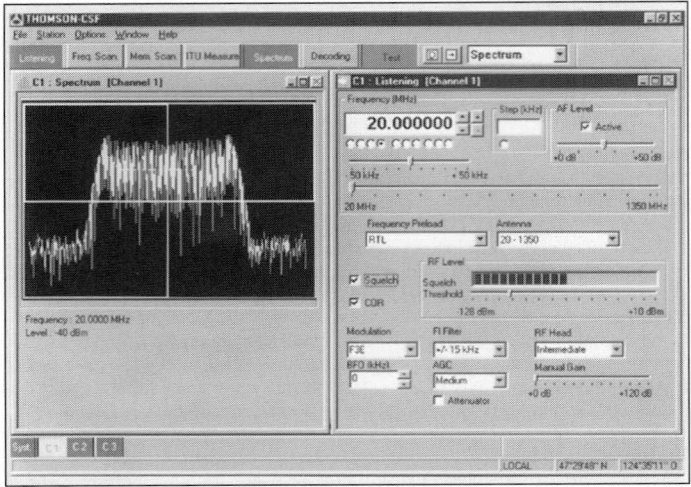

TRC 2000's Windows-based display format

Status

Over time, the TRC 2000 digital receiver is reported as having been procured.

Contractor

Thales Communications, RadioSurveillance and COMINT Systems Unit, Gennevilliers, France.

VERIFIED

TRC 6100 digital direction-finder

Type

0.3 to 3,000 MHz band digital direction-finder.

Description

The modular, tri-service TRC 6100 digital direction-finder is designed for electronic warfare system applications and utilises multichannel parallel processing and optimised algorithms (including Watson Watt and correlative matrix) to handle modern transmission techniques such as frequency hopping, free channel search and burst. System flexibility is enhanced by the equipment's ability to make use of a range of tactical and integrated antenna types. TRC 6100 is controlled by means of an external personal computer and offers a number of display formats including polar histogram, frequency versus azimuth and waterfall. Other system features comprise:

- a 'very wide' instantaneous bandwidth
- software control of direction-finding functions
- high protection efficiency in the Ops environment (high IP 3 and 'very low' noise figure)
- electromagnetic compatibility protection
- automatic communications type classification (extraction algorithm)
- 'easy to use' Windows NT man/machine interface
- compact architecture
- interferometry or super resolution (with multipath and co-channel impunity).

Status

As of May 2001, the status of the TRC 6100 direction-finder was uncertain.

Specifications

Frequency: 0.3-3,000 MHz
DF accuracy/reliability: 0.5 RMS class
Scan rate: up to 10 GHz/s

Contractor

Thales Communications, RadioSurveillance and COMINT Systems Unit, Gennevilliers, France.

VERIFIED

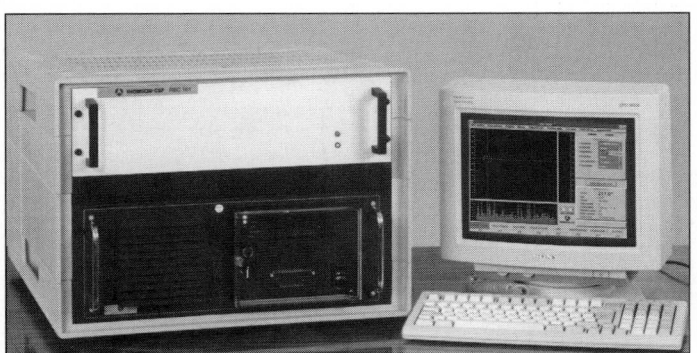

The TRC 6100 digital direction-finder 0044242

VKP 4000 wideband Direction-Finding (DF) system

Type

0.5 to 30 MHz wideband DF system.

Description

The VKP 4000 DF system is designed to provide simultaneous, continuous wideband monitoring and surveillance of all radio activity within its frequency range. The system offers 'reliable' interception of low probability of intercept signals (such as burst, frequency hopping and chirp), detection and analysis of unknown signals and retrieval of known networks that have changed their frequencies and call signs. The architecture has a modular configuration that starts with simultaneous surveillance of an 800 kHz real-time bandwidth (expandable to 1.6 or 2.4 MHz if required) within its 0.5 to 30 MHz frequency range. Within this real-time bandwidth, the system intercepts all active transmitters with signal detection and DF on active emitters being performed in parallel (for example, 19,200 Fast Fourier Transform (FFT) subchannels in a 2.5 MHz real-time band). Larger frequency ranges can be handled via the use of a scanning facility that operates at speeds of up to 150 MHz/s. The equipment comprises an Adcock, Watson Watt or interferometric DF antenna, analogue and digital sections and an operator workstation. Of these, the analogue section houses an analogue interface, wideband and analogue-to-digital converters and a data synchroniser. The digital section incorporates an FFT processor and polyphase filter bank, a DF processor and a post processing unit. The operator's workstation houses a former DaimlerChrysler Aerospace (now part of the European Aeronautic, Defence and Space (EADS) Co) E 2000 receiver (see separate entry) and a computer. VPK 4000's manufacturer also notes that the equipment can be customised to meet specific requirements with real bandwidths of between 400 KHz and 2.4 MHz.

Status

As of early 2001, Jane's sources were reporting the VKP 4000 DF system as having entered operational service during 1997.

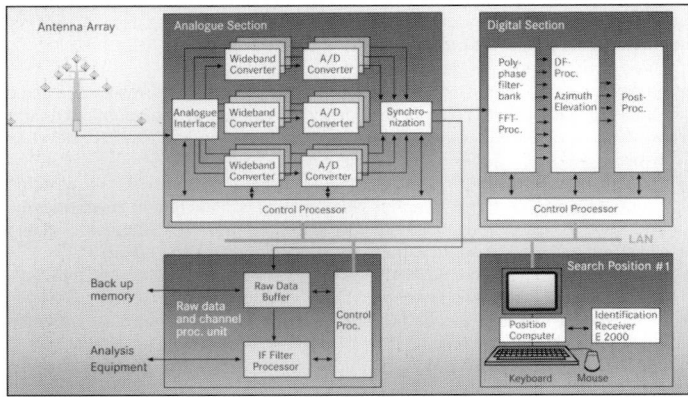

A block diagram of the VKP 4000 wideband DF system 0009344

The VKP 4000's operator position showing a typical spectrogram display 0009345

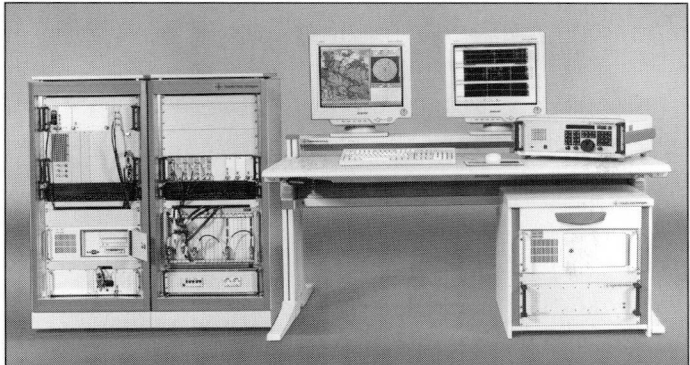

The VKP 4000 wideband DF system   0044249

Specifications
Frequency coverage: 0.5-30 MHz
Effective bandwidth: 2.4 MHz modularity (divided into 3 800 kHz sub-bands for simultaneous surveillance)
Signal buffer: video buffer (standard - for replay of last 60 s of display)
DF mode: interferometric (small and wide aperture) or Watson Watt
Resolution: 19,200 125 Hz wide subchannels (standard mode - 250, 500 or 1,000 Hz spacing selectable)
Operator interface: Windows-based
Data presentation: 3 windows (200 × 1,000 pixels/window); azimuth spectrogram; bearing histogram; power spectrogram; zoom window

Contractor
European Aeronautic, Defence and Space (EADS) Co Systems and Defence Electronics, Ulm, Germany.

UPDATED

Weasel III Electronic Support (ES)/ELectronic INTelligence (ELINT) system

Type
Land-based ES/ELINT system.

Description
The Weasel III ES/ELINT equipment is a highly automated, open architecture system that can accommodate a range of search and analysis subsystems. In terms of search, 100 per cent probability of intercept interferometer and amplitude comparison direction-finding packages are offered, while the analysis subsystem can be configured to accommodate a variety of receiver types including Thales' Corvus III channelised receiver, narrow or wideband superheterodyne receivers, compressive (microscan) receivers and customer-supplied equipment. In terms of system function, the Weasel III architecture offers automated signal detection, analysis, identification, recording and reporting. Control modes comprise master/ slave, integrated network and training and there is an automatic built-in test facility. Analysis modes include intrapulse and statistical and the architecture is described as being 'fully interoperable' with Thales' electronic warfare command and control systems.

Status
As of March 2001, the Weasel III ES/ELINT architecture was reported as being available. As of the cited date, Weasel-type systems are understood to have been procured.

An exterior view of the Weasel III ES/ELINT architecture showing an array of search, analysis, radio relay and combat net radio antennas

Specifications
Search subsystem
Frequency: 2-18 GHz (extension to 40 GHz as an option)
Array: 8 port switched amplitude comparison system
Bearing accuracy: 2.5° RMS
Sensitivity: –62 dBmi
Dynamic range: 60dB
Analysis subsystem - Corvus option
Frequency: 0.4-18 GHz (extension to 40 GHz as an option)
Antenna: spinning dish
Receiver type: channelised (1 GHz instantaneous bandwidth split into 29 channels)
Bearing accuracy: 1° RMS
Sensitivity: –90 dBmi

Contractor
Thales Sensors, Crawley, UK.

VERIFIED

Wideband Search, Location and Classification (WSLC) system

Type
0.5 to 30 MHz band automatic detection, location and classification system.

Description
The MRCM (a joint venture between European Aeronautic, Defence and Space (EADS) EWATION and South Africa's Grintek EWATION) WSLC system is a real-time, 0.5 to 30 MHz band (including low probability of intercept signals) automatic detection, location and classification system for use against emitters located in a predefined geographic area. WSLC functions automatically via an architecture that includes wideband Direction-Finders (DF - including the MRD 4000 unit), wide and narrowband receivers (including the wideband MRR 4000W unit) and the MCL 4000 classification subsystem. System features include:
- simultaneous surveillance and Direction-Finding (DF) on up to 19,200 channels with a total bandwidth of up to 2.4 MHz (subchannel raster of 125 Hz)
- determination and real-time processing of up to 2.4 Mio DF data/s with an instrumented accuracy of 0.5°
- additional scanning facility for surveillance of larger frequency bands (with a scanning speed of up to 300 MHz/s)
- automatic new signal classification
- 'reliable' interception from emission start (first bit/sign)
- demodulation, classification, decoding and production of relevant emission content
- simultaneous multi-operator operation

The MCL 4000 signal classification subsystem that is used in the WSLC architecture (MRCM0
2002/0098320

The MRD 4000W wideband receiver that is used in the WSLC architecture (MRCM) 2002/0098321

- wide DF antenna compatibility (Adcock, interferometric, Wullenwever and the like)
- emission location by means of either a single station location algorithm or by DF base.

Status

As of late 2000, the WSLC system was reported as being in service with the German armed forces.

Specifications

Frequency range: 0.5-30 MHz
Real-time bandwidth: 2-6 MHz (MRR 4000W - scaleable to 30 MHz); up to 2.4 MHz (MRD 4000)
Scanning speed: up to 300 MHz/s (MRD 4000)
Bearing accuracy: 0.7° RMS (MRD 4000 - instrumented); <1° RMS (MRD 4000 - with DF antenna)
Min signal duration: 1 ms (MRD 4000)
Filterbank resolution: 125, 250, 500 Hz and 1 kHz (MRD 4000 - programmable)
DF mode: interferometric (small and wide aperture mode) or Watson Watt (MRD 4000)
Signal filtering: digital down converters (MRR 4000W); signal buffer for predetection recording (MRR 4000W)
Segmentation/classification modules: data protocols; modulation mode; parameters (MCL 4000)
Data production/analysis extensions: decoders; demodulators; equaliser (MCL 4000)

Contractor

MRCM (an EADS/Grintek EWATION joint venture), Ulm, Germany/Pretoria, South Africa

NEW ENTRY

ISRAEL

CDF 1500 surveillance and emitter location system

Type

20 to 1,200 MHz spectral surveillance and emitter location system.

Description

Comprising two line-replaceable units, CDF 1500 utilises commercial-off-the-shelf hardware and software (together with proprietary digital signal processing techniques) to provide spectral surveillance and precision emitter location against sources operating within the 20 to 1,200 MHz frequency band. Functionally, a combination of a fast synthesised receiver and digital processing are used to detect activity and signals of interest. Direction-finding is executed concurrently with the search function and makes use of an interferometer technique combined with a rapidly deployable, wide aperture antenna. Search modes include band sweep and step search and the equipment is supported by an integrated wood processing package, digital voice recording and a range of analysis and reporting tools. The CDF 1500 workstation features a UNIX operating system running on a 90 MHz Pentium processor. The package has 32 MBytes of memory, a 1 GBytes hard disc and an X Windows/MOTIF graphic user interface. System options comprise:

- an extended frequency range (0.5 to 30 MHz (including single site location capability) and 1,200 to 2,000 MHz)
- a 'mobile sensor' package made up of a quick erection antenna, a magnetic heading indicator and a Global Positioning System sensor

The CDF 1500 surveillance and emitter location system 0044250

- a multi-operator architecture making use of a local area network and bridge/routers for remote workstation connection
- a multisensor architecture using 2,400 to 19,200 bits/s radio and/or landline, synchronous and/or asynchronous connectivity
- a low-profile, vehicle-mounted antenna for covert operation.

Status

As of the second quarter of 2000, the CDF 1500 surveillance and precision emitter location system was reported as being available.

Specifications

Frequency coverage: 20-1,200 MHz
Operating modes: DF histogram (10 frequencies); step (256 frequencies); sweep (16 frequency sub-bands)
Screening filters: geographic lockout (256 frequencies)
Field of view: 360° (instantaneous)
DF accuracy: <1° RMS
DF resolution: 0.1°
DF sensitivity: 2 µV/m (20-150 MHz); 4 µV/m (150-500 MHz); 10 µV/m (500-1,200 MHz)
Polarisation: ±45° vertical
Modulation: any communications signal with passband
Signal duration: 4 µs (min)
Noise figure: 8 dB (typical)
Frequency resolution: 100 Hz
Dynamic range: 80 dB
Temperature: −30 to +60°C (system, operating); +5 to +45°C (workstation, operating)
Altitude: 4,500 m (max operating)
Wind load: 110 km/h (operating)

Contractor

NICE Systems Ltd, Tel Aviv.

VERIFIED

··

CR-2740A ELectronic INTelligence (ELINT) system

Type

Ground-based 0.5 to 18 GHz band ELINT system.

Description

The 0.5 to 18 GHz ground-based CR-2740A ELINT system is described as being an automatic tactical and strategic equipment that is designed to operate in a dense electronic environment and scans the azimuth and frequency domains according to a 'sophisticated' scan table. The system incorporates high-gain, rotating dish antennas, superheterodyne receivers and is reported to be capable of analysing 'exotic' signals such as biphase, chirp and multipulse. CR-2740A is normally vehicle-mounted.

Status

Over time, the CR-2740A ELINT system is reported as having been procured by the Israeli Defense Forces and 'other customers'.

Contractor

Elisra Electronic Systems Ltd (a member of the Elisra Group), Bene Beraq.

UPDATED

EL/K-7035 COMmunications INTelligence (COMINT) system

Type
All-platform, communications band COMINT system.

Description
The all-platform EL/K-7035 COMINT system is designed to intercept, monitor, locate, analyse and report on communications traffic in the 20 to 500 MHz frequency range. The system is based around a multi-task workstation that is designated as the 'Basic Post' (BP) and is equipped with two to four COMINT receivers, two to four control units, two to four dual-channel tape recorders, a ruggedised computer, a built-in test routine, an optional intermediate frequency panoramic display and an optional time code generator reader, all mounted in a shock absorbing 48 cm (19 in) rack. Using this a building block, system configurations possible with EL/K-7035 include architectures that feature:

- a supervisor's console (similar to the BP)
- two to five BP-based operator consoles
- a system controller and mass storage unit
- a radio frequency and antenna distribution subsystem
- a Direction-Finding (DF) subsystem that is remotely controlled from the operator consoles and incorporates an array of up to eight antennas
- a plotter position
- a communications data link
- an S-250 or S-280 military standard shelter to accommodate the architecture's equipment and operators.

The equipment's modularity further allows it to control the same system resource from a number of operator consoles, vary the role of individual consoles according to specific requirements, integrate two EL/K-7035 systems together and configure the architecture and its shelter types for ground-based, shipboard and airborne applications.

In general terms, EL/K-7035 offers independent automatic search and scan and access to any available DF capability at each operator position, the ability to operate its receivers manually and a system wide monitoring and operator direction facility for a mission supervisor. Other operating functions comprise:

Search and acquisition
EL/K-7035 is able to scan up to 500 channels (excluding protected ones) every second, across its full frequency range, when searching for activity. Detected signals are automatically monitored and evaluated.

Preset task monitoring
So employed, the EL/K-7035's scan strategy is based on up to 64 preset taskings. As in the search and acquisition function, detected signals are automatically monitored and evaluated. When undertaking preset roles, the system can make use of additional receivers to provide monitoring and recording support for high priority jobs.

Localising
EL/K-7035's localising capability is based on automatic or manual DF, with acquired data being used for subsequent emitter location analysis.

Analysis
EL/K-7035 includes dedicated software that is used for off-line analysis of signal activity and DF results data held on the system's hard drive. Optional equipment is also available for re-transcription of selected communications for later analysis, study and/or editing.

Data transfer
An optional microwave datalink subsystem is available for the transmission of data to a remote monitoring station or stations.

Training
The EL/K-7035 architecture includes simulation software to allow the equipment to be used for operator training.

Status
As of the second quarter of 2000, the all-platform EL/K-7035 COMINT system was reported as being available.

Specifications
COMINT subsystem
Frequency range: 20-500 MHz
Frequency resolution: 1 kHz
Demodulation: AM and FM
Scan rate: up to 500 steps/s
Probability-of-detection: 95% (25 kHz bandwidth, 11 dB C/N)
Probability of false alarm: 10^{-4} (25 kHz bandwidth, 11 dB C/N)
Preset channels: 64 per receiver
Protected channels: 96 per receiver
Selectable frequency search ranges: 8
System noise figure: 8 dB
System sensitivity: −97 dBm (FM, 17 dB S/N, 100 kHz bandwidth, 30 kHz DF); −104 dBm (FM, 17 dB S/N, 20 kHz bandwidth, 6 kHz DF)
DF subsystem
Frequency range: 20-500 MHz (platform and configuration dependent)
Frequency resolution: 1 kHz
Demodulation: AM and FM
DOA accuracy: platform and configuration dependent
Signal duration: 300 ms (min)
Polarisation: vertical
Bearing display resolution: 0.1°
System sensitivity: −105 dBm (10 kHz IF bandwidth)
EMC: MIL-STD-461 compliant

Contractor
Elta Electronics Industries Ltd (a subsidiary of Israel Aircraft Industries Ltd), Ashdod.

VERIFIED

EL/K-7036 tactical COMmunications INTelligence (COMINT) system

Type
2 to 500 MHz tri-service COMINT system.

Description
EL/K-7036 is described by its manufacturer as being the basic building block for its latest generation, tri-service, mobile or fixed-site COMINT systems. It can operate as a stand-alone equipment or be integrated into a larger COMINT/electronic warfare architecture and is noted as being suitable for strategic intelligence, battlefield, internal security and 'early alert' applications. Elta states that the system makes use of the 'flexibility and maintainability' of the IBM compatible personal computer family (in a stand-alone or distributed processing configuration) and offers an automatic, three-level screening process to both focus system resources on target signals that meet pre-defined parameters and draw its operator's attention to networks of 'potential importance'. Equipment operation is by means of a menu-driven man-machine interface, with the various menus being accessed using a mouse or keyboard function keys. System functions include:

- a monitoring capability that makes use of an auxiliary monitoring receiver
- activity report generation for data hand-off to tactical in-theatre forces or a central site
- activity searches of predefined frequency bands and geographic areas
- audio recording and data input for data accumulation and analysis purposes
- automatic rejection of recognised signals of little or no interest
- direction-finding analysis to ensure the accuracy of direction-finding results and network analysis
- map displays showing lines-of-bearing against a geographical map background

The interior of an EL/K-7035 COMINT system application   2000/0080488

The EL/K-7036 tactical COMINT system's man/machine interface   2000/0081436

The EL/K-7036 tactical COMINT system as applied to an armoured personnel carrier chassis   2000/0081435

- 'very fast' scanning of known signals-of-interest to ensure a high probability of intercept whatever the duration of the transmission.

Other system features include:

- the use of filtering algorithms in both the frequency and geographic domains to support the system's probability of intercept value and to ensure that only relevant signals are queued for further processing
- an open, modular architecture to facilitate system adaptation and expansion.

Status
As of the second quarter of 2000, the EL/K-7036 tactical COMINT system was reported as being available and 'field-proven'.

Specifications
Interception
Frequency range: 2-30 MHz (0.5-2 MHz option) and 20-500 MHz (500-1,000 MHz option)
Modulation types: AM, CW, FM and SSB
IP3: +3 dBm (20-500 MHz band); +20 dBm (2-30 MHz band)
IF bandwidth: 0.5, 3.2, 6 and 16 kHz (2-30 MHz band); 10, 20, 50 and 100 kHz (20-500 MHz band); up to 1,000 kHz (optional)
IF rejection: 80 dB (20-500 MHz band); 100 dB (2-30 MHz band)
Frequency step: 10 Hz (2-30 MHz band); 10-1,000 Hz (20-500 MHz)
Image rejection: 80 dB (20-500 MHz band); 100 dB (2-30 MHz band)
Fast-scan: 100 frequency cells/s (2-30 MHz band); 500 frequency cells/s (20-500 MHz band)
Direction finding
Frequency range: 2-30 MHz (0.5-2 MHz option) and 20-500 MHz (500-1,000 MHz option)
Modulation types: AM, CW, FM and SSB
Response time: <200 ms (20-500 MHz band); <500 ms (2-30 MHz band)
Accuracy: 1.5° RMS (20-500 MHz band); 2° RMS (2-30 MHz band)

Contractor
Elta Electronics Industries Ltd (a subsidiary of Israel Aircraft Industries Ltd), Ashdod.

VERIFIED

GES-210E mobile radar detection and analysis system

Type
C- through J-band (0.5 to 18 GHz sub-band) mobile radar detection and analysis system.

Description
GES-210E is based on Elisra's AES-210E airborne electronic support/intelligence system (see separate entry) and is designed for ground-mobile or fixed-site applications. A baseline GES-210E system comprises a high-gain, spinning parabolic antenna (with a static direction-finding array as an option); a multichannel, receiver/processor and a display/control console. Of these, the receiver processor incorporates a multibandwidth superheterodyne receiver, an instantaneous frequency measuring receiver and a single processor. Parametric data is supplied to the operator in the form of a colour graphic situational display

and acquired data is stored on magnetic media. GES-210E can operate as a stand-alone equipment as well as being interfaced with a jamming subsystem or a local command and control network. The equipment can also be operated by remote control.

Status
As of January 2002, Elisra was reported to have delivered GES-210E type equipment to an as yet unidentified customer during September 1997.

Specifications
Frequency range: 0.5-18 GHz
Sensitivity: up to −90 dBm
Bearing accuracy: 1° RMS (typical)

Contractor
Elisra Electronic Systems Ltd (a member of the Elisra Group), Bene Beraq.

UPDATED

MTI Electronic Warfare (EW) system antennas and arrays

Type
Family of EW system antennas and arrays.

Description
Israeli contractor MTI Technology and Engineering Ltd produces a range of specialist EW system antennas and arrays, the known details of which are as follows:

MT-6005
MT-6005 is a half-space Direction-Finding (DF) array that is suitable for armoured personnel carrier applications and incorporates 21 horn antennas. Each array element has a 3 dB, 10° bandwidth with the entire assembly providing 210° coverage in azimuth. An armoured frame is used to house the MT-6005 unit and the assembly is completed by a sidelobe suppression horn that is mounted on top of the array housing.

Specifications
DF antenna
Frequency: 8.5-17 GHz (7.5-18 GHz option)
Antenna type: pyramidal horn with thin 45° slant polariser
Polarisation: 45° linear slant
2 dB beamwidth: 9° ± 1° (azimuth); >10° (elevation)
Gain: 18 dBi (min)
Sidelobe: −20 dB (max)
VSWR: 2:1 (max)
Weight: 1.75 kg
Dimensions (L × W × H): 700 × 255 × 110 mm
Sidelobe suppression antenna
Frequency: 8.5-17 GHz
Polarisation: 45° linear slant
Azimuth coverage: 200° (at max sidelobe level of the DF antenna)
Elevation beamwidth: >10°
Gain: >6 dBi
VSWR: 3:1 (max)

MT-6010
The 700 MHz to 18 GHz band MT-6010 interferometric array covers its frequency range in four system specific sub-bands and is designed for installation on both sides of a high-speed (up to Mach 0.9) subsonic airborne platform. The assembly consists of 10 flush-mounted antenna subsets (including interferometer, reference and side-looking units) and establishes target bearing by comparing the phase signals detected at three inputs within an 80° field of view. MT-6010 complies with MIL-E-5400R.

Specifications
Band 4 interferometric antenna
Frequency: 12-18 GHz
Antenna type: corrugated horn
Polarisation: 45° linear slant
Boresight gain: 7 dBi (min)
VSWR: <2:1
Amplitude matching: 2.5 dB (max over ± 45° in azimuth for all 3 antennas)
Phase tracking: 16° PTP, 5° RMS (side antennas over ± 45° in azimuth with reference to centre antenna)
Weight: 4.5 kg
Dimensions (W × H × D): 225 × 240 × 175 mm

MT-6015
Designed to withstand 'high' hydrostatic pressures, the MT-6015 circular DF array comprises eight horn antennas that are covered by an integral cylindrical 45° slant linear polariser and radome. MT-6015 complies with MIL-A-23836 and MIL-E-16400G.

Specifications
Frequency: 8-18 GHz
Polarisation: 45° linear slant
Coverage: 360°
Number of horns: 8
Gain: 7-11 dBi
3 dB beamwidth: 30° (elevation, nominal); 45° (azimuth, nominal)
Crossover: 2-5 dB down (1st); 6-20 dB down (2nd)
VSWR: 3:1
Dimensions (Ø × H): 160 × 150 mm (including connector plates)

MT-6020
Specifically designed for a 'special purpose, high accuracy, circular, amplitude comparison DF system', MT-6020 incorporates an array of 60 horn antennas. The assembly weighs 65 kg and is compliant with MIL-E-16400. The system's monopulse gain tracking value does not deviate by more than 2 dB from the mean curve between the second crossover points. In terms of peak gain tracking, no two antennas within the system differ by more than a maximum of 2 dB at peak gain at particular frequencies within their particular frequency range. MT-6020 is 45 cm high and has a diameter of 140 cm.

Specifications

	Low band	Mid band	High band
Frequency	2-4 GHz	4-8 GHz	8-18 GHz
Number of system elements	12	12	36
Electrical squint	1.5° (max)	1.5° (max)	0.7° (max)
Element gain	11 dBi (min)	12 dBi (min)	20 dBi (min)
Second crossover level below peak	6.5-19 dB	7.5-17 dB	7-13 dB
A3 sidelobe level	18 dB	20 dB	22 dB
V/H ratio (between second crossover)	2 dB (max)	2 dB (max)	2 dB (max)
Elevation beamwidth	30° (nominal)	25° (nominal)	20° (nominal)

MT-6025
MT-6025 is described as being applicable to 'very accurate, circular, amplitude comparison DF systems' and covers the 2 to 18 GHz frequency range in three system specific sub-bands. The assembly incorporates 60 horns and can be supplied to cover three 120° sectors. MT-6025 weighs 78 and has a monopulse gain tracking value that does not deviate by more than 2 dB from the mean curve between the second crossover points. In terms of peak gain tracking, no two antennas within the system differ by more than a maximum of 1.5 dB at peak gain at particular frequencies within their particular frequency range. MT-6025 is 38 cm high and has a diameter of 255 cm.

Specifications

	Low band	Mid band	High band
Frequency	2-4 GHz	4-8 GHz	8-18 GHz
Number of system elements	12	12	36
Polarisation	45° linear	45° linear	45° linear
VSWR	2:1 (max)	1.7:1 (max)	2:1 (max)
Electrical squint	1.5° (max)	1.5° (max)	0.7° (max)
Element gain	15 dBi (min)	13 dBi (min)	16 dBi (min)
Second crossover level below peak	9-17 dB	9-17 dB	6-13 dB
A3 sidelobe level	18 dB	20 dB	22 dB
V/H ratio (between second crossover)	1 dB (max)	1 dB (max)	1.5 dB (max)
Elevation beamwidth	30° (nominal)	25° (nominal)	20° (nominal)

MT-6030
The MT-6030 dual polarised horn is described as being a 'special' antenna that is capable of performing beam scanning within the confines of a single unit. Operating over octave bands within the 250 MHz to 18 GHz frequency range, MT-6030 is a compact, quad-ridged device with each ridge being fed separately at four orthogonal inputs into a square waveguide. By means of a radio frequency network (consisting of a number of 180° hybrids and delay lines), the horn's beam can be switched over 110° in azimuth. Possible feeding networks for the MT-6030 unit include dual and circular polarised multibeam horns, a dual polarised monopulse feed for a parabolic dish and an electronic conical scan feed.

Specifications
Frequency: 250 MHz-18 GHz
Instantaneous bandwidth: 1 octave
Polarisation: dual or circular
Gain: in excess of 8 dBi (nominal)
Number of beams: 3
Azimuth coverage: >110°
Input impedance: 50 Ω
VSWR: 3:1 (max)
Length: 0.7 wavelength (low end)
Aperture: 0.6 × 0.6 wavelengths (low end)

MT-7005
MT-7005 is a broadband, omnidirectional antenna that operates within the 100 kHz to 18 GHz frequency range and is designed for installation on top of submarine masts. As such, the assembly is made up of three elements, as follows:
- a two part microwave array that comprises two vertically polarised biconical antennas that are covered by a 45° slant polariser and a radome. Of the two antennas, the lower one covers the 2 to 8 GHz band while the upper element operates in the 8 to 18 GHz frequency range
- sub-microwave stubs that consist of two short monopoles, one of which, encircles the microwave antenna radome (in a 45° slant helix shape to minimise mutual interference) with appropriate active matching circuits
- a radio frequency module that incorporates limiters, amplifiers and detectors for the system's various reception channels.

MT-7005's gain varies monotonically as a function of frequency and it is claimed that the device suffers no sudden changes in swept gain measurement curves when both vertical and horizontal polarised. MT-7005 complies with MIL-E-16400.

Specifications
Microwave antenna
Polarisation: 45° slant
Azimuth coverage: omnidirectional
Azimuth ripple: ± 3 dB
3 dB elevation beamwidth: 20° (nominal)
Gain at matched polarisation: −1 to 3 dB (8-18 GHz); −3 to 4 dB (2-18 GHz)
VSWR: 3:1 (max)
Sub-microwave active stubs
Frequency response: 0.1-30 MHz (band 1); 30-1,000 MHz (band 2)
Polarisation: 45° slant (band 1); vertical (band 2)
Azimuth coverage: omnidirection, ± 3 dB ripple
Elevation coverage: −10 to +50°
Antenna factor: −20 dB (min)
VSWR: 2:1 (max)
Radio frequency module - microwave amplifier
Frequency range: 2-8 GHz (band 1); 8-18 GHz (band 2)
RF gain: 30 dB (bands 1 and 2, min)
Noise figure: 8 dB (bands 1 and 2, max)
Radio frequency module - sub-microwave amplifier
Frequency range: 0.1-30 MHz (band 1); 30-1,000 MHz (band 2)
RF gain: 20 dB (band 2, min); 30 dB (band 1, min)
Noise figure: 4dB (bands 1 and 2, max)

MT-7010
Designed for installation on top of a submarine mast, the MT-7010 wideband omnidirectional array comprises a stack of two vertically polarised biconical antennas that are covered by a 45° slant polariser and a radome. Here, the lower deck covers the 2 to 18 GHz frequency range with the upper segment operating in the 8 to 18 GHz band. Like MTI's MT-7005 unit, MT-7010 features monotonical antenna gain. The assembly complies with MIL-E-16400.

Specifications
Polarisation: 45° slant
Azimuth coverage: omnidirectional
Azimuth ripple: ± 3 dB
3 dB elevation beamwidth: 20° (nominal)
Gain at matched polarisation: −1 to 3 dB (8-18 GHz); −3 to 4 dB (2-18 GHz)
VSWR: 3:1 (max)
Dimensions (Ø × H): 16 × 25 cm (Ø = mounting base)

MT-7015/-7020
Designed for use in submarine instantaneous DF systems, MT-7015 is a wide bandwidth corrugated horn that covers the 8 to 18 GHz frequency range while MT-2070 is a cavity backed spiral that operates in the 2 to 8 GHz band. Both units are sealed to withstand 'high' hydrostatic pressures. Both MT-7015 and MT-7020 comply with MIL-E-16400.

Specifications

	MT-7015	MT-7020
Frequency range	8-18 GHz	2-8 GHz
Polarisation	45° linear slant	Circular
VSWR	3:1 (max)	3:1 (max)
Gain	9 dB (nominal)	
Azimuth/elevation half-power beamwidth	45° (nominal)	75° (± 25°)
Amplitude tracking between horns	1 dB	± 2 dB
Dimensions (Ø)	120 mm	90 mm

MT-7025
The 2 to 18 GHz band MT-2075 is designed for submarine electronic support applications and comprises a stack of two biconical antennas covered by a polariser and a radome. Of the two antennas used, the lower element is 45° slant linear polarised and covers the 2 to 8 GHz band while the upper antenna is circularly polarised and operates within the 8 to 18 GHz range. MT-7025 complies with MIL-E-16400.

Specifications
Polarisation: 45° linear slant (2-8 GHz); circular (8-18 GHz)
Azimuth coverage: omnidirectional
Azimuth ripple: ± 3 dB (2-18 GHz, max)
3 dB elevation bandwidth: 15° (min)
Gain: −2 dBi (8-18 GHz, min); −5 dBi (2 GHz, min)
VSWR: 3:1 (2-18 GHz, max)
Dimensions (Ø × H): 260 × 265 mm (Ø = mounting base)

MT-2035
The MT-2035 blade antenna is described as covering the 30 to 90 MHz frequency band 'instantaneously' and as being suitable for both high-power, instantaneous jamming and frequency-hopping communications applications. The following specification data relates to the antenna's performance as measured on a large ground plane.

Specifications
Frequency: 30-90 MHz
Instantaneous bandwidth: 60 MHz
Gain: 2 dBi (30 MHz, min)
Impedance: 50 Ω (nominal)
VSWR: 3:1 (max)
Power handling: 100 W (CW)
Polarisation: vertical
Azimuth pattern: omnidirectional
Weight: 25 kg
Dimensions: (H × W): 1 m × 84 cm (at base)

MT-2040
The 20 to 500 MHz MT-2040 monopole blade antenna is designed for amplitude and phase DF applications operating in 'severe, subsonic airborne environments'. The following specification data relates to the antenna's performance as measured on a large ground plane.

Specifications
Frequency: 20-500 MHz
Polarisation: vertical
VSWR: 2.5:1 (100-500 MHz); 3.5:1 (20-100 MHz)
Impedance: 50 Ω (input)
Gain: −42 dBi (20 MHz); −27 dBi (30 MHz); −19 dBi (60 MHz); −12 dBi (90 MHz); −2 dBi (140 MHz); 1 dBi (500 MHz); 1,5 dBi (400 MHz); 3 dBi (220 MHz)
Azimuth coverage: omnidirectional
3 dB elevation coverage: 30° (max)
Phase tracking: ± 4° (max)
Weight: 1.8 kg
Height: 36 cm

MT-2045
The 500 to 1,000 MHz MT-2045 blade antenna is designed for amplitude and phase DF applications operating in 'severe, subsonic airborne environments'. The following specification data relates to the antenna's performance as measured on a large ground plane.

Specifications
Frequency: 500-1,000 MHz
Polarisation: vertical
VSWR: 2.5:1 (max)
Impedance: 50 Ω (input)
Gain: 4 dBi (nominal)
Azimuth coverage: omnidirectional
3 dB elevation coverage: 30° (min)
Phase tracking: ± 2° (max)
CW power handling: 100 W
Weight: 400 g
Height: 12 cm

Status
As of the second quarter of 2000, antennas of the type described were reported as being available.

Contractor
MTI Technology and Engineering Ltd, Rosh Ha'ayin.

VERIFIED

NiceCall™ digital voice logger

Type
Compact digital voice logging system.

Description
The NiceCall™ compact digital voice logger is designed to record the audio output of telephones and telephone exchanges, radios and intercom systems in a variety of civilian and paramilitary applications. System options include:
• analogue off-hook detection

The NiceCall™ digital voice logger   0044251

• DTMF detection and query
• caller identification detection (on Bellcore GR-30 lines)
• four channel synchronised playback
• noise reduction with tone removal
• external IRIG-B time sychronisation.

Status
As of the second quarter of 2000, the NiceCall™ digital voice logger was reported as being available.

Specifications
Frequency range: 300-3,400 MHz
Signal-to-noise ratio: 40 dB (64 kbytes sampling)
Audio outputs: built-in speaker, headset jack; RCA jack (external tape recorder or speaker)
Display: 180 × 240 pixel graphic liquid crystal display
Input gain control: automatic gain control (fast attack/slow release)
Voice processing: ADPCM 16
Recording channels: up to 24 analogue input channels (optional beep tones)
Voice storage: 250 h (on disk), 500 h (DAT DDS2)
User interface: front panel function keys, LAN or modem
Access control: password
Dimensions (W × D × H): 390 × 395 × 155 mm
Power requirement: 115/220 V AC (60/50 Hz, 150 W)
Environmental: 5-40°C (operating)

Contractor
NICE Systems Ltd, Tel Aviv.

VERIFIED

....

NiceFix™ air/harbour traffic management direction-finder

Type
100 to 400 MHz air/harbour traffic management direction-finder.

Description
The NiceFix™ direction-finder is an interferometer-based, Very/Ultra High Frequency (VHF/UHF), wide aperture direction-finder that is designed for military and civilian air and harbour traffic management. The system utilises commercial-off-the-shelf hardware and software and makes use of proprietary circuit cards, algorithms and digital processing techniques. It detects and displays all

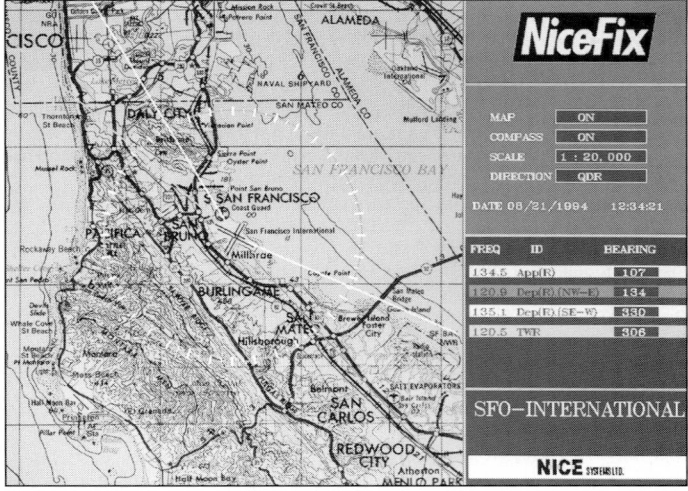

The NiceFix™ system's display format   0044252

communications traffic in the relevant frequencies and the introduction of an optional processing card facilitates real-time recording of audio communications directly to the system's hard disk. Standard NiceFix™ systems incorporate an automatic triangulation capability and can handle up to 64 channels simultaneously. The system's man/machine interface is menu driven with operator functions and direction-finding vectors being displayed digitally and graphically in colour. NiceFix™ also features built-in test.

Status

As of the first quarter of 2000, the NiceFix™ DF system was reported as being available.

Specifications

Frequency coverage: 100-150 MHz (VHF air band); 156-162 MHz (VHF marine band); 225-400 MHz (UHF air band)
Field of view: 360° (instantaneous)
DF accuracy: <1° RMS (operational)
DF resolution: 0.1°
DF sensitivity: 4 µV/m (100-400 MHz, typical)
Polarisation: ±45°
Modulation: any communications signal within passband
Signal duration: 10 µs (min)
Bearing type: auto triangulation, QDR, QDM, QTE and QWJ
Dimensions: 3 m (antenna diameter)
Weight: 36 kg (antenna)
Power source: 115/230 V AC (47-440 Hz, 290 W)
Temperature: –30 to +60°C (system, operating); 0 to 50°C (processor, operating)
Altitude: 4,500 m (max operating)
Wind load: 110 km/h (max operating)

Contractor

NICE Systems Ltd, Tel Aviv.

VERIFIED

Nicelog™ and NiceCLS™ telephony logging systems

Type

Computerised voice telephony and incoming/outgoing call logging systems.

Description

Nicelog™ and NiceCLS™ provide telephony logging capabilities for a range of applications including air traffic control and civilian, paramilitary and governmental security systems. Of the two, Nicelog™ is a Computer Telephony Integrated (CTI) digital voice logging system, while NiceCLS™ acts as a CTI call logging system that stores all incoming and outgoing call information in a searchable database. Nicelog™ provides access controlled logging, archiving and retrieval capabilities for between 16 and 'thousands' of channels of analogue/digital telephone and radio communications activity. Using the two systems, authorities can:

- retrieve call data by date, time, extension, caller identification or other parameters such as customer and/or transaction identification
- identify calls made by specific individuals regardless of station or extension used
- selectively record significant calls
- provide quality monitoring of selected calls
- integrate voice data with other information to provide a comprehensive activity review.

The Nicelog™ digital voice logging system
0044253

Status

As of the second quarter of 2000, both the Nicelog™ and NiceCLS™ systems were reported as being available.

Specifications

Recording channels: 16-92 analogue/digital input channels per logger; digital interface for E1, T1, PCM and Nortel TCM extensions, ISDN (BRI, PRI and proprietary), DASS2 or DPNSS; digital recording without digital-to-analogue conversion; optional warning tone during analogue line recording.
On-line voice storage: 1 or 2 DDS2 or DDS3 format DAT drives; 128-8,760 channel-h on-line (hard drive); up to 1,370 channel-h per DDS2 DAT cassette (compression rate dependent); up to 4,200 channel-h per DDS3 DAT cassette (compression rate dependent); up to 21,000 h on-line with 6 DAT option; remote tape server (using dedicated LAN segment); database for voice information and retrieval
Search time: instantaneous (hard disk); 2 min (DAT cassette)
Voice processing: A-Law (at 64 Kbytes PCM); DTMF/CLID detection; CCITT G.762 (at 16 and 32 Kbytes ADPCM); optional 5.6, 6.4, 7.2 and 8 Kbytes advanced compression algorithm; sampling in accordance with CCITT G.711 and G.712
Frequency range: 300-3,400 Hz
Signal-to-noise ratio: >40 dB (64 Kbytes sampling)
Playback capabilities: multiple user playback; playback loop; playback while recording; noise reduction algorithm; random access; synchronised (up to 4 channels)
Audio output: 4-12 channels; line (external speaker or headphones); telephone (active connection); voice on LAN (sound card playback or speaker)
Graphical user interface: Windows 3.11; Windows 95; Windows NT; UNIX
Dimensions (W × D × H): 483 × 520 × 185 mm
Power requirement: 115/220 V AC (60/50 Hz, 300 W)
Environmental: 5-40°C

Contractor

NICE Systems Ltd, Tel Aviv

VERIFIED

RAF-5100 super resolution Direction-Finding (DF) system

Type

Super resolution DF system.

Description

The RAF-5100 DF system is designed to overcome multipath and co-channel interference and make use of Rafael developed super resolution algorithms. The equipment is fully remotely controlled, offers 'powerful' reporting capabilities and extracts data (including direction of arrival, power level and centre frequency) from each intercepted wave front in real time. RAF-5100 is also noted as being available with both standard off-the-shelf and customised antenna arrays that include omnidirectional, directional, narrow aperture and wide aperture types.

Status

As of this edition, the status of the RAF-5100 DF system was uncertain.

Contractor

Rafael Electronic Systems Division, Haifa.

VERIFIED

TC/DF COMmunications INTelligence (COMINT) and Direction-Finding (DF) system

Type

Self-contained COMINT and DF system.

Description

TC/DF is described by its manufacturer as being a fully operational, self-contained COMINT and DF system that is suitable for installation in armoured personnel carriers, shelters, fixed-sites, ships, aircraft or van type wheeled vehicles. The system automatically detects signals in the 1 to 1,000 MHz frequency range, measures their direction of arrival, analyses them and records the processed data. TC/DF can operate as a stand-alone item, form part of a larger DF master/slave network or be incorporated into an integrated electronic warfare system.

Status

As of this edition, TC/DF was reported as being available.

Specifications

Frequency range: 1-1,000 MHz (1-3,000 MHz or sub-bands as options)
Measurement time: up to tens of thousands of channels/s

The TC/DF COMINT and DF system in van-mounted configuration
0009347

Outputs/peripherals: printers, audio recorders; RS-232, I/O to integrated EW/C³I systems
Input power: AC/DC (configuration dependent)

Contractor
Tadiran Electronic Systems Ltd, Holon.

VERIFIED

TDF 1200 communications band Direction-Finding (DF) system

Type
Communications band DF system.

Description
The TDF 1200 communications band DF system is described as being a multichannel, wideband equipment that operates within the 20 to 1,000 MHz frequency range (20 to 3,000 MHz as an option). The system is noted as being self-contained, as using the interferometer DF principle, as being suitable for ground-based mobile or fixed-site applications and as being able to operate as a stand-alone equipment or as part of a DF 'master-slave' network. Tadiran claims that TDF 1200's scanning rates facilitate its handling of 'extremely' dense electromagnetic environments that contain frequency hopping, burst and other types of agile emitters. TDF 1200 is further noted as being suitable for naval installations and as being a part of Tadiran's Electronic Warfare Integrated System (EWIS - see separate entry).

Status
As of this edition, the TDF 1200 communications band DF system was reported as being 'fully operational'.

Specifications
Frequency range: 20-1,000 MHz (20-3,000 MHz option)
Scan rate: 1 GHz/s
Instantaneous bandwidth: 4 MHz
Accuracy: < 1° RMS (instrumented); < 1.5° RMS (field, typical)
Response time: 1 ms (fast mode); 1, 5, 10 and 50 ms (histogram mode)

Contractor
Tadiran Electronic Systems Ltd, Holon.

VERIFIED

TDF 2020 communications band Direction-Finding (DF) system

Type
Communications band DF system.

Description
The TDF 2020 communications band DF system covers the 10 kHz to 30 MHz and 20 to 3,000 MHz frequency bands and can be configured for stationary, mobile, sheltered and transportable ground applications. The system can operate as a

stand-alone equipment or as part of a DF 'master-slave' network. TDF 2020 makes use of the weighted beam-forming DF technique and offers scan rates that facilitate the accurate determination of the direction of fixed-frequency, frequency hopping and burst emitters.

Status
As of this edition, the TDF 2020 DF system was reported as being 'fully operational'.

Specifications
Frequency range: 10 kHz-30 MHz and 20-3,000 MHz
Scan speed: up to 200 channels/s (10 kHz-30 MHz band); up to 1,000 channels/s (20-3,000 MHz band)
Instantaneous bandwidth: 340 kHz
Min signal duration: 5 ms
Accuracy:

Band	Fixed/sheltered	Mobile
10 kHz-30 MHz	1.5° RMS	3° RMS
20-3,000 MHz	1° RMS	2.5° RMS

Contractor
Tadiran Electronic Systems Ltd, Holon.

VERIFIED

TWR-1000 receiver/detector

Type
20 to 1,000 MHz wideband receiving system.

Description
The TWR-1000 wideband receiving system comprises a WideBand Receiver (WBR), an embedded (in the equipment's computer) digital signal processing-based WideBand Detector (WBD) and a system computer and is designed for COMmunications INTelligence (COMINT) and/or electronic warfare applications. As such, its manufacturer claims that it is able to handle 'all modern communications arenas', including 'exotic' emitter types such as burst and frequency hoppers. Processed data from the system takes the form of activity vectors that include all detected channels within the selected frequency slice. The system computer uses these vectors as the basis for data presentation and/or additional analysis. Subsequent to processing, TWR-1000 derived information is formatted for use on standard Windows NT personal computers or in open architecture systems. Other features include:
• 'very' high scan and detection rates
• low false alarm rate
• high probability of detection
• high dynamic range
• integration compatibility with a range of COMINT systems
• field-proven detection algorithms
• ruggedised design.
Alongside TWR-1000, Tadiran Electronics Systems has developed a wideband, multichannel direction-finding system that incorporates a coherent four-channel receiver, a dedicated digital signal processor and controller cards (embedded in the system computer), a direction-finding array and a system computer. Display

The TWR-1000 radio frequency front end   0044256

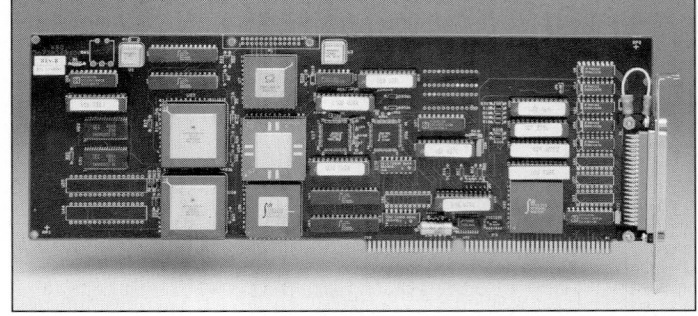

One of the TWR-1000 wideband receiving system's processing boards
0044255

formats include graphic maps, direction versus frequency, direction versus time, polar and direction of arrival. Other system features comprise:

- parallel digital processing
- high accuracy interferometric emitter location using a wide opening aperture antenna
- the ability to handle fixed-frequency, frequency hopping, burst and other types of agile emitter operating in dense electromagnetic environments
- the ability to integrate with a wide range of communications intelligence systems
- field-proven direction-finding and emitter location algorithms
- a built-in software simulator for operator training
- user-friendly man/machine interface running on Windows NT
- a built-in test capability.

Status
As of this edition, the TWR-1000 wideband receiving system and its associated wideband, multichannel direction-finder were reported as being available.

Specifications
TWR-1000
Frequency: 20-1,000 MHz (20-3,000 MHz option)
Channel resolution: 25 kHz
Scan speed: 40,000 channels/s
Instantaneous bandwidth: 4 MHz
Probability of detection: >99% (at 10 dB SNR)
False alarm rate: <4 × 10 ~ 4
Noise figure: <13 dB
Dynamic range: 67 dB (97 dB with optional AI.C)
Dimensions (W × D × H): 48 × 56 × 13 cm (WBR)
Weight: 25 kg (WBR)
Environmental: suitable for naval and ground-mobile applications

Wideband multichannel direction-finder
Frequency: 20-1,000 MHz (20-3,000 MHz option)
Channel resolution: 25 kHz
Scan speed: 40,000 channels/s
Instantaneous bandwidth: 4 MHz
Coverage: ±10° RMS (elevation); 360° (azimuth)
Polarisation: vertical
Accuracy: <1° (instrumented); <1.5° (typical)
Response time: 1 µs (fast direction-finding mode)
Detected signal modulations: analogue and digital
Environment: suitable for military applications

Contractor
Tadiran Electronic Systems Ltd, Holon.

VERIFIED

ITALY

ELT/888 Electronic Support (ES) system

Type
Tri-service ES system.

Description
The ELT/888 ES system can be configured for ground-based, shipboard and airborne applications and features a range of antenna types and interface to facilitate such installations. System functions include:
- radar and countermeasures signal interception
- the simultaneous analysis of multiple emissions

The ELT/888 ES system

- real-time emitter classification and identification
- emitter direction-finding
- digital recording and printing of parametric and analysis data.

All ELT/888's functions are fully automatic while its incorporation of high sensitivity receivers is claimed to ensures high probability of intercept. Measurement accuracy is ensured by the use of statistical processing and data collected by the system can be transferred to an associated Electronic Warfare Analysis Centre (EWAC). The EWAC provides facilities for deeper analysis of data produced by the ELT/888 architecture, multisensor system data integration and the compilation of databases.

Status
As of this edition, the ELT/888 ES system was reported as having been procured. Alongside the baseline system, Elettronica is also understood to have 'completed developed of' and produced the 'enhanced' ELT/888(V)2G and 'third generation' ELT/888(V)2E variants.

Contractor
Elettronica SpA, Rome.

VERIFIED

NORWAY

RL1 laser warning receiver

Type
Multiplatform laser warning system.

Description
The RL1 laser-warning receiver detects radiation from single and multiple pulse laser range-finders and target markers. It is suitable for installation on armoured fighting vehicles, mobile weapon systems and command posts. It can also be fitted to helicopters, ships and any type of fixed installations.

The detector unit has a field of view in azimuth of 360° and can also detect radiation coming from above. It incorporates five silicon photodiode detectors, four of which are mounted horizontally (providing overlapping 135° sector cover) with the fifth facing vertically upwards. Here, coverage is of the complete hemisphere down to 22.5° below the horizon. The indicator unit gives the approximate direction of the laser source by means of Light Emitting Diodes (LEDs) mounted in a circle. Eight 45° sectors are indicated. A ninth LED mounted in the centre of the circle indicates radiation coming from above.

RL1 detects radiation within the 0.6 to 1.1 µm near infra-red band, covering the most common types of pulsed lasers currently used in range-finders and target markers.

The receiver will give an acoustic alarm consisting of a pulsed audio signal in the crew's intercom when a laser pulse is detected. The duration of the alarm is two seconds for a single laser pulse. When more than one pulse is detected, the alarm remains on as long as laser pulses are being received.

Status
As of this edition, the status of the RL 1 laser warning receiver was uncertain.

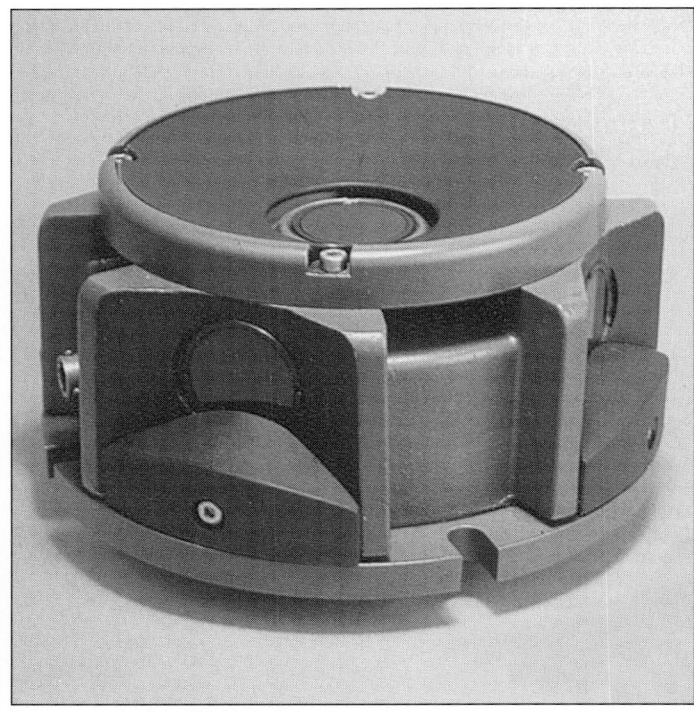

Detector unit of the RL1 laser warning system

Specifications
Detection unit
Number of detectors: 4 horizontal; 1 vertical
Type of detector: PIN photodiode
Optical bandwidth: 0.66-1.1 μm
False alarm rate: <10⁻³/h
Dimensions: 180 mm diameter × 80 mm high
Weight: 3 kg
Indicator Unit
Resolution: 45° (horizontal and vertical)
Number of sectors: 8 horizontal; 1 vertical
Display brightness control (manual): adjustable
Duration, acoustic alarm: 2 s
Duration, display: 8 s
Operating voltage: 20-32 V (24 V nominal)
Dimensions: 120 × 120 × 60 mm
Weight: 1.25 kg

Contractor
Simrad Optronics A/S, Etterstad.

VERIFIED

POLAND

MUR-20 ELectronic INTelligence (ELINT) and Electronic Support (ES) system

Type
C- through J-band (0.5 to 18 GHz sub-band) ground-mobile ELINT and ES system.

Description
According to Jane's sources, the MUR-20 system is a fully automated, ground-mobile, 'frontline' (5 to 10 km behind the forward edge of battle) tactical ELINT and ES sensor that is able to detect, localise and identify threat emitters operating in the 0.5 to 18 GHz frequency range. The system can be operated as a stand-alone equipment or integrated into a three to four unit network. MUR-20 is operated by a crew of three (ELINT operator, ES operator and a driver-technician) and is installed on an eight-wheeled SPR-22B armoured chassis that provides its crew with ballistic and nuclear/biological/chemical protection. The system incorporates dedicated ELINT (upper) and ES antenna arrays and is equipped with individual ELINT and ES workstations, each of which is provided with a single Barco MPRD 9639 and two Barco MPRD 9651 ruggedised colour displays. Onboard communications are understood to comprise a datalink and two tactical communications radios (likely to be Thomson-CSF/Radmor TRC-9500s) and there is a Przemyslowy Instytut Telekomunikacji (PIT - Telecommunications Research Institute) UNS combined Global Positioning System/inertial navigation system for accurate platform location. MUR-20 is also noted as being equipped with a chassis/antenna array auto-levelling subsystem together with an integral threat library that is reported to have been developed by Poland's Military Academy of Technology. When carrying the MUR-20 payload, the SPR-22B is credited with a maximum road speed and range of 80 km/h and 400 to 500 km respectively. SPR-22B is also noted as having a fording depth of 1.2 m and, as of this edition, it is thought likely that PIT is developing a lighter weight version of the MUR-20 for installation aboard a Tatrapan 6 × 6 all-terrain vehicle or a BRDM-2 Zbik armoured reconnaissance vehicle.

Status
As of this edition, Jane's sources were suggesting that MUR-20 development began during 1993 and that the system was field tested by the Polish armed forces during 1999. The Polish Army is understood to have a requirement for between 16 and 20 such equipments, with the first production delivery thought to have taken place during early 2000. As of January 2000 (and alongside the national application), PIT was noted as offering an export variant of the system (possibly designated as the MOSTAR-2 system) for a then year unit price of US$5 to US$7 million. Jane's sources also suggest that the MUR-20's current radar ES subsystem may be complemented and/or superceded by a communications intelligence capability at a later stage in the sensor's development.

Specifications
Frequency range: 0.5-18 GHz (ELINT and ES subsystems - extendable to 40 GHz)
Azimuth coverage: 360° (ES antenna)
Elevation coverage: −2° to +7° (ELINT antenna); −15° to +30° (ES antenna)
Sensitivity: −40 to −50 dBm (ES subsystem); −57 to −80 dBm (ELINT subsystem)

Contractor
Przemyslowy Instytut Telekomunikacji, Warsaw.

VERIFIED

RUSSIAN FEDERATION AND ASSOCIATED STATES (CIS)

85V6 Orion ELectronic INTelligence (ELINT) station

Type
B- to J-band (0.4 to 18 GHz sub-band) ELINT station.

Description
The 85V6 ELINT station is designed to detect, localise, identify and classify land-based, seaborne and airborne radar emitters together with the provision of an electronic environment monitoring capability for use around industrial centres and air- and sea ports. The system's basic mode of operation is circular scanning of the environment with automatic signal processing and data delivery. When fitted with data transmission equipment, individual Orion stations can deliver information to a user every five to ten seconds. Additional operating modes allow an operator to control the system manually with signal parameters being measured automatically or on operator command and with signal time and frequency characteristics being displayed to the operator. Signal processing is by means of fast Fourier transform with direction-finding being executed using the monopulse technique. The Orion system is mounted on a six wheeled truck chassis and draws electrical power from an associated diesel generator or from the civilian grid. Other system features (according to Jane's sources) include:
- false trajectory data elimination by means of software filtering
- automatic target tracking
- built-in test
- an integral 1,024 mode threat library
- provision for manual antenna pointing.

The Orion-1 variant of the system is designed for use in a triangulation network and is noted as featuring an enhanced antenna array (for more accurate direction-finding) and a frequency range of 0.1 to 40 GHz.

Status
As of the second quarter of 2000, the described 85V6 Orion ELINT station variants were reported as being available.

Specifications
Frequency coverage: 0.4-18 GHz (Orion); 0.1-40 GHz (Orion-1)
Frequency measurement accuracy: 1 MHz
Equivalent sensitivity: −145 dB/W
Azimuth measurement accuracy: 1° (over 2.5 GHz); 2-5° (below 2.5 GHz)
Observation zone: 0-20° (elevation); 360° (azimuth)
Minimum azimuth observation period: 2 s
Target capacity: 30-50 (designated to user); 300 (angle tracked)
Pulse measurement accuracy: 0.1 μs (pulse duration - not confirmed); 1 μs (pulse repetition period - not confirmed)
Detection range: 400 km (min - not confirmed)
Scan rate: 180°/s (azimuth - not confirmed)

Contractors
Spets-Radio Research and Production Enterprise, Belgorod.
Ulyanovsk Mechanical Plant State Enterprise, Ulyanovsk.

VERIFIED

....

Land-based SIGnals INTelligence (SIGINT) systems

Description
The following lists (primarily by NATO reporting name) known RFAS ground-based signals intelligence and electronic support systems. Inclusion here does not signify current operational use.

Bar Brick — vehicle-mounted radar intercept system
Bee Hive — ground-based tower-mounted intercept system
Big Ear — an Electronic Support Measures (ESM) system
Box Top — High Frequency Direction-Finding (HF/DF) system
Chuck Luck — twin antenna intercept system
Crab Pot — ground-based intercept system
Fix Series — a family of fixed and semi-mobile DF systems (see following)
Front Plate — vehicle-mounted intercept system with single dish antenna
Full House — HF/DF system, held at army level
Grid Shield — ground-based intercept system
Krug — a large HF/VHF DF system (see later)
Lone Box — intercept system
Loop Series — a family of HF/DF systems; Loop 3 is a trailer-mounted tactical system
Moon — a four-tower fixed HF/DF array
Pie Dish — radar intercept dish antenna
Pole Dish — single dish, jeep or man-portable tripod-mounted ELectronic INTelligence (ELINT) system with a 25 km range. Information processing time is 5 to 7 minutes. Russian designation is NRS-1

The Viewpoint electronic emission and detection system mounted on a Ural 375D truck (Jane's Intelligence Review)

The Twin Box intercept and DF system shown here mounted on a GAZ-66 truck

Prayer Wheel — mast-mounted communications intercept system
Quad Spring — quadruple antenna intercept system, probably V/UHF band (30 MHz to 1 GHz)
Ramona — a SIGINT system
Rib Cone — vehicle-mounted communications intercept and DF system
Ring Two — a trailer-mounted tactical HF/DF system
Small Cross — vehicle-mounted Adcock-type antenna DF system
Soft Ball — an ESM system
Spaced Loop — a fixed system consisting of a pair of square loop DF antennas
Spike Square — mast-mounted DF system with eight dipole antennas
SPZ/SPZM — man-portable radio and radar DF systems respectively, for use by special forces. The SPZM has been exported
Square Four — an array of four box-shaped DF antennas
Square Nose — mast-mounted intercept system
Square Pick — quadruple-mounted communication intercept system
Stick Tree — mast-mounted communication intercept system
Swing Box — trailer-mounted intercept system
Tall Rods — mobile DF system
Thick Eight — a fixed DF array of vertically polarised HF dipole antennas
Tube Tree — mast-mounted intercept and DF system with a complex of antennas
Turn Series — range of vehicle-mounted communications intercept and DF systems (Turn Cut, Turn Pole, Turn Spike, Turn Twist)
Twin Box — mobile intercept and DF system, with telescopic antenna (see later)
Viewpoint — ESM system
Krug SIGINT System
The Krug SIGINT system is a large, land-based HF/VHF band (3 to 300 MHz) DF system which equips many of the large permanently sited ESM sites. The main purpose of these stations is to monitor tactical radio traffic used by NATO for

ground and airborne forces. The early Krug systems, produced in the 1950s from a German Second World War design, covered frequencies of about 6 to 20 MHz, but the design has been continuously updated since then. The present antenna array is circular, approximately 100 m in diameter and consists of 120 equally spaced broadband vertically polarised antennas. Accuracy of the system is claimed to be better than 0.5°. The system has been further improved by the addition of an inner ring of dipoles that increases coverage into the VHF band (30 to 300 MHz). Over time, a large number of Krug stations have been set up in the RFAS and Eastern Europe, together with sites in Cuba and Vietnam.

Fix Series
The Fix family consists of a number of direction-finding systems in the VHF and HF (3 to 30 MHz) frequency ranges. One of the most widely used of these is the Fix-24 which is a circular array of 24 vertical monopoles, placed at 15° intervals in a ring approximately 150 m in diameter. In some stations a second ring of 24 monopoles has been placed inside the outer ring to increase the frequency coverage. Although fulfilling much the same function as Krug, it is a much less complex array and the accuracy and sensitivity is reported to be considerably less. Other Fix systems are Fix-4, Fix-6 and Fix-8, all of which are VHF/DF equipments.

Twin box
Twin Box, mounted on a GAZ-66 truck as shown in the picture, is a signature equipment of the divisional reconnaissance battalions. It is used for radar intercept and direction-finding. The system illustrated is the army version, there is also a variant in service with the air force. Twin Box is now being mounted on an MT-LB chassis for greater mobility and protection.

VERIFIED

Okhota ELectronic INTelligence (ELINT) system

Type
D- through J-band (1 to 18 GHz sub-band) ELINT system.

Description
Designed to 'monitor the radio frequency emissions of air and ground targets' in the 1 to 18 GHz frequency range, the automated Okhota ELINT system is described as being suitable for fixed-site, vehicular or shipboard applications and, in baseline form, comprises:
- an antenna array
- an analogue frequency-time signal converter
- a first-stage digital processor
- control and processing personal computers
- built-in test, training (integral signal simulation) and 'registration' subsystems.

Additionally, if the station is vehicle mounted, it is further equipped with:
- digital and voice communications subsystems
- an electrical power supply
- an onboard 'life support' subsystem
- an antenna raising and lowering subsystem.

Looking at some of these elements in more detail, the system's antenna assembly is described as incorporating an eight-port reception array, wideband, low noise amplifiers, a wideband frequency conversion unit and an electronic switching device. The first-stage digital processor is noted as being used to measure signal parameters and establish bearings on target emitters. Overall system functions comprise:
- high-speed, wideband signal detection
- measurement of 'basic' signal characteristics
- the establishment of emitter bearing
- emitter identification
- classification of detected emitters by type or other predetermined criteria.

Status
As of the second quarter of 2000, the Okhota ELINT system was reported as being available.

Specifications
Frequency range: 1-18 GHz
Sensitivity: 120-130 dB/W (unmodulated signals)
Panoramic observation bandwidth: 500 MHz
Frequency measurement accuracy: 0.5-1 MHz
Measurement accuracy: 0.1 µs (against pulses of up to 12 µs duration); 1 µs (against pulses of over 12 µs duration and pulse repetition period)
Dynamic range: at least 60 dB (one signal - instantaneous); at least 90 dB (selectable)
Direction-finding accuracy: 3° or better (emitting signals (as a continuous sequence of pulses) or emitting quasi-continuous signal); 5° or better (emitting burst of pulses)
Control: semi-automatic (via keyboard)
Equipment-antenna separation: up to 25 m
Power supply: 26 VDC (built-in generator); 220 VAC (50 Hz); 6ST190A battery (integral twin battery installation)

Contractors
Spets-Radio Research and Production Enterprise, Belgorod.
Ulyanovsk Mechanical Plant State Enterprise, Ulyanovsk.

VERIFIED

SOUTH AFRICA

Grintek EWATION communications band Direction-Finding (DF) systems

Type
Family of communications band DF systems.

Description
Grintek EWATION produces a range of communications band DF systems, the known details of which are as follows:

GDF30n series
GDF30n series equipments make use of Grintek EWATION active crossed loop/monopole antennas that are deployed in seven-element L-type and six-element pentagon arrays according to the specific sub-variant. GDF30n series equipments are remotely controlled (via a local area network) from a Man/Machine Interface (MMI) that runs under the Windows NT operating system. Display options include azimuth and elevation histograms, rosette DF, Ionospheric model and map. Frequency, mode and intermediate frequency bandwidth selection are performed using the associated receiver's control panel. Here, the receiver used is the 100 kHz to 30 MHz frequency band the Grintek EWATION GRX30n multichannel unit. GRX30n can be configured with up to eight channels, each of which is a high sensitivity receiver that incorporates digital signal processing. Other system features include:

- narrow and medium bandwidth measurements
- medium to wide aperture antenna configurations
- interception of low and high angle of arrival signals
- single site location using an integrated Ionospheric model
- generation of a quality factor for each DF measurement
- selectable integration times
- integrated receiver calibration
- built-in test to module level.

As of this edition, three GDF30n series variants have been identified, as follows:

GDF30n3
GDF30n3 is a three-channel equipment that is designed for restricted site or mobile/semi-mobile applications and employs a single crossed loop/monopole antenna or an active ferrite array. DF is performed using the Watson-Watt algorithm. Functionally, the equipment's antenna array is connected to a receiver subsystem with amplitude and phase measurements being performed by digital signal processors on each channel. A DF algorithm and system calibration data are used to calculate the DF result, which is displayed in real time. GDF30n3 can produce line of bearing measurement for ground-wave signals only.

GDF30n6
GDF30n6 is a six-channel configuration that makes use of a six-element pentaganol antenna array that is deployed over a 50 m open space. GDF30n6 is suitable for static installations (including single site location) and applications where high DF accuracy is required.

GDF30n7
GDF30n7 is a seven-channel equipment that employs a seven-element L-type antenna array that is deployed across an 85 m aperture. GDF30n7 is designed for static installations (including single site location) and is claimed to provide 'superior' accuracy and 'improved' ambiguity resolution.

Specifications

	GDF30n3	GDF30n6	GDF30n7
No of channels:	3	6	7
Monitoring channels:	1	1	1
Accuracy (instrumented):	2° RMS (3-30 MHz); 3° RMS (1.5-3 MHz)	1.5° RMS (3-30 MHz); 2.5° RMS (1.5-3 MHz)	1° RMS (3-30 MHz); 1.5° RMS (1.5-30 MHz)
DF results rate (scanning, BW=3 kHz):	50 DFs/s	50 DFs/s	50 DFs/s
DF results rate (scanning, BW=50 kHz, 10% occupancy):	5,000 DFs/s	2,500 DFs/s	2,500 DFs/s
DF algorithm:	Watson-Watt (ground-wave)	Interferometric	interferometric
User interface:	Windows NT, SVGA, 1,024 × 768	Windows NT, SVGA, 1,024 × 768	Windows NT, SVGA, 1,024 × 768

GDF3000w5
GDF3000w5 is wide-band DF system that comprises a mobile or static, five-element antenna array, an antenna band switching unit, a GDF3000w5 radio frequency front-end, a digital DF processor subsystem, an MMI (with control), an optional slaved demodulator receiver and an optional narrowband DF server. Other system features include:

- a frequency hopping DF capability
- a 20-3,000 MHz scan window
- an instantaneous 10 MHz bandwidth

- angle of arrival against frequency, spectrum, time DF (signal activity against frequency) and time waterfall displays
- remote interfacing
- an MMI running under Windows NT
- built-in test.

Specifications
DF system parameters
No of DF channels: 5
Monitoring channels: slaved narrowband receiver
Instrumented accuracy: 1.5° RMS (100-3,000 MHz); 2.5° RMS (20-100 MHz)
Spectrum surveillance
Scan rates: >10GHz/s (multiples of 10 MHz)
Frequency resolution: 12.5 kHz
Signal acquisition time: 80 µs
Wideband DF measurement
DF result rate: 10,000 DFs/s (100 µs/DF, BW=10 MHz, 10% occupancy)
Equivalent DF scan rate: 1 GHz/s
Instantaneous DF bandwidth: 1.25/2.5/5/10 MHz
Frequency resolution: 12.5 kHz
DF algorithm: interferometric
General
User interface: Windows NT, SVGA, 1,024 x 768
Sensitivity (10 MHz BW): a -110 dBm RF signal (CW) will produce a detectable output on the wideband receiver display when integrated over 10 ms
Data interface: LAN (TCP/IP) and RS-232C
Power supply: 110/240 V AC (50-400 Hz)

Status
As of this edition, Grintek EWATION communications band DF systems of the types described were reported as being available.

Contractor
Grintek EWATION, Pretoria.

UPDATED

Grintek EWATION SIGnals INTelligence (SIGINT) receivers

Type
Family of SIGINT receivers.

Description
As of this edition, Grintek EWATION is thought to be supporting a family of narrow and wideband SIGINT receivers that are suitable for channel scanning, monitoring, real-time spectrum analysis, signal detection and spectrum surveillance applications. As of the given edition, six such receivers have been identified, the known details of which are as follows:

GRX30n
The 100 kHz to 30 MHz GRX30n narrowband receiver is a multichannel monitoring equipment that can be configured to contain up to eight channels, each of which consists of a high-resolution, high-sensitivity receiver and a digital signal processor for signal demodulation. The receiver is fully remotely controlled (via a serial or local area network interface) and its Man/Machine Interface (MMI) software runs on computer workstations under the Windows NT operating system.

R30
The 5 kHz to 30 MHz R30 narrowband receiver is a high-resolution, high-sensitivity monitoring equipment that employs digital signal processing to facilitate selectable intermediate frequency filtering and digital demodulation. The receiver's front panel allows 'easy, loose standing' operation where operator assisted tuning is preferred. Remote control interfaces are provided as standard.

R3000
The 20 to 3,000 MHz R3000 narrowband receiver employs digital frequency synthesis to facilitate stable resolution and fast tuning. The equipment is fully controllable via serial or local area network interfaces and is able to receive and demodulate both voice and data signals.

S30
The 1.5 to 30 MHz S30 wideband receiver is a search and detection equipment that has a 500 kHz instantaneous bandwidth (at 625 Hz resolution) and an effective scan rate of 50 MHz/s. The unit's display allows the operator to view up to 55 channels (with 9 kHz spacing) and a peak select facility allows it to fast tune to any selected active channel.

S3000
The 20 MHz to 3 GHz S3000 wideband receiver offers a 10 MHz instantaneous bandwidth (at 12.5 kHz resolution) and has a real-time activity display that can handle up to 800 adjacent channels. An effective scan rate of 15 GHz/s allows the receiver to detect frequency hoppers with a hop rate of up to 500 hops/s and a scan window can be set up to search for activity within a specific band while still displaying the set's 10 MHz instantaneous bandwidth.

Specifications: Grintek EWATION SIGnals INTelligence (SIGINT) receivers

Narrowband receivers

	GRX30n	R30	R3000
Frequency range:	100 kHz-30 MHz	5 kHz-30 MHz	20-3,000 MHz
No of channels:	Up to 8 (max)	1	1
Frequency resolution:	1 Hz	1 Hz	100 Hz (30 KHz IF bandwidth)
Demodulation:			
CW	A1A/B and A2A/B	A1A/B and A2A/B	A1A
AM	A3E, 2A and H2B/E	A3E and H2A/B/E	A3A
SSB	J2A, J3E, R2A and	J2A, J3E, R2A and	J3E and H3E
SSB	R3E	R3E	B8E
ISB	B8E	B8E	F3E and G3E
FM	F3E	F3E	
FSK	F1A/B	F3C (optional)	
Fax	F3C (1 channel)		
Sensitivity:	≤−110 dBm (3 kHz 10 dB S/N, =0.707 uV, AM demodulator	≤−110 dBm (3 kHz 10 dB S/N, =0.707 uV, AM demodulator	−110 dBm (30 kHz 20 dB S/N, =0.707 uV, FM 10 kHz; ≤−112 dBm (6 kHz 12 dB S/N, =1.0 uV, AM 1 kHz
Image rejection:	≥80 dB (IF 1/2)	≥80 dB (IF 2); ≥90 dB (IF 1)	≥80 dB
IP3:	+24 dBm	+30 dBm	+15 dBm
IF filters:	22 DSP filters (150 Hz-50 kHz, phase linear)	19 DSP filters (150 Hz-16 kHz, phase linear)	12 DSP filters (0.3-300 kHz)
Notch filters:	None	none	7 filters (± 40 dB rejection)
Channel scan rate:	100 channels/s	100 channels/s	100 channels/s; 1,000 channels/s (option)
IF rejection:	≥80 dB	≥90 dB	≥80 dB
Antenna input:	2:1	2.5:1	<3:1
IF output frequency:	100 kHz (BW=6 or 50 kHz)	30 kHz (BW=16 kHz)	70 MHz (BW=20 MHz)
Power supply:	110/240 V AC (50-400 Hz, 270 VA)	115/230 V AC (47-420 Hz, 56 VA)	110/240 V AC (50-400 Hz, 35 VA, 24 V DC, 50 W)
Temperature:	−10 to +55°C (operating); −10 to +70°C (storage)	0 to +55°C (operating); −30 to +70°C (storage)	−10 to +55°C (operating); −30 to +70°C (storage)
Weight:	13.2 kg + 1.1 kg/channel	8.8 kg	8.8 kg
Dimensions (w × h × d):	482 × 178 ×515 mm	217 × 134 ×429 mm	217 × 134 ×429 mm

Wideband receivers

	S30	S3000	X3000
Frequency range:	1.5-30 MHz	20-3,000 MHz	20-3,000 MHz
Bandwidth/ resolution (instantaneous window):	500 kHz (625 Hz resolution)	10 MHz (12.5 kHz resolution)	10 MHz (12.5 kHz resolution)
Effective scan speed (scan window):	50 MHz/s	15 GHz/s (25 kHz resolution)	72 GHz/s (25 kHz resolution)
Zoom function (instantaneous window):	62.5/125/250 kHz	1.25/2.5/5/10 MHz	1.25/2.5/5/10 MHz
Sensitivity:	−113 dBm RF signal level (CW) will produce a spectrum display	−110 dBm RF signal level (CW) will produce a wideband receiver display (integrated over 10 ms)	−110 dBm RF signal level (CW) will produce a wideband receiver display (integrated over 10 ms)
A/D spurious free dynamic range:	70 dB	70 dB	70 dB
A/D converter:	12 bit	12 bit	12 bit
A/D sampling rate:	2.56 million samples/s	51.2 million samples/s	51.2 million samples/s
FFT (instantaneous):	1,024 point complex	1,024 point complex	1,024 point complex
FFT (scan window):	1,024 point complex	512 point complex	512 point complex
Image rejection:	>80 dB	>80 dB	>80 dB
Output baseband IF:	640 kHz (BW=500 kHz)	12.8 MHz (BW=10 MHz)	12.8 MHz (BW=10 MHz)
User interface:	RS-232C	high-speed synchronous serial interface	high-speed synchronous serial interface
Temperature:	−10 to +55°C (operating); −30 to +70°C (storage)	−10 to +55°C (operating); −30 to +70°C (storage)	−10 to +55°C (operating); −30 to +70°C (storage)
Weight:	7.5 kg	9.2 kg	15 kg
Dimensions (w × h × d):	217 × 134 × 429 mm	217 × 134 × 429 mm	482 × 134 × 420 mm

X3000

The 20 to 3,000 MHz X3000 wideband receiver is an ultra-fast scanning unit that is based on the S3000 equipment. Capable of an equivalent scanning speed of 72 GHz/s, one of X3000's primary design goals is the detection of very fast frequency-hopping emitters. Other system features include:

- instantaneous frequency, amplitude, band scanning and time waterfall displays
- selectable detection algorithms
- a spurious free dynamic range
- zoom functions for frequency domain analysis
- built-in test
- computer control under the Windows NT operating system
- interfaces for hand-off receivers and direction-finding equipment.

Status

As of this edition, the precise status of the described family of SIGINT receivers was uncertain.

Contractor

Grintek EWATION, Pretoria.

UPDATED

GSY 1450 COMmunications INTelligence (COMINT) system

Type

Single-operator COMINT system.

Description

The single-operator GSY 1450 COMINT system is described as being a stand-alone, tri-service equipment that is suitable for the detection, monitoring, decoding and recording of voice, teletype and facsimile signals in the 5 kHz to 3,000 MHz frequency range. The architecture comprises:

- an S30 wideband spectrum surveillance receiver
- an S3000 wideband spectrum surveillance receiver
- a narrowband R30 monitoring receiver (slaved to the display cursor)
- a narrowband R3000 monitoring receiver (slaved to the display cursor)
- a PD1000 panoramic display (displaying the spectral content of the monitoring receiver signals)
- a digital audio recorder
- a ruggedised personal computer system controller
- a W41PC data decoding card
- a GPU1000 power filter unit
- a GRD1300 antenna power supply unit.

System features include:

- instantaneous, wideband search, detection and display
- spectral display of signal content
- monitoring and demodulation of voice signals
- analysis and decoding of data signals
- digital audio recording
- signal database management
- a Windows-based man/machine interface
- optional interfaces for integration with direction-finders and electronic countermeasures systems.

Status

As of this edition, the GSY 1450 single-operator COMINT system was reported as being available.

Specifications

Frequency range: 5 kHz-1.5 MHz (R30 receiver only), 1.5-30 MHz (search/monitor) and 20-3,000 MHz (search/monitor)

Detection sensitivity: −113 dBm RF signal level (CW) will produce a detectable output on the wideband receiver display

Wideband scan rate: 50 MHz/s (1.5-30 MHz sub-band, 500 kHz BW, 625 Hz FFT resolution); 15 GHz/s (20-3,000 MHz sub-band, 10 MHz BW, 12.5 kHz FFT resolution)

Frequency resolution (zoom function): down to 1.56 kHz (20-3,000 MHz sub-band, wideband mode); down to 78.125 Hz (1.5-30 MHz sub-band, wideband mode)

Demodulation: A1A/B, A2A/B, A3E, B8E, F1A/B/E, H2A/B/E, J2A/E, R2A and R3E (1.5-30 MHz sub-band); A3E and F3E (20-3,000 MHz sub-band)

Power supply: 230 V AC (single-phase, 50 Hz)

Power consumption: <1.5 kVA

Temperature: −10 to +55°C (operating); −30 to +70°C (storage)

Contractor
Grintek EWATION, Pretoria.

UPDATED

Laser Warning System for Combat Vehicles (LWS-CV)

Type
Laser warning system for combat vehicle applications.

Description
Avitronics' LWS-CV system provides combat personnel with threat identification and a direction-finding indication of laser range-finders, designators and lasers used for missile guidance purposes. As of this edition, LWS-CV is available with one of two Laser Warning System (LWS) sensors (designated as the LWS-200 and LWS-300 respectively), both of which are electrically compatible and use the same LWS control/display unit for data processing. Here, commonality allows for ease of installation and system upgrading. The display integrated into LWS controller consists of three, seven segment, Liquid Crystal Displays (LCD) centred in a circle of Light Emitting Diodes (LED). On detection of a threat, an audio alarm is generated and the angle of arrival is indicated by the illumination of the appropriate LED and in numerical form on the LCD displays. LWS-CV can be configured for stand-alone operation or interfaced with an existing onboard system (via its LWS display/control unit) to facilitate turret slewing and activation of countermeasures in an automatic or semi-automatic mode. Four LWS sensors are required to provide 360° coverage in azimuth and the system is quoted as having a false alarm rate of less than one every 16 operational hours.

Status
As of November 2000, LWS-CV was reported as being available.

Specifications
LWS-200/-300 sensors
Wavelength coverage: 0.5-1.8 μm (LWS-300); 0.6-1.8 μm (LWS-200)
Threat coverage: erbium glass, GaAs, NdYAG, raman sifted NdYAG and ruby (LWS-200); erbium glass, doubled NdYAG, GaAs, NdYAG, raman sifted NdYAG and ruby (LWS-300)
AOA accuracy: 11° RMS (LWS-200, azimuth) 15° RMS (LWS-300 azimuth)
Spatial coverage (azimuth): 360° (LWS-300, 90° max per sensor); 360° (LWS-200, 99° per sensor)
Spatial coverage (elevation): 60° (LWS-200 and LWS-300)
Probability of intercept: >99% (LWS-200 and LWS-300, for single pulse)
Weight: 0.8 kg (single sensor LWS-200); 1.2 kg (single sensor LWS-300)
Dimensions: 103 × 86 × 64 mm (LWS-200); 115 × 90 × 76 mm (LWS-300)
LWS display/control unit
Interfaces: RS-422 (MIL-STD-1553, RS-232 and RS-485 bus options)
Power requirement: +28 V DC (715 mA, 25 W)
Weight: 2.5 kg
Dimensions: 188 × 89 × 131 mm

Contractor
Avitronics (Pty) Ltd (a Saab and Grintek company), Centurion.

UPDATED

The LWS-CV system configured with LWS-200 sensors   0044257

SPAIN

EN/UYQ-100 V3 Electronic Support (ES)/Electronic INTelligence (ELINT) processing system

Type
Tactical ES/ELINT processing system.

Description
The EN/UYQ-100V3 processor is designed to analyse data derived from external ES/ELINT receivers. The system is controlled by one to three operators, co-operating in the classification of detected radar signals. With the help of more than 80 commands, UYQ-100V3 is able to process data in real- or delayed-time, or from an intermediate frozen situation. Other modes allow the generation and updating of technical and tactical map files, as well as the classification and printing of results. The system can be integrated with a number of applications including data processing ground stations and the airborne Thomson-CSF DETEXIS Syrel ES collection pod.

Status
As of this edition, the EN/UYQ-100 V3 ES/ELINT processor was reported as having been procured by the Spanish Air Force.

Specifications
Architecture: 68030 and VME multiprocessor system
Cartographic display: 1,024 dots and 1,024 vectors in a 48 cm colour 1,280 × 1,024 pixels graphic terminal
Sight histogram display: 36 cm colour, 640 × 480 pixels
Conversational and tabular display: 36 cm colour, 640 × 480 pixels
Max number of radar sights: 10,000
Max number of located radar sights: 200 tracks
Technical memory: 2,000 elements (expandable)
Tactical file: 500 elements (expandable)

Contractor
INDRA DTD, Madrid.

VERIFIED

The EN/UYQ-100 ES/ELINT processing equipment

TURKEY

Armoured Tactical Direction-Finding Vehicle (ATDFV)

Type
COMmunications INTelligence (COMINT) and DF system.

Description
ASELSAN's computer controlled, modular architecture ATDFV has been developed to intercept, direction-find, monitor and record hostile communications traffic in combat environments where mobility and armour protection are required. Mounted aboard 'light' armoured vehicles, the ATDFV mission system is noted as being able to operate under 'harsh' environmental conditions and as being able to be set up for operations in a 'short' time span. The system's DF function covers the 20 to 1,200 MHz frequency band, makes use of the interferometric principle and is

claimed to be effective against frequency-hopping and burst emitters. Other system features include:

- High/Very High/Ultra High Frequency (HF/VHF/UHF - 3 MHz to 3 GHz band) monitoring and recording capabilities
- netted functionality for location fixing
- use of the company's DFINT-3T communications band DF system (see separate entry) to provide the system's 'basic' intercept, DF and analysis functions within the specified frequency band
- an external or integral Global Positioning System receiver
- a communications capability built around the ASELSAN 9600 VHF (30 to 300 MHz band) frequency hopping radio
- an audio tape recorder
- integral power generation and air conditioning facilities.

With regard to the specific ATDFV DFINT-3T application, it should be noted that the equipment and its associated communications system are deployable in order to facilitate the setting up of optimised DF sites in terrain that the host vehicle cannot negotiate.

Status

As of May 2001, ATDFV was reported as being in production and service.

Specifications

DFINT-3T subsystem
Frequency range: 20-1,200 MHz
DF accuracy: less than 3° RMS
LOB resolution: 0.1°
Frequency (FFT) resolution: 500 Hz-160 kHz
DF receiver stability: 1 ppm
Demodulation: AM and FM

Contractor

ASELSAN Inc, Microwave and Systems Technologies Division, Ankara.

NEW ENTRY

....

ASELSAN Field Artillery Meteorology System (AFAMS)

Type

Upper air meteorological data acquisition, processing and dissemination system.

Description

ASELSAN's AFAMS is a mobile, upper air, meteorological data acquisition, processing and dissemination system that provides field artillery units with the meteorological information needed to maximise battery effectiveness. Functionally, AFAMS measures temperature, humidity, wind direction and velocity by means of a balloonborne radiosonde that is automatically tracked by a radiotheodolite system, using radio direction finding techniques. Other system features include:

- a 'user friendly' man/machine interface
- real-time data and message presentation
- graphical data presentation
- storage of all acquired sounding data
- a simulation mode
- a built-in test equipment facility
- meteorological message formats that are compatible with international standard agreements (including STANAG and WMO)
- a hydraulically controlled levelling system
- 'enhanced' tube loading to facilitate the loading/unloading of helium tubes and a hydrogen generator
- MIL-STD-461C and -810D compliant.

Status

As of May 2001, the AFAMS system was reported as being in production and service.

Specifications

Pressure: 3-1,060 mbar (± 1 mbar accuracy)
Temperature: −90 to +60°C (± 0.5°C accuracy)
Relative humidity: 0-100% (3% accuracy)
Windspeed: 0-180 m/s (2 m/s accuracy)
Wind Direction: 0-360° (1° accuracy)
Tracking range: 160 km
Tracking height: 30 km
Deployment time: <30 min
Power requirement: 24 V DC (vehicle alternator); 220 V AC (dual generator)

Contractor

ASELSAN Inc, Microwave and System Technologies Division, Ankara.

NEW ENTRY

....

DFINT-3 communications band Direction-Finding (DF) and intelligence system

Type

Tactical communications band DF and intelligence system.

Description

ASELSAN describes DFINT-3 as being a fully computer-controlled, interferometer-based DF and intelligence system that operates in the 20 to 1,000 MHz frequency band. As such, the equipment is integrated aboard a 4 × 4 vehicle that tows a single axle trailer-mounted Alternating Current (AC) generator. For silent operation, DFINT-3 makes use of an integral battery bank. Functionally, the system is able to locate and track targets in collaboration with other stations (using a frequency modulated Very High Frequency (VHF - 30 to 300 MHz band) radio or wire links for inter-communication) and is capable of unmanned operation when used in its fully automatic mode. Utilising digital signal processing, DFINT-3 is claimed to have a band scan speed that is sufficient to intercept and direction-find against state-of-the-art frequency-hopping and burst emitters. Demonstrated field accuracy is quoted as being better than 2° RMS with a Line Of Bearing (LOB) resolution of 0.1°. DFINT-3 is further noted as being 'user friendly' menu-driven and as having a man/machine interface that incorporates a keypad (including soft keys), a tracker ball and a colour monitor. Data presentation formats are interactive, graphical or list-based and include:

- simultaneous, real-time LOB versus frequency and amplitude (versus frequency in DF mode)
- an on-screen, digital panoramic display and polar LOB plot for target analysis and DF step mode
- simultaneous display of up to 16 targets on a LOB versus time graph for moving target detection.

DFINT-3 is also reported as incorporating two additional monitoring receivers, two dual-track tape recorders, an electronic compass and an integral Global Positioning System receiver.

Status

As of May 2001, the DFINT-3 communications band DF and intelligence system was reported as having been in service since 1995.

Contractor

ASELSAN Inc, Microwave and System Technologies Division, Ankara.

VERIFIED

The DFINT-3 tactical communications band DF and intelligence system
2001/0100385

....

DFINT-3A tactical communications band Direction-Finding (DF) and intelligence system

Type

Tactical communications band DF and intelligence system.

Description

ASELSAN describes DFINT-3A as being a fully computer-controlled, interferometer-based DF and intelligence system that operates in the 20 to

1,000 MHz frequency band. As such, the equipment is integrated aboard a 4 × 4 vehicle and can be operated using external power sources. For silent operation, DFINT-3A makes use of an integral battery bank. Functionally, the system is able to locate and track target transmitters in collaboration with other stations (using a frequency modulated Very High Frequency (VHF - 30 to 300 MHz band) radio or wire links for inter-communication) and is capable of unmanned operation when used in its fully automatic mode. Utilising digital signal processing, DFINT-3A is claimed to have a band scan speed that is sufficient to intercept and direction-find against state-of-the-art frequency-hopping and burst emitters. Demonstrated field accuracy is quoted as being better than 2° RMS with a Line Of Bearing (LOB) resolution of 0.1°. DFINT-3A is further noted as being 'user friendly' menu-driven and as having a man/machine interface that incorporates a keypad (including soft keys), a tracker ball and a colour monitor. Data presentation formats are interactive, graphical or list-based and include:

- simultaneous, real-time LOB versus frequency and amplitude (versus frequency in DF mode)
- an on-screen, digital panoramic display and polar LOB plot for target analysis and DF step mode
- simultaneous display of up to 16 targets on a LOB versus time graph for moving target detection.

DFINT-3A is also reported as incorporating two additional monitoring receivers, two dual-track tape recorders, an electronic compass and an integral Global Positioning System receiver.

Status
As of May 2001, the DFINT-3A communications band DF and intelligence system is reported as having been in service since 1995.

Specifications
Frequency range: 20-1,000 MHz
DF accuracy: ≤ 2° RMS
LOB resolution: 0.1°
Frequency (FFT) resolution: 156 Hz-40 kHz (selectable)
Hopper performance: >500 hops/s
Scan speed: >500 MHz/s
DF receiver stability: 1 ppm
Monitoring receiver: 2 units
Tuning step: 100 Hz
IF bandwidths: up to 5 IF bandwidths between 3.2 kHz-1 MHz
Demodulation: AM, CW, FM and SSB
Frequency stability: 1 ppm
Audio recording: 2 dual-track units
Recording modes: voice operated, carrier operated or manual
Monitor: SVGA colour, 640 × 480 resolution
Printer: 80 columns, continuous form
Power: 220 V AC (50 Hz) 1Ø generator or integral 24 V DC battery bank
Temperature: −30 to +50°C (operating); −40 to +60°C (storage)
Humidity: 90% (non-condensing)

Contractor
ASELSAN Inc, Microwave and System Technologies Division, Ankara.

VERIFIED

The DFINT-3A tactical communications band DF and intelligence system
2001/0100386

DFINT-3T Direction-Finding (DF) system

Type
Very/Ultra High Frequency (V/UHF - 30 MHz to 3 GHz) band DF system.

Description
The tri-service DFINT-3T system is transportable DF equipment that can intercept, take a bearing on and analyse V/UHF band communications traffic. Emitter location is achieved via collaboration with other DFINT systems (with intercommunication being by radio or landline) and the equipment can handle 'state-of-the-art' emitter types including frequency hoppers and burst transmitters. The system can be deployed with a standard lightweight, high-gain antenna, sub-band arrays and/or a range of application specific antennas that include disguised low-profile, airborne and heliborne types. In rough terrain land-based applications, the environmentally ruggedised DFINT-3T equipment can be deployed by a two-man team and the system presents data in either graphical or list formats. Examples of available display types include:

- simultaneous Line Of Bearing (LOB), versus frequency and amplitude, versus frequency graphics (with equipment operating in DF scan mode)
- on-screen, digital panoramic spectrum and polar LOB plot (for monitoring or fast scanning of targets with equipment operating in DF step mode)
- location fixing on a digital map.

DFINT-3T uses the interferometric DF technique and has a digital signal processing-based modular architecture, together with integral software drivers for communications radios, data terminals/modems (for radio and line networking), a Global Positioning System receiver, an electronic compass and monitoring receivers.

Status
As of May 2001, the DFINT-3T DF system was reported as having been in service and production since 1997.

Specifications
Frequency range: 20-1,200 MHz
DF accuracy: ≤ 2° RMS
LOB resolution: 0.1°
Frequency (FFT) resolution: 500 Hz - 160 kHz (selectable)
Hopper performance: >1,000 hops/s
Scan speed (with DF): >1,000 MHz/s
DF receiver stability: 1 ppm
Demodulation: AM and FM (SSB optional)
Monitor: 26 cm (10.4 in); 640 × 480 pixel; 8 colour
Power: 24 V DC or 110/220 V AC or both
Weight: 9.8 kg (antenna); 19.7 kg (system)
Temperature range: −20 to +50°C (operating); −40 to +70°C (storage)
Humidity: 100%
Environmental: MIL-STD-810E
Antenna mast: 6.25 m; 12 kg (5 sections)
Power generator: 8.8 kg (dry weight); 220 V AC (50 Hz, 450 V A)
GPS receiver: external or built-in (optional accessory)

Contractor
ASELSAN Inc, Microwave and System Technologies Division, Ankara.

UPDATED

The DFINT-3T portable direction-finding system   2001/0100390

UKRAINE

Colchuga Electronic Support (ES) system

Type
Ground-mobile air defence ES system.

Description
The Colchuga air defence ES system is described as being installed aboard a two vehicle (sensor and power generation (16 kW) trucks) convoy and is operated by a seven man crew (a figure that includes three system operators). Of the cited system

A general view of the sensor vehicle used in the Colchuga ES system (Jane's/IDR)
2002/0083266

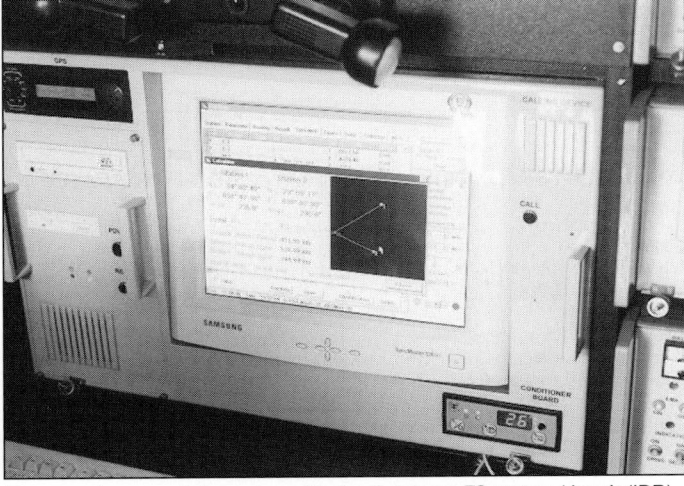

The operator console format used in the Colchuga ES system (Jane's/IDR)
2002/0083267

vehicles, the sensor truck incorporates a 'control and processing station' that makes use of a personal computer-based console running a Windows 98 operating system. Externally, the vehicle is equipped with a folding, mechanically scanned antenna assembly that makes use of distinct sub-arrays to cover specific sub-bands within the system's overall 0.1 to 18 GHz frequency band coverage. Functionally, four Colchuga sensor stations are netted together to determine target position by triangulation and time of arrival techniques. Within the net, the individual stations are typically located 80 km apart and Global Positioning System data is used for accurate geolocation. Colchuga's manufacturer claims that the system can give warning of the take-off of hostile aircraft at ranges that 'exceed the distances of modern radar systems' and can provide notification of target numbers and direction of flight. Target identity can be deduced from analysis of the characteristics of the detected emitter.

Status
As of the second half of 2000, Topaz was actively promoting the Colchuga ES system.

Specifications
Frequency range: 0.1-18 GHz
Range: 600 plus km (max detection and localisation)
Coverage: simultaneous narrowband (1-5°) and wideband (45°) sectors over the 360° arc in azimuth
Power requirement: 8 kW
Track capacity: up to 30 targets simultaneously (single azimuth)

Contractor
Joint Stock Company Topaz, Donetsk.

NEW ENTRY

UNITED KINGDOM

MS series ELectronic INTelligence (ELINT)/ Electronic Support (ES) systems

Type
Range of ELINT/ES equipments.

Description
Over time, TMD Technologies (formerly Thorn Microwave Devices) has produced the MS33XX range of receivers. Identified examples of the series comprise the MS3363 narrowband and the MS3365 wideband units, the known details of which are as follows:

MS3363/5 Integrated Microwave Receivers
MS3363/5 are high-performance, swept superheterodyne 0.5 to 18 GHz receivers with simultaneous data capture in time and frequency domains. The units are fitted with an integral pulse analyser providing the functionality previously only found in multibox solutions. They are designed for all tactical and strategic signal monitoring applications, in either stand-alone mode or in conjunction with analysis software. It provides real-time fine grain analysis in a cost-effective small volume unit. The units are fully ruggedised and have an inbuilt base cooling plate for high reliability.

Operation is straightforward from the man/machine interface or by remote control via GPIB or Ethernet. To aid the operator in fast set on, predefined settings can be placed in memory for instant recall. The pulse analyser provides time stamping of all received pulses with a resolution of 10 ns. Pulse Descriptor Words (PDW) for each of up to 128,000 pulses is available from the receiver. These PDWs and the statistical analysis of them can also be displayed on the receivers main display, with cursors for ease of measurement. Qualification of pulses is also available, to eliminate unwanted pulses from the pulse analysis.

While the system is a single-box solution to most requirements, the internal construction comprises separate stand-alone modules, of standard size and control architecture. This building block approach enables TMD Technologies to offer customised ELectronic INTelligence (ELINT)/Electronic Support (ES) installations by providing integrated modules as required. This modular approach allows systems to be modified or upgraded as technology advances or as threat scenarios change. Currently ELINT/ES stations can be configured to operate anywhere in the range 0.5 MHz to 40 GHz, with bandwidths of 20 to 1,000 MHz, depending on individual requirements. A recent development to this system adds facilities that support analysis of microwave communications systems. TMD also notes that it is involved in 'some of the latest' digital signal processing with respect to its ELINT/ES products.

Contractor
TMD Technologies Ltd, Hayes.

UPDATED

Tactical Surveillance Receiver Module (TSRM)

Type
Building block for a tactical intercept system.

Description
The TSRM is a compact module covering the 20 to 1,350 MHz frequency range. It is designed to be the building block of a modern tactical intercept system. The

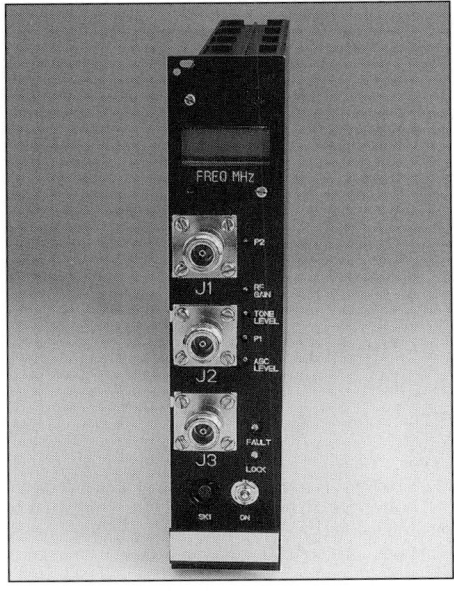

BAE System's Tactical Surveillance Receiver Module

TSRM is small enough to be stacked nine to a 48 cm (19 in) receiver tray and is programmed via a standard IEEE-488 interface. A manual reversionary control mode is available using a keypad which plugs into the unit's front panel. The equipment is capable of both fixed and swept frequency operation, as well as being able to be programmed to perform complex frequency monitoring strategies.

The TSRM is a fully tunable, synthesised superhet design with a front end tracking preselector. This preselection offers full band performance in sensitivity and selectivity. Up to five separate antennas may be connected to the front panel, allowing up to five different frequency bands to input simultaneously without the need for external radio frequency switching or reconnection. As standard, the TSRM has a baseband output and front panel control of automatic gain control threshold and signal level setting. Digital and audio outputs are available as options.

Status
As of August 2001, the status of the TSRM was uncertain.

Specifications
Frequency range: 20-1,350 MHz
Frequency stability: =1 part in 10^{-7} (with external reference signal)
IF bandwidth: 10 kHz; 50 kHz; 300 kHz
Baseband BW: 0.1-1.1 MHz
Noise figure: 17 dB
Gain: 70 dB ± 9 dB
MTBF: >10,000 h
Temperature: 0 to +40°C
Weight: 3.8 kg
Dimensions: 380 × 190 × 45.3 mm

Contractor
BAE Systems Avionics - Sensor Systems Division, Stanmore.

UPDATED

UNITED STATES OF AMERICA

AN/MSQ-103 Teampack Electronic Support (ES) system

Type
Radar band ES system.

Description
AN/MSQ-103 is a transportable ES system that is designed to collect, identify and locate ground-based radar emitters operating (according to Jane's sources) in the 500 MHz to 40 GHz frequency range. As of March 2001, three variants of the equipment (MSQ-103A, -103B and -103C) have been identified that operate within the 500 MHz to 40 GHz frequency range and provide field support for division level formations. MSQ-103 is housed in a ballistically protected shelter that can be carried by either the M-35 utility truck or the M-1015 tracked vehicle. If required, the Teampack architecture can be configured for installation aboard light armoured vehicles, jeep type and other forms of combat vehicle. MSQ-103 was originally developed as a replacement for the US Army's AN/MLQ-24 system and incorporates computer processing, a wideband datalink and secure voice communications subsystem. Jane's sources suggest that the MSQ-103A was an upgrade of the basic system that included improved reliability, performance and crew protection together with the ability to be netted. In its MSQ-103C variant, the system was given an unspecified extension in frequency coverage, the ability to handle frequency agile emitters, automatic north seeking and nuclear, biological and chemical warfare protection for its operator crew.

Status
As of March 2001, the AN/MSQ-103 ES system was understood to have been in service with the US Army.

Contractor
Systems & Electronics Inc, St Louis, Missouri.

UPDATED

AN/PRD-12 Direction-Finding (DF) system

Type
Man-portable DF system.

Description
The 500 kHz to 500 MHz band AN/PRD-12 (ES1200) man-portable direction-finder is a dual-channel system that is claimed to offer 'high accuracy, speed and operational flexibility'. It can be used as part of a network or in a stand-alone mode and has, as its primary target, battlefield command and fire-control communications links. When networked, bearing and co-ordination data is passed

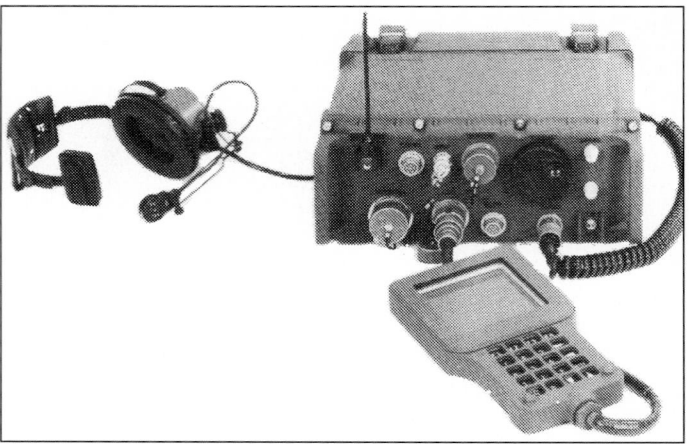

The AN/PRD-12 lightweight man-portable direction-finding system

to other stations via a separate manpack radio. The baseline equipment weighs approximately 27 kg and can be carried by two men (individual receiver-processor and antenna subsystem loads). A hand-held operator-control keyboard is attached by cable to the receiver-processor and processed data is displayed on a high-resolution alphanumeric display.

Status
As of this edition, the AN/PRD-12 portable DF system was reported as having been procured by the US Army and Marine Corps (designated as Top Hunter). By the end of 1993, over 300 PRD-12 equipments were understood to have been delivered.

Contractor
TRW, Sunnyvale, California.

VERIFIED

AN/PRD-13(V) intercept and Direction-Finding (DF) system

Type
Man-portable communications band intercept and DF system.

Description
AN/PRD-13(V) is a man-portable communications band intercept and DF system that covers the 0.1 to 2,000 MHz frequency range and is designed for use by special forces. Key components of the baseline system comprise:
- the 45 to 1,400 MHz MA-118 DF/intercept antenna
- the 2 to 500 MHz MA-308 hand-held directional antenna
- the 1 to 45 MHz MA-551E DF antenna
- the 1 to 1,400 MHz MA-713 monitoring antenna
- the MD-403 manpack receiver/processor.

Of these, the MA-118 and MA-551E operate alone or in tandem to provide DF coverage within the 1 to 1,400 MHz band while the hand-held MA-308 unit is used to localise nearby transmitters. In the monitoring role, the MA-713 whip aerial can be used alone or teamed with the MA-308 DF unit. Other antenna options can be applied to the system to configure it for shipboard, airborne or HF (3 to 30 MHz) skywave applications.

Baseline PRD-13(V) is operable by a single person and can be set up in less than five minutes. It can maintain an active signals list of up to 400 signals and measures centre frequency, bandwidth, time first seen, per cent time active, DF bearing and signal strength. A directed search capability is included which allows threshold and other parameters to be independently set for each of up to 400 channels. The MD-403 receiver/processor incorporates a range of interfaces (serial remote control, tape recorder, headphones, external power and input) and the system's man/machine interface. The processor uses single channel interferometry to

The MA-551E (left)/MA-118 (centre) antennas and man/machine interface (right) used in the AN/PRD-13 DF/intercept system

establish DF bearing and an HF band DF histogram display is provided for use in weak signal or adverse propagation conditions with simplex nets. The equipment is noted as being able to function for up to 24 hours on a single battery and there are solar panel, rechargeable Ni/Cd battery and cassette recorder options available.

Status

As of early 2001, the versions of the AN/PRD-13(V) intercept and DF system were reported as having been procured by a customer base that includes the Australian Army and the US Special Operations Command. The PRD-13(V)2 variant forms the basis of the US Army's production standard AN/MLQ-40(V)2 Ground Prophet Block I electronic support system (see separate entry).

Specifications

Baseline system
Frequency: 0.1-1,200 MHz (intercept); 1-1,400 MHz (DF)
DF resolution: 1°
DF accuracy: 3° RMS (typical)
Coverage: 0-60° (elevation); 360° (azimuth)
Polarisation: vertical
Demodulation: AM, CW, FMn, FMw, LSB, USB
Operating modes: DF/PAN (manual); automatic/semi-automatic band sweep (9 selectable bands); measurement (BW, centre frequency, SNR, time statistics); signal list (log up to 400); pass list (avoid up to 400); channel scan (up to 400); set up; BIT
Operating temperature: −20 to +50°C
Altitude: up to 15,240 m
Power consumption: 3.9 W (typical)
Power input: 10-16 V DC
Weight: 19 kg

Contractor

Titan Systems Corporation - Delfin Systems Division, Santa Clara, California.

UPDATED

....

AN/TRQ-32A(V)2 Teammate intercept and Direction-Finding (DF) system

Type

Ground-mobile communications band intercept and DF system.

Description

The AN/TRQ-32A(V)2 Teammate communications band intercept and DF system is a ground-mobile equipment that is designed to support tactical formations. Mounted on an M-1097 chassis, the system provides High, Very High and Ultra High Frequency communications intercept and Very High Frequency DF. TRQ-32A (V)2 features two operator positions and an interface unit to connect it with other intelligence and electronic warfare sensors.

Status

As of this edition, the AN/TRQ-32A(V)2 intercept and DF system was reported as having been procured by the US Army.

Contractor

Raytheon Electronic Systems, Fort Wayne, Indiana.

VERIFIED

The AN/TRQ-32A mobile intercept and DF system

AN/TSQ-114 Trailblazer intercept and Direction-Finding (DF) system

Type

Radio communications intercept and DF system.

Description

The AN/TSQ-114 Trailblazer intercept and DF system is a local or remotely controlled, tactical, ground-based communications intercept and DF system. It is designed to provide real-time intelligence and combat information to commanders at divisional level and below. Frequency coverage is understood to be in the 0.5 to 150 MHz frequency range with DF available across the 20 to 80 MHz sub-band. The latest known iteration of the system is designated as the AN/TSQ-114B(V)2. The baseline Trailblazer architecture consists of two Master Control Sets (MCS) and three remote slave sets that a linked via an Ultra High Frequency (UHF - 0.3 to 1 GHz) datalink. Other system elements include the AN/UYK-19B computer, the RT-524A/VRC transceiver, the R-442A/VRC radio receiver, the RT-1167/ARC-164(V) transceiver, the AN/UNH-17A recorder/reproducer set, the CRC/1 recorder/reproducer, the AN/UYQ-10 plasma display set and the R-1444/UR radio receiver. Of these, the MCSs are also netted with an associated reporting facility, with each unit houses two operators who are provided with independent and duplicate capabilities. A system-to-operator interface provides test, voice link, data storage, intercept, DF display and reporting functions. Additional features provide the operator with a system status monitoring capability via a visual/aural caution and warning subsystem. that gives visual and aural alerts. The system's electronics and operator positions are contained in a modified, ballistically-protected S-280 shelter that is mounted on a dedicated signals intelligence/electronic warfare M1015 tracked chassis. The antenna assembly used is mounted on a 15 m hydraulic/pneumatic, quick-erecting mast that is attached to the shelter roof. A support trailer contains a 30 kW generator for power supplies. It is believed that subsequent to its original development, the Trailblazer system has been modified so that all five vehicles within a system can be used as an MCS rather that the two dedicated platforms included in the baseline architecture.

Status

As of this edition, the AN/TSQ-114 Trailblazer intercept and DF system is reported as having been procured by the US Army. Jane's sources were further suggesting that, over time, the TSQ-114B(V)2 variant has been deployed in Germany and South Korea.

Contractor

TRW, Sunnyvale, California.

VERIFIED

A Trailblazer vehicle deployed in the field

....

AN/TSQ-138 COMmunications INTelligence (COMINT) system

Type

COMINT and Direction-Finding (DF) system.

Description

AN/TSQ-138 is an improved variant of the AN/TSQ-114A Trailblazer for use at division level and below. It is designed for COMINT processing and DF and is mounted in an S-280 shelter that is installed n an M1015 tracked chassis. It is interoperable with the EH-60A electronic warfare helicopter for DF. A number of improvements thought to have been implemented include an enhanced self-location capability, digital temporary storage recorder and provisions for interfacing with other integrated electronic warfare systems.

Status
As of this edition, the AN/TSQ-138 COMINT and DF system was reported as having been supplied to the US Army.

Contractor
TRW, Sunnyvale, California.

VERIFIED

AN/TSQ-199 Enhanced TRACKWOLF intercept/emitter location system

Type
Transportable intercept and emitter location system.

Description
Enhanced TRACKWOLF is a fully automated, man transportable, redesign of the AN/TSQ-152(V) system that provides collection and direction-finding capabilities against High Frequency (HF - 3 to 30 MHz) band emitters including 'advanced' modems, 'packet' radios and low probability-of-intercept types. Enhanced TRACKWOLF's full spectrum delay captures the entire signal of interest (including its turn-on) and provides 'high-accuracy' DF. 'High-performance' hardware is claimed to provide the 'maximum achievable dynamic range', thereby ensuring optimum performance in a dense environment. A common set of sensor equipment performs all signal acquisition, collection, geolocation and analysis functions. The system's wideband digital architecture allows it to rapidly map unknown signal environments and to transition from target development to automatic signal exploitation in a matter of hours. AN/TSQ-199 can be set up quickly, is powered by a range of commercial and military power sources and is mounted in transit cases (decreasing its deployment lift requirements by over 95 per cent when compared with the original TRACKWOLF system). It is transportable by commercial air, military air, High-Mobility Multipurpose Wheeled Vehicle (HMMWV) or any available vehicle for early insertion into a crisis area.

Status
As of this edition, the AN/TSQ-199 intercept and emitter location system was reported as having been procured by the US Army.

Contractor
Raytheon Electronic Systems, Falls Church, Virginia.

VERIFIED

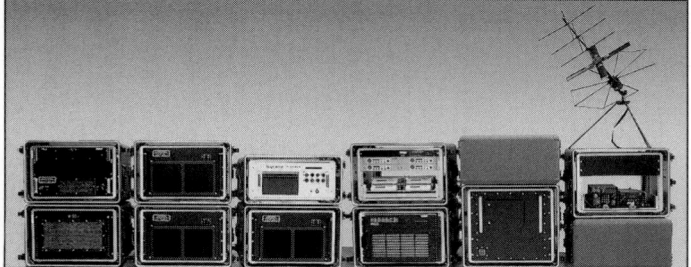

The elements which go to make up the TSQ-199 Enhanced TRACKWOLF intercept system

AN/ULQ-16(V) pulse analyser

Type
Microwave frequency automatic or manual pulse analyser.

Description
AN/ULQ-16(V) is a precision automatic or manual pulse analyser capable of fine grain analysis. It is suitable for land-based, shipborne or airborne applications and may be connected to the video output of any microwave receiver having a 160 MHz intermediate frequency. Two versions have been produced: (V)1 for rack mounting applications and (V)2 for Air Transport Racking standard tray mounting. Operationally the two systems are identical. The equipment offers automatic analysis, time-base, delayed time-base, single sweep and falling raster operating modes, is menu-driven and provides real-time displays of the cited operational modes. The built-in library is loaded from a data recorder and can be edited from the keypad provided or from an external keyboard. Measurements provided by ULQ-16(V) are pulse rate (pulse repetition frequency/interval), pulse-width, pulse amplitude, scan time/rate, illumination time and bandwidth. Available options include larger displays, multichannel processing and pulse-by-pulse frequency analysis.

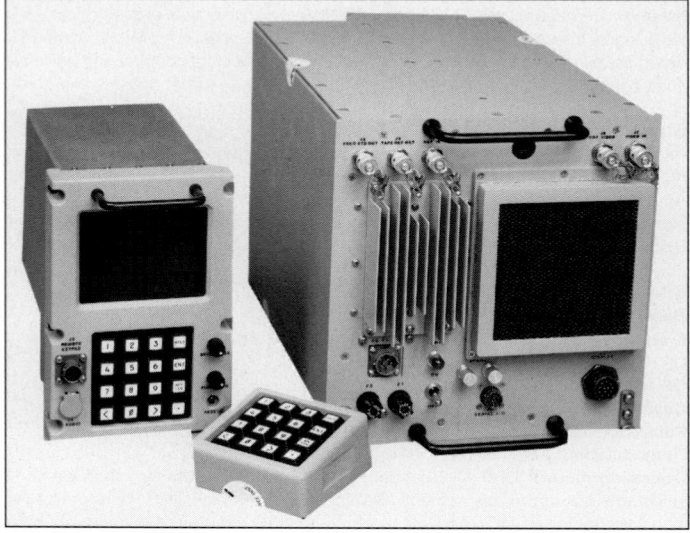

The AN/ULQ-16(V)2 pulse analyser

In the specific US Navy (USN) EP-3E Airborne Reconnaissance Integrated Electronics Suite (ARIES) II, context, the ULQ-16(V) pulse analysers installed in the aircraft's crew stations 8, 10, 11, 12 and 20 have been upgraded as part of the the platform's overall Sensor System Improvement Program (SSIP). Here, the modification involves:
- replacement of the existing IP-1159A/A and FR-185/U (XAN-3) pulse analysis units
- the introduction of dual channel, real-time, video inputs
- the introduction of new circuitry in the architecture's signal data processor (the CP-1499 modification)
- the introduction of the 23 cm (9 in) EI-1700 high-resolution display
- the introduction of the EI-1400 control display unit
- the introduction of processor software upgrades.

Introduction of the described ULQ-16(V) modifications are understood to reduce the EP-3E ARIES II's overall weight by 3.2 kg per analyser (?)

Status
As of September 2001, over 900 ULQ-16(V) pulse analysers are reported as having been procured by the USN.

Contractors
Condor Systems Inc, San Jose, California.
Northrop Grumman Corporation, Electronic Systems, Buffalo, New York.

UPDATED

CDR-3100 receiver series

Type
Family of communications band receivers.

Description
The CDR-3100 family of Very Low/High Frequency (VLF/HF - 3 to 30 kHz and 3 to 30 MHz) receivers incorporates digital signal processing microchip technology and is designed to solve many of the typical communications, surveillance and direction-finding problems encountered in air- and shipboard applications. CDR-3100 series receivers have digitally tuned preselectors, 250 programmable channels, 51 selectable intermediate frequency bandwidths (ranging from 100 Hz to 18 kHz) and integral diagnostics. When a hardware failure is encountered, a message is displayed to the operator that identifies the system module that needs to be replaced.

Status
As of this edition, the status of the CDR-3100 receiver family was uncertain.

Contractor
Cubic Communications Inc, San Diego, California.

VERIFIED

Communications Direction-Finding (DF) system

Type
Modular DF system.

Description
Raytheon Systems' COMMunications Direction-Finding (COMM DF) system is a full-function emitter acquisition, direction-finding and reporting system. The system

The Raytheon Systems' COMM DF modular DF system

provides high performance and accurate line of bearing measurements on operator selected targets of interest. COMM DF has three components: the sensor package, the antenna array and an operator workstation. All components feature an open architecture VME-based hardware design and a UNIX/X-Windows-based software design. The sensor package incorporates a single-board computer, an A/D converter circuit card assembly, an antenna switch, a flexible datalink interface and two nanomin receivers (see separate entry) and operates over the 20 to 1,500 MHz frequency spectrum. Operating at 28 V DC, the sensor package can interface with a number of inertial navigation systems. As an option, it can incorporate and control up to four monitoring receivers. The UNIX-based workstation (selectable by the user) includes a VME datalink/audio distribution circuit card and operationally proven interface software. The datalink (selectable by the user) operates using RS-422, RS-232 (at 512 Kbytes/s or 1.544 Mbytes/s) or RS-170 (standard broadcast video).

Specifications
Frequency range: 20-1,500 MHz
DF accuracy: <1° RMS using typical VHF/UHF antenna array
DF sensitivity: −116 dBm (10 dB S/N in 6.4 KHz bandwidth)
DF output: line of bearing, quality factor, other technical parameters
Audio output: 6 audio channels
DF throughput: 8-10 bearings/s

Contractor
Raytheon Electronic Systems, Falls Church, Virginia.

VERIFIED

Condor Systems demodulators, receivers and tuners

Type
Family of microwave demodulators, receivers and tuners.

Description
Over time, Condor Systems has developed a range of specialist signals intelligence demodulators, receivers and tuners that are described as being characterised by wide instantaneous bandwidths, fast tuning and RS-422, Ethernet, Versanet or digital parallel bus control. Available options have included extended dynamic range, multiple channels and customised control and packaging. Examples of the range include the following:

CS-2400 receiver
The CS-2400 receiver covers the 0.5 to 18 GHz frequency band (with 1 kHz steps) and offers up to four amplitude and phase matched channels. The equipment's Intermediate Frequency (IF) output is given as 4.5 GHz (with 2 GHz bandwidth) with tuning speed being noted as being 1 ms. CS-2400 also incorporates a high-speed fibre optic interface.

CS-4002 instantaneous frequency measuring receiver
CS-4002 covers the 2 to 18 GHz frequency range in 2 to 6 GHz and 6 to 18 GHz sub-bands. The equipment is noted as having a sensitivity value of −60 dBm, as

generating time of arrival flags and as providing a high-speed parallel output (2 million PWD/s). CS-4002 measures 22 × 39 × 58 cm and was formerly designated as the ESSI DR-4002.

MD-128-03 modulator
The MD-128-03 analysis demodulator is a plug-in, multifunction demodulator that is designed for use with the Condor Systems CY-128-03 demodulator mainframe. As such, the device accepts a 160 MHz Intermediate Frequency (IF) input and provides selectable and fixed IF outputs at both 21.4 and 160 MHz. Eight selectable IF bandpass filters and IF attenuation are provided together with LOG, LIN (MGC/AGC), frequency modulated video and audio outputs. MD-128-03 also contains a digitised IF pan (LOG/LIN) and a digitised linear instantaneous frequency monitor. Of these, the IF pan is self-centering.

Specifications
IF section
IF input: 160 MHz (75 MHz bandwidth)
IF outputs: 21.4 MHz (narrowband filtered); 160 MHz (pre- and post-filtered)
IF bandwidths: 0.1, 0.5, 2, 5, 10, 20, 30 or 40 MHz (selectable)
IF attenuation: 0-60 dB (1 dB steps)
Resolution: 1 dB
Noise figure: 22 dB
Impedance: 50 Ω
VSWR: 2:1
Demodulator section
Outputs: display video, frequency modulated, linear, LOG, normal or stretched audio and selected video
Video bandwidths: 0.05, 0.25, 1, 2.5, 5, 10, 15 MHz and 'max selectable'
Output impedance: 50 Ω
IF pan video: LOG/LIN video (peak detected and digitised)
IF pan bandwidth: 100 kHz or 1 MHz (selectable)
IF pan sweep rate: 5-25 Hz
FM/LIN/LOG video
Dynamic range: 20 dB (FM/LIN); 50 dB (LOG)
Linearity: ±2 dB (LOG/LIN)
Rise time: 30 ns (FM/LOG/LIN)
DC output: 0.1-1 V (LOG/LIN)
Gain mode: MGC/AGC AVG (LIN)
Gain control: 50 dB (LIN, starting at 10 dB above MDS)
Output linearity bandwidth: 50 MHz (FM)
Output: 2 V AC p-p (FM)
General
Audio output impedance: 8 Ω
Audio output level: 2.5 V AC p-p
Power: 43 W (provided by CY-128)
Temperature: −20 to +60°C (non-operating): 0 to +50°C (operating)
Altitude: SL-3,658 m (operating); SL-13,716 m (non-operating)
Dimensions (h × w × l): 193 × 381 × 437 mm
Weight: 4.5 kg

TN-118 microwave tuner
TN-118 is a fully synthesised, airborne qualified, superheterodyne tuner that covers the 0.5 to 18 GHz frequency band (with 100 kHz steps). IF output is given as 400 MHz (200 MHz bandwidth) with 160 MHz (50 MHz bandwidth) as an option. The equipment's frequency range can also be extended to 0.1 to 18 GHz if required and RS-422 or US MILitary STanDard (MIL-STD)-1553B control interfaces are a further option. TN-118 measures 19 × 13 × 50 cm and was formerly designated as the Watkins-Johnson TN-118A.

TN-340 microwave tuner
TN-340 covers the 0.5 to 18 GHz frequency range (with 100 kHz steps) and is frequency locked while scanning (1 MHz steps). The equipment features automatic local oscillator alignment and a high-speed parallel digital control bus. IF outputs are given as 160 MHz (60 MHz bandwidth) and 1 GHz (500 MHz bandwidth). TN-340 weighs 16 kg and measures 19 × 14 × 50 cm. This equipment was formerly designated as the ESSI ER-3400.

TN-613 microwave tuner
TN-613 is a 0.5 to 18 GHz, internally synthesised microwave tuner that is designed for communications intelligence and ELINT applications. IF outputs are given as 70 MHz (50 MHz bandwidth), 160 MHz (75 MHz bandwidth) and 1 GHz (250 MHz bandwidth) while phase noise is noted as being <1.8°. TN-613 weighs 15 kg and measures 19 × 15 × 52 cm.

TN-618-15 microwave tuner
The TN-618-15 is a fully synthesised microwave tuner that covers the 0.5 to 18 GHz frequency band in 1 kHz steps. As such, it is functionally and software compatible with previous TN-618 variants and offers the following performance and feature enhancements:
- automatic alignment of YIG devices at switch-on
- a lower noise figure and an improved dynamic range
- reduced gain variation
- 1 kHz tuning steps
- a hard-wired phase lock indicator to support faster hopping in pulse processing systems

TN-618-15 also includes a built-in scan demodulator in order to digitise scan video in support of the C-144 and C-244 resource controllers while an integral microprocessor controller provides scan and set on capabilities without the need for continual updating by the system controller.

Specifications

RF input frequency: 0.5-18 GHz
1 GHz IF performance
Noise figure: 13 dB (max)
Bandwidth: 500 MHz (6 dB)
Gain: 8.5 ±2.5 over frequency and temperature
1 dB comp dynamic range: 95 dB (min, 100 dB typical)
1 dB comp input intercept: −7 dBm (min)
3rd order dynamic range: 65 dB (min, 75 dB typical)
3rd order input intercept: −4 dBm (min, +9 dBm typical)
Group delay: 3 ns (p-p type over centre, 450 MHz of passband)
160 MHz performance
Noise figure: 14 dB (max)
Bandwidth: 65 MHz (min)
Gain: 11 ±3 over frequency and temperature
1 dB comp dynamic range: 92 dB (min)
1 dB comp input intercept: −10 dBm (min)
3rd order dynamic range: 63 dB (min)
3rd order input intercept: −4 dBm (min)
Group delay: 5 ns (p-p type over centre, 40 MHz of passband)
General
VSWR: 2:1 (IF output, max, 50 Ω system); 2.5:1 (RF output, max, 50 Ω system)
Image rejection: 70 dB (min)
Frequency stability vs temperature: $\pm1 \times 10^{-9}$ (0-50°C, max); $\pm5 \times 10^{-9}$ (−30 to +55°C, max)
Frequency stability vs time: 1×10^{-9}/day (max)
Tuning step size: 1 kHz
Scan time: 30 ms (min, no hand-breaks)
Integrated phase noise: 10° RMS (max, 100 Hz-20 MHz)
LO reradiation: −80 dBm (max)
Power requirement: 115 V AC (±10%, 47-440 Hz); 230 V AC (±10%, 47-440 Hz)
Power consumption: 130 W (max)
Temperature: −30 to +55°C (operating); −54 to +71°C (storage)
Altitude: up to 13,716 m (max temperature +32°C)
Dimensions (h × w × l): 196 × 148 × 523 mm (ATR Style A1D chassis)
Weight: 15 kg (max)

Status

During the period May to August 2001, Condor Systems were promoting the demodulators, receivers and tuners described here.

Contractor

Condor Systems Inc, San Jose, California.

UPDATED

Condor Systems specialist antennas

Type

Family of specialist signals intelligence gathering antennas.

Description

Over time, Condor Systems has developed and produced a range of antennas that are suitable for tri-service Direction-Finding (DF), search and surveillance applications. Known details of a selection of such antennas are as follows:
AS-105-02
The AS-105-02 antenna assembly is a DF antenna that covers the 0.4 to 18 GHz frequency range and is designed for use in an aircraft-mounted pod that acts as a radome. A brush type Direct Current (DC) motor rotates the turntable on which the equipment's antennas (a 0.4 to 2 GHz horizontally polarised log periodic unit and a 2 to 18 GHz assembly that comprises a slant 45° polarised, log periodic feed located at the focus of a CoSeCant squared (CSC^2) shaped reflector) are mounted. The assembly's radio frequency output is routed through a dual-rotary joint to output connectors on the pedestal.

Specifications

Frequency: 0.4-18 GHz
Gain/beamwidth:

Frequency	Gain *	Azimuth beamwidth **	Elevation beamwidth ***
0.4-0.5 GHz	0.5 dBiL	90° ****	75°
0.5-1 GHz	0.5 dBiL	85°	75°
1-2 GHz	4 dBiL	80°	75°
2 GHz	4 dBiL	70°	75°
2-4 GHz	10 dBiL	24°	CSC^2 to 20°
4-8 GHz	15.5 dBiL	12.4°	CSC^2 to 20°
8-12 GHz	20 dBiL	6.7°	CSC^2 to 20°
12-18 GHz	21.5 dBiL	4.6°	CSC^2 to 20°

Key
* Nominal value ** Max value *** Typical value **** Beamwidth between 0.4 and 0.5 GHz is typically less than 75°
Squint: <10% of beamwidth or 1°, whichever is the larger
Impedance: 50 Ω
VSWR: 3.5:1 (max)
Polarisation: horizontal (0.4-2 GHz sub-band); 45° slant linear (2-18 GHz sub-band)

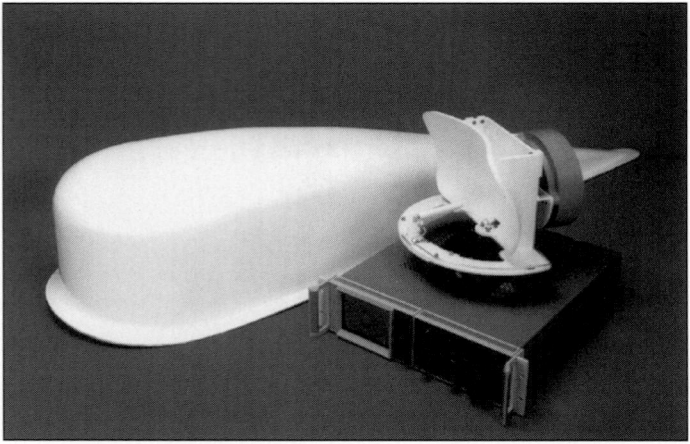

The CS-8056 DF antenna system

Control modes: manual, sector scan and spin
Azimuth positioning range: 0-359.9° (manually adjustable)
Sector scan centre: continuously adjustable through 360°
Sector scan width: 4-350° (adjustable)
Sector scan rate: adjustable
Sector scan direction: bidirectional
Spin rate: 0-200 rpm (adjustable)
Direction of spin: clockwise
Spin rate accuracy: within 5% of the command value
Power: 26 V AC (resolver, 95 mA (max), 400 Hz); 28 V DC (motor, 13 A (max))
Temperature: −54 to +55°C (operating, with optional heaters); −54 to +71°C (non-operating)
Altitude: up to 9,144 m
Dimensions: 105 × 787 mm (Vibration mount); 487 mm (height of antenna assembly with pedestal); 607 mm (antenna assembly Ø)
Weight: 10 kg (vibration mount assembly); 23 kg (cover, antenna and pedestal)

AS-105-03
The AS-105-03 antenna assembly is a DF antenna that covers the 0.5 to 18 GHz frequency range and is designed for use on aircraft. A brush type DC motor rotates the turntable on which the equipment's antennas (a 0.5 to 2 GHz vertically polarised log periodic unit and a 2 to 18 GHz assembly that comprises a horizontally polarised, log periodic feed located at the focus of a CSC^2 shaped reflector) are mounted. The assembly's radio frequency output is routed through a dual-rotary joint to output connectors on the pedestal.

Specifications

Frequency: 0.5-18 GHz
Gain/beamwidth:

Frequency	Gain *	Azimuth beamwidth **	Elevation beamwidth ***
0.5 GHz	2 dBiL	80°	75°
0.85 GHz	4 dBiL	65°	75°
2 GHz	8 dBiL	50°	75°
6 GHz	17.5 dBiL	11.5°	CSC^2 to 20°
10 GHz	21 dBiL	6.5°	CSC^2 to 20°
15 GHz	22.5 dBiL	4.5°	CSC^2 to 20°
18 GHz	23 dBiL	4°	CSC^2 to 20°

Key
* Nominal value ** Max value *** Typical value
Squint: <7% of beamwidth or 1°, whichever is the larger
Impedance: 50 Ω
VSWR: 3.5:1 (max)
Polarisation: horizontal (2-18 GHz sub-band); vertical (0.5-2 GHz sub-band)
Control modes: manual, sector scan and spin
Azimuth positioning range: 0-359.9° (manually adjustable)
Sector scan centre: continuously adjustable through 360°
Sector scan width: 4-350° (adjustable)
Sector scan rate: adjustable
Sector scan direction: bidirectional
Spin rate: 0-200 rpm (adjustable)
Direction of spin: clockwise
Spin rate accuracy: within 5% of the command value
Power: 26 V AC (resolver, 95 mA (max), 400 Hz); 28 V DC (motor, 13 A (max))
Temperature: −54 to +55°C (operating, with optional heaters); −54 to +71°C (non-operating)
Altitude: up to 9,144 m
Dimensions (h × Ø): 487 × 607 mm
Weight: 23 kg

AS-105-18
The AS-105-18 antenna assembly is a DF antenna that covers the 0.5 to 18 GHz frequency range and is designed for use on aircraft. A brush type DC motor rotates the turntable on which the equipment's antennas (a low band, horizontally polarised, log periodic unit and a slant 45° polarised, log periodic feed located at the focus of a CSC^2 shaped reflector) are mounted. The assembly's radio frequency output is routed through a dual-rotary joint to output connectors on the pedestal.

Specifications

Frequency: 0.5-18 GHz
Gain/beamwidth:

Frequency	Gain *	Azimuth beamwidth **	Elevation beamwidth ***
0.5 GHz	0.5 dBiL	65°	60°
0.85 GHz	4 dBiL	60°	50°
2 GHz	8 dBiL	35°	50°
6 GHz	17.5 dBiL	8.5°	CSC² to 20°
10 GHz	21 dBiL	4.7°	CSC² to 20°

Key
* Typical values at the pedestal output ** Nominal value *** Nominal value
Squint: <7% of beamwidth or 1°, whichever is the larger
Impedance: 50 Ω
VSWR: 3.0:1 (over 80% of band)
Polarisation: horizontal (low band); 45° slant linear (high band)
Control modes: manual, sector scan and spin
Azimuth positioning range: 0-359.9° (manually adjustable)
Sector scan centre: continuously adjustable through 360°
Sector scan width: 4-350° (adjustable)
Sector scan rate: adjustable
Sector scan direction: bidirectional
Spin rate: 0-200 rpm (adjustable)
Direction of spin: clockwise or anti-clockwise
Spin rate accuracy: within 5% of the command value
Power: 26 V AC (resolver, 95 mA (max), 400 Hz); 28 V DC (motor, 3 A (max))
Temperature: −54 to +55°C (operating, with optional heaters); −54 to +71°C (non-operating)
Altitude: up to 9,144 m
Dimensions (h × Ø): 487 × 607 mm
Weight: 23 kg

AS-106-21

The AS-106-21 antenna assembly is a DF antenna that covers the 0.5 to 18 GHz frequency range and is designed for use on aircraft. A brush type DC motor rotates the turntable on which the equipment's antennas (a pair of 0.5 to 2 GHz, circularly polarised, spiral antenna elements (whose outputs are combined) and a slant 45° polarised, log periodic feed located at the focus of a CSC² shaped reflector) are mounted. The assembly's radio frequency output is routed through a dual-rotary joint to output connectors on the pedestal.

Specifications

Frequency: 0.5-18 GHz
Gain/beamwidth:

Frequency	Gain *	Azimuth beamwidth **	Elevation beamwidth ***
0.5 GHz	−6 dBiL	90°	90°
1 GHz	−4 dBiL	65°	80°
2 GHz	0 dBiL	30°	70°
6 GHz	10 dBiL	20°	CSC² to 20°
10 GHz	12 dBiL	15°	CSC² to 20°
15 GHz	18 dBiL	10°	CSC² to 20°
18 GHz	20 dBiL	4°	CSC² to 20°
12-18 GHz	21.5 dBiL	4.6°	CSC² to 20°

Key
* Nominal values at the pedestal output ** Nominal value *** Typical value
Squint: <7% of beamwidth or 1°, whichever is the larger (at 0.5 GHz, squint is typically ±5°)
Impedance: 50 Ω
VSWR: 3.5:1 (max)
Polarisation: right hand circular (0.5-2 GHz sub-band); 45° slant linear (2-18 GHz sub-band)
Control modes: manual, sector scan and spin
Azimuth positioning range: 0-359.9° (manually adjustable)
Sector scan centre: continuously adjustable through 360°
Sector scan width: 4-350° (adjustable)
Sector scan rate: adjustable
Sector scan direction: bidirectional
Spin rate: 0-200 rpm (adjustable)
Direction of spin: clockwise (when installed under an aircraft)
Spin rate accuracy: within 5% of the command value
Power: 26 V AC (resolver, 95 mA (max), 400 Hz); 28 V DC (motor, 3 A (max))
Power dissipation: 70 W (typical)
Temperature: −54 to +55°C (operating, with optional heaters); −54 to +71°C (non-operating)
Altitude: up to 9,144 m
Dimensions (h × Ø): 381 × 432 mm
Weight: 12.5 kg

AS-162-06

The AS-162-06 antenna assembly is a DF antenna that covers the 1 to 18 GHz frequency range and is designed for use in ground or shipboard applications. A brush type DC motor rotates the turntable on which the equipment's antenna (a slant 45° polarised, log periodic feed located at the focus of a shaped reflector that provides a 'fairly constant' beamwidth) are mounted. The assembly's radio frequency output is routed through a single rotary joint and its base unit is so designed as to facilitate the installation of a radome and additional optional circuitry.

Specifications

Frequency: 1-18 GHz
Gain/beamwidth:

Frequency	Gain *	Azimuth beamwidth **	Elevation beamwidth ***
1 GHz	12.3 dBi	22.5°	26°
2 GHz	16.6 dBi	11.5°	29°
4 GHz	20 dBi	6.3°	19°
8 GHz	23.2 dBi	3.2°	22°
12 GHz	25.5 dBi	2.1°	21°
18 GHz	27 dBi	1.4°	18°

Key
* Typical values measured at pedestal output with slant polarisation ** Typical value *** Typical value
Impedance: 50 Ω
VSWR: 3.5:1 (max)
Polarisation: 45° slant linear (1-18 GHz band)
Control modes: point, sector scan and spin
Point azimuth positioning range: 0-359.9° (adjustable)
Sector scan centre: continuously adjustable through 360°
Sector scan width: 4-350° (in 4° increments)
Sector scan rate: adjustable
Sector scan direction: bidirectional
Spin rate: 0-200 rpm (adjustable)
Direction of spin: clockwise
Spin rate accuracy: within 5% of the command value
Power: 26 V AC (resolver, 95 mA (max), 400 Hz); 28 V DC (motor, 7 A (max))
Temperature: −54 to +55°C (operating, with optional heaters); −54 to +71°C (non-operating)
Altitude: up to 3,048 m
Dimensions (h × Ø): 1,181 × 813 mm
Weight: 52.3 kg

CS-8050

The baseline CS-8050 assembly is a three axis, carry-on, signals intelligence antenna that covers the 1 to 18 GHz frequency band, features an integral radome and can be controlled at ranges of up to 1 km, using fibre optic control lines.

Specifications

Frequency coverage: 1-18 GHz (1-18 GHz (61 cm dish) or 2-18 GHz (31 cm dish) sub-bands)
Coverage: −9.9 to +90° (elevation); ±99° (polarisation); ±200° (azimuth - mechanical stop limits)
Control: via C-1830 antenna control unit
Dimensions (h × d): 1.1 × 0.9 m
Weight: 41 kg
Options: 18-40 GHz coverage with 31 cm dish; integrated preamplifiers; 18-40 GHz downconverters

CS-8056

The CS-8056 assembly is described as being a 'compact' DF antenna system that covers the 0.5 to 18 GHz frequency band, makes use of the AS-106 antenna and the C-8101 control/display unit and is designed for airborne ('small' aircraft), shipboard or ground-mobile applications.

Specifications

Frequency coverage: 0.5-18 GHz
Polarisation: circular (0.5-2 GHz sub-band); slant linear (2-18 GHz sub-band)
DF accuracy: 2° RMS (2-18 GHz); 5° RMS (0.5-2 GHz)
Beamshape: CSC² to 20° elevation
Control: via C-8101 antenna control and display unit
Options: integrated omnidirectional antenna; integrated preamplifiers; ruggedised pedestal; aerodynamic radome; 2° RMS low-band precision DF

Status

During the period May to August 2001, Condor Systems were promoting the antenna units described here.

Contractor

Condor Systems Inc, San Jose, California.

UPDATED

CS-2040 Mini Hawk receiving system

Type

Manual ELectronic INTelligence (ELINT) receiver system.

Description

The CS-2040 Mini Hawk receiver is based on Condor Systems' Hawk equipment. The CS-2040 system covers 0.5 to 18 GHz and features multiple superheterodyne timer/demodulator channels (TN-618 tuner/MD-128 demodulator) and a single operator control and display unit (C-244). The C-244 control unit features a 16 cm colour liquid crystal display with keyboard tuning knob and operator defined

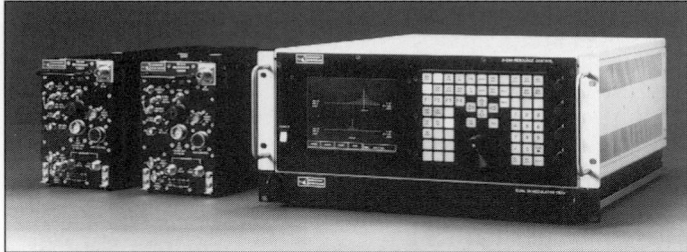

The CS-2040 Mini Hawk ELINT receiving system

parameter encoders. The TN-618 tuner covers the 0.5 to 18 GHz frequency band and provides an instantaneous Intermediate Frequency (IF) bandwidth of 500 MHz. The MD-128 is a full-function analysis demodulator with selectable IF filters, attenuators, LOG/LIN video outputs and built in IF pan/Linear Instantaneous Frequency Measurement (LIFM) display modules. Operational features include 40 GHz frequency extension and control/display of spinning direction-finding antennas. Radio frequency and intermediate frequency pan display modes are available.

Status
As of August 2001, the CS-2040 Mini-Hawk receiver system was understood to be available.

Contractor
Condor Systems Inc, San Jose, California.

VERIFIED

CS-5060 ELectronic INTelligence (ELINT)/Electronic Support (ES) system

Type
Automatic ELINT/ES system.

Description
CS-5060 is a fully automatic, instantaneous frequency measuring ELINT/ES system that acquires, identifies and locates non-communication signals in the 0.5 to 18 GHz frequency range (expandable to 40 GHz). The system is designed for use in stand-alone or multiplatform, networked configurations. Key features include:
- wideband intercept and processing providing high probability of intercept and modern modulation signal handling
- high-sensitivity acquisition, Direction-Finding (DF) and geolocation for extended range operation against emitter backlobes and low-power radars
- geolocation by means of line of bearing and time difference of arrival
- lightweight and low power
- ruggedised for use in airborne, mobile and shipborne applications
- 2° RMS direction-finding accuracy
- X-Windows/Motif graphical user interface.

Status
As of August 2001, the CS-5060 ELINT/ES system was understood to be available.

Specifications
Frequency: 0.5-18 GHz (0.5-40 GHz optional)
Sensitivity: −85 dBm
Max pulse density: 1,024 mpps
DF accuracy: 2° RMS

Contractor
Condor Systems Inc, San Jose, California.

VERIFIED

CS-5060 ELINT/ESM system

DF-18110 Direction-Finding (DF) system

Type
Wideband microwave and millimetre-wave DF system.

Description
The DF-18110 DF system is a wideband equipment which can detect incoming signals within the frequency range of 18 to 110 GHz. A transparent window with a 120° azimuthal span and a splash reflector are designed to receive the incoming signals without a rotary joint. The combination of five square pyramid horns and associated ortho-mode transducers provides the detection capability of omnipolarisation, low polarisation and full waveguide bandwidth signals over the equipment's entire frequency range. The DF-18110 antenna can be attached to the roof of a vehicle.

Status
As of January 2002, DF-18110 was understood to be a live programme.

Specifications
Frequency range: 18-110 GHz (0.1-18 GHz optional extension)
Antenna gains: 23-31 dBi average
Polarisation: omnipolarisation (H, V, LS, RS, LHCP, RHCP)
Weight: 40 kg approx
Dimensions (D × H): 686 × 889 mm

Contractor
Telestar (MM-Wave Technology), Ontario, California.

UPDATED

A general view of the DF-18110B DF system   0009352

A general view of the DF-18110B DF system showing the antenna radomes used   0009353

Digital Multimedia Watchdog recording/processing systems

Type
Digital recording and processing systems.

Description
Digital Multimedia Watchdog enables intelligence and law enforcement professionals to monitor, record and process all types of communications used by criminal and intelligence targets. Operating as a multichannel passive monitoring system, Watchdog automatically collects, digitises, stores and indexes all incoming voice, fax and modem data as well as supporting virtually any telephone based input including standard analogue, T1, E1 and ISDN. Data from sources such as cellular phones, pagers, receivers, microphones and video may also be input into the system. Watchdog's front end digital signal processing differentiates between voice and data signal types, allowing both to be processed from a

The Raytheon Systems' family of Digital Multimedia Watchdog recording and processing systems

common input source. Encoding techniques are automatically adapted to accommodate high-speed data signals so that maximum integrity is preserved. Watchdog demodulates fax and modem data (up to V.32 bis and V.34) and nearly all proprietary fax protocols.

Standard operator functions include: live monitor, digital record, playback; demodulation of fax/modem files; network printing and on-screen displays; report generation and database queries involving calls and other Signal Related Information (SRI). Input channel number; start and stop time of call; signal type; contact and caller identification and incoming or outgoing DTMF (digits dialled) are all included in the SRI. Information retrieval is managed through the database. Commercially based software simplifies transcription and report generation.

All Watchdog products are built around a flexible architecture using commercial off-the-shelf VME card sets and industry standard PCs and networks. This enables the products to be easily embedded into existing military intelligence systems. Various archive media are available which support high-capacity, low-cost storage.

Watchdog's Suite of Systems
Watchdog's suite includes Timberwolf, Coyote and Jackal. Each of these gives users complete functionality through the use of standard watchdog application software.

Timberwolf
Timberwolf is designed for large, centralised monitoring facilities and can support up to 200 simultaneous users and 1,000 inputs. Its open architecture easily supports new technology insertion, and its local/Wide Area Network (WAN) interconnection allows workstations, recording units and storage to be configured for today and grow as requirements evolve.

Coyote
Coyote is designed for small sites and quick response to crisis situations. It has the capacity to handle up to 60 input lines, 12 simultaneous users and up to 2,250 hours of online storage. When configured with the embedded database server, Coyote is ideal for store-and-forward and remote operations. Coyote may be set up and left unattended allowing administrative functions and audio collection via the WAN to support intelligence operations. Coyote operates stand alone and easily scales to the larger Timberwolf system.

Jackal
The portable Jackal is the smallest member of the Watchdog family. The system accepts up to 16 input lines or single T1/E1, four simultaneous users and up to 1,350 hours of online storage. As with Coyote, Jackal is designed for store-and-forward and remote intelligence operations via the WAN. Jackal offers full watchdog functionality through its embedded file server and database management system. Online audio and call detail storage is included in the system's convenient tower configuration.

Status
As of this edition, the status of the equipment types described was uncertain.

Specifications
Analogue interfaces: 600Ω or 15 kΩ balanced (selectable); switch contact closure; microphone 7 kHz bandwidth; audio 3.5 kHz bandwidth
Digital interfaces: T1, E1, ISDN; digital tape, AES/EBU & SPDIF
Communication interfaces: TCP/IP; Ethernet, FDDI, ATM; X.25; RS-232, SCSI-2; secure communications
Storage: magneto-optical disk, digital tape, optical disk; RAID; jukebox; hierarchical storage management
Physical: standard 48 cm rack mountable; tower configuration 43 × 27 × 66 cm; customised packaging options available; 110 or 220 V AC

Contractor
Raytheon Electronic Systems, Falls Church, Virginia.

VERIFIED

Digital receiver Intermediate Frequency (IF) processor cards

Description
The 6U VME digital receiver building blocks, when combined with the appropriate radio frequency converter, perform a variety of systems functions including search at speeds up to 20 GHz/s scan rate, recognition, direction-finding and collection (demodulation). Signal types include standard amplitude and frequency modulation/pulse modulation/continuous wave, cellular radio, pulsed signals and low probability of intercept signals. This card set provides a flexible, open architecture solution to many systems' needs using common hardware that is reconfigured using different software loads. Other product variations include higher packaging density (eight channels per CCA) and IEEE-1101.2 conduction cooled cards for severe airborne environments.

Contractor
Raytheon Electronic Systems, Falls Church, Virginia.

VERIFIED

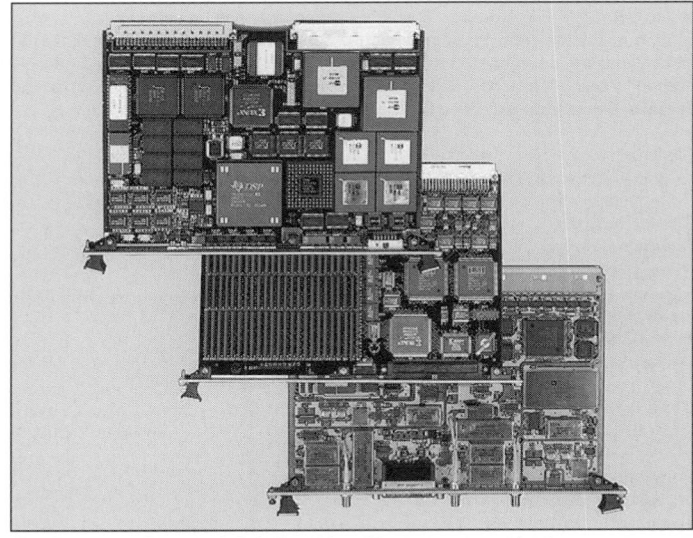

The Raytheon Systems' digital receiver IF processor card set

HH1307 Direction-Finding (DF) antenna

Type
Portable intercept and DF equipment.

Description
The HH1307 hand-held DF antenna is a wide frequency coverage, multiple feed disk radio direction-finding antenna designed for homing applications over the 2 to 520 MHz or 100 to 1,000 MHz frequency ranges. It is a self-contained unit requiring only a radio receiver with an amplitude modulation demodulation capability. A single carrying case will accommodate this compact, folding antenna, together with any small receiver.

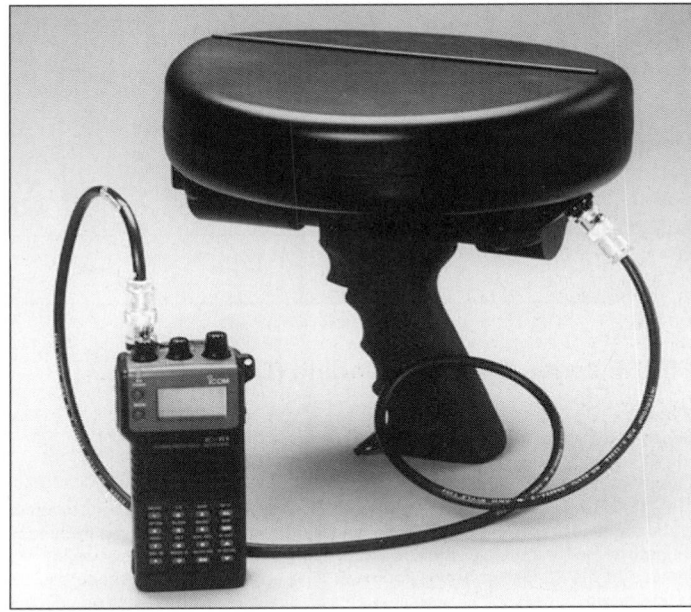

VERIFIED   *The hand-held HH1307 DF antenna*

The triple mode of this system allows the operator to monitor the target, home in (course mode) and pinpoint (fine mode) while on the move. Power is by a single 9 V alkaline battery which lasts approximately 30 days in the monitor mode, and six to eight hours in DF modes.

Contractor
Cubic Communications Inc/OAR DF Products, San Diego, California.

VERIFIED

MA 1320 Direction-Finding (DF) antenna

Type
Lightweight DF antenna.

Description
MA 1320 is a 0.5 to 1,300 MHz DF antenna array that is designed for use in mobile or man-portable DF systems. The array comprises a four-element loop with a vertical sense antenna combined with a four-element Adcock. Polarisation is vertical and the entire 15 × 30 × 30 cm system weighs 2.95 kg. Bearing accuracy is quoted as being typically 7° RMS. The MA 1320M variant differs from the baseline model in being supplied with a flux gate compass.

Status
As of this edition, the MA 1320 DF antenna was understood to be available.

Specifications
Frequency range: 0.5-1,300 MHz (2-1,000 MHz option)
Azimuth coverage: 360°
Antenna type: 4-element loop array with vertical sense antenna (HF/VHF subband); 4-element monopole Adcock array (UHF sub-band)
Bearing accuracy: 7° RMS (typical); 10° RMS (max)
Power: 11.5-20 V DC (supplied through DF processor, 250 mA for VHF, 310 mA for UHF)
DF sensitivity (typical, 15 kHz IF bandwidth): 2 µV/M (200 and 500 MHz); 3 µV/M (100 MHz); 4 µV/M (20 MHz); 6 µV/M (10 MHz); 10 µV/M (1,000 MHz); 12 µV/M (5 MHz); 30 µV/M (2 MHz); 100 µV/M (0.5 MHz)
Polarisation: vertical
Impedance: 50 Ω (nominal)
Temperature: −40 to +60°C (operating); −40 to +70°C (storage)
Humidity: 95% RH (MIL-STD-810D (507.2))
Dimensions (h × w × d): 15.3 × 30.5 × 30.5 cm
Weight: 3 kg

Contractor
Cubic Communications Inc/OAR DF Products, San Diego, California.

UPDATED

The MA1320 lightweight mobile DF antenna

MMDF series Direction-Finding (DF) systems

Type
Series of DF/surveillance systems.

Description
The MMDF series of direction-finding/surveillance systems is designed to detect and determine angles of arrival of incoming signals in the 0.1 to 40 GHz frequency range. The antenna systems provide a minimum 15° elevation plane beamwidth via the use of the EL-PAR reflector approach. The antenna size and shape can be reconfigured to meet specific requirements on dimensions, frequency range, gain and beamwidths.

Status
As of January 2002, MMDF series equipments were reported as having been procured.

Specifications

Model	MMDF0518	MMDF0118	MMDF0140
Frequency range	0.5-18 GHz	0.1-18 GHz	0.1-40 GHz
Gain	5-24 dBi	3-24 dBi	3-25 dBi
3dB beamwidth			
Azimuth	55-1.5°	70-1.5°	70-1.5°
Elevation	15°	15°	15°
Velocity	1-200 rpm	1-200 rpm	1-200 rpm
VSWR (average)	2.5:1	2.5:1	2.5:1

Contractor
Telestar (MM-Wave Technology), Ontario, California.

VERIFIED

Model 4400 Direction-Finding (DF) processor/receiver

Type
DF processor and receiver.

Description
The Model 4400 DF processor/receiver covers, dependent on antenna configuration, the 0.5 to 1,300 MHz frequency range and is designed for use in vehicular, airborne, fixed station and shipboard applications. The equipment features an embedded central processing unit; a bitmapped graphics display; a standard PC parallel printer port; a compatible keyboard interface and a VGA monitor port. The system also incorporates remote control and monitoring functions to allow for unattended remote operation.

Status
As of this edition, the Model 4400 DF receiver/processor was reported as being available.

Specifications
Frequency range: 0.5-1,300 MHz (DF); 0.1-2,036 MHz (monitor)
Bearing resolution: 1°
Azimuthal coverage: 360°
Receiver selectivity: 2.4 kHz at −6 dB, 4.5 kHz at −60 dB (SSB, CW)
Dimensions: 13 × 28 × 26 cm
Weight: 6 kg
Environmental range: 0 to +50°C

Contractor
Cubic Communications Inc/OAR DF Products, San Diego, California.

VERIFIED

The Model 4400 DF processor/receiver

Nanomin receivers

Type
Miniature communications band intercept receivers.

Description
Nanomin receivers (single- or dual-channel) combine a modular design, broad tuning range, small size, low power dissipation and light weight. They have the capability of intercepting signals over the frequency range of 20 to 1,500 MHz and demodulating Amplitude Modulation (AM), narrowband frequency modulation (FM), wideband FM and continuous wave signals. Multiple Intermediate Frequency (IF) bandwidths accommodate single or multiple channel voice and data, telemetry and pulse reception.

Single- and dual-channel Nanomin receivers

Flexibility to accommodate diverse system requirements is an integral design feature. Receiver control, via the serial databus, is a significant advantage in multiple receiver applications. Receiver audio, video and IF outputs support all types of signal analysis and exploitation from intelligence listing to fine grain spectral analysis. The single channel version is 52.9 mm (height) × 123.5 mm (width) × 113.4 mm (length), with a weight of 1.1 kg.

The dual-channel, differentially coherent version of the receiver is available for Direction-Finding (DF) applications, where the IF outputs are amplitude and phase matched to support all types of DF processing and analysis. Dimensions of the dual-channel version are 108.4 × 123.5 × 113.4 mm, with a weight of 2.1 kg.

The Nanomin product line has been expanded to include the following options:
- frequency extension to cover 0.5 to 1,500 MHz
- dual-channel with HF
- wideband IF output at 21.4 and 19.2 MHz to support 'Digital Receiver' applications
- fast tuning (<200 μs)

Other product variations include a 20 to 500 MHz search receiver capable of supporting first syllable detection search systems and a 150 to 4,000 MHz pulse receiver that presorts pulsed signals based on characteristics such as pulse repetition interval and pulsewidth.

Status
As of this edition, Nanomin receivers were reported as having been installed aboard a number of US and international airborne platforms.

Contractor
Raytheon Electronic Systems, Falls Church, Virginia.

VERIFIED

Northrop Grumman Corporation Emitter Feature Extractor (EFE)

Type
Radio Frequency (RF) emitter pulse train analyser.

Description
The Northrop Grumman Emitter Feature Extractor (EFE) sorts complex RF signal environments and measures and identifies emitters in that environment. EFE connects to the video output from any microwave (narrow or wideband) receiver system. The system supports electronic intelligence and electronic support measures signal processing; training for electronic warfare equipment operators and RF test verification for laboratory or range situations. EFE can be used as a fine grain analysis tool when connected to a receiver system or spectrum analyser.

The system provides Pulse Repetition Interval (PRI) measurement, pulse-width, amplitude, scan type, beamwidth, scan rate and emitter identification. Measurements include fine grain intra-frame intervals (up to 64 levels of stagger) and accommodates PRI and pulse-width agile threats. Measured parameters and pulse buffers can be recorded in a magnetic removable flash memory. Data are also plotted in many combinations for real-time parameter-to-parameter analysis.

The system's identification capability is designed to support positive identification even in the most ambiguous situations. The identification library can be developed on a PC and downloaded into EFE; manually entered from the keyboard, or can be based upon actual detections saved from previous encounters. EFE incorporates an advanced PRI de-interleaving algorithm that separates multiple complex pulse trains even in highly corrupted signal environments.

The 6U VME bus, two card set can be either integrated into a receiver system or used stand alone with control via a standard personal computer. The stand-alone EFE is in a ruggedised chassis suitable for land-, sea- or air-based platforms.

Status
As of September 2001, the Northrop Grumman EFE unit was reported as having been procured by the US Navy, the Royal Air Force and Royal Navy.

The EFE pulse train analyser

Specifications
Max pulse density: >0.5 Mpps
Processing capability: up to 500 emitters simultaneously
Power: 240 V AC
Dimensions: 470 × 254 × 191 mm
Weight: 14.5 kg

Contractor
Northrop Grumman Corporation, Electronic Systems, Buffalo, New York.

UPDATED

Prophet Electronic Attack (EA) and Electronic Support (ES) architecture

Type
Battlefield Electronic Attack (EA) and Electronic Support (ES) architecture.

Description
The US Army's Prophet battlefield EA/ES architecture replaces the service's now abandoned Intelligence and Electronic Warfare Common Sensor (IEWCS) programme. As such, the effort encompasses the Prophet Air, Prophet Control and Prophet Ground, the known details of which are as follows:

Prophet Air
As of February 2000, Prophet Air subsystem was described as being an unmanned aerial vehicle mounted EA/ES capability that was being developed as a replacement for the US Army's existing EH-60A Quickfix electronic warfare helicopter. As of the cited date, the system was being developed in two phases, with the 20 MHz to 2 GHz Block I EA/ES capability scheduled to enter service during US Fiscal Year (FY) 2006. While incorporating an EA capability, Block I Prophet Air's primary role is understood to be that of the detection, identification and location of conventional and low probability of intercept battlefield emitters. Prophet Air Block II envisages a system capable of detecting signals at frequencies below 20 MHz and above 2 GHz. While not confirmed, it is thought that Block II frequency coverage goal is 0.5 to 40 GHz and that the application could take the form of either a preplanned improvement to the Block I receiver system or a completely new design. As of the given date, Engineering and Manufacturing Development (EMD) of the Prophet Air Block II configuration was scheduled to begin during US FY2005.

Prophet Control
As of February 2000, Jane's sources were suggesting that a putative Prophet Control network control and data processing and dissemination tool was scheduled to undergo operational test and evaluation during US FY2007, with a production decision following in FY2008.

Prophet Ground
As with the Prophet Air subsystem, development of Prophet Ground is being undertaken in phases. As of mid 2000, Interim Prophet Ground Block I was being described as a 20 MHz to 2 GHz band ES capability that was to be developed for both man-pack and High-Mobility Multipurpose Wheeled Vehicle (HMMWV) - mounted applications. At this time, Jane's sources were describing the overall architecture as incorporating a Titan Systems Corporation Delfin Systems Division (formerly Delfin Systems) direction-finding package (made up of Titan's AN/PRD-13(V)2 equipment (see separate entry) and a 6 m high mast-mounted antenna assembly), two monitoring receivers, a SINCGARS communications radio, a Global Positioning System receiver/processor, an electrical power subsystem and a KVH Industries fibre-optic gyro-based tactical navigation system. Set up and tear down times of five and 'less than' three minutes were being quoted for the system, which, in its vehicle-mounted form, was to be single load transportable by a C-130 class aircraft. The man-pack application was being developed initially for the US Army's 82nd Airborne Division and as of February 2000, Interim Prophet Ground Block I was scheduled to enter full-scale production during US FY2001/2002 (see Status). As of the cited date, a Prophet Ground Block II variant was scheduled to enter production during US FY2003. As originally described, the Block II configuration adds an EA capability, tactical internet connectivity, an improved man/machine interface, an increase in frequency coverage and the ability to handle 'exotic' emitters such as frequency hoppers. According to Jane's sources, Interim Prophet

Ground Block I meets approximately 80 per cent of the overall Prophet ES requirement; provides direction-finding accuracies of 10° RMS when stationary and 22.5° RMS on the move and can demodulate amplitude and frequency-modulated, continuous wave and single sideband signals.

Status

On 2 July 1999, the then Delfin Systems was awarded a then year US$5,523,000 increment from an existing then year US$9,142,951 cost-plus, fixed-fee contract covering the integration and development of seven equipments in support of the Interim Prophet Ground Block I ES architecture. At the time of the increment's announcement, the effort was described as being primarily concerned with the design and implementation of a High-Mobility, Multipurpose Wheeled Vehicle (HMMWV) PRD-13(V)2 direction-finding (see previously) application, as being divided between production facilities at San Diego, California (65 per cent workshare) and Santa Clara, California (35 per cent) and as being scheduled for completion on 31 July 2000. The programme's contracting activity was the US Army's Communications and Electronics Command, Fort Monmouth, New Jersey. Elsewhere in the programme, as yet unconfirmed sources were suggesting that two Prophet Ground ES systems were delivered to Fort Polk, Louisiana during May 2000 for use in the Joint Contingency Force Advanced Warfighting Experiment (JCFAWE), with a further eight being allocated for evaluation by two brigade combat teams. May 2000 is also reported to have seen the start of developmental testing of the dismounted Prophet Ground ES configuration, with final developmental testing of the mounted variant being noted during the following June.

On 12 June 2001, the Titan Systems Corporation's Delfin Systems Division (Santa Clara, California) was awarded a then year figure of US$6 million as part of an estimated then year US$58,328,451 firm, fixed-price contract covering the procurement of 83 production standard Prophet Ground Block I systems (designated as the AN/MLQ-40(V)2) together with training, fielding and test support services. At the time of the announcement, six first article test vehicles were to be acquired during 2001, with the remaining 77 equipments to be procured by means of the exercising of annual options to purchase (based on proposed range quantities and associated pricing) up to a maximum of 83 systems. Work on the effort was to take place at facilities at Santa Clara, California (5 per cent workshare) and Melbourne, Florida (95 per cent) and at the time of the award's announcement, was scheduled for completion by 11 June 2008. The programme's contracting activity was the US Army's Communications and Electronics Command, Fort Monmouth, New Jersey. According to as yet (July 2001) unconfirmed sources, the described MLQ-40(V)2 deal involved Titan in installing its PRD-13(V)2 direction-finder aboard government furnished HMMWVs, with the first article units being scheduled to have been delivered by June 2002. Following acceptance, these equipments would be refurbished and issued to two US Army Brigade combat teams stationed at Fort Lewis, Washington State by September 2002. If carried through, full rate production of the 77 remaining systems would be delivered during the period May 2003 to 2005.

Contractor

Titan Systems Corporation - Delfin Systems Division, Santa Clara, California.

UPDATED

Raytheon Electronic Systems Direction-Finding (DF) system

Type
Modular DF system.

Description
Raytheon Electronic Systems' modular DF system is an emitter location equipment that features an open architecture VME-based design, and provides fast and accurate direction-finding for ground and airborne applications.

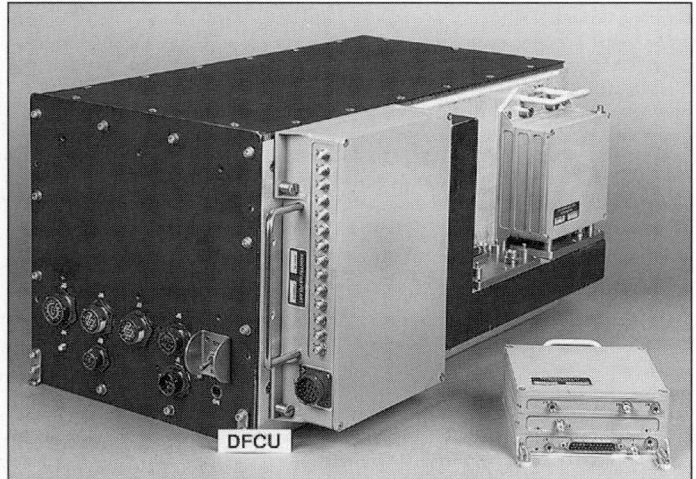

The Raytheon Electronic Systems' modular DF system

At the heart of the system is the DF control unit which uses Raytheon Systems dual 20 to 1,500 MHz Nanomin receiver front ends, which are modular receivers and which are claimed to be the smallest in the industry. Interfaced to these are A/D converters, digital decimating down converter and DF processing and interface boards, all residing on a standard VME bus. The VME-based architecture includes four slots for future expansion, and allows for easy upgrades to system capabilities.

The modular DF system design utilises the MUSIC super resolution DF algorithm and provides high performance and accurate line of bearing measurements, even in the presence of severe co-channel interference. A low-noise Nanomin-based receiving system ensures DF sensitivity to the weakest signals. Digital high-speed processing enables quick response time. Pulsed radar signals and low probability of intercept signals can also be handled with optional VME processing boards.

Specifications
Frequency range: 20-1,500 MHz
DF accuracy: <1° RMS for fine DF using typical airborne VHF/UHF array
DF sensitivity: −116 dBm (10 dB S/N in 6.4 kHz bandwidth)
DF output: line of bearing, quality factor, frequency, time, in/out of geosort limits, emitter location (calculated in workstation)
DF throughput: coarse DF >30 DF/s; fine DF>10 DF/s

Contractor
Raytheon Electronic Systems, Falls Church, Virginia.

VERIFIED

Raytheon Electronic Systems intelligence workstation

Type
Compact SIGnals INTelligence (SIGINT) workstation.

Description
Raytheon Electronic Systems' rugged, compact workstation is specially designed for ground and airborne applications. Its software was developed to support the control of SIGINT assets and the processing of signal intelligence. It supports the SUN (UNIX) operating system and is capable of operating a variety of UNIX applications in either stand-alone or in an Ethernet environment.

Ideally suited where space is limited, the compact workstation integrates a 6U VME SPARC central processing unit with a colour active matrix, 264 mm Liquid Crystal Display (LCD) and hard disk. The 1 Gbyte disk canister is ruggedised and readily removable for container storage of classified material.

The current Force CPU-2CE SPARC board provides 28 Mips of processing power and 16 to 32 Mbytes of random access memory, enough for the most memory intensive applications. A single-slot, Sbus LCD frame buffer controls the active matrix thin film transistor LCD display, offering 512 colours. SPARC station 2 input/output interfaces are available at the rear panel through circular MIL connectors, while configurability and expandability are maintained through standard off-the-shelf Sbus modules from third-party vendors.

Status
As of this edition, equipment of the type described was reported as having been introduced into service during 1994.

Contractor
Raytheon Electronic Systems, Falls Church, Virginia.

VERIFIED

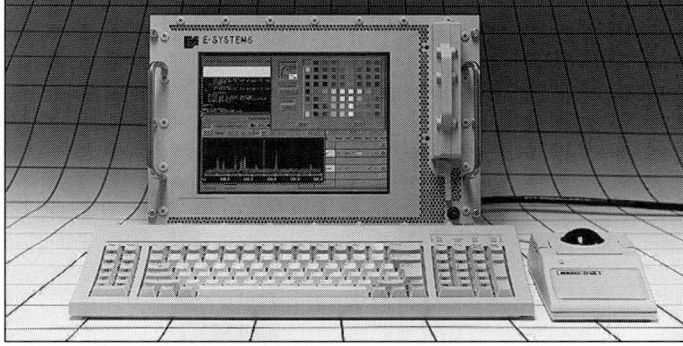

The Raytheon Electronic Systems' SIGINT workstation

SCP-2960 pulse analyser

Type
Pulse analysis system.

Description
The SCP-2960 pulse analyser is described as providing the 'wide bandwidth, digitising accuracy, processing speed, large memory capacity and interactive

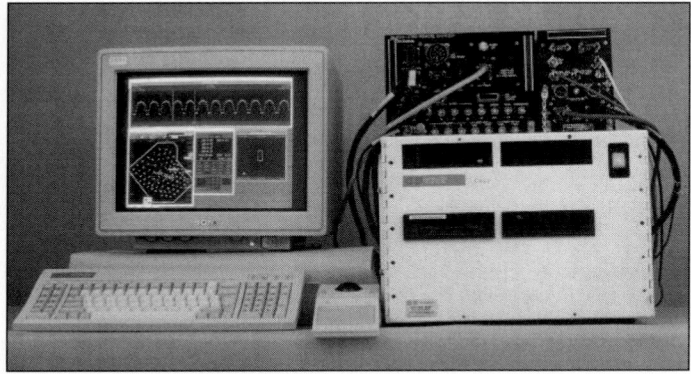

The SCP-2960's remote digitiser, operator control panel and display together with an external computer

display capability required to analyse complex, high-density electronic intelligence/ Electronic Support Measures (ESM) environments'. It is a third-generation, fine grain, real-time equipment that is based on the former Andrew SciCom SCP-2760 equipment (see following) and features a flexible architecture to facilitate firmware and hardware modification and upgrading. SCP-2960 incorporates three modular line-replaceable units namely, a remote digitiser unit, an operator control panel and a display unit. Designed to be 'highly flexible and user-friendly', the device makes use of soft-keys and menus to assist the operator with equipment set up and signal analysis. SCP-2960 can be automatically operated from an external system computer (through an IEEE-488 bus or 802.3 Ethernet local area network) and can function as either a pulse analyser (using the hardware sorting capabilities) or as a pulse digitiser for ESM applications.

Status
Over time, the SCP-2960 pulse analyser is noted as having been supplied to US and NATO customers. Before its acquisition by Condor Systems, SCP-2960 was an Andrew SciComm product.

Contractor
Condor Systems Inc, San Jose, California.

VERIFIED

SCR-7204 communications monitoring receivers

Type
Family of communications monitoring receivers.

Description
The baseline SCR-7204 family of receivers covers the 20 to 1,200 MHz frequency range (optionally extendable to 2,500 MHz), with up to six selectable Intermediate Frequency (IF) bandwidths. The receiver provides fine tuning resolution (less than 10 Hz when required) and a local oscillator with high spectral purity. The unit is constructed in a 9 cm high half-rack wide package and uses a computer-controlled 'soft-key' front panel that allows any new or optional features to be added as needed. SCR-7204 equipments provide operator or computer control of all receiver functions and can scan or be fixed tuned. When used in conjunction with the former Andrew SciComm's SCD-7304 IF spectrum display unit (see following), receivers of the type can provide digitally refreshed radio frequency panoramic displays as well as the usual IF panoramic formats. Up to 144 internal memory locations provide for automatic memory scanning and/or frequency lockouts. Automatic acquisition of signals, under automatic or manual control, is provided. Specific frequency ranges of the three basic SCR-7204 models are 20 to 500 MHz for the SCR-7204D; 500 to 1,200 MHz for the SCR-7204E and 20 to 1,200 MHz for the SCR-7204DE. Other identified variants include the SCR-7204AD, the SCR-7204AF, the SCR-7204FX, the SCR-7028 and the COM-LOG equipments. Of these, the 20 to 2,600 MHz SCR-7204FX model is described as featuring a 'robust' front end and is described as being suitable for dense environment applications including airborne operations and surveillance of large metropolitan areas. The 0.1 to 500 MHz SCR-7204AD model is designed to support remote, multichannel direction-finding applications (using an SCR-7204 series 'slave') while the SCR-7028 variant, which uses a full rack-width chassis to incorporate additional signal processing, provides data clock recovery and internal IF-to-tape conversion in addition to the normal SCR-7204 features. The 0.1 to 2,600 MHz COM-LOG communications monitoring and activity logging receiver teams an SCR-7204 series receiver with a Personal Computer (PC) and, so configured, can download specific search scenarios from the PC and initiate spectrum surveillance. Here, the emphasis is on detecting and logging time-tagged signals and provides a falling raster graphics display.

Status
The US Army is understood to have awarded the then Andrew SciComm a mid-1993 contract for the supply of 74 SCR-7204AF receivers for use in the service's TROJAN signals intelligence (?) programme. As of mid-2001, the status of

the SCR-7204 series was uncertain due to Andrew SciComm's acquisition by Condor Systems.

Specifications
SCR-7204AF
Frequency: 0.1-2,600 MHz
Coverage: 360° (azimuth)
Sensitivity: –120 dBm
Bandwidth: 12 selectable
Dimensions: 559 × 216 × 89 mm
Weight: 9.5 kg

Contractor
Condor Systems Inc, San Jose, California.

VERIFIED

SI-8614 Nanoceptor

Type
Subminiature general purpose surveillance receiver.

Description
The BAE Systems North America 762 × 1,270 × 279 mm SI-8614 Nanoceptor device is described as being a 'high performance', subminiature, general purpose surveillance receiver that consists of a tuner that covers the 20 to 3,000 MHz frequency range. The receiver incorporates low noise synthesisers (to facilitate a 'high' dynamic range), fine tuning resolution (100 Hz) and a combination of tracking preselectors and sub-octave filters to reject unwanted, out-of-band signals. Other system features include:

- six intermediate frequency bandwidths within the range 10 kHz to 25 MHz
- a 12 dB noise figure
- a less than 2.5 W power requirement
- amplitude and frequency modulation detection modes
- Windows-based control software (WJ-RCS) compatible.

Status
As of August 2001, the SI-8614 Nanoceptor general purpose surveillance receiver was understood to be available.

Contractor
BAE Systems North America - Aerospace Electronics, Gaithersburg, Maryland.

NEW ENTRY

SI-9111

Type
10 kHz to 31.4 MHz band tuner.

Description
The SI-9111 device is described as being a two-channel, 'direct' tuner with a frequency coverage of 10 kHz to 31.4 MHz in a single-slot 6U VME form factor. The equipment's 'direct' approach utilises wideband baseband circuitry to process the spectrum instead of 'traditional' frequency conversion. Here, the radio frequency input is first routed to a five-band preselector that divides the spectrum into five segments. These filters can be specified to include a selection of multi-octave and sub-octave elements. A low-pass and anti-alias filter is included to reject spurious responses due to the sampling process. This latter component is designed to provide a minimum of 100 dB of alias rejection for external Analogue/Digital (A/D) sampling at 80 Msps. The tuner provides 45 dB of front-end gain control in one dB steps. In addition, post amplifier gain control is provided for A/D converter optimisation. A dither circuit is included that can be turned on or off by software command. Other system features include:

- a 10 dB noise figure
- a +90 dB output at IP2 and a +45 dB one at IP3
- 64 dB radio frequency gain control
- a 'C' source code driver.

Status
As of August 2001, the SI-9111 tuner was understood to be available.

Contractor
BAE Systems North America - Aerospace Electronics, Gaithersburg, Maryland.

NEW ENTRY

SI-9137

Type
20 to 3,000 MHz band tuner.

Description
The SI-9137 is described as being a dual-channel, wideband, tuner converter that is packaged in a single-slot 6U VME (160 × 233 mm) form factor. Each tuner channel independently provides conversion of 20 to 3,000 MHz band radio frequency signals to two intermediate frequency outputs centred on 16.25 MHz and 70 MHz, with an instantaneous bandwidth of 25 MHz. A high-stability, 10 MHz reference oscillator (with front-mounted SMA outputs) provides precise frequency control and each of the tuner channels is 'completely' shielded from the electromagnetic/radio frequency interference environment within the VME chassis. An onboard microprocessor provides the control interface between the VME backplane and the various digital signals to be tuned, sets the attenuation level and monitors the status of the tuner module. Other system features include:
- a 25 MHz intermediate frequency bandwidth
- VME bus register-based control
- output passband group delay within ±70 ns
- output passband flatness within ±1 dB
- gain adjustment in 1 dB steps from 20 to 50 dB
- provision of a 'C' source code driver.

Status
As of August 2001, the SI-9137 tuner was understood to be available.

Contractor
BAE Systems North America - Aerospace Electronics, Gaithersburg, Maryland.

NEW ENTRY

SI-9250

Type
0.5 to 18 GHz band tuner.

Description
Designed to extend the frequency coverage of 'any' Very/Ultra High Freqency (VHF/UHF - 30 MHz to 3 GHz) receiver system, the SI-9250 tuner is described as being a single-slot, 6U, VME equipment that accepts 0.5 to 18 GHz input signals. Signals below 2 GHz are routed directly to the Intermediate Frequency (IF) output, while those above the 2 GHz threshold are filtered with sub-octave preselection before being converted to an output IF centred at 1.6 MHz. Other system features include:
- an 880 MHz IF bandwidth
- compatibility with 'most' V/UHF receivers
- the ability to be configured for multichannel, phase-coherent operation
- Windows-based control software (WJ-RCS) compatibility.

Status
As of August 2001, the SI-9250 tuner was understood to be available.

Contractor
BAE Systems North America - Aerospace Electronics, Gaithersburg, Maryland.

NEW ENTRY

SI-9460

Type
Programmable digital signal processor.

Description
The SI-9460 is described as being a 'complete', self-contained, programmable, Digital Signal Processor (DSP) that incorporates an 8 million signals/s digitiser, a programmable digital downconverter and a 32-bit, floating point TMS320C6701 processor. As such, it provides audio and video reconstruction and digital data outputs and can be 'tightly coupled' to BAE Systems Miniceptor family of receivers (SI-9460 is also noted as being capable of integration with 'other' receiver types). Other system features include:
- the ability to act as a programmable DSP platform for user's custom algorithms such as code for spectral analysis, private voice systems, digital modulation, modulation recognition and double demodulation
- integral 'standard' surveillance receiver functionality
- 22 bandwidths within the 200 Hz to 1.23 MHz range
- digital data outputs (audio, custom algorithm data, I/Q, magnitude/phase and video)
- support for a 21.4 MHz input.

Status
As of August 2001, the SI-9460 digital signal processor was understood to be available.

Contractor
BAE Systems North America - Aerospace Electronics, Gaithersburg, Maryland.

NEW ENTRY

SP-103

Type
Automated signal processor.

Description
The Condor Systems SP-103 wideband signal processor is described as being an automated equipment that provides real-time analysis and identification of radar signals for a 'wide range' of tri-service Electronic Support (ES) and ELectronic INTelligence (ELINT) applications. When used with the company's receivers and antennas, SP-103 also functions as a system controller. Here, the equipment controls the tuner (that is executing a frequency search scenario) and a rotating Direction-Finding (DF) antenna (that provides Angle Of Arrival (AOA) measurements). An operator interface software package (operating on a Unix/X-Windows based workstation) provides all the necessary displays and controls needed to operate the device.

SP-103 can be integrated with new or existing superheterodyne receivers and accepts a 1 GHz (500 MHz bandwidth) Intermediate Frequency (IF) input. Its automated processing output consists of parametric emitter characterisation reports that can be displayed to the operator and/or transmitted to an external user. Here, SP-103 'precisely' measures and characterises emitter frequency, pulse repetition interval and agility, pulse-width and amplitude, time-first-seen, time-last-seen, scan period and scan type. The processor also associates the measured and characterised parameters of the emitter with a customer-furnished internal library of known radars to determine emitter identity.

When used in an ES/ELINT system with a rotating DF capability, SP-103 performs DF processing using 'advanced' automatic AOA measurement software. To support detailed emitter analysis and/or emitter identification ambiguity resolution, SP-103 provides its operator with 'complete, interactive analysis capabilities'. Here, the equipment provides snapshot collection and real-time displays 'without the need for dedicated display and control units'. Other system features include:
- 'precision' frequency and time of arrival measurement for pulsed and continuous wave emitters
- manual ELINT analysis display software tools
- receiver control functions that support set on and search receiver tuning
- an optional line of bearing geolocation measurement capability for single moving platform or multiple netted systems
- automated built-in test and diagnostics.

Status
As of May 2001, Condor Systems was actively promoting the SP-103 signal processor.

Specifications
IF input
Centre frequency: 1 GHz
Bandwidth: 500 MHz
IF processing bandwidth: 20, 50, 100, 250 or 500 MHz (selectable)
Video bandwidth: 5 or 20 MHz (selectable)
Max input burst pulse rate: 2 Mpps
Signal characterisation/pulse measurement
Frequency resolution: 125 kHz
Frequency accuracy: 1 MHz RMS
FMOP recognition: flag
PRF range: 100-500,000 pps
PRI range: 2 μs-10 ms
PRI types recognised/characterised: constant, continuous wave, jitter (±10%), pulse groups and stagger (up to 32 levels)
Pulse-width
Resolution: 40 ns
Range: 50 ns-2.62 ms
Accuracy: 40 ns RMS
Amplitude
Range: 55 dB
Accuracy: 2 dB RMS
Scan
Rate range: 0.05-100 Hz
Accuracy: 2% RMS
Types: bidirectional, circular, complex, conical, sector and steady
General
Active emitters: up to 1,024
Emitter library size: 10,000
Built-in test signal: IF pulse generator
Temperature: −30 to +55°C (operating); −51 to +71°C (storage)
Power consumption: 250 W
Operating voltage: 115/230 V AC (47-440 Hz)
Standard interface: IEEE 402.3 Ethernet
Optional interfaces: blanking, GPS (1 pps), MIL-STD-1553B, RS-232/-433 and synchronised reference

Contractor
BAE Systems North America - Aerospace Electronics, Gaithersburg, Maryland.

NEW ENTRY

Dimensions (h × w × d): 217 × 260 × 458 (519 with chassis extension) mm
Weight: 16.3 kg

Contractor
Condor Systems Inc, San Jose, California.

NEW ENTRY

SP-2060 pulse processor

Type
Microwave frequency precision pulse analyser.

Description
SP-2060 measures radar signal parameters, providing fast, accurate results. Operation is automatic and gives the operator a display of signal activity with identification of specific radar types. A continuous log of activity is stored in memory for post-mission analysis. Requiring only a single video input, it is compatible with any type of radar receiver. Two models of control/display units are available. The C-1500 control/display unit has a 23 cm display and fits into a standard 48 cm rack. The C-1400 control/display with a 13 cm display is available for aircraft installations or where space and weight are limited. Both models have built-in floppy disk drives.

Status
SP-2060 is an upgraded version of AN/ULQ-16(V) (see separate entry) and, as of mid-2001, was understood to be available and as having been selected by the US Navy for installation aboard maritime patrol aircraft.

Specifications
Weight: 17 kg
Dimensions (h × w × d): 27 × 27 × 47 cm

Contractor
Condor Systems Inc, San Jose, California.

VERIFIED

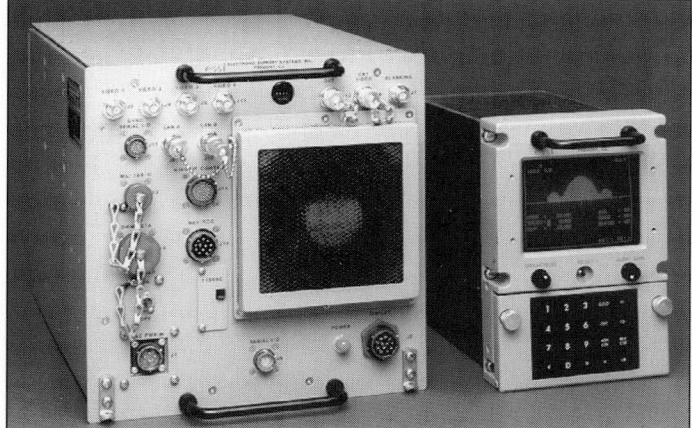

The SP-2060 pulse processor with associated control/display unit

Tactical Electronic Support (ES) system modules

Type
Family of modules for the construction of tactical ES systems.

Description
Raytheon Electronic Systems' family of ES modules is designed to facilitate fast delivery of customer systems which are tailored to meet specific requirements. Each module within the range incorporates flexible signal recognition (based on modulation type, data rate, signal content and structure) and direction-finding (using correlated interferometry) capabilities. Currently available modules incude:
10 MHz wide digital staring channels
These staring channels (each of which has a 10 MHz spectrum) are designed to provide high discernability of Signals Of Interest (SOI) in a dense environment. Each 10 MHz spectrum is delayed to allow for effective signal recognition and Direction-Finding (DF). Specifically, the capability analyses the radio frequency spectrum, assigns recognition resources to candidate SOIs and identifies up to approximately 500 signal types by modulation characteristics and internal data pattern.
High-speed acquisition scanner
In the high-speed acquisition scanner architecture, a single 10 MHz acquisition receiver is time-shared over a wide Radio Frequency (RF) range. High-speed tuning and Fast Fourier Transforms allow the receiver to scan 1 GHz of RF with revisit

A typical tactical ES system constructed from Raytheon Electronic Systems' family of ES modules

intervals of less than 25 ms. Other system features include customised spectrum analysis for each scanned RF range; use of a constant false alarm approach to speed detection decisions; SOI prioritisation and a DF calculation for each assigned signal.
Narrowband search and monitoring
Narrowband search is available for the HF (3 to 30 MHz) and VHF/UHF (30 to 1,000 MHz) frequency bands. Features of the architecture include:
- Directed search mode with automatic frequency revisit based on operator selection with optimised bandwidth
- General search mode using a step search over a designated range using selectable bandwidths with automatic or operator-cued assignment of recognition/collection resources
- Recognition/collection/DF mode using the resources available to the described delayed channel and scanning architectures
- Manual receiver mode for operator monitoring and optional panoramic spectral displays.

Contractor
Raytheon Electronic Systems, Falls Church, Virginia.

VERIFIED

TC-5100 series Direction-Finding (DF) systems

Type
Radio communications monitoring and DF systems.

Description
The TC-5100 series of direction-finders has been developed to conform with the US Army requirements for the 'AirLand Battle 2000' plan. The equipment provides monitoring, intercept and bearing information of signals over the 0.01 MHz to 2.6 GHz frequency range. Information is displayed as a three-digit light emitting diode display and a circular compass point.
The system employs a small broadband slot antenna with a shape and size that allows for rapid deployment and redeployment without the need to assemble and reassemble any of the elements. The antenna is designed to be mast mounted on a vehicle and lowered during movement to a height of only a few centimetres above the vehicle or shelter roof-line. In certain types of shelter, the antenna can be lowered below the roof-line and still provide bearing information while the vehicle is

The TC-5134-1A 0.5-1,300 MHz DF antenna mounted on the roof of a small van

moving. The antennas are small discs, varying in diameter from 56 to 86 cm and measuring less than 15 cm in height. They can be easily mounted or concealed on a variety of mobile ground, shipborne or airborne platforms. The single 0.5-1,300 MHz TC-5134-1A DF antenna can be employed in mobile roles. The antenna requires no assembly when deployed and provides DF during movement.

Status
As of this edition, the status of the TC-5100 series was uncertain.

Specifications
Frequency range: 0.1 MHz-2.6 GHz (special designs from 10 kHz to microwave are available)
Processing accuracy: ± 1°
Processing resolution: 0.1° (1° displayed)

Contractor
Tech Comm Inc, Coral Springs, Florida.

UPDATED

TCI/BR Model 9050 signal classifier

Type
HF/VHF/UHF-band (3 MHz to 1 GHz) digital signal classifier.

Description
The Model 9050 digital signal classifier measures signal modulations and modulation parameters of communications signals in the HF, VHF and UHF frequency bands. Its digital signal processing-based components and specialised algorithms are designed to achieve fast, reliable and fully automatic operation. The equipment accepts Intermediate Frequency (IF) inputs at different IF frequencies/bandwidths generated by customer-supplied or optional TCI furnished receivers. System features include a minimum signal duration of 80 to 320 ms; a frequency resolution of 3 to 12 Hz; 200 kHz, 455 kHz, 1.4 MHz and 21.4 MHz IF input frequencies; selectable - 10 or 0 dBm IF input levels and a greater than 70 dB instantaneous dynamic range. The unit can be operated in stand-alone mode from its local control panel, remotely from an optional tasking terminal or as an integrated part of a multinode network. Available with or without optional TCI receivers, Model 9050 can be supplied as a 20 kg, 22 × 48 × 51 cm rack-mountable unit or installed in a transit case for transportable applications.

Status
As of this edition, the status of the Model 9050 signal classifier was uncertain.

Contractor
TCI/BR (formerly Technology for Communications International), Sunnyvale, California.

UPDATED

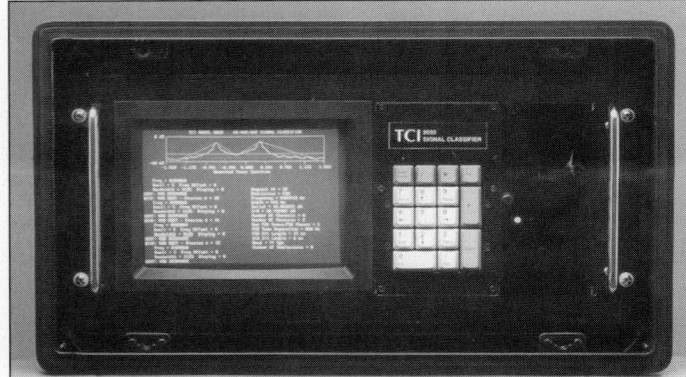

The TCI/BR Model 9050 digital signal classifier

TCI Model 802 High Frequency (HF) Direction-Finding (DF)/Single Station Location (SSL) systems

Type
Family of 0.3 to 30 MHz DF/SSL systems.

Description
The TCI Model 802 series of HF DF/SSL systems cover the 0.3 to 30 MHz frequency range with all system configurations based on the TCI 8060 digital DF processor. A specific system configuration within the family is designated as the Model 802-N-n where 'N' is the number of DF receiver channels and 'n' is the number of elements of the DF antenna array. This latter item is selected on the basis of the particular signal intercept/frequency coverage/deployment

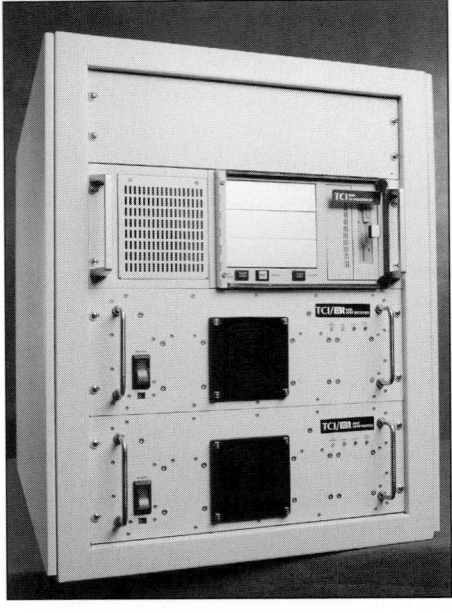

The TCI Model 802 HF DF/SSL system
0044258

requirement and can take the form of Monopole (whip) and crossed-loop elements arranged in linear or circular arrays. The TCI 8060 processor can also accommodate complex circular arrays such as the FLR-9, Pusher and TCI 410.

The TCI 8060 processor permits multichannel DF receiver operation without the need for perfect receiver matching. This is achieved through a technique of calibration and software-based compensation for Radio Frequency (RF) path mis-match. When using one DF receiver channel per antenna array element (Nn), no RF switching is necessary and a major source of DF errors is claimed to be eliminated. The TCI 8400 HF receiver has been developed specifically for 802 systems but a wide range of good quality receivers can be used.

The TCI 8060 digital DF processor uses a Personal Computer (PC) as the basic computer platform. A modified interferometry technique is used for simple arrays, with Wave Front Analysis (WFA) being employed for more complex arrays. Fast Fourier Transform processing (using modern digital signal processing techniques) and components replaces traditional I-Q demodulators. Signals are processed with a single intermediate frequency filter (3 kHz bandwidth). The operator may then isolate a signal affected by co-channel interference by reducing bandwidth with software-controlled function keys. A technique of high-speed data sampling (cuts) and cluster analysis is used. The operator can adjust the DF dwell-time, or a number of successful cuts, to handle low duty cycle signals, such as Morse or single sideband.

All Model 802 systems have SSL capabilities as both the azimuth and elevation angles of arrival are measured. The TCI 820 vertical ionosphere sounder (see separate entry) is used to obtain real-time ionospheric height data necessary for SSL range calculations. Installations for fixed and mobile stations are available and system options include:
- delayed DF (frequency and time filtering) to resolve co-channel interference
- super resolution for enhanced separation of multiple, closely spaced signals
- selectable sequential DF processing or interleaving (providing simultaneous service to multiple users via time sharing).

Model 802 systems are fully automatic and are controlled remotely. A net of DF stations can provide emitter location results by triangulation or SSL directly to a signal surveillance and monitoring system. Operational displays (with colour VGA) include:
- polar (with azimuth and elevation angles of arrival)
- histograms
- tabular DF/SSL summaries
- signal spectrum
- digital map.

Status
Model 802 system production is reported to have begun during the early 1990s and as of this edition, several configurations are understood to have been procured. Identified system configurations comprise:
Model 802-2-9
Two-channel DF receiver operating with a nine-whip element antenna array. Forms part of AN/TRD-27 Trackwolf equipment.
Model 802-9-9
Fixed-site configuration incorporating the Model 820 sounder, a nine-whip element antenna array and a nine-channel DF receiver.
Model 802-24-24
Fixed-site configuration incorporating the Model 820 sounder, a 24-monopole circular antenna array and a 24-channel DF receiver section (under TCI 8060 control)
As of November 2000, Jane's sources were suggesting that the Model 802 programme was live.

Specifications
Frequency range: 0.3-30 MHz (1.5-30 MHz standard)
Modulation rates: all
DF bandwidth: 100 Hz-3 kHz (100 Hz increments)

Signal duration: 50 ms (min, 24-element array, DAA = 100 ms)
Data collection speed: 10 m s/cut
DF accuracy: 0.1° RMS (instrumented); 1.6° RMS (standard frequency range, antenna dependent)
Sensitivity: 3 kHz (BW); 10 dB (SNR)
DF integration time: 1-10 s (selectable)

TCI 8400 receiver
Frequency range: 0.3-30 MHz
Noise figure: 14 dB
Tuning: 1 ms (max, 1 Hz increments)
Dynamic range: 120 dB
IF rejection: 80 dB
Image rejection: 80 dB
Intermodulation distortion: + 30 dBm (IP3); + 75 dBm (IP2)

Contractor
TCI/BR (formerly Technology for Communications International), Sunnyvale, California.

UPDATED

TCI Model 820 vertical ionospheric sounder

Type
Automatic vertical ionospheric sounder system.

Description
As part of a high-frequency Direction-Finding (DF)/Single Site Location (SSL) system, the fully automatic Model 820 vertical ionospheric sounder provides real-time ionospheric measurements that enable the DF system to calculate target emitter location without the need for multisite triangulation. The equipment comprises a vertical incidence Model 9046 Chirpsounder® transceiver, an 8020 DSP processor (with a built-in VGA display and keyboard) and a cross-delta antenna. The system can operate in a stand-alone mode or be controlled via data/voice interfaces (landlines or radio) which link it to a remote terminal such as those used in the TCI Model 802 DF/SSL system.

Functionally, the sounder sweeps the required frequency range, takes measurements and uses proprietary algorithms to calculate ionospheric height and frequency data for use in SSL DF calculations. Whilst designed to interface directly with TCI's Model 802 DF/SSL system, the Model 820 can be easily modified for use with other DF/SSL equipments. It is available in a standard equipment rack format for fixed site installations or a transit case for tactical deployments.

Status
As of November 2000, Jane's sources were suggesting that the Model 820 sounder programme was live and as having been a production item since the early 1990s.

Contractor
TCI/BR (formerly Technology for Communications International), Sunnyvale, California.

UPDATED

The TCI Model 820 ionospheric sounder shown in its transit case configuration but without its display and keyboard

TCI Model 8000 series workstation equipments

Type
Family of Direction-Finding (DF) and monitoring system workstation equipments.

Description
The Model 8000 series workstation equipments are designed for the rapid and accurate remote/local co-ordination of multisite communications band DF and monitoring systems. Such equipments utilise off-the-shelf components with hardware in a typical series server/workstation including an Intel Pentium central processing unit; a 2.5 Gbyte hard drive; a 1.44 Mbyte floppy drive; a compact disc - read only memory; 8 RS-232 I/F ports; a colour graphic display and a keyboard. The standard operating system is Windows NT with the option of implementing customer specified software if required. Details of the series members are as follows:

Signal collection workstation (Model 8070)
The model 8070 signal collection workstation is designed for use with TCI's signals intelligence collection, DF/Single Site Location (SSL) and reporting systems. It shares the same architecture as (and is compatible with) the company's Model 8013 DF/SSL operator and Model 8014 collection supervisor workstations. The equipment consists of a PC/AT processor, a digital signal processing board and up to four High Frequency receivers. The Model 8070 runs on a Windows NT operating system.

The PC/AT employed includes a VGA monitor and an Ethernet Local Area Network (LAN) interface board. The monitoring receivers used are miniaturised and are equipped with digital demodulators (Model 8014). The signals from two of these receivers can be displayed simultaneously. Multiple tasking is facilitated by a screen design that employs separate screens with multiple windows. The standard collection screen has windows for Signal Display and DF Request (SD/DFR),

The PC/AT configuration used in the Model 8021 and 8070 signal collection position

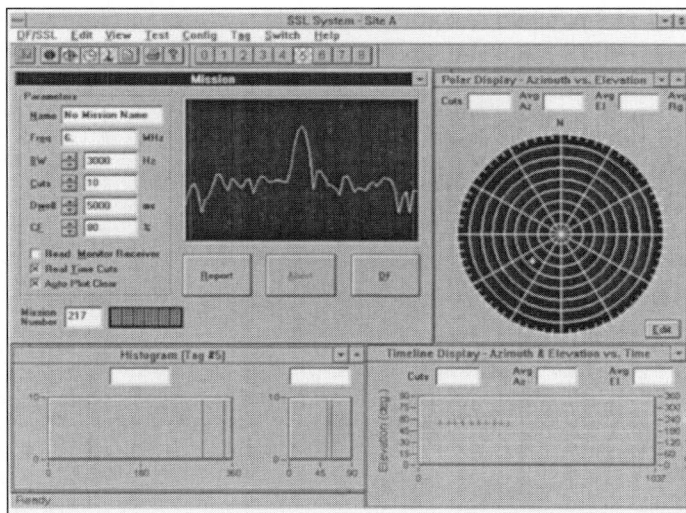

A typical display generated by the Model 8013 HF/DF workstation

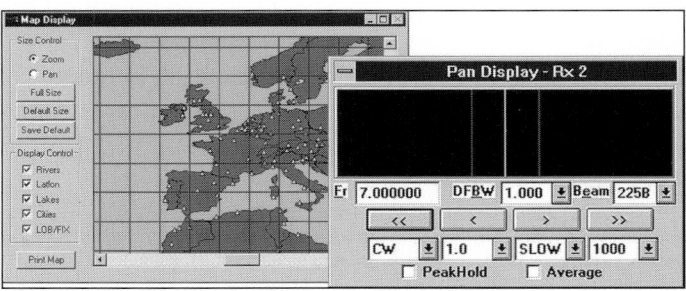

Two of the display formats available on the Model 8070 signal collection workstation

0052057

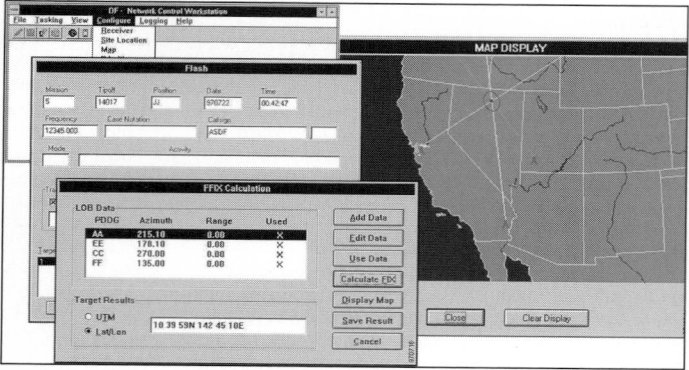

Display formats generated by the Model 8015 command centre workstation

0052058

Collection/Results (CR) and Status (S). Of these, the SD/DFR window displays the 6 kHz passband of two receivers with colour-coded bandwidth markers to help the operator correctly tune the receiver to initiate an automatic DF request. This window also allows the initiation of operator-selected DF and emitter location requests. The CR window facilitates control of up to 4 receivers, the creation of collection files (including header, operator entries, radio-printer collection and DF results) and can be customised to meet a user's specific collection requirements. The S window displays the real-time operational status of assigned LAN nodes, priority levels, the identification of the particular collection position and the local time.

The Model 8070 equipment is available in transportable, mobile or fixed station applications. The PC/AT used is also available in a number of configurations that include rack-mounted, ruggedised commercial, laptop and Mil-Spec standard.

High Frequency (HF) DF operator workstation (Model 8013)

The Model 8013 HF band (3 to 30 MHz) DF workstation is a modular design which incorporates off-the-shelf components and is designed for use in a package made up of TCI proprietary software, antennas and the 1.5 to 30 MHz Model 8174 digital receiver. Functionally, the equipment can automatically scan up to 20 frequencies, initiate Line Of Bearing (LOB) and single site location tasks and hand-off results to DF equipment such as TCI's Model 800 series systems. For local DF requests, LOBs can be displayed on polar plots and azimuth and elevation histograms. The Model 8013 can also assist the Model 8021 and 8074 HF band monitoring positions in the location of difficult signals.

Command centre workstation (Model 8015)

The Model 8015 command centre workstation is designed to control and co-ordinate DF functions at all stations in a multisite DF network using Ethernet connection. System functions include DF operation tasking; the direction of data collection; the reception of data from remote sites and its analysis to determine emitter locations and the production, storage, display and distribution of fixed result reports generated throughout the network. The Model 8015 features a geographic map system against which DF results can be displayed; built-in Check Target analysis functions (for DF site performance tracking) and a relational database management system for the storage of results. Various centre to remote station communications links can be employed including cellular and landline telephone and modem.

Communications server (Model 8011)

The Model 8011 communications server is designed to provide flexible and high-speed data communication between remote DF stations and network access to all DF assets within the system. It supports industry-standard protocols (TCP/IP, IPX/SPX and NETBEU); up to 4 Ethernet cards; up to 32 RS-232/422 serial connections and wireless connections using HF band and cellular technology. With optional routing features, the equipment can also provide wide area network and LAN access for all system workstations. Model 8011 can be configured for a rack-mounted PC/AT chassis.

Status

As of this edition, the status of equipments of the types described was uncertain.

Contractor

TCI/BR (formerly Technology for Communications International), Sunnyvale, California.

UPDATED

VISTA surveillance and target acquisition system

Type

Transportable surveillance and target acquisition equipment.

Description

VISTA (Visual Interactive Surveillance and Target Acquisition) is a small and compact transportable system that provides the field commander with computer-assisted reception and analysis of communications band signals traffic, with optional direction-finding. The system can be rapidly deployed, signals of interest catalogued in a minimum of time and information transferred by telephone lines, UHF, microwave link or satellite link. Designed for mobility, flexibility and

VISTA surveillance and target acquisition system

0009354

expandability, VISTA covers the frequency ranges 10 kHz to 30 MHz and 20 to 2,400 MHz.

The VISTA system features:

- the system software uses an application of Microsoft Windows 95, which allows use of interactive windows for system control. Advanced radio frequency monitoring enables signals to be acquired from a visual search of a wideband spectrum display, or from receiver sweep and scan modes
- the system can be netted for database transfers during set up, hand-off of signals of interest, and co-ordinated operations in the direction-finding mode
- VISTA receivers can also be remoted from the system controller over telephone lines, radio frequency modems or via UHF, microwave or satellite
- a variety of receivers can be selected depending on customer requirements
- system architecture is expandable with the addition of receivers, remote stations, operator positions, and more sophisticated software.

Status

As of this edition, the VISTA surveillance and target acquisition system was reported as having been procured.

Specifications

Frequency range: 10 kHz-30 MHz; 20-2,400 MHz
Power in Operating modes: AM, FM, CW, USB, LSB, ISB
Programmable channels: 250
Sweep/scan rate: up to 100/s
Tuning time: 3 ms typical

Contractor

Cubic Communications Inc, San Diego, California.

VERIFIED

WJ-860X Miniceptor™ surveillance receivers

Type

Family of miniature surveillance receivers.

Description

BAE Systems WJ-860X series of miniature surveillance receivers are designed for applications where high performance with a low power requirement is desirable. BAE Systems further claims that the series' small size (46 × 17 × 27 cm), fast tuning speed and high-level interfaces makes it 'ideal' for multiple receiver systems. As of this edition, the WJ-860X series comprises the WJ-8604, WJ-8604A, WJ-8607, WJ-8607A/B/C and WJ-8609 units. System features include:

- 20 to 3,000 MHz tuning range (with extensions as options)
- A worst case 300 µs tuning speed
- Five selectable intermediate frequency bandwidths within the 3.2 kHz to 40 MHz range (3.2 kHz to12 MHz for the WJ-8604A and WJ-8607A/B/C units; 0.5 MHz to 40 MHz for the WJ-8609A)
- A tuneable baseband output with up to 4 MHz of bandwidth
- 'smart' sweep and step algorithms
- a high-speed serial interface (RS-232, -422 or -485) that supports data rates of up 230.4 kbaud (WJ-8604A and WJ-8607A/B/C units)
- low phase noise
- less than 19 W power requirement
- Windows-based control software available as an option.

The WJ-8607A Miniceptor communications band intercept receiver

Status

As of August 2001, the described range of WJ-860X miniature surveillance receivers was understood to be available.

Specifications

WJ-8607A
Frequency: 20-512 MHz
Tuning resolution: 100 Hz (synthesised)
Max RF input: +20 dBm (without damage)
Noise figure: 8 dB (5-512 MHz, preselector off); 12 dB (20-512 MHz, preselector on); 15 dB (with frequency extender)
Receiver tuning speed: 300 μs max (from receipt of last data byte of binary frequency message to within 10 kHz of final frequency)
Sweep tuning speed: 200 μs (25 kHz steps)
Detection modes: AM, CW, FM, Pulse standard, SSB (option)
Power requirement: 12 V DC
Operating temperature: −25 to +55°C
Options: WJ-8607/IFBW IF bandwidth unit; WJ-8607/SSB single sideband unit; WJ-8607/WBO wideband output unit; WJ-8607A/FE frequency extender (512-2,000 MHz); WJ-8607A/HPIL Hewlett Packard interface loop; WJ-8607A/DSO digital scan output unit
Weight: 2.48 kg (standard); 3.18 kg (with frequency extender)
Dimensions: 38.1 × 165.1 × 266.7 mm (standard); 38.1 × 165.1 × 339 mm (with frequency extender)

Contractor

BAE Systems North America - Aerospace Electronics, Gaithersburg, Maryland.

VERIFIED

WJ-8611 surveillance receiver

Type

Communications band surveillance receiver.

Description

WJ-8611 is a half-rack, fully synthesised, general purpose, digital surveillance receiver which covers the 2 to 1,000 MHz frequency range in 10 Hz steps. Key features include a digital Intermediate Frequency (IF) section (15 IF filters from 200 Hz to 200 kHz); amplitude and frequency modulated, continuous wave, single sideband and independent sideband detection modes; low phase noise; high linear radio frequency performance; built-in tracking preselection; 200 channel memory scan and RS-232C and IEEE-488.2 interfaces for remote control.

The WJ-8611 general purpose digital surveillance receiver

Status

As of August 2001, the WJ-8611 surveillance receiver was understood to be available.

Specifications

Frequency: 2-1,000 MHz
Tuning resolution: 10 Hz
Noise figure: 12 dB (max)
Preselection: 20% BW (with tracking filter)
Detection modes: AM, CW, FM (all BWs); LSB, USB (3.2 kHz BW); ISB (6.4 kHz BW)
Power requirement: 90-264 V AC, 48-440 Hz
Operating temperature: 0 to +50°C
Dimensions: 133.4 × 209.6 × 133.4 mm (excl control knobs/connectors)
Weight: 6.8 kg

Contractor

BAE Systems North America - Aerospace Electronics, Gaithersburg, Maryland.

VERIFIED

WJ-8615P general purpose receiver

Type

Compact general purpose receiver.

Description

The 2 to 1,600 MHz band WJ-8615P general purpose receiver is a half-rack (8.89 cm), fully synthesised equipment that features:
- a 'high' dynamic range
- amplitude/frequency modulated, continuous wave, independent sideband (optional) and pulse detection modes
- an optional, built-in, tracking preselector
- IEEE-488 bus compatible (talk/listen)
- front panel and remote control
- front panel setting hand-off to other receivers
- up to five intermediate frequency bandwidths (3.2 kHz to 8 MHz).

Status

As of August 2001, the WJ-8615P general purpose receiver was understood to be available.

Contractor

BAE Systems North America - Aerospace Electronics, Gaithersburg, Maryland.

VERIFIED

WJ-8621 compact receiver

Type

20 to 2,700 MHz band compact receiver.

Description

The WJ-8621 compact receiver is packaged in a single-slot, C-size, VXI (VME bus eXtension for Instrumentation) module and features:
- low phase noise frequency synthesisers
- a preselector-equipped front end
- 100 Hz tuning resolution
- a 'high' dynamic range
- a +10 dBm typical third order intercept value
- seven selectable intermediate frequency bandwidths (from 3.2 kHz to 12 MHz)
- a built-in sub-octave preselector
- a tunable baseband output
- amplitude/frequency-modulated, continuous wave and upper/lower sideband detection modes
- multichannel, phase-coherent configurability for direction-finding and beam-forming applications
- a 12.5 MHz wideband intermediate frequency output
- a built-in reference oscillator
- a front panel-mounted RS-232 auxiliary control port
- WJ-RCS control software compatibility.

Status

As of August 2001, the WJ-8621 compact receiver was understood to be available.

Contractor

BAE Systems North America - Aerospace Electronics, Gaithersburg, Maryland.

VERIFIED

WJ-8629A VXI (VMEbus eXtensions for Instrumentation formats) receiver

Type
Software definable VXI receiver.

Description
The 20 to 2,700 MHz band WJ-8629A is a variant of the WJ-8629A VXI receiver that integrates a software, downloadable, digital signal processor with the baseline hardware. The receiver can process Intermediate Frequency (IF) bandwidths of up to 1.23 MHz and allows the user to upload custom bandwidth, demodulation, analysis and decoding algorithms directly into the unit. WJ-8629A comes preloaded with standard IF and demodulation algorithms and features:

- 25 (including five user defined) IF bandwidth filters (from 200 Hz to 1.23 MHz)
- amplitude/frequency modulated, continuous wave, frequency shift keying and independent/upper sideband demodulation modes as standard
- compatibility with the WJ-SDK software developers kit
- multichannel, phase-coherent configurability for direction-finding and beam-forming applications
- a front panel-mounted RS-232 auxiliary control port
- WJ-RCS control software compatibility.

Status
As of August 2001, the software definable WJ-8629A VXI receiver was understood to be available.

Contractor
BAE Systems North America - Aerospace Electronics, Gaithersburg, Maryland.

VERIFIED

WJ-8629 VXI (VMEbus eXtensions for Instrumentation formats) receiver

Type
20 to 2,700 MHz VXI receiver.

Description
The WJ-8629 receiver is packaged in a single-slot, C-size, VXI package and features:

- digitally signal-processed intermediate frequency and demodulators
- a typical +10 dBm third order intercept value
- 15 selectable, digital intermediate frequency bandwidth filters (from 200 Hz to 200 kHz)
- digital audio, video, direct analogue-to-digital samples or I & Q data available over the VXI interface
- front panel digital data outputs that are configurable for interface to a TI C40 comport or a Motorola 56000 family processor
- multichannel, phase-coherent configurability for direction-finding and beam-forming applications
- a 12.5 MHz wideband intermediate frequency output
- a front panel-mounted RS-232 auxiliary control port
- WJ-RCS control software compatibility.

Status
As of August 2001, the WJ-8629 VXI receiver was understood to be available.

Contractor
BAE Systems North America - Aerospace Electronics, Gaithersburg, Maryland.

VERIFIED

WJ-8633

Type
20 to 2,700 MHz band VXI receiver.

Description
The WJ-8633 VXI receiver is described as being a general purpose, high-performance receiver that utilises a 160 MHz Intermediate Frequency (IF) to provide nine IF bandwidths within the 250 kHz to 80 MHz range. The device is housed in a single-slot, C-size VME module and is further noted as combining receiver control directly with a standard instrumentation and computing bus. Here, the approach is claimed to add 'significant' system capability while reducing the complexity of integration. Other system features include:

- 100 Hz frequency resolution
- a typical +5 dBm 3rd-order input intercept point
- a typical 12 dB noise figure
- switchable radio frequency preamplification
- the ability to be configured for multichannel, phase-coherent operation (local oscillator inputs/outputs, two WJ-8633s being grouped together to form a two channel coherent system)

- VXI message-based controllability
- an integral reference oscillator
- compatibility with Windows-based control software (WJ-RCS).

Status
As of August 2001, the WJ-8633 VXI receiver was understood to be available.

Contractor
BAE Systems North America - Aerospace Electronics, Gaithersburg, Maryland.

NEW ENTRY

WJ-8634 surveillance/monitoring receiver

Type
Communications band surveillance and monitoring receiver.

Description
WJ-8634 is a fully synthesised, general purpose surveillance and monitoring receiver which covers the 20 to 1,000 MHz frequency range and is packaged in a single-slot, C-size VXI (VMEbus eXtensions for Instrumentation) module for high density/high integration level applications. Key features of the receiver include its size; a −5 dBm third order intercept capability; tracking preselector filtering, low phase noise and VXI message-based control. Operating modes comprise fixed-frequency manual, sweep and step functions and the receiver is available in a number of configurations comprising the standard 20 to 1,000 MHz option together with two frequency extended models (covering the 0.5 to 1,000 MHz and 20 to 2,400 MHz bands respectively), a narrowband variant and a wideband configuration. The narrowband model supports four Intermediate Frequency BandWidths (IFBW) between 3.2 and 100 kHz when operating in amplitude/ frequency modulation, continuous wave, IFT and single sideband detection modes. The wideband option supports IFBWs between 300 kHz and 12 MHz when operating in the amplitude/frequency modulation detection modes.

Status
As of August 2001, the WJ-8634 surveillance receiver was understood to be available.

Specifications
Frequency: 20-1,000 MHz (standard configuration)
Tuning resolution: 100 Hz (standard configuration)
Noise figure (standard configuration): 10 dB max (below 500 MHz); 12 dB max (above 500 MHz)
Tuning time (to within 1 kHz of final frequency): 2 ms (sweep); 19 ms (20-1,000 MHz manual); 24 ms (0.5-30/1,000-2,400 MHz bands)
Demodulation: AM, CW, FM, IFT and SSB (narrowband); AM and FM (wideband)
Operating temperature: −20 to +55°C
Dimensions: 233.7 × 30.5 × 340.4 mm
Weight: <2.73 kg

Contractor
BAE Systems North America - Aerospace Electronics, Gaithersburg, Maryland.

VERIFIED

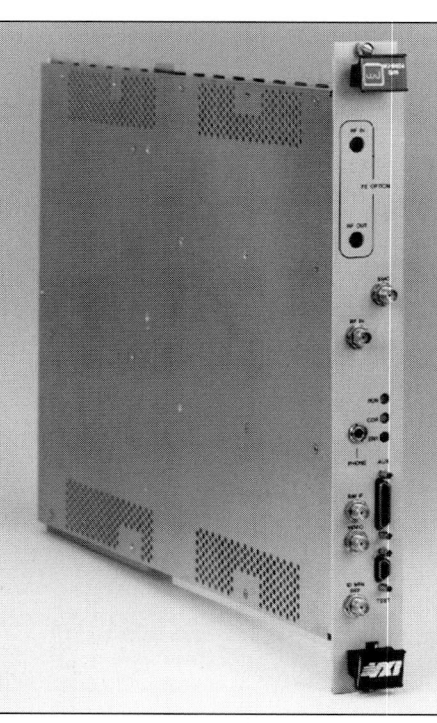

The WJ-8634 general purpose surveillance and monitoring receiver

WJ-8635 surveillance receiver

Type
20 to 1,000 GHz band general purpose surveillance receiver.

Description
The WJ-8635 general purpose surveillance receiver is a modular equipment that is available in narrow- and wideband configurations. System features include:
- 0.5 to 1,000 MHz and 20 to 2,400 MHz optional frequency extensions
- four selectable intermediate frequency bandwidths (3.2 to 100 kHz narrowband, 300 kHz to 12 MHz wideband)
- tunable narrowband baseband output
- narrowband amplitude/frequency modulated, continuous wave and independent/single sideband and wideband amplitude/frequency modulated demodulation
- 'smart' scan and step algorithms
- RS-232 or -422 system control
- WJ-RCS control software compatibility.

Status
As of August 2001, the WJ-8635 general purpose surveillance receiver was understood to be available.

Contractor
BAE Systems North America - Aerospace Electronics, Gaithersburg, Maryland.

VERIFIED

WJ-8654 surveillance receiver

Type
Miniature surveillance receiver.

Description
The WJ-8654 surveillance receiver is a small (96.52 cm³) general purpose surveillance receiver covering the 20 to 1,000 MHz frequency band. Its small size, low weight (0.96 kg) and low power consumption makes it ideal for portable narrowband surveillance applications where weight and power are crucial. WJ-8654 features low phase noise frequency synthesisers, a tunable narrowband baseband output, four selectable intermediate frequency bandwidths (3.2 to 100 kHz narrowband, 300 kHz to 12 MHz wideband), 'smart' step and scan algorithms, multiple serial interface support and an accurate tuning resolution of 100 Hz. A high-performance tracking preselector filters incoming radio frequency signals and rejects undesired out-of-band signals. If required, the WJ-8654's frequency range can be extended to cover the 0.5 to 1,000 MHz, 0.5 to 2,400 MHz or 20 to 2,400 MHz sub-bands. The receiver also features a low-power, 'sleep' operating mode and is able to demodulate narrowband AM, CW, FM, IFT and SSB or wideband AM and FM signals.

Status
As of August 2001, the WJ-8654 receiver was understood to be available.

Contractor
BAE Systems North America - Aerospace Electronics, Gaithersburg, Maryland.

VERIFIED

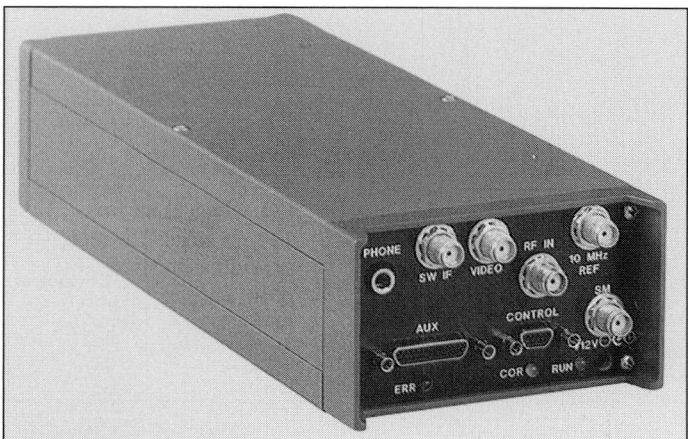

The WJ-8654 miniature surveillance receiver

WJ-8655A

Type
20 to 2,700 MHz band surveillance receiver.

Description
The WJ-8655A is described as being a 'high performance', subminiature, general purpose surveillance receiver that covers the 20 to 2,700 MHz frequency range. BAE Systems claims that the WJ-8655A's 'extremely' small size and 'low' power consumption make it 'ideal' for portable surveillance applications where 'weight and power reduction are critical'. Other system features include:
- six intermediate frequency bandwidths within the range 10 kHz to 25 MHz
- a −96 dBc/Hz phase noise figure value (at 20 kHz offset)
- a 12 dB noise figure value
- a less than 3 W power requirement
- sub-octave and tracking preselectors
- Windows-based control software (WJ-RCS) compatible.

Status
As of August 2001, the WJ-8655A surveillance receiver was understood to be available.

Contractor
BAE Systems North America - Aerospace Electronics, Gaithersburg, Maryland.

NEW ENTRY

WJ-8710A surveillance receiver

Type
Surveillance and monitoring receiver.

Description
WJ-8710A is a fully synthesised, general purpose digital receiver that is designed for the surveillance and monitoring of radio frequency communications in the 5 kHz to 30 MHz band with 1 Hz tuning resolution. The receiver is operated by one of two selectable interfaces. With the exception of audio output level and remote-control mode selection, all receiver parameters are controllable and accessible via an RS-232 remote interface. In lieu of the RS-232 interface, the operator can use a Carrier Sense Multiple Access (CSMA) with collision detection interface with a limited instruction set. The CSMA controls WJ-8710A by using a command protocol similar to several popular consumer receivers. Selection of the active interface is via an internal switch setting or by front panel entry.

Status
As of August 2001, the WJ-8710A receiver was understood to be available.

Specifications
Frequency range: 5 kHz-30 MHz
Tuning resolution: 1 Hz
Dynamic range: +30 dBm (3rd order intercept, typical)
IF bandwidths: 66 (up to 16 kHz)
Internal reference stability: better than 0.7 ppm (0-50°C)
Detection modes: AM, CW, FM, ISB, LSB, SAM and USB
Power requirement: 12 V DC
Dimensions: 6 × 19 × 29 cm (approx)

Contractor
BAE Systems North America - Aerospace Electronics, Gaithersburg, Maryland.

VERIFIED

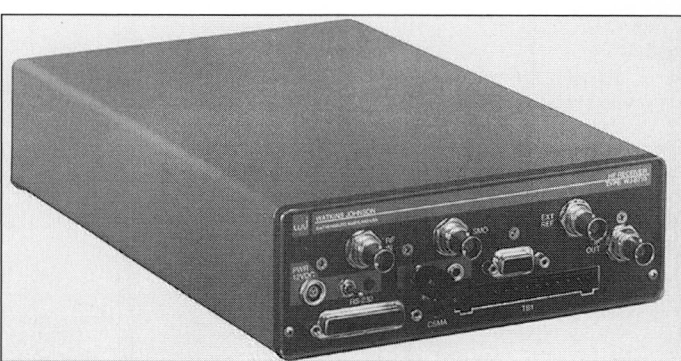

The WJ-8710A surveillance receiver

WJ-8711A surveillance/monitoring receiver

Type
Digital communications band surveillance and monitoring receiver.

Description
WJ-8711A is a fully synthesised, general purpose surveillance and monitoring receiver which covers the 5 kHz to 30 MHz frequency range and combines

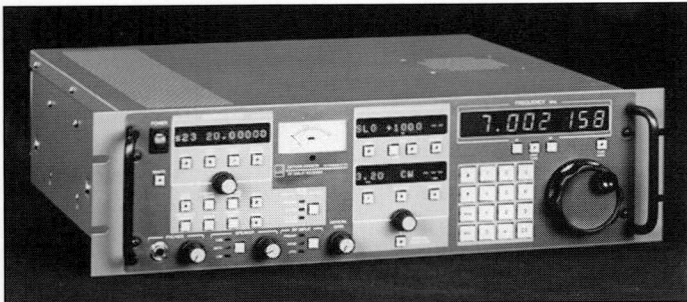

The WJ-8711A general purpose surveillance and monitoring receiver

analogue with digital signal processing to maximise performance while keeping down cost. Key features of the receiver include +30 dBm third order interception; digital filtering (66 intermediate frequency bandwidths up to 16 kHz); fast-scanning with 100 memory channels; easily readable light emitting diode front panel displays; 100 memory channels; three variable automatic gain control decay settings; noise blanking; passband tuning; internal, switchable preamplifier and attenuator; standard remote interfaces and built-in self-test. A digital data output and a sub-octave preselector are available options.

Status

As of August 2001, the WJ-8711A surveillance/monitoring receiver was understood to be available.

Specifications

Frequency: 5 kHz-30 MHz
Tuning resolution: 1 Hz
Noise figure: 11 dB max (preamplifier engaged); 14 dB max (baseline)
Detection modes: AM, CW, FM, ISB, LSB, SAM, USB
Power requirement: 97-253 V AC, 47-440 Hz
Operating temperature: 0 to +50°C
Altitude: 7,315 m (operating), 15,240 m (non-operating)
Weight: 6.78 kg
Dimensions: 133.6 × 482.6 × 508 mm

Contractor

BAE Systems North America - Aerospace Electronics, Gaithersburg, Maryland.

VERIFIED

WJ-8712A surveillance/monitoring receiver

Type
Digital communications band surveillance and monitoring receiver.

Description
WJ-8712A is a fully synthesised, general purpose surveillance and monitoring receiver which covers the 5 kHz to 30 MHz frequency band. Effectively, the equipment is the WJ-8711A receiver without the former's front panel and optimised for remote control via a range of interface options. Other system features include 100 memory channels, noise blanking, built-in test, tunable notch filtering, internally switchable preamplifier and attenuator, selectable roofing filters and a remote control capability for dial-up collection.

Status
As of August 2001, the WJ-8712A surveillance and monitoring receiver was understood to be available.

Specifications
Frequency: 5 kHz-30 MHz
Tuning resolution: 1 Hz
Noise figure: 11 dB max (preamplifier engaged); 14 dB max (baseline)
Dynamic range: +30 dBm (3rd order intercept, typical)
IF bandwidths: 66 (up to 16 kHz)

The WJ-8712A general purpose surveillance and monitoring receiver

Detection modes: AM, CW, FM, ISB, LSB, SAM, USB
Operating temperature: 0 to +50°C
Altitude: 7,315 m (operating); 15,240 m (non-operating)
Weight: <5.5 kg
Dimensions: 88.9 × 209.6 × 508 mm

Contractor

BAE Systems North America - Aerospace Electronics, Gaithersburg, Maryland.

VERIFIED

WJ-8721 surveillance/monitoring receiver

Type
Communications band surveillance and monitoring receiver.

Description
WJ-8721 is a fully synthesised, general purpose surveillance and monitoring receiver which covers the 5 kHz to 30 MHz frequency band and is packaged in a single-slot, C-size VXI (VMEbus eXtensions for Instrumentation) module for high-density/high integration level applications. Key features include its size; +30 dBm third order intercept capability; digital filtering (66 intermediate frequency bandwidths up to 16 kHz); amplitude/frequency modulated, continuous wave and upper/lower/independent sideband detection modes; VXI message-based control; high-density packaging (up to 12 receivers in a single VXI chassis); built-in self-test, 100 memory channels and sub-octave preselection/digital outputs (including in-phase and quadrature) as standard.

Status
As of August 2001, the WJ-8721 surveillance and monitoring receiver was understood to be available.

Specifications
Frequency: 5 kHz-30 MHz
Tuning resolution: 1 Hz
Noise figure: 11 dB (preamplifier engaged), 14 dB (baseline)
Detection modes: AM, CW, FM, ISB, LSB, USB
Operating temperature: 0 to +50°C
Altitude: 7,315 m (operating); 15,240 m (non-operating)
Power consumption: 21 W (typical)
MTBF: >14,000 h
Dimensions: 233.7 × 30.5 × 335.3 mm
Weight: <2.26 kg

Contractor
BAE Systems North America - Aerospace Electronics, Gaithersburg, Maryland.

VERIFIED

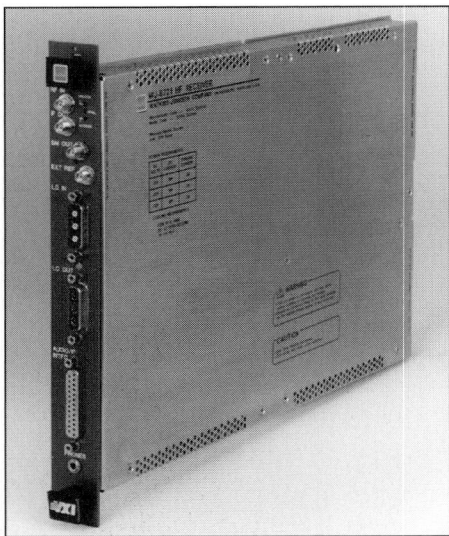

The WJ-8721 general purpose surveillance and monitoring receiver

WJ-8991/SYS Independent Collection Equipment (ICE)

Type
Manpack intercept and Direction-Finding (DF) system.

Description
The WJ-8991 ICE package is a complete manpack intercept and direction-finding system. It is easily transported by a single person and weighs less than 23 kg. It

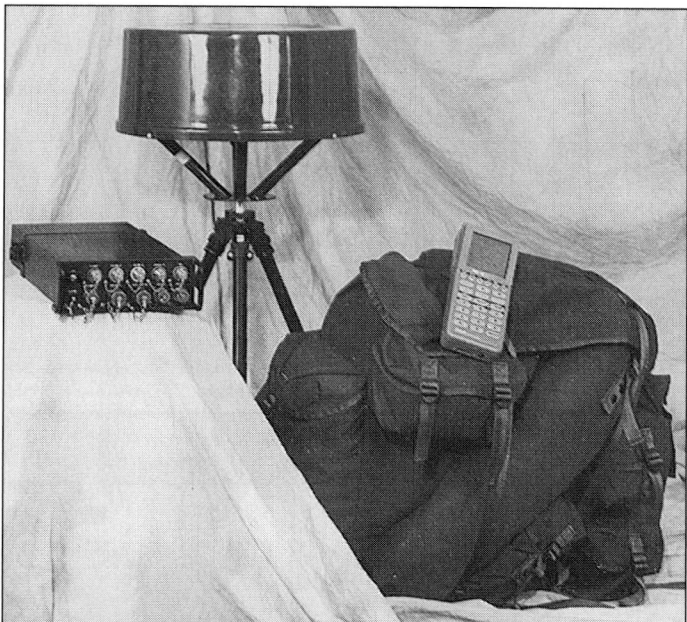

The WJ-8991/SYS manpack ICE system

provides DF and intercept capabilities from 20 to 1,200 MHz. Available options allow customisation to particular requirements. The basic WJ-8991 system consists of:

- WJ-8996-1 receiver/processor.
- WJ-9887 DF antenna.
- WJ-8996/HHC hand-held controller.
- tripod.
- audio headset.
- system cables.

WJ-8991 is compact enough to fit into an ALCE-type backpack. It operates as a stand-alone intercept DF post or as an element in a DF net intended to conduct emitter location operations. System options include the WJ-8991/ASU antenna switch, the WJ-9896 monopole antenna array, the 10.7 m high WJ-9887 antenna mast, the WJ-8996/AC power supply, the WJ-8996/WCS user interface software package, the WJ-8996/NBC laptop computer and 20 to 2,000 MHz frequency coverage when using the noted WJ-9896 antenna.

Status

As of August 2001, the WJ-8991A/SYS package was understood to be available.

Specifications

Frequency range: 20-1,200 MHz; 1,200-2,000 MHz with reduced performance
Accuracy: 3° RMS (antenna on mast); 5° RMS (antenna on tripod)
Sensitivity: <10μV/m typical
System weight: <23 kg

Contractor

BAE Systems North America - Aerospace Electronics, Gaithersburg, Maryland.

VERIFIED

WJ-8996-1 Direction-Finding (DF) processor

Type

20 to 2,000 MHz band DF processor.

Description

The correlative-vector WJ-8996-1 DF processor is a lightweight, man-portable equipment to incorporate four complete radio frequency channels to facilitate modulation-independent acquisition of short duration signals. Other system features include:

- antenna-type flexibility
- tunability down to 0.5 MHz
- a low power requirement
- digital signal processing
- ruggedisation to MIL-STD-810C.

Status

As of August 2001, the WJ-8996-1 DF processor was understood to be available.

Contractor

BAE Systems North America - Aerospace Electronics, Gaithersburg, Maryland.

VERIFIED

WJ-9012A High Frequency (HF) Direction-Finding (DF) system

Type

1 to 30 MHz HF DF system.

Description

WJ-9012A is an open architecture, modular HF DF system that is built around the IEEE-1155 VMEbus eXtensions for Instrumentation (VXI) standard and makes use of commercial-off-the-shelf hardware. The equipment's man/machine interface utilises 'field-proven' software and can be 'grown' to accommodate the Very High and Ultra High Frequency bands if required. Other system features include:

- N-channel design (eight channels typical)
- resolution of co-channel interference
- super resolution algorithms
- use of the WJ-8721 VXI standard receiver
- built-in test
- networking options
- variable antenna configurations.

Status

As of August 2001, the WJ-9012A HF DF system was understood to be available.

Contractor

BAE Systems North America - Aerospace Electronics, Gaithersburg, Maryland.

VERIFIED

WJ-9104A digital tuner

Type

20 to 2,000 MHz band digital tuner.

Description

The WJ-9104A digital tuner offers up to eight radio frequency channels that can be remotely configured for either independent operation or a phase-coherent direction-finding where channels share common Local Oscillators (LO). In either mode, each channel provides a digitised, 10 MHz instantaneous bandwidth with 12-bit precision. If required, 2 and 25 MHz bandwidths are available as options, while system control is executed via an SCSI-2 small computer system interface. A direct tuning interface allows for precision triggering and timing. By changing the LO distribution scheme, WJ-9104A can support up to four dual channels or a combination of dual and independent channels. Other system features include:

- tunability across the 10 to 2,600 MHz frequency range
- 80 dB digital and 85 dB analogue spur-free dynamic ranges
- a 60 ms tuning speed.

Status

As of August 2001, the WJ-9104A digital tuner was understood to be available.

Contractor

BAE Systems North America - Aerospace Electronics, Gaithersburg, Maryland.

VERIFIED

WJ-9104B digital tuner

Type

20 to 3,000 MHz band digital tuner.

Description

The WJ-9104B digital tuner offers up to eight radio frequency channels that can be remotely configured for either independent operation or a phase-coherent direction-finding where channels share common Local Oscillators (LO). In either mode, each channel provides a digitised, 10 MHz instantaneous intermediate frequency bandwidth that is sampled at 25.6 MHz with 14-bit precision. If required, 2 MHz and 25 MHz bandwidths are available as options, while system control is executed via an SCSI-2 small computer system interface. A direct tuning interface allows for precision triggering and timing. The WJ-9104B is under 32 kg in weight and can be configured with either an AC or a DC power supply and is packaged in a single, standard, full-rack chassis. Other system features include:

- 'low' phase and amplitude mismatch
- 85 dB spur-free dynamic ranges
- a 60 ms tuning speed.

Status

As of August 2001, the WJ-9104B digital tuner was understood to be available.

Contractor

BAE Systems North America - Aerospace Electronics, Gaithersburg, Maryland.

VERIFIED

WJ-9107 VXI (VMEbus eXtensions for Instrumentation formats) tuner

Type
800 to 1,000 MHz and 1,700 to 2,000 MHz band VXI tuner.

Description
The WJ-9107 VXI tuner is designed to receive cellular telephone and PCS signals and consists of independently housed dual-channel tuner (WJ-9107/DTM) and dual-tuned synthesiser (WJ-9107/DLO) single-slot, C-size, VXI modules. Each WJ-9107/DTM converter channel can digitise a 25 MHz bandwidth to facilitate signal processing while the WJ-9107/DLO module provides synthesised tuning over the cellular, cordless, wireless data and PCS bands. Other system features include:

- seven band preselection
- a 75 to 80 dB spur-free dynamic range
- 1 MHz tuning resolution
- 12 bit analogue-to-digital conversion
- a 'multichannel ready' configuration.

Status
As of August 2001, the WJ-9107 VXI tuner was understood to be available.

Contractor
BAE Systems North America - Aerospace Electronics, Gaithersburg, Maryland.

VERIFIED

WJ-9119A and WJ-9119A-1 VXI (VMEbus eXtensions for Instrumentation formats) tuning units

Type
0.1 to 32 MHz band VXI tuning units.

Description
The WJ-9119A and WJ-9119A-1 VXI tuners provide a 95 dB instantaneous, spur-free dynamic range in 4 MHz (WJ-9119A) or 8 MHz (WJ-9119A-1) bandwidths. Each device comprises radio frequency tuning and local oscillator synthesiser sections, both of which are packaged as single-width, C-size, VXI modules. Of the two, WJ-9119A interfaces with Hewlett-Packard's E1430A analogue-to-digital converter, while the WJ-9119A-1 interfaces with the company's E1437 converter. Direct and frequency converted paths optimise performance for 'any' input frequency and the two tuners incorporate 'special' circuits and components, including a proprietary BAE Systems North America mixer. Other system features include:

- 250 kHz tuning resolution
- phase and amplitude stability between channels
- built-in test circuitry.

Status
As of August 2001, the WJ-9119A and WJ-9119A-1 VXI tuners were understood to be available.

Contractor
BAE Systems North America - Aerospace Electronics, Gaithersburg, Maryland.

VERIFIED

WJ-9127 VXI (VMEbus eXtensions for Instrumentation formats) tuner

Type
0.1 to 32 MHz band VXI tuner.

Description
The WJ-9127 VXI tuner provides a 100 dB instantaneous, spur-free dynamic range in a 4 MHz bandwidth and interfaces with the Hewlett-Packard E1430A analogue-to-digital converter for both single and multichannel applications. Two WJ-9127 tuners can be housed in a single C-size VXI module and the use of direct and frequency converted paths optimises performance for 'any' input frequency. The unit includes a proprietary BAE Systems North America mixer and other system features include:

- two channels per 6U VXI slot
- VXI register-based control
- phase and amplitude stability between channels
- built-in test circuitry
- WJ-9119/LO module compatibility.

Status
As of August 2001, the WJ-9127 VXI tuner was understood to be available.

Contractor
BAE Systems North America - Aerospace Electronics, Gaithersburg, Maryland.

VERIFIED

WJ-9128A VXI (VMEbus eXtensions for Instrumentation formats) tuner

Type
0.1 to 32 MHz band VXI tuner.

Description
The WJ-9128A VXI tuner comprises radio frequency tuner and local oscillator synthesiser sections, each of which is housed in a single-width, C-size VXI module. The equipment incorporates direct and frequency converted baseband tuning (to maximise performance for 'any' frequency input) and is equipped with a proprietary BAE Systems North America mixer. Other system features include:

- a minimum 88 dB spur-free dynamic range
- 2 MHz bandwidth
- an internal 14 bit analogue-to-digital converter
- built-in test circuitry
- a 'high-speed' fibre-optic digital output.

Status
As of August 2001, the WJ-9128A VXI tuner was understood to be available.

Contractor
BAE Systems North America - Aerospace Electronics, Gaithersburg, Maryland.

VERIFIED

WJ-9482-1 demodulator unit

Type
Digital demodulator unit.

Description
WJ-9482-1 makes use of digital signal processing and surface-mounted technology to achieve precision demodulation of received signals in a compact, price competitive package. The unit accepts 140 MHz (160 MHz as option) Intermediate Frequency (IF) inputs, demodulates low- to high-rate pulse code modulated signals and provides a decoded symbol data output in word parallel format. An input automatic gain control circuit allows operation with a −30 to −10 dBm analogue input. The use of digital processing allows WJ-9482-1 to implement any IF bandwidth between 1 and 56 MHz with BAE Systems standard set of filters providing 96 bandwidths within the specified range. Other system features include selectable absolute or differential encoding; flexible symbol bit mapping; Ethernet remote control and built-in test.

Status
As of August 2001, the WJ-9482-1 demodulator was understood to be available.

Specifications
IF input: 140 MHz (160 MHz option)
IF bandwidths: 96 selectable (from 1 to 56 MHz)
Input level: −30 to −10 dBm
Input signal rate: 1 to 40 Mbaud
Noise figure: 22 dB (max)
Gain control: manual or automatic
Demodulation modes: BPSK, QPSK, SQPSK, 8 PSK and 16 QAM
Operating temperature: 0 to +50°C
Altitude: 4,572 m (max)
Weight: 9.05 kg (max)
Dimensions: 44.4 × 483 × 503 mm (excl connectors and handles)

Contractor
BAE Systems North America - Aerospace Electronics, Gaithersburg, Maryland.

VERIFIED

The WJ-9482-1 demodulator unit

WJ-9488 digital sub-band tuner

Type
Digital sub-band tuner.

Description
The WJ-9488 digital sub-band tuner is described as incorporating digital signal processing, application-specific integrated circuitry and surface-mount technology. Functionally, the unit provides precision tuning, filtering, decimation and gain control facilities. In addition to power-of-two decimations (applied commensurate with the selected bandwidth), the device performs fractional resampling to facilitate 'almost continuous' adjustment of the output rate. A precision time tagging facility is included to support direct connection to Datolite and SDN data streams. Other system features include:
- a 12-bit digital input
- selectable 8-, 10- or 12-bit digital outputs
- input sample rates of between 0.1963125 and 54.4 megasamples/s (configuration dependent)
- 1 Hz tuning resolution (or 2:32 of the input sample rate)
- 21 selectable bandwidths as standard.

Status
As of August 2001, the WJ-9488 digital sub-band tuner was understood to be available.

Contractor
BAE Systems North America - Aerospace Electronics, Gaithersburg, Maryland.

VERIFIED

WJ-9548 demultiplexer

Type
Digital Frequency-Division Multiplex (FDM) demultiplexer.

Description
WJ-9548 is a compact multichannel tunable FDM demultiplexer incorporating the accuracy and efficiency of a digital signal processing approach. It combines analogue and digital processing techniques in a scheme that enhances the performance relative to demultiplexer implementations that are purely analogue or digital. Its modular design enables WJ-9548 to be configured easily as a 6-, 12-, 18- or 24-channel unit and to be tailored to meet specific system requirements. It accepts up to four 20 MHz analogue FDM basebands and connects them in a non-blocking fashion to any one of the independently tunable channel demodulators. A buffered version of each baseband input is also provided as an output allowing multiple units to access the same basebands.

Control of WJ-9548 can be performed either locally (via the front panel liquid crystal display and keypad controls) or remotely using the standard IEEE-488 interface. A variety of other remote interfaces is available as drop-in, alternative options. All operator selectable parameters (except headphone volume control) are accessible and controllable over the remote control interface. This capability also includes programmable scan strategies. A built-in test feature, capable of detecting circuit faults to module level, can also be initiated remotely.

Two WJ-9548 units, mounted side-by-side, fit into a standard 48 cm rack, occupying only 9 cm of vertical rack space. The data and control architecture allows up to eight units to be stacked in a master/slave configuration and function as a single unit. A total of 192 FDM channels can be independently mapped into the various pulse code modulation and audio outputs supported by the individual units.

Status
As of August 2001, the WJ-9548 demultiplexer was understood to be available.

The WJ-9548 FDM demultiplexer

Contractor
BAE Systems North America - Aerospace Electronics, Gaithersburg, Maryland.

VERIFIED

WJ-9887 Direction-Finding (DF) antenna

Type
Low-profile DF antenna.

Description
The WJ-9887 DF antenna is intended for applications where a low-profile, lightweight and compact antenna array is a requirement. Covering the Very/Ultra High Frequency (V/UHF - 30 MHz to 1 GHz) frequency band, WJ-9887 is designed to be used with four channel, vector-correlation DF systems such as the WJ-8996-1 equipment.

Status
As of August 2001, the WJ-9887 DF antenna was understood to be available.

Contractor
BAE Systems North America - Aerospace Electronics, Gaithersburg, Maryland.

VERIFIED

WJ-9896 Direction-Finding (DF) array

Type
2 to 30 MHz band DF array.

Description
The WJ-9896 DF array is used with the WJ-8996-1 receiver/DF processor and consists of four antenna elements, a calibration switch and coaxial radio frequency cabling. The system is described as being effective against both ground and skywave signals and the antenna elements used are collapsible monopoles. A single BNC coaxial cable connects each of the array elements to a switch box, with the entire assembly being set out in a 4.26 m² square. The array baseline can be enlarged to enhance signal collection at the lower end of the equipment's frequency range and its manufacturer recommends deploying the system's antenna elements at least 75 m away from any obstruction for optimum performance. When folded for storage, each WJ-9896 monopole measures 0.6 × 0.3 × 0.3 m.

Status
As of August 2001, the WJ-9896 DF array was understood to be available.

Contractor
BAE Systems North America - Aerospace Electronics, Gaithersburg, Maryland.

VERIFIED

ZS-1015 COMmunications INTelligence (COMINT) workstation

Type
Integrated COMINT workstations for tri-service applications.

Description
The ZS-1015 integrated COMINT workstation covers task specific bands within the 0.5 to 3,000 MHz frequency range and is designed for both tactical and strategic applications. The equipment's basic architecture is software intensive in order to allow it to perform multiple functions within a single unit. System capabilities include communications band signal search and detection; direction-finding; emitter location (with data presented on a digital geographic map display); signal monitoring; signal parameter measurement and analysis; Digital Audio Recording and Playback (DARP) and audio signals analysis. All acquired data is entered and stored in an integrated relational database. ZS-1015 operating modes are as follows:
Continuous scan
The equipment scans operator selected frequency ranges in specific step sizes. The signal environment is presented via a tabular display of frequency, bearing and signal status. There is also provision for simultaneous polar and spectrum display formats.
Preset frequency scan
As continuous scan but with the operator specifying individual frequencies rather than ranges of frequencies.
Monitor/direction-finding
A single frequency mode with a line of bearing histogram display of all the emitters detected on the selected frequency.

Frequency/bearing

Scans an operator selected frequency range (as in the continuous scan mode) with results being graphically displayed in a frequency (vertical axis) versus bearing (horizontal axis) format.

ZS-1015 is available in a number of configurations, the known details of which are as follows:

ZS-1015(F)

Ground-based, fixed-site systems equipped with either omnidirectional or high gain/high direction-finding accuracy antenna arrays.

ZS-1015(M)

Mobile system for installation in disguised vehicles with concealed antennas. Configured for operation of all functions while the host vehicle is moving.

ZS-1015(S)

Shipboard systems for communications band electronic support. Equipped with situation monitor screen displays and an interface to a ship's combat information centre.

ZS-1015(T)

Mobile, tactical, transportable systems for installations in shelters, military tracked vehicles or High-Mobility Multipurpose Wheeled Vehicles (HMMWV).

Status

As of January 2002, ZS-1015 series COMINT workstations were understood to be available.

Specifications

Frequency range: 1.5-1,000 MHz (ZS-1015(T) option - Spectrum Scanning (SS)/Monitoring (MO)/Measurement (Me)/DF); 20-1,000 MHz (ZS-1015(F)/(M)/ (S)/(T) baseline - SS/Mo/Me/DF); 100-1,000 MHz (ZS-1015(S) option - SS/Mo/ Me/DF)

Scanning/detection speed: up to 3,000 channels/s

DF technique: correlative interferometry

DF accuracy: 2° RMS (ZS-1015(F)/(S)/(T) - typical; 3° RMS (ZS-1015(M) - typical)

DF measurement speed: 4 ms (ZS-1015(F)/(M)/(T) - fast DFmax); 200 ms (all - normal DF max)

Remote control: RS-232C or other interface

Emitter location: by triangulation of integrated DF measurements from up to 4 stations; automatic time tagging for unambiguous location of multiple emitters received simultaneously at same frequency in receiver bandwidth

Digital map display: full colour display of digital vector or bit-maps (raster scan) with emitter location overlaid on map display

Demodulation: AM/CW/FM/LSB/USB

DARP: up to 4 record/playback channels stored as files on hard disk; up to 72 channels hours capacity; simultaneous record and playback (on different channels); instantaneous replay of stored records; 88.2 kbytes/s sampling speed (multimedia standard); VOX feature for recording only when audio signal active; automatic time tagging of audio record

Audio signal analysis tools: spectral analysis; time analysis; variable speed playback (constant pitch); loop playback

Report/file management: operator aids for development of datacards for each emitter; datacard data (including audio files) stored in object-orientated relational database

Contractor

Zeta (an Integrated Defense Technologies company), Morgan Hill, California.

UPDATED

LAND-BASED ACTIVE AND PASSIVE COUNTERMEASURES SYSTEMS AND DEFENSIVE AIDS SUITES (DAS)

BULGARIA

Cactus stationary site jamming system

Type
Stationary site jamming system.

Description
The Cactus system is designed to record and jam spurious information emissions from stationary and semi-stationary sites within the 0.1 to 120 kHz range, together with information (electrical and magnetic field components) that has penetrated the power supply network. The equipment's transmitter subsystem can also be used for noise masking jamming within special designation cable lines while its receiver can be used for radio control operations within the 0.3 to 150 kHz frequency band.

Status
As of this edition, the Cactus system was reported as having been procured by the Bulgarian Army.

Specifications
Transmitter
Frequency range: 0.1-120 kHz
Capacity: 100 W
Jamming type: barrage
Effective protection area: 2,500 m^2
Equivalent noise power in the net: 1 W
Power supply: 220 V (50-60 Hz)
Power consumption: 200 W
Receiver
Frequency range: 0.3-150 kHz

A general view of the Cactus stationary site jamming system
0009378

Operator equipment used with the Cactus stationary site jamming system
0009379

Sensitivity: 10 µV (SNR = 10/1)
Work modes: SSB physical
Frequency/level indication: digital
Power supply: 12 V

Contractor
Kintex, Sofia.

VERIFIED

Lilia expendable communications band jammers

Type
Family of 1.5 to 120 MHz expendable communications band jammers.

Description
Lilia series expendable jammers cover the 1.5 to 120 MHz frequency band and are delivered via 152 mm artillery pieces or the 122 mm BM-21 rocket system. The equipment is described as being effective against frequency modulated and single side band transceivers, together with frequency agile and spread spectrum units. Kintex notes that the Lilia series offers precision in jammer location (irrespective of terrain type) and round the clock usage. Lilia is a standard munition that requires no special handling and has a storage life of up to 10 years.

Status
As of this edition, Lilia series jammers were reported as having been procured by the Bulgarian Army.

Specifications
Frequency range: 1.5-120 MHz
Firing range: 4-30 km
Range accuracy: ≥700 m
Operating period: 1 h (continuous)
Temperature range: −40 to +55°C
Storage period: 10 years

Contractor
Kintex, Sofia.

VERIFIED

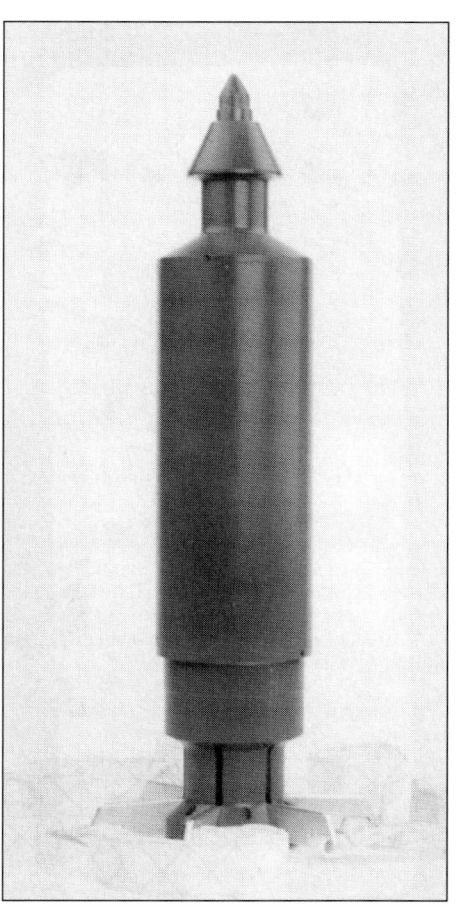

The Lilia jammer is an expendable artillery or rocket system delivered expendable device
0009376

Starshel 122 expendable communications band jammers

Type
Family of 20 to 100 MHz, artillery delivered, expendable communications band jammers.

Description
Aimed at fixed-frequency and hopping tactical radios, the 122 mm Starshel 122 expendable, artillery delivered, barrage jammer family is compatible with the D-30 and M-30 towed howitzers and the 2C1 self-propelled weapon and comprises the VRS5, VRS5L, VRS6, VRS6L, VRS-463 and VRS-463L rounds. Within the range, details of weapon compatibility, propellant charge and battery types are as follows:

Round	Artillery system	Propellant charge	Battery type
VRS5	D-30 or 2C1	full	self-activating
VRS5L	D-30 or 2C1	reduced	lithium
VRS6	D-30 or 2C1	full	self-activating
VRS6L	D-30 or 2C1	reduced	lithium
VRS-463	M-30	full	self-activating
VRS-463L	M-30	reduced	lithium

Status
As of this edition, Starshel 122 expendable jammer rounds were reported as having been procured by the Bulgarian Army.

Specifications
Frequency range: 20-100 MHz
Coverage: >700 m
Operating time: >1 h
Type of jamming: barrage
Delivery system calibre: 122 m
Firing range: 6,000 m (min); as defined for the weapon type (max)

Contractor
Kintex, Sofia.

VERIFIED

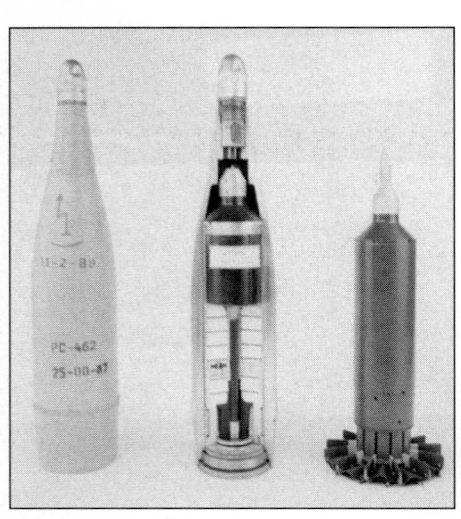

A general view of members of the Starshel 122 expendable jammer family
0009375

Starshel 152 expendable communications band jammers

Type
Family of 20 to 100 MHz, artillery-delivered, expendable communications band jammers.

Description
Aimed at fixed-frequency and hopping tactical radios, the 152 mm Starshel 152 expendable, artillery-delivered, barrage jammer family is compatible with the D-20 and ML-20 towed howitzers and the 2C3M self-propelled weapon and comprises the VRS-546, VRS-546L, VRS-546U and VRS-546UL rounds. Within this range, details of propellant charge and battery types are as follows:

Round	Propellant charge	Battery type
VRS-546	full	self-activating
VRS-546L	full	lithium
VRS-546U	reduced	self-activating
VRS-546UL	reduced	lithium

Status
As of this edition, Starshel 152 expendable jammer rounds were reported as having been procured by the Bulgarian Army.

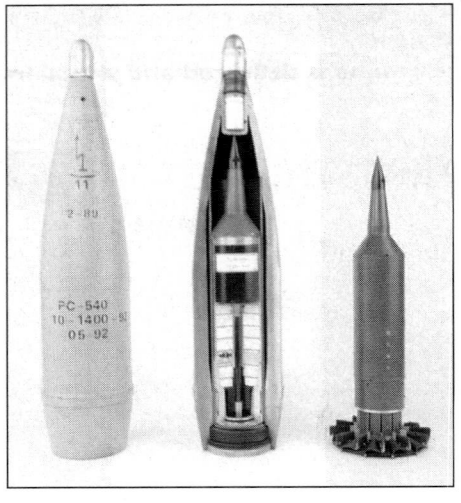

A general view of members of the Starshel 152 expendable jammer family
0009374

Specifications
Frequency range: 20-100 MHz
Coverage: >700 m
Operating time: >1 h
Type of jamming: barrage
Delivery system calibre: 152 mm
Firing range: 6,000 m (min); as defined for the weapon type (max)

Contractor
Kintex, Sofia.

VERIFIED

SHTURETS manpack communications band jammer

Type
Manpack communications band radio transmitter jammer.

Description
SHTURETS is a manpack barrage radio jamming transmitter with remote control. It is designed to create barrage jamming against fixed frequency or frequency agile radio communication transmissions. Components of the jammer are as follows:
- transmitter
- directional antenna
- non-directional antenna
- mobile set
- Very High Frequency (VHF — 30 to 300 MHz) transceiver
- remote-control panel
- remote-control executive device.

Three operational versions can be configured from the above modules:
Manpack 1
Utilises the described transmitter and non-directional antenna modules.
Manpack 2
Utilises the described transmitter, directional antenna and non-directional antenna modules.
Mobile
Each of these variants is equipped with remote control (by radio, using the described VHF transceiver, remote control panel and remote control executive

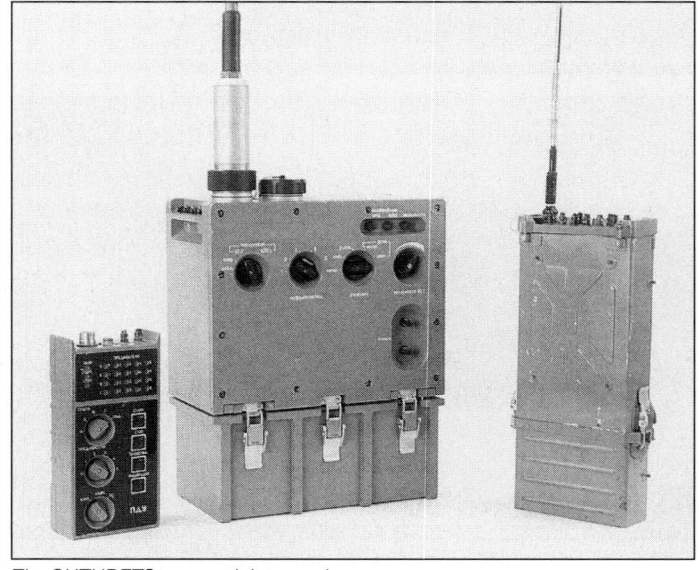

The SHTURETS manpack barrage jammer

device modules) or by wire (using the described remote control panel and executive device modules) and the necessary cabling. The set is battery powered and a charging device for the battery packs is supplied. It can also be powered by a vehicle power supply.

Status
As of this edition, the SHTURETS communications jammer was reported as having been procured by the Bulgarian Army.

Specifications
Frequency range: 20-100 MHz
Sub-ranges: 5 pcs
Range: 700 m (flagpole antenna); 1,200 m (directional antenna)
Operating time: 3 h (battery); continuous (vehicle power supply)
Remote control by radio: 8 km
Number of simultaneously controlled jammers: 20

Contractor
Kintex, Sofia.

VERIFIED

CHINA, PEOPLE'S REPUBLIC

BM/DJG 8715 jamming system

Type
Mobile I/J-band (8 to 20 GHz) radar jamming system.

Description
BM/DJG 8715 is a ground-based jammer that is aimed at airborne missile guidance, missile seekers, navigation and terrain-following radars. The architecture incorporates one Electronic Support (ES) station with up to eight Electronic CounterMeasures (ECM) sites, all of which are integrated by means of datalinks to ensure automatic control, direction of jamming and feedback of ECM data. BM/DJG 8715 is stated to feature a wide frequency coverage; automatic classification and identification of threat radars; monopulse auto-angular tracking with 'high' direction-finding accuracy; multithreat, multibearing, multifrequency jamming; the ability to counter 'exotic' emitters and a high value, travelling wave tube-based jamming output. A variety of ES displays are provided at the ES station, including polar, alphanumeric and an air track (showing the graphical background of the operations zone). Both the ES and the ECM stations are mounted on self-propelled vehicles to ensure high mobility.

Status
As of this edition, BM/DJG 8715 is thought to have been procured by the Chinese Army.

Specifications
ESM Station
Frequency coverage: 8-18 GHz
IFM accuracy: 5-12 MHz RMS
Azimuth coverage: 360°
DF accuracy: 5-8°
Measurable pulse-width: 0.1-100 µs
ECM Station
Frequency coverage: 8-16 GHz
IFM accuracy: 10 MHz RMS
AFC spotting accuracy: <2 MHz
Angular tracking accuracy: 1° RMS
Output power: 2 × 100 W (CW) min

Contractor
Southwest China Research Institute of Electronic Engineering (SWIEE), Chengdu, Sichuan province.

VERIFIED

Model 970 radar jamming system

Type
Mobile I/J-band (8 to 20 GHz) radar jamming system.

Description
Model 970 is a mobile, ground-based radar jammer that is primarily designed to protect high value ground targets from air attack. The system operates in I/J-band and uses noise-modulated 'blanket' jamming to produce a continuous interference sector on the screen of an aircraft's surveillance radar. It can also be used to jam air-launched missile guidance radars, offers a range of jamming modes and can be operated either automatically or manually. Normally, a number of Model 970

The Type 970 radar jammer trailer showing the roof-mounted antenna assembly

equipments are placed some 3 to 5 km from the protected target and are used in conjunction with a target indicating radar. Functionally, the system measures the bearing, frequency and antenna rotation speed of the hostile radar and when used in conjunction with a pulse analyser, is able to establish parameters such as emitter pulse-width and pulse repetition frequency. The equipment consists of a number of units including an antenna feed, a transmitter, coarse and fine receivers, servoes, a display unit and a power supply. It is installed in a trailer that incorporates a roof-mounted antenna assembly.

Status
As of this edition, the Model 970 radar jammer was reported as having been procured by the Chinese Army.

Specifications
Frequency: 8-12 GHz
Receiver sensitivity: better than −76 dBm
Power output: better than 120 W; a 200 W output transmitter is also available.

Contractor
China National Electronics Export and Import Corporation, Beijing.

VERIFIED

FRANCE

GALIX self-protection system

Type
Self-protection system for Armoured Fighting Vehicles (AFV).

Description
Aimed primarily at AFV applications, the GALIX self-protection system is described as offering a 'complete response' to a range of threats that includes the use of a range of nine, 80 mm visual and multispectral smoke, Infra-Red missile decoy, tear gas, warning rounds and illuminating rounds (designated as the GALIX 4 self-protection smoke, GALIX 6 IR decoy, GALIX 7 illuminating, GALIX 13 multiband smoke, GALIX 15 tear gas, GALIX 16 test, GALIX 17 practice smoke, GALIX 18 practice self-protection smoke and GALIX 19 stun grenades) A typical GALIX architecture consists of a firing panel and launch tubes of which, the former controls the latter and is located within the host vehicle. The preloaded launch tubes are installed on the host vehicle's turret or chassis, with their quantity and orientation being determined as a function of the vehicle operational role. The GALIX system can be integrated with threat warning devices, offers manual and automatic operating modes and can be configured for naval as well as land vehicle installation.

Status
As of September 2000, Jane's sources were reporting that the GALIX AFV self-protection system was in production and that it had been procured by the armies of France (installed aboard Leclerc Main Battle Tank - MBT), Italy (Ariete MBT and 8 × 8 Centauro tank destroyer/armoured car), Saudi Arabia (8 × 8 Piranha light armoured vehicle), Sweden (CV90 series and Strv 122 MBT) and the United Arab Emirates (Leclerc MBT and variants). GALIX is also noted as forming part of Giat's *Kit Basique de ContreMesures (KBCM)* AFV defensive aids system. Comprising a man/machine interface/processor unit, an IR jammer, missile launch and laser warning subsystems, a GALIX firing unit's KBCM is reported to have been tested on a French Army 6 × 6 AMX-10RC armed reconnaissance vehicle.

Specifications

Round:	GALIX 4	GALIX 6	GALIX 7	GALIX 13	GALIX 15	GALIX 17	GALIX 18	GALIX 19
Type:	SPSR	IRDR	IIIR	MBSR	TGR	PSR	PSPSR	SGR
Length:	250 mm	300 mm	500 mm	400 mm	274 mm	400 mm	250 mm	265 mm
Weight:	3.6 kg	2.4 kg	4.8 kg	5.1 kg	1.7 kg	5.1 kg	1.8 kg	3.0 kg

Key
IIIR Illumination Round **IRDR** IR Decoy Round **MBSR** MultiBand Smoke Round **PSPSR** Practice Self-Protection Smoke Round **PSR** Practice Smoke Round **SGR** Stun-Grenade Round **SPSR** Self-Protection Smoke Round **TGR** Tear Gas Round

Contractors

Giat Industries, Versailles Cedex.
Etienne LACROIX, Muret Cedex.

UPDATED

GERMANY

Maske series multispectral screening grenades

Type

Family of multispectral screening smoke grenades for armoured fighting vehicle applications.

Description

As of this edition, known details of the Maske multispectral screening grenade family are as follows:

Maske 66/76/81 EL RP/C

Available in 66, 76 and 81 mm calibres, the Maske 66/76/81 EL RP/C grenade has been jointly developed by BUCK and the Swiss contractor SM Swiss Ammunition Enterprise and contains a bimodular red phosphorus and carbon-based payload that is designed to maximise its effectiveness in the Infra-Red (IR) spectrum.

Specifications

Maske 76 EL RP/C
IR screening time: 60 s
Ejection range: 60 m
Weight: 1.8 kg
Dimensions (Ø × L): 76 × 248 mm

Maske 66/76/81 ST RP/RP

Available in 66, 76 and 81 mm calibres, the Maske 66/76/81 ST RP/RP grenade contains a bimodular red phosphorus payload that is designed to provide screening in both the visible light and IR portions of the electromagnetic spectrum.

Specifications

Maske 76 ST RP/RP
Deployment time: within 1 s
IR screening time: 30-40 s
Ejection range: 40 m
Weight: 1.2 kg
Dimensions (Ø × L): 76 × 180 mm

Status

As of this edition, the Swiss Army was reported as having procured the Maske 76 EL RP/C grenade (under the designation 7.6 cm Nb Pat 95 el Zü) for use on its Leopard II main battle tanks. As originally scheduled, the Swiss Army was to have received approximately 165,000 Nb Pat 95 el Zü rounds by mid-2001. For its part, the Maske 76 ST RP/RP grenade was introduced in mid-1999 and has been adopted by the German Army as the DM53A1. DM53A1 replaces the service's existing DM53 visible light-only round and as of the given edition, the Maske 66/76/81 ST RP/RP grenade was noted as being NATO approved.

Contractor

BUCK Neue Technologie GmbH (a Rheinmetall AG subsidiary), Bad Reichenhall

VERIFIED

MUltifunctional Self-protection System (MUSS)

Type

Integrated Defensive Aids Suite (DAS) for Armoured Fighting Vehicle (AFV) applications.

Description

The baseline MUSS AFV DAS comprises missile and laser warning sensors, a DAS computer/controller/display, a multispectral (visible and Infra-Red (IR) smoke, IR flares and chaff) pyrotechnics launcher and an active IR jammer. Of these, the missile warning subsystem is a modified Lenkflugkörpersysteme (LFK) AN/AAR-60 MILDS® sensor (designated as P-MILDS) while the laser warner used is a variant of LFK's Common Opto-electronic Laser Detection System (COLDS) equipment that covers the 0.4 to 1.7 μm band. P-MILDS differs from the baseline system in incorporating new optics and filters, a new image amplifier, new algorithms and new packaging. To provide 360° coverage on a Main Battle Tank (MBT), four MILDS/COLDS sensor head packages are mounted on the vehicle's turret in such a way as to cover individual 96° sectors. The system's two, four by four, 76 or 81 mm pyrotechnic launchers are also turret mounted with the DAS computer/controller/display and a pyrotechnic launcher control box being located inside the vehicle's hull. Subsystem intercommunication is by means of RS-422 interfaces and the active IR jammer is sourced by Israel Aircraft Industries. MUSS is intended for upgrade with an active/passive laser decoying variant of the COLDS system. It can fire a mixed decoy salvo and can be integrated into its host vehicle's fire and control systems if required. Alongside LFK, the MUSS development team includes BUCK Neue Technologie GmbH and Krauss-Maffei Wegmann.

Status

Work on the MUSS programme is reported to have been started during 1998 and entered Phase I technology demonstration trials during June 1999. Here, the technology demonstrator incorporated P-MILDS and COLDS sensors, a control personal computer and a pyrotechnic launcher rig. As of this edition, Phase II MUSS trials (with the system fitted to a German Leopard II MBT) were reported as having begun during January 2000 and as having been scheduled for completion during the following August. If successful, Phase III of the MUSS programme would see the system installed on German Army AFVs from 2002 onwards. It is further understood that MUSS is to be tested by Switzerland (Leopard II MBT) and Turkey (Leopard I MBT).

Contractor

Lenkflugkörpersysteme GmbH (prime), Munich.

VERIFIED

INTERNATIONAL

BLB 20 expendable jammer

Type

Hand-emplaced expendable barrage jammer.

Description

The 20 W BLB 20 expendable barrage jammer operates within the 20 to 110 MHz frequency range and is intended to prevent communications across all or part of this band. The system is claimed to be able to neutralise 'all' communications (including frequency hoppers) and can also be used for training personnel in how to maintain radio communications in a dense electronic warfare environment. BLB 20 is described as being 'compact and lightweight' and offers continuous barrage, intermittent transmission and sliding frequency range operating modes. A data fill device is used to preprogramme the equipment which, if required, can be remotely controlled via radio link.

Status

As of May 2001, development of the BLB 20 expendable jammer was reported as having been completed.

The BLB 20 hand-laid expendable jammer

Specifications
Frequency range: 20-110 MHz
Output power: 20 W
Signal bandwith: programmable from 100 kHz to 90 MHz
Delayed activation: up to 4 h
Antenna: omnidirectional 1.8 m high
Autonomy: 3 h (typical)
Dimensions: 260 × 260 × 100 mm
Weight: 4.3 kg (incl battery and antenna)

Contractor
Thales Communications, RadioSurveillance and COMINT Systems Unit, Gennevilliers, France.

VERIFIED

CBJ-40 radar jammer

Type
Lightweight battlefield radar jammer.

Description
CBJ-40 is a lightweight, battlefield radar detector-jammer that is designed for use by ground forces or installation in unmanned aerial vehicles and helicopters. It is described as being able to counter battlefield surveillance, low-altitude air surveillance, artillery spotting and counterbattery pulsed, continuous wave and coherent radars. In ground force applications, the device appears to comprise a single transceiver/processing unit that is tripod or mast mounted. System features include a broad coverage antenna array; simultaneous processing of radar groups into separate sub-bands; denial and deception modulations; instantaneous set on; full programmability and high probability of intercept, receiver sensitivity and output values.

Status
As of April 2001, development of the CBJ-40 radar jammer was understood to have been completed.

Specifications
Frequency range: D- to G-band (1-6 GHz) or G- to J-band (6-18 GHz sub-band) - customer selectable
Volume: 26 litres
Weight: 27.5 kg (excl tripod)

Contractor
Thales Airborne Systems, Elancourt, France.

VERIFIED

The CBJ-40 battlefield radar jamming system
0009373

Cerberus Armoured Fighting Vehicle (AFV) Defensive Aids Subsystem (DAS)

Type
AFV DAS suite.

Description
Cerberus is an AFV DAS suite that incorporates up to 12 Avimo laser detector arrays, a Thales (formerly Helio) FVS 25 control/display unit and various

The FVS 25 control/display unit used in the Cerberus AFV DAS suite

arrangements of Thales' (formerly Helio) FVG 66 or FVG 76 grenade dischargers. Of these, the FVS 25 control/display unit features audio and visual threat alarms together with threat identification and threat bearing determination. The unit's grenade discharger control subsystem incorporates two discharger selection switches (one for individual discharger control and the other for salvo control), a mode selector and a firing/test button. Selectable operating modes comprise Test, Salvo, Auto, All and Single. When using the Single, Salvo and All options, discharger control is manual. The Auto setting automatically fires a salvo of eight grenades on activation of a system detector array, giving a threat facing 180° smokescreen within 2 seconds. A Test facility allows for the diagnosis of faults in the control box, discharger control cables and the individual dischargers together with a grenade fuze safety check. The FVG 66/FVG 76 dischargers are modular in design and can be installed singly or in groups according to customer requirement. Each discharger is equipped with a radio frequency interference protection device to prevent accidental grenade discharge by radio signal.

Status
As of January 2001, the Cerberus AFV DAS suite was reported as being in production and service.

Contractor
Thales Optronics, Belvedere, UK.

UPDATED

CICADA series communications jammers

Type
Family of 0.525 to 3,000 MHz communications band jammers.

Description
Based on the former DaimlerChrysler Aerospace's experience with the HUMMEL jamming system (see separate entry), the MRCM (a joint venture between European Aeronautic, Defence and Space (EADS) EWATION and South Africa's Grintek EWATION) CICADA family is aimed at communications emitters operating in the 0.525 to 3,000 MHz frequency band and is described as incorporating the following features:

- an overall performance that is optimised via matched receiver sensitivity to jamming range
- automatic, computer controlled jamming sequences
- use of state-of-the-art, high-power, broadband, liquid-cooled power amplifiers
- a look-through capability
- a prioritised, predefined target frequency operating mode
- operator controlled deception jamming
- a broadband barrage jamming mode for use against multiple, frequency-hopping emitters
- an automatic or operator-controlled electronic support function
- programmed channel protection
- 'comprehensive' built-in test
- local or remote control facilities
- a countermeasures tasking and reporting facility during active jamming functions
- 'unique' signal detection circuitry designed to 'guarantee' immunity from strong signal interference, a high probability of intercept and a low false alarm rate
- fast time division multiplexed function for the 'effective' jamming of multiple targets
- a Windows standard man/machine interface.

CICADA system components are mounted in 48 cm (19 in) racks and the operator's console and can be installed in tri-service transportable shelters (including units mounted on cross-country capable trucks), armoured personnel carriers (with integral power generation) and all-terrain vehicles. Looking at the system's power amplifier design in more detail, the CICADA unit is described as offering an output power of up to 10 kW and as featuring enhanced reliability and

The TOR communications jamming system (MRCM)   0044260

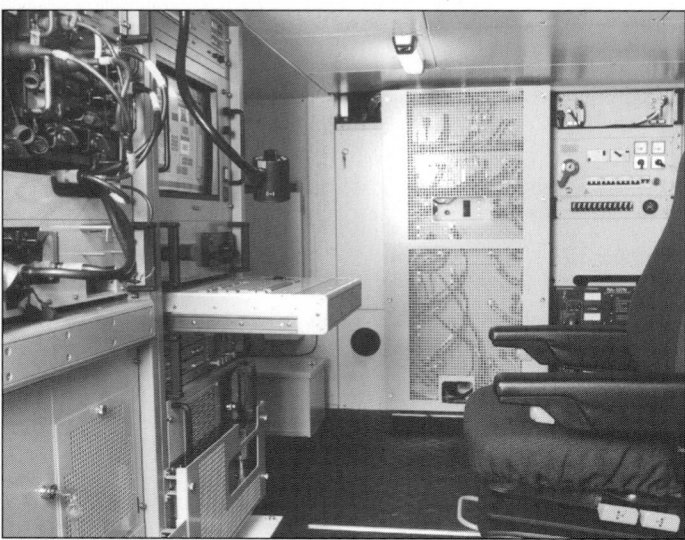

The interior of the TOR jamming shelter (MRCM)   2002/0098335

mean time between failure values through its use of liquid cooling and solid-state components. The equipment is also noted as featuring a 'soft failure' mode (ensures graceful degradation of output power rather than system shut-down following module failure), electronically switched harmonic filters, integral transmit/receive switches and mismatch/overheating protection. CICADA is further supported by a software-based mission planning tool that can be used by individual jammers or an operations centre that is controlling a number of jamming stations.

Status

CICADA jammers (in the form of the DaimlerChrysler Aerospace SGS 2000 series equipment) have been in production since 1997. As of June 2001, an all-terrain vehicle-mounted CICADA variant (designated as the Tor system) has been procured by the Norwegian armed forces. Here, Jane's sources suggest that the specific variant incorporates a software mission planning tool that was jointly developed by DaimlerChrysler and the Norwegian contractor Teleplan AS. The same sources also suggest that CICADA-type receiver and direction-finding subsystems form the basis of the German Army's EUKOP Very/Ultra High Frequency (V/UHF – 30 MHz to 1 GHz) ground-mobile surveillance system procurement, which was launched during September 1998. Installed in a Fuchs armoured personnel carrier, EUKOP is noted as being able to function while on the move.

Specifications

Frequency range: 0.525-3,000 MHz
Spot jamming: up to 16 frequencies
Jamming modulation: FM (effective against AM, CW, FM, FSK and SSD emitters)
Modulation signals: adapted signals against amplitude, frequency or phase modulated communications links
Barrage jamming bandwidth: up to 240 MHz (simultaneous generation of up to 16 separate broadband segments)

A vehicle-mounted application of the CICADA communications band jamming system (MRCM)   2002/0098315

Receive modes: A1A, A1B, A2A, A2B, A3E, F1A, F1B, F1C, F3E, H3E, J3E(USB/LSB) and R3E
Output power: up to 10 kW (into 50 Ω load)
Harmonic attenuation: built-in harmonic filters
Target frequency adaptation: several times/s (frequency range dependent)
Frequency change time: <100 μs
Search speed: 250 MHz/s (V/UHF range – 1,000 MHz/s option)
Transmit modes: A1B, A3E, F1B, F1C, F3E and J3E (USB/LSB) (for deception/burn-through)
Antennas: log periodic dipole or wideband vertically/horizontally polarised units (frequency dependent – wideband units for jamming 'on-the-move')
Jamming efficiency: up to 16 active radio nets simultaneously
Temperature: –25 to +55°C (operating); –40 to +70°C (storage)

Contractors

MRCM (an EADS/Grintek EWATION joint venture), Ulm, Germany/Pretoria, South Africa.

UPDATED

HUMMEL communications band jammer

Type

Automatic communications band jamming system.

Description

The HUMMEL jamming system is an automatic, response equipment that is designed for use against voice and datalinks, operating in the 20 to 80 MHz band. It functions as a multichannel jammer which is capable of attacking up to 10 individual voice/data channels simultaneously, using up to six jamming modulations per channel. The baseline HUMMEL system works with a complementary 1 to 180 MHz electronic support measures equipment that identifies targets and selects appropriate jamming modulations. A computer-controlled transceiver is at the heart of the system, whose operating modes are as follows:

- transmission and reception with manual control
- automatic single channel operation as responding jamming transmitter
- multichannel fixed-frequency operation
- search operation with multichannel jamming transmissions.

HUMMEL is normally mounted in an armoured fighting vehicle (with integral power supply) such as the Fuchs six-wheel armoured personnel carrier. In such an installation, the system is operated by a crew of three made up of a driver, a system operator and a deception/communications operator. A Fuchs-based installation is reported to have a maximum output of 2 kW to utilise a high-gain monopulse transceiver antenna array and be able to operate while in motion. The HUMMEL system can be operated as a stand-alone equipment or in a remote-control mode reporting to a control centre. In the latter, all targeting and modulation selection data are passed to the jammer via a datalink or (if stationary) wire. At the particular jammer, such data are input automatically without the need for operator intervention.

In addition to the above, the jammer can be installed in a 2 × 2 m shelter or be carried on customer furnished vehicles. This version uses a log-periodic dipole on a telescopic mast. The power generator is carried on a separate trailer. The jammer shelter can be helicopter transported as a single load.

The HUMMEL jamming system installed in a Fuchs armoured personnel carrier

Status

As of July 2001, Jane's sources report that HUMMEL variants have been supplied to the armies of Germany, Spain and, possibly, Netherlands. Of these, the Spanish equipments are shelter-mounted while, those operated by Germany are vehicle-mounted and described as being 'revised'. In this context, sources were suggesting during early 1999 that 16 German HUMMELs were to be the subject of a Combat Effectiveness Improvement (CEI) programme. Here, use would be made of technology from the MRCM CICADA jammer series system (see separate entry) and the so called 'Kampfwertsteigerung' HUMMEL would incorporate a new mast-mounted (12 m) wideband direction-finding antenna, emitter classification and identification capabilities, an onboard emitter parameter database and an integral mission planning tool designated as KESS

Contractor

European Aeronautic, Defence and Space (EADS) Co Systems and Defence Electronics, Ulm, Germany.

UPDATED

RHINO communications band detector-jammer

Type

10 kHz to 30 MHz band detector-jammer system.

Description

RHINO is a transportable, shelter-mounted, 10 kHz to 30 MHz band detector-jammer system that has a maximum 1 kW output and is designed for use against communications networks. The system employs two types of antennas (12 m whip and V-slope) that allow it to handle both ground and skywave transmissions. The equipment can be used as either an integrated element of an Electronic CounterMeasures (ECM)/Electronic Support (ES) system or as a stand-alone unit. RHINO functions as a spot jammer that counters hostile signals traffic on a frequency/channel on a time basis. Multichannel jamming is facilitated by a time-sharing regime with the system as a whole making use of high speed receivers, transmitters and antenna tuning. Here, the tuner unit can be preprogrammed with up to 1,024 channels. Alongside conventional signals traffic, RHINO is also claimed to be effective against frequency-agile, burst and frequency-hopping transmitters. In the case of frequency hoppers, the system can cope with a hop rate of up to a few tens of hops/s and is most effective in countering such emitters when it is teamed with a suitable ES subsystem. In terms of system control, RHINO can be operated in local or remote modes. In the local control configuration, RHINO stands alone with the operator performing search, interception and jamming activities within the system's shelter. In remote mode, one or more unattended RHINO stations operate under the control of a computerised jamming control centre. Command data is passed between stations by means of an ultra-high frequency radio link, the whip antenna and tuning unit for which can be located up to 100 m from a particular system shelter.

Status

Over time, the RHINO detector-jammer system is reported as having been procured by a number of European NATO members, with whom it was noted as having been used in the 'ES, ECM/ES and communications' roles.

The RHINO HF band detector-jammer system

Specifications

Frequency: 10 kHz-30 MHz (receiver); 1.5-3 MHz (transmitter)
Output: 125 W; 250 W; 500 W; 1 kW (PEP/average)
Jamming options: manual or automatic
Demodulations: AM; CW; FM, FSK, ISB; SSB (USB/LSB)

Contractor

Thales International, Chieti Scalo, Italy.

VERIFIED

Squirrel communications band detector-jammer

Type

Land-based medium power communications band detector-jammer system.

Description

Squirrel is a High Frequency (HF - 3 to 30 MHz) detector-jamming system that is similar to Thales' RHINO equipment (see separate entry). It is designed for shelter and transportable applications and has a 400 W output. It can be controlled either locally or by remote control. Several antenna types can be used to jam hostile ground or skywave communications traffic. Squirrel is also equipped with fast electronic support units to facilitate its use as a stand-alone equipment or as part of an integrated electronic warfare system. Sequential multichannel jamming, using time-sharing techniques for the best power efficiency, is provided by very fast automatic, fully silent antenna tuning.

Contractor

Thales International, Chieti Scalo, Italy.

VERIFIED

TRC 274 communications jammer

Type

1 MHz to 3 GHz digital, multirange communications jamming system.

Description

The modular TRC 274 digital jamming system is designed to neutralise military, paramilitary and civilian communication links and networks operating in the 1 MHz to 1 GHz (extension up to 3 GHz if required) frequency range. The equipment

The TRC 274 communications jamming system   0044259

comprises electronic countermeasures and support subsystems and is capable of intercepting and monitoring signals, establishing their frequencies and setting up an optimum jamming strategy based on integral emitter and jamming mode databases. TRC 274 can operate as a stand-alone unit or as part of an integrated electronic combat architecture and can identify targets via its own electronic support subsystem or from cues supplied by external battlefield electronic warfare equipments.

The system incorporates automatic jamming functions and makes use of programmable algorithms to generate the best output for the particular emitter type (fixed-frequency, frequency hopper or burst) in the particular environment (high- or low-density network). For attacks on frequency hoppers, TRC 274 makes use of what its manufacturer terms the CHIRP technique, that is, a 'smart' compromise between barrage jamming and the follower principle that is optimised for use in dense High and Very High Frequency environments. CHIRP is independent of hopping speed and emitter numbers and incorporates spectrum management to avoid friendly communications fratricide. Other system features include a Windows NT man/machine interface, 100 W to 1 kW amplifier output power (other values available as options) and emitter specific jamming and deception modes.

Status
As of May 2001, the status of the TRC 274 communications jamming system was uncertain.

Contractor
Thales Communications, RadioSurveillance and COMINT Systems Unit, Gennevilliers, France.

VERIFIED

ISRAEL

Electronic Warfare Integrated System (EWIS)

Type
Integrated, ground-based EW system.

Description
The 0.5 to 18 GHz EWIS architecture comprises COMmunications INTelligence (COMINT) and Direction-Finding (DF), communications jamming, electronic support/ELectronic INTelligence (ELINT), electronic countermeasures and command and control subsystems. Functionally, the architecture is described as being capable of rapid spectrum scanning (using wide-band receivers), monitoring, DF, location-finding, signal classification, digital audio recording and jamming. The system is further noted as being able to handle frequency hopping, burst and 'other' agile signal types and as being suitable for fixed-site, ground-mobile, armoured personnel carrier-mounted and shelterised applications.

Status
As of this edition, the EWIS architecture was reported as being 'fully operational' with 'many' customers worldwide.

Specifications

	COMINT/DF HF band	COMINT/DF V/UHF band	COMJAM HF band	COMJAM V/UHF band
Frequency:	0.15-30 MHz	20-1,000 MHz (20-3,000 MHz option)	1.5-30 MHz	20-1,000 MHz (20-3,000 MHz option)
Scan rate:	100 MHz/s	1 GHz/s		
DF accuracy:	<1° RMS (instrumented); 1.5° RMS (field, typical)	<1° RMS (instrumented); 1.5° RMS (field, typical)		
Power output:			1-2 kW	1 kW
Jamming modes:			signal initiated, fast sequential and broadband	signal initiated, fast sequential and broadband

Contractor
Tadiran Electronic Systems Ltd, Holon.

VERIFIED

EL/K-7000MT communications band jamming systems

Type
Family of communications band jamming systems.

Description
The EL/K-7000MT family of modular communications band jamming systems is designed to 'effectively block' military, paramilitary and civilian communications links and is noted as being suitable for ground-based (fixed and mobile), airborne and maritime applications. System features include:
- 'very fast' broadband receivers for signals intercept and jammer look-through
- digital signal processing
- a modular architecture that facilitates customising systems to meet specific customer requirements
- multiple jamming modes including sequential, signal initiated, barrage and deception
- stand-alone operation or integration into a larger communications intelligence, direction-finding and jamming system
- integral mission-planning tools to aid the selection of jamming sites and the evaluation of jammer effectiveness against target communications networks and links
- built-in test
- jamming output optimisation relative to specific targets
- protection of friendly communications channels
- anti-frequency hopping capability.

Status
As of late 1999, the EL/K-7000MT family of modular communications band jamming systems was understood to be available.

Specifications
Frequency range: 1 MHz - 1 GHz (optional extensions available)
Output power: 200 W, 500 W, 1 kW or higher
Frequency switching time: <100 microseconds
Jamming/reception modulations: AM, CW, FM, FSK, PSK and SSB
Jamming bandwidth: narrow/broad/selectable
Jamming modes: manual/automatic
Antenna types: directional/omni-directional (selectable as per specific application requirement)
Reception bandwidth: 10, 20, 50 or 100 kHz (options available)
Number of search frequency bands: 8 (typical)
Number of preset scan channels: 50 (typical)
Number of protected channels: 100 (typical)

Contractor
Elta Electronic Industries Ltd (a subsidiary of Israel Aircraft Industries Ltd), Ashdod.

VERIFIED

NS-9005G radar jammer

Type
Ground-based radar jammer.

Description
The ground-based NS-9005G radar jammer is primarily a stand-off equipment that can also be used in self-defence and/or escort applications. Functionally, the system analyses multitarget signals and manually/automatically generates an

appropriate response. NS-9005G make use of a 'unique' Windows NT™ operator console and a transmission chain that includes high-power amplifiers and mechanically or electronically steerable antenna arrays.

Status
As of January 2002, NS-9005G was understood to be available.

Contractor
Elisra Electronic Systems Ltd (a member of the Elisra Group), Bene Beraq.

UPDATED

Rattler jammer system

RAJ 101 radar detector-jammer

Type
Mobile land-based radar detection and jamming system.

Description
RAJ 101 is a ground-based radar jammer that covers a 180° sector and is effective at ranges of up to 30 km. It offers 'high' effective radiated power, a wide frequency band and is capable of dealing with three threats simultaneously via the use of time-sharing techniques. In acquisition mode, RAJ 101 measures frequency, direction, pulse repetition frequency, pulse-width and amplitude in order to detect, identify and locate hostile radars. The information collected can also be used for analysis. The equipment's jammer subsystem operates in fully automatic or semi-automatic mode. Effectiveness is achieved by concentrating the transmitted power in either spot frequency or barrage modes and reaching a high jam/signal ratio. Frequency and direction are set up to prevent hostile radars from detecting and locating friendly forces. The system is completely self-contained, is vehicle mounted and is able to operate and travel in extremely rugged terrain.

Status
As of this edition, the RAJ 101 radar jamming system is understood to have been procured by the Israel Defence Forces.

Contractor
Rafael Electronic Systems Division, Haifa.

VERIFIED

The RAJ 101 radar jamming system mounted on an armoured personnel carrier

Rattler radar jammer

Type
Multiplatform radar jamming system.

Description
Rattler is a jamming system that is designed for use in ground, naval and airborne applications, as either part of an overall Electronic CounterMeasures (ECM) system or in a stand-alone configuration. The system can operate as both standoff and stand-in and is designed to jam up to three surveillance, search and tracking radars simultaneously using time-share techniques. The system consists of four main units: a low-power microwave source, a high-power wideband amplifier, power supply and a control unit. Output power is believed to be greater than 400 W in spot or barrage mode and the equipment (which is ruggedised and MIL-STD-5400 and -16400 compliant) covers the 2 to 18 GHz frequency band. Up to 16 Rattler systems can be connected on the same 488 databus and an Electronic Support (ES) system, enabling a full system to jam 48 frequencies under the control of a single ES/ECM suite. Remote control is provided and can operate the system from a distance of up to 200 m. Range of the jammer is believed to be 20 to 30 km. When an emitter is received and identified as a threat, Rattler goes into operation, either automatically through an existing computer, or manually via its control unit. Voltage

controlled oscillator sources determine the jamming frequencies which are produced by the low-power microwave jamming source. The low-power outputs are transferred to the amplifier and wideband power is transmitted to jam the enemy radar.

Contractor
Rafael Electronic Systems Division, Haifa.

VERIFIED

Tactical Automatic Communications Jamming System (TACJS)

Type
2 to 1,000 MHz tactical communications jamming system.

Description
TACJS is an automated, power-managed, 2 to 1,000 MHz, look-through, communications jamming system that is designed for installation in an armoured personnel carrier or in a shelter configuration aboard a high-mobility wheeled vehicle. Functionally, the system offers operator initiated, automatic, cyclic multifrequency, time-shared jamming or multifrequency, signal initiated jamming that is based on activity within the targeted communications net. Several TACJSs

The TACJS communications jamming system as applied to an armoured personnel carrier
0010136

can be netted to an electronic warfare headquarters or to a communications intelligence system to provide a combat integrated power multiplier. TACJS hardware comprises broadband antennas, solid-state amplifiers, transmit/receive switches and high-speed receivers/activity detectors. System control is by means of a computer that also facilitates the use of a Windows-driven, man/machine interface.

Status
As of this edition, the TACJS communications jamming system was understood to be available.

Specifications
Frequency range: 2-30, 20-500 or 20-1,000 MHz
Monitored/detected signal types: AM/CW/FM/SSB
Power output: 0.5-2 kW (frequency dependent)
Outputs/peripherals: printers; audio recorder; RS-232; I/O to integrated EW/C³I system
Input power: AC/DC (configuration dependent on host vehicle supply or towed generator configuration)
Installation: APC or shelter mounted on high-mobility wheeled vehicle

Contractor
Tadiran Electronic Systems Ltd, Holon.

UPDATED

ITALY

ELT/D-1000 communication band detector-jammer systems

Type
Family of tactical detection and jamming systems.

Description
The ELT/D-1000 series is a family of microprocessor-controlled detector-jammers that operate in the Very High Frequency (VHF - 30 to 300 MHz) and Ultra High Frequency (UHF - 0.3 to 1 GHz) frequency bands. They can be installed on a variety of ground, naval and airborne platforms. The use of a distributed processing approach allows a very high frequency setting speed and stand-alone operation. The jamming subsystem features either manual (or local) and slaved (integrated into a system and linked to a host computer) modes. There are two variants, the ELT/D-1000A that covers the VHF/UHF bands and the ELT/D-1000B for VHF. The systems are modular, fully solid-state and have a capability for single or multichannel jamming. The system consists of two main elements, as follows:
- the control and exciter section is composed of a control unit and noise generator, a digital bus and interfaces for external sources/remote operation, a fast synthesised exciter and a local control panel

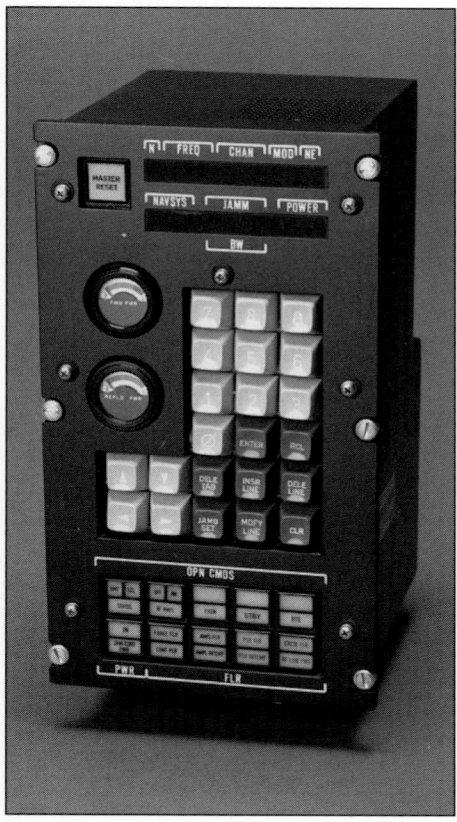

The ELT/D-1000 local control panel

- the power amplifier section includes the power supply and radio frequency power amplifier. The latter is a fully solid-state, wideband low distortion amplifier. It features total protection against overloads, is self-cooled and incorporates comprehensive built-in test. This section may also be used for Amplitude and Frequency Modulation (AM/FM) transmitters and for electromagnetic interference tests.

In manual mode, the system can jam a single channel or multiple channels in a pseudo-random switched sequence. Manual mode involves the following parameter settings:
- frequency tables (emissions of interest)
- jammed frequencies (designated in the frequency table)
- type of modulation (continuous wave, AM, FM)
- modulating signal type (noise, swept tones and a combination of noise and tones).

In the slaved mode, the parameter settings are performed by a host computer, which also manages the look-through function. Parameter changes require the exchange of only a minimum amount of information between the local microprocessor and the host. The built-in processing and data storing capability provides a quick reaction to situation changes, ensuring effective real-time jamming response. Weight of the system is 126 kg for the ELT/D-1000A and 114 kg for the ELT/D-1000B.

Status
As of this edition, the ELT/D-1000 communications detector-jammer family was reported as being fully developed.

Contractor
Elettronica SpA, Rome.

VERIFIED

IGS-5 Electronic Support (ES)/Electronic CounterMeasures (ECM) system

Type
Transportable automatic ES/ECM sensor system.

Description
IGS-5 is a transportable, modular, automatic system that is capable of detection and analysing the prevailing electromagnetic scenario. It is a member of the RQH-5(V) family that is designed for ground-based, naval and airborne applications. The equipment is housed in a standard air conditioned shelter and consists of an antenna group, an instantaneous frequency measuring receiver, a video processor unit, a digital signal processor, a processor and graphics generator and a system console (38 cm (15 in) monochrome or colour display and multifunction keyboard). Other system features include:
- wide frequency coverage and high sensitivity
- high-accuracy direction-finding measurement with a very high-accuracy optional module
- real-time automatic extraction of pulsed and continuous wave signals
- full analysis and tracking of all incoming radar signals (scanning or not, known or unknown)
- automatic track correlation identification, classification and threat evaluation using a very wide identification library
- computer generated interactive display modes
- a man/machine computer assisted interface.

The IGS-5 ECM capability is noted as using high accuracy frequency tuning and a pencil beam antenna to maximise its power-on-target value. Targeting accuracy is maintained by an antenna-pointing passive tracking subsystem and the capability is further noted as incorporating an integral growth facility.

Status
As of this edition, the status of the IGS-5 ES/ECM system was uncertain.

Contractor
Elettronica SpA, Rome.

VERIFIED

RUSSIAN FEDERATION AND ASSOCIATED STATES (CIS)

Introduction
Entries here relate to equipment produced by, exported by and/or used by the armed forces of the Russian Federation and the independent states of Armenia, Azerbaijan, Belarus, Georgia, Kazakhstan, Kyrgyzstan, Moldova, Tajikistan, Turkmenistan, Ukraine and Uzbekistan. Taken together, these 12 entities form the Commonwealth of Independent States economic grouping.

VERIFIED

Expendable communications jammer

Type
Rocket delivered communications jamming rounds.

Description
Developed in co-operation with a Bulgarian manufacturer (Kintex?), this equipment is reported to be designed to disrupt communications in the 1.5 to 120 MHz frequency range and takes the form of a payload for a modified 122 mm bombardment rocket. The launch/transporter vehicle associated with the programme is described as being a six-wheel cross-country type.

Specifications
Calibre: 122 mm
Length of rocket: 3,026 mm
Rocket weight: 66 kg
Range: 4.55-18.3 km
Number of rockets per transporter: 7
Jammer duration: 1 h
Temperature range: −40 to +50°C

Contractor
Splav State Research and Production Enterprise, Tula.

VERIFIED

Gazetchik anti - Anti-Radiation Missile (ARM) system

Type
Anti-ARM system.

Description
The Gazetchik anti-ARM system is designed to protect friendly radar emitters from attack by ARMs. The equipment is made-up of a stand-alone detector, active Radio Frequency (RF) decoys, passive countermeasures dispensers and an interface with the radar or radars being defended. Functionally, the detector unit alerts the system to the approach of an incoming weapon, a warning that activates a basket of responses comprising interruption of the protected emitter's transmissions, transmission of RF decoy signals on the protected emitter's operating frequency and the firing of passive decoys from the equipment's chaff and aerosol launchers. Gazetchik is reported to be available in a number of variants and has an automatic operating mode if required. Power for the system is drawn from the emitter being protected.

Status
As of the first quarter of 2000, Gazetchik anti-ARM system was understood to be available.

Specifications
Coverage: up to 90° (elevation); 360° (azimuth)

Contractor
All-Russian Radio Engineering Research Institute, Moscow.

VERIFIED

SPN series radar jammers

Type
Family of ground-based radar jamming systems.

Description
As of this edition, two SPN series equipments have been positively identified, the known details of which are as follows:

SPN-2
The five-man SPN-2 radar jammer is designed to protect ground forces and 'small size' installations and is described as being effective against airborne pulsed side-looking (SL), air-to-surface weapon control (ASWC), navigation and terrain-following (TF) radars. The equipment comprises antenna, system control and power generation vehicles, all of which make use of the URAL-4320 or KamAZ-4310 chassis. Of the three, the antenna vehicle is equipped with:
- an array assembly that comprises multibeam transmission and reception antennas and a sidelobe compensation unit
- a 'sensing' unit
- a frequency determination and reproduction unit
- an analysis and control subsystem
- prime and multichannel output power amplifiers
- a computer.
The control vehicle features:
- a system control console
- a data transmission unit

The SPN-2 radar jamming system showing the architecture's antenna vehicle in the foreground

- communication radios
- an automatic 'monitoring and registration' subsystem
- a signal simulation unit.
The system's power generation vehicle houses a 60 kW diesel generator and an 'industrial electrical network' connector and each SPN-2 architecture carries sufficient spares, tools and accessories as to make it self-sufficient in the field.

Functionally, SPN-2 can operate as a stand-alone unit or as part of a jamming network with overall control being exercised from a central control station. Using its two beam output, SPN-2 is claimed to be effective against frequency-agile and 'slow carrier wave tuning' radars and as being able to handle up to two SL or navigation/TF or up to six WC (one SL or navigation/TF radar plus three WC radars per beam) simultaneously. The equipment offers narrow (10 by 45°) and wide (45 by 45°) radiation patterns and can generate continuous noise (masking) and noise repeater jamming modulations. In terms of its effectiveness against WC emitters, the system is claimed to be able to prevent the detection of an up to 10 m² radar cross section target at ranges of between 10 and 15 km. SPN-2's operating frequency is given as 'within 2 cm', which is taken to mean within the J-band (10 to 20 GHz).

Specifications
Total output power: 1,100 W
Receiver sensitivity: 90 dB/W
Detection range: 70-80 km ('dete−mination of SL/ASWC radar class'); 130-150 km (detection of SL, ASWC and navigation/TF radars)
Angular limits: 0-360° (azimuth); -2.5 to +45° (elevation, narrow radiation pattern); −7.5 to +78° (elevation, wide radiation pattern)
Accuracy of angular co-ordinate follow-up: 1.4° (elevation); 3.75° (azimuth)
Accuracy of mid-band frequency determination on received signals: 3.5 MHz (max)
Repeater noise delay: 10 μs (max)
Continuous operation: 24 h (max)
Power supply: 220 V AC (400 Hz); 380 V AC (50 Hz)
Power consumption: 50 kW
Temperature: −50 to +40°C (operating)

SPN-4
Like SPN-2, the computer controlled SPN-4 radar jammer is designed to protect ground forces and 'small size' installations and is described as being effective against airborne pulsed side-looking (SL), air-to-surface weapon control (ASWC), navigation and terrain-following (TF) radars. The equipment comprises antenna, system control and power generation vehicles, all of which make use of the KamAZ-4310 cross-country chassis. Of the three, the antenna vehicle is equipped with:
- an array assembly that comprises multibeam transmission and reception antennas and a sidelobe compensation unit
- a 'sensing' unit
- a frequency determination and reproduction unit
- an analysis and control subsystem
- a built-in test subsystem
- a power amplifier.
The control vehicle features:
- a system control console
- a data transmission, 'combat operation control' and 'registration' unit
- communication radios
- an air conditioning unit.
The system's ED60-T230P-1RAM1 power generation vehicle houses a 60 kW diesel generator and an 'industrial electrical network' connector and each SPN-4 architecture carries sufficient spares, tools and accessories as to make it self-sufficient in the field.

Functionally, SPN-4 can operate as a stand-alone unit or as part of a jamming network with overall control being exercised from a central control station. The equipment automatically searches for targets, identifies them as 'friend or foe', classifies their type (SL, ASWC or navigation/TF), measures their parameters (carrier frequency, pulse duration, pulse repetition frequency and the 'rate of change in the pulse sequence envelope'), prioritises them and generates an appropriate output. SPN-4 jamming signals are described as having 'predetermined structures' within the system's 24-beam, 'Simultaneous Operating Sector' (SOS). Overall, SPN-4 has three operating modes as follows:

Scan
In which, the system determines the target emitter's bearing.

Suppression
In which, the equipment identifies target type, selects targets, automatically tracks targets and jams targets.

Simulation
In which, the system generates synthetic data for crew training.

SPN-4's operating frequency range is given as 'within 3 cm', which is taken to mean the junction of the I- (8 to 10 GHz) and J- (10 to 20 GHz), bands.

Specifications
Total output power: 1,250-2,500 W
Receiver sensitivity: 90 dB/W
Range: 30-50 km (navigation/TF radar jamming); 40-60 km (SL radar jamming); 80 km (signal parameter and radar class determination, min); 150 km (ASWC radar detection, min)
Angular limits: 0-360° (azimuth); −2.5 to +45° (elevation)
SOS width: 45° (azimuth and elevation)
Power supply: 220 V AC (400 Hz); 380 V AC (50 Hz)
Power consumption: 50 kW
Temperature: −50 to +40°C (operating)

Status
As of the second quarter of 2000, the SNP-2 and SNP-4 radar jammers were understood to be available.

Contractors
SPN-2
Gradient Research Institute, Rostov-on-Don.
Kvant State Production Enterprise, Novgorod.
SPN-4
Gradient Research Institute, Rostov-on-Don.
Bryansk Electromechanical Plant, Bryansk.

VERIFIED

....

TShU 1-7 Infra-Red (IR) jamming system

Type
IR jamming system for Armoured Fighting Vehicle (AFV) applications.

Description
TShU 1-7 is designed to provide AFVs (such as the T-72 and T-80 main battle tanks) and fixed-site targets with protection from anti-tank missiles such as TOW, HOT, Milan, Dragon, Cobra and AT-3. A vehicle fit comprises two 280 × 350 × 350 mm IR radiator units (weighing 30 to 35 kg each); a 280 × 350 × 120 mm power supply/control unit (10 to 15 kg) and a 100 × 70 × 50 mm control panel (0.3 to 0.6 kg). In an AFV application, the TShU 1-7 radiators are mounted on the vehicle's turret and can also be used as illuminators for night vision devices. In countermeasures mode, the system is described as offering variable jamming modulations and the ability to counter several types of threat missile simultaneously. Specified system life is given as 1,000 hours with a mean time between failure value of 250 hours. The TShU 1-7 radiation source is noted as having a 50 hour life. While not confirmed, TShU 1-7 may also form part of the Shtora-1 automatic optical jamming AFV defensive aids subsystem. If this reading is correct, Shtora-1 teams TShU 1-7 with a 0.6 to 1.1 μm band, 360 × −5/+25° field of view laser warner and a grenade launching package. The grenade subsystem has a 12 round capacity, uses launch tubes set at an elevation of 12° and can generate an aerosol smoke cloud in less than 3 seconds.

Contractor
Elers-Electron Ltd, Moscow.

VERIFIED

SOUTH AFRICA

GSY 1800 Electronic Warfare (EW) system

Type
Mobile tactical EW system for communications band signal detection, analysis, Direction-Finding (DF) and countermeasures.

Description
GSY1800 has been developed to provide Electronic Support (ES), Electronic CounterMeasures (ECM) and EW command and control support for ground forces. Looking at these various functions in more detail, the ESM capability covers the 1 MHz to 1 GHz (extension to 3 GHz as an option) band and includes spectrum surveillance, search, fixed task, DF and best point calculation functions. The fixed

task role includes signal intercept, translation, data documentation and classification while best point calculations comprise line of bearing measurement, triangulation and single site location. All derived data is stored in an integral database with individual databases being networked between stations by radio. The ECM subsystem can be used for deception and jamming (covering the 1 MHz to 0.5 GHz (extension to 3 GHz as an option) band) and each GSY1800 station is noted as being able to undertake up to 4 hours of continuous silent running with all onboard equipment functioning. Direction-finding coverage is quoted as being across the 1 MHz to 1 GHz band (up to 3 GHz as an option) and recharge time (using an onboard generator) is given as 8 hours. The system vehicle is air-conditioned and incorporates a radio communications facility.

Status
As of this edition, the GSY 1800 EW system was thought to be available.

Contractor
Grintek EWATION, Pretoria.

UPDATED

SWEDEN

Tele-Weapon System (TWS) 80

Type
Ground mobile jamming system.

Description
TWS 80 is a ground mobile jamming system comprising Direction-Finding (DF) and jamming vehicles and a wheeled Electronic Warfare (EW) centre system controller. Prime contractor is Ericsson Microwave Systems with Thomson Racal Defence providing the DF subsystem and DMW the jamming chain. Functionally, a number of TWS 80 DF and jamming vehicles are controlled from a single EW centre, which is also equipped with surveillance receivers for monitoring applications. No technical details have been publicised.

Status
As of this edition, the TWS 80 jamming system is reported as having been procured by the Swedish Army and to have entered service during 1985. The system is

The tracked DF and jamming vehicle/trailer combination used in the TWS 80 system (MS/Swedish Army)

The EW control vehicle used in the TWS 80 system (MS/Swedish Army)

further understood to have been upgraded with new computers and a new DF subsystem. An additional requirement is noted as being the incorporation of a TWS 80 type system into an armoured chassis.

Contractor
Ericsson Microwave Systems AB, Mölndal (prime contractor).

UPDATED

TURKEY

JAMINT-3 tactical communications band jamming system

Type
Tactical Very/Ultra High Frequency (V/UHF - 20 to 500 MHz sub-band) communications jamming system.

Description
Mounted in a 4 × 4 vehicle and an associated single-axle trailer, the JAMINT-3 V/UHF communications jamming system is designed to intercept, analyse and if instructed by an associated command centre, jam hostile communications channels. The equipment incorporates an integral air conditioning unit to facilitate operations in extreme (hot and cold) climatic conditions and the system's power generator is housed in the previously noted trailer. An integral pneumatic antenna mast is provided to facilitate rapid deployment and/or tear down. JAMINT-3 makes use of a log periodic antenna array (horizontally or vertically polarised and capable of providing 360° coverage in azimuth) and other system features include:

- a high degree of rough terrain mobility (via the use of a 4 × 4 chassis)
- two-man set up/tear down
- user friendly, menu-driven operation
- a computer controlled, modular system architecture
- built-in self test
- sequential jamming of multiple targets
- a voice or recorded message 'spoofing' facility
- adjustable bandwidth barrage jamming
- an integral panoramic display receiver to facilitate radio frequency environment analysis
- an updateable threat list facility
- 'quick learning, easy-to-operate' design
- secure voice and data communications links.

Status
As of May 2001, the JAMINT-3 communications band jamming system was reported as being in service.

Specifications
Frequency range: 20-500 MHz
Tuning step: 100 Hz
Receiver bandwidths: user selectable

Detection modes: AM, CW and FM
Audio recording: dual-track
Recording modes: carrier operated, manual or voice operated
Jamming bandwidths: adjustable
Modulation types: AM, AM + FM, CW and FM
Modulation waveforms: audio memory, audio recorder, microphone, noise and tone
Jamming modes: barrage, sequential and spot
Jamming types: continuous, look-through and signal initiated
Output power: 25-500 W (adjustable)
Threat list: non-volatile, mission reprogrammable and can be prioritised
Primary power: 220 V AC (50 Hz)
Temperature: −30 to +50°C (operating); −40 to +60°C (storage)
Humidity: 90% (non-condensing)

Contractor
ASELSAN Inc, Microwave and System Technologies Division, Ankara.

VERIFIED

JAMINT-4S tactical communications band jamming system

Type
1 to 30 MHz band tactical communications jamming system.

Description
JAMINT-4S is a tactical communications jamming system that operates in the 1 to 30 MHz frequency band and comprises a truck-mounted operator/equipment shelter and a power generation/system accessories trailer. Functionally, the equipment detects and analyses hostile communications traffic and jams those channels specified by a command centre. The JAMINT's operator/equipment shelter is air-conditioned and the system can be operated via a man/machine interface in the shelter or by remote control, as required. In the remote control mode, one or more JAMINT stations operate under the control of a central jamming command station. Other system features include:

- two-man set up and tear down
- a computer controlled, modular architecture
- user-friendly, menu-driven operation
- built-in self-test
- ground and sky wave jamming capabilities
- sequential jamming of multiple targets
- a voice or recorded message 'spoofing' facility
- adjustable bandwidth barrage jamming
- a panoramic, radio frequency environment analysis display
- an updateable threat list
- colour graphic and list format data displays
- a secure voice and data communications subsystem
- 'quick to learn and operate' systems design
- environmental ruggedisation.

The JAMINT-3 communications band jamming system
2001/0100383

The JAMINT-4S tactical communications jamming system
2001/0100384

Status

As of May 2001, JAMINT-4S communications jamming system was reported as having entered production and service during 1997.

Specifications

Frequency range: 1-30 MHz (ES); 1.5-30 MHz (ECM)
Tuning step: 1 Hz
Receiver bandwidths: user selectable (up to 12 BW)
Detection modes: AM, CW, FM, FSK, SSB, (ISB, LSB and USB)
Audio recording: dual track
Recording modes: voice operated, carrier operated or manual
Jamming bandwidth: adjustable
Modulation types: AM, AM + FM, CW, FM, FSK, SSB, (ISB, LSB and USB)
Modulation waveforms: audio memory, audio recorder, microphone, Morse and noise and tone in any combination
Jamming modes: barrage, sequential and spot
Jamming types: continuous, look-through and signal initiated
Output power: 25-1,000 W (adjustable)
Threat list: non-volatile (updateable and allows for prioritisation)
Remote control: by laptop computer (all functions)
Primary power: 220/380 V AC (50 Hz)
Temperature range: −30 to +50°C (operating); −40 to +60°C (storage)
Humidity: 90% (non-condensing)

Contractor

ASELSAN Inc, Microwave and System Technologies Division, Ankara.

UPDATED

UNITED KINGDOM

Multiband Screening Grenade Mk 4

Type

Multiband screening grenade for Armoured Fighting Vehicle (AFV) applications.

Description

The Pains-Wessex Multiband Screening Grenade Mk 4 is designed to generate a visual and Infra-Red (IR) smoke screen for the protection of AFVs. The round is compatible with UK/US 66 mm jackplug contact grenade dischargers and produces a smoke that is opaque to thermal imagers, optical sights, passive image intensifiers and laser range-finders. Each Mk 4 grenade contains two fast rise time, ground-burning smoke pots that ignite instantaneously on launch and are ejected to a nominal range of 35 m from the launch vehicle. In still air conditions, a salvo of 12 Mk 4 grenades will produce a screen capable of obscuring a main battle tank within four seconds.

Status

As of the second quarter of 2000, the Multiband Screening Grenade Mk 4 was reported as being available.

Specifications

IR band coverage: 3-5 µm and 8-14 µm
Salvo screen area: 110° in azimuth × 5 m high (min height value)
Duration: 20 s (IR, nominal); 25 s (visual, nominal)
Windspeed: up to 24 km/h
Operating voltage: 1.5V (min)
Operating current: 1 A (min)
Weight: 590 g (payload); 840 g (complete round)
Dimensions (Ø × L): 66 × 215 mm

Contractor

Pains-Wessex Ltd, High Post.

VERIFIED

....

Rampart countermeasures system

Type

Chaff, infra-red decoy, smoke and tethered obstruction countermeasures system.

Description

The Rampart countermeasures system is designed to protect high value targets such as airfields and missile batteries against air attacks and is claimed to be the world's first passive defence system so applied. It is based on a number of individual firing units which are radio controlled by a compact transmitter located at a convenient central position. Any combination of defensive measures can be activated instantly at any time and at ranges of up to 15 km.

A complete Rampart system is inexpensive, requires a minimum of personnel and is at immediate readiness for action at all times. It provides protection against manned aircraft, laser- and TV-guided missiles by the rapid emission of smoke and

The firing unit used in the Rampart countermeasures system

against aircraft and missile radars by the use of chaff decoys. It also offers a unique defence against low-level aircraft attack by the quick release of a mass of 'Skysnare' airborne tethered obstacles that provide an obstruction and which can remain aloft indefinitely. A series of Skysnare balloons is placed around a target area that, because of their kite-like design, will hold station even in the lightest wind. Attacking aircraft will be subject to severe weapon aiming and delivery problems and will be forced to climb to an altitude that will expose them to active defence systems. 'Skynet' is a larger balloon reaching its deployment height of 1,000 m in 6 minutes. It can be deployed further and higher from the defended area than Skysnare, precluding terrain masking by the attacker during his approach.

Firing stations are portable, can be solar powered if appropriate and are each capable of firing rocket decoys (chaff or infra-red), smoke and Skynet Skysnare obstructions. Rapid or slow burning smoke is available to give both immediate area coverage and sustained coverage thereafter. The Skysnare system consists of a balloon, 100 m of Kevlar cable, case, ground anchor, a gas cylinder and pyrotechnic inflation and release. It is at continuous readiness and can be fully deployed within two minutes of initiation by radio signal. The whole system creates an effective deterrent in less than 40 seconds. The equipment can be further enhanced by the incorporation of a Large Area Smoke Screening subsystem (LASS) which is fully automatic and, in response to commands from the Rampart control unit, will generate smoke to counter visual or infra-red sensors for up to 90 minutes. Wallop is also noted as having developed a Rampart Mk II variant that has been designed to meet NATO requirements. This version can be mobile and has a new infra-red screening smoke back-up to complement the existing decoy features.

Status

As of this edition, the Rampart countermeasures system was reported as having been procured by at least two customers.

Contractor

Wallop Defence Systems (a division of Flight Refuelling Ltd), Middle Wallop.

VERIFIED

....

Type S373 Electronic CounterMeasures (ECM) system

Type

Radar frequency surveillance and jamming system.

Description

The BAE Systems Avionics - Sensor Systems Division (formerly GEC-Marconi) Type S373 ECM system is a multiband radar jamming system that incorporates an autonomous surveillance subsystem for the detection, localisation and analysis of

The S373 mobile ECM system showing its 'double-bubble' antenna radome in the deployed position

target radar emitters. The equipment makes use of a twin, horn-fed, parabolic dish reflector assembly that is housed in a 'double-bubble' radome and is deployed (using hydraulic jacks) from the roof of the system vehicle. These antennas can be turned at rates of up to 200 rpm (by means of slab torque motors) and can scan sectors of between 5 and 180° in azimuth (with a positional accuracy of <0.1°) and between −5 and +8° in elevation. The assembly's movement in elevation is driven by electrical actuators, with the required angular setting being selected remotely. Antenna scan speed and positions in azimuth/elevation are displayed on a console in the system's one ton, flatbed truck-mounted operator/equipment shelter. Each of the equipment's jamming sub-bands has its own high-power travelling wave tube transmitter feeding its own antenna, with each transmitter incorporating a digitally controlled variable frequency oscillator, a driver stage and modulator with an output amplifier stage. A comprehensive range of modulation types and patterns is provided including amplitude/wideband frequency modulation (sinusoidal, linear swept or noise), pulse modulation (multiple pulse-widths and duty cycles) and combinations of the three.

Status
Over time, the British Army is reported as having procured the Type S373 ECM system.

Contractor
BAE Systems Avionics - Sensor Systems Division, Stanmore (product support).

UPDATED

UNITED STATES OF AMERICA

AN/TLQ-32 ARM-D anti-radiation missile decoy

Type
Decoy system against anti-radar missiles.

Description
AN/TLQ-32 ARM-D is a miniature radar transmitter designed to protect radars in the field from Anti-Radiation Missiles (ARMs) which are guided by homing in on the radar's own transmission signals. The ARM-D provides protection to the radar by emulating the transmission characteristics of the host radar, thereby deceiving and confusing the incoming missile. Features of ARM-D include its capability to

The ITT Gilfillan AN/TLQ-32 ARM-D decoy

emulate frequency-agile radars; 360° coverage; protection of both the radar and the decoy assets against ARMs; lightweight fibre optic interface between the radar and decoy emitter groups and low prime power operation. It also features rugged, lightweight modular packaging, extensive built-in test capability and rapid set up and tear down. It is claimed that the decoy can be transported by two people with individual decoys being deployable within 15 minutes. In operational use, three decoys are allocated to each radar system. The surveillance decoys are designed to be capable of protecting the radar site from multiple missile launches, whether simultaneous or consecutive.

Status
The AN/TLQ-32 ARM-D was selected by the US Air Force (USAF) in March 1989, with a then year US$11.6 million contract for two 'first article' examples being awarded during the following September. Testing of these began in May 1992 and full-scale production of 14 systems to protect USAF AN/TPS-75 radars began in December 1992. During 1996, additional TLQ-32 systems were delivered to the US Air National Guard.

Contractor
ITT Gilfillan, Van Nuys, California.

VERIFIED

AN/VLQ-8A Infra-Red (IR) countermeasures set

Type
IR countermeasures system for land vehicles.

Description
The AN/VLQ-8A IR countermeasures set is designed to provide immediate protection to armoured fighting vehicles against a wide variety of second-generation, IR-guided anti-tank missiles. The system comprises a small, lightweight, robust transmitter that incorporates an integral Electronic Control Unit (ECU). The ECU contains the operator signal and built-in test circuitry while, overall, VLQ-8A's optical features are claimed to provide optimum field of view and wavelength coverage without compromising night vision covertness.

Status
The AN/VLQ-8A is reported as being operationally qualified on the US Army's M1A1/M1A2 Abrams main battle tank and M2/3 Bradley fighting vehicle. Over 1,000 equipments of this type are understood to have been delivered to the US Army.

Specifications
Dimensions: 406 × 267 × 254 mm
Weight: 11.4 kg

Contractor
BAE Systems North America - Information and Electronic Warfare Systems, Nashua, New Hampshire.

UPDATED

MGARJS mobile radar jamming system

Type
Network of Electronic Support (ES) and Electronic CounterMeasures (ECM) subsystems.

Description
The Mobile Ground-to-Air Radar Jamming System (MGARJS) is designed to provide electronic warfare protection support for high-value targets and installations. In particular, the system (which can be integrated into an air defence system) provides surveillance, acquisition and analysis of airborne radar systems; directed ECM against such emitters and radar track correlations. MGARJS consists of mobile stations strategically located around high-value installations, with the exact number and role of particular stations being tailored to the mission requirement. Multiple ECM subsystems provide the necessary tracking and jamming capabilities and are under the direct command and control of the ES subsystem which detects, identifies and tracks threat emitters. Overall, MGARJS is configured to minimise activation time and is supported by a mobile field maintenance station which can also act as a communications relay node if so required.

Status
As of this edition, at least 20 MGARJS systems were reported as having been delivered to customers.

Specifications
ES subsystem
Frequency coverage: 0.35-18 GHz
Measurement/bearing accuracy: 5 MHz RMS/1-2° RMS (8-18 GHz)

Azimuth coverage: 360°
Pulse-width/PRF range: 50 ns-50 μs (±100 ns accuracy)/3-20,000 μs (500 ns RMS accuracy)
Pulse handling capability: 1 Mpps (approx)
Automatic emitter correlation: simultaneous geolocation/tracking of up to 100 targets by correlation with air defence radar tracks
Automatic analysis capability: 1,000 separate emitters
Library: up to 3,000 emitters (0.2-1.2 s identification time)
ECM subsystem
Frequency: 8-18 GHz
Measurement/tracking accuracy: 1 MHz RMS/1° RMS
Output: 250 kW ERP per polarisation (2 per ECM; other parameters as per ES subsystem)
Jamming modes: 19 noise and deception modes (min)
Weight: <1,814 kg

Contractor
Alliant Defense Electronic Systems, Clearwater, Florida.

VERIFIED

PACJAM portable countermeasures system

Type
Portable communications band jamming equipment.

Description
PACJAM is a lightweight, portable, high-power communications band jammer with operator selectable modulations and power output. It uses discone and log-periodic antennas. The system intercepts and disrupts hostile communications using priority preset frequencies with automatic look-through. It is ruggedised for low-intensity conflict, special operations and uses battery power for remote operating capability. The receiver/transmitter unit weighs 13.6 kg, with the batteries and case adding a further 16 kg. The system is adaptable to unmanned aerial vehicle operations. PACJAM system features include:
- manual or automatic mode operation with 512 programmable frequencies, 10 scan bands and programmable look-through
- wideband Frequency Modulation (FM), narrowband (FM), amplitude modulation, continuous wave and noise modulation capabilities
- pulsed FM transmission scheme to extend battery life
- up to three hours intermittent jamming
- presettable target and priority frequencies
- selectable output power
- discone or log-periodic antenna for full bandwidth operation
- available options include 12/24 V operation, RS-232 interface, 200 W output 100 to 500 MHz.

Status
As of this edition, the PACJAM portable communications band jamming system was reported as having been procured by an as yet unidentified US agency.

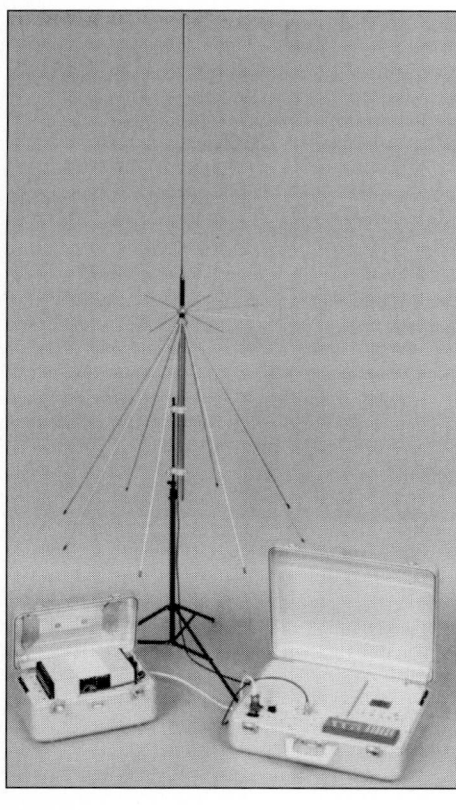

The PACJAM portable countermeasures system

Specifications
Receiver
Frequency range: 100 kHz-2,036 MHz
Presets: 400 channels in scan mode
Scan rate: 20 channels/s
Modulation: WB/NB FM, AM, CW, USB, LSB
Jammer
Frequency range: 20-500 MHz
RF output (nominal): high power 100 W CW ±1.5 dB; low power 20 W CW ±1.5 dB
Autojam look-through: 40 ms with 1-2 s jam time
Modulation: WB/NB FM noise, pseudo-random AM, external WB/NB FM, PTT

Contractor
BAE Systems North America, Lansdale, Pennsylvania.

VERIFIED

Shortstop Electronic Protection System (SEPS)

Type
Family of countermeasures systems for use against proximity-fuzed mortar/artillery threats.

Description
The SEPS is a programmable and responsive Radio Frequency (RF) countermeasure that is designed to predetonate electronic proximity-fuzed battlefield munitions such as artillery shells, mortar rounds and rockets. Functionally, SEPS equipments use an integral passive detection capability to detect emitting threats. Once a signal is detected, control algorithms establish whether it is hostile, with the SEPS RF transmission subsystem responding accordingly. SEPS units are noted as being able to create a protection zone that is the size of 'multiple (American) football fields' and as of August 2001, five SEPS variants have been identified, the known details of which are as follows:

AN/GLQ-16
AN/GLQ-16 is a stand-alone, ground-mounted, second-generation SEPS variant that weighs 21.8 kg and requires a 28 V DC (800 W peak) power supply. The GLQ-16 unit is 24 cm high by 40 cm long by 40.4 cm deep.

AN/PLQ-7
AN/PLQ-7 is a second-generation, manpack SEPS variant that weighs 11.3 kg and is powered by three BA 5590/U batteries that give it a minimum operating period of 8 hours. The PLQ-7 unit is 17.3 cm high by 29 cm long by 45.5 cm deep.

AN/VLQ-9
AN/VLQ-9 was a first-generation, vehicle-mounted SEPS variant that was developed under a Quick Reaction Capability (QRC) programme.

AN/VLQ-10
Like AN/VLQ-9, AN/VLQ-10 was a first-generation, vehicle-mounted, QRC SEPS variant.

AN/VLQ-11
AN/VLQ-11 is a second-generation, vehicle-mounted SEPS variant that weighs 22.7 kg and requires a 28 V DC (800 W peak) power supply. The VLQ-11 unit is 26.7 cm high by 40 cm long by 40.4 cm deep.
Jane's sources suggest that the second-generation SEPS equipments make use of positional data derived from the Global Positioning System.

Status
The AN/VLQ-9 and AN/VLQ-10 QRC SEPS variants were produced to meet an urgent operational need during the 1991 Gulf War. In July 1994, the system's developer — the Whittaker Electronic Systems (now Condor Systems Electronic Systems Division - CSESD) — was awarded a then year US$10.9 million US Army contract to supply an additional nine systems (36 QRC models having already been delivered) by the end of 1996. This batch of SEPS units is reported to have comprised three manpack, three tripod-mounted and three vehicle-mounted equipments. In late 1995/early 1996, a number of the original QRC equipments were reported to have been deployed to Bosnia in support of the US NATO Implementation FORce (IFOR) contingent. Elsewhere in the world, Jane's sources were suggesting that the second-generation AN/VLQ-11 SEPS unit was deployed by the US Army in South Korea during 1999. Here, the primary threat was viewed as being North Korea's 240 mm artillery rocket. On 1 May 2000, CSESD announced that it had been awarded a then year US$16,983,200 modification to an existing firm, fixed-price contract (DAAB07-98-C-H005) that covered the supply of 92 SEPS (including antennas and two units as spares) to the US military. At the time of the announcement, work on the effort was to take place at Simi Valley, California and was scheduled for completion by 30 April 2002. The efforts's contracting activity was the US Army's Communications and Electronics Command, Fort Monmouth, New Jersey. As of September 2000, Jane's sources were reporting that Condor and UK contractor Hunting Engineering were contemplating a Shortstop marketing and development agreement that would include work on an anti-frequency modulated continuous wave proximity fuze capability. On 9 March 2001, CSESD announced that it had negotiated a then year US$2.1 million SEPS research and development contract covering analysis of 'additional and new threat capabilities'

and the development of 'new and/or improved countermeasures'. The effort was to be applicable to the AN/GLQ-16, AN/VLQ-11 and AN/PLQ-7 configurations and was billed as being the first in a series of planned, multiyear, SEPS research and development projects. At the time of the announcement, CSESD noted that the March 2001 contract was the then culmination of more than then year US$ 70 million that the US Army and US Marine Corps had invested in SEPS development and production. It further noted that over 6,000 rounds had been fired against the SEPS during its developmental and operational test programmes and that since

1999, more than 220 SEPS units had been produced or ordered, with a further 45 to be booked during June/July 2001. Of this figure, four production examples had been delivered to the US Marine Corps for 'unique testing' which, as of the given date, was continuing.

Contractor
Condor Systems Electronic Systems Division, Simi Valley, California.

UPDATED

NAVAL SIGNALS INTELLIGENCE (SIGINT), ELECTRONIC SUPPORT AND THREAT-WARNING SYSTEMS

AUSTRALIA

PRISM III Electronic Support (ES) receiver

Type
Naval ES surveillance receiver.

Description
The PRISM III lightweight naval ES system comprises antenna and signal processing units with a dedicated operator console as an option. As such, the equipment automatically detects, direction-finds and classifies emitters operating in the microwave frequency band and provides immediate warning of potential threat emitters. PRISM III features a radar emitter library that can be 'easily' programmed with specific and general radar types and provides the necessary interface between PRISM and other onboard countermeasures systems such as chaff and/or active offboard decoys. Other system features include:
- 'easy' integration with a host vessel's combat, tactical data, weapon, sensor and countermeasures systems
- a distributed array pulse processing architecture
- 'advanced' pulse de-interleaving algorithms for 'fast' target identification
- removable media storage of the radar emitter library and operational software
- 'extensive' built-in test.

Readers should note that, as of January 2001, PRISM III's manufacturer had not verified the provided specification data.

Status
As of the given date, Jane's sources were reporting that naval PRISM ES system applications included the Royal Australian Navy's 'Freemantle' class large patrol craft (15 ships in class - T133 PRISM variant) and the same service's 'Huon' class coastal minehunters (four ships, two building - PRISM III variant).

Specifications
Frequency range: 2-18 GHz (in sub-bands, 0.5-2 GHz sub-band as an option)
Receiver type: channelised crystal video; IFM
Frequency resolution: 1 MHz
Frequency accuracy: better than 5 MHz RMS
Tracking sensitivity: better than −55 dBm (band edge/sub-band crossovers); better than −60 dBm (in band)
Dynamic range: 50 dB (min)
DF accuracy: better than 5° RMS (8-18 GHz); better than 7° RMS (2-8 GHZ)
Emitter POI: 100 % within 1 s
Emitter library: 1,000 emitter modes (expandable)
Weight: 31 kg (signal processing unit); 36 kg (antenna assembly)

Contractor
BAE Systems Australia, Elizabeth, South Australia.

UPDATED

The PRISM III antenna (left) and signal processing units   0044609

Warrlock Direction-Finding (DF) system

Type
1 to 500 MHz DF system.

Description
Warrlock is a 1 to 500 MHz (expandable upwards to 1 GHz) DF system that is suitable for shipborne, land-mobile or fixed-site applications. The equipment can

be configured for one or two operators and, in its baseline, single operator configuration, comprises a broadband, active, DF array (comprising up to eight individual antenna elements); an antenna multiplexer; a local DF receiver/ processor (incorporating a wideband receiver module and a DF processor) and a remote display/control unit with an optional laptop/desktop remote controller. Other system features are reported to include:
- high system sensitivity
- fast, programmable scan revisit regimes
- a time bearing display with a review facility for multiple frequencies
- built-in test
- easy to programme DF calibration
- screen-based calibration facilities that allow the operator to correct site and/or structure-based DF errors
- ±70° display viewing angle

Status
As of late 2000, 39 examples of the Warrlock DF system were reported as having been procured by the Royal Australian Navy. As of May 2001, CEA was understood to be actively promoting the Warrlock DF system.

Specifications:
Frequency: 1-1,000 MHz (omni); 2 MHz-1 GHz (DF)
Antenna type: broadband active dipoles (up to 8, proprietary phase/amplitude correlative DF technique)
Antenna sensitivity: better than 20 μV/m (1-300 MHz); better than 40 μV/m (300-500 MHz)
Antenna polarisation: vertical
Antenna array directivity: omnidirectional in azimuth to ±3 dB
1dB compressed field strength: ≥2 V/m
Receiver frequency control: 10 Hz tuning resolution
DF accuracy: 2° RMS (VHF); 3° RMS (UHF); 5-15° RMS (HF - antenna configuration dependent)
Input signal protection: >30 V/m
Input/output interfaces: CEA multidrop serial bus; customer nominated bus; IEEE 488; RS-232/RS-485; serial DCS
Power: 80-250 V AC (2 A, max)
Ship data: M-Type transmitter (heading only) or synchro (heading, roll and/or pitch)
Display screen: >10:1 (contrast ratio); >140° (viewing angle); 640 × 350 pixel (resolution)
Temperature: −25 to +70°C (operating)
Dimensions (h × w × l): 56 × 64 × 140 mm (antenna splitter); 80 × 140 × 140 mm (M-type converter unit); 90 × 425 × 450 mm (receiver/processor); 170 (h) × 105 (d) mm (omni-antenna - 170 (h) × 220 (d) mm with false earth); 210 × 140 mm (display screen); 210 × 400 × 80 mm (display/controller); 260 × 120 × 240 mm (antenna multiplexer); 305 × 62 × 342 mm (DF dipole)
Weight: 0.38 kg (antenna splitter); 0.5 kg (omni-antenna); 1 kg (M-type converter unit); 1.4 kg (DF dipole - per unit); 3 kg (display/controller); 6 kg (antenna multiplexer); 9 kg (receiver/processor)

Contractor
CEA Technologies Pty Ltd, Canberra, Australian Capital Territory

UPDATED

CANADA

AN/SLQ-501 CANEWS Electronic Support (ES) system

Type
Shipborne radar detection and analysis system.

Description
Specifically designed to perform real-time detection, analysis, identification, classification and warning of both hostile/friendly platforms and missiles, the CANEWS architecture comprises four major subsystems:
- an instantaneous frequency measuring antenna and receiver subsystem which covers the desired frequency bands and provides a very high probability of intercept
- a data transfer unit incorporating a high-speed digital processor
- twin processors which carry out the real-time analysis and supervisory function, and which provide an interface with the ship command and control system
- a display console which provides the operator interface

Built-in test equipment is also provided.

In operation, a main library holds data on numerous radar modes and signatures. Analysis of the detected radio frequency signals identifies the radar type and platform and provides a confidence level indication. In addition, the operator can

The CANEWS antenna installation on the Canadian frigate Halifax

build a tactical library of threat radar signatures to meet immediate mission requirements. While undertaking many processes automatically (to reduce operator workload and fatigue), AN/SLQ-501 provides facilities for manual control if required. One operator only is required to control the complete ES system. Data is processed automatically and displayed clearly in a number of different information formats. The manual pulse analyser enables the operator to undertake additional complex analyses. He can also expand a selected frequency band to interrogate dense parts of the spectrum more closely. Alongside the baseline configuration (and as of 1999), Lockheed Martin Canada was also developing a CANEWS 2 configuration that mated the system's existing antennas and receivers with a new processor, new software and an enhanced tactical display format.

Status
As of this edition, SLQ-501 was reported as being installed aboard 'Iroquois' class destroyers and 'Halifax' class frigates of the Canadian Navy. With regard to the CANEWS 2 effort, Phase 1B of the programme was completed during 1997, at which time an interim Advanced Development Model (ADM) had been delivered. During 1999, Lockheed Martin Canada was under contract to develop a fully capable ADM (CANEWS 2 Phase II). Elsewhere in the programme, Software Kinetics is known to have been awarded a C$32 million plus CANEWS 2 software contract during December 1998.

Contractor
Lockheed Martin Canada, Montreal, Quebec.

VERIFIED

CHINA, PEOPLE'S REPUBLIC

BM/HZ 8610 Electronic Support (ES) system

Type
Shipborne ES system.

Description
BM/HZ 8610 is an ES equipment that provides direction-finding and analysis of threat radar equipment. It can operate in conjunction with both passive and active electronic countermeasures systems and provides maximum prior warning and threat identification. The system is suitable for warships of all types and sizes. BM/HZ 8610 possesses a sophisticated radar signal processing capability and provides both tabular and graphic displays to the operator. The system's claimed direction-finding accuracy and acquisition capability facilitate it use as a cueing tool for radars, optical systems and passive fire-control systems.

Status
Over time, the BM/HZ 8610 ES system is reported as having been installed aboard 'Huangfen', 'Shanghai II' and 'Soju' class fast attack craft and 'Hainan' class large patrol craft operated by the North Korean Navy.

Specifications
Frequency coverage: 2-8 GHz; 7.5-18 GHz
Azimuth coverage: 360°
Elevation coverage: −10 to +30°
Frequency accuracy: 5 MHz RMS
DF accuracy: 2.5 RMS

Sensitivity: better than −70 dBWi
Dynamic range: 40 dB
Pulsewidth range: 0.1-99.9 µs
Environmental signal density: 10^5 pps
Radar data store capability: 500-1,000 modes

Contractor
Southwest China Research Institute of Electronic Equipment, Chengdu, Sichuan province.

VERIFIED

GERMANY

Common Opto-electronic Laser Detection System (COLDS)

Type
Multiplatform laser warning system.

Description
The COLDS laser warning sensor is described as being a lightweight, electromagnetic interference resistant sensor that is primarily designed for naval applications and determines threat direction, type and coding. The equipment comprises multiple wavelength sensor heads, an electronics unit (connected to the sensor heads by fibre-optic cable) and an RS-422 interface between the system and its host vessel's Combat Management System (CMS). To compliment the passive COLDS sensor, Lenkflugkörpersysteme (LFK) has also developed an active subsystem to provide a laser designation countermeasures capability. Here, the passive COLDS sensor detects a laser threat and measures its Pulse Repetition Frequency (PRF) and azimuth angle. The active subsystem then projects a 'high intensity' decoy spot 'several hundreds of meters' from the target vessel that matches the threat signal's PRF and is within the incoming missile seeker's field of view. At windspeeds of greater than 8 on the Beaufort scale, the decoy spot is accompanied by a laser light-absorbing smokescreen and the laser source used is mounted in such a way as to provide 360° in azimuth and ± 60° in elevation. Threat data is handed off from the passive sensor to the laser projector via the host vessel's CMS, which is also used to feedback optical data from the projector to ensure that the decoy signal's PRF remains within a 'few' microseconds of that of the threat designator. LFK has also proposed a variant of the described active/passive COLDS laser decoying system for use on armoured fighting vehicles.

Status
As of the second quarter of 2000, Jane's sources were reporting that eight passive COLDS systems had been procured by the navies of Canada (one system), Germany (three systems installed aboard 'Oste' class intelligence gathering vessels) and Finland (four systems installed on ocean-going patrol craft). As of June 1999, the system was being proposed for installation aboard German 'K 130' class corvettes and had been tested by the United Arab Emirates for possible use aboard its 'Lewa 1' class fast attack craft and 'Lewa 2' class corvettes. As of the same date, sea trials of the active/passive COLDS laser decoying system were anticipated aboard a German Navy vessel already equipped with the COLDS passive sensor. Jane's sources also suggest that a 10.6 µm capable COLDS variant was tested by the US Army's Tank Automotive Command during the early 1990s and that COLDS technology was licenced to US contractor Tracor (now BAE Systems North America) during 1995.

Specifications
COLDS passive sensor
Waveband: 0.4-1.7 µm (2-6 µm and 5-12 µm options)
Angular coverage: ± 45° (elevation); 360° (azimuth)
Angular resolution: 1.6° or 3° (azimuth)
Dynamic range: >80 dB (signal voltage)
Weight: 50 kg (Band 1 (0.5-1.1 µm) sensor head); 85 kg (Bands I and II (1.2-1.7 µm) sensor head)

Contractor
Lenkflugkörpersysteme GmbH, Munich.

VERIFIED

INTERNATIONAL

ALTESSE shipborne alert and awareness monitoring Electronic Warfare (EW) system

Type
Shipborne alert and awareness monitoring EW system.

Description
ALTESSE is a naval alert and awareness monitoring EW system that is suitable for both surface ship and submarine applications. As such, the system provides a

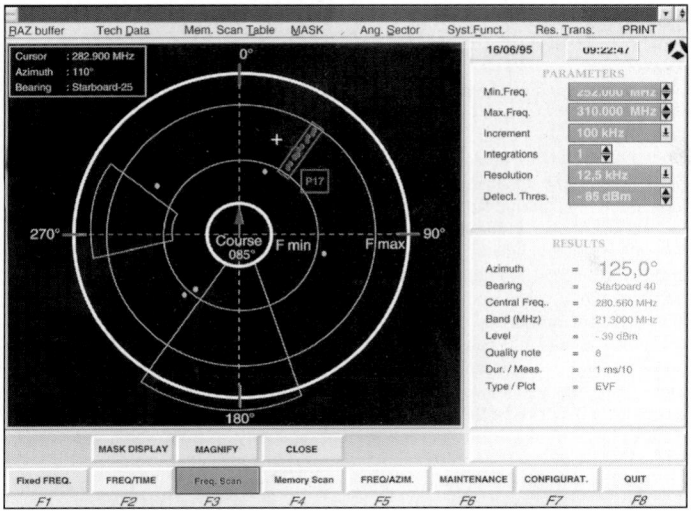

The ALTESSE system's Windows NT man/machine interface   0044611

The ANT 184-A antenna unit used in the ALTESSE shipborne alert and awareness monitoring EW system
0044612

communications intelligence collection capability, warnings of communications band signals activity and a tactical situation monitoring facility. ALTESSE is described as having a high scan rate, the ability to measure the bearing of a signal in a 'short' space of time and as being effective against 'modern threats' (such as frequency hopping and burst emitters) operating in a dense electromagnetic environment. The equipment utilises a user-friendly Windows NT man/machine interface and can be interfaced with its host ship's combat system if required. Baseline frequency coverage is 20 to 500 MHz with optional extensions down to 1 MHz and up to 3,000 MHz being available. Alongside its use in the communications bands, ALTESSE can also be used for electronic intelligence collection against radars operating at the lower end of the frequency spectrum. Thales Communications also notes that the system can be 'associated' with its TRC 274 High/Very High/Ultra High Frequency jammer and is compatible with Thales Airborne Systems' DR 3000 radar band electronic support equipment (see separate entries).

Status
As of October 2001, the ALTESSE naval EW system was reported as having been 'sea proven' by the French Navy and as being installed aboard Saudi Arabia's 'Arriyad' class frigates (three building as of March 2001).

Specifications
Frequency range: 20-500 MHz (extensions down to 1 MHz and up to 3,000 MHz as options)
Bearing accuracy: 1° RMR (operational)
Combat system interface: Ethernet; RS-232C
Antenna type: ANT 184-A (20-3,000 MHz); ANT 206 (1-3,000 MHz)
Antenna dimensions (h × w): 1.3 × 0.5 m (ANT 184-A); 2 × 1.5 m (ANT 206)

Contractor
Thales Communications, RadioSurveilllance and COMINT Systems Unit, Gennevilliers, France.

UPDATED

ARBR 17 Electronic Support (ES) system

Type
Shipborne radar detection and analysis system.

Description
The ARBR 17 ES suite is a radar detection, analysis and threat warning system that analyses the measured radar parameters and uses them to determine emitter type and identity. The equipment uses a mainmast-mounted omnidirectional antenna and two eight-port directional assemblies mounted on the same mast but slightly below. The omnidirectional element of this array provides initial detection and threat frequency and carries out primary analysis, while the directional units provide more accurate bearing and frequency information. The antenna assemblies contain preamplifiers for improved performance.

Status
During the period February to July 2001, Jane's sources were reporting ARBR 17 as being installed aboard the following classes of warship:
France   • the frigate *La Fayette*
    • 'Cassard' (two ships in class - ARBR 17B variant), 'Georges Leygues' (seven ships) and 'Suffren' (two ships) class destroyers
    • 'Floréal' class frigates (six ships)
Morocco   • 'Floréal' class frigates (two ships building as of April 2001)

Contractor
Thales Airborne Systems, Elancourt, France.

UPDATED

Cutlass Electronic Support (ES) systems

Type
Family of shipborne ES systems.

Description
The Cutlass series of equipment consists of a range of computer-controlled ES systems, primarily intended for use aboard ships, but capable of deployment on other land-, sea- or air-based platforms. Designed for operation in very dense signal environments, the equipments receive signals in the 1 to 18 GHz frequency range, measure their parameters, compare these with those in a preprogrammed radar library and display the information within 1 second. The Electronic Warfare (EW) operator is presented with a tabular display for threat identity and threat evaluation and a tactical display giving a pictorial representation of the radio frequency environment. Selected digital outputs can be sent to other local systems and hard-copy printout of the intercepted radar is also available. The tabular display can indicate 150 prioritised intercepts. The central processor used houses a library containing the parameters of up to 2,000 radars.

Cutlass ES systems are wide open in bearing and frequency (they do not employ sweep techniques) and are claimed to offer near 100 per cent probability of intercept. They are modular equipments and can be configured to suit installation aboard a wide range of ship classes. As of May 2001, the current variant is thought to be the Cutlass B1 (the Cutlass family comprising the E and B1) which utilises an Instantaneous Frequency Measuring (IFM) receiver and a 32-element antenna array to provide bearing measurement by phase analysis techniques. This antenna also provides a radio frequency output for instantaneous frequency measurement. Cutlass ES equipments can be integrated with active radar jammers (such as Thales' Cygnus or Scorpion units) and other ships' systems, such as a tactical data system.

The Cutlass E's antenna array is mounted on the host ship's masthead

Status

During the period January to September 2001, Jane's sources were reporting Cutlass ES systems as being installed aboard the following warships and classes of warship:

Algeria	• the landing ships logistic *Kalaat Beni Hammad* and *Kalaat Beni Rached*
Brazil	• the aircraft carrier *Minas Gerais* (Cutlass B1)
	• 'Niteroi' class frigates (six ships in class - Cutlass B1BW)
Denmark	• 'Niels Juel' (three ships) and 'Thetis' (four ships) class frigates
	• 'Falster' class minelayers (two ships)
Egypt	• 'October' (five ships) and 'Ramadan' (six ships) class fast attack craft - missile
Estonia	• the frigate *Admiral Pitka*
Germany	• 'Albatros' (10 ships - Cutlass B1 as part of the Octopus electronic warfare suite with Thales Scorpion radar jammer) and 'Tiger' (five ships - Cutlass B1 as part of the Octopus electronic warfare suite with Thales Scorpion radar jammer) class fast attack craft - missile
Kenya	• 'Nyayo' class fast attack craft - missile (two ships)
Kuwait	• the fast attack craft - missile *Al Sanbouk* and *Istiqlal*
Malaysia	• 'Musytari' class offshore patrol vessels (two ships)
Nigeria	• 'Vosper Thornycroft Mk 9' class corvettes (two ships)
Oman	• the patrol ship *Al Mabrukah* (Cutlass variant unknown)
Turkey	• 'Barbaros' class frigates (four ships - Cutlass as part of an electronic warfare suite with the Thales Scorpion radar jammer)
	• 'Kilic' (three ships, four building as of March 2001) and 'Yildiz' (two ships) fast attack craft - missile
United Arab Emirates	• 'Ban Yas' class fast attack craft - missile (six ships - Cutlass variant unknown)

Contractor

Thales Sensors, Crawley, UK.

UPDATED

···

Cutlass Type 242 Electronic Support (ES) system

Type

Shipborne ES system.

Description

The Cutlass Type 242 is a lightweight and compact shipborne ES equipment. The system is designed for installations where space, weight, heat generation and power consumption must be kept to a minimum. It receives radar emissions and provides instantaneous bearing and frequency measurement over 360° azimuth. The received signals are sorted by means of high-speed automatic processors to enable comparison with a threat library. Detailed information on the radar environment is presented to the operator in a variety of formats. Selected data on threats received can be routed automatically, under simple operator command, to the Action Information Organisation and to countermeasures systems using proven equipment interfaces.

Status

During the period January to August 2001, Jane's sources were reporting the Cutlass Type 242 ES system as being installed aboard the following warships and classes of warship:

Bangladesh	• the frigate *Bangabandhu*
Canada	• 'Protecteur' class replenishment oilers (two ships in class - Cutlass Type 242 designated as SLQ-504 in Canadian service)
Oman	• 'Dhofar' class fast attack craft - missile (four ships)

Contractor

Thales Sensors, Crawley, UK.

UPDATED

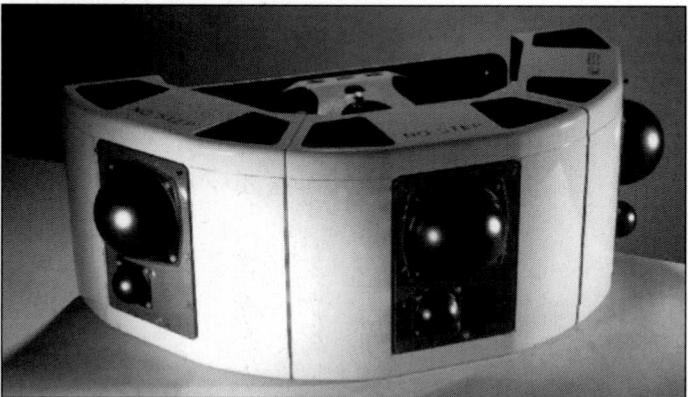

The antenna assembly used in the Cutlass Type 242 ESM system   0009368

DR 2000 Electronic Support (ES) receiver

Type

Shipborne radar surveillance system. French Navy designation ARBR 16.

Description

DR 2000 (Mk 1, Mk 2 or Mod 3) -DALIA is an ES system for use in seaborne, ground-based or airborne applications. Three versions have been developed: the DR 2000S for surface ship/ground station use, the DR 2000U for submarines and the DR 2000A for airborne fitment. In all cases the basic system consists of a radar intercept receiver (DR 2000) and an analyser (DALIA 1000 Mk 1 or Mk 2), plus the necessary antenna units. The DR 2000 intercept receiver is built up of six Direction-Finding (DF) antennas and one omni-antenna, and a processing control/display console. It provides a virtual 100 per cent probability of detection over the complete 360° of azimuth. The equipment carries out passive search, detection of all pulse and continuous wave signals and gives an instantaneous audible and visual alert. The DR 2000 Mk 2 and Mod 3 are improved versions with better sensitivity and coverage of the frequency bands from D to J and a DF accuracy of 5° RMS. The DALIA analyser provides alarm analysis and identification facilities through a library of 1000 radar modes and parameters, which is easily reprogrammable. It provides high automatic selectivity in dense radar environments, with a high analysis accuracy.

Status

During the period January to April 2001, Jane's sources were reporting ARBR 16/DR 2000 ES systems as being in service aboard the following warships/submarines and classes of warship/submarine:

ARBR 16

France	• 'Tourville' class destroyers (two ships-in-class)
	• 'D'Estienne d'Ovres' class frigates (nine ships)
Turkey	• 'Burak' class frigates (six ships)

DR 2000

Argentina	• the anti-submarine warfare capable submarine *Salta*
Colombia	• 'Pijao' class general purpose submarines (two boats)
Ecuador	• 'Manta' (two ships) and 'Quito' (three ships) class fast attack craft - missile
Germany	• 'Ensdorf' class coastal minesweepers (five ships)
	• 'Kulmbach' class coastal minehunters (five ships)
Greece	• 'Anninos' (four ships), 'Laskos' (nine ships), 'Super Vita' (three building, four proposed as of February 2001) and 'Votsis' (six ships) class fast attack craft - missile
	• 'Armatolos' class large patrol craft (two ships)
	• 'Niki' class gunboats (five ships)
Indonesia	• the fast attack craft - missile *Badik* and *Keris*
	• 'Singa' class large patrol craft (four ships - DR 2000S3 with DALIA analyser)
Malaysia	• 'Perdana' class fast attack craft - missile (four ships)
Peru	• 'Velarde' class fast attack craft - missile (six ships)
Tunisia	• 'Combattante IIIM' class fast attack craft - missile (three ships)
Turkey	• 'Atilay' class anti-submarine warfare capable submarines (six boats - DR 2000 suggested as an alternative to Thales Sensors' Sealion or Porpoise systems)
Venezuela	• 'Sabalo' class anti-submarine warfare capable submarines (two boats)

Contractor

Thales Airborne Systems, Elancourt, France.

UPDATED

···

DR 3000S/U Electronic Support (ES) system

Type

ES receiver for surface vessels and submarines. French Navy designations ARBR 21 and ARUR 13 (see following).

Description

Naval DR 3000 is a modular tactical ES system that is designed for shipboard applications and offers threat detection, Direction-Finding (DF), identification and target designation of enemy radars operating in the C- through J-bands (0.5 to 20 GHz). It monitors the tactical situation in a given area and performs electronic intelligence functions. Two versions of the equipment are available, the DR 3000S for surface vessels and the DR 3000U for submarines. A third version (the DR 3000A) is available for fitment in fixed- and rotary-wing aircraft. The system uses gallium arsenide technology and is of modular and lightweight design. A wide range of options can be added to the basic unit to tailor the system to the specific requirements of the end-user. It has a high-sensitivity range that covers the breadth of the frequency spectrum, with a detection capability of nearly 100 per cent. DF accuracy is understood to be in the order of 3°. Highly automated operation includes the use of artificial intelligence and allows for very short reaction times. A technical data gathering capacity is proposed as an option for the DR 3000 equipments. As of April 2001, known naval variants of the modular DR 3000 system are as follows:

DR 3000 CLASSIC

DR 3000 CLASSIC is an Engineering Change Product (ECP) DR 3000 variant that was launched during 2000 and incorporates enhanced (when compared with the baseline architecture) signal analysis and data processing capabilities, a new antenna design and a new Man/Machine Interface (MMI). As such, DR 3000 CLASSIC is designed for 'small size platform' applications.

DR 3000 COMPACT

DR 3000 COMPACT is a lightweight ES system for use on vessels such as fast attack craft. As such, COMPACT employs a 40 kg amplitude DF antenna unit and a 57 kg receiver/processor.

DR 3000 HIGH CLASS

DR 3000 HIGH CLASS is an Engineering Change Product (ECP) DR 3000 variant that was launched during 2000 and incorporates enhanced (when compared with the baseline architecture) signal analysis and data processing capabilities, a new antenna design and a new Man/Machine Interface (MMI). As such, DR 3000 HIGH CLASS is designed for 'capital warship' applications.

DR 3000 LEADER CLASS

DR 3000 LEADER CLASS is an Engineering Change Product (ECP) DR 3000 variant that was launched during 2000 and incorporates enhanced (when compared with the baseline architecture) signal analysis and data processing capabilities, a new antenna design and a new Man/Machine Interface (MMI). As such, DR 3000 LEADER CLASS is designed for installation aboard surface ships ranging in size from corvettes to 'light' frigates.

DR 3000S1

DR 3000S1 incorporates a 40 kg amplitude DF antenna unit, a 77 kg receiver, a 47 kg processor and a 75 kg control/display console.

DR 3000S1X

DR 3000S1X comprises one 40 kg amplitude and two 49 kg phase DF antenna units, the 47 kg processor, the 77 kg receiver and the 75 kg control/display console. The equipment utilises interferometry for DF.

DR 3000S2

DR 3000S2 is generally similar to the S1 configuration with the exception that it uses two 60 kg amplitude DF antenna units. As with S1X, interferometry is used for DF.

DR 3000U

The submarine configured DR 3000U incorporates (according to the specific application) a 47 kg amplitude DF antenna unit, a periscope-mounted amplitude DF or omnidirectional radar warning antenna, a 47 kg processor, the 77 kg receiver, a 57 or 140 kg receiver/processor and the 75 kg control/display console. In French Navy service, DR 3000U is designated as ARUR 13.

Status

During the period February to July 2001, Jane's sources were reporting ARBR-21/ARUR-13/DR 3000 ES systems as being installed aboard the following warships and classes of warship/submarine:

ARBR 21
France
- the aircraft carrier *Charles de Gaulle*
- the 'La Fayette' class frigates *Aconit*, *Courbet*, *Guepratte* and *Surcouf*

ARUR 13
France
- 'Le Triomphant' class nuclear-powered ballistic missile submarines (two boats in class, two building as of July 2001)
- 'L'Inflexible M4' class nuclear-powered ballistic missile submarines (two boats)
- 'Rubis Amethyste' class nuclear-powered attack submarines (six boats)

DR 3000
Greece
- Batch 2 'Pirpolitis' class large patrol craft (four building as of March 2001)
Indonesia
- 'Kakap' (four ships - DR 3000S1) and 'Todak' (two ships, two building as of February 2001 - DR 3000S1) class large patrol craft
Kuwait
- 'Um Almaradim' class fast patrol craft - missile (eight ships - DR 3000 Compact)
Oman
- 'Qahir' class corvettes (two ships - DR 3000S1)
- 'Al Bushra' class offshore patrol vessels (three ships - DR 3000S1)
Pakistan
- 'Tariq' class frigates (six ships - being retrofitted with DR 3000S1X as of March 2001)
Qatar
- 'Barzan' class fast patrol craft - missile (four ships - DR 3000S1)
Saudi Arabia
- 'Arriyad' class frigates (three building as of March 2002 - DR 3000S2)
Taiwan
- 'Kang Ding' class frigates (six ships - DR 3000 subvariant unknown)

As of late 2000, Thales was reporting overall sales of 80 plus DR 3000 systems worldwide and as of the same date, as yet unconfirmed Jane's sources were suggesting a Colombian procurement of DR 3000S1. Elsewhere in the world, 5 February 2002 saw Thales Airborne Systems announce that it had signed a then year €26.3 million contract with respect to the supply of 11 DR 3000-based Shipborne Integrated Electronic Warfare Systems (SIEWS) to the Finnish Navy. Here, the systems were to be used to outfit six 'Squadron 2000' and four 'Rauma' class warships, with the remaining equipment to be used in a shore-based installation. As described at the time of the announcement, SIEWS incorporated both a DR 3000 application and a central EW computer and was intended to provide threat warning together with self-defence (via control of onboard active/passive countermeasures subsystems), targeting and intelligence collection capabilities. Again at the time of the announcement, work on the programme was scheduled for completion by the end of 2005.

Specifications

Generic DR 3000
Frequency: C- to J-bands
Azimuth coverage: 360°
Elevation coverage: -10 to +45°
Sensitivity: -68 dBm
Dynamic range: up to 60 dB
Radar modes: 12,000 in library

Contractor

Thales Airborne Systems, Elancourt, France.

UPDATED

DR 4000S/U Electronic Support (ES) system

Type

Shipborne radar surveillance system.

Description

The DR 4000 is an Instantaneous Frequency Measuring (IFM) ES system with frequency coverage from C- to J-band. Received data is displayed to the operator (in graphic and alphanumeric formats) on a three-colour cathode ray tube display. Symbols, alphanumeric data and interface comply with NTDS standards. The display shows the tactical situation in the upper part and intercept parameters below. DR4000 can be configured for airborne (DR4000A), surface ship (DR4000S) and submarine (DR4000U) applications. Overall, the system can be interfaced with decoy dispensers or jammers and automatically controls their operation. It is understood that the system combines crystal video and IFM receiver technology to create a system which offers threat warning, surveillance, emitter identification, electronic intelligence, and automatic electronic warfare system control capabilities. The antenna assembly used with DR 4000S includes two Direction-Finding (DF) arrays (2 × 6 DF channels), one omni-antenna (frequency channel) and modules housing radio frequency amplifiers. The two sets of DF antennas correspond to two complementary frequency bands, H to J (6 to 20 GHz - system specific high band) and C to G (0.5 to 6 GHz - system specific low band). The installation can be at the masthead, around the mast or on both sides of the bridge or superstructure. The processing unit, either frequency band designation or instantaneous frequency measurement, has reprogramming capabilities with de-interleaving, analysis and identification. The DR 4000U antenna array consists of an omnidirectional unit for frequency measurements and two concentric sets of six antennas, one set for the H to J high band for DF and one set for the C to G low band. The DR 4000U provides a very high intercept probability on even short single pulses, and is very sensitive, using crystal video and IFM techniques on both the omnidirectional and DF channels. Bearing accuracy is about 5°. The situation is displayed in raw video on a colour display giving azimuth and amplitude, and in symbols which are distinguished in colour and shape. The three colour code is used to distinguish between different degrees of threat. The operator can select alphanumeric presentation to provide an overall activity display, or can select all data concerning a specific signal, together with analysis results.

The DR 4000 IFM ES system's operators console

Status

During the period January to March 2001, Jane's sources were reporting DR 4000 as being installed aboard the following classes of warship/submarine:

Brazil   • 'Tupi' class anti-submarine warfare capable submarines (four boats in class - DR 4000U)

Saudi Arabia   • 'Madina' class frigates (four ships - DR 4000S)

Specifications

Dimensions: 415 × 300 mm (h × Ø, omni-antenna); 425 × 480 mm (h × Ø, DF antenna); 316 × 360 × 520 mm (IFM processor); 590 × 1,100 × 1,543 mm (console)

Weight: 9 kg (omni-antenna); 2 × 22 kg (DF antenna); 25 kg (IFM processor); 310 kg (console)

Contractor

Thales Airborne Systems, Elancourt, France.

UPDATED

FL 1800 U Electronic Support (ES) and Direction-Finding(DF) system

Type

Radar ES and DF system for submarine applications.

Description

The FL 1800 U ES and DF system is a fully automatic, modular equipment that is designed for submarine applications. It incorporates the USK 800/4 DF sensor (see separate entry) and provides intercept, automatic analysis, automatic DF classification and display capabilities against 'all signals' within its frequency range. Processed, analysed and classified signal data is presented on a tactical picture of the radar environment with emitter classification being by means of library comparison. FL 1800 U generates automatic visual and 'external' warnings if a detected signal exhibits certain characteristics (such as lock-on) and the system's display features a snapshot mode. Here, emitter track data can be frozen for review after the ES sensor has been lowered. FL 1800 U also offers the ability to transfer selected data from itself to its host's combat system. In terms of the USK 800 series sensors, the equipment's manufacturer notes that they can be integrated with 'all common' periscopes or electro-optic masts and that individual sensor heads can combine omnidirectional and DF antennas, ES electronic modules, a global positioning system navigation capability and a very high/ultra high frequency communications capability in a single unit. The company believes that such an approach obviates the need for submarines to be equipped with a dedicated ES mast and enables the boat's ES capability to be available immediately after its periscope/electro-optic mast breaks the surface, thereby facilitating simultaneous optical and electromagnetic search and detected target correlation.

Status

As of early 2001, Jane's sources were reporting that the FL 1800U radar ES and DF system was mandated for installation aboard German (four boats building) and

The USK /4 DF sensor used in the FL 1800 U submarine ESM and DF system

Italian (two building) Type 212A anti-submarine warfare capable attack submarines.

Contractor

European Aeronautic, Defence and Space (EADS) Co Systems and Defence Electronics, Ulm, Germany.

UPDATED

MAIGRET II

Type

Shipboard communications and radar band Electronic Support (ES) system.

Description

The MRCM (a joint venture between European Aeronautic, Defence and Space (EADS) EWATION and South Africa's Grintek EWATION) MAIGRET II shipboard communications and radar band ES system is designed to provide a vessel's commander with a 'highly effective', real-time, tactical threat picture. Additionally, the system is configured to capture all relevant intelligence data for analysis in quasi real-time or offline. Acquired data is stored and managed within a single, tactical electronic warfare database. Other system features include:

• fusion of radar and communications band data into a single picture
• electronic and communications intelligence expandable
• single or multiple operator functionality
• a modular communications band antenna array that is mounted on a single mast section

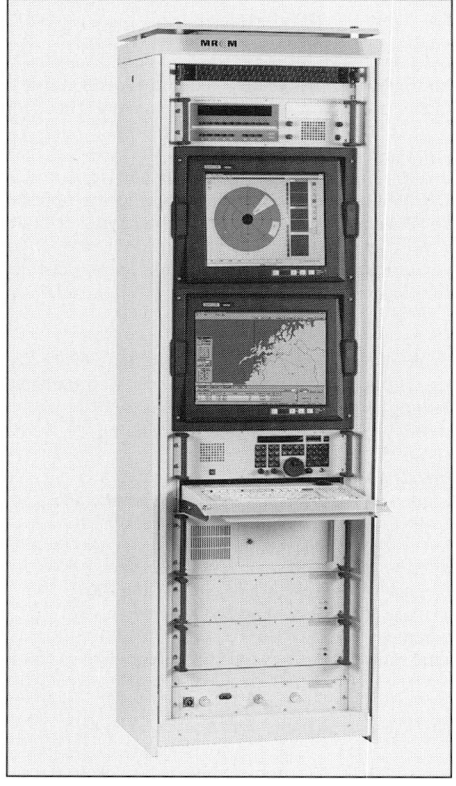

The MAIGRET II MMI (MRCM)
2002/0098317

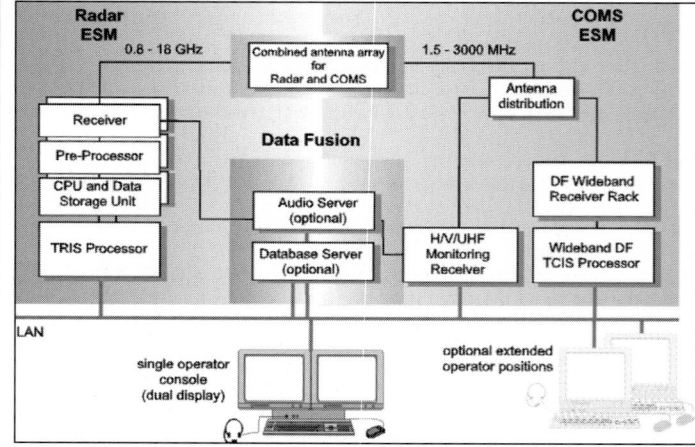

A schematic showing the layout of the MAIGRET II shipboard radar/communications band ES system (MRCM)   2002/0098318

Status
As of late 2000, the MAIGRET II shipboard radar/communications band ES equipment was reported as being in development.

Specifications
Frequency coverage: 10 kHz-3,000 MHz and 0.8-18 GHz (18-40 GHz option)
DF frequency coverage: 20-3,000 MHz (1.5-20 MHz option) and 0.8-18 GHz (18-40 GHz option)
Analysis: COMINT, ELINT and ES
Audio recording: digital format on audio server
MMI: Windows NT based

Contractor
MRCM (an EADS/Grintek EWATION joint venture), Ulm, Germany/Pretoria, South Africa

NEW ENTRY

MAIGRET communications band Electronic Support (ES) system

Type
Naval tactical communications band ES system.

Description
The compact, modular, MAIGRET surveillance system is a reconnaissance equipment that is designed to provide a rapid representation of the radio communications scenario. Its tactical and strategic sensors are especially designed for surface ship and submarine applications. As such, the architecture consists of subsystems with passive functions for interception and monitoring, direction-finding and tactical exploitation. The system capabilities are comparable with those of the SIGMA system (see separate entry).

Status
As of 2000, the MAIGRET communications band ES system was reported as having been procured by a number of navies (including that of a NATO member) around the world.

Specifications
Frequency range: 1-1,000 MHz or sub-bands
Search speed: 125 channels/s (max HF band); 40,000 channels/s (max V/UHF band)
Probability of intercept: 99%
DF modes: Watson Watt and interferometer
False alarm rate: 10^{-6}

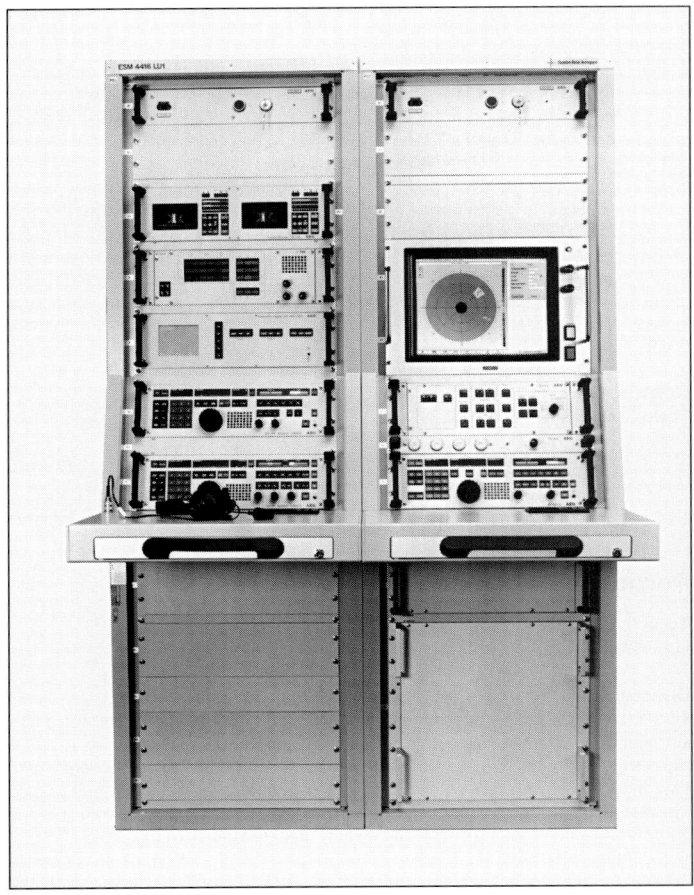

The MAIGRET naval ES system   0009370

Computer workstation: PC; AT compatible, class 486
Operator interface: graphic based on Windows

Contractor
European Aeronautic, Defence and Space (EADS) Co Systems and Defence Electronics, Ulm, Germany.

UPDATED

Manta Electronic Support (ES) system

Type
Submarine radar detection and analysis system.

Description
Manta is an ES system suitable for submarine applications with or without an electronic warfare mast. So installed, Manta intercepts, analyses, classifies and identifies radars in the 2 to 18 GHz range. It is designed as a wideband modular equipment that enables the most cost-effective system to be built up to match the operational requirement for a particular class of submarine. Different antenna modules provide options for extended frequency coverage and high-resolution targeting, while the digital processing modules can extend the pulse density and library capacities. The Manta ES system can be interfaced with the central data handling and the output data may be recorded on either magnetic tape and/or hard copied. Using wide open ES techniques, the system provides 360° coverage and 100 per cent probability of intercept. Intercepted signals are automatically analysed and identified by reference to a comprehensive library of known radar types. Processed data is immediately displayed to the operator in both cartesian and alphanumeric formats on two interchangeable colour displays. The display console can be either a dedicated display or interface with a multifunction console. Manta uses a central management computer to control subsystems, each of which, contains microprocessors for local processing tasks.

Status
As of March 2001, Jane's sources were reporting the Manta ES system as being installed aboard the following classes of submarine:

Spain	• 'Delfin' class anti-submarine warfare capable submarines (four boats in class)
	• 'Galerna' class anti-submarine warfare capable submarines (four boats - Manta E)
Sweden	• 'Gotland' class anti-submarine warfare capable submarines (three boats - Manta S)

Contractor
Thales Sensors, Crawley, UK.

UPDATED

Outfits UAP(1), UAP(3) and UAP(4) Radar Electronic Support (RES) systems

Type
Family of submarine RES systems.

Description
The Outfit UAP series of RES systems has been developed to satisfy the RES requirements of the Royal Navy's (RN) Submarine Flotilla. Outfit UAP equipments

A general view of the Outfit UAP submarine RES system's human-computer interface that shows the equipment's integral onboard trainer

feature identical operator consoles and 88 per cent component commonality between variants. The systems feature Thales Sensors' (formerly Racal) proprietary, extended window addressable memory-based SADIE signal analysis processor that is described as being capable of 'full' pulse-by-pulse analysis on up to 500,000 pulses/s 'without performance degradation'. UAP equipments feature a 'highly ergonomic' human-computer interface that facilitates computer-assisted manual signals analysis and other 'advanced' functions. A fully representative onboard trainer is included in the architecture, as are deployed antenna assemblies that are optimised to suit the mast architecture of particular classes of submarine.

Status

During the period March to September 2001, Jane's sources were reporting Outfit UAP RES systems as installed aboard or mandated for the following classes of submarine:

UK • 'Vanguard' class nuclear-powered ballistic missile submarines (four boats in class - Outfit UAP(3) variant)
• 'Astute' (three building, two proposed as of April 2001 - Outfit UAP(4) variant), 'Swiftsure' (five boats - Outfit UAP(1) variant) and 'Trafalgar' (seven boats - Outfit UAP(1) variant) class nuclear-powered attack submarines

As of late 1999, a UAP enhancement programme was under consideration. Here, the intention was to address technical harmonisation questions between the Outfit UAP family and other RES systems in RN service.

Contractor

Thales Sensors, Crawley, UK.

UPDATED

Outfit UAA(2) Electronic Support (ES) system

Type

Surface ship ES system.

Description

UAA(2) is a radar signal analysis and direction-finding system that incorporates instantaneous frequency measuring receivers. The intercept and direction-finding antenna array is a series of radome-covered horns normally mounted under a masthead radar. UAA(2) is an upgrade of the earlier UAA(1) system that incorporates solid-state electronics, an increase in the number of available warner channels, an improved threat library and greater resistance to interference signals.

Status

As of April 2001, Jane's sources were suggesting that the UAA(2) ES system was likely to remain in service aboard the Royal Navy destroyers *Cardiff*, *Glasgow* and *Newcastle* until their decommissioning during the period 2007 (*Cardiff* and *Newcastle*) to 2009 (*Glasgow*).

Contractor

Thales Sensors, Crawley, UK.

UPDATED

Outfit UAT Electronic Support (ES) system

Type

Surface ship ES system.

Description

The Outfit UAT ES system comprises an eight-element antenna array, a receiver/processor and an operator's console. The equipment can be integrated into its host vessel's command system and features an integrated receiver subsystem and system-wide software control. The operator console incorporates two display screens showing, respectively, the overall electronic environment and alphanumeric tabular data on detected emitters.

Status

During the period March to August 2001, Jane's sources were reporting Outfit UAT ES variants as being installed aboard the following Royal Navy warships and classes of warship:

UK • the amphibious assault ship-helicopter HMS *Ocean* (Outfit UAT(1) variant)
• 'Invincible' class aircraft carriers (three ships in class - Outfit UAT(5) variant)
• Batch 2 and 3 'Type 42' class destroyers (eight ships - Outfit UAT(5) variant)
• 'Broadsword' class frigates (five ships - Outfit UAT(6) retrofit programme)
• 'Duke' class frigates (15 ships, one building as of August 2001 - Outfit UAT(1) aboard Batch 2 and 3 vessels, Outfit UAT(7) retrofitted to Batch 1 ships)

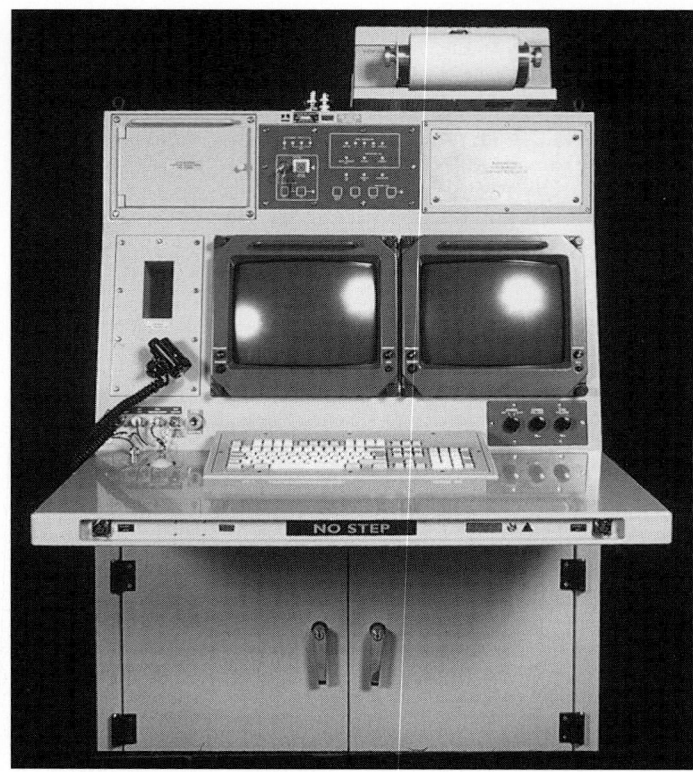

The Outfit UAT operator console

• 'Albion' class amphibious transport docks (two ships - Outfit UAT(12) variant)
• 'Fort Victoria' class replenishment oilers (two ships - unknown Outfit UAT variant replaces original Outfit UAG ES system ship's fit)

Alongside the noted seagoing Outfit UAT variants, Outfit UAT(2) and Outfit UAT(3) are the designations given to the system when used in shore installations. Identified Outfit UAT contracting activity includes the following:

April 1990
Initial contract for 12 Outfit UAT systems for installation aboard 'Invincible' class aircraft carriers (two equipments) and 'Duke' class frigates (eight). The remaining pair of systems were for installation at shore establishments.

April 1996
A then year £36 million award for 16 Outfit UAT shipsets (including systems for retrofit aboard 'Type 42' class destroyers as part of the Outfit UAA(2) Mid-Life Update effort).

Mid-1997
A then year £27 million award for nine Outfit UAT equipments for installation aboard eight 'Duke' class frigates and the aircraft carrier HMS *Invincible*.

In addition, Thales Sensors is known to have supplied BAE Systems with three Outfit UAT systems for installation aboard HMS *Albion*, HMS *Bulwark* and HMS *Ocean*. The value of this contract is believed to have been in the region of then year £6 million. It should also be noted that Outfit UAT processing and display technology forms the basis of a Thales upgrade package that has been developed for those of the company's Sceptre ES systems that are installed aboard 'ANZAC' class frigates of the Royal Australian and Royal New Zealand Navies. Here, Thales Sensors was awarded a then year £3 million contract for the programme during early 2001, with, at the time of the announcement, the first installation of the upgraded system (designated as the Centaur equipment) being scheduled to take place during early 2002.

Contractor

Thales Sensors, Crawley, UK.

UPDATED

Porpoise Electronic Support (ES) equipment

Type

Submarine ES system.

Description

The Porpoise ES system is a submarine version of the Cutlass system and is a fully automatic equipment that provides 360° arc in azimuth. The system receives signals in the 2 to 18 GHz frequency range, measures their parameters and compares these with those contained in a preprogrammed radar threat library. Processing of the radar emitter signals is carried out against the library to give the operator an alphanumeric or graphic display of identification in threat significance order. Intercepted signals are preamplified in the mast unit before being passed to the processing and analysis equipment inside the hull. Bearing data are extracted using amplitude comparison. Amplitude, pulse-width, frequency and time are combined in a single digital word before being passed to the processor unit in the operator's console. There the pulse trains of the different radars are de-interleaved and identified from the library information. The library carries up to 2,000 emitter modes. Porpoise is capable of being integrated with its host boat's fire-control and communications systems and may also be integrated with periscope-mounted radar warning equipments. The system can also provide alert warnings when prime threats (such as helicopter or maritime surveillance radars) reach a preprogrammed danger level. The Porpoise antenna is a compact six port system giving good bearing accuracy and may be mounted on either hull penetrating or non-hull penetrating masts. It is built of titanium to reduce weight and overcome corrosion and is pressure resistant to 60 bar.

Status

As of March 2001, Jane's sources were citing the Porpoise ES system as a potential option aboard Turkish 'Atilay' class anti-submarine warfare capable submarines (six boats in class as of the given date). When supplied to the Royal Navy (for use on 'Oberon' class submarines), the equipment was known as Outfit UAJ(1).

Contractor

Thales Sensors, Crawley, UK.

UPDATED

Sceptre Electronic Support (ES) systems

Type

Family of shipborne radar intercept and analysis systems covering all classes of naval vessels.

Description

Sceptre is a family of shipboard wideband modular ES systems designed for specific operational requirements within any prevailing space and weight limitations. The system provides automatic intercept, analysis, classification and identification of all radar emissions. A wide range of antenna and receiver options is offered to increase frequency coverage and sensitivity, as well as to enhance bearing and frequency resolution and passive targeting capabilities. Optional digital processing modules are available to extend the pulse density and library handling capabilities. Output data can be stored either on magnetic tape or in hard copy format and the system can be interfaced with any central data handling system.

Using wide open ES antenna technology, Sceptre provides 360° coverage and 100 per cent probability of intercept. Intercepted signals are automatically analysed and identified by reference to a comprehensive library of known radar types. Processed data is immediately displayed to the ES operator in both cartesian and alphanumeric formats. Sceptre is designed to interface with the ship's combat system, supplying a comprehensive set of emitter details and threat alerts to any requested standard. A processor unit designed for use with multifunction consoles is available.

Over time, a number of Sceptre variants have been proposed/developed, the known details of which are as follows:

Sceptre X

Sceptre X is a high-performance ES system that detects, measures and identifies radars operating in the 2 to 18 GHz frequency band. Coverage of other bands is available. All detected signals are rapidly measured and sorted to provide identification on each radar. Radars are automatically identified by matching the pulse train data with known radar data held in the emitter library. Immediate warning is given of any threat radar and the system will provide automatic control of chaff or other self-protection equipment. Data may be recorded for intelligence gathering purposes. Sceptre X is compact and of low weight and can be integrated with the ship's command suite. It features one-man operation; a multicolour display; programmable threat, mission and operator libraries; 100 per cent probability of intercept; instantaneous detection over 360°; and tracking of all emitters detected.

Sceptre XL

Sceptre XL is a high-capability ES system operating in the 2 to 18 GHz frequency band. Extensions to cover higher and lower bands are available. All detected signals are rapidly measured and sorted into pulse trains. These are identified automatically by matching the measured data with known radar data held in a software library. Warning of threat and initiation of countermeasures is very fast to enhance ship survival against attack. Sceptre XL possesses high sensitivity and dynamic range, real-time displays and enhanced manual analysis. It also features one-man operation; a real-time, multicolour display; comprehensive programmable threat, mission and operator library facilities; continuous 360° detection and tracking; and 100 per cent probability of intercept.

Status

As of the period March to August 2001, Jane's sources were reporting that a Sceptre ES system variant was installed aboard 'ANZAC' class frigates of the Royal Australian (five ships in class, five building as of August 2001) and Royal New Zealand (two ships) Navies. As of early 2001, the same sources were indicating that Thales had received a then year £3 million contract to upgrade the ANZAC Sceptre systems to Centaur standard. Here, the upgrade was understood to centre on processing and display technology from the company's Outfit UAT system (see

separate entry), with the first Centaur configured equipment being scheduled for installation during the first half of 2002.

Contractor

Thales Sensors, Crawley, UK.

UPDATED

Sealion Electronic Support (ES) system

Type

2 to 18 GHz band radar ES and threat warning system.

Description

The Sealion ES and threat warning system is designed primarily for naval surface and subsurface applications and comprises a main antenna assembly, a search periscope-mounted, omnidirectional, threat warning antenna (as its description suggests, specific to submarine applications), Receiver-Processor-Digitiser (RPD) and Pulse Train Analyser (PTA) units and a low-volume, low-weight Man/Machine Interface (MMI). Of these, the main antenna assembly consists of two omnidirectional antennas, an optional Global Positioning System antenna and an eight port (four active) Direction-Finding (DF) array. The system switches between the elements of the DF array on a pulse-by-pulse basis and when fitted aboard a submarine, the entire assembly can be configured for installation on penetrating or non-penetrating masts. The equipment's receiver incorporates very large scale integrated circuitry and is claimed to be able to handle 'complex' radio frequency environments. Emitter identification is by means of real-time library comparison, with the system being equipped with main and auxiliary parameter libraries that are loaded from an integral disk drive and by the operator respectively. The MMI used is ruggedised and incorporates a high resolution monitor on which data are displayed in tabular or graphic formats. Data shown include emitter identity, bearing and threat significance. The equipment's processor automatically tracks up to 200 emitters and provides a continuous output on their status. Acquired data can be recorded onto magnetic disk for post-mission analysis if required.

Status

As of the period February to March 2001, Jane's sources were reporting the Sealion ES system as being installed aboard the following submarines and classes of submarine:

Denmark • 'Narhvalen' (two boats in class) and 'Tumleren' (three boats) class anti-submarine warfare capable submarines

Norway • 'Ula' class anti-submarine warfare capable submarines (six boats)

Turkey • the 'Preveze' class *Canakkale* and *Gur*
• 'Atilay' class anti-submarine warfare capable submarines (six boats - Sealion as an alternative to the Thales DR 2000 or Porpoise ES systems)

Sealion is also noted as having been used in at least one surface ship application, in which context, Thales announced on 18 March 2001 that it was supplying a Sealion variant to the Bahrain Navy in a deal noted as being valued at then year 'several' million pounds.

Specifications

Frequency range: 2-18 GHz
Bearing accuracy: 2.25° RMS
Sensitivity: −65 dBmi
Dynamic range: 60 dB
Track capacity: 200-500
Library: up to 10,000 emitter modes
Coverage: −10 to +45° (elevation); 360° (azimuth)
Pulse density: up to 1 Mpps
Dimensions (h × w(Ø) × d): 280 × 4,807 × 437 mm (MMI); 600 × 600 × 480 mm (RPD); 630 × 460 mm (main antenna assembly, penetrating mast); 725 × 460 mm (main antenna assembly, non-penetrating mast)
Weight: 14.5 kg (RPD and PTA units); 50 kg (main antenna assembly - surface ship application); 72 kg (main antenna assembly - submarine application); 110 kg (MMI)

Contractor

Thales Sensors, Crawley, UK.

UPDATED

Shiploc Radar Warning Receiver (RWR)

Type

Miniature shipborne RWR.

Description

Shiploc is a navalised variant of the airborne Sherloc RWR that provides threat warning across the D- through J-bands (1 to 20 GHz). It is designed for use on smaller surface ships such as fast attack and patrol craft. It has a high sensitivity and provides instantaneous and accurate threat identification. Software controlled, Shiploc is fully reprogrammable during the mission and will carry out target

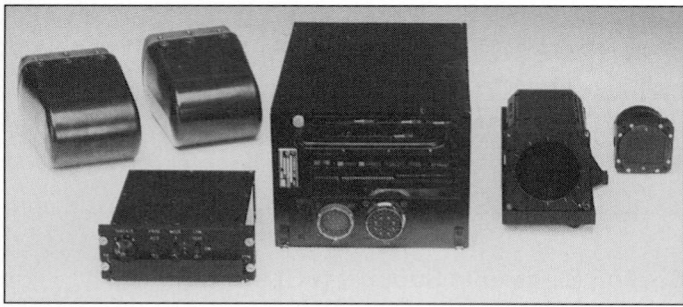

The Shiploc radar warning receiver

designation for countermeasures equipment such as chaff launchers. Display of threat information can be either on a Cathode Ray Tube (CRT) or a two colour Liquid Crystal Display (LCD) or a combined CRT/LCD display giving the direction, type and level of threat and the measured radar parameters.

Status
As of April 2001, Thales Airborne Systems was reporting that the Shiploc RWR was no longer in production and that over time, the equipment had been installed aboard Saudi Arabian 'Al Jawf' class coastal minehunters (three ships in class as of March 2001).

Contractor
Thales Airborne Systems, Elancourt, France.

UPDATED

SPS-N 5000 Electronic Support (ES) system

Type
ES system for surface ship and submarine applications.

Description
SPS-N 5000 has been designed as a compact, modular, surface/subsurface ES system. It has been conceived for tactical deployment (surveillance) at fixed sites or on mobile platforms. Its technique comprises an instantaneous wide-open receiving system covering the radar frequency range 2 to 18 GHz, with a 0.5 to 40 GHz extension capability possible. SPS-N 5000 is based on the latest technology and reliable processes for most currently known signals in dense scenarios. It consists of four major subsystems:
- a six-port direction-finding monopulse USK 800/6 sensor (see separate entry)
- a wideband receiver (RX 5000)
- a signal data processor (SDP 5000) for analysis and classification
- an operator workstation (MMI 5000).

Apart from automatic detection, analysis and classification, the system may also be operated manually. SPS-N 5000's role is to support its host platform, its command systems and weapons with the objective of enhancing the success of its own actions. For out-of-area missions the capacity of the threat library can be expanded. The basic version of SPS-N 5000 is a stand-alone system interfaced to a combat information centre.

Status
As of July 2001, the status of the SPS-N 5000 naval ES system was uncertain.

Contractor
European Aeronautic, Defence and Space (EADS) Co Systems and Defence Electronics, Ulm, Germany.

UPDATED

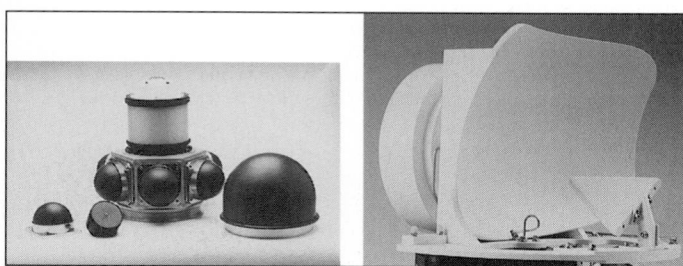

The antenna assembly used in the SPS-N 5000 ES system

USK 800 sensor family

Type
Family of communications and radar band intercept and Direction-Finding (DF) sensors for naval applications.

Description
USK 800 series sensors are tailored to specific customer requirements and are 'mixed and matched' from a range of subcomponents that comprises:

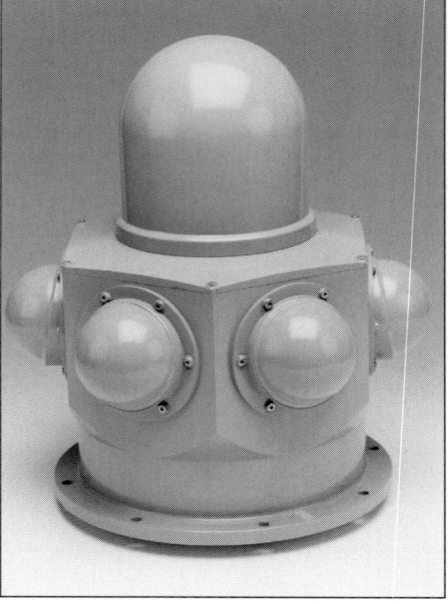

The USK 800/6 sensor head used in the SPS-N 5000 naval ES system
0044615

The periscope electronic support measures/ communications/DF array used in the USK 800/4 application
0044614

The internal arrangement of two multifunction assemblies used in the USK 800 range
0044613

- an L1/L2 Global Positioning System navigation capability
- a 2 to 18 GHz omnidirectional intercept capability
- a four/eight channel DF capability within the 2 to 18 GHz frequency range
- a communications capability within the 225 to 400 MHz frequency range.

Such equipments are suitable for both surface ship and submarine applications and are reported to be wide open in azimuth and frequency when being used for both DF and omnidirectional collection. Other system features include:
- high sensitivity
- a wide dynamic range
- small radar reflection cross-sections

- modular architecture
- light weight and compact dimensions
- pressure resistant
- ISO 9001 compliant.

Status

As of July 2001, USK 800 series sensors were reported as having been procured by a number of navies around the world.

Contractor

European Aeronautic, Defence and Space (EADS) Co Systems and Defence Electronics, Ulm, Germany.

UPDATED

ISRAEL

C-Pearl Electronic Support (ES) system

Type

Tri-service ES system.

Description

C-Pearl is a 'compact, lightweight' ES system that can be 'easily' installed in fixed-wing aircraft, surface ships and submarines. It is designed for the automatic detection and identification of radar threats in complex electromagnetic environments and is capable of identifying a range of signal types that includes chirp, frequency-agile signals and stagger/jitter pulse repetition intervals. It evaluates signal parameters, and carries out data processing and extraction. The main C-Pearl subsystems are a platform configured antenna assembly and a receiver/operator console. Of these, the core antenna assembly used is a miniaturised element that contains Instantaneous Frequency Measurement (IFM) and Instantaneous Direction-Finding (IDF) arrays and an IDF receiver. The IFM receiver, data processing and display are packaged in the single receiver/console.

Status

As of March 2001, Jane's sources were reporting that the six frigates making up the Royal Australian Navy's 'Adelaide' class were scheduled to be retrofitted with the C-Pearl ES system from 2003 onwards. Elsewhere in the world, the equipment may have been supplied to the Indian Navy (not confirmed as of August 2001) and as of May 2001, Rafael was understood to be actively promoting the sensor.

Specifications

Frequency coverage: 2-18 GHz (0.5-2 GHz optional)
Spatial coverage: 40° (elevation, typical); 360° (azimuth)
Sensitivity: −60 dBm
Measurement accuracy: 1° (bearing); 1.5 MHz (frequency); 2° (bearing, 2-4 GHz range)
Instantaneous dynamic range: 60 dB
Dimensions: 25 (Ø) × 25 (h) cm or 25 (Ø) × 76 (h) cm or 38 (w) × 38 (d) × 31 (h) cm (antenna assembly); 61 (w) × 76 (d) × 132 (h) cm (receiver/operator console)
Weight: 35 kg (antenna assembly); 125 kg (console)

Contractor

Rafael Electronic Systems Division, Haifa.

UPDATED

A general view of one of the antenna configurations (less radome) used with the tri-service C-Pearl ES system (Rafael)
2002/0114373

NAval TActical Communications intelligence System (NATACS) 2000

Type

0.3 to 1,000 MHz band COMmunications INTelligence (COMINT) and Direction-Finding (DF) system.

Description

The NATACS 2000 naval COMINT and DF system is described as being optimised to handle frequency agile emitters operating in an 'extremely' dense electromagnetic environment. The equipment makes use of fast scanning, wideband receivers and is noted as being able to perform 'rapid' spectrum scanning, DF and location finding, signal classification and digital audio recording functions. NATACS 2000 is further designed for data fusion with information generated by a radar electronic support system.

Status

As of this edition, the NATACS 2000 naval COMINT and DF system was reported as being 'fully operational'.

Specifications

Frequency range: 0.3-30 MHz and 20-1,000 MHz sub-bands (20-3,000 MHz option)
Scan rate: 100 MHz/s (0.3-30 MHz sub-band, COMINT and DF functions); 1 GHz/s (20-1,000 MHz sub-band, COMINT and DF functions)
DF accuracy: <1° RMS (20-1,000 MHz sub-band, instrumented); 1.5° RMS (20-1,000 MHz sub-band, at sea, typical); optional (0.3-30 MHz sub-band)

Contractor

Tadiran Electronic Systems Ltd, Holon.

UPDATED

NS-9003A-V2 Electronic Support (ES) system

Type

2 to 18 GHz shipboard ES system.

Description

The NS-9003A-V2 shipboard ES system is designed to receive, analyse and identify signals in the 2 to 18 GHz frequency band and is claimed to offer 100 per cent probability of intercept throughout the 360° arc in azimuth. The equipment's man/machine interface is noted as including a 'unique' Windows NT™ operator console and interfaces are provided to link NS-9003A-V2 with onboard electronic countermeasures systems, decoy launchers, command and control systems and 'other shipboard devices'. The NS-9003A-V1 configuration is described as being essentially similar to that of the NS-9003A-V2 with the exception of offering a 'reduced' direction-finding accuracy of 5° RMS.

Status

Over time, the NS-9003A-V2 shipboard ES system is reported as having been installed aboard 'several classes of ship' operated by a number of 'classified' customers. As of January 2002, NS-9003A-V1 and -V2 were reported as being available.

Specifications

NS-9003A-V2
Frequency coverage: 2-18 GHz (0.5-40 GHz option)
Frequency accuracy: 2 MHz
Azimuth coverage: 360°
DF accuracy: 2° RMS
Pulse density: 1 Mpps
POI: 100% (claimed)
Sensitivity: −65 to −75 dBm
Dynamic range: 60 dB
Signal modulation: frequency, PRI and PW

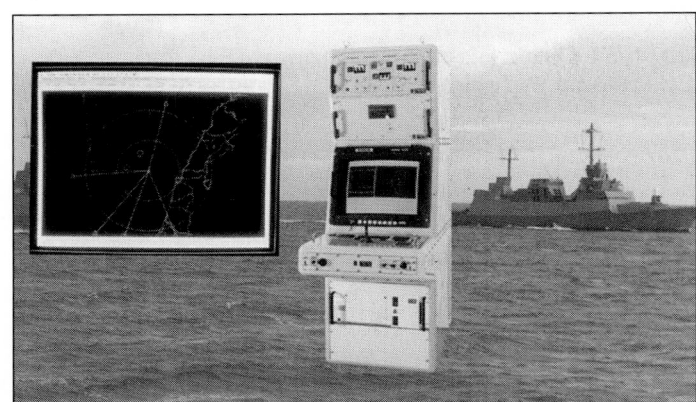

The operator workstation used in the NS-9003A-V2 naval radar ES system
0044616

CW: full parameter measurement
Scan rate: simultaneous measurement of all detected emitters
Data recording/replay: environmental data, emitter parameters and pulse data (digital)
Radar library: over 10,000 modes
Display formats: activity (Cartesian and polar), tabular (alphanumeric), interrogation (ELINT) and geographic map (option)
Interfaces: includes countermeasures (active and passive, integral provision), C³I, radar, navigation and GPS
Environment: MIL-STD-2036 compliant

Contractor
Elisra Electronic Systems Ltd (a member of the Elisra Group), Bene Beraq.

UPDATED

SES-210/E ELectronic INTelligence (ELINT) and Electronic Support (ES) system

Type
0.5 to 18 GHz naval ELINT and ES system.

Description
SES-210/E is a lightweight, naval ELINT and ES system that comprises a display/control package (located in the ship's combat information centre and comprising a colour cathode ray tube display, keyboard and processor), a receiver/processor unit and, in baseline form, 0.7 to 2 GHz and 2 to 18 GHz band mast-mounted antenna assemblies. In more detail, the equipment's reception chain incorporates both digital instantaneous frequency measurement and superheterodyne receivers, while direction-finding is by means of the monopulse amplitude technique. Functionally, the system automatically:

- detects naval, ground-based and airborne radar signals
- measures the electronic parameters and angle of arrival of each signal received
- identifies detected emitters by means of comparison with an integral emitter parameter file
- sorts 'exotic' emitters
- computes emitter location by means of triangulation (monopulse amplitude direction-finding)
- records intercepted media on magnetic media for further analysis.

Data are displayed to the operator in selectable polar, Cartesian and alphanumeric formats and the SES-210/E system also features acoustic threat warning and provision for integration with an onboard countermeasures suite. Here, overall system control is exercised manually, semi-automatically or automatically via the host platform's ES console.

Status
As of January 2002, the SES-210/E ELINT/ES system was reported as being available.

Specifications
Frequency range: 0.5-18 GHz (0.5-40 GHz option)
Frequency resolution: 500 kHz
Frequency accuracy: 3 MHz RMS
Reception sensitivity: −65 dBm (pulse); −85 dBm (CW)
DF accuracy: 8° (3° option)
Data recording: environmental data logging; onboard playback
Interfaces: interface provision for radar blanking, gyro equipment, weapon systems (option) and countermeasures systems (active and passive)
Power consumption: 500 W

Contractor
Elisra Electronic Systems Ltd (a member of the Elisra Group), Bene Beraq.

UPDATED

TIMNEX series Electronic Support (ES) and ELectronic INTelligence (ELINT) system

Type
Family of combined naval ES and ELINT systems.

Description
TIMNEX ES/ELINT systems are described as being designed for the detection, location, identification and analysis of radar emitters. Such equipments are claimed to offer short response times and a 'high', omnidirectional, probability of detection in 'dense' environments. As of May 2001, the latest identified member of the family was the 2 to 18 GHz band TIMNEX II subsurface configuration which, as of the given date, was described as being a fully integrated or stand-alone ES/ELINT system that comprised main and miniature antenna assemblies, a processing cabinet and a display/control console and offered a range of capabilities that included:

- claimed 100 per cent probability of instantaneous Direction-Finding (DF) and frequency measurement (using a channelised receiver)

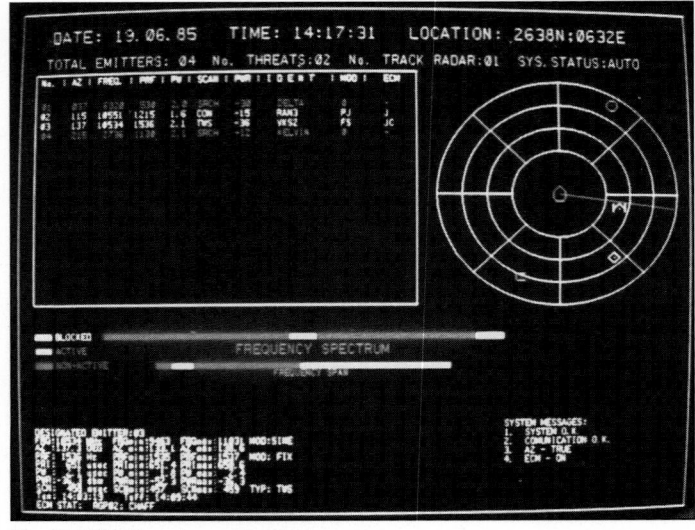

The TIMNEX II work monitor display

- automatic operation and threat analysis
- automatic signal identification/library correlation
- computer or operator controlled ELINT analysis
- video and digital data recording
- real time processing of up to 64 active emitters simultaneously
- automatic threat, high amplitude and continuous wave alarms
- automatic identification of predefined targets.

Looking at some of the equipment's subsystems in more detail, the TIMNEX II main antenna assembly is noted as being designed to withstand hydrostatic pressures that are in excess of 50 atmospheres and as comprising the following modules:

- a 2 to 18 GHz band omni-assembly that takes into account radome effects on signal path
- an array of eight cavity backed Archimedes DF antennas covering the 2 to 8 GHz band
- an array of eight corrugated conical horn DF antennas covering the 8 to 18 GHz band
- an integral radio frequency receiver front end (the OMNI module)
- an integral amplitude monopulse instantaneous DF module
- an integral self-test module that splits test signals in order to inject them into the antenna mast's receiver channel inputs and acts as a calibration tool for the equipment's DF channels.

The system's channelised receiver is reported as being continuously open across the equipment's frequency band and as being configured for automatic or operator initiated monitoring of individual frequency and channel blocking elements. In terms of processing, TIMNEX II is noted as making use of a processor network to perform pulse separation and sorting and as using multiple processing to increase data rates and capacities. The system's operator console features two display screens (characterised as activity and work monitors) and a keyboard and trackball. Of the two displays, the activity monitor presents a graphical picture of a detected source while the work monitor shows an alphanumeric display of summary data, environment information, system messages and operation menus. In its integrated configuration, TIMNEX II is combined with the host vessel's combat system and facilitates multiple use (command and control, ES/ELINT and sonar) of its operator console. As options, the system can also be interfaced with external displays and onboard fire- and torpedo-control systems.

Status
Over time, Jane's sources suggest that TIMNEX 4CH(V) ES and ELINT systems have been installed aboard Chinese 'Han' class submarines (4CH(V)2); Ecuadorean 'Gal' class submarines (4CH(V)1); Israeli 'Dolphin' (4CH(V)2) and 'Gal' (4CH(V)1) class submarines and Taiwanese 'Hai Lung' class submarines (4CH (V)2).

Contractor
Elbit Systems Ltd, Haifa.

UPDATED

ITALY

CO-NEWS communications band Electronic Support (ES) system

Type
Shipborne communications monitoring and analysis system.

Description
CO-NEWS (COmmunications Naval Electronic Warfare System) is a naval ES system which provides full monitoring and analysis of communications in the VHF/UHF (30 MHz to 1 GHz) band. The system is completely automated to achieve very

The CO-NEWS antenna assembly

high probability of intercept and to handle high-density traffic. It comprises a lightweight antenna assembly, (normally mounted at the masthead and consisting of three co-planar dipole arrays and a switching matrix) and an operator console located below decks. The equipment is modular in design to cater for expansion and for additional units. A baseline configuration comprises:

- a surveillance receiver which automatically scans the communications frequency bands
- a direction-finding receiver that automatically takes the bearing and performs the technical analysis of intercepted emissions.
- a computer which manages the sensors, as well as the processing and presentation of the collected data.
- an operator console with displays and a 'command' keyboard
  CO-NEWS automatically carries out continuous surveillance of the communications band, technical analysis and bearing of the intercepted emissions and real-time data processing and presentation. The intercepted signals can also be demodulated for monitoring purposes. The system operates automatically under the supervision of an operator who can also override the system's automatic operation to solve ambiguous situations or to monitor the emissions.

Status
Over time, the CO-NEWS ES system is reported as having been procured.

Contractor
Elettronica SpA, Rome.

VERIFIED

RQN-5C Electronic Support (ES)/ELectronic INTelligence (ELINT) system

Type
Shipboard ES/ELINT system.

Description
As of July 2001, RQN-5C was the latest identified variant of Elettronica's RQN-5 shipboard ES/ELINT architecture. System features include:

- a compact antenna group that, on ships of a suitable size, can incorporate an optional fine direction-finding antenna (E-mode unit) for 'very' high accuracy direction-finding and 'superior' sensitivity within steerable azimuth sectors
- instantaneous and omnidirectional coverage of the C- through J-band (5 to 20 GHz) frequency range with an upward extension to K-band (20 to 40 GHz) as an available option
- 'high' sensitivity receivers for both signal analysis and direction-finding
- a 'high accuracy', full band, Instantaneous Frequency Measurement (IFM) facility.
- an optional IFM receiver module to enhance frequency measurement accuracy
- fully automated ES functionality (including automatic real-time extraction and analysis and tracking of all intercepted known and unknown emitters)
- claimed near 100 per cent probability of extraction on a single scan
- automatic warning and identification (by programmable library comparison) of known threats and emitters
- automatic warning of unknown lock-ons and suspected continuous wave threats

- computer aided ELINT-type analysis (including all types of MOP) of individual emitters (with data recording)
- an automatic 'fingerprinting' facility on previously analysed emitters
- the ability to automatically drive active and passive countermeasures systems
- standard interfaces to facilitate integration with a wide range of peripherals
- configuration modularity to facilitate installation aboard ships ranging in size from offshore patrol vessels to frigates and system growth.

Status
Over time, RQN-5 ES/ELINT architecture variants are reported as having been procured by a 'number' of navies around the world.

Contractor
Elettronica SpA, Rome.

UPDATED

JAPAN

Japanese naval Electronic Support (ES)/threat warning systems

Description
Japan began to develop indigenous naval ES/threat warning systems during the 1960s and has gone on to produce a range of equipments which are currently in service. Known details of such systems are as follows:

NOLR-5
The NEC NOLR-5 is an ES system which, according to Jane's sources, has been installed aboard 'Chikugo' class frigates and 'Minegumo' class training ships of the Japanese Maritime Self-Defence Force (JMSDF).

NOLR-6
The NEC NOLR-6 system is a multi-variant ES system that is reported to have been installed aboard the JMSDF destroyers *Tachikaze* and *Takatsuki* and the same service's 'Asagiri' (NOLR-6C on seven ships), 'Hatsuyuki' (NOLR-6C on nine ships) and 'Yamagumo' class destroyers together with 'Abukuma' (NOLR-6C), 'Ishikari' (NOLR-6C) and 'Yubari' (NOLR-6C) class frigates.

NOLR-8
NOLR-8 is noted as being the ES system that has been fitted to the JMSDF destroyers *Asayuki*, *Setoyuki* and *Yamagiri*.

NOLR-9
NOLR-9 is reported to be the ES system that has been installed aboard the JMSDF destroyer *Kikuzuki*.

OLR-9
The Fujitsu OLR-9 is understood to be a multi-variant threat warning system that has been fitted to JMSDF 'Hatakaze' (OLR-9B), 'Haruna' and 'Shirane' (OLR-9B) class destroyers.

ZLR series threat warners
Over time, Jane's sources suggest that ZLR series threat warners have been installed aboard JMSDF 'Harushio' (ZLR 3-6), 'Oyashio' (ZLR-7) and 'Yuushio' (ZLR 3-6) class submarines.

VERIFIED

RUSSIAN FEDERATION AND ASSOCIATED STATES (CIS)

Introduction

The following entries represent equipment produced by and/or exported by the Russian Federation, Armenia, Azerbaijan, Belarus, Georgia, Kazakhstan, Kyrgyzstan, Moldova, Tajikistan, Turkmenistan, Ukraine and Uzbekistan. Of these, the Russian Federation, Azerbaijan, Kazakhstan and Ukraine have operational navies. Taken together, these 12 entities form the Commonwealth of Independent States economic grouping.

VERIFIED

RFAS Electronic Support (ES) and threat warning systems

Description
The following equipments are identified by their NATO reporting names or, where known, their RFAS designation.

Bald Head
As of this edition, Bald Head was reported as being the ES and radar warning system that was fitted to the Russian nuclear-powered attack submarine *Tula*. As

such, it may have been used as a passive targeting aid for the P-20L Ametiste subsurface-to-surface anti-shipping missile.

Bell Slam

Bell Slam is reported to be an E- through H-band (2 to 8 GHz) threat warner that has been derived from the Bell Tap ES system (see following). Over time, Jane's sources report that the Bell Slam system has been installed aboard the Indian destroyers *Rajput*, *Rana* and *Ranjit*, the Romanian destroyer *Marasesti* and the Russian cruisers *Admiral Golovko* and *Kerch*.

Bell Tap

Bell Tap is described as being a G- through J-band (possibly 5 to 18 GHz sub-band) threat warning and ES system with a bearing accuracy of approximately 2.5°. Over time, Jane's sources suggest that the Bell Tap system has been installed aboard the following ship classes:

Algeria	'Nanuchka II' class corvettes
India	the destroyers *Rajput, Rana* and *Ranjit* together with the same country's 'Durg' class corvettes
Libya	'Nanuchka II' class corvettes
Russia	the cruiser *Admiral Golovko* and Type 1234 ('Nanuchkia I/Burya' class) corvettes

Brick Group

The Brick Group family is a series of ES and threat-warning systems that appear to be designed for submarine applications where they are frequently co-located with Snoop Group radar installations. Over time, identified members of the family have included the Brick Plug, Brick Pulp, Brick Split and Clay Brick sub-variants and Jane's sources suggest that members of the family have been installed aboard the following submarine classes:

Algeria	'Kilo' class Anti-Submarine Warfare (ASW)-capable submarines
Poland	the ASW-capable submarine *Orzel*
Romania	the ASW-capable submarine *Delfinul 521*
Russia	the submarine *K 395* and the same country's Type 641B ('Tango/Som' class - Brick Group or Squid Head) ASW-capable and Type 671RTN ('Victor III/Schuka' class - Brick Spit and Brick Pulp) nuclear-powered attack submarines.

Cross Loop

Cross Loop is reported to be a Direction-Finding (DF) system that, as of this edition, was reported as having been installed aboard Algerian 'Mourad Rais' class frigates and 'Nanuchka II' class corvettes and Russian Type 12660 ('Gorya' class) ocean-going minehunters.

Kursk

Given the NATO Reporting Name Bell Nip, Kursk is described as being a set-on receiver for chaff launchers and as incorporating both I/J- (8 to 20 GHz) and J/K-band (10 to 40 GHz) antennas arrays. As of this edition, Jane's sources were suggesting that the Kursk system has been installed aboard the Russian aircraft carrier *Admiral Gorshkov* and Russian Type 1144/1144.1/1144.2 ('Kirov/Orlan' class) battle cruisers.

MP-401 Start

Given the NATO Reporting Name Bell Shroud, Start is reported to be an upgrade of the Watch Dog B radar intercept and warning system and is frequently teamed with the Bell Squat jammer. As of this edition, Jane's sources were suggesting that the Start system has been installed aboard the following ship classes:

Poland	the destroyer *Warszawa*
Russia	the destroyer *Admiral Chabanenko*, the frigate *Yastreb*, the amphibious transport dock *Mitrofan Moskalenko* and the same country's Type 61/61M ('Kashin/Modified Kashin' class) destroyers and Type 956/956A ('Sovremenny/Sarvich-Sarya' class) destroyers and Type 1135/1135M/ 1135MP ('Krivak I - Burevestnik/II/III - Nerey' class) frigates
Ukraine	the frigate *Hetman Sagaidachny* and the same country's 'Krivak I' class frigates

MP-404

Given the NATO Reporting Name Rum Tub, MP-404 is described as being a surface ship ES antenna group in which four arrays are normally disposed to cover 90° sectors of the 360°. As of this edition, Jane's sources were suggesting that MP-404 antenna groups have been installed aboard the Russian cruiser *Kerch* and the same country's Type 1164 ('Slava/Atlant' class) cruisers.

MRKP-60

MRKP-60 is a combined radar warning and active radar navigation system that is designed for submarine applications. The equipment comprises an inboard electronics/display unit (housing the active radar's transceiver and the threat warner's receiver and analyser) and a mast-mounted, combined active/passive antenna array. MRKP-60 offers 24 threat warning and eight active radar channels and weighs 0.7 tonnes. It can be brought into action in approximately 5 minutes and, as of this edition, MRKP-60 was thought to be available through Russia's Rosvoorouzhenie export agency.

Nakat

Given the NATO Reporting Name Stop Light, Nakat is reported to be a broadband submarine ES system that in its Stop Light B configuration, incorporates an Instantaneous Frequency Measuring (IFM) capability. As of this edition, Jane's sources were suggest that the Nakat system has been installed aboard the following submarine classes:

Bulgaria	'Romeo' class general-purpose submarines
India	'Foxtrot' class general purpose submarines
North Korea	'Romeo' class general purpose submarines
Libya	'Foxtrot' class general-purpose submarines
Poland	'Foxtrot' class general-purpose submarines
Russia	Type 641 ('Foxtrot' class) general-purpose submarines
Ukraine	The submarine *Zaporizya*
Yugoslavia	'Heroj' and 'Sava' class general-purpose submarines

Nakat-M

Given the NATO Reporting Name Watch Dog, Nakat -M is described as being a wide-open ES system that covers the D- through H-band (1 to 8 GHz) frequency band. Nakat-M is understood to utilise two antenna arrays and in its Watch Dog B variant, is thought to incorporate an IFM capability. As of this edition, Jane's sources were suggesting that the Nakat-M system has been installed aboard the following ship classes:

Algeria	'Mourad Rais' class frigates
Bulgaria	The frigate *Smeli* and 'Letyashti' class corvettes
Cuba	'Koni' class frigates
Indonesia	'Kapitan Patimura' class corvettes
North Korea	The frigate *823* and the same country's 'Najin' class frigates (Chinese RW-23 version (NATO Reporting Name Jug Pair) of Nakut-M)
Libya	'Koni' class frigates
Lithuania	'Grisha' III class frigates
Romania	The destroyer *Marasesti* and the same country's 'Tetal/Improved Tetal' class frigates, 'Zborul' class corvettes, 'Cossar' class minelayers/mine warfare support ships and 'Croitor' class logistic support ships
Russia	The ocean-going minesweeper *Yurka*, the training ship *Gangut*, the submarine depot ship *Volga* and the same country's Type 133.1M ('Parchim II' class) frigates, Type 264M ('T-58' class) patrol ships, Type 266 ('Yurka/Rubin' class) ocean-going minesweepers, Type 503M/R ('Alpinist class) intelligence collection/research ships, Type 887 ('Smolny' class) training ships and Type 1124/1124P/1124M/1124EM ('Grisha I/II/III/V - Albatros' class) frigates
Ukraine	'Grisha I/II/V' class frigates
Vietnam	'Petya' class frigates

Park Lamp

Park Lamp is reported to be a submarine DF system which, as of this edition, is reported as having been installed aboard the Russian submarines *K 395* and *Tula* together with the same country's Type 667B/667BDR/667BDRM ('Delta I - Murena/Delta III - Kalmar/Delta IV - Delfin' class) and Type 941 ('Typhoon/Akula' class) nuclear ballistic missile submarines.

Quad Loop

Quad Loop is described as being a submarine DF and monitoring system that, as of this edition, was reported as having been installed aboard the following submarine classes:

Iran	'Kilo' class ASW-capable submarines
Poland	The ASW-capable submarine *Orzel*
Romania	the ASW-capable submarine *Delfinul 521*
Russia	Type 636/877/877K/877M ('Kilo/Kilo4B - Vashavyanka' class) and Type 641B ('Tango/Som' class) ASW-capable submarines

Rim Hat

Over time, Jane's sources suggest that the Rim Hat radar warning system has been fitted aboard Russian Type 941 ('Typhoon/Akula' class) nuclear ballistic missile submarines, Russian Type 945A ('Sierra II/Kondor' class) class nuclear attack submarines, Russian Type 949A ('Oscar II/Antyey' class) cruise missile submarines and Russian Type 971/971M ('Akula I - Bars/II' class) nuclear attack submarines.

Squid Head

Squid Head is desribed as being a submarine DF and intercept system that, as of this edition, is reported as having been installed aboard the following submarine classes:

People's Republic of China	'Kilo' class ASW-capable submarines
India	'Sindhughosh' class ASW-capable submarines
Iran	'Kilo' class ASW-capable submarines
Russia	Type 641B ('Tango/Som' class) and Type 877/877K/877M 'Kilo/Kilo4B - Vashavyanka' class) ASW-capable submarines

VERIFIED

SOUTH AFRICA

Shrike Electronic Support (ES) System

Type

Compact naval radar ES system

Description

The Shrike ES system is described as being a low cost, lightweight naval ES equipment that is designed for installation aboard 'small ships', support vessels and submarines and comprises an antenna assembly, an operator console and a processor unit. The use of high speed parallel processing is claimed to facilitate Shrike's operation in 'extremely dense' electromagnetic environments while the systems man/machine interface is noted as providing an automated basic analysis function that establishes emitter frequency versus azimuth and frequency versus amplitude readings. Parameter analysis of detected emitters is available in prioritised text form and full analysis and 'other' auxiliary functions can be selected via an operator's trackball. Other system features include:

- an over-the-horizon detection capability
- user definable threat libraries

- pulse Doppler emitter detection
- instantaneous frequency measurement
- monopulse direction-finding
- built-in test
- recording and playback facilities

System options include interfaces for a laser warner, UV sensors and/or electronic countermeasures equipment.

Status

As of this edition, the Shrike ES system was reported as being available and, according to Jane's sources, possibly selected for retrofit aboard South African 'Warrior' class fast attack craft.

Specifications

Frequency range: 2-18 GHz (0.7-2GHz and 18-40 GHz options)
Direction-finding accuracy: 5-6° RMS
Coverage: ± 45° (elevation); 360° (azimuth)
Pulse density capability: > 3 million pulses/s (up to 54 levels of stagger)
Frequency resolution: 10 MHz
Sensitivity: –60 dBm
Amplitude resolution: 1 dB
Pulse-width range: 0.1-50 μs
Power drain: 170 W

Contractor

Avitronics (Pty) Ltd (Maritime) (a Saab and Grintek company), Tokai.

UPDATED

SPAIN

BLQ-355 Electronic Support (ES) system

Type

E- through J-band (2 to 20 GHz) radar ES system for submarine applications.

Description

BLQ-355 is designed to detect (with 100 per cent probability of intercept), identify and track radar emitters for both tactical and strategic intelligence purposes. The system's architecture comprises a six port, cavity-backed spiral antenna array/ digital instantaneous frequency measuring receiver assembly, a digital processor, a dedicated operator console and a Unix-based application software package. System functions include scanning for and the detection of active radar emitters; signal analysis; parameter measurement and data recording for future analysis. An intra-pulse analysis module is an available option and BLQ-355 as a whole is claimed to have very high reliability, thereby reducing maintenance at the 'preventative and corrective' levels. System reliability is achieved via the use of commercial-off-the-shelf components and a three level (initialisation, operator ordered and continuous) built-in test routine. Emitter identification is by means of comparison with a 3,000 mode main library and a 256 mode operator library. The system is able to play back recordings of detected emitters and pulses, a feature that its manufacturer claims is a useful tool for 'pop-up' searching. Immunity from Continuous Wave (CW) interference is ensured via the use of a CW-proofed, digital, logarithmic video amplifier.

Status

As of this edition, the BLQ-355 ES system was reported as having been installed aboard Spanish 'Galerna/Agosta' class submarine together with Type 209 boats operated by 'other' navies.

Contractor

INDRA DTD, Madrid.

VERIFIED

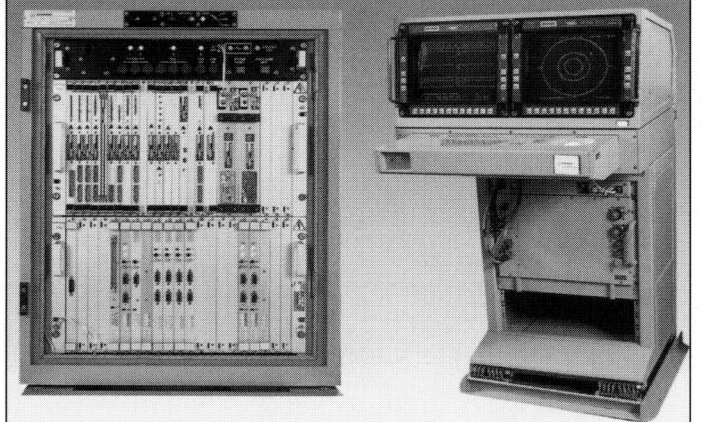

A composite view showing the display console and inboard electronics used in the BLQ-355 submarine ES system   1996

SWEDEN

EWS-905 Electronic Support (ES) system

Type

Shipborne radar ES system.

Description

This shipborne system is designed to identify targets detected by the vessel's surveillance radar. The antenna turns automatically towards the direction indicated (using information passed from the ship's tactical display) and acquires and analyses threat signals. Signals received from a sector around the designated bearing are analysed and parameters representing bearing, frequency band, pulse interval, pulsewidth and scan time are displayed. The parameters are compared with a threat library and the target is identified.

Status

Over time, the EWS-905 ES system is reported as having been installed aboard 'Hugin/Kaparen' class fast attack craft of the Swedish Navy.

Contractor

Saab Dynamics AB, Linköping.

VERIFIED

UNITED KINGDOM

Mentor 2000 Electronic Support (ES) system

Type

Shipborne radar intercept and analysis system.

Description

Mentor 2000 is a surveillance and threat warning equipment that can be configured for installation in warships ranging in size from patrol craft to aircraft carriers. The system is designed to operate in dense and demanding environments where it provides passive detection and identification of complex emitters. At the heart of Mentor 2000 is a signal processing chain that employs commercial-off-the-shelf processors and Ada software. The equipment can be operated automatically with up to 8,000 library modes and alarm channels. Simultaneous tracking of up to 500 radars is possible with new intercepts displayed in approximately 0.6 second.

Status

As of August 2001, four examples of the earlier Mentor 1 and 2 series equipments are known to have been sold. Of these, two Mentor 1 equipments have been installed on the Royal Navy's 'Fort Victoria' class fleet replenishment vessels (service designation Outfit UAG) with a further pair of Mentor 2 systems being fitted to 'Lekiu' class frigates of the Malaysian Navy. In the 'Fort Victoria' class installation, Outfit UAG has been replaced by a variant of Outfit UAT (see separate entry). According to BAE Systems sources, Mentor was a live programme during 2001.

Contractor

BAE Systems Avionics - Sensor Systems Division, Stanmore.

UPDATED

Outfit UCB(1) Electronic Warfare (EW) control processor

Type

EW system integration processor.

Description

UCB(1) is a knowledge-based processor unit which is designed to integrate the available EW assets aboard its host platform, reduce the picture compilation and soft kill allocation loads on the platform's EW director and provide an enhanced EW system interface with the host vessel's command system. Royal Navy EW assets which will be integrated using UCB(1) include the Type 675(2) onboard radar jammer, the Outfit UAT electronic support receiver, various applications of the Seagnat decoy launcher, the Outfit DLH active offboard decoy, the Outfit DLF passive offboard radar decoy and the Outboard communications intelligence system. UCB(1) makes extensive use of commercial-off-the-shelf hardware and housekeeping software, has its operational software written in Ada and employs a dedicated operator console.

Status

During March 1995, the then GEC-Marconi Stanmore (now BAE Systems Avionics - Sensor Systems Division) was awarded a then year £17 million contract to supply the Royal Navy with 14 UCB(1) processor ship sets. As originally planned, these were to be installed aboard Batch 2 and 3 'Type 42' class destroyers and all three of its 'Invincible' class aircraft carriers, with the first UCB(1) package scheduled for delivery during 1998. During 1999, Jane's sources were suggesting that the equipment had been successfully sea trialled aboard HMS *Exeter*. In the 'Type 42' class destroyer application, Outfit UCB(1) interfaces with the ship's ADAWS Mod 1 command system.

Contractor

BAE Systems Avionics - Sensor Systems Division, Stanmore.

UPDATED

UNITED STATES OF AMERICA

AN/BRD-7 Direction-Finding (DF) system

Type
Submarine DF system.

Description
The AN/BRD-7 system is described as being a high-performance equipment that provides a DF capability over a wide frequency range.

Status
Over time, approximately 100 AN/BRD-7 DF equipments are reported as having been procured by the US Navy. As of the April 2001, Jane's sources were reporting the system as having been installed aboard the service's 'Los Angeles' class nuclear-powered attack submarines.

Contractor
BAE Systems North America - Information and Electronic Warfare Systems, Nashua, New Hampshire.

UPDATED

The US Navy 'Los Angeles' class nuclear-powered attack submarines are fitted with the AN/BRD-7 DF system

AN/SRS-1 combat Direction-Finding (DF) system

Type
Signal exploitation system.

Description
SRS-1 is a shipboard, digital DF and signals exploitation system that has been procured by the US Navy. It is based on a distributed architecture concept that enhances its growth potential. It incorporates high-speed signal processing for demodulation, analysis and signal recognition of a wide variety of both conventional and spread spectrum types.

Status
Over time, the AN/SRS-1 combat DF system is reported as having been procured by the US Navy. In terms of programme contracting activity, Sanders is known to have received a then year US$56 million production contract for seven complete SRS-1A systems and three SRS-1 upgrade kits from the USN during February 1997. This award is also understood to have included options on a further 16 systems for delivery through to 2002. A complete SRS-1A system comprises inboard equipment racks together with deck-edge and mast-mounted antenna

arrays. At least 22 SRS-1 systems have been delivered since production began in 1986.

Contractor
BAE Systems North America - Information and Electronic Warfare Systems, Nashua, New Hampshire.

UPDATED

AN/SSQ-108 Outboard

Type
Shipborne countermeasures detection and analysis system.

Description
The AN/SSQ-108 Outboard system is a follow-on of the earlier AN/SSQ-72 equipment and is designed to provide technical and tactical intelligence data so that the force commander can analyse enemy dispositions and intentions for Over-The-Horizon (OTH) detection and identification of surface ships for targeting purposes. As of this edition, two SSQ-108 variants (designated as SSQ-108(V)1 and SSQ-108(V)2) have been identified. Of these, the SSQ-108(V)1 configuration consists of the AN/SRD-19A equipment, the AN/SLR-16A countermeasures receiver, the OK-324/SYQ System Supervisor Station (SSS), a local monitoring station and tactical intelligence communications package. SSQ-108(V)2 adds the AN/SLR-23 automated narrow-band acquisition system and an OK-324/SYQ modification kit. The SSS integrates and controls Outboard sensors as well as external datalinks and communication to provide information processing and technical dissemination. The system uses hull and mast-mounted antenna arrays.

Status
Over time, the Outboard detection and analysis system has been reported as having been installed aboard American 'Spruance' class destroyers and British 'Invincible' class aircraft carriers and 'Broadsword' class frigates. In British service, the system is designated as Outfit UAD or UAK (installation and specific role dependent) while US usage is, over time, thought to have involved installations aboard approximately 36 destroyers and cruisers. In the American application, the system is reported as being used to provide signals intelligence, early warning and, possibly, over-the-horizon targeting for US Navy aircraft carrier battle groups. During February 1995, UK contractor BAE Systems announced that it had been appointed the prime UK subcontractor on a BAE Systems North America (formerly Sanders) - led Anglo-American Outboard upgrade programme that was designated as the Co-operative OutBoard Logistic Upgrade (COBLU) effort.

Contractor
BAE Systems North America - Information and Electronic Warfare Systems, Nashua, New Hampshire.

UPDATED

AN/WLQ-4(V) SIGnals INTelligence (SIGINT) system

Type
Submarine SIGINT detection and analysis system.

Description
The AN/WLQ-4(V) SIGINT system is an automated, modular signal collection equipment that facilitates the identification and location of unknown radar and communications emitters. It incorporates a network of mini-computers and

An AN/WLQ-4 SIGINT system under test

microprocessors, data from which is correlated with information received from satellite sensors. The system is part of Sea Nymph, a highly classified US Navy (USN) programme. WLQ-4(V) system has a number of key features, including:

- automatic search, acquisition and signal processing
- automatic logging book-keeping and reporting
- semi-automatic correlation of real-time measured data with input from an external system
- 400,000 lines of AN/UYK-44 source code
- 50,000 lines of executable code in 40 microprocessors
- a significant growth capability to handle new threats.

Status

As of this edition, the AN/WLQ-4(V) SIGINT system was reported as having been installed aboard USN 'Seawolf', 'Sturgeon' and 'Virginia' class nuclear-powered attack submarines. Of these, the WLQ-4(V)1 'Seawolf' application is reported to have been the subject of a then year US$49 million development and production contracts.

Contractor

General Dynamics Information Systems and Technology business group, Mountain View, California.

VERIFIED

··

AN/WLR-1H(V) Electronic Support (ES) systems

Type

ES and Direction-Finding (DF) system.

Description

The 0.5 to 20 GHz band AN/WLR-1H(V) ES and DF system (up to and including the WLR-1H(V)5 configuration - see following) is described as making use of electronically tuned superheterodyne receivers and as being designed for the high-speed acquisition, analysis, identification and direction-finding of high-frequency Radio Frequency (RF) signals. As such, the architecture is said to combine the long-range and high-sensitivity of the earlier AN/WLR-1G equipment with independently computer-controlled receivers, automatic signal analysis/threat warning and automatically tasked monopulse DF. Key features of the architecture include:

- a distributed, multiprocessor architecture
- fully automated signal processing
- a dedicated DF processor
- a 300 active signal tracking capacity
- automatic threat identification (based on an integral 80 threat mode library)
- prioritised DF and scan measurement on a prioritised basis
- antenna dictated monopulse, spinning or interferometer direction-finding
- elevation angle correction of DF azimuth
- separate port and starboard antenna arrays
- real-time calibration
- use of an emitter classification/platform correlation algorithm (with a tape loaded database)
- tactical and analytical operator interfaces (alphanumeric, panoramic, analysis and DF analogue display options)
- built-in test.

During the mid- to late 1990s, Wide Band Systems (WBS) Inc was awarded a contract to upgrade the performance of the WLR-1H(V)5 equipment, with the new configuration being given the designation WLR-1H(V)7 (as of January 2002, Jane's

sources had identified -1H(V)3, -1H(V)5, -1H(V)6 and -1H(V)7 system variants). As yet unconfirmed sources suggest that WBS WLR-1H(V)7 architecture includes a channelised superheterodyne receiver, company wide- and narrow-band instantaneous frequency measuring receivers, dual-Pentium processors, two 51 cm (20 in) flat panel display units and a 'dedicated' antenna. Received signals are said to be passed from the antenna to the system's workstation via a fibre-optic link and the system as a whole is reported as weighing approximately 91 kg (as against 1,361 kg plus for the 22 line replaceable unit -1H(V)5 configuration), as saving 3 kW of power when compared with its predecessor and as having an improved time of arrival precision measurement value of to within half of a nanosecond. The equipment's graphical user interface is based on the Windows NT operating system and its manufacture claims that it is capable of near 100 per cent probability of intercept in environments with pulse densities of greater than four million pulses/s.

Status

As of January to November 2001, Jane's sources were reporting WLR-1 and WLR-1H(V) receiver systems as being installed aboard the following warships and classes of warship:

Australia	• the destroyer *Brisbane* (WLR-1H(V))
Brazil	• 'Para' class frigates (four ships in class - WLR-1)
South Korea	• 'Gearing (Fram 1)' class destroyers (five ships - WLR-1)
Mexico	• 'Gearing (Fram 1)' class destroyers (two ships - WLR-1)
Thailand	• the frigate *Pin Klao* (WLR-1)
	• the training ship *Makut Rajakumarn* (WLR-1)
Turkey	• 'Tang' class general purpose submarines (two boats - WLR-1)
USA	• the aircraft carrier *Enterprise* (WLR-1H(V))

- the 'Los Angeles' class nuclear attack submarines *Columbia*, *Greenville* and *Cheyene* (WLR-1H(V))
- the command ships *Coronado* (WLR-1) and *La Salle* (WLR-1)
- 'Kitty Hawk/John F Kennedy' class aircraft carriers (three ships - WLR-1H(V) - WLR-1H(V)7 planned for retrofit aboard the *John F Kennedy* - see following)
- 'Nimitz' class aircraft carriers (eight ships in service, one building, one proposed - WLR-1H(V) - WLR-1H(V)7 aboard the *Eisenhower* and the *Harry S Truman* and planned for the *Ronald Reagon* - see following)
- 'Spruance' class destroyers ('some' of 22 ships - WLR-1)
- 'Hamilton/Hero' class high endurance cutters (US Coast Guard - 12 ships - WLR-1C)

In terms of specific contracting activity, WBS Inc is understood to have received a then year US$1.3 million WLR-1H(V)7 contract during 1999, with the system having undergone sea trials aboard the aircraft carriers the USS *Eisenhower* (June 1999) and the USS *Truman*. As of 1999, unconfirmed sources were suggesting that the USN was acquiring 11 WLR-1H(V)7 equipments, with a possible overall procurement of 36. On 1 April 2001, WBS announced that it had completed delivery of two WLR-1H(V)7 equipments for installation aboard the aircraft carriers the USS *John F Kennedy* and the USS *Reagan*. By the end of 1994, over 100 WLR-1 systems of all types were reported as having been produced.

Specifications

excluding the WLR-1H(V)7 configuration
Frequency range: 0.55-20 GHz
Dynamic range: 60 dB
Automatic analysis: Frequency, PRI, pulse-width, angle of arrival, scan type, scan period, pulse amplitude, beamwidth
DF modes: Automatic, manual, back-up
Signal tracking: 300 emitters
Library: 80 threats, 300 radars, 1,500 radar modes, 150 platforms
Alarms: Audio/visual

Contractors

Condor Systems Inc, San Jose, California.
Wide Band Systems Inc (WLR-1H(V)7), Neshanic Station, New Jersey.

UPDATED

··

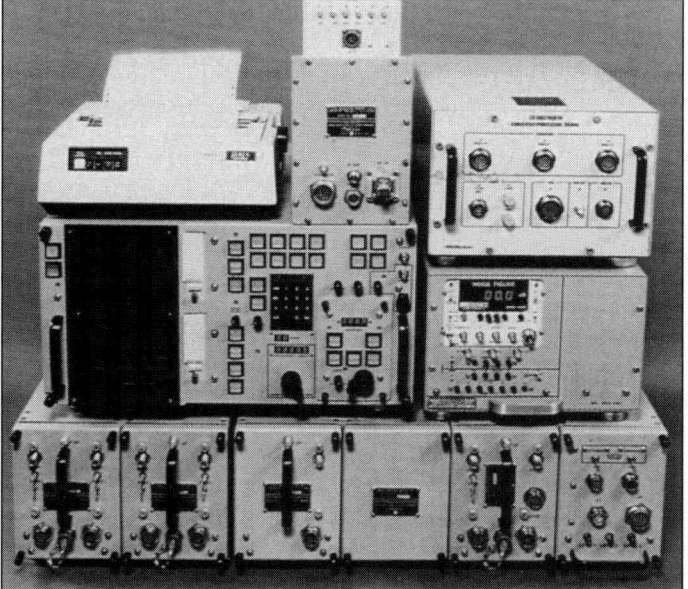

The WLR-1H(V) ES and DF system

AN/WLR-8(V) Electronic Warfare (EW) receiver system

Type

50 MHz to 18 GHz tactical radar detection and analysis system.

Description

ANWLR-8 is a solid-state, tactical EW and surveillance receiver designed for installation in both surface ships and submarines of the US Navy (USN). The system is of modular construction and provisions are made for operation in conjunction with numerous types of direction-finding or omni-antennas together with a wide range of optional peripheral equipment to provide comprehensive electronic support measures facilities. WLR-8 is compatible with NTDS (Navy Tactical Data System) and similar action information automation systems. The system can be expanded in frequency/signal handling capability (by 'simple' hardware additions and/or software changes) and makes use of a PSP-300 digital computer for system control, automatic signal acquisition/analysis and file processing and a PSP-200

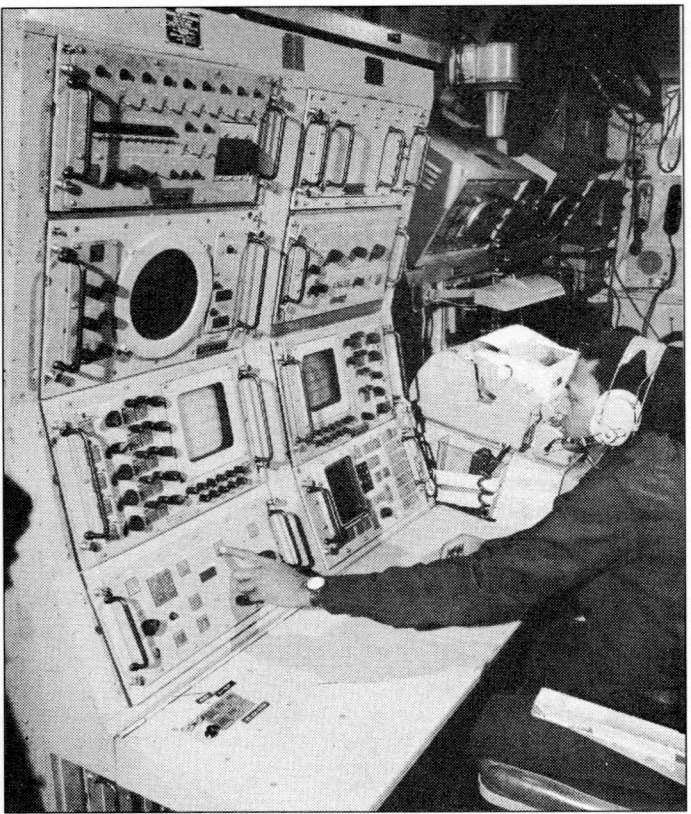

The operating console used with the AN/WLR-8(V) tactical EW receiver

microcomputer for hardware control functions. Digital techniques are employed throughout WLR-8 and its operational facilities include:

- automatic measurement of signal direction of arrival
- signal classification and recognition
- sequential or simultaneous scanning over a wide frequency range
- signal activity detection for threat warning
- analysis of signal parameters such as frequency, pulse repetition frequency, modulation, pulsewidth, amplitude and scan rate
- logging of signal parameters for display to operator(s), and printout of hard copy to teletype or printer
- extensive built-in test equipment
- directed priority searches of specific frequency segments.

Direct reporting to onboard computers, such as NTDS, permits response times in the ms range with minimal operator involvement. A two-trace Cathode Ray Tube (CRT) is provided for display purposes, and this can be supplemented by an optional five-trace panoramic display for presentation of signal activity data. Another CRT display is incorporated if WLR-8 is used with automatic or manual direction-finding antenna systems.

Status

As of the first quarter of 2000, AN/WLR-8 systems were reported as being installed aboard the USN's dry deck shelter nuclear-powered attack submarine USS *Kamehameha* and the same service's 'Ohio' class nuclear ballistic missile submarines (WLR-8(V)5) and 'Los Angeles' nuclear-powered attack submarines (WLR-8(V)2/6).

Contractor

General Dynamics Information Systems and Technology business group, Mountain View, California.

VERIFIED

AR-700A Electronic Support (ES) and Direction-Finding (DF) systems

Type

Family of multiplatform ES and DF systems.

Description

The AR-700A (V) ES/DF system is a modular design that can be configured for installation aboard surface ships, submarines and aircraft. It covers the 0.5 to 18 GHz frequency band with an optional extension to 40 GHz. The wide open monopulse DF and parallel instantaneous frequency receiver approach has no channelisation or band stepping which can degrade probability of intercept and short-pulse performance. The ASP-2000 signal processor is at the heart of the system, where it processes pulse data at up to 3 Mpps and handles an extremely wide range of pulse repetition frequencies and signal modulations. The ASP-2000 drives a colour graphics display and can also be interfaced to a variety of weapon control or mission avionics systems. Known AR-700A subvariants are as follows:

AR-700A(V)1

The AR-700A (V)1 provides ES and DF capabilities over the frequency range 2 to 18 GHz with a system sensitivity of –60dBm; monopulse DF accuracy of 5° RMS; frequency measurement accuracy of 4 MHz; 1 second reaction time and pulse processing of 1.5 Mpps. The antenna assembly can be packaged in a single unit for installation on fast patrol boats, frigates, larger warships and submarines.

AR-700A(V)2

This version augments the (V)1 with precision DF accuracy of 2° RMS and an added frequency coverage of 0.5 to 2 GHz (18 to 40 GHz optional). In the submarine version, AR-700 (V)2 uses a small periscope antenna unit incorporating an omnidirectional antenna providing instantaneous DF accuracy of 5° RMS together with an ES antenna providing precision 2° RMS DF accuracy with sensitivity for over-the-horizon targeting.

In surface ship installations, the (V)2 uses an integrated instantaneous and precision DF capability in two separated, lightweight monopulse receiver units. These are designed to minimise topside weight and to allow flexible installation. Dense signal environments of up to 3 Mpps can be handled. An optional phase-to-phase interferometer DF subsystem offering 0.7° RMS is also available.

AR-700A(V)3

Addition of an Electronic Warfare Operator's Console (EWOC) to the (V)2 configuration results in the (V)3. The EWOC is an ultra-high-performance workstation with software designed specifically for advanced electronic warfare applications. Capabilities include real-time operation and advanced graphics presentations. Multiple graphics windows provide text and tabular information, geographical plots, colour-coded threat-correlated spectral displays, waterfall displays and pulse analysis displays. The EWOC's relational database and tactical situation displays are designed to allow ship- or north-oriented map presentations with overlaid ESM and/or radar data for correlation and detailed situation evaluation. Optional video windows can display infra-red or TV sensor images.

Alongside the described AR-700A variants and the earlier AR-700 system (reported to have been fitted with an ASP-32 signal processor and digital instantaneous frequency measuring receiver), ARGOSystems appears to have used the technology to create a series of dedicated submarine systems, designated as the AR-700SF, AR-700-S5 and the AR-740. Of these, the AR-740 is described as covering the 0.5 to 26 GHz frequency range and having an interferometric DF capability for missile targeting.

Status

During the period February to May 2001, Jane's sources identified AR-700 series ES, threat warning and DF systems as being in service aboard the following ship and submarine classes:

Australia	• 'Collins' class ASW-capable submarines (six boats in class, AR-740)
Egypt	• 'Improved Romeo' class ASW-capable submarines (six boats, AR-700-S5)
Greece	• 'Glavkos' class ASW-capable submarines (eight boats, AR-700-S5) and 'Hydra' class frigates (four ships, AR-700 as part of APECS II?)
India	• 'Shishumar' class ASW-capable submarines (four boats (two building), AR-700 or Sea Sentry)
South Korea	• 'Okpo' class frigates (three ships, AR-700 as part of APECS II)
Netherlands	• 'Walrus' class ASW-capable submarines (four boats, AR-700) and 'Karel Doorman' class frigates (eight ships, AR-700 as part of APECS II)
Norway	• 'Oslo' class frigates (three ships, AR-700)
Portugal	• 'Comandante João Belo' class frigates (three ships, as part of APECS II)
Sweden	• the minelayer *Carlskrona* (AR-700), the minelayer/support ship *Visborg* (AR-700) and the same country's 'Vastergotland' (four boats, AR-700-S5 or CS-3701) class ASW-capable submarines and 'Norrkoping' class fast attack craft (six ships, AR-700 or Susie)
Taiwan	• 'Hai Lung' class ASW-capable submarines (two boats, AR-700SF)

Contractor

Condor Systems Inc, San Jose, California.

UPDATED

AU-506 submarine Electronic Support (ES) antenna

Type

Communications band submarine ES system antenna.

Description

Identified as covering the 3 to 1,200 MHz frequency range (divided into High, Very High and Ultra High Frequency (HF/VHF/UHF) sub-bands), the AU-506 submarine ES system antenna provides both Direction-Finding (DF) and signal acquisition outputs. Each of the three noted sub-bands is provided (in parallel) with its own output, with each specific output containing sine, cosine and omnidirectional data to facilitate arctangent (Watson-Watt) or vector matching DF processing. AU-506 can be used with its manufacturer's own narrow and wideband processing equipment, other manufacturer's equipment or as an upgrade replacement for existing arrays. The unit is designed to survive the hydrodynamic flow, hydrostatic pressure, loading, vibration, shock and wave slap conditions associated with submarine operations. A variant that offers an integrated COMmunications and ELectronic INTelligence (COMINT/ELINT) capability is designated as the AU-506A.

Status

As of this edition, the AU-506 communications band ES array was understood to be 'proven'.

Specifications

Frequency range: 3 -1,200 MHz (divided into HF, VHF and UHF sub-bands)
DF accuracy*: 2° RMS (nominal, clear site, vertical polarisation, −5 to +10° elevation, 40 dB SNR, averaged over the operating frequency after system calibration); 3° RMS (max, clear site, vertical polarisation, −5 to +10° elevation, 40 dB SNR, averaged over the operating frequency after system calibration)
Outputs: DF and acquisition outputs for each sub-band
Sensitivity:

Sub-band	Frequency	Nominal field strength **
HF	4 MHz	60 dBμV/m
HF	24 MHz	30 dBμV/m
HF	32 MHz	25 dBμV/m
HF	50 MHz	15 dBμV/m
VHF	50 MHz	30 dBμV/m
VHF	60-90 MHz	25 dBμV/m
VHF	100 MHz	15 dBμV/m
VHF	125-500 MHz	10 dBμV/m
UHF	500-650 MHz	20 dBμV/m
UHF	650-1,000 MHz	10 dBμV/m
UHF	1,200 MHz	20 dBμV/m

Temperature: −28 to +65°C (operating); −48 to +71°C (storage)
MTBF: 15,000 h (min)
Hydrostatic pressure: 1,000 psi (max)
Weight: 107 kg (incl pressure-bearing radome, nominal)
Dimensions (Ø × H): 51 × 81 cm
* DF accuracy degraded by cosine roll-off for elevations in the +10 to +45° range. VHF/UHF accuracy is degraded by no more than a nominal 4° RMS for standard waves (45° elevation, 45° linear polarisation).
** Nominal field strength required to produce 10 dB (S+N)/N in a 3 kHz intermediate frequency bandwidth at the antenna outputs.

Contractor

Southwest Research Institute, San Antonio, Texas.

VERIFIED

Condor Systems naval Electronic Support (ES) and SIGnals INTelligence (SIGINT) systems

Type

Family of naval ES and SIGINT systems.

Description

US contractor Condor Systems has, over time, developed a range of naval ES and SIGINT systems that has included the following:

AR-900

Designed for both surface ship and submarine applications, the 2 to 18 GHz band AR-900 ES and Direction-Finding (DF) system comprises an antenna unit (surface ship, subsurface or miniaturised for use on optronic masts), a receiver/processor unit and Windows-NT operator's workstation. System features include:

- 2 to 18 GHz receiver calibration and built-in test
- a wide open amplitude monopulse DF array
- 'successful' integration with a number of combat system types
- 2 to 18 GHz receiver calibration and built-in test
- an integrated printer/read/write CD/removable hard drive package
- integral training simulator and mission playback facility
- optional frequency extensions (1 to 2 and 18 to 40 GHz) and integrated Global Positioning System (GPS) and satellite communications antennas
- claimed 100 per cent probability of intercept with interfering signal protection.

Frequency range: 2-6 GHz and 6-18 GHz sub-bands
Processing sensitivity: −65 dBm (both sub-bands)
Azimuth coverage: 360° (both sub-bands)
Instantaneous DF measurement accuracy: 2° RMS (surface ship application, 6-18 GHz sub-band); 3° RMS (subsurface application, 6-18 GHz sub-band); 3.5° RMS (surface ship application, 2-6 GHz sub-band); 4° RMS (subsurface application, 2-6 GHz sub-band)
Display resolution: 1°
Dynamic range: 70 dB
Signal types: agile frequency (±10%), continuous wave, conventional pulse trains, jittered PRI (±10%), pulse-Doppler, pulse modulation and staggered PRI (2-16 positions)
Frequency measurement displayed resolution: 1 MHz (both sub-bands)

Frequency measurement accuracy: 2 MHz RMS (2-6 GHz sub-band); 3 MHz RMS (6-18 GHz sub-band)
Pulse-width measurement range: 50 ns-230 μs
Amplitude measurement range: >60 dB
PRI measurement range: 2-20,000 μs (50-500,000 Hz)
Scan types: bidirectional, circular, conical and omnidirectional
Polarisations received: horizontal, LHCP, slant linear, RHCP and vertical
System alarms: amplitude, continuous wave, steady illumination and threat
Threat library capacity: 10,000 emitter modes
Number of signals tracked: 500
System reaction time: 1 s (max)
Pulse density capacity: 1 Mpps
Temperature: −28 to +65°C (operating, antenna assembly); 0 to +50°C (operating, inboard equipment)
Power: 250 W (operator's workstation); 550 W (receiver/processor - does not include 110 W requirement for anti-condensation heaters, antenna assembly draws power from the system receiver/processor)
Dimensions (h × w × d): 300 (h) × 178 (d) mm (miniature antenna assembly); 526 (h) × 320 (d) mm (surface ship antenna assembly); 668 (h) × 320 (d) mm (subsurface antenna assembly); 721 × 490 × 635 mm (operator's workstation); 940 × 559 × 356 mm (receiver/processor)
Weight: 6.8 kg (miniature antenna assembly); 14.5 kg (surface ship antenna assembly); 27.2 kg (receiver/processor); 29 kg (subsurface antenna assembly); 31.7 kg (operator's workstation)

CS-3701

Described as being a tactical ES and radar surveillance system that is suitable for tri-service applications, the 2 to 18 GHz CS-3701 equipment comprises a monopulse Direction-Finding (DF) antenna assembly, a receiver, a signal processor and an operator workstation. Of these, the monopulse DF antenna assembly incorporates (from top to bottom) Global Positioning System, 6 to 18 GHz DF and 2 to 6 GHz DF antenna arrays and a preamplifier/calibration generator unit. The two DF arrays are described as being interferometers and as generating both omni- and DF data. The equipment's receiver incorporates phase DF 'autohet' and frequency measurement units and produces digitised bearing angle, frequency, amplitude, time of arrival and pulse-width values that are combined to produce pulse descriptor words for each signal received. The system's signal processor accepts these pulse descriptors, de-interleaves them to form pulse trains and compares them with an emitter library for identification. Signal reports produced by the processor are passed to the operator workstation via an Ethernet local area network. The CS-3701's operator workstation incorporates an operator-configurable, multiWindows format tactical display unit, a keyboard and a Windows-NT operating system. Other system features include:

- long-range, precision, over-the-horizon targeting
- multimode radar report merging
- multipath and reflection processing
- 2 to 18 GHz receiver calibration and built-in test
- a built-in training simulator
- PDW processing for high duty cycle signal environments
- tunable, 2 to 18 GHz, continuous wave notch filters for the system's omni- and DF channels
- real-time pulse analysis displays
- claimed 100 per cent probability of intercept with interference signal rejection.

Frequency coverage: 2-18 GHz (0.5-2 GHz and 18-40 GHz frequency extension options)
Azimuth coverage: 360°
Pulse environment: up to 1 Mpps
Threat library: >10,000 emitter modes
Tracking capacity: up to 500 signals simultaneously
Automatic processing sensitivity: −65 dBm
Tangential signal sensitivity: −70 dBm
Dynamic range: 60 dB (instantaneous processing)
DF accuracy: 2° RMS (across the complete dynamic range)
Frequency measurement accuracy: 3 MHz
Pulse-width measurement: 50 ns (min)
Options: 0.5-18 GHz superheterodyne receiver; enhanced sensitivity

CS-5060

The 0.5 to 18 GHz CS-5060 electronic intelligence system is described as being suitable for tri-service applications and as offering automatic emitter intercept, identification, DF and geolocation facilities. The equipment comprises an antenna assembly, an SP-103 signal processor, a TN-618 microwave tuner and a workstation. Other system features include:

- operational and technical report generation
- high-gain, directional antenna to facilitate long-range detection
- modulated signal sorting, characterisation and association
- precision frequency and time measurement
- a VME-based open architecture
- a Windows-based man/machine interface
- integral pulse analysis and PDW data recording for post-mission analysis.

Frequency range: 0.5-18 GHz
Frequency accuracy: 1 MHz RMS
Instantaneous bandwidth: 500 MHz
IF bandwidths: 20, 50, 100, 250 or 500 MHz
Video bandwidths: 5 and 20 MHz
Automatic DF accuracy: 2° RMS
PRI range: 2 μs-10 ms

TOA resolution: 5 ns
PRI accuracy: 50 ns
Pulse-width range: 50 ns-2.62 ms
Pulse-width resolution: 40 ns
Pulse-width accuracy: 40 ns RMS
Scan rate range: 0.05-100 Hz
Scan rate accuracy: 2% RMS
Scan types: bidirectional, complex, conical, circular and steady
Pulse throughput: 2 Mpps (burst)
Power: 125 W (tuner); 200 W (antenna assembly); 250 W (processor)
Dimensions (w × h × d): 147 × 198 × 559 mm (tuner); 269 × 218 × 457 mm (processor); 483 (d) × 737 (h) mm (antenna assembly)
Weight: 14.9 kg (tuner); 16.3 kg (processor); 31.8 kg (antenna assembly)

Status
As of May 2001, the systems cited here were being actively promoted by Condor Systems. As of the given date, Jane's sources were suggesting that a CS-3701 variant was an 'alternative' ES system aboard Swedish 'Vastergotland' class anti-submarine warfare attack submarines and that the equipment had been selected for use aboard Australian 'Collins' class anti-submarine warfare attack submarines (fourth and fifth of class only), Norwegian 'Fridtjof Nansen' class frigates and Swedish 'Visby' class corvettes.

Contractor
Condor Systems Inc, San Jose, California

UPDATED

Phoenix Electronic Support (ES) system

Type
Submarine radar ES system.

Description
Phoenix has been designed for use in submarines to provide automatic identification and bearing of intercepted radar emissions in the 2 to 18 GHz frequency range. The system comprises an antenna array and an instantaneous frequency measuring receiver that provides precise measurement of the threat radar operating frequency. A detailed display is provided to the operator, giving all significant parameters of the incoming signals. The system is modular in concept and can be configured to meet individual customer requirements. Both manual and fully automatic systems and analysis capabilities are available. The fully automatic version is known as Phoenix IV.

Status
As of March 2001, Jane's sources were reporting that the Phoenix ES system was in service aboard the New Zealand frigate *Canterbury*.

Contractor
Condor Systems Inc, San Jose, California.

UPDATED

Type 18 periscope-mounted Automatic Direction-Finding (ADF) system

Type
Submarine Electronic Support (ES) and ADF system.

Description
The Type 18 ES and ADF system is noted as being fitted to the search periscopes of US Navy 'Seawolf', 'Sturgeon' and 'Los Angeles' class nuclear-powered attack submarines. As such, the equipment incorporates a mast-mounted antenna assembly (located above the periscopes gyro-stabilised optics), radio frequency receivers, a digital processor and a display.

Status
As of this edition, at 24 Type 18 ES and ADF systems were reported as having been procured by the US Navy.

Contractor
BAE Systems North America, Lansdale, Pennsylvania.

VERIFIED

WBR-2000 Electronic Support (ES) System

Type
2 to 18 GHz band tactical radar ES system.

Description
Functionally, the WBR-2000 ES system makes use of a combined omni-directional and direction-finding antenna assembly to detect radar signals within the 2 to 18

GHz frequency band. These are then passed to a receiver/processor where the frequency and angle-of-arrival of each received pulse is measured and digitised. This information is forwarded to a signal processor that associates separated pulses from a particular emitter, measures pulse train modulation characteristics and generates an emitter characterisation report. This is compared with the system's integral emitter parameter library in order to identify the detected radar and the platform/platforms with which it is normally associated. A computer workstation is used to store processed data and display information to the equipment's operator. Other system features include:
- a fully open architecture
- claimed 100 percent probability-of-intercept in a dense, continuous wave-type environment
- tactical displays with operator access to all parametric, signal environment, threat and built-in test data
- the detection, identification and characterisation of complex emitter types
- provision for Ethernet, NTDS, RS-232 and RS-422 interfaces
- an interactive, user-friendly, single operator, control display format
- built-in test and maintenance functions
- integral expansion capabilities

Status
As of the first quarter of 2000, the WBR-2000 ES system was reported as being 'available, deployed and operational'. Within the US military, the equipment forms the basis of the Bobcat carry-on ES package that has been procured for use aboard US Special Operations Command's (SOCOM) Mk V Fast Patrol and 'Cyclone' class coastal patrol boats based at Tampa, Florida. Identified contracting activity relating to the Bobcat programme is as follows:
November 1998
On 4 November 1998, Sensytech Inc announced that it had been awarded a then year US$6.3 million contract covering the supply of Bobcat ES systems for use aboard SOCOM Mk V and 'Cyclone' class patrol craft. As originally setout, delivery of the equipment was scheduled to begin during the second half of 1999 and to be completed during 2000.
May 1999
On 19 May 1999, Sensytech Inc announced that it had been awarded a then year US$3.2 million contract covering the supply of additional Bobcat ES systems for use on SOCOM's Mk V and 'Cyclone' class patrol boats. This award was the exercise of an option contained in the original November 1998 Bobcat procurement contract.
August 1999
On 24 August 1999, Sensytech Inc announced that it had been awarded a then year US$1.8 million contract covering the supply of a third tranche of Bobcat ES systems for use aboard SOCOM Mk V and 'Cyclone' class patrol craft.

Specifications
WBR-2000
Frequency: 2-18 GHz (0.5-2 GHz and 18-40 GHz options)
Sensitivity: -65 dBm (-85 dBm option)
DF accuracy: 10° RMS (2° and 5° RMS options)
Dynamic range: 60 dB
Probability-of-intercept: 100%
TOA accuracy: 30 ns
PRI range: 500 ns-33 ms
Minimum pulsewidth: 50 ns

Contractor
Sensytech Inc, Newington, Virginia.

VERIFIED

WBR-2500 Electronic Support (ES) system

Type
Automatic, tactical, submarine radar ES system.

Description
Functionally, the WBR-2500 ES system utilises a combined omni-directional and directional antenna assembly to receive signals within the 2 to 18 GHz frequency band. These are then passed to a receiver/processor unit that measures and digitises their frequency and angle-of-arrival. The derived information is next forwarded to a signal processor where separated pulses from a particular emitter are re-associated, pulse train modulation characteristics are established and emitter characterisation reports generated. Emitter identification and platform association is achieved via the matching of individual characterisation reports to parametric data in the system's library. WBR-2500 incorporates a computer workstation capable of both storing and displaying data and other system features include:
- a full open architecture
- claimed 100 probability-of-intercept of a single pulse in a dense continuous wave environment
- tactical displays with operator access to all parametric, signal environment and built-in test data
- the detection, identification and characterisation of complex emitter types
- Ethernet, NTDS, RS-232 and RS-422 interfaces
- an interactive, user friendly, single operator, control display format

- built-in test and maintenance functions
- an integral expansion capability

Status

As of the first quarter 2000, the WBR-2500 ES system was reported as being available.

Specifications

Frequency: 2-18 GHz (0.5-2 GHz and 18-40 GHz options)
Sensitivity: -65 dBm (-85 dBm option)
DF accuracy: 10° RMS (2° and 5° RMS options)
Dynamic range: 60 dB
POI: 100%
TOA accuracy: 30 ns
PRI range: 500 ns-33 ms
Minimum pulsewidth: 50 ns

Contractor

Sensytech Inc, Newington, Virginia.

VERIFIED

WBR-3000 ELectronic INTelligence (ELINT)/ Electronic Support (ES) system

Type

Automated ELINT/ES system.

Description

The WBR-3000 ELINT/ES system is designed to 'quickly' identify all 2 to 18 GHz band signals within a particular environment with 100 percent probability-of-intercept. Additionally, the equipment is able to perform 'high' sensitivity background searches and signals analysis functions with the 0.5 to 18 GHz frequency range. WBR-3000 can be configured with one or more 'interactive, user friendly' workstations and other system features include:

- a full open architecture
- a fine grain signals analysis capability
- tactical displays with operator access to all parametric, signal environment, threat and built-in test data
- the detection, identification and characterisation of complex emitter types
- Ethernet, NTDS, RS-232 and RS-422 interfaces
- built-in test and maintenance functions
- an integral expansion capability

Status

As of the first quarter of 2000, the WBR-3000 ELINT/ES system was reported as being available.

Specifications

	Wide-band	Narrow-band
Frequency:	2-18 GHz (0.1-2 GHz option)	0.5-18 GHz (18-40 GHz option)
Frequency accuracy:	4 MHz RMS	1 MHz RMS
Sensitivity:	−65 dBm	−85 dBm
DF accuracy:	7° RMS	2° RMS
Dynamic range:	60 dB	75 dB
POI:	100%	scan dependant
TOA accuracy:	30 ns	30 ns
PRI range:	500 ns-33 ms	500 ns-33 ms
Minimum pulsewidth:	50 ns	50 ns

Contractor

Sensytech Inc, Newington, Virginia.

VERIFIED

NAVAL ACTIVE AND PASSIVE COUNTERMEASURES SYSTEMS AND DEFENSIVE AIDS SUITES (DAS)

AUSTRALIA

NULKA active missile decoy system

Type

Rocket-launched hovering decoy system.

Development

The I/J-band (8 to 20 GHz) NULKA (NATO designation Mk 234 Mod 1) device is an active missile decoy that provides effective all-weather self-protection for naval vessels against anti-ship missiles. It can be used as part of a multilayer defence system or for stand-alone ship protection. Information on the threat is provided by the ship's electronic support measures system or other equipment and NULKA uses this information to calculate the optimum launch time and trajectory for the decoy. The system allows for automatic or operator designation of a missile threat and, upon designation of a particular threat, will respond rapidly by launching an autonomous airborne decoy. Before launch the system calculates the optimum decoy flight trajectory for the mission and programmes that trajectory data into the decoy's flight control unit. With its programmable and controllable flight path, the rocket hovers and positions itself to provide a more attractive target for the threat missile. The decoy payload is provided by Sippican Inc and is optimised to provide effective protection to both large and small surface ships.

In more detail, the NULKA hovering rocket decoy air vehicle is held in an hermetically sealed canister that acts as a lifetime storage container, as well as the launch tube for the decoy. It is propelled by a solid fuel rocket motor. Control of the decoy's flight is achieved by a thrust control mechanism that acts on the motor's efflux, and a spin control unit mounted on top of the decoy. After preflight programming from the launcher processor, the decoy's flight trajectory is determined by a digital flight control unit mounted immediately above the rocket motor. The combination of thrust and flight control enables successful launches to be made in severe sea state and high wind conditions. The hovering flight characteristics of the decoy vehicle permit the effective use of the Sippican payload with wide area coverage, thus enabling one decoy to counter multiple threats. Once launched, the decoy operates autonomously and following its stored flight commands, moves away from the ship at its preprogrammed height and speed to present an alternative and more attractive target to incoming missiles.

The NULKA payload comprises a 13.2 kg, 81 by 15 cm mid-body electronics section and modular transceiver antenna subassemblies. The electronics section is compartmentalised (for reasons of electromagnetic interference management) into receiver and transmitter units. The antenna design used contributes to the system's overall gain and provides sufficient angular coverage to engage multiple anti-ship missile attacks. The receiver unit detects and amplifies all in-band signals and (using programmable signal processing) rejects own-force emitter signals. Each subassembly derives power from a high-reliability thermal battery. As noted previously, the broadband horn antennas used are modular and feature high levels of isolation. The receiver unit makes use of hybrid microwave circuitry and features built-in protection from high power inputs and high gain with 'state-of-the-art' gain stability. An associated Fire-Control System (FSC) allows NULKA to be installed and operated on ships that are not fitted with combat and fire-control systems as well as providing a back-up aboard ships that are so equipped. The FCS accepts the minimum input data required via an automatic interface with an electronic support system or a manual input from an operator. With its own processor, the FCS manages the launching of decoys from multiple launchers located around the ship.

Status

NULKA is a collaborative Australian/American programme. According to Jane's sources, the NULKA decoy completed full-scale engineering development during December 1993 and in May 1995, the US Navy (USN), on behalf of the two nation NULKA Joint Project Office, awarded an engineering and manufacturing development contract that covered the supply of 13 prototype decoys. During September 1996, the USN designated a NULKA-capable Mk 36 decoy launcher variant as the Mk 53 (see separate entry) and in June 1997, the Royal Australian Navy (RAN) was authorised to place an initial NULKA production contract (valued at then year A\$112 million) that included 52 rounds for the USN. Developmental and operational testing of the device (aboard the USS *Peterson*) took place during the summer of 1998 and in January 1999, NULKA (and its associated Mk 53 launch system) was assessed as being 'potentially effective and suitable' for operational use by the USN and limited fleet introduction of the capability was recommended with the proviso that additional follow-on test and evaluation take place. As a result, a production contract for 11 Mk 53 launchers was issued during February 1999, with the USN eventually intending to install the system aboard 'Ticonderoga' class cruisers (27 ships in class as of April 2001), 'Arleigh Burke' (Flights I and II) class destroyers (28 ships as of April 2001) and 'San Antonio' class amphibious transport docks (four building and eight projected as of April 2001).

For its part, the RAN installed a prototype NULKA system aboard HMAS *Melbourne* during June 1997 and in the following May, nine NULKA rounds were launched during land trials at the Woomera test range and the system fitted aboard the *Melbourne* was upgraded to a representative production standard. So equipped, the frigate participated in the RIMPAC 98 exercise and became the first

A general view of the NULKA decoy when deployed

warship to deploy operationally with NULKA when it joined the UN's multinational Maritime Interception Force in the Persian Gulf during the summer of 1999. In RAN service, NULKA is installed (as a retrofit) aboard its 'Adelaide' (six ships as of March 2001) and 'ANZAC' (three in service and five building as of March 2001) class frigates, with (as of February 2000) HMAS *Canberra*, *Melbourne* and *Newcastle* being reported as having been equipped with the system. While not confirmed, it is likely that New Zealand's pair of 'ANZAC' frigates have provision for the retrofitting of the NULKA system.

Elsewhere in the world, Jane's sources report that the Canadian Navy signed a then year A\$18 million contract covering the supply of NULKA systems for installation aboard 'Iroquois' class destroyers (four ships as of March 2001) on 5 June 1997. Forming part of the type's Tribal Class Update and Modernisation Project (TRUMP), NULKA replaces the AN/ULQ-6 radar jammer aboard 'Iroquois' class ships, the first two of which were noted as having had the NULKA system installed by February 2000. As of March 2001, the same sources were suggesting that NULKA was also going to be fitted aboard the 12 frigates that made up the Canadian Navy's 'Halifax' class.

Contractors

BAE Systems, Elizabeth, South Australia.
Sippican Inc, Marion, Massachusetts (decoy payload).

UPDATED

CANADA

AN/SLQ-503 RAMSES Electronic CounterMeasures (ECM) system

Type

Shipborne radar noise and deception repeater jammer system.

Description

The AN/SLQ-503 Reprogrammable Advanced Multimode Shipborne ECM System (RAMSES) is a modular naval electronic countermeasures system designed principally to provide protection for ships against I/J-band (8 to 20 GHz) radar threats. It is an integrated responsive jammer capable of long-range jamming of search radars and the deception of missiles once launched. The system's normal mode of operation is to be connected to and controlled by the ship's combat control system, thus removing the need for a dedicated system operator. It is also capable of being connected directly to and controlled by the ship's electronic support measures system to provide an integrated electronic warfare solution. RAMSES operates in I/J-band and is capable of high-pulse and continuous wave effective radiated power. Multimode jamming operation is possible with multitarget handling capability. The system is reprogrammable and power management is provided. The antennas are fully trainable (port and starboard each through 180°) and are stabilised for roll and pitch.

The RAMSES antenna installation on HMCS Halifax

Status

Over time, the AN/SLQ-503 ECM system is reported as having been installed aboard 'Iroquois' class destroyers and 'Halifax' class frigates of the Canadian Forces' Maritime Command.

Contractor

Lockheed Martin Canada Inc, Montreal, Quebec.

VERIFIED

DENMARK

SKWS (Soft Kill Weapon System) decoy launching system

Type

Surface-ship decoy launching system.

Description

TERMA Elektronik's Seagnat-compatible, mortar-type, SKWS can be configured for small- (DL-6T launcher unit) or medium- to large-ship (DL-12T launcher unit) applications and, according to Jane's sources, is capable of supporting the Mk 214 seduction and Mk 216 distraction Seagnat chaff rounds together with the Mk 245 Giant, Perfect, Pirate and Talos Infra-Red (IR) seduction devices, the Chimera dual-mode chaff and IR decoy and the Launched Expendable Acoustic Device (LEAD) torpedo decoy. The system is operated from the host ship's Combat Management System (CMS) which also supplies the necessary threat, sensor and operator data needed to calculate the optimal engagement proposal in a multithreat situation.

Looking at the system elements in more detail, the DL-6T and DL-12T units are multitube, fixed azimuth angle launchers that have port and starboard firing angles

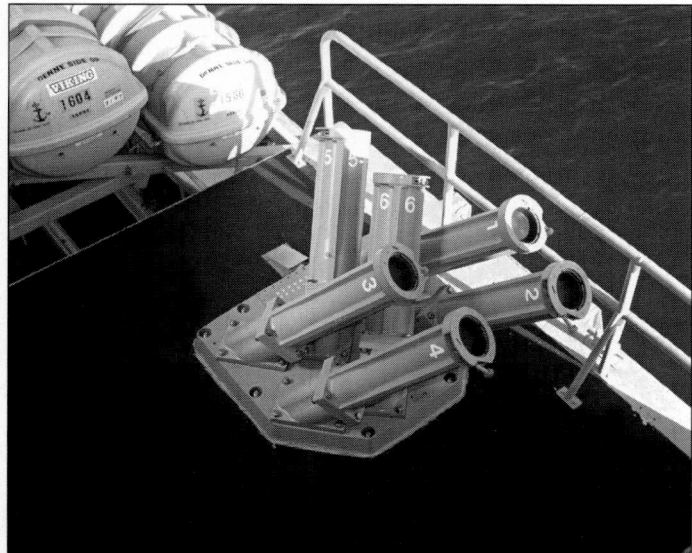

The six-barrelled DL-6T launcher unit   0009364

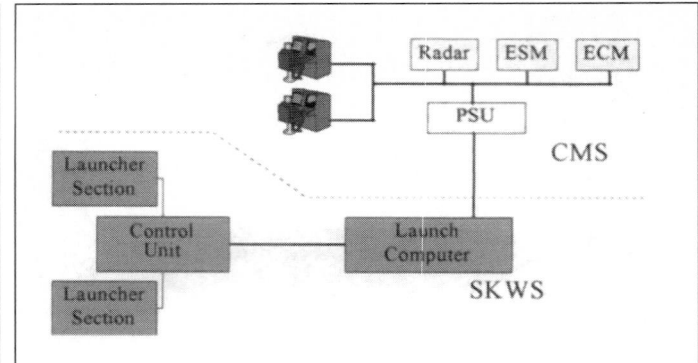

A block diagram showing the CMS to SKWS interface   0009363

The 12-barrelled DL-12T launcher unit   0009365

(in relation to the host vessel's direction of travel) of 20, 40 and 135° for the DL-6T and 10, 40, 60 and 135° for the DL-12T. The firing angles used are based on modelling undertaken by the Royal Danish Defence Research Institute and the individual tubes have a 45° angle of elevation in order to facilitate the discharge of both mortar and rocket munitions. Both types of launcher unit incorporate an intelligent measuring/scanning system for ammunition detection and a submunitions counter.

Two or four, 20 round capacity, ready service lockers are used for munitions storage with each locker being equipped with a sprinkler system (water nozzle and shut-off valve) for safety purposes. The equipment's power supply converts the input voltage (115 V AC) to system voltage (28 V DC), with other voltage options being available. The SKWS control unit has a graphic system status display that shows data such as ammunition type, remaining ammunition in the tube (IR-Giant round), loaded/empty status, proposed launch and tube limit.

The system's launch control computer is a workstation type unit that incorporates a fast microprocessor and software engagement algorithms that are used to calculate the optimum firing scheme for a particular threat. Such calculations are based on threat type, wind, 'own ship' and launcher loading data in combination with stored information concerning threat, 'own ship' and decoy characteristics. Stored data can be accessed and updated via the CMS and the firing schemes generated incorporate platform manoeuvre advice. The launch computer interfaces with a system control unit to initiate decoy launch and to establish individual tube contents and overall system status. SKWS offers the following operating modes:

Manual

Where the operator selects the tubes to be fired using the CMS or the SKWS control unit.

Semi-automatic

Where threats are assigned from the CMS using automatic designation (radar fast target alerts or electronic support measures alarm/electronic warfare lines) or via the operator who can select a proposed firing scheme.

Automatic

Where threats are assigned automatically from the CMS, the optimum response proposal is initiated and the operator notified.

Threat Evaluation and Weapon Assignment (TEWA)

Available where the CMS is able to support computer assisted or automatic, multiweapon, multithreat engagements, the TEWA mode assigns threats and requests proposals for specific threats or manoeuvre combinations.

Status

The first production decoy launching system was delivered to the Danish Navy (for installation aboard a 'Willemoes' class fast attack craft) during the second quarter of 1989. During the period February to March 2001, Jane's sources were reporting that DL-6T configured SKWS systems were in service aboard Danish 'Falster' class minelayers (two ships in class) and 'Flyvefisken' class multirole warships (14 ships), while DL-12T configured systems were installed aboard the country's 'Niels Juel' (three ships) and 'Thetis' (four ships) class frigates. As of the given date (and looking to the future), SKWS manufacturer TERMA was understood to be considering modifying the system to support the BAE Systems Siren active offboard decoy (see separate entry) and as implementing it alongside legacy Mk 137 launchers as a way of addressing the anti-missile/anti-torpedo requirement. Here, the proposal would increase the number of available launch barrels per beam to 24 and would incorporate a specific Anti-Submarine Warfare (ASW) interface to allow the host vessel's ASW officer to control the launch of torpedo decoys and co-ordinate their use with other measures such as manoeuvre and/or towed decoy deployment.

Contractor
TERMA Elektronik AS (Naval and Communications Systems Division), Tästrup.

UPDATED

FRANCE

LACROIX naval countermeasures

Type
Shipborne expendable munitions.

Description
LACROIX manufactures a large range of expendable munitions that are used in launchers such as Dagaie/Sagaie (both developed in collaboration with Matra Défense Equipements & Systemes) and Philax (with CelsiusTech). These various countermeasures operate in seduction, distraction and seduction/dissimulation modes using both chaff and infra-red mortar modules and chaff rockets. The company's electromagnetic decoys are normally made of aluminised glass fibre chaff with a rapid bloom time and cover the I/J-band (8 to 20 GHz). Its infra-red decoy rounds make use of a composition covering the 3 to 5 and 8 to 14 µm bands. LACROIX also manufactures pyroacoustic ammunition for anti-submarine warfare. This consists of both bubbles cloud material and acoustic noise generators to provide an increase of ambient noise and mask the ship by false echoes.

Status
As of this edition, LACROIX's range of naval countermeasures was reported as being available.

Contractor
Etienne LACROIX, Muret Cedex.

VERIFIED

LACROIX infra-red decoys in action

Sidewind countermeasures management system

Type
Shipborne countermeasures management system.

Description
Sidewind is a countermeasures management system that interfaces with the Dagaie Mk 2 decoy launcher (see separate entry), generates optimised responses to identified threats and provides a tactical situation display. When linked with an Electronic Support Measures (ESM) system and a jammer, Sidewind can act as the core of an onboard Electronic Warfare (EW) suite. At the heart of the system is a digital computer that incorporates two 68020 processors, a 48 cm (19 in) colour monitor and an alphanumeric keyboard. One of the processors co-ordinates the electronic warfare tactics and the other interfaces with the ship's sensors and command system. Functionally, Sidewind receives tracks from the ship's ESM, determines the nature of the threat, provides a threat assessment and weapons

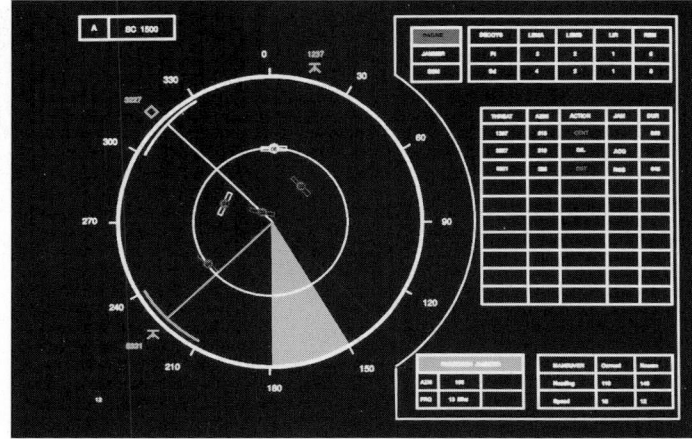

The Sidewind countermeasures management system's display

allocation function. It informs the ship's EW co-ordinator as to the most threatening track, best combination of resources to defend the ship, status of hard and soft kill resources, possible interference between hard kill and soft kill, and the best course to steer. Threat response is completely automatic, with a veto available to the EW co-ordinator. At the Sidewind console, a bearing versus time variable range display shows the known characteristics of the threat, availability of countermeasures, ship course and speed and course to steer to free blind arcs or to reduce radar cross-section. The bearing/time display also shows the position of decoys once deployed, the mode of operation, status of the ship's jammer, windspeed and direction. Up to 10 threats can be handled simultaneously. The operator can also interrogate Sidewind to provide a preview facility based on current course and speed of known tracks. The system's processor, monitor and keyboard can be separated to meet installation requirements. Co-ordination between soft kill and hard kill tactics can be provided by the Cuirasse system. A version of Sidewind that incorporates the Vampir ML 11 infra-red detection system, the Dagaie Mk 2 decoy launcher and the Sadral surface-to-air missile, Cuirasse carries out threat evaluation, weapon assignment and hard/soft kill co-ordination and incorporates an additional processing unit to handle its increased data requirements.

Status
As of the first quarter of 2000, the Sidewind countermeasures management system was reported as having been installed aboard 'Barzan' class patrol craft of the Qatar Navy.

Contractor
Matra Défense Equipements & Systemes (a European Aeronautic Defence and Space (EADS) Co component), Les Ulis.

UPDATED

GERMANY

BUCK chaff and Infra-Red (IR) decoy rounds

Type
Range of naval decoy rounds.

Description
BUCK Neue Technologie manufactures a range of naval chaff and IR decoy rounds, the known details of which are as follows:

DM 19 DUERAS
Designed for use with the SCLAR launch system (see separate entry), DM 19 is a chaff rocket that can be used for seduction, distraction and/or confusion. The munition is electronically fuzed and NATO qualified (Stock No 1340-12-314-7364)

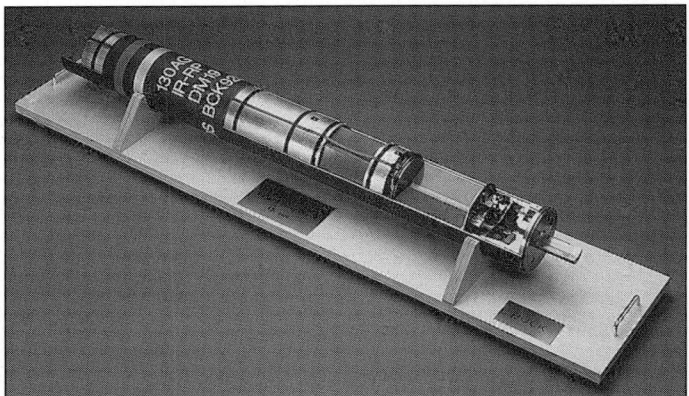

The BUCK 130 mm DM 19 GIANT/Mk 245 Mod O IR decoy round

Specifications
Weight: 31.3 kg
Length: 1,752 mm

DM 19 GIANT

Designed for use with the Mk 36/Super Rapid Bloom Offboard Countermeasures (SRBOC) and Seagnat launchers, GIANT is an IR seduction decoy round that is equipped with a five submunition payload that is designed to 'walk off' incoming anti-shipping missiles. It is claimed to offer high radiant intensity, a large effective area and long duration. GIANT is targeted against first, second and third-generation IR seekers and is NATO qualified (Stock No 1329-12-232-1749, designation Mk 245 Mod O).

Specifications
Number of submunitions: 5
Submunition firing rate: 0.1 s (min)
Band coverage: 3-5 μm, 4.1-4.5 μm and 8-14 μm
Weight: 3 kg (submunition); 21 kg (complete round)
Dimensions (Ø × l): 130 × 1,208 mm

DM 29 IRRAS

Designed for use with the SCLAR launch system (see separate entry), DM 29 is a multiwavelength IR decoy rocket that can be used for seduction, distraction and/or confusion. The munition is electronically fuzed and NATO qualified (Stock No 1340-12-323-0293)

Specifications
Weight: 31.3 kg
Length: 1,752 mm

Status
As of the first quarter of 2000, the DM 19 DUERAS and DM 29 IRRAS rocket munitions were understood to be in service with a number of navies who use the SCLAR launch system. The Mk 245 Mod O/DM 19 GIANT IR decoy round was reported as having been procured by the American (evaluation quantity), Australian (evaluation quantity), British, Danish, Netherlands, German, Portuguese and Spanish navies. In terms of specific contracting activity, Buck is known to have received a UK Mk 245 Mod O procurement contract during August 2001.

Contractor
BUCK Neue Technologie GmbH (a subsidiary of Rheinmetall AG), Neuenburg.

UPDATED

INTERNATIONAL

ARBB 32 radar jammer

Type
Shipborne noise and deception radar jammer.

Description
The ARBB 32 jammer was developed during the late 1960s/early 1970s and has been fitted to a number of ships of the French Navy in its ARBB 32B production configuration. It has been superseded by the ARBB 33.

Status
As of February 2001, Jane's sources were reporting the ARBB 32 radar jammer as being installed aboard the following warships and classes of warship:
France   • the 'Georges Leygues' class destroyers *Georges Leygues*, *La Motte-Picquet*, *Latouche-Treville* and *Primauguet*
    • 'Tourville' class destroyers (two ships in class)

Contractor
Thales Airborne Systems, Elancourt, France.

UPDATED

....

ARBB 33 radar jammer

Type
Shipborne noise and deception radar jammer.

Description
ARBB 33 is a high-performance jammer, conceived and manufactured for the French Navy, which follows on from the ARBB 32 equipment. It is designed for installation on board a variety of surface ships, has a multiple threat capability and is able to counter up to four threats simultaneously. ARBB 33 is capable of jamming all modern threats including target designation, pulsed fire-control and active missile seeker radars. It offers a wide range of jamming modes including continuous noise, pulsed noise, cover pulse jamming, synchronous and

The port side ARBB 33 installation aboard a French warship

asynchronous false echoes and range gate pull-off. The system can be alerted initially by a threat detection equipment, but also possesses an autonomous detection subassembly. It provides 360° coverage in azimuth by means of electronically switched antennas for both detection and jamming. This approach is claimed to improve reaction times against threats on any bearing. Overall, ARBB 33 is divided into four elements, as follows:
• two canisters, port and starboard, each one comprising a transmitter, a receiver and their electronically switched antennas. The latter are circular phased-arrays for reception and transmission.
• a technical cabinet
• a control and display console.
Of these, the transceiver canisters are divided into transmission and reception units. The system operates in H-, I- and J-bands (6 to 20 GHz) and covers two 180° sectors. Output power is more than 100 kW and the system can handle two to four (option) threats simultaneously.

Status
As of the period March to July 2001, Jane's sources were reporting the ARBB 33 radar jammer as being installed aboard the following warships and classes of warship:
France   • the aircraft carrier *Charles de Gaulle* (twin ARBB 33 installation)
    • 'Cassard' (two ships in class) and 'Suffren' (two ships) class destroyers
    • 'La Fayette' class frigates (five ships - provision for ARBB 33)
Qatar   • 'Barzan' class fast patrol craft - missile (four ships)
As of April 2001, ARBB 33 was reported as being no longer in production.

Specifications
Frequency: H-, I- and J-bands (6-20 GHz)
Azimuth coverage: 2 sectors of 180°
Reaction time: <0.5 s
Detection sensitivity: better than −50 dBm
Bearing accuracy: better than 5° RMS
Dimensions: 1.2 × 2.2 m (canister); 1.8 × 0.6 × 0.78 m (cabinet)
Weights: 400 kg (cabinet); 500 kg (canister);

Contractor
Thales Airborne Systems, Elancourt, France.

UPDATED

....

Cygnus radar jammer

Type
Shipborne or land-based radar noise and deception jamming system.

Description
The Cygnus radar jammer is a ship- or land-based equipment operating in the I- or J-bands (8 to 10 GHz or 10 to 20 GHz). Originally designed for integration with the Cutlass electronic support system, Cygnus uses both responsive noise and deception jamming to provide an effective jamming capability against 'all types' of radar including early warning, target acquisition and missile guidance. The jammer has a narrow beamwidth (giving an effective radiated power of 300 kW) and is kept

The antenna assembly used in the Cygnus jammer (MS/RRDS)

on target by a built-in, interferometer type, passive tracking system in both azimuth and elevation. Radio Frequency (RF) signals received by the tracking antennas are also used as the basis of transmitted RF and modulation and power management is under the control of a processor that also controls the tracking procedure. Types of modulation sequence available include range gate pull-off and false target generation.

Status

As of the period January to September 2001, Jane's sources were reporting the Cygnus radar jamming system as being installed aboard the following warships and classes of warship:

Algeria	• the landing ships logistic *Kaleet Beni Hammad* and *Kaleet Beni Rached*
Bahrain	• the 'Ahmad El Fateh' class fast attack craft - missile *Abdul Rahman Al Fadel* and *Al Taweelah*
Brazil	• 'Niteroi' class frigates (six ships in class - Cygnus or Elebra SLQ-1 radar jammer)
Egypt	• 'Ramadan' class fast attack craft - missile (six ships)
Kenya	• 'Nyayo' class fast patrol craft - missile (two ships)
Kuwait	• the fast attack craft - missile *Istiqlal*
Qatar	• 'Damsah' class fast attack craft - missile (three ships)
United Arab Emirates	• 'Muray Jib' class corvettes (two ships)

In terms of recent contracting activity, Egypt placed an order for additional Cygnus jammers during March 1995.

Contractor

Thales Sensors, Crawley, UK.

UPDATED

··

Dagaie Decoy Launching System (DLS)

Type

Shipborne chaff and Infra-Red (IR) DLS.

Description

The Dagaie DLS is an automatic dispenser for IR decoy flares, chaff grenades and chaff rockets. The system accepts threat data from a variety of sources such as radar, Electronic Support (ES) electro-optical sensors and so on. This information is used with ship's speed and heading to compute the optimum countermeasures deployment. The system's launcher unit comprises 10 replaceable munitions modules, each of which is loaded with either IR/centroid chaff or rocket launcher module chaff. The ammunition weight is 50 kg for the centroid 'suitcase' and 20 kg for the rocket unit. The firing sequence runs automatically and is triggered on a missile alarm from a variety of sources (radar, electronic support or electro-optic) thus providing a very short reaction time. The firing direction is optimised in accordance with the threat bearing, wind speed and direction, ship heading and speed. The launching of radar and IR decoys is so arranged that the most advanced mixed-guidance missile cannot discriminate between decoy and target.

Dagaie installations take the form of a double mounting for ships over 1,000 tonnes or a single mounting for smaller vessels and comprise:
- one or two trainable mountings capable of accommodating 10 'suitcase' or rocket launcher module munitions. Each suitcase is loaded with IR or chaff projectiles and the rocket launcher module is loaded with three chaff rockets, including chaff projectiles. The range of existing ammunition enables decoys to be adapted to a variety of countermeasures scenarios
- one or two servo units
- one data processing unit to compute firing data
- one power supply unit
- one supervision unit used to operate the system and display the loading state of the mounting
- one manoeuvre indicator that displays the proposed ship's manoeuvre in order to extend the protection time as long as possible without renewing the decoy.

For either the single or double installation, a reduced size mounting carrying six suitcases can be supplied.

Dagaie Mk 2

The Dagaie Mk 2 DLS is an improved version of the Mk 1 that differs from its predecessor in offering:
- easier integration of the equipment with a central processing system (NTDS)
- centralised operation in conjunction with other Electronic Warfare (EW) equipment
- use of a new generation of ammunition (including the REM chaff rocket) requiring new firing rules and consequently, more sophisticated software
- an increase in operational characteristics via the use of a new medium-range electromagnetic ammunition (REM) which allows the system to cover, alone or associated with a jammer, the whole gambit of EW actions possible during an attack sequence (dilution, centroid seduction, dissimulation seduction and range gate pull off seduction).

Physically, Dagaie Mks 1 and 2 are similar, with the main differences being in the latter's computer and supervision unit. Dagaie Mk 1 can be upgraded to Mk 2 standard by replacing the former's supervision unit and upgrading the system's data processor. Overall, modifications include the replacement of the unit's acquisition and processor racks, 'minor modifications' to its ancillaries' rack and, according to Jane's sources, the replacement of the Mk 1's central 'suitcases' with a pair of three-barrel launch modules for the REM chaff rocket.

Status

As of the period January to April 2001, Jane's sources were reporting that Dagaie DLSs were in service aboard the following warships and warship classes:

Argentina	• 'Almirante Brown' class destroyers (four ships in class)
	• 'Drummond' (three ships) and 'Espora' (five ships, one building as of January 2001) class frigates
Bahrain	• 'Al Manama' class corvettes (two ships)
	• 'Ahmad El Fateh' class fast attack craft - missile (four ships)
Colombia	• 'Almirante Padilla' class frigates (four ships)
France	• 'Cassard' (two ships), 'Georges Leygues' (seven ships - Dagaie Mk 1 or 2), 'Suffren' (two ships) and 'Tourville' (two ships - Dagaie replaces Syllex DLS) class destroyers
	• 'D'Estienne d'Orves' (nine ships), 'Floreal' (six ships - Dagaie Mk 2) and 'La Fayette' (five ships - Dagaie Mk 2) class frigates
Indonesia	• 'Todak' (two ships) and 'Singa' (four ships) class patrol craft
Italy	• 'Maestrale' class frigates (eight ships)
South Korea	• 'Okpo' class destroyers (three ships - Dagaie Mk 2)
Kuwait	• the fast attack craft - missile *Al Sanbouk* and *Istiqlal*
	• 'Um Almaradin' class fast patrol craft - missile (eight ships - Dagaie Mk 2)
Malaysia	• 'Kasturi' class corvettes (two ships)

The Dagaie Mk 2 countermeasures dispenser

Morocco	• the frigate *Lieutenant Colonel Errhamani*
	• 'Floreal' class frigates (two building as of April 2001 - Dagaie Mk 2)
Peru	• the cruiser *Almirante Grau*
Qatar	• 'Barzan' (four ships - Dagaie Mk 2) and 'Damsah' (three ships) class fast patrol craft - missile
Saudi Arabia	• 'Arriyad' (three building as of March 2001 - Dagaie Mk 2) and 'Madina' (four ships) class frigates
Taiwan	• 'Kang Ding' class frigates (six ships)
Thailand	• 'Rattanakosin' class corvettes (two ships)
Tunisia	• 'Combattante IIIM' class fast patrol craft - missile (three ships)
Turkey	• 'Burak' class frigates (six ships)
United Arab Emirates	• 'Murray Jib' class corvettes (two ships)
	• 'Ban Yas' (six ships) and 'Mubarraz' (two ships) class fast patrol craft - missile

Specifications

Resetting time for 90°: <2 s
Medium aiming speed: 1.5 rad/s
Average reaction time: <4 s (reaction time is the time lapse between reception of 'missile alarm' information and the end of IR firing)
Traverse: 360°
Power consumption: 1 kVA (normal); 5 kVA (max)
Launcher weight: 450 kg (6 munitions module capacity); 500 kg (10 munitions module capacity)
Launcher height: 75 cm (6 munitions module capacity); 95 cm (10 munitions module capacity)

Contractor

European Aeronautic, Defence and Space (EADS) Company - Matra Systèmes and Information , Les Ulis, France.

UPDATED

FL 1800S Electronic Support and CounterMeasures (ES/ECM) system

Type

Shipborne radar detection, analysis and jamming system.

Description

The FL 1800S shipboard ES/ECM system is understood to cover the 0.5 to 18 GHz frequency range, as offering 360° cover against all threats operating within its operating band and as incorporating full threat detection, analysis and jamming capabilities. It can handle several threats simultaneously, even when these differ considerably in frequency and/or direction of arrival. FL 1800S's antenna system consists of an omnidirectional unit, four Direction-Finding (DF) arrays and two ECM units, with the omni and DF arrays being installed on the host vessel's mast. Although the equipment's operation is primarily automatic, a below-decks operator's console provides system control and displays information on tabular and tactical cathode ray tube screens. This console is intended to be operated by one person but can accommodate two operators if necessary. The architecture's ECM subsystem consists of a tracking processor, mode modulators and multibeam antenna assemblies. The FL 1800S ECM segment is able to generate threat specific outputs and has a look-through facility to both check jammer effectiveness and update the ES picture. The effective radiated power of the jammer subsystem is described as being 'sufficient' to mask its host vessel's radar cross-section. FL 1800S is connected to the CIC (Combat Information Centre) via NTDS interfaces and sends data to and receives messages from the CIC.

Alongside the baseline system, the former DaimlerChrysler Aerospace (see following) is reported to have developed an FL 1800S Stage II equipment that was

first installed as a prototype aboard an 'S 143A' class patrol craft during 1993. The Stage II architecture features enhancements in the areas of signal processing, analysis and classification, the man/machine interface used and jammer performance. A nine sub-band channelised receiver has been incorporated and provides 'extremely' accurate frequency measurement and 'continuous wave resistance'. The system's ability to determine absolute pulse amplitude (using a 'special' calibration process) allows it to generate alarm messages within its host vessel's CIC if a defined pulse amplitude threshold value is exceeded by a designated emitter. Two fully integrated superheterodyne receivers have been incorporated for pulse analysis and to provide a fingerprinting capability. FL 1800S Stage II remains wide open at all times and its performance is noted as remaining stable even in 'extremely dense and complex' environments. According to Jane's sources, the Stage II ECM subsystem incorporates two travelling wave tube-driven Multiple Beam Array Transmitters (MBAT) that feature folded pillbox lens beam-forming and are capable of generating both noise and deception outputs. FL 1800S Stage II also incorporates a built-in test routine together with expanded capacity threat libraries to facilitate onboard reprogramming for out of area missions.

Status

As of early 2001, the FL 1800S ES/ECM system was reported as being installed aboard 'Lütjens' class destroyers, 'Brandenburg' (Stage II), 'Bremen' (to be upgraded to Stage II standard) and 'Sachsen' (Stage II) class frigates and 'Gepard' class fast attack craft (to be upgraded to Stage II standard) of the German Navy. The cited 'Bremen' and 'Gepard' class FL 1800S enhancement programme is reported to have been launched during 1995 when the then DaimlerChrysler Aerospace (now part of the European Aeronautic, Defence and Space (EADS) Co) received a then year US$122.2 million contract to upgrade their existing systems to Stage II standard. As of the given date, work on the upgrade was originally scheduled for completion by the end of August 2000 and the FL 1800S ES/ECM suite was also mandated for installation aboard the German Navy's future 'K130' class corvettes. As of February 2001, five 'K130' class vessels were scheduled for delivery during the period 2005 to 2008.

Contractor

European Aeronautic, Defence and Space (EADS) Co Systems and Defence Electronics, Ulm, Germany.

UPDATED

Janet radar jammer

Type

Shipborne noise and deception radar jammer.

Description

Janet is a naval electronic countermeasures system designed to jam locked on missile radar seekers as well as surveillance radars operating in the H-, I- and J-bands (6 to 20 GHz). It operates against pulsed radars, which may use fixed or staggered frequency, frequency agility or pulse compression transmission modes and can function as a noise jammer, a deception jammer or a two-mode jammer. Janet is of modular design and comprises up to four modules, each of which is a complete and autonomous jammer with its own two-horn receive antenna for tracking; radar detector; computer/signal generator and two Travelling Wave Tubes (TWTs) and directional transmit antennas. One TWT/directional antenna combination is used for noise jamming and the other for deception jamming. This enables the system to jam simultaneously two radar threats arriving from different directions. Each jammer module can also jam two signals if these are located close to each other. The system's receiver is used for accurate target bearing and frequency measurement following an alert from an associated radar warning receiver. The computer controls the signal generator and selects the type of

The German Navy's 'Brandenburg' class frigates (Brandenburg shown here) are fitted with the FL 1800S ES/ECM system

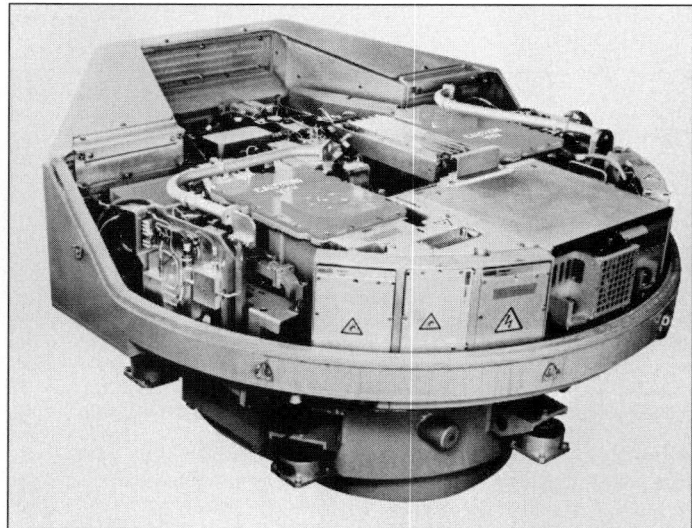

The Janet jammer's antenna and transmitter assembly

modulation used. This high flexibility in jamming modes programming and selection provides a variety of noise signals (including continuous or pulsed noise) as well as dual-mode (noise/deception), deception and repeater signals. The system has been designed to store received target signals in order to recreate identical ones for deception purposes. Since Janet is a processor-controlled combined noise and deception jammer, it provides 'smart' jamming to combat modern electronic counter-countermeasures equipped radars. With its modular concept of from one to four modules, it can be installed on many types of ships. Each module has an overall diameter of 1,000 mm, a height of 270 mm and weighs 250 kg.

Status
As of March 2001, Jane's sources were reporting the Janet radar jammer as being installed aboard the four frigates that make up Saudi Arabia's 'Madina' class.

Contractor
Thales Airborne Systems, Elancourt, France.

UPDATED

NATO Seagnat decoy system

Type
Anti-ship missile decoy system.

Development
The NATO Seagnat project was a joint development between the governments of USA, UK and Denmark. It entered production in 1986.

Description
A joint American, British and Danish development (with German and Norwegian support), NATO Seagnat is a 130 mm calibre, offboard, chaff and Infra-Red (IR) countermeasures capability that is designed to decoy hostile Radio Frequency (RF) and IR guided anti-ship missiles away from their intended targets. Over time, German contractor BUCK (Mk 245 Mod O IR round), US contractor Hycor and the UK companies Chemring (Mk 214 RF seduction/Mk 216 RF distraction rounds and TALOS IR round), Hunting Engineering (launchers) and Racal Defence Electronics (control unit) have been associated with the programme. Seagnat decoy rounds are compatible with the US Mk 36/Super Rapid Bloom Offboard Countermeasures (SRBOC) launcher.

Status
Over time, the Seagnat system is understood to have been implemented by the American ('Arleigh Burke' (Flts I and II) class destroyers and 'Tarawa' and 'Wasp' class amphibious assault ships); Australian ('ANZAC' class frigates); Brazilian ('Broadsword' class frigates); British (the helicopter carrier HMS *Ocean* (Outfit DLJ launcher), the aviation training ship HMS *Argus* (Outfit DLB) and the same service's 'Invincible' class aircraft carriers (Outfit DLJ launcher), 'Type 42' class destroyers (Outfit DLB), 'Duke' (Outfit DLB) and 'Broadsword' (Outfit DCH) class frigates, 'Albion' class assault ships (Outfit DLJ), 'Fearless' class amphibious warfare vessels (Outfit DLB) and 'Fort Victoria' class fleet replenishment ships); Danish (the frigate *Beskytteren* and the same service's 'Niels Juel' and 'Thetis' class frigates, 'Flyvefisken' class multirole warships, 'Willemoes' class fast attack craft and 'Falster' class minelayers) and Portuguese ('Vasco Da Gama' class frigates) navies.

VERIFIED

New Generation Decoy System (NGDS)

Type
Shipboard Electronic Warfare (EW) and Anti-Torpedo (AT) countermeasures decoy system.

Description
NGDS is a fully automatic countermeasures system that is designed to protect all types of ships from anti-shipping missile and torpedo attack. As such, the system is

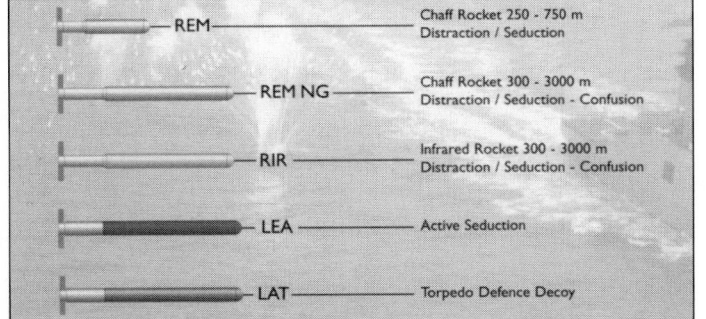

A schematic showing a selection of the munitions that are compatible with the NGDS system 0009362

A general view of the twin-axis launcher used in the NGDS EW and AT countermeasures system
0009361

configured to be effective against both Radio Frequency (RF) and Infra-Red (IR) guided missiles (singly or in combination) together with acoustic, self- or wire-guided torpedoes. An NGDS installation comprises up to four twin-axis launchers, a data processing unit, a supervisory unit and a manoeuvre indicator. Of these, the system's launcher unit can accommodate four rocket launcher modules with each module being loaded with three RF, IR or AT munitions. Individual modules can be RF, IR or AT configured with each launcher being controlled in train and elevation by means of short reaction time electric drives, with movement being standardised irrespective of the type of munition being used.

The NGDS data processing unit acts as the system's electrical power source and computes the EW and AT tactics and firing data necessary to counter incoming threats. The supervisory unit is used to operate the system and display its status, while the manoeuvre unit displays suggested ship manoeuvres that will maximise the efficiency of the response. Further flexibility is provided by the ability of a two launcher NGDS configuration to be coupled with a pair of Dagaie Mk 2 launchers (see separate entry) to provide a passive, centroid seduction capability. NGDS munition can be deployed within a 200 to 3,000 m range envelope.

NGDS is intended to transcend the simple decoy launcher role and provide a truly effective self-defence capability. It can be operated as a stand-alone system on smaller vessels or be fully integrated into its host's combat management system aboard larger ships, operating as one of its major self-defence modules. In both cases, NGDS can perform EW command and control functions together with the co-ordination of EW and AT operations. When used against anti-shipping missiles, the system offers confusion, dilution, distraction/active centroid seduction (when used alone) and dump seduction (when used with an onboard jammer) modes. In the AT role, NGDS is noted as being capable of decoying and jamming threats even in their terminal attack phase.

Status
As of early 2001, Jane's sources were reporting that France's planned pair of 'Horizon' class anti-air warfare frigates would be the first ships to be fitted with the NGDS decoy launcher.

Specifications
Launcher height: 1,300 mm
Launcher sweep radius: 1,250 mm
Launcher weight: 600 kg
Munition capacity: 12 (rockets per launcher)
Resetting time: <1.5 s (for 90°)

Contractor
European Aeronautic, Defence and Space (EADS) Company - Matra Systèmes and Information , Les Ulis, France.

UPDATED

Sabre Electronic Warfare (EW) system

Type
Integrated surface ship EW system.

Description
Sabre is an integrated surface ship EW system that comprises:
- a lightweight, Electronic Support (ES) subsystem, mast-mounted antenna unit that includes the necessary reception arrays, Radio Frequency (RF) protection, filtering and preamplification elements
- a masthead ES processor assembly that incorporates microwave receiver, RF filtering and control and RF to digital conversion circuitry
- a below decks ES equipment cabinet that houses the subsystem's digital instantaneous direction-finding and pulse train analysis circuitry together with the subsystem to ship's command system interface

A schematic showing the elements making up the Sabre surface ship EW system

0017688

- port and starboard jammer assemblies, each of which comprises a stabilised platform, a steerable dish antenna, a reception array, a cooling unit, phased-array transmission heads and a techniques generator
- two inboard electronic countermeasures 'equipment units'
- an operator console.

Looking at some of these elements in more detail, the Sabre's ES subsystem covers the 0.5 to 18 GHz frequency band and provides instantaneous interception, analysis and classification of surveillance, acquisition and targeting threat radars. The jamming subsystem covers the 7.5 to 18 GHz frequency range and is capable of tracking and jamming multiple threats. The Techniques Generator (TG) used comes from the same technology base as the Shrike TG Thales has developed for the Royal Air Force's Nimrod MRA.4 maritime patrol and Airborne STand-Off Radar (ASTOR) aircraft and is digital radio frequency memory based. The subsystem's steerable antennas are used to counter threats such as monopulse seekers where polarisation diversity is required. Sabre is designed to support the following operational functions:

Air defence and self-protection
Here, the equipment's ES subsystem provides early detection of and a fast response to, potential and terminal threats. Acquired parametric data is passed to the jamming subsystem in such a way as to allow the jamming subsystem's output to be both timely and optimised to maximise platform survivability. The ES subsystem also facilitates tactical awareness via the generation of threat warnings, cues and tracking data for the ship's combat direction system.

Surveillance
In the surveillance role, Sabre gathers parametric data on all intercepted threats and presents the information to the system's operator position in a range of formats. The system allows for in-depth operator analysis of received data and can record it for post-mission study

Area operations
In the area operations role, the data gathered by the Sabre system is shared with associated platforms, such as ships and helicopters, via its host ship's combat direction system.

Status
As of October 2001, the Sabre EW system had been selected for installation aboard 'de Zeven Provincien' class frigates of the Royal Netherlands Navy (four building as March 2001). Readers should also note that Thales (formerly Racal) has previously used the designation Sabre to describe a 0.5 to 18 GHz ES system that the company has supplied to Denmark for use aboard 'Flyvefisken' class multirole combat vessels. As of the given date, this Danish Sabre system is understood to remain in service.

Specifications
ES subsystem
Frequency range: 0.5-18 GHz (0.5-6 GHz and 6-18 GHz sub-bands; 0.1-40 GHz coverage as option)
Bearing accuracy: 2° (2-18 GHz); 4° (0.5-2 GHz)
Sensitivity: –62 dBmi (typical)
Pulse density: 10^6 pps

Jamming subsystem
Frequency range: 7.5-18 GHz
ERP: up to 180 kW
Threat handling capacity: 6/180° sector
Jamming techniques: cross polarisation; false targets; range-gate; pull-off/pull-in; scan rate modulation; swept, cover and pulsed noise

Contractor
Thales Sensors, Crawley, UK.

UPDATED

Sagaie Decoy Launching System (DLS)

Type
Naval DLS.

Description
The Sagaie DLS is a fully automatic, passive countermeasures system designed to protect medium and large ships from anti-ship missiles. It will provide protection against missiles guided by ElectroMagnetic (EM) or Infra-Red (IR) seekers, or any combination of the two, even when the missiles are attacking simultaneously over the entire horizon. The system can operate in conjunction with Dagaie and provides long-range defence, by confusion and dilution effects against, enemy target designating radars and close-in defence by distraction and seduction against acquisition and tracking systems in missile seekers. Here, Sagaie supports Dagaie so that the latter can maximise its centroid deception capability. Sagaie is fully automatic from the reception of a missile threat alarm originating from any of the surveillance systems (radar, IR, optical or electronic support measures) and will optimise the use of decoys in a very short reaction time. The equipment can be used alone (confusion and distraction) or jointly with a jammer (dilution, substitution after concealment, substitution after deception) by firing of substitution decoys. The installation comprises:
- one or two fully stabilised launchers loaded with 10 rockets in containers and trainable in azimuth and elevation
- one or two servo units
- one or two aiming and maintenance units
- a processing unit which computes the rocket launching sequence and provides information for the supervision unit
- a supervision unit with controls and status displays
- an interface unit which provides for operation with any tactical data or target designation system.

Sagaie can fire either EM or IR decoy rocket type ammunition or both. The rocket is packaged in a waterproof launcher container. This container is used for handling, transport and storage and when secured to the mounting, ensures guidance of the rocket at the moment of launch.

Status
As of the period February to March 2001, Jane's sources were reporting the Sagaie DLS in service aboard the following warships and classes of warship:
Brazil • the aircraft carrier *Sao Paulo* (ex *Foch*)
France • the aircraft carrier *Charles de Gaulle* (four launchers)
• 'Cassard' (two ships in class, two launchers per ship) and 'Suffren' (two ships, two launchers per ship) class destroyers
Italy • 'De La Penne' (two ships, two launchers per ship) class destroyers
Peru • the cruiser *Almirante Grau* (one Sagaie and two Dagaie DLSs)

Specifications
Launcher
Height: 2,000 mm
Sweeping radius: 1,600 mm
Weight (empty): 1,600 kg

The Sagaie DLS

Number of rockets: 10
Container
Length: 1,800 mm
Diameter: 330 mm
Weight: 80 kg

Contractor
European Aeronautic, Defence and Space (EADS) Company - Matra Systèmes and Information , Les Ulis, France.

UPDATED

Salamandre radar jammer family

Type
Shipborne multithreat jammer using multibeam array transmitters.

Description
Based on the experience gained with the ARBB 33, the Salamandre electronic warfare system has been designed to protect fighting ships and land-based sites against surveillance and targeting radars, fire-control radars and missile electromagnetic seekers. Key features of the system include a very high accuracy in determining the direction of emitters, a transmitter with a phased-array of mini-travelling wave tubes and electronic beam-steering. Salamandre is capable of:

- detection of threats; measurement of characteristics; fast alert; identification; classification; tracking; data transmission to the combat system and display of the corresponding information on a console
- activation of jamming, either automatically after detection or when ordered by the combat system or an operator. Salamandre offers a wide range of jamming modes.

The modular design of Salamandre allows its installation on board vessels up to 6,000 tonnes and the system is available in a number of variants as follows:

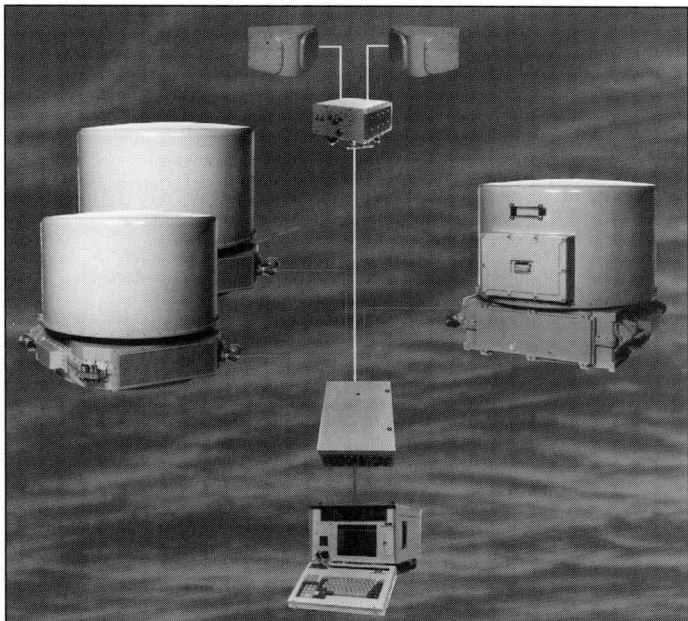

The layout of the Salamandre B3 jamming system showing the equipment's reception antenna (top), three jamming canisters and its control unit (bottom)

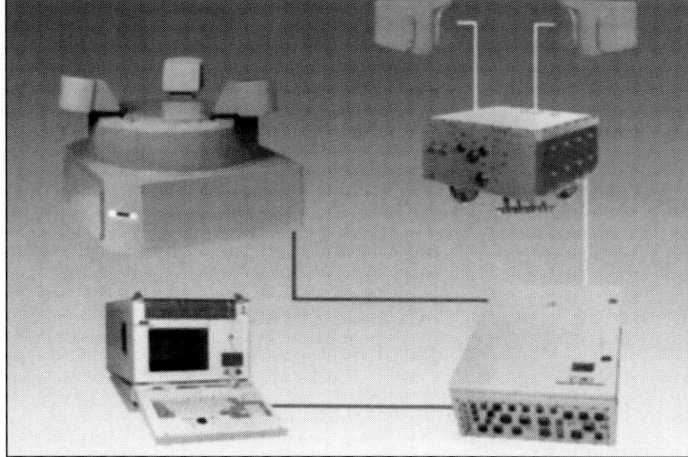

A schematic showing the layout of the single transmitter unit equipped Salamandre B1
0009357

In French Navy service, both Salamandre B2 (shown here) and B4 appear to be designated as ARBB 36
0009358

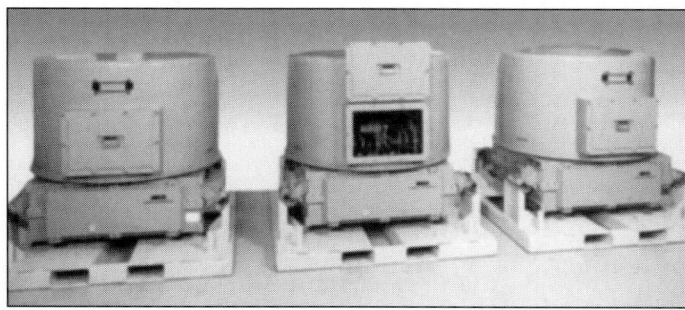

A close-up of the three transmitter units used in the Salamandre B3 jamming system
0009359

In French Navy service, both Salamandre B2 and B4 (shown here) appear to be designated as ARBB 36
0009360

Salamandre B1
Salamandre B1 is designed for patrol craft applications and features a single transmitter unit to facilitate installation aboard such relatively small ships.

Salamandre B2
Salamandre B2 is optimised for installation aboard corvettes and frigates. In French Navy service the system is designated as ARBB 36.

Salamandre B3
Salamandre B3 is optimised for installation aboard destroyers and is described as being the most powerful member of the family. It incorporates three transmitter assemblies, each of which covers 120° in azimuth.

Salamandre B4
Salamandre B4 is an alternative designation for the French Navy's ARBB 36 system.

Status
During the period February to March 2001, Jane's sources were reporting Salamandre series radar jammers as being installed aboard or mandated for the following warships and classes of warship:

France • the 'Georges Leygues' class destroyers *Dupleix*, *Jean de Vienne* and *Montcalm* (ARBB 36)

Saudi Arabia • 'Arriyad' class frigates (three building as of March 2001 - twin Salamandre B2 installation)

Specifications
Frequency: H- through J-band (6-20 GHz)
Coverage: 360° azimuth
Jamming modes: continuous noise, pulsed noise, cover-pulse jamming, synchronous false echoes, asynchronous false echoes, gate pull-off, dual mode

Contractor
Thales Airborne Systems, Elancourt, France.

UPDATED

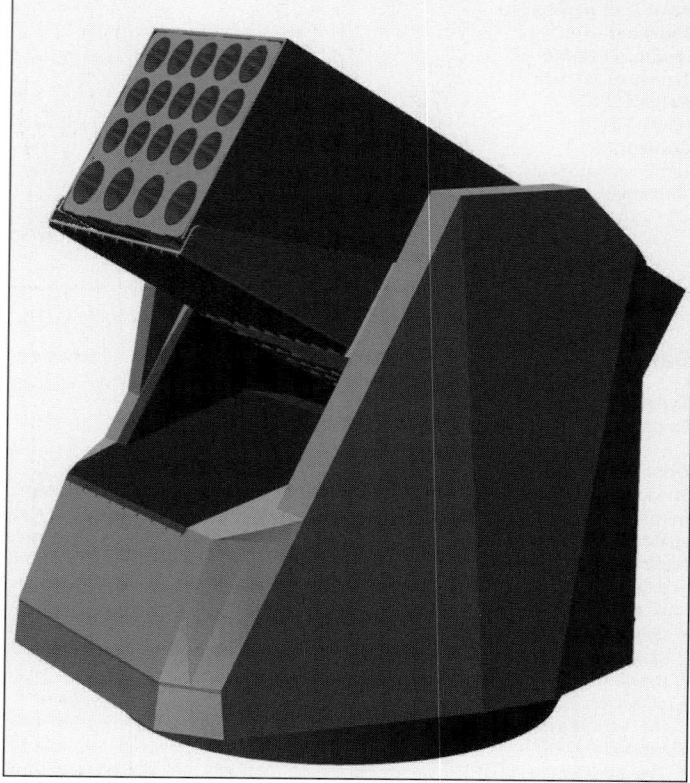

A general view of the low radar cross section launcher unit used in the SCLAR-H decoy launching system (Alenia Marconi Systems) 2002/0083797

SCLAR Mk 2 rocket launching system

Type
Shipborne countermeasures launching system.

Description
SCLAR Mk 2 is a development of the baseline 105 mm SCLAR system and was originally designed and manufactured for use by the Italian Navy. SCLAR Mk 2 employs the same rocket launcher as its predecessor but with launch data being processed by a digital multiprocessing computer, based on the Elsag designed ESA 24 microprocessor. The system is defined as multipurpose, since it is designed for a range of tasks including the decoying of incoming missiles, confusion of hostile radars, illumination or high explosive bombardment. Role change is by means of munition selection and a 'very large' set of decoying patterns can be stored in the system's digital computer. These are recalled and automatically executed as required and there is provision for reprogramming. Physically, SCLAR Mk 2 consists of a control console (which requires a single operator), a data processing unit, one or two rocket launchers, one or two power supply cabinets and one or two local control panels. In addition, the system requires alarm loading and acoustic warning units for each launcher. An optional illuminating launching remote-control panel is also available. Two remotely controlled rocket launchers can be driven by the SCLAR Mk 2 computer, which positions them in azimuth and elevation and performs all the necessary logic and computation. This includes rocket pattern programming, azimuth and elevation computation for the accurate launch of each rocket to the programmed burst point and automatic rocket selection and firing. The system is reported as being able to accommodate a range of 105/118 mm munitions that includes the Buck DM 19 DUERAS chaff and DM 29 IRRAS Infra-Red (IR) decoys; the SNIA 105 LR-C chaff distraction (12 km maximum range), 105 LR-I illumination (4 km range), 105 MR-C chaff seduction (4.7 km maximum range) and 105 MR-IR IR seduction rounds and Wallop Defence Systems 105 mm chaff and IR decoy (3 to 5 µm and 8 to 14 µm bands) rounds.

As of August 2001, the latest SCLAR variant to be identified was the SCLAR-H. Featuring Otobreda-developed launcher sets, SCLAR-H control system manufacturer Alenia Marconi Systems describes the architecture as offering a 'very short' reaction time, the ability to 'fully' co-ordinate with other onboard electronic warfare assets and automatic selection of the 'most appropriate' decoy type for the particular threat. The launcher unit used is able to accommodate munitions with calibres that range from 102 to 130 mm and can be configured with between 16 and 20 launch tubes arranged in four rows. In its standard configuration, SCLAR-H is further noted as having 'reduced' radar cross section launch units and as being capable of launching long-range chaff and Infra-Red (IR) decoy rockets together with radio frequency and IR mortar munitions for close range platform protection. As of early 2001, SCLAR-H was slated for installation aboard the Italian Navy's Nuova Unita Maggiore (NUM - New Major Unit) - the multirole, air capable *Luigi Einaudi* (provisional name) - and its 'Horizon' (two ships) and 'Soldati' (four ships) class frigates.

Status
As of the period January to April 2001, Jane's sources were reporting SCLAR variants as being installed aboard the following warships and classes of warship:

Argentina • 'Almirante Brown' class destroyers (four ships in class, two SCLAR launchers per ship)
Ecuador • 'Esmeraldas' class corvettes (six ships, one SCLAR launcher per ship)
Germany • the small ammunition ship *Westerwald* (two SCLAR launchers)
• 'Brandenburg' class frigates (four ships, two SCLAR launchers per ship)
• 'Luneburg' class small repair ships (two ships, two SCLAR launchers per ship)
Italy • the aircraft carrier *Giuseppe Garibaldi* (two SCLAR launchers)
• the command cruiser *Vittorio Veneto* (two SCLAR launchers)
• 'Audace' class destroyers (two ships, two SCLAR launchers per ship)
• 'Lupo' (four ships, two SCLAR launchers per ship) and 'Maestrale' (eight ships, two SCLAR launchers per ship) class frigates
• 'Artigliere' class fleet patrol ships (four ships, two SCLAR launchers per ship)
Nigeria • the frigate *Arudu* (two SCLAR launchers)
Peru • 'Carvajal' class frigates (four ships, two SCLAR launchers per ship)
Venezuela • 'Modified Lupo' class frigates (six ships, two SCLAR launchers per ship)

Specifications
SCLAR Mk 2
Training: ±150°
Max slewing speed: 30°/s (elevation); 60°/s (azimuth)
Elevation: −1 to +54°
No of rockets: 20 per launcher
Firing rate: approx 1 rocket/s

Contractor
Alenia Marconi Systems (an Alenia-BAE Systems joint venture), Chelmsford, UK/ Rome, Italy.

UPDATED

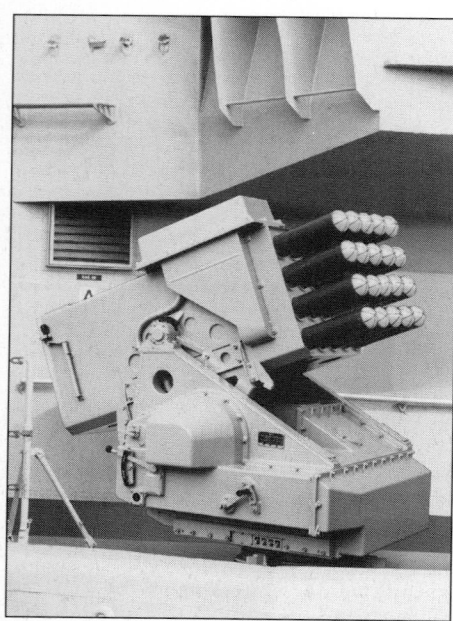

The SCLAR multipurpose naval rocket launcher

Scorpion Electronic CounterMeasures (ECM) system

Type
Shipborne noise and deception jamming system.

Description
The 7.5 to 18 GHz band Scorpion radar jammer is designed to counter both long-range search target acquisition and missile seeker radars operating in search and

The Scorpion jammer/Cutlass ES array used with the German Navy's Octopus ECM suite (MS/Racal)

lock-on modes and is claimed to be able to counter salvos of sea-skimming missiles over a wide angle and a wide frequency range. The system is lightweight, high power and uses 'state-of-the-art' jamming techniques. It has undergone trials on board ships in the Baltic and North Sea and has shown itself capable of dealing with multiple targets and 'highly sophisticated' airborne radars. It can be integrated with Racal's Cutlass and Type 242 Electronic Support (ES) systems together with ES equipments from other suppliers. When teamed with the Cutlass ES system, the latter's library can incorporate customer defined jamming response codes for each known threat. Functionally, detected threats are prioritised and segregated from the general signals environment by Scorpion's integral pulse repetition interval tracking circuitry. Here, received pulse trains are de-interleaved to establish time of arrival data for use in the system's pulse prediction circuits. The data are then fed through a techniques generator to a fast tuning voltage controlled oscillator, whose output is fed to a high-power amplifier for transmission. Scorpion can generate 10 types of jamming modulation including continuous and burst noise, cover pulse noise, synchronised railing and false targets. Overall, the system can handle up to eight threats simultaneously, with each target being treated to a time domain managed mix of the 10 jamming modulations. Scorpion is stabilised in all three axes and incorporates a twin antenna array to facilitate the transmission of wide beam multiple target and narrow beam high effective radiated power outputs.

Status

As of the period February to August 2001, Jane's sources were reporting the Scorpion radar jammer as being installed aboard the following warships and classes of warship:

Bangladesh • the frigate *Bangabandhu*
Denmark • 'Thetis' class frigates (four ships in class)
Germany • 'Albatros' (10 ships) and 'Tiger' (five ships) class fast attack craft - missile
Oman • 'Dhofar' class Fast attack craft - missile (four ships)
Turkey • 'Barbaros' class frigates (four ships)

By the mid-1990s, the then Racal (now Thales Sensors) was reporting worldwide sales of over 30 Scorpion radar jammers.

Contractor

Thales Sensors, Crawley, UK.

UPDATED

....

Shield Decoy Launching System (DLS)

Type

Surface ship DLS.

Description

Over time, three generations of the Shield DLS (designated as Shield I (UK Royal Navy Outfit DLB), Shield II (UK Royal Navy Outfit DLE) and Shield III) have been fielded. Of these, Shield I is described as being an essentially manually operated architecture that incorporated fixed launcher units and a stand-alone control system and made use of the N5 BroadBand Chaff (BBC) rocket (see following).

Shield II is noted as being equipped to fire the P6 Infra-Red (IR) and P8 chaff rounds (see following), as introducing a rule-based control system and as having provision for a ship's combat system automatic control interface. Shield III is said to incorporate a new central control unit (based around Motorola 68040 microprocessors), an 'updated' man/machine interface (incorporating plasma screen touch-sensitive technology), 'additional' operating modes and an architecture that is based on an Ethernet local area network.

Looking at Shield II in more detail, the system takes the form of a Z80 processor-controlled architecture that incorporates multiple launcher units, a command module, launcher control modules, loader control units, a bridge module, emergency power modules and payload rockets. Of these, the system's launcher units can be configured to offer 6, 12, 18 or 24 barrels, with the individual barrels being arranged in parallel or angled at elevations of up to 30°. The launch unit configuration is further subdivided into small ship (fast attack craft or smaller), corvette or frigate type applications. Here, the small ship configuration is described as making use of two crossed, six-barrel launchers that are arranged to fire 30° off the host vessel's bow and 70° off its stern. The corvette application can take the form of two crossed, nine-barrel launchers, while the frigate configuration incorporates up to four straight, six-barrel launch units arranged in pairs to port and starboard. Here, the two units were trained at 30° and 110° respectively, with each pair of launchers being supported by a launcher control module.

The system's command module is normally located in the host vessel's operations area and provides system control when in manual mode, together with 'routine' monitoring of system function. The architecture's bridge module indicates the current load status and initiates payload firing when the system is in local mode. The loader control units are described as being safety devices that allow the system's firing circuit to be isolated during loading. Payload deployment can be initiated automatically, semi-automatically or manually. Here, the automatic mode fires seduction decoys in response to cues (threat type, ship's heading/speed, relative wind, ship's pitch and roll and threat bearing/distance speed) from the host vessel's command and weapons control system. If the number of rounds available is insufficient to counter the detected threat, an automatic reload instruction is generated. In semi-automatic mode, Shield permits operator choice of distraction or seduction modes and operator initiation. The barrels to be fired are selected automatically, with fuze times being based on windspeed and direction. In manual mode, the system operator inputs the necessary cueing data into the system (via a keypad), decides on the use of distraction or seduction rounds and/or an appropriate mix of the two and initiates firing. Known Shield payloads are as follows:

N5 BBC chaff rocket

The N5 BBC chaff rocket is noted as having manually set fuzing and as being applicable to at least the Shield I and II configurations.

P6 IR decoy round

The P6 IR decoy is reported as being an IR decoy mortar round that incorporates seven submunitions that operate in the 3 to 5 µm and 8 to 13 µm bands and have individual burn times of between 15 and 20 seconds. Submunition deployment begins at a distance of 50 to 200 m from the launch point and lasts for up to 15 seconds.

P8 chaff rocket

Jane's sources describe the P8 chaff rocket as having a 7.5 to 8.5 kg payload that is effective against both vertically and horizontally polarised emitters and as featuring electronically set fuzing to optimise chaff cloud deployment at ranges of between 50 and 2,500 m from the launch point.

Status

As of the period January to April 2001, Jane's sources were reporting the Shield DLS as being in service aboard the following warships and warship classes:

Brazil • the aircraft carrier *Minas Gerais*
• 'Niteroi' class frigates (six ships in class, two Shield II (?) launchers per ship)
• 'Inhauma' class corvettes (four ships, two Shield II (?) launchers per ship)

The Shield tactical decoy system

Canada
- the replenishment oiler *Protecteur* (provision for four Shield II launchers)
- 'Iroquois' class destroyers (four ships, four Shield II launchers per ship)
- 'Halifax' class frigates (12 ships, four Shield II launchers per ship)

Pakistan
- the destroyer *Nazim* (two Shield launchers)

Portugal
- the replenishment oiler *Berrio* (two Shield launchers)

Singapore
- 'Victory' class corvettes (six ships, two Shield launchers per ship)
- 'Fearless' class offshore patrol vessels (12 ships, two Shield III launchers per ship)
- 'Endurance' class amphibious transport docks/landing ships tank (four ships, two Shield III launchers per ship)

UK
- the forward repair ship *Diligence* (two Shield launchers)
- the transport oiler *Oakleaf* (two Shield launchers)
- the landing ships logistic *Sir Bedivere* (two Shield launchers) and *Sir Galahad* (four Shield launchers)
- 'Castle' class offshore patrol vessels (two ships, two Shield or two Corvus launchers per ship)
- 'Fort Victoria' class replenishment oilers (two ships, four Shield or four Seagnat launchers per ship)
- 'Rover' class small replenishment oilers (three ships, two Shield or two Corvus launchers per ship)
- 'Appleleaf' class transport oilers (three ships, two Shield or two Corvus launchers per ship)
- 'Fort Grange' class combat stores ships (two ships, two Shield or two Corvus launchers per ship)

Specifications

Shield II

Dimensions (h × w × d): 90 × 260 × 160 mm (bridge control module); 309 × 442 × 392 mm (command and launch modules); 320 × 390 × 180 mm (loader control unit); 1.25 × 1.1 × 1.8 m (straight, six-barrel launcher); 1.7 × 1.8 × 1.8 m (crossed, six-barrel launcher); 2 × 1.7 × 1.9 m (crossed, nine-barrel launcher)

Weight: 15 kg (command module and loader control unit); 18 kg (bridge control module); 20 kg (launcher control module); 400 kg (crossed, six- and nine-barrel launchers); 425 kg (straight, six-barrel launcher)

Contractor

Alenia Marconi Systems (an Alenia-BAE Systems joint venture), Chelmsford, UK/ Rome, Italy.

UPDATED

Type 675(2) radar band detector-jammer

Type
Shipboard radar jamming system.

Description
Type 675(2) is an I/J-band (8 to 20 GHz) shipboard, point and area defence radar jamming system that has been developed for the Royal Navy and is designed to

The Type 675(2) operator's console (MS/Thales)

The Type 675(2) deck shelter installation as applied to Type 42 destroyers (MS/Thales)

counter and confuse surveillance and missile homing radars via the use of 'selective' techniques. The Type 675(2) system has two antenna assemblies (to ensure uninterrupted 360° coverage), each of which incorporates direction-finding receiving and transmission arrays. Antenna movement is controlled by mechanical steering and received signals are passed from the antenna assembly to a wideband receiver for processing and identification. A high-power transmitter subsystem provides a jamming capability against multiple targets simultaneously, with mode selection being based on a 'comprehensive' integral electronic countermeasures library. Overall system control is by means of distributed microprocessors and a central microcomputer processor and Type 675(2) interfaces with its host vessel's systems. The jammer's main equipment cabinets are situated below decks or installed in custom deck containers depending on platform type. The system is completed by an operator console that is equipped with a plasma display panel.

Status
Over time, approximately 20 Type 675(2) equipments have been procured by the Royal Navy (RN) and installed aboard 'Invincible' class aircraft carriers (three ships in class as of August 2001 - internal installation), 'Type 42' class destroyers (eight Batch 2 and 3 ships as of April 2001 - deck shelter installation) and 'Broadsword' class frigates (four (?) Batch 3 ships as of April 2001 - internal installation). During October 1997, the then Racal (now Thales Sensors) was awarded a then year £2 million RN contract covering the supply of at least one additional Type 675(2) jammer.

Contractor
Thales Sensors, Crawley, UK.

UPDATED

ISRAEL

DESEAVER Mk II decoy system

Type
Decoy control and launching system.

Description
DESEAVER is a decoy control and launching system that responds to the threats in the combat scenario, taking all factors into account. Acting on data received and processed by the ship's command and control system and/or its sensors, the system launches a barrage of expendable decoys with precise priority and timing. Using an array of fixed or fast manoeuvring stabilised rocket launchers, backed by advanced processing and control units, DESEAVER delivers payloads at accurate time intervals and according to specified anti-missile doctrines and guidelines. The system is adaptable to a broad range of vessels, from patrol boats to frigates. It can

be configured to integrate with any existing onboard equipment. Functionally, DESEAVER performs the following tasks:

- evaluation of the threat tactical picture and selection of the appropriate passive countermeasures
- operation in conjunction with the Electronic Warfare (EW) defence system and within the architecture of the ship's overall combat system
- deployment procedure computation
- recommendation of preferred ship's course and speed as part of the anti-missile defence procedure
- launching of defence decoys
- control of up to six types of EW rockets (chaff and infra-red)
- resource allocation and rocket launching according to selected programmes
- deployment of tactical decoys
- providing the commanding officer with a display of the threat and decoy environment.

DESEAVER is designed to be operated from a single console. It operates in three modes:

- automatic — evaluates tactical picture, selects threats to be engaged and generates countermeasures response programmes
- semi-automatic — functions automatically requiring operator approval by single key activation for initiating rocket launching
- manual — most functions performed automatically but operator approval is required at two stages: designation of targets and initiation of rocket launch.

The launching capabilities of DESEAVER offer comprehensive coverage under complete control. The system controls three stabilised azimuthal launchers that accommodate a total of 72 rockets. The stabilised rocket launching unit coverage is 340° in azimuth. After installation, azimuth coverage is ±170° relative to a preset heading line. In elevation, coverage is according to the selected launch angle, from +15 to +55° relative to earth.

Status

Over time, Jane's sources have suggested that the DESEAVER system has been installed aboard 'Eilat' class corvettes and 'Hetz' class fast attack craft of the Israeli Navy. It should also be noted that there is a DESEAVER Mk I configuration that makes use of an array of fixed launchers and incorporates commercial-off-the-shelf hardware and software. As of January 2002, a DESEAVER Mk I application is understood to have been supplied to a 'non-disclosed' customer.

Contractor

Elbit Systems Ltd, Haifa.

UPDATED

Integrated Decoy System (IDS)

Type

Naval decoy, decoy launcher and system control package.

Description

The Manor IDS package comprises a 'complete range' of decoy rounds (Manor's BT-4/1, Heatrap, LRCR and MRCR devices (see separate entry) plus the Super Rapid Bloom Offboard Countermeasures (SRBOC) range of munitions (see separate entry), a computerised decoy and launcher control system, a rotating, trainable launcher unit and provision for a torpedo decoy round (such as the Manor LESCUT device - see separate entry) if required. As such, the package is noted as offering 'three layers of defence' against missile, radar and if appropriately configured, torpedo threats. Other system features include 'precise' decoy location, optimised threat responses, 'no' installation constraints, 'no' deck penetration and 'improved' (better decoy placement and reduced recoil) SRBOC performance.

Status

As of May 2001, Jane's sources were reporting that Manor was actively promoting the IDS package.

Specifications

Azimuth position range: ±125° or more
Angular velocity: 60°/s
Acceleration: up to 230°/s²
Launcher weight: 500 kg (empty); 920 kg (max loaded)

Contractor

Manor Expendable Decoys Directorate (a Rafael subsidiary), Haifa.

NEW ENTRY

Manor naval decoy systems

Type

Family of naval decoy launchers and munitions.

Description

The Manor naval decoy system family is designed to provide a layered defence against radar emitters and Infra-Red (IR) sensors. It comprises two launch system modules and four munitions, the known details of which are as follows:

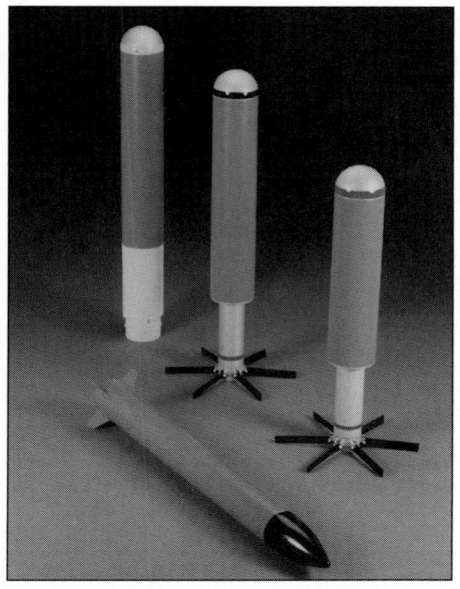

The Manor Heatrap (back row, left), BT-4/1 (back row, centre), MRCR (back row, right) and LRCR (foreground) naval decoy rounds   0009356

BT-4/1

BT-4/1 is a seduction chaff rocket that is designed to protect ships of up to 5,000 tons against radar guided anti-shipping missiles that have achieved lock-on.

Heatrap

Heatrap is a rocket-powered Infra-Red (IR) seduction/distraction round that incorporates a high- and low-IR band decoy flare. Once ignited, Heatrap's decoy payload is parachute supported.

LRCR

LRCR is a long-range, tactical confusion chaff rocket that is designed to protect ships of all sizes against surface and airborne search, targeting and missile radars before the start of the attack sequence.

LRCR launch system module

An LRCR launch system module comprises a 3 kg fire-control unit and two 216 × 1,250 mm, 12 kg launch tubes. A computerised decoy controller and a programmable electronic timer (installed in the decoy round) provide automatic threat/scenario optimisation. Overall, the module is described as being lightweight, watertight and suitable for installation on all sizes of ship. As of May 2001, it was thought likely that elements from the LRCR launch system module were incorporated in Manor's Integrated Decoy System (IDS) package (see separate entry).

MRCR

MRCR is a distraction chaff rocket that is designed to protect all sizes of ship against radar-guided anti-shipping missiles before target lock-on. MRCR is claimed to be able to defeat a bracket of missiles that are attacking at the same time from different directions.

MRCR/BT-4/1/Heatrap launch system module

An MRCR/BT-4/1/Heatrap launch system module comprises a 3 kg fire-control unit, two 5 kg wiring boxes and four, 13 kg, 270 × 365 × 1,020 mm, six tube launchers. A computerised decoy controller and a programmable electronic timer (installed in the decoy round) provide automatic threat/scenario optimisation. Overall, the module is described as being lightweight, watertight and suitable for installation on all sizes of ship. As of May 2001, it was though likely that elements from the MRCR/BT-4/1/Heatrap launch system module were incorporated in Manor's Integrated Decoy System (IDS) package (see separate entry).

Status

As of May 2001, Jane's sources were reporting that Manor was actively promoting the BT-4/1, Heatrap, LRCR and MRCR decoy rounds.

Specifications

	BT-4/1	Heatrap	LRCR	MRCR
Deployment range	up to 500 m	up to 500 m	>12 km	up to 1,500 m
Deployment altitude	50-100 m	50-100 m	900 m	100-200 m
Decoy persistence	up to 3 min	20-30 s	>10 min	up to 5 min
Rocket dimensions	115 × 920 mm	115 × 920 mm	89 × 920 mm	115 × 780 mm
Rocket/payload weight	10.1/5.2 kg	10.1/3.0 kg	9.4/1.3 kg	8.1/3.2 kg
Measured radiation		up to 1,500 W/str		
Measured RCS	6,000-8,000 m²		ship size	2,500-3,500 m²

Contractor

Manor Expendable Decoys Directorate (a Rafael subsidiary), Haifa.

UPDATED

MBAT/RAN-1010 and -1020 radar jamming transmitters

Type
Multibeam array radar jamming transmitters.

Description
MBAT/RAN-1010 and -1020 are MultiBeam Array Transmitters (MBAT) that feature 'ultra-high speed' electronic beam steering and the capability for jamming several signals simultaneously. So configured, MBAT jamming systems can provide protection against simultaneous threats from a number of different directions in a multimissile attack situation. The use of lens technology for a beam-forming network enables the equipment to provide a wide frequency beamwidth, instantaneous azimuth coverage and a 'very high' effective radiated power. MBAT/RAN series equipments can form part of a complete electronic warfare suite or can be used to upgrade an existing system. Options available include a wide/narrow beamwidth in elevation, a roll and pitch stabiliser and dual inputs for simultaneous pulse-on-pulse jamming in any direction. The MBAT/RAN family has been designed, built and tested to MIL-STD-16400 and RAN-1020 differs from RAN-1010 in its incorporation of the dual pulse-on-pulse jamming input option as standard. The equipment comprises two transmitters, each of which covers 360° in azimuth and can instantaneously jam multiple threats coming from different directions.

Status
As of May 2001, Jane's sources were reporting that Rafael was actively promoting the MBAT/RAN-1020 jamming transmitter.

Specifications
Frequency coverage: 7.5-18 GHz
ERP: 70-80 dBm (without stabilisation); 83-86 dBm (with stabilisation)
Instantaneous azimuth coverage: 360°
Beamwidth: 6° min (azimuth); 25° (elevation)
Number of beams: 32
Beam switching time: 250 ns
Transponder set on: 1 µs (to within 1 MHz)

Contractor
Rafael Electronic Systems Division, Haifa.

VERIFIED

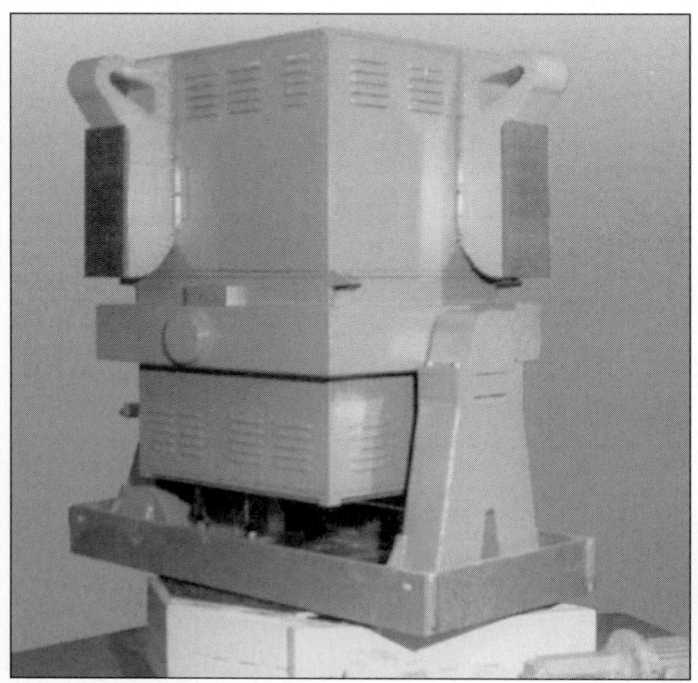

A stabilised 360° MBAT from Rafael's MBAT/RAN series of equipments

NS-9003A/9005 Electronic Warfare (EW) suite

Type
Integrated naval Electronic Support/Electronic CounterMeasures (ES/ECM) suite.

Description
NS-9003A/9005 is an integrated naval EW suite that receives and identifies over-the-horizon signals and uses a multibeam array transmitter to counter identified threats. Instantaneous frequency and direction-finding measurements are performed by the ES subsystem on a pulse-by-pulse basis with 'special' techniques being employed to analyse exotics such as chirp, frequency agile and staggered Pulse Repetition Frequency (PRF) signals. The suite's ECM function is automatically activated as soon as a threat signal has been identified against the

system's integral threat library. Tailored transmission techniques are used to counter individual threats and the suite's ECM subsystem architecture is such as to allow it to engage up to 16 threats simultaneously. System control is executed via a single operator console that is equipped with a 48 cm (19 in) colour graphic display, a colour alphanumeric display and a video monitor. In addition to its ES/ECM functions, the NS-9003A/9005 system is noted as offering a number of tactical electronic intelligence features. Here, the range includes intrapulse analysis, PRF analysis, frequency profile analysis and radar scan pattern analysis. Data logging is by means of a removable hard disk and a printer and the equipment as a whole complies with the MIL-E-16400G and MIL-STD-2036 environmental standards.

Status
Over time, the NS-9003A/9005 EW suite is understood to have been procured for installation aboard 'Eilat' class corvettes of the Israeli Navy. As of late 2000, NS-9003A/9005 was reported as being available.

Specifications
ES subsystem
Instantaneous frequency range: 2-18 GHz (0.5-40 GHz option)
Frequency resolution: 2 MHz
Instantaneous azimuth coverage: 360°
DF accuracy: 2° RMS
Environment density: > 500,000 pulses/s
Detection probability: 100% (claimed)
Sensitivity: −65 to −75 dBm
Dynamic range: > 60 dB
Pulse-width range: 0.1 to CW
CW: full parametric measurement
Scan rate: simultaneous analysis of all detected emitters
Data recording: environmental and emitter pulse data
Identification library: up to 5,000 sets of emitter parameters
ECM subsystem
Frequency range: I/J-band (8-20 GHz, 6-18 GHz option)
Transmitter ERP: 75 dBm (higher ERP as option)
ECM techniques: noise, deception and combined modes
ECM response modes: automatic, library selected and operator defined

Contractor
Elisra Electronic Systems Ltd, Bene Beraq.

UPDATED

NS-9005 radar jammer

Type
Shipborne noise and deception radar jammer.

Description
NS-9005 is a compact, wideband naval jamming system covering the 7.5 to 18 GHz frequency band. It is designed to provide self-protection for various types of vessels and includes both noise and deception techniques. It is controlled by an advanced digital techniques generator. NS-9005 is intended to operate in conjunction with an automatic, computerised Electronic Support (ES) system and can cope with several simultaneous threats, each with its own optimised electronic countermeasures programme. It is effective against frequency-agile and spread-spectrum signals and has a look-through-while-jam capability. The system can generate spot, barrage or swept noise along with deception modulations, with the latter including range/angle gate pull-off and false target options. It uses a 100 per

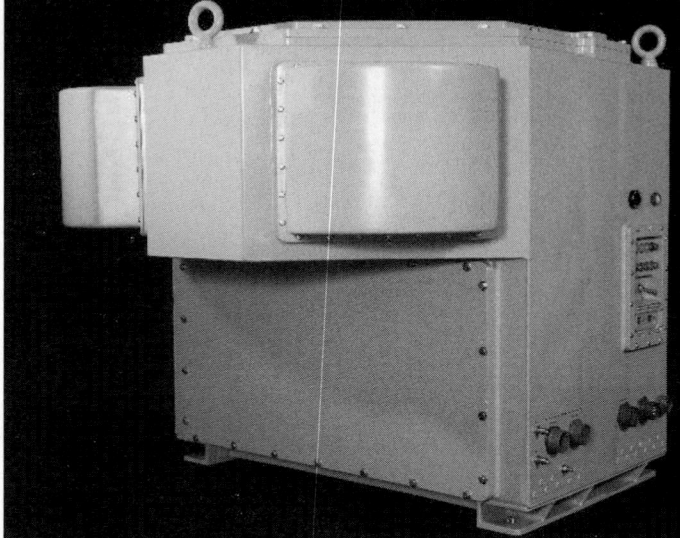

The MBAT unit used in the NS-9005 system

cent duty cycle power amplifier and a steerable antenna for transmission of the jamming signals. If the foreseen threat environment requires simultaneous handling of multiple incoming threats arriving from different directions and a high average effective radiated power, NS-9005 can be integrated with the Elisra's MultiBeam Array Transmitter (MBAT). An integrated electronic warfare suite consisting of the NS-9005 and the NS-9003A ES system is designated as the NS-9003A/9005 equipment (see separate entry).

Status
Over time, the NS-9005 radar jammer is understood to have been procured (as part of the NS-9003A/9005 suite) for installation aboard 'Eilat' class corvettes of the Israeli Navy.

Contractor
Elisra Electronic Systems Ltd (a member of the Elisra Group), Bene Beraq.

UPDATED

Rafael torpedo decoy systems

Type
Torpedo decoy series.

Description
As of late 2000, Rafael's Ordnance Systems Division has produced at least two torpedo decoy systems, the known details of which are as follows:

LESCUT
Developed in collaboration with BAE Systems North America, LESCUT is described as being a 'flexible, reactive decoy' that can be fired from both surface ships (using a Mk 36/Super Rapid Bloom Offboard Countermeasures (SRBOC) rocket/mortar or a pneumatic launcher) or submarines. The device is noted as making maximum use of commercial-off-the-shelf components and as being able to overcome 'all known' acoustic counter-countermeasures employed by 'modern' torpedoes. LESCUT is further described as being able to react to simultaneous threats and as incorporating the capabilities of Rafael's SCUTTER decoy (see following).

SCUTTER
SCUTTER is described as being a self-propelled, expendable decoy that is capable of providing 360° protection for submarines against acoustic homing, active and/or passive torpedo threats. Functionally, the device is launched immediately after a torpedo alert has been sounded and moves autonomously to an operating position, 'discerns' the threat and generates and transmits deception signals. The threat torpedo is lured into attacking and reattacking the SCUTTER round until its power source is exhausted, at which point it sinks and self-destructs. Other system features include prelaunch test and setting free launch and the ability to be fired from a standard submarine signal ejector.

Status
As of May 2001, Jane's sources were reporting that the LESCUT torpedo decoy was being actively promoted.

Specifications
Operating depth: 10-300 m
Operating time: 5-10 min
Weight 7.8 kg
Dimensions (Ø × l): 101 × 1,020 mm

Contractor
Rafael Ordnance Systems Division, Haifa.

VERIFIED

SEWS/RAN-1110 shipboard Electronic Warfare (EW) suite

Type
Surface ship EW suite.

Description
Based on the C-Pearl Electronic Support (ES) system (see separate entry) and the Shark countermeasures MultiBeam Array Transmitter (MBAT), the SEWS/RAN-1110 EW suite is designed for installation aboard a range of ships that includes patrol vessels, corvettes, frigates and destroyers. As such, the equipment includes an ES antenna assembly, an instantaneous direction-finding receiver, a radio frequency head, an instantaneous frequency measuring receiver, a signal processor and identification computer, an operator console, a power management computer, repeater and transponder units, dual trackers and technique generators and two MBAT units. Overall, the SEWS/RAN-1110 suite is modular, can be interfaced with onboard combat and data systems, meets the MILitary Electronics (MIL-E) 16400 standard and is operated by a single individual. Other system features include 180° coverage by each of the MBATs, the ability to undertake coincidence jamming in the direction of an identified threat, immediate beam positioning, power management and real-time signal detection, analysis and identification.

Status
As of May 2001, Jane's sources were reporting that Rafael was actively promoting the SEWS/RAN-1110 EW suite.

Specifications
Single pulse POI: 100%
Sensitivity: better than −65 dBm (over the 1-18 GHz frequency range)
ERP: 100-500 kW

Contractor
Rafael Electronic Systems Division, Haifa.

UPDATED

SHARK/RAN-1101 Electronic CounterMeasures (ECM) system

Type
Shipborne ECM system.

Description
An operational shipborne ECM system with high effective radiated power, SHARK/RAN-1101 is capable of jamming and deceiving a large number of multi-directional threats simultaneously using power management techniques. It has been designed for fitment on board vessels such as corvettes, frigates and destroyers. The SHARK/RAN-1101 system consists of two MultiBeam Array Transmitters (MBAT) each covering 180° in azimuth. These transmitters enable immediate positioning of a transmitter beam (capable of coincidence jamming) in the direction of the received threat. An optional single MBAT covering 360° in azimuth is available for smaller vessels. The transponder unit includes a group of voltage-controlled oscillators generating Radio Frequency (RF) signals at the frequencies and bandwidths matching the relevant ECM technique. The tracking unit (controlled by a power management computer) tracks any threat against which jamming is directed. The trackers are fed with the pulse descriptor and their outputs are the anticipated gates in time and the anticipated frequencies for the techniques generator. The latter implements the desired techniques against any threat using signal processors in accordance with instructions received from the power management computer. The latter manages the active response and efficiently allocates resources against threats. It is built around a high-speed microprocessor and uses a high-level software language. The principal features of SHARK/RAN-1101 include:

- simultaneous coincidence jamming
- high transmitting power
- fast beam switching
- ultra-short and accurate frequency set on transponder
- long RF memory
- advanced trackers and techniques generator
- power management.

Status
As of the first quarter of 2000, Jane's sources were suggesting that the SHARK/RAN-1101 ECM system was installed aboard 'Victory' class corvettes of the Singapore Navy.

Contractor
Rafael Electronic Systems Division, Haifa.

VERIFIED

The MBAT used in the SHARK/RAN-1101 ECM system   0009355

ITALY

Nettuno naval Electronic Warfare (EW) systems

Type
Family of shipborne integrated radar detection and countermeasures systems.

Description
Nettuno series EW systems are designed for use aboard major warships such as large frigates, destroyers and aircraft carriers. A Nettuno installation provides integrated radar Electronic Support/CounterMeasures (ES/ECM), while an independent ES section with a dedicated operator console is available as an option. A secondary electronic intelligence function is available without impact on the suite's primary functions. Nettuno is claimed to provide 100 per cent probability of intercept and a 'very high' level of automation ensures immediate and automatic library-based emitter identification and tracking. Real-time preprocessing is achieved via the use of a multimicro-preprocessor that incorporates proven (in computer simulations and during sea trials) algorithms. Processing is computer controlled and employs structured modular software using a high-level language. The architecture is fully and interactively integrated with its host vessel's command and control and decoy launcher systems. All system functions are controlled by a single operator working at a multimode synthetic display console with an 'enhanced' man/machine interface. Roll-stabilised multibeam direction-finding antennas provide high bearing accuracy and feature very high sensitivity, wide dynamic range logarithmic amplification and completely digital processing. Nettuno's ECM subsystem provides self-protection via the generation of environment adaptive, instant-by-instant automatic jamming against simultaneous multiple threats. If required, both the ship's command and control system and the Nettuno operator can, if necessary, override the suite's automatically selected jamming option. The subsystem's antennas are roll stabilised and consist of electronically steered linear arrays that provide a narrow high-gain beam. Jamming implementation follows automatic instructions provided by the ES section and by the power management computer. High effective radiated power is achieved by using a transmitter design employing paralleled matched high-power Elettronica-designed travelling wave tubes.

As of late 2000, the latest identified member of the Nettuno family was the Nettuno 2100/4100 equipment. Divided into the Nettuno 2100 ES and the Nettuno 4100 ECM subsystems, the suite comprises:

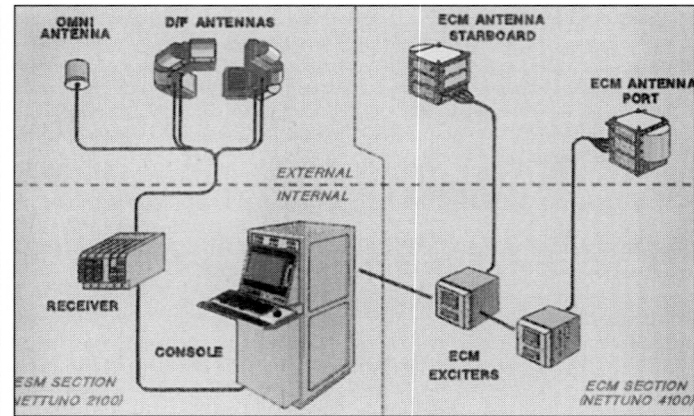

A schematic showing the Nettuno 2100/4100 configuration (Elettronica)

2002/0059923

- an array of monopulse direction-finding antennas (2100 subsystem)
- an omni-antenna assembly (2100)
- port and starboard jamming antenna units (with active phased-array modules - 4100)
- port and starboard jamming signal exciter units (4100)
- an operator console (2100)
- a receiver unit (2100).

Overall, the Nettuno 2100/4100 configuration is described as being fully integrated, as being 'totally' based on solid-state components and as having the ability to incorporate externally sourced ES systems such as Thales Airborne Systems DR 3000 (see separate entry) if required. System operating modes comprise:

Online
- automatic, real-time surveillance providing low probability of intercept emitter detection, automatic emitter analysis and identification and electronic intelligence functions
- automatic reaction (with operator override).

Offline
- analysis of collected data
- operator training
- library loading/editing
- built-in test/maintenance.

Other system features include:
- the ability to track, analyse and identify 'large numbers' of emitters operating in a 'very dense' environment
- 'very accurate' signal parameter measurement (including modulation on pulse)
- a data collection facility for post mission processing and single emission fine and statistical analysis
- digital radio frequency memory generated noise and deception jamming modulations
- electronic beam steering (electronically stabilised against ship movement)
- built-in test down to module/card level
- 'easy' onboard integration and installation (no waveguides).

A first-generation Nettuno display and control console

Status
As of the first quarter of 2000, Nettuno ES/ECM suites (bearing the service designation SLQ-732?) were reported as being installed aboard the Italian aircraft carrier *Giuseppe Garibaldi* and the same country's 'Audace' and 'De La Penne' class destroyers together with the Spanish aircraft carrier *Principe de Asturias* and the same country's 'Santa Maria' class frigates *Santa Maria*, *Victoria* and *Numancia*. In Spanish service, Nettuno is known as Nettunel. As of late 2000, the Nettuno 2100/4100 configuration was being actively promoted.

Specifications
Nettuno 2100/4100
Environmental: MIL-STD-2036 compliant
Netunno 2100 ES subsystem
Frequency range: 1-20 GHz
Spatial coverage: 40° (elevation); 360° (azimuth)
POI: 100% (nominal)
DF accuracy: suitable for co-operative fixing
Pulse density: up to 1 Mpps
Analysis/identification: automatic (traditional and exotic emitter types)
MOP detection: amplitude, frequency and phase
Tracking channels: >100
Nettuno 4100 ECM subsystem
Frequency range: 6-20 GHz
Spatial coverage: 40° (elevation); 360° (azimuth)
ERP: 'adequate' for large ship protection
Sensitivity: 'adequate' for sidelobe jamming
ECM response: multithreat jamming capability

Contractor
Elettronica SpA, Rome.

A first-generation Nettuno EW suite's ECM antenna installation

UPDATED

TQN-2BB

Type
H- through J-band (6 to 20 GHz) shipboard radar jammer.

Description
As of July 2001, the H- through J-band TQN-2BB equipment was the latest identified variant of Elettronica's TQN-2 shipboard radar jamming architecture. As such, TQN-2BB is described as being a modular, high power system that comprises:

- one or two antenna units
- one or two exciters
- one or two transmitters
- a 'TQN cabinet'

Looking at these various elements in more detail, the TQN-2BB antenna unit comprises a high gain, parabolic Electronic CounterMeasures (ECM) transmission antenna, together with four (?) direction-finding arrays that provide target azimuth and elevation data to the jammer's pair of target tracking receivers. The whole assembly is mounted on a two axis (azimuth and elevation), electromechanically aimed pedestal. The noted tracking receivers are housed in the 'TQN cabinet' which also contains the system's ECM processor and servo units/electronics. Functionally, target tracking is undertaken by the ECM processor on the basis of tracking error data received from the two tracking receivers. The TQN-2BB transmitter unit makes use of 'high efficiency' travelling wave tubes to produce 'high' Radio Frequency (RF) output power with 'limited' electrical power consumption. Transmitter cooling is by means of ambient air circulation or connection with an air conditioning system and makes use of multiple stage heat exchangers and low noise blowers. The equipment's exciter takes the form of a fast tuning source that incorporates a 'sophisticated' modulator. As such, it is capable of generating both noise and deception ECM signals and following (in RF) frequency agile radars.

Functionally, the TQN-2BB's antenna unit can accept and combine signals from up to two transmitters, with the most complete system configuration (two exciters, two transmitters, two antenna units and the 'TQN cabinet') facilitating the radiation of the highest power level signal against one or two threats coming from the same direction. Alternatively, the output can be split (one transmitter serving each of two antenna units) to counter up to four threats coming from two separate directions. Elettronica also notes that the system can be configured with a single transmitter feeding two antenna units. Here, the transmitter's output is directed to the antenna unit that has the best 'view' of the identified threat at any given time. Such an arrangement eliminates 'blind spots' created by obstacles such as the host vessel's superstructure. The company further reports that TQN-2BB can be 'fully' integrated with its RQN-5 or Nettuno 2100 electronic support systems (see separate entries) or externally sourced equipments such as Thales Airborne System's DR 3000 system (see separate entry).

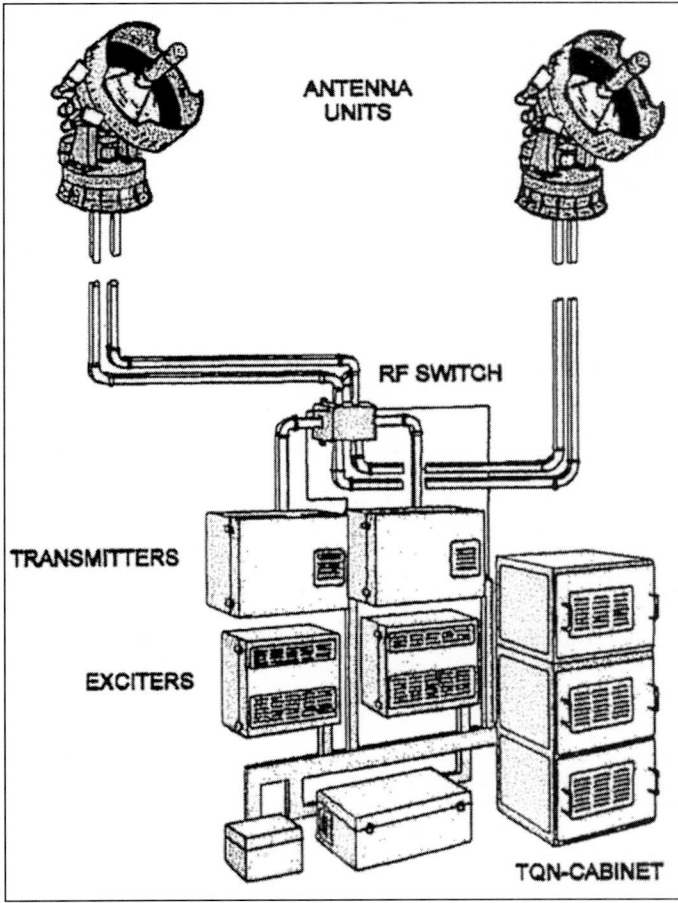

A schematic showing the full capability TQN-2BB configuration (Elettronica)
2002/0114014

Status
As of early 2001, Jane's sources were suggesting that TQN-2 variants were in service aboard 'Delhi' class destroyers and 'Brahmaputra' and 'Godavari' class frigates of the Indian Navy together with 'Laksamana' class corvettes of the Malaysian Navy. As of July 2001, Elettronica noted that TQN-2BB was 'already operational in various countries'.

Specifications
Frequency range: H- through J-band (6-20 GHz)
Spatial coverage: 40° (elevation); 360° (azimuth)
ERP: adequate to protect large ships
Sensitivity: adequate for sidelobe jamming
ECM response: single or dual threat jamming per exciter with the full TQN-2BB configuration (two exciters, two transmitters, two antenna units and the 'TQN cabinet') offering two fully independent beam outputs, with each beam able to engage two threats simultaneously
ECM programmes: includes modulated AM and/or FM noise programmes, pulse modes and combined pulse and noise programmes

Contractor
Elettronica, Rome.

NEW ENTRY

JAPAN

Japanese naval Electronic CounterMeasures (ECM) systems

Japan began to develop indigenous naval ECM systems during the 1960s and has gone on to produce a range of equipments which are currently in service. Known details of these systems are as follows:
NOLQ-1
NOLQ-1 is described as being an Electronic Warfare (EW) suite which is produced by the Mitsubishi Electric Corporation and comprises an electronically scanning electronic support measures subsystem, two Fujitsu OLT-2 deception jammers and the American SRBOC decoy launcher. Jane's sources suggest that NOLQ-1 has been installed aboard 'Haruna', 'Hatakaze', 'Shirane', and 'Tachikaze' (two ships) class destroyers of the Japanese Maritime Self-Defence Force (JMSDF).
NOLQ-2
NOLQ-2 is noted as being a Mitsubishi naval EW suite that is reported to have been installed aboard JMSDF 'Kongou' and 'Marasame' class destroyers.
OLT-2
The Fujitsu OLT-2 is described as being a deception radar jammer with automatic direction-finding. According to Jane's sources, the unit forms part of the NOLQ-1 suite and has been fitted as a stand-alone equipment aboard JMSDF 'Ishikari' class frigates.
OLT-3
The Fujitsu OLT-3 is reported to be a radar jamming system which (according to Jane's sources) has been fitted to JMSDF 'Asagiri', 'Hatsuyuki', 'Tachikaze' and 'Tackatsuki' class destroyers together with 'Yubari' class frigates.

VERIFIED

RUSSIAN FEDERATION AND ASSOCIATED STATES (CIS)

The following entries detail naval Electronic CounterMeasures (ECM) systems produced by and/or exported by the Russian Federation and the independent states of Armenia, Azerbaijan, Belarus, Georgia, Kazakhstan, Kyrgyzstan, Moldova, Tajikistan, Turkmenistan, Ukraine and Uzbekistan. Of these, the Russian Federation, Azerbaijan, Kazakhstan and Ukraine have operational navies. Taken together, these 12 entities form the Commonwealth of Independent States economic grouping.

VERIFIED

MP-401S naval Electronic CounterMeasures (ECM) system

Type
Shipboard threat warning, ECM and decoy launcher control system.

Description
MP-401S is a surface ship Electronic Warfare (EW) suite which provides warning of radar threats, spot and barrage noise jamming modulations (against both air and

surface emitters) and control of decoy launchers. The suite requires a five person operating crew and incorporates receiver, active jamming, control, electromagnetic compatibility, built-in test and power supply switching subsystems. The equipment's receiver subsystem can cover four frequency bands while the jammer is directed against a single band. MP-401S is categorised as a Group 58 Communications, Detection and Coherent Radiation equipment (Class 5865 Electronic Countermeasures).

Status

As of August 2001, the status of the MP-401S EW suite was uncertain.

Specifications

Bearing error on illuminating radar: 7-30°
Power consumption: 30 kW
Weight: 6,100 kg (incl spares and accessories)
Volume: 55 m²

VERIFIED

MP-403 Gorzuf radar jamming system

Type

Surface ship radar jamming system. NATO reporting name Side Globe.

Description

The MP-403 system is described as being a radar noise jammer that covers the C- (0.5 to 1 GHz), E/F- (2 to 4 GHz) and I/J- (8 to 20 GHz) bands and is suitable for installation aboard large surface ships. MP-403 incorporates multiple, mechanically scanning antennas and is noted as offering 360° coverage in azimuth.

Status

As of May 2001, the MP-403 radar jamming system was reported as being in service aboard the Russian cruisers *Kerch*, *Marshal Ustinov*, *Moskva* (ex *Slava*) and *Varyag* (ex *Chervona Ukraina*) and the Ukrainian cruiser *Ukraina* (ex *Admiral Lobov*).

UPDATED

MP-405 naval Electronic CounterMeasures (ECM) system

Type

Shipboard threat warning ECM and decoy launcher control system.

Description

MP-405 is described as being a surface ship threat warning, active jamming and decoy (PK-16 system) control system. Operated by a two person crew, the equipment comprises receiver, active jamming and control subsystems and can generate repeater, masking, spot and barrage jamming modulations. Its receiver subsystem covers four frequency bands while its jamming element utilises directional antennas and can be configured to cover up to three system specific bands. Each jamming band makes use of an individual jammer head. MP-405 is classified as a Group 58 Communications, Detection and Coherent Radiation equipment (Class 5865 Electronic Countermeasures).

Status

As of August 2001, the status of the MP-405 threat warning, ECM and decoy control system was uncertain.

Specifications

Coverage: 15° (azimuth-jamming subsystem); 30° (elevation-jamming subsystem); 360° (receiver subsystem)
Weight: 2,000 kg (receiver subsystem and 1 jammer head)

VERIFIED

MP-407 naval Electronic CounterMeasures (ECM) system

Type

Shipboard threat warning, ECM and decoy launcher control system.

Description

Designed for use aboard medium- to large-size surface ships, MP-407 is an electronic warfare suite, which provides threat warning, active jamming and decoy launcher (PK-2 and PK-16 systems) control. Operated by a three-man crew, the system comprises receiver, active jamming and control subsystems and can generate repeater, masking, deception, spot and barrage jamming modulations.

The MP-407 receiver subsystem covers four frequency bands with the jammer element utilising directional antennas to target two. The system is designated as a Group 58 Communications, Detection and Coherent Radiation equipment (Class 5865 Electronic Countermeasures).

Status

As of August 2001, the status of the MP-407 threat warning, ECM and decoy control system was uncertain.

Specifications

Coverage: 15° (azimuth-jamming subsystem); 30° (elevation-jamming subsystem); 360° (receiver subsystem)
Weight: 6,500 kg

VERIFIED

PK-2 decoy launcher

Type

Shipboard decoy launching system.

Description

PK-2 (ПК-2) is a surface ship decoy launcher system that comprises two ZIF-121 launcher units and a Tertsiya fire-control package. System payloads comprise the 43.7 kg, 1,105 × 36 mm TSP-47 chaff/infra-red decoy, the 40.1 kg, 1,105 × 37.5 mm TST-47 visual/electro-optic and the 47.1 kg, 1,105 × 38.5 mm TSO-47 visual/electro-optic rounds. The system requires a five person operating crew and is classed as a Group 58 Communications, Detection and Coherent Radiation equipment (Class 5865 Electronic Countermeasures).

Status

As of May 2001, the PK-2 decoy launcher was reported as being installed aboard the following warships and warship classes:

China
• 'Sovremenny' class destroyers (two ships in class - 2 × PK-2 and 8 × PK-10 decoy launchers)

India
• 'Delhi' class destroyers (three ships (three projected) - 2 × PK-2)
• 'Talwar' class frigates (four building - 2 × PK-2)

Russia
• the aircraft carrier *Admiral Kuznetsov* (ex *Tbilisi* - 4 × PK-2 and 10 × PK-10)
• the battle cruisers *Admiral Nakhimov* (ex *Kalinin*) and *Pyotr Velikiy* (ex *Yuri Andropov* - both 2 × 'twin' PK-2)
• the cruisers *Kerch* (2 × PK-2), *Marshal Ustinov* (2 × PK-2), *Moskva* (ex *Slava* - 2 × PK-2) and *Varyag* (ex *Chervona Ukraina* - 2 × PK-2)
• the destroyer *Admiral Chabanenko* (2 × PK-2 and 8 × PK-10)
• 'Sovremenny/Sarych' (four ships - 2 × PK-2 and 8 × PK-10) and 'Udaloy/Fregat' (seven ships - 2 × PK-2 and 8 × PK-10) class destroyers

Specifications

Calibre: 140 mm
Ammunition load: 100 rds/launcher
Weight: 15,000 kg
Temperature range: −40 to +50°C

UPDATED

PK-10 decoy launcher

Type

Shipboard decoy launching system.

Description

PK-10 is a 120 mm calibre surface ship decoy launching system which incorporates a remote control console and between two (small ship fit) and 16 (large ship fit) launchers. The launch units can be fitted with a drive mechanism to facilitate loading and can be configured to meet specific customer azimuth and elevation training requirements. Munitions are thought to include the SK-50 chaff/ infra-red decoy (9.1 kg payload), the SOM-50 infra-red seduction (7.2 kg payload) and the SR-50 chaff (11 kg payload) rounds. PK-10 is classed as a Group 58 Communications, Detection and Coherent Radiation equipment (Class 5865 Electronic Countermeasures).

Status

As of May 2001, PK-10 decoy launchers were reported as being installed aboard the following warships and classes of warship:

China
• 'Sovremenny' class destroyers (two ships in class - 8 × PK-10 and 2 × PK-2 decoy launchers)

India
• 'Kora' class corvettes (three ships, one building - 4 × PK-10)

Russia
• the aircraft carrier *Admiral Kuznetsov* (ex *Tbilisi* - 10 × PK-10 and 4 × PK-2)
• the destroyers *Admiral Chabanenko* (8 × PK-10 and 2 × PK-2) and *Neustrashimy* (8 × PK-10 and 2 × PK-16)

A general view of a PK-10 decoy launcher as installed aboard an Indian Navy corvette (Richard Scott)
2002/0083794

Poland	• the destroyer *Warszawa* (4 × PK-16)
	• 'Gornik' class corvettes (four ships - 2 × PK-16)
Romania	• the destroyer *Marasesti* (2 × PK-16)
	• 'Tetal/Improved Tetal' class corvettes (four/two ships - 2 × PK-16)
	• 'Zborul' class corvettes (three ships - 2 × PK-16)
Russia	• the destroyer *Neustrashimy* (2 × PK-16 and 8 × PK-10)
	• the landing platform dock *Mitrofan Moskalenko* (4 × PK-16 and 16 × PK-10)
	• the fast attack patrol hydrofoil *Vladimirets* (2 × PK-16)
	• 'Grisha/Albatros' (27 ships - 2 × PK-16 and 4 × PK-10) and 'Krivak' (15 ships - 4 × PK-16 and 10 × PK-10) class frigates
	• 'Dergach/Sivuch' (two ships - 2 × PK-16 or 2 × PK-10) and 'Parchim II' (11 ships - 2 × PK-16) class frigates
	• 'Tarantul III/ Molnya' class corvettes (24 ships - 2 × PK-16 or 4 × PK-10)
	• 'Pauk I/Molnya' (17 ships - 2 × PK-16 or 4 × PK-10) and 'Svetlyak' (27 ships - 2 × PK-16) class fast attack patrol craft
	• 'Matka/Vekhr' class fast attack hydrofoils (three - 2 × PK-16)
	• 'Gorya' class ocean-going minehunters (two ships - 2 × PK-16)
	• 'Aist/Dzheyran' class amphibious warfare hovercraft (seven - 2 × PK-16)
Syria	• 'Osa II' class fast attack craft (eight ships - ? × PK-16)
Ukraine	• the frigates *Hetman Sagaidachny* (4 × PK-16) and *Mikolaiv* (ex *Bezukoriznenny* - 4 × PK-16 or 10 × PK-10)
	• the patrol craft *Khmelnitsky* (ex *MPK 116* - 2 × PK-16 or 4 × PK-10)
	• the auxiliary *Slavutich* (2 × PK-16)
	• 'Grisha' class frigates (three ships - 2 × PK-16)
	• 'Pauk I/Molnya' class patrol craft (three ships (border guard) - 2 × PK-16 or 4 × PK-10)
Vietnam	• 'Tarantul' class corvettes (four ships - 2 × PK-16)

Specifications

Calibre: 82 mm
Firing intervals: 20 to 100 s (single shot/burst — automatic mode)
Rate of fire: 1-2 rds/s (burst mode)
Number of projectiles in burst: 2-3
Range: 200-1,800 m

UPDATED

RFAS naval Electronic CounterMeasures (ECM) systems

Alongside the equipments already described, a number of other RFAS naval ECM systems have been identified by their NATO reporting names. Known details of these are as follows:

Flat Track

Flat Track is described as a surface ship ECM system derived from the Wine Flask equipment with increased frequency coverage to counter anti-shipping missiles. As of May 2001, four Flat Track systems were reported as being installed aboard the Russian aircraft carrier *Admiral Kuznetsov*.

Wine Flask

Over time, Wine Flask has been variously described as being a surface ship barrage/deception radar jamming system that utilises a four element antenna array and as a surface ship electronic warfare 'intercept' equipment. Whatever its role, as of May 2001, Jane's sources were reporting that the Wine Flask system was in service aboard the Russian aircraft carrier *Admiral Kuznetsov* (four systems) and the same country's battle cruisers *Admirals Nakhimov* and *Velikiy* (four systems per vessel).

UPDATED

• the landing platform dock *Mitrofan Moskalenko* (16 × PK-10 and 4 × PK-16)
• 'Sovremenny/Sarych' (four ships - 8 × PK-10 and 2 × PK-2) and 'Udaloy/Fregat' (seven ships - 8 × PK-10 and 2 × PK-2) class destroyers
• 'Grisha/Albatros' (27 ships - 4 × PK-10 and 2 × PK-16) and 'Krivak' (15 ships - 10 × PK-10 or 4 × PK-16) class frigates
• 'Dergach/Sivuch' (two ships - 2 × PK-10 or 2 × PK-16), 'Nanuchka/Veter' (16 ships - 4 × PK-10) and 'Tarantul III/Molnya' (24 ships - 4 × PK-10 or 2 × PK-16) class corvettes
• 'Pauk I/Molnya' class fast attack patrol craft (17 ships - 4 × PK-10 or 2 × PK-16)

Ukraine
• the cruiser *Ukraina* (ex *Admiral Lobov* - 12 × PK-10)
• the frigate *Mikolaiv* (ex *Bezukoriznenny* - 10 × PK-10 or 4 × PK-16)
• the patrol craft *Khmelnitsky* (ex *MPK 116* - 4 × PK-10 or 2 × PK-16)
• 'Pauk I/Molnya' class fast attack patrol craft (three ships (border guard) - 4 × PK-10 or 2 × PK-16)

Contractor

Institute of Applied Physics, Novosibirsk.

UPDATED

PK-16 decoy launcher

Type

Shipboard decoy launching system.

Description

PK-16 is a surface ship decoy launching system that comprises two PK-16 launch units, a remote control console and a power rectifier. Loading is manual and the system has manual/automatic single round or automatic burst fire operating modes. It is classified as a Group 58 Communications, Detection and Coherent Radiation equipment (Class 5865 Electronic Countermeasures).

Status

During the period January to May 2001, Jane's sources were reporting the PK-16 decoy launcher as being in service aboard the following warships and classes of warship:

Algeria	• 'Moourad Rais' class frigates (three ships in class - 2 × PK-16)
Bulgaria	• the frigate *Smeli* (ex *Delfin* - 2 × PK-16)
	• the corvette *Mulniya* (2 × PK-16)
	• 'Reshitelni' class corvettes (two ships - 2 × PK-16)
Cuba	• the corvette *321* (2 × PK-16)
India	• the corvette *Sindhudurg* (2 × PK-16)
	• 'Rajput' class destroyers (five ships - 4 × PK-16)
	• 'Abhay' (four ships - 2 × PK-16), 'Khukri' (four ships - 2 × PK-16) and 'Veer' (12 ships (one building, three projected) - ? × PK-16) class corvettes
Indonesia	• 'Kapitan Patimura' class corvettes (16 ships - 2 × PK-16)
North Korea	• 'Hainan' class large patrol craft (six ships - 2 × PK-16)

TC series naval decoy rounds

Type

Family of decoy rounds for use with the PK-2 (ПK-2) launcher.

Description

The Novosibirsk-based Institute of Applied Physics' TC family of naval decoy rounds is designed for use with the Russian 140 mm PK-2 launcher unit (see separate entry) and is reported as offering effective protection for 'large and medium displacement' warships. TC rounds operate as confusion and/or distraction devices and are noted as being suitable for storage in 'tropical' conditions. As of May 2001, three family members had been identified, the known details of which are as follows:

TCП-47

The TCП-47 chaff round can be used as a confusion and/or distraction device and is described as being effective across a 'wide range of wavelengths'.

Specifications

Length: 1,105 mm
Weight: 7.73 kg (payload); 36 kg (complete round)

TCO-47

TCO-47 is an electro-optical masking round that features a dual pyrotechnic and laser reflecting payload and is designed for use in the distraction role.

Specifications

Length: 1,105 mm
Weight: 8.6 kg (pyrotechnic payload); 38.5 kg (complete round)

TCT-47

TCT-47 is an Infra-Red (IR) decoy round that is designed for use in the distraction role.

Specifications

Length: 1,105 mm
Weight: 8.6 kg (IR payload); 38.5 kg (complete round)

Status

As of May 2001, Jane's sources report that the TC family of naval decoy rounds was being actively promoted.

Contractor

Institute of Applied Physics, Novosibirsk.

NEW ENTRY

SPAIN

SCR-390(V) Electronic Support (ES) and jamming system

Type

Shipboard communications band ES and jamming system.

The mast-top antenna array associated with SCR-390(V)

The main computer/display unit used in the SCR-390(V) communications band ES/ECM system

Description

SCR-390(V) is designed to detect, locate, monitor, classify and analyse Very/Ultra High Frequency (V/UHF — 30 MHz to 1 GHz) signals traffic and, with the edition of a countermeasures subsystem, jam it. The equipment is described as being modular and as featuring a single operator position for both its active and passive functions. The man/machine interface used is noted as offering an 'interactive graphical interface'. System functions handled at the workstation include electromagnetic environment monitoring, interpretation of intercepted transmissions, their recording and, 'in certain cases', their 'modification'. SCR-390(V)'s countermeasures subsystem generates noise jamming and can be used to transmit 'spoof' traffic, while an automatic transmission modulation type classifier can be added to the ESM subsystem to provide a communications intelligence capability. Interfaces to the host ship's control and navigation systems are provided and there is a built-in test function. SCR-390(V) is also described as being a 'totally national' product.

Status

As of the first quarter of 2000, the SCR-390(V) communications band ES and jamming system was reported as being in service aboard 'several' frigates of the Spanish Navy.

Contractor

INDRA DTD, Madrid.

VERIFIED

SLQ-380(V) Electronic Warfare (EW) system

Type

Scalable, shipboard radar Electronic Support (ES) and jamming system.

Description

The scalable SLQ-380(V) naval EW system is described as being an outgrowth of INDRA's experience with the Acrux, Alderbaran, Canapus, Deneb and Nettunel countermeasures programmes and can be configured as follows:

Basic ESM SLQ-380(V)
Incorporating a spiral reception antenna array and a wide-open receiver, Basic ESM SLQ-380(V) is described as being suitable for ES and radar warning functions aboard patrol craft.

Advanced ESM SLQ-380(V)
Advanced ES SLQ-380(V) features I/J-band (8 to 20 GHz) multibeam antennas for targeting chaff and flare decoy launchers, close-in weapons systems and/or jammers.

Full ESM/ELectronic INTelligence (ELINT) SLQ-380(V)
Full ES/ELINT SLQ-380(V) comprises either the Basic or Advanced ESM variants combined with a spinning Direction-Finding (DF) antenna and a superheterodyne receiver to provide a narrowband fine analysis capability.

ESM/Electronic CounterMeasures (ECM) SLQ-380(V)
ES/ECM SLQ-380(V) adds a jamming subsystem (featuring mechanical pointing and selectable radiated power) to either the Basic or Advanced ESM variants of the system.

Full EW Suite SLQ-380(V)
The full EW Suite variant of the system combines the Advanced ES capability with a 'full' jamming and deception capability and 'sophisticated' multithreat electronic pointing.

In terms of SLQ-380(V)'s scalability, INDRA stresses the importance of the design and characteristics of the system's DF group, frequency group and preprocessor set to the process. Overall system features include:

- C/D- (0.5 to 2 GHz) and E/J-band (2 to 20 GHz) frequency coverage
- de-interleaving and characterisation of detected emissions
- emitter identification by library comparison (3,000 to 6,000 mode main library, 256 mode operator library)
- pulse-by-pulse and intra-pulse analysis
- claimed 99 per cent probability-of-intercept and low false alarm rate
- the ability to operate in a 2 Mpps plus environment
- better than 3 MHz RMS frequency accuracy over the system's full frequency range
- continuous Wave (CW) interference immunity via the use of CW-proof digital logarithmic video amplifiers
- automatic ESM/ECM hand-off in less than 25 ms
- three level built-in test
- Ada language programmes running on a UNIX operating system
- a high resolution operator display
- data recording and playback facilities
- an integrated operator training routine
- a multithreat countermeasures capability using independent port and starboard jamming assemblies
- digital radio frequency memory-based countermeasures techniques
- power management
- the use of high-power, fast steering phase-array technology.

Status

As of the first quarter of 2000, SLQ-380(V) variants were reported as having been procured for use aboard 'F-100' class frigates, amphibious assault ships, patrol boats and fleet logistic tankers of the Spanish Navy.

Contractor

INDRA DTD, Madrid.

VERIFIED

Spanish naval Electronic Warfare (EW) programmes

Description

Traditionally, Spanish land, sea and air EW systems have been procured offshore (most notably from France and Italy) or built under licence (mainly from Italy's Elettronica). More recently, Spain has developed an indigenous EW capability (centred around the national contractor INDRA DTD) and has undertaken a number of development programmes. The known details of those efforts directed at naval applications are as follows:

Acrux

Acrux appears to be aimed at the development of submarine Electronic Support (ES) equipment.

Aldebaran

Aldebaran is a Spanish Ministry of Defence research and development programme that is aimed at the development of a modular EW suite that is suitable for a range of applications such as frigates, corvettes and patrol boats. As of the first quarter of 2000, the Aldebaran effort appears to have resulted in the production of the SLQ-380(V) scalable EW system (see separate entry).

Canopus

An electronic countermeasures system developed by Inisel (now INDRA DTD). It is designed as an upgrade for the Spanish Navy's 'Descubierta' class frigates where it replaces the Elettronica ELT/311 and ELT/511 jamming systems.

Deneb

Deneb is an interface package to allow outputs from the Elettronica ES systems installed aboard 'Descubierta' class frigates to be integrated with equipment such as Canopus.

Elios

Elios is an identification and operation database management system for land and sea applications.

Elinath

Elinath is a communications intercept and jamming system.

Kochab

Kochab is a controller for decoy launchers that is designed for retrofit into existing systems.

VERIFIED

UNITED KINGDOM

Barricade series decoy systems

Type

Family of naval chaff and Infra-Red (IR) decoy launching systems.

Description

Wallop Defence Systems' family of Barricade naval chaff and IR decoy launching systems includes the Barricade Mk III, the SuperBarricade and Ultrabarricade equipments. Known details of these are as follows:

Barricade Mk III

Barricade Mk III is designed to provide a 'lightweight, reliable and cost effective' decoy launching capability for 'large' fast attack craft, mine warfare vessels and

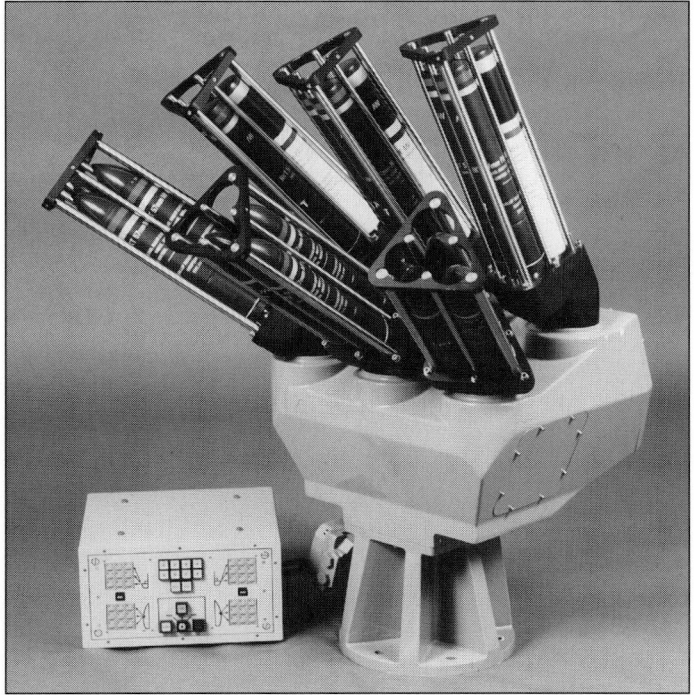

The Barricade Mk III decoy launcher and control unit

SuperBarricade firing 102 mm distraction rocket

'larger stealth' ships. A complete system incorporates multiple launch units, safety switches, a bridge display unit, a tactical computer and a control unit. Of these, the tactical computer provides automatic system control; instructs the launchers to fire distraction, dump and/or centroid decoy patterns; generates optimum centroid evasion manoeuvres and Radar Cross Section (RCS) minimisation data for the launch platform and interfaces with the host vessel's electronic support, sensor, active jamming and combat systems. The microprocessor-based control unit facilitates manual control of the system (if required) and provides a stored programme capability and launcher status displays. Barricade Mk III makes use of lightweight, low RCS launch units that fire Wallop Defence Systems' 57 mm, spin-stabilised, chaff, IR decoy (3 to 5 µm and 8 to 14 µm bands), illuminating, maroon and flare rocket munitions. In the seduction role, the countermeasures rounds are effective out to ranges of 80 m (4 seconds reaction time), a figure that rises to up to 600 m in dump mode (6 seconds reaction time). In the distraction role, such rounds are effective at ranges of up to 1,000 m (9 seconds reaction time), a figure that doubles (up to 2,000 m) when they are used in confusion mode (17 seconds reaction time). Overall, the system is noted as having a low reaction time and a multiple firing capability without the need to reload.

Specifications
Weight: 12.5 kg (control unit); 95 kg (launcher)
Dimensions   (h × w × d):   183 × 335 × 330 mm   (control   unit); 1,096 × 520 × 760 mm (launcher)

SuperBarricade

The SuperBarricade system is a multidirectional, 102 mm calibre equipment that provides a fast, computer-assisted countermeasures reaction to anti-shipping missile threats. A complete system incorporates multiple, 12 barrel (18 barrel option) launch units, safety switches, a bridge display unit, a tactical computer and a control unit. Of these, the tactical computer provides automatic system control; instructs the launchers to fire distraction, dump and/or centroid decoy patterns; generates optimum centroid evasion manoeuvres and Radar Cross Section (RCS) minimisation data for its launch platform and interfaces with the host vessel's electronic support, sensor, active jamming and combat systems. The microprocessor-based control unit facilitates manual control of the system (if required) and provides a stored programme capability and launcher status displays. SuperBarricade is offered with a computer-based training option and is able to fire both chaff and IR decoy, spin-stabilised, rocket munitions. Such munitions are effective in the seduction/centroid role at ranges of up to 150 m, in the dump role at ranges of up to 800 m, in the distraction role at ranges of up to 1,200 m and in the confusion role at ranges of up to 2,000 m.

Specifications
Weight: 12.5 kg (control unit); 150 kg (launcher)
Dimensions   (h × w × d):   183 × 335 × 330 mm   (control   unit); 1,275 × 1,500 × 840 mm (launcher)

Ultrabarricade

Developed jointly by Wallop and South African contractor Avitronics Maritime (a Saab/Grintek company), Ultrabarricade comprises four, fully trainable, fully stabilised, 102 mm 'stealthy' rocket launcher units and a fully automated system controller that interfaces with its host vessel's sensors and command system using FDDI, Ethernet or RS-422 busses. The system controller offers manual, automated and remote control options and automatically calculates and implements appropriate soft-kill solutions for each threat encountered. Ultrabarricade is compatible with a wide range of payloads that include radio frequency and IR seduction, dump, distraction and confusion decoys, illumination rockets, small calibre artillery rockets, torpedo decoys, laser and optical obscurant rounds, active offboard decoys and, with some modification, point defence surface-to-air missiles. The artillery rocket option can also be used in the missile defence role when it is interfaced with target designation and tracking systems. The system's decoy rounds are described as being effective at ranges of up to 80 m in the seduction role (4 seconds reaction time), at up to 600 m in the dump role (6 seconds reaction

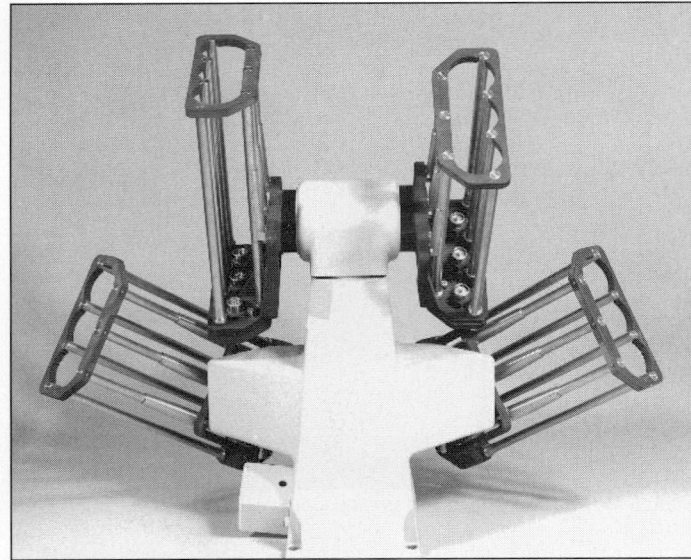

A SuperBarricade launcher assembly

time), at up to 1,000 m in the distraction role (9 seconds reaction time) and at up to 12,000 m in the confusion role (17 to 60 seconds reaction time).

Specifications
System controller
Power requirement: 115 V (single phase, 60 Hz)
Weight: 25 kg
Dimensions (h × w × d): 267 × 483 × 500 mm
Launcher
Coverage: 70° (elevation); 270° (azimuth)
Slew rate: 2 s (full arcs)
Pointing accuracy: 0.1°
Power requirement: 230 V (3-phase, 60 Hz)
Recoil capability: 10 kN
Weight: 450 kg (launch unit without rockets); 600 kg (payload)

Status
As of the period January to April 2001, Jane's sources were reporting that Barricade, SuperBarricade and Ultrabarricade decoy systems had been selected for use on or where installed aboard the following warships and classes of warship:

Algeria	• the landing ships logistic *Kalaat Beni Hammad* and *Kallat Beni Rachad* (Barricade system)
Australia	• 'Huon' class coastal minehunters (four ships in class, two building as of January 2002 - two SuperBarricade launchers per ship)
Bahrain	• 'Al Riffa' class fast attack craft - gun (two ships, one Barricade launcher per ship)
Chile	• 'Prat' class destroyers (three ships, two Barricade launchers per ship)
	• 'Leander' class frigates (three ships, Barricade systems)
Croatia	• the fast attack craft - missile *Sibenik* (two Barricade launchers)
	• 'Kralj' class corvettes (one ship, one building as of March 2001 - two Barricade launchers per ship)
Finland	• the corvette *Karjala* (Barricade system)
Greece	• 'Hunt' class coastal minehunters/minesweepers (two ships, two Barricade Mk III launchers per ship)
India	• 'Brahmaputra' (one ship, one building as of May 2001 - two SuperBarricade launchers per ship) and 'Godavari' (three ships, two SuperBarricade launchers per ship) class frigates
Italy	• 'Minerva' class corvettes (eight ships, two Barricade launchers per ship)
Kenya	• 'Nyayo' class fast attack craft - missile (two ships, two Barricade launchers per ship)
Malaysia	• 'Lekiu' class frigates (two ships, two SuperBarricade launchers per ship)
Oman	• the patrol ship *Al Mabrukah* (Barricade system)
	• the landing ship logistic *Nasr Al Bahr* (Barricade system)
	• 'Qahir' class corvettes (two ships, two Barricade launchers per ship)
	• 'Al Bushra' class offshore patrol vessels (three ships, Barricade system)
	• 'Dhofar' class fast attack craft - missile (four ships, two Barricade launchers per ship)
South Africa	• 'Meko A-200' class frigates (four building as of March 2001 - two SuperBarricade or Ultrabarricade launchers per ship)
UK	• 'Type 42 (Batches 1 and 2)' class destroyers (seven ships, Barricade as an IR decoy supplement)
	• 'Hunt' class coastal minehunters/minesweepers (11 ships, two Barricade Mk III launchers (Outfit DLK) per ship)
	• 'Sandown' class coastal minehunters/single role minehunters (11 ships, one building as of March 2001 - two Barricade launchers per ship when on deployment)

USA	• 'Cyclone' class coastal patrol ships (13 ships as of April 2001, to be reduced - two Mk 52 and/or Barricade Mk III launchers per ship)
Yugoslavia	• 'Split/Kotor' class frigates (three ships, two Barricade launchers per ship)
	• 'Koncar' class fast patrol craft - missile (five ships, two Barricade launchers per ship)

Contractor
Wallop Defence Systems (a division of Flight Refuelling Ltd), Middle Wallop

UPDATED

Chemring naval countermeasures

Chemring Countermeasures (a division of Pains Wessex Ltd) produces a range of radio frequency and Infra-Red (IR) naval countermeasures rounds. Examples of the family are as follows:
CHIMERA
CHIMERA is a 130 mm calibre, dual-mode (chaff and IR) munition that is compatible with both Seagnat and Super Rapid Bloom Offboard Countermeasures (SRBOC) launchers. The device is designed to protect 'lower signature warships' (conventional vessels up to light frigate size and larger 'stealthy' ships) and features a 'new design', centre-burst chaff payload together with a four sequential airburst IR capability that is claimed to be effective in both the long and short IR bands. The IR component of the munition is further claimed to be effective against 'all' seeker types including imaging and two colour devices. As of the first quarter of 2000, CHIMERA is known to have been procured by the Royal Danish Navy for use aboard its 'Flyvefisken' class patrol vessels and 'Niels Juel' class corvettes. In Danish service, CHIMERA is fired from a modified Terma Elektronik Soft Kill Weapon System (SKWS) launcher.

Specifications
Burst range: typically 40 m (at 45° launch angle)
Burst height: typically 30 m (at 45° launch angle)
Flight time to burst: typically 1.5 s
Burst sequence: 1st IR cloud/chaff (1-1.5 s); 2nd IR cloud; 3rd IR cloud (approx 7.4 s); 4th IR cloud
Chaff RCS: up to 8,000 m^2 (I-band (8-10 GHz), horizontally polarised)
Chaff frequency bands: I/J-band (8-20 GHz - frequency cuts can be made to customer requirements)
IR output: >1.5 kW/sr mean (3-5 μm band); >3.5 kW/sr mean (8-14 μm band)
IR duration: >20 s
Length: 1,222 mm
Calibre: 130.2 mm
Net explosive weight: 1.5 kg
Overall weight: 18.5 kg

Mark 33 (MOD 212) seduction chaff round
The Mark 33 chaff round is a 113 mm calibre, mortar-launched munition which is designed for use with the SAFAK and 113 mm Rapid Bloom Offboard Countermeasures (RBOC) launchers. It is noted as being effective in giving cover to small- to medium-sized vessels and as of 2000, was noted as being in service with NATO customers.
Mark 33 (MOD 212) IR decoy
The Mark 33 IR round is a 113 mm calibre, mortar-launched munition which provides an output in both the long and short IR wavebands. It is designed to be fired from SAFAK and 113 mm RBOC launchers. As with the Mark 33 chaff round, the Mark 33 IR decoy provides cover for small- to medium-sized ships and as of 2000, was reported as being in service with NATO customers.
Mark 36 seduction chaff round
The Mark 36 chaff round is a 130 mm calibre, mortar-launched munition that can be fired from all 130 mm Seagnat and SRBOC launchers. It features a fast bloom

A Mk 36 SRBOC launcher loaded with PIRATE IR decoy rounds 0050686

time and a large radar cross section value and is designed for the protection of medium- to large-sized vessels. As of 2000, the device was understood to be in service with a number of navies around the world.

PIRATE IR decoy

PIRATE is a 130 mm calibre, mortar-launched munition which provides a ship-like signature in both the 3 to 5 and 8 to 14 µm IR wavebands and can be fired from all 130 mm Seagnat and Super Rapid Bloom Offboard Countermeasures (SRBOC) launchers. It combines air and surface emitters to generate a high intensity, long duration, large area output close to the sea's surface. PIRATE can be used in both distraction and seduction modes and is claimed to be effective against a wide range of IR seeker types including imaging units. As of 2000, the device was reported as being fully qualified and as being in service with a number of navies (including NATO forces) around the world.

PW216 distraction chaff decoy

The PW216 chaff round is a 130 mm calibre, rocket launched munition that can be fired from all 130 mm Seagnat and SRBOC launchers. It features variable range (out to 3 km) and a barometric height sensor for optimum chaff deployment at all selected ranges. As of 2000, PW216 was reported as being in service with several navies (including NATO forces) around the world.

TALOS

Standing for Thermal Anti-missile Launched Offboard Seduction, TALOS is a walk-off IR decoy round that is compatible with the Seagnat and Mk 36 SRBOC decoy launching systems. It is interchangeable with the existing NATO Mk 245 munition and is described as offering 'enhanced performance characteristics and materials' when compared with the latter device. As of 2000, TALOS was the subject of a then year £3 million plus Royal Navy procurement.

Alongside its own range of munitions, Chemring Countermeasures have (over time) produced chaff and IR decoy payloads for a number of international systems including Breda's 105 mm SCLAR equipment, Bofor's 57 mm system, Buck's DUERAS and Silver Dog, Matra's Dagaie and Sagaie equipments, Wallop Defence Systems' Barricade and Super Barricade systems, Rafael's Beamtrap and Saab's ELMA equipment. The company is also known to be developing a new range of anti-submarine warfare decoy payloads and vehicles.

Contractor

Chemring Countermeasures/Pains Wessex Ltd, Salisbury.

VERIFIED

....

Corvus decoy launcher

Type

Shipborne chaff decoy launching system.

Description

Corvus (UK service designation Outfit ULC) is a simple, lightweight, quick reaction system for the self-defence of surface vessels against surface-to-surface and air-to-surface missiles. Chaff dispensing rockets are fired to form a radar decoy screen around the vessel. The system can be installed in any surface vessel (other than patrol craft) and its simplicity of construction makes it particularly suitable for installation in destroyers and frigates. The complete system comprises two multibarrelled rocket launchers, a launcher control and firing panel and chaff rockets.

Two modes of operation are available for the system. The modes differ in the azimuth angle at firing and the range at which chaff is released. The modes of operation are:

- distraction decoy
- centroid decoy.

A firing and control panel can be ideally situated in the combat information centre. The maximum loading of 16 chaff rockets provides protection against three missile attacks before reloading. Any member of the combat information centre crew can control and operate the system.

A cylindrical rotating structure carries eight launching tubes mounted in two sets of three (one above the other) and crossed at 90° in azimuth. Two further tubes are set above this arrangement and are aligned midway between the other tubes, all at a fixed elevation of 30°. A deck-mounted pedestal supports the rotating structure on its training bearing and also houses a self-contained electrical power conversion unit for the control circuits and the associated electrical equipment. The training drive consists of a gearbox driven by a reversible motor, while braking is achieved by an electromagnetic brake. A slip-ring positioning unit, located within the pedestal, controls the training of the launcher to preselected bearings. The normal limits of the training arc for the launcher are between 60 and 120° but, as the lower sets of rocket barrels are angled about the centreline of the rotating structure, the arc covered by the rockets is considerably greater. Within these normal limits, fixed firing bearings for the launcher, at 15° intervals, can be selected by completing the appropriate circuits at the control panel. Training stops and a buffer assembly are incorporated to prevent over-stressing and twisting of the cable bight in the event of an accidental overrun. The launchers can be fitted in any position on No 1 deck or above, providing adequate space is available for slewing and loading, and there is a clear firing arc of approximately 160° in azimuth and 30° elevation.

Status

During the period January to May 2001, Jane's sources were reporting that the Corvus decoy launching system was in service aboard the following warships and classes of warship:

The eight-barrel launcher used in the Corvus decoy launching system

Argentina	• 'Hercules' class destroyers (two ships in class)
	• 'Drummond' class frigates (three ships, three proposed as of January 2001 - Dagaie (see separate entry) or Corvus decoy launchers fitted)
Bangladesh	• the frigate *Umar Farooq*
	• 'Leopard' class frigates (two ships)
Chile	• 'Prat' class destroyers (three ships, two Corvus launch units per ship)
	• 'Leander' class frigates (three ships, two Corvus launch units per ship)
Ecuador	• 'Leander' class frigates (two ships, two Corvus launch units per ship)
India	• the aircraft carrier *Viraat* (two Corvus launch units)
Indonesia	• 'Ahmad Yani' (six ships, two Corvus launch units per ship) and 'Fatahillah' (three ships, two Corvus launch units per ship) class frigates
Pakistan	• the frigates *Badr* and *Shahjahan* (two Corvus launch units per ship)
	• 'Leander' class frigates (two ships, two Corvus launch units per ship)
Portugal	• the replenishment oiler *Berro* (two Corvus launch units)
UK	• 'Appleleaf' class transport oilers (three ships, two Corvus or two Shield (see separate entry) launch units per ship)
	• 'Rover' class small transport oilers (three ships, two Corvus and two Shield launch units per ship)
	• 'Fort Grange' class combat stores ships (two ships, two Corvus or two Shield launch units per ship)

Readers should also note that the French Navy has employed a Corvus-type system (designated as Syllex) and that during the period February to March 2001, Jane's sources were reporting it as being installed aboard the service's helicopter carrier *Jeanne d'Arc* and destroyers *de Grasse* and *Tourville* (two launchers per ship).

Specifications

Bearing arc: 15-165° (tube axis, max); 60-120° (launcher axis, normal)
Number of tube guides: 4 (right-hand spiral)
Tube dimensions (Ø × l): 195 mm × 1.6 m
Weight: 25 kg (empty tube); 42 kg (off-launcher equipment); 582 kg (launcher unit)

Contractor

BAE Systems Marine, Barrow-in-Furness.

UPDATED

....

Siren active offboard decoy

Type

Naval active offboard decoy system.

Description

The 28 kg, 130 by 1,700 mm Siren round is an offboard decoy that is designed to counter radar-guided anti-ship missiles. It consists of an 'intelligent' electronic decoy round, a 130 mm calibre multibarrelled launcher (see following) and a microprocessor-based fire-control unit. As such, the system interfaces with the ship sensors to give fully automatic, semi-automatic or manual activation, with the fire-control unit providing the link with the host vessel's other onboard defence systems. In its primary control role, the fire-control unit utilises its own microprocessor to select the optimum launcher, programme the correct jamming technique, choose the best deployment position and initiate the firing sequence. Manual operation can be carried out via a keyboard and slave controllers are available for local control of individual launchers. Once launched (and clear of its

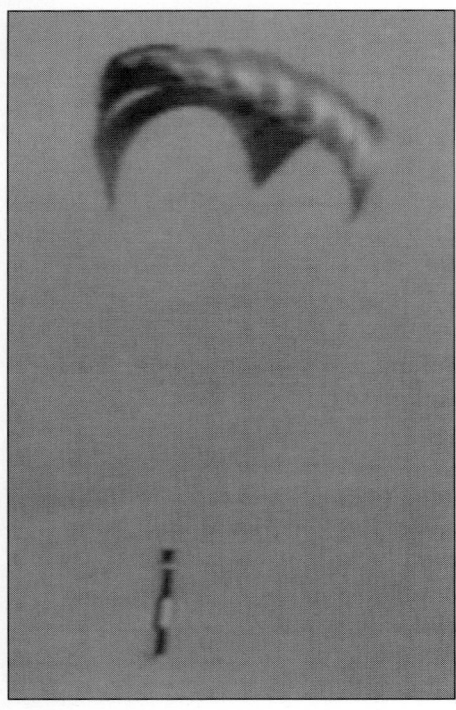

The Mk 251 Siren offboard decoy

host vessel), the Siren round radiates a powerful jamming signal to seduce the missile away from the ship while slowly descending on a deployed parasail. The system can be adapted to almost any warship design (from a patrol boat to a capital ship) and can also be fitted to naval auxiliaries and merchant ships. The decoy itself consists of a low *g* rocket, the noted parasail-type parachute, a transmitter, a receiver, antennas, control electronics and a battery source. With regard to its launch system, Siren's manufacturer considers two launchers as being adequate to protect smaller vessels, with four being needed to provide effective cover for larger ships. Alongside the Royal Navy's Outfit DLH (itself an update of the Outfit DLB/DLJ(2) system - see following), Siren can be launched from any standard 130 mm Seagnat barrel and other types of 130 mm calibre launchers. For ships with Seagnat, Siren shares the same magazines and ready use ammunition stowage with the system's chaff and infra-red rounds. For other applications, a dedicated magazine locker and launcher package is available.

Status

In July 1994, the then GEC-Marconi (now BAE Systems Avionics - Sensor Systems Division) was awarded a then year £80 million contract covering the supply of 21 system control units and approximately 720 examples of a Royal Navy specific Siren variant (designated as the Mk 251 Active Decoy Round - ADR). Originally scheduled to enter service during April 2000, technical difficulties with its payload's ability to withstand the stress of launch are understood to have delayed the Mk 251 ADR's service qualification and as of early 2001, to have put back its service entry until September 2001, at which time, seven ships were expected to be Mk 251 ADR capable, with each vessel carrying 20 such devices. As of the given date, Jane's sources went on to suggest that completion of the programme required the procurement of a further 20 Outfit DLB/DLJ(2) to DLH conversion kits and sufficient additional ADR rounds to complete the service's Mk 251 war arsenal and that once in service, the Outfit DLH launcher would be modified so that its lower pair of barrels could be angled in steps between 25° and 45° to facilitate decoy deployment at lower altitudes and at closer ranges to the host vessel. It should also be noted that France's Thales Airborne Systems (formerly Thomson-CSF DETEXIS) is a subcontractor on the Siren programme and is understood to have at one time proposed a system variant as a solution to the French Navy's Leurre Electromagnétique Actif (Active Electromagnetic Decoy - LEA) requirement.

Contractors

BAE Systems Avionics - Sensor Systems Division, Stanmore, UK.
Thales Airborne Systems, Elancourt (France).

UPDATED

UNITED STATES OF AMERICA

AN/SLY-2(V) Advanced Integrated Electronic Warfare System (AIEWS)

Type
Integrated shipboard EW system.

Description
AN/SLY-2(V) AIEWS is intended to replace the AN/SLQ-32(V) radar warning and jamming system (see separate entry) aboard US Navy (USN) surface ships during

the early part of the 21st century. When fully implemented, AIEWS is intended to incorporate Electronic Support (ES), Electronic Attack (EA), Infra-Red Search and Track (IRST) and IR jamming capabilities that are fully integrated with a host vessel's combat information system. As of early 2001, it is expected that AIEWS applications will take the form of full capability or ES/IRST only systems according to perceived individual platform requirements.

Status
The USN awarded two then year US$75,000 AIEWS study contracts to industry during December 1995. These were completed in September 1996 and were followed by a formal Request For Proposals (RFP) which was issued during the summer of 1997. This document covered Increment 1 of the AIEWS programme and addressed the ES and IRST elements of the system. Elements included are understood to have comprised requirements for precision ES, specific emitter identification, high processing throughput, advanced display equipment and decoy integration. Two contractor teams (led by the then Raytheon Systems (now Raytheon Electronic Systems) and Lockheed Martin) responded and in January 1998, it was announced that the Lockheed Martin led consortium (see contractors) had been awarded a then year US$66.5 million contract covering the delivery of 12 Low-Rate Initial Production (LRIP) systems (with options for up to 40 production examples) during the period US Fiscal Year (FY) 2001 to 2004. Ships slated to receive AIEWS included 'Arleigh Burke' and 'DD-21' class destroyers, 'Ticonderoga' class cruisers, 'Enterprise' and 'Nimitz' class aircraft carriers and 'San Antonio/Wasp/Whidbey Island' class amphibious warfare ships. AIEWS Increment 2 covers the required EA and IR jamming capabilities and was scheduled to be addressed during 1999 following the completion of a countermeasures Advanced Technology Demonstration (ATD) that was being undertaken by Northrop Grumman. If followed through, AIEWS Increment 2 will see the production of five EMD and seven LRIP systems that may be based on the ATD work or may be bid competitively. During July 2000, SLY-2(V) passed its critical design review and as of the given date, the first low-rate initial production AIEWS ES capability was scheduled to be installed aboard the 'Arleigh Burke' class destroyer *DDG-91* during the autumn of 2003. Elsewhere in the programme, testing of the first prototype SLY-2(V) antenna array began during July 2000 and as of the cited date, the first AIEWS below-deck cabinet was scheduled for delivery during the early part of 2001 for integration with the latest version of the Aegis combat system.

Contractors
Lockheed Martin Naval Electronics and Surveillance Systems - Syracuse (prime and team member developing the SLY-2(V)'s ES antenna arrays, pulse-sorting hardware and processing software).
Northrop Grumman Electronic Sensors and Systems Sector - Defensive Systems Division (team member developing the SLY-2(V)'s ES antenna arrays, pulse-sorting hardware and processing software).
SenSyTech (team member developing the SLY-2(V)'s ES antenna arrays, pulse-sorting hardware and processing software).
DSR (control and processing software and displays - DSR is an associated AIEWS contractor to the USN).
Northrop Grumman Electronic Systems sector (Increment 2 ATD contractor).

UPDATED

Advanced Programmable Electronic Countermeasures System (APECS) II

Type
Shipborne radar detection and jamming system.

Description
APECS II is a modular, 0.5 to 18 GHz band Electronic Support and CounterMeasures (ES/ECM) system that is designed for use on a range of vessel types (fast patrol boats to destroyer) and is intended to counter 'most known' radar threats. It is a fully automatic system that identifies and jams threat emitters using instantaneous frequency measuring receivers and microprocessor-controlled jammers. The jamming array used is a phased-array assembly and high-speed switching enables up to 16 different threats to be jammed simultaneously. Overall, APECS II consists of four subsystems, as follows:

- a mast-mounted ES antenna assembly comprising two preamplified, monopulse Direction-Finding (DF) arrays
- a system processing equipment package made up of ES receivers, DF processors, ES signal processors, an ECM controller and a Power Management/ Techniques Generator (PM/TG)
- an operator's console with an 'ergonomically designed' colour monitor and keyboard with touch-sensitive overlay. It is designed for fully automatic operation
- port and starboard ECM transmitter units with internally mounted transmission arrays.

Of these various elements, the ES subsystem is based on either the Condor Systems AR-700 or AR-900 equipments. The antenna configuration used consists of two monopulse, three-band receiver units (with a total of 24 antennas) and an omnidirectional antenna. This latter array provides a preamplified Radio Frequency (RF) output to the system's digital instantaneous frequency measuring receivers while the DF monopulse arrays provide RF to the DF receiver inputs. The receiving and processing units allow pulse-by-pulse measurements of RF, pulse-width, amplitude, direction of arrival and time of arrival over the complete 0.5 to 18 GHz frequency range. Each of the suite's ECM transmission units offer transponder and repeater capabilities over a 180° azimuth sector. Phased-array, polarisation-diverse transmitters are employed and can generate 60 separate, overlapping beams to provide 360° jamming coverage for the ship. The two transmitters are controlled by the PM/TG unit located in the system's equipment rack. This unit receives emitter data on a pulse-by-pulse basis, controls the transmitter (including frequency and beam switching) and implements the appropriate techniques. An important capability of the design is time-shared management of a frequency memory loop that enables the ECM subsystem to handle multiple, simultaneous, agile and jittered threats. Both transmitters are stabilised in pitch and roll to maximise 'on-target' jamming strength. In all, the ECM subsystem can simultaneously jam up to 16 threat signals (in any combination of direction and frequency) with 'optimum' power and technique management. An additional eight signals are tracked in standby, waiting for assignment. Up to eight frequency-agile and/or pulse repetition interval agile signals can be handled simultaneously. The PM/TG uses a variety of noise and deception techniques. The high effective radiated power is the same for Continuous Wave (CW) jamming because of the use of CW travelling wave tubes. The APECS II operator's console integrates multiple, separate displays into a multiwindow, stand-alone system with pseudo real-time displays (including pan and plan position indicator). The colour graphic units used are 48 cm (19 in) raster displays with 1,280 × 1,024 pixel resolution.

Status

As of the first quarter of 2000, Condor Systems (and the system's originator, ARGOSystems) were reported as having sold more than 20 integrated electronic warfare systems to five navies around the world. As of February/March 2001, Jane's sources had identified APECS II installations aboard Dutch 'Karel Doorman' class frigates (eight ships in class), Greek 'Hydra' class frigates (four ships) and Portuguese 'Commandante João Belo' (three ships) and 'Vasco da Gama' class frigates (three ships).

Specifications

ES frequency coverage: 0.5-18 GHz
Sensitivity: up to −67 dBm
Detection range: 370 km ships; 93 km aircraft
DF accuracy: 4° RMS; 2° RMS fine DF
Automatic tracking: 500 active signals in dense signal environment
Signal identification speed: <1s
ECM azimuth coverage: 360° using a 60-beam jammer (1 beam per 6° azimuth sector)
Radiated power: 180 kW nominal mid-band
ECM frequency coverage: 7.5-18 GHz
Jammer capacity: up to 16 threats simultaneously

Contractor

Condor Systems Inc, San Jose, California.

UPDATED

....

AN/SLQ-32(V) Electronic Warfare (EW) system

Type

Shipboard radar threat detection, analysis and jamming system.

Description

AN/SLQ-32(V) series EW systems are the US Navy's (USN) standard radar threat detection, analysis and jamming suites which were originally developed as a replacement for the AN/WLR-1 receiver/AN/ULQ-6 radar jammer combination aboard the service's surface ships. System development took the form of a multiyear USN industry programme that produced two competing solutions; the Hughes AN/SLQ-31 and Raytheon SLQ-32. As a deployed system, SLQ-32(V) is modular (see following) and employs lens-feed, multibeam reception antennas for all signals except for those in the lowest frequency bands. The antenna technology used consists of an array of receivers fed through coaxial cables by a stripline, multibeam, parallel-plate lens. This arrangement provides a set of individual, contiguous, high-gain outputs (all of which exist simultaneously) that individually benefit from the full gain of the available aperture. Within the system, each array offers more than an octave of frequency coverage. As noted previously, SLQ-32(V) is a modular equipment which is (or has been) deployed in five distinct variants, as follows:

SLQ-32(V)1

SLQ-32(V)1 is a wide-open (in angle and frequency) Electronic Support (ES) system that covers radar threats within the US Band 3 frequency range. It also provides cues and control facilities for onboard decoy launchers and, if fitted, the active Sidekick radar jammer (see following). The system comprises processing and display equipment, inboard equipment racks and two multibeam antenna/receiver assemblies, each of which is equipped with a 180° semi-omni sense antenna (for instantaneous frequency measurement) and two 90° Direction-Finding (DF) lens arrays. A special purpose digital processing subsystem (designated as the Presorter and incorporating DF/frequency correlation and digital tracking elements) correlates coarse frequency and amplitude data which, together with time of arrival, is used to create a pulse descriptor word. This is then catalogued (by frequency and angle cell) and stored in the emitter file memory of the digital tracker. If three or more pulses meeting a specific angle/frequency signature are received within a programmable (up to 32 ms) time span, the system's AN/UYK-19

One of the AN/SLQ-32(V)2 EW system's antenna arrays
(Stefan Terzibaschitsch)

central computer is informed that a new emitter has been detected. This in turn triggers a command for the digital tracker to provide additional pulses for full-scale analysis. From this, the threat signal's pulse repetition frequency, scan type, scan rate and frequency are established in order to allow the system to identify the emitter against a software mode library. Having completed the identification process, the equipment generates appropriate alert signals and commands for the available countermeasures systems.

SLQ-32(V)2

SLQ-32(V)2 adds early warning, indentification and DF capabilities against anti-shipping missile launch and seeker radars to the system's basic Band 3 coverage. The increase in capability is achieved via the introduction of two additional receiver chains, namely a US Band 2 subsystem within the existing antenna/receiver assemblies and a Band 1 chain which utilises four spiral antennas located on the ship's yardarm in such a way as to provide 360° cover. Like the Band 3 chain, the Band 2 subsystem features two DF receivers and lens arrays (each covering 90° sectors) and an associated 180° sense antenna. Aside from the described hardware changes, the SLQ-32(V)2's central computer is provided with an enhanced memory to cope with the additional data flow generated by the increase in system coverage. SLQ-32(V)2 can also be interfaced with the Sidekick radar jammer, the addition of which changes the system's designation to SLQ-32(V)5 within the US Navy.

SLQ-32(V)3

SLQ-32(V)3 adds the Active Electronic CounterMeasures (AECM) subsystem to the (V)2's ES and decoy cueing functions. To meet the AECM's hardware requirements, additional inboard equipment racks house eight high voltage power supplies for Travelling Wave Tube (TWT) sources, a transponder, a digital switching unit and a techniques generator. Externally, the system's receiver/lens array assemblies are enlarged to house four (two per assembly) Band 3 transmission arrays and their associated electronics. Additionally, the (V)3 introduces roll stabilisation for its external reception/transmission assemblies. In the countermeasures role, SLQ-32(V)3 identifies the available jamming technique most appropriate to the identified threat and hands-off parametric data to the AECM's techniques generator. Here, an appropriate waveform is synthesised which, in turn, is used to modulate voltage controlled oscillators in the transponder and its associated drive unit. The derived output is then directed to the appropriate input in the appropriate transmission lens array to ensure maximum radiated power in the direction of the threat. The system's central computer (which incorporates a further increase in memory capacity) manages the AECM's ouput in such a way as to enable it to engage a number of threats simultaneously and the subsystem has semi-automatic (operator initiated jamming) or automatic (system initiated jamming) operating modes.

SLQ-32(V)4

SLQ-32(V)4 is the (V)3 capability optimised for use aboard aircraft carriers. Hardware additions include the introduction of a fibre optic interface for data communication between the port and starboard reception/transmission assemblies.

SLQ-32(V)5

As noted previously, the addition of the Sidekick radar jammer to the SLQ-32(V)2 ES system results in the US Navy's SLQ-32(V)5 electronic countermeasures/ES suite. The Sidekick jammer itself makes use of port and starboard transmitter assemblies each of which contains two high-gain, multibeam, Rotman lens antenna arrays which provide 90° sector coverage fore and aft. An amplification train comprising 12 moderate power TWT's is used to maximise the equipment's effective radiated power and its ability to counter multiple threats simultaneously and at any azimuth angle. Sidekick is also described as offering a full techniques repertoire and as utilising modified SLQ-32(V)3 software. While it is noted as being able to fully integrate with both SLQ-32(V)1 and (V)2, Raytheon is also understood to be able to supply an alternative set on receiver/ES system for use with Sidekick in applications where no SLQ-32 system is available.

Readers should also note that the SLQ-32A(V) designation has been introduced for new build (and, possibly, upgraded) systems, reflecting the scope of the architectural and processing advances introduced into the equipment over time. Among these is a new software package from COMPTEK Amherst Systems that was introduced during 1995. SLQ-32's Phase E processing update is further understood to provide a baseline for the US Navy's next generation Advanced Integrated Electronic Warfare System (AIEWS) with which the service hopes to replace SLQ-32 in the first decade of the next century.

Status
As of this edition, over 360 SLQ-32 systems were reported as having been delivered to the USN. Known contracting activity is as follows:

September 1987/ January 1988	Raytheon was awarded two SLQ-32(V) production contracts with a combined then year value of US$80 million
May 1988	Raytheon was awarded a then year US$70 million contract covering the supply of 25 SLQ-32(V) systems
June 1990	Hughes (now part of Raytheon Electronic Systems) was awarded a then year US$29 million SLQ-32(V) second-source production contract covering the supply of eight systems for the USN and one for Taiwan. After producing one more equipment of the type, Hughes withdrew from the programme
December 1993/ January 1994	The USN awarded two production contracts covering the supply of SLQ-32A(V)2 systems
Summer 1998	Raytheon was awarded a then year US$9.9 million contract covering the 'restoration' of the first seven of a projected 30 in-service SLQ-32(V)2 systems. This award was also understood to contain options covering the upgrading of an eighth SLQ-32(V)2 and a single SLQ-32(V)3 model. As originally planned, delivery of the refurbished and 'improved' systems was scheduled to begin in July 1999 and end during October 2001

Over time, SLQ-32(V) EW systems are reported as having being installed aboard the following ship classes:

Australia	'Adelaide' class frigates (SLQ-32C)
Bahrain	the frigate Sabha (SLQ-32(V)2)
Greece	'Kimon' and 'Kidd' class destroyers (both SLQ-32(V)5?) and 'Epirus' class frigates (SLQ-32(V)2)
Mexico	'Knox' class frigates (SLQ-32(V)2)
Saudi Arabia	'Badr' class corvettes and 'Al Siddiq' class fast attack craft (both SLQ-32(V)1)
Taiwan	a single 'Anchorage' class landing ship (SLQ-32(V)1) and the same country's 'Cheng Kung' (Taiwanese produced Chang Feng IV variant of SLQ-32(V)2 with Sidekick jammer) and 'Knox' (SLQ-32(V)2) class frigates
Thailand	the aircraft carrier Chakri Naruebet (SLQ-32(V)3) and the same country's 'Knox' class frigates (SLQ-32(V)2)
Turkey	'Gaziantep' and 'Tepe' class frigates (both SLQ-32(V)2)
USA	the aircraft carrier USS Enterprise (SLQ-32(V)4); the command ship La Salle (SLQ-32(V)3); the mine warfare vessel Inchon (SLQ-32(V)3) and the same country's 'Nimitz' and 'Kitty Hawk/ John F Kennedy' class aircraft carriers (both SLQ-32(V)4); 'Ticonderoga' class cruisers (SLQ-32(V)3); 'Arleigh Burke (Flts I/II)' (SLQ-32(V)2 or (V)3 in seven ships), 'Arleigh Burke (Flt IIA)' (SLQ-32(V)3) and 'Spruance' (SLQ-32(V)2) class destroyers; 'Oliver Hazard Perry' class frigates (SLQ-32(V)2); 'Blue Ridge' class command ships (SLQ-32(V)3); 'Wasp' class amphibious assault ships (SLQ-32(V)3); 'Austin' class amphibious transport docks (SLQ-32(V)1); 'Anchorage' (SLQ-32(V)1) and 'Whidbey Island/Harpers Ferry' (SLQ-32(V)1 or (V)2) class dock landing ships; 'Jumboised Cimarron' class oilers (SLQ-32(V)1); 'Sacramento' and 'Supply' class fast combat support ships (both SLQ-32(V)3) and 'Famous' class medium endurance cutters (SLQ-32(V)2)

Contractor
Raytheon Electronic Systems, Goleta, California.

VERIFIED

Automated Launch of EXpendables (ALEX) system

Type
Automated Decoy Launching System (DLS).

Description
Capable of operating as a stand-alone equipment or as part of an integrated electronic warfare suite, the ALEX DLS links a decoy launching subsystem to the host vessel's electronic support, anemometry and navigation systems to create an automated offboard countermeasures capability. The ALEX system is described as being modular and as being able to be configured to meet specific customer installation and control (manual, semi-automatic and automatic) requirements. System modules comprise:
- the 112 mm Mk 135/Rapid Bloom Offboard Countermeasures (RBOC) II or 130 mm Super RBOC Mk 137 launcher units
- Launch Alert Horns (LAH)
- a Launcher Electronics Unit (LEU)

- a System Processor (SP)
- a Ship Manoeuvre Indicator (SMI)
- a Bridge Control Panel (BCP)
- a Master Control Panel (MCP)
- a Master Power Panel (MPP)
- an Uninterruptable Power Source (UPS)

Functionally, an ALEX system makes use of one or more deck-mounted launchers that are under the control of one or more LEUs. Each LEU receives commands from the SP and operator interface and control of the system is accomplished via the use of fixed action buttons and/or a touchscreen on the MCP. The MCP also displays menus, inventories, modes, ship input data status, system configuration and built-in test fault alerts. The SMI also interfaces with the SP and provides an audio alarm when a recommended course-to-steer forms part of the selected tactic against a particular threat. The data and power circuits for all the ALEX system's modules are routed through the MPP, which can also be used as a controller via its circuit breakers. An LAH is mounted in the vicinity of each of the system's launchers and sounds at predetermined intervals when the architecture is in the 'enable fire' mode. The UPS provides conditioned AC power and will continue to supply back-up power (without operator intervention) in the event of a ship's power system failure. The BCP is the ALEX system's manual alternative to the MCP and incorporates safety, arming and fire-control switches together with status indicators. Transfer of system control from the MCP to the BCP is at the operator's discretion. The SP also houses the system's operational software in a non-volatile memory. When the architecture is in automatic or semi-automatic mode, this software package utilises data from the host vessel's electronic support, anemometry and navigation systems to determine the optimum response to the tactical situation using expendable decoys and ship manoeuvres.

Status
As of early 2001, Jane's sources were reporting that customers for the ALEX DLS included the navies of Egypt, Greece (for installation aboard four projected Super Vita fast attack craft - missile), Saudi Arabia and Turkey. As of May 2001, ALEX was noted as being actively promoted.

Contractor
Sippican Inc, Hycor Products Group, Marion, Massachusetts.

UPDATED

Direct Coupled Jamming/Surveillance System (DCJ/ SS)

Type
Shipborne command, control and communications system for voice and data communications jamming and surveillance.

Description
DCJ/SS is a command, control and communications countermeasures system intended to disrupt enemy voice and data communications with simultaneous jamming and surveillance over the frequency range 100 to 500 MHz. This is achieved by directly coupling two monitor receivers and two 400 W jammer combinations (receiver/exciter/transmitter) in the basic system. The system can be used in a shipborne, airborne and ground-based role and is modularised for easy expansion to up to 14 monitor receivers (2 to 1,100 MHz); up to 10 receiver/exciter combinations; up to 10 high-power amplifiers with power combiner options and a direction-finding system. In a typical system, configured with multiple high-powered amplifiers, the DCJ/SS can operate in any of four modes, from manual to fully automatic. Collection and jamming operations can be performed simultaneously. DCJ/SS can also simulate a full range of threat signals, jamming techniques and effective radiated power levels for training purposes.

Status
As of this edition, a DCJ/SS variant was reported as having been procured by the Royal Navy.

Contractor
BAE Systems North America - Integrated Defense Solutions, Austin, Texas.

UPDATED

Hycor naval decoy rounds

Sippican Inc's Hycor Products Group produces a series of naval chaff and Infra-Red (IR) decoy rounds, the known details of which are as follows:

CHAFFSTAR II decoy cartridge
CHAFFSTAR II (NATO Mk 214 equivalent) is a rapid blooming, seduction chaff cartridge which is compatible with the Rapid Bloom Offboard Countermeasures (RBOC) II launcher. Hycor claims that a single cartridge of this type will provide a sufficiently large chaff cloud to protect a ship of up to frigate size without, in many cases, special manoeuvring.

Specifications
Frequency coverage: 2-20 GHz (single spot frequency, multiple frequencies or full-band coverage)
Chaff type: rapid blooming aluminised glass
RCS: up to 5,000 m² (8-18 GHz sub-band); up to 12,000 m² (single frequency)
Payload volume: 7,700 cm³
Muzzle velocity: 60 m/s
Weight: 8.2 kg (payload); 15.5 kg (complete round)
Dimensions (Ø × l): 112 × 1,067 mm

GEMINI decoy cartridge
GEMINI is a combined chaff and IR decoy cartridge that is compatible with the RBOC II and US Mk33/34 launch systems. Functionally, the device generates a large radar cross section chaff cloud (sufficient to mask a small vessel such as a fast patrol boat) while simultaneously deploying a co-located, parachute supported IR decoy flare.

Specifications
Frequency coverage: 2-20 GHz (single frequency, multiple frequencies or full-band coverage)
Chaff type: aluminised glass
Chaff volume: 1,550 cm³
IR decoy type: proprietary formulation
Burn time: 30 s (min)
Weight: 1.5 kg (chaff payload); 5.2 kg (complete round)
Dimensions (Ø × l): 112 × 434 mm

HIRAM II decoy cartridge
HIRAM (Hycor IR Anti-Missile) II is an IR decoy cartridge that is compatible with the RBOC II launcher and is designed to ignite on impact with the sea's surface, producing a 2 m high flame source that acts as an attractive target for IR seekers.

Specifications
Fuel type: liquid
Burn time: 45 s (min)
Flotation method: inflatable collar
Payload volume: 7,500 cm³
Weight: 4.7 kg (payload); 15.9 kg (complete round)
Dimensions (Ø × l): 112 × 1,220 mm

LOROC decoy cartridge
LOROC (LOng-Range Offboard Chaff) is a chaff distraction decoy that is compatible with the RBOC II launcher. The device is designed to be used in conjunction with seduction rounds and generates a ship-sized radar echo at ranges of between 1 and 4.5 km from its launch point.

Specifications
Frequency coverage: 2-20 GHz (single frequency, multiple frequencies or full-band coverage)
Chaff type: all available types
Chaff volume: 2,043 cm³
RCS: up to 2,000 m² (8-18 GHz sub-band); up to 5,000 m² (single frequency)
Weight: 2.3 kg (payload); 14.74 kg (complete round)
Dimensions (Ø × l): 112 × 1,067 mm

Ship-Launched Acoustic Decoy (SLAD)
Hycor's SLAD round is available in 112 and 130 mm diameter variants that are compatible with the RBOC II and SRBOC launchers respectively. On impacting the sea's surface, the SLAD round dispenses a payload assembly and a flotation device. The former (suspended from the float by a tether) now transmits a high-power, wideband acoustic signal to jam incoming acoustic torpedo seekers. At a preset point, the device automatically switches to the generation of a 'ship-like' acoustic signature that is designed to seduce acoustic torpedoes away from their target.

Super CHAFFSTAR decoy cartridge
Super CHAFFSTAR (NATO Mk 214 equivalent) is a rapid bloom, seduction chaff decoy which is compatible with the Mk 36/Super Rapid Bloom Offboard Countermeasures (SRBOC) launch system. Hycor claims that the chaff cloud generated is sufficiently large to obviate frequently the need for ship manoeuvres.

Specifications
Frequency coverage: 2-20 GHz (single frequency, multiple frequencies or full-band coverage)
Chaff type: rapid blooming aluminised glass
Chaff volume: 12,500 cm³
RCS: up to 10,000 m² (8-18 GHz); up to 20,000 m² (single frequency)
Muzzle velocity: 65 m/s
Weight: 12.7 kg (payload); 22.2 kg (complete round)
Dimensions (Ø × l): 130 × 1,220 mm

Super GEMINI decoy cartridge
Super GEMINI is a combined chaff and IR decoy which is compatible with the SRBOC launcher. Functionally, the device generates a large radar echo chaff cloud together with a parachute supported IR decoy flare.

Specifications
Frequency coverage: 2-20 GHz (single frequency, multiple frequencies or full-band coverage)

Chaff type: aluminised glass
Chaff volume: 8,000 cm³
IR decoy type: proprietary formulation
Burn time: 30 s (min)
Weight: 8.2 kg (chaff payload); 20 kg (complete round)
Dimensions (Ø × l): 130 × 1,200 mm

Super HIRAM III decoy cartridge
Super HIRAM III is an IR decoy cartridge which is compatible with the SRBOC launcher. Functionally, the device ignites on impact with the sea's surface to create a 2.5 m high flame source that is claimed to match the IR signature of a large ship.

Specifications
Fuel type: liquid
Burn time: 45 s (min)
Flotation method: inflatable collar
Payload volume: 10,000 cm³
Weight: 6.4 kg (payload); 22 kg (complete round)
Dimensions (Ø × l): 130 × 1,220 mm

Super LOROC decoy cartridge
Super LOROC is a distraction chaff cartridge that is designed to complement seduction rounds and is compatible with the SRBOC launcher. The device can create chaff clouds at ranges of between 1 and 4.5 km from its launch point.

Specifications
Frequency coverage: 2-20 GHz (single frequency, multiple frequencies or full-band coverage)
Chaff type: all available types
Chaff volume: 1,850 cm³
RCS: up to 5,000 m² (8-18 GHz sub-band); up to 20,000 m² (single frequency)
Weight: 6.4 kg (payload); 22 kg (complete round)
Dimensions (Ø × l): 130 × 1,200 mm

Super Walk-Off IR (SWOIR) decoy
The SWOIR decoy is a seduction IR round that is mechanically and electrically compatible with the SRBOC launcher. After launch, SWOIR produces a number of small IR clouds that are progressively further away from the launch vessel and are designed to 'walk' the incoming missile off its target. At the end of this sequence, SWOIR produces a longer duration 'keeper' IR cloud to complete capture of the missile's seeker. Hycor claims that the output from a single SWOIR round is sufficient to match the radiant intensity of a 'large' ship.

Specifications
IR material: 'special IR material' (pyrophoric?)
Burn time: 15 s (min)
Weight: 23 kg (complete round)
Dimensions (Ø × l): 130 × 1,200 mm

Status
As of May 2001, Jane's sources were reporting that Sippican was actively promoting the described decoy rounds.

Contractor
Sippican Inc, Hycor Products Group, Marion, Massachusetts.

UPDATED

Mk 53 Decoy Launching System (DLS)

Type
Naval DLS.

Description
The Mk 53 DLS is a US adaptation of the Mk 36/Super Rapid Bloom Offboard Countermeasures (SRBOC) launcher that is designed to provide an automated, fast response self-defence capability for surface ships. The system can accommodate active and passive radio frequency and infra-red decoys (including the NULKA/Mk 234 active offboard device and the Mks 214, 216 and 245 rounds) and can be interfaced with both the AN/SLQ-32(V) Electronic Support (ES) and the AN/SLY-2(V) Advanced Integrated Electronic Warfare System (AIEWS) architectures. Alongside the host vessel's electronic warfare capability, the Mk 53 DLS interfaces with its synchronisation systems, wind sensing equipment and electrical power system. A typical Mk 53 installation comprises:
- a Mk 24 Mod 1 Decoy Launch Processor (DLS)
- four (Mk 53 Mod 1) or six (Mk 53 Mod 4) Mk 174 Mod 1 Processor/Power Supply (PPS) units
- four Mk 137 Mod 7 launcher units (six passive round mortar barrels and two Mk 234 decoy modules per unit)
- a Mk 164 Bridge Control Panel (BCP).
Of these, the DLS hosts the system's operational software and provides its primary control and interface capability. Functionally, it accepts operator commands and threat data from the host vessel's ES system. On the basis of such information, it determines the most tactically advantageous placement for the Mk 234 active offboard decoys relative to its host platform and selects the appropriate decoy

tactic for the identified threat, selects the launcher or launchers to be used, initialises the PPS units and authorises decoy launch. The equipment incorporates digital interfaces for up to six launchers, the operator (or operators) and an optional data recording device. Windspeed and direction and own ship's direction, heading, speed, roll and pitch information is handed off to the DLS in either analogue or digital (via the operator's interface) formats. The PPS units communicate with the launchers and receive launch commands from the DLS or, in the case of passive decoys, from the BCP. Each PPS is located inboard (within a 20 m cable's length of its respective launcher) and serves as the two-way communications link between the DLS and the system's decoy rounds. The Mk 137 Mod 7 launcher unit is created via the addition of a two round Mk 234 decoy module to the existing Mk 137 unit.

Status

As of this edition, the Mk 53 Mod 4 DLS was identified as having been mandated for installation aboard US Navy 'San Antonio' class amphibious transport docks.

Contractor

Sippican Inc, Marion, Massachusetts.

VERIFIED

Rapid Bloom Offboard Countermeasures (RBOC) Decoy Launching System (DLS)

Type

Shipborne chaff and Infra-Red (IR) flare DLS.

Description

According to Jane's sources, the baseline RBOC DLS system comprises up to four, 112 mm, six barrelled, Mk 135 launcher units, a Mk 164 Mod O bridge launcher control unit, a Mk 158 Mod O master launcher control unit and Mk 4 Mod O or Mk 5 Mod O ready lockers with a maximum capacity of 40 rounds per locker. When configured with two launcher units, the RBOC system carries the US designation Mk 34 while the four launcher unit version is designated as the Mk 33. RBOC compatible munitions include the CHAFFSTAR chaff, GEMINI combined chaff/Infra-Red (IR) decoy, HIRAM II IR decoy, LOROC chaff and Mk 171 Mod O chaff rounds (see *Hycor naval decoy rounds* entry). The RBOC II DLS is an improved version of the Mk 33/Mk 34 system that can fire both standard RBOC cartridges and longer format rounds that are better suited to the protection of larger ships. The number of launchers per ship can be tailored to the needs of the particular vessel, as can the cartridge load-out. Mk 33/34 RBOC and RBOC II systems can be upgraded to Automated Launch of EXpendables (ALEX) standard (see separate entry) via the introduction of the Automatic Round IDentification (ARID) package, a new power supply and a new control subsystem. Thus configured, the system integrates data from the host vessel's electronic support receiver and its wind and navigation sensors to provide an automatic, optimised response to a specific threat. Also available as new build equipment, the ALEX configuration can, if required, be supported by the IBM-compatible, personal computer-based Tactical Analysis WorkStation (TAWS) and the ALEX training simulator. TAWS is an engagement analysis tool which allows for the creation of 'expert' tactics while the

associated training simulator offers tutorial, operator training and engagement analysis modules.

Status

Over time, RBOC DLSs have been reported as having been installed aboard the following warships:

Brazil	• 'Para' class frigates (Mk 33 application)
China	• 'Jianghu I' class frigates (Mk 33 application)
South Korea	• 'Alligator' class landing ships and 'Pae Ku' class fast attack craft (Mk 33 application)
Singapore	• 'Sea Wolf' class fast attack craft
Taiwan	• 'Cheng Kung' class frigates (Taiwanese-produced Kung Fen RBOC variant)
Thailand	• the training ship *Makut Rajakumarn* (Mk 34 application?) and the same country's 'Chon Buri' fast attack craft (Mk 33 application?)

Specifications

RBOC II launcher
Weight: 125 kg
Dimensions (l × w × h): 1,242 × 400 × 700 mm

Contractor

Sippican Inc, Hycor Products Group, Marion, Massachusetts.

VERIFIED

Super Rapid Bloom Offboard Countermeasures (SRBOC) Decoy Launching System (DLS)

Type

Shipborne chaff and Infra-Red (IR) decoy flare DLS.

Description

The Mk 36 SRBOC system is a deck-mounted, mortar-type DLS that is used to launch an array of decoy munitions against 'a variety' of threats. As such, the equipment is controlled from the host ship's bridge and/or Combat Information Centre (CIC) and is dependent on cueing data provided by the vessel's threat detection and analysis subsystems. A generic Mk 36 SRBOC system comprises the following:

- a minimum of two (port and starboard) 130 mm Mk 137 launcher units
- a Mk 158 Mod 1 or 2 master launch control panel
- a Mk 164 Mod 1 or 2 bridge launch control panel
- a Mk 160 Mod 1 power supply for each Mk 137 launcher unit within the particular system configuration
- a minimum of two ready service lockers (one for each Mk 137 launcher within the particular system configuration).

Of these, the deck-mounted Mk 137 launcher unit comprises two parallel rows of three launch tubes per row with the individual tubes being mounted at angles of 45° and/or 60°. The firing circuits employ electromagnetic induction to initiate the propellant charges in the equipment's decoy rounds. The Mk 160 power supply converts the host ship's AC power to DC for use by the system's firing circuits and safe lights. As noted previously, the Mk 164 control panel is bridge-mounted and provides firing circuit controls and a system status display while the Mk 158 unit provides similar capabilities within the host vessel's CIC. System flexibility is enhanced by SBROC's ability to handle munitions designed for use with the Seagnat launcher system (see separate entry) alongside its own family of decoy rounds. The system is also understood to be upgradable to an Automated Launch

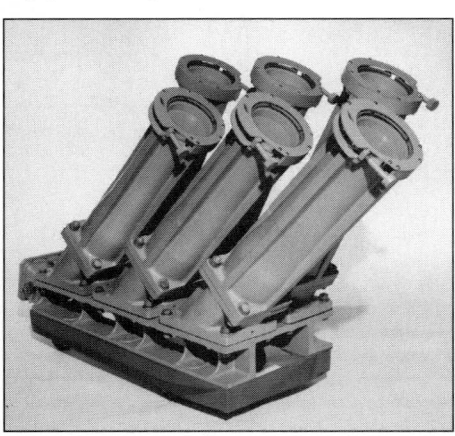

The RBOC II launcher unit

The RBOC Mk II's master launcher controller

A Mk 137 SRBOC launcher loaded with Super CHAFFSTAR cartridges

of EXpendables (ALEX - see separate entry) standard. This is achieved via the introduction of an automatic round identification capability together with new power supply and control subsystems. So configured, the equipment can integrate threat, windspeed and navigation data to generate an automatic, optimised countermeasures response. Over time, SRBOC has been configured to a number of 'Mod' standards for different ship installations, the known details of which are as follows:

Mod	No of launchers	No of lockers (capacity)	Control configuration (bridge/CIC)
Mod 1	2	2 (40 rounds)	Mk 164 panel/integrated with AN/SLQ-32(V) and Mk 158 panel
Mod 2	4	4 (80 rounds)	Mk 164 panel/integrated with AN/SLQ-32(V) and Mk 158 panel
Mod 5	2	2 (40 rounds)	Mk 164 panel/integrated with AN/LQ-32(V)
Mod 6	4	4 (80 rounds)	Mk 164 panel/integrated with AN/SLQ-32(V)
Mod 8	8	8 (160 rounds)	Mk 164 panel/Mk 158 panel
Mod 9	4	4 (80 rounds)	Mk 164
Mod 10	6	6 (120 rounds)	Mk 164
Mod 11	4	4 (140 rounds)	Mk 164
Mod 12	6	6 (210 rounds)	Mk 164

Status

Over time, the following ships and classes of ship have been identified as having been equipped with the MK 36 decoy launching system:

Australia	• 'Perth' class destroyers and 'Adelaide' and 'ANZAC' (Mk 36 Mod 1) class frigates
Bahrain	• the frigate *Sabha*
Bangladesh	• the frigate *Osman*
Belgium	• 'Wielingen' class frigates
Canada	• 'Improved Provider' class replenishment oilers
China	• 'Luhu' class destroyers and 'Jiangwei I/II' class frigates
Ecuador	• 'Leander' class frigates
Egypt	• 'Knox' and 'Oliver Hazard Perry' class frigates
Germany	• 'Lutjens' class destroyers and 'Bremen' and 'Sachsen' class frigates
Greece	• 'Kidd' and 'Kimon' class destroyers and 'Elli', 'Epirus' and 'Hydra' (Mk 36 Mod 2) class frigates
Japan	• the training ship *Shimayuki* and the same country's 'Asagiri', 'Haruna', 'Hatakaze', 'Hatsuyuki', 'Kongou', 'Murasame', 'Shirane', 'Takatsuki' and 'Tachikaze' class destroyers; 'Abukuma' and 'Shikari/Yuubari' class frigates and 'PG 01' class fast attack hydrofoils
South Korea	• 'Ulsan' class frigates and 'Po Hang' class corvettes
Mexico	• 'Knox' class frigates
Netherlands	• the fast combat support ships *Amsterdam* and *Zuiderkruis* and the same country's 'De Zeven Provincien' class destroyers; 'Karel Doorman', 'Jacob Van Heemskerck', 'Kortenaer' and 'Tromp' class frigates and 'Rotterdam' class amphibious warfare ships
New Zealand	• the military sealift ship *Charles Upham* and the same country's 'Broad-Beam Leander' class frigates
Pakistan	• the replenishment oilers *Moawin* and *Nasr* and the same country's 'Tariq' class frigates (Corvus or Mk 36)
Portugal	• 'Commandante Joao Belo' and 'Vasco Da Gama' class frigates
Saudi Arabia	• 'Badr' class corvettes; 'Al Siddiq' class fast attack craft and 'Al Jawf' class coastal minehunters
Spain	• the aircraft carrier *Principe De Asturias*, the attack transport *Arago*, the fleet logistic tanker *Patino* and the same country's 'Balearas' (Mk 36 Mod 2), 'Descubierta', 'F 100' (Mk 36 Mod 2) and 'Santa Maria' (Mk 36 Mod 1/2) class frigates and 'Galicia' class amphibious transport docks
Taiwan	• a single 'Anchorage' class amphibious warfare vessel and the same country's 'Knox' class frigates
Thailand	• 'Knox' class frigates
Turkey	• 'Barbaros' (Mk 36 Mod 1), 'Gaziantep', 'Tepe' and 'Yavuz' (Mk 36 Mod 1) class frigates and 'Dogan', 'Kilic' and 'Yildiz' class fast attack craft
UAE	• 'Kortenaer' class frigates
UK	• 'Type 42' class destroyers (Outfit DLB or Mk 36 and Corvus)
USA	• the aircraft carrier USS *Enterprise*, the command ships *Coronado* and *La Salle*, the mine warfare vessel *Inchon* and the same country's 'Kitty Hawk/John F Kennedy' and 'Nimitz' class aircraft carriers; 'Ticonderoga' class cruisers; 'Arleigh Burke (Flts I/II)' (Mk 36 Mod 12), 'Oliver Hazard Perry' and 'Spruance' class frigates; 'Blue Ridge' class command ships; 'Tarawa' and 'Wasp' class amphibious assault ships; 'Austin' class amphibious transport docks; 'Anchorage' and 'Whidbey Island/Harpers Ferry' class dock landing ships; 'Jumboised Cimarron' class oilers; 'Sacramento' and 'Supply' class fast combat support ships; 'Hamilton/ Hero' class high endurance cutters and 'Famous' class medium endurance cutters

Specifications

Calibre: 130 mm
Launcher tube muzzle velocity: 75 m/s
Power requirment: 24 V DC (operating, emergency batteries); 28 V DC (normal operation); 440 V AC (60 Hz, single phase)
Weight: 9.1 kg (Mk 164); 13.6 kg (Mk 158); 209 kg (Mk 137)
Dimensions (l × w × h): 260 × 200 × 100 mm (Mk 164); 310 × 480 × 200 mm (Mk 158); 1,600 × 460 × 900 mm (Mk 137)

Contractors

Over time, the SRBOC system is reported as having been produced by a number of contractors including:

United Defense, Armament Systems Division, Louisville, Kentucky.
Sippican Inc, Hycor Products Group, Marion, Massachusetts.

VERIFIED

AIRBORNE SIGNALS INTELLIGENCE (SIGINT), ELECTRONIC SUPPORT AND THREAT WARNING SYSTEMS

Introduction

This section covers those passive airborne electronic warfare systems and equipments that are used for SIGINT collection, electronic support and to provide warning of radio frequency, laser and missile threats. Within the SIGINT field, the emphasis is on those systems that are used to collect and process data on communications emitters (known respectively as COMmunications and ELectronic INTelligence — COMINT and ELINT) aboard manned and unmanned air vehicles.

VERIFIED

AUSTRALIA

ALR-2002 Radar Warning Receiver (RWR)

Type
RWR for fixed- and rotary-wing aircraft.

Description
The ALR-2002 is claimed to be the first Australian developed RWR that meets the operational requirements of the Australian armed forces and as such, has been promoted for installation aboard the service's F/RF-111 strike/reconnaissance aircraft, F/A-18 tactical fighters and C-130J-30 transports, together with the Australian Army's CH-47D and S-70A-9 helicopters. The system's modular design is intended to facilitate the development of a family of applications that utilise a common set of hardware and software modules. Accordingly, the five port ALR-2002A has been developed to fit the 'unique antenna configuration' of the Royal Australian Air Force's F/RF-111 fleet, while the ALR-2002B variant is tailored to the service's F/A-18 fighters. For its part, the ALR-2002D is understood to be optimised for installation aboard the Australian Army's S-70A-9 helicopters. BAE Systems Australia notes that the three described variants have 75 per cent commonality in their hard and software modules (representing a 'significant' saving for the Australian Defence Forces in terms of initial development and through life support costs) and that ALR-2002 as a whole is 'readily adaptable' to a range of platforms that includes fast jets, transports and commercial aircraft.

Status
According to Jane's sources, an ALR-2002 concept demonstration programme was 'successfully' completed during 1993 and was followed by the award of a then year A$ 20 million contract covering Full-Scale Engineering Development (FSED) of the ALR-2002A variant during April 1997. This was followed by a then year A$7 million ALR-2002D FSED award that was announced during May 1999. During July 2001, the same sources were reporting that BAE Systems Australia and Tenex Defence Systems had been awarded concurrent, 12 month duration Initial Design Activity (IDA) contracts with respect to Project Air 5416 (otherwise known as Project 'Echidna'). As of the given date, 'Echidna' was designed to create a common Electronic Warfare Self-Protection (EWSP) suite for installation aboard Australian Army CH-47D Chinook and S-70A-9 Black Hawk helicopters and Royal Australian Air Force C-130J-30 Hercules transport and F-111C/G strike aircraft. Forming Phase 1, Stage 2 of the 'Echidna' effort, award of the described IDAs was preceded by Phase 1, Stage 1 activity that is described as centering around the development of a PrePlanned Product Improvement (P³I) programme for the ALR-2002 RWR. If successfully completed (scheduled for the end of 2004), the described ALR-2002 P³I effort was noted as placing the equipment at the heart of the various 'Echidna' applications. As of the given date, it was envisaged that the type specific Black Hawk EWSP system would achieve its initial operating capability (aboard a minimum of six aircraft) by the end of 2005, with that for the F-111 starting installation a year earlier in 2004. As of August 2001, the status of the proposed ALR-2002B variant for the F/A-18 was uncertain.

Contractor
BAE Systems Australia, Elizabeth, South Australia.

UPDATED

CHILE

DM/A-104 Radar Warning Receiver (RWR)

Type
Airborne RWR.

Description
DM/A-104 is a wideband RWR for helicopters and combat aircraft that provides instantaneous detection of threat radar emitters. The system consists of four orthogonal spiral antennas (each connected to wideband crystal video receiving channels) that provide full 360° coverage in the 2 to 18 GHz frequency range, thereby assuring the detection of most search, acquisition and fire-control radars.

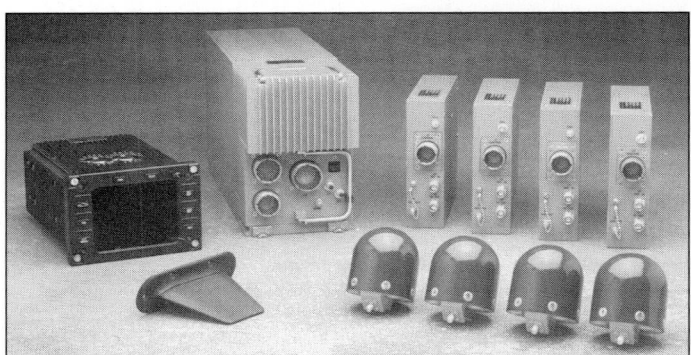

The DM/A-104 radar warning receiver

A powerful digital processor and advanced de-interleaving software ensure a real-time automatic radar sorting and evaluation. A cockpit cathode ray tube display presents to the pilot the most dangerous detected threat radars. In addition, an audio warning system is sent to the aircraft intercom system. The threat itself is classified and the pilot is informed whether his aircraft is being targeted by a Surveillance Radar (SR), an acquisition radar (AQ) or a fire-control radar in lock on mode (LK), or whether the transmission is from a continuous wave radar.

DM/A-104 can be interfaced with the DM/A-202 chaff/flare dispensing system, to provide an automatic self-protection capability, as well as onboard systems for blanking of self-emitted signals. It features a compact and modular design that does not demand a great deal of aircraft space, and can be retrofitted easily on any combat aircraft. It is designed for a high level of reliability and maintainability with a complete built-in test capacity for online diagnosis.

Status
As of this edition, the status of the DM/A-104 RWR was uncertain.

Specifications
Frequency coverage: 2-18 GHz (4 sub-bands); C/D-band (0.7-1.3 GHz)
Detection: pulse, CW, pulse Doppler
Sensitivity: pulse −50 dBm; C/D-band −50 dBm
DF accuracy: better than 10° RMS

Contractor
DTS, Santiago.

VERIFIED

Itata ELectronic INTelligence (ELINT) system

Type
Airborne ELINT system.

Description
The Itata ELINT system is a development by DTS and is a high-sensitivity electronic intelligence gathering system that can detect, locate and measure the parameters of emissions from search, acquisition and fire-control radars. Itata consists of a fully programmable superheterodyne receiver, a digital pulse analyser and a high-gain, wideband, rotating dish antenna which provides 360° coverage and bearing information to within a few degrees accuracy. Although intended primarily for light transport type aircraft, the equipment can also be installed in shipborne or ground vehicle configurations.

The receiver operates over the 30 MHz to 18 GHz frequency range with coverage divided into six system specific sub-bands. It can be used in either a wide open mode over the complete frequency range or in a selective mode over a single band. After detection of a transmission of interest, the receiver automatically locks onto it and measures its frequency and other parameters. Digitised data of each intercepted signal can be recorded automatically for subsequent analysis.

Status
As of this edition, Jane's sources were reporting that the ITATA ELINT system was installed aboard three Beech 99A Petrel Beta aircraft operated by the Chilean Air

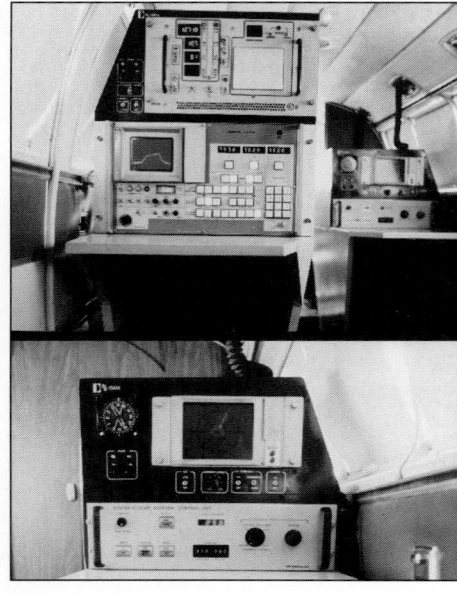

The band selection/ analysis (top) and antenna control/ direction-finding display workstations used in the Itata ELINT system

Force's Escuadrilla de Guerra Electrónica/2nd Grupo de Aviación which (as of the given edition) was believed to be based at Iquique.

Specifications

Frequency range: 0.03-18 GHz in 6 bands
Sweep modes: multiband; total or independent programmable sub-bands
single band; total or independent programmable sub-bands
manual
Sensitivity: −83 dBm
Dynamic range: 70 dB
Pulse-width range: 0.05-24 μs
PRF resolution: 0.25 μs
PRF range: 0.092 to 12.83 kHz
Azimuth coverage: 360°
Polarisation: circular
Azimuth beamwidths: 8° E/F-band; 1.8° J-band

Contractor

DTS Ltd, Santiago.

VERIFIED

CHINA, PEOPLE'S REPUBLIC

BM/KJ 8602 Radar Warning Receiver (RWR)

Type

RWR for combat aircraft.

Description

BM/KJ 8602 is a radar warning receiver designed for tactical and other combat aircraft. It consists of a digital signal analyser, a cathode ray tube display unit, control box, six receivers, four direction-finding antennas and an omni-antenna. It is stated to be a wideband unit capable of dealing with multiple threats from all pulse and continuous wave type radars, with automatic sorting and identification and automatic audio alarm and recording. It can operate with electronic countermeasures units and chaff/flare dispensers for self-protection. Flexible reprogramming is claimed, with a low weight and built-in test facilities.

Status

As of this edition, the BM/KJ 8602 RWR was reported as having been procured by the Chinese Air Force. It is perhaps worth noting that during the mid-1990s, Jane's sources were pointing up the strong physical resemblance between the BM/KJ 8602 RWR and the Israeli SPS-1000 equipment.

Specifications

Frequency coverage: 0.7-1.4 GHz and 2-18 GHz
Azimuth coverage: 360°
Elevation coverage: −30 to +30°
Number of radars: 16 threat radar signals can be sorted, identified and displayed simultaneously
Response time: 1 s
DF accuracy: 15° RMS
Power consumption: <120 VA
Overall weight: 20 kg

Contractor

Southwest China Research Institute of Electronic Equipment, Guanxian, Sichuan.

UPDATED

BM/KZ 8608 ELectronic INTelligence (ELINT) system

Type

Airborne ELINT detection and analysis system.

Description

BM/KZ 8608 is an airborne ELINT system developed by the Southwest China Research Institute. It is designed to detect, identify, analyse and locate land-based or shipborne radar emitters with a high probability of intercept, high sensitivity and accurate measurement of parameters. It consists of five main parts; antennas, superheterodyne receiver, instantaneous frequency measuring receiver, processor and display unit.

Although very few technical details have been released it is claimed to have a wide frequency coverage, long range, the ability to operate in dense electromagnetic environments, automatic signal identification and an emitter fixing capability. No information is available on the size of the processor library.

Status

Over time, the BM/KZ 8608 ELINT system is understood to have been installed on at least one Xian Y-8 (An-12) aircraft of the Chinese Air Force. So equipped, the Y-8 is apparently designated as the EY-8. It is perhaps worth noting that during the mid-1990s, Jane's sources were suggesting that BM/KZ 8608 is a derivative of the Israeli CR-2800 system.

Specifications

Frequency range: 1-18 GHz
Frequency accuracy: 5 MHz
Azimuth coverage: 360°
Bearing accuracy: 5° (1-8 GHz); 3° (8-18 GHz)
Sensitivity: −100 dBW
Dynamic range: 50 dB
Signal density: 200,000 pps
PRF range: 100 Hz-20 kHz
PRF accuracy: ±1% (100 Hz-2 kHz); ±2% (2-20 kHz)
Pulse-width range: 0.1-99.9 μs
Power supply: 28 V DC; 115 V AC (400 Hz)

Contractor

Southwest China Research Institute of Electronic Equipment, Chengdu, Sichuan.

UPDATED

FRANCE

SAMIR Infra-Red (IR) missile detector

Type

Airborne missile detector.

Description

SAMIR (*Système d'Alerte Missile Infra Rouge* - also known as the DDM system) is designed to provide aircraft with automatic detection of ground-to-air and air-to-air missile threats. It locates the missile in flight and feeds processed threat data to the aircraft defensive aids system in real time so that the appropriate countermeasures can be initiated. The equipment makes use of passive IR detection techniques and

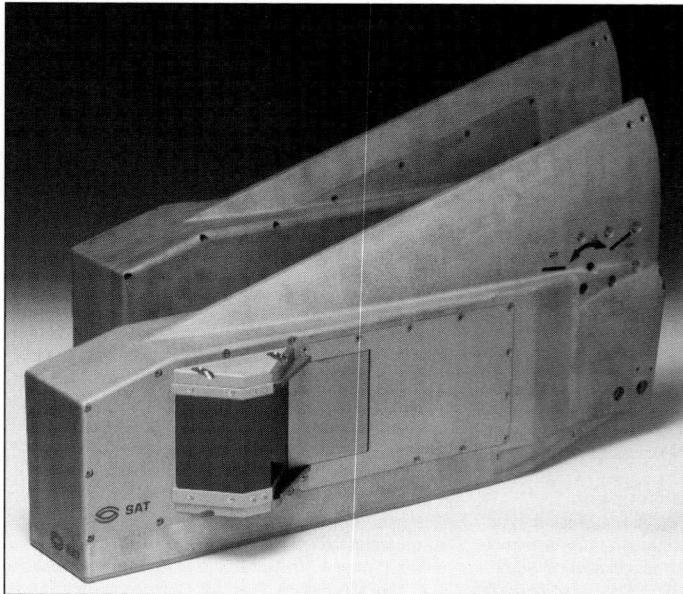

SAMIR 2000 for the Mirage 2000

SAMIR PRIME for the Rafale multirole combat aircraft

is reported to be able to function in severe countermeasures environments and detect 'all' types of missiles (particularly those associated with very short range air defence weapon systems). It features 'advanced' embedded algorithms that classify threats, reject parasitic sources and generate an 'extremely high' probability of detection to 'low' false alarm ratio. Performance is further enhanced by the use of a multispectral mosaic detector array to provide IR signature discrimination. In terms of its construction, SAMIR is made-up of two component modules (an electro-optical head and a signal processing unit) that can be packaged to meet specific platform size and weight requirements. Candidate platforms are listed as including combat, transport, VIP transport and rotary-winged types. Complete angular coverage is achieved via the use of multiple electro-optical heads, with each head providing 180° sector coverage. SAMIR is a Matra British Aerospace Dynamics France (now part of MBDA Missile Systems) - Sagem co-development that, as of August 2001, had been identified as having been produced in two configurations, the known details of which are as follows:

SAMIR 2000
The SAMIR 2000 system variant is designed specially for use on the Dassault Mirage 2000 fighter and features aft-facing, MAGIC air-to-air missile launch rail-mounted electro-optic head/processor assemblies. When applied to the Mirage 2000, SAMIR 2000 is understood to be an 'extension' of the aircraft's SPIRALE countermeasures dispensing system (see separate entry).

SAMIR PRIME
The SAMIR PRIME variant is designed for use on the Dassault Rafale multirole fighter aircraft where it forms part of the type's SPECTRA defensive aids suite (see separate entry).

Status
Full-scale development of the SAMIR/DDM airborne missile detector is understood to have begun during 1985. Flight testing of the system began in 1988, with qualification being achieved during 1990. An initial SAMIR 2000 variant production contract was awarded to the then Matra British Aerospace Dynamics France during October 1994. A second production contract was awarded in 1995 with regard to the SAMIR PRIME model. SAMIR is also understood to have been selected for evaluation by the Japanese Air Self-Defence Force. As of July 2001, SAMIR applications were reported as being in service aboard French Air Force Mirage 2000s and the first production Rafale combat aircraft delivered to the French Navy.

Specifications
Angular coverage: 180°
Locating accuracy: better than 2°
Weight: 3.6 kg (signal processor); 5.6 kg (electro-optic head)
Interface bus: MIL-STD-1553B, Digibus, RS-422, multitrack capability

Contractors
MBDA Missile Systems, Vélizy-Villacoublay.
SAGEM SA, Defence & Security Division, Paris.

VERIFIED

GERMANY

MILDS® AN/AAR-60 missile detection system

Type
Passive missile warning system.

Description
The Missile Launch Detection System (MILDS®) AAR-60 is a passive imaging sensor device that is optimised to detect the solar blind spectral band radiation

The sensor head line-replaceable unit used in the MILDS® AAR-60 passive missile warning system
0052549

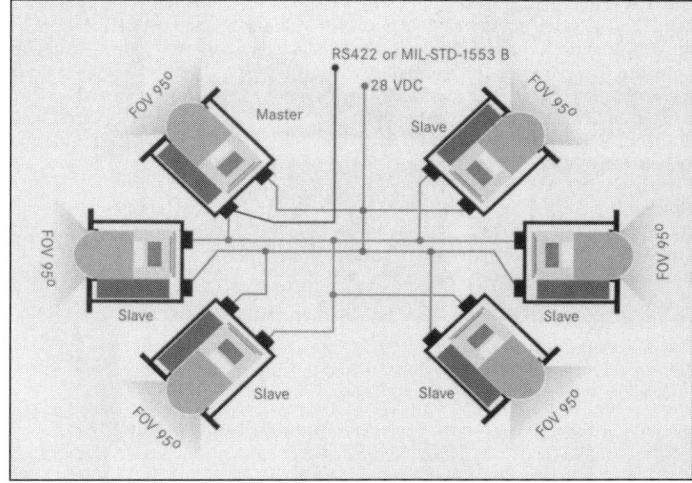

A schematic showing a MILDS® AAR-60 system configured with six sensor heads
0017839

signature that is emitted from an approaching hostile missile exhaust plume. As such, the system is designed to detect threat missiles with the maximum of warning time. The principle components of the MILDS® AAR-60 are an array of one to six identical sensor heads that incorporate entrance optics, filters and imaging processors.

MILDS® AAR-60's 'unique' imaging sensor/sensors allow it to track Ultra-Violet (UV) sources, classify them, determine their angle of attack and prioritise them. Thereafter, the system can initiate threat appropriate countermeasures and warn the host platform's pilot so that he can begin evasive manoeuvres. A combination of high spatial resolution and 'advanced' temporal processing is claimed to increase MILDS ® AAR-60's probability of detection while, at the same time, 'virtually eliminating' false alarms.

The system's sensor heads are self-contained line-replaceable units that require no additional hardware. In a multihead array, one sensor unit acts as a master with the remainder being slaved to the head. Only the master head communicates with the host platform's electronic warfare system computer and countermeasures dispensing system. This being a purely software function, the hardware used in the master and slave heads is identical. Each MILDS® AAR-60 sensor head generates a fully processed signal and the system as a whole can be installed as a stand-alone package (initiating a CounterMeasures Dispensing System (CMDS) and driving its own display) or be fully integrated into an existing threat warning system or a directed infra-red countermeasures unit (reporting through existing displays and communications paths).

Status
As of early 2001, MILDS® AAR-60 was reported as having been in serial production, to have been test flown more than 200 times and to have been qualified for installation aboard the multinational NH90 naval/transport and Franco-German Tiger battlefield attack helicopters. On the NH90, MILDS®AAR-60 is teamed with Thales Airborne Systems' radar and laser Threat Warning Equipment (TWE - see separate entry) and a Matra BAe Dynamics CMDS. As of June 2000, Germany was understood to have a requirement for 214 MILDS® AAR-60 systems with which to equip its tactical transport NH90 and Tiger helicopters.

Specifications
Field of view (Elevation × Azimuth): 95 × 360° (4 sensor heads); 180 × 360° (6 sensor heads)
IFOV: <1°
AOA: 2.5° (ref, mounting surface)
Threat analysis capability: 8 simultaneous threats
Power consumption: <15 W (LRU)
Interfaces: RS-422; optional MIL-STD-1553B; discrete lines
MTBF: >9,600 h (LRU)
Weight: <2 kg (sensor)
Dimensions (l × w × h): 108 × 107 × 120 mm (sensor head excl mounting flange and connectors)

Contractor
LFK-Lenkflugkörpersysteme GmbH, Munich.

UPDATED

INTERNATIONAL

AN/ALR-68A(V)3 radar warning system

Type
Airborne radar warning system.

Description
Northrop Grumman's (formerly Litton's) AN/ALR-68A(V)3 equipment is a digital radio-frequency threat warning receiver that has been developed specifically for use by Germany's Luftwaffe (Air Force). As such, it is described as being a wide-open, crystal video equipment that is field-programmable and provides in-cockpit threat parameter programming and data hand-off to other onboard countermeasures systems. ALR-68A(V)3 is equipped with a digital threat processor that makes use of software that is unique to the German threat scenario and is manufactured 'in association' with the former German contractor DaimlerChrysler Aerospace (now a part of the European Aeronautics, Defence and Space (EADS) Co).

Status
As of early 2001, ALR-68A(V)3 was being reported as being in service with the Luftwaffe and as being installed aboard the service's C.160 transport, F-4F interceptor and RF-4F reconnaissance aircraft.

Contractors
EADS Systems and Defence Electronics, Ulm, Germany.
Northrop Grumman Electronic Sensors and Systems Sector - Defensive Systems
  Division, Rolling Meadows, Illinois, USA.

UPDATED

ASTAC ELectronic INTelligence (ELINT) system

Type
Airborne tactical ELINT system.

Description
The *Analyseur de Signaux TACtiques (ASTAC)* electronic reconnaissance system consists of an internally or pod-mounted airborne sensor package and an associated ground processing station. It is intended to perform detection, identification and localisation of any radar type in a very dense environment. A datalink between the pod application and the ground station enables a very rapid build-up of the electronic order of battle of the observed area.

The main characteristics of the system are a very wide frequency coverage, wide instantaneous bandwidth, high sensitivity, high discriminating power and high direction measurement accuracy by interferometer. The system is fully automatic, fully reprogrammable and possesses a very high-speed processing capability of up to 20 radars/s. It can process pulse modulated radar with pulse repetition internal diversity or agility, radio frequency diversity or agility, as well as pulse compression, Continuous Wave (CW) and interrupted CW systems.

ASTAC uses two wideband compressive surface acoustic wave receivers. One receiver is used to obtain a very precise measurement of the radar frequency and the two together can handle frequency-agile emitters. The system uses interferometer phase-measuring antenna arrays to determine the azimuth of any threat emitter operating within its 0.5 to 18 GHz (with 18 to 40 GHz as an option) frequency range. When packaged as a pod, ASTAC can store acquired data in an onboard recording subsystem as well as transmit it to its associated ground station using an Ultra High Frequency (UHF - 300 MHz to 1 GHz) datalink. When installed on a two-seat aircraft, the ASTAC system can be configured to display to the back seater in tabular and liquid crystal formats. In such an installation, a keyboard is provided for the operator to interface with the equipment.

Status
As of October 2001, the ASTAC ELINT system was reported as having been installed aboard French Air Force C-160 Gabriel (internal installation), DC-8 SARIGUE NG (internal) and Mirage F1-CR tactical reconnaissance (pod) aircraft together with Japanese Air Self-Defence Force RF-4EJ tactical reconnaissance platforms (pod). Elsewhere, Jane's sources report the acquisition of podded ASTAC applications by the Hellenic and Taiwanese Air Forces. Here, the Hellenic Air Force is reported to have placed an order for up to five ASTAC pods during the first half of 1999 and to have been testing the system's compatibility with the RF-4E reconnaissance aircraft during the summer of 2000. For its part, Taiwan is understood to have procured an undisclosed number of podded ASTAC systems for carriage by its Mirage 2000 aircraft (part of the 'Flying Dragon' programme) during the spring of 2001. This deal is further reported as including at least one ASTAC ground station. Looking at the Japanese application, the specific ASTAC application is understood to be known locally as the TACtical Electronic Reconnaissance (TACER) system, with Mitsubishi acting as the national prime contractor on the effort. The French Air Force is known to have used ASTAC operationally over Bosnia and Thales notes that the system has been demonstrated aboard NATO F-16 aircraft. In this context, Jane's sources have reported that a demonstration that used an F-16B as host aircraft and was flown in support of the Greek procurement showed podded ASTAC to be able to detect, identify and track (via their radars) nine frigate-sized surface combatants in a naval

An ASTAC ELINT pod mounted on a French Air Force Mirage F1-CR reconnaissance aircraft   1996

exercise area measuring 56 × 56 km from a range of 170 km. The first ASTAC trial aboard an F-16 is noted as having taken place during 1993. Outside the various cited programmes, a podded ASTAC application is also known to have been proposed to Germany for use aboard Tornado reconnaissance aircraft.

Specifications
Podded configuration
Frequency coverage: 0.5-18 GHz (18-40 GHz option)
Location accuracy: <1% of the distance between the ASTAC aircraft and the target emitter (aircraft flying at a speed and altitude of M0.9 and 12,192 m respectively)
Direction-finding accuracy: 0.5-1°
Instantaneous area coverage: 164 km² (aircraft flying at a speed and altitude of M0.9 and 12,192 m respectively)
Emitter location time: 2-3 min (111 km range, aircraft flying at a speed and altitude of M0.9 and 12,192 m respectively)
Antenna coverage: ±20° (elevation); 120° (azimuth, lateral antenna, 0.5-4 GHz sub-band); 240° (azimuth, forward antenna, 2/4-18 GHz sub-band)
Parameter measurement

	Range	Resolution	Accuracy
Frequency	0.5-18 GHz	0.125 MHz	0.2 MHz RMS
	18-40 GHz	0.125 MHz	0.4 MHz RMS
PRI	3-32 µs	1 µs	±1.5 µs
	32 µs-16 ms	0.125 µs	±0.25 µs
			<10 ns (technical analysis mode)
Pulse-width	0.1 µs-2 ms	62.5 ns	60 ns + 5% RMS
Pulse amplitude	50 dB	0.33 dB	
Spectrum width	0,125-10 MHz	125 kHz	200 MHz
Antenna rotation period	1-20 s	0.1 s	0.1 s RMS
TOA	0-120 s	0.125 µs	

Power supply: 2 kVA
Weight: 400 kg
Dimensions (Ø × l): 0.4 × 4.1 m

Contractor
Thales Airborne Systems, Elancourt, France.

UPDATED

Carapace threat warning system

Type
Radar threat warning system.

The elements making up the Carapace threat warning system as applied to the two-seat F-16B   1996

Description

Carapace is a version of the passive detection element of the EWS-16 system for the F-16 aircraft. The main part of the system is a hybrid receiver system that includes instantaneous frequency measuring, crystal video and superheterodyne receivers plus an interferometric direction-finding array. The system probably covers the C to K frequency bands (0.5 to 40 GHz) and is designed to detect, identify and localise all modern threats with great accuracy. It is able to analyse, on a pulse-to-pulse basis, local emitters accurately at long range in severe electronic countermeasures environments and offers an electronic support measures capability. Accuracy of direction-finding is classified but is probably of the order of 1°.

Status

As of October 2001, the Carapace threat warning system was reported as being in service aboard F-16 aircraft of the Belgian Air Force. Thales Airborne Systems notes that Carapace-equipped F-16s of the Belgian Air Force flew combat missions during 1999 as part of Operation 'Allied Force'.

Contractor

Thales Airborne Systems, Elancourt, France.

VERIFIED

DR 3000A Electronic Support (ES) system

Type

Airborne version of the DR 3000 ES system series.

Description

The DR 3000A ES system is the airborne variant of the tri-service DR 3000 ES system series and is described as being suitable for maritime patrol applications. The equipment is described as comprising six Direction-Finding (DF) and intercept antenna assemblies, a processor and a display unit. DR 3000A is claimed to offer a 'very high' probability of detection factor over the complete 360° arc in azimuth with 'high' sensitivity across the C- through J-band (0.5 to 20 GHz) frequency band. Emitter identification is based on 'efficient' de-interleaving ('even in very dense electromagnetic environments'), 'accurate' parameter measurement and artificial intelligence techniques. The system is designed to provide warning, surveillance, ELectronic INTelligence (ELINT) and targeting data and can have its baseline ELINT and DF performance enhanced via a range of add-on modules. DR 3000A's weight (including processor, display and antennas) is given as less than 80 kg.

Status

As of October 2001, Jane's sources were reporting the DR 3000A ES system as having been procured for installation aboard Atlantique (Pakistan), Fokker F.27 maritime patrol aircraft and Embraer EMB 145 airborne early warning and control platforms (Greece - two of four aircraft delivered during late September 2001).

Contractor

Thales Airborne Systems, Elancourt, France.

UPDATED

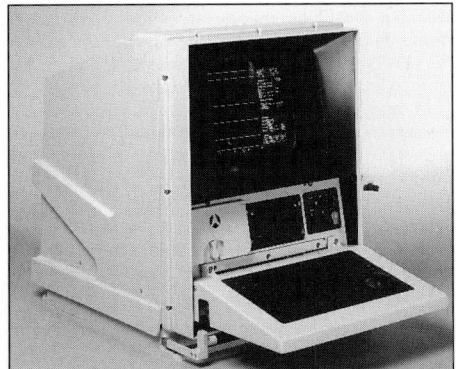

The display console used in the DR 3000A ES system

Enhanced Radar Warning Equipment (ERWE) II

Type

C- through J-band (0.5 to 20 GHz) airborne threat warning system.

Description

The 0.5 to 20 GHz ERWE II threat warning system is produced jointly by the European Aeronautic, Defence and Space (EADS) Co Systems and Defence Electronics and Northrop Grumman Electronic Sensors and Systems Sector - Defensive Systems Division and is designed to detect and analyse threat radar signals, provide aural and visual threat warnings and hand off data to other onboard countermeasures equipments. System features include:

- a multireceiver and processor architecture with a wide, instantaneous bandwidth

- a 'high sensitivity', narrowband receiver
- a dedicated C/D-band (0.5 to 2 GHz) receiver
- two MIL-STD databus interfaces
- discrete interfaces for look-through management
- embedded maintenance facilities
- a reprogrammable threat library
- an optional data recording facility
- a 'terminal threat' operating mode that makes use of 'smart' analysis algorithms for threat identification and classification
- a 'bypass' operating mode that provides a high probability of intercept capability against scanning emitters, real-time signal indications and data recording.

Status

As of July 2001, the ERWE II threat warning system is understood to have been installed aboard Luftwaffe and Marineflieger Tornado interdiction and strike aircraft.

Contractors

EADS Systems and Defence Electronics, Ulm, Germany.
Northrop Grumman Electronic Sensors and Systems - Defensive Systems Division, Rolling Meadows, Illinois, USA.

UPDATED

Kestrel Electronic Support (ES)/ELectronic INTelligence (ELINT) system

Type

Airborne ES/ELINT system.

Description

The Kestrel ES/ELINT system (Royal Navy designation Orange Reaper) receives and processes pulse and continuous wave radar emissions within (in baseline form) the 0.5 to 20 GHz frequency band. Instantaneous bearing and frequency measurement are provided over the full 360° arc in azimuth. Functionally, Kestrel employs six antenna modules that are mounted around the host platform's airframe or in wingtip pods. Each module contains 'high-sensitivity' receivers and derived radio frequency and video outputs are fed into a Parameter Measurement Unit (PMU) for processing. The PMU incorporates superheterodyne receivers, undertakes digital instantaneous frequency measurement, establishes the time of arrival, direction of arrival, frequency, pulse repetition frequency, pulse-width and amplitude of received signals and digitises the processed data for onward transmission to the system's Data Processing Unit (DPU). For its part, the DPU performs 'high pulse density' de-interleaving, analysis of 'complex' pulse repetition intervals, scan types and frequency agilities and emitter identification. It is programmed before flight with a mission data package that includes an emitter library for automatic radar recognition. Emitter identification is achieved via library comparison using a software library with a capacity of 'at least' 2,000 radar modes. Kestrel also incorporates a radar warning function that is capable of displaying threats within 1 second of receipt and interfaces with an onboard defensive aids suite (active radar jamming and chaff/infra-red decoy flare dispensing subsystems) if available. Intelligence gathered during a mission can be recorded for post-flight analysis if required.

Kestrel offers a number of tactical and ELINT display options (including fine grain measurement of 'agile and complex radar signatures') that are dependent on the type of aircraft in which the system is installed and the role it is being used for. As an alternative, analysed data can be transmitted to the host platform's mission computer or an aircraft recorder by means of a MILitary STanDard (MIL-STD) -1553B databus. Other system features include:

- an over-the-horizon detection capability
- an in-flight reprogrammable auxiliary emitter mode library
- claimed 'near' 100 per cent probability of intercept against continuous wave and pulsed signals in 'very dense' environments
- continuous and operator initiated built-in test (no first line test equipment required)
- the ability to be integrated into an active/passive airborne early warning architecture.

While not confirmed by its manufacturer, Jane's sources suggest that the specific Orange Reaper application has a frequency coverage of 0.6 to 18 GHz.

Status

As of May 2001, Kestrel is reported as having been supplied to Denmark (for use on Lynx helicopters) and in its Orange Reaper form, to the Royal Navy (RN) for installation aboard 44 Merlin HM Mk 1 anti-submarine warfare helicopters. As of the given date, Thales Sensors were noting that Orange Reaper (together with other defensive aids systems) had been integrated with (but not fitted to) the RN's Sea King AEW Mk 7 helicopters.

Specifications

Baseline Kestrel configuration
Frequency coverage: 0.5-20 GHz (20-40 GHz option)
Receiver types: log video amplifier; digital log video amplifier; instantaneous frequency measuring; continuous wave superheterodyne
Coverage: 45° (elevation); 360° (azimuth)
Polarisation: all linear and one hand of circular

The elements making up the Kestrel ES system 0009390

Direction-finding accuracy: better than 5° RMS

Warning/identification: all pulse types including continuous wave, interrupted continuous wave, pulse-Doppler, 3-D low probability of intercept, jitter/stagger/agile, pulse compression; frequency-agile radars and unknown emitters

Emitter library storage: at least 2,000 modes (main library)

Display: high brightness, colour cathode ray tube with computer controlled, software programmable symbology

Emitter displays: up to 400 pages of 20 emitters per page, each showing tote format parametric data and emitter identity

Range bearing displays: 25 highest priority emitters showing lethality, type and track identification

Expanded track information: showing full parametric data and ranges for a single track

BITE: continuous monitoring of key system parameters; provision of 'on request' fault isolation for more than 95 per cent of the system; provision of DOA quality check

Additional features: true or relative bearing display presentation; processor controlled audio alarms; operator initiated emitter hand-off; operator controlled 'specialised' computer modes; an ELINT data recording interface

Weight: 55 kg (display configuration dependent)

Power consumption: 1 kVA

Interfaces: serial and parallel interfaces for jammers, chaff/flare systems, displays, telemetry links and recorders

MTBF: over 800 h (demonstrated to MIL-STD-781C)

Environment: MIL-E-5400 Class II

Contractor

Thales Sensors, Crawley, UK.

UPDATED

MIR-2 Electronic Support (ES) system

Type

ES system for helicopters, fixed-wing aircraft and small naval vessels.

Description

The MIR-2 ES system (Royal Navy designation Orange Crop) covers the C- through J-band (0.5 to 18 GHz sub-band) frequency range and features a digital receiver around which a fully solid-state wideband system has been built. It incorporates

A close up of one of the forward hemisphere MIR-2 antenna assemblies on a Royal Navy Sea King helicopter (Martin Streetly)

lightweight antennas and a cockpit display which includes a very compact solid-state light emitting diode presentation of signal intercept data. Six antenna packages (which can be flush or externally mounted) are fitted on the aircraft with their boresights at 60° intervals in azimuth. Each package consists of two cavity-backed spiral antennas and a radio frequency unit.

The pulse receiver processes the received signals and presents the information on the control indicator unit. Signals are displayed to the operator by frequency band, bearing and amplitude on the solid-state activity display. The total frequency coverage is divided into four frequency ranges with three amplitude levels indicated in each band at each bearing. The bearing scale is divided into 10° steps over a full 360°. Strobes are provided to select signals by frequency band and bearing for pulse analysis and for accurate bearing measurement. For pulse analysis, signals are selected in 10° steps and pulse repetition frequency is displayed on a digital indicator.

The Royal Navy's Orange Crop application has been updated (the MA and PA programme) to improve its continuous wave detection capability and visual and aural threat warning of illumination by pulse radars. Both enhancements are effective across the system's full frequency range and the equipment fitted to UK Lynx helicopters is, as of May 2001, scheduled to be integrated into the aircraft's central tactical system.

Status

Over time, MIR-2 is reported as having been procured by the Brazilian (Lynx helicopters) and UK (Lynx and Sea King helicopters) navies. The system has also been fitted to Royal Air Force (RAF) Hercules C Mk 1 transport aircraft serial numbers XV201, XV203, XV204, XV206 and XV213 under the designation Orange Blossom. As of early 2000, photographic evidence suggests that aircraft XV206 had retained its Orange Blossom capability and was in service as a special forces support platform. Usually reliable sources suggest that XV206 was still in service as of January 2001. As of May 2001, a total of 264 MIR-2 systems are reported to have been supplied to customers around the world.

Specifications

Frequency range: 0.5-18 GHz
Azimuth cover: 360°
Pulse-width: 0.15-10 μs
PRF range: 0.1-10 kHz
System weight: 47.7 kg

Contractor

Thales Sensors, Crawley, UK.

UPDATED

MWS-20 missile warning system

Type

Active missile approach warner for helicopter (MWS-20 H) and transport aircraft (MWS-20 TA) applications.

Description

MWS-20 is an active pulse-Doppler missile approach warner that comprises a transceiver/processor unit, four antennas and a cockpit display/control box. It is designed for the protection of helicopters and fixed-wing transport aircraft and performs direction of arrival, time to impact and missile range, speed and bearing calculations on passive, active and semi-active surface-to-air, air-to-air, anti-shipping and anti-radiation missiles that use radar, infra-red, laser, fibre optic or wire guidance. The equipment makes use of active technology in order to minimise false alarm rates while providing an all weather capability and insensitivity to decoys. MWS-20 can also exercise automatic or semi-automatic control over a decoy dispenser subsystem. The system incorporates built-in test and makes use of miniaturisation to reduce its weight and volume. The use of application specific integrated circuitry and solid-state transmitter technology is claimed to enhance the systems reliability.

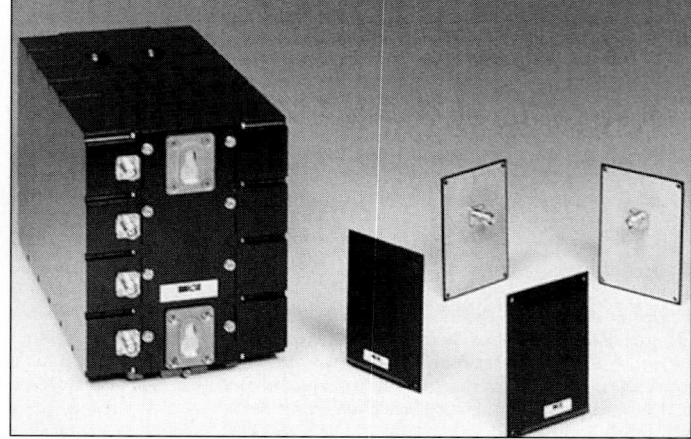

The MWS-20 missile warning system 0009385

Status

As of April 2001, the MWS-20 missile warning system was reported as being in production and service. Platforms equipped with or mandated to be equipped with the system include helicopters operated on behalf of France's special forces, French Army's Horizon radar surveillance helicopter and French Air Force's Cougar combat search and rescue helicopter and C-130H/H-30 (eight aircraft) and C-160 (12) transport aircraft.

Specifications

MWS-20H
Angular coverage: 360°
Power consumption: 600 V A
Weight: 10 kg

Contractor

Thales Airborne Systems, Elancourt, France.

UPDATED

Naval Aviation COMmunications INTelligence (COMINT) system

Type

High to Ultra High Frequency (HF to UHF) band COMINT system for maritime patrol aircraft.

Description

The HF to UHF band Naval Aviation COMINT System (NACS) is primarily designed for use aboard maritime patrol aircraft and provides real-time technical analysis and decoding of intercepted voice and data communications traffic. NACS subsystems are computer controlled to facilitate overall system control, monitoring report management and data display. A mission planning capability allows NACS to be preprogrammed to undertake specific collection cycle scenarios and the system features both in-flight and ground-based data processing (with the former being executed in real time). NACS's open architecture allows for the incorporation of enhancements such as extended frequency coverage, technical analysis channels, monitoring facilities and a direction-finding capability. Alongside its primary usage, NACS can also be configured for surface ship and submarine applications.

Status

As of October 2001, the status of the NACS system was uncertain.

Contractor

Thales Communications, RadioSurveillance and COMINT Systems Unit, Gennevilliers, France.

VERIFIED

Phalanger Electronic Support (ES)/ELectronic INTelligence (ELINT) system

Type

Airborne ES/ELINT payload.

Description

Phalanger is a new-generation ES/ELINT payload for airborne platforms such as unmanned aerial vehicles, helicopters or light multipurpose aircraft. Based on phase interferometry and digital receiver techniques, it is designed to offer high performance combined with 'minimal' weight, volume and power consumption. The use of interferometry allows for 'accurate' direction-finding while that of a digital receiver is used to facilitate long-range signal detection, 'superior' parameter acquisition and 'very short' acquisition times. Radar tracks can be preprocessed and Phalanger can either deliver track data for real-time display and analysis or record information for post-mission download. An associated battlefield tactical workstation can take the form of a stand-alone laptop computer or a

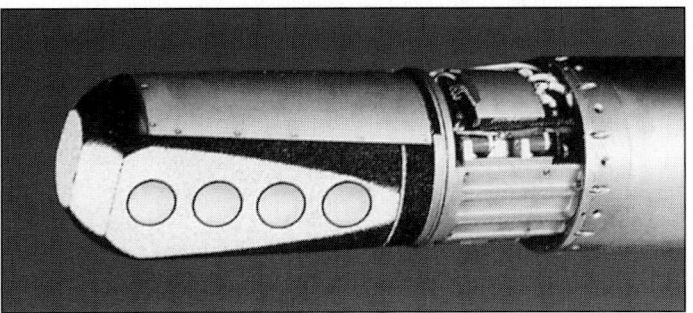

A close-up of the antenna head used in the Phalanger ES/ELINT system
0009392

'compatible ground workstation'. Data display formats include emitter maps and analysis histograms.

Status

As of April 2001, the Phalanger ES/ELINT system was reported as having been validated by the French Ministry of Defence in ground and flight trials.

Specifications

Frequency range: 2-18 GHz (standard); 0.5-40 GHz (option)
FOV: 360°
Accuracy: 1° class (bearing); 2 MHz class (frequency)
Power consumption: <300 W
Volume: 20 dm³
Weight: <20 kg (complete system)
Dimensions (Ø × l): 0.20 × 0.90 m

Contractor

Thales Airborne Systems, Elancourt, France.

VERIFIED

Serval Radar Warning Receiver (RWR)

Type

Airborne RWR.

Description

The Serval RWR is a channelised crystal video detector that provides the pilot with warning when his aircraft is illuminated by surface or airborne threat radars. Frequency coverage is E- through J-band (2 to 20 GHz). Displayed information comprises threat direction, level and identification. The latter is obtained by comparison of the received signal with a threat library contained in the system. The display unit is a panel-mounted cathode ray tube where several threats can be displayed simultaneously. At the same time, an audio alarm is generated in the pilot's headset. Four antennas are used, located at the aircraft wingtips and vertical fin. They are connected to an analogue and digital processing unit. Thales Airborne Systems is also understood to have developed a second-generation Serval RWR under the designation *Serval Nouvelle Génération-Distance* (New Generation-Distance - NG-D). Serval NG-D adds range data to the existing bearing, amplitude and identification capabilities.

Status

As of April 2001, Serval was reported as being the standard RWR installed on the Mirage 2000 multirole combat aircraft. With regard to Serval NG-D, Thales Airborne Systems notes that it has completed development of the necessary algorithms and has tested these in what it terms 'complex and demanding' threat scenarios. As of the given date, the Serval RWR was noted as being no longer in production.

Contractor

Thales Airborne Systems, Elancourt, France.

VERIFIED

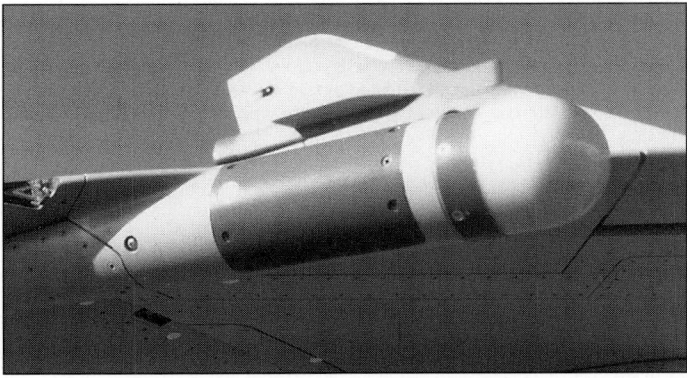

A close-up of one of the wingtip antennas used in the Serval RWR system
0009380

Sherloc Radar Warning Receiver (RWR)

Type

RWR for aircraft and helicopters.

Description

Sherloc is an airborne crystal video RWR that is intended for fighter/attack aircraft and helicopters. It is designed for instantaneous detection and identification of radar emitters in complex and dense signal environments. Sherloc is also designed to integrate data from laser and missile warning receivers and to control a decoy

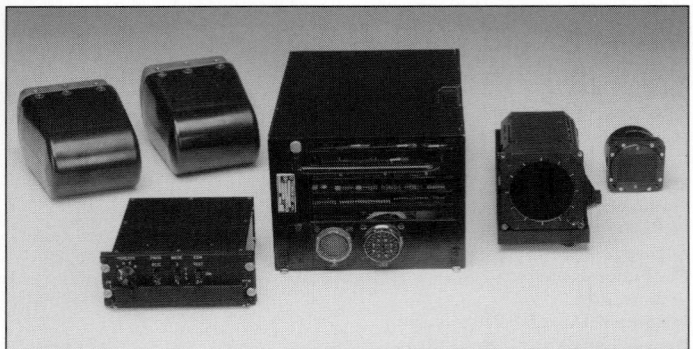

The Sherloc RWR combat aircraft configuration. The display on the extreme right is a smaller unit for helicopters

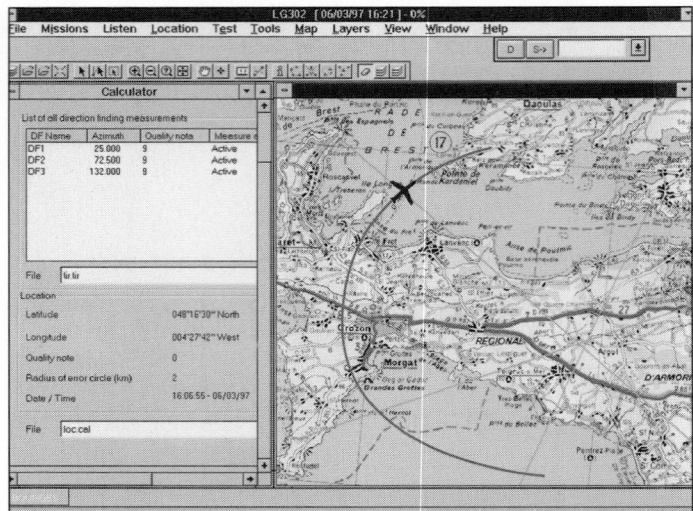

The SAS MMI display showing a detected transmitter location   0052548

dispenser. This provides the crew with comprehensive information regarding the overall threat environment of current and future missions.

The basic system covers D- through J-bands (1 to 20 GHz) with extensions available at both ends of the range to cover the C- through K-band (0.5 to 40 GHz) frequency range. A high-speed analogue and digital processor is incorporated with a threat library of several hundred modes that can be readily reprogrammed on the flight line. The library can be further extended.

The system is claimed as having 'very high' sensitivity, is of small size and weighs only 13 kg. It is capable of detecting all types of pulse radar (from low to high pulse repetition frequency) as well as continuous wave illuminators. Sherloc can be fitted as a new installation or can be retrofitted to nearly any type of aircraft. Radar data are presented to the pilot in the form of alphanumeric symbology on a liquid crystal display. These symbols correspond to the various signal identities and their actual position on the display indicates signal bearing and approximate range. An alternative small size light emitting diode display is available. In both cases eight threat emitters can be presented simultaneously to the pilot. Sherloc also provides an electronic support measures function for unknown threats. An aural warning is also incorporated and an optional synthetic voice is available. This latter function is particularly useful on helicopters where the pilot is not able to look at the display for long periods. The latest known Sherloc variant (which incorporates an instantaneous frequency measurement capability) is designated as Sherloc F.

Status
As of April 2001, the Sherloc RWR was reported as being in production and service. Outside France, the system is noted as having been procured by four air forces for installation aboard Mirage 50 and F.1 fast jets and Dauphin and Super Puma helicopters. In French service, the baseline Sherloc system is reported as having been fitted to French Air Force C-135F, C-160 (see following) and DC-8 aircraft. The latest version of the equipment (designated as Sherloc-F) is (or can be) installed aboard the French Navy's Super Etendard strike fighter, the French Air Force's C-130 and C-160 transport aircraft and Cougar helicopter and the Eurocopter Panther and Super-Puma helicopters. In the C-130/C-160 Sherloc-F context, Jane's sources suggest that the equipment forms part of the electronic warfare suite that the French Air Force ordered in early 2000 for installation aboard eight C-130H/H-30 and 12 C-160 transport aircraft. Here, the suite was designated as the *Système d'AutoProtection (SAP)* and teamed the Sherloc RWR with an MWS-20 missile warning system and an MBDA Missile Systems' Alkan chaff and infra-red decoy flare dispensing subsystem. SAP installation aboard the C-130 aircraft is noted as having been undertaken by French contractor Sogerma with that for the C-160 aircraft being carried out by Atelier Industriel de l'Aeronautique at Clemont-Ferrand. Readers are cautioned that over time, the use of Sherloc in the SAP system may have been superseded by that of Thales' TDS-TA equipment and the SAP suite may have been re-badged as the company's SPS-TA architecture.

Specifications
Frequency coverage: D/J-bands (1-20 GHz), optional extension to C/K-band (0.5-40 GHz)
Receiver: crystal video and IFM
DF accuracy: better than 10°
Total weight: 10 kg (depending on display system used)

Contractor
Thales Airborne Systems, Elancourt, France.

UPDATED

Spectrum Airborne Surveillance (SAS) system

Type
Airborne 300 kHz to 3,000 MHz COMmunications INTelligence (COMINT) system.

Description
The 300 kHz to 3,000 MHz SAS COMINT system is designed for the real-time interception, location, monitoring, analysis and recording of 'all types' of voice communications and digital data emitters. It can be installed in a variety of airframes (ranging from small, single engine types to business jets) and can be configured with one or more operator workstations as required. Offering a 'wide variety' of demodulation and decoding routines, the SAS system makes use of a direction-finding antenna package that comprises either an array of sabre antennas or a circular arrangement of elements mounted in a radome. Other system features include:

- an open architecture, digital technology and Fast Fourier Transform processing
- the ability to integrate and merge data from optical and infra-red sensors with its own
- data recording for post-mission debriefing and analysis
- a high scan rate to facilitate the processing of mobile and cellular digital radio systems
- a user-friendly Man/Machine Interface (MMI) running Windows NT
- system tasking according to mission type.

Status
As of May 2001, the SAS system is reported as having been installed aboard at least one F406 Vigilant aircraft and was thought to be available.

Specifications
Frequency coverage: 300 kHz-3,000 MHz (monitoring); 20 MHz-3,000 MHz (direction-finding)
DF accuracy: <1.5° RMS
Scan rate: 40-2,000 MHz
Demodulation: A1A, A1B, A2A, A2B, A3E, B8E, F1A, F1B, F1C, F3E, F7B, H3E, J2A, J2B, J3E and R3E
Sensitivity: 3 µV/m (typical operation, 10 MW at 180 km)
Power supply: 115/230 V AC
Power consumption: 300 W
Dimensions (h × l × w): 134 × 485 × 520 mm (48 cm 3U standard rack and personal computer)
Weight: 45 kg

Contractor
Thales Communications, RadioSurveillance and COMINT Systems Unit, Gennevilliers, France.

VERIFIED

Syrel ELectronic INTelligence (ELINT) system

Type
Pod-mounted airborne ELINT system.

Description
Syrel is a fully automatic electronic reconnaissance pod that is pylon mounted under the fuselage of the fighter. It is designed primarily for tactical penetration missions at medium or low altitudes. It acquires and automatically records data relating to the identification and location of ground-based electronic systems. It is intended to provide reliable information of early warning, search and acquisition,

The Syrel ELINT pod mounted on Mirage F.1

ground controlled interception and fire-control radars associated with anti-aircraft missiles or artillery.

In normal operations, the aircraft will carry out direction-finding and position fixing of transmitters by flying along a path and taking a series of bearings to establish the position of the emitter. The pod is 3.57 m long × 0.42 m diameter and weighs 265 kg. It has antennas at both front and rear, with receiver units, amplifier and recorders in the centre section. A downlink enables real-time information to be fed to a ground station. The pylon houses a cooling system that has a ram-air intake in the leading edge. The system's high processing speed is assisted by thick and thin film microwave circuit assemblies on ceramic substrates. Thales Airborne Systems also produces an associated ground station.

Status
Over time, the Syrel ELINT system is reported as having been procured by the Spanish and 'other' air forces.

Contractor
Thales Airborne Systems, Elancourt, France.

VERIFIED

TDS-FA radar warning system

Type
Radar warning and electronic warfare system management equipment.

Description
The Threat Detection System for Fighter Aircraft (TDS-FA) is a compact (10 kg), user reprogrammable radar warning receiver that is designed for the upgrading of existing self-protection suites. The equipment is described as being lightweight and as being able to detect continuous wave, pulsed and pulse-Doppler emitters operating in dense environments. Thales Airborne Systems identifies TDS-FA's key features as being a 'very short' reaction time and 'highly dependable' emitter identification that is based on the use of instantaneous frequency measurement technology and a claimed 100 per cent probability of intercept. Alongside its warning function, TDS-FA can exercise control over onboard jammers and/or countermeasures dispensing systems and can be integrated with its host platform's weapon system via serial link and/or multiplex bus interfaces.

Status
As of April 2001, the TDS-FA radar warning system was reported as being in full-scale production for installation aboard 110 Mirage F.1 aircraft of the French Air Force (the Aigle (Eagle) programme) where it is teamed with the PAJ-FA jamming pod (which is itself the subject of a mid-life upgrade programme). Alongside the Mirage F.1 application, three other variants of the TDS-FA system have been developed that are (respectively) compatible with the Mirage III/V/50, the F-5 and the MiG-21 types. As of the given date, the various TDS-FA configurations were noted as having been validated in flight tests.

Contractor
Thales Airborne Systems, Elancourt, France.

VERIFIED

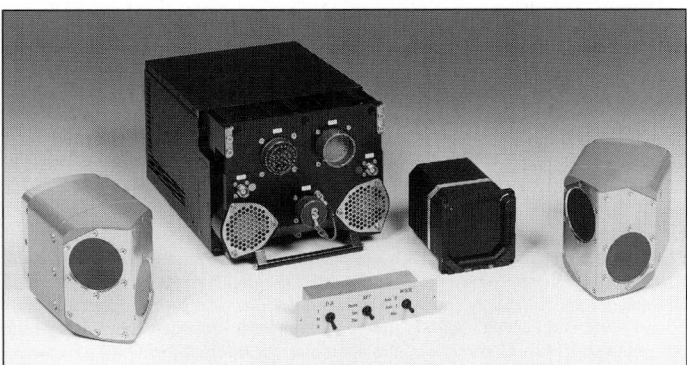

The components making up the TDS-FA configuration that is being fitted to French Air Force Mirage F.1 aircraft   1996

TDS-H Radar Warning Receiver (RWR)

Type
RWR for helicopter applications.

Description
The TDS-H RWR is a user programmable, database oriented system that is described as having a 'very wide' frequency range and as making use of full-band, instantaneous frequency measuring technology. The system is able to detect continuous wave, pulse and pulse-Doppler emitters operating in 'very dense' electromagnetic environments. Alongside its warning function, TDS-H is noted as offering 'smart' management of an onboard countermeasures dispensing system and as being able to interface with a missile warning system to create a self-protection suite.

Status
As of April 2001, the TDS-H RWR was reported as being in production and service. Identified applications the whole of the French Army's helicopter fleet (including the Gazelle, Puma and Super Puma types) together with the Polish Sokol multirole helicopter. As of the cited date, the French Army applications were understood to be operational.

Contractor
Thales Airborne Systems, Elancourt, France.

VERIFIED

TDS-TA Radar Warning Receiver (RWR)

Type
RWR for transport aircraft applications.

Description
The Threat Detection System for Transport Aircraft (TDS-TA) RWR is described as being a user programmable, database oriented equipment that covers a 'very wide' frequency range and makes use of full-band instantaneous frequency measurement. TDS-TA is further noted as being able to detect continuous wave, pulse and pulse-Doppler emitters that are operating in 'very' dense electromagnetic environments. Alongside its threat warning function, TDS-TA is noted as being able to offer 'smart' management of an onboard countermeasures dispensing system and as being able to interface with a missile warning system to create a self-protection suite.

Status
As of April 2001, the TDS-TA RWR was reported as being in production for installation aboard French Air Force C-130 and C-160 transport aircraft.

Contractor
Thales Airborne Systems, Elancourt, France.

VERIFIED

Threat Warning Equipment (TWE)

Type
Combined radar and laser warning system.

Description
TWE comprises radar and laser warning receivers, a Central Processing Unit (CPU), a library module, a symbol generator, a multifunction display, radar antennas, laser heads and a control box. Coverage is E- through K-band (2 to 40 GHz) in the radar domain and bands I and II for lasers. The radar warning subsystem is capable of instantaneous frequency measurement and can handle pulse, pulse-Doppler and continuous wave emitters. The CPU processes received radar and laser signals, interfaces with the library module for threat identification, maintains electromagnetic compatibility and manages a self-protection suite (decoys and missile approach/launch warners) if required. The system generates a colour threat display that shows information on threat type, bearing and danger level. The visual format is backed up by an audio warning which sounds each time a radar or laser threat is detected. TWE also allows crew members to call up data lists from the library module during flight in order to change priorities for different phases of a mission. The system also incorporates a built-in test routine and because of its design, is claimed not to require a test bench for routine maintenance.

Status
TWE has been developed for installation aboard the Tigre HAC (*Hélicoptère Anti-Char*)/HAP (*Hélicoptère d'Appui et de Protection*) and UHT (*Unterstützungs Hubschrauber Tiger*) battlefield escort/anti-tank helicopters, which are on order for the French and German armies respectively. Alongside being the basis of the Electronic Warfare System (EWS) that has been selected for use on all versions of the described Tiger battlefield helicopter, TWE is also mandated for the NH90 tactical transport helicopter, where it is teamed with the MILDS® AN/AAR-60 passive missile warner and a countermeasures dispensing system. The then DaimlerChrysler Aerospace received the first Tiger EWS system production contract (covering 160 systems (out of a total requirement of 427) and valued at then year US$47 million) during March 2000.

Specifications
Frequency: 2-40 GHz (radar); bands I and II (laser)
Coverage: ±45° (elevation); 360° (azimuth)

Radar threats supported: continuous wave, pulse and pulse-Doppler
Accuracy: better than 10° RMS (angular); 20 MHz (frequency)
Library: 2,000 radar/laser mode class
Power consumption: 200 W
Weight: <15 kg
Dimensions: 199 × 194 × 350 mm (CPU)

Contractors

Thales Airborne Systems, France (radar warning).
European Aeronautic, Defence and Space (EADS) Deutschland - LFK-Lenkflugkörpersysteme GmbH, Germany (laser warning).

UPDATED

ISRAEL

AES-210E Electronic Support/ELectronic INTelligence (ES/ELINT) system

Type
Airborne ES/ELINT system.

Description
AES-210E is primarily an airborne ES/ELINT system that is designed for installation aboard a variety of platforms engaged in maritime patrol, overland surveillance and/or ELINT gathering. The system also acts as a radar warning receiver. Functionally, AES-210E detects, measures the parameters of and identifies ground-based, shipboard and airborne radar emitters and calculates their position. Processed data is displayed to the operator via a colour graphic situation display. The system also provides the host vehicle's pilot with audio and visual threat warnings.

Status
As of this edition, AES-210E manufacturer Elisra reported that the system was 'in service' aboard 'fixed- and rotary-winged aircraft' operated by 'classified' customers. According to Jane's sources, a variant of the AES-210E system (possibly designated as AES-210I) has been selected by the Royal Australian Navy for installation aboard its 16 S-70B-2 and 11 SH-2G(A) helicopters and a variant of the system may be the ES equipment fitted to Indian Coastguard Service Do 228 surveillance aircraft. AES-210 is also noted as being carried by the Argentine Navy's six S-2T maritime patrol aircraft.

Specifications
Frequency range: 0.5-18 GHz (optional extension up to 40 GHz available)
Azimuth coverage: 360°
DF accuracy: 3° (typical fine DF); 7° (typical coarse DF)
Identification library: over 1,000 emitter modes (can be updated in flight)

Contractor
Elisra Electronic Systems Ltd, Bene Beraq.

UPDATED

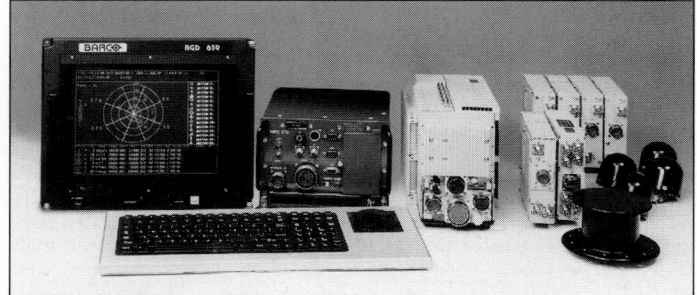

The elements that go to make up the AES-210E ES/ELINT system

2000/0069824

CDF-3001 COMmunications INTelligence (COMINT) and Direction-Finding (DF) system

Type
Airborne COMINT and DF system.

Description
CDF-3001 is a self-contained airborne system that is designed to acquire, monitor and establish the Direction Of Arrival (DOA) of tactical radio transmissions in the 20 to 500 MHz frequency band. The equipment utilises a Windows-driven man/machine interface and databases and features fast radio frequency spectrum scanning and an audio and control subsystem that facilitates signals recording and interfacing with the host platform. DOA measurement is automatic.

The CDF-3001 COMINT and DF system antenna array as applied to a UH-1 helicopter   0009391

Status
As of this edition, the CDF-3001 COMINT and DF system was reported as being available.

Specifications
Frequency range: 20-500 MHz (1.5-1,000 MHz option)
Monitored/detection signal types: AM/FM
Outputs/peripherals: printer; audio recorder; RS-232; I/O to platform navigation and command and control subsystems
Input power: AC or DC
Platform types: fixed-wing or helicopter

Contractor
Tadiran Electronic Systems Ltd, Holon.

VERIFIED

EL/K-7032 COMmunications INTelligence (COMINT) system

Type
Airborne COMINT system.

Description
EL/K-7032 is an airborne COMINT system designed for electronic surveillance and interception of signals within the 20 to 500 MHz frequency range. This is performed by local or remote digitally controlled communications receivers and features high-speed switching between bands to allow very fast and reliable coverage of the complete spectrum. Computer-controlled automatic operation enables a minimum number of individual operators to carry out multitask COMINT missions.

Status
As of this edition, K-7032 was understood to form the COMINT subsystem aboard Chile's Boeing 707-based Phalcon airborne early warning and surveillance aircraft.

Specifications
Frequency range: 20-500 MHz
Modes: AM, FM (CW and SSB optional)
Frequency resolution: 1 kHz (10 Hz optional)

Contractor
Elta Electronics Industries Ltd (a subsidiary of Israel Aircraft Industries Ltd), Ashdod.

VERIFIED

EL/L-8300 Electronic Support Measures (ESM)/ SIGnals INTelligence (SIGINT) system

Type
Airborne ESM/SIGINT system.

Description
Somewhat confusingly, Elta has used the EL/L-8300 designation to describe both an airborne strategic SIGINT suite and a maritime patrol ESM system. In its strategic SIGINT guise, L-8300 is a multi-operator system which has been noted as incorporating the EL/L-8312A ELectronic INTelligence (ELINT), EL/K-7032 COMmunications INTelligence (COMINT) and EL/L-8350 control and analysis subsystems. The EL/L-8351 ELINT training simulator, EL/L-8352 ELINT data analysis facility and EL/L-8353 tactical ground station have also been reported as supporting the described airborne segment. According to the same brief, SIGINT EL/L-8300 has a detection range of 450 km when being flown at a typical operating altitude aboard a Boeing 707 host platform. In a later release, Elta describes SIGINT EL/L-8300's ELINT subsystem as covering the 0.5 to 18 GHz (0.03 to 40 GHz as an option) frequency range and offering high probability of intercept in dense environments, instantaneous frequency measurement and an electronic order of battle analysis capability. The suite's COMINT subsystem is noted as being an acquisition, exploration and monitoring capability which covers the 20 to 1,000 MHz (2 to 1,500 MHz as an option) frequency band and features both wide and narrowband direction-finding. The equipment's airborne 'command station' is noted as providing data integration, report generation/mission support, threat warning and air-to-surface communications facilities. While not confirmed, logic points to the described ELINT subsystem forming the basis of the ESM EL/L-8300 system.

Status
As of this edition, Jane's sources suggest that an upgraded version of the SIGINT EL/L-8300 architecture forms the basis of the electromagnetic spectrum hardware used in the SIGMA mission suite that is installed in Spain's Boeing 707 SIGINT and electro-optic surveillance platform. As an ESM, EL/L-8300 variants have been selected for use on Australian Lockheed Martin AP-3C Orion (designated as ALR-2001 Odyssey) and Singaporean Fokker Enforcer Mk 2 maritime patrol aircraft. The Australian programme is a joint effort between Elta and Australian industry and involves equipment being installed in 19 aircraft. A further variant (designated as EL/L-8300 UK) has been selected as the ESM system for the Royal Air Force's Nimrod MRA Mk 4 maritime patrol aircraft.

Contractor
Elta Electronics Industries Ltd (a subsidiary of Israel Aircraft Industries Ltd), Ashdod.

VERIFIED

Operator positions for the EL/L-8300 SIGINT suite as installed in a Boeing 707 host aircraft

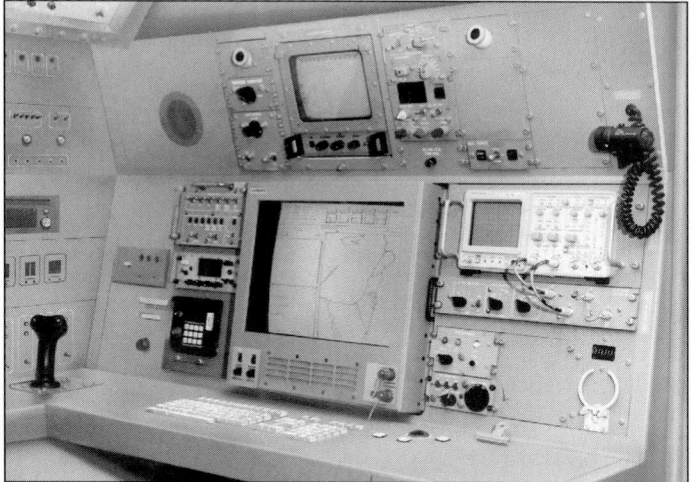
An operator console option used in the EL/L-8300 maritime patrol ESM system

EL/M-2160 Missile Approach Warning (MAW) system

Type
Active MAW system.

Description
EL/M-2160 is a solid-state, pulse-Doppler MAW that can be configured for fighter, helicopter and transport aircraft applications. The system comprises a Transceiver/Processing Unit (TPU) and Radio Frequency (RF) head and between four and six antennas (with the number depending on the platform type and coverage required). A control panel is available as an option. Overall system features include all-weather operation; missile detection during boost, sustain and post burnout flight phases; time-to-impact calculation; timing data generation for chaff/flare release; effective function over the complete flight envelope and built-in test/fault isolation.

Status
As of this edition, EL/M-2160 was in service and a miniaturised and upgraded variant is known to have been developed. Jane's sources suggest the EL/M-2160 MAWs have been installed on Danish C-130s, French and German C-160s, Israeli F-16C/Ds, Netherlands C-130Hs, F27s and F60s and a variety of helicopter types including the Bell 212/412 series and the MIL Mi-17. EL/M-2160 can also be supplied as part of a 'complete' integrated self-protection system that includes a countermeasures dispensing system and the necessary installation kit.

Specifications
Dimensions: 508 × 254 × 216 mm (TPU)
Weight: 20 kg (miniaturised/upgraded variant); 32 kg (original (?) TPU and RF head)
Power consumption: 400 W
Interfaces: 1553B/PPD/RS-232/RS-422 and/or discrete lines

Contractor
Elta Electronics Industries Ltd (a subsidiary of Israel Aircraft Industries Ltd), Ashdod.

VERIFIED

The EL/M-2160 pulse Doppler missile approach warner

Guitar 350 missile warning system

Type
Passive electro-optical missile warning system.

Description
Guitar 350 is a lightweight, passive, autonomous warning system that detects missiles by sensing the electro-optic signatures of their exhaust plumes. The system provides audio and visual alarms, giving sufficient time for the activation of the host platform's defensive systems. The equipment has been developed for the protection of fixed- and rotary-wing aircraft and is based on a patented sensor. The device automatically controls an infra-red countermeasures system and emits no detectable electromagnetic or electro-optical signals. Guitar 350 is designed to detect surface-to-air and air-to-air missiles with 'extremely low' false alarm rates. The system is further noted as covering 'all possible' attack angles and is claimed to offer high discrimination against background clutter. Guitar 350 is also

Rafael's Guitar 350 missile warning system is designed for both helicopter and fixed-wing applications   0009904

described as being compatible with cockpit displays and as being able to withstand adverse environmental conditions.

Status

As of this edition, the Guitar 350 missile warning system is understood to have been test flown.

Specifications

Spatial coverage: 120° (elevation); 360° (azimuth)
Warning time: 4-6 s
Weight: <15 kg
Power consumption: <200 W
Interface: MIL-STD-1553B

Contractor

Rafael Electronic Systems Division, Haifa.

VERIFIED

LWS-20V-2 laser warning system

Type

Airborne laser warner.

Description

As of January 2002, LWS-20V-2 was the latest identified variant of Elisra Electronic System's LWS-20 laser warning system. Designed for stand-alone operations or integration into a multithreat warning system (such as Elisra's SPS-65V - see separate entry), the stand-alone LWS-20V-2 configuration comprises four sensor heads, a Laser Warning Analyser (LWA), a system control panel and a pilot's display. Functionally, the four sensor heads receive and detect pulsed laser signals and send the detected pulses to the LWA. Here, each detected pulse is characterised by its angle of arrival (for direction-finding purposes), relative time of arrival and amplitude. Threat identification is by means of library comparison. The LWA is noted as incorporating a 'powerful' military 32-bit microprocessor with complementary metal oxide semiconductor-based random access memory for temporary data storage and electronically erasable programmable read only memory for combat software and programmable tables. Identified threats are presented in alphanumeric and symbolic formats and as threat types, azimuths and lethalities. Audio warnings are also generated within the host platform's intercommunications system. The LWS-20V-2's combat software is based on 'proven self-protection' code and the system as a whole incorporates built-in test of its front end, preprocessor, input/output interface and central processor unit memory.

The LWS-20 laser warning system

Status

As of January 2002, Jane's sources were reporting that LWS-20 variants had been selected or procured for installation aboard S-70B-2 (?) and SH-2G(A) helicopters of the Royal Australian Navy, Canadian Forces CH-146 Griffon helicopters (as part of the SPS-65V integrated self-protection system) and Heersflieger (German Army Aviation) CH-53G helicopters (again as part of an SPS-65V application).

Specifications

Power source: 28 V DC (MIL-STD-704)
Power consumption: 120 W
Interfaces: active electronic countermeasures; MUX bus (option); onboard laser source blanking; RS-422
Environmental: to MIL-E-5400 Class IB
Weight: 0.2 kg (control panel); 0.6 kg (sensor head - each); 1.4 kg (display unit); 2.5 kg (LWA); 6.5 kg (complete system - stand-alone configuration)

Contractor

Elisra Electronic Systems Ltd (a member of the Elisra Group), Bene Beraq.

UPDATED

Passive Airborne Warning System (PAWS)

Type

Passive missile warning system.

Description

The PAWS passive missile warning system is described as being a lightweight, Infra-Red (IR) missile launch and approach warning system that is designed for both fixed- and rotary-winged applications. 'Specified, characterised and integrated' by Elisra, the system comprises of an Elisra-sourced processor and four or six staring IR sensors produced by Elop Electronic Industries. As such, PAWS can operate as an independent equipment or be integrated into an electronic warfare suite. Functionally, the system detects the threat missile's IR signature and tracks it. Having detected a threat, it determines whether or not the missile specifically threatens its host platform, provides an 'accurate' readout of its approach direction and estimates the time to intercept. PAWS can also select an appropriate 'narrowbeam or flare' countermeasures response and activate it automatically. The equipment is claimed to have a 'very low' false alarm rate in a 'high' clutter environment with its host platform executing 'violent' manoeuvres. As noted earlier, PAWS provides both launch and approach warning and has a multithreat warning capability. Elisra further notes that PAWS can be integrated with the SPS-65V threat warner, the LWS-20 laser warner, a countermeasures dispenser system and a directional IR jammer.

Status

As of January 2002, PAWS was noted as being 'fully developed'.

Specifications

Power source: 28 V DC (115 V, 400 Hz, MIL-STD-704D)
Power consumption: 200 W
Environmental: MIL-E-5400 Class IB
Interfaces: MIL-STD-1553B MUX bus; RS-422; transputerlink
Dimensions: 120 × 120 × 230 mm (sensor head); 203 × 389 × 127 mm (processor, short ½ ATR format)
Weight: 2.5 kg (sensor head - each); 10 kg (processor)

Contractor

Elisra Electronic Systems Ltd (a member of the Elisra Group), Bene Beraq.

UPDATED

The elements that make-up the PAWS IR missile warning system (Elisra)
2002/0131094

SPS-20V-2 self-protection system

Type
Airborne radar warning system.

Description
The wideband SPS-20V-2 is the latest identified configuration of Elisra Electronic Systems' SPS-20V airborne radar warning system. As such, it is designed to fit 'existing airborne installations' and to detect and display pulsed radar threats operating within the 0.7 to 18 GHz frequency range. Continuous wave radar threats are detected by a dedicated high-sensitivity receiver. A high-sensitivity 8 cm (3 in) display unit provides an alphanumeric presentation of the type, coarse angle of arrival, relative lethality and status of the analysed radar threats.

SPS-20V-2 features a digital signal analyser that carries out the data processing and interfacing tasks and contains a reprogrammable emitter library file. Functionally, analysed threats are displayed on the display unit screen together with audio warnings that are transmitted to the host platform's pilot via its intercommunication system. The displayed threat representations on the screen allow the pilot to establish the azimuth and emitter type of up to 16 top lethal threats. The displayed symbol and two additional arrow heads indicate the relative threat lethality. Up to 30 different symbols can be programmed to represent specific threat types. Additional alerting capability for missile launch status is also provided. As well as in stand-alone applications, SPS-20V forms part of Elisra's SPS-65V multisensor self-protection system (see separate entry).

Status
As of July 2001, Jane's sources were reporting SPS-20V variants as having been selected or procured for use aboard Canadian CH-146 Griffon helicopters (SPS-65V application), German CH-53G helicopters (SPS-65V application) and upgraded Romanian MiG-21 (Lancer programme) and MiG-29 (Sniper programme) multirole fighters.

Specifications
Frequency coverage: 0.7-1.3 GHz (CD pulse); 2-8 GHz (DJ pulse - 1/2 sub-bands)
Sensitivity: −38 dBm (DJ pulse); −40 dBm (CD pulse)
Threat sorting parameters: PRI, stagger, jitter, PW, conical scan modulation, angle of arrival
Weights: 0.2 kg (control unit); 0.4 kg (single-channel receiver - 2 used); 1.1 kg (dual-channel receiver - 2 used); 1.2 kg (display unit); 1.4 kg (analyser); 1.63 kg (optional CW receiver)

Contractor
Elisra Electronic Systems Ltd (a member of the Elisra Group), Bene Beraq.

UPDATED

The SPS-20V-2 self-protection system

SPS-45V self-protection system

Type
Integrated airborne self-protection system.

Description
SPS-45V is described as being an airborne self-protection system that integrates the SPS-20V pulse radar threat warning receiver ('low band' to 18 GHz coverage) with a superheterodyne SRS-25 receiver that is used to detect continuous wave, high pulse repetition frequency and low-effective radiated power emitters. Elisra further notes that the LWS-20 laser warner can be incorporated into the architecture as an option and that all identified threats are displayed to the host platform's pilot via an 8 cm (3 in) display unit. Here, the pilot is presented with a graphical representation of threat type, angle of arrival, relative lethality and status. SPS-45V is also reported as being 'completely' threat reprogrammable, using a portable memory loader/verifier to load or unload operational software and emitter tables (for the equipment's integral emitter library) 'in the field'.

Functionally, SPS-45V's four element, orthogonal, cavity-backed spiral antenna array receives pulsed signals within the 2 to 18 GHz frequency range and hands them off to the equipment's dual-band SPS-20V receiver. The video output from the SPS-20V is then feed to an analyser that tags each pulse with its angle of arrival,

amplitude, pulsewidth and time of arrival. The analyser's 80386 Central Processing Unit (CPU) processes the characterised pulses, de-interleaves pulse trains and identifies the threat emitter by means of library comparison. 'Low band' pulsed signals are also noted as being detected and characterised to provide 'additional' information and threat status identification.

For its part, the SRS-25 is connected to the system's antenna array via the SPS-20V receiver to facilitate signal direction-finding. Input circuitry allows selection of one of the four antenna elements, with any detected signal being passed through a preliminary filter, mixed with an output signal from a voltage controlled oscillator and passed to an intermediate frequency section. The signal is then amplified, filtered and detected. The video signal is transferred to video circuitry for digitisation and final analysis/identification is performed by an 80386 microprocessor. The processed data is then passed to the analyser CPU (via a serial datalink) where it is integrated into a common SPS-20V/SRS-25 threat file. Analysed threats are displayed both graphically and in the form of audio warnings generated in the host platform's intercommunications system. In terms of its graphic output, SPS-45V can generate read outs representing the azimuth, lethality and type of up to 16 'top' threat emitters simultaneously. In all, up to 250 different symbols can be programmed to represent specific threat types. Additionally, the architecture has a missile launch status capability. System control is by means of a pushbutton control unit that allows the platform's pilot to power up the equipment, activate an automatic self-test routine, define modes of operation and delete preprogrammed threats from the display options if required. As a final point, Elisra notes that both the SPS-20V and SRS-25 subsystems incorporate their own microprocessors with electronically erasable programmable read-only and random access memories. Each processor works autonomously, using its own emitter and mission tables.

Status
As of January 2002, SPS-45V was understood to be available.

Specifications
Frequency: low band −18 GHz
Received signals: CW, low EPR, high PRF and pulsed
Coverage: up to full azimuth (depending on antenna installation)
Communications channels: RS-232/-422; TTL in/out
Threat sorting: angle of arrival, conical scan modulation, frequency, jitter, PRI, PW and stagger
Audio: composite, missile launch and 'new guy'
Display: alphanumeric and 'special' symbology
Interfaces: 1553 MUX bus (optional), CMDS, missile warner, radar blanking, radar jammer and simplified blanking centre
Supply voltage: 28 V DC (MIL-STD-704)
Environment: MIL-E-5400 Class II qualified
Weight: 3.8 kg (SRS-25); 7.5 kg (SPS-20V)

Contractor
Elisra Electronic Systems Ltd (a member of the Elisra Group), Bene Beraq.

NEW ENTRY

SPS-65V integrated self-protection system

Type
Integrated radar and laser warning system.

Description
The SPS-65V integrated radar and laser warning system is a lightweight equipment that is claimed to offer 'outstanding sensitivity' and as being capable of intercepting and analysing a 'wide range of known and anticipated emissions' from 'the battlefield'. As such, SPS-65V comprises three subsystems, as follows:

- the SPS-20V low-volume, lightweight radar warning system that detects, processes and displays threat emitters operating within the 0.7 to 18 GHz frequency band
- the 6.5 to 18 GHz band SRS-25 superheterodyne receiver that detects 'modern' continuous wave, high pulse repetition frequency and low effective radiated power radar emitters
- the LWS-20 laser warning receiver that detects, identifies and locates the emitters associated with laser-guided weapon systems.

An 8 cm (3 in) display unit provides an alphanumeric readout of the type, relative bearing, relative lethality and status of up to 16 analysed threats. In addition, audio warnings are presented to the host platform's pilot via his intercommunications system. SPS-65V can be further augmented by an automatically or manually initiated CounterMeasures Dispensing System (CMDS). The equipment's software and emitter tables can be field-loaded into its integral emitter library 'in seconds' and are claimed to provide 'complete and immediate adaptability to changing threat environments'. The system's analyser records flight events and downloads then to a tape cassette for post-mission playback. When incorporating the SPS-20V-2 pulsed radar threat warner, SPS-65V is designated as the SPS-65V-2 equipment.

Status
As of July 2001, Jane's sources were reporting that the Heersflieger (German Army Aviation) and Canadian Forces had procured SPS-65V variants for installation aboard CH-53G and CH-146 Griffon helicopters respectively. In more detail, the

The SPS-65V integrated radar and laser warning system (Elisra) 2002/0131093

Heersflieger procurement is understood to have involved at least 20 SPS-65V shipsets. Here, three shipsets are understood to have been acquired during 1997, with the remaining 17 (valued at then year US\$10 million) during 1998. As of June 2001, Elisra was reporting that it was supplying Canada with 17 SPS-65V systems in a two phase deal valued at then year C\$25 million. Phase one of the effort involved 10 systems (together with installation and system integration support) with phase two seeing the delivery of the remaining seven shipsets, together with soft- and hardware system 'upgrades'. As of May 2001, Jane's sources were suggesting that Canada was acquiring the V-2 variant of the SPS-65V architecture and that the Canadian Forces had a possible requirement for a further 15 such systems. In the CH-53G application, SPS-65V is reported to be teamed with an Israeli sourced CMDS, while the Griffon fit is reported as being integrated with the AN/AAR-47 missile approach warner and the AN/ALE-39 or -47 CMDS. As of January 2002, the preceding was neither confirmed nor denied by Elisra.

Specifications
Frequency: 0.7-18 GHz (SPS-20V); 6.5-18 GHz (SRS-25)
Power supply: 28 V DC
Environmental: MIL-STD-5400 Class II
Weight: 2.5 kg (four sensor head LWS-20 application); 4 kg (SRS-25); 7.5 kg (SPS-20V)

Contractor
Elisra Electronic Systems Ltd (a member of the Elisra Group), Bene Beraq.

UPDATED

SPS-1000V-5 radar warning system

Type
Radar threat warning system.

Description
SPS-1000V-5 is an airborne self-protection system offering high performance and reliability. The system integrates two subsystems:
- a basic SPS radar warning equipment which detects the 'classic' radars from low to K-band (40 GHz)
- a superheterodyne receiver subsystem which provides for detection and unambiguous identification of modern continuous wave, high pulse repetition frequency, low effective radiated power and pulse-Doppler radars.

Overall, the system is threat reprogrammable and makes use of a portable field loader to upload updated emitter tables and download recorded mission data. Analysed threat data is presented to the pilot on an 8 cm (3 in) display unit that provides a graphic representation of threat type, angle of arrival, relative lethality

The SPS-1000V-5 self-protection system

and threat status. These displayed threat representations allow the pilot to establish the azimuth (with respect to aircraft heading) and emitter type of up to 16 top lethal threats. In addition, audio warnings are provided on the host platform's intercommunication system. Up to 80 different symbols can be preprogrammed to represent specific threat types. An additional alert capability for missile launch status is provided. The system's mode of operation is controlled from a front panel single control unit that enables the pilot to power-up the system, activate its self-test procedure, delete preprogrammed threats from the display, define high- or low-altitude emitter priorities and activate recording of threat data during a mission.

Status
As of early 2001, Jane's sources were suggesting that SPS-1000V variants had been selected for installation aboard Royal Australian Air Force C-130H transport aircraft and formed part of Israel Aircraft Industries' F-16 Avionics Capability Enhancement (ACE) upgrade package. As of the autumn of 1999, Jane's sources were pointing to SPS-1000V-5 being bid for retrofit to South Korean F-4 and F-5 combat aircraft and noting that the Chinese BM/KJ-8602 radar warning receiver bore a 'strong resemblance' to the SPS-1000V system.

Specifications
Frequency coverage: existing radar frequencies (millimetric-wave optional)
Receiver signals: pulse, CW, High PRF, Pulse Doppler
Weights: superheterodyne receiver subsystem 4 kg; analyser and CD receiver 13 kg; Pilot controller and display 2.4 kg; antennas and amplifier/detectors 8 kg

Contractor
Elisra Electronic Systems Ltd (a member of the Elisra Group), Bene Beraq.

VERIFIED

SRS-25 superheterodyne receiver

Type
Airborne detection receiver.

Description
SRS-25 is an airborne superheterodyne receiver that forms part of the SPS-65V integrated helicopter self-protection system (see separate entry). It detects Continuous Wave (CW), high Pulse Repetition Frequency (PRF) and medium effective radiated power radars, performs frequency measurement and direction-finding on detected signals and undertakes 'sophisticated' signal processing operations. Its primary function is to detect and process those CW radar and high PRF signals that are not detected by conventional radar warning systems. As such, SRS-25 scans the frequency range, measures frequency, calculates the angle of arrival, analyses emitter parameters and provides data for display. The SPS-45 integrated system combines the SRS-25 receiver with the SPS-20V self-protection system (see separate entry) and is intended primarily for the retrofit and upgrading market.

Status
As of July 2001, SRS-25 receivers formed part of the SPS-65V multisensor threat warning systems procured for installation aboard Canadian Forces CH-146 Griffon and Heersflieger (German Army Aviation) CH-53G helicopters.

Specifications
Frequency: 6.5-18 GHz
Received signals: CW, high PRF, medium ERP
Reception coverage: ±30° (elevation); 360° (azimuth)
Frequency accuracy: ±10 MHz
Frequency resolution: 2 MHz
Dimensions: 101 × 122 × 250 mm
Weight: 4 kg

Contractor
Elisra Electronics Systems Ltd (a member of the Elisra Group), Bene Beraq.

VERIFIED

STRATUS COMmunications INTelligence (COMINT) system

Type
Balloonborne COMINT system.

Description
STRATUS is a balloonborne COMINT system that covers the 20 to 1,200 MHz frequency range and can remain airborne for up to 4 weeks of continuous operations. It is based on field-proven techniques and subsystems and Rafael's long-term experience with balloonborne systems.

   The system consists of an aerostat equipped with a remotely controlled, customer specified sensor payload; a tether and a ground support system which includes mooring equipment, computers and operators' consoles. Once the

STRATUS aerostat and ground station

balloon reaches the designated altitude, data are collected and passed to the ground station where they are processed and stored in an information bank. The STRATUS system is also understood to be able to hand-off data to radar sites directly from its mission payload.

Status
As of July 1999, the Israel Defence Force was being reported as having confirmed its use of the STRATUS system in support of its south Lebanese 'security zone'.

Contractor
Rafael Electronic Systems Division, Haifa.

UPDATED

ITALY

ALR-733

Type
Family of airborne C- through J-band (0.5 to 20 GHz) Electronic Support (ES) systems.

Description
Elettronica's ALR-733 series of radar band ES systems is designed for maritime patrol (fixed-wing aircraft and helicopters), airborne early warning and air-to-ground surveillance applications and provides automatic ES surveillance and computer aided, ELectronic INTelligence (ELINT) - type analysis capabilities, with the latter being under operator control. A 'core' ALR-733 system comprises:

- an omnidirectional antenna assembly
- six to eight 'basic' Direction-Finding (DF) antennas
- a receiver unit
- a processor unit.

Available options include:
- up to four fine DF antennas (providing sector coverage)
- an 'E-mode' fine DF antenna (steerable over 360° in azimuth and designed for passive over-the-horizon targeting)
- a solid-state pilot's radar warning receiver display
- a data-recorder
- expanded frequency coverage using additional millimetre-wave antennas.

Functionally, ALR-733's automatic ES surveillance mode provides:
- a 'fast and fully automatic' description of the electromagnetic environment (in which, emitters are detected, analysed, identified, tracked and displayed)
- automatic detection of intentional frequency modulation on pulse signals
- automatic identification and warning of 'high priority' emitters.

Other system features include:
- nominal 100 per cent probability of intercept
- monopulse DF
- real-time signal analysis (including modulation on pulse).

As of May 2001, two ALR-733 subvariants had been identified, the known details of which are as follows:

ALR-733(V)2
ALR-733(V)2 is designed for small to medium sized maritime patrol aircraft and helicopters and incorporates six DF antennas, an omni-antenna assembly, a floppy disk drive, a receiver unit and a processor unit. The equipment covers the 2 to 18 GHz frequency band (extendable to 0.6 to 40 GHz if required) and is described as incorporating the family's ES surveillance and ELINT-type analysis operating modes. Typical platform applications for the system are noted as including maritime patrol variants of the Fokker F-27 and ATR-42 airliners and the anti-submarine/surface warfare AB 212 helicopter.

ALR-733(V)4
ALR-733(V)4 has been specifically developed for installation aboard the naval variant of the multinational NH-90 helicopter and comprises an omnidirectional antenna assembly, six basic and one to four fine DF antennas, receiver and processor units and a data-recorder. The equipment is believed to cover the 2 to 18 GHz frequency band in baseline form and offers the ALR-733 family's already described ES surveillance and ELINT-type analysis operating modes.

Status
As of May 2001, ALR-733 series ES systems were understood to be available.

Contractor
Elettronica SpA, Rome.

NEW ENTRY

ELT/156, 156X, 158 series Radar Warning Receivers (RWRs)

Type
Family of RWRs.

Description
The ELT/156 series of RWRs is designed to provide rapid non-ambiguous warning of illumination by hostile search/ tracking radars and, depending on variant, is applicable to light fighter/ground attack aircraft, high-performance combat aircraft and battlefield attack helicopters. Known details of the identified family members are as follows:

ELT/156(V)
The ELT/156(V) RWR is designed for light strike aircraft and helicopter applications and provides all-round azimuth and elevation coverage together with a threat bearing/range presentation on a cockpit-mounted dual-mode bright display. An audio warning facility is also provided and the equipment can generate simultaneous warnings of multiple threats. ELT/156(V) comprises four antennas, two radio frequency heads, a signal processor, a display unit and a control panel, weighs 10.5 kg and if fully computer controlled. The system's software is reprogrammable.

Specifications
Azimuth coverage: 360°
Weight: 0.15 kg (antenna); 0.6 kg (control panel); 1 kg (radio frequency head); 1.9 kg (display); 5.2 kg (signal processor); 10.5 kg (total system)

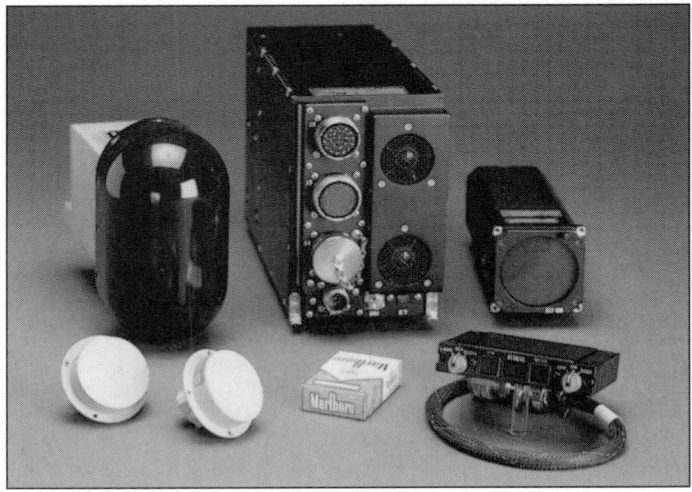

The ELT/156X RWR showing antennas, processor, control panel and display

The ELT/158 radar warning receiver (l to r: processor, wingtip antennas, display, rear antenna)

ELT/156X(V)

The 17.7 kg ELT-156X(V) is a wideband, miniature crystal video RWR whose frequency coverage includes the millimetric band. It features distributed bit-slice multiprocessing and an electronically erasable programmable read-only memory. ELT/156X(V) is flight line re-programmable and features a dual-mode (raw/synthetic) cockpit display. Here, the raw mode emitter vectors give the pilot an immediate picture of the surrounding environment, while the synthetic display provides threat identity and helps the pilot to explore transient gaps in air defence networks during penetration missions. The 20.1 kg ELT/156X(V)2 variant is designed for helicopter applications and makes use of four, 1.7 kg, single channel radio frequency head/antenna units.

ELT/158

The ELT/158 RWR is designed for high-performance combat aircraft applications and integrates with its host's onboard countermeasures equipment. The system provides 'high-speed, unambiguous' threat identification; a 'high' probability of intercept value intercept probability; full digital processing and 360° azimuth coverage. Visual threat warnings are provided on a dual-mode (raw/synthetic) cockpit display, together with aural warning on the aircraft intercom system.

Status

As of this edition, identified ELT/156-158 series RWR applications were reported as comprising the Alenia/Aermacchi/Embraer AMX strike fighter (ELT/156(V)), the Agusta A 129 Mangusta (ELT/156-05 (lot 2) or ELT/156(V)2 (lot 3)) and the EH Industries EH 101 (ELT/156X(V)2 on Italian Navy aircraft).

Contractor

Elettronica SpA, Rome.

VERIFIED

ALR-73x series equipments form part of the RQH-5(V) technology family

ELT/263 Electronic Support (ES) system

Type

ES system for small- and medium-size aircraft.

Description

The ELT/263 airborne ES system is designed principally for use in maritime aircraft such as the Beech 200T and can be configured for a range of other types including the Guardian, Casa C212, Learjet 35A, F27 Maritime, Bandeirante, Piaggio P166, and Britten-Norman Maritime Defender. The surveillance of coastal water is its main function, with detection, analysis and identification of emission in the E- to J-bands (2 to 20 GHz) being performed together with bearing measurement. Emitter location is by means of triangulation and the data gathered by the system can be transferred to external users (such as ground stations, naval units and patrol aircraft radar operators/tactical navigators) and/or printed out for post-mission analysis on an optional printer. The ELT/263 system comprises four direction-finding arrays, an omnidirectional antenna set, direction-finding and instantaneous frequency measuring receivers, a display and control console, a radar warning processor and a radar warning display unit.

Status

As of this edition, the ELT/263 ES system is reported as having been procured. Elettronica notes that the system is no longer in production.

Contractor

Elettronica SpA, Rome.

VERIFIED

RQH-5(V) Electronic Support (ES)/ELectronic INTelligence (ELINT) system

Type

Multi-application ES/ELINT system.

Description

RQH-5(V) is a family of systems that can meet electonic warfare requirements ranging from threat detection and analysis to electronic intelligence. The various configurations and options allow the system to be tailored to meet specific requirements. All RQH-5(V) series configurations can be integration with other onboard equipments and can be used for targeting. The small size and low weight of RQH-5(V) components make the system readily adaptable to current airborne, naval and ground installations with a minimum of effort. The system can be operated after 'minimal' instruction and is designed for 'maximum operating time with minimum servicing'. RQH-5(V) equipments are broadband (0.65 to 18 GHz, with an optional extension to 40 GHz) and entirely automatic. They provide real-time automatic extraction, analysis and tracking of all incoming radar signals together with parametric data on their pulse and intrapulse structure. Such systems can also function as ELINT sensors, providing fine frequency, jitter, stager, pulse repetition interval, histogram, antenna pattern and amplitude data. In list mode, the display used can show data on up to 200 emitters (16 per page) and the architecture as a whole can operate without *a priori* knowledge of the electromagnetic scenario, without significant reductions in performance.

The basic modules used to create RQH-5(V) systems comprise an antenna group, direction-finding and Instantaneous Frequency Measuring (IFM) receivers and the data extractor. Of these, the antenna group includes one omnidirectional

unit and either four or eight Direction-Finding (DF) antennas. Various types of DF and fine DF antennas that cover different frequency ranges are available, with each antenna unit incorporating its associated circuitry. The DF receiver used is a wide open, omnidirectional and instantaneous unit that uses the amplitude comparison monopulse techniques. An eight port configuration is normally adopted, but it can also use a four element DF antenna subsystem for radar warning receiver applications. The RQH-5(V)'s wide open IFM receiver features 'high' sensitivity/probability of detection, 'fast' response times and a 'very high' instantaneous dynamic range. Two versions (FR-6 and FR-7) are available, providing different levels of frequency accuracy. The data extractor unit features a multiprocessor structure that is specifically designed for real-time applications and includes a range of standard input/output interfaces. The resulting automatic data extraction process acquires information on up to 20 new emitters every 60 ms, even in a completely unknown environment.

RQH-5(V) configurations offer a number of graphic and alphanumeric display modes that include frequency and direction of arrival, tabular lists, emitter characteristics, true or relative bearing, frequency, frequency-agile deviation, pulse repetition frequency, jitter and stagger, emitter identity and threat level, emitter scan period and scan type. Tactical, panoramic and geographic (ES, radar and navigation) modes are also available. The architecture's integral emitter identification library has a 3,000 mode capacity and the operator can input known data and preassigned threat and confidence level information into the memory. RQH-5(V) systems can be further expanded via the addition of the following options:

- a high accuracy DF package that consists of a multiple beam antenna system and a crystal video, amplitude sectorial, monopulse receiver. Several multibeam flat arrays, covering different frequency ranges, are available, providing different instantaneous fields-of-view and 'very high' DF accuracy
- a K-band (20 to 40 GHz), high-gain, steerable antenna and a formatter unit (½Air Transport Racking Short standard). This provides high-sensitivity detection and direction of the emitter signals
- a fine analysis ELINT receiver, which gives supplementary information on the active extracted emitter, including the presence of simultaneous multiple RF or intrapulse modulations.

RQH-5(V) configurations currently available include the ALR-733 and -735 ES/ELINT variants and the 26 kg, single unit ALR-741 ESM. Of these, ALR-741 is described as being a 'full feature' system despite its compact nature while all three systems are noted as having a synthetic omni-antenna configuration option.

Status

As of this edition, RQH-5(V) ES/ELINT system variants were reported as having been procured. Jane's sources suggest that the ALR-735(V)3 configuration has been selected for installation aboard the Italian Navy's eight anti-submarine/anti-surface vessel warfare and four airborne early warning EH 101 helicopters. As of October 2000, Jane's sources were suggesting that an ALR-733 variant was mandated for the Meteor's proposed Mirach 2000 Falco unmanned aerial vehicle.

Contractor

Elettronica SpA, Rome.

UPDATED

Smart Guard COMmunications INTelligence (COMINT) system

Type

Intelligence monitoring system for COMINT purposes.

Description

The Smart Guard COMINT system is designed for intelligence monitoring purposes of the Very/Ultra High Frequency (V/UHF - 30 MHz to 1 GHz)

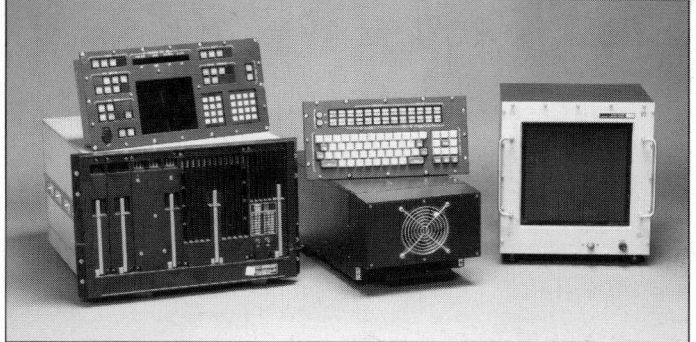

The Smart Guard basic ESM subsystem

communications bands. Although it is mainly intended for airborne applications, its compact and lightweight construction allows it to be installed on ground vehicles and ships. The main features of the system are modularity, coverage of the VHF and UHF bands, high speed and intercept capability and high Direction-Finding (DF) accuracy. Equipment of this type can be configured as follows:

Basic Electronic Support (ES) system
A basic ES system using the equipment's modules consists of a search and analysis receiver and a monitoring subsystem. In this configuration, the system facilitates the search for and intercept of communications through a computer-controlled operation. Search is carried out over the complete bandwidth or on specified sub-bands or channels. On operator control, the receiver stops on a specific channel, so allowing the demodulation and analysis of the particular channel. A continuous monitoring of up to eight channels is formed by up to eight remotely controlled receivers and associated recorders. The frequency tuning of each monitoring receiver is set automatically by the system computer. Voice and associated data are recorded for subsequent analysis. In this configuration the system is manned by a single operator.

Smart Guard
In the Smart Guard configuration, the basic ES system is augmented by a direction-finding and fixing subsystem. The latter adds the capability to measure the Direction Of Arrival (DOA) of the communications transmissions and to determine their location. DOA measurements are performed by a specific dual-channel superheterodyne receiver controlled by the system computer, and are obtained through an interferometric measurement. In addition to DOA, the frequency value and signal strength are measured for each specific channel. An interface with a navigation system provides the actual position of the platform in such a way that for each specific emission a set of data is stored. This contains DOA, frequency value, signal strength and platform position. The fixing of a specific communication emission is performed by a dedicated computer and associated software algorithms by using the DOA and platform position data of a specific channel. At least two significant DOA measurements are required for a fixing computation. In the Smart Guard configuration, two operators are required, one of whom controls the ES subsystem with the other handling emitter location fixing. A ground-based retrieval and analysis system is available as an option. Integration of the Smart Guard COMINT system with the ELT/888 electronic intelligence system makes it possible to monitor the full electromagnetic environment and to identify and locate all hostile weapon systems.

Status
As of this edition, the status of the Smart Guard architecture was uncertain.

Contractor
Elettronica SpA, Rome.

VERIFIED

JAPAN

Japanese airborne SIGnals INTelligence (SIGINT), Electronic Support (ES) and threat warning systems

Description
Over time, the Japanese defence electronics industry has produced a range of airborne SIGINT, ES and threat warning systems, the known details of which are as follows:

J/ALR-2
Jane's sources describe J/ALR-2 as being part of, or the entire mission suite that was installed in two SIGINT configured Japanese Air Self-Defence Force (JASDF) YS-11 transport aircraft during the early 1990s.

J/APQ-1
J/APQ-1 is thought to be an active radar missile approach warner that the Mitsubishi Electronic Corporation has developed for use aboard JASDF Mitsubishi F-15J aircraft. Installed in a fairing on the aircraft's starboard fin, the system is said to utilise a 20 cm antenna and is reported to generate both visual and audio threat alerts and to be able to automatically activate an onboard countermeasures dispensing system when a threat is confirmed. J/APQ-1 is noted as having entered service during 1992, a date that, as of this edition, is not confirmed.

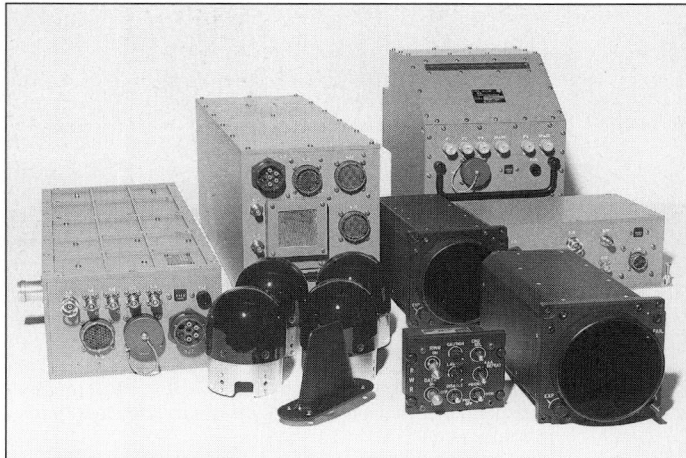

The J/APR-6 radar warning receiver that is installed on the Mitsubishi F-4EJ Kai

J/APR-2
The J/APR-2 Radar Warning Receiver (RWR) was a first generation equipment that was installed aboard JASDF Mitsubishi F-4EJ fighter and RF-4E reconnaissance aircraft.

J/APR-3
J/APR-3 is the RWR that is fitted to JASDF Mitsubishi F-1 close-support aircraft.

J/APR-4
The J/APR-4 Radar Warning Receiver (RWR) is installed aboard early production JASDF Mitsubishi F-15J interceptor aircraft and is described as being able to process multiple threats simultaneously in a dense electromagnetic environment. The equipment is noted as utilising a reprogrammable digital computer and a daylight viewing display. Data presentation is in alphanumeric and graphic formats and J/APR-4 can interface with other onboard electronic warfare systems such as the J/ALQ-8 radar jammer.

J/APR-4A
As its designation suggests, the J/APR-4A RWR is an upgraded version of the J/APR-4 equipment that is installed aboard late production JASDF Mitsubishi F-15J/DJ interceptor/operational training aircraft. The system incorporates multiple, high-speed microprocessors and is reprogrammable.

J/APR-5
The J/APR-5 RWR is described as being the J/APR-4A equipment repackaged for retrofit aboard JASDF Mitsubishi RF-4E reconnaissance aircraft.

J/APR-6
Like J/APR-5, the J/APR-6 RWR is described as being a J/APR-4A equipment repackaged for installation aboard JASDF Mitsubishi F-4EJ Kai reconnaissance aircraft.

EP-3 SIGINT suite
Japanese contractors NEC and the Mitsubishi Electronic Corporation are reported to have developed the 30 MHz to 20 GHz SIGINT mission system that is installed aboard Japanese Maritime Self-Defence Force's (JMSDF) Kawasaki EP-3 electronic reconnaissance aircraft.

HR-108
Jane's sources suggest that the HR-108 equipment is the ES system installed aboard JMSDF Mitsubishi SH-60J anti-submarine warfare helicopters.

Contractors
Mitsubishi Electronic Corporation, Tokyo (J/APQ-1 and EP-3 SIGINT suite).
NEC Corporation, Tokyo (EP-3 SIGINT suite).
Tokimec Inc, Tokyo (J/APR-2, -3, -4/4A, -5 and -6).

VERIFIED

RUSSIAN FEDERATION AND ASSOCIATED STATES (CIS)

RFAS airborne SIGnals INTelligence (SIGINT), electronic Support (ES) and threat warning systems

Description
Over time, the defence electronic industries of the members of the Commonwealth of Independent States economic grouping (Armenia, Azerbaijan, Belarus, Georgia, Kazakhstan, Kyrgystan, Moldova, the Russian Federation, Tajikistan, Turkmenistan, Ukraine and Uzbekistan) have produced a range of airborne SIGINT, ES and threat warning systems, the known details of which are as follows:

Koob-3M
Koob-3M is reported to be the SIGINT system that was installed aboard the Mikoyan MiG-25RBK reconnaissance/strike aircraft.

Landysh
Landysh is an internally mounted SIGINT system that is installed aboard the Sukhoi Su-24MP electronic warfare aircraft.

A close-up of the antennas associated with as yet unidentified RWR (possibly SPO-2, left) and terrain bounce jamming (right) subsystems fitted to a Tupolev Tu-95K cruise missile launcher (Martin Streetly)   1996

LIP
LIP is reported to be an active (pulse-Doppler?) radar missile approach warning system that has been used on a range of fixed- and rotary-wing platforms including the Kamov KA-29, the MIL Mi-24 and the Sukhoi Su-25.

LO-6 Beryoza
Beryoza is described as being a Radar Warning Receiver (RWR) that was installed aboard late production Mikoyan MiG-25 series aircraft as an alternative to the Sirena-3M (see following) system. This may be an alternative designation for the SPO-15 equipment (see following).

LO-81 Fantasmagoria-B
Fantasmagoria-B is a pod-mounted emitter location system that is used in support of anti-radiation missiles and as a general electronic reconnaissance tool aboard Sukhoi Su-24M strike aircraft. LO-81 is understood to have replaced the earlier LO-80 Fantasmagoria-A system.

LO-82 Mak-UL
Mak-UL is described as being the ultra-violet missile approach warning system that is installed aboard Sukhoi Su-24 series strike aircraft.

Pastil
Pastil is the designation given to the 1.2 to 18 GHz RWR that is installed aboard the Sukhoi Su-39 close-support aircraft. Credited to the Central Scientific Institute for Radiological Measurement (the TSNITI), the system is reported to be effective against continuous wave, pulse and pulse-Doppler radars and as being able to function as a stand-alone unit or as part of a defensive aids suite.

SG-1
SG-1 is reported to be the RWR fitted to Mikoyan MiG-27 series aircraft.

Shar-25
Shar-25 is the SIGINT system that was installed in the Mikoyan MiG-25RBF reconnaissance/strike aircraft.

Sirena-3
Sirena-3 is the designation given to an RWR family that has been installed on a wide range of aircraft types that includes the MIL Mi-24, the Mikoyan MiG-21PFM/PFS/MF, MiG-25 (Sirena-3M variant, 360° field of view) and the Sukhoi Su-25K.

SPO-2 Sirena-2
SPO-2 is the RWR that was installed aboard Tupolev Tu-95/-95K strategic bombers/cruise missile carriers.

SPO-10
SPO-10 is described as being an H- through J-band (6 to 20 GHz) RWR that was installed on early production Mikoyan MiG-29 aircraft.

SPO-15
SPO-15 is the designation given to a series of RWRs that are fitted to a range of aircraft that includes the MIL Mi-24, Mikoyan MiG-29 series (SPO-15LM variant, 360° coverage), Sukhoi Su-24 series and Sukhoi Su-27 series.

SPO-23
According to Jane's sources, the G- through J-band (4 to 18 GHz) SPO-23 RWR is a replacement for the SPO-15 and comprises two or four, four-beam azimuth antennas; up to four High-Frequency (HF) units (for the azimuth arrays); a receiver/processor unit and a cockpit indicator. Available options include a wide-angle azimuth and two elevation antennas together with two additional HF units to support the new elevation arrays. SPO-23 is quoted as offering 360° coverage in azimuth together with ±30° in elevation. Bearing accuracy is given as being 10° and the system contains a 128 mode integral threat library. Equipment weight is between 18 and 30 kg depending on the specific aircraft installation.

SRS-4A, -4B and -4V
The SRS-4 series is a family of SIGINT systems that were installed aboard the Mikoyan MiG-25RB reconnaissance/strike aircraft.

SRS-9 Virzh
SRS-9 is the SIGINT system that was installed in the Mikoyan MiG-25RBT reconnaissance/strike aircraft.

Tangazh
The Tangazh (aeronautical pitch) system is a pod-mounted 'radio monitoring' (electronic intelligence) equipment that is carried on the centre line stores station of the Sukhoi Su-24MR reconnaissance aircraft.

Yaguar
Yaguar is the targeting system that was installed aboard the Mikoyan MiG-25BM defence suppression aircraft. The primary armament of such aircraft was the Kh-58U anti-radiation missile

VERIFIED

SOUTH AFRICA

Airborne Laser Warning System (ALWS)

Type
Airborne laser warning system.

Description
Avitronics' ALWS provides threat identification and direction-finding indications of laser range-finders, designators and missile guidance systems. The equipment is designed to interface with existing onboard radar warning receiver/electronic support host systems and is available with one of three sensor models (the LWS-200, LWS-300 or LWS-400). All the noted sensor models utilise the same laser warning controller for data processing and interfacing with the host Electronic Warfare (EW) system. On detection of a threat, audio and visual alarms are generated via the host EW architecture or, if the ALWS is installed as a stand-alone equipment, via a dedicated, Avitronics-sourced, ALWS threat display and control unit. Of the three types of sensor, the LWS-300 and LWS-400 models are form-fit compatible and all three types of unit are electronically compatible (facilitating sensor upgrade if required). ALWS's 'broad coverage' of the laser spectrum is claimed to ensure detection of 'all known threats' and the system incorporates a programmable threat library that is field-loaded via the host EW architecture.

Status
As of November 2000, ALWS series laser warners were reported as being available.

Specifications

	LWS-200	LWS-300	LWS-400
Wavelength coverage	0.6-1.8µm	0.5-1.8µm	05-1.8 and 2-12 µm.
Threat coverage	Ruby, GaAs, NdYAG, Raman shifted NdYAG and Erbium lass lasers	as LWS-200 plus doubled NdYAG lasers	as LWS-300 plus CO_2 lasers
AOA accuracy	11° RMS (AZ)	15° RMS (AZ)	15° RMS (AZ)
Spatial coverage	60° (EL); 360° (AZ, 99° per sensor)	60° (EL); 360° (AZ, 90° per sensor)	40° (EL, 2-12 µm); 60° EL, 0.5-1.8 µm); 360° (AZ, 90° per sensor)
POI	>99% (single pulse)	>99% (single pulse)	>99% (single pulse)
Dimensions	103 × 86 × 64 mm	115 × 90 × 76 mm	107 × 90 × 76 mm
Weight	0.8 kg (per sensor)	1.2 kg (per sensor)	1.2 kg (per sensor)

Contractor
Avitronics (Pty) Ltd (a Saab and Grintek company), Centurion.

UPDATED

The ALWS configured with LWS-300 sensors   0017841

Electronic Surveillance Payload (ESP)

Type
Electronic Support (ES) payload for Unmanned Aerial Vehicle (UAV) applications.

Description
The Avitronics ESP is an ES payload that has been developed for installation aboard the Kentron Seeker UAV. It is based on the company's Emitter Location System (ELS - see separate entry) and features an improved probability of intercept value over that of its predecessor. The system comprises a single box acquisition and analysis receiver/controller unit and an interferometric antenna array that is mounted under the air vehicle's nose. The architecture is designed to operate as a

stand-alone ES capability that is fully integrated with its host vehicle and its associated ground station. Functionally, the airborne segment of the payload acquires and analyses radar signals prior to downloading then (via a datalink) to a remote ground terminal that both displays the received data and provides system control. ESP is designed primarily for electronic order of battle compilation and incorporates an onboard mini flash data recorder. Other system features include:
- low mass and volume
- intrapulse channel switching to provide a single pulse Direction-Finding (DF) capability
- the use of combined phase and amplitude comparison DF techniques
- a pulse Doppler radar handling capability.

Status

As of November 2000, the Avitronics ESP was reported as being available.

Specifications

Frequency coverage: 0.5-18 GHz
Frequency resolution: 1 MHz
Instantaneous bandwidth: 80 MHz (narrowband) or 1 GHz
DF accuracy: 1° RMS class (above 2 GHz); 3.5° RMS (at 700 MHz)
FOV: 70° (elevation, fully calibrated); 240° (azimuth, in three sectors, fully calibrated)
Antennas: phase amplitude matched
Weight: 6 kg (antenna array); 12 kg (receiver/controller unit)
Dimensions: 343 × 127 × 193 mm (receiver/controller unit)

Contractor

Avitronics (Pty) Ltd (a Saab and Grintek company), Centurion.

UPDATED

Emitter Location System (ELS)

Type

ELS adjunct to radar warning and electronic support systems.

Description

The Avitronics' ELS consists of a single box receiver/controller together with one or more interferometric antenna arrays. It is designed to be fully integrated with a host radar warning or electronic support system in order to enhance the latter's Direction-Finding (DF) capability. The architecture provides:
- acquisition, analysis and precision DF against search, tracking and fire-control radar emitters
- determination of bearings and parametric data for signals designated by the host system
- frequency measurement of designated emitters
- an autonomous, low probability of intercept emitter detection operating mode
- detailed emitter data sets for electronic support/intelligence analysis
- a pulse Doppler radar handling capability.

Status

As of November 2000, the Avitronics' ELS was reported as being available.

Specifications

Frequency coverage: 2-18 GHz (0.5-2 GHz option)
Frequency resolution: 2 MHz (narrowband mode)
Instantaneous bandwidth: 80 MHz or 1 GHz
DF accuracy: better than 1° RMS
FOV: 70° (elevation, per antenna array, fully calibrated); 120° (azimuth, per antenna array, fully calibrated)
Antennas: phase and amplitude matched
Weight: 5 kg (2-18 GHz antenna array); 7 kg (0.5-18 GHz antenna array); 14 kg (ELS controller)
Dimensions: 250 × 90 × 110 mm (2-18 GHz antenna array); 343 × 127 × 193 mm (ELS controller); 500 × 80 × 180 mm (0.5-18 GHz antenna array)

Contractor

Avitronics (Pty) Ltd (a Saab and Grintek company, Centurion.

UPDATED

MAW-200 Missile Approach Warning (MAW) system

Type

Integrated or stand-alone ultra-violet MAW.

Description

The MAW-200 MAW comprises a system controller and four sensor heads that utilise what their manufacturer describes as a 'unique' optical design (including 'state-of-the-art' filter technology, purpose-designed image intensifiers and photon-counting, focal-plane array processors) in order to facilitate threat detection at long ranges. The system also makes use of a distributed, hierarchical, data processing

The sensor head used in the MAW-200 missile approach warning system
0009901

architecture to optimise the utilisation of information in real time. Functionally, digitising and preprocessing functions are performed at detector level, with acquired data being passed to a sensor head digital signal processor unit. Here, equalisation, segmentation and feature extraction operations are carried out on multiple targets, with the derived spatial and temporal feature information being handed-off to the system controller. At this stage, spatial data is integrated with real-time inertial navigation system information (to compensate for platform movement, attitude and altitude) before the controller executes pattern recognition algorithms to identify detected threats with a minimum false alarm rate. Other system features include:
- accurate direction-finding
- neural network feature extraction and identification algorithms
- lightweight, low power, 'no cooling', skin-mounted sensor heads
- seamless tracking and hand-over between sensor heads.

Status

As of January 2000, the MAW-200 MAW system was reported as being 'in production' and as having been field tested against 'various' missiles. The equipment was further reported as being available during the following November.

Specifications

Power consumption: 28 V DC (0.6 A per sensor head)
FOV: 94° (conical, per sensor head)
DF resolution: better than 5°
Spatial coverage: 360° (azimuth, four sensors)
False alarm rate: maximum of 1 false alarm in every 2 h of operation in a high clutter environment (typically, better than 1 false alarm in 5 operational flying hours)
Warning probability: >99%
Threat capacity: up to at least 8 targets simultaneously
Weight: 2.5 kg (controller); 3.1 kg (sensor head)
Dimensions: 223 × 98 × 152 mm (controller); 230 × 130 × 130 mm (sensor head)

Contractor

Avitronics (Pty) Ltd (a Saab and Grintek company), Centurion.

UPDATED

MultiSensor Warning System (MSWS)

Type

Integrated warning and self-protection system.

Description

MSWS provides tactical aircraft with a complete warning capability for self-protection purposes. This capability includes radar warning, laser warning (using the LWS-400 subsystem) and missile approach warning (using the MAW-200 subsystem).The system consists of four front end receivers, four laser warning sensor heads, four missile approach warning sensor heads, an Electronic Warfare (EW) controller, a threat display and control unit and four 0.7 to 40 GHz spiral antennas. Functional capabilities of the system include:
- radar warning function catering for pulse Doppler and continuous wave radars in high pulse density environments. The function is complemented by an instantaneous frequency measurement module internal to the EW control unit
- man/machine interface via a colour multifunction display
- interface to and control of chaff and flare dispensing systems for the automatic dispensing of chaff and infrared flares upon threat detection
- interface to and control of (growth option) an active electronic countermeasures system offering the automatic activation of the electronic countermeasures system upon threat detection
- avionics system interface (growth option) via a 1553 bus

The elements that make up the MSWS system   0017842

- laser warning function providing threat warning with 360° azimuth direction-finding indication of laser illuminations
- missile approach warning function providing warning of approaching missiles, with 360° coverage via four passive sensors.

The MSWS architecture provides for a variety of sensors to be integrated and managed by the system, so allowing the user to upgrade the system in a phased and planned way as the battlefield requirements, operational experience and cost considerations dictate. It allows for easy and effective upgrading by the addition of components and subsystems, rather than major upgrade programmes. Generic design and low unit count allows easy installation in aircraft ranging from helicopters to fighters.

Status
As of November 2000, the MSWS was reported as being available.

Specifications
Radar warning
Frequency coverage: 0.7-18 GHz (CW signals); 0.7-40 GHz (pulsed signals)
Direction-finding: 10°-12° RMS (pulsed signals, 2-40 GHz sub-band)
Spatial coverage: 90° (elevation or spherical); 360° (azimuth or spherical)
Pulse density capability: >2.5 Mpps
Frequency resolution: 10 MHz
Laser warning
Wavelength coverage: 0.5-1.8 µm and 2-12 µm
AOA accuracy: 15° RMS (azimuth)
Spatial cover: 40° (elevation, 2-12 µm sub-band); 60° (elevation, 0.5-1.8 µm sub-band); 360° (azimuth, 90° per sensor head)
Laser threat coverage: CO_2, doubled NdYAG, Erbium glass, GaAs, NdYAG, Raman shifted NdYAG and Ruby
Laser threat types: beam riders, designators and range-finders
POI: >99%
Missile warning
Operating frequency: solar blind UV band
DF resolution: better than 5°
FOV: 94° (conical, per sensor)
Spatial coverage: 360° (azimuth, four sensor heads)
False alarm rate: max of 1 false alarm in 2 h operating in a high clutter environment (typically, better than 1 false alarm in 5 operational flying hours)
Probability of warning: >99%
Threat capacity: up to at least 8 targets simultaneously
Dimensions/weight: 107 × 90 × 76 mm/1.2 kg (LWS-400 sensor head); 110 × 110 × 67.5 mm/0.7 kg (0.7-40 GHz band spiral antenna); 128 × 127 × 120 mm/2.2 kg (threat display/control unit); 176 × 45 × 158 mm/3 kg (front-end receiver); 230 × 130 × 130 mm/3.1 kg (MAW-200 sensor head); 343 × 127 × 193 mm/14 kg (EW controller)

Contractor
Avitronics (Pty) Ltd (a Saab and Grintek company), Centurion.
UPDATED

....

RWS-50 radar warning system

Type
Radar warning system.

Description
The RWS-50 radar warning system comprises four antenna heads, two dual-channel detector-amplifiers, a control processor and a multifunction colour display/control unit and has been specifically designed to meet the needs of customers who require a minimum warning capability that can be subsequently upgraded as budgetary and operational considerations dictate. As such, the system provides tactical aircraft with a comprehensive radar warning capability for self-protection purposes. This capability can be extended to include laser and missile approach warning. The system includes an interface for the automatic or manual control of a countermeasures dispensing system and its integral threat library can be loaded via a memory loading unit on the flight line. RWS-50's generic

The elements that make up the baseline RWS-50 radar warning system
0017843

design is such as to allow it to be installed in aircraft types ranging from helicopters to fighters and available upgrade options comprise:
- an omni- or full direction-finding 0.7 to 1.4 GHz band capability
- a frequency extension of 0.7 to 40 GHz (in one antenna)
- an instantaneous frequency measuring capability
- a MIL-STD-1553B databus interface
- a laser warning capability
- a missile approach warning subsystem interface
- spherical coverage.

System upgrading is further facilitated by the use of the same controller unit throughout (with changes being executed via the introduction of printed circuit boards, Radio Frequency (RF) modules and/or new software within the unit), retention of the baseline receivers (with changes to or the addition of RF components) and minimum airframe structural changes.

Status
As of November 2000, the RWS-50 radar warning system was reported as being available.

Specifications
Frequency coverage: 2-18 GHz (basic configuration)
DF accuracy: 10° RMS
Spatial coverage: 90° (elevation); 360° (azimuth)
Power requirement: 28 V DC (140 W)
Dimensions/weights: 110 × 70 × 70 mm/0.4 kg (2-18 GHz spiral antennas); 128 × 127 × 120 mm/2.2 kg (display/control unit); 176 × 45 × 158 mm/3 kg (dual-channel detector-amplifiers); 343 × 127 × 193 mm/14 kg (control processor)

Contractor
Avitronics (Pty) Ltd (a Saab and Grintek company), Centurion.
UPDATED

_____

SPAIN

AMES Electronic Support (ES) system

Type
2 to 18 GHz airborne ES system.

Description
The modular AMES ES system is designed to be a compact, lightweight, low cost equipment that is suitable for airborne and 'small' surface vessel applications. In its 'Basic' form (AMES can be supplied in 'Basic' and 'Advanced' configurations with the latter adding a fine DF subsystem to the baseline architecture), the system comprises an array of four spiral Direction-Finding (DF) antenna/channelised receiver assemblies and a digital system/signal processor. Covering the 2 to 18 GHz frequency band, the 'Basic' configuration scans for and detects activity in a series of system specific bands that are programmed according to mission type. Signals analysis is undertaken by the system operator and received data can, if desired, be recorded for post-mission use. For customers requiring an electronic intelligence capability, AMES can be enhanced to such a standard via the addition of superheterodyne receivers, an intrapulse analysis package and the 'Advanced' configuration's fine DF capability.

Status
Over time, the AMES radar ES system has been reported as having been produced for 'light aircraft' applications for 'foreign' naval customers.

Contractor
INDRA DTD, Madrid.
VERIFIED

EN/ALR-300V1 Radar Warning Receiver (RWR)

Type
Airborne RWR.

Description
The EN/ALR-300V1 is a fully programmable RWR that is made up four Direction-Finding (DF) spiral antennas plus four channelised DF receivers (E- to J-band, 2 to 20 GHz), a blade antenna, a C/D-band (0.5 to 2 GHz) receiver, a processor, a control unit and an azimuth indicator display. It features a flight line reprogrammable threat library and its main features include:

- unique identification of all detected emitters by alphanumeric symbology on the cathode ray tube display
- voice-synthesised messages in accordance with the lethality of the detected emitters
- a recording capability of up to 100 emitters during flight time.

Radar data is presented to the operator via alphanumeric symbology on a high-brightness display. The position of the symbols on the display indicates lethality and bearing of the detected signal, as well as identifying the threat emitter. Symbol blinks and audio alarms indicate high-priority radar types. As an option, the EN/ALR-300V1 may incorporate DF capabilities in the C/D-band.

Specifications
Frequency coverage: C-J in 5 sub-bands
Radars detected: all pulsed and CW radars
Receiver type: channelised crystal video
Azimuth coverage: 360°
DF accuracy: better than 12° RMS
Number of radars simultaneously displayed: 15
Weight: 20 kg

Contractor
INDRA DTD, Madrid.

VERIFIED

EN/ALR-300V2 Radar Warning Receiver (RWR)

Type
Airborne RWR.

Description
EN/ALR-300V2 is a computer-controlled RWR that has been adopted for universal combat aircraft use within the Spanish Air Force. The equipment is an upgrade of the EN/ALR-300V1 system that features a superheterodyne/digital instantaneous frequency measuring receiver, a new power supply and new operating software to improve its performance. As an option, the receiver may incorporate direction-finding capabilities in the C/D-band (0.5 to 2 GHz). The system can include auxiliary equipment for library generation, loading/unloading its electronically erasable programmable read-only memories, maintenance purposes and software support. Threat data is presented to the operator via alphanumeric symbology on a high-brightness display. The position of the symbols on the display indicates the lethality and bearing of the detected signal, as well as identifying the threat emitter. Symbol blinks and audio alarms indicate high-priority radar types.

Specifications
Frequency coverage: C-J in 4 sub-bands
Azimuth coverage: 360°
Radars detected: all pulse and CW radars
DF accuracy: better than 12° RMS
Number of radars simultaneously displayed: 15

The EN/ALR-300V2 radar warning receiver

Receiver: channelised crystal video and superheterodyne DIFM
Symbology: programmable
Weight: 38 kg

Contractor
INDRA DTD, Madrid.

VERIFIED

EN/ALR-310 Radar Warning Receiver (RWR)

Type
Airborne RWR.

Description
EN/ALR-310 is a fully programmable RWR that incorporates a digital computer to provide emitter identification in complex signal environments. It is a lightweight, 360° coverage system that includes crystal video receiver technology. ALR-310 is noted as being suitable for fighter, light strike/attack and helicopter applications. It is fitted with a programmable emitter library and provides single identification of all detected emitters by alphanumeric symbology on a high-brightness cathode ray tube, ANVIS compatible display.

Status
As of this edition, the EN/ALR-310 RWR was reported as having been procured by the Spanish Ministry of Defence for use on helicopters.

Specifications
Frequency coverage: 0.7-1.5 GHz, and E/J-band in 2 bands
Radars detected: pulsed and CW
Azimuth coverage: 360°
DF accuracy: better than 15° RMS
Number of radars simultaneously displayed: 15
Symbology: programmable
Receiver: crystal video
Processor memory: EEPROM
Weight: 20 kg

Contractor
INDRA DTD, Madrid.

VERIFIED

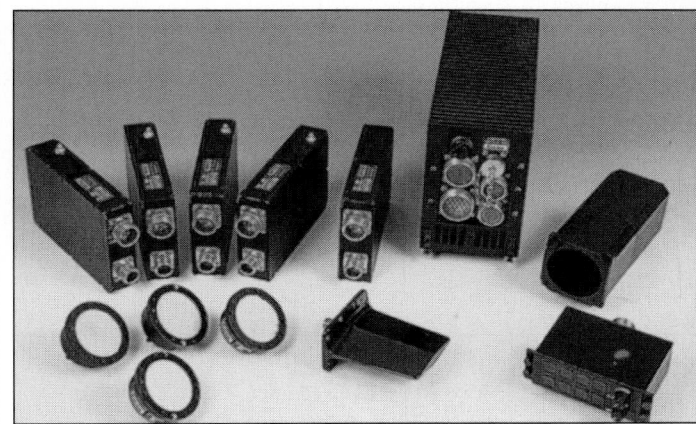

The EN/ALR-310 radar warning receiver

SIGMA SIGnals INTelligence (SIGINT) system

Type
Multispectral airborne SIGINT system.

Description
INDRA's SIGMA modular airborne SIGINT system incorporates multiple radar and communications band sensors and is designed to monitor and analyse dense tactical electromagnetic environments, detect and identify emitters and signals traffic and locate emitters. SIGMA makes use of a core INDRA developed software package that controls the systems' hardware, manages, processes and analyses received data and updates an associated signal/emitter type database. Overall, the SIGMA system employs a modular and configurable architecture that is built around a range of standard elements that includes an operator processor console, use of the Ada programming language, a UNIX-based operating system, an Ethernet local area network and standard interfaces. So configured, SIGMA applications can be 'mixed and matched' to suit specific customer requirements. Flexibility is further enhanced by the architecture's ability to accommodate both INDRA developed radar and communications band receiver equipment and

offshore sourced electromagnetic and electro-optic sensors. Alongside the described airborne elements of the system, SIGMA also incorporates a ground-based operational support facility.

Status
As of this edition, a 'basic small aircraft' SIGMA configuration was reported as having been produced, while a larger, more capable application has been installed aboard a Spanish Air Force Boeing 707 SIGINT aircraft that, according to Jane's sources, is operated by the service's Torrejon de Ardoz-based Escuadron 408. Thought to have been previously designated as the SCAPA system, this latter configuration makes use of the described INDRA developed software package and standard component suite mated to Elta Electronics-sourced electromagnetic sensors and a Tamam-sourced long-range, stabilised, electro-optical observation system.

Contractor
INDRA DTD, Madrid.

VERIFIED

SOCCAM Electronic Support (ES) and Direction-Finding (DF) system

Type
Airborne, Very/Ultra High Frequency (V/UHF — 30 MHz to 1 GHz) band ES and DF system.

Description
INDRA's SOCCAM ESM and DF system is described as being a modular, platform adaptive, communications surveillance and monitoring equipment that makes use of a compact, low cost ESM receiver that incorporates a DF capability. Designed for both military and paramilitary applications, SOCCAM system features include:
- digital processing
- operator-based signals analysis
- multiple, automatically controlled receivers
- programmable (by threat and mission type), system specific, frequency search sub-bands
- standard cartographic applications
- digital audio recording
- an Ethernet local area network
- Ada/UNIX-based ruggedised workstations.

SOCCAM makes extensive use of commercial-off-the-shelf components and the airborne element of the equipment is complemented by a ground-based support centre. The system is also noted as being suitable for land-based vehicle and surface applications.

Status
As of this edition, the SOCCAM communications band ES and DF system was reported as having been produced for installation aboard CASA C-212 and Fokker F-27 aircraft of an as yet unidentified customer air force and navy.

Contractor
INDRA DTD, Madrid.

VERIFIED

SWEDEN

BOW-21

Type
E- through J-band (2 to 20 GHz) Radar Warning Receiver.

Description
The Saab Avionics BOW-21 RWR comprises:
- four (six as an option) Receiver Front end Units (RFU)
- a Low Band Antenna (LBA)
- a Receiver Processor Unit (RPU).

Of these, the baseline RFU incorporates an antenna assembly, a Radio Frequency (RF) preamplifier, filters and a microcontroller. Optionally, such units can be equipped with a K-band (20 to 40 GHz) antenna and front end and/or an interferometric antenna array and a phase discriminator. The LBA takes the form of a dedicated blade antenna for use with the system's C- to D-band (0.5 to 2 GHz) coverage option, while the RPU includes receivers, a pulse processor, an RWR/ Electronic Warfare (EW) computer and host platform interfaces. In its baseline version, the RPU incorporates a number of empty board slots that can be used to implement the equipment's spherical coverage (as against toroidal coverage in its baseline, four RFU configuration), C- to D-band coverage and/or digital receiver options. Of the various cited RPU sub-units, the receivers installed take the form of narrow and wideband units. Of these, the system's narrowband receiver is noted as being a four channel device with a 'full' monopulse capability. The local oscillator

used is a high speed synthesiser (facilitating optimised searches based on library information) while the device's precision synthesiser and narrowband Digital Frequency Discriminator (DFD) provide it with 'excellent' frequency accuracy and resolution. The BOW-21 wideband receiver takes the form of a four channel amplitude monopulse device that incorporates a 'high-resolution' DFD and switchable filters to handle 'interoperability' effects.

In terms of processing, the RPU's pulse processor is described as being a 'unique' but 'proven' design that functions in real time and makes use of all 'primary and derived parameters' in the correlation (pulse to burst and burst to emitter) process. The system's RWR/EW computer is a commercial-off-the-shelf device that is a single board equipment that incorporates 'several' serial channels, Flash memory, Ethernet and two PCI interfaces for standard or customised input/output. The unit makes use of an industry-standard VxWorks real-time operating system, undertakes emitter identification, passive ranging and threat evaluation and incorporates spare capacity to facilitate control of an onboard defensive aids suite. Secondary functions include built-in test control, mission data recording and emitter simulation for training purposes. The BOW-21 RPU sub-units are Versa Modula Eurocard (VME) compliant and the unit as a whole is housed in a 1 Air Transport Racking (ATR) Short sized casing. With regard to data display, the system can either generate an output for display on integrated cockpit displays (using a MILitary StanDard (MIL-STD) -1553 databus) or a dedicated threat warning display unit, as required.

Status
As of May 2001, Jane's source's were suggesting the BOW-21 development was launched during the mid-1990s to fulfill a requirement for a second-generation threat warning system for Sweden's JAS 39 Gripen next-generation combat aircraft. Saab Avionics reports that 'extensive' flight testing of a number of system configurations was undertaken before the then CelsiusTech Electronics (now part of the Saab group) was awarded a full-scale development and production contract for the system during the second half of 1999. Here, Jane's sources suggest that this contract formed part of the larger, November 1999, then year SKr1.2 billion (then year US$127 million) award received by Saab AB Gripen in respect of the final development of the Gripen's EWS 39 defensive aids suite (see separate entry). Saab Avionics also notes that a BOW-21 variant has been selected for use in Germany's IDS Tornado mid-life upgrade programme.

Specifications
Frequency range: 2-20 GHz (0.5-2 GHz and 20-40 GHz bands as options)
Coverage: ±45° (elevation, −5 dB, baseline configuration); ±90° (elevation, option); 360° (azimuth)
DF accuracy: 1° RMS (interferometric option); 7° RMS (baseline configuration, narrow and wideband)
RF accuracy: 1 MHz (narrowband); 5 MHz (wideband)
Dynamic range: 75 dB
Pulse density capability: 2 Mpps
Tracked emitters: 500
Emitter modes in library: 10,000
Reaction time: 1 s (max)
A/c interfaces: MIL-STD-1553B; RS-422
Power: 3 × 115 V (50 V A - RFU); 3 × 115 V (500 V A - RPU)
Cooling: conduction (RFU) and forced air (RPU)
Dimensions (l × l × d): 44 × 114 × 113 mm (LBA); 125 × 125 × 150 mm (RFU - excl installation specific casing); 256 × 194 × 387 mm (RPU)
Weight: 0.4 kg (LBA); 2.5 kg (RFU); 15 kg (RPU)

Contractor
Saab Avionics AB, Stockholm.

UPDATED

TURKEY

Personal Computer-based Loader/Unloader (PCLU)

Type
PCLU.

Description
STM's PCLU is designed to load electronic warfare and avionic system mission data files aboard combat aircraft such as the F-16 and to download and analyse information recorded during a mission. System features include:
- use of an Intel Pentium processor-based IBM compatible personal computer running Windows 95™
- 16 Mbytes of random access memory
- a removable 810 Mbyte hard disk (optional increase in capacity available)
- a 3.5 in, 1.44 Mbyte floppy disc drive
- a 24 cm, 640 x 480, 256 colour active matrix TFT colour liquid crystal display
- an 86 key dust- and shower-proof keyboard
- an integrated micro module pointing device
- an external PIO/SIO cable
- an NiMH battery and battery charger
- more than one protocol on a single system
- MIL-STD-1553 and RS-422 interface ports

- two PCMCIA slots for systems programmed via Type II or Type III PCMCIA memory cards
- verification of loaded data
- the ability to monitor and modify data (as a panel function) at any memory location
- optional secure software for data transfer via modem
- optional on-aircraft data loading and unloading via MIL-STD-1553 databus
- ruggedisation for field operations
- MIL-STD-461C and -810E compliance.

Status

As of this edition, STM PCLU configurations were reported as being utilised in support of F-16 multirole combat aircraft and Sea Hawk helicopters. STM is also known to have designed and developed a PCLU for the Turkish Air Force.

Specifications

Data loading time: 61 s
Data unloading time: 35 s
Capacity: >770 Mbyte
Temperature: −20 to +55°C (operating); −40 to +70°C (non-operating)
Humidity: 5-95% RH (non-condensing)
Power input: AC/DC combined adaptor (90-264 V AC (50-400 Hz), 20-32 V DC)
Weight: 7 kg

Contractor

STM Defense Technologies Engineering Inc, Ankara.

VERIFIED

UNITED KINGDOM

ARI 18241 Radar Homing and Warning Receiver (RHWR) system

Type

RHWR system for Tornado strike and interceptor aircraft. UK service designations ARI 18241/1 and ARI 18241/2.

Description

ARI 18241 is the standard radio frequency threat warning system installed on UK standard Tornado strike and Air Defence Variant (ADV) aircraft. The ARI 18241/1 variant equips the ADV and differs from the strike Tornado's ARI 18241/2 system in incorporating the higher level of direction-finding capability necessary for the interception role. Covering the C- through J-band (0.5 to 20 GHz) frequency range, ARI 18241 can detect and identify threats up to, and beyond, the radar horizon. The high-accuracy direction-finding is obtained by the use of interferometers.

The system uses a digital processor to accept the parametric data provided from the receivers (including frequency, pulse repetition interval, pulse-width and scan characteristics) and to compare these data with a comprehensive emitter library. Identified emitters are then passed to the cockpit display units. The onboard displays provide pilot and navigator with both tabular and graphic presentations of threats, as well as the overall radar environment.

Status

As of August 2001, ARI 18241/1 standard RHWRs were reported as being installed aboard Tornado F Mk 3 interceptors of the Royal (as of April 2001, 118 aircraft) and Italian (as of October 2000, 24 aircraft) air forces, together with Tornado Air Defence Variant (ADV) interceptors of the Royal Saudi Air Force (as of October 2000, 22 aircraft). As of the given date, ARI 18241/2 standard equipments were fitted to strike and reconnaissance Tornados operated by the UK (as of April 2001,

68 Tornado GR Mk 1, 24 GR Mk 1B, 64 GR Mk 4 and 1 GR Mk 4B aircraft) and Saudi Arabia (as of October 2000, 76 strike and 12 reconnaissance Tornado InterDiction and Strike (IDS) aircraft). In February 1989, the then Marconi Electronic Systems (now Part of BAE Systems Avionics - Sensor Systems Division) was awarded a contract to upgrade all ARI 18241 equipments in service with the Royal Air Force. As of the given date, delivery of these modified systems is understood to have been completed.

Contractor

BAE Systems Avionics - Sensor Systems Division, Stanmore.

UPDATED

Aware Radar Warning Receiver (RWR)

Type

Airborne RWR. UK service designation ARI 23491/1.

Description

The AWARE RWR is a modular system that can, by means of subassembly or line-replaceable unit substitution, be upgraded to electronic support measures standard. The baseline model is AWARE-3 that covers the E- through J-band (2 to 20 GHz) frequency range and is intended for helicopter and light tactical fixed-wing applications. The system detects and identifies pulse, pulse-Doppler, Continuous Wave (CW) and interrupted CW signals. An immediate identification of the threat is provided and the particular threat type is identified.

Identification is achieved by comparison with a stored threat library. The library, threat priorities and electronic countermeasures interfaces are under software control with the necessary data being loaded prior to a mission, allowing rapid updating of threats and priorities on a mission-by-mission basis. Additionally, security of system and intelligence data are enhanced by the ability to clear the library and software. The system operates in peak pulse densities of several hundred thousand pulses/s, radiated by many simultaneous emitters. The AWARE display is heading compensated with computer-controlled symbology. An extensive symbol library can be programmed with symbols of the user's choice.

The system consists of four planar spiral antennas, a hand-portable programme loading unit, two dual-channel crystal video receivers, a fast Instantaneous Frequency Measurement (IFM) receiver and a signal processor. Control of the system can be achieved either as an autonomous system (where a control unit and a 76.2 mm display are employed) or as an integral part of the mission management system, with control being exercised via a MIL-STD-1553B standard interface. All line-replaceable units are convection cooled. The four antennas are mounted in mutually orthogonal azimuth directions. The signals received are passed to the two dual-channel crystal video receivers. These receivers provide the preamplification, detection and compression of the video signals for input to the signal processor. They also provide controlled radio frequency outputs to the IFM receiver. The output of the IFM receiver is presented to the signal processor as a digital number. The signal processor determines the direction of arrival and time of arrival of the signals. The results of these operations, when combined with output of the IFM receiver, allow de-interleaving of the incoming signals to be performed. Once de-interleaved and characterised, the signals are classified by comparing them with a stored library of known emitters. Following classification, the source of emission and its range and bearing may be displayed in simplified plan form on a 76.2 mm monitor, or passed to the aircraft mission management system via a monitor, or via an MIL-STD-1553B databus.

Status

As of August 2001, AWARE-3 is noted as being installed aboard Gazelle and Lynx Mk 7/9 helicopters of the British Army Air Corps and SH-14D Lynx helicopters of the Royal Netherlands Navy. While not confirmed, it is possible that the AWARE-3 RWR is called Rewarder in UK service.

The elements making up the ARI 18241/1 RHWR system fitted to ADV Tornado aircraft

The forward hemisphere AWARE-3 antennas as applied to the British Army Air Corps' Gazelle helicopters (MS/BAE Systems)

Specifications

Frequency coverage: 2-18 GHz
Emission detection: pulse, pulse-Doppler, CW
Bearing accuracy: better than 10° RMS
Weight: 13 kg

Contractor

BAE Systems Avionics - Sensor Systems Division, Stanmore.

UPDATED

Hermes Electronic Support (ES) system

Type

Airborne ES system.

Description

Hermes is an electronic warfare surveillance system for the interception and analysis of radar signals. It can function as both an ES system and a threat warner and is equipped with a superheterodyne receiver for long-range signal detection. Digital processing is used to sort measured parameters that are then compared with those in an integral library to identify the emitter. Precise direction-finding accuracy is based on the use of phased-arrays. Separate antennas are used for high and low band reception in the C to J frequency bands (0.5 to 20 GHz). The standard display used with the system is a colour raster monitor that can be configured to show both graphic and tabular format data. Signals can be prioritised so that only the most relevant signals are displayed to the operator. The system can be controlled either from a standard keyboard or from an ergonomically designed infra-red programmable touch panel, with integral menu symbology.

Status

As of August 2001, BAE Systems Avionics is thought to be supporting the Hermes ES systems fitted to the Indian Navy's 15 Westland Sea King Mk 42B anti-submarine warfare helicopters. Whether or not the equipment is installed on the service's eight Mk 42 and three Mk 42A anti-submarine warfare helicopters remains uncertain.

Specifications

Frequency coverage: C- to J-bands (0.5-20 GHz)
Azimuth coverage: 360°
Elevation coverage: ±45°
Antenna type: reduced size spiral
Emitter library: more than 1,000 modes
Signal types detected: pulse, CW, ICW, RF agile, PRI agile, jammers
Weight: 110 kg (approx value)

Contractor

BAE Systems Avionics - Sensor Systems Division, Stanmore.

UPDATED

HOstile Fire INdicator (HOFIN) system

Type

Airborne shockwave passive warning system.

Description

The MS Instruments HOFIN system is a passive warning equipment that is designed to alert a helicopter pilot that his aircraft is under fire. The first indication is an audible warning that draws the pilot's attention to a visual display that shows the general direction of the threat. This enables evasive action to be taken according to the current operational situation. The complete system (which weighs less than 6 kg) comprises a sensor array, a computer unit and an indicator unit. Of these, the sensor array is mounted beneath the helicopter and is aligned along its longitudinal axis. It detects the shockwave front generated by the projectile and converts the impulsive pressure change to electrical signals. The electrical signals are fed to the computer unit for further processing. The equipment's computer unit is housed in a ⅜ Air Transport Racking (ATR) standard short case and is normally mounted in the equipment bay of the aircraft. This unit processes the signals provided by the sensor array and computes the general direction of the hostile fire. The computer unit generates the 1 second audible warning signal, which is fed into the intercom system of the helicopter and also drives the indicator unit. The HOFIN indicator unit is housed in an instrument case (which is normally located on the instrument panel) and consists of a circular red display which is divided into eight 45° octants. When displaying information to the pilot, four adjacent octants are illuminated for 5 seconds. The illuminated arc rotates in 45° steps and shows the source of the hostile fire relative to the longitudinal airframe axis. The indicator unit also houses the operating controls for the system and consists of the display dimmer control, power on/off switch, system test switch, and the reset switch.

Status

As of May 2001, HOFIN was reported as having been supplied to the defence forces of Canada, Italy, the UK and a number of other countries.

Contractor

MS Instruments plc, Bromley.

VERIFIED

PVS 2000 Missile Approach Warning (MAW) system

Type

Active pulse-Doppler MAW system.

Description

The PVS 2000 MAW uses a low-power pulse-Doppler radar to detect the approach of missiles from the rear and to initiate countermeasures automatically. The equipment has been optimised for the low-level strike aircraft (such as the Tornado GR Mk 4 and Harrier GR Mk 5/7) and is capable of detecting small high-speed missiles in a dense radar clutter environment. Variants of the system, giving all-round detection of missiles at greater ranges, are in development for other aircraft types and roles, including helicopters, fighters, transports and land applications. Functionally, PVS 2000 transmits from its main antenna and phase compares the received echoes with internal reference pulses to identify returns with a Doppler shift. The hardware and software design discriminates between the real positive and the clutter negative Doppler signals. To obviate false alarms from clutter sources, a receiver-only forward-looking antenna is used to collect interference data that is then eliminated by software system algorithms.

PVS 2000 consists of five units:

- a transmit/receive unit containing a low-power solid-state transmitter and receiver, an integral antenna, analogue processing and digital conversion circuitry. The unit measures 15 × 25 cm and weighs 5.5 kg. It is designed for installation within 1 m of the main antenna
- a signal processing unit which houses all the digital signal processing necessary for detection and analysis of signal returns. This unit weighs 7 kg and measures 327 × 125 × 160 mm
- an active array main antenna unit and radome designed for installation in an area of the aircraft tail with a clear view aft. This unit measures 21 × 10 cm diameter and weighs 0.55 kg
- a cockpit control unit which provides on/off control and incorporates a built-in test status indicator lamp
- a small blade antenna mounted on any suitable site close to the main antenna but with a relatively clear view forward. This unit measures 150 × 100 × 50 mm and weighs less than 0.5 kg.

PVS 2000 pulse-Doppler technology also provides the basis for the Advanced MIssile Detection System (AMIDS) configuration that BAE Systems and Italy's Elettronica are developing for use as part of the Eurofighter/Typhoon multirole fighter aircraft's Defensive Aids SubSystem (DASS). As reported by Jane's sources during June 2001, the Eurofighter/Typhoon AMIDS is described as incorporating transmitter and receiver/processor line-replaceable units, one rear-facing and two forward-facing antenna assemblies and two low-noise amplifiers. So configured, the Eurofighter/Typhoon AMIDS is also noted as having the ability to be integrated with a passive missile launch warner for enhanced threat detection.

Specifications

AMIDS
Dimensions: 380 × 160 × 190 mm (transmitter); 480 × 160 × 190 mm (receiver/processor)
Weight: 2 kg (individual antenna and amplifier units); 13 kg (transmitter); 18 kg (receiver/processor)

Status

Over time, the PVS 2000 pulse-Doppler MAW system is reported as having been installed on Harrier GR Mk 5 and Mk 7 strike aircraft of the UK's Royal Air Force. With regard to the technology's AMIDS application, the then Marconi Electronic Systems (now a part of BAE Systems Avionics - Sensor Systems Division) was awarded a then year £16 million contract to develop the AMIDS MAW for the Eurofighter/Typhoon's Defensive Aids SubSystem (DASS - see separate entry) during April 1992.

Contractor

BAE Systems Avionics - Sensor Systems Division, Portsmouth.

UPDATED

The transceiver and processing units of the PVS 2000 MAW

Series 1220 laser warning system

Type
Multirole laser warning system.

Description
BAE Systems Avionics' Sensor Systems Division has developed a laser warning system that is suitable for use on a variety of aircraft, armoured fighting vehicles and ships. It is claimed to combine high dynamic range, low noise and a low false alarm rate. The system is designed to give visual and aural warning of a hostile laser emission at a range of several kilometres. The Series 1220 Laser Warning Receiver (LWR) incorporates 'sophisticated' parameter measurement features and a threat library, enabling non-ambiguous identification of 'most' military laser types. Series 1220 is an expandable system that is based on a ½ Air Transport Racking (ATR) standard short electronics unit that can accommodate a range of plug-in modules. This enables an appropriate system to be configured for 'any' platform and to be integrated with other defensive equipment. On most medium-sized platforms (combat aircraft, helicopters and armoured fighting vehicles) two sensor heads are required, mounted in such a way as to provide an unobstructed field of view. A processor can output raw data (electronic intelligence mode) or threat classification/identification and internal data storage can be provided. Control and display functions may be stand-alone or may be integrated with a radar warning receiver. The system is capable of semi-autonomous activation of countermeasures and optional modules provide additional functions, such as wavelength analysis or band extensions.

Status
Over time, Series 1220 equipment is noted as having been 'extensively' ground and flight tested 'internationally'. Trials on a Chieftain main battle tank have also been 'sucessfully' completed using a system variant able to detect far infra-red carbon dioxide lasers. Series 1220 equipments are also reported to have been sold to unspecified customers in Europe and Asia.

Specifications
Spectral range: 0.35-1.1 µm (basic system); 1-1.8 µm/8-11 µm (options)
Spectral resolution: 15-100 nm, using optional wavelength discriminator module, yielding unambiguous identification of the most common laser types (ruby, GaAs, Nd:YAG)
Angular resolution: ±22.5° in azimuth (15 or 10° optional)
Field of view: 360° azimuth; 55° elevation (typically −15 to +40° for land vehicles; other elevation biases available)
Mass: sensor head 0.4 kg; electronics unit 5 kg; control panel 0.4 kg; display panel 0.4 kg
Dimensions: sensor head 50 mm diameter × 25 mm high; electronics unit 127 × 194 × 323 mm (½ ATR short)
Data interfaces: RS-422 or MIL-STD-1553B

Contractor
BAE Systems Avionics - Sensor Systems Division, Stanmore.

UPDATED

The Series 1220 laser warning system

Series 1223 laser warning system

Type
Multirole laser warning system.

Description
The Series 1223 laser warning is a high dynamic range development of the Series 1220 equipment and is optimised for rapid declaration of laser beam-rider threats. The equipment is suitable for tri-service applications and can function as a stand-alone item or as a fully integrated part of a defensive aids suite. The system requires a minimum of two semi-cylindrical sensors per installation in order to guarantee a high probability of intercept. The normal configuration incorporates four sensor heads of the type illustrated. The Series 1223 laser warner is form fit compatible with an AN/AVR-2A(V) installation (see separate entry) and offers angle of arrival information rather than a quadrant of detection reading.

The sensor head used in a normal configuration Series 1223 laser warning system   0009906

Status
The Series 1223 laser warner is understood to have been selected for use in the Helicopter Integrated Defensive Aids Suite (HIDAS — see separate entry) that is to be installed on the British Army Air Corps' Apache AH Mk 1 battlefield attack helicopter.

Specifications
System field of view: ±45° (elevation); 360° (azimuth)
Spectral range: 0.4-1.7 µm (designators, range-finders and trackers); 0.8-1.1 µm (beamriders)
Power supply: 28 V DC (95 W, nominal)
Dimensions (h × w × d): 176 × 144 × 170 mm (standard sensor head)
Weight: 1.5 kg (standard sensor head)

Contractor
BAE Systems Avionics - Sensor Systems Division, Stanmore.

UPDATED

Sky Guardian 200 Radar Warning Receiver (RWR)

Type
Airborne RWR and electronic support measures system for helicopters and fixed-wing aircraft.

Description
Sky Guardian 200 is a radar warning receiver that uses digital processors to give accurate analysis of threats and utilises crystal video technology to provide high probability of intercept. Functionally, Sky Guardian 200 measures the parameters of the intercepted signal, which are then passed to a software controlled digital processor that analyses and identifies radars. Information is presented to the aircrew by alphanumeric presentation on a cathode ray tube display, on a common function display, or on a head-up display. Sky Guardian 200 will control a range of

The Sky Guardian 200 radar warning system

countermeasures including jammers, chaff dispensers and infra-red flares. Other system features include:

- a wide frequency range - units are available for all bands from C to J (0.5 to 20 GHz)
- detection of pulse, Continuous Wave (CW) and interrupted CW radars
- an interface for the host platform's radar
- signals identified as threats trigger visual and audible cockpit alarms, they also enable countermeasures to be initiated where appropriate
- a full direction-finding capability
- display persistence for fleeting intercepts
- data recording for playback and analysis.

Status

Over time, the Sky Guardian 200 RWR is noted as having been procured by Austria (for use on the J35Ö interceptor), the British Army Air Corps (limited fit aboard Lynx helicopters), Oman (Hawk 103/203 aircraft), the Royal Air Force (C-130 (trials installation), Jaguar and VC-10 fixed-wing types and Chinook and Puma helicopters) and the Royal Navy (Sea Harrier aircraft). In addition, the system is thought to have been fitted aboard Thai AV-8Ss, Indonesian F-5E/Fs and Hawk 63As operated by the United Arab Emirates. A Sky Guardian 200 variant is also understood to have been mandated as the 'RWR-of-choice' for the Aero Vodochody L-159 light combat aircraft.

Specifications

Frequency coverage: E-J (2-20 GHz) in 3 sub-bands; (C/D- (0.5-2 GHz) and K- (20-40 GHz) bands optional)
Azimuth coverage: 360° instantaneous
DF accuracy: better than 10° on all bands and options
Response time: <1 s
Multiprocessor memory: 500 kbyte RAM/PROM/EEPROM (100% growth capacity)
Processor operations rate: 5 Mops
Emitter library storage: In EEPROM accommodating over 400 emitters with 2,500 modes
Weight: 17 kg

Contractor

BAE Systems Avionics - Sensor Systems Division, Stanmore.

UPDATED

··

Sky Guardian 2000 Radar Warning Receiver (RWR)

Type
Lightweight airborne RWR.

Description
The Sky Guardian 2000 RWR is configured for rapid response in high-density environments, with the processing power to identify and display multiple threat signals in less than 1 second. The system can operate as a stand-alone RWR or be integrated with an active radar jammer and/or other onboard countermeasures subsystems. Sky Guardian 2000 can also accept and process data from a laser warning receiver and provides a common display for radar and laser threat information. The system has been designed from the outset to act as the core of an integrated electronic warfare system and can accommodate a defensive aids suite management module within its receiver/processor if required. Overall, Sky Guardian 2000 consists of four antennas, a control unit, a display unit and a ½ Air Transport Racking (ATR) standard-sized receiver/processor. The system's operational features include:

- long-range detection of airborne and surface threat radars
- display persistence for fleeting intercepts
- emitter identification with selectable display formats
- built-in recording for post-flight analysis
- effective and efficient use of countermeasures
- low band targeting options
- provision for future threat expansion
- the ability to be integrated into a multifunction cockpit control and display system.

The Sky Guardian 2000 radar warning system   0009905

Status
As of August 2001, the Sky Guardian 2000 RWR was reported as having been selected for use on the British Army Air Corp's Apache AH Mk 1 battlefield attack (as part of the type's Helicopter Integrated Defensive Aids Suite (HIDAS) – see separate entry) and the Royal Air Force's Merlin HC Mk 3 battlefield support helicopters.

Specifications
Frequency coverage: E-J (2-20 GHz) bands (C/D- (0.5-2 GHz) and K- (20-40 GHz) band options)
Frequency measurement: pulse and CW/ICW
Pulse density: >1 Mpps
Azimuth coverage: 360° instantaneous
DF accuracy: better than 10°
Polarisation: dual-polarisation option
Emitter library: over 4,000 emitter modes (expandable)

Contractor
BAE Systems Avionics - Sensor Systems Division, Stanmore.

UPDATED

UNITED STATES OF AMERICA

AN/AAR-44, AAR-44A and AAR-44(V) Infra-Red (IR) missile warning receivers

Type
Family of airborne IR missile warning receivers.

Description
As of this edition, the AN/AAR-44 IR missile warning receiver has been identified as appearing in three variants, respectively designated as the AAR-44, AAR-44A and AAR-44(V). AAR-44 systems are designed to provide a long-range, multithreat search and verification capability while continuously tracking already detected threats. Specific AAR-44/44A features include:

- scanning IR technology
- demonstrated pinpoint threat resolution for directional countermeasures steering
- functionality over the entire flight envelope
- variable sensor configurations for hemispheric or spherical fields-of-view
- multispectral discriminators for background and countermeasures rejection
- long-range detection to maximise the time available to counter threats
- full compatibility with EMCON operations
- full system operation with or without databus integration
- claimed high probability of detection and low false alarm rate.

AAR-44A differs from AAR-44 by virtue of its use of 'current technology' throughout the system, the availability of a Directional IR CounterMeasures (DIRCM) interface and its ability to be reprogrammed on the flight line. Jane's sources suggest that AAR-44(V) is a collaborative venture between BAE Systems Canada and Raytheon Electronic Systems that features:

- multicolour IR detection technology for 'positive' missile warning with a 'minimum' false alarm rate
- multiple, simultaneous threat detection
- resistance to decoy flares
- ±135° by 360° coverage in elevation and azimuth per sensor head
- better than 1° angular detection
- an integral laser-pointing growth path.

Status
As of this edition, AAR-44 IR missile warning systems are field proven, available and in service. The updated AAR-44A is also described as being field proven and

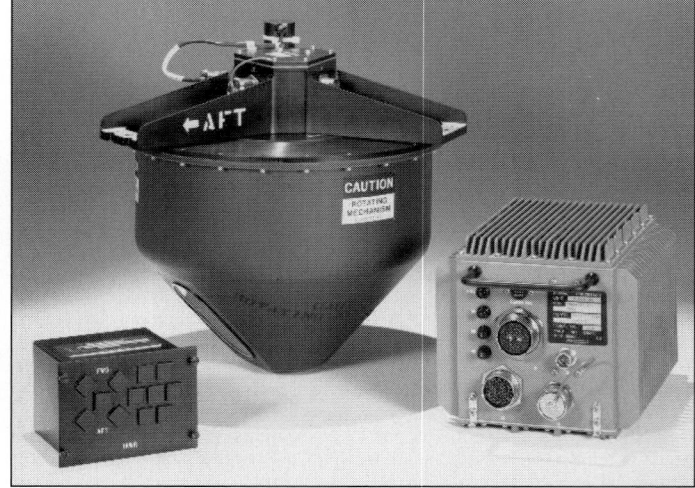

The AAR-44A infra-red warning system

available and is noted as having entered service during 1999/2000. As of June 2000, AAR-44(V) was noted as having been 'extensively' field tested.

Specifications

AAR-44
Technology: scanning IR
Display provision: crew visual, audio alerts and external countermeasures command
Compatibility: MIL-STD-1553B
Power: 115 V AC (400 Hz, 3-phase, 180 W - typical operating mode); +28 V DC (120 W - typical operating mode)
Temperature: –54 to +90°C (storage); –54 to +71°C (operating)
Altitude: 0-15,240 m (storage and operating)
Weight: 1.12 kg (control/display unit); 8.82 kg (processor); 17.91 kg (sensor heads)
Dimensions: 102 × 142 × 150 mm (control/display unit); 200 × 212 × 252 mm (processor); 400 × 369 mm (individual sensor head)

AAR-44A
Technology: scanning IR
Display provision: crew visual, audio alerts and external countermeasures command
Compatibility: MIL-STD-1553B
Power: 115 V AC (400 Hz, 3-phase, 155 W - typical operating mode)
Temperature: –54 to +90°C (storage); –54 to +71°C (operating)
Altitude: 0-15,240 m (storage and operating)
Weight: 1.13 kg (control/display unit); 8.27 kg (processor); 18.57 kg (sensor heads)
Dimensions: 102 × 142 × 150 mm (control/display unit); 200 × 212 × 252 mm (processor); 400 × 369 mm (individual sensor head)

AAR-44(V)
Detection range: beyond lethal range
Altitude: up to 13,716 m
Coverage: ±135° x 360° (single sensor head)
Cueing accuracy: better than 1° (angle of arrival)
Reliability: in excess of 1,000 h
Interfaces: MIL-STD-1553B; RS-232; RS-422
Power: 28 V DC (1.1 A); 115 V AC (1 A, 400 Hz)

Contractor
BAE Systems Canada, Mason, Ohio.
Raytheon Electronic Systems (AAR-44(V)), Goleta, California.

UPDATED

....

AN/AAR-47(V) missile warning system

Type
Passive warning system against infra-red homing missiles.

Description
Functionally, the AN/AAR-47 missile warning system detects Surface-to-Air Missiles (SAM) that are fired at helicopters and low altitude/slow-flying, fixed-wing aircraft and automatically provides a countermeasures response (via an AN/ALE-39, -40 or -47 countermeasures dispenser system) and audio/visual threat warnings to its host platform's crew. As such, the system comprises a Computer Processor (CP), a Control Indicator Unit (CIU) and four to six Optical Sensor Converter (OSC) heads. Of these, the OSC array is hard-mounted on the host platform's outer skin in such a way as to provide all-round coverage, with overlap between the individual heads to avoid blanking. Each OSC detects any SAM engine exhaust plume that is within its field-of-view and hands-off the acquired data to the system CP for processing. The CP analyses received data from the OSCs individually as an array and automatically triggers an appropriate countermeasures response to any perceived threat. Multidirectional threats are prioritised for countermeasures sequencing. For its part, the CIU displays the incoming direction

The AN/AAR-47(V) missile warning set

Aft facing AN/ AAR-47(V) OSCs as installed on a Royal Air Force C-130 transport aircraft (Tim Ripley) 1996

of the highest priority threat to assist with tactical manoeuvre responses. AAR-47 also features self-initiated built-in test and as of January 2002, was understood to be available in three configurations that are designated as the AAR-47(V), AAR-47(V)1 and AAR-47(V)2. Of these, AAR-47(V) is the original baseline configuration, AAR-47(V)1 is an upgraded version of AAR-47(V) (see Status) and AAR-47(V)2 is the AAR-47(V)1 configuration with the addition of a laser warning capability. As of 2001, US aircraft/helicopters that were fitted with the AAR-47(V) threat warner were reported as including the AH-1W, C-5, C-17, C-130E/H, C-141, CH-46, CH-53D, DHC-7, HC-130N/P, HH/SH-60, KC-130, MH-47E, MH-53J, MH-60, P-3 and UH-1N. Outside the US, AAR-47(V) systems have been acquired by Australia, Canada and the UK.

Status
Over time, Jane's sources suggest that AAR-47(V) has been manufactured by both Sanders (now BAE Systems North America - Information and Electronic Warfare Systems) and Alliant Techsystems. Of these, the then Sanders is understood to have received orders from the US Navy for at least 750 equipments for installation aboard helicopters and lower speed fixed-wing aircraft. For its part, Alliant Techsystems has received orders for in excess of 1,100 systems (the first of which was delivered in August 1993) to meet the needs of the USAF, the US Army and a number of allied air arms (see previously). During September 1996, the US Navy's Naval Air Systems Command awarded Alliant Techsystems' Alliant Defense Electronics Systems (ADES) subsidiary (now the Alliant Integrated Defense Company - Florida Operations) a then year US$2.8 million contract covering the design, development and manufacture of 60 sets of backplane and microprocessor circuit cards with which to upgrade the AAR-47(V)'s CP. This effort was designed to address limitations identified in the system's threat declaration and non-threat rejection capabilities, as well as increase available software memory. So upgraded, the AAR-47(V) is understood to become the AAR-47(V)1. On 10 December 1998, the then ADES was awarded two contracts covering, in the first instance, the design and development of upgrade kits and a flight-line tester for the AAR-47(V)'s OSCs (valued at then year US$12 million) and in the second, production of the previously described CP backplanes and microprocessor circuit cards (valued at then year US$6 million). Both awards contained production options. On 8 November 1999, it was announced that the ADES had been awarded the Clinton Presidency's Hammer Award for its leadership of a government-industry team that was estimated to be going to save the US taxpayer approximately then year US$240 million over the lifetime of the AAR-47(V) upgrade effort. It should be further noted that, alongside the AAR-47(V) threat warning system itself and appropriate flight-line testers, Alliant produces an AAR-47(V) sensor simulator unit.

Specifications
Coverage: 360° (azimuth - 6 OSCs)
Power supply: 28 V DC (4 OSC AAR-47(V) system - 4 W per OSC, 59 W for the CP)
Operational altitude: up to 5,240 m (AAR-47(V))
Dimensions: 120 × 200 mm (OSC - AAR-47(V)); 203 × 257 × 204 mm (CP - AAR-47(V))
Weight: 1.5 kg (individual OSC - AAR-47(V)); 7.9 kg (CP - AAR-47(V))

Contractors
Alliant Integrated Defense Company - Florida Operations, Clearwater, Florida.

UPDATED

AN/AAR-54(V) Passive Missile Approach Warning System (PMAWS)

Type
Passive electro-optic missile warning system.

Description
Northrop Grumman and the US Department of Defense (DoD) have developed and tested a PMAWS that can be used on a wide variety of aircraft and ground vehicles. The system is available to provide internal advanced missile warning for tactical and transport aircraft, helicopters and armoured fighting vehicles. It is also used to provide identification, missile tracking information and target cueing to Directed Infra-Red CounterMeasures (DIRCM) systems. Because of the adaptive design of AN/AAR-54(V), all applications can use common hardware and software.

The fine 1° angle of arrival discrimination capability of AAR-54(V) provides greatly reduced false alarm rates as well as detection ranges nearly double that of existing ultra-violet systems. Passive time to intercept is another system feature providing optimum cueing to countermeasures dispensers. AAR-54(V) provides all-weather, all-altitude operation while protecting against multiple simultaneous engagements in dense clutter environments.

The system consists of wide field of view, high resolution ultra-violet sensors and a modular electronics unit. From one to six sensors can be utilised, providing up to full spherical coverage. Full in-flight built-in test and fault isolation to a single sensor or electronics line-replaceable unit provides level-2 maintenance.

Status
As of June 2001, Northrop Grumman, the US DoD and the UK Ministry of Defence are noted as having completed AN/AAR-54(V) design verification testing. This trials programme is understood to have included several hundred live fire demonstrations with the warner installed on QF-4 drones, a cable car test rig and ground vehicles. The system has also been integrated into and demonstrated with a DIRCM system and the AN/ALQ-131 electronic countermeasures system. In April 1995, the UK's Ministry of Defence and the US Special Operations Command awarded Northrop Grumman a US$35 million engineering, manufacturing and development/production contract for the AN/AAQ-24(V) DIRCM system which includes AAR-54(V). In this application, the equipment provides missile detection, lethal missile declarations and fine angle of arrival hand-off to the DIRCM. AAQ-24(V) is scheduled to be installed on over 15 types of UK and US fixed-wing aircraft and helicopters.

Outside the AAQ-24(V) application, AAR-54(V) is known to have been procured by the Australian Navy (SH-2G(A)) and the air forces of Australia (Boeing 737 'Wedgetail' airborne early warning and control aircraft), Denmark (Mid-Life Update (MLU) F-16s), Germany (C.160 transport aircraft), Japan (H-60), the Netherlands (AS 532U2, CH-47D and MLU F-16s), Portugal (C-130H Hercules) and Norway (MLU F-16s). Of these, the Netherlands' helicopter application involves 13 CH-47D and 17 AS 532U2 aircraft, while the Portuguese fit teams a six sensor AAR-54(V) configuration with the AN/ALE-40 countermeasures dispensing system and TERMA's Electronic Warfare Management Unit to create a defensive aids suite. According to Jane's sources, the F-16 MLU application involves integration of the AAR-54(V) sensor into TERMA's Pylon Integrated Dispenser System (PIDS), a configuration that was first flight tested during 1998.

Contractor
Northrop Grumman Electronic Systems sector - Defensive Systems Division, Rolling Meadows, Illinois.

UPDATED

AN/ALQ-78 Electronic Support (ES) system

Type
Airborne ES system.

Description
ALQ-78 is used aboard the Lockheed Martin P-3C Orion maritime patrol aircraft as its ES sensor. The antenna is carried on a pylon under the inner wing. It automatically detects and measures the characteristics and bearings of intercepted radar signals of anti-submarine and electronic warfare interest. The measured parameters and bearing of the intercepted signals are supplied to the aircraft central data processing system for evaluation, recording and presentation on the aircraft displays.

The system uses a high-speed rotating antenna and a scanning, superheterodyne receiver for acquisition of signals in specific frequency bands of particular interest to the P-3C. Operation is to a great extent automatic, based on parametric data. The countermeasures set normally operates in an omnidirectional search mode. When a radar signal of interest is acquired and analysed, ALQ-78 automatically initiates a direction-finding routine. The signal data is processed by the central data computer and formatted for readout on a multipurpose display.

Status
ALQ-78 is understood to have been in service aboard Lockheed Martin P-3C Orion patrol aircraft since 1969 and as of October 2001, is scheduled to be replaced aboard US Navy aircraft by a variant of the AN/ALR-66 system (see separate entry). Outside the US, ALQ-78 has been built under licence (by Mitsubishi) for use on Japanese P-3Cs and is reported to have been supplied to Netherlands for use on its

ALQ-78's detection antenna array is mounted in a pod beneath the P-3C's port wing (Martin Streetly)

Orion maritime patrol aircraft. Sources suggest that the ALQ-78A variant of the basic design was the original ES system fitted aboard Update II-5 and III P-3Cs.

Contractor
BAE Systems North America - Threat Warning and Defense Systems, Yonkers, New York.

UPDATED

AN/ALQ-128 Radar Warning Receiver (RWR)

Type
Airborne RWR.

Description
AN/ALQ-128 is a multimode RWR that forms part of the Tactical Electronic Warfare System (TEWS) installed on US Air Force (USAF) F-15 aircraft. Here, it is teamed with the AN/ALE-45 countermeasures dispensing system, the AN/ALQ-135 radar jammer and the AN/ALR-56 RWR. Few details have been released regarding this equipment, but it is known that recent updating changes have been included measures to modify its software to increase the number and type of threats that the equipment can recognise. The equipment probably covers the area beyond the upper J-band (10 to 20 GHz) limit of the ALR-56 receiver.

Status
As of this edition, most of the USAF's AN/ALQ-128 requirements were reported as having been met. Production of the system is now complete.

Contractor
Raytheon Electronic Systems, Fort Wayne, Indiana.

VERIFIED

AN/ALQ-131 receiver/processor

Type
Receiver/processor for the AN/ALQ-131 jamming system.

Description
The ALQ-131 receiver/processor is a self-contained, single modular package that fits within the ALQ-131 Block II jamming pod. It enhances the operation of this electronic countermeasures system by maximising its jamming capability and effectiveness against a multiple radar threat environment. This is accomplished through the concept of power management.

The equipment incorporates a wideband, frequency-agile, double-conversion, superheterodyne receiver together with a crystal video receiver for low-band coverage. The module has a self-contained processor performing automatic signal sorting and threat identification.

The system receivers conduct a signal search within a prescribed frequency range, under control of the processor. When a signal representing a threat is

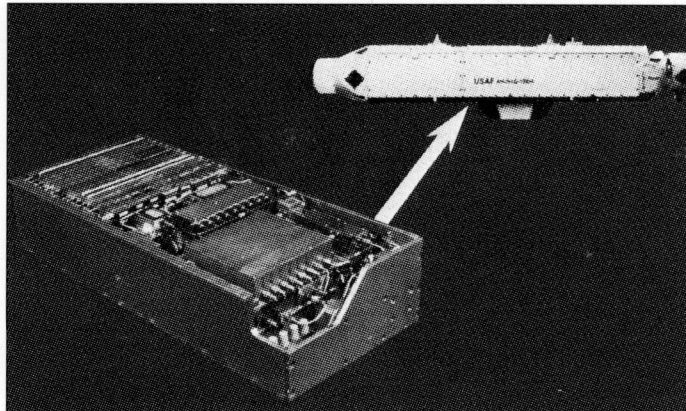

BAE Systems' AN/ALQ-131 receiver/processor showing its location in ALQ-131 jamming pod

acquired, it is analysed for parameter values, formatted and used to control the operation of the jammers. The jamming energy is applied with optimum timing and is better concentrated within the emitter's intermediate radio frequencies and servo bandwidths. The concept of power management thereby assures that energy waste and beaconing, (normally inevitable in conventional jamming) is minimised while facilitating jamming against multiple, simultaneous threats. A look-through feature of the receiver/processor permits continuous surveillance of the radar environment while jamming is in process.

Status
As of May 2001, the ALQ-131 receiver/processor module was noted as being in service with (or on order for) the USAF together with the air forces of Bahrain, Belgium, Egypt, Japan, Netherlands, Norway and Portugal. In terms of identified 1999 programme activity, 3 November 1999 saw the then Lockheed Martin Fairchild Systems (Yonkers, New York - now BAE Systems North America - Threat Warning and Defense Systems) being awarded a then year US$15,737,150 firm, fixed-price contract (F09603-99-C-0417) covering the supply of 39 ALQ-131 receiver/ processors to Egypt. At the time of the announcement, this Foreign Military Sales effort was scheduled for completion on 1 October 2001 with the USAF's Warner Robins Air Logistics Center, Robins Air Force Base, Georgia acting as the programme's contracting activity. As of May 2001, the latest identified programme activity saw BAE Systems North America - Threat Warning and Defense Systems being awarded a then year US$6,180,000 firm, fixed-price Foreign Military Sales contract covering the supply of 12 ALQ-131 receiver/ processor modules for installation in ALQ-131 radar jamming pods operated by the Bahrain Air Force. Announced on 16 February 2001, this effort was originally scheduled for completion by the end of December 2001 and its contracting activity was the USAF's Warner Robins Air Logistics Center, Robins Air Force Base, Georgia.

Contractor
BAE Systems North America - Threat Warning and Defense Systems, Yonkers, New York.

VERIFIED

AN/ALQ-142 Electronic Support (ES) system

Type
Airborne ESM surveillance system.

Description
AN/ALQ-142 is a lightweight ES system that provides surveillance in the 2 to 25 GHz frequency band for US Navy SH-60B Anti-Submarine Warfare (ASW)/anti-surface warfare helicopter. It acts as the airborne front end of a total ship/air Electronic Warfare (EW) system, and is datalinked to the ship's AN/SLQ-32(V) EW system and correlated with the onboard database to enhance system response. It is also suitable for other helicopters, marine patrol aircraft and small vessels (such as fast patrol boats). The system consists of three variants configured for ASW operations, surface ship surveillance and missile targeting and airborne area surveillance and missile targeting. Each system has been designed to fulfil a specific mission role. Intercept and direction-finding are performed over 360° in azimuth, with elevation coverage tailored to Light Airborne MultiPurpose System (LAMPS) requirements. Four Rotman-lens array antenna units (each providing 90° azimuth coverage) are located fore and aft on the helicopter with the receiver processor centrally located on board.

The ASW operations variant provides real-time identification and bearing of surveillance radars and their associated platforms. The variant makes use of the basic ALQ-142 helicopter configuration to which, modules can be added for increased performance. Sorting techniques are used to detect, identify and analyse all electromagnetic radiation from aircraft and surface vessels in the tactical area. It has a high probability of intercept on a single scan of a hostile radar at extended ranges. Threat bearing is measured on each pulse, and emitter identification accomplished by comparing measured data with parameters stored

in the AN/AYK-14 system computer. Emitter identification and bearing are displayed on the shipboard SLQ-32(V) operator console. Intercept and direction-finding are performed over 360° in azimuth with elevation tailored to the LAMPS requirements.

The surface targeting variant is formed by the addition of receiver modules. This allows surveillance and missile targeting from a small surface ship at ranges beyond the radar horizon, including the detection of long-range cruise missiles.

The maritime patrol variant provides additional frequency coverage for assessment of enemy forces from the air. Airborne area surveillance and missile targeting use the receivers in the ALQ-142 to identify emissions at extended ranges, to measure its bearing accurately and provide targeting information for air-to-sea and surface-to-surface missile systems.

An improved version, known as the ALQ-142(I), which incorporates a more sensitive and accurate interferometer, is being developed by the US Navy. Raytheon is also understood to have considered an electronic intelligence version of the Raytheon Beech Model 350 aircraft that incorporates a derivative of the ALQ-142.

Status
As of this edition, at least 200 AN/ALQ-142 ES systems are reported as having been produced.

Contractor
Raytheon Electronic Systems, Goleta, California.

VERIFIED

AN/ALQ-156(V) missile warning system

Type
Airborne pulse Doppler radar for missile detection.

Description
The AN/ALQ-156(V) missile detection system has been developed to provide a means of automatically triggering the ejection of decoys from an aircraft in response to an attack by heat-seeking and radar-directed missile(s). The system is compatible with the AN/ALE-39, -40, -47 and M-130 dispensers and comprises a radar transceiver, a control indicator and between two and four standard antennas, depending on the type of aircraft being equipped (see following). At the heart of the capability is the equipment's pulse-Doppler radar that creates a continuous protection ring around the aircraft and senses incoming missiles. This results in automatic triggering of a countermeasures dispenser (normally the M-130 unit - see separate entry) to eject the appropriate decoy. ALQ-156(V) evaluates the threat by comparison of the closing rates and other parameters, and is stated to be capable of operating in close proximity to the ground without detriment to the performance. The use of a microprocessor and a programmable digital signal processor permits rapid adaptation of the ALQ-156(V) from a helicopter to a fixed-wing aircraft configuration and it can be used in conjunction with other systems such as laser or radar warning receivers and radio frequency decoys.

As of this edition, two variants of the basic ALQ-156(V) design have been identified, the known details of which are as follows:

AN/ALQ-156(V)1
Weighing approximately 22.5 kg, the ALQ-156(V)1 missile approach warner comprises a pair of AS-3149/ALQ-156(V) antennas, an RT-1220/ALQ-156(V) transceiver and a C-10131(V) control indicator. Of these, the equipment's transceiver is noted as incorporating built-in test and the system as a whole is supported by the TS-3609 bench test set and the Maintenance Assistance Module (MAM).

AN/ALQ-156(V)2
Weighing approximately 25.3 kg, the ALQ-156(V)2 missile approach warner comprises four AS-3650/ALQ-156(V) circular horn antennas, an RT-1220A/ALQ-156(V) transceiver and a C-10131(V) control indicator. Like the (V)1 iteration, ALQ-156(V)2 is supported by the TS-3609 bench test set and the MAM.

A later model, ALQ-156A has been developed for the US Navy for tactical aircraft. It comprises a transceiver, a buffer storage unit and associated antennas and provides complete spherical cover around the aircraft. The buffer storage unit is a computer/controller that allows the system to communicate with onboard avionics systems. ALQ-156A integrates radar warning receiver and deceptive electronic countermeasures data to allow 'smart' control of decoys and flares. Flight line

The ALQ-156(V) missile detection set

reprogrammability adjusts the system responses to meet changing threats and theatre operational needs. Enhancements to ALQ-156A include a digital clutter filter to replace the analogue type and this more than doubles the effective range of detection.

Status

Over time, ALQ-156(V) is understood to have been procured for use aboard US Army aviation assets (applicable to the CH-47D (ALQ-156(V)1) and EH-60A (ALQ-156(V)2) helicopters and the C-23B and RC-12 fixed-wing aircraft (both ALQ-156(V)2)) and US Air Force C-130 transports. A development programme was initiated in 1986 to adapt the system for supersonic applications. Sanders received a then year US$31 million contract in October 1988 for 12 engineering development models of ALQ-156A for various US Navy tactical aircraft applications. Deliveries commenced in April 1991 and have been completed. In early 1990, an externally mounted pod configuration of ALQ-156A, installed on a QF-100 fighter drone, was tested successfully against live missile firings. On 23 December 1999, Sanders was awarded a then year US$1,964,260 increment as part of a firm, fixed-price then year US$7,100,000 Foreign Military Sales contract that covered the supply of 17 ALQ-156(V)2/(V)3T missile warners (together with a spares, documentation and in-country training package) for installation aboard CH-47D helicopters of the Elliniki Aeroporia Stratou (Hellenic Army Aviation). At the time of the increment announcement, work on the effort was scheduled for completion on 30 June 2001 and its contracting activity was the US Army's Communications and Electronics Command, Fort Monmouth, New Jersey. As of January 2000, Sanders noted that ALQ-156(V) equipments have been flown 'successfully' for more than 4,000 hours in a 'combat environment'.

Specifications

Weight: 20.5 kg (ALQ-156A); 22.5 kg (ALQ-156(V)1); 25.3 kg (ALQ-156(V)2)
Dimensions: 518 × 259 × 193 mm (transceiver)
Volume: 0.028 m³
Accuracy: ±4.5°
MTBF: 300 h
Power: 360 W(ALQ-156A); 425 W (ALQ-156(V))

Contractor

BAE Systems North America - Information and Electronic Warfare Systems, Nashua, New Hampshire.

UPDATED

AN/ALQ-210( ) Electronic Support (ES) system

Type
Multi-application (ES) system.

Description
Designed for fixed- and rotary-winged electronic intelligence, anti-submarine/surface warfare and airborne early warning applications, Lockheed Martin's AN/ALQ-210( ) ES system is described as combining a multibandwidth phase, frequency and amplitude measuring receiver with digital receiver technology and as comprising a commercial-off-the-shelf receiver/processor unit and four antenna modules. ALQ-210( ) employs a Versa Modular Eurocard (VME) - based open architecture and its combination of digital receiver technology and 'sophisticated' signal processing techniques is claimed to facilitate the detection of moving and stationary targets to an order of magnitude that 'far exceeds' that of 'classical' direction-finding approaches. Other noted system features include:

- claimed high probability of intercept in both open ocean and 'dense' littoral environments
- 'rapid' targeting solutions against short on-time radars
- 'unambiguous' emitter identification for 'high-confidence fratricide avoidance'
- a 'fast' reaction time and a threat mode change detection capability for self protection

The ALQ-210( ) ES system has been selected for installation aboard the USN's MH-60R Strikehawk helicopter, the first LRIP example of which is shown here (Sikorsky)
2002/0114483

The system's threat identification and mode determination capabilities are reported as being based on 'operationally proven' algorithms and digital parametric accuracy.

Status

As of January 2002, Jane's sources were reporting AN/ALQ-210( ) as having been selected for installation aboard the US Navy's (USN) MH-60R Strikehawk helicopter. The same sources also note that first of four Low Rate Initial Production (LRIP) MH-60Rs (USN serial number 166402) made its maiden flight on 19 July 2001.

Specifications

RF bandwidth: up to 1,000 MHz (instantaneous)
Processor: digital signal processing plus 200 MHz Power PC™
MTBF: > 3,000 h
Volume: 24.8 litres (receiver/processor); 21.6 litres (individual antenna module)
Weight: 40.4 kg

Contractor

Lockheed Martin Systems Integration - Owego, New Jersey.

NEW ENTRY

AN/ALQ-217 Electronic Support (ES) system

Type
Airborne ES system.

Description
Jane's sources describe the Lockheed Martin Systems Integration - Owego/Anaren Microwave AN/ALQ-217 ES system as comprising port and starboard active front-ends (amplifiers), four (port, starboard, forward-facing and aft-facing) antenna heads and a receiver/processor. This latter item incorporates a Lockheed Martin SP-103A Power PC 603e/704e/740™ single-board computer. Other system features include threat mode variance detection, non-developmental front-ends and a Versa Modular Eurocard (VME) - based open architecture with commercial-off-the-shelf processing. Within the Hawkeye 2000 Airborne Early Warning (AEW) aircraft application, ALQ-217 interfaces with the type's mission computer and tactical displays and is claimed as offering a wider frequency coverage, improved emitter classification and system reliability/maintainability and a 55 per cent reduction in installed weight when compared with the E-2C variant of the Hawkeye's AN/ALR-73 passive detection system.

Status

As of January 2002, Jane's sources were suggesting that 1999 saw a consortium of then Lockheed Martin Federal Systems (subsequently Lockheed Martin Systems Integration - Owego) and Anaren Microwave Inc awarded a then year US$33 million, five-year duration contract covering the supply of 21 ALQ-217 ES systems for installation aboard US Navy Hawkeye 2000 AEW aircraft. As of the given date, the same sources were reporting that Northrop Grumman was working on a selective emitter identification capability for the ALQ-217, development of which was scheduled for completion by mid-2003.

Specifications

MTBF: >1,400 h
Volume: 0.05 m³ (receiver/processor); 0.1 m³ (active front-ends/antennas); 0.15 m³ (complete system)
Weight: 42 kg (receiver/processor and tray); 44 kg (active front-ends/antennas)

Contractors

Lockheed Martin Systems Integration - Owego, New York.
Anaren Microwave Inc, Syracuse, New York.

UPDATED

AN/ALR-56A/C Radar Warning Receiver (RWR)

Type
Airborne RWR.

Description
The ALR-56 RWR is used with the AN/ALQ-135 Internal Countermeasures Set jamming equipment to form the Tactical Electronic Warfare System (TEWS) of the F-15 fighter aircraft.

The basic ALR-56 system incorporates the R-1867 processor/low-band receiver, the R-1866 high-band receiver, the IP-1164 display, the C-9429 immediate action control unit, the PP-6968 power supply, a TEWS controller and an antenna array. In more detail, the processor and low-band receiver unit contains three major sections: a single-channel low-band superheterodyne receiver, a dual-channel Intermediate Frequency (IF) section and a processor.

The low-band receiver is electronically tuned under control of the processor. The dual-channel IF section operates with either the high-band dual-channel receiver or

The wingtip antenna pod used in the Danish C-130 ALR-69(V) installation (Martin Streetly)

The starboard hemisphere ALR-69(V) antenna installation fitted to Netherlands' Fokker 60 transport aircraft (Martin Streetly)

warning and direction-finding on continuous wave signals, accurate frequency measurements on pulse signals for ambiguity resolution, threat antenna scan and rate analysis, and frequency set on for jammer power management and blanking functions.

Outside the US, Jane's sources suggest the Danish contractor Radartronic A/S has reconfigured the ALR-69(V) system for a specific Danish Army helicopter application. Reported as being designated as the 'ALR-DK' or 'Mini-69' system, the repackaging effort is understood to have created an equipment that weighs (in baseline form) 13.6 kg and comprises an azimuth indicator, prime and auxiliary control panels, a signal processor, frequency selective and C/D-band (500 MHz to 1 GHz) receivers, four amplifier-detectors and an antenna array. When applied to helicopters, the frequency selective and C/D-band receivers are omitted. Size and weight reduction measures are said to include a redesigned control panel, new circuit boards and antennas and repackaged amplifier-detectors.

Status

As of July 2001, ALR-69(V) was no longer in production. Since 1978, Jane's sources suggest that more than 3,500 systems have been delivered to the US Air Force (USAF) and off-shore customers. Over time, the equipment has been fitted to A-10, F-4(D?), F-16A/B, Fokker 60, AC/C/MC-130 and MH-53J(?) aircraft. Of these, F-16 applications include examples operated by Bahrain, Denmark, Egypt, Netherlands, Norway and the USAF. Other known installations include those aboard Danish C-130H and Dutch Fokker 60/C-130H transport aircraft. The cited 'ALR-DK/Mini-69' variant is reported as being installed aboard Danish AS 550 C2 Fennec helicopters. Jane's sources also suggest that ALR-69(V) has been 'continually' updated and that during 1993, US contractor Henderson Industries was awarded contracts (valued at then year US$32 million plus) covering the supply of over 3,000 ALR-69(V) upgrade kits for the USAF and 'other' customers. On 27 August 2001, the Raytheon Systems Company (Goleta, California) was awarded a then year US$25,950,520 firm, fixed-price contract covering the design, development, test and production of a situational awareness upgrade for the ALR-69(V) RWR. At the time of the announcement, the programme was divided into Engineering, Manufacturing and Development (EMD) and production phases, with the EMD effort being initially tailored to produce a 'core' capability. Thereafter, software upgrade options/increments would be used to enhance the 'core' capability. As of late August 2001, work on the effort was scheduled for completion by the end of September 2004, with the USAF's Warner-Robins Air Logistics Center (Robins Air Force Base, Georgia) acting as the programme's contracting activity. Solicitations for this contract were garnered via the World Wide Web, with negotiations being completed by the end of August 2001.

Contractor

Northrop Grumman Electronic Systems sector - Defensive Systems Division, Rolling Meadows, Illinois.

UPDATED

AN/ALR-73 Passive Detection System (PDS)

Type
Airborne passive Electronic Support (ES) system.

Description
The AN/ALR-73 PDS is an airborne ES equipment developed for the Northrop Grumman E-2C airborne early warning aircraft. It is an improved version of and successor to the AN/ALR-59. Because of the extensive nature of the updating, the new model has been given its own US Army/Navy (AN) designation. The primary operational roles assigned to the E-2C consist of surveillance of airborne and surface forces, early warning of hostile aircraft and the exercise of real-time control over carrierborne tactical aircraft. ALR-73 is intended to augment the airborne early warning, command and control role of the E-2C by enhancing its threat detection and identification capabilities. It is a completely automatic, computer-controlled, superheterodyne receiver/processing system that communicates directly with the E-2C's central processor. The design of the system was motivated by four major considerations:

- very high probability of intercept in dense environments
- automatic system operation
- high reliability
- ease of maintenance.

Features of the ALR-73 which are related to its intercept probability performance include four quadrant 360° antenna coverage; four independently controlled receivers; dual processor channels and a digital closed loop rapid-tuned local oscillator. Other features concerned with automated system operation are low false alarm report rate; automatic overload logic an adaptive hardware AYK-14 computer and a degraded operation operating mode.

The PDS system can measure Direction Of Arrival (DOA), frequency, pulse-width and amplitude and Pulse Repetition Interval (PRI) simultaneously. Scan rate information is also available if called for by the central processor. Special emitter tags can also be provided. The PDS detects and analyses electromagnetic radiations within the microwave spectrum. It sends emitter data reports (pulse-width, PRI, DOA, frequency, pulse amplitude and special tags) to the E-2C's central processor via the PDS data processor. The central processor performs the identification function. The PDS immediately reports new emitters to the central processor. It eliminates redundant data on emitters after a programmable period of time, thus significantly reducing the data rate to the central processor. The PDS, a multiband, parallel scan, mission programmable, superheterodyne receiving system, covers the frequency range in four bands through step sweeping. Programmable frequency bands and dwell time permit very rapid surveillance of priority threat bands. Non-priority bands are also monitored, but at a reduced rate. Probability of intercept is increased without sacrificing sensitivity through the detection of both real and image sidebands.

Installation of the ALR-73 on an existing aircraft required a functional and mechanical modularity in the system partitioning to facilitate installation on the airframe. Antenna packages located in the four quadrants of the aircraft provide 360° azimuth coverage. The large phase tolerance of the binary beam minimises the sensitivity of the DOA system to radome distortion and aircraft reflection. All four bands in the normal mode of operation scan through their respective frequency limits independently and simultaneously.

Activity indications may be obtained on any of the four bands. Following an indication of activity, permitted dwell time is increased and processing of the intercepted signal is started. Dual signal processing circuits allow intercepted signals in any two of the four bands to be processed simultaneously.

The signal processor is a special purpose logic processor which performs pulse train separation, direction-finding correlation, band tuning and timing and built-in test logic functions. The signal processor also contains the input/output circuitry necessary for the computer to communicate with both the signal processor and the aircraft central processor. The AYK-14 provides the system control function, data storage and formatting. Variable frequency coverage, along with variable dwell times and processing times, provide a means for optimising probability of intercept for any given theatre.

The local oscillator system consists of three units: an Instantaneous Frequency Measuring (IFM)/Local Oscillator (LO) generator, LO amplifier and LO power divider. The system is unique in that it is extremely fast and accurate. The key unit in the LO system is the IFM receiver which samples the LO frequency and converts it

The ALR-73 PDS provides the Northrop Grumman E-2C airborne early warning aircraft with a passive detection capability (MS/Northrop Grumman)

to a digital tuning command. This closed loop system permits frequency measurement accuracy while being largely insensitive to environmental variations. The local oscillator signals from the IFM/LO are amplified by the LO amplifiers to overcome the long LO cable losses and are then routed, via the power divider, to the various receiver front ends. Most recently, ALR-73 has been upgraded with solid-state in place of travelling wave tube amplifiers. The system is also noted as having been successfully installed and integrated aboard a C-130 aircraft.

Status

As of early 2001, at least 181 ALR-73 systems were reported as having been delivered to the US Navy (USN), France, Japan, Singapore and Taiwan. Another six such systems are understood to have been delivered during 1996 and in late 1998, the USN announced that it was considering the acquisition of a new electronic support system to replace ALR-73 aboard the Hawkeye 2000 variant of the E-2C. As of December 1999, this proposal had crystalised into the procurement of 21 examples of the Lockheed Martin Federal Systems AN/ALQ-217 system (see separate entry) for installation on the service's first tranche of new-build Hawkeye 2000s.

Contractor

Northrop Grumman Electronic Systems sector - Defensive Systems Division, Rolling Meadows, Illinois.

UPDATED

AN/ALR-76 Electronic Support (ES) system

Type
Airborne ES system.

Description

AN/ALR-76 is an ES system that was originally developed as a successor to the AN/ALR-47 in the US Navy's S-3B aircraft upgrade programme. It detects and processes electromagnetic signals in the microwave frequency region, with particular emphasis on radar transmissions of short duration. This includes tracking, classification, location and identification of emitters in dense environments. The system combines ES and radar warning functions in a single system and consists of two sets of spiral antennas, two multibandwidth receivers and a signal comparator. Audio alerts are provided, as are outputs for a countermeasures dispenser. Emitter identification, location and parametric data from the system are made available over a digital interface for sensor integration and display. ALR-76 makes use of an array of fixed antennas that drive 'high-sensitivity' receivers to provide a 'high-accuracy' amplitude direction-finding capability. The broadband spiral antennas installed provide five-octave coverage in a single unit, with each antenna being a planar cavity-backed device with an integral broadband balun feed and a protective radome. Looking at the equipment's receiver chain, each of its multibandwidth receivers features a wide bandwidth for signal acquisition together with a narrow bandwidth for signal analysis and interference processing. Dual-channel operation provides for full 360° instantaneous azimuth coverage. The signal comparator measures the characteristics of the video outputs from the four receiver channels simultaneously and provides an instantaneous 360° azimuth field of view and a monopulse direction-finding capability. All emitter parametric data is digitally encoded. Signal processing and system control functions are performed by computer programmes contained in a pulse processor and a general purpose Control and Correlation Processor (CCP), both of which are contained within the signal comparator unit. These two processors have an aggregate memory capacity of 448,000 words. Functionally, the pulse processor performs real-time sorting and tracking functions in very dense environments while the CCP facilitates the loading and storage of emitter identification libraries and also provides processed emitter data to the external interface. Total weight of the system, including eight spiral antennas, two receivers, one signal comparator and the antenna cabling is 61 kg.

The wingtip ALR-76 antenna assembly as applied to a US Navy EP-3E Aries II land-based SIGINT aircraft (Martin Streetly)

Status

As of October 2001, ALR-76 was installed aboard US Navy S-3B anti-submarine/anti-surface warfare and EP-3E Aries II signals intelligence aircraft.

Contractor

Lockheed Martin Systems Integration - Owego, Owego, New York.

UPDATED

AN/ALR-81(V) microwave receiver system

Type
Microwave receiver system.

Description

The Condor Systems AN/ALR-81(V) receiver system is a fully synthesised, microprocessor controlled, multiconfiguration equipment that is designed for general search, collection, signal analysis and recording tasks. The ALR-81(V) architecture is claimed to provide a 'high degree' of flexibility that has, over time, 'yielded several different versions of the system'. In baseline configuration (designated as the ALR-81(V)2), the equipment comprises the C-11701/ALR-81(V) Control/Display/Demodulator (CDD) unit and the TN-613/ALR-81(V) tuner. Of these, the C-1170 CDD unit includes a multitrace, digitally refreshed, spectrum/Intermediate Frequency (IF) pan display together with a 'complete analysis' demodulator. As such, the C-11701 requires 133 mm (5.25 in) of vertical rack space and features an integral MILitary STanDard (MIL-STD) -1553B remote control interface (Institute of Electrical and Electronics Engineers (IEEE) -488 as an option), multiple IF bandwidth filter selection and LOG/LIN/ frequency modulation detection modes. For its part, the TN-613 tuner is described as being fully synthesised and as being able to cover the system's 0.5 to 18 GHz frequency range in 'precision' 100 kHz steps. TN-613 is further claimed to offer 'low phase-noise combined with wide, instantaneous, IF bandwidth'. ALR-81(V) features a built-in test regime that facilitates field verification of performance (including memory, communications links and oscillator/synthesiser phase lock checks) and fault diagnosis and incorporates a high-resolution, vector graphics display. Here, the provision is noted as offering a flicker-free signal display 'regardless of the receiver scan or digital data update rates'. Spectrum information is provided over the equipment's complete frequency range and takes the form of combined, multitrace, band scan and IF pan displays with alphanumeric annotation. In 'Manual Tune' mode, ALR-81(V) provides its operator with a split-screen display that offers a frequency band overview with a high-resolution IF pan output. Condor claims that this split-screen format 'provides highly interactive and easy-to-interpret controls for the operator to tune signals-of-interest'. Jane's sources also suggest that ALR-81(V) is usually integrated with one or more AS-135 spinning direction-finding antenna assemblies, each of which, contains direction-finding and millimetric detection arrays and a downconverter.

Status

As of August 2001, an AN/ALR-81(V) microwave receiver system variant was in service aboard the US Navy's EP-3E Airborne Reconnaissance Integrated Electronics Suite (ARIES) II land-based signals intelligence aircraft.

Specifications
ALR-81(V)

Frequency: 0.5-18 GHz (extendable to 40 GHz with an appropriate downconverter)
Noise figure: 16 dB (typical, 0.5-8 GHz band, 23°C ambient); 19 dB (typical, 8-18 GHz band, 23°C ambient); 23 dB (max, 18.5 GHz, 23°C ambient)
Frequency accuracy: 1×10^{-9}/day and $\pm 10^{-9}$ over 0 to +50°C ($\pm$18 Hz at 18 GHz)
Displayed frequency resolution: to 100 kHz (sector limits to 1 MHz)
Phase noise: -70 dBc/Hz (typical, 100 Hz offset); -72 dBc/Hz (typical, 1,000 Hz offset); -82 dBc/Hz (typical, 10,000 Hz offset); -104 dBc/Hz (typical, 100,000 Hz offset); -122 dBc/Hz (noise floor); <1.8° RMS (max, integrated from 100 Hz to 20 MHz)
Frequency step: 100 kHz (synthesised tuned frequency displayed to operator)
Scan time: 30 ms (min, 0.5-8 GHz and 8-18.5 GHz sub-bands); 60 ms (min, 0.5-18.5 GHz band)
Image rejection: 70 dBm (min)

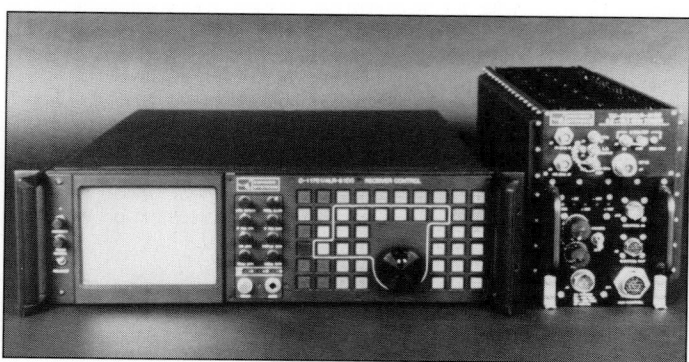

The AN/ALR-81(V) microwave receiving system

LO reradiation: −90 dBm (max); −100 dBm (typical)
Dynamic range: >90 dB (min, 1 dB compression, 1 MHz reference bandwidth)
IF bandwidths: 0.5 MHz (standard); 5 MHz (standard); 20 MHz (standard); 36 MHz (standard - bandwidths from 0.5 to 50 MHz available, plug-in field changeable IF filter modules, software filter identification for correct display format, gain bandwidth normalised IF filter modules for constant noise power output)
IF differential group delay: <35 ns (36 MHz filter, measured across a 1 dB bandwidth); <40 ns (20 MHz filter, measured across a 1 dB bandwidth); <85 ns (5 MHz filter, measured across a 1 dB bandwidth); <200 ns (0.5 MHz filter, measured across a 1 dB bandwidth)
IF outputs (demodulator): 70 MHz (post-filtered, 50 ±10% MHz bandwidth); 140 MHz (75 ±10% MHz bandwidth); 160 MHz (75 ±10% MHz bandwidth)
Wideband IF output (tuner): 1 GHz (prefiltered, 250 MHz min bandwidth)
Video outputs: LIN, LOG, narrowband FM, wideband FM 1, wideband FM 2 and selected video (any of the preceding)
Video levels: 0.2-2 V (LIN - into 75 Ω over a min 20 dB input power range, LIN gain control adjusts 20 dB window over a min 50 dB range; LOG - into 75 Ω over a min 70 dB input power range); ±1 V (into 75 Ω, gain scaled to the selected IF bandwidth)
Audio output: 600 Ω (to headset, audio output derived from selected video via 20 kHz low pass filter)
Digital control: fibre optic bus or RS-422 serial interface for data transfer between units
Temperature: 0 to +55°C (operating)
Operating power: 90-140 V AC or 160-240 V AC (47-440 Hz)
Power consumption: <115 W (tuner); <120 W (C-11701)
Dimensions: 133 × 483 × 559 mm (C-11701); 198 × 125 × 493 mm (tuner)
Weight: <14.5 kg (tuner); <21.3 kg (C-11701)
AS-135 antenna assembly
Frequency coverage: 0.5-18 GHz (400 MHz-0.5 GHz and 18-40 GHz options)
Azimuth beamwidth: 2-80° (typical)
Gain: +4 to +26 dBi (typical)
Beamshape: cosec^2 to 30° elevation (2-18 GHz frequency range)
Control: via a PE-105 pedestal electronics unit
Options: radome; integrated preamplifiers; vibration isolators
Dimensions (h × d): 48 × 61 cm
Weight: 23 kg

Contractor

Condor Systems Inc, San Jose, California.

UPDATED

AN/ALR-87 threat warning system

Type

C- through J-band (0.5 to 20 GHz) airborne threat warning system.

Description

The AN/ALR-87 airborne threat warning system is a replacement for the AN/ALR-46(V) equipment and is described as being able to handle both 'identified and postulated' threat environments. The equipment is based around a digital instantaneous frequency measuring receiver, is microprocessor controlled and consists of four high-band antennas, four quadrant receivers, a low-band antenna, a low-band receiver and power supply, a processor/receiver unit, an azimuth indicator and an indicator-control unit. ALR-87 is fully programmable and features an in-flight data recording capability together with interfaces for onboard active and passive countermeasures subsystems.

Status

As of this edition, ALR-87 is no longer in production. Jane's sources suggest that the Swiss Air Force has procured approximately 200 examples of the system for use on F-5E/F and Mirage IIIS aircraft. Of these, the Mirage IIIS has been withdrawn from service.

Specifications

Frequency range: 0.5-20 GHz
Processors: 3 MIL-STD-1750A microprocessors
Memory: 128 k (EEPROM); 162 k (RAM)
Mission data recorder: 1 M × 17 dynamic RAM
Interfaces: MIL-STD-1553B (avionics and electronic warfare bus); RS-232; RS-422
Power: 580 W
Weight: 0.13 kg (low-band antenna); 0.45 kg (4 × high-band antennas); 0.68 kg (indicator-control unit); 1.36 kg (azimuth indicator); 5.45 kg (low-band receiver/power supply); 10 kg (4 × quadrant receivers); 11.82 kg (processor/receiver)
Dimensions: 46 × 130 × 109 mm (indicator-control unit); 51 × 70 (Ø) mm (4 × high-band antennas); 82 × 82 × 222 mm (azimuth indicator); 102 × 127 × 76 mm (low-band antenna); 127 × 193 × 371 mm (processor/receiver); 175 × 155 × 41 mm (4 × quadrant receivers); 254 × 152 × 102 mm (low-band receiver/power supply)

Contractor

Litton (a wholly owned Northrop Grumman subsidiary), San Jose, California.

UPDATED

AN/ALR-93(V)1 Radar Warning Receiver (RWR) and Electronic Warfare (EW) suite controller

Type

C- through J-band (0.5 to 20 GHz) airborne RWR and EW suite controller.

Description

AN/ALR-93(V)1 is described as being a computerised RWR and EW suite controller that provides automatic detection and display of radio frequency signals and features a 'robust' tri-receiver (amplified crystal video, Instantaneous Frequency Measuring (IFM) and superheterodyne) architecture. It is designed to operate in 'dense' and 'complex' environments (with 'near' 100 per cent probability of intercept), generate aural and visual threat warnings and be integrated (if required) with a host platform's control and display system. A data recording capability is incorporated and an associated Software Support Facility (SSF) is used to replay recorded data for post-mission analysis. The SSF also facilitates user reprogramming of the equipment's integral emitter library. ALR-93(V)1 is further supported by a 'complete' manufacturer's integrated logistics package. Here, the service is described as offering 'complete' in-country support for both the system's hard- and software.

Status

As of late 2000, ALR-93(V)1 was being offered as part of the multicontractor Advanced Self-Protection Integrated Suite (ASPIS) EW system. At the given date, ASPIS comprised ALR-93(V)1 teamed with the Northrop Grumman/European Aeronautic, Defence and Space (EADS) Co AN/AAR-60 missile launch warner, the BAE Systems North America - Integrated Defense Systems AN/ALE-47 countermeasures dispensing system and the Raytheon Systems AN/ALQ-187 pulse/continuous wave radar jammer. Identified ALR-93(V)1 applications comprise Greek F-16C/D multirole fighters (as part of the ASPIS system), 'A-4M' fighter-bombers operated by a 'Latin American' country (confirmed by Jane's sources as Argentina's A-4AR Fighting Hawk multirole combat aircraft), CN-253 platforms operated by an 'Asian' country and Taiwan's Indigenous Defensive Fighter (IDF).

Specifications

Frequency: C/D-band (0.5-2 GHz) and E- through J-band (2-20 GHz)
Receiver types: amplified crystal video, IFM and superheterodyne
Radar types: CW, frequency agile, jitter/stagger, low probability of intercept, pulse, pulse compression, pulse Doppler and pulse repetition interval agile/frequency agile
Direction-finding accuracy: 15° RMS (E- through J-band); omnidirectional (C/D-band)
Emitter library storage: 1,800 modes
Programmability: both OFP and emitter library are contained on EEPROM
Software: 'C' high-order language
Power: <190 W
Interface options: discretes; MIL-STD-1553B (2 × dual redundant); RS-232C; RS-422
Warning: aural and visual
BIT: background; operated initiated (>95% LRU fault detection); maintenance initiated (>95% SRU fault detection)
MTBF: 525 h
Environmental: MIL-E-5400 (flight hardware)
Weight: <27.2 kg

Contractor

Northrop Grumman Electronic Systems sector - Defensive Systems Division, Rolling Meadows, Illinois.

UPDATED

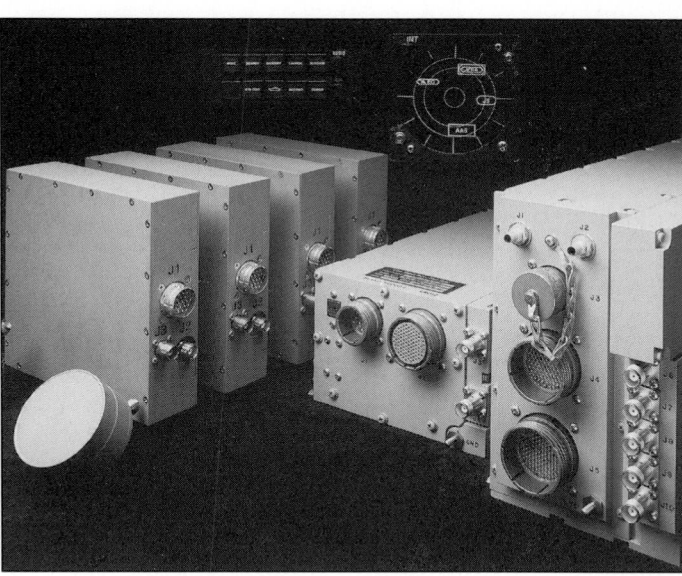

The ALR-93(V)1 RWR and EW suite controller

AN/APR-39A(V) threat warning systems

Type
Family of radar signal detecting and countermeasures suite control systems.

Description
The AN/APR-39A(V) threat warner was originally developed as a digital replacement for the analogue AN/APR-39(V) Radar Warning Receiver (RWR - see separate entry). As of July 2001, four APR-39A(V) variants have been identified, the known details of which are as follows:

AN/APR-39A(V)1
Produced by BAE Systems North America - Threat Warning and Defense Systems (with Northrop Grumman as second source), the AN/APR-39A(V)1 digital threat warning system is designed to provide timely and accurate detection and location of the threat emitters associated with anti-aircraft artillery and air-to-air and surface-to-air missiles. As described by LAS and the US Army, APR-39A(V)1 comprises a CP-1597/APR-39A(V) or CP-1597A/APR-39A(V) digital signal processor, two R-2218/APR-39(V) H- through M-band (6 to 100 MHz) receivers, pairs of AS-3548A/APR-39(V) and AS-3549A/APR-39(V) spiral antenna/detectors, a single AS-2890/APR-39(V) C/D-band (0.5 to 2 GHz) blade antenna, an IP-1150A/APR-39(V) radar signal indicator and a C-11308/APR-39A(V) detecting set control unit. Of these, the equipment's signal processor incorporates a main system control microprocessor, a video processor, display and articulated audio units, a multiplexer, a C/D-band receiver, a power supply and a User Data Module (UDM). The H- through M-band crystal video receivers are noted as providing continuous coverage across their entire frequency range, while the C/D-band receiver provides 'correlated' data on the status of detected emitters. In terms of system function, the equipment is Night Vision Goggles (NVG) compatible and can be 'fully' reprogrammed via the UDM. Here, all threat data is stored on electrically erasable, read-only memory chips, with the system processor's threat responses being changed via the simple expedient of slotting in a newly reprogrammed module. UDM removal and installation can be accomplished at unit level or during intermediate maintenance.

AN/APR-39A(V)2
Produced by Northrop Grumman, AN/APR-39A(V)2 is described as being a combined radar signal detector and countermeasures suite controller that is suitable for lightweight airborne applications including, among others, fixed- and

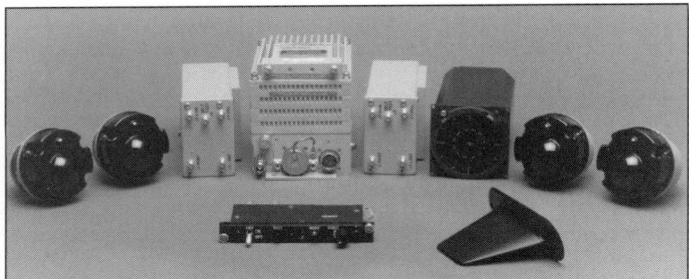

The line-replaceable units that make up the AN/APR-39A(V)1 threat warning system

The line-replaceable units that make up the AN/APR-39A(V)2 threat warning and countermeasures control system   0009909

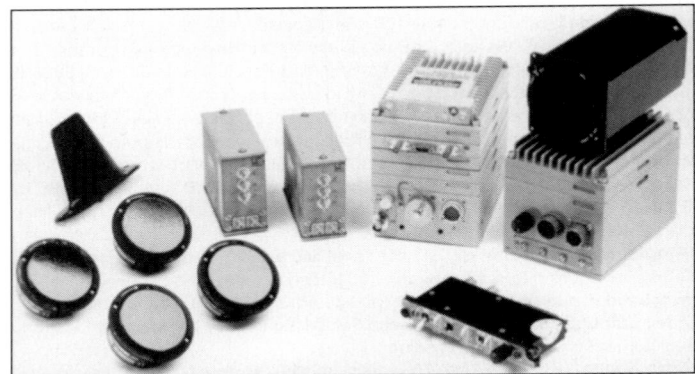

The line-replaceable units that make up the AN/APR-39A(V)3 threat warning system when it is configured for use with a MIL-STD-1553B databus   0009910

rotary-wing aircraft, tilt rotors and transports. The equipment covers the C/D- (0.5 to 2 GHz) and E- through K-band (2 to 40 GHz) frequency ranges and is understood to comprise two R-2390/APR-39A(V) dual-channel radar receivers, pairs of AS-4130/APR-39A(V) and AS-4131/APR-39A(V) antenna-detectors, a CP-1895/APR-39A(V) digital processor, an IP-1150A/APR-39(V) radar signal indicator, a C-11308/APR-39A(V) detecting set control unit and an AS-2890/APR-39(V) C/D-band blade antenna. Other system features include:

- automatic identification of threat types, bearing, mode and lethality
- 100 per cent use of Ada software language
- flight line threat library reprogrammability
- audio and graphic video threat alerts
- 360° coverage in azimuth
- interfaces for multispectral warning and semi-automatic/automatic countermeasures system control.

AN/APR-39A(V)3
Produced by BAE Systems North America (with Northrop Grumman as second source), the AN/APR-39A(V)3 threat warner is described as being specifically designed for helicopter and 'light' fixed-wing applications where the platforms are operating at low level ('nap-of-the-earth' type mission profiles). As marketed by Northrop Grumman, APR-39A(V)3 covers the C/D- (0.5 to 2 GHz) and E- through J-band (2 to 20 GHz) frequency ranges and comprises a digital signal processor, two crystal video receivers, four E- through J-band spiral antennas, a C/D-band blade antenna, an NVG compatible 7.6 cm (3 in) display unit and a control unit. Looking at some of these elements in more detail, the equipment's signal processor incorporates its main system control microprocessor, a video processor, display and articulated audio units, a multiplexer, a C/D-band receiver, a power supply and a reprogrammable UDM. The warner's C/D- and E- through J-band antennas are described as being 'off-the-shelf' units while the R-1838, dual-channel crystal video receivers used are noted as being 'US Government standard' items.

AN/APR-39A(VE)
Northrop Grumman describes AN/APR-39A(VE) as being a country specific variant of APR-39A(V)1/(V)3 technology that has been supplied to the Royal Norwegian Air Force.

Status
As of July 2001, the known status of various of the cited APR-39A(V) variants was as follows:

AN/APR-39A(V)1
Jane's sources suggest that over time, the US military has procured approximately 5,000 APR-39A(V)1 threat warners with additional units being exported. Within the US Army, the equipment is (or has been) applicable to the AH-1F, AH-64A/D, CH-47D, MH-47E, UH-60A/L, UH-60Q, MH-60K and OH-58C/D helicopters. Northrop Grumman also reports that APR-39A(V)1 is suitable for installation aboard AH-1S attack helicopters, C-130 transport aircraft, F-5A fast jets and UH-1H utility helicopters.

AN/APR-39A(V)2
As of mid-2000, APR-39A(V)2 is understood to have successfully completed US Navy/US Marine Corps (USMC) operational evaluation and to have been mandated

Specifications

	AN/APR-39A(V)1	AN/APR-39A(V)2	AN/APR-39A(V)3
Frequency range:	0.5-2 GHz/6-100 GHz	0.5-2 GHz/2-40 GHz	0.5-2 GHz/2-20 GHz
Power:	28 V DC/58 W	28 V DC/200 W	28 V DC/58 W
Software:		Ada (100%)	
Display:	bright light/NVG compatible	bright light/NVG compatible	bright light/NVG compatible
Audio:		synthetic voice	
Databus:	MIL-STD-1553B (optional); RS-422	MIL-STD-1553B; RS-422	MIL-STD-1553B (optional); RS-422
Interfaces:	E-O warners; missile launch detectors; radar jammers; CW warners	AAR-47; ALE-47; AVR-2/2A; ALQ-136; ALQ-162; ASM-687; mission data recorder; RRT	E-O warners; missile launch detectors; radar jammers; CW warners; CMDSs
Volume:	7,194 cm³	12,701 cm³	6,000 cm³
Weight:	7 kg	15.9 kg	8 kg

for installation aboard USMC AH-1Z, CH-46, CH-53 and UH-1Z helicopters, V-22 tilt rotors and KC-130 fixed-wing tanker aircraft. Northrop Grumman further notes that the equipment is also suitable for use on A-109, AH-1W, AH-64, CH/MH-3, CH-47, Cougar, EH-60, EH 101, HH-60 and UN-1N helicopters, C-130 transport aircraft, F-5 fast jets and RC-12 fixed-wing signals intelligence platforms. In terms of recent system production, the then Litton Advanced Systems (now part of Northrop Grumman) was awarded a then year US$9.8 million modification to an existing APR-39A(V)2 indefinite delivery/quantity contract (DAAB07-96-D-D601) on 13 July 2000. At the time of the contract modification announcement, work on the effort was to be split between the then Litton's facilities at San Jose, California (52 per cent workshare), College Park, Maryland (31 per cent) and Lansdale, Pennsylvania (17 per cent) and was scheduled for completion on 2 September 2002. The programme's contracting activity was the US Army's Communications and Electronics Command, Fort Monmouth, New Jersey.

AN/APR-39A(V)3

As of mid-2000, Jane's sources suggest that approximately 600 APR-39A(V)3 threat warners have been supplied to customers around the world. LAS notes that the equipment is suitable for installation aboard A-109, AH-1S, AH-64, B-105, CH-47D, Ecureuil, HH-60, Lynx, OH-58D AHIP, SH-70, UH-1H and UH-60 helicopters, C-130 transport aircraft and F-5A fast jets.

Contractors

BAE Systems North America - Threat Warning and Defense Systems, Yonkers, New York.

Northrop Grumman Electronic Systems sector - Defensive Systems Division, Rolling Meadows, Illinois.

UPDATED

AN/APR-39B(V) threat warning system

Type

Airborne threat warning system.

Description

The AN/APR-39B(V) threat warning system is described as representing a 'significant' increase in performance over the AN/APR-39A(V) series systems (see separate entry) and differs from its predecessors in adding a single line-replaceable unit to the overall architecture together with a number of 'plug and play' substitutes for existing APR-39A(V) components. Benefits derived are claimed to include improved ambiguity resolution, airborne intercept/pulse-Doppler radar handling, sensitivity, pulse density handling, continuous wave (including direction of arrival establishment) and processing capabilities. Alongside the described changes, APR-39B(V) equipments can be fitted with a number of optional enhancements, the known details of which are as follows:

1553B Communications Interface Unit (CIU)
Provides remote terminal functions and is capable of transmitting all system track file, display file and status data.

Situational Awareness Display (SAD)
The APR-39B(V) SAD features a datalink capability together with the ability to load electronic order of battle data (via a PCMCIA card) and record mission data (using the same card). The unit displays threat data on a digital map (synchronised with Global Positioning System (GPS) information) and provides a 3-D view ahead of the aircraft together with 'intervisibility' between the threat site and the host platform.

Articulated Audio Unit (AAU)
The APR-39B(V) AAU is a 'plug and play' replacement for the baseline audio card that requires no A-Kit modifications or software changes during its installation. The module is designed to improve the system's audio clarity and can be reprogrammed with any 'message or sound' in 'any language'. An editing station allows the user to set up own or prerecorded voice, digitised voice or sound and/or tone warnings.

Mission Data Recorder (MDR)
The APR-39B(V) MDR is a 'plug and play' add-on module that automatically captures emitter parameters, time tags each emitter intercept and is operationally transparent to both the host aircraft's crew and overall system operation. Acquired data can be downloaded on the flight line (using a laptop computer) and there is an optional GPS interface to facilitate the logging of platform location and heading at the time of intercept.

Serial Input/Output Unit (SIOU)
The APR-39B(V) SIOU provides a real-time RS-232C bus output.

APR-39B(V) equipments integrate 'fully' with external laser and missile warning systems and the described package can be used to upgrade APR-39A(V)1 and APR-39A(V)3 threat warners to APR-39B(V)1 and APR-39B(V)3 standard respectively.

Status

As of 2000, the then Litton Advanced systems (now a part of Northrop Grumman) is understood to have received a then year US$3 million contract from the Royal Norwegian Air Force covering the supply of 25 APR-39B(V) kits with which to upgrade the service's APR-39A(VE) threat warning systems. This effort is known as the 'Viking' programme. Elsewhere, Jane's sources suggest that the APR-39A(V)1 equipments fitted to US Army AH-64D attack helicopters may be configured with the APR-39B(V)'s optional 1553B CIU and that the Royal Netherlands Air Force has

procured sufficient APR-39B(V)2 systems to equip its fleet of 17 AS 532U2 and 13 CH-47D helicopters.

Contractor

Northrop Grumman Electronics Systems sector - Defensive Systems Division, Rolling Meadows, Illinois.

UPDATED

AN/APR-39(V) Radar Warning Receiver (RWR)

Type

Airborne RWR.

Description

The APR-39(V)1 RWR equipment provides automatic warning of emitters in the E, F, G, H, I and most of J radar bands (2 to 18 GHz), as well as the appropriate portions of C- and D-bands (0.5 to 2 GHz - BAE Systems North America neither confirms nor denies the quoted frequency bands). It is intended for use on either fixed-wing aircraft or helicopters. The equipment provides indications of bearing, identity and the mode of operation of detected signals with the acquired data being displayed on a cockpit indicator. Proportional pulse repetition frequency of displayed signals and alarm tones are presented to the crew via an integral audio warning subsystem.

The basic APR-39(V)1 system comprises two dual-video receivers, four spiral cavity-backed antennas, one blade antenna, an indicator unit, a comparator and a control unit. Without cables and brackets, this weighs 3.63 kg.

An updated version, APR-39(V)2 has been produced by BAE Systems North America. In this variant, the comparator is replaced by a CM-480/APR-39(V) digital processor. This performs signal sorting, identification of emitters, bearing computation and character generation for the presentation of threat details in alphanumeric form on the cockpit display. The unit weighs 6.5 kg and incorporates an adaptive noise threshold and angle-gate, programmable pulse repetition interval filters, and coded emitter outputs. A 19 kword programmable read-only memory/random access memory is provided.

Status

APR-39(V)1 was designed and developed by E-Systems (now part of Raytheon Electronic Systems) while the (V)2 variant was originally a Lockheed Martin product. As of October 2001, several thousand APR-39(V) equipments had been produced, 230 of which have been supplied to Germany for use on the PAH-1 anti-tank helicopter. Within the US Army, APR-39(V) RWRs are applicable to EH-60A and RC-12 Special Electronic Mission Aircraft (SEMA). Overall, the system is noted as having been installed in 15 types of aircraft and patrol/fast attack craft.

Contractors

BAE Systems North America - Threat Warning and Defense Systems, Yonkers, New York.

UPDATED

The elements making up the APR-39(V)1 RWR

AN/APR-44(V) Radar Warning Receiver (RWR)

Type

Airborne RWR.

Description

AN/APR-44(V) is a small, 'power-frugal', lightweight radar warning receiver designed to detect specific radar signals with omnidirectional coverage. Both visual and aural warnings of continuous wave threat radar signals are provided. The system is available in three configurations, (V)1, (V)2 and (V)3, for different frequency band coverage. The basic system comprises monopole antennas, a receiver and a control unit. The radio frequency circuitry, including switches, is a microwave integrated circuit module. False alarms are eliminated by special pulse rejection circuitry.

The (V)1 system employs an R-2097 low-band receiver covering the 6 to 10 GHz band, the (V)2 version an R-2098 high-band receiver covering 10 to 20 GHz band,

The AN/APR-44(V)3 radar warning system

while the (V)3 system employs both the R-2097 and R-2098. Other coverages are available.

Status
As of this edition, the AN/APR-44(V) RWR was reported as having been procured by the US Army and the US Marine Corps. In US Army service, the system was noted as being applicable to MH-47E, MH-60K, OH-58D and 'enhanced' UH-60A helicopters and RC-12 fixed-wing Special Electronic Mission Aircraft (SEMA).

Contractor
BAE Systems North America - Aerospace Electronics, Lansdale, Pennsylvania.

UPDATED

AN/APR-46A(V)1 receiver system

Type
Radar frequency surveillance system.

Description
The APR-46A(V)1 system is a high-performance, ruggedised receiver that covers the 30 MHz to 18 GHz frequency range. It incorporates control and display facilities which are claimed to make it ideal for threat warning/avoidance applications where high probability of intercept is required. In more detail, the system controller allows an operator to programme distinct frequency bands of interest and optimise these bands for prioritised probability of intercept. Many different scenarios can be programmed, together with control of the equipment's analysis display traces that may be tuned by a marker and bandwidth controls.

APR-46A(V)1 provides superheterodyne sensitivity together with fast scanning of all programmed scenarios. Two tuners' intermediate frequency outputs are routed to the demodulator that accomplishes both channels of demodulation and houses the central computer electronics for both tuners. Outputs are routed to a cathode ray tube display to create an eight trace digitally refreshed display for all scan and analysis activity. The user-control interfaces for the system allow all operator programming and control. The standard system uses two omnidirectional antennas for the 0.5 to 18 GHz spectrum and allows selection of these antenna inputs. A separate antenna is used for the 0.03 to 0.5 GHz region.

The eight-trace display is divided into six traces for scan activity and two for any selected region based on marker location, with adjustable bandwidths for each analysis trace. The internal memory cells allow the operator to programme up to two regions of frequency activity in each trace and assign a priority to each independent region. When compared with the baseline APR-46A, APR-46A(V)1 adds a monopulse Direction-Finding (DF) capability for the 0.5 to 18 GHz region, together with an angle of arrival polar display, a remote control facility (via a MIL-STD-1553B databus) and emitter identification processing.

Status
As of August 2001, AN/APR-46A/A(V) variants were reported as having been installed aboard MC-130 Combat Talon special forces transport aircraft of the US Air Force.

The AN/APR-46A(V) receiving system

Specifications
Frequency range: 0.03-18 GHz
DF accuracy: 5° RMS (0.5-18 GHz frequency range)
Emitter PRF: CW to 100 kHz
Emitter pulse-width: 200 ns to CW

Contractor
Condor Systems Inc, San Jose, California.

VERIFIED

AN/APR-48A radar frequency interferometer

Type
Lightweight airborne passive radar receiver and threat detection system.

Description
AN/APR-48A provides 360° continuous emitter (including early warning, ground targeting, counter battery and airborne types) target detection, identification and target azimuth for a range of air vehicles including the AH-64D Longbow Apache, the OH-58D Kiowa Warrior and the WAH-64D Apache AH Mk 1 helicopters. Such applications are mast mounted and incorporate a four element interferometer. A four element coarse Direction-Finding (DF) array with a 360° Field Of View (FOV) is used for initial signal acquisition, while the specific AH/WAH-64D application features a four element, long baseline interferometer (with a rotating 90° FOV) to enable it to provide fine DF cuts. The system is also noted as being able to look through a helicopter's rotor disc to detect targets at lower elevations than itself. Total system weight is given as 13.4 kg.

APR-48A has been designed for both helicopter and fixed-wing applications (including unmanned aerial vehicle) and can operate as a stand-alone system or as part of a target acquisition system. Here, it is used to cue electro-optic or Radio Frequency (RF) sensors. In the AH-64D application, the equipment is claimed to significantly reduce the aircraft's exposure time to threats, thereby increasing survivability and overall weapon system lethality. The equipment also lends itself to suppression of enemy air defences and reconnaissance tasks as was demonstrated during the US Army's APR-48A Initial Operational Test and Evaluation programme that took place during 1995. During this effort, the system is noted as proving itself capable of detecting radars in a dense countermeasures environment and providing continuous situational awareness even in extremely dense air defence emitter environments.

APR-48A is further reported as making use of 'advanced' packaging techniques that include the use of an ultra-lightweight, second-generation wideband antenna design, integrated gallium arsenide radio frequency components, Very Large Scale Integrated (VLSI) monopulse, parameter measurement technology and VLSI input/output chip design. The receiver used is a four channel, wide instantaneous bandwidth, superheterodyne equipment that employs delay line descriminators at

The APR-48A passive radar receiver and threat detection system is installed on the AH-64D battlefield attack helicopter   0017845

a high intermediate frequency and is typical of instantaneous frequency measuring systems. The equipment's manufacturer is also understood to have developed system upgrades that reduce APR-48A's weight, volume and power requirements for use in air defence, ship and fixed-wing applications while maintaining current system performance. The system may also be used as the primary sensor for its host platform's defensive survivability suite.

Status

According to Jane's sources, the then Loral Federal Systems - Owego (subsequently Lockheed Martin Owego and now part of Lockheed Martin Systems Integration - Owego) was awarded a then year US$5.1 million contract during the first quarter of 1996 that covered the supply of an initial tranche of APR-48A equipments for installation aboard US Army AH-64D battlefield attack helicopters. During May 1996, the then Lockheed Martin Owego received a then year US$3.4 million, six unit APR-48A contract from the UK as part of its WAH-64D Apache AH Mk 1 programme. In January 1998, the then Lockheed Martin Owego was awarded a then year US$94.4 million multiyear production contract covering the supply of 207 prime and 16 spare APR-48A systems for use on US Army AH-64Ds. During the same year, the company is further understood to have received a then year US$30 million contract covering the supply of 62 prime and spare APR-48As for use on British Apache AH Mk 1 helicopters. Here, deliveries are reported as having begun during December 1998 and as originally being scheduled for completion during February 2003. Elsewhere in the world, Jane's sources suggest that Both Israel and Netherlands have procured APR-48A with the two requesting 12 and up to 30 systems respectively during November 1999.

Contractor

Lockheed Martin Systems Integration - Owego, Owego, New York.

UPDATED

AN/APR-50 Defensive Management System (DMS)

Type
Aircrew threat awareness system.

Description
The AN/APR-50 DMS is installed aboard US Air Force (USAF) B-2 strategic bombers and is, according to Jane's sources, designed to present aircrews with a pre-programmed picture of hostile emitter locations around their aircraft, overlaid with updates based on signals intercepted by the platform's antenna farm. APR-50 is further described as being able to perform precision interferometric emitter location; as featuring a distributed architecture and as incorporating a wideband, channelised receiver chain. Readers should be aware that the system's manufacturer will neither confirm nor deny the accuracy of the foregoing system description.

Status
As of October 2001, the AN/APR-50 DMS was reported as being operational aboard USAF B-2 strategic bombers. According to Jane's sources, the US Department of Defense's Directorate of Operational Test and Evaluation (DOT & E) noted that the baseline version of APR-50 "failed to provide aircrew with timely, concise threat indications in all circumstances". Software changes designed to address such shortcomings are reported to have completed testing during April 1998 and to have been incorporated into the operational B-2 fleet (the APR-50 systems installed on Block 30 B-2 aircraft being described as being "fully capable in Bands 1 to 4"). Additional software upgrades were noted as being scheduled for implementation during 2001-2002.

Contractor
Lockheed Martin Systems Integration - Owego, Owego, New York.

UPDATED

AN/ASQ-213 HARM Targeting System (HTS)

Type
Radar emitter detection, location, identification and data processing system.

Description
The Raytheon Systems ASQ-213 HTS is a pod-mounted equipment that weighs 40.8 kg and is installed aboard US Air Force F-16CJ Suppression of Enemy Air Defences (SEAD) aircraft. It is designed to detect, identify and locate radar emitters and provide the data needed to calculate appropriate targeting and launch parameters for the AGM-88 High Speed Anti-Radiation (HARM) missile. As of this issue, the equipment exists in two configurations, designated as Lot I and Lot II. Lot I systems are operational and have been the subject of at least five software releases (Release 5 being under test during 1995). According to Jane's sources, the Lot II configuration (when compared with the Lot I system) incorporates an enhanced capability against low frequency emitters, increased processing power, better target resolution and improvements in the system's ability to handle multiple targets. On the F-16CJ, the ASQ-213 pod is carried on the aircraft's starboard chin station.

When installed on the F-16CJ SEAD aircraft, the ASQ-213 HTS allows the AGM-88 HARM missile to be fired in its most effective 'range-known' mode (USAF)   0053500

Status
ASQ-213 was developed and procured under a special access ('black') programme and was declassified (to 'secret/no foreigners' status) during August 1993. Lot II development is reported to have begun during US Fiscal Year (FY) 1996 and in 1997, ASQ-213 was publicly displayed for the first time as a precursor to possible export. Lot II flight-testing began during the summer of 1998 and as of this edition, the configuration is scheduled to enter service during US FY2000 in the form of a retrofit package for the existing Lot I pods. As of July 1998, the USAF was noted as having received 112 ASQ-213 HTS pods with a further 22 scheduled for delivery during US FYs 1998 and 1999. As of this edition, Jane's sources were suggesting that the USAF was planning to further upgrade ASQ-213 with improved range-finding algorithms and the ability to use targeting data transmitted via the Joint Tactical Information Distribution System (JTIDS). At the time of the report (September 1999), the programme to integrate the described JTIDS data reception capability into ASQ-213 (known as the 'R7' configuration) was scheduled to start during US FY 2005. In its export configuration, ASQ-213 is designated as the HARM Targeting System (Export) or HTS(E).

Specifications
Dimensions: 20 cm (Ø); 1.4 m (length)
Weight: 41 kg

Contractor
Raytheon Electronic Systems, Tucson, Arizona.

UPDATED

AN/AVR-2A(V) laser detecting set

Type
Airborne laser detecting set.

Description
The AN/AVR-2A(V) laser detecting set detects, identifies and characterises optical signals over 360° around the aircraft together with the provision of laser threat warning. It consists of four sensor units and an interface unit comparator. AVR-2A (V) interfaces with all variants of the AN/APR-39(V)/A(V) series radar signal detecting set to function as an integrated radar and laser warning receiver system.

Status
As of this edition, AN/AVR-2 series equipments were reported as having been procured by the US Army and Marine Corps. In February 1990, the then Hughes Danbury (now part of Raytheon Electronic Systems) was awarded a multiyear AVR-2 production contract. By mid-1995, approximately 700 systems are reported

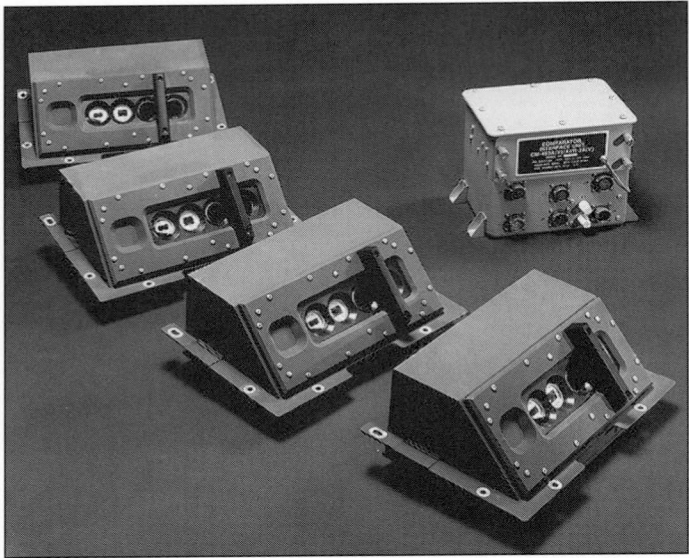

The four sensor heads and the interface unit which go to make up the AVR-2A (V) laser warner

to have been delivered to the US Army with a further 800 plus on order. As of the cited edition, AVR-2 series warners were understood to be mandated for use on US AH-1F, AH-1W, AH-64(AVR-2A), HH-60H, MH-47E(AVR-2A), MH-60K(AVR-2A), OH-58D(AVR-2A), and UH-1N helicopters. The system has also been proposed for use on the RAH-66, SH-60B, SH-60F and the V-22 air vehicles. Alongside AVR-2, Raytheon Electronic Systems is also known to have developed the MINiature LAser Warning Sensor (MINLAWS). MINLAWS comprises six sensor heads (three for each of two spectral bands) and a laser-processing interface unit and is reported as having been tested by the US Air Force.

Contractor
Raytheon Electronic Systems, Danbury, Connecticut.

VERIFIED

AN/AYR-1 Electronic Support (ES) system

Type
Airborne ESM sensor.

Description
AYR-1 is an ES system which is designated for the upgrade of US Air Force (USAF), French Air Force and NATO E-3 Airborne Warning And Control System (AWACS) aircraft. It is designed to carry out passive interception, identification and analysis of radar and radio signals. The extra processing units required to analyse and classify radar returns, plus associated equipment, are understood to add some 855 kg to the overall weight of the aircraft. Jane's sources suggest that AYR-1 is based on the one time Boeing subsidiary ARGOSystems' AR-1080 precision ESM equipment. AR-1080 is described as incorporating fast-tuning synthesisers and multiple baseline interferometer technology. Boeing claims that the system can offer direction-finding accuracies measured in tenths of a degree.

Status
The Electronic Systems Division of Boeing's Defense & Space Group has received a US$300 million contract to develop and integrate the AYR-1 ESM system into the E-3. During 1995, AYR-1 systems for the first 24 USAF/NATO aircraft were noted as

The nose and side antenna assemblies for the AYR-1 ESM as installed on a JE-3C trials aircraft

The aft facing antenna assembly for the AYR-1 ESM as installed on a JE-3C trials aircraft

being in production at the Boeing Facility in Corinth, Texas. Recent identified AYR-1 contracting activity includes:
- **15 November 1996**
  A then year US$38,579,590 increase to an existing Boeing AYR-1 contract that covered the supply of a fourth system and two laboratory kits for use in the USAF's E-3 Block 30/35 upgrade programme.
- **27 January 1997**
  A then year US$32,400,000 award to Boeing covering the supply of four AYR-1 systems for use on French E-3F aircraft. The first of these was installed during June 1999 with the full, four aircraft installation programme scheduled (as of this edition) to be completed by the end of December 2000.

As of this edition, AYR-1 is understood to be operational with the USAF and NATO.

Contractor
Boeing Defense & Space Group, Seattle, Washington.

VERIFIED

AN/USD-9(A) Improved Guardrail V COMmunications INTelligence (COMINT) system

Type
Airborne signals intelligence system.

Description
Improved Guardrail V is a US Army airborne COMINT system that intercepts, locates and classifies target systems and transmits data to ground processors to provide real-time intelligence information. The USD-9(A) system consists of a transportable ground-based system and RC-12D aircraft carrying remotely controlled mission equipment. Improved Guardrail V is reported to operate in three frequency bands: 20 to 75 MHz, 100 to 150 MHz and 350 to 450 MHz. The concept behind Guardrail is that data streams from multiple receivers are microwave downlinked to a ground-based integrated processing facility housed in four connected 15.24 m long trailers. In these, the operators can remotely tune and monitor the radios mounted in the aircraft. Time of arrival algorithms incorporated into the direction-finding software allow operators to determine the location of hostile transmissions.

Status
Two Improved Guardrail V systems were delivered to the US Army during 1984-85. As of this edition, the system is understood to be in service with the US Army's 15th Military Intelligence Battalion (Aerial Exploitation) at Robert Gray Army Airfield, Fort Hood, Texas. As of this edition, this unit's Improved Guardrail V architecture was scheduled to be replaced by the Guardrail Common Sensor No 2/Guardrail 2000 system (see separate entry) during 2000/2001.

Contractor
TRW Systems and Information Technology, Sunnyvale, California.

VERIFIED

AN/USD-9(B)/(C)/(V)D Guardrail Common Sensor SIGnals INTelligence (SIGINT) systems

Type
Airborne SIGINT system.

Description
The Guardrail Common Sensor system is the US Army's latest Guardrail variant and differs from its predecessors in that it provides a full SIGINT capability in a single airborne platform. This has been achieved by adding the Advanced

The RC-12K mission aircraft used in the USD-9(B) GRCS #4 system

QuickLook (AQL) ELectronic INTelligence (ELINT) and Communications High Accuracy Airborne Location System (CHAALS) subsystems into an evolved Improved Guardrail V COMmunications INTelligence (COMINT) architecture. Other aircraft upgrades included pooled receivers and modifications to the RC-12 host airframe so that it can accommodate an increase in mission system payload. Work has also been done to improve the system's ground-based Integrated Processing Facility (IPF) and Guardrail's interoperability with other US tri-service SIGINT assets. Of the two new airborne mission suite subsystems, AQL provides ELINT collection, emitter location (using the Time Difference Of Arrival (TDOA) technique) and target processing capabilities while CHAALS, as its name suggests, provides high-accuracy communications emitter location using both TDOA and differential Doppler techniques. As of this edition, there are four standards of Guardrail Common Sensor USD-9 deployed or scheduled for deployment.

AN/USD-9(C) Guardrail Common Sensor System Number 1 (GRCS #1)
Fielded in early 1995, GRCS #1 is the latest USD-9 system to be delivered and is a full capability architecture which includes some ELINT software processing improvements. The mission aircraft for GRCS #1 is the RC-12N.

AN/USD-9(V)D Guardrail Common Sensor System Number 2/Guardrail 2000 (GRCS #2)
GRCS #2 (also known as Guardrail 2000), which was rolled out during February 2000, retains the ground elements of GRCS #1 (upgraded to allow the system to operate in less dense signals environments with less than four IPF vans) and introduces a revised airborne mission suite and mother ship capability. In terms of the mission suite, GRCS #2 replaces the existing GRCS COMINT and direction-finding subsystems with wideband digital tuners and processors (developed for the USAF Senior Smart programme) and the existing CHAALS system with the newer Communications High Accuracy Location System (CHALS)-X equipment. Other improvements include the introduction of detailed ELINT/COMINT databases, an embedded operator training capability and improved airborne platform onboard processing and software. The mother ship capability will be provided by the specialised RC-12Q aircraft which, as well as carrying the GRCS #2 mission suite, will be equipped with a data correlator and a Direct Airborne Satellite Relay subsystem. The combination of these two items will allow GRCS #2 to operate outside the footprint of the US's Defense Satellite Communications System with the Q aircraft acting as a direct satellite relay for the system's RC-12P mission aircraft.

AN/USD-9(B) Guardrail Sensor System Number 3(-) (GRCS #3(-))
The GRCS #3(-) RC-12H airborne platform, as initially deployed, was equipped with the AN/USD-9(A) Improved Guardrail V mission suite and lacked (but had some provision for) the GRCS AQL and CHALS subsystems. GRCS #3(-) was upgraded to a full capability GRCS system during 1996.

AN/USD-9(B) Guardrail Common Sensor System Number 4 (GRCS #4)
GRCS #4 was the first full capability (COMINT, AQL and CHAALS) to be fielded. The GRCS #4 mission aircraft is the RC-12K.

Status
As of this edition, GRCS #1 was reported as being deployed with the 224th Military Intelligence Battalion based at Hunter Army Airfield, Georgia with RC-12N mission aircraft. GRCS #2 was scheduled to be deployed with the 15th Military Intelligence Battalion at Fort Hood, Texas during 2000/2001. GRCS #3(-) was located in South Korea where it was being operated by the 3rd Military Intelligence Battalion (Aerial Exploitation) with RC-12H mission aircraft. GRCS #3(-) was brought up to full GRCS#4 standard during 1996. GRCS #4 was located in Germany and was being operated by the 1st Military Intelligence Battalion (Aerial Exploitation) with RC-12K mission aircraft.

Contractor
TRW Systems and Information Technology, Sunnyvale, California (prime).

UPDATED

CS-2010 Hawk SIGnals INTelligence (SIGINT) and Direction-Finding (DF) system

Type
Airborne SIGINT and DF system.

Description
CS-2010 is a multiple operator, pooled resource SIGINT and DF system that covers the 20 MHz to 40 GHz frequency band. The equipment includes microwave tuners, demodulators, 20 MHz to 1 GHz upconverters, 18 to 40 GHz downconverters, and

The CS-2010 Hawk SIGINT and DF system

spinning DF antenna subsystems. System elements are maintained by the system as a pool of resources and are shared by the operators. Operators 'acquire' resources from the pool as needed and release them back to the pool when no longer required. Operators interface with CS-2010 through a Resources Control Console (RCC) consisting of a resource control unit, a keyboard and a pair of high-resolution raster scan display units. The RCC provides the operator with control of the pooled resource elements and provides him with real-time displays of both antenna and receiver functional outputs.

CS-2010 features a variety of operational modes including common band scan, analysis and DF. Common band scan is a display synthesised from the radio frequency scan spectral data produced by all 'unused' tuners in the system resource pool (that is the display is produced by parallel scanning all tuners which have not been 'acquired' by an operator for use in analysis of DF operation). The analysis displays (normal and expanded) provide the operator with radio frequency, Intermediate Frequency (IF) pan displays and LIFM for receiver channels 'acquired' by the operator from the resource pool. The operator has control over five independent traces representing up to five independent receiver channels. The DF displays include digitally generated polar, inverse polar and rising raster rectilinear. These support the manual DF capabilities of the system.

The system is built around VersaNet, a time domain multiple access bus developed by Condor for ELINT system integration. VersaNet distributes control information between systems elements and carries digitised display and audio information data generated in the system elements to the RCC. VersaNet enables additional resources such as tuners, demodulators, resource control units and pedestal electronic units, to be added to the system as required. The principle tuner in the family is the TN-618 microwave unit which features very low phase (<1° RMS), a 500 MHz wide IF and 10 kHz step size over the 0.5 to 18 GHz frequency range. Block upconverters and downconverters can extend the available coverage. The tuner also includes a built-in single bandwidth demodulator. An external full function demodulator with eight IF bandwidths, IF pan and LOG/LIN/frequency modulated video inputs is also available.

Status
As of August 2001, the CS-2010 Hawk airborne SIGINT and DF system was reported as being available.

Contractor
Condor Systems Inc, San Jose, California.

VERIFIED

CS-5550

Type
Airborne Electronic Support (ES) system.

Description
The Condor Systems CS-5550 airborne ES system includes six wideband spiral antenna units, four radio frequency heads, a high-gain Direction-Finding (DF) antenna assembly, a Radio Frequency (RF) distribution unit, a TN-613 tuner, an SP-150 processor, a removable hard drive and an operator workstation. Overall, the architecture is designed to provide situational awareness, signal analysis and signal recording facilities. Other system features include:
- automatic emitter identification
- a 10,000 mode emitter library
- wideband amplitude monopulse and narrowband (using a high-gain directional antenna) precision DF
- 'precision' parameter measurement

- 'high' throughput with frequency/azimuth mapping
- dual channel functionality
- a flat panel colour display
- a Windows-NT or Unix-based operator workstation.

Status
As of May 2001, Condor Systems was actively promoting the CS-5550 airborne ES system.

Specifications
Frequency range: 0.5-18 GHz (narrowband); 2-18 GHz (wideband)
Sensitivity: −60 dBm (wideband); −90 dBm (narrowband at 18 GHz)
Signal types processed: complex, continuous wave, jitter, normal, pulse-Doppler and pulse stagger
Parameters measured: amplitude, beamwidth, bearing, frequency, illumination time, pulse repetition frequency/interval, pulsewidth and scan
Digital signal recording: pulse descriptor words (recording and playback, up to 20,000 pulses)
Displays: analysis (bearing vs frequency, frequency vs time, polar DF, time vs amplitude and time vs time), built-in test status, intercept and library edit
Interfaces: audio out, Ethernet LAN, RS-232 and video out
Temperature: −55 to +55°C (antennas, RF distribution unit and wideband receiver); 0 to +55°C (processor)
Power: 30 W (wideband receiver); 35 W (RF distribution unit); 80 W (high-gain DF antenna); 140 W (tuner); 390 W (processor)
Dimensions (h × w × d): 914 × 137 × 213 mm (wideband receiver and RF distribution unit); 114 (Ø) × 102 (d) mm (wideband spiral antenna unit); 198 × 147 × 533 mm (tuner); 267 × 274 × 483 mm (processor); 483 (Ø) × 353 (h) mm (high-gain DF antenna assembly)
Weight: 0.2 kg (wideband spiral antenna unit); 4.1 kg (wideband receiver); 4.5 kg (RF distribution unit); 9.1 kg (high-gain DF antenna assembly); 16.3 kg (tuner); 21.4 kg (processor)

Contractor
Condor Systems Inc, San Jose, California.

NEW ENTRY

....

Emitter Location System (ELS)

Type
Airborne radar detection and location system.

Description
Raytheon Electronic Systems, under contract to the former DaimlerChrysler Aerospace, has developed an Emitter Location System (ELS) for use on the Tornado Electronic Combat and Reconnaissance (ECR) aircraft. The ELS detects, identifies and locates radar emissions through the use of a high probability of intercept system. It features multi-octave frequency coverage, phase

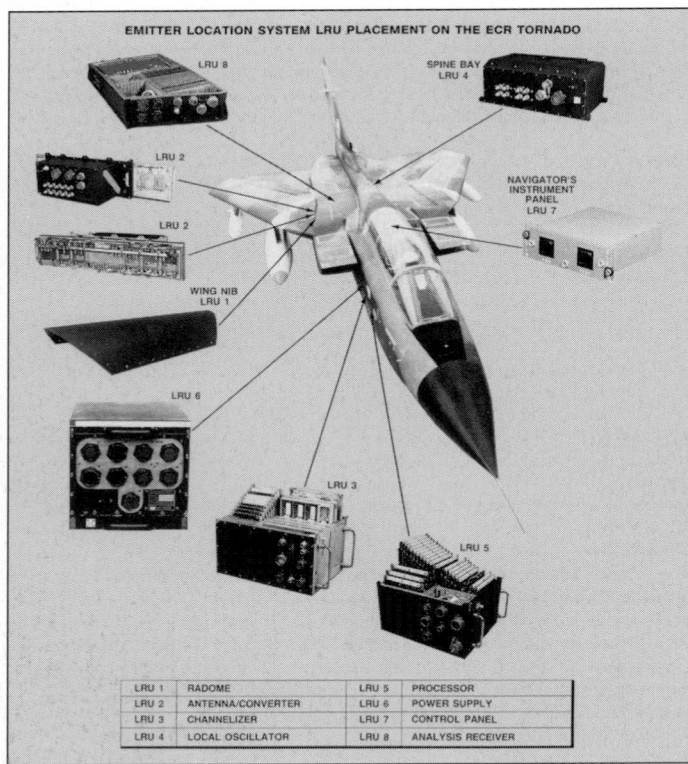

A schematic showing the layout of the ELS suite aboard the ECR Tornado aircraft

interferometric antenna arrays for precision direction-finding, passive ranging channelised receivers and a multiple 1750A digital processor.

The ELS operates across the radio frequency spectrum for all primary surface-to-air and air-to-air threats. Data acquired by the system are transmitted (via a MIL-STD-1553B databus) to the tactical displays of both crew members. Threat assessments can then be made and the appropriate countermeasures activated. Threats that have been located by the ELS can also be communicated to follow-on forces through an operational/data interface. This can then be used for threat avoidance or suppression operations.

The design uses a surface acoustic wave channeliser in a cued analysis receiver configuration. In this configuration, the delay lines allow the channeliser to measure the frequency of the incoming signal and (using a fast settling local oscillator) tune and cue the correct narrowband receiver for subsequence analysis and pulse report generation. The channeliser provides cued analysis contiguous high signal selectivity across a wide instantaneous bandwidth. The basic channeliser design consists of multiple parallel surface acoustic wave filter banks, each bank having multiple signal channels.

High-speed real-time emitter analysis is processed for positive threat identification. In addition, accurate direction-finding and passive ranging is computed within the processor, correctly locating and displaying large numbers of emitters. The processed information can then be used to cue weapons and can be configured to steer jamming pods.

The configuration of the ELS in the Tornado consists of eight individual Line-Replaceable Units (LRUs) which, together with the forward-looking infra-red sensor, weigh about 205 kg. The two antenna/radio frequency converter units are mounted conformal with the aircraft's left and right wing roots. The remaining units are distributed within the spin, shoulder and gun bays. Operator control is provided by the ELS control panel and the Tornado's TV/TABS display.

Supporting the ELS in the German Tornados is a defensive aids package comprising a Litton/former DaimlerChrysler Aerospace radar warning receiver, a former DaimlerChrysler Aerospace/Elta countermeasures pod and a SaabTech BOZ 100 chaff/flare dispenser. The Italian ECR Tornados will include a radar warning receiver and radar jammer from Elettronica and the SaabTech dispenser.

Status
As of this edition, 70 ELS systems have been procured with 35 going to the German ECR programme, 16 to the Italian ECR programme and 19 to a common spares holding. Within the Luftwaffe, the ECR Tornado is operated by the 321st and 322nd *Staffeln of Jagdbombergeschwader 32* based at Lechfeld. Aircraft from this *Geschwader* have been detached to Italy in support of UN/NATO operations in Bosnia. Within the Italian Air Force, the service's 16 ECR aircraft (local designation Tornado IT ECR) are being issued to the *155° Gruppo Caccia Bombardieri/50° Stormo Caccia Bombardieri* which is based at Piacenza-San Damiano. The *155°Gruppo* received its first IT ECR aircraft on 27 February 1998. Both German and Italian ECR aircraft are reported as having flown defence suppression missions during NATO's 1999 vintage Operation 'Allied Force'.

Contractor
Raytheon Electronic Systems, Tucson, Arizona.

UPDATED

....

ES5000 SIGnals INTelligence (SIGINT) system

Type
Airborne SIGINT system.

Description
The ES5000 is a combined communications and electronic intelligence system that can be installed in most types of aircraft. It performs signal interception, direction-finding and analysis, data collection, and location of 2 to 1,200 MHz band communication emitters and 0.02 to 18 GHz band non-communication emitters. Information gathered by the airborne system is transmitted by microwave link to a ground control station. The airborne SIGINT equipment can be controlled either by operators on board the aircraft for independent, special purpose SIGINT missions with preflight mission planning, or remotely from the ground analysis centre with no SIGINT operators on board the aircraft.

On a typical mission, one or two airborne collection centres communicate by microwave datalink with the ground centre. One ES5000 can perform signal intercept, direction-finding and emitter location, producing emitter line of bearing data. A two aircraft configuration can perform instantaneous emitter location on short burst transmissions. Both systems pass data to the ground analysis centre where it is analysed, combined and mapped.

Advanced capabilities of the ES5000 include:
- remote and autonomous 'standoff' SIGINT direction-finding
- accurate and rapid communication and non-communication emitter characterisation
- improved detection and location of short duration signals by automatic system tasking
- increased COMmunications INTelligence (COMINT) accuracy by the use of wide aperture direction-finding antenna arrays
- high operating speeds through multiple receiver channels combined with high-speed synthesis, switching and measurement
- wideband ELectronic INTelligence (ELINT) intercept, up to 500 MHz instantaneous bandwidth

In Egyptian service, the ES5000 SIGINT system is installed aboard Raytheon Beech 1900C aircraft as shown here

- choice of automatic collection or manual analysis mode
- modulation independence for any type of signal modulation
- analysis of fused COMINT and ELINT data
- database storage and display of emitter map location and signal data.

Status
As of this edition, Jane's sources were suggesting that an ES5000 system variant constitutes the mission suite installed aboard four Egyptian Air Forces Beech 1900C SIGINT configured aircraft.

Specifications
COMINT subsystem
Frequency range: 2-1,200 MHz
Instantaneous coverage: 500 MHz
Field of view: primary 120° each side; secondary ±30° front and back
DF accuracy: primary ≤2° RMS; secondary ≤3°
Signal polarisation: vertical ±20°
Depression angle: +2 to −10°
DF rate: fine DF 4/s; coarse 12/s
ELINT subsystem
Frequency range: 0.02-18 GHz
Field of view: 360°
DF accuracy: 1-3° RMS
Antenna spin rate: 200 rpm
System measurements: frequency, PRI/PRF, pulse-width, amplitude, scan time, DF, time of arrival, intrapulse
PRI range: 2-10,000 µs
Agilities recognised: pulsed, 16-level stagger, CW, interval pulse, complex
Active emitters: 1,024
Emitter library size: 10,000 modes
Microwave link
Frequency: S-band
Max range: 350 km for 10,000 m altitude

Contractor
TRW System and Information Technology, Sunnyvale, California.

VERIFIED

....

EW-1017 surveillance system

Type
Airborne electronic surveillance equipment.

Description
As an airborne electronic surveillance system, EW-1017 automatically acquires and identifies emissions within the C- to J-bands (0.5 to 20 GHz). The system is also designed to receive and identify all those emissions illuminating the aircraft (including short bursts) in particular when it is operating in very dense signal environments. The warning of possible danger is given both visually and aurally on a display unit while preferential scan is used to ensure immediate recognition of possible lethal threats.

EW-1017 consists of antenna arrays, receiver, a processor system, an electronic support measures operator interactive display subsystem and a pilot's display. Broadband spiral antennas are used to provide omnidirectional coverage which, together with their separate multiband receivers, are mounted in pods on each wingtip. This location drastically limits aircraft 'shadowing' and the proximity of the receiver cuts signal losses to a minimum. Angular bearing of the emissions is determined by using selected pairs of antennas.

The hybrid superheterodyne receiver offers high probability of acquisition, high sensitivity, frequency accuracy and a high degree of frequency selection and selectivity. A broad bandwidth is used in the acquisition mode to obtain the initial intercept with a narrow bandwidth then being used for accurate bearing measurement and analysis. To ensure processing capability in highly dense signal conditions a high-speed digital computer performs the data processing functions, supplemented by microprocessors. This enables the receivers to scan the frequency band continuously on a reprogrammable basis, so that conventional, continuous wave and agile signals are processed for identification. For special applications the receiver can be interfaced via a smart post processor unit to a centralised tactical display and control system.

An interactive display subsystem provides a full range of operator facilities to manage and optimise collection. It also provides a readily accessible real-time

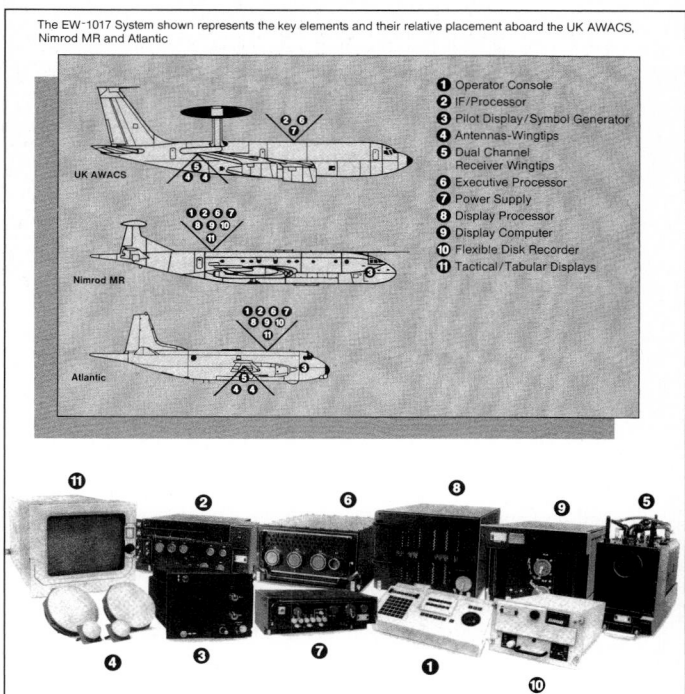

The EW-1017 system

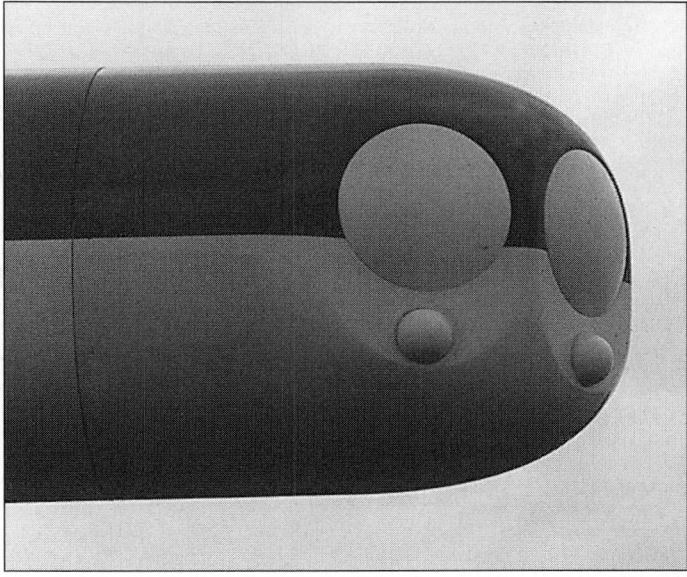

A close-up of the antenna arrangement at the forward end of the EW-1017's wingtip antenna housings (Martin Streetly)

emitter and platform library storage and analysis capability, facilitated by a modern data management system. A control/display unit allows the operator to monitor and control the automatic surveillance function to resolve possible ambiguities and evaluate and use to the best advantage the data displayed.

Status
As of October 2001, EW-1017 was reported as being in service with the German Naval Air Arm (Breguet Atlantic maritime patrol aircraft) and the Royal Air Force (British Aerospace Nimrod MR Mk 2 maritime patrol and Boeing E-3D Sentry AEW Mk 1 airborne warning and control system aircraft). In UK service, the system is designated as ARI.18240 'Yellowgate'. During April 1999, Racal Defence Electronics (now Thales Sensors) announced that it had been awarded an approximately then year £5 million contract that covered the upgrading of the 'Yellowgate' Electronic Support (ES) systems installed aboard the UK's Sentry AEW Mk 1 fleet. Jane's sources reported that the upgrade focuses on the enhancement of the equipment's reliability/maintainability, together with the introduction of Racal's commercial-off-the-shelf 'Melinda' signal processor. Here, the technology used is described as being similar to that employed in the company's Outfit UAT naval ES system and, as of this issue, the 'Yellowgate' upgrade was expected to remain operationally viable until 2025.

Contractor
BAE Systems North America - Threat Warning and Defense Systems, Yonkers, New York.

UPDATED

LR-100 warning and surveillance receiver

Type
Lightweight radar band receiver.

Description
LR-100 is a lightweight radar signal receiver covering the 2 to 18 GHz frequency band. This receiver has been built to provide precision radar warning, electronic support measures and electronic intelligence in a system with a total installed weight (including receiver, antennas, cables and brackets) of less than 23 kg. LR-100 is a complete, two-channel interferometer receiver with a 500 MHz bandwidth and VME-based processor. Options are available for frequency extensions of 70 to 200 MHz and 18 to 40 GHz. Phase and amplitude calibration signals are injected at the antenna to achieve precision angle and location measurements. This highly integrated receiver was built as a commercial product to save cost yet achieves a 1,500 hour mean time between failure value via ruggedised industrial components.

Emitter identification and location is digitally passed to the host vehicle for warning, display, analysis and recording. The only support equipment needed is a Windows compatible PC. Help and maintenance manuals are built into Windows-based software. User-defined receiver functions and identification parameters are programmed with the same software tool and the information can be graphically displayed aboard a host vehicle or via datalink (for example, unmanned aerial vehicle applications). LR-100 can be modified in real time for special receiver modes or directed tuning.

The open architecture of LR-100 supports multiple interfaces on a VME backplane (RS-422, RS-232, MIL-STD-1553, Ethernet). The receiver will operate with any antenna configuration to provide amplitude (<15°) or interferometer (<1.0°) direction-finding accuracy. In addition to standard azimuth or azimuth/elevation arrays, the system's antennas can be tailored for any ground, ship, pod or airborne application.

Status
As of July 2001, the LR-100 receiver system is known to have been installed and tested on an aerostat (located at Gulfport, Mississippi), the RQ-4A Global Hawk and RQ-5A Hunter Unmanned Aerial Vehicles (UAV - see following), a Mk V patrol boat (operating from Tampa, Florida) and a High Mobility Multipurpose Wheeled Vehicle (at Patuxent River, Maryland). Of the two UAV programmes, the sensor was installed aboard a US Army RQ-5A Hunter during 1997 and was used to detect radars on the Melrose bombing range (near Cannon Air Force Base, New Mexico) during a January 1998 US Air Force (USAF) UAV Battle Laboratory experiment. Here, the UAV handed off LR-100 detected emitter data directly to two F-16 fighter-bombers using the Improved Data Modem (IDM) as the transmission medium. In the case of the RQ-4A, the Global Hawk air vehicle that was deployed to Australia during April/May 2001 was equipped with an LR-100 receiver that was used to cue the vehicle's other onboard sensors and to demonstrate a surface ship tracking capability. Elsewhere, the system is reported to have been the subject of a mid 1990s 'submarine system' contract (which, as originally scheduled, equipment to be delivered during November 1996 for sea trials during 1997) and has been selected for installation aboard New Zealand's SH-2G(NZ) naval helicopters. Readers are cautioned that the following specification data is interim and should be treated with caution.

Specifications
Frequency: 2-18 GHz (70-200 MHz and 18-40 GHz options)
Field of view: ±45° (elevation); 360° (azimuth)
Accuracy: 0.78° RMS (interferometric function); 2 MHz (frequency accuracy); <15° (amplitude DF function)
Dynamic range: 60 dB
Dimensions (l × w × h): 429 × 216 × 218 mm (receiver/processor)
Weight: 15.9 kg (receiver/processor); <22.7 kg (total system)

Contractor
Northrop Grumman Electronic Systems sector - Defensive Systems Division, Rolling Meadows, Illinois.

UPDATED

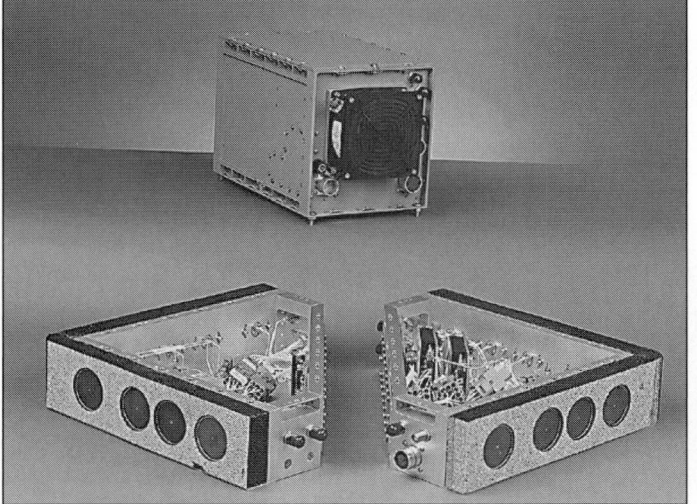

The LR-100 receiver/processor unit shown with four quadrant interferometer arrays

LT-500 emitter targeting system

Type
Passive precision interferometric Radio Frequency (RF) emitter targeting system.

Description
The fully militarised LT-500 Emitter Targeting System (ETS) is a joint development by the then Litton Advanced Systems (now part of Northrop Grumman) and TRW Avionics. Designed to achieve the goal of balancing performance, cost and reliability in a single receiver/processor electronics package, its primary application is passive precision RF targeting for modern air-to-ground weapons via retrofit and upgrade of existing tactical fighter aircraft. This cost-effective implementation of a totally modular 'building block' architecture (SEM-E size electronics modules weighing 34 kg total without antennas) can be easily tailored to match specific emitter tracking/targeting mission requirements. The LT-500 ETS may be installed with a variety of interferometer antenna array configurations and is operated via a MIL-STD-1553 control/display interface. An embedded, 32-bit JIAWG compliant, RISC processor hosts the Ada operational flight programme. Employing patented interferometric direction-finding and ranging techniques, LT-500 achieves very rapid situation awareness, emitter identification and precision geolocation in dense electromagnetic environments. The LT-500 ETS is compatible with either internal or pod installation on board the F-15, F-16, F/A-18 or similar tactical aircraft. Because of its modularity, it may be easily adapted to a wide variety of other airborne, shipboard or ground-based platforms.

Status
As the Precision Direction-Finding (PDF) system, LT-500 type technology is understood to have completed a successful series of flight trials aboard an F-15 aircraft as part of a US Air Force defence suppression demonstration/validation programme.

Contractor
Northrop Grumman Electronic Systems sector - Defensive Systems Division, Rolling Meadows, Illinois.

UPDATED

Passive Ranging SubSystem (PRSS)

Type
Passive Electronic Support (ES) radar location system.

Description
PRSS is a flight-proven, digital receiver that is designed to provide very quick and accurate real-time emitter location and autonomous ES cueing (radar detection, identification and location) in dense electronic environments. The system utilises self-scaling Doppler and self-resolving, long baseline, interferometer techniques (which are claimed to offer an order of magnitude improvement over traditional direction-finding systems) for single aircraft emitter location and features an open system, modular VME architecture that is based on commercial standards. Such modularity ensures that the system is scaleable via the introduction of additional plug-in modules.

Status
As of October 2001, PRSS development for tactical aircraft applications was reported as having been completed and the equipment is noted as forming the basis of the AN/ALQ-210 and AN/ALQ-217 electronic support systems.

Contractor
Lockheed Martin Systems Integration - Owego, Owego, New York.

UPDATED

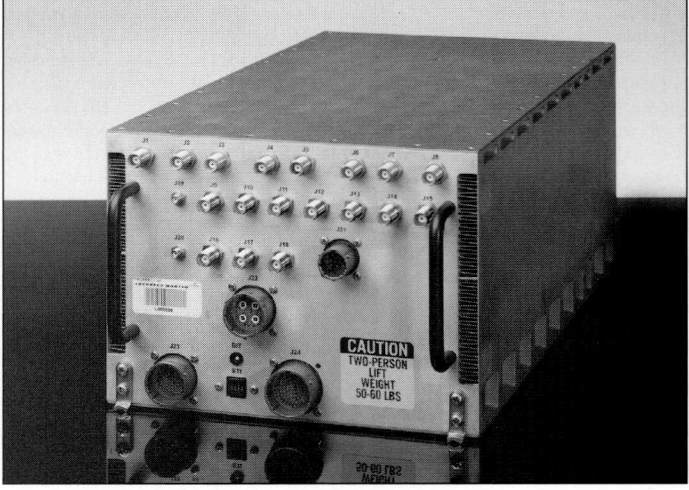

The Lockheed Martin PRSS digital receiver   0017846

Raytheon Remote-Controlled Signals intelligence System (RCSS)

Type
SIGINT system for aircraft.

Description
RCSS is an airborne remotely controlled system for electronic surveillance. It consists of a number of radio receivers and antennas operated by remote radio control from a ground control facility that is equipped with the necessary processing and display equipment. It is intended to be carried in small and large platforms and the system configuration can be easily tailored to the particular aircraft type. Onboard equipment would normally consist of radar detection units covering the frequency band 0.5 to 18 GHz, radio monitoring receivers covering 0.1 MHz to 18 GHz and radio direction-finding systems covering 1.5 to 1,500 MHz. RCSS is designed to operate with various datalink systems including line of sight and satellite configurations. The system can also be operated in a manned configuration.

Status
As of this edition, the RCSS architecture was reported as having been installed aboard a number of aircraft types.

Contractor
Raytheon Electronic Systems, Falls Church, Virginia.

VERIFIED

AIRBORNE ACTIVE AND PASSIVE COUNTERMEASURES SYSTEMS AND DEFENSIVE AIDS SUITES (DAS)

This section covers those airborne electronic warfare systems that are either basically active in operation or act in co-operation with such systems. This includes complete Electronic Warfare (EW) suites, jamming systems, radar warning receivers, laser warning receivers, electro-optical warning equipment, electromagnetic decoys, chaff and infrared flare decoys, expendables and decoy launchers. It provides information on the sensors, processing systems and displays for all types of ECM systems.

VERIFIED

BULGARIA

AJ-XX communications band jammer

Type
The AJ-XX wideband communications jamming system is of modular construction and is designed for both piloted and unmanned aerial vehicle applications.

Status
As of this edition, the status of the AJ-XX system was uncertain.

Specifications
Frequency range: 1.5-1,200 MHz
Weight: 1.5 kg (1 module)
Dimensions (l × w × h): 160 × 220 × 50 mm (1 module)
Output: 30 W
Power supply: 24 W DC
Operating time: 24 h
Temperature: −40 to +50°C (operating)
Modulation: noise

Contractor
Kintex, Sofia.

VERIFIED

CHILE

DM/A-202 CounterMeasures Dispensing System (CMDS)

Type
Airborne CMDS for helicopter and fixed-wing applications.

Description
DM/A-202 is a self-protection system for helicopters and combat aircraft that provides chaff and/or infra-red flares to break the lock of radar and infra-red guided missiles. It consists of five line-replaceable units, including the cockpit control unit and four launching units containing the chaff and flare cartridges. The control unit includes a bright display that shows the amount of chaff and flare cartridges remaining and a mode selector, whereby the pilot can select one of four different launching sequences.

Each launching unit comprises an easily removable magazine, allowing quick reloading of the cartridges, as well as the associated firing circuits. A typical configuration consists of 108 chaff cartridges and 54 flare cartridges contained in the four launching units. The system can be expanded easily to handle up to eight launching units.

Flexibility and ease of operation have been achieved by the use of a fast and reprogrammable microprocessor that guarantees adaptability of the system to changing tactical situations. Simplicity of operation has also been enhanced by the installation of control switches on the aircraft throttle and/or stick, allowing the pilot to operate the system without taking his hands away from these controls (the HOTAS – Hands On Throttle And Stick concept). A safety switch is incorporated in the system so that all the stores can be jettisoned in an emergency. Modes of operation are manual, semi-automatic, automatic and emergency.

To avoid reducing the aircraft operational load-carrying capacity, the launching units are normally attached externally to the rear fuselage or to the sides of the ventral and wing pylons. Internal or semi-recessed installation can also be adopted depending on the aircraft configuration and available space.

The DM/A-202 also provides an interface with the DM/A-104 radar warning receiver for full aircraft self-protection.

Status
Over time, the DM/A-202 CMDS is reported as having been procured.

Contractor
DTS, Santiago.

VERIFIED

EWPS-100 Electronic Warfare (EW) system

Type
Airborne self-protection EW system.

Description
EWPS-100 has been developed to protect helicopters and combat aircraft from present and future radar-controlled weapon systems. The system operates over the 0.7 to 18 GHz frequency band and provides a low-cost and effective answer to operational requirements in the air-to-air and air-to-ground roles. Based on experience gained on various electronic support measures and electronic countermeasures systems, an integrated EW system architecture has been developed using modular techniques, field-proven software and hardware building blocks. The EWPS integrates the DM/A-104 radar warning receiver, the DM/A-202 chaff/flare dispenser and the DM/A-401 self-protection jammer.

The main features of the EWPS-100 are a high probability of threat interception, high sensitivity, initiation of countermeasures, power management in time and frequency and control of chaff/flare cartridge dispensing.

Specifications
Frequency coverage: 0.7-18 GHz
Detection: pulse; pulse-Doppler, CW
Sensitivity: −50 dBm
Max pulse density: 500,000 pps
Threat display: 16 simultaneously

Status
Over time, the EWPS-100 EW system is reported to have been procured.

Contractor
DTS, Santiago.

VERIFIED

The DM/A-202 CMDS

The EWPS-100 self-protection system

CHINA, PEOPLE'S REPUBLIC

BM/KG 8601/8605/8606 radar jammers

Type
Family of self-protection noise and repeater radar jammers.

Description
Over time, the Southwest China Research Institute has developed a range of airborne, self-protection, radar jamming systems for fixed-wing aircraft applications. Of these, the BM/KG 8605 is noted as operating in the I/J-band (8 to 20 GHz) and as being a smart noise jammer that produces a hybrid output that incorporated elements of both noise and deception modulations. The BM/KG 8606 equipment is reported as operating within the I-band (8 to 10 GHz) and as making use of orthogonal and dual circularly polarised jamming techniques. Both BM/KG 8605 and KG 8606 are described as being configured for integration with chaff/flare dispensers. BM/KG 8601 type repeater jammers are reported as operating in the E/F- (2 to 4 GHz) and G/H- (4 to 8 GHz) bands and as being suitable for strike and fighter/bomber aircraft applications. BM/KG 8601 repeater jammers are noted as having a high power output, minimal repeater delay times, threat management through radio frequency channelling, a wide antenna coverage and multijamming techniques.

Contractor
Southwest China Research Institute of Electronic Equipment, Chengdu, Sichuan.

VERIFIED

The units making up the BM/KG 8601 repeater jammer

GT-1 CounterMeasures Dispensing System (CMDS)

Type
Airborne CMDS.

Description
GT-1 is a CMDS that is suitable for both fixed-wing and helicopter applications. It provides chaff dispensing countermeasures in the 2 to 18 GHz band and Infra-Red (IR) countermeasures in the 1.5 to 5 μm wavelength. The complete system consists of a programme controller, an operations control, dispensers and cartridges. It is intended to be interconnected with an appropriate radar warning receiver. The system is able to dispense 68 chaff cartridges or 32 IR cartridges. For salvo firing, the GT-1 can be controlled for the number of payloads in a salvo, interval between salvos and number of salvos. Other options are continuous dispensing, random dispensing, double or single cartridge dispensing and jettison. Two dispensers must be employed for chaff and IR cartridges respectively, as chaff and IR cartridges cannot be mixed in the same dispenser.

Contractor
China National Import and Export Corporation, Beijing.

VERIFIED

DENMARK

AN/ALE-40 digital sequencer switch

Type
CounterMeasures Dispensing System (CMDS) sequencer switch unit.

Description
TERMA's digital, solid-state, enhanced sequencer switch assembly is an autonomous, line-replaceable unit that is designed for use with the ALE-40 CMDS.

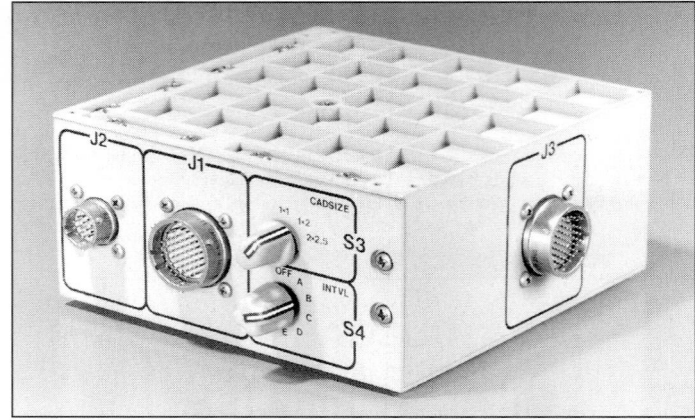

The TERMA ALE-40 digital sequencer switch   0017847

On command from the cockpit control unit, the sequencer switch routes payload dispensing signals to the breech-plate firing pins of the dispenser assemblies, which, in turn, execute the sequential firing of chaff, flares and advanced decoy countermeasures payloads. The sequencer can be set for single or double firing at one of eight, preprogrammed burst intervals from 5 ms upwards. Fast firing is available down to the squib limit of 15 ms or less. Other system features include:
- automated misfire correction
- 'full' built-in test
- a demonstrated mean time between failure value of more than 18,000 hours.

Alongside the described sequencer unit, TERMA has developed an advanced dispenser variant that is equipped with a serial bus interface that facilitates full, intelligent, real-time operation with advanced CMDSs. Unit features include:
- digitally controlled communications
- the ability to service 16 or 32 multiple and/or grouped dispenser station locations to provide full simultaneous symmetrical/asymmetrical dispensing
- automated payload coding for four categories of eight payload types per category
- positive, weight-off-wheels squib power control for enhanced safety.

Status
As of this edition, TERMA ALE-40 digital sequencer switches were reported as having been procured for use in CMDSs operated by the air forces of Belgium (F-16 aircraft), Denmark (C-130 and F-16 aircraft), Germany (C-160 and F-4 aircraft), Netherlands (C-130, F-16 and Fokker 27/60 aircraft), Norway (F-16 aircraft), Portugal (C-130 aircraft) and the US (Air National Guard F-16 aircraft).

Contractor
TERMA Elektronik AS, Lystrup.

VERIFIED

Electronic Combat Integrated Pylon System (ECIPS)

Type
Pylon mounting system for electronic combat equipment.

Description
ECIPS is the generic designation for a number of weapons pylon-mounted F-16 electronic combat systems (including the AN/ALQ-162 radar jammer). As such, ECIPS applications retain all the conventional weapon capabilities of their host pylons, are capable of carrying electronic countermeasures pods on special missions and do not impact on weapon-pylon separation/delivery or existing aircraft wiring. System control is normally by means of TERMA's Electronic Warfare Management System (EWMS - see separate entry).

Status
As of this edition, the Danish Air Force was reported as having procured 62 ECIPS systems for use on F-16 aircraft.

Contractor
TERMA Industries Grenaa SA, Grenaa.

VERIFIED

Electronic Warfare Management System (EWMS)

Type
Control system for the co-ordination, integration and operation of electronic warfare self-protection subsystems installed in fighters, transport aircraft and helicopters. In US service, the EWMS is designation as AN/ALQ-213(V).

Description
EWMS was originally developed in collaboration with the Danish Air Force in order to reduce the pilot's workload aboard the service's Lockheed Martin F-16s and to

Elements of the TERMA EWMS showing, from left to right, the EWPI, the EWMU and the Electronic Warfare AGE panel

improve the operational effectiveness of the defensive aids fit (AN/ALE-40 dispenser, AN/ALR-69 Radar Warning Receiver (RWR) and AN/ALQ-162 radar jammer) carried by such aircraft. Within the installation, all existing cockpit countermeasures system controls and indicators (with the exception of the RWR display) are replaced by two units: a computer-based Electronic Warfare Management Unit (EWMU) and an Electronic Warfare Prime Indicator (EWPI). The fit is completed by a ground access and data loading panel designated as the Electronic Warfare AGE Panel (EWAP). Subsequent to this original programme, EWMS has gone on to be used on a number of aircraft types that have been equipped with a range of threat warners, dispensers and jammers.

Functionally, self-protection programmes are set up, controlled and initiated via EWMU software driven menus in manual, semi-automatic or automatic modes according to pilot choice. Any combination of chaff, Infra-Red (IR) decoy flare and jammer can be selected in order to achieve the most effective counter to the threat at hand. Optimisation of the response is achieved through the use of what is termed Electronic Combat Adaptive Processing (ECAP) which allows EWMS to analyse the incoming threat and automatically select (and initiate if in automatic mode) the most effective combination of available countermeasures. While not originally featured in the system, ECAP is now an operational part of it. EWMS can be operated 'hands on' using a four position thumb switch on the control stick together with bump and elbow switches which the pilot can operate without taking his hand off the throttle. This 'hands on' feature includes selection of subsystem data, selection of countermeasures programmes and countermeasures initiation. EWMS is also fully night vision compatible.

As noted earlier, EWMS is able to control a wide range of RWRs, dispensers, jammers and missile warning systems (including IR and ultra-violet types). TERMA notes that the system has growth potential including increased computer capacity, control of new countermeasures subsystems (including towed decoys, directed IR countermeasures and laser warners) and the introduction of a Tactical Threat Display (TTD), 3-D audio warning and the Tactical Data Equipment (TDE) data loader. Of these, the TTD will be offered as a replacement for the system's existing combination of EWPI and RWR indicator and is intended to provide the pilot with 'all necessary information' from his onboard self-protection systems. The 3-D audio warning capability provides the pilot with a real-time audio indication of threat direction and is being developed to give an indication of threat range. The TDE is a commercial-off-the-shelf equipment that links up to a personal computer workstation and is designed to facilitate up- and downloading of data.

Status
As of this edition, EWMS was reported as being installed aboard the following aircraft types in the following countries:

Country	Aircraft type/types	Start date	Programme status
Australia	F-111C	2000-	continuing
Belgium	F-16A/B Mid-Life Upgrade (MLU*)	1995	continuing
Denmark	C-130H and F-16A/B	1992	completed
Germany	C-160**	1992	continuing upgrade
Netherlands	AS-532 Cougar** (1999 start date/ programme continuing); C-130H (1994/ programme completed); CH-47** (1998/ continuing), F-16A/B MLU (1994/ completed), Fokker F.27 (1992/completed) and Fokker F.60U (1994/completed)		
Norway	F-16A/B MLU	1995	completed
Portugal	C-130H (1996/completed) and F-16A/D (1999/continuing)		
USA	A-10 (1997/continuing), F-16C/D* (1997/ continuing) and HH-60G*** (1997/continuing)		

* Includes control of the F-16 Tactical Reconnaissance System (TRS)
** System includes or likely to include the TDE unit
*** New-build aircraft only

Contractor
TERMA Elektronik AS, Lystrup.

VERIFIED

Modular Countermeasures Pod (MCP)

Type
Countermeasures dispenser pod.

Description
MCP is, as its name suggests, a modular system which can be configured to accommodate chaff and infra-red decoy flare dispensers, radar warning receivers and active jamming system antennas and/or missile approach warning sensors according to specific customer requirements. A typical configuration is noted as being able to accommodate up to seven dispenser modules in a 2,270 × 381 × 434 mm pod shell. Empty, such a system would weigh in at approximately 65 kg. System control is normally by means of TERMA's Electronic Warfare Management System (EWMS - see separate entry).

Status
As of this edition, the MCP countermeasures pod was reported as having been supplied to Netherlands for use on Fokker 60U transport aircraft.

Contractors
TERMA Industries Grenaa SA, Grenaa.
TERMA Elektronik AS (EWMS and digital dispenser sequencers), Lystrup.

VERIFIED

The TERMA MCP modular dispensing pod

Pylon Integrated Dispenser Station (PIDS)

Type
Pylon mounting system for countermeasures dispensers and missile approach warning sensors.

Description
TERMA's PIDS system takes the form of a Lockheed Martin F-16 underwing stores pylon that, in baseline PIDS form, has been modified to accommodate the AN/ALE-40 or -47 CounterMeasures Dispensing System (CMDS) or Chemring Chaffblock payload module (see separate entries) without impacting on the station's ability to carry munitions or drop tanks. In the PIDS+ configuration, the CMDS capability is augmented by a missile warning capability.

PIDS stores pylons can be mounted on the F-16's stations 3 and/or 7 and system control is either by means of the standard ALE-40/-47 cockpit control panel or TERMA's Electronic Warfare Management System (EWMS – see separate entry). PIDS systems are compatible with the F-16's A, B, C and D models and the PIDS+ configuration can accommodate a range of missile warning systems that includes the AN/AAR-54(V), AN/AAR-57, Guitar and MILDS® AN/AAR-60 systems (see separate entries). Three sensor heads are fitted to each PIDS+ pylon with a port/

VERIFIED | *The TERMA PIDS system installed on an F-16*

A PIDS+ CMDS/missile approach warning configured stores pylon mounted on an F-16   0017848

starboard, two pylon/six sensor head installation providing virtually 360° cover in azimuth. Stores are carried on an MAU-12 rack that is attached to the bottom of the PIDS pylon and can be configured for both air-to-air missiles (AIM-9 and AIM-120) and air-to-surface munitions. TERMA also notes that baseline PIDS pylons can be upgraded to PIDS+ configuration if required. System advantages are noted as being:

- improved chaff blooming when compared to fuselage-mounted CMDSs
- optimum location (rear quadrant) for decoy flare dispensing
- up to 200 per cent increase in chaff capacity when compared with conventional F-16 CMDS provision.

Status

As of this edition, at least 600 baseline PIDS pylons are reported as having been supplied to the air forces of Belgium (72 known units), Denmark (66 known units), Netherlands (108 known units) and the US (401 known units for the Air Force Reserve and Air National Guard with options on a further 300 equipments). As of the given edition, development of the PIDS+ configuration was reported as being a multinational programme that was being sponsored by the air forces of Denmark, Netherlands and Norway together with the 'participation' of the US Air National Guard. As originally scheduled, flight trials of the PIDS+ configuration were scheduled to begin during the spring of 1998 with production following during the autumn of the same year. It should also be noted that as of late 1998, TERMA was noted as working on the F/A-18 Pylon Accommodated Self-protection System (PASS) which was intended to provide the Hornet with a PIDS+ type capability. As far as can be ascertained, the PASS concept can accommodate at least three CMDS magazines together with an AN/AAR-54 (V), AN/ARR-57, AN/AAR-60 or 'other' missile warning subsystem.

Specifications

	PIDS	PIDS+
Length	2.6 m	2.8 m
Body Width	23 cm	26 cm
Height	0.38 m	0.38 m
Weight	161 kg	171 kg

Contractor

TERMA Industries Grenaa SA, Grenaa.

UPDATED

FRANCE

BEL expendable radar jammer

Type

Miniaturised expendable radar jamming decoy.

Description

BEL is an expendable, very wideband coherent jammer for self-protection of aircraft against missiles and tracking radars. It presents a credible electromagnetic signature with regard to velocity, direction and range. BEL is, therefore, able to counter continuous wave, pulsed and pulsed-Doppler radar threats, in particular those used in active and semi-active missile seeker heads, either air-to-air or ground-to-air and including radars with monopulse direction-finding or other counter-countermeasures. Dimensions of the BEL cartridge are compatible with current-generation decoy dispensers. The design of this compact device has been made possible by research and development in the area of microwave micro-electronics, such as the design of gallium arsenide monolithic microwave integrated circuits.

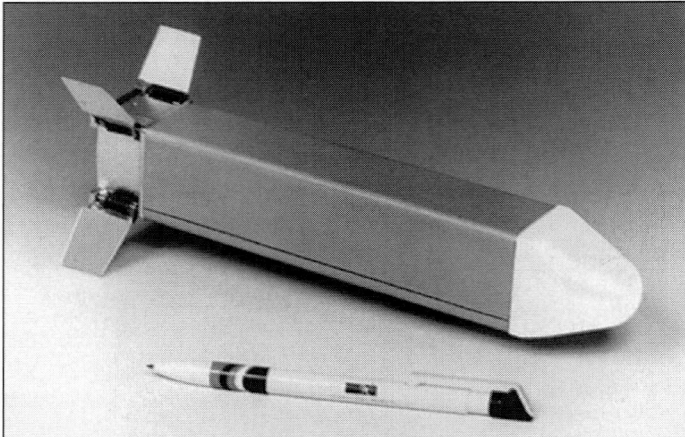

The BEL expendable jammer   0009389

Status

As of early 2001, the BEL expendable radar jammer was reported as having entered limited initial rate production.

Contractors

Thales Airborne Systems, Elancourt.
Etienne LACROIX, Muret.

UPDATED

Integrated CounterMeasures Suite (ICMS)

Type

Internally mounted Electronic Warfare (EW) suite for the Mirage 2000.

Description

Thales Airborne Systems and MBDA Missile Systems (formerly Matra BAe Dynamics France) have developed an integrated, internally mounted EW suite for the Mirage 2000 aircraft that is designated as the Integrated CounterMeasures Suite (ICMS). ICMS features a central interface and management unit, three radio frequency receivers, two detector-jammers, a CounterMeasures Dispensing System (CMDS) and, if required, an optional MBDA Missile Systems infra-red missile warning receiver. Of these, the system's receiver chain comprises a variant of the Serval Radar Warning Receiver (RWR - see separate entry), a superheterodyne receiver (for the detection of continuous wave, pulse compressed and low-power pulse-Doppler emitters) and a nose-mounted receiver/processor (for the detection of missile command links). The detector-jammer package consists of high- and low-frequency units that are aimed at airborne/surface-to-air and surface-to-air threats respectively. The use of detector-jammers in the suite's jamming chain, allows ICMS to continue functioning should its basic RWR capability fail. The equipment's CMDS takes the form of the Spirale system (see separate entry).

Status

As of early 2000, the ICMS defensive aids suite was reported as being in production for installation aboard Mirage 2000-5 aircraft and as having been supplied to the air forces of Greece, Qatar and Taiwan. As of early 2001, a variant of the system was being developed for Mirage 2000-9 aircraft procured by the United Arab Emirates. While not confirmed by the manufacturer, Jane's sources suggest that this latter iteration is the product of an industrial consortium made up of Thales Airborne Systems, MBDA Missile Systems and Elettronica and incorporates a number of

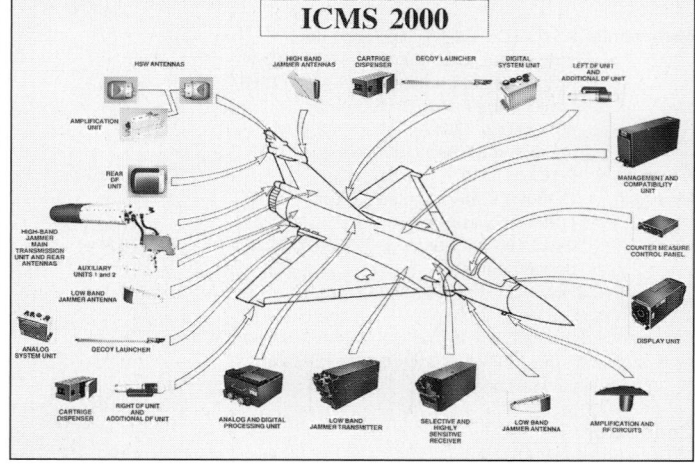

The layout of the ICMS system aboard the Mirage 2000 (MS/Thales)

advanced features including electronically steered jamming antennas and a very high degree of direction-finding accuracy.

Contractors

Thales Airborne Systems, Elancourt.
MBDA Missile Systems, Vélizy-Villacoublay.

UPDATED

Leurre Electromagnétique Actif (LEA) active radar decoy

Type

Air-launched active expendable radar jammer.

Description

The Thales Airborne Systems/MBDA Missile Systems (formerly Matra BAe Dynamics France) LEA active radar decoy (also known as the SPIDER device) is designed to counter threats such as active coherent missile homing heads and monopulse trackers. In more detail, the device is described as being an expendable, pseudo-repeater, 'mini' jammer that is programmable in flight and is compatible with 'current' countermeasures dispensing systems. LEA's payload is battery powered and makes use of monolithic microwave integrated circuitry, a gallium arsenide amplifier and a 'high' degree of integration to facilitate its installation in the limited volume of the LEA flight body.

Status

As of early 2001, the LEA active radar was reported as having shown itself capable of jamming pulse-Doppler radars during flight trials.

Contractors

Thales Airborne Systems, Elancourt.
MBDA Missile Systems, Vélizy-Villacoublay.

UPDATED

The Thales Airborne Systems/MBDA Missile Systems LEA active radar decoy
0017852

PYROTRONICS chaff, Infra-Red (IR) and Electro-Optic (EO) decoys

Type

Family of chaff, IR and EO decoy cartridges and payload modules.

Description

PYROTRONICS (a 50/50 joint venture between SNPE and LACROIX) produces a range of chaff, IR decoy flare and EO countermeasures cartridges, payload modules and tracking flares for drone and towed targets. The company is able to define (in terms of signature, characterisation and effectiveness modelling), design and produce countermeasure cartridges that are compatible with a wide range of dispenser systems produced by Alkan, SaabTech, Matra British Aerospace (BAe) Dynamics (a Matra-BAE Systems joint venture), BAE Systems North America and

Vinten. As of this edition, known details of the company's ISO 9001 compatible product range are as follows:

Barrette

Barrette is a combined dispenser/decoy system that consists of a number of electrically initiated cartridges that are fixed at their base to a metal frame that provides a mechanical and electrical interface to the host aircraft. The frame design is such as to ensure a fully sealed firing circuit, vibration resistance and minimum recoil forces. Barrette systems have a storage life of 5 to 8 years, can be operated at temperatures of −55 to +90°C and are effective up to an altitude of 24,384 m.

EO cartridges

PYROTRONICS produces EO cartridges for ALE-40/-47 (LEO S12) and 60 mm (LEO 685) dispensers.

Specifications

	LEO S12	LEO 685
Dimensions (CS/L)	1 × 2 × 8 in	60(Ø) × 150 mm
Weight	600 g	800 g
Payload weight	160 g	220 g
Ejection velocity	>20 m/s	30 m/s
Storage life	5-8 years	5-8 years
Temperature	−55 to +90°C	−55 to +90°C
Altitude	24,384 m (max)	24,384 m (max)
Application	fighter	fighter

HELIR

HELIR is a 72 round (JUNON chaff and VERDITE IR cartridges) payload module that is designed for use in the Alkan ELIPS dispenser system.

Specifications

Dimensions (l × h × w): 243.5 × 173.6 × 125 mm
Weight: 1.3 kg (unloaded); 6.5 kg (loaded)

MUCALIR

MUCALIR is a 72 round (JUNON chaff and VERDITE IR cartridges) payload module that is designed for use in the Matra BAe Dynamics France SAPHIR dispenser system.

Specifications

Dimensions (l × h × w): 243.5 × 165.6 × 125 mm
Weight: 1.3 kg (unloaded); 6.5 kg (loaded)

MICLIR

MICLIR is an integrated, 30 shot, chaff/IR decoy cartridge (JUNON and VERDITE types) module that is both stackable (up to three) and reloadable. PYROTRONICS claims that the module can be controlled by any available sequencer or cockpit control unit.

Specifications

Dimensions (l × h × w): 290 × 80 × 190 mm
Weight: 2 kg (empty); 4.2 kg (loaded with 30 IR cartridges)

SIDEMIR

SIDEMIR is a missile launch simulator that is designed to exercise missile approach warning systems and comprises a towable launcher, an MF29 independent firing system and a number of SIDEMIR simulators. This latter item weighs approximately 1 kg before launch, features a braking system and has a drop zone of no more than 500 m in diameter. Resetting can be achieved in less than 5 minutes.

Status

As of this edition, the described PYROTRONICS product line was understood to be available.

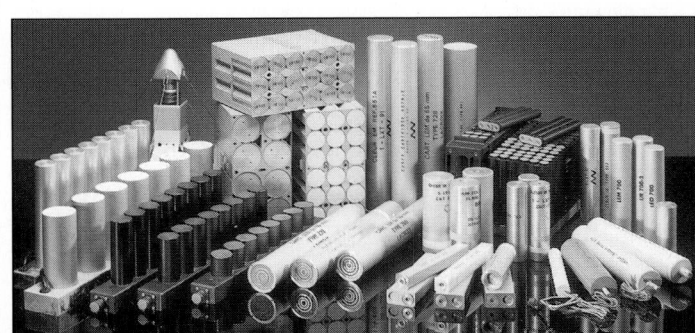

PYROTRONICS produces a wide range of airborne chaff, EO and IR decoy cartridges
0009383

Specifications: PYROTRONICS chaff, Infra-Red (IR) and Electro-Optic (EO) decoys
Chaff applications

	B 677	B 785	LEM 852
Operating principle	8 independent LEM 677 decoys	8 independent LEM 785 decoys	8 independent LEM 887 decoys
Dimensions	52.5 × 170 × 440 mm	52.5 × 170 × 440 mm	52.5 × 233 × 440 mm
Weight	3,000 g	3,800 g	3,100 g
Payload weight	4 × 30 g (per LEM 677)	130 g (per LEM 785)	235 g (per LEM 887)
Chaff dispersion	successive payload ejections at preset intervals	airflow	Z dispersion
Ejection velocity	=35 m/s	30 m/s	30 m/s

IR applications

	B 733	B 750	LIR 842	LIR 853
Operating principle	8 independent LIR 733 decoys	8 independent LIR 750 decoys	8 independent LIR 595 decoys	7 independent LIR 659 decoys
Dimensions	52.5 × 170 × 440 mm	52.5 × 78 × 440 mm	52.5 × 116 × 440 mm	75 × 170 × 580 mm
Weight	3,500 g	1,700 g	2,300 g	6,600 g
Payload weight	110 g (per LIR 733)	80 g (per LIR 750)	110 g (per LIR 595)	330 g (per LIR 659)
Ejection velocity	30 m/s	30 m/s	30 m/s	20-35 m/s

Chaff cartridges

PYROTRONICS produces chaff cartridges for AN/ALE-39 (designated as LEM 657); AN/ALE-40/-47 (LEM 622); BOH (LEM 700); 19 mm (JUNON 1 and 3); 26 mm (LEM 26-1); 40 mm (LEM 677, 785 and 887) and 55 mm (LEM 707DS, 651 and 728) dispensers.

Specifications

	JUNON 1	JUNON 3	LEM 26-1	LEM 407DS
No of initiators	1	3	1	2
Dimensions	19 × 145 × 56 mm	19 × 145 × 56 mm	26(Ø) × 77 mm	55(Ø) × 375 mm
Weight	220 g	220 g	130 g	1,650 g
Payload weight	6 × 30 g	60 g	80 g	4 × 70 g
Operating principle	delayed dispersion with semi-Doppler content	dispersion at ejection	1 payload	2 independent decoys (each comprising 4 successive payload ejects at preset intervals)
Chaff dispersion				Central burster charge + ¼ section shells
Ejection velocity	30 m/s	30 m/s	20-35 m/s	=15 m/s
Storage life	5 years	5 years	5 years	5-8 years
Temperature	−40 to +75°C	−40 to +75°C	−40 to +75°C	−55 to +90°C
Altitude	6,096 m (max)	6,096 m (max)		24,384 m (max)
Application	helicopter	helicopter	towed target/ target drone	fighter

	LEM 622	LEM 651	LEM 657	LEM 677
No of initiators	1	3	1	1
Dimensions	1 × 2 × 8 in	55(Ø) × 375 mm	36(Ø) × 148 mm	40(Ø) × 150 mm
Weight	390 g	1,700 g	230 g	280 g
Payload weight	180 g	3 × 70 g	140 g	4 × 30 g
Operating principle	dual payload (1 instantaneous/ 1 delayed release)	3 independent decoys (each comprising 3 successive payload ejections at preset intervals)	3 successive payload ejections at preset intervals	successive payload ejections at preset intervals
Chaff dispersion	Shutter	central burster charge + ¼ section shells	central burster charge with ½ section shells	central burster charge with ½ section shells
Ejection velocity	>10 m/s	≥15 m/s	≥20 m/s	>35 m/s
Storage life	5-8 years	5-8 years	5-8 years	5-8 years
Temperature	−55 to +90°C	−55 to +90°C	−55 to +90°C	−55 to +90°C
Altitude	24,384 m (max)	24,384 m (max)	24,384 m (max)	24,384 m (max)
Application	various	fighter	various	fighter

	LEM 700	LEM 728	LEM 785	LEM 887
No of initiators	1	1	1	1
Dimensions	40(Ø) × 200 mm	55(Ø) × 375 mm	40(Ø) × 150 mm	40(Ø) × 213 mm
Weight	450 g	1,700 g	320 g	450 g
Payload weight	4 × 30 g	9 × 70 g	130 g	235 g
Operating principle	sequential payload ejection	9 successive payload ejections at preset intervals	delayed burst	single ejection
Chaff dispersion		central burster charge + ¼ section shells	packs	Z burst
Ejection velocity	20-30 m/s	20 m/s	30 m/s	30 m/s
Storage life	5-8 years	5-8 years	5-8 years	5-8 years
Temperature	−55 to +90°C	−55 to +90°C	−55 to +90°C	−55 to +90°C
Altitude	24,384 m (max)	24,384 m (max)	24,384 m (max)	24,484 m (max)
Application	helicopter	transport/maritime patrol	helicopter	fighter

Contractor

SAS PYROTRONICS Contremesures LACROIX/SNPE, Muret.

IR decoy cartridges
PYROTRONICS produces IR decoy cartridges for ALE-39 (designated as LIR 658); ALE-40/-47 (JASPE 2, LIR R11 and LIR 623); BOH (LIR 700-2 and 700-3); 19 mm (VERDITE 1, 1.S and 3); 26 mm (LIR 26-1); 40 mm (Amethyste LS and LIR 595, 733 and 893-B); 55 mm (LIR 407, 698, 730, 760 and 1010) and 60 mm (LIR 659, 659-10 and 684) dispensers.

Specifications

	AMETHYSTE LS	JASPE 2	LIR R11	LIR 26-1
No of initiators	1	1	1	1
Dimensions	40(Ø) × 150 mm	1 × 1 × 8 in	1 × 1 × 8 in	26(Ø) × 77 mm
Weight	300 g	210 g	200 g	125 g
Payload weight	240 g	2 × 75 g	150 g	75 g
Operating principle		sequential ejection		adapted IR output
Ejection velocity	30 m/s	>25 m/s	>25 m/s	20-30 m/s
Storage life	5-8 years	5-8 years	5-8 years	5 years
Temperature	−55 to +90°C	−55 to +90°C	−55 to +90°C	−40 to +75°C
Altitude	24,384 m (max)	24,384 m (max)	24,384 m (max)	
US designation			M 206	
Application	fighter	helicopter	fighter/transport	towed target/ target drone

	LIR 407	LIR 595	LIR 623	LIR 658
No of initiators	2	1	1	1
Dimensions	55(Ø) × 375 mm	40(Ø) × 100 mm	1 × 2 × 8 in	36(Ø) × 148 mm
Weight	1,500 g	200 g	350 g	250 g
Payload weight	780 g	110 g	235 g	145 g
Ejection velocity	20 m/s	30 m/s	>20 m/s	30 m/s
Storage life	5-8 years	5-8 years	5-8 years	5-8 years
Temperature	−55 to +90°C	−55 to +90°C	−55 to +90°C	−55 to +90°C
Altitude	24,384 m (max)	24,384 m (max)	24,384 m (max)	24,384 m (max)
US designation			MJU 7A/B	MJU 8
Application	Fighter	transport	fighter/transport	fighter

	LIR 659	LIR 659-10	LIR 684	LIR 698
No of initiators	1	1	1	3
Dimensions	60(Ø) × 150 mm	60(Ø) × 150 mm	60(Ø) × 150 mm	55(Ø) × 375 mm
Weight	800 g	800 g	800 g	1,500 g
Payload weight	330 g	330 g	330 g	680 g
Operating principle	Cartridge integral with C 814 magazine	cartridge integral with C 810 magazine		
Ejection velocity	20-35 m/s	20-35 m/s	20-35 m/s	≥20 m/s
Storage life	5-8 years	5-8 years	5-8 years	5-8 years
Temperature	−55 to +90°C	−55 to +90°C	−55 to +90°C	−55 to +90°C
Altitude	24,384 m (max)	24,384 m (max)	24,384 m (max)	24,384 m (max)
Application	Fighter	fighter	fighter	transport/maritime patrol

	LIR 700-2	LIR 700-3	LIR 730	LIR 733
No of initiators	2	3	2	1
Dimensions	40(Ø) × 200 mm	40(Ø) × 200 mm	55(Ø) × 375 mm	40(Ø) × 150 mm
Weight	360 g	400 g	1,600 g	300 g
Payload weight	90 g (per payload)	65 g (per payload)	780 g	200 g
Operating principle	2 payloads	3 payloads		
Ejection velocity	20-35 m/s	20-35 m/s	≥20 m/s	30 m/s
Storage life	5-8 years	5-8 years	5-8 years	5-8 years
Temperature	−55 to +90°C	−55 to +90°C	−55 to +90°C	−55 to +90°C
Altitude	24,384 m (max)	24,384 m (max)	24,384 m (max)	24,384 m (max)
Application	Helicopter	helicopter	fighter	helicopter

	LIR 760	LIR 893-B	LIR 1010
No of initiators	4	1	2
Dimensions	55(Ø) × 375 g	40(Ø) × 213 g	55(Ø) × 375 mm
Weight	1,700 g	480 g	1,700 g
Payload weight	780 g	180 g	780 g
Ignition delay		short	
Ejection velocity	≥20 m/s	20-35 m/s	≥20 m/s
Storage life	5-8 years	5-8 years	5-8 years
Temperature	−55 to +90°C	−55 to +90°C	−55 to +90°C
Altitude	24,384 m (max)	24,384 m (max)	24,384 m (max)
Application	fighter	helicopter	fighter

	VERDITE 1	VERDITE 1.S	VERDITE 3
Dimensions	19 × 145 × 56 mm	19 × 145 × 56 mm	19 × 145 × 56 mm
Payload weight	50 g	55 g	50 g
Operating principle	3 decoys (Ø19 mm)	3 decoys (Ø19 mm)	3 decoys (Ø19 mm) with successive ejection
Ejection velocity	30 m/s	30 m/s	30 m/s
Storage life	5 years	5 years	5 years
Temperature	−40 to +75°C	−40 to +75°C	−40 to +75°C
Altitude	6,096 m (max)	6,096 m (max)	6,096 m (max)
Application	helicopter	helicopter	helicopter

UPDATED

SPECTRA Electronic Warfare (EW) system

Type
Integrated airborne EW suite.

Description
The Spectra EW suite is a collaborative effort between Thales Airborne Systems and MBDA Missile Systems and is the first such system to have been developed in France that covers the electromagnetic, laser and Infra-Red (IR) domains. The architecture is described as making use of a range of 'sophisticated' techniques that include interferometry, digital frequency memory, electronic scanning, multispectral IR detection, artificial intelligence, monolithic microwave integrated circuitry (on gallium arsenide substrates) and very high-speed integrated circuitry. The Spectra system includes a phased-array, flight-tested radar jamming transmitter and is installed in 10 internal locations aboard the Rafale multirole combat aircraft. The suite's various elements are integrated via a dedicated electronic warfare databus and a central system processor. If desired, Spectra can be configured for external carriage. Within the development consortium, MBDA Missile Systems provides the SAMIR/DDM IR missile launch detector (see separate entry) and the LCM countermeasures dispensing system. This latter item comprises four cartridge dispenser modules ('built into the airframe') and four internally mounted chaff dispensers. For its part, Thales Airborne Systems is responsible for system integration and the electromagnetic warning, radar jamming and laser warning subsystems. Dassault Aviation has been responsible for integrating the suite into the Rafale weapon system.

Status
The SPECTRA programme was launched during 1990, with the first prototype system being delivered during 1993. Flight trials (aboard a Mystère 20 testbed and Rafale aircraft M02) started during 1994. As of July 2001, the SPECTRA EW suite was reported as being in 'mass production' and as being 'operational' aboard the first production Rafale aircraft.

Contractors
Thales Airborne Systems, Elancourt
MBDA Missile Systems, Vélizy-Villacoublay.

VERIFIED

The layout of SPECTRA's components within the Rafale airframe

GERMANY

BUCK array Infra-Red (IR) decoy flares

Type
Family of multiwavelength IR decoy flares.

Description
BUCK has developed a range of IR decoy flares that are designed to protect large, fixed-wing aircraft and helicopters. Functionally, each cartridge dispenses a large number of mini-payloads that together form a large, dual-wavelength target for a missile's IR seeker head. According to Jane's sources, the technology used is based on the company's naval IR decoy work (see separate entry) and can be tuned during manufacture to meet specific spectral and temporal requirements, including rise time and burn duration. Buck array IR decoy flares are currently packaged in 1 × 2 × 8 in square format cartridges and are suitable for use in dispensers such as the AN/ALE-40, AN/ALE-47 and SAPHIR units.

Status
According to Jane's sources, BUCK began array flare development in response to a 1986 German helicopter requirement. This programme was subsequently abandoned and the concept was resurrected in January 1994 to provide Luftwaffe C.160 Transall transport aircraft operating over Bosnia with improved protection against the IR missile threat. An array flare (designated as DM69) for this application was placed in production during 1995 and has been followed by the improved DM69A1 round which is understood to have been flight trialled during September 1996. During 1995, the US Navy (USN) is understood to have undertaken a series of BUCK array flare ground tests together with flight trials using a P-3 Orion maritime patrol aircraft as the dispenser platform. In October 1996, a BUCK array flare designated as RP-12 became the subject of a formal US Foreign Comparative Test (FCT) programme that involved BUCK in the supply of 525 rounds for FCT flight trials aboard a C-17 transport aircraft. BUCK is also understood to have developed array flares for fast jet applications and the NH90 helicopter programme.

Specifications
DM69A1
Temperature: –54 to +71°C (operating)
Hazard classification: UN0093, 1.3G
Qualification: MIL-STD-810E
Standard packaging: 60 pcs
Dispensers: AN/ALE-40, AN/ALE-47, SAPHIR, MYRIAD or similar
Dimensions: 254 × 508 × 2,032 mm (1 × 2 × 8 in)
Weight: approximately 350 g

Contractor
BUCK Neue Technologien GmbH (a Rheinmetall AG subsidiary), Neuenburg.

UPDATED

INTERNATIONAL

Alkan CounterMeasures Dispensing Systems (CMDS)

Type
Family of airborne CMDSs.

Description
Alkan has developed a range of CMDSs that can accommodate either chaff or infra-red flare cartridges, or a combination of the two. The cartridges are arranged in interchangeable, easily handled magazines. Typically, a single dispenser will accommodate five to seven dispenser modules, each of which houses a magazine containing, for instance, either 1,840 mm chaff or infra-red or NATO 2.54 × 2.54 cm² square format cartridges. The dispenser's electronic management system (whether or not connected to a radar warning receiver) initiates software programmed firing sequences, permanently manages the inventory of available cartridges and provides the necessary information to a cockpit control unit that displays the status of the complete equipment.

Status
Over time, Alkan has produced a range of CMDSs for aircraft as diverse as the Mirage III and V, the Mirage F.1, the Mirage 2000, the MiG-21, the Super Etendard

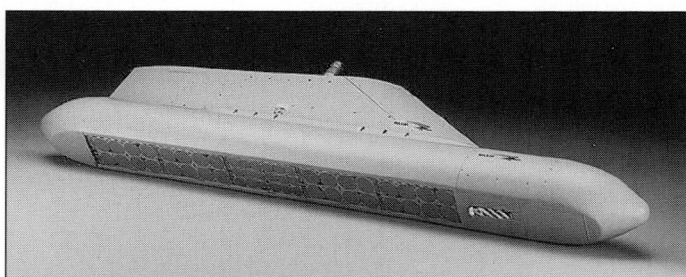

Alkan's chaff/flare dispensing pod for the Super Etendard naval strike fighter

The Mirage F1 gondola installation for the Alkan Type 5020 dispenser unit

and the Jaguar. Of these, the system designed for Jaguar light strike aircraft of the French Air Force includes two dispensers under each wing in a conformal installation near the aircraft fuselage. Each dispenser contains seven modules (Type 5020). The same module is used as the 'heart of a complete self-protection system' for the Mirage F.1 and Mirage 2000 aircraft types. Another example of an integrated installation is the Type 5013 that is designed for retrofit of the Mirage III/5 aircraft. This unit comprises four modules and is installed in the rear part of the aircraft fuselage. The Alkan Type 5080 pod is designed to fit either to the jet assisted take-off points on the MiG-21 or any 36 cm standard armament hard point and as of July 2001, was reported as having been procured for installation aboard Finnish MiG-21 interceptors. Other applications of the same concept include the Type 5081 dispensing pod for the Super Etendard and the CADMIR dispenser pylon for the Mirage 2000. As of July 2001, the Types 5020, 5080 and 5081 CMDSs were thought to be available.

Contractor
R Alkan & Cie (an MBDA Missile Systems subsidiary), Valenton, France.

UPDATED

A Barem/Barax self-protection jammer pod installed on a French Air Force Mirage F1 (MS/Thales Airborne Systems)

AN/ALQ-119 radar jammer upgrade kit

Type
Enhancement package for existing AN/ALQ-119 jamming pods.

Description
The European Aeronautic, Defence and Space (EADS) Co's Systems and Defence Electronics business unit's AN/ALQ-119 update kit consists of radio frequency modules, logic modules, wiring and software for the upgrade of existing pods so that they can cope with modern airborne and ground-based threats. The primary enhancements offered comprise:
- continuous frequency coverage
- improved repeater techniques
- receiver capability
- extended noise capability
- new jamming techniques
- improved techniques programming capability
- intelligent interface to radar warning receiver
- improvements in maintainability.

Status
As of July 2001, the EADS Systems and Defence Electronics' ALQ-119 upgrade kit was thought to have been procured.

Contractor
European Aeronautic, Defence and Space (EADS) Co Systems and Defence Electronics, Ulm, Germany.

UPDATED

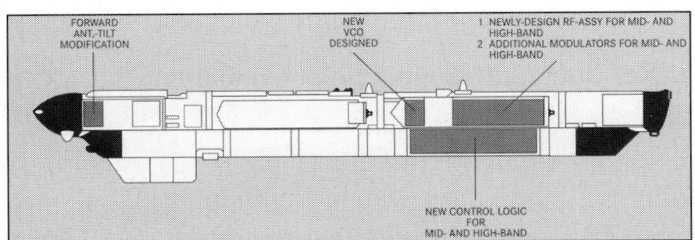

The elements making up the EADS Systems and Defence Electronics' ALQ-119 upgrade package

Barem/Barax detector-jammer

Type
Pod-mounted radar band detector-jammer.

Description
Barem/Barax is an 85 kg, pod-mounted radar band detector-jammer that is designed for tactical/strike aircraft applications. It is equipped with a travelling wave tube and provides instantaneous coverage over a wide frequency band. It can detect, identify and simultaneously jam multiple threats from a variety of ground-to-air and air-to-air pulse-Doppler and continuous wave radars. The equipment has an extensive memory capacity and a modular software design that is reprogrammable to cater for future threat developments. Thales Airborne Systems also notes that Barem/Barax can be equipped with a techniques generator which deals with the most modern coherent radar systems.

The M2.0 capable Barem/Barax system consists of reception and transmission antennas, a receiver and a transmitter. Threats are detected by the antennas and analysed by the receiver against a wide range of radar parameters. The ultra-wideband travelling wave tube amplified transmitter uses noise and deception

modes to jam the radar in less than 1 second. Operation is fully automatic. Two threats can be countered at the same time at the aircraft front or rear. Barem/Barax is further noted as incorporating an integrated superheterodyne receiver that automatically adjusts to 'differing environments', can be reprogrammed and provides flight reports for situation monitoring in quasi real time. Barem/Barax also provides an electronic support function and has been developed using much of the experience gained with the earlier Remora system. For internal installation, the equipment can be packaged into two units with a combined weight of 65 kg. The Barem/Barax pod shell has a diameter of 0.16 m and is 3.45 m long.

Status
As of spring/summer 2001, pod-mounted Barem/Barax detector-jammer applications were reported as having been procured by the French and 'other' air forces together with the French Navy for use on its Super Etendard Modernisé strike aircraft. During the mid-1990s, Barem/Barax was proposed as a potential upgrade for the Sukhoi Su-22 strike aircraft and the system is described as being 'combat proven'. When 'modernised', the Barem/Barax detector-jammer is designated as the PAJ-FA system (see separate entry).

Contractor
Thales Airborne Systems, Elancourt, France.

VERIFIED

Caiman radar jammer

Type
Airborne noise and deception radar jamming system.

Description
Caiman is an airborne radar jamming pod that was probably developed from the earlier Alligator system. It is intended for aircraft fulfilling electronic warfare support missions, operating to protect groups of similar aircraft in ground attack roles by jamming surveillance and target designation radars. Within the pod are two jammers (each with its own receiver) and fore and aft receiver antennas. Radiated power output from each jammer is 750 W in the I-band (8 to 10 GHz) or 3 kW in the D-band (1 to 2 GHz). The jamming signal bandwidth is programmable before the mission. Three modes of operation are available: in manual mode the pilot can select any one of three predetermined frequencies at will; in semi-automatic mode a signal indicates to the pilot when a transmission is being received so that he can take the appropriate action and in fully automatic mode the jammers are automatically activated when such a signal is received.

The jamming system is an autonomous pod with ram air entering the unit through an annular intake to drive a power turbine and provide cooling. It can be installed either underwing or on a fuselage pylon.

Status
As of April 2001, the Caiman noise and deception radar jammer was reported as being no longer in production.

Remora (wing station) and Caiman (centreline) jamming pods mounted on a Mirage 2000

Specifications
Dimensions (l × Ø): 5.95 m × 0.41 m
Weight: 550 kg

Contractor
Thales Airborne Systems, Elancourt, France.

VERIFIED

CORAIL CounterMeasures Dispensing System (CMDS)

Type
Airborne CMDS.

Description
CORAIL is a radar and optronic CMDS that is designed for use on various versions of the Mirage F1 fighter. It can also be applied to other types of aircraft. The system can be housed conformally or in an external 'gondola'.

Specifications
Dimensions: 2,616 × 163 × 218 cm
Weight: 129.7 kg
Capacity: 2 × 56-126 cartridges

Status
As of July 2001, the CORAIL CMDS is reported as having been procured by the French Air Force for installation aboard Mirage F1-CR/-CT combat aircraft. The CORAIL CMDS has been in service since 1992 and is noted as having been used by the French Air Force in operations over Bosnia.

Contractor
MBDA Missile Systems, Vélizy-Villacoublay, France.

UPDATED

As applied to the Mirage F1, the Corail countermeasures dispenser is mounted in underwing gondolas
0017849

A close-up of the Corail countermeasures dispensing system mounted in the underwing gondolas of a Mirage F1 combat aircraft
0017850

ECLAIR-M CounterMeasures Dispensing System (CMDS)

Type
Airborne CMDS.

Description
The ECLAIR-M CMDS is designed to supplement the Mirage 2000's SPIRALE CMDS (see separate entry) by increasing the latter's decoy capacity. As such, the ECLAIR-M system comprises a mechanical framework, six dispenser racks (each capable of holding eight flares or 18 chaff cartridges) and an electronics unit. When ECLAIR-M is loaded with flares, it quadruples the SPIRALE system's flare capacity. To free up available hard-points, the ECLAIR-M assembly is mounted in the Mirage 2000's drag-chute bay. The electronics unit used is drawn from the next-generation Rafale multirole combat aircraft's CMDS. A serial datalink connects ECLAIR-M to the SPIRALE system's digital system unit in order to facilitate the latter's control of the former.

Status
As of July 2001, the ECLAIR-M CMDS was reported as being in production for 'all versions' of the Mirage 2000 in French Air Force service and the Mirage 2000-9 variant. As of the given date, 260 examples were noted as being on order.

Contractor
MBDA Missile Systems, Vélizy-Villacoublay, France.

UPDATED

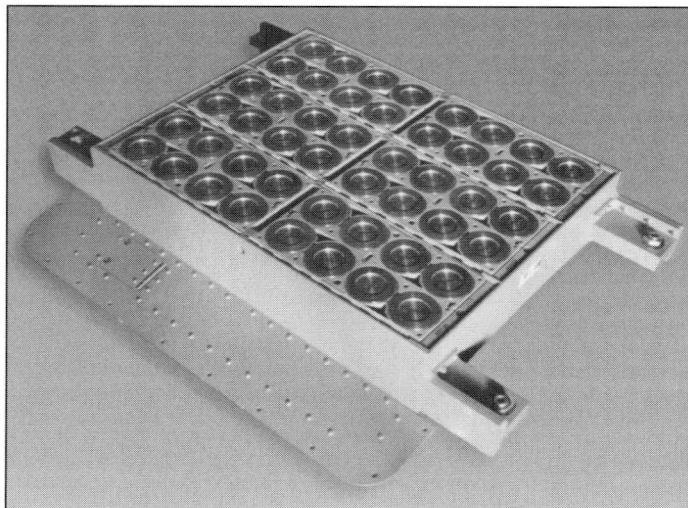

The ECLAIR-M CMDS
0017851

ELIPS CounterMeasures Dispensing System (CMDS)

Type
CMDS for helicopter applications.

Description
The ELIPS CMDS has been specifically designed for helicopter applications and incorporates two to four dispenser modules (each capable of holding 24 square format payloads) and a cockpit control unit which is compatible with night vision goggles.

Status
As of July 2001, the ELIPS CMDS was reported as being in service aboard Eurocopter Cougar Mk 2 and Panther helicopters operated by customers in 'several countries' around the world. The equipment's manufacturer is also noted as having entered into an agreement with UK contractor Chemring concerning the development of ELIPS-specific Chemring Chaff and Flare Block payload modules. As of the given date, ELIPS was further understood to be available and to be the

The ELIPS CMDS is designed specifically for use on helicopters

subject of a development programme that was scheduled to produce a 'new generation' ELIPS-NG configuration by 2002.

Contractor

R Alkan & Cie (an MBDA Missile Systems subsidiary), Valenton, France.

UPDATED

Eurofighter Defensive Aids SubSystem (EuroDASS)

Type

Integrated Electronic Warfare (EW) suite.

Description

Created specifically for use aboard the four-nation Eurofighter/Typhoon multirole combat aircraft, the EuroDASS EW suite is designed to provide its host platform with protection against air-to-air and surface-to-air threats. As such, 360° cover is provided with the system detecting and evaluating threats before automatically activating an appropriate countermeasures response without pilot intervention. According to Jane's sources, the suite incorporates:

- a Defensive Aids Computer (DAC)
- an integrated, radar band Electronic Support and CounterMeasures (ES/ECM) package
- a Missile Approach Warner (MAW)
- a Laser Warning Receiver (LWR)
- an associated CounterMeasures Dispensing System (CMDS).

In more detail, the DAC provides system control and allocates system resources, while the ES/ECM package provides the aircraft's pilot with 'immediate' warnings of potential threats, analyses received signals and generates 'strategic' information for co-ordination with that from the aircraft's radar and infra-red search and track system. Once cued by the ES capability, the suite's ECM capability is described as providing the 'optimum number' of available onboard and offboard jamming 'techniques and resources' necessary to defeat the detected threat. Among other elements, this capability includes a multiple active towed decoy installation.

The EuroDASS suite's MAW is the BAE Systems Avionics/Elettronica active, pulse-Doppler Advanced MIssile Detection System (AMIDS - see separate entry) and is described as making 'extremely efficient' use of the host aircraft's prime power supply. The LWR is a UK only fit and provides both a warning of and the direction of arrival of detected laser threats. The CMDS used is understood to incorporate a chaff capability that is similar in format to the Swedish BOL system (see separate entry) and while integrated with the main EuroDASS suite, is not part of it. The system, as a whole, features built-in test and has been optimised for ease of maintenance.

Status

As of August 2001, a three-member consortium (comprising BAE Systems Avionics in the UK, Elettronica in Italy and INDRA in Spain) was producing the EuroDASS EW suite. According to Jane's sources, the EuroDASS consortium was awarded a then year £200 million development contract for the Eurofighter/Typhoon EW suite during March 1992. At this time, workshare within the consortium was noted as being approximately 60 per cent for the UK arm of the effort and 40 per cent for the Italian one. The same sources suggest that May 1998 saw the award of a then year £170 million Eurofighter/Typhoon DASS production investment contract. During June 2001, the EuroDASS consortium announced that it had been awarded a then year £300 million plus contract covering the supply of 103 DASS suites for installation aboard Eurofighter/Typhoon production aircraft. Under the deal, the EuroDASS partners were mandated to deliver DASS line-replaceable items to their home countries, with, at the time of the announcement, deliveries scheduled to begin during the first quarter of 2002. While not confirmed (and as of August 2001), the remaining Eurofighter/Typhoon partner (Germany) was expected to rejoin the EuroDASS programme after abandoning a national Eurofighter/Typhoon EW solution that it had pursued throughout the 1990s. Here, a programme re-entry fee of then year US$158.5 million was being quoted.

Contractors

BAE Systems Avionics - Sensor Systems Division, Stanmore and Portsmouth, UK (prime)
Elettronica SpA, Rome, Italy.
INDRA DTD, Madrid, Spain.

UPDATED

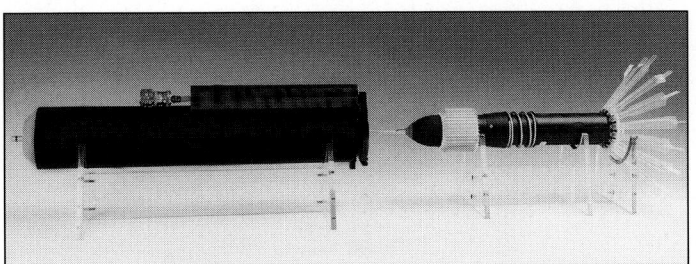

The active towed radar decoy used in the EuroDASS integrated EW suite

0017857

EWS-16 self-protection system

Type

Internal self-protection system for the F-16 aircraft.

Description

EWS-16 is an integrated system that consists of a threat warning system and an active jammer. The threat warning system can detect, identify and localise all modern threats with great accuracy. All threats can be identified in less than 1 second without any ambiguity, even in dense electromagnetic environments. The active jammer features high radiated power and can counter pulsed and continuous wave radars. Analysis of threats is on a pulse-by-pulse basis, with very accurate identification and priority assessment, a clear pilot interface and full integration with the weapon system. EWS-16 is fully programmable and makes use of a separate threat library. The electronic support measures subsystem can accurately locate ground-based radars and can record data during flight. Features of the EWS-16 system include:

- a management and compatibility unit based on a 32-bit processor
- a crystal video receiver and a high-speed wideband superheterodyne receiver
- an instantaneous wideband direction-finding interferometer
- an instantaneous frequency measuring receiver
- real-time spectral analysis processing
- a plug-in mission report module
- a multiple-threat jammer employing a high-power transmitter
- synergy between jammer and decoy dispenser.

Status

As of April 2001, a variant of the EWS-16's warning subsystem (designated as Carapace) was reported as being in service aboard Belgian F-16 aircraft (see separate entry). As of September 2000, sources were suggesting that the Thales Electronic Warfare (EW) suite then on order for installation aboard Turkish F-16s was based on the EWS-16 system. As of the given date, the programme was understood to involve then year US$108 million and US$82 million awards to Thales and Turkish contractor ASELSAN respectively. In the first instance, the Thales award covered the supply of system hardware, while the ASELSAN contract involved the company in system integration and the development of an EW support centre that would facilitate in-country reprogramming. As of February 2001, usually reliable sources were reporting that Turkey may have cancelled the described contract with Thales, following a diplomatic dispute between itself and France. As of October 2001, such a cancellation had not been confirmed.

Contractor

Thales Airborne Systems, Elancourt, France.

UPDATED

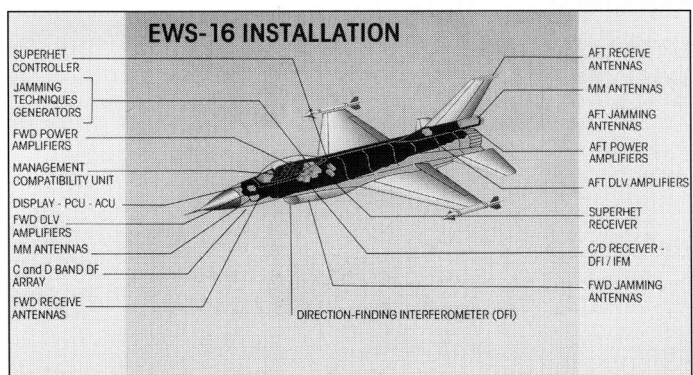

The general layout of the EWS-16 system as applied to the F-16

Modular Self-Protection System (MSPS)

Type

Airborne Electronic Warfare (EW) suite.

Description

MSPS is a lightweight, easy to install EW suite that is designed for retrofit aboard existing aircraft. It combines the Sherloc radar warning receiver with the Barem/Barax radar band detector-jammer and an Alkan Type 5081 podded decoy system (see separate entries). System capabilities include wide frequency coverage, reprogrammability, a combined crystal video/superheterodyne receiver subsystem and fast data transmission between subsystems to facilitate automatic responses to detected threats. MSPS type systems can be installed on a variety of fixed- and rotary-wing aircraft and the suite's jamming subsystem can be installed internally if a sufficiently large space envelope is available within the host airframe. The system can be augmented with add-ons such as laser and missile warners and support jamming equipment and has an overall weight of 80 kg when mounted internally, a figure that rises to 100 kg if its radar jamming subsystem is pod-mounted.

Thales Airborne Systems MSPS package has been installed on French Navy Super Etendard Modernisé strike fighters as part of a type upgrade programme   0017853

Status

As of April 2001, the MSPS EW suite was reported as having been installed aboard French Navy Super Etendard Modernisé strike aircraft.

Contractor

Thales Airborne Systems, Elancourt, France.

VERIFIED

PAJ-FA detector-jammer

Type

H- through J-band (6 to 20 GHz) pod-mounted detector-jammer.

Description

PAJ-FA is an airborne, pod-mounted, detector-jammer system that is designed to counter fire-control, target designation and missile seeker radars. PAJ-FA incorporates digital radio frequency memory technology and offers power-managed amplitude modulation, barrage and spot noise, 'clutter', combination, false target and range/velocity gate pull-off jamming modes. Other system features include:

- a twin travelling wave tube transmission chain
- full software control
- a user programmable mission library

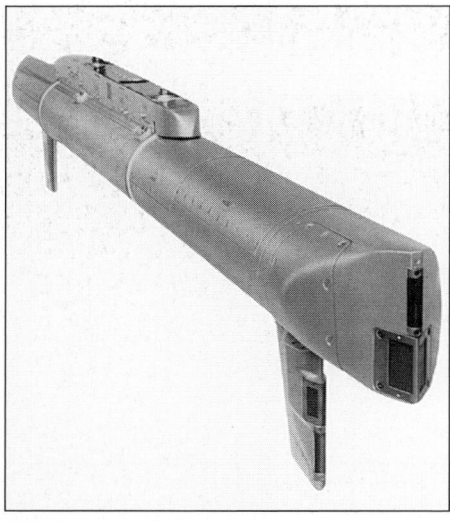

The PAJ-FA detector-jammer
0009384

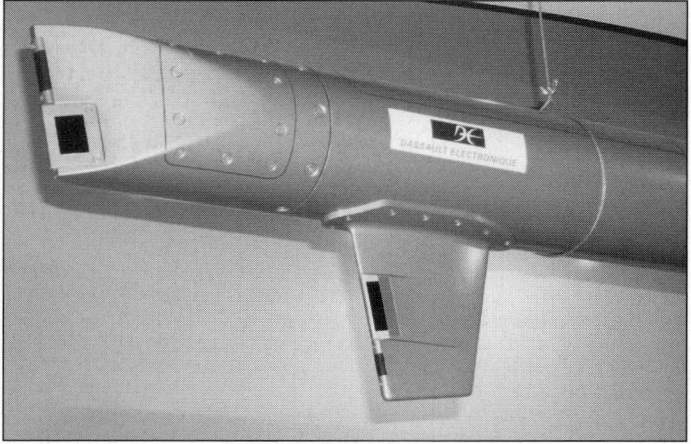

A close-up of the forward end of the PAJ-FA detector-jammer pod (Martin Streetly)   0017854

- prioritisation of detected threats
- the ability to track and jam multiple threats on separate channels
- fore and aft transmission arrays
- output matched to threat polarisation
- built-in test
- generation of flight reports for post-mission analysis
- M2.0 flight speed compatibility.

PAJ-FA can be used as a stand-alone system or as part of a self-protection system where it exercises control over a countermeasures dispensing system.

Status

As of April 2001, PAJ series detector-jammers were reported as having been procured by France and Spain. Of these, French PAJ systems (also known as the Barem/Barax equipment before 'modernisation') have, over time, been applied to Jaguar, Mirage III and F.1 and Super Etendard Modernisé aircraft. The Spanish application appears to be flown on Mirage F.1 aircraft and is described as being a country-specific variant that has been manufactured in co-operation with Spain's defence electronics industry. According to Jane's sources, the system received a new techniques generator during the 1990s and in the French Mirage F.1 application, is 'fully' integrated with the type's TDS-FA radar warning receiver.

Specifications

Frequency coverage: 6-20 GHz
Power consumption: 1 kW
MTBF: >250 h
Weight: 85 kg
Dimensions (Ø × l): 0.16 × 3.42 m

Contractor

Thales Airborne Systems, Elancourt, France.

VERIFIED

PHIMAT CounterMeasures Dispensing System (CMDS)

Type

Pod-mounted chaff dedicated CMDS.

Description

PHIMAT is a pod-mounted, chaff dedicated CMDS that comprises dispenser and control units. The dispenser is a 3.6 m long × 108 mm diameter tube containing the chaff packs, ejection mechanism and drive electronics. The complete system weighs 105 kg and can carry 200 chaff packs, with manual or automatic triggering. PHIMAT can generally be adapted to all types of modern aircraft, particularly those with weapon stations capable of taking the R550 Magic or Sidewinder missiles.

Status

Over time, approximately 300 PHIMAT CMDS pods have been produced for carriage aboard Jaguar strike aircraft of the British and French air forces, French Air Force Mirage F1s and Mirage 2000s, Super Etendards of the French Navy and as an option on Tornados of the Royal Air Force. As of July 2001, the PHIMAT CMDS was in service but no longer in production.

Contractor

MBDA Missile Systems, Vélizy-Villacoublay, France.

UPDATED

French Air Force personnel loading a PHIMAT expendables pod fitted to a Jaguar strike aircraft (MS/MBDA)

SAPHIR CounterMeasures Dispensing System (CMDS)

Type
CMDS for helicopter applications.

Description
The SAPHIR CMDS is made up of two to eight cartridge dispensers (each containing 24 to 72 cartridges) and a control unit that enables the pilot to choose from 12 different dispensing programmes. These scenarios (together with the constantly available survival programme) have the following parameters:

- the number of cartridges per salvo
- the time interval between salvos and the number of salvos in the programme.

These parameters are stored in an erasable programmable read-only memory and, in addition to stand-alone applications, SAPHIR can be incorporated into an overall countermeasures system and/or weapons system. So configured, SAPHIR delivers optimised real-time decoying programme parameters, based on data supplied, from the countermeasures system or the navigation system. The equipment features automatic, semi-automatic and manual triggering modes and its software is optimised for rotary-winged applications. As of July 2001, MBDA Missile Systems was noting that the SAPHIR system existed in -A, -B and -M variants (see following).

Status
Over time, the SAPHIR CMDS is reported as having been procured for installation aboard French military Cougar, Ecureuil, Gazelle, Lynx and Puma helicopters. As of July 2001, the SAPHIR-M variant was reported as being the 'latest' variant of the system and as featuring modular cartridge launchers and software that was 'specifically optimised for use on new-generation helicopters'. SAPHIR-M was further noted as having been selected for use on the NH-90 and Tiger helicopters from 2004 onwards. Elsewhere, existing SAPHIR variants are described as having been used operationally during the 1991 Gulf War and more recent operations in the Balkans.

Contractor
MBDA Missile Systems, Vélizy-Villacoublay, France.

UPDATED

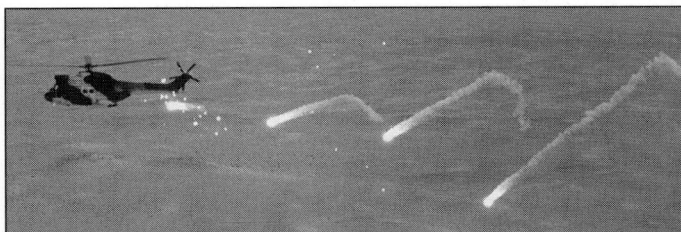

A French Puma helicopter firing infra-red decoy flares from its SAPHIR CMDS during the 1991 Gulf War (MS/MBDA)

....

Shrike airborne Techniques Generator (TG)

Type
Airborne TG application.

Description
The Shrike TG that has been developed by Thales Sensors is, in the first instance, aimed at the Royal Air Force's (RAF) Nimrod MRA Mk 4 maritime reconnaissance aircraft programme. It is based on work that Thales has undertaken in collaboration with the UK's former Defence Evaluation and Research Agency on the Advanced Jamming System technology demonstration programme and is described by its manufacturer as being a 'third generation' equipment that in the 'large aircraft' context, is in the 'same category' as the TG being installed aboard the US Air Force's next generation F-22 air superiority fighter. As applied to the Nimrod MRA Mk 4, the equipment is used to drive a Fibre-Optic Towed radar Decoy (FOTD) variant of Raytheon Systems AN/ALE-50 device (see separate entry) and is mounted in a standard 48 cm (19 in) rack, incorporates a digital radio-frequency memory and is equipped with integral threat parameter and countermeasures technique software libraries. The particular application also sees the TG interfaced with and cued by an AN/ALR-56M(V) radar warning receiver (see separate entry). Functionally, the system generates a coherent response and is able to function in noise, deception and repeater modes. Thales claims that when compared with existing towed decoy systems, the described combination of its TG and the Raytheon FOTD is able to triple the number of simultaneous threats able to be countered in a single decoy deployment.

Status
Under a £10 million plus contract from Lockheed Martin, Thales Sensors has developed and fabricated the Shrike TG for use in the Nimrod MRA Mk 4 programme. Under this award, Thales was scheduled to deliver a Shrike integration model to Lockheed Martin during September 1998 for a 12 month trial programme. As of May 2001, Thales reported that it was in the process of delivering Shrike for installation in the Nimrod MRA Mk 4 systems integration rig. If successful, all 21

Thales' Shrike TG is a central component of the defensive aids suite installed aboard the UK's Nimrod MRA Mk 4 maritime patrol aircraft (BAE Systems)
2002/0116877

Nimrod MRA Mk 4 aircraft are to be fitted with production examples of the device. Subsequent to this Nimrod-related award, June 2001 saw Thales announce that it had received an approximately then year £6 million contract to supply the Shrike TG into the UK's Airborne STandOff Radar (ASTOR) programme.

Alongside the two contracted efforts, Thales and Lockheed Martin are understood to have agreed to collaborate on the use of the Shrike TG in future Lockheed Martin large aircraft defensive aid system bids and as of May 2001, the company was understood to be 'talking' to Raytheon about the continuance of the ALE-50/Shrike relationship. Targeted applications are understood to include transports such as the C-130J, maritime patrol aircraft, air-to-air refuelling tankers and 'high value' systems such as the E-3 Airborne Warning And Control System (AWACS). Thales is further believed to be studying repackaging the system for fast jet applications and Shrike technology is known to have been used in the company's Sabre naval electronic warfare system (see separate entry).

Contractor
Thales Sensors, Crawley, UK.

UPDATED

....

Sky Buzzer towed decoy

Type
Active towed radar jamming and decoying system.

Description
The European Aeronautic, Defence and Space (EADS) Co's Systems and Defence Electronics business unit's Sky Buzzer is an active, towed, radar jamming and decoying system that is designed to protect both large bodied aircraft and fighters. Functionally, the system detects radar threats and generates and transmits jamming signals to induce angular errors in the tracking loops of airborne and ground-based target trackers and active missile seekers. The equipment is described as being effective against a wide range of radar types (including monopulse equipments) and can be configured with parachute and winch recoverable transmitter flight bodies for use on fighter and large body aircraft respectively. EADS notes that Sky Buzzer's effectiveness is based on a high transmitted power value, a broad frequency range (covering 'all relevant' threat systems), techniques to prevent the flight body being hit by an incoming missile and a range of countermeasures modulations for use against continuous wave, pulse and pulse-Doppler radars. In the device's proposed Tornado strike aircraft application, the Sky Buzzer capability is packaged in a modified BOZ countermeasures dispensing system pod. Here, the pod shell is used to house fore and aft reception antenna arrays, two decoy launchers and an electronics payload that is thought to comprise a radio frequency front end, an amplitude/frequency modulation unit, a Digital Radio Frequency Memory (DRFM) transponder, an onboard transmitter, an RF-to-optic conversion unit and a high voltage power supply. The system's reusable flight body features fore and aft radomes, four fixed stabilising fins, four variable 'pop-up' flaps and a mid-body array of cooling vanes. The device's payload comprises a travelling wave tube amplifier, a power-

The Sky Buzzer towed decoy system utilises both fixed vanes and spring-loaded flaps to stabilise its transmitter flight body
0017856

A rear view of a BOZ countermeasures dispensing system pod that has been modified for Sky Buzzer carriage and launch (EADS)   2002/0018294

conditioning unit and an optic-to-RF conversion unit. Flight body-to-launcher connection is by means of a broadband fibre-optic cable (40 to 150 m long when fully deployed) and explosive squibs are used to launch the flight body and to sever the tow cable before flight body recovery. EADS claims that the system's flight bodies can be refurbished for future use within 1 to 2 hours of recovery.

Status

As of July 2001, Sky Buzzer was reported as having been 'successfully' flight-tested on a range of aircraft (including the F-4 and Tornado and at speeds of up to M1.4) and was initially scheduled to have become 'commercially available' at the end of 1998. EADS notes that the flight trial programme with the Tornado involved single aircraft and multiship formations and that deployment of the decoy had no effect on the type's flight behaviour. Towing manoeuvres included pop-ups, 'S' shapes, spirals and figures-of-eight. When activated, the decoy is described as having being 'instantly effective' in single and multiple threat environments. As of the cited date, Jane's sources were suggesting that a Sky Buzzer variant was to be procured by the German Air Force for use on its Tornado strike and electronic combat/reconnaissance aircraft subject to funding availability.

Contractor

European Aeronautic, Defence and Space (EADS) Co Systems and Defence Electronics, Ulm, Germany.

UPDATED

SPIRALE CounterMeasures Dispensing System (CMDS)

Type

Airborne CMDS.

Description

SPIRALE is the CMDS installed on the Mirage 2000 and is a major component of the type's ICMS 2000 electronic warfare suite. It is internally mounted and

SPIRALE dispensers on the lower surfaces of a Mirage 2000 multirole combat aircraft

comprises four dispenser tubes (located in the aircraft's wing roots) and two cartridge dispensers in the fuselage. The system can also be configured to include two DDM/SAMIR missile launch detectors (see separate entry) with the sensors being installed in the aircraft's Magic air-to-air missile launch rails. Capacity is 16 infra-red or electro-optical cartridges and 112 electromagnetic chaff packs. A further system option is the ECLAIR-M complementary cartridge dispenser (see separate entry) that can be fitted to increase the available onboard decoy payload.

Status

The SPIRALE CMDS entered service aboard the Mirage 2000 at the end of 1987. As of July 2001, SPIRALE was reported as forming an integral part of the ICMS 2000 integrated countermeasures system and as being in service with the air forces of France and five other nations. As of the given date, the system was further noted as having been ordered by Greece (Mirage 2000-5 Mk 2 aircraft) and the United Arab Emirates (Mirage 2000-9 aircraft) and as having been used 'extensively' on Mirage 2000 aircraft engaged in multinational operations over the Balkans.

Contractor

MBDA Missile Systems, Vélizy-Villacoublay, France.

UPDATED

Spirit CounterMeasures Dispensing System (CMDS)

Type

CMDS for the Transall C-160 and other large aircraft.

Description

Spirit is a countermeasures system for the dispensing of both chaff and infra-red high-volume cartridges in the quantities necessary to protect a large turboprop-powered airframe. To this end, system capacity is up to 168 flares and 42 chaff cartridges housed in dispensers scabbed on to the sides of the aircraft's fuselage. System control is by means of a night vision goggles-compatible control unit that offers automatic, semi-automatic or manual operation.

Status

As of July 2001, the Spirit CMDS was reported as having been procured for installation aboard French Air Force C-160s and 'other' applications (thought to be French Air Force C-130s). With regard to the C-160 application, Spirit is understood to interface with a Thales Airborne Systems radar warning receiver and an Elta missile approach warner. The Spirit CMDS is reported as having entered service with the French Air Force during 1994 and as of July 2001, was understood to be available.

Contractor

R Alkan & Cie (an MBDA Missile Systems subsidiary), Valenton, France.

UPDATED

Infra-red decoy flares being fired from Alkan's Spirit dispenser system (MS/Alkan)

SPS-H and SPS-TA self-protection systems

Type

Family of airborne Defensive Aids Suites (DAS)

Description

Thales Airborne Systems has developed a family of integrated DAS systems that are designed for helicopter (designated as SPS-H) and fixed-wing transport aircraft (SPS-TA) applications. Taking these in the order given, the SPS-H DAS is available in V1 and V2 configurations. Both architectures incorporate an optimised MWS-20 missile approach warner variant (see separate entry), 'SPS system processing' and a countermeasures dispensing system, with the SPS-HV2 adding the TDS-H Radar

Warning Receiver (RWR - see separate entry), an optional laser warner and an optional infra-red jammer and/or radar jammer to the core system. Of the two, SPS-HV1 is described as being a 'switch-on and forget system' that requires no human intervention and is claimed to offer both an 'extremely low' false alarm rate and 'high-level' system performance that has been demonstrated in an 'extensive' test programme. The SPS-HV2 is noted as being a 'complete turnkey solution' that is able to protect helicopters from both active and passive threats. 'Full electronic warfare situational awareness' is provided via the system's processing. As with SPS-H, SPS-TA can be configured in V1 and V2 variants that make use of similar equipment fits to those described for the SPS-H iterations (SPS-TAV2 making use of the TDS-TA RWR - see separate entry) and offer similar capability packages that are tailored to the needs of fixed-wing transport aircraft such as the C-130 and the C-160.

Status

As of April 2001, SPS series equipments were reported as being in production and service. As of the given date, SPS-H was noted as being in service aboard French Army Cougar helicopters, while 'particularly complete configurations' of the SPS-H and SPS-TA were reported as being in production for installation aboard French Air Force Cougar combat search and rescue helicopters and C-130/C-160 transport aircraft respectively.

Contractor

Thales Airborne Systems, Elancourt, France.

VERIFIED

SYCOMOR CounterMeasures Dispensing System (CMDS)

Type

Airborne CMDS.

Description

SYCOMOR is a CMDS which has been developed for the Mirage F1 and is packaged in either a 2.95 m long externally mounted pod or a 2.5 m conformal 'gondola'. Each gondola has three chaff dispensing tubes and seven magazines, housing four decoy cartridges per magazine. Each pod has the capacity of two gondolas. Although designed for the Mirage F1, it can be adapted to other types of aircraft.

Status

Over time, the SYCOMOR CMDS has been procured for use aboard the Mirage F1 combat aircraft operated by a number of air forces around the world. The SYCOMOR CMDS entered service during 1984.

Contractor

MBDA Missile Systems, Vélizy-Villacoublay, France.

UPDATED

Sycomor countermeasures pod under Mirage F1

Tornado Self-Protection Jammer (TSPJ)

Type

Pod-mounted, self-protection radar jammer for fighter aircraft.

Description

TSPJ is the third-generation of a line of pod-mounted jamming technology that is designed to provide fighter aircraft with electronic self-protection against radar threats. As such, the equipment is a generic system that contains no threat specific hardware and can generate a 'large repertoire' of jamming techniques for use against continuous wave, pulse and pulse-Doppler emitters. At the heart of the capability is a European Aeronautic, Defence and Space (EADS) Co Systems and Defence Electronics business unit-developed Digital Radio Frequency Memory

The TSPJ self-protection radar jamming pod

(DRFM) that is used for deceiving and jamming modern coherent radars. The equipment's DRFM incorporates both the RF and memory sections needed to digitise and precisely store signals of interest and the system's techniques generator. Alongside its capability against coherent radars, TSPJ is noted as being effective against non-coherent types and as being fully software reprogrammable. The system can function as an autonomous unit or be controlled by a radar warning receiver system and has been designed for ease of maintenance.

Status

According to Jane's sources, TSPJ is a co-operate venture between the EADS Systems and Defence Electronics business unit and Israel's Elta Electronics that was successfully flight tested in Germany (three month long trial) and the United States during 1997. Deliveries of TSPJ pods to the Luftwaffe (German Air Force) began during 1998, with the first examples being issued to the Tornado electronic combat/reconnaissance aircraft-equipped Jagdbombergeschwader (Fighter-Bomber Wing) 32 at Lechfeld. In Luftwaffe service, TSPJ replaces the Cerberus III pod-mounted radar jammer and for its part in the programme, Elta Electronics is understood to have received a US$156 million plus contract for the supply of four prototype TSPJ pods, 60 production units and a supportive integrated logistics package.

Contractors

European Aeronautic, Defence and Space (EADS) Co Systems and Defence Electronics, Germany.
Elta Electronics Ltd (a subsidiary of Israel Aircraft Industries Ltd), Israel.

UPDATED

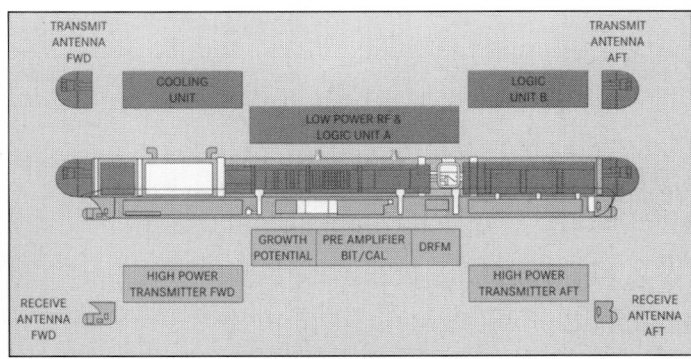

The layout of the TSPJ self-protection jammer

In Luftwaffe service, the TSPJ system replaces the earlier Cerberus III radar jamming pod shown here (Martin Streetly) 0017858

Unmanned Aerial Vehicle (UAV) communications band jamming payload

Type
UAV communications band jamming payload.

Description
Thales Communications has developed a compact, lightweight communications band jamming payload for UAVs that utilises digital techniques to counter a range of 'modern' emitters that includes free channel search, frequency hopping and burst types. Functionally, the payload is activated in the mission area with the best jamming coverage being achieved via the use of an associated, deployable antenna that is integrated with the host vehicle. Thales further notes that this military specified equipment can also be integrated with air vehicles other than UAVs.

Status
As of early 2001, the Thales Communications UAV communications band jamming payload was reported as having been successfully test flown aboard a Dragon UAV.

Specifications
Dimensions (l × h × w): 370 × 220 × 210 mm
Weight: 12.5 kg

Contractor
Thales Communications, RadioSurveillance and COMINT Systems Unit, Gennevillers, France.

VERIFIED

Thales' communications band jamming payload for UAVs   0017855

Vicon 70 CounterMeasures Dispensing System (CMDS)

Type
Lightweight pod-mounted CMDS.

Description
Vicon 70 is a modular, lightweight, pod-mounted CMDS that uses components of the Vicon 78 system to provide cost-effective self-protection against radar threats and heat-seeking missiles. It is designed for helicopters and fixed-wing aircraft where structural and other constraints preclude the use of airframe-mounted dispensers. The baseline architecture is configured for two dispensers, a number that can be increased to up to eight by adding modules to the pod. Vicon 70 modules can also be used to house radar and missile approach warners and the system as a whole can be installed on any fixed- or rotary-wing aircraft type equipped with weapons racks or fuselage/wing pylons with 36 cm NATO attachments. The firing trajectory of the Vicon 70's dispenser modules can be

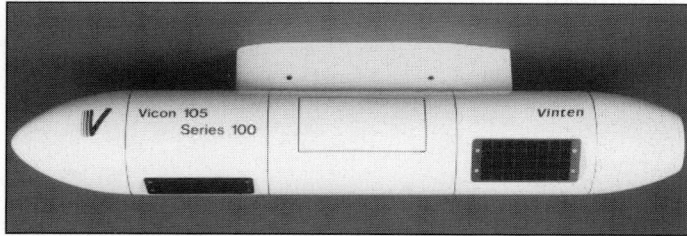

The Vicon 70 CMDS pod

preset to optimise the effectiveness of the chaff and infra-red flares. Manual or fully automatic dispensing versions of the system have been developed with the automatic configuration interfacing directly with a radar warning receiver.

Status
As of March 2001, the Vicon 70 pod shell was being seen by Thales Optronics (formerly Vinten) as a 'bus' capable of accommodating countermeasures or reconnaissance equipment, with an emphasis on the latter.

Specifications
Dimensions (l × Ø): 2,705 mm × 356 mm
Weight: 210 kg fully loaded with 8 dispensers

Contractor
Thales Optronics, Bury St Edmunds, UK.

UPDATED

Vicon 78 CounterMeasures Dispensing Systems (CMDS)

Type
Family of airborne lightweight CMDS.

Description
The Vicon 78 CMDS series is a family of advanced lightweight chaff and Infra-Red (IR) decoy dispensing equipments which are designed for use on high-performance combat aircraft, tactical transports, maritime patrol aircraft, helicopters and Unmanned Aerial Vehicles (UAV). Vicon 78 systems can be fitted as original equipment in new build airframes or as retrofit equipments and can be configured for manual, semi-automatic and fully automatic operation. Of these, fully automatic function is achieved by interfacing the dispensing system to a threat detection sensor. In automatic mode, the dispensing system programme is initiated by the Radar Warning Receiver (RWR) or missile approach warner. In semi-automatic mode the programme to be dispensed is downloaded from the threat-warning receiver and initiated by the crew. In manual mode the crew can dispense preset chaff and/or flare programmes. Vicon 78 series systems are offered with a range of dispenser configurations that range from multiple modules for large transport/maritime patrol aircraft applications to a single dispenser arrangement for small aircraft or Unmanned Aerial Vehicles (UAV). The architecture's dispenser module can accept interchangeable, six round chaff or flares payload magazines; can be configured to house expendable jammers, towed decoys and other format expendables and can be mounted as internal, semi-recessed or podded units. As of October 2001, identified members of the Vicon 78 CMDS family comprised:

Vicon 78 Series 200
The Vicon 78 Series 200 has been supplied to BAE Systems for its Sea Harrier programme. The Series 200 CMDS is a two-module system, controlled by a salvo control processor unit and manually operated by the pilot. Vicon 78 Series 200 equipments have undergone electromagnetic compatibility testing by the UK's

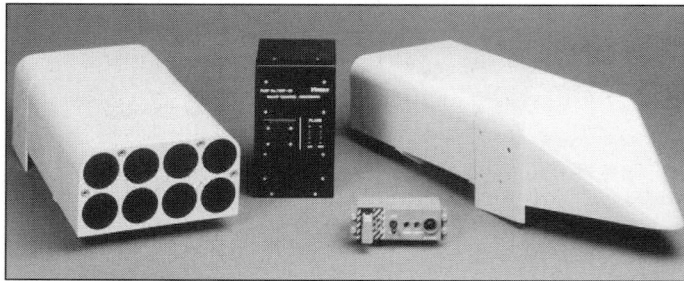

The Vicon 78 Series 210 airborne decoy dispensing system was used by the Royal Air Force on Tornado F Mk 3 aircraft during the 1990/1991 Gulf War

The Vicon 78 Series 420 countermeasures dispenser has been supplied for installation aboard British Army Air Corps Gazelle and Lynx helicopters

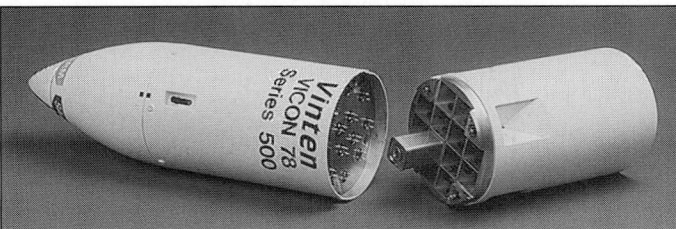

The Vicon 78 Series 500 system has been developed for light aircraft, helicopters and UAVs

The Vicon 78 Series 455 dispenser system

former Defence Evaluation and Research Agency and have been accepted for service by the UK Ministry of Defence. Within the Vicon 78 Series 200 range, Series 203 is a four dispenser module system which is controlled by a two-box processor and offers a fully automatic operating mode when interfaced with an RWR. Series 205 is equipped with a dispenser module capable of accommodating both AN/ALE-39 and AN/ALE-40 style cartridges, while Series 210 is optimised to fire 55 mm twin-shot IR decoy flares.

Vicon 78 Series 300

The Vicon 78 Series 300 was originally developed for the BAE Systems Hawk 100 and 200 series aircraft. In the Hawk, the Series 300 CMDS is a digitally controlled, fully automatic two-dispenser system that interfaces (via an RS-422 port) to either the Thales Sensors Prophet or the BAE Systems Sky Guardian RWRs. The Series 300 system has been designed with the capability to control up to 16 dispensers, with each dispenser able to be configured to hold up to 64 chaff or flare payloads. Each dispenser can accept interchangeable chaff and flare magazines. When power is applied, the system will identify the number and type of expendables in each dispenser and display the count of chaff and flare cartridges on the Cockpit Control Unit (CCU). A continuous built-in test facility is executed by the system; in the event of failure the fail indications are displayed on the CCU.

The Series 300 CMDS has been designed for use in any RWR/CMDS system that has a serial data interface capability. The RWR/CMDS interface has been achieved successfully with Northrop Grumman (formerly Litton), BAE Systems and Thales Sensors (formerly Racal) equipments and agreements are in place to interface with RWRs from other manufacturers. Vicon 78 CMDS also have the ability to interface to missile approach warners and missile launch detectors.

Vicon 78 Series 400

The 400 Series incorporates the electronics of the 300 Series, which have been improved and miniaturised, together with the eight-cartridge 16-shot 55 mm dispenser format of the Series 210. The system uses a single miniature Line-Replaceable Unit (LRU) Chaff and Flare Dispenser Control Unit (CFDCU), which can be either cockpit or remote mounted and provides the system control, processing and programme loading functions. Vicon 78 Series 400 is installed on Royal Air Force (RAF) and Italian Tornado F Mk 3 interceptors.

Vicon 78 Series 420

Derived from the Series 400 equipment, the Series 420 CMDS has been configured to accept standard square format NATO payload cartridges. The system can control up to four 8 × 4 in dispenser modules together with two BOL chaff dispensers. The Series 420 CMDS has been optimised for helicopter applications.

Vicon 78 Series 455

The Series 455 CMDS is an advanced, lightweight system that is suitable for helicopter, fast jet and transport/maritime patrol aircraft applications. The system's low mass and reduced number of LRUs facilitates its use in upgrade/retrofit programmes and its integration into new build aircraft. Series 455 can control up to 24 countermeasures dispensers and two BOL chaff dispensers. Dispensers offered come in a range of formats that can accommodate a spread of payload formats, including the new 6 × 5 and 8 × 4 combined load modules. Series 455 is night vision goggle compatible; has a number of different threat sensor interfaces and can be integrated via a MIL-STD-1553B databus. The Vicon 78 Series 455 CMDS has been ordered for use on the AeroVodochody L-159, the RAF's Nimrod MRA Mk 4, the British Army's WAH-64D Apache battlefield attack helicopter, the Belgian Army's A109BA helicopters and a number of other international new build and retrofit programmes. The Vicon 78 Series 456 is a Series 455 variant that has been specifically configured for the Lead-In Fighter Trainer (LIFT) model of the BAE Systems Hawk trainer.

Vicon 78 Series 500

The Series 500 lightweight stand-alone dispenser uses the proven electronics of the Series 300 system, reconfigured to provide a simple and cost-effective countermeasure dispensing solution for light aircraft, helicopters and UAVs. This 12-shot dispenser has a simple mechanical and electrical interface and can be operated from the cockpit, or in the case of a UAV via a radio frequency link. This system can be configured to interface to a missile approach warner.

Status

As of March 2001, identified (by Jane's sources) Thales Optronics (formerly Vinten) Vicon 78 dispenser applications were reported as follows:

Vicon 78 Series 200	•Sea Harrier FRS Mk 51 interceptor
Vicon 78 Series 203	•export CASA 101 training aircraft (interfacing with the AN/ALR-80(V) RWR) •CN-235 transport aircraft (interfacing with the AN/ALR-85 RWR)
Vicon 78 Series 210	•fully approved modification for Royal Air Force Tornado interceptors post 1990/91
Vicon 78 Series 300	•export Hawk 100/200 series training/light fighter aircraft
Vicon 78 Series 400	•UK and Italian Tornado F. Mk 3 interceptors
Vicon 78 Series 420	•British Army Air Corps Gazelle and Lynx helicopters
Vicon 78 Series 455	•Nimrod MRA Mk 4 maritime patrol aircraft •AeroVodochody L159 multirole light combat aircraft •export C-130 transport aircraft •WAH-64D Apache battlefield attack helicopter •Agusta A109BA helicopter •Denel Oryx helicopter •MIL Mi-24 battlefield assault helicopter
Vicon 78 Series 456	•BAE Systems LIFT Hawk

Contractor

Thales Optronics, Bury St Edmunds, UK.

UPDATED

ISRAEL

Advanced Digital Dispensing System (ADDS)

Type

Airborne CounterMeasures Dispensing System (CMDS).

Description

ADDS is a computer-controlled, threat-adaptive CMDS that is designed to protect aircraft from both ground and air threats by dispensing decoy payloads. The system is configured for high-performance aircraft, helicopters, transports and maritime patrol aircraft.

The ADDS system consists of a cockpit control and display unit (which provides interface to the crew), a programmer (handling threat-adaptive processing and the optimisation of the system's dispensing programmes), a dispenser (which interfaces with and controls one payload module) and a payload module.

ADDS can control and dispense chaff, flare, radio frequency and future types of expendable payloads by means of programmes that adapt to the specific threat and the engagement parameters, in automatic, semi-automatic and manual modes. The dispensing programmes are user programmable.

Special design allows ADDS to fire dual-chaff cartridges to double the number of onboard chaff stores on each mission. In addition, the system can dispense multiple payloads simultaneously to provide a multispectral response or a stronger decoy signal when necessary. ADDS incorporates improvements derived from combat experience and hundreds of installations on a wide variety of aircraft, including the A-4, AH-64, CH-53, F-4, F-5, F-15, F-16, Mi-17, Mi-24, Super Puma and others. A fully integrated logistics support package is provided.

The units making up the ADDS CMDS Advanced dispensing system for fighters, helicopters and transport aircraft

Status

As of this edition, ADDS is reported to be in service with 10 air forces around the world. While neither confirmed nor denied by Rokar, Jane's sources suggest that this number includes the Israel Air Force.

Contractor

Rokar International Ltd, Jerusalem.

VERIFIED

Airborne Self-Protection Suite (ASPS)

Type

Airborne defensive aids suite.

Description

Elisra describes its ASPS as being a 'state-of-the-art, self-protection suite for fighter aircraft' that comprises active (SPJ) and passive (SPS) subsystems and makes use of a range of 'proven and new' technologies that include monolithic microwave integrated circuitry, application specific integrated circuitry and multiminiature travelling wave tubes. For its part, the architecture's SPS subsystem is described as providing wide- and narrow-band reception (using 'various reception techniques'), a 'full' signal parameter measurement capability and a selectable operating bandwidth to enhance sensitivity and signal separation. Data processing and analysis is performed using multiple processors and 'special' gate array modules. The system's SPJ component is described as offering a high effective radiated power value, wideband transmission and reception coverage, wide angular coverage, the ability to cope with a multithreat emitter environment and 'synergetic coupling' with the SPS subsystem. SPJ features are noted as including a multichannel architecture, an autonomous threat acquisition receiver, 'full' integration with the host platform's avionic system and the use of 'compact and modular' line replaceable units. Available countermeasures techniques include Doppler, range, amplitude modulation and noise modes and the system as a whole provides a pilot's threat display (threat azimuth (with respect to aircraft heading) and type), a missile launch alert facility and the ability to display via a central multifunction display unit (using a MILitary STanDard (MIL-STD) - 1553B data bus) if required. System control is by means of front panel switches on a cockpit control unit.

Status

As of January 2002, ASPS was reported as being in production and service aboard Israeli F-15I strike aircraft. The system is also noted as having been selected for installation aboard Israeli F-16I multirole fighter.

Specifications

SPS subsystem
Frequency coverage: existing radar frequencies
Signals reception: CW, high PRF, pulse and pulse-Doppler
Reception coverage: full arc in azimuth
Communications channels: MIL-STD-1553B; RS-422
Data processing: multiple autonomous processors
Field loading capability: portable field loader (emitter table and operational software)
Recording capability: flight data (downloaded via the field loader)
Audio warnings: composite, missile launch and 'new guy'
Display: CRT type (graphic representations of emitter type, reception angle, relative lethality and threat status)
Interfaces: radar and CMDS
Environmental: MIL-E-5400 Class II qualified

Contractor

Elisra Electronic Systems Ltd (a member of the Elisra Group), Bene Beraq.

UPDATED

Elisra's ASPS has been selected for installation aboard Israeli F-15I and F-16I aircraft (Elisra)
2002/0131095

AIRMOR Defensive Aids Suite (DAS)

Type

Helicopter and transport aircraft DAS.

Description

The AIRMOR DAS is described as being an 'add-on' countermeasures system for use on helicopters and transport aircraft that comprises:
- an AN/AAR-60 ultra-violet missile warner (see separate entry)
- management Control and Control Display Units (MCU and CDU)
- a fixed-source Infra-Red (IR) jammer
- a CounterMeasures Dispensing System (CMDS).

Of these, the Lenkflugkörpersysteme GmbH-sourced AAR-60 is noted as being an off-the-shelf sensor that is integrated into AIRMOR via an RS-422 serial link. While AAR-60 is the suite's primary sensor, radar and laser warners can be accommodated if required. Integration of such equipment is achieved via an RS-442 interface or an MIL-STD-1553B databus.

The AIRMOR MCU and CDU are produced by Israel Military Industries (IMI). Of the two, the MCU receives inputs from the suite's sensor(s) and the host aircraft's avionics (navigational, weapons, altimeter and air speed data), integrates and analyses the received information and activates an optimal (direction, intervals and expendable type and quantity) countermeasures response. The CDU is described as being a state-of-the-art man/machine interface that provides the host platform's crew with 'situational awareness analysis'. If required, AIRMOR can accommodate two CDUs.

The AIRMOR fixed-source IR jammer is noted as being an 'off-the-shelf, battle-proven' device that is controlled via the suite's MCU. The CMDS used is IMI's off-the-shelf SAMP unit (see separate entry), with subsystem control (manual or automatic) being exercised via the suites MCU. In 'maximum configuration', the AIRMOR SAMP application can accommodate 480 expendables of up to four different types. Here, compatible payloads include IMI's CG-17 chaff cartridge, the FG-3, FG-6 and FG-9 IR decoy flares and the MultiBlu IR decoy flare (see IMI expendables entry).

Status

As of January 2002, Jane's sources were suggesting that Turkey had selected an AIRMOR DAS configuration for a helicopter application, with any finalised negotiated contract having a potential value of then year US$105 million.

Contractor

Israel Military Industries Ltd, Ramat Hasharon.

UPDATED

EL/L-8222 radar jammer

Type

Airborne radar jamming pod.

Description

EL/L-8222 is described as being an advanced self-protection system that features wide frequency coverage; a high power, broadband output; selectable countermeasures techniques flightline reprogrammability (threat parameters, techniques and mission data); manual or automatic operation; integrated ram air/liquid cooling; low aerodynamic drag; modular construction; built-in test and a high mean time between failure value.

Status

As of this edition, the EL/L-8222 radar jamming pod is reported to have been procured by a number of world air forces for use on A-4, F-5, F-16, F-111, Jaguar, Kfir, MiG-21 and Tornado combat aircraft.

Contractor

Elta Electronics Industries Ltd (a subsidiary of Israel Aircraft Industries Ltd), Ashdod.

VERIFIED

Elta's L-8222 radar jamming pod under test aboard an F-16

EL/L-8233 Integrated Self-Defence System (ISDS)

Type
ISDS for lightweight fighter applications.

Description
The EL/L-8233 modular ISDS has been specifically designed to meet the operational needs and volume constraints of small, lightweight fighter aircraft. The modular nature of the suite's architecture enables it to be applied as an integrated suite, a MIL-STD-15533B databus-controlled system or as individual, stand-alone subsystems. As applied to the F-5E/F, EL/L-8233 comprises a Radar Warning Receiver (RWR), a mini-Continous Wave (CW) repeater-jammer and a countermeasures dispenser. Known details of these subsystems are as follows:

RWR
When integrated into the EL/L-8233 ISDS, the system's RWR is responsible for communications between the host aircraft's avionics and the elements making up the ISDS. The equipment is noted as being effective against pulse, CW, pulse-Doppler and exotic emitters and as having a short cycle time for its complete frequency range. Other features include emitter identification (by frequency and time domain characteristics); field reprogrammability; dispenser initiation and data recording.

CW repeater jammer
The EL/L-8233 ISDS jammer is designed to provide a self-defence deception capability against pulse-Doppler radars and the CW illuminators used with semi-active missile systems. Operating modes include velocity gate pull off and angle deception and the unit features power management to enable it to handle two different threat types simultaneously. Other features include an integral receiver capability; field reprogrammability; low power consumption; self-cooling and the option of a back up control unit to allow for manual initiation. Like other elements of the EL/L-8233 suite, the CW repeater jammer can function as a stand-alone item.

Countermeasures dispenser
The EL/L-8233 countermeasures dispenser is described as being a modular, computerised, threat adaptive system that automatically optimises its response to threats according to type and engagement geometry.

Status
As of the second quarter of 2000, EL/L-8233 ISDS programme appeared to be active.

Contractor
Elta Electronics Industries Ltd (a subsidiary of Israel Aircraft Industries Ltd), Ashdod.

VERIFIED

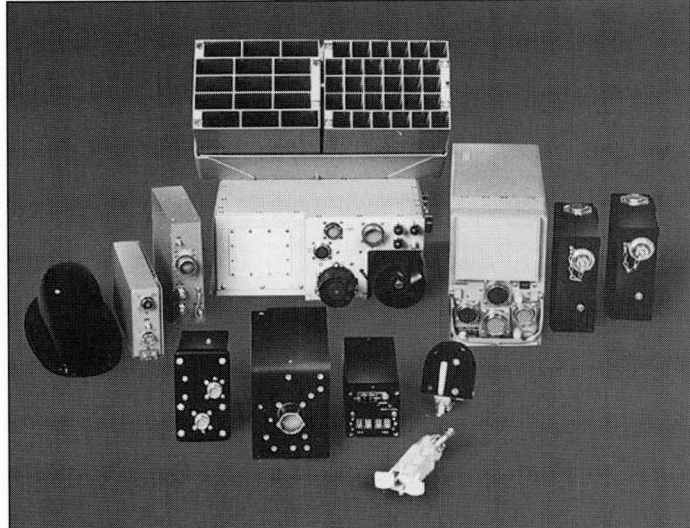

The jamming, warning and dispenser subsystems which make up the L-8233 defensive aids suite

EL/L-8240 Electronic CounterMeasures (ECM) system

Type
Airborne radar warning and jamming system.

Description
EL/L-8240 is an internally mounted self-protection ECM system designed principally for advanced combat aircraft. It consists of a fully integrated radar warning and radar jamming/deception equipment and can be adapted for installation on any combat aircraft. It provides wide frequency and angular coverage, with a short response time and a high transmitted power in the jamming mode. A substantial number of jamming and deception techniques are included. Total weight is 136 kg.

Status
As of this edition, Jane's sources were suggesting that the EL/L-8240 ECM system was installed aboard Batch 3 F-16C and D aircraft of the Israeli Air Force.

Contractor
Elta Electronics Industries Ltd (a subsidiary of Israel Aircraft Industries Ltd), Ashdod.

VERIFIED

Israel Military Industries (IMI) chaff and Infra-Red (IR) decoy flares

Type
Family of chaff cartridges/payloads and IR decoy flares.

Description
Over time, IMI has produced a range of chaff cartridges/payloads and IR decoy flares, the known details of which are as follows:

CG-17
CG-17 is a RR-170A/AL form compatible chaff cartridge that is designed for use in SAMP 60/120/240 and AN/ALE-40/47 dispenser systems. Its payload covers the 2 to 20 GHz frequency band and ejection is by means of a BBU-35/B squib.

Chaff package H/G
Chaff package H/G covers the 2 to 20 GHz frequency band and is designed for area saturation and corridor screening. On fighters, the package is carried as a store on a bomb pylon, while in larger aircraft (such as transports), it is dispensed via conveyor.

FG-3
FG-3 is an IR decoy flare that is designed for use on the F-16 fighter and the AH-64 battlefield attack helicopter. It is a form, fit and function equivalent of the US M-206 cartridge and is compatible with the SAMP 60/120/240, M-130 and ALE-40/47 dispensers.

Specifications
Dimensions: 25 × 25 × 205 mm
Payload weight: 120 g
Ejection velocity: 20 m/s (min)

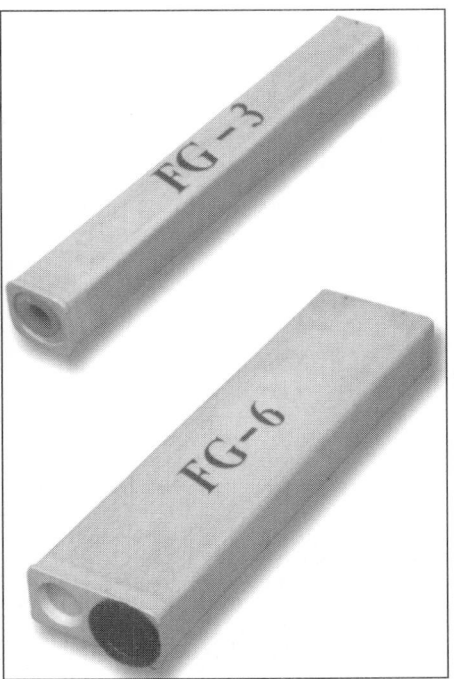

The FG-3 and FG-6 devices form part of IMI's range of IR decoy flares
0017860

IMI's MULTI-BLU IR decoy flare showing its three foil wrapped payload pellets
0017859

Rise time: 0.25 s (max)
Burn time: >3 s
Radiation intensity: >10 kW/STR
Safety and initiation: yes
Squib type: M-796 or BBU-35
Pellet: magnesium/PTFE cartridge

FG-6

FG-6 is an IR decoy flare that is designed for use on the F-4E and F-16 fighter aircraft. It is a form, fit and function equivalent of the US MJU-7B cartridge and is compatible with SAMP 60/120/240 and ALE-40 dispensers.

Specifications

Dimensions: 25 × 51 × 205 mm
Payload weight: 300 g
Ejection velocity: 20 m/s (min)
Rise time: 0.25 s (max)
Burn time: >3.5 s
Radiation intensity: >20 kW/STR
Safety and initiation: yes
Squib type: BBU-36/B
Pellet: magnesium/PTFE cartridge

FG-8

FG-8 is a round format, aerodynamically stabilised IR decoy flare that is designed for use against air-to-air and surface-to-air IR guided threats and is compatible with existing dispenser systems. IMI claims that FG-8 provides a longer period of effective protection than that offered by a similarly sized, non-stabilised cartridge. The company also noted that the device has an immediate response time and a decreased decoy-to-host angular separation velocity. FG-8 utilises a magnesium flare pellet and a PTFE cartridge. As of this edition, FG-8's status is uncertain.

FG-9

FG-9 is an IR decoy flare that is designed for use on the F-15 fighter aircraft. It is a form, fit and function equivalent of the US MJU-10B cartridge and is compatible with the SAMP 60/120/240 and AN/ALE-45 dispensers.

Specifications

Dimensions: 64 × 51 × 205 mm
Payload weight: 840 g
Ejection velocity: 17 m/s (min)
Rise time: 0.25 s (max)
Burn time: >3.5 s
Radiation intensity: >60 kW/STR
Safety and initiation: yes
Squib type: BBU-36/B
Pellet: magnesium/PTFE cartridge

MULTI-BLU

MULTI-BLU is a multipayload IR decoy flare that is designed for helicopter applications. The device is compatible with the ALE-40 and -47 dispensers together with IMI'S MPMN and SAMP 60 systems. MULTI-BLU's payload ejection sequence can be customised to meet specific customer requirements. The device is also noted as being form compatible with the US M-206 cartridge.

Specifications

Dimensions: 25 × 25 × 205 mm
Weight: 220 g
Payload weight: 3 × 30 g
Ejection velocity: 20-55 m/s
Ejection times: as per customer requirements
Temperature: −40 to +60°C (operating)
Squib type: M769 or BBU-35

Status

As of this edition, devices of the types described were thought to be available.

Contractor

Israel Military Industries Ltd, Ramat Hasharon.

VERIFIED

***

Long Star Electronic Warfare (EW) system

Type

Airborne EW system for attack helicopter applications.

Description

The Long Star EW system is a lightweight, modular architecture that offers detection, location and support-jamming capabilities against radar and communication link emitters. It is designed for rapid installation aboard and removal from helicopter platforms. It comprises Electronic Support Measures (ESM) and instantaneous frequency measuring receivers, an identification processor, a power management unit, an operator console, an ESM/instantaneous direction-finding antenna assembly and a multibeam array transmitter. Overall, the

Rafael's Long Star EW system is designed for applications such as the AH-64 battlefield attack helicopter
0017861

system is quoted as being suitable for electronic intelligence gathering and suppression of enemy air defences missions alongside its support-jamming role. Within the system's jamming subsystem, modularity allows the transmitter chain to be customised for specific missions.

Status

As of early 2000, the Long Star EW system was understood to be available and to have been offered to meet a late 1990 US Army requirement for an airborne Army Support Jamming (ASJ) system. The ASJ bid was made in collaboration with Northrop Grumman.

Contractor

Rafael Electronic Systems Division, Haifa.

VERIFIED

***

SAMP series CounterMeasures Dispensing Systems (CMDS)

Type

Family of airborne CMDS.

Description

SAMP series CMDS are modular equipments that can be configured for fast jet, helicopter, tactical transport and maritime patrol retrofit and new-build applications. Available system configurations are understood to be as follows:

SAMP 60

A typical SAMP 60 system comprises a firing control unit, a control and operating panel and two 30 round Infra-Red (IR) decoy flare/chaff cartridge magazines. The equipment is designed to provide protection against both IR and radar guided air-to-air and surface-to-air missiles and features automatic payload identification. The control and operating panel is night vision goggle compatible and offers a payload inventory display; mode selection (Radar Warning Receiver (RWR), pickle and trigger); dispenser programme scrolling; a Built-In Test (BIT) display and push-button firing.

SAMP 120

A typical SAMP 120 fast jet system comprises a firing control unit, a control and operating panel and four 30 round IR decoy flare/chaff cartridge magazines. The system can accommodate 2 to 20 GHz RR-170 chaff cartridges together with M-206 or MJU-7B IR decoy flares and can be activated from the control and operating panel or via the pilot's Hands On Throttle And Stick (HOTAS) controller. Other system features include firing events based on payloads/bursts/salvos; in-line downloading of firing programmes; on- and offline programming and periodic/on request BIT and diagnostics. SAMP 120 incorporates up to 16 dispense, escape or jettison programmes and offers manual (pilot selects programmes and initiates them), semi-automatic (the onboard RWR selects firing programmes which are then initiated by the pilot) and automatic (the onboard RWR selects and initiates firing programmes) operating modes.

SAMP 240

A typical SAMP 240 large aircraft system includes a firing control unit and eight IR decoy flare/chaff cartridge magazines that can accommodate up to 240 chaff rounds or up to 120 2 × 1 in IR decoy flares. The system is controlled from the flight deck and the equipment features the same operating programmes as SAMP 120 together with multiple operating modes.

Status

As of this edition, a SAMP 60 supply and installation contract was noted as having been procured for an as yet unidentified retrofit programme during 1997.

Contractor

Israel Military Industries Ltd, Ramat Hasharon.

VERIFIED

SPJ-20 self-protection radar jammer

Type
Airborne self-protection radar jammer.

Description
The SPJ-20 self-protection jammer is described as being 'powerful but compact', as being designed to provide protection against ground and airborne, radar-guided, fire-control systems and as being suitable for both fixed- and rotary-winged applications. System features include:
- installation of an acquisition receiver and an Electronic CounterMeasures (ECM) exciter in a single line replaceable unit (no separate radar warning receiver required)
- system support of up to two multiport pulse/Continuous Wave (CW) transmitters
- wideband operation
- ECM techniques for use against pulsed, CW and pulse-Doppler threat emitters
- interfaces to external avionic system and a countermeasures dispenser system
- an integral, field programmable threat library
- in-flight data recording

Of these various elements, the SPJ-20's acquisition receiver is noted as being a wideband, Voltage Controlled Oscillator (VCO) - based, 'fast' instantaneous frequency measuring/superheterodyne device that performs amplitude monopulse direction-finding together with threat identification and classification. For its part, the system's transmitter takes the form of a four tube, multimode (pulse and CW), mini-travelling wave tube cluster with its output being combined in phase and amplitude. This combined output signal can be directed to any combination of outport ports on a pulse-by-pulse basis. Overall, SPJ-20's ECM resources include a 'fast' VCO-based noise transponder, pulse and CW repeater channels, generators and trackers and Doppler, noise and amplitude jamming techniques. SPJ-20 mission support equipment includes a laptop personal computer-based mission loader/verifier, a pre-flight message generator and a post-mission data analyser.

Status
As of January 2002, SPJ-20 airborne radar jammer was understood to be available.

Specifications
Frequency: 2-18 GHz (receive); 6-17.5 GHz (transmit)
Power: 28 V DC (240 W - each transmitter); 115 V AC (3-phase, 3,000 W - each transmitter)
Interfaces: MIL-STD-1553B; RS-232/-422
Environmental: MIL-E-5400T Class 2X
Weight: 12 kg (acquisition receiver/ECM exciter): 30 kg (each transmitter)

Contractor
Elisra Electronics Systems Ltd (a member of the Elisra Group), Bene Beraq.

UPDATED

The acquisition receiver/ECM exciter (right) and transmitter units used in the SPJ-20 self-protection jammer (Elisra)   2000/0081734

SPJ-40 self-protection radar jammer

Type
Airborne self-protection radar jammer.

Description
The SPJ-40 self-protection radar is designed for fighter aircraft applications and its major features include:
- wideband transmission and reception coverage
- a 'high' effective radiated power value
- wide angular transmission and reception coverage
- a 'high'-sensitivity value
- the ability to handle a multithreat emitter environment

- a multichannel architecture
- synergistic coupling with a countermeasures dispensing system
- an autonomous threat acquisition receiver (with a performance that is 'commensurate with that of a high-end radar warning receiver')
- full integration with its host platform's avionics system
- compact and modular line-replaceable units.

Status
As of January 2002, the SPJ-40 self-protection radar jammer was noted as forming the active countermeasures subsystem in Elisra's ASPS defensive aids suite (see separate entry).

Contractor
Elisra Electronic Systems Ltd (a member of the Elisra Group), Bene Beraq.

UPDATED

Tactical Air-Launched Decoy (TALD)

Type
Unmanned, air-launched passive or active radar decoy system.

Description
TALD is an unmanned, air-launched, gliding decoy vehicle that can be equipped with a passive Luneberg lens radar cross-section multiplier and a wideband active radar repeater or a chaff dispenser. The vehicle is designed for multiple carriage on standard ejection racks and is compatible with a wide range of combat aircraft that includes the A-4, AV-8B, F-4, F-16, F/A-18, Kfir and Mirage. Functionally, the vehicle can be programmed to follow a variety of flight profiles and manoeuvres can be launched singly or in multiples at high and low altitudes (including toss modes). Once manufactured, the vehicle is reported to require no ground maintenance and can be programmed on the flight line before a mission.

Status
TALD has been used operationally and as of this edition, was reported as having been procured by both the Israeli Air Force and the US Navy (USN). In US service, the device is designated as the ADM-141.

Specifications
Launch modes: high/low altitude; low level toss; high speed toss
Launch speed: up to M0.95
Launch altitude: up to 12,192 m
Range: up to 124 km
Manoeuvres: preprogrammed
RF payload: active/passive; chaff dispenser (alternative)
Go/No-Go test: 15 s (preflight preparation)
Mission programming: 30 s (preflight preparation)
Weight: 181.4 kg

Contractor
Israel Military Industries Ltd, Ramat Hasharon.

VERIFIED

ITALY

Apex radar jammer

Type
Pod-mounted radar deception jamming system.

Description
The Apex radar deception jamming system makes use of the existing ELT/555 pod shell and is claimed to be effective against all types of fire-control radars. The equipment is described as being effective against pulsed and continuous wave emitters, as having an 'extended' frequency range and as weighing 140 kg. Apex uses fast digital radio frequency memory technology to facilitate coherent response techniques and the system is flight line programmable. Apex is noted as being suitable for aircraft such as the F-5, Hawk 100/200, MB-339 and Mirage F.1.

The Apex radar jamming pod (MS/Elettronica)

Status

As of this edition, the Apex radar jammer was reported as having been 'fully developed'.

Contractor

Elettronica SpA, Rome.

VERIFIED

ARIES Electronic Warfare (EW) system

Type

Integrated airborne EW system.

Description

ARIES is a modular airborne EW system for tactical support and training. It consists of two subsystems, the radar band ARIES A and the communications band Smart Guard/Fast Jam package. As a whole, the system provides tactical ELectronic INTelligence (ELINT)/COMmunications INTelligence (COMINT), tactical surveillance, Electronic CounterMeasures (ECM) support and EW training capabilities as follows:

Tactical ELINT/COMINT
- intercept, fine analysis and direction-finding of radar emissions
- intercept, direction-finding, location, demodulation and monitoring of communications emissions
- collection of intercepted data.

Tactical surveillance
- continuous monitoring of the radar and communications frequency bands
- alerts to changes in the electromagnetic environment
- recognition of unexpected, new or priority emissions.

ECM support
- escort ECM to protect aircraft penetrating a hostile air defence environment by simultaneous jamming of search, acquisition and tracking radars and communications channels
- self-protection when employed in tactical missions.

Training
- familiarising radio operators with the use of electronic counter-countermeasures techniques
- training EW operators in the most effective jamming techniques and procedures
- training aircrews in target approach and attack tactics under real ECM conditions
- field evaluation of EW deployment doctrines.

    The data collected by the aircraft in the ELINT/COMINT and surveillance roles can be transmitted to the ground over secure datalink for near real-time analysis.

Status

As of this edition, more than 60 ARIES systems were reported as having been supplied to air forces around the world since the suite first entered service in 1980.

Contractor

Elettronica SpA, Rome.

VERIFIED

The ARIES modular airborne EW system is designed to provide tactical support and training for air assets for tactical support training

ELT/553(V)2, ELT/554 and ELT/558 series radar jammers

Type

Family of self-protection radar jamming systems.

Description

The ELT/55x family is a series of active self-protection equipments that are targeted against radars, continuous wave illuminators and missile radar seekers. They provide high Effective Radiated Power (ERP), advanced multithreat engagement capability, diversified deception programmes, wide angular cover and short reaction times. They are designed to be integrated with the aircraft Electronic CounterMeasures (ECM) and weapon system or, in some cases, can be used as stand-alone equipment. They have a look-through-while-jam capability and are effective against frequency-agile sources. Known details of the identified members of the family are as follows:

ELT/553(V)2

ELT/553(V)2 is described as being a second generation self-protection system that covers the E to J frequency band (2 to 20 GHz). It provides high ERP over a very large radio frequency bandwidth, an 'advanced' multithreat engagement capability, diversified and easily reprogrammable deception programmes, wide angular coverage and fast reaction. It is intended for internal installation in combat aircraft and helicopters and features low weight and reduced volume. The system can be employed as a stand-alone equipment or can be integrated with radar warning/electronic support equipments, such as the ELT/156X threat warner.

ELT/554

ELT/554 is a lightweight (under 20 kg) deception jammer that is derived from the ELT/553 family. It is specifically conceived for installation in combat helicopters to whose operational roles and constraints the system has been designed. High ERP in J-band (10 to 20 GHz) and ample antenna coverage ensure necessary short self-screening range.

ELT/558

This system is also derived from the ELT/553 family and is designed as the ECM section of a fully integrated electronic defence suite (radar warning receivers and other avionics) for high-performance combat aircraft.

Status

As of this edition, equipment of the types described were reported as having been procured. More specifically, Jane's sources were suggesting that the ELT/553(V)2 had been selected for installation aboard Brazilian AMX strike aircraft and that the ELT/554 unit was fitted to Agusta A 129 Mangusta battlefield attack helicopters.

Contractor

Elettronica SpA, Rome.

VERIFIED

The ELT/558 self-protection radar jammer

ELT/555 and ELT/555B radar warning and jamming systems

Type

Family of airborne radar warning and deception jamming systems.

Description

ELT/555 is a pod-housed airborne deception jammer and warning system that is designed for the self-protection of modern attack and fighter aircraft from enemy anti-aircraft systems (surface-to-air and air-to-air and anti-aircraft artillery). The system is completely self-sufficient as regards cooling and primary power and makes no demands whatsoever on the carrier aircraft's power supply. The philosophy adopted for the ELT/555 system ensures that the jammer will automatically tailor its reaction to the type of hostile radiation detected. The system operates against both pulsed and Continuous Wave (CW) radar emissions and presents the following salient operational features: simultaneous multithreat reaction; very high resistance to electronic counter-countermeasures (including frequency agility); full automation; optimised threat deception programmes; multiband frequency coverage (H- through J-band, 6 to 20 GHz)) and look-through-while-jam capability.

The equipment is claimed to be effective against all types of fire-control radars and active or semi-active homing missiles. Various deception techniques are employed and multiple target engagement is possible.

Airspeeds of up to Mach 1.1 at sea level and Mach 1.5 at 12,192 m were tested during the qualification of the pod. The central body section and tail module are completely clean, except for the two bottom radomes and stabilising fins: the latter, located toward the rear of the main body section, serve to increase the safety envelope during jettison. The pod can be jettisoned by the pilot and is provided with a self-destruction device. Self-destruction occurs at a presettable interval after the jettison command. The device can be installed or removed at flight line level.

The pod is provided with a cooling system and self-sufficient power generating system that supplies the electric power necessary to the pod's equipment. The prime mover of the power generating system consists of a Ram Air Turbine (RAT) which drives a four-pole permanent magnet generator. The RAT also drives the compressor of the pod's cooling system, which operates on a closed-circuit fluid recycling principle.

There are forward- and rear-facing CW transmission antennas on the undersurface of the pod body and nose and tail radomes house antennas for transmission and reception of pulse signals and reception of CW signals.

The ELT/555 pod measures 300 cm in length, with a body diameter of 27 cm (34 cm RAT maximum) and a weight of 150 kg. Types of aircraft nominated by Elettronica as capable of using the ELT/555 include the F-5, Mirage III, Alpha Jet, MB 326, MB 339, Jaguar, MiG-21, Phantom, Mirage V and F1, Hawk and Saab 105.

A new version of the system (designated as ELT/555B) is also available and incorporates improved deception jamming techniques, a more powerful processing capability and flight line reprogrammability.

Status
As of this edition, ELT/555 series equipments were reported as having been procured. The ELT/555B variant was reported as having become 'operational' during 1995.

Contractor
Elettronica SpA, Rome.

VERIFIED

JAPAN

Japanese airborne radar and communications band jamming systems

Description
Known examples of Japanese airborne radar and communications band jamming systems are as follows:
J/ALQ-6
J/ALQ-6 is a radar jammer that is understood to have been installed aboard Japanese Air Self-Defence Force (JASDF) F-1 and F-4 combat aircraft.
J/ALQ-7
J/ALQ-7 is understood to be a communications band jamming system that was developed by NEC, under a then year US$20 million contract that was awarded during the early part of 1989.
J/ALQ-8
J/ALQ-8 is reported to be the radar jamming system that is installed aboard JASDF F-15J interceptors. The equipment is understood to operate in three frequency bands (1 to 4, 4 to 8 and 7.5 to 18 GHz) and to be interfaced with the J/APR-4A radar warning receiver (see separate entry).

VERIFIED

NORWAY

Samovar radar jammer

Type
Pod-mounted radar jammer.

Description
Samovar is a pod-mounted radar jammer built by Kongsberg Gruppen in collaboration with the Norwegian Defence Research Establishment. It is a programmable responsive noise jammer (operating in the 8 to 16 GHz band) and is fitted to F-5 and F-16 aircraft to provide standoff jamming protection for F-16s launching Penguin Mk 3 anti-ship missiles. The system automatically identifies and responds to threats (via a multiprocessor control system) and generates Continuous Wave (CW), interrupted CW and pulse outputs. It is a velocity controlled oscillator-based, liquid-cooled pod that can be integrated with radar warning receivers and chaff/flares, if required for self-protection. The pod is 2.75 m long, 0.254 m diameter and weighs about 150 kg.

Status
Over time, the Royal Norwegian Air Force is reported to have procured up to 22 Samovar radar jamming pods with which to equip F-5 and F-16 aircraft.

The Samovar jamming pod mounted on the centreline of a Norwegian Air Force F-16 (Martin Streetly)

Contractor
Kongsberg Gruppen AS, Kongsberg.

VERIFIED

RUSSIAN FEDERATION AND ASSOCIATED STATES (CIS)

RFAS active and passive airborne countermeasures systems

The following entry provides known details of active and passive airborne countermeasures systems produced by, exported by and/or in service with the air forces of the Russian Federation and the independent states of Armenia, Azerbaijan, Belarus, Georgia, Kazakhstan, Kyrgyzstan, Moldova, Tajikistan, Turkmenistan, Ukraine and Uzbekistan. Taken together, these 12 entities form the Commonwealth of Independent States economic grouping.
Akatsiya
Akatsiya is reported as being the mission equipment that is installed aboard the Mi-8PPA battlefield Electronic Warfare (EW) helicopter.

APP-50MA/APP-50MR
Family of CounterMeasures Dispensing Systems (CMDS) for medium to large aircraft such as the Antonov An-22, the Sukhoi Su-24 and -27 and the Tupolev Tu-142 and -160.

ASO-2 Series
Family of CMDSs that are suitable for a range of aircraft including the Mil Mi-17 and Mi-24D (ASO-2V version) helicopters and Sukhoi Su-25K (ASO-2V) assault aircraft.

ASO-3
The CMDS installed on the Mil Mi-8MT helicopter.

A close-up of the L-166 fixed-source IR jammer and the port side ASO-2 CMDS fitted to a MIL Mi-24 battlefield attack helicopter of the Czech Air Force (Martin Streetly)

0017862

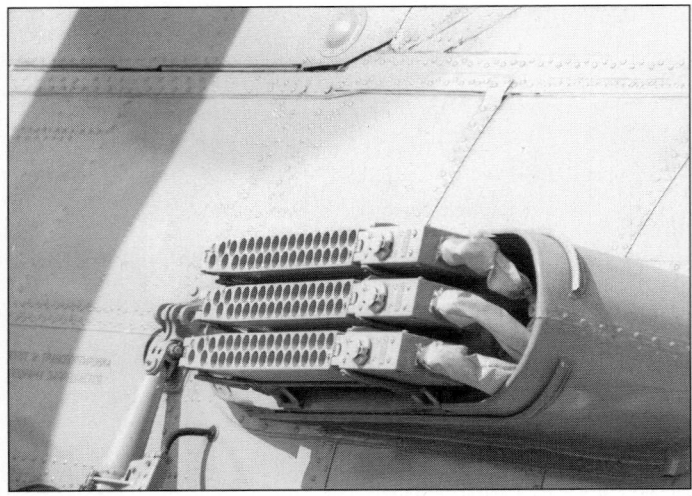

A close-up of an ASO-3 CMDS fitted to a Mil Mi-8MT helicopter

BVP-30-26M
The Infra-Red (IR) decoy flare dedicated CMDS fitted to the Mikoyan MiG-29 series. Comprises two 26 mm by 30 round capacity magazines mounted in the aircraft's fin root extensions.

Gardeniya
Family of internally mounted and podded radar and communications band jammers. The series is understood to include an E/F-band (2 to 4 GHz), 73 kg system for use on the Sukhoi Su-27P and the **Gardeniya-IFUE** jammer that is mounted in the Mil Mi-17PG battlefield EW helicopter. Here, the equipment is reported as being able to jam up to eight H/I-band (6 to 10 GHz) pulsed, Continuous Wave (CW) or 'quasi CW' threat emitters within a 25° (azimuth) × 12° (elevation) sector.

Ikebana
Credited to the Kaluzhsky Nauchno-Issledovatelskiy Institut Radioteckhnicheskikh Sistemi (KNIIRS) organisation, Ikebana is described as being the mission suite installed aboard the Mi-8MTI battlefield EW helicopter. An 'improved' variant of the system is also reported as being installed aboard the Mi-8TPI EW helicopter.

KS418E
KS418E is reported as being a pod-mounted EW system for use on the Su-24M, a replacement for the SPS-161 equipment and as a product of the KNIIRS organisation.

L-005S Sorbtsya
The L-005S Sorbtsya is described as being a wingtip pod-mounted EW self-protection system for the Su-27 that is a product of the KNIIRS organisation.

L-028K
A CMDS for medium to large aircraft applications such as the Ilyushin Il-76, Sukhoi Su-24 and -27 and Tupolev Tu-22.

L-145VE Khibiny
L-145VE Khibiny is described as being an EW self-protection system that is designed for installation aboard the Su-32 and is a product of the KNIIRS organisation.

L-166 Series
Family of fixed-source Infra-Red (IR) jammers that are described as being suitable for use on helicopters such as the Kamov Ka-25 and Mil Mi-8, -17, -24 and -28 together with American types such as the AH-1, the UH-60 and the CH-46. Within the series, the Mi-24 is noted as being fitted with the **L-166V-11E** variant while the **L-166B-1A** model is configured for the Mil Mi-8 series. L-166B-1A comprises a 365 × 463 mm transmitter-radiator unit and a 95 × 70 × 45 mm cockpit control unit. The system's mean time between failure value is quoted as being 250 hours with IR source life being 50 hours. The equipment is designed to protect against a range of IR-guided missiles including Sidewinder, IR Falcon, MICA, Strela 2M, Redeye and Chaparral.

Landysh
Emitter detection, location, analysis, identification, classification and jamming suite which appears to be capable of integration with the **Fasol** (String Bean), **Los** (Moose) and **Mimoza** (Mimosa) active radar jammers and is installed aboard the Sukhoi Su-24MP electronic warfare aircraft.

LO-80 Fantasmagoria-A
Emitter location and classification pod that is used with anti-radiation missiles such as the Kh-31P and Kh-58 weapons aboard the Sukhoi Su-24M strike aircraft. As of August 2001, LO-80 is likely to have been superseded by the LO-81 system (see following).

LO-81 Fantasmagoria-B
Emitter location and classification pod that is used with the Kh-58 anti-radiation missile aboard the Sukhoi Su-24M.

An SPS-141 EW pod carried beneath the wing of an Su-25 close-support aircraft (Piotr Butowski)   2002/0114484

MSP-25 Omul
MSP-25 Omul is a podded electronic countermeasures system that appears to be a deception equipment that is carried on the port and starboard outer wing stations of the Su-39 close support aircraft. In such an installation, one pod is used to receive and process radar threats (including the emitters associated with the Hawk and Patriot Surface-to-air missile (systems) while the second acts as a transmitter. In other applications, the Omul pod appears to be able to be configured for both reception and fore and aft transmission. As applied to the Su-39, system coverage is noted as being 360° in azimuth with the exception of a 15° half angle on either side of the aircraft's centreline.

Pole
The Pole system is described as being the radar jammer that is installed aboard the Mi-8PP battlefield EW helicopter.

Shokogruz IR jammer
The Shokogruz IR jammer is a modulated source IR jamming system that is mounted in the tailcone of the Su-39 close support aircraft. So installed, the system is credited as being able to protect its host platform at up to and including 95 per cent engine thrust levels. Over and above this, IR protection is provided by a mixture of the Shokogruz system and IR decoy flares.

Smalta-V
Smalta-V is the radar jamming system installed aboard the Mi-8SMV battlefield EW helicopter and is a product of the KNIIRS organisation. Jane's sources suggest that KNIIRS has further developed the system to the single operator Smalta-PGE standard. Smalta-PGE is noted as being designed to counter air-defence, airborne fire-control and anti-aircraft missile seeker radars.

SNOP
As of August 2001, SNOP was understood to be an IR jamming system, the status of which was uncertain.

Sorbtsya-S
Sorbtsya-S is described as being a 'wide spectrum' active jamming system that is reported as being designed for installation aboard the Su-34 and -35 aircraft and possibly intended for co-operative jamming and/or the provision of jamming cover for non-electronically protected aircraft within a formation.

SPS-63/-66/-68
While not confirmed, it is thought possible that the SPS-63, -66 and -68 systems are the radar jammers installed aboard the Mi-17P EW helicopter. Here, the platform's three jamming subsystems cover the B- (250 to 500 MHz), D- (1 to 2 GHz) and F- (3 to 4 GHz) bands and offer 30° (azimuth) × 12° (elevation) and 120° (azimuth) × 30° (elevation) transmission arcs in the B/D- and F-bands respectively. D- through F-band threat signals can be detected within a 180° (azimuth) × 30° (elevation) sector.

SPS-130/140 Gvozdika
First-generation (?) self-protection jammers used on Mikoyan MiG-25 and -27 series aircraft.

SPS-141
Reported as being a product of the KNIIRS organisation, SPS-141 is understood to be an EW self-protection system that is suitable for internal and podded installations and as of August 2001, was reported as being in service aboard 'export aircraft' such as the MiG-21, MiG-23, MiG-25, Su-22 and Su-25. Jane's sources further report that KNIIRS was promoting an SPS-141 upgrade kit at the August 2001 MAKS 2001 airshow. Described as taking the form of a 'module-for-module' replacement kit, the enhanced SPS-141 was also noted as being effective against 'quasi CW', high pulse repetition frequency and track-while-scan radars.

SPS-150 Lyutik/SPS-151
Self-protection jammer fitted to the Mikoyan MiG-25RB. Jane's sources describe the SPS-151 EW system as being a product of the KNIIRS organisation.

SPS-160 Geran/SPS-161
Family of self-protection jammers installed on Sukhoi Su-24 (**Geran-F**) and Su-27 series aircraft. Jane's sources describe the SPS-161 EW system as being a product of the KNIIRS organisation.

SPS-170 Series

Family of self-protection jammers that have been associated with the Sukhoi Su-34 and -35 aircraft. Jane's sources describe the SPS-171 EW system as being a product of the KNIIRS organisation.

Tangazh (Aeronautical pitch)

Podded signals intelligence collection system that forms part of the Sukhoi Su-24MR's BKR-1 **Shtyk** (Bayonet) reconnaissance suite.

UV-3A

Comprising a 'Control Console' (CC), a Programme Setting Unit (PSU), a Control-Information Unit (CIU), a Safety Switch (SS) and Cartridge Firing Units (CFU), the UV-3A CMDS is compatible with both 26 and 50 mm circular munitions and can accommodate between eight and 512 dependent on configuration. The system's CFUs can be mounted internally and externally, with the latter making use of the UV26S.8750-0 ejection module that can accommodate 16 50 mm decoys. As of August 2001, a UV-3A CMDS variant was understood to be installed on the MiG-31 interceptor.

Specifications

Number of available preset programmes: over 50,000
Number of memorised programmes that can be selected operationally: 8
Variable parameters per programme: 5
Number of cartridges per salvo: 1-8
System cartridge capacity: 8-512
Salvo interval: 0.025 to 16 s
Programme duration: 28 min (max)
Programme parameter/cartridge data storage: up to 1 year
Power consumption: 0.5 kVA (115 V AC, 400 Hz); 1.2 kW (24-30 V DC)
Weight: 0.35 kg (SS); 0.6 kg (CC); 5.5 kg (PSU); 7.4 kg (32 × 26 mm cartridge CFU); 9.5 kg (8 × 50 mm cartridge CFU); 9.8 kg (CIU)
Dimensions: 94 × 90 × 66 mm (SS); 130 × 384 × 172 mm (32 × 26 mm cartridge CFU); 130 × 384 × 285 mm (8 × 50 mm cartridge CFU); 146 × 64 × 104 mm (CC); 172 × 373 × 214 mm (CIU); 186 × 221 × 212 mm (PSU)

UV-26

Comprising a 'Control Console' (CC), a Preset Programme Unit (PPU), a Control Unit (CU), a Safety Switch (SS), distributors and Cartridge Firing Units (CFU), the UV-26 CMDS is compatible with both 100 g 26 mm and 1 kg 50 mm circular munitions and can accommodate between eight and 512 dependent on configuration. The number of distributors used is system configuration dependent and such devices are included in the architecture when both 26 mm and 50 mm CFUs are fitted. Each distributor can handle two 32 × 26 mm cartridge and two to four 16 × 50 mm cartridge CFUs. As of August 2001, a 192 cartridge capacity UV-26 CMDS variant was understood to be installed on the Sukhoi Su-39 close support aircraft.

Specifications

Number of available ejection programmes: 575
Number of preset 'quick-to-select' programmes: 3
Variable parameters per programme: 3
Number of cartridges per salvo: 1-8
System cartridge capacity: 8-512
Salvo interval: 0.05 to 8 s
Programme duration: 120 s (max)
Power consumption: 50 W (24-30 V DC); up to 200 VA (115 V AC, 400 Hz)
Weight: 0.35 kg (SS); 0.53 kg (PPU); 1.2 kg (CC); 1.37 kg (distributors); 4.23 kg (CU); 6.5 kg (32 × 26 mm cartridge CFU); 15.5 kg (16 × 50 mm cartridge CFU)
Dimensions: 66 × 94 × 90 mm (SS); 128 × 722 × 62 mm (32 × 26 mm cartridge CFU); 134 × 64 × 105 mm (PPU); 146 × 64 × 176 mm (CC); 150 × 308 × 95 mm (CU); 248 × 260 × 110 mm (CU); 255 × 1,063 × 74 mm (16 × 50 mm cartridge CFU)

UP-P1/UV-P2

Series of CMDSs that are suitable for medium to large aircraft applications such as the Ilyushin Il-76 and Sukhoi Su-24 and Su-27 series aircraft.

UPDATED

SOUTH AFRICA

GSY1500 communications band jamming system

Type

Airborne, ground-based or shipborne communications band jamming system.

Description

The main purpose of the GSY1500 communications band jamming system is the disruption and/or jamming of identified communications signals in the 20 to 500 MHz frequency range. Because of its modular design, GSY1500 can be mounted in an aircraft (such as a helicopter or transport aircraft), a naval vessel or in a static or mobile ground-based installation. GSY1500 operates in the frequency range 20 to 100 MHz, 100 to 500 MHz or both, dictated by the frequency range of the power amplifiers. Depending on the antenna subsystems selected, the power

Airborne deployment of the GSY1500 system

amplifier outputs can be arranged to provide continuous coverage over the full frequency range.

GSY1500 consists of a wideband fast-setting receiver, radio frequency synthesiser power amplifiers covering 20 to 100 and 100 to 500 MHz, a countermeasure generator unit and a man/machine interface consisting of a graphic display and computer unit with keyboard. A wide range of directional or omnidirectional antennas is available. It can be controlled by a single operator using the man/machine interface. The system can be pretasked to perform intelligent (automatic) jamming on certain identified channels, once they are detected during an operation. It can be either operated in this mode or manually controlled by the operator. In the latter mode the operator is able to intercept other communication channels. Features of the GSY1500 include the following:

- look-through capability to monitor continued target presence
- selected, optimised counter modulation for various target signal types
- configurations for both time division multiplex jamming (signals are jammed with full power in sequential order), or freqency division multiplex jamming (signals on different frequencies are jammed simultaneously, sharing the available power between targets)
- jamming of up to 20 prioritised target frequencies, preprogrammed in memory
- effective against fixed frequency voice and data
- selection of different output power settings
- spurious suppression of unwanted harmonics and spurious signals to protect own communications and limit unwanted radiation.

Other than control of the system, the man/machine interface provides display of system status and detected activities in signal bands and channels, prediction of jamming effectiveness based on analysis of ground and air communication links, pretasking of system, mission log analysis and comprehensive built-in test facilities. Software flexibility can accommodate specific customer requirements.

Status

As of this edition, the GSY 1500 communications band jamming system was reported as having been procured by the South African Air Force and as likely to be supported by the Grintek EWATION organisation.

Specifications

Frequency range: 20-100 MHz or 100-500 MHz or both
Output power: 500 or 1,000 W (20-400 MHz); 400 or 800 W (400-500 MHz)
Counter modulation types: gauss noise at 0.75 kHz or 3.3 kHz bandwidths, mixed voices, tones, FSK, recorded audio, external audio
Counter modulation bandwidth: AM 6 kHz; FM 3.5, 7.5, 12.5, 50 or 75 kHz
Antenna options: log-periodic, discone, whip or high-power blade
Scanning speed: 80 channels/s
System reaction time: 10 ms
No of pretasked channels: 20

UPDATED

GSY1501 airborne communications Electronic Warfare (EW) system

Type

Airborne communications band Electronic Support (ES) and jamming system.

Description

GSY1501 is a comprehensive airborne communications EW system that is suitable for installation on passenger aircraft, cargo carriers, business jets or similar types. The system provides a complete capability enabling the detection, interception, direction-finding, recording and disruption (jamming or deception) of enemy command and control communication networks. The system provides coverage of the frequency spectrum from 20 to 1,000 MHz and can be extended to operate from 1.5 MHz. Because of the modular approach in design, the GSY1501 can be

A general view of the operator stations associated with the GSY1501 airborne ES/countermeasures system

adapted to accommodate any suitable number of operators and additional capabilities can be added as required.

The GSY1501 system consists of a number of subsystems as follows:

Radio Frequency (RF) reception and distribution
A 20 to 500 MHz and 500 to 1,000 MHz blade antenna with omnidirectional response in the azimuth plane provides reception over the full frequency range. An antenna distribution unit distributes the received Radio Frequency (RF) signals to the receivers in the system, and a blanking unit protects the platform antennas during active transmission.

Spectrum scanning
Scanning receivers perform scanning of the spectrum in the band as configured by the operators.

Direction-Finding (DF)
A 7-channel interferometer direction-finder determines the line of bearing of a signal in the 20 to 500 MHz frequency band (optionally to 1,000 MHz). The DF is fed by a uniquely designed antenna array with wide aperture to provide high accuracy.

System control
ES and countermeasures system control units consist of a number of processors which control the scanning receivers and direction-finder, power amplifiers, synthesisers and counter modulation generators.

Operator workstations
The number of operator workstations is configurable depending on the available space and platform limitations. A workstation consists of a monitoring receiver, digital voice recorder, spectral display unit, operator interface unit and intercom control unit. The operator interface is an environmentally hardened PC with a colour flat panel display and a conventional PC keyboard.

Jamming
30 to 100 and 100 to 500 MHz high-power antennas capable of transmitting more than 1 kW of RF power. RF jamming synthesisers, each having three independent RF channels capable of amplitude and frequency modulation. Counter modulation generators that generate the base-band signals from a digital source and are used as modulation sources for the jamming signal. Power amplifiers covering 20 to 100 and 100 to 500 MHz, capable of generating up to 1 kW of RF power.

Status
As of this edition, Jane's sources suggest that a variant of the GSY1501 communications EW system was installed aboard the South African Air Force's Boeing 707-344C stand-off communications jamming aircraft '1421/621'. As of October 2000, the platform was operated by the service's No 60 Squadron and was likely to have been supported by the Grintek EWATION organisation.

Specifications
ES capability
Frequency range: 20-1,000 MHz (1.5-30 MHz optional)
Frequency scan rate: 1.5-30 MHz at 0.3 MHz/s at 3 kHz channel width; 20-100 MHz 4-16 GHz/s
Demodulation (fixed frequencies): AM, FM
DF capability
Frequency range: 20-500 MHz (1.5-30 and 500-1,000 MHz optional)
DF accuracy (instrumental): 1.5° RMS
DF agility: 40 bearings/s
ECM capability
Frequency range: 30-500 MHz
Output power: 20-400 MHz 500 W or 1,000 W RMS; 400-500 MHz 400 or 800 W RMS
No of target nets: up to 3 simultaneous (fixed frequencies per frequency band)
Counter modulation types: gauss noise (various bandwidths), mixed voice, swept tones, FSK
Deception: acts as normal radio with AM and FM transmitter
Jamming synthesiser agility: setting within 10 μs on any frequency

UPDATED

SPAIN

TARAN communications and navigation system countermeasures suite

Type
Airborne communications and navigation system countermeasures suite.

Description
TARAN is a multi-operator, communications, command and control countermeasures system that fuses tactical communications countermeasures with a C/D-band (0.5 to 2 GHz) capability that is aimed primarily at radio navigation aids and Identification Friend-or-Foe (IFF) systems. Within the system, a 'low-band' console handles the communications frequencies while a 'high band' workstation is dedicated to C/D-band activity. A third console is used to exercise overall system control. So configured, TARAN is capable of detecting and jamming communications band, navigation aid and IFF signals traffic.

TARAN has been developed according to MIL standards and has a modular architecture to facilitate system growth as required. Functionally, the suite is capable of scanning for, intercepting and analysing signals and can record and replay those that are considered to be of interest. The jamming function is automatic with target frequency selection being made in-flight or prior to the start of the mission. Environment data is displayed to the operators in the form of tables that can be manipulated using keyboard-driven screen menus. Recorded data can be downloaded for post-mission analysis and there is a computer-based system remote control facility.

Status
As of this edition, Jane's sources were suggesting that the TARAN airborne communications and navigation system countermeasures suite was installed aboard one or more of the three Dassault TM.11 (Falcon 20D/E) aircraft operated by the Spanish Air Force's Escuadron 408 (based at Torrejon de Ardoz).

Contractor
INDRA DTD, Madrid.

VERIFIED

A close-up of one of the workstations used in the TARAN communications and navigation system countermeasures suite

SWEDEN

BO series CounterMeasures Dispensing Systems (CMDS)

Type
Family of airborne CMDS.

Description
Swedish contractor Saab Avionics produces a range of CMDS, the known details of which are as follows:

BOP/A
The 31 kg (empty) BOP/A CMDS is made up of beam-shaped launch assemblies that accommodate six, two round capacity magazine modules and are compatible with 55 mm diameter chaff and Infra-Red (IR) decoy flare (up to three per round) cartridges. The system ejects its payloads downwards and towards the rear of the host platform and is loaded from below. BOP/A is designed for ease and rapidity of installation and a typical fighter fit incorporates two launch assemblies. For larger airframes (such as transport aircraft), a four launch assembly installation is used. In the case of fighter applications, BOP/A can also be installed conformally to

minimise drag. Maximum reload time is given as being 60 seconds and individual BOP/A launch assemblies communicate with their host aircraft's Electronic Warfare (EW) system via an integral electronics unit and an RS-422 serial datalink. The system can be fully integrated into an onboard countermeasures suite and an optional IR sensor can be mounted in the rear of the launch assemblies to provide confirmation of IR decoy flare ignition.

BOP/AT

The BOP/AT CMDS is designed to be scab mounted on the top of the fuselages of fighter aircraft. So installed, the equipment ejects expendables upwards and forwards, an arrangement that, in particular, allows IR decoy flares to develop their full spectrum characteristics close in to the launch vehicle, thereby maximising their effectiveness. System features include:

- an up to nine magazine system capacity, with each magazine capable of accommodating up to 24 payload cartridges
- a high level of system automation (including automatic adjustment of ignition pulses to specific payload cartridge ignition characteristics)
- payload type and quantity identification
- databus system control
- support of nested sequencing and random dispensing

The magazine format used in the BOP/AT and AX CMDSs
2000/0068926

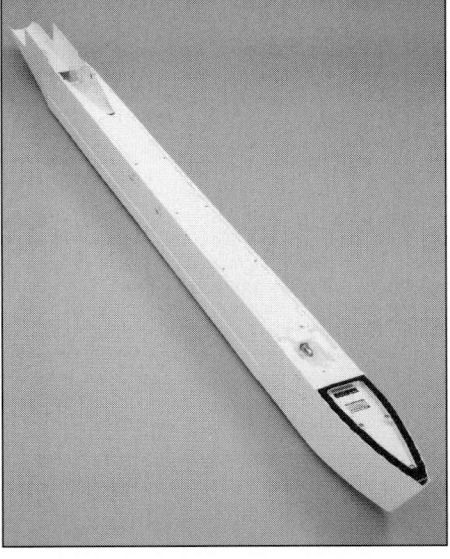

The BOP/A CMDS launch assembly

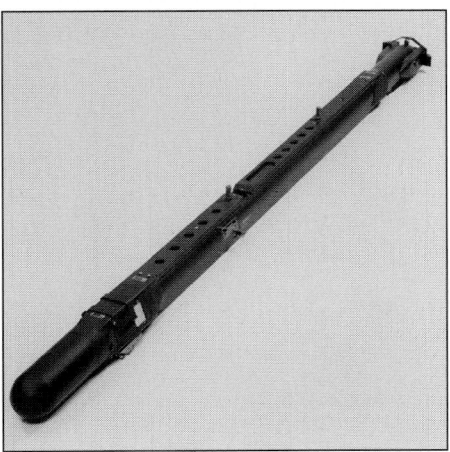

A BOL CMDS mounted in the rear of a LAU-7/-138 missile launch rail and configured for use by the Royal Air Force (MS/SaabTech)

- automatic ejection of failed payloads and the immediate initiation of a replacement round
- capacity for integration with a missile approach warner to create a 'complete' defence against IR and radar guided missiles.

BOP/AX

The BOP/AX CMDS is designed to be scab mounted on the sides of the fuselages of, typically, transport aircraft. So installed, the equipment can eject payloads upwards, downwards and forwards. By fitting double or quadruple installations, BOP/AX can be configured to dispense payloads in 'virtually' any direction required, thereby allowing the equipment's control circuitry to select the most effective ejection sequence for any detected threat. System features include:

- an up to nine magazine system capacity, with each magazine capable of accommodating up to 24 payload cartridges
- a high level of system automation (including automatic adjustment of ignition pulses to specific payload cartridge ignition characteristics)
- payload type and quantity identification
- databus system control
- support of nested sequencing and random dispensing
- automatic ejection of failed payloads and the immediate initiation of a replacement round
- capacity for integration with a missile approach warner to create a 'complete' defence against IR and radar guided missiles.

BOP/B

The 7.2 kg (including magazine) BOP/B CMDS is designed specifically for fighter aircraft applications and can incorporate up to eight dispenser modules, with each

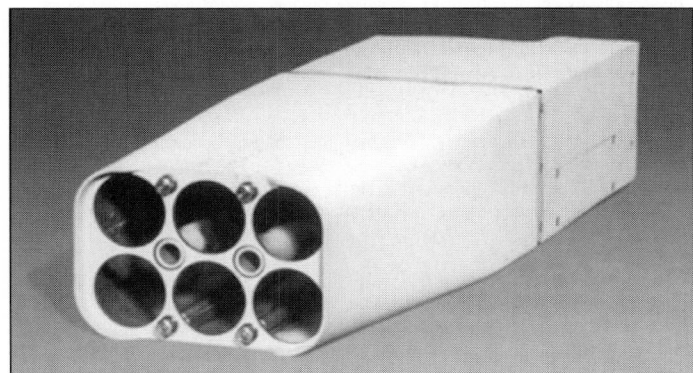

A BOP/B CMDS dispenser module

A BOZ 102 dispenser mounted on an Italian Tornado strike aircraft (Martin Streetly)

The BOL CMDS as applied to the LAU-128 launch rail when carried by the US Navy's F-14 Tomcat interceptor
0017863

module being able to accommodate 15 MJU-7, six MJU-10 or six 55 mm calibre cartridges. The system's dispenser modules each comprise a magazine, a breech plate and an electronics unit and feature breech plate coding pins to identify the type of payload loaded into the particular magazine. As a whole, BOP/B offers a wide range of mounting options and can handle the latest generation of active offboard/towed expendables in addition to conventional chaff and IR decoy flare payloads.

BOP/C

Jane's sources suggest that the 5 kg (8.1 kg with magazine) BOP/C designation had been given to an internally mounted CMDS that can be configured to accept single type or mixed loads drawn from a range of payload formats that include 25 × 25 × 210 mm (up to 40), 51 × 25 × 210 mm (up to 20) or 51 × 64 × 210 mm rectangular cartridges or 36 or 40 mm diameter cylindrical munitions.

BOL

The BOL CMDS is designed to be installed in the rear of a range of air-to-air missile launch rails that include the LAU-7 (known as the LAU-138 with BOL installed), LAU-127, LAU-128, LAU-129 and flight refuelling Common Rail Launcher (CRL) units. So installed, the system comprises an electromechanical dispenser assembly that fits into the rear of the host launch rail (which also features a revised and rearranged missile seeker cooling system) and can accommodate both chaff (up to 160 RR-184 packages) or IR countermeasures payloads mounted in plastic holding frames. Functionally, these are ejected one by one into the airflow aft of the launcher where, in the case of the IR payloads, they ignite or, if chaff, break up to form the required cloud formation. Saab Avionics maintains that the BOL electromechanical dispensing method is more precise than pyrotechnic ejection and offers advantages in terms of safety on the ground and maintainability. If required, the BOL dispensing mechanism can be installed conformally where no suitable launch rail is available. Saab Avionics has developed system test and chaff loading equipments to support BOL installations.

BOZ 3

The BOZ 3 CMDS is a subsonic, pod-mounted chaff dispenser that was originally developed for use by the Swedish Air Force and has subsequently gone on to be adapted for 'electronic aggressor' training work. The system has manual and automatic initiation modes and can be used to generate break-lock and corridor chaff clouds.

BOZ 100

BOZ 100 series CMDS take the form of a pod-mounted system that was originally developed for the Swedish Air Force and has subsequently been widely exported. As such, the equipment is capable of break-lock and corridor functions at sub- and supersonic speeds. BOZ pods have comprehensive EW system interfaces, are microprocessor controlled, have reprogrammable memories and are suitable for use during deep-penetration strike and reconnaissance sorties together with the provision of EW support.

Status

Over time, the BOP/A CMDS has been installed on Viggen aircraft of the Swedish Air Force, with BOP/B systems being fitted to J 35J Draken interceptors and JAS 39 Gripen multirole combat aircraft of the same service. The BOL CMDS has been selected for use on the four nation Eurofighter Typhoon multirole combat aircraft (CRL installation) and has been/is being procured by the air forces of Sweden (in service on the JA 37 Viggen and selected for the JAS 39 Gripen), the UK (in service on the Harrier GR Mk 7 and the Tornado F Mk 3) and the US (F-15 aircraft with BOL installed in the LAU-128 launch rail - see following). Elsewhere, the US Navy has introduced a LAU-128/BOL installation aboard its F-14 Tomcat interceptors. The BOZ 3 training CMDS is reported to be in service with a 'number' of countries and the civilian contractor FR Aviation, while the BOZ 100 CMDS pod is operational with the air forces of Germany (BOZ 101 on Tornado strike aircraft), Italy (BOZ 102 on Tornado strike aircraft), Saudi Arabia (unidentified variant on Tornado strike aircraft), Sweden (as the BOX 9) and the UK (BOZ 107 on Tornado strike aircraft together with unidentified variants on Nimrod MR Mk 2 and R Mk 1 maritime reconnaissance and signals intelligence aircraft). Outside air force use, the system is also noted as being in service aboard Super Etendard strike aircraft of the French Navy. Of these various BOZ 100 series applications, the Tornado procurement is

probably the largest, with approximately 700 units having been supplied into this trinational programme. In terms of the F-15 programme, US contractor BAE Systems North America - Integrated Defense Solutions was awarded a then year US$18,495,236 not to be exceeded, firm, fixed-price contract on 28 September 2001 that covered the production of BOL Group-B kits and modified LAU missile launchers for use aboard F-15 aircraft. Specifically, the award (which at the time of its announcement, was scheduled for completion by April 2002) related the qualification, initial production test and production dispenser systems for 99 F-15 aircraft plus spares. The US Air Force's Aeronautical Systems Center, Wright-Patterson Air Force Base, Ohio was the programme's contracting activity.

Specifications

BOP/AT
Cartridge capacity: 9 magazines of 24 × 1 × 1 in, 12 × 2 × 1 in, 6 × 2 × 2.5 in, 16 × Ø36 mm or 10 × Ø40 mm payloads; a mix of the foregoing up to a total of 9 magazines
Reloading time: 3 min (max)
Control signals: 1 dispense signal (discrete input, +28 V DC); MIL-STD-1553B or RS-485 (databus)
Safety signals: MASS (+28 VDS); safety pin (option)
Weight: 2.5 kg (single empty magazine); 40 kg (dispenser structure)
Dimensions (l × w × h): 2,650 × 240 × 355 mm
BOP/AX
Cartridge capacity: 9 magazines of 24 × 1 × 1 in, 12 × 2 × 1 in, 6 × 2 × 2.5 in, 16 × Ø36 mm or 10 × Ø40 mm payloads; a mix of the foregoing up to a total of 9 magazines
Reloading time: 3 min (max)
Control signals: 1 dispense signal (discrete input, +28 V DC); MIL-STD-1553B or RS-485 (databus)
Safety signals: MASS (+28 VDS); safety pin (option)
Weight: 2.5 kg (single empty magazine); 40 kg (dispenser structure)
Dimensions (l × w × h): 2,650 × 355 × 240 mm
BOL
Control signal: RS-485 serial datalink or single +28 V discrete signal
Reload time: <1 min
Power supply: 115 V (400 Hz)
Loaded weight: 19 kg (dispenser only); 56.5 kg (LAU-138)

Contractor

Saab Avionics AB, Stockholm.

UPDATED

....

BO2D expendable towed decoy

Type

Expendable towed broadband repeater decoy.

Description

Saab Avionics, in collaboration with Sweden's Armaments Authority, has developed a pyrotechnically ejected, expendable, towed, broadband repeater decoy that has been given the designation BO2D. Weighing less than 2 kg, BO2D is a solid-state device that makes use of 'advanced' microwave integrated circuitry, is compatible with 55 mm countermeasures dispensing systems (such as Saab Avionics BOP and BOZ systems – see separate entries) and provides protection for fighter aircraft against active and semi-active missiles with Radio Frequency (RF) sensors. Described as being an H/I/J-band (6 to 20 GHz) broadband, high-gain repeater with a frequency modulation capability, the BO2D utilises internal battery power, communicates with its host aircraft via its towline and is noted as being all solid-state with 'advanced' microwave integrated circuitry. Once deployed, the

The BOL CMDS as applied to the LAU-128 launch rail when carried by the US Air Force's F-15 aircraft
0017864

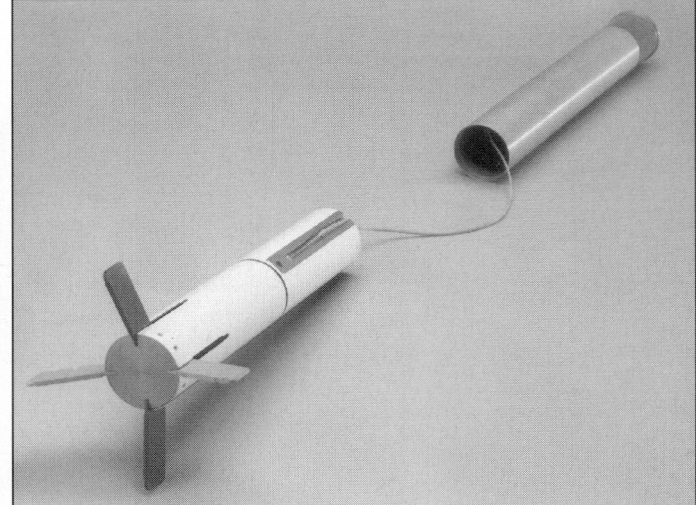

A general view showing the BO2D flight body, towline and launch sleeve
0017865

device's RF output can be switched on and off and different countermeasures techniques selected, with the necessary commands being communicated to the flight body through the towline. While not fully confirmed, the BO2D flight body appears to house thermal and standby batteries, control logic, a mode library, communications circuitry, reception and transmission antennas, a modulator and signal and power amplifiers. After use, this fin-stabilised unit is cut adrift using a pyrotechnic cutter.

Status

Work on the BO2D programme began in 1990/91 and as of late 2000, system development is reported to have been completed during 1997. Prototypes are understood to have been evaluated by the Swedish Air Force with the intention of applying an operational system to Gripen and Viggen combat aircraft.

Contractor

Saab Avionics AB, Stockholm.

UPDATED

....

Electronic Warfare Core System (EWCS)

Type

Integrated Electronic Warfare (EW) suite for the JAS 39 Gripen multirole fighter aircraft.

Description

The Saab Avionics EWCS has been developed specifically for the JAS 39 Gripen multirole fighter aircraft and is described as being a modular, Versa Modular Eurocard (VME) - based, open architecture that can be configured to meet specific customer requirements. The suite's manufacturer further notes that its modularity facilitates both future upgrading and the addition of new functions, thereby maintaining the system's over time ability to handle an 'ever changing' EW environment. As described in July 2001, the baseline EWCS comprises Radar Warning/Electronic Support (RW/ES) and internally mounted jamming subsystems and is made up of seven line-replaceable units (four wingtip units, one fin pod unit, one forward transmitter unit and an EW central unit). So configured, the suite's RW/ES subsystem covers the E- through J-band (2 to 20 GHz), with its internal jamming element operating within the H- through J-band (6 to 20 GHz) spectrum. In both cases, radio frequency coverage can be expanded to meet specific customer requirements and the EWCS as a whole can be integrated with missile and laser warners, towed decoys, countermeasures dispensers and external jamming pods. When required to, the EWCS controls expendables via interaction with a stores management unit and the system as a whole communicates with its host platform's avionics via a MILitary STanDard (MIL-STD) -1553B databus. The EWCS is designed to interact with and, in most cases, be controlled by its host platform's Systems Computer (SC). The SC also provides the July 2001 configuration suite with mission data (such as the necessary threat and countermeasures response libraries) and is further used to process EWCS generated data for recording, display and audio warning purposes. The system's application software includes emitter identification, estimation of emitter location, dynamic threat analysis and countermeasures management code. Saab Avionics claims that an EWCS suite with 'all options incorporated' provides its host with self-protection and escort jamming capabilities in both air-to-air and air-to-ground scenarios.

Contractor

Saab Avionics AB, Stockholm.

UPDATED

The Saab Avionics EWCS for the Gripen multirole fighter aircraft can be complemented by an external jamming pod as shown here (Saab Avionics)
2002/0017866

Erijammer 200

Type

Airborne self-protection radar jammer.

Description

Erijammer 200 is a high-performance jammer pod for self-protection of combat aircraft. It is an automatic, multimode jammer effective against pulse and continuous wave radar threats. It consists of a pod-mounted jammer with built-in antennas and electronics, and a cockpit-mounted control unit. Protection is accomplished by the transmission of jamming signals when informed of radar threats by the built-in receiver. The jamming modes provided are:

- several range deception programmes using advanced range gate pull-off techniques
- angle deception with high modulation sensitivity for conical scan radars
- velocity deception with high suppression of unwanted signals for maximum deceptive action
- noise jamming with frequency modulated noise spectrum automatically locked to incoming signals.

Signal analysis and response is computer-controlled with optimal preflight setting of certain jamming parameters via a control panel in the pod.

The pod can be adapted easily to a variety of aircraft types. The mounting has been chosen for flexible installation and maintenance and it is ideal for retrofit purposes. There are a number of options available, including an external Electronic CounterMeasures (ECM) control feature which can control other ECM units on the aircraft such as a chaff launcher.

While the Erijammer 200 is intended primarily for use against ground threats, it also gives protection against radar guided missiles.

Status

As of July 2001, Erijammer 200 (service designation U 22) was reported as being in service aboard AJS 37 Viggen reconnaissance and attack aircraft of the Swedish Air Force.

Specifications

Frequency coverage: extended octave coverage in H-, I- and J-bands
Output power: 300 W
Noise bandwidth: 50-250 MHz
Antennas: elliptically polarised horn types
Speed: M1.0+
Pod dimensions: 0.425 × 4.25 m
Weight: 350 kg

Contractor

Saab Avionics AB, Stockholm.

UPDATED

The Erijammer 200 pod under the wing of an AJS 37 Viggen

....

Erijammer A110

Type

6.8 to 10.5 GHz tactical and training radar jammer.

Description

Based on the proven Responsive Electronic Warfare Training System (REWTS) Erijammer A100/ALQ-503 system (see separate entry), Erijammer A110 is a pod-

The Erijammer A110 dual-use radar jamming pod installed on a JA 37 Viggen of the Swedish Air Force's F17 Wing 0017867

mounted, dual-use system that can be used as a training tool 'without compromising' its operational capability. The switch between training tool and operational system is achieved via the loading ('in seconds') of software appropriate to the particular mission. The Erijammer A110 system comprises the A110 jamming pod itself, cockpit Control and Display Units (CU/DU) and a ground-based Mission Planning Workstation (MPW). Of these, the A110 pod is carried on standard NATO 36 cm (14 in) and 76 cm (30 in) hard points and incorporates an electronic support measures capability. The CU and the MPW incorporate a personal computer interface. The DU offers an 84 × 112 mm display area. The MPW uses a standard Pentium-equipped desktop and runs a Windows-based/A110 specific software operating system.

Status
Series deliveries of the A110 radar jamming pod to the Swedish Air Force began during February 1998.

Specifications
Frequency coverage: 6.8-10.5 GHz
Input power: 115 V; 400 Hz; 2.5 kVA; 28 V DC; 6 A (A110 pod)
Dimensions: 178 × 134 × 134 mm (DU); 250 × 145 × 103 mm (CU)
Weight: <4 kg (CU and DU); <230 kg (A110 pod)

Contractor
Saab Avionics AB, Stockholm.

UPDATED

UNITED KINGDOM

Ariel Towed Radar Decoy (TRD)

Type
Airborne TRD.

Description
The fully programmable Ariel TRD is designed to counter monopulse radars, semi-active missiles and home-on-jam weapon systems using angle deception. The system can be configured for large aircraft and fast jet applications with host platforms that are large enough to accommodate a winch mechanism being able to recover the device in flight. Platforms unable to do so may utilise a decoy recovery option whereby the device is jettisoned before landing and deploys a parachute to ensure a soft landing before ground recovery. Ariel installations may be within the host's platform airframe, located in a pylon-mounted pod or scabbed on to its fuselage skin.

The TRD flight body incorporates pre- and final travelling wave tube amplifiers, transmission antennas and a power-conditioning unit. Functionally, the system utilises the host platform's radar warning receiver to cue an onboard techniques generator. Data from the techniques generator are passed to the TRD via a decoy interface module and an optical fibre link incorporated in the tow cable. The TRD then transmits the countermeasures signal created by the techniques generator. As an alternative, the Ariel TRD can be equipped with its own dedicated receiver, techniques generator and power supply unit for operation in a stand-alone mode. As of August 2001, the Ariel TRD is noted as having been flight proven at speeds ranging from 278 km/h to M1.2.

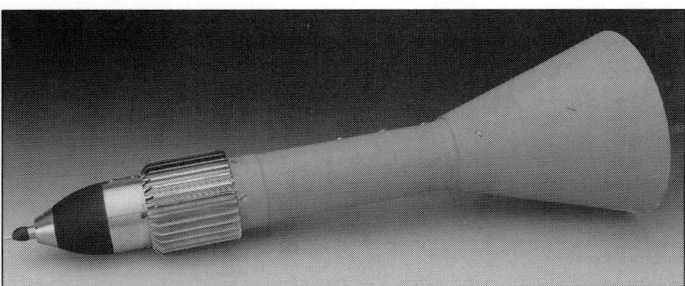

A fast jet Ariel TRD flight body configuration

Status
Over time, variants of the Ariel TRD are reported as having been procured by the Royal Air Force (RAF). According to Jane's sources, identified RAF applications comprise the Tornado F Mk 3 intercept (TRD launched from a modified SaabTech BOZ dispenser pod) and the Nimrod MR Mk 2 (BOZ pod installation evaluated). The same sources suggest that the device has also been evaluated by the US Air Force's Special Operations Command aboard an MC-130 special forces transport/support aircraft. BAE Systems sources were reporting that the Ariel TRD programme was live during 2001.

Specifications
Frequency coverage: H- through J-bands (6-20 GHz)
Spatial coverage: spherical
Techniques: noise, repeater and advanced
Cooling: air

Contractor
BAE Systems Avionics - Sensor Systems Division, Stanmore.

UPDATED

Chemring chaff and Infra-Red (IR) countermeasures

Type
Range of chaff and IR countermeasures cartridges.

Description
Chemring Countermeasures (a division of Chemring Group plc) is the UK's design authority for and prime supplier of airborne, offboard, expendable countermeasures and produces a range of fixed-wing and helicopter applicable chaff and IR decoy cartridges, the known details of which are as follows:

Chemring chaff (Radio Frequency - RF) countermeasures
The Chemring Countermeasures chaff product range includes:
- standard format and customised RR129 and RR170 type chaff cartridges that are compatible with a wide range of dispensers that includes the AN/ALE-29, -40, -45 and -47 equipments together with systems from Alkan, eastern European manufacturers, Matra British Aerospace (BAe) Dynamics, MES, Rokar, SaabTech, TERMA and Vinten
- flat-packs
- payloads for the SaabTech BOL and BOZ and Matra BAe Dynamicas Phimat dispensers
- Chaff Block payloads.

Of these, the BOL product takes the form of a chaff cassette, 160 of which form the dispenser's payload. Functionally, the cassettes are dispensed via a drive mechanism that is controlled by the host aircraft's defensive aids system and are designed in such a way as to ensure effective break lock via rapid bloom in close proximity to the dispensing platform. In terms of eastern European dispenser work, Chemring notes that its 26 mm chaff cartridges have been qualified and supplied for use aboard MiG-29 fighters and Mi-24 helicopters operated by 'unspecified customers'. The company's Chaff Block payloads form part of its Modular Expendable Block System (MEBS) product range (see following) and are designed to maximise chaff payloads within a particular dispenser by means of platform optimised chaff dispersion.

IR countermeasures
The Chemring Countermeasures IR decoy cartridge product range includes:
- formats that are compatible with the AN/ALE-29, -39, -40, -45, -47 and M-130 dispensers together with equipments produced by Alkan, eastern European manufacturers, Matra BAe Dynamics, MES, Rokar, SaabTech, TERMA and Vinten
- next-generation, advanced decoy flares for the Royal Air Force (RAF)

A selection of airborne chaff and IR decoy cartridge formats produced by Chemring Countermeasures

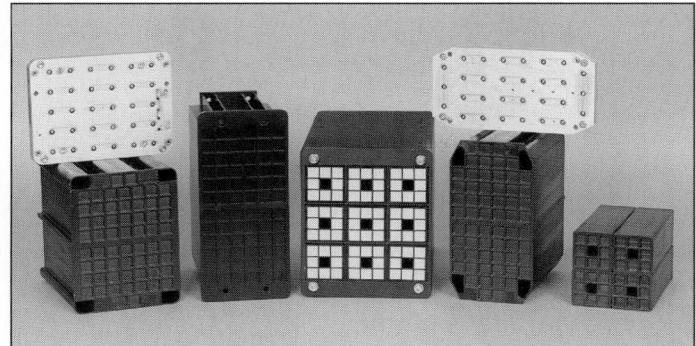

A general view of Chemring's MEB range showing, from left to right, the M147, Vinten Vicon 78 Series 455 (8 × 4 format), Hawk, ALE-47 and TERMA configurations
0055012

- the 55 mm N04 Mk 1 IR decoy flare for the Italian and, possibly, Royal Saudi Air Forces
- a 60 mm IR decoy flare for use on a Mirage 2000 variant
- a 55 mm IR decoy payload for the multinational Eurofighter/Typhoon combat aircraft.

In general, Chemring's IR decoy products are described as making use of 'high-grade materials, pressed consolidation, composition formulae, pellet profiles and specific deployment techniques' to optimise their output characteristics to specific customer requirements. The company goes on to note that it can offer spectrally matched, aerodynamic, kinetic and multishot solutions for use against the 'latest' target discrimination and flare rejection techniques. Chemring's IR decoy payloads are complemented by a range of impulse cartridges (including the PW11 and PW21 types) that the company claims offer great ejection velocity consistency over a wider temperature range than their BBU series equivalents. Looking at some of the specific applications cited, the next-generation advanced IR decoy flare for the RAF uses the company's proprietary PW118/218 Mk 3 flare and PW55 impulse cartridge as a basis, while its 26 mm IR payloads have been qualified and supplied for use aboard MiG-29 fixed-wing and Mi-24 rotary-wing applications. Known specification data for a selection of Chemring Countermeasures' IR decoy flares is as follows:

MEBS

MEBS is a family of preloaded and tested chaff and IR decoy flare block munitions that are designed to exploit the digital technology inherent in Alkan, BAE Systems North America - Integrated Defense Systems, Matra BAe Dynamics, MES, Rokar, TERMA and Vinten countermeasures dispensing systems. The MEBS concept allows the number of available salvoes to be increased without any increase in the number of dispensers and achieves maximum utilisation of the available magazine space via the use of optimised payloads.

Chaff (RF) MEBS

Chemring's chaff MEBSs take into account the redundant dipoles that have been proven to saturate chaff clouds generated by conventional cartridges and are optimised to maximise the radar cross section generated by the available payload. Typically, the company's RF MEBSs contain 60 preloaded chaff payloads that are fitted with tested BBU35B or PW11 squibs. Such an arrangement effectively doubles the number of chaff shots (from 30 to 60) available from a conventional RR170 magazine configuration, without any host aircraft modifications. Examples of Chemring RF MEBS applications/potential applications include:

- supplied systems for use on Australian, Czech (L159 light attack aircraft), NATO, New Zealand and RAF aircraft, with those fitted to C-130 transports and F-16 fighters being operational
- a qualified and flight tested RF MEBS (developed in conjunction with Joyce Loebl) for the M-147 (upgraded M-130) helicopter dispenser system
- an eight shot 'mini-block' application for use (in the first instance) on the British Army's Apache AH Mk 1 battlefield attack helicopter. Chemring has identified Kuwait, Singapore and the United Arab Emirates as potential customers for this RF MEBS variant
- development of an 80 shot capability for the SaabTech BOP dispenser that is targeted at the multinational Nordic helicopter programme
- development of RF MEBS applications for the Alkan ELIPS NG and Matra BAe Dynamics Saphir dispensers.

Known specification data for a range of Chemring RF MEBSs is as follows:

IR MEBS

Within Chemring IR MEBSs, the flare payload size (13.5 mm nominal) and performance are harmonised with the specific platform's signature requirements. In the case of helicopter applications, IR MEBSs are claimed to double (typically 60 off) the payload count over a dispenser equipped with conventional 1 × 1 in² format flares. MEBS flare output characteristics can be tailored to suit specific dispenser installations, with 'tuned' PW11 squib variants providing the required ejection velocities. Chemring notes that its IR MEBSs have been 'extensively' trialled in South Africa (using the Oryx transport helicopter as the test platform) and have shown themselves to be as effective as its 'larger flare counterparts', as well as providing sufficient salvoes as to facilitate pre-emptive and reactive deployments during a single mission. Elsewhere, the system has been evaluated aboard a Netherlands CH-47 Chinook transport helicopter and has been the subject of 'extensive' operational evaluations in Italy and the UK. Chemring also points out that it has completed design and trials of a larger MEBS IR payload (17.5 mm nominal) for light, fast jet attack aircraft such as the BAE Systems Hawk, AMX International AMX, EADS Alpha Jet and the Aerovodychody L159. As of 2000, initial target applications here included South African Hawk and Italian AMX aircraft. Known specification data for a range of Chemring IR MEBSs applications is as follows:

MEBS future concepts

As of 2000, Chemring was working on the development of mixed payload (chaff and flare or, in the pure IR context, MTV and spectral flares) MEBSs together with

Designation	CS/Ø	Length	Impulse cartridge	Typical CMDS
PW26 Mk 1	26.5 mm Ø	80 mm	PW26 Mk 1	BB11-30-26M
PW36 Mk 1	36 mm Ø	148 mm	PW36 Mk 1 CCU63B	AN/ALE-29, -39 and -47
PW40 Mk 1	40 mm Ø	150 mm	PW40 Mk 1	Alkan and Matra BAe Dynamics variants
PW50 Mk 1	50 mm Ø	200 mm	PW50 Mk 1	Gorizant
PW55	55 mm Ø	375 mm	PW106 integral	BOP/BOZ and Vinten
PW60	60 mm Ø	150 mm	PW60 Mk 1	Alkan and Matra BAe Dynamics variants
PW118 Mk 3	25 × 25 mm	205 mm	PW11 Mk 1 fitted	Alkan, AN/ALE-40/47, M-130, Matra BAe Dynamics, MES, Rokar, TERMA and Vinten
PW218 Mk 3	25 × 50 mm	205 mm	PW21 Mk 1	Alkan, AN/ALE-40/47, M-130, Matra BAe Dynamics, MES, Rokar, TERMA and Vinten
PW228 Mk 1	50 × 64 mm	205 mm	PW21 Mk 3	AN/ALE-47

Key
CS/Ø Cross section/diameter **CMDS** Countermeasures dispensing system

Nomenclature	Nominal payload cross section	Payloads per side/ per dispenser	CMDS type
RF MEB 65/30	13.5 × 13.5 mm	30/60	AN/ALE-47, MES, TERMA, Rokar and Vinten
RF MEB 84/32	13.5 × 13.5 mm	32/64	Vinten 456
RF MEB 147/30	13.5 × 13.5 mm	30/60	Joyce Loebl M-147
RF MEB BOP/40	13.5 × 13.5 mm	40/80	SaabTech BOP
RF MEB ELNG/36	13.5 × 13.5 mm	36/72	Alkan ELIPS NG
RF MEB SAP/36	13.5 × 13.5 mm	36/72	Matra BAe Dynamics Saphir
RF MEB Mini	11.5 × 11.5 mm	8 (per dispenser)	Vinten 456

Key
CMDS Countermeasures dispensing system

Nomenclature	Nominal payload cross section	Payloads per side/ per dispenser	CMDS type
IR MEB 65/30	13.5 × 13.5 mm	30/60	AN/ALE-47, MES, TERMA, Rokar and Vinten
IR MEB 66/26	17.5 × 17.5 mm	26/52	Vinten 456 *
IR MEB 84/32	13.5 × 13.5 mm	32/64	Vinten 456 **
IR MEB 147/30	13.5 × 13.5 mm	30/60	Joyce Loebl M-147
IR MEB ELNG/36	13.5 × 13.5 mm	36/72	Alkan ELIPS NG
IR MEB SAP/36	13.5 × 13.5 mm	36/72	Matra BAe Dynamics Saphir

Key
* Hawk application ****** Helicopter application **CMDS** Countermeasures dispensing system

platform IR signature studies for a range of types including the C-130 and G.222 transport aircraft and F-16, MiG-21 and MiG-23 fast jets. Here, the aim was to enhance the platform's flare count over that for the existing MJU7 type decoy flare dispensers and complement the RF MEBSs already in service on some of the cited platforms.

Status

As of 2000, Chemring Countermeasures activity included:

- development and production of the RAF's next generation of IR decoy flares and full-scale development/initial series production of the IR decoy flare for the multinational Eurofighter/Typhoon combat aircraft
- introduction of the M-147 RF MEBS application on UK Chinook, Lynx, Puma and Sea King helicopters
- introduction of an ALE-47/Vinten dispenser RF MEBS application on UK C-130 and Canberra PR Mk 9 fixed-wing aircraft
- use of an ALE-47 dispenser RF MEBS application on Australian, Portuguese and New Zealand C-130s
- completion of IR MEBS trials on the South African Oryx transport helicopter
- completion of PW118 Mk 3 IR decoy flare trials on South African C-130s
- completion of IR MEBS trials on a Royal Netherlands Air Force Chinook transport helicopter
- completion of IR and RF MEBS trials on an Mi-24 battlefield attack/assault helicopter
- completion of RF MEBS trials on a British Army WAH-64D Apache AH Mk 1 battlefield attack helicopter
- product trials on the Aerovodychody L159 (RF MEBS), Italian HH-3F helicopters (IR and RF MEBS) and UK Chinook, Lynx, Puma and Sea King helicopters (all IR MEBS)

Contractor

Chemring Countermeasures (a division of Chemring Group plc), Salisbury.

UPDATED

Helicopter Integrated Defensive Aids Suite (HIDAS)

Type

Integrated DAS for helicopter applications.

Description

The BAE Systems Avionics HIDAS helicopter DAS system is mixed and matched from a range of subsystems that comprise radar, missile and laser warners, offboard decoys and radio frequency and IR countermeasures. System control is exercised by a DAS management module within the suite's Sky Guardian 2000 Radar Warning Receiver (RWR – see separate entry). Functionally, this module correlates threat data received by the DAS sensors against an integral, 4,000-mode software library. Having identified the threat, the unit declares the threat to the crew, determines the optimum response to the particular threat and automatically initiates it with a maximum of 2 seconds from initial threat detection. DAS to aircraft mission system communication is by means of a 1553B databus with threat data (within a situational awareness picture) and/or system status reports being shown on the host's multifunction displays or a dedicated Electronic Warfare (EW) display. If required, HIDAS can be restricted to semi-automatic operation and a manual override is always available. Aside from the dispenser system, HIDAS's countermeasures capability takes the form of the Apollo radar jammer (see separate entry) and the Directional IR CounterMeasures (DIRCM – see separate entry) system. HIDAS is flight line reprogrammable and a smartcard PCM port can be provided on the system's cockpit control unit to facilitate the loading of preflight messages and the recording of in-flight mission data. As an alternative, both these functions can be performed by the host aircraft's systems using the 1553B databus. If required, system control can be exercised via the aircraft's mission computer and the use of soft controls on its multifunction displays.

Status

As of August 2001, a BAE Systems Avionics HIDAS configuration was mandated for installation aboard the British Army Corps' next-generation Apache AH Mk 1 battlefield attack helicopter. Aside from the Sky Guardian 2000 RWR, this architecture will feature BAE Systems Avionic's Series 1223 laser warner (see separate entry), a Vinten's Vicon 78 Series 455 countermeasures dispenser (see separate entry) and the BAE Systems North America - Information and Electronic Warfare Systems (formerly Sanders) AN/AAR-57 passive missile approach warner. Future integration of an active frequency jammer and a DIRCM subsystem are understood to be options.

Contractor

BAE Systems Avionics - Sensor Systems Division, Stanmore.

UPDATED

Skyshadow radar jammer

Type

Autonomous airborne radar jamming system. UK service designation ARI 23246/1.

Description

The pod-mounted Skyshadow radar jammer is a combat-proven system that is carried by InterDiction and Strike (IDS) Tornado aircraft operated by the British and Saudi air forces. The prime contract was awarded to the then GEC-Marconi Defence Systems (now part of BAE Systems Avionics' - Sensor Systems Division) which had full responsibility for overall design, development and testing, plus software design and integration. Skyshadow makes use of high-power travelling wave tube amplifiers (with a dual-mode capability) to generate deceptive and continuous wave modulations together with voltage-controlled oscillators that are equipped with full frequency, varactor-tuned Gunn diodes. A Thales Sensors set on receiver is fitted with system signal processing employing BAE Systems hard and software.

Functionally, Skyshadow is a fully autonomous system that can jam multiple simultaneous threats. Its dual receivers and jamming transmitters are connected to a dedicated central processor that identifies the threat, establishes priorities and decides optimum jamming strategy. Once a threat has been identified, classified and given a priority, Skyshadow automatically sets the jamming frequency and optimum modulation technique to counter the enemy radar. The system is software programmable to allow modification of the threat library and will easily accommodate changes in operating environment or mission requirement. The equipment digitises receiver outputs which are then passed on to the processor. The processor maintains a file of all active emitters detected by the system and identifies emissions not previously detected. It identifies radars and, for those designated as threats, it selects the optimum response and jams. The 3.35 m long by 38 cm diameter Skyshadow pod has fore and aft radomes and is capable of countering multiple ground and air threats, including surveillance, missile and airborne radar types. Automatic power management is provided and modular construction probably allows for variations in the pod's configuration for differing operational missions.

Status

As of August 2001, the Skyshadow radar jammer was reported as being in service aboard British and Saudi Tornado IDS aircraft. During the late 1980s/early 1990s, the then GEC-Marconi is understood to have carried out a Skyshadow upgrade programme to improve the system's ability to deal with the 'latest' threat environment. A feature of this upgrade was intercommunication between the Skyshadow pod and the IDS Tornado's ARI 18241/2 radar homing and warning receiver (which has itself been upgraded – see separate entry). Thus modified, the two systems can operate as a federated suite, thereby improving their overall operational effectiveness.

Contractor

BAE Systems Avionics - Sensor Systems Division, Stanmore.

UPDATED

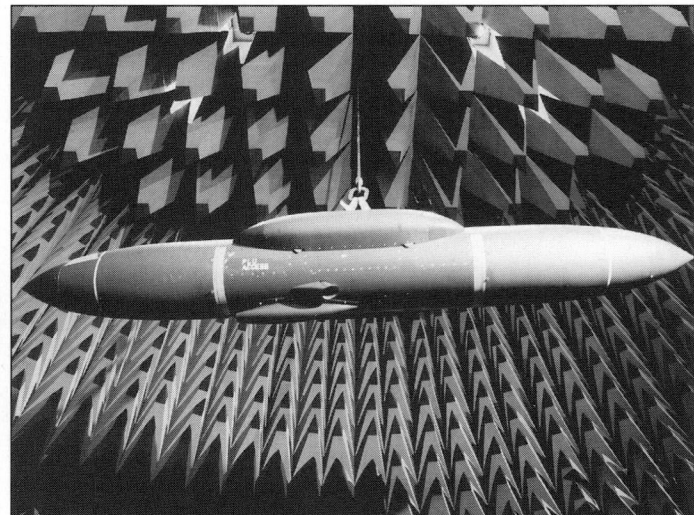

A general view of the radar and laser warning elements of the BAE Systems HIDAS system shown here together with the equipment's cockpit control unit and dedicated electronic warfare display   0009908

The Skyshadow ECM pod is a standard installation on British and Saudi Arabian Tornado IDS aircraft

Dimensions	**H.S.1**		**H.S.2**		**H.S.4**
cross section	25 × 25 mm		25 × 50 mm		50 × 64 mm
length	205 mm		205 mm		205 mm
Weight	200 g		350 g		1.20 kg
Ejection weight	150 g		300 g		1 kg
Ejection velocity	25-55 m/s		25-55 m/s		20-30 m/s
Time to effective output	<0.2 s		<0.25 s		<0.25s
Peak output	>10 kWsr		>20 kWsr		>50 kWsr
Duration	>3 s		>3.5 s		>3.55 s
Impulse cartridge	Integral W1 cart Impulse No.8 Mk 1 (US - M796 or BBU 35B)		Integral W1 cart Impulse No.1 Mk 1 (US - BBU 36B)		Integral W1 cart Impulse No.12 Mk 1
US equivalent/UK designation	M-206/118		MJU-7B/218W1 Mk 2		MJU-10/228
Temperature range	−40 to +70°C		−40 to +70°C		−40 to +90°C
Shelf life (magazine condition)	5 years		5 years		5 years
Net explosive content			270 g		730 g
Dispenser	AN/ALE-40, M-130 and Vicon 78		AN/ALE-40/-45 and Vicon 78		AN/ALE-45/-47 and Vicon 78

Dimensions	**H.S.6 Mk II**	**CART CM 55 mm No.1 Mk 1**	**CART CM 55 mm No.2 Mk 1**
Ø	36 mm	55 mm	55 mm
length	148 mm	375 mm	200 mm
Weight	280 g	1.65 kg	785 g
Ejected weight	200 g	2 × 650 g	610 g
Ejection velocity	20-45 m/s	20-30 m/s	
Time to effective output	<0.2 s	<0.25 s	
Peak output	according to type	>30 kWsr	
Duration	according to type	>3.5 s	
Impulse cartridge	CCU 63B	integral M106	integral M106
US equivalent/UK designation		MJU-8A-A/B/none	None/Cart CM 55 mm No 1 Mk 1
Temperature range	−40 to +70°C	−40 to +70°C	−54 to +90°C
Shelf life (magazine condition)	6 years	6 years	6 years
Net explosive content	150 g	950 g	550 g
Dispenser	AN/ALE-29/-39	BOZ 107 and BOP/Z	ML Aviation Eurofighter/Typhoon flare dispenser

PW Defence decoy flares

Type
Range of Infra-Red (IR) decoy flares.

Description
PW Defence Ltd manufactures a wide range of IR decoy flares for use on a variety of fixed-wing aircraft and helicopters. All PW Defence flares are fitted with safety and initiating mechanisms to ensure aircraft safety in compliance with NATO STANAG 3525 (DEFSTAN 13-27). Known details of identified members of the company's IR decoy flare product line are as in the table above:

Status
As of March 2001, PW Defence IR decoy flares were noted as having been supplied to the UK Royal Air Force/Royal Navy, NATO countries, Saudi Arabia and Australia and as having been evaluated by a number of Pacific Rim countries.

Contractor
PW Defence Ltd, Amesbury/Draycott.

UPDATED

....

Zeus integrated Defensive Aids Suite (DAS)

Type
Integrated airborne radar warning and jamming system. UK service designation ARI 23333/1.

Description
Zeus is a compact integrated DAS that has been designed by BAE Systems with US company Northrop Grumman as chief subcontractor. As such, it provides a complete defensive capability that consists of a radar receiver and a jammer. It can be fitted as a stand-alone Radar Warning Receiver (RWR), with the RWR processor providing power management of the jammer. This unit also interfaces with the host platform's countermeasures dispensing system. The system's receiver chain makes use of instantaneous frequency measuring and fast superheterodyne techniques and is able to intercept and measure the characteristics of all radar-controlled systems likely to be a threat to the aircraft. Parameters measured include direction of arrival, time of arrival, frequency, pulse repetition interval, pulse-width, amplitude and scan interval and rate. Types of radar threats identified include anti-aircraft artillery, surface-to-air missile, air-to-air missile and airborne intercept radar emitters. Having measured the threat parameters, the receiver passes the signal to the Zeus digital processor that identifies the radar type, displays it and gives an audio warning. The architecture's jamming subsystem is controlled directly by the receiver's digital processor and the transmitter chains have alternative configurations to meet specific tasks. Signals from the processor set all the necessary parameters to allow the transmitter chains to jam both pulse and continuous wave radars, while control features ensure that home-on-jam

weapons are not allowed enough time to acquire the aircraft. A range of different jamming modes are available, with individual installation being tailored to the operational requirements of the particular customer. Typically, a full Zeus system (receiver and jammer) weighs approximately 118 kg, although the actual weight depends on the system's exact configuration and the number of transmitter chains used.

Status
As of this edition, the Zeus integrated defensive aids system was reported as being installed aboard Royal Air Force (RAF) Harrier GR Mk 7 strike aircraft. Installations are also noted as having been designed for various other types of aircraft including the AV-8B, F-16, F/A-18 and JAS-39. As of December 2000, Jane's sources were suggesting that Zeus was likely to be replaced in RAF service by a new, 'off-the-shelf' fast-jet defensive aids system.

Specifications
Frequency coverage: C-J band (0.5-20 GHz)
Azimuth/elevation cover: 360° azimuth; ±45° elevation
Types of jammers: noise, VGPO, deception, co-operative
Response time: 1 s (typical)
Weight: 118 kg

Contractors
BAE Systems, Stanmore, UK.
Northrop Grumman Corporation, Electronics & Systems Integration Division, Rolling Meadows, Illinois, USA.

UPDATED

A composite illustration showing the elements that make up the Zeus integrated DAS

UNITED STATES OF AMERICA

Advanced Self-Protection Integrated Suite (ASPIS)

Type
'Mix and match' airborne Electronic Warfare (EW) suite for fixed-wing and helicopter applications.

Description
ASPIS is an airborne EW warfare system developed by an industrial consortium that comprises Northrop Grumman Electronic Systems sector - Defensive Systems Division, BAE Systems North America - Integrated Defense Systems and Raytheon Electronic Systems. It utilises subsystems from each of the contractors' existing product ranges and 'mixes and matches' equipment to create customer specific solutions. ASPIS systems can be grown by upgrading existing subsystems, adding new ones and/or upgrading its software. The suite's inherent flexibiliy also allows it to be installed in a wide variety of aircraft. As of December 2001, the basic building blocks used to create an ASPIS suite comprised:
- Northrop Grumman's AN/ALR-93(V)1 Radar Warning Receiver (RWR)
- Raytheon Electronic System's AN/ALQ-187 radar jamming system
- BAE Systems North America - Integrated Defense Systems' AN/ALE-47 CounterMeasures Dispensing System (CMDS)

Status
As of early 2001, the ASPIS architecture was understood to be available. In February 1993, it was announced that an ASPIS team made up of the then Litton Advanced Systems (now part of Northrop Grumman), Raytheon E-Systems (now part of Raytheon Electronic Systems) and Tracor (now BAE Systems North America - Integrated Defense Systems) had been selected to supply a defensive aids suite for use on F-16s of the Hellenic Air Force. The specific application comprised the then Litton's ALR-93(V)1 RWR, Raytheon's DIAS radar jammer (a country specific variant of the contractor's ALQ-187 system) and BAE Systems' ALE-47 CMDS. During the following October, the then Litton is reported to have ordered 33 DIAS jammers and 80 ALE-47 CMDSs from its partners to service this contract. Deliveries of the application began during mid-1994 with a prototype suite starting flight trials in March 1996. Delivery of production ASPIS systems to Greece is noted as having begun during August 1996.

Contractors
BAE Systems North America - Integrated Defense Solutions - Integrated Survivability Systems, Austin, Texas.

Northrop Grumman Electronic Systems sector - Defensive Systems Division, Rolling Meadows, Illinois.

Raytheon Electronic Systems, Goleta, California.

UPDATED

Alloy Surfaces Infra-Red (IR) decoys

Type
Family of cartridges and payloads.

Description
Alloy Surfaces Co Inc (a subsidiary of the UK's Chemring Group) produces a range of airborne expendable payloads, the known details of which are as follows:

AN/ALE-50(V) Infra-Red (IR) decoy payload
As of mid-2000, Alloy Surfaces was understood to be supplying the payload element for an IR variant of Raytheon Systems' AN/ALE-50(V) towed decoy (see separate entry). As such, the specific variant is described as making use of the standard ALE-50 launch unit, as weighing 2.27 kg and as measuring 292 × 64 mm. As yet unconfirmed sources suggest that the application is a pre-emptive rather than reactive measure and that it is to be flight-tested, deployed from an AV-8B testbed.

ASD-118L
The 25 × 25 × 203 mm 'special materials' ASD-118L IR decoy flare is described as having been specially designed for a Royal Air Force helicopter application.

MJU-27A/B
MJU-27A/B is described as being a 146 × 36 mm IR decoy round that is compatible with the US Navy's (USN) AN/ALE-39 CounterMeasures Dispensing System (CMDS). The device is noted as weighing 0.39 kg and as having a 2 to 3 second burn time.

MJU-49/B
Noted as being a 'special material' (pyrophoric) IR decoy for use by USN helicopters, MJU-49/B is described as being a 0.39 kg, 36 (Ø) × 146 mm device that is compatible with the AN/ALE-39 CMDS. Burn time is given as being 2 to 3 seconds.

MJU-50/B
MJU-50/B is described as being a 'special materials' (pyrophoric), 0.34 kg, 25 × 25 × 203 mm, covert Advanced Strategic and Tactical Expendable (ASTE) IR decoy round that is compatible with AN/ALE-40/-47 series CMDS. Burn time is given as being 2 to 3 seconds.

MJU-51/B
MJU-51/B is noted as being a 'special materials' (pyrophoric), 0.73 kg, 25 × 51 × 203 mm, covert ASTE IR decoy round that is compatible with AN/ALE-40/-47 series CMDS. Burn time is given as being 2 to 3 seconds.

MJU-52/B
MJU-52/B is described as being the 0.05 kg, 83 × 70 × 10 mm IR decoy payload used in the BOL CMDS. It is noted as being a pre-emptive device that provides 'continuous' protection for its host platform.

XM211
XM211 is described as being a 'special materials' (pyrophoric), 0.34 kg, 25 × 25 × 203 mm IR decoy round that is associated with the AN/ALQ-212(V) directed IR countermeasures system (see separate entry) and has a burn time of 2 to 3 seconds. The device is reported as having been developed as part of the US Army's Advanced IR CounterMeasures Munitions (AIRCMM) programme.

Status
As of July 2001, the various described Alloy Surfaces IR decoys are understood to be either available or under development. Previously identified Alloy Surfaces expendables include the ASD-7, ASD-118 and MJU-46/B IR decoy rounds and the combined IR and chaff RR191 cartridge. Identified contracting activity for the period 1999/2001 is as follows:

13 July 1999
On 13 July 1999, the US Government announced its intention to award Alloy Surfaces a sole source contract covering the supply of 44,054 (22,027 as an option) MJU-49/B IR decoy flares.

17 May 2000
Alloy Surfaces was awarded a then year US$9,946,444 firm, fixed-price delivery order (on a basic ordering agreement) that covered the supply of 90,720 MJU-27A/B and 28,800 MJU-49/B IR decoys (together with associated data) to the US Navy (USN). At the time of this order's announcement, work on the effort was to be undertaken at Alloy Surfaces' facility at Chester Township, Pennsylvania and was scheduled for completion by the end of November 2001. The programme's contracting activity was the USN's Naval Surface Warfare Center, Crane Division, Crane, Indiana.

17 July 2001
Alloy Surfaces was awarded a then year US$6,679,648 firm, fixed-price job order under a basic ordering agreement for 266,791 MJU-52/B IR decoy payloads (for use in BOL dispensers mounted in LAU-138A/A missile launch rails), tooling and associated data. At the time of the announcement, work on the programme was to take place at Alloy Surface's Chester Township, Pennsylvania facility and was scheduled for completion by the end of March 2002. The effort's contracting activity was the US Navy's Naval Surface Warfare Center, Crane Division, Crane, Indiana.

Contractor
Alloy Surfaces Co Inc, Chester Township, Pennsylvania.

UPDATED

AN/AAQ-4/8 Infra-Red (IR) CounterMeasures (IRCM) systems

Type
Internally mounted airborne IRCM systems.

Description
AN/AAQ-4 is an internally mounted reprogrammable system that electronically modulates a sapphire IR source to produce a highly effective jamming signal against heat-seeking missiles. AN/AAQ-8 is a second-generation system that is capable of operation in a supersonic environment. Mounted in an aerodynamically faired pod, it has been extensively deployed on both fighter and transport aircraft. The pod can be configured with a ram air turbine to provide protection independent of aircraft prime power and cooling resources. AAQ-8 has been upgraded on several occasions since its initial deployment in 1972.

Status
Over time, the AN/AAQ-4 IRCM system is noted as having been installed aboard a number of US Air Force helicopters and fixed-wing combat and transport aircraft. As of December 2001, systems of the types described were reported as being in service aboard F-5E and Mirage F.1 fighter aircraft.

Contractor
Northrop Grumman Electronic Systems sector - Defensive Systems Division, Rolling Meadows, Illinois.

UPDATED

AN/AAQ-24(V) Directional Infra-Red CounterMeasures (DIRCM) system

Type
Directable IRCM system. UK service designation ARI 18246.

Description
The AN/AAQ-24(V) Nemesis Anglo-American DIRCM system is designed to provide transport aircraft and helicopters with an enhanced defence against IR-guided missiles. The equipment comprises Northrop Grumman's AAR-54(V) passive missile warning sensor, a Boeing-sourced high-resolution, 256 × 256 staring array fine tracking subsystem, a control indicator unit, a system processor (Northrop Grumman sourced), modulators, power supplies and up to four, four-axis, steerable tracking/jamming turrets. UK contractor BAE Systems is responsible for the construction of these turrets (which are derived from technology developed for the TIALD designation system) together with the system's pointing and tracking subsystem, power supplies and control unit. The split between UK and US work share is understood to be approximately equal. Functionally, Nemesis detects the threat missile's exhaust plume using the AAR-54(V) which, in turn, hands off the bearing data to the steering subsystem. With the turret (or turrets) suitably aligned, the tracking subsystem establishes a target box in which the threat is calculated to appear and maintains turret alignment in this box for the duration of the engagement. The initial jamming source is a high-power, modulated arc lamp with an upgrade path to the use of a laser source already established.

Status
The Northrop Grumman/Boeing/BAE Systems Nemesis consortium was awarded a then year US$271 million fixed-price development contract by the UK Ministry of Defence during May 1995. It is understood that the USAF's Special Operations Command is taking up an initial batch of 60 systems for installation on MC-130E/H and AC-130H/U aircraft. In the UK, an undisclosed number are to be supplied to the Royal Air Force/British Army Air Corps for use on 14 types of fixed-wing aircraft and helicopters. As of July 2001, AAQ-24(V) is understood to have undergone its first live fire trials at the White Sands Missile Range in New Mexico during the early summer of 1998. UK trials with the system have centred on a Sea King testbed helicopter while a 1998 US test round made use of an NC-130E aircraft. As of the given date, it was also noted that a laser equipped Nemesis configuration had been selected for installation aboard Australian Boeing 737-700 Wedgetail airborne early warning and control aircraft and the CV-22 tilt-rotor transport. Elsewhere, Jane's sources report that AAQ-24(V) was tested with Northrop Grumman's Viper solid-state, diode-pumped laser source during August 1999. Here, the modified system was test-flown aboard a US Army UH-60A helicopter at the service's Aviation Technology Test Center at Fort Rucker, Alabama. During the effort, the testbed helicopter is reported to have made approximately 100 trial runs during the course of a 10 hour, five flight programme. Northrop Grumman further notes that the Viper source can be integrated into the production AAQ-24(V) system as an upgrade. Jane's sources also suggest that Northrop Grumman is developing a miniaturised AAQ-24(V) derivative under the designation WANDA. Here, the system is reported as incorporating a laser source and as operating in the 1 to 2 and 3 to 5 μm bands.

Specifications
Dimensions: 121 × 169 mm (missile warning sensor); 176 × 146 × 48 mm (control indicator); 330 × 50 mm (laser unit); 348 × 124 × 193 mm (power supply/modulator-helicopter application); 356 × 356 mm (transmitter-helicopter application); 411 × 658 mm (transmitter – fixed-wing application); 439 × 190 × 193 mm (processor); 533 × 223 × 230 mm (power supply – fixed-wing application); 534 × 304 × 190 mm (modulator – fixed-wing application)

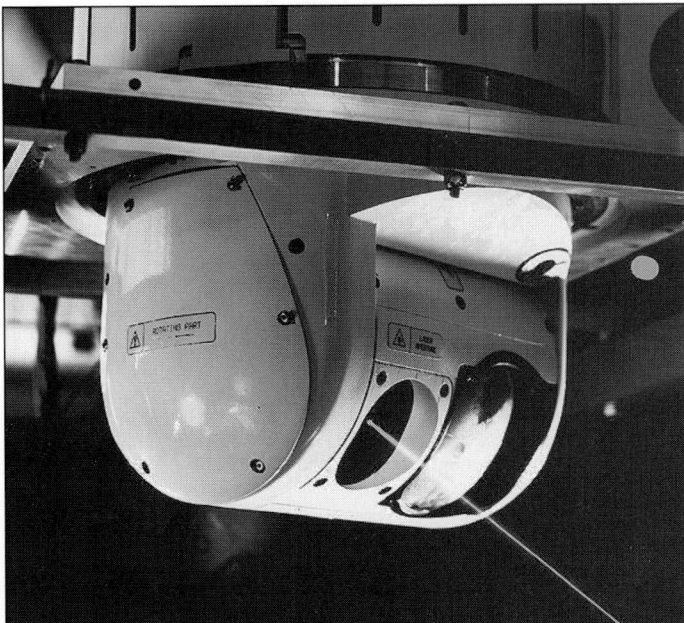

The AAQ-24(V) Nemesis helicopter application uses steerable tracking jamming turrets of the type shown here

Weight: 55.7 kg (helicopter application without laser); 60.5 kg (helicopter application with laser); 219.3 kg (fixed-wing application without laser); 228.8 kg (fixed-wing application with laser)
Transmitter power: 0.6 kW (fixed-wing application without laser); 0.96 kW (fixed-wing application with laser); 2.58 kW (helicopter application without laser); 3 kW (helicopter application with laser)

Contractors
Northrop Grumman Corporation, US (prime, system installation, missile launch detector and system processor).
Boeing , US (staring fine tracking sensor).
BAE Systems, UK (turrets, pointing and tracking subsystem, power supplies, control unit and modelling)

UPDATED

AN/ALE-38/41 CounterMeasures Dispensing Systems (CMDS)

Type
Airborne chaff dedicated CMDSs.

Description
The AN/ALE-38 (US Air Force) and AN/ALE-41 (US Navy) equipments are high-capacity bulk chaff CMDSs. They employ dispensing techniques that provide continuous dipole dispersal and instantaneous bloom for laying chaff corridors. They can also be used for aircraft self-protection and can be turned on automatically by the radar warning receiver. Precut dipoles are sandwiched between two wraps of mylar film with six 22.7 kg rolls of this composite being carried in each pod.

Status
As of this edition, AN/ALE-38 and -41 CMDSs were reported as having been procured.

Contractor
BAE Systems North America - Integrated Defense Solutions, Austin, Texas.

UPDATED

AN/ALE-39B CounterMeasures Dispensing System (CMDS)

Type
Airborne CMDS.

Description
The AN/ALE-39B CMDS is an upgrade of the ALE-39A system and is designed to protect tactical aircraft from missile and radar-directed anti-aircraft gun threats. It is capable of accommodating up to three types of expendable payloads (chaff, Infra-Red (IR) flares and offboard expendable (POET or GEN-X) jammers), loaded in any combination of multiples of 10. All three types of payload can be dispensed manually (single payload) or automatically in accordance with preset programmes. The dispensing function can be initiated by the pilot (or the weapon systems operator in the case of the F-14). The system is also capable of accepting dispense commands from aircraft warning receivers.

Status
Over time, the AN/ALE-39 CMDS is noted as having been installed aboard A-6, AV-8B, EA-6B, ES-3A, F-14, F/A-18, S-3B and SH-2F/G aircraft of the US Navy, US Marine Corps and other air forces/naval air arms. As of this edition, more than

On the AV-8B, multiple ALE-39 dispensers are mounted in the upper section of the aircraft's rear fuselage (MS/BAE Systems)

1,000 ALE-39 systems were reported as having been delivered since the device's initial production release during October 1973. ALE-39 is no longer in production.

Contractor
BAE Systems North America - Integrated Defense Solutions, Austin, Texas.

UPDATED

AN/ALE-40 CounterMeasures Dispensing System (CMDS)

Type
Airborne CMDS.

Description
Originally developed for use on the McDonnell Douglas F-4 fighter-bomber, the AN/ALE-40 CMDS has gone on to be configured for a range of aircraft and mounting options. A typical F-4 installation comprises four dispensers, a chaff/flare programmer and a Cockpit Control Unit (CCU). One dispenser is mounted on each side of the inboard armament pylons. A dispenser consists of a mounting plate to provide attachment to the pylon structure, an aerodynamic nose fairing for drag reduction and a detachable magazine for carrying the chaff cartridge. Each dispenser accommodates 30 RR-170 chaff cartridges for a total of 60 cartridges per pylon and 120 per aircraft. The outboard dispensers can carry 15 MJU-7/B Infra-Red (IR) flares. System drag is comparable to a Sidewinder missile and launcher. No weapons are displaced. Loaded weight is under 60 kg. Examples of the range of ALE-40 configurations that have been developed are as follows:

ALE-40(V)4,5,6
ALE-40(V)4,5.6 is an internal, flush-mounted ALE-40 configuration that was designed for installation aboard F-16 multirole combat aircraft. The architecture consists of a CCU, a chaff/flare programmer, a sequencer switch, an ElectroMagnetic Interference (EMI) filter and two dispensers modules. These later items are located in the aircraft's rear fuselage and can accommodate 30 RR-170 chaff cartridges or 15 MJU-7/B IR flares.

ALE-40(V)7,8,9
The ALE-40(V)7,8,9 CMDS configuration is a semi-internal application that was developed for use on the F-5E/F light fighter/trainer aircraft. The architecture consists of CCU, a chaff/flare programmer, a sequencer switch, an EMI filter and a dispenser module housing containing two dispenser magazines. Each magazine can accommodate RR-170 chaff cartridges or 15 MJU-7/B IR flares. As applied to the F-5, the ALE-40(V)7,8,9 CMDS is mounted in the aircraft's belly near the left-wing root.

ALE-40(V)10
The ALE-40(V)10 CMDS configuration was designed for internal installation aboard the A-10 close support aircraft.

ALE-40(V)11
The ALE-40(V)11 CMDS configuration was designed for internal flush mounting aboard the A-7D strike aircraft.

ALE-40(N)
The ALE-40(N) CMDS configuration is a skin-mounted application that was developed for installation on Netherlands' NF-5 light fighters. The architecture consists of a cockpit control box, a chaff/flare programmer and two dispensers modules. Each dispenser can accommodate 30 RR-170 chaff units, 30 M206 flares or 15 MJU-7 flares and the system's loaded weight is noted as being less than 30 kg. As well as being applied to NF-5 aircraft, ALE-40(N) was also flight tested on the F-104 interceptor.

Other known ALE-40 variants have, over time, included configurations designed for installation on Jaguar (scabbed), Harrier GR Mk3 (airbrake installation), Hunter and Mirage aircraft.

Status
As of this edition, the AN/ALE-40 CMDS was reported as being in service but no longer in production. Examples of F-16 ALE-40 installation executed over time have included aircraft operated by the air forces of Bahrain, Belgium (ALE-40XX which

interfaces with the Carapace threat warning system), Denmark, Egypt, Indonesia, South Korea, Netherlands, Norway, Portugal, Thailand and Turkey (Peace Oryx I aircraft only).

Contractor
BAE Systems North America - Integrated Defense Solutions, Austin, Texas.

UPDATED

AN/ALE-43(V) CounterMeasures Dispensing System (CMDS)

Type
Airborne chaff-dedicated CMDS.

Description
The AN/ALE-43(V) CMDS is designed for internal installation (ALE-43(V)3) or pod (ALE-43(V)1) mounting. In the external pod installation, the chaff cutter assembly is mounted on a structural bulkhead at the rear of the pod centre section. This assembly consists of a drive motor, clutch/brake unit, cutting mechanism and supporting framework. The cutting mechanism contains a rubber platen roller and three cutter rollers. The cutter rollers are designed with blades that yield three specific combinations of dipole lengths as commanded. In the internal installation the cutter assembly is attached directly to the aircraft structure.

For either installation, the system consists of a chaff roving supply, chaff cutter assembly and a cockpit programmer (the C-11393/ALE-43(V) unit). The chaff supply is made of metallised glass roving packages contained within a chaff hopper. The chaff is dispensed when rovings are drawn simultaneously from each of the roving supply packages. Each roving passes through a guide tube and is drawn between the platen roller and the cutter rollers for dipole cutting. The cut dipole lengths of chaff are then discharged at the exit that opens into the airstream.

A control unit allows the pilot to select dispensing programmes and to start and stop dispensing sequences. The pod has mounting provisions for standard NATO 36 and 76 cm stores suspension hardware and can be mounted conveniently on existing stores stations.

Status
As of this edition, the AN/ALE-43(V) CMDS is reported as having been procured.

Specifications

	ALE-43(V)1	ALE-43(V)3
Dimensions	Ø482 mm × 3.37 m	403 × 338 × 310 mm
Weight	154 kg (empty)	36 kg (empty)
Power requirement	115/208 V (400 Hz)	115/208 V (400 Hz)
Chaff supply	18 kg RR-179/AL roving bundle	18 kg RR-179/AL or special roving bundle
Chaff payload	144 kg (8 rovings)	up to 225 kg (up to 9 rovings)
Dispensing rate	216 g/s (per 8 rovings)	27 g/s (per roving)
Modes	1-9 s (pulse-on/off at 1 s intervals); 660 s (continuous)	1-9 s (pulse-on/off at 1 s intervals); up to 880 s (continuous)
Cutter rollers	3 (in-flight selectable)	3 (in-flight selectable)

Contractor
Alliant Defense Electronics Systems, Clearwater, Florida.

VERIFIED

The ALE-43(V)1 chaff cutter dispensing pod

AN/ALE-40 CMDS payload modules

AN/ALE-45 CounterMeasures Dispensing System (CMDS)

Type
Automatic airborne CMDS.

Description

The AN/ALE-45 CMDS is a microprocessor-controlled countermeasures dispenser system that has been developed specifically for installation aboard the F-15 aircraft. It responds automatically to threat notification from the radar warning receiver, a tail warning set or the pilot. The system consists of a programmer assembly and four dispensing switch/twin magazine assemblies. Of these, the programmer assembly houses the central processor and the input/output circuitry. An operational flight programme (incorporating the dispenser programmes and other preprogrammed data defining the functional operation of the system) is accessed via a front panel connector. The ALE-45 software is designed to provide automatic priority selection of dispensing programmes based on computations of threat sources from the radar warning receiver, or the pilot and aircraft flight data. The dispensing programme parameters are selectable, including payload class, burst count and interval and payload count and interval. The four dispensing switch assemblies are identical and interchangeable. ALE-45 can accommodate RR-170 and/or RR-180 chaff cartridges, MJU-7 and MJU-10 flare cartridges and other payload types of similar form factor.

Status

As of this edition, the AN/ALE-45 CMDS was reported as being in service aboard F-15 aircraft flown by the US and other air forces.

Contractor

BAE Systems North America - Integrated Defense Solutions, Austin, Texas.

UPDATED

··

AN/ALE-47 CounterMeasures Dispensing System (CMDS)

Type

Airborne threat adaptive CMDS.

Description

The AN/ALE-47 CMDS is a threat-adaptive, software-programmable equipment that is capable of dispensing chaff, flares and active radio frequency expendables. It has been designed primarily to replace existing AN/ALE-39 systems used by the US Navy (USN) and the US Air Force's (USAF) AN/ALE-40. As such, ALE-47 applications are constructed from the following modules:

- a 953 × 146 × 173 mm cockpit control unit that weighs less than 1.82 kg, consumes 23 W of power and provides aircrew/system interface, inventory/built-in test indication and system mode/inhibit functions
- a 216 × 170 × 256 mm F-16 compatible dispenser/magazine assembly that weighs less than 2.27 kg when empty (between 10.16 and 12.34 kg when loaded) and provides payload housing/firing, magazine-to-sequencer identification and 'smart' payload interface functions. The ALE-47 F-16 compatible dispenser/magazine assembly accepts a total of five different magazine types and up to 32 such assemblies can be accommodated within a single system/host platform
- a 250 × 258 × 167 mm F-18 compatible dispenser/magazine assembly that weighs less than 2.27 kg (between 10.85 and 12.78 kg when loaded) and provides payload housing/firing, magazine-to-sequencer identification and 'smart' payload interface functions. The ALE-47 F-18 compatible dispenser/magazine assembly accepts a total of five different magazine types and up to 32 such assemblies can be accommodated within a single system/host platform
- a 328 × 227 × 118 mm helicopter dispenser/magazine assembly that weighs less than 2.27 kg when empty (between 10.16 and 12.34 kg when loaded) and provides payload housing/firing, magazine-to-sequencer identification and 'smart' payload interface functions. The ALE-47 helicopter dispenser/magazine assembly accepts a total of five different magazine types and up to 32 such assemblies can be accommodated within a single system/host platform
- a 953 × 146 × 159 mm programmer unit that weighs less than 2.27 kg, consumes 20 W of power and provides threat adaptive processing, threat evaluation, dispensing strategy, built-in test and payload management and CMDS-Electronic Warfare (EW) system/air data computer link functions. The

The AN/ALE-47 countermeasures dispensing system

ALE-47 programmer is described as being a system option that provides 'additional functions' and can be supplied as a VME card if required
- a 655 × 118 × 927 mm safety switch that weighs less than 0.45 kg and provides the safety interlock for the system's firing circuits
- a 734 × 158 × 169 mm sequencer switch that weighs less than 1.82 kg, consumes 5 W of power and provides the dispenser firing source together with payload inventory and automatic misfire detection/correction functions. Each ALE-47 sequencer can manage up to two dispenser/magazine assemblies and up to 16 sequencer switches can be installed aboard a particular host platform.

ALE-47 can be retrofitted into any aircraft (tactical, cargo, rotary-wing) that is capable of carrying the ALE-39, ALE-40 or M-130 CMDSs. Functionally, the system has three operating modes, namely:
- manual which provides six preprogrammed, cockpit selectable, dispensing sequences/programmes
- semi-automatic which selects the optimum expendable(s) and programme (based on the aircraft radar warning receiver and other sensor systems and aircraft avionics inputs) to be implemented by aircrew switch action
- automatic that evaluates sensor and avionics inputs, selects the appropriate response, then automatically initiates the dispenser programme without any action from the aircrew.

ALE-47 can accept customised dispenser/magazine assemblies (external conformal, special payload form factor and/or special internal shape) and is claimed to be compatible with glass cockpit aircraft, 'advanced' next-generation expendable payloads and dropsondes, and chemical detection units in addition to its range of countermeasures cartridges and active expendables. As already suggested, ALE-47 can be integrated with a wide range of EW systems as outlined in Table 1.

Table 1: 'Proven' ALE-47 - EW system integration

RWR	MWR	LWR	Jammers
AN/ALR-56M	AN/AAR-47	AN/AVR-2A	AN/ALQ-126
AN/ALR-66	AN/AAR-54	LWS-20	AN/ALQ-136
AN/ALR-67	AN/AAR-57		AN/ALQ-165
AN/ALR-69	AN/AAR-60		AN/ALQ-167
AN/ALR-93	AN/ALQ-156		AN/ALQ-214
AN/APR-39A(V)2	EL/M-2160		
AES 210			
ALR-2002			
ELT/156			
LR-100			
Sky Guardian 2000			

Key
RWR Radar Warning Receiver *MWR* Missile Warning Receiver *LWR* Laser Warning Receiver

Status

As of November 2001, approximately 2,000 ALE-47 systems are understood to have been delivered to customers in 19 countries since 1993. In more detail, over 1,500 ALE-47 systems are noted as having been delivered (for installation aboard at least 21 different aircraft types) by mid-1997 and, during the following May, the then Marconi Aerospace Defense Systems (now BAE Systems North America - Integrated Defense Systems) was awarded a then year US$19,770,688 contract covering production of ALE-47 CMDS Lots 5 through 7 (approximately 952 shipsets). Lots 5/7 ALE-47s were noted as being applicable to C-5, C-17, C-130, C-141, E-8C, F-16, F/A-18, HH-60H, MH-47E, MH-53J, MH-60K/L, P-3, UH-3D, VH-60N and V-22 aircraft of the USAF, Army, Marine Corps and Navy. This contract may also have included ALE-47 systems for Finland, Malaysia, Switzerland and Taiwan. Other identified programme activity includes the type's selection for use on Royal Air Force Merlin HC Mk 3 transport helicopters and its installation aboard a number of Australian C-130H transport aircraft. During December 1999, Jane's sources were reporting that Greek contractor Miltech Hellas were to produce 50 examples of the equipment under licence for installation aboard two squadrons of Greek Air Force A-7E fighter-bombers stationed at Araxos Air Base. As of October 2000, Jane's sources were suggesting that work was in hand on the development of an ALE-47 variant that would be suitable for unmanned aerial vehicle applications. ALE-47 contracting activity identified during the first 11 months of 2001 comprised the following:

May 2001
Northrop Grumman Field Support Services Inc (Jacksonville, Florida) was awarded a then year US$6,347,620 firm, fixed-price delivery order covering the manufacture and installation of components (including 104 ALE-47 CMDSs) aboard F-14B and D aircraft. At the time of the announcement, work on the effort was scheduled for completion by the end of December 2002, with the programme's contracting activity being the USN's Naval Air Systems Command, Naval Air Station, Patuxent River, Maryland.

29 August 2001
BAE Systems North America - Integrated Defense Solutions (Austin, Texas) was awarded a then year US$7,175,000 (estimated) firm, fixed-price, indefinite delivery/quantity contract covering the provision of AN/ALM-288 support equipment for the ALE-47 CMDS. At the time of its announcement, this effort was noted as being in support of a US Foreign Military Sales programme, with BAE Systems North America's subsidiary Rokar International undertaking 80 per cent of the necessary work at its facilities in Jerusalem and at 'other locations' in Israel. The USAF's Warner-Robins Air Logistics Center (Robins Air Force Base, Georgia) was this programme's contracting activity.

15 September 2001
Symetrics Industries (Melbourne, Florida) was awarded a then year US$16,800,000 (estimated) firm, fixed-price contract covering the supply of ALE-47 control display

units, programmers, safety switches, sequencers and dispenser modules for use in systems installed aboard a range of aircraft types including the C-17, C-130 and F-16. Part of this award was in support of a US Foreign Military Sales programme for the Netherlands and at the time of the announcement, work on the effort was scheduled for completion 13 months from receipt of order. The USAF's Warner-Robins Air Logistics Center (Robins Air Force Base, Georgia) was this programme's contracting activity.

20 September 2001

BAE Systems North America - Integrated Defense Solutions (Austin, Texas) was awarded a then year US$17,467,000 (maximum) firm, fixed-price, indefinite delivery/quantity contract covering the supply of up to 492 control display units, up to 1,096 programmers, up to 784 safety switches, up to 2,294 sequencers and up to 3,846 dispenser modules for use in ALE-47 CMDSs. The USAF's Warner-Robins Air Logistics Center (Robins Air Force Base, Georgia) was this programme's contracting activity.

14 November 2001

BAE Systems North America - Integrated Defense Solutions (Austin Texas) was awarded a then year US$5,406,850 firm, fixed-price contract covering the provision of 'modified ALE-47 Aircraft Survivability Equipment (ASE)' components for use on 131 US Army CH-47D transport helicopters. At the time of the announcement, work on the effort was scheduled for completion by 30 June 2003, with the programme's contracting activity being the US Army's Aviation and Missile Command, Redstone Arsenal, Alabama.

Contractor

BAE Systems North America - Integrated Defense Solutions - Integrated Survivability Systems, Austin, Texas.

UPDATED

AN/ALE-50(V) Advanced Airborne Expendable Decoy (AAED)

Type

Airborne towed decoy.

Description

The AN/ALE-50(V) AAED system consists of an integrated launcher/controller and towed decoys. The power supply and decoy monitor/control electronics are in the launch controller. The launcher typically holds three decoys and can be customised to fit the required aircraft. Decoys come in a sealed canister that includes a payout reel. Carried internally, the decoy is towed behind the aircraft when deployed and protects the aircraft against radio frequency homing missiles by seducing them away from the host aircraft. The system performs over the entire flight envelope of high-performance aircraft, interfacing with the aircraft over a standard databus. ALE-50(V) has also been integrated into the AN/ALQ-184(V) radar jamming system (see separate entry) with the resultant equipment being designated as the ALQ-184(V)9. Raytheon is also developing an Infra-Red (IR) decoy payload for the system (see Alloy Surfaces IR decoys entry), as well as a retrievable, Fibre-Optic Towed Decoy (FOTD) variant for use in the UK's Nimrod MRA Mk 4 maritime patrol aircraft programme. The company is also thought to have demonstrated an AL-50(V) type millimetre wave capability.

Status

As of May 2001, the ALE-50(V) radio frequency AAED was reported as being operational on US Air Force (USAF) F-16s, Air National Guard F-16s (AN/ALQ-184(V)9 pod) and USAF B-1B strategic bombers, and as having been mandated for use aboard the Royal Air Force's Nimrod MRA Mk 4 (see above). Elsewhere, the US Navy is acquiring a limited number of ALE-50(V) systems for test and contingency use on the F/A-18E/F aircraft and the device is under consideration for installation aboard the American Global Hawk unmanned reconnaissance vehicle. In terms of contract activity, Raytheon was awarded an initial ALE-50(V) multiyear production contract in December 1996. Production contract increases (valued at then year US$60.5 million) were sanctioned in late 1997 and early 1998 and in May 1998, the company was awarded a further then year US$5 million to produce 50 ALE-50(V) launchers and 150 magazines for the

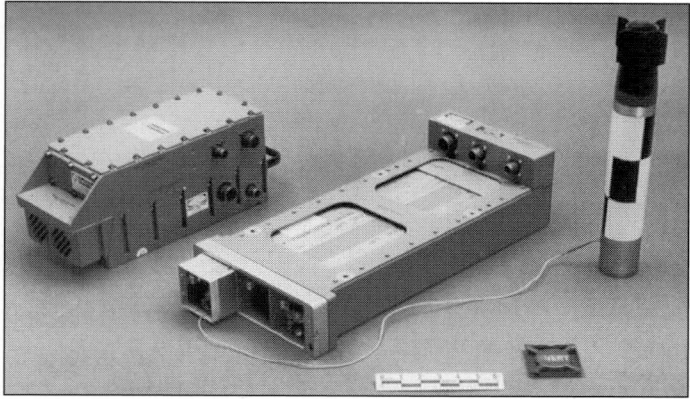

AN/ALE-50 launch controller, launcher and representative decoy

B-1 bomber. Production deliveries of ALE-50(V) systems began in January 1998, by which time, Raytheon had produced its 1,000th ALE-50(V) decoy round. Radio frequency ALE-50(V) decoys were first used operationally during NATO's 1999 vintage Operation 'Allied Force' during which the device was flown on both USAF F-16s and B-1Bs. With regard to the latter type's use of the system during the campaign, Jane's sources suggest that during the course of 100 plus sorties flown against targets in Kosovo and Serbia, approximately 125 surface-to-air missiles were fired at participating B-1s. During these engagements, ALE-50(V) decoys were regularly deployed and are credited with saving individual aircraft on between five and 10 occasions. Identified programme activity during the period 1999-2001 comprises the following:

15 April 1999

The then Raytheon Company (Goleta, California) was awarded a then year US$6,500,000 firm, fixed-price contract covering accelerated ALE-50(V) production in support of Operation 'Allied Force'. At the time of the described announcement, then year US$4,875,000 of the cited total had been obligated and the programme's contracting activity was the USAF's Aeronautical Systems Center, Wright-Patterson Air Force Base, Ohio.

4 August 1999

The then Raytheon Company (Goleta, California) was awarded a then year US$41,225,194 modification to an existing firm, fixed-price contract (F33657-96-C-0036-P00024) that covered the supply of 1,596 ALE-50(V) towed decoys, 39 spare launcher/controllers, 62 spare magazines and 39 isolation racks in support of system installations aboard F-16 aircraft. At the time of the announcement, work on the effort was scheduled for completion on 28 February 2001, with the USAF's Aeronautical Systems Center (Wright-Patterson Air Force Base, Ohio) acting as the programme's contracting activity.

13 June 2000

Raytheon Electronic Systems (Goleta, California) was awarded a then year US$33,914,660 modification to an existing firm, fixed-price contract that covered the supply of 97 ALE-50(V) mass models for US Navy trials; 289 1 × 2 launcher/controllers, 203 magazines and 33 fault isolation testers for use on or with F-16 aircraft and 50 1 × 4 dual compatible launchers, 146 magazines, test equipment and warranties applicable to the B-1 bomber. At the time of the announcement, work on the effort was scheduled for completion on 30 July 2002, with the USAF's Warner Robins Air Logistics Center (Robins Air Force Base, Georgia) acting as the programme's contracting activity.

6 March 2001

Raytheon Electronic Systems (Goleta, California) was awarded a US$5,406,710 modification to an existing firm, fixed-price contract (ALE-50(V) production programme, Lot 5) that covered the supply of 57 1 × 4 dual compatible ALE-50(V) launchers (five as spares) and 166 1 × 4 ALE-50(V) operational magazines (10 as spares) for use on B-1B strategic bombers. At the time of the announcement, work on the effort was scheduled for completion by the end of February 2003, with the USAF's Warner Robins Air Logistics Center (Robins Air Force Base, Georgia) acting as the programme's contracting activity.

5 April 2001

Raytheon Electronic Systems (Goleta, California) was awarded a US$54,577,779 modification to an existing firm, fixed-price contract that covered the supply of 2,477 ALE-50(V) decoys, 111 1 × 2 ALE-50(V) launcher/controllers and 80 ALE-50(V) magazines (plus associated warranties) for use aboard USAF F-16 aircraft. At the time of the announcement, work on the effort was scheduled for completion by the end of October 2003, with the USAF's Warner Robins Air Logistics Center (Robins Air Force Base, Georgia) acting as the programme's contracting activity.

Contractor

Raytheon Electronic Systems, Goleta, California.

UPDATED

AN/ALQ-99(V) Tactical Jamming System (TJS)

Type

Airborne tactical smart noise jamming and electronic surveillance system.

Description

AN/ALQ-99(V) is the tactical jamming and electronic surveillance suite that has been installed aboard US Navy (USN) EA-6B and US Air Force (USAF) EF-111A countermeasures aircraft respectively. In the EA-6B configuration, the system comprises up to five externally mounted transmitter pods, a System Integrated Receiver (SIR) subsystem, a system computer and workstations for two or three operators. In more detail, the transmitter pods house two transmitters; fore and aft, high gain, directional or electronically steerable (dependent on model) transmission antennas; a tracking receiver and a ram-air turbine to provide the necessary electrical power. The SIR system utilises reception antennas mounted in fairings on either side of the aircraft's vertical tail surface and in the 'American football' housing on the top of the fin. Overall, the system offers automatic, semi-automatic and manual operating modes. In automatic mode, the system computer sorts received signals, selects the appropriate jamming response and activates it. In the semi-automatic mode, classified threat data is displayed to the system operators who select and initiate the appropriate response to particular threats. In manual mode, each operator monitors a preselected area of the spectrum, identifies threats and activates responses. When not functioning as a jammer, the ALQ-99(V)'s SIR subsystem is frequently used as a passive electronic intelligence gathering tool.

A close-up of one of the external noise jamming pod configurations used in ALQ-99(V) when it is applied to the EA-6B (Martin Streetly)

Over time, Jane's sources suggest that the EA-6B configured ALQ-99(V) system has gone through a number of iterations, the known details of which are as follows:

AN/ALQ-99(V)
The original ALQ-99(V) model fitted to Standard EA-6B aircraft and able to cover system specific frequency bands 1/2 (30 MHz to 1 GHz), 4 (approximately 500 MHz to 1.05 GHz) and 7 (approximately 2.6 to 3.5 GHz). Now obsolete and out of service.

AN/ALQ-99A(V)
Upgraded equipment installed aboard late production Standard and EXtended CAPability (EXCAP) EA-6Bs. ALQ-99A(V) differed from its predecessor in being able to cover system specific frequency bands 1/2,4,5/6 (approximately 1.1 to 2.7 GHz), 7,8 (approximately 4.3 to 7 GHz) and 9 (approximately 7 to 10 GHz). Other new features included new bands 5/6 and 9 travelling wave tube power sources; new pod configurations for bands 5/6, 8 and 9; new transmitter exciters and the introduction of a frequency sector operating mode. Now obsolete and out of service.

AN/ALQ-99B(V)
ALQ-99B(V) is described as being generally similar to ALQ-99A(V) but with improvements to system electronics to improve reliability and maintainability. Now obsolete and out of service.

AN/ALQ-99C(V)
Described as being a minor enhancement of ALQ-99B(V) with an improved automatic jamming cabability. Now obsolete and out of service.

AN/ALQ-99D(V)
Deployed on Improved CAPability (ICAP) 1 EA-6B aircraft, ALQ-99D(V) is reported to have incorporated digitally tuned receivers and fully integrated computer control for the jamming subsystem. Now obsolete and out of service.

AN/ALQ-99F(V)
Believed to be the current standard equipment installed on ICAP 2 EA-6Bs, ALQ-99F(V) is understood to incorporate a new universal exciter, electronically steerable transmission antennas, an AN/AYK-14 system computer and an improved SIR system. As of this edition, ALQ-99F(V) was understood to be in service.

Alongside the described USN ALQ-99 variants, the highly automated, single operator ALQ-99E(V) was developed for use in the USAF's EF-111A electronic combat aircraft. ALQ-99E(V)'s operational capabilities are understood to be similar to those of the naval equipments. As of this edition, the EF-111A has been withdrawn from service.

Status
ALQ-99(V) variants are understood to be in service aboard USN and Marine Corps EA-6B ICAP 2 Block 82 and 89/89A aircraft. The US military has now decided to standardise on the EA-6B as its multiservice jamming platform and the ICAP 2 Block 89A configuration forms the starting point for the new, multiservice ICAP 3 aircraft. Under development by a consortium of Northrop Grumman (prime and incorporating team members Litton and PRB Associates) and BAE Systems North America - Information and Electronic Warfare Systems, the ICAP 3 configuration incorporates new band 1/2/3 and band 9/10 transmitters, an enhanced universal exciter, a 'reactive' S1R group (the Litton LR-700 equipment with a top end frequency of 18 to 20 GHz), full TJS-AN/USQ-113(V) communications jammer integration, new operator graphical interfaces, improved connectivity and situational awareness and a 'selective reactive' narrowband jamming capability for use against frequency agile threats.

Contractors
Over time, a large number of contractors have been involved in the ALQ-99(V) programme. Identified participants include:
AIL Systems Inc, New York (prime contractor for ALQ-99(V) and ALQ-99E(V)).
Raytheon Electronic Systems, Goleta, California.

Northrop Grumman Corporation, Electronic & Systems Integration Division.
BAE Systems North America - Aerospace Electronics, Lansdale, Pennsylvania.
Lockheed Martin Systems Integration - Owego, Owego, New York.
Astronautics Corporation of America, Milwaukee, Wisconsin.
Motorola Inc, Scottsdale, Arizona.
Smiths Industries, Grand Rapids, Michigan.

UPDATED

AN/ALQ-126B Deception Electronic CounterMeasures (DECM) system

Type
Airborne radar deception countermeasures system.

Description
AN/ALQ-126 is a radar DECM system that was originally developed by the US Navy's (USN) Naval Air Systems Command for its Tactical Air Electronic Warfare programme. As such, the system was designed to provide wider frequency coverage (D through I-band - 1 to 10 GHz) than its predecessor (the AN/ALQ-100 system) and was initiated in response to new air and surface threats. The initial production version was designated as the ALQ-126A and was widely used by the USN aboard aircraft such as the A-4, A-6, A-7 and F-4. The latest version is the ALQ-126B which again increases frequency coverage and incorporates a digital instantaneous frequency measuring receiver, improved deception techniques and updated construction, packaging and cooling arrangements. The equipment also includes a distributed, microprocessor control system to improve signal processing and enable it to be programmed on the flight line. Covering the 2 to 18 GHz frequency band, ALQ-126B is capable of generating a variety of jamming modulations that include inverse conical scanning, range gate pull off, swept square wave and main lobe blanking. ALQ-126B operates either autonomously or as part of an integrated weapon system made up of the AN/ALR-45F or AN/ALR-67 radar warning equipment, the AN/ALQ-162 continuous wave radar jammer and the AGM-88, AIM-120, Phoenix and Sparrow missiles. ALQ-126B is compatible with the A-4, A-6, A-7, F-4, F-14 and F/A-18 aircraft types.

The AN/ALQ-126B equipment

ALQ-126B is virtually a standard fit on the F/A-18 multirole combat aircraft (MS/Sanders)

Status

As of May 2001, the AN/ALQ-126B DECM system was reported as being in service. Over time, the USN is noted as having awarded Sanders ALQ-126B production contracts to the value of nearly US$500 million. Outside the US, the equipment is installed in Australian F/A-18, Canadian CF-18, Kuwaiti F/A-18, Malaysian F/A-18 and Spanish EF-18 aircraft. Recent export activity includes the procurement of ALQ-126B sets for use in Italian AN/ALQ-164 countermeasures systems (see separate entry) and aboard Kuwaiti (42 sets delivered by late 1996) and Malaysian (four sets ordered during 1995 at a then year price of US$4.6 million) F/A-18 Hornets. At least 1,200 ALQ-126B systems have been deployed worldwide. As of early 2001, the latest identified contract relating to the ALQ-126B programme was announced on 19 April 2001 and comprised a then year US$19,008,600 indefinite delivery/quantity fixed-price award to Teledyne Electronic Technologies (Rancho Cordova, California) in respect of the repair and testing of ALQ-126B high band travelling wave tubes. At the time of the announcement, work on the effort was to be undertaken at Rancho Cordova and was scheduled for completion by the end of May 2006. The programme's contracting activity was the USN's Naval Surface Warfare Center, Crane, Indiana.

Specifications

ALQ-126B
Frequency coverage: 2-18 GHz
Dimensions: 412 × 269 × 610 mm
Weight: 86 kg
Power: 3 kVA (max); 115/200 V AC (3-phase, 400 Hz)

Contractor

BAE Systems North America - Information and Electronic Warfare Systems, Nashua, New Hampshire.

UPDATED

AN/ALQ-131 radar jammer

Type

Airborne pod-mounted radar jamming system.

Description

The AN/ALQ-131 radar jammer is an automatic, modular self-protection system. It was designed to provide advanced broadband coverage against all types of modern Red, Blue and Grey radar-guided weapons. ALQ-131 is carried externally on a variety of front-line, high-performance aircraft. Internal installations are also available. Over time, it has been certified on a variety of aircraft, including the F-16, F-15, F-111, F-4, A-7, A-10, Harrier and C-130.

The modular design of the pod structure and electronic assemblies, plus its central computer software architecture, enables the ALQ-131 system to adapt quickly to a broad spectrum of Electronic Warfare (EW) applications. This feature proved very helpful in the 1990/91 Gulf War where ALQ-131s provided over 48 per cent of the USAF tactical aircraft EW self-protection. The latest Block II version experienced no combat losses in over 12,000 combat sorties flown during that campaign. Increased effectiveness can be achieved by incorporating various mission modules, including missile warning systems, offboard and towed countermeasures dispensers and advanced technique generators. This capability allows maximum flexibility in countering all threat types.

The basic structural elements of the pod are modular canisters that provide structural support, cooling and environmental protection. Each canister is an I-beam structure that also serves as a cold plate. Both sides of the I-beam form equipment bays into which functional equipment modules are mounted. The modules can also be mounted in lower equipment bays on the bottom surface of the I-beam. Using common mounting techniques, each canister can accommodate

several equipment modules that can be removed directly from the bays without disassembling the pod. The system is 2.83 m long and its weight ranges from 260 to 324 kg.

An innovative feature of the ALQ-131 is a high reliability, self-contained Freon-to-ram air cooling system that has no moving parts and requires no electrical power for operation. Each pod also includes a transition/air module, forward and aft antennas and an optional ram air turbine for applications requiring prime power not available from the aircraft.

The functional organisation of the system is centred around the Interface and Control (I/C) module that contains a programmable digital computer that acts as the system controller. The modules required for a given configuration are connected to the I/C by a digibus that carries all sensor and control data. A memory loader/verifier allows operational flight and mission specific programme software to be loaded into the pod on the flight line in less than 15 minutes. The I/C module also contains a digital waveform generator that can permit up to 48 simultaneous waveforms for deception modulation. When any countermeasures technique requires a deception waveform, the latter's values are transmitted to the onboard equipment via the waveform distribution bus.

Maintenance of the ALQ-131 is based on the pod's Centrally Integrated Test System (CITS) which provides a comprehensive functional check of system operation, both in flight and on the ground. During flight the CITS continuously monitors the operational status of the equipment, including repeater channel modulation, high voltage for the travelling wave tubes, noise power output, primary bus voltage and the integrity of the computer memory. I-level maintenance is provided by the AN/ALM-256 I-Level Support Equipment (ILSE). The ILSE automatically provides 100 per cent fault detection and 98 per cent fault isolation.

The pod is a software reprogrammable system that allows a tactical commander to tailor the pod's responses to his mission requirements. Using ALQ-131's ability to be flight line reprogrammed, mission specific data can be created in response to threat changes and, using a memory loader verifier, loaded into the pod's digital computer. Another important capability of the system is its self-contained power management feature. This is included in a receiver/processor module that detects radar threats, measures their key parameters and performs weapon type and operational mode identification. This information is then used by the I/C module to select optimum jamming techniques automatically and tailor their parameters for countering all detected threats, thereby permitting specific tailored jamming techniques to be applied only when needed. Additionally, numerous threats can be countered simultaneously, each with independent techniques, using the receiver processor's pulse-repetition interval tracking capability.

Status

As of late 2000, more than 560 Block I and 460 Block II AN/ALQ-131 radar jamming pods were reported as having been delivered to the USAF. In addition, over 320 pods have been procured by the air forces of Bahrain (6 Block II pods), Belgium, Egypt (49(?) Block I and 42 Block II), Israel, Japan, Netherlands (26 Block I and 79 Block II), Norway (16 Block I), Pakistan (21 Block I and (?) Block II) Portugal, Singapore and Thailand. A programme to update the original configuration Block I system to the latest Block II version is underway. Recent activity here includes the 25 May 1999 announcement of a then year US$15,761,936 firm, fixed-price contract to upgrade 16 Norwegian ALQ-131 Block I pods up to Block II standard. The Norwegian deal was scheduled for completion by the end of May 2001 with the USAF's Warner Robins Air Logistics Center, Robins Air Force Base, Georgia acting as the programme's contracting activity. Elsewhere in the world, Northrop Grumman is also known to have been awarded a then year US$19.6 million contract during April 1999 that covered the supply of ALQ-131 kits to the Mitsubishi Electric Corporation for final assembly and testing. Mitsubishi has been licence building ALQ-131 for the Japanese Air Self-Defence Force since 1991.

In October 1993, the USAF completed testing of ALQ-131 modified with an AN/ALQ-153 active Missile Warning System (MWS) and an AN/ALE-47 CounterMeasures Dispenser (CMD) installed on an F-16 aircraft. In April 1994, ALQ-131 was demonstrated with a fully integrated AN/AAR-54(V) passive missile warning system and an ALE-47 countermeasure dispenser installed, while August 1998 is reported to have seen an ALQ-131 pod/'high-powered' fibre-optic towed decoy combination tested aboard a C-130 aircraft. Northrop Grumman has also proposed using the ALQ-131 pod shell as a bus which can incorporate a customer specific electronics suite drawn from elements including active countermeasures, threat warners, electro-optic reconnaissance systems and radar.

Contractor

Northrop Grumman Electronic Systems sector - Defensive Systems Division, Baltimore, Maryland.

UPDATED

AN/ALQ-135 radar jamming system

Type

Airborne radar jamming system.

Description

The AN/ALQ-135 Internal Countermeasures Set (ICS) is a component of the Tactical Electronic Warfare System for the US Air Force (USAF) F-15 Eagle fighter. It is installed in varying configurations in the F-15A, C, E and S models and operates with the AN/ALR-56 radar warning system, the AN/ALQ-128 radar warning receiver and the AN/ALE-45 countermeasures dispenser in the type's Tactical Electronic

A close-up of an ALQ-131 jamming pod mounted on a USAF A-10 aircraft (MS/Northrop Grumman)

The ALQ-135 ICS incorporates miniaturised component technology as shown here on one of its circuit boards

Warfare System (TEWS). ICS is an internally mounted system that features high power transmitters and integral jamming management and provides automatic jamming of threat pulsed and continuous wave radars. The equipment's computer management system (20 microprocessors) enables ALQ-135 to adapt automatically to changes in hostile transmissions. In its original 1970s configuration, ALQ-135 was made up of Line-Replaceable Units (LRU - plus associated waveguides and a four-antenna array) and incorporated one Band 1 and two Band 2 transmitters. Over the years, the system has evolved to match the development of the F-15 and changes in the threat. While maintaining commonality with the original system and support electronics, ALQ-135 has been updated to include 'full' band coverage and 'effective' techniques flexibility. Development of a new Band 3 transmitter, housed in two boxes (transceiver/processor and power amplifier), was initiated in 1983. Frequency coverage of Band 3 overlaps that of Band 2 and this has made it possible for the current Band 1 and 2 transmitters to be combined. The new low-band transmitter is known as the Band 1.5 unit and is half the size of the Bands 1 and 2 transmitters. The F-15E ALQ-135 configuration incorporates Band 1.5 and Band 3 transmitters while that for the F-15C is understood to make use of Bands 1, 2 and 3 transmitters.

Status

Initial development funding for the AN/ALQ-135 ICS was made available during August 1974 and led to a then year US$25 million production contract in September 1975. This has been followed by further awards covering the procurement of more than 1,750 sets. In addition to the USAF, ALQ-135 is fitted to F-15 aircraft operated by Israel (F-15A and C aircraft) and Saudi Arabia (F-15C and S). In April 1993, Northrop Grumman was awarded a then year US$173 million contract for the Lot 4 buy of ALQ-135, consisting of 60 Band 3 sets and associated contractor support. In September 1994, the company received a then year US$44 million contract for Band 3 subsystems followed by a further then year US$141 million award in March 1995 covering a further tranche of 60 units. Lot 9 deliveries (including Band 3 units) are reported to have begun during 1997, the same year that Saudi contractor Advanced Electronics Co Ltd was awarded a then year SR6 million contract covering the supply of power supplies for ALQ-135 Band 1.5 transmitters. The ALQ-135 Band 1.5 unit underwent initial flight testing during 1998 and a USAF/ Northrop Grumman team is reported to have successfully completed installation of the first operational ALQ-135 Band 1.5 transmitter aboard a USAF F-15E strike aircraft during the fourth quarter of 2000. Congressional approval for the modification of 170 ALQ-135 suites to Band 1.5/3 configuration is understood to have been given during November 1998 and, as of November 2001, identified ALQ-135 contract activity during the period 1999 to 2001 comprised:

22 December 1999

Northrop Grumman (Rolling Meadows, Illinois) was awarded a then year US$33,368,614 modification to an existing firm, fixed-price contract that covered the supply of 15 ALQ-135 Band 1.5 transmitters. At the time of the announcement, work on the effort was scheduled for completion by 30 December 1999, with the USAF's Aeronautical Systems Center (Wright-Patterson Air Force Base (AFB), Ohio) acting as the programme's contracting activity.

9 January 2001

As of the given date, Northrop Grumman (Rolling Meadows, Illinois) announced that it had been awarded a then year US$30 million contract covering the supply of 17 ALQ-135 Band 1.5 transmitters for installation aboard USAF F-15E strike aircraft. At the time of the announcement, work on the effort was scheduled for completion by the end of December 2001.

30 March 2001

Northrop Grumman (Rolling Meadows, Illinois) was awarded a then year US$6,424,735 modification to an existing firm, fixed-price contract that covered the supply of 31 ALQ-135 Band 3 transmitters. The USAF's Aeronautical Systems Center (Wright-Patterson AFB, Ohio) acted as the programme's contracting activity.

25 September 2001

The Northrop Grumman Corporation (Rolling Meadows, Illinois) was awarded a then year US$10,652,565 firm, fixed-price contract covering the supply of 19 control receivers for ALQ-135 ICSs together with the 'necessary' materials and non-

recurring engineering work required to overcome 'existing obsolescence problems' with the system. At the time of the announcement, work on the effort was scheduled for completion by the end of October 2002 and the award incorporated an option that allowed for the purchase of 10 addition control receivers if required. This programme's contracting activity was the USAF's Warner-Robins Air Logistics Center (Robins AFB, Georgia).

11 December 2001

The Northrop Grumman Corporation (Rolling Meadows, Illinois) was awarded a then year US$65,111,239 firm, fixed-price contract covering the supply of 23 ALQ-135 Band 1.5 transmitters applicable to the F-15E strike aircraft. At the time of the announcement, work on the effort was scheduled for completion by the end of December 2002, with the USAF's Aeronautical Systems Center (Wright-Patterson AFB, Ohio) acting as the programme's contracting activity.

Contractor

Northrop Grumman Electronic Systems sector - Defensive Systems Division, Rolling Meadows, Illinois.

UPDATED

....

AN/ALQ-136(V) radar jammer

Type

Airborne radar jammer for helicopter applications.

Description

AN/ALQ-136(V) series jammers are designed to protect helicopters from radar-guided weapons. The system consists of three main units; a control unit, two spiral antennas (one each for transmit and receive) and a transmitter/receiver. System features include a fully automatic transponder system, automatic countermeasures technique selection, simultaneous handling of multiple threats, software reprogrammability and high receiver sensitivity. When the aircraft is illuminated by a hostile radar, the jammer automatically analyses the received pulses, compares them against its threat library, assigns them a priority and then provides the most appropriate countermeasures response. The design of ALQ-136(V)2 traces its lineage back to the combat-proven ALQ-136(V)1/5. Deployed on the US Army's fleet of Apache and Cobra attack helicopters, ALQ-136(V) demonstrated high levels of reliability and effectiveness as individual systems accumulated over 300 hours of continuous operating time during Operation Desert Storm. ALQ-136(V) can handle multiple threats, is internally mounted and interfaces with other onboard systems via a 1553B databus.

Status

As of this edition, the AN/ALQ-136(V) radar jammer was reported as being in US Military service. More than 1,400 systems have been delivered to the US Army. Within this service, ALQ-136(V) is applicable to the AH-IF (ALQ-136(V)1), AH-64A/D (ALQ-136(V)5), EH-60A (ALQ-136(V)2), MH-47E (ALQ-136A(V)2) and MH-60K (ALQ-136A(V)2) helicopters together with RC-12 series (ALQ-136(V)2) fixed-wing Special Electronic Mission Aircraft (SEMA). Other platforms configured to accept the system include MH-53J Pave Low helicopters and the AH-IW Super Cobra and it is known that both Greece and Saudi Arabia have procured ALQ-136(V) jammers. In March 2000, ITT announced that it had been awarded a then year US$7.5 million contract to supply the US Army with 31 High Power Remote Transmitters (HPRT) for installation aboard MH-47E and MH-60K helicopters where they form a jamming suite alongside the ALQ-136A(V)2 pulse radar jammer and the AN/ALQ-162 continuous wave jammer. The described HPRT award is noted as following-on from an initial contract awarded during 1998.

Contractor

ITT Avionics, Clifton, New Jersey.

VERIFIED

The ALQ-136(V)2 jammer system

AN/ALQ-144A(V) Infra-Red CounterMeasures (IRCM) system

Type
Airborne IRCM system.

Description
ALQ-144A(V) is the latest configuration of the electrically powered ALQ-144(V) IRCM system that provides medium-sized air vehicles with protection against IR guided missiles. It is an omnidirectional equipment that consists of a transmitter assembly and an Operator Control Unit (OCU). Of these, the transmitter assembly is made up of a chassis that supports a cylindrical IR source surrounded by a modulation screen. ALQ-144A(V) features built-in test, is supported by the system dedicated AN/ALM-178 test set and has appeared in three variants designated as ALQ-144A(V)1, -144A(V)3 and -144A(V)5. Of these, the 12.5 kg ALQ-144A(V)1 is used on utility, observation and special mission helicopters, the 12.7 kg ALQ-144A(V)3 on attack helicopters and the ALQ-144A(V)5 on naval helicopters. Both ALQ-144A(V)1 and -144A(V)3 make use of the T-1360A(V)1/ALQ-144A(V) transmitter assembly and differ from one another in that the former employs the C-10280/ALQ-144(V) OCU while the latter makes use of the C-9576/ALQ-136(V) unit. The C-9576/ALQ-136(V) OCU features a 25 pin connector assembly (as opposed to the C-10280's nine pin assembly) and can control both the AN/ALQ-136(V)1/5 series of radar jammers (see separate entry) and the AN/ALQ-144A(V)3 or ALQ-144A(V)3 IRCM systems. The ALQ-144A(V)5 system makes use of a phase lock configuration of two transmitters that are electrically phased and are controlled via a single OCU that offers a choice of nine different jamming programmes.

Status
Over time, in excess of 5,000 ALQ-144(V)/A(V) equipment are reported to have been supplied to the US military and offshore customers. ALQ-144(V)/A(V) manufacturer BAE Systems North America - Information and Electronic Warfare Systems (formerly Sanders) is known to have received contracts worth nearly US$100 million to produce upgrade kits to bring ALQ-144(V)1 and (V)3 equipments up to ALQ-144A(V)1 and (V)3 standard. As of mid 2000, ALQ-144A(V)5 is noted as being or having been used aboard US Navy/Marine Corps SH-2F/G, SH-3D, SH-60B/R and VH-60, helicopters, while the US Army states that ALQ-144A(V)1 is applicable to its EH-60A, MH-60K, OH-58D and UH-60A/L platforms. Within that same service, ALQ-144A(V)3 is applicable to AH-1F and AH-64A/D attack helicopters together with AH-1W platforms operated by the US Marine Corps. Recent identified programme activity includes the delivery of 28 ALQ-144 IRCM jammers to Turkey during April 1998 (with a further 200 systems following during 1999/2000) and a July 1999 then year US$4.8 million contract covering the supply of 133 ALQ-144A(V)1, 3 and 5 jammers to the US military. Identified ALQ-144(V)/-144A(V) programme activity during 1999-2001 comprised the following:

24 February 1999
Sanders was awarded a firm, fixed-price contract (with a not-to-exceed then year value of US$30 million) covering the supply of 400 ALQ-144A(V)1/(V)3 or -144A(V)5 IRCM systems, 100 associated AN/ALM-178 test sets, support equipment, 'essential' spares, maintenance training and engineering and technical field support for Foreign Military Sales of ALQ-144A(V) equipments to Israel, Netherlands and Taiwan. At the time of the contract's announcement, work on the effort was scheduled for completion on 19 February 2000 and the programme's contracting activity was the US Army's Communications and Electronics Command, Fort Monmouth, New Jersey.

26 May 2000
Sanders was awarded a then year US$28,076,727 (with all options exercised) firm, fixed-price, indefinite delivery/quantity contract covering the supply of spares for an estimated 6,366 ALQ-144(V) and ALQ-144A(V) IRCM systems. At the time of the award's announcement, work on the effort was scheduled for completion by 25 May 2005 and the programme's contracting activity was the US Army's Communications and Electronics Command, Fort Monmouth, New Jersey.

12 July 2001
BAE Systems North America - Information and Electronic Warfare Systems was awarded a then year US$6 million delivery order (part of firm, fixed-price contract DAAB07-99-D-B605) covering the procurement of 300 maintenance work order kits for the installation and conversion of 300 US Navy basic ALQ-144(V) transmitters to T-1360A(V)/ ALQ-144A(V) standard. At the time of the announcement, work on the effort was to be performed at BAE Systems' Nashua, New Hampshire facility and was scheduled for completion on 30 May 2002. The programme's contracting activity was the US Army's Communications and Electronics Command, Fort Monmouth, New Jersey.

Specifications
Power: 1,675 W (max); 28 V DC
MBTF: 300 h (min demonstrated); >600 h (achieved)
Dimensions: 2.8 × 14.7 × 13.4 cm (OCU); 32 × 34 cm (transmitter)
Weight: 0.45 kg (OCU); 12 kg (transmitter)

Contractors
BAE Systems North America - Information and Electronic Warfare Systems, Nashua, New Hampshire.

UPDATED

AN/ALQ-151(V)2 QUICKFIX Electronic Warfare (EW) system

Type
Airborne intercept and countermeasures system.

Description
The AN/ALQ-151(V)2 QUICKFIX system is a direction-finding, intercept and countermeasures jamming suite that is installed aboard US Army EH-60 special electronic mission helicopters. The system can interface with all other tactical army aircraft using a secure communications link and also interfaces with an operations centre. As of this edition, the ALQ-151(V)2 variant of the system is installed on EH-60A aircraft.

Status
During 1984, the then Tracor Aerospace (now BAE Systems North America) won a contract to install and integrate AN/ALQ-151(V)2 in the EH-60A. Total contract value was then year US$100 million which included necessary aircraft configuration changes and integration of an extensive aircraft survivability equipment suite. Delivery of the full batch of 66 systems was completed during 1989 and as of this edition, ALQ-151(V)2 was scheduled to have been completely withdrawn from service by the end of 2005.

Specifications
Frequency range: 2-80 MHz (DF and jamming)
Range: line of sight
Signal types: AM, FM, CW, SSB
Bandwidth: 8, 30 or 50 kHz
Power output: 500 W

Contractors
TRW Systems and Information Technology (ALQ-151(V)2 QUICKFIX mission suite), Sunnyvale, California.

VERIFIED

The transmitter used in the ALQ-144(V)/A(V) IRCM system

The US Army's EH-60A special electronic mission helicopter is fitted with the ALQ-151(V)2 QUICKFIX countermeasures suite

AN/ALQ-155(V) electronic warfare power management system

Type
Integration system for the AN/ALR-46 radar warning receiver and the AN/ALT-28 jammer.

Description
The AN/ALQ-155(V) countermeasures power management system forms part of the defence avionics system installed aboard US Air Force (USAF) B-52H strategic bombers. It provides integral set on receivers for each jamming transmitter, plus increased effective radiated power density through accurate frequency set on. The system is a power-managed evolution for the ALT-28(V) set providing automated hand-off from the AN/ALR-46 radar warning receiver with near instantaneous jammer response. It is computer-managed and field programmable. The system contains automatic frequency control in all modes and a wide variety of electronic countermeasures techniques that are automated, semi-automated or manual. A 12 transmitter upload capability is provided.

A variety of improvements have, over time, been incorporated in ALQ-155(V), including frequency agility against multiple threats, pulse-repetition interval trackers, cover pulse jamming techniques, false target generation through pseudo-random noise, coherent and incoherent jamming and downlink jamming. The system also incorporates a hybrid instantaneous frequency measuring receiver and central receiver capability, programmable noise optimisation, increased pulse-up power for continuous wave to pulse operations, electronically steerable antenna system compatibility, and compatibility with the AN/ALQ-117 deceptive I/J-band (8 to 20 GHz) jammer.

Status
Total AN/ALQ-155(V) design, development and production contracts since 1975 exceed US$160 million in value, with approximately 300 examples of the equipment being known to have been delivered to the USAF. On 2 October 2000, Northrop Grumman announced that it had been awarded a then year US$3.5 million contract covering the procurement of long-lead materials for the upgrading of the ALQ-155(V) sets installed aboard the B-52H strategic bomber and funding for the effort through its preliminary design review. Characterised as a 'low-cost, low-risk' programme, the upgrade involved the integration of non-developmental items from Northrop Grumman's AN/ALQ-135 and AN/ALQ-165 radar jammers into the existing ALQ-155(V) receivers and transmitters. So configured, the 'new' ALQ-155(V) will be a form-fit replacement for its predecessor, offer increased reliability and a five fold increase in output power and provide 'advanced' countermeasures techniques for use against targets operating in 'multiple' frequency bands. Despite the launch of this effort, as of July 2001 usually reliable sources were reporting that Boeing had issued a request for interest in a potentially more 'robust' replacement for the ALQ-155(V) in order to keep the B-52's low- and mid-band countermeasures capability viable until *circa* 2040. As described (and if funded), requests for proposals for such equipment would be released during US Fiscal Years (FY) 2002/2003, with up to three Engineering and Manufacturing Development (EMD) contracts being awarded during US FY 2004. Following completion of the initial phase of a four year EMD effort, the USAF would select one contractor to complete the work. Thereafter, the service would award a six year long production contract covering the supply of jammers to equip a fleet of 76 aircraft. As of the given date, the full 11 year effort was noted as having a potential value of then year US$600 million.

Contractor
Northrop Grumman Electronic Systems sector - Defensive Systems Division, Rolling Meadows, Illinois.

UPDATED

The dual heat exchanger element of the ALQ-155(V) power management set for USAF B-52 bomber

AN/ALQ-157 Infra-Red CounterMeasures (IRCM) system

Type
Airborne jamming system against infra-red homing missiles.

Description
The AN/ALQ-157 is designed to provide protection against infra-red heat-seeking missiles and is being manufactured for use on large troop-carrying helicopters and transport aircraft. For large helicopter applications, two transmitters are installed (one on each side of the sail) for unobstructed protection in azimuth. Transport aircraft applications utilise transmitters mounted on either side of the fuselage. ALQ-157 consists of four basic subsystems: transmitters, a control power supply, electromagnetic interference filter assembly and a pilot's control indicator. A switch on the control power supply allows in-flight selection of any one of five preprogrammed jamming codes. Additional codes can be preprogrammed as new threats are defined. The microprocessor also directs all operational sequences of the system. The system ensures full-time protection against threats, easy access to all components and employs built-in test circuits to perform operational readiness tests automatically. Weight of the complete system is 100 kg.

Status
As of this edition, the AN/ALQ-157 IRCM system was reported as being installed aboard helicopters and fixed-wing aircraft operated by the US Navy, US Air Force, US Marine Corps and the National Guard. It has also fitted to a number of Royal Air Force transport helicopters and fixed-wing aircraft and Italian Army helicopters.

Contractor
Lockheed Martin Information Systems, Electro-Optical Systems Division, Pomona, California.

VERIFIED

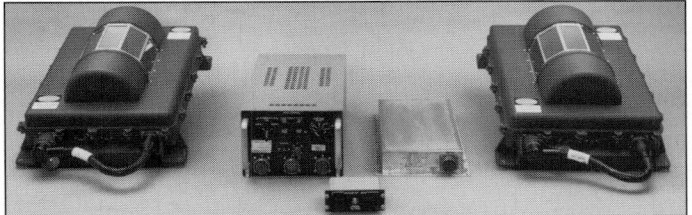

A display of AN/ALQ-157 LRUs showing two of the system's transmitters and its EMI filter, pilot control indicator and control power supply

AN/ALQ-161A Electronic CounterMeasures (ECM) system

Type
Airborne ECM system for the B-1 strategic bomber.

Description
The B-1's AN/ALQ-161A defensive avionics system is intended to counter surface-to-air missile, anti-aircraft and air-to-air missile fire-control radars and to degrade by noise jamming early warning and ground controlled interception radars. The system incorporates and controls a large quantity of jamming transmitters and antennas. In addition to the jamming hardware, a sophisticated control system managed by a network of digital computers is employed. This can jam signals from many radars simultaneously. The numerous jamming chains are deployed around the periphery of the B1-B to jam signals in the 200 MHz to 25 GHz frequency range which are coming from any direction. Integrated with the jamming control system is a network of separate receiving antennas, receivers and processors that act as the ears of the system. By means of this receiving system, new signals can be picked up, identified and then jammed (with optimised jamming techniques) in a fraction of a second. The receiver subsystem continuously monitors the environment and incorporates jamming look-through.

The ALQ-161A system is made up of 108 Line-Replaceable Units (LRUs) which, exclusive of cabling, displays and controls, weigh approximately 2,359 kg and consume about 120 kW of power in the all-out jamming mode. Most of the LRUs have volumes of between 0.02 and 0.06 m³ and weigh about 17.4 to 36.3 kg. A system operator has displays and controls by which the system can be monitored and directed. The computer performs the power management role based on threats sensed by the radio frequency sensor subsystem and prestored information about threat types. Terminal threat jammers, phased-array antennas and other hardware are provided by subcontractors. Since the system was originally deployed, a new frequency band capability and an integrated Tail Warning Function (TWF) have been added. The purpose of the latter is to detect the presence of approaching missiles in the tail region and initiate a defensive countermeasure. The TWF is a pulse-Doppler radar and shares the ECM subsystem's transmitter/antenna assembly. A separate receive antenna interfaces with the TWF receiver/processor located in the tailfin area. The TWF is capable of operating in a stand-alone mode, independent of the rest of the system. ALQ-161A also features a monitoring network that is integrated with a central integrated test

Units of the ALQ-161 defensive avionics suite displayed in front of a B-1B aircraft

system. This network automatically monitors and reports any electronic warfare system degradation or computer failures. The system independently routes signals around failed or damaged components to maintain full jamming capability against high-priority threats.

Status

As of this edition, AN/ALQ-161A contracts worth over US$2 billion have been awarded to AIL and its subcontractors, since the initial development began in the 1970s. After the cancellation and subsequent reinstatement of the B-1B programme, delivery of the first line-replaceable units began in February 1984 and was followed by a three year long flight test programme. The production phase of the programme was completed in December 1987.

There have been problems with the ALQ-161A which AIL, working in conjunction with the USAF, notes have mostly been overcome. In this context, AIL is known to have been authorised to retrofit the TWF subsystem into ALQ-161A during 1991. TWF deliveries began in September of that year and are now complete. Following further upgrade studies, Jane's sources suggest that the USAF is augmenting ALQ-161A with an AN/ALE-50 towed decoy fit comprising two, four round dispensers. As of October 1998, the service was proceeding with ALE-50 installations on seven B-1B aircraft with the last example scheduled for completion in March 1999. Identified ALQ-161A programme activity during 2000 and 2001 was as follows:

16 May 2000
AIL Systems (Deer Park, New York) was awarded a then year US$6,133,457 firm, fixed-price contract covering repair of components used in the ALQ-161A ECM system. At the time of this award's announcement, work on the effort was scheduled for completion by the end of April 2001 and its contracting activity was the USAF's Warner Robins Air Logistics Center, Robins Air Force Base (AFB), Georgia.

30 August 2001
The Northrop Grumman Corporation (Rolling Meadows, Illinois) was awarded a then year US$19,996,297 firm, fixed-price contract covering the supply of 42 ALQ-161A 'components' for use on USAF B-1B strategic bombers. At the time of the announcement, the effort was structured into a basic 12-month contract period with four annual options and was scheduled to be undertaken by Northrop Grumman personnel at the USAF's Warner-Robins Air Logistics Center, Robins AFB, Georgia (also the programme's contracting activity).

16 November 2001
AIL Systems (Deer Park, New York) was awarded a then year US$5,587,609 firm, fixed-price contract covering the repair of 'various' ALQ-161A components. At the time of the announcement, the effort was scheduled to be completed by the end of May 2002 and its contracting activity was the USAF's Warner-Robins Air Logistics Center, Robins AFB, Georgia.

Contractor

AIL Systems Inc, Deer Park, New York.

UPDATED

AN/ALQ-162(V) radar jammer

Type

Airborne radar jamming system.

Description

AN/ALQ-162(V) (also known as Shadowbox) is a single Weapon-Replaceable Assembly (WRA) that incorporates transmitter, receiver/processor, user data memory and an antenna module. It is a continuous wave, chopped repeater jammer that can operate autonomously or be interfaced with other onboard electronic warfare equipments. It is packaged in a single unit that weighs 18.6 to

19 kg (see following) and has a volume of less than 0.01 m³. The system provides self-protection against radar threats by continuously scanning the threat signal environment, identifying emitters and then generating specific countermeasures against prioritised threats.

Of the various WRA modules, the receiver/processor (which breaks out as 12 shop-replaceable assemblies) performs the system's reception and processing functions, test and tactical system programming, modulation techniques generation, system time gating and transmitter control. It contains all the low-power elements, system radio frequency video processing and analysis circuits, central processing unit, memory, control logic and modulation generators. The ALQ-162(V) transmitter consists of three assemblies and houses high- and low-voltage power supplies and inverters, the travelling wave tube amplifier and equaliser, modulator and cooling fan. The equipment's user data memory module is a single printed circuit assembly (with a programmable read-only memory) and provides a reprogrammable data bank for system control parameters, threat tables, threat priorities and modulation techniques. The system's antenna module consists of two vertically polarised antennas, a coaxial switch or power divider and interconnecting coaxial cables. The antennas provide nominal coverage of ±60° in azimuth and ±30° in elevation. ALQ-162(V) will also interface with additional antennas or cross-polarisation antenna subsystems if required.

Known details of specific ALQ-162(V) subvariants are as follows:

AN/ALQ-162(V)2
Weighing approximately 19 kg, ALQ-162(V)2 comprises two AS-3554 antennas, an RT-1377A/ALQ-162(V) transceiver/processor WRA and a C-11080/ALQ-162(V)2 control unit. The equipment is reprogrammable via its user data module and is supported by the SM-756/APR-44(V) radar signal simulator.

AN/ALQ-162(V)3
Weighing approximately 19 kg, ALQ-162(V)3 comprises two AS-3554 antennas and the RT-1377A/ALQ-162(V) transceiver/processor WRA. The equipment is reprogrammable via its user data module and is supported by the SM-756/APR-44(V) radar signal simulator.

AN/ALQ-162(V)4
Weighing approximately 18.6 kg, ALQ-162(V)4 comprises a single AS-3554 antenna and the RT-1377A/ALQ-162(V) transceiver/processor WRA. The equipment is reprogrammable via its user data module and is supported by the SM-756/APR-44(V) radar signal simulator.

Shadowbox II
Northrop Grumman has developed an upgraded ALQ-162(V) under the designation Shadowbox II. Here, the new configuration adds pulse-Doppler jamming to the baseline capability in order to counter the lock on of hostile tracking/fire-control radars and active radar seekers. Shadowbox II is the same size as the original ALQ-162(V), a goal that was achieved via the use of monolithic microwave and application specific circuitry and microwave power module technology. As is the case with its predecessor, Shadowbox II can be installed internally or in a pod.

Status

As of February 2001, approximately 700 ALQ-162(V) radar jammers were understood to be in service with (or selected by) 10 armed services around the world. Within the American military, ALQ-162(V) is applicable to EH-60A (ALQ-162(V)2), MH-47E (ALQ-162(V)4) and MH-60K (ALQ-162(V)3) helicopters of the US Army together with the same service's RC-12N and P fixed-wing signals intelligence aircraft (ALQ-162(V)3). Outside the US, ALQ-162(V) is understood to be fitted to F/A-18 aircraft operated by the air forces of Canada, Denmark, Kuwait and Spain, together with Danish C-130 and F-16 aircraft and AV-8Bs flown by the naval air arms of Italy and Spain. In these two latter applications, ALQ-162(V) forms part of the AN/ALQ-164 system (see separate entry). Following test flights of Shadowbox II aboard a US Navy F/A-18, the system is noted as having been sold to the Royal Norwegian (29 units) and Royal Danish (4 units) Air Forces in a then year US$ 28.9 million US Foreign Military Sales (FMS) deal that was identified during June 2000. At the time this contract was announced, it was thought that both countries were intending to install Shadowbox II aboard their F-16 aircraft and it was further understood that the contract included an option for an FMS sale to Egypt, with the system slated for use aboard that country's AH-64 battlefield attack helicopters. Here, the cited option was converted into a firm order when, on 12 February 2001, it

The ALQ-162(V) radar jamming system

was announced that Northrop Grumman had been awarded a then year US$15 million FMS contract covering the supply of an unspecified quantity of 'enhanced' ALQ-162(V)6 radar jammers for installation aboard Egyptian AH-64Ds.

Contractor
Northrop Grumman Electronic Systems sector - Defensive Systems Division, Rolling Meadows, Illinois.

UPDATED

AN/ALQ-164 Deception Electronic CounterMeasures (DECM) system

Type
Airborne pod-mounted radar jamming system.

Description
AN/ALQ-164 is an airborne, reprogrammable, multimode, power managed, pod-mounted deception jamming system that is effective against both continuous wave and pulse radars. The pod integrates the capabilities of two radar jamming systems, the AN/ALQ-126B and AN/ALQ-162(V) units (see separate entries). It can operate autonomously or be integrated with a radar warning receiver and other avionics on board the aircraft. Within the system, ALQ-126B employs automatic search and multiple jamming techniques for each identified threat, selecting the most effective approach for the threat priority. ALQ-162 uses automatic search and jamming to counter the effectiveness of continuous wave threats. ALQ-164 is intended for fitment on the centreline pylon of the AV-8B close-support aircraft and has applications to a wide range of tactical and transport aircraft.

Status
As of this edition, ALQ-164 is noted as having been installed on AV-8B aircraft operated by the US Marine Corps and the navies of Italy (then year US$47.1 million procurement with four ALQ-126B units being ordered for use in Italian ALQ-164 systems during 1995) and Spain.

Specifications
Power: 3 kVA; 28 V DC (at 1.5 A)
Dimensions: 0.4 × 0.4 × 2 m
Weight: 188 kg

Contractor
BAE Systems North America - Information and Electronic Warfare Systems, Nashua, New Hampshire.

UPDATED

The ALQ-164 DECM pod

AN/ALQ-165 Airborne Self-Protection Jammer (ASPJ)

Type
Airborne internally mounted electronic countermeasures system.

Development
The AN/ALQ-165 ASPJ was designed as the next-generation electronic countermeasures system for internal installation in the US Navy's (USN) F/A-18C/D and F-14D aircraft. It was also designed for use in US Air Force (USAF) F-16 fighters and was intended to be available in a pod configuration for the US Marine Corps AV-8B. When ASPJ development first commenced, it was a joint effort between the USN and the USAF. In January 1990, the latter service withdrew from the programme citing overall budget limitations and the need to cut the number of its tactical fighter squadrons. The USN continued the programme until December

The ASPJ configuration selected by South Korea for use on its F-16 aircraft

ASPJ has been flown operationally aboard US Marine Corps F/A-18 aircraft

1992, when it too withdrew from the production effort that was then formally terminated. ASPJ has since gone on to be reappraised for USN use and has been exported (see Status).

Description
ASPJ incorporates a range of state-of-the-art technologies including applications specific integrated circuits, monolithic microwave integrated circuits, gate arrays, Travelling Wave Tubes (TWTs) and progressive packaging techniques. It can be electrically programmed on the flight line and its built-in test and modular plug-in design permits rapid replacement of assemblies at the operational level without external test equipment. Weight of the system is between 91 and 150 kg, depending on configuration and it occupies a space of 0.06 m³.

ALQ-165 has the ability to select automatically the best jamming techniques to use against any given threat, based on the system's own computer data and real-time information on the specific threat signal from the receiver/processor. The computer software can be modified to accommodate new threats as they arise. The equipment covers the frequency range in two bands, a coverage that can be expanded if required.

The system's transmitters can jam a large number of threats simultaneously over various ranges and in different modes. The computer selects the power and duty cycle criteria. An augmented version, with an additional transmitter power-booster is also available. The transmitters are of a dual-mode type with parallel pulse and Continuous Wave (CW) TWTs. The receiver can handle conventional pulsed, high-duty cycle pulsed, complex waveform and CW transmission, with the appropriate signal processing, threat evaluation and prioritisation accomplished automatically.

Three ASPJ configurations were originally envisaged; a 'standard' five Line-Replaceable Unit (LRU) format for F-16 and F/A-18 type aircraft, a seven LRU 'augmented' system for the F-14 and F-16 and a podded application for the AV-8B. The 'augmented' system incorporated two additional boxes to boost the equipments rear hemisphere ouput while the podded iteration was a repackaging of the 'standard' format and was intended for carriage on the AV-8Bs centreline stores station.

Status
The ITT Avionics/Northrop Grumman ASPJ Joint Venture was awarded a US$376 million Product Verification (PV) contract in August 1987. The full-scale development phase under which 12 development systems were delivered and tested, has been completed. Subsequently, the system has undergone 8,000 hours of ground testing and 500 hours of flight testing. The PV programme required the delivery of six systems in Phase 1 and 14 additional systems in Phase 2 during 1989-90. The system completed PV developmental flight testing in July 1991. PV operational flight tests began in August 1991 and were scheduled for completion in 1992. Both contractors were producing 50 ASPJ systems each,

under a Lot 1 initial production contract awarded in September 1989. In the second competitive procurement of the programme, the USN awarded ITT Avionics a US$38 million Lot 2 low-rate production contract to produce 12 ASPJ systems together with a US$51 million contract to Westinghouse for 24 systems. Delivery of the Lot 2 systems was scheduled to be completed by May 1994.

In mid-December 1992, the USN formally terminated the ALQ-165 programme on the basis that the system, according to the US Department of Defense, was 'not operationally suitable' to meet the service's requirements. The ASPJ Joint Venture team considers this to have been a politically driven decision. In a rather bizarre turn of events, the USN has gone on to outfit the Marine Corps F/A-18C/D aircraft of VMFAs -224, -251 and -533 with the system for operations over Bosnia during 1995 and 1996. In May 1996, ASPJ was further approved by the USN for unrestricted fleet use on the F-14D aircraft following the completion of an F-14 ASPJ Follow-On Test and Evaluation programme. The F-14D/ASPJ combination made its first carrier deployment (aboard USS *Carl Vinson*) during the same month and in the following August, the US Congress appropriated and authorised funding for additional ASPJ production (36 systems) for use on USN F/A-18C/D aircraft during US Fiscal Year 1997. Since its initial deployment in 1995, ASPJ has logged more than 40,000 operational flight hours on US F/A-18C/D and F-14D aircraft flying over Bosnia and the Persian Gulf.

ALQ-165 is approved for export sales and in September 1994, the Finnish Air Force signed a contract with the ASPJ Joint Venture to procure ASPJ for its F/A-18 aircraft. This was followed three months later by a decision from the Swiss Air Force to fit the ASPJ to its F/A-18 aircraft.

Finland's first ALQ-165 was delivered in November 1995 and the country has taken up a contractual option to procure additional equipments of the type. The first Swiss system was delivered during September 1996. On 24 January 1997, the ASPJ Joint Venture announced a third export sale, this time to South Korea. Under a contract valued at more than US$100 million, South Korea is procuring the system for use on its F-16 fleet. Like other export customers, Korea is obtaining its ASPJ hardware in a direct commercial sale with system software and integration being acquired via the US Foreign Military Sales procedure. In December 1997, the Northrop Grumman/ITT ALQ-165 Joint Venture was awarded a then year US$36 million contract to build 30 additional ALQ-165s for the USN and US Marine Corps. On 26 February 1999, the Republic of Korea Air Force accepted its first ASPJ-modified F-16. During October 2000, South Korea awarded the ALQ-165 joint venture team a follow-on contract for additional ASPJ jammers with which to equip its F-16s.

Contractors

Northrop Grumman Electronic Sensors and Systems Sector - Defensive Systems Division, Baltimore, Maryland.
ITT Avionics, Clifton, New Jersey.

UPDATED

AN/ALQ-167(V) radar jamming system

Type
Airborne noise and deception radar jamming system.

Description
AN/ALQ-167(V) is a modular noise and deception jamming system that is available in pod or internal configurations. The system is completely self-contained and provides all the receiving, processing and transmitting functions required to jam radar-directed fire-control systems. It's operating parameters can be preset on the ground or, with the addition of an RS-422 serial bus, controlled and updated in real time during a mission. Using the same bus format, ALQ-167(V) can also be directly interfaced with an onboard electronic support system and the system can be fitted with a crystal video receiver or a crystal video/superheterodyne receiver system if required. The addition of frequency memory loops or digital radio frequency memories enables the equipment to handle 'advanced' threats.

ALQ-167(V) provides continuous fore and aft coverage in the 0.25 to 40 GHz frequency range using multiple high-power travelling wave tube transmitters. These provide high-power pulse and continuous wave jamming. The system uses dedicated microprocessors for threat identification, countermeasures mode control and overall system control. Noise jamming modes available include continuous, spot, intermittent spot, swept amplitude modulation, continuous barrage, intermittent barrage noise swept amplitude modulation, swept noise and fast set on spot noise. For deception jamming, ALQ-167(V) provides multiple frequency repeater, repeater swept amplitude modulation, velocity gate stealer, combination of velocity gate stealer and repeater swept amplitude modulation, chirp gate stealer, narrowband repeater noise, random Doppler and noise Doppler.

Over time, examples of ALQ-167(V) configurations deployed include D-band (1 to 4 GHz) noise-only, E/F-band (2.8 to 3.8 GHz sub-band) noise-only, G- through I-band (5 to 6 GHz and 8 to 10 GHz sub-bands) noise and deception and J-band (12.5 to 17.5 GHz sub-band) noise and deception variants. Of these, the cited higher frequency configurations were responsive, facilitating, for example, the production of false Doppler returns. Effective radiated power for the described configurations included a value of 4 kW at I-band (8 to 10 GHz) and 2 kW at G-band (4 to 6 GHz).

Status
As of March 2001, ALQ-167(V) is understood to be in service with a number of operators including NATO's Multiservice Electronic Warfare Support Group, the

When applied to Royal Navy Lynx HAS Mk 3 helicopters, ALQ-167(V) is known as the Yellow Veil system (Michael J Gething)

civilian countermeasures training providers Phoenix Air and FR Aviation, the UK Royal Navy and the US Navy. As of late 1990s, more than 100 Condor Systems-sourced ALQ-167(V) systems were reported to have been in service worldwide. It is also worth noting that, over time, the US Navy, the UK's Royal Navy and the South Korean Air Force have all made use of ALQ-167(V) variants in an operational (as against training) role. Of these, the Royal Navy programme involved a dual-band (2 to 4 GHz and 6.5 to 11 GHz) pod configuration (designated as the Yellow Veil system) that was taken into service during 1990/1991 for use on Lynx HAS Mk 3 and Sea King helicopters, while usually reliable sources suggest that the South Korean programme also involved a podded application and was as recent as the 'late 1990s'.

Contractors

Condor Systems Electronic Systems Division, Simi Valley, California.
Rodale Electronics Inc, Hauppauge, New York.

UPDATED

AN/ALQ-172(V) Electronic CounterMeasures (ECM) system

Type
Airborne ECM system for bomber and C-130 special operations aircraft.

Description
AN/ALQ-172(V) is a combat proven ECM system that is installed aboard US Air Force (USAF) B-52 strategic bombers, Special Operations MC-130E/H Combat Talon I and II and AC-130H/U gunships. System design features include full automation, multiband coverage, simultaneous multiple threat recognition and jamming, digital computer control, advanced jamming techniques, high effective radiated power, threat reprogrammability, high-gain array antenna, threat warning display, dual MIL-STD-1553B databus interfaces and extensive built-in test capabilities. With the ability to jam multiple pulse, pulse-Doppler and continuous wave threats simultaneously, ALQ-172(V)2 (see following) can also counter monopulse radars. Organic logistics support and test equipment are in place to maintain the ALQ-172(V) system. An automated diagnostics system allows flight line crews to run a system self-test and obtain probable cause of failure down to the shop-replaceable unit level in minutes. As of August 2001, three variants of the basic design had been fielded, with the individual models being designated as ALQ-172(V)1, (V)2 and (V)3. Of these, ALQ-172(V)1 and (V)2 are fully automatic systems with the two variants differing from one another in the (V)2's use of phased-array antenna technology. ALQ-172(V)3 incorporates in-flight reprogrammability, extended frequency coverage (including both 'high' and 'low' band) and an improved threat handling capability.

ITT's ALQ-172(V) serves as the common ECM system of the USAF fleet of C-130 aircraft (MC-130E/H and AC-130H/U). The picture is of an MC-130E Combat Talon aircraft

Status

As of November 2001, the AN/ALQ-172(V) ECM system was reported as being in operational service. Starting in January 1993, ITT Avionics was awarded contracts totalling nearly then year US$70 million in connection with the development of the ALQ-172(V)3 system for use on the USAF's AC-130H gunship. In December 1995, ITT received a then year US$16 million award covering the initial phase of (V)3 production which was followed by a then year US$27 million production contract in July 1996. An additional then year US$10 million production option was exercised in March 1997. In the Spring of 1995, the then Chrysler Technologies (now part of Raytheon) was awarded a then year US$8.8 million contract covering the development of an ALQ-172(V)3 installation kit for the AC-130H. Elsewhere in the US inventory, over 6,000 ALQ-172(V)1 and (V)2 jammers are reported to have been supplied to the service over a 10 year period. ALQ-172(V)1 is noted as being installed aboard USAF AC-130U and MC-130E and H aircraft. On 11 September 1998, ITT Avionics was awarded a then year US$17,811,880 contract covering upgrading of ALQ-172(V)1 (ECP-93) standard equipments with expanded memories, additional channels and flight-line reprogrammability. Three years later (on 7 August 2001), ITT was awarded a then year US$13,437,583 (ceiling) firm, fixed-price, time and materials contract covering the upgrading of ALQ-172(V)2 systems installed aboard B-52H strategic bombers. Specifically, the contract required delivery of 12 aircraft installation kits, five spare kits, an upgrade kit for an associated AN/ALM-252 'hot' mock-up, an upgrade kit for an associated integrated support station, an upgrade kit for an associated semi-automatic test station, an upgrade kit for an associated weapon system trainer and 'associated data'. At the time of the announcement, work on the effort was scheduled for completion by the end of August 2003 and the programme's contracting activity was the USAF's Warner-Robins Air Logistics Center, Robins Air Force Base (AFB), Georgia. Elsewhere in the programme, 21 September 2001 saw the then Litton Systems (now part of Northrop Grumman) being awarded a then year US$5,007,800 firm, fixed-price contract covering the supply of four line items for use as spares for the USAF's ALQ-172(V) inventory. The USAF's Warner-Robins Air Logistics Center, Robins AFB, Georgia acted as the effort's contracting activity.

Contractor

ITT Avionics, Clifton, New Jersey.

UPDATED

AN/ALQ-178(V) radar warning and electronic countermeasures suite

Type

Airborne radar warning and jamming system.

Description

ALQ-178(V) is an integrated radar warning and active countermeasures suite for tactical fighter aircraft, such as the F-16. An earlier version of the system was known as RAPPORT III. The system utilises a central programmable computer for data analysis and system control, with independent microprocessors to direct the Radar Warning Receiver (RWR), display, jamming and countermeasures dispensing functions. The wideband superheterodyne RWR continuously scans the threat radar environment. Detected signals are de-interleaved, identified as to radar type and displayed to the pilot in a legible, unambiguous format. The power management algorithm optimally matches the countermeasures to the RWR's constantly changing threat picture. Separate directional right/left, forward and aft jammers are used for maximum spatial coverage. The jammer's output power is accurately set for maximum power output at each victim radar's frequency to maximise effective radiated power and jammer effectiveness. The jammer frequency range provides coverage of the entire threat band, while the jammer power is sufficient to counter these radars throughout their lethal area. The countermeasures dispenser is automatically controlled, including the pilot's display and cues, to provide optimum response to the threat between active jamming and chaff dispensing.

Although the ALQ-178(V) consists of an RWR, an active jammer and countermeasures control, the RWR can be installed as an independent threat warning system (controlling a countermeasures dispensing system), with the jammer added later. The latest derivative of the ALQ-178(V) family incorporates technology advances and state-of-the-art packaging to increase system capability significantly. Enhanced performance is derived from the introduction of an agile channel RWR receiver; digital radio frequency memory technology; agile jamming channels; distributed high speed/capacity processors and precision direction-finding.

A follow-on to ALQ-178(V), combined with the AN/ALR-56M or other RWR, is known as the AN/ALQ-202(V) autonomous jammer (see separate entry).

Status

ALQ-178(V)3 was selected by Turkey in January 1989 for its F-16 aircraft and, in October 1989, it was announced that Turkey had awarded a contract worth then year US$325 million for supply of systems, test and support and ancillary equipment. Deliveries ran from October 1991 to early 1996. Lockheed Martin (the then ALQ-178 contractor - see following) formed the MIKrodalgo Electronik Sistmeleri (MIKES) joint venture to produce, in the first instance, the ALQ-178(V)3 in Turkey after initial deliveries from the USA. The ALQ-178(V)1 is fitted to F-16s of the Israel Air Force with the threat warning subsystem being built under licence in Israel by Elisra. Jane's sources suggest that the Israeli application may cover aircraft up to batch 3 F-16C/Ds and excludes the country's F-16D-30 aircraft. ALQ-178(V)1 and (V)3 differ in the former incorporating a set of crystal video direction-finding receivers that have been eliminated from the latter. In the latest known move relating to the Turkish programme, the Turkish Air Force announced its decision to procure 80 MIKES assembled ALQ-178(V)5 systems for installation on its second batch of F-16 fighters during October 1998. Of the 80 systems acquired, 20 were configured as receive-only equipments and were not be fitted with the ALQ-178(V)'s jamming subsystem.

Contractor

BAE Systems North America - Threat Warning and Defensive Systems, Yonkers, New York.

VERIFIED

AN/ALQ-184(V) radar jammer

Type

Airborne pod-mounted radar jamming system.

Description

The AN/ALQ-184 Electronic CounterMeasures (ECM) pod is designed to provide electronic protection against surface-to-air missiles, radar-directed gun systems and airborne interceptors. As an upgrade of AN/ALQ-119(V), the new system increases effective radiated power, reduces ECM response time and improves reliability. Programmable operations with a digital microprocessor-controlled system provide mission flexibility.

ALQ-184(V) retains the same dimensions as ALQ-119(V), provides contiguous sub-band frequency coverage and is able to generate transponder, repeater and noise jamming modes. The continued updating of the system has resulted in the ALQ-184(V)5 standard. Earlier models of the system are understood to have been upgraded to this standard. The key to ALQ-184(V) function is the Raytheon E-Systems multibeam system that uses Rotman lenses, high-gain antennas, medium-power mini-Travelling Wave Tubes (TWTs), crystal video receivers and

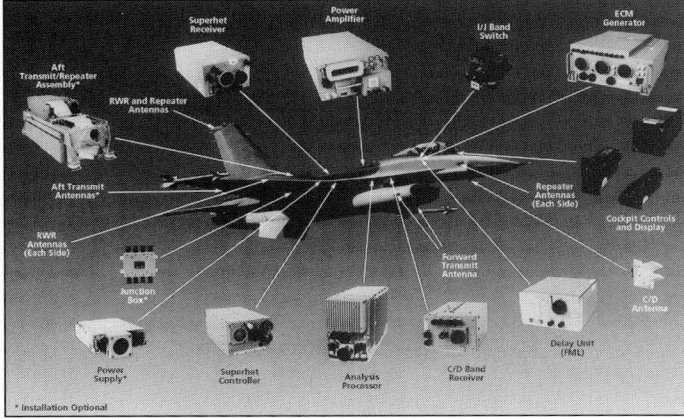

The elements which make up BAE Systems' ALQ-178(V) ECM system

An ALQ-184(V) jamming pod being installed on a USAF F-16

signal processors. In the transponder and noise modes an internally generated signal is selected from the voltage control oscillator assembly. This signal is modulated by the techniques generator and fed through switches and mini-TWT amplifiers to the correct antenna array for retransmission. Repeater loop delay is reduced as the antennas, Radio Frequency (RF) switches, channelisers and Direction-Finding (DF) receivers are located in proximity to minimise RF path length between receiver and transmitter arrays. ALQ-184(V) does not transmit until preset pulse-count thresholds (programmed by sub-band) are exceeded. Overall, ALQ-184(V) offers the following features:

- instantaneous RF signal processing wide open in angle and frequency
- DF on every received signal, independent of frequency
- selective directional high-power effective radiated power against multiple emitters
- full 100 per cent duty cycle transmit capability
- ECM techniques custom-designed for maximum effectiveness against engaged emitters
- rapid retrodirective automatic ECM response.

Functionally, once a received signal is determined to be a threat (by comparing the signal parameters to those in the library), ALQ-184 determines the correct response, based on preprogrammed rules. It automatically initiates the necessary countermeasures in real time. A pod transmitter comprising identical transmit assemblies accomplishes both forward and aft ECM coverage. RF drive signals are amplified by solid-state predriver amplifiers, connected to corresponding beam switches forward and aft and to the transmitter assemblies. Each transmitter provides a single combiner mid/high-band channel. The transmit switches select the transmission angle (or beam) to be transmitted. Rotman lenses provide the correct phasing and feed the medium-power mini-TWTs to the antenna array elements. In the repeater mode, the signal is retrodirectively retransmitted to the threat radar with the selected appropriate deceptive modulation(s). ALQ-184(V) radar jamming pods can be configured as two- ('mid/high') or three-band ('low' and 'mid/high') systems. Of these, the two-band configuration weighs 210 kg while the three-band variant reaches the 289 kg mark. Mean time between failure values for the two configurations are 150 hours and 80 hours respectively. As of this edition, Raytheon Systems has also incorporated its AN/ALE-50(V) towed radar decoy (see separate entry) into the three-band ALQ-184(V) configuration. Designated as the ALQ-184(V)9, this application incorporates a four-shot ALE-50 launch module at the rear of its lower gondola structure.

Status
As of this edition, the AN/ALQ-184(V) ECM pod was reported as being in service with the US Air Force (USAF) and the US Air National Guard (including 10 ALQ-184(V)9 units) and has been used operationally during the 1990–91 Gulf War and more recently over Bosnia, Kosovo/Serbia and Iraq. ALQ-184 achieved initial operating capability in 1988 and as of this edition, at least 850 ALQ-184(V) pods were reported as having been procured by the USAF. Outside America, Taiwan is thought to be acquiring at least 80 ALQ-184 pods for use on its F-16 aircraft. On 10 April 2000, Raytheon Electronic Systems was awarded a then year US$5,437,361 modification to an existing cost-plus, fixed-fee contract that covered the supply of 298 modification kits designed to upgrade the ALQ-184(V)7 and (V)8 ECM pods used on USAF A-10, F-15 and F-16 aircraft to ALQ-184(V)11 standard. At the time of the modification's announcement, 149 of these kits were intended for depot installation, with the remainder to be field installed. Work on the effort was scheduled for completion by the end of September 2001 and the programme's contracting activity was the USAF's Warner Robins Air Logistics Center, Robins Air Force Base, Georgia.

Contractor
Raytheon Electronic Systems, Goleta, California.

UPDATED

The ALQ-187 countermeasures system

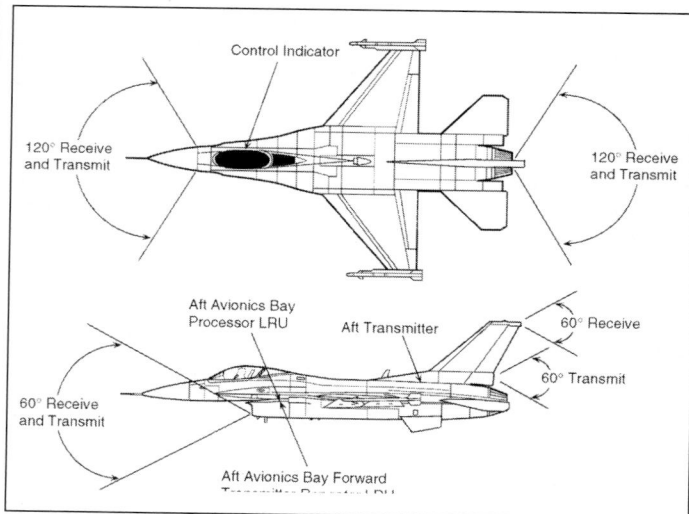

A schematic showing the layout of the ALQ-187 system when applied to the F-16
0017868

Status
As of this edition, a country specific variant of AN/ALQ-187 (designated as DIAS) was reported as having been procured by Greece for installation aboard A-7, F-4E, F-16 and RF-4E aircraft.

Specifications
Frequency range: 6.5-18 GHz
Jamming modulations: bandwidth selectable spot noise; countdown; Doppler false target; false scan/false lobe; inverse gain; keeper; range false target, range gate pull off, swept audio, swept spot noise; time sequence; velocity gate pull off and WPWO
Weight: 7 kg

Contractor
Raytheon Electronic Systems, Goleta, California.

UPDATED

AN/ALQ-187 detector-jammer

Type
Airborne 6.5 to 18 GHz band detector-jammer.

Description
The AN/ALQ-187 is an internally mounted, fully automatic, radar detector-jammer that operates in the 6.5 to 18 GHz frequency band and can be integrated with an onboard radar warning receiver and countermeasures dispensing system to create a defensive aids suite. ALQ-187 comprises a system control processor, a forward transmitter/repeater, an aft transmitter, a cockpit control/indicator and reception and transmission antenna arrays. System options are a data recorder and an integrated control/indicator for applications where a stand-alone unit is not required. Of these various subsystems, the forward transmitter/repeater receives, amplifies, modulates and transmits radio frequency signals in the forward hemisphere while the cockpit control/indicator unit displays system 'go/no-go' status, system faults, threat environment data and transmitter status. The system's transmit and receive antenna arrays both cover the full 6.5 to 18 GHz frequency range. Functionally, this software programmable system automatically detects radar threats and selects an appropriate response according to whether the threat is a ground-based, shipboard or airborne continuous wave, pulse or pulse-Doppler emitter. ALQ-187 is flight line reprogrammable and makes use of power management techniques to ensure maximum jamming effectiveness.

AN/ALQ-202 autonomous radar jammer

Type
Airborne autonomous radar jamming system.

Description
The AN/ALQ-202 radar jammer is described by its manufacturer as being an 'advanced, autonomous, Electronic CounterMeasures (ECM)' equipment that is designed for internal installation aboard F-16, F/A-18 and other types of combat aircraft. The system provides automatic and prioritised ECM responses to what is termed the full spectrum of airborne and ground-based radar threats. ALQ-202 is fully integrated with its host's avionic systems including its Radar Warning Receiver (RWR), fire-control radar, CounterMeasures Dispensing System (CMDS) and mission computer. The system interfaces used are designed for non-interference in order to facilitate fully compatible operational performance of all onboard avionics. Integration with the CMDS allows for automated and co-ordinated active and passive countermeasure responses and ALQ-202 can operate as a stand-alone unit or be interfaced with a range of RWRs including the AN/ALR-56M, AN/ALR-67 and AN/ALR-69 units (see separate entries).

ALQ-202 incorporates an internal Jammer Support Receiver (JSR) that provides a rapid and independent threat detection capability against pulse and continuous wave emitters. Functionally, the JSR automatically scans through a programmed frequency range of interest with the system as a whole determining direction of

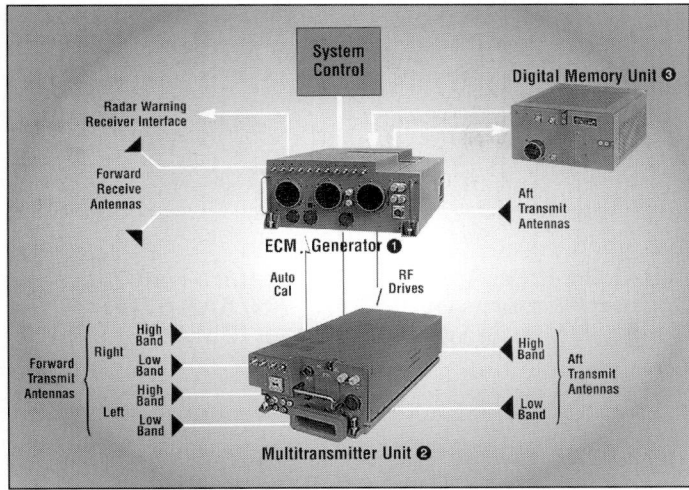

A schematic showing the elements that make up the ALQ-202 autonomous radar jamming system

0009915

arrival for each identified threat and selecting an automatically initiated, threat optimised, directional response from a multilevel list of prioritised ECM techniques. ALQ-202 continually searches for new threats and updates system information on those previously detected. The system utilises an extensive library of proven deception and denial techniques and the jamming output is of the transponder and repeater types (including continuous and time-gated responses).

The ALQ-202 system comprises an ECM generator, a MultiTransmitter Unit (MTU) or units, a Digital Radio Frequency Memory (DRFM) and system control unit. The ECM generator houses the JSR and generates the necessary jamming techniques while the DRFM is provided to facilitate an additional techniques capability. The MTUs are equipped with latest technology travelling wave tubes to improve efficiency and reliability and reduce the thermal noise level to ensure non-interference and compatibility with other onboard systems.

Contractor
BAE Systems North America - Threat Warning and Defensive Systems, Yonkers, New York.

VERIFIED

AN/ALQ-204(V) directional infra-red jammer (LANCIR)

Type
Airborne self-protection jamming system for use against Infra-Red (IR) guided missiles.

Description
LANCIR is an upgrade version of Lockheed Martin's previously produced/deployed ALQ-204(V) MATADOR IR CounterMeasures (IRCM) system. Comprising steerable transmitters, a power supply, a control unit, an EMI filter and a controller, LANCIR upgrades MATADOR to a directional IRCM configuration that provides higher peak jamming radiant intensity levels for additional and 'more advanced' IR missile threat protection. LANCIR maintains MATADOR's US Federal Aviation Administration certification lineage and is specifically designed for Head-of-State/VIP and civilian aircraft applications. LANCIR is offered as a MATADOR upgrade kit or as a stand-alone new-build system and can be integrated with 'any of today's' imaging missile warning systems that are capable of providing accurate angle of arrival data.

Status
As of this edition, the status of the LANCIR IRCM system was uncertain.

Contractor
Lockheed Martine Information Systems, Electro-Optical Systems Division, Pomona, California.

VERIFIED

AN/ALQ-204(V) Infra-Red (IR) jammer (MATADOR)

Type
Self-protection jamming system against IR guided missiles.

Description
AN/ALQ-204(V) (which is also known as MATADOR) is a modular IR countermeasures system that has been offered in 11 different configurations to suit particular airframe applications. The system is suitable for all types of jet transport

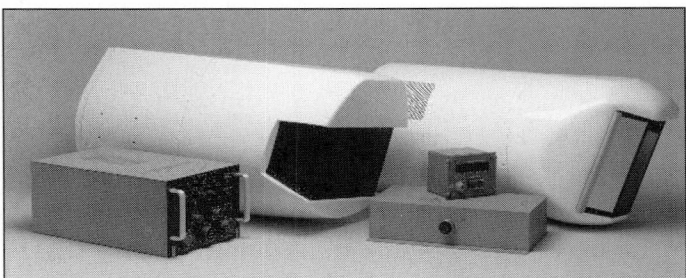

ALQ-204(V) system components

aircraft with unsuppressed engines and provides 360° azimuth coverage. The basic equipment consists of multiple transmitters, a controller unit and an operator's controller. Transmitters are electronically synchronised by the controller unit that controls and monitors up to two transmitters. The operator's controller (which is common to all configurations) controls from 1 to 12 transmitters and incorporates a system status display. Each transmitter contains a 4 to 12 kW IR source that emits pulsed radiation to combat multi-IR guided missiles. Preprogammed multithreat jamming codes are provided, selectable on the operator's control unit and all new codes can be entered as required to cope with new threats. All MATADOR hardware is US Federal Aviation Administration certified for civilian and commercial aircraft installation and operation.

Status
As of this edition, AN/ALQ-204(V)/MATADOR series equipments were reported as being in service aboard at least 20 Head of State/VIP platforms. Designated as the ALQ-204(V)1 in US Air Force service, current operational installations include Boeing 707 and 747 (including US 'Air Force One') aircraft, Lockheed Martin L-1011s, Airbus A340s, BAe-146s (UK Queen's flight) and Gulfstream IVs.

Contractor
Lockheed Martin Information Systems, Electro-Optical Systems Division, Pomona, California.

VERIFIED

AN/ALQ-211 Suite of Integrated Radio Frequency Countermeasures (SIRFC)

Type
Fully integrated passive and active electronic combat system for rotary-wing and special operations aircraft.

Description
The AN/ALQ-211 SIRFC system's wide frequency range, high-precision direction-finding, threat geolocation and electronic countermeasures response capabilities provide threat warning and self-protection to aircrews. SIRFC's multispectral (radio frequency, laser and infra-red) sensor data fusion capability provides comprehensive situational assessment and awareness that provides aircrews with a real-time picture of the air defence threats on the battlefield. The same information is transmitted off board (via datalinks), to ground commanders and other airborne platforms. ALQ-211 provides an embedded capability directly supportive of battlefield digitisation initiatives. The system allows aviators to evade or defeat a diversity of airborne and ground-based threats and has been designed to enhance the survivability and lethality of rotary-wing and special operations aircraft well into the 21st century. SIRFC's in-flight reprogrammability provides the

The type of real-time picture of the battle area expected to be produced by the SIRFC system

SIRFC makes extensive use of SEM-E modules and application specific and monolithic microwave integrated circuitry

combat pilot with the capability to meet changing threat scenarios instantaneously. MIL-STD-1553B and high-speed databus structures facilitate system integration and pilot interface. ALQ-211 also features extensive built-in test and the system makes use of standard processors and incorporates the extensive use of both application specific and monolithic microwave integrated circuits. Packing techniques include the use of Standard Electronics Modules (SEM-E) and composite materials. The complementary use of these technologies allows SIRFC to achieve its weight requirement of approximately 45 kg while increasing system performance. ALQ-211's flexible design provides for installation on a variety of platforms with the AH-64D, MH-47E, MH-60K and 'some modernised' UH-60 helicopters and the CV-22 tilt-rotor air vehicle scheduled to be fitted with the equipment.

Status

In September 1990, the US Army awarded ITT Avionics a then year US$18 million cost plus incentive fee contract for the advanced development of the AN/ALQ-211 SIRFC system. The SIRFC advanced development model completed successful flight tests in December 1993. SIRFC is reported to have met and in most cases exceeded, its performance criteria during this phase of the programme. In July 1994, ITT Avionics was awarded a then year US$54.2 million contract by the US Army for the engineering and manufacturing development of ALQ-211. The 48-month contract was captured in a competitive solicitation by ITT Avionics. During 1995, situational awareness algorithms central to SIRFC were demonstrated on a Black Hawk helicopter as part of a US Army Advanced Technology Demonstration programme. During 1997, ITT was awarded a then year US$25.2 million contract by Bell Helicopter Textron to integrate ALQ-211 on the CV-22 Osprey. Here, SIRFC will fulfill the role of CV-22 sensor fusion processor and is intended to fuse and integrate multispectral, on and off board, sensor-derived data and provide the pilot with a real-time, comprehensive picture of the battlefield. ITT delivered the first ALQ-211 Engineering and Development Model (EDM) to the US Army in August 1998, with a second EDM unit being supplied during the following October. On 23 March 1999, SIRFC made its maiden flight aboard an AH-64D at Boeing's Mesa, Arizona plant, with system flight testing (42 sorties) taking place during the period July to October 1999. Elsewhere in the world, August 2000 saw Australia procure an ALQ-211 system for use in the Australian-US Advanced Integrated Aircraft Survivability Technology (AIAST) programme while in the following October, the Boeing Company selected ITT Avionics to provide an ALQ-211-based Integrated Survivability System (ISS) for the US Army's next-generation Comanche attack and reconnaissance helicopter. Here, the ISS is built around the ALQ-211(V)3 radar warning receiver and as scheduled at the time of its announcement, the ISS programme's engineering and manufacturing development phase was to encompass the provision of 14 ISS suites made of the ALQ-211(V)3, a laser warning receiver system and a point chemical detector. For ALQ-211(V)3 development, ITT is understood to be teamed with BAE Systems North America. As of May 2001, the latest identified programme activity was ITT Avionic's receipt of a then year US$24,314,292 time and materials contract covering the provision of engineering and support services (including studies and evaluation work) for the ALQ-211(V) effort. At the time of the announcement, work on the contract was scheduled for

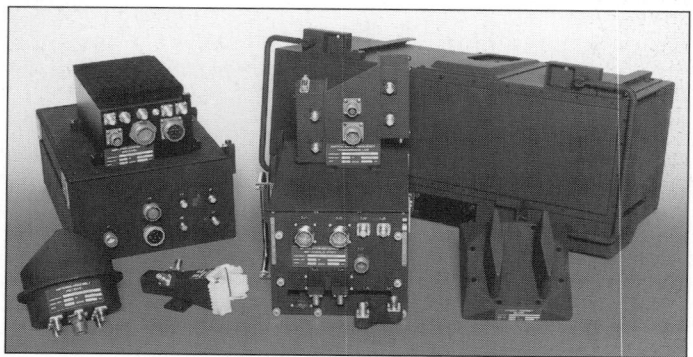

The elements making up the ALQ-211 SIRFC system   0017869

completion by 19 April 2006, with its contracting activity being the US Army's Communications-Electronics Command, Fort Monmouth, New Jersey.

Contractor

ITT Avionics, Clifton, New Jersey.

UPDATED

AN/ALQ-212(V) Advanced Threat Infra-Red CounterMeasures (ATIRCM) — AN/AAR-57(V) Common Missile Warning System (CMWS)

Type

Directable IRCM system.

Description

The AN/ALQ-212(V) system is an integrated missile warning, IR jamming and CounterMeasures Dispensing System (CMDS) suite that comprises:

• a missile warning subsystem Electronic Control Unit (ECU)
• four AN/AAR-57(V) variant CMWS sensor heads
• two steerable jamming heads
• a Jamming Head Control Unit/Processor (JHCU/P)
• a laser jamming source
• an AN/ALE-47 CMDS sequencer unit
• a system specific number of CMDS dispenser modules.

Functionally, ALQ-212(V) generates a co-ordinated, multispectral response from its directional jamming heads and CMDS dispenser modules when cued by its integral missile warning subsystem. This latter element takes the form of a variant of the AAR-57(V) CMWS that was also designed for tri-service use on US military helicopters, tilt-rotor and fixed-wing aircraft together with the UK's WAH-64D and Nimrod MRA Mk 4 platforms. Within ALQ-212(V), the AAR-57(V) subsystem detects the launch of threat missiles and generates accurate direction of arrival data (to 'within a few degrees') with which to cue the suite's jamming heads. Once targeted, each jamming head focuses a modulated, deceptive waveform, jamming beam onto the seeker of the approaching threat missile. ALQ-212(V) is designed to handle 'all currently operational' IR guided missiles and has integral growth potential to maintain its effectiveness against future threats.

Status

A prototype AN/ALQ-212(V) jamming subsystem underwent live fire trials during 1994 and in September 1995, the US Army awarded an industrial team led by the then Sanders (now BAE Systems North America - Information and Electronic Warfare Systems) a then year US$98 million ALQ-212(V) suite and AAR-57(V) CMWS engineering and development contract. Of four years duration, this award included the delivery of six ALQ-212(V) suites and 50 examples of the AAR-57(V) CMWS. ALQ-212(V) is reported to have passed its preliminary design review in June 1996 and in the following September, demonstrated its ability to detect, track and counter IR guided missiles during a test programme at the US Army's White Sands Missile Range (new Mexico). ALQ-212(V)'s critical design review is understood to have taken place during February 1997 and, on 28 October 1999, the then Sanders was awarded a modification to an existing cost-sharing contract (DAAB07-95-C-D606) that covered contract restructuring of the ALQ-212(V)'s engineering and manufacturing development programme. At the time of the announcement, the estimated cumulative value of DAAB07-95-C-D606 was put at then year US$171,783,577, with the US Army's Communications and Electronics

A general view of the ALQ-212(V) ATIRCM jamming head and one of its associated AAR-57 CMWS sensors (BAE Systems North America)
2002/0059921

Command (Fort Monmouth, New Jersey) acting as the programme's contracting activity. During the period 4 to 24 April 2001, both the ATIRCM jammer and the CMWS were subject to a series of ten live fire tests at the White Sands Missile Range, during which, the equipment was faced with single and multiple missiles fired at them from differing angles and ranges. BAE Systems North America reports that this trials round (a 'major' phase of the ALQ-212(V)'s Development Test/ Operational Test (DT/OT) programme) was completed 'successfully'. As of mid 2000, Jane's sources were suggesting that the system's first operational application was likely to be the US Army's AH-64 attack helicopter. With regard to the AAR-57(V) CMWS, BAE Systems suggests that the system is potentially applicable to US Air Force A-10, C-5, C-17, C-130, C-141, CV-22, F-15E and F-16 aircraft; US Army AH-64D, CH/MH-47, MH-60K/L, OH-58D and UH/EH-60 helicopters and US Navy/Marine Corps AH-1W, AV-8B, CH-60, F/A-18, MV-22, SH-60R and UH-1N aircraft. As noted previously, the system has also been selected for use on the UK's Nimrod MRA Mk 4 and WAH-64D platforms.

Specifications
AN/AAR-57(V)

LRU	No	Power	Dimensions	Weight
ECU	1	281 W	13 (h) × 23 (w) × 28 (d) cm	8.5 kg
Sensor head	up to 6	10 W (without anti-ice); 26 W (with anti-ice)	08 (Ø) × 11 (d) cm	1.4 kg (each)

ALQ-212(V)

LRU	No	Power	Dimensions	Weight
ECU	1	327 W	13 (h) × 23 (w) × 33 (d) cm	8.2 kg
Sensor head	4	111 W	10 (Ø) × 10 (d) cm	1.4 kg (each); 5.6 kg (total fit)
JHCU/P	1	330 W	20 (h) × 18 (w) × 23 (d) cm	5.9 kg
Jamming head	2	1,550 W	36 (h) × 23 (w) × 23 (d) cm	14.5 kg (each); 29 kg (total fit)
Laser source	1	400 W	18 (h) × 18 (w) × 46 (d) cm	9.5 kg

Contractors

BAE Systems North America - Information and Electronic Warfare Systems, Nashua, New Hampshire (prime).

UPDATED

....

AN/ALQ-214(V) Integrated Defensive Electronic CounterMeasures (IDECM) Radio Frequency CounterMeasures (RFCM) system

Type
Airborne deception countermeasures system.

Description

The AN/ALQ-214(V) IDECM RFCM system is a US Navy (USN) lead, joint USN/US Air Force (USAF) programme which is designed to provide a range of aircraft types with next-generation protection from RF threats. The system's primary application is the F/A-18E/F carrierborne, multirole combat aircraft where it is teamed with a radar warning receiver, the AN/AAR-57 Common Missile Warning System (CMWS) and the Advanced Strategic/Tactical Expendable (ASTE) Infra-Red (IR) decoy flare to create a total defensive aids package. Here, system design objectives include improved situational awareness through the use of the CMWS and the fusion of *a priori* and real-time information; improved IR threat countering through the use of the kinetic ASTE round and 'smart' dispensing routines and a counter RF capability which can defeat radars which incorporate features such as monopulse angle tracking, signal coherency and man-in-the-loop tracking.

The IDECM RFCM system comprises a Techniques Generator (TG), an Independent WideBand Repeater (IWBR) and AN/ALE-55 Fibre Optic Towed

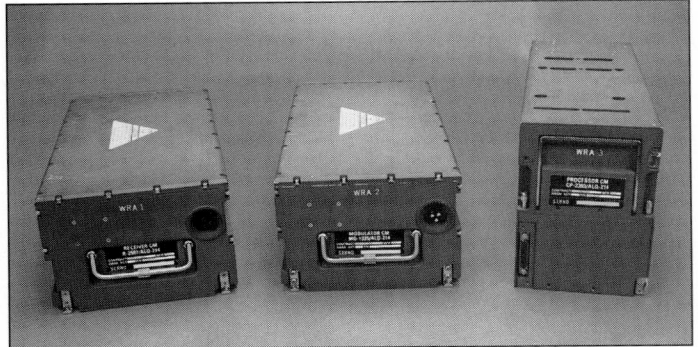

The three weapon-replaceable assemblies that make up the ALQ-214's TG

0017870

The ALE-55 FOTD used in the IDECM RFCM system

Decoys (FOTD). Both the TG and the IWBR interface with the FOTD through the decoy dispensing system. Electronic CounterMeasures (ECM) techniques are synthesised in the TG and transduced to optical frequencies for transmission to the FOTD via its tow line. Within the FOTD, the optical data is converted back to RF format for amplification and transmission. The IWBR provides an alternative source of ECM by passing the threat signal (as received on the host aircraft) to the FOTD.

As noted above, the IDECM RFCM system is mandated for the USN F/A-18E/F and the USAF B-1B and is under consideration for application to the F-15 aircraft. Additionally, the IDECM RFCM system is understood to be under study for possible application to a wider range of types including the USAF's AC/MC-130, F-16 and U-2 aircraft. As installed in the F/A-18E/F, the IDECM RFCM system makes maximum use of the existing AN/ALE-50 and AN/ALQ-165 Group A provisioning, with the described TG and IWBR functions being implemented in four weapon-replaceable assemblies; a receiver, modulator, processor and signal conditioning unit. It should also be noted that onboard transmitters are an option for the Hornet application.

Status

In November 1996, the contractor team of BAE Systems North America - Information and Electronic Warfare Systems (formerly Sanders) and ITT Avionics announced that BAE Systems (as prime) had been awarded a then year US$26.8 million, five year duration, IDECM RFCM system Engineering and Manufacturing Development (EMD) contract. Here, the contractor team was mandated to deliver and integrate 15 TGs (developed by ITT Avionics), 50 FOTDs and 50 FOTD mass models (developed by BAE Systems) for use in development flight testing. Under a contract option, work was also to be done on a common, high-powered Towed Decoy System design, development and test effort for the USAF, together with a B-1B architecture study to determine how the IDECM RFCM system could support the type's Defensive System Upgrade programme. During the period 14 to 19 April 1999, BAE Systems undertook the first ALE-55 FOTD airborne fast deployments (using Learjet and Saab Draken testbed aircraft) and in the following May, the company delivered the first six EMD FOTD devices. On 22 June 1999, the US Navy successfully completed the first flight test of the IDECM RFCM with an FOTD using an F/A-18D avionics test aircraft that had been equipped with F/A-18E/F systems. During this flight, the FOTD was reported as having been successfully deployed and the IDECM system as a whole as correctly detecting, identifying and radiating against prioritised threats. On 24 November 1999, the IDECM team delivered Block 4 IDECM software packages to Boeing's F/A-18E/F integration laboratory, BAE Systems' B-1B Defensive Systems Upgrade Programme (DSUP) integration laboratory and to the USN's Naval Air Weapon Center's Weapons Division Laboratory. The Block 4 standard software was described as forming the 'foundation' of the iteration that will be used during the summer 2000 IDECM operational evaluation effort. During September 2000, ITT announced that ALQ-214(V) had successfully completed its USN Operational Assessment test and was scheduled to begin Operational Evaluation during 2001. On 22 March 2001, BAE Systems North America - Information and Electronic Warfare Systems was awarded a then year US$59,003,220 firm, fixed-price contract covering the fabrication, assembly, test and delivery of six low-rate initial production examples of the ALQ-214(V) RFCM unit and 30 ALE-55 FOTD. At the time of the announcement, work on the effort was scheduled for completion by the end of November 2003, with the programme's contracting activity being the USN's Naval Air Systems Command (Naval Air Station Patuxent River, Maryland).

Specifications

WRA	Size (m³)	Weight (kg)	Power (W)
Receiver	0.01	14.8	395
Modulator	0.01	14.1	234
Processor	0.01	15.6	444
FOTD		4.5	

Contractors

BAE Systems North America - Information and Electronic Warfare Systems, Nashua, New Hampshire (prime).

ITT Avionics, Clifton, New Jersey (team partner).

Ortel Corporation, Alhambra, California (linear fibre-optic technology).

UPDATED

Apollo Defensive Aids Suite (DAS)

Type

Off-the-shelf DAS for transport aircraft.

Description

The Apollo DAS is an early- to mid-1990s vintage 'quick fix' defensive aids package for Lockheed Martin C-130 operators that used off-the-shelf US and/or locally sourced subsystems and was obtainable, subject to State Department approval, under the US Foreign Military Sales programme. A typical Apollo package was customer specific and usually incorporated radar and/or missile warners (including the AN/AAR-44 and -47, AN/ALQ-156, AN/APR-39(V) and AN/ALR-56M equipments) together with a dispenser subsystem (typically, AN/ALE-40 or an ALE-47 variant). The programme contractor (the then Lockheed Martin Aircraft Service (LMAS) Company) designed the particular suite around the selected subsystems, installed it and fabricated, where necessary, items such as radomes, sensor head fairings and equipment racking. Individual Apollo installation efforts took from three weeks to three months to complete depending on complexity and the urgency of the requirement.

Status

The Apollo effort is understood to have begun during 1992 and by July 1993, Jane's sources suggest that LMAS had completed then year US$12 million worth of Apollo work involving 10 aircraft installations and the provision of 11 conversion kits for a single customer for local installation. As of this edition, Jane's sources were reporting that Apollo defensive aids suites had been fitted to aircraft from the following countries:

Australia	Four C-130H aircraft fitted with an Apollo suite made up of a Radar Warning Receiver (RWR), a Missile Approach Warner (MAW) and a chaff/infra-red decoy flare dispenser. Contract valued at US$10 million and reported to have been completed during 1994.
France	France may have been the recipient of the cited 11 Apollo conversion kits supplied to a customer during 1993.
Norway	A percentage of Norway's fleet of C-130H transport aircraft are reported to have been fitted with an Apollo suite which incorporates the ALE-40 dispenser system.
Sweden	Two Swedish Air Force C-130 aircraft have been fitted with an Apollo suite made up of an APR-39(V) RWR, the ALQ-156 MAW and a six magazine ALE-40 dispenser subsystem. Work carried out during the early 1990s.

A number of Norwegian C-130H aircraft have been fitted with an Apollo DAS (MS/LAS)

As of this edition, the latest identified Apollo programme activity included the suite's inclusion in a Lockheed Martin bid to upgrade five C-130B and 10 C-130H transports of the Hellenic Air Force and the introduction of the (V)2/(V3T) model of the ALQ-156 missile approach warner into the package.

UPDATED

BAE Systems North America chaff cartridges/ payloads and Infra-Red (IR) decoy flares

Type

Family of chaff cartridges/payloads and IR decoy flares.

Description

As of July 2001, BAE Systems North America - Integrated Defense Solutions was producing a range of chaff cartridges/ payloads and IR decoy flares, the known details of which are as follows:

M-206

The M-206 IR decoy flare is described as employing an extruded flare pellet, a lightweight extruded aluminium outer case and moulded plastic inert components and as offering a 1.25 s duration 'high' initial output with a regressive output thereafter. M-206 makes use of an electrically initiated M796 impulse cartridge and is compatible with AN/ALE-40, AN/ALE-45, AN/ALE-47, M-130 and 'similar' dispenser systems. Over time, the M-206 decoy flare has been carried by the A-10, AH-64, C-5, C-130, CH-47, F-15, RC-12 and RU-21 aircraft types.

Altitude envelope: sea level-15,240 m
Automatic ignition temperature: >149°C
Operating temperature: −10 to +77°C
Shelf life: >5 years
Dimensions: 25 × 25 × 208 mm
Weight: 200 g

Mk 46

Compatible with the AN/ALE-29A, AN/ALE-39, AN/ALE-47 and 'similar' dispenser systems, the Mk 46 IR decoy flare makes use of an electrically initiated CCU-63 impulse cartridge and an extruded flare pellet. The device has a specified burn time, pull-wire ignition and 'performs at all operational altitudes'. Over time, the Mk 46 decoy flare has been carried by the A-4, A-6, A-7, AV-8A, CH-46, CH-47, CH-53, F-4, F-14 and UH-1 aircraft types.

Automatic ignition temperature: >149°C
Operating temperature: −18 to +74°C
Shelf life: >5 years
Dimensions: 36 (Ø) × 148 mm
Weight: 273 g

MJU-7A/B and MJU-7/B

Capable of being tailored to specific requirements, the basic MJU-7 IR decoy flare employs an electrically initiated BBU-36 impulse cartridge, an extruded flare pellet and parasitic or interrupted type ignition. The cartridge is compatible with the AN/ALE-29A, AN/ALE-39, AN/ALE-47 and 'similar' dispenser systems and (over time) has been carried by the C-130, F-4, F-5, F-15, F-16, F-104 and 'other types' of NATO aircraft.

Automatic ignition temperature: >149°C
Operating temperature: −10 to +93°C
Shelf life: >5 years
Dimensions: 25 × 52 × 208 mm
Weight: 370 g

MJU-10/B

The MJU-10/B IR decoy flare makes use of an electrically initiated BBU-36 impulse cartridge and a pressed flare pellet and is claimed to have a 'high' initial output, a 'standard' burn time and a 'rapid' rise time. The device's output can be tailored to 'unique' operational requirements and it is compatible with the AN/ALE-40, AN/ALE-45, AN/ALE-47 and 'similar' dispenser systems. Over time, the MJU-10/B IR decoy flare has been carried by the C-17, F-15 and 'similar' types of 'high performance' aircraft.

Operating temperature: −18 to +93°C
Shelf life: >5 years
Dimensions: 52 × 65 × 208 mm
Weight: 1,020 g

MJU-47/B and MJU-48/B ASTE

The MJU-47/B and -48/B Advanced Strategic and Tactical Expendable (ASTE) IR decoy flares make up a 'multicomponent' decoy set that is designed to be effective against 'sophisticated' IR seekers. The devices are compatible with the AN/ALE-47 dispenser system and are noted as being suitable for all 'US fighter aircraft'.

Operating temperature: −18 to +74°C
Dimensions: 52 × 65 × 208 mm and 25 × 52 × 208 mm

RR-129/AL

Described as being the US Navy's (USN) standard tactical countermeasure radar reflector, the RR-129/AL chaff cartridge is described as covering the 2 to 6 and 8 to 10 GHz frequency bands using half-wavelength, aluminium-coated, glass monofilament fibre dipoles. The cartridge makes use of the electrically sequenced CCU-41/B impulse cartridge and is claimed to offer rapid bloom, multiple broadband frequency protection, no bird-nesting, 'high' reliability and an 'excellent' operational Radar Cross Section (RCS). RR-129/AL can be configured with specific dipole lengths to meet specific operational requirements and is compatible with the AN/ALE-29A, AN/ALE-37A, AN/ALE-39, AN/ALE-42, AN/ALE-47 and 'similar' dispenser systems. Aircraft types that have, over time, carried the cartridge include the A-4, A-6, A-7, AH-1, AV-8A, CH-46, CH-53, F-4, F-14, F/A-18 and UH-1.

Dimensions: 36 (Ø) × 148 mm
Weight: 134 g

RR-155/AL, RR-156/AL, RR-163/AL, RR-167/AL, RR-171/AL and RR-172/AL

The RR-155/AL, -156/AL, -163/AL, -167/AL, -171/AL and -172/AL roll chaff configurations are described as being standard USN and USAF formats that are claimed to offer a 'demonstrated' saturation capability, 'continuous' dipole dispersal, an 'excellent' operational RCS, 'instantaneous' bloom, multiple broadband frequency protection and 'high' reliability. Of the various configurations, RR-171/AL is described as covering the 2 to 6 and 10 to 20 GHz frequency bands

Dimensions: 32 (Ø) cm
Weight: 23.6 kg (typical)

RR-170

Described as being the US Air Force's standard tactical countermeasure radar reflector, the RR-170 chaff cartridge is reported as covering the 2 to 6 and 10 to 20 GHz frequency ranges using half-wavelength, aluminium-coated, glass monofilament fibre dipoles. The cartridge makes use of the electrically sequenced BBU-35 impulse cartridge and is claimed to offer rapid bloom, multiple broadband frequency protection, no bird-nesting, 'high' reliability and an 'excellent' operational RCS. RR-170 can be configured with specific dipole lengths to meet specific operational requirements and is compatible with the AN/ALE-45, AN/ALE-47 and 'similar' dispenser systems. The M-1 RR-170 variant is used by the US Army for helicopter protection. Aircraft types that have, over time, carried the cartridge include the A-7, A-10, F-4, F-5, F-15 and F-16.

Dimensions: 25 × 25 × 208 mm
Weight: 121 g

RR-179

RR-179 bundle chaff is described as covering the 0.03 to 60 GHz frequency band and as being made up of aluminium glass filaments in a continuous strand of 3,000 fibres. Here, eight, 18 kg rovings making up the payload of the AN/ALE-43 dispenser. RR-179 is claimed to offer a chaff corridor creation capability together with 'instantaneous' bloom, 'high' reliability, an 'excellent' operational RCS and continuous dipole dispersal. Over time, RR-179 has been carried by the EA-6, EC-24, F/A-18, NC-135 and SH-3 aircraft types.

Dimensions: 27 (Ø) × 31 cm
Weight: 19 kg

RR-180 Dual

The RR-180 Dual chaff cartridge is reported as offering 'double the payload capacity within the same aircraft space' and makes use of 'superfine' glass filaments. It can be configured with specific dipole lengths to meet specific operational requirements and is claimed to offer 'high' reliability, multiple broadband frequency protection, an 'excellent' operational RCS, rapid bloom and no bird-nesting. RR-180 Dual makes use of an electrically sequenced BBU-48 dual impulse cartridge that responds to separate firing signals and is compatible with the AN/ALE-40, AN/ALE-45, AN/ALE-47 and 'similar' dispenser systems.

Dimensions: 25 × 25 × 208 mm
Weight: 127 g

RR-184

Manufactured under licence from UK contractor the Chemring Group, the RR-184 payload is designed for use with the non-pyrotechnic BOL chaff dispensing system. RR-184 can be supplied with specific dipole lengths to meet specific operational requirements.

Dimensions: 84 × 71 × 8 mm
Weight: 39 g

Status

As of November 2001, BAE Systems North America - Integrated Defense Solutions - Integrated Survivability Systems was reported as being an active manufacturer of chaff cartridges and IR decoy flares. On 21 July 1999, BAE Systems North America - Integrated Defense Solutions (formerly Marconi Aerospace Defense Systems) was awarded a then year US$4,058,520 base year increment of a then year US$11,160,930 firm, fixed-price contract covering the supply of 53,368 MJU-10/B IR decoy flares. At the time of the increment's announcement, work on the effort was to have been undertaken at the company's East Camden, Arkansas facility and

was scheduled for completion on 28 February 2002. The programme's contracting activity was the US Army's Armament, Munitions and Chemical Command, Rock Island, Illinois. Subsequently, 27 August 2001 saw the company being awarded a then year US$7,986,600 increment of a then year US$15,973,200 firm, fixed-price contract covering the supply of a total of 234,900 MJU-7A/B IR decoy flares. At the time of the announcement, work on the particular tranche was scheduled for completion by 29 March 2003, with the US Army's Operations Command (Rock Island, Illinois) acting as the programme's contracting activity.

Contractor

BAE Systems North America - Integrated Defense Solutions - Integrated Survivability Systems, Austin, Texas.

UPDATED

Challenger Infra-Red CounterMeasures (IRCM) system

Type

Airborne IR self-protection system.

Description

Challenger is an IRCM system for use on small attack, utility and observation helicopters. It is a low-cost, lightweight, low-power system contained in an omnidirectional jammer assembly that is driven by a compact controller and power supply. The installation can comprise single- or dual- jammer heads, giving a wide variety of installation options. The Challenger series can also be fitted in fixed-aircraft apertures or deployed on retractable platforms.

Installation consists of one or two External Transmitter Units (ETUs) configured to cover up to 360° in azimuth, a Control Power Supply (CPS – which can power one or two transmitters with a selectable multithreat jamming code), a pilot control indicator, an electronic interface unit to provide an interface between the CPS and each ETU and an electromagnetic interference filter.

Status

Over time, the Challenger IRCM system is reported as having been procured by the Royal Navy and the Japanese Ground Self-Defence Force.

Contractor

Lockheed Martin Information Systems, Electro-Optical Systems Division, Pomona, California.

VERIFIED

A dual Challenger IRCM installation on a Royal Navy Lynx helicopter

DASS 2000 Electronic Warfare (EW) suite

Type

Defensive aids suite.

Description

DASS 2000 is a turnkey, fully integrated defensive aids suite that can be configured to meet specific customer requirements and is applicable to a wide range of platform types. The suite's long-range, multispectral threat detection and identification capabilities provide the aircrew with full situational awareness through sensor fusion and ensures the optimisation of countermeasure responses. Such capabilities are claimed to ensure platform survivability at all stages of the mission.

DASS 2000 resources typically include a radar warning receiver, missile warning system, laser warning system, chaff/flare dispensers, electronic countermeasures jammer, towed radar decoy and infra-red countermeasures. BAE Systems will

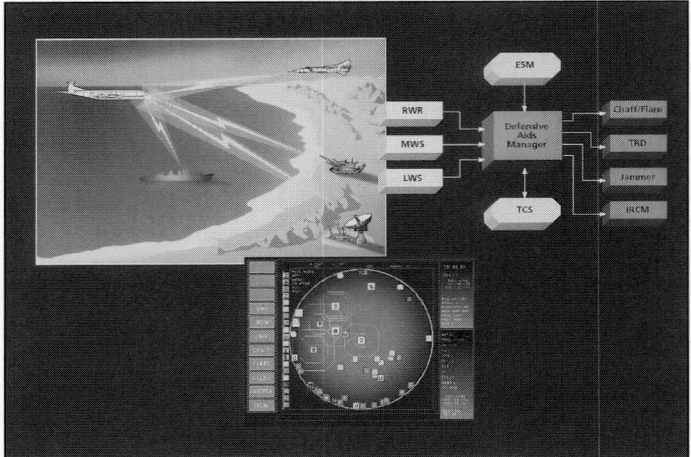

DASS 2000 is intended to provide maritime patrol with an effective, integrated EW capability

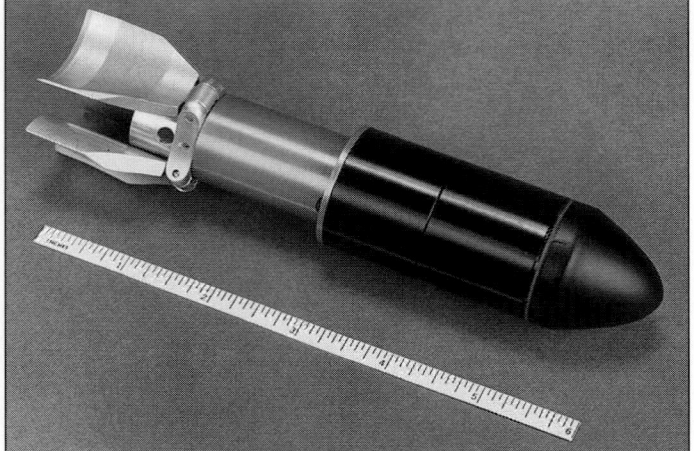

The GEN-X expendable active decoy

adapt the configuration to user requirements, integrate it into the platform ensuring full compatibility and provide a full range of training and logistics support.

DASS 2000 features user-friendly, interactive functions that are principally centred on a full colour display. This facilitates threat assessment together with an automatic countermeasures response.

Status
As of October 2001, the UK's Ministry of Defence was procuring a DASS 2000 type EW suite for use on the Nimrod MRA Mk 4 maritime patrol aircraft. Jane's sources suggest that the specific Nimrod application incorporates a BAE Systems North America AN/ALR-56M radar warning receiver (see separate entry), a fibre optic variant of Raytheon's AN/ALE-50 active towed radar decoy (see separate entry) and Thales Shrike onboard techniques generator (see separate entry). Sources also suggest that a DASS 2000 variant has been proposed for use on C-130H and -130J transport aircraft.

Contractor
BAE Systems North America - Threat Warning and Defensive Systems, Yonkers, New York.

UPDATED

....

Defendir Infra-Red (IR) jamming systems

Type
Family of IR jamming systems for rotary- and fixed-wing aircraft.

Description
Defendir is a family of systems that can be used for small and large helicopter applications, as well as turboprop/jet transport fixed-wing aircraft. They can be supplied in various configurations including an aft-mounted jammer for aft protection only; an aft-mounted jammer with side windows for aft and side protection; a nose-mounted jammer for forward protection only and aft and nose jammers for all-round protection.

Status
As of this edition, the status of the Defendir IR jamming system series was uncertain.

Contractor
Lockheed Martin Information Systems, Electro-Optical Systems Division, Pomona, California.

VERIFIED

....

GENeric eXpendable (GEN-X)

Type
Aircraft-dispensed expendable radar deception jammer.

Description
The GEN-X RTE-1489/ALE cartridge is a small, active expendable countermeasures cartridge that is ejected from aircraft and emits radar-like signals to lure a radar-guided missile from its intended target. It is a follow-on from the Primed Oscillator Expendable Transponder (POET) system with a wider frequency range and is dispensed from aircraft using equipment such as the AN/ALE-39 or AN/ALE-47. GEN-X can also be used with an AN/ALE-40 dispenser that is equipped with a round slot magazine module. The GEN-X decoy is programmable and features a broadband antenna and wide frequency coverage. It has a high

gallium arsenide monolithic microwave integrated circuit content. The decoy is dispensed by an impulse cartridge and, after ejection, extends three small fins for stability. Its lithium thermal battery ignites to provide power, the receiver locks on to the threat radar signal and a deception signal is generated by the GEN-X techniques generator.

Status
Raytheon is understood to have completed advanced development of GEN-X by the end of 1986 and to have been awarded a full-scale development contract for the device during September 1987. Full-scale development deliveries were completed in October 1990 and successful technical and operational evaluations have since been completed. The first production contract for 7,000 units was awarded in May 1992, with a further contract for 3,416 decoys following in April 1993 with a third (for 1,440 rounds) being placed in October 1994. It is thought likely that the US Navy's total GEN-X procurement (11,856 rounds) had been delivered by the end of 1996. As of this edition, the only known export order for GEN-X has come from the UK's Ministry of Defence, which is procuring an undisclosed number of rounds for use on Royal Navy Sea Harrier F/A Mk 2 aircraft.

Specifications
Length: 15 cm
Diameter: 36 cm
Weight with cartridge: 0.45 kg
Length with fin extended: 17 cm

Contractor
Raytheon Electronic Systems, McKinney, Texas.

VERIFIED

....

Improved Self-Defence System (ISDS) Infra-Red CounterMeasures (IRCM)

Type
IRCM system.

Description
Developed as an improvement on BAE Systems North America - Information and Electronic Warfare Systems' Self-Defence System, ISDS is a fuselage or pylon-

The ISDS IRCM self-defence system

mounted equipment that is designed to provide protection against IR homing anti-aircraft missiles. The system comprises a new multiband IRCM transmitter, an electronic control unit and an operational control unit. System configuration varies according to aircraft type with a typical installation incorporating an electronic control unit/transmitter package for each engine. A single, cockpit-mounted operational controller can control up to four transmitters. ISDS has also been flight tested in a ram-air turbine-powered pod configuration. Within the system as a whole, the individual transmitters weigh 29.5 kg; the electronic control unit, 2.3 kg and the 146 × 125 × 57 mm operational control unit, 1.56 kg.

Status

As of this edition, BAE Systems was reported as having sold in excess of 150 ISDS systems worldwide. Platforms fitted with the system are noted as being both turboprop and jet powered.

Contractor

BAE Systems North America - Information and Electronic Warfare Systems, Nashua, New Hampshire.

UPDATED

Kilgore Infra-Red (IR) decoy flares

Type

Family of air-launched IR decoy flares.

Description

The Kilgore Flares Company (a Chemring Countermeasures subsidiary) produces a range of air-launched IR decoy flares, the known details of which are as follows:

Designation	Size	Weight	Launcher compatibility
KC-001/003	36 × 148 mm	280 g	AN/ALE-29/-39
KC-002	25 × 50 × 205 mm	350 g	AN/ALE-40/-45
KC-004	55 × 375 mm	1.65 kg	BOP 300/ BOZ 107
KC-005	25 × 25 × 205 mm	200 g	AN/ALE-40/ M-130
KC-006	50 × 64 × 205 mm	1.2 kg	AN/ALE-40/-47
M-206	25 × 25 × 205 mm	200 g	AN/ALE-40/-45 and M-130
MJU-7A/B	25 × 52 × 205 mm	370 g	AN/ALE-40/-45/-47
MJU-8A/B	36 × 148 mm	273 kg	AN/ALE-29A/-39/-47
MJU-10/B	52 × 65 × 205 mm	1.2 kg	AN/ALE-40/-45
Mk 46	36 × 148 mm	273 g	AN/ALE-29A/-39/-47

Status

As of November 2001, IR decoy flares of the types described were understood to be available and in service. Identified contracting activity during the period 1999 to 2001 is as follows:

29 July 1999

The then Alliant Techsystems' Kilgore operation was awarded a then year US$15,918,291 firm, fixed-price contract covering the supply of 440, 340 M-206 IR decoy flares and a then year US$5,095,668 increment from a then year US$14,013,087 firm, fixed-price contract covering the supply of a total of 78,552 MJU-10/B IR decoy flares. At the time of the announcements, work on the two efforts were scheduled for completion on 31 December 2002 and 28 February 2002 respectively, with the US Army's Armament, Munitions and Chemical Command (Rock Island, Illinois) acting as the contracting activity on both programmes.

27 August 2001

The Kilgore Flare Company was awarded a then year US$6,072,948 increment from a then year US$12,186,046 firm, fixed-price contract covering the supply of a

A KC-004 IR decoy flare following ignition   0017871

total of 156,000 MJU-7A/B IR decoy flares. At the time of the announcement, work on the particular tranche was scheduled for completion on 29 March 2003, with the US Army's Operations Support Command (Rock Island, Illinois) acting as the programme's contracting activity.

Contractor

Kilgore Flares Company (a Chemring Countermeasures subsidiary), Toone, Tennessee.

UPDATED

M130 CounterMeasures Dispensing System (CMDS)

Type

General purpose airborne CMDS.

Description

The M130 airborne general purpose CMDS is a lightweight countermeasures system that has been developed using AN/ALE-40 technology for the US Army. The system is designed for applications such as the AH-1, CH-47, OH-58 and UH-1 and is of modular construction to facilitate flexibility in its operational configuration. The modules used consist of a cockpit control unit and electronics module, one or more dispensers, cabling and aircraft adaptors. The cabling and adaptors are unique to the aircraft type, whereas the other modules are common among all aircraft types. In terms of payloads, the M130 CMDS is compatible with the M1 chaff cartridge and the M206 Infra-Red (IR) decoy flare. Each dispenser module can accommodate 30 payloads of a single type with the complete system incorporating one or two dispensers. A dual dispenser, 60 payload configuration weighs 21.8 kg and M130 can be directly interfaced with a passive IR missile warner. The M1 chaff cartridge and M206 IR decoy flare have a nominal 25 mm square configuration, are 210 mm in length and have a form factor that is identical to (and functionally interchangeable with) the RR-170A/AL chaff cartridge used in the AN/ALE-40 CMDS.

Status

As of this edition, the M130 CMDS was reported as being in service. Within the US Army, the system is applicable to the AH-IF (chaff only), AH-64A/D (chaff only), CH-47D (flare only), EH-60A (chaff and flare), MH-47E (chaff and flare), MH-60K (chaff only) and UH-60A/L/Q (chaff only) helicopters together with fixed-wing RC-12 signals intelligence aircraft (chaff and flare). Over time, M130 CMDSs are known to have been exported with Greece's mid-1999 procurement of four systems for use on AH-64A attack helicopters being an example.

Contractor

BAE Systems North America - Integrated Defense Solutions, Austin, Texas.

UPDATED

M130 Threat Adaptive Countermeasures Dispenser System (TACDS)

Type

Airborne countermeasures dispensing system.

Description

The M130 TACDS is a mechanical replacement for the M130 helicopter dispenser system that utilises solid-state microprocessor technology to provide threat adaptive, automatically programmed chaff, infra-red decoy flare and advanced expendable dispensing sequences. The system's electronics module contains an embedded processor which accepts inputs from the host aircraft's radar, missile and laser warners and avionics system, compares the received data to integral threat response files and selects the optimum dispensing programme. Overall, M130 TACDS offers the following in-flight selectable operating modes:

- manual — selection and activation (via a cockpit control unit) of one of six, preprogrammed dispensing programmes
- semi-automatic — threat warning and system prompted selection of the optimum expendable/sequence to counter the specific threat
- automatic — automatic threat analysis followed by response selection and activation without crew intervention.

M130 TACDS is designed to be easily retrofitted in existing M130 installations. The upgrading process can be partial or whole, consisting of implementation of the TACDS digital sequencer only, the digital sequencer and the cockpit control unit only or the complete system. The equipment communicates with other subsystems via a range of buses (such as RS-422 or 1553 A/B) and incorporates a built-in test routine that offers continuous automatic and crew initiated fault finding to system replaceable unit level. Available load/mission mixes include five payload types per magazine; 15 magazine mixes; multiple, simultaneous firing pulses and extremely short burst intervals. The system also features automatic misfire detection and correction; user developed, flight line loadable response/mission specific dispensing files and integral growth potential to counter future threats.

Status

As of this edition, the M130 TACDS was reported as having been procured by an 'international customer' for a UH-60 application.

Contractor

BAE Systems North America - Integrated Defense Solutions, Austin, Texas.

UPDATED

Martin Electronics Infra-Red (IR) decoy flares

Type

Family of IR decoy flares.

Description

As of mid-1999, Florida-based contractor Martin Electronics was known to be producing a range of IR decoy flares, the known details of which are as follows:

Designation	Size	Weight	Shelf life	Launcher compatibility
M206	25 x 25 x 205 mm	180 g	7 yrs	AN/ALE-40, -45, -47 & M130
MJU-7A/B	25 x 52 x 205 mm	370 g	9 yrs	AN/ALE-40, -45 & -47

Status

On 21 July 1999, Martin Electronics was awarded contracts covering the production of M206 and MJU-7A/B IR decoy flares. Taking these in the order given, the M206 award took the form of a then year US$4,124,518 base year increment from an existing then year US$12,457,854 firm, fixed-price contract covering the supply of 293,560 M206 IR decoy flares. At the time of the increment's announcement, work on the effort was scheduled for completion on 31 December 2002 and its contracting activity was the US Army's Armament, Munitions and Chemical Command, Rock Island, Illinois. For its part, the MJU-7A/B award took the form of a then year US$6,224,400 base year increment from an existing then year US$14,796,288 firm, fixed-price contract covering the supply of 296,400 MJU-7A/B IR decoy flares. At the time of the increment's announcement, work on the effort was scheduled for completion on 31 December 2002 and its contracting activity was the US Army's Armament, Munitions and Chemical Command, Rock Island, Illinois.

Contractor

Martin Electronics, Perry, Florida.

NEW ENTRY

Quick Reaction Contract (QRC) 83-05 Infra-Red CounterMeasures (IRCM) system

Type

Airborne IRCM system.

Description

QRC 83-05 originated in an internally mounted IRCM system that was installed aboard US Air Force EB-66 electronic warfare aircraft deployed to Southeast Asia. The system has been reconfigured for helicopter applications and updated with multithreat capabilities. Subsequently, it has been fitted to USAF MH-53 helicopters. In the helicopter configuration, the dual transmitter provides protection on bands I and II without engine suppressors. The system electronically modulates a vistal caesium infra-red source to produce a highly effective jamming system.

Status

As of December 2001, the operational status of the QRC 83-05 IRCM system was uncertain.

Contractor

Northrop Grumman Electronics sector - Defensive Systems Division, Rolling Meadows, Illinois.

UPDATED

Semiconductor Infra-red Laser Countermeasures (SILC) system

Type

Airborne Infra-Red (IR) self-protection system.

Description

SILC is an all-laser IR CounterMeasures (IRCM) system that is designed for use on airborne platforms that range in size from the smallest helicopter to large jet transport aircraft such as the Boeing 747. As such, SILC utilises multiband semiconductor lasers that are claimed to yield 'the smallest yet highest performing IRCM technology available today'. Each SILC transmitter provides hemispherical field of regard protection with a two transmitter ship-set providing full spherical protection for the host aircraft. The SILC pointer-tracker is claimed to provide the 'fastest timelines achievable', while the system's lasers generate the high levels of jamming power required to ensure rapid, simultaneous, multithreat protection. SILC can also be configured to integrate with 'any of todays' missile warning systems that are capable of providing accurate angle of arrival data.

Status

As of this edition, the status of the SILC IRCM system was uncertain.

Contractor

Lockheed Martin Information Systems, Electro-Optical Systems Division, Pomona, California.

VERIFIED

Survivability Augmentation for Transport INstallation (SATIN)

Type

Electronic self-protection system for large transport aircraft.

Description

SATIN is a minimum modification kit that protects large transport or tanker aircraft against air-to-air and surface-to-air missiles. SATIN installations accommodate a variety of electronic warfare equipment including chaff and flare dispensing systems, radar warning receivers, missile detection systems and infra-red jammers. SATIN configurations are tailored to meet customer requirements and specific aircraft applications. The degree of protection provided by a particular SATIN suite depends on the type and quantity of systems selected for installation. A fully equipped suite is effective against infra-red homing missiles, radar-guided missiles and radar-aimed and anti-aircraft artillery.

Status

Since the implementation of the SATIN programme, Lockheed Martin has produced C-130 and C-5 defensive systems for foreign and domestic military agencies. SATIN systems may be incorporated on new production aircraft or as a field retrofit kit for in-service aircraft.

Contractor

Lockheed Martin Aeronautical Systems, Marietta, Georgia.

VERIFIED

The SATIN EW systems provide countermeasures protection for aircraft such as the C-130

UST-107 Command and Control Warfare (C²W) systems

Type

0.5 to 2,000 MHz C²W system.

Description

The UST-107 C²W system is described as providing 'reliable, low cost' Electronic Support (ES), Electronic Attack (EA) and radio communications capabilities in the 0.5 to 2,000 MHz frequency range. The system is intended for fixed-station, shipboard, ground-mobile and airborne applications and can be operated autonomously or via Ethernet, MIL-STD-1553B or RS-232 interfaces using its integral Windows operating programme. The UST-107 architecture consists of multiple 95V-1 direct conversion surveillance receivers, 95E-1 direction conversion jamming exciters and power amplifiers that function under the control of a ruggedised, laptop computer-based man/machine interface that runs Windows NT. The receiver/exciter units are installed in a ruggedised VME chassis in such a way as to allows system components to be tailored to meet particular mission needs and upgraded when new technology becomes available. Scanning algorithms are contained in a flash memory that is held on the VME central processor unit card and are initiated and monitored by the Windows operating programme. The equipment's receivers and exciters tune across the 5 kHz to 2,000 MHz range, facilitating UST-107's use in 'virtually any jamming mission in any theatre of operations'. System operating modes as are follows:

Communications mode

In communications mode, UST-107 acts as a voice and data transceiver linking two or more stations. As such, the equipment provides:

- simplex, half-duplex and full-duplex operation
- single- or two-channel (independent sideband) function
- compatibility with Voice Privacy™ and ANDVT encryption
- compatibility with single- and multitone data modems
- amplitude/frequency and independent/single sideband modulation/demodulation
- manual, channelised operation
- fixed frequency or hopping functionality

ES/Surveillance mode

In ES/surveillance mode, UST-107 offers two submodes, as follows:

Acquisition mode

In which the system detects signals of varying amplitude while avoiding false responses due to noise.

Scan mode

In which the system records the time on and time off for each detected emitter together with a sample of the received audio.

Reactive jamming mode

In reactive jamming mode, UST-107 offers:

- programmable jammer acquisition, timing, waveforms and other system parameters for each frequency or frequency band under attack
- optimised jamming to prevent excessive jamming against a single target in the presence of multiple targets

- prioritised jamming that ensures EA against low priority targets only when high priority emitters are off air

Pre-emptive jamming mode

UST-107 provides a 'blind' pre-emptive jamming mode that is designed to jam weak signals or to deny an entire frequency band by time multiplexing the jamming signal.

Status

As of this edition, the UST-107 C²W system was reported as being under development.

Specifications

EA capability

Frequency range: 5 kHz - 2,000 MHz
Jamming modulation: programmable wave file
Amplifier output power: 100 W
Frequency switching: 500 µs
Elementary jamming duration: 1 ms - 2 min (selectable)
Jamming look-through: 1 ms - 2 min (selectable)
Look-through duration: <1 ms (two antenna systems)

ES capability

Frequency range: 5 kHz - 2,000 MHz
Frequency resolution: 1 Hz
Tuning time: 500 µs (typical figure with AGC disabled and under VME control)
Frequency stability: ±0.1 ppm (with electronic adjustment of the internal standard)
Frequency accuracy: ±1 ppm (0°C to +55°C)
Noise figure: 12 dB (20-1,000 MHz antenna, typical); 14 dB (0-30 MHz antenna, typical)
Modulation modes: AM, CW, FM, ISB and SSB
Demodulation modes: AM, CW, FM, I/Q, ISB and SSB
2nd order intercept: +60 dBm (typical, ≥30 MHz); +80 dBm (typical, <30 MHz)
3rd order intercept: −3 dBm (typical, >500 MHz); +7 dBm (30-500MHz); +25 dBm (typical, <30 MHz)

Contractor

Rockwell Collins, Cedar Rapids, Iowa.

UPDATED

RADAR AND ELECTRONIC WARFARE SIMULATION AND TRAINING SYSTEMS

RADAR AND ELECTRONIC WARFARE SIMULATION AND TRAINING SYSTEMS

Introduction

This section deals exclusively with those radar and Electronic Warfare (EW) simulation and training systems used in the military sphere, whether land-based, shipborne or airborne. It does not cover those systems used for ship or aircraft handling, or for any civil applications, such as air traffic control. A comprehensive survey of all military training and simulation systems is provided in *Jane's Simulation and Training Systems.*

VERIFIED

AUSTRALIA

Simulated System Maintenance Trainer (SSMT)

Type
Maintenance trainer simulation system.

Description
SSMT is a large classroom training device for training users or maintainers by simulation of an equipment system or subsystem. Applications include aircraft systems such as electronic support measures, avionics, flight controls or hydraulics.

SSMT is specifically designed to allow fault finding of a specific range of faults by use of procedure or free play (not bound by procedures). The simulation allows operation of the system in normal mode, as with the real system. Actual or replicated controls are part of the trainer and the simulation is supported by dynamic interactive video and sound.

A typical system consists of four main assemblies: two panels with system components mounted or represented and audibly simulated, a control console to allow instructor fault setting and location of the video screen and a small console to mount and drive major moving controls. In the case of an aircraft trainer this can be enlarged to replicate a complete cockpit.

Status
Over time, the aircraft trainer variant (SAMT - Simulated Aircraft Maintenance Trainer) of this system is thought to have been procured by the armed forces of Australia, Canada and an as yet unidentified Asian country.

Specifications
Size: 2.43 × 1.52 m artworked panels mounted ergonomically
Power: electrical, in an air conditioned environment
Weight: 500 kg per unit, average
Design: computer software simulation controlling simulation system hardware using digital and analogue input/output and supported by video and sound systems.

Contractor
BAE Systems Australia, Elizabeth, South Australia.

UPDATED

A representative SSMT installation

BRAZIL

AV-ST2 training simulator

Type
Simulator equipment for the EDT-FILA fire-control system.

Description
AV-ST2 has been developed to meet a Brazilian Army requirement for an advanced training simulator for the EDT-FILA fire-control system. As such, it complements the first level ST1 simulator that is embedded in the EDT-FILA's software. AV-ST2 comprises a single equipment console which houses a radar/TV tracker display unit (identical to that used in the EDT-FILA system), an operator's monitor and keyboard, intercommunications elements, system controls and a printer. Functionally, AV-ST2 generates scenario-driven aircraft imagery for the TV tracker and radar responses (including interference) for the radar display. A total of 42 programmed exercises is available, each of which contains up to eight fixed-wing, helicopter, remotely piloted vehicle and/or air-to-ground missile targets. Scenarios include platform self-defence countermeasures while target velocities and direction of flight are variable. Variability is also built in to the visibility and light level of the TV display. For debrief purposes, AV-ST2 provides a chronological evaluation of trainee actions during an exercise.

Status
Over time, AV-ST2 is noted as having been procured by the Brazilian Army, with which it became operational during 1991.

Contractor
AVIBRAS Indústria Aerospacial SA, Sao José dos Campos, Brazil.

VERIFIED

BULGARIA

Radar jamming simulators

Type
Radar jamming simulation and training systems.

Description
Kintex radar jamming simulators are designed to train radar operators in how to detect and track real targets in the presence of simulated noise, chaff and pulse jamming, recognise jamming types and operate anti-jamming equipment. They provide active noise; modulated and unmodulated jamming by two sources; chaff cloud; asynchronous; interpulse and multiresponse jamming. Simulator-to-radar connection is by means of two video outputs (with positive and negative polarity) or a 10, 20, 24.6 or 30 MHz intermediate frequency output.

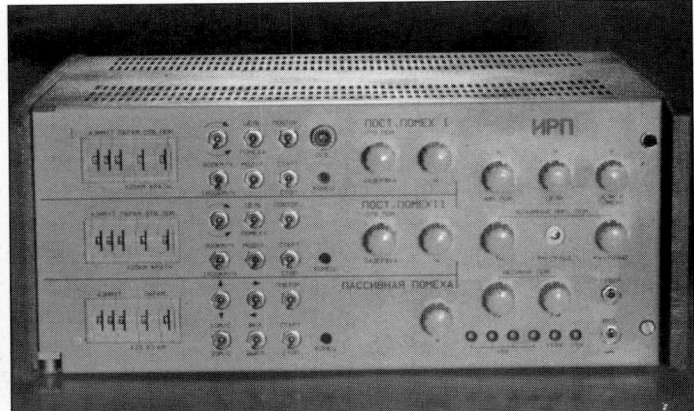

Kintex's radar jamming simulator equipment   0010003

Status

Over time, equipment of the type is understood to have been procured by the Bulgarian Army.

Specifications

Weight: 22 kg
Simulated target characteristics
Flight speed: 900 and 1,800 km/h
Mean initial azimuth: 7° intervals (max - set digitally)
Course parameters: 25 km intervals (set digitally)
Chaff simulation characteristics
Wind velocity: 10 and 20 km/h
Mean initial azimuth: 7° intervals (max - set digitally)
Course parameters: 25 km intervals (set digitally)
Power supply: 220 V AC; 50-60 Hz
MTBF: >300 h
Temperature range: 5 to 40°C

Contractor

Kintex, Sofia.

VERIFIED

···

Roubin ground radar simulator

Type

Lightweight ground radar simulator.

Description

Roubin is a ground radar system that simulates the radiation pattern of various types of radars operating in the 3 cm waveband. It consists of a tripod-mounted simulator main unit, a battery and connecting cable.

Status

As of August 2001, the status of the Roubin ground radar simulator was uncertain.

Specifications

Range: 1-60 km
Radiated pulse power: 5-7 kW
Pulse duration: 0.15-0.25 µs
PRF: 500, 1,000, 1,500 and 2,000 Hz
Continuous operation time: 24 h
Power supply: 12 V
Consumption: 60 kW
Weight: 29 kg

Contractor

Kintex, Sofia.

VERIFIED

The Roubin ground radar simulator

···

Universal general purpose shipborne radar simulator

Type

Naval and maritime radar simulator and trainer.

Description

The universal simulator is designed for mass training of operators on civilian and military maritime surveillance radars. The system consists of an instructor's panel,

students' panels, a 16-bit PC and software with test programmes for the operator's initial preparation. Up to 20 students can take part in the training at any one time, each station being equipped with indicators, controls and radar simulation devices.

The system provides for simultaneous simulation of 80 moving targets, up to 15 navigation signs, active and passive jamming, navigation and identification. It provides training in both airborne and surface target identification and tracking, noise situation analysis and data processing, as well as identifying the various types of jamming and the methods used to counter them.

Status

Over time, Kintex's universal shipborne radar simulator is understood to have been procured by the Bulgarian armed forces.

Specifications

Simulation types (simultaneous): coastline; active/passive jamming; identification signals; >15 pcs (ship targets - individual or groups); >15 pcs (navigation signals); 80 pcs (moving targets - different speeds and directions); different types of shipborne radar
Power supply: 3 × 380 V
Consumption: 20 W (for 20 operator stations)

Contractor

Kintex, Sofia.

VERIFIED

———————————————————————————————————

CANADA

Airborne Electronic Warfare Trainer (AEWT)

Type

Integrated Electronic Warfare (EW) training system.

Description

AEWT is an integrated electronic warfare training system that is designed to provide 'comprehensive' EW training for modern air defence radar sensors and networks and can be installed aboard either a business jet or a transport aircraft. Covering the A- to J- frequency bands (0.03 to 18 GHz sub-band), AEWT is a modular architecture that is claimed to have been 'optimised to give the greatest EW capability within a single, compact system solution'. As such, the system can be installed aboard smaller aircraft 'where a conventional solution using a large collection of black boxes would not be feasible'. Identified platform applications range from space and prime power limited business jets to larger aircraft when the operational requirement is for a system that has high effective radiated power and duty cycle values and operating costs are 'of less concern'.

Status

As of February 2001, AEWT was reported as being 'under consideration' for procurement by potential North American and Asian customers. As of the summer of 2001, AEWT was understood to be a live programme.

Specifications

Frequency range: 0.03-18 GHz (standard); 0.03-40 GHz (option)
Instantaneous bandwidth: 63,500 MHz
Pulse-width: 50 ns (min); 2,000 µs to CW (max)
ECM transmitter: travelling wave tube or solid state
Transmitter power: 200 W (CW); 1 kW (pulsed)
Transmitter antenna type: reflector/horn/dipole
ECM receiver: scanning DF superheterodyne; DIFM/DLVA
Sensitivity: −60 dBm
Dynamic range: 60 dB
ECM techniques: deception and denial (techniques and sequences in both cases)

Contractor

Telemus Inc, Kanata, Ontario.

UPDATED

···

AN/ALM-507 simulator

Type

End-to-end receiver test set.

Description

ALM-507 was developed for on-aircraft testing of Electronic Support Measures (ESM) receivers fitted to a variety of maritime and reconnaissance aircraft. The unit can be used as an onboard electronic warfare training simulator, as well as for first and second line maintenance. The system consists of a small portable set and a remote-control panel. The user sits at the ESM sensor station inside the aircraft and, using a portable control panel, initiates the injection of simulated emitter signals into the aircraft antennas. Through diagnostic routines, the unit can provide identification of equipment failure to a specific antenna at a specific frequency range. During testing, the set can vary a number of parameters: frequency range, angle of arrival, pulse-width, pulse repetition interval and prime power (AC or DC). The system is aimed at systems such as AN/ALR-47, AN/ALR-66 and AN/ALR-76. Equipment can be configured for amplitude or phase comparison systems.

Status
As of January 2002, ALM-507 was no longer an 'active' Telemus product line but is understood to have been procured by the Canadian Forces Air Command in support of the CP-140 Aurora maritime patrol aircraft.

Contractor
Telemus Inc, Kanata, Ontario.

UPDATED

Excalibur Electronic Warfare (EW) test and training simulators

Type
Family of EW and radar threat training and test simulators.

Description
Excalibur's family of Radio Frequency (RF), video and software based EW simulation, test and training systems is designed to address current defence requirements, can be used in a wide range of land, sea and airborne applications and provides what their manufacturer terms sophisticated, high-fidelity emitter simulations. The company's TS series RF systems covers the 0.5 to 40 GHz (extendable to 96 GHz if required) frequency band and all Excalibur simulators utilise user-friendly, Windows-based ThreatBuilder operator interfaces and modular architectures. The company's RF systems offer full amplitude and phase direction of arrival simulation and continuous wave; pulse-Doppler and noise modules are available if required. By using common modules across the entire family, Excalibur's RF and video systems can be configured for a variety of roles that range from man-portable flight-line testers to multichannel laboratory test and evaluation, high power transmitting and multiposition training systems. As of August 2001, known details of Excalibur sourced EW and radar threat simulation systems were as follows:

HPTS
HPTS is a threat simulation system that is designed to operate on a training or test range with live aircraft. It employs any one of the TS series simulators as an exciter for a high-power amplifier or transmitter. Transmitter output is radiated via an antenna towards the test aircraft that attempts to defeat the threat simulation using a variety of techniques. HPTS employs a number of subsystems and techniques to track the aircraft under test and to evaluate the pilot's response. These include identification friend-or-foe, infra-red and visual trackers; digital video capture for post engagement analysis and intelligent mode changes in response to pilot action as detected by electronic countermeasures evaluation receivers.

Using combinations of dynamically allocated complex and background emitters, any TS series chassis can support hundreds of simulated threats. Very large simulators can be created by linking together any number and combination of TS series systems via their Ethernet interfaces. This allows for staged procurement,

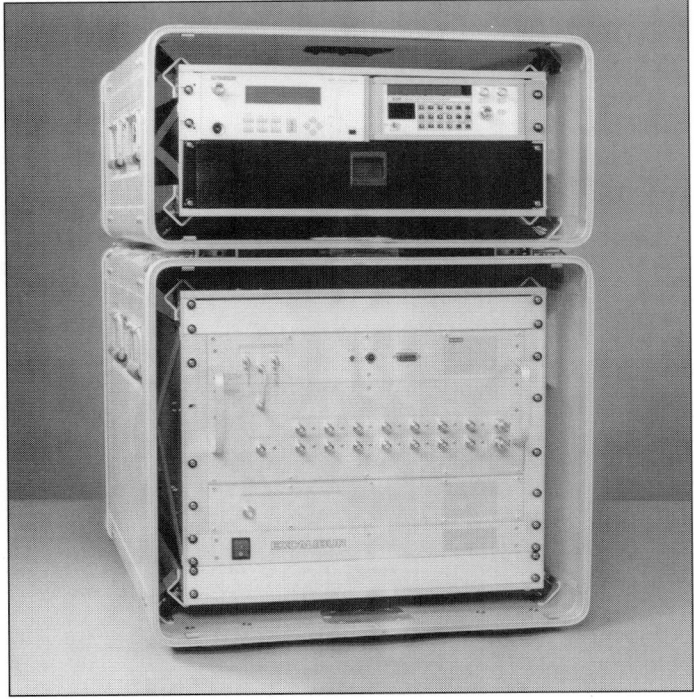

The TS-200 simulator with its associated calibration unit   0017872

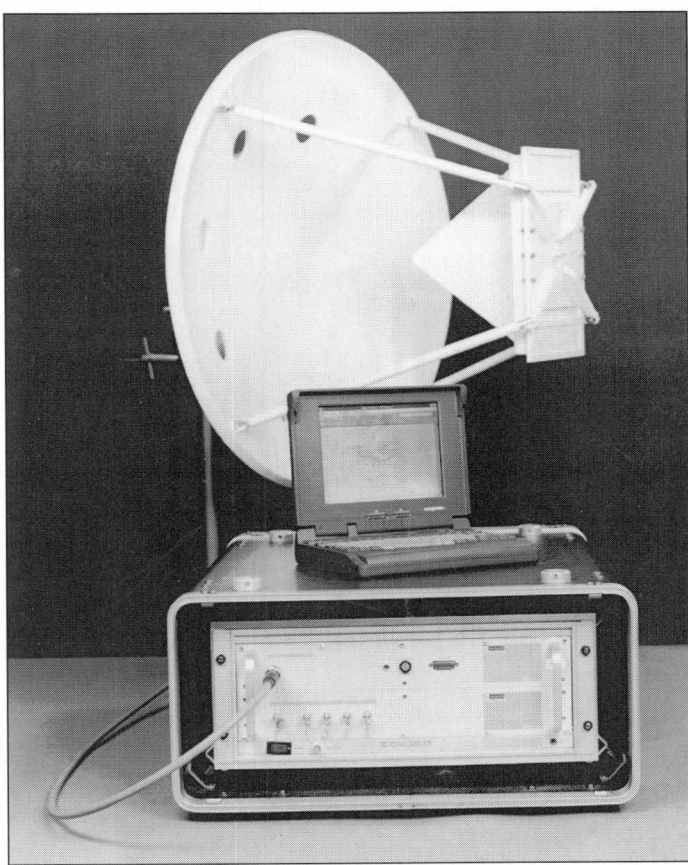

The TS-300 simulator with its associated antenna subsystem   0017873

multiple roles, ease of expansion and minimal spares holdings. Overall TS series features include synthesised and/or digitally tuned oscillators; 0.5 to 40 GHz coverage (with options available up to 96 GHz); frequency resolution to 0.4 Hz for synthesised sources; up to 5 million digital pulses/s per simulator unit; up to 770,000 different pulses/s per microwave channel; dynamic, 3-D scenarios with large numbers of players and programmed emitters; the ability to trade any complex emitter for eight background emitters at any time; multichannel amplitude and phase direction of arrival capabilities (including the simulation of spinning direction-finding antennas) and multichannel video outputs and user-configurable digital interfaces in addition to microwavable outputs.

TS-100
Reported as being designed for research and development and test and evaluation purposes, the rack-mounted TS-100 equipment is noted as:
- operating within the 0.5 to 18 GHz or 18 to 40 GHz frequency bands
- supporting pulse densities of more than 3 Mpps
- generating from 16 to an 'unlimited' number of threats
- generating agile, continuous wave, pulse agile, pulse-Doppler and 'other' types of waveform
- offering agile, continuous wave, stable and 'other' types of modulation
- generating both coupled and radiated radio frequency outputs
- being able to emulate both electronic warfare receivers and pulse analysers.

TS-200
TS-200 equipments are rack-mounted systems that are designed for high-capacity applications and are typically configured with more than one chassis. Each chassis used can support up to 80 complex emitters and four broadband microwave channels.

TS-300
Reported as being designed for research and development and test and evaluation purposes, the rack- or desktop/workstation-mounted TS-300 equipment is noted as:
- operating within the 0.5 to 18 GHz or 18 to 40 GHz frequency bands
- supporting pulse densities of more than 1 Mpps
- generating from 32 to an 'unlimited' number of threats
- generating agile, continuous wave, pulse agile, pulse-Doppler and 'other' types of waveform
- offering agile, continuous wave, stable and 'other' types of modulation
- generating both coupled and radiated radio frequency outputs
- being able to emulate both electronic warfare receivers and pulse analysers.

Virtual Integrated EW Simulator (VIEWS)
The VIEWS EW training and simulation package is a comprehensive software product that utilises an off-the-shelf Personal Computer (PC) to produce graphical emulations of a range of EW equipments that includes receivers, pulse analysers and direction-finding displays. Emulated EW equipment is simulated in real time with complex emitters and scenarios being built using ThreatBuilder, a user-friendly, Windows-based scripting tool that incorporates Wizards and extensive

online help. VIEWS may be employed as either a stand-alone, one-station trainer or as a multiplace trainer using multiple PCs linked together via Ethernet or a local area network.

Status

Over time, TS series simulators are reported as having been procured for civilian contractors, the Brazilian Air Force (TS-100), the Canadian Forces, the French Ministry of Defence (TS-300), NATO's Command, Control and Communications Agency and Multiservice Electronic Warfare Support Group (MEWSG), the Royal Netherlands Navy, the Royal Norwegian Navy (TS-100), the Republic of Korea's Navy (TS-100) and the US Navy (four TS-200 RF systems and nine TS-200 video systems). As of the summer of 2001, the TS-100 and TS-300 systems were understood to be live programmes.

Contractors

Excalibur Systems Ltd, Kanata, Ontario, Canada.
Excalibur Defense Systems Corporation, Orlando, Florida.

UPDATED

....

INtegrated Electronic Warfare Trainer (INEWT)

Type
Mobile naval EW simulator

Description

INEWT is an integrated naval EW training solution that combines dedicated Electronic CounterMeasures (ECM) and radar signal simulators in a single architecture. Telemus claims that the system provides the user with an 'advanced' and 'cost-effective' solution that can be programmed to simulate a 'wide variety' of radar/ECM signal and 'total' anti-shipping scenarios (with the latter using an array of high power transmitters and antennas). The INEWT architecture comprises:

- an ECM signal simulator
- a radar signal simulator
- an Electronic Support (ES) receiver and pulse analyser
- a target tracking subsystem
- a common set of high-power continuous wave and pulse transmitters and antennas.

Of these various elements, the ECM and threat simulators are connected to a set of common travelling wave tube amplifier transmitters and an array of transmit and receive antennas. Normally, these (together with the ES receiver subsystem and antenna steering/stabilisation electronics) are housed in a self-contained, mobile International Standards Organisation S-280 shelter. Functionally, INEWT is designed to exercise active and passive naval radio frequency sensors, that are used in surface warfare and the air defence environment. When working with active radars, the system facilitates the training of operators in the identification of countermeasures effects and the use of Electronic Counter-CounterMeasures (ECCM) facilities within the sensor. In the passive role, INEWT generates threat emitter signatures for display on the ES system and the complete system is noted as being capable of generating dynamic scenarios that can contain simple to complex signatures as required. The INEWT architecture is also supported by an integrated radar signal analysis and recording capability.

Status

Over time, INEWT has been reported as having been procured by an as yet unidentified Asian navy. As of the summer of 2001, INEWT was understood to be a live programme.

Specifications
Frequency range: 0.5-18 GHz
Pulse-width: <50 ns (min); >2,000 μs or CW (max)
Transmitter type: travelling wave tube and solid state
Transmitter power: 200 W CW (per band); 1 kW pulsed (per band)
Transmitter antenna type: reflector array/horn
Antenna mounting: fixed or stabilised mount
ES receiver type: DIFM/DLVA; scanning DF superheterodyne
Instantaneous bandwidth: 500 MHz
Modes: linear/logarithmic
Sensitivity: −60 dBM
Dynamic range: 60 dB
ECM techniques: deception (techniques and sequences); denial (techniques and sequences)
Radar modes: acquisition/targeting; active radar seeker; multimode/electronic scan; surveillance/air early warning/surface search; tracking/fire-control/illuminator
GUI: Windows 95/NT; X-Windows
Real-time capability: custom instruction set processor-based techniques
Shelter (W × H × L): 2.4 × 2.4 × 6 m ISO S-280
Power: internal motor generator or ship's power
Training support: operator/maintenance training; EW/ECM instructor training

Contractor
Telemus Inc, Kanata, Ontario.

UPDATED

TActical CounterMeasures Evaluation Trainer (TACMET)

Type
Electronic CounterMeasures (ECM) simulators/training radar jammer.

Description

TACMET is a radar jammer and ECM simulator that can be 'easily' programmed to simulate a wide variety of ECM scenarios, jammers and their ECM waveforms. Currently available systems represent the culmination of 'prudent and continuous' insertion of new technology in order to provide customers with 'advanced and cost-effective' solutions. Although an ECM system, TACMET is primarily used for radar Electronic Counter-CounterMeasures (ECCM) Test and Evaluation (T&E) or radar ECM training. TACMET's basic capabilities are claimed to 'meet or exceed' the performance of many operational ECM systems in a 'one-on-one' engagement and to optimise the system for the training and T&E roles. Performance can be derated (under computer control) to simulate 'less capable' ECM systems. Telemus notes that considerable effort has been expended on making the control of TACMET equipments 'easy and intuitive' and the programming of ECM techniques clear to the user. This is achieved via the use of an interactive, custom MS-Windows™ graphic user interface programme. Here, the sequence of windows and menus allows the system to be 'easily' operated in a number of modes. TACMET equipments are supplied as complete turnkey packages that include vehicle or shelter installation, training, maintenance and infrastructure support.

Status
As of early 2001, TACMET was reported as having been procured by customers in North America and Asia.

Specifications
Frequency range: 0.5-18 GHz
Instantaneous bandwidth: 100, 200 or 500 MHz
Pulse-width: <50 ns (min); >838 μs or CW (max)
ECM transmitter: travelling wave tube or solid state
Transmitter power: 200 W CW TWTs (standard); 1 kW and 10 kW pulsed (optional)
Transmitter antenna type: reflector/horn/dipole
ECM receiver type: DIFM/DLVA
Sensitivity: −50 dBm
Dynamic range: >60 dB
ECM techniques: angle denial (AM, inverse gain, pulse and swept AM); arbitrary waveform generator (modulation on pulse); noise denial (barrage, blink, burst, spot and swept); range deception (hooks, multiple false target and RGPO/I); range denial (CW and noise); synchronised range and velocity deception (R/VGPO/I); velocity deception (VGPO/I); velocity denial (velocity noise).

Contractor
Telemus Inc, Kanata, Ontario.

UPDATED

GERMANY

Radar and navigational aid simulator (RASI)

Type
Radar and navigational aid simulator.

Description

RASI is suitable for operational training on any type of radar, as well as other applications such as navigation aids and ship handling. It can be used to train both

The instructor's console of the RASI system

naval and mercantile marine personnel and consists of an instructor station, students' cubicles, the necessary electronics and a line data input station. The latter unit is used for generating any required exercise area, updating existing exercise areas, and generating new types of own and other ships. Each student's cubicle contains a radar set, a slave radar, navigational instruments, a chart table, communications unit, indicators and a steering unit.

Status
Over time, identified RASI system procurements are reported as including a single example for the German Navy, four for German maritime training establishments and one for a customer in Taiwan.

Contractor
STN ATLAS Elektronik GmbH (a BAE Systems-Rheinmetall AG joint venture), Bremen.

VERIFIED

STN ATLAS Electronic Maintenance Trainer (EMT)

Type
Training system for maintenance personnel.

Description
EMT is designed to provide maintenance and repair training for service personnel working with radar and electronic warfare systems. The system uses either single or multiple consoles linked to an instructor's station. Each console consists of a workstation with graphics screen, text display and disk memory connected to peripherals such as printers, plotters and scanners, as well as a signal generator to produce the required spectra. The system is capable of providing modulated signals from 0 to 18 MHz and digital signals with Transistor/transistor logic/Emitter coupled logic. Spectra and number of signal sources can be specified by the user. All training data are stored on disk and can be used to introduce new parameters, tasks or projected fault occurrences.

Status
Over time, the EMT training system is reported as having been procured by the German Air Force.

Contractor
STN ATLAS Elektronik GmbH (a BAE Systems-Rheinmetall AG joint venture), Bremen.

VERIFIED

INTERNATIONAL

Air Navigation Trainer (ANT)

Type
Air navigation training simulator

Description
Alenia Marconi System's ANT system is an entry level radar navigation training tool which can be rescaled to meet specific student/trainer ratios. It incorporates radar and multifunction touchscreen displays as well as exercise preparation and control facilities. High-resolution colour graphics are used with the formats being fully reconfigurable to match specific navigation instrument fits. Radar system features reproduced by ANT include azimuth and elevation polar response; receiver gain and noise control; pulse-length; antenna tilt and refraction, sea clutter, earth curvature, shadowing and edge enhancement landmass effects. Available instructor facilities include group or single student one-to-one training configurations together with a full capability hand-held terminal for 'over-the-shoulder' instruction.

The ANT simulator

Status
Over time, the ANT simulator is reported as having been procured by the UK's Royal Air Force.

Contractor
Alenia Marconi Systems (an Alenia-BAE Systems joint venture), Chelmsford, UK/ Rome, Italy.

UPDATED

MESA Electronic Warfare (EW) simulators

Type
Family of interactive electronic warfare simulators.

Description
MESA is a family of EW simulators covering the full range of configurations from digital to hardware-in-the-loop simulations. As of early 2001, the MESA family included digital, hardware-in-the-loop, onboard training and ground-based instruction equipments. Specific application details of these various configurations are as follows:

MESA-D digital version
- design and analysis of EW systems
- EW mission planning and analysis
- training of operational personnel.

MESA-H hardware-in-the-loop version
- assessment of EW system performance.

MESA-O onboard training version
- EW training during flight.

MESA-T ground-based training version
- ground-based training and instruction tool.
  The use of a common baseline configuration for these various equipments, results in a high level of coherence, not only for design and analysis but also for the final assessment of systems. The modular architecture of MESA is based on the use of unclassified generic models (defence systems, airborne, seaborne and ground-based EW systems, operational scenarios, terrain configurations) fed from a separate classified database containing all model parameters. In its digital version, MESA is an autonomous and portable simulator operating under UNIX. The hardware-in-the-loop version can be supplied with either radio frequency environment generators or adapted to specific instrumentation.

Status
As of early 2001, MESA systems were reported as having been procured by the armed forces of 'several' countries around the world.

Contractor
Thales Airborne Systems, Elancourt, France.

VERIFIED

SS2932 COMSIM Mk II communications simulator

Type
Communications and Electronic Warfare (EW) simulator.

Description
COMSIM Mk II is a tactical communications and EW simulator that is designed to provide a realistic training environment for a range of communications/EW personnel including trainee operators, system managers and staff officers. At the heart of the architecture is a computer-controlled propagation matrix. User radios are connected to the system and radio frequency propagation conditions are simulated according to the notional position of the radios within the exercise area. The architecture comprises a central control console (with map display screen and equipment rack) and a number of radio stations to which the student's radios are connected. All exercises are carried out with a high degree of realism based on any mapped exercise area. The system's control computer calculates the effective propagation loss between all radio stations and sets the attenuators in the propagation matrix accordingly. The exercise controller changes the positions of radio stations as required by moving the symbol over the display using his mouse. He also specifies the positions of hostile Electronic CounterMeasures (ECM) or electronic support measures stations in a similar manner. Using a graphical user interface, he controls the execution of the complete exercise.

As an option, COMSIM can incorporate a number of ECM scenario generators or jammers and this allows either another instructor to inject ECM into a radio net, or an EW student to train for the ECM controller role. Flexible audio monitoring facilities provide both control consoles with up to 10 configurable audio channels each. This allows all or selected radios or nets to be monitored. Facilities for radio transmission logging and multichannel audio recording are optionally available for post-exercise debrief and game analysis. In static configuration, COMSIM is installed in two cabinets, one engineering control console and one exercise control position.

Status
Over time, the COMSIM system is reported as having been procured by the armed services of a number of countries. The equipment is no longer in production.

Specifications
Exercise coverage area: 80 × 80 km
Displayed map scales: 80, 26.7 and 8 km across the screen
Height matrix: 500 m granularity
Height resolution: 1 m

Contractor
Thales Defence, Wells, UK.

UPDATED

Tornado Electronic Warfare (EW) simulator module

Type
EW module for German Air Force Tornado simulators.

Description
The EW module (comprising a PE 3252 computer and three 48 cm (19 in) electronics racks) is an addition to the baseline Tornado strike aircraft flight simulators operated by the Luftwaffe (German Air Force). Such appliques facilitate aircrew training under 'full' EW conditions, with an instructor pilot selecting appropriate scenarios. Within the simulation, the platform's defensive aids systems react to 'received' radar emissions within the field of view of a real-world Tornado's overall antenna array. With the module primarily simulating discrete radar sources, terrain feature interference is generated by a dedicated submodule designated as the Digital Radar Land Mass System. A dedicated processor is used to transfer data from the EW module to the simulator and over time, 'considerable' improvements have been made to the module's instructor console so that it can display complete scenarios.

Status
Over time, equipment of the type described is reported as having been installed on the flight simulators operated by each of the Luftwaffe's Tornado Wings. Industrial activity connected with the described EW module has been co-ordinated by the former DaimlerChrysler Aerospace AG (now a component of the European Aeronautic, Defence and Space (EADS) Co).

Contractor
EADS - Systems and Defence Electronics, Ulm, Germany.

UPDATED

ISRAEL

NS-9002E Electronic Warfare (EW) simulator

Type
Vehicle-mounted radar EW simulator.

Description
NS-9002E is an EW simulation system that can be used for system evaluation, maintenance and operator training. It is entirely self-contained and is capable of

The NS-9002E EW simulator

generating any known electromagnetic environment, both at digital and Radio Frequency (RF) levels. Moreover, NS-9002E can provide a real-time simulation of any received environment and also dynamically adapts to aircraft manoeuvres. The system generates up to 256 independent emitters in the 0.5 to 18 GHz range, with an extended range available. Each emitter is independently programmable for a variety of parameters and trajectories. An emitter library can be prepared and stored for any number of radars. The NS-9002E features multi-emitter generation in an extremely dense environment (up to 2 Mpps) and RF transmission through antennas or direct coupling. The receiving part of NS-9002E enables the evaluation of the electronic countermeasures system in the EW suite.

Specifications
Frequency coverage: 0.5-18 GHz (extended range optional)
Power output: 30 dBm ERP (solid state); higher ERP with TWT amplifiers on request
Dynamic range: 60 dB
Spurious: −60 dBc
Frequency accuracy: 1 MHz (0.5-2 GHz); 2 MHz (2-18 GHz)
Frequency repeatability: ±0.5 MHz (0.5-8 GHz); ±1 MHz (8-18 GHz)

Contractor
Elisra Electronic Systems Ltd (a member of the Elisra Group), Bene Beraq.

UPDATED

RANEWS Electronic Warfare (EW) simulator

Type
EW simulator.

Description
The RANEWS system is designed to simulate engagements between anti-ship/anti-aircraft radar homing missiles and target ships/aircraft in an active and passive EW environment. System features include real radio frequency signal transmission (real time, power and frequency) to the seeker under test; electronically controlled target positioning; the ability to handle complex scenarios; a wideband capability and modular design to facilitate system growth and maintainability.

The target wall in the RANEWS EW simulator
0010005

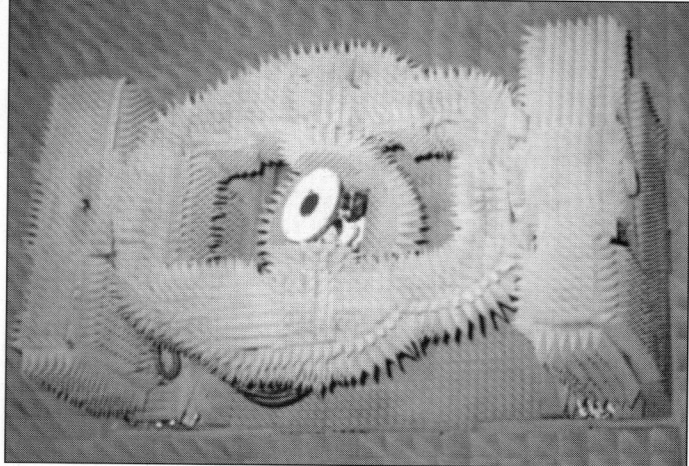

A generic seeker installed in the RANEWS EW simulator   0010006

Status
As of January 2002, the status of the RANEWS was uncertain.

Contractor
Rafael Electronic Systems Division, Haifa.

VERIFIED

NATO

NATO training and simulation equipment

Type
Range of simulation and training equipment operated by NATO.

Description
Over time, NATO's multinational Multiservice Electronic Warfare Support Group (MEWSG) has been equipped with the NEWVAN and TRACSVAN systems for land-based and land-based/shipboard applications respectively. The known details of these two systems are as follows:

NEWVAN
NEWVAN is designed to offer EW simulation and training across the 2 to 1,000 MHz frequency band. It is housed in a vehicle-mounted shelter and incorporates a towed generator for autonomous operation. The NEWVAN surveillance subsystem provides surveillance, intercept, technical analysis and direction-finding facilities, with fast scanning techniques used to permit sampling of the complete band every second. In terms of its jamming capability, the architecture incorporates an integral subsystem together with a control facility for up to 25 remote UnAttended Jammer (UAJ) units. Each of these UAJs is understood to operate in the 30 to 89 MHz range and to have an output up to 10 W Equivalent Isotropically Radiated Power (EIRP). Jamming is possible on up to three subfrequencies simultaneously on a time-sharing basis. For the integral capability, up to 10 subfrequencies can be jammed simultaneously using, again, time-sharing. Jamming modulations include

MEWSG's TRACSVAN can be deployed on land or aboard ship (MS/MEWSG)

MEWSG's latest land-based asset is the NEWVAN system shown here (Martin Streetly)

frequency shift keying noise, 'white' noise, fixed and variable audio tone and prerecorded 'spoof' traffic. Output is believed to be of the order of 750 W EIRP and is aimed at amplitude and frequency modulated, single sideband and continuous wave communications equipment. The NEWVAN system entered service during July 1993.

TRACSVAN
TRACSVAN consists of three vans and is designed principally for maritime and air defence training and exercise work. TRACSVAN capabilities include a 0.5 to 18 GHz radar simulator, 0.2 to 17.2 GHz jamming, electronic support measures coverage over the microwave and communications band and 3 MHz to 1 GHz communications band jamming. Each system is housed in a 6 m container and is operated by a three-man crew. During the mid- to late 1990s, MEWSG planned to upgrade the TRACSVAN system so that it would be able to simulate up to eight radar emitters (including frequency and pulse-compression types) and 'look-through' its communications band jamming output. As of August 2001, it was not clear as to whether or not this upgrade has taken place.

VERIFIED

SOUTH AFRICA

MRSS-200 mini radar signal simulator

Type
Portable radar signal simulator.

Description
MRSS-200 is a portable, microprocessor-controlled unit that transmits simulated radar signals for flight line testing and maintenance of aircraft electronic warfare systems. The system is capable of generating tuneable (1.6 MHz stops) microwave signals in the 0.5 to 18 GHz frequency band. Each signal generated can be either pulse, pulse-Doppler or continuous wave modulated. All signals are transmitted via self-contained antennas within the unit. Alternately, a front panel jack is provided for connecting the MRSS directly to an antenna port of a radar warning receiver or electronic countermeasures system using a flexible coaxial cable. MRSS can be either held by the operator or be tripod mounted for remote operation. Threat programming is executed on a PC and is loaded to a PCMCIA card. This is then inserted into the MRSS-200 for transmission. Such an approach is claimed to offer a flexible and secure threat data control environment. Several tripod-mounted units can be operated simultaneously via remote control to simulate a multi-emitter environment with spatial separation of emission.

Status
Over time, the MRSS-200 radar simulator is reported as having been procured by South Africa's armed forces and a number of international customers.

Specifications
Dimensions: 550 × 160 × 360 mm
Weight: 15 kg

Contractor
Avitronics (Pty) Ltd (a Saab and Grintek company), Centurion, Pretoria.

VERIFIED

The MRSS-200 in use

SPAIN

SIGEL Electronic Warfare (EW) simulator

Type
EW training system.

Description
SIGEL is a simulator designed for training pilots and EW specialists. The equipment can simulate all radio frequency signals of interest, analyse them and generate the appropriate countermeasures. The system's main elements comprise a simulation module, a Radio Frequency (RF) module and an intercom system that allows users to listen to the audio signals generated. The simulation module, which can be enlarged if required and is linked to the RF module by a local area network, is understood to have one instructor and four student positions. The system's man/machine interface uses a Windows environment in order to minimise learning time and reduce the possibility of human error. The simulation module used is a modified version of one developed by Thales Airborne Systems and can create, manage and maintain parametric databases of emitter types and the EW equipment installed aboard Spanish Air Force aircraft. SIGEL's simulation module also generates operational scenarios, using target and sensor data input by the instructor. The modes and operating sequences of the various systems modelled all appear in a realistic fashion, as do mobile platforms (own and enemy aircraft, ships or ground vehicles). The architecture's RF module generates the radar signals (including frequency-agile and other complex waveforms) and analyses them. It may also operate as an electronic intelligence system and can, with the addition of an appropriate interface, be used to activate real-world EW equipment. The module has been improved several times since its introduction, mainly via increases in its analysis and signal generation performance. SIGEL uses Hewlett-Packard computers (running Unix) as the instructor and student workstations and, in part, the RF frequency module. The software used was developed by INDRA DTD and architecture is noted as being able to be configured to represent ground-based or naval platforms.

Status
Over time, the SIGEL EW simulator is reported as having been procured by the Spanish Air Force.

Contractor
INDRA DTD, Madrid.

VERIFIED

SWEDEN

Erijammer A100 Responsive Electronic Warfare Training System (REWTS)

Type
Airborne jamming pod for training and countermeasures support.

Description
Erijammer A100 is a manually or automatically computer-controlled pod for electronic counter-countermeasures and Electronic CounterMeasures (ECM) training of air defence fighters and anti-aircraft radar operators. It provides the operator with a high degree of situational awareness via a combination of the system's integral radar warning receiver, set on receiver and jammer look-through mode. Over 45 smart noise, advanced range, velocity and angle deception modes are available. The pod is also capable of providing simulation of missile seeker radars. Single and/or multithreat capability is provided by a frequency memory loop, the set on receiver and selectable bandwidths.

The pod is entirely self-contained and requires only power from the carrier aircraft. The system is controlled by an ECM operator and features an electronically erasable, programmable read-only memory. A100 is 100 per cent programmable in flight. The analysis and subsequent jamming on incoming signals combined with 360° antenna coverage with selectable high- and low-gain antennas, gives the system unique flexibility in aggressor flying and cost-effective training.

The Erijammer A100 training pod mounted on a Swiss PC-9 aircraft   0010007

A close up of the A100 installation on the PC-9 (Saab Avionics)   2002/0054969

Status
As of July 2001, Erijammer A100 was in service with the Canadian, Swedish and Swiss air forces. In Canadian service, A100 is designated as ALQ-503.

Specifications
Frequency coverage: 6.8-10.5 GHz
Output power: 350 W, ERP 1-2 or 10 kW
Dimensions: 3.2 × 0.42 m
Weight: 210 kg
Speed: M0.20 to M1.0+
Lug mounting: 36 and 76 cm NATO standard
Cooling: ram-air

Contractor
Saab Avionics AB, Stockholm.

UPDATED

MI 745 training radar jammer

Type
G/H- (4 to 8 GHz) and I/low J-band (8 to 12 GHz) ground-mobile or airborne training radar jammer.

Description
MI 745 is a radar jammer that is designed to train anti-aircraft system operators in how to work in a countermeasures polluted environment and to act as a radar evaluation and test tool. The equipment comprises receiver, output generator and power supply subsystems together with +46 dBm power amplifiers, an array of transmit/receive antennas and a Windows-based laptop PC for system control. The system can be set up in a master-slave configuration with the slave stations being controlled from a central station via an RS-232 interface.

The MI 745's output generator subsystem incorporates four discrete units providing bidirectional transmissions in both the noted frequency bands. Effective radiated power (using a 40 W power amplifier and 22 dB gain horn transmission antennas) is given as 6.3 kW. The receiver subsystem is a microscanner type and operates across seven adjacent 40 MHz frequency slots. Having detected a signal, its carrier frequency is determined (in 1 MHz slots) together with its amplitude and Pulse Repetition Frequency (PRF), with the acquired data being passed to the

The MI 745 training radar jammer with its transmit/receive antenna array deployed

The MI 745 training radar jammer in travelling configuration

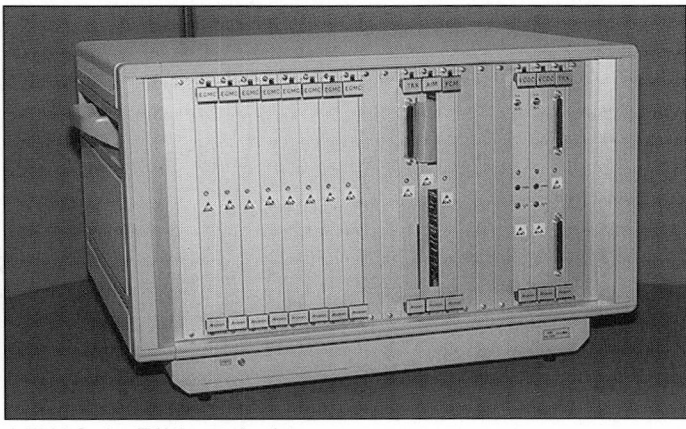

A 7000 Series EW threat simulator

control PC. Detected signals are revisited once every microsecond to check for changes in amplitude and PRF. The reception array can be aligned to the system's direction of transmission or can take the form of left/right bearing antennas. Best bearing accuracy is quoted as being 5°. The standard system's output is frequency modulated (noise and two sawtooth), amplitude modulated (noise and fluctuating frequency) and/or pulse modulated (fast train, slow train, triggered squall and synchronised) and is usually initiated on reception of the second detected pulse from the target radar. MI 745 can be customised (in terms of frequency coverage (2 to 4 GHz and 12 to 18 GHz options) and jamming mode/type) according to customer requirements.

Status
As of August 2001, MI 745 is reported to be available and to have been in service with the Swedish armed forces since 1995. The system is also noted as having been successfully tested against a number of different radar systems and as being supplied to civilian electronic warfare training contractor Saab Nyge Aero AB for use on the company's Learjet 35 and MU-2 mission aircraft.

Specifications
Frequency: G/H- (5.1-5.9 GHz sub-band) and I/low J-band (8.5-10.3 GHz sub-band)
Resolution: 1 MHz
Generator power output: 100 mW
Receiver sensitivity: <−60 dBm
MTBF: >2,000 h
Operating temperature: 0-70°C
Altitude: 12,000 m (max)

Contractor
μ-data/Microdata Innovation AB, Solna.

VERIFIED

UNITED KINGDOM

7000 series Electronic Warfare (EW) simulators

Type
Static mode EW threat simulators.

Description
7000 series EW simulators are available with digital or Radio Frequency (RF) outputs. The digital output card produces a standard 96-bit (plus time of arrival) word RF pulse descriptor that can be input into most electronic support measures processor databusses. It is intended to provide a simple to use test and evaluation facility for receiver processor systems or pulse de-interleaver algorithm developers. A single page emitter form-based software package is all that is needed to describe fully a radar's parametrics and to control the simulator.

Status
As of July 2001, 7000 series simulators were reported as having been procured by customers in Europe, the Middle East, Asia, the UK and the US. A 2 to 18 GHz RF variant (designated as RSS 7000 and capable of generating up to 16 static signals) is understood to have been acquired by customers in Australia (1998 sale), Egypt, Greece, Sweden and the US.

Specifications
Signals: 1-256
RF range: 0.5-18 GHz
Agile: PRI/F, RF and PW/D
Radar types: AI, target acquisition, AAA, surface search, air search, missile guidance, AEW, IFF

RF types: stable, hopper, random, chirp, Bi/quad-phase
PRI types: CW/pulse-Doppler, stable, jitter, stagger, bursts, grouped, switcher, slider, synchronised
Scan types: lock-on (+LORO), omni, circular, sector, raster, conical (+COSRO), helical, spiral, nodding

Contractor
EW Simulation Technology Ltd, Aldershot.

VERIFIED

7500 series threat simulators

Type
Range of Electronic Warfare (EW) threat simulators for system training, test and evaluation.

Description
The 7500 series is made up of real-time EW threat simulators that produce signal stimulus to drive external EW receivers. The product line-up includes the digital output Digital Electronic Warfare Simulator, video VTG and Radio Frequency (RF) (or optionally Intermediate Frequency (IF)) RSS simulators and the portable 7500/P equipment. The 7500 is built in a modular fashion from an industry standard VME bus 48 cm chassis. This chassis can be duplicated as many times as is needed to achieve the signal density or number of RF channels required. The digital video, IF or RF modules all fit, either separately or together, into the same single chassis. The RF module incorporates a high tuning speed Digital Tuning Oscillator (DTO) and other components needed to produce a single port RF output. A multiport card direction-finding option requires an additional half-height chassis for the attenuator stack. The DTO can be replaced, or added to, by a highly stable, phase coherent, synthesised RF channel. Standard chirp circuitry on the RF source controller card can be replaced to produce even more complex frequency modulation patterns including phase shift keying, frequency shift keying, pulse amplitude modulation and on/off keying.

7500 series simulators are operator controlled by a standard PC (desktop or laptop) via an intuitive Graphical User Interface (GUI) under Microsoft Windows.

A 7500 Series RSS simulator

The VME bus embedded real-time scenario engine power control unit (M680x0) operates under OS/9 real-time executive. All software is designed as easily portable code while the scenario engine runs on the PC or 680x0 to enable a future port into more powerful platforms (Intel Pentium, Power PC, DEC Alpha, SG MIPS) to be added. An embedded microprocessor pulse generator design enables the already impressive scan types to be added to via a simple software upgrade. 7500 equipments are designed to have high mean time between failure and low mean time to repair values, with each system module being field replaceable by the user. An automatic full system built-in test software programme identifies the faulty line replacement unit and maintenance charts advise a course of action to repair the system. A recent addition to the range is the portable 7500/P that covers the 2 to 18 GHz frequency range and can simulate up to 40 multiplexed emitters. The equipment features a user-friendly GUI software package for ease of scenario creation, definition and execution. 7500/P is also described as being easily reconfigured and/or upgraded because of its modular design.

Status

As of July 2001, the 7500 series customer base was reported as including EW system houses and the navies, armies and air forces of a 'wide range' of countries in Europe, the Americas, the Middle East and Asia. Sales during 1997 are understood to have included units for customers in Norway, Sweden and the UK. Procurements of 7500 series equipments during 1998 included those by customers in the UK and involved a system variant capable of generating angle of arrival direction-finding outputs. As of the given date, EW Simulation Technology is known to have developed an upgrade package for the 7500 series.

Specifications

Signals: 1-4,096 (complex)
RF range: 50 MHz-40 GHz
PRI range: 500 ns-800 μs (typical)
PW range: 25 ns-1.6 μs (typical)
RF output power: 0 dBm-200 kW (typical)
System control: PC with Windows NT
Number of RF channels: 32 (wideband)
Agile: PRI/F, RF and PW/D
Radar types: AI, target acquisition, AAA, surface search, air search, missile guidance, AEW, IFF
RF types: multibeam, stable, hopper, random, drift, chirp, Bi/quad-phase, FSK, PSK
PRI types: CW/pulse-Doppler, stable, jitter, stagger, bursts, grouped, switcher, slider, wobbulated, synchronised, PPM
Scan types: electronic, lock-on, omni, circular, sector, raster, conical (+COSRO), helical, spiral, nodding, TWS, lobe switching (+LORO)

Contractor

EW Simulation Technology Ltd, Aldershot.

VERIFIED

8000 series radar threat simulators

Type

Family of radar threat simulators for training, test and evaluation applications.

Description

Operating within the 50 MHz to 40 GHz frequency band, EW Simulation Technology's 8000 series radar threat simulators make use of the 'latest' integrated technologies to generate 'complex and accurate' radar signals. As such, the equipments are available in a variety of modular configurations that range from small portable units to large, multisource, multichannel applications. The equipment's manufacturer notes its belief in the 8000 series' ability to meet 'all' electronic warfare system test, evaluation and training requirements, either in the classroom or in an 'onboard' environment. Other system features include:

- a Pentium personal computer simulation controller
- C++ software
- Microsoft Windows 2000© functionality
- a VME64 bus architecture
- a 100 Mbps Ethernet control link
- an embedded PowerPC and VxWorks© operating system
- a six degrees of freedom receiver
- a real-time simulation engine
- dynamic updating of emitter parameters
- employment of live threat databases
- use of Director© dynamic scenario and Director/Lt Static test builders
- Microsoft Excel©-based pattern data entry
- a Microsoft Access© database engine
- a log of lost pulses due to collision
- scenario event file logging
- VCR-style control buttons
- in excess of 24 h game times
- pulse timing synchronisation and digital pulse descriptor outputs
- transportable 12U 48 cm (19 in) cabinets as standard
- automatic built-in test fault isolation to line replaceable unit level
- unattended radio frequency calibration
- a health monitoring regime

- optional maps and terrain masking
- rainfall effect amplitude jitter.

Status

As of August 2001, 8000 series radar threat simulators were reported as having been supplied to customers in the UK and the Middle and Far East.

Specifications

RF source/DF ports
Frequency coverage: 50 MHz-40 GHz
Frequency resolution: 1 MHz or 0.25%
Dynamic range: >90 dB
Noise: >90 dB/MHz
Spurious level: <−50 dBc
Harmonic level: <−60 dBc
General: 512,000 pattern points (azimuth, elevation and frequency cuts); AOA (amplitude), phase or DTOA DF options; independent port patterns; multiple RF source configurations; external fast or slow (GPIB) synthesiser; internal, fast tuning DTO or synthesiser; modular banded operation
Platforms
absolute or relative movement; altitude up to 30,408 m; curved earth modelling; flightpath (defined by waypoints); independent or convoyed platforms; movement over 3,704 km (X and Y); placement to 1 m (X, Y and Z); speeds of up to 3,704 km/h; straight or curved motion; targeted ('follow me') motion; turn rates of up to 180°/s; X, Y, Z and roll and pitch motion
Digital pulse generation
Number of emitters: up to 4,096 complex types
Scan amplitude calculation period: every 100 μs
General: 10 ns step AMOP, FMOP and PMOP; modular DPG card architecture; real-time geometry and path loss calculations; scan-to-pulse train synchronisation; simultaneous AMOP, FMOP or PMOP; unrestricted emitter agility
Emitters
Pulse density: in excess of 0.8 Mpps (for each RF source)
PRI range: 1 μs-800 ms
PRI resolution: 10 ns (0.1 ns Hi-Fi)
Pulse-width range: 25-50 ns and continuous wave
Pulse-width resolution: 25 ns
Modulations: agile, burst, cycler, discrete, doublet and triplet, drift, dwell, exponential, group, jitter, periodic, sawtooth, sinusoidal, stable, stagger, switcher, synchronised, triangular, user defined and wobble
Stagger positions: 4,096 (with 64,000 pulse repeats)
Jitter: uniform or Gaussian 1-99%
Scan patterns: circular, bidirectional raster, bidirectional sector, conical, electronic, helical, lobing, lock-on, multibeam, nodding, spiral, stable, track-while-scan, unidirectional raster, unidirectional sector and user defined
Scan rates: 0.01-200 Hz
Electronic beam dwell period: 100 μs-1 s
Antenna beam patterns: cos X, cos² X, cosec² X, cosine array, cosine taper, fan, isotropic, pencil, sin X/X and user defined
Antenna beamwidth: 0.5-40°
Beamwidth resolution: 0.1°
Antenna coverage: ±90° (elevation); ±180° (elevation)
DF antenna pattern modulation range: 64 dB

Contractor

EW Simulation Technology Ltd, Aldershot.

NEW ENTRY

Chameleon I series Electronic CounterMeasures (ECM) simulators

Type

Range of ECM simulators used for system training, test and evaluation.

Description

The Chameleon I series are ECM simulators which are designed for radar electronic counter-countermeasures testing and radar operator training. They are Digital Radio Frequency Memory (DRFM) technology-based real-time ECM equipments that produce jamming techniques and real target echo signal stimulus to stimulate the victim radar. The product line-up includes simulators producing single or dual channel outputs. User selection of a victim radar is made via a graphical amplitude timeline display that sets a frequency and amplitude threshold acceptance window. An all new (single channel) variant operating in the 7.5 to 18 GHz band operates with a fully automatic set on receiver that does not require an acceptance window as it automatically selects each victim radar as it appears.

Chameleon I is claimed to provide a cost-effective solution for ECM assessment in a variety of applications. The unit's design features enable realistic land-based, airborne or shipboard situations to be created in a mobile installation. Chameleon I equipments are built in a modular fashion on an industry standard VME bus 48 cm chassis. The system's single/twin DRFM and direct digital synthesiser modules all fit into this chassis. Chameleon I units can also be packaged in a full Air Transport Racking standard box form factor.

Chameleon I systems are operator controlled by a standard PC (desktop or laptop) running a graphical user interface under Microsoft Windows. All

A Chameleon I series ECM simulator

parameters needed to describe the various jamming techniques are entered via simple dialogue boxes. The Chameleon series features jamming technique daisy chaining to form a repeating sequence of different jamming modes. The real-time techniques engine uses an embedded VME bus power control unit (M680x0) which operates under the OS/9 real-time executive. All software is designed as easily portable code to facilitate transfer into more powerful platforms. An embedded microprocessor techniques generator enables new ECM types to be added via a simple software upgrade. As a whole, Chameleon I is designed to be both reliable and maintainable with each module being field replaceable by the user. An automatic built-in test software programme identifies the faulty line-replaceable unit and maintenance charts advise a course of action to repair the system.

Status

As of Ju;y 2001, Chameleon I systems were reported as having been procured by a number of armed services in Australia, Europe, the Americas and Asia (including Indonesia and Singapore).

Specifications

RF range: 1-18 GHz
Dynamic range: better than 50 dB
Bandwidth: 400 MHz (instantaneous)
Sensitivity: −55 dBm
ECM: noise (spot, barrage, swept, burst, blink) RGPO, VGPO, co-ordinated RGPO/VGPO, multiple false targets, range/velocity false targets, inverse gain, swept AM, recirculation/CW

Contractor

EW Simulation Technology Ltd, Aldershot.

VERIFIED

Chameleon II Electronic CounterMeasures (ECM) simulator

Type
ECM simulator

Description

Primarily designed for hardware-in-the-loop, high-fidelity applications (such as anechoic chamber system test and evaluation and radar performance testing), Chameleon II is an enhanced version of EW Simulation Technology's Chameleon I system (see separate entry), that incorporates two 8-bit amplitude Digital Radio Frequency Memory (DRFM) channels. Chameleon II DRFM features include:

- 1,000 µs memory depth and 16 ns delay resolution
- up to eight memory files
- a read/write interface to facilitate the creation of customer specific RF outputs
- ±50 MHz Doppler at 1 Hz resolution
- RF amplitude control and a programmable system threshold
- continuous wave operation and a recirculation mode

Other overall system features include:

- radar target modelling and ECM signal generation
- an instantaneous frequency measuring analysis receiver
- an internal VME bus structure
- a built-in test facility
- VxWorks™ real-time processing and Power PC™ implementation
- 48 cm (19 in) rack mounting
- Windows 2000™ graphic user interface software
- user defined ECM libraries
- 'fill-in-the-blanks' ECM techniques
- simulated targets with variable radar cross section modelling.

System options include radar target and clutter modelling.

Status

As of July 2001, the Chameleon II ECM simulator was reported as having been procured by customers in Australia (two), France and Switzerland.

Specifications
Frequency coverage: 1-18 GHz (continuous operation)
Bandwidth: 400 MHz (instantaneous)
Sensitivity: −55 dBm
Dynamic range: ≥60 dB
Output power: 0 dBm
Power requirement: 110/240 V AC
ECM techniques: amplitude modulation; co-ordinated RGPO/VGPO; inverse gain; noise (blink, burst, spot and swept); range/frequency false targets; RGPO; VGPO

Contractor
EW Simulation Technology Ltd, Aldershot.

UPDATED

Chemring Countermeasures Radar Cross Section (RCS) profiler

Type
RCS data capture and analysis system.

Description

Chemring Countermeasures (a division of PW Defence Ltd) has developed a portable RCS profiler package which can be integrated into a tracking radar to capture and analyse RCS data generated by an illuminated target object. While its primary purpose is the establishment of chaff cloud RCS, the package can also be used to measure the performance of on- and offboard electronic countermeasures systems. Functionally, the system records intermediate frequency in-phase and quadrature data at the radar pulse repetition frequency level. In a typical chaff analysis application (measurement of RCS amplitude and Doppler), the radar's transmit/receive triggers are used as reference to generate a series of range gates relative to the target. Up to 32 of these gates can be created and can be positioned ahead of and/or to the rear of the target. The radar's elevation and azimuth signals are also recorded so that target track may be plotted. Both real-time and subsequent analysis displays (using a wide range of presentation formats) are available.

Status

Over time, the Chemring Countermeasures RCS profiler package is noted as having been used in a number of NATO and UK chaff trials in conjunction with a number of European fire-control radars.

Contractor

Chemring Countermeasures (a division of PW Defence Ltd), Salisbury.

UPDATED

EES series threat simulators

Type
Range of Radio/Intermediate Frequency (RF/IF) threat simulators.

Description

BAE Systems Defence Training Solutions (DTS - formerly Reflectone UK Ltd) produces a range of RF/IF threat simulators for use in Electronic Warfare (EW) and research and development roles. The series comprises the EES 400, 500 and 600 families, the known details of which are as follows:

EES 400 series

EES 400 simulators are based on RISC processing and offer multiple threat simulations (up to 1,500) for superheterodyne receiver systems operating in the 0.1 to 18 GHz (optional extension to 40 GHz if required) frequency range. Inputs can be at IF or RF level and the receiver's tuner is used for real-time selection of threat descriptors. These are used to generate the appropriate RF signature which is injected into the receiver's IF or RF port. The equipment also offers amplitude comparison and scanning direction-finding antenna interfaces, as well as time and direction of arrival.

EES 500 series

EES 500 equipments are generally similar to the EES 400 series and are optimised to match required signal densities. The equipments can also be linked in real time to other simulators. Ada software is used for emission simulation in both combat and test/performance proving scenarios.

EES 600 series

EES 600 series equipments are configured for EW system validation and acceptance testing work. RF formats generated include those associated with multiple continuous wave, pulse and exotic emitter types (up to and including multimode formats) and modelling real-time air/land engagements, terrain masking, blind arcs and network site screening in relation to target aircraft altitude. As with the EES 500 series, EES 600 equipments can be connected to other simulators.

Status

Over time, EES series equipments are reported as having been procured by the UK's Ministry of Defence (EES 400/500) and former Defence Evaluation and Research Agency (EES 600).

Contractor

BAE Systems Defence Training Solutions (DTS), Bristol.

UPDATED

Electronic Combat Modelling and Evaluation System (ECMES)

Type

Electronic combat simulation system.

Description

ECMES is a digital simulation of a total electronic combat system for use in design and assessment of performance/ operational effectiveness. It is a pulse by pulse, event-driven model where each equipment is simulated at a functional level. ECMES was designed as a modular system so that, if required, a user may increase the level of detail for specific subsystems by adding new program modules. The architecture also incorporates a facility by which, with simple changes to the ECMES framework, completely independent system models can interact with the rest of ECMES via defined interfaces. A redisplay facility is now included which offers a 'video playback' for the online graphical analysis, the data being logged during an execution run. The ECMES program consists of a model framework and a set of user-configurable simulation modules to represent the electronic engagement and determine the effectiveness of various forms of electronic support/countermeasures, electronic counter-countermeasures and radar or infra-red tracking techniques. It provides an environment model through which the electromagnetic interaction between a large number of independent, user-defined platforms can be modelled at a pulse by pulse level.

Status

Over time, ECMES systems are reported as having been procured by a number of research and training establishments in Europe and North America.

Contractor

IBM Global Services, Farnborough.

VERIFIED

Electronic Warfare Evaluation System (EWES)

Type

Electronic Support Measures (ESM) simulation and training system.

Description

EWES is an ESM simulation system for use in system design and assessment. It consists of an integrated suite of computer programs that can be used to assess the performance of modern ESM systems in a controlled and objective manner. The simulation is carried out at a pulse by pulse level, providing an unrivalled ability to analyse system performance in detail. Analysis is aided by reports that are internally generated by the simulation programs and integrated graphical analysis facilities. The programs are all independent (transferring data by means of disk files), thereby allowing users to stimulate the models with their own trials data and use their own special purpose analysis routines if required.

The EWES Scenario Generator (EWSG) can represent up to 6,000 platforms and a total of up to 12,000 emitters. The generation and propagation of each electromagnetic pulse is modelled in detail, including effects caused by the environment, terrain masking and ground reflection. It provides a powerful and flexible method of describing the parametric, scan and antenna definitions of each emitter and records the arrival of the electromagnetic pulses at a sensor in the scenario.

The EWES Processor Model (EWPM) can be stimulated with data from EWSG, the EWES receiver model or the users' own data. EWPM is a powerful and generic model that attempts to de-interleave the received signals to form tracks, to analyse those tracks and to classify them against known emitters from a library.

The EWES suite provides full life cycle support for the evaluation of ground-based, shipborne and airborne ESM systems and has found many uses in research and development, EW equipment design and training.

Status

Over time, EWES systems are reported as having been procured by a number of research, development and training establishments in Europe, North and South America and the Middle East.

Contractor

IBM Global Services, Farnborough.

VERIFIED

Electronic Warfare Training System – Radar (EWTS-R)

Type

Electronic Warfare (EW) simulation and training system.

Description

EWTS-R is a flexible computer-based facility designed specifically to allow operators to develop and maintain the skills necessary to provide effective EW performance.

High-performance commercially available workstations, communicating via a high-speed local area network, ensures that the EWTS-R meets the most stringent needs of the EW training requirement. The facility's workstations are fully equipped with colour displays giving graphic and alphanumeric readouts, audio networks and a range of interfaces for emulated EW control.

Generic and customer-specific EW equipments are emulated in the software and stored at the instructor's position. At system set up, the downloading of the relevant emulation software to individual student workstations provides realistic equipment displays and controls. Software is written in Ada for maximum portability, ease of maintenance and subsequent enhancement. The system can be expanded to meet changing requirements such as additional student workstations, new EW equipments and/or additional processing power, with minimal changes to the structure of system software or hardware architecture.

EWTS simulation options already available, or in development, include electronic support measures signal characterisation/classification and threat environment analysis; electronic countermeasures (including barrage, swept, spot, continuous wave noise, cover pulse, false target, amplitude modulated blinking and range/velocity/angle gate pull off jamming modulations); radar warning receiver (control, position fixing and air intercept), electronic intelligence (pulse and pulse chain characterisation, classification and scan pattern recognition); communications intelligence (data gathering and reports, training, communications band direction-finding and jamming and resource allocation).

Status

Over time, EWTS-R systems are understood to have been procured by a number of research and training establishments throughout Europe and North America.

Contractor

IBM Global Services, Farnborough.

VERIFIED

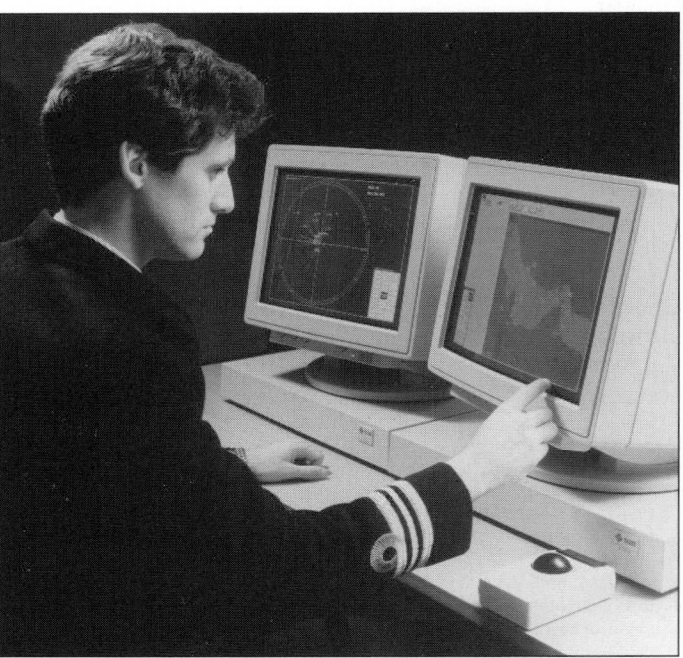

The EWTS-R training system

Epsilon™ Radar Cross-Section (RCS) analysis tool

Type

RCS prediction and analysis tool.

Description

Roke Manor Research has developed Epsilon™ for the RCS prediction and analysis of land, sea and air platforms. True parallel computing on low-cost, parallel systems permits large and complex targets to be modelled quickly and effectively. Epsilon™ exploits direct-input geometry from Computer Aided Design/ Engineering (CAD/CAE) files to provide considerable savings in both time and cost of modelling. Epsilon™ utilises the PATRAN (MacNeal Schwendler Co Ltd) neutral file system as an interface to the raw surface geometry model in the designer's

CAD/CAE system. When the geometry model is read into Epsilon, the scattered field is calculated. This comprises specular, multiple-bounce and re-entrant reflections (which arise from ducts, cavities and intakes) to provide an accurate solution. Multiple scattering and shading effects are automatically generated and require no operator intervention. Integrating the individual contributions to the scattered field from all the illuminated parts of the target model gives its RCS at a particular aspect to the observing radar(s), making the simulation as near to real life as possible.

Status

As of December 2001, the Epsilon™ RCS prediction and analysis tool was understood to be a live programme and had been billed as 'probably the world's most widely used RCS prediction code'.

Contractor

Roke Manor Research Ltd, Romsey.

UPDATED

....

Flexible Attrition Model (FAM)

Type

Land/air battle simulator.

Description

FAM is a simulation model for representing the land/air battle. It simulates the interaction between ground-based defence assets and the aircraft's defensive aids system. It operates at mission level allowing many-on-many interactions to be investigated. FAM is highly modular being based on object oriented design principles, allowing a library of alternative modules to be selected and used (radar warning receivers, electronic countermeasures pods and other avionics subsystems being examples). Full ground radar coverage and digital terrain is incorporated and the system operates under UNIX on most graphics workstations.

Status

Over time, FAM systems are reported as having been procured by a number of research and training establishments in Europe and North America.

Contractor

IBM Global Services, Farnborough.

VERIFIED

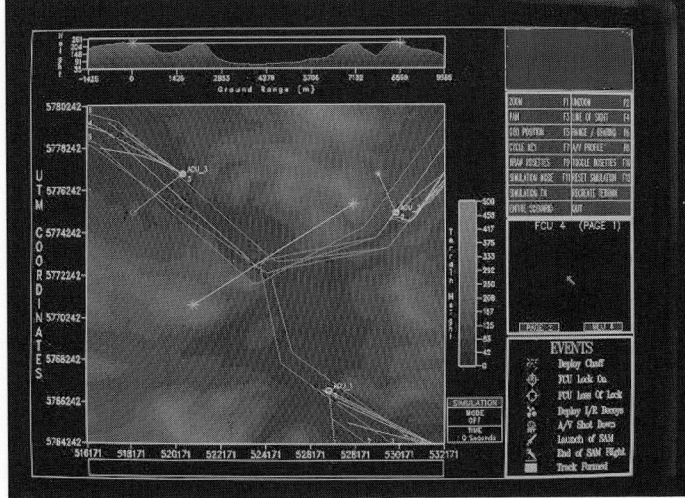

The FAM screen picture

....

Mobile Electronic warfare and Radar Test System (MERTS)

Type

Turnkey radar and Electronic Warfare (EW) test and evaluation system.

Description

The MERTS mobile radar and EW test and evaluation system is a vehicle-mounted (Mercedes Unimog four-wheel drive chassis), turnkey equipment that covers the 2 to 18 GHz frequency band and incorporates EW Simulation Technology's Chameleon Electronic CounterMeasures (ECM) simulator (see separate entry) and its RSS7500 radar threat generator as primary subsystems. Other equipment features include:

- 'full mobility'
- an onboard, diesel-powered, 18 kW, AC generator

- a stabilised antenna system (comprising three parabolic reflectors with 'high-power' horn feeds) with television camera control (alignment and tracking)
- use of 'high power' travelling wave tube amplifiers
- adjustable receive and transmit polarisation
- use of an instantaneous frequency measuring analysis receiver and digital radio frequency memory technology
- a fully stabilised azimuth and elevation positioner
- integrated radar threat/ECM software
- emitter/platform/scenario databases
- the generation of up to 16 threats simultaneously as standard
- 'fill-in-the-blanks' ECM techniques
- Windows™ GUI software
- user defined ECM libraries
- an air-conditioned, International Standards Organisation compatible operator/equipment shelter
- 'full' Chameleon and RSS7500 product features.

Status

As of August 2001, a MERTS application was understood to be in service with the Indonesian armed forces and to have been used in an operational role against Australian forces engaged in peacekeeping operations in East Timor.

Specifications

Frequency: 2-18 GHz (other bands available)
Bandwidth: 400 MHz (ECM subsystem DRFM)
Sensitivity: −55 dBm (ECM subsystem receiver)
Dynamic range: >60 dB (ECM subsystem)
ERP: up to 200 kW (radar threat subsystem)
Antenna gain: up to +30 dBi
ECM techniques: amplitude modulation, blink/burst/spot/swept noise, co-ordinated RPGO/VGPO, inverse gain, range/frequency false targets, RGPO and VGPO

Contractor

EW Simulation Technology Ltd, Aldershot.

NEW ENTRY

....

Modular Electronic Warfare (EW) training suite

Type

Suite of stand-alone/networked software programmes for use in EW training courses.

Description

The Dundridge College EW training centre has developed a suite of software programmes which are designed for use in advanced EW operator and aircrew training and run on standard IBM compatible PCs using Windows 3.X, Windows 95 or Windows NT. The suite architecture is such as to allow individual tuition at single workstations or networked activity using multiple PCs to create an interactive training environment. To support the training packages, instructor-led instruction and interactive training modules are available on a wide range of EW and related topics including radar, communications, signals intelligence, Electronic Support Measures (ESM), electronic countermeasures/counter-countermeasures, electro-optics, expendables, suppression of enemy air defences, EW tactics and mission planning. The described software suite includes the following modules:

EW database
The EW database stores information on EW equipment, associated platforms and weapons. It can be supplied with unclassified data that can be added to by the customer using their own classified/unclassified data.

Electronic Order of Battle (EOB)
The EOB simulation is designed for ELectronic INTelligence (ELINT) operator training using a generic ESM display which can be tailored to represent equipment operated by the customer. While in the EOB simulation, operators can access the EW database to identify detected emitters.

Recognised Air and Surface Picture (RASP)
The RASP module combines radar data with ELINT data derived from the EOB simulation. Two ELINT operators can be networked with a RASP co-ordinator to permit real-time training in ELINT data gathering and the co-ordination of remote sensing sites for emitter location.

Defensive Aids SubSystem (DASS)
The DASS simulation allows students to plan an EW mission and to 'fly' through a realistic threat scenario using a preflight message they have designed and programmed themselves. A digital terrain database is used to provide realistic terrain masking effects and by combining the DASS with ELINT data derived from the EOB and RASP modules, realistic EW wargames can be run.

Status

As of 2001, Dundridge College was understood to be providing EW training to a number of armed forces around the world. The establishment is noted as offering residential EW courses (basic to advanced) and specialist training in areas such as EW database management and equipment programming. In early 1996, it was awarded a three year contract to support Royal Air Force EW training with wargaming simulations and expertise. As of the given date, readers should note that Dundridge College is understood to be a part of Alenia Marconi Systems (an

Alenia-BAE Systems joint venture). Previously, the establishment was a subsidiary of GEC-Marconi Radar and Defence Systems Ltd's Simulation and Training Division.

VERIFIED

OnBoard Trainer (OBT) 7500

Type
Digital output Electronic Warfare (EW) threat simulator for operator training on board ship or submarine.

Description
OBT is part of the EW Simulation Technology 7500 series of EW simulators. It is designed as a ruggedised unit so that it can withstand the harsh environment found on board naval platforms. It generates digital data sets as RF pulse descriptors to mimic the live environment which surrounds the host platform at sea. Because these pulse descriptors are intended for direct injection into the Electronic Support Measures (ESM) digital data processor, the OBT fully emulates the surrounding electromagnetic environment, antenna system and receiver front end. Environment models are built into the software and systematic receiver models handled by dedicated hardware, both of which avoid 'negative training'. A single field-replaceable module carries out the specific ESM emulation function thereby allowing the unit to be easily reconfigured to handle a wide range of receiver systems. EW Simulation Technology now has experience of over 10 different receiver/processor combinations.

The unit either fits into the ESM receiver or console racks or is carried on board in a shockproof box as a stand-alone trainer. In both instances the OBT is built into a compact self-enclosed industry standard VME bus-based 48 cm chassis. The OBT is wired for expandability to accept extra pulse density and can be fitted on site by trained maintenance personnel. Additional chassis can be added at a later date to expand the unit.

The OBT is controlled by an embedded microprocessor card on which runs the six degrees of freedom simulation engine software. Scenarios are normally generated at a central location (such as a training establishment) and then distributed to the fleet as controlled exercises on industry standard Flash memory cards. The EW instructor can optionally edit these exercises or create his own on board ship. A rugged hand-held terminal acts as the controller of the OBT where space is limited. Elsewhere, the equipment is controlled via a portable rugged laptop computer. The operator interacts with the OBT through a simple menu and option list user interface on the hand-held terminal or an intuitive Microsoft Windows-based graphical user interface control programme on the PC.

Status
As of July 2001, OBT 7500 was reported as having been procured by the Royal Navy and the armed forces of a number of European (including the Netherlands Navy) and Asian countries. According to Jane's sources, EW Simulation Technology had delivered 50 OBT units by the end of 1998.

Specifications
Frequency coverage: 0.05-18 GHz (fixed, agile, hopper, jitter, chirp emitter types)
PRI: 0.5 µs-0.8 s (fixed, stagger, jitter, periodic, switcher, grouped, burst, user defined types)
Pulse-width: 25 ns - CW (fixed, agile, stagger, jitter, CW, user defined types)
Scan: 0.01-100 Hz (lock on, circular, sector, raster, spiral, conical, Palmer, nodding, helical, lobing, TWS, electronic, synchronised multibeam types)
Beamwidth: 1-40°

Contractor
EW Simulation Technology Ltd, Aldershot.

UPDATED

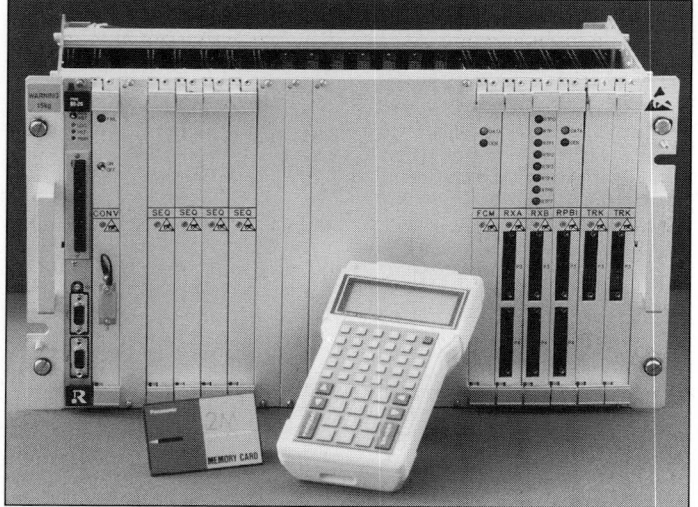

The OBT EW threat simulator

PROTEUS air defence simulator

Type
Simulation and design tool for air defence systems.

Description
PROTEUS is a simulator that is used as a design tool for integrated air defence systems. It can be used to assist with design issues and to evaluate overall system performance against a set of standard measures of effectiveness. PROTEUS represents multiple ground-based radar systems, communication networks, command post processing and terrain effects. Command post functions include multiradar tracking, correlation and recognised air picture distribution. Identification friend-or-foe systems are also represented. PROTEUS is currently being used by UK and US government agencies and also by air defence system manufacturers. It is available on SUN and DEC workstations. It is a readily usable design tool with good graphics displays and simple man/machine interface.

Status
Over time, PROTEUS systems are reported as having been procured by a number of research and training establishments in Europe and North America.

Contractor
IBM Global Services, Farnborough.

VERIFIED

RBOT radar trainer

Type
Radar training system.

Description
RBOT is an advanced radar picture compilation training system. Written in Ada, the RBOT system is built around high-performance SUN Sparc processing stations connected (via an Ethernet local area network) into general purpose student consoles. Each console is equipped with a trackball, keyboard and special keypad. The workstation's high-resolution screen emulates the radar's Level Plan Display. The RBOT architecture incorporates 16 student consoles and two instructor positions. Students may be trained individually or consoles may be linked in units of four to represent the air, surface and Link 11 consoles on a Royal Netherlands Navy frigate. In this mode, students will act as a team on a common radar picture. Track data may be passed between 'frigates' by simulated datalink. An intercom system allows operators to communicate with one another and with the instructors. Using an advanced and highly flexible Windows-based graphical user interface, the instructors are able to define and control scenarios. Student performance is automatically monitored and recorded by the system.

Status
RBOT has been developed by a joint Dutch/UK team. Over time, the system is reported as having been procured by the Royal Netherlands Navy.

Contractor
IBM Global Services, Farnborough (UK contractor).

VERIFIED

The RBOT radar training system (MS/IBM Global Services)

Talisman II Electronic Warfare (EW) training system

Type
Classroom EW training system.

Description
Talisman II is a self-contained EW classroom training system, custom designed to take new students step-by-step through the electronic support measures/ electronic intelligence operator learning process. The system can be readily

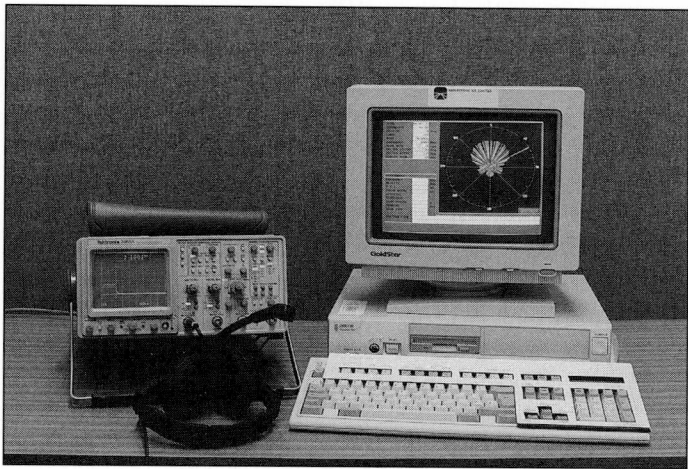

The Talisman II classroom student console

extended to provide advanced training on operational in-service EW equipment. It provides practical reinforcement of theory and maximises the effectiveness of any EW training source. Talisman II is suited to operators at a variety of levels, from new recruits undergoing introductory exercises, to the experienced operator going through operational conversion.

Training is carried out on individual consoles under instructor control. The consoles provide a wide range of computer-based displays and controls which are representative of modern EW systems. Exercise scenarios are held on magnetic tape and are reusable and upgradeable. Comprehensive debrief facilities are also provided. Up to 10 consoles may be included in the network. Exercise scenarios are prepared by the customer's instructional staff and are held on hard disk or magnetic tape. These are reusable or can be amended or updated. Full emitter programming provides Radio Frequency (RF), pulse-width, pulse repetition frequency, stagger/jitter parameters, scan type, effective radiated power, and antenna characteristics. Moving emitter and EW platforms are also included, together with programmable terrain and propagation effects. Software is provided to allow efficient management of the emitter database. For advanced training the core instructor system can also drive an RF signal generator for cable-coupled injection of signals into operational in-service EW equipments, either classroom-based or within an RF interference screened portable container.

Portable RF Simulator

This has been designed as an extension to the Talisman family and is programmed by scenarios created on similar lines to the Talisman core system using a ruggedised rack-mounted SBC workstation with data being saved to floppy disk. The system consists of two rugged, man-portable, splashproof cabinets plus antenna dishes according to the frequency range required. The main features of the simulator are:

- provides live added realism to 'on-the-job' EW training for electronic support measures and radar system operators
- provides electronic countermeasures/jamming training for radar operators
- operates over octave and multi-octave bandwidths
- provides 16 pulsed emitters simultaneously or one Continuous Wave (CW) emitter or one electronic countermeasures signal
- fully programmable
- provides frequency agility
- includes pulse repetition interval jitter and stagger modes
- noise or pulse jamming
- 200 W (CW) and 2 kW travelling wave tube amplifiers
- variety of antenna options.

Status

Over time, a Talisman E RF cable-coupled variant of Talisman II is reported to have been delivered to the UK's Ministry of Defence during 1995. This equipment is also noted as having been procured by the British Army for 'advanced' electronic support measures training.

Specifications

Talisman II
Scenario characteristics
Area: 1,000 km × 1,000 km × 99,999 m
Resolution: 0.1 km (range); 1 m (height)
Platform speed: up to 7,408 km/h
Climb/descent: up to 50,000 m/s
Event resolution: 1 s
Number of platforms: 256
Number of emitters per platform: 9
Emitter characteristics
Frequency range: 0.5-18 GHz (0.5-2, 2-4, 4-8, 8-12, 12 and 18 GHz)
Accuracy: ±1 MHz (±50 Hz at RF)
Simultaneous emitters: 96
Emitter characteristics: AOA, jitter, on/off, peak amplitude, PRF/PRI, pulse-width, scan pattern rate and stagger
Polarisation: horizontal, vertical and 43 or 133° from vertical
Scan types: circular, conical sector, fixed, helical, interleaved raster, Palmer dynamic, Palmer raster and spiral

Antenna patterns: 15 user defined beam shapes with 255 amplitude × 255 azimuth steps per emitter; programmable back and sidelobes

Contractor

BAE Systems Defence Training Solutions (DTS), Bristol.

UPDATED

Warfare communications and Electronic Warfare (EW) modelling simulation

Type

Communications and EW modelling software package.

Description

The Warfare software package is designed to model the performance of communication radio systems operating across the 30 MHz to 38 GHz plus frequency range and their vulnerability to countermeasures. In terms of radio performance prediction, the package models point to point and wide area applications (using fixed and mobile ground-to-ground and ground-to-air transmitters) together with natural interference, network traffic flows and vulnerability to interception and jamming. All propagation algorithms used are noted as having been internationally validated and verified for the frequency range described.

Within the specific EW field, Warfare is able to model own intercept and jamming (including wideband, frequency-hopping units) assets in order to predict performance against known or deduced hostile communications links. This includes a numerical assessment of the percentage of links within the target net disrupted. The limitations of jamming, direction-finding and signals intelligence systems can be assessed with the data being displayed graphically. In order to overcome perceived equipment weaknesses, the package also includes tools to optimise the location and parameters of available EW assets. Outside its modelling role, Warfare can be directly connected to EW systems and used as a mission management tool. Here, the package is used to collate hostile traffic to establish the enemy's communications architecture. The operator uses this data to run simulations designed to establish the best countermeasures option and to configure and initiate available jammer assets accordingly.

Alongside the communications bands, Warfare includes ground-based radar modules with which to plan the deployment of tactical air control, anti-armour and air defence systems. As a whole, the package runs under Windows 95/NT on a standard Pentium-based PC and includes an ODBC interface for databases such as ORACLE and Microsoft ACCESS. Other databases and command information system interfacing can be configured for specific applications and simulation results can be exported to the ARC/INFO system. Terrain data used is derived from the DTED NATO standard or other sources. Matching overlays of digitised maps or satellite images allow the operator to relate terrain to tactically important features. As an optional extra, Advanced Topographic Development & Images (ATDI) can provide the Image Cartography System Map Server tool that provides the ability to import a wide range of geographic data formats and generate new terrain models from digitised maps if no other data on the target region is available.

Status

Over time, the Warfare system is reported as having been procured by the British, Finnish, French and Israel armies, the French Air Force and Navy, the Royal Air Force and a number of intelligence services around the world. Within the British Army procurement, the baseline system is understood to have been complemented by an EW training variant while the French Army is understood to have used the system as an EW mission management tool.

Contractor

Advanced Topographic Development & Images Ltd, Crawley.

VERIFIED

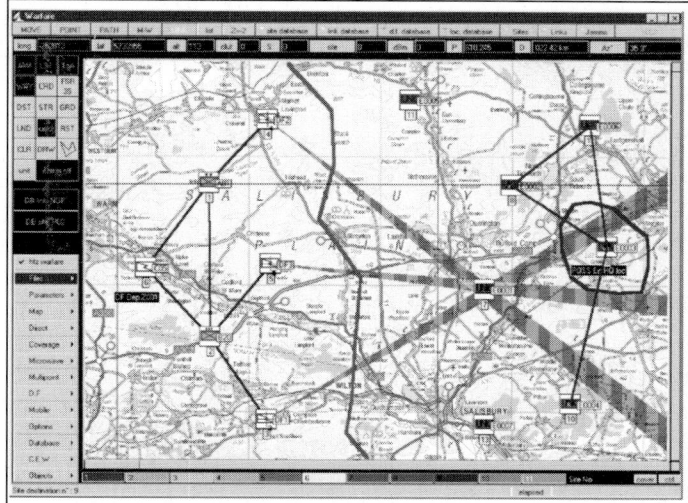

One of the display formats generated by the Warfare simulation   0017875

Weapon System Model (WSM)

Type

Weapon system simulation system.

Description

WSM is a flexible and generic weapon system engagement model that can be used to support concept and system studies, system integration and subsystem trade-offs and procurement assessment. The system is equipped with digital terrain and culture databases, using digital terrain elevation data/digital feature analysis data and/or digital chart of the world data. Key model elements provided are a six degrees of freedom weapon body; propulsion and retardation; autopilot and instrumentation; millimetric-wave, infra-red and laser radar seekers; guidance laws and navigation systems. The model elements provided by the WSM allow complex weapons, such as cruise missiles and precision guided munitions to be represented quickly and simply. The model framework allows for significant enhancement of existing elements and the addition of new ones at relatively low cost. The system has an advanced graphical user interface for model definition, results analysis and presentation. Extensive use is made of forms and direct graphical inputs to provide a flexible and intuitive tool. Generic libraries of weapon components, complete weapons and targets can be defined, which can then be deployed in any number of scenarios by simple selection and configuration for use. The comprehensive analysis facilities include scenario preview and playback; data tabulation; statistical analysis; data plotting (2-D and 3-D); and animation of the weapon body, weapon flyout and target close-up.

Status

Over time, the WSM system is reported as having been procured by an arm of the UK's Ministry of Defence.

Contractor

IBM Global Services, Farnborough.

VERIFIED

UNITED STATES OF AMERICA

0-1843(V)/ULQ-21(V) waveform controller oscillator

Type

Low-cost radar jammer for training and testing.

Description

The 0-1843(V)/ULQ-21(V) waveform controller oscillator is designed for use in ground-based, shipborne and airborne applications where it produces low-power radio frequency jamming signals over the frequency range 0.5 to 18 GHz. The unit can be pod-mounted for carriage on high-speed tactical aircraft or rack-mounted for internal carriage on aircraft or drones. Items such as a high-power amplifier, antenna and digital radio frequency memory can be added to the system to form a deployable Electronic CounterMeasures (ECM) system. The unit is microprocessor-controlled and can serve as the system controller over other subassemblies in the ECM system. The unit can store up to 16 technique/parameter sets in electronically erasable, programmable read-only memory for a preplanned response. It can also be controlled via an RS-232 interface. As a whole, the ULQ-21(V) system is capable of generating various types of noise and deception techniques, including continuous noise, blinking noise, noise with swept amplitude modulation, swept noise, repeater with swept amplitude modulation, range gate stealer, range gate stealer with swept amplitude modulation and range gate stealer with swept amplitude modulation hold. ULQ-21(V) can also be combined with a velocity deception techniques generator and set on receiver to provide Doppler and fast set on techniques.

Status

Over time, the 0-1843(V)/ULQ-21(V) oscillator is reported as having formed the basis of a number of in-service countermeasures training systems.

Contractor

Rodale Electronics Inc, Hauppauge, New York.

UPDATED

AFEWES simulator

Type

Laboratory Electronic Warfare (EW) evaluation simulator for the US Air Force.

Description

The US Air Force's (USAF) Air Force EW Evaluation Simulator (AFEWES) is a major service test facility that is operated and maintained by Lockheed Martin Aeronautical Systems. The Electronic CounterMeasures (ECM) equipment on all major USAF aircraft normally passes through AFEWES several times before operational testing, test flight confirmation, or deployment. It is also used by the US

Army, US Navy and friendly foreign governments. AFEWES is a secure laboratory where EW hardware is evaluated in a simulated threat environment. The threat simulators are specifically designed transmitters/receivers with simulated antenna patterns and target aircraft. Signal levels are continuously adjusted to simulate the effects of range, aircraft movement, antenna scanning and other factors occurring in actual combat environments. AFEWES can simulate ECM systems so that techniques and concepts can be evaluated. AFEWES has been in existence for over 40 years and has been upgraded on numerous occasions to incorporate new threats, new technology and new EW areas. AFEWES hardware resources include:

- individual radar and threat system simulators, hardware systems, complete with operational consoles and Radio Frequency (RF) receivers, designed to simulate the actual circuitry of specific threat weapon systems
- a multiple emitter generator which simultaneously simulates multiple, dynamic radar signatures at numerous scenario locations to test (stimulate) EW receivers and processors
- a communications/datalink simulator which determines RF jamming effects on voice and datalink communications
- a jammer techniques simulator which generates actual or conceptual ECM techniques at RF
- an infra-red simulator - provides the necessary elements and environments to evaluate Infra-Red (IR) missile seeking heads and IR countermeasures with or without the associated background.

Status

Over time, AFEWES was reported as being US Department of Defense high-level architecture compliant and as coming under the control of the USAF's 412th Test Wing (Edwards Air Force Base, California).

VERIFIED

Advanced Multiple Environment Simulator (AMES) series Electronic Warfare (EW) simulators

Type

Family of multiple environment EW simulators.

Description

AMES series equipments are modular EW threat simulation systems that are capable of expansion and customisation. They take the form of a dynamic and static Radio Frequency (RF) simulators that are able to generate a radar threat environment of over 2,000 'complex' emitters. As of December 2001, the AMES series was understood to centre on the AMES II and III architectures, the known details of which are as follows:

AMES II
AMES II system features include:
- realistic simulation of 'sophisticated' high-density radar signal environments
- a frequency coverage of 10 MHz to 96 GHz in selectable sub-bands
- modelling of up to 2,048 software programmable independent complex emitters
- use of Ada software running a Windows™ environment
- system control via a host DEC/Alpha computer
- support of up to 64 RF sources
- 3-D scan computed for each pulse
- 3-D amplitude or phase angle of arrival simulation computation on each pulse
- 3-D dynamic scenarios with frame freeze/playback facilities
- DMA map, EWIR and KILTING databases
- beacon range, multipath and ducting propagation effects
- simulation of polarisation effects
- closed loop receiver/jammer operation
- interactive controls and colour displays
- a simulated radar warning receiver display
- 500 fully programmable receiver platforms
- simulation of aircraft cabling and antenna losses
- full programmability of threat (missiles, aircraft, vehicles and ships) movement
- terrain masking and map facilities
- digitised entry of scan, agility and receiver antenna patterns
- automatic frequency, power and phase calibration
- data snapshot recording (simulator and SUT)
- the generation of data reduction/correlation reports.

AMES III
AMES III system features include:
- geometry computation on a pulse by pulse basis
- generation of a radar threat environment containing over 2,000 'complex' emitters
- in-pulse sculpturing
- precision frequency and pulse repetition interval
- the modeling of non-collocated, multiple receivers with independent flight paths, geometries, times, delays and phases
- clock drift for each emitter modeled
- programmable pulse sculpturing with user-defined shapes, amplitude ranges of up to 40 dB and minimum segment times of 12.5 ns
- real-time scenarios for all platforms that include real-time geometry on pulse; azimuth/elevation updating for every pulse; 'high', SUT compatible frame rates; external, real-time platform control and up to 2,000 individual sites/platforms
- Windows NT™ man-machine interface

- multi-user, multi-project, multi-library databases
- use of client/server SQL commercial databases
- a direction-finding capability that can be interfaced with 'any' receiver
- automatic PFM/UDF sweep testing
- query reports with customisable filter and sorting criteria
- test automation that includes a multiple scenario sequenced batch mode with recording
- on-board radar/jammer coupling
- the simulation of 'complex' emitters.

Status

Over time, over 80 AMES II series simulators are reported as having been procured by customers in Australia, Europe, the Middle East, the Pacific region and the US (tri-service applications).

Specifications

AMES II

Frequency characteristics

Coverage: 0.5-18 GHz (in selectable communications, 0.5-2 GHz, 2-6 GHz, 6-18 GHz, 2-18 GHz, 7-12 GHz and millimetre wave sub-bands - 10 MHz-96 GHz as option)

Accuracy: ±1 MHz (0.6-6 GHz range); ±3 MHz (6-18 GHz range)

Resolution: 4 kHz (0.1-18 GHz range - synthesiser); 50 kHz (0.1-2 GHz - VCO); 100 kHz (2-6 GHz range - VCO); 200 kHz (6-18 GHz - VCO)

Chirp: 20 MHz/s-500 MHz/ms

Switch time: <300 ns (synthesiser); 1 ms (VCO)

Dynamic range: 90 dB

Spurious/harmonics: −60 dBc

Modulations: agile, chirp, coherent, phase coded, slider, stable, stagger and simultaneous synchronised beam

Pulse characteristics

PRI range: 2-100,000 ms

PRI resolution: 1.6 ns

PRI modulations: agile, exponential, formula patterns, jitter, parabolic, ramp, sinusoidal, stable, stagger/switcher, trapezoid, triangular, uniform or Gaussian random and user-defined

PW range: 0.05-3,000 ms

PW resolution: 12.5 ns

PW rise/fall time: 20 ns (standard)

Control rise/fall time: UMOP (optional)

PW modulations: agile, exponential, formula patterns, jitter, parabolic, ramp, sinusoidal, stable, stagger/switcher, trapezoid, triangular, uniform or Gaussian random and user-defined

Emitter antenna simulation characteristics

Antenna patterns: cosecant², cosine, cosine with pedestals, fan, isotropic, omni, sin X/X and user defined

Mainlobe beamwidth: 0.5-22.5° (in 0.1° steps)

BW modulation range: 0-60 dB (in 0.25 dB steps)

Number of sidelobes: 0-99 pairs

Sidelobe level: 0 to −60 dB (in 0.25 dB steps)

Sidelobe position: 0 to ±179° (relative to main lobe)

Pattern resolution: up to 22.5° (in 0.02° steps); up to 360° (in 0.35° steps)

Polarisation: circular (right or left hand), horizontal, rotating, slant (right or left) and vertical

Emitter scan characteristics

Scan motion: bidirectional (raster and sector), circular, circular/sector, conical, electronic, helical, lobe on receive only, lobe switching, nodding, orthogonal, random, spiral, steady, unidirectional (raster and sector) and user-defined

Dynamic range: 0-60 dB (in 0.25 dB steps)

Sector width: 0.5-180° (in 0.1° steps)

Scan rate: 0.00102 Hz-1 kHz (in 0.000073 Hz steps)

Scan coverage: ±90° (elevation): ±180° (azimuth)

Emitter synchronisation

Synchronised modulation patterns: multiple step/dwell scan/pulse synchronisation; multiple synchronised pulse transmissions; multiple synchronised RF modulations; multiple synchronised scan modulations; scan/pulse, RF/scan, pulse/RF and scan/pulse/RF; step/dwell scan/pulse synchronisation

AOA models

Amplitude: 4 ports per RF source (additional ports as an option)

Phase interferometer: 4 ports per sector (additional ports as an option)

Monopulse: 4 integrated amplitude/phase channels (sum, delta azimuth, delta elevation and reference)

DTOA: number of ports optional; 1-256 ns delay/port

Rotating: up to 1,000 rpm

General

External control: Ethernet, RS-232, RS-422 and MIL-STD-1553

Co-ordinates: latitude/longitude; X/Y

Gaming area: 1,852 × 1,852 km

Platform performance: altitude, heading, pitch, roll, speed, time and yaw

Event timing resolution: 1-50 Hz

Terrain modeling: contour line terrain maps, DMA terrain maps, ducting, filled contour line terrain maps, multipath, radar horizon, spherical earth, terrain masking and user-defined maps

AMES III

DF capability

Amplitude angle of arrival: 1 dB

Phase angle of arrival: 4°

Long baseline interferometry: up to 20 m

Differential time of arrival: 1 ns

Amplitude modulation on pulse

Number of segments: 1 million shareable (maximum)

Segment duration: 12.5 ns (minimum)

Segment dwell: 3 ms (maximum)

Amplitude dynamic range: 40 dB

Amplitude resolution: 0.2 dB

Modulation blanking on pulse

Number of segments: 1 million shareable (maximum)

Segment duration: 12.5 ns (minimum)

Amplitude dynamic range: 90 dB

Phase shift key

Number of segments: 1 million shareable (maximum)

Segment duration: 25 ns (minimum)

Segment dwell: 3 ms (maximum)

Phase: bi and quad

General

PRI resolution: 0.1 ns

Clock synchronised PRI resolution: 0.1 ns

Emitter synchronisation resolution: 0.1 ns

Programmable frequency drift: down to 0.75 fs/0.18192 ms

Transmit antenna: (sin X/X); circular; circular with pedestal (10 or 20%); cosecant² (fan-horizon and vertical); cosine/cosine²; cosine with pedestal (10 or 20%)/ cosine² with pedestal (10 or 20%); gaussian; hyperbolic; interpolation (variable resolution down to 0.02°); multi-cut 3-D scan (plus principal cuts); omni (horizontal, vertical and isotropic); tapered hyperbolic (10, 50 or 80%) triangular; triangular with pedestal; truncated cosine/truncated cosine² and user defined (symmetrical, asymmetrical and multiple cuts)

Electronic scan: beam widening/amplitude warping (off-boresight effects); dynamic multiple target tracking (unique waveforms); frequency-amplitude/ beamwidth warping; independent electronic/mechanical motions and subscan around tracked targets (track pattern(s) plus search pattern)

Sensitivity profile (in addition to threshold): azimuth-elevation/average frequency profile and sensitivity degradation per beam

Mode changes (defensive order of battle): autonomous or C4I across emitters and platforms; formula-based PRI changes and state controller transitions on altitude, blockage, closing velocity, dwell, FOV, time and range

Environment: atmosphere; ducting; Faraday rotation (missiles); multipath (sea state, reflectivity map and omni-reflectors); terrain masking and weather

Contractor

Northrop Grumman Electronic Systems sector - Amherst Systems, Buffalo, New York.

UPDATED

··

AN/ALQ-188A(V) Electronic CounterMeasures (ECM) system

Type

Airborne ECM training pod.

Description

ALQ-188A(V) is a low-cost threat ECM simulator for aircrew training. The system is a modular design that is based on the AN/DLQ-3C(V) system. It provides a maximum effective radiated power of 2 kW over the frequency range 8.6 to 10.2 GHz. ALQ-188A(V) is capable of producing over 30 technique combinations and is programmable from an RS-232 source. The technique types include noise (spot, barrage or swept spot), velocity gate pull off, narrowband repeater noise, pseudo-random noise, random Doppler, range gate pull off, cover pulse, false target, and Amplitude Modulation (AM) (blink, random blink, fixed AM, swept square wave, sequenced AM). The equipment's antennas are circularly polarised with coverage of ±30° for low gain and ±15° for high gain.

Status

Over time, ALQ-188A(V) is reported to have been cleared for use on A-10, F-15, F-16, T-33 and Learjet 35/36 aircraft.

The ALQ-188A(V) countermeasures pod

Specifications

Dimensions: 64 × 277 × 25 cm
Weight: 136 kg max
Temperature/altitude: −55 to +71°C; sea level to 15,240 m
Vibration: 20-2,000 Hz, 33.4 g average
EMI: MIL-STD-461/462
Transients: MIL-STD-704
Power: 115 V AC, 400 Hz 3-phase, 5 A/phase

Contractor

USAF Quick Reaction Contract Support Center Northrop Grumman Corporation, Illinois.

VERIFIED

AN/APM-427 Radar Warning Receiver (RWR) test set

Type

Radar simulator for RWR testing.

Description

The AN/APM-427 RWR test set provides threat simulation of ground-to-air and air-to-ground radars for the complete testing of RWRs. The equipment provides free-space radiation of Radio Frequency (RF) pulses and Continuous Wave (CW) power to stimulate aircraft RWRs. The total system concept tests for correct assimilation, analysis and display of Electronic Warfare (EW) radar threats by simulating the emission characteristics of the radar systems. The system consists of five units: three pulse RF units, one CW RF unit and one battery charger unit. Each of the three pulse units contains three independent sources which may be pulse and amplitude modulated to simulate the scan parameters of particular radar threats. The RF sources radiate at different frequencies within the C/D- and J-bands (0.5 to 2 GHz and 10 to 20 GHz respectively). Threats may be built manually and stored in non-volatile memory or recalled by a number from a preprogrammed threat library. When using the library, the operator can transmit a predetermined group of threats in any order. Pulse trains of up to 256 independent Pulse Repetition Intervals (PRIs) may be operated on each RF source or superimposed on a single source to produce very complex pulse trains. The CW unit produces a non-modulated CW threat in the H- and I-bands (6 to 8 GHz and 8 to 10 GHz respectively). APM-427 is powered by rechargeable batteries with battery operation exceeding 1.5 hours continuous. Output power is continually monitored and displayed during operation. An RS-232 external interface is available for remote operation. Under programme control the operator is able to display and edit threat parameters. Programmable features include modes of operation, pulse-widths, PRI, scan pattern simulation, depth of modulation, scan rate and complex pulse trains. APM-427 offers three operating modes as follows:

- automatic transmission of a group of preprogrammed radar threats
- construction of a threat scenario via the combination of preprogrammed threats
- manual construction of threat scenarios via a 'user-friendly' display prompts and keyboard programming

Status

Over time, more than 500 APM-427 systems are reported to have been procured by the US military and off-shore customers.

Contractor

AAI Corporation, Hunt Valley, Maryland.

VERIFIED

The five sub-units that make up the APM-427 simulator system

AN/FPS-127 height-finder

Type

Simulator to provide height-finding radar capability.

Description

AN/FPS-127 is designed to add height-finding capability to the AN/MPS-T9 or other early warning/acquisition radars. It is a replica class emitter/receiver/

The FPS-127 height-finder radar simulator

processor simulator operating within the 2.7 to 2.9 GHz frequency range. The equipment is mounted on a mobile elevation over azimuth platform and consists of:

- a 1 MW dual pulse-width radar transmitter, which is adjustable in frequency across the band and is locked in frequency to the receiver local oscillator
- a radar receiver with automatic frequency control
- a synchroniser
- a dual quadrature digital moving target indicator
- a constant speed evaluation nodding equipment
- a high-gain parabolic E-band antenna
- an azimuth positioning system
- a remote-control panel.

Status

As of early 2001, the FPS-127 simulator was reported as being in service with the US Air Force at Nellis Air Force Base, Nevada.

Contractor

Metric Systems (an Integrated Defense Technologies company), Fort Walton Beach, Florida.

UPDATED

AN/MPQ-T3A simulator

Type

Multiple Anti-Aircraft Artillery (AAA)/Surface-to-Air Missile (SAM) threat radar simulator.

Description

AN/MPQ-T3A includes two different H/I-band (6 to 10 GHz) radar transmitters, an H-band (6 to 8 GHz) guidance illuminator, I- and J-band (8 to 10 GHz and 10 to 20 GHz respectively) radar transmitters and a J-band (10 to 20 GHz) guidance transmitter. These are used to simulate threat AAA and SAM radars for aircrew training. The T3-A system has full automatic track capabilities for each radar transmitter and can simulate any H-, I- and J-band threat simultaneously, using up to four of its six transmitters. Effective radiated power is stated to be 150 kW. H/I-band radar capabilities include pulse compression and pulse-Doppler modes. Each of the radar transmitters feature computer programmed pulse repetition

The MPQ-T3-A multiple AAA/SAM threat radar simulator system

frequency, pulse-width and frequency parameters, providing versatile threat simulation capabilities. MPQ-T3A can accept and/or provide slaving information, performing parallax correction on incoming and outgoing data. System instrumentation includes provision for digital data recording, radar video recording and playback. The system is configured in a 12 m long fifth-wheel type trailer suitable for towing over unimproved roads. The antenna pedestals are mounted on jacks for rapid lowering onto the trailer for transport. Outriggers and trailer jacks are included for stable operation on uneven surfaces.

Status
As of early 2001, the MPQ-T3A simulator was reported as being in service with the US Air Force at Mountain Home Air Force Base (AFB), Idaho, Nellis AFB, Nevada, Seymour Johnson AFB, North Carolina and Yukon AFB, Alaska and at ranges in Germany (Polygone), South Korea (Pil-Sung) and the UK (Spadeadam).

Contractor
Metric Systems (an Integrated Defense Technologies company), Fort Walton Beach, Florida.

UPDATED

The MSQ-T13 radar simulator

processing, Digital Moving Target Indication (DMTI), non-synchronous pulse suppression and plan position indicator and B-scope displays. The control radar subsystem provides a two-channel, monopulse tracking and control capability that features DMTI, non-synchronous pulse suppression A- and B-scope displays and a bit processor for automatic detection and lock-on. The subsystem emits a continuous wave (klystron) target illuminator signal alongside the pulse (magnetron) tracking output. Real-time data processing provides 3-D target information and performance monitoring. Data processing has been modified to include radar and signal processing, bit processing, three-pulse canceller DMTI, and five-pulse canceller DMTI for the track radar channel. MSQ-T13 can be fitted with a retrofit package that includes a pulse repetition frequency generator, multibeam acquisition receivers, acquisition track DMTIs, a constant false alarm rate bit processor, transmitters and associated hardware.

Status
As of early 2001, the MSQ-T13 simulator was reported as being in service with the US Air Force at Nellis Air Force Base, Nevada.

Contractor
Metric Systems (an Integrated Defense Technologies company), Fort Walton Beach, Florida.

UPDATED

AN/MPS-T9 radar simulator

Type
Early warning radar simulator.

Description
AN/MPS-T9 is an early warning/ground control intercept radar threat simulator. The system operates in both the E and F frequency bands (2 to 4 GHz). The T9 radar is housed in two mobile electronic shelters that are designated as the Operations Van (OV) and the Antenna Van (AV). A multibeam antenna, mounted on a rotating structure, determines the target positions while scanning in the azimuth plane. The OV contains the system displays, controls and communications circuits, all of which are housed in a console/rack format while the AV houses the system's power distribution unit, digital moving target indicator, compressor/dehydrator, slipring assembly, built-in test equipment, identification friend-or-foe interrogator and six transmitter/modulator subsystems with their respective receivers. Each transmitter/modulator subsystem independently transmits 1 MW of peak power at its selected frequency. The transmitters cover the frequency band from 2.7 to 3.1 GHz.

Status
As of early 2001, the MPS-T9 simulator was reported as being in service and deployed in the US at Naval Air Station Fallon, Nevada and Nellis Air Force Base, Nevada.

Contractor
Metric Systems (an Integrated Defense Technologies company), Fort Walton Beach, Florida.

UPDATED

AN/MSQ-T43(V) simulator

Type
Threat radar system simulator.

Description
The AN/MSQ-T43(V) modular threat emitter has been designed to simulate missile launching and/or anti-aircraft gun threat radar systems for aircrew training. It is a computer embedded, in-band tracking radar that consists of a pallet-mounted antenna/transmitter/pedestal group and an operations group housed in an S-280

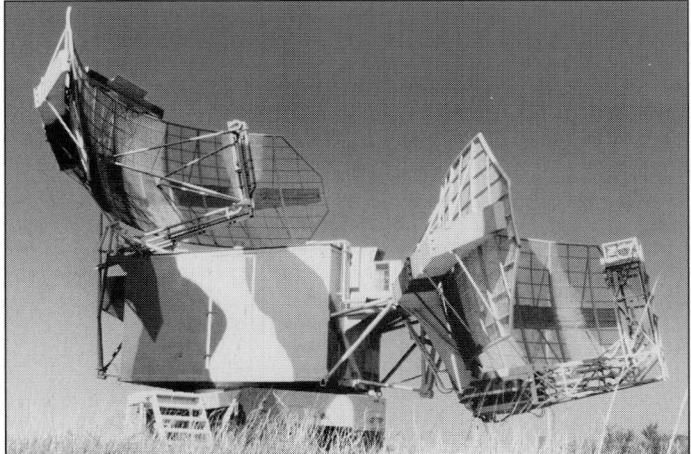

The MPS-T9 early warning radar simulator

AN/MSQ-T13 radar simulator

Type
Surface-to-Air Missile (SAM) acquisition and control radar simulator.

Description
AN/MSQ-T13 is a shelter-mounted, SAM acquisition and control radar simulator that incorporates two transmitters that generate simultaneous outputs and provide overlapping upper and lower beams. The acquisition system provides signal

The MSQ-T43(V) modular threat emitter

shelter. The shelter-mounted operations group includes a computer processor that is used for all operation and control of the system as well as diagnostic routines for self-test purposes. Weight of the pedestal group is 3,045 kg while that of the shelter group is 2,270 kg. MSQ-T43(V) operates at real threat radio frequency power and features automatic radar tracking in the threat band, track-on-jam, and better than ±32 mrad angle tracking.

Status

Over time, the MSQ-T43(V) simulator is reported as having been procured by the US Air Force, the US Navy and 'other agencies' for test and training purposes.

Contractor

Condor Systems Inc, San Jose, California.

VERIFIED

....

AN/MST-T1(V) Mini-MUTES Electronic Warfare (EW) training system

Type

EW training system for USAF aircrew.

Description

The AN/MST-T1(V) Mini-MUTES is an EW training system which simulates radar emissions in the 2 to 18 GHz frequency band. It allows aircrews on training flights to experience dense multiple electronic threats that may be encountered in an integrated air defence system. The architecture consists of a Master Control Group (MCG) and from one to five Remote Emitter Units (REU). The MCG is manned by a single operator who sets up the operational parameters of the system, monitors the scenario as it is played out and monitors the status of the REUs. The latter can be located up to 80 km from the MCG and communicates through normal telephone systems or other links. There are four configurations of the unmanned preprogrammed REU, each of which permits up to 10 threat signals to be generated at realistic power levels, frequencies and other signal characteristics. Each REU is able to:

- automatically acquire and track a co-operative aircraft and direct emitter signals
- undertake multiple identification friend-or-foe acquisition/tracking and pointing/slaving modes
- record electronic countermeasures and aircrew response to the simulated radar emissions with corresponding flight path data
- process noise jamming detection, range gate pull off detection, velocity gate pull off detection, and amplitude modulation detection
- send status, pointing and tracking data, electronic countermeasures receiver data and fault and warning messages.

Status

Mini-MUTES was originally developed by General Dynamics for threat simulation and 33 were built and delivered. In 1990 the Harris Corporation won a competition follow-on contract to build three new MCGs and 13 REUs. In September 1992, Harris was awarded a further then year US$10 million contract to upgrade the MST-T1(V) threat simulation capability. On 10 August 1998, Harris announced that the US Air Force's Sacramento Air Logistics Center (McClellan Air Force Base, California) had awarded it a then year US$49.9 million contract with regard to the Mini-MUTES Modification Program (M³P). Under M³P, Harris took responsibility for the system's 'total' performance and augmented the architecture's capabilities in the areas of threat capacity, new technology insertion and user interface improvements in order to maintain its viability beyond 2015. At the time of the announcement, Harris was reporting that a total of 20 MST-T1(V) MCGs and 77 REUs were operational at nine locations worldwide. As of the summer of 2001, MST-T1(V) was understood to be a live programme.

Contractor

Harris Corporation - Government Communication Systems Division, Melbourne, Florida.

UPDATED

The MST-T1(V) training system

AN/TPQ-45 simulator

Type

Ground-based mobile threat simulator.

Description

The AN/TPQ-45 Aircraft Survivability Equipment Trainer (ASET IV) uses mobility and interactive simulation to provide advanced air defence threat training. Within each ASET IV system, five High-Mobility Multipurpose Wheeled Vehicles (HMMWV) are equipped as threat systems with a sixth functioning as the architecture's Command, Control and Communications (C³) centre. Of the five threat system vehicles, two are anti-aircraft artillery analogues, with a further pair replicating infra-red guided Surface-to-Air Missile (SAM) systems. The remaining threat vehicle is equipped as a radio frequency SAM emitter simulator. TPQ-45 can operate autonomously, without instrumentation, on any range. The system also has a common interface for operation on instrumented ranges.

Status

Over time, AN/TPQ-45 is reported as having been procured by the US Army.

Contractor

Sierra Research (an Integrated Defense Technologies Inc company), Buffalo, New York.

VERIFIED

An AN/TPQ-45 threat simulator 'convoy'

....

AN/TPT-T1(V) simulator

Type

UnManned Threat Emitter (UMTE) simulator.

Description

AN/TPT-T1(V) is an unmanned threat emitter system that simulates various surface-to-air missile system radars and anti-aircraft radars. It provides tracking radar and missile guidance signals. Each system consists of a master controller and up to five unmanned remote emitters that may be deployed in areas during live fire air-to-ground training exercises. They are directed by a system operator to provide realistic threat signals to aircrews who are undergoing training. The operator is positioned in a control centre that, for safety purposes, is located well

The TPT-T1(V) unmanned threat emitter simulator

away from the target. The system is configured to interface with a rangemaster computer system to allow the entire mission to be controlled from this computer. Provision for manual operation is provided for ranges without this capability.

Status

Over time, Sierra Research is reported to have supplied the US Air Force (USAF) and the Air National Guard with five AN/TPT-T1(V) systems. Of these, units are understood to have been allocated to the USAF's Nellis Air Force Base, Nevada, the service's training facilities in the Yukon and the Air National Guard's range at Volk Field, Wisconsin.

Contractor

Sierra Research (an Integrated Defense Technologies Inc company), Buffalo, New York.

VERIFIED

ASCOT IV radar simulator

Type

Airspace control and operations training radar simulator.

Description

The ASCOT (AirSpace Control and Operations Trainer) IV system is a portable, personal computer-based radar simulator that is capable of emulating 'most' military and civilian surveillance and precision approach radars. It can be used as a stand-alone simulator/operations trainer for a single unit (radar site/air defence or Air Traffic Control (ATC) operations centre with numerous sensor inputs) or for modelling a large integrated system (air defence/ATC control region(s) in a large-scale exercise scenario).

Capable of unlimited scenario time lengths using a rolling buffer, ASCOT IV can generate and display up to 1,024 plots per sweep or frame rate, including any combination of preprogrammed and/or dynamically generated plots. Tabular and situation displays (36 cm and 43 cm video monitors) provide detailed data for simulator operator control and input via mouse or keyboard. Tabular display features include 24-track data blocks providing aircraft heading, altitude, speed, bank angle, position, Identification Friend-or-Foe (IFF)/Selective Identification Feature (SIF), remaining fuel and electronic countermeasures status. Each operator can display the Fire-Control System (FCS) 'radar picture' of fighter aircraft, providing the capability to display, target and fire. The FCS display provides positions from navigation aids and other aircraft (including threats), aircraft weapons loading, countdown, firing range and kill probability. Prioritised target range, heading, altitude and speed for up to five threats within the fighter's FCS azimuth/range are also displayed.

Simulator operators can generate or take control of any dynamic or preprogrammed track during a mission. Over 100 aircraft types and their parameters may be preprogrammed into the ASCOT simulator database, providing scenario flexibility and responsiveness. Custom flight routes (eight routes with up to 12 gates per route) and formations (eight options with up to six aircraft/formations) are also available. Mechanical and electronic jamming can be simulated from either preprogrammed or dynamic simulator operator-controlled tracks. Operators can use their own on-site equipment to interrogate and read out the simulated IFF/SIF responses (Modes 1, 2, 3A, 3C and 4) generated by ASCOT IV. When operating under EMCON conditions, the ASCOT IV simulator can also be used for operational equipment checkout, eliminating the need for turning on the radar transceiver.

To facilitate training, exercise scenario 'pilot-to-controller' voice communications scripts for simulator operators may be preprogrammed using the ASCOT IV. Simulator operators or students may review previous exercises and missions using the system's record/playback feature. Terrain masking, weather/ground clutter, antenna rotation rate and simulation-over-live video are also standard features.

The ASCOT IV 'effectively' integrates with operational equipment via the PLEXSYS IOP2 proprietary interface, providing radar and IFF data to systems

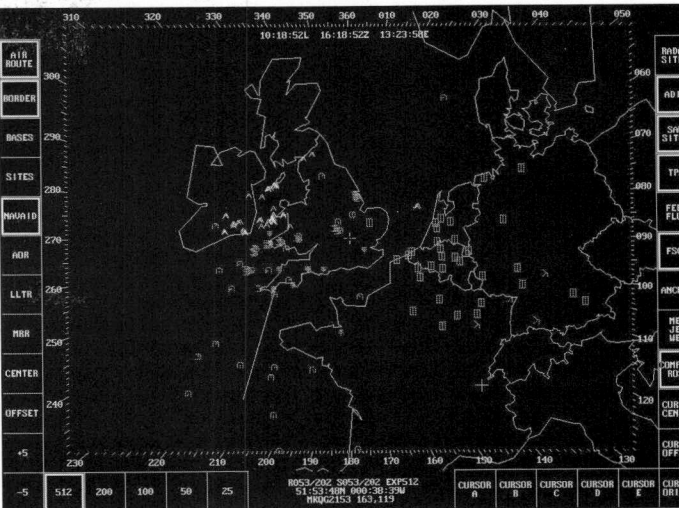

A typical ASCOT IV display 2000/0068928

requiring similar or dissimilar formats/protocols. Individual ASCOT simulators can be linked together in a local area network to support local exercises/training or into a wide area network in support of multi-unit/region scenarios/distributed interactive simulations. ASCOT IV incorporates exercise editor and video mapping features in order to provide the ability to generate preprogrammed training scenarios and situation display maps. The situation display provides simultaneous displays of radar video, target symbology, alphanumeric characters and vector graphics. Up to five dynamic cursors and up to 20 map layers are available. Each map layer provides individually programmable and selectable geopolitical map data that is customised for the particular scenario. Preset display ranges of 46, 93, 185, 370 and 948 km are provided and may be adjusted as desired using ± 9 km increments. For automated systems, an optional symbology editor permits generation of customised symbology. Online help screens for ASCOT IV's features provide immediate assistance and 'significantly' reduce learning times.

Status

Over time, the ASCOT IV radar simulator is reported as having been procured by customers in Bangladesh, Botswana, Canada, Egypt, Saudi Arabia, the United Arab Emirates and the US. In the US, the system has been used (under contract) to support three Joint System Training Exercises annually.

Contractor

PLEXSYS Interface Products Inc, Portland, Oregon.

VERIFIED

BAE Systems ground-based radar simulators

Type

Full power radar threat simulation systems.

Description

A TActical Steerable Emitter Threat Simulator (TASETS) system comprises a group of emitters, steerable pedestals/antennas and pedestal interface units. Each emitter is programmed with the parameters of a specific threat anti-aircraft artillery or surface-to-air missile system. Interface units process antenna steering data from an outside pointing reference to control positioning of the antenna pedestal. Each emitter can be paired against any desired aircraft in the training area. TASETS systems can be provided for installation on Tactical Aircrew Combat Training System/Air Combat Manoeuvring Instrumentation (TACTS/ACMI) ranges and remotely controlled from the display and debriefing system facility. Steering information can also be received from Identification Friend-or-Foe (IFF) trackers, optical tracking systems, full-scale threat simulators or any tracking radar. For training areas without ACMI, a fully self-contained TASETS can be provided with an IFF tracker or tracking radar to provide antenna pointing information. Transmitter options include solid-state or hard-tube modulator simulators with pulse group flexibility, or line type modulator simulators with one or two fixed pulse-widths. RS 565M3(CW) simulators with a Continuous Wave (CW) output power of up to 2 kW can also be provided. Other options include pedestal interface units to process antenna steering data, and pedestal/antenna combinations to replicate accurate threat antenna patterns.

RS 565M series simulators

Four versions of the RS 565M series radar simulators have been developed. RS 565M(V) models simulate pulse threat radars while the RS 565M3(CW) model simulates CW threat radars. Units can be provided at various frequencies and power levels through the 2.7 to 18 GHz frequency band and 40 kW to 1 MW peak power outputs. The pulse emitters provide high application flexibility and capability through the use of solid-state and hard-tube modulators. These modulators permit the generation of clean, well defined pulse shapes at any pulse-width. Increased flexibility is achieved through variable Pulse Repetition Frequency (PRF). PRFs

The ASCOT IV radar simulator 2000/0068929

A general view of one of BAE System's TASETS simulators   0017876

from 100 Hz to 4 kHz can be preset in the field and selected by the operator. Higher PRFs can be provided if required. The RS 565M3(CW) emitter consists of one unit containing the high-voltage power supply and a second unit containing the transmitter. Up to 2 kW of CW power can be produced by this emitter. Units can be provided at any frequency for which compatible klystrons are available.

RS 575 simulator

The RS 575 line type modulator emitter is designed for applications where the pulse flexibility of the RS 565M(V) is not required. RS 575 is capable of full-power emulation of a variety of threat radar signals to support aircrew training or test and evaluation missions. Various frequency and power combinations are from 2.7 to 18 GHz and with a peak power of 40 to 500 kW. The RS 575 is especially suitable for emulating the single pulse trains associated with surface-to-air missile and anti-aircraft artillery radar tracking signals. They can be installed on land ranges, remote-controlled boats or on ship hulks for use as a target radar for live anti-radiation missile firings.

Status

Over time, TASETS systems of types described are reported as having been procured by US and 'international' customers.

Contractor

BAE Systems North America - Technical Services, Fort Walton Beach, Florida.

UPDATED

....

BAE Systems simulation systems

Type

Range of radar and electronic warfare simulators and associated equipment.

Description

Over time, BAE Systems North America has produced a variety of threat simulator systems including full power Radio Frequency (RF) emitters, RF and communications jammers, remote-control systems, RF recorders, radar environment simulators and associated electronic equipment. Full power emitters include solid-state, hard-tube, or line-type modulators at power levels up to 1 MW. Ground-based and airborne training jammers produce the hostile radar and communications jamming threats faced by combat aircrews. Ground-based jamming systems are built into mobile shelters for remote location use. The same jamming components are available in an airborne configuration for training of both surface and air combatants. The Radar Environment Simulator produces a dense RF environment from communications bands to millimetre-wave radar bands. With a dynamic scenario generator and computer simulation, the Radar Environment Simulator produces a real-time simulation of a threat scenario. The computer simulations are interfaced with the Tactical Environment Simulation for development, evaluation and training.

Exotic Threat Emitter System (ETES)

ETES is a radar threat simulator system that is capable of the very latest, most complex and sophisticated radar simulations. It is well suited for generating extremely complex signals and outputs up to four overlapping radar signals.

Radar Emitter Simulation (RES) System 32

The System 32 RES provides simulated radar signals to Electronic Warfare (EW) receivers in test and development. The system outputs 32 separate emitters, multiplexed through four RF sources. Each emitter can simulate complex radars with up to 1,024 levels (variations) in pulse-width, frequency and amplitude (scan). This output consists of actual RF pulses that reproduce, in high fidelity, all the characteristics of a radar pulse, including frequency, pulse repetition interval, pulse-width, scan and direction of arrival. All parameters are independently

programmable. Each signal output is independent of any other signal being output simultaneously. This radar threat simulator is ideally suited for generating a dense radar environment.

Tactical Environment Simulation (TES)

TES is a software package specifically designed to network multiple manned simulators in a real-time tactical warfare environment. TES duplicates the mission environment challenging one or more player stations to execute tasks while exposed to representative tactical stimuli. Interactive software incorporates all players, as well as ground- and airborne-based threat defence systems, effecting realistic mission oriented decisions, actions and workload. Powerful 3-D graphics provide a real-time, out-of-the-cockpit display of the underlying terrain, targets and cultural features as well as both friendly and hostile aircraft. Realistic Hands-On-Throttle-And-Stick (HOTAS) controls may be used to provide both primary flight control as well as sensor, weapons and countermeasures controls. TES is readily networked to one or more high-fidelity dome simulators or helmet-mounted displays, with auxiliary player stations providing an interactive supplementary control function. In a stand-alone mode, the player station can function as a rapidly configurable part-task trainer or as a real-time performance analysis tool.

Threat generation hardware

The Radar Emitter Simulation System and the Exotic Threat Emitter System are used to support development and testing of EW receivers. The two simulators can be used in conjunction with each other and controlled by the same software. This provides coverage of both dense and complex radar signals. The BAE Systems-produced software controlling both simulators provides realistic radar simulation. By changing signal parameters, the radar threat output realistically matches the change in radar environment that an EW receiver would experience in an actual electronic combat situation.

Status

Over time, equipment and software of the types described are thought to have seen operational service.

Contractor

BAE Systems North America - Technical Services, Fort Walton Beach, Florida.

UPDATED

....

Combat Electromagnetic Environment SIMulator (CEESIM) series

Type

Family of Radio Frequency (RF) signal environment simulators for test, evaluation, training support and maintenance of air-/shipborne Electronic Warfare (EW) warning systems, Electronic CounterMeasures (ECM) systems and Electronic Support Measures (ESM) systems.

Description

CEESIM is a design and integration tool for use in the development of radar warning receivers and other types of EW equipment. CEESIM also provides realistic environments for use in EW operational trainers. The CEESIM family of simulators is based on a modular design that allows unique customer requirements to be readily satisfied with existing off-the-shelf hardware modules. CEESIM has the ability to generate from 16 to 8,192 simultaneous threat signals over the frequency range 1 MHz to 95 GHz. CEESIM can also be configured as a small low-power flight line simulator, an outdoor range threat emitter or as an anechoic/laboratory simulator. CEESIM is a family of different configurations – each addressing the unique requirements of a specific test application (for example, laboratory, anechoic chamber, test-training range, and so on). All configurations are based on

The CEESIM Model 128 RF simulator

the same internal architecture and utilise a common user-friendly software interface. Two examples of CEESIM for laboratory use are:

CEESIM Model 128

The Model 128 provides simultaneous, real-time stimulation of amplitude, Butler phase and interferometer phase direction of arrival systems in three contiguous bands covering 0.5 to 18.5 GHz with up to −10 dBm output power. Harmonic and spurious suppression is −60 dBc and frequency accuracy is better than 2 MHz. CEESIM 128 is designed and packaged to generate up to 128 simultaneous emitters, located on any of 128 simultaneous platforms, all of which are fully dynamic, allowing six degrees of freedom. All platforms are independently programmable for 3-D motion, velocity and location, with an update rate of 1 Hz. Multipath effects may be incorporated into the simulation on command from the operator. All emitters are independently programmable in pulse-width, Pulse Repetition Interval (PRI), jitter, chirp, Frequency Modulation On Pulse (FMOP), BP Shift Keying (BPSK), frequency, Effective Radiated Power (ERP) and scan combinations.

CEESIM Model 256

The Model 256 provides real-time stimulation for amplitude direction-finding systems using eight multiplexed RF sources covering 0.5 to 18 GHz, with 0 dBm output power across the band. Additionally, 16 Continuous Wave (CW)/RF channels are available across the 0.5 to 18.5 GHz frequency range. The CW channels have scan, ERP, range loss and direction of arrival modulation capabilities to increase RF signal density while ensuring a low pulse drop-out rate on the multiplexed channels. Harmonic and spurious suppressions are −60 dBc and frequency accuracy is better than 1 MHz.

The CEESIM Model 256 generates up to 256 simultaneous complex emitters, located on any of 256 simultaneous platforms, all of which are fully dynamic, allowing six degrees of freedom. All platforms are independently programmable for 3-D motion, velocity and location, with an update rate of 16 Hz. All emitters are independently programmable in pulse-width, PRI, jitter, chirp, FMOP, BPSK, frequency, ERP and scan combinations.

Status

Over time, CEESIM series simulators are reported as having been procured for use in laboratory, anechoic chamber and test range applications worldwide. Specific EW systems identified as having been tested using CEESIM include the AN/ALR-46, AN/ALR-47, AN/ALR-56, AN/ALR-59, AN/ALR-66 AN/ALR-67, AN/ALR-69, AN/ALR-73, AN/ALR-74, AN/ALR-87, AN/ALQ-99, AN/ALQ-126B, AN/ALQ-135, AN/ALQ-142, AN/ALQ-161, AN/ALQ-165, AN/APR-39, AN/APR-50, AN/ASQ-171, AN/SLQ-32 and INEWS equipments.

Contractor

Northrop Grumman Electronic Systems sector - Amherst Systems, Buffalo, New York.

UPDATED

CrossJam 2000 radar jammer

Type

Low-cost Radio Frequency (RF) jammer for training and testing.

Description

CrossJam 2000 is a family of modular radar jammers. Used in a podded or internally mounted installation, it can be operated from 0.5 to 18 GHz by varying the RF source. It is available in configurations ranging from a basic noise jammer to a sophisticated deception jammer that can include a fast set on receiver or digital RF memory. The system is designed to simulate numerous self-protection and escort jamming systems. It provides an airborne jammer source for training fighter pilots to intercept a jammer-equipped target. It can also be configured to train ground control intercept and acquisition radar operators in a jamming environment. CrossJam 2000 can be used with most high-power amplifiers and has been configured in the 100 to 500 W range. The AN/ALQ-167 pod shell used for the podded version is certified on more aircraft than any other pod. The system can be configured internally for aircraft such as the Lear, Falcon and Challenger business jets. It can also be packaged for mounting in other locations such as the gun bays of an F-5 aircraft. The full-up version is capable of generating 'all' significant electronic countermeasures techniques 'in use today'. Like third-generation combat jammers, the CrossJam 2000 is able to 'stack' techniques, giving an almost unlimited number of combinations and variations. Techniques include continuous and blinking noise, range and velocity deception (both separately and in combination) and fast set on techniques. CrossJam 2000 is used in radar and missile scenarios for airborne, open air testing. The internal version has been used extensively in hardware-in-the-loop and chamber testing. The CrossJam 2000 is also a credible combat jammer for the air-to-air arena.

Status

Over time, the CrossJam 2000 radar jammer is reported as having been installed aboard a number of 'electronic aggressor' Learjet aircraft operated by US civil contractor Phoenix Air Group Inc.

Contractor

BAE Systems North America - Technical Services, Fort Walton Beach, Florida.

UPDATED

CrossRES 2000 radar environment simulator

Type

High-fidelity, real-time target simulation.

Description

BAE Systems has developed and manufactured Radar Environment Simulators (RES) that inject complex peacetime or wartime environments into air surveillance radars.These simulations are used to train radar operators and evaluate radar operational readiness in an environment that includes large numbers of aircraft, active jamming, passive jamming (chaff), weather and ground clutter. RES systems consist of a set of equipment that enables an operator to define a scenario and to control and monitor its execution. Control may be exercised through either a local RES control console or the radar's own console via a specified digital interface (Ethernet or other serial link). Scenario data is used in conjunction with timing and control information received from the radar to synthesise signals that simulate the selected environment and to inject these into the radar at its intermediate frequency. Simulated radar returns are mixed with live return signals received from the real radar environment to allow the radar to carry out its primary air defence surveillance mission while performing training operations. The latest version of BAE Systems' RES equipment family uses digital signal synthesis to create up to 512 simultaneous targets. The standard Ethernet interface permits RES sited at different locations to be networked together to provide joint training at multiple radar sites. A Windows based graphical user interface running on commercial workstations provides the user with full control of scenario generation, system administration, maintenance, and diagnostic routines.

Status

Over time, BAE Systems RES systems are reported as having been supplied to a number of NATO customers for use with air defence and surveillance radars.

Contractor

BAE Systems North America - Technical Services, Fort Walton Beach, Florida.

UPDATED

Flightline electronic warfare Simulator (FLSim)

Type

Flightline Electronic Combat (EC) system simulator.

Description

FLSim is the lowest cost member of the Amherst System's family of EC simulators and takes the form of a ruggedised, miniature (0.057 m³) simulator system that is designed to assess EC system mission readiness and GO/NO GO system status in free space. It can also be used to verify the integrity of preflight mission data and user-defined files. Signals security can be maintained via direct coupling to the system under test (using antenna hoods and cabling) and programming is accomplished via a commercial-off-the-shelf (COTS) workstation and plug-in PCMCIA cards that can each hold in excess of 4,000 emitter definitions and test sequences. FLSim is operated by a single person and comprises two main components: a support station and a radiating unit. The support station (which can be used to maintain several radiating units) is a COTS computer and is used to define the scenarios and emitters using the standard CEESIM (see separate entry) graphical user interface. The generated data is transferred to the radiating unit using the previously noted PCMCIA cards. The radiating unit is operated via a hand-held remote controller in an automatic or manual mode. Testing can be accomplished using a walk-around method or by direct coupling. Multiple radiating units can be linked together (using a standard fibre optic interface) to provide a combined, co-ordinated stimulus for advanced open air testing. Weighing under 41 kg (configuration dependent), FLSim has an output power of +35 dBm and can be supplied with emitter libraries.

Status

As of December 2001, FLSim was understood to be a live programme.

Contractor

Northrop Grumman Electronic Systems sector - Amherst Systems, Buffalo, New York.

NEW ENTRY

HIgh DEnsity Signal Simulator (HIDESS)

Type

2 to 18 GHz band threat radar simulator.

Description

The Computer Science and Applications (CSA) HIDESS threat radar simulator is described as being a dual-cabinet, four-channel equipment that can simulate up to 64 emitters, each of which has its own high resolution clock source. Providing a 'fully asynchronous, real-world' emulation of 'various' threat radar systems,

HIDESS is designed primarily for laboratory use, but can be operated 'in the field' when teamed with a high-power antenna radiating unit. The equipment's frequency range can be extended to cover the 0.5 to 2 GHz and 33 to 38 GHz sub-bands if required and it is claimed as being able to simulate 'all known' radar pulse-widths, pulse repetition intervals, scan types, antenna patterns and modulation modes. Equally, HIDESS is noted as being able to accommodate 'as many' antenna ports as the particular test article needs for signal injection. The use of ESCAPE2000 operational and WINCAL 2000 calibration software is noted as providing 'user friendly' GUI's for system set up, control and calibration. HIDESS also features 'integral flexibility' to facilitate expansion and upgrading in terms of channel numbers, emitters, antenna output ports and so on.

CSA goes on to report that HIDESS forms the basis of its Beambox and Briefcase simulators. Of these, Beambox is described as being a low cost, single-channel sub-set of HIDESS that operates in the 2 to 18 GHz frequency range and is designed for laboratory signal injection use. As such, Beambox is noted as offering the same capabilities as HIDESS in terms of pulse-widths, pulse repetition intervals, scan types, antenna patterns and modulation types. For its part, the CSA Briefcase simulator is noted as being functionally the same as Beambox but packaged in such a way as to provide a laptop computer-controlled portable capability. The Briefcase output can take the form of a signal directly injected into the test item or routed into a high-power radiating unit for radiation mode testing. The HIDESS ESCAPE2000 operating software is also used in both the Beambox and Briefcase applications.

Status

As of July 2001, CSA was reporting that the baseline for the HIDESS simulator was established during the 1990s and that following its acquisition of the Antekna electronic warfare product line during 1999, it has completed development of the architecture. As of the given date, CSA was expecting to deliver its first HIDESS system during December 2001, with the launch customer being the Royal Danish Air Force. As of the summer of 2001, sources were suggesting that Beambox and Briefcase were both live programmes.

Contractor

Computer Science and Applications Inc, Shalimar, Florida.

NEW ENTRY

....

KOR simulation systems

Type

Family of Digital Radio Frequency Memories (DRFM), Radar Environment Simulators (RES) and Electronic CounterMeasures/Target (ECM/T) simulators.

Description

KOR Electronics produces a range of DRFMs, synthesised, Digital Signal Processing (DSP)-based and DRFM-based RESs and DRFM-based ECM/T simulators. In terms of the company's DRFM line, as of October 2001, KOR was known to have produced the following models:

Model 1027

The Model 1027 DRFM is described as being a four-bit, pod-mounted unit that offers an 800 MHz instantaneous bandwidth together with range and velocity capabilities. It is a self-contained unit that includes Radio Frequency (RF) and power supply units.

Model 1030

The Model 1030 DRFM is described as being a one-bit, pod-mounted unit that offers a 450 MHz instantaneous bandwidth together with internal techniques. It is a self-contained unit that includes Radio Frequency (RF) and power supply units.

Model 1058

The Model 1058 DRFM is described as being an eight-bit, laboratory unit that offers a 500 MHz instantaneous bandwidth together with range, velocity, multiple Doppler, range extent and memory access capabilities.

Model 1177

The Model 1177 DRFM is described as being a one-bit, pod-mounted unit that offers a 500 MHz instantaneous bandwidth together with internal techniques. It is a self-contained unit that includes RF and power supply units.

Model 1225

The Model 1225 DRFM is described as being a three-bit miniaturised unit that, as of October 2001, was under development for airborne, pod and unmanned aerial vehicle applications. The unit is noted as providing 10-12 MHz of instantaneous bandwidth, together with what are described as 'internal techniques'. Model 1225 is further described as being a self-contained unit that includes RF and power supplies.

Model 1232

The Model 1232 DRFM is described as being an eight-bit laboratory unit that offers a ≥500 MHz instantaneous bandwidth (eventual goal, 700 MHz), together with range and velocity capabilities, multiple Doppler, range extent and memory access.

Status

As of October 2001, KOR Electronics' range of DRFM's, RESs and ECM/T simulators was understood to be available. Systems known to have been tested using KOR synthesised RESs include the Eurofighter, the German Tornado Self-Protection Jamming (TSPJ) pod, the F/A-18 combat aircraft, the Electronic Combat and Reconnaissance (ECR) Tornado and the Captor fire-control radar.

The CSA HIDESS threat radar simulator (CSA)   2002/0113332

The CSA Beambox threat radar simulator (CSA)   2002/0113331

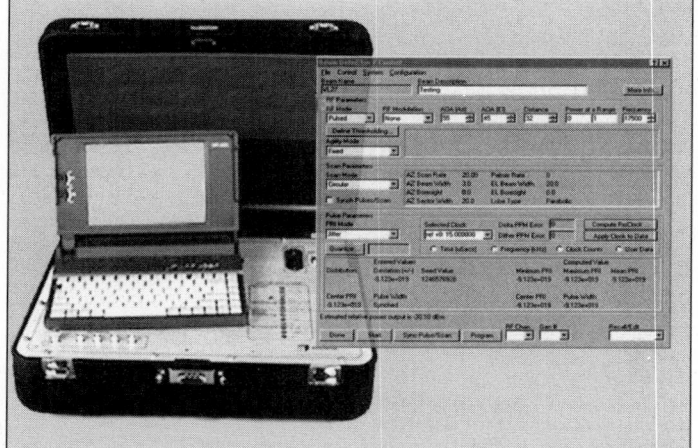

The CSA Briefcase threat radar simulator (CSA)   2002/0113330

Specifications: KOR simulation systems

(1)   DRFMs

	Model 1027	Model 1030	Model 1058	Model 1177
Bandwidth				
Operating	9-9.8 GHz and 9.5-10.3 GHz	3.07-3.53 GHz	100 GHz	3.05-3.55 GHz
Instantaneous	>800 MHz	>460 MHz	up to >500 MHz	>500 MHz
Dynamic range				
Input	−55 to −35 dBm	−20 to +15 dBm	−55 to +5 dBm	−23 to +15 dBm
Output (min)	0 dBm	≥+5 dBm	5 dBm	≥+10 dBm
Flatness	±2 dBm	±3 dBm	±3 dBm	±2 dBm
VSWR (in/out)	≤2:1	≤2:1	≤2:1	<2:1
Noise				
LO leakage (output)	22 dBc	>10 dBc	>40 dBc	>10 dBc
Noise floor (DRFM active)	>40 dBc	>25 dBc	<−110 dBc/Hz	>25 dBc
Noise floor (DRFM inactive)	<80 dBm	<−80 dBm	<−150 dBc/Hz	<−80 dBm
Spurious (in-band worst case)	>22 dBc	>6 dBc	>32-40 dBc	>6 dBc
Spurious (in-band typical)	>25 dBc	>9 dBc	>45 dBc	>10 dBc
Out-of-band	>30 dBc	>45 dBc	>30 dBc	>45 dBc
Timing - Pipeline mode				
PW (min input)	50 ns	100 ns	50 ns	100 ns
PW (max input)	CW	CW	CW	CW
PW (output)	50 ±4.4 ns	100 ±3.7 ns	50 ±4 ns	100 ±8 ns
PW (step size)	4.4 ns	3.7 ns	4 ns	8 ns
Min delay (insertion)	=90 ns	≤−95ns	<120 ns	≤90 ns
Max delay	582 µs	720 µs	2.1 µs	63 µs
Delay resolution	4.4 ns	3.7 ns	4 ns	8 ns
Timing - Stretched pulse mode				
PW (min input)	120 ns	<100 ns	100 ns	<100 ns
PW (max input)	582 µs	>720 µs	2.1 µs	65 µs
PW (output)	175.75 ns-9.999 s	100s-9.999 s or CW	160 ns-9.999 s or CW	100s-9.999 s or CW
PW (step size)	4.4 ns	29.7 ns	32 ns	8 ns
Min delay (insertion)	=90 ns	≤95 ns	<120 ns	≤95 ns
Max delay	2.3 µs	1.9 µs	2.1 µs	1.9 µs
Delay resolution	4.4 ns	29.7 ns	32 ns	8 ns
Timing - Multiple false targets				
PW (min input)	120 ns	<100 ns	100 ns	<100 ns
PW (max input)	582 µs	>720 µs	2.1 µs	>65 µs
PW (output)	120 ±35.5 ns	<100 ±3.7 ns	100 +32 ns	<100 ±32 ns
PW (step size)	4.4 ns	29.7 ns	4 ns	8 ns
Min delay (insertion)	=90 ns	≤90 ns	<120 ns	≤90 ns
Max delay	2.3 µs	1.9 µs	2.1 µs	1.9 µs
Delay resolution	4.4 ns	29.7 ns	32 ns	8 ns
Timing - Targets				
Number	1-255	1-255	1-255	1-255
Min spacing	input PW +35.5 ns	input PW +29.7 ns	input PW +32 ns	input PW +32 ns
Max spacing	2.3 µs	1.9 µs	2.1 µs	1.9 µs
Spacing step size	35.5 ns	29.7 ns	32 ns	32 ns
Doppler				
Range	−300 MHz to +300 kHz	−300 kHz to +300 kHz (option)	−2 MHz to +2 MHz	−300 kHz to +300 kHz (option)
Resolution	=19.5 Hz	≤25 Hz (option)	<0.1 Hz	≤10 Hz (option)
Power				
Nominal	<300 W	100 W	<300 W	75 W
Input voltage	25 V DC	23 V DC	110-240 V AC (46-400 Hz)	23 V DC
Max input current	15 A	5 A	4.0 A	2 A
Environmental				
Operating temperature	−20 to +50°C	−40 to +71°C	0 to +50°C	−40 to +71°C
Storage temperature	−54 to +71°C	−54 to +71°C	−20 to +71°C	−54 to +71°C
Altitude	SL to 15,240 m	SL to 15,240 m	SL to 3,048 m	SL to 12,192 m
Cooling	forced air	forced air	forced air	forced air
Dimensions (h × w × d)	170 × 157 × 483 mm	184 × 146 × 470 mm	400 × 483 × 711 mm	160 × 119 × 311 mm
Weight	13.6 kg	9.1 kg	<36.3 kg	6.4 kg

	Model 1225	Model 1232
Bandwidth		
Operating	8.6-9.8 GHz	baseband to 100 GHz
Instantaneous	>1,200 MHz	up to >500 MHz
Dynamic range		
Input	−55 to +10 dBm	−55 to +5 dBm
Min output	−30 to +10 dBm (0.25 dB steps)	5 dBm
Flatness	<±2 dBm	<±3 dBm
VSWR (in/out)	=2.0:1	≤2.0:1
Noise		
LO leakage (output)	<−17 dBc	>40 dBc
Noise floor (DRFM active)	>40 dBc	<−110 dBc Hz
Noise floor (DRFM inactive)	<80 dBm	<−159 dBc Hz
Spurious (in-band worst case)	>17 dBc	>32 dBc (>40 dBc for most configurations)
Spurious (in-band typical)	>22 dBc	45 dBc
Out-of-band	>60 dBc	>30 dBc (without additional filters); >60 dBc (with additional filters)
Timing - Pipeline mode		
PW (min input)	20 ns	50 ns
PW (max input)	CW	CW

Specifications: KOR simulation systems (cont)

	Model 1225	Model 1232
PW (output)	20 (±4.4) ns	50 (±2) ns
PW (step size)	=4.4 ns	2 ns
Min delay (insertion)	=90 ns	<120 ns
Max delay	128-582 µs (TBD)	>16.5 ms
Delay resolution	=4.4 ns	2 ns
Timing - Stretched pulse mode		
PW (min input)	=140 ns	100 ns
PW (max input)	128-582 µs (TBD)	>10 ms
PW (output)	100-255 ms	160 ns-9.999s or CW
PW (step size)	=4.4 ns	64 ns
Min delay (insertion)	=80 ns	≤120 ns
Max delay	2.3 ms	2.1 ms
Delay resolution	=4.4 ns	2 ns
Timing - Multiple false targets		
PW (min input)	=140 ns	100 ns
PW (max input)	582 µs	>10 ms
PW (output)	=140 (± 35.5) ns	100 (±2) ns
PW (step size)	=4.4 ns	2 ns
Min delay (insertion)	=90 ns	<120 ns
Max delay	2.3 ms	>16.5 ms
Delay resolution	4.4 ns	2 ns
Timing - Targets		
Number	1-225	1-225
Min spacing	2.3 ms	>10 ms
Spacing step size	35.5 ns	2 ns
Doppler		
Range	–300 to +300 kHz	–2,000 to +2,000 kHz
Resolution	=19.5 Hz	<0.1 Hz
Power		
Nominal	<80 W	<250 W
Input voltage	23 V DC	110-240 V AC (46-400 Hz)
Max input current	4.5 A (max)	3.5 A (max at 110 V AC)
Environmental		
Operating temperature	–40 to +71°C	0 to +50°C
Storage temperature	–54 to +71°C	–20 to +71°C
Altitude	SL to 12,192 m	SL to 3,048 m
Cooling	forced air	forced air
Dimensions (h × w × d)	160 × 119 × 311 mm	400 × 483 × 711 mm
Weight	6.8 kg	<27.2 kg

(1) RESs	DRFM-based	DSP-based	Synthesised
Scenario			
Targets (in environment)		up to 500	up to 250
Targets (in-beam)		up to 20	up to 8
Jammers (in environment)		up to 12	up to 12
Jammers (in-beam)		up to 4	up to 4
Chaff (in environment)		up to 12 bundles or corridors	up to 12 bundles or corridors
Chaff (in-beam)		up to 4 bundles or corridors	up to 4 bundles or corridors
Weather clutter (in environment)		up to 4 features	up to 4 features
Weather clutter (in-beam)		up to 2 features	up to 2 features
Pilot-in-the-loop aircraft		up to 16	
Terrain clutter (ground/sea)			downloadable 360° clutter definition
			dynamic mainlobe, sidelobe and altitude line return
Airborne			
Number of scenarios		50 (min)	
Duration of scenarios		2 h (min)	
Target fidelity			
Target types		>15 definable	
Mean RCS value	0.001-10,000 m2	0.001-10,000 m2	0.001-10,000 m2
3-D RCS patterns	±90° (AZ); ±180° (EL); –60 to +60 dB	±90° (AZ); ±180° (EL); –60 to +60 dB	±90° (AZ); ±180° (EL); –60 to +60 dB
Scintillation	Swerling cases 0-4	Swerling cases 0-4	Swerling cases 0-4
Target modulation	user definable; variable modulations based on aspect angle		user definable; variable modulations based on aspect angle
Geometry modelling	6-DOF	6-DOF	6-DOF
Jamming assets	CW and wide/narrowband noise	combination of active/passive	combination of active/passive
Radar fidelity			
Antenna types		mechanical, electronic or combination	mechanical, electronic or combination
Narrowbeam antenna pattern			±30° at 0.1° resolution
Gain range/resolution			>60 dB at 0.25 dB resolution
Antenna update rate			>1 kHz
Waveforms	All within DRFM	pulse, phase coded and linear/ non-linear FM	Pulse (50 ns to CW), Q/BPSK phase coded (2,048 chips max), linear/non-linear FM and support for monopulse, FM ranging and LPI
ECCM modes		supported	Supported
Range		1.9-1,852 km	30.5 m-1,852 km (0.9 m resolution)

Specifications: KOR simulation systems (cont)

(1) RESs

	DRFM-based	DSP-based	Synthesised
Radio frequency fidelity			
Dynamic range	>120 dB		>120 dB
Amplitude resolution	0.25 dB		0.25 dB
RF on/off isolation	>100 dB		>120 dB
Doppler range	±2 MHz		±2 MHz
Doppler resolution	<0.1 Hz		<0.1 Hz
Spurious (worst case in-band)	32 to >60 dBc		>60 dBc
Spurious (typical in-band)	40 to >65 dBc		>65 dBc
Return delay	150 ns-2.1 µs at 4 ns resolution		
Delay accuracy	4 ns		
Monopulse phase accuracy			±5°
Range accuracy			±0.61 m
Phase noise (1 Hz bandwidth)			−80 dBc/Hz (100 Hz); −100 dBc/Hz (1 kHz); −120 dBc/Hz (10 kHz); −140 dBc/Hz (100 kHz)
Noise floor			−140 dBm/Hz
Environmental			
Operating temperature	0 to +40°C	0 to +40°C	0 to +40°C
Non-operating temperature	−20 to +60°C	−20 to +60°C	−20 to +60°C
Operating altitude	SL to 3,048 m	SL to 3,048 m	SL to 3,048 m
Non-operating altitude (air shipment)	SL to 12,192 m	SL to 12,192 m	SL to 12,192 m
Cooling	forced air	forced air	forced air
Input voltage	110-240 V AC (45-65 Hz)	110-240 V AC (45-65 Hz)	110-240 V AC (45-65 Hz)
Dimensions (h × w × d)	configuration dependent × 483 × 914 mm	~213 × 483 × 914 mm	~213 × 483 × 914 mm

(2) ECT/T simulators (typical example)

Scenario	
Targets	user defined
Jammers	user defined
Target fidelity	
Mean RCS value	0.001-10,000 m²
Scintillation	Swerling cases 0-4
Target modulation	user definable with downloadable patterns for JEM, RSM or other modulations
Radio frequency fidelity	
Waveform	CW, FM, phase coded and pulse. All waveforms within the DRFMs operating band will be successfully stored and retransmitted
Dynamic range	>120 dB
Amplitude resolution	0.25 dB
RF on/off isolation	>100 dB
Doppler range	±2 MHz
Doppler resolution	<0.1 Hz
Spurious (worst case in-band)	32 to >60 dBc
Spurious (typical in-band)	40 to >65 dBc
Return delay	150 ns-2.1 µs at 4 ns resolution
Delay accuracy	4 ns
Standard coherent and pulse techniques	
Range deception	RGPO/RGPI, range false targets, cover pulse and random range targets
Velocity deception	VGPO/VGPI and velocity false targets
Co-ordinated	Synchronised R/VGPO and unsynchronised R/VGPO
Masking	range bin masking, velocity bin masking, asynchronous pulse repetition masking, combined range and velocity bin masking and intrapulse phase modulation
Standard continuous wave and noise techniques	
CW	fixed, swept, click and random CW
Noise	fixed, responsive, swept, random click and gated spot and fixed barrage
Environmental	
Operating temperature	0 to +40°C
Non-operating temperature	−20 to +60°C
Operating altitude	SL to 3,048 m
Non-operating altitude (air shipment)	SL to 12,192 m
Cooling	forced air
Input voltage	110-240 V AC (45-65 Hz)
Dimensions (h × w × d)	Configuration dependent × 483 × 914 mm

Contractor

KOR Electronics Inc, Garden Grove, California.

UPDATED

....

Metric Advanced Threat Emitter System (ATES)

Type

Multiband transmitter simulator group.

Description

Metric's ATES consists of emitters (see Specifications) in the C- (850 to 942 MHz sub-band), E/F- (2.1 to 2.4 GHz sub-band) and H/I/J- (7.7 to 8.5 GHz, 8.3 to 8.9 GHz, 9.5 to 10.5 GHz and 9.9 to 10.2 GHz sub-bands) frequency ranges. The emitters have peak powers of up to 300 kW (in C-band) and agile Pulse Repetition Frequencies (PRF) of up to 100 kHz (in the 9.9 to 10.2 GHz band). ATES offers horizontal, vertical and circular polarisation options and each transmitter within the system is housed in an air-conditioned S-280 shelter that incorporates the particular equipment's frequency synthesiser, tunable klystron, klystron modulator, PRF generator, high voltage power supply, heat exchangers and 'all' supporting subsystems.

The Model 3383C transmitter used in Metric's ATES

Status

Over time, ATES architectures are reported to have been deployed at Naval Air Station China Lake, California in the US and at the UK's electronic warfare range at Spadeadam, Northumberland.

Specifications

	Model 3383A (850-942 MHz)	Model 3383B (2.1-2.4 GHz)	Model 3383C (9.9-10.2 GHz)
RF peak power	300 kW	150 kW	62 kW
Max duty cycle	2.5%	2.4%	8.5%
PRF	1-17,000 pps agile	1-30,000 pps agile	1-100,000 pps agile
RF modulation	0-8 MHz chirp	0-5 MHz chirp	optional
Antenna type	2 back-to-back sectional parabolas	sectional parabola	CSEC^2 sectional parabola
Gain	22 dB	30 dB	34 dB
Polarisation	1 horizontal 1 vertical	Horizontal	circular

Contractor

Metric Systems (an Integrated Defense Technologies company), Fort Walton Beach, Florida.

UPDATED

Mobile Reprogrammable Emitter Simulator (MRES)

Type

Mobile reprogrammable emitter simulator for test and training range applications.

Description

The Amherst Systems' MRES is described as being able to radiate threat signals at target aircraft or ships within a 161 km radius of itself. Target aircraft/ship movement is monitored by radar (or other range time/space position indicator system) and is tracked by pedestal-mounted MRES antennas in order to achieve a high and effective radiated power value in any direction within the specified radius. All threat and electronic countermeasures signals are monitored and measured using an integrated signal measurement subsystem, thereby facilitating realistic threat reactivity and after-action debriefing and analysis. MRES features emitter and scenario compatibility with other Amherst radio frequency-based simulators and as such, is claimed to generate signals with the same fidelity as the company's full-scale laboratory equipments. 'All known' electronic warfare emitters can be simulated 'to the highest levels of accuracy' and multiple radiating sites can be networked and operated using IADS tactics. Here, all system functions can be remotely controlled (including 'full' emitter reprogrammability in real-time).

Status

As of December 2001, a MRES was reported as being in production for the US Navy's Naval Air Warfare Center (NAWC) at Patuxent River, Maryland. NAWC is further noted as having procured an Amherst Reprogrammable Emitter Simulator. Specifically identified MRES contracting activity during 2001 is as follows:

22 June 2001
Amherst Systems was awarded a then year US$6,948,956 modification to a previously awarded firm, fixed-price contract (N00019-99-C-1472) covering the exercising of a procurement option for a single MRES and two supporting Combat

Electromagnetic Environment Simulators (CEESIM). At the time of the announcement, work on the effort was scheduled for completion by the end of May 2003, with the USN's Naval Air Systems Command, Patuxent River, Maryland acting as its contracting activity.

20 September 2001
Amherst Systems was awarded a then year US$6,300,000 modification to a previously awarded firm, fixed-price contract (N00019-99-C-1472) covering the exercising of a procurement option for a single MRES. At the time of the announcement, work on the effort was scheduled for completion by the end of September 2003, with the USN's Naval Air Systems Command, Patuxent River, Maryland acting as its contracting activity.

Contractor

Northrop Grumman Electronic Systems sector - Amherst Systems, Buffalo, New York.

NEW ENTRY

Mobile Threat Emitter System (MoTES)

Type

Unmanned threat emitter simulator.

Description

Sierra Technologies' MoTES is an unmanned mobile emitter that simulates a variety of surface-to-air missile and anti-aircraft artillery radars. It provides authentic replication of tracking radar and missile guidance signals for the 'realistic and cost-effective' training of aircrews. MoTES' mobility features facilitate speedy relocation and the system incorporates a remote-control capability. To enhance mobility, the deployment and tear down time is reduced to less than 30 minutes. MoTES controls are compatible with UMTE so that a combined mobile/stationary system is possible.

Status

Over time, two MoTES simulators are reported as having been procured by the US Air National Guard.

Contractor

Sierra Research (an Integrated Defense Technologies company), Buffalo, New York.

VERIFIED

The MoTES threat emitter simulator   0017878

Model 8102 Smart Crow Electronic Counter-CounterMeasures (ECCM)/Electronic CounterMeasures (ECM) training system

Type

Internally mounted airborne electronic warfare training system.

Description

Smart Crow is an internally mounted modular countermeasures system capable of generating ECM techniques with radiated power in excess of 1 kW effective radiated power over the 8 to 11 GHz frequency range. Control of jamming modes and parameters is carried out by a cockpit control indicator or a computer. A full range of jamming modulations and a velocity deception capability are included and a built-in set on receiver connected to antennas with integral amplification is used to counter frequency-agile radars. A number of ECM modes are available, as follows:

- spot/barrage noise with or without countdown blink or SAM
- range or velocity gate stealer with or without SAM
- random Doppler
- hold out and hook
- pseudo-random noise
- pseudo-random noise with countdown blink
- repeater swept amplitude modulation
- repeater, single or multiple frequency
- narrowband repeater noise.

Status
Over time, 11 Smart Crow systems are known to have been procured.

Specifications
Frequency range: 8-11 GHz
Transmitter output power: 200 W typical
System ERP: 1-2 kW
Total weight: 115 kg

Contractor
Rodale Electronics Inc, Hauppauge, New York.

VERIFIED

The SLQ-32 OTD

Northrop Grumman AN/SLQ-32 environment generator

Type
Environment simulation device to provide stimulus for the AN/SLQ-32 electronic warfare system.

Description
The Environment Generator (EG) creates a simulated digital environment for the AN/SLQ-32 system. It connects to the AN/UYK-19 processor bus of the SLQ-32 system and emulates its entire front end (including the Radio Frequency (RF) environment) on a pulse-by-pulse basis. The EG is used to stimulate the actual onboard system, laboratory versions of the system or its training systems. The SLQ-32 EG can create a wide range of realistic RF environments with signals being processed by an embedded VME processor to emulate the functions of the antennas, receivers, amplifiers, presorters and digital tracking unit in the SLQ-32. This allows execution of the operational software without the expensive hardware assets necessary to stimulate the system using RF signals, thus providing for software support activities, system testing and operator training.

Status
Over time, the SLQ-32 EG is reported as having been procured by the Australian and US navies.

Contractor
Northrop Grumman Electronic Systems sector - PRB Systems, Goleta, California.

UPDATED

Northrop Grumman AN/SLQ-32 operator training device

Type
Operator training device for the AN/SLQ-32 Electronic Warfare (EW) system.

Description
The Operator Training Device (OTD) provides the capability to train EW operators with consoles identical in appearance to those of the operational AN/SLQ-32 system. Low cost is maintained via the use of commercial non-military specifications equipment. The unit contains a commercial-off-the-shelf microprocessor emulating the console with a commercial AN/UYK-19 processor allowing the OTD to directly execute the SLQ-32's software. The device provides all actual operator functions including displays, console lamps, light emitting diode indications, rotary/mechanical switches and realistic emitter audio. EW environment stimulus for training exercises is provided through Comptek's environment generator, Comptek's Training/Readiness Assessment Device (TRAD) or through scenarios within the SLQ-32 training software. Operator training can be accomplished either in a stand-alone format or as part of a combat direction system. External interfaces are provided to allow OTD to interact with other devices or programmes such as data recording tools or test analysis tools.

Status
Over time, the SLQ-32 OTD is reported as having been procured by the Australian and US navies.

Contractor
Northrop Grumman Electronic Systems sector - PRB Systems, Goleta, California.

UPDATED

Northrop Grumman AN/SLQ-32 operator training laboratory

Type
Operator training for the AN/SLQ-32 Electronic Warfare (EW) system.

Description
The AN/SLQ-32 training laboratory is a comprehensive training environment for shipboard EW operators. The laboratory allows multiple operators to train in parallel, experiencing a realistic environment in which multiple SLQ-32 systems are operating. Cost-effective training is realised with a single instructor teaching multiple operators at one time. The operators are trained using realistic electromagnetic environment scenarios from the Comptek environment generator while running any version of the SLQ-32 system operational software in the Comptek Operator Training Device. This allows training with the exact software configurations in use on board ship. In addition to console operations, the system provides for exercises with employment of decoys and Naval Tactical Data System interfaces. The laboratory is available in two-, four- or six-seat versions. Individuals can be exercised as multiple operators on board the same ship or as individuals within a large force involving multiple ships in the same electromagnetic environment. This permits the operators to be trained in fleet and joint force tactics. The scenarios may be simple or complex and are designed to increase in

Northrop Grumman's SLQ-32 Operator Training Laboratory

complexity as exercises progress. The instructor may modify the environment scenario in real time by injecting events that modify the exercise.

Contractor
Northrop Grumman Electronic Systems sector - PRB Systems, Goleta, California.

UPDATED

Northrop Grumman AN/SLQ-32 Software Support and Training System (SSTS)

Type
System to support software verification and validation and to support operator training for the AN/SLQ-32 Electronic Warfare (EW) system.

Description
SSTS is a multipurpose system designed to provide the full stand-alone capability to validate threat libraries and perform operator training for the AN/SLQ-32 EW system. The SSTS provides the capability to execute the SLQ-32 operational software in real time without any of the operational hardware. The SSTS incorporates:
- a Northrop Grumman environment generator to provide the radio frequency environment to stimulate the software
- Northrop Grumman's Operator Training Device (OTD) to execute the software and support AN/SLQ-32 controls and displays
- a Northrop Grumman emitter feature generator to verify the environment and compare to SLQ-32 responses
- a standard personal computer for equipment control purposes
- computer peripherals to perform tape, disk and hard copy input/output

The system supports the essential process of verifying the SLQ-32 emitter library and the proper operation of emitter identification with new intelligence information. Individual emitters are simulated in the environment generator while effectiveness and accuracy are verified using the current operational software and library in the OTD. When not in use for verification, the SSTS can be used to train shipboard operators with realistic scenarios from the environment generator.

Status
Over time, the SLQ-32 SSTS is reported as having been procured by the Australian Navy.

Contractor
Northrop Grumman Electronic Systems sector - PRB Systems, Goleta, California.

UPDATED

The SLQ-32 SSTS

Northrop Grumman AN/SLQ-32 Training/Readiness Assessment Device (TRAD)

Type
Simulation and training device to support the AN/SLQ-32 Electronic Warfare (EW) system.

Description
TRAD is a microprocessor-based device that provides environment stimulus and operator response assessment for training SLQ-32 operators. The TRAD allows student operators to participate in EW training and allows instructors to verify and evaluate reactions of student operators in real time. The instructor can use this

The SLQ-32 TRAD

information to interface more effectively with continuing training. The device is an open system consisting of the Comptek environment generator, a data extractor module, a video generator and a scenario generation/control module; all housed in a single portable VME chassis. TRAD control software that runs in any PC DOS Windows environment is included in the system. TRAD interfaces with both operational SLQ-32 systems and shore-based training devices. TRAD features graphical user interfaces, realistic radio frequency diverse environments, real-time non-intrusive data extraction and interleaved emitter video for realistic audio and operator pulse analysis. The PC running the control software interfaces to TRAD via a standard RS-232 port.

Status
Over time, the SLQ-32 TRAD is reported as having been procured by the US Navy.

Contractor
Northrop Grumman Electronic Systems sector - PRB Systems, Goleta, California.

UPDATED

Northrop Grumman Real-time Infra-red/electro-optic Scene Simulator (RISS)

Type
Multispectral Infra-Red (IR) and Electro-Optical (EO) scene simulator.

Description
Northrop Grumman's RISS is a high fidelity, physics-based scene generation system that can simulate the 1 to 14 μm IR spectrum (with coverage of the 0.1 to 3 μm ultra-violet band as a planned future enhancement) and includes hardware and software components that provide real-time, frame-by-frame, high-fidelity IR imagery for applications such as hardware-in-the-loop testing of forward-looking IR sensor, IR missile warning and IR missile seeker systems. Identified RISS subsystems comprise:

The RISS display showing modelling and database development imagery. The A-10 aircraft image (right hand side of screen) is part of a 10 μm thermal scene (Northrop Grumman)   2002/0102905

Modelling and scenario development
The architecture's model builder application is integrated with MultiGen CreatorPro to create 3-D object and terrain databases, with the individual databases being created in OpenFlight format with extensions for IR attribution. The model toolkit application facilitates the importation of measured signature data together with attributed IR models from signature prediction codes such as SPF, SIRRM/SPURC, PRISM, RadTherm and SPIRITS. Atmospheric and weather modelling is performed using MODTRAN. For its part, the scenario builder used makes use of situation display and event sequencing tools to create 'complex' test scenarios. Trajectory models such as BLUEMAX, ESAMS, DISAMS and TRAP can be used interactively to script player movement and all the various modelling and scenario development applications feature 'extensive' Graphical User Interface (GUI) support.

Scene generation
The RISS's controller application provides a GUI-based test control and visualisation environment while the scene generator software used performs real-time scene generation that is synchronised to the sensor unit under test. Physics-based scene radiance calculations are performed on a frame-by-frame basis at rates and sizes of up to 500 MHz and up to 1,024 × 1,024 pixels respectively. Scene rendering can be hosted on either the SGI InfiniteReality or RISS scene rendering subsystems. The architecture's digital scene output can be formatted to drive commercial optical projection devices as well as a Universal Programmable Interface (UPI - see following).

Scene Rendering Subsystem (SRS)
The RISS SRS is a real-time rendering engine that is designed and manufactured by Northrop Grumman and is based on parallel arrays of PowerPC processing modules that are hosted on custom VME circuit card assemblies. As such, it is claimed to facilitate higher fidelity, more flexible renderings than other visual scene generation systems. System features include one to six colours at 16-bit resolution, synchronisation with sensors that are operating at higher than visual frame rates (30 to 60 Hz), non-symmetric frame size and deterministic anti-aliasing.

UPI
The RISS UPI is a real-time image processing subsystem that provides an interface between a real-time image generator and sensor system hardware, scene projection device or display hardware. Its primary function is to provide image post processing for functions such as IR/EO sensor emulation, direct signal injection formatting and scene projector processing. The emulation of by-passed sensor components is by means of user programmable convolution, image blur, image warping and noise modelling.

External control synchronisation
The RISS architecture can be controlled by an external simulator or simulation, a facility that supports closed loop simulation in missile seeker testing, IR countermeasures evaluation and multispectral simulation test applications. Protocols (supported via custom interface code) include SCRAMNet, Internet, Gagabit local area network and ATM.

Non real-time scene generation
Software emulations of the SRS and UPI are available that are fully compatible with real-time databases. Uses include algorithm development and sensor prototyping and an API is provided so that scene generation can be embedded in a customer's simulation environment.

Status
Over time, RISS is reported as having been procured by the US Air Force (USAF – as part of the AN/AAR-44 Missile Warning Receiver Integration Support Station) and as being the basis of Northrop Grumman IR sensor simulators delivered to the USAF's Flight Test Center (Edwards Air Force Base, California) and the US Navy's Naval Air Warfare Center – Aircraft Division (Naval Air Station Patuxent River, Maryland). Other users include the European Aeronautic, Defence and Space (EADS) Company and as of September 2001, a RISS system/s was due for delivery to a Japanese customer during 2002.

Contractor
Northrop Grumman Electronic Systems sector - Amherst Systems, Buffalo, New York.

UPDATED

Northrop Grumman Reprogrammable Emitter Simulator (RES)

Type
Reprogrammable emitter simulator for test and training range applications.

Description
Derived from the US Navy's (USN) Remote Emitter System (located at Point Lookout, Maryland), Northrop Grumman's RES is a Combat Electromagnetic Environment SIMulator (CEESIM – see separate entry) - based configuration that is capable of radiating threat signals at target aircraft or ships within a 161 km radius of the transmitter site. Target movements are monitored by radar and are tracked by pedestal-mounted RES antennas to achieve a high effective radiated power in any direction within the 161 km range. Transmitted signals are monitored and measured via an integrated verification system. RES systems can be supplied as turnkey packages that include remote-control computers, datalinks, signal generation and modulation units, site construction, transmitters, antennas, radomes, pedestal-tower integration and signal verification systems. Northrop

A general view of a Northrop Grumman RES site   0010010

Grumman notes that such turnkey applications range in price from US$1 million to US$5 million depending on specific configuration and site requirements. The company claims that the described equipment enables the user to create a realistic, remotely controlled, programmable, multisource integrated air defence system analogue within the range. Further, the use of distributed simulation protocols (high-level architecture and distributed interactive simulation) allows systems to be linked to other assets located worldwide in a co-ordinated test or training exercise.

Status
As of December 2001, a Northrop Grumman's RES was reported as being in service at the USN's Atlantic Test Range.

Contractor
Northrop Grumman Electronic Systems sector - Amherst Systems, Buffalo, New York.

UPDATED

Pico-AMES threat simulator

Type
Portable threat simulator.

Description
Pico-AMES is a low-cost, modular, desktop Electronic Warfare (EW) simulator which is fully compatible with Comptek's AMES II system and can generate up to 64 complex emitters operating in the 0.5 to 18 GHz frequency range. A dynamic scenario simulation capability allows the user to generate combinations of static and/or dynamic scenarios containing pulse, pulse-Doppler and continuous wave radars for an amplitude Angle Of Arrival (AOA) EW system. Pico-AMES can be configured as part of a host system (using an Ethernet interface into a local or wide area network) and features automatic calibration and built-in test. Its AOA amplitude simulation has up to 10 ports while its virtual memory environment open architecture is designed to facilitate modular expansion. The equipment uses fast tuning voltage controlled oscillator sources and there is a synthesised phased lock option if required. The system's scenario control facility includes online editing; dynamic formats with up to 64 emitters (in increments of 16) on up to 64 mobile platforms; latitude/longitude or X/Y co-ordinates; a 1,852 × 1,852 km gaming area; variable platform speed, altitude, heading, roll/pitch and yaw/time; variable event timing resolution (1 to 10 Hz) and spherical, earth, contour line terrain map, radar horizon and user-defined terrain map terrain modelling.

Status
Over time, Pico-AMES simulators are reported as having been procured by customers in Canada, Europe and the US.

Specifications
Frequency: 0.5-18 GHz (0.5-2 GHz and 2-18 GHz sub-bands – mm wave expansion option)
Dynamic range: 90 dB
Modulations: agile, chirp, coherent, phase coded, stable, stagger, slider, simultaneous synchronised beams
PRI range/resolution: 2-100,000 μs/1.6 ns
PRI modulations: agile, formula patterns, jitter, stable, stagger, ramp, uniform or Gaussian random

PW range/resolution: 0.05-3,000 µs/50 ns
PW rise/fall time: 20 ns
PW modulation: as PRI
Scan characteristics: bidirectional sector and raster, circular, circular/sector, conical, electronic, helical, lobe on receive, lobe switching, nodding, orthogonal, scan motion steady, spiral, random, user-defined, unidirectional sector and raster
Scan coverage: ±90° (elevation); ±180° (azimuth)

Contractor
Northrop Grumman Electronic Systems sector - Amherst Systems, Buffalo, New York.

UPDATED

Portable Combat Electromagnetic Environment Simulator (CEESIM)

Type
Portable Electronic Combat (EC) system simulator.

Description
Portable CEESIM is a compact, low-cost system that can be used throughout all phases of an EC system's lifecycle where high-fidelity, limited density simulation is required. Applications range from the laboratory to free space radiating field tests and when combined with an EC system, portable CEESIM can also be used for classroom and onboard training. The equipment can be carried by a single person and comprises three main components: Control; Digital Generation and Radio Frequency Generation Subsystems (CS, DGS and RFGS respectively). Looking at these in more detail, the CS incorporates a control computer (containing the programmes and data needed to define the simulated environment) and an operator workstation to access the system's graphical user interface. The DGS receives commands and data from the CS and translates the environment into real-time digital pulse commands that are fed in turn to the RFGS. This latter subsystem converts the received digitised data into real-time RF pulses and applies them to the system under test using direct injection or free space radiation as appropriate. Unlike larger versions of CEESIM (see separate entry), portable CEESIM's DGS and RFGS reside in the same chassis. The CS can comprise an external workstation or an embedded processor and a PC. When using an embedded processor, the system can be accessed via a lap- or desktop PC or a network link. Portable CEESIM can provide audio, digital, intermediate frequency, RF and video stimuli and is claimed to be able to simulate all known types of pulse repetition intervals and scan patterns. A maximum of 128 simultaneous, high-fidelity emitters (including pulse-on-pulse, pulse-on-Continuous Wave (CW) and CW-on-CW signals) can be generated within the 0.5 to 18 GHz frequency range. Each of the two available RF channels can provide up to 12 ports (each of which has a unique amplitude angle of arrival) for direct injection testing. Free space radiation is supported by an optional travelling wave tube amplifier gating/grid control that provides programmable duty cycle thresholding.

Status
Over time, portable CEESIM is reported as having been procured by 'number' of US and offshore customers.

Specifications
Frequency range: 0.5-18 GHz (0.05-0.5 GHz and 18-40 GHz options)
Dimensions (W × L × H): 445 × 533 × 356 mm
Weight: 29.5-38.6 kg (configuration dependent)
Power: 110/220 V AC; 50-400 Hz

Contractor
Northrop Grumman Electronic Systems sector - Amherst Systems, Buffalo, New York.

UPDATED

Pulse-Doppler/high-power Airborne Emitter System (AES)

Type
Airborne intercept radar threat simulators.

Description
Metric's pulse-Doppler (alternately designated low-power)/high-power AES equipments are pod-mounted threat emitter simulators that can mimic the output of a variety of airborne intercept radars. Using a standard 2.86 × 0.47 m pod shell, both systems are microprocessor-controlled and provide up to eight threat programmes that can be selected from a remote location. Each programme activates the system and selects from a 32 mode menu to provide the desired emitter parameters. Antenna position and rate can be changed according to preflight programming. Following selection of a particular programme, mode execution is accomplished via time basis or antenna position preprogramming.

Status
As of early 2001, Metric's pulse-Doppler/high-power AES equipment was reported as having been procured by users in France/Germany (the Polygon range), the UK (the Spadeadam, Northumberland range) and the US (Eglin Air Force Base (AFB), Florida, Eielson AFB, Alaska, Mountain Home AFB, Idaho and Nellis AFB, Nevada ; Naval Air Stations China Lake and Point Mugu, California and the Fallon, California and Pinecastle, Florida ranges).

Specifications

	high-power AES	pulse-Doppler/low-power AES
Frequency	7.8-17.5 GHz (magnetron selectable)	8-18 GHz (magnetron selectable)
Power	150 kW	200 W
Pulse-width	0.2-1.2 µs	0.1 µs - CW
Duty cycle	0.001	100%
PRF	5,000 pps max	1 pps - CW
Antenna gain	26 dB (nominal)	26 dB (nominal)
Antenna rates	10-70°/s	10-70°/s
Sector width	120°	120°
Boresight angles	−30°, 0°, +30°	−30°, 0°, +30°

Contractor
Metric Systems (an Integrated Defense Technologies company), Fort Walton Beach, Florida.

UPDATED

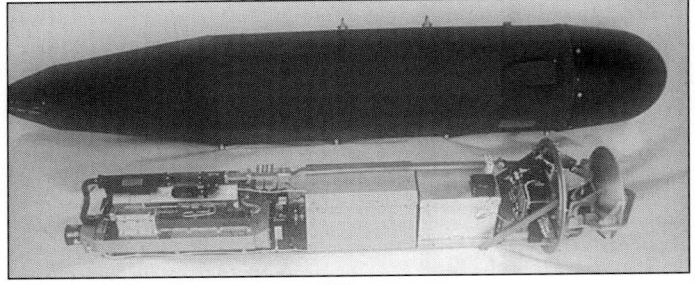
Metric's pulse-Doppler/low-power AES

Metric's high-power AES

Radio Frequency Simulator (RFS)

Type
6 to 18 GHz band Radio Frequency (RF) missile simulator.

Description
Specifically developed to meet a US Air Force (USAF) requirement, Computer Science and Applications' (CSA) RFS is described as being designed for the non-destructive, laboratory testing of active, passive or semi-active, RF terminally guided, missile systems. As such, the architecture is understood to take the form of an eight channel system that is built around a 15 m anechoic chamber, using a curved antenna wall that incorporates 384 target and 42 background emitter antennas with a 52 × 54° field of view to the missile. Facilities for true and synthetic line of sight simulation are provided within the 6 to 18 GHz frequency range and the architecture as a whole is claimed to offer 'realistic' models of the missile under test, targets, RF phenomena and countermeasures and a 'high accuracy' simulation of the weapon's real-world performance from launch to impact.

Status
The described RFS was developed as a joint project between CSA, the USAF's 46th Test Wing (Eglin Air Force Base, Florida), an on-site support contractor and 'several other' subcontractors who provided administrative, assembly and analysis support for the project. Within this consortium, CSA designed the system and built and tested all the required prototype assemblies before turning over fabrication to Vitro Services (now a part of BAE Systems North America). Fabricated assemblies were returned to CSA for final installation, hardware/software integration and system testing. Elsewhere in the programme, Carco Electronics supplied the necessary flight motion simulator, while KOR Electronics provided digital RF memories. The complete RFS was assembled and installed at Eglin's Guided

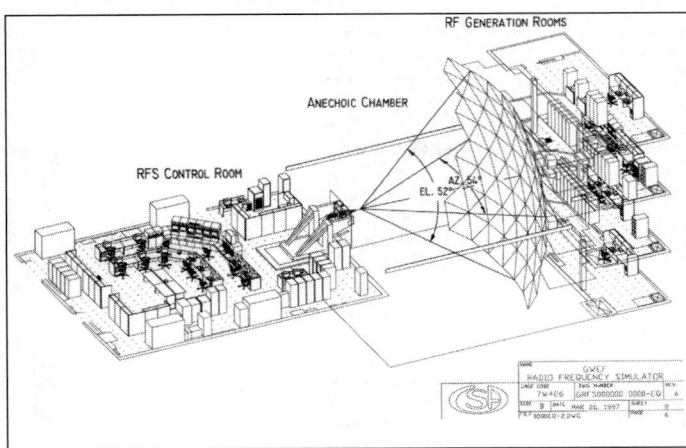

A schematic showing the layout of the GWEF at Eglin AFB in Florida (Computer Science and Applications)   2002/0113336

Weapons Evaluation Facility (GWEF) during the period 1992 to 1997 and achieved full operational capability during May 1997. At this time, its primary purpose was to provide 'state-of-the-art' simulation for continuation simulation and testing of weapons such as the AIM-120 Advanced Medium Range Air-to-Air Missile (AMRAAM).

Contractor
Computer Science and Applications Inc, Shalimar, Florida.

NEW ENTRY

....

Simulator for Electronic Combat Training (SECT)

Type
Electronic combat trainer.

Description
SECT is a software-based generic simulator developed for the US Air Force (USAF). It consists of six student stations, with 'glass-cockpit', graphical-based equipment suites that are reconfigurable and expandable using embedded software tools. Realistic training is provided through representative airborne displays and simulation of 200 fully interactive threat emitters (including electronic countermeasures) which encompass the full gamut of the electronic combat scenario. Each threat type has been provided with a detailed, high-fidelity tactics model. Threats can be integrated into a complete air defence network simulation. Sophisticated voice recognition technology has also been employed in each student station for independent student communication control of 'own ship' manoeuvres and evaluation of student crew co-ordination tasks.

SECT is controlled and operated by a single two-place instructor console, providing a comprehensive software mediated performance monitoring and feedback system. The instructor selects which suite of mission equipment will be active and presented at any one of the student stations. The students are immersed in a synthetic environment that can be self-paced instruction or real-time mission simulation, including visual and aural cues that depict realistic combat situations.

The simulator covers all types of missions, including strategic and covert operations, standoff jamming and electronic intelligence gathering. It is intended that realistic interactive training be given with the student stations simulating representative simulated electronic combat positions in the aircraft. Using SECT, a new mission can be developed and presented for real-time mission rehearsal within 72 hours of receipt of the new specifications. An associated training systems support centre is also noted as providing low-cost life cycle support for organic software upgrades, the maintenance of student records and new lesson generation.

Status
Over time, SECT is reported as having been procured by the USAF. As of January 2002, SECT appeared to have been a live programme.

Contractor
AAI Corporation, Hunt Valley, Maryland.

UPDATED

....

Signal Measurement System (SMS)

Type
Transportable, high-fidelity, signal monitoring system.

Description
SMS is a signal monitoring system that is designed to verify and validate the electronic combat test environment. The equipment measures the output (at radio frequency, video or digital level) of threat simulators, measures the response of the radar and/or electronic countermeasures system, correlates radar and

countermeasure responses and analyses countermeasures techniques. SMS is made up of input, receiver and signal processor subsystems (plus a control computer), each of which uses commercially available hardware supported by application specific signal analysis software. Looking at these elements in more detail, the input subsystem is application specific with an open-air range configuration utilising a broadband antenna to sample the environment. An anechoic chamber application utilises multiple antennas distributed around the test chamber. The reception element incorporates tuneable narrowband and wideband receivers with the latter capability utilising a number of instantaneous frequency measuring receivers to cover the system's 0.5 to 18 GHz (18 to 40 GHz option) frequency range.

SMS signal processing is based on Northrop Grumman Amherst System's Receiver Processor System and incorporates a processor chain that executes proprietary signal processing software that has been specifically designed to de-interleave corrupted pulse trains (missing pulses, multipath effects and incorrectly parametised pulses). A multiprocessor architecture is used to achieve throughput rates in excess of 1 million pulses/s and to simplify maintenance and evolutionary improvement. Individual processors within the chain function in a concurrent, asynchronous fashion to offer true pulse by pulse processing and a high tolerance of corrupted data. The modular design of the subsystem supports expansion to an arbitrarily large number of processors to meet evolving system requirements and the replacement of existing processors by more powerful boards if required.

Status
As of December 2001, SMS was reported as having been procured by the US Air Force.

Contractor
Northrop Grumman Electronic Systems sector - Amherst Systems, Buffalo, New York.

UPDATED

....

Threat Emitter Simulation System (TESS)

Type
Low-cost threat radar simulator for training and testing.

Description
TESS is a programmable threat simulator that is tunable in flight and is used to simulate threat radar and air-to-surface missile emissions for training weapon systems operators. The system can be housed in either a standard AN/ALQ-167 pod shell or standard 48 cm racks for internal carriage. TESS comprises a transmitter assembly, magnetrons, antennas, associated Radio Frequency (RF) components, interface assembly and low-voltage power supply. Both internal and external installations utilise identical assemblies. For the podded configuration, these are mounted on an electronic tray that slides into the pod. In flight the operator can choose from 15 programmed scenarios that feature a variety of frequency and scan patterns. Closed-loop remote magnetron tuning aided by a frequency measurement device prevents excessive drift.

TESS is designed to allow for maximum system flexibility while requiring a minimum of component replacement between bands. For the lower bands (7.8 to 8.5 GHz and 8.5 to 9.6 GHz) only a change of magnetrons is required. For the upper band (12 to 13.2 GHz and 14 to 15.2 GHz) the magnetron and associated RF components are replaced. Mechanical packaging of the TESS allows rapid replacement of components for either band change or maintenance functions. The TESS pod has been designed and tested to meet MIL-E-5400, Class 1A environment and MIL-STD-461/462 for electromagnetic interference. The pod is qualified for carriage on several commercial and military aircraft at speeds up to 1,389 km/h calibrated airspeed or M1.3.

Status
Over time, TESS is reported as having been procured by the Canadian Forces Maritime Command and the Royal Navy.

Contractor
BAE Systems North America - Technical Services, Fort Walton Beach, Florida.

UPDATED

....

Transportable Combat Electromagnetic Environment Simulator (CEESIM)

Type
Radio Frequency (RF) signal environment simulator.

Description
Transportable CEESIM is designed to provide military users with a transportable RF signal simulator for flight line or onboard Electronic Warfare (EW) system check out. It can also be used as a development tool for EW systems that require a high-fidelity, high-density, pulse environment during their design, integration, test and/or validation. Northrop Grumman Amherst claims that transportable CEESIM matches the signal fidelity of other members of its CEESIM family (see separate entries) in a portable and application flexible package. Transportable CEESIM unit

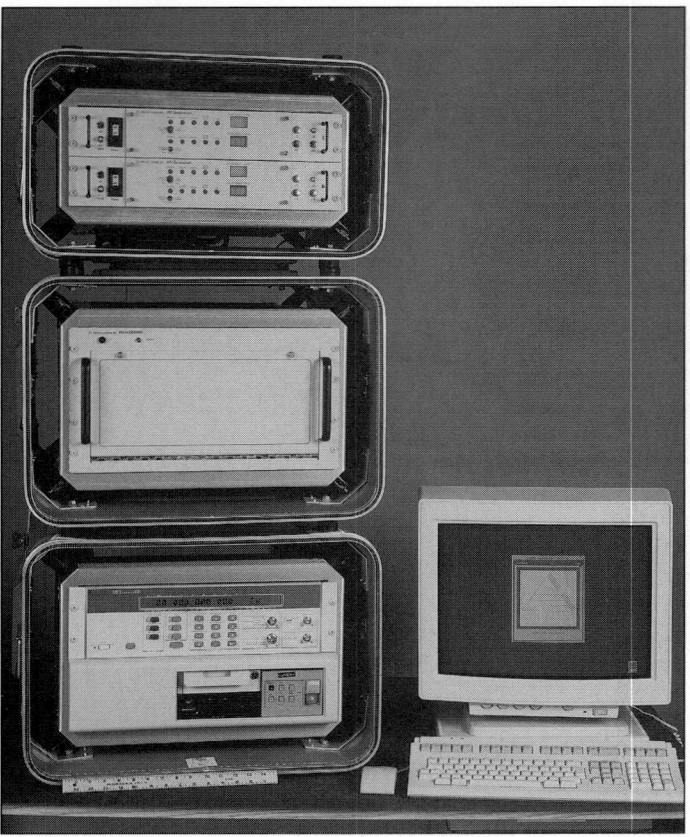

The Transportable Combat Electromagnetic Environment Simulator (CEESIM)

price ranges between US$250,000 and US$1 million depending on specific configuration. Other system features include:

- Online emitter and scenario editing
- Generation of up to 1,024 simultaneous, real-world emitters on 512 simultaneous platforms
- A +10 dB output power
- Quadphase modulation on pulse
- A 70 dB scan/effective radiated power/range loss dynamic range
- A full complement of complex scan types
- 0.05 to 40 GHz frequency coverage
- Optional amplitude and phase Angle Of Arrival (AOA) modulations. AOA simulations are accommodated via the introduction of a single RF AOA subsystem and its associated digital hardware. Northrop Grumman Amherst notes that amplitude, interferometer phase and Butler matrix phase AOA simulations have been delivered to customers worldwide
- A user-friendly, intuitive Windows operator interface that allows the operator to input all parameters associated with scenario events, emitters and platforms. The interface also provides multiple display/reporting options and flexible run-time scenario adaptability
- A digital architecture that can support easy, low-cost expansion to up to 32 RF channels
- Component mounting in shock isolated, environmentally sealed cases for transport between sites.

Status

Over time, transportable CEESIM is reported as having been procured for a number of US and international laboratory, anechoic chamber and range applications. EW equipment known to have tested using transportable CEESIM include the AN/BLD-1, AN/BRD-7, AN/WLR-1H, AN/WLR-8/-8 HPI and AN/WLQ-4 equipments.

Contractor

Northrop Grumman Electronic Systems sector - Amherst Systems, Buffalo, New York.

UPDATED

Turbo Crow Responsive Electronic Warfare Training System (REWTS)

Type

EW training and countermeasures support suite.

Description

The Turbo Crow suite comprises radar jamming, radar simulation and monitoring/control subsystems. It is a dual-role training and tactical support system that can be installed in commercial jets such as the Learjet and/or similarly sized military types.

A typical Turbo Crow operator's station (MS/Rodale)

The radar jamming subsystem offers noise, smart noise, deception, cover pulse, multiple false target, Doppler, co-ordinated and combination modes and is capable of simultaneous multiband operation. The system's hardware can be tailored to specific airframes and can take the form of internally and/or pod-mounted (AN/ALQ-167 or A100 pod shell types) units. A hybrid option is also available which splits the capability between an inboard techniques generator, controller, and external podded amplifier/antenna units.

The radar simulation subsystem allows the suite to simulate a wide range of airborne and ground-based search, tracking, missile guidance and fire-control radars. The capability is built around the Rodale Digital Airborne Radar Threat Simulator (DARTS) equipment and can be operated at the same time as the radar jamming subsystem. DARTS modes comprise circular, steady, sector centred and sector off-centre scan patterns; stable, jitter and stagger pulse repetition frequencies and adjustable pulse-widths, delays and frequencies. The computerised monitoring/control subsystem is described as being at the heart of the Turbo Crow system and allows the operator full real-time control over the suite and the engagement scenario of the moment. In the training role, mission data is recorded for post mission analysis and debriefing while the system can be upgraded to provide an electronic support measures/electronic intelligence capability for tactical applications. An integral databank of emitter parameters is incorporated in the facility.

Status

Over time, an example of the Turbo Crow suite is understood to have been procured by the US commercial EW training provider Corporate Jets.

Specifications

Frequency range: 0.85-18 or 0.2-18 GHz (jamming subsystem in 3 or 4 system specific bands respectively); 7.8-8.5, 8.6-9.6, 12-13.2 and 14-15.2 GHz (DARTS subsystem); 0.4-18 GHz (ESM/ELINT capability - extendable to 40 GHz)
Jammer output power: typically 200 or 400 W
DARTS output power: min 100 kW
ESM/ELINT sensitivity: −70dBm
DF accuracy: 3-5°

Contractor

Rodale Electronics Inc, Hauppauge, New York.

UPDATED

US Navy (USN) Electronic Warfare (EW) simulation systems

Type

Series of EW simulation systems developed by the USN.

Description

Over time, the USN has, in concert with industry, developed a range of EW simulation systems that are designed for use on training and research and development applications. Known details of such equipments are as follows:

AN/AST-6(V) pod

The AN/AST-6(V) supersonic pod is a computer-controlled and programmable threat emitter pod and is a replacement for the earlier AN/AST-4(V). The system's digital cockpit control box controls up to 12 pods on a single aircraft. Each pod transmits up to 99 different Anti-Ship Cruise Missile (ASCM) threat emission

The AN/AST-6(V) threat emitter pod

The AN/ULQ-21(V) digital countermeasures system

sequences. Programmable variables include scan simulations, pulse modes, pulse repetition frequencies, pulse-widths, frequency tuning and frequency jitter. Scenarios can be programmed to transition from launch platform search, missile launch and final attack mode in time frames that represent actual ASCM threats.

AN/AST-7(V)

AN/AST-7(V) is a podded threat emitter simulator system that is installed in an AN/ALQ-167 type pod shell. It covers H- to K-bands (6 to 40 GHz) with pulse-width and frequency values factory preset. Peak power is given as 125 kW, with pulse-width and pulse repetition frequency settings ranging from 0.1 to 1.3 μs and 1 to 6,000 pulses/s respectively.

AN/DLQ-3C(V)

AN/DLQ-3C(V) is a modular, analogue, radar jammer that was originally designed for use in aerial targets. Over time, it has now been upgraded and expanded so that it generates 21 techniques ranging from simple continuous noise to complex deception techniques in the frequency range 0.4 to 18 GHz. It can be configured for use in AN/ALQ-167(V) type pod shells, aircraft, laboratories, vans, land-based sites, and tactical and quick reaction capability applications.

AN/DPT-1(V)

AN/DPT-1(V) is a variable frequency radar threat simulator that has a peak power of 70 kW and operates in the H-, I- or J-bands (6 to 20 GHz overall) depending on the mission profile. Over time, it has been installed in AQM-127A, MQM-8 and BQM-34S drones together with the AN/AST-4 pod.

AN/DPT-2(V)

AN/DPT-2(V) is a miniaturised version of the AN/DPT-1(V) system that, over time, has been installed in the BQM-74C and TDU-34 target systems and has been used in anti-radiation missile firing exercises and for specialised electronic support system test and training. It has a peak power of 20 kW and operates in one of two frequency bands (8 to 10 GHz or 10 to 20 GHz) depending on the missile profile for the target.

AN/ULQ-21(V)

AN/ULQ-21(V) is a digital countermeasures set that is a replacement for the AN/DLQ-3C(V) system. It is capable of simulating present and projected threats and is designed to meet a wide range of developmental test and training requirements. The system includes an automatic set on noise capability, a digital radio frequency memory, terrain bounce and non-adaptive swept polarisation techniques together with the full range of AN/DLQ-3C(V) jamming modes. Baseline ULQ-21(V) operates in the 5 to 11 GHz range (expandable to 1 to 40 GHz) and can be installed in an AN/ALQ-167 type pod shell if required.

AN/UPT-2

AN/UPT-2 is a radar threat simulator that, over time, has been installed in BQM-34S, MQM-8 and AQM-127 drones. The system operates in the H-, I- and J-bands (6 to 20 GHz overall), has a peak power of 150 kW and offers four selectable pulse-widths together with variable pulse repetition frequencies (in the 200-5,000 pulses/s range).

TREE series

The TREE series of threat simulation products are special, high-power ECM pods. They were initially designed to provide a threat environment for surface weapons test and training at sea. Their use has now been expanded to provide threat ECM for airborne weapons. These pods have a radiated power that is equivalent to that of operational stand-off equipments and feature special technique transmitter add-ons (such as the TREE FOX exciter and the TREE BASS transponder) to produce realistic self-protection and escort jamming techniques. By way of example, the TREE HAWK and TREE FOX systems offer 'advanced' deception techniques for use against surveillance, tracking and fire-control radars. Other special purpose ECM devices (such as the AN/ASR-101 ECM pod), when used with the TREE FOX exciter, provide a realistic battle scenario for operator training and shipboard system testing.

ECM jammers

The Basic ECM Environment Simulators (BEES) and portable shore-based jammers are installed in towers located near land-based weapon system development and training sites. These devices produce advanced techniques for realistic test and training. Protective ECM is also produced by airborne and shipborne threat simulation dispensers, launchers, and their expendable countermeasures payloads.

Status

Over time, equipments of the types described are known to have been taken into service. Identified contracting activity that related to the overall programme during 2001 comprises a then year US$21,653,423 firm, fixed-price, indefinite delivery/indefinite-quantity contract that was awarded to KOR Electronics (Garden Grove, California) on 7 March 2001 with respect to the supply of not more than 101 miniaturised, I/J-band (8 to 20 GHz) digital radio frequency memories for use in ULQ-21(V) digital countermeasures sets. At the time of the announcement, work on the effort was scheduled for completion by the end of March 2007 and the effort's contracting activity was the USN's Naval Air Warfare Center Weapons Division, Naval Air Station (NAS) Point Mugu, California.

Contractor

US Naval Air Warfare Center, Weapons Division, NAS Point Mugu, California.

UPDATED

TABULATIONS

World submarine radar/electronic warfare fits
World surface ship radar fits
World surface ship electronic warfare fits

The following tabulations are designed to provide readers with ready-reckoner type information on the radar and electronic countermeasures fits of the world's submarine and surface ship inventories. All vessels included are identified by their country of service and class name, except where they are a single unit. Here, they are identified by name or pennant number. Numbers of vessels in service (as of this edition) are also noted. Within the submarine data, no attempt is made to differentiate between boat roles. Within the surface ship material, entries are divided into the broad categories of aircraft carrier, cruiser, destroyer, frigate, corvette and patrol craft. Vessels operated by or originating from the Russian Federation and associated states are, for brevity's sake, referred to by the class names given to them by NATO rather then their national Project designations. Readers seeking more specific information on particular vessels or ship classes are referred to *Jane's Fighting Ships*.

WORLD SUBMARINE RADAR/ELECTRONIC WARFARE FITS

Class name (no)	SSR	NR	ESM/RW	Decoy
ALGERIA				
'Kilo' (2)	Snoop Tray		Brick Group	
ARGENTINA				
'Salta' (2)		Calypso II	DR 2000	
'Santa Cruz' (2)		Calypso IV	Sea Sentry III	
AUSTRALIA				
'Collins' (6)		T.1007	AR-700 series	SSE
'Oberon' (3)	T.1006		Mavis	SSDE
BRAZIL				
'Humaita' (3)		T.1006	UA-4	
'Tupi' (4)		Calypso III	DR series	
Tacantins (1)		Calypso III	DR series	
BULGARIA				
'Romeo' (2)	Snoop Plate		Stop Light	
CANADA				
'Oberon' (3)		T.1006 or Furuno 1831	Guardian Star	
CHILE				
'Oberon' (2)		T.1006	ESM/RW	
'Thomson' (2)	Calypso II		DR 2000	
CHINA				
200 (1)		Snoop Plate		
320 (1)	SSR		T.921A	
351 (1)	Snoop Plate and Snoop Tray			
'Han' (5)	Snoop Tray		T.921A	
'Kilo' (4)	Snoop Tray		Squid Head or Brick Pulp	
'Ming' (13)	Snoop Tray			
'Romeo' (65)	Snoop Plate or Snoop Tray			
Xia (1)	Snoop Tray		T.921A	
COLUMBIA				
'Pijao' (2)	Calypso II		DR 2000	
CUBA				
'Foxtrot' (3)	Snoop Tray		Stop Light	
DENMARK				
'Narhvalen' (2)	TERMA SSR		Sea Lion	
'Tumleren' (3)	TERMA SSR		Sea Lion	
ECUADOR				
'Type 209' (2)	Calypso series			
EGYPT				
'Romeo' (4)	Snoop Plate		AR-700 series	
FRANCE				
'Agosta' (4)	DRUA 33		ARUD	
'Daphné' (3)	Calypso series			
'Le Triomphant' (3)	Dassault SSR		DR 3000	
'L'Inflexible M4' (5)		DRUA 33	DR 3000	
'Rubis Améthyste' (6)	Kelvin Hughes SSR		DR 3000	
GERMANY				
'Type 205' (2)	Calypso II		ESM/RW	
'Type 206/206A' (17)	Calypso II		DR 2000	
'Type 212' (4)	T.1007		FL 1800U	TAU 2000
GREECE				
'Glavkos' (8)	Calypso II		AR-700 series	
INDIA				
'Foxtrot' (6)	Snoop Tray		Stop Light	
'Kilo' (8)		Snoop Tray	Squid Head	
'Shishumar' (4)	Calypso series		ARGO ESM or Sea Sentry RW	
INDONESIA				
'Cakra' (2)	Calypso series		DR 2000	

Class name (no)	SSR	NR	ESM/RW	Decoy
IRAN				
'Kilo' (3)	Snoop Tray		Squid Head and Quad Loop	
ISRAEL				
'Gal' (3)	SSR		Timnex 4	
ITALY				
'Improved Sauro' (4)	Officine Galileo SSR/NR		Elettronica ESM/RW	
'Sauro' (4)	Officine Galileo SSR/NR		Elettronica ESM/RW	
JAPAN				
'Harushio' (7)	ZPS 6		ZLR 3/6	
'Improved Harushio' (4)	ZPS 6 or 7		ZLR 7	
Yaeshio (1)	ZPS 4		ZLR 3/6	
'Yuushio' (10)	ZPS 6		ZLR 3/6	
KOREA (North)				
'Romeo' (22)	Snoop Plate and Snoop Tray		T.921A	
KOREA (South)				
'Chang Bogo' (7)		NR	ARGO ESM/RW	
LIBYA				
'Foxtrot' (4)	Snoop Tray		Stop Light	
NETHERLANDS				
'Walrus' (4)	ZW 07		AR-700 series	
NORWAY				
'Modernised Kobben' (6)	T.1007		ARGO RW	
'Ula' (6)	T.1007		Sea Lion	
PAKISTAN				
'Hangor' (4)	DRUA 31		ARUD	
'Hashmat' (3)	DRUA 33		ARUD	
PERU				
'Abtao' (2)		NR	ESM/RW	
'Casma' (6)	Calypso series		ESM/RW	
POLAND				
'Foxtrot' (2)	Snoop Tray		Stop Light	
Orzel (1)	Snoop Tray		Brick Group and Quad Loop	
PORTUGAL				
'Albacora' (3)	T.1007		ESM/RW	
ROMANIA				
Delfinul 521 (1)	Snoop Tray		Brick Group and Quad Loop	
RUSSIAN FEDERATION AND ASSOCIATED STATES (CIS)				
'Akula I/II' (17)	Snoop Half or Snoop Pair		Rim Hat	
'Charlie II' (2)	Snoop Tray		Stop Light, Brick Spit/ Brick Pulp and Park Lamp	
'Delta I/II/III/IV' (30)	Snoop Tray		Brick Pulp/Group and Park Lamp	
'Echo II' (2)	Snoop Tray/ Slab and Front Door (FCR)		Stop Light, Brick Pulp (or Squid Head) and Quad Loop	
'Foxtrot' (6)	Snoop Plate or Snoop Tray		Stop Light	
'Kilo' (24)	Snoop Tray		Brick Pulp or Squid Head and Quad Loop	
'Oscar I/II' (15)	Snoop Half or Snoop Pair		Bald Head or Rim Hat	
'Severodvinsk' (2)	SSR		ESM/RW	
'Sierra I/II' (4)	Snoop Pair		Bald Head, Park Lamp and Rim Hat	
'Tango' (16)	Snoop Tray		Brick Group or Squid Head and Quad Loop	
'Typhoon' (6)	Snoop Pair		Rim Hat and Park Lamp	
'Victor I/II/III' (31)	Snoop Tray		Brick Group and Park Lamp	
'Yankee Notch' (3)	Snoop Tray		Brick Group and Park Lamp	

Class name (no)	SSR	NR	ESM/RW	Decoy
SOUTH AFRICA				
'Daphne' (3)	Calypso II		ARUD	
SPAIN				
'Delfin' (4)	DRUA 31 or		Manta	
	DRUA 33A			
'Galerna' (4)	DRUA 33C		Manta	
SWEDEN				
'Gotland' (3)		TERMA NR	Manta	
'Näcken' (3)		TERMA NR	ARGO ESM	
'Sjöormen' (3)		TERMA NR		
'Västergötland' (4)		TERMA NR	AR-700 series	
TAIWAN				
'Guppy II' (2)	SSR		WLR-1 or -3	
'Hai Lung' (2)	ZW 06		AR-700 series and	
			Timnex 4	
TURKEY				
'Atilay' (6)	SSR		DR 2000 or Porpoise	
'Guppy IIA/III' (7)	SSR			
'Preveze' (4)	SSR		Porpoise	
'Tang' (2)	BPS-12		WLR-1	
UK				
'Swiftsure' (5)		T.1006	UAP	T.2066/2071
'Trafalgar Batch 1' (7)		T.1007	UAP	T.2066/2071
'Vanguard' (4)		T.1007	UAP	T.2066/2071
USA				
'Benjamin Franklin' (2)	BPS-15		WLR-8 and -10	Emerson
	(SS/N/FCR)			Mk 2
'Los Angeles' (58)	BPS-15A or -16		WLR-1H, -8(V)2 and	Emerson
	(SS/N/FCR)		-10	Mk 2
Narwhal (1)	BPS-14		WLQ-4	Emerson
	(SS/N/FCR)			Mk 2
'Ohio' (18)	BPS-15A		WLR-8(V)5 and-10	Emerson
	(SS/N/FCR)			Mk 2
'Seawolf' (3)		BPS-16	WLQ-4(V)1	WLY-1
'Sturgeon' (18)	BPS-14 or -15		WLQ-4(V)	Emerson
	(SS/N/FCR)			Mk 2
VENEZUELA				
'Cabalo' (2)		TERMA NR	DR 2000	
YUGOSLAVIA (Federal Republic of)				
'Heroj' (2)	Snoop Group		Stop Light	
'Sava' (2)	Snoop Group		Stop Light	

Key
SSR Surface Search Radar; **NR** Navigation Radar; **FCR** Fire-Control Radar; **ESM/RW** Electronic Support Measures/Radar Warning; **Decoy** Torpedo countermeasures

VERIFIED

WORLD SURFACE SHIP RADAR FITS

AIRCRAFT CARRIERS (Including helicopter carriers)

Class name (no)	A/SSR	NR	FCR
BRAZIL			
Minas Gerais (1)	AWS-4 and SPS-40B	ZW 06	SPG-34
FRANCE			
Charles de Gaulle (1)	DRBJ-11B and DRBV-15C	Racal Decca 1229	Arabel
Jeanne D'Arc (1)	DRBV-22D	DRBN-34A	DRBC-32A
INDIA			
Viraat (1)	DA 05 and T.996	T.1006	T.904
ITALY			
Giuseppe Garibaldi (1)	SPS-52C,-702, -774 and -768	SPN-728 and -749(V)	SPG-74 and -75
RUSSIAN FEDERATION AND ASSOCIATED STATES (CIS)			
Admiral Gorshkov (1)	Plate Steer, Sky Watch and Strut Pair	Palm Frond	Bass Tilt, Cross Sword, Front Door and Kite Screech
Admiral Kuznetsov (1)	Sky Watch and Strut Pair	Palm Frond	Cross Sword and Hot Flash
SPAIN			
Principe de Asturias (1)	SPS-52C/D and SPS-55		RAN-12L, RTN-11X and VPS-2
THAILAND			
Chakri Naruebet (1)	SPS-52C and -64		STIR
UK			
'Invincible' (3)	T.992, 996 and 1022	T.1006 or 1007	T.909 or 909(1)
Ocean (1)	T.996	T.1007	
USA			
Enterprise (1)	Mk 23 and SPS-48E, -49(V)5 and -67	Furuno 900 and SPS-64(V)9	Mk 95
'Kitty Hawk'/'John F Kennedy' (3)	Mk 23 and SPS-48E, -49(V)5 and -67	Furuno 900 and SPS-64(V)9	Mk 95
'Nimitz' (9)	Mk 23 and SPS-48E, -49(V)5 and -67	Furuno 900 and SPS-64(V)9	Mk 95

Key
A/SSR Air/Surface Search Radar; **SSR** Surface Search Radar; **NR** Navigation Radar; **FCR** Fire-Control Radar

UPDATED

CRUISERS

Class name (no)	A/SSR	NR	FCR
ITALY			
Vittorio Veneto (1)	SPS-52C, -702 and -768	SPS-748	SPS-55C, -70 and -74
PERU			
Aguirre (1)	DA 02, LW 02 and ZW 03	Racal Decca 1226	M25 and 45
Almirante Grau (1)	DA 08 and LW 08	Racal Decca 1226	STIR and WM 25
RUSSIAN FEDERATION AND ASSOCIATED STATES (CIS)			
'Kara' (5)	Flat Screen, Head Net A and Top Sail	Don Kay or Palm Frond	Bass Tilt, Head Light, Owl Screech, Pop Group and Top Dome
'Kirov' (4)	Strut Pair, Top Pair, Top Plate and Top Steer	Palm Frond	Bass Tilt, Cross Sword, Eye Bowl, Hot Flash, Kite Screech, Pop Group and Top Dome
'Slava' (3)	Top Pair, Top Plate and Top Steer	Palm Frond	Bass Tilt, Front Door, Kite Screech, Pop Group and Top Dome
USA			
'California' (2)	SPS-48E, -49(V)5 and -67(V)1	SPS-64(V)9	SPG-51D and -60D
'Ticonderoga' (27)	SPS-49(V)7, -49(V)8	SPS-64(V)9	SPG-62 and SPQ-9A

Key
A/SSR Air/Surface Search Radar; **SSR** Surface Search Radar; **NR** Navigation Radar; **FCR** Fire-Control Radar

VERIFIED

DESTROYERS

Class name (no)	A/SSR	NR	FCR
ARGENTINA			
'Almirante Brown' (4)	DA 08A and ZW 06	Racal Decca 1226	STIR
'Hercules' (2)	T.965P and 992Q	T.1006	T.909
AUSTRALIA			
'Perth' (3)	SPS-40C, -52C and -67(V)		SPG-51C
BRAZIL			
Mariz E Barros (1)	SPS-10 and -40		Mk 25 Mod 3
CANADA			
'Iroquois' (4)	DA 08 and LW 08	Pathfinder	STIR 1.8
CHILE			
'Prat' (4)	EL/M-2228S and T.227M, 992Q/R, 965M and 966	T.978 or 1006	T.901 and 902 and EL/M-2221
CHINA			
'Luda I/II' (15)	Eye Shield, Knife Rest, Rice Screen, Sea Tiger and Square Tie	Fin Curve and Racal Decca 1290	Castor II, Rice Lamp, Sun Visor B, T.347G and Wasp Head
'Luhu' (3)	ESR 1, God Eye and Sea Tiger	Racal Decca 1290	Castor II, Rice Lamp and T.347G
FRANCE			
'Cassard' (2)	DRBJ-11B and DRBV-26C	DRBN-34A	DRBC-33A and SPG-51C
'Georges Leygues' (7)	DRBV-15A, -26 and -51C	Racal Decca 1226	DRBC-32E and -33A and Vega
'Suffren' (2)	DRBI-23 and DRBV-15A	Racal Decca 1226	DRBC-33A and DRBR-51
'Tourville' (3)	DRBV-26 and -51B	Racal Decca 1226	DRBC-32D
GERMANY			
'Lütjens' (3)	SPS-10, -40 and-52		SPG-51 and -60 and SPQ-9
GREECE			
'Kimon' (4)	SPS-10D/F, -39, -40D and -64	LN 66	SPG-51D and -53A
INDIA			
'Delhi' (4)	DA 05 and LW 08	Bharat NR	Bass Tilt
'Rajput' (5)	Big Net A and Head Net C	Don Kay	Bass Tilt, Drum Tilt, Owl Screech and Peel Group
IRAN			
'Babr' (2)	SPS-10B and -29C	LN 66	Mk 25
ITALY			
'Audace' (2)	SPS-52C, -774, -768 and SPQ-2D	SPN-748	SPG-51 and -76
'De La Penne' (2)	SPS-52C, -702,-774 and -768	SPN-748	SPG-51D and -76
JAPAN			
'Asagiri' (8)	OPS-14C, -24 and -28C		
'Haruna' (2)	OPS-11C and-28	OPN-11	
'Hatakaze' (2)	OPS-11C and -28B and SPS-52C		SPG-51C and Melco 2-21
'Hatsuyuki' (2)	OPS-14B and -18		
'Kongou' (4)	OPS-28C/D and SPY-1D		SPG-62
'Marasame' (6)	OPS-24 and -28D	OPS-20	
'Shirane' (2)	OPS-12 and -28	OFS-2D	WM 25
'Tachikaze' (3)	OPS-16 and -28 and SPS-52B		SPG-51
'Takatsuki' (3)	OPS-11B and -17		Mk 35
'Yamagumo' (3)	OPS-11 and -17		Mk 35
KOREA (South)			
'Gearing Fram I/II' (7)	SPS-10, -37 and -40		Mk 25
MEXICO			
'Gearing Fram I' (2)	SPS-10, -29 and -40	LN 66	Mk 12/22
PAKISTAN			
'Gearing Fram I' (3)	SPS-10 and -40	Racal Decca 1226	Mk 25
POLAND			
Warszawa (1)	Big Net and Head Net C	SRN 207 and 7453	Bass Tilt, Owl Screech and Peel Group

Class name (no)	A/SSR	NR	FCR
ROMANIA			
Marasesti (1)	Strut Curve and SSR	Spin Trough	Drum Tilt and Hawk Screech
RUSSIAN FEDERATION AND ASSOCIATED STATES (CIS)			
'Sovremenny' (19)	Plate Steer, Top Plate and Top Steer	Palm Frond	Bass Tilt, Front Dome and Kite Screech
'Udaloy' (10)	Strut Pair and Top Plate	Palm Frond	Bass Tilt, Eye Bowl, Kite Screech and Sword
TAIWAN			
'Wu Chin I/II' (6)	EL/M-1040 and SPS-10, -40 and -58		HR 76, Mk 25 and RTN-10X
'Wu Chin III' (7)	DA 08		STIR and W-160
TURKEY			
'Carpenter Fram I' (2)	SPS-10 and -40		Mk 35
'Gearing Fram I' (3)	SPS-10 and -40		Mk 25
UK			
'Type 42' (12)	T.996 and 1022	T.1006 or 1007	T.909 or 909(1)
USA			
'Arleigh Burke Flts I/II' (28)	SPS-67(V)3 and SPY-1D	SPS-64(V)9	SPG-62
'Arleigh Burke Flts IIA' (4)	SPS-67(V) and SPY-1D(V)		SPG-62
'Kidd' (4)	SPS-48E, -49(V)5 and -55	SPS-64	SPG-51D and SPQ-9A
'Spruance' (31)	Mk 23 and SPS-40B, -49(V) and -55	SPS-64(V)9	Mk 95, SPG-60 and SPQ-9A

Key
A/SSR Air/Surface Search Radar; **SSR** Surface Search Radar; **NR** Navigation Radar; **FCR** Fire-Control Radar

VERIFIED

FRIGATES

Class name (no)	A/SSR	NR	FCR
ALGERIA			
'Mourad Rais' (3)	Strut Curve	Don 2	Drum Tilt, Hawk Screech and Pop Group
ARGENTINA			
'Drummond' (3)	DRBV-51A	Racal Decca 1226	DRBC-32E
'Espora' (6)	DA 05	Racal Decca 1226	
AUSTRALIA			
'ANZAC' (8)	SaabTech A/SSR and SPS-49(V)8	Atlas 9600	SaabTech FCR
'Adelaide' (6)	SPS-49 and-55		Mk 92 Mod 2 and SPG-60
'River' (2)	Atlas 8600 and LW 02		M22
BANGLADESH			
'Leopard' (2)	T.965 and 993	T.978 and 1007	T.275
Osman (1)	Eye Shield and Square Tie	Fin Curve	
Umar Farooq (1)	T.965 and 993	T.978	T.275 and 278
BELGIUM			
'Wielingen' (3)	DA 05	Raytheon 1645	WM 25
BRAZIL			
'Broadsword' (4)	T.967 and 968	T.1006	T.911
'Inhauma' (4)	AWS-4	T-1007	RTN-10X
'Ipara' (4)	SPS-10C and -40B	LN 66	Mk 35
'Niteroi' (6)	AWS-2, RAN-20S and ZW 06		RTN-10X
CANADA			
'Annapolis' (2)	SPS-502 and -503	Sperry 127E	SPG-515
'Halifax' (12)	Sea Giraffe and SPS-49(V)5	T.1007	STIR 1.8
'Improved Restigouche' (2)	SPS-10 and -503	Sperry 127E	SPG-515
CHILE			
Almirante Williams (1)	AWS-1 and SNW 10	Racal Decca 1629	M4/3 and SGR 102
'Leander' (4)	T.965, 966, 992Q and 994	T.1006	T.903 and 904

Class name (no)	A/SSR	NR	FCR
CHINA			
'Jianghu I' (27)	MX 902, Rice Screen and Square Tie	Don 2, Fin Curve and Racal Decca NR	Rice Lamp and Sun Visor
'Jianghu III/IV' (3)	MX 902 and Square Tie	Fin Curve	Rice Lamp and Sun Visor B
'Jiangwei' (6)	Knife Rest A	Racal Decca 1290	Fog Lamp, Rice Lamp and Sun Visor
COLUMBIA			
'Almirante Padilla' (4)	Sea Tiger		Castor II
CUBA			
'Koni' (3)	Strut Curve	Don 2	Drum Tilt, Hawk Screech and Pop Group
DENMARK			
'Niels Juel' (3)	AWS 5, SaabTech SSR and TRS		SaabTech FCR and Mk 95
'Thetis' (4)	AWS-6 and TERMA SSR		SaabTech FCR
ECUADOR			
'Leander' (2)	T.994 and 996	T.1006	T.903 and 904
EGYPT			
'Descubierta' (2)	DA 05 and ZW 06		WM 25
'Jianghu' I (2)	Eye Shield, T.765 and Square Tie	Racal Decca 1290A	Fog Lamp
'Knox' (2)	SPS-10, -40 and -67	LN 66	SPG-53A
FRANCE			
Commandant Bory (1)	DRBV-22 and Racal 1226		DRBC-32C
'D'Estienne d'Orves' (17)	DRBV-51A	Racal Decca 1226	DRBC-32E
'Floréal' (6)	DRBV-21A	Racal Decca 1229	
'La Fayette' (6)	DRBV-15C	Racal Decca 1229	Castor IIJ
GERMANY			
'Brandenburg' (4)	LW 08 and SMART	Raypath	STIR 180
'Bremen' (8)	DA 08 and TRS	Officine Galileo NR	STIR and WM 25
'Type 124'	SMART, SPY-1D and TRS		APAR
GREECE			
'Elli' (6)	LW 08 and ZW 06		STIR and WM 25
'Epirus' (3)	SPS-10F and -40D		SPG-53
'Hydra' (4)	DA 08 and MW 08	APAR and Racal Decca 2690	STIR
INDIA			
'Godavari' (6)	Head Net C, LW 08 and ZW 06		Drum Tilt, Muff Cob, Pop Group and Seaguard
'Nilgiri' (6)	LW 08, T.965M and ZW 06	Racal Decca 1226	M44 and 45 and T.903 and 904
'Petya II' (4)	Slim Net	Don 2	Hawk Screech
INDONESIA			
'Ahmad Yani' (6)	DA 05 and LW 03	Racal Decca 1229	M44 and 45
'Fatahillah' (3)	DA 05	Racal Decca 1229	WM 28
'Samadikun' (4)	SPS-4, -5D and -6E	Racal Decca 1226	SPG-52
'Tribal' (3)	T.965 and 993	Racal Decca 978	T.903
IRAN			
'Alvand' (3)	AWS-1 and Racal 1226	Racal Decca 629	Sea Hunter
ITALY			
'Alpino' (2)	SPS-12 and -702(V)3	SPN-748	EMPAR and SPG-70
'Artigliere' (4)	SPQ-712 and SPS-774	SPN-703	SPG-70 and -74
'Lupo' (4)	SPQ-2F and SPS-702 and -774	SPN-748	SPG-70 and -74
'Maestrale' (8)	SPS-702 and -774	SPN-703	SPG-74 and -75
JAPAN			
'Abukuma' (6)	OPS-14C and -28		
'Chikugo' (11)	OPS-14 and -16		
'Yubari' (2)	OPS-19B and -28C		
KOREA (North)			
'Najin' (2)	Pot Head and Square Tie	Pot Drum	Drum Tilt
KOREA (South)			
'Okpo' (3)	MW 08 and SPS-49(V)	DTR-92	STIR 180
'Ulsan' (9)	DA 05, S1810 and ZW 06	SPS-10C	ST1802 and WM 28

Class name (no)	A/SSR	NR	FCR
LIBYA			
'Koni' (2)	Plank Shave and Strut Curve	Don 2	Drum Tilt, Hawk Screech and Pop Group
LITHUANIA			
'Grisha III' (2)	Strut Curve	Don 2	Bass Tilt and Pop Group
MALAYSIA			
'Lekiu' (2)	DA 08 and Sea Giraffe		S1802
MEXICO			
'Bronstein' (2)	SPS-10F and -40D	LN 66	Mk 35
NETHERLANDS			
'Jacob van Heemskerck' (2)	LW 08, Scout and SMART		STIR 180 and 240
'Karl Doorman' (8)	LW 08 and SMART	Racal Decca 1226	STIR
'Kortenaer' (4)	LW 08 and ZW 06		STIR
'LCF' (2)	APAR, Scout and SMART		
'Tromp' (2)	Signaal A/SSR	Racal Decca 1226	SPG-51C and WM 25
NEW ZEALAND			
'ANZAC' (2)	SaabTech A/SSR and SPS-49(V)8	Atlas 9600	SaabTech FCR
'Broad Beam Leander' (2)	LW 08 and T.993	T.1006	TR 76
NORWAY			
'Oslo' (4)	AWS-9	Racal Decca 1226	SaabTech FCR and Mk 95
OMAN			
'Qahir' (2)	MW 08	T.1007	DRBV-51C and STING
PAKISTAN			
'Leander' (2)	T.966 and 993	T.1006	T.904
'Tariq' (6)	DA 08 and T.992R	T.1006	T.91C
PERU			
'Meliton Carvajal' (4)	RAN-10S and -11L/X	Officine Galileo NR	RTN-10X and -20X
PORTUGAL			
'Baptista de Andrade' (4)	AWS-2	Racal Decca 316P	Pollux
'Commandante Joao Belo' (4)	DRBV-22A and -50	T.1007	DRBC-31D
'Joao Coutinho' (6)	Kelvin Hughes A/SSR	Racal Decca 1226	SPG-34
'Vasco Da Gama' (3)	DA 08 and MW 08	T.1007	STIR
ROMANIA			
'Tetal' (4)	Strut Curve	Nayada	Drum Tilt and Hawk Screech
RUSSIAN FEDERATION AND ASSOCIATED STATES (CIS)			
'Grisha II/III/IV' (71)	Half Plate B, Strut Curve and Strut Pair	Don 2	Bass Tilt, Muff Cob and Pop Group
'Krivak I/II/III' (24)	Head Net C and Top Plate	Don 2, Don Kay, Palm Frond, Peel Cone and Spin Trough	Bass Tilt, Kite Screech, Owl Screech and Pop Group
'Neustrashimy' (2)	Top Plate	Palm Frond	Cross Sword and Kite Screech
'Parchim II' (12)	Cross Dome	Kivach II or Nayada	Bass Tilt
'Petya II' (3)	Slim Net A	Don 2	Hawk Screech
SAUDI ARABIA			
'La Fayette' (2)	DRBV-26C and Sea Tiger	Racal Decca 1226	Castor II
'Madina' (4)	Sea Tiger	Racal Decca 1226	Castor IIB and DRBC-32
SPAIN			
'Baleares' (5)	SPS-10 and -52B	Pathfinder	RAN-12L, SPG-51C and -53B and VPS-2
'Descubierta' (6)	DA 05	ZW 06	WM 22/41 and 25
'Santa Maria' (6)	SPS-49(V)5 and -55	Raytheon 1650 and SPS-67	Mk 92, RAN-30L, STING and VPS-2
SYRIA			
'Petya III' (2)	Slim Net	Don 2	Hawk Screech
TAIWAN			
'Cheng Kung' (7)	Chang Bei and SPS-49(V)5 and -55		Mk 92 Mod 6 and STIR
'Kang Ding' (6)	DRBV-26D and Triton G		Castor IIC
'Knox' (12)	SPS-10, -40B and -67	LN 66	SPG-53A/D/F

Class name (no)	A/SSR	NR	FCR
THAILAND			
'Chao Phraya' (4)	Eye Shield and Square Tie	Racal Decca 1290	Rice Lamp and Sun Visor
'Knox' (2)	SPS-10, -40B and -67	LN 66	SPG-53A/D/F
'Naresuan' (2)	LW 08 and T.360	SPS-64(V)5	STIR and T.374G
'Tapi' (2)	LW 04 and SPS-53E		WM 22/61
TURKEY			
'Barbaros' (2)	AWS-9	Racal Decca 2690	Contraves TMX, Seaguard and STIR
'Berk' (2)	SPS-10, and -40	Racal Decca 1226	Mk 34
'Tepe' (8)	SPS-10, -40B and -67	LN 66	SPG-53A/D/F
'Yavuz' (4)	DA 08	Racal Decca 1226	Seaguard, STIR and WM 25
UK			
'Broadsword' (11)	T.967 and 968	T.1006 or 1007	T.910 and 911
'Type 23' (16)	T.996(1)	T.1007	T.911
USA			
'Oliver Hazard Perry' (45)	SPS-49(V)4/5 and -55	Furuno	Mk 92 and STIR
URUGUAY			
'Commandant Rivière' (3)	DRBV-22A	Racal Decca 1226	DRBC-32C
VENEZUELA			
'Modified Lupo' (6)	RAN-10S and SPQ-2F		RTN-10X and-20X
VIETNAM			
'Petya' (5)	Strut Curve	Don 2	Hawk Screech
YUGOSLAVIA (Federal Republic of)			
'Split'/'Kotor' (3)	Strut Curve	Don 2 or Palm Frond	SaabTech FCR, Drum Tilt, Owl Screech and Pop Group

Key
A/SSR Air/Surface Search Radar; **SSR** Surface Search Radar; **NR** Navigation Radar; **FCR** Fire-Control Radar

UPDATED

CORVETTES

Class name (no)	A/SSR	NR	FCR
ALGERIA			
'Djebel Chinoise' (3)	Racal Decca 1226		
'Nanuchka II' (3)	Square Tie	Don 2	Muff Cob and Pop Group
ECUADOR			
'Esmeraldas' (6)	RAN-10S	Officine Galileo NR	RTN-10X
FINLAND			
'Turunmaa' (2)	TERMA SSR	ARPA	WM 22
GREECE			
'Niki' (5)	TRS 3001		
INDIA			
'Abhay' (4)	Cross Dome	Pechora	Bass Tilt
'Durg' (3)	Square Tie	Don 2	Muff Cob and Pop Group
'Khukri' (6)	Cross Dome and Plank Shave	Bharat 1245	Bass Tilt
'Veer' (12)	Plank Shave	Pechora	Bass Tilt
INDONESIA			
'Kapitan Patimura' (16)	Strut Curve	TSR 333	Muff Cob
IRAN			
'Bayandor' (2)	SPS-6C	Raytheon 1650	Mk 36
IRAQ			
'Assad' (2)	RAN-12L/X	SPN-703	RTN-10X
ISRAEL			
'Eilat' (3)	EL/M-2218S and SPS-55		EL/M-2221
ITALY			
'Minerva' (8)	SPS-774	SPN-728(V)2	SPG-76

Class name (no)	A/SSR	NR	FCR
KOREA (North)			
'Sariwon'/'Tral' (5)	Pot Head and Don 2	T.351	
KOREA (South)			
'Dong Hae' (4)	SPS-64		WM 28
'Po Hang' (25)	SPS-64 and ST1810		S1802 and WM 28
MALAYSIA			
'Assad' (2)	RAN-12L/X	T.1007	RTN-10X
'Kasturi' (2)	DA 08	Racal Decca 1226	WM 22
POLAND			
'Gornik' (4)	Plank Shave	Kivach and Pechora	Bass Tilt
'Orkan' (3)	NUR-27XA	SRN-443XTA	Bass Tilt
ROMANIA			
'Democratia' (4)		Don 2	
'Zborul' (3)	Plank Shave	Spin Trough	Bass Tilt
SAUDI ARABIA			
'Badr' (4)	SPS-40B and -55		Mk 92
SINGAPORE			
'Victory' (6)	Sea Giraffe	T.1007	EL/M-2221X
SWEDEN			
'Göteborg' (4)	Sea Giraffe	PN 612	SaabTech FCR
'Stockholm' (2)	Sea Giraffe	PN 612	SaabTech FCR
THAILAND			
'Khamronsin' (3)	AWS-4	Racal Decca 1226	
'Rattanakosin' (2)	DA 05 and ZW 06	Racal Decca 1226	WM 25/41
UNITED ARAB EMIRATES			
'Murray Jib' (2)	Sea Giraffe	Racal Decca 1226	SaabTech FCR and DRBV-51C
VIETNAM			
'Tarantul I' (2)	Plank Shave	Pechora	Bass Tilt

Key
A/SSR Air/Surface Search Radar; **SSR** Surface Search Radar; **NR** Navigation Radar; **FCR** Fire-Control Radar

UPDATED

PATROL CRAFT

Class name (no)	A/SSR	NR	FCR
ALGERIA			
'Osa I/II' (11)	Square Tie		Drum Tilt
BAHRAIN			
'Ahmad El Fateh' (4)	Sea Giraffe	Racal Decca 1226	SaabTech FCR
'Al Manama' (2)	Sea Giraffe	Racal Decca 1226	SaabTech FCR
'Al Riffa' (2)	SaabTech SSR	Racal Decca 1226	
BANGLADESH			
'Durbar' (5)	Square Tie		
'Durdharsha' (5)	Square Tie		
'Huchuan' (8)	T.753		
'Shaheed' (9)	Pot Head and Skin Head		
CUBA			
'Osa I/II' (14)	Square Tie		Drum Tilt
'Turya' (9)	Pot Drum		Muff Cob
DENMARK			
'Flyvefisken' (14)	AWS-6, TERMA SSR and TRS	Furuno	SaabTech FCR
'Willemoes' (10)	SaabTech A/SSR	TERMA NR	SaabTech FCR
EGYPT			
'Hainan' (8)	Pot Head and Skin Head		
'October' (6)	S810		ST802
'Osa I' (5)	Kelvin Hughes A/SSR	Racal Decca 916	Drum Tilt
'Ramadan' (6)	S810 and 820		ST802
'Shershen' (6)	Pot Drum		Drum Tilt
FINLAND			
'Helsinki' (4)	SaabTech SSR	ARPA	SaabTech FCR
'Rauma' (5)	SaabTech SSR	ARPA	SaabTech FCR

Class name (no)	A/SSR	NR	FCR
GERMANY			
'Albatros' (10)	WM 27 (SS/FCR)	Officine Galileo NR	
'Gepard' (10)	WM 27 (SS/FCR)	Officine Galileo NR	
'Tiger' (16)	Triton	Officine Galileo NR	Castor
GREECE			
'Jaguar' (4)	Racal Decca 1226		
'La Combattante II' (4)	Triton	Racal Decca 1226	Pollux
'La Combattante IIA' (4)	Triton	Officine Galileo NR	
'La Combattante III' (10)	Triton	Racal Decca 1226	Castor II and Pollux
'Nasty' (4)	Racal Decca 1226		
INDONESIA			
'Dagger' (4)	Racal Decca 1226		WM 28
'Kakap' (4)	Racal Decca 2459	T.1007	
'PB 57' (4)	Scout	T.1007	
'Sibarau' (8)	Racal Decca 916		
'Singa' (4)	Racal Decca 2459 and Scout		WM 22
IRAN			
'Hudong' (10)	SR-47A	FM 1070	Rice Lamp
'Kaman' (10)	WM 28 (SS/FCR)	Racal Decca 1226	
ISRAEL			
'Aliya' (2)	Neptune		RTN-10X
'Hetz' (5)	Neptune		EL/M-2221
'Reshef' (6)	Neptune		RTN-10X
'Super Dvora' (14)		Raytheon NR	
ITALY			
'Cassiopea' (4)	SPS-702(V)2	SPN-748(V)2	SPG-70
'Sparviero' (6)	Officine Galileo SSR and SPQ-701		SPG-70
KENYA			
'Madaraka' (3)		Racal Decca 1226	RTN-10X
'Nyayo'	AWS-4	Racal Decca 1226	ST802
KOREA (North)			
'Chaho' (62)	Pot Head		
'Chong-Jin' (52)	Skin Head		
'Hainan' (6)	Pot Head		
'Kimjin' (65)	Furuno		
'Komar'/'Sohung' (16)	Square Tie		
'Ku Song'/'Sin Hung/ 'modified Sin Hung' (103)	Skin Head		
'Osa I'/'Huangfen' (12)	Square Tie		Drum Tilt
'P 6/Shantou/Sinpo'/'Sinnam' (36)	Skin Head and Furuno		
'Shanghai II' (12)	Pot Head and Skin Head		
'SO 1' (19)	Pot Head	Don 2	
'Soju' (15)	Square Tie		Drum Tilt
'Taechong I/II' (13)	Pot Head		Drum Tilt
'TB 11PA'/'TB 40A' (25)	Furuno		
'Yongdo' (50)	Furuno		
KOREA (South)			
'Pae Ku' (8)	HC 75 and SPS-58		SPG-50 and W-120
Pae Ku 51 (1)	Raytheon 1645		SPG-50
'Sea Dolphin' (87)	Raytheon 1645		
KUWAIT			
'Combattante I' (8)	MRR	Racal Decca NR	
Istiqlal (1)	S810	Racal Decca 1226	SaabTech FCR
'Simmoneau Star' 'Naja' (12)	Furuno		
LIBYA			
'Osa II' (12)	Square Tie		Drum Tilt
MALAYSIA			
'Handalan' (4)	SaabTech SSR	Racal Decca 616	SaabTech FCR
'Musytari' (2)	DA 05	Racal Decca 1226	SaabTech FCR
'Perdana' (4)	Triton	Racal Decca 616	Pollux
MEXICO			
'Holzinger' (4)	SPS-64(V)6	Nucleus	
'Uribe' (6)	Racal Decca 1226		
MOROCCO			
'Lazaga' (4)	ZW 06		WM 25
'Okba' (2)	Racal Decca 1226		
NEW ZEALAND			
'Moa' (4)	Racal Decca 916		

Class name (no)	A/SSR	NR	FCR
NORWAY			
'Hauk' (14)	Racal Decca 1226 (SS/NR)		
'Storm' (8)	Racal Decca 1226		WM 26
OMAN			
'Al Bushra' (3)	T.1007		
'Al Waafi' (2)	Racal Decca 1226	Racal Decca 1229	
'Dhofar' (4)	AWS-4 and -6	Racal Decca 1226	SaabTech FCR
'Seeb' (4)	Racal Decca 1226		
PAKISTAN			
'Haibat' (4)	Pot Head		
'Huangfen' (4)	Square Tie		
'Shanghai II' (3)	Skin Head		
PERU			
'Velarde' (6)	Triton	Racal Decca 1226	Castor II
POLAND			
'Modified Obluze' (8)			Drum Tilt
'Puck' (7)	Square Tie		Drum Tilt
PORTUGAL			
'Albatroz' (6)	Racal Decca 316P		
'Cacine' (10)	T.1007		
QATAR			
'Barzan' (4)	Racal Decca 1226		
'Damsah' (3)	Triton	Racal Decca 1226	Castor II
'Vita' (4)	MRR	T.1007	STING
ROMANIA			
'Epitrot' (12)	Pot Drum		Drum Tilt
'Huchuan' (20)	T.753		
'Osa I' (6)	Square Tie		Drum Tilt
'Shanghai' (16)	Don 2		
RUSSIAN FEDERATION AND ASSOCIATED STATES (CIS)			
'Matka' (6)	Plank Shave		Bass Tilt
'Molnya'/'Pauk II' (35)	Peel Cone and Positive-E	Kivach and Pechora	Bass Tilt
'Muravey' (14)	Peel Cone		Bass Tilt
'Svelyak' (27)	Peel Cone	Palm Frond	Bass Tilt
'Zhuk' (38)	Spin Trough		
SAUDI ARABIA			
'Al Siddiq' (9)	SPS-55		Mk 92
'Halter' (17)	SPS-64		
SINGAPORE			
'Fearless' (12)	EL/M-2228X (SS/FCR)		
'Sea Wolf' (6)		Racal Decca NR	WM 28/5
'Swift' (4)	Racal Decca 626		WM 26
'Vosper Type A' (3)	Racal Decca 626		
'Vosper Type B' (3)	Racal Decca 626		WM 26
SOUTH AFRICA			
'Minister' (9)	Triton		RTN-10X
SPAIN			
'Serviola' (4)	Racal Decca 2459	Racal Decca 2690	
SWEDEN			
'Dalarö' (3)	TERMA 610		
'Hugin' (12)	TERMA SSR		SaabTech FCR
Inshore Patrol Craft (12)	Racal Decca 914		
'Norrköping' (12)	Sea Giraffe		SaabTech FCR
SYRIA			
'Osa I/II' (12)	Square Tie		Drum Tilt
'Zhuk' (8)	Spin Trough		
TAIWAN			
'Hai Ou' (49)	LN 66		R 76
THAILAND			
'Chon Buri' (3)	ZW 06		WM 22/61
'PGM 71' (7)	Racal Decca SSR		
'Prabparapak' (3)	Kelvin Hughes SSR		WM 28/5
'Ratcharit' (3)		Racal Decca NR	WM 25
'Sattahip' (6)	Racal Decca SSR		

Class name (no)	A/SSR	NR	FCR
TUNISIA			
'Bizerte' (3)	DRBN-31		
'La Combattante IIIM' (3)	Triton		Castor II
'Modified Shanghai II' (3)	Pot Head		
'Shanghai II' (2)	Skin Head		
'Tazarka' (2)	Racal Decca 916		
TURKEY			
'Dogan' (8)	Racal Decca 1226		WM 28/41
'Turk' (12)	Racal Decca SSR		
'Yildiz' (5)	AWS-6 and MW 08	Racal Decca 1226	Contraves TMX and STING
UNITED ARAB EMIRATES			
'Ardhana' (6)	Racal Decca 1226		
'Ban Yas' (6)	Sea Giraffe	Racal Decca 1226	SaabTech FCR
'Mubarraz' (2)	Sea Giraffe	Racal Decca 1226	SaabTech FCR
URUGUAY			
'Cape' (2)		SPS-64	
'Vigilante' (3)	Racal Decca 1226		
VENEZUELA			
'Constitucion' (6)			RTN-10X and SPQ-2D
VIETNAM			
'Osa II' (8)	Square Tie		Drum Tilt
'Shershen' (16)	Pot Drum		Drum Tilt
'Turya' (5)	Pot Drum		Muff Cob
YUGOSLAVIA (Federal Republic of)			
'Koncar' (5)	Racal Decca 1226		SaabTech FCR
'Osa I' (5)	Square Tie		Drum Tilt

Key

A/SSR Air/Surface Search Radar; **SSR** Surface Search Radar; **NR** Navigation Radar; **FCR** Fire-Control Radar

UPDATED

WORLD SURFACE SHIP ELECTRONIC WARFARE FITS

AIRCRAFT CARRIERS (Including helicopter carriers)

Class name (no)	Decoy	ESM/RW	Jammer	Suite
BRAZIL				
Minas Gerais (1)	Shield	SLR-2		
FRANCE				
Charles de Gaulle (1)	Sagaie	ARBR-21	ARBB-33	
Jeanne D'Arc (1)	Syllex	ARBR-16 and ARBX-10		
ITALY				
Giuseppe Garibaldi (1)	SCLAR			Nettuno
RUSSIAN FEDERATION AND ASSOCIATED STATES (CIS)				
Admiral Gorshkov (1)	PK-2	Boll Nip		Bell Thump, Cage Pot, Foot Ball and Wine Flask
Admiral Kuznetsov (1)	PK-2 and 10			Ball Shield A and B, Flat Track, Foot Ball and Wine Flask
SPAIN				
Principe de Asturias (1)	SRBOC			Nettunel
UK				
'Invincible' (3)	DLJ	UAF	T.675(2)	
Ocean (1)	DLJ and Seagnat	UAT	T.675(2)	
USA				
Enterprise (1)	SRBOC			SLQ-32(V)4
'Kitty Hawk'/'John F Kennedy' (3)	SRBOC			SLQ-32(V)4
'Nimitz' (9)	SRBOC			SLQ-32(V)4

Key
Decoy Chaff, infra-red decoy, electro-optic decoy and active offboard decoy launching system; **ESM/RW** Electronic Support Measures/Radar Warning; **Suite** Combined electronic countermeasures/ESM system.

Note
In the RFAS entries, where there is doubt as to whether a system is an ECM or an ESM equipment, all onboard systems are logged under the *Suite* heading

UPDATED

CRUISERS

Class name (no)	Decoy	ESM/RW	Jammer	Suite
ITALY				
Vittorio Veneto (1)	SCLAR	SLR-4	SLQ-B and -C	
PERU				
Almirante Grau (1)	Dagaie and Sagaie			
RUSSIAN FEDERATION AND ASSOCIATED STATES (CIS)				
'Kara' (3)	PK-2 and -10	Bell Tap and Bell Slam		Bell Clout, Rum Tub and Side Globe
'Kirov' (4)	PK-2			Bell Bash, Bell Nip, Foot Ball, Half Cup, Rum Tub, Side Globe and Wine Flask
'Slava' (3)	PK-2 and -10			Rum Tub and Side Globe
USA				
'California' (2)	SRBOC			SLQ-32(V)3
'Ticonderoga' (27)	SRBOC			SLQ-32(V)3

Key
Decoy Chaff, infra-red decoy, electro-optic decoy and active offboard decoy launching system; **ESM/RW** Electronic Support Measures/Radar Warning; **Suite** Combined electronic countermeasures/ESM system.

Note
In the RFAS entries, where there is doubt as to whether a system is an ECM or an ESM equipment, all onboard systems are logged under the *Suite* heading

VERIFIED

DESTROYERS

Class name (no)	Decoy	ESM/RW	Jammer	Suite
ARGENTINA				
'Almirante Brown' (4)	Dagaie and SCLAR		Sphinx-Scimitar	
'Hercules' (2)	Corvus	RCM-2	RDL-257	
AUSTRALIA				
'Perth' (3)	SRBOC and Nulka	WLR-1H	ULQ-6	
BRAZIL				
Mariz E Barros (1)		WLR-1C and -3A	ULQ-6B	
CANADA				
'Iroquois' (4)	Shield	SLQ-501	SLQ-503 or ULQ-6	
CHILE				
'Prat' (4)	Barricade and Corvus	UA-8 and -9	T.667	
CHINA				
'Luda I/II' (15)		RW-23/1		
'Luhu' (3)	SRBOC		BM 8610	
FRANCE				
'Cassard' (2)	Dagaie	ARBR-17B and Saigon	ARBB-33	
'Georges Leygues' (7)	Dagaie	ARBR-17	ARBB-32B and -36	
'Suffren' (2)	Sagaie	ARBR-17	ARBB-33	
'Tourville' (3)	Syllex	ARBR-16	ARBB-32	
GERMANY				
'Lütjens' (3)	SRBOC			FL 1800S
GREECE				
'Kimon' (4)	SRBOC	SLQ-32(V)2		
INDIA				
'Delhi' (4)		Ajanta	TQN-2	
'Rajput' (5)	PK-16	Bell Tap		Bell Clout, Bell Shroud, Bell Slam, Bell Squat and Top Hat
IRAN				
'Babr' (2)		WRL-1	ULQ-6	
ITALY				
'Audace' (2)	SCLAR			Nettuno
'De La Penne' (2)	Sagaie			Nettuno
JAPAN				
'Asagiri' (8)	SRBOC	NOLR-6C and -8	OLT-3	
'Haruna' (2)	SRBOC	OLR-9		NOLQ-1
'Hatakaze' (2)	SRBOC	OLR-9B		NOLQ-1
'Hatsuyuki' (12)	SRBOC	NOLR-6 and -8	OLT-3	
'Kongou' (4)	SRBOC			NOLQ-2
'Marasame' (6)	SRBOC			NOLQ-2
'Shirane' (2)	SRBOC	OLR-9B		NOLQ-1
'Tachikaze' (3)	SRBOC	NOLR-6	OLT-3	NOLQ-1
'Takatsuki' (3)	SRBOC	NOLR-6C and -9	OLT-3	
'Yamagumo' (3)		NOLR-6		
KOREA (South)				
'Gearing Fram I/II' (7)		WJ.1140 and WRL-1		
MEXICO				
'Gearing Fram I' (3)		WRL-1		
PAKISTAN				
'Gearing Fram I' (3)	Shield			APECS II
POLAND				
Warszawa (1)	PK-16			Bell Shroud and Bell Squat
ROMANIA				
Marasesti (1)	PK-16	Watch Dog		
RUSSIAN FEDERATION AND ASSOCIATED STATES (CIS)				
'Sovremenny' (19)	PK-2 and-10			Foot Ball and Half Cup
'Udaloy' (10)				Bell Squat, Foot Ball, Half Cup and Wine Glass
TAIWAN				
'Wu Chin I/II' (6)	Kung Fen	WRL-1 and -3	ULQ-6	Chang Feng II
'Wu Chin III' (7)	Kung Fen			Chang Feng III

Class name (no)	Decoy	ESM/RW	Jammer	Suite
TURKEY				
'Carpenter Fram I' (2)		WRL-1		
'Gearing Fram I' (3)	SCLAR	WRL-1 and -3	ULQ-6	
UK				
'Type 42' (12)	DLB	UAT	T.675(2)	
USA				
'Arleigh Burke Flts I/II' (28)	SRBOC and Seagnat	SLQ-32(V)2		SLQ-32(V)3 and (V)5
'Arleigh Burke Flt IIA' (4)	SRBOC			SLQ-32(V)3
'Kidd' (4)	SRBOC			SLQ-32(V)5
'Spruance' (31)	SRBOC	WRL-1		SLQ-32(V)5

Key
Decoy Chaff, infra-red decoy, electro-optic decoy and active offboard decoy launching system; **ESM/RW** Electronic Support Measures/Radar Warning; **Suite** Combined electronic countermeasures/ESM system.
Note
In the RFAS entries, where there is doubt as to whether a system is an ECM or an ESM equipment, all onboard systems are logged under the *Suite* heading

VERIFIED

FRIGATES

Class name (no)	Decoy	ESM/RW	Jammer	Suite
ALGERIA				
'Mourad Rais' (3)	PK-16	Cross Loop and Watch Dog		
ARGENTINA				
'Drummond' (3)	Dagaie	DR 2000	Alligator	
'Espora' (6)	Dagaie	RQN-3B	TQN-2X	
AUSTRALIA				
'ANZAC' (8)	SRBOC and Nulka	Sceptre and Telegon		
'Adelaide' (6)	SRBOC and Nulka	SLQ-32	Israeli jammer	
'River' (2)		Elettronica ESM		
BANGLADESH				
Osman (1)	SRBOC	Watch Dog		
BELGIUM				
'Wielingen' (3)	SRBOC	DR 2000		
BRAZIL				
'Broadsword' (4)	Seagnat			
'Inhauma' (4)	Shield	Cutlass B1	SDR-7 and SLQ-1	
'Ipara' (4)	RBOC	WLR-1 and -6	ULQ-6	
'Niteroi' (6)	Shield and local system	Cutlass and local system	Cygnus and SLQ-1	
CANADA				
'Annapolis' (2)	SRBOC	SLQ-501 and intercept		
'Halifax' (12)	Shield	SLQ-501 and intercept	SLQ-503	
'Improved Restigouche' (2)	SRBOC	SLQ-501 and intercept	ULQ-6	
CHILE				
Almirante Williams (1)		WLR-1		
'Leander' (4)	Barricade and Corvus	UA-8 and-9	T.668	
CHINA				
'Jianghu I' (27)	RBOC	Jug Pair and Watch Dog		
'Jianghu III/IV' (3)		Elettronica RW	Elettronica jammer	
'Jiangwei' (6)	SRBOC	RWD 8	NJ81-3	
COLUMBIA				
'Almirante Padilla' (4)	Dagaie	ARGO RW	Scimitar	
CUBA				
'Koni' (3)	PK-16	Watch Dog		
DENMARK				
'Niels Juel' (3)	Seagnat	Cutlass		
'Thetis' (4)	Seagnat	Cutlass	Scorpion	

Class name (no)	Decoy	ESM/RW	Jammer	Suite
ECUADOR				
'Leander' (2)	Corvus and SRBOC	UA-8 and -9	T.667 and 668	
EGYPT				
'Descubierta' (2)			Elettronica jammer	
'Jianghu I' (2)		Triton	Elettronica jammer	
'Knox' (2)	SRBOC	SLQ-32(V)2		
FRANCE				
Commandant Bory (1)	Dagaie	ARBR-16		
'D'Estienne d'Orves' (17)	Dagaie	ARBR-16		
'Floréal' (6)	Dagaie	ARBR-17		
'La Fayette' (6)	Dagaie	ARBR-21	ARBB-33	
GERMANY				
'Brandenburg' (4)	SCLAR			FL 1800S
'Bremen' (8)	SRBOC			FL 1800S
GREECE				
'Elli' (6)	SRBOC	Elettronica RW		
'Epirus' (3)	SRBOC			SLQ-32(V)2/Sidekick
'Hydra' (4)	SRBOC	Telegon		APECS II
INDIA				
'Godavari' (6)	Super Barricade	Ajanta	TQN-2	
'Niligiri' (6)		Ajanta and Telegon	T.667	
INDONESIA				
'Ahmad Yani' (6)	Corvus	UA-8, -9 and -13		
'Fatahillah' (3)	Corvus	Susie I		
'Samadikun' (4)		WLR-1C		
'Tribal' (3)	Corvus			
IRAN				
'Alvand' (3)		RDL-2		
ITALY				
'Alpino' (2)	SCLAR			Elettronica suite
'Artigliere' (4)	SCLAR			Elettronica suite
'Lupo' (4)	SCLAR	SLR-4	SLQ-D	
'Maestrale' (8)	SCLAR	SLR-4	SLQ-D	
JAPAN				
'Abukuma' (6)	SRBOC	NORL-6	OLT-3	
'Chikugo' (11)		NORL-5		
'Yubari' (2)	SRBOC	NORL-6C	OLT-3	
KOREA (North)				
'Najin' (2)		Jug Pair and Watch Dog		
KOREA (South)				
'Okpo' (3)	Dagaie			APECS II
'Ulsan' (9)	SRBOC			
LIBYA				
'Koni' (2)		Watch Dog		
LITHUANIA				
'Grisha III'	PK-2	Watch Dog		
MALAYSIA				
'Lekiu' (2)	Super Barricade	Mentor	Scimitar	
MEXICO				
'Bronstein' (2)		WLR-3	ULQ-6	
NETHERLANDS				
'Jacob van Heemskerck' (2)	SRBOC			Ramses
'Karl Doorman' (8)	SRBOC			APECS II
'Kortenaer' (4)	SRBOC			Ramses
'LCF' (2)	SRBOC			APECS II
'Tromp' (2)	SRBOC			Ramses
NEW ZEALAND				
'ANZAC' (2)	Seagnat	SRBOC	Sceptre and Telegon	
'Broad Beam Leander' (2)	SRBOC			ARGO suite
NORWAY				
'Oslo' (4)		AR-700 series		
OMAN				
'Qahir' (2)	Barricade	DR 3000		

Class name (no)	Decoy	ESM/RW	Jammer	Suite
PAKISTAN				
'Leander' (2)	Corvus	UA-8, -9 and -13	T.668	
'Tariq' (6)	Corvus	DR 3000		
PERU				
'Meliton Carvajal' (4)	SCLAR	Elettronica ESM		
PORTUGAL				
'Commandante Joao Belo' (4)	SRBOC	ARBR-10 and AR-700 series		
'Vasco Da Gama' (3)	SRBOC			APECS II
ROMANIA				
'Tetal' (4)	PK-16	Watch Dog		
RUSSIAN FEDERATION AND ASSOCIATED STATES (CIS)				
'Grisha II/III/IV' (71)	PK-10 and -16	Watch Dog		
'Krivak I/II/III' (24)	PK-10 and -16	Bell Shroud		Bell Squat and Half Cup
'Neustrashimy' (2)	PK-10 and -16	Half Hat		Foot Ball and Half Cup
'Parchim II' (12)	PK-16	Watch Dog		
'Petya II' (3)		Watch Dog		
SAUDI ARABIA				
'La Fayette' (2)	Dagaie	DR 3000		
'Madina' (4)	Dagaie	DR 4000	Janet	
SPAIN				
'Balearas' (5)	SRBOC	Deneb	Canopus	
'Descubierta' (6)	SRBOC			Deneb
'Santa Maria' (6)	SRBOC			Nettunel
TAIWAN				
'Cheng Kung' (7)	Kung Fen			Chang Feng IV
'Kang Ding' (6)	Dagaie			Chang Feng IV
'Knox' (12)	SRBOC			SLQ-32(V)2/Sidekick
THAILAND				
'Chao Phraya' (4)	T.945 GPJ	T.923	T.981	
'Knox' (2)	SRBOC			SLQ-32(V)5
'Naresuan' (2)	T.945 GPJ			Elettronica suite
TURKEY				
'Barbaros' (2)	SRBOC	Cutlass	Scorpion	
'Berk' (2)		WLR-1		
'Tepe' (8)	SRBOC	SLQ-32(V)2	Sidekick?	
'Yavuz' (4)	SRBOC			Ramses?
UK				
'Broadsword' (11)	DLB and Seagnat	UAA(2) and UAT	T.675(2)	
'Type 23' (16)	DLB and Seagnat	UAF and UAT		
USA				
'Oliver Hazard Perry' (45)	SRBOC			SLQ-32(V)5
URUGUAY				
'Commandant Rivière' (3)		ARBR-16		
VENEZUELA				
'Modified Lupo' (6)	SCLAR	NS-9003		
VIETNAM				
'Petya' (5)		Watch Dog		
YUGOSLAVIA (Federal Republic of)				
'Split'/'Kotor' (3)	Barricade			

Key

Decoy Chaff, infra-red decoy, electro-optic decoy and active offboard decoy launching system; **ESM/RW** Electronic Support Measures/Radar Warning; **Suite** Combined electronic countermeasures/ESM system.

Note

In the RFAS entries, where there is doubt as to whether a system is an ECM or an ESM equipment, all onboard systems are logged under the *Suite* heading

VERIFIED

CORVETTES

Class name (no)	Decoy	ESM/RW	Jammer	Suite
ALGERIA				
'Nanuchka II' (3)	PK-16	Bell Tap and Cross Loop		
ECUADOR				
'Esmeraldas' (6)	SCLAR			Elettronica suite
FINLAND				
'Turunmaa' (2)	Barricade	ARGO RW		
INDIA				
'Abhay' (4)	PK-16			
'Durg' (3)	PK-16	Bell Tap		
'Khukri' (6)	PK-16	Ajanta P		
INDONESIA				
'Kapitan Patimura' (16)	PK-16	Watch Dog		
IRAQ				
'Assad' (2)	Italian decoy	INS-3	TQN-2	
ISRAEL				
'Eilat' (3)	Desaver		RAN-1010?	NS-9003/5
ITALY				
'Minerva' (8)	Barricade			Elettronica suite
KOREA (South)				
'Dong Hae' (4)	Protean			
'Po Hang' (25)	Protean and SRBOC			
MALAYSIA				
'Assad' (2)	Italian decoy	INS-3	TQN-2	
'Kasturi' (2)	Dagaie	Rapids	Scimitar	
POLAND				
'Gornik' (4)	PK-16			
'Orkan' (3)	PK-12 and -32			
ROMANIA				
'Zborul' (3)	PK-16	Watch Dog		
SAUDI ARABIA				
'Badr' (4)	SRBOC	SLQ-32(V)		
SINGAPORE				
'Victory' (6)	LRCR and Shield	Israeli ESM	SHARK/RAN-1101?	
SWEDEN				
'Göteborg' (4)	Philax	AR-700 series		
'Stockholm' (2)	Philax	AR-700 series		
THAILAND				
'Rattanakosin' (2)	Dagaie			
UNITED ARAB EMIRATES				
'Murray Jib' (2)	Dagaie	Cutlass	Cygnus	
VIETNAM				
'Tarantul' I (2)	PK-16			

Key
Decoy Chaff, infrared decoy, electro-optic decoy and active offboard decoy launching system; **ESM/RW** Electronic Support Measures/Radar Warning; **Suite** Combined electronic countermeasures/ESM system.
Note
In the RFAS entries, where there is doubt as to whether a system is an ECM or an ESM equipment, all onboard systems are logged under the *Suite* heading

VERIFIED

PATROL CRAFT

Class name (no)	Decoy	ESM/RW	Jammer	Suite
BAHRAIN				
'Ahmad El Fateh' (4)	Dagaie	RDL-2	Cygnus	
'Al Manama' (2)	Dagaie	Cutlass	Cygnus	
'Al Riffa' (2)	Barricade	RDL-2		
DENMARK				
'Flyvefisken' (14)	Seagnat	Sabre	Cygnus	
'Willemoes' (10)	Seagnat	Racal ESM		
EGYPT				
'October' (6)	Protean	Cutlass		
'Ramadan' (6)	Protean	Cutlass	Cygnus	
FINLAND				
'Helsinki' (4)	Philax	ARGO RW		
'Rauma' (5)	Philax	Matilda		
GERMANY				
'Albatros' (10)	Hot/Silver Dog			Octopus
'Gepard' (10)	Hot/Silver Dog			FL 1800
'Tiger' (16)	Wolke chaff			Octopus
GREECE				
'La Combattante IIA' (4)	Wolke chaff			
'La Combattante III' (10)	Wegmann decoy			
INDONESIA				
'PB 57' (4)	Dagaie			
'Singa' (4)	Dagaie	DR 2000 and Telegon		
ISRAEL				
'Aliya' (2)				NS-9003/5
'Hetz' (5)				NS-9003/5
'Reshef' (6)				NS-9003/5
KOREA (North)				
'Hainan' (6)	PK-16	Chinese RW		
'Osa I/Huangfen' (12)		Chinese RW		
'Soju' (15)		Chinese RW		
KOREA (South)				
'Pae Ku' (8)	RBOC			
KUWAIT				
'La Combattante I' (8)	Dagaie	DR 3000	Salamandre	
Istiqlal (1)	Dagaie	Cutlass		
MALAYSIA				
'Handalan' (4)		Susie		
'Musytari' (2)		Cutlass		
NORWAY				
'Hauk' (14)		ARGO RW		
OMAN				
'Al Bushra' (3)	Barricade	DR 3000		
'Dhofar' (4)	Barricade	T.242	Scorpion	
PERU				
'Velarde' (6)		DR 2000		
QATAR				
'Damsah' (3)	Dagaie	Cutlass	Cygnus	
'Vita' (4)	Dagaie	DR 3000	Salamandre	
RUSSIAN FEDERATION AND ASSOCIATED STATES (CIS)				
'Matka' (6)	PK-16	Clay Brick		
'Molnya'/'Pauk II' (35)	PK-10 and -16	Brick Plug and Half Hat		
'Svelyak' (27)	PK-16			
SAUDI ARABIA				
'Al Siddiq' (9)	SRBOC	SLQ-32(V)		
SINGAPORE				
'Fearless' (12)	Shield	NS-9010C?		
'Sea Wolf' (6)	RBOC	TDF-205?		Elettronica suite
SOUTH AFRICA				
'Warrior' (9)	ACDS	Grinaker RW	Rattler	

Class name (no)	Decoy	ESM/RW	Jammer	Suite
SWEDEN				
'Hugin' (12)		Swedish RW		
'Norrköping' (12)	Philax	AR-700 series		
		and Susie		
TAIWAN				
'Hai Ou' (49)	Israeli decoy	WD-2A		
THAILAND				
'Chon Buri' (3)	US chaff	Elettronica RW		
'Prabparapak' (3)		RDL-2		
'Ratcharit' (3)		RDL-2		
TUNISIA				
'La Combattante IIIM' (3)	Dagaie			
TURKEY				
'Dogan' (8)	SRBOC	Susie		
'Yildiz' (5)	SRBOC	Cutlass		
UNITED ARAB EMIRATES				
'Ban Yas' (6)	Dagaie	Cutlass		
'Mubarraz' (2)	Dagaie	Cutlass	Cygnus	
YUGOSLAVIA (Federal Republic of)				
'Koncar' (5)	Barricade			

Key
Decoy Chaff, infra-red decoy, electro-optic decoy and active offboard decoy launching system; **ESM/RW** Electronic Support Measures/Radar Warning; **Suite** Combined electronic countermeasures/ESM system.

Note
In the RFAS entries, where there is doubt as to whether a system is an ECM or an ESM equipment, all onboard systems are logged under the *Suite* heading

VERIFIED

CONTRACTORS

The following listing is designed to provide readers with known details of contractor addresses, telephone, facsimile and telex numbers, e-mail addresses and web sites that are relevant to the entries in the main body of this yearbook. Where an address discrepancy exists between the text pages and this list, the latter (which is compiled after the main text) should be taken as the most up-to-date. Telephone and facsimile numbers are given as per the international dialling convention. To help us maintain the correctness of this listing, please notify change of address details to the Editor, *Jane's Radar and Electronic Warfare Systems*, Sentinel House, 163 Brighton Road, Coulsdon, Surrey CR5 2YH, UK. While every care is taken in the compilation of this material, Jane's Information Group Ltd can take no responsibility for its accuracy.

AUSTRALIA

BAE Systems Australia
Industrial Contractors Area
North Gate DSTO Salisbury
East Avenue (off Belchambers Road)
Elizabeth
South Australia 5112
Tel: (+61 8) 84 80 88 88
Fax: (+61 8) 84 80 88 32
WWW: http://www.baesystems.com

CEA Technologies Pty Ltd
Head office
59-65 Gladstone Street
Fyshwick ACT 2609
Tel: (+61 2) 62 13 00 00 (switchboard)
     (+61 2) 62 13 00 01 (enquiries)
Fax: (+61 2) 62 13 00 13
e-mail: cea@cea.com.au
WWW: http://www.cea.com.au

US operation
CEA Technologies Inc
Suite 131
4901 Morena Boulevard
San Diego
California 92117
Tel: (+1 853) 490 51 44
Fax: (+1 853) 490 51 30

BELGIUM

Belgium Advanced Technology Systems SA
Parc Industriel de Recherches Sart Tilman
Avenue Des Noisetiers
B-4031 Angeleur
Tel: (+32 4) 367 08 88
Fax: (+32 4) 367 13 14

BRAZIL

AVIBRAS Aeroespacial SA
Rodovia dos Tamoios
KM 14
CX Postal 278
CEP 12300-000
Jacareí
Tel: (+55 1) 23 55 60 00
Fax: (+55 1) 23 51 62 77

BULGARIA

Kintex Share Holding Co
PO Box 209
66 James Baucher Str
BG-1407 Sofia
Tel: (+359 2) 66 23 11
Fax: (+359 2) 65 81 91; 65 81 01
Telex: 23 243/22 471

CANADA

Array Systems Computing Inc
1120 Finch Avenue, 7th Floor
Toronto
Ontario M3J 3H7
Tel: (+1 416) 736 09 00
Fax: (+1 416) 736 47 15
WWW: http://www.array.ca

BAE Systems Canada Inc
Kanata facility
415 Legget Drive
PO Box 13330
Kanata
Ontario K2K 2B2
Tel: (+1 613) 592 65 00
Fax: (+1 613) 592 74 27

Excalibur Systems Ltd (a Sierra Research/Integrated Defense Technologies business unit)
Canadian operation
50 Hines Road
Kanata
Ontario K2K 2M5
Tel: (+1 613) 591 60 00
Fax: (+1 613) 591 60 01
e-mail: info@excalibur.com
sales@excalibur.com
WWW: http://www.excalibur.com

US operation
3505 Lake Lynda Drive
Suite 103
Orlando
Florida 32817
Tel: (+1 407) 249 83 84
Fax: (+1 407) 277 35 81
e-mail: info@excalibur.com
sales@excalibur.com
WWW: http://www.excalibur.com

Litton Systems Canada (a subsidiary of the Northrop Grumman Corporation)
25 City View Drive
Toronto
Ontario M9W 5A7
Tel: (+1 416) 249 12 31
WWW: http://www.littoncanada.com

Telemus Inc
88 Hines Road
Kanata
Ontario K2K 2T8
Tel: (+1 613) 592 22 88
Fax: (+1 613) 592 88 55
WWW: http://www.telemus.com

CHILE

DTS (Desarrollo de Tecnología y Sistemas) Ltda
Rosas 1444
Santiago
Tel: (+56 2) 397 10 00
Fax: (+56 2) 397 10 01
e-mail: info@dts.cl
WWW: http://www.dts.cl

CHINA, PEOPLE'S REPUBLIC

China National Electronics Import and Export Corporation (CEIEC)
Electronics Building
A23 Fuxing Road
Beijing 100036
Tel: (+86 10) 68 29 61 09; 68 29 62 10
Fax: (+86 10) 68 21 23 52; 68 22 39 07
e-mail: ceiec@ceiec.com.cn
WWW: http://www.ceiec.com.cn

East China Research Institute of Electronic Engineering (ECRIEE)
Box 9023
Hefei
Anhui 230031
Tel: (+86 55) 15 56 44 33
Fax: (+86 55) 15 56 29 98

CZECH REPUBLIC

Omnipol a.s.
Electronics and communications department
Nekázanka 11
CZ-112 21 Praha 1
Tel: (+420 2) 24 01 11 20
Fax: (+420 2) 24 01 22 40
e-mail: omni25@omnipol.cz
WWW: http://www.omnipol.cz

DENMARK

TERMA A/S
Headquarters
Hovmarken 4
DK-8520 Lystrup
Tel: (+45) 87 43 60 00
Fax: (+45) 87 43 60 01
e-mail: terma.hq@terma.com
WWW: http://www.terma.com

Naval and Communications Division
Mårkærvej 2
DK-2630 Tåstrup
Tel: (+45) 43 52 15 13
Fax: (+45) 43 52 57 58
e-mail: terma.ncs@terma.com

Terma Industries
Fabrikvej 1
DK-8500
Grenaa
Tel: (+45) 86 32 19 88
Fax: (+45) 86 32 14 48
e-mail: terma.ind@terma.com

FRANCE

Etienne LACROIX Tous Artifices SA
6 boulevard de Joffrey
BP 213
F-31607 Muret Cédex
Tel: (+33 5) 61 56 65 00; 61 56 64 31
Fax: (+33 5) 61 51 42 77; 61 56 64 26
WWW: http://www.etienne-lacroix.com

SAS PYROTRONICS Contremesures LACROIX/SNPE
6 boulevard de Joffrey,
BP 213
F-31607 Muret Cédex
Tel: (+33 5) 61 56 64 31
Fax: (+33 5) 61 56 64 26

GERMANY

Buck Neue Technologien GmbH (a Rheinmetall W&M GmbH subsidiary)
Hans-Buck Strasse 1
D-79395 Neuenburg
Tel: (+49 7631) 70 20
Fax: (+49 7631) 702 70
e-mail: info@buck-tech.com

C Plath GmbH Nautish-Elektronische Technik
Gotenstrasse 18
D-20097 Hamburg
Tel: (+49 40) 23 73 40
Fax: (+49 40) 23 73 41 73
e-mail: c.plath@plath.de
WWW: http://www.plath.de

Rohde & Schwarz GmbH & Co KG
(street address)
Mühldorfstrasse 15
D-81671 München
(postal address)
Postfach 80 14 69
D-81614 München
Tel: (+49 89) 412 90
Fax: (+49 89) 412 91 21 64
WWW: http://www.rohde-schwarz.com

SEL Defense Systems GmbH
Headquarters
Ostendstrasse 3
D-75175 Pforzheim
Tel: (+49 7231) 150 (Information)
e-mail: info@sel.de
WWW: http://www.sel.de

STN ATLAS Elektronik GmbH (a BAE Systems - Rheinmetall DeTec AG joint venture company)
Sebaldsbrücker Heerstrasse 235
D-28305 Bremen
Tel: (+49 421) 45 70 (Headquarters)
Fax: (+49 421) 457 29 00 (Headquarters)
e-mail: marketing@stn-atlas.de
WWW: http://www.stn-atlas.de

INDIA

Bharat Electronics Ltd
Head office
2nd Floor, Shankaranaraya Building
25 MG Road

Karnataka 560001
Bangalore
Tel: (+91 80) 558 38 51; 559 50 01; 559 50 17
Fax: (+91 80) 558 36 75; 558 49 11
e-mail: imd@bel-india.com
WWW: http://www.bel-india.com

INTERNATIONAL

Alenia Marconi Systems NV
Italian headquarters
Via Tiburtina Km 12,400
I-00131 Rome
Tel: (+39 06) 415 01; 419 71; 41 88 31
Fax: (+39 06) 413 14 36; 413 11 33; 413 10 91

UK headquarters
Eastwood House
Glebe Road
Chelmsford
Essex CM1 1QW
Tel: (+44 1245) 70 27 02
Fax: (+44 1245) 70 27 00
WWW: http://www.aleniamarconisystems.com

EURO-ART Advanced Radar Technology GmbH
(street address)
Dogmagstrasse 11
D-80807 München
(postal address)
PO Box 46 07 05
D-80915 München
Tel: (+49 89) 63 65 87 09
Fax: (+49 89) 63 65 87 13
e-mail: Euroart@jetnet.de

European Aeronautic, Defence and Space (EADS) Company - Matra Systèmes and Information (formerly Matra Défense Equipments and Systems)
Head office
6 rue Dewoitine
BP 14
F-78143 Vélizy-Villacoublay Cédex
France
Tel: (+33 1) 34 63 70 00
Fax: (+33 1) 34 63 70 70
WWW: http://www.matra-msi-com

Les Ulis operation
ZA de Courtaboef 2-6
avenue des Tropiques
F-91943 Les Ulis Cédex
France
Tel: (+33 1) 69 86 85 00
Fax: (+33 1) 69 07 03 70

European Aeronautic, Defence and Space (EADS) Company - Systems and Defence Electronics (former DaimlerChrysler Aerospace AG business units)
(street address)
Wörthstrasse 85
D-89077 Ulm
Germany
(postal address)
D-89070 Ulm
Germany
Tel: (+49 731) 39 20
Fax: (+49 731) 392 33 93
WWW: http://www.eads-nv.com

Air Systems Division
Wörthstrasse 85
D-89070 Ulm
Germany
Tel: (+49 731) 392 54 16
Fax: (+49 731) 392 41 08
WWW: http://www.eads-nv.com

Former SI Sicherungstechnik operation
(street address)
Landshuter Strasse 26
D-85716 Unterschleissheim
Germany
(postal address)
Postfach 1661
D-85705 Unterschleissheim
Germany
Tel: (+49 893) 17 90
Fax: (+49 893) 179 22 19
WWW: http://www.eads-nv.com

Ewation GmbH (a European Aeronautic, Defence and Space (EADS) Company - Systems and Defence Electronics business unit)
Wörthstrasse 85
D-89077 Ulm
Germany
Tel: (+49 731) 39 20
Fax: (+49 731) 392 33 93
WWW: http://www.eads-nv.com

GTDAR
La Clef de Saint Pierre
1 boulevard Jean Moulin
BP No 7
F-78966 Elancourt Cédex
Tel: (+33 1) 34 59 68 88
Fax: (+33 1) 34 59 70 93

LFK-Lenkflugkörpersysteme GmbH (a European Aeronautic, Defence and Space (EADS) Company - Systems and Defence Electronics business unit)

Headquarters
(street address)
Landshuter Strasse 26
D-85716 Unterschleissheim
Germany
(postal address)
Postfach 1661
D-85705 Unterschleissheim
Germany
Tel: (+49 893) 17 90
Fax: (+49 893) 179 22 19
WWW: http://www.eads-nv.com

Ottobrunn operation
(street address)
Willy Messerschmitt Strasse
D-85521 Ottobrunn
Germany
(postal address)
Postfach 801149
D-81663 München
Germany
Tel: (+49 89) 60 70
Fax: (+49 89) 60 72 64 81

MBDA Missile Systems
Headquarters
11 The Strand
London WC2N 5RJ
UK
Tel: (+44 20) 74 51 60 00
Fax: (+44 20) 74 51 60 01
WWW: http://www.mbda.net

Sites
20/22 rue Grange Dame Rose
BP 150
F-78141 Vélizy-Villacoublay Cédex
France
Tel: (+33 1) 34 88 30 00
Fax: (+33 1) 34 88 22 88
Via Tiburtina
Km. 12,400
I-00131 Rome
Italy
Tel: (+39 06) 415 01
Fax: (+39 06) 413 14 36

Alkan
rue de 8 Mai 1945
BP 23
F-94460 Valenton Cédex
Tel: (+33 1) 45 10 86 00
Fax: (+33 1) 43 89 10 61
Telex: 266 126 F
e-mail: alkan@compuserve.com

SOSTAR GmbH
D-88039 Friedrichshafen 1
Germany
Tel: (+49 7545) 846 13
Fax: (+49 7545) 858 88

Thales

Thales Airborne Systems (Thales Systèmes Aeroportes SA)
2 avenue Gay-Lussac
F-78851 Elancourt Cédex
France
Tel: (+33 1) 34 81 60 00
Fax: (+33 1) 30 66 79 66
e-mail: customer.service.tas@fr.thalesgroup.com
WWW: http://www.thales-airbornesystems.com

Thales Air Defence
7/9 rue des Mathurins
F-92221 Bagneux Cédex
France
Tel: (+33 1) 40 84 40 00
Fax: (+33 1) 40 84 33 81

Thales Communications
160 boulevard de Valmy
BP 82
F-92704 Colombes Cédex
France
Tel: (+33 1) 41 30 30 00
Fax: (+33 1) 41 30 33 57

Radiosurveillance and communications intelligence systems unit
66 rue de Fossé-Blanc
BP 156
F-92231 Gennevilliers Cédex
France
Tel: (+33 1) 46 13 20 00
Fax: (+33 1) 46 13 21 63

Thales Defence Information Systems
Wookey Hole Road
Wells
Somerset BA5 1AA
UK
Tel: (+44 1749) 68 24 84
Fax: (+44 1749) 68 23 43
e-mail: tdis.marketing@thalesgroup.com
WWW: http://www.thales-defence.co.uk

Thales International
Via E Mattei 20
I-66013 Chiete Scalo
Italy
Tel: (+39 0871) 56 94 89
Fax: (+39 0871) 56 94 67

Thales Naval France
18 avenue Maréchal Juin
F-92366 Meudon-la-Forêt Cédex
France
Tel: (+33 1) 39 45 55 00
Fax: (+33 1) 39 45 55 46 (sales)

Thales Nederland
PO Box 42
NL-7550 GD Hengelo Ov
Netherlands
Tel: (+31 74) 248 81 11
Fax: (+31 74) 242 59 36
e-mail: info@thales.nl
WWW: http://www.thales-nederland.com

Thales Optronics
Crabtree Manorway North
Belvedere
Kent DA17 6AY
UK
Tel: (+44 20) 83 19 75 00
Fax: (+44 20) 83 19 75 26
e-mail: info@helio-ltd.com
WWW: http://helio-ltd.com
Lisieux Way
Taunton
Somerset TA1 2JZ
UK
Tel: (+44 1823) 33 10 71
Fax: (+44 1823) 27 44 13
e-mail: sales@avimo.co.uk
WWW: http://www.avimo.co.uk

Vicon House
Western Way
Bury St Edmunds
Suffolk IP33 3SP
UK
Tel: (+44 1284) 75 05 99
Fax: (+44 1284) 75 05 98
e-mail: sales@wvintenltd.com
WWW: http://www.vintenltd.com

Thales Sensors
Manor Royal
Crawley
West Sussex RH10 2PZ
UK
Tel: (+44 1293) 52 87 87
Fax: (+44 1293) 54 28 18

Thales Raytheon Systems
Head office
1-5 Avenue Carnot
91300 Massy
France
Tel: (+33 1) 69 75 50 00
e-mail: first name.last name@ thalesraytheon-fr.com
WWW: http://thalesraytheon.com

ISRAEL

BAE Systems ROKAR International Ltd
Science Based Industry Campus,
Mount Hotzvim
PO Box 45049
IL-91450 Jerusalem
Tel: (+972 2) 532 98 88
Fax: (+972 2) 582 25 22

Elbit Systems Ltd
Advanced Technology Centre
PO Box 539
IL-31053 Haifa
Tel: (+972 4) 831 53 15
Fax: (+972 4) 855 00 02
e-mail: marcom@elbit.co.il
WWW: http://www.elbit.co.il

Elisra Electronic Systems Ltd (a member of the Elisra Group)
48 Mivtza Kadesh Street
IL-51203 Bene Beraq
Tel: (+972 3) 617 55 22
Fax: (+972 3) 617 58 50
e-mail: marketing@elisra.com
WWW: http://www.elisra.com

Elta Electronics Industries Ltd (a subsidiary of Israel Aircraft Industries Ltd)
100 Yitzhak Hanassi Boulevard
PO Box 330
IL-77102 Ashdod
Tel *(switchboard)*: (+972 8) 857 22 22
Tel *(radar)*: (+972 8) 857 22 32
Tel *(EW)*: (+972 8) 857 29 20
Tel *(AEW & C)*: (+972 8) 857 26 18
Fax *(radar)*: (+972 8) 857 29 70
Fax *(EW)*: (+972 8) 857 29 73
Fax *(AEW & C)*: (+972 8) 857 21 10
WWW: http://www.elta-iai.com

Israel Military Industries Ltd
Rocket Systems Division
PO Box 1044/6604
IL-47100 Ramat Hasharon
Tel: (+972 8) 924 26 84
Fax: (+972 8) 925 28 96
WWW: http://www.imi-israel.com

Manor Expendable Decoys Directorate (a Rafael subsidiary)
PO Box 2250-M1
Haifa 31021
Tel: (+972 4) 879 26 88
Fax: (+972 4) 879 43 92

NICE Systems Ltd
Headquarters
8 Hapnina Street
PO Box 690
IL-43107 Ra'anana
Tel: (+972 9) 775 37 77
Fax: (+972 9) 743 42 82
e-mail: info@nice.com
WWW: http://www.nice.com

Rafael
Electronic Systems Division
PO Box 2250/80
IL-31021 Haifa
Tel: (+972 4) 879 20 02
Fax: (+972 4) 879 40 93

Ordnance Systems Division
PO Box 2250/20
IL-31021 Haifa
Tel: (+972 4) 879 47 02
Fax: (+972 4) 879 50 27

Tadiran Electronic Systems Ltd (a member of the Elisra Group)
29 Hamerkava Street
PO Box 150
IL-58101 Holon
Tel: (+972 3) 557 74 41
Fax: (+972 3) 556 45 36
e-mail: mkt@tadsys.com
WWW: http://www.tadsys.com

ITALY

Alcatel Italia SpA
via Trento 30
I-20059 Vimercate
Milan
Tel: (+39 02) 39 68 61
WWW: http://www.alcatel.il

Alenia Difesa Avionic Systems and Equipment Division – Officine Galileo
Via A Einstein 35
I-50013 Campi Bisenzio-Firenza
Tel: (+39 055) 895 01
Fax: (+39 055) 895 06 00
Telex: 570126 GALILE I

Elettronica SpA
Via Tiburtina Valeria Km 13,700
I-00131 Rome
Tel: (+39 06) 415 41
Fax: (+39 06) 415 49 24
e-mail: eltia@tin.it

FIAR (a Finmeccanica subsidiary)
Via GB Grassi 93
I-20156 Milan
Tel: (+39 02) 357 90
Fax: (+39 02) 33 40 09 81

GEM Elettronica Srl
Via Amerigo Vespucci 9
PO Box 280
I-63039 San Benedetto Del Tronto (AP)
Tel: (+39 0735) 590 51
Fax: (+39 0735) 59 05 40; 59 05 80

ITALTEL
Via A di Tocqueville 13
I-20154 Milan
Tel: (+39 02) 438 81
Fax: (+39 02) 43 88 52 20
WWW: http://www.italtel.it

Oerlikon Contraves SpA
Via Afile 102
I-00131 Rome
Tel: (+39 06) 436 11
Fax: (+39 06) 438 30 41

JAPAN

Mitsubishi Electric Corporation
Head office
2-2-3 Marunouchi Chiyoda-ku
Tokyo 100-8310
Tel: (+81 3) 32 18 23 46
Fax: (+81 3) 32 18 24 31
e-mail: prd.prdesk@hq.melco.jp
WWW: http://www.mitsubishielectric.com

NEC Corporation
Headquarters
7-1 Shiba 5-chome
Minato-ku
Tokyo 108-8001
Tel: (+81 3) 34 54 11 11
Fax: (+81 3) 37 98 15 10/11/12
e-mail: webmaster@nec.co.jp
WWW: http://www.nec-global.com

TOKIMEC Inc
Headquarters
2-16-46 Minami-Kamata
Ohta-ku
Tokyo 144-8551
Tel: (+81 3) 37 32 21 11
Fax: (+81 3) 37 36 02 61
e-mail: corp-comm@tokimec.co.jp
WWW: http://www.tokimec.co.jp

KOREA, SOUTH

Daewoo Telecom Ltd
Special Projects Division
275-6 Yangjae Dong
Socho-Gu
PO Box 187
Seoul
Tel: (+82 2) 589 28 94
Fax: (+82 2) 589 28 99

NORWAY

Ericsson Radar A/S
Isebakkeveien 49
N-1788 Halden
Tel: (+47) 69 21 41 00
Fax: (+47) 69 21 41 88
WWW: http://www.ericsson.no

POLAND

PIT (Przemyslowy Instytut Telekomunikacji)
(Telecommunications Research Institute)
PL-04-051 Warsaw
ul Pollgonowa 30
Tel/Fax: (+48 2) 28 10 23 81

RUSSIAN FEDERATION

All-Russian Radio Engineering Institute
22B Pochtovaya Street
Moscow 107082
Tel: (+7 095) 267 66 04
Fax: (+7 095) 265 60 38

Defence Systems
62nd Spasonalivkovsky Pereulok
Moscow 117909
Tel: (+7 095) 238 72 31; 16 87
Fax: (+7 095) 238 18 65; 77 42

Institute of Applied Physics
1/1 Arbuzov Street
Novosibirsk 630117
Russia
Tel: (+7 3832) 32 18 50
Fax: (+7 3832) 32 18 56
Telex: 133061 KLIUZ

Kuntsevo Design Bureau
29 Vereyskaya Street
Moscow 121357
Tel: (+7 095) 440 12 38
Fax: (+7 095) 443 72 72

Lianozovo Electromechanical Plant
110 Dmitrovskoe Shosse
Moscow 127411
Tel: (+7 095) 485 15 22
Fax: (+7 095) 485 15 63

Lira Design Bureau
110 Dmitrovskoe Shosse
Moscow 127411
Tel: (+7 095) 484 00 90
Fax: (+7 095) 485 01 09

Long Range Radio Communication Research Institute Production Complex Joint Stock Company
12/11 Bukhvostov Street
Moscow 107258
Tel: (+7 095) 963 50 13
Fax: (+7 095) 962 10 02

Mints Radio Engineering Institute
10-12 Vosmogo Marta Street
Moscow 125083
Tel: (+7 095) 212 42 83
Fax: (+7 095) 212 10 81; 214 06 62

Moscow Radio Engineering Plant
29 Vereyskaya Street
Moscow 121357
Tel: (+7 095) 443 72 80; 444 50 54
Fax: (+7 095) 443 71 40

MRI AGAT
54 Malaya Gruzinskaya Street
Moscow 123557
Tel: (+7 095) 253 09 96
Fax: (+7 095) 253 65 44
Telex: 346719 Agat

Nauchno Issledovatelskiy Institut Priborostroyeniya VEGA-M
(Scientific Research Institute for Instrument Engineering VEGA-M)
34 Kutuzov Prospekt
Moscow 121170
Tel: (+7 095) 249 76 10
Fax: (+7 095) 148 79 96
Telex: +412268 VEGA SU

Nitel Joint Stock Company
37 Gagarin Avenue
Nizhni Novgorod 603009
Tel: (+7 8312) 65 51 59
Fax: (+7 8312) 65 50 19

Nizhegorodsky Radio Engineering Research Institute
7 Shaposhnikov Street
Nizhni Novgorod 603600
Tel: (+7 8312) 65 00 69
Fax: (+7 8312) 65 79 88

Phazotron-NIIR Joint Stock Company
1 Electrichesky Pereulok
Moscow 123557
Tel: (+7 095) 253 56 13
Fax: (+7 095) 253 04 95

Pravdinsk Radio Relay Equipment Plant
34 Gorky Street
Pravdinsk District
Balakhna 606406
Nizhni Novgorod Region
Tel: (+7 8314) 42 39 99
Fax: (+7 8314) 42 64 56

RADAR MMS
37 Novoselkovskaya
St Petersburg 197349
Tel: (+7 812) 393 96 00
Fax: (+7 812) 394 40 00

Radio Measuring Instrument Plant
2 Karacharovskoye Shosse
Murom 602200
Vladimir Region
Tel: (+7 0923) 42 01 82
Fax: (+7 0923) 42 02 25

Rosoboronexport (formerly Rosvoorouzhenie)
21 Gogolevsky Boulevard
Moscow 119865
Tel: (+7 095) 202 66 03
Fax: (+7 095) 202 45 94
WWW: http://www.rusarm.ru

Spets-Radio Research and Production Enterprise Joint Stock Company
4 Promyshlennaya Street
Belgorod 308023
Tel: (+7 0722) 34 22 72
Fax: (+7 0722) 34 76 82

State Unitary Enterprise 'State Moscow Plant Salyut'
6 Plekhanov Street
Moscow 111123
Tel: (+7 095) 176 11 39
Fax: (+7 095) 176 08 39

STRELA Research Institute
6 Gorky Street
Tula 300002
Russia
Tel: (+7 0872) 77 05 60
Fax: (+7 0872) 34 11 04

SOUTH AFRICA

Avitronics (Pty) Ltd (a Saab - Grintek joint venture company)
PO Box 8492
Centurian 0046
Tel: (+27 12) 672 60 00
Fax: (+27 12) 672 62 22
e-mail: avitronics@grintek.com
WWW: http://www.avitronics.co.za

Avitronics (Pty) Ltd (Maritime) (a Saab - Grintek joint venture company)
PO Box 30189
Tokai 7966
Tel: (+27 21) 705 62 95
Fax: (+27 21) 705 31 15
e-mail: maritime@grintek.com
WWW: http://www.avitronics.co.za

Grintek Ewation (formerly Grintek System Technologies (GST) - a Grintek - European Aeronautics, Defence and Space (EADS) Company Deutschland joint venture)
(street address)
13 De Havilland Crescent
Persequor Technopark
Pretoria
(postal address)
PO Box 912-561
Silverton 0127
Tel: (+27 12) 421 62 00
Fax: (+27 12) 349 13 08
e-mail: gst@grintek.com
WWW: http://www.grintek.com

Reutech Systems (Pty) Ltd
42 James Crescent
PO Box 35
Halfway House
Midrand 1685
Tel: (+27 11) 652 55 55
Fax: (+27 11) 805 31 90
e-mail: info@reutech.co.za
WWW: http://www.reutech.co.za

SPAIN

INDRA
Head office
Velázquez 132
E-28006 Madrid
Tel: (+34 91) 396 33 00
Fax: (+34 91) 396 31 31
WWW: http://www.indra.es

SWEDEN

Ericsson Microwave Systems AB
SE-431 84 Mölndal
Tel: (+46 31) 747 00 00
Fax: (+46 31) 747 17 27
WWW: http://www.ericsson.com/microwave
e-mail: defenseinfo@emw.ericsson.se

Försvarets Forskningsanstalt (FOA)
(National Defence Research Establishment, Department of Sensor Technology)
Olaus Magnus väg 42
PO Box 1165
SE-581 11 Linköping
Tel: (+46 13) 37 80 00
WWW: http://www.foa.se

Microdata Innovations AB (μ.data)
Köpenhamnsgaten 4
PO Box 1063
S-164 25 Kista
Tel: (+46 8) 477 77 70
Fax: (+46 8) 477 77 80
WWW: http://www.microdata.se

Saab Avionics AB
Head office
Torshamnsgatan 32C
Kista
SE-164 84 Stockholm
Tel: (+46 8) 757 30 00
Fax: (+46 8) 752 81 72
e-mail: info@avionics.saab.se
WWW: http://www.avionics.saab.se

SaabTech Systems AB
Head office
Nettovägen 6
Jakobsberg
SE-175 88 Järfälla
Tel: (+46 8) 58 08 40 00
Fax: (+46 8) 58 03 22 44
e-mail: info@systems.saab.se
WWW: http://www.saab.se/saabtechsystems

SWITZERLAND

Oerlikon Contraves AG
Birchstrasse 155
Postfach CH-8050 Zürich
Tel: (+41 1) 316 22 11
Fax: (+41 1) 311 31 54
WWW: http://www.oerlikoncontraves.com

TURKEY

ASELSAN Inc
(street address)
Mehmet Akif Ersoy Mah 16
cadde No 16
Macunköy
TR-06370 Ankara
(postal address)
PO Box 101
Yenimahalle
TR-06172 Ankara
Tel: (+90 312) 385 19 00
Fax: (+90 312) 354 13 02; 354 26 29
WWW: http://www.aselsan.com.tr

STM Defense Technologies Engineering Inc
Oguzlar 1
cadde 35
sokak no 28
Balgat
TR-06520 Ankara
Tel: (+90 312) 286 11 11
Fax: (+90 312) 285 86 66

UNITED KINGDOM

Advanced Topographic Development & Images Ltd (ATDI)
15 Kingsland Court
Three Bridges Road
Crawley
West Sussex RH10 1HL
Tel: (+44 1293) 52 20 52
Fax: (+44 1293) 52 25 21
e-mail: atdi@atdi.co.uk
WWW: http://www.atdi.co.uk

BAE Systems Avionics - Sensor Systems Division
Christopher Martin Road
Basildon
Essex SS14 3EL
Tel: (+44 1268) 52 28 22
Fax: (+44 1268) 88 31 40

Crewe Toll
Ferry Road
Edinburgh EH5 2XS
Tel: (+44 131) 332 24 11
Fax: (+44 131) 332 06 90

The Grove
Warren Lane
Stanmore
Middlesex HA7 4LY
Tel: (+44 20) 89 54 23 11
Fax: (+44 20) 84 20 39 90

Browns Lane
The Airport
Portsmouth
Hampshire PO3 5PH
Tel: (+44 2392) 22 60 00
Fax: (+44 2392) 22 75 95

BAE Systems DTS (Defence Training Systems - formerly Reflectone UK Ltd)
Building 20A1
Southmead Road
Filton
Bristol BS34 7RP
Tel: (+44 117) 936 44 38
Fax: (+44 117) 936 44 62

BAE Systems Marine
Barrow in Furness
Cumbria LA14 1AF
Tel: (+44 1229) 82 33 66
Fax: (+44 1229) 87 40 00

Chemring Countermeasures (a division of Pains Wessex Ltd)
High Post
Salisbury
Wiltshire SP4 6AS
Tel: (+44 1722) 41 16 11
Fax: (+44 1722) 42 87 92
e-mail: info@chemringcm.com
WWW: http://www.chemringcm.com

EW Simulation Technology (EWsT) Ltd
Unit 1
The Royston Centre
Lynchford Road
Ash Vale
Hampshire GU12 5PQ
Tel: (+44 1252) 51 29 51
Fax: (+44 1252) 51 24 28
e-mail: info@ewst.co.uk
WWW: http://www.ewst.co.uk

Australian operation
EWsT Australia Pty Ltd
17 Cumberland Avenue
Nowra
New South Wales 2541
Tel: (+61 2) 44 22 81 60
Fax: (+61 2) 44 22 77 33
e-mail: info@ewst.com.au

IBM Global Services
Meudon House
Meudon Avenue
Farnborough
Hampshire GU14 7NB
Tel: (+44 1252) 80 55 55
Fax: (+44 1252) 80 60 01

Kelvin Hughes Ltd (a part of the Naval and Marine Division of Smiths Industries Aerospace)
Central office
New North Road
Hainault
Ilford
Essex IG6 2UR
Tel: (+44 20) 85 00 10 20
Fax: (+44 20) 85 59 85 35
Telex: 884934
WWW: http://www.kelvinhughes.co.uk

M/A-Com Ltd
Humphrys Road
Woodside Estate
Dunstable
Bedfordshire LU5 4SX
Tel: (+44 1582) 47 12 00
Fax: (+44 1582) 47 22 77

MS Instruments plc
Electron House
Farwig Lane
Bromley
Kent BR1 3RE
Tel: (+44 20) 82 90 02 00
Fax: (+44 20) 84 64 65 96
e-mail: Sales@msinstruments.co.uk
WWW: http://www.msinstruments.co.uk

PW Defence Ltd (formerly Pains Wessex Ltd)
Sales and marketing
Unit 6
Minton Distribution Park
London Road
Amesbury
Wiltshire SP4 7EN
Tel: (+44 1980) 62 46 71
Fax: (+44 1980) 62 57 30

Factory and administration
Wilne Mill
Draycott
Derbyshire DE72 3QJ
Tel: (+44 1332) 87 24 75
Fax: (+44 1332) 87 30 46

Raytheon Systems Ltd
Head office
80 Park Lane
London W1K 7TR
Tel: (+44 20) 75 69 55 00
Fax: (+44 20) 75 69 55 91
WWW: http://www.raytheon.co.uk

IFF/MSSR operation
The Pinnacles
Elizabeth Way
Harlow
Essex CM19 5BB
Tel: (+44 1279) 42 68 62
Fax: (+44 1279) 41 04 13

Roke Manor Research Ltd
Roke Manor
Romsey
Hampshire SO51 0ZN
Tel: (+44 1794) 83 30 00
Fax: (+44 1794) 83 34 33
e-mail: info@roke.co.uk
WWW: http://www.roke.co.uk

TMD Technologies Ltd (formerly Thorn Microwave Devices (TMD) Ltd)
Swallowfield Way
Hayes
Middlesex UB3 1DQ
Tel: (+44 20) 85 73 55 55
Fax: (+44 20) 85 69 18 39
WWW: http://www.tmd.co.uk

Wallop Defence Systems (a Cobham plc company)
Flight Refuelling Ltd
Wallop Defence Systems Division
Craydown Lane
Middle Wallop
Stockbridge
Hampshire SO20 8DX
Tel: (+44 1264) 78 14 56
Fax: (+44 1264) 78 20 84
WWW: http://www.cobham.com

UNITED STATES OF AMERICA

AAI Corporation (a United Industrial Corporation subsidiary)
Corporate headquarters
124 Industry Lane
Hunt Valley
Maryland 21030-0126
Tel: (+1 410) 666 14 00
WWW: http://www.aaicorp.com

Alliant Integrated Defense Company - Florida Operations
PO Box 4648
Clearwater
Florida 33758-4648
Tel: (+1 727) 572 19 00
Fax: (+1 727) 572 21 80
WWW: http://www.atk.com

Alloy Surfaces Co Inc (a Chemring Countermeasures subsidiary)
121 N Commerce Drive
Chester Township
Pennsylvania 19014
Tel: (+1 610) 497 79 79
Fax: (+1 610) 494 72 50
e-mail: contact@alloysurfaces.com
WWW: http://www.alloysurfaces.com

Anaren Microwave Inc
Corporate offices
6635 Kirkville Road
East Syracuse
New York 13057
Tel: (+1 315) 432 89 09
Web: http://www.anaren.com

BAE Systems North America
Corporate headquarters
1601 Research Boulevard
Rockville
Maryland 20850
Tel: (+1 301) 738 40 00

Advanced Systems
1 Hazeltine Way
Greenlawn
New York 11740-1600
Tel: (+1 516) 262 70 00
Fax: (+1 516) 262 80 02
WWW: http://www.asd.marconi-na.com

Aerospace Electronics
700 Quince Orchard Road
Gaithersburg
Maryland 20878-1794
Tel: (+1 301) 948 75 50
Fax: (+1 301) 921 94 79
WWW: http://www.signalsurveillance.com

305 Richardson Road
Lansdale
Pennsylvania 19446
Tel: (+1 215) 996 20 00
Fax: (+1 215) 996 20 88

Information and Electronic Warfare Systems
65 Spit Brook Road
PO Box 868
Nashua
New Hampshire 03061-0868
Tel: (+1 603) 885 22 22;885 28 17; 885 43 21; 885 57 77
Fax: (+1 603) 885 28 16; 885 36 55
WWW: http://www.sanders.com

Information and Electronic Warfare Systems - Countermeasures business area
95 Canal Street
Nashua
New Hampshire 03061
Tel: (+1 603) 885 38 24
Fax: (+1 603) 885 61 09

Information and Electronic Warfare Systems - Information dominance systems
Pope Technical Park
65 River Road
Hudson
New Hampshire 03051
Tel: (+1 603) 885 27 18
Fax: (+1 603) 885 77 01

Infrared Imaging Systems
2 Forbes Road
Lexington
Massachusetts 02421
Tel: (+1 781) 863 33 00
Fax: (+1 781) 863 33 34

Integrated Defense Solutions
6500 Tracor Lane
Austin
Texas 78725-2070
Tel: (+1 512) 929 28 84
Fax: (+1 512) 929 23 12
WWW: http://www.ids.na.baesystems.com

Reconnaissance and Surveillance Systems
300 Robbins Lane
Syosset
New York 11794
Tel: (+1 516) 349 22 00

Technical Services
Industrial Parkway
557 Mary Esther Cut Off
Fort Walton Beach
Florida 32548
Tel: (+1 850) 244 77 11
Fax: (+1 850) 244 77 60
e-mail: info@ts-technicalservices.com
WWW: http://www.ts-technicalservices.com

Threat Warning and Defense Systems
1 Ridge Hill
Yonkers
New York 10710-5598
Tel: (+1 914) 964 25 07
Fax: (+1 914) 964 07 49

Cincinnati Electronics (a BAE Systems Canada subsidiary)
7500 Innovation Way
Mason
Ohio 45040-9699
Tel: (+1 513) 573 61 00; 80 05 43 82 20
Fax: (+1 513) 573 67 41
e-mail: sales@cinele.com
WWW: http://www.cinele.com

Comptek Federal Systems Inc (a Logicon Inc business unit)
2732 Transit Road
Buffalo
New York 14224-2523
Tel: (+1 716) 677 40 70
Fax: (+1 716) 677 00 14

Advanced Systems Division
96-10 23rd Avenue
East Elmhurst
New York 11369-1230
Tel: (+1 718) 565 23 00
Fax: (+1 718) 565 02 88

Computer Science and Applications Inc
2 Clifford Drive
Shalimar
Florida 32579
Tel: (+1 850) 651 49 91
Fax: (+1 850) 651 28 16
WWW: http://www.antekna.com

802 Park Drive
Warner Robins
Georgia 31088
Tel: (+1 478) 329 09 70
Fax: (+1 478) 329 09 35

Condor Systems Inc
2133 Samaritan Drive
San Jose
California 95124
Tel: (+1 408) 371 95 80; (+1 703) 415 05 77 (Condor Systems East)
Fax: (+1 408) 371 95 89
WWW: http://www.condorsys.com

Electronic Systems Division
996 Flower Glen Street
Simi Valley
California 93065
Tel: (+1 805) 584 82 00
Fax: (+1 805) 527 83 32
WWW: http://www.condoresd.com

Cubic Communications Inc
9535 Waples Street
San Diego
California 92121
Tel: (+1 858) 643 58 00
Fax: (+1 858) 643 58 03
WWW: http://www.cubic.com/cci

DRS Electronic Systems Inc
200 Professional Drive
Gaithersburg
Maryland 20879
Tel: (+1 301) 921 81 00
Fax: (+1 301) 977 61 58
e-mail: marketing@drs-esq.com
WWW: http://www.drs.com

EDO Corporation Electronic Systems Group (formerly AIL Systems Inc)
455 Commack Road
Deer Park
New York 11729-4591
Tel: (+1 516) 595 33 10
Fax: (+1 516) 595 53 66
WWW: http://www.nycedo.com/edocorp

General Dynamics Information Systems and Technology business group
100 Ferguson Drive
PO Box 7188
Mountain View
California 94039
Tel: (+1 650) 966 20 00
Fax: (+1 650) 966 34 01

Harris Corporation - Government Communication Systems Division
2400 Palm Bay Road NE
Palm Bay
Florida 32905
Tel: (+1 321) 727 69 63
Fax: (+1 321) 727 45 00
e-mail: govt@harris.com
WWW: http://www.harris.com

Honeywell
Corporate headquarters
101 Columbia Road
Morristown
New Jersey 07962
Tel: (+1 973) 455 20 00
Fax: (+1 973) 455 48 07
WWW: http://www.honeywell.com

ITT Avionics
100 Kingsland Road
Clifton
New Jersey 07014-1993
Tel: (+1 973) 284 01 23
Fax: (+1 973) 284 41 22
WWW: http://www.ittavionics.com

ITT Gilfillan
7821 Orion Avenue
PO Box 7713
Van Nuys
California 91406-7713
Tel: (+1 818) 988 26 00
Fax: (+1 818) 901 24 35
WWW: http://www.gilfillan.itt.com

Kilgore Flares Company (a Chemring Countermeasures subsidiary)
155 Kilgore Drive
Toone
Tennessee 38381
Tel: (+1 901) 658 52 31
Fax: (+1 901) 658 41 73
WWW: http://www.atk.com

KOR Electronics Inc
11958 Monarch Street
Garden Grove
California 92841
Tel: (+1 714) 898 82 00
Fax: (+1 714) 895 75 26
WWW: http://www.korelectronics.com

L-3 Telemetry-East
47 Friends Lane
PO Box 328
Newtown
Pennsylvania 18940-0328
Tel: (+1 215) 968 42 71
Fax: (+1 215) 968 32 14
e-mail: Sales/Mktg@TE.L-3.com
WWW: http://www.l-3com.com

Lockheed Martin

Aeronautics Company
Lockheed Martin Aeronautics Company - Marietta Operations
86 South Cobb Drive
Marietta
Georgia 30063
Tel: (+1 770) 494 44 11
WWW: http://www.lmaeronautics.com

Systems Integration sector
Lockheed Martin Naval Electronics and Surveillance Systems (NE & SS) - Surface
*   Systems*
199 Borton Landing Road
Moorestown
New Jersey 08057-0927
Tel: (+1 609) 722 50 00
WWW: http://ness.external.lmco.com/ss

Lockheed Martin NE & SS - Syracuse
Electronics Parkway
Liverpool
New York 13088
Syracuse
New York 13221-4840
Tel: (+1 315) 456 01 23
WWW: http://www.lockheedmartin.com/syracuse

Lockheed Martin Systems Integration - Owego
1801 State Route 17C
Owego
New York 13827-3998
Tel: (+1 607) 751 20 00
Fax: (+1 607) 751 32 59
WWW: http://www.owego.com

Metric Systems (an Integrated Defense Technologies company)
645 Anchors Street
Fort Walton Beach
Florida 32548
Tel: (+1 850) 302 30 00
WWW: http://www.metricsys.com

Northrop Grumman Electronic Systems sector
Sector headquarters
1580-A West Nursery Road
Linthicum
Maryland 21090
Tel: (+1 410) 765 10 00
Fax: (+1 410) 993 87 71
WWW: http://www.northgrum.com

PO Box 451
Baltimore
Maryland 21203
Tel: (+1 410) 765 27 00
Fax: (+1 410) 993 66 98
WWW: http://www.sensors.northgrum.com

Defensive Systems Division
600 Hicks Road
Rolling Meadows
Illinois 60008
Tel: (+1 847) 259 96 00
Fax: (+1 847) 506 79 81

Amherst Systems
30 Wilson Road
Buffalo
New York 14221-7082
Tel: (+1 716) 631 06 10
Fax: (+1 716) 631 06 29
WWW: http://www.amherst.com

Norden Systems (CT)
10 Norden Place
PO Box 5300
Norwalk
Connecticut 06856
Tel: (+1 203) 852 49 14
Fax: (+1 203) 852 78 58

Norden Systems (NY)
65 Marcus Drive
Melville
New York 11747
Tel: (+1 631) 719 46 00

PRB Systems
44 Castilian Drive
Goleta
California 93117
Tel: (+1 805) 685 45 71
Fax: (+1 805) 685 78 53

Sperry Marine
1070 Seminole Trail
Charlottesville
Virginia 22901
Tel: (+1 434) 974 20 00
Fax: (+1 434) 974 22 59
WWW: http://www.litton-marine.com

Burlington House
118 Burlington Road
New Malden
Surrey KT3 4NR
Tel: (+44 20) 83 29 20 00
Fax: (+44 20) 83 29 24 15

PLEXSYS Interface Products Inc
PO Box 301459
Portland
Oregon 97294-9459
Tel: (+1 503) 251 04 55
Fax: (+1 503) 251 04 59
WWW: http://www.plexsysipi.com

Raytheon Electronic Systems
2000 East El Segundo Boulevard
PO Box 902
El Segundo
California 90245
Tel: (+1 310) 616 10 22
WWW: http://www.raytheon.com

One South Los Carneros
Goleta
California 93117-3187
Tel: (+1 805) 967 55 11
Fax: (+1 805) 964 81 15
WWW: http://www.raytheon.com

100 Wooster Heights Road
Danbury
Connecticut 06810-7589
Tel: (+1 800) 797 28 72
Fax: (+1 203) 797 59 58
e-mail: info@hdos.hac.com
WWW: http://www.raytheon.com

1010 Production Road
Fort Wayne
Indiana 46808
Tel: (+1 219) 429 60 00
Fax: (+1 219) 429 46 55
WWW: http://www.raytheon.com

1151 E Hermans Road
Tucson
Arizona 85706-9367
WWW: http://www.raytheon.com

7700 Arlington Boulevard
Falls Church
Virginia 22046-2900
Tel: (+1 703) 849 15 67
Fax: (+1 703) 280 46 27
WWW: http://www.raytheon.com

Rockwell Collins
Corporate headquarters
400 Collins Road North East
Cedar Rapids
Iowa 52498
Tel:(+1 319) 295 10 00 (switchboard)
Fax: (+1 319) 295 54 29
e-mail: collins@collins.rockwell.com (military/government customers)
mktgsvc@collins.rockwell.com (all other customers)
WWW: http://www.collins.rockwell.com

Rodale Electronics Inc
20 Oser Avenue
Hauppauge
New York 11788
Tel: (+1 631) 231 00 44
Fax: (+1 631) 231 13 45
e-mail: jbclement@earthlink.net
WWW: http://www.rodaleelectronics.com

Sandia National Laboratories
(street address)
1515 Eubank South East
Albuquerque
New Mexico 87123
(postal address)
PO Box 5800
Albuquerque
New Mexico 87185

SenSyTech Inc
Corporate headquarters
8419 Terminal Road
PO Box 1430
Newington
Virginia 22122-1430
Tel: (+1 703) 550 70 00
Fax: (+1 703) 550 74 70
WWW: http://www.sensytech.com

Sierra Research (an Integrated Defense Technologies company)
485 Cayuga Road
Buffalo
New York 14225
Tel: (+1 716) 631 62 00
Fax: (+1 716) 631 78 49
WWW: http://www.sierra-idt.com

Sippican Inc
Electronic Warfare Division
7 Barnabas Road
Marion
Massachusetts 02738
Tel: (+1 508) 748 11 60
Fax: (+1 508) 748 17 18
e-mail: countermeasures@sippican.com
WWW: http://www.sippican.com

Hycor Products Group
7 Barnabas Road
Marion
Massachusetts 02738
Tel: (+1 508) 748 11 60
Fax: (+1 508) 748 68 94
e-mail: hycorgroup@sippican.com
WWW: http://www.sippican.com

Southwest Research Institute
PO Box Drawer 28510
San Antonio
Texas 78228-0510

Systems & Electronics Inc (an Engineered Support Systems Inc (ESSI) company)
Headquarters
201 Evans Lane
MS 4361
St Louis
Missouri 63121-1126
Tel: (+1 314) 553 49 01/10
Fax: (+1 314) 553 49 49
e-mail: moreinfo@seistl.com
WWW: http://www.seistl.com

TCOM L.P.
7115 Thomas Edison Drive
Columbia
Maryland 21046
Tel: (+1 410) 312 24 00
Fax: (+1 410) 312 24 55
e-mail: aerostat@tcomlp.com
WWW: http://www.tcomlp.com

TechComm Inc
3650 Coral Ridge Drive, Suite 106
Coral Springs
Florida 33065
Tel: (+1 954) 341 11 11
Fax: (+1 954) 341 67 87
e-mail: techcomm@gate.net

Technology for Communications International/BR Communications (TCI/BR)
47300 Kato Road
Fremont
California 94538
Tel: (+1 510) 687 61 00
Fax: (+1 510) 687 61 01
WWW: http://www.tcibr.com

Telephonics Corporation (a Griffon company)
Command Systems Division
815 Broad Hollow Road
Farmingdale
New York 11735-3940
Tel: (+1 631) 755 73 65
Fax: (+1 631) 755 76 44
e-mail: csdprodsupt@telephonics.com
WWW: http://www.telephonics.com

Telestar Corporation
1461 South Balboa Avenue
Ontario
California 91761
Tel: (+1 909) 923 69 99
Fax: (+1 909) 923 68 98
WWW: http://www.telestarcorporation.com

Titan Systems Corporation - Delfin Systems Division
Headquarters
3000 Patrick Henry Drive
Santa Clara
California 95054
Tel: (+1 408) 748 12 00
WWW: http://www.delfinsystems.com

TRW Systems and Information Technology Group
One Federal Systems Park Drive
Fairfax
Virginia 22033
Tel: (+1 703) 803 54 98
Fax: (+1 703) 803 47 46
WWW: http://www.trw.com

Wide Band Systems Inc
110 Woodfern Road, Building R
PO Box 550
Neshanic Station
New Jersey 08853
Tel: (+1 908) 369 64 14
Fax: (+1 908) 369 64 30
e-mail: Marketing@widebandsystems.com
WWW: http://www.widebandsystems.com

Zeta (an Integrated Defense Technologies company)
17680 Butterfield Boulevard
Morgan Hill
California 95037
Tel: (+1 408) 852 08 00
Fax: (+1 408) 852 08 01
e-mail: zeta@zeta-idt.com
WWW: http://www.zeta-idt.com

YUGOSLAVIA, FEDERAL REPUBLIC

Yugoimport SDPR
2 Bulevar umetnosti
PO Box 89
YU-11070 Belgrade
Tel: (+381 113) 11 27 43
Fax: (+381 113) 24 87 91
e-mail: sdpr@yugoimport.co.yu
WWW: http://www.yugoimport.co.yu

INDEXES

Manufacturers' Index

Lianozovsky Electromechanical Plant
76N6S air defence radar ... 100
96L6E 3-D surveillance radar ... 34

Lira Design Bureau
1L117 surveillance radar ... 32
76N6 multifunction radar ... 99
76N6S air defence radar ... 100
96L6E 3-D surveillance radar ... 34

Litton Advanced Systems
AN/ALR-87 threat warning system ... 505
AN/APR-39A(V)1/2/3 EW system ... 506
AN/APR-39B(V) threat warning system ... 506

Litton Data Systems
AN/GYQ-51 Advanced Tracking System (ATS) ... 286
AN/UPX-24(V) IFF system ... 288

Litton Guidance & Control Systems
AN/APX-101 airborne transponder ... 284
AN/APX-109(V) interrogator/transponder ... 285

Litton Marine Systems, Decca Division
Surveillance and navigation radars ... 306

Litton Systems Canada Ltd
APS-504(V) series radars ... 199

Lockheed Martin Aeronautical Systems
AFEWES simulator ... 592
SATIN transport defensive system ... 572

Lockheed Martin Canada Inc
AN/SLQ-501 CANEWS ESM system ... 417
AN/SLQ-503 RAMSES ECM system ... 439

Lockheed Martin Electro-Optical Systems
AN/ALQ-157 countermeasures system ... 559
AN/ALQ-204(V) (LANCIR) jammer ... 565
AN/ALQ-204(V) (Matador) jammer ... 565
Challenger IRCM system ... 569
Defendir IR jamming systems ... 570
SILC laser countermeasures ... 572

Lockheed Martin Missiles & Fire-Control
AN/APG-78 Longbow radar ... 265

Lockheed Martin Mission Systems
Silent Sentry passive detection system ... 65

Lockheed Martin Naval Electronics and Surveillance Systems
Aerostat-borne surveillance radar ... 235
AIEWS EW system ... 462
AIRSTAR–Airborne Surveillance and Target Acquisition Radar ... 221
AN/APG-67 multimode airborne radar ... 261
AN/APS-125/-138/-139/-145 airborne radar ... 225
AN/MPS-39 multiple object tracking radar ... 310
AN/PPQ-2 PSTAR battlefield radar ... 116
AN/SLY-2(V) AIEWS EW system ... 462
AN/SPG-60 radar ... 195
AN/SPY-1 multifunction radar ... 176
AN/TPS-59(V) tactical radar ... 57
AN/TPS-73 ATC radar ... 312
COBRA counterbattery radar ... 79
Mk 92 fire-control system ... 197
MMSR MultiMission Surveillance Radar ... 177

Lockheed Martin Ocean, Radar & Sensor Systems
AN/FPS-6 height-finding radar ... 51
AN/FPS-8 search radar ... 51
AN/FPS-88 surveillance radar ... 52
AN/FPS-117 air defence radar ... 53
AN/FPS-118 OTH radar ... 54
AN/FPS-124(V) surveillance radar ... 55
LAADS radar system ... 64
TRACKSTAR low-altitude air defence system ... 66

Lockheed Martin Systems Integration
AN/ALQ-99(V) Tactical Jamming System (TJS) ... 554
AN/ALQ-210() ES system ... 498
AN/ALQ-217 Electronic Support system ... 498
AN/ALR-67(V)3&4 countermeasures system ... 502
AN/ALR-76 ESM system ... 504
AN/APR-48A Radar Frequency Interferometer (RFI) ... 508
AN/APR-50 threat emitter location system ... 509

Lockheed Martin Tactical Defense Systems
LAIRS–Advanced Imaging Radar System ... 234
STacSAR radar ... 235

Long Range Radio Communication Research Institute
Dunai-3U ABM and tracking radar ... 36
Rezonans surveillance radar ... 42

M

M/A-COM Ltd
ARI 5983 IFF transponder ... 277
Outfit RRB IFF receiver ... 278

Manor Expendable Decoys
Integrated decoy system ... 451
Naval decoy systems ... 451

Marconi Italiana SpA
Creso surveillance radar ... 212

Martin Electronics
IR decoy flares ... 572

MBDA Missile Systems
Corail countermeasures ... 526
Eclair-M CMDS ... 526
Integrated CounterMeasures Suite (ICMS) ... 520
LEA active radar jammer ... 521
Phimat chaff decoy ... 528
SAMIR IR missile warner ... 470
SAPHIR countermeasures ... 528
SPECTRA EW system ... 524
SPIRALE countermeasures ... 530
Sycomor countermeasures ... 531

Measuring Instruments Research Institute (NIIP), Novosibirsk
9S15MTZ target acquisition radar ... 97
9S18M1E target acquisition radar ... 98
64N6E target acquisition radar ... 99

Metric Systems Corporation
AES Airborne Emitter System ... 608
AN/FPS-127 height-finder simulator ... 594
AN/MPQ-T3-A simulator ... 594
AN/MPS-T9 radar simulator ... 595
AN/MSQ-T13 radar simulator ... 595
ATES Advanced Threat Emitter System ... 603
Falcon II surveillance radar ... 177

Microdata Innovation AB
MI 745 training radar jammer ... 584

Mints Radio Engineering Institute
Daryal ABM and tracking radar ... 35
Dnepr ABM radar ... 35
Don-2N ABM and tracking radar ... 36

Mitsubishi Electric Corporation
Airborne radar ... 249
EW equipment ... 485
J/FPS-3 air defence radar ... 28
OPS series shipborne search radar ... 156

Moscow Radio Engineering Plant
Kama-N trajectory measuring radar ... 305

Motorola Inc, Government Electronics Group
AN/ALQ-99(V) Tactical Jamming System (TJS) ... 554
SLAMMR side-looking radar ... 239

MRCM Products
CICADA series jammers ... 403
DFS2000 man/machine interface ... 340
MAIGRET II naval ESM system ... 422
MRD2000 LH DF system ... 345
MRD4008 DF system ... 345
Polygon MRD30w3/n DF system ... 346
Polygon MRD3000w5 DF system ... 347
SIGMA II signals intercept and analysis ... 350
Telegon MRD 1920 DF system ... 353
WSLC Wideband Search, Location and Classification . 357

MS Instruments plc
HOFIN HOstile Fire INdicator ... 492

MTI Technology and Engineering
EW system antennas and arrays ... 360

N

Nanjing Marine Institute
Hai Ying air surveillance radar ... 125

Nanjing Research Institute of Electronics Technology
146-1 target indication radar ... 7
YLC-4 surveillance radar ... 12

NEA Lindberg A/S
Mk91 fire-control system radar ... 183

NEC Corporation
EW equipment ... 485
J/FPS-2 3-D air defence radar ... 28
J/TPS-102 air defence radar ... 29
OPS series shipborne search radar ... 156

NICE Systems Ltd
CDF 1500 surveillance and emitter location ... 358
NiceCall digital voice logger ... 362
NiceFix traffic management DF ... 362
NiceLog/NiceCLS logging systems ... 363

NIIP Nauchno-Issledovatelskiy Institut Priborostroyeniya, Zhukovsky
N011/014 fire-control radars ... 252
Osa multimode radar ... 252
Zaslon fire-control radar ... 254

NITEL JSC
55G6-1 Nebo 3-D surveillance radar ... 33
Furgon early-warning radar ... 36
Lena early-warning radar ... 37
Oborona-14 early-warning radar ... 38

Nizhegorodsky Radio Engineering Research Institute
55G6-UE Nebo-U 3-D surveillance radar ... 33
Delta surveillance radar ... 35
Furgon early-warning radar ... 36
Lena early-warning radar ... 37
Oborona-14 early-warning radar ... 38
Protivnik-GE surveillance radar ... 41
Struna-1 surveillance radar ... 42

Northern Telecom Europe, Radio and Microwave Division
APAR–Active Phased Array Radar ... 129

Northrop Grumman Amherst Systems Inc
AMES II/III EW simulators ... 592
CEESIM simulators ... 598
Flightline EW simulator (FLSim) ... 599
MRES Mobile Reprogrammable Emitter Simulator ... 604
Pico-AMES threat simulator ... 607
Portable CEESIM ... 608
RES Reprogrammable Emitter Simulator ... 607
RISS infra-red simulator ... 606
SMS Signal Measurement Systems ... 609
Transportable CEESIM ... 609

Northrop Grumman Corporation
AIEWS EW system ... 462
AN/AAQ-4/8 IR countermeasures ... 550
AN/AAQ-24(V) IR countermeasures ... 551
AN/AAR-54(V)–(PMAWS) missile warner ... 496
AN/ALQ-99(V) Tactical Jamming System (TJS) ... 554
AN/ALQ-131 ECM pod ... 556
AN/ALQ-135 jamming system ... 556
AN/ALQ-155(V) EW power management system ... 559
AN/ALQ-162(V) jammer ... 560
AN/ALQ-165 self-protection jammer ... 561
AN/ALR-66 series RWR/ESM systems ... 500
AN/ALR-67(V)2 countermeasures system ... 501
AN/ALR-68A(V)3 radar warning system ... 472
AN/ALR-69(V) RWR ... 502
AN/ALR-73 detection system ... 503
AN/ALR-93(V)1 RWR/ESM system ... 505
AN/APG-66 multimode airborne radar ... 260
AN/APG-68(V) airborne radar ... 261
AN/APG-77 airborne radar ... 265
AN/APG-78 Longbow airborne radar ... 265
AN/APG-80 AESA radar ... 266
AN/APN-241 airborne radar ... 221
AN/APQ-164 multimode radar ... 223
AN/APY-1/2 AWACS radar ... 229
AN/APY-3 multimode SLAR ... 231
AN/SLY-2(V) AIEWS EW system ... 462
AN/SPS-40 radar ... 170
AN/SPS-58/65 naval radar ... 174
AN/TPS-43 tactical 3-D radar ... 56
AN/TPS-63 surveillance radar ... 59
AN/TPS-70 tactical radar ... 59
AN/TPS-75 tactical 3-D radar ... 61
AN/ULQ-16(V) pulse analyser ... 376
AN/ZPQ-1 TESAR ... 233
ARSR-4 air route surveillance radar ... 61
ASPIS–Self-Protection Integrated Suite ... 550
ASR-9 airport surveillance radar ... 313
ASR-12 airport surveillance radar ... 313
Emitter Feature Extractor (EFE) ... 383
Enhanced Radar Warning Equipment (ERWE) II ... 473
F-35 AESA radar ... 267
MESA radar ... 237
Monopulse SSR system ... 292
QRC 83-05 IRCM system ... 572
Tactical UAV radar ... 239
TESAR Tactical Endurance Synthetic Aperture Radar ... 233
Zeus integrated DAS ... 549

Northrop Grumman Norden Systems
AN/APG-76 multimode radar ... 264
AN/APS-130 multimode radar ... 226
AN/APY-3 multimode SLAR ... 231
AN/APY-6 surface surveillance radar ... 233
AN/SPQ-9B fire-control radar ... 196
AN/SPS-40 radar ... 170
B-52 airborne radar ... 234

Jane's Radar and Electronic Warfare Systems 2002-2003

www.janes.com

Alphabetical index

To help users of this title evaluate the published data, *Jane's Information Group* has divided entries into three categories.

[N] NEW ENTRY Information on new equipment and/or systems appearing for the first time in the title.

[V] VERIFIED The editor has made a detailed examination of the entry's content and checked its relevancy and accuracy for publication in the new edition to the best of his ability.

[U] UPDATED During the verification process, significant changes to content have been made to reflect the latest position known to *Jane's* at the time of publication. Items in italics refer to entries which have been deleted from this edition with the relevant page numbers from last year.

NOTES